[second edition]

# Principles of Biology

Robert J. Brooker
University of Minnesota - Minneapolis

Eric P. Widmaier
Boston University

Linda E. Graham
University of Wisconsin - Madison

Peter D. Stiling
University of South Florida

PRINCIPLES OF BIOLOGY, SECOND EDITION

Published by McGraw-Hill, a business unit of The McGraw-Hill Companies, Inc., 1221 Avenue of the Americas, New York, NY 10020. 

This book is printed on acid-free paper.

1 2 3 4 5 6 7 8 9 0 LWI 21 20 19 18 17

ISBN 978–1–259–87512–0
MHID 1–259–87512–1

Senior Vice President, Products & Markets: *G. Scott Virkler*
Vice President, General Manager, Products & Markets: *Marty Lange*
Vice President, Content Production & Technology Services: *Betsy Whalen*
Managing Director: *Lynn Breithaupt*
Brand Manager: *Michelle Vogler*
Product Developer: *Elizabeth M. Sievers*
Executive Marketing Manager: *Patrick E. Reidy*
Director, Content Production: *Linda Avenarius*
Program Manager: *Angela FitzPatrick*
Content Project Manager (Print): *Jayne Klein*
Content Project Manager (Media): *Christina Nelson*
Senior Buyer: *Sandy Ludovissy*
Senior Designer: *David W. Hash*
Cover Image: © *Chase Dekker Wild-Life Images/Moment/Getty Images*
Content Licensing Specialists: *Carrie Burger/Shannon Manderscheid*
Compositor: *MPS Limited*
Typeface: *10/12 STIX MathJax Main*
Printer: *LSC Communications*



**Library of Congress Cataloging-in-Publication Data**

Brooker, Robert J., author. | Widmaier, Eric P., author. | Graham, Linda E., 1946- author. | Stiling, Peter D., author.
Principles of biology/Robert J. Brooker, University of Minnesota-Minneapolis, Eric P. Widmaier, Boston University, Linda E. Graham, University of Wisconsin-Madison, Peter D. Stiling, University of South Florida.
Second edition. | New York, NY : McGraw-Hill Education, [2018] | Includes index.
LCCN 2016035146 | ISBN 9781259875120 (alk. paper)
LCSH: Biology—Textbooks.
LCC QH308.2 .B75 2018 | DDC 570—dc23 LC record available at https://lccn.loc.gov/2016035146



mheducation.com/highered

# Brief Contents

# About the Authors

## Robert J. Brooker

Rob Brooker received his Ph.D. in genetics from Yale University in 1983. At Harvard, he studied lactose permease, the product of the *lacY* gene of the *lac* operon. He continued working on transporters at the University of Minnesota, where he is a Professor in the Department of Genetics, Cell Biology, and Development At the University of Minnesota, Dr. Brooker teaches undergraduate courses in biology and genetics. In addition to many other publications, he has written two undergraduate genetics texts: *Genetics: Analysis & Principles*, Sixth Edition, copyright 2018, and *Concepts of Genetics*, Second Edition, copyright 2015; and he is the lead author of *Biology*, Fourth Edition, copyright 2017, all published by McGraw-Hill Education.

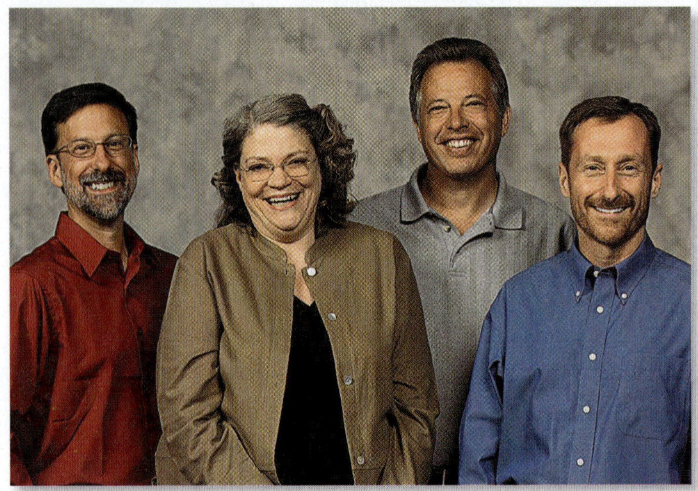

Left to right: Eric Widmaier, Linda Graham, Peter Stiling, and Rob Brooker

## Eric P. Widmaier

Eric Widmaier received his Ph.D. in 1984 in endocrinology from the University of California at San Francisco. His research focuses on the control of body mass and metabolism in mammals, the hormonal correlates of obesity, and the effects of high-fat diets on intestinal cell function. Dr. Widmaier is currently Professor of Biology at Boston University, where he teaches undergraduate courses in human physiology, comparative physiology, and endocrinology and recently received the university's highest honor for excellence in teaching. Among other publications, he is a coauthor of *Vander's Human Physiology: The Mechanisms of Body Function*, Fourteenth Edition, copyright 2017; and *Biology*, Fourth Edition, copyright 2017, both published by McGraw-Hill Education.

## Linda E. Graham

Linda Graham received her Ph.D. in botany from the University of Michigan, Ann Arbor. Her research explores the evolutionary origin of algae land-adapted plants, focusing on their cell and molecular biology as well as ecological interactions with microbes. Dr. Graham is now Professor of Botany at the University of Wisconsin–Madison. She teaches undergraduate courses in microbiology and plant biology. She is the coauthor of, among other publications, *Algae*, Third Edition, copyright 2015, a major's textbook on algal biology; and *Plant Biology*, Third Edition, copyright 2015, both published by LJLM Press. She is also a coauthor of *Biology*, Fourth Edition, copyright 2017, published by McGraw-Hill Education.

## Peter D. Stiling

Peter Stiling obtained his Ph.D. from University College, Cardiff, Wales, in 1979. Subsequently, he became a postdoctoral fellow at Florida State University and later spent two years as a lecturer at the University of the West Indies, Trinidad. During this time, he began photographing and writing about butterflies and other insects, which led to publication of several books on local insects. Dr. Stiling is currently a Professor of Biology at the University of South Florida at Tampa. His research interests include plant-insect relationships, parasite-host relationships, biological control, restoration ecology, and the effects of elevated carbon dioxide levels on plant–herbivore interactions. He teaches graduate and undergraduate courses in ecology and environmental science as well as introductory biology. He has published many scientific papers and is the author of *Ecology: Global Insights and Investigations*, Second Edition, copyright 2015, and is coauthor of *Biology*, Fourth Edition, copyright 2017, both published by McGraw-Hill Education.

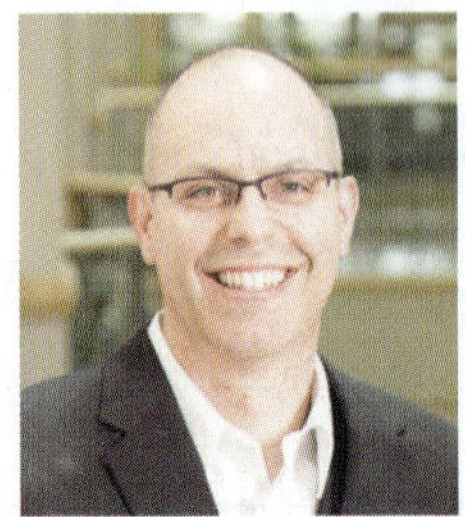

© Ian J. Quitadamo, Ph.D.

## Ian Quitadamo, Lead Digital Author

Ian Quitadamo is a Professor with a dual appointment in Biological Sciences and Science Education at Central Washington University in Ellensburg, Washington. He teaches introductory and majors biology courses and cell biology, genetics, and biotechnology, as well as science teaching methods courses for future science teachers and interdisciplinary content courses in alternative energy and sustainability. Dr. Quitadamo was educated at Washington State University and holds a BA in biology, Masters degree in genetics and cell biology, and an interdisciplinary Ph.D. in science, education, and technology. Previously a researcher of tumor angiogenesis, he now investigates the behavioral and neurocognitive basis of critical thinking and has published numerous studies of factors that improve student critical thinking performance. He has received the Crystal Apple award for teaching excellence, led multiple initiatives in critical thinking and assessment, and is active nationally in helping transform university faculty practice.

## A Note about *Principles of Biology* . . .

A recent trend in science education is the phenomenon called "flipping the classroom." This phrase refers to the idea that some of the activities that used to be done in class are now done out of class, and vice versa. For example, instead of spending the entire class time lecturing about textbook and other materials, some of the class time is spent engaging students in various activities, such as problem solving, working through case studies, and designing experiments. This approach is called active learning. For many instructors, the classroom has become more learner-centered rather than teacher-centered. A learner-centered classroom provides a rich environment in which students can interact with each other and with their instructors. Instructors and fellow students often provide formative assessment—immediate feedback that helps each student understand if his or her learning is on the right track.

What are some advantages of active learning? Educational studies reveal that active learning usually promotes greater learning gains. In addition, active learning often focuses on skill development rather than the memorization of facts that are easily forgotten. Students become trained to "think like scientists" and to develop a skill set that enables them to apply scientific reasoning.

A common concern among instructors who are beginning to try out active learning is that they think they will have to teach their students less material. However, this may not be the case. Although students may be provided with online lectures, "flipping the classroom" typically gives students more responsibility for understanding the textbook material on their own. Along these lines, *Principles of Biology* is intended to provide students with a resource that can be effectively used out of the classroom. Several key pedagogical features include the following:

- **Focus on Core Concepts:** Although it is intended for majors in the biological sciences, *Principles of Biology* is a shorter textbook that emphasizes core concepts. Twelve principles of biology are enunciated in Chapter 1 and those principles are emphasized throughout the textbook with specially labeled figures. An effort has also been made to emphasize some material in bulleted lists and numbered lists, so students can more easily see the main points.
- **Learning Outcomes:** Each section of every chapter begins with a set of learning outcomes. These outcomes help students understand what they should be able to do if they have mastered the material in that section. Certain learning outcomes, labeled as **SCISKILLS**, emphasize experimental skills needed in the study of biology. Skills such as *analyze data, form hypotheses, make predictions, make calculations,* are skills that scientists generally perform and students majoring in biology should practice.
- **Formative Assessment:** When students are expected to learn textbook material on their own, it is imperative that they be given regular formative assessments so they can gauge whether or not they are mastering the material. Formative assessment is a major feature of this textbook and is bolstered by McGraw-Hill Connect®—a state-of-the-art digital assignment and assessment platform. In *Principles of Biology*, formative assessment is provided in multiple ways.

    1. Each section of every chapter ends with multiple-choice questions.
    2. Most figures have concept check questions so students can determine if they understand the key points in the figure.
    3. End-of-chapter questions continue to provide students with feedback regarding their mastery of the material.
    4. Further assessment tools are available in Connect. Question banks, Test banks, and Quantitative Question banks can be assigned by the professor. McGraw-Hill SmartBook® allows for individual study as well as assignments from the professor.

- **Quantitative Analysis:** Many chapters have a subsection that emphasizes quantitative reasoning, an important skill for careers in science and medicine. In these subsections, the quantitative nature of a given topic is described, and then students are asked to solve a problem related to that topic.
- **BioConnections and Evolutionary Connections:** To help students broaden their understanding of biology, two recurring features are BioConnections and Evolutionary Connections. BioConnections are placed in key figure legends in each chapter and help students relate a topic they are currently learning to another topic elsewhere in the textbook, often in a different unit. Evolutionary Connections provide a framework for understanding how a topic in a given chapter relates to evolution, the core unifying theme in biology.

- New **BioTIPS:** In Connect, the digital partner to this textbook, we have a new feature called **BioTIPS**, which is intended to help students refine problem-solving skills. Most of the **BioTIPS** are called out with icons in the textbook, but additional **BioTIPS** are included in the SmartBook. The **BioTIPS** themselves are accessed through links in SmartBook. **BioTIPS** will focus on 11 strategies that will help students solve problems:

  1. ***Make a drawing***.
  2. ***Compare and contrast***.
  3. ***Relate structure and function***.
  4. ***Sort out the steps in a complicated process***.
  5. ***Propose a hypothesis***.
  6. ***Design an experiment***.
  7. ***Predict the outcome***.
  8. ***Interpret data***.
  9. ***Use statistics***.
  10. ***Make a calculation***.
  11. ***Search the literature.***

**BioTIPS** will provide students with practice at applying these problem-solving strategies.

Overall, the pedagogy of *Principles of Biology* has been designed to foster student learning. Instead of being a collection of "facts and figures," *Principles of Biology* is intended to be an engaging and motivating textbook in which formative assessment allows students to move ahead and learn the material in a productive way.

## Content Changes to the Second Edition

The author team of *Principles of Biology* is fully committed to keeping the content up to date; the second edition has five new chapters that reflect modern trends in the field. They are intended to achieve three goals:

- **Prepare Students for Careers in Modern Biology:** Chapters 11, 16, and 24 are concerned with the topics of Non-coding RNAs, Epigenetics, and Microbiomes, respectively. The emerging importance of these areas in the field of medicine is dramatic. It's difficult to pick up a newspaper and not see a story that concerns at least one of these areas and its impact on human health. For example, researchers are now studying how the manipulation of certain non-coding RNAs may be used as therapeutic tools to treat diseases such as cancer. Similarly, these same topics have broad importance in the fields of agriculture, biotechnology, and environmental science.
- **An Emphasis on Systems Biology:** The first edition of *Principles of Biology* already had an emphasis on systems biology and trying to relate topics in biology to its evolutionary foundation. In the second edition, we have added a new chapter to the animal unit (Chapter 41) that explores how the whole body responds to a major challenge to homeostasis (hemorrhage). This allows students to appreciate how various organs and organ systems work together as a larger integrated system—the animal body.
- **Impact on Society:** Not only do we want to help our students learn biology and prepare them for careers in this field, we also want them to appreciate their roles as citizens of the world. Chapter 46 pulls together many of the key topics involving the impact of humans on the environment, thereby making students aware of current and future problems. This chapter may inspire some students to pursue a career in ecology or environmental science, and may encourage others to educate the public regarding the negative effects that humans have had on the environment and ways to evoke positive changes.

To make room for this material and other updated material, some chapters have been streamlined and combined, and obsolete methods have given way to new techniques described in these new chapters. The major content changes that have occurred in the second edition are summarized below.

**Chapter 1 An Introduction to Biology.** Has a new section on the adaptations that have occurred during the evolution of polar bears.

**Chapter 4 Evolutionary Origin of Cells and their General Features.** This chapter now begins with a section on the evolutionary origin of cells.

**NEW Chapter 11 The Expression of Genetic Information via Genes II: Non-coding RNAs.** This "first of its kind chapter" recognizes the great importance of non-coding RNAs in biology and devotes an entire chapter to this topic. The author team feels it is long overdue.

**NEW Chapter 16 Transmission of Genetic Information from Parents to Offspring II: Epigenetics, Linkage, and Extranuclear Inheritance.** Due to the rapidly expanding topic of epigenetics, the inheritance chapter in the first edition has been split into two chapters. Chapter 16, which is the second chapter devoted to inheritance, has four sections on epigenetics and also includes the topics of linkage and extranuclear inheritance.

**Chapter 18 Genetic Technologies: How Biologists Study Genes and Genomes.** Has a new section on CRISPR-Cas technology, which is used to introduce mutations into genes.

**Chapter 19 Evolution of Life I: How Populations Change from Generation to Generation.** The Evolution unit has been reorganized so that it now begins with a description of the basic mechanisms that underlie evolutionary change.

**Chapter 22 The History of Life on Earth and Human Evolution.** The topic of human evolution has been moved from the Diversity unit to the Evolution unit. The description of human evolution has been greatly expanded, and has new topics including how humans are still evolving and the level of genetic variation in modern human populations.

**Chapter 23 Diversity of Microbial Life: Archaea, Bacteria, Protists, and Fungi.** This chapter on the diversity of prokaryotic and eukaryotic microbial life has been heavily revised to integrate material previously covered in separate chapters. Newer concepts of phylogenetic diversification of these groups have been incorporated into evolutionary tree diagrams.

This revision provides several pedagogical advantages. A focus on microbial diseases of humans and crops, a continuing thread through coverage of bacteria, and also protists and fungi, reveals greater pathogen diversity than students may previously have realized. The diversity of technological applications involving microbes, previously described in several separate places, has now been aggregated at the end of the chapter. Long important in terms of food or antibiotic production, microbial applications are now taking on new relevance to the fields of environmental pollution control and renewable biofuels.

Finally, by integrating fundamental aspects of four microbial groups, Chapter 23 now provides the broad diversity background necessary to comprehend microbiomes, a topic of vast medical, ecological, and technological importance that is presented in a new chapter.

**NEW Chapter 24 Microbiomes: Microbial Systems on and Around Us.** This entirely new chapter integrates information about the occurrence of microbes (archaea, bacteria, protists, and fungi) within complex organism-gene systems known as microbiomes, a major frontier of biological sciences. The new chapter expands the much briefer and scattered introductions to symbiotic relationships between microbes, plants, and animals presented in the first edition. New Chapter 24 begins by linking basic information on microbial life provided in Chapter 23 with important environments in which microbiomes occur: physical environments such as oceans, ice, and soils, and biotic environments that include the bodies of humans and agricultural plants. The new chapter then focuses on genetic methods that microbiologists use to comprehend and compare Earth's microbiomes. This helps students to review and extend basic genetics presented earlier in the text and understand important applications of genetic and genomic technologies. The new chapter also focuses on evolutionary and diversity aspects of microbiomes that are key to fostering agricultural production and human health, thereby connecting students to previous text chapters describing fundamental principles of evolutionary biology.

**Chapter 25 Plant Evolution and Diversity.** New information about the evolutionary history of plants has been incorporated to maximize currency, without increasing complexity or level of detail. A new **BioTIPS** feature, which aims to foster student understanding of experimental design in the scientific process, has been developed for the popular Feature Investigation on *Cannabis* secondary metabolites, of high societal significance.

**Chapter 26 Invertebrates: The Vast Array of Animal Life Without a Backbone.** Our animal classification as depicted in Figure 26.2 has been reworked and redrawn to reflect the position of the Ctenophora or comb jellies, as the earliest diverging animal clade. Additional photographs have also been added to Figure 26.12 to illustrate the polyp and medusa form of cnidarians. We have also included a new multi-part figure, Figure 26.32, to illustrate the different echinoderm classes.

**Chapter 27 Vertebrates: Fishes, Amphibians, Reptiles and Mammals.** The material on primates and human evolution has been moved to Chapter 22, The History of Life on Earth and Human Evolution.

**Chapter 28 An Introduction to Flowering Plant Form and Function.** A **BioTIPS** feature designed to help students interpret graphical quantitative information has been developed for the Feature Investigation, which focuses on leaf structural variation.

**Chapter 29 How Flowering Plants Sense and Interact with Their Environments.** Some new images have been incorporated.

**Chapter 30 How Flowering Plants Obtain and Transport Nutrients.** Some new images have been incorporated.

**Chapter 31 How Flowering Plants Reproduce and Develop.** A new **BioTIPS** feature, based on the Feature Investigation about flower blooming, not only fosters student ability to interpret graphical quantitative information, but also leads them to make additional calculations to answer new questions about the topic.

**Chapter 32 General Features of Animal Bodies, and Homeostasis as a Defining Principle of Animal Biology.** This chapter now includes a section entitled "Principles of Homeostasis of Internal Fluids," which has been moved here from later in the Animal Unit where it was previously covered (former Chapter 38). New "Test Yourself" questions and several improved figures and new concept checks have been added.

**Chapter 33 Neuroscience I: The Structure, Function, and Evolution of Nervous Systems.** New figures, including an electron micrograph of a cross section through a nerve, have been added, while several existing figures have been modified with additional labeling or text boxes to improve clarity. Numerous SCISKILLS features have been incorporated throughout the section-opening learning outcomes.

**Chapter 34 Neuroscience II: How Sensory Systems Allow Animals to Interact with the Environment.** Numerous subheadings are now interspersed in the chapter to help the reader navigate through difficult passages and to help the instructor and student organize the readings. Numerous SCISKILLS have been incorporated throughout the section-opening learning outcomes, and new concept checks have been added. Throughout the chapter, material has been updated to reflect key new research, particularly with respect to olfaction and balance.

**Chapter 35 How Muscles and Skeletons are Adaptations for Movement, Support, and Protection.** In addition to numerous SCISKILLS features, a new conceptual question has been added and several figures have been improved for even greater clarity.

**Chapter 36 Circulatory and Respiratory Systems: Transporting Solutes and Exchanging Gases.** The former chapters on circulation and respiration (chapters 36 and 37) have now been merged into one cohesive chapter that covers both topics in a fully integrated way. As one example, a new table has been added that covers the relationship between an animal's body mass and various respiratory parameters; this table now parallels a similar one that was present in the former Circulatory System chapter that described the relationship between body mass and circulatory features. As with other chapters, numerous SCISKILLS features, figure modifications, and assessments have added or updated. A new figure depicting human bronchioles in health and disease has also been added.

**Chapter 37 Digestive and Excretory Systems Help Maintain Nutrient, Water, and Energy Balance and Remove Waste Products from Animal Bodies.** The former chapters on digestion and nutrition, and the excretory system, have now been integrated into one chapter. The combined focus is now on nutrient processing and energy balance and the elimination of soluble wastes. Numerous text boxes and figure labels have been adjusted in the artwork to enhance understanding. The advantages and disadvantages of generating a particular type of nitrogenous waste are now elaborated. SCISKILLS features have been added to all sections. Two new concept checks and Bioconnection features have been added.

**Chapter 38 How Endocrine Systems Influence the Activities of all Other Organ Systems.** Several text boxes, labels and figure legends have been modified for additional detail to improve understanding. SCISKILLS features have been added to each section, and the text has been updated to reflect modern research in endocrinology.

**Chapter 39 The Production of Offspring: Reproduction and Development.** The Impact on Public Health section has been reorganized with numerous subheadings for clarity. SCISKILLS have been added, as has a new Bioconnections question. Certain key figures have been updated or modified for clarity.

**Chapter 40 Immune Systems: How Animals Defend Against Pathogens and Other Dangers.** The opening section is now reorganized with subheadings for clarity, and includes discussions of some important animal diseases. Additional new subheadings also break up complex text throughout the chapter. SCISKILLS and a new test question have been added, and key figures have been improved for clarity or detail, or updated (such as latest figures on the number of people living with HIV/AIDS as of today).

**NEW Chapter 41 Integrated Responses of Animal Organ Systems to a Challenge to Homeostasis.** This new chapter integrates the functions of all organ systems found in animals, using a challenge to homeostasis (hemorrhage) as the central theme. It introduces ten new figures and a new table covering topics such as baroreceptors, chemoreceptors, Starling forces and many others, all in the context of an integrated response to a large homeostatic insult.

**Chapter 43: Population Growth and Species Interactions.** Two new **BioTIPS** questions have been added to better familiarize students with mark-recapture analyses and competition and resource utilization. The material on human population growth has been moved to Chapter 46. There are three new conceptual and collaborative questions.

**Chapter 44: Communities and Ecosystems: Ecological Organization of Large Scales.** Chapter 44 has been reworked to include a discussion of both community and ecosystem ecology together in the same chapter. We have combined chapters 45 and 46 from the first edition. However, the material on biogeochemical cycles has been moved to Chapter 46.

**Chapter 45: How Climate Affects the Distribution of Species on Earth.** This chapter uses elements of Chapter 43: Ecology and the Physical Environment, from the first edition and expands on them. In the first section, 45.1, Climate, we show what causes global temperature and precipitation differentials across the Earth. In the next section, 45.2, Major Biomes, we describe and illustrate the major biomes on Earth.

**NEW Chapter 46: The Age of Humans.** This is a new chapter. We begin by introducing the concept of a new geological era, the Anthropocene, and then discuss the effects of humans on natural systems. We start with an examination of human population growth, which continues in an upward trend. Next, we explain how humans are contributing to climate change via global warming. This is followed by section 46.3, Pollution and Human Influences on Biogeochemical Cycles. In this section we describe human influences on the carbon, water, phosphorous, and nitrogen cycles from the burning of fossil fuels, the use of chemical fertilizers and pesticides, and other factors. This can lead to biomagnification, as explained next in section 46.4. One of the biggest effects of humans is habitat destruction and in section 46.5 we detail the effects of deforestation and agriculture on wildlife loss. In section 46.6, Overexploitation, we discuss the effects of overhunting and overfishing on land mammals, whales, birds, fishes, and plants. Lastly, in section 46.7, Invasive Species, we consider the many and varied effects of deliberate and accidental plant and animal introductions on native wildlife via competition, predation and parasitism.

**Chapter 47: Biodiversity and Conservation Biology.** We have updated Table 47.1, which provides details of the world's ecosystem services. The material on causes of extinction and loss of biodiversity has been moved to chapter 46. However, section 47.3, Conservation Strategies, has been expanded to include new material and figures on crisis ecoregions and "last of the wild" in addition to megadiversity countries and biodiversity hot spots.

# Guiding You Through *Principles of Biology*

*Principles of Biology* and its online assets have been carefully crafted to help students work efficiently and effectively through the material in the course, making the most of their study time. This *Guiding You Through Principles of Biology* section explains how students can use the text and Connect® to help them succeed in majors biology.

## EMPHASIZING SKILLS DEVELOPMENT AND PROBLEM SOLVING

### Skills Development

At the beginning of each section, **Learning Outcomes** inform students of concepts they should understand. New to the second edition are skills-based Learning Outcomes. Labeled as **SCISKILLS**, these Learning Outcomes are specific to the skills students will acquire when mastering the material and provide a specific understanding of how such skills may be assessed. SCISKILLS is a mental action such as analyze data, form hypotheses, make predictions, or perform calculations. These are skills scientists generally perform and students should practice.

The emphasis on skills development continues in the **Feature Investigations**. Feature Investigations provide a complete description of experiments, including data analysis, so students can understand how experimentation leads to an understanding of biological concepts.

The **Quantitative Analysis** feature helps develop analytical skills. This feature walks through biological concepts that have a quantitative component. The Crunching the Numbers provides a sample problem to test understanding.

**5.2 Fluidity of Membranes**

**Learning Outcomes**

1. Describe the fluidity of membranes.
2. **SCISKILLS ▸** Predict how fluidity will be affected by changes in lipid composition.
3. **SCISKILLS ▸** Analyze the results of experiments indicating that certain membrane proteins can diffuse laterally within the membrane.

Let's now turn our attention to the dynamic properties of membranes. Although a membrane provides a critical interface between a cell and its environment, it is not a solid, rigid structure. Rather, biological membranes exhibit properties of **fluidity,** which means that individual molecules remain in close association yet have the ability to readily move within the membrane. In this section, we will examine the fluid properties of biological membranes.

**Membranes Are Semifluid**

Though membranes are often described as fluid, it is more appropriate to say they are **semifluid,** because movements of lipids and membrane proteins occur in only two dimensions. In a fluid

**Quantitative Analysis**

**RECEPTORS HAVE A MEASURABLE AFFINITY FOR THEIR LIGANDS**

In general, the binding and release between a ligand and its receptor are relatively rapid, and therefore an equilibrium is reached when the rate of formation of new ligand•receptor complexes equals the rate at which existing ligand•receptor complexes dissociate:

$$k_{on}\,[\text{Ligand}][\text{Receptor}] = k_{off}\,[\text{Ligand•Receptor complex}]$$

Rearranging,

$$\frac{[\text{Ligand}][\text{Receptor}]}{[\text{Ligand•Receptor complex}]} = \frac{k_{off}}{k_{on}} = K_d$$

$K_d$ is called the **dissociation constant** between a ligand and its receptor. The $K_d$ value is inversely related to the affinity between the ligand and receptor. A low $K_d$ value indicates that a receptor has a high affinity for its ligand.

Let's look carefully at the left side of this equation and consider what it means. At a ligand concentration where half of the receptors are bound to a ligand, the concentration of the ligand•receptor complex equals the concentration of receptor that doesn't have ligand bound. At this ligand concentration, [Receptor] and [Ligand•Receptor complex] cancel out of the equation because they are equal. Therefore, at a ligand concentration where half of the receptors have bound ligand

$$K_d = [\text{Ligand}]$$

When the ligand concentration is above the $K_d$ value, most of the receptors are likely to have ligand bound to them. In contrast, if

### Problem Solving

A new feature will help students develop their problem solving skills. **BioTIPS**, which stands for **T**opic, **I**nformation, and **P**roblem Solving **S**trategy is a feature available in Connect. Icons appearing throughout the book indicate the textual material supporting the **BioTIPS** online. These solved problems follow a consistent pattern in which students are given advice on how to solve problems in biology using different types of problem solving strategies. These strategies include: Make a drawing; Compare and contrast; Relate structure and function; Sort out the steps in a complicated process; Propose a hypothesis; Design an experiment; Predict the outcome; Interpret data; Use statistics; Make a calculation; and Search the literature.

products of cellular metabolism must be released from cells before they reach toxic levels. For example, a transporter removes lactic acid, a by-product of muscle cells during exercise. Other transporters, which are involved with ion transport, play an important role in regulating internal pH and controlling cell volume. Transporters tend to be much slower than channels. Their rate of transport is typically 100 to 1,000 ions or molecules per second.

Transporters are named according to the number of solutes they bind and the direction in which they transport those solutes (Figure 5.15).

- **Uniporters** bind a single ion or molecule and transport it across the membrane.
- **Symporters** bind two or more ions or molecules and transport them in the same direction.
- **Antiporters** bind two or more ions or molecules and transport them in opposite directions.

BIO TIPS ONLINE

Active Transport Is the Movement of Solutes Against a Gradient

As mentioned, active transport is the movement of a solute across a membrane against its concentration gradient—that is, from a region of low concentration to higher concentration. Active transport is energetically unfavorable and requires an input of energy. Two general types are found:

- **Primary active transport** involves the functioning of a

A biological question related to chapter content is posed. The **BioTIPS** then walks the student through the process of answering the question. First, they help the student identify the topic of the question—what is really being asked in the question? Then they help the student collect information that was presented in the chapter that is related to the question. Finally, they help the student settle on one or more strategies that can be followed to answer the question. The answers are provided to complete the problem solving process.

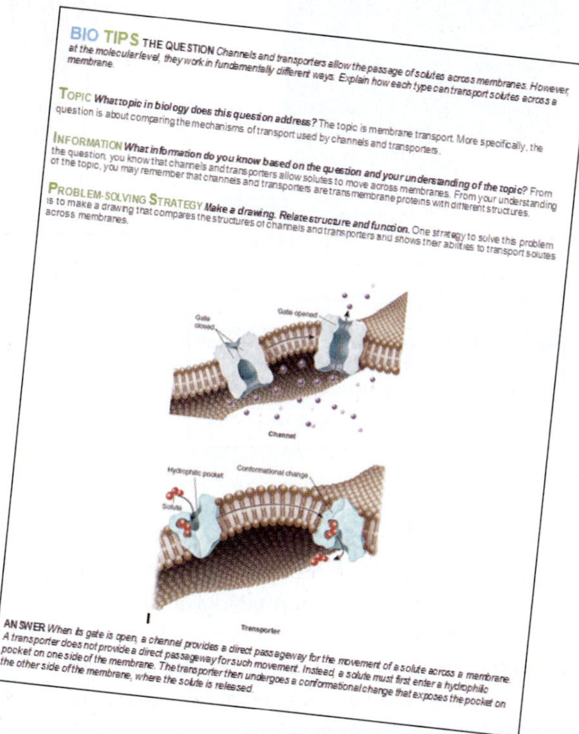

BIO TIPS **THE QUESTION** *Channels and transporters allow the passage of solutes across membranes. However, at the molecular level, they work in fundamentally different ways. Explain how each type can transport solutes across a membrane.*

TOPIC **What topic in biology does this question address?** The topic is membrane transport. More specifically, the question is about comparing the mechanisms of transport used by channels and transporters.

INFORMATION **What information do you know based on the question and your understanding of the topic?** From the question, you know that channels and transporters allow solutes to move across membranes. From your understanding of the topic, you may remember that channels and transporters are transmembrane proteins with different structures.

PROBLEM-SOLVING STRATEGY **Make a drawing. Relate structure and function.** One strategy to solve this problem is to make a drawing that compares the structures of channels and transporters and shows their abilities to transport solutes across membranes.



**ANSWER** *When its gate is open, a channel provides a direct passageway for the movement of a solute across a membrane. A transporter does not provide a direct passageway for such movement. Instead, a solute must first enter a hydrophilic pocket on one side of the membrane. The transporter then undergoes a conformational change that exposes the pocket on the other side of the membrane, where the solute is released.*

## USING STUDENT USAGE DATA TO MAKE IMPROVEMENTS

To help guide the revision for the second edition, student usage data and input were used, derived from thousands of SmartBook® users of the first edition. SmartBook "heat maps" provided a quick visual snapshot of chapter usage data and the relative difficulty students experienced in mastering the content. These data directed the authors to evaluate text content that was particularly challenging for students. These same data were also used to revise the SmartBook questions.

- If the data indicated that the subject was more difficult than other parts of the chapter, as evidenced by a high proportion of students responding incorrectly to the questions, the authors revised or reorganized the content to be as clear and illustrative as possible by rewriting the section, providing additional examples or revised figures to assist visual learners, etc.
- In other cases, one or more of the SmartBook questions for a section was not as clear as it might be or did not appropriately reflect the content in the chapter. In these cases the *question*, rather than the text, was edited.

Below is an example of one of the heat maps. The color-coding of highlighted sections indicates the various levels of difficulty students experienced in learning the material, topics highlighted in red being the most challenging for students.

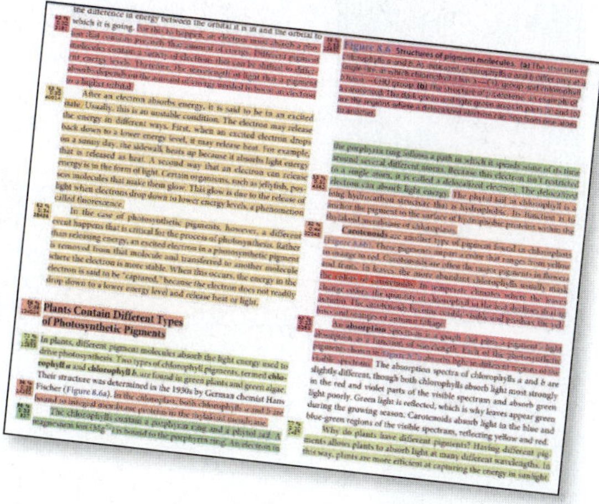

Plants Contain Different Types of Photosynthetic Pigments

# MAKING CONNECTIONS

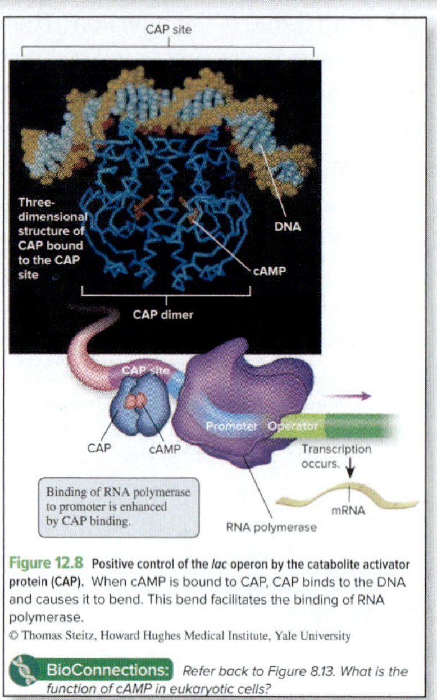


**Figure 12.8** **Positive control of the *lac* operon by the catabolite activator protein (CAP).** When cAMP is bound to CAP, CAP binds to the DNA and causes it to bend. This bend facilitates the binding of RNA polymerase.
© Thomas Steitz, Howard Hughes Medical Institute, Yale University

**BioConnections:** *Refer back to Figure 8.13. What is the function of cAMP in eukaryotic cells?*

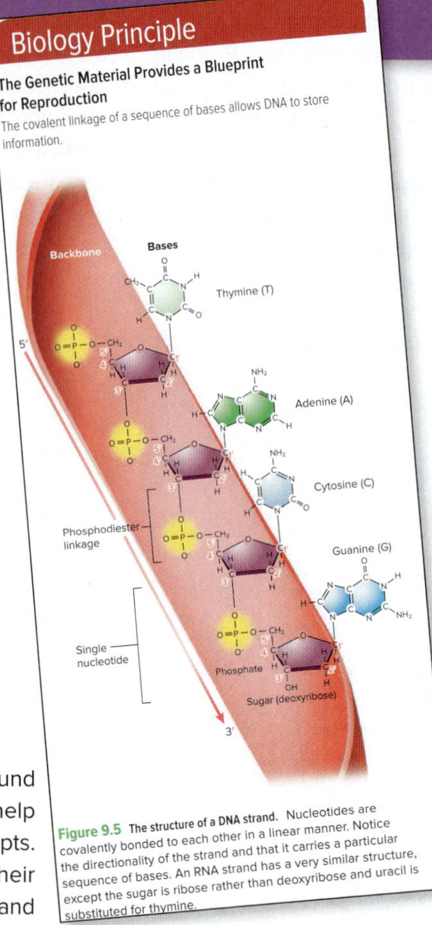


**Figure 9.5** **The structure of a DNA strand.** Nucleotides are covalently bonded to each other in a linear manner. Notice the directionality of the strand and that it carries a particular sequence of bases. An RNA strand has a very similar structure, except the sugar is ribose rather than deoxyribose and uracil is substituted for thymine.

***Principles of Biology*** are introduced in Chapter 1 and are then threaded throughout the entire textbook. This is achieved in two ways. First, the principles are highlighted in selected figures in which the specific principle is illustrated.

In addition, a Conceptual Question at the end of each chapter is directly aimed at exploring a particular Biology Principle related to the content of the chapter.

***BioConnections*** BioConnections are questions found in selected figure legends in each chapter that help students make connections between biological concepts. BioConnections help students understand that their study of biology involves linking concepts together and building on previously learned information. Answers to the BioConnections are found in Appendix B.

## Conceptual Questions

1. What is the difference between inducible and repressible operons?
2. Transcriptional regulation often involves a regulatory protein that binds to a segment of DNA and a small effector molecule that binds to the regulatory protein. Do the following terms apply to a regulatory protein, a segment of DNA, or a small effector molecule? (a) repressor, (b) inducer, (c) operator, (d) corepressor, (e) activator
3. **PRINCIPLES** A principle of biology is that the genetic material provides a blueprint for reproduction. Explain how gene regulation is an important mechanism for reproduction and sustaining life.

***Unit openers*** serve two purposes. They allow the student to see the big picture of the unit. In addition, the unit openers draw attention to the principles of biology that will be emphasized in that unit.

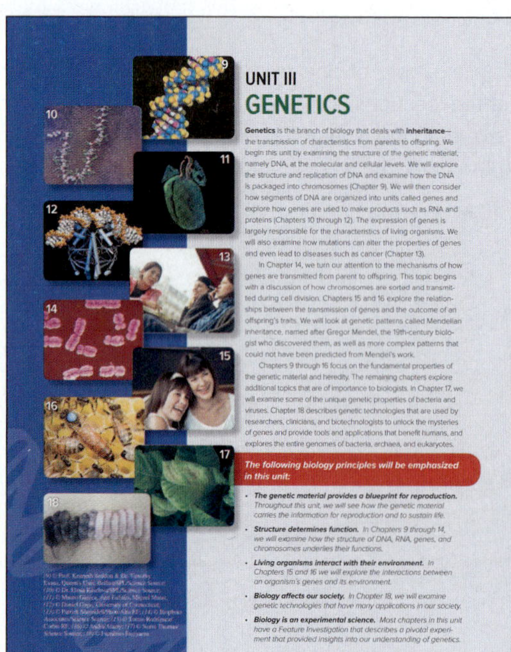


***The following biology principles will be emphasized in this unit:***

- ***The genetic material provides a blueprint for reproduction.*** *Throughout this unit, we will see how the genetic material carries the information for reproduction and to sustain life.*
- ***Structure determines function.*** *In Chapters 9 through 14, we will examine how the structure of DNA, RNA, genes, and chromosomes underlies their functions.*
- ***Living organisms interact with their environment.*** *In Chapters 15 and 16 we will explore the interactions between an organism's genes and its environment.*
- ***Biology affects our society.*** *In Chapter 18, we will examine genetic technologies that have many applications in our society.*
- ***Biology is an experimental science.*** *Most chapters in this unit have a Feature Investigation that describes a pivotal experiment that provided insights into our understanding of genetics.*

# Strengthen Problem Solving Skills and Key Concept Development with Connect®

## Problem Solving Skills

### Detailed Feedback in Connect®

Learning is a process of iterative development, of making mistakes, reflecting, and adjusting over time. The question and test banks in Connect® for *Principles of Biology,* Second Edition, are more than direct assessments; they are self-contained learning experiences that systematically build student learning over time.

For many students, choosing the right answer is not necessarily based on applying content correctly; it is more a matter of increasing their statistical odds of guessing. A major fault with this approach is students don't learn how to process the questions correctly, mostly because they are repeating and reinforcing their mistakes rather than reflecting and learning from them. To help students develop problem solving skills, all higher level Blooms questions in Connect are supported with hints, to help students focus on important information for answering the question. After submitting an answer, the student is given detailed feedback that walks through the problem solving process, using Socratic questions in a decision tree-style framework to scaffold learning. Each step models and reinforces the learning process.

The feedback for each higher level Blooms question (Apply, Analyze, Evaluate) follows a similar process: Clarify Question, Gather Content, Choose Answer, Reflect on Process.

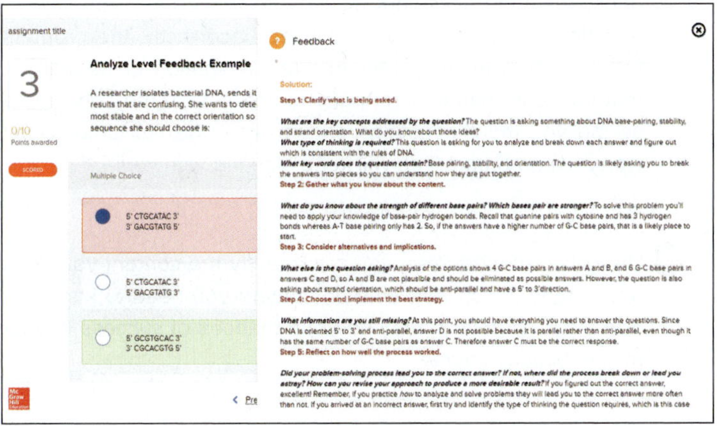


## Key Concept Development

### SmartBook with Learning Resources

To help students understand key concepts, SmartBook® for *Principles of Biology,* Second Edition, is enhanced with Learning Resources. Based on student usage data, derived from thousands of SmartBook users of the first edition, concepts that proved more challenging for students are supported with Learning Resources to enhance the textbook presentation. Learning Resources, such as animations and tutorials, are indicated in Smartbook adjacent to the textbook content. If a student is struggling with a concept based on his/her performance of the SmartBook questions, the student is given an option to review the Learning Resource or the student can click on the Learning Resources at any time.

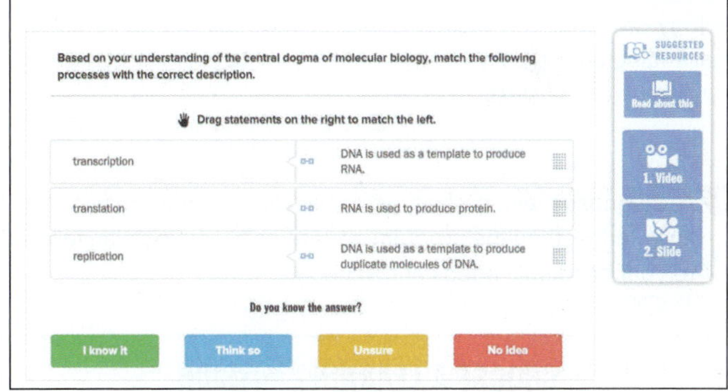


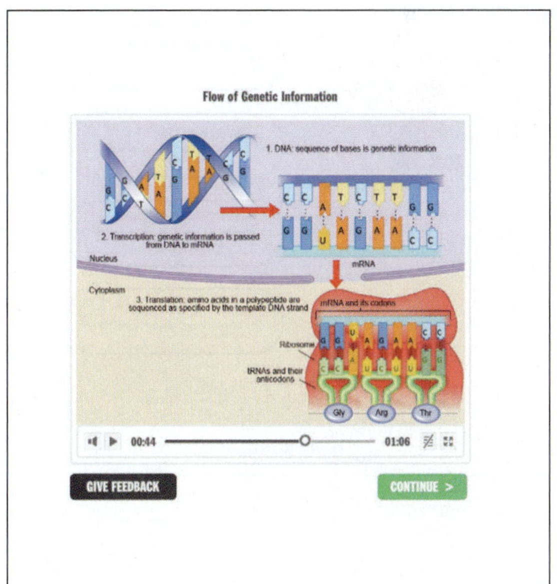

# Using Connect® and *Principles of Biology*, Second Edition

*Principles of Biology,* Second Edition, and its online assets have been carefully crafted to help professors and students work efficiently and effectively through the material in the course, making the most of instructional and study time.

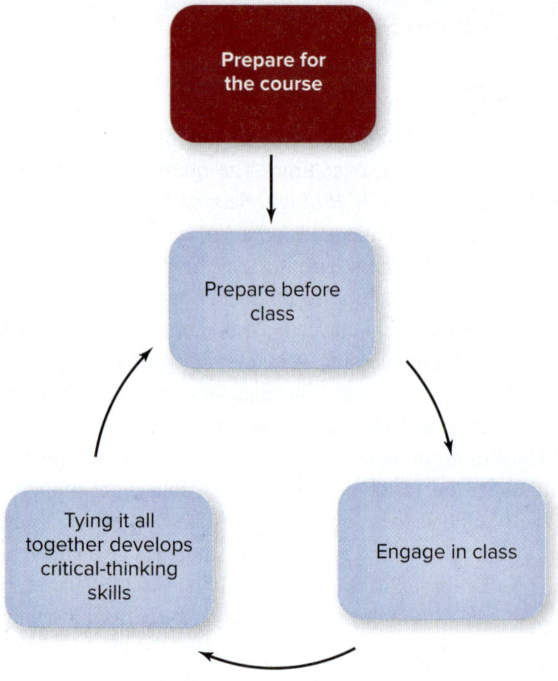


## Prepare for the Course

Many biology students struggle the first few weeks of class. Many institutions expect students to start majors biology having a working knowledge of basic chemistry and cellular biology. *LearnSmart Prep* is now available in Connect. Professors can assign modules in LearnSmart Prep to help students get up to speed on core concepts, or students can access LearnSmart Prep directly through the LearnSmart Prep link.

Mc Graw Hill Education | LEARNSMART PREP®

*LearnSmart Prep* is an adaptive learning tool designed to increase student success and aid retention through the first few weeks of class. Using this digital tool, majors biology students can master some of the most fundamental and challenging principles of biology before they begin to struggle in the first few weeks of class.

1. A diagnostic establishes your baseline comprehension and knowledge; then the program generates a learning plan tailored to your academic needs and schedule.

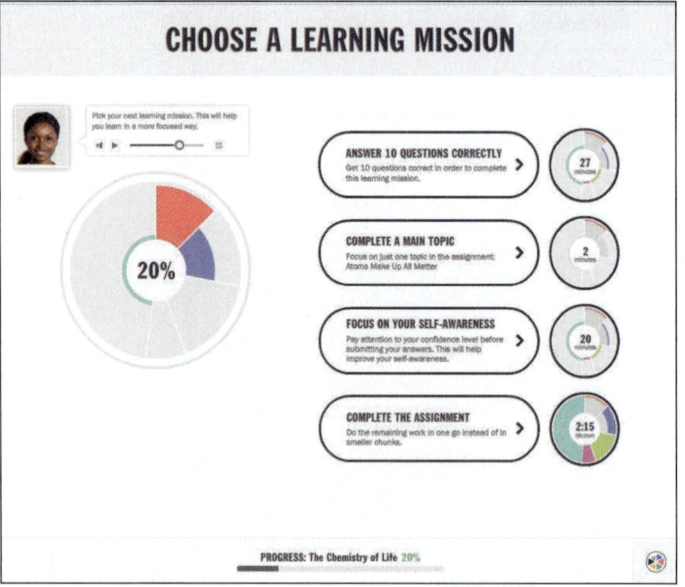


2. As you work through the learning plan, the program asks you questions and tracks your mastery of concepts. If you answer questions about a particular concept incorrectly, the program will provide a learning resource (ex. animation or tutorial) on that concept, then ensure that you understand the concept by asking you more questions. Didn't get it the first time? Don't worry—*LearnSmart Prep* will keep working with you!

3. Using *LearnSmart Prep*, you can identify the content you don't understand, focus your time on content you need to know but don't, and therefore improve your chances of success in your majors biology course.

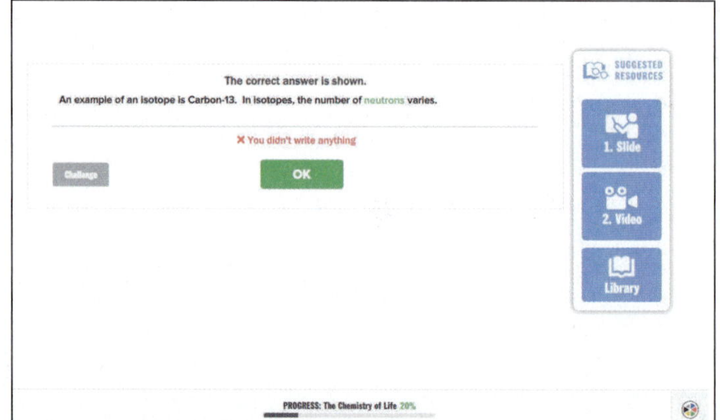

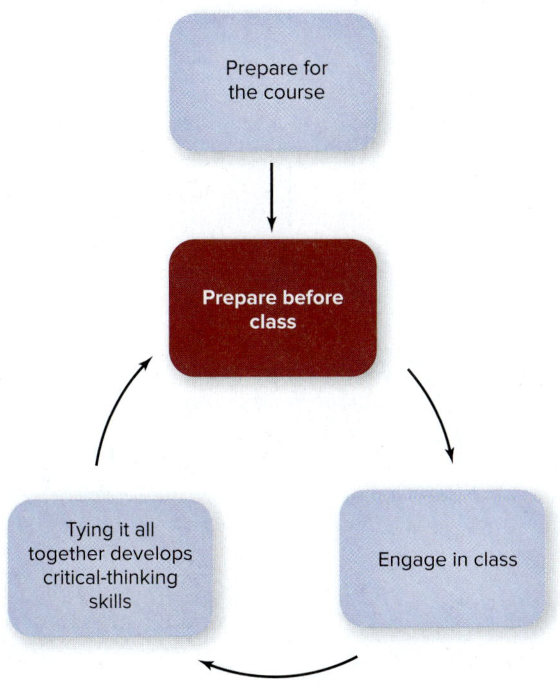


## Prepare Before Class

Students who are most successful in college are those who have developed effective study skills and who use those skills before, during, and after class.

Students can maximize time in class by previewing the material before stepping into the lecture hall. *Principles of Biology*, Second Edition, is available in two formats: the printed text as well as the online SmartBook. Students can use either of these options to preview the material before lecture. Becoming familiar with terminology and basic concepts will allow students to follow along in class and engage in the content in a way that allows for better retention.

Professors can help students prepare for class by making preclass assignments. SmartBook assignments are effective for introducing terminology and general concepts.

SmartBook provides a personalized, adaptive reading experience.

Powered by an intelligent diagnostic and adaptive engine, **SmartBook** facilitates the reading process by identifying what content a student knows and doesn't know through adaptive assessments.

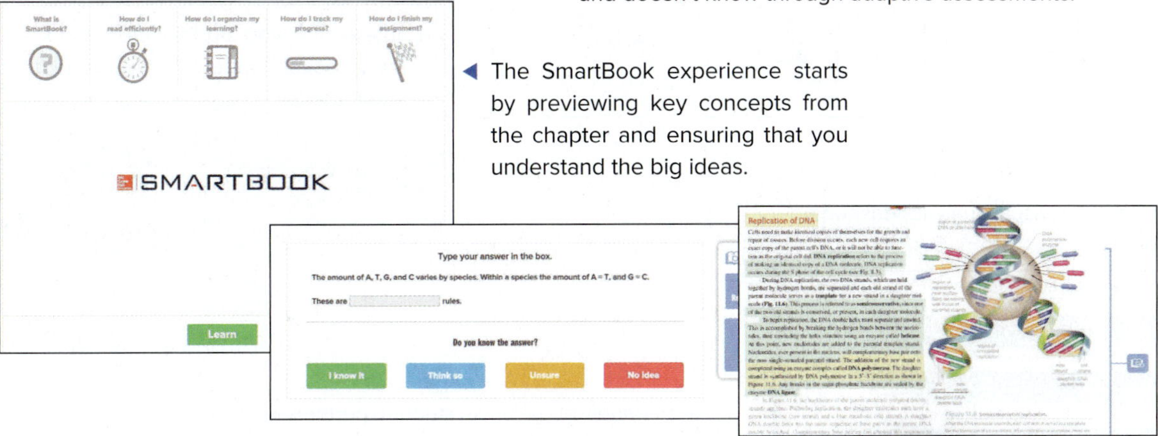


◀ The SmartBook experience starts by previewing key concepts from the chapter and ensuring that you understand the big ideas.

▲ SmartBook asks you questions that identify gaps in your knowledge. The reading experience then continuously adapts in response to the assessments—highlighting the material you need to review based on what you don't know.

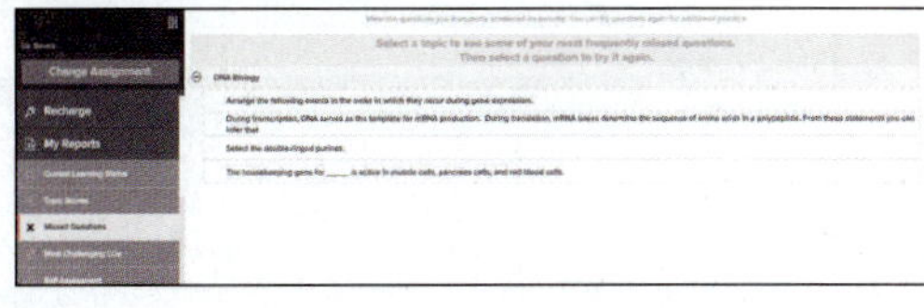

▲ The reports in SmartBook help identify topics where you need more work.

## Engage in Class

**McGraw-Hill Connect**® provides online presentation, assignment, and assessment solutions. It connects students with the tools and resources they'll need to achieve success. A robust set of questions and activities is presented in the Question Bank and a separate set of questions to use for exams is presented in the Test Bank. Instructors can edit existing questions and author entirely new problems. They can track individual student performance—by question, assignment, or in relation to the class overall—with detailed grade reports.

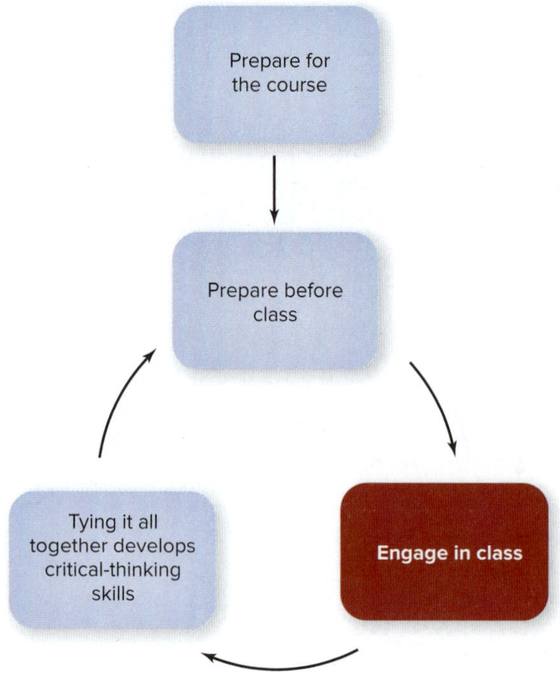


1 Preclass assignments to help students engage in the content during class.

Assignments are accessed through Connect and could include homework assignments, quizzes, SmartBook assignments, and other resources.

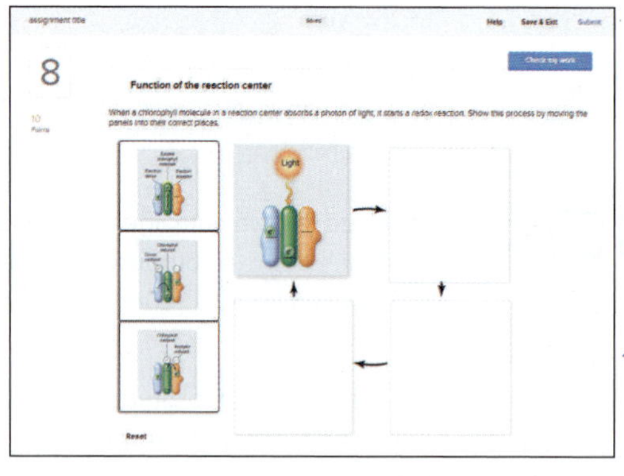


Interactive and traditional questions help assess students' knowledge of the material.

2 Connect Insight is Connect's visual analytics dashboard for instructors and students.

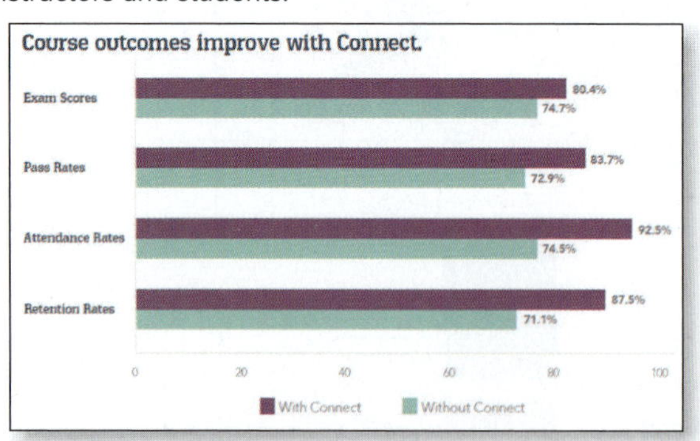


Provides at-a-glance student performance on assignments. Instructors can use the information for a just-in-time approach to teaching.

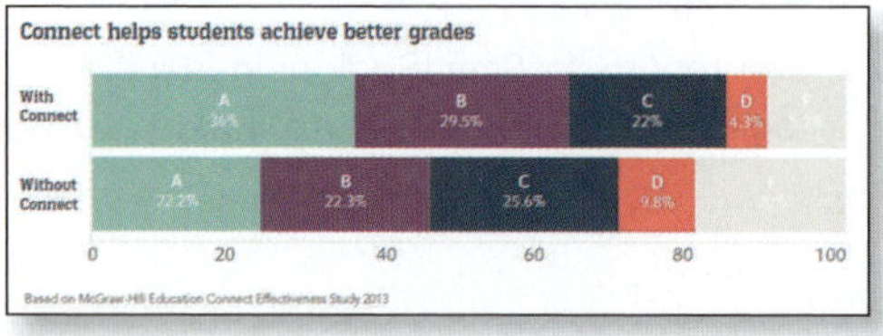


Presents data that empower students to improve performance that is efficient and effective.

## Tying It All Together

Follow up class with assessment that helps students develop critical-thinking skills. Set up assignments from the various assessment banks in Connect.

The Question and Test Banks contain higher order critical-thinking questions that require students to demonstrate a more in-depth understanding of the concepts—instructors can quickly and easily filter the banks for these questions using higher level Blooms tags.

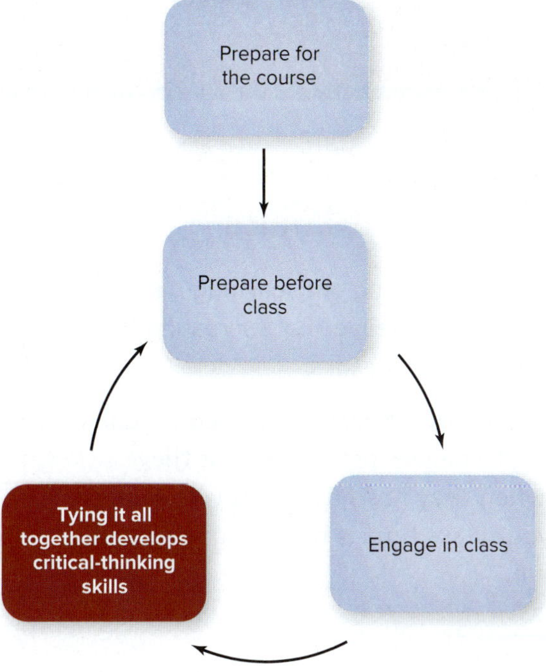


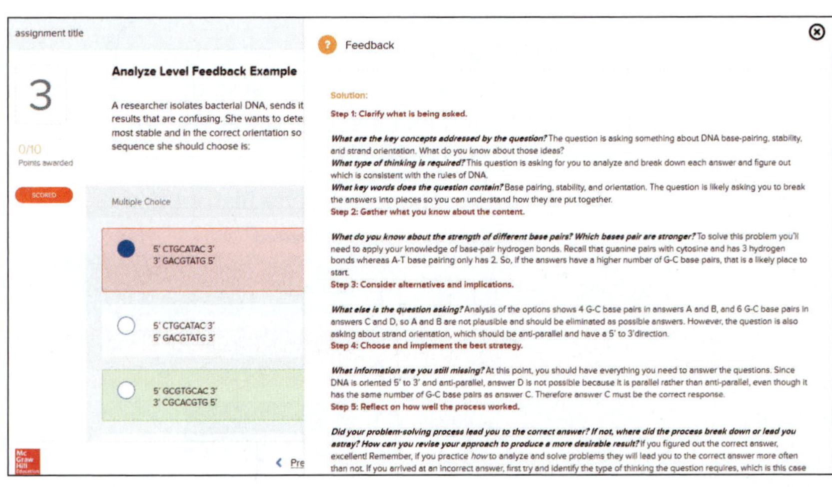


◀ **Detailed Feedback** All higher level Blooms questions that involve problem solving contain detailed feedback in Connect. The feedback walks students through the steps of the problem-solving process and helps them evaluate their scientific-thinking skills.

Many chapters also contain a **Quantitative Question Bank**. ▶ These are more challenging algorithmic questions, intended to help your students practice their quantitative reasoning skills. Hints and guided solution options step students through a problem.

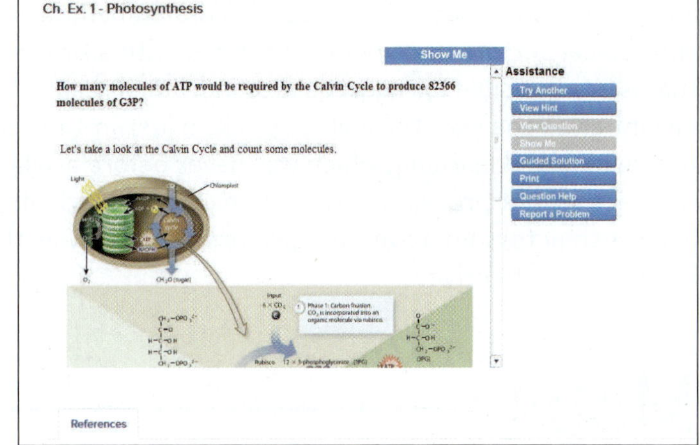

Required=Results

©Getty Images/iStockphoto

## McGraw-Hill Connect®
## Learn Without Limits

Connect is a teaching and learning platform that is proven to deliver better results for students and instructors.

Connect empowers students by continually adapting to deliver precisely what they need, when they need it, and how they need it, so your class time is more engaging and effective.

73% of instructors who use **Connect** require it; instructor satisfaction **increases** by 28% when **Connect** is required.

### Connect's Impact on Retention Rates, Pass Rates, and Average Exam Scores

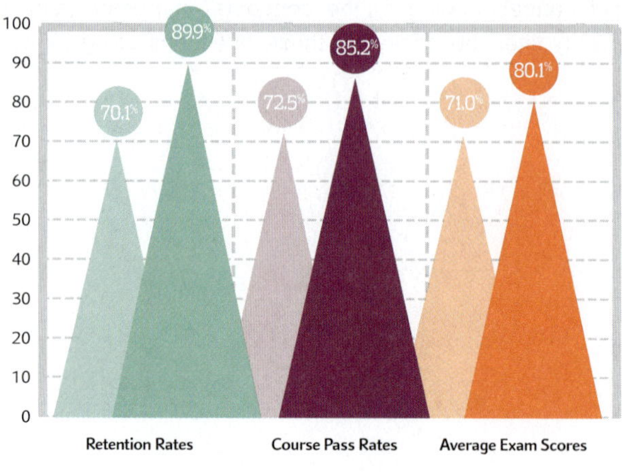


Using **Connect** improves retention rates by **19.8%**, passing rates by **12.7%**, and exam scores by **9.1%.**

# Analytics

### Impact on Final Course Grade Distribution

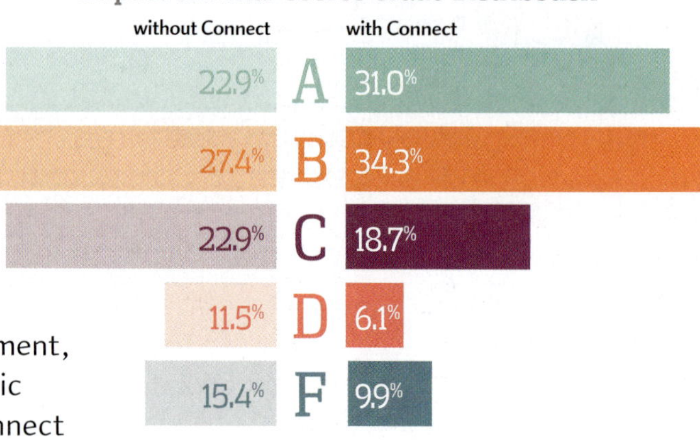


## Connect Insight®

Connect Insight is Connect's new one-of-a-kind visual analytics dashboard—now available for both instructors and students—that provides at-a-glance information regarding student performance, which is immediately actionable. By presenting assignment, assessment, and topical performance results together with a time metric that is easily visible for aggregate or individual results, Connect Insight gives the user the ability to take a just-in-time approach to teaching and learning, which was never before available. Connect Insight presents data that empowers students and helps instructors improve class performance in a way that is efficient and effective.

Students can view their results for any **Connect** course.

# Mobile

Connect's new, intuitive mobile interface gives students and instructors flexible and convenient, anytime–anywhere access to all components of the Connect platform.

# Adaptive

THE **ADAPTIVE READING EXPERIENCE** DESIGNED TO TRANSFORM THE WAY STUDENTS READ

More students earn **A's** and **B's** when they use McGraw-Hill Education **Adaptive** products.

## SmartBook®

Proven to help students improve grades and study more efficiently, SmartBook contains the same content within the print book, but actively tailors that content to the needs of the individual. SmartBook's adaptive technology provides precise, personalized instruction on what the student should do next, guiding the student to master and remember key concepts, targeting gaps in knowledge and offering customized feedback, and driving the student toward comprehension and retention of the subject matter. Available on tablets, SmartBook puts learning at the student's fingertips—anywhere, anytime.

Over **8 billion questions** have been answered, making McGraw-Hill Education products more intelligent, reliable, and precise.

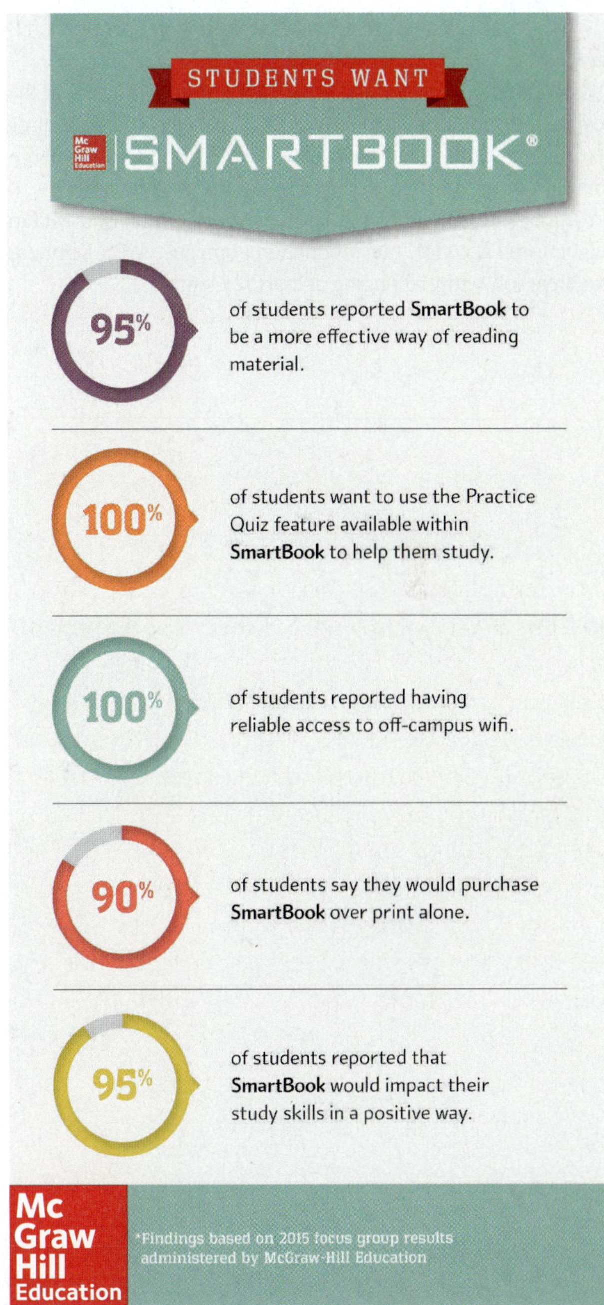


www.mheducation.com

# Acknowledgments

The lives of most science-textbook authors do not revolve around an analysis of writing techniques. Instead, we are people who understand science and are inspired by it, and we want to communicate that information to our students. Simply put, we need a lot of help to get it right.

Editors are a key component that help the authors modify the content of their book so it is logical, easy to read, and inspiring. The editorial team for *Principles of Biology*, Second Edition, has been a catalyst that kept this project rolling. The members played various roles in the editorial process. Justin Wyatt, Brand Manager for Majors Biology, did an outstanding job of overseeing the development of this new text. Elizabeth Sievers, Lead Product Developer, has been the master organizer. Liz's success at keeping us on schedule is greatly appreciated.

We would also like to acknowledge our copy editor, Deb DeBord, for keeping our grammar on track.

Another important aspect of the editorial process is the actual design, presentation, and layout of materials. It's confusing if the text and art aren't near each other or if a figure is too large or too small. We are indebted to the tireless efforts of Jayne Klein, Content Project Manager, and David Hash, Senior Designer, at McGraw-Hill Education. Likewise, our production company, MPS Limited, did an excellent job with the paging and art revisions.

We would like to acknowledge the ongoing efforts of the superb marketing staff at McGraw-Hill Education. Special thanks to Patrick Reidy, Executive Marketing Manager, Life Sciences, for his ideas and enthusiasm for this book.

Other staff members at McGraw-Hill Education have ensured that the authors and editors were provided with adequate resources to achieve the goal of producing a superior textbook. These include Scott Virkler, Senior Vice President, Products & Markets; Marty Lange, Vice President, General Manager, Products & Markets; Lynn Breithaupt, Managing Director for Life Science.

We would like to thank the digital authors and subject matter experts who helped in the development of the digital assets in Connect that support *Principles of Biology*, Second Edition. Ian Quitadamo, Central Washington University, serves as the Lead Digital Author for Majors Biology, directing the development plan for the digital content. Cynthia Dadmun serves as the Lead for the Connect digital content team, and Johnny El-Rady, University of South Florida, serves as the Lead for the SmartBook development team.

*The authors are grateful for the help, support, and patience of their families, friends, and students: Deb, Dan, Nate, and Sarah Brooker; Maria, Rick, and Carrie Widmaier; Jim, Michael, and Melissa Graham; and Jacqui, Zoe, Leah, and Jenna Stiling.*

# CONTENTS



## UNIT I Chemistry

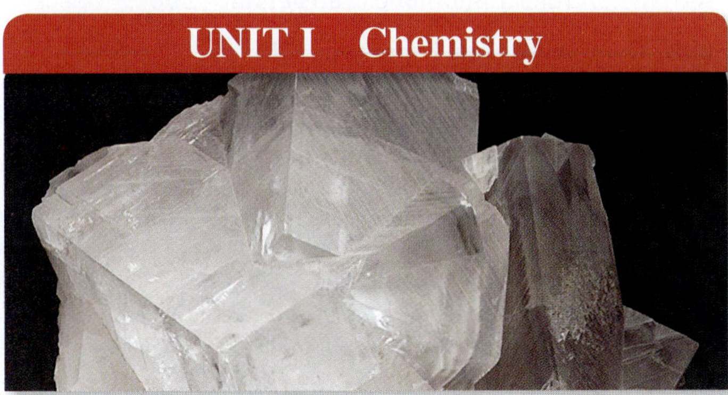

© Dr. Parvinder Sethi



## UNIT II Cells

© Dennis Kunkel Microscopy, Inc./Phototake

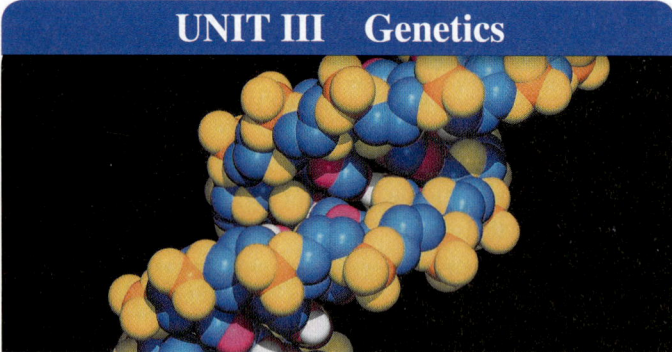

© Mark Dadswell/Getty Images

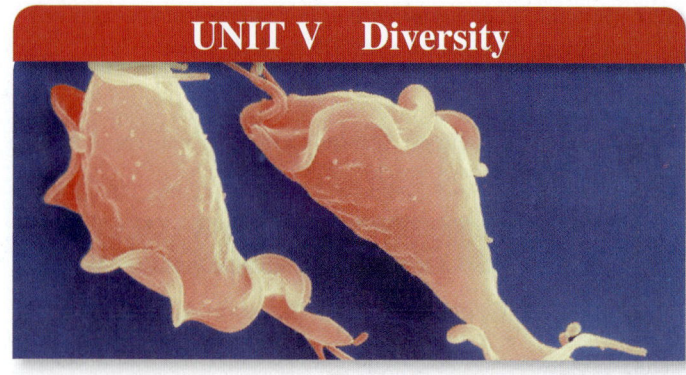

© John Swithinbank/agefotostock

© John Rowley/Getty Images RF

## CHAPTER 32

### General Features of Animal Bodies, and Homeostasis as a Key Principle of Animal Biology 671



## CHAPTER 33

### Neuroscience I: The Structure, Function, and Evolution of Nervous Systems 692



## CHAPTER 34

### Neuroscience II: How Sensory Systems Allow Animals to Interact with the Environment 721



## CHAPTER 35

### How Muscles and Skeletons Are Adaptations for Movement, Support, and Protection 742



## CHAPTER 36

### Circulatory and Respiratory Systems: Transporting Solutes and Exchanging Gases 757



## CHAPTER 37

### Digestive and Excretory Systems Help Maintain Nutrient, Water, and Energy Balance and Remove Waste Products from Animal Bodies 790

# UNIT VIII Ecology

# 1 An Introduction to Biology

© micheldenijs/Getty Images RF

**The polar bear (*Ursus maritimus*).** The polar bear exhibits many characteristics that are adaptations to living in the cold Arctic climate and hunting seals.

## Chapter Outline



**Biology** is the study of life. The diverse forms of life found on Earth provide biologists with an amazing array of organisms to study. In many cases, the investigation of living things leads to discoveries with far-reaching benefits. Certain ancient civilizations, such as the Greeks, Romans, and Egyptians, discovered that the bark of the white willow tree (*Salix alba*) could be used to fight fever. Chemists determined that willow bark contains a substance called salicylic acid, which led to the development of the related compound acetylsalicylic acid, more commonly known as aspirin (Figure 1.1). Today, aspirin is not only taken for fever and pain relief but is also recommended for the prevention of heart attacks and strokes.

As a more recent example, researchers determined that the venom from certain poisonous snakes contains a chemical that lowers blood pressure in humans. By analyzing that chemical, scientists have developed drugs called ACE inhibitors that treat high blood pressure (Figure 1.2).

These are just a couple of the many discoveries that make biology an intriguing discipline. The study of life not only reveals the fascinating characteristics of living species but also leads to the development of medicines and research tools that benefit the lives of people.

To make new discoveries, biologists view life from many different perspectives. What is the composition of living things? How is life organized? How do organisms reproduce? Sometimes the questions posed by biologists are fundamental and even philosophical in nature. How did living organisms originate? Can we live forever? What is the physical basis for memory? Can we save endangered species?

Future biologists will continue to make important advances. Biologists are scientific explorers looking for answers to some of life's most enduring mysteries. Unraveling these mysteries presents an exciting challenge to the best and brightest minds. The rewards of a career in biology include the excitement of forging into uncharted territory, the thrill of making discoveries that can improve the health and lives of people, and the goal of trying to preserve the environment and protect

endangered species. For these and many other compelling reasons, students seeking challenging and rewarding careers may wish to choose biology as a lifelong pursuit.

In this chapter, we will begin our survey of biology by examining the basic principles that underlie the characteristics of all living organisms. We then take a deeper look at the process of evolution and how it explains the unity and diversity that we observe among living and extinct species. Finally, we will explore the general approaches that scientists follow when making new discoveries.

**Figure 1.1 The white willow (*Salix alba*) and aspirin.** Modern aspirin, acetylsalicylic acid, was developed after analysis of a chemical found in the bark of the white willow tree.


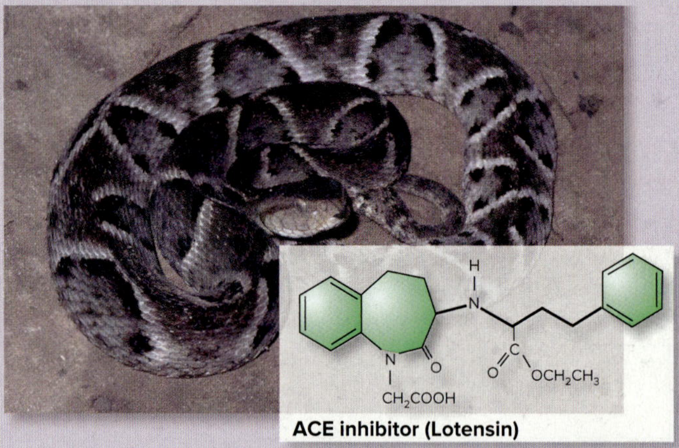


**Figure 1.2 The Brazilian arrowhead viper and inhibitors of high blood pressure.** Originally found in the venom of the Brazilian arrowhead viper (*Bothrops jararaca/jararacussa*), angiotensin-converting enzyme (ACE) inhibitors, including benazepril (Lotensin, chemical structure shown here), are commonly used to treat high blood pressure.


## 1.1 Principles of Biology and the Levels of Biological Organization

### Learning Outcomes

1. Describe the principles of biology.
2. Explain how life can be viewed at different levels of biological complexity.

Because biology is the study of life, a good way to begin a biology textbook is to distinguish living organisms from nonliving objects. At first, the distinction might seem obvious. A person is alive, but a rock is not. However, the distinction between living and nonliving may seem less obvious when we consider microscopic entities. Is a bacterium alive? What about a virus or a chromosome? In this section, we will examine the principles that underlie the characteristics of all forms of life and explore other broad principles in biology. We will then consider the levels of organization that biologists study, ranging from atoms and small molecules to very large geographical areas.

### The Study of Life Has Revealed a Set of Unifying Principles

In the course of studying a vast number of species, biologists have learned that a set of principles applies to all fields of biology. Twelve broad principles are described in Figure 1.3. The first eight principles are often used as criteria for defining the basic features of life. You will see these 12 principles at many points as you progress through this textbook. In particular, we will draw your attention to them at the beginning of each unit, and we will refer to them within particular figures in Chapters 2 through 47. It should be noted that the principles of biology are also governed by the laws of chemistry and physics, which are discussed in Chapters 2, 3, and 6.

**Principle 1: Cells are the simplest units of life.** The term **organism** can be applied to all living things. Organisms maintain an internal order that is separated from the environment (Figure 1.3a). The simplest unit of such organization is the cell, which we will examine in Unit II. One of the foundations of biology is the **cell theory,** which states that (1) all organisms are composed of one or more cells, (2) cells are the smallest units of life, and (3) new cells come from pre-existing cells by cell division. Unicellular organisms are composed of one cell, whereas multicellular organisms such as plants and animals contain many cells. In plants and animals, each cell has an internal order, and the cells within the organism have specific arrangements and functions.

**Principle 2: Living organisms use energy.** The maintenance of organization requires energy. Therefore, all living organisms acquire energy from the environment and use that energy to maintain their internal order. Cells carry out a variety of chemical reactions that are responsible for the breakdown of energy-yielding nutrients. Such

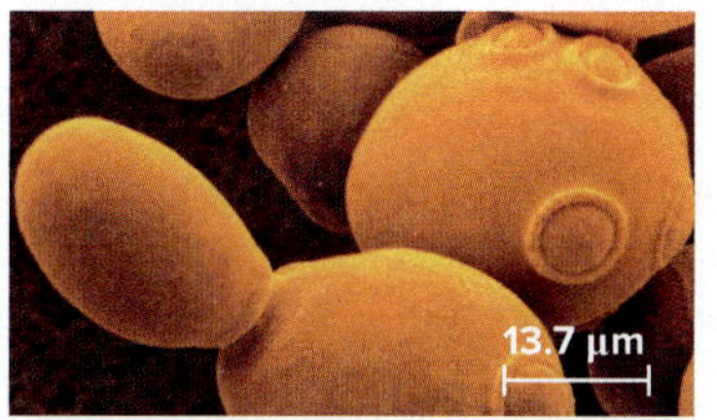


**(a) Cells are the simplest units of life:** Organisms maintain an internal order. The simplest unit of organization is the cell. Yeast cells are shown here.

**(b) Living organisms use energy:** Organisms need energy to maintain internal order. These algae harness light energy via photosynthesis. Energy is used in chemical reactions collectively known as metabolism.

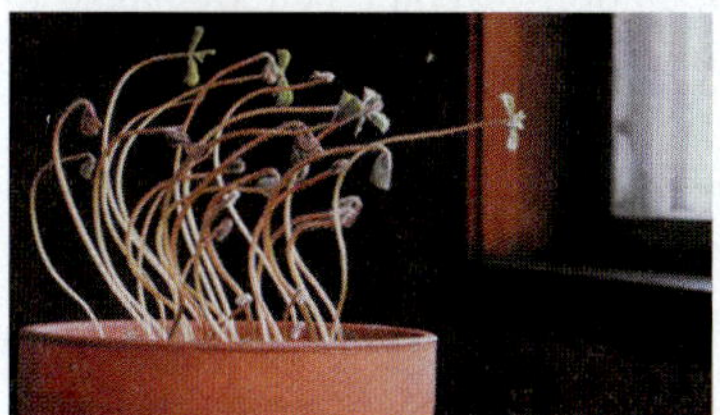

**(c) Living organisms interact with their environment:** Organisms respond to environmental changes. These plants are growing toward the light.

**(d) Living organisms maintain homeostasis:** Organisms regulate their cells and bodies, maintaining relatively stable internal conditions, a process called homeostasis. For example, this bird maintains its internal body temperature on a cold day.

**(e) Living organisms grow and develop:** Growth produces more or larger cells, whereas development produces organisms with a defined set of characteristics.

**(f) The genetic material provides a blueprint for reproduction:** To sustain life over many generations, organisms must reproduce. Due to the transmission of genetic material, offspring tend to have traits like their parents.

**(g) Populations of organisms evolve from one generation to the next:** Populations of organisms change over the course of many generations. Evolution results in traits that promote survival and reproductive success.

**(h) All species (past and present) are related by an evolutionary history:** The three mammal species shown here share a common ancestor, which was also a mammal.

**(i) Structure determines function:** In the example seen here, webbed feet (on ducks) function as paddles for swimming. Nonwebbed feet (on chickens) function better for walking on the ground.

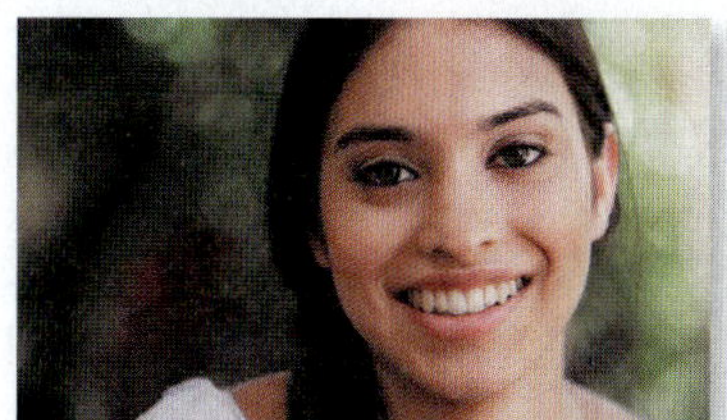

**(j) New properties of life emerge from complex interactions:** Our ability to see is an emergent property due to interactions among many types of cells in the eye and neurons that send signals to the brain.

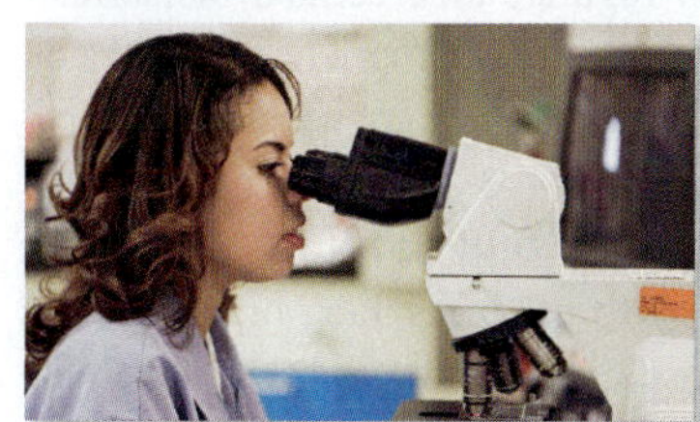

**(k) Biology is an experimental science:** The discoveries of biology are made via experimentation, which leads to theories and biological principles.

**(l) Biology affects our society:** Many discoveries in biology have had major effects on our society. For example, biologists developed Bt-corn, which is resistant to insect pests and is widely planted by farmers.

**Figure 1.3 Twelve principles of biology.** The first eight principles are often used as criteria for defining the basic features of life. Note: The 12 principles described here were modeled after the themes and core competencies described in *Vision and Change in Undergraduate Biology,* a report that was published in 2009 and organized by the American Association for the Advancement of Science. *Vision and Change* proposed five themes. We have divided them into 10 principles to make them more accessible to beginning biology students. The five *Vision and Change* themes are related to our principles in the following manner: (1) evolution (principles g and h); (2) structure and function (principle i); (3) information flow, exchange, and storage (principles e and f); (4) pathways and transformations of energy and matter (principles b, c, and d); (5) systems (principles a and j). The last two principles are modeled after two core competencies described in *Vision and Change*: ability to apply the process of science (principle k) and ability to understand the relationship between science and society (principle l).



**BioConnections:** *Look ahead to Figure 4.15. Which of these principles is this figure emphasizing?*

reactions often release energy in a process called **cellular respiration.** The energy may be used to synthesize the components that make up individual cells and living organisms. Chemical reactions involved with the breakdown and synthesis of cellular molecules are collectively known as **metabolism.** Plants, algae, and certain bacteria directly harness light energy to produce their own nutrients in the process of **photosynthesis** (Figure 1.3b). They are the primary producers of food on Earth. In contrast, some organisms, such as animals and fungi, are consumers—they must use other organisms as food to obtain energy.

**Principle 3: Living organisms interact with their environment.** To survive, living organisms must interact with their environment, which includes other organisms they may encounter. All organisms must respond to environmental changes. In the winter, many species of mammals develop a thicker coat of fur that protects them from the cold temperatures. Plants respond to changes in the angle of the Sun. If you place a plant in a window, it will grow toward the light (Figure 1.3c).

**Principle 4: Living organisms maintain homeostasis.** Although life is a dynamic process, living cells and organisms regulate their cells and bodies to maintain relatively stable internal conditions, a process called **homeostasis** (from the Greek, meaning to stay the same). The degree to which homeostasis is achieved varies among different organisms. For example, most mammals and birds maintain a relatively stable body temperature in spite of changing environmental temperatures (Figure 1.3d), whereas reptiles and amphibians tolerate a wider fluctuation in body temperature. By comparison, all organisms continually regulate their cellular metabolism so that nutrient molecules are used at an appropriate rate and new cellular components are synthesized when they are needed.

**Principle 5: Living organisms grow and develop.** All living organisms grow and develop. **Growth** produces more or larger cells, which usually results in an increase in size and weight. Multicellular organisms, such as plants and animals, begin life at a single-cell stage (for example, a fertilized egg) and then undergo multiple cell divisions to develop into a complete organism with many cells. Among unicellular organisms such as bacteria, new cells are relatively small, and they increase in volume by the synthesis of additional cellular components. **Development** is a series of changes in the state of a cell, a tissue, an organ, or an organism, eventually resulting in organisms with a defined set of characteristics (Figure 1.3e).

**Principle 6: The genetic material provides a blueprint for reproduction.** All living organisms have a finite life span. To sustain life, organisms must **reproduce,** or generate offspring (Figure 1.3f). A key feature of reproduction is that offspring tend to have characteristics that greatly resemble those of their parent(s). How is this possible? All living organisms contain genetic material composed of **deoxyribonucleic acid (DNA),** which provides a blueprint for the organization, development, and function of living things. During reproduction, a copy of this blueprint is transmitted from parent to offspring. DNA is **heritable,** which means that offspring inherit DNA from their parents.

As discussed in Unit III, **genes,** which are segments of DNA, govern the characteristics, or traits, of organisms. Most genes are transcribed into a type of **RNA (ribonucleic acid)** molecule called messenger RNA (mRNA) that is then translated into a **polypeptide** with a specific amino acid sequence. A **protein** is composed of one or more polypeptides. The structures and functions of proteins are largely responsible for the traits of living organisms.

**Principle 7: Populations of organisms evolve from one generation to the next.** The first six characteristics of life, which we have just considered, apply to individual organisms over the short run. Over the long run, another universal characteristic of life is **biological evolution,** or simply **evolution,** which refers to a heritable change in a population of organisms from generation to generation. As a result of evolution, populations become better adapted to the environment in which they live. For example, the long snout of an anteater is an adaptation that enhances its ability to obtain food, namely ants, from hard-to-reach places (Figure 1.3g). Over the course of many generations, the fossil record indicates that the long snout occurred via biological evolution in which modern anteaters evolved from populations of organisms with shorter snouts.

In many chapters of this textbook, you will find a subsection called Evolutionary Connections, which focuses on the evolutionary aspects of the chapter's material.

**Principle 8: All species (past and present) are related by an evolutionary history.** Principle 7 considers evolution as an ongoing process that happens from one generation to the next. Evidence from a variety of sources, including the fossil record and DNA sequences, also indicates that all organisms on Earth share a common ancestry. For example, the three species of mammals shown in Figure 1.3h shared a common ancestor in the past, which was also a mammal. We will discuss evolutionary relationships further in Section 1.2.

**Principle 9: Structure determines function.** In addition to the preceding eight characteristics of life, biologists have identified other principles that are important in all fields of biology. The principle that structure determines function pertains to very tiny biological molecules as well as very large biological structures. For example, at the microscopic level, a cellular protein called actin naturally assembles into structures that are long filaments. The function of these filaments is to provide support and shape to cells. At the macroscopic level, let's consider the feet of different birds (Figure 1.3i). Aquatic birds have webbed feet that function as paddles for swimming. By comparison, the feet of nonaquatic birds are not webbed and are better adapted for grasping food, perching on branches, and running along the ground. In this case, the structure of a bird's feet, webbed versus nonwebbed, is a critical feature that affects their function.

**Principle 10: New properties of life emerge from complex interactions.** In biology, when individual components in an organism interact with each other or with the external environment to create novel structures and functions, the resulting characteristics are called **emergent properties.** For example, the human eye is composed of many different types of cells that are organized to sense incoming light and transmit signals to the brain (Figure 1.3j). Our ability to see is an emergent property of this complex arrangement of different cell types.

**Principle 11: Biology is as an experimental science.** Biology is an inquiry process. Biologists are curious about the characteristics of living organisms and ask questions about those characteristics. For example, a cell biologist may wonder why a cell produces a specific protein when it is confronted with high temperature. An ecologist may ask herself why a particular bird eats insects in the summer and seeds in the winter. To answer such questions, biologists typically gather additional information and ultimately form a hypothesis, which is a proposed explanation for a natural phenomenon. The next stage is to design one or more experiments to test the validity of a hypothesis (Figure 1.3k).

Like evolution, experimentation is such a key aspect of biology that many chapters of this textbook include a Feature Investigation—an actual research study that showcases the experimental approach.

**Principle 12: Biology affects our society.** The influence of biology is not confined to textbooks and classrooms. The work of biologists has far-reaching effects in our society. For example, biologists have discovered drugs that are used to treat many different human diseases. Likewise, biologists have created technologies that have many uses. Examples include the use of microorganisms to make medical products, such as human insulin, and the genetic engineering of crops to make them resistant to particular types of insect pests (Figure 1.3l).

## Living Organisms Are Studied at Different Levels of Organization

The organization of living organisms can be analyzed at different levels of biological complexity, starting with the smallest level of organization and progressing to levels that are physically much larger and more complex. Figure 1.4 depicts a scientist's view of the levels of biological organization.

1. **Atoms:** An **atom** is the smallest unit of an element that has the chemical properties of the element. All matter is composed of atoms.
2. **Molecules and macromolecules:** As discussed in Unit I, atoms bond with each other to form **molecules.** Many smaller molecules bonded together to form a large polymer is called a **macromolecule.** Carbohydrates, proteins, and nucleic acids (DNA and RNA) are important macromolecules found in living organisms.
3. **Cells:** Molecules and macromolecules associate with each other to form larger structures such as cells. A **cell** is surrounded by a membrane and contains a variety of molecules and macromolecules. As noted earlier, a cell is the simplest unit of life.
4. **Tissues:** In the case of multicellular organisms such as plants and animals, many cells of the same type associate with each other to form **tissues.** An example is muscle tissue.
5. **Organs:** In complex multicellular organisms, an **organ** is composed of two or more types of tissue. For example, the heart is composed of several types of tissue, including muscle, nervous, and connective tissue.
6. **Organism:** All living things can be called **organisms.** Biologists classify organisms as belonging to a particular **species,** which is a related group of organisms that share a distinctive form and set of attributes in nature. The members of the same species are closely related genetically. In Units VI and VII, we will examine plants and animals at the level of cells, tissues, organs, and complete organisms.
7. **Population:** A group of organisms of the same species that occupy the same environment is called a **population.**
8. **Community:** A biological **community** is an assemblage of populations of different species that live in the same environment. The types of species found in a community are determined by the environment and by the interactions of species with each other.
9. **Ecosystem:** Researchers may extend their work beyond living organisms and also study the physical environment. Ecologists analyze **ecosystems,** which are formed by interactions between a community of organisms and its physical environment. Unit VIII considers biology from populations to ecosystems.
10. **Biosphere:** The **biosphere** includes all of the places on the Earth where living organisms exist. Life is found in the air, in bodies of water, on the land, and in the soil.

## 1.1 Reviewing the Concepts

- Biology is the study of life. Discoveries in biology help us understand how life exists, and they have many practical applications, such as the development of drugs to treat human diseases (Figures 1.1, 1.2).
- Eight principles underlie the characteristics that are common to all forms of life. All living things (1) are composed of cells as their simplest unit; (2) use energy; (3) interact with their environment; (4) maintain homeostasis; (5) grow and develop; and (6) have genetic material for reproduction. Also, (7) populations of organisms evolve from one generation to the next and (8) are connected by an evolutionary history (Figure 1.3).
- Additional important principles of biology are that (9) structure determines function; (10) new properties emerge from complex interactions; (11) biology is an experimental science; and (12) biology influences our society.
- Living organisms can be viewed at different levels of biological organization: atoms, molecules and macromolecules, cells, tissues, organs, organisms, populations, communities, ecosystems, and the biosphere (Figure 1.4).

## 1.1 Testing Your Knowledge

1. The wing of a bird, the wing of an insect, and the wing of a bat have similar shapes. Which principle of biology does this observation pertain to?
   a. Living organisms use energy.
   b. Living organisms maintain homeostasis.
   c. Structure determines function.
   d. Populations of organisms evolve from one generation to the next.
   e. All of the above are correct.

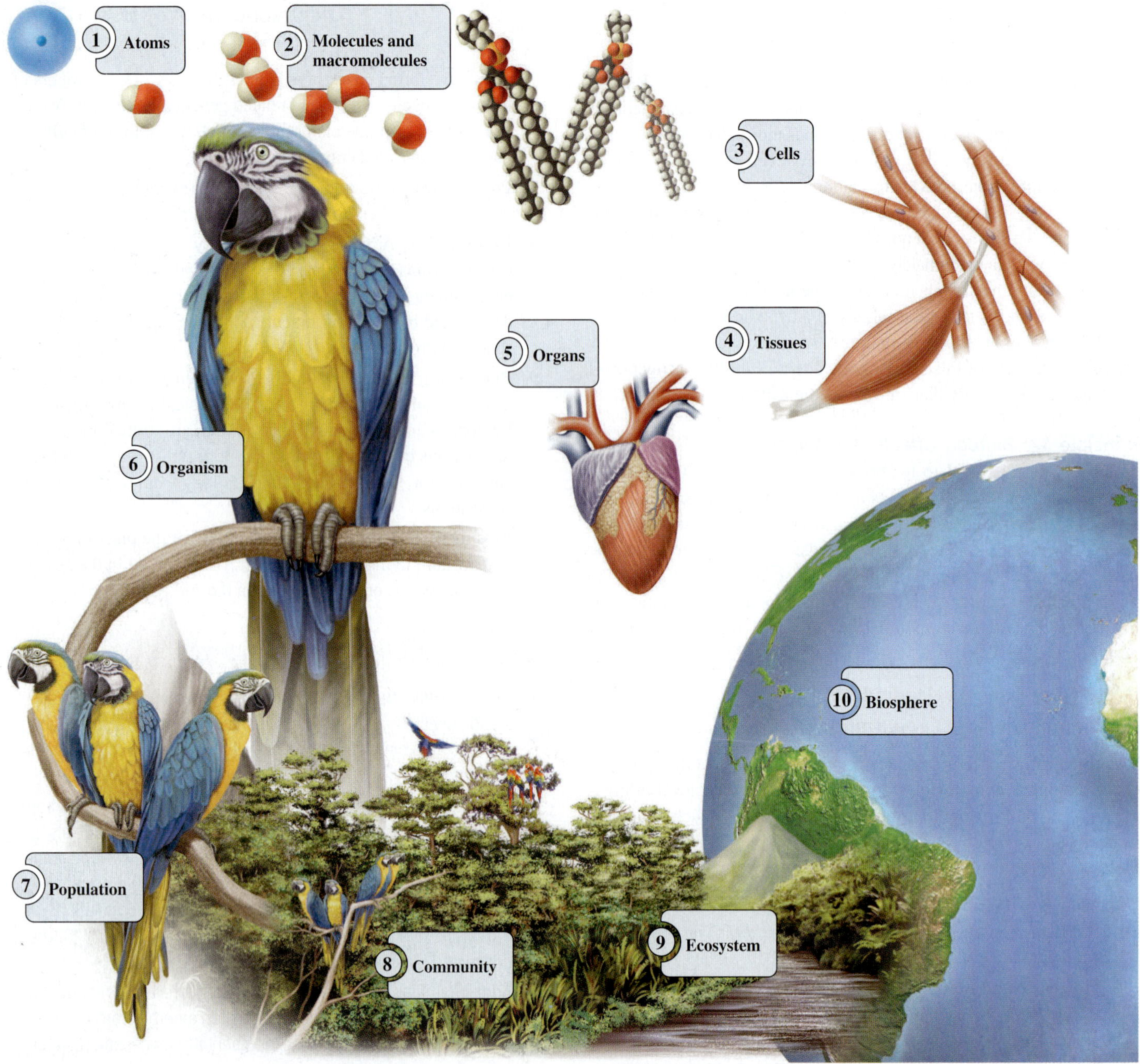


**Figure 1.4** The levels of biological organization.

**Concept Check:** *At which level of biological organization would you place a herd of buffalo?*

2. Which of the following is the most complex level of biological organization?
   a. organism b. tissue c. community d. population

## 1.2 Unity and Diversity of Life

### Learning Outcomes

1. Explain the two basic mechanisms by which evolutionary change occurs: vertical descent with mutation and horizontal gene transfer.
2. Outline how organisms are classified (taxonomy).
3. Describe how evolution accounts for unity and diversity in biology.

Unity and diversity are two words that often are used to describe the living world. As we have seen, all modern forms of life display a common set of characteristics that distinguish them from nonliving objects. In this section, we will explore how this unity of common traits is rooted in the phenomenon of biological evolution. Life on Earth is united by an evolutionary past in which modern organisms have evolved from populations of pre-existing organisms.

Evolutionary unity does not mean that organisms are exactly alike. The Earth has many different types of environments, ranging from tropical rain forests to salty oceans, from hot and dry deserts to cold mountaintops. Diverse forms of life have evolved in ways that help them prosper in the different environments the Earth has to offer. In this section, we will begin to examine the unity and diversity that exist within the biological world.

## Modern Forms of Life Are Connected by an Evolutionary History

Life began on Earth as primitive cells about 3.5–4 billion years ago (bya). Since that time, populations of living organisms have undergone evolutionary changes that ultimately gave rise to the species we see today. Understanding the evolutionary history of species can provide key insights into an organism's structure and function, because evolutionary change involves modifications of characteristics in pre-existing populations. Over long periods of time, populations may change so that structures with a particular function may become modified to serve a new function. For example, the wing of a bat is used for flying, and the flipper of a dolphin is used for swimming. Evidence from the fossil record indicates that both structures were modified from a front limb that was used for walking in a pre-existing ancestor (Figure 1.5).

Evolutionary change occurs by two mechanisms: vertical descent with mutation and horizontal gene transfer. Let's take a brief look at each of these mechanisms.

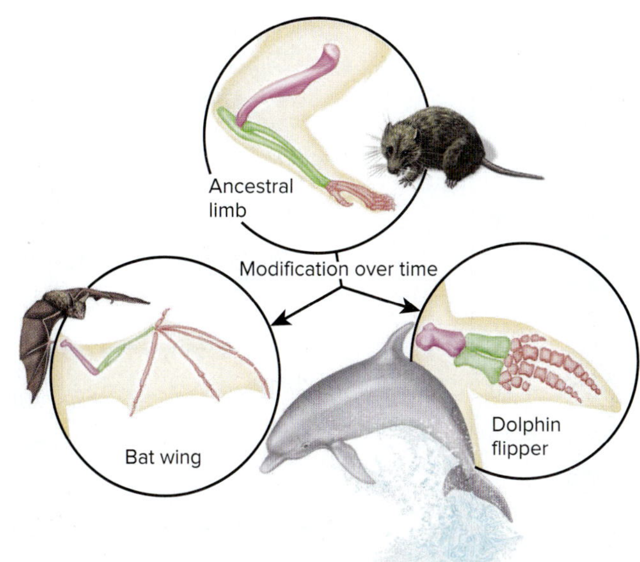


**Figure 1.5 An example showing a modification that has occurred as a result of biological evolution.** The wing of a bat and the flipper of a dolphin are modifications of a limb that was used for walking in a pre-existing ancestor.

**Concept Check:** *Among mammals, give two examples of how the tail has been modified and has different purposes.*

**Vertical Descent with Mutation** The traditional way to study evolution is to examine a progression of changes in a series of ancestors. Such a series is called a **lineage.** Biologists have traditionally depicted such evolutionary change in a diagram like the one shown in Figure 1.6, which shows a portion of the lineage that gave rise to modern horses. In this mechanism of evolution, called **vertical evolution,** new species evolve from pre-existing ones by the accumulation of **mutations,** which are heritable changes in the genetic material of organisms. But why would some mutations accumulate in a population and eventually change the characteristics of an entire species? One reason is that a mutation may alter the traits of organisms in a way that increases their chances of survival and reproduction. When a mutation causes such a beneficial change, the frequency of the mutation may increase in a population from one generation to the next, a process called **natural selection.** This topic is discussed in Units IV and V. Evolution also involves the accumulation of neutral changes that do not benefit or harm a species, and evolution sometimes involves rare changes that may be harmful.

With regard to the horses shown in Figure 1.6, the fossil record has revealed adaptive changes in various traits such as size and tooth morphology. The first horses were the size of dogs, whereas modern horses typically weigh more than a half ton. The teeth of *Hyracotherium* were relatively small compared with those of modern horses. Over the course of millions of years, horse teeth have increased in size, and a complex pattern of ridges has developed on the molars. How do evolutionary biologists explain these changes in horse characteristics? They can be attributed to natural selection in which changing global climates favored the survival and reproduction of horses with certain types of traits. Over North America, where much of horse evolution occurred, large areas changed from dense forests to grasslands. Horses with genetic variation that made them larger were more likely to escape predators and travel greater distances in search of food. The changes seen in horses' teeth are consistent with a dietary shift from eating tender leaves to eating grasses and other vegetation that are more abrasive and require more chewing.

**Horizontal Gene Transfer** The most common way for genes to be transferred is in a vertical manner. This can involve the transfer of genetic material from a mother cell to daughter cells, or it can occur via gametes—sperm and egg—that unite to form a new organism. However, as discussed in Chapter 21, genes are sometimes transferred between organisms by **horizontal gene transfer**—a process in which an organism incorporates genetic material from another organism without being the offspring of that organism. In some cases, horizontal gene transfer can occur between members of different species. For example, you may have heard in the news media that resistance to antibiotics among bacteria is a growing medical problem. This can occur by the transfer of an antibiotic resistance gene from one bacterial species to another via horizontal gene transfer.

Traditionally, biologists have described evolution using diagrams such as that in Figure 1.6, which depict the vertical evolution of species over a long timescale. In this view, all living organisms evolved from a common ancestor, resulting in a "tree of life" that could describe the evolution that gave rise to all modern species. Now that we understand the great importance of horizontal gene transfer in the evolution of life on Earth, biologists have re-evaluated the concept of evolution as it occurs over time. Rather than a tree of life, a more appropriate way to view the unity of living organisms is to describe it as a "web of life" (as discussed in Chapter 21; look ahead to Figure 21.12), which accounts for both vertical evolution and horizontal gene transfer.

## The Classification of Living Organisms Allows Biologists to Appreciate the Unity and Diversity of Life

As biologists discover new species, they try to place them into groups based on their evolutionary history. This is a difficult task because researchers estimate that the Earth has between 5 and 50 million different species! The rationale for categorization is usually based

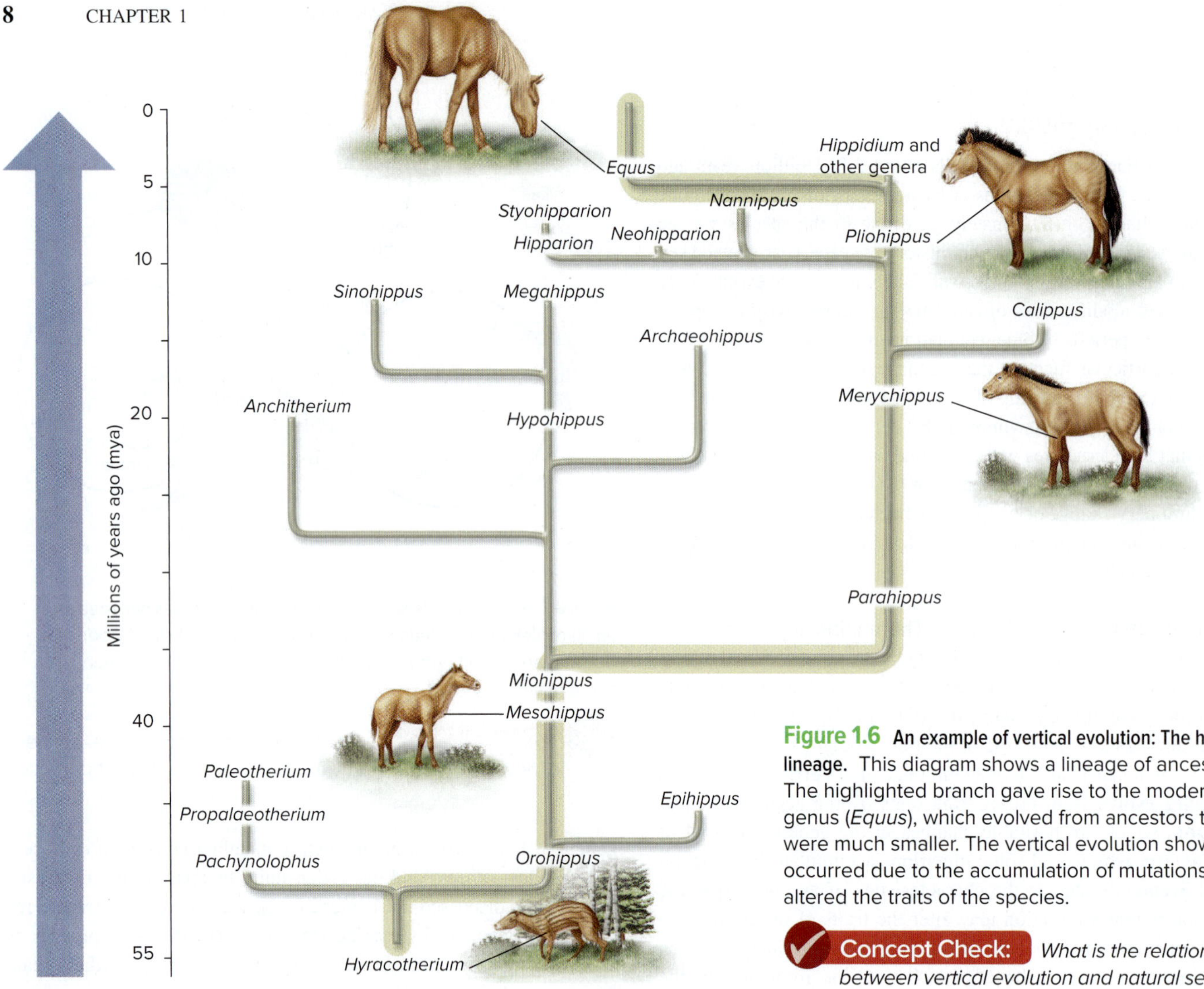

**Figure 1.6 An example of vertical evolution: The horse lineage.** This diagram shows a lineage of ancestors. The highlighted branch gave rise to the modern horse genus (*Equus*), which evolved from ancestors that were much smaller. The vertical evolution shown here occurred due to the accumulation of mutations that altered the traits of the species.

**Concept Check:** *What is the relationship between vertical evolution and natural selection?*

on vertical evolution. Species with a recent common ancestor are grouped together, whereas species whose common ancestor was in the very distant past are placed into different groups. The grouping of species is termed **taxonomy.**

Let's first consider taxonomy on a broad scale. From an evolutionary perspective, all forms of life can be placed into three large categories, or **domains,** called **Bacteria, Archaea,** and **Eukarya** (Figure 1.7). Bacteria and archaea are microorganisms that are also termed **prokaryotic** because their cell structure is relatively simple. At the molecular level, bacterial and archaeal cells show significant differences in their compositions. By comparison, organisms in domain Eukarya are termed **eukaryotic;** they have larger cells with internal compartments that serve various functions. A defining distinction between prokaryotic and eukaryotic cells is that eukaryotic cells have a **nucleus** in which the genetic material is surrounded by a membrane. The organisms in domain Eukarya are divided into seven broad categories called **supergroups.**

Taxonomy involves multiple levels in which particular species are placed into progressively smaller and smaller groups of organisms that are more closely related to each other evolutionarily. Such an approach emphasizes the unity and diversity of different species. As an example, let's consider clownfish, which are found in the Indian and Pacific Oceans and are popular among saltwater aquarium enthusiasts (Figure 1.8). Several species of clownfish have been identified.

One species of clownfish, which is orange with white stripes, has several common names, including Ocellaris clownfish. The broadest grouping for this clownfish is the domain, Eukarya, followed by progressively smaller divisions, from supergroup (Opisthokonta) to kingdom (Animalia) and eventually to species. In the animal kingdom, clownfish are part of a phylum, Chordata, the chordates, which is subdivided into classes. Clownfish are in a class called Actinopterygii, which includes all ray-finned fishes. The common ancestor that gave rise to ray-finned fishes arose about 420 million years ago (mya). Actinopterygii is subdivided into several smaller orders. The clownfish are in the order Perciformes (bony fish). The order is, in turn, divided into families; the clownfish belong to the family of marine fish called Pomacentridae, which are often brightly colored. Families are divided into genera (singular, genus). The genus *Amphiprion* is composed of 28 different species; these are various types of clownfish. Therefore, the genus contains species that are very similar to each other in form and have evolved from a common (extinct) ancestor that lived relatively recently on an evolutionary timescale.

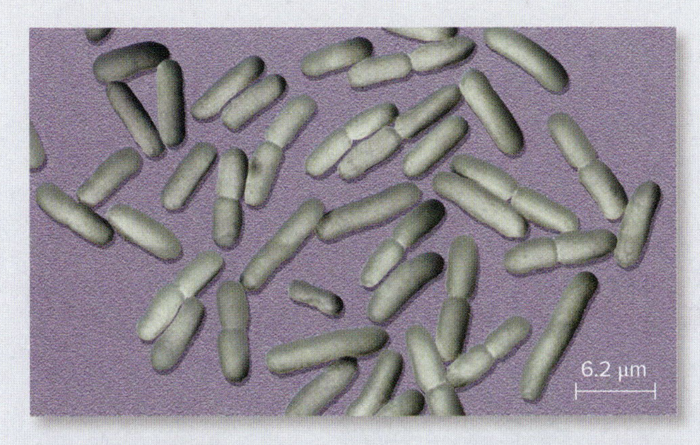


**(a) Domain Bacteria: Mostly unicellular prokaryotes that inhabit many diverse environments on Earth**

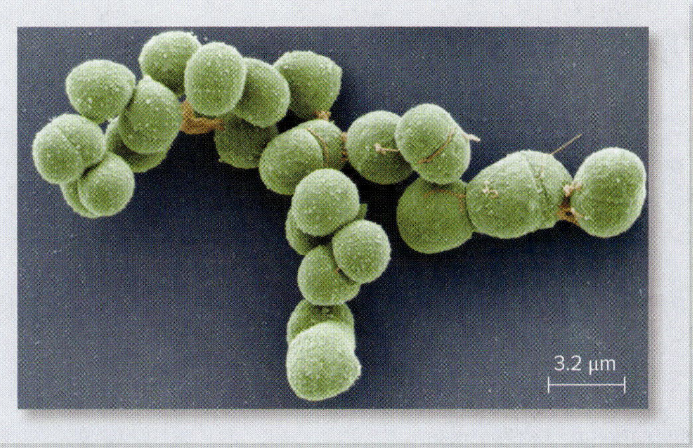


**(b) Domain Archaea: Unicellular prokaryotes that often live in extreme environments, such as hot springs**

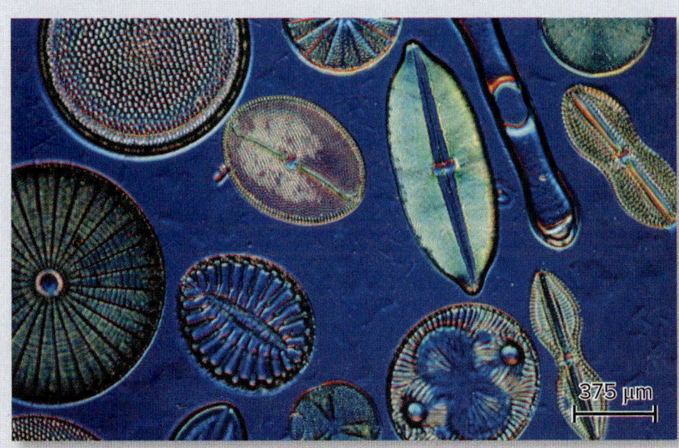


**Protists:** Mostly unicellular and some multicellular organisms that are now subdivided into seven broad groups based on their evolutionary relationships

**Plants:** Multicellular organisms that can carry out photosynthesis

**Fungi:** Unicellular and multicellular organisms that have a cell wall but cannot carry out photosynthesis; fungi usually survive on decaying organic material

**Animals:** Multicellular organisms that usually have a nervous system and are capable of locomotion; they must eat other organisms or the products of other organisms to live

**(c) Domain Eukarya: Unicellular and multicellular organisms having cells with internal compartments that serve various functions**

**Figure 1.7** **The three domains of life.** **(a)** Bacteria and **(b)** Archaea are domains consisting of prokaryotic cells. The third domain, **(c)** Eukarya, comprises species that are eukaryotes.

*(a)* © BSIP/agefotostock RF; *(b)* © Eye of Science/Science Source; *(c Protists)* © Jan Hinsch/Getty Images; *(c Plants)* © Kent Foster/Science Source; *(c Fungi)* © Carl Schmidt-Luchs/SPL/Science Source; *(c Animals)* © Ingram Publishing/agefotostock RF

*Look ahead to Figure 21.1. Are fungi more closely related to plants or animals?*

| Taxonomic group | Ocellaris clownfish is found in | Approximate time when the common ancestor for this group arose | Approximate number of modern species in this group | Examples |
|---|---|---|---|---|
| **Domain** | Eukarya | 2,000 mya | > 5,000,000 |  |
| **Supergroup** | Opisthokonta | > 1,000 mya | > 1,000,000 |  |
| **Kingdom** | Animalia | 600 mya | > 1,000,000 |  |
| **Phylum** | Chordata | 525 mya | 65,000 |  |
| **Class** | Actinopterygii | 420 mya | 30,000 |  |
| **Order** | Perciformes | 80 mya | 7,000 |  |
| **Family** | Pomacentridae | ~ 40 mya | 360 |  |
| **Genus** | *Amphiprion* | ~ 9 mya | 28 |  |
| **Species** | *ocellaris* | < 3 mya | 1 |  |

**Figure 1.8** Taxonomic classification of the clownfish.

*Why is it useful to place organisms into taxonomic groupings?*

Biologists use a two-part description, called **binomial nomenclature,** to provide each species with a unique scientific name. The scientific name of the Ocellaris clownfish is *Amphiprion ocellaris*. The first word is the genus, and the second word is the specific epithet, or species descriptor. By convention, the genus name is capitalized, whereas the specific epithet is not. Both names are italicized. Scientific names are usually Latinized, which means they are made similar in appearance to Latin words. The origins of scientific names are typically Latin or Greek, but they can come from a variety of sources, including a person's name.

## EVOLUTIONARY CONNECTIONS

### The Study of Evolution Allows Us to Appreciate the Unity and Diversity Among Different Species

As described earlier in Figure 1.3a-f, living organisms have a unifying set of features, because all species evolved from a common set of ancestors. Evolution also results in adaptations to specific environments, which accounts for the wonderful diversity that is observed in the living world. An **adaptation** is a characteristic in a species that is the result of natural selection. In our Evolutionary Connections subsections, we will often explore the characteristics of species to appreciate how evolution has resulted in certain types of adaptations.

Your textbook cover provides an interesting and striking example of an evolutionary connection. Polar bears (*Ursus maritimus*) live in one of Earth's most extreme environments, the Arctic Circle. Their evolutionary history can be traced back to ancestral populations of brown bears in Siberia and Alaska that migrated northward and adapted to living in and around the Arctic Ocean. **Figure 1.9** shows a simplified evolutionary tree that relates the 8 bear species that are alive today. (Evolutionary trees are explained in Chapter 21.) The polar bear is most closely related to the brown bear (*Ursus arctos*).

As you might expect, certain adaptations in the polar bear allow them to withstand cold temperatures.

- **Fur:** Except for the tip of the nose, polar bears are entirely covered in fur, which has a very thick undercoat and is denser than the coats of other bears. The outer hairs are covered in oil, which helps to insulate polar bears when they swim in cold, Arctic water.
- **Fat:** Adult polar bears have a layer of fat that can be 10 cm (~4 inches) thick, which also provides insulation from the cold.
- **Large size; small ears and tail:** Polar bears are one of the largest carnivores on Earth. Large size is a common adaptation for many species in cold habitats. Large animals have a lower surface area/volume (SA/V) ratio, which allows for less heat loss through the skin. Similarly, body parts that stick out also affect the SA/V ratio. The polar bear's small ears and tail have less surface area than large ears or tails, and thereby prevent heat loss.

The food of bears depends on their environment. Most bear species eat diets that are rich in nuts and berries. In contrast, the main diet of polar bears consists of different species of seals. However, they will eat vegetation, such as kelp and berries, when seals are not available. Polar bears show several adaptations that allow them to be successful hunters of seals.

- **Sharp teeth and claws:** The teeth of polar bears are consistent with their diet and evolutionary history. Polar bears have long, sharp canines and a row of sharp incisors across the front for grasping prey. Their molars can grind vegetation, but not as well as the other bear species. Polar bears have claws that are curved and sharp for catching and holding prey.
- **Tapered body shape:** Polar bears have an unusual body shape for a bear. They have a large rump and smaller shoulders that lead up to a long, slender neck and skull. This shape makes them more streamlined for rapid swimming and helps them keep their head above the surface for spotting seals.
- **Large paws:** Polar bear paws are enormous! They can be up to 30 cm (12 inches) wide. These large paws act like paddles, helping the polar bear get more power out of every stroke.
- **Hollow, white hairs:** As mentioned, polar bear fur insulates them from the cold, but it also has other characteristics that are helpful for swimming and catching seals. Each hair is hollow and acts like a tiny float, which makes the polar bear more buoyant. In addition, the white color of their fur is thought to camouflage them on snow and ice, thereby making them less visible to their prey.

Taken together, we can see how the evolution of polar bear populations resulted in adaptations that distinguish them from other bears. Natural selection has favored traits that allow the polar bear to withstand the harsh Arctic climate and to be effective hunters of seals.

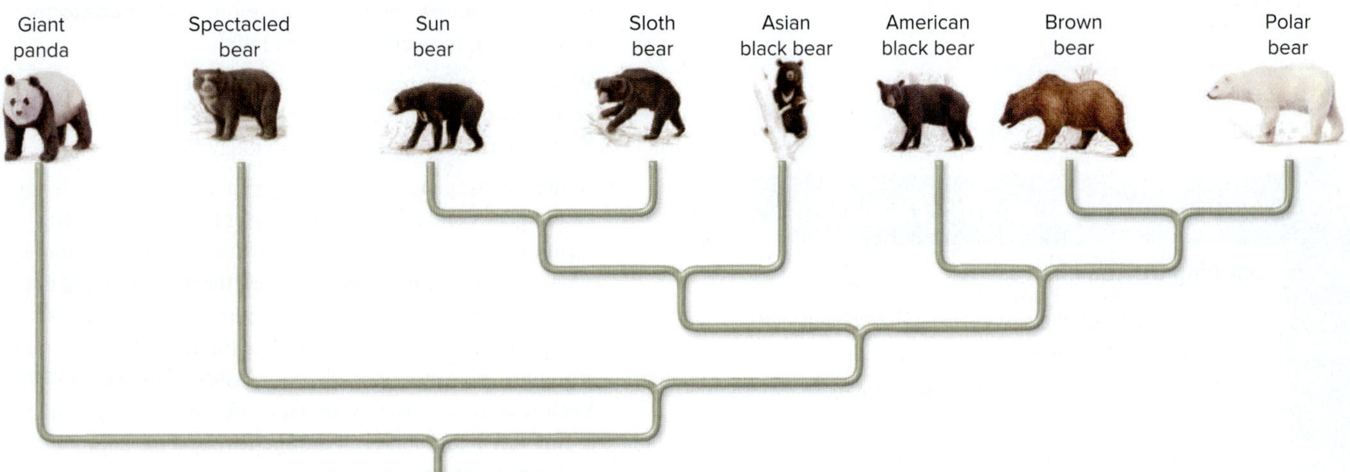


**Figure 1.9** **Bear evolution.** An evolutionary tree that shows the relationships among the 8 living bear species: giant panda (*Ailuropoda melanoluca*), spectacled bear (*Tremarctos ornatus*), sun bear (*Helarctos malayanus*), sloth bear (*Melursus ursinus*), Asian black bear (*Ursus thibetanus*), American black bear (*Ursus americanus*), brown bear (*Ursus arctos*), and polar bear (*Ursus maritimus*).

## 1.2 Reviewing the Concepts

- Changes in species often occur as a result of modification of preexisting structures (Figures 1.5).
- During vertical evolution, mutations in a lineage alter the characteristics of species. Individuals with greater reproductive success are more likely to contribute certain mutations to future generations, a process known as natural selection. Over the long run, this process alters species and may produce new species (Figure 1.6).
- Horizontal gene transfer is the transfer of genes from one organism to another organism that is not its offspring. It may involve the transfer of genes between different species. Along with vertical evolution, it is an important process in biological evolution, producing a web of life.
- Taxonomy is the grouping of species according to their evolutionary relatedness to other species. Going from broad to narrow groups, each species is placed into a domain, supergroup, kingdom, phylum, class, order, family, and genus (Figures 1.7, 1.8).
- Evolution explains the unity and diversity among different species. Throughout this textbook, subsections entitled Evolutionary Connections will relate a specific topic in a given chapter to evolution (Figure 1.9).

## 1.2 Testing Your Knowledge

1. Which of the following is an example of horizontal gene transfer?
   - **a.** the transmission of genes from a mother cell to a daughter cell during cell division
   - **b.** the transmission of a mutant gene from a father to his daughter
   - **c.** the transfer of an antibiotic-resistance gene from one bacterial species to another
   - **d.** all of the above
   - **e.** both a and b
2. Which of the following is the broadest group?
   - **a.** phylum
   - **b.** kingdom
   - **c.** class
   - **d.** species
   - **e.** genus

## 1.3 Biology as a Scientific Discipline

### Learning Outcomes

1. Explain how researchers study biology at different levels, ranging from molecules to ecosystems.
2. **SCISKILLS ▶** Distinguish between discovery-based science and hypothesis testing.
3. **SCISKILLS ▶** Describe the steps of the scientific method, also called hypothesis testing.

What is science? Surprisingly, the definition of science is not easy to state. Most people have an idea of what science is, but actually articulating that idea proves difficult. In biology, we can define **science** as the observation, identification, experimental investigation, and theoretical explanation of natural phenomena.

Science is conducted in different ways and at different levels. Some biologists study the molecules that compose life, and others try to understand how organisms survive in their natural environments. Experimentally, they often focus their efforts on **model organisms**—organisms studied by many different researchers so that they can compare their results and determine scientific principles that apply more broadly to other species, including humans. Examples include *Escherichia coli* (a bacterium), *Saccharomyces cerevisiae* (a yeast), *Drosophila melanogaster* (a fruit fly), *Caenorhabditis elegans* (a nematode worm), *Mus musculus* (a mouse), and *Arabidopsis thaliana* (a flowering plant). Model organisms offer experimental advantages over other species. For example, *E. coli* is a very simple organism that can be easily grown in the laboratory. By focusing their work on a model organism, researchers can gain a deeper understanding of such species.

In this section, we will begin by examining how biologists generally follow a standard approach, called the **scientific method,** to test their ideas. We will explore how scientific knowledge makes predictions that can be experimentally tested. Even so, not all discoveries are the result of researchers following the scientific method. Some discoveries are simply made by gathering new information. As described earlier in Figures 1.1 and 1.2, the characterization of many living organisms has led to the development of important medicines. In this section, we will also consider how researchers often set out on fact-finding missions aimed at uncovering new information that may eventually lead to modern discoveries in biology.

### Biologists Investigate Life at Different Levels of Organization

In Figure 1.5, we examined the various levels of biological organization. The study of these different levels depends not only on the scientific interests of biologists but also on the tools available to them.

- The study of organisms in their natural environments is a branch of biology called **ecology,** which considers populations, communities, and ecosystems (**Figure 1.10a**).
- Researchers may examine the structures and functions of plants and animals, which are disciplines called **anatomy** and **physiology,** respectively (**Figure 1.10b**).
- With major advances in microscopy in the 20th century, **cell biology,** which is the study of cells, became an important branch of biology and remains so today (**Figure 1.10c**).
- In the 1970s, genetic tools became available for studying single genes and the proteins they encode. This genetic technology enabled researchers to study individual molecules, such as proteins, in living cells and thereby spawned the field of **molecular biology.** Together with biochemists and biophysicists, molecular biologists focus their efforts on the structure and function of the molecules of life (**Figure 1.10d**). Such researchers want to understand how biology works at the molecular and even atomic levels. This approach is called **reductionism**—reducing complex systems to simpler components as a way to understand how the system works.
- More recently, scientists have invented new tools that allow them to study groups of genes and groups of proteins. **Systems biology** is aimed at understanding how emergent

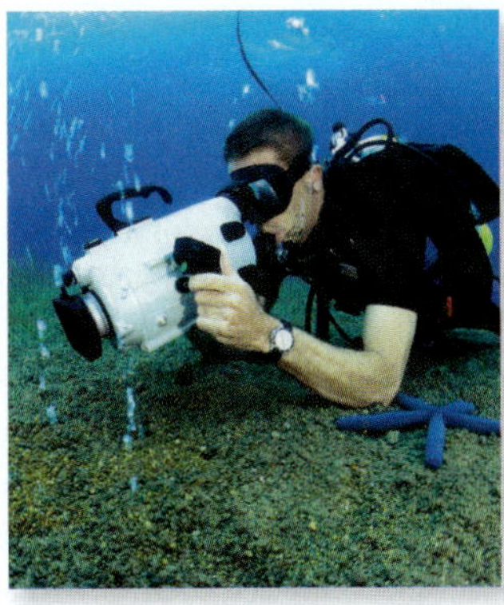

Ecologists study species in their native environments.

**(a) Ecology—population/community/ecosystem levels**

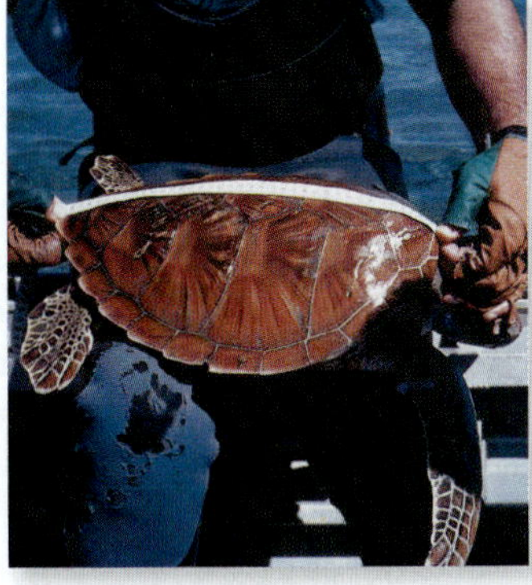

Anatomists and physiologists study how the structures of organisms are related to their functions.

**(b) Anatomy and physiology—tissue/organ/organism levels**

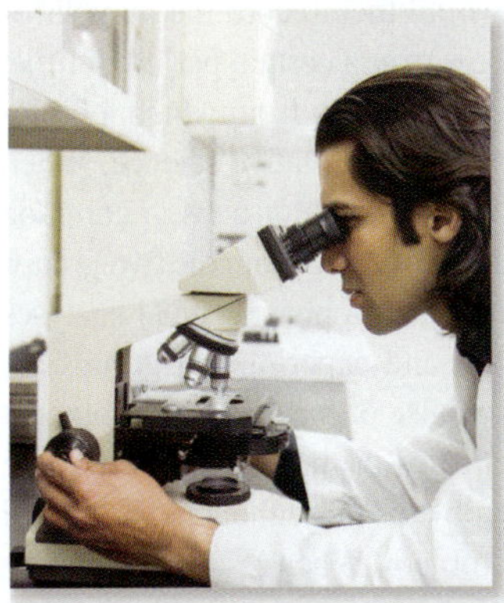

Cell biologists often use microscopes to learn how cells function.

**(c) Cell biology—cellular levels**

Molecular biologists and biochemists study the molecules and macromolecules that make up cells.

**(d) Molecular biology—atomic/molecular levels**

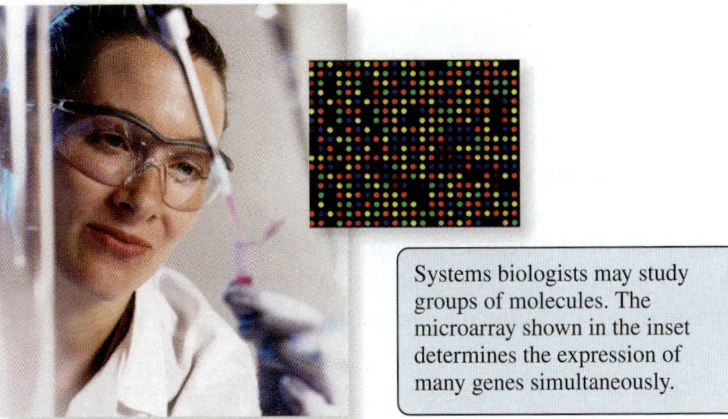

Systems biologists may study groups of molecules. The microarray shown in the inset determines the expression of many genes simultaneously.

**(e) Systems biology—all levels, shown here at the molecular level**

**Figure 1.10** **Biological investigation at different levels.**

*(a)* © Purestock/SuperStock RF; *(b)* © Diane Nelson; *(c)* © Erik Isakson/Blend Images RF; *(d)* © Northwestern, Shu-Ling Zhou/AP Photo; *(e)* © Andrew Brookes/Corbis; *(e inset)* © Alfred Pasieka/Science Source

properties arise from complex interactions. At the molecular or cellular level, systems biology may involve the investigation of groups of proteins with a common purpose (Figure 1.10e). For example, a systems biologist may conduct experiments that try to characterize an entire cellular process, which is driven by dozens of different proteins.

## A Hypothesis Is a Proposed Idea, Whereas a Theory Is a Broad Explanation Backed by Extensive Evidence

Let's now consider the process of science. In biology, a **hypothesis** is a proposed explanation for a natural phenomenon. It is a proposition based on previous observations or experimental studies. For example, with knowledge of seasonal changes, you might hypothesize that maple trees drop their leaves in the autumn because of the shortened amount of daylight. An alternative hypothesis might be that the trees drop their leaves because of colder temperatures. In biology, a hypothesis requires more work by researchers to evaluate its validity.

A useful hypothesis must make **predictions—**expected outcomes that can be shown to be correct or incorrect. In other words, a useful hypothesis is **testable.** If a hypothesis is incorrect, it should be **falsifiable,** which means that it can be shown to be incorrect by additional observations or experimentation. Alternatively, a hypothesis may be correct, so further experiments will not disprove it. In such cases, we would say that the researcher has failed to reject the hypothesis. Even so, in science, a hypothesis is never really proven but rather always remains provisional. Researchers accept the possibility that perhaps they have not yet conceived of the correct hypothesis. After many experiments, biologists may conclude their hypothesis is consistent with known data, but they should never say the hypothesis is proven.

By comparison, the term **theory,** as it is used in biology, is a broad explanation of some aspect of the natural world that is substantiated by a large body of evidence. Biological theories incorporate observations, hypothesis testing, and the laws of other disciplines such as chemistry and physics. Theories are powerful because they allow us to make many predictions about the properties of living organisms. As an example, let's consider the theory that DNA is the genetic material and that it is organized into units called genes. An overwhelming body of evidence has substantiated this theory. Thousands of living species have been analyzed at the molecular level. All of them have been found to use DNA as their genetic material and to express genes that produce the proteins that lead to their characteristics. This theory makes many valid predictions. For example, certain types of mutations in genes are expected to affect the traits of organisms. This prediction has been confirmed experimentally. Similarly, this theory predicts that genetic material is copied and transmitted from parents to offspring. By comparing the DNA of parents and offspring, this prediction has also been confirmed. Furthermore, the theory explains the observation that offspring resemble their parents. Overall, two key attributes of a scientific theory are

1. consistency with a vast amount of known data and
2. the ability to make many correct predictions.

The meaning of the term theory is sometimes muddled because the word is used differently depending on the situation. In everyday language, a "theory" is often viewed as little more than a guess. For example, a person might say, "My theory is that Professor Simpson did not come to class today because he went to the beach." However, in biology, a theory is much more than a mere guess. A theory is an established set of ideas that explains a vast amount of data and offers valid predictions that can be tested. Theories are viewed as **knowledge,** which is the awareness and understanding of information.

## Discovery-Based Science and Hypothesis Testing Are Scientific Approaches That Help Us Understand Biology

The path that leads to an important discovery is rarely a straight line. Rather, scientists ask questions, make observations, ask modified questions, and may eventually conduct experiments to test their hypotheses. The first attempts at experimentation may fail, and new experimental approaches may be needed. To suggest that scientists follow a rigid scientific method is an oversimplification of the process of science. Scientific advances often occur as scientists dig deeper and deeper into a topic that interests them. Curiosity is the key phenomenon that sparks scientific inquiry. How is biology actually conducted? As discussed next, researchers typically follow two general types of approaches: discovery-based science and hypothesis testing.

**Discovery-Based Science** The collection and analysis of data without having a preconceived hypothesis is called **discovery-based science,** or simply discovery science. Why is discovery-based science carried out? The information gained from discovery-based science may lead to the formation of new hypotheses and, in the long run, may have practical applications that benefit people. Researchers, for example, have identified and begun to investigate newly discovered genes within humans without already knowing the function of the gene they are studying. The goal is to gather additional clues that may eventually allow them to propose a hypothesis that explains the gene's function. Discovery-based science often leads to hypothesis testing.

**Hypothesis Testing** In biological science, the scientific method—also known as **hypothesis testing—**is usually followed to formulate and test the validity of a hypothesis. This strategy may be described as a five-step method:

1. Observations are made regarding natural phenomena.
2. These observations lead to a hypothesis that tries to explain the phenomena. A useful hypothesis is one that is testable because it makes specific predictions.
3. Experimentation is conducted to determine if the predictions are correct.
4. The data from the experiment are analyzed.
5. The hypothesis is considered to be consistent with the data, or it is rejected.

The scientific method is intended to be an objective way to gather knowledge. As an example, let's return to our scenario of maple trees dropping their leaves in autumn. By observing the length of daylight throughout the year and comparing those data with the time of the year when leaves fall, one hypothesis might be that leaves fall in response to a shorter amount of daylight (**Figure 1.11**). This hypothesis makes a prediction—exposure of maple trees to shorter amounts of daylight will cause their leaves to fall. To test this prediction, researchers would design and conduct an experiment.

How is hypothesis testing conducted? Although hypothesis testing may follow many paths, certain experimental features are common to this approach. First, data are often collected in two parallel manners. One set of experiments is done on the **control group,** while another set is conducted on the **experimental group.** In an ideal experiment, the control and experimental groups differ by only one factor. For example, an experiment can be conducted in which two groups of trees are observed, and the only difference between their environments is the length of time they receive light each day. To conduct such an experiment, researchers would grow small trees in a greenhouse, where they could keep other factors such as temperature, water, and nutrients the same between the control and experimental groups, while providing them with differing amounts of daylight. In the control group, the number of hours of light provided by lightbulbs would be kept constant each day, whereas in the experimental group, the amount of light each day would become progressively shorter to mimic seasonal light changes. The researchers would then record the number of leaves dropped by the two groups of trees over a certain period of time.

The result of experimentation is a set of data from which a biologist tries to draw conclusions. Biology is a quantitative science. When experimentation involves control and experimental groups, a common form of analysis is to determine if the data collected from the two groups are truly different from each other. Biologists apply statistical analyses to their data to determine if the control and experimental groups are likely to be different from each other because of the single variable that differs between the two groups. A statistics primer is provided at the website for this textbook, and several problems throughout the book will ask you to apply them to a biological problem. When results are statistically significant, the differences between the control and experimental data are not likely to have occurred as a matter of random chance.

In the example shown in Figure 1.11, the trees in the control group dropped far fewer leaves than did those in the experimental group. Statistical analysis can determine if the data collected from the two greenhouses are significantly different from each other. If the two sets of data are found not to be significantly different, the hypothesis is rejected. Alternatively, if the differences between the two sets of data are significant, as shown in Figure 1.11, biologists conclude that the hypothesis is consistent with the data and, therefore, cannot be rejected. A hallmark of science is that valid experiments are **repeatable,** which means that similar results are obtained when the experiment is conducted on multiple occasions. For our example in Figure 1.11, the data would be valid only if the experiment was repeatable.

As described next, discovery-based science and hypothesis testing are often used together to learn more about a particular scientific topic. As an example, let's look at how both approaches led to successes in the study of the disease called cystic fibrosis.

## The Study of Cystic Fibrosis Provides Examples of Discovery-Based Science and Hypothesis Testing

Let's consider how biologists made discoveries related to the disease cystic fibrosis (CF), which affects about 1 in every 3,500 Americans. Persons with CF produce abnormally thick and sticky mucus that obstructs the lungs and leads to life-threatening lung infections. The

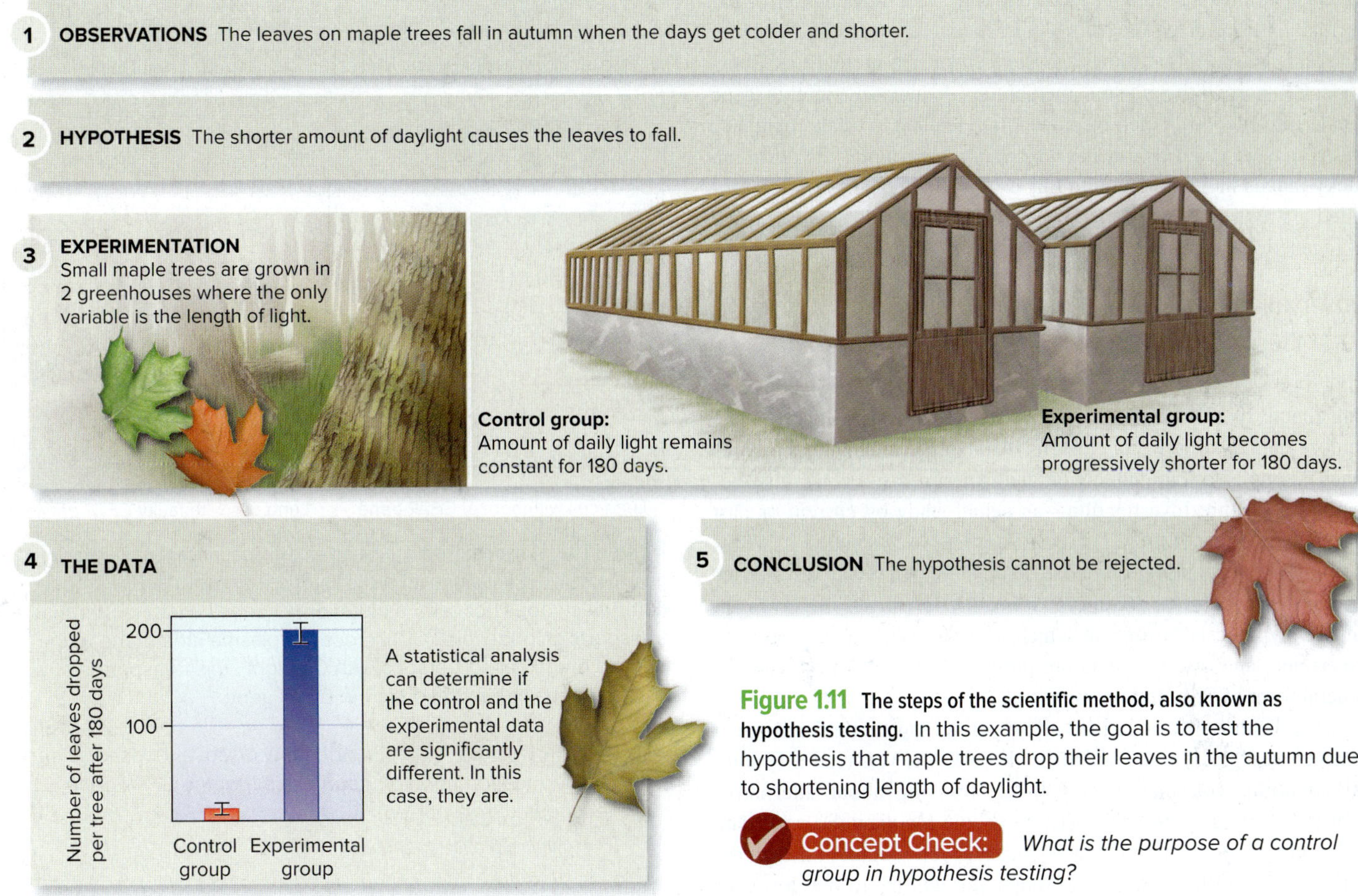

1 OBSERVATIONS The leaves on maple trees fall in autumn when the days get colder and shorter.
2 HYPOTHESIS The shorter amount of daylight causes the leaves to fall.
3 EXPERIMENTATION
Small maple trees are grown in 2 greenhouses where the only variable is the length of light.
Control group:
Amount of daily light remains constant for 180 days.
Experimental group:
Amount of daily light becomes progressively shorter for 180 days.
4 THE DATA
Number of leaves dropped per tree after 180 days
200
100
Control group
Experimental group
A statistical analysis can determine if the control and the experimental data are significantly different. In this case, they are.
5 CONCLUSION The hypothesis cannot be rejected.


**Figure 1.11 The steps of the scientific method, also known as hypothesis testing.** In this example, the goal is to test the hypothesis that maple trees drop their leaves in the autumn due to shortening length of daylight.

**Concept Check:** *What is the purpose of a control group in hypothesis testing?*

thick mucus also blocks ducts in the pancreas, which prevents the digestive enzymes this organ produces from reaching the intestine. Without these enzymes, the intestine cannot fully absorb proteins and fats, which can cause malnutrition. Persons with this disease may also experience liver damage because the thick mucus can obstruct the liver. On average, people with CF in the United States currently live into their late 30s. Fortunately, as more advances have been made in treatment, their life expectancy has steadily increased.

Because of this disease's medical significance, many scientists are conducting studies aimed at gaining greater information about the underlying cause of CF. The hope is that knowing more about the disease may lead to improved treatment options and perhaps even a cure. As described next, discovery-based science and hypothesis testing have been critical to gaining a better understanding of CF.

**The *CFTR* Gene and Discovery-Based Science** In 1935, American physician Dorothy Andersen determined that cystic fibrosis is a genetic disorder. Persons with CF have inherited two faulty *CFTR* genes, one from each parent. (We now know this gene encodes a protein named the cystic fibrosis transmembrane regulator, abbreviated CFTR.) In the 1980s, researchers used discovery-based science to identify this gene. Their search for the *CFTR* gene did not require any preconceived hypothesis regarding the function of the gene. Rather, they used genetic strategies similar to those described in Chapter 18. Research groups headed by Lap-Chee Tsui, Francis Collins, and John Riordan identified the *CFTR* gene in 1989.

The discovery of the gene made it possible to devise diagnostic testing methods to determine if a person carries a faulty *CFTR* gene. In addition, the characterization of the *CFTR* gene provided important clues about its function. Researchers observed striking similarities between the *CFTR* gene and other genes that were already known to encode proteins called transport proteins, which function in the transport of substances across membranes. Based on this observation, as well as other kinds of data, the scientists hypothesized that the function of the normal *CFTR* gene is to encode a transport protein. In this way, the identification of the *CFTR* gene led them to conduct experiments aimed at testing a hypothesis of its function.

**The *CFTR* Gene and Hypothesis Testing** Researchers considered the characterization of the *CFTR* gene along with other studies showing that patients with the disorder have an abnormal regulation of salt balance across their plasma membranes. They hypothesized that the normal *CFTR* gene encodes a transport protein that functions in the transport of chloride ions ($Cl^-$) across the membranes of cells (Figure 1.12). This hypothesis led to experimentation in which they tested normal cells and cells from CF patients for their ability to transport $Cl^-$. The CF cells were found to be defective in chloride transport. In 1990, scientists successfully transferred the normal

*CFTR* gene into CF cells in the laboratory. The introduction of the normal gene corrected the cells' defect in chloride transport. Overall, the results showed that the *CFTR* gene encodes a protein that transports $Cl^-$ across the plasma membrane. A mutation in this gene causes it to encode a defective transporter, leading to a salt imbalance that affects water levels outside the cell, which explains the thick and sticky mucus in CF patients. In this example, hypothesis testing provided a way to evaluate a hypothesis about how a disease is caused by a genetic change.

## Observation and Experimentation Form the Core of Biology

Biology is largely about the process of discovery. Therefore, a recurring theme of this textbook is how scientists design experiments, analyze data, and draw conclusions. Although each chapter contains many examples of data collection and experiments, a consistent element is a Feature Investigation—an actual study by current or past researchers. Some of these involve discovery-based science, in which biologists collect and interpret data in an attempt to make discoveries that are not hypothesis driven. Most Feature Investigations, however, involve hypothesis testing, in which a hypothesis is stated and the experiment and resulting data are presented. Figure 1.11 shows the general form of Feature Investigations.

The Feature Investigations allow you to appreciate the connection between science and scientific theories. As you read a Feature Investigation, you may find yourself thinking about alternative approaches and hypotheses. Different people can view the same data and arrive at very different conclusions. As you progress through the experiments in this textbook, we hope you will try to develop your own skills at formulating hypotheses, designing experiments, and interpreting data.

## Science Is a Social Discipline

Finally, it is worthwhile to point out that biology is a social as well as a scientific discipline. Different laboratories often collaborate on scientific projects. After performing observations and experiments, biologists communicate their results in different ways. Most importantly, papers are submitted to scientific journals. Following submission, papers usually undergo a **peer-review process** in which other scientists who are experts in the area evaluate the paper and make suggestions regarding its quality. As a result of peer review, a paper is accepted or rejected for publication, or the authors of the paper may be given suggestions for how to revise the work or conduct additional experiments before it is acceptable for publication.

Another social aspect of research is that biologists often attend meetings where they report their most recent work to the scientific community. They comment on each other's ideas and work, eventually shaping together the information that builds into scientific theories over many years. As you develop your skills at scrutinizing experiments, it is helpful to discuss your ideas with other people, including fellow students and faculty members. Importantly, you do not need to "know all the answers" before you enter into a scientific discussion. Instead, a more realistic way to view science is as an ongoing and never-ending series of questions.

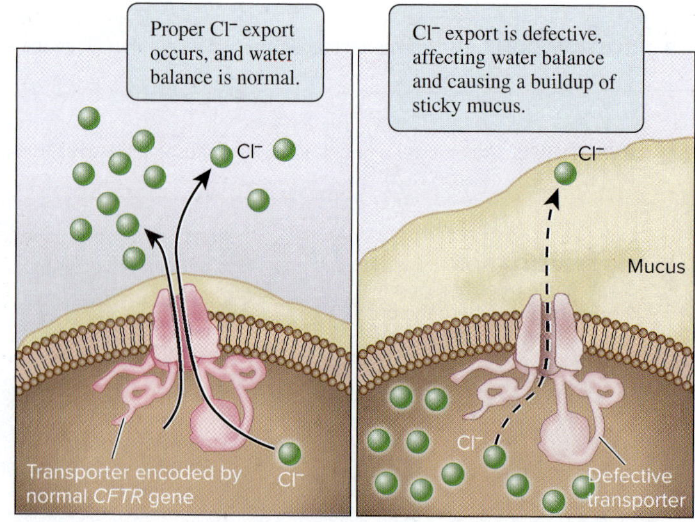


**Figure 1.12 A hypothesis suggesting an explanation for the function of a gene defective in patients with cystic fibrosis.** The normal *CFTR* gene (left), which does not carry a mutation, encodes a transporter that transports chloride ions ($Cl^-$) across the plasma membrane to the outside of the cell. In persons with CF (right), this transporter is defective due to a mutation in the *CFTR* gene.

**Concept Check:** *Explain how discovery-based science helped researchers to hypothesize that the CFTR gene encodes a transporter.*

## 1.3 Reviewing the Concepts

- Biological science is the observation, identification, experimental investigation, and theoretical explanation of natural phenomena. Biologists study life at different levels, ranging from ecosystems to the molecular components in cells (Figure 1.10).
- A hypothesis is a proposal to explain a natural phenomenon. A useful hypothesis makes a testable prediction. A biological theory is a broad explanation that is substantiated by a large body of evidence.
- Discovery-based science is an approach in which researchers conduct experiments and analyze data without a preconceived hypothesis.
- The scientific method, also called hypothesis testing, is a series of steps to formulate and test the validity of a hypothesis. The experimentation often involves a comparison between control and experimental groups (Figure 1.11).
- The study of cystic fibrosis provides an example in which both discovery-based science and hypothesis testing led to key insights into the nature of the disease (Figure 1.12).
- Biology is a social discipline in which scientists often work in teams. To be published, a scientific paper is usually subjected to a peer-review process in which other scientists evaluate the paper and make suggestions regarding its quality. Advances in science often occur when scientists gather and discuss their data.

## 1.3 Testing Your Knowledge

1. Which of the following is a level of study in biology?
   a. studying organisms in their native environments
   b. studying parts of organisms, such as the heart of a frog
   c. studying cells
   d. studying specific molecules within cells
   e. all of the above
2. Which of the following is *not* a step in the scientific method?
   a. Observations lead to a hypothesis that tries to explain a phenomenon.
   b. Experimentation is conducted to determine if the predictions of a hypothesis are correct.
   c. The data from the experiment are analyzed.
   d. The hypothesis is considered to be consistent with the data, or it is rejected.
   e. All of the above are steps in the scientific method.
3. A key difference between the scientific method and discovery-based science is that
   a. the scientific method does not require a hypothesis.
   b. discovery-based science does not require a hypothesis.
   c. the scientific method does not require experimentation.
   d. discovery-based science does not require experimentation.

# Assess and Discuss

## Test Yourself

1. A bird maintains a relatively stable internal body temperature on a cold day. This is an example of
   a. adaptation. c. metabolism. e. development.
   b. evolution. d. homeostasis.
2. Populations of organisms change over the course of many generations. Many of these changes result in increased survival and reproduction. This phenomenon is
   a. evolution. c. development. e. metabolism.
   b. homeostasis. d. genetics.
3. A biologist is studying the living organisms in a valley in western Colorado and their interactions with the environment. She is studying
   a. an ecosystem. c. the biosphere. e. a population.
   b. a community. d. a viable landmass.
4. Which of the following is an example of horizontal gene transfer?
   a. the transmission of an eye color gene from father to daughter
   b. the transmission of a mutant gene causing cystic fibrosis from father to daughter
   c. the transmission of a gene conferring pathogenicity (the ability to cause disease) from one bacterial species to another
   d. the transmission of a gene conferring antibiotic resistance from a mother cell to its two daughter cells
   e. all of the above
5. The scientific name for humans is *Homo sapiens*. The name *Homo* is the _____ to which humans are classified.
   a. kingdom c. order e. species
   b. phylum d. genus
6. The underlying factor that explains the unity and diversity of modern species is
   a. energy. c. homeostasis. e. all of the above.
   b. evolution. d. systems biology.
7. By observing certain desert plants in their native environment, a researcher proposes that they drop their leaves to conserve water. This is an example of
   a. a theory. c. a prediction. e. an experiment.
   b. a law. d. a hypothesis.
8. In science, a theory should
   a. be equated with knowledge.
   b. be supported by a substantial body of evidence.
   c. provide the ability to make many correct predictions.
   d. do all of the above.
   e. b and c only.
9. Conducting research without a preconceived hypothesis is called
   a. discovery-based science. d. a control experiment.
   b. the scientific method. e. none of the above.
   c. hypothesis testing.
10. What is the purpose of using a control group in scientific experiments?
    a. A control group allows the researcher to practice the experiment first before actually conducting it.
    b. A researcher can compare the results in the experimental group and control group to determine if a single variable is causing a particular outcome in the experimental group.
    c. A control group provides the framework for the entire experiment so that the researcher can recall the procedures that should be conducted.
    d. A control group allows the researcher to conduct other experimental changes without disturbing the original experiment.
    e. All of the above are correct.

## Conceptual Questions

1. Of the first eight characteristics of life described in Figure 1.3, which apply to individuals and which apply to populations?
2. Explain how it is possible for evolution to result in unity among different species yet also produce amazing diversity.
3. **PRINCIPLES** In your own words, describe the 12 principles of biology that are detailed at the beginning of this chapter.

## Collaborative Questions

1. Discuss whether or not you think that theories in biology are true. Outside of biology, how do you decide if something is true?
2. The polar bear has evolved characteristics that make it a successful hunter of seals. Make a list of other examples in which one species has evolved characteristics that affect its interactions with a different species.

# Online Resource

**connect.mheducation.com**

© Dr. Parvinder Sethi

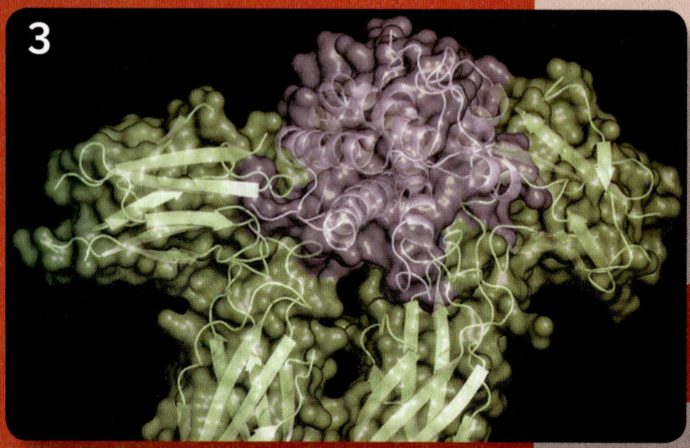


© Daniel Gage, University of Connecticut

# UNIT I
# CHEMISTRY

Living organisms are composed of chemicals, which are altered via chemical reactions. These reactions occur between atoms and molecules and may require or release energy. Chemical reactions and interactions between molecules play a role in virtually all aspects of a cell's activities. In order to understand how living organisms function, grow, develop, behave, and interact with their environments, we first need to understand some basic principles of atomic and molecular structure. In addition, it is necessary to understand the forces that allow atoms and molecules to interact with each other. We begin this unit with an overview of inorganic chemistry—that is, the nature of atoms and molecules, with the exception of those that contain rings or chains of carbon. Such carbon-containing molecules form the basis of organic chemistry and are covered in Chapter 3.

## *The following biology principles are emphasized in this unit:*

- ***Living organisms use energy:*** *We will see how the chemical energy stored in certain bonds in molecules such as sugars and fats can be released and used by living organisms to perform numerous functions that support life, including growth, digestion, and locomotion.*
- ***Structure determines function:*** *As described in Chapters 2 and 3, the three-dimensional structure of molecules is critical to their ability to interact with and influence the function of other specific molecules.*
- ***The genetic material provides a blueprint for reproduction, and all species (past and present) are related by an evolutionary history:*** *Nucleic acids, the basis of inherited genetic material, are first introduced in Chapter 3.*
- ***New properties emerge from complex interactions:*** *You will learn in this unit how atoms and simple molecules are connected together to create more complex molecules with new biological properties.*

# The Chemical Basis of Life I: Atoms, Molecules, and Water

2

© Dr. Parvinder Sethi

Crystals of sodium chloride (NaCl), a molecule composed of two elements.

## Chapter Outline



Upon leaving home and beginning college, an 18-year-old student forgoes the usual well-balanced diet he was accustomed to at home in favor of large amounts of chips, crackers, pizza, burgers, and fries. By the time he graduates, he discovers at a routine physical exam that his blood pressure is a little high. One common denominator of his college diet has been the large amount of salt in his favorite foods. Table salt is a molecule made up of two elements, or atoms: sodium and chlorine. In ordinary amounts, this molecule, symbolized NaCl, is vital for all living organisms. In excess, however, it can cause numerous health problems, including, in some susceptible people, high blood pressure. An understanding of what atoms and molecules are, and how they are important for life and health, is a critical component of understanding biology.

Biology—the study of life—is founded on the principles of chemistry. All living organisms are a collection of atoms and molecules bound together and interacting with each other through the forces of nature. Throughout this textbook, we will see how chemistry can be applied to living organisms as we discuss the components of cells, the functions of proteins, the flow of nutrients in plants and animals, and the evolution of new genes. This chapter lays the groundwork for understanding these and other concepts. We begin with an overview of the structure of atoms. We next explore the various ways that atoms combine with other atoms to create molecules, looking at the different types of chemical bonds between atoms, how these bonds form, and how they determine the structures of molecules. We then examine chemical reactions between molecules. The water molecule is then highlighted, including the properties that make it a crucial component of living organisms. We conclude with a discussion about the importance of pH in the watery fluids of living organisms.

## 2.1 Atoms

### Learning Outcomes

1. Describe the general structure of atoms.
2. Define orbital and electron shell.
3. Relate atomic structure to the periodic table of the elements.
4. **SCISKILLS ▶** Quantify atomic mass using units such as daltons and moles.
5. Explain how a single element may exist in two or more forms, called isotopes.
6. List the elements that make up most of the mass of all living organisms.

All organisms are composed of **matter,** which is anything that contains mass and occupies space (mass, then, is the amount of matter in any object). Matter may exist in any of three states: solid, liquid, or gas. **Atoms** are the smallest functional units of matter that form all chemical substances and ultimately all organisms; they cannot be further broken down into other substances by ordinary chemical or physical means. Atoms, in turn, are composed of different types of smaller, subatomic particles. When two or more atoms are bonded together, a **molecule** is formed. In this section, we will explore the physical properties of atoms so that we can understand how atoms combine to form molecules of biological importance.

### Atoms Are Composed of Subatomic Particles

The chemicals within living organisms are composed of many different types of atoms. Each specific type of atom—nitrogen, hydrogen, oxygen, and so on—is called an **element** (or chemical element), which is defined as a pure substance made up of only one kind of atom. If you hold in your hand an ounce of pure gold, you are holding an element composed of many, many gold atoms.

Three subatomic particles—**protons ($p^+$), neutrons ($n^0$),** and **electrons ($e^-$)**—are found within atoms (Figure 2.1). The protons and neutrons are confined to a very small volume at the center of an atom, the **atomic nucleus,** whereas the electrons are found in regions at various distances from the nucleus. In most atoms, the numbers of protons and electrons are identical, but the number of neutrons may vary. Each of the subatomic particles has a different electric charge:

- Protons have one unit of positive charge.
- Electrons have one unit of negative charge.
- Neutrons are electrically neutral.

Like charges repel each other, and opposite charges attract each other. It is the opposite charges of the protons and electrons that create an atom—the positive charges in the nucleus attract the negatively charged electrons.

Because the protons are located in the atomic nucleus, the nucleus has a net positive charge equal to the number of protons it contains.

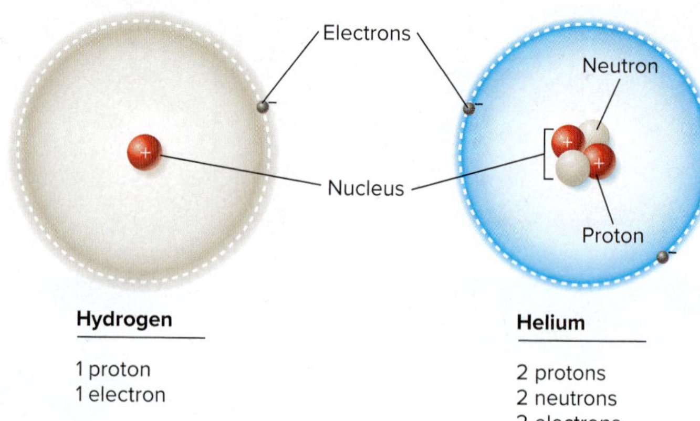


**Figure 2.1 Diagrams of two simple atoms.** These are models of the two simplest atoms, hydrogen and helium. The nucleus consists of protons and neutrons, whereas electrons are found outside the nucleus. Note: In all figures of atoms, the sizes and distances are not to scale. For example, the radius of a typical atom is roughly 10,000 times the radius of its nucleus. If an atomic nucleus were the size of a soccer ball, then the entire atom would have a radius of about 2 miles!

The entire atom has no net electric charge, however, because the number of negatively charged electrons around the nucleus is equal to the number of positively charged protons in the nucleus.

### Electrons Occupy Orbitals Around an Atom's Nucleus

At one time, scientists visualized an atom as a mini-solar system, with the nucleus being the Sun and the electrons traveling in clearly defined orbits around it. Diagrams of the two simplest atoms—hydrogen and helium—which have the smallest numbers of protons, were shown in Figure 2.1. This model of the atom is now known to be an oversimplification, because electrons do not actually orbit the nucleus in a defined path like planets around the Sun. However, this depiction of an atom remains a convenient way to diagram atoms in two dimensions.

For complex reasons associated with the physics of subatomic particles, it is impossible to precisely predict the exact location of a given electron. We can only describe the region of space surrounding the atomic nucleus in which there is a high probability of finding that electron. These regions are called **orbitals.** A better model of an atom, therefore, is a central nucleus surrounded by cloudlike orbitals. An orbital can contain a maximum of two electrons. Consequently, any atom with more than two electrons must contain additional orbitals.

An **electron shell** is composed of one or more orbitals. The orbitals of a given electron shell have a characteristic amount of energy. **Energy** is the capacity to do work or effect a change. In biology, we often refer to various types of energy, such as light energy, mechanical energy, and chemical energy. Electrons have kinetic energy, or the energy of moving matter. Atoms with progressively more electrons have additional electron shells that are at greater and greater distances from the nucleus. These shells are numbered, with shell number 1 being closest to the nucleus. Electron shells may contain one or more orbitals, with each orbital containing up to two electrons. The first, innermost electron shell of all atoms has room for only two electrons in one orbital. The second electron shell has

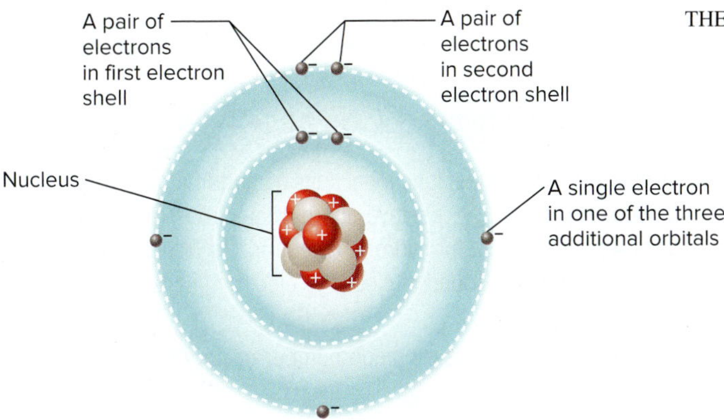


**Simplified depiction of a nitrogen atom (7 electrons; 2 electrons in first electron shell, 5 in second electron shell)**

**Figure 2.2 Diagram showing the two electron shells of a nitrogen atom.** An atom's shells fill up one by one with electrons. The first, innermost shell fills up first with two electrons. Subsequent shells gain one electron at a time until they become filled. Heavier elements contain additional shells.

four orbitals holding up to four pairs of electrons, or eight electrons altogether.

Electrons are found in orbitals of varying shapes, such as spherical or propeller-type shapes. However, scientists often use simplified diagrams when depicting an atom without showing individual orbitals within an electron shell. Figure 2.2 illustrates an example of such a depiction of a nitrogen atom. An atom of this element has seven protons and seven electrons. Two electrons fill the first shell, leaving five electrons for the second, outer shell. Two of these fill one orbital and are shown as a pair of electrons in the second shell. The other three electrons in the second shell are found singly in each of the three other available orbitals. The diagram in Figure 2.2 makes it easy to see whether the outer shell of an atom is full.

Most atoms, like nitrogen, have outer shells that are not completely filled with electrons. Atoms that have unfilled electron shells tend to share, lose, or gain electrons to fill their outer shell. Those electrons in the outermost shell are called the **valence electrons.** As you will learn shortly, such electrons allow atoms to form chemical bonds with each other, in which two or more atoms become joined together to create a new substance.

## Each Element Has a Unique Number of Protons

Each chemical element has a specific and unique number of protons in its nucleus that distinguishes it from another element. The number of protons in an atom is its **atomic number.** For example, hydrogen, the simplest atom, has an atomic number of 1, corresponding to its single proton. Magnesium has an atomic number of 12, corresponding to its 12 protons. With the exception of ions, which are described later, the number of protons is identical to the number of electrons in a given atom. Therefore, the atomic number is also equal to the number of electrons in the atom, resulting in a net electric charge of zero.

Appendix A shows the periodic table of the elements, which arranges the known elements according to their atomic number and electron shells. A one- or two-letter symbol is used as an abbreviation for each element. The rows (known as "periods") indicate the number of electron shells. For example, hydrogen (H) has one shell, lithium (Li) has two shells, and sodium (Na) has three shells. The columns (called "groups") indicate the numbers of electrons in the outer shell. The outer shell of lithium (Li) has one electron, beryllium (Be) has two, boron (B) has three, and so forth. This organization of the periodic table tends to arrange elements based on similar chemical properties. For example, magnesium (Mg) and calcium (Ca) each have two electrons in their outer shell, so these two elements tend to combine with many of the same other elements. The similarities of elements within a group occur because they have the same number of electrons in their outer shells, and this gives them similar chemical bonding properties.

## Atoms Have a Small but Measurable Mass

Atoms are extremely small and thus have very little mass. Protons and neutrons are nearly equal in mass, and each has more than 1,800 times the mass of an electron. Because of their tiny size relative to protons and neutrons, the mass of the electrons in an atom is ignored in calculations.

The **atomic mass** indicates an atom's mass relative to the mass of other atoms. By convention, the most common type of carbon atom, which has six protons and six neutrons, is assigned an atomic mass of exactly 12. On this scale, a hydrogen atom has an atomic mass of 1, indicating that it has 1/12 the mass of a carbon atom. A magnesium atom, with an atomic mass of 24, has twice the mass of a carbon atom.

Atomic mass is measured in units called daltons. One **dalton (Da)** equals 1/12 the mass of a carbon atom. Therefore, the most common type of carbon atom has an atomic mass of 12 Da. Because atoms such as hydrogen have a small mass, but atoms such as carbon have a larger mass, 1 gram (g) of hydrogen contains more atoms than 1 g of carbon. As first described by Italian physicist Amedeo Avogadro, 1 **mole** (mol) of any substance contains the same number of atoms as there are atoms in exactly 12 g of carbon. Twelve grams of carbon equals 1 mol of carbon, and 1 g of hydrogen equals 1 mol of hydrogen. One mol of any element contains the same number of atoms—$6.022 \times 10^{23}$, which is called Avogadro's number. For example, 12 g of carbon contains $6.022 \times 10^{23}$ atoms, and 1 g of hydrogen, whose atoms have 1/12 the mass of a carbon atom, contains $6.022 \times 10^{23}$ atoms.

## Isotopes Vary in Their Number of Neutrons

Many elements can exist in multiple forms, called **isotopes,** that differ in the number of neutrons they contain. For example, the most abundant form of the carbon atom, $^{12}C$, contains six protons and six neutrons, and thus has an atomic number of 6 and an atomic mass of 12 Da. The superscript placed to the left of $^{12}C$ is the sum of the protons and neutrons. The rare carbon isotope $^{14}C$, however, contains six protons and eight neutrons. Although $^{14}C$ has an atomic number of 6, it has an atomic mass of 14 Da. Nearly 99% of the carbon in living organisms is $^{12}C$. Consequently, the average atomic mass of carbon is slightly greater than 12 Da because of the existence of a small amount of heavier isotopes. This explains why the atomic masses given in the periodic table do not add up exactly to whole numbers.

Isotopes of an atom have similar chemical properties but may have very different physical properties. For example, many isotopes found in nature are inherently unstable. Such unstable isotopes are called **radioisotopes,** because they lose energy by emitting

**Table 2.1 Chemical Elements Essential for Life in Many Organisms**

| Element | Symbol | % Human body mass | % All atoms in human body | | |
|---|---|---|---|---|---|
| **Most abundant in living organisms (approximately 95% of total mass)** | | | | | |
| Oxygen | O | 65 | 25.5 | | |
| Carbon | C | 18 | 9.5 | | |
| Hydrogen | H | 9 | 63.0 | | |
| Nitrogen | N | 3 | 1.4 | | |
| **Element** | **Symbol** | **Element** | **Symbol** | **Element** | **Symbol** |
| **Mineral elements (less than 1% of total mass)** | | **Trace elements (less than 0.01% of total mass)** | | | |
| Calcium | Ca | Boron | B | Manganese | Mn |
| Chlorine | Cl | Chromium | Cr | Molybdenum | Mo |
| Magnesium | Mg | Cobalt | Co | Selenium | Se |
| Phosphorus | P | Copper | Cu | Silicon | Si |
| Potassium | K | Fluorine | F | Tin | Sn |
| Sodium | Na | Iodine | I | Vanadium | V |
| Sulfur | S | Iron | Fe | Zinc | Zn |

subatomic particles and/or radiation, which converts them to a stable form. At the very low amounts found in nature, radioisotopes usually pose no serious threat to life, but exposure of living organisms to high amounts of radioactivity can result in the disruption of cellular function, cancer, and even death.

### All Living Organisms Are Largely Composed of Four Elements

The elements oxygen, carbon, hydrogen, and nitrogen account for the majority of atoms in living organisms (**Table 2.1**), typically making up about 95% of an organism's mass. Much of the oxygen and hydrogen occur in the form of water, which accounts for approximately 60% of the mass of most animals and up to 95% or more in some plants. Carbon is a major building block of all living matter, and nitrogen is a vital element in all proteins. Note in Table 2.1 that although hydrogen accounts for about 63% of the atoms in the body, it makes up only a small percentage (9%) of the mass of the human body. That is because the atomic mass of hydrogen is so much lighter than that of heavier elements such as oxygen.

Other important elements in living organisms include the **mineral elements,** which make up less than 1% of total mass. Calcium and phosphorus, for example, are important constituents of the skeletons and shells of animals. In addition, all living organisms require **trace elements.** These elements are present in extremely small quantities but are still essential for normal growth and function. For example, iron plays an important role in the transport of oxygen in the blood of many animals.

## 2.1 Reviewing the Concepts

- Atoms are the smallest functional units of matter. Atoms are composed of protons ($p^+$, positive charge), electrons ($e^-$, negative charge), and (except for hydrogen) neutrons ($n^0$, electrically neutral). Electrons are found in cloudlike electron shells around the nucleus (Figures 2.1, 2.2).
- Each element contains a unique number of protons—its atomic number.
- The atomic mass scale indicates an atom's mass relative to the mass of other atoms.
- Many atoms exist as isotopes, which differ in the number of neutrons they contain. Some isotopes are unstable radioisotopes and emit radiation.
- Oxygen, carbon, hydrogen, and nitrogen account for the majority of atoms in organisms. Living organisms require mineral and trace elements that are essential for growth and function (Table 2.1).

## 2.1 Testing Your Knowledge

1. Refer to the periodic table of the elements in Appendix A. Which element has 15 electrons and three electron shells?
   **a.** O **b.** P **c.** Na **d.** Cl **e.** Ar
2. How many electrons would be required to fill the outer electron shell of fluorine (F)?
   **a.** 0 **b.** 1 **c.** 8 **d.** 7 **e.** 2

## 2.2 Chemical Bonds and Molecules

### Learning Outcomes

1. Compare and contrast the types of chemical bonds and atomic interactions that lead to the formation of molecules.
2. Explain the concept of electronegativity and how it contributes to the formation of polar and nonpolar covalent bonds.
3. Describe how a molecule's shape is important for its ability to interact with other molecules.

The linkage of atoms with other atoms serves as the basis for life and gives life its great diversity. Two or more atoms bonded together make up a molecule. For example, two oxygen atoms can combine to form an oxygen molecule, represented as $O_2$. This representation is called a **molecular formula.** It consists of the chemical symbols for all of the atoms that are present (here, O for oxygen) and a subscript that indicates how many of those atoms are present in the molecule (in this case, two). The term **compound** refers to a molecule composed of two or more different elements. Examples include water ($H_2O$), with two hydrogen atoms and one oxygen atom, and the sugar glucose ($C_6H_{12}O_6$), which has 6 carbon atoms, 12 hydrogen atoms, and 6 oxygen atoms.

One of the most important features of compounds is their emergent properties. This means that the characteristics of a compound differ greatly from those of its constituent elements. Let's consider sodium as an example. Pure sodium (Na) is a soft, silvery metal that reacts violently with water to produce NaOH and hydrogen gas. By comparison, the compound that results from combining sodium with the yellow-green gas chlorine ($Cl_2$) is table salt (NaCl). NaCl is a white, relatively hard crystal (as seen in the chapter-opening photo) that dissolves in water. Thus, the properties of sodium in a compound can be dramatically different from its properties as a pure element.

The atoms in molecules are held together by chemical bonds. In this section, we will examine the different types of chemical bonds, how these bonds form, and how they determine the structures of molecules.

## Covalent Bonds Are Formed When Atoms Share Electrons to Fill Their Outer Shells

**Covalent bonds,** in which atoms share a pair of electrons, can occur between atoms whose outer shells are not full. A fundamental principle of chemistry is that atoms tend to be most stable when their outer shells are filled with electrons. Figure 2.3 shows this principle as it applies to the formation of hydrogen fluoride (HF), a molecule with many important industrial and medical applications. The outer shell of a hydrogen atom is full when it contains two electrons, though a hydrogen atom has only one electron. The outer shell of a fluorine atom is full when it contains eight electrons, though a fluorine atom has only seven electrons in its outer shell. When HF is made, the two atoms share a pair of electrons, which spend time in the outer electron shells of both atoms. This allows both of the outer shells of those atoms to be full. Covalent bonds are strong chemical bonds, because the shared electrons behave as if they belong to each atom.

Chemists sometimes depict molecules with a **structural formula** in which each covalent bond is represented by a line indicating a pair of shared electrons. For example, HF is diagrammed as

H—F

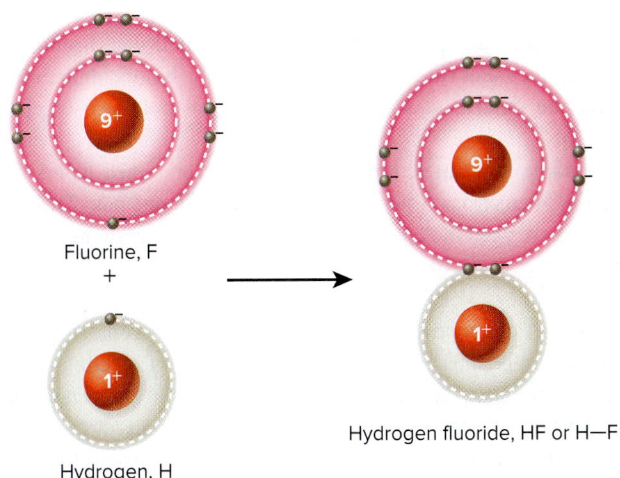


**Figure 2.3 The formation of covalent bonds.** In covalent bonds, electrons from the outer shell of two atoms are shared with each other in order to complete the outer shells of both atoms. This simplified illustration shows hydrogen forming a covalent bond with fluorine. The number of protons in the nuclei is indicated.

A molecule of water ($H_2O$) can be diagrammed as

H—O—H

The structural formula of water indicates that the oxygen atom is covalently bonded to two hydrogen atoms.

Each atom forms a characteristic number of covalent bonds, which depends on the number of electrons required to fill the outer shell. The atoms of some elements important for life, notably carbon, form more than one covalent bond and become linked simultaneously to two or more other atoms. Figure 2.4 shows the number of covalent bonds formed by the four most abundant atoms found in the molecules of living organisms—hydrogen, oxygen, nitrogen, and carbon.

For many types of atoms, their outermost shell is full when they contain eight electrons, an octet. The **octet rule** states that many atoms are most stable when they have eight electrons in their outermost electron shell. This rule applies to most atoms found in living organisms, including oxygen, nitrogen, carbon, phosphorus, and sulfur. These atoms form a characteristic number of covalent

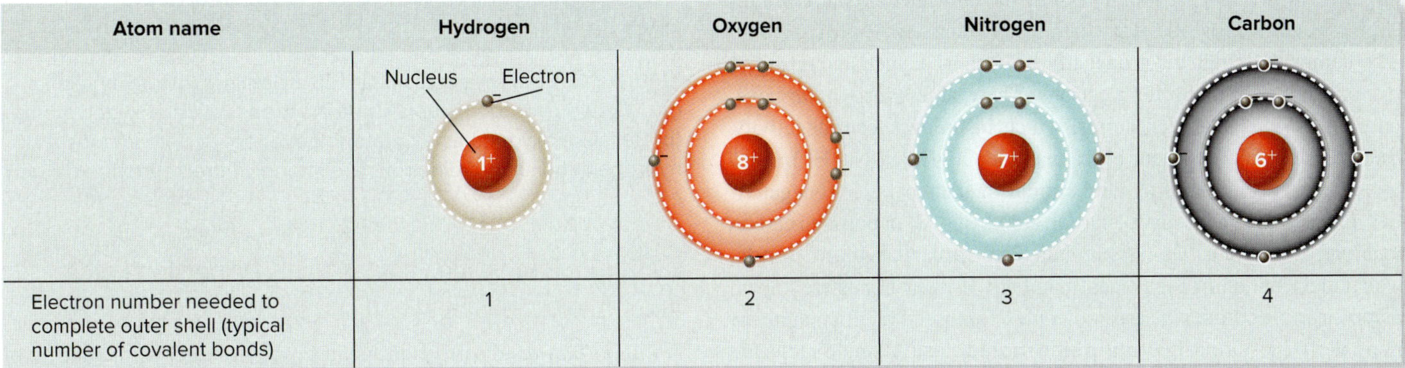


| Atom name | Hydrogen | Oxygen | Nitrogen | Carbon |
|---|---|---|---|---|
| Electron number needed to complete outer shell (typical number of covalent bonds) | 1 | 2 | 3 | 4 |

**Figure 2.4 The number of covalent bonds formed by common elements found in living organisms.** These elements form different numbers of covalent bonds due to the number of electrons in their outer shells.

bonds to make an octet in their outermost shell (see Figure 2.4). However, the octet rule does not always apply. For example, hydrogen has an outermost shell that can contain only two electrons, not eight.

In some molecules, a **double bond** occurs when atoms share two pairs of electrons (four electrons) rather than one pair. This is the case for an oxygen molecule ($O_2$), which can be diagrammed as

$$O{=}O$$

Another common example occurs when two carbon atoms form bonds in compounds. They may share one pair of electrons (single bond) or two pairs (double bond), depending on how many other covalent bonds each carbon forms with other atoms. In rare cases, carbon can even form triple bonds, in which three pairs of electrons are shared between two atoms.

## Covalent Bonds May Be Polar or Nonpolar

Some atoms attract shared electrons more readily than do other atoms. The **electronegativity** of an atom is a measure of its ability to attract electrons in a bond with another atom. Generally, electronegativity is greater in atoms that have more protons (and therefore more positive charges) but similar numbers of electron shells; thus, fluorine is more electronegative than oxygen, which is more electronegative than nitrogen (refer to Appendix A). As atoms get much larger, however, electronegativity decreases, because the distance between the nucleus (where the positive charges reside) and the outer electron shell increases. Thus, fluorine is more electronegative than chlorine.

When two atoms with different electronegativities form a covalent bond, the shared electrons are more likely at any time to be closer to the nucleus of the atom of higher electronegativity than to the atom of lower electronegativity. Such bonds are called **polar covalent bonds,** because the distribution of electrons around the nuclei creates a polarity, or difference in electric charge, across the molecule. $H_2O$ is a classic example of a molecule containing polar covalent bonds. Because oxygen is much more electronegative than hydrogen, the shared electrons tend to be pulled closer to the oxygen nucleus than to either of the hydrogens, resulting in a higher probability of an electron being nearer the oxygen atom at any given instant. This unequal sharing of electrons gives the molecule a region of partial negative charge (indicated by the Greek letter δ and a minus sign, $\delta^-$) and two regions of partial positive charge ($\delta^+$) (**Figure 2.5**).

Atoms with high electronegativity, such as oxygen and nitrogen, have a strong attraction for electrons. These atoms form polar covalent bonds with hydrogen atoms, which have low electronegativity. Examples of polar covalent bonds include O—H and N—H. In contrast, bonds between atoms with similar or identical electronegativities, for example, between two carbon atoms (C—C) or between carbon and hydrogen atoms (C—H) are called **nonpolar covalent bonds.** Molecules containing significant numbers of polar bonds are known as **polar molecules,** whereas molecules composed predominantly of nonpolar bonds are called **nonpolar molecules.** A single molecule may have different regions with nonpolar bonds and polar bonds. As we will explore later, the physical characteristics of polar and nonpolar molecules, especially their solubility in water, are quite different.

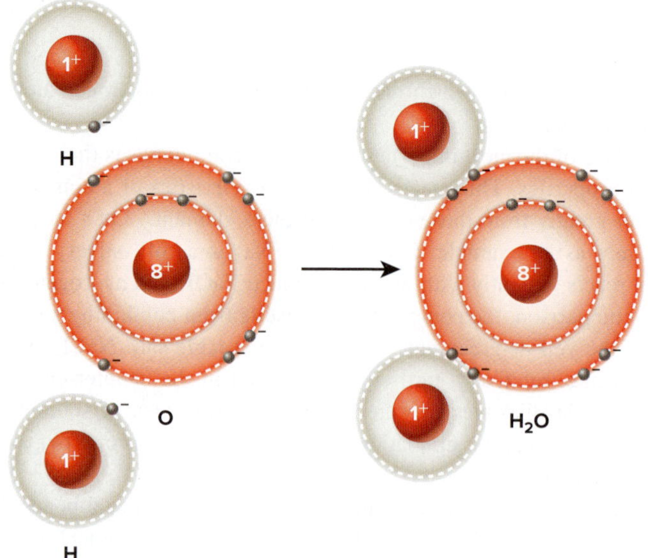


(a) In $H_2O$, electrons are shared between an oxygen atom and two hydrogen atoms.

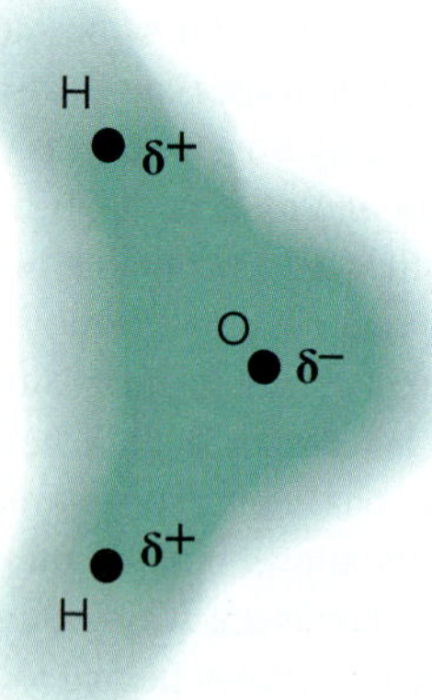


(b) The shared electrons in $H_2O$ are more likely to be near the oxygen atom at any moment, as shown in this model of a water molecule. The shading represents a rough approximation of the predicted electron densities between the three atoms. This gives oxygen a partial negative charge ($\delta^-$) and each hydrogen a partial positive charge ($\delta^+$).

**Figure 2.5** **Polar covalent bonds in $H_2O$ molecules.** **(a)** In a molecule of $H_2O$, two hydrogen atoms share electrons with an oxygen atom. **(b)** Because oxygen has a greater electronegativity than does hydrogen, the shared electrons are strongly attracted to the oxygen.

## Hydrogen Bonds and van der Waals Dispersion Forces Allow Interactions Between and Within Molecules

An important result of certain polar covalent bonds is the ability of one molecule to loosely associate with another molecule through a weak interaction called a **hydrogen bond.** A hydrogen bond forms when a hydrogen atom from one polar molecule becomes electrically

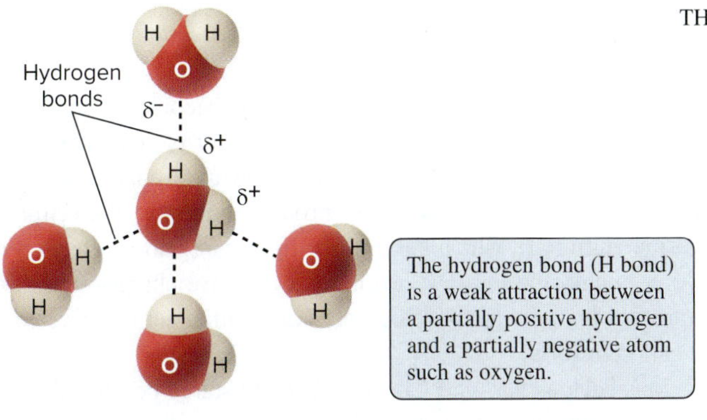

Hydrogen bonds
δ−
δ+
δ+
H
O
The hydrogen bond (H bond) is a weak attraction between a partially positive hydrogen and a partially negative atom such as oxygen.
Hydrogen bonds between $H_2O$ molecules


**Figure 2.6** **Example of hydrogen bonds.** Hydrogen bonds are important because they allow for interactions between different molecules or interactions of atoms within a molecule. This example depicts hydrogen bonds (shown as dashed lines) between $H_2O$ molecules. For simplicity, only a few partial charges are indicated. In this diagram, the atoms are depicted as solid shapes, which represent the outer electron shell (this is called a space-filling model for an atom).

attracted to an electronegative atom, such as an oxygen or nitrogen atom, in another polar molecule. Hydrogen bonds, like those between $H_2O$ molecules, are represented by dashed or dotted lines in diagrams to distinguish them from covalent bonds (**Figure 2.6**). A single hydrogen bond is very weak. The strength of a hydrogen bond is only a small percentage of the strength of the polar covalent bonds linking the hydrogen and oxygen within a molecule of $H_2O$.

Hydrogen bonds can also occur within a single large molecule. Large molecules may have many hydrogen bonds within their structure. Collectively, many hydrogen bonds add up to a strong force that helps maintain the three-dimensional structure of a molecule.

In contrast to the cumulative strength of many hydrogen bonds, the weakness of individual bonds is also important. When an interaction between two molecules involves relatively few hydrogen bonds, such interactions tend to be weak and readily broken. The reversible nature of hydrogen bonds allows molecules to interact and then to become separated again. For example, small molecules may bind to proteins called enzymes via hydrogen bonds. **Enzymes** are molecules that catalyze many biologically important chemical reactions in cells. After binding to an enzyme, a small molecule may be converted to a different molecule, which is then released.

Temporary attractive forces that are even weaker than hydrogen bonds also form between molecules. These **van der Waals dispersion forces** arise because electrons occupy orbitals in a random, probabilistic way, as mentioned previously. At any moment, the electrons in the outer shells of the atoms in a nonpolar molecule may be evenly distributed or unevenly distributed. In the latter case, a fleeting electrical attraction may arise with other nearby molecules. As with hydrogen bonds, the collective strength of these transient attractive forces between molecules can be quite strong.

## Ionic Bonds Involve an Attraction Between Positive and Negative Ions

Atoms are electrically neutral because they contain equal numbers of negative electrons and positive protons. If an atom or a molecule gains or loses one or more electrons, it acquires a net electric charge and becomes an **ion** (**Figure 2.7a**). For example, when a sodium atom (Na),

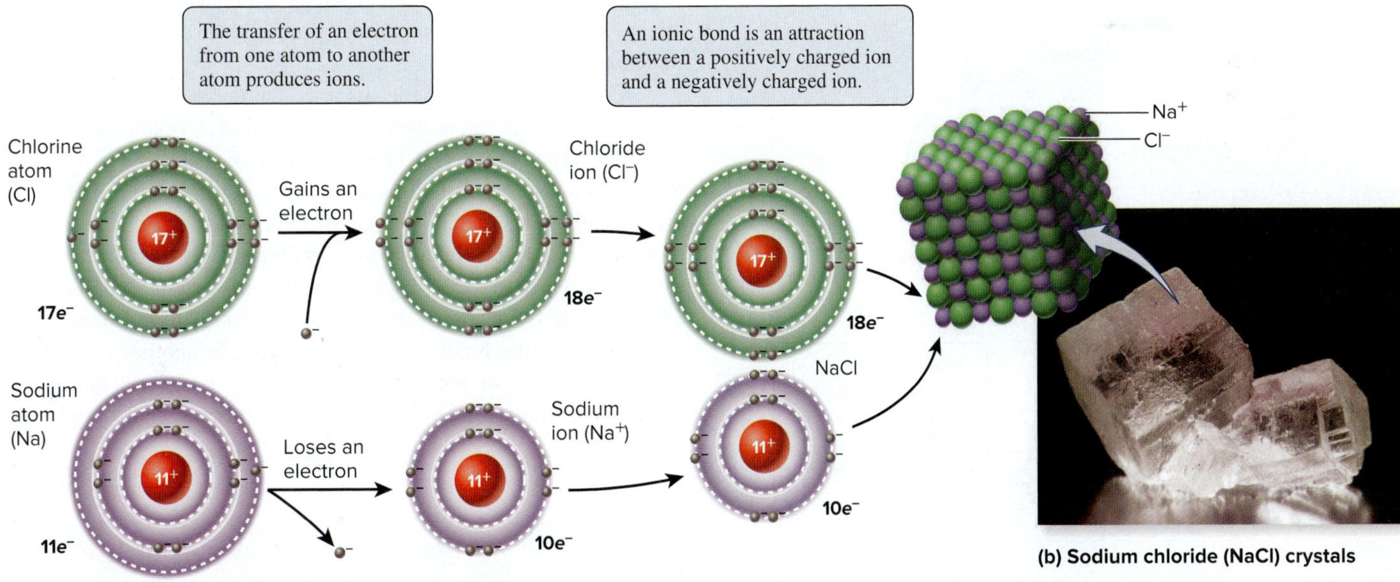


**Figure 2.7** **Ionic bonding in table salt (NaCl).** **(a)** When a sodium atom loses an electron, a sodium ion is formed, and when a chloride ion gains an electron, a chloride ion is formed. Note: Electrons do not exist free in solution; when an electron is removed from one atom, it must be transferred to some other atom. **(b)** In a salt crystal, a lattice is formed in which the positively charged sodium ions ($Na^+$) are attracted to negatively charged chloride ions ($Cl^-$).

*(b)* © Charles D. Winters/Science Source

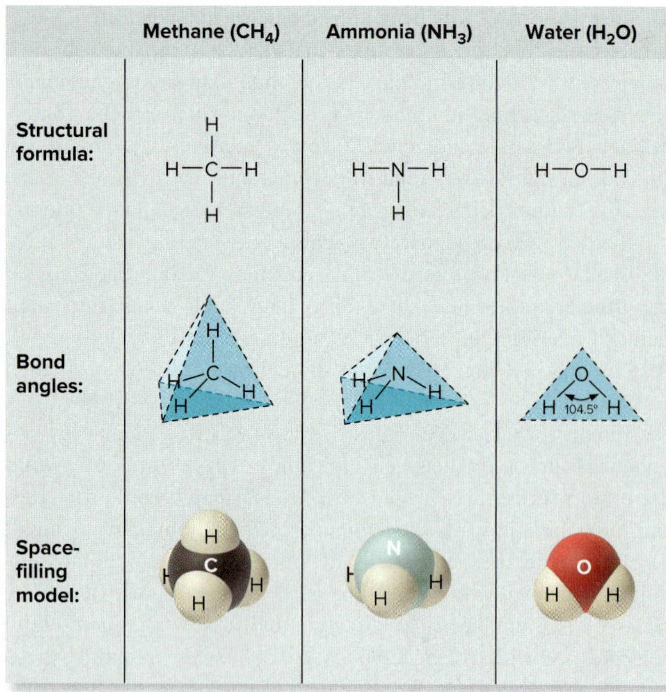


**Figure 2.8** **Shapes of molecules.** Molecules may assume different shapes, depending on the types of bonds between their atoms. The angles between groups of atoms are well defined. For example, in water at room temperature, the angle formed by the covalent bonds between the two hydrogen atoms and the oxygen atom is approximately 104.5°.

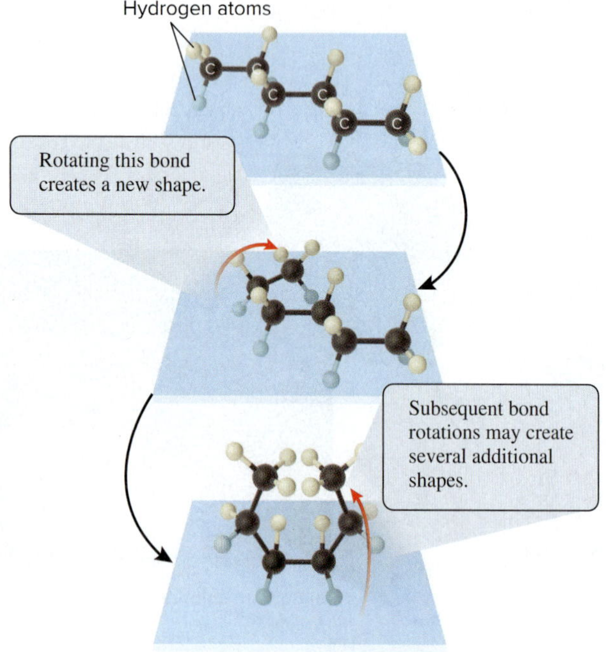


**Figure 2.9** **Shape changes in molecules.** A single molecule may assume different three-dimensional shapes without breaking any of the covalent bonds between its atoms, as shown for a six-carbon molecule. Hydrogen atoms above the blue plane are shown in white; those below the blue plane are blue.

which has 11 electrons, loses 1 electron, it becomes a sodium ion ($Na^+$) with a net positive charge (the superscript indicates the net charge of an ion). Ions that have a net positive charge are called **cations.** A sodium ion still has 11 protons, but only 10 electrons. A chlorine atom (Cl), which has 17 electrons, can gain an electron and become a chloride ion ($Cl^-$) with a net negative charge—it still has 17 protons but now has 18 electrons. Ions with a net negative charge are called **anions.**

Ions such as $Na^+$ and $Cl^-$ are relatively stable because the outer electron shells are full. A sodium atom has one electron in its third (outermost) shell. If it loses this electron to become $Na^+$, it no longer has a third shell, and the second shell, which is full, becomes its outermost shell (see Figure 2.7). Alternatively, a Cl atom has seven electrons in its outermost shell. If it gains an electron to become a chloride ion ($Cl^-$), its outer shell becomes full with eight electrons. Some atoms can gain or lose more than one electron. For instance, a calcium atom, which has 20 electrons, loses 2 electrons to become a calcium ion, depicted as $Ca^{2+}$.

An **ionic bond** occurs when a cation binds to an anion. The right side of Figure 2.7a shows an ionic bond between $Na^+$ and $Cl^-$ to form NaCl, or common table salt. NaCl can form crystals in which the cations and anions form a regular array. Figure 2.7b shows a NaCl crystal, in which the sodium and chloride ions are held together by ionic bonds.

## Molecules May Change Their Shapes

When atoms combine, they form molecules with various three-dimensional shapes, depending on the arrangements and numbers of bonds between their atoms. As an example, let's consider the arrangements of covalent bonds in a few simple molecules, including $H_2O$ (Figure 2.8). These covalent bonds create defined angles between the atoms. This gives molecules very specific shapes, as shown in the three examples of Figure 2.8.

Within certain limits, the shape of a molecule can change without breaking its covalent bonds. As illustrated in Figure 2.9, a molecule of 6 carbon atoms and 14 hydrogen atoms can assume different shapes as a result of rotations around various covalent bonds. The three-dimensional, flexible shape of molecules contributes to their biological properties. Imagine the possible variety of shape changes that could exist in very large molecules of dozens, hundreds, or even thousands of atoms, as typically occur in living organisms. In such cases, the binding of one molecule to another may affect the shape of one or both of the molecules; it is often this shape change that triggers the biological activity of a molecule in a living organism.

## 2.2 Reviewing the Concepts

- Two or more atoms bonded together make up a molecule. A compound is a molecule composed of two or more different elements.
- Atoms tend to form bonds that fill their outer shell with electrons. Covalent bonds, in which atoms share electrons, are very strong chemical bonds (Figures 2.3, 2.4).

- The electronegativity of an atom is a measure of its ability to attract electrons in a bond with another atom. When two atoms with different electronegativities combine, they form a polar covalent bond. Polar molecules, such as $H_2O$, are largely composed of polar bonds, and nonpolar molecules are composed predominantly of nonpolar bonds (Figure 2.5).
- One molecule may loosely associate with another molecule through weak interactions such as hydrogen bonds and van der Waals dispersion forces (Figure 2.6).
- If an atom or a molecule gains or loses one or more electrons, it acquires a net electric charge and becomes an ion. The strong attraction between two oppositely charged ions forms an ionic bond (Figure 2.7).
- The three-dimensional, flexible shape of molecules allows them to interact and contributes to their biological properties (Figures 2.8, 2.9).

## 2.2 Testing Your Knowledge

1. Which of the following is a compound?
   a. Na
   b. $Na^+$
   c. $O_2$
   d. HF
   e. C
2. How many additional covalent bonds are available for each carbon in a C═C double bond?
   a. 0
   b. 1
   c. 2
   d. 3
   e. 4

## 2.3 Chemical Reactions

### Learning Outcome

1. Relate the concept of a chemical reaction to the concept of chemical equilibrium.

### Chemical Reactions Create New Compounds from Elements or Other Compounds

A **chemical reaction** occurs when one or more substances are changed into other substances by the making or breaking of chemical bonds. This can happen when two or more elements or compounds combine to form a new compound, or when one compound breaks down into two or more molecules.

Chemical reactions share many similar properties:

- They require a source of energy (heat) for molecules to move and encounter each other.
- Many times, they must be catalyzed. A **catalyst** is an agent that speeds up the rate of a chemical reaction. An enzyme is an example of a catalyst.
- They tend to proceed in a particular direction but eventually reach a state of equilibrium.
- They usually occur in watery environments.

To understand what we mean by direction and equilibrium, let's consider a chemical reaction between methane (a component of natural gas) and oxygen. When a single molecule of methane reacts with two molecules of oxygen, one molecule of carbon dioxide and two molecules of water are produced:

$$\underset{\text{(methane)}}{CH_4} + \underset{\text{(oxygen)}}{2\,O_2} \rightleftharpoons \underset{\text{(carbon dioxide)}}{CO_2} + \underset{\text{(water)}}{2\,H_2O}$$

As it is written here, $CH_4$ and $O_2$ are the starting materials, or **reactants,** and $CO_2$ and $H_2O$ are the **products.** The bidirectional arrows indicate that this reaction can proceed in both directions. Whether a chemical reaction is likely to proceed in a forward (left-to-right) or reverse (right-to-left) direction depends on changes in free energy, which you will learn about in Chapter 6. If we began with only $CH_4$ and $O_2$, the forward reaction would be very favorable. The reaction would produce a large amount of $CO_2$ and $H_2O$. If the products are not converted to other molecules, chemical reactions eventually reach **chemical equilibrium,** in which the rate of the formation of products equals the rate of the formation of reactants; in other words, the concentrations of products and reactants stay the same. In the case of the reaction involving $CH_4$ and $O_2$, this equilibrium occurs when nearly all of the reactants have been converted to products.

In biological systems, many reactions do not have a chance to reach chemical equilibrium. For example, the products of a reaction may immediately be converted within a cell to different substances through a second reaction. In this case, the reactants will continue to form new products until all the reactants are used up.

One common feature of chemical reactions in living organisms is that many reactions occur in watery environments. Such chemical reactions involve reactants and products that are dissolved in water. Next, we examine the properties of this amazing liquid and its importance to life.

## 2.3 Reviewing the Concepts

- A chemical reaction occurs when one or more substances are changed into different substances. All chemical reactions eventually reach chemical equilibrium, unless the products of the reaction are continually removed.

## 2.3 Testing Your Knowledge

1. In the reaction $CO_2 + H_2O \rightleftharpoons H_2CO_3$, chemical equilibrium will occur when
   a. the rate of $CO_2$ and $H_2O$ formation equals the rate of $H_2CO_3$ formation.
   b. the reactants are completely converted to products.
   c. the products are completely converted to reactants.
   d. the products are converted to other substances by a second reaction.
   e. either c or d occurs.

## 2.4 Properties of Water

### Learning Outcomes

1. Define solute and solvent.
2. Compare and contrast hydrophilic and hydrophobic substances.
3. **SCISKILLS ▶** Calculate the molarity of a solution, and explain its meaning.
4. Describe the three states of $H_2O$.
5. Make a list of the roles of water that are critical for the survival of living organisms.

The bodies of all organisms are composed largely of water; most of the cells in an organism's body are not only filled with water but also are surrounded by it. Up to 95% of the weight of certain plants comes from water. In humans, typically 60–70% of body weight is from water. The brain is roughly 70% water, blood is about 80% water, and the lungs are nearly 90% water. Even our bones are about 20% water! In addition, water is an important liquid in the surrounding environments of living organisms. For example, vast numbers of species live in watery environments.

Thus far in this chapter, we have considered the features of atoms and molecules and the nature of bonds and chemical reactions between atoms and molecules. In this section, we will turn our attention to issues related to the liquid properties of living organisms and the environment in which they live. Most of the chemical reactions that occur in nature involve molecules that are dissolved in water, including those reactions that happen inside the cells of living organisms and in the spaces that surround them.

In this section, we will examine the properties of chemicals that influence whether they dissolve in water and consider how biologists quantify the amounts of dissolved substances. In addition, we explore some of the special properties of water that make it a vital component of living organisms and their environments.

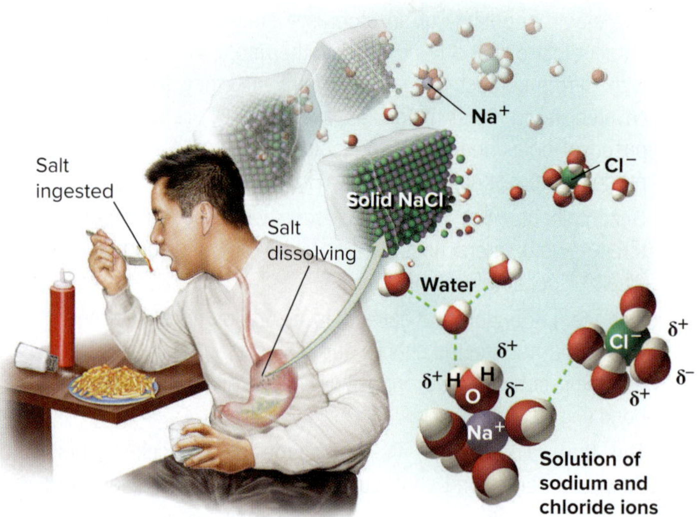


**Figure 2.10** **Table salt (NaCl crystals) dissolving in water.** The ability of water to dissolve sodium chloride crystals depends on the electrical attraction between the polar water molecules and the charged sodium ($Na^+$) and chloride ions ($Cl^-$). Water molecules surround each ion as it becomes dissolved. For simplicity, the partial charges are indicated for only two water molecules.

### Ions and Polar Molecules Readily Dissolve in Water

Substances dissolved in a liquid are known as **solutes,** and the liquid in which they are dissolved is the **solvent.** In all living organisms, the solvent for chemical reactions is water. Solutes dissolve in a solvent to form a **solution.** Solutions made with water are called **aqueous solutions.** To understand why a substance dissolves in water, we need to consider the chemical bonds in the solute molecule and those in water. As discussed earlier, the covalent bonds linking the two hydrogen atoms to the oxygen atom in a molecule of $H_2O$ are polar. Therefore, the oxygen in $H_2O$ has a slight negative charge, and each hydrogen has a slight positive charge. To dissolve in water, a substance must be electrically attracted to $H_2O$ molecules. For example, table salt (NaCl) is a solid, crystalline substance because of the strong ionic bonds between positive sodium ions ($Na^+$) and negative chloride ions ($Cl^-$). When a crystal of sodium chloride is placed in water, the partially negatively charged oxygens of water molecules are attracted to the $Na^+$, and the partially positively charged hydrogens are attracted to the $Cl^-$ (**Figure 2.10**). Clusters of $H_2O$ molecules surround the ions, allowing the $Na^+$ and $Cl^-$ to separate from each other and enter the water—that is, to dissolve.

Generally, molecules that contain ionic and/or polar covalent bonds dissolve in water. Such molecules are said to be **hydrophilic,** which literally means water-loving. In contrast, molecules composed predominantly of carbon and hydrogen are relatively insoluble in water, because carbon-carbon and carbon-hydrogen bonds are nonpolar covalent bonds. These molecules do not have partial positive and negative charges and, therefore, are not attracted to $H_2O$ molecules. Such molecules are **hydrophobic,** meaning water-fearing. Oils are a familiar example of hydrophobic molecules. Try mixing vegetable oil with water and observe the result. The two liquids separate into an oil layer and a water layer, with very little oil dissolving in the water.

### Some Molecules Have Both Hydrophilic and Hydrophobic Regions

Molecules that have both hydrophilic regions at one or more sites and hydrophobic regions at other sites are called **amphipathic** (or amphiphilic, from the Greek for both loves). When mixed with water, long, amphipathic molecules may aggregate into spheres called **micelles,** with their polar (hydrophilic) regions at the surface of the micelle, where they are attracted to the surrounding water molecules. The nonpolar (hydrophobic) ends are oriented toward the interior of the micelle (**Figure 2.11**). Such an arrangement minimizes the interaction between $H_2O$ molecules and the nonpolar ends of the amphipathic molecules, which face inward. Nonpolar molecules can dissolve in the central nonpolar regions of these clusters and thus can exist in an aqueous environment in far higher amounts than would otherwise be possible

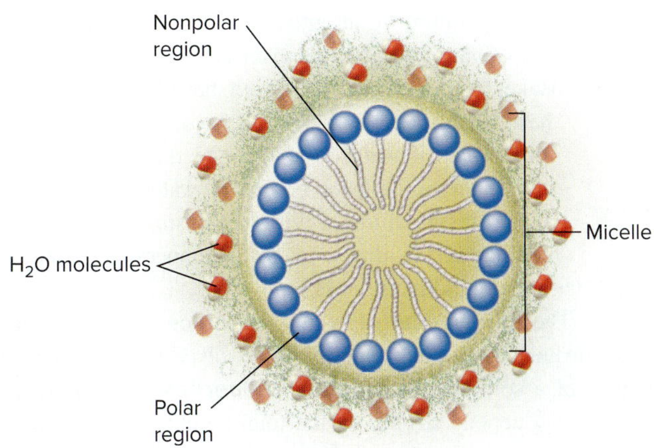


**Figure 2.11** **The formation of micelles by amphipathic molecules.** In water, amphipathic molecules tend to arrange themselves so their nonpolar regions are directed away from water molecules, while their polar regions are directed toward water molecules and can form hydrogen bonds with them.

**Concept Check:** *When oil dissolves in detergent micelles, where is the oil found?*

due to their low solubility in water. Familiar examples of amphipathic molecules are those in detergents, which can form micelles that help to dissolve oils and nonpolar molecules found in dirt. The detergent molecules in soap have polar and nonpolar ends. Oils on your skin dissolve in the nonpolar regions of the detergent, and the polar ends help the detergent rinse off in water, taking the oil with it.

In addition to micelles, amphipathic molecules may form structures consisting of double layers of molecules called bilayers. Such bilayers have two hydrophilic surfaces facing outward on both sides, in contact with water, and a hydrophobic interior facing away from water. As you will learn in Chapter 5, bilayers play a key role in cell membrane structure (look ahead to Figure 5.1).

## The Amount of a Dissolved Solute per Unit Volume of Liquid Is Its Concentration

Solute **concentration** is defined as the amount of a solute dissolved in a unit volume of solution. For example, if 1 g of NaCl were dissolved in enough water to make 1 liter (L) of solution, we would say that the solute concentration is 1 g/L.

A comparison of the concentrations of two different substances on the basis of the number of grams per liter of solution does not directly indicate how many molecules of each substance are present. For example, let's compare 10 g each of glucose ($C_6H_{12}O_6$) and sodium chloride (NaCl). Because the individual molecules of glucose have more mass than those of NaCl, 10 g of glucose contains fewer molecules than 10 g of NaCl. Therefore, another way to describe solute concentration is according to the moles of dissolved solute per volume of solution. To make this calculation, we must know three things: the amount of dissolved solute, the molecular mass of the dissolved solute, and the volume of the solution.

## Quantitative Analysis

### CONCENTRATIONS OF MOLECULES IN SOLUTION CAN BE DEFINED BY MASS AND MOLES

The **molecular mass** (also referred to as molecular weight) of a molecule is equal to the sum of the atomic masses of all the atoms in the molecule. For example, using Appendix A as a guide and rounding off to the nearest whole number, glucose ($C_6H_{12}O_6$) has a molecular mass of 180 ($[6 \times 12] + [12 \times 1] + [6 \times 16] = 180$). As mentioned earlier, a mole of a substance is the amount of the substance in grams equal to its atomic or molecular mass; thus, 1 mol of glucose is 180 g of glucose. The **molarity** of a solution is defined as the number of moles of a solute dissolved in 1 L of solution. A solution containing 180 g of glucose dissolved in enough water to make 1 L is a 1 **molar** solution of glucose (1 mol/L). By convention, a 1 mol/L solution is usually written as 1 M, where the capital M stands for molar and is defined as mol/L. If 90 g of glucose (half its molecular mass) were dissolved in enough water to make 1 L, the glucose concentration would be 0.5 mol/L, or 0.5 M.

The concentrations of solutes dissolved in the fluids of living organisms are usually much less than 1 M. Many have concentrations in the range of millimoles per liter (1 mM = 0.001 M = $10^{-3}$ M), and others are present in even smaller concentrations—micromoles per liter (1 μM = 0.000001 M = $10^{-6}$ M) or nanomoles per liter (1 nM = 0.000000001 M = $10^{-9}$ M), or less.

**Crunching the Numbers:** A typical concentration of glucose in the blood of many animals is around 5.5 mM. At that concentration, how many grams of glucose are present in 1 L of blood?

## $H_2O$ Exists in Three States

$H_2O$ is an abundant compound on Earth that exists in all three states of matter—solid (ice), liquid (water), and gas (water vapor). At the temperatures found over most regions of the planet, $H_2O$ is found primarily as a liquid in which the weak hydrogen bonds between molecules are continuously being formed, broken, and formed again as $H_2O$ molecules move due to heat energy. If more heat is added, the speed at which $H_2O$ molecules move increases, and hydrogen bonds are broken more readily; under such conditions, some molecules of $H_2O$ are able to escape into the gaseous state, becoming water vapor. If the temperature falls, $H_2O$ molecules move more slowly and consequently hydrogen bonds are broken less frequently. Larger and larger clusters of $H_2O$ molecules are formed, until at 0°C water freezes into a crystalline matrix—ice. The $H_2O$ molecules in ice exist in a much more orderly and highly structured arrangement than in the liquid state (**Figure 2.12**).

Ice: $H_2O$ molecules are less likely to move apart due to decreased heat energy. Hydrogen bonds are more stable, resulting in an orderly array of molecules.
150 ml
Liquid water: $H_2O$ molecules are in rapid motion, and hydrogen bonds continually break and re-form.


**Figure 2.12** **Structure of liquid water and ice.** In its liquid form, the hydrogen bonds between water molecules continually form, break, and re-form, resulting in a changing arrangement of molecules from instant to instant. At temperatures at or below its freezing point, $H_2O$ forms a crystalline matrix called ice. In this solid form, hydrogen bonds are more stable.

Changes between the solid, liquid, and gaseous states of $H_2O$ involve an input or a release of energy. For example, when energy is supplied to raise the temperature of water to 100°C, it boils; that is, it changes from the liquid to the gaseous state—a process called vaporization. The heat required to vaporize 1 mol of any substance at its boiling point is called the substance's **heat of vaporization.** For water, this value is very high because of the hydrogen bonds between the molecules. It takes more than five times as much heat to vaporize water than it does to raise the temperature of water from 0°C to 100°C. In contrast, energy is released when water freezes to form ice. A substance's **heat of fusion** is the amount of heat energy that must be withdrawn or released from a substance to cause it to change from the liquid to the solid state. For water, this value is also high.

Another important feature for living organisms is that water has a very high **specific heat,** defined as the amount of heat energy required to raise the temperature of 1 g of a substance by 1°C (or conversely, the amount of heat energy that must be lost to lower the temperature by 1°C). A high specific heat means that it takes considerable heat to raise the temperature of water. A related concept is **heat capacity,** which refers to the amount of heat energy required to raise the temperature of an entire object or substance. A large beaker of pure water has a greater heat capacity than a small beaker of pure water, but both have the same specific heat because both contain the same substance. These properties of water contribute to the relatively stable temperatures of large bodies of water compared with inland temperatures. Living organisms have evolved to function best within a range of temperatures consistent with the liquid phase of $H_2O$.

The temperatures at which a solution freezes or vaporizes are influenced by the amounts of dissolved solutes. These are examples of a solution's **colligative properties,** defined as those properties that depend strictly on the total number of dissolved solutes, not on the specific type of solute. Pure water freezes at 0°C and vaporizes at 100°C. Addition of solutes to water lowers its freezing point below 0°C and raises its boiling point to above 100°C. The presence of large amounts of solutes partly explains why the oceans do not freeze when the temperature falls below 0°C, whereas freshwater lakes and ponds do.

## Water Performs Many Important Tasks in Living Organisms

As discussed previously, water is the primary solvent in the fluids of all living organisms. Water permits atoms and molecules to interact in ways that would be impossible in their nondissolved states.

In addition to acting as a solvent, water serves many other remarkable functions that are critical for the survival of living organisms. For example, $H_2O$ molecules participate in many chemical reactions of this general type:

$$R_1—R_2 + H—O—H \rightarrow R_1—OH + H—R_2$$

R is a general symbol used in this case to represent a group of atoms. In this equation, $R_1$ and $R_2$ are distinct groups of atoms. On the left side, $R_1—R_2$ is a compound in which the groups of atoms are connected by a covalent bond. To be converted to products, a covalent bond is broken in each reactant, $R_1—R_2$ and H—O—H; next, OH and H (from $H_2O$) form covalent bonds with $R_1$ and $R_2$, respectively. Reactions of this type are known as **hydrolysis reactions** (from the Greek *hydro,* meaning water, and *lysis,* meaning to break apart), because water is used to break apart another molecule (Figure 2.13a). As discussed in Chapter 3 and later chapters, many large molecules are broken down into smaller, biologically important units by hydrolysis reactions.

Another feature of water is that it is incompressible—its volume does not significantly decrease when subjected to high pressure. This has biological importance for many organisms that use water to provide force or support. For example, water supports the bodies of some invertebrates, and it provides turgidity (stiffness) and support for plants (Figure 2.13b).

Water is also the means by which waste compounds are eliminated from an animal's body (Figure 2.13c). In mammals, for example, the kidneys filter out soluble waste products derived from the breakdown of proteins and other compounds. The filtered products remain in solution in a watery fluid, which eventually becomes urine and is excreted.

## Biology Principle

### Living Organisms Maintain Homeostasis

Healthy organisms have a normal supply of water that maintains such homeostatic functions as body temperature, ion balance, and waste levels.

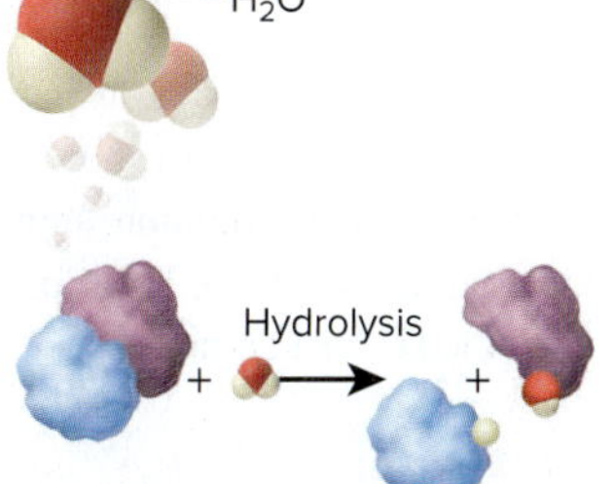


**(a) Water participates in chemical reactions.**

**(b) Water provides support.** The plant on the right is wilting due to lack of water.

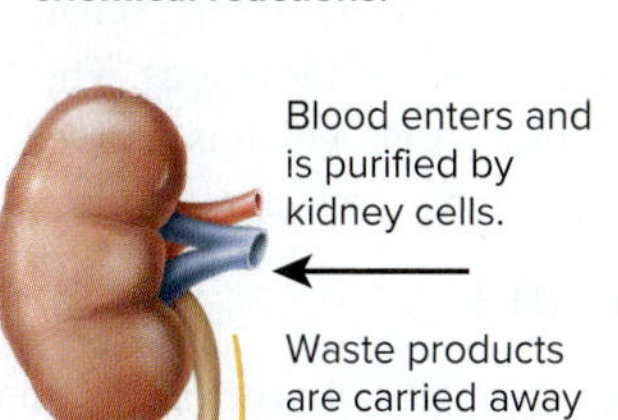

Blood enters and is purified by kidney cells.
Waste products are carried away in the watery urine.

**(c) Water is used to eliminate soluble wastes.**

**(d) Evaporation helps some animals dissipate body heat.**

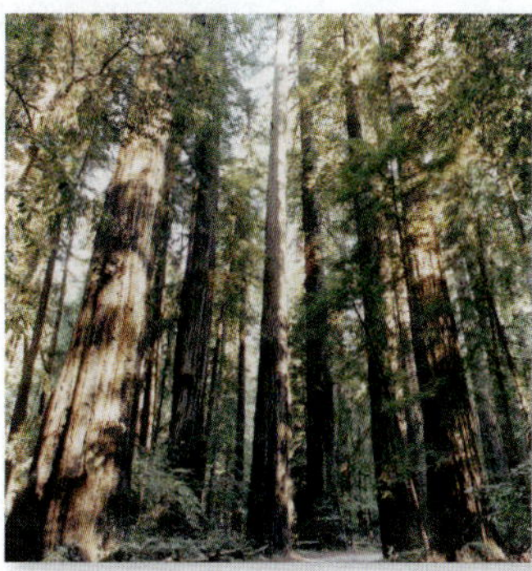

**(e) The cohesive force of water molecules aids in the movement of water against gravity in plants.**

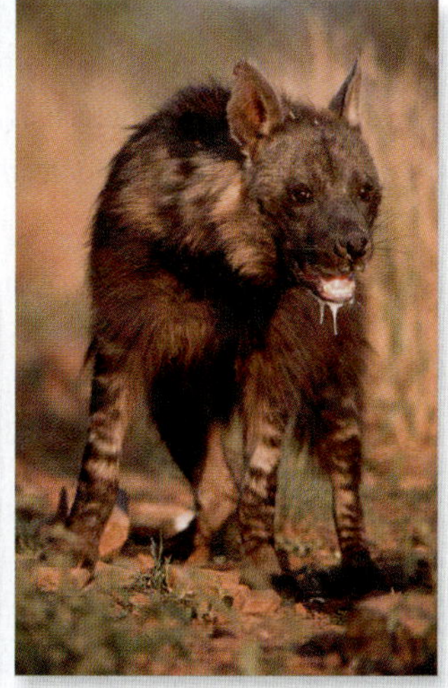

**(f) Water in saliva serves as a lubricant during—or as shown here, in anticipation of—feeding.**

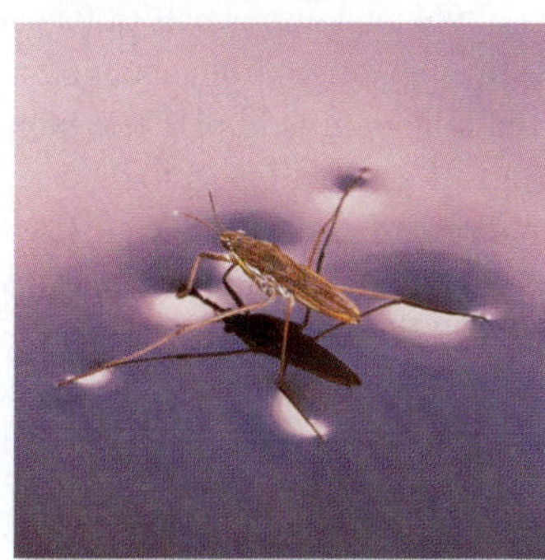

**(g) The surface tension of water explains why this water strider doesn't sink.**

**Figure 2.13** **Some of the amazing functions of water.** In addition to acting as a solvent, water serves many crucial functions in nature.

*(b)* © Aaron Haupt/Science Source; *(d)* © Chris McGrath/Getty Images; *(e)* © Dana Tezarr/Getty Images; *(f)* © Anthony Bannister/Gallo Images/Corbis; *(g)* © Hermann Eisenbeiss/Science Source

Recall from our discussion of water's properties that it takes considerable energy in the form of heat to convert water from a liquid to a gas. This feature has great biological significance. Although we are familiar with the phenomenon of boiling water being converted to water vapor, water can also vaporize into the gaseous state even at ordinary temperatures. This process is known as **evaporation.** To understand this, imagine that in any volume of water at any temperature, some vibrating water molecules have higher energy than others. Those with the highest energy break their hydrogen bonds and escape into the gaseous state. During this process, energy in the form of heat is released into the environment. Evaporation is an important mechanism by which many animals cool themselves on hot days (Figure 2.13d).

The hydrogen-bonding properties of water affect its ability to form droplets and to adhere to surfaces. The phenomenon of water molecules attracting each other is called **cohesion.** Water exhibits strong cohesion due to hydrogen bonding. Cohesion aids in the movement of water through the vessels of plants (Figure 2.13e). A property similar to cohesion is **adhesion,** which is the ability of water to be attracted to, and thus adhere to, a surface that is not electrically neutral. Water tends to cling to surfaces to which it can hydrogen bond, such as a paper towel; paper is comprised of compounds such as cellulose that have an abundance of partial negative and positive charges, giving paper the ability to form hydrogen bonds. In organisms, the adhesive properties of water allow it, for example, to coat the surfaces of the digestive tract of animals and act as a lubricant for the passage of food (Figure 2.13f).

**Surface tension** is a measure of the attraction between molecules at the surface of a liquid. In the case of water, the attractive force between hydrogen-bonded water molecules at the interface between water and air is what causes water to form droplets. The surface water molecules attract each other into a configuration (roughly that of a sphere) that reduces the number of water molecules in contact with air. You can see this by slightly overfilling a glass with water; the water forms an oval shape above the rim. Likewise, surface tension allows certain insects, such as water striders, to walk on the surface of a pond without sinking (Figure 2.13g).

## 2.4 Reviewing the Concepts

- Water is the solvent for most chemical reactions in all living organisms. A solute dissolves in water to form a solution. Solute concentration refers to the amount of a solute dissolved in a unit volume of water (Figure 2.10).
- Molecules with ionic and polar covalent bonds generally are hydrophilic, whereas nonpolar molecules are hydrophobic. Amphipathic molecules have hydrophilic and hydrophobic regions (Figure 2.11).
- $H_2O$ exists as ice, liquid water, and water vapor (gas) (Figure 2.12).
- The colligative properties of water depend on the number of dissolved solutes.
- Water's high heat of vaporization and high heat of fusion make it a very stable liquid.

- $H_2O$ molecules participate in many chemical reactions in living organisms, including hydrolysis reactions. In living organisms, water provides support, is used to eliminate wastes, dissipates body heat, and serves as a lubricant (Figure 2.13).

## 2.4 Testing Your Knowledge

1. The amount of heat required to convert 1 mol of $H_2O$ from liquid to gas is called water's
   a. specific heat.
   b. heat of vaporization.
   c. heat capacity.
   d. heat of fusion.
   e. cohesion.
2. The molecular formula of table sugar (sucrose) is $C_{12}H_{22}O_{11}$. What is its molecular mass?
   a. 45
   b. 182
   c. 67
   d. 342
   e. 91

## 2.5 pH and Buffers

### Learning Outcomes

1. **SCISKILLS ▶** Explain how $H_2O$ has the ability to ionize into hydroxide ions ($OH^-$) and hydrogen ions ($H^+$), and calculate the concentration of hydrogen and hydroxide ions at a given pH.
2. Give examples of how buffers maintain a stable environment in an animal's body fluids.

### Hydrogen Ion Concentrations Are Changed by Acids and Bases

Pure water has the ability to ionize to a very small extent into **hydroxide ions** ($OH^-$) and hydrogen ions ($H^+$). In pure water, the concentrations of $H^+$ and $OH^-$ are both $10^{-7}$ mol/L, or $10^{-7}$ M. An inherent property of water is that the product of the concentrations of $H^+$ and $OH^-$ is always $10^{-14}$ M at 25°C. Therefore, in pure water, $[H^+][OH^-] = [10^{-7}\text{ M}][10^{-7}\text{ M}] = 10^{-14}$ M. (The brackets around the symbols for the hydrogen and hydroxide ions indicate concentration.)

When certain substances are dissolved in water, they may release or absorb $H^+$ or $OH^-$, thereby altering the relative concentrations of these ions. Substances that release hydrogen ions in solution are called **acids.** Two examples are hydrochloric acid and carbonic acid:

$$\underset{\text{(hydrochloric acid)}}{HCl} \rightarrow H^+ + \underset{\text{(chloride ion)}}{Cl^-}$$

$$\underset{\text{(carbonic acid)}}{H_2CO_3} \rightleftharpoons H^+ + \underset{\text{(bicarbonate ion)}}{HCO_3^-}$$

Hydrochloric acid is called a **strong acid** because it completely dissociates into $H^+$ and $Cl^-$ when added to water (which is why the arrow is not bidirectional in this reaction). By comparison, carbonic acid is a **weak acid** because some of it remains in the $H_2CO_3$ state when dissolved in water (note the bidirectional arrows).

Compared with an acid, a **base** has the opposite effect when dissolved in water—it binds hydrogen ions in solution. This can occur in different ways. Some bases, such as sodium hydroxide (NaOH), release $OH^-$ when dissolved in water:

$$\underset{\text{(sodium hydroxide)}}{NaOH} \rightarrow \underset{\text{(sodium ion)}}{Na^+} + OH^-$$

When a base such as NaOH increases the $OH^-$ concentration, some of the hydrogen ions bind to these hydroxide ions to form $H_2O$. Therefore, increasing the $OH^-$ concentration decreases the $H^+$ concentration.

### pH Is a Measure of the $H^+$ Concentration of a Solution

The addition of acids and bases to water can greatly change the $H^+$ and $OH^-$ concentrations over a very broad range. Therefore, scientists use a log scale to describe the concentrations of these ions. The $H^+$ concentration is expressed as the solution's **pH,** which is defined as the negative logarithm to the base 10 of the $H^+$ concentration.

$$pH = -\log_{10}[H^+]$$

To understand what this equation means, let's consider a few examples. A solution with an $H^+$ concentration of $10^{-7}$ M has a pH of 7. A solution in which the pH is 7 is said to be neutral because $[H^+]$ and $[OH^-]$ are equal. A solution in which $[H^+] = 10^{-6}$ M has a pH of 6. A solution at pH 6 is said to be **acidic,** because it contains more $H^+$ ions than $OH^-$ ions. Note that as the acidity increases, the pH decreases. A solution with a pH above 7 is considered to be **alkaline.** Figure 2.14 shows the pH values of some familiar fluids. Keep in mind that each change of one pH unit represents a 10-fold difference in $H^+$ concentration.

Why is pH of importance to biologists? The answer lies in the observation that $H^+$ and $OH^-$ can readily bind to many kinds of ions and molecules. For this reason, the pH of a solution can affect

- the shapes and functions of molecules,
- the rates of many chemical reactions,
- the ability of two molecules to bind to each other, and
- the ability of ions or molecules to dissolve in water.

Due to the various effects of pH, many biological processes function best within a very narrow range of pH, and even small shifts can have a negative effect. In living cells, the pH ranges from about 6.5 to 7.8 and is carefully regulated to avoid major shifts in pH. The blood of the human body has a normal range of about pH 7.35 to 7.45 and is therefore slightly alkaline. Certain diseases, such as kidney disease, can change blood pH by a few tenths of a unit. When this happens, the enzymes in the body that are required for normal metabolism can no longer function optimally, leading to additional illness. As described next, living organisms have buffers to help prevent such changes in pH.

### Buffers Minimize Fluctuations in pH

What factors might alter the pH of an organism's fluids? In plants, external factors such as acid rain and other forms of pollution can decrease the pH of water entering the roots. In animals, exercise generates lactic acid, and certain disease states can raise or lower the pH of blood.

Organisms have several ways to cope with changes in pH. In mammals, for example, the kidneys secrete acidic or alkaline

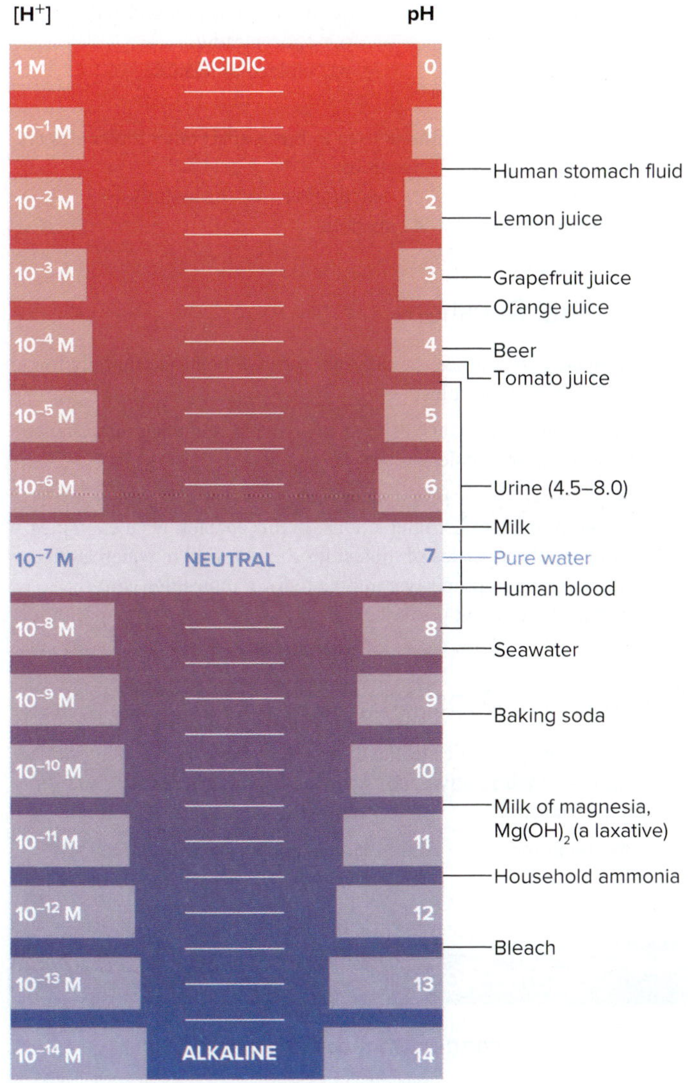


**Figure 2.14** The pH scale and the relative acidities of common substances.

**Concept Check:** *What is the $OH^-$ concentration at pH 8?*

compounds into the blood when the blood pH becomes imbalanced. A very common mechanism by which pH balance is regulated in diverse organisms involves the actions of acid-base buffers. A **buffer** is a pair of substances that minimizes fluctuations in the pH of the fluids of living organisms. A buffer is composed of a weak acid and its related base. One such buffer is carbonic acid ($H_2CO_3$) and bicarbonate ions ($HCO_3^-$), called the bicarbonate pathway, which functions to keep the pH of an animal's body fluids within a narrow range:

$$CO_2 + H_2O \rightleftharpoons \underset{\text{(carbonic acid)}}{H_2CO_3} \rightleftharpoons H^+ + \underset{\text{(bicarbonate)}}{HCO_3^-}$$

This buffer can work in both directions. For example, if the pH of an animal's body fluids were to increase (that is, the $H^+$ concentration decreased), the bicarbonate pathway would proceed primarily from left to right. $CO_2$ would combine with water to make $H_2CO_3$, and then the $H_2CO_3$ would dissociate into $H^+$ and $HCO_3^-$. This would increase the $H^+$ concentration and thereby decrease the pH. Alternatively, when the pH of an animal's blood decreases (that is, the $H^+$ concentration increases), this pathway proceeds primarily in the reverse direction. Bicarbonate combines with $H^+$ to make $H_2CO_3$, which then dissociates to $CO_2$ and $H_2O$. This process removes $H^+$ from the blood, restoring it to its normal pH, and the $CO_2$ is eliminated by the lungs or other structures. Many buffers exist in nature. Buffers found in living organisms function most efficiently at the normal range of pH values that occur in that organism.

## 2.5 Reviewing the Concepts

- The pH of a solution is its hydrogen ion concentration. Alkaline solutions have a pH higher than 7, and acidic solutions have a pH lower than 7 (Figure 2.14).
- Buffers minimize pH fluctuations in the fluids of living organisms.

## 2.5 Testing Your Knowledge

**1.** Refer to the following reaction occurring in a solution of water:

$$CO_2 + H_2O \rightleftharpoons H_2CO_3 \rightleftharpoons H^+ + HCO_3^-$$

If additional $HCO_3^-$ was added to this solution, what would happen?

**a.** The rate of $H^+$ production would increase.
**b.** The rate of $CO_2$ production would increase.
**c.** The forward reaction (left to right) would be favored.
**d.** The pH of the solution would decrease.
**e.** The solution would become more acidic.

**2.** The pH of a solution is defined as

**a.** the logarithm of the $H^+$ concentration.
**b.** the negative logarithm to the base 10 of the $H^+$ concentration.
**c.** the concentration of hydrogen ions in a solution.
**d.** the inverse of the concentration of hydrogen ions in a solution.
**e.** the negative logarithm to the base 10 of the $OH^-$ concentration.

# Assess and Discuss

## Test Yourself

1. ____________________ make up the nucleus of an atom.
   a. Protons and electrons
   b. Protons and neutrons
   c. Protons, neutrons, and electrons
   d. Neutrons and electrons
   e. Electrons

2. Living organisms are composed mainly of which atoms?
   a. calcium, hydrogen, nitrogen, and oxygen
   b. carbon, hydrogen, nitrogen, and oxygen
   c. hydrogen, nitrogen, oxygen, and helium
   d. carbon, helium, nitrogen, and oxygen
   e. carbon, calcium, hydrogen, and oxygen

3. The ability of an atom to attract electrons in a bond with another atom is termed its
   a. hydrophobicity.
   b. electronegativity.
   c. solubility.
   d. valence.
   e. van der Waals dispersion force.
4. Hydrogen bonds differ from covalent bonds in that
   a. covalent bonds can form between any type of atom, but hydrogen bonds form only between H and O.
   b. covalent bonds involve sharing of electrons, and hydrogen bonds involve the complete transfer of electrons.
   c. covalent bonds result from equal sharing of electrons, but hydrogen bonds involve unequal sharing of electrons.
   d. covalent bonds involve sharing of electrons between atoms, but hydrogen bonds are the result of weak attractions between a hydrogen atom of a polar molecule and an electronegative atom of another polar molecule.
   e. covalent bonds are weak bonds that break easily, but hydrogen bonds are strong links between atoms that are not easily broken.
5. During the formation of ions
   a. the atomic numbers of an anion and a cation both increase by 1.
   b. an anion loses an electron and a cation gains an electron.
   c. an anion gains an electron and a cation loses an electron.
   d. an anion and a cation both gain an electron.
   e. an anion and a cation both lose an electron.
6. Chemical reactions in living organisms
   a. may continue to completion if the products are removed.
   b. often require a catalyst to speed up the process.
   c. are in many cases reversible.
   d. occur in liquid environments, such as water.
   e. may exhibit all of the above.
7. Solutes that easily dissolve in water are said to be
   a. hydrophobic.
   b. hydrophilic.
   c. nonpolar.
   d. aqueous.
   e. micelles.
8. The molecular mass of glucose is about 180 g/mole. If 45 g of glucose is dissolved into a final volume of 0.5 L of water, what is the molarity of the solution?
   a. 0.125 M
   b. 0.25 M
   c. 0.5 M
   d. 1.0 M
   e. 2.0 M
9. The sum of the atomic masses of all the atoms of a molecule is its
   a. atomic weight.
   b. molarity.
   c. molecular mass.
   d. concentration.
   e. molecular formula.
10. Reactions in which water is used to break apart other molecules are known as __________ reactions.
   a. hydrophilic
   b. hydrophobic
   c. dehydration
   d. anabolic
   e. hydrolysis

## Conceptual Questions

1. Compare and contrast the different types of bonds commonly found in biological molecules.
2. What is the significance of molecular shape, and what may change the shape of molecules?
3. **PRINCIPLES** A principle of biology is that new properties emerge from complex interactions. In this chapter, what examples of emergent properties of molecules can you find in which atoms with one type of property combine to form molecules with completely different properties?

## Collaborative Questions

1. Discuss the properties of the three subatomic particles of atoms, including their relative mass and charge, and where they exist within the atom.
2. Discuss properties of water that make it possible for life to exist.

# The Chemical Basis of Life II: Organic Molecules

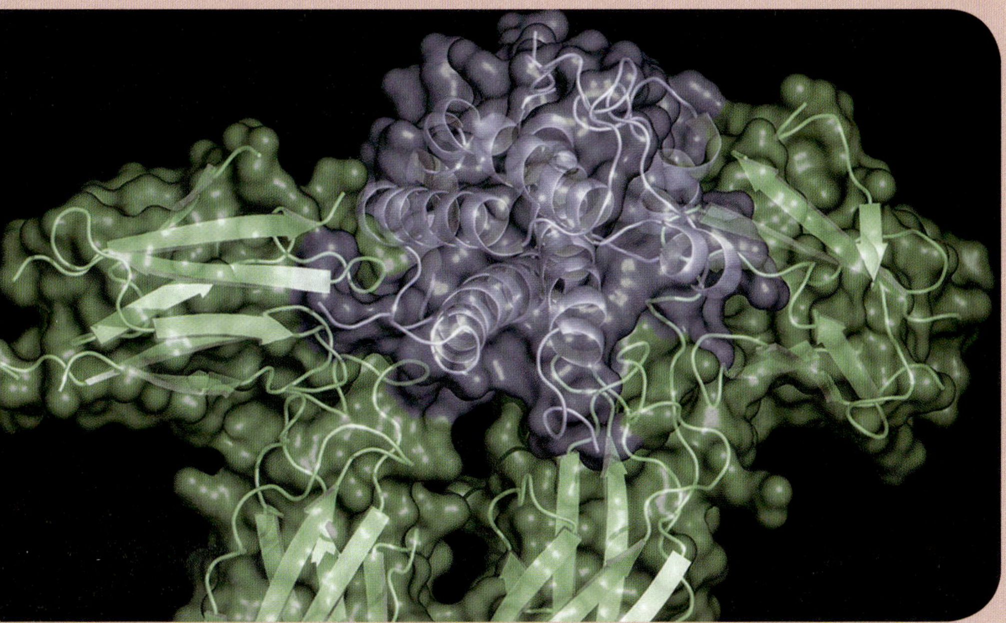

© Daniel Gage, University of Connecticut

**A model showing the structure of two proteins that have bonded together.** A protein is a type of organic macromolecule.

## Chapter Outline



Of the countless possible molecules that can be produced from the known elements in nature, certain types contain carbon and are found in all forms of life. These carbon-containing molecules are collectively referred to as **organic molecules,** so named because they were first discovered in living organisms. The science of carbon-containing molecules is known as organic chemistry.

Prior to 1828, some scientists accepted a concept called "vitalism," which held that organic molecules were created by, and therefore imparted with, a vital life force that was contained within living organisms. Consequently, there was no rationale for attempting to synthesize such molecules, because they were believed to arise only through the intervention of mysterious qualities associated with life. However, in 1828, German researcher Friedrich Wöhler discovered that he had synthesized the organic molecule urea, a common waste product in animals. Wöhler's great achievement paved the way for a revolution in our understanding of the link between chemistry and biology, a discipline known as **biochemistry.**

Organic molecules include many lipids and also large, complex compounds called **macromolecules,** such as carbohydrates, proteins, and nucleic acids. Macromolecules differ from lipids in being long strings of repeating units of smaller molecules bonded together. In this chapter, we survey the structures of these molecules and examine their chief functions. We begin with the element whose chemical properties are fundamental to the formation of biologically important molecules: carbon. This element provides the atomic scaffold on which life is built.

# 3.1 The Carbon Atom and Carbon-Containing Molecules

**Learning Outcomes**

1. Explain the properties of carbon that make it the chemical basis of all life.
2. Describe the variety and chemical characteristics of common functional groups of organic compounds.

In this section, we will examine the bonding properties of carbon that create biologically important organic molecules with distinct functions and shapes.

## Carbon Is an Essential Element for Life

A key property of the carbon atom is its ability to form multiple covalent bonds with other atoms, including other carbon atoms. Carbon has four electrons in its outer (second) shell, and it requires eight electrons, or four additional electrons, to fill that shell (Figure 3.1). In living organisms, carbon atoms most commonly form covalent bonds with other carbon atoms and with hydrogen, oxygen, nitrogen, and sulfur atoms. Bonds between two carbon atoms, between carbon and oxygen, or between carbon and nitrogen can be single or double, or in the case of certain C≡C and C≡N bonds, triple. In addition, carbon bonds may occur in configurations that are linear, ringlike, or highly branched, creating even more molecular diversity. As a result of these properties of carbon bonding, a vast number of organic compounds may be formed from only a small number of chemical elements.

As you learned in Chapter 2, bonds between atoms with similar electronegativities—for example, between two carbon atoms (C—C) or between carbon and hydrogen atoms (C—H)—are called nonpolar covalent bonds. **Hydrocarbons** are molecules with predominantly or entirely C—H and C—C bonds, and consequently are hydrophobic and poorly soluble in water. In contrast, when carbon forms polar covalent bonds with more electronegative atoms, such as oxygen or nitrogen, the molecule assumes regions of partial negative and partial positive charges (refer back to Figure 2.5). As a result, such a molecule is hydrophilic and much more soluble in water, due to its electrical attraction to polar water molecules. The ability of carbon to form both polar and nonpolar bonds (Figure 3.2) contributes to its presence in an astonishing variety of biologically important molecules.

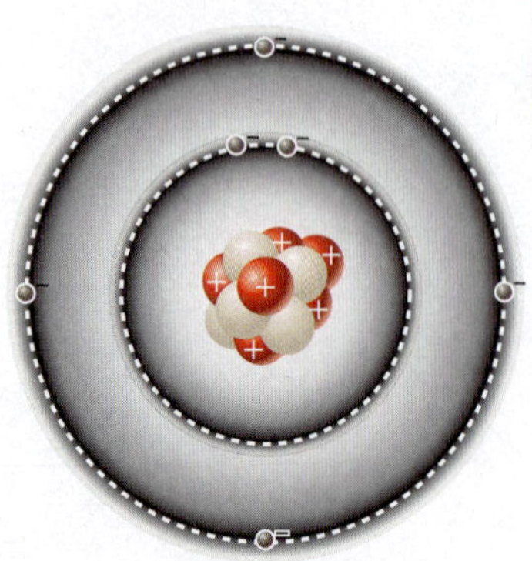

The electrons in the second shell are available for forming four covalent bonds.
The carbon atom


**Figure 3.1 Model of the carbon atom.** Carbon atoms have only four electrons in their outer (second) electron shell, which allows carbon to form four covalent bonds.

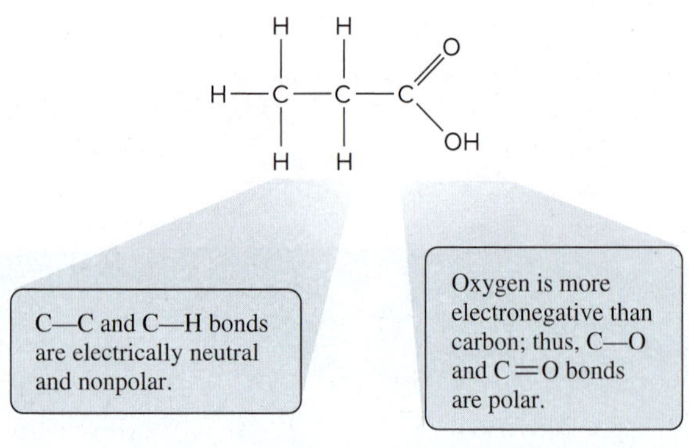


**Figure 3.2 Nonpolar and polar bonds in an organic molecule.** Carbon can form both nonpolar and polar bonds, and both single and double bonds, as shown here for the molecule propionic acid, a common food preservative.

Another feature of carbon that is important to living organisms is that carbon bonds are stable within the large range of temperatures associated with life. This property arises in part because the carbon atom is very small relative to most other atoms. Therefore, the distance between carbon atoms forming a C—C bond is short. Shorter bonds tend to be stronger and more stable than longer bonds between two large atoms. For this reason, carbon bonds are compatible with what we observe about life-forms today; namely, living organisms can inhabit environments with a range of temperatures, from the Earth's frigid, icy poles to the superheated water of deep-sea vents.

## Carbon Atoms May Be Arranged into Functional Groups That Have Specific Properties

Aside from the simplest hydrocarbons, most organic molecules and macromolecules contain **functional groups**—groups of atoms with characteristic chemical structures and properties. Each type of functional group exhibits similar chemical properties in all molecules in which it occurs. For example, the amino group ($NH_2$) acts as a base. At the pH found in living organisms, amino groups readily bind $H^+$ to become $NH_3^+$, thereby removing $H^+$ from an aqueous solution and increasing the pH. As discussed later in this chapter, amino groups are widely found in proteins and certain other types of organic molecules. Table 3.1 describes examples of functional groups found in many different types of organic molecules.

### 3.1 Reviewing the Concepts

- Carbon can form up to four covalent bonds (polar or nonpolar) with other atoms (Figures 3.1, 3.2). A vast number of organic compounds can be formed from carbon and a relatively few other elements.
- Carbon bonds are stable at the different temperatures associated with life.
- Organic compounds may contain functional groups of atoms with characteristic properties (Table 3.1).

**Table 3.1 Some Biologically Important Functional Groups That Bond to Carbon**

| Functional group* (with shorthand notation) | Formula | Examples of where they are found | Properties |
|---|---|---|---|
| Amino ($-NH_2$) | $R-NH_2$ (R—N bonded to two H) | Amino acids | Weakly basic (can accept $H^+$); polar; forms part of peptide bonds |
| Carbonyl† ($-CO$) | | Steroids, waxes, and proteins | Polar; highly chemically reactive; forms hydrogen bonds |
| Ketone | $R-C(=O)-R'$ | | |
| Aldehyde | $R-C(=O)-H$ | | |
| Carboxyl ($-COOH$) | $R-C(=O)-OH$ | Amino acids, fatty acids | Acidic (gives up $H^+$ in water); forms part of peptide bonds |
| Hydroxyl ($-OH$) | $R-OH$ | Steroids, alcohol, carbohydrates, some amino acids | Polar; forms hydrogen bonds with water |
| Methyl ($-CH_3$) | $R-CH_3$ (R—C bonded to three H) | May be attached to DNA, proteins, and carbohydrates | Nonpolar |
| Phosphate ($-PO_4^{2-}$) | $R-O-P(=O)(-O^-)-O^-$ | Nucleic acids, ATP, phospholipids | Polar; weakly acidic and thus negatively charged at typical pH of living organisms |
| Sulfate ($-SO_4^-$) | $R-O-S(=O)(=O)-O^-$ | May be attached to carbohydrates, proteins, and lipids | Polar; negatively charged at typical pH |
| Sulfhydryl ($-SH$) | $R-SH$ | Proteins that contain the amino acid cysteine | Polar; forms disulfide bridges in many proteins |

*This list contains some of the functional groups that are most important in biology. However, many more functional groups have been identified by biochemists. R and R′ represent the remainder of the molecule.

†A carbonyl group is C=O. In a ketone, the carbon forms covalent bonds with two other carbon atoms. In an aldehyde, the carbon is linked to a hydrogen atom.

## 3.1 Testing Your Knowledge

1. Carbon has an atomic number of ________ and has ______ electrons in its outer electron shell. (Refer back to Chapter 2 for help with atomic numbers.)
   a. 4 and 6
   b. 6 and 4
   c. 4 and 2
   d. 12 and 4
   e. 12 and 6

2. Hydrocarbons
   a. are molecules containing predominantly carbon, hydrogen, nitrogen, and oxygen.
   b. typically are not very soluble in water.
   c. are created when two atoms of very different electronegativities form bonds.
   d. are molecules primarily composed of bonds between atoms of similar electronegativities.
   e. both b and d.

## 3.2 Synthesis and Breakdown of Organic Molecules

### Learning Outcome

1. Diagram how small organic molecules are assembled into larger ones by dehydration reactions and how hydrolysis reactions can reverse this process.

Many organic molecules are relatively small, but some are extremely large macromolecules composed of many thousands of atoms. Some important macromolecules found in living cells are formed by linking together many smaller molecules called **monomers** (meaning one part) and are thus also known as **polymers** (meaning many parts). The linkage of monomers occurs via a type of chemical reaction called a **dehydration reaction,** so named because a water molecule is released when the monomers are linked together (Figure 3.3a). Notice that the length of a polymer may be extended again and again by additional dehydration reactions. Some polymers can reach great lengths by this mechanism.

Polymers can be broken down into their constituent monomers by **hydrolysis reactions**. The term hydrolysis refers to the observation that a water molecule is used to break, or lyse, the linkage that holds monomers together (Figure 3.3b). Therefore, the formation of polymers in organisms is generally reversible; once formed, a polymer can later be broken down.

## 3.2 Reviewing the Concepts

- Polymers are large macromolecules built up by dehydration reactions, in which individual monomers combine with each other. Polymers are broken down into monomers by hydrolysis reactions (Figure 3.3).

## 3.2 Testing Your Knowledge

1. Formation of polymers from monomers typically involves
   a. removal of a molecule of water.
   b. addition of a molecule of water.
   c. a hydrolysis reaction.
   d. a dehydration reaction.
   e. both a and d.

## 3.3 Overview of the Four Major Classes of Organic Molecules Found in Living Cells

### Learning Outcome

1. Compare and contrast the structures and functions of carbohydrates, lipids, proteins, and nucleic acids.

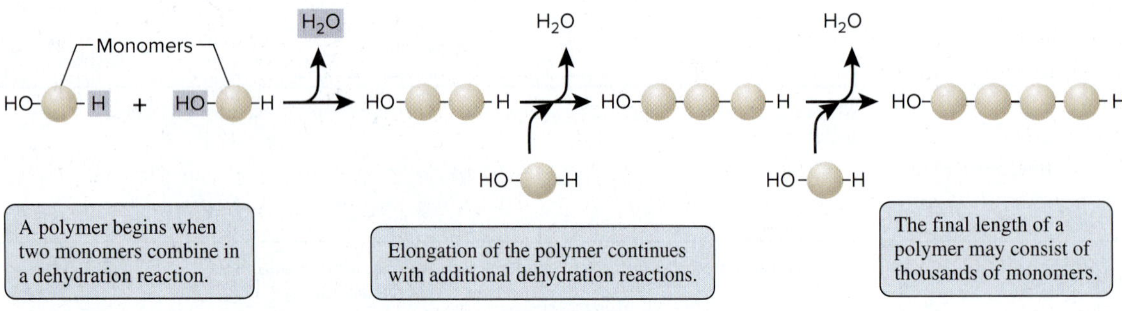

Monomers
HO- -H + HO- -H
$H_2O$
HO- -H
A polymer begins when two monomers combine in a dehydration reaction.
Elongation of the polymer continues with additional dehydration reactions.
The final length of a polymer may consist of thousands of monomers.


**(a) Polymer formation by dehydration reactions**

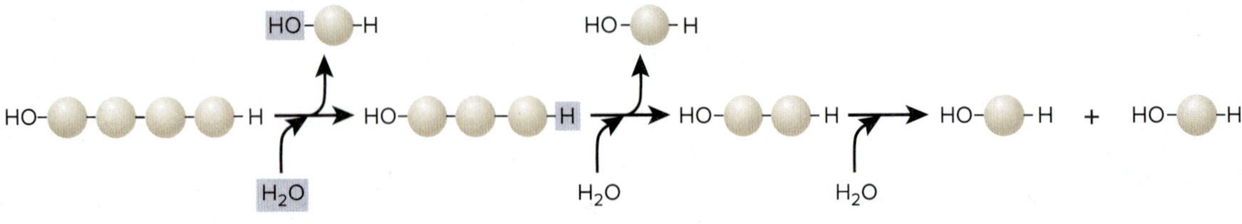


**(b) Breakdown of a polymer by hydrolysis reactions**

**Figure 3.3 Formation and breakdown of polymers.** **(a)** Monomers combine to form polymers in living organisms by dehydration reactions, in which a molecule of $H_2O$ is removed each time a new monomer is added to the growing polymer. **(b)** Polymers can be broken down into their constituent monomers by hydrolysis reactions, in which a molecule of $H_2O$ is added each time a monomer is released.

**Table 3.2 A Comparison of the Four Types of Organic Molecules Found in Living Organisms**

| Type | Structure | Key Functions | Examples |
|---|---|---|---|
| Carbohydrates | The general formula is $C_n(H_2O)_n$, where $n$ is a whole number. | Simple carbohydrates are broken down to make ATP, which is used as a source of energy. Larger carbohydrates store energy or may play a structural role, as in plant cell walls. Some carbohydrates function as molecular tags, allowing recognition of specific cells and molecules. | Simple sugars, such as glucose; larger polymers, such as starch and cellulose |
| Lipids | Lipids are nonpolar molecules that are primarily composed of carbon and hydrogen, with some oxygen. | Lipids are a key part of cell membranes and function as hormones and in energy storage; in animals, they act as insulators and shock absorbers. | Phospholipids, estrogen, testosterone, triglycerides |
| Proteins | A polypeptide is a structural unit composed of a linear sequence of amino acids. A protein is a functional unit composed of one or more polypeptides. | Proteins play a key role in cell structure and carry out a diverse array of cellular functions; for example, there are proteins involved with gene expression and regulation, motor proteins, defense proteins, cell-signaling proteins, metabolic enzymes, structural proteins, and transporters. | Look ahead to Table 3.3. |
| Nucleic acids | A nucleic acid is a linear sequence of nucleotides; DNA is double-stranded. | DNA stores genetic information in units called genes. RNA is made from DNA and provides access to that information. | DNA and RNA |

By analyzing the cells of many different species, researchers have determined that all forms of life have organic molecules and macromolecules that fall into four broad categories, based on their chemical and biological properties: carbohydrates, lipids, proteins, and nucleic acids. Table 3.2 outlines the general structures and functions of these molecules and provides some examples. In the next sections, we will examine them in greater detail.

## 3.3 Reviewing the Concepts

- The four major classes of organic molecules are carbohydrates, lipids, proteins, and nucleic acids (Table 3.2).

## 3.3 Testing Your Knowledge

1. Which of the following classes of organic molecules are important for storing energy?
   a. nucleic acids (DNA and RNA)
   b. proteins
   c. carbohydrates
   d. lipids
   e. both c and d

## 3.4 Carbohydrates

### Learning Outcomes

1. Distinguish among different types of carbohydrate molecules, including monosaccharides, disaccharides, and polysaccharides.
2. Relate the functions of plant and animal polysaccharides to their structure.

**Carbohydrates** are composed of carbon, hydrogen, and oxygen atoms. Most of the carbon atoms in a carbohydrate are linked to another carbon atom, a hydrogen atom, and an —OH functional group. However, other functional groups, such as $—NH_2$ and —COOH groups, are also found in certain carbohydrates. As discussed next, sugars are simple carbohydrates, whereas polysaccharides are large macromolecules.

### Monosaccharides and Disaccharides Are Simple Carbohydrates

The simplest carbohydrates are the monomers known as **monosaccharides** (from the Greek, meaning single sugars). The most common types contain five carbons, called pentoses,

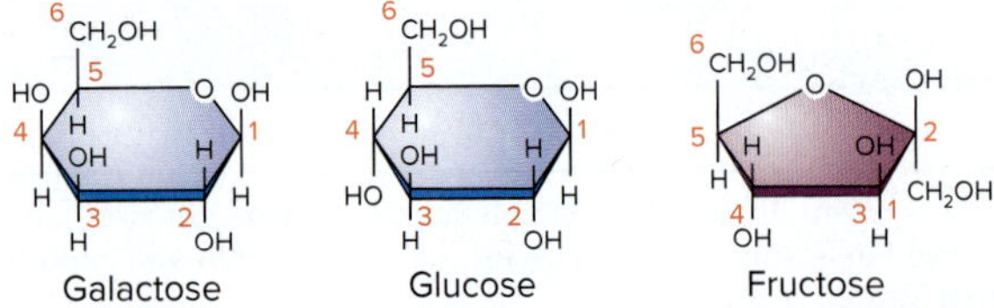


**Figure 3.4 Monosaccharide structures.** Glucose and galactose differ in the position of the —OH group attached to carbon atom number 4. Fructose has two carbon atoms outside the main ring structure.

or six carbons, called hexoses. Important pentoses are ribose ($C_5H_{10}O_5$) and the closely related deoxyribose ($C_5H_{10}O_4$), which are part of RNA and DNA molecules, respectively (described later in this chapter). The most common hexoses are galactose, glucose, and fructose (**Figure 3.4**). Note in Figure 3.4 that all three hexoses have the same chemical formula ($C_6H_{12}O_6$), but the atoms are arranged differently within the molecules. Molecules with identical formulas but different structures are called **isomers.** The hexoses may exist in a linear structure, or more typically in a ring. The carbon atoms in the ring are numbered by convention, as shown in Figure 3.4. The ring is formed with an oxygen atom; in glucose, the oxygen forms a linkage between carbons 1 and 5. The hydrogen atoms and the —OH groups may lie above or below the plane of the ring structure; this is one feature that distinguishes one isomer from another.

Of these three hexoses, glucose plays an especially important role in the cellular activities of living organisms. For example, glucose is very water-soluble and thus circulates in the blood or body fluids of animals, where it can be transported across cell membranes. Once inside a cell, enzymes can break down glucose into smaller molecules, releasing energy that was stored in the chemical bonds of glucose. This energy is then stored in the bonds of another molecule, called adenosine triphosphate, or ATP (see Chapter 6), which, in turn, provides an energy source for a variety of cellular processes.

Monosaccharides can join together by a dehydration reaction to form **disaccharides** (meaning two sugars), which are carbohydrates composed of two monosaccharides. A familiar disaccharide is sucrose, or table sugar, which is composed of the monomers glucose and fructose (**Figure 3.5**). Sucrose is the major transport form of sugar in plants. The linking together of most monosaccharides involves the removal of an —OH group from one monosaccharide and a hydrogen atom from the other, giving rise to a molecule of $H_2O$ and covalently bonding the two sugars together through an oxygen atom. The bond formed between two sugar molecules by such a dehydration reaction is called a **glycosidic bond.**

## Polysaccharides Are Carbohydrate Polymers

When many monosaccharides are linked together to form long polymers, they form **polysaccharides** (meaning many sugars). In living organisms, polysaccharides include

- **starch,** found in plant cells,
- **glycogen,** present in certain types of animal cells,
- **cellulose,** found in the cell walls of plant cells,
- **peptidoglycans,** found in the cell walls of certain bacteria,
- **chitin,** found in cell walls of fungi and the exoskeletons of arthropods, and
- **glycosaminoglycans,** polysaccharides found in connective tissue and surrounding cells in animals.

Starch, glycogen, and cellulose are composed of thousands of glucose molecules linked together in long chains, differing in the extent of branching and the orientation of monomers along the chain (**Figure 3.6**). The sugar monomers within other polysaccharides may have nitrogen-containing groups, such as amino groups, attached to them.

## Biology Principle

### Living Organisms Use Energy

The chemical energy stored in the bonds of glucose molecules can be harnessed by living organisms. This energy can be used to perform numerous functions that support life, including the synthesis of new molecules, growth, digestion, locomotion, and many others.

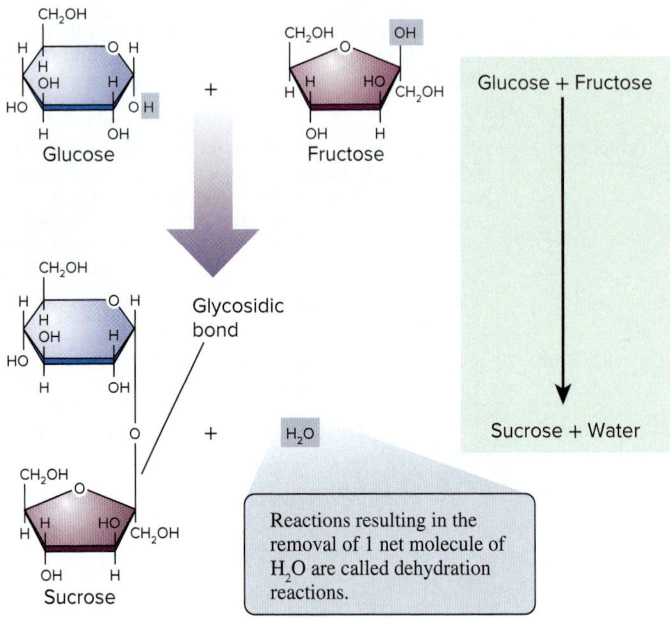


**Figure 3.5 Formation of a disaccharide.** Two monosaccharides can bond to each other to form a disaccharide, such as sucrose, by a dehydration reaction.

**Concept Check:** *What type of reaction is the reverse of the one shown here, in which a disaccharide is broken down into two monosaccharides?*

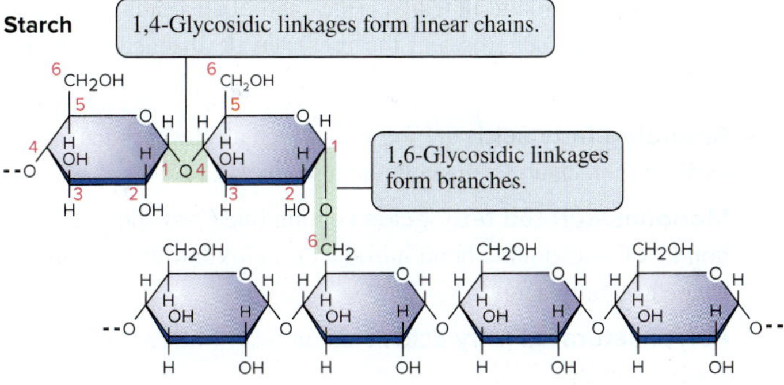


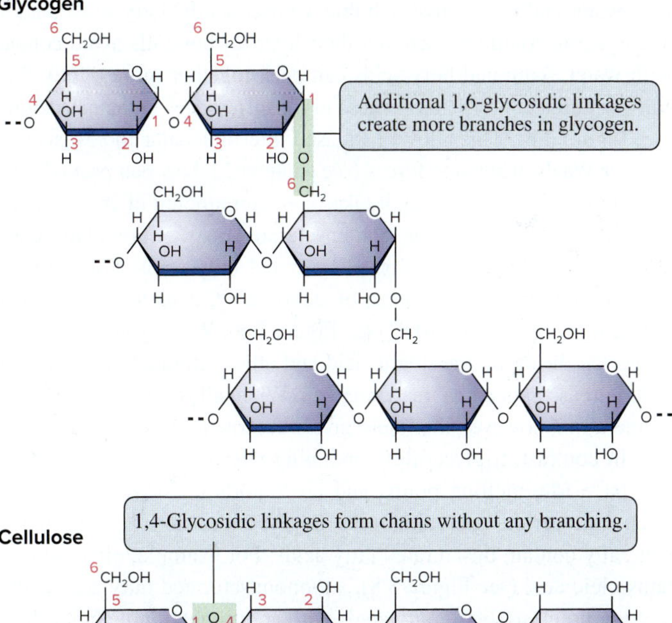


Cellulose

1,4-Glycosidic linkages form chains without any branching.

**Figure 3.6 Polysaccharides that are polymers of glucose.** These polysaccharides differ in their extent of branching. Note: In cellulose, the bonding arrangements cause every other glucose to be inverted with respect to its neighbors.

**BioConnections:** *Look ahead to Figures 4.34 and 4.35 for the role of cellulose in plant cell structure and to Figures 28.10 and 28.11 for its role in plant growth. Considering the amount of plant life on Earth, what might you conclude about the abundance of cellulose on the planet?*

The bonds that form in polysaccharides are between specific carbon atoms of each molecule. In starch and glycogen, the bonds form between carbons 1 and 4, and between 1 and 6. The high degree of branching in glycogen contributes to its solubility in animal tissues, such as muscle. This is because the extensive branching creates a more open structure, in which many hydrophilic —OH functional groups have access to water and can hydrogen-bond with it. Starch is less branched, making it less soluble.

Starch and glycogen are used to store energy in cells. Like disaccharides, these polysaccharides can be hydrolyzed to yield monosaccharides, which are broken down to provide the energy to make ATP. Starch and glycogen are an efficient means of storing energy for those times when a plant or an animal cannot obtain sufficient energy from its environment or diet for its metabolic requirements.

Cellulose, peptidoglycans, chitin, and glycosaminoglycans, by contrast, play a structural rather than energy-storing role. Cellulose has a linear arrangement of carbon-carbon bonds and no branching (see Figure 3.6). The linear arrangement allows vast numbers of hydrogen bonds to form between cellulose molecules, which stack together in sheets and provide great strength to structures like plant cell walls. The hydrogen bonds form between —OH groups on carbons 3 and 6.

Peptidoglycans are found in the cell walls of certain bacteria; they consist of sugars and amino acids bonded together in a strong, lattice-like arrangement that gives strength and rigidity to the cell wall. Chitin is a tough, structural polysaccharide that forms the external skeleton of insects and crustaceans (for example, shrimp and lobsters) as well as the cell walls of fungi. Glycosaminoglycans are found abundantly in cartilage—the tough, fibrous material in bone and certain other animal structures. Glycosaminoglycans are also abundant in the extracellular matrix that provides a structural framework surrounding many of the cells in an animal's body (look ahead to Figure 4.31). The nitrogen in these polysaccharides allows for additional hydrogen bonding and thus contributes to the strength of structures containing them.

## 3.4 Reviewing the Concepts

- Carbohydrates are composed of carbon, hydrogen, and oxygen atoms. Cells can break down glucose, an important carbohydrate, releasing energy that is then stored in the bonds of ATP.
- Carbohydrates include monosaccharides, disaccharides, and polysaccharides. Some polysaccharides store energy in cells. Others may serve a support or structural function (Figures 3.4, 3.5, 3.6).

## 3.4 Testing Your Knowledge

1. Glucose, galactose, and fructose are examples of
   a. polysaccharides.
   b. hexoses.
   c. isomers.
   d. disaccharides.
   e. both b and c.
2. __________ is a storage polysaccharide commonly found in the cells of animals.
   a. Glucose
   b. Sucrose
   c. Glycogen
   d. Starch
   e. Cellulose

## 3.5 Lipids

### Learning Outcomes

1. List the different classes of lipid molecules important in living organisms.
2. Diagram the structure of a triglyceride, and explain how it is formed and how its structure is affected by the presence of saturated and unsaturated fatty acids.
3. Explain why some fats are solid at room temperature and others are liquid.
4. Discuss how triglycerides function as energy-storage molecules.
5. Explain why phospholipids form a bilayer when dissolved in water.
6. Describe the chemical nature of steroids, and give an example of their biological importance.

**Lipids** are hydrophobic molecules composed mainly of hydrogen and carbon atoms, and some oxygen. The defining feature of lipids is that they are nonpolar and therefore poorly soluble in water. Lipids include triglycerides, phospholipids, steroids, and waxes.

### Triglycerides Are Made from Glycerol and Fatty Acids

**Triglycerides** (commonly referred to as fats) consist of a glycerol molecule linked to three fatty acids (**Figure 3.7**). Glycerol is a three-carbon molecule with one —OH group bonded to each carbon. A fatty acid is a chain of carbon and hydrogen atoms with a carboxyl group (—COOH) at one end. Each of the —OH groups in glycerol is linked to the —COOH group of a fatty acid by the removal of a molecule of $H_2O$ by a dehydration reaction.

The fatty acids found in triglycerides and other lipids may differ with regard to their lengths and the presence or absence of double bonds (**Figure 3.8**).

- **Saturated fatty acids** are those in which all the carbons in the hydrocarbon chain form single bonds (C—C).
- **Monounsaturated fatty acids** contain one C═C double bond. A C═C double bond introduces a kink into the linear shape of a fatty acid.
- **Polyunsaturated fatty acids** contain two or more C═C double bonds.

The hydrocarbon tail of a fatty acid does not form hydrogen bonds with water and is very hydrophobic. Consequently, fatty acids tend to aggregate in solutions such that their hydrocarbon tails avoid contact with water. Saturated fatty acids can pack together more tightly than can unsaturated fatty acids because of the differences in their structures (see Figure 3.8). This allows increased intermolecular forces, such as van der Waals dispersion forces (see Chapter 2), between packed saturated fatty acids. As a result, it takes a greater amount of energy (heat) to melt saturated fatty acids than unsaturated ones. Animal fats generally contain a high proportion of saturated fatty acids. For example, beef fat contains high amounts of stearic acid, a saturated fatty acid with a melting point of 70°C (see Figure 3.8). When you heat a hamburger on the stove, the stearic acid and other saturated fats melt, and liquid grease appears in the frying pan. When allowed to cool to room temperature, however, the liquid grease returns to its solid form.

In contrast, triglycerides containing unsaturated fatty acids usually have low melting points and are liquids at room temperature. Such triglycerides are called oils. Triglycerides derived from plants generally contain unsaturated fatty acids. For example, olive oil contains oleic acid (see Figure 3.8), a monounsaturated fatty acid with a melting point of 16°C. Fatty acids with additional double bonds have even lower melting points; for example, linoleic acid has two double

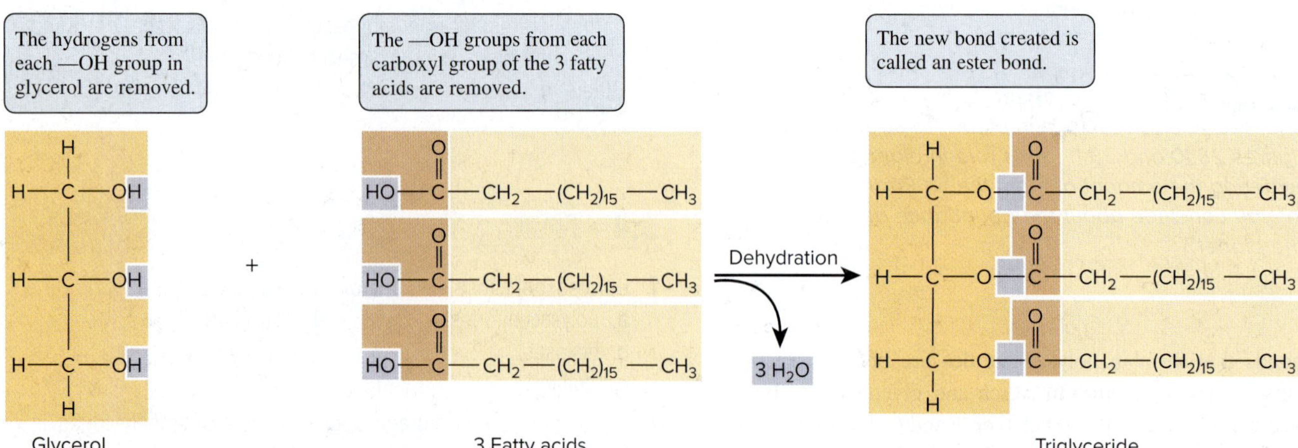


**Figure 3.7 The formation of a triglyceride.** The formation of a triglyceride occurs via three dehydration reactions in which fatty acids are bonded to glycerol. Note in this figure that a common shorthand notation is used for depicting fatty acids, in which a portion of the $CH_2$ groups is illustrated as $(CH_2)_n$, where $n$ in this example is 15. Thus, this would be a fatty acid with 18 carbons.

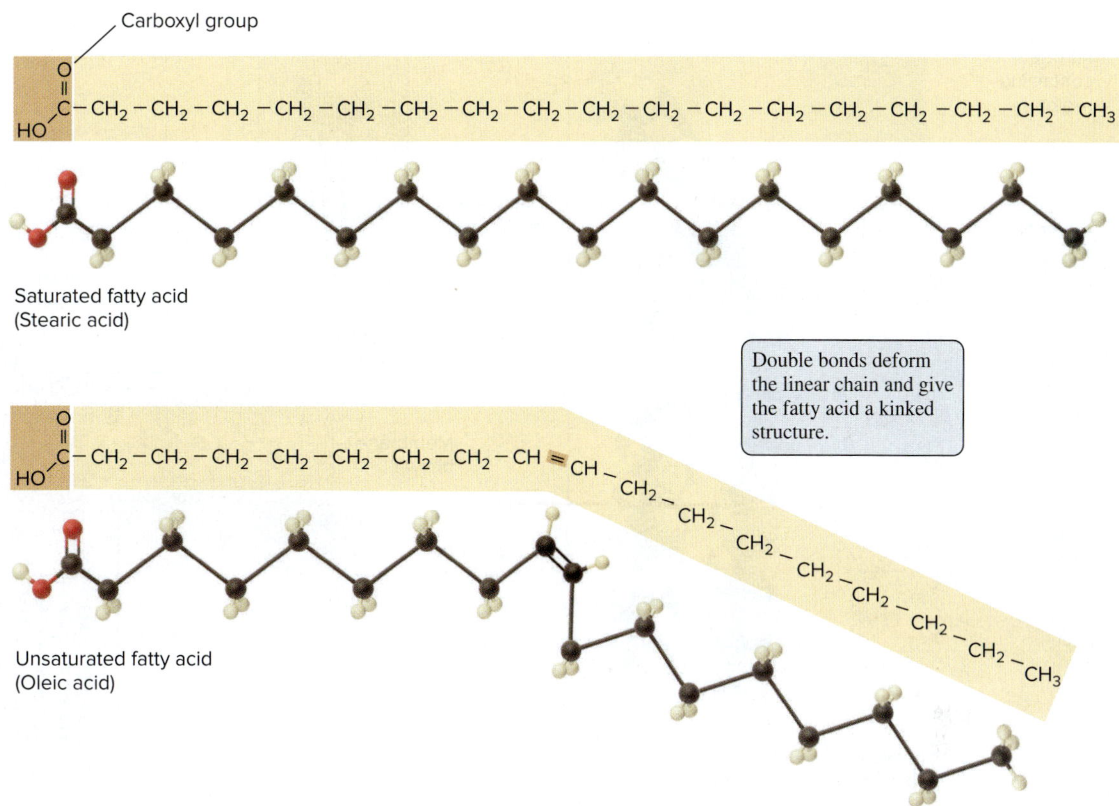


**Figure 3.8 Examples of fatty acids.** Fatty acids are hydrocarbon chains with a carboxyl functional group at one end and either no double-bonded carbons (saturated) or one or more double bonds (unsaturated). Stearic acid, for example, is an abundant saturated fatty acid in animals, whereas oleic acid is an unsaturated fatty acid found in plants. Note that the presence of a C═C double bond introduces a kink into the shape of oleic acid. As a consequence, unsaturated fatty acids are not able to pack together as tightly as saturated fatty acids.

bonds and melts at –5°C. Safflower and sunflower oils contain large amounts of linoleic acid.

Like starch and glycogen, triglycerides are important for storing energy. The hydrolysis of triglycerides releases their fatty acids, which can be metabolized to provide energy to make ATP. Certain organisms, most notably mammals, have the ability to store large amounts of energy by accumulating triglycerides.

## Phospholipids Are Amphipathic Lipids

**Phospholipids** are similar in structure to triglycerides, but with one important difference. The third —OH group of glycerol is linked to a phosphate group instead of a fatty acid. In most phospholipids, a small polar or charged nitrogen-containing molecule is attached to this phosphate (Figure 3.9a). The glycerol backbone, phosphate group, and charged molecule constitute a polar (hydrophilic) head at one end of the phospholipid, whereas the two fatty acids are nonpolar (hydrophobic) tails at the opposite end. Recall from Chapter 2 that molecules with polar and nonpolar regions are called amphipathic molecules.

In water, phospholipids become organized into a double layer of molecules called a bilayer, with their polar heads interacting with the $H_2O$ molecules and their nonpolar tails facing the interior, where they avoid contact with water. As you will learn in Chapter 5, this bilayer arrangement of phospholipids is critical for determining the structure of cell membranes, as shown in Figure 3.9b.

## Steroids Contain Ring Structures

**Steroids** have a distinctly different chemical structure from that of the other types of lipid molecules discussed thus far. One class of steroids contains an —OH group at a particular carbon; these are called sterols and they are found in fungi, plants, and animals (Figure 3.10). Cholesterol, for example, is found in the cell membranes of animals, where it contributes to membrane structure and function. It is also found in numerous animal cells, where it serves as a precursor for the synthesis of steroid hormones. These include estrogens and androgens, which are necessary for animal reproduction.

## Waxes Are Complex Lipids That Prevent Water Loss from Organisms

Many plants and animals produce lipids called **waxes** that are secreted onto their surface, such as the leaves of plants and the cuticles of insects. Although any wax may contain hundreds of different compounds, all waxes contain one or more hydrocarbons and long structures that resemble a fatty acid attached by its carboxyl group to another long hydrocarbon chain. Most waxes are very nonpolar and therefore exclude water, providing a barrier to water loss. They may also be used as structural elements in colonies, like those of bees, where beeswax forms the honeycomb of the hive.

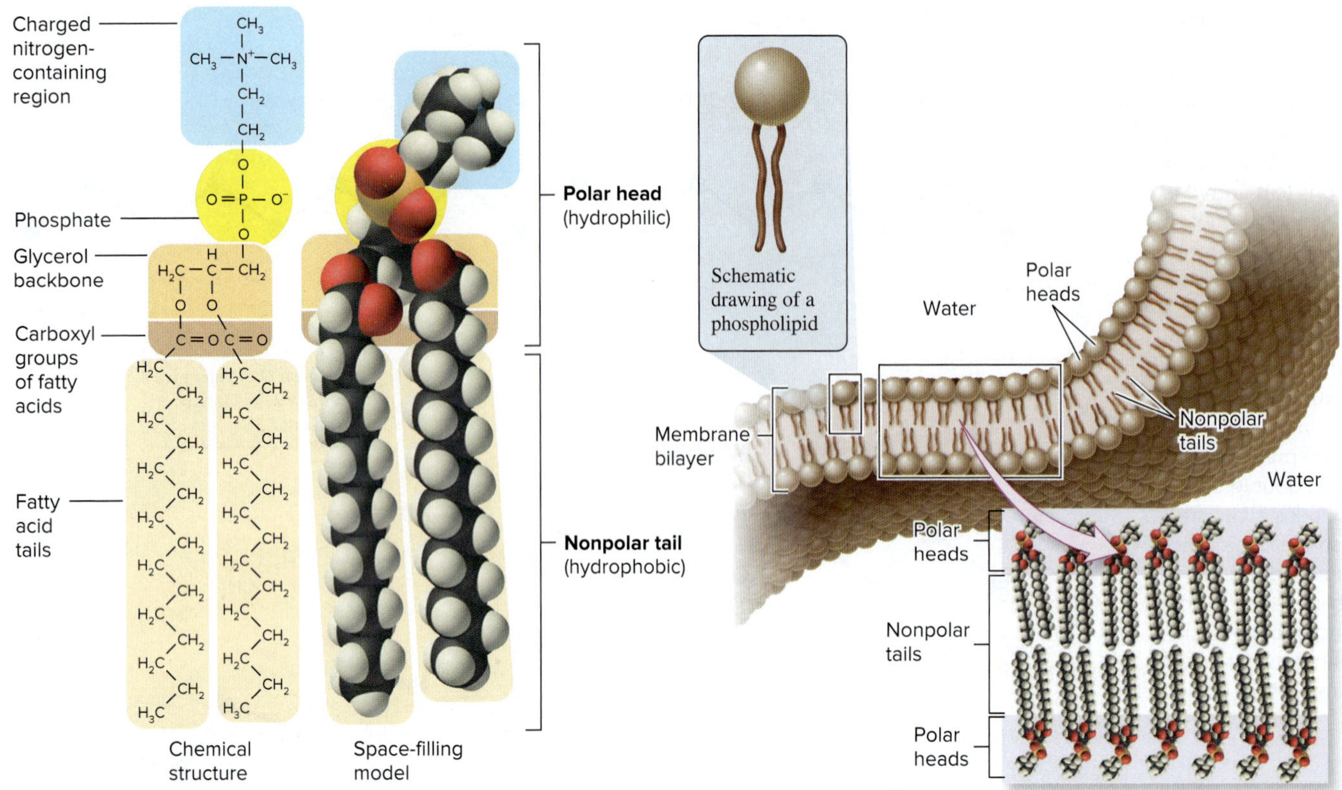


**Figure 3.9** **Structure of phospholipids.** **(a)** Chemical structure and space-filling model of phosphatidylcholine, a common phospholipid found in living organisms. Phospholipids contain both polar and nonpolar regions, making them amphipathic. The fatty acid tails are nonpolar. The rest of the molecule is polar. **(b)** Arrangement of phospholipids in a biological membrane, such as the plasma membrane that encloses cells. The hydrophilic region of the phospholipid faces the watery environment, whereas the hydrophobic regions associate with each other in the interior of the membrane, forming a bilayer.

**Concept Check:** *When water and oil are added to a test tube, the two liquids form two separate layers (think of oil and vinegar in a bottle of salad dressing). If a solution of phospholipids was added to a mixture of water and oil, where would the phospholipids dissolve?*

**BioConnections:** *For a more detailed view of the components of cell membranes, including the phospholipid bilayer, look ahead to Figure 5.1.*

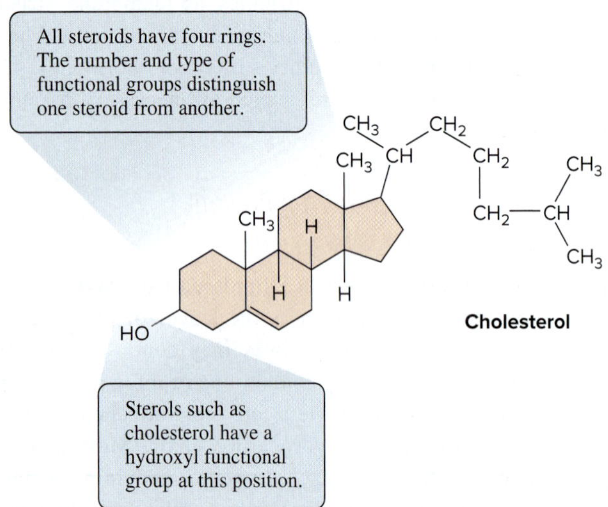

All steroids have four rings. The number and type of functional groups distinguish one steroid from another.
Cholesterol
Sterols such as cholesterol have a hydroxyl functional group at this position.


**Figure 3.10** **Structure of a common steroid.** All steroids share a common ring structure, shown here. One class of steroids is the sterols, such as cholesterol.

## 3.5 Reviewing the Concepts

- Lipids are nonpolar and very insoluble in water. Major classes of lipids include triglycerides, phospholipids, steroids, and waxes.
- Triglycerides are formed by bonding glycerol with three fatty acids. In a saturated fatty acid, all the carbons are linked by single covalent bonds. Unsaturated fatty acids contain one or more C═C double bonds (Figures 3.7, 3.8).
- Phospholipids contain a polar head and two nonpolar tails, making them amphipathic (Figure 3.9).
- Steroids are constructed of four fused rings of carbon atoms. Sterols are a type of steroid and include cholesterol (Figure 3.10).
- Waxes are nonpolar and exclude water, and they often are found as protective coatings on leaves and the surfaces of some animals' bodies.

## 3.5 Testing Your Knowledge

1. Hydrolysis of a triglyceride yields
   a. 3 glycerols and 3 fatty acids.
   b. 1 glycerol and 3 fatty acids.
   c. 3 glycerols and 1 fatty acid.
   d. 3 glycerols and 6 fatty acids.
   e. a shorter triglyceride.
2. Which of the following are highly amphipathic?
   a. phospholipids
   b. triglycerides
   c. steroids
   d. waxes
   e. all of the above

## 3.6 Proteins

### Learning Outcomes

1. Give examples of the general functions that are carried out by different proteins.
2. Describe how amino acids are joined to form a polypeptide.
3. Explain the four levels of protein structure.
4. Outline the factors that determine protein shape and function.
5. Define protein domain.

**Proteins** are macromolecules composed of carbon, hydrogen, oxygen, nitrogen, and small amounts of other elements, such as sulfur. They play critical roles in nearly all life processes (Table 3.3).

### Amino Acids Are the Building Blocks of Proteins

Proteins are polymers of **amino acids.** All amino acids contain a carbon atom, called the α-carbon, that is linked to an amino group (—$NH_2$) and a carboxyl group (—COOH). When an amino acid is dissolved in water at neutral pH, the —$NH_2$ group acts as a base and accepts a hydrogen ion, becoming positively charged. The —COOH group acts as an acid; it loses a hydrogen ion and is negatively charged at neutral pH. The α-carbon also is linked to a hydrogen atom and a side chain, which is given a general designation R.

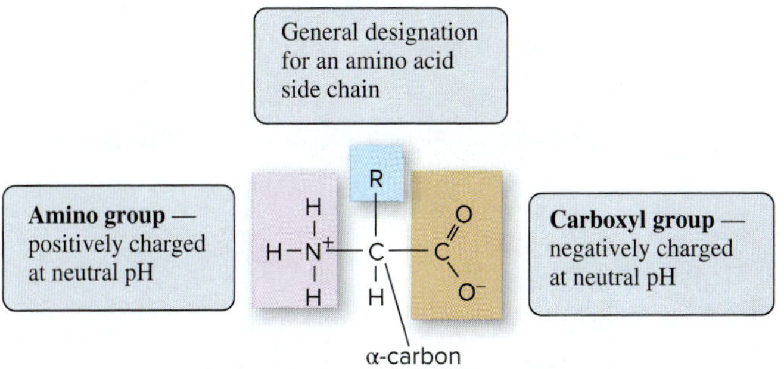


The 20 amino acids found in proteins are distinguished by their side chains (Figure 3.11). The amino acids are categorized by the properties of their side chains, whether nonpolar, polar and uncharged, or polar and charged. The varying structures of the side chains are critical features of protein structure. The arrangement and chemical features of the side chains cause proteins to fold and adopt their three-dimensional shapes. In addition, certain amino acids and their side chains may be critical in protein function. For example, amino acid side chains within specific regions of enzymes are essential for enzymes to catalyze chemical reactions.

Amino acids are joined together by a dehydration reaction that links the carboxyl group of one amino acid to the amino group of another (Figure 3.12a). The covalent bond formed between a carboxyl group and an amino group is called a **peptide bond.** When multiple amino acids are joined by peptide bonds, the resulting molecule is called a **polypeptide** (Figure 3.12b). The backbone of the polypeptide in Figure 3.12 is highlighted in yellow. The amino acid side chains project from the backbone. When two or more amino acids are linked together, one end of the resulting molecule has a free amino group. This is the amino end, also called the **N-terminus.** The other end of the polypeptide, called the carboxyl end, or **C-terminus,** has a

**Table 3.3 Major Categories and Functions of Proteins**

| Category | Functions | Examples |
|---|---|---|
| Proteins involved in gene expression and regulation | Make mRNA from a DNA template; synthesize polypeptides from mRNA; regulate genes | RNA polymerase assists in synthesizing RNA from DNA. |
| Motor proteins | Initiate movement | Myosin provides the contractile force of muscles. |
| Defense proteins | Protect organisms against disease | Antibodies help destroy bacteria or viruses. |
| Metabolic enzymes | Increase rates of chemical reactions important in energy balance | Hexokinase is an enzyme involved in glucose metabolism. |
| Cell-signaling proteins | Enable cells to communicate with each other and to sense the environment | Notch proteins coordinate growth of cells in developing animals. |
| Structural proteins | Support and strengthen structures | Actin provides shape to the cytoplasm of plant and animal cells. Collagen gives strength to tendons. |
| Transporters | Mediate movement of solutes across membranes | Glucose transporters move glucose from outside cells to inside cells, where it can be used for energy. |

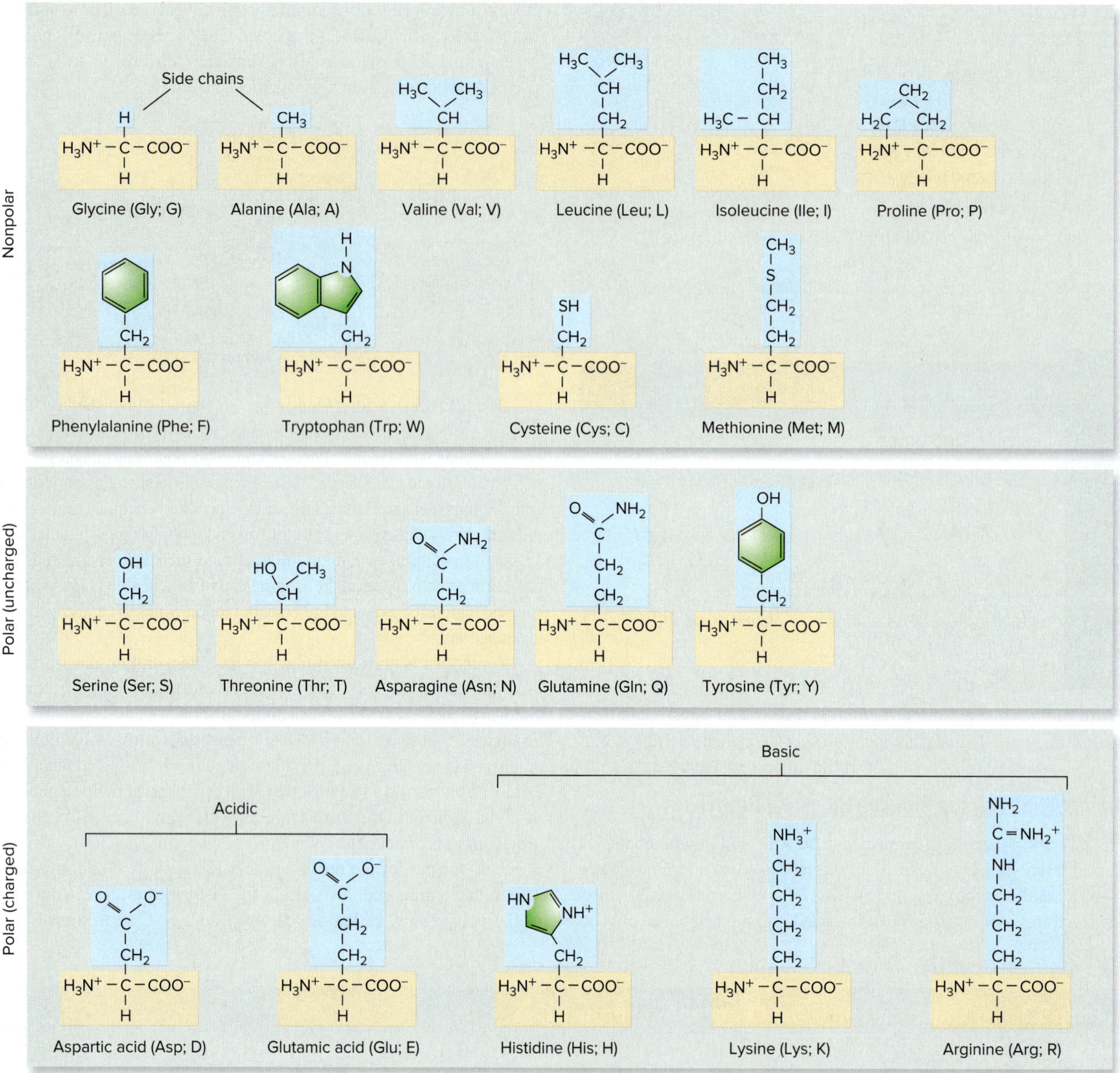


**Figure 3.11 The 20 amino acids found in living organisms.** Amino acids have different chemical properties (for example, nonpolar versus polar) due to the nature of their different side chains. These properties contribute to the differences in the three-dimensional shapes and chemical properties of proteins, which, in turn, influence their biological functions. Note: Tyrosine has both polar and nonpolar characteristics and is listed in just one category for simplicity. The common three-letter and one-letter abbreviations for each amino acid are shown in parentheses.

free carboxyl group. As shown in Figure 3.12c, amino acids within a polypeptide are numbered sequentially from the N-terminus to the C-terminus.

The term polypeptide refers to a structural unit composed of a linear sequence of amino acids. A protein is a functional unit composed of one or more polypeptides that have folded and twisted into a precise three-dimensional shape. Many proteins also have carbohydrates (glycoproteins) or lipids (lipoproteins) attached at certain amino acids; these modifications impart unique functions to such proteins.

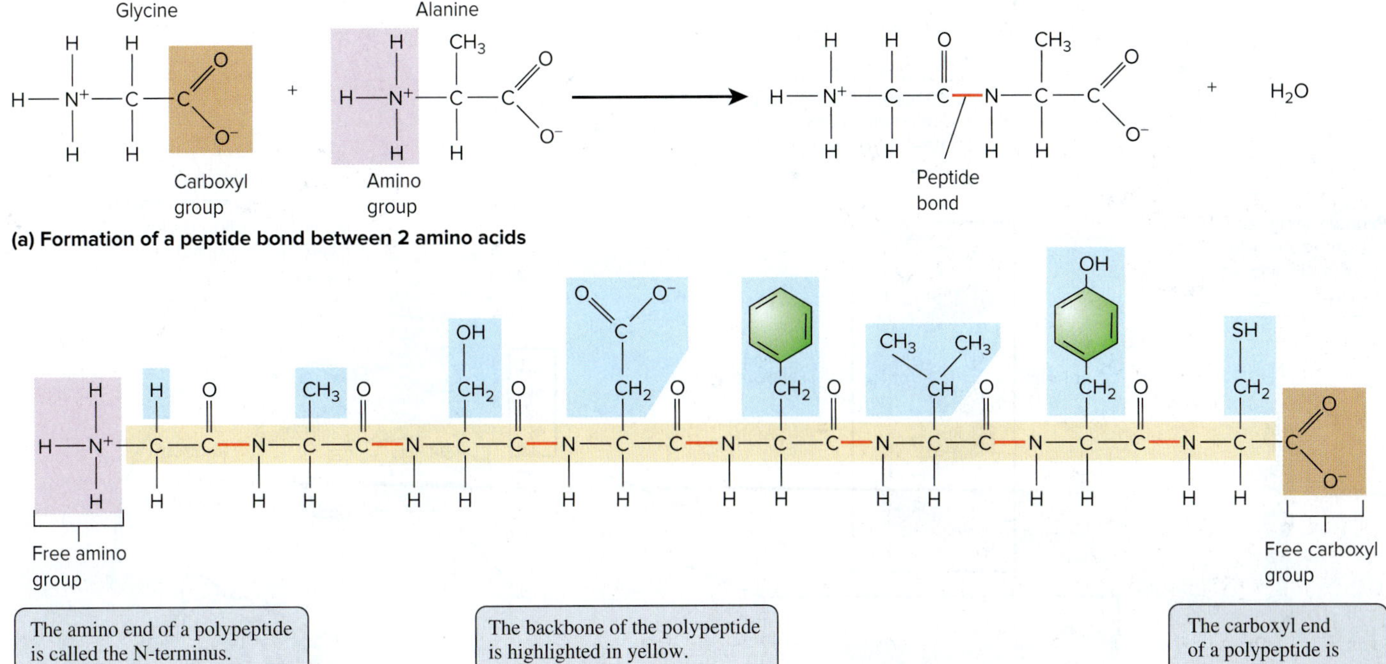


(b) Polypeptide—a linear chain of amino acids

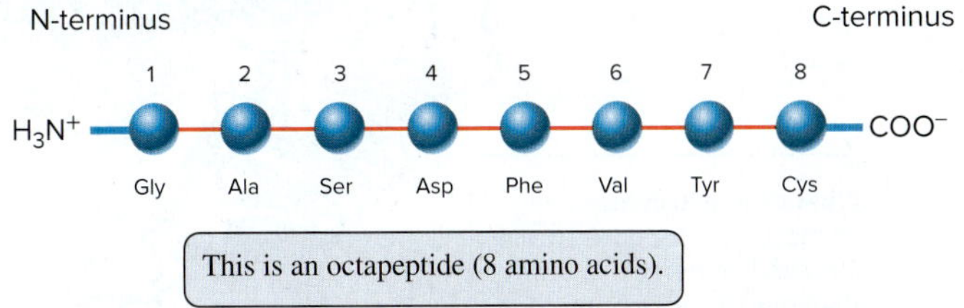


(c) Numbering system of amino acids in a polypeptide

**Figure 3.12** **The mechanism of polypeptide formation.** Polypeptides are polymers of amino acids. They are formed by linking amino acids via dehydration reactions to make peptide bonds. Every polypeptide has an amino end, or N-terminus, and a carboxyl end, or C-terminus.

**Concept Check:** *How many $H_2O$ molecules would be produced in making a polypeptide that is 72 amino acids long?*

## Proteins Have a Hierarchy of Structure

Scientists view protein structure at four progressive levels: primary, secondary, tertiary, and quaternary, shown schematically in Figure 3.13. Each higher level of structure depends on the preceding levels. For example, changing the primary structure may affect the secondary, tertiary, and quaternary structures.

**Primary Structure** The **primary structure** (see Figure 3.13) of a protein is its amino acid sequence, from beginning to end. The primary structures of proteins are determined by genes. As we will explore in Chapter 10, genes carry the information for the production of polypeptides with specific amino acid sequences.

**Secondary Structure** The amino acid sequence of a polypeptide, together with the principles of chemistry and physics, cause a protein to fold into a more compact structure. Amino acids can rotate around bonds within a polypeptide. Consequently, proteins are flexible and can fold into a number of shapes, just as a string of beads can be twisted into many configurations. Folding can be irregular, or certain regions can have a repeating folding pattern called a protein's **secondary structure.** The two basic types of a protein's secondary structure are the α helix and the β pleated sheet.

In an **α helix,** the polypeptide backbone forms a repeating helical structure that is stabilized by hydrogen bonds along the length of the backbone. As shown in Figure 3.13, the hydrogen linked to a nitrogen atom forms a hydrogen bond with an oxygen atom, which in turn is double-bonded to a carbon atom. These hydrogen bonds occur at regular intervals within the polypeptide backbone and cause the backbone to twist into a helix.

In a **β pleated sheet,** regions of the polypeptide backbone lie parallel to each other. Hydrogen bonds between a hydrogen linked to a nitrogen atom and a double-bonded oxygen form between these adjacent, parallel regions. When this occurs, the polypeptide backbone adopts a repeating zigzag, or pleated, shape.

Secondary structure contributes to the great strength of certain proteins, including those found in plant cell walls; the proteins that make up the silk webs of spiders; and collagen, the chief component of cartilage in vertebrate animals.

Some regions along a polypeptide do not assume an α helix or β pleated sheet conformation and do not have a secondary structure. These are sometimes called random coiled regions.

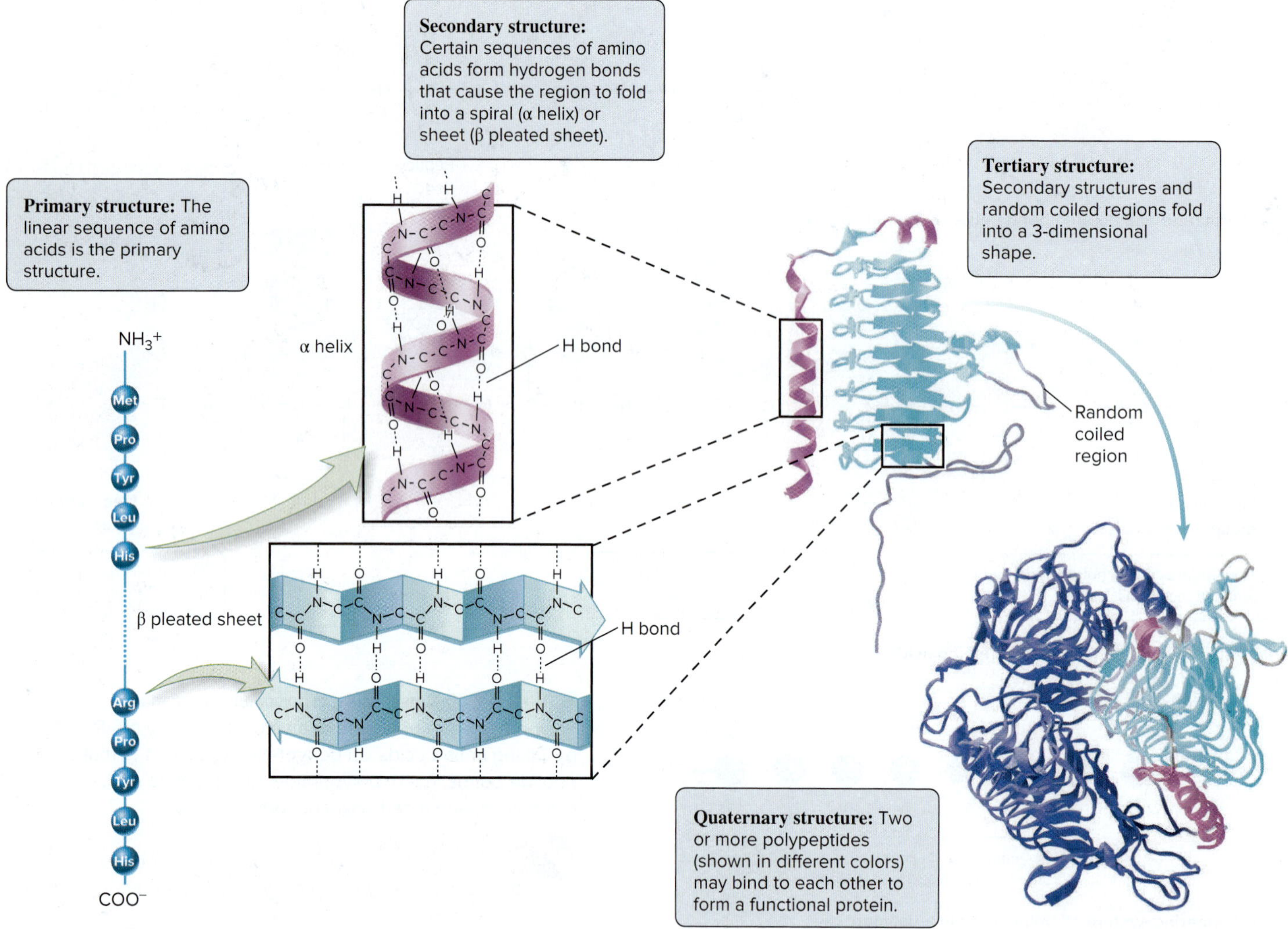


**Figure 3.13** **The hierarchy of protein structure.** The R groups are omitted for simplicity.

**Tertiary Structure** As the secondary structure of a polypeptide becomes established due to the particular primary structure, additional side chains of amino acids interact. The polypeptide folds and refolds upon itself to assume a complex three-dimensional shape—its **tertiary structure** (see Figure 3.13). Tertiary structure includes all secondary structures plus any other interactions involving amino acid side chains. For some proteins, the tertiary structure is the final structure. However, as described next, other proteins are composed of two or more polypeptides.

**Quaternary Structure** Most proteins are composed of two or more polypeptides, which each adopt a tertiary structure and then assemble with each other (see Figure 3.13). The individual polypeptides are called **protein subunits.** Subunits may be identical polypeptides or they may be different. When proteins consist of more than one polypeptide, they are said to have **quaternary structure.** A common example is the oxygen-binding protein called hemoglobin, found in the red blood cells of vertebrate animals. As you will learn later (look ahead to Figure 37.8), four protein subunits combine to form one hemoglobin protein. Each subunit can bind a single molecule of oxygen; therefore, each hemoglobin protein can carry four molecules of oxygen in the blood.

## Protein Structure Is Determined by Several Factors

Several factors determine the way proteins adopt their secondary, tertiary, and quaternary structures. The following factors are critical for protein folding and stability:

- *Hydrogen bonds*—The large number of weak hydrogen bonds within a polypeptide and between polypeptides adds up to a collectively strong force that promotes protein folding and stability.
- *Ionic bonds and other polar interactions*—Some amino acid side chains are positively or negatively charged. Positively charged side chains may bind to negatively charged side chains via ionic bonds.
- *Hydrophobic effect*—Some amino acid side chains are nonpolar. As a protein folds, the hydrophobic amino acids are likely to

be found in the center of the protein, minimizing contact with water. Some proteins have stretches of nonpolar amino acids that anchor them in the hydrophobic portion of membranes.

- *van der Waals dispersion forces*—Atoms within molecules have temporary attractions for each other if they are an optimal distance apart. This weak attraction is termed the van der Waals dispersion force, and it contributes to tertiary and quaternary structure (see Chapter 2).
- *Disulfide bridges*—The side chain of the amino acid cysteine (see Figure 3.11) contains a sulfhydryl group (—SH), which can react with an —SH group in another cysteine side chain. The resulting bond is a **disulfide bridge,** which links the two amino acid side chains together (—S—S—). Disulfide bridges are covalent bonds that can occur within a polypeptide or between different polypeptides.

The first four factors are also important in the ability of different proteins to interact with each other. Many cellular processes involve steps in which two or more different proteins interact with each other. For this to occur, proteins must bind to each other. Such binding is usually very specific. The shape of one protein may permit it to precisely fit into a region of another protein (Figure 3.14). Such protein-protein interactions are important in building complicated cellular structures and in processes that provide organization and function to cells.

## Biology Principle

### New Properties Emerge from Complex Interactions

This principle of biology is apparent at the protein level. The primary sequence of proteins determines their final three-dimensional structures. Compare this with the chapter-opening depiction of two real proteins and the intermediate levels of protein structure shown in Figure 3.13. It is the three-dimensional shape of different proteins that determines their ability to interact with other molecules, including other proteins.

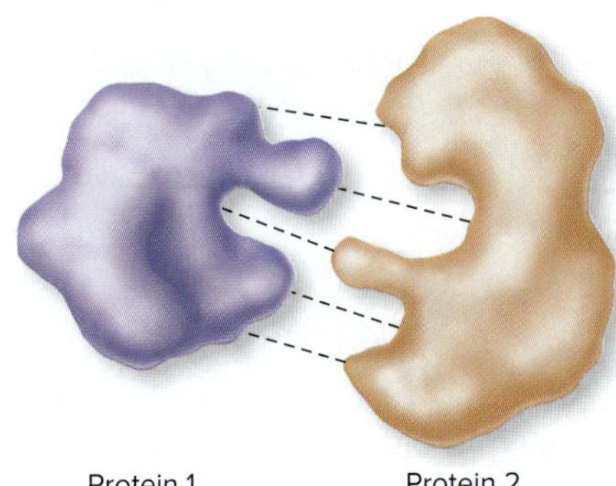


**Figure 3.14 Protein-protein interaction.** Two different proteins may interact with each other due to hydrogen bonding, ionic bonding, the hydrophobic effect, and van der Waals dispersion forces. Interaction is also facilitated by their respective three-dimensional shapes.

## FEATURE INVESTIGATION

### Anfinsen Showed That the Primary Structure of Ribonuclease Determines Its Three-Dimensional Structure

Prior to the 1960s, the mechanisms by which proteins assume their three-dimensional structures were not understood. Scientists hypothesized that correct folding required unknown cellular components or that ribosomes, the cellular structures in which polypeptides are synthesized, somehow shaped proteins as they were being made. American biochemist Christian Anfinsen, however, postulated that the amino acid sequence of proteins is the primary factor that determines how proteins fold into their proper conformation. He hypothesized that proteins spontaneously assume their most stable conformation based on the principles of chemistry and physics (Figure 3.15).

To test this hypothesis, Anfinsen studied ribonuclease, an enzyme that degrades RNA molecules. Anfinsen began with purified ribonuclease. The important point is that other cellular components were not present. Biochemists had already determined that ribonuclease has four disulfide bridges (S—S) between eight cysteine amino acids. Anfinsen exposed ribonuclease to a chemical called β-mercaptoethanol, which breaks S—S bonds, and to urea, which disrupts hydrogen bonds (H bonds) and ionic bonds. This treatment caused ribonuclease to be denatured, that is, to become unfolded. Following this treatment, he measured the ability of the enzyme to degrade RNA molecules. The enzyme had lost essentially all of its ability to degrade RNA. Therefore, Anfinsen concluded that when ribonuclease is unfolded or denatured, it is no longer functional.

The key step in this experiment came when Anfinsen removed the urea and β-mercaptoethanol from the solution. Because these molecules are much smaller than ribonuclease, removing them from the solution could be accomplished with a technique called size-exclusion chromatography. In size-exclusion chromatography, solutions are poured on top of a glass column of beads and allowed to flow down the column to an open collection port at the bottom. The beads in the column have microscopic pores that trap small molecules like urea and mercaptoethanol but permit large molecules such as ribonuclease to pass down the length of the column and out the collection port.

Using size-exclusion chromatography, Anfinsen was able to separate ribonuclease from β-mercaptoethanol and urea. He then allowed the purified enzyme to sit in water for up to 20 hours, after which he retested the ribonuclease for its ability to degrade RNA. The result revolutionized our understanding of proteins. The activity of the ribonuclease was almost completely restored! Therefore, even in the complete absence of any cellular factors or ribosomes, an unfolded protein can refold into its correct, functional structure.

**Figure 3.15** Anfinsen's experiments with ribonuclease, demonstrating that the primary structure of a polypeptide plays a key role in protein folding.

**HYPOTHESIS** Within their amino acid sequence, proteins contain all the information needed to fold into their correct, three-dimensional shapes.

**KEY MATERIALS** Purified ribonuclease, RNA, denaturing chemicals, size-exclusion columns.

| | Experimental level | Conceptual level |
|---|---|---|
| **1** Incubate purified ribonuclease in test tube with RNA, and measure its ability to degrade RNA. | | 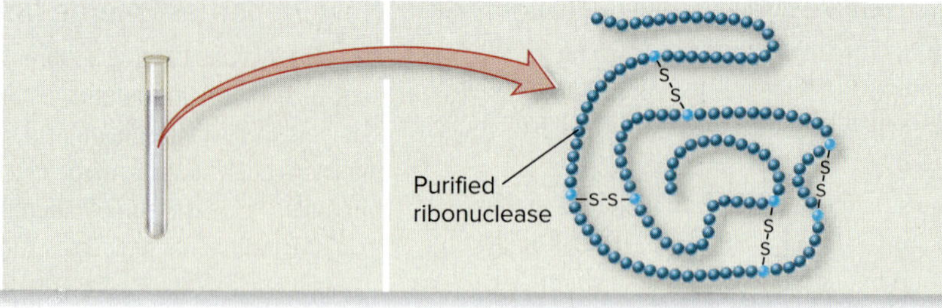 Purified ribonuclease; S-S bonds. Numerous H bonds and ionic bonds (not shown) and 4 S—S bonds. Protein is properly folded. (For simplicity, the three-dimensional shape is not shown; see Panel 3 for a computer model of the true structure.) |
| **2** Denature ribonuclease by adding β-mercaptoethanol (breaks S—S bonds) and urea (breaks H bonds and ionic bonds). Measure its ability to degrade RNA. | 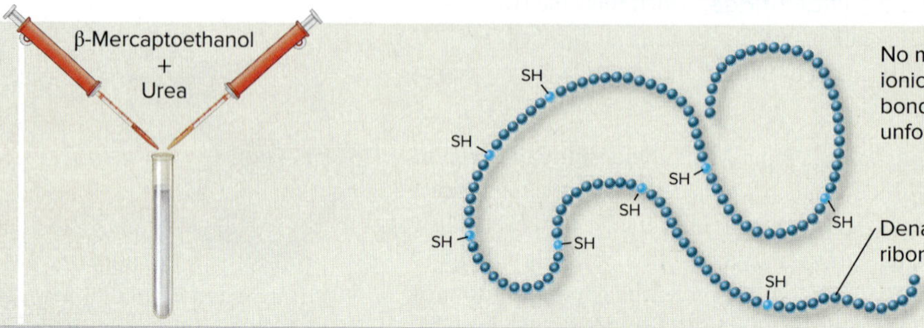 β-Mercaptoethanol + Urea | SH groups. No more H bonds, ionic bonds, or S—S bonds. Protein is unfolded. Denatured ribonuclease |
| **3** Pour mixture from step 2 into a size-exclusion chromatography column. Beads in the column trap β-mercaptoethanol and urea, whereas ribonuclease flows to the bottom. Collect ribonuclease in a test tube. Allow ribonuclease to sit for up to 20 hours and then measure its ability to degrade RNA. | 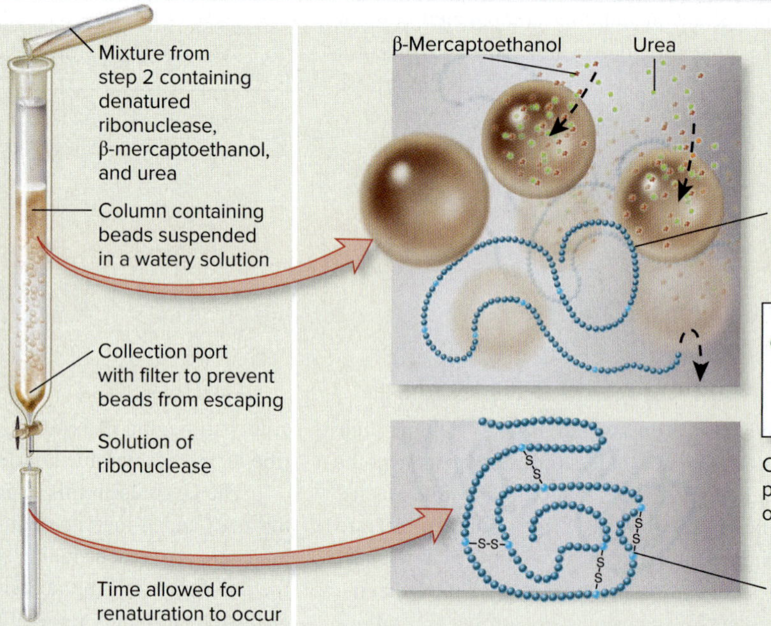 Mixture from step 2 containing denatured ribonuclease, β-mercaptoethanol, and urea; Column containing beads suspended in a watery solution; Collection port with filter to prevent beads from escaping; Solution of ribonuclease; Time allowed for renaturation to occur | β-Mercaptoethanol; Urea. Beads have microscopic pores that trap β-mercapto-ethanol and urea, but not ribonuclease. Denatured ribonuclease. Computer model of properly folded structure of ribonuclease. Renatured ribonuclease |

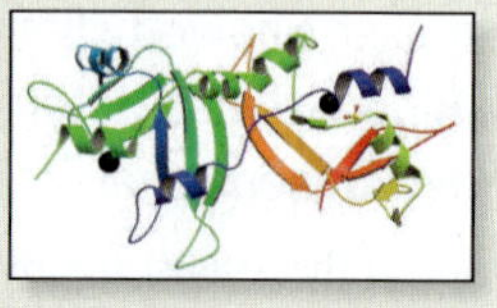

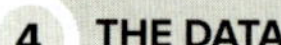

**4 THE DATA**

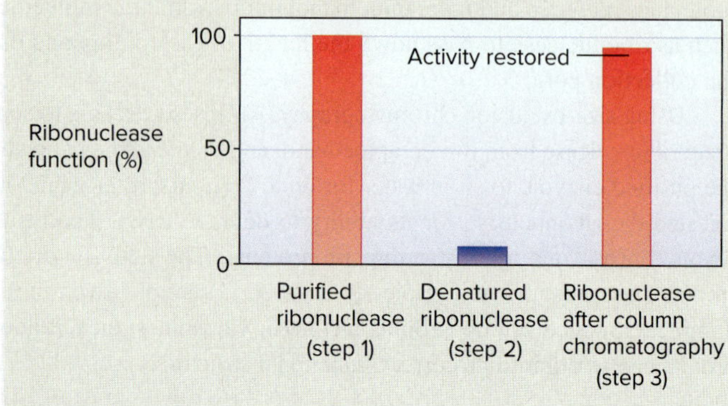


**5 CONCLUSION** Certain proteins, like ribonuclease, can spontaneously fold into their final, functional shapes without assistance from other cellular structures or factors. (However, as described in the text, this is not true of many other proteins.)

**6 SOURCE** Haber, E., and Anfinsen, C. B. 1961. Regeneration of enzyme activity by air oxidation of reduced subtilisin-modified ribonuclease. *Journal of Biological Chemistry* 236:422–424.

*(Computer model of properly folded structure of ribonuclease)* Image from the RCSB PDB (www.pdb.org) of PDB ID 1A2W (Liu, Y., Hart, P.J., Schlunegger, M.P., Eisenberg, D. (1998) *PNAS*, 95: 3437–3442)

Researchers have since learned that ribonuclease's ability to refold into its functional structure does not occur for all proteins. Some proteins do require enzymes and other proteins to assist in their proper folding. Even so, Anfinsen's experiments provided compelling evidence that the primary structure of a polypeptide is a key determinant of a protein's tertiary structure, an observation that allowed him to share in the Nobel Prize in Chemistry in 1972.

*Experimental Questions*

1. What hypothesis was Anfinsen testing?
2. Why did Anfinsen use urea and β-mercaptoethanol in his experiments?
3. **SCISKILLS ▸** Explain the result that was crucial to the discovery that the tertiary structure of ribonuclease may depend entirely on the primary structure.

## EVOLUTIONARY CONNECTIONS

### Proteins Contain Functional Domains

Modern research has revealed that many proteins have a modular design. This means that portions within proteins, called **domains,** have distinct structures and functions. These domains have been duplicated during evolution, so the same kind of domain may be found in many different proteins. When the same domain is found in different proteins, the domain has the same three-dimensional tertiary structure and performs a function that is characteristic of that domain.

As an example, **Figure 3.16** shows two members of a family of related proteins, known as nuclear receptors, which travel into the nucleus and play critical roles in regulating how certain genes are turned on in animal cells. These types of proteins are involved in animal development, reproduction, metabolism, and homeostasis. Nuclear receptors contain four or more domains within their structure:

- One domain found in nuclear receptors is a ligand-binding domain. These receptors are activated by a ligand, which is a molecule that binds to the receptor. For the two examples in Figure 3.16, the ligands are the steroid hormones estrogen and testosterone (a type of androgen), and the nuclear receptors are called the estrogen receptor and the androgen receptor, respectively. Though the ligand-binding domains in these two nuclear receptors have similar structures, slight structural differences allow one to bind estrogen and the other to bind testosterone (see the insets in Figure 3.16).
- A second domain, called the DNA-binding domain, binds to DNA once the receptor is activated by its ligand. The DNA-binding domains of these two receptors are sufficiently similar to enable both of them to bind to DNA, but dissimilar enough that they bind to different genes. Therefore, estrogen and testosterone have different effects, because they regulate different sets of genes.

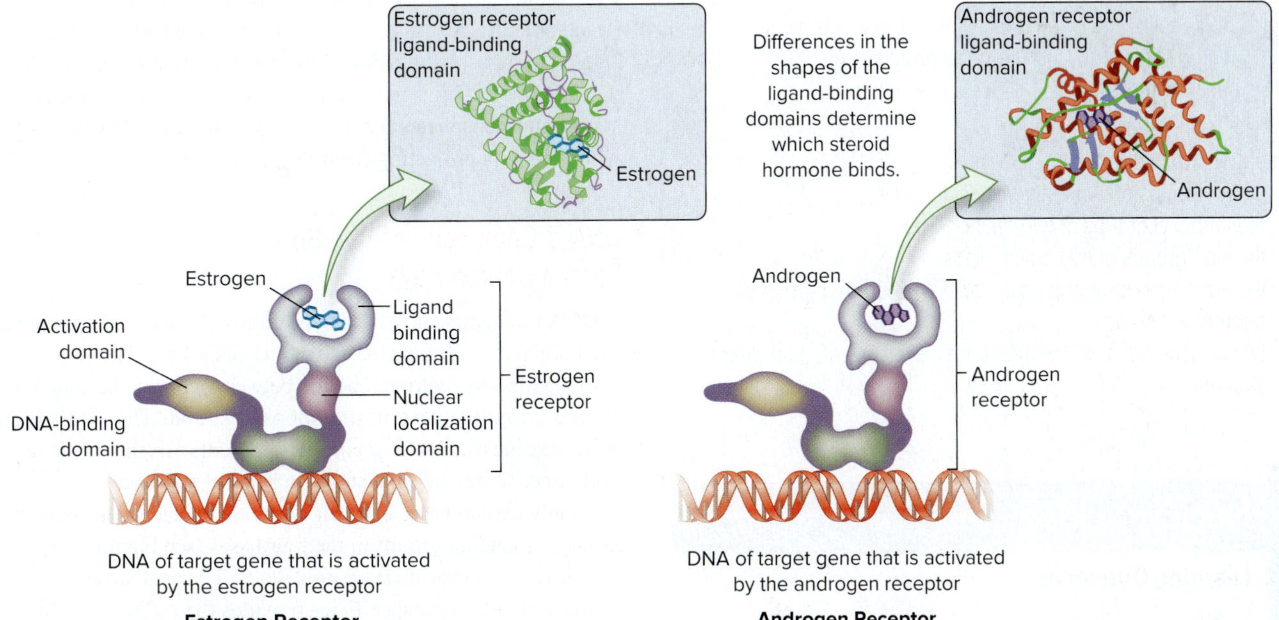


**Figure 3.16** **The domain structures of two related proteins called nuclear receptors.** The schematic drawings do not depict the actual three-dimensional structures of the estrogen and androgen receptors but rather emphasize that both receptors are composed of a series of different domains. The insets show the actual structures of the ligand-binding domains for both receptors. Note: The estrogen and androgen receptors are each composed of two identical polypeptides. For simplicity, only one polypeptide is shown here.

- Another domain, called a nuclear localization domain, facilitates the movement of the protein into the cell nucleus, where the DNA is located.
- Once the protein binds to DNA, a fourth domain, called the activation domain, activates the transcription of the target gene.

Overall, a nuclear receptor is a single protein with multiple domains, each with a unique function.

## 3.6 Reviewing the Concepts

- Proteins are macromolecules that play critical roles in almost all life processes. The proteins of all living organisms are composed of the same set of 20 amino acids, corresponding to 20 different side chains (Figure 3.11, Table 3.3).
- Amino acids are joined by peptide bonds to form polypeptides. A protein is a functional unit composed of one or more polypeptides that have folded into three-dimensional shapes (Figure 3.12).
- The four levels of protein structure are primary (its amino acid sequence), secondary (α helices or β pleated sheets), tertiary (folding to assume a three-dimensional shape), and in most cases quaternary. Domains are units of structure found in many different proteins where they serve similar functions, such as binding to DNA (Figures 3.13, 3.14, 3.15, 3.16).

## 3.6 Testing Your Knowledge

1. Which of the following is the primary structure of a protein?
   a. β pleated sheet
   b. its amino acid sequence
   c. its three-dimensional shape
   d. its protein subunits
   e. its random coils
2. A peptide bond is formed between
   a. amino groups of two amino acids.
   b. the carboxyl groups of two amino acids.
   c. the side groups of two amino acids.
   d. the amino group of one amino acid and carboxyl group of another amino acid.
   e. two polypeptides, which then form a protein with quaternary structure.

## 3.7 Nucleic Acids

### Learning Outcomes

1. Describe the three components of a nucleotide.
2. Distinguish between the structures of DNA and RNA.
3. Explain how certain bases pair with others in DNA.

**Nucleic acids** are responsible for the storage, expression, and transmission of genetic information. The expression of genetic information in the form of specific proteins determines whether an organism is a human, a fungus, an oak tree, or a bacterium. In this section, we survey the general features of DNA and RNA, which are discussed in greater detail in Chapter 9.

### Nucleotides Are the Building Blocks of DNA and RNA

The two classes of nucleic acids are **deoxyribonucleic acid (DNA)** and **ribonucleic acid (RNA).** DNA molecules store genetic information coded in the sequence of their building blocks. RNA molecules are involved in decoding this information into instructions for linking a specific sequence of amino acids to form a polypeptide.

Like other macromolecules, DNA and RNA are polymers consisting of linear sequences of repeating monomers. Each monomer, known as a **nucleotide,** has three components:

- A phosphate group
- A pentose sugar (deoxyribose or ribose)
- A single or a double ring of carbon and nitrogen atoms known as a **base** (Figure 3.17)

The nucleotides in DNA contain the five-carbon sugar **deoxyribose.** Four different nucleotides are present in DNA, corresponding to the four different bases that can be linked to deoxyribose. The **purine** bases, **adenine (A)** and **guanine (G),** have fused double rings of carbon and nitrogen atoms, and the **pyrimidine** bases, **cytosine (C)** and **thymine (T),** have a single-ring structure (Figure 3.17).

Nucleotides are covalently linked to each other to form a DNA strand or an RNA strand. The carbon atoms of the sugar are numbered 1′–5′. The phosphate groups link the 3′ carbon of one nucleotide to the 5′ carbon of the next (see Figure 3.17). The linkages between sugars and phosphates form the backbone of a DNA or RNA strand with the bases projecting from the backbone.

### DNA Is Composed of Two Strands That Form a Double Helix

A DNA molecule consists of two strands of nucleotides coiled around each other to form a **double helix** (Figure 3.18). The two strands are held together by hydrogen bonds between a purine base in one strand and a pyrimidine base in the opposite strand. The ring structure of each base lies in a flat plane perpendicular to the sugar-phosphate backbone, somewhat like steps on a spiral staircase.

Only certain bases can pair with others, due to the location of the hydrogen-bonding groups in the four bases (see Figure 3.18). In DNA, an A in one strand is always paired with a T in another strand, and a G is paired with a C. This specificity provides the mechanism for duplicating and transferring genetic information (see Chapter 9). Two hydrogen bonds are formed between adenine and thymine (A-T pairing), and three are formed between guanine and cytosine (G-C pairing).

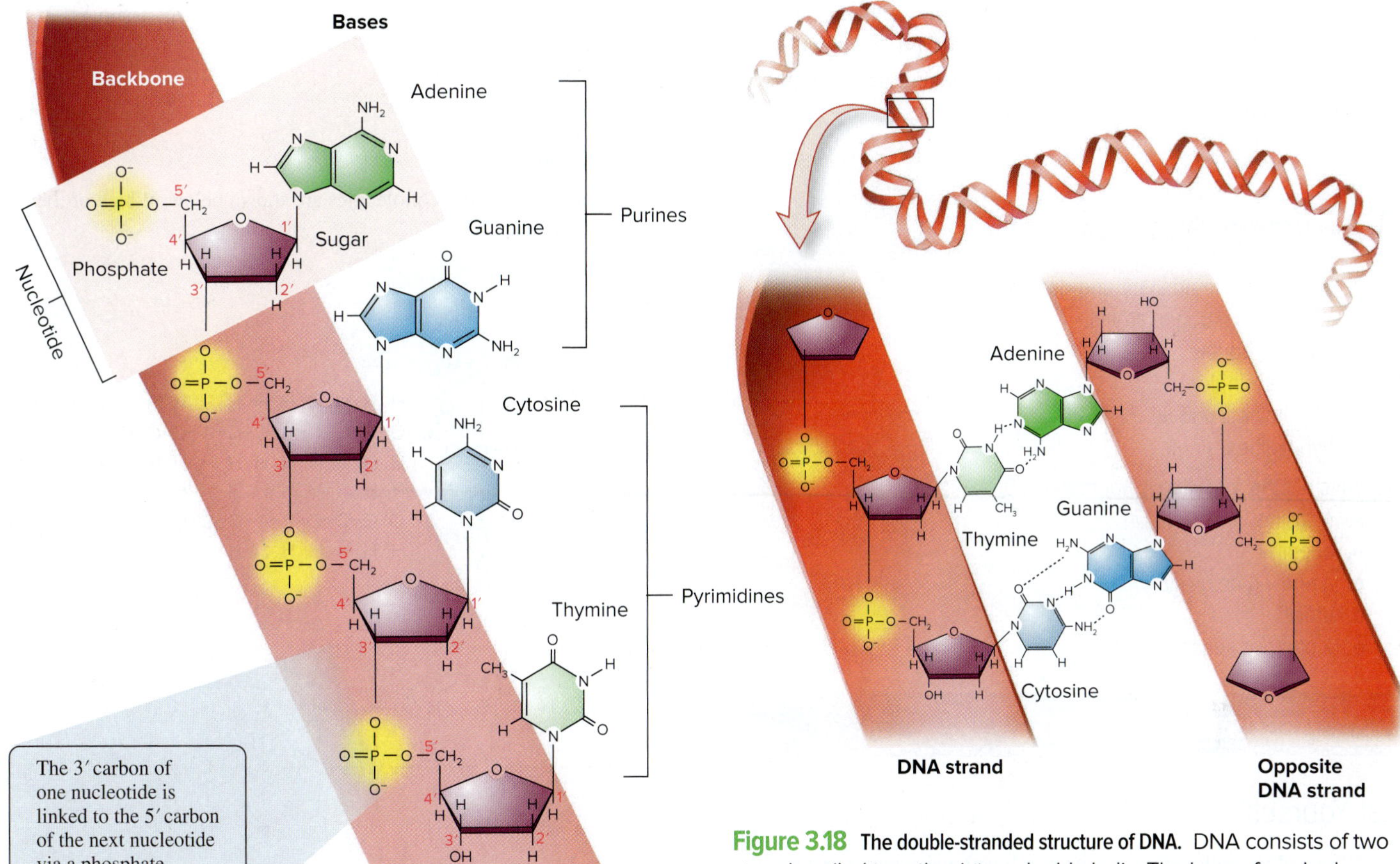


**Figure 3.17 Structure of a DNA strand.** Nucleotides are linked to each other to form a strand of DNA. The four bases found in DNA are shown. A strand of RNA is similar except the sugar is ribose, and uracil is substituted for thymine.

**Figure 3.18 The double-stranded structure of DNA.** DNA consists of two strands coiled together into a double helix. The bases form hydrogen bonds (dashed lines) in which A pairs with T, and G pairs with C.

**Concept Check:** *If the sequence of bases in one strand of a DNA double helix is known, can the base sequence of the opposite strand be predicted?*

## RNA Is Usually Single Stranded

RNA structure differs in only a few respects from DNA. Except in some viruses, RNA consists of a single strand of nucleotides rather than a double-stranded structure. In RNA, the sugar in each nucleotide is **ribose** instead of deoxyribose. Also, the pyrimidine base thymine in DNA is replaced in RNA with the pyrimidine base **uracil (U)** (Figure 3.19). The other three bases—adenine, guanine, and cytosine—are found in both DNA and RNA. Certain forms of RNA called messenger RNA (mRNA), ribosomal RNA (rRNA), and transfer RNA (tRNA) are responsible for using the information contained in DNA, that is, within genes, in order to make polypeptides. This topic will be discussed in Unit III.

Base (uracil)

**Figure 3.19 Uracil, a pyrimidine base found only in RNA.** Uracil forms hydrogen bonds with adenine.

## 3.7 Reviewing the Concepts

- Nucleic acids (DNA and RNA) are responsible for the storage, expression, and transmission of genetic information. DNA and RNA consist of repeating monomers known as nucleotides composed of a phosphate group, a five-carbon sugar, and a base (Figure 3.17).
- DNA consists of two strands of nucleotides coiled to form a double helix, held together by hydrogen bonds between a purine base in one strand and a pyrimidine base in the opposite strand. DNA molecules store genetic information coded in the sequence of their monomers (Figure 3.18).

- RNA consists of a single strand of nucleotides, which include the base uracil. RNA molecules are involved in decoding information in DNA into instructions for making polypeptides (Figure 3.19).

## 3.7 Testing Your Knowledge

**1.** _______ include adenine and guanine and have_______ rings of carbon and nitrogen atoms.
  **a.** Purines, double
  **b.** Purines, single
  **c.** Pyrimidines, double
  **d.** Pyrimidines, single

**2.** Which of the following is *not* found in DNA?
  **a.** cytosine
  **b.** adenine
  **c.** a phosphate group
  **d.** uracil
  **e.** a pentose sugar

## Assess and Discuss

### Test Yourself

1. Which of the following is a common functional group found in organic molecules?
   a. a hydrocarbon
   b. a ketone
   c. an isomer
   d. a carbon atom
   e. an oxygen atom
2. Dehydration reactions may produce which of the following?
   a. monomers
   b. glycogen
   c. glucose
   d. amino acids
   e. glycerol
3. The versatility of carbon to serve as the backbone for a variety of molecules is due in part to
   a. the ability of carbon atoms to form four covalent bonds.
   b. the fact that carbon usually forms ionic bonds with many different atoms.
   c. the abundance of carbon in the environment.
   d. the ability of carbon to form single, double, and even triple bonds with many different types of atoms.
   e. both a and d.
4. Which of the following bonds is/are nonpolar?
   a. C—C
   b. C—O
   c. C—H
   d. both a and b
   e. both a and c
5. _________ is a polysaccharide characterized by an unbranched arrangement and a high degree of hydrogen bonding.
   a. Glucose
   b. Starch
   c. Wax
   d. Cellulose
   e. Glycogen
6. All _________ contain four rings of carbon in their structure.
   a. fatty acids
   b. steroids
   c. triglycerides
   d. proteins
   e. nucleotides
7. The structures of three molecules are shown below. Which of them are amphipathic?

(a)

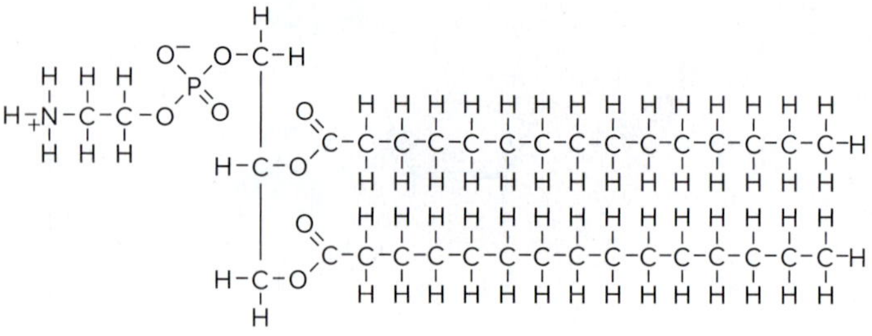

(b)

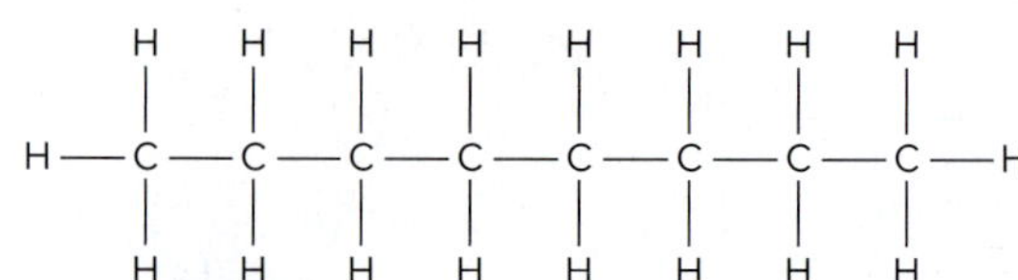

(c)

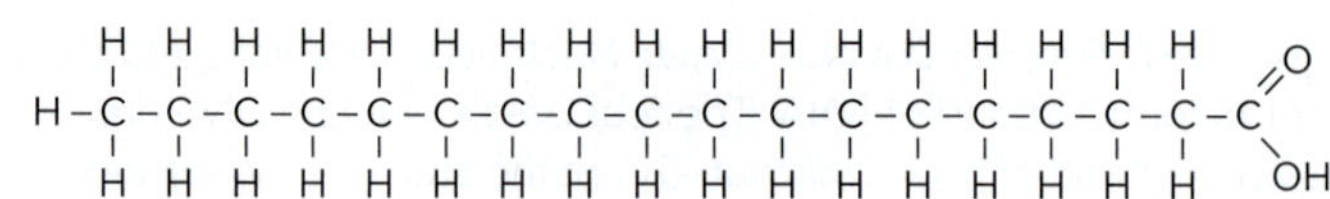

   a. (a) only
   b. (b) only
   c. (c) only
   d. (b) and (c)
   e. (a) and (c)
8. The monomers of proteins are __________, and these are linked by polar covalent bonds commonly referred to as _______________ bonds.
   a. nucleotides, peptide
   b. polypeptides, peptide
   c. dipeptides, amino
   d. amino acids, peptide
   e. monosaccharides, glycosidic
9. A(n) ________ is a modular portion of a protein with a particular structure and function.
   a. peptide bond
   b. domain
   c. phospholipid
   d. amino acid
   e. monosaccharide

10. A DNA molecule contains 30% G. What percentage is A?
    a. 30%
    b. 20%
    c. 15%
    d. 10%
    e. none of the above

## Conceptual Questions

1. Imagine that a protein's primary structure could somehow be altered; would this affect its tertiary structure?
2. Explain the difference between saturated and unsaturated fatty acids, and how the structural differences between them contribute to differences in their properties.
3. **PRINCIPLES** A principle of biology is that structure determines function. What does this mean for organic molecules such as carbohydrates, lipids, and proteins?

## Collaborative Questions

1. Discuss the structural and functional differences between different types of carbohydrates.
2. Discuss some of the general functions of proteins in organisms.

## Online Resource

**connect.mheducation.com**

SMARTBOOK® SmartBook® is the first and only adaptive reading experience designed to change the way students read and learn.

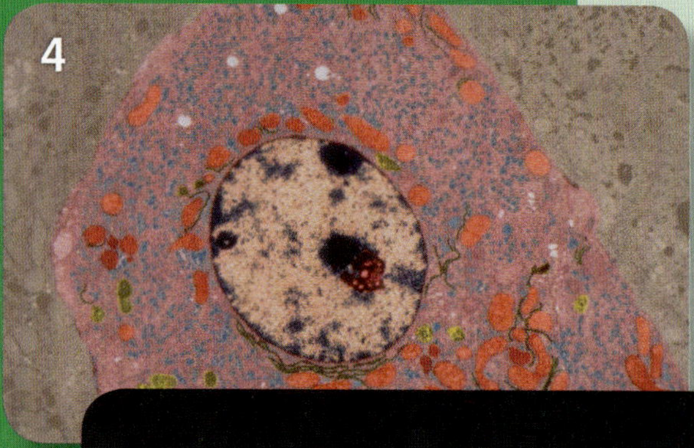

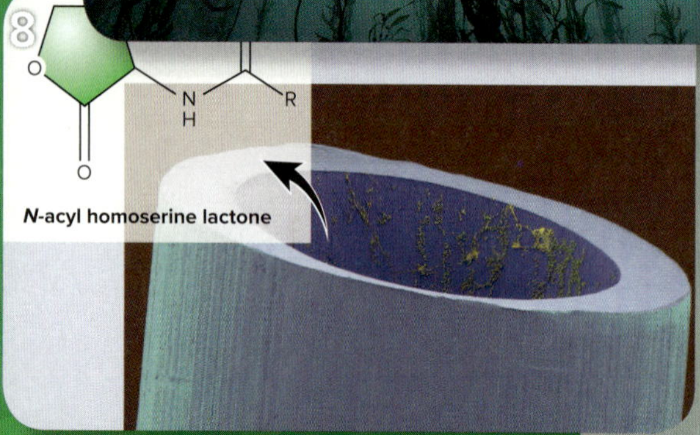

# UNIT II
# CELLS

Cell biology is the study of life at the cellular level. Although cells are the simplest units of life, they are also wonderfully complex and interesting, providing information about all living things. In this unit, Chapter 4 begins with a description of the evolutionary origin of cells and then provides an overview of their structures and functions. Chapter 5 focuses on the structure of cell membranes and the transport of substances into and out of the cell. We will also examine how cells are held together by cell junctions. Chapters 6 and 7 are largely devoted to metabolism—the sum of the chemical reactions in a cell or an organism. Chapter 6 explores the topic of energy and considers how enzymes facilitate chemical reactions in a cell. We will also examine the pathways for carbohydrate breakdown and how those pathways are used to make an energy intermediate called adenosine triphosphate (ATP). Chapter 7 describes how carbohydrates are made via photosynthesis by capturing the energy of sunlight. Finally, Chapter 8 considers how cells interact with their environment and with each other. We will examine how cells respond to signals, either those that come directly from their environment or those that are made by other cells.

***The following biology principles are emphasized in this unit:***

- ***Cells are the simplest units of life.*** *Throughout this unit, we will examine how life exists at the cellular level.*
- ***Living organisms use energy.*** *Chapters 6 and 7 address how cells utilize and store energy contained within organic molecules such as glucose.*
- ***Living organisms interact with their environment.*** *Chapter 8 largely focuses on the plasma membrane of the cell and how it provides an interface between the cell and its environment.*
- ***Structure determines function.*** *Throughout this unit, we will see many examples in which the structure of proteins determines their cellular functions.*
- ***Biology is an experimental science.*** *Most chapters in this unit have a Feature Investigation describing a pivotal experiment that provided insights into the workings of cells.*

4

5

6

7

8

*N*-acyl homoserine lactone

# Evolutionary Origin of Cells and Their General Features

# 4

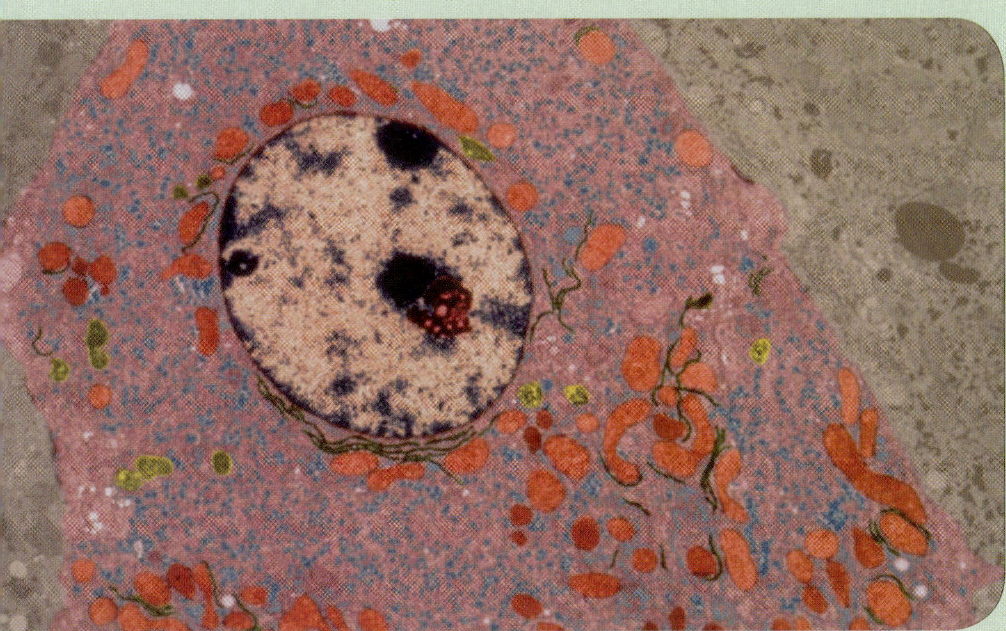

© Dennis Kunkel Microscopy, Inc./Phototake

**A cell from the human liver.** The cell in this micrograph was viewed by a technique called transmission electron microscopy (TEM), which is described in this chapter. The micrograph was artificially colored using a computer to enhance the visualization of certain cell structures.

## Chapter Outline



Cancer is a disease characterized by uncontrolled cell growth. Cancer cells form tumors that invade nearby parts of the body and may spread to distant parts via the bloodstream. About 1 in 4 Americans will eventually die from cancer. In the U.S., approximately 1.6 million people were diagnosed with cancer during 2012, and about 600,000 died from the disease. Though such numbers are sobering, there is some reason for optimism. Over the past 10 years, the death rate from cancer has consistently decreased by over 1 percent per year. This decrease can be attributed to a variety of factors, including lifestyle changes (for example, fewer people smoking cigarettes) and improved treatment methods.

Topics such as cancer are within the field of **cell biology**—the study of individual cells and their interactions with each other. Researchers in this field want to understand the basic features of cells and apply that knowledge in the treatment of diseases such as cystic fibrosis, sickle cell disease, and lung cancer.

The idea that organisms are composed of cells originated in the mid-1800s. German botanist Matthias Schleiden studied plant material under the microscope and was struck by the presence of many similar-looking compartments, each of which contained a dark area. Today we call those compartments cells and the dark area the nucleus. Schleiden speculated in 1838 that cells are living entities and that plants are aggregates of cells arranged according to definite laws. In 1839, German physiologist Theodor Schwann extended Schleiden's hypothesis to animals. About two decades later, German biologist Rudolf Virchow proposed that *omnis cellula e cellula*, or "every cell originates from another cell." This idea arose from his research, which showed that diseased cells divide to produce more diseased cells.

The **cell theory,** which is credited to both Schleiden and Schwann with contributions from Virchow, has three parts.

1. All living organisms are composed of one or more cells.
2. Cells are the smallest units of life.
3. New cells come only from pre-existing cells by cell division.

In this chapter, we will begin our exploration of cells with a description of hypotheses of how living cells arose on Earth. We then consider the general features of cell structure and function. Later chapters in this unit will examine certain aspects of cell biology in greater detail.

## 4.1 Origin of Living Cells on Earth

### Learning Outcomes

1. Outline the four overlapping stages that are hypothesized to have led to the origin of living cells.
2. List various hypotheses about how complex organic molecules formed.
3. Explain the concept of an RNA world and describe how it could have evolved into a DNA/RNA/protein world.

As described in Chapter 1, cells are the simplest unit of life. Living cells are complex collections of molecules and macromolecules. DNA stores genetic information, RNA acts as an intermediary in the process of protein synthesis and plays other important roles, and proteins form the foundation for the structure and activities of living cells. Life as we know it requires this interplay between DNA, RNA, and proteins for its existence and perpetuation. On modern Earth, every living cell is made from a pre-existing cell.

But how did life get started? Because DNA, RNA, and proteins are the central players in the enterprise of life, scientists who are interested in the origin of cells have focused much of their attention on the formation of these macromolecules and their building blocks, namely, nucleotides and amino acids. To understand the origin of life, we can view the process as occurring in four overlapping stages:

- **Stage 1:** Nucleotides and amino acids were produced prior to the existence of cells.
- **Stage 2:** Nucleotides became polymerized to form RNA and/or DNA, and amino acids became polymerized to form proteins.
- **Stage 3:** Polymers became enclosed in membranes.
- **Stage 4:** Polymers enclosed in membranes acquired properties that are associated with living cells.

Researchers have followed a variety of experimental approaches to determine how life may have begun, including the synthesis of organic molecules in the laboratory without the presence of living cells or cellular material. This work has led researchers to propose a variety of hypotheses regarding the origin of cells. In this section, we examine the origin of life at each of these stages and consider a few scientific viewpoints that wrestle with the question "How did life on Earth begin?"

### Stage 1: Organic Molecules Formed Prior to the Existence of Cells

Let's begin our inquiry into the first stage of the origin of life by considering how nucleotides and amino acids may have been made prior to the existence of living cells. In the 1920s, the Russian biochemist Alexander Oparin and the Scottish biologist J. B. S. Haldane independently proposed that organic molecules, such as nucleotides and amino acids, arose spontaneously under the conditions that occurred on early Earth. According to this hypothesis, the spontaneous appearance of organic molecules produced what they called a "primordial soup," which eventually gave rise to living cells.

The conditions on early Earth, which were much different from today, may have been more conducive to the spontaneous formation of organic molecules. Current hypotheses suggest that organic molecules, and eventually macromolecules, formed spontaneously. This is termed prebiotic (before life) or abiotic (without life) synthesis. These slowly forming organic molecules accumulated because there was little free oxygen gas, so they were not spontaneously oxidized, and there were as yet no living organisms, so they were also not metabolized. The slow accumulation of these molecules in the early oceans over a long period of time formed what is now called the **prebiotic soup.** The formation of this medium was a key event that preceded the origin of life.

Though most scientists agree that life originated from the assemblage of nonliving matter on early Earth, the mechanism of how and where these molecules originated is widely debated. Many intriguing hypotheses have been proposed, which are not mutually exclusive. A few of the more widely debated ideas are the reducing atmosphere hypothesis, the extraterrestrial hypothesis, and the deep-sea vent hypothesis.

**Reducing Atmosphere Hypothesis** Based largely on geological data, many scientists in the 1950s proposed that the atmosphere on early Earth was rich in water vapor ($H_2O$), hydrogen gas ($H_2$), methane ($CH_4$), and ammonia ($NH_3$). These components, along with a lack of atmospheric oxygen ($O_2$), produce a reducing atmosphere because methane and ammonia readily give up electrons to other molecules, thereby reducing them. Such oxidation-reduction reactions, or redox reactions, are required for the formation of complex organic molecules from simple inorganic molecules.

In 1953, American chemist Stanley Miller, a student in the laboratory of the physical chemist Harold Urey, was the first scientist to use

## Biology Principle

### Biology Is an Experimental Science

By conducting experiments, researchers were able to demonstrate the feasibility of the synthesis of organic molecules prior to the emergence of living cells.

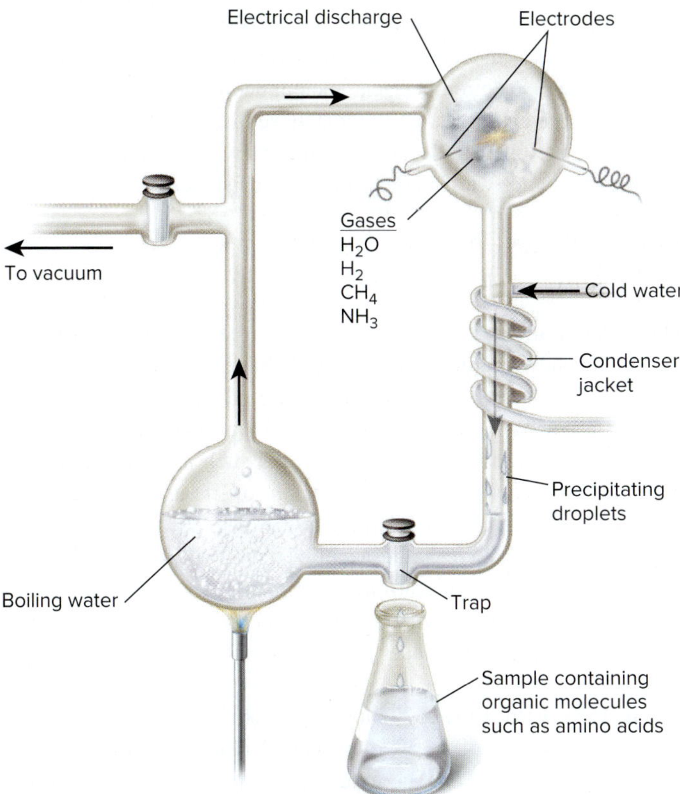


**Figure 4.1 Testing the reducing atmosphere hypothesis for the origin of life—the Miller and Urey experiment.**

**Concept Check:** *With regard to the origin of life, why are biologists interested in the abiotic synthesis of organic molecules?*

experimentation to test whether the prebiotic synthesis of organic molecules is possible. His experimental apparatus was intended to simulate the conditions on early Earth that were postulated in the 1950s (Figure 4.1). Water vapor from a flask of boiling water rose into another chamber containing hydrogen gas ($H_2$), methane ($CH_4$), and ammonia ($NH_3$). Miller inserted two electrodes that sent electrical discharges into the chamber to simulate lightning bolts. A condenser jacket cooled some of the gases from the chamber, causing droplets to form that fell into a trap. He then took samples from this trap for chemical analysis. In his first experiments, he observed the formation of hydrogen cyanide (HCN) and formaldehyde ($CH_2O$). Such molecules are precursors of more complex organic molecules. These precursors also combined to make larger molecules such as the amino acid glycine. At the end of 1 week of operation, 10–15% of the carbon had been incorporated into organic compounds. Later experiments by Miller and others demonstrated the formation of sugars, a few types of amino acids, lipids, and nitrogenous bases found in nucleic acids (for example, adenine).

In a study published in 2011, researchers analyzed samples that Miller had preserved from a 1958 experiment in which he used a mixture of $CH_4$, $NH_3$, hydrogen sulfide ($H_2S$), and carbon dioxide ($CO_2$). For unknown reasons, Miller had not analyzed what products were made in this experiment. When these preserved samples were analyzed using modern technology, they were found to contain 23 different amino acids and 4 amines (another type of organic molecule), more organic compounds than seen in Miller's classic experiments.

Why were these studies important? The work of Miller and Urey was the first attempt to apply scientific experimentation to our quest to understand the origin of life. Their pioneering strategy showed that the prebiotic synthesis of organic molecules is possible, although it could not prove that it really happened that way. In spite of the importance of these studies, critics of the so-called reducing atmosphere hypothesis have argued that Miller and Urey were wrong about the composition of early Earth's environment. More recently, many scientists have suggested that the atmosphere on early Earth was not reducing, but instead was a neutral gaseous mixture consisting mostly of carbon monoxide (CO), carbon dioxide ($CO_2$), nitrogen gas ($N_2$), and $H_2O$. These newer ideas are derived from studies of volcanic gas, which has much more $CO_2$ and $N_2$ than $CH_4$ and $NH_3$, and from the observation that ultraviolet (UV) radiation destroys $CH_4$ and $NH_3$, so these molecules would have been short-lived on early Earth, which had high levels of UV radiation. Nevertheless, since the experiments of Miller and Urey, many newer investigations have shown that organic molecules can be made under a variety of conditions. For example, organic molecules can be made prebiotically from a neutral gaseous mixture composed primarily of CO, $CO_2$, $N_2$, and $H_2O$.

**Extraterrestrial Hypothesis** Many scientists have argued that sufficient organic molecules may have been present in meteorites, material from asteroids and comets that reached the surface of early Earth. A significant proportion of meteorites belong to a class known as carbonaceous chondrites. Such meteorites may contain a substantial amount of organic carbon, including amino acids and nucleic acid bases. Based on this observation, some scientists have postulated that such meteorites could have transported a significant amount of organic molecules from outer space to early Earth.

Opponents of this hypothesis argue that most of this material would have been destroyed by the intense heating that accompanies the passage of large bodies through the atmosphere and their subsequent collision with the surface of the Earth. Though some organic molecules are known to reach the Earth via such meteorites, the degree to which heat would have destroyed many of the organic molecules remains a matter of controversy.

**Deep-Sea Vent Hypothesis** In 1988, German lawyer and organic chemist Günter Wächtershäuser proposed that key organic molecules may have originated in deep-sea vents, which are cracks in the Earth's surface where superheated water rich in metal ions and hydrogen sulfide ($H_2S$) mixes abruptly with cold seawater. These vents release hot gaseous substances from the interior of the Earth at temperatures in excess of 300°C (572°F). Supporters of this hypothesis propose that biologically important molecules may have been formed in the temperature gradient between the extremely hot vent water and the cold water that surrounds the vent (Figure 4.2a).

Experimentally, the temperatures within this gradient are known to be suitable for the synthesis of molecules that form components of biological molecules. For example, the reaction between iron and $H_2S$ yields pyrites and $H_2$ and has been shown to provide the energy necessary for the reduction of $N_2$ to $NH_3$. Nitrogen is an essential component of both nucleic acids and amino acids—the molecular building blocks of life. But $N_2$, which is found abundantly on Earth, is chemically inert, so it is unlikely to have given rise to life. The conversion of $N_2$ to $NH_3$ at deep-sea vents may have led to the production of amino acids and nucleic acids.

Interestingly, complex biological communities are found in the vicinity of modern deep-sea vents. Various types of fish, worms, clams, crabs, shrimp, and bacteria are found in significant abundance in those areas (Figure 4.2b). Unlike most other forms of life on our planet, these organisms receive their energy from chemicals in the vent and not from the Sun. In 2007, American scientist Timothy Kusky and colleagues discovered 1.43 billion-year-old fossils of deep-sea microbes near ancient deep-sea vents. This study provided more evidence that life may have originated on the bottom of the ocean. However, debate continues as to the primary way that organic molecules were made prior to the existence of life on Earth.

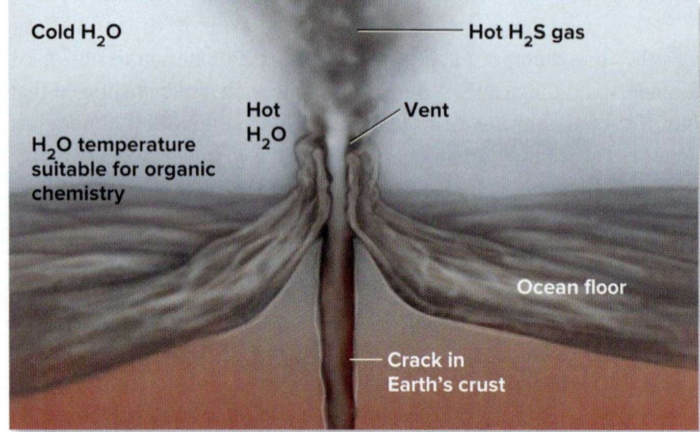


(a) Deep-sea vent hypothesis

(b) A deep-sea vent community

**Figure 4.2 The deep-sea vent hypothesis for the origin of life.**
**(a)** Deep-sea vents are cracks in the Earth's surface that release hot gases such as hydrogen sulfide ($H_2S$). This heats the water near the vent and results in a gradient between the very hot water adjacent to the vent and the cold water farther from the vent. The synthesis of organic molecules can occur in this gradient. **(b)** Photograph of a biological community near a deep-sea vent, which includes giant (orange) tube worms, crabs, and eels.
*(b)* © Emory Kristof/National Geographic Creative

**Concept Check:** *What properties of deep-sea vents made them suitable for the prebiotic synthesis of molecules?*

## Stage 2: Organic Polymers May Have Formed on the Surface of Clay or in Water

The preceding three hypotheses provide reasonable mechanisms by which small organic molecules could have accumulated on early Earth. Scientists hypothesize that the second stage in the origin of life was a period in which simple organic molecules polymerized to form more complex organic polymers such as DNA, RNA, or proteins. Most ideas regarding the origin of life assume that polymers with at least 30–60 monomers are needed to store enough information to make a viable genetic system. Because hydrolysis competes with polymerization, many scientists have speculated that the synthesis of polymers did not occur in a watery prebiotic soup, but instead took place on a solid surface or in evaporating tidal pools.

In 1951, Irish X-ray crystallographer John Bernal first suggested that the prebiotic synthesis of polymers took place on clay. In his book *The Physical Basis of Life,* he wrote that "clays, muds, and inorganic crystals are powerful means to concentrate and polymerize organic molecules." Many clay minerals are known to bind organic molecules such as nucleotides and amino acids. Experimentally, many research groups have demonstrated the formation of nucleic acid polymers and polypeptides on the surface of clay, given the presence of monomer building blocks. During the prebiotic synthesis of RNA, the purine bases of the nucleotides could have interacted with the silicate surfaces of the clay. Cations, such as $Mg^{2+}$, bound the nucleotides to the negative surfaces of the clay, thereby positioning the nucleotides in a way that promoted bond formation between the phosphate of one nucleotide and the ribose sugar of an adjacent nucleotide. In this way, polymers such as RNA may have formed.

Though the formation of polymers on clay remains a reasonable hypothesis, studies by American chemist Luke Leman and his colleagues English chemist Leslie Orgel and Iranian-American chemist M. Reza Ghadiri indicate that polymers can also form in aqueous solutions, which is contrary to popular belief. Their work in 2004 showed that carbonyl sulfide, a simple gas present in volcanic gases and deep-sea vent emissions, can bring about the formation of peptides from amino acids under mild conditions in water. These results indicate that the synthesis of polymers could have taken place in the prebiotic soup.

### Stage 3: Cell-Like Structures May Have Originated When Polymers Were Enclosed by a Boundary

The third stage in the origin of living cells is hypothesized to be the formation of a boundary that separated the polymers such as RNA from the environment. The term **protobiont** is used to describe an aggregate of prebiotically produced molecules and macromolecules that acquired a boundary, such as a lipid bilayer, that allowed it to maintain an internal chemical environment distinct from that of its surroundings. What characteristics make protobionts possible precursors of living cells? Scientists envision the existence of four key features:

1. A boundary, such as a membrane, separated the internal contents of the protobiont from the external environment.
2. Polymers inside the protobiont contained information.
3. Polymers inside the protobiont had catalytic functions.
4. The protobionts eventually developed the capability of self-replication.

Protobionts were not capable of precise self-reproduction like living cells but could divide to increase in number. Such protobionts are thought to have exhibited simple metabolic pathways in which the structures of organic molecules were changed. In particular, the polymers inside protobionts must have gained the catalytic ability to link organic building blocks to produce new polymers. This would have been a critical step in the process that eventually provided protobionts with the ability to self-replicate. According to this scenario, metabolic pathways became more complex, and the ability of protobionts to self-replicate became more refined over time. Eventually, these structures exhibited the characteristics that we attribute to living cells. As described next, researchers have hypothesized that protobionts may have existed as coacervates or liposomes.

**Coacervates** In 1924, Alexander Oparin hypothesized that living cells evolved from **coacervates,** droplets that form spontaneously from the association of charged polymers such as proteins, carbohydrates, or nucleic acids surrounded by water. Their name derives from the Latin *coacervare,* meaning to assemble together or cluster. Coacervates measure 1–100 μm (micrometers) across, are surrounded by a tight skin of water molecules, and possess osmotic properties (Figure 4.3a). This skin of water allows the selective absorption of simple molecules from the surrounding medium.

Enzymes trapped within coacervates can perform primitive metabolic functions. For example, researchers have made coacervates containing the enzyme glycogen phosphorylase. When glucose-1-phosphate was made available to these coacervates, it was taken up into them, and starch was produced. The starch merged with the wall of the coacervates, and they increased in size and eventually divided into two. When the enzyme amylase was included, the starch was broken down to maltose, which was released from the coacervates.

**Liposomes** As a second possibility, protobionts may have resembled **liposomes**—vesicles surrounded by a lipid bilayer (Figure 4.3b). When certain types of lipids are dissolved in water, they spontaneously form liposomes. As discussed in Chapter 5, lipid bilayers are selectively permeable (look ahead to Figure 5.7), and some liposomes can even store energy in the form of an electrical gradient. Such liposomes can discharge this energy in a neuron-like fashion, showing rudimentary signs of excitability, which is characteristic of living cells.

In 2003, Danish chemist Martin Hanczyc, American chemist Shelly Fujikawa, and Canadian-American biologist Jack Szostak showed that clay can catalyze the formation of liposomes that grow and divide, a primitive form of self-replication. Furthermore, if RNA was on the surface of the clay, the researchers discovered that liposomes that enclosed RNA were formed. These experiments are compelling because they showed that the formation of membrane vesicles containing RNA molecules is a plausible explanation of the emergence of cell-like structures based on simple physical and chemical properties.

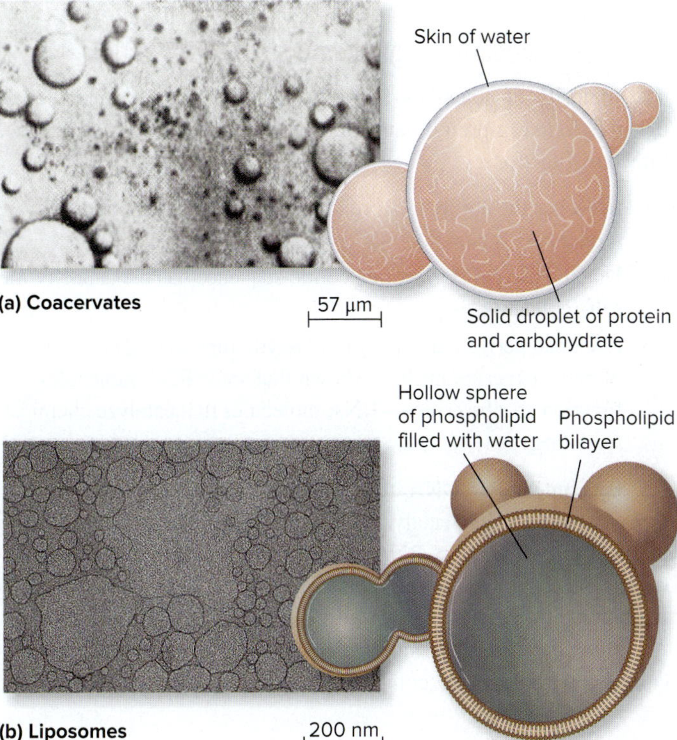


**Figure 4.3 Protobionts and their lifelike functions.** Primitive cell-like structures such as coacervates or liposomes could have given rise to living cells. **(a)** A micrograph and illustration of coacervates, which are droplets of protein and carbohydrate surrounded by a skin of water molecules. **(b)** An electron micrograph and illustration of liposomes. Each liposome is made of a phospholipid bilayer surrounding an aqueous compartment.

*(a)* Source: A.I. Oparin. From: *The Origin of Life,* New York: Dover, 1952; *(b)* © Mary Kraft

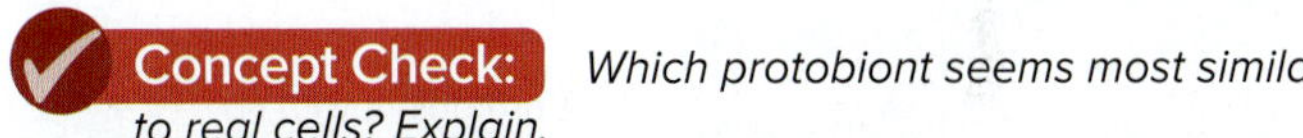

**Concept Check:** *Which protobiont seems most similar to real cells? Explain.*

**BioConnections:** *Refer back to Figure 3.9. What is the physical/chemical reason why phospholipids tend to form a bilayer?*

## Stage 4: Cellular Characteristics May Have Evolved via Chemical Selection, Beginning with an RNA World

The majority of scientists favor RNA as the first macromolecule that was found in protobionts. Unlike other polymers, RNA exhibits three key functions:

1. RNA has the ability to store information in its nucleotide sequence.
2. Due to base pairing, its nucleotide sequence has the capacity for self-replication.
3. RNA can perform a variety of catalytic functions. The results of many experiments have shown that some RNA molecules function as **ribozymes**—RNA molecules that catalyze chemical reactions.

By comparison, DNA and proteins are not as versatile as RNA. DNA has very limited catalytic activity, and proteins are not known to undergo self-replication. RNA can perform functions that are characteristic of proteins and, at the same time, can serve as genetic material with replicative and informational functions.

How did the RNA molecules that were first made prebiotically evolve into more complex molecules that produced cell-like characteristics? Researchers propose that a process called chemical selection was responsible. **Chemical selection** occurs when a chemical within a mixture has special properties or advantages that cause it to increase in number relative to other chemicals in the mixture. (As we will discuss in Chapter 19, natural selection is a similar process except that it describes changes in a population of living organisms over time due to survival and reproductive advantages.) Chemical selection results in **chemical evolution,** in which a population of molecules changes over time to become a new population with a different chemical composition.

Scientists speculate that initially the special properties enabling certain RNA molecules to undergo chemical selection were their ability (1) to self-replicate and (2) to perform other catalytic functions. As a way to understand the concept of chemical selection, let's consider a hypothetical scenario showing two steps of chemical selection. The first step of Figure 4.4 shows a group of protobionts that contain RNA molecules that were made prebiotically. RNA molecules inside these protobionts can be used as templates for the prebiotic synthesis of complementary RNA molecules. Such a process of self-replication, however, would be very slow because it would not be catalyzed by enzymes in the protobiont. In a first step of chemical selection, the sequence of one of the RNA molecules

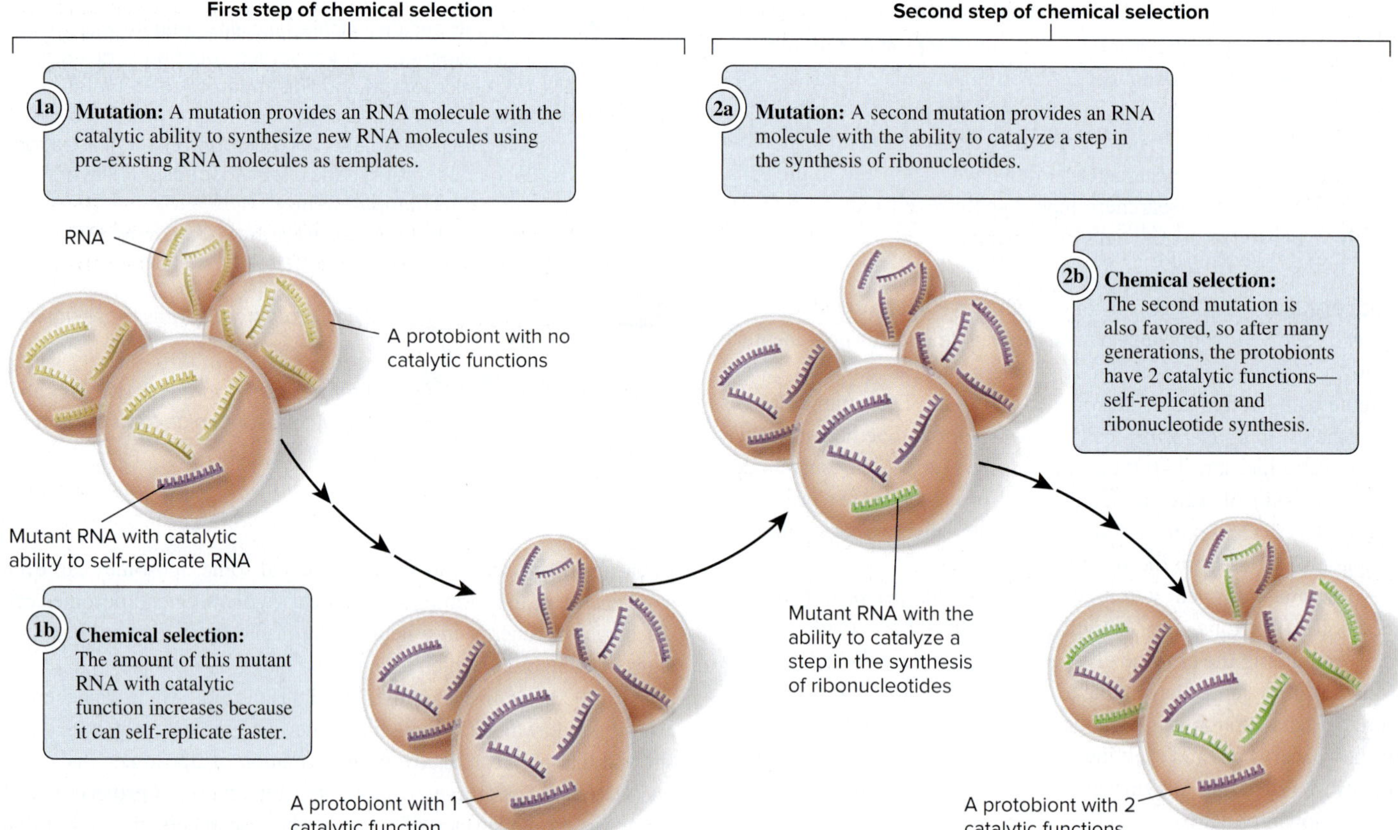

First step of chemical selection
Second step of chemical selection
1a Mutation: A mutation provides an RNA molecule with the catalytic ability to synthesize new RNA molecules using pre-existing RNA molecules as templates.
2a Mutation: A second mutation provides an RNA molecule with the ability to catalyze a step in the synthesis of ribonucleotides.
RNA
A protobiont with no catalytic functions
2b Chemical selection: The second mutation is also favored, so after many generations, the protobionts have 2 catalytic functions—self-replication and ribonucleotide synthesis.
Mutant RNA with catalytic ability to self-replicate RNA
1b Chemical selection: The amount of this mutant RNA with catalytic function increases because it can self-replicate faster.
Mutant RNA with the ability to catalyze a step in the synthesis of ribonucleotides
A protobiont with 1 catalytic function
A protobiont with 2 catalytic functions


**Figure 4.4 A hypothetical scenario illustrating the process of chemical selection.** This figure shows a two-step scenario. In the first step, RNAs that can self-replicate are selected, and in the second step, RNAs with the ability to catalyze a step in ribonucleotide synthesis are selected.

*What is meant by the term chemical selection?*

has undergone a mutation that gives it the catalytic ability to attach nucleotides together, using other RNA molecules as a template. This protobiont would have an advantage over the others because it would be capable of faster self-replication of its RNA molecules. Over time, due to this enhanced rate of replication, protobionts carrying such RNA molecules would increase in number compared with the others. Eventually, the group of protobionts shown in the figure contains only this type of catalytic RNA.

In the second step of chemical selection (Figure 4.4, right side), a second mutation in an RNA molecule produces a catalytic function that promotes the synthesis of ribonucleotides, the building blocks of RNA . For example, a hypothetical ribozyme may catalyze the attachment of a base to a ribose, thereby catalyzing one of the steps necessary for making a ribonucleotide. This protobiont would not rely solely on the prebiotic synthesis of ribonucleotides, which is a very slow process. Therefore, the protobiont having the ability to both self-replicate and catalyze a step in the synthesis of ribonucleotides would have an advantage over a protobiont that could only self-replicate. Over time, the faster rate of self-replication and ribonucleotide synthesis would cause an increase in the numbers of the protobionts with both functions.

The **RNA world** is a hypothetical period on early Earth when both the information needed for life and the catalytic activity of living cells were contained solely in RNA molecules. In this scenario, lipid membranes enclosing RNA exhibited the properties of life due to RNA genomes that were copied and maintained through the catalytic function of RNA molecules. Scientists envision that, over time, mutations occurred in these RNA molecules, occasionally introducing new functional possibilities. Chemical selection would have eventually produced an increase in complexity in these cells, with RNA molecules accruing activities such as the ability to link amino acids together into proteins and other catalytic functions.

But is an RNA world a plausible scenario? Remarkably, scientists have been able to perform experiments in the laboratory that can select for RNA molecules with a particular function. American biologists David Bartel and Jack Szostak conducted the first study of this type in 1993 in which they selected for RNA molecules with the catalytic ability to link nucleotides together. After 10 rounds of chemical selection, they obtained a collection of RNA molecules that had catalytic activity that was 3 million times higher than their original random collection of molecules! Like the work of Miller and Urey, Bartel and Szostak showed the feasibility of another phase of the prebiotic process that led to life. In this case, chemical selection resulted in chemical evolution. The results showed that chemical selection can change the functional characteristics of a group of RNA molecules over time by increasing the proportion of those molecules with enhanced function.

## The RNA World Was Superseded by the Modern DNA/RNA/Protein World

Assuming that an RNA world was the origin of life, researchers have asked, "Why and how did the RNA world evolve into the DNA/RNA/protein world we see today?" The RNA world may have been superseded by a DNA/RNA world or an RNA/protein world before the emergence of the modern DNA/RNA/protein world. Let's now consider the advantages of a DNA/RNA/ protein world as opposed to the simpler RNA world and explore how this modern biological world might have come into being.

**Information Storage** RNA can store information in its base sequence. If so, why did DNA take over that function, as is the case in modern cells? During the RNA world, RNA had to perform two roles: information storage and the catalysis of chemical reactions. Scientists have speculated that the incorporation of DNA into cells would have relieved RNA of its role of storing information, thereby allowing RNA to perform a variety of other functions. For example, if DNA stored the information for the synthesis of RNA molecules, such RNA molecules could bind cofactors, have modified bases, or bind peptides that might enhance their catalytic function. Cells with both DNA and RNA would have had an advantage over those with just RNA, and so they would have been selected. Another advantage of DNA is its stability. Compared with RNA, DNA strands are less likely to spontaneously break.

A second issue is how DNA came into being. Scientists have proposed that an ancestral RNA molecule had the ability to make DNA using RNA as a template. This function, known as reverse transcription, is described in Chapter 17 in the discussion of retroviruses.

**Metabolism and Other Cellular Functions** The emergence of proteins as catalysts may have been a great benefit to early cells. Due to the different chemical properties of the 20 amino acids, proteins have vastly greater catalytic ability than do RNA molecules, again providing a major advantage to cells that had both RNA and proteins. In modern cells, proteins have taken over most, but not all, catalytic functions. In addition, proteins can perform other important tasks. For example, cytoskeletal proteins carry out structural roles, and certain membrane proteins are responsible for the uptake of substances into living cells.

How would proteins have come into being in an RNA world? Chemical selection experiments have shown that RNA molecules can catalyze the formation of peptide bonds and even attach amino acids to primitive tRNA molecules. Similarly, modern protein synthesis still involves a central role for RNA in the synthesis of polypeptides. First, mRNA provides the information for a polypeptide sequence. Second, tRNA molecules act as adaptors for the formation of a polypeptide chain. And finally, ribosomes containing rRNA provide a site for polypeptide synthesis. Furthermore, rRNA within the ribosome acts as a ribozyme to catalyze peptide bond formation. Taken together, the analysis of translation in modern cells is consistent with an evolutionary history in which RNA molecules were instrumental in the emergence and formation of proteins.

## 4.1 Reviewing the Concepts

- Life on Earth is hypothesized to have occurred in four overlapping stages. The first stage involved the synthesis of organic molecules to form a prebiotic soup. Possible scenarios of how this occurred are the reducing atmosphere, extraterrestrial, and deep-sea vent hypotheses (Figures 4.1, 4.2).

- The second stage was the formation of polymers from simple organic molecules. This may have occurred on the surface of clay or in water. The third stage occurred when polymers became enclosed in structures called protobionts that separated them from the external environment (Figure 4.3).
- In the fourth stage, polymers enclosed in membranes acquired properties of cells, such as self-replication and other catalytic functions, via chemical selection (Figure 4.4).
- In the hypothesized period called the RNA world, the first living cells used RNA for both information storage and catalytic functions.
- The RNA world was eventually superseded by the modern DNA/RNA/protein world.

## 4.1 Testing Your Knowledge

1. Place the following four stages for the origin of life in their correct order:
   1. Nucleotides became polymerized to form RNA and/or DNA, and amino acids became polymerized to form proteins.
   2. Nucleotides and amino acids were produced prior to the existence of cells.
   3. Polymers enclosed in membranes acquired cellular properties.
   4. Polymers became enclosed in membranes.

   **a.** 1, 2, 3, 4 **c.** 2, 1, 4, 3
   **b.** 1, 3, 2, 4 **d.** 2, 3, 4, 1
2. Which of the following is *not* a hypothesis to explain the origin of the first organic molecules on Earth?
   **a.** the reducing atmosphere hypothesis
   **b.** the extraterrestrial (meteorite) hypothesis
   **c.** the chemical selection hypothesis
   **d.** the deep-sea vent hypothesis
3. Many evolutionary biologists hypothesize that the macromolecule that had informational and catalytic function and arose first in primordial cells was
   **a.** DNA. **d.** carbohydrates.
   **b.** RNA. **e.** both a and c.
   **c.** proteins.

## 4.2 Microscopy

### Learning Outcomes

1. **SCISKILLS ▸** Explain the three key parameters in microscopy: resolution, contrast, and magnification.
2. **SCISKILLS ▸** Compare and contrast the different types of light and electron microscopes and their uses.

Before we examine the general features of modern cells, we will first consider the **microscope,** which is a magnification tool that enables researchers to visualize the structure and inner workings of cells. A **micrograph** is an image taken with the aid of a microscope. The first compound microscope—a microscope with more than one lens—was invented in 1595 by Zacharias Jansen of Holland. In 1665, an English biologist, Robert Hooke, studied cork under a primitive compound microscope he had made. Hooke coined the word cell, derived from the Latin word *cellula*, meaning small compartment, to describe the structures he observed.

As described next, three important parameters in microscopy are resolution, contrast, and magnification:

- **Resolution:** the ability to observe two adjacent objects as distinct from one another. It is a measure of the clarity of an image. For example, a microscope with good resolution enables a researcher to distinguish two adjacent chromosomes as separate objects, which would appear as a single, blurry object under a microscope with poor resolution.
- **Contrast:** the ability to visualize a particular cell structure based on how different it looks from an adjacent structure. Staining the cellular structure of interest with a dye can make viewing much easier. The application of stains, which selectively label individual components of the cell, greatly improves contrast. However, staining should not be confused with colorization. Many of the micrographs shown in this textbook are colorized, or artificially colored, to emphasize certain cellular structures, such as different parts of a cell (see the chapter opener, for example). In colorization, particular colors are added to micrographs with the aid of a computer.
- **Magnification:** the ratio between the size of an image produced by a microscope and its actual size. For example, if the image size is 100 times larger than its actual size, the magnification is designated 100×. Depending on the quality of the lens and illumination source, every microscope has an optimal range of magnification before objects appear too blurry to be readily observed.

Microscopes are categorized into two groups based on the source of illumination. A **light microscope** utilizes light for illumination, whereas an **electron microscope** uses a beam of electrons for illumination. Very good light microscopes resolve structures that are as close as 0.2 μm (micron, or micrometer) from each other. Resolution is improved when the illumination source has a shorter wavelength. A major advance in microscopy occurred in 1931, when Max Knoll and Ernst Ruska invented the first electron microscope. Because the wavelength of an electron beam is much shorter than that of visible light, the resolution of the electron microscope is far better than that of any light microscope. The resolution limit is typically around 2 nm (nanometers), which is about 100 times better than that for the light microscope. Figure 4.5 shows the range of resolving powers of the electron microscope, light microscope, and unaided eye and compares them with various cells and cell structures.

Over the past several decades, technological advances have made light microscopy a powerful research tool. Improvements in lens technology, microscope organization, sample preparation, sample illumination, and computerized image processing have enabled

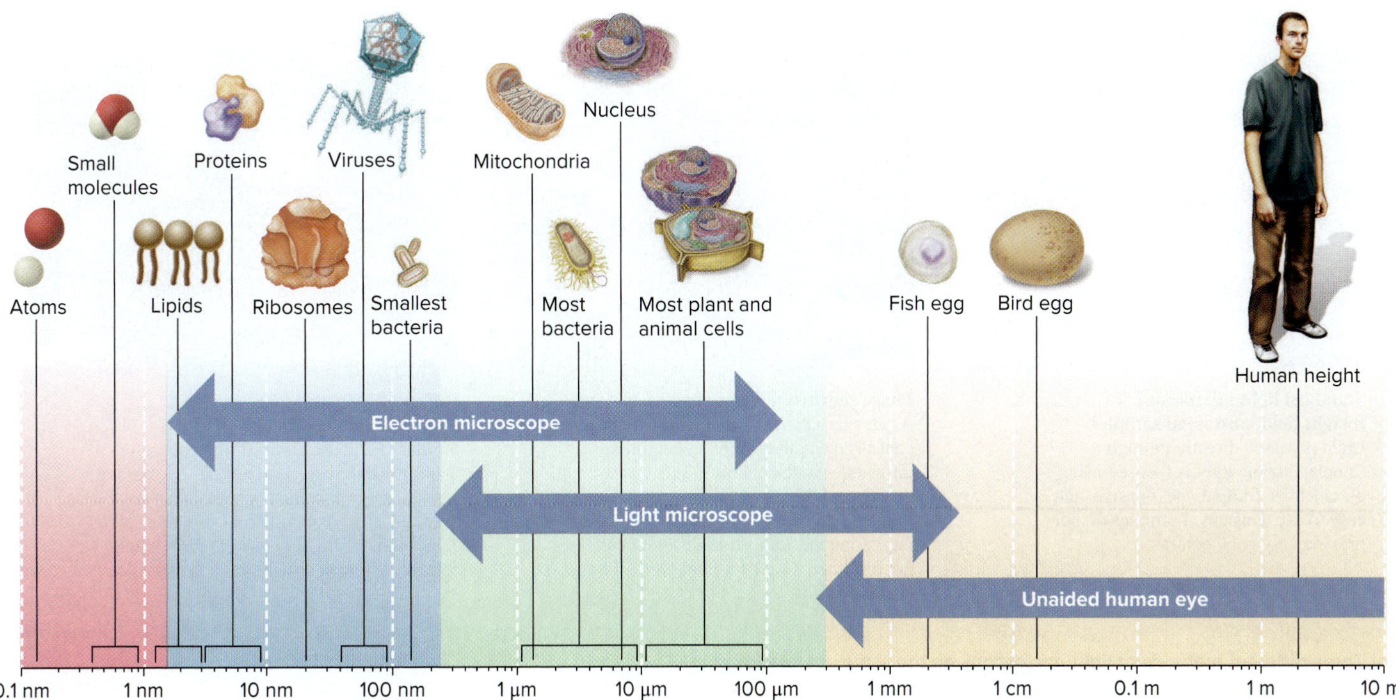


**Figure 4.5 A comparison of the sizes of various chemical and biological structures, and the resolving power of the unaided human eye, light microscope, and electron microscope.** The scale at the bottom is logarithmic to accommodate the wide range of sizes in this drawing.

*Which type of microscopy would you use to observe a virus?*

researchers to invent different types of light microscopes, each with its own advantages and disadvantages (Figure 4.6).

Similarly, improvements in electron microscopy occurred during the 1930s and 1940s, and by the 1950s, the electron microscope was playing a major role in advancing our understanding of cell structure. Two general types of electron microscopy have been developed: transmission electron microscopy and scanning electron microscopy. In **transmission electron microscopy (TEM),** a beam of electrons is transmitted through a biological sample. To provide contrast, the sample is stained with a heavy metal, which binds to certain cellular structures such as membranes. The sample is then adhered to a copper grid and placed in a transmission electron microscope. When the beam of electrons strikes the sample, some of them hit the heavy metal and are scattered, while those that pass through without being scattered are focused to form an image on a photographic plate or screen (Figure 4.7a). The metal-stained regions of the sample that scatter electrons appear as darker areas, due to reduced electron penetration. TEM provides a cross-sectional view of a cell and gives the best resolution compared with other forms of microscopy. However, such microscopes are expensive and cannot be used to view living cells.

**Scanning electron microscopy (SEM)** is used to view the surface of a sample. A biological sample is coated with a thin layer of heavy metal, such as gold or palladium, and then is exposed to an electron beam that scans its surface. Secondary electrons are emitted from the sample, which are detected and create an image of the three-dimensional surface of the sample (Figure 4.7b). SEM is not usually used to view living cells.

## 4.2 Reviewing the Concepts

- Three important parameters in microscopy are magnification, resolution, and contrast.
- A light microscope utilizes light for illumination, whereas an electron microscope uses an electron beam. Transmission electron microscopy (TEM) provides the best resolution of any form of microscopy, and scanning electron microscopy (SEM) produces an image of a three-dimensional surface (Figures 4.5, 4.6, 4.7).

## 4.2 Testing Your Knowledge

1. A researcher uses a dye that specifically stains the cell nucleus. When viewing cells under the microscope, the nucleus is easier to see because of improved
   a. resolution.
   b. contrast.
   c. magnification.
   d. resolution and contrast.
   e. resolution and magnification.

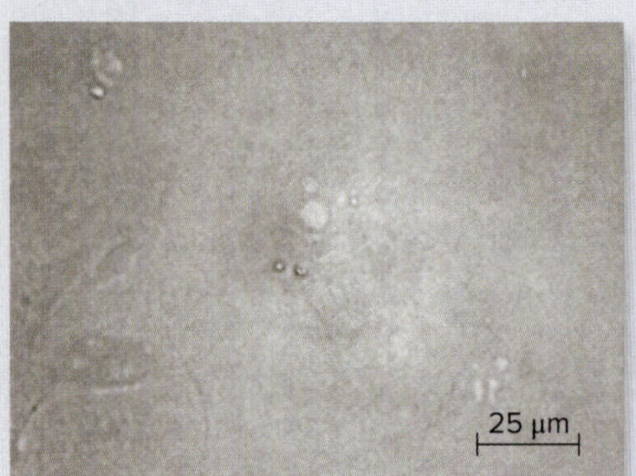


**Standard light microscopy (bright field, unstained sample).**
Light is passed directly through a sample, and the light is focused using glass lenses. Simple, inexpensive, and easy to use. Unstained samples do not provide very good contrast.

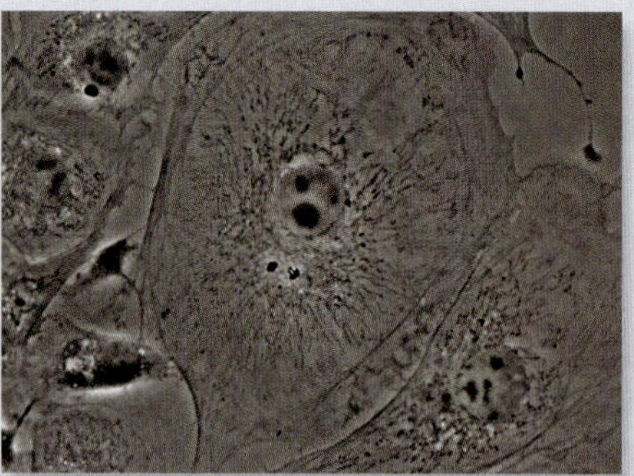

**Phase contrast microscopy.**
As an alternative to staining, this microscope controls the path of light and amplifies differences in the phase of light transmitted or reflected by a sample. The dense structures appear darker than the background, thereby improving the contrast in different parts of the specimen. Can be used to view living, unstained cells.

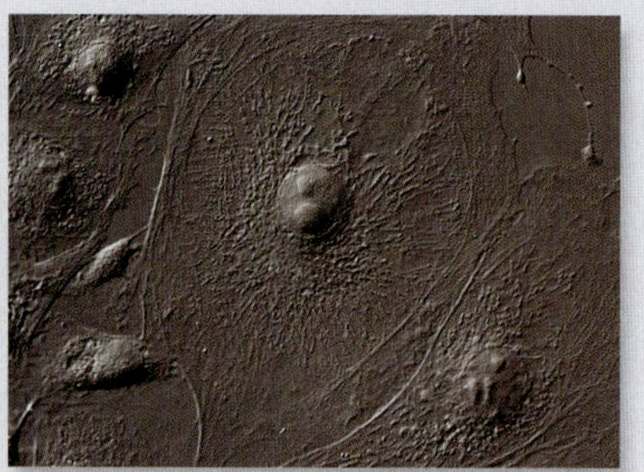

**Differential interference contrast (Nomarski) microscopy.**
Similar to a phase contrast microscope in that it uses optical modifications to improve contrast in unstained specimens. Can be used to visualize the internal structures of cells and is commonly used to view whole cells or large cell structures such as nuclei.

**(a) Three different methods of light microscopy on the same unstained sample**

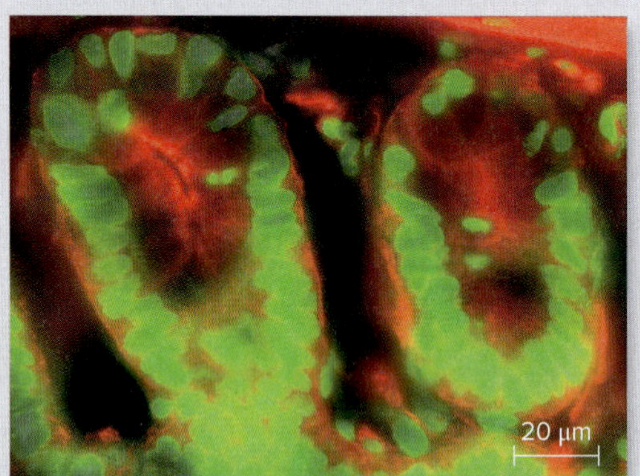


**Standard (wide-field) fluorescence microscopy.**
Fluorescent molecules specifically label a particular type of cellular protein or organelle. A fluorescent molecule absorbs light at a particular wavelength and emits light at a longer wavelength. This microscope has filters that illuminate the sample with the wavelength of light that a fluorescent molecule absorbs, and then only the light that is emitted by the fluorescent molecules is allowed to reach the observer. To detect their cellular location, researchers often label specific cellular proteins using fluorescent antibodies that bind specifically to a particular protein.

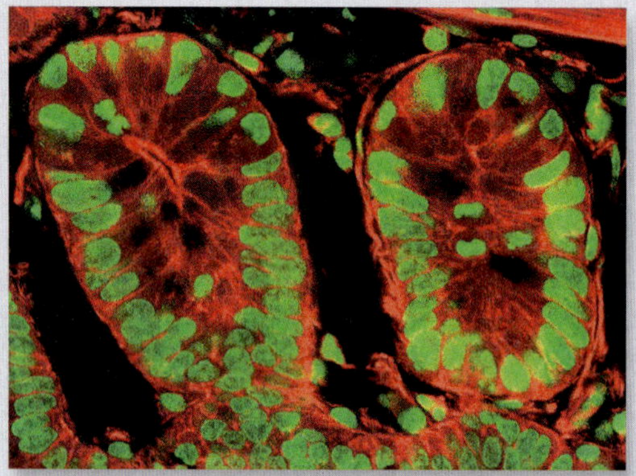

**Confocal fluorescence microscopy.**
Uses lasers that illuminate various points in the sample. These points are processed by a computer to give a very sharp focal plane. In this example, this process is used in conjunction with fluorescence microscopy to view fluorescent molecules within a cell.

**(b) Two different methods of fluorescence microscopy**

**Figure 4.6** **Examples of light microscopy.** **(a)** These micrographs compare three microscopic techniques on the same unstained sample of endothelial cells that line the interior surface of arteries in the lungs. **(b)** These two micrographs compare standard (wide-field) fluorescence microscopy with confocal fluorescence microscopy on the same stained sample of cells. The sample is a section through a mouse intestine, showing two villi—projections from the small intestine that are described in Chapter 37. In this sample, the nuclei are stained green, and the actin filaments (discussed later in this chapter) are stained red.

*(a-b)* Images courtesy of Michael Davidson at Florida State University and Molecular Expressions

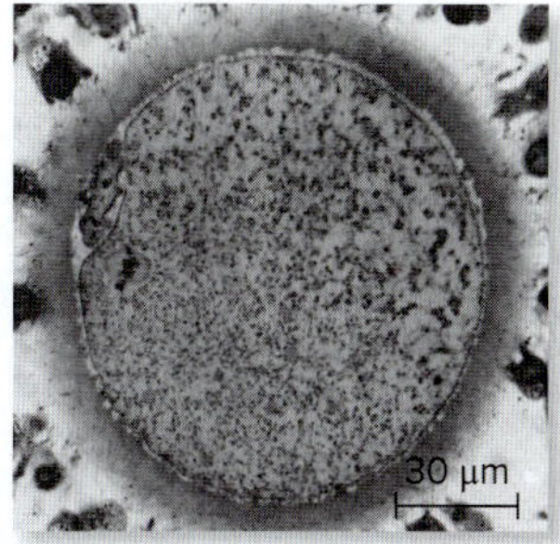


(a) Transmission electron micrograph (TEM)

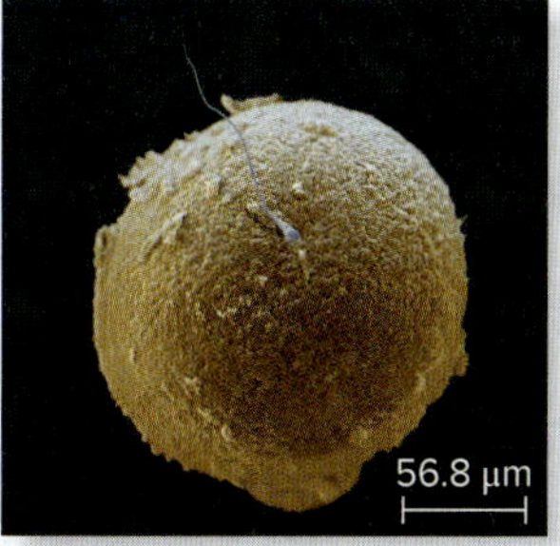


(b) Scanning electron micrograph (SEM)

**Figure 4.7 A comparison of transmission and scanning electron microscopy.** **(a)** Section through a developing human egg cell, observed by TEM, shortly before it was released from an ovary. **(b)** An egg cell, with an attached sperm, was coated with a heavy metal and observed via SEM. This SEM is colorized.

*(a)* © Don W. Fawcett/Science Source; *(b)* © Eye of Science/Science Source

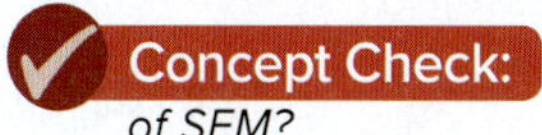


*What is the primary advantage of SEM?*

2. Which of the following statements regarding transmission electron microscopy (TEM) and scanning electron microscopy (SEM) is/are correct?
   a. In TEM, electrons pass through the sample, whereas in SEM secondary electrons are emitted from the sample.
   b. TEM gives better resolution than SEM.
   c. SEM gives a three-dimensional view of the surface of a sample.
   d. All of the above are correct statements.
   e. Only statements a and b are correct.

## 4.3 Overview of Cell Structure and Function

### Learning Outcomes

1. Compare and contrast the general structural features of prokaryotic and eukaryotic cells.
2. Explain how the proteome underlies the structure and function of cells.
3. **SCISKILLS ▶** Analyze how cell size and shape affect the ratio between surface area and volume.

Cell structure and function are primarily determined by four factors:

1. **Matter:** As discussed in Chapters 2 and 3, the matter found in living organisms is composed of atoms, molecules, and macromolecules. Each type of cell synthesizes a unique set of molecules and macromolecules that contribute to cell structure.
2. **Energy:** Energy is needed to produce molecules and macromolecules and to carry out many cellular functions (discussed in Chapters 6 and 7).
3. **Organization:** A cell is not a haphazard bag of components. The molecules and macromolecules that constitute cells are found at specific sites. For example, if we compare the structure of a muscle cell in two different humans, or two muscle cells within the same individual, we would see striking similarities in their overall structures. All living cells have the ability to build and maintain their internal organization. Proteins often bind to each other in much the same way that building blocks snap together. These types of **protein-protein interactions** create intricate cell structures and facilitate processes in which proteins interact in a consistent series of steps.
4. **Information:** Cell structure and function require instructions. These instructions are found in the blueprint of life, namely, the genetic material (DNA), which is discussed in Unit III. Every species has a distinctive **genome,** the entire complement of its genetic material. Likewise, each living cell has a copy of that genome. This information is passed from cell to cell and from parent to offspring to yield new generations of cells and new generations of life. The **genes** within each species' genome contain the information to produce cellular proteins with particular structures and functions.

In this section, we will explore the general features of cells and examine how the genome contributes to cell structure and function.

### Prokaryotic Cells Have a Relatively Simple Structure

Based on cell structure, all forms of life can be placed into two categories called prokaryotes and eukaryotes. We will first consider **prokaryotic cells,** which have a relatively simple structure. The term comes from the Greek *pro* and *karyon*, which means before a kernel—a reference to the kernel-like appearance of what would later be named the cell nucleus. Prokaryotic cells lack a membrane-enclosed nucleus.

From an evolutionary perspective, the two categories of organisms that have prokaryotic cells are **bacteria** and **archaea.** Both types are microorganisms, which are relatively small, with cell sizes that usually range between 1 and 10 μm in diameter. Bacteria are abundant throughout the world, being found in soil, water, and even our digestive tracts. Most bacterial species are not harmful to humans, and they play vital roles in ecology. However, some species are pathogenic, meaning they cause disease. Archaea are also widely found throughout the world, though they are less common than bacteria and often occupy extreme environments such as hot springs and deep-sea vents.

Figure 4.8 shows a typical bacterial cell. A few key features include the following:

- **Plasma membrane:** a double layer of phospholipids and embedded proteins that form a barrier between the cell and its external environment

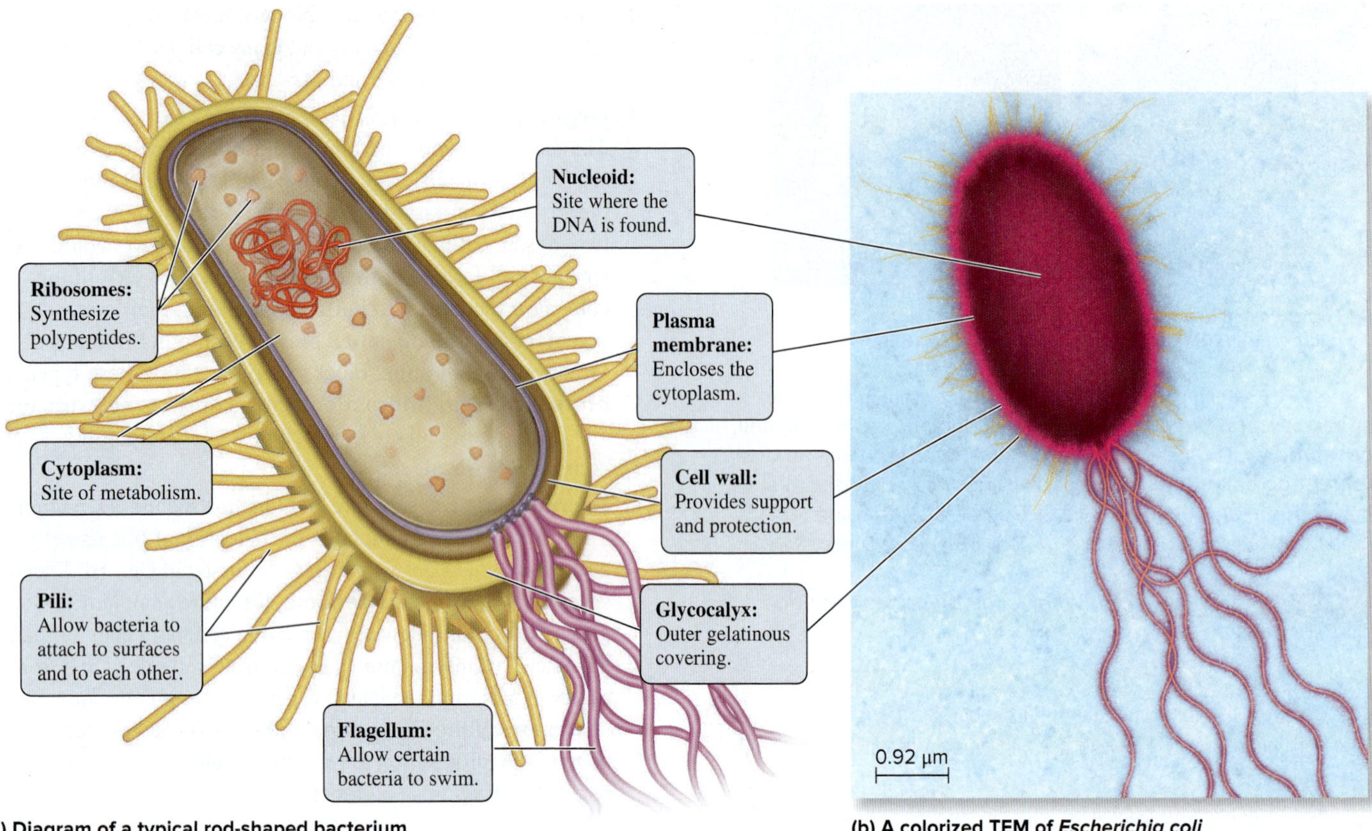


**(a) Diagram of a typical rod-shaped bacterium** **(b) A colorized TEM of *Escherichia coli***

**Figure 4.8** **Structure of a typical bacterial cell.** Prokaryotic cells, which include bacterial and archaeal cells, lack internal compartmentalization.



- **Cytoplasm:** the region of the cell contained within the plasma membrane
- **Nucleoid:** location where the genetic material (DNA) is found. It is not a membrane-bound compartment.
- **Ribosomes:** cell components involved in polypeptide synthesis
- **Cell wall:** a relatively rigid structure that supports and protects the plasma membrane and cytoplasm. The cell-wall composition varies widely among prokaryotic cells but commonly contains peptides and carbohydrates. The cell wall, which is relatively porous, allows most nutrients in the environment to reach the plasma membrane.
- **Glycocalyx:** an outer viscous covering surrounding many species of bacteria. The glycocalyx traps water and helps protect bacteria from drying out. Certain strains of bacteria that invade animals' bodies produce a very thick, gelatinous glycocalyx called a capsule that may help them avoid being destroyed by the animal's immune (defense) system or may aid in the attachment to cell surfaces.
- **Pili** (singular, pilus): hairlike projections that allow cells to attach to surfaces and to each other
- **Flagella** (singular, flagellum): long appendages that provide prokaryotic cells with a way to move, also called motility

## Eukaryotic Cells Are Compartmentalized by Internal Membranes to Create Organelles

Aside from bacteria and archaea, all other species are **eukaryotes** (from the Greek, meaning true nucleus), which include protists, fungi, plants, and animals. Paramecia and algae are types of protists; yeasts and molds are types of fungi. Figure 4.9 illustrates the morphology of a typical animal cell. Eukaryotic cells have a nucleus, where most of the genetic material (DNA) is housed. A nucleus is a type of **organelle—**a membrane-bound compartment with its own unique structure and function. In contrast to prokaryotic cells, eukaryotic cells exhibit extensive **compartmentalization,** which means they have many membrane-bound organelles that separate the interior of the cell into different regions. Compartmentalization allows a cell to carry out specialized chemical reactions in different places.

Some general features of cell organization, such as a nucleus, are found in nearly all eukaryotic cells. Even so, the shape, size, and organization of cells vary considerably among different species and

## Biology Principle

### Cells Are the Simplest Units of Life

A cell, such as the animal cell illustrated here, is the smallest unit that satisfies all of the characteristics of living organisms. These characteristics are discussed in Chapter 1 (refer back to Figure 1.3).

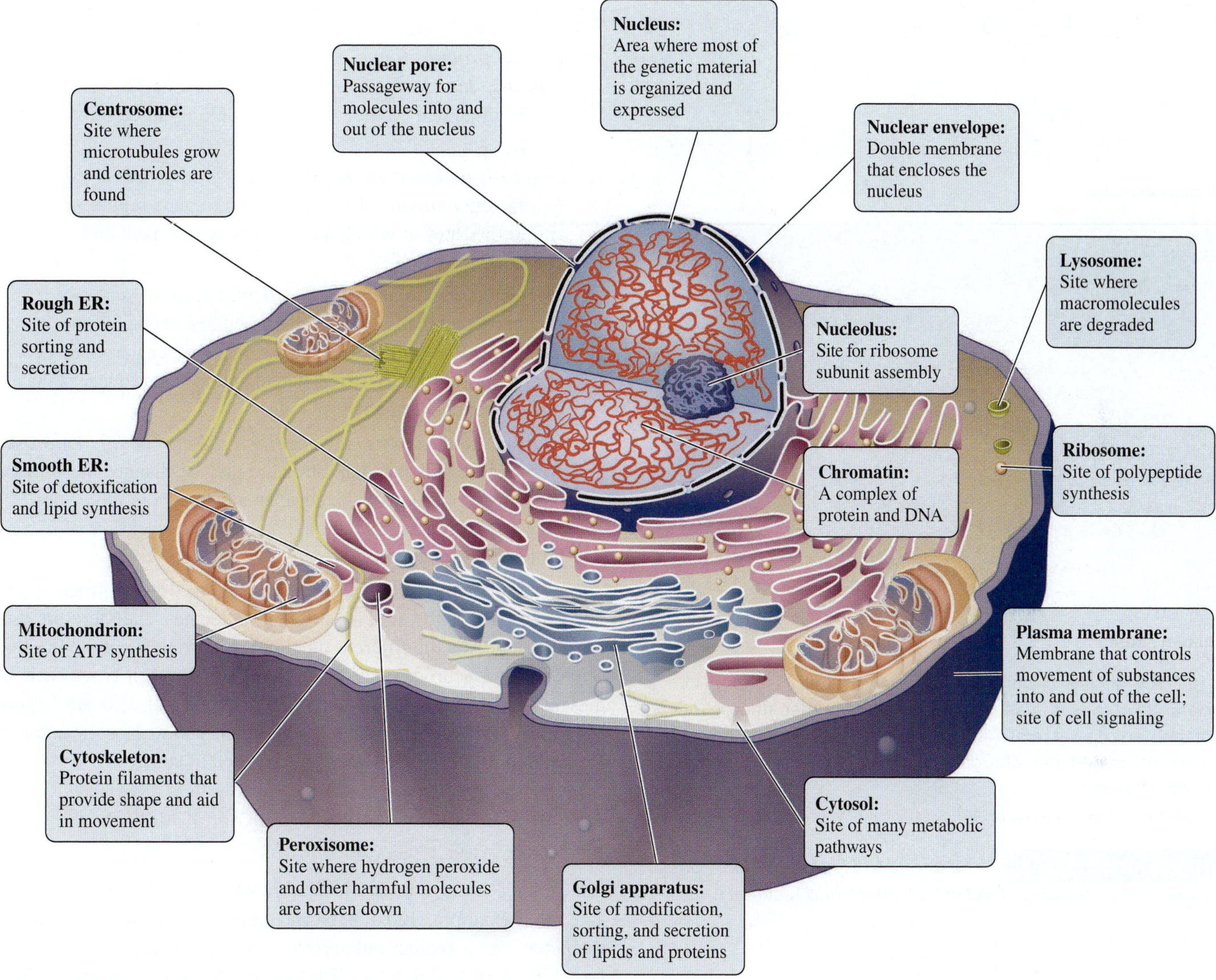


**Figure 4.9** General structure of an animal cell.

even among different cell types of the same species. For example, micrographs of a human skin cell and a human neuron show that, although these cells contain the same types of organelles, their overall morphologies are quite different (Figure 4.10).

Plant cells have a collection of organelles similar to those in animal cells (Figure 4.11). Additional structures found in plant cells but not animal cells include chloroplasts, a central vacuole, and a cell wall, which are discussed later in this chapter.

### The Characteristics of a Cell Are Largely Determined by the Proteins It Makes

Many organisms, such as animals and plants, are multicellular, meaning that a single organism is composed of many cells. However, as we saw in Figure 4.10, the cells of a multicellular organism are not all identical. For example, your body contains skin cells, neurons, muscle cells, and many other types. An intriguing

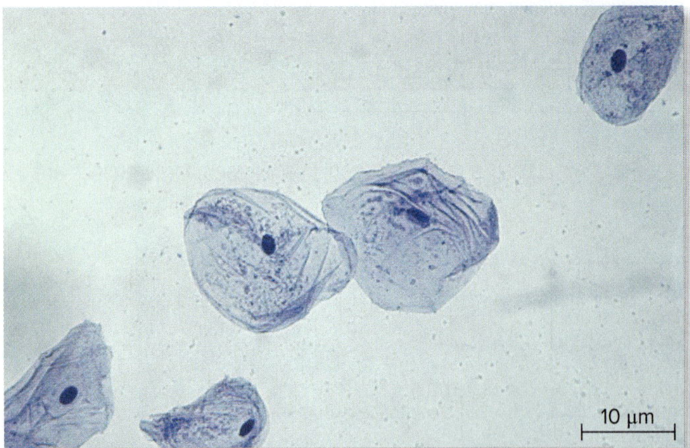


**(a) Human skin cell**

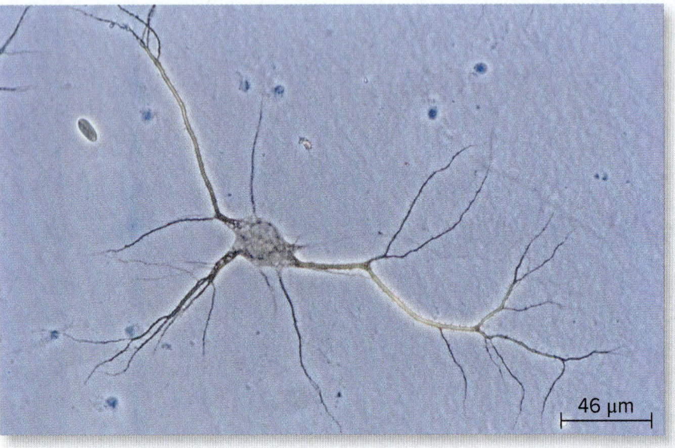


**(b) Human neuron**

**Figure 4.10 Variation in morphology of eukaryotic cells.** Light micrographs of **(a)** a human skin cell and **(b)** a human neuron (a cell of the nervous system). Although these cells have the same genome and same types of organelles, their general morphologies are quite different.

*(a)* © Ed Reschke/Getty Images; *(b)* © Eye of Science/Science Source

BioConnections: *Look ahead to Figure 12.17. How does alternative splicing affect cell structure and function?*

question, therefore, is how does a single organism produce different types of cells?

To answer this question, we need to consider the distinction between a cell's genome and its proteome. Recall that the genome constitutes all types of genetic material, namely DNA, which contains many different genes. Most genes encode the production of polypeptides, which assemble into functional proteins. However, only a subset of genes is expressed (transcribed into mRNA) in any given cell. Due to **differential gene regulation** (discussed in Chapter 12), every cell expresses a unique set of mRNAs. These mRNAs are then translated to produce a specific set of polypeptides that form proteins. The **proteome** is defined as the complete protein composition of a cell or an organism. The set of proteins that is made by a given cell type is largely responsible for determining the characteristics of that cell.

The set of proteins made in one cell type is not the same as that made in a different cell type. As an example, let's consider skin cells and neurons—two cell types with dramatically different organization and structure (see Figure 4.10). In any individual, the genes in a human skin cell are identical to those in a human neuron. However, their protein compositions are quite different for the following reasons:

1. *Certain proteins found in skin cells may not be produced in neurons, and vice versa.* As described in Chapter 12, genes can be regulated so they are turned on only in certain cell types.
2. *Skin cells and neurons may produce the same protein but in different amounts.* This is also due to gene regulation and to the rates at which a protein is synthesized and degraded.
3. *The amino acid sequences of particular proteins can vary in skin cells and neurons.* As discussed in Chapter 12, the messenger RNA (mRNA) from a single gene can produce two or more polypeptides with different amino acid sequences via a process called alternative splicing.
4. *Skin cells and neurons may alter their proteins in different ways.* After a protein is made, its structure may be changed in a variety of ways. These include the covalent attachment of molecules, such as phosphate and carbohydrates, and the cleavage of a protein to a smaller size.

These four phenomena enable skin cells and neurons to produce different sets of proteins, that is, different proteomes, and therefore different structures with different functions. Likewise, the proteomes of skin cells and neurons differ from those of other cell types such as muscle and liver cells. Ultimately, the proteomes of cells are largely responsible for producing the traits of organisms, such as the color of the iris of a person's eye, which is composed of cells.

## Surface Area and Volume Are Critical Parameters That Affect Cell Sizes and Shapes

As we have seen, a common feature of most cells is their small size. For example, most bacterial cells are about 1–10 μm in diameter, and a typical eukaryotic cell is 10–100 μm in diameter (see Figures 4.5 and 4.10). Though some exceptions are known, such as an ostrich egg, small size is nearly a universal characteristic of cells. In general, large organisms attain their large sizes by having more cells, not by having larger cells. For example, the various types of cells found in an elephant and a mouse are roughly the same sizes. However, an elephant has many more cells than a mouse.

Why are cells usually small? One key factor is the interface between a cell and its extracellular environment, which is the plasma membrane. For cells to survive, they must import substances across their plasma membranes and export waste products. If the internal volume of a cell is large, it will require a greater amount of nutrient uptake and waste export. The rate of transport of substances across the plasma membrane, however, is limited by

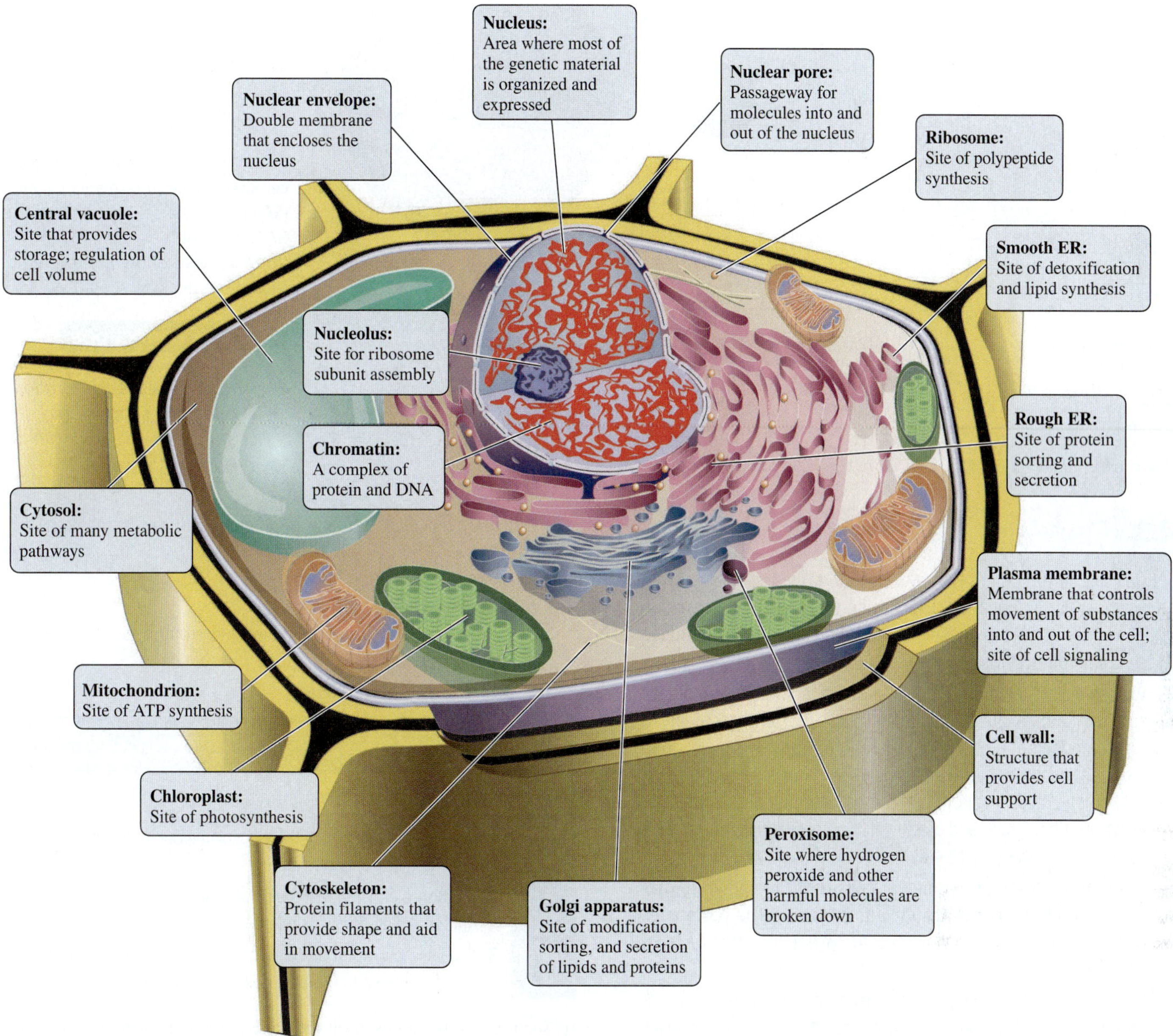
Nucleus:
Area where most of the genetic material is organized and expressed
Nuclear envelope:
Double membrane that encloses the nucleus
Nuclear pore:
Passageway for molecules into and out of the nucleus
Ribosome:
Site of polypeptide synthesis
Central vacuole:
Site that provides storage; regulation of cell volume
Smooth ER:
Site of detoxification and lipid synthesis
Nucleolus:
Site for ribosome subunit assembly
Rough ER:
Site of protein sorting and secretion
Chromatin:
A complex of protein and DNA
Cytosol:
Site of many metabolic pathways
Plasma membrane:
Membrane that controls movement of substances into and out of the cell; site of cell signaling
Mitochondrion:
Site of ATP synthesis
Cell wall:
Structure that provides cell support
Chloroplast:
Site of photosynthesis
Peroxisome:
Site where hydrogen peroxide and other harmful molecules are broken down
Cytoskeleton:
Protein filaments that provide shape and aid in movement
Golgi apparatus:
Site of modification, sorting, and secretion of lipids and proteins

**Figure 4.11** **General structure of a plant cell.** Plant cells lack lysosomes and centrioles. Unlike animal cells, plant cells have an outer cell wall; a large central vacuole that functions in storage, digestion, and cell volume; and chloroplasts, organelles that carry out photosynthesis.

**Concept Check:** *What are the functions of the cell structures and organelles that are (1) found in animal cells but not plant cells and (2) found in plant cells but not animal cells?*

its surface area. Therefore, a critical issue for sustaining a cell is the surface area/volume (SA/V) ratio. This concept is illustrated in Figure 4.12, which considers a simplified case in which cells are spherical. As cells get larger, the surface area of their plasma membrane increases with the square of the radius ($A = 4\pi r^2$), whereas the volume increases with the cube of the radius ($V = 4/3\pi r^3$). Therefore, as the radius of the cell gets larger, the SA/V ratio gets smaller. Biologists hypothesize that most cells are small because a high SA/V ratio is needed for cells to sustain an adequate level of nutrient uptake and waste export.

One way for cells to partially overcome the dilemma posed by the SA/V ratio is to have elongated and irregularly shaped surfaces. For example, look back at Figure 4.10. The skin cell is roughly spherical, whereas the neuron is very elongated and has an irregularly shaped surface. If a skin cell and a neuron cell had the same internal volume, the neuron would have a much higher SA/V ratio.

BIO TIPS ONLINE

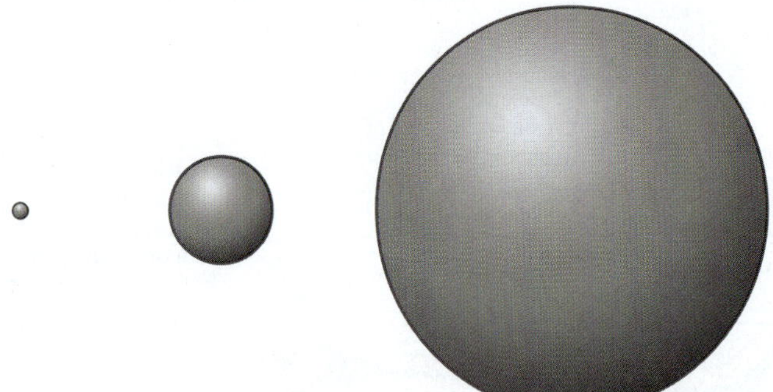

| | | | |
|---|---|---|---|
| **Radius** (μm): | 1 | 10 | 100 |
| **Surface area** ($\mu m^2$) ($A = 4\pi r^2$): | 12.6 | ~1260 | ~124,600 |
| **Volume** ($\mu m^3$): ($V = \frac{4}{3}\pi r^3$) | 4.2 | ~4200 | ~4,200,000 |
| **Surface area/volume ratio:** | 3.0 : 1 | 0.3 : 1 | 0.03 : 1 |

**Figure 4.12 Relationship between cell size and the surface area/volume (SA/V) ratio.** As cells get larger, the SA/V ratio gets smaller. Note: The three spheres shown here are not drawn precisely to scale.

**BioConnections:** *Look ahead to Figure 36.15. How does the SA/V ratio affect the shapes of structures involved with gas exchange?*

## 4.3 Reviewing the Concepts

- Cell structure is determined by four factors: matter, energy, organization, and information. Every living organism has a genome. The genes within the genome contain the information to produce cells with particular structures and functions.
- We can classify all forms of life into two categories based on cell structure: prokaryotic and eukaryotic cells. Bacteria and archaea are composed of prokaryotic cells with a relatively simple structure that lacks a membrane-enclosed nucleus. Structures in prokaryotic cells include the plasma membrane, cytoplasm, nucleoid, ribosomes, and a cell wall. Some prokaryotic cells also have a glycocalyx (Figure 4.8).
- Differences in the proteomes of cells result in cells with dramatically different organization and structure. The proteomes of cells are largely responsible for producing the traits of organisms.
- Eukaryotic cells are compartmentalized into organelles and contain a nucleus that houses most of the DNA. The surface area/volume (SA/V) ratio is thought to limit cell size (Figures 4.9, 4.10, 4.11, 4.12).

## 4.3 Testing Your Knowledge

1. Which of the following structures would *not* be found outside the plasma membrane of a prokaryotic cell?
   a. cell wall d. pili
   b. glycocalyx e. flagella
   c. nucleoid
2. A key difference between prokaryotic and eukaryotic cells is that only eukaryotic cells have
   a. a plasma membrane.
   b. genetic material (DNA).
   c. flagella.
   d. many types of membrane-bound organelles.
   e. all of the above.
3. Within a multicellular organism, differences in cell types usually arise from
   a. differences in their genomes.
   b. differences in their proteomes.
   c. differences in both their genomes and proteomes.
   d. none of the above.

## 4.4 The Cytosol

### Learning Outcomes

1. Identify the location of the cytosol in a eukaryotic cell, and describe its general functions.
2. Compare and contrast the three types of protein filaments that make up the cytoskeleton.
3. Explain how motor proteins interact with microtubules or actin filaments to promote cellular movements.

In Section 4.3, we focused on the general features of prokaryotic and eukaryotic cells. In the rest of this chapter, we will survey the various compartments of eukaryotic cells with a greater emphasis on structure and function. Figure 4.13 compares four different regions in both an animal and a plant cell. We will start with the **cytosol** (shown in yellow), the region of a eukaryotic cell that is outside the membrane-bound organelles but inside the plasma membrane. The other regions of the cell, which we will examine later in this chapter, include the interior of the nucleus (blue), the endomembrane system (purple and pink), and the semiautonomous organelles (orange and green). As in prokaryotic cells, the term cytoplasm refers to the region enclosed by the plasma membrane. This includes the cytosol and the organelles.

### Synthesis and Breakdown of Molecules Occur in the Cytosol

**Metabolism** is defined as the sum of the chemical reactions by which cells produce the materials and utilize the energy necessary to sustain life. Although many steps of metabolism also occur

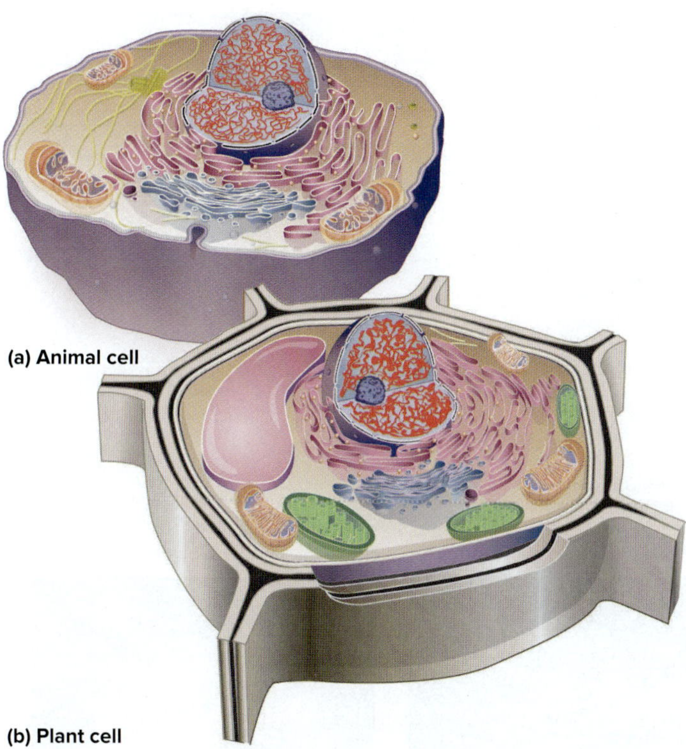


**Figure 4.13** **Compartments within (a) animal and (b) plant cells.** The cytosol, which is outside the organelles but inside the plasma membrane, is shown in tan. The membranes of the endomembrane system are shown in purple (nuclear envelope and plasma membrane), pink (endoplasmic reticulum and vacuole), blue (Golgi apparatus), and dark purple (peroxisome). The interior of the nucleus is blue. Semiautonomous organelles are shown in orange (mitochondria) and green (chloroplasts).

in cell organelles, the cytosol is a central coordinating region for many metabolic activities of eukaryotic cells. Metabolism often involves a series of steps called a metabolic pathway. Each step in a metabolic pathway is catalyzed by a specific **enzyme**—a protein that accelerates the rate of a chemical reaction. In Chapter 6, we will examine the properties of enzymes and consider a few important metabolic pathways that occur in the cytosol and cell organelles.

Some pathways involve the breakdown of a molecule into smaller components, a process termed **catabolism.** Such pathways are needed by the cell to utilize energy and to generate molecules that provide the building blocks for constructing macromolecules. Conversely, other pathways are involved in **anabolism**—the synthesis of molecules and macromolecules. For example, polysaccharides are made by linking sugar molecules. To make proteins, amino acids are covalently connected to form a polypeptide, using the information within an mRNA, a process called translation (see Chapter 10). Translation occurs on ribosomes, which are found in various locations in the cell. Some ribosomes may float freely in the cytosol, others are attached to the outer membrane of the nuclear envelope and endoplasmic reticulum membrane, and still others are found within the mitochondria or chloroplasts.

## The Cytoskeleton Provides Cell Shape, Organization, and Movement

The **cytoskeleton** is a network of three different types of protein filaments: **microtubules, intermediate filaments,** and **actin filaments** (Table 4.1). Each type is constructed from many protein monomers. The cytoskeleton is a striking example of protein-protein interactions. The cytoskeleton is found primarily in the cytosol as well as in the nucleus along the inner nuclear membrane. Let's first consider the structure of cytoskeletal filaments and their roles in the construction and organization of cells. Later, we will examine how they are involved in cell movement.

1. **Microtubules:** Microtubules are long, hollow, cylindrical structures about 25 nm in diameter composed of protein subunits called α- and β-tubulin. The assembly of tubulin to form a microtubule results in a polar structure with a plus end and a minus end (see Table 4.1). Microtubules grow only at the plus end but can shorten at either the plus or minus end. A single microtubule can oscillate between growing and shortening phases, a phenomenon termed **dynamic instability.** This phenomenon is important in many cellular activities, including the sorting of chromosomes during cell division.

The sites where microtubules form within a cell vary among different types of organisms. Animal cells that are not in the act of dividing contain a single structure near their nucleus called the **centrosome,** also called a microtubule-organizing center (see Table 4.1). Within the centrosome are the **centrioles,** a conspicuous pair of structures arranged perpendicular to each other. In animal cells, microtubule growth typically starts at the centrosome in such a way that the minus end is anchored there. In contrast, most plant cells and many protists lack centrosomes and centrioles. Microtubules are created at many sites that are scattered throughout a plant cell. In plants, the nuclear membrane appears to function as a microtubule-organizing center.

Microtubules are important for cell shape and organization. Organelles such as the Golgi apparatus are attached to microtubules. In addition, microtubules are involved in the organization and movement of chromosomes during mitosis and in the orientation of cells during cell division, events we will examine in Chapter 14.

2. **Intermediate filaments:** Intermediate filaments are another class of cytoskeletal filament found in the cells of many but not all animal species. Their name is derived from the observation that they are intermediate in diameter between actin filaments and microtubules. Intermediate filament proteins bind to each other in a staggered array to form a twisted, ropelike structure with a diameter of approximately 10 nm (see Table 4.1). They function as tension-bearing fibers that help maintain cell shape and rigidity. Intermediate filaments tend to be relatively permanent. By comparison, microtubules and actin filaments readily lengthen and shorten in cells.

Several types of proteins form intermediate filaments. Keratins are intermediate filaments in skin, intestinal, and kidney cells, where they are important for determining cell shape and providing mechanical

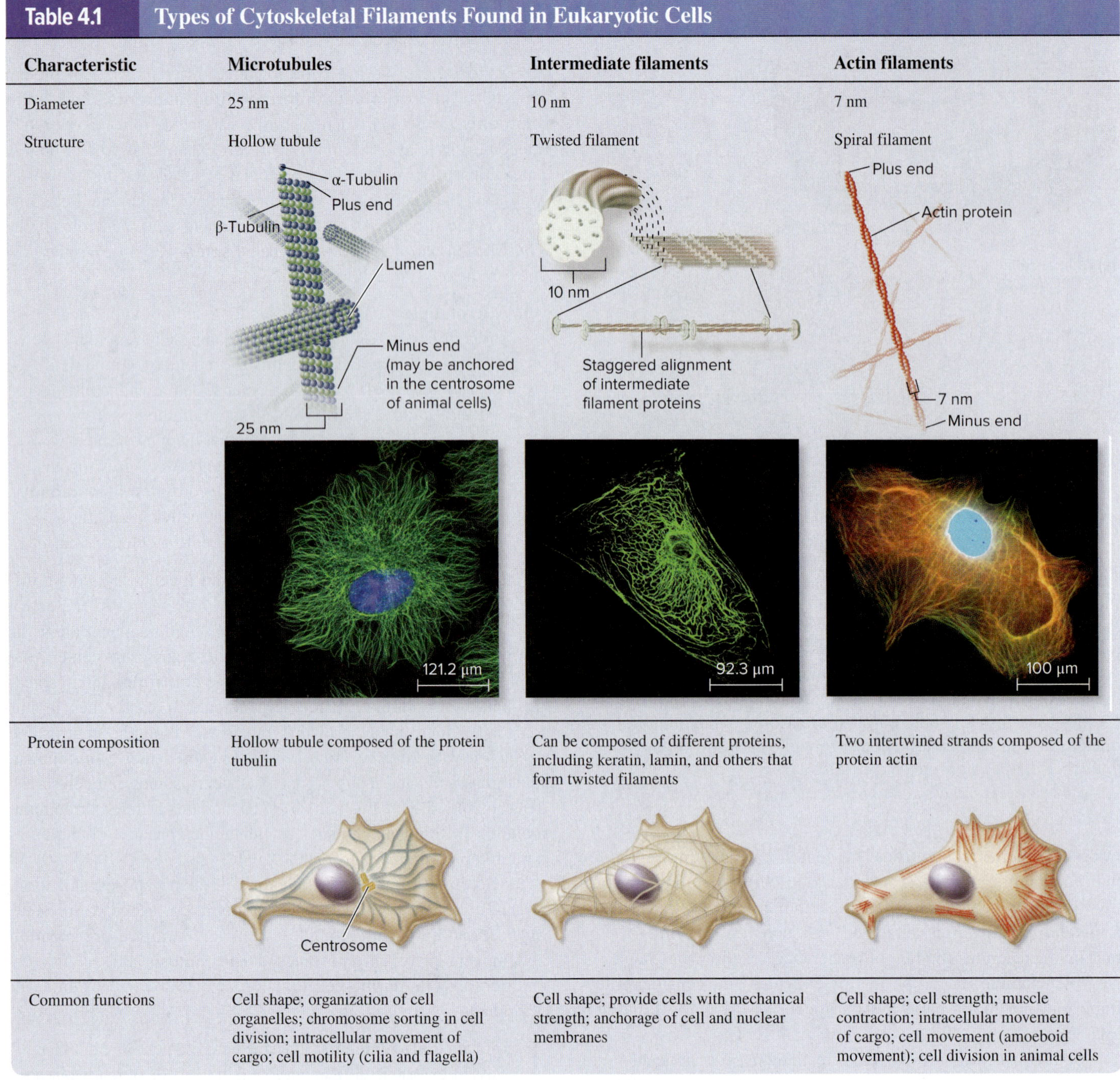


**Table 4.1 Types of Cytoskeletal Filaments Found in Eukaryotic Cells**

| Characteristic | Microtubules | Intermediate filaments | Actin filaments |
|---|---|---|---|
| Diameter | 25 nm | 10 nm | 7 nm |
| Structure | Hollow tubule | Twisted filament | Spiral filament |
| Protein composition | Hollow tubule composed of the protein tubulin | Can be composed of different proteins, including keratin, lamin, and others that form twisted filaments | Two intertwined strands composed of the protein actin |
| Common functions | Cell shape; organization of cell organelles; chromosome sorting in cell division; intracellular movement of cargo; cell motility (cilia and flagella) | Cell shape; provide cells with mechanical strength; anchorage of cell and nuclear membranes | Cell shape; cell strength; muscle contraction; intracellular movement of cargo; cell movement (amoeboid movement); cell division in animal cells |



strength. They are also a major constituent of hair and nails. As discussed later, some intermediate filaments are found inside the cell nucleus.

3. **Actin filaments:** Actin filaments are also known as **microfilaments,** because they are the thinnest cytoskeletal filaments. They are long, thin fibers approximately 7 nm in diameter. Like microtubules, actin filaments have plus and minus ends, and they are dynamic structures in which each strand grows at the plus end by the addition of actin monomers (see Table 4.1). This assembly process produces a fiber composed of two strands of actin monomers that spiral around each other.

Despite their thinness, actin filaments play a key role in cell shape and strength. They tend to be highly concentrated near the plasma membrane. In many types of cells, actin filaments support the plasma membrane and provide shape and strength to the cell. The sides of actin filaments are often anchored to other proteins near the plasma membrane, which explains why they are typically found there. The plus ends grow toward the plasma membrane and play a key role in cell shape and movement.

## Motor Proteins Interact with Cytoskeletal Filaments to Promote Movements

**Motor proteins** are a category of proteins that use ATP as a source of energy to promote various types of movements. As shown in Figure 4.14a, a motor protein consists of three domains:

- **Head:** the site where ATP binds and is hydrolyzed to adenosine diphosphate (ADP) and inorganic phosphate ($P_i$)
- **Hinge:** the site that bends in response to ATP binding and hydrolysis. This bending is what causes movement to occur.
- **Tail:** an elongated region that is attached to other proteins or to other kinds of cellular molecules

To promote movement, the head region of a motor protein interacts with a cytoskeletal filament—a microtubule or actin filament (Figure 4.14b). When ATP binds and is hydrolyzed, the motor protein attempts to "walk" along the filament. The head of the motor protein is initially attached to a filament. To move forward, the head detaches from the filament, cocks forward, binds to another point further along on the filament, and cocks backward. To picture how this works, consider the act of walking and imagine that the ground is a cytoskeletal filament, your leg is the head of the motor protein, and your hip is the hinge. To walk, you lift your leg up, you move it forward, you place it on the ground, and then you cock it backward (which propels you forward). This series of events is analogous to how a motor protein moves along a cytoskeletal filament.

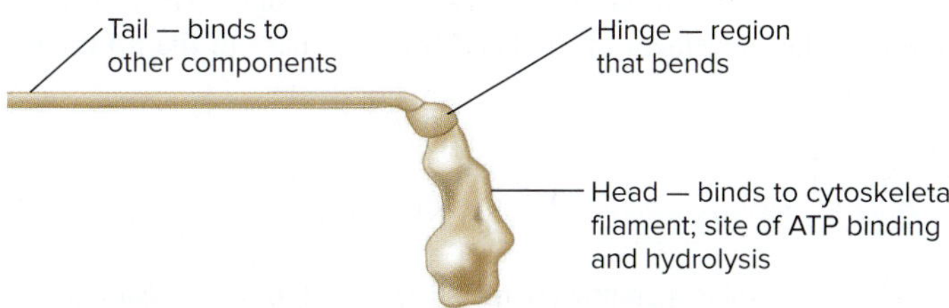


**(a) Three-domain structure of myosin, a motor protein**

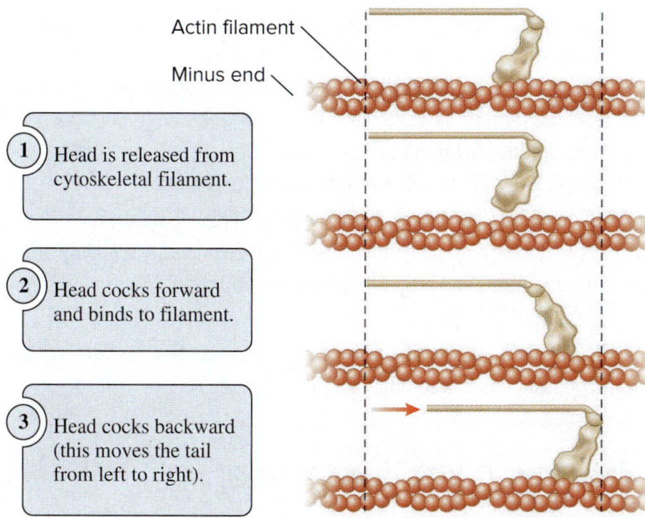


**(b) Movement of a motor protein along a cytoskeletal filament**

**Figure 4.14 Motor proteins and their interactions with cytoskeletal filaments.** The example illustrated here is the motor protein myosin (discussed in Chapter 35), which interacts with actin filaments. **(a)** Three-domain structure of myosin. **(b)** Conformational changes in a motor protein that allow it to walk along a cytoskeletal filament.

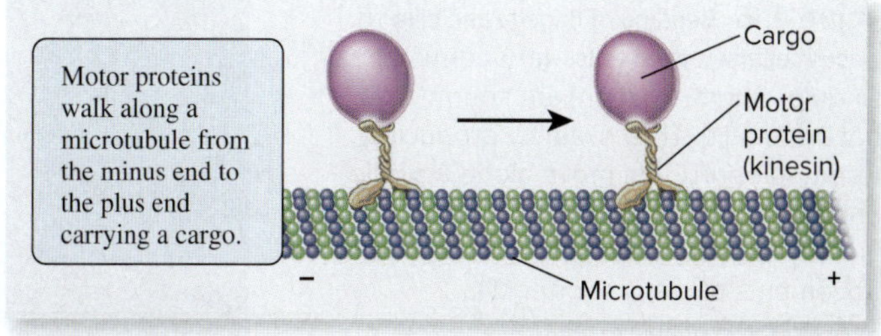


**(a) Motor protein moves**

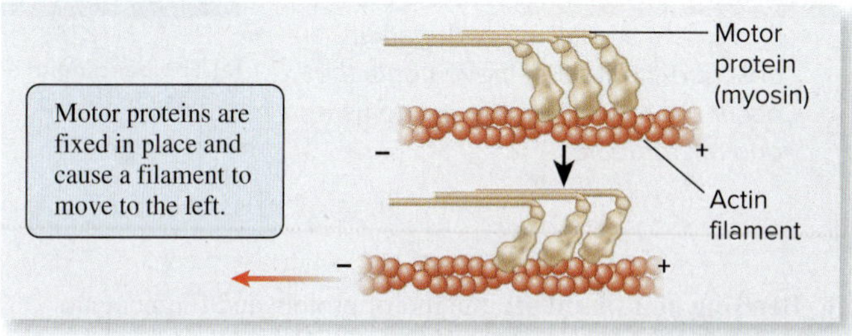


**(b) Filament moves**

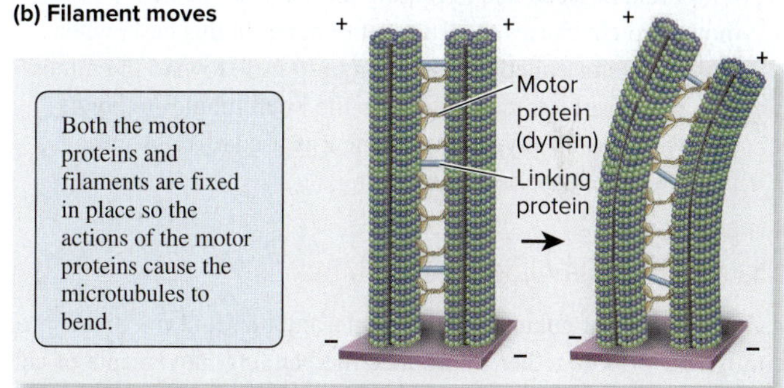

Both the motor proteins and filaments are fixed in place so the actions of the motor proteins cause the microtubules to bend.
Motor protein (dynein)
Linking protein
+
–


**(c) Filaments bend**

**Figure 4.15 Three ways that motor proteins and cytoskeletal filaments cause movement.**

**BioConnections:** *Look ahead to Figure 35.8. Which of these three types of movements occurs during muscle contraction?*

Cells utilize the actions of motor proteins to promote three different kinds of movements:

1. **Movement of cargo:** the filament is fixed in place and the motor protein moves a cargo from one location to another. In the example shown in Figure 4.15a, the tail region of a motor protein called kinesin is attached to a cargo, such as a membrane vesicle containing proteins.
2. **Movement of a filament:** the motor protein is fixed in place. The action of the motor protein moves the filament (Figure 4.15b). In this example, the motor protein myosin is moving an actin filament. This type of movement occurs during muscle contraction, which is described in Chapter 35.

**Figure 4.16 Bending of flagella and cilia causes cellular movements. (a)** Spermatozoa (singular, spermatozoon) are sperm cells that are motile. They swim by producing repeated bends that move along a single, long flagellum. **(b)** Ciliated protozoa such as this paramecium swim by the bending movements of many shorter cilia.
*(b)* © SPL/Science Source

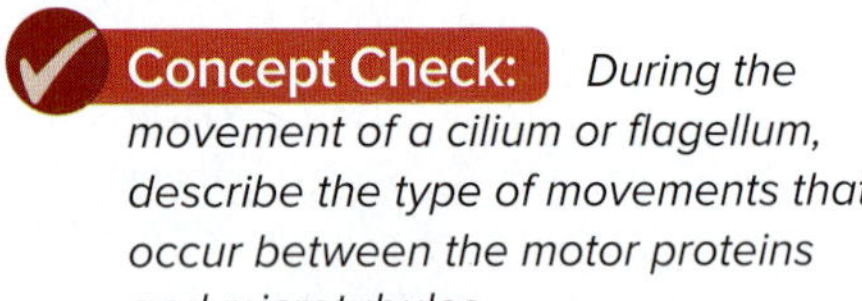

**Concept Check:** *During the movement of a cilium or flagellum, describe the type of movements that occur between the motor proteins and microtubules.*

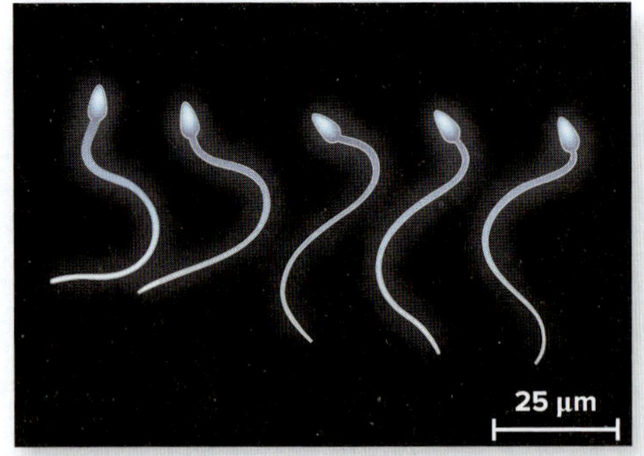


**(a) The bending of a sperm flagellum from the base to the tip**

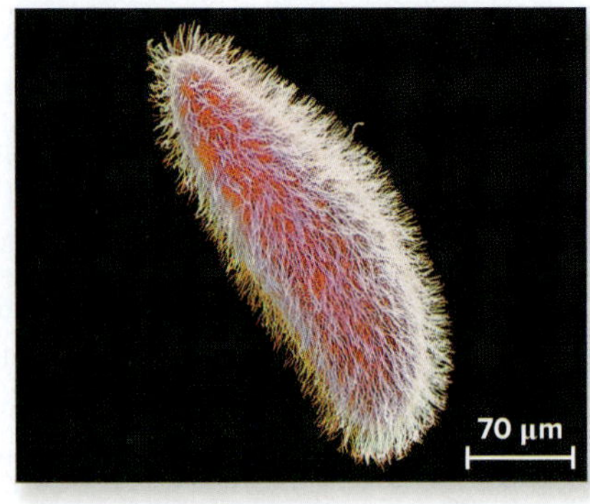


**(b) Paramecium with many cilia**

3. **Bending of a filament:** the motor protein and filament are both fixed in place due to linking proteins, so the action of the motor protein causes the filament to bend. In this case, when motor proteins called dynein attempt to walk toward the minus end, they exert a force that causes the microtubules to bend (Figure 4.15c). This type of movement is used by cilia and flagella, which are described next.

## Cilia and Flagella Allow Cells to Move

In certain kinds of eukaryotic cells, microtubules and the motor protein dynein work together to produce the bending movements of cell appendages called **flagella** and **cilia** (singular, flagellum and cilium). Flagella are usually longer than cilia, and they are typically found singly or in pairs. Both flagella and cilia produce movement by generating bends that move along the length and push backwards against the surrounding fluid. A sperm cell generates bends alternatively in each direction, which begin at the head and move (propagate) toward the tip of the flagellum (Figure 4.16a). By comparison, cilia are often shorter than flagella and tend to cover all or part of the surface of a cell. Protists such as paramecia may have hundreds of cilia, which are spaced closely together and bend in a coordinated fashion to propel the organism through the water (Figure 4.16b).

Despite their differences in length, flagella and cilia have the same internal structure called the **axoneme.** The axoneme contains microtubules, the motor protein dynein, and linking proteins (Figure 4.17). In the cilia and flagella of most eukaryotic organisms, the microtubules form an arrangement called a 9 + 2 array. The outer nine are doublet microtubules, which are composed of a partial microtubule attached to a complete microtubule. Each of the two central microtubules consists of a single microtubule. Radial spokes project from the outer doublet microtubules toward the central pair. The microtubules in flagella and cilia emanate from **basal bodies,** which are anchored to the cytoplasmic side of the plasma membrane. At the basal body, the microtubules form a triplet structure. Much like the centrosome of animal cells, the basal bodies provide a site for microtubules to grow.

The movement of both flagella and cilia involves the propagation of a bend, which begins at the base of the structure and proceeds toward the tip (see Figure 4.16a). The bending occurs because dynein is activated to walk toward the minus end of the microtubules. ATP hydrolysis is required for this process. However, the microtubules and dynein are not free to move relative to each other because of linking proteins. Therefore, instead of dyneins freely walking along the microtubules, they exert a force that bends the microtubules (see Figure 4.15c). The dyneins at the base of the flagellum or cilium are activated first, followed by dyneins that are progressively closer to the tip. The movement of the bend from the base to the tip propels the cell.

## 4.4 Reviewing the Concepts

- The cytosol is a central coordinating region for many metabolic activities of eukaryotic cells, including polypeptide synthesis (Figures 4.13).
- The cytoskeleton is a network of three different types of protein filaments: microtubules, intermediate filaments, and actin filaments. Microtubules are important for cell shape, organization, and movement. Intermediate filaments help maintain cell shape and rigidity. Actin filaments support the plasma membrane and play a key role in cell strength, shape, and movement (Table 4.1).
- Motor proteins interact with cytoskeletal filaments and promote three types of movements, which include walking along the filament, moving the filament, and bending the filament (Figures 4.14, 4.15).
- The structure and function of cilia and flagella provide certain types of cells with motility (Figures 4.16, 4.17).

## 4.4 Testing Your Knowledge

1. The cytosol is the region of a eukaryotic cell that is
   a. inside the plasma membrane.
   b. inside the plasma membrane but outside the membrane-bound organelles.
   c. inside the cell nucleus.
   d. inside the plasma membrane but outside the nucleus.
   e. inside the organelles.

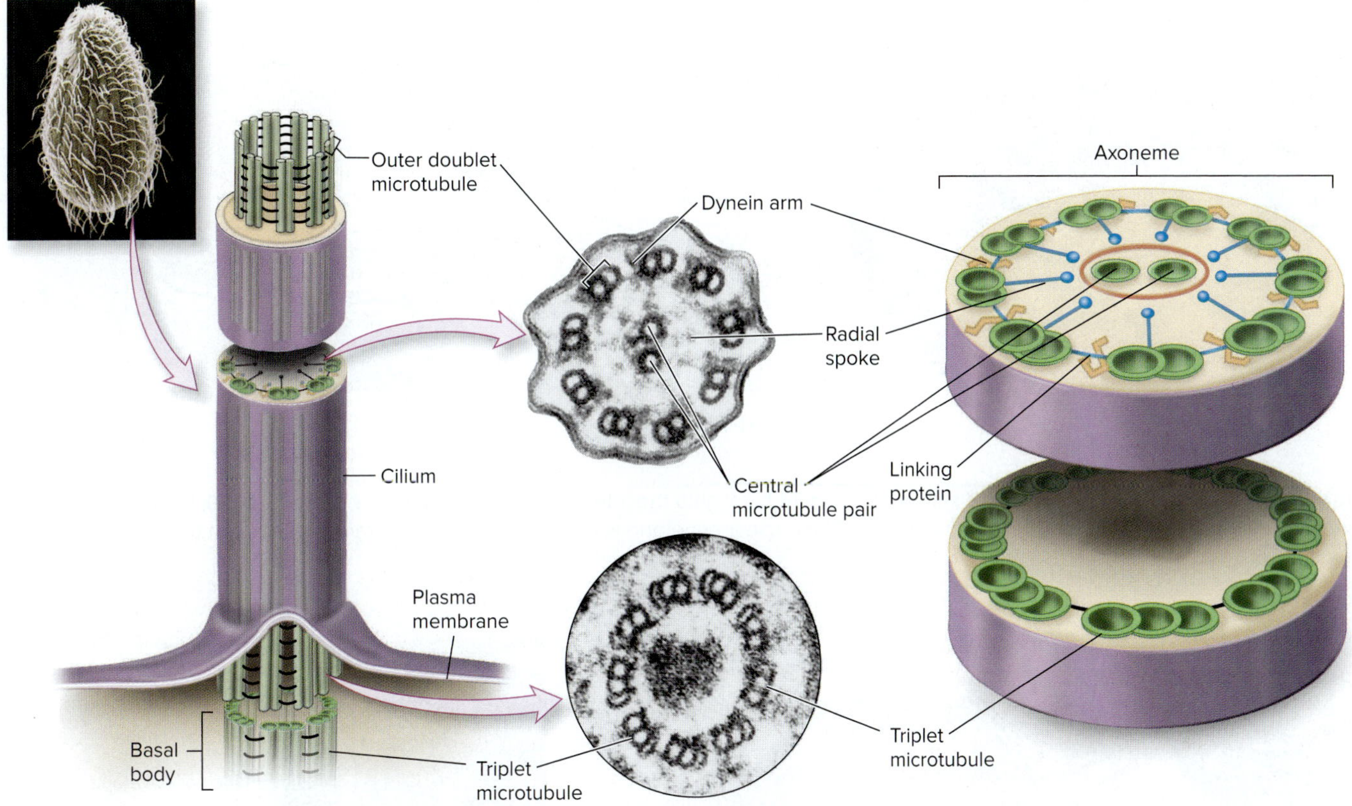


**Figure 4.17 Structure of a eukaryotic cilium or flagellum.** The structure of a cilium of a protist, *Tetrahymena thermophila* (see inset), consists of a 9 + 2 arrangement of nine outer doublet microtubules and two central microtubules. This structure is anchored to the basal body, which has nine triplet microtubules, in which three microtubules are fused together. Note: The structure of the basal body is very similar to centrioles in animal cells.
*(top left)* © Aaron J. Bell/Science Source; *(top & bottom middle)* © Dr. William Dentler/University of Kansas

**BioConnections:** *Look ahead to Figure 23.15. What are some different uses of cilia and flagella among protists?*

2. Which of the following would *not* be caused by a motor protein?
   a. movement of a cargo along a filament
   b. breakage of a filament
   c. movement of a filament
   d. bending of a filament
   e. none of the above
3. Which of the following is a key difference between cilia and flagella?
   a. On a single cell, cilia are more numerous than flagella.
   b. Flagella are usually longer than cilia.
   c. The internal axoneme structure is very different between cilia and flagella.
   d. Cilia contain microtubules but flagella contain actin.
   e. Both a and b are correct.

## 4.5 The Nucleus and Endomembrane System

### Learning Outcomes

1. Describe the structure and organization of the cell nucleus.
2. Outline the structures and general functions of the components of the endomembrane system.
3. Distinguish between the rough endoplasmic reticulum and the smooth endoplasmic reticulum.

In Chapter 2, we learned that the nucleus of an atom contains protons and neutrons. In cell biology, the term **nucleus** has a different meaning. It is an organelle found in eukaryotic cells that contains most of the cell's genetic material. A small amount of genetic material is also found outside the nucleus, in mitochondria and chloroplasts.

The membranes that enclose the nucleus are part of a larger network of membranes called the **endomembrane system.** This system includes not only the nuclear envelope, which encloses the nucleus, but also the endoplasmic reticulum, Golgi apparatus, lysosomes, vacuoles, and peroxisomes. The prefix *endo* (from the Greek, meaning inside) originally referred only to these organelles and internal membranes. However, we now know that the plasma membrane is also part of this integrated membrane system (Figure 4.18). We will consider the major functions of the plasma membrane in Chapters 5 and 8.

Some of the membranes of the endomembrane system, such as the outer membrane of the nuclear envelope and the membrane of the endoplasmic reticulum, have direct connections to one another. Other organelles of the endomembrane system pass materials to each other via **vesicles**—small, membrane-enclosed spheres. In this section, we examine the nucleus and survey the structures and functions of the organelles and membranes of the endomembrane system.

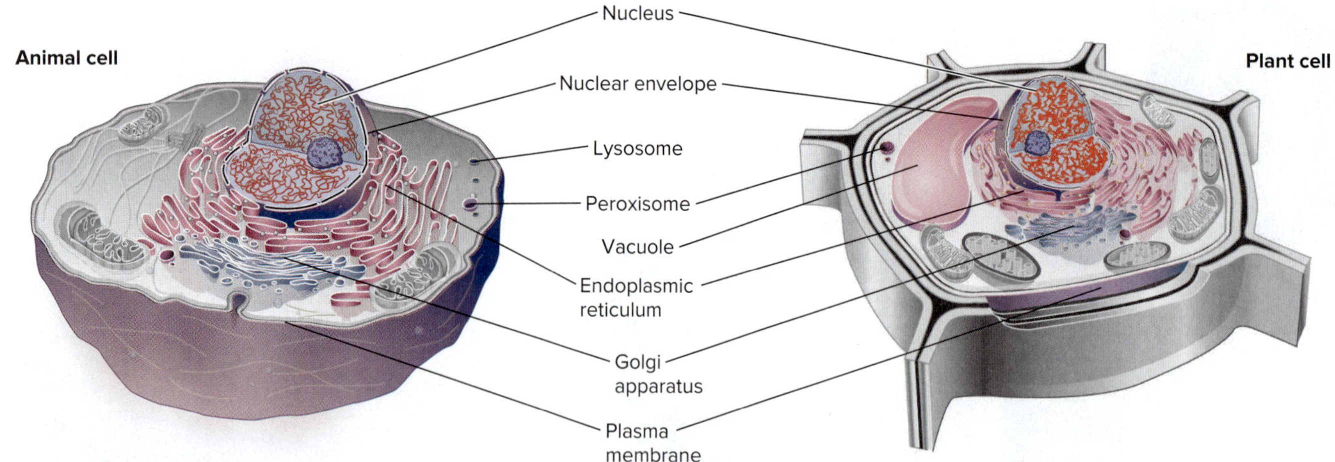


**Figure 4.18 The nucleus and endomembrane system.** This figure highlights the internal compartment of the nucleus (blue), the membranes of the endomembrane system (purple, blue, and pink). The nuclear envelope is part of the endomembrane system, but the interior of the nucleus is not.

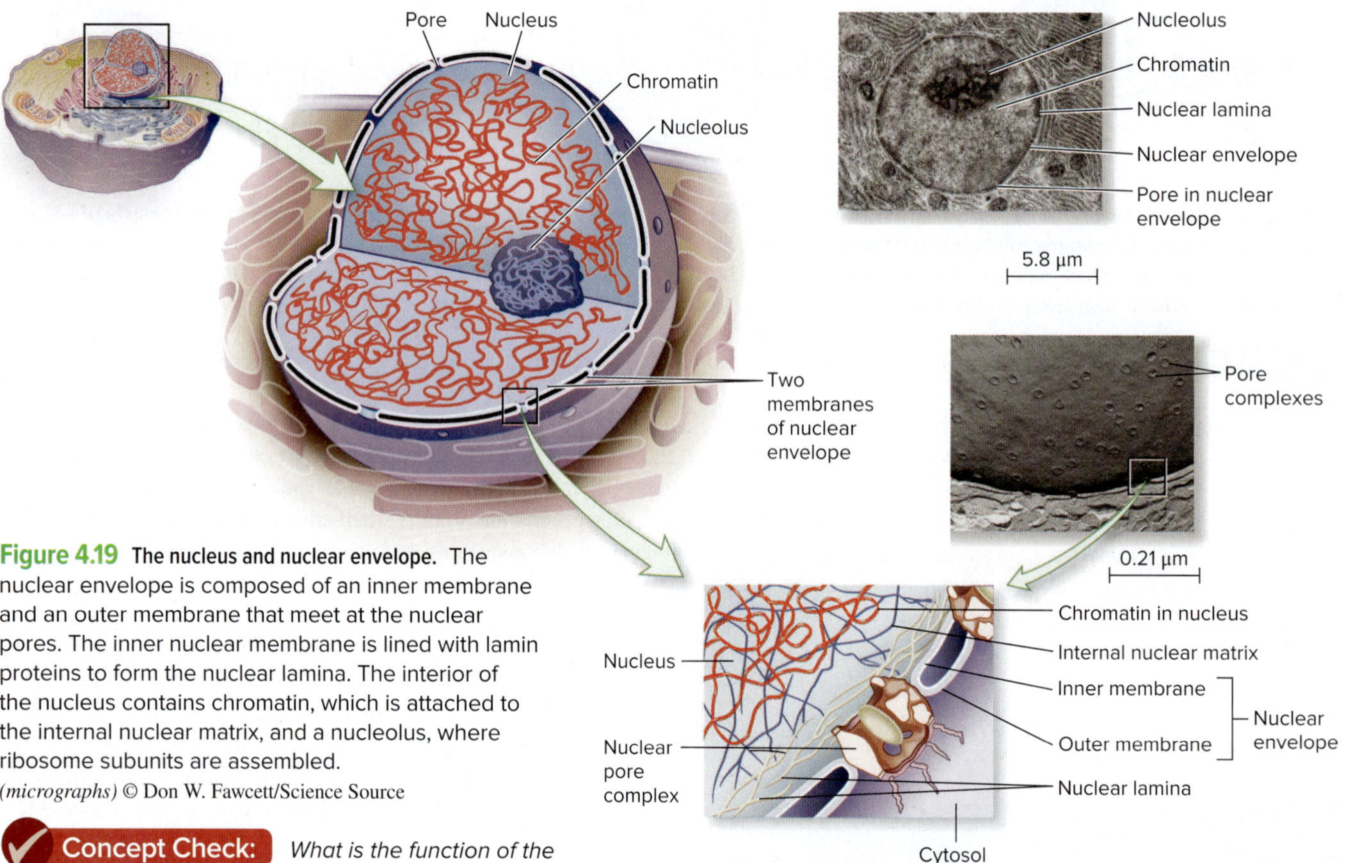


**Figure 4.19 The nucleus and nuclear envelope.** The nuclear envelope is composed of an inner membrane and an outer membrane that meet at the nuclear pores. The inner nuclear membrane is lined with lamin proteins to form the nuclear lamina. The interior of the nucleus contains chromatin, which is attached to the internal nuclear matrix, and a nucleolus, where ribosome subunits are assembled.
*(micrographs)* © Don W. Fawcett/Science Source

**Concept Check:** *What is the function of the nuclear lamina and the internal nuclear matrix?*

## The Eukaryotic Nucleus Contains Chromosomes

The nucleus is the compartment that is enclosed by a double-membrane structure termed the **nuclear envelope** (Figure 4.19). **Nuclear pores** are formed where the inner and outer nuclear membranes make contact with each other. Nuclear pore complexes provide a passageway for the movement of molecules and macromolecules into and out of the nucleus. Although cell biologists view the nuclear envelope as part of the endomembrane system, the materials within the nucleus are not.

Inside the nucleus are the chromosomes and a filamentous network of proteins called the nuclear matrix. Each **chromosome** is composed of genetic material, namely DNA, and many types of proteins that help to compact the chromosome to fit inside the nucleus. An example are histone proteins, described in Chapter 9. The complex formed between DNA and such proteins is termed **chromatin.** The **nuclear matrix** consists of two parts: the nuclear lamina, which is composed of intermediate filaments that line the inner nuclear membrane, and an internal nuclear matrix, which is connected to the lamina and fills the interior of the nucleus. The nuclear matrix organizes the chromosomes within the nucleus. Each chromosome is located in a distinct, nonoverlapping **chromosome territory,** which is visible when cells are exposed to dyes that label specific types of chromosomes (Figure 4.20).

The primary function of the nucleus is the protection, organization, replication, and expression of the genetic material. These topics are discussed in Unit III. Another important function is the assembly of ribosome subunits—cellular structures involved in producing polypeptides during the process of translation. The assembly of ribosome subunits occurs in the **nucleolus** (plural, nucleoli), a prominent region in the nucleus. A ribosome is composed of two subunits: one small and one large (look ahead to Chapter 10, Table 10.3). Each subunit contains one or more RNA molecules and several types of proteins. Most of the RNA molecules that are components of ribosomes are made in the vicinity of the nucleolus. By comparison, the ribosomal proteins are produced in the cytosol and then imported into the nucleus through the nuclear pores. The ribosomal proteins and RNA molecules then assemble in the nucleolus to form the ribosomal subunits. Finally, the subunits exit through the nuclear pores into the cytosol, where they are needed for protein synthesis.

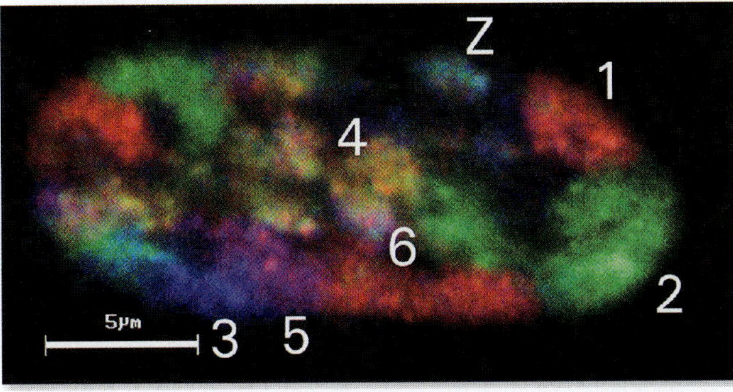


**Figure 4.20** **Chromosome territories in the cell nucleus.** Chromosomes from a chicken were labeled with chromosome-specific probes. Seven types of chicken chromosomes are stained with a different dye. Each chromosome occupies its own distinct, nonoverlapping territory within the cell nucleus.
Courtesy of Felix A. Habermann

**BioConnections:** *Look ahead to Figure 14.7. What happens to chromosome territories during cell division?*

## The Endoplasmic Reticulum Initiates the Sorting of Some Proteins and Carries Out Metabolic Functions

The **endoplasmic reticulum (ER)** is a network of membranes that form flattened, fluid-filled tubules, or **cisternae** (Figure 4.21). The terms endoplasmic (Greek, for in the cytoplasm) and reticulum (Latin, for little net) refer to the location and shape of this organelle when viewed under a microscope. The term **lumen** describes the internal space of an organelle. The ER membrane, which is continuous with the outer nuclear membrane, encloses a single compartment called the **ER lumen.** There are two distinct, but continuous types of ER: rough ER and smooth ER.

**Rough ER** The outer surface of the **rough endoplasmic reticulum (rough ER)** is studded with ribosomes, giving it a bumpy appearance when viewed under the microscope. Rough ER performs a few key functions:

- **Protein sorting:** Proteins that are destined for the ER, Golgi apparatus, lysosomes, vacuoles, plasma membrane, or outside of the cell are first directed to the rough ER. This topic is described later in Section 4.7.
- **Insertion of membrane proteins:** Membrane proteins with segments that span the membrane are initially inserted into the rough ER membrane as the protein is being synthesized.
- **Glycosylation:** An important function of the rough ER is the attachment of carbohydrates to proteins and lipids. This process is called **glycosylation.**

**Smooth ER** The **smooth endoplasmic reticulum (smooth ER),** which lacks ribosomes, functions in diverse processes:

- **Metabolism:** The extensive network of smooth ER membranes provides an increased surface area for enzymes that play important metabolic roles. In liver cells, enzymes in the smooth ER detoxify many potentially harmful organic molecules, including barbiturate drugs and ethanol (alcohol). These enzymes convert hydrophobic toxic molecules into more hydrophilic molecules, which are easily excreted from the body. Chronic alcohol consumption, as occurs in alcoholics, leads to a greater amount of smooth ER in liver cells, which increases the rate of alcohol breakdown. This explains why people who consume alcohol regularly must ingest more alcohol to experience its effects.
- **Storage of $Ca^{2+}$:** The smooth ER contains calcium pumps that transport $Ca^{2+}$ into the ER lumen. The regulated release of $Ca^{2+}$ into the cytosol is involved in many vital cellular processes, including muscle contraction in animals.
- **Lipid synthesis and modification:** The smooth ER is the primary site for the synthesis of phospholipids, which are the main lipid component of cell membranes. In addition, enzymes in the smooth ER are necessary for certain modifications of the lipid cholesterol that are needed to produce steroid hormones such as estrogen and testosterone.

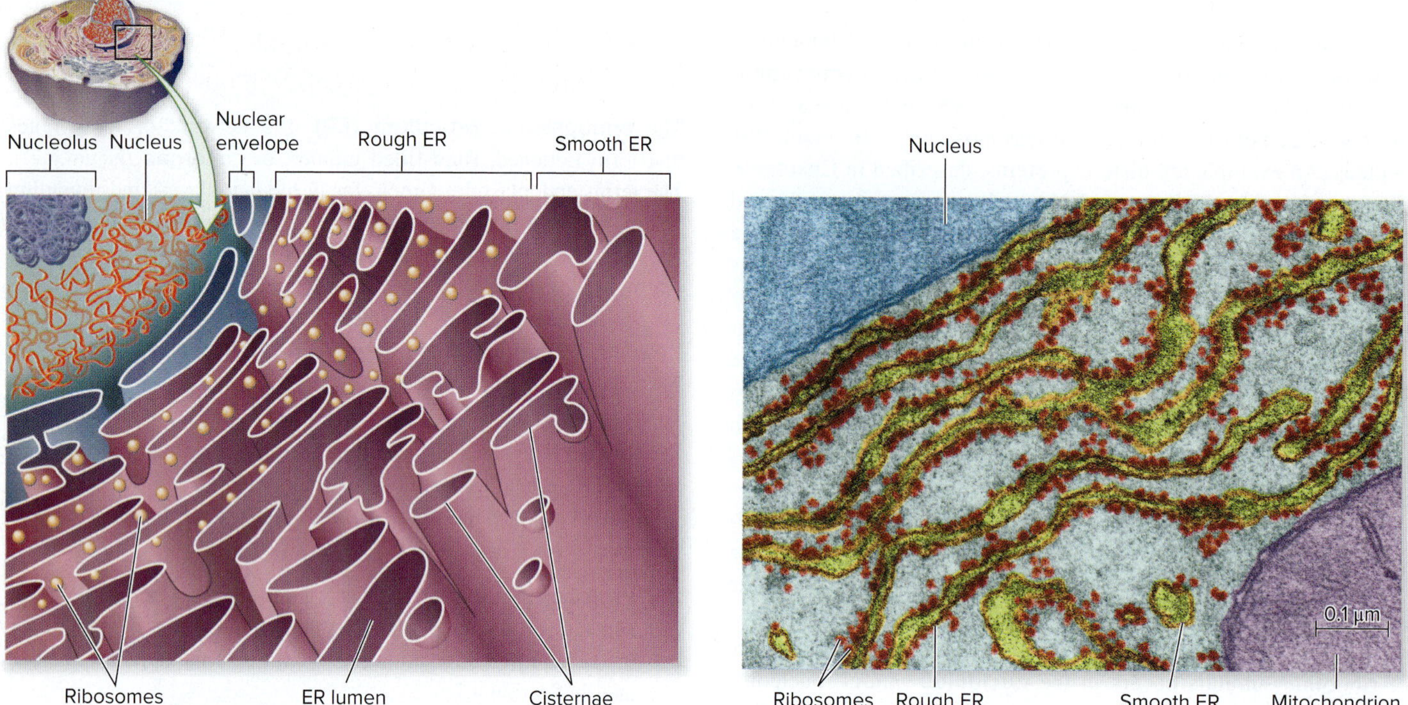


**Figure 4.21** **Structure of the endoplasmic reticulum.** (Left side) The endoplasmic reticulum (ER) is composed of a network of flattened tubules called cisternae that enclose a continuous ER lumen. The rough ER is studded with ribosomes, whereas the smooth ER lacks ribosomes. The rough ER is continuous with the outer nuclear membrane. (Right side) In this colorized TEM of the ER, the lumen is colored yellow and the ribosomes are red.

*(right)* © Dennis Kunkel Microscopy, Inc./Phototake

## The Golgi Apparatus Directs the Processing, Sorting, and Secretion of Cellular Molecules

The **Golgi apparatus** (also called the Golgi body, Golgi complex, or simply Golgi) was discovered by the Italian microscopist Camillo Golgi in 1898. It consists of a stack of flattened membranes, each of which encloses a single compartment called a **cisterna** (plural cisternae) (**Figure 4.22**). The Golgi compartments are named according to their orientation in the cell. The *cis* Golgi is near the ER membrane, the *trans* Golgi is closest to the plasma membrane, and the medial Golgi is found in the middle. Two models have been proposed to explain how materials move through the Golgi apparatus:

- **Vesicular transport model:** Materials are transported between the Golgi cisternae via membrane vesicles that bud from one compartment in the Golgi (for example, the *cis* Golgi) and fuse with another compartment (for example, the medial Golgi).
- **Cisternal maturation model:** Vesicles from the ER fuse to form a cisterna at the *cis* face; the cisterna that was previously at the *cis* face becomes a medial cisterna. Likewise, this moves medial cisternae toward the *trans* face. The cisterna at the *trans* face is lost due to the export of vesicles from its surface.

Further research will be needed to determine if either or both of these models are correct.

The Golgi apparatus performs three overlapping functions:

1. **Protein sorting:** As discussed in Section 4.7, the Golgi plays a role in the sorting of several categories of proteins.
2. **Processing:** Enzymes in the Golgi apparatus process, or modify, certain proteins and lipids. As mentioned earlier, carbohydrates can be attached to proteins and lipids in the endoplasmic reticulum. Glycosylation continues in the Golgi. For this to occur, a protein or lipid is transported via vesicles from the ER to the *cis* Golgi. Most of the glycosylation occurs in the medial Golgi. A second type of processing event is **proteolysis,** whereby enzymes called **proteases** cut proteins into smaller polypeptides. For example, the hormone insulin is first made as a large precursor protein termed proinsulin. In the Golgi apparatus, proinsulin is packaged with proteases into vesicles. The proteases cut out a portion of the proinsulin to create a smaller insulin molecule that is a functional hormone.
3. **Secretion:** The Golgi apparatus packages different types of materials into **secretory vesicles** that fuse with the plasma membrane, thereby releasing their contents outside the cell. Proteins destined for secretion are synthesized into the ER, travel to the Golgi, and then are transported by vesicles to the plasma membrane (see Figure 4.22). The vesicles then fuse with the plasma membrane and are secreted to the outside of the cell. The entire route is called the **secretory pathway.**

## Lysosomes Are Involved in the Intracellular Digestion of Macromolecules

We now turn to another organelle of the endomembrane system, **lysosomes,** which are small organelles found in animal cells that break down macromolecules. Lysosomes contain many **acid hydrolases,** which are hydrolytic enzymes that use a molecule of water to break a covalent bond. As described in Chapter 3

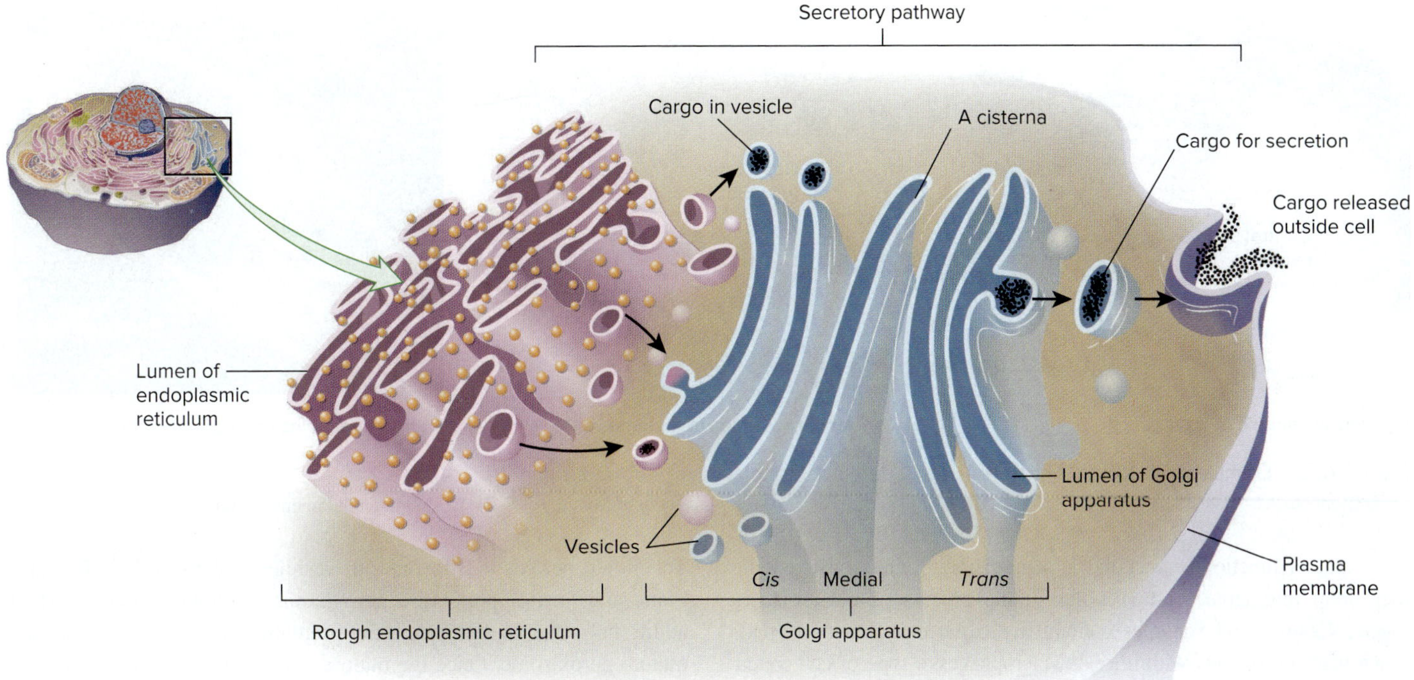


**Figure 4.22** **The Golgi apparatus and secretory pathway.** The Golgi is composed of stacks of membranes that enclose separate compartments. Transport to and from the Golgi occurs via membrane vesicles. Vesicles bud from the ER and go to the Golgi, and vesicles from the Golgi fuse with the plasma membrane to release cargo to the outside. The pathway from the ER to the Golgi to the plasma membrane is termed the secretory pathway.

**Concept Check:** *If we consider the Golgi apparatus as three compartments (cis, medial, and trans), describe the compartments that a protein travels through to be secreted.*

(refer back to Figure 3.3b), this type of chemical reaction is called hydrolysis:

$$R_1\text{—}R_2 + H_2O \xrightarrow{\text{Acid hydrolase}} R_1\text{—}OH + H\text{—}R_2$$

The acid hydrolases in a lysosome function optimally at an acidic pH. The fluid-filled interior of a lysosome has a pH of approximately 4.8. If a lysosomal membrane breaks, releasing acid hydrolases into the cytosol, the enzymes are not very active because the cytosolic pH is neutral (approximately pH 7.0) and buffered. This prevents significant damage to the cell from lysosome breakage.

Lysosomes contain many different types of acid hydrolases that break down carbohydrates, proteins, lipids, and nucleic acids. This enzymatic function enables lysosomes to break down complex materials. One function of lysosomes involves the digestion of substances that are taken up from outside the cell via a process called endocytosis (look ahead to Chapter 5, Figure 5.21).

## Vacuoles Function in Storage, Regulation of Cell Volume, and Degradation

**Vacuoles** are prominent organelles in plant cells, fungal cells, and certain protists. The term vacuole (Latin, for empty space) came from early microscopic observations of these compartments. We now know that vacuoles are not empty but instead contain fluid and sometimes even solid substances. Most vacuoles are made from the fusion of many smaller membrane vesicles. In animal cells, vacuoles tend to be smaller and are more commonly used to temporarily store materials or transport substances. In animals, such vacuoles are sometimes called storage vesicles.

The functions of vacuoles are extremely varied, and they differ among cell types and even environmental conditions. The best way to appreciate vacuole function is to consider a few examples. Mature plant cells usually have a large **central vacuole** that occupies 80% or more of the cell volume (**Figure 4.23a**). The central vacuole serves two important purposes. First, it stores a large amount of water, enzymes, and inorganic ions such as calcium, as well as other materials, including proteins and pigments. Second, it performs a space-filling function. The central vacuole exerts a pressure on the cell wall, called turgor pressure. If a plant becomes dehydrated and this pressure is lost, a plant will wilt. Turgor pressure is important in maintaining the structure of plant cells and the plant itself, and it helps to drive the expansion of the cell wall, which is necessary for growth.

Certain species of protists use vacuoles to maintain cell volume. Freshwater organisms such as the alga *Chlamydomonas reinhardtii* have small, water-filled **contractile vacuoles** that expand as water enters the cell (**Figure 4.23b**). Once they reach a certain size, the vacuoles fuse with the plasma membrane, expelling their contents to the exterior of the cell (look ahead to Figure 5.11). This mechanism is necessary to remove excess water that continually enters the cell by diffusion across the plasma membrane.

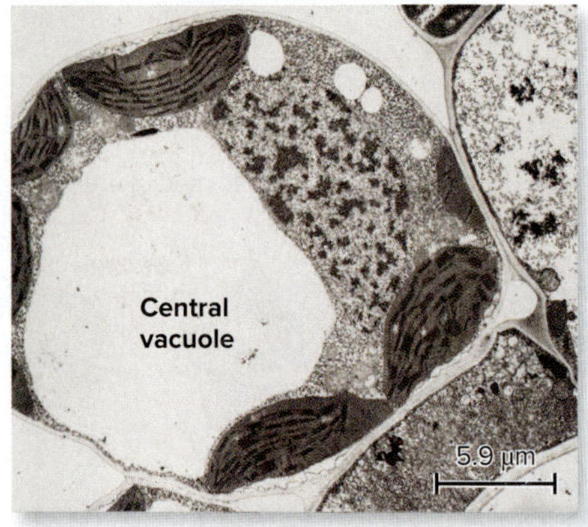


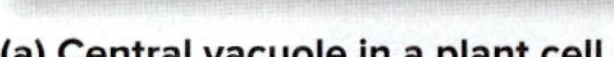

(a) Central vacuole in a plant cell

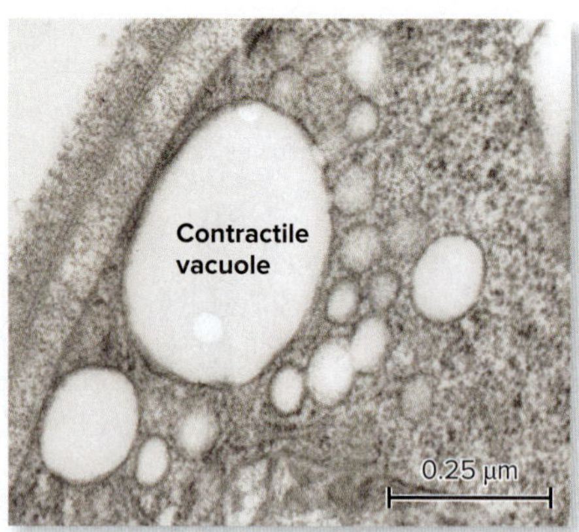


(b) Contractile vacuoles in an algal cell

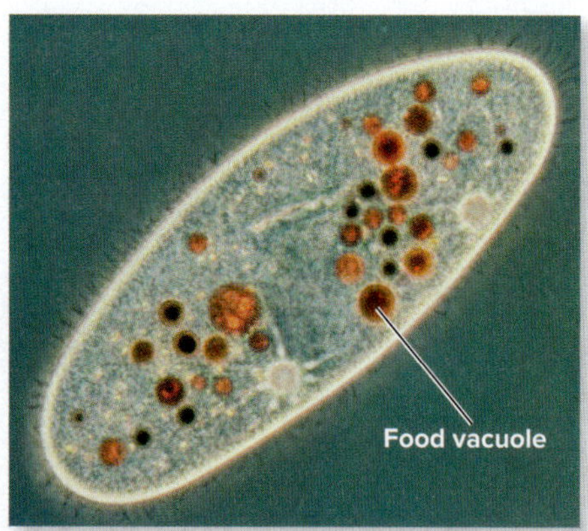


(c) Food vacuoles in a paramecium

**Figure 4.23** **Examples of vacuoles.** These are TEMs. Part (c) is colorized.

*(a)* © Biophoto Associates/Science Source; *(b)* Courtesy Dr. Peter Luykx, Biology, University of Miami; *(c)* © Dr. David Patterson/Science Source

Another function of vacuoles is degradation. Some protists engulf their food into large food vacuoles in the process of phagocytosis (Figure 4.23c). Food vacuoles contain hydrolytic enzymes that break down macromolecules within food. Macrophages, a type of cell found in animals' immune systems, engulf bacterial cells into phagocytic vacuoles, which then fuse with lysosomes, where the bacteria are destroyed.

## Peroxisomes Catalyze Detoxifying Reactions

**Peroxisomes,** discovered by Christian de Duve in 1965, are small organelles found in all eukaryotic cells. Peroxisomes consist of a single membrane that encloses a fluid-filled lumen. A typical eukaryotic cell contains several hundred of them.

Peroxisomes catalyze a variety of chemical reactions, including some reactions that break down organic molecules and others that are biosynthetic. In mammals, large numbers of peroxisomes are found in liver cells, where toxic molecules accumulate and are broken down. A common by-product of the breakdown of toxins is hydrogen peroxide, $H_2O_2$:

$$\underset{\text{(toxin)}}{RH_2} + O_2 \longrightarrow R + H_2O_2$$

Hydrogen peroxide has the potential to be highly toxic. In the presence of metals such as iron ($Fe^{2+}$), which are found naturally in living cells, $H_2O_2$ can be broken down to form a hydroxide ion ($OH^-$) and a molecule called a hydroxide free radical ($\cdot OH$):

$$Fe^{2+} + H_2O_2 \longrightarrow Fe^{3+} + OH^- + \cdot OH \text{ (hydroxide free radical)}$$

The $\cdot OH$ is highly reactive and can damage proteins, lipids, and DNA. Therefore, it is beneficial for cells to break down $H_2O_2$ in an alternative manner that does not form a $\cdot OH$. Peroxisomes contain an enzyme called **catalase** that breaks down hydrogen peroxide to make water and oxygen gas (hence the name peroxisome):

$$2\,H_2O_2 \xrightarrow{\text{Catalase}} 2\,H_2O + O_2$$

A general model for peroxisome formation is shown in Figure 4.24, though the details may differ among animal, plant, and fungal cells. To initiate peroxisome formation, vesicles bud from the ER membrane and form a premature peroxisome. Following the import of additional proteins and lipids, the premature peroxisome becomes a mature peroxisome. Once the mature peroxisome has formed, it may then divide to further increase the number of peroxisomes in the cell.

## 4.5 Reviewing the Concepts

- The primary function of the nucleus is the organization and expression of the genetic material. A second important function is the assembly of ribosomal subunits in the nucleolus (Figures 4.18, 4.19, 4.20).
- The endomembrane system includes the nuclear envelope, endoplasmic reticulum (ER), Golgi apparatus, lysosomes, vacuoles, peroxisomes, and plasma membrane. The rough endoplasmic reticulum (rough ER) plays a key role in the initial sorting of proteins. The smooth endoplasmic reticulum (smooth ER) functions in metabolic processes such as detoxification, carbohydrate metabolism, accumulation of calcium ions, and synthesis and modification of lipids. The Golgi apparatus performs three overlapping functions: processing, protein sorting, and secretion. Lysosomes degrade macromolecules and help digest substances taken up from outside the cell (endocytosis) (Figures 4.21, 4.22).
- Types of vacuoles include central vacuoles, contractile vacuoles, and food vacuoles (Figure 4.23).
- Peroxisomes catalyze a variety of chemical reactions, including those involved with the breakdown of toxic molecules such as hydrogen peroxide, and typically contain enzymes involved in the metabolism of fats and amino acids. Peroxisomes are made via budding from the ER, followed by maturation and division (Figure 4.24).

## 4.5 Testing Your Knowledge

1. Which of the following organelles is surrounded by a double membrane?
   a. nucleus
   b. endoplasmic reticulum
   c. Golgi apparatus
   d. lysosome
   e. peroxisome

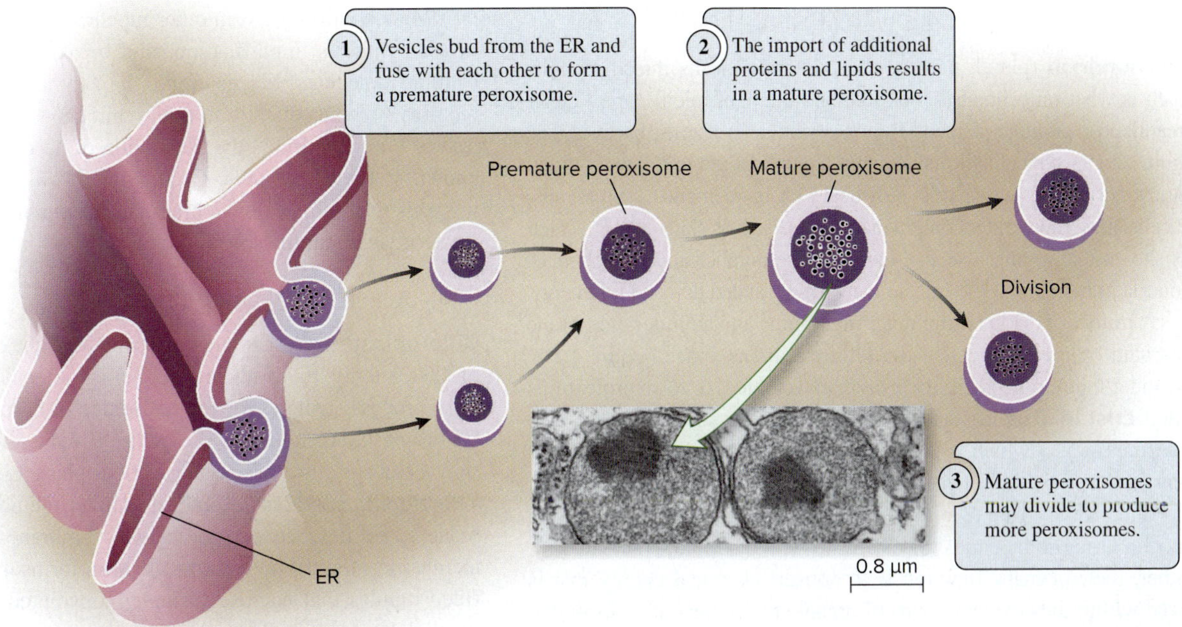

1 Vesicles bud from the ER and fuse with each other to form a premature peroxisome.
2 The import of additional proteins and lipids results in a mature peroxisome.
Premature peroxisome
Mature peroxisome
Division
3 Mature peroxisomes may divide to produce more peroxisomes.
ER
0.8 µm


**Figure 4.24** **Formation of peroxisomes.** The inset is a TEM of mature peroxisomes.
© Don W. Fawcett/ Science Source

2. The secretory pathway occurs in which of the following orders?
   a. Golgi to ER to plasma membrane to outside of cell
   b. ER to Golgi to plasma membrane to outside of cell
   c. plasma membrane to ER to Golgi to outside of cell
   d. ER to plasma membrane to Golgi to outside of cell
   e. none of the above
3. A characteristic of lysosomes is
   a. they have a low internal pH.
   b. they contain different acid hydrolases.
   c. they hydrolyze organic molecules into small components.
   d. they are surrounded by a single membrane.
   e. all of the above.

## 4.6 Semiautonomous Organelles

### Learning Outcomes

1. Describe the structures and general functions of mitochondria and chloroplasts.
2. **SCISKILLS ▸** Discuss the evidence for the endosymbiosis theory.

We now turn to those organelles in eukaryotic cells that are considered semiautonomous—mitochondria and chloroplasts. These organelles grow and divide, but they are not completely autonomous because they depend on other parts of the cell for their internal components (**Figure 4.25**). For example, most of the proteins found in mitochondria are imported from the cytosol. In this section, we survey the structures and functions of the semiautonomous organelles in eukaryotic cells and consider their evolutionary origins. In Chapters 6 and 7, we will explore the functions of mitochondria and chloroplasts in greater depth.

## Biology Principle

### Living Organisms Use Energy

Chloroplasts capture light energy and synthesize organic molecules. Mitochondria break down organic molecules and make ATP, which is used as an energy source to drive many different cellular processes.

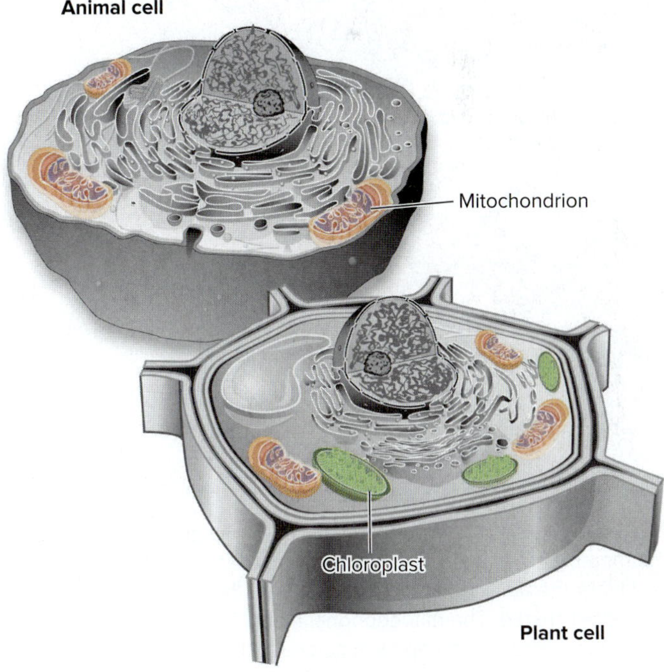


**Figure 4.25** **Semiautonomous organelles.** These are the mitochondria (top and bottom cell) and chloroplasts (bottom cell).

## Mitochondria Supply Cells with Most of Their ATP

**Mitochondrion** (plural, mitochondria) literally means thread granule, which is what mitochondria look like under a light microscope—either threadlike or granular-shaped. They are similar in size to bacteria (see Figure 4.8). A typical cell may contain a few hundred to a few thousand mitochondria. Cells with particularly heavy energy demands, such as muscle cells, have more mitochondria than others, such as skin cells. Research has shown that regular exercise increases the number and size of mitochondria in human muscle cells to meet the expanded demand for energy.

A mitochondrion has an outer membrane and an inner membrane separated by a region called the intermembrane space (Figure 4.26). The inner membrane is highly invaginated (folded) to form projections called **cristae.** The cristae greatly increase the surface area of the inner membrane, which is the site where ATP is made. The compartment enclosed by the inner membrane is the **mitochondrial matrix.**

The primary role of mitochondria is to make ATP. Even though mitochondria produce most of a cell's ATP, they do not create energy. Rather, their primary function is to convert chemical energy that is stored within the covalent bonds of organic molecules into a form that can be readily used by cells. Covalent bonds in sugars, fats, and amino acids store a large amount of energy. The breakdown of these molecules into simpler molecules releases energy that is used to make ATP. Many proteins in living cells use ATP as a source of energy to carry out their functions, such as muscle contraction, the uptake of nutrients, cell division, and many other cellular processes.

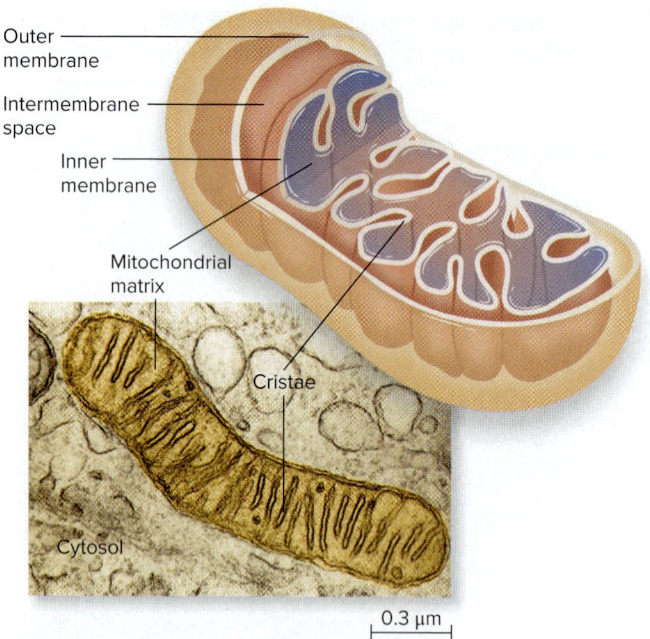


**Figure 4.26 Structure of a mitochondrion.** This figure emphasizes the membrane organization of a mitochondrion, which has an outer and inner membrane. The invaginations of the inner membrane are called cristae. The mitochondrial matrix lies inside the inner membrane. The micrograph is a colorized TEM.



**Concept Check:** *What is the advantage of having a highly invaginated inner membrane?*

Mitochondria perform other functions as well. They are involved in the synthesis, modification, and breakdown of several types of cellular molecules. For example, the synthesis of certain hormones requires enzymes that are found in mitochondria. Another interesting role of mitochondria is the generation of heat in specialized fat cells known as brown fat cells. Groups of brown fat cells serve as heating pads that help to revive hibernating animals and protect sensitive areas of young animals from the cold.

## Chloroplasts Carry Out Photosynthesis

**Chloroplasts** are organelles that capture light energy and use some of that energy to synthesize organic molecules such as glucose. This process, called **photosynthesis,** is described in Chapter 7. Chloroplasts are found in the cells of plants and algae. Figure 4.27 shows the structure of a typical chloroplast. Like a mitochondrion, a chloroplast contains an outer and inner membrane. An intermembrane space lies between these two membranes. A third system of membranes, the **thylakoid membrane,** forms many flattened, fluid-filled tubules that enclose a single, convoluted compartment called the thylakoid lumen. These tubules tend to stack on top of each other to form a structure called a **granum** (plural, grana). The **stroma** is the compartment of the chloroplast that is enclosed by the inner membrane but outside the thylakoid membrane.

Chloroplasts are a specialized version of plant organelles that are more generally known as **plastids.** All plastids are derived from unspecialized **proplastids.** The various types of plastids are distinguished by their synthetic abilities and the types of pigments they contain:

- Chloroplasts, which carry out photosynthesis, contain the green pigment chlorophyll. The abundant number of chloroplasts in the leaves of plants gives them their green color.

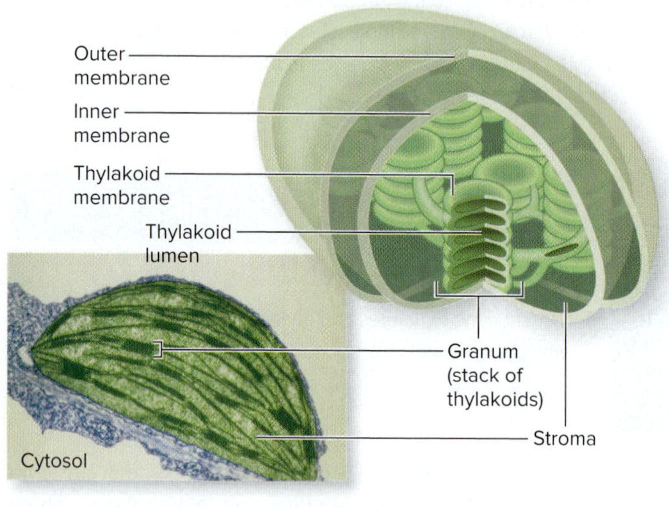


**Figure 4.27 Structure of a chloroplast.** Like a mitochondrion, a chloroplast is enclosed in a double membrane. In addition, it has an internal thylakoid membrane system that forms flattened compartments. These compartments stack on each other to form grana. The stroma is located inside the inner membrane but outside the thylakoid membrane. This micrograph is a colorized TEM.

- Chromoplasts, a second type of plastid, function in synthesizing and storing the yellow, orange, and red pigments known as carotenoids. Chromoplasts give many fruits, vegetables, and flowers their colors. In autumn, the chromoplasts also give many leaves their yellow, orange, and red colors.
- Leucoplasts typically lack pigment molecules. An amyloplast is a leucoplast that synthesizes and stores starch. Amyloplasts are common in underground structures such as roots and tubers.

## Mitochondria and Chloroplasts Contain Their Own Genetic Material and Divide by Binary Fission

The chromosomes found in mitochondria and chloroplasts are referred to as the **mitochondrial genome** and **chloroplast genome,** respectively, whereas the chromosomes found in the nucleus of the cell constitute the **nuclear genome.** Like bacteria, the genomes of most mitochondria and chloroplasts are composed of a single circular chromosome. Compared with the nuclear genome, they are very small. For example, the human genome has approximately 22,000 different protein-encoding genes, whereas the human mitochondrial genome has only about a dozen protein-encoding genes.

Mitochondria and chloroplasts increase in number via **binary fission,** or splitting in two. Figure 4.28 illustrates the process for a mitochondrion. The mitochondrial chromosome, which is found in a region called the nucleoid, is duplicated, and the organelle divides into two separate organelles. Mitochondrial and chloroplast division are needed to maintain a full complement of these organelles when cell growth occurs following cell division. In addition, environmental conditions may influence the sizes and numbers of these organelles. For example, when plants are exposed to more sunlight, the number of chloroplasts in leaf cells increases.

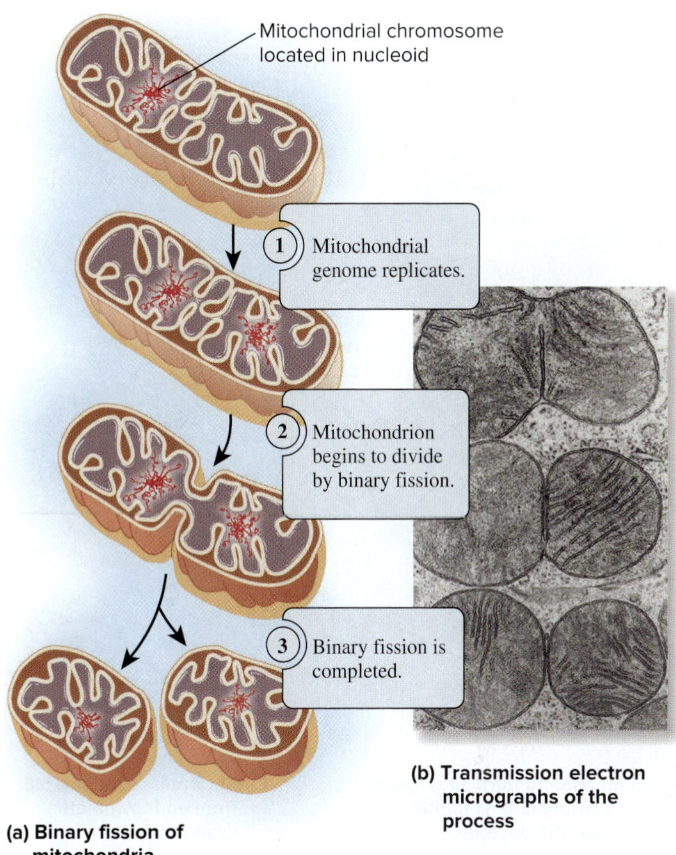


**Figure 4.28** **Division of mitochondria by binary fission.**
*(b)* © Don W. Fawcett/Science Source

**BioConnections:** *Look ahead to Figure 17.8. How is this process similar to bacterial cell division, and how is it different?*

# EVOLUTIONARY CONNECTIONS

## Mitochondria and Chloroplasts Are Derived from Ancient Symbiotic Relationships

The observation that mitochondria and chloroplasts contain their own genetic material may seem puzzling. Perhaps you might think that it would be simpler for a eukaryotic cell to have all of its genetic material in one place—the nucleus. The distinct genomes of mitochondria and chloroplasts can be traced to their evolutionary origin, which involved an ancient symbiotic association.

A symbiotic relationship occurs when two different species live in direct contact with each other. **Endosymbiosis** describes a symbiotic relationship in which the smaller species—the symbiont—actually lives inside the larger species. In 1883, Andreas Schimper proposed that chloroplasts were descended from an endosymbiotic relationship between cyanobacteria (a bacterium capable of photosynthesis) and eukaryotic cells. In 1922, Ivan Wallin also hypothesized an endosymbiotic origin for mitochondria.

In spite of these interesting ideas, the question of endosymbiosis was largely ignored until the discovery that mitochondria and chloroplasts contain their own genetic material. In 1970, the issue of endosymbiosis as the origin of mitochondria and chloroplasts was revived by Lynn Margulis in her book *Origin of Eukaryotic Cells*. During the 1970s and 1980s, the advent of molecular genetic techniques allowed researchers to analyze genes from mitochondria, chloroplasts, bacteria, and eukaryotic nuclear genomes. Researchers discovered that genes in mitochondria and chloroplasts are very similar to bacterial genes. Likewise, mitochondria and chloroplasts are strikingly similar in size and shape to certain bacterial species. These observations provided strong support for the **endosymbiosis theory,** which proposes that mitochondria and chloroplasts originated from bacteria that took up residence within a primordial eukaryotic cell (Figure 4.29). Over the next 2 billion years, the characteristics of these intracellular bacterial cells gradually changed to those of a mitochondrion or chloroplast. The origin of eukaryotic cells is discussed in more detail in Chapter 22.

A symbiotic relationship is beneficial to one or both species. According to the endosymbiosis theory, this relationship provided eukaryotic cells with useful cellular characteristics. Chloroplasts, which were derived from cyanobacteria, have the ability to carry out photosynthesis. This benefits plant cells by giving them the ability to use the energy from sunlight. By comparison, mitochondria are thought to have been derived from a different type of bacteria known as purple bacteria, or alphaproteobacteria. In this case, the endosymbiotic relationship enabled eukaryotic cells to synthesize greater amounts of ATP. How the relationship would have been beneficial to

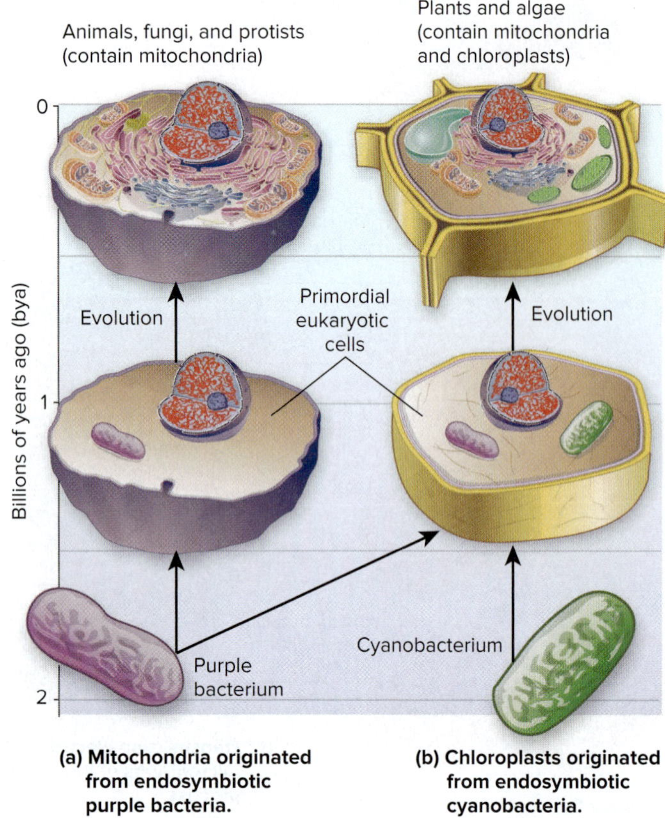


(a) Mitochondria originated from endosymbiotic purple bacteria.

(b) Chloroplasts originated from endosymbiotic cyanobacteria.

**Figure 4.29** **A simplified view of the endosymbiosis theory.** **(a)** According to the endosymbiosis theory, modern mitochondria were derived from purple bacteria, also called alphaproteobacteria. Over the course of evolution, their characteristics changed into those found in mitochondria today. **(b)** A similar phenomenon occurred for chloroplasts, which were derived from cyanobacteria (blue-green bacteria), bacteria that are capable of photosynthesis.

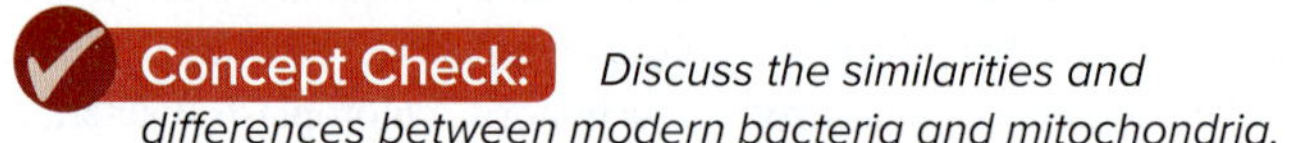
Concept Check: *Discuss the similarities and differences between modern bacteria and mitochondria.*

a cyanobacterium or purple bacterium is less clear, though the cytosol of a eukaryotic cell may have provided a stable environment with an adequate supply of nutrients.

During the evolution of eukaryotic species, many genes that were originally found in the genomes of the primordial purple bacteria and cyanobacteria have been transferred from the organelles to the nucleus. This has occurred many times throughout evolution, so modern mitochondria and chloroplasts have lost most of the genes that still exist in present-day purple bacteria and cyanobacteria. Some researchers speculate that the movement of genes into the nucleus makes it easier for the cell to control the structure, function, and division of mitochondria and chloroplasts. In modern cells, hundreds of different proteins that make up these organelles are encoded by genes that have been transferred to the nucleus. These proteins are made in the cytosol and then taken up into mitochondria or chloroplasts. We will discuss this topic next.

## 4.6 Reviewing the Concepts

- Mitochondria and chloroplasts are considered semiautonomous because they grow and divide, but they still depend on other parts of the cell for their internal components (Figure 4.25).
- Mitochondria produce most of a cell's ATP, which is utilized by many proteins to carry out their functions. Other mitochondrial functions include the synthesis, modification, and breakdown of cellular molecules and the generation of heat in specialized fat cells (Figure 4.26).
- Chloroplasts, which are found in the cells of plants and algae, carry out photosynthesis (Figure 4.27).
- Mitochondria and chloroplasts contain their own genetic material and divide by binary fission (Figure 4.28).
- According to the endosymbiosis theory, mitochondria and chloroplasts originated from bacteria that took up residence in early eukaryotic cells (Figure 4.29).

## 4.6 Testing Your Knowledge

1. Most of the ATP in a eukaryotic cell is made in the
   a. cytosol.
   b. endoplasmic reticulum.
   c. mitochondrion.
   d. chloroplast.
   e. peroxisome.
2. Chloroplasts are found in
   a. plants.
   b. algae.
   c. fungi.
   d. both a and b.
   e. all of the above.

## 4.7 Protein Sorting to Organelles

### Learning Outcome

1. List which categories of proteins are sorted cotranslationally and which are sorted post-translationally.

As we have seen, eukaryotic cells contain a variety of membrane-bound organelles. Each protein that a cell makes usually functions within one cellular compartment or is secreted from the cell. How does each protein reach its appropriate destination? For example, how does a mitochondrial protein get sent to the mitochondrion rather than to a different organelle such as a lysosome? In eukaryotes, most proteins contain short stretches of amino acid sequences that direct them to their correct cellular location. These sequences are called **sorting signals.** Each sorting signal is recognized by specific cellular components that facilitate the proper routing of that protein to its correct location.

Most eukaryotic proteins begin their synthesis on ribosomes in the cytosol using mRNA, which contains the information for polypeptide synthesis (**Figure 4.30**). The cytosol provides amino acids, which are used as building blocks to make these proteins during translation. Cytosolic proteins lack any sorting signal, so they remain within the cytosol. By comparison, the synthesis of proteins destined for the ER, Golgi, lysosomes, vacuoles, or secretory

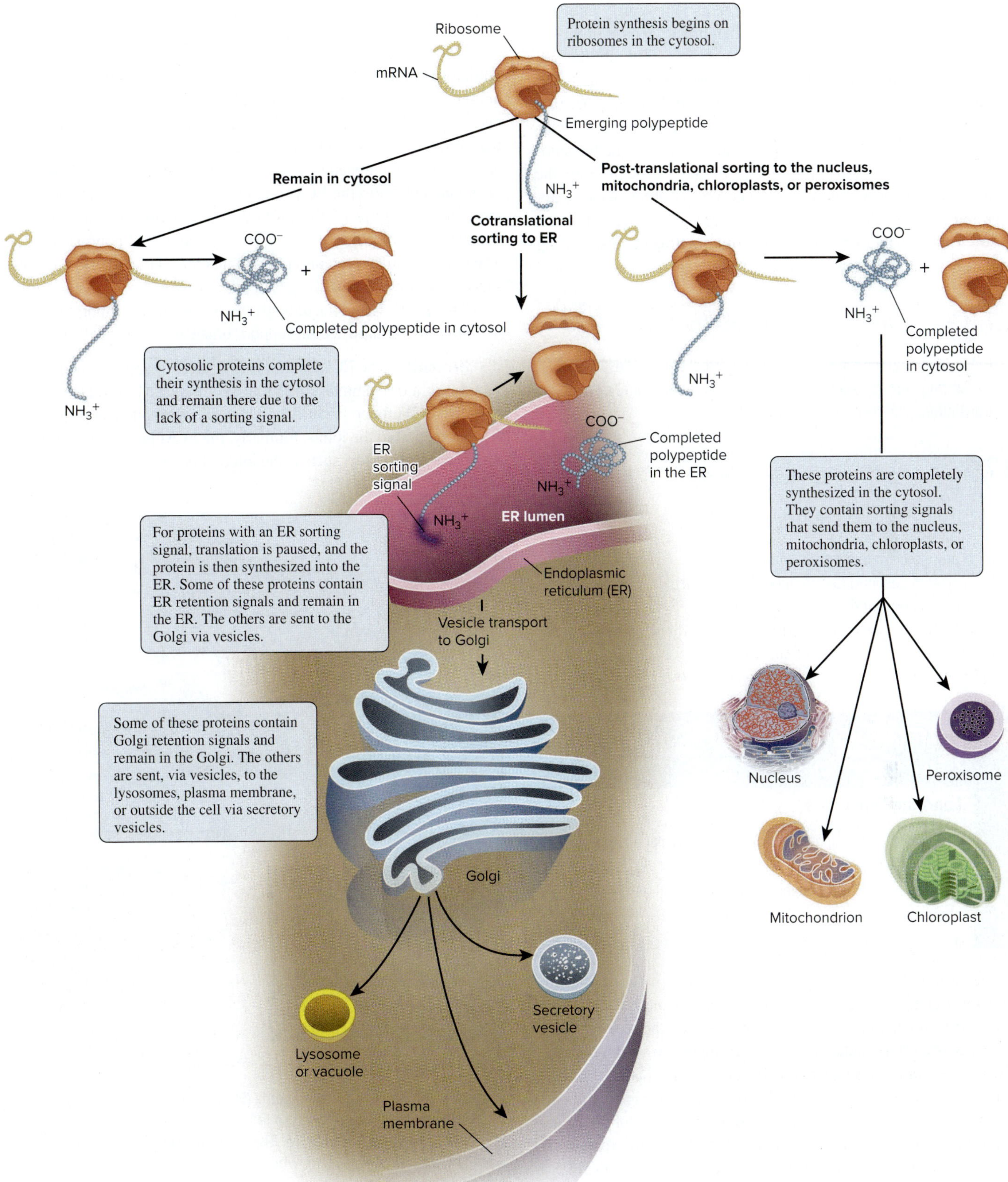


Ribosome

mRNA

Emerging polypeptide

Protein synthesis begins on ribosomes in the cytosol.

Some of these proteins contain Golgi retention signals and remain in the Golgi. The others are sent, via vesicles, to the lysosomes, plasma membrane, or outside the cell via secretory vesicles.

Golgi

Nucleus

Peroxisome

Mitochondrion

Chloroplast

Secretory vesicle

Lysosome or vacuole

Plasma membrane

**Figure 4.30** **Three pathways for protein sorting in a eukaryotic cell.** Proteins remain in the cytosol, are sorted into the ER (cotranslational sorting), or are sorted into other organelles after the protein is completely made (post-translational sorting).

vesicles begins in the cytosol and then halts temporarily until the ribosome has become bound to the ER membrane. After this occurs, translation resumes and the polypeptide is synthesized into the ER. Some proteins remain in the ER, and the rest are sent to the Golgi apparatus. This process is called **cotranslational sorting** because the first step in the sorting process begins while translation is occurring. These proteins eventually move to new locations via the endomembrane system. Finally, the uptake of most proteins into the nucleus, mitochondria, chloroplasts, and peroxisomes occurs after the protein is completely made (that is, completely translated) in the cytosol. This is called **post-translational sorting** because sorting does not happen until translation is finished.

## 4.7 Reviewing the Concepts

- Eukaryotic proteins are sorted to their correct cellular destination via sorting signals. These can occur via cotranslational or post-translational sorting (Figure 4.30).

## 4.7 Testing Your Knowledge

1. Which of the following types of protein is sorted post-translationally?
   a. a protein destined for the cytosol
   b. a protein destined for the endoplasmic reticulum
   c. a protein destined for the lysosome
   d. a protein destined for the mitochondrion
   e. a protein that is secreted from the cell

# 4.8 Extracellular Matrix and Plant Cell Walls

### Learning Outcomes

1. Explain the functional roles of the extracellular matrix (ECM) in animals.
2. Outline the major structural components of the ECM of animals.
3. Describe the structure and function of plant cell walls.

A large portion of an animal or a plant consists of a network of material that is secreted from cells and forms a complex meshwork outside of the plasma membrane. In animals, this is called the **extracellular matrix (ECM),** whereas plant cells are surrounded by a **cell wall.** The ECM and cell walls are a major component of certain parts of animals and plants, respectively. For example, bones and cartilage in animals are composed largely of ECM, and the woody portions of plants are composed mostly of cell walls. Although the cells within wood eventually die, the cell walls they have produced remain and provide a rigid structure that supports the plant for years or even centuries.

In this section, we begin by examining the structure and role of the ECM in animals, focusing on the functions of the major ECM components. We then explore the structure and functions of the cell wall of plant cells.

## The Extracellular Matrix in Animals Supports and Organizes Cells and Plays a Role in Cell Signaling

Unlike the cells of bacteria, archaea, fungi, and plants, the cells of animals are not surrounded by a rigid cell wall that provides structure and support. Instead, animal cells secrete materials that form an ECM that provides support and helps to organize cells. Certain animal cells are completely embedded within an extensive ECM, whereas other cells may adhere to the ECM on only one side. Figure 4.31 illustrates the general features of the ECM and its relationship to cells. The major macromolecules of the ECM are proteins and polysaccharides. The most abundant proteins are those that form large fibers. The polysaccharides give the ECM a gel-like character.

The ECM found in animals performs many important roles, including strength, structural support, organization, and cell signaling.

- **Strength:** The ECM is the "tough stuff" of animals' bodies. In the skin of mammals, the strength of the ECM prevents tearing. The ECM found in cartilage resists compression and provides protection to the joints. Similarly, the ECM protects the soft parts of the body, such as the internal organs.

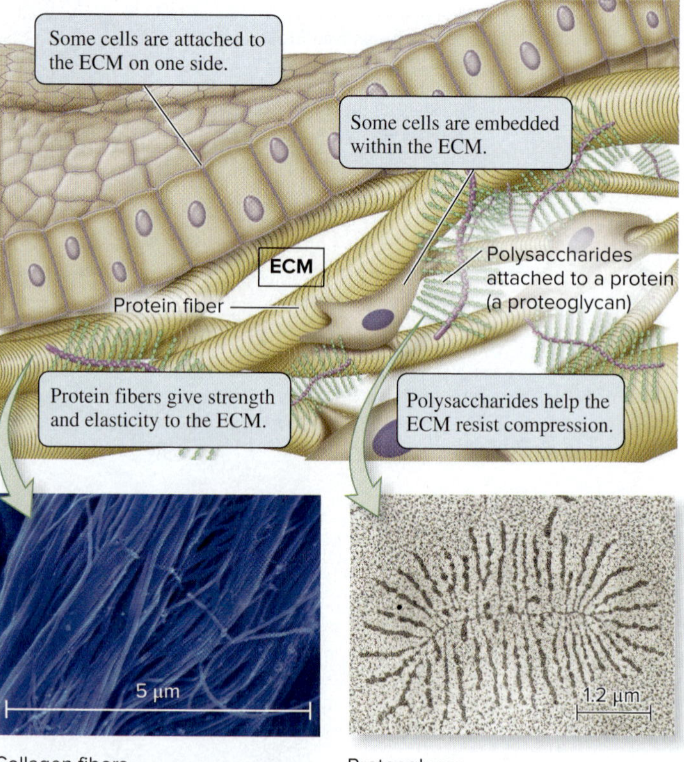


**Figure 4.31** **The extracellular matrix (ECM) of animal cells.** The micrograph (SEM) at the bottom left shows collagen fibers, a type of protein fiber found in the ECM. The micrograph (TEM) at the bottom right shows a proteoglycan, which consists of polysaccharides attached to a protein.

*(left)* © Dennis Kunkel Microscopy, Inc./Phototake; *(right)* Courtesy of Dr. Joseph Buckwalter/University of Iowa

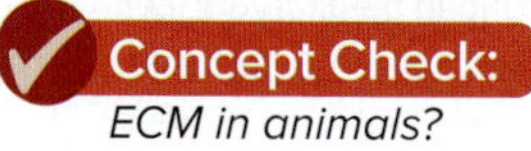

**Concept Check:** *What are the four functions of the ECM in animals?*

- **Structural support:** The bones of many animals are composed primarily of ECM. Skeletons not only provide structural support but also facilitate movement via the functioning of attached muscles.
- **Organization:** The attachment of cells to the ECM plays a key role in the proper arrangement of cells throughout the body. In addition, the ECM binds many body parts together, such as tendons to bones.
- **Cell signaling:** A less obvious role of the ECM is cell signaling. One way that cells in multicellular organisms sense their environment is via changes in the ECM.

## Adhesive and Structural Proteins Are Major Components of the ECM of Animals

In the 1850s, German biologist Rudolf Virchow suggested that all extracellular materials are made and secreted by cells. Around the same time, biologists realized that gelatin and glue, which are produced by the boiling of animal tissues, contain a common fibrous substance. This substance was named **collagen** (from the Greek, meaning glue-producing). Since that time, experimental techniques in biochemistry, microscopy, and biophysics have enabled scientists to probe the structure of the ECM. We now understand that the ECM contains a mixture of several different components, including proteins such as collagen.

The proteins found in the ECM are grouped into adhesive proteins, such as fibronectin and laminin, and structural proteins, such as collagen and elastin (**Table 4.2**).

**Adhesive Proteins** Fibronectin and laminin have multiple binding sites that bind to other components in the ECM, such as protein fibers and polysaccharides. These same proteins also have binding sites for receptors on the surfaces of cells. Therefore, adhesive proteins are so named because they adhere ECM components together and to the cell surface. They provide organization to the ECM and facilitate the attachment of cells to the ECM.

**Structural Proteins and Tensile Strength** Structural proteins, such as collagen, form large fibers in the ECM (see Figure 4.31). A key function of collagen is to impart tensile strength, which is a measure of how much stretching force a material can bear without tearing apart. Collagen provides high tensile strength to many parts of an animal's body. It is the main protein found in bones, cartilage, tendons, skin, and the lining of blood vessels and internal organs. In mammals, more than 25% of the total protein mass consists of collagen, much more than any other protein. Approximately 75% of the protein in mammalian skin is composed of collagen. Leather is largely a pickled and tanned form of collagen.

**Table 4.2 Proteins in the ECM of Animals**

| General type | Example | Function |
|---|---|---|
| **Adhesive** | Fibronectin | Connects cells to the ECM and helps to organize components in the ECM |
| | Laminin | Connects cells to the ECM |
| **Structural** | Collagen | Forms large fibers and interconnected fibrous networks in the ECM; provides tensile strength |
| | Elastin | Forms elastic fibers in the ECM that can stretch and recoil |

## Biology Principle

### Structure Determines Function

Two structural features—the tendency of the elastin protein to adopt a compact conformation and the covalent crosslinking of multiple elastin proteins—result in fibers with the characteristic of elasticity.

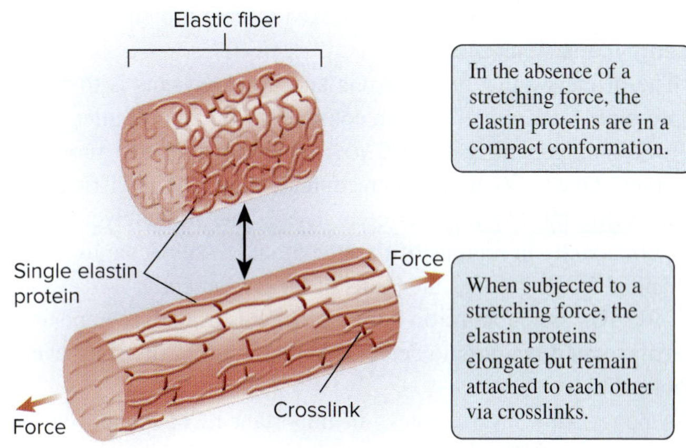

Elastic fiber
In the absence of a stretching force, the elastin proteins are in a compact conformation.
Single elastin protein
Force
When subjected to a stretching force, the elastin proteins elongate but remain attached to each other via crosslinks.
Crosslink
Force


**Figure 4.32** **Structure and function of elastic fibers.** Elastic fibers are made of elastin, one type of structural protein found in the ECM surrounding animal cells.

**Concept Check:** *Suppose you started with an unstretched elastic fiber and treated it with a chemical that breaks the crosslinks between adjacent elastin proteins. What would happen when the fiber was stretched?*

**Structural Proteins and Elasticity** Elasticity is needed in regions of the body such as the lungs and blood vessels, which regularly expand and return to their original shape. In these places, the ECM contains elastic fibers composed primarily of the protein **elastin** (**Figure 4.32**). Elastin proteins form many covalent crosslinks to make a fiber with remarkable elastic properties. In the absence of a stretching force, each protein tends to adopt a compact conformation. When subjected to a stretching force, however, the compact proteins elongate, with the covalent crosslinks holding the fiber together. When the stretching force has ended, the proteins naturally return to their compact conformation. In this way, elastic fibers behave much like a rubber band, stretching under tension and snapping back when the tension is released.

## Animal Cells Also Secrete Polysaccharides into the ECM

Polysaccharides are the second major component of the ECM of animals. As discussed in Chapter 3, polysaccharides are polymers of sugars. Among vertebrates, the most abundant types of polysaccharides in the ECM are **glycosaminoglycans (GAGs).** These macromolecules are long, unbranched polysaccharides containing a repeating disaccharide unit (Figure 4.33a). GAGs are highly negatively charged molecules that tend to attract positively charged ions and water. The majority of GAGs in the ECM are linked to core proteins, forming **proteoglycans** (Figure 4.33b).

The primary function of GAGs and proteoglycans is to resist compression. Once secreted from cells, these macromolecules form a gel-like component in the ECM. How is this gel-like property important? Due to its high water content, the ECM is difficult to compress and thereby protects cells. GAGs and proteoglycans are found abundantly in regions of the body that are subjected to harsh mechanical forces, such as the joints of the human body. Two examples of GAGs are chondroitin sulfate, which is a major component of cartilage, and hyaluronic acid, which is found in the skin, eyes, and joint fluid.

Among many invertebrates, an important ECM component is **chitin,** a nitrogen-containing polysaccharide. Chitin forms the hard, protective outer covering (called an exoskeleton) of insects, such as crickets and grasshoppers, and of crustaceans, such as lobsters and shrimp. As these animals grow, they must periodically shed this rigid outer layer and secrete a new, larger one—a process called molting (look ahead to Figure 26.8).

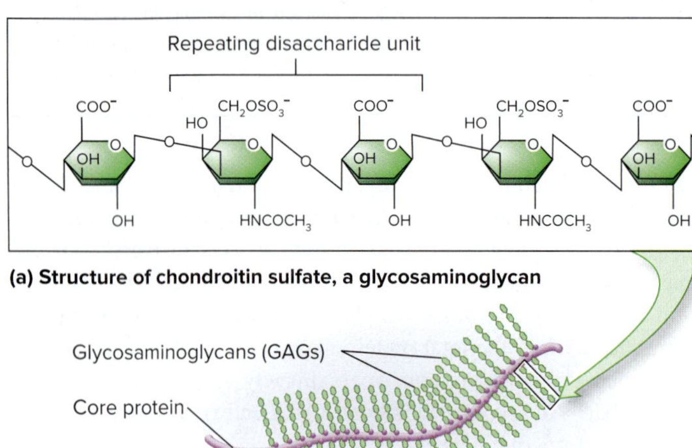


**(a) Structure of chondroitin sulfate, a glycosaminoglycan**

**(b) General structure of a proteoglycan**

**Figure 4.33 Structures of glycosaminoglycans and proteoglycans.** These macromolecules are found in the ECM, which is located outside of animal cells. **(a)** Glycosaminoglycans (GAGs) are composed of repeating disaccharide units. They range in length from several dozen to 25,000 disaccharide units. The GAG shown here is chondroitin sulfate, which is a component of cartilage. **(b)** Proteoglycans are composed of a long, linear core protein with many GAGs attached. Note that each GAG is typically 80 disaccharide units long, but only a short chain of sugars is shown in this illustration.

**Concept Check:** *What structural feature of GAGs and proteoglycans give them a gel-like character?*

## The Cell Wall of Plants Provides Strength and Resistance to Compression

Let's now turn our attention to the cell walls of plants. Plant cells are surrounded by a cell wall, a protective layer that forms outside of the plasma membrane. Like animal cells, the cells of plants are surrounded by material that provides tensile strength and resistance to compression. The cell walls of plants, however, are usually thicker, stronger, and more rigid than the ECM found in animals. Plant cell walls provide rigidity for mechanical support and play a role in the maintenance of cell shape and the direction of cell growth. As discussed in Chapter 5, the cell wall also inhibits expansion when water enters the cell, thereby preventing the cell from bursting.

The cell walls of plants are composed of a primary cell wall and a secondary cell wall (Figure 4.34). These walls are named according to the timing of their synthesis—the primary cell wall is made before the secondary cell wall.

**Primary Cell Wall** During cell division, the **primary cell wall** develops between two newly made daughter cells. It is usually very flexible and allows new cells to increase in size. The following are the main components of the primary cell wall.

- **Cellulose:** The main macromolecule of the plant cell wall is **cellulose,** a polysaccharide made of repeating molecules of glucose attached end to end. These glucose polymers associate with each other via hydrogen bonding to form microfibrils that provide great tensile strength (Figure 4.35). The mechanism of cellulose synthesis is described in Chapter 28 (look ahead to Figure 28.11).
- **Hemicellulose:** Hemicellulose (see Figure 4.34) is another linear polysaccharide, with a structure similar to that of cellulose, but it contains sugars other than glucose in its structure and usually forms thinner microfibrils.
- **Glycans:** Polysaccharides with branching structures are also important in cell-wall structure. The crosslinking glycans bind to cellulose and provide organization to the cellulose microfibrils.
- **Pectins:** Highly negatively charged polysaccharides, such as pectins, attract water and have a gel-like character. They provide the cell wall with the ability to resist compression.

**Secondary Cell Wall** The **secondary cell wall** is synthesized and deposited between the plasma membrane and the primary cell wall (see Figure 4.34) after a plant cell matures and has stopped increasing in size. It is made in layers by the successive deposition of cellulose microfibrils and other components. Whereas the primary wall structure is relatively similar in nearly all cell types and plant species, the structure of the secondary cell wall is more variable. Some plant cells

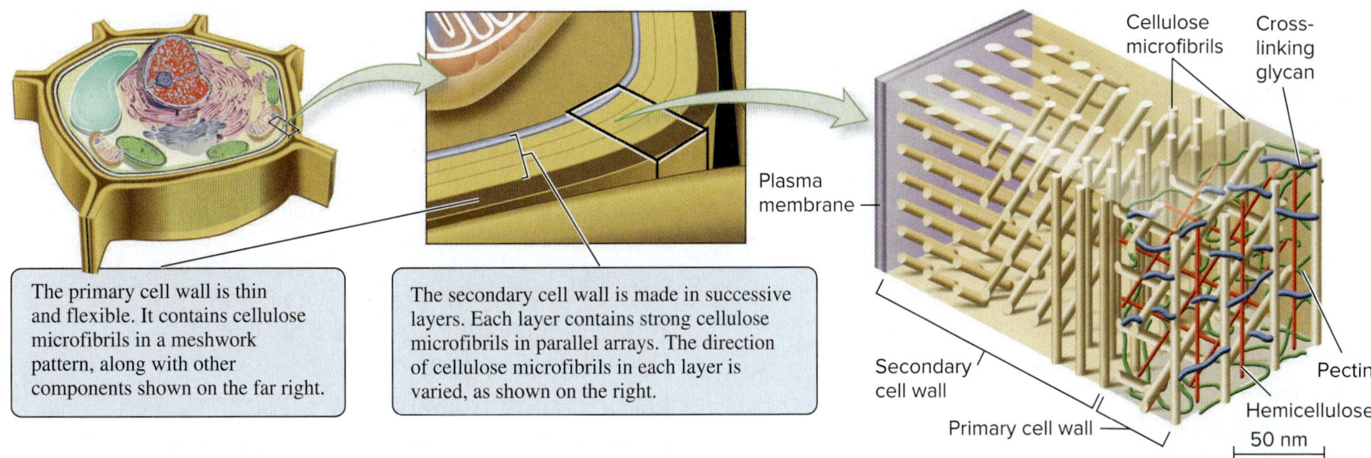


**Figure 4.34 Structure of the cell wall of plant cells.** The primary cell wall is relatively thin and flexible. It contains cellulose (tan), hemicellulose (red), crosslinking glycans (blue), and pectin (green). The secondary cell wall, which is produced only by certain plant cells, is made after the primary cell wall and is synthesized in successive layers.

*With regard to cell growth, what would happen if the secondary cell wall was made too soon?*

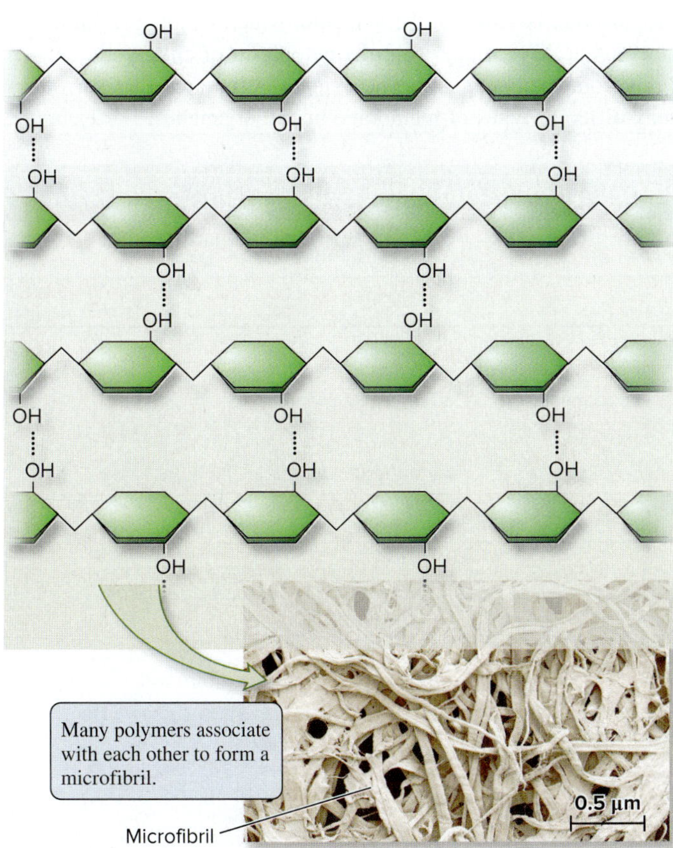


**Figure 4.35 Structure of cellulose, the main macromolecule of the primary cell wall.** Cellulose is made of repeating glucose units linked end to end that hydrogen-bond to each other to form microfibrils (SEM).



have no secondary cell wall; leaf cells that are involved in photosynthesis lack a secondary wall, allowing light to enter the cells more readily. In contrast, other cells have a thick secondary cell wall that contains components that are not found in the primary cell wall. For example, phenolic compounds called lignins, which are found in the woody parts of plants, are very hard and impart considerable strength to the secondary wall structure.

## 4.8 Reviewing the Concepts

- The extracellular matrix (ECM) is a network of material that forms a complex meshwork outside of animal cells; the cell wall is a similar component of plant cells.
- Proteins and polysaccharides are the major constituents of the ECM in animals. These materials are involved in strength, structural support, organization, and cell signaling (Figure 4.31).
- Adhesive proteins, such as fibronectin and laminin, help cells adhere to the ECM. Structural proteins include collagen fibers, which provide tensile strength, and elastic fibers, which allow regions of the body to stretch (Table 4.2, Figure 4.32).
- Glycosaminoglycans (GAGs) are polysaccharides of repeating disaccharide units that give a gel-like character to the ECM of animals. Proteoglycans consist of a long, linear core protein with attached GAGs (Figure 4.33).
- Plant cells are surrounded by a cell wall composed of primary and secondary cell walls. The primary cell wall is composed largely of cellulose. The secondary cell wall is synthesized and deposited between the plasma membrane and the primary cell wall and is often thick and rigid (Figures 4.34, 4.35).

## 4.8 Testing Your Knowledge

1. The function of the extracellular matrix (ECM) in animals is
   a. to provide strength.
   b. to provide structural support.
   c. to organize cells and other body parts.
   d. cell signaling.
   e. all of the above.
2. The protein found in the ECM of animals that provides tensile strength and resistance to tearing when stretched is
   a. elastin. c. collagen. e. fibronectin.
   b. cellulose. d. laminin.
3. A major component of the plant cell wall is
   a. elastin. c. collagen. e. fibronectin.
   b. cellulose. d. laminin.

## 4.9 Systems Biology of Cells: A Summary

### Learning Outcomes

1. Outline the differences in complexity among bacteria, animal cells, and plant cells.
2. Describe how a eukaryotic cell can be viewed as four interacting systems: the nucleus, cytosol, endomembrane system, and semiautonomous organelles.

We will conclude this chapter by reviewing cell structure and function from a perspective called **systems biology,** the study of how new properties of life arise by complex interactions of its components. The system being studied can be anything from a metabolic pathway to a cell, an organ, or even an entire organism. In this section, we view the cell as a system. First, we compare prokaryotic and eukaryotic cells as systems, then examine the four interconnected parts that make up the system that is the eukaryotic cell.

### Bacterial Cells Are Relatively Simple Systems, Whereas Eukaryotic Cells Are a System of Four Interacting Parts

Bacterial cells are relatively small and lack the extensive internal compartmentalization characteristic of eukaryotic cells (**Table 4.3**). On the outside, bacterial cells are surrounded by a cell wall, and many species have flagella. Animal cells lack a cell wall, and only certain cell types have flagella or cilia. Like bacteria, plant cells also have cell walls, but their chemical composition is different from that of bacterial cells. Plant cells rarely have flagella.

As mentioned earlier in this chapter, the cytoplasm is the region of the cell enclosed by the plasma membrane. Ribosomes are found in the cytoplasm of all cell types. In bacteria, the cytoplasm is a single compartment. The bacterial genetic material, usually a single chromosome, is found in the nucleoid, which is not surrounded by a membrane. By comparison, the cytoplasm of eukaryotic cells is highly compartmentalized. The cytosol is the area that surrounds many different types of membrane-bound organelles. For example,

**Table 4.3 A Comparison of Cell Complexity Among Bacterial, Animal, and Plant Cells**

| Structures | Bacteria | Animal cells | Plant cells |
|---|---|---|---|
| **Structures on the cell surface** | | | |
| Cell wall* | Present | Absent | Present |
| Flagella/cilia | Flagella sometimes present | Cilia or flagella present on certain cell types | Rarely present† |
| Plasma membrane | Present | Present | Present |
| **Interior structures** | | | |
| Cytoplasm | Usually a single compartment inside the plasma membrane | Composed of membrane-bound organelles that are surrounded by the cytosol | Composed of membrane-bound organelles that are surrounded by the cytosol |
| Ribosomes | Present | Present | Present |
| Chromosomes and their location | Typically one circular chromosome per nucleoid; a nucleoid is not a separate compartment | Multiple linear chromosomes in the nucleus; nucleus is surrounded by a double membrane; mitochondria also have chromosomes | Multiple linear chromosomes in the nucleus; nucleus is surrounded by a double membrane; mitochondria and chloroplasts also have chromosomes |
| Endomembrane system | Absent | Present | Present |
| Mitochondria | Absent | Present | Present |
| Chloroplasts | Absent | Absent | Present |

*Note that the biochemical composition of bacterial cell walls is very different from that of plant cell walls.
†Some plant species produce sperm cells with flagella, but flowering plants produce sperm within pollen grains that lack flagella.

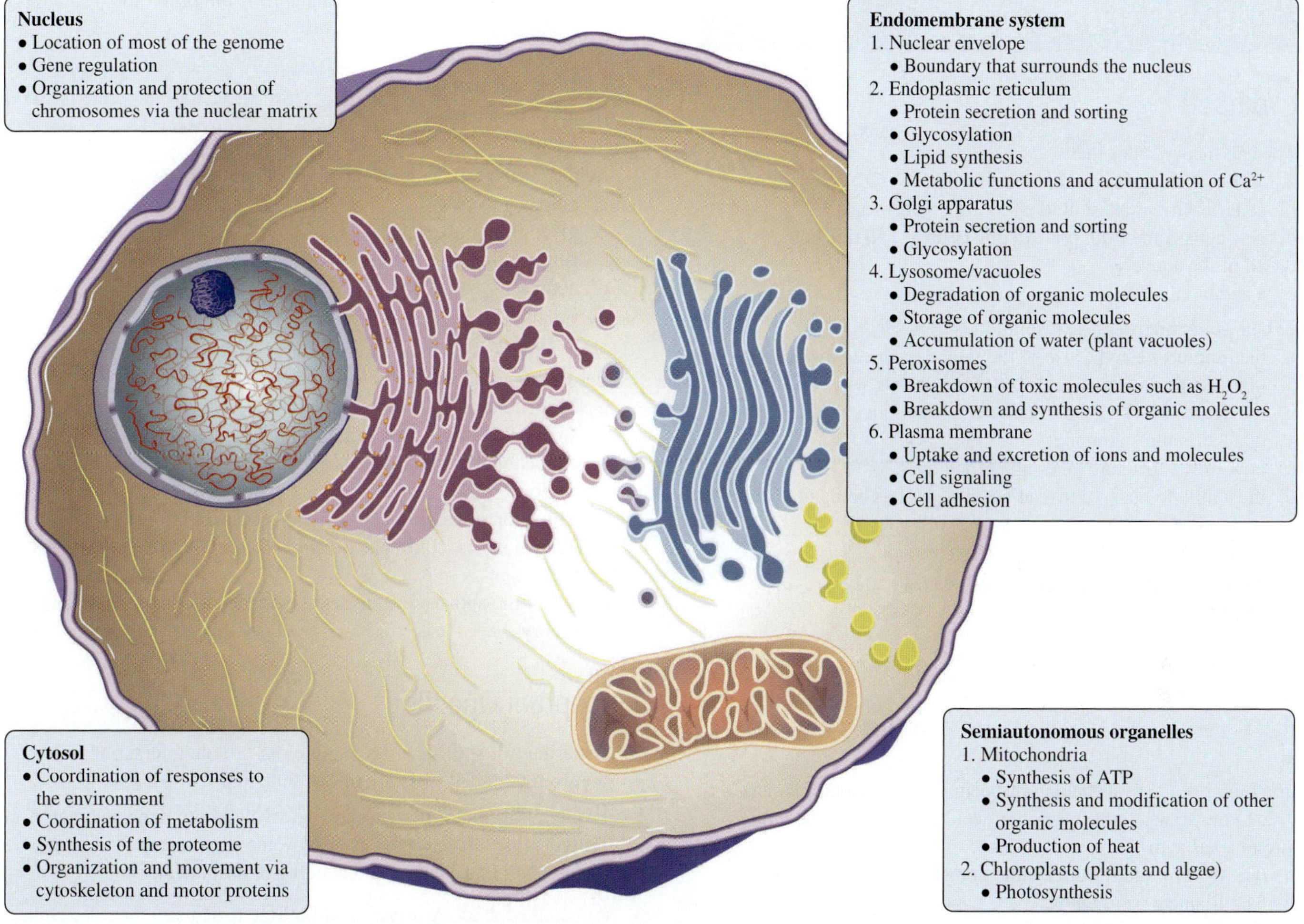

Nucleus
• Location of most of the genome
• Gene regulation
• Organization and protection of chromosomes via the nuclear matrix
Endomembrane system
1. Nuclear envelope
• Boundary that surrounds the nucleus
2. Endoplasmic reticulum
• Protein secretion and sorting
• Glycosylation
• Lipid synthesis
• Metabolic functions and accumulation of $Ca^{2+}$
3. Golgi apparatus
• Protein secretion and sorting
• Glycosylation
4. Lysosome/vacuoles
• Degradation of organic molecules
• Storage of organic molecules
• Accumulation of water (plant vacuoles)
5. Peroxisomes
• Breakdown of toxic molecules such as $H_2O_2$
• Breakdown and synthesis of organic molecules
6. Plasma membrane
• Uptake and excretion of ions and molecules
• Cell signaling
• Cell adhesion
Cytosol
• Coordination of responses to the environment
• Coordination of metabolism
• Synthesis of the proteome
• Organization and movement via cytoskeleton and motor proteins
Semiautonomous organelles
1. Mitochondria
• Synthesis of ATP
• Synthesis and modification of other organic molecules
• Production of heat
2. Chloroplasts (plants and algae)
• Photosynthesis


**Figure 4.36 The four interacting parts of eukaryotic cells: Nucleus, cytosol, endomembrane system, and semiautonomous organelles.** An animal cell is illustrated here.

eukaryotic chromosomes are found in the nucleus, which is surrounded by a double membrane. In addition, all eukaryotic cells have an endomembrane system and mitochondria, and plant cells also have chloroplasts.

We can view a eukaryotic cell as a system with four interacting parts: the interior of the nucleus, the cytosol, the endomembrane system, and the semiautonomous organelles (**Figure 4.36**). These four regions have their own structure and organization, while also playing a role in the structure and organization of the entire cell.

## 4.9 Reviewing the Concepts

- Systems biology is the study of how new properties of life arise by complex interactions of its components. In systems biology, the cell is viewed in terms of its structural and functional connections, rather than its individual molecular components.
- Prokaryotic and eukaryotic cells differ in their levels of organization. In eukaryotic cells, four parts—the nucleus, cytosol, endomembrane system, and semiautonomous organelles—work together to produce dynamic organization (Table 4.3, Figure 4.36).

## 4.9 Testing Your Knowledge

1. Which of the following could be found in a eukaryotic cell but not a bacterium?
   **a.** cytoplasm
   **b.** cell wall
   **c.** lysosome
   **d.** plasma membrane
   **e.** genetic material (DNA)
2. The region of a eukaryotic cell that coordinates metabolism and plays a major role in cell movements is
   **a.** the nucleus.
   **b.** the cytosol.
   **c.** the endomembrane system.
   **d.** the semiautonomous organelles.
   **e.** both a and c.

# Assess and Discuss

## Test Yourself

1. The cell theory states that
   a. all living things are composed of cells.
   b. cells are the smallest units of living organisms.
   c. new cells come from pre-existing cells by cell division.
   d. all of the above.
   e. a and b only.
2. When using microscopes, resolution is
   a. the ratio between the size of the image produced by the microscope and the actual size of the object.
   b. the degree to which a particular structure looks different from other structures around it.
   c. how well a structure takes up certain dyes.
   d. the ability to observe two adjacent objects as being distinct from each other.
   e. the degree to which the image is magnified.
3. A spherical cell has a radius of 34 μm. What is its surface area/volume ratio?
   a. 0.088
   b. 0.12
   c. 11.3
   d. 55.7
   e. 127
4. If a motor protein was held in place and a cytoskeletal filament was free to move, what type of motion would occur when the motor protein was active?
   a. The motor protein would walk along the filament.
   b. The filament would move.
   c. The filament would bend.
   d. All of the above would happen.
   e. Only b and c would happen.
5. Each of the following is part of the endomembrane system except
   a. the nuclear envelope.
   b. the endoplasmic reticulum.
   c. the Golgi apparatus.
   d. lysosomes.
   e. mitochondria.
6. Which of the following is *not* a feature of the nucleus?
   a. bounded by a double membrane
   b. nuclear envelope contains nuclear pores
   c. contains a nucleolus
   d. contains microtubules
   e. has chromosomes located in chromosome territories
7. Functions of the smooth endoplasmic reticulum include
   a. detoxification of harmful organic molecules.
   b. metabolism of carbohydrates.
   c. protein sorting.
   d. all of the above.
   e. a and b only.
8. The central vacuole in many plant cells is important for
   a. storage.
   b. photosynthesis.
   c. structural support.
   d. all of the above.
   e. a and c only.
9. Which of the following would *not* be found outside of animal cells?
   a. collagen
   b. glycosaminoglycans
   c. fibronectin and laminin
   d. actin
   e. elastin
10. Which of the following observations would *not* be considered evidence for the endosymbiosis theory?
    a. Mitochondria and chloroplasts have genomes that resemble smaller versions of bacterial genomes.
    b. Mitochondria, chloroplasts, and bacteria all divide by binary fission.
    c. Mitochondria, chloroplasts, and bacteria all have ribosomes.
    d. Mitochondria, chloroplasts, and bacteria all have similar sizes and shapes.
    e. All of the above are considered evidence for the endosymbiosis theory.

## Conceptual Questions

1. Describe two specific ways that protein-protein interactions are involved with cell structure or cell function.
2. Explain how motor proteins and cytoskeletal filaments interact to promote three different types of movements: movement of cargo, movement of a filament, and bending of a filament.
3. **PRINCIPLES** A principle of biology is that structure determines function. Explain how the invaginations of the inner mitochondrial membrane are related to mitochondrial function.

## Collaborative Questions

1. Discuss the roles of the genome and proteome in determining cell structure and function.
2. Discuss and draw the structural relationship among the nucleus, the rough endoplasmic reticulum, and the Golgi apparatus.

# Online Resource

# Membranes: The Interface Between Cells and Their Environment

# 5

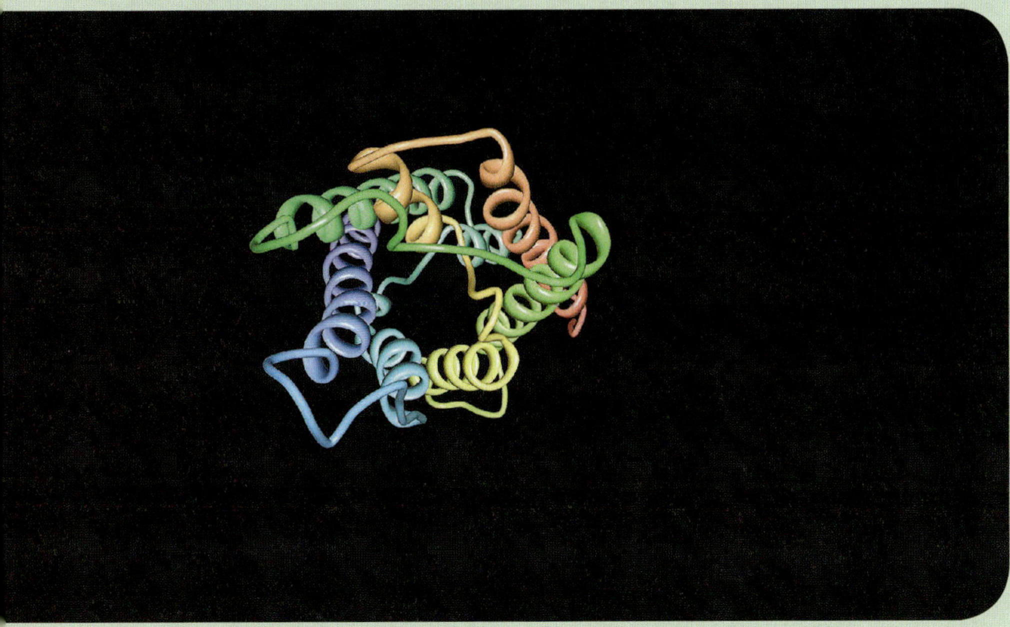

Image from the RCSB PDB (www.pdb.org) of PDB ID 1H6I (B.L. De Groot, A. Engel, H. Grubmuller, (2001) *FEBS Lett*, 504:206)

**A model for the structure of aquaporin.** This protein, found in the plasma membrane of many cell types, such as red blood cells and plant cells, forms a channel that allows the rapid movement of water molecules across the membrane. The central pore allows water molecules to cross the membrane.

## Chapter Outline



A membrane called the **plasma membrane** separates the internal contents of a cell from its external environment. With such a role, you might imagine that the plasma membrane would be thick and rigid. Remarkably, the opposite is true. All cell membranes, including the plasma membrane, are thin (typically 5–10 nm) and even somewhat fluid. It would take 5,000–10,000 membranes stacked on top of each other to equal the thickness of a piece of paper! Despite their thinness, membranes are impressively dynamic structures that effectively maintain the separation between a cell and its surroundings.

All cells have a plasma membrane that encloses the cytoplasm, and eukaryotic cells have internal membranes that form organelles (see Chapter 4). Both types are also called **biological membranes.** The plasma membrane and internal membranes provide an interface to carry out many vital cellular activities (**Table 5.1**). In this chapter, we will begin by considering the components that provide the structure and fluid properties of membranes and then explore how they are synthesized. Next, we will examine one of a membrane's primary functions—membrane transport. Biological membranes regulate the traffic of substances into and out of the cell and its organelles. This occurs via simple diffusion, transport proteins, intercellular channels, exocytosis, and endocytosis. Finally, we will consider how adjacent cells are held together by cell junctions.

**Table 5.1 Important Functions of Biological Membranes**

| Function |
|---|
| Selective uptake and export of ions and molecules |
| Cell compartmentalization |
| Protein sorting |
| Anchoring of the cytoskeleton |
| Production of energy intermediates such as ATP |
| Cell signaling |
| Cell and nuclear division |
| Adhesion of cells to each other and to the ECM |

## 5.1 Membrane Structure

### Learning Outcomes

1. Describe the fluid-mosaic model of membrane structure.
2. Identify the three different types of membrane proteins.

The two primary components of membranes are phospholipids, which form the basic matrix of a membrane, and proteins, which are embedded in the membrane or loosely attached to its surface. A third component is carbohydrates, which may be attached to membrane lipids and proteins. In this section, we will be mainly concerned with the organization of these components to form a biological membrane and how they are important in the overall function of membranes.

### Biological Membranes Are a Mosaic of Lipids, Proteins, and Carbohydrates

Figure 5.1 shows the biochemical organization of a membrane, which is similar in composition among all living organisms. The framework of the membrane is the **phospholipid bilayer,** which consists of two layers of phospholipids. Recall from Chapter 3 that phospholipids are **amphipathic** molecules (refer back to Figure 3.9). They have two hydrophobic (water-fearing) nonpolar tails and a hydrophilic (water-loving) polar head. The nonpolar tails of lipids are found in the interior of the membrane, and the polar heads are on the surface. Biological membranes also contain proteins, and most membranes have carbohydrates attached to lipids and proteins. Overall, the membrane is considered a mosaic of lipid, protein, and carbohydrate molecules. The membrane structure illustrated in Figure 5.1 is referred to as the **fluid-mosaic model,** originally proposed by S. Jonathan Singer and Garth Nicolson in 1972. As discussed later, the membrane exhibits properties that resemble a fluid because lipids and proteins can move relative to each other within the membrane.

Half of a phospholipid bilayer is termed a **leaflet.** Each leaflet faces a different region. For example, the plasma membrane contains a leaflet that faces the cytosol (cytosolic leaflet) and a leaflet that faces the extracellular environment (extracellular leaflet) (see Figure 5.1). With regard to lipid composition, the two leaflets are highly asymmetric, with certain types of lipids more abundant in one leaflet than the other. A striking asymmetry occurs with glycolipids—lipids with carbohydrate attached. These are found primarily in the extracellular leaflet. The carbohydrate portion of a glycolipid is attached to the polar head and protrudes into the extracellular environment. The types of carbohydrates on the surfaces of cells often allow cells to recognize each other, a phenomenon called cell surface recognition.

### Proteins Associate with Membranes in Three Ways

Although the phospholipid bilayer forms the basic foundation of biological membranes, the protein component carries out many key functions. Some of these functions were considered in Chapter 4. Later in this chapter, we will explore how membrane proteins are involved in transporting ions and molecules across membranes and in the adhesion of cells to each other. In later chapters, we will examine how membrane proteins are responsible for other functions, including ATP synthesis (Chapter 6), photosynthesis (Chapter 7), and cell signaling (Chapter 8).

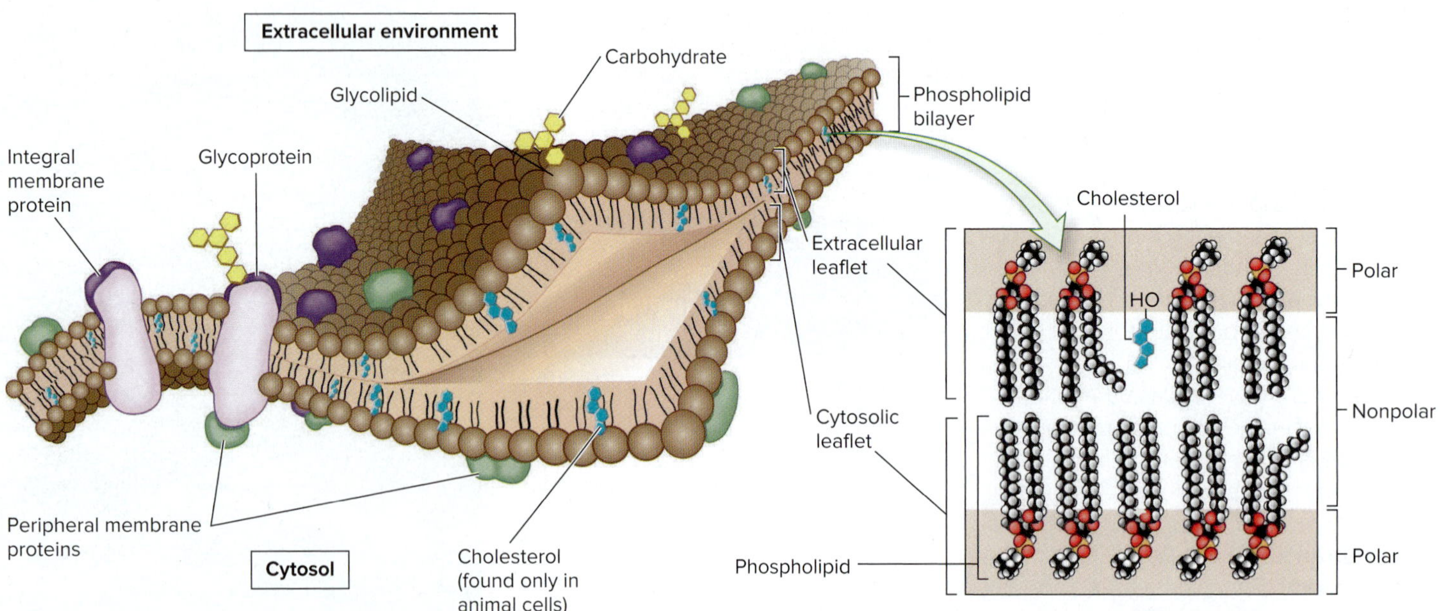


**Figure 5.1 Fluid-mosaic model of membrane structure.** The basic framework of a plasma membrane and other biological membranes is a phospholipid bilayer. Proteins may span the membrane and may be bound on the surface to other proteins or to lipids. Proteins and lipids that have covalently bound carbohydrates are called glycoproteins and glycolipids, respectively. The inset shows nine phospholipids and one cholesterol molecule (blue) in a bilayer, and it emphasizes the two leaflets and the polar and nonpolar regions of the bilayer.

The three types of membrane proteins are transmembrane proteins, lipid-anchored proteins, and peripheral membrane proteins. Transmembrane proteins and lipid-anchored proteins are called **integral membrane proteins,** because a portion of them is integrated into the hydrophobic region of the membrane. Each of the three types has a different way of associating with a membrane (**Figure 5.2**).

1. **Transmembrane proteins** span or traverse the membrane from one leaflet to the other. Transmembrane segments within these proteins have stretches of nonpolar amino acids that are inserted into the hydrophobic interior of the phospholipid bilayer, making it possible for a portion of the protein to traverse the membrane. In most transmembrane proteins, each transmembrane segment is folded into an α helix structure. Such a segment is stable in a membrane because the nonpolar amino acids interact favorably with the nonpolar tails of the lipid molecules.
2. **Lipid-anchored proteins** are proteins that associate with a membrane because they have a lipid molecule that is covalently attached to an amino acid side chain within the protein. The lipid tail is inserted into the hydrophobic portion of the membrane, thereby keeping the protein firmly attached to the membrane.
3. **Peripheral membrane proteins** are proteins that are noncovalently bound to regions of transmembrane membrane proteins that project out from the membrane or to the polar head groups of phospholipids (see Figure 5.1). Peripheral membrane proteins are typically bound to the membrane by hydrogen bonds or ionic bonds (or both).

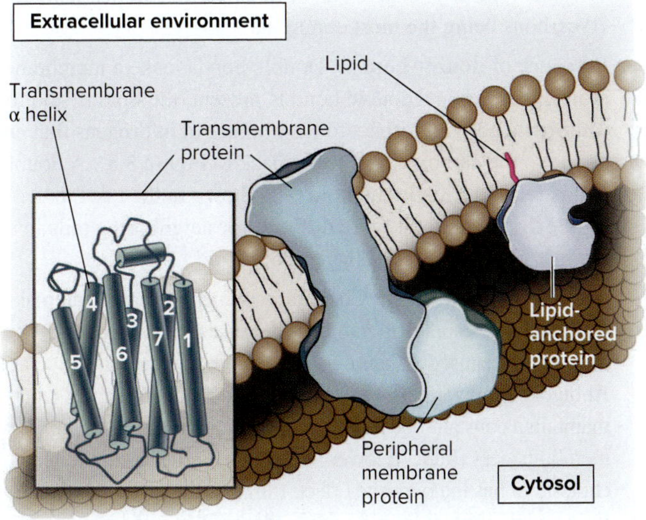


**Figure 5.2** **Types of membrane proteins.** Integral membrane proteins are of two types: transmembrane proteins and lipid-anchored proteins. Peripheral membrane proteins are noncovalently bound to the hydrophilic regions of integral membrane proteins or to the polar head groups of lipids. Inset: The protein bacteriorhodopsin contains seven transmembrane segments, depicted as cylinders, in an α helix structure. Bacteriorhodopsin is found in halophilic (salt-loving) archaea.

## 5.1 Reviewing the Concepts

- A plasma membrane separates a cell from its surroundings. Biological membranes provide interfaces for carrying out vital cellular functions (Table 5.1).
- The accepted model of membranes is the fluid-mosaic model, and its basic framework is the phospholipid bilayer. Biological membranes also contain proteins, and some membranes have attached carbohydrates (Figure 5.1).
- The three main types of membrane proteins are transmembrane proteins, lipid-anchored proteins, and peripheral membrane proteins. Transmembrane proteins and lipid-anchored proteins are classified as integral membrane proteins (Figure 5.2).

## 5.1 Testing Your Knowledge

1. Which of the following is *not* a characteristic of a biological membrane?
   a. contains a bilayer of phospholipids
   b. contains proteins that are inserted into the membrane
   c. contains a high percentage of water molecules
   d. has carbohydrates attached to lipids and proteins
   e. has asymmetric leaflets
2. A membrane protein that is noncovalently attached to a protein or a lipid is
   a. a transmembrane protein.
   b. a lipid-anchored protein.
   c. a peripheral membrane protein.
   d. a glycoprotein.
   e. all of the above.

# 5.2 Fluidity of Membranes

### Learning Outcomes

1. Describe the fluidity of membranes.
2. **SCISKILLS ▶** Predict how fluidity will be affected by changes in lipid composition.
3. **SCISKILLS ▶** Analyze the results of experiments indicating that certain membrane proteins can diffuse laterally within the membrane.

Let's now turn our attention to the dynamic properties of membranes. Although a membrane provides a critical interface between a cell and its environment, it is not a solid, rigid structure. Rather, biological membranes exhibit properties of **fluidity,** which means that individual molecules remain in close association yet have the ability to readily move within the membrane. In this section, we will examine the fluid properties of biological membranes.

## Membranes Are Semifluid

Though membranes are often described as fluid, it is more appropriate to say they are **semifluid,** because movements of lipids and membrane proteins occur in only two dimensions. In a fluid

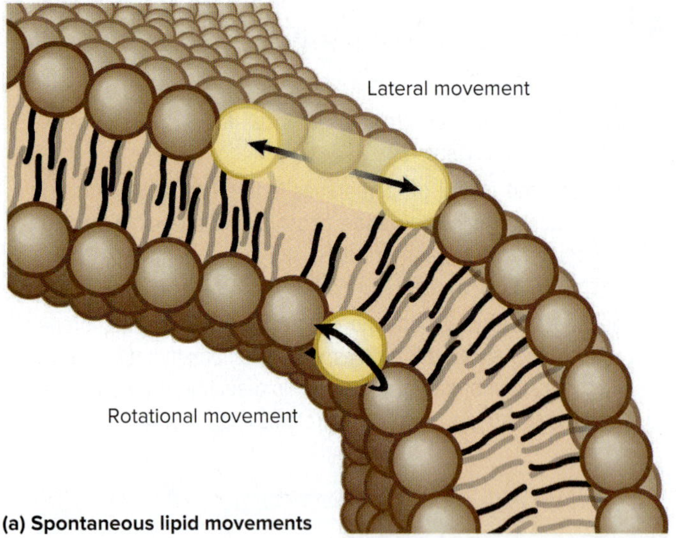

(a) Spontaneous lipid movements

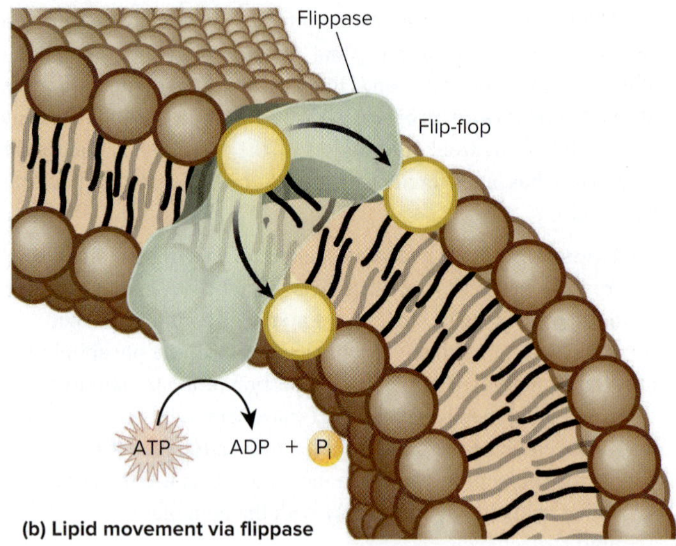

(b) Lipid movement via flippase

**Figure 5.3** Semifluidity of the lipid bilayer. **(a)** Spontaneous movements in the bilayer. Lipids can rotate (that is, move 360°) and move laterally (for example, from left to right in the plane of the bilayer). **(b)** Flip-flop does not happen spontaneously, because the polar head group would have to pass through the hydrophobic region of the bilayer. Instead, the enzyme flippase uses ATP to flip phospholipids from one leaflet to the other.

*In an animal cell, how can changes in lipid composition affect membrane fluidity?*

substance, molecules can move in three dimensions. By comparison, most lipids can rotate freely around their long axes and move laterally within the membrane leaflet (Figure 5.3a). This type of motion is considered two-dimensional, which means it occurs within the plane of the membrane. Because rotational and lateral movements keep the nonpolar tails within the hydrophobic interior, such movements are energetically favorable. At 37°C, a typical lipid molecule exchanges places with its neighbors about $10^7$ times per second, and it can move several micrometers per second. At this rate, a lipid can traverse the length of a bacterial cell (approximately 1 μm) in only 1 second and the length of a typical animal cell in 10 to 20 seconds.

In contrast to rotational and lateral movements, the "flip-flop" of lipids from one leaflet to the opposite leaflet does not occur spontaneously. Flip-flop is energetically unfavorable because the polar head of a phospholipid has to travel through the hydrophobic interior of the membrane. How are lipids moved from one leaflet to the other? The transport of lipids between leaflets requires the action of the enzyme flippase, which requires energy input in the form of ATP (Figure 5.3b).

Although most lipids diffuse rotationally and laterally within the plane of the lipid bilayer, researchers have discovered that certain types of lipids in animal cells tend to strongly associate with each other to form structures called lipid rafts. As the word raft suggests, a **lipid raft** is a group of lipids that float together as a unit within a larger sea of lipids. Lipid rafts have a lipid composition that differs from the surrounding membrane. For example, they usually have a high amount of cholesterol. In addition, lipid rafts may contain unique sets of lipid-anchored proteins and transmembrane proteins. The functional importance of lipid rafts is the subject of a large amount of current research. Lipid rafts may play an important role in endocytosis (discussed later in this chapter) and cell signaling.

## Lipid Composition Affects Membrane Fluidity

The biochemical properties of phospholipids affect the fluidity of the phospholipid bilayer. These include the following:

- **Length of the nonpolar tails:** Shorter tails are less likely to interact with each other, which makes the membrane more fluid. The tails typically range from 14 to 24 carbon atoms, with 18 to 20 carbons being the most common.
- **Presence of double bonds:** Double bonds make a membrane more fluid. When a double bond is present, the lipid is said to be **unsaturated** with respect to the number of hydrogens that are bound to the carbon atoms (refer back to Figure 3.8). A double bond creates a kink in the lipid tail (shown in the inset to Figure 5.1), making it more difficult for neighboring tails to interact and making the bilayer more fluid.
- **Presence of cholesterol:** Cholesterol, which is found in animal cells, tends to stabilize membranes because it is a short, rigid molecule (see inset to Figure 5.1). Its effects depend on temperature. At higher temperatures, such as those observed in mammals that maintain a constant body temperature, cholesterol makes the membrane less fluid. At lower temperatures, such as icy water, cholesterol has the opposite effect. It makes the membrane more fluid and prevents it from freezing. Plant cell membranes contain phytosterols that resemble cholesterol in their chemical structure.

An optimal level of bilayer fluidity is essential for normal cell function, growth, and division. If a membrane is too fluid, which may occur at higher temperatures, it can become leaky. However, if a membrane becomes too solid, which may occur at lower temperatures, the functioning of membrane proteins will be inhibited. How can organisms cope with changes in temperature? The cells of many species adapt to changes in temperature by altering the lipid composition of their membranes. For example, when the water temperature drops, the cells of certain fish will incorporate more cholesterol in their membranes, making the membrane more fluid. If a plant cell is exposed to high temperatures for many hours or days, it will alter its lipid composition to have longer tails and fewer double bonds, which will make the membrane less fluid.

## Many Transmembrane Proteins Can Rotate and Move Laterally, but Some Are Restricted in Their Movement

Like lipids, many transmembrane proteins may rotate and laterally move throughout the plane of a membrane. Because transmembrane proteins are larger than lipids, they move within the membrane at a much slower rate. Flip-flop of transmembrane proteins does not occur, because the proteins also contain hydrophilic regions that project out from the phospholipid bilayer, and it would be energetically unfavorable for the hydrophilic regions of membrane proteins to pass through the hydrophobic portion of the phospholipid bilayer.

In 1970, Larry Frye and Michael Edidin conducted an experiment that verified the lateral movement of transmembrane proteins (**Figure 5.4**). Mouse and human cells were mixed together and exposed to agents that caused them to fuse with each other to produce mouse-human cell hybrids. Some cells were cooled to 0°C, while others were incubated at 37°C before being cooled. Both sets of cells were then exposed to fluorescent antibodies that became specifically bound to a mouse transmembrane protein called H-2. The fluorescent label was observed with a fluorescence microscope. If the cells were maintained at 0°C, a temperature that greatly inhibits lateral movement, the fluorescence was seen on only one side of the fused cell. However, if the cells were incubated for several hours at 37°C and then cooled to 0°C, the fluorescence was distributed throughout the plasma membrane of the fused cell. This occurred because the higher temperature allowed the lateral movement of the H-2 protein throughout the fused cell.

Unlike the example shown in Figure 5.4, not all transmembrane proteins are capable of rotational and lateral movement. Depending on the cell type, 10–70% of membrane proteins may be restricted in their movement. Transmembrane proteins may be bound to components of the cytoskeleton, which restricts the proteins from moving (**Figure 5.5**), or may be attached to molecules that are outside the cell, such as the interconnected network of proteins that forms the extracellular matrix (ECM) of animal cells (see Chapter 4, Figure 4.31).

## Biology Principle

### Biology Is an Experimental Science

This experiment verified that membrane proteins can diffuse laterally within the plane of the lipid bilayers.

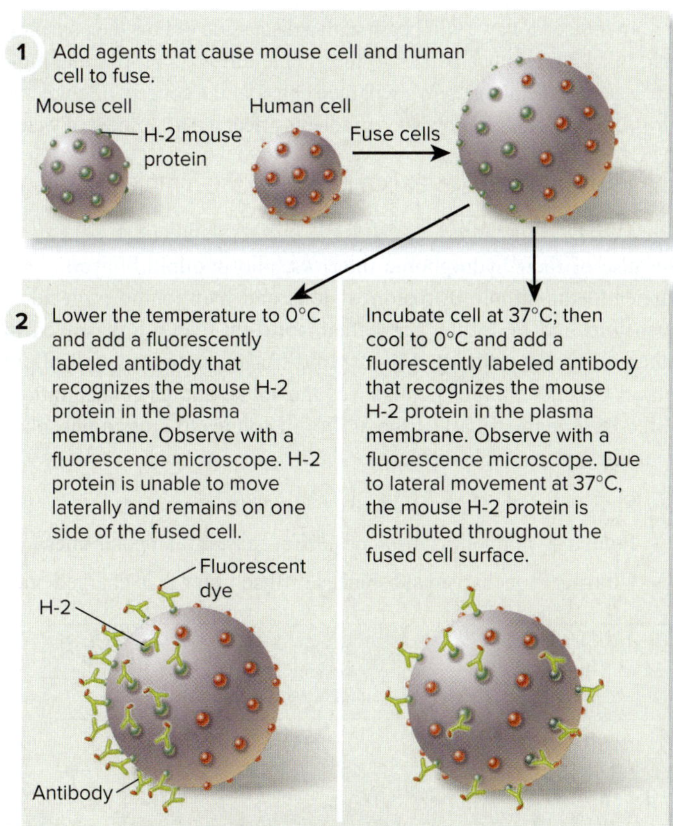


**Figure 5.4 A method to measure the lateral movement of membrane proteins.**

**Concept Check:** *Explain why the H-2 proteins were found on only one side of the cell when the cells were incubated at 0°C.*

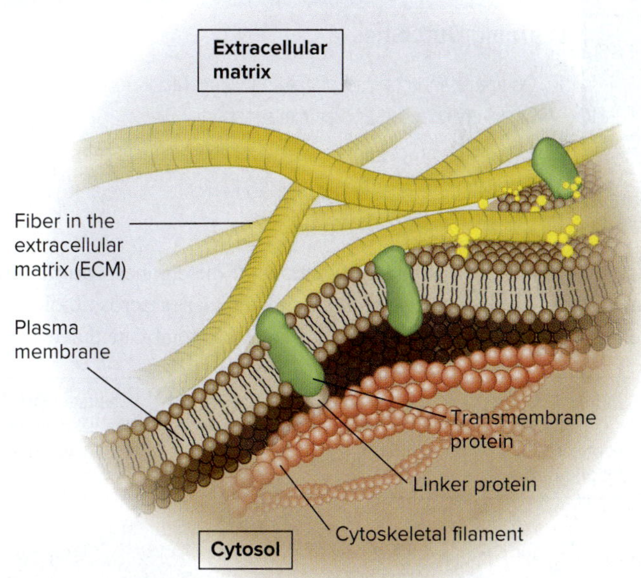


**Figure 5.5 Attachment of transmembrane proteins to the cytoskeleton and ECM of an animal cell.** Some transmembrane proteins have regions that extend into the cytosol and are anchored to large cytoskeletal filaments via linker proteins. Being bound to these large filaments restricts the movement of these proteins. Similarly, some transmembrane proteins are bound to large, immobile fibers in the ECM, which restricts their movements.

## 5.2 Reviewing the Concepts

- Membrane fluidity is essential for normal cell function, growth, and division. Lipids and many proteins can move rotationally and laterally, but the flip-flop of lipids from one leaflet to the opposite does not occur spontaneously. Some membrane proteins are restricted in their movements (Figures 5.3, 5.4, 5.5).
- The chemical properties of phospholipids—such as tail length and the presence of double bonds—and the amount of cholesterol affect the fluidity of membranes.

## 5.2 Testing Your Knowledge

1. Which of the following lipid movements would *not* occur spontaneously in a membrane?
   a. lateral movement (side-to-side)
   b. rotational movement
   c. flip-flop from one leaflet to the other
   d. both a and c
   e. all of the above
2. Which of the following changes would make a membrane more fluid?
   a. decrease the temperature
   b. increase the percentage of lipids with short tails
   c. decrease the percentage of lipids with double bonds in their tails
   d. increase the percentage of glycoproteins
   e. increase the percentage of glycolipids

## 5.3 Overview of Membrane Transport

### Learning Outcomes

1. Compare and contrast simple diffusion, facilitated diffusion, passive transport, and active transport.
2. Explain the process of osmosis and how it affects cell structure.
3. **SCISKILLS ▸** Predict the direction of water movement in response to a solute gradient.

We now turn to one of the key functions of membranes, **membrane transport**—the movement of ions and molecules across biological membranes. All cells are enclosed by a plasma membrane that exhibits **selective permeability,** allowing the passage of some ions and molecules but not others. As a protective envelope, its structure ensures that essential molecules such as glucose and amino acids enter the cell, metabolic intermediates remain in the cell, and waste products exit. The selective permeability of the plasma membrane allows the cell to maintain a favorable internal environment.

Substances can move directly across a membrane in three general ways (**Figure 5.6**).

- **Simple diffusion** occurs when a substance moves from a region of high concentration to a region of lower concentration. Some substances can move directly through a biological membrane via simple diffusion.
- In **facilitated diffusion,** a transport protein provides a passageway for a substance to diffuse across a membrane. Simple and facilitated diffusion are examples of **passive transport—**the transport of a substance across a membrane that does not require an input of energy.
- A third mode of transport, called **active transport,** moves a substance from an area of low concentration to one of higher concentration with the aid of a transport protein. Active transport requires an input of energy from a source such as ATP.

In this section, we will begin with a discussion of how the phospholipid bilayer presents a barrier to the simple diffusion of ions and molecules across membranes. We will then consider the concept of gradients across membranes and how such gradients affect the movement of water.

### The Phospholipid Bilayer Is a Barrier to the Simple Diffusion of Hydrophilic Solutes

Because of their hydrophobic interiors, phospholipid bilayers are a barrier to the simple diffusion of ions and hydrophilic molecules. Such ions and molecules are called **solutes;** they are dissolved in water, which is a **solvent.** The rate of diffusion across a phospholipid bilayer depends on the chemistry of the solute and its concentration. Three factors greatly affect the ability of solutes to cross a phospholipid bilayer by simple diffusion:

- **Size:** small substances diffuse faster than larger ones
- **Polarity:** nonpolar substances diffuse faster than polar ones
- **Charge:** noncharged substances diffuse faster than charged ones

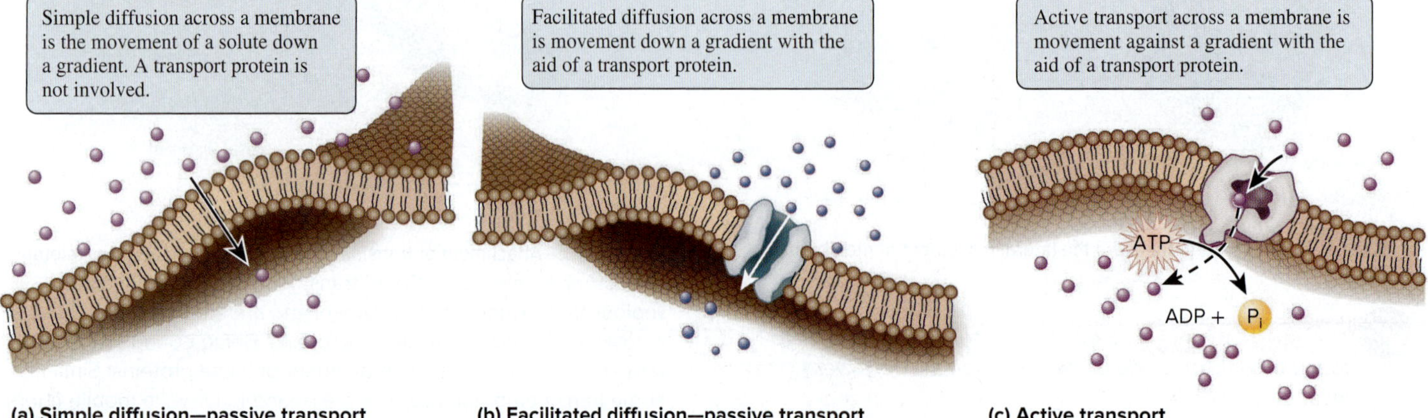


**Figure 5.6** Three general types of membrane transport.

Figure 5.7 compares the relative permeabilities of various solutes through an artificial phospholipid bilayer that does not contain any proteins or carbohydrates. Gases, such as $O_2$ and $CO_2$, and a few small, uncharged molecules, such as ethanol, can readily diffuse across the bilayer. However, the permeability of ions and larger polar molecules, such as sugars, is relatively low, and the permeability of macromolecules, such as proteins and polysaccharides, is even lower.

## Cells Maintain Gradients Across Their Membranes

A hallmark of living cells is their ability to maintain a relatively constant internal environment that is distinctively different from their external environment. Solute gradients are formed across the plasma membrane and across internal membranes. When we speak of a **transmembrane gradient,** or **concentration gradient,** we mean the concentration of a solute is higher on one side of a membrane than the other. Transmembrane gradients of solutes are a universal feature of all living cells. For example, immediately after you eat a meal containing carbohydrates, a higher concentration of glucose is found outside your cells than inside; this is an example of a chemical gradient (Figure 5.8a).

Gradients involving ions have two components: electrical and chemical. An **electrochemical gradient** is a dual gradient with both electrical and chemical components (Figure 5.8b). It occurs with solutes that have a net positive or negative charge. For example, let's consider a gradient involving $Na^+$. An electrical gradient can exist in which the amount of net positive charge outside a cell is greater than inside. In Figure 5.8b, an electrical gradient is due to differences in the amounts of different types of ions across the membrane, including sodium, potassium, and chloride ($Na^+$, $K^+$, and $Cl^-$). At the same time, a chemical gradient—a difference in $Na^+$ concentration across the membrane—can exist in which the concentration of $Na^+$ outside is greater than inside. Thus, the $Na^+$ electrochemical gradient is composed

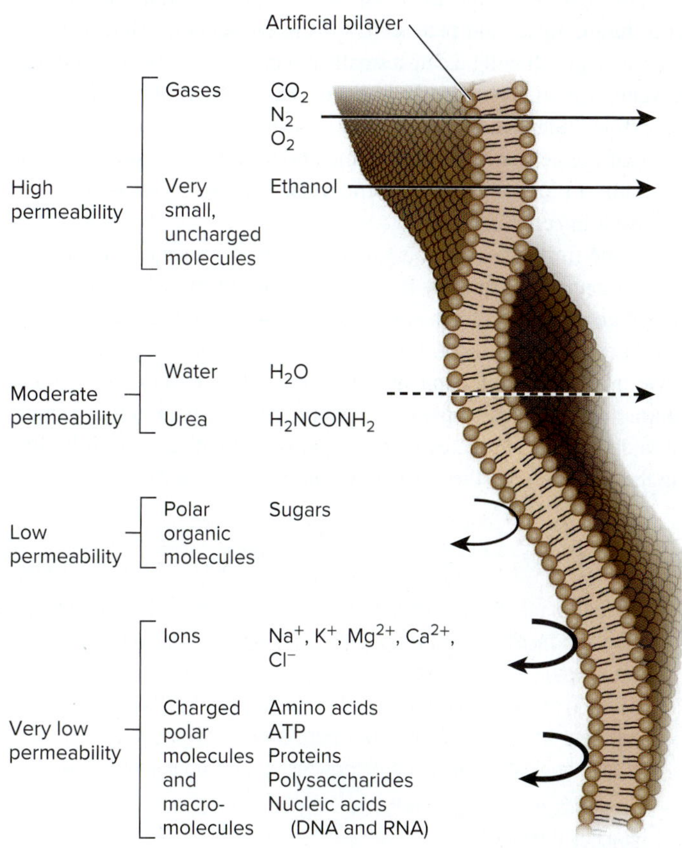


**Figure 5.7 Relative permeability of an artificial phospholipid bilayer to a variety of solutes.** Solutes that easily penetrate are shown with a straight arrow that passes through the bilayer. The dotted line indicates solutes that have moderate permeability. The remaining solutes shown at the bottom are relatively impermeable.

**BioConnections:** *Which amino acid, described in Chapter 3 (refer back to Figure 3.11), would you expect to cross an artificial membrane more quickly, leucine or lysine?*

**Concept Check:** *Which molecule would you expect to pass through a phospholipid bilayer more quickly, methanol ($CH_3OH$) or methane ($CH_4$)?*

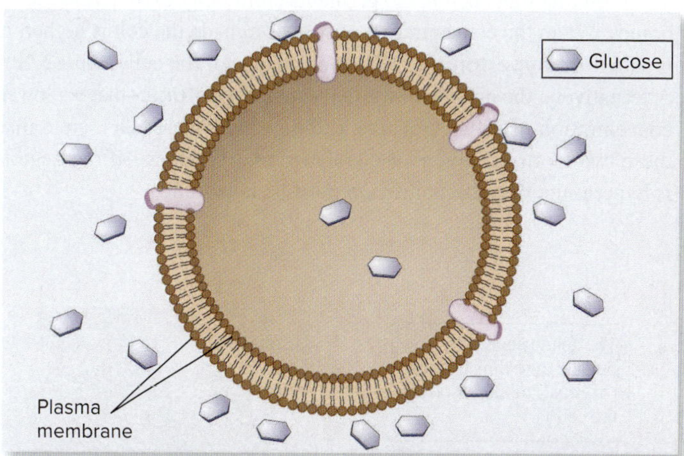


**(a) Chemical gradient for glucose—a higher glucose concentration outside the cell**

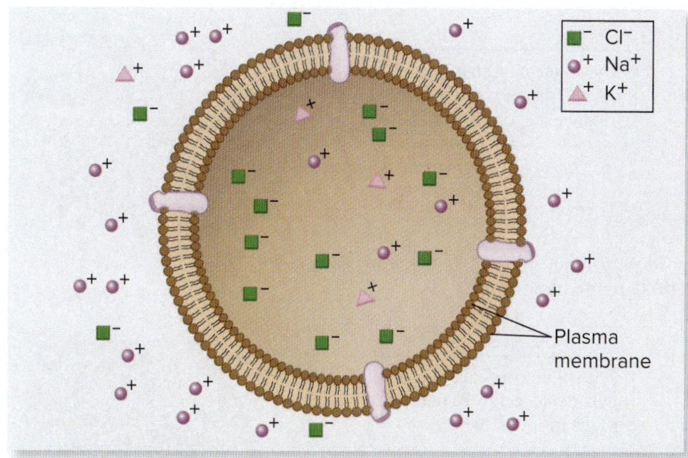


**(b) Electrochemical gradient for $Na^+$—more positive charges outside the cell and a higher $Na^+$ concentration outside the cell**

**Figure 5.8 Gradients across cell membranes.**

**BioConnections:** *Look ahead to Figure 33.9. What types of ion gradients are important for the conduction of an action potential across the plasma membrane of a neuron?*

of both an electrical gradient due to charge differences across the membrane and a chemical gradient for $Na^+$.

One way to view the transport of solutes across membranes is to consider how the transport process affects the pre-existing gradients across membranes. Passive transport tends to decrease the magnitude of a pre-existing gradient. It is a process that is energetically favorable and does not require an input of energy (see Figure 5.6a, b). By comparison, active transport produces a chemical gradient and/or electrochemical gradient. The formation of a gradient requires an input of energy.

## Osmosis Is the Movement of Water Across Membranes to Balance Solute Concentrations

Let's now turn our attention to how gradients affect the movement of water across membranes. When the concentrations of dissolved substances (solutes) on both sides of the plasma membrane are equal, the two solutions are said to be **isotonic** (Figure 5.9a). However, we have also seen that transmembrane gradients commonly exist across membranes. When the concentration of solutes outside the cell is higher, it is said to be **hypertonic** relative to the inside of the cell (Figure 5.9b). Alternatively, the outside of the cell can be **hypotonic**—have a lower concentration of solutes relative to the inside (Figure 5.9c). Note that these two terms are always used relative to each other—if one region is hypertonic, the adjacent region must be hypotonic.

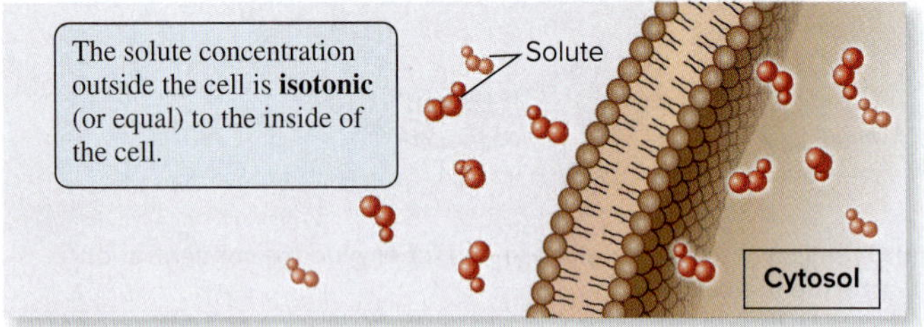


**(a) Outside isotonic**

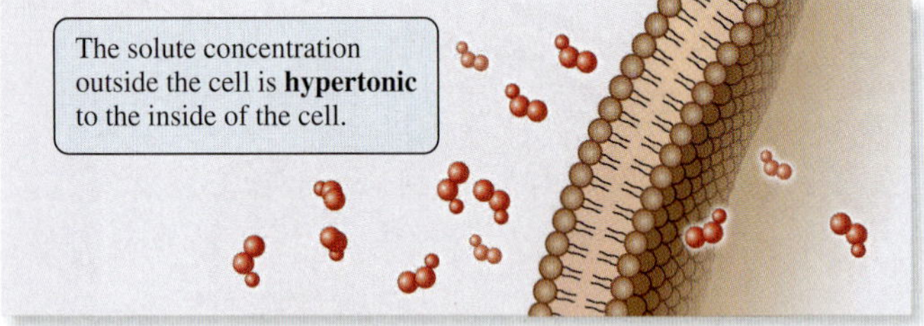


**(b) Outside hypertonic**

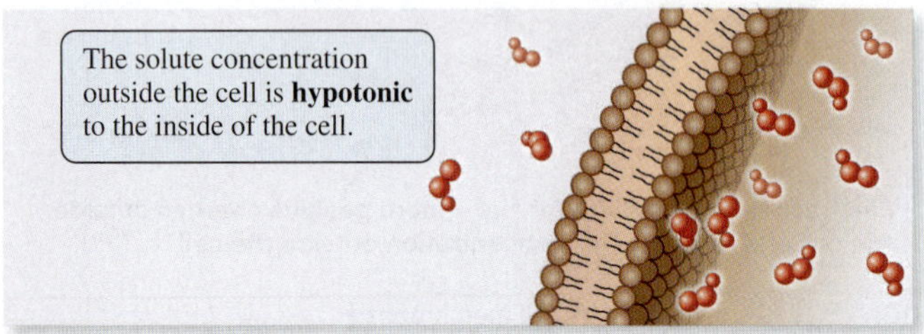


**(c) Outside hypotonic**

**Figure 5.9** Relative solute concentrations outside and inside cells.

If solutes cannot readily move across the membrane, water will move and tend to balance the solute concentrations. In this process, called **osmosis,** water moves across a membrane from the hypotonic compartment (with a lower solute concentration) into the hypertonic compartment (with a higher solute concentration). Animal cells, which are not surrounded by a rigid cell wall, must maintain a balance between the extracellular and intracellular solute concentrations; they are isotonic. Animal cells contain a variety of transport proteins that sense changes in cell volume and allow the necessary movements of solutes across the membrane to prevent osmotic changes and maintain normal cell shape. However, if an animal cell is placed in a hypotonic medium, water will enter the cell and thereby equalize solute concentrations on both sides of the membrane. In extreme cases, a cell may take up so much water that it ruptures, a phenomenon called osmotic lysis (Figure 5.10a). Alternatively, if an animal cell is placed in a hypertonic medium, water will exit the cell via osmosis and equalize solute concentrations on both sides of the membrane, causing it to shrink in a process called crenation.

How does osmosis affect cells with a rigid cell wall, such as bacteria, fungi, algae, and plant cells? If the extracellular fluid is hypotonic, a plant cell will take up a small amount of water, but the cell wall prevents osmotic lysis from occurring (Figure 5.10b). Alternatively, if the extracellular fluid surrounding a plant cell is hypertonic, water will exit the cell and the plasma membrane will pull away from the cell wall, a process called **plasmolysis.** If plasmolysis is severe, it can result in cell death.

Some freshwater microorganisms, such as amoebae and paramecia, are found in extremely hypotonic environments where the external solute concentration is always much lower than the concentration of solutes in their cytosol. Because of the great tendency for water to move into the cell by osmosis, such organisms contain one or more contractile vacuoles to prevent osmotic lysis. A contractile vacuole takes up water from the cytosol and periodically discharges it by fusing the vacuole with the plasma membrane (Figure 5.11).

## 5.3 Reviewing the Concepts

- Biological membranes exhibit selective permeability. Simple diffusion and facilitated diffusion, both examples of passive transport, occur when a solute moves from a region of high concentration to a region of lower concentration. Passive transport does not require an input of energy. Active transport is the movement of a substance against a gradient and requires energy input (Figure 5.6).
- The phospholipid bilayer is relatively impermeable to many substances (Figure 5.7).
- Living cells maintain an internal environment that is separated from their external environment. Transmembrane gradients, in which the concentration of a solute is higher on one side of a membrane than the other, are established across the plasma membrane and across organellar membranes (Figure 5.8).
- In the process of osmosis, water diffuses through a membrane from a solution that is hypotonic (lower concentration of dissolved particles) into a solution that is hypertonic (higher concentration of dissolved particles). Solutions with identical concentrations are isotonic. Some cells have contractile vacuoles to eliminate excess water (Figures 5.9, 5.10, 5.11).

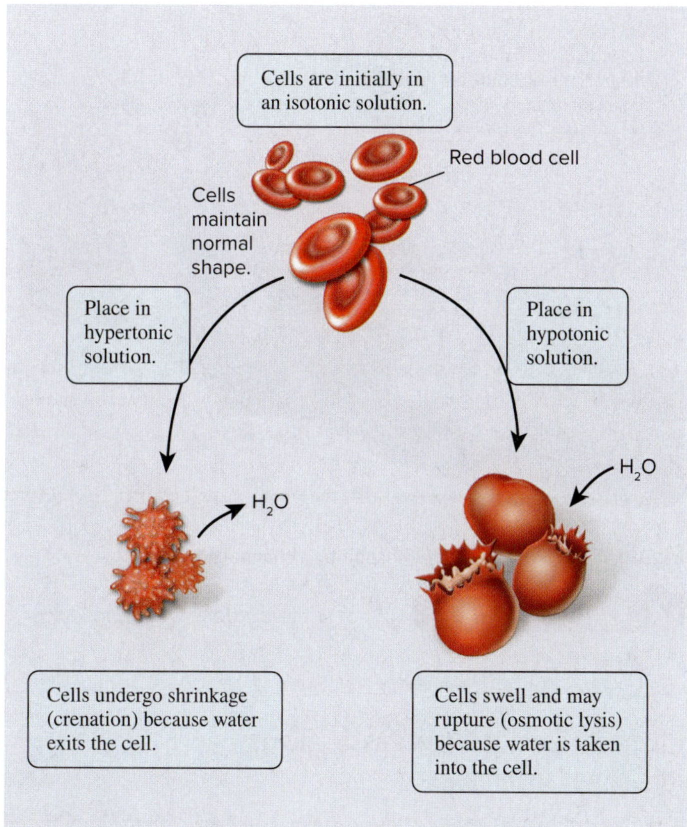


**(a) Osmosis in animal cells**

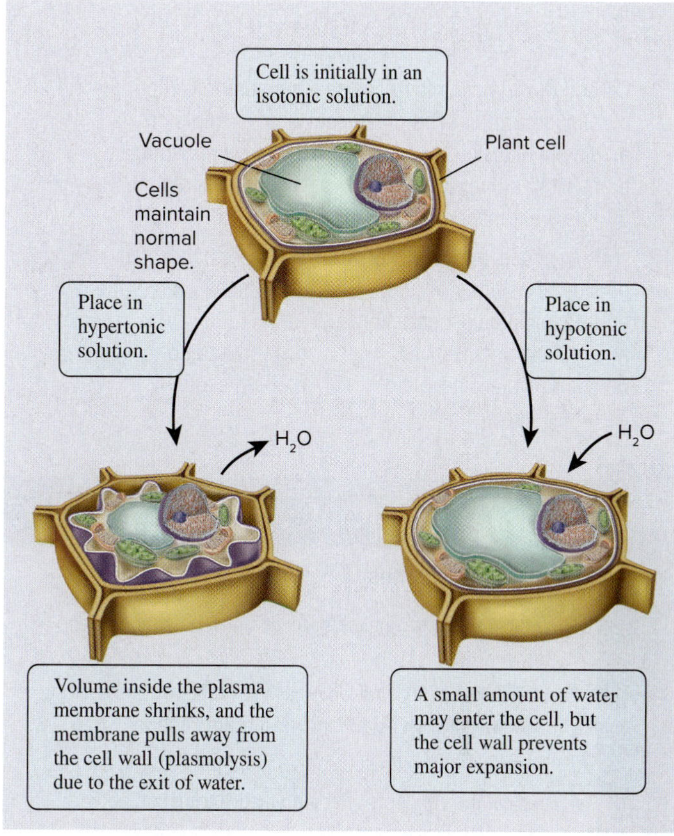


**(b) Osmosis in plant cells**

**Figure 5.10** **The phenomenon of osmosis.** **(a)** In cells that lack a cell wall, such as animal cells, osmosis may promote cell swelling and possible rupture (osmotic lysis) or shrinkage (crenation). **(b)** In cells that have a rigid cell wall, such as plant cells, a hypotonic medium causes only a minor amount of expansion, whereas a hypertonic medium causes the plasma membrane to pull away from the cell wall.

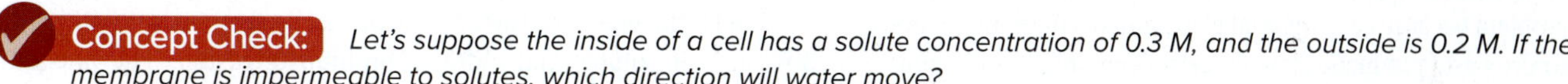

**Concept Check:** *Let's suppose the inside of a cell has a solute concentration of 0.3 M, and the outside is 0.2 M. If the membrane is impermeable to solutes, which direction will water move?*

## Biology Principle

### Living Organisms Maintain Homeostasis

In this example, the paramecium maintains a relatively constant internal volume by using contractile vacuoles to remove excess water.

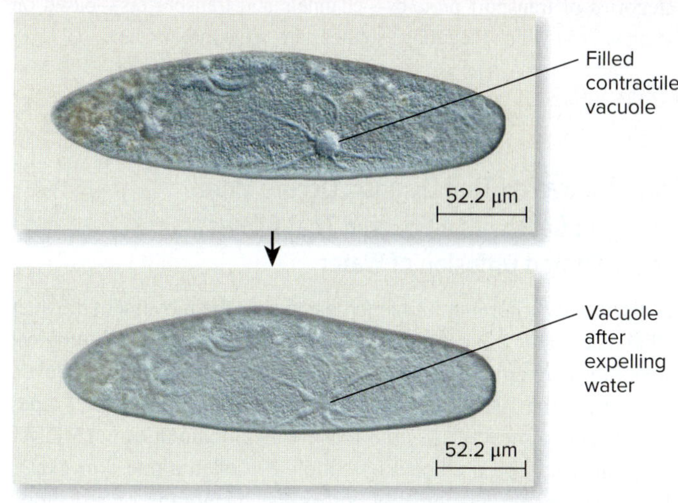


**Figure 5.11** **The contractile vacuole in *Paramecium caudatum*.** In the upper photo, a contractile vacuole is filled with water from radiating canals that collect fluid from the cytosol. The lower photo shows the cell after the contractile vacuole has fused with the plasma membrane (which would be above the plane of this page) and has released the water from the cell.

## 5.3 Testing Your Knowledge

1. Carbon dioxide can move across a membrane without the aid of a transport protein. This movement is an example of
   a. simple diffusion.
   b. facilitated diffusion.
   c. passive transport.
   d. active transport.
   e. both a and c.
2. Let's suppose the inside of a cell is hypertonic relative to the outside. If osmosis occurs, _______ will move ______ the cell to equalize solute concentrations.
   a. solute molecules, into
   b. solute molecules, out of
   c. water, into
   d. water, out of
   e. both a and d

## 5.4 Transport Proteins

### Learning Outcomes

1. Outline the functional differences between channels and transporters.
2. **SCISKILLS** ▶ Analyze the results of Agre, and explain how they indicated that aquaporin is a water channel.
3. Compare and contrast uniporters, symporters, and antiporters.
4. Explain the concepts of primary active transport and secondary active transport.
5. Describe the structure and function of pumps.

Because the phospholipid bilayer is a physical barrier to the diffusion of most hydrophilic molecules and ions, cells can separate their internal contents from the external environment. However, this barrier also poses a potential problem because cells must take up nutrients from the environment and export waste products. How do cells overcome this dilemma? Over the course of millions of years, species have evolved a multitude of **transport proteins**—transmembrane proteins that provide passageways for the movement of ions and hydrophilic molecules across the phospholipid bilayer. Transport proteins play a central role in the selective permeability of biological membranes. In this section, we will examine the two categories of transport proteins—channels and transporters—based on the manner in which they move solutes across the membrane.

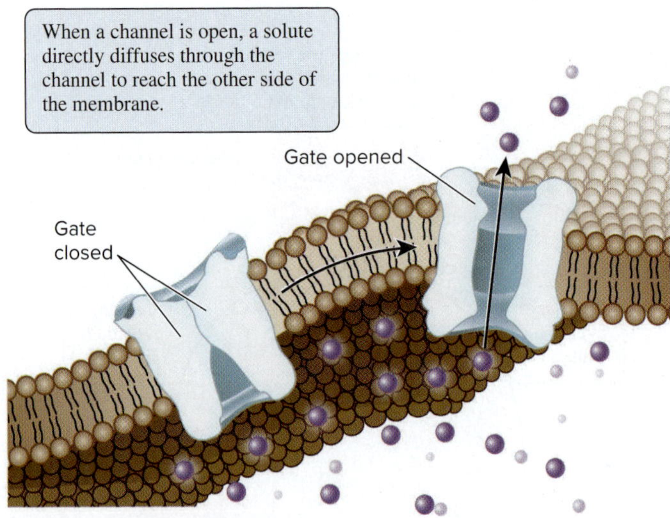

When a channel is open, a solute directly diffuses through the channel to reach the other side of the membrane.
Gate opened
Gate closed


**Figure 5.12** Mechanism of transport by a channel protein.

*What is the purpose of gating?*

### Channels Provide Open Passageways for Solute Movement

The term **channel** refers to an open passageway for the facilitated diffusion of ions or molecules across a membrane. The type of channel shown in Figure 5.12 is a transmembrane protein. Solutes move directly through this channel to get to the other side. When a channel is open, the transmembrane movement of solutes can be extremely rapid, up to 100 million ions or molecules per second!

Most channels are **gated,** which means they open to allow the diffusion of solutes and close to prohibit diffusion. The phenomenon of gating allows cells to regulate the movement of solutes. For example, gating may involve the direct binding of a molecule to the channel protein itself. These gated channels are controlled by the noncovalent binding of small molecules—called ligands—such as hormones or neurotransmitters. The ligands are often important in the transmission of signals between neurons and muscle cells or between two neurons.

## FEATURE INVESTIGATION

### Agre Discovered That Osmosis Occurs More Quickly in Cells with a Channel That Allows the Facilitated Diffusion of Water

As discussed earlier in this chapter, osmosis is the movement of water to balance solute concentrations. Water can slowly cross biological membranes by simple diffusion through the phospholipid bilayer. However, in the 1980s, researchers discovered that certain cell types allow water to move across the plasma membrane at a much faster rate than would be predicted by simple diffusion alone. For example, water moves very quickly across the membrane of red blood cells, which causes them to shrink and swell in response to changes in extracellular solute concentrations (see Figure 5.10a). Likewise, bladder and kidney cells, which play a key role in regulating water balance in the bodies of vertebrates, allow the rapid movement of water across their membranes. Based on these observations, researchers speculated that certain cell types might have channel proteins in their plasma membranes that enable the rapid movement of water.

One approach to characterizing a new protein is to first identify a protein based on its relative abundance in a particular cell type and then attempt to determine the protein's function. This rationale was applied to the discovery of proteins that allow the rapid movement of water across membranes. Peter Agre and his colleagues first identified a protein that was abundant in red blood cells and kidney cells but not found in high amounts in many other cell types. Though they initially did not know the function of the protein, its physical structure was similar to other proteins that were already known to function as channels. They named this protein CHIP28, which stands for channel-forming integral membrane protein with a molecular mass of 28,000 daltons. During the course of their studies, they also identified and isolated the gene that encodes CHIP28.

In 1992, Agre and his colleagues conducted experiments to determine if CHIP28 functions in the transport of water across membranes (**Figure 5.13**). Because they already had isolated the gene that

**Figure 5.13** **The discovery of water channels (aquaporins) by Agre and colleagues.**

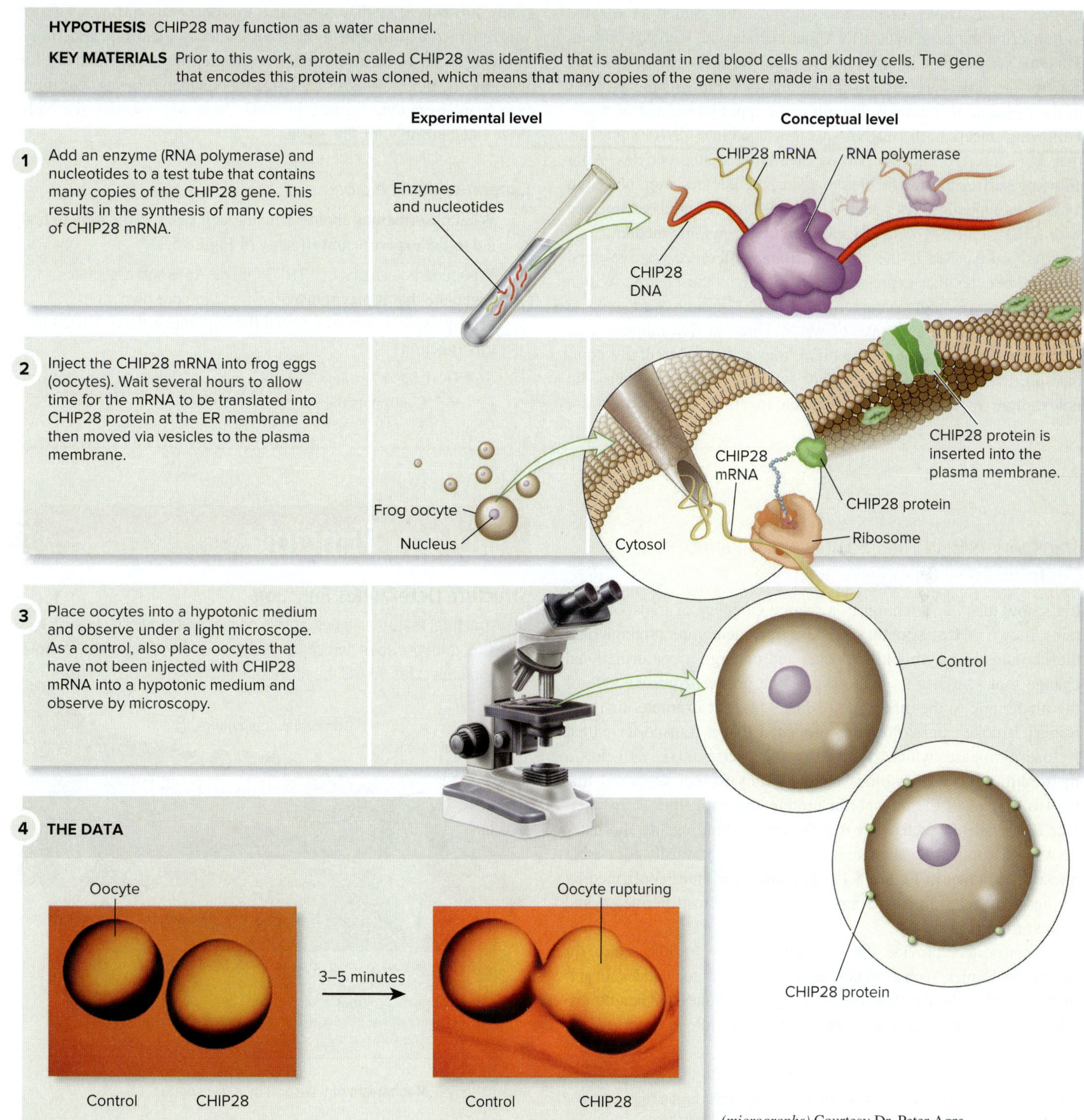


*(micrographs)* Courtesy Dr. Peter Agre

5 **CONCLUSION** The CHIP28 protein, now called aquaporin, allows the rapid movement of water across the membrane.

6 **SOURCE** Preston, G. M., Carroll, T. P., Guggino, W. B., and Agre, P. 1992. Appearance of Water Channels in *Xenopus* Oocytes Expressing Red Cell CHIP28 Protein. *Science* 256: 385–387.

encodes CHIP28, they could make many copies of this gene in a test tube (in vitro) using gene cloning techniques (see Chapter 18). Starting with many copies of the gene in vitro, they added an enzyme to transcribe the gene into mRNA that encodes the CHIP28 protein. This mRNA was then injected into frog oocytes, chosen because frog oocytes are large, are easy to inject, and lack pre-existing proteins in their plasma membranes that allow the rapid movement of water. Following injection, the mRNA was translated into CHIP28 proteins that were inserted into the plasma membrane of the oocytes. After allowing sufficient time for this to occur, the oocytes were placed in a hypotonic medium. As a control, oocytes that had not been injected with CHIP28 mRNA were also exposed to a hypotonic medium.

As you can see in the data, a striking difference was observed between oocytes that expressed CHIP28 versus the control. Within minutes, oocytes that contained the CHIP28 protein were seen to swell due to the rapid uptake of water. Three to five minutes after being placed in a hypotonic medium, they actually lysed! By comparison, the control oocytes did not swell as rapidly, and they did not rupture even after 1 hour. Taken together, these results are consistent with the hypothesis that CHIP28 functions as a channel that allows the facilitated diffusion of water across the membrane. Many subsequent studies confirmed this observation. Later, CHIP28 was renamed **aquaporin** to indicate its newly identified function of allowing water to diffuse through a channel in the membrane. More recently, the three-dimensional structure of aquaporin was determined (see chapter-opening figure). In 2003, Agre was awarded the Nobel Prize in Chemistry for this work.

*Experimental Questions*

1. What observations about particular cell types in the human body led to the experimental strategy of Figure 5.13?
2. What characteristics of CHIP28 made Agre and associates speculate that it may transport water? In your own words, briefly explain how they tested the hypothesis that CHIP28 has this function.
3. **SCISKILLS ▸** Explain how the results of the experiment of Figure 5.13 support the proposed hypothesis.

## Transporters Bind Their Solutes and Undergo Conformational Changes

Let's now turn our attention to a second category of transport proteins known as **transporters.*** These transmembrane proteins bind their solutes in a hydrophilic pocket and undergo a conformational change that switches the exposure of the pocket from one side of the membrane to the other side (**Figure 5.14**). For example, in 1995, Robert Brooker and colleagues proposed that a transporter called lactose permease, which is found in the bacterium *E. coli*, has a hydrophilic pocket that binds lactose. They further proposed that the two halves of the transporter protein come together at an interface that moves in such a way that the lactose-binding site alternates between an outwardly accessible pocket and an inwardly accessible pocket, as shown in Figure 5.14. This idea was later confirmed by studies that determined the structure of the lactose permease and related transporters.

Transporters provide the principal pathway for the uptake of organic molecules, such as sugars, amino acids, and nucleotides. In animals, they also allow cells to take up certain hormones and neurotransmitters. In addition, many transporters play a key role in export. Waste

* Transporters are also called carriers. However, this term is misleading because transporters do not physically carry their solutes across the membrane.

## Biology Principle

### Structure Determines Function

Two structural features—a hydrophilic pocket and the ability to switch back and forth between two conformations—allow transporters to move ions and molecules across the membrane.

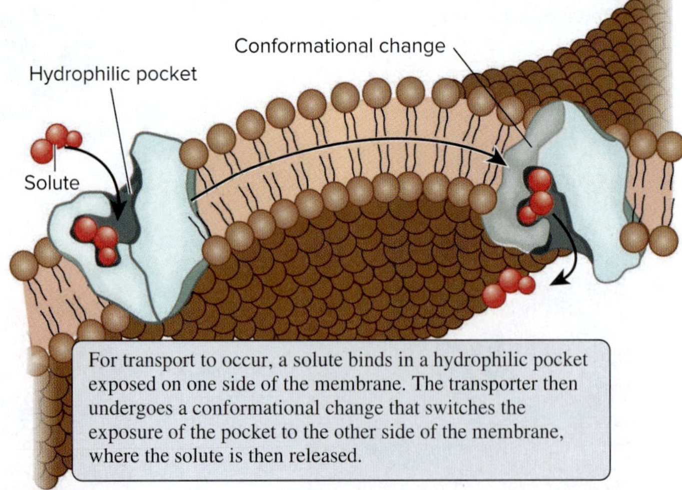

Conformational change
Hydrophilic pocket
Solute
For transport to occur, a solute binds in a hydrophilic pocket exposed on one side of the membrane. The transporter then undergoes a conformational change that switches the exposure of the pocket to the other side of the membrane, where the solute is then released.


**Figure 5.14** Mechanism of transport by a transporter.

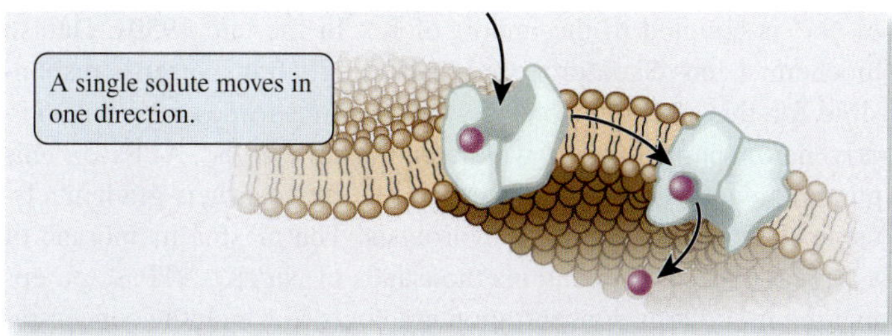


(a) Uniporter

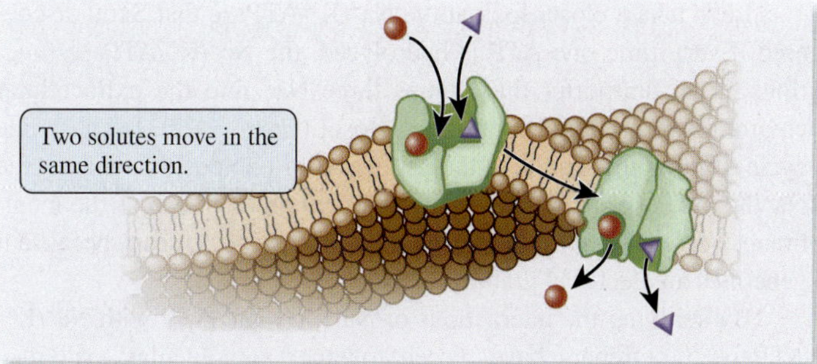


(b) Symporter

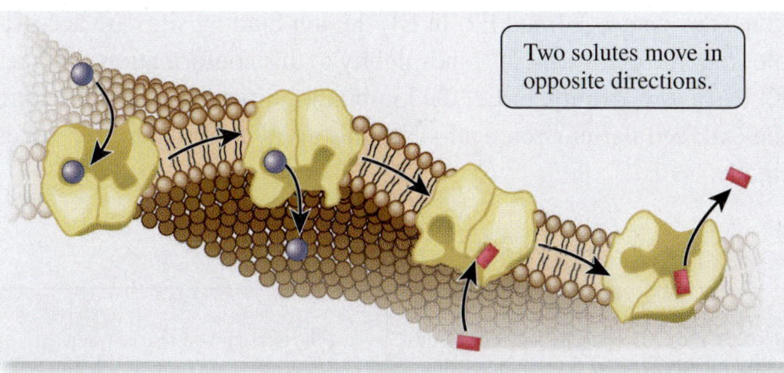


(c) Antiporter

**Figure 5.15** Types of transporters based on the direction of transport.

products of cellular metabolism must be released from cells before they reach toxic levels. For example, a transporter removes lactic acid, a by-product of muscle cells during exercise. Other transporters, which are involved with ion transport, play an important role in regulating internal pH and controlling cell volume. Transporters tend to be much slower than channels. Their rate of transport is typically 100 to 1,000 ions or molecules per second.

Transporters are named according to the number of solutes they bind and the direction in which they transport those solutes (Figure 5.15).

- **Uniporters** bind a single ion or molecule and transport it across the membrane.
- **Symporters** bind two or more ions or molecules and transport them in the same direction.
- **Antiporters** bind two or more ions or molecules and transport them in opposite directions.

BIO TIPS ONLINE

## Active Transport Is the Movement of Solutes Against a Gradient

As mentioned, active transport is the movement of a solute across a membrane against its concentration gradient—that is, from a region of low concentration to higher concentration. Active transport is energetically unfavorable and requires an input of energy. Two general types are found:

- **Primary active transport** involves the functioning of a **pump—**a type of transporter that directly uses energy to transport a solute against a concentration gradient. Figure 5.16a shows a pump that uses energy in the form of ATP to transport $H^+$ against a gradient. Such a pump establishes an $H^+$ electrochemical gradient across a membrane.
- **Secondary active transport** involves the use of a pre-existing gradient to drive the active transport of another

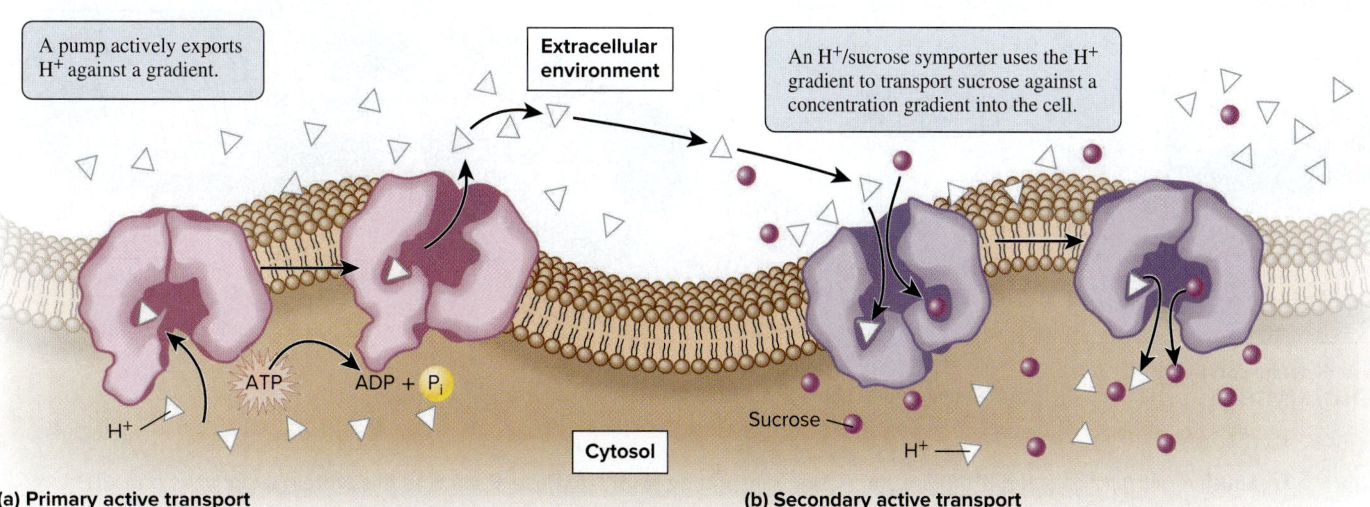

A pump actively exports H+ against a gradient.
Extracellular environment
An H+/sucrose symporter uses the H+ gradient to transport sucrose against a concentration gradient into the cell.
ATP
ADP + Pi
H+
Sucrose
H+
Cytosol


(a) Primary active transport

(b) Secondary active transport

**Figure 5.16** **Types of active transport.** **(a)** During primary active transport, a pump directly uses energy, in this case from ATP, to transport a solute against a concentration gradient. The pump shown here uses ATP to establish an $H^+$ electrochemical gradient. **(b)** Secondary active transport via symport involves the use of this gradient to drive the active transport of a solute, such as sucrose.

solute. For example, an $H^+$/sucrose symporter uses an $H^+$ electrochemical gradient, established by a pump, to move sucrose against its concentration gradient (Figure 5.16b). In this case, only sucrose is actively transported. Hydrogen ions move down their electrochemical gradient via the symporter.

Symporters enable cells to actively import nutrients against a gradient. These proteins use the energy stored in the electrochemical gradient of $H^+$ or $Na^+$ to power the uphill movement of organic solutes such as sugars, amino acids, and other needed molecules. Therefore, with symporters in their plasma membrane, cells can scavenge nutrients from the extracellular environment and accumulate them to high levels within the cytoplasm. $H^+$/solute symporters are more common in bacteria, fungi, algae, and plant cells, because $H^+$ pumps are found in their plasma membranes. In animal cells, a pump that exports $Na^+$ maintains a $Na^+$ gradient across the plasma membrane. $Na^+$/solute symporters are prevalent in animal cells.

## ATP-Driven Ion Pumps Generate Ion Electrochemical Gradients

The phenomenon of active transport was discovered in the 1940s based on the study of the transport of sodium ions ($Na^+$) and potassium ions ($K^+$). In animal cells, the concentration of $Na^+$ is lower inside the cell than outside, whereas the concentration of $K^+$ is higher inside the cell than outside. After analyzing the movement of these ions across the plasma membrane of muscle cells, neurons, and red blood cells, researchers determined that the export of $Na^+$ is coupled to the import of $K^+$. In the late 1950s, Danish biochemist Jens Skou proposed that a single transporter is responsible for this phenomenon. He was the first to describe an ATP-driven ion pump, which was later named $Na^+/K^+$-ATPase. This pump can actively transport $Na^+$ and $K^+$ against their gradients by using the energy from ATP hydrolysis. The plasma membrane of a typical animal cell contains thousands of $Na^+/K^+$-ATPase pumps that maintain large concentration gradients in which the concentration of $Na^+$ is higher outside the cell and the concentration of $K^+$ is higher inside the cell.

Let's take a closer look at the $Na^+/K^+$-ATPase that Skou discovered. Every time one ATP is hydrolyzed, the $Na^+/K^+$-ATPase functions as an antiporter that pumps three $Na^+$ into the extracellular environment and two $K^+$ into the cytosol (Figure 5.17a). Because one cycle of pumping results in the net export of one positive charge, the $Na^+/K^+$-ATPase also produces an electrical gradient across the membrane. For this reason, it is called an **electrogenic pump,** because it generates an electrical gradient.

By studying the interactions of $Na^+$, $K^+$, and ATP with $Na^+/K^+$-ATPase, researchers have pieced together a molecular road map of the steps that direct the pumping of ions across the membrane (Figure 5.17b). $Na^+/K^+$-ATPase can alternate between two conformations, designated E1 and E2. In E1, the ion-binding sites are accessible from the cytosol—$Na^+$ binds tightly to this conformation, whereas $K^+$ has a low affinity. In E2, the ion-binding sites are accessible from the extracellular environment—$Na^+$ has a low affinity, and $K^+$ binds tightly.

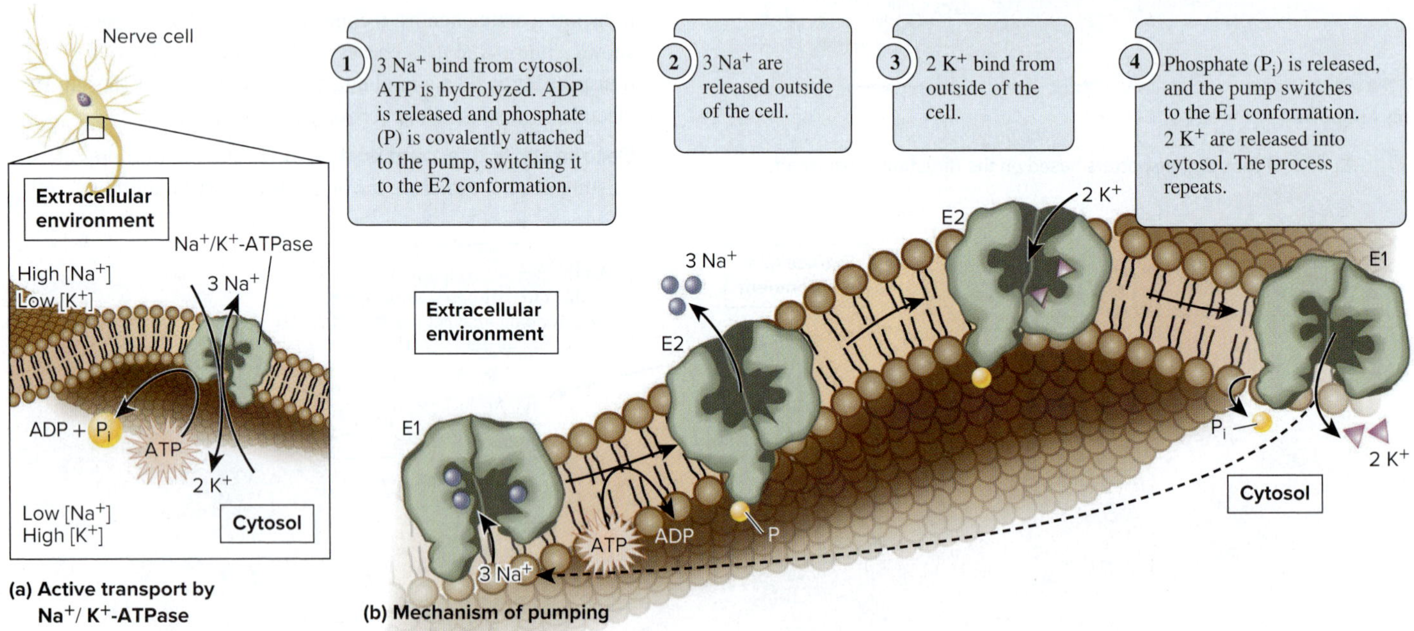


**Figure 5.17** **Structure and function of $Na^+/K^+$-ATPase.** **(a)** Active transport by $Na^+/K^+$-ATPase. Each time this protein hydrolyzes one ATP molecule, it pumps out three $Na^+$ and pumps in two $K^+$. **(b)** Pumping mechanism. This figure illustrates the protein conformational changes between E1 and E2. As this occurs, ATP is hydrolyzed to ADP and phosphate. During the process, phosphate is covalently attached to the protein but is released after two $K^+$ bind.

The pumping mechanism of $Na^+$/ $K^+$-ATPase occurs as follows (Figure 5.17b):

1. Three $Na^+$ bind to the E1 conformation from the cytosol. ATP is hydrolyzed to ADP and phosphate. Temporarily, the phosphate is covalently bound to the pump, an event called phosphorylation. The pump switches to the E2 conformation.
2. The three $Na^+$ are released into the extracellular environment, because they have a lower affinity for the E2 conformation.
3. Two $K^+$ bind from the outside.
4. The binding of two $K^+$ causes the release of phosphate. The pump switches back to the E1 conformation. Because the E1 conformation has a low affinity for $K^+$, the two $K^+$ are released into the cytosol. $Na^+$/$K^+$-ATPase is now ready for another round of pumping.

$Na^+$/$K^+$-ATPase is a critical ion pump in animal cells because it maintains $Na^+$ and $K^+$ gradients across the plasma membrane. Many other types of ion pumps are also found in the plasma membrane and in organellar membranes. Ion pumps play the primary role in the formation and maintenance of ion gradients that drive many important cellular processes (**Table 5.2**). ATP is commonly the source of energy to drive ion pumps, and cells typically use a substantial portion of their ATP to keep them working. For example, neurons use up to 70% of their ATP just to operate ion pumps! As discussed in Chapter 33, neurons need to maintain $Na^+$ and $K^+$ gradients so that they can conduct signals known as action potentials.

**Table 5.2 Important Functions of Ion Electrochemical Gradients**

| Function | Description |
|---|---|
| Transport of ions and molecules | Symporters and antiporters use $H^+$ and $Na^+$ gradients to take up nutrients and export waste products (see Figure 5.16). |
| Production of energy intermediates | In the mitochondrion and chloroplast, $H^+$ gradients are used to synthesize ATP. |
| Osmotic regulation | Animal cells control their internal volume by regulating ion gradients between the cytosol and extracellular fluid. |
| Neuronal signaling | $Na^+$ and $K^+$ gradients are involved in conducting action potentials, the signals transmitted by neurons. |
| Muscle contraction | $Ca^{2+}$ gradients regulate the ability of muscle fibers to contract. |
| Bacterial swimming | $H^+$ gradients drive the rotation of bacterial flagella. |

## 5.4 Reviewing the Concepts

- Two classes of transport proteins are channels and transporters.
- Channels form an open passageway for the facilitated diffusion of solutes across the membrane. One example is aquaporin, which allows the movement of water. Most channels are gated, which allows cells to regulate the movement of solutes (Figures 5.12, 5.13).
- Transporters, which tend to be slower than channels, bind their solutes in a hydrophilic pocket and undergo a conformational change that switches the exposure of the pocket to the other side of the membrane. They can be uniporters, symporters, or antiporters (Figures 5.14, 5.15).
- Primary active transport involves pumps that directly use energy to generate a solute gradient. Secondary active transport uses a pre-existing gradient (Figure 5.16).
- $Na^+$/$K^+$-ATPase is an electrogenic pump that uses energy in the form of ATP to transport ions across the membrane (Figure 5.17, Table 5.2).

## 5.4 Testing Your Knowledge

1. Which of the following is *not* a characteristic of channels?
   a. They may be gated.
   b. They form an open passageway for the movement of solutes.
   c. They are usually faster than transporters.
   d. They bind their solutes in a hydrophilic pocket, much like an enzyme.
   e. They are transmembrane proteins.
2. Let's suppose $Na^+$/$K^+$-ATPase was in an environment in which $Na^+$ was inside the cell, ATP was inside the cell, but no $K^+$ was outside the cell. Under these conditions, $Na^+$/$K^+$-ATPase would be stuck in which conformation?
   a. E1 with phosphate covalently bound
   b. E2 with phosphate covalently bound
   c. E1 without phosphate covalently bound
   d. E2 without phosphate covalently bound
   e. either a or c

## 5.5 Intercellular Channels

### Learning Outcome

1. Outline the structure and function of gap junctions and plasmodesmata.

Thus far, we have learned that cells produce many types of transport proteins that allow the transport of solutes across the plasma membrane. Some transport solutes into the cell and others facilitate the export of solutes. The cells of multicellular organisms also may have **intercellular channels** that allow the direct movement of substances between adjacent cells. In this section, we will examine gap junctions

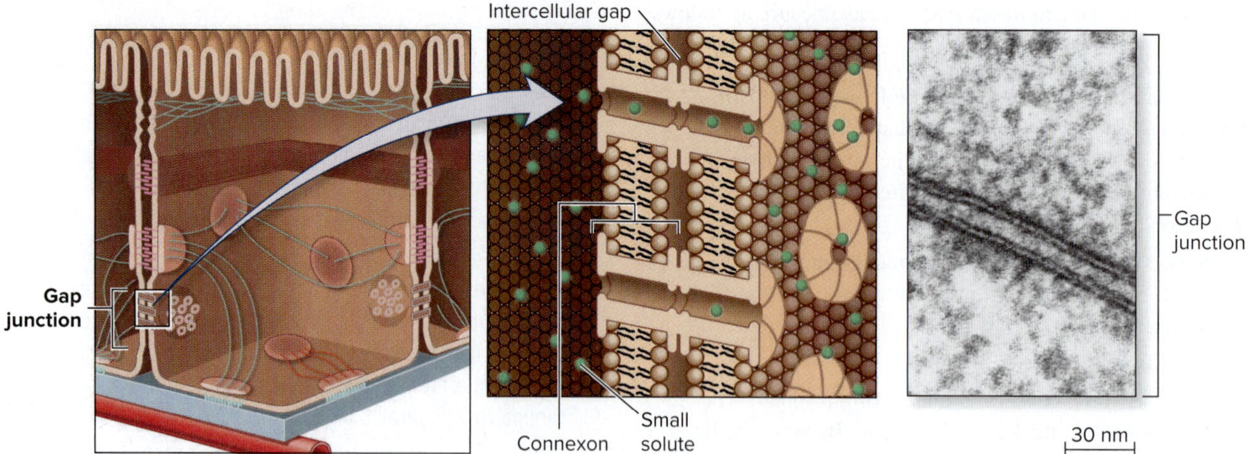


**Figure 5.18 Gap junctions between adjacent cells that line the intestine.** Gap junctions form intercellular channels that allow the passage of small solutes with masses less than 1,000 Da. One connexon consists of six proteins called connexins. Two connexons align to form an intercellular channel. The micrograph shows a gap junction, which is composed of many connexons, between intestinal cells.
*(photo)* © Don W. Fawcett/Science Source

that often occur between adjacent animal cells and plasmodesmata that occur between plant cells.

## Gap Junctions Between Animal Cells Provide Passageways for Intercellular Transport

A type of junction found between animal cells is called a **gap junction,** because a small gap occurs between the plasma membranes of cells connected by these junctions (Figure 5.18). Gap junctions are abundant in tissues and organs where cells need to communicate with each other. For example, cardiac muscle cells, which cause your heart to beat, are interconnected by many gap junctions. Because gap junctions allow the passage of ions, electrical changes in one cardiac muscle cell are easily transmitted to an adjacent cell that is connected via gap junctions. This is needed for the coordinated contraction of cardiac muscle cells.

In vertebrates, gap junctions are composed of many integral membrane proteins called connexins. Invertebrates have a structurally similar protein called innexin. Six connexin proteins in one vertebrate cell form a channel called a **connexon.** A connexon in one cell aligns with a connexon in an adjacent cell to form an intercellular channel (see middle inset to Figure 5.18). The term gap junction refers to a cluster of many connexons that are close to each other in the plasma membrane and form many intercellular channels.

Gap junction channels allow the passage of ions and small molecules, including amino acids, sugars, and signaling molecules such as $Ca^{2+}$ and cAMP, between cells. In this way, gap junctions allow adjacent cells to share metabolites and directly signal each other. However, gap junction channels are too small to allow the passage of RNA, proteins, or polysaccharides. Therefore, cells that communicate via gap junctions still maintain their own distinctive sets of macromolecules.

## Plasmodesmata Are Channels Connecting the Cytoplasm of Adjacent Plant Cells

In 1879, Eduard Tangl, a Russian botanist, observed intercellular connections in the seeds of the strychnine tree and hypothesized that the cytoplasm of adjacent cells is connected by ducts in the cell walls. He was the first to propose that direct cell-to-cell communication integrates the functioning of plant cells. The ducts or intercellular channels that Tangl observed in plant cells are now known as **plasmodesmata** (singular, plasmodesma).

Plasmodesmata are functionally similar to gap junctions in animal cells because they are open pores that allow the passage of ions and molecules between the cytosol of adjacent plant cells. However, the structure of plasmodesmata is quite different from that of gap junctions. As shown in Figure 5.19, the plasma membrane of one cell is continuous with the plasma membrane of the adjacent cell, which forms a pore that permits the diffusion of molecules from the cytosol of one cell to the cytosol of the other. In addition to a cytosolic connection, plasmodesmata also have a central tubule, called a desmotubule, connecting the smooth ER membranes of adjacent cells.

Plasmodesmata can change the size of their opening among closed, open, and dilated states. In the open state, they allow the passage of ions and small molecules, such as sugars and ATP. In this state, plasmodesmata play a similar role to gap junctions between animal cells. Plasmodesmata tend to close when a plant is wounded. The closure of plasmodesmata between adjacent cells helps to prevent the loss of water and nutrients from the wound site.

Unlike gap junctions between animal cells, plasmodesmata can dilate to allow the passage of macromolecules and even viruses between adjacent plant cells. Though the mechanism of dilation is not well understood, the wider opening of plasmodesmata is important for the passage of proteins and mRNA during plant development.

## 5.5 Reviewing the Concepts

- Gap junctions are intercellular channels composed of connexons, which permit the direct passage of ions and small molecules between adjacent animal cells (Figure 5.18).
- The plasma membranes and ER of adjacent plant cells are connected via plasmodesmata that allow the passage of ions and molecules between the cytosol of adjacent plant cells (Figure 5.19).

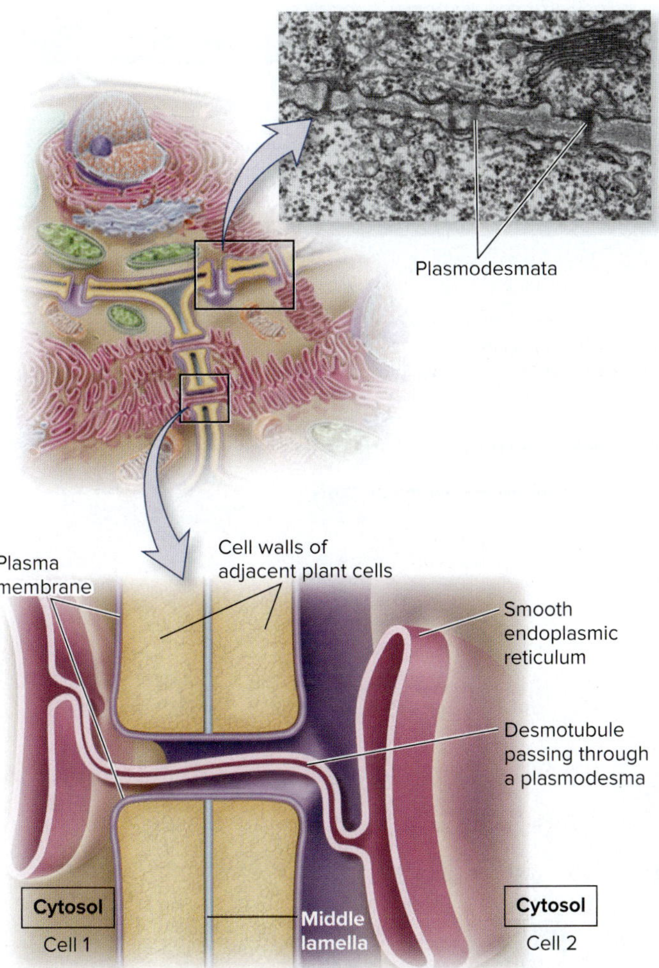


**Figure 5.19** **Structure of plasmodesmata.** Plasmodesmata are cell junctions connecting the cytosol of adjacent plant cells, allowing water, ions, and molecules to pass from cell to cell. At these pores, the plasma membrane of one cell is continuous with the plasma membrane of an adjacent cell. In addition, the smooth ER from one cell is connected to that of the adjacent cell via a desmotubule.



**BioConnections:** *Look ahead to Figure 30.17. How do plasmodesmata play a role in the movement of nutrients through a plant root?*

## 5.5 Testing Your Knowledge

1. Which of the following is *not* a characteristic of gap junctions?
   a. They allow the passage of small molecules between adjacent cells.
   b. They are composed of membrane proteins called connexins, which form connexons.
   c. They are found between adjacent plant cells.
   d. They are found between adjacent animal cells.
   e. All of the above are correct.

## 5.6 Exocytosis and Endocytosis

### Learning Outcomes

1. Describe the steps in exocytosis and endocytosis.
2. Compare and contrast three types of endocytosis.

We have seen that most small substances are transported via channels and transporters, which provide a passageway for the movement of ions and molecules directly across the membrane. Eukaryotic cells have two other mechanisms, exocytosis and endocytosis, to transport larger molecules such as proteins and polysaccharides, and even very large particles. Both mechanisms involve the packaging of the transported substance, sometimes called the cargo, into a membrane vesicle or vacuole. **Table 5.3** describes some examples.

**Exocytosis** During **exocytosis,** material inside the cell is packaged into vesicles and then excreted into the extracellular environment (**Figure 5.20**). These vesicles are usually derived from the Golgi apparatus. As the vesicles form, a specific cargo is loaded into their interior and a protein coat forms around the emerging vesicle. The assembly of coat proteins on the surface of the Golgi membrane causes the bud to form. Eventually, the bud separates from the membrane to form a vesicle. After the vesicle is released, the coat is shed. Finally, the vesicle fuses with the plasma membrane and releases the cargo into the extracellular environment.

**Endocytosis** During **endocytosis,** the plasma membrane invaginates, or folds inward, to form a vesicle that brings substances into the cell. Three types of endocytosis are receptor-mediated endocytosis, pinocytosis, and phagocytosis.

- In **receptor-mediated endocytosis,** a receptor in the plasma membrane is specific for a given cargo (**Figure 5.21**). Receptor-mediated endocytosis allows cells to take up cargo that may not be very concentrated in the extracellular environment, such as cholesterol. Cargo molecules binding to their specific receptors stimulate many receptors to aggregate, and then the coat proteins bind to the membrane. The protein coat causes the membrane to invaginate and form a vesicle. Once it is released into the cell, the vesicle sheds its coat. In most cases, the vesicle fuses with an internal membrane organelle, such as a lysosome, and the receptor releases its cargo. Then, the cargo may be directly released into the cytosol, or it may be digested into simpler building blocks before release.
- **Pinocytosis** (from the Greek, meaning cell-drinking) involves the formation of membrane vesicles from the plasma membrane as a way for cells to internalize the extracellular fluid. This allows cells to sample the extracellular solutes. Pinocytosis is particularly important in cells that are actively involved in nutrient absorption, such as cells that line the intestine in animals.
- **Phagocytosis** (from the Greek, meaning cell-eating) involves the formation of an enormous membrane vesicle called a phagosome, which engulfs a large particle such as a

**Table 5.3 Examples of Exocytosis and Endocytosis**

| Exocytosis | Description |
|---|---|
| Hormones | Certain hormones, such as insulin, are composed of polypeptides. To exert its effect, insulin is secreted into the bloodstream via exocytosis from cells of the pancreas. |
| Digestive enzymes | Digestive enzymes that function in the lumen of the small intestine are secreted via exocytosis from cells of the pancreas. |
| **Endocytosis** | **Description** |
| Uptake of vital nutrients | Many important nutrients are very insoluble in the bloodstream. Therefore, they are bound to proteins in the blood and then taken into cells via endocytosis. Examples include the uptake of lipids (bound to low-density lipoprotein) and iron (bound to transferrin protein). |
| Root nodules | Nitrogen-fixing root nodules found in certain species of plants, such as legumes, are formed by the endocytosis of bacteria. After endocytosis, the bacterial cells are contained within a membrane-enclosed compartment in the root nodules. |
| Immune system | Cells of the immune system, known as macrophages, engulf and destroy bacteria via phagocytosis. |

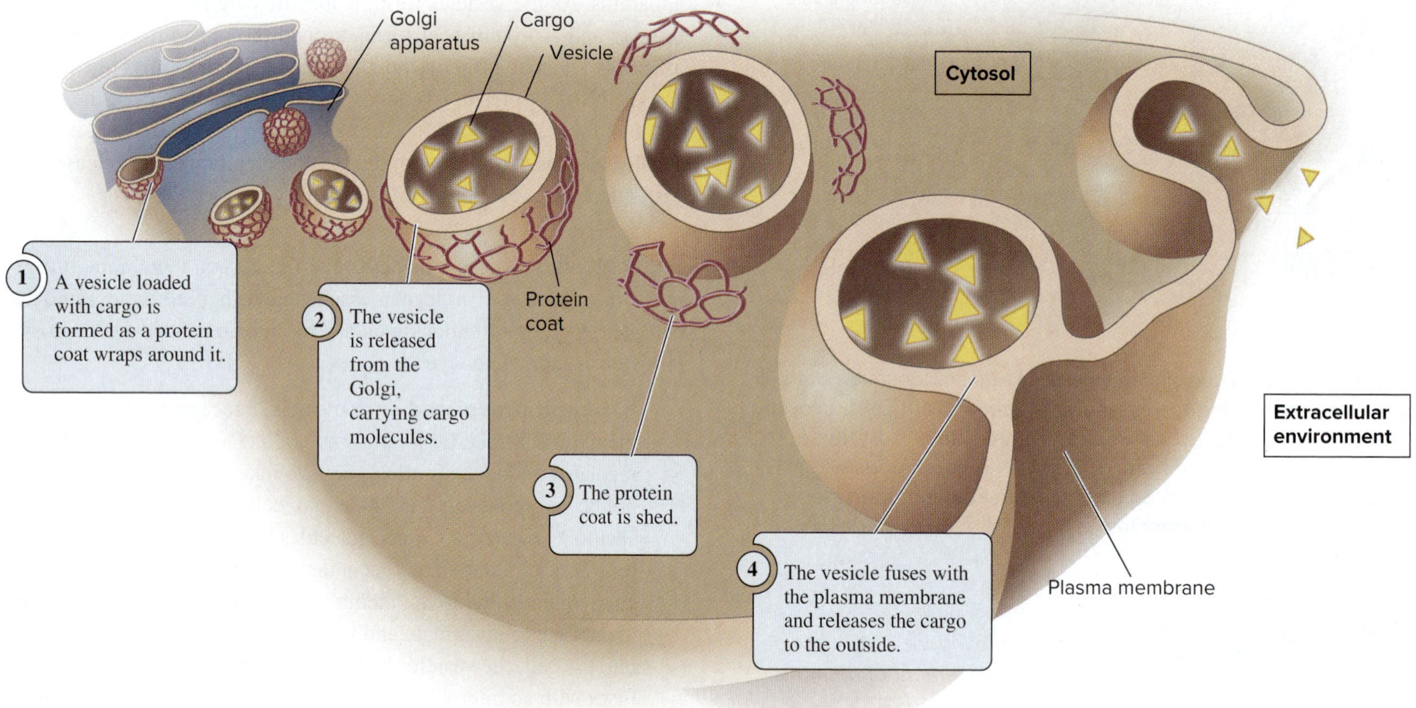


**Figure 5.20** Exocytosis.

*What is the function of the protein coat?*

bacterium. Only certain kinds of cells, termed phagocytes, can carry out phagocytosis. For example, macrophages, which are cells of the immune system in mammals, kill bacteria via phagocytosis. Macrophages engulf bacterial cells into phagosomes. Once inside the cell, the phagosome fuses with a lysosome, and the digestive enzymes within the lysosome destroy the bacterium.

## 5.6 Reviewing the Concepts

- In eukaryotes, exocytosis and endocytosis are used to transport large molecules and particles. Exocytosis is a process in which material inside the cell is packaged into vesicles and excreted into the extracellular environment. During endocytosis, the plasma membrane folds inward to form a vesicle that brings

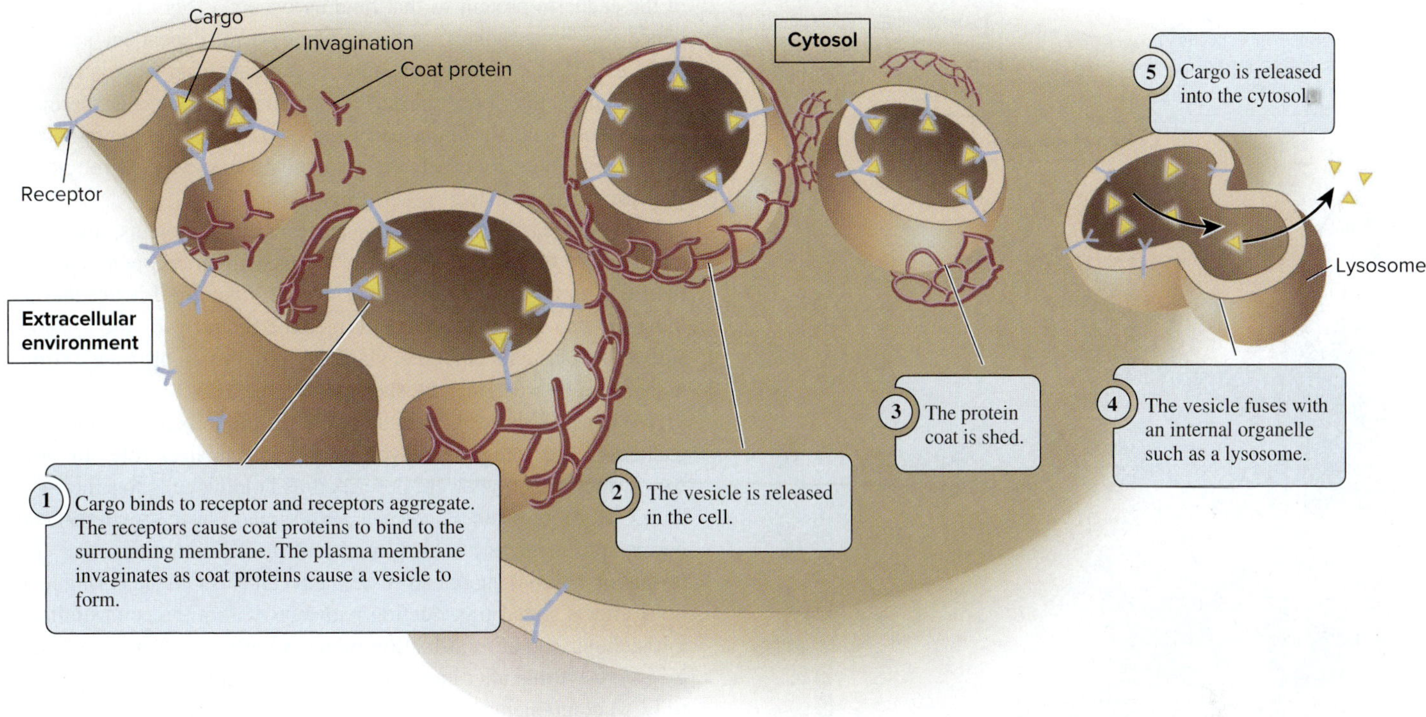


**Figure 5.21** Receptor-mediated endocytosis.

substances into the cell. Receptor-mediated endocytosis, pinocytosis, and phagocytosis are forms of endocytosis (Figures 5.20, 5.21, Table 5.3).

## 5.6 Testing Your Knowledge

1. A form of endocytosis that involves the formation of an enormous intracellular vesicle is called
   a. receptor-mediated endocytosis.
   b. pinocytosis.
   c. phagocytosis.
   d. all of the above.
   e. both a and b.

## 5.7 Cell Junctions

### Learning Outcome

1. Outline the structure and function of anchoring junctions and tight junctions.

To become a multicellular organism, cells within the body must be linked to each other. In Section 5.5, we considered junctions that form intercellular channels that allow the movement of solutes between adjacent cells. In this section, we will explore junctions that function to physically adhere animal cells to each other or to the extracellular matrix (ECM).

### Anchoring Junctions Link Animal Cells to Each Other and to the Extracellular Matrix (ECM)

Electron microscopy allows researchers to explore the types of junctions that occur between adjacent cells and between cells and the extracellular matrix (ECM), collectively called **anchoring junctions.** These junctions are particularly common in parts of the body where the cells are tightly connected and form linings. An example is the layer of cells that line the small intestine. Anchoring junctions keep intestinal cells tightly adhered to one another, thereby forming a strong barrier between the lumen of the intestine and the blood. A key component of anchoring junctions that form the actual connections are integral membrane proteins called **cell adhesion molecules (CAMs).** Two types of CAMs are cadherins and integrins.

Anchoring junctions are grouped into four main categories, according to their functional roles and their connections to cellular components. Figure 5.22 shows these junctions between cells of the mammalian small intestine.

- **Adherens junctions** connect cells to each other via cadherins. In many cases, these junctions are organized into bands around cells. On the cytosolic side of the plasma membrane, adherens junctions bind to cytoskeletal filaments called actin filaments described in Chapter 4 (refer back to Table 4.1).
- **Desmosomes** also connect cells to each other via cadherins. They are spotlike points of intercellular contact that rivet cells together. Desmosomes are connected to cytoskeletal filaments called intermediate filaments described in Chapter 4 (refer back to Table 4.1).

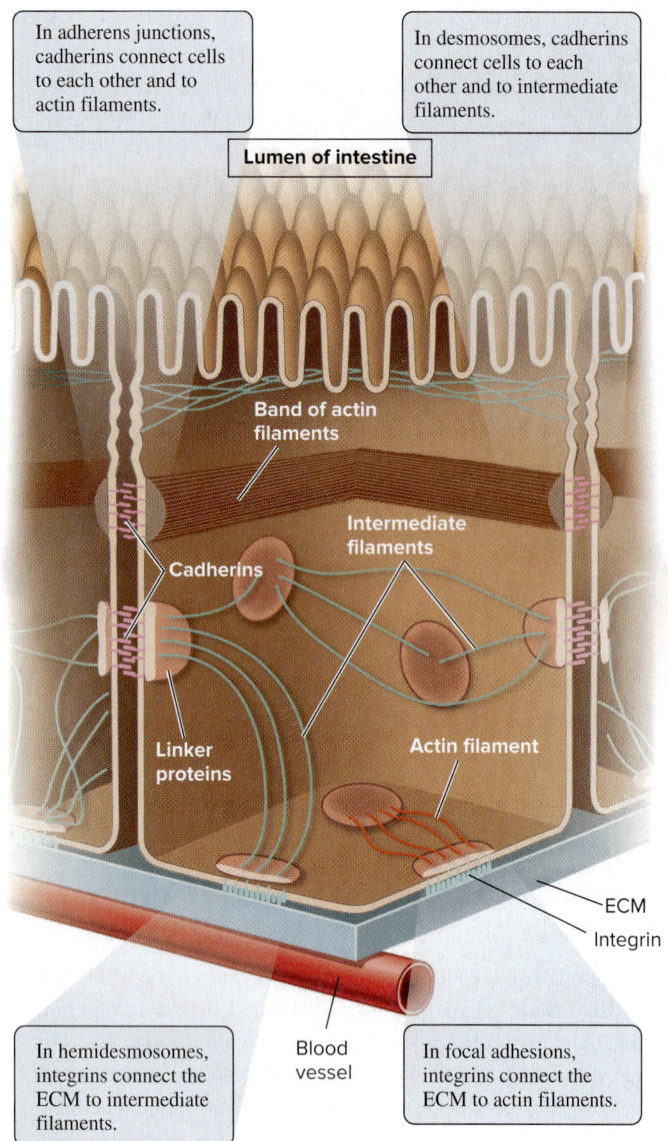


**Figure 5.22** **Types of anchoring junctions.** This figure shows these junctions in three adjacent intestinal cells. The tops of these cells face the lumen of the intestine, whereas the bottoms are adjacent to the ECM and face the blood.

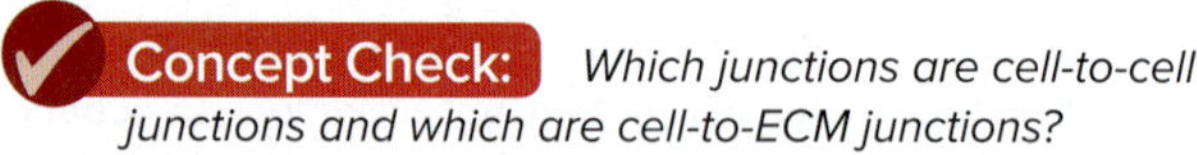

*Which junctions are cell-to-cell junctions and which are cell-to-ECM junctions?*

- **Hemidesmosomes** connect cells to the ECM via integrins. Like desmosomes, they interact with intermediate filaments.
- **Focal adhesions** also connect cells to the ECM via integrins. On the cytosolic side of the plasma membrane, focal adhesions bind to actin filaments.

Let's now consider the molecular components of anchoring junctions. As mentioned, **cadherins** are CAMs that create cell-to-cell junctions. Two cadherin proteins, each in adjacent cells, bind to each other to promote cell-to-cell adhesion (see Figure 5.22). This binding requires the presence of calcium ions ($Ca^{2+}$), which change the conformation of the cadherin protein so that binding occurs. (This calcium dependence is where cadherin gets its name—$Ca^{2+}$-dependent adhering molecule.) The two identical cadherin proteins form what is called a dimer. On the interior of the cell, linker proteins connect cadherins to actin or intermediate filaments of the cytoskeleton. This promotes a more stable interaction between two cells because their strong cytoskeletons are connected to each other.

**Integrins,** a group of cell-surface receptor proteins, are a second type of CAM, one that creates connections between cells and the ECM. Integrins do not require $Ca^{2+}$ to function. Like cadherins, integrins also bind to actin or intermediate filaments on the cytosolic side of the plasma membrane, via linker proteins, to promote a strong association between the cytoskeleton and the ECM.

When CAMs were first discovered, researchers imagined that cadherins and integrins played only a mechanical role. In other words, their functions were described as holding cells together or to the ECM. More recently, however, experiments have shown that cadherins and integrins are also important in cell communication. The formation or breaking of cell-to-cell and cell-to-ECM anchoring junctions affects signal transduction pathways within the cell. Similarly, intracellular signal transduction pathways affect cadherins and integrins in ways that alter intercellular junctions and the binding of cells to ECM components.

Abnormalities in CAMs such as integrins are associated with the ability of cancer cells to metastasize, that is, to move to other parts of the body. CAMs are critical for keeping cells in their correct locations. When they become defective due to cancer-causing mutations, cells lose their proper connections with the ECM and adjacent cells and may move to other parts of the body.

## Tight Junctions Prevent the Leakage of Materials Across Animal Cell Layers

In animals, **tight junctions** are a second type of junction, one that forms a tight seal between adjacent cells, thereby preventing material from leaking between cells. As an example, let's consider the intestine. The cells that line the intestine form a sheet that is one cell thick. One side faces the intestinal lumen, and the other faces the ECM and blood vessels (**Figure 5.23**). Tight junctions are formed between these cells.

Tight junctions are made by integral membrane proteins, called occludin and claudin, that form interlaced strands in the plasma membrane (see inset in Figure 5.23). These strands of proteins, each in adjacent cells, bind to each other, thereby forming a tight seal between cells. Tight junctions are not mechanically strong like anchoring junctions, because they do not have strong connections with the cytoskeleton. Therefore, adjacent cells that have tight junctions also have anchoring junctions to hold the cells in place.

Tight junctions perform several important functions. Let's consider a few examples.

- Tight junctions between intestinal cells prevent the leakage of materials from the lumen of the intestine into the blood, and vice versa.
- Tight junctions help maintain the polarity of intestinal cells by preventing the lateral diffusion of integral membrane proteins between the apical side (which faces the lumen of the intestine)

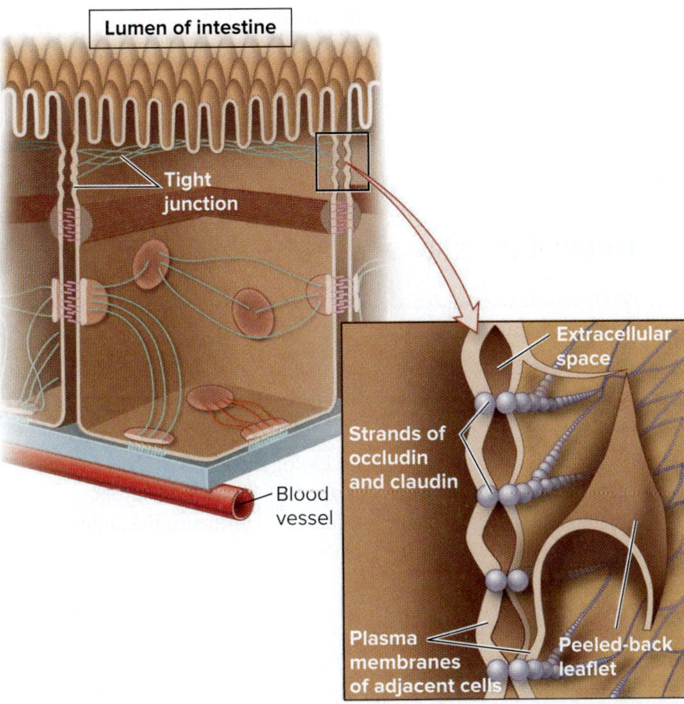


**Figure 5.23 Tight junctions between adjacent intestinal cells.** In this example, tight junctions form a seal between cells. The inset shows the interconnected network of occludin and claudin that forms the tight junction.

**Concept Check:** *What do you think is the role of tight junctions in the layers of your skin?*

**BioConnections:** *Look ahead to Figure 37.8. What problems could arise if tight junctions did not connect the cells that line your small intestine?*

and the basolateral side (which faces a blood vessel). For example, proteins involved with receptor-mediated endocytosis are restricted to the apical side, and proteins involved with exocytosis are located at the basolateral side. Thus, intestinal cells are able to take up nutrients from the intestinal lumen and export them into the bloodstream, a phenomenon called **transepithelial transport.**

- Tight junctions prevent microbes from entering the body. In mammals, the skin on the exterior of the body and the lining of the digestive tract are formed from interconnected cells that have tight junctions. Some pathogenic microorganisms, such as those that cause certain forms of diarrhea, are able to cause infection by disrupting tight junctions.

## 5.7 Reviewing the Concepts

- Anchoring junctions attach cells to each other and to the ECM. The four types of anchoring junctions are adherens junctions, desmosomes, hemidesmosomes, and focal adhesions (Figure 5.22).
- Key components of anchoring junctions are integral membrane proteins called cell adhesion molecules (CAMs). Cadherins and integrins are two types of CAMs. Cadherins link cells to each other, whereas integrins link cells to the ECM. On the cytosolic side of the plasma membrane, CAMs bind to actin or intermediate filaments.
- Tight junctions between cells, composed of occludin and claudin, prevent the leakage of materials across a layer of cells (Figure 5.23).

## 5.7 Testing Your Knowledge

1. Anchoring junctions that adhere adjacent animal cells to each other are
   a. adherens junctions.
   b. desmosomes.
   c. hemidesmosomes.
   d. focal adhesions.
   e. both a and b.

# Assess and Discuss

## Test Yourself

1. Which of the following statements best describes the chemical composition of biological membranes?
   a. Biological membranes are bilayers of proteins with associated lipids and carbohydrates.
   b. Biological membranes are composed of two layers: one layer of phospholipids and one layer of proteins.
   c. Biological membranes are bilayers of phospholipids with associated proteins and carbohydrates.
   d. Biological membranes are composed of equal numbers of phospholipids, proteins, and carbohydrates.
   e. Biological membranes are composed of lipids with proteins attached to the outer surface.
2. Which of the following events in a biological membrane would *not* be energetically favorable and therefore *not* occur spontaneously?
   a. the rotation of phospholipids
   b. the lateral movement of phospholipids
   c. the flip-flop of phospholipids to the opposite leaflet
   d. the rotation of membrane proteins
   e. the lateral movement of membrane proteins
3. Let's suppose an insect, which doesn't maintain a constant body temperature, was exposed to a shift in temperature from 60°F to 80°F. Which of the following types of membrane changes would be the most beneficial to help this animal cope with the temperature shift?
   a. increase in the number of double bonds in the nonpolar tails of phospholipids
   b. increase in the length of the nonpolar tails of phospholipids
   c. decrease in the amount of cholesterol in the membrane
   d. decrease in the amount of carbohydrate attached to membrane proteins
   e. decrease in the amount of carbohydrate attached to phospholipids
4. Carbohydrates of the plasma membrane
   a. are bonded to a protein or lipid.
   b. are located on the outer surface of the plasma membrane.
   c. can function in cell surface recognition.
   d. all of the above.
   e. a and c only.

5. Which of the following movements would *not* be an example of passive transport?
   a. the movement of water through aquaporin
   b. the intercellular transport of molecules via gap junctions
   c. the transport of $Na^+$ and $K^+$ via $Na^+/K^+$-ATPase
   d. the diffusion of $CO_2$ through a phospholipid bilayer
   e. All of the above are examples of passive transport.
6. The tendency for $Na^+$ to move into the cell can be due to
   a. the higher numbers of $Na^+$ outside the cell, resulting in a chemical concentration gradient.
   b. the net negative charge inside the cell, attracting the positively charged $Na^+$.
   c. the attractive force of $K^+$ inside the cell, pulling $Na^+$ into the cell.
   d. all of the above.
   e. a and b only.
7. Let's suppose the solute concentration inside the cells of a plant is 0.3 M and outside is 0.2 M. If we assume that the solutes do not readily cross the membrane, which of the following statements best describes what will happen?
   a. The plant cells will lose a lot of water, and the cells will crenate.
   b. The plant cells will lose a little water, and the cells will expand.
   c. The plant cells will take up a lot of water, and the cells will undergo osmotic lysis.
   d. The plant cells will take up a little water, and the plasma membrane of the cells will slightly push against the cell wall.
   e. Both a and b are correct.
8. What features of a membrane are a major contributor to its selective permeability?
   a. phospholipid bilayers
   b. transport proteins
   c. glycolipids on the outer surface of the membrane
   d. peripheral membrane proteins
   e. both a and b
9. What is the process in which solutes are moved across a membrane against their concentration gradient?
   a. simple diffusion
   b. facilitated diffusion
   c. osmosis
   d. passive transport
   e. active transport
10. Large particles can be brought into the cell by
   a. facilitated diffusion.
   b. active transport.
   c. phagocytosis.
   d. exocytosis.
   e. all of the above.

## Conceptual Questions

1. With your textbook closed, draw and describe the fluid-mosaic model of membrane structure.
2. Describe two different ways that integral membrane proteins associate with a membrane. How do peripheral membrane proteins associate with a membrane?
3. **PRINCIPLES** A principle of biology is that living organisms interact with their environment. Discuss how lipid bilayers, channels, transporters, and cell junctions influence the ability of cells to interact with their environment.

## Collaborative Questions

1. Proteins in the plasma membrane are often the target of medicines. Discuss why you think this is the case. How would you determine experimentally that a specific membrane protein was the target of a drug?
2. With regard to bringing solutes into the cell across the plasma membrane, discuss the advantages and disadvantages of simple diffusion, facilitated diffusion, active transport, and endocytosis.

# How Cells Utilize Energy

# 6

**Exercise and mitochondria.** Regular exercise increases the number of mitochondria in muscle cells to meet the higher energy demands.

## Chapter Outline



The human body adapts to environmental and behavioral changes. Many research studies have shown that regular exercise increases the number of mitochondria in muscle cells. Why is this change beneficial? During exercise, muscle cells need more energy. Mitochondria play a key role in the production of ATP, which is used to power a variety of processes, including muscle contraction. By increasing the number of mitochondria, muscle cells have a greater capacity to generate ATP. Exercise is a powerful inducer of mitochondrial biogenesis. Recent research has even shown that exercise may increase the number of mitochondria in brain cells. Therefore, exercise may not only enhance muscle function but also may improve brain function.

We begin this chapter with a discussion of energy and how it is used by living organisms to power a variety of processes. Much of our discussion will center on **chemical reactions,** which are processes in which one or more substances are changed into other substances (see Chapter 2). Every living cell continuously performs thousands of chemical reactions to sustain life. We will examine what factors control the direction of a chemical reaction and what determines its rate, paying particular attention to the role of enzymes.

**Metabolism** is the sum total of all chemical reactions that occur within a cell or an organism. Metabolism also refers to a specific set of chemical reactions occurring at the cellular level. For example, biologists may speak of sugar metabolism or fat metabolism. In this chapter, we will consider metabolism at the cellular level, examining how chemical reactions are often coordinated with each other in metabolic pathways. We will largely focus on a group of chemical reactions that involve the breakdown of carbohydrates, namely, the sugar glucose. As you will learn, mitochondria play a key role in this process.

## 6.1 Energy and Chemical Reactions

### Learning Outcomes

1. Define energy, and distinguish between potential and kinetic energy.
2. State the first and second laws of thermodynamics, and discuss how they relate to living things.
3. Explain how the change in free energy determines the direction of a chemical reaction and how chemical reactions eventually reach a state of equilibrium.
4. Distinguish between exergonic and endergonic reactions.
5. Describe how cells use the energy released by ATP hydrolysis to drive endergonic reactions.

Two general factors govern the fate of a given chemical reaction in a living cell: its direction and rate. In this section, we will begin by examining the interplay of energy and the concentration of reactants as they govern the direction of a chemical reaction. You will learn that cells use energy-intermediate molecules, such as ATP, to drive chemical reactions in a desired direction.

### Energy Exists in Different Forms

To understand why a chemical reaction occurs, we first need to consider **energy,** which is the ability to promote change or do work. Physicists often consider energy in two general forms: kinetic energy and potential energy (Figure 6.1). **Kinetic energy** is energy associated with movement, such as the movement of a baseball bat from one location to another. By comparison, **potential energy** is the energy that a substance or an object possesses due to its structure or location, such as the location of an arrow when a bow is drawn. The energy contained within covalent bonds in molecules is a type of potential energy called **chemical potential energy.** The breaking of those bonds is one way that energy can be released, which is then used by living cells to perform various functions. **Table 6.1** summarizes various forms of energy that are common in biological systems.

**(a) Kinetic energy**

**(b) Potential energy**

**Figure 6.1 Examples of energy. (a)** Kinetic energy, such as swinging a bat, is energy associated with motion. **(b)** Potential energy is stored energy, as in a bow that is ready to fire an arrow.
*(a)* © moodboard/Corbis RF; *(b)* © amanaimages/Corbis RF

An important factor in biology is the ability of energy to be converted from one form to another. The study of energy transfers and transformations is called **thermodynamics.** Physicists have determined that four laws govern thermodynamics, the following two of which we'll study here:

1. The **first law of thermodynamics**—Energy cannot be created or destroyed. However, energy can be transferred from one place to another and can be transformed from one type to another (as when, for example, chemical energy is transformed into heat).

**Table 6.1 Types of Energy That Are Important in Biology**

| Energy type | Description | Biological example |
|---|---|---|
| Light | Light is a form of electromagnetic radiation that is visible to the eye. The energy of light is packaged in photons. | During photosynthesis, light energy is captured by pigments in chloroplasts (described in Chapter 7). Ultimately, this energy is used to produce organic molecules. |
| Heat | Heat is the transfer of kinetic energy from one object to another or from an energy source to an object. In biology, heat is often viewed as kinetic energy that can be transferred due to a difference in temperature between two objects or locations. | Many organisms, including humans, maintain their bodies at a constant temperature. This is achieved, in part, by chemical reactions that generate heat. |
| Mechanical | Mechanical energy is the energy possessed by an object due to its motion or its position relative to other objects. | In animals, mechanical energy is associated with movements due to muscle contraction such as walking. |
| Chemical potential | Chemical potential energy is potential energy stored in the electrons of molecules. When bonds are broken and rearranged, energy may be released. | The covalent bonds in organic molecules, such as glucose and ATP, store large amounts of energy. When bonds are broken in larger molecules to form smaller molecules, the chemical energy that is released can be used to drive cellular processes. |
| Electrical/ion gradient | The movement of charge or the separation of charge can provide energy. Also, a difference in ion concentration across a membrane constitutes an electrochemical gradient, which is a source of potential energy. | During a stage of cellular respiration called oxidative phosphorylation (described in this chapter), an $H^+$ gradient provides the energy to drive ATP synthesis. |

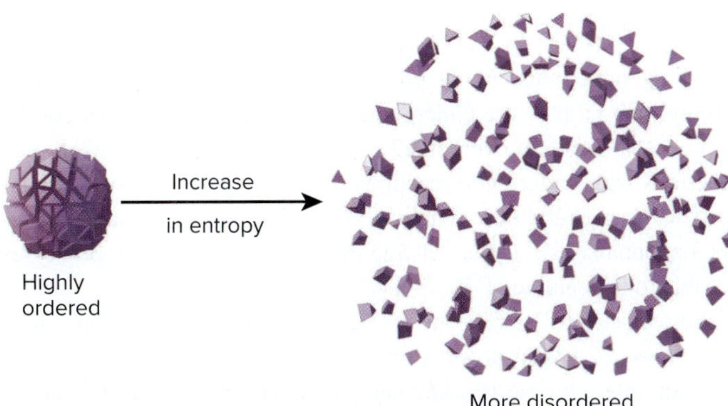


**Figure 6.2** **Entropy, a measure of the disorder of a system.** An increase in entropy means an increase in disorder.

**Concept Check:** *Which do you think has more entropy, an NaCl crystal at the bottom of a beaker of water or the same beaker of water after the $Na^+$ and $Cl^-$ in the crystal have dissolved in the water?*

2. The **second law of thermodynamics**—Any energy transfer or transformation from one form to another increases the degree of disorder of a system, called **entropy** (Figure 6.2). Entropy is a measure of the randomness of molecules in a system. When a physical system becomes more disordered, the entropy increases. As the energy becomes more evenly distributed, that energy is less able to promote change or do work.

## The Change in Free Energy Determines the Direction of a Chemical Reaction

Energy is required for many cellular processes, including chemical reactions, cellular movements such as those occurring in muscle contraction, and the maintenance of cell organization. To understand how organisms use energy, we need to distinguish between the energy that can be used to promote change or do work (usable energy) and the energy that cannot (unusable energy):

$$\text{Total energy} = \text{Usable energy} + \text{Unusable energy}$$

Why is some energy unusable? The main culprit is entropy. As stated by the second law of thermodynamics, energy transfers or transformations involve an increase in entropy, a degree of disorder that cannot be harnessed in a useful way. The total energy is termed **enthalpy (*H*),** and the usable energy—the amount of available energy that can be used to promote change or do work—is called the **free energy (*G*).** The letter *G* is in recognition of American physicist J. Willard Gibbs, who proposed the concept of free energy in 1878. The unusable energy is the system's entropy (*S*). Gibbs proposed that these three factors are related to each other in the following way:

$$H = G + TS$$

where *T* is the absolute temperature in kelvins (K). Because our focus is on free energy, we can rearrange this equation as

$$G = H - TS$$

A critical issue in biology is whether a process does or does not occur spontaneously. For example, will glucose be broken down into carbon dioxide and water? Another way of framing this question is to ask "Is the breakdown of glucose a spontaneous reaction?" A spontaneous reaction or process is one that occurs in a particular direction without being driven by an input of energy. The key way to evaluate if a chemical reaction is spontaneous is to determine the free-energy change that occurs as a result of the reaction:

$$\Delta G = \Delta H - T\Delta S$$

where the symbol Δ (the Greek letter delta) indicates a change, such as before and after a chemical reaction. If a chemical reaction has a negative free-energy change ($\Delta G < 0$), the products have less free energy than the reactants, and therefore free energy is released during product formation. Such a reaction is said to be **exergonic.** Exergonic reactions are spontaneous. Alternatively, if a reaction has a positive free-energy change ($\Delta G > 0$), requiring the addition of free energy, it is termed **endergonic.** An endergonic reaction is not a spontaneous reaction.

If $\Delta G$ for a chemical reaction is negative, the reaction favors the conversion of reactants to products, whereas a reaction with a positive $\Delta G$ favors the formation of reactants. Chemists have determined free-energy changes for a variety of chemical reactions, which allows them to predict their direction. As an example, let's consider **adenosine triphosphate (ATP),** which is a molecule commonly used to drive cellular processes. ATP is broken down to adenosine diphosphate (ADP) and inorganic phosphate ($HPO_4^{2-}$, abbreviated $P_i$). Because water is used to remove a phosphate group, chemists refer to this reaction as the hydrolysis of ATP (Figure 6.3). In converting 1 mole of ATP to 1 mole of ADP and $P_i$, $\Delta G$ equals –7.3 kcal/mol. Because this is a negative value, the reaction strongly favors the formation of products.

## Chemical Reactions Eventually Reach a State of Equilibrium

Even when a chemical reaction is associated with a negative free-energy change, not all of the reactants are converted to products. The reaction reaches a state of **chemical equilibrium** in which the rate of formation of products equals the rate of formation of reactants. Let's consider the generalized reaction

$$a\text{A} + b\text{B} \rightleftharpoons c\text{C} + d\text{D}$$

where A and B are the reactants, C and D are the products, and *a*, *b*, *c*, and *d* are the number of moles of reactants and products. An equilibrium occurs, such that

$$K_{eq} = \frac{[\text{C}]^c[\text{D}]^d}{[\text{A}]^a[\text{B}]^b}$$

where $K_{eq}$ is the equilibrium constant. Each type of chemical reaction has a specific value for $K_{eq}$. When $K_{eq}$ is greater than 1, the reaction favors the formation of products; when it is less than 1, the reaction favors the formation of reactants.

Biologists make two simplifying assumptions when determining values for equilibrium constants. First, the concentration of water does not change during the reaction, and the pH remains constant

## Biology Principle

### Living Organisms Use Energy

The hydrolysis of ATP is an exergonic reaction that is used to drive many cellular reactions.

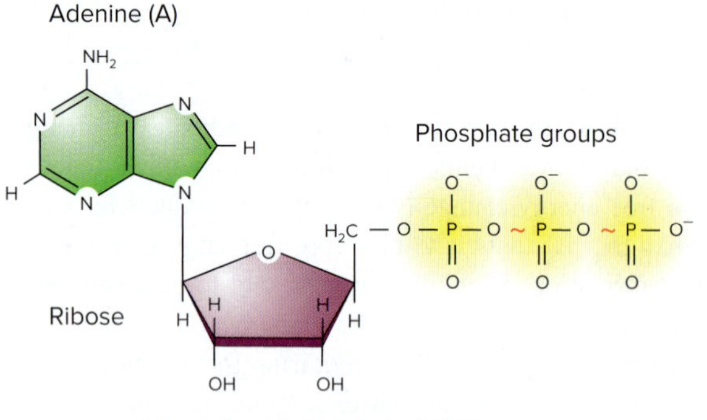


**Figure 6.3** **The hydrolysis of ATP to ADP and $P_i$.** As shown in this figure, ATP has a net charge of –4, and ADP and $P_i$ are shown with net charges of –2 each. When these compounds are shown in chemical reactions with other molecules, the net charges are also indicated. Otherwise, these compounds are simply designated ATP, ADP, and $P_i$. At neutral pH, $ADP^{2-}$ dissociates to $ADP^{3-}$ and $H^+$.

at pH 7. The equilibrium constant under these conditions is designated with a prime symbol, $K_{eq}'$. If water is one of the reactants, as in a hydrolysis reaction, it is not included in the chemical equilibrium equation. As an example, let's consider the chemical equilibrium for the hydrolysis of ATP.

$$ATP^{4-} + H_2O \rightleftharpoons ADP^{2-} + P_i^{2-}$$

$$K_{eq}' = \frac{[ADP][P_i]}{[ATP]}$$

Experimentally, the value for $K_{eq}'$ for this reaction has been determined to be approximately 1,650,000 M. Such a large value indicates that the equilibrium greatly favors the formation of products—ADP and $P_i$.

## Cells Use ATP to Drive Endergonic Reactions

In living organisms, many processes require the addition of free energy; that is, they are endergonic and do not occur spontaneously. How is this problem overcome? One strategy is to couple exergonic reactions with endergonic reactions. If an exergonic reaction is coupled with an endergonic reaction, the endergonic reaction will proceed spontaneously if the net free-energy change for both processes combined is negative.

As discussed later in this chapter, the breakdown of sugars, such as glucose, is a vital process for living organisms to utilize energy. For example, glucose is a key energy source for your brain. Certain steps in glucose metabolism are endergonic and are coupled to ATP hydrolysis. An example is described next:

$$\text{Glucose} + \text{Phosphate}^{2-} \longrightarrow \text{Glucose-6-phosphate}^{2-} + H_2O \qquad \Delta G = +3.3\ \text{kcal/mol}$$

$$ATP^{4-} + H_2O \longrightarrow ADP^{2-} + P_i^{2-} \qquad \Delta G = -7.3\ \text{kcal/mol}$$

Coupled reaction:

$$\text{Glucose} + ATP^{4-} \longrightarrow \text{Glucose-6-phosphate}^{2-} + ADP^{2-} \qquad \Delta G = -4.0\ \text{kcal/mol}$$

The first reaction, in which phosphate is covalently attached to glucose, is endergonic and by itself is not spontaneous. The second, the hydrolysis of ATP, is exergonic. If the two reactions are coupled, the net free-energy change for both reactions combined is negative ($\Delta G$ = –4.0 kcal/mol); the coupled reaction is exergonic. In the coupled reaction, a phosphate is directly transferred from ATP to glucose, in a process called **phosphorylation.** This coupled reaction proceeds spontaneously because the net free-energy change is negative. Exergonic reactions, such as the breakdown of ATP, are commonly coupled to chemical reactions and other cellular processes that would otherwise be endergonic.

## 6.1 Reviewing the Concepts

- The fate of a chemical reaction is determined by its direction and rate.
- Energy, the ability to promote change or do work, exists in many forms. According to the first law of thermodynamics, energy cannot be created or destroyed, but it can be converted from one form to another. The second law of thermodynamics states that energy transfers or transformations involve an increase in entropy (Figures 6.1, 6.2, Table 6.1).
- Free energy is the amount of available energy that can be used to promote change or do work. Spontaneous or exergonic reactions, which release free energy, have a negative free-energy change, whereas endergonic reactions have a positive free-energy change (Figure 6.3).
- Chemical reactions proceed until they reach a state of chemical equilibrium, where the rate of formation of products equals the rate of formation of reactants.
- Cells couple exergonic reactions, such as the hydrolysis of ATP, to endergonic reactions, with the result that endergonic reactions proceed spontaneously.

## 6.1 Testing Your Knowledge

1. When a reaction or process is exergonic,
   a. the free-energy change is negative.
   b. it is spontaneous.
   c. it does not need added energy to proceed.
   d. the entropy usually increases.
   e. all of the above are correct.
2. Cells often use ATP hydrolysis to
   a. decrease their water content.
   b. drive endergonic reactions.
   c. prevent exergonic reactions.
   d. achieve chemical equilibrium.
   e. do all of the above.

## 6.2 Enzymes

### Learning Outcomes

1. Explain how enzymes increase the rates of chemical reactions by lowering the activation energy.
2. Describe how enzymes bind their substrates with high specificity and undergo induced fit.
3. **SCISKILLS ▶** Analyze the velocity of chemical reactions, and evaluate the effects of competitive and noncompetitive inhibitors.
4. Explain how factors, such as nonprotein molecules or ions, temperature, and pH, influence enzyme activity.

For most chemical reactions in cells to proceed at a rapid pace, a catalyst is needed. A **catalyst** is an agent that speeds up the rate of a chemical reaction without being permanently changed or consumed by it. In living cells, the most common catalysts are **enzymes,** which are proteins. The term was coined in 1876 by a German physiologist, Wilhelm Kühne, who discovered trypsin, an enzyme in pancreatic juice that is needed for the digestion of food proteins. In this section, we will explore how enzymes work.

### Enzymes Increase the Rates of Chemical Reactions

If a chemical reaction has a negative free-energy change, the reaction will be spontaneous; it will tend to proceed in the direction of reactants to products. Although thermodynamics governs the direction of an energy transformation, it does not determine the rate of a chemical reaction. For example, the breakdown of the molecules in gasoline to smaller molecules is an exergonic reaction. Even so, we could place gasoline and oxygen in a container and nothing much would happen (provided it wasn't near a flame). If we came back several days later, we would expect to see the gasoline still sitting there. Perhaps if we came back in a few million years, the gasoline would have been broken down. On a timescale of months or a few years, however, the chemical reaction would proceed very slowly.

In living cells, the rates of enzyme-catalyzed reactions typically occur millions of times faster than the corresponding uncatalyzed reactions. A dramatic example is the enzyme catalase, which catalyzes the breakdown of hydrogen peroxide ($H_2O_2$) into water and oxygen. Catalase speeds up this reaction $10^{15}$-fold faster than the uncatalyzed reaction!

Why are catalysts necessary to speed up a chemical reaction? Chemical reactions between molecules involve bond breaking and bond forming. The process of either breaking or forming a covalent bond may begin with the straining or stretching of one or more bonds in the starting molecule(s), and/or it may involve the positioning of two molecules so that they interact with each other properly. Enzymes help to facilitate these types of events. As an example, let's consider the reaction in which ATP is used to phosphorylate glucose:

$$\text{Glucose} + \text{ATP}^{4-} \longrightarrow \text{Glucose-6-phosphate}^{2-} + \text{ADP}^{2-}$$

For this reaction to occur between glucose and ATP, the molecules must collide in the correct orientation and possess enough energy so the chemical bonds can be changed. As glucose and ATP get close together, the electrons in the outer shells of their atoms repel each other. To overcome this repulsion, an initial input of energy, called the **activation energy,** is required (Figure 6.4). Activation

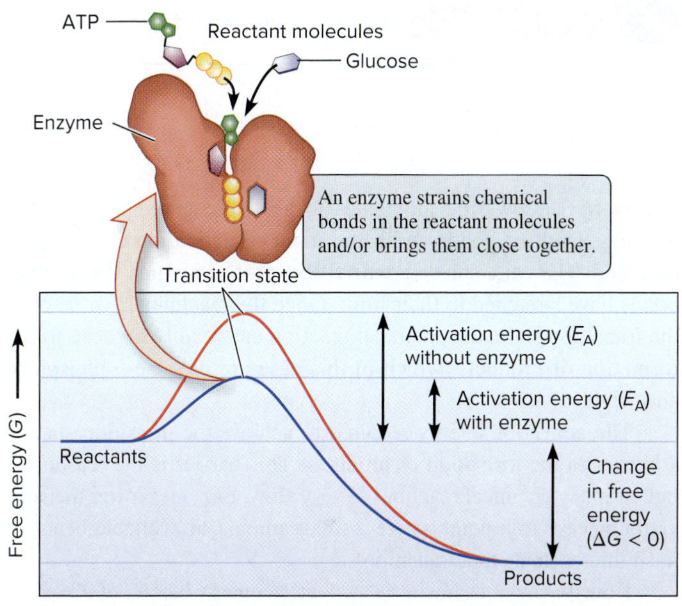


**Figure 6.4** **Activation energy of a chemical reaction.** This figure depicts an exergonic reaction. The activation energy ($E_A$) is needed for molecules to achieve a transition state. One way that enzymes lower the $E_A$ is by straining chemical bonds in the reactants so less energy is required to attain the transition state. A second way is by binding two reactants so they are close to each other and in a favorable orientation.

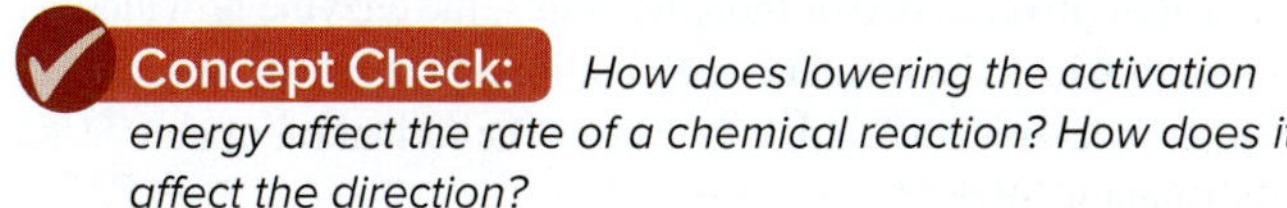

**Concept Check:** *How does lowering the activation energy affect the rate of a chemical reaction? How does it affect the direction?*

## Biology Principle

### Structure Determines Function

A key function of enzymes is their ability to bind their substrates with high specificity. This is due to the structure of the enzyme's active site.

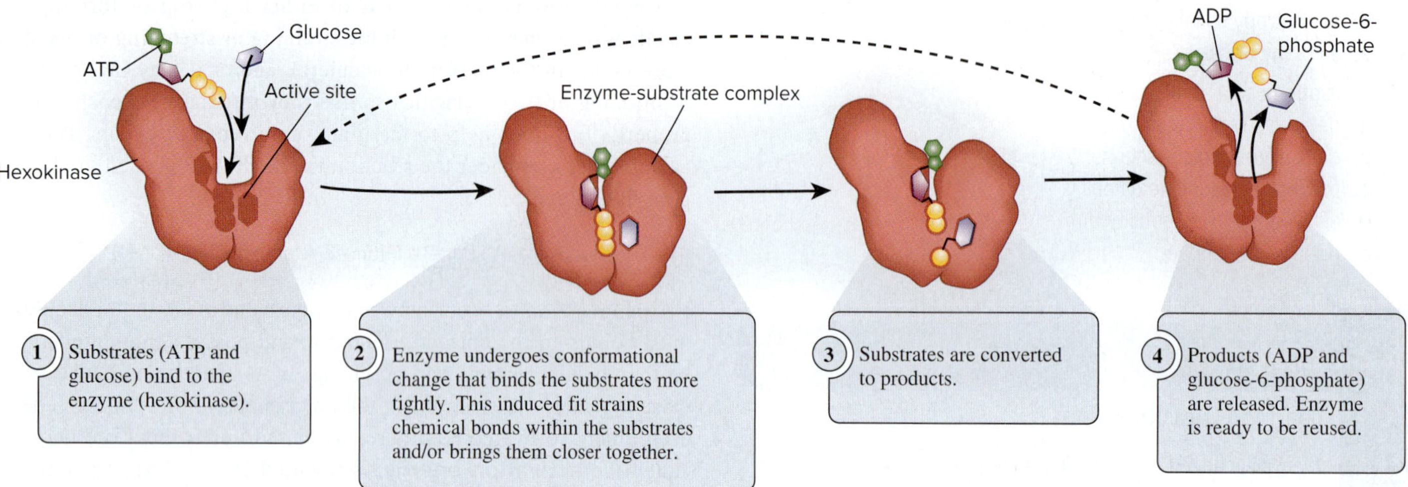

Glucose
ATP
Active site
Hexokinase
Enzyme-substrate complex
ADP
Glucose-6-phosphate
1 Substrates (ATP and glucose) bind to the enzyme (hexokinase).
2 Enzyme undergoes conformational change that binds the substrates more tightly. This induced fit strains chemical bonds within the substrates and/or brings them closer together.
3 Substrates are converted to products.
4 Products (ADP and glucose-6-phosphate) are released. Enzyme is ready to be reused.


**Figure 6.5 The steps of an enzyme-catalyzed reaction.** The example shown here involves the enzyme hexokinase, which binds glucose and ATP. The products are glucose-6-phosphate and ADP, which are released from the enzyme.

*During which step is the activation energy lowered?*

energy ($E_A$) allows the molecules to get close enough to cause a rearrangement of bonds. With the input of activation energy, glucose and ATP can achieve a **transition state** in which the original bonds have stretched to their limit. Once the reactants have reached the transition state, the chemical reaction can readily proceed to the formation of products, which in this case are glucose-6-phosphate and ADP.

The activation energy required to achieve the transition state is a barrier to the formation of products. This barrier is the reason the rate of many chemical reactions is very slow. Enzymes lower the activation energy to a point where a small amount of available heat can push the reactants to a transition state.

How do enzymes lower the activation energy barrier of chemical reactions? The following are two common mechanisms:

- **Straining reactants:** Enzymes are large proteins that bind relatively small reactants. When bound to an enzyme, the bonds in the reactants can be strained, thereby making it easier for them to achieve the transition state (see Figure 6.4).
- **Positioning reactants close together:** When a chemical reaction involves two or more reactants, the enzyme provides sites in which the reactants are positioned very close to each other in an orientation that facilitates the formation of new covalent bonds.

### Enzymes Recognize Their Substrates with High Specificity and Undergo Conformational Changes

Thus far, we have considered how enzymes lower the activation energy of a chemical reaction and thereby increase its rate. Let's consider some other features of enzymes that enable them to serve as effective catalysts in chemical reactions. The **active site** is the location in an enzyme where the chemical reaction takes place. The **substrates** for an enzyme are the reactant molecules that bind to an enzyme at the active site and participate in the chemical reaction. For example, hexokinase is an enzyme whose substrates are glucose and ATP (**Figure 6.5**). The binding between an enzyme and substrate produces an **enzyme-substrate complex.**

A key feature of nearly all enzymes is their ability to bind their substrates with a high degree of specificity. For example, hexokinase recognizes glucose but does not recognize other similar sugars, such as fructose and galactose, very well. In 1894, German chemist Emil Fischer proposed that the recognition of a substrate by an enzyme resembles the interaction between a lock and key: Only the right-sized key (the substrate) will fit into the keyhole (active site) of the lock (the enzyme). Further research revealed that the interaction between an enzyme and its substrates also involves movements or conformational changes in the enzyme itself. As shown in Figure 6.5 step 2, these conformational changes cause the substrates to bind more tightly to the enzyme, a phenomenon called **induced fit,** which was proposed by American biochemist Daniel Koshland in 1958. Only after this induced fit takes place does the enzyme catalyze the conversion of reactants to products. Induced fit is a key phenomenon that lowers the activation energy.

## Quantitative Analysis

### ENZYME FUNCTION IS INFLUENCED BY SUBSTRATE CONCENTRATION AND BY INHIBITORS

The degree of attraction between an enzyme and its substrate(s) is called the **affinity** of the enzyme for its substrate(s). Some enzymes recognize their substrates with very high affinity, which means they have a strong attraction for their substrates. Such enzymes bind their substrates even when the substrate concentration is relatively low. Other enzymes recognize their substrates with lower affinity; the enzyme-substrate complex is likely to form only when the substrate concentration is higher.

Let's consider how biologists analyze the relationship between substrate concentration and enzyme function. In the experiment of Figure 6.6a, tubes labeled A, B, C, and D each contained 1 μg of enzyme but they varied in the amount of substrate that was added. This enzyme recognizes a single substrate and converts it to a product. The samples were incubated for 60 seconds, and then the amount of product in each tube was measured. The velocity or rate of the chemical reaction is expressed as the amount of product produced per second. As we see in Figure 6.6a, the velocity increases as the substrate concentration increases but eventually reaches a plateau. Why does the plateau occur? At high substrate concentrations, nearly all of the active sites of the enzyme are occupied with substrate, so increasing the substrate concentration further has a negligible effect. At this point, the enzyme is saturated with substrate, and the velocity of the chemical reaction is near its maximal rate, called its $\boldsymbol{V_{max}}$.

Figure 6.6a also helps us understand the relationship between substrate concentration and velocity. The $\boldsymbol{K_M}$ is the substrate concentration at which the velocity is half its maximal value. The $K_M$ is also called the Michaelis constant in honor of German biochemist Leonor Michaelis, who carried out pioneering work with Canadian biochemist Maud Menten on the study of enzymes. The $K_M$ is a measure of the substrate concentration required for a chemical reaction to occur. An enzyme with a high $K_M$ requires a higher substrate concentration to achieve a particular reaction velocity than does an enzyme with a lower $K_M$.

For an enzyme-catalyzed reaction, we can view the formation of product as occurring in two steps: (1) binding or release of substrate and (2) formation of product:

$$E + S \rightleftharpoons ES \rightarrow E + P$$

where $E$ is the enzyme, $S$ is the substrate, $ES$ is the enzyme-substrate complex, and $P$ is the product.

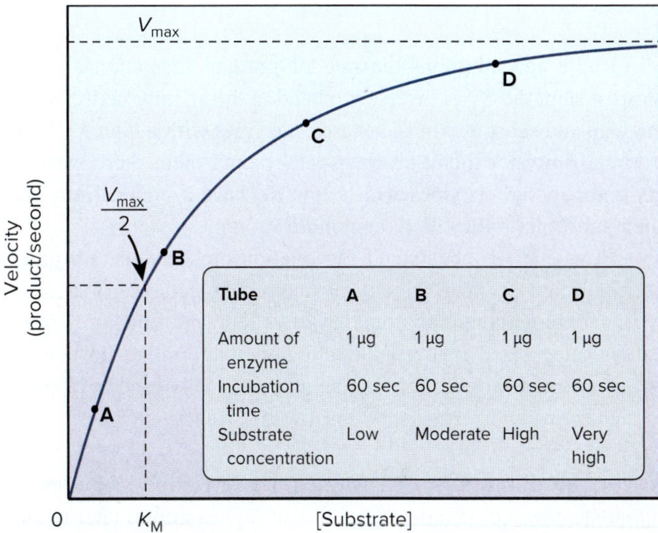


(a) Reaction velocity in the absence of inhibitors

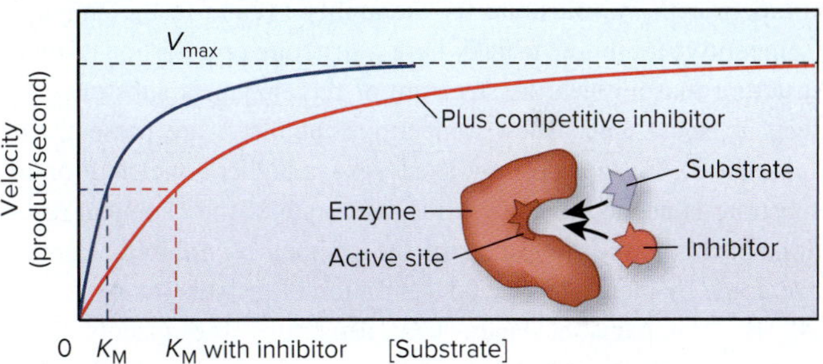


(b) Competitive inhibition

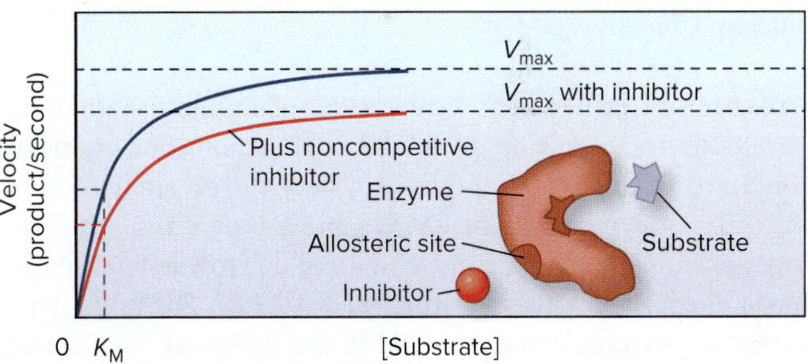


(c) Noncompetitive inhibition

**Figure 6.6 The relationship between velocity and substrate concentration in an enzyme-catalyzed reaction, and the effects of inhibitors. (a)** In the absence of an inhibitor, the maximal velocity ($V_{max}$) is achieved when the substrate concentration is high enough to be saturating. The $K_M$ value is the substrate concentration at which the velocity is half the maximal velocity. **(b)** A competitive inhibitor binds to the active site of an enzyme and raises the $K_M$ for the substrate. **(c)** A noncompetitive inhibitor binds to an allosteric site outside the active site and lowers the $V_{max}$ for the reaction.

**Concept Check:** *Enzyme A has a $K_M$ of 0.1 mM, whereas enzyme B has a $K_M$ of 1.0 mM. They both have the same $V_{max}$. If the substrate concentration was 0.5 mM, which reaction—the one catalyzed by enzyme A or B—would have the higher velocity?*

If the second step—the rate of product formation—is the slowest step, the $K_M$ is inversely related to the affinity between the enzyme and substrate. In such cases, enzymes with a high $K_M$ have a low affinity for their substrates—they bind them more weakly. By comparison, enzymes with a low $K_M$ have a high affinity for their substrates—they bind them more strongly.

Now that we understand the relationship between substrate concentration and the velocity of an enzyme-catalyzed reaction, we can explore how inhibitors may affect enzyme function. These can be categorized as reversible inhibitors that bind noncovalently to an enzyme or irreversible inhibitors that usually bind covalently to an enzyme and permanently inactivate its function.

**Reversible Inhibitors** Cells often use reversible inhibitors to modulate enzyme function. We will consider examples later in our discussion of glucose metabolism. **Competitive inhibitors** are molecules that bind noncovalently to the active site of an enzyme and inhibit the ability of the substrate to bind. Such inhibitors compete with the substrate for the ability to bind to the enzyme. Competitive inhibitors usually have a structure or a portion of their structure that mimics the structure of the enzyme's substrate. As seen in Figure 6.6b, when competitive inhibitors are present, the apparent $K_M$ for the substrate increases—a higher concentration of substrate is needed to achieve the same rate of the chemical reaction. In this case, the effects of the competitive inhibitor can be overcome by increasing the concentration of the substrate.

By comparison, Figure 6.6c illustrates the effects of a **noncompetitive inhibitor.** This type of inhibitor lowers the $V_{max}$ for the reaction without affecting the $K_M$. A noncompetitive inhibitor binds noncovalently to an enzyme at a location outside the active site, called an **allosteric site,** and inhibits the enzyme's function.

**Irreversible Inhibitors** Irreversible inhibitors usually bind covalently to an enzyme to inhibit its function. For example, some irreversible inhibitors bind covalently to an amino acid at the active site of an enzyme, thereby preventing it from catalyzing a chemical reaction. An example of an irreversible inhibitor is diisopropyl phosphorofluoridate (DIFP). DIFP is a type of nerve gas that was developed in the 1930s as a chemical weapon. This molecule binds covalently to the enzyme acetylcholinesterase, which prevents the proper communication between neurons and muscle cells.

Irreversible inhibition is not a common way for cells to control enzyme function. Why do cells usually control enzymes via reversible inhibitors? The answer is that a reversible inhibitor allows an enzyme to be used again, when the inhibitor concentration is lowered. Being able to reuse an enzyme is energy-efficient. In contrast, irreversible inhibitors permanently inactivate an enzyme, thereby preventing its further use.

**Crunching the Numbers:** As we have seen, the presence of a competitive inhibitor raises the $K_M$ for an enzyme-catalyzed reaction. The $K_M$ value in the presence of the inhibitor is called the apparent $K_M$ (abbreviated $K_M^{app}$). The relationship between the $K_M$ and $K_M^{app}$ is given by the following equation:

$$K_M^{app} = K_M\left(1 + \frac{[I]}{K_i}\right)$$

where $K_i$ is a measure of the affinity between the enzyme and the inhibitor and [I] is the inhibitor concentration. As an example, let's consider hexokinase, which was described in Figure 6.5. The $K_M$ for glucose for a particular hexokinase found in humans is 0.15 mM. A structurally related sugar called 2-deoxyglucose acts as a competitive inhibitor with a $K_i$ of 0.3 mM. What would be the $K_M^{app}$ for glucose if the 2-deoxyglucose concentration was 1.0 mM?

## Additional Factors Influence Enzyme Function

Enzymes, which are proteins, sometimes require the following nonprotein molecules or ions to carry out their functions:

- **Prosthetic groups** are small molecules that are permanently attached to the surface of an enzyme and aid in enzyme function.
- **Cofactors** are usually inorganic ions, such as $Fe^{3+}$ or $Zn^{2+}$, that temporarily bind to the surface of an enzyme and promote a chemical reaction.
- **Coenzymes** are organic molecules that temporarily bind to an enzyme and participate in the chemical reaction but are left unchanged after the reaction is completed. Some of these coenzymes are synthesized by cells, but many of them are taken in as dietary vitamins by animal cells.

The ability of enzymes to increase the rate of a chemical reaction is also affected by their environment. In particular, the temperature, pH, and ionic conditions play an important role in the proper functioning of enzymes. Most enzymes function maximally in a narrow range of temperature and pH. For example, many human enzymes work best at 37°C (98.6°F), which is the normal body temperature. If the temperature is several degrees above or below an enzyme's optimum temperature due to infection or environmental causes, the function of many enzymes is greatly inhibited (Figure 6.7). Very high temperatures may denature a protein, causing it to unfold and lose its three-dimensional shape, thereby inhibiting its function.

Enzyme function is also sensitive to pH. Certain enzymes in the stomach function best at the acidic pH found in this organ. For example, pepsin is a protease—an enzyme that digests proteins into peptides—that is released into the stomach. The optimal pH for pepsin function is around pH 2.0, which is extremely acidic. By comparison, many cytosolic enzymes function optimally at a more neutral pH, such as pH 7.2, which is the pH normally found in the cytosol of human cells. If the pH was significantly above or

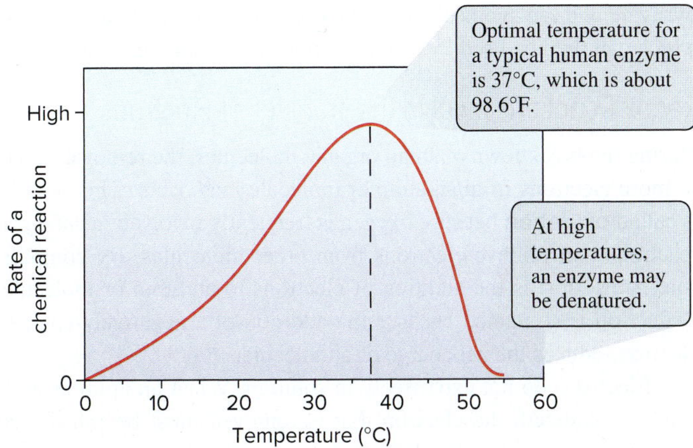


**Figure 6.7** **Effects of temperature on a typical human enzyme.** Most enzymes function optimally within a narrow temperature range. Many human enzymes function best at 37°C— normal body temperature.

below this value, enzyme function would be decreased for cytosolic enzymes.

## 6.2 Reviewing the Concepts

- Enzymes are proteins that speed up the rate of a chemical reaction by lowering the activation energy ($E_A$) needed to achieve a transition state (Figure 6.4).
- Enzymes recognize reactant molecules, also called substrates, with a high specificity. Conformational changes cause substrates to bind more tightly to enzymes, called induced fit (Figure 6.5).
- Each enzyme-catalyzed reaction exhibits a maximal velocity ($V_{max}$). The $K_M$ is the substrate concentration at which the velocity of the chemical reaction is half of the $V_{max}$ value. Competitive inhibitors raise the apparent $K_M$ for the substrate, whereas noncompetitive inhibitors lower the $V_{max}$ (Figure 6.6).
- Enzymes sometimes need nonprotein molecules (prosthetic groups, cofactors, and coenzymes) in order to function, and other factors, including temperature and pH, can affect the proper functioning of enzymes. (Figure 6.7)

## 6.2 Testing Your Knowledge

1. With regard to their effects on a chemical reaction, the main effect of enzymes is to
   a. lower the activation energy and thereby increase the rate of the reaction.
   b. raise the activation energy and thereby increase the rate of the reaction.
   c. lower the activation energy and thereby make the free energy more negative.
   d. raise the activation energy and thereby make the free energy more positive.
   e. both a and c.
2. An inhibitor raises the $K_m$ for an enzyme but has no effect on the $V_{max}$. This inhibitor probably binds to
   a. the active site.
   b. an allosteric site.
   c. the substrate.
   d. all of the above.
   e. both a and b.

## 6.3 Overview of Metabolism and Cellular Respiration

### Learning Outcomes

1. Describe the concept of a metabolic pathway, and distinguish between catabolic and anabolic reactions.
2. Explain how catabolic reactions are used to generate building blocks to make larger molecules and to produce energy intermediates.
3. Define redox reaction.
4. Compare and contrast three ways that metabolic pathways are regulated.
5. Outline the four metabolic pathways needed to break down glucose to $CO_2$.

In the previous sections, we examined the underlying factors that govern chemical reactions and explored the properties of enzymes. In living cells, chemical reactions are coordinated with each other and often occur in a series of steps called a **metabolic pathway,** with each step catalyzed by a different enzyme (**Figure 6.8**). These pathways are categorized according to whether the reactions lead to the breakdown or synthesis of substances. **Catabolic reactions** result in the breakdown of larger molecules into smaller ones. Such reactions are often exergonic. By comparison, **anabolic reactions** involve the synthesis of larger molecules from smaller precursor molecules. These reactions are usually endergonic and, in living cells, must be coupled to an exergonic reaction. In this section, we will survey the general features of catabolic and anabolic reactions and explore the ways in which metabolic pathways are controlled. Also, we will begin to examine cellular respiration, a process that involves four metabolic pathways.

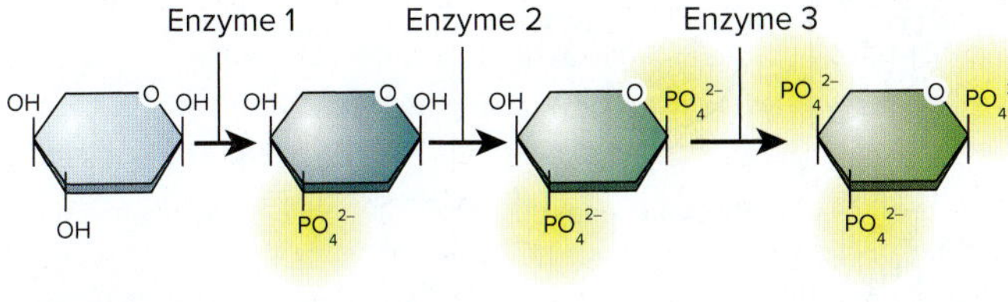


**Figure 6.8** **A metabolic pathway.** In this metabolic pathway, three different enzymes catalyze the attachment of phosphate groups to various positions on a sugar molecule, beginning with a starting substrate and ending with a final product.

## Catabolic Reactions Recycle Organic Building Blocks and Produce Energy Intermediates Such as ATP

Catabolic reactions result in the breakdown of larger molecules into smaller ones. Such catabolic reactions have two advantages.

**Recycling of Organic Building Blocks** One reason for the breakdown of macromolecules is to recycle their organic molecules, which are used as building blocks to construct new molecules and macromolecules. For example, polypeptides, which make up proteins, are composed of a linear sequence of amino acids. When a protein is improperly folded or is no longer needed by a cell, the peptide bonds between the amino acids in the protein are broken by enzymes called proteases. This generates amino acids that can be used in the construction of new proteins.

$$\text{Protein} \xrightarrow{\text{Proteases}} \rightarrow \rightarrow \rightarrow \rightarrow \rightarrow \rightarrow \rightarrow \rightarrow \text{Many individual amino acids}$$

**Breakdown of Organic Molecules to Obtain Energy** A second reason for the breakdown of macromolecules and smaller organic molecules is to obtain energy that is used to drive endergonic processes in the cell. Covalent bonds store a large amount of energy. However, when cells break covalent bonds in organic molecules such as glucose, they do not directly use the energy released in this process. Instead, the released energy is stored in **energy intermediates,** molecules such as ATP, which are directly used to drive endergonic reactions in cells.

As an example, let's consider the breakdown of glucose into two molecules of pyruvate. As discussed later, the breakdown of glucose to pyruvate involves a catabolic pathway called glycolysis. Some of the energy released during the breakage of covalent bonds in glucose is harnessed to synthesize ATP. Glycolysis involves a series of steps in which covalent bonds are broken and rearranged. This process produces molecules that readily donate a phosphate group to ADP, thereby producing ATP. For example, phosphoenolpyruvate has a phosphate group attached to pyruvate. Due to the arrangement of bonds in phosphoenolpyruvate, this phosphate bond is unstable and easily broken. Therefore, the phosphate can be transferred from phosphoenolpyruvate to ADP, which is used to synthesize ATP:

Unstable phosphate bond

Phosphoenolpyruvate ($O^-$, C=O, CO–P, $CH_2$) + ADP → (Pyruvate kinase) → ATP + Pyruvate ($O^-$, C=O, C=O, $CH_3$)

$\Delta G = -7.5$ kcal/mol

This is an exergonic reaction and therefore favors the formation of products. In this step of glycolysis, the breakdown of an organic molecule, namely phosphoenolpyruvate, results in the formation of pyruvate and the synthesis of an energy intermediate molecule, ATP, which can then be used by a cell to drive endergonic reactions.

## Redox Reactions Involve the Transfer of Electrons

During the breakdown of small organic molecules, the removal of one or more electrons from an atom or molecule may occur. This process is called **oxidation** because oxygen is frequently involved in chemical reactions that remove electrons from other molecules. By comparison, **reduction** is the addition of electrons to an atom or molecule. Reduction is so named because the addition of a negatively charged electron reduces the net charge of a molecule.

Electrons do not exist freely in solution. When an atom or molecule is oxidized, the electron that is removed must be transferred to another atom or molecule, which becomes reduced. This type of reaction is termed a **redox reaction,** which is short for a reduction-oxidation reaction. As a generalized equation, an electron may be transferred from molecule A to molecule B as follows:

$$\text{A}e^- + \text{B} \rightarrow \underset{\text{(oxidized)}}{\text{A}} + \underset{\text{(reduced)}}{\text{B}e^-}$$

As shown on the right side of this reaction, A has been oxidized (that is, had an electron removed), and B has been reduced (that is, had an electron added). In general, a substance that has been oxidized has less energy, whereas a substance that has been reduced has more energy.

During the oxidation of organic molecules such as glucose, the electrons are used to produce energy intermediates such as NADH (Figure 6.9). In this process, an organic molecule has been oxidized, and **$NAD^+$ (nicotinamide adenine dinucleotide)** has been reduced to NADH. Cells use NADH in two common ways. First, as discussed later, the oxidation of NADH is a highly exergonic reaction that can be used to make ATP. Second, NADH can donate electrons to other organic molecules and thereby energize them. Such energized molecules can more readily form covalent bonds. Therefore, NADH is often needed in anabolic reactions that involve the synthesis of larger molecules through the formation of covalent bonds between smaller molecules.

## Metabolic Pathways Are Regulated in Three General Ways

The regulation of metabolic pathways is important for a variety of reasons. Catabolic pathways are regulated so organic molecules are broken down only when they are no longer needed or when the cell requires energy. During anabolic reactions, regulation ensures that a cell synthesizes molecules only when they are needed. The regulation of catabolic and anabolic pathways occurs at the genetic, cellular, and biochemical levels.

**Gene Regulation** Enzymes are proteins that are encoded by genes. One way that cells control metabolic pathways is via gene regulation. For example, if a bacterial cell is not exposed to a particular sugar in its environment, it will turn off the genes that encode the enzymes that are needed to break down that sugar. Alternatively, if the sugar becomes available, the genes are switched on. Chapter 12 examines the steps of gene regulation in detail.

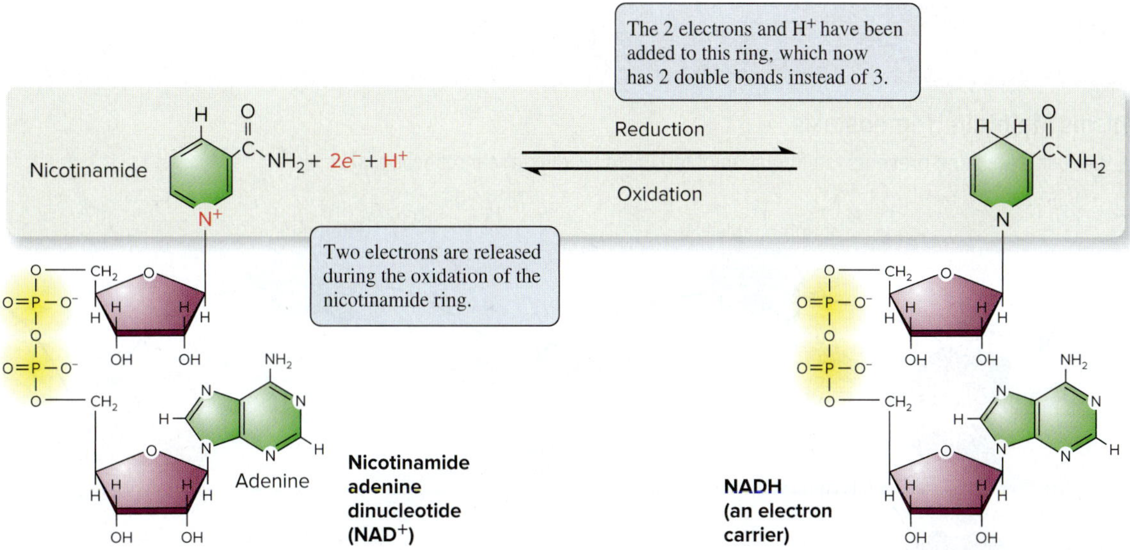


**Figure 6.9 The reduction of $NAD^+$ to produce NADH.** $NAD^+$ is composed of two nucleotides: one with an adenine base and one with a nicotinamide base. The oxidation of organic molecules releases electrons that can bind to $NAD^+$ (and along with a hydrogen ion) result in the formation of NADH. The two electrons and $H^+$ are incorporated into the nicotinamide ring. Note: The actual net charges of $NAD^+$ and NADH are –1 and –2, respectively. They are designated $NAD^+$ and NADH to emphasize the net charge of the nicotinamide ring, which is involved in oxidation-reduction reactions.

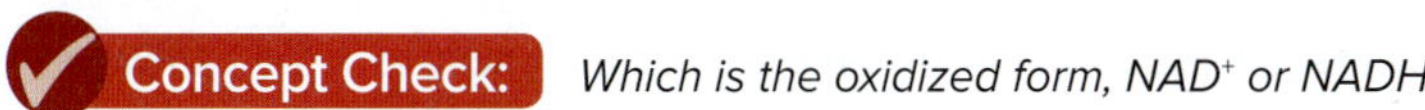

*Which is the oxidized form, $NAD^+$ or NADH?*

**Regulation via Cell-Signaling Pathways** Metabolic reactions are also coordinated at the cellular level. Cells integrate signals from their environment and adjust their metabolic pathways to adapt to those signals. As discussed in Chapter 8, cell-signaling pathways often lead to the activation of protein kinases—enzymes that covalently attach a phosphate group to target proteins. For example, when people are frightened, they secrete a hormone called epinephrine into their bloodstream. This hormone binds to the surface of muscle cells and stimulates an intracellular pathway that leads to the phosphorylation of specific enzymes involved in carbohydrate metabolism. These activated enzymes promote the breakdown of carbohydrates, an event that supplies the frightened individual with more energy. Epinephrine is sometimes called the "fight-or-flight" hormone because the added energy prepares an individual to either stay and fight or run away quickly.

**Biochemical Regulation** A third and very prominent way that metabolic pathways are controlled is at the biochemical level. In this case, the noncovalent binding of a molecule to an enzyme directly regulates its function. As discussed earlier, one form of biochemical regulation involves the binding of molecules, such as competitive or noncompetitive inhibitors (see Figure 6.6). An example of noncompetitive inhibition is a type of regulation called **feedback inhibition,** in which the product of a metabolic pathway inhibits an enzyme that acts early in the pathway, thus preventing the overaccumulation of the product (**Figure 6.10**).

Many metabolic pathways use feedback inhibition as a form of biochemical regulation. In such cases, the enzyme being inhibited has two binding sites. One site is the active site, where the reactants are converted to products. In addition, enzymes controlled by feedback inhibition also have an allosteric site, where a molecule can bind noncovalently and affect the function of the active site. The binding of a molecule to an allosteric site causes a conformational change in the enzyme that inhibits its catalytic function. For example, the conformational change may alter the structure of the active site so that reactant(s) cannot bind. Allosteric sites are often found in the enzymes that catalyze the early steps in a metabolic pathway. Such allosteric sites typically bind molecules that are the products of the metabolic pathway. When the products bind to these sites, they inhibit the function of these enzymes, thereby preventing the formation of too much product.

**Rate-Limiting Step** Cell-signaling and biochemical regulation are important, rapid ways to control chemical reactions in a cell. For a metabolic pathway composed of several enzymes, which enzyme in a pathway should be controlled? In many cases, a metabolic pathway has a **rate-limiting step,** which is the slowest step in a pathway. If the rate-limiting step is inhibited or enhanced, such changes will have the greatest influence on the formation of the product of the metabolic pathway. Rather than affecting all of the enzymes in a metabolic pathway, cell-signaling and biochemical regulation are often directed at the enzyme that catalyzes the rate-limiting step. This is an efficient, rapid way to control the amount of product of a pathway.

## The Breakdown of Glucose Occurs in Four Stages

**Cellular respiration** is a process by which living cells obtain energy from organic molecules and release waste products. A primary aim of cellular respiration is to make adenosine triphosphate (ATP).

## Biology Principle

### Living Organisms Maintain Homeostasis

In this example, feedback inhibition prevents the formation of too much product of a metabolic pathway.

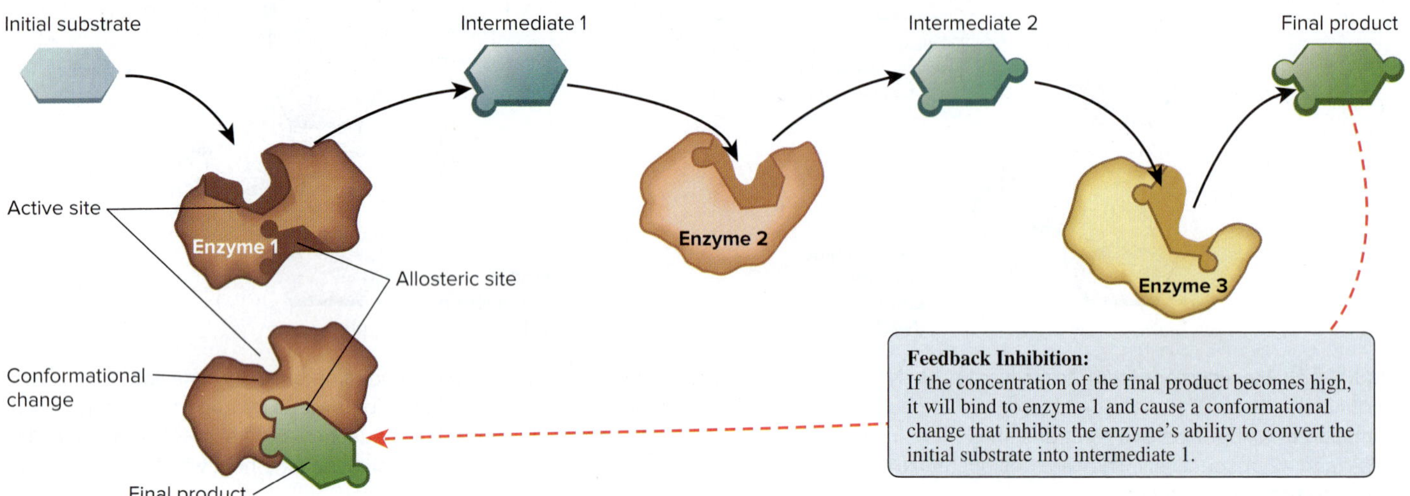


**Figure 6.10** **Feedback inhibition.** In this process, the final product of a metabolic pathway inhibits an enzyme that functions in the pathway, thereby preventing the overaccumulation of the product.

---

When oxygen ($O_2$) is used, this process is termed **aerobic respiration.** During aerobic respiration, $O_2$ is used, and carbon dioxide ($CO_2$) is released via the oxidation of organic molecules. When we breathe, we inhale the oxygen needed for aerobic respiration and exhale $CO_2$, a by-product of the process. For this reason, the term respiration has a second meaning—the act of breathing.

Different types of organic molecules, such as carbohydrates, proteins, and fats, are used as energy sources to drive aerobic respiration. In the rest of this chapter, we will largely focus on the use of glucose as an energy source for cellular respiration, as described in the following reaction:

$$C_6H_{12}O_6 + 6\ O_2 \rightarrow 6\ CO_2 + 6\ H_2O \qquad \Delta G = -686 \text{ kcal/mol}$$
(Glucose)

Certain covalent bonds within glucose store a large amount of energy. When glucose is broken down via oxidation, ultimately to $CO_2$ and water, a tremendous amount of free energy is released (–685 kcal/mol). Some of the energy is lost as heat, but much of it is used to make three energy intermediates: ATP, NADH, and $FADH_2$. This process involves four metabolic pathways: (1) glycolysis, (2) the breakdown of pyruvate, (3) the citric acid cycle, and (4) oxidative phosphorylation (Figure 6.11):

1. **Glycolysis:** In glycolysis, glucose (a compound with six carbon atoms) is broken down to two pyruvate molecules (with three carbons each), producing a net energy yield of two ATP molecules and two NADH molecules. The two ATP are synthesized via **substrate-level phosphorylation,** which occurs when an enzyme directly transfers a phosphate from an organic molecule to ADP. In eukaryotes, glycolysis occurs in the cytosol.
2. **Breakdown of pyruvate to an acetyl group:** The two pyruvate molecules enter the mitochondrial matrix, where each one is broken down to an acetyl group (with two carbons each) and one $CO_2$ molecule. For each pyruvate broken down via oxidation, one NADH molecule is made by the reduction of $NAD^+$.
3. **Citric acid cycle:** Each acetyl group is incorporated into an organic molecule, which is later oxidized to liberate two $CO_2$ molecules. One ATP, three NADH, and one $FADH_2$ are made in this process. Because there are two acetyl groups (one from each pyruvate), the total yield is four $CO_2$, two ATP via substrate-level phosphorylation, six NADH, and two $FADH_2$. This process occurs in the mitochondrial matrix.
4. **Oxidative phosphorylation:** The NADH and $FADH_2$ made in the three previous stages contain high-energy electrons that can be readily transferred in a redox reaction to other molecules. Once removed from NADH or $FADH_2$, these high-energy electrons release some energy, and through an electron transport chain, that energy is harnessed to produce an $H^+$ electrochemical gradient. In **chemiosmosis,** energy stored in the $H^+$ electrochemical gradient is used to synthesize ATP from ADP and $P_i$. The overall process of electron transport and ATP synthesis is called

# Biology Principle

## Living Organisms Use Energy

Molecules such as glucose store a large amount of energy. The breakdown of glucose is used to make energy intermediates, such as ATP molecules, which drive many types of cellular processes.

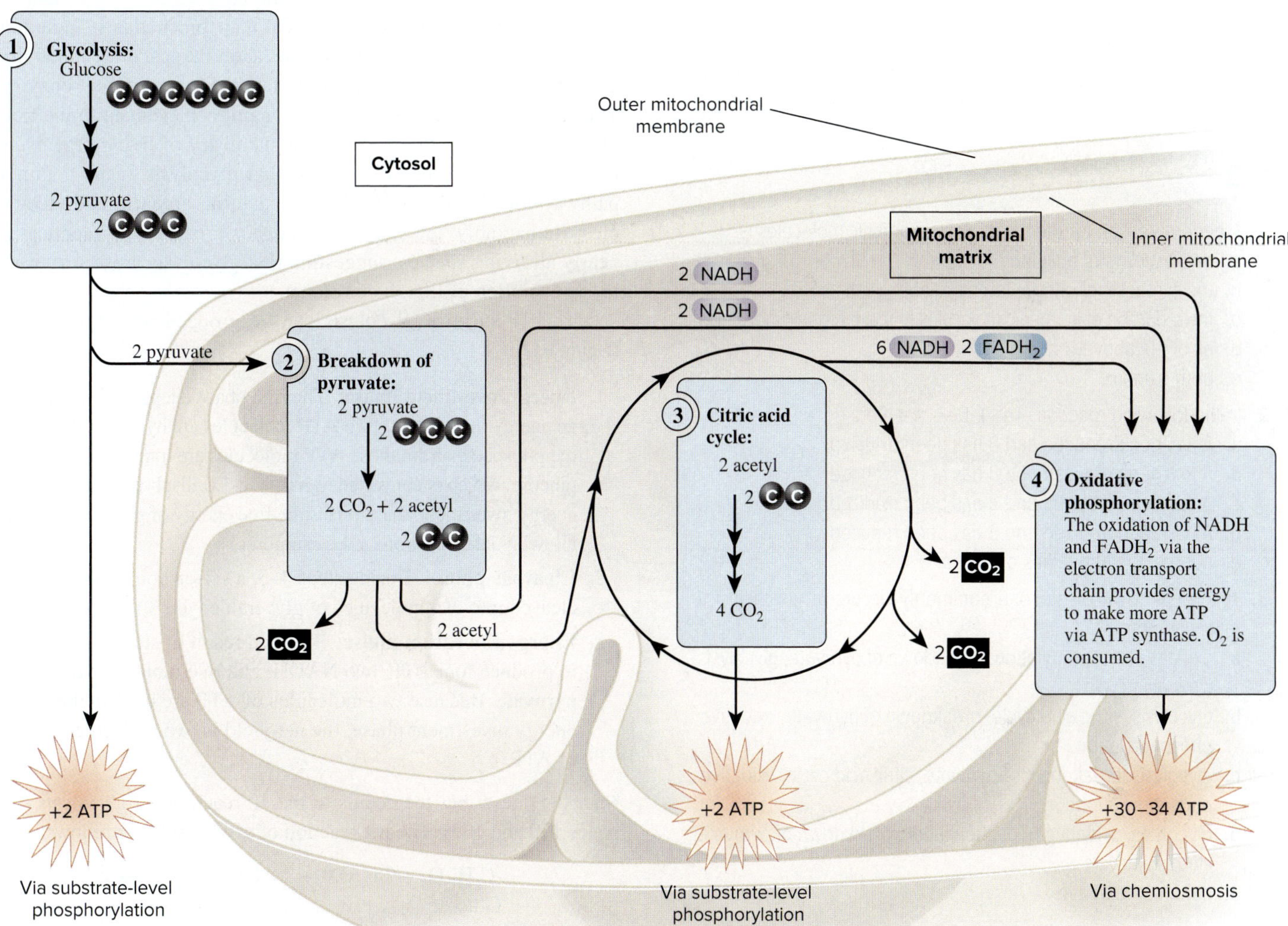


**Figure 6.11** **An overview of cellular respiration.** The 30–34 ATP molecules produced via chemiosmosis is the maximum number possible. As described later in this chapter, mitochondria may use NADH, $FADH_2$, and the $H^+$ gradient for purposes other than ATP synthesis.

oxidative phosphorylation because NADH or $FADH_2$ have been oxidized and ADP has become phosphorylated to make ATP. Approximately 30 to 34 ATP molecules can be made via oxidative phosphorylation.

In eukaryotes, oxidation phosphorylation occurs along the **cristae,** which are invaginations of the inner mitochondrial membrane. The invaginations greatly increase the surface area of the inner membrane, thereby increasing the amount of ATP that can be made. In bacteria and archaea, oxidative phosphorylation occurs along the plasma membrane because the proteins that carry out this process are located there.

## 6.3 Reviewing the Concepts

- Metabolism is the sum of the chemical reactions in a living organism. Metabolic pathways consist of chemical reactions that occur in steps and are catalyzed by specific enzymes (Figure 6.8).
- Catabolic reactions involve the breakdown of larger molecules into smaller ones. These reactions recycle organic molecules that are used as building blocks to make new molecules. The organic molecules are also broken down to make energy intermediates such as ATP. Anabolic reactions require an input of energy to synthesize larger molecules and macromolecules.

- Some chemical reactions are redox reactions in which electrons are transferred from one molecule to another. These can be used to make energy intermediates such as NADH (Figure 6.9).
- Metabolic pathways are controlled by gene regulation, cell-signaling, and biochemical regulation. An example of biochemical regulation is feedback inhibition (Figure 6.10).
- Cells obtain energy via cellular respiration, which involves the breakdown of organic molecules and the export of waste products. With regard to glucose, cellular respiration is a process involving four metabolic pathways: glycolysis, pyruvate breakdown, citric acid cycle, and oxidative phosphorylation (Figure 6.11).

## 6.3 Testing Your Knowledge

1. A general reason for the catabolism of organic molecules is
   a. the recycling of building blocks.
   b. the synthesis of energy intermediates, such as ATP.
   c. the reduction of inorganic molecules to make organic molecules.
   d. all of the above.
   e. both a and b.
2. In the following reaction, $Ae^- + B \rightarrow A + Be^-$,
   a. A has been reduced and B has been oxidized.
   b. A has been reduced and B has been reduced.
   c. A has been oxidized and B has been oxidized.
   d. A has been oxidized and B has been reduced.
   e. none of the above has occurred.
3. The breakdown of glucose commonly occurs in which of the following orders?
   a. oxidative phosphorylation, breakdown of pyruvate, glycolysis, citric acid cycle
   b. glycolysis, citric acid cycle, breakdown of pyruvate, oxidative phosphorylation
   c. glycolysis, breakdown of pyruvate, citric acid cycle, oxidative phosphorylation
   d. glycolysis, oxidative phosphorylation, breakdown of pyruvate, citric acid cycle
   e. none of the above

## 6.4 Glycolysis

### Learning Outcomes

1. Outline the three phases of glycolysis, and identify the net products.
2. Describe the series of enzymatic reactions that constitute glycolysis.
3. **SCISKILLS ▶** Explain the underlying basis for the use of positron-emission tomography to detect tumors.

Thus far, we have examined the general features of the four metabolic pathways involved in the breakdown of glucose. We will now turn our attention to a more detailed understanding of the pathways for glucose metabolism, beginning with glycolysis.

### Glycolysis Is a Metabolic Pathway That Breaks Down Glucose to Pyruvate

**Glycolysis** (from the Greek *glykos*, meaning sweet, and *lysis*, meaning splitting) involves the breakdown of glucose, a simple sugar, into two molecules of a compound called pyruvate. This process can occur in the presence of oxygen, that is, under aerobic conditions, and it can occur in the absence of oxygen. During the 1930s, the efforts of several German biochemists, including Gustav Embden, Otto Meyerhof, and Jacob Parnas, determined that glycolysis involves 10 steps, each catalyzed by a different enzyme. The elucidation of these steps was a major achievement in the field of **biochemistry**—the study of the chemistry of living organisms. Researchers have since discovered that glycolysis is the common pathway for glucose breakdown in bacteria, archaea, and eukaryotes. Remarkably, the steps of glycolysis are virtually identical in nearly all living species, suggesting that glycolysis arose very early in the evolution of life on our planet.

The 10 steps of glycolysis can be grouped into three phases (Figure 6.12):

1. **Energy investment phase:** The first phase (steps 1–3) involves an energy investment. Two ATP molecules are hydrolyzed, and the phosphates from those ATP molecules are transferred to glucose, which is converted to fructose-1,6-bisphosphate. The energy investment phase raises the free energy of glucose, thereby allowing later reactions to be exergonic.
2. **Cleavage phase:** During steps 4–5, a six-carbon molecule is cleaved into two molecules of glyceraldehyde-3-phosphate.
3. **Energy liberation phase:** The end result of steps 6–10 is to produce four ATP, two NADH, and two molecules of pyruvate. Because two molecules of ATP are used in the energy investment phase, the net yield is two molecules of ATP.

Figure 6.13 describes the details of the 10 reactions of glycolysis (see pages 132 and 133). The net reaction of glycolysis is as follows:

$$\underset{\text{Glucose}}{C_6H_{12}O_6} + 2\,NAD^+ + 2\,ADP^{2-} + 2\,P_i^{2-} \rightarrow$$

$$\underset{\text{Pyruvate}}{2\,CH_3(C{=}O)COO^-} + 2\,H^+ + 2\,NADH + 2\,ATP^{4-} + 2\,H_2O$$

**Regulation of Glycolysis** How do cells control glycolysis? The rate of glycolysis is regulated by the availability of substrates, such as glucose, and by feedback inhibition. A key control point involves the enzyme phosphofructokinase, which catalyzes the third step in glycolysis (see Figure 6.13, step 3), the step believed to be the slowest, or rate-limiting, step. When a cell has a sufficient amount of ATP, feedback inhibition occurs. At high concentrations, ATP binds reversibly to an allosteric site in phosphofructokinase, causing a conformational change that renders the enzyme functionally inactive. This prevents the further breakdown of glucose, thereby inhibiting the overproduction of ATP. Alternatively, if ADP binds to this enzyme, its function is stimulated, thereby increasing the rate of glycolysis.

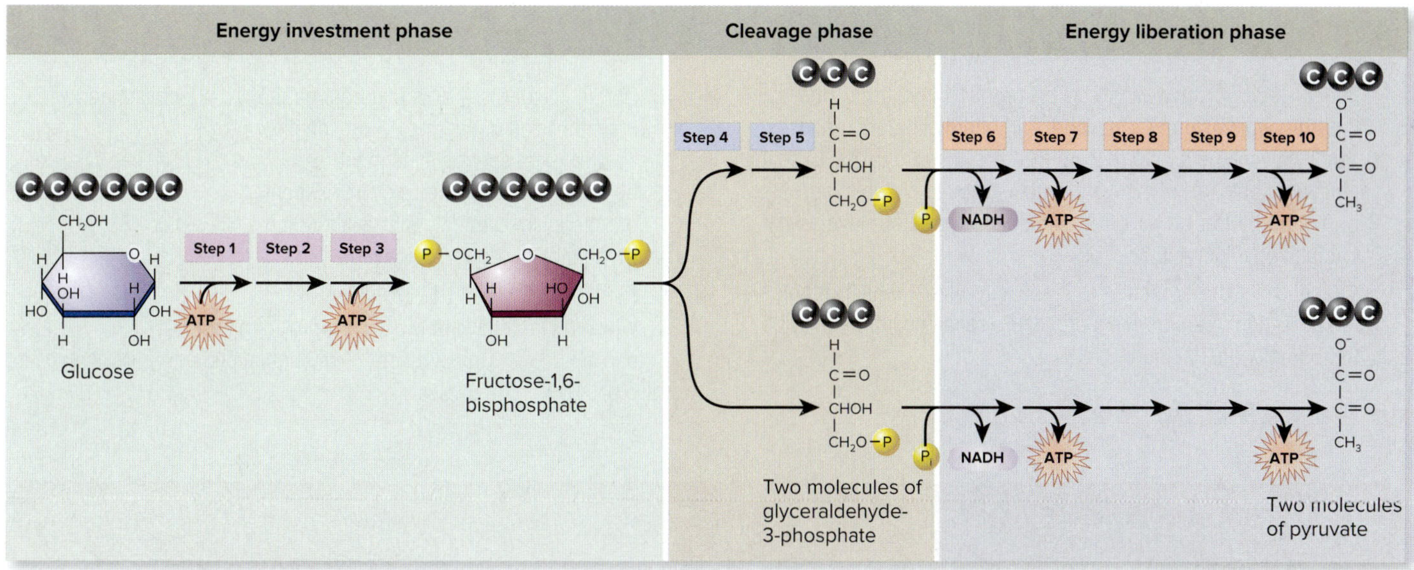


**Figure 6.12** Overview of glycolysis.

**Concept Check:** *Explain why the three phases are named the energy investment phase, the cleavage phase, and the energy liberation phase.*

**BioConnections:** *Look ahead to Chapter 35. With regard to oxygen needs, what is an advantage of glycolytic muscle fibers?*

## Cancer Cells Usually Exhibit High Levels of Glycolysis

In 1931, the German physiologist Otto Warburg discovered that certain cancer cells preferentially use glycolysis for ATP production, in contrast to healthy cells, which mainly generate ATP from oxidative phosphorylation. This phenomenon, termed the Warburg effect, is very common among different types of tumors. The Warburg effect is the basis for the detection of cancer via a procedure called positron-emission tomography (PET). In this technique, patients are injected with a radioactive glucose analogue called $[^{18}F]$-fluorodeoxyglucose (FDG). FDG is taken up by high-glucose-using cells such as cancer cells. A scanner detects regions of the body that accumulate high amounts of FDG, which are visualized as bright spots on the PET scan.

**Figure 6.14** shows a PET scan of a patient with lung cancer. The bright regions next to the arrows are tumors that show abnormally high levels of glycolysis. This result occurred because certain genes that encode enzymes involved with glycolysis have an increased expression in cancer cells. Research has shown that the enzymes of glycolysis are overexpressed in approximately 80% of all types of cancer, including lung, skin, colon, liver, pancreatic, breast, ovarian, and prostate cancers. The three enzymes of glycolysis whose overexpression is most commonly associated with cancer are glyceraldehyde-3-phosphate dehydrogenase, enolase, and pyruvate kinase (shown in Figure 6.13, steps 6–10). In many cancers, all 10 glycolytic enzymes are overexpressed!

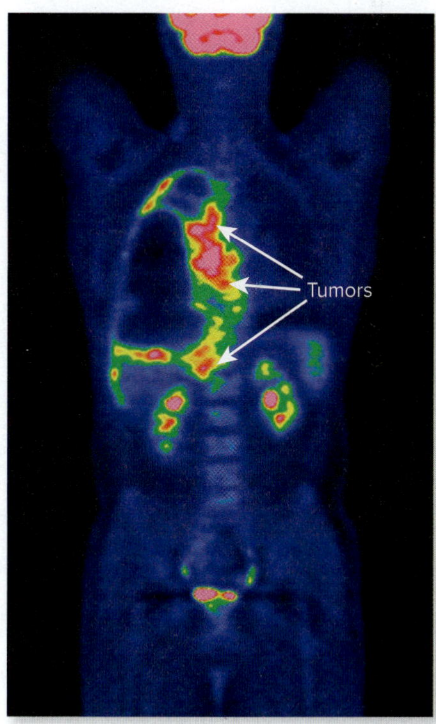


**Figure 6.14** **A PET scan of a patient with lung cancer.** The bright regions in the lungs are tumors (see arrows). Organs such as the brain, which are not cancerous in this case, appear bright because they perform high levels of glucose metabolism. Also, the kidneys and bladder appear bright because they filter and accumulate FDG. (Note: FDG is taken up by cells and converted to FDG-phosphate by hexokinase, the first enzyme in glycolysis. However, because FDG lacks an —OH group, it is not metabolized further. Therefore, FDG-phosphate accumulates in metabolically active cells.)
© ISM/Phototake

## 6.4 Reviewing the Concepts

- The 10 steps of glycolysis can be grouped into three phases: the energy investment, cleavage, and energy liberation phases. During glycolysis, which occurs in the cytosol, glucose is split into two molecules of pyruvate, with a net yield of two molecules of ATP and two of NADH. ATP is made by substrate-level phosphorylation (Figures 6.12, 6.13).
- Cancer cells preferentially carry out glycolysis, which enables the detection of tumors via a procedure called positron-emission tomography (PET) (Figure 6.14).

## 6.4 Testing Your Knowledge

1. During glycolysis, ATP is used during the ______ phase, and ATP is synthesized during the ______ phase.
   - a. cleavage, energy investment
   - b. energy liberation, energy investment
   - c. energy investment, energy liberation
   - d. energy investment, cleavage
   - e. energy liberation, cleavage
2. Which of the following is *not* a product of glycolysis?
   - a. ATP
   - b. NADH
   - c. pyruvate
   - d. $CO_2$
   - e. both a and b

# Biology Principle

### Living Organisms Maintain Homeostasis

To maintain a relatively constant level of ATP in the cytosol, the activity of phosphofructokinase is controlled by feedback inhibition. A high level of ATP inhibits its function.

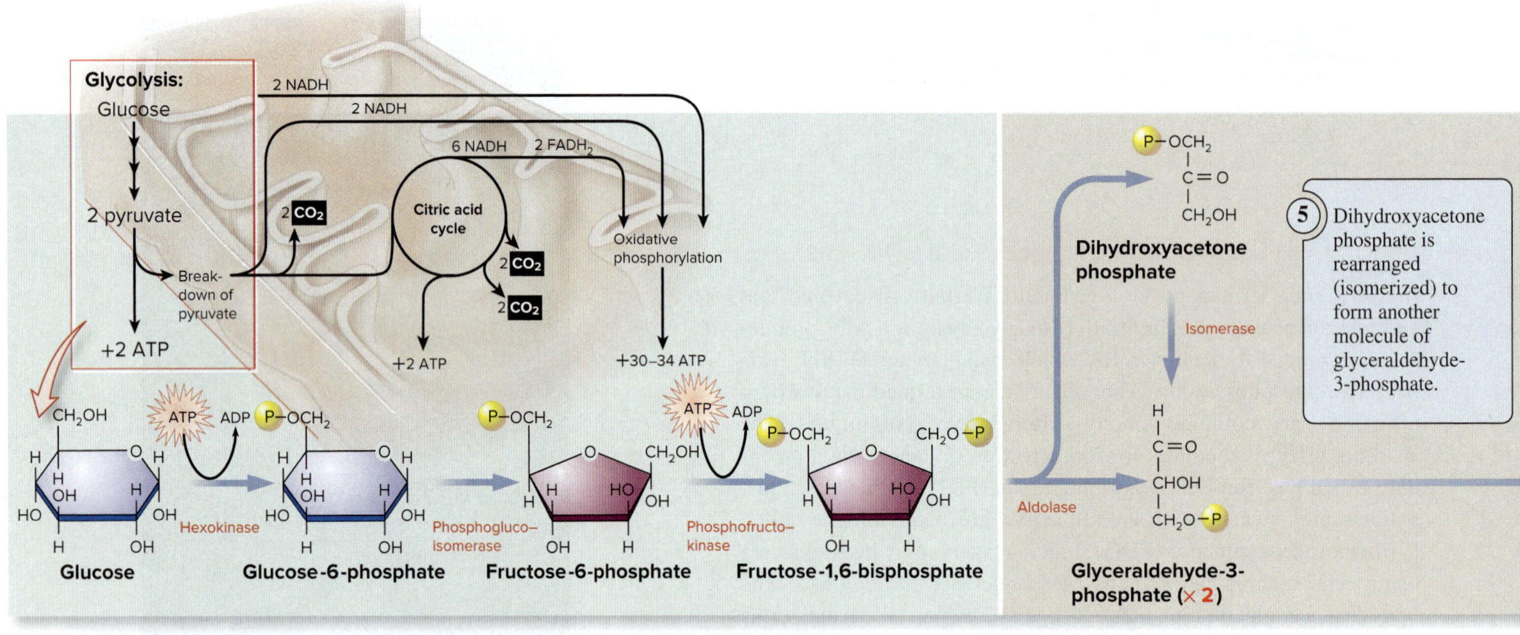


1. A phosphate is transferred from ATP to glucose to make glucose-6-phosphate, which is more easily trapped in the cell than glucose.

2. The structure of glucose-6-phosphate is rearranged to fructose-6-phosphate.

3. A second phosphate is transferred from ATP to fructose-6-phosphate to make fructose-1,6-bisphosphate.

4. Fructose-1,6-bisphosphate is cleaved into dihydroxyacetone phosphate and glyceraldehyde-3-phosphate.

**Figure 6.13 A detailed look at the steps of glycolysis.** The pathway begins with a 6-carbon molecule (glucose) that is eventually broken down into 2 molecules (pyruvate) that contain 3 carbons each. The notation × 2 in the figure indicates that 2 of these 3-carbon molecules are produced from each glucose molecule.

*Which of these organic molecules donate a phosphate group to ADP during substrate-level phosphorylation?*

## 6.5 Breakdown of Pyruvate

### Learning Outcome

1. Describe how pyruvate is oxidized after entering the mitochondrial matrix.

In eukaryotes, glycolysis produces pyruvate in the cytosol, which is then transported into the mitochondria. Once in the mitochondrial matrix, pyruvate molecules are broken down (oxidized) by an enzyme complex called pyruvate dehydrogenase (**Figure 6.15**). A molecule of $CO_2$ is removed from pyruvate, and the remaining acetyl group is attached to an organic molecule called coenzyme A (CoA) to produce acetyl CoA. (In chemical equations, CoA is depicted as CoA—SH to emphasize how the SH group participates in the chemical reaction.) During this process, two high-energy electrons are removed from pyruvate and transferred to $NAD^+$, which together with $H^+$ produce a molecule of NADH. For each pyruvate, the net reaction is as follows:

$$\underset{\text{Pyruvate}}{{}^{-}O{-}\overset{O}{\overset{\|}{C}}{-}\overset{O}{\overset{\|}{C}}{-}CH_3} + \underset{\text{CoA}}{CoA{-}SH} + NAD^+ \rightarrow$$

$$\underset{\text{Acetyl CoA}}{CoA{-}S{-}\overset{O}{\overset{\|}{C}}{-}CH_3} + CO_2 + NADH$$

The acetyl group is attached to CoA via a covalent bond to a sulfur atom. The hydrolysis of this bond releases a large amount of free energy, making it possible for the acetyl group to be transferred to other organic molecules. As described in Section 6.6, the acetyl group attached to CoA enters the citric acid cycle.

## 6.5 Reviewing the Concepts

- Once in the mitochondrial matrix, pyruvate is broken down to $CO_2$ and an acetyl group that becomes attached to coenzyme A (CoA). NADH is made during this process (Figure 6.15).

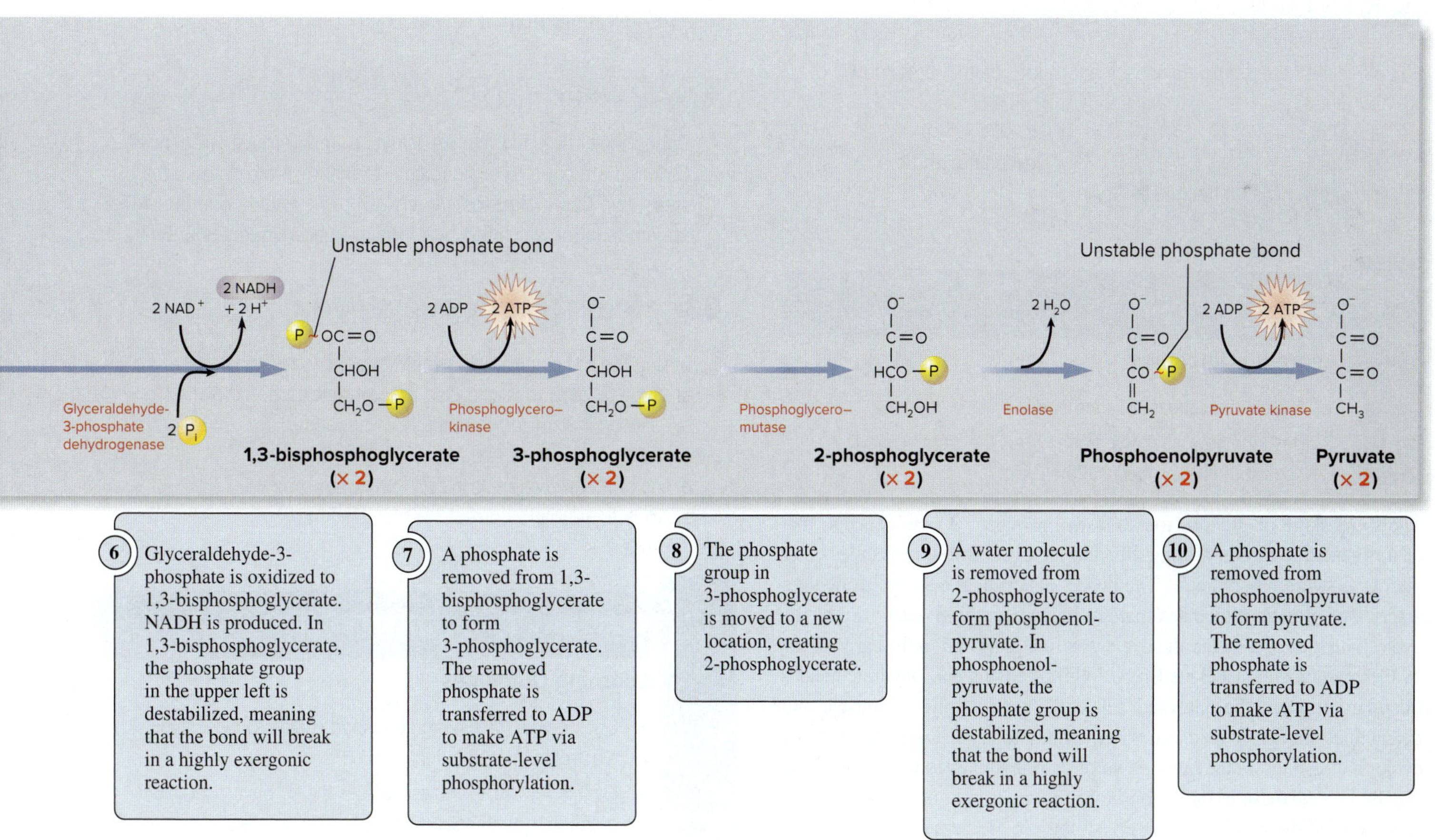

Unstable phosphate bond
2 NAD⁺
2 NADH + 2 H⁺
Glyceraldehyde-3-phosphate dehydrogenase
2 Pi
1,3-bisphosphoglycerate (× 2)
2 ADP
2 ATP
Phosphoglycero-kinase
3-phosphoglycerate (× 2)
Phosphoglycero-mutase
2-phosphoglycerate (× 2)
2 H₂O
Enolase
Unstable phosphate bond
Phosphoenolpyruvate (× 2)
2 ADP
2 ATP
Pyruvate kinase
Pyruvate (× 2)
6 Glyceraldehyde-3-phosphate is oxidized to 1,3-bisphosphoglycerate. NADH is produced. In 1,3-bisphosphoglycerate, the phosphate group in the upper left is destabilized, meaning that the bond will break in a highly exergonic reaction.
7 A phosphate is removed from 1,3-bisphosphoglycerate to form 3-phosphoglycerate. The removed phosphate is transferred to ADP to make ATP via substrate-level phosphorylation.
8 The phosphate group in 3-phosphoglycerate is moved to a new location, creating 2-phosphoglycerate.
9 A water molecule is removed from 2-phosphoglycerate to form phosphoenol-pyruvate. In phosphoenol-pyruvate, the phosphate group is destabilized, meaning that the bond will break in a highly exergonic reaction.
10 A phosphate is removed from phosphoenolpyruvate to form pyruvate. The removed phosphate is transferred to ADP to make ATP via substrate-level phosphorylation.


**Figure 6.13** ***continued***

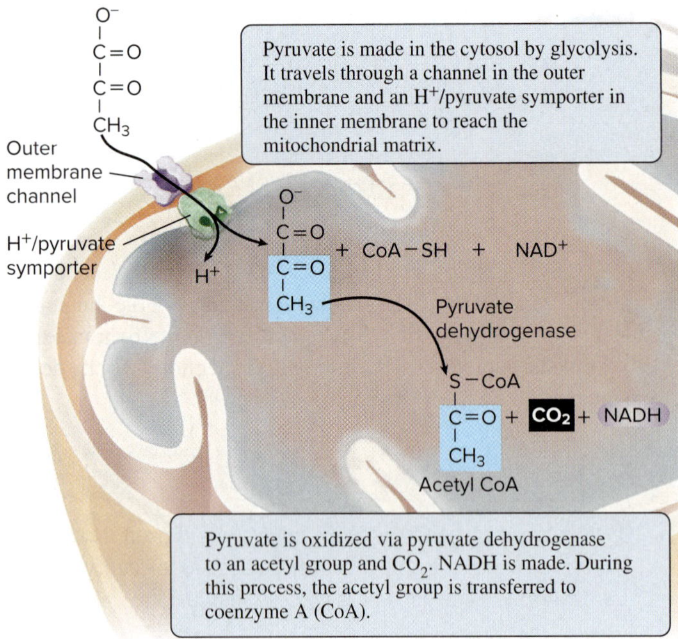

Pyruvate is made in the cytosol by glycolysis. It travels through a channel in the outer membrane and an H+/pyruvate symporter in the inner membrane to reach the mitochondrial matrix.
Outer membrane channel
H+/pyruvate symporter
H+
+ CoA−SH + NAD+
Pyruvate dehydrogenase
S−CoA
C=O + CO2 + NADH
CH3
Acetyl CoA
Pyruvate is oxidized via pyruvate dehydrogenase to an acetyl group and CO2. NADH is made. During this process, the acetyl group is transferred to coenzyme A (CoA).


**Figure 6.15** Breakdown of pyruvate and the attachment of an acetyl group to CoA.

## 6.5 Testing Your Knowledge

1. Which of the following is *not* a product of the breakdown of pyruvate?
   a. ATP
   b. NADH
   c. acetyl group (attached to CoA)
   d. $CO_2$
   e. both a and b

## 6.6 Citric Acid Cycle

### Learning Outcomes

1. Explain the concept of a metabolic cycle.
2. Describe how an acetyl group enters the citric acid cycle, and list the net products of the cycle.

The third stage of glucose metabolism introduces a new concept, that of a **metabolic cycle.** During a metabolic cycle, some molecules enter the cycle while others leave. The process is cyclical because it involves a series of organic molecules that are regenerated with each turn of the cycle. The idea of a metabolic cycle was first proposed in the early 1930s by German biochemist Hans Krebs. While studying carbohydrate metabolism in England, he analyzed cell extracts from pigeon muscle and determined that citric acid and other organic molecules participated in a cycle that resulted in the breakdown of carbohydrates to $CO_2$. This cycle is called the **citric acid cycle,** or the Krebs cycle, in honor of Krebs, who was awarded the Nobel Prize in Physiology or Medicine in 1953.

An overview of the citric acid cycle is shown in Figure 6.16. In the first step of the cycle, the acetyl group (with two carbons) is removed from acetyl CoA and attached to oxaloacetate (with four carbons) to form citrate (with six carbons), also called citric acid. During the cycle, two $CO_2$ molecules are released, and three molecules of NADH, one molecule of $FADH_2$, and one molecule of guanine triphosphate (GTP) are made. The GTP, which is made via substrate-level phosphorylation, is used to make ATP. After a total of eight steps, oxaloacetate is regenerated so the cycle can begin again, provided that acetyl CoA is available. Figure 6.17 shows a more detailed view of the citric acid cycle (see page 136). For each acetyl group attached to CoA, the net reaction of the citric acid cycle is as follows:

$$\text{Acetyl CoA} + 2\,H_2O + 3\,NAD^+ + FAD + GDP^{2-} + P_i^{2-} \rightarrow$$
$$\text{CoA—SH} + 2\,CO_2 + 3\,NADH + FADH_2 + GTP^{4-} + 3\,H^+$$

**Regulation of the Citric Acid Cycle** How is the citric acid cycle controlled? The rate of the cycle is largely regulated by the availability of substrates, such as acetyl CoA and $NAD^+$, and by feedback inhibition. The three steps in the cycle that are highly exergonic are those catalyzed by citrate synthase, isocitrate dehydrogenase, and $\alpha$-ketoglutarate dehydrogenase (see Figure 6.17). Each of these steps is rate-limiting under certain circumstances, and the way that each enzyme is regulated varies among different species. Let's consider an example. In mammals, NADH and ATP act as feedback inhibitors of isocitrate dehydrogenase, whereas $NAD^+$ and ADP act as activators. In this way, the citric acid cycle is inhibited when NADH and ATP levels are high, but it is stimulated when $NAD^+$ and ADP levels are high.

## 6.6 Reviewing the Concepts

- During the citric acid cycle, an acetyl group is removed from acetyl CoA and attached to oxaloacetate to make citrate. In a series of steps, two $CO_2$ molecules, three NADH, one $FADH_2$, and one ATP are made, after which the cycle begins again (Figures 6.16, 6.17).

## 6.6 Testing Your Knowledge

1. During the citric acid cycle, what happens to carbon?
   a. Organic carbon is released as inorganic carbon dioxide.
   b. Carbon is oxidized to make NADH and $FADH_2$.
   c. Carbon dioxide is used to make ATP.
   d. All of the above are correct.
   e. Only a and b are correct.

## 6.7 Oxidative Phosphorylation

### Learning Outcomes

1. Outline how the electron transport chain produces an $H^+$ electrochemical gradient.
2. Explain how ATP synthase utilizes the $H^+$ electrochemical gradient to synthesize ATP.
3. **SCISKILLS ▸** Analyze and interpret the results of the experiment that showed that ATP synthase is a rotary machine.

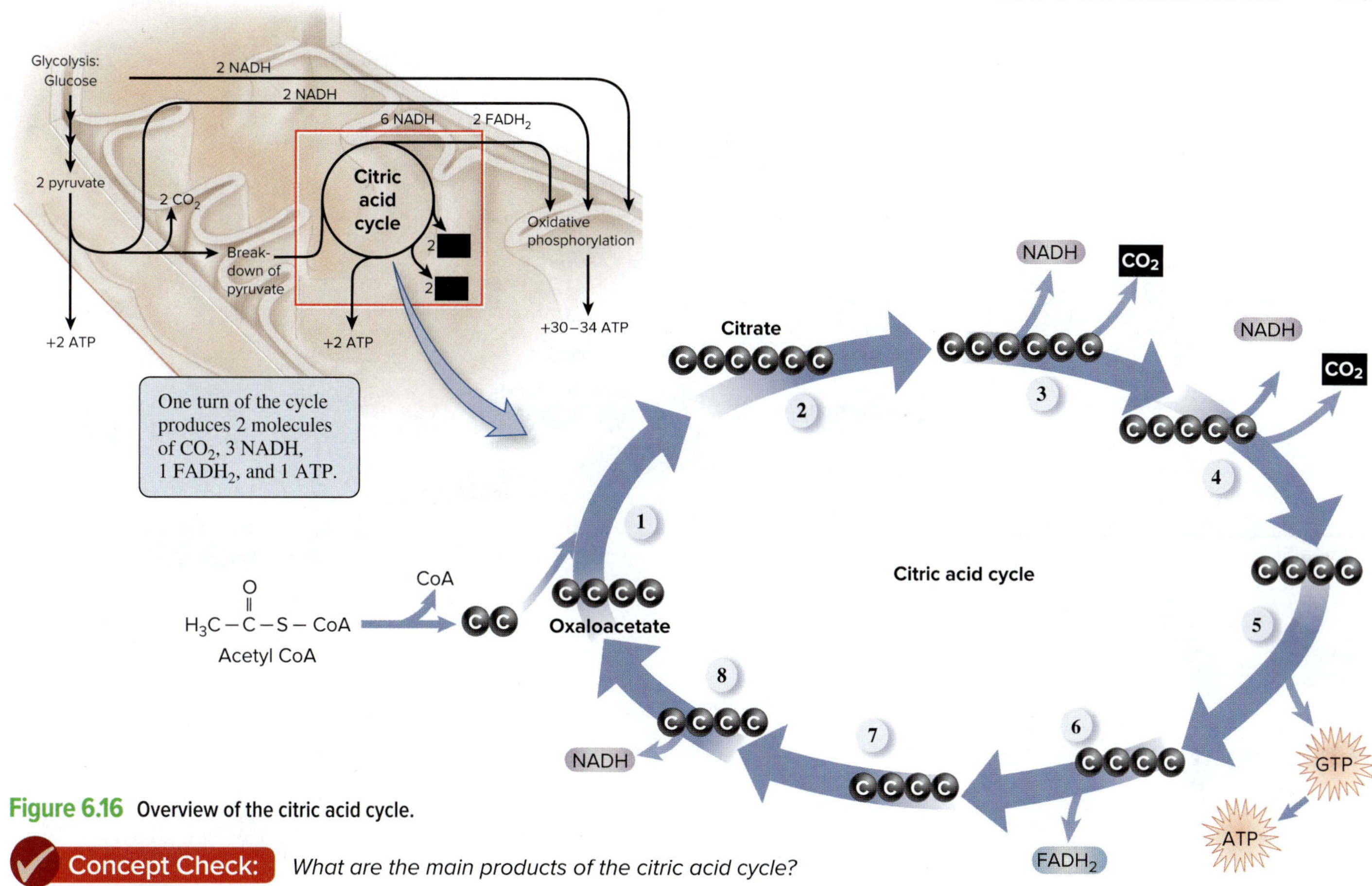


**Figure 6.16** Overview of the citric acid cycle.

**Concept Check:** *What are the main products of the citric acid cycle?*

During the first three stages of glucose metabolism, the oxidation of glucose yields 6 molecules of $CO_2$, 4 molecules of ATP, 10 molecules of NADH, and 2 molecules of $FADH_2$. Let's now consider how high-energy electrons are removed from NADH and $FADH_2$ to produce more ATP. This process is called **oxidative phosphorylation.** As mentioned earlier, the term refers to the observation that NADH and $FADH_2$ have had electrons removed and have thus become oxidized, and ATP is made by the phosphorylation of ADP. In this section, we will examine how the oxidative process involves the electron transport chain, whereas the phosphorylation of ADP occurs via ATP synthase.

## The Electron Transport Chain Establishes an Electrochemical Gradient

The **electron transport chain (ETC)** consists of a group of protein complexes and small organic molecules embedded in the inner mitochondrial membrane. These components are referred to as an electron transport chain because electrons are passed from one component to the next in a series of redox reactions (**Figure 6.18**). Most members of the ETC are protein complexes (designated I–IV) that have prosthetic groups. For example, cytochrome oxidase contains two prosthetic groups, each with an iron atom. The iron in each prosthetic group can readily accept and release an electron. One member of the ETC, ubiquinone (Q), is not a protein but is instead a small organic molecule that can accept and release an electron.

The red line in Figure 6.18 shows the path of electron flow. The electrons, which are originally located on NADH or $FADH_2$, are transferred to components of the ETC. The electron path is a series of redox reactions in which electrons are transferred to components with increasingly higher electronegativity. As discussed in Chapter 2, electronegativity is a measure of an atom's ability to attract electrons. At the end of the chain is oxygen, which is the most electronegative component and the final electron acceptor. The ETC is also called the **respiratory chain** because the oxygen we breathe is used in this process.

NADH and $FADH_2$ donate their electrons at different points in the ETC. Two high-energy electrons from NADH are first transferred one at a time to NADH dehydrogenase (complex I). They are then transferred to ubiquinone (Q), cytochrome *b-c*$_1$ (complex III), cytochrome *c*, and cytochrome oxidase (complex IV). The final electron acceptor is $O_2$. By comparison, $FADH_2$ transfers electrons to succinate reductase (complex II), then to ubiquinone and the rest of the chain.

As shown in Figure 6.18, some of the energy that is released during the movement of electrons is used to pump $H^+$ across the inner mitochondrial membrane into the intermembrane space. This active transport establishes a large **$H^+$ electrochemical gradient,** in which the concentration of $H^+$ is higher outside of the matrix than inside and an excess of positive charge exists outside the matrix. Because hydrogen ions consist of protons, the $H^+$ electrochemical gradient is also called the **proton-motive force.**

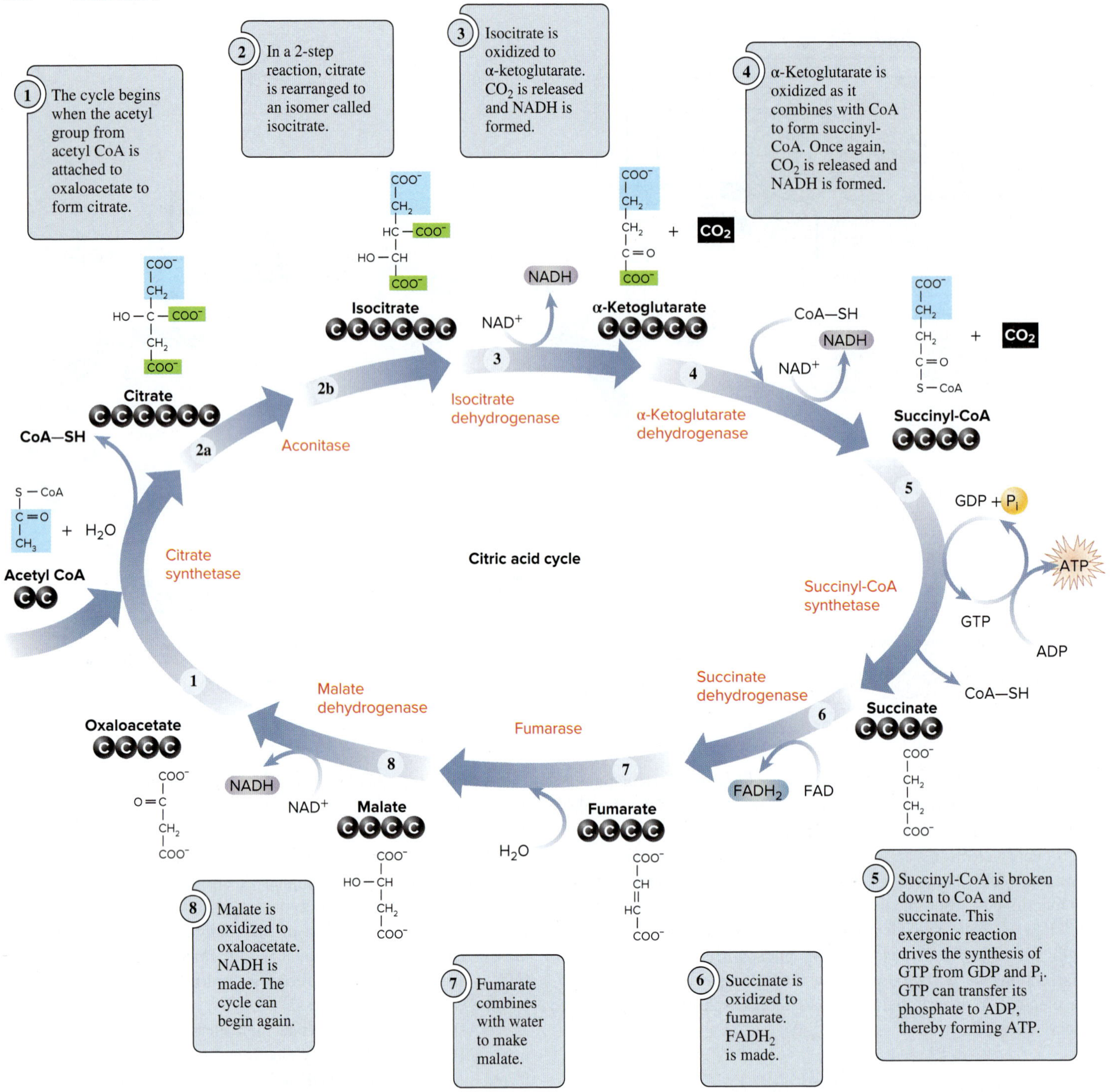

1 The cycle begins when the acetyl group from acetyl CoA is attached to oxaloacetate to form citrate.
2 In a 2-step reaction, citrate is rearranged to an isomer called isocitrate.
3 Isocitrate is oxidized to α-ketoglutarate. $CO_2$ is released and NADH is formed.
4 α-Ketoglutarate is oxidized as it combines with CoA to form succinyl-CoA. Once again, $CO_2$ is released and NADH is formed.
Citrate
CoA—SH
Acetyl CoA
+ $H_2O$
Citrate synthetase
Aconitase
Isocitrate
NAD+
NADH
Isocitrate dehydrogenase
α-Ketoglutarate
+ $CO_2$
α-Ketoglutarate dehydrogenase
CoA—SH
NADH
NAD+
Succinyl-CoA
+ $CO_2$
Citric acid cycle
GDP + $P_i$
ATP
GTP
ADP
Succinyl-CoA synthetase
CoA—SH
Succinate
Succinate dehydrogenase
$FADH_2$
FAD
Fumarate
Fumarase
$H_2O$
Malate
Malate dehydrogenase
NAD+
NADH
Oxaloacetate
5 Succinyl-CoA is broken down to CoA and succinate. This exergonic reaction drives the synthesis of GTP from GDP and $P_i$. GTP can transfer its phosphate to ADP, thereby forming ATP.
6 Succinate is oxidized to fumarate. $FADH_2$ is made.
7 Fumarate combines with water to make malate.
8 Malate is oxidized to oxaloacetate. NADH is made. The cycle can begin again.


**Figure 6.17 A detailed look at the steps of the citric acid cycle.** The blue boxes indicate the location of the acetyl group, which is oxidized at step 6. (It is oxidized again in step 8.) The green boxes indicate the locations where $CO_2$ molecules are removed.

NADH dehydrogenase, cytochrome *b*-$c_1$, and cytochrome oxidase are $H^+$ pumps. The movement of electrons along the ETC is an exergonic process—it releases energy. The binding of electrons to certain proteins of the ETC enables them to actively transport $H^+$ out of the matrix into the intermembrane space against the $H^+$ electrochemical gradient.

Chemicals that inhibit the flow of electrons along the ETC can have lethal effects. For example, one component of the ETC, cytochrome oxidase (complex IV), is inhibited by cyanide. The deadly effects of cyanide ingestion occur because the ETC is shut down, preventing cells from making enough ATP for survival.

## ATP Synthase Makes ATP via Chemiosmosis

The second event of oxidative phosphorylation is the synthesis of ATP by an enzyme called **ATP synthase.** The $H^+$ electrochemical gradient across the inner mitochondrial membrane is a source of potential energy. How is this energy used? The passive flow of $H^+$ back into the matrix is an exergonic process. The lipid bilayer

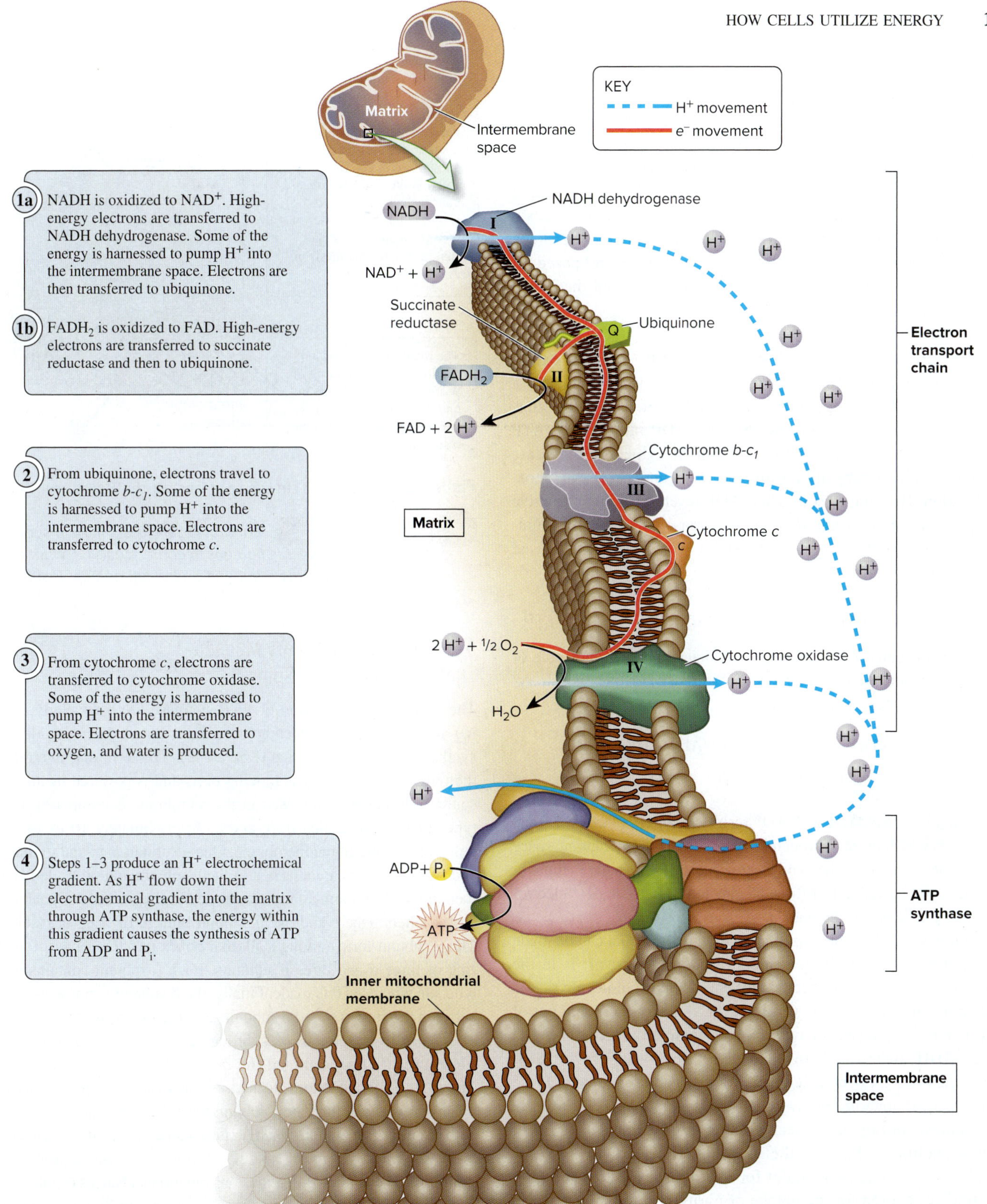


**Figure 6.18 Oxidative phosphorylation.** This process consists of two distinct events: the electron transport chain (ETC) and ATP synthesis. The ETC oxidizes, or removes, electrons from NADH or $FADH_2$ and pumps $H^+$ across the inner mitochondrial membrane. In chemiosmosis, ATP synthase uses the energy in this $H^+$ electrochemical gradient to phosphorylate ADP, thereby synthesizing ATP. In this figure, an oxygen atom is represented as $\frac{1}{2}$ $O_2$ to emphasize that the ETC reduces oxygen when it is in its molecular ($O_2$) form.

**Concept Check:** *Can you explain the name of cytochrome oxidase? What might be another appropriate name?*

is relatively impermeable to $H^+$. However, $H^+$ can pass through the membrane-embedded portion of ATP synthase. The flow of $H^+$ causes conformational changes in ATP synthase that result in the synthesis of ATP from ADP and $P_i$ (see bottom of Figure 6.18). This is an example of an energy conversion: Energy in the form of an $H^+$ electrochemical gradient is converted to chemical potential energy in ATP. The synthesis of ATP that occurs as a result of pushing $H^+$ across a membrane is called chemiosmosis (from the Greek *osmos*, meaning to push). The theory behind it was proposed by Peter Mitchell, a British biochemist who was awarded the Nobel Prize in Chemistry in 1978.

**Regulation of Oxidative Phosphorylation** How is oxidative phosphorylation controlled? This process is regulated by a variety of factors, including the availability of ETC substrates, such as NADH and $O_2$, and by the ATP/ADP ratio. When ATP levels are high, ATP binds to a subunit of cytochrome oxidase (complex IV), thereby inhibiting the ETC and oxidative phosphorylation. By comparison, when ADP levels are high, oxidative phosphorylation is stimulated for two reasons: (1) ADP stimulates cytochrome oxidase, and (2) ADP is a substrate that is used (with $P_i$) to make ATP.

## NADH Oxidation Makes a Large Proportion of a Cell's ATP

For each molecule of NADH that is oxidized and each molecule of ATP that is made, the two chemical reactions of oxidative phosphorylation can be represented as follows:

$$NADH + H^+ + {}^1\!/\!_2\, O_2 \rightarrow NAD^+ + H_2O$$

$$ADP^{2-} + P_i^{2-} \rightarrow ATP^{4-} + H_2O$$

The oxidation of NADH to $NAD^+$ results in an $H^+$ electrochemical gradient in which more hydrogen ions are in the intermembrane space than are in the matrix. The synthesis of one ATP molecule is thought to require the movement of three to four ions down their $H^+$ electrochemical gradient into the matrix.

When we add up the maximal amount of ATP that can be made by oxidative phosphorylation, most researchers agree it is in the range of 30 to 34 ATP molecules for each glucose molecule that is broken down to $CO_2$ and $H_2O$. However, the maximum amount of ATP is rarely achieved for two reasons. First, although 10 NADH and 2 $FADH_2$ are available to make the $H^+$ electrochemical gradient across the inner mitochondrial membrane, a cell uses some of these molecules for anabolic pathways. For example, NADH is used in the synthesis of organic molecules such as glycerol (a component of phospholipids). Second, the mitochondrion may use some of the $H^+$ electrochemical gradient for other purposes. For example, the gradient is used for the uptake of pyruvate into the matrix via an $H^+$/pyruvate symporter (see Figure 6.15). Therefore, the actual amount of ATP synthesis is usually a little less than the maximum number of 30 to 34. Even so, when we compare the amount of ATP that is made by glycolysis (2), the citric acid cycle (2), and oxidative phosphorylation (30–34), we see that oxidative phosphorylation provides a cell with a much greater capacity to make ATP.

The nonmembrane-embedded portion consists of 1 ε, 1 γ, 1 δ, 3 α, and 3 β subunits. Movement of $H^+$ between a *c* subunit and the *a* subunit causes the γ subunit to rotate. The rotation, in 120° increments, causes the β subunits to progress through a series of 3 conformational changes that lead to the synthesis of ATP from ADP and $P_i$.

The membrane-embedded portion consists of a ring of 9–12 *c* subunits, 1 *a* subunit, and 2 *b* subunits. $H^+$ moves between a *c* subunit and the *a* subunit.

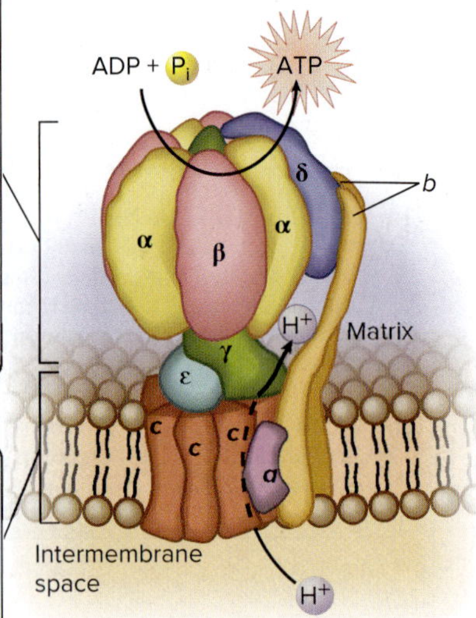


**Figure 6.19** The subunit structure and function of ATP synthase.

**Concept Check:** *Look at Figure 6.20. If the β subunit in the front center of this figure is in conformation 2, what are the conformations of the β subunit on the left and the β subunit on the back right?*

## ATP Synthase Is a Rotary Machine That Makes ATP as It Spins

The structure and function of ATP synthase are particularly intriguing and have received much attention over the past few decades. ATP synthase is a rotary machine (**Figure 6.19**). The term machine refers to the observations that it has moving parts and is powered by an energy source. The region embedded in the membrane is composed of three types of subunits called *a*, *b*, and *c*. Approximately 10 to 14 *c* subunits form a ring in the membrane. One *a* subunit is bound to this ring, and two *b* subunits are attached to the *a* subunit and protrude from the membrane. The nonmembrane-embedded subunits are designated with Greek letters. One ε and one γ subunit bind to the ring of *c* subunits. The γ subunit forms a long stalk that pokes into the center of another ring of three α and three β subunits. Each β subunit contains a catalytic site where ATP is made. Finally, the δ subunit forms a connection between the ring of α and β subunits and the two *b* subunits.

When hydrogen ions pass through a narrow channel at the contact site between a *c* subunit and the *a* subunit, a conformational change causes the γ subunit to turn clockwise (when viewed from the intermembrane space). Each time the γ subunit turns 120°, it changes its contacts with the three β subunits, which, in turn, causes the β subunits to change their conformations. How do these conformational changes promote ATP synthesis? The answer is that the conformational changes occur in a way that favors ATP synthesis and release. As shown in **Figure 6.20**, the conformational changes in the β subunits happen in the following order:

- Conformation 1: ADP and $P_i$ bind with good affinity.
- Conformation 2: ADP and $P_i$ bind very tightly, which strains chemical bonds so that ATP is made.
- Conformation 3: ATP (and ADP and $P_i$) bind very weakly, and ATP is released.

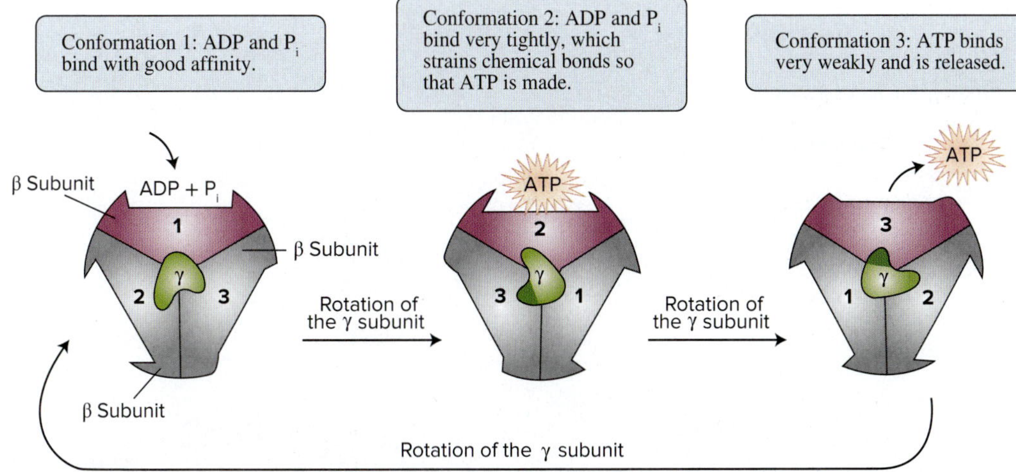


**Figure 6.20 Conformational changes that result in ATP synthesis.** For simplicity, the α subunits are not shown. This drawing emphasizes the conformational changes in the β subunit shown at the top. The other two β subunits also make ATP. All three β subunits alternate between three conformational states due to their interactions with the γ subunit.

Each time the γ subunit turns 120°, it causes a β subunit to change to the next conformation. After conformation 3, a 120° turn by the γ subunit returns a β subunit back to conformation 1, and the cycle of ATP synthesis can begin again. Because ATP synthase has three β subunits, each subunit is in a different conformation at any given time.

American biochemist Paul Boyer proposed the concept of a rotary machine in the late 1970s. In his model, the three β subunits alternate between three conformations, as described previously. Boyer's original idea was met with great skepticism, because the concept that part of an enzyme could spin was very novel, to say the least. In 1994, British biochemist John Walker and his colleagues determined the three-dimensional structure of the nonmembrane-embedded portion of the ATP synthase. The structure revealed that each of the three β subunits had a different conformation—one with ADP bound, one with ATP bound, and one without any nucleotide bound. This result supported Boyer's model. In 1997, Boyer and Walker shared the Nobel Prize in Chemistry for their work on ATP synthase. As described in the Feature Investigation, other researchers subsequently visualized the rotation of the γ subunit.

# FEATURE INVESTIGATION

## Yoshida and Kinosita Demonstrated That the γ Subunit of ATP Synthase Spins

In 1997, Masasuke Yoshida, Kazuhiko Kinosita, and colleagues set out to experimentally visualize the rotary nature of ATP synthase (**Figure 6.21**). The membrane-embedded region of ATP synthase can be separated from the rest of the protein by treatment of mitochondrial membranes with a high concentration of salt, releasing the portion of the protein containing one γ, three α, and three β subunits. The researchers adhered the resulting $\gamma\alpha_3\beta_3$ complex to a glass slide so the γ subunit was protruding upward. Because the γ subunit is too small to be seen with a light microscope, the rotation of the γ subunit cannot be visualized directly. To circumvent this problem, the researchers attached a large, fluorescently labeled actin filament to the γ subunit via linker proteins. The fluorescently labeled actin filament is very long compared to the γ subunit and can be readily seen with a fluorescence microscope.

Because the membrane-embedded portion of the protein is missing, you may be wondering how the researchers could get the γ subunit to rotate. The answer is they added ATP. Although the normal function of ATP synthase is to make ATP, it can also hydrolyze ATP. In other words, ATP synthase can run backwards. As shown in the data in Figure 6.21, when the researchers added ATP, they observed that the fluorescently labeled actin filament rotated in a counterclockwise direction, which is opposite to the direction that the γ subunit rotates when ATP is synthesized. Actin filaments were observed to rotate for more than 100 revolutions in the presence of ATP. These results convinced the scientific community that ATP synthase is a rotary machine.

### *Experimental Questions*

1. **SCISKILLS ▸** The components of ATP synthase are too small to be visualized by light microscopy. For the experiment of Figure 6.21, how did the researchers observe the movement of ATP synthase?
2. **SCISKILLS ▸** In the experiment of Figure 6.21, what observation did the researchers make that indicated ATP synthase is a rotary machine? What was the control of this experiment? What did it indicate?
3. **SCISKILLS ▸** Were the rotations seen by the researchers in the data of Figure 6.21 in the same direction as expected in the mitochondria during ATP synthesis? Why or why not?

**Figure 6.21** Yoshida and Kinosita provide evidence that ATP synthase is a rotary machine.



**HYPOTHESIS** ATP synthase is a rotary machine.

**KEY MATERIALS** Purified complex containing 1 $\gamma$, 3 $\alpha$, and 3 $\beta$ subunits.

**Experimental level** | **Conceptual level**

**1** Adhere the purified $\gamma\alpha_3\beta_3$ complex to a glass slide so the base of the $\gamma$ subunit is protruding upward.

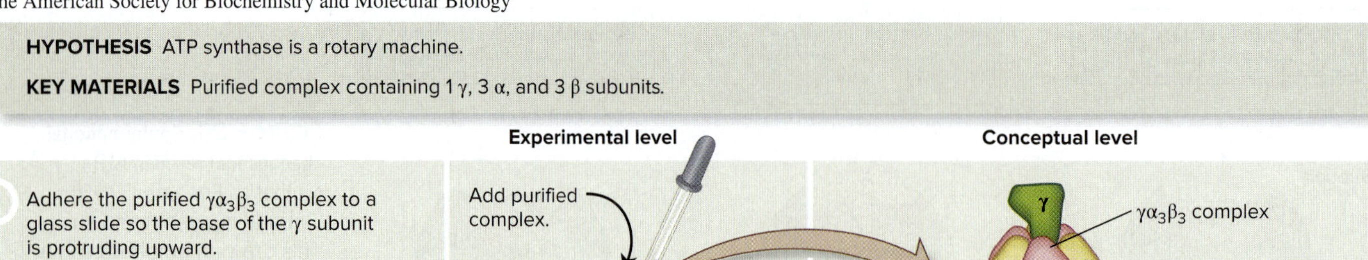


**2** Add linker proteins and fluorescently labeled actin filaments. The linker protein recognizes sites on both the $\gamma$ subunit and the actin filament.

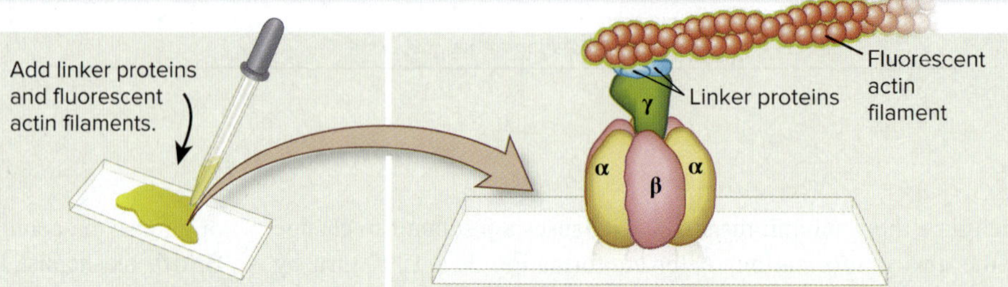


**3** Add ATP. As a control, do not add ATP.

**4** Observe under a fluorescence microscope. The method of fluorescence microscopy is described in Chapter 4.

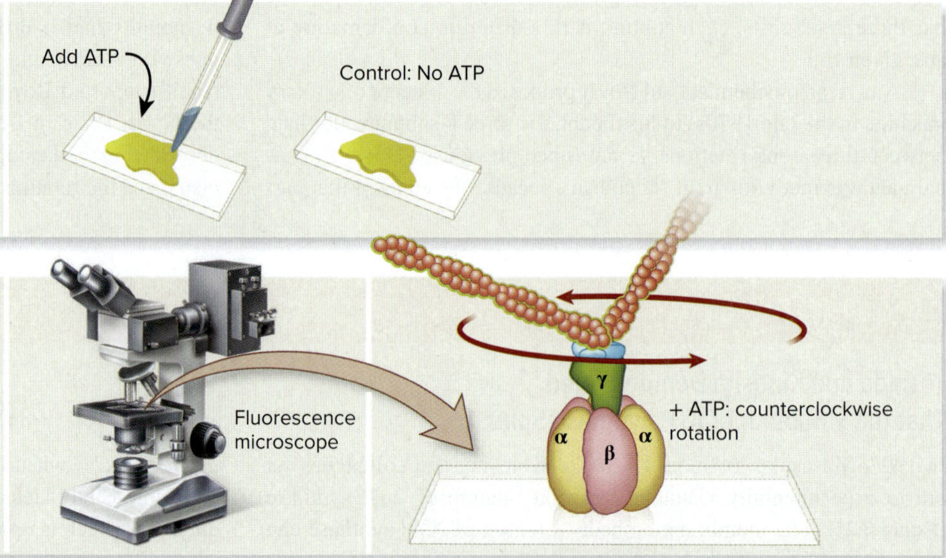


**5** **THE DATA**

Results from step 4:

| ATP | Rotation |
|---|---|
| No ATP added | No rotation observed. |
| ATP added | Rotation was observed as shown below. This is a time-lapse view of the rotation in action. |

Row 1

Row 2

6 **CONCLUSION** The γ subunit rotates counterclockwise when ATP is hydrolyzed. It would be expected to rotate clockwise when ATP is synthesized.

7 **SOURCE** Noji, H., Yasuda, R., Yoshida, M., and Kinosita, K. 1997. Direct observation of the rotation of $F_1$-ATPase. *Nature* 386: 299–303.

## 6.7 Reviewing the Concepts

- Oxidative phosphorylation involves two events: (1) The electron transport chain (ETC) oxidizes NADH or $FADH_2$ and generates an $H^+$ electrochemical gradient, and (2) this gradient is used by ATP synthase to make ATP via chemiosmosis (Figures 6.18).
- ATP synthase is a rotary machine. The rotation is triggered by the passage of $H^+$ through a channel between a *c* subunit and the α subunit that causes the γ subunit to spin, resulting in three conformational changes in the β subunits that promote ATP synthesis (Figures 6.19, 6.20).
- Yoshida and Kinosita demonstrated rotation of the γ subunit by attaching a fluorescently labeled actin filament to the γ subunit via linker proteins and observing its movement during the hydrolysis of ATP (Figure 6.21).

## 6.7 Testing Your Knowledge

1. The electrons that travel down the electron transport chain come from
   a. NADH.
   b. $FADH_2$.
   c. $O_2$.
   d. ATP.
   e. either a or b.
2. The source of energy that directly drives the synthesis of ATP during oxidative phosphorylation is
   a. the oxidation of NADH.
   b. the oxidation of glucose.
   c. the oxidation of pyruvate.
   d. the $H^+$ gradient.
   e. the reduction of $O_2$.
3. The spinning of the γ subunit of ATP synthase is caused by
   a. the oxidation of NADH.
   b. the synthesis of ATP.
   c. the hydrolysis of ATP.
   d. the movement of $H^+$ through ATP synthase.
   e. either c or d.

## 6.8 Connections Among Carbohydrate, Protein, and Fat Metabolism

### Learning Outcome

1. Explain how carbohydrate, protein, and fat metabolism are interconnected.

When you eat a meal, it usually contains not only carbohydrates (including glucose) but also proteins and fats. These molecules are broken down by some of the same enzymes involved with glucose metabolism. By using the same pathways for the breakdown of sugars, amino acids, and fats, cellular metabolism is more efficient because the same enzymes can be used for the breakdown of different starting molecules.

As shown in **Figure 6.22**, proteins and fats can enter into glycolysis or the citric acid cycle at different points. Proteins are first acted

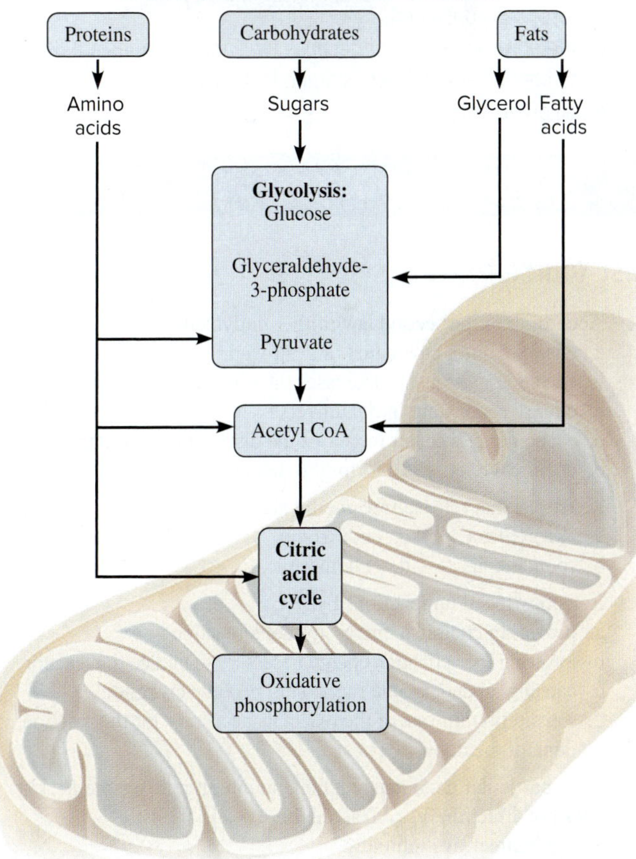


**Figure 6.22** **Integration of carbohydrate, protein, and fat metabolism.** Breakdown products of proteins and fats can be used as fuel for cellular respiration, entering the same pathway used to break down carbohydrates.

*(photo)* © Ernie Friedlander/Cole Group/Getty Images RF

**Concept Check:** *What is a cellular advantage of integrating protein, carbohydrate, and fat metabolism?*

on by enzymes, either in digestive juices or within cells, that cleave the bonds connecting individual amino acids. Breakdown products of some amino acids can enter at later steps of glycolysis, or an acetyl group can be removed from certain amino acids and become attached to CoA and then enter the citric acid cycle. Other amino acids can be modified and enter the citric acid cycle.

Fats are typically broken down to glycerol and fatty acids. Glycerol can be modified to glyceraldehyde-3-phosphate and enter glycolysis. Fatty acid tails have two carbon acetyl units removed, which bind to CoA and enter the citric acid cycle.

## 6.8 Reviewing the Concepts

- Proteins and fats can enter into glycolysis or the citric acid cycle at different points (Figure 6.22).

## 6.8 Testing Your Knowledge

1. The advantage of connecting metabolic pathways for carbohydrate, protein, and fat metabolism is that
   a. more ATP is made.
   b. more NADH is made.
   c. less $O_2$ is consumed.
   d. the efficiency is higher because the same enzymes are used to break down different starting materials.
   e. all of the above.

# Assess and Discuss

## Test Yourself

1. According to the second law of thermodynamics,
   a. energy cannot be created or destroyed.
   b. each energy transfer decreases the disorder of a system.
   c. energy is constant in the universe.
   d. each energy transfer increases the disorder of a system.
   e. chemical energy is a form of potential energy.
2. Reactions that release free energy are
   a. exergonic. c. endergonic. e. both a and b.
   b. spontaneous. d. endothermic.
3. Enzymes speed up reactions by
   a. providing chemical energy to fuel a reaction.
   b. lowering the activation energy necessary to initiate the reaction.
   c. causing an endergonic reaction to become exergonic.
   d. substituting for one of the reactants necessary for the reaction.
   e. none of the above.
4. For a particular chemical reaction, an inhibitor raises the $K_M$ but does not affect the $V_{max}$. This inhibitor
   a. is a competitive inhibitor.
   b. is a noncompetitive inhibitor.
   c. binds to the active site of the enzyme.
   d. binds to an allosteric site of the enzyme.
   e. is a competitive inhibitor and binds to the active site of the enzyme.
5. When NADH is converted to $NAD^+ + H^+$, NADH has been
   a. reduced. c. oxidized. e. methylated.
   b. phosphorylated. d. decarboxylated.
6. Which of the following occurs in the cytosol of eukaryotes?
   a. glycolysis
   b. breakdown of pyruvate to an acetyl group
   c. citric acid cycle
   d. oxidative phosphorylation
   e. all of the above
7. The net products of glycolysis are
   a. 6 $CO_2$, 4 ATP, and 2 NADH.
   b. 2 pyruvate, 2 ATP, and 2 NADH.
   c. 2 pyruvate, 4 ATP, and 2 NADH.
   d. 2 pyruvate, 2 GTP, and 2 $CO_2$.
   e. 2 $CO_2$, 2 ATP, and glucose.
8. The ability to diagnose tumors using [$^{18}$F]-fluorodeoxyglucose (FDG) is based on the phenomenon that most types of cancer cells exhibit higher levels of
   a. glycolysis. d. oxidative phosphorylation.
   b. pyruvate breakdown. e. all of the above.
   c. citric acid metabolism.
9. ATP is made via chemiosmosis during
   a. glycolysis. d. oxidative phosphorylation.
   b. the breakdown of pyruvate. e. all of the above.
   c. the citric acid cycle.
10. Certain drugs act as ionophores that cause the mitochondrial membrane to be highly permeable to $H^+$. How would such drugs affect oxidative phosphorylation?
    a. Movement of electrons down the ETC would be inhibited.
    b. ATP synthesis would be inhibited.
    c. ATP synthesis would be unaffected.
    d. ATP synthesis would be stimulated.
    e. Both a and b are correct.

## Conceptual Questions

1. Describe the mechanism and purpose of feedback inhibition in a metabolic pathway.
2. What causes the rotation of the γ subunit of ATP synthase? How does this rotation promote ATP synthesis?
3. **PRINCIPLES** A principle of biology is that living organisms maintain homeostasis. How is glucose breakdown regulated to maintain homeostasis? What would be some potentially harmful consequences if glucose metabolism was not regulated properly?

## Collaborative Questions

1. Discuss how life can maintain its order in spite of the second law of thermodynamics. Are we defying this law?
2. Read more about PET scans from other sources. Which types of cancers are most easily detected by this procedure and which types are not? Is the ability to detect cancer via a PET scan related to the degree of hypoxia (low oxygen levels) in a tumor?

# Online Resource

**connect.mheducation.com**

SMARTBOOK® SmartBook® is the first and only adaptive reading experience designed to change the way students read and learn.

# How Cells Capture Light Energy via Photosynthesis

© Steven Trainoff Ph.D./Getty Images RF

**Algae in the open ocean.** Algae, such as these giant kelp, are photosynthetic organisms.

## Chapter Outline



Nearly all of the oxygen in every breath you take is made by the abundant plant life, algae (see chapter-opening photo), and cyanobacteria on Earth. We often think of photosynthesis in terms of terrestrial plants such as grasses and trees. Even so, you may be surprised to learn that nearly half of the oxygen produced globally is made by phytoplankton—a diverse mixture of aquatic and photosynthetic microorganisms that include cyanobacteria and protists such as algae. Global warming may decrease phytoplankton populations in oceans and lakes, thereby decreasing oxygen levels on a global scale.

**Photosynthesis** is the process in which the energy from light is captured and used to synthesize glucose and other organic molecules. Nearly all living organisms rely on photosynthesis for their nourishment, either directly or indirectly. Photosynthesis is also responsible for producing the oxygen that makes up a large portion of the Earth's atmosphere. Therefore, all aerobic organisms rely on photosynthesis for cellular respiration.

We begin this chapter with an overview of photosynthesis as it occurs in green plants and algae. We will then explore the two stages of photosynthesis in more detail. In the first stage, called the **light reactions,** light energy is absorbed by pigments and converted to chemical energy in the form of two energy intermediates: ATP and NADPH. During the second stage, known as the **Calvin cycle,** ATP and NADPH are used to drive the synthesis of carbohydrates. We will conclude with a consideration of the variations in photosynthesis that occur in plants existing in hot and dry conditions.

## 7.1 Overview of Photosynthesis

### Learning Outcomes

1. Write the general equations that represent the process of photosynthesis.
2. Diagram how photosynthesis powers the biosphere.
3. Describe the general structure of chloroplasts.
4. Explain how photosynthesis occurs in two stages: the light reactions and the Calvin cycle.

In the mid-1600s, a Flemish physician, Jan Baptista Van Helmont, conducted an experiment in which he transplanted the shoot of a young willow tree into a bucket of soil and allowed it to grow for 5 years. After this time, the willow tree had added 164 pounds to its original weight, but the soil had lost only 2 ounces. Van Helmont correctly concluded that the willow tree did not get most of its nutrients from the soil. From this work, he also hypothesized that the mass of the tree came from the water he had added over the 5 years. This hypothesis was partially correct, but we now know that $CO_2$ from the air also is a major contributor to the growth and mass of plants.

Photosynthesis is carried out by certain species of bacteria, algae, and plants. Most photosynthetic organisms use $H_2O$ for photosynthesis, although some bacteria use hydrogen sulfide ($H_2S$) instead of water. When water is used, a general equation for photosynthesis is

$$CO_2 + 2\ H_2O + \text{Light energy} \rightarrow CH_2O + O_2 + H_2O$$

When the carbohydrate produced is glucose ($C_6H_{12}O_6$), we multiply each side of the equation by 6 to obtain

$$6\ CO_2 + 12\ H_2O + \text{Light energy} \rightarrow C_6H_{12}O_6 + 6\ O_2 + 6\ H_2O$$
$$\Delta G = +685\ \text{kcal/mol}$$

In this redox reaction, $CO_2$ is reduced during the formation of glucose, and $H_2O$ is oxidized during the formation of $O_2$. Notice that the free-energy change required for the production of 1 mole of glucose from carbon dioxide and water is a whopping +685 kcal/mol! As we learned in Chapter 6, endergonic reactions are driven forward by coupling the reaction with an exergonic process that releases free energy. In this case, the energy from sunlight ultimately drives the synthesis of glucose.

In this section, we will survey the general features of photosynthesis as it occurs in green plants and algae. Later sections will examine the various steps in this process.

### Photosynthesis Powers the Biosphere

The term **biosphere** refers to the places on the surface of the Earth and in the atmosphere where living organisms exist. Organisms can be categorized as autotrophs and heterotrophs.

- **Autotrophs** sustain themselves by producing organic molecules from inorganic sources. For example, plants can make glucose (an organic molecule) from inorganic $CO_2$ and $H_2O$.

  **Photoautotrophs** are autotrophs that use light as a source of energy to make organic molecules. These include plants, algae, and some bacterial species such as cyanobacteria.
- **Heterotrophs** must consume food—organic molecules from their environment—to sustain life. Most species of bacteria and protists, as well as all species of fungi and animals, are heterotrophs.

Life in the biosphere is largely driven by the photosynthetic power of plants, algae, and cyanobacteria. The existence of most species relies on a key energy cycle that involves the interplay between organic molecules (such as glucose) and inorganic molecules, namely $O_2$, $CO_2$, and $H_2O$ (**Figure 7.1**). Photoautotrophs make a large proportion of the Earth's organic molecules via photosynthesis, using light energy, $CO_2$, and $H_2O$. During this process, they also produce $O_2$. To supply their energy needs, both photoautotrophs and heterotrophs metabolize organic molecules via cellular respiration. As described in Chapter 6, cellular respiration generates $CO_2$ and $H_2O$ and is used to make ATP. The $CO_2$ is released into the atmosphere and can be reused by photoautotrophs to make more organic molecules such as glucose. In this way, an energy cycle between photosynthesis and cellular respiration sustains life on our planet.

## Biology Principle

### Living Organisms Use Energy

Photosynthetic species capture light energy and store it in organic molecules, which are used by photosynthetic and nonphotosynthetic organisms as energy sources.

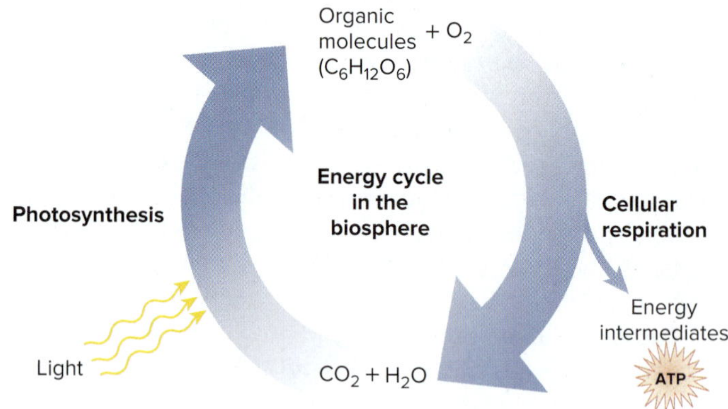


**Figure 7.1 An important energy cycle between photosynthesis and cellular respiration.** Photosynthesis uses light, $CO_2$, and $H_2O$ to produce $O_2$ and organic molecules. The organic molecules are broken down to $CO_2$ and $H_2O$ via cellular respiration to supply energy in the form of ATP; $O_2$ is reduced to $H_2O$.

**Concept Check:** *Which types of organisms carry out cellular respiration? Is it heterotrophs, autotrophs, or both?*

## In Plants and Algae, Photosynthesis Occurs in the Chloroplast

**Chloroplasts** are organelles found in plant and algal cells that carry out photosynthesis. These organelles contain **chlorophyll,** which is a pigment that gives plants their green color. All green parts of a plant contain chloroplasts and can perform photosynthesis, although the majority of photosynthesis in most species of plants occurs in the leaves (**Figure 7.2**). The tissue in the internal part of the leaf, called the **mesophyll,** contains cells with chloroplasts. For photosynthesis to occur, the mesophyll cells must receive light as well as obtain water and carbon dioxide. The water is taken up by the roots of the plant and is transported to the leaves by small veins. Carbon dioxide gas enters the leaf, and oxygen exits via pores called **stomata** (singular, stoma or stomate).

A chloroplast contains an outer and inner membrane, with an intermembrane space between the two. A third membrane, the thylakoid membrane, contains pigment molecules, including chlorophyll. The thylakoid membrane forms many flattened, fluid-filled tubules called **thylakoids,** which enclose a single, convoluted compartment known as the **thylakoid lumen.** Thylakoids stack on top of each other to form a structure called a **granum** (plural, grana). The **stroma** is the fluid-filled region of the chloroplast between the thylakoid membrane and the inner membrane (see Figure 7.2).

## Photosynthesis Occurs in Two Stages: Light Reactions and the Calvin Cycle

How does photosynthesis take place? The process of photosynthesis occurs in two stages called the light reactions and the Calvin cycle. The term photosynthesis is derived from the association between these two stages: Photo refers to the light reactions that capture the energy from sunlight needed for the synthesis of carbohydrates that occurs in the Calvin cycle. The light reactions take place at the thylakoid membrane, and the Calvin cycle occurs in the stroma (**Figure 7.3**).

The light reactions involve an amazing series of energy conversions, starting with light energy and ending with chemical energy that is stored in the form of covalent bonds. The light reactions produce three chemical products: ATP, NADPH, and $O_2$. ATP and NADPH are energy intermediates that provide the needed energy and electrons to make carbohydrates during the Calvin cycle. Like NADH, **NADPH (nicotinamide adenine dinucleotide phosphate)** is an electron carrier that can accept two electrons. Its structure differs from NADH by the presence of an additional phosphate group. The structure of NADH is described in Chapter 6 (refer back to Figure 6.9).

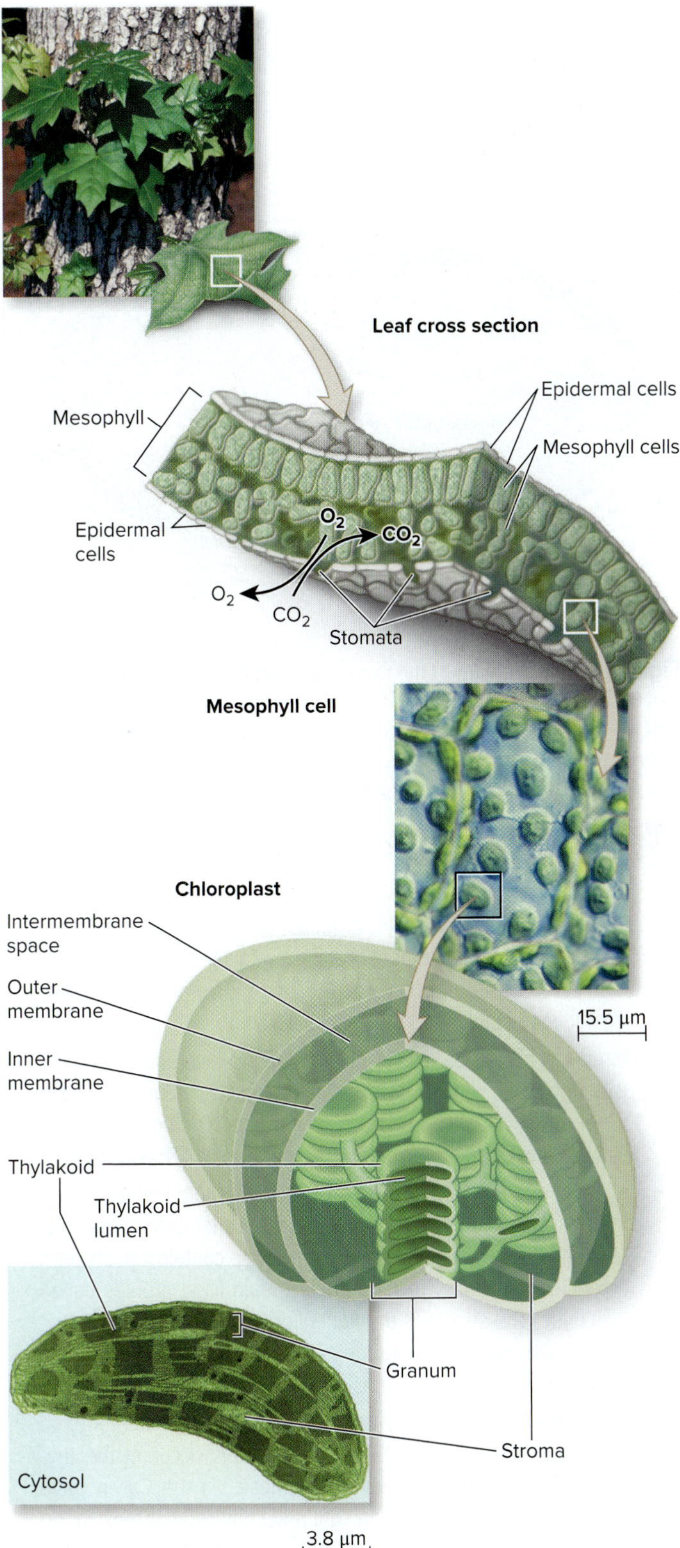


**Figure 7.2 Leaf organization.** Leaves are composed of layers of cells. The epidermal cells are on the outer surface, both top and bottom, with mesophyll cells sandwiched in the middle. The mesophyll cells contain chloroplasts and are the primary sites of photosynthesis in most plants.
*(top)* © Photoshot; *(middle)* © Biophoto Associates/SPL/Science Source; *(bottom)* © Science Source

**BioConnections:** *Look ahead to Figure 30.23. How many guard cells make up a stoma (plural, stomata)?*

### 7.1 Reviewing the Concepts

- Photosynthesis is the process by which plants, algae, and cyanobacteria capture light energy to make energy intermediates and synthesize carbohydrates.
- During photosynthesis, $CO_2$, $H_2O$, and energy are used to make carbohydrates and $O_2$.

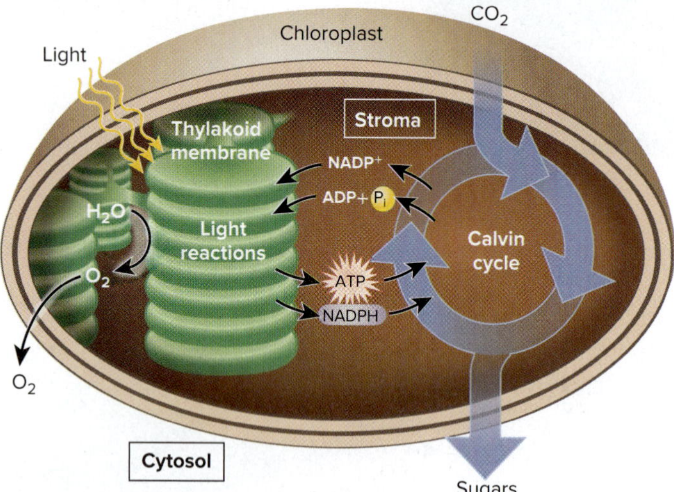


**Figure 7.3 An overview of the two stages of photosynthesis: Light reactions and the Calvin cycle.** The light reactions, through which ATP, NADPH, and $O_2$ are made, occur at the thylakoid membrane. The Calvin cycle, in which enzymes use ATP and NADPH to incorporate $CO_2$ into carbohydrate, occurs in the stroma.

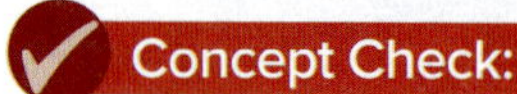

*Can the Calvin cycle occur in the dark?*

- Heterotrophs must obtain organic molecules in their food, whereas autotrophs make organic molecules from inorganic sources. Photoautotrophs use the energy from light to make organic molecules.
- An energy cycle occurs in the biosphere in which photosynthesis uses light, $CO_2$, and $H_2O$ to make organic molecules, and the organic molecules are broken back down to $CO_2$ and $H_2O$ via cellular respiration to supply energy in the form of ATP (Figure 7.1).
- In plants and algae, photosynthesis occurs within chloroplasts—organelles with an outer membrane, inner membrane, and thylakoid membrane. The thylakoid membrane contains chlorophyll molecules and forms many flattened, fluid-filled tubules called thylakoids. The stroma is the fluid-filled region between the thylakoid membrane and inner membrane. In plants, the leaves are the major site of photosynthesis (Figure 7.2).
- The light reactions of photosynthesis capture light energy to make ATP, NADPH, and $O_2$. These reactions occur at the thylakoid membrane. Carbohydrate synthesis via the Calvin cycle uses ATP and NADPH from the light reactions and happens in the stroma (Figure 7.3).

## 7.1 Testing Your Knowledge

1. A heterotroph is an organism that
   a. can make organic molecules from inorganic sources.
   b. cannot make organic molecules starting with inorganic molecules.
   c. cannot make any organic molecules.
   d. must consume autotrophs or other heterotrophs to survive.
   e. both b and d.
2. If you were traveling from the cytosol to the thylakoid lumen, how many membranes would you have to cross?
   a. 0 c. 2 e. 4
   b. 1 d. 3
3. The products of the light reactions of photosynthesis include
   a. NADPH. c. $O_2$. e. only a, b, and c.
   b. ATP. d. carbohydrate.

## 7.2 Reactions That Harness Light Energy

### Learning Outcomes

1. Describe the general properties of light.
2. Explain how pigments absorb light energy, and list the types of pigments found in plants and green algae.
3. Outline the steps in which photosystems II and I capture light energy and produce $O_2$, ATP, and NADPH.
4. Describe the process of cyclic photophosphorylation in which only ATP is made.

According to the first law of thermodynamics discussed in Chapter 6, energy cannot be created or destroyed, but it can be transferred from one place to another and transformed from one form to another. During the light reactions of photosynthesis, energy in the form of light is transferred from the Sun, nearly 93 million miles away, to a pigment molecule in a photosynthetic organism such as a plant. What follows is an interesting series of energy transformations in which light energy is transformed into electrochemical energy and then into energy stored within energy intermediates.

In this section, we will explore this series of transformations, collectively called the light reactions of photosynthesis. We begin by examining the properties of light and then consider the features of chloroplasts that allow them to capture light energy. The remainder of this section focuses on how the light reactions of photosynthesis generate three important products: ATP, NADPH, and $O_2$.

### Light Energy Is a Form of Electromagnetic Radiation

Light is essential to support life on Earth. Light is a type of electromagnetic radiation, so named because it consists of energy in the form of electric and magnetic fields. Electromagnetic radiation travels as waves caused by the oscillation of the electric and magnetic fields. The **wavelength** is the distance between the peaks in a wave pattern. The **electromagnetic spectrum** encompasses all possible wavelengths of electromagnetic radiation, from relatively short wavelengths (gamma rays) to much longer wavelengths (radio waves) (**Figure 7.4**). Visible light is the range of wavelengths detected by the human eye, commonly between 380 and 740 nm. As discussed later, visible light provides the energy to drive photosynthesis.

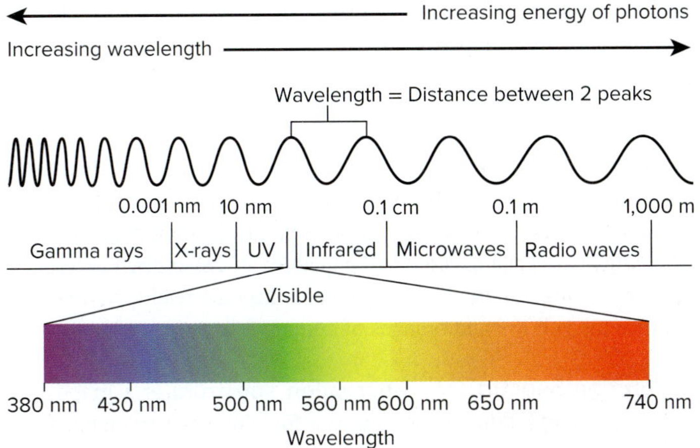


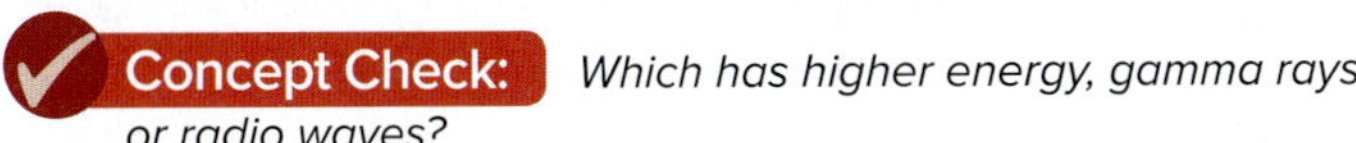


**Figure 7.4** **The electromagnetic spectrum.** The bottom portion of this figure emphasizes visible light—the wavelengths of electromagnetic radiation visible to the human eye. Light in the visible portion of the electromagnetic spectrum drives photosynthesis.

**Concept Check:** *Which has higher energy, gamma rays or radio waves?*

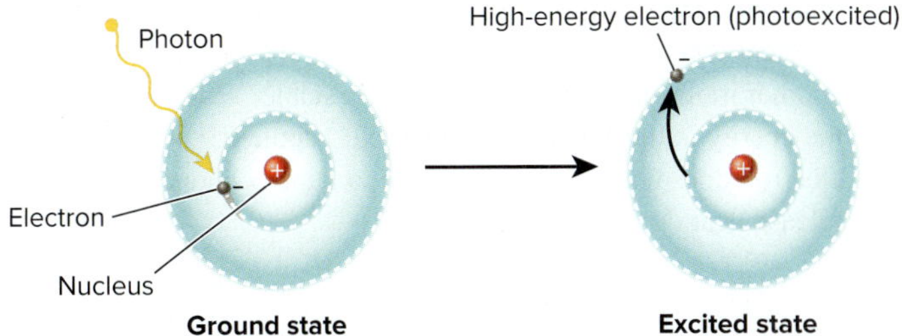


**Figure 7.5** **Absorption of light energy by an electron.** When a photon of light of the correct amount of energy strikes an electron, the electron is boosted from the ground (unexcited) state to a higher energy level (an excited state). When this occurs, the electron occupies an orbital that is farther away from the nucleus of the atom. At this farther distance, the electron is held less firmly and is considered unstable.

**Concept Check:** *Describe three ways that a photoexcited electron can become more stable.*

Light also has properties that are characteristic of particles. Albert Einstein proposed that light is composed of discrete particles called **photons**—massless particles traveling in a wavelike pattern and moving at the speed of light (about 300 million m/sec). Each photon contains a specific amount of energy. An important difference between the various types of electromagnetic radiation, shown in Figure 7.4, is the amount of energy found in the photons. Shorter-wavelength radiation carries more energy per unit of time than longer-wavelength radiation. For example, the photons of gamma rays carry more energy than those of radio waves.

The Sun radiates the entire spectrum of electromagnetic radiation, but the atmosphere prevents much of this radiation from reaching the Earth's surface. For example, the ozone layer forms a thin shield in the upper atmosphere, protecting life on Earth from much of the Sun's ultraviolet (UV) radiation. Even so, a substantial amount of electromagnetic radiation reaches the Earth's surface. The effect of light on living organisms is critically dependent on the energy of the photons that reach them. The photons found in gamma rays, X-rays, and UV rays have very high energy. When molecules in cells absorb such energy, the effects can be devastating. Such radiation can cause mutations in DNA and even lead to cancer. By comparison, the energy of photons found in visible light is much milder. Molecules can absorb this energy in a way that does not cause damage. Next, we will consider how molecules in living cells absorb the energy within visible light.

## Pigments Absorb Light Energy

When light strikes an object, one of three things happens. First, light may simply pass through the object. Second, the object may change the path of light toward a different direction. A third possibility is that the object may absorb the light. The term **pigment** is used to describe a molecule that can absorb light energy. When light strikes a pigment, some of the wavelengths of light energy are absorbed, while others are reflected. For example, leaves look green to us because they reflect radiant energy of the green wavelength. Various pigments in the leaves absorb the energy of other wavelengths. At the extremes of color reflection are white and black. A white object reflects nearly all of the visible light energy falling on it, whereas a black object absorbs nearly all of the light energy. This is why it is coolest to wear white clothes on a sunny, hot day.

What do we mean when we say that light energy is absorbed? In the visible spectrum, light energy may be absorbed by boosting electrons to higher energy levels (Figure 7.5). The location in which an electron is found is called its orbital. Electrons in different orbitals possess different amounts of energy. For an electron to absorb light energy and be boosted to an orbital with a higher energy, it must overcome the difference in energy between the orbital it is in and the orbital to which it is going. For this to happen, an electron must absorb a photon that contains precisely that amount of energy.

After an electron absorbs energy, it is said to be in an excited state. Usually, this is an unstable condition. The electron may release the energy in different ways.

- An excited electron may release heat. For example, on a sunny day, the sidewalk heats up because it absorbs light energy that is released as heat.
- An electron can release energy in the form of light. Certain organisms, such as jellyfish, possess molecules that make them glow. This glow is due to the release of light when electrons drop down to lower energy levels, a phenomenon called fluorescence.
- In the case of photosynthetic pigments, a different event happens that is critical for the process of photosynthesis. Rather

than releasing energy, an excited electron in a photosynthetic pigment is removed from that molecule and transferred to another molecule where the electron is stable. When this occurs, the energy in the electron is said to be "captured," because the electron does not readily drop down to a lower energy level and release heat or light.

## Plants Contain Different Types of Photosynthetic Pigments

Plant cells that carry out photosynthesis synthesize different pigment molecules that absorb the light energy used to drive photosynthesis.

**Chlorophylls** Two types of chlorophyll pigments, termed **chlorophyll *a*** and **chlorophyll *b*,** are found in green plants and green algae (Figure 7.6a). In the chloroplast, both chlorophylls *a* and *b* are bound to integral membrane proteins in the thylakoid membrane, which traverse the membrane.

The chlorophylls contain a porphyrin ring and a hydrocarbon tail, also called a phytol tail. A magnesium ion ($Mg^{2+}$) is bound to the porphyrin ring. An electron in the porphyrin ring follows a path in which it spends some of its time around several different atoms. Because this electron isn't restricted to a single atom, it is called a delocalized electron. The delocalized electron can absorb light energy. The phytol tail in chlorophyll is a long hydrocarbon that is hydrophobic. Its function is to anchor the pigment to the surface of hydrophobic proteins within the thylakoid membrane of chloroplasts.

CHO in chlorophyll *b*
$CH_3$ in chlorophyll *a*
Porphyrin ring
Phytol tail

**(a) Chlorophylls *a* and *b***

**(b) β-Carotene (a carotenoid)**

**Figure 7.6 Structures of pigment molecules. (a)** The structure of chlorophylls *a* and *b*. As indicated, chlorophylls *a* and *b* differ only at a single site, at which chlorophyll *a* has a $—CH_3$ group and chlorophyll *b* has a —CHO group. **(b)** The structure of β-carotene, an example of a carotenoid. The dark green and light green areas in parts **(a)** and **(b)** are the regions where a delocalized electron can hop from one atom to another.

**Carotenoids** These pigments impart a color that ranges from yellow to orange to red (Figure 7.6b). Carotenoids are often the major pigments in flowers and fruits. In leaves, the more abundant chlorophylls usually mask the colors of carotenoids. In temperate climates where the leaves change colors, the quantity of chlorophyll in the leaf declines during autumn. The carotenoids become readily visible and produce the yellows and oranges of autumn foliage.

**Absorption Spectra** An **absorption spectrum** is a graph that plots a pigment's light absorption as a function of wavelength. Each of the photosynthetic pigments shown in Figure 7.7a absorbs light in

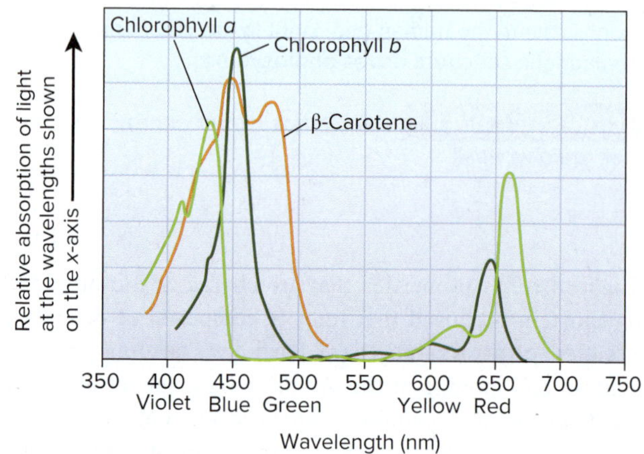


**(a) Absorption spectra**

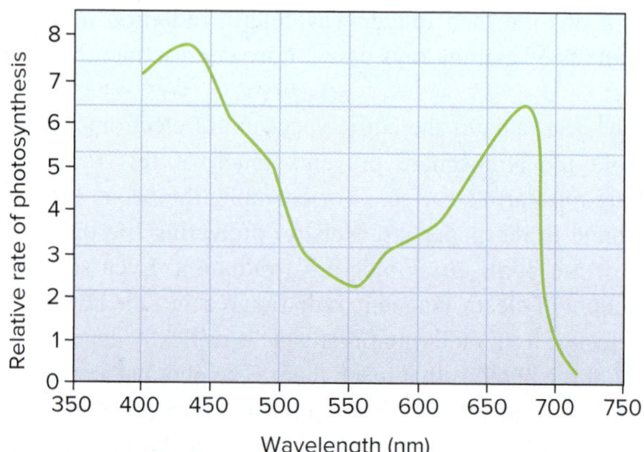


**(b) Action spectrum**

**Figure 7.7 Properties of pigment function: absorption and action spectra. (a)** These absorption spectra show the absorption of light by chlorophyll *a*, chlorophyll *b*, and β-carotene. **(b)** An action spectrum of photosynthesis depicting the relative rate of photosynthesis in green plants at different wavelengths of light.

**Concept Check:** *What is the advantage of having different pigment molecules?*

different regions of the visible spectrum. The absorption spectra of chlorophylls *a* and *b* are slightly different, though both chlorophylls absorb light most strongly in the red and violet-blue parts of the visible spectrum and absorb green light poorly. Carotenoids absorb light in the blue and blue-green regions of the visible spectrum, reflecting yellow and red.

Why do plants have different pigments? Having different pigments allows plants to absorb light at many different wavelengths. In this way, plants are more efficient at capturing the energy in sunlight. This phenomenon is highlighted in an **action spectrum,** which plots the rate of photosynthesis as a function of wavelength (Figure 7.7b). The highest rates of photosynthesis in green plants correlate with the wavelengths that are strongly absorbed by the chlorophylls and carotenoids. Photosynthesis is poor in the green region of the spectrum, because these pigments do not readily absorb this wavelength of light.

## Photosystems II and I Work Together to Produce ATP and NADPH via Linear Electron Flow

As noted previously, photosynthetic organisms have the unique ability not only to absorb light energy but also to capture that energy in a stable way. Let's now consider how chloroplasts capture light energy. The thylakoid membranes of the chloroplast contain two distinct complexes of proteins and pigment molecules called **photosystem I (PSI)** and **photosystem II (PSII)** (Figure 7.8). Photosystem I was discovered before photosystem II, but photosystem II is the initial step in photosynthesis.

**Events Within Photosystem II** As described in steps 1a and 1b of Figure 7.8, light excites electrons in pigment molecules, such as chlorophylls, which are located in a region of PSII called the light-harvesting complex. Rather than releasing their energy in the form

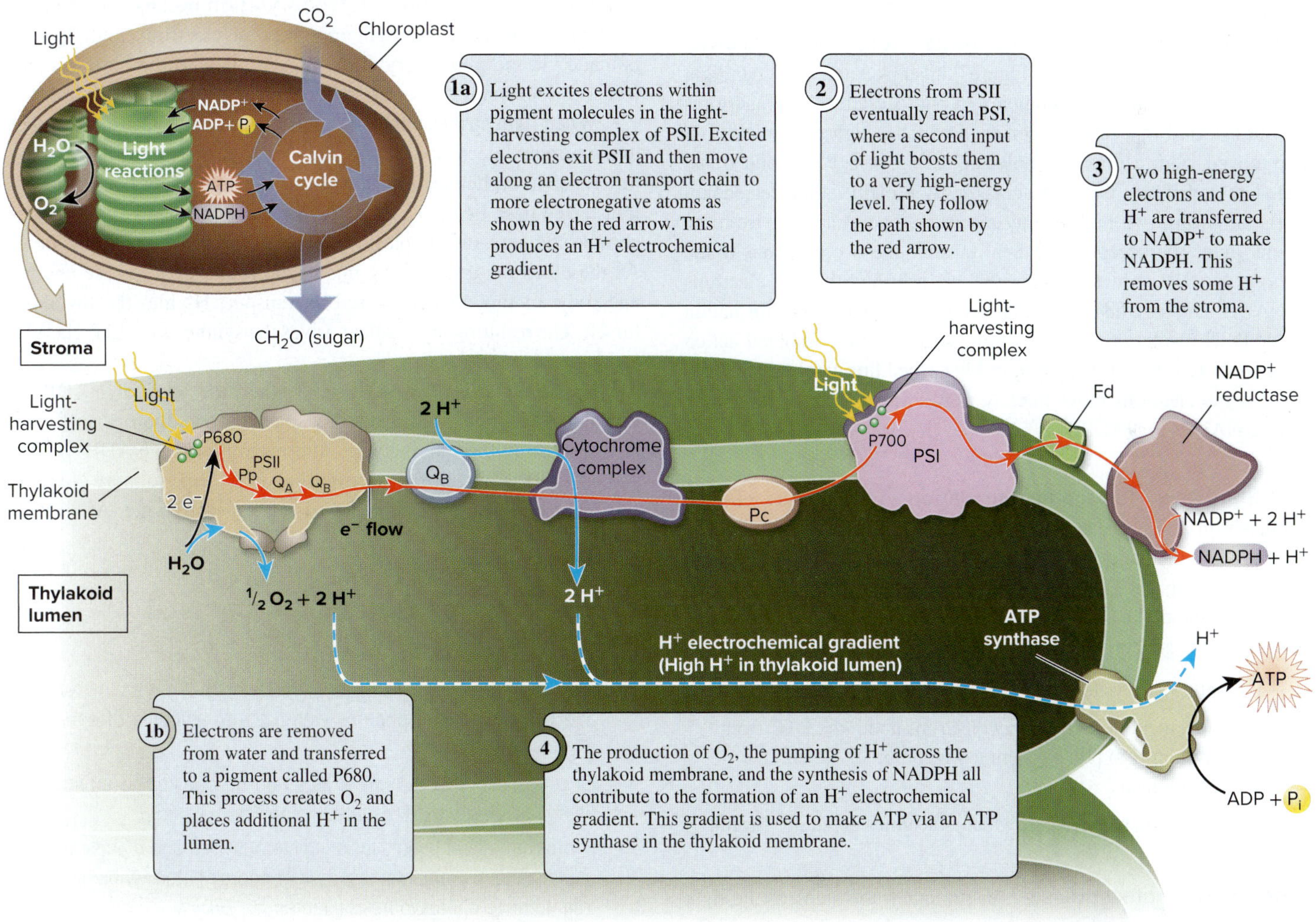

**1a** Light excites electrons within pigment molecules in the light-harvesting complex of PSII. Excited electrons exit PSII and then move along an electron transport chain to more electronegative atoms as shown by the red arrow. This produces an $H^+$ electrochemical gradient.

**2** Electrons from PSII eventually reach PSI, where a second input of light boosts them to a very high-energy level. They follow the path shown by the red arrow.

**3** Two high-energy electrons and one $H^+$ are transferred to $NADP^+$ to make NADPH. This removes some $H^+$ from the stroma.

**1b** Electrons are removed from water and transferred to a pigment called P680. This process creates $O_2$ and places additional $H^+$ in the lumen.

**4** The production of $O_2$, the pumping of $H^+$ across the thylakoid membrane, and the synthesis of NADPH all contribute to the formation of an $H^+$ electrochemical gradient. This gradient is used to make ATP via an ATP synthase in the thylakoid membrane.

**Figure 7.8 The synthesis of ATP, NADPH, and $O_2$ by the concerted actions of photosystems II and I.** The movement of electrons from photosystem II to photosystem I to NADPH is called linear electron flow.

*Are ATP, NADPH, and $O_2$ produced in the stroma or in the thylakoid lumen?*

of heat, the excited electrons begin to follow a path shown by the red arrow.

Initially, the excited electrons move from a pigment molecule called P680 in PSII to other electron carriers called pheophytin (Pp), $Q_A$, and $Q_B$. PSII also oxidizes water, which generates $O_2$ and adds $H^+$ into the thylakoid lumen. The electrons released from the oxidized water molecules are used to replenish the electrons that exit PSII via $Q_B$.

**Electron Transport Chain** After a pair of electrons reaches $Q_B$, each one enters an **electron transport chain (ETC)**—a series of electron carriers—located in the thylakoid membrane. From $Q_B$, an electron goes to a cytochrome complex; then to plastocyanin (Pc), a small protein; and then to photosystem I. Along its journey from photosystem II to photosystem I, the electron releases some of its energy at particular steps and is transferred to the next component that has a higher electronegativity. The energy released is harnessed to pump $H^+$ into the thylakoid lumen.

**Photosystem I and NADPH Synthesis** A key role of photosystem I is to make NADPH (see Figure 7.8, steps 2 and 3). When light strikes the light-harvesting complex of photosystem I, this energy is also transferred to a reaction center in which a high-energy electron is removed from a pigment molecule, designated P700, and transferred to a primary electron acceptor. A protein called ferredoxin (Fd) can accept two high-energy electrons, one at a time, from the primary electron acceptor. Fd then transfers the two electrons to the enzyme $NADP^+$ reductase. This enzyme transfers the two electrons to $NADP^+$ and together with an $H^+$ produces NADPH. The formation of NADPH results in fewer $H^+$ in the stroma. The combined action of photosystem II and photosystem I is termed **linear electron flow** because the electrons move linearly from PSII to PSI and ultimately reduce $NADP^+$ to NADPH.

**ATP Synthesis** **Photophosphorylation** is the process of ATP synthesis in chloroplasts. The term refers to the idea that light is used to drive the phosphorylation of ADP to make ATP. As shown in Figure 7.8 (step 4), ATP is synthesized as $H^+$ flow through ATP synthase from the thylakoid lumen into the stroma. An $H^+$ electrochemical gradient is generated in three ways:

- The splitting of water, which places $H^+$ in the thylakoid lumen
- The movement of high-energy electrons along the ETC from photosystem II to photosystem I, which pumps $H^+$ into the thylakoid lumen
- The formation of NADPH, which consumes $H^+$ in the stroma

**Products of Photosynthesis** In summary, the steps of the light reactions of photosynthesis produce three chemical products: $O_2$, NADPH, and ATP:

1. $O_2$ is produced in the thylakoid lumen by the oxidation of water by photosystem II. Two electrons are removed from water, which produces 2 $H^+$ and $^1/_2$ $O_2$. The two electrons are transferred to $P680^+$ molecules.
2. NADPH is produced in the stroma using high-energy electrons that originate in the light-harvesting complex of photosystem II and are boosted a second time in photosystem I. Two high-energy electrons and one $H^+$ are transferred to $NADP^+$ to produce NADPH.
3. ATP is produced in the stroma via ATP synthase that uses an $H^+$ electrochemical gradient.

## PSI Produces Only ATP via Cyclic Electron Flow

The mechanism of harvesting light energy described in Figure 7.8 is called linear electron flow because it is a linear process. This electron flow produces ATP and NADPH in roughly equal amounts. However, as we will see later, the Calvin cycle uses more ATP than NADPH. How can plant cells avoid making too much NADPH and not enough ATP? In 1959, Polish-born American plant physiologist Daniel Arnon discovered a pattern of electron flow that is cyclic and generates only ATP (**Figure 7.9**). This process is called **cyclic photophosphorylation** because (1) the path of electrons is cyclic, (2) light energizes the electrons, and (3) ATP is made via the phosphorylation of ADP. Due to the path of electrons, the mechanism is also called **cyclic electron flow.**

When light strikes photosystem I, high-energy electrons are sent to the primary electron acceptor and then to ferredoxin (Fd). The key difference in cyclic electron flow is that the high-energy electrons are transferred from Fd to $Q_B$. From $Q_B$, the electrons then go to the cytochrome complex, then to plastocyanin (Pc) and back to photosystem I. As the electrons travel along this cyclic route, they release energy, and some of this energy is used to transport $H^+$ into the thylakoid lumen. The resulting $H^+$ gradient drives the synthesis of ATP via ATP synthase.

Cyclic electron flow is favored when the level of $NADP^+$ is low and NADPH is high. Under these conditions, there is sufficient NADPH to run the Calvin cycle. Alternatively, when $NADP^+$ is high and NADPH is low, linear electron flow is favored, so more NADPH can be made. Cyclic electron flow is also favored when ATP levels are low.

## 7.2 Reviewing the Concepts

- Light is a form of electromagnetic radiation that travels in waves and is composed of photons with discrete amounts of energy (Figure 7.4).
- Electrons can absorb light energy and be boosted to a higher energy level—an excited state (Figure 7.5).
- Photosynthetic pigments include chlorophylls *a* and *b* and carotenoids. These pigments absorb light energy in the visible spectrum to drive photosynthesis (Figures 7.6, 7.7).
- During linear electron flow, electrons from photosystem II follow a pathway along an electron transport chain (ETC) in the thylakoid membrane. This pathway generates an $H^+$ gradient that is used to make ATP. In addition, light energy striking photosystem I (PSI) boosts electrons to a very high-energy level that allows the synthesis of NADPH (Figure 7.8).

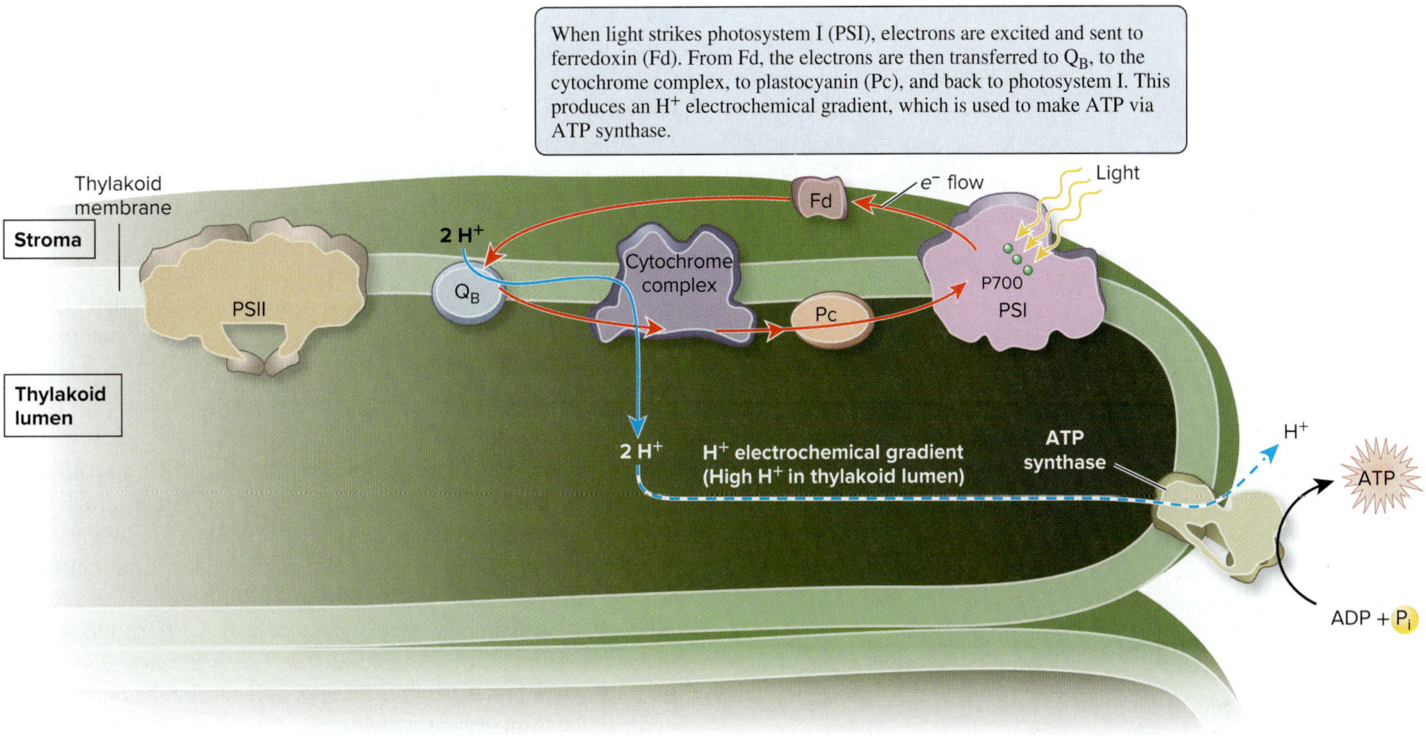

**Figure 7.9 The synthesis of only ATP by cyclic photophosphorylation.** In this process, electrons follow a cyclic path that is powered by photosystem I (PSI). This produces an $H^+$ electrochemical gradient, which is then used to make ATP by ATP synthase.

*Why is using cyclic photophosphorylation an advantage to a plant over using only linear electron flow?*

- During cyclic photophosphorylation, electrons are excited in PSI and flow from ferredoxin (Fd) through the ETC and back to PSI. This cyclic electron route produces an $H^+$ gradient that is used to make ATP (Figure 7.9).

## 7.2 Testing Your Knowledge

1. When a photoexcited electron returns to the ground state, which of the following will *not* occur?
   a. It could give off light.
   b. It could absorb a photon of light.
   c. It could give off heat.
   d. All of the above are possible.
2. During linear electron flow, the primary function of photosystem I (PSI) is to make
   a. ATP.
   b. NADPH.
   c. $O_2$.
   d. glucose.
   e. all of the above.
3. What is the purpose of cyclic photophosphorylation, which involves photosystem I but *not* photosystem II?
   a. production of ATP
   b. production of NADPH
   c. production of both ATP and NADPH
   d. production of $O_2$
   e. production of ATP and $O_2$

## 7.3 Molecular Features of Photosystems

### Learning Outcomes

1. Describe the molecular mechanism in which photosystem II absorbs light energy and produces $O_2$.
2. Diagram how the energy of an electron changes as it moves from PSII to $NADP^+$.

The previous section provided an overview of how chloroplasts absorb light energy and produce ATP, NADPH, and $O_2$. As you have learned, two photosystems—PSI and PSII—play critical roles in two aspects of photosynthesis. First, both PSI and PSII absorb light energy and capture that energy in the form of excited electrons. Second, PSII oxidizes water, thereby producing $O_2$. In this section, we will take a closer look at how these events occur at the molecular level.

### Photosystem II Captures Light Energy and Produces $O_2$

PSI and PSII have two main components: a light-harvesting complex and a reaction center. Figure 7.10 shows how these components function in PSII. In 1932, American biologist Robert Emerson and an undergraduate student, William Arnold, discovered the **light-harvesting complex** in the thylakoid membrane. It is composed

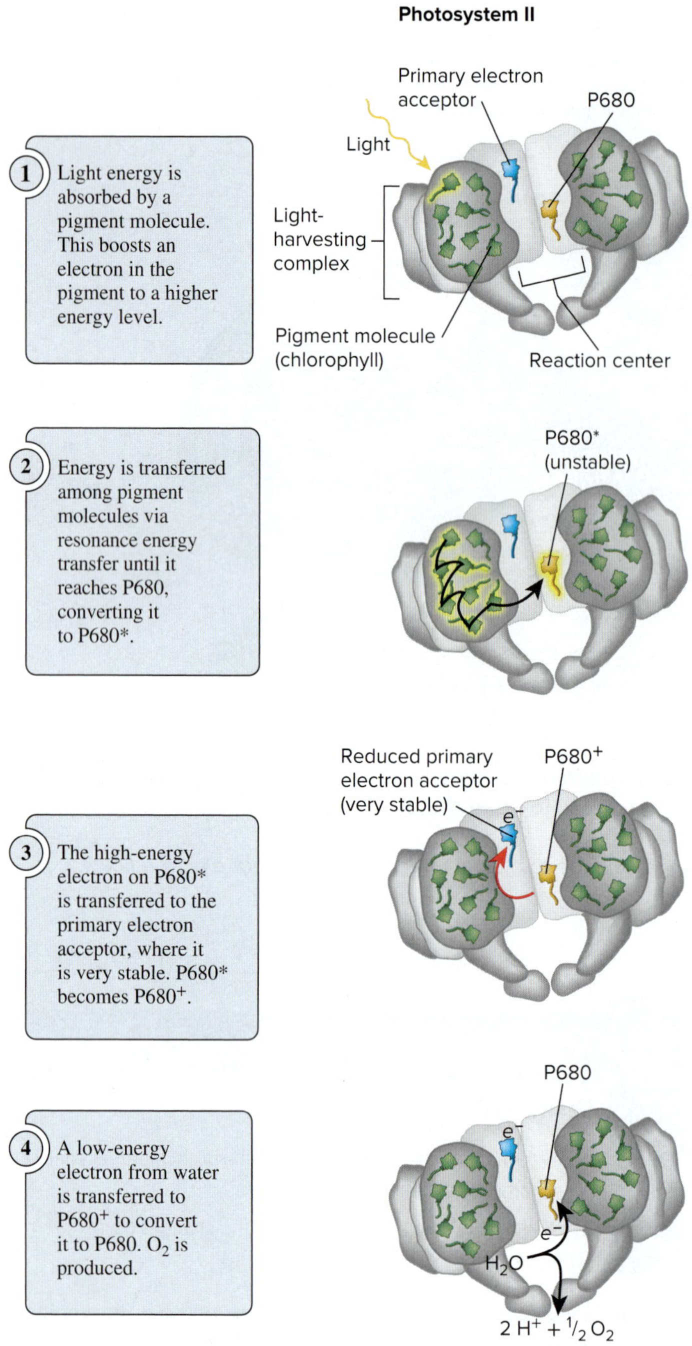

**Photosystem II**

(1) Light energy is absorbed by a pigment molecule. This boosts an electron in the pigment to a higher energy level.

(2) Energy is transferred among pigment molecules via resonance energy transfer until it reaches P680, converting it to P680*.

(3) The high-energy electron on P680* is transferred to the primary electron acceptor, where it is very stable. P680* becomes $P680^+$.

(4) A low-energy electron from water is transferred to $P680^+$ to convert it to P680. $O_2$ is produced.

**Figure 7.10 A closer look at how photosystem II harvests light energy and oxidizes water.** Note: Two electrons are released during the oxidation of water, but they are transferred one at a time to $P680^+$. The primary electron acceptor is a pigment molecule named pheophytin; see Figure 7.8.

of several dozen pigment molecules that are anchored to transmembrane proteins. The role of the complex is to directly absorb photons of light. When a pigment molecule absorbs a photon, an electron is boosted to a higher energy level. As shown in Figure 7.10, the energy (not the electron itself) is transferred to adjacent pigment molecules by a process called **resonance energy transfer.** The energy may be transferred among multiple pigment molecules until it is eventually transferred to a special pigment molecule designated P680, which is located within a site in PSII called the reaction center, where redox reactions take place. The P680 pigment is so named because it can directly absorb light at a wavelength of 680 nm. However, P680 is more commonly excited by resonance energy transfer from a chlorophyll pigment in the light-harvesting complex. In either case, when an electron in P680 is excited, it is designated P680*.

Unlike the pigments in the light-harvesting complex that undergo resonance energy transfer, P680* can actually release its high-energy electron and become $P680^+$.

$$P680^* \rightarrow P680^+ + e^-$$

One role of the reaction center is to carry out a redox reaction in which the high-energy electron from P680* is transferred to another molecule, where the electron is more stable. This molecule is called the **primary electron acceptor** (see Figure 7.10). The transfer of the electron from P680* to the primary electron acceptor is remarkably fast. It occurs in less than a few picoseconds! (One picosecond equals one-trillionth of a second, or $10^{-12}$ sec.) Because this occurs so quickly, the excited electron does not have much time to release its energy in the form of heat or light.

After the primary electron acceptor has received this high-energy electron, the light energy has been captured and can be used to perform cellular work. As discussed earlier, the work it performs is to synthesize the energy intermediates ATP and NADPH (see Figure 7.8).

Let's now consider what happens to $P680^+$, which has given up its high-energy electron. After $P680^+$ is formed, it is necessary to replace the electron so that P680 can function again. Therefore, another role of the reaction center is to replace the electron that is removed when P680* becomes $P680^+$. This missing electron of $P680^+$ is replaced with a low-energy electron from water (see Figure 7.10).

$$H_2O \rightarrow 2\,H^+ + \tfrac{1}{2}\,O_2 + 2\,e^-$$

$$2\,P680^+ + 2\,e^- \rightarrow 2\,P680$$

(from water)

The oxidation of water results in the formation of oxygen gas ($O_2$), which many organisms use for cellular respiration. Photosystem II is the only known protein complex that can oxidize water, resulting in the release of $O_2$ into the atmosphere.

## Electrons Vary in Their Energy Level as They Travel from Photosystem II to Photosystem I to $NADP^+$

In 1960, Robin Hill and Fay Bendall proposed that photosynthesis involves two events in which light is absorbed. According to their model, known as the **Z scheme,** an electron proceeds through a series of energy changes during photosynthesis (**Figure 7.11**). The Z refers to the zigzag shape of this energy curve. Based on our modern understanding of photosynthesis, we now know these events involve increases and decreases in the energy of an electron as it moves from photosystem II through photosystem I to $NADP^+$ during linear electron flow.

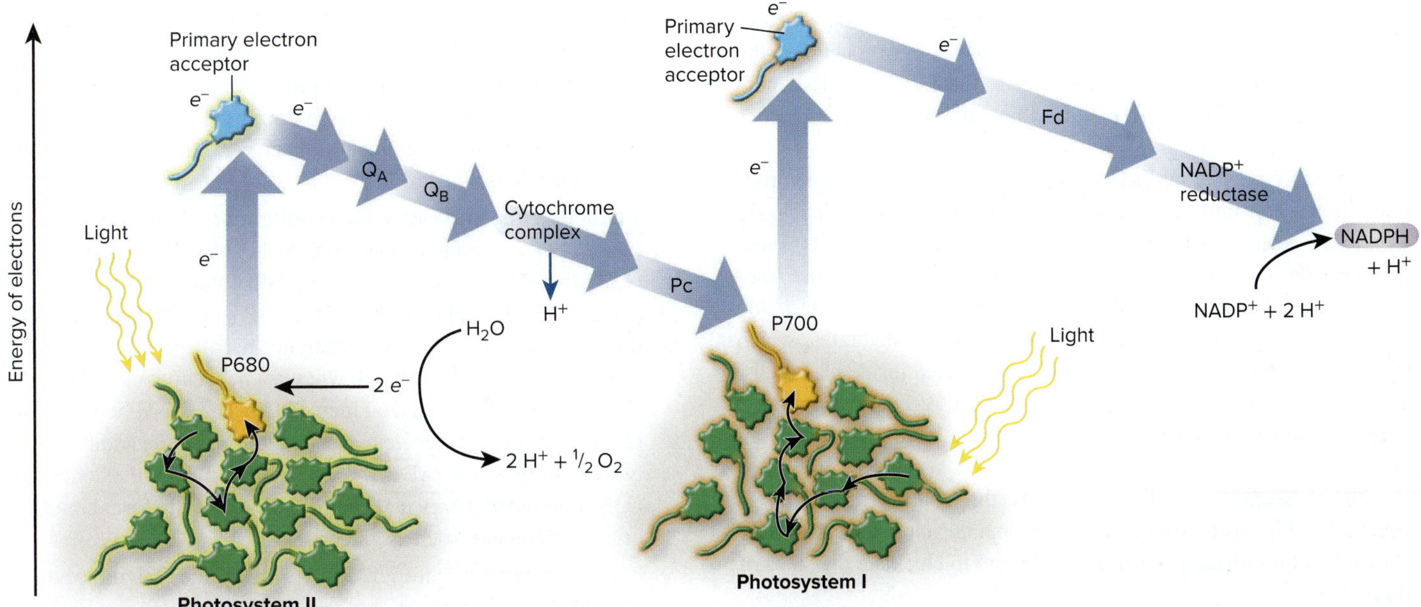


**Figure 7.11** **The Z scheme, showing the energy of an electron moving from photosystem II to NADP⁺.** The oxidation of water releases two electrons that travel one at a time from photosystem II to NADP⁺. As seen here, the input of light boosts the energy of the electron twice. At the end of the pathway, two electrons are used to make NADPH.

**Concept Check:** *During its journey from photosystem II to NADP⁺, at what point does an electron have the highest amount of energy?*

- An electron on a nonexcited pigment molecule in the light-harvesting complex of photosystem II has the lowest energy.
- In photosystem II, light boosts such an electron to a much higher energy level.
- As the electron travels from photosystem II to photosystem I, some of the energy is released.
- The input of light in photosystem I boosts the electron to an even higher energy than it attained in photosystem II.
- The electron releases a little energy before it is eventually transferred to NADP⁺.

## 7.3 Reviewing the Concepts

- In the light-harvesting complex of photosystem II (PSII), pigment molecules absorb light energy that is transferred to P680 in the reaction center via resonance energy transfer. A high-energy electron from P680* is transferred to a primary electron acceptor. An electron from water is then used to replenish the electron lost from P680*. The oxidation of water produces $O_2$ (Figure 7.10).
- The Z scheme proposes that an electron absorbs light energy twice, at both PSII and PSI, losing some of that energy as it flows along the ETC before it is transferred to NADP⁺ (Figure 7.11).

## 7.3 Testing Your Knowledge

1. After a chlorophyll pigment in the light-harvesting complex of PSII absorbs a photon of light, the
   a. excited electron is transferred to the reaction center.
   b. excited electron is transferred to the electron transport chain.
   c. energy is transferred to P680 via resonance energy transfer.
   d. energy is transferred to the primary electron acceptor via resonance energy transfer.
   e. energy is released as heat, which is absorbed by the primary electron acceptor.
2. After P680* becomes P680⁺, it is converted back to P680
   a. via resonance energy transfer.
   b. by receiving an electron from the light-harvesting complex.
   c. by receiving an electron from the primary electron acceptor.
   d. by receiving an electron from water.
   e. by receiving an electron from $O_2$.

## 7.4 Synthesizing Carbohydrates via the Calvin Cycle

### Learning Outcomes

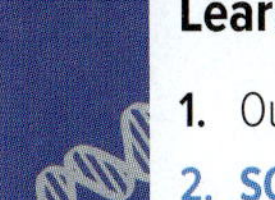

1. Outline the three phases of the Calvin cycle.
2. **SCISKILLS ▸** Analyze the results of Calvin and Benson, and explain how they identified the components of the Calvin cycle.

In the previous sections, we learned how the light reactions of photosynthesis produce $O_2$ and two energy intermediates, ATP and NADPH. We will now turn our attention to the second phase of photosynthesis—the Calvin cycle—in which ATP and NADPH are used to make carbohydrates. The Calvin cycle consists of a series of steps that occur in a metabolic cycle. In plants and algae, it occurs in the stroma of chloroplasts. In cyanobacteria, the Calvin cycle occurs in the cytoplasm where the enzymes that catalyze each of the steps are located.

The Calvin cycle takes $CO_2$ from the atmosphere and incorporates the carbon into organic molecules, namely carbohydrates. As mentioned earlier, carbohydrates are critical for two reasons. First, they provide the precursors to make the organic molecules and macromolecules of nearly all living cells. The second key reason is the storage of energy. The Calvin cycle produces carbohydrates, which store energy. These carbohydrates are accumulated inside plant cells. The stored carbohydrates are used as a source of energy. Similarly, when an animal consumes a plant, it uses the carbohydrates as an energy source.

In this section, we will examine the three phases of the Calvin cycle. We will also explore the experimental approach of Melvin Calvin and his colleagues that enabled them to elucidate the steps of this cycle.

## The Calvin Cycle Incorporates $CO_2$ into Carbohydrate

The Calvin cycle, also called the Calvin-Benson cycle, was determined by chemists Melvin Calvin and Andrew Adam Benson and their colleagues in the 1940s and 1950s. This cycle requires a massive input of energy. For every 6 $CO_2$ molecules that are incorporated into a carbohydrate such as glucose ($C_6H_{12}O_6$), 18 ATP molecules are hydrolyzed and 12 NADPH molecules are oxidized:

$$6\,CO_2 + 12\,H_2O \rightarrow C_6H_{12}O_6 + 6\,O_2 + 6\,H_2O$$

$$18\text{ ATP} + 18\,H_2O \rightarrow 18\text{ ADP} + 18\,P_i$$

$$12\text{ NADPH} \rightarrow 12\text{ NADP}^+ + 12\,H^+ + 24\,e^-$$

Although biologists commonly describe glucose as a product of photosynthesis, glucose is not directly made by the Calvin cycle. Instead, molecules of glyceraldehyde-3-phosphate (G3P), which are products of the Calvin cycle, are used as starting materials for the synthesis of glucose and other molecules, including sucrose. After glucose molecules are made, they may be linked together to form a polymer of glucose called starch, which is stored in the chloroplast for later use. Alternatively, the disaccharide sucrose may be made and transported out of the leaf to other parts of the plant.

The Calvin cycle is divided into three phases: carbon fixation, reduction and carbohydrate production, and regeneration of ribulose bisphosphate (RuBP) (Figure 7.12).

1. **Carbon fixation:** During **carbon fixation,** $CO_2$ is incorporated into RuBP, a five-carbon sugar. The term fixation means that the carbon has been removed from the atmosphere and fixed into an organic molecule that is not a gas. More specifically, the product of the reaction is a six-carbon intermediate that immediately splits in half to form two molecules of 3-phosphoglycerate (3PG). The enzyme that catalyzes this step is named RuBP carboxylase/oxygenase, or **rubisco.** It is the most abundant protein in chloroplasts and perhaps the most abundant protein on Earth! This observation underscores the massive amount of carbon fixation that happens in the biosphere.
2. **Reduction and carbohydrate production:** In the second phase, ATP is used to convert 3PG to 1,3-bisphosphoglycerate (1,3-BPG). Next, electrons from NADPH reduce 1,3-BPG to glyceraldehyde-3-phosphate (G3P). G3P is a carbohydrate with three carbon atoms. The key difference between 3PG and G3P is that 3PG forms a C—O bond, whereas the analogous carbon in G3P has a C—H bond (see Figure 7.12). The C—H bond occurs because the G3P molecule has been reduced by the addition of two electrons from NADPH. Compared with 3PG, the bonds in G3P store more energy and enable G3P to readily form larger organic molecules such as glucose.

   As shown in Figure 7.12, only some of the G3P molecules are used to make glucose or other carbohydrates. Phase 1 begins with 6 RuBP molecules and 6 $CO_2$ molecules. By the end of the second phase, 12 G3P molecules have been made but only 2 of these G3P molecules are used in carbohydrate production. As described next, the other 10 G3P molecules are needed to keep the Calvin cycle turning by regenerating RuBP.
3. **Regeneration of RuBP:** In the third phase of the Calvin cycle, a series of enzymatic steps converts the 10 G3P molecules into 6 RuBP molecules, using 6 molecules of ATP. After the RuBP molecules are regenerated, they serve as acceptors for $CO_2$, thereby allowing the cycle to continue.

As we have just seen, the Calvin cycle begins by using carbon from an inorganic source (that is, $CO_2$) and ends with organic molecules that will be used by the plant to make other molecules. You may be wondering why $CO_2$ molecules cannot be directly linked to form these larger molecules. The answer lies in the number of electrons that are located around carbon atoms. In $CO_2$, the carbon atom is considered electron poor. Oxygen is a very electronegative atom that monopolizes the electrons it shares with other atoms. In a covalent bond between carbon and oxygen, the shared electrons are closer to the oxygen atom.

By comparison, in an organic molecule, the carbon atom is electron rich. During the Calvin cycle, ATP provides energy and NADPH donates high-energy electrons, so the carbon originally in $CO_2$ has been reduced. The Calvin cycle combines less electronegative atoms with carbon atoms so that C—H and C—C bonds are formed. This allows the eventual synthesis of larger organic molecules, including glucose, amino acids, and so on. In addition, the covalent bonds within these molecules are capable of storing large amounts of energy.

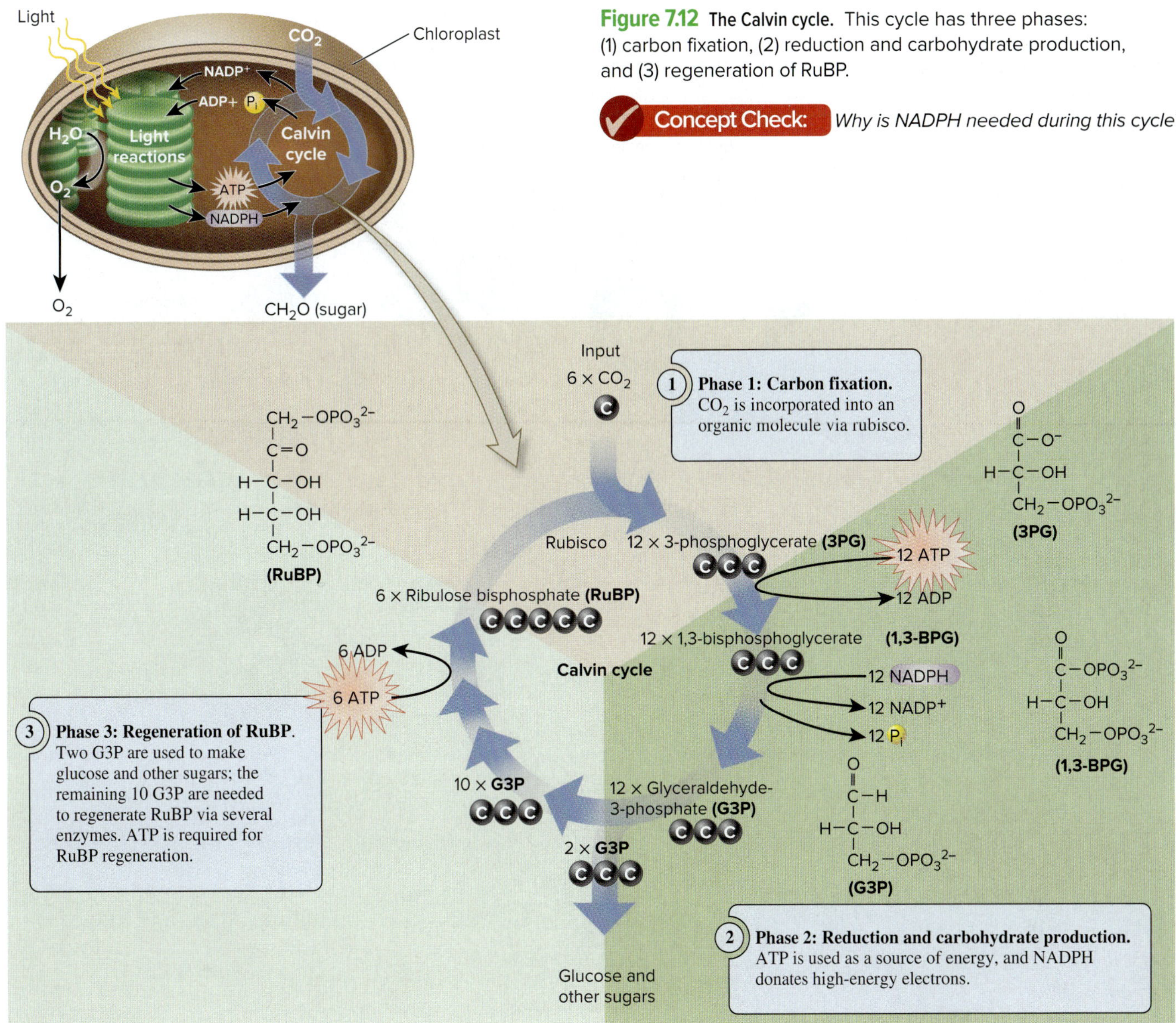


**Figure 7.12** **The Calvin cycle.** This cycle has three phases: (1) carbon fixation, (2) reduction and carbohydrate production, and (3) regeneration of RuBP.

**Concept Check:** *Why is NADPH needed during this cycle?*

# FEATURE INVESTIGATION

## The Calvin Cycle Was Determined by Isotope-Labeling Methods

The steps in the Calvin cycle involve the conversion of one type of molecule to another, eventually regenerating the $CO_2$ acceptor, RuBP. In the 1940s and 1950s, Calvin and his colleagues used $^{14}C$, a radioisotope of carbon, to label and trace molecules produced during the cycle (**Figure 7.13**). They injected $^{14}C$-labeled $CO_2$ into cultures of the green alga *Chlorella pyrenoidosa* grown in an apparatus called a "lollipop" (because of its shape). The *Chlorella* cells were given different lengths of time to incorporate the $^{14}C$-labeled carbon, ranging from fractions of a second to many minutes. After a specific incubation period, the cells were abruptly placed into a solution of alcohol to inhibit enzymatic reactions, thereby stopping the cycle.

**Figure 7.13** The determination of the Calvin cycle using $^{14}C$-labeled $CO_2$ and paper chromatography.

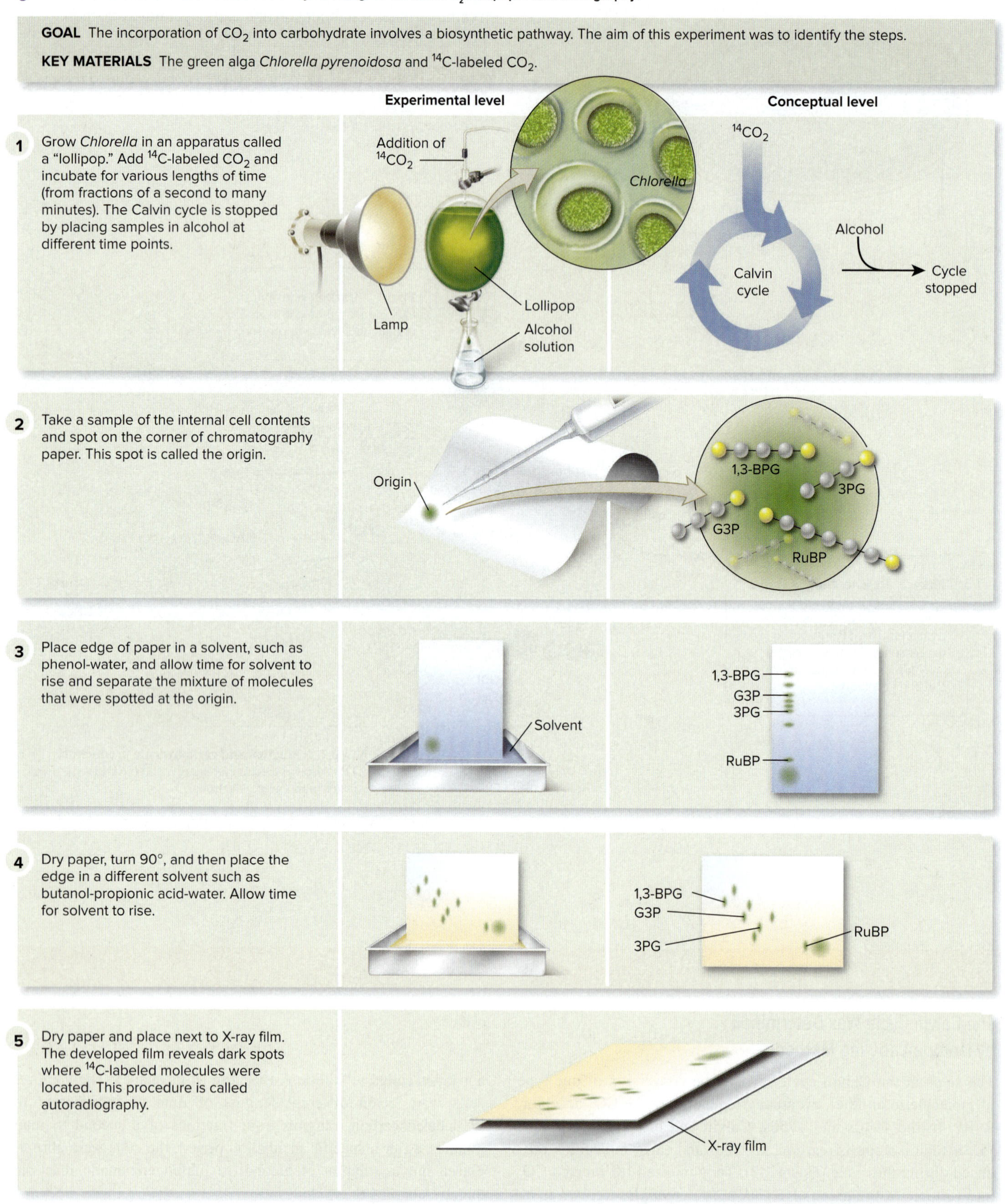

GOAL The incorporation of $CO_2$ into carbohydrate involves a biosynthetic pathway. The aim of this experiment was to identify the steps.
KEY MATERIALS The green alga *Chlorella pyrenoidosa* and $^{14}C$-labeled $CO_2$.
Experimental level
Conceptual level
1 Grow *Chlorella* in an apparatus called a "lollipop." Add $^{14}C$-labeled $CO_2$ and incubate for various lengths of time (from fractions of a second to many minutes). The Calvin cycle is stopped by placing samples in alcohol at different time points.
Addition of $^{14}CO_2$
Chlorella
Lamp
Lollipop
Alcohol solution
$^{14}CO_2$
Calvin cycle
Alcohol
Cycle stopped
2 Take a sample of the internal cell contents and spot on the corner of chromatography paper. This spot is called the origin.
Origin
1,3-BPG
3PG
G3P
RuBP
3 Place edge of paper in a solvent, such as phenol-water, and allow time for solvent to rise and separate the mixture of molecules that were spotted at the origin.
Solvent
1,3-BPG
G3P
3PG
RuBP
4 Dry paper, turn 90°, and then place the edge in a different solvent such as butanol-propionic acid-water. Allow time for solvent to rise.
1,3-BPG
G3P
3PG
RuBP
5 Dry paper and place next to X-ray film. The developed film reveals dark spots where $^{14}C$-labeled molecules were located. This procedure is called autoradiography.
X-ray film

**6 THE DATA***

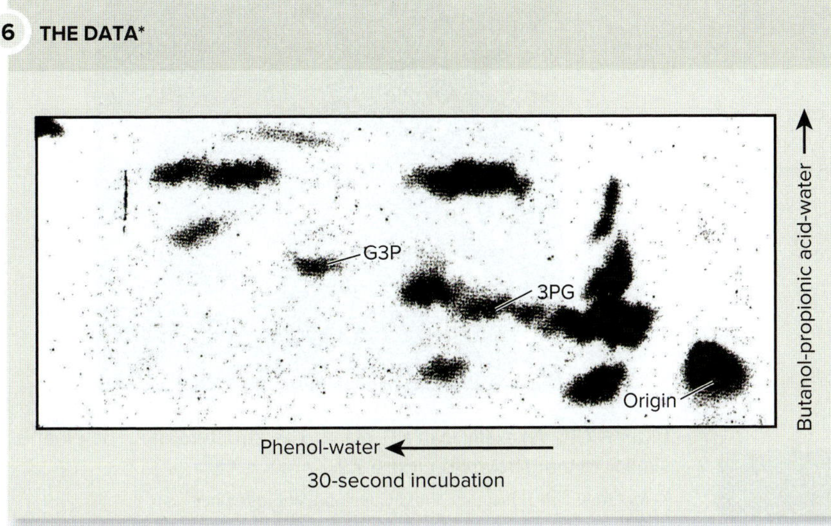


*An autoradiograph from one of Calvin's experiments.

*(image)* M. Calvin, "The path of carbon in photosynthesis," *Nobel Lecture, December 11, 1961,* 618–644, Fig. 4. © The Nobel Foundation

**7 CONCLUSION** The identification of the molecules in each spot elucidated the steps of the Calvin cycle.

**8 SOURCE** Calvin, M. December 11, 1961. "The path of carbon in photosynthesis." Nobel Lecture.

The researchers separated the newly made radiolabeled molecules by a variety of methods. The most commonly used method was two-dimensional paper chromatography. In this approach, a sample containing radiolabeled molecules was spotted onto a corner of the paper at a location called the origin. The edge of the paper was placed in a solvent, such as phenol-water. As the solvent rose through the paper, so did the radiolabeled molecules. The rate at which they rose depended on their structures, which determined how strongly they interacted with the paper. This step separated the mixture of molecules spotted onto the paper at the origin.

The paper was then dried, it was turned 90°, and then the edge was placed in a different solvent, such as butanol-propionic acid-water. Again, the solvent rose through the paper (in a second dimension), thereby separating molecules that may not have been adequately separated during the first separation step. After this second separation step, the paper was dried and exposed to X-ray film, a procedure called autoradiography. Radioactive emission from the $^{14}C$-labeled molecules caused dark spots to appear on the film.

The pattern of spots changed depending on the length of time the cells were incubated with $^{14}C$-labeled $CO_2$. When the incubation period was short, only molecules that were made in the first steps of the Calvin cycle were seen—3-phosphoglycerate (3PG) and 1,3-bisphosphoglycerate (1,3-BPG). Longer incubations revealed molecules synthesized in later steps—glyceraldehyde-3-phosphate (G3P) and ribulose bisphosphate (RuBP).

A challenge for Calvin and his colleagues was to identify the chemical nature of each spot. They achieved this by a variety of chemical methods. For example, a spot could be cut out of the paper, the molecule within the paper was washed out or eluted, and then the eluted molecule was subjected to the same procedure that included a radiolabeled molecule whose structure was already known. If the unknown molecule and known molecule migrated to the same spot in the paper, this indicated they were likely to be the same molecule. During the late 1940s and 1950s, Calvin and his coworkers identified all of the $^{14}C$-labeled spots and the order in which they appeared. In this way, they determined the series of reactions of what we now know as the Calvin cycle. For this work, Calvin was awarded the Nobel Prize in Chemistry in 1961.

### *Experimental Questions*

1. What was the purpose of the study conducted by Calvin and his colleagues?
2. **SCISKILLS ▸** In Calvin's experiment shown in Figure 7.13, why did the researchers use $^{14}C$? Why did they examine samples at several different time periods? How were the different molecules in the samples identified?
3. **SCISKILLS ▸** Explain the results of Calvin's study.

## 7.4 Reviewing the Concepts

- The Calvin cycle is composed of three phases: carbon fixation, reduction and carbohydrate production, and regeneration of ribulose bisphosphate (RuBP). In the cycle, ATP supplies the energy, and NADPH supplies the high-energy electrons that drive the incorporation of $CO_2$ into carbohydrate (Figure 7.12).
- Calvin and Benson determined the steps in the Calvin cycle by isotope-labeling methods in which the products of the Calvin cycle were separated by chromatography (Figure 7.13).

## 7.4 Testing Your Knowledge

1. During the Calvin cycle, what is used to reduce carbon?
   a. NADH
   b. NADPH
   c. ATP
   d. GTP
   e. $O_2$
2. What happens during the phase of the Calvin cycle called carbon fixation?
   a. $CO_2$ is incorporated into an organic molecule.
   b. $CO_2$ is released into the atmosphere.
   c. $CO_2$ is taken into the leaves via the stomata.
   d. Carbon is reduced.
   e. RuBP is regenerated.

## 7.5 Variations in Photosynthesis

### Learning Outcomes

1. Explain the process of photorespiration.
2. Describe how $C_4$ and CAM plants avoid photorespiration and conserve water.

Certain environmental conditions such as temperature, water availability, and light intensity alter the way in which the Calvin cycle operates. In this section, we will begin by examining how hot and dry conditions may reduce the output of photosynthesis. We will then explore two adaptations that allow certain plant species to conserve water and help maximize photosynthetic efficiency in such environments.

### Photorespiration Decreases the Efficiency of Photosynthesis

In the previous section, we learned that rubisco adds a $CO_2$ molecule to an organic molecule, RuBP, to produce two molecules of 3-phosphoglycerate (3PG):

$$\text{RuBP} + CO_2 \rightarrow 2\ \text{3PG}$$

For most species of plants, the incorporation of $CO_2$ into RuBP is the only way for carbon fixation to occur. Because 3PG is a three-carbon molecule, these plants are called **$C_3$ plants.** An example of a $C_3$ plant is an oak tree (**Figure 7.14**). About 90% of the plant species on Earth are $C_3$ plants.

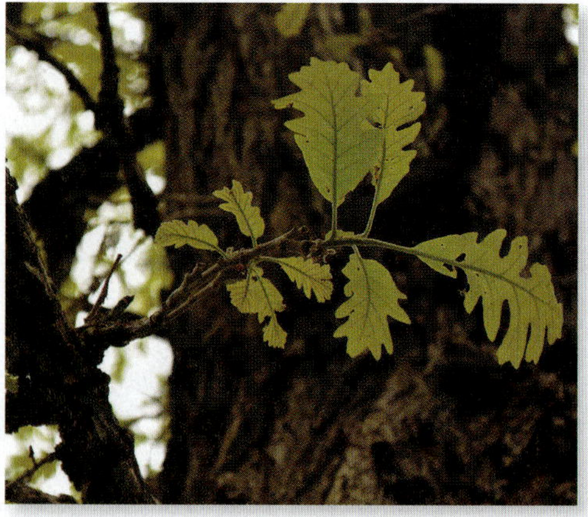

**Figure 7.14 An example of a $C_3$ plant.** The structure of an oak leaf is similar to that shown in Figure 7.2.



The active site of rubisco can also add $O_2$ to RuBP, although its affinity for $CO_2$ is over 10-fold better than that for $O_2$. Even so, when $CO_2$ levels are low and $O_2$ levels are high, rubisco adds an $O_2$ molecule instead of $CO_2$ to RuBP. This produces only one molecule of 3PG and a two-carbon molecule called phosphoglycolate:

$$\text{RuBP} + O_2 \rightarrow \text{3PG} + \text{Phosphoglycolate}$$

The phosphoglycolate is then dephosphorylated to glycolate, which is released from the chloroplast. In a series of several steps, the two-carbon glycolate is eventually oxidized in peroxisomes and mitochondria to produce an organic molecule plus a molecule of $CO_2$:

$$\text{Phosphoglycolate} \rightarrow \text{Glycolate} \rightarrow \rightarrow \text{Organic molecule} + CO_2$$

This process, called **photorespiration,** uses $O_2$ and liberates $CO_2$. Photorespiration is considered wasteful because it releases $CO_2$, thereby limiting plant growth.

Photorespiration is more likely to occur when plants are exposed to a hot and dry environment. To conserve water, the stomata of the leaves close, inhibiting the uptake of $CO_2$ from the air and trapping the $O_2$ that is produced by photosynthesis. When the level of $CO_2$ is low and $O_2$ is high, photorespiration is favored. If $C_3$ plants are subjected to hot and dry environmental conditions, as much as 25–50% of their photosynthetic work is reversed by the process of photorespiration.

Why do plants carry out photorespiration? The answer is not entirely clear. One common view is that photorespiration may have a protective advantage. On hot and dry days when the stomata are closed, $CO_2$ levels within the leaves fall and $O_2$ levels rise. Under these conditions, highly toxic oxygen-containing molecules such as free radicals may be produced that could damage the plant. Therefore, plant biologists have hypothesized that the role of photorespiration may be to protect the plant against the harmful effects of such toxic molecules by consuming $O_2$ and releasing $CO_2$.

## EVOLUTIONARY CONNECTIONS

### $C_4$ and CAM Plants Have Evolved a Mechanism to Minimize Photorespiration

Certain species of plants have evolved adaptations to minimize photorespiration. In the early 1960s, Hugo Kortschak discovered that the first product of carbon fixation in sugarcane is not 3GP but instead is a molecule with four carbon atoms. Species such as sugarcane are called **$C_4$ plants** because of this four-carbon molecule.

**$C_4$ Plants with Mesophyll and Bundle-Sheath Cells** Some $C_4$ plants have a unique leaf anatomy that allows them to minimize photorespiration (**Figure 7.15**). An interior layer in the leaves of many $C_4$ plants has a two-cell organization composed of mesophyll cells and bundle-sheath cells. $CO_2$ and $O_2$ from the atmosphere enter the mesophyll cells via stomata. Once inside, the enzyme PEP

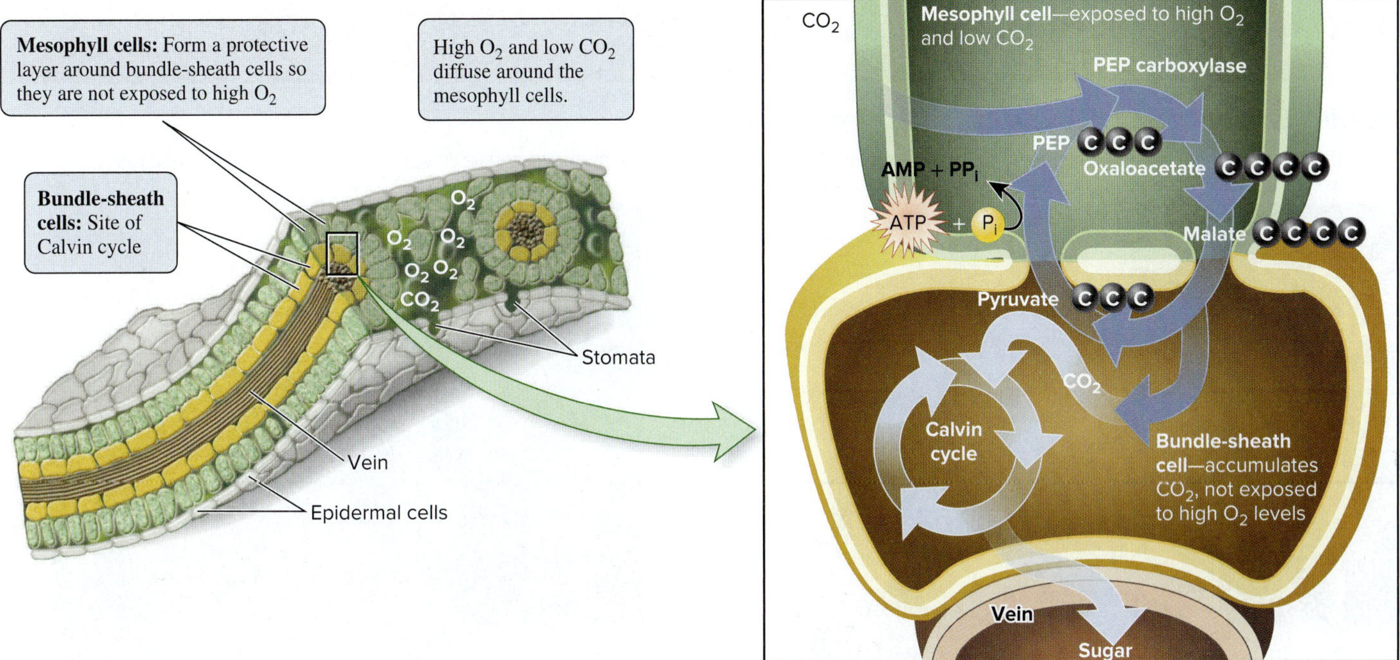


**Figure 7.15** **Leaf structure and its relationship to the $C_4$ cycle.** $C_4$ plants have mesophyll cells, which initially take up $CO_2$, and bundle-sheath cells, where much of the carbohydrate synthesis occurs. Compare this leaf structure with the structure of $C_3$ leaves shown in Figure 7.2.

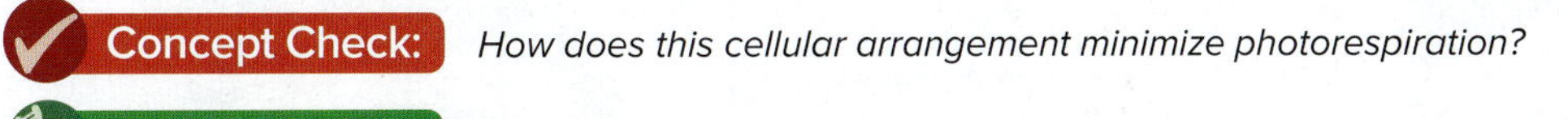


*How does this cellular arrangement minimize photorespiration?*

BioConnections: *Look ahead to Figure 30.20. How do plants transport the water needed for photosynthesis into their leaves?*

carboxylase adds $CO_2$ to phosphoenolpyruvate (PEP), a three-carbon molecule, to produce oxaloacetate, a four-carbon molecule. PEP carboxylase does not recognize $O_2$. Therefore, unlike rubisco, PEP carboxylase does not promote photorespiration when $CO_2$ is low and $O_2$ is high. Instead, PEP carboxylase fixes only $CO_2$.

As shown in Figure 7.15, oxaloacetate is converted to the four-carbon molecule malate, which is transported into the bundle-sheath cell. Malate is then broken down into the three-carbon compound pyruvate and $CO_2$. The pyruvate returns to the mesophyll cell, where ATP is used to convert it to PEP that binds $CO_2$, and the cycle in the mesophyll cell can begin again. The $CO_2$ enters the Calvin cycle in the chloroplasts of the bundle-sheath cells. Because the mesophyll cell supplies the bundle-sheath cell with a steady supply of $CO_2$, the concentration of $CO_2$ remains high in the bundle-sheath cell. Also, the mesophyll cells shield the bundle-sheath cells from high levels of $O_2$. This strategy minimizes photorespiration, which requires low $CO_2$ and high $O_2$ levels to happen.

Which is better—being a $C_3$ or a $C_4$ plant? The answer is that it depends on the environment. In warm and dry climates, $C_4$ plants have an advantage. During the day, they can keep their stomata partially closed to conserve water. Furthermore, they minimize photorespiration. Examples of $C_4$ plants are sugarcane, crabgrass, and corn. In cooler climates, $C_3$ plants have the edge because they use less energy to fix $CO_2$. Recall that $C_4$ plants use ATP to regenerate PEP from pyruvate (see Figure 7.15), which $C_3$ plants do not have to expend.

**CAM Plants That Separate Carbon Fixation and the Calvin Cycle in Time** We have just learned that certain $C_4$ plants prevent photorespiration by providing $CO_2$ to the bundle-sheath cells, where the Calvin cycle occurs. This mechanism spatially separates the processes of carbon fixation and the Calvin cycle. Another strategy followed by other $C_4$ plants, called **CAM plants,** separates these processes in time. CAM stands for crassulacean acid metabolism, because the process was first studied in members of the plant family Crassulaceae. Most CAM plants are water-storing succulents such as cacti, bromeliads (including pineapple), and sedums. To avoid water loss, CAM plants keep their stomata closed during the day and open them at night, when it is cooler and the relative humidity is higher.

**Figure 7.16** compares CAM plants with the other type of $C_4$ plants we considered in Figure 7.15. Photosynthesis in CAM plants occurs entirely within mesophyll cells, but the synthesis of a $C_4$ molecule and the Calvin cycle occur at different times. During the night, the stomata of CAM plants open, thereby allowing the entry of $CO_2$ into mesophyll cells. $CO_2$ is joined with PEP to form the four-carbon molecule oxaloacetate. This is then converted to malate, which accumulates during the night in the central vacuoles of the cells. In the morning, the stomata close to conserve moisture. The accumulated malate in the mesophyll cells leaves the vacuole and is broken down to release $CO_2$, which then drives the Calvin cycle during the daytime.

# Biology Principle

### Species Evolve from One Generation to the Next

The adaptations shown here have evolved to help plants living in hot and dry environments to conserve water and minimize photorespiration.

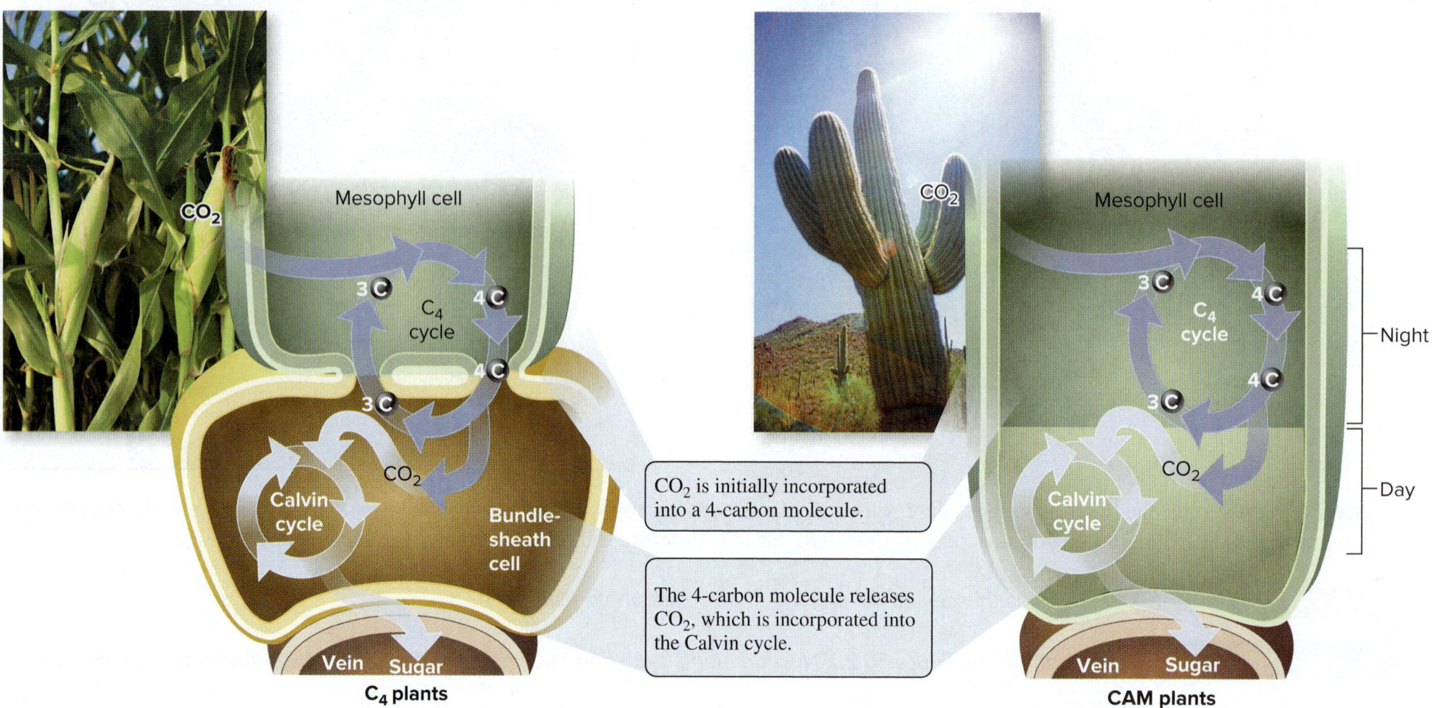


**Figure 7.16** **A comparison of $C_4$ and CAM plants.** $C_4$ plants are so named because the first organic product of carbon fixation is a four-carbon molecule. Using this definition, CAM plants are a type of $C_4$ plant. CAM plants, however, do not separate the functions of making a four-carbon molecule and the Calvin cycle into different types of cells. Instead, they make a four-carbon molecule at night that is broken down to release $CO_2$ during the day that drives the Calvin cycle.

*(left)* © Wesley Hitt/Getty Images; *(right)* © John Foxx/Getty Images RF

*What are the advantages and disadvantages for $C_3$, $C_4$, and CAM plants?*

## 7.5 Reviewing the Concepts

- For $C_3$ plants, the incorporation of $CO_2$ into RuBP to make 3PG, a three-carbon molecule, is the only way for carbon fixation to occur. Photorespiration occurs when the level of $O_2$ is high and $CO_2$ is low, which happens under hot and dry conditions. During this process, some $O_2$ is used and $CO_2$ is liberated. Photorespiration is inefficient because it reverses the incorporation of $CO_2$ into an organic molecule (Figure 7.14).
- Some $C_4$ plants avoid photorespiration because the $CO_2$ is first incorporated, via PEP carboxylase, into a four-carbon molecule, which is pumped from mesophyll cells into bundle-sheath cells. This maintains a high concentration of $CO_2$ in the bundle-sheath cells, where the Calvin cycle occurs. The high $CO_2$ concentration minimizes photorespiration (Figure 7.15).
- CAM plants, a type of $C_4$ plant, minimize photorespiration by fixing $CO_2$ into a four-carbon molecule at night and then running the Calvin cycle during the day with their stomata closed to reduce water loss (Figure 7.16).

## 7.5 Testing Your Knowledge

1. Under hot and dry conditions, why does the level of $O_2$ within leaves become higher but the level of $CO_2$ becomes lower?
   - **a.** because the stomata are closed
   - **b.** because the stomata are open
   - **c.** because there is insufficient water to drive photosynthesis
   - **d.** because the light reactions stop working
   - **e.** because the Calvin cycle stops working
2. An evolutionary adaptation to minimize photorespiration is found in
   - **a.** $C_3$ plants.
   - **b.** $C_4$ plants with mesophyll and bundle-sheath cells.
   - **c.** CAM plants that separate carbon fixation and the Calvin cycle in time.
   - **d.** all of the above.
   - **e.** b and c only.

# Assess and Discuss

## Test Yourself

1. The water necessary for photosynthesis
   a. is split into $H_2$ and $O_2$.
   b. is directly involved in the synthesis of carbohydrates.
   c. provides the electrons to replace lost electrons in photosystem II.
   d. provides $H^+$ needed to synthesize G3P.
   e. does none of the above.
2. In PSII, P680 differs from the pigment molecules of the light-harvesting complex in that it
   a. is a carotenoid.
   b. absorbs light energy and transfers that energy to other molecules without the transfer of electrons.
   c. transfers an excited electron to the primary electron acceptor.
   d. transfers an electron to $O_2$.
   e. acts as an ATP synthase to produce ATP.
3. The cyclic electron flow that occurs via photosystem I produces
   a. NADPH.
   b. oxygen.
   c. ATP.
   d. all of the above.
   e. a and c only.
4. During linear electron flow, the high-energy electron from P680
   a. eventually moves to $NADP^+$.
   b. becomes incorporated into water molecules.
   c. is pumped into the thylakoid space to drive ATP production.
   d. provides the energy necessary to split water molecules.
   e. falls back to the low-energy state in photosystem II.
5. During the first phase of the Calvin cycle, carbon dioxide is incorporated into ribulose bisphosphate (RuBP) by
   a. oxaloacetate.
   b. rubisco.
   c. RuBP.
   d. malate.
   e. G3P.
6. The NADPH produced during the light reactions is used during
   a. the carbon fixation phase, which incorporates carbon dioxide into an organic molecule of the Calvin cycle.
   b. the reduction phase, which produces carbohydrates in the Calvin cycle.
   c. the regeneration of RuBP of the Calvin cycle.
   d. all of the above.
   e. a and b only.
7. The majority of the G3P produced during the reduction and carbohydrate production phase is used to produce
   a. glucose.
   b. ATP.
   c. RuBP to continue the cycle.
   d. rubisco.
   e. all of the above.
8. Photorespiration
   a. is the process in which plants use sunlight to make ATP.
   b. is an inefficient way plants can produce organic molecules and in the process use $O_2$ and release $CO_2$.
   c. is a process that plants use to convert light energy to NADPH.
   d. occurs in the thylakoid lumen.
   e. is the normal process of carbohydrate production in cool, moist environments.
9. Photorespiration is minimized in $C_4$ plants because
   a. these plants separate the formation of a four-carbon molecule from the rest of the Calvin cycle in different cells.
   b. these plants carry out only anaerobic respiration.
   c. the enzyme PEP carboxylase functions to maintain high $CO_2$ concentrations in the bundle-sheath cells.
   d. all of the above.
   e. a and c only.
10. Plants commonly found in hot and dry environments that carry out carbon fixation at night are
   a. oak trees.
   b. $C_3$ plants.
   c. CAM plants.
   d. all of the above.
   e. a and b only.

## Conceptual Questions

1. What are the two stages of photosynthesis? What are the key products of each stage?
2. What is the function of NADPH in the Calvin cycle?
3. **PRINCIPLES** A principle of biology is that living organisms use energy. At the level of the biosphere, what is the role of photosynthesis in the utilization of energy by living organisms?

## Collaborative Questions

1. Discuss the advantages and disadvantages of being a heterotroph or a photoautotroph.
2. Biotechnologists are trying to genetically modify $C_3$ plants to convert them to $C_4$ or CAM plants. Why would this be useful? What genes might you introduce into $C_3$ plants to convert them to $C_4$ or CAM plants?

# Online Resource

# 8 How Cells Communicate with Each Other and with the Environment

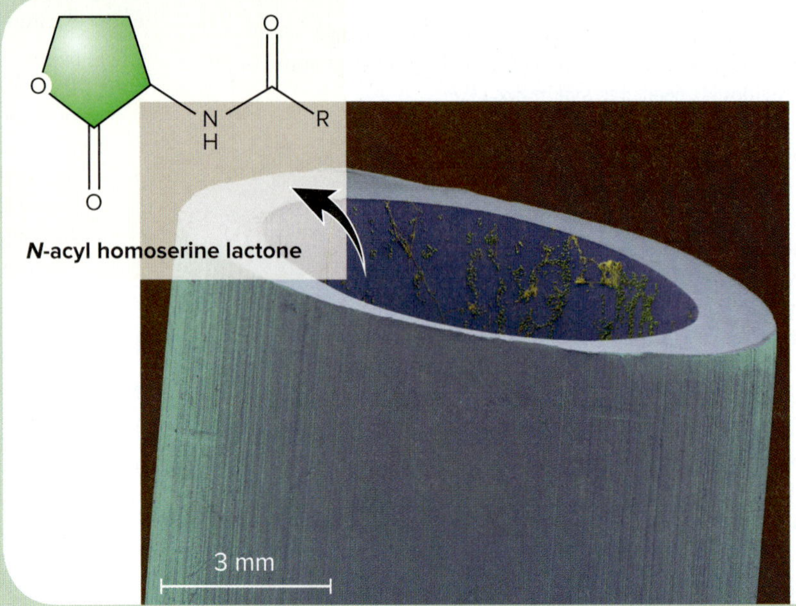


© Dennis Kunkel Microscopy, Inc./Phototake

**A colorized SEM showing the formation of a biofilm on a catheter**. Inside the blue catheter is a biofilm (colored in tan and orange), which is composed of microorganisms embedded within a matrix. The inset shows the structure of *N*-acyl homoserine lactone, which is a signaling molecule made by certain bacterial species that allows them to communicate with each other and form a biofilm.

## Chapter Outline



It wasn't supposed to happen. Philippe, age 37, died 5 weeks after a fairly routine surgical procedure. Following the procedure, Philippe had a catheter (a small, flexible tube) inserted into one of his blood vessels that was needed to deliver fluids and drugs. The catheter was in place for several weeks. The cause of death was sepsis, a bacterial infection of the blood. The likely source of the bacteria was a biofilm—an aggregate of microorganisms that adhere to a surface, such as the inside of a catheter (see chapter-opening photo). A biofilm develops when the attached cells secrete polymers that promote adhesion to a surface and cause the cells to become embedded within a matrix. The characteristics of microorganisms within biofilms are very different from those that exist as independent cells. In particular, bacteria within biofilms typically exhibit dramatic resistance to antibiotics and are able to evade the immunological defenses of their host. Compared with independent bacterial cells, those that are formed within a biofilm can be deadly.

The formation of biofilms is one example of a response that involves **cell communication**—the process through which cells detect, interpret, and respond to signals in their environment. A **signal** is an agent that influences the properties of cells. *N*-acyl homoserine lactones are signaling molecules that are made and secreted by certain bacterial species and promote biofilm formation (see inset of chapter-opening photo).

In this chapter, we will examine how cells detect environmental signals and how they produce signals that enable them to communicate with other cells. Communication at the cellular level involves not only receiving and sending signals but also interpreting them. For this to occur, a signal must be recognized by a cellular protein called a **receptor.** When a signal and a receptor interact, the receptor changes shape—or conformation—thereby changing the way the receptor interacts with cellular factors. These interactions eventually lead to some type of response in the cell. We will begin the chapter with the general features of cell communication and then discuss the main ways in which cells receive, process, and respond to signals sent by other cells. As you will learn, cell communication involves an amazing diversity of signaling molecules and cellular proteins that are devoted to this process.

## 8.1 General Features of Cell Communication

### Learning Outcomes

1. List two general reasons cells need to respond to signals.
2. Compare and contrast the five ways that cells communicate with each other based on the distance between them.
3. Outline the three-stage process of cell signaling.

All living organisms, including bacteria, archaea, protists, fungi, plants, and animals, conduct and require cell communication to survive. Cell communication, also known as **cell signaling,** involves both incoming and outgoing signals. For example, on a sunny day, cells can sense their exposure to ultraviolet (UV) light—a physical signal—and respond accordingly. In humans, UV light acts as an incoming signal, and one way that we respond to it is to synthesize melanin, a protective pigment that helps to prevent the harmful effects of UV radiation. In addition, cells produce outgoing signals that influence the behavior of neighboring cells. Plant cells, for example, make hormones that influence the pattern of cell elongation so the plant grows toward light. Cells of all living organisms both respond to incoming signals and produce outgoing signals. Cell communication is a two-way street.

In this section, we will begin by considering why cells need to respond to signals. We will then examine various forms of signaling that are based on the distance between the cells that communicate with each other. Finally, we will examine the main steps that occur when a cell is exposed to a signal and elicits a response to it.

### Cells Detect and Respond to Signals from Their Environment and from Other Cells

Before getting into the details of cell communication, let's take a general look at why cell communication is necessary.

**1. Responding to Changes in the Environment** Changes in the environment are a persistent feature of life, and living cells are continually faced with alterations in temperature and availability of nutrients, water, and light. A cell may even be exposed to a toxic chemical in its environment. Being able to respond to change at the cellular level is called a **cellular response.** As an example, let's consider the response of a yeast cell to glucose in its environment (Figure 8.1). Some of the glucose acts as a signaling molecule that binds to a receptor and causes a cellular response. In this case, the cell responds by increasing the number of glucose transporters needed to take glucose into the cell and by increasing the number of metabolic enzymes required to utilize glucose once it is inside. The cellular response allows the cell to use glucose efficiently.

**2. Cell-to-Cell Communication** Cells need to communicate with each other—a type of cell communication called **cell-to-cell communication.** In one of the earliest experiments demonstrating cell-to-cell communication, Charles Darwin and his son Francis

## Biology Principle

### Living Organisms Interact with Their Environment

In this example, cell signaling allows yeast cells to respond to the presence of glucose in their environment.

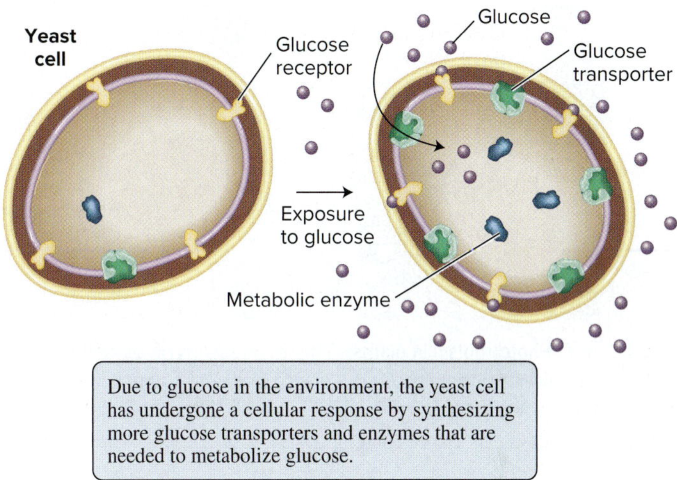


**Figure 8.1** **Response of a yeast cell to glucose.** When glucose is absent from the extracellular environment, the cell is not well prepared to take up and metabolize this sugar. However, when glucose is present, some of that glucose binds to receptors in the membrane, which leads to changes in the amounts and properties of glucose transporters and metabolic enzymes so the cell can readily use glucose.

**Concept Check:** *What is the signaling molecule in this example?*

studied phototropism, the phenomenon in which plants grow toward light (Figure 8.2). The Darwins observed that the actual bending occurs in a zone below the growing shoot tip. They concluded that a signal must be transmitted from the growing tip to lower parts of the shoot. Later research revealed that the signal is a molecule called auxin, which is transmitted from cell to cell. A higher amount of auxin accumulates on the nonilluminated side of the shoot. It binds to a receptor that promotes cell elongation on that side of the shoot only, thereby causing the shoot to bend toward the light source.

### Cell-to-Cell Communication Occurs Between Adjacent Cells and Between Cells That Are Long Distances Apart

Organisms have a variety of mechanisms to achieve cell-to-cell communication. The mode of communication depends, in part, on the distance between the cells. One way to categorize cell signaling is by the manner in which the signal is transmitted from one cell to another. Signals are relayed between cells in five common ways, all of which involve a cell that produces the signal and a target cell that receives it (Figure 8.3).

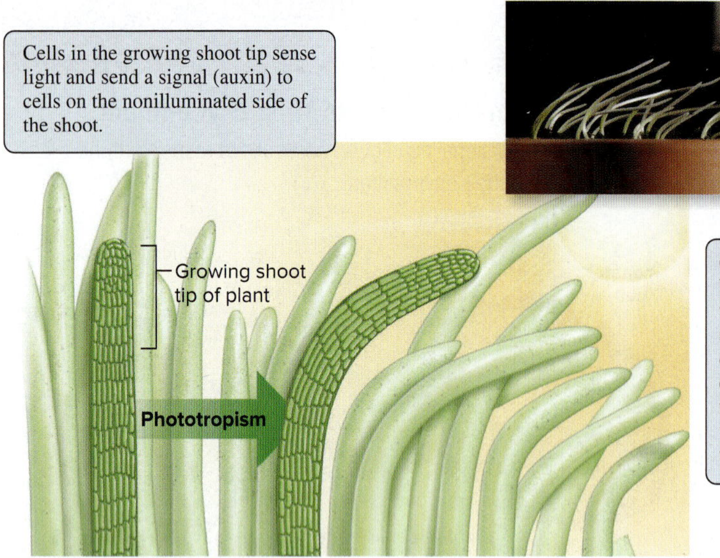

Cells in the growing shoot tip sense light and send a signal (auxin) to cells on the nonilluminated side of the shoot.
Growing shoot tip of plant
Phototropism
Cells located below the growing tip receive this signal and elongate, thereby causing a bend in the shoot. In this way, the tip grows toward the light.


**Figure 8.2** **Phototropism in plants.** This process involves cell-to-cell communication that leads to a shoot bending toward light just beneath its growing tip.



**1. Direct Intercellular Signaling** In a multicellular organism, cells adjacent to each other may have contacts, called intercellular channels, that enable them to pass ions, signaling molecules, and other materials between the cytosol of one cell and the cytosol of another (Figure 8.3a). For example, cardiac muscle cells, which cause your heart to beat, have intercellular channels called gap junctions that allow the passage of ions needed for the coordinated contraction of these cells.

**2. Contact-Dependent Signaling** Some molecules are bound to the surface of cells and provide a signal to other cells that make contact with the surface of that cell (Figure 8.3b). In this case, one cell has a membrane-bound signaling molecule that is recognized by a receptor on the surface of another cell. This occurs, for example, when neurons (cells of the nervous system) grow and make contact with other neurons. This is important for the formation of the proper connections between neurons.

**3. Autocrine Signaling** In autocrine signaling, a cell secretes signaling molecules that bind to receptors on its own cell surface or on neighboring cells of the same cell type, stimulating a response (Figure 8.3c). One purpose of autocrine signaling is for groups of cells to sense cell density. When cell density is high, the concentration of autocrine signals is also high. In some cases, such signals will inhibit further cell growth and thereby limit cell density.

**4. Paracrine Signaling** In paracrine signaling, a cell secretes a signaling molecule that does not affect the cell secreting the signal but instead influences the behavior of target cells in close proximity (Figure 8.3d). Paracrine signaling is typically of short duration. Usually, the signal is broken down too quickly to be carried to other parts of the body and affect distant cells. A specialized form of paracrine signaling occurs in the nervous systems of animals. Neurotransmitters—molecules made in neurons—are released at the end of the neuron and traverse a narrow space called the synaptic cleft (see Chapter 33, Figure 33.12). The neurotransmitter then binds to a receptor in a target cell.

**5. Endocrine Signaling** In contrast to the previous mechanisms of cell signaling, endocrine signaling occurs over relatively long distances (Figure 8.3e). In both animals and plants, molecules involved in long-distance signaling are called **hormones.** They usually last longer than signaling molecules that are used in autocrine and paracrine signaling. In vertebrate animals, endocrine signaling involves the secretion of hormones into the bloodstream, which may affect virtually all cells of the body, including those that are far from the cells that secrete the signaling molecules. In vascular plants, hormones move through the vascular system as well as through adjacent cells. Some hormones are even gases that diffuse into the air. Ethylene, a gas given off by plants, plays a variety of roles, one of which is to accelerate the ripening of fruit.

## Cells Usually Respond to Signals via a Three-Stage Process

Up to this point, we have learned that signals influence the behavior of cells in close proximity or at long distances, interacting with receptors to elicit a cellular response. What events occur when a cell encounters a signal? In most cases, the binding of a signaling molecule to a receptor causes the receptor to activate a signal transduction pathway, which then leads to a cellular response. Figure 8.4 diagrams the three common stages of cell signaling: receptor activation, signal transduction, and a cellular response.

**Stage 1: Receptor Activation** In the initial stage, a signaling molecule binds to a receptor in the target cell, causing a conformational change in the receptor that activates its function. In most cases, the activated receptor initiates a response by causing changes in a series of proteins that collectively forms a signal transduction pathway, as described next.

**Stage 2: Signal Transduction** During signal transduction, the initial signal is converted—or transduced—to a different signal inside the cell. This process is carried out by a group of proteins that form a **signal transduction pathway.** These proteins undergo a series of changes that may result in the production of intracellular signals.

**Stage 3: Cellular Response** Signals cause cells to respond in several different ways. Figure 8.4 shows three common categories of proteins that are controlled by cell signaling: enzymes, structural proteins, and transcription factors.

- Many signaling molecules exert their effects by altering the activity of one or more enzymes. For example, certain hormones

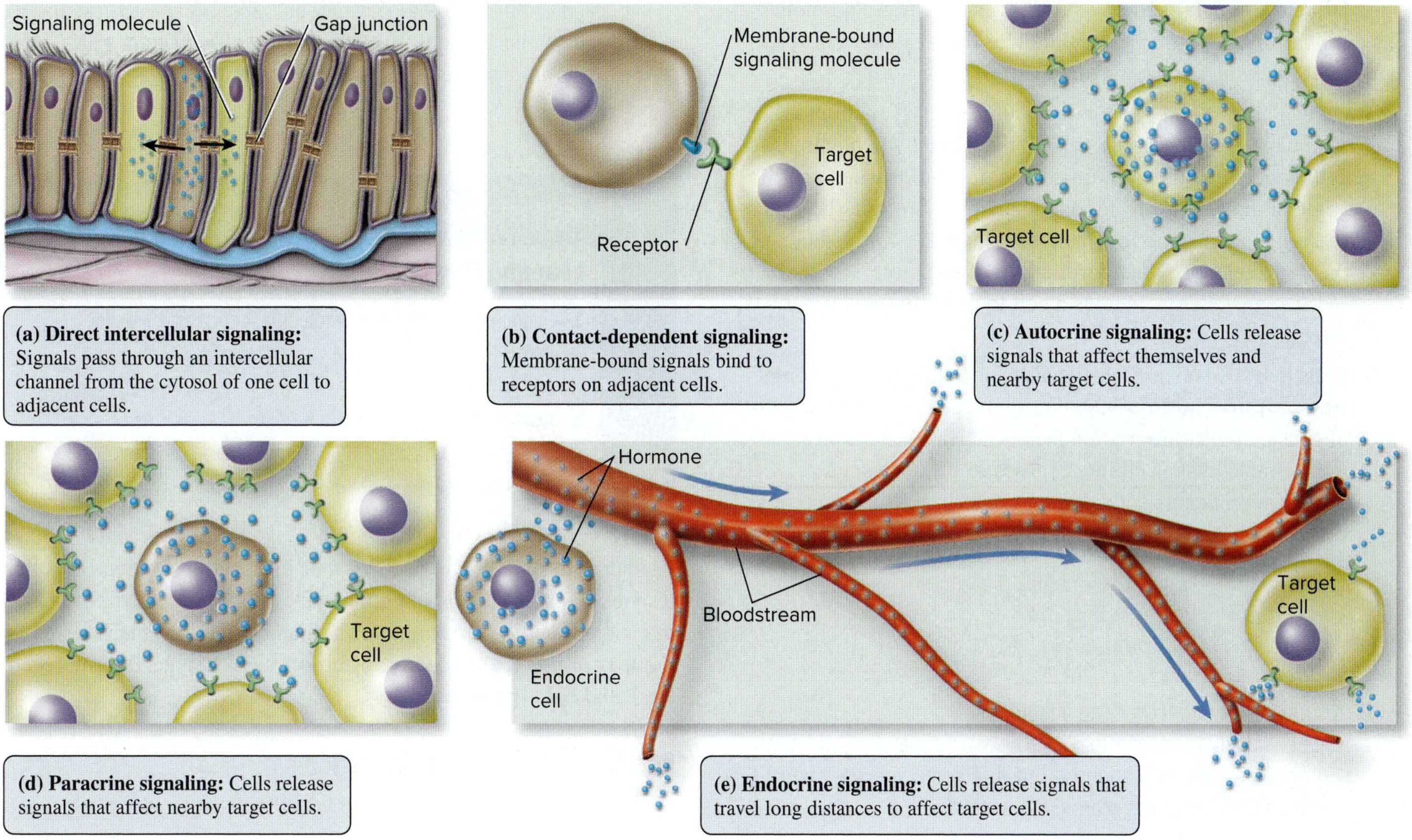

Signaling molecule
Gap junction

**(a) Direct intercellular signaling:** Signals pass through an intercellular channel from the cytosol of one cell to adjacent cells.

Membrane-bound signaling molecule
Target cell
Receptor

**(b) Contact-dependent signaling:** Membrane-bound signals bind to receptors on adjacent cells.

Target cell

**(c) Autocrine signaling:** Cells release signals that affect themselves and nearby target cells.

Target cell

**(d) Paracrine signaling:** Cells release signals that affect nearby target cells.

Hormone
Bloodstream
Endocrine cell
Target cell

**(e) Endocrine signaling:** Cells release signals that travel long distances to affect target cells.

**Figure 8.3** Types of cell-to-cell communication based on the distance between cells.

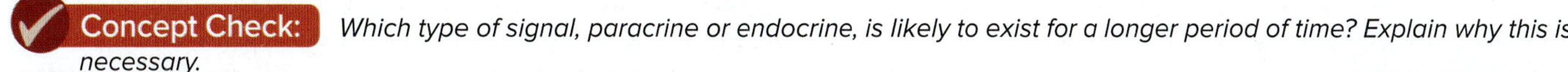

Concept Check: *Which type of signal, paracrine or endocrine, is likely to exist for a longer period of time? Explain why this is necessary.*

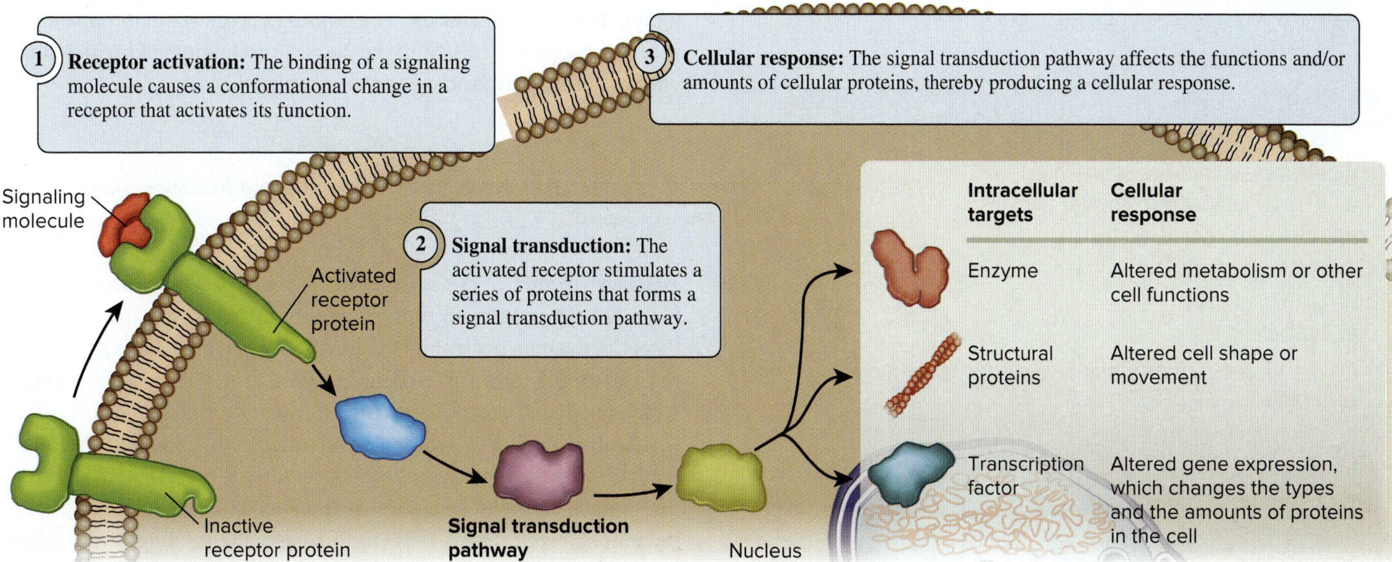

1 Receptor activation: The binding of a signaling molecule causes a conformational change in a receptor that activates its function.
3 Cellular response: The signal transduction pathway affects the functions and/or amounts of cellular proteins, thereby producing a cellular response.
Signaling molecule
Activated receptor protein
2 Signal transduction: The activated receptor stimulates a series of proteins that forms a signal transduction pathway.
Intracellular targets
Cellular response
Enzyme
Altered metabolism or other cell functions
Structural proteins
Altered cell shape or movement
Transcription factor
Altered gene expression, which changes the types and the amounts of proteins in the cell
Inactive receptor protein
Signal transduction pathway
Nucleus


**Figure 8.4** The three stages of cell signaling: Receptor activation, signal transduction, and a cellular response. Note: The examples described in this chapter involve signaling molecules that activate receptors. Some signaling molecules inhibit receptor function.

Concept Check: *For most signaling molecules, explain why a signal transduction pathway is necessary.*

provide a signal that the body needs energy. These hormones activate enzymes that are required for the breakdown of molecules such as carbohydrates.

- Cells also respond to signals by altering the functions of structural proteins in the cell. For example, when animal cells move during embryonic development or when an amoeba moves toward food, signals play a role in the rearrangement of actin filaments, which are components of the cytoskeleton. The coordination of signaling and changes in the cytoskeleton enables a cell to move in the correct direction.
- Cells may also respond to signals by affecting the function of **transcription factors**—proteins that regulate the transcription of genes. Some transcription factors activate gene expression. For example, when cells are exposed to sex hormones, transcription factors activate genes that change the properties of cells, which can lead to changes in the sexual characteristics of organisms.

## 8.1 Reviewing the Concepts

- A signal is an agent that can influence the properties of cells. A signal binds to a receptor to elicit a cellular response. Cell signaling enables cells to respond to environmental changes and to communicate with each other (Figures 8.1, 8.2).
- Cell-to-cell communication varies in the mechanism and distance that a signal travels. Signals are relayed between cells in five common ways: direct intercellular, contact-dependent, autocrine, paracrine, and endocrine signaling (Figure 8.3).
- Cell communication is usually a three-stage process involving receptor activation, signal transduction, and a cellular response. A signal transduction pathway is a group of proteins that convert an initial signal to a different signal inside the cell (Figure 8.4).

## 8.1 Testing Your Knowledge

1. A general reason for cell signaling is
   a. to respond to environmental changes.
   b. cell-to-cell communication.
   c. protein degradation.
   d. all of the above.
   e. both a and b.
2. A type of cell signaling that involves signaling molecules that travel long distances is
   a. direct intercellular.
   b. contact-dependent.
   c. autocrine.
   d. paracrine.
   e. endocrine.
3. Which of the following is the correct order for the three stages of cell signaling?
   a. receptor activation → cellular response → signal transduction
   b. receptor activation → signal transduction → cellular response
   c. signal transduction → receptor activation → cellular response
   d. signal transduction → cellular response → receptor activation
   e. cellular response → signal transduction → receptor activation

## 8.2 Receptor Activation

### Learning Outcomes

1. Explain how a signaling molecule, or ligand, binds to its receptor and causes a conformational change that activates the receptor.
2. **SCISKILLS ▶** Calculate the affinity, measured as a dissociation constant, that a receptor has for its signaling molecule, or ligand.

In this section, we will take a closer look at receptors and how they interact with signaling molecules. Our focus will be on receptors that respond to chemical signaling molecules. Other receptors, discussed in Units VI and VII, respond to mechanical motion (mechanoreceptors), pressure (baroreceptors), temperature change (thermoreceptors), and light (photoreceptors).

### Receptors Bind to Specific Signals and Undergo Conformational Changes

The ability of cells to respond to a signal usually requires precise recognition between the signal and its receptor. In many cases, the signal is a molecule, such as a steroid or a protein, which binds to the receptor. A signaling molecule binds to a receptor in much the same way that a substrate binds to the active site of an enzyme, as described in Chapter 6 (refer back to Figure 6.5). The signaling molecule, which is called a **ligand,** binds noncovalently to the receptor molecule with a high degree of specificity. The binding occurs when the ligand and receptor happen to collide in the correct orientation and with enough energy to form a **ligand•receptor complex.**

$$[\text{Ligand}] + [\text{Receptor}] \underset{k_{off}}{\overset{k_{on}}{\rightleftharpoons}} [\text{Ligand•Receptor complex}]$$

Brackets [ ] refer to concentration. The value $k_{on}$ is the rate at which binding occurs. After a complex forms between the ligand and its receptor, the noncovalent interaction between a ligand and its receptor remains stable for a finite period of time. The term $k_{off}$ is the rate at which the ligand•receptor complex falls apart or dissociates.

Unlike enzymes, which convert their substrates into products, receptors do not usually alter the structure of their ligands. Instead, the ligands alter the structure of their receptors, causing a conformational change (Figure 8.5). In this case, the binding of the ligand to its receptor changes the receptor in a way that activates its ability to initiate a cellular response. Because the binding of a ligand to its receptor is a reversible process, the ligand and receptor will also dissociate. Once the ligand is released, the receptor is no longer activated.

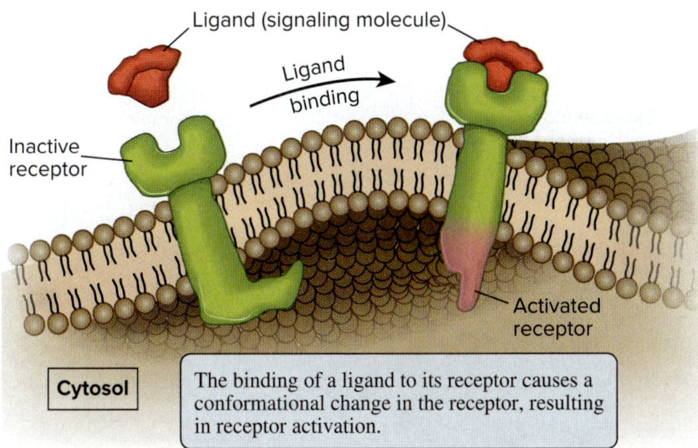


**Figure 8.5** Receptor activation.

**BioConnections:** *Look back at Figure 6.5. How is the binding of a ligand to its receptor similar to the binding of a substrate to an enzyme? How are they different?*

# Quantitative Analysis

## RECEPTORS HAVE A MEASURABLE AFFINITY FOR THEIR LIGANDS

In general, the binding and release between a ligand and its receptor are relatively rapid, and therefore an equilibrium is reached when the rate of formation of new ligand•receptor complexes equals the rate at which existing ligand•receptor complexes dissociate:

$$k_{on}\,[\text{Ligand}][\text{Receptor}] = k_{off}\,[\text{Ligand•Receptor complex}]$$

Rearranging,

$$\frac{[\text{Ligand}][\text{Receptor}]}{[\text{Ligand•Receptor complex}]} = \frac{k_{off}}{k_{on}} = K_d$$

$K_d$ is called the **dissociation constant** between a ligand and its receptor. The $K_d$ value is inversely related to the affinity between the ligand and receptor. A low $K_d$ value indicates that a receptor has a high affinity for its ligand.

Let's look carefully at the left side of this equation and consider what it means. At a ligand concentration where half of the receptors are bound to a ligand, the concentration of the ligand•receptor complex equals the concentration of receptor that doesn't have ligand bound. At this ligand concentration, [Receptor] and [Ligand•Receptor complex] cancel out of the equation because they are equal. Therefore, at a ligand concentration where half of the receptors have bound ligand

$$K_d = [\text{Ligand}]$$

When the ligand concentration is above the $K_d$ value, most of the receptors are likely to have ligand bound to them. In contrast, if the ligand concentration is substantially below the $K_d$ value, most receptors will not be bound by their ligand. The $K_d$ values for many different ligands and their receptors have been experimentally determined. How is this information useful? It allows researchers to predict when a signaling molecule is likely to cause a cellular response. If the concentration of a signaling molecule is far below the $K_d$ value, a cellular response is not likely because relatively few receptors will form a complex with the signaling molecule.

**Crunching the Numbers:** Type I diabetes is a human disease in which the pancreas is unable to make sufficient amounts of the hormone insulin. The disorder often begins in childhood, but its onset may occur in adulthood. Type I diabetes affects over 1 million Americans.

After a nondiabetic person eats a meal, a proper amount of insulin is secreted into the bloodstream. Insulin binds to the insulin receptor found on the surfaces of certain cell types, such as muscle cells, and promotes the uptake of glucose from the blood into the cells. The insulin receptor on skeletal muscle cells has a $K_d$ of approximately 0.2 nM. Keisha, who does not have type I diabetes, eats a meal and her blood insulin level rises to 0.45 nM. Ryan has type I diabetes. If he is not given insulin by injection, his blood insulin level following a meal rises to only 0.06 nM. For both Keisha and Ryan, is the percentage of insulin receptors that have insulin bound under these conditions above or below 50%? Why does Ryan have to take regular insulin injections?

## 8.2 Reviewing the Concepts

- The binding of a signaling molecule, also called a ligand, to a receptor is usually very specific and alters the conformation of the receptor (Figure 8.5).
- A ligand binds to a receptor with an affinity that is measured as a dissociation constant, or $K_d$ value. The $K_d$ value is inversely related to the affinity between the ligand and its receptor.

## 8.2 Testing Your Knowledge

1. When a ligand (signaling molecule) binds to a receptor,
   **a.** the ligand is converted to a product.
   **b.** the ligand forms a covalent bond with the receptor.
   **c.** the receptor undergoes a conformational change.
   **d.** the receptor is rapidly degraded.
   **e.** the receptor phosphorylates the ligand.
2. A receptor has a $K_d$ for its ligand of 50 nM. At a ligand concentration of 100 nM,
   **a.** a small percentage of receptors will have ligand bound to them.
   **b.** 50% of the receptors will have ligand bound to them.
   **c.** most receptors will have ligand bound to them.
   **d.** all receptors will have ligand bound to them.
   **e.** none of the above.

## 8.3 Cell Surface Receptors

### Learning Outcome

1. Compare and contrast the three general types of cell surface receptors.

Most signaling molecules are either small, hydrophilic molecules or large molecules that do not readily pass through the plasma membrane of cells. Such extracellular signals bind to **cell surface receptors**—receptors embedded in the plasma membrane. A typical cell has dozens or even hundreds of different cell surface receptors that enable the cell to respond to different kinds of extracellular signaling molecules. By analyzing the functions of cell surface receptors from many different organisms, researchers have determined that most fall into one of three categories: enzyme-linked receptors, G-protein-coupled receptors, and ligand-gated ion channels.

**1. Enzyme-Linked Receptors** Receptors known as **enzyme-linked receptors** are found in the cells of all living species. Many human hormones bind to this type of receptor. For example, when insulin binds to an enzyme-linked receptor in muscle cells, it enhances the ability of those cells to use glucose. Enzyme-linked receptors typically have two important domains: an extracellular domain, which binds a signaling molecule, and an intracellular domain, which has a catalytic function (Figure 8.6a). When a signaling molecule binds to the extracellular domain, a conformational change is transmitted through the membrane-embedded portion of the protein that affects the conformation of the intracellular catalytic domain. In most cases, this conformational change causes the intracellular catalytic domain to become functionally active.

Most types of enzyme-linked receptors function as **protein kinases,** enzymes that transfer a phosphate group from ATP to a specific amino acid in a protein (Figure 8.6b). For example, tyrosine kinases attach phosphate to the amino acid tyrosine, whereas serine/threonine kinases attach phosphate to the amino acids serine and threonine. In the example shown in Figure 8.6b, the catalytic domain of the receptor is inactive when no signaling molecule is present. However, when a signal binds to the extracellular domain, the catalytic domain becomes activated. Under these conditions, the receptor may phosphorylate itself, or (as shown in Figure 8.6b) it may phosphorylate intracellular proteins. The attachment of a negatively charged phosphate changes the structure of a protein and can thereby alter its function. Later in this chapter, we will explore how this event leads to a cellular response, such as the activation of enzymes that affect cell function.

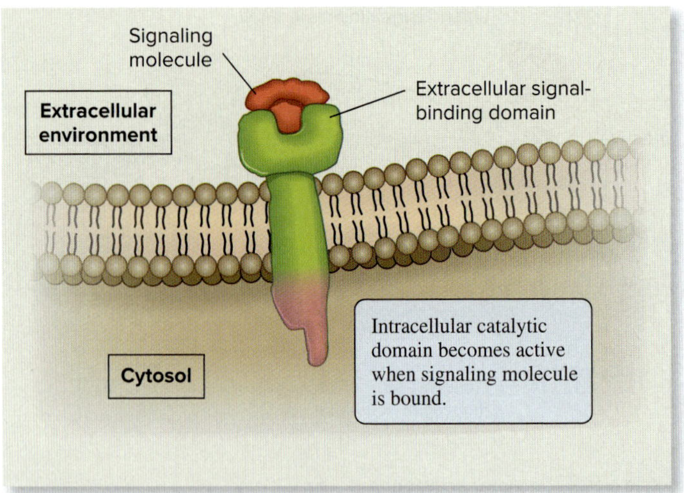


(a) Structure of an enzyme-linked receptor

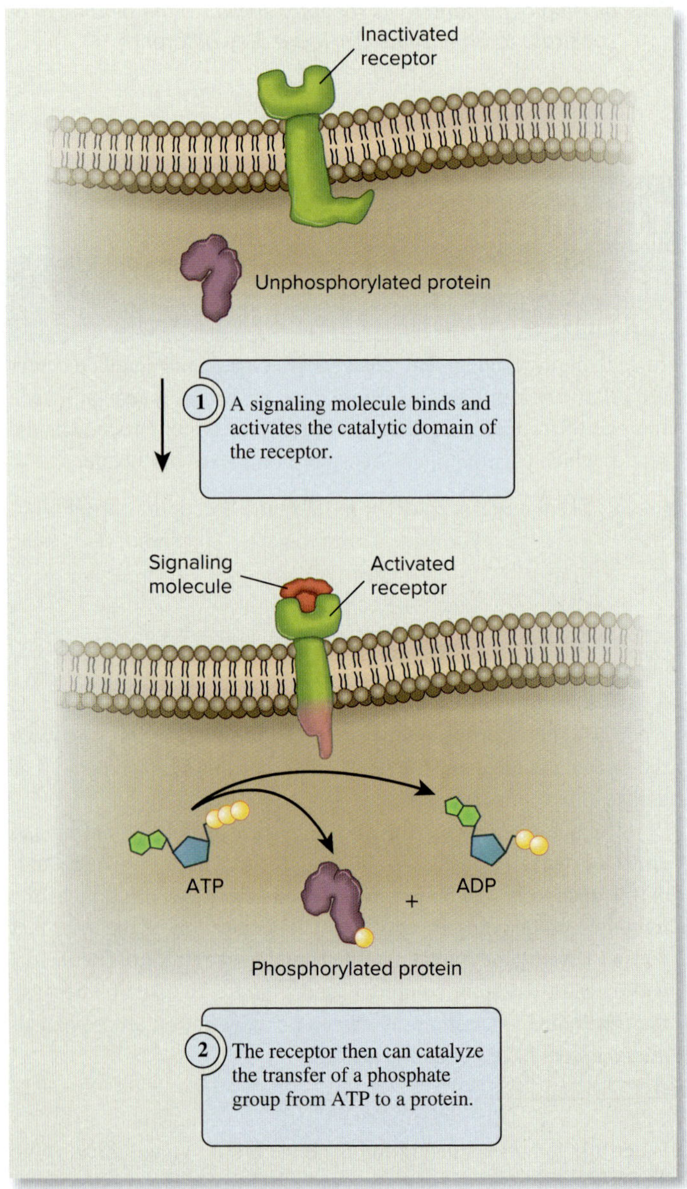


(b) A receptor that functions as a protein kinase

**Figure 8.6** Enzyme-linked receptors.

**Concept Check:** *Based on your understanding of ATP as an energy intermediate, is the phosphorylation of a protein via a protein kinase an exergonic or endergonic reaction? What is the purpose of phosphorylation?*

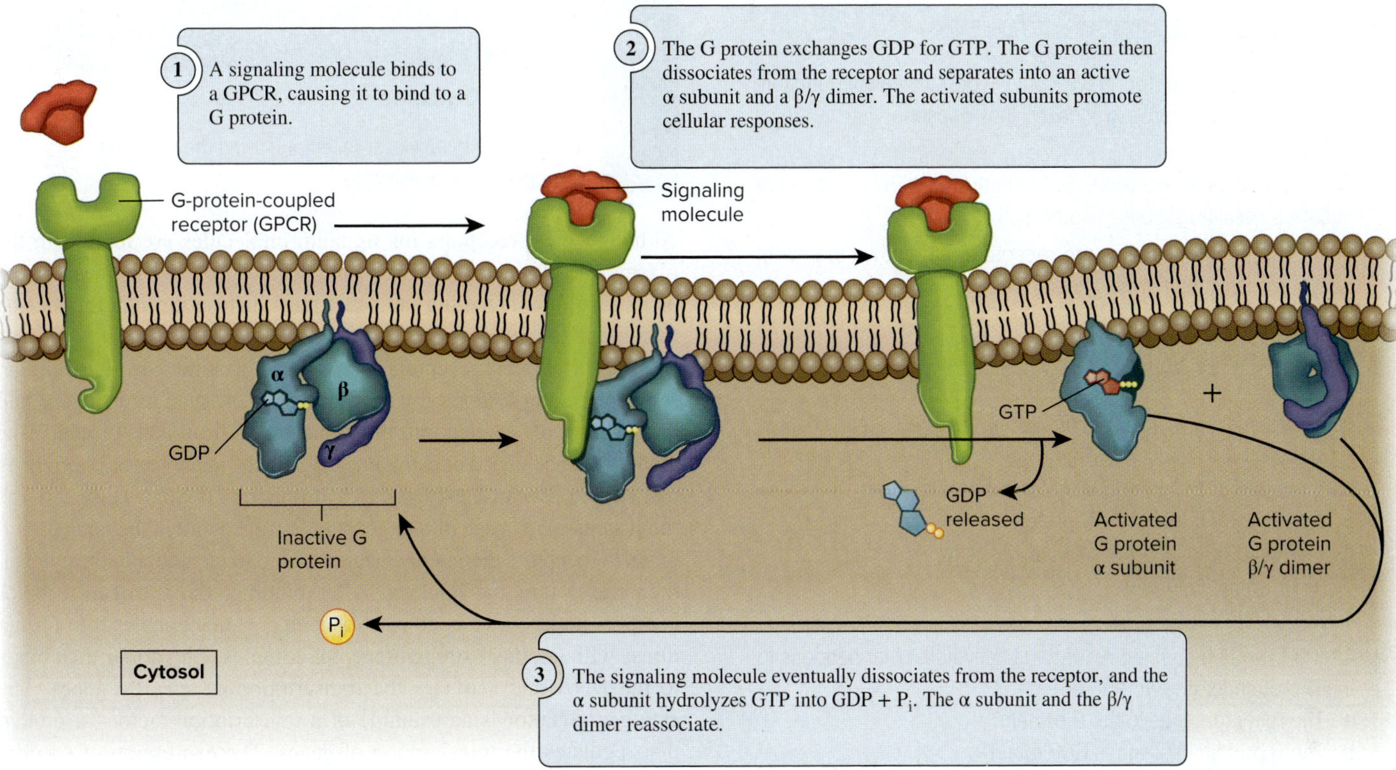


**Figure 8.7** The activation of G-protein-coupled receptors (GPCRs) and G proteins.

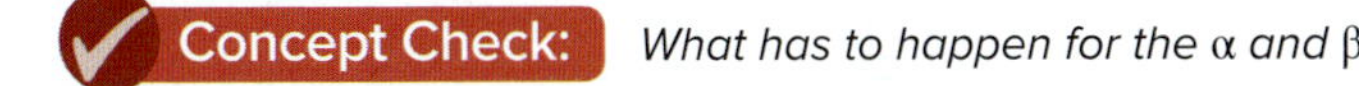

*What has to happen for the α and β/γ subunits of the G protein to reassociate with each other?*

**2. G-Protein-Coupled Receptors** Receptors called **G-protein-coupled receptors (GPCRs)** are found in the cells of all eukaryotic species and are particularly common in animals. GPCRs typically contain seven transmembrane segments that wind back and forth through the plasma membrane. The receptors interact with intracellular proteins called **G proteins,** which are so named because of their ability to bind guanosine triphosphate (GTP) and guanosine diphosphate (GDP). GTP is similar in structure to ATP except it has guanine as a base instead of adenine.

Figure 8.7 shows how a GPCR and a G protein interact. At the cell surface, a signaling molecule binds to a GPCR, resulting in a conformational change that activates the receptor, causing it to bind to a G protein. The G protein, which is a lipid-anchored protein, releases GDP and binds GTP instead. GTP binding changes the conformation of the G protein, causing it to dissociate into an α subunit and a β/γ dimer. Later in this chapter, we will examine how the α subunit interacts with other proteins in a signal transduction pathway to elicit a cellular response.

**3. Ligand-Gated Ion Channels** As described in Chapter 5 (refer back to Figure 5.12), ion channels are proteins that allow the diffusion of ions across biological membranes. **Ligand-gated ion channels** are a third type of cell surface receptor found in the plasma membrane of animal, plant, and fungal cells. When signaling molecules (ligands) bind to this type of receptor, the channel opens and allows the flow of ions through the membrane, changing the concentration of the ions in the cell (Figure 8.8). In animals, ligand-gated ion channels are important in the transmission of signals between neurons and muscle cells and between two neurons.

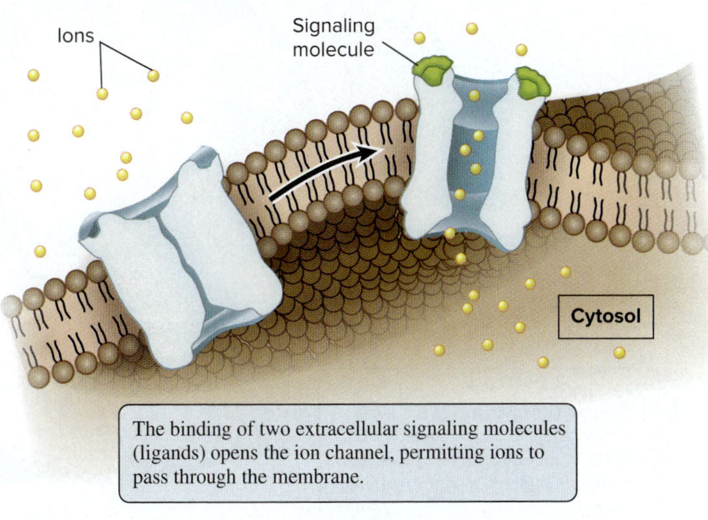

Ions
Signaling molecule
Cytosol
The binding of two extracellular signaling molecules (ligands) opens the ion channel, permitting ions to pass through the membrane.

**Figure 8.8** The function of a ligand-gated ion channel.

## 8.3 Reviewing the Concepts

- Enzyme-linked receptors have some type of catalytic function. Many of them are protein kinases that phosphorylate proteins (Figure 8.6).
- G-protein-coupled receptors (GPCRs) interact with G proteins to initiate a cellular response (Figure 8.7).
- Ligand-gated ion channels are receptors that allow the flow of ions across cell membranes (Figure 8.8).

## 8.3 Testing Your Knowledge

1. When a signaling molecule binds to an enzyme-linked receptor, it causes
   a. the receptor to undergo a conformational change.
   b. the receptor to bind to a G protein.
   c. the receptor to become catalytically active.
   d. an ion channel to open.
   e. both a and c.
2. After a G-protein-coupled receptor is activated, what happens to the G protein for it to be active?
   a. The receptor binds to the G protein.
   b. The G protein exchanges GDP for GTP.
   c. The G protein dissociates into $\alpha$ and $\beta/\gamma$ subunits.
   d. All of the above happen.
   e. None of the above happens.

## 8.4 Intracellular Receptors

### Learning Outcome

1. Describe intracellular receptors, using the estrogen receptor as an example.

Although most receptors for signaling molecules are located in the plasma membrane, some are found inside the cell. In these cases, an extracellular signaling molecule must diffuse through the plasma membrane to gain access to its receptor.

In vertebrates, receptors for steroid hormones are intracellular. As discussed in Chapter 39, steroid hormones, such as estrogens and androgens, are secreted into the bloodstream from cells of endocrine glands. The behavior of estrogen is typical of many steroid hormones (**Figure 8.9**). Because estrogen is hydrophobic, it can diffuse through the plasma membrane of a target cell. Estrogen enters the nucleus and binds to receptors there. Other steroids bind to receptors in the cytosol, then travel into the nucleus. After binding, the estrogen•receptor complex undergoes a conformational change that enables it to form a dimer with another estrogen•receptor complex. The dimer then binds to the DNA and activates the transcription of specific genes. The estrogen receptor is an example of a transcription factor—a protein that regulates the transcription of genes. The expression of specific genes changes cell structure and function in a way that results in a cellular response.

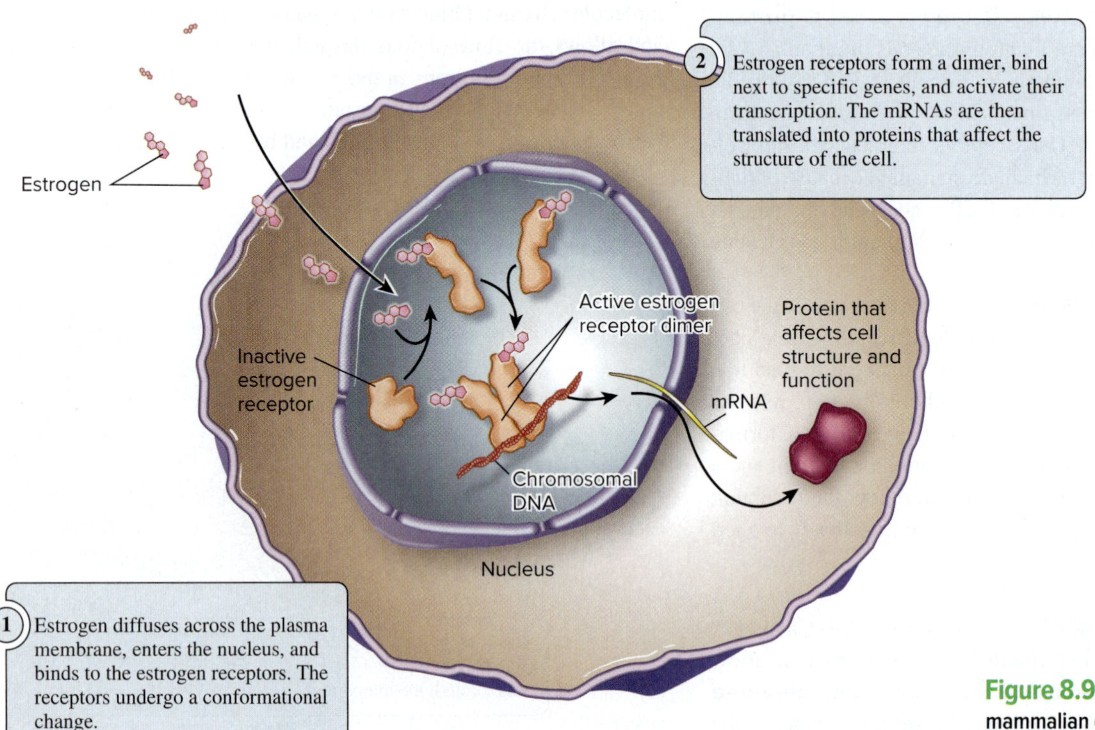

Estrogen
Inactive estrogen receptor
Active estrogen receptor dimer
Protein that affects cell structure and function
mRNA
Chromosomal DNA
Nucleus
2 Estrogen receptors form a dimer, bind next to specific genes, and activate their transcription. The mRNAs are then translated into proteins that affect the structure of the cell.
1 Estrogen diffuses across the plasma membrane, enters the nucleus, and binds to the estrogen receptors. The receptors undergo a conformational change.


**Figure 8.9 Estrogen receptor in mammalian cells.** The estrogen receptor is an example of an intracellular receptor.

## 8.4 Reviewing the Concepts

- Although most receptors involved in cell signaling are found on the cell surface, some receptors, such as the estrogen receptor, are found inside the cell and are termed intracellular receptors (Figure 8.9).
- The estrogen receptor functions as a transcription factor that activates specific genes.

## 8.4 Testing Your Knowledge

1. Prior to the binding of estrogen, the estrogen receptor is located
   a. in the plasma membrane.
   b. in the cytosol.
   c. in the nucleus.
   d. in the ER.
   e. none of the above.

## 8.5 Signal Transduction and Cellular Response via an Enzyme-Linked Receptor

### Learning Outcomes

1. Describe the signal transduction pathway for signaling molecules that bind to a receptor tyrosine kinase.
2. Explain how the signal transduction pathway elicits a cellular response.

We will now turn our attention to the intracellular events that enable a cell to respond to a signaling molecule that binds to a cell surface receptor. In this section, we will examine a signal transduction pathway that is controlled by an enzyme-linked receptor.

### Receptor Tyrosine Kinases Activate Signal Transduction Pathways Involving a Protein Kinase Cascade That Alters Gene Transcription

**Receptor tyrosine kinases** are a category of enzyme-linked receptors that are found in all animals. As an example, Figure 8.10 describes a simplified signal transduction pathway for epidermal growth factor (EGF). A **growth factor** is a signaling molecule that promotes cell division. EGF is responsible for stimulating epidermal cells, such as skin cells, to divide. In vertebrate animals, EGF is secreted from endocrine cells, travels through the bloodstream, and binds to a receptor tyrosine kinase, which is located on target cells and called the EGF receptor. For receptor activation to occur, each EGF receptor subunit binds a molecule of EGF. The binding of EGF causes the subunits to dimerize and phosphorylate each other on tyrosines within the receptors, which is why they are named receptor tyrosine kinases.

Following receptor activation, the general parts of the signal transduction pathway are as follows: (1) Relay proteins activate a protein kinase cascade; (2) the protein kinase cascade phosphorylates proteins in the cell such as transcription factors; and (3) the phosphorylated transcription factors stimulate gene transcription.

**1. Relay Proteins** The phosphorylated form of the EGF receptor is first recognized by a relay protein, causing it to bind another relay protein in the signal transduction pathway. The second relay protein then binds to a third relay protein called Ras. The binding between the second relay protein and Ras causes Ras to release GDP and bind GTP. The GTP form of Ras is the active form.

**2. Protein Kinase Cascade** The relay proteins activate additional proteins in the signal transduction pathway that form a **protein kinase cascade.** In the example shown in Figure 8.10, this cascade involves the sequential activation of three protein kinases. The first protein kinase is activated by Ras. It then phosphorylates and thereby activates the second protein kinase, which then phosphorylates the third protein kinase.

**3. Activation of Transcription Factors and the Cellular Response** The phosphorylated form of the third protein kinase enters the nucleus and phosphorylates specific transcription factors. Once these transcription factors are phosphorylated, they promote the transcription of genes that encode proteins that cause cell division. After these proteins are made, the cell is stimulated to divide.

Growth factors such as EGF cause a rapid increase in the expression of many genes in mammals, perhaps as many as 100. As discussed in Chapter 13, growth factor–signaling pathways are often involved in cancer. Mutations that cause proteins in these pathways to become hyperactive result in cells that divide uncontrollably. BIO TIPS ONLINE

## EVOLUTIONARY CONNECTIONS

### Receptor Tyrosine Kinases Are Found in Choanoflagellates and Animals

The choanoflagellates are a group of unicellular and colonial protists (Figure 8.11). The term choanoflagellate means collared flagellate. Choanoflagellates are composed of ovoid or spherical cells with a single flagellum surrounded by a collar of 30–40 microvilli. Movement of the flagellum creates water currents that propel free-swimming choanoflagellates through the water and trap bacteria and other food particles against the collar, where they are engulfed. Similarly, sponges, which are very simple animals, contain cells called choanocytes (Figure 8.11c). These cells line the sponges' inner passages and fan their flagella to move seawater and food particles. Similarities in the genomes of choanoflagellates and animals indicate that animals are descended from a choanoflagellate species (see Chapter 26).

Receptor tyrosine kinases are a category of enzyme-linked receptors that are found in all animals and in choanoflagellates. However, they are not found in other eukaryotic species or in bacteria and archaea. (Bacteria have receptor histidine kinases, and all eukaryotes

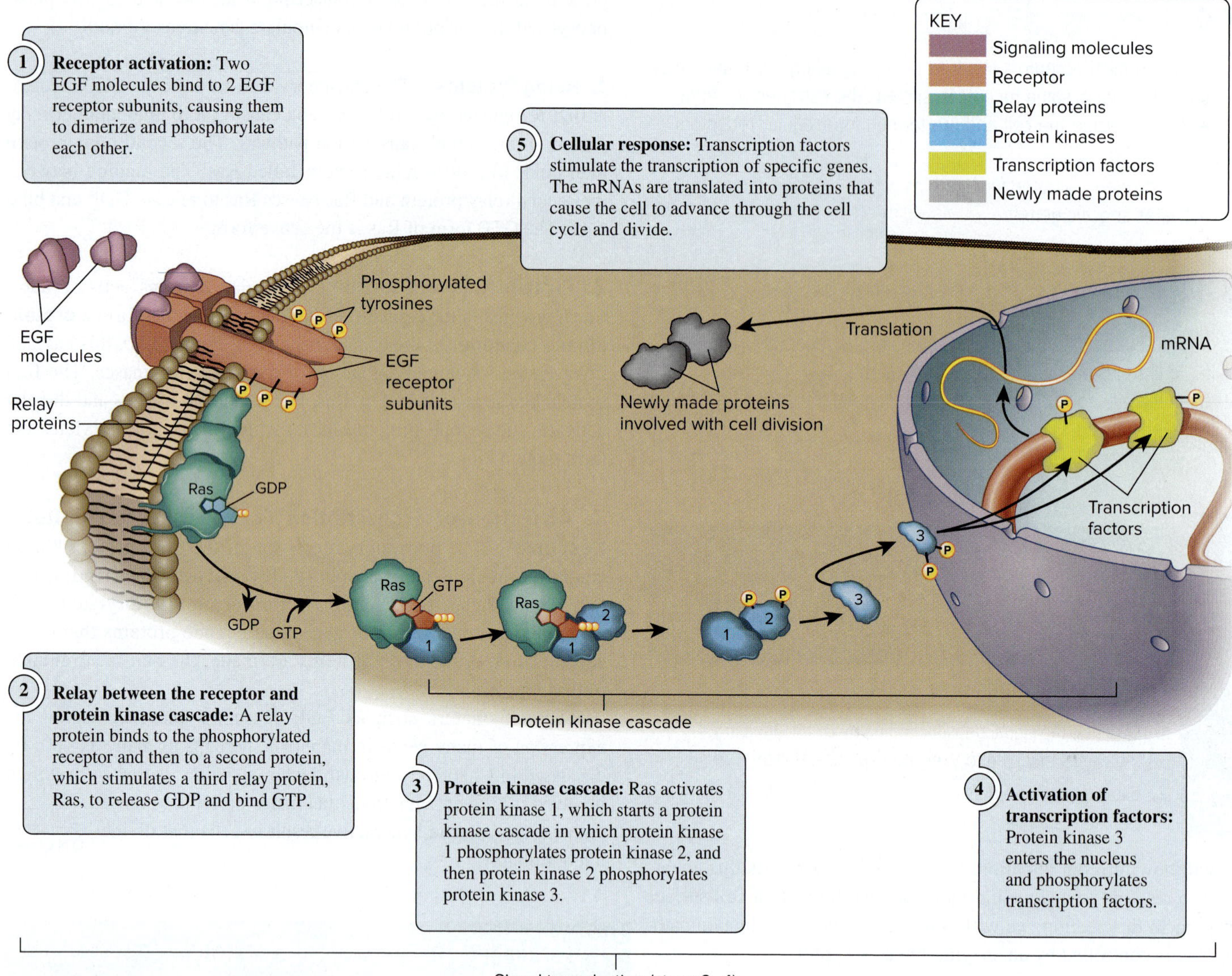


**Figure 8.10** The epidermal growth factor (EGF) pathway that promotes cell division.

**BioConnections:** *Look ahead to Figures 13.9 and, in particular, 13.10. Certain mutations alter the structure of the Ras protein so it will not hydrolyze GTP. Such mutations cause cancer. Explain why.*

have receptor serine/threonine kinases.) The human genome contains about 60 different genes that encode receptor tyrosine kinases that recognize various types of signaling molecules such as hormones. The primordial gene that gave rise to the first receptor tyrosine kinase is presumed to have arisen in a choanoflagellate species that was an ancestor to sponges and other animals. This explains why genes that encode receptor tyrosine kinases are found only in these eukaryotic species, but not others. Researchers have speculated that this form of cell communication may have been an important feature in the evolution of multicellular animals.

## 8.5 Reviewing the Concepts

- Signal transduction pathways influence whether or not a cell will divide. An example is the pathway that is stimulated by epidermal growth factor (EGF). EGF binds to a receptor tyrosine kinase, which is a type of enzyme-linked receptor (Figure 8.10).
- Receptor tyrosine kinases are found only in choanoflagellates and animals, an observation consistent with the hypothesis that animals evolved from a choanoflagellate species (Figure 8.11).

## Biology Principle

### All Species (Past and Present) Are Related by an Evolutionary History

Animals evolved from an ancestral choanoflagellate.

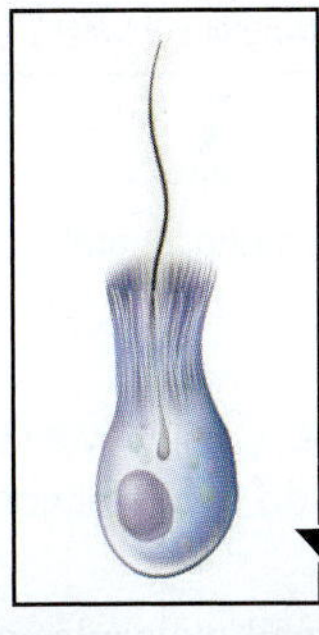

**Figure 8.11** **Choanoflagellates and sponges.** **(a)** A unicellular choanoflagellate (*Monosiga brevicollis*). **(b)** A colonial choanoflagellate (*Proterospongia* sp.). **(c)** A sponge (*Ulosa stuposa*), which is a very simple animal. The inset depicts choanocytes within the body of a sponge.

*(a, b)* Mark J. Dayel (2012) Choanoflagellates and animal multicellularity, http://www.dayel.com/choanoflagellates (visited on 03/26/2013); *(c)* © Wilfried Bay Nouailhat

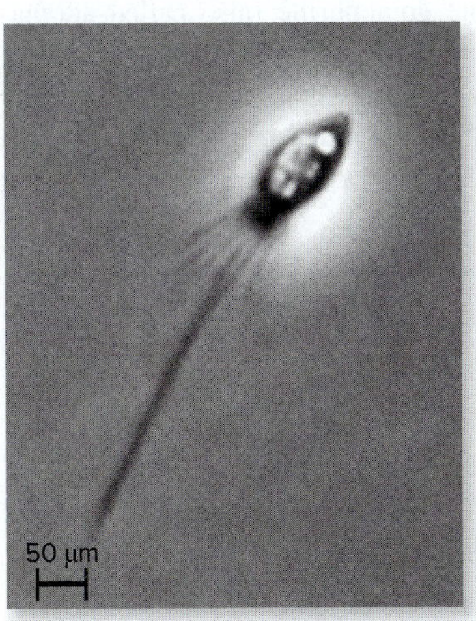


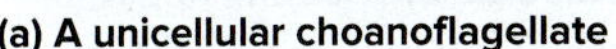

**(a) A unicellular choanoflagellate**

**(b) A multicellular choanoflagellate**

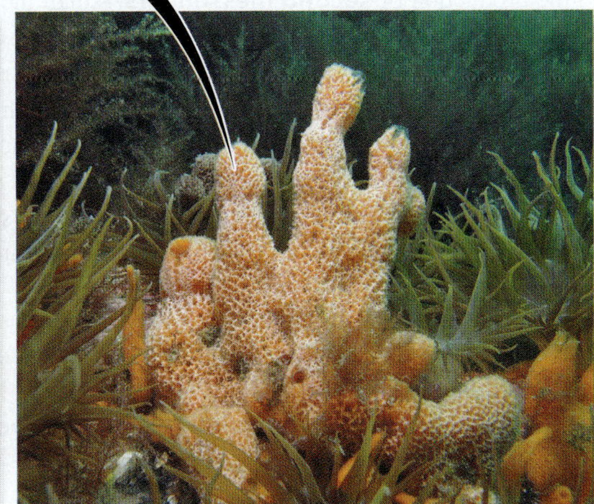

**(c) A sponge**

## 8.5 Testing Your Knowledge

1. Upon binding of EGF, the EGF receptor first activates
   a. relay proteins.
   b. a protein kinase cascade.
   c. transcription factors.
   d. proteins directly involved with cell division.
   e. none of the above.
2. One way to turn off the EGF pathway is for the Ras protein (a relay protein) to hydrolyze GTP to GDP and $P_i$. If a mutation resulted in a Ras protein that could bind GTP but could not hydrolyze GTP to GDP and $P_i$, how would this affect the EGF signal transduction pathway?
   a. The pathway would be inhibited, and cell division would be inhibited.
   b. The pathway would be inhibited, and cell division would be accelerated.
   c. The pathway would be kept on longer than it should, and cell division would be inhibited.
   d. The pathway would be kept on longer than it should, and cell division would be accelerated.
   e. The pathway would not be affected.

## 8.6 Signal Transduction and Cellular Response via a G-Protein-Coupled Receptor

### Learning Outcomes

1. Describe the signal transduction pathway for signaling molecules that bind to a G-protein-coupled receptor (GPCR).
2. Explain the advantages of the production of second messengers such as cAMP.

In this section, we will examine a pathway that is controlled by a G-protein-coupled receptor (GPCR). As you will learn, this pathway involves the production of intracellular signals called second messengers.

### Second Messengers Such as Cyclic AMP Are Key Components of Many Signal Transduction Pathways

Extracellular signaling molecules that bind to cell surface receptors are sometimes referred to as first messengers. After first messengers bind to receptors such as GPCRs, many signal transduction pathways

**Figure 8.12** **The synthesis and breakdown of cyclic AMP.** Cyclic AMP (cAMP) is a second messenger formed from ATP by adenylyl cyclase, an enzyme in the plasma membrane. cAMP is inactivated by the action of an enzyme called phosphodiesterase, which converts cAMP to AMP.

lead to the production of **second messengers—**small molecules or ions that relay signals inside the cell. The signals that result in second messenger production often act quickly, in a matter of seconds or minutes, but their duration is usually short. Therefore, such signaling typically occurs when a cell needs a quick and short cellular response. Cells use several different types of second messengers, and more than one type may be used at the same time. We will focus our discussion of second messengers on cyclic AMP, which is described next.

**Production of cAMP** Mammalian and plant cells make several different types of G protein α subunits. One type of α subunit binds to **adenylyl cyclase,** an enzyme in the plasma membrane. This interaction stimulates adenylyl cyclase to synthesize **cyclic adenosine monophosphate** (**cyclic AMP,** or **cAMP**) from ATP (Figure 8.12). cAMP is an example of a second messenger.

**Signal Transduction Pathway and Cellular Response** Let's explore a signal transduction pathway in which the G-protein-coupled receptor recognizes the hormone epinephrine (also called adrenaline). This hormone is sometimes called the fight-or-flight hormone. Epinephrine is produced when an individual is confronted with a stressful situation and helps the individual deal with a perceived threat or danger by staying and fighting or by fleeing. The following steps occur at the cellular level.

1. Epinephrine binds to a GPCR, which promotes the dissociation of the α subunit from the G protein (Figure 8.13).
2. The α subunit then activates adenylyl cyclase, which catalyzes the production of cAMP from ATP.
3. One effect of cAMP is to activate protein kinase A (PKA), which is composed of four subunits: two catalytic subunits that phosphorylate specific cellular proteins, and two regulatory subunits that inhibit the catalytic subunits when they are bound to each other. cAMP binds to the regulatory subunits of PKA. The binding of cAMP separates the regulatory and catalytic subunits, which allows each catalytic subunit to be active.

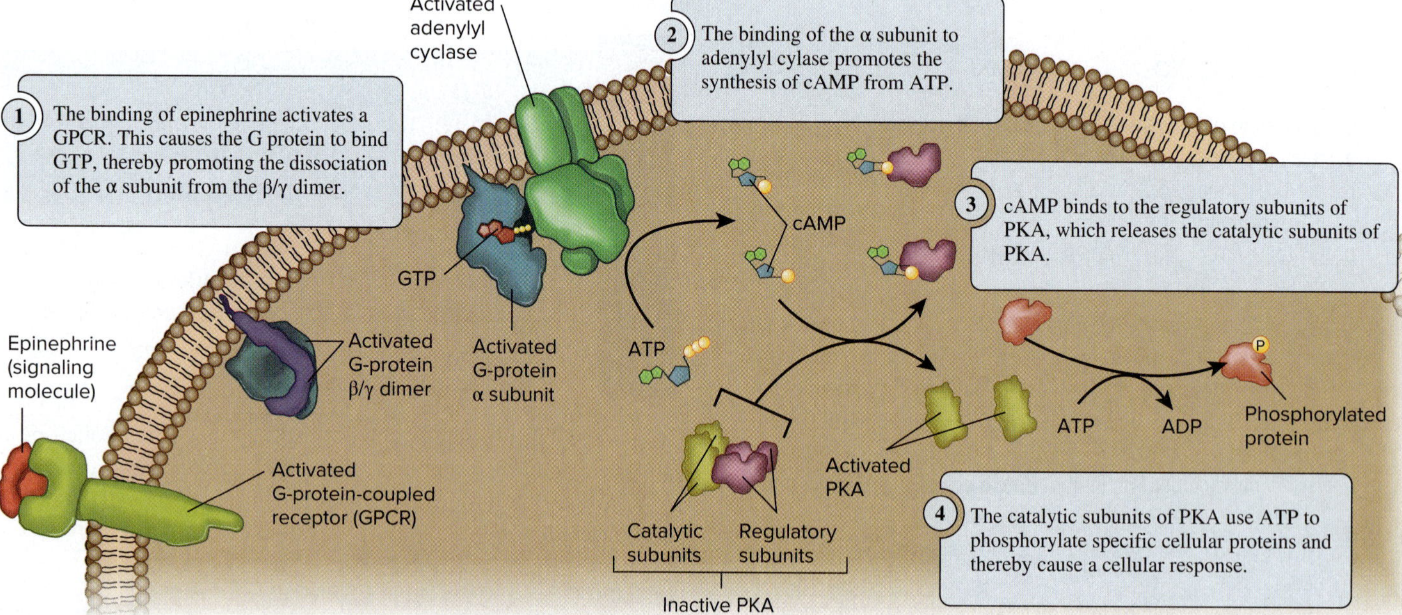


**Figure 8.13** **A signal transduction pathway involving cAMP.** The pathway leading to the formation of cAMP and subsequent activation of PKA, which is mediated by a G-protein-coupled receptor (GPCR).

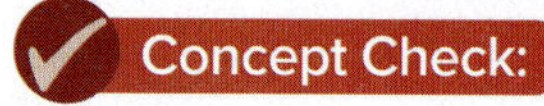


*In this figure, which part is the signal transduction pathway, and which is the cellular response?*

4. The catalytic subunit of PKA phosphorylates specific cellular proteins such as enzymes, structural proteins, and transcription factors, thus affecting cell structure and function.

As a specific example of a cellular response, Figure 8.14 shows how a skeletal muscle cell responds to elevated levels of epinephrine. When PKA becomes active, it phosphorylates two enzymes: phosphorylase kinase and glycogen synthase. Both of these enzymes are involved with the metabolism of glycogen, which is a polymer of glucose used to store energy. When phosphorylase kinase is phosphorylated, it becomes activated. The function of phosphorylase kinase is to phosphorylate another enzyme in the cell called glycogen phosphorylase, which then becomes activated. This enzyme causes glycogen breakdown by phosphorylating glucose units at the ends of a glycogen polymer, which releases individual glucose-phosphate molecules from glycogen:

$$\text{Glycogen}_n + P_i \xrightarrow{\text{Glycogen phosphorylase}} \text{Glycogen}_{n-1} + \text{Glucose-phosphate}$$

where $n$ is the number of glucose units in glycogen.

When PKA phosphorylates glycogen synthase, the function of this enzyme is inhibited rather than activated (see Figure 8.14). The function of glycogen synthase is to make glycogen. Therefore, the effect of cAMP is to prevent glycogen synthesis.

Taken together, the effects of epinephrine in skeletal muscle cells are to stimulate glycogen breakdown and inhibit glycogen synthesis. This provides these cells with more glucose molecules, which they can use for the energy needed for muscle contraction. In this way, the individual is better prepared to fight or flee.

**Reversal of the Cellular Response** When a signaling molecule is no longer produced and its level falls, a larger percentage of the receptors are not bound by their ligands. When a ligand dissociates from the GPCR, the GPCR becomes deactivated. Intracellularly, the α subunit hydrolyzes its GTP to GDP and $P_i$, and the α subunit and β/γ dimer reassociate to form an inactive G protein (see Figure 8.7, step 3). The level of cAMP decreases due to the action of an enzyme called **phosphodiesterase,** which converts cAMP to AMP (see Figure 8.12).

As the cAMP level falls, the regulatory subunits of PKA release cAMP, and the regulatory and catalytic subunits reassociate, thereby inhibiting PKA. Finally, enzymes called **protein phosphatases**

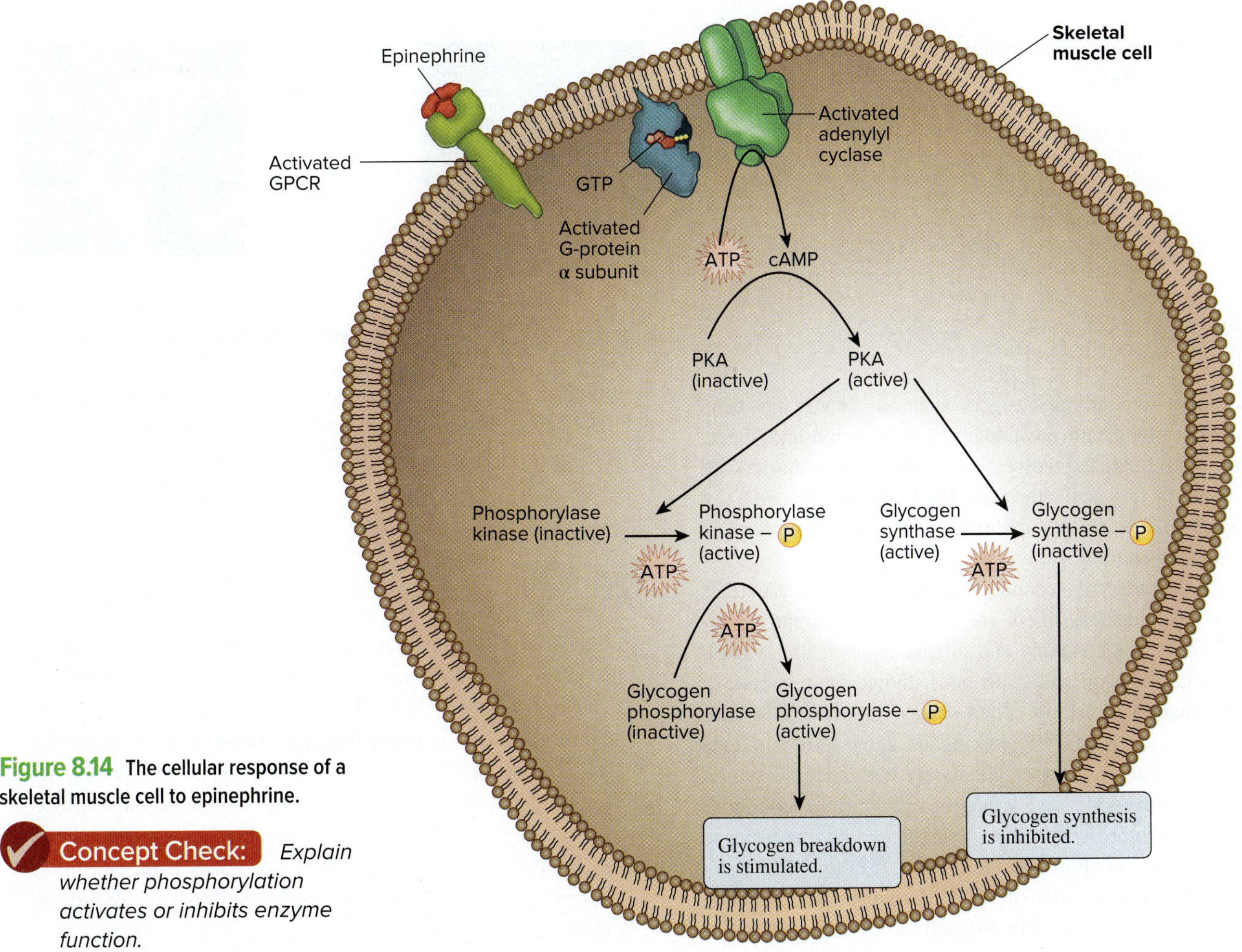


**Figure 8.14 The cellular response of a skeletal muscle cell to epinephrine.**

**Concept Check:** *Explain whether phosphorylation activates or inhibits enzyme function.*

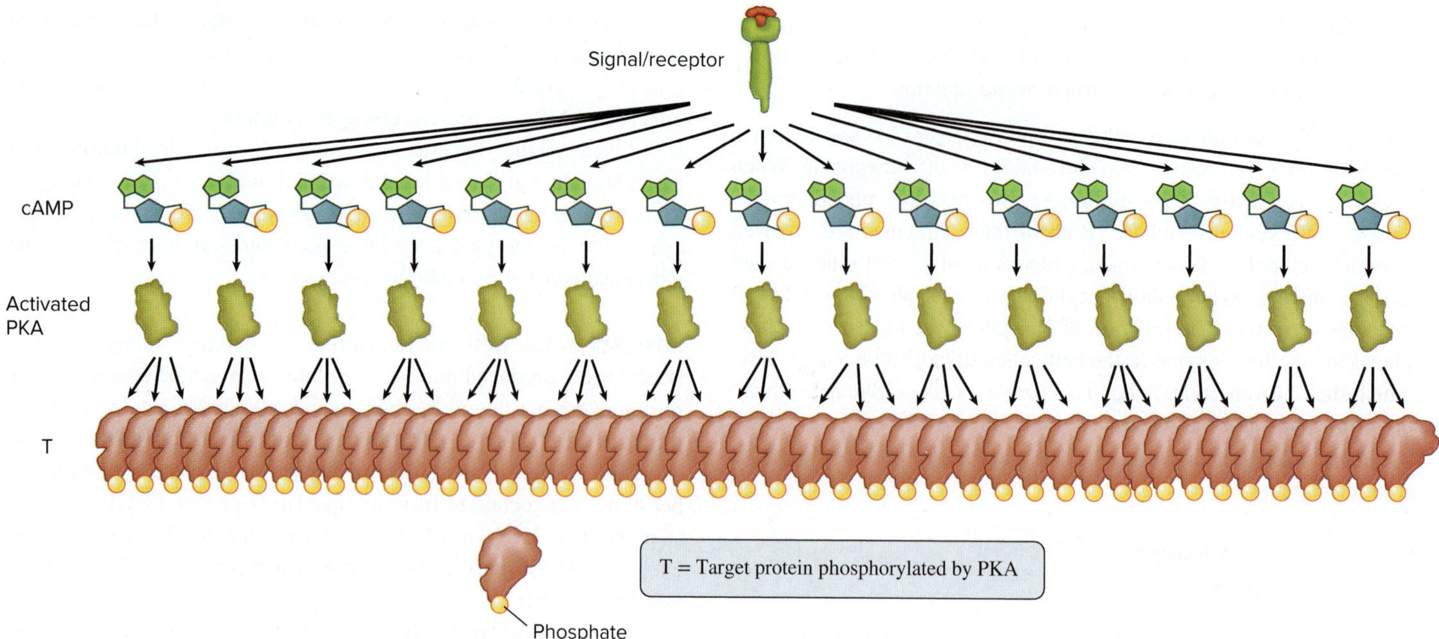


**Figure 8.15 Signal amplification.** An advantage of second messengers is the amplification of a signal. In this case, a single signaling molecule leads to the phosphorylation of many target proteins, perhaps hundreds or thousands.

*In the case of signaling molecules such as hormones, why is signal amplification an advantage?*

are responsible for removing phosphate groups from proteins, which reverses the effects of PKA:

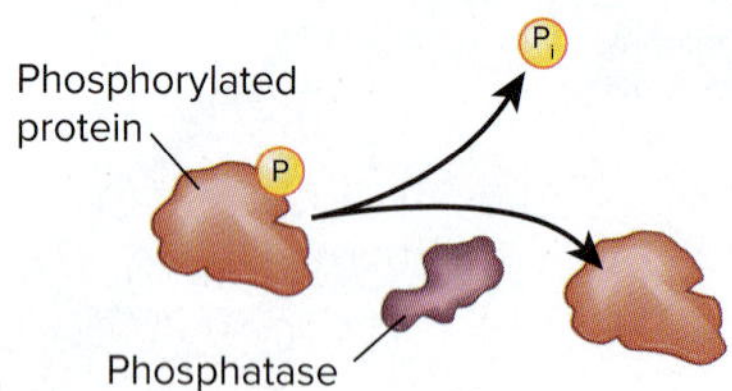


## The Main Advantages of Second Messengers Are Amplification and Speed

The production of second messengers such as cAMP has been found to have two important advantages: amplification and speed. Amplification of the signal involves the synthesis of many cAMP molecules, which in turn activate many PKA proteins (Figure 8.15). Likewise, each PKA protein phosphorylates many target proteins in the cell to promote a cellular response.

A second advantage of second messengers such as cAMP is speed. Because second messengers are relatively small and water-soluble, they can be made rapidly and diffuse through the cytosol. For example, Brian Bacskai and colleagues studied the response of neurons to a signaling molecule called serotonin, which is a neurotransmitter that binds to a GPCR. In humans, a low serotonin level is believed to play a role in depression, anxiety, and other behavioral disorders. To monitor cAMP levels, neurons grown in a laboratory were injected with a fluorescent protein that changes its fluorescence when cAMP is made. As shown in the drawing in Figure 8.16, such cells made a substantial amount of cAMP within 20 sec after the addition of serotonin, which quickly diffused through the cell.

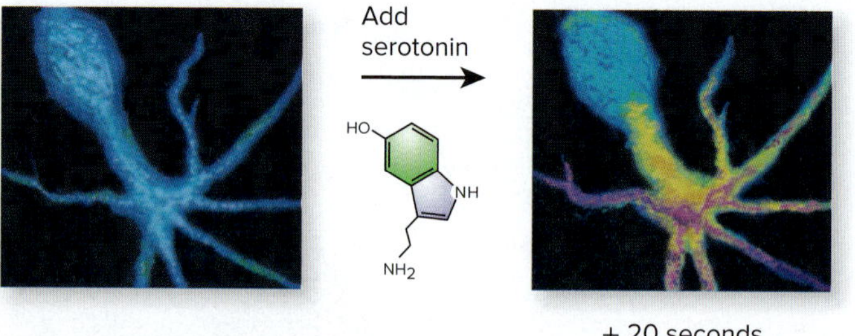


**Figure 8.16 The rapid speed of cAMP production.** The schematic drawing on the left shows a neuron prior to its exposure to serotonin, a signaling molecule; the drawing on the right shows the same cell 20 sec after exposure. Blue indicates a low level of cAMP, yellow is an intermediate level, and red/purple is a high level.

## 8.6 Reviewing the Concepts

- Second messengers, such as cAMP, play a key role in signal transduction pathways, such as those that occur via G-protein-coupled receptors (Figures 8.12, 8.13).
- An example of a pathway that uses cAMP is found in skeletal muscle cells responding to elevated levels of epinephrine, the fight-or-flight hormone. Epinephrine enhances the function of enzymes that increase glycogen breakdown and inhibits enzymes that cause glycogen synthesis (Figure 8.14).
- Second messenger pathways amplify the signal and occur with great speed (Figures 8.15, 8.16).

## 8.6 Testing Your Knowledge

**1.** The enzyme that is responsible for the synthesis of cAMP is
- **a.** the α subunit of a G protein.
- **b.** adenylyl cyclase.
- **c.** protein kinase A.
- **d.** phosphodiesterase.
- **e.** phosphatase.

**2.** The synthesis of cAMP can amplify the signal caused by a first messenger because
- **a.** cAMP diffuses rapidly through the cell.
- **b.** many cAMP molecules are made.
- **c.** many PKAs are activated, which phosphorylate many target proteins.
- **d.** all of the above.
- **e.** both b and c.

# Assess and Discuss

## Test Yourself

1. An agent that allows a cell to respond to changes in its environment is termed
   - a. a cell surface receptor.
   - b. an intracellular receptor.
   - c. a structural protein.
   - d. a signal.
   - e. a cellular response.
2. When a cell secretes signaling molecules that bind to receptors on neighboring cells as well as the cell itself, this is called ___________ signaling.
   - a. direct intercellular
   - b. contact-dependent
   - c. autocrine
   - d. paracrine
   - e. endocrine
3. Which of the following does *not* describe a typical cellular response to signaling molecules?
   - a. activation of enzymes within the cell
   - b. change in the function of structural proteins, which determine cell shape
   - c. alteration of levels of certain proteins in the cell by changing the level of gene expression
   - d. change in a gene sequence that encodes a particular protein
   - e. All of the above are examples of cellular responses.
4. A receptor has a $K_d$ for its ligand of 50 nM. This receptor
   - a. has a higher affinity for its ligand compared to a receptor with a $K_d$ of 100 nM.
   - b. has a higher affinity for its ligand compared to a receptor with a $K_d$ of 10 nM.
   - c. will be mostly bound by its ligand when the ligand concentration is 100 nM.
   - d. must be an intracellular receptor.
   - e. both a and c.
5. Which of the following is *not* an example of a cell surface receptor?
   - a. an enzyme-linked receptor
   - b. a G-protein-coupled receptor
   - c. a ligand-gated ion channel
   - d. the estrogen receptor
   - e. All of the above are cell surface receptors.
6. ___________ bind(s) to receptors inside cells.
   - a. Estrogen
   - b. Epinephrine
   - c. Epidermal growth factor
   - d. All of the above
   - e. None of the above
7. The EGF receptor functions as
   - a. a receptor tyrosine kinase.
   - b. a G-protein-coupled receptor.
   - c. a ligand-gated ion channel.
   - d. a transcription factor.
   - e. none of the above.
8. Which of the following events does *not* happen via the EGF receptor pathway?
   - a. The binding of EGF activates the protein kinase activity of the receptor.
   - b. After EGF binding, the receptor is autophosphorylated on serine side chains.
   - c. Relay proteins are activated after autophosphorylation.
   - d. Relay proteins turn on a protein kinase cascade.
   - e. The protein kinase cascade leads to the phosphorylation and activation of transcription factors.
9. Small molecules, such as cAMP, that relay signals within the cell are called
   - a. first messengers.
   - b. ligands.
   - c. G proteins.
   - d. second messengers.
   - e. transcription factors.
10. The benefit of second messengers in signal transduction pathways is
    - a. an increase in the speed of a cellular response.
    - b. duplication of the ligands in the system.
    - c. amplification of the signal.
    - d. all of the above.
    - e. a and c only.

## Conceptual Questions

1. What are the two general reasons that cells need to communicate?
2. What are the three stages of cell signaling? What stage does not occur when the estrogen receptor is activated?
3. **PRINCIPLES** A principle of biology is that living organisms interact with their environment. Discuss how cell signaling helps organisms to interact with their environment.

## Collaborative Questions

1. Discuss and compare several different types of cell-to-cell communication. What are some advantages and disadvantages of each type?
2. Discuss the advantages and disadvantages of second messengers.

# Online Resource

**connect.mheducation.com**

# UNIT III
# GENETICS

**Genetics** is the branch of biology that deals with **inheritance**—the transmission of characteristics from parents to offspring. We begin this unit by examining the structure of the genetic material, namely DNA, at the molecular and cellular levels. We will explore the structure and replication of DNA and examine how the DNA is packaged into chromosomes (Chapter 9). We will then consider how segments of DNA are organized into units called genes and explore how genes are used to make products such as RNA and proteins (Chapters 10 through 12). The expression of genes is largely responsible for the characteristics of living organisms. We will also examine how mutations can alter the properties of genes and even lead to diseases such as cancer (Chapter 13).

In Chapter 14, we turn our attention to the mechanisms of how genes are transmitted from parent to offspring. This topic begins with a discussion of how chromosomes are sorted and transmitted during cell division. Chapters 15 and 16 explore the relationships between the transmission of genes and the outcome of an offspring's traits. We will look at genetic patterns called Mendelian inheritance, named after Gregor Mendel, the 19th-century biologist who discovered them, as well as more complex patterns that could not have been predicted from Mendel's work.

Chapters 9 through 16 focus on the fundamental properties of the genetic material and heredity. The remaining chapters explore additional topics that are of importance to biologists. In Chapter 17, we will examine some of the unique genetic properties of bacteria and viruses. Chapter 18 describes genetic technologies that are used by researchers, clinicians, and biotechnologists to unlock the mysteries of genes and provide tools and applications that benefit humans, and explores the entire genomes of bacteria, archaea, and eukaryotes.

***The following biology principles will be emphasized in this unit:***

- ***The genetic material provides a blueprint for reproduction.*** *Throughout this unit, we will see how the genetic material carries the information for reproduction and to sustain life.*
- ***Structure determines function.*** *In Chapters 9 through 14, we will examine how the structure of DNA, RNA, genes, and chromosomes underlies their functions.*
- ***Living organisms interact with their environment.*** *In Chapters 15 and 16 we will explore the interactions between an organism's genes and its environment.*
- ***Biology affects our society.*** *In Chapter 18, we will examine genetic technologies that have many applications in our society.*
- ***Biology is an experimental science.*** *Most chapters in this unit have a Feature Investigation that describes a pivotal experiment that provided insights into our understanding of genetics.*

# The Information of Life: DNA and RNA Structure, DNA Replication, and Chromosome Structure

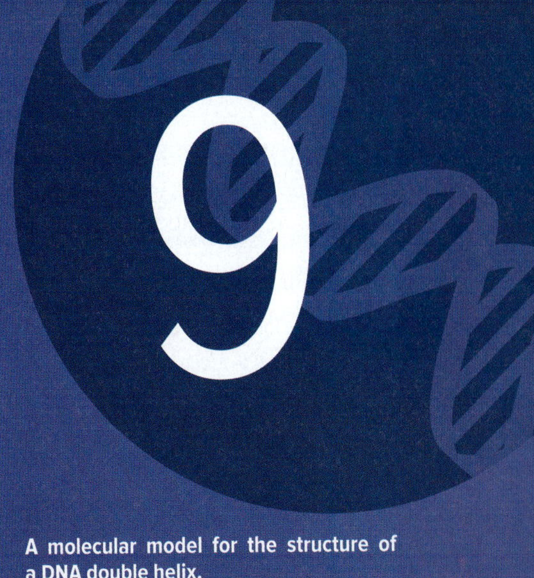


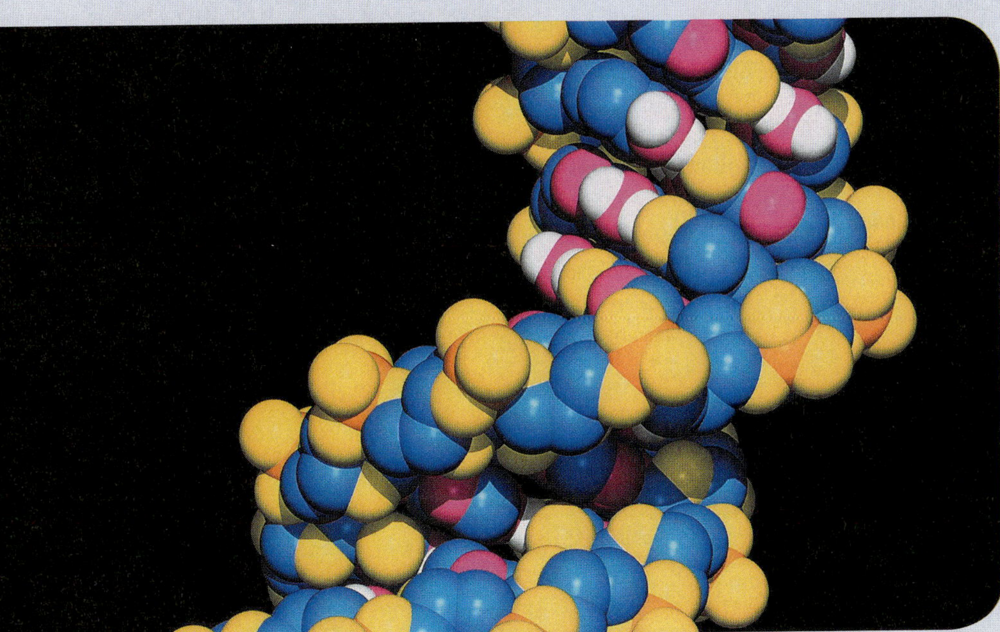

© Prof. Kenneth Seddon & Dr. Timothy Evans, Queen's Univ. Belfast/SPL/Science Source

A molecular model for the structure of a DNA double helix.

## Chapter Outline



It's an age-old and messy problem: dog owners who do not clean up after their pets. It can be particularly annoying to people living in the city, where dogs are taken for a walk in public areas. Recently, some owners of apartment buildings have taken action. They are relying on DNA in the unscooped feces to identify the dog and its negligent owner. Here's how it works. Dog owners from certain apartment buildings in New York City must have the DNA from their dogs analyzed. Dogs' cheeks are swabbed for DNA and sent to a company that analyzes the DNA and keeps the information in a DNA registry. Small samples from unscooped feces are sent to the same company and analyzed. Because the DNA of each dog is unique, the company tries to find a match between a sample and a dog in the DNA registry. If a match is found, the culprit is identified!

**Deoxyribonucleic acid,** or **DNA,** is the genetic material that provides the blueprint to produce an organism's traits. Researchers have determined that all living organisms, including bacteria, archaea, protists, fungi, plants, and animals, use DNA as their genetic material. Viruses are not considered to be living organisms, because they do not fulfill all of the criteria described in Chapter 1 (refer back to Figure 1.3a–h). Some viruses use DNA as their genetic material, while others use RNA.

We begin our survey of genetics by examining DNA at the molecular level. Once we understand how DNA works at this level, it becomes easier to see how the function of DNA controls the properties of cells and ultimately the characteristics of unicellular and multicellular organisms.

To a large extent, our understanding of genetics comes from our knowledge of the molecular structure of DNA. In this chapter, we will begin by considering some classic experiments that provided evidence that DNA is the genetic material. We will then survey the molecular features of DNA, which allows us to appreciate how DNA can store information and be accurately copied. We will explore the details of how DNA strands are used as templates for the synthesis of new DNA strands. We will also consider the components of ribonucleic acid (RNA), which show striking similarities to those of DNA. Finally, we will examine the molecular composition of chromosomes, where DNA is found.

## 9.1 Properties and Identification of the Genetic Material

### Learning Outcomes

1. List the four key criteria that the genetic material must fulfill.
2. **SCISKILLS ▶** Analyze the results of the experiments that identified DNA as the genetic material.

DNA carries the genetic instructions for the characteristics of all living organisms. In the case of multicellular organisms such as plants and animals, the information stored in the genetic material enables a fertilized egg to develop into an embryo and eventually into an adult organism. To fulfill its role, the genetic material must meet the following key criteria:

1. **Information:** The genetic material must contain the information necessary to construct an entire organism.
2. **Replication:** The genetic material must be accurately copied.
3. **Transmission:** After it is replicated, the genetic material can be passed from parent to offspring. It also must be passed from cell to cell during the process of cell division.
4. **Variation:** Differences in the genetic material must account for the known variation within each species and among different species.

How was the genetic material discovered? The quest to identify the genetic material began in the late 1800s, when a few scientists postulated that living organisms possess a blueprint that has a biochemical basis. In 1883, August Weismann and Karl Nägeli championed the idea that a chemical substance exists within living cells that is responsible for the transmission of traits from parents to offspring. During the next 30 years, experimentation along these lines centered on the behavior of **chromosomes,** the cellular structures that we now know contain the genetic material. Taken literally, chromosome comes from the Greek words *chromo* and *soma*, meaning colored body, which refers to the observation of early microscopists that chromosomes are easily stained by colored dyes. By studying the transmission patterns of chromosomes from cell to cell and from parent to offspring, researchers were convinced that chromosomes carry the determinants that control the outcome of traits. The two main components of chromosomes are proteins and DNA. As we will examine in this section, experimentation revealed that the DNA component is the genetic material of living organisms.

### Griffith's Bacterial Transformation Experiments Indicated the Existence of a Genetic Material

Studies in microbiology were important in developing an experimental strategy to identify the genetic material. In the late 1920s, an English microbiologist, Frederick Griffith, studied a type of bacterium known then as pneumococci and now classified as *Streptococcus pneumoniae*. Some strains of *S. pneumoniae* secrete a polysaccharide capsule, and other strains do not. When streaked on petri plates containing solid growth media, capsule-secreting strains have a smooth colony morphology. Those strains unable to secrete a capsule have a colony morphology that looks rough. In mammals, smooth strains of *S. pneumoniae* may cause pneumonia and other symptoms. In mice, such infections are usually fatal because the capsule prevents the mouse's immune system from killing the bacteria. By comparison, rough strains are not pathogenic because the immune system can destroy them.

As shown in Figure 9.1, Griffith injected live and/or heat-killed bacteria into mice and then observed whether or not the bacteria caused them to die. He investigated the effects of two strains of *S. pneumoniae:* type S (for smooth) and type R (for rough).

- **Step 1:** When injected into a live mouse, the type S strain killed the mouse. The capsule made by a type S strain prevents the mouse's immune system from killing the bacterial cells. Following the death of the mouse, many type S bacteria were found in the mouse's blood.
- **Step 2:** When type R bacteria were injected into a mouse, the mouse survived, and after several days, living bacteria were not found in the live mouse's blood because the mouse's immune system was able to destroy bacteria that had no capsule.

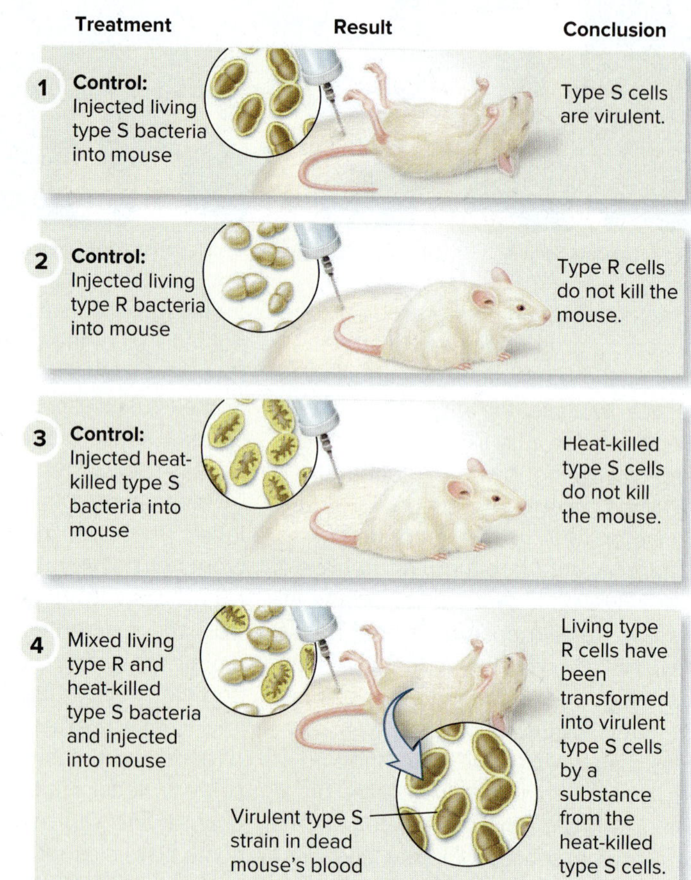

Treatment
Result
Conclusion
1
Control:
Injected living type S bacteria into mouse
Type S cells are virulent.
2
Control:
Injected living type R bacteria into mouse
Type R cells do not kill the mouse.
3
Control:
Injected heat-killed type S bacteria into mouse
Heat-killed type S cells do not kill the mouse.
4
Mixed living type R and heat-killed type S bacteria and injected into mouse
Virulent type S strain in dead mouse's blood
Living type R cells have been transformed into virulent type S cells by a substance from the heat-killed type S cells.


**Figure 9.1 Griffith's experiments showing that genetic material can be transferred from one bacterium to another.** Note: To determine if a mouse's blood contained live bacteria, a sample of blood was applied to solid growth media to see if smooth or rough bacterial colonies would form.

**BioConnections:** *Look ahead to Figure 17.10. How does bacterial transformation play a role in the transfer of genes, such as antibiotic-resistance genes, from one bacterial species to another?*

- **Step 3:** In a follow-up to these results, Griffith also heat-killed the smooth bacteria and then injected them into a mouse. As expected, the mouse survived.
- **Step 4:** A surprising result occurred when Griffith mixed live type R bacteria with heat-killed type S bacteria and then injected them into a mouse—the mouse died. The blood from the dead mouse contained living type S bacteria! How did Griffith explain these results? He postulated that a substance from dead type S bacteria transformed the type R bacteria into type S bacteria. Griffith called this process **transformation,** and he termed the unidentified material responsible for this phenomenon the "transformation principle."

Let's consider what these observations mean with regard to the four criteria for the genetic material: information, replication, transmission, and variation. According to Griffith's results, the transformed bacteria had acquired the information (criterion 1) to make a capsule from the heat-killed cells. For the transformed bacteria to proliferate and thereby kill the mouse, the substance conferring the ability to make a capsule must be replicated (criterion 2) and then transmitted (criterion 3) from mother to daughter cells during cell division. Finally, Griffith already knew that variation (criterion 4) existed in the ability of his strains to produce a capsule (S strain) or not produce a capsule (R strain). The experiment of Figure 9.1, step 4, was consistent with the idea that some genetic material from the heat-killed type S bacteria had been transferred to the living type R bacteria and provided those bacteria with a new trait. At the time of his studies, however, Griffith could not determine the biochemical composition of the transforming substance.

## FEATURE INVESTIGATION

### Avery, MacLeod, and McCarty Used Purification Methods to Reveal That DNA Is the Genetic Material

In the 1940s, American physician Oswald Avery and American biologists Colin MacLeod and Maclyn McCarty were also interested in the process of bacterial transformation. During the course of their studies, they realized that Griffith's observations could be used as part of an experimental strategy to biochemically identify the genetic material. They asked the question "What substance is being transferred from the dead type S bacteria to the live type R bacteria?"

To answer this question, Avery, MacLeod, and McCarty needed to purify the general categories of substances found in living cells. They used established biochemical procedures to purify classes of macromolecules, such as proteins, DNA, and RNA, from the type S streptococcal strain. Initially, they discovered that only the purified DNA could convert type R bacteria into type S. To further verify that DNA is the genetic material, they performed the investigation outlined in Figure 9.2. They purified DNA from the type S bacteria and

**Figure 9.2** The Avery, MacLeod, and McCarty experiments that identified DNA as Griffith's transformation principle—the genetic material.

**HYPOTHESIS** A purified macromolecule from type S bacteria, which functions as the genetic material, will be able to convert type R bacteria into type S.

**KEY MATERIALS** Type R and type S strains of *Streptococcus pneumoniae*.

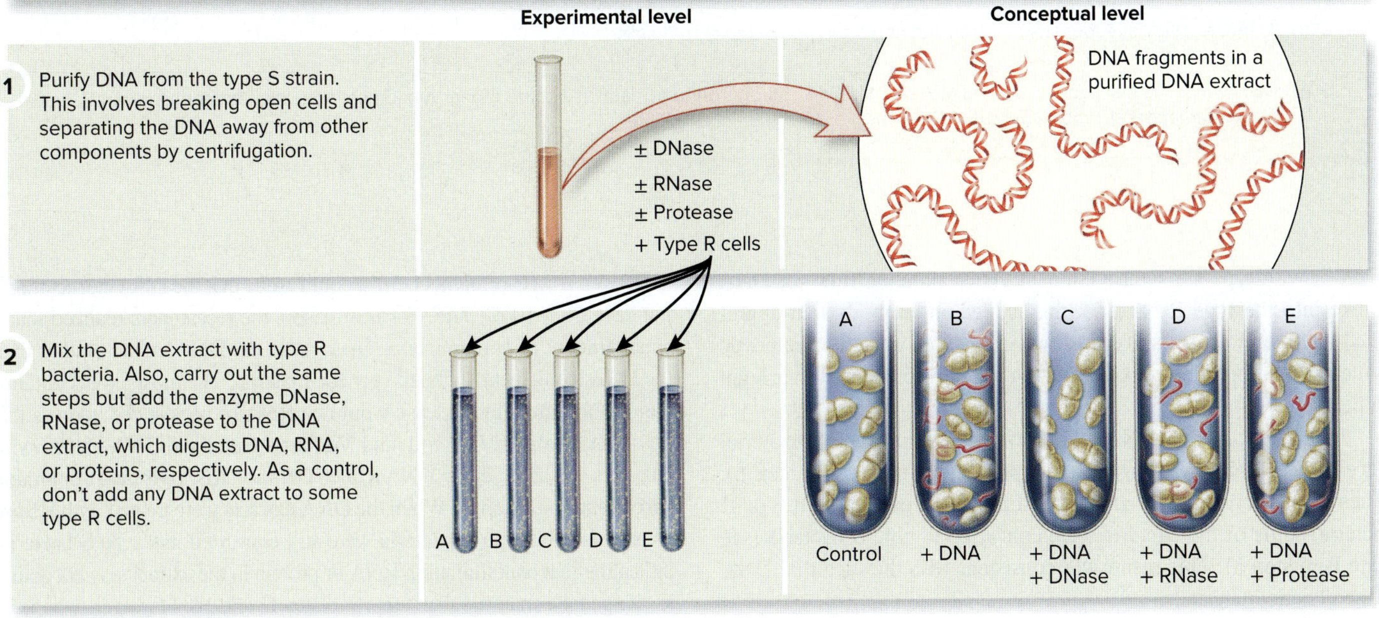

| | Experimental level | Conceptual level |
|---|---|---|
| 1 Purify DNA from the type S strain. This involves breaking open cells and separating the DNA away from other components by centrifugation. | ± DNase<br>± RNase<br>± Protease<br>+ Type R cells | DNA fragments in a purified DNA extract |
| 2 Mix the DNA extract with type R bacteria. Also, carry out the same steps but add the enzyme DNase, RNase, or protease to the DNA extract, which digests DNA, RNA, or proteins, respectively. As a control, don't add any DNA extract to some type R cells. | A B C D E | A: Control; B: + DNA; C: + DNA + DNase; D: + DNA + RNase; E: + DNA + Protease |

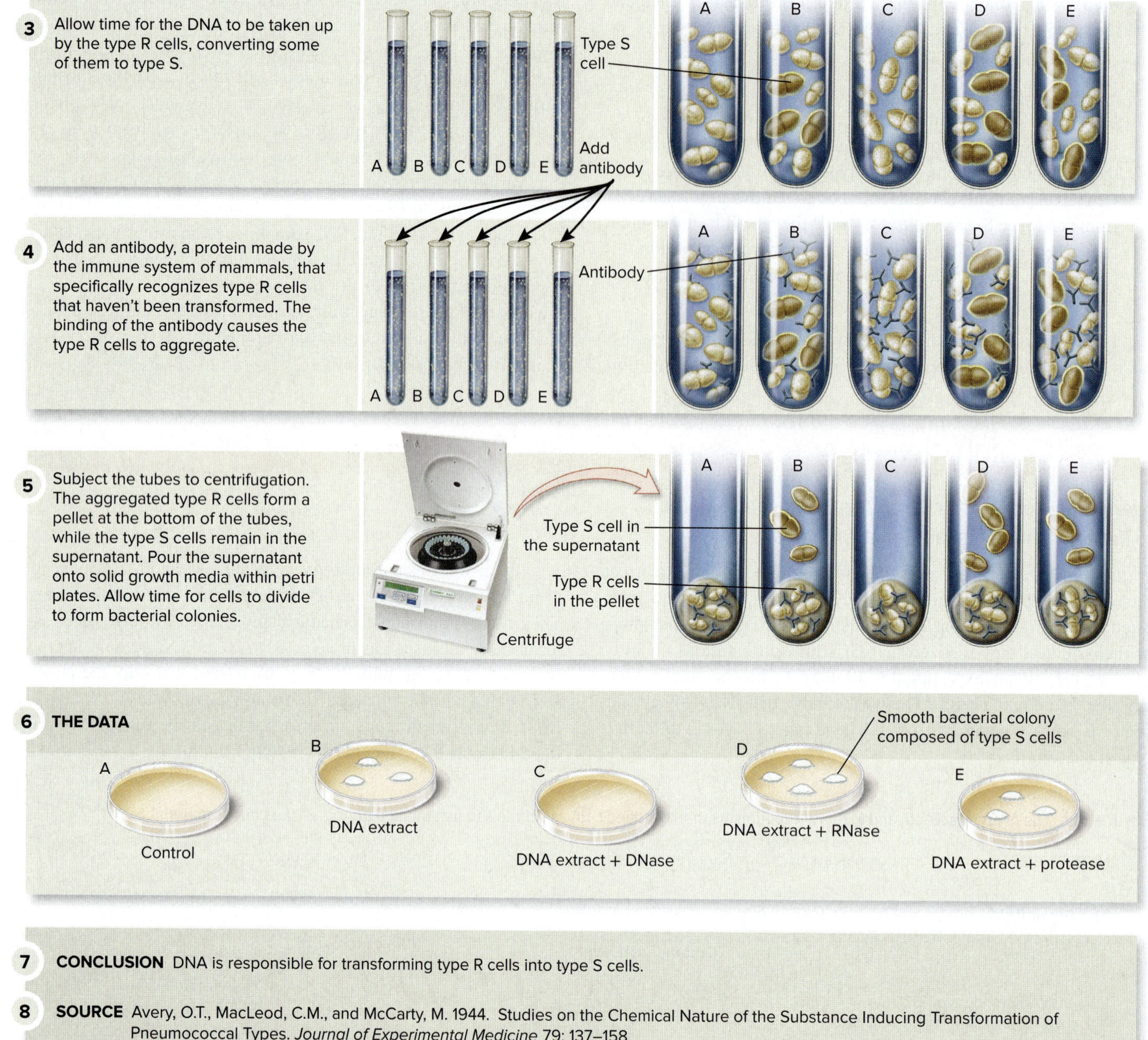

3 Allow time for the DNA to be taken up by the type R cells, converting some of them to type S.
A B C D E
Type S cell
Add antibody
A B C D E
4 Add an antibody, a protein made by the immune system of mammals, that specifically recognizes type R cells that haven't been transformed. The binding of the antibody causes the type R cells to aggregate.
Antibody
5 Subject the tubes to centrifugation. The aggregated type R cells form a pellet at the bottom of the tubes, while the type S cells remain in the supernatant. Pour the supernatant onto solid growth media within petri plates. Allow time for cells to divide to form bacterial colonies.
Type S cell in the supernatant
Type R cells in the pellet
Centrifuge
6 THE DATA
A
Control
B
DNA extract
C
DNA extract + DNase
D
DNA extract + RNase
Smooth bacterial colony composed of type S cells
E
DNA extract + protease


7 **CONCLUSION** DNA is responsible for transforming type R cells into type S cells.

8 **SOURCE** Avery, O.T., MacLeod, C.M., and McCarty, M. 1944. Studies on the Chemical Nature of the Substance Inducing Transformation of Pneumococcal Types. *Journal of Experimental Medicine* 79: 137–158.

mixed it with type R bacteria (step 1). After allowing time for DNA uptake into the type R bacteria, they added an antibody that aggregated any nontransformed type R bacteria, which were then removed by centrifugation. The remaining bacteria were incubated overnight on petri plates.

As a control, if no DNA extract was added, no type S bacterial colonies were observed on the petri plates (see plate A in step 6). When the researchers mixed their S strain DNA extract with type R bacteria, some of the bacteria were converted to type S bacteria (see plate B in step 6). This result was consistent with the idea that DNA is the genetic material.

Even so, a careful biochemist could argue that the DNA extract might not have been 100% pure. For this reason, the researchers realized that a small amount of contaminating material in the DNA extract, rather than the DNA, could actually be the genetic material. The most likely contaminating substances in this case would be RNA or protein. To address this possibility, Avery, MacLeod, and McCarty treated the DNA extract with enzymes that digest either DNA (called **DNase**), RNA (**RNase**), or protein (**protease**) (see step 2). When the DNA extracts were treated with RNase or protease, the type R bacteria were still converted into type S bacteria, indicating that contaminating RNA or protein in the extract was not acting as the genetic material (see step 6, plates D and E). However, when the

DNA extract was treated with DNase, it lost the ability to convert type R bacteria into type S bacteria (see plate C). Taken together, these results were consistent with the idea that DNA is the genetic material.

*Experimental Questions*

1. **SCISKILLS ▸** Avery, MacLeod, and McCarty worked with two strains of *Streptococcus pneumoniae* to determine the biochemical identity of the genetic material. Explain the characteristics of the *S. pneumoniae* strains that made them particularly well suited for such an experiment.
2. What is a DNA extract?
3. **SCISKILLS ▸** In the experiment of Avery, MacLeod, and McCarty, what was the purpose of using the DNase, RNase, and protease if only the DNA extract caused transformation?

## 9.1 Reviewing the Concepts

- The genetic material carries information to produce the traits of organisms. It is accurately replicated and transmitted from cell to cell and parent to offspring. The genetic material carries differences that explain the variation among different organisms.
- Griffith's work with type S and type R bacteria indicated the existence of a genetic material, which he called the transformation principle (Figure 9.1).
- Avery, MacLeod, and McCarty used biochemical methods to show that DNA is the genetic material (Figure 9.2).
- All living organisms use DNA as the genetic material but some viruses use RNA.

## 9.1 Testing Your Knowledge

1. In the work of Griffith, the mixture of heat-killed S strain bacteria and living R strain bacteria resulted in the death of the mouse. The correct interpretation of these results is that
   a. the S strain bacteria came back to life.
   b. genetic material was transferred from the heat-killed S strain to the R strain.
   c. heat-killed S strain bacteria can kill mice.
   d. R strain bacteria can kill mice.

## 9.2 Nucleic Acid Structure

### Learning Outcomes

1. Outline the structural features of DNA at five levels of complexity.
2. Describe the structure of nucleotides, a DNA strand, and the DNA double helix.

DNA and its molecular cousin, RNA, are known as **nucleic acids,** polymers consisting of nucleotides, which are responsible for the storage, expression, and transmission of genetic information. This term is derived from the discovery of DNA by Friedrich Miescher in 1869. He identified a novel phosphorus-containing substance from the nuclei of white blood cells found in waste surgical bandages. He named this substance nuclein. As the structure of DNA and RNA became better understood, they were found to be acidic molecules, which means they release hydrogen ions ($H^+$) in solution and have a net negative charge at neutral pH. Thus, the name nucleic acid was coined.

DNA is a very large macromolecule composed of smaller building blocks. As shown in **Figure 9.3**, we can consider the structural features of DNA at different levels of complexity:

1. **Nucleotides** are the building blocks of DNA.
2. A **strand** of DNA is formed by the covalent linkage of nucleotides in a linear manner.

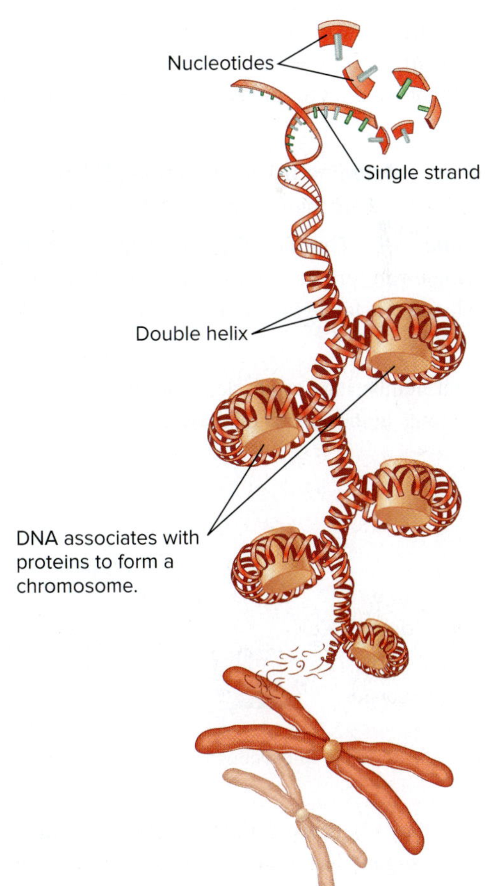


**Figure 9.3** Levels of DNA structure to create a chromosome.

3. Two strands of DNA hydrogen-bond with each other to form a **double helix**—two DNA strands that are twisted together to form a structure that resembles a spiral staircase, or helix.
4. In living cells, DNA is associated with an array of different proteins to form chromosomes. The association of proteins with DNA organizes the long strands into a compact structure.
5. A **genome** is the complete complement of an organism's genetic material. For example, the genome of most bacteria is a single circular chromosome, whereas eukaryotic cells have DNA in their nucleus, mitochondria, and chloroplasts.

The first two levels of complexity will be the focus of this section. Level 3 will be explored in Section 9.3, level 4 will be discussed in Section 9.6, and level 5 is examined in Chapter 18.

## Nucleotides Contain a Phosphate, a Sugar, and a Base

A nucleotide has three components: a phosphate group, a pentose (five-carbon) sugar, and a nitrogen-containing base. The base and phosphate group are attached to the sugar molecule (Figure 9.4). The nucleotides in DNA and RNA contain different sugars:

- **Deoxyribose** is found in DNA.
- **Ribose** is found in RNA.

Five different bases are found in nucleotides, although any given nucleotide contains only one base. The five bases are subdivided into two categories due to differences in their structures (see Figure 9.4):

- **Purine** bases: **Adenine (A)** and **guanine (G)** have a double-ring structure and are found in both DNA and RNA.
- **Pyrimidine** bases: **Thymine (T), uracil (U),** and **cytosine (C)** have a single-ring structure. Cytosine is found in both DNA and RNA. Thymine is found only in DNA, whereas uracil is found only in RNA.

A conventional numbering system describes the locations of carbon and nitrogen atoms in the sugars and bases:

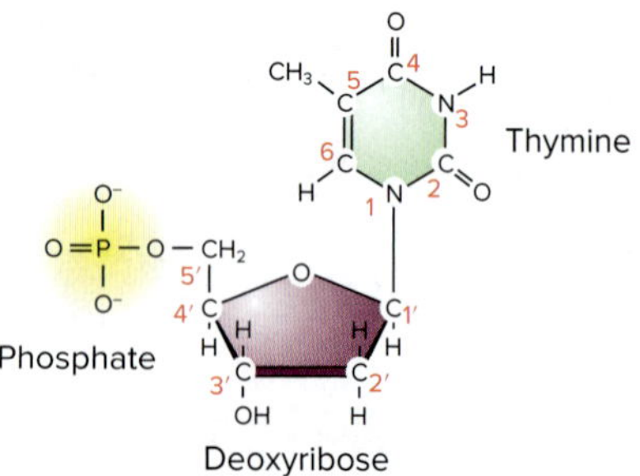


The prime symbol (′) is used to distinguish the numbering of carbons in the sugar. The atoms in the ring structures of the bases are not given the prime designation. The sugar carbons are designated 1′ (that is, "one prime"), 2′, 3′, 4′, and 5′, with the

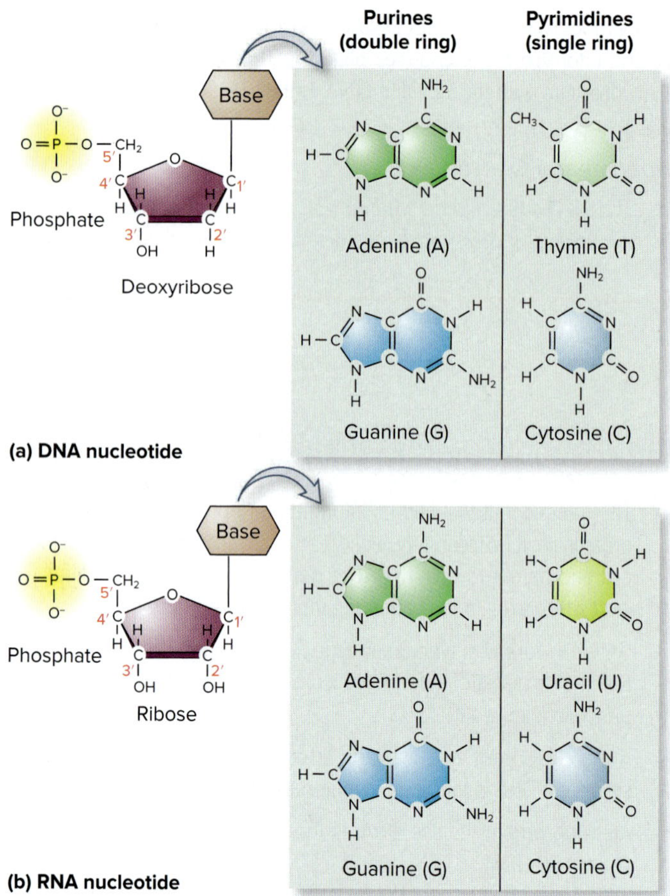


**Figure 9.4** Nucleotides and their components.

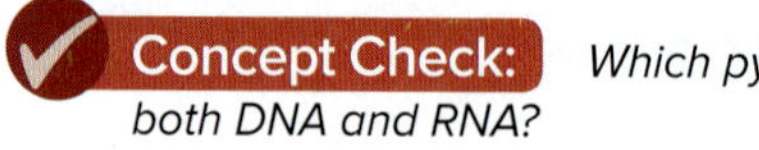
**Concept Check:** *Which pyrimidine(s) is (are) found in both DNA and RNA?*

carbon atoms numbered in a clockwise direction starting with the carbon atom to the right of the ring oxygen atom. The fifth carbon is outside the ring. A base is attached to the 1′ carbon atom, and a phosphate group is attached at the 5′ position. Compared with ribose, deoxyribose lacks a single oxygen atom at the 2′ position; the prefix deoxy- (meaning without oxygen) refers to this missing atom.

## A Strand Is a Linear Linkage of Nucleotides with Directionality

The next level of nucleotide structure is the formation of a strand of DNA or RNA in which nucleotides are covalently attached to each other in a linear fashion. Figure 9.5 depicts a short strand of DNA with four nucleotides. The key structural features of a DNA strand are as follows:

- Nucleotides are linked together by covalent bonds—called phosphoester bonds—between phosphorus and oxygen. A sugar molecule in one nucleotide is attached to a phosphate group in the next nucleotide. Another way of viewing this

## Biology Principle

### The Genetic Material Provides a Blueprint for Reproduction

The covalent linkage of a sequence of bases allows DNA to store information.

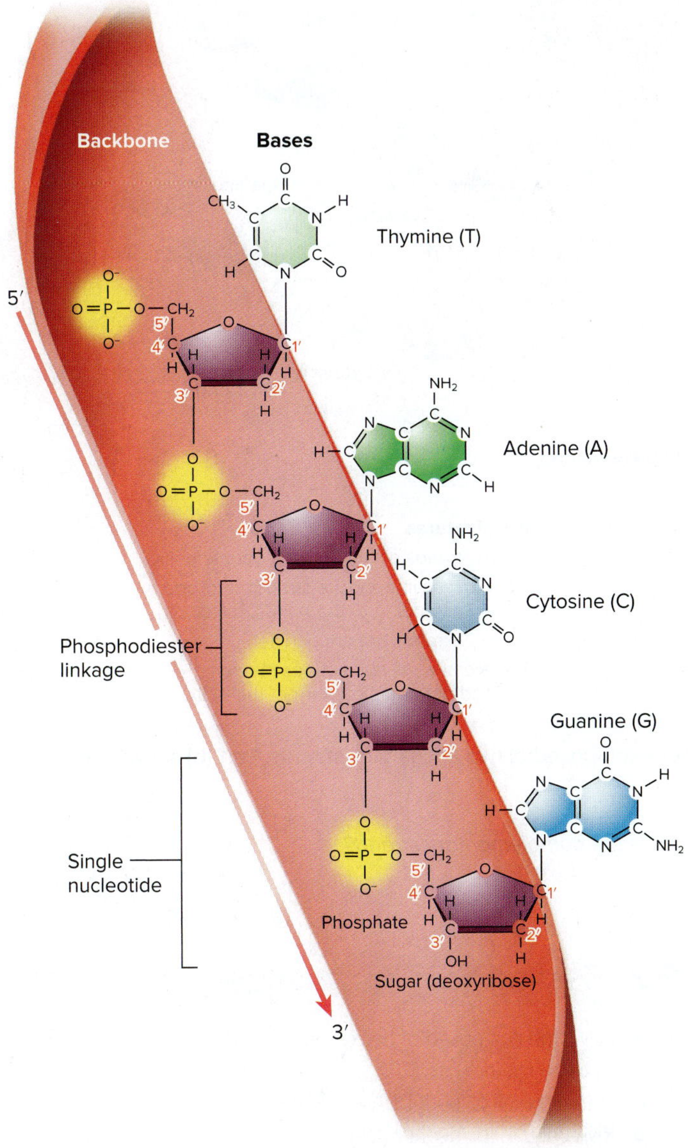


**Figure 9.5 The structure of a DNA strand.** Nucleotides are covalently bonded to each other in a linear manner. Notice the directionality of the strand and that it carries a particular sequence of bases. An RNA strand has a very similar structure, except the sugar is ribose rather than deoxyribose and uracil is substituted for thymine.

**Concept Check:** *What is the difference between a phosphoester bond and a phosphodiester linkage?*

linkage is to notice that a phosphate group connects two sugar molecules. From this perspective, the linkage in DNA strands is called a **phosphodiester linkage,** which has two phosphoester bonds.

- The phosphates and sugar molecules form the **backbone** of a DNA or RNA strand, and the bases (thymine, adenine, cytosine, and guanine) project from the backbone.
- A strand has a directionality based on the orientation of the sugar molecules within that strand. In Figure 9.5, the direction of the strand is said to be 5′ to 3′ when going from top to bottom. The 5′ end of a DNA strand has a phosphate group, and the 3′ end has an —OH group. This strand is abbreviated 5′–TACG–3′ to indicate its directionality.

## DNA Has a Repeating, Antiparallel Helical Structure Formed by the Complementary Base Pairing of Nucleotides

The DNA double helix has several distinguishing features (Figure 9.6a):

- DNA is a double-stranded structure with the sugar-phosphate backbone on the outside and the bases on the inside. It is a helical structure, which means it is cylindrically spiral.
- The double helix is stabilized by hydrogen bonding between the bases in opposite strands to form **base pairs.** Because hydrogen bonds are relatively weak, they can be easily broken and rejoined.
- Base pairing is specific. An adenine (A) in one strand forms two hydrogen bonds with a thymine (T) in the opposite strand, or a guanine (G) forms three hydrogen bonds with a cytosine (C) (Figure 9.6b). This specificity is called the **AT/GC rule.**
- Due to the AT/GC rule, the base sequences of two DNA strands are complementary to each other. For example, if one strand has the sequence of 5′–GCGGATTT–3′, the opposite strand must be 3′–CGCCTAAA–5′.
- One complete turn of the double helix is 3.4 nm in length and comprises about 10 base pairs.
- With regard to their 5′ and 3′ directionality, the two strands of a DNA double helix are **antiparallel.** If you look at Figure 9.6, one strand runs in the 5′ to 3′ direction from top to bottom, whereas the other strand is oriented 3′ to 5′ from top to bottom.
- As shown in Figure 9.7, two grooves, called the major groove and the minor groove, spiral around the double helix. This figure shows a space-filling model in which the atoms are depicted as spheres. It emphasizes the surface of DNA. The sugar-phosphate backbone is on the outermost surface of the double helix. The indentations where the atoms of the bases make contact with the surrounding water are termed grooves. The **major groove** occurs where the DNA backbones of the two strands are farther apart, whereas the **minor groove** is where they are closer together.

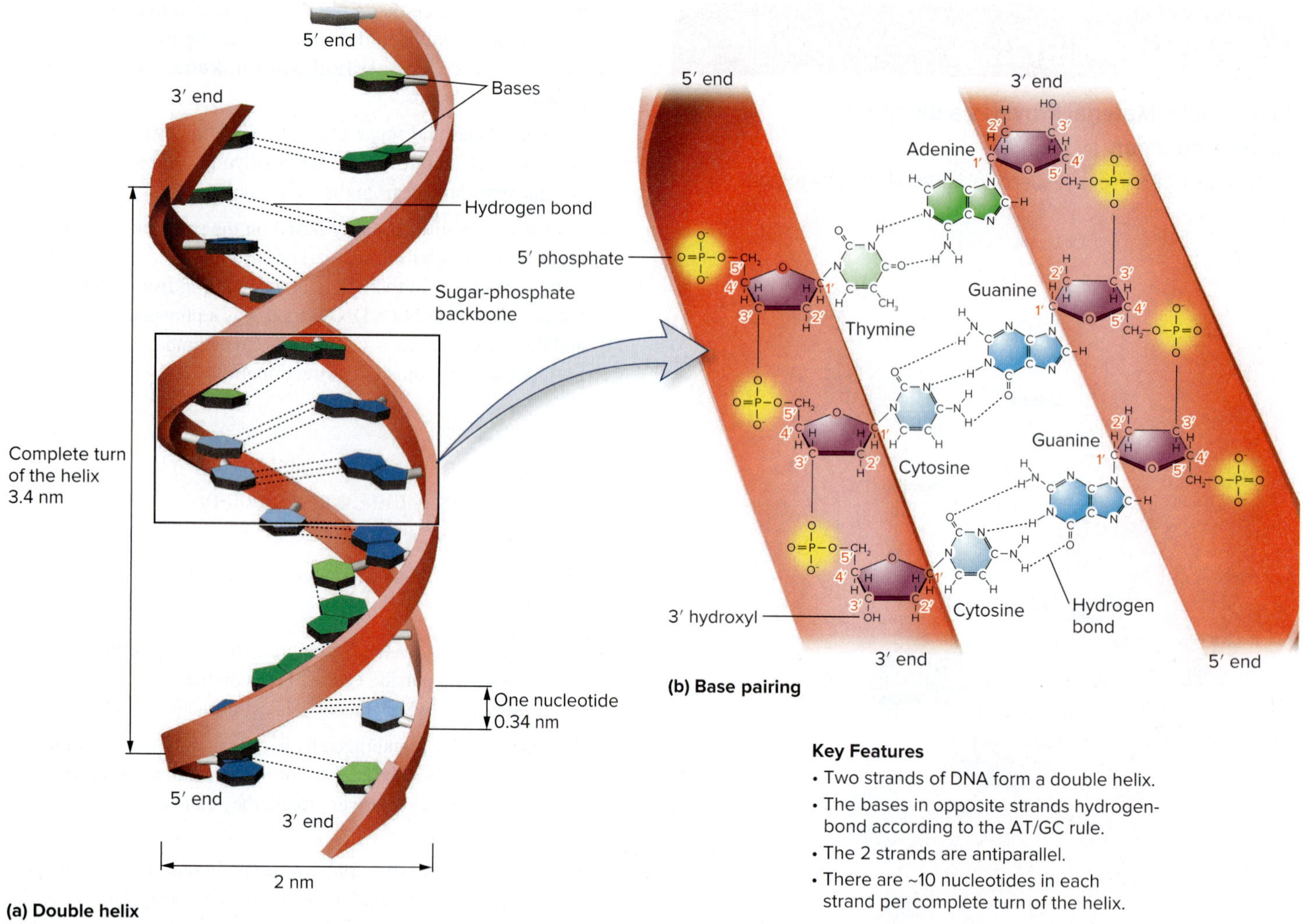

(a) Double helix

(b) Base pairing

**Key Features**

- Two strands of DNA form a double helix.
- The bases in opposite strands hydrogen-bond according to the AT/GC rule.
- The 2 strands are antiparallel.
- There are ~10 nucleotides in each strand per complete turn of the helix.

**Figure 9.6 Structure of the DNA double helix.** As seen in part **(a)**, DNA is a helix composed of two antiparallel strands. Part **(b)** shows the AT/GC base pairing that holds the strands together via hydrogen bonds.

**Concept Check:** *If one DNA strand is 5′–GATTCGTTC–3′, what is the complementary strand?*

## 9.2 Reviewing the Concepts

- DNA is composed of nucleotides, which are covalently linked to form DNA strands. Two DNA strands are held together by hydrogen bonds between the bases to form a double helix. DNA associates with various proteins to form a chromosome (Figure 9.3).
- Nucleotides are composed of a phosphate, a sugar, and a base. The sugar can be deoxyribose (DNA) or ribose (RNA). The purine bases are adenine and guanine, and the pyrimidine bases are thymine (DNA only), uracil (RNA only), and cytosine. The atoms in a nucleotide are numbered in a conventional way (Figure 9.4).
- A strand of DNA (or RNA) has a directionality due to the orientation of the sugar molecules (Figure 9.5).
- DNA is a double helix in which the DNA strands are antiparallel and obey the AT/GC rule (Figures 9.6, 9.7).

## 9.2 Testing Your Knowledge

1. A nucleotide composed of deoxyribose, phosphate, and adenine could be found in
   a. DNA only. c. both DNA and RNA.
   b. RNA only. d. neither DNA nor RNA.
2. The terms 5′ and 3′ refer to the directionality of a DNA strand based on
   a. the orientation of phosphate molecules in the backbone.
   b. the orientation of sugar molecules in the backbone.
   c. the direction in which the bases project from the backbone.
   d. the orientation of base pairing.
3. The strands of a DNA double helix are held together by base pairing between
   a. A to G and C to T. c. A to T and G to C.
   b. A to C and G to T. d. A to A, T to T, G to G, and C to C.

## Biology Principle

### Structure Determines Function

The major groove provides a binding site for proteins that control the expression of genes.

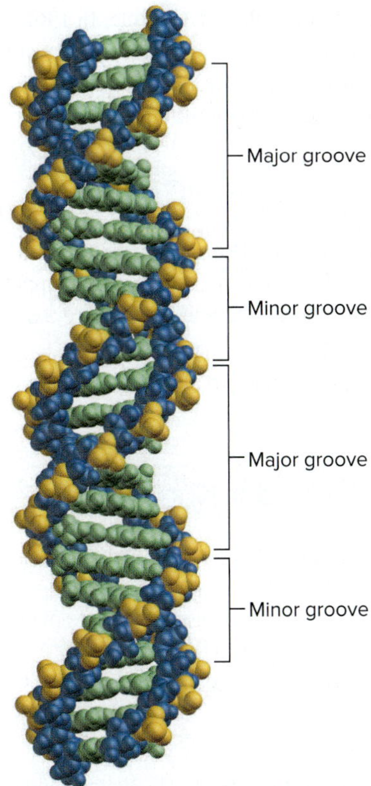


**Figure 9.7 A space-filling model of the DNA double helix.** In the sugar-phosphate backbone, sugar molecules are shown in blue, and phosphate groups are yellow. The backbone is on the outermost surface of the double helix. The atoms of the bases, shown in green, are more internally located within the double-stranded structure. Notice the major and minor grooves that are formed by this arrangement.

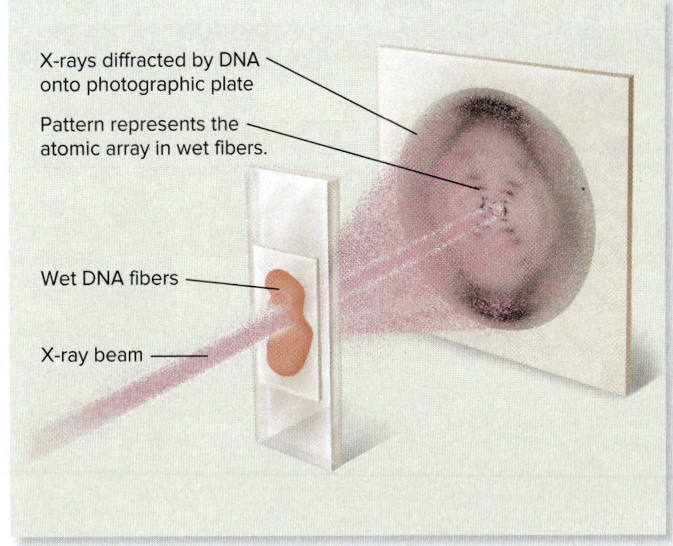


**Figure 9.8 Rosalind Franklin's X-ray diffraction of DNA fibers.** The exposure of wet DNA fibers to X-rays causes the X-rays to be scattered and the pattern of scattering is related to the position of the atoms in the DNA fibers.

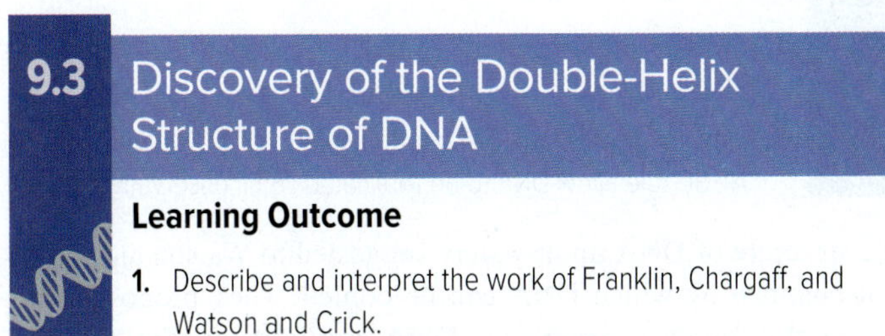

## 9.3 Discovery of the Double-Helix Structure of DNA

### Learning Outcome

1. Describe and interpret the work of Franklin, Chargaff, and Watson and Crick.

Now that we have an understanding of DNA structure, let's take a look back at the methods that were used to deduce its structure. X-ray diffraction was a key experimental tool that led to the discovery of the DNA double helix. When a substance is exposed to X-rays, the atoms in the substance cause the X-rays to be scattered (Figure 9.8). If the substance has a repeating structure, the pattern of scattering, known as the diffraction pattern, is related to the structural arrangement of the atoms causing the scattering. The diffraction pattern is analyzed using mathematical theory to provide information regarding the three-dimensional structure of the molecule. Rosalind Franklin, a British biophysicist working in the 1950s in the same laboratory as Maurice Wilkins, was a gifted researcher who made marked advances in X-ray diffraction techniques involving DNA. The diffraction pattern of DNA fibers produced by Franklin suggested a helical structure with a diameter that is relatively uniform and too wide to be a single-stranded helix. In addition, the pattern provided information regarding the number of nucleotides per turn and was consistent with a 2-nm (nanometer) spacing between the strands in which a purine (A or G) bonds with a pyrimidine (T or C).

Another piece of information that proved to be critical for the determination of the double-helix structure came from the studies of Austrian-born biochemist Erwin Chargaff. In 1950, Chargaff analyzed the base composition of DNA that was isolated from many different species. His experiments consistently showed that the amount of adenine in each sample was similar to the amount of thymine, and the amount of cytosine was similar to the amount of guanine (Table 9.1).

In the early 1950s, more information was known about the structure of proteins than that of nucleic acids. American chemist Linus Pauling correctly proposed that some regions of proteins fold into a structure known as an α helix. To determine the structure of the α helix, Pauling built large models by linking together simple ball-and-stick units. In this way, he could see if atoms fit together properly in a complicated three-dimensional structure. This approach is still widely used today, except researchers now construct three-dimensional models on computers.

**Table 9.1 Base Composition of DNA from a Variety of Organisms as Determined by Chargaff**

| | Percentage of bases | | | |
|---|---|---|---|---|
| Organism | Adenine | Thymine | Guanine | Cytosine |
| *Escherichia coli* (bacterium) | 26.0 | 23.9 | 24.9 | 25.2 |
| *Streptococcus pneumoniae* (bacterium) | 29.8 | 31.6 | 20.5 | 18.0 |
| *Saccharomyces cerevisiae* (yeast) | 31.7 | 32.6 | 18.3 | 17.4 |
| Turtle | 28.7 | 27.9 | 22.0 | 21.3 |
| Salmon | 29.7 | 29.1 | 20.8 | 20.4 |
| Chicken | 28.0 | 28.4 | 22.0 | 21.6 |
| Human | 30.3 | 30.3 | 19.5 | 19.9 |

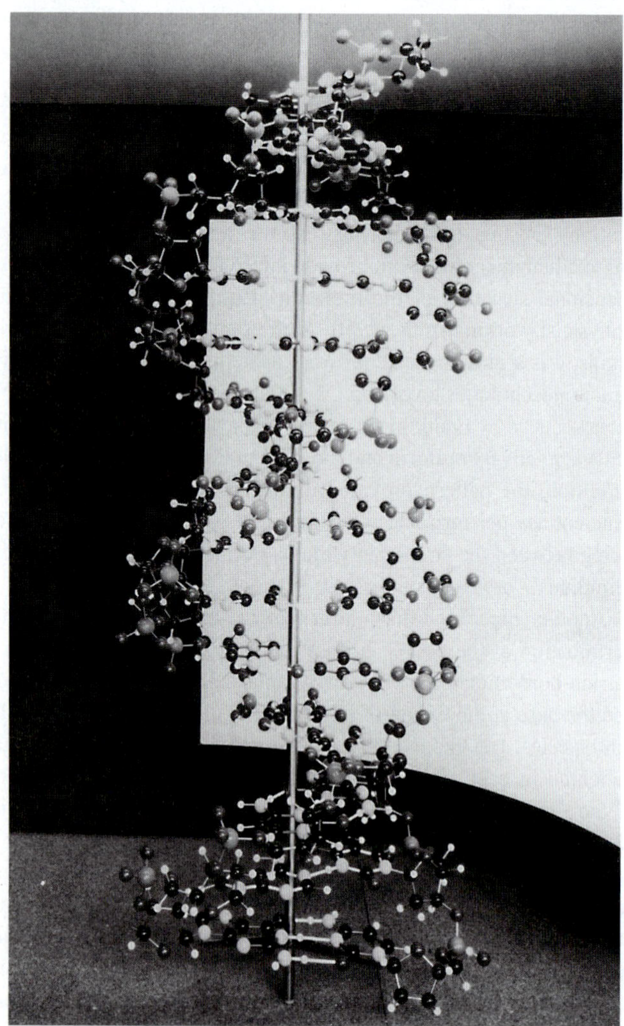

**Figure 9.9 Original ball-and-stick model of the DNA double helix proposed by Watson and Crick.**


Using the ball-and-stick approach, American biologist James Watson and English biologist Francis Crick, in collaboration with Wilkins, set out to determine the structure of DNA. They assumed that nucleotides are linked together in a linear fashion and that the chemical linkage between two nucleotides is always the same. Figure 9.9 shows a ball-and-stick model that helped them deduce the structure. In a 1953 article in the journal *Nature*, Watson and Crick proposed that the structure of DNA is a double helix in which the DNA strands are antiparallel and obey the AT/GC rule. In 1962, Watson, Crick, and Wilkins were awarded the Nobel Prize in Physiology or Medicine. Unfortunately, Rosalind Franklin had died before this time, and the Nobel Prize is awarded only to living recipients.

## 9.3 Reviewing the Concepts

- The X-ray diffraction data of Franklin, the biochemical data of Chargaff, and the ball-and-stick model approach of Pauling helped reveal the structure of DNA (Figure 9.8, Table 9.1).
- Watson and Crick determined that DNA is a double helix in which the DNA strands are antiparallel and obey the AT/GC rule (Figure 9.9).

## 9.3 Testing Your Knowledge

1. The data of Chargaff suggested that in DNA the amount of A = T and G = C. This conclusion was based on
   a. a determination of the base sequences of opposite DNA strands.
   b. the composition of viral DNA.
   c. a comparison of base composition of several different species.
   d. both b and c.
2. To determine the structure of DNA, Watson and Crick used
   a. the X-ray diffraction data of Franklin.
   b. the biochemical data of Chargaff.
   c. the ball-and-stick model approach of Pauling.
   d. all of the above.

## 9.4 Overview of DNA Replication

### Learning Outcomes

1. **SCISKILLS ▶** Describe and interpret the experiments of Meselson and Stahl.
2. Explain how the double-stranded structure of DNA and the AT/GC rule allow DNA to be replicated semiconservatively.

The structure of DNA immediately suggested to Watson and Crick a mechanism by which DNA can be copied. They proposed that during this process, known as **DNA replication,** the original DNA strands are used as templates for the synthesis of new DNA strands. In this section, we will look at an early experiment that helped to determine the mechanism of DNA replication and then examine the structural characteristics that enable a double helix to be faithfully copied.

## Meselson and Stahl Considered Three Proposed Mechanisms of DNA Replication

Researchers in the late 1950s considered three different models for the mechanism of DNA replication (Figure 9.10). In all of these models, the two newly made strands are called the **daughter strands,** and the original strands are the **parental strands.**

1. **Semiconservative mechanism** (Figure 9.10a): In this model, the double-stranded DNA is half conserved following the replication process. The new double-stranded DNA contains one parental strand and one daughter strand.
2. **Conservative mechanism** (Figure 9.10b): Following DNA replication, the original arrangement of parental strands is completely conserved, and the two newly made daughter strands are also together following replication.
3. **Dispersive mechanism** (Figure 9.10c): After DNA replication is completed, segments of parental DNA and newly made DNA are interspersed in both strands.

In 1958, Matthew Meselson and Franklin Stahl devised an experimental approach to distinguish among these three mechanisms. An important feature of their research was the use of isotope labeling. Nitrogen, which is found in DNA, occurs in a common light ($^{14}N$) form and a rare heavy ($^{15}N$) form. Meselson and Stahl studied DNA replication in the bacterium *Escherichia coli*. They grew *E. coli* cells for many generations in a medium that contained only the heavy ($^{15}N$) form of nitrogen (Figure 9.11). This produced a population of bacterial cells in which all of the DNA was labeled with the heavy isotope. Then they switched the bacteria to a medium that contained only the light ($^{14}N$) form as its nitrogen source. The cells were allowed to divide, and samples were collected after one generation (that is, one round of DNA replication), two generations, and so on. Because the bacteria were doubling in a medium that contained only $^{14}N$, all of the newly made DNA strands were labeled with light nitrogen, whereas the original strands remained labeled with the heavy form.

Meselson and Stahl then used centrifugation to separate DNA molecules based on differences in density. Samples were placed on the top of a solution that contained a salt gradient, in this case, cesium chloride (CsCl). A double helix containing all heavy nitrogen has a higher density and will travel closer to the bottom of the gradient. By comparison, if both DNA strands contained $^{14}N$, the DNA would have a lower density and remain closer to the top of the gradient. If one strand contained $^{14}N$ and the other strand contained $^{15}N$, the DNA would be half-heavy and have an intermediate density, ending up near the middle of the gradient. The DNA within the gradient was observed by exposing the gradient to UV light.

After one cell doubling (that is, one round of DNA replication), all of the DNA exhibited a density that was half-heavy or intermediate density (Figure 9.11, step 5). These results are consistent with both the semiconservative and dispersive models. In contrast, the conservative mechanism predicts two different DNA bands: one of high density and one of low density. Because the DNA was found in a single half-heavy band after one doubling, the conservative model was disproved.

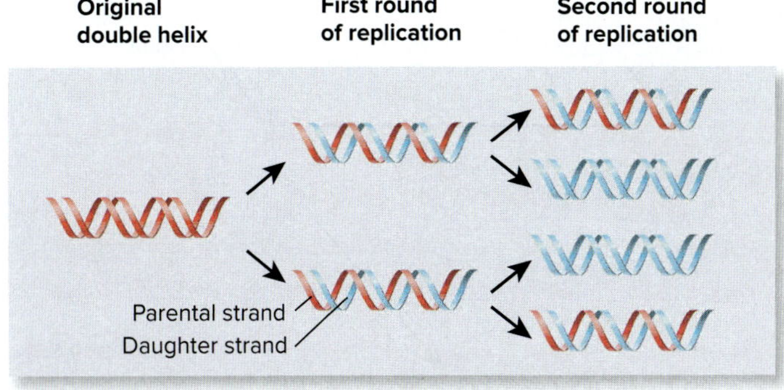


**(a) Semiconservative mechanism. DNA replication produces DNA molecules with 1 parental strand and 1 newly made daughter strand.**

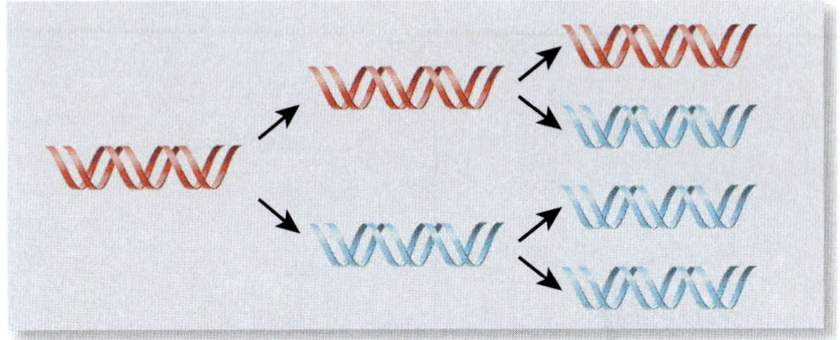

**(b) Conservative mechanism. DNA replication produces 1 double helix with both parental strands and the other with 2 new daughter strands.**

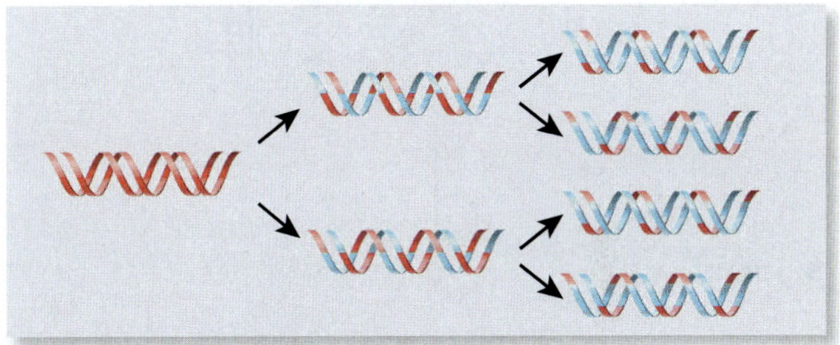

**(c) Dispersive mechanism. DNA replication produces DNA strands in which segments of new DNA are interspersed with the parental DNA.**

**Figure 9.10 Three proposed mechanisms for DNA replication.** The strands of the original (parental) double helix are shown in red. Two rounds of replication are illustrated, with the daughter strands shown in blue.

After two cell doublings, both light DNA and half-heavy DNA bands were observed. This result was also predicted by the semiconservative mechanism of DNA replication, because a parental strand that was light would produce all light DNA, whereas a parental strand that was heavy would produce half-heavy DNA (see Figure 9.10a). However, in the dispersive mechanism, all of the DNA strands would have been $^{1}/_{4}$ heavy after two generations. This mechanism predicts that the heavy nitrogen would be evenly dispersed among four double helices, each strand containing $^{1}/_{4}$ heavy nitrogen and $^{3}/_{4}$ light nitrogen (see Figure 9.10c). This prediction did not agree with the data.

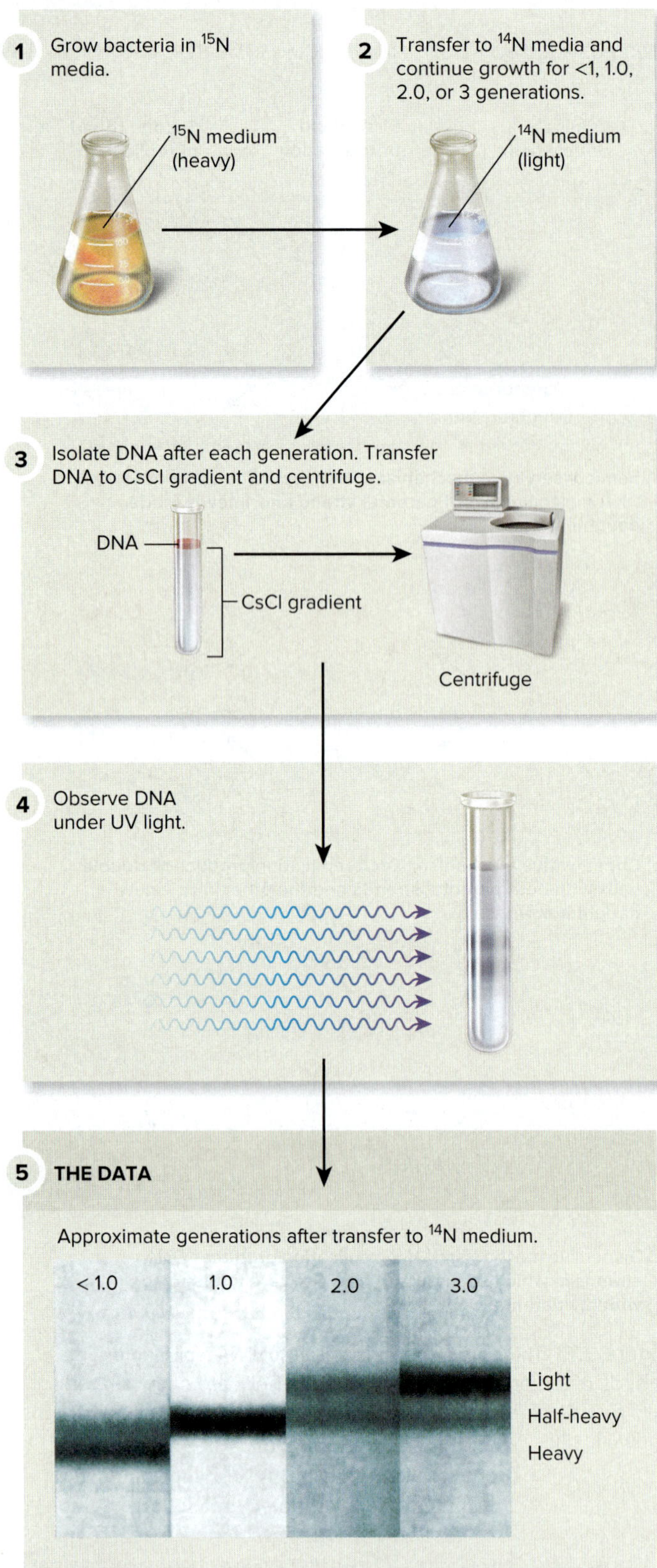


**Figure 9.11** **The Meselson and Stahl experiment showing that DNA replication is semiconservative.**

*(5)* © M. Meselson, F. Stahl (1958), "The replication of DNA in *Escherichia coli*," *PNAS*, 44(7): 671-682, Fig. 4a

**Concept Check:** *If this experiment was conducted for four rounds of DNA replication (that is, four generations), what would be the expected fractions of light DNA and half-heavy DNA according to the semiconservative model?*

Taken together, the results of the Meselson and Stahl experiment are consistent only with a semiconservative mechanism for DNA replication.

## DNA Replication Proceeds According to the AT/GC Rule

As originally proposed by Watson and Crick, DNA replication relies on the complementarity of DNA strands according to the AT/GC rule. During the replication process, the two complementary strands of DNA separate and serve as **template strands** for the synthesis of daughter strands of DNA (**Figure 9.12a**). After the double helix has separated, individual nucleotides have access to the template strands in a region called the replication fork. First, individual nucleotides hydrogen-bond to the template strands according to the AT/GC rule: An adenine (A) in one strand bonds with a thymine (T) in the opposite strand, or a guanine (G) bonds with a cytosine (C). Next, a covalent bond is formed between the phosphate of one nucleotide and the sugar of the previous nucleotide. The end result is that two double helices are made that have the same base sequence as the original DNA molecule (**Figure 9.12b**). This is a critical feature of DNA replication, because it enables the replicated DNA molecules to retain the same information (that is, the same base sequence) as the original molecule. In this way, DNA has the remarkable ability to direct its own duplication.

## 9.4 Reviewing the Concepts

- Meselson and Stahl used $^{15}N$- and $^{14}N$-isotope labeling methods to show that DNA is replicated by a semiconservative mechanism in which the product of DNA replication is one original strand and one new strand (Figures 9.10, 9.11).
- During DNA replication, the double-stranded DNA separates and each strand serves as a template for the synthesis of daughter strands. Nucleotides hydrogen-bond to the template strands according to the AT/GC rule: An adenine (A) in one strand bonds with a thymine (T) in the opposite strand, or a guanine (G) bonds with a cytosine. The result of DNA replication is two double helices with the same base sequence as the original DNA (Figure 9.12).

## 9.4 Testing Your Knowledge

1. According to Meselson and Stahl's procedure, what would be the predicted results after four generations according to the semiconservative model?
   **a.** $^{1}/_{4}$ half-heavy and $^{3}/_{4}$ light
   **b.** $^{1}/_{8}$ half-heavy and $^{7}/_{8}$ light
   **c.** $^{1}/_{16}$ half-heavy and $^{15}/_{16}$ light
   **d.** $^{1}/_{32}$ half-heavy and $^{31}/_{32}$ light
2. What key features of DNA structure allow it to be replicated?
   **a.** double stranded and 10 base pairs per turn
   **b.** double stranded with major and minor grooves
   **c.** double stranded and the AT/GC rule
   **d.** AT/GC rule and 10 base pairs per turn

## Biology Principle

### Structure Determines Function

A double-stranded structure that obeys the AT/GC rule underlies the function of DNA replication.

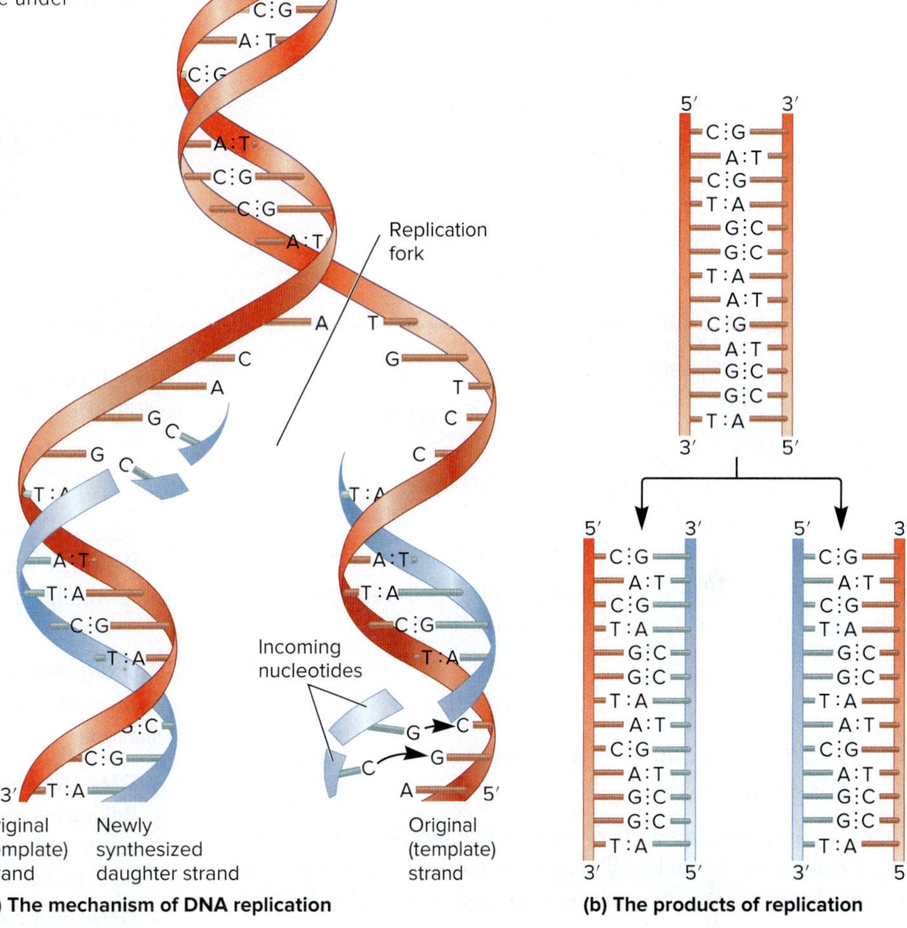


**Figure 9.12** **DNA replication according to the AT/GC rule.** **(a)** The mechanism of DNA replication as originally proposed by Watson and Crick. As we will see in Section 9.5, the synthesis of one newly made strand (the leading strand on the left side) occurs in the direction toward the replication fork, whereas the synthesis of the other newly made strand (the lagging strand on the right side) occurs in small segments away from the fork. **(b)** DNA replication produces two copies of DNA with the same sequence of bases as the original DNA molecule.

## 9.5 Molecular Mechanism of DNA Replication

### Learning Outcomes

1. Explain how the synthesis of new DNA strands begins at an origin of replication.
2. Describe the functions of helicase, topoisomerase, single-strand binding protein, primase, and DNA polymerase at the replication fork.
3. Outline the key differences between the synthesis of the leading and lagging strands.
4. List three reasons why DNA replication is very accurate.

Thus far, we have examined the general mechanism of DNA replication, known as semiconservative replication, and seen how DNA synthesis obeys the AT/GC rule. In this section, we will examine the details of DNA replication as it occurs inside living cells. As you will learn, several different proteins are needed to initiate DNA replication and allow it to proceed quickly and accurately.

### DNA Replication Begins at an Origin of Replication

An **origin of replication** is a site within a chromosome that serves as a starting point for DNA replication. At the origin, the two DNA strands unwind (Figure 9.13). DNA replication proceeds outward from two **replication forks,** a process termed **bidirectional replication.** A bacterial chromosome is relatively small and usually circular and has a single origin of replication. Bidirectional replication starts at the origin of replication and proceeds until the new strands meet on the opposite side of the chromosome. Eukaryotes have larger chromosomes that are linear. They require multiple origins of replication so the DNA can be replicated within a reasonable length of time. The newly made strands from each origin eventually make contact with each other to complete the replication process.

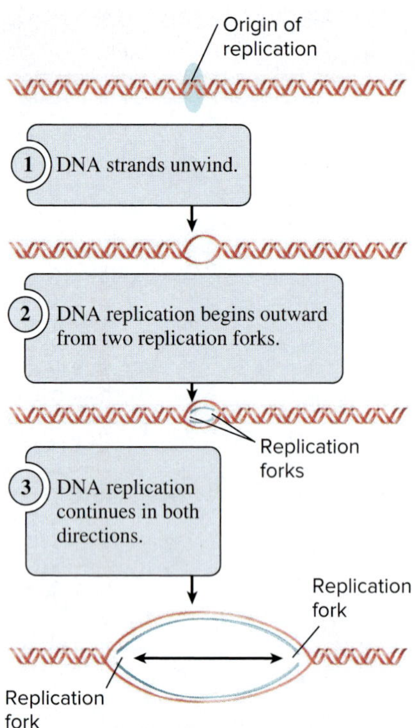


**Figure 9.13 The bidirectional replication of DNA.** DNA replication proceeds in both directions from an origin of replication.

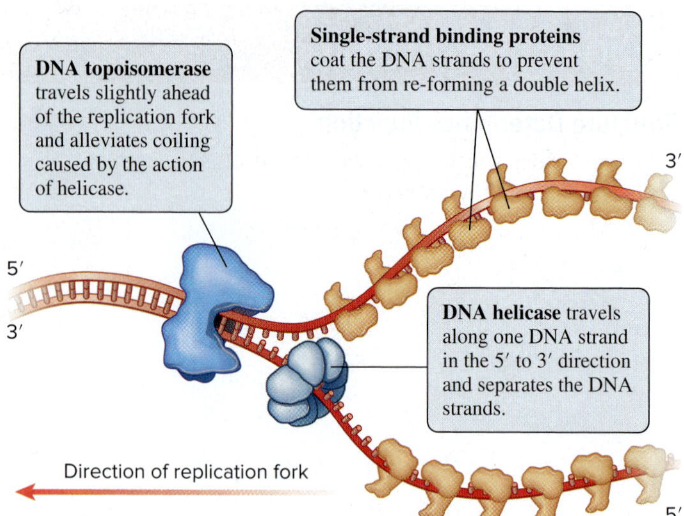


**Figure 9.14 Proteins that facilitate the formation and movement of the replication fork.**

## DNA Replication Requires the Action of Several Different Proteins

In all living species, a set of several different proteins is involved in DNA replication. An understanding of the functions of these proteins is critical to explaining the replication process at the molecular level.

**Helicase, Topoisomerase, and Single-Strand Binding Proteins: Formation and Movement of the Replication Fork** To act as a template for DNA replication, the strands of a double helix must separate, and the resulting replication fork must move. As mentioned, an origin of replication serves as a site where this separation begins. Each fork then moves outward from the origin. DNA helicase, DNA topoisomerase, and single-strand binding proteins are responsible for fork formation and movement:

- **DNA helicase:** At each fork, DNA helicase binds to one DNA strand and travels in the 5′ to 3′ direction (Figure 9.14). It uses energy from ATP to break hydrogen bonds between base pairs and thereby separates the DNA strands. This keeps the fork moving forward.
- **DNA topoisomerase:** The action of DNA helicase can cause knots (called supercoils) to form just ahead of the replication fork. These knots are alleviated by another enzyme called DNA topoisomerase.
- **Single-strand binding proteins:** After the two DNA strands have separated, they must remain that way so they can act as templates to make complementary daughter strands. The function of single-strand binding proteins is to coat both of the single strands of template DNA and prevent them from re-forming a double helix.

**DNA Polymerase and Primase: Synthesis of DNA Strands** Two enzymes are needed to synthesize DNA strands during DNA replication.

- **DNA polymerase:** This enzyme is responsible for covalently linking nucleotides together to form DNA strands. The name indicates that this enzyme makes a polymer of DNA nucleotides.
- **DNA primase:** DNA polymerase is unable to begin DNA synthesis on a bare template strand. A different enzyme called **DNA primase** is required if the template strand is bare. DNA primase makes a complementary **primer,** which is actually a short segment of RNA, typically 10 to 12 nucleotides in length that starts, or primes, the process of DNA replication.

Let's first consider the catalytic properties of DNA polymerase, and then we will examine how DNA strands are made at the replication fork. The structure of DNA polymerase resembles a human hand with the DNA threaded through it (Figure 9.15a). As DNA polymerase slides along the DNA, individual nucleotides with three phosphate groups, called **deoxynucleoside triphosphates,** hydrogen-bond to the exposed bases in the template strand according to the AT/GC rule (Figure 9.15b). At the catalytic site, DNA polymerase breaks a bond between the first and second phosphate and then attaches the resulting nucleotide with one phosphate group (a deoxynucleoside monophosphate) to the 3′ end of a growing strand via a phosphoester bond. The breakage of the covalent bond that releases pyrophosphate ($PP_i$) is an exergonic reaction that provides the energy to covalently connect adjacent nucleotides. The pyrophosphate is broken down to two phosphates.

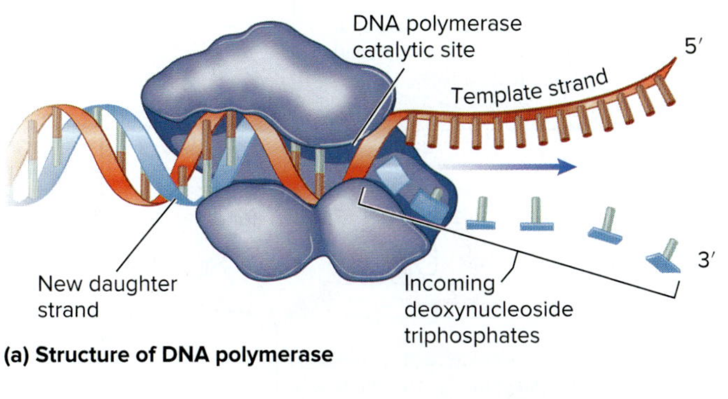

(a) Structure of DNA polymerase

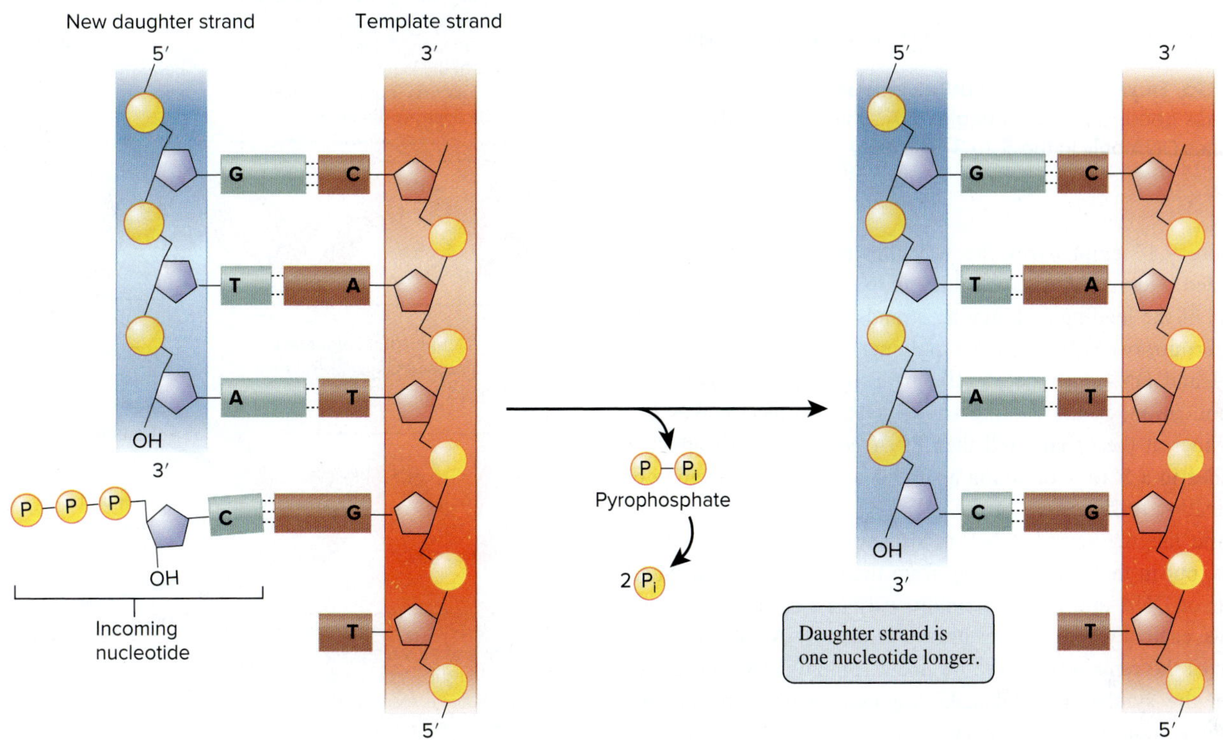

(b) Chemistry of DNA replication

**Figure 9.15** **Enzymatic synthesis of DNA.** **(a)** Structure of DNA polymerase. **(b)** DNA polymerase breaks the bond between the first and second phosphate in an incoming deoxynucleoside triphosphate, causing the release of pyrophosphate. This provides the energy to form a covalent bond between the resulting deoxynucleoside monophosphate and the previous nucleotide in the growing strand. The pyrophosphate is broken down to two phosphates.

The rate of DNA synthesis is truly remarkable. In bacteria, DNA polymerase synthesizes DNA at a rate of 500 nucleotides per second, whereas eukaryotic species make DNA at a rate of about 50 nucleotides per second.

DNA polymerase has two additional enzymatic features that affect how DNA strands are made. First, DNA polymerase is unable to begin DNA synthesis on a bare template strand. However, if a DNA strand or an RNA primer is already attached to a template strand, DNA polymerase can elongate such a pre-existing strand by making DNA (Figure 9.16a). A second feature of DNA polymerase is that once synthesis has begun, it can synthesize new DNA only in a 5′ to 3′ direction (Figure 9.16b).

## Leading and Lagging DNA Strands Are Made Differently

Let's now examine how new DNA strands are made at replication forks. DNA replication occurs near the opening that forms each replication fork (Figure 9.17, step 1). The synthesis of a strand begins with an RNA primer (depicted in yellow), and the new DNA is made in the 5′ to 3′ direction. The manner in which the two daughter strands are synthesized is strikingly different. One strand, called the **leading strand,** is made in the same direction that the fork is moving. The leading strand is synthesized as one long, continuous molecule.

By comparison, the other daughter strand, termed the **lagging strand,** is made differently. Because DNA polymerase can synthesize DNA only in the 5′ to 3′ direction (see Figure 9.16), the lagging strand is made as a series of small fragments that are subsequently connected to each other to form a continuous strand. These DNA fragments are known as **Okazaki fragments,** after Japanese molecular biologists Reiji and Tsuneko Okazaki, who initially discovered them in the late 1960s. The synthesis of Okazaki fragments occurs away from the direction that the replication fork is moving. For example, notice that the Okazaki fragments seen in Figure 9.17a, steps 2 and 3 are synthesized from left to right. As shown in Figure 9.17a, step 4, the RNA primer is eventually removed, and adjacent Okazaki fragments are connected to each other to form a continuous strand of DNA.

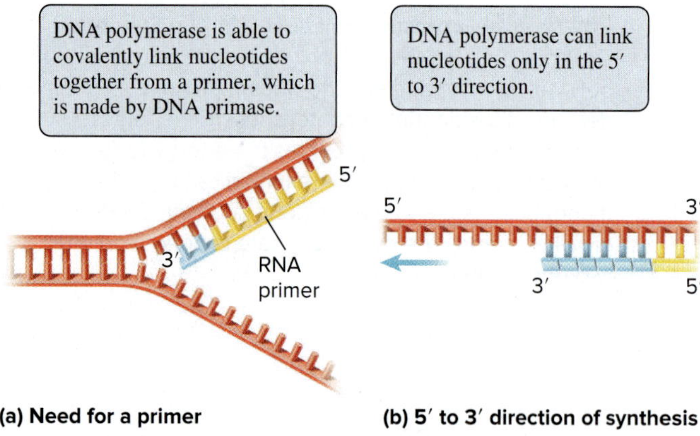


**Figure 9.16** **Enzymatic features of DNA polymerase. (a)** DNA polymerase needs a primer to begin DNA synthesis, and **(b)** it synthesizes DNA only in the 5′ to 3′ direction.

Figure 9.17 emphasizes the way that new DNA strands are synthesized. Figure 9.18 also includes the proteins involved in the synthesis of the leading and lagging strands in *Escherichia coli*. In this bacterium, two different DNA polymerases, called DNA polymerase I and DNA polymerase III, are primarily responsible for DNA replication. In the leading strand, DNA primase makes one RNA primer at the origin, and then DNA polymerase III attaches nucleotides in a 5′ to 3′ direction as it slides toward the opening of the replication fork.

In the lagging strand, DNA is also synthesized in a 5′ to 3′ direction, but in the direction away from the replication fork. Short segments of DNA are made discontinuously as a series of Okazaki fragments, each of which requires its own primer. DNA polymerase III synthesizes the remainder of the fragment (Figure 9.18, step 2). To complete the synthesis of Okazaki fragments within the lagging strand, three additional events occur: the removal of the RNA primers, the synthesis of DNA in the area where the primers have been removed, and the covalent joining of adjacent fragments of DNA (Figure 9.18, steps 3 and 4). The RNA primers are removed by DNA polymerase I, which digests the linkages between nucleotides in a 5′ to 3′ direction and fills in the vacant region with DNA. However, once the DNA has been completely filled in, a covalent bond is missing between the last nucleotide added by DNA polymerase I and the first DNA nucleotide in the adjacent Okazaki fragment. An enzyme known as **DNA ligase** catalyzes the formation of a covalent bond between these two DNA fragments to complete the replication process in the lagging strand (Figure 9.18, step 4).

## DNA Replication Is Very Accurate

Although errors can happen during DNA replication, permanent mistakes are extraordinarily rare. For example, during bacterial DNA replication, only 1 mistake is made per 100 million nucleotides. Biologists use the term high fidelity to refer to a process that occurs with relatively few mistakes. Three factors explain such a remarkably high fidelity for DNA replication:

1. Hydrogen bonding between A and T or between G and C is more stable than between mismatched pairs.

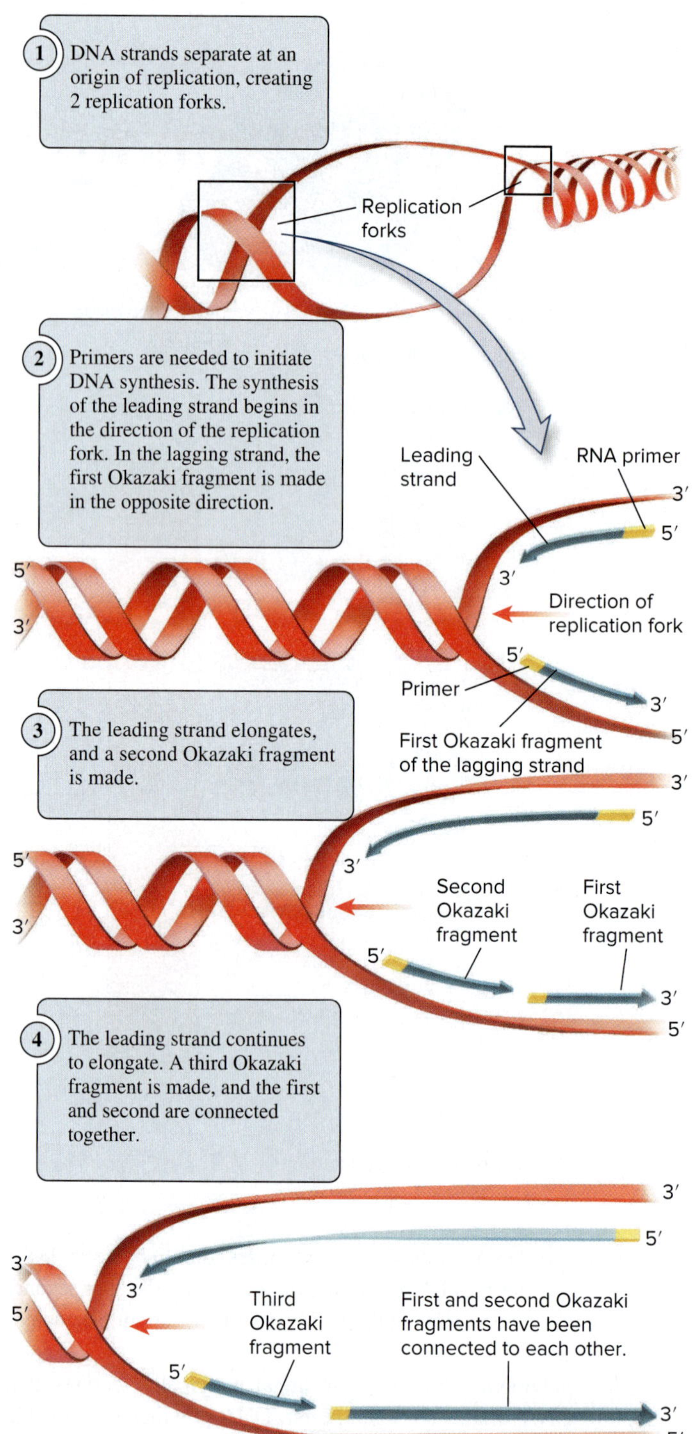


**Figure 9.17** **Synthesis of new DNA strands.** The separation of DNA at the origin of replication produces two replication forks that move in opposite directions. New DNA strands are made near the opening of each fork. The leading strand is made continuously in the same direction the fork is moving. The lagging strand is made as small pieces (Okazaki fragments) in the opposite direction. Eventually, these small pieces are connected to each other to form a continuous lagging strand.

**Concept Check:** *Which strand, the leading or lagging strand, is made discontinuously in the direction opposite the movement of the replication fork?*

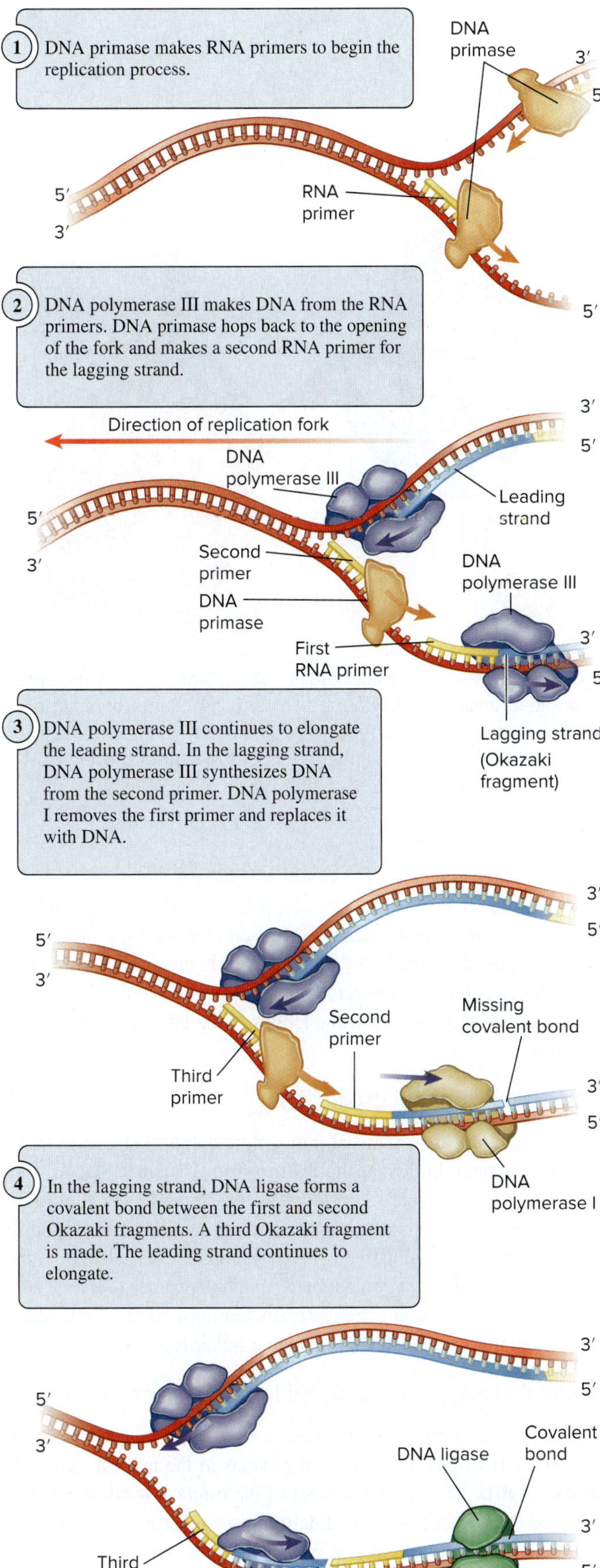

1 DNA primase makes RNA primers to begin the replication process.
DNA primase
3′
5′
5′
3′
RNA primer
5′
2 DNA polymerase III makes DNA from the RNA primers. DNA primase hops back to the opening of the fork and makes a second RNA primer for the lagging strand.
Direction of replication fork
3′
5′
DNA polymerase III
Leading strand
5′
3′
Second primer
DNA primase
DNA polymerase III
First RNA primer
3′
5′
Lagging strand (Okazaki fragment)
3 DNA polymerase III continues to elongate the leading strand. In the lagging strand, DNA polymerase III synthesizes DNA from the second primer. DNA polymerase I removes the first primer and replaces it with DNA.
3′
5′
5′
3′
Second primer
Missing covalent bond
Third primer
3′
5′
DNA polymerase I
4 In the lagging strand, DNA ligase forms a covalent bond between the first and second Okazaki fragments. A third Okazaki fragment is made. The leading strand continues to elongate.
3′
5′
5′
3′
DNA ligase
Covalent bond
3′
5′
Third primer


**Figure 9.18** **Proteins involved with the synthesis of the leading and lagging strands in *E. coli*.**

2. The active site of DNA polymerase is unlikely to catalyze bond formation between adjacent nucleotides if a mismatched base pair is formed.
3. DNA polymerase can identify a mismatched nucleotide and remove it from the daughter strand. This event, called **proofreading,** occurs when DNA polymerase detects a mismatch and then reverses its direction and digests the linkages between nucleotides at the end of a newly made strand in the 3′ to 5′ direction. Once it passes the mismatched base and removes it, DNA polymerase then continues to synthesize DNA in the 5′ to 3′ direction.

## 9.5 Reviewing the Concepts

- DNA synthesis occurs bidirectionally from an origin of replication. The synthesis of new DNA strands happens near each replication fork (Figure 9.13).
- DNA helicase separates DNA strands, single-strand binding proteins keep them separated, and DNA topoisomerase alleviates coiling ahead of the fork (Figure 9.14).
- Deoxynucleoside triphosphates bind to the template strands according to the AT/GC rule. DNA polymerase recognizes these deoxynucleoside triphosphates and attaches a deoxynucleoside monophosphate to the 3′ end of a growing strand (Figure 9.15).
- DNA polymerase requires a primer and synthesizes new DNA strands only in the 5′ to 3′ direction (Figure 9.16).
- The leading strand is made continuously, in the same direction the fork is moving. The lagging strand is made in the opposite direction from that of the replication fork as short Okazaki fragments that are synthesized and connected together (Figure 9.17).
- DNA primase makes one RNA primer in the leading strand and multiple RNA primers in the lagging strand. In *E. coli*, DNA polymerase III extends these primers with DNA, and DNA polymerase I removes the primers when they are no longer needed and fills in with DNA. DNA ligase connects adjacent Okazaki fragments in the lagging strand (Figure 9.18).
- DNA replication is very accurate because (1) hydrogen bonding according to the AT/CG rule is more stable than mismatched pairs, (2) DNA polymerase is unlikely to catalyze bond formation if a mismatched base pair is formed, and (3) DNA polymerase carries out proofreading.

## 9.5 Testing Your Knowledge

1. Which protein is needed to form a replication fork and keep it moving?
   **a.** DNA helicase
   **b.** topoisomerase
   **c.** single-strand binding protein
   **d.** all of the above
2. Which of the following is *not* a property of DNA polymerase?
   **a.** It cannot begin DNA synthesis on a bare template strand.
   **b.** It can elongate a DNA strand from a primer or pre-existing DNA strand.
   **c.** It can synthesize DNA in the 3′ to 5′ direction.
   **d.** It can synthesize DNA in the 5′ to 3′ direction.

3. Okazaki fragments are made during the synthesis of the ______ strand. They are connected together by ______.
   a. lagging, DNA ligase
   b. lagging, helicase
   c. leading, DNA ligase
   d. leading, helicase

## 9.6 Molecular Structure of Eukaryotic Chromosomes

### Learning Outcomes

1. Describe the structure of nucleosomes and the 30-nm fiber, and how the 30-nm fiber forms radial loop domains.
2. Outline the various levels of compaction that lead to a metaphase chromosome.

We now turn our attention to the structure of eukaryotic chromosomes. A typical eukaryotic chromosome contains a single, linear, double-stranded DNA molecule that may be hundreds of millions of base pairs in length. If the DNA from a single set of human chromosomes was stretched from end to end, it would be over 1 meter in length! By comparison, most eukaryotic cells are only 10–100 μm (micrometers) in diameter, and the cell nucleus is typically about 2–4 μm in diameter. Therefore, to fit inside the nucleus, the DNA in a eukaryotic cell must be folded and compacted by a staggering amount.

The term chromosome is used to describe a discrete unit of genetic material. For example, a human somatic cell contains 46 chromosomes. By comparison, the term chromatin has a biochemical meaning. **Chromatin** is the complex of DNA and proteins that makes up eukaryotic chromosomes. As described in Chapter 11, it may also contain noncoding RNA. Chromosomes are very dynamic structures that alternate between tight and loose compaction states. In this section, we will focus on two issues of chromosome structure. First, we will consider how chromosomes are compacted and organized within the cell nucleus. Then we will examine the additional compaction that is necessary to produce the highly condensed chromosomes that occur during cell division.

### DNA Wraps Around Histone Proteins to Form Nucleosomes

Eukaryotic DNA is first compacted by wrapping itself around a group of proteins called **histones.** As shown in Figure 9.19, a repeating structural unit of chromatin is the **nucleosome,** which is 11 nm in diameter at its widest point. Each nucleosome is composed of 146 or 147 bp (base pairs) of DNA wrapped around a core of eight histone proteins called a histone octamer. Each octamer contains two molecules of four types of histone proteins: H2A, H2B, H3, and H4. Histone proteins contain a large number of positively charged amino acids, namely lysine and arginine. The negative charges found in the phosphates of DNA are attracted to the positive charges on histone proteins. The amino terminal tail of each histone protein protrudes from the histone octamer. As discussed in Chapter 12, these tails can be covalently modified and play a key role in gene regulation.

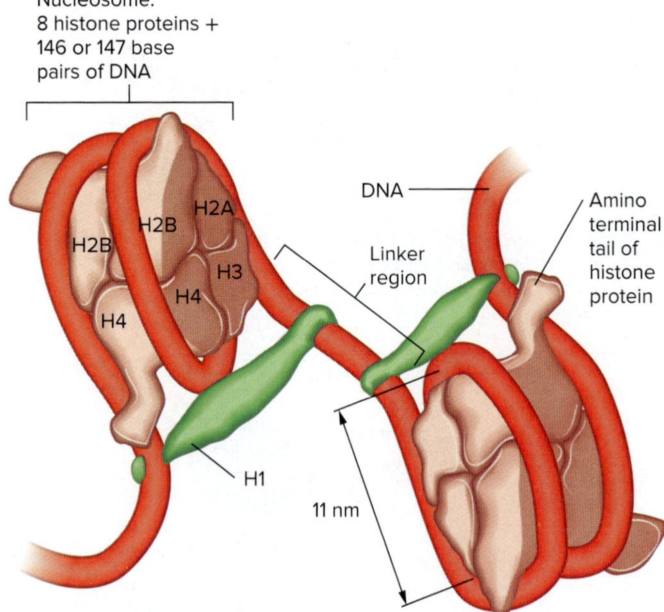


**Figure 9.19 Structure of a nucleosome.** A nucleosome is composed of double-stranded DNA wrapped around an octamer of histone proteins. A linker region connects two adjacent nucleosomes. Histone H1 is bound to the linker region, as are other proteins not shown in this figure.

The nucleosomes are connected by linker regions of DNA that vary in length from 20 to 100 bp, depending on the species and cell type. A particular histone named H1 is bound to the linker region, as are other types of proteins (see Figure 9.19). In electron micrographs, the overall structure of connected nucleosomes resembles beads on a string. This structure shortens the length of the DNA molecule about sevenfold.

### Nucleosomes Form a 30-nm Fiber

Nucleosome units are organized into a more compact structure that is 30 nm in diameter, known as the **30-nm fiber** (Figure 9.20a). Histone H1 and other proteins are important in the formation of the 30-nm fiber, which shortens the nucleosome structure another sevenfold. In one model for the 30-nm fiber, linker regions are variably bent and twisted, with little direct contact between nucleosomes (Figure 9.20b). The 30-nm fiber forms an irregular, fluctuating structure with stable nucleosome units connected by bendable linker regions.

### Chromatin Loops Are Anchored to the Nuclear Matrix

A third level of compaction involves interactions between the 30-nm fibers and a filamentous network of proteins in the nucleus called the **nuclear matrix.** This matrix consists of the **nuclear lamina,** which is composed of protein fibers that line the inner nuclear membrane, and an internal nuclear matrix that is connected to the lamina and fills the interior of the nucleus (refer back to Figure 4.19). The nuclear matrix is involved in the compaction of the 30-nm fiber by participating in the formation of **radial loop domains.** These loops, often 25,000–200,000 bp in size, are anchored to the nuclear matrix (Figure 9.21).

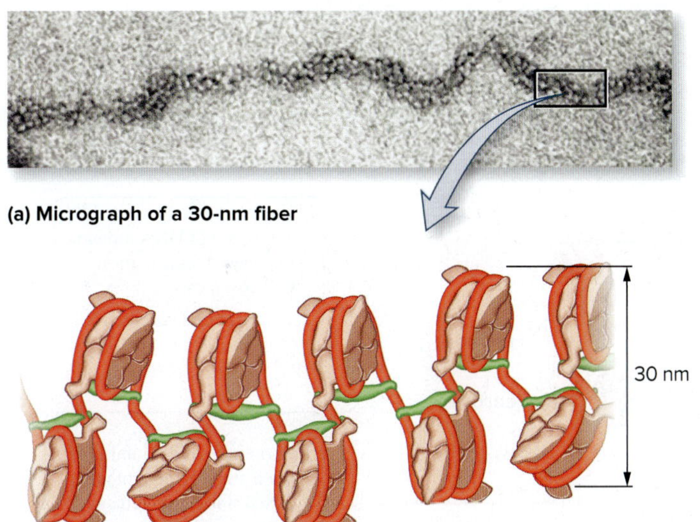


**Figure 9.20** **The 30-nm fiber.** **(a)** A photomicrograph of the 30-nm fiber. **(b)** In this model, the linker DNA forms a bendable structure with little contact between adjacent nucleosomes.
*(a)* © Dr. Barbara A. Hamkalo

Protein fiber of the nuclear matrix
30-nm fiber
Protein that binds to the 30-nm fiber and to a protein fiber of the nuclear matrix
Gene
Radial loop domain
Gene
Gene

**Figure 9.21** **Attachment of the 30-nm fiber to a protein fiber to form a radial loop domain.**

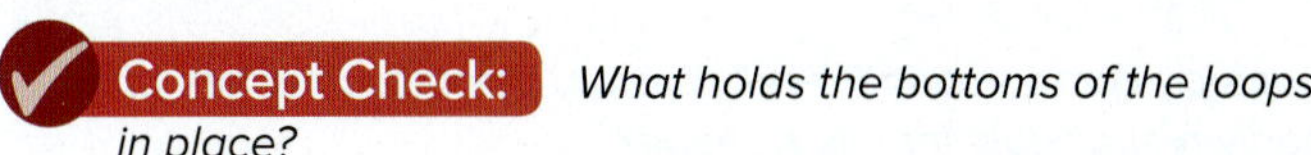

*What holds the bottoms of the loops in place?*

The compaction level of chromosomes in the cell nucleus is not completely uniform. This variability can be seen with a light microscope and was first observed by the German cytologist E. Heitz in 1928. He coined the term **heterochromatin** to describe the highly compacted regions of chromosomes. The less condensed regions are known as **euchromatin,** which is the form of chromatin in which the 30-nm fiber forms radial loop domains. In heterochromatin, these radial loop domains are compacted even further. In nondividing cells, most chromosomal regions are euchromatic, and these are the regions where genes are usually located. By comparison, the centromeric and telomeric regions are often heterochromatic and may not contain genes.

## During Cell Division, Chromosomes Undergo Maximum Compaction

When cells prepare to divide, the chromosomes become even more compacted or condensed. Each chromosome becomes entirely compacted into heterochromatin. This aids in their proper alignment during metaphase, which is a stage of eukaryotic cell division described in Chapter 14. Figure 9.22 illustrates the levels of compaction that contribute to the formation of a metaphase chromosome. DNA in the nucleus is always compacted by the formation of nucleosomes, 30-nm fibers, and radial loop domains (Figure 9.22a–d). In euchromatin, the 30-nm fibers are arranged in radial loop domains that are relatively loose, meaning that a fair amount of space occurs between the 30-nm fibers (Figure 9.22d). The average width of such loops is about 300 nm.

By comparison, heterochromatin involves a much tighter packing of the loops, so little space is between the 30-nm fibers (Figure 9.22e). Heterochromatic regions tend to be wider, about 700 nm. When cells prepare to divide, all of the euchromatin becomes highly compacted into heterochromatin, which greatly shortens the chromosomes. For a metaphase chromosome that contains two copies of the DNA (Figure 9.22f), the width averages about 1,400 nm, but a metaphase chromosome is much shorter than the same chromosome in the nucleus of a nondividing cell.

## 9.6 Reviewing the Concepts

- Chromosomes carry the genetic material. Chromatin is the name given to the complex of DNA, proteins, and noncoding RNA that makes up chromosomes.
- In eukaryotic chromosomes, the DNA is wrapped around histone octamers to form nucleosomes. Nucleosomes are further compacted into 30-nm fibers (Figures 9.19, 9.20).
- A third level of compaction of eukaryotic chromosomes involves the formation of radial loop domains in which the 30-nm fibers are anchored to the nuclear matrix to form euchromatin. In heterochromatin, the loops are more closely packed together (Figure 9.21).
- When cells prepare to divide, chromosomes become even more condensed. They are entirely heterochromatic (Figure 9.22).

## 9.6 Testing Your Knowledge

1. A nucleosome is composed of
   a. 146 bp or 147 bp of DNA.
   b. an octamer of histone proteins.
   c. nuclear matrix proteins.
   d. both a and b.

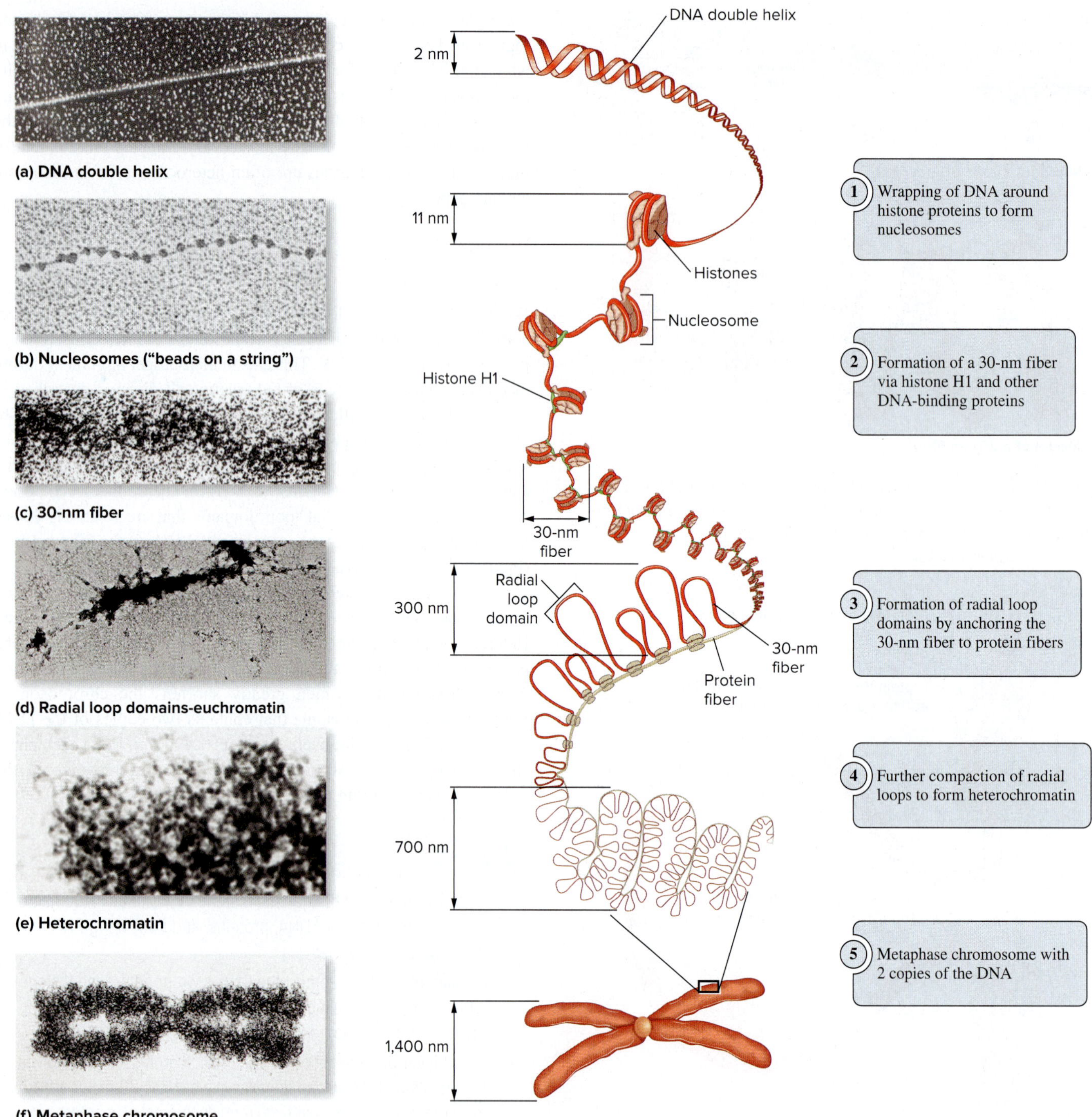


**Figure 9.22** The steps in eukaryotic chromosomal compaction leading to the metaphase chromosome.

*(a)* © Science Source; *(b)* © Dr. Barbara A. Hamkalo; *(c)* Courtesy Dr. Jerome B. Rattner, Cell Biology and Anatomy, University of Calgary; *(d)* Micrograph courtesy of Dr. James R. Paulson; *(e, f)* © Peter Engelhardt/Department of Virology, Haartman Institute

**Concept Check:** *After they have replicated and become compacted in preparation for cell division, chromosomes are often shaped like an X, as in part (f) of this figure. Which proteins are primarily responsible for this X shape?*

**BioConnections:** *Look ahead to Figure 14.7. Why do you think it is necessary for the chromosomes to become compact in preparation for cell division?*

**2.** Which of the following is the correct order for the progression of chromatin compaction seen in metaphase chromosomes?
  **a.** radial loop domains, further compaction of radial loop domains, nucleosomes, 30-nm fiber
  **b.** radial loop domains, further compaction of radial loop domains, 30-nm fiber, nucleosomes
  **c.** nucleosomes, 30-nm fiber, radial loop domains, further compaction of radial loop domains
  **d.** 30-nm fiber, nucleosomes, radial loop domains, further compaction of radial loop domains

## Assess and Discuss

### Test Yourself

1. What is/are the main component(s) of chromosomes?
   a. DNA only
   b. RNA only
   c. proteins only
   d. DNA and proteins
   e. DNA and RNA
2. Considering the components of a nucleotide, what component is always different when comparing nucleotides in a DNA strand to an RNA strand?
   a. phosphate group
   b. pentose sugar
   c. base
   d. both b and c
   e. a, b, and c
3. Which of the following equations would be appropriate when considering DNA base composition?
   a. %A + %T = %G + %C
   b. %A = %G
   c. %A = %G = %T = %C
   d. %A + %G = %T + %C
4. If the sequence of a segment of DNA is 5′–CGCAACTAC–3′, what is the appropriate sequence for the opposite strand?
   a. 5′–GCGTTGATG–3′
   b. 3′–ATACCAGCA–5′
   c. 5′–ATACCAGCA–3′
   d. 3′–GCGTTGATG–5′
5. Of the following statements, which is correct when considering the process of DNA replication?
   a. New DNA molecules are composed of two completely new strands.
   b. New DNA molecules are composed of one strand from the old molecule and one new strand.
   c. New DNA molecules are composed of strands that are a mixture of sections from the old molecule and sections that are new.
   d. None of the above are correct.
6. Meselson and Stahl were able to demonstrate semiconservative replication in *E. coli* by
   a. using radioactive isotopes of phosphorus to label the old strand and visually determining the relationship of old and new DNA strands.
   b. using different enzymes to eliminate old strands from DNA.
   c. using isotopes of nitrogen to label the DNA and determining the relationship of old and new DNA strands by density differences of the new molecules.
   d. labeling viral DNA before it was incorporated into a bacterial cell and visually determining the location of the DNA after centrifugation.
7. The difference in the synthesis of the leading and lagging strands is the result of which of the following?
   a. DNA polymerase replicates only the leading strand.
   b. The two template strands are antiparallel, and DNA polymerase makes DNA only in the 5′ to 3′ direction.
   c. The lagging strand is the result of DNA breakage due to UV light.
   d. The cell does not contain enough nucleotides to make two complete strands.
8. The protein that separates DNA strands at the replication fork is
   a. DNA polymerase.
   b. DNA ligase.
   c. DNA topoisomerase.
   d. DNA helicase.
   e. DNA primase.
9. A nucleosome is
   a. a dark-staining body composed of RNA and proteins found in the nucleus.
   b. a protein that helps organize the structure of chromosomes.
   c. another word for a chromosome.
   d. a structure composed of DNA wrapped around eight histones.
   e. the short arm of a chromosome.
10. The conversion of euchromatin into heterochromatin involves
   a. the formation of more nucleosomes.
   b. the formation of less nucleosomes.
   c. greater compaction of loop domains.
   d. less compaction of loop domains.
   e. both a and c.

### Conceptual Questions

1. What are the four criteria that the genetic material must fulfill? What was Griffith's contribution to the study of DNA, and why was it important?
2. What are the key features of DNA that allow it to be replicated? How does the replication process affect the ability of DNA to store information?
3. **PRINCIPLES** A principle of biology is that structure determines function. Discuss how the structure of DNA underlies different aspects of its function.

### Collaborative Questions

1. **SCISKILLS ▶** Some bacterial strains exhibit resistance to killing by antibiotics. For example, certain strains of bacteria are resistant to the drug tetracycline, whereas other strains are sensitive to this antibiotic. Describe an experiment you would carry out to demonstrate that tetracycline resistance is an inherited trait carried in the DNA of the resistant strain.
2. **SCISKILLS ▶** How might you provide evidence that DNA is the genetic material in mice?

## Online Resource

# 10 The Expression of Genetic Information via Genes I: Transcription and Translation

© Dr. Elena Kiseleva/SPL/Science Source

**An electron micrograph of many ribosomes in the act of translating an mRNA molecule into polypeptides.** The mRNA is colored in red. Many ribosomes (colored in blue) are attached to this mRNA molecule. Polypeptides (colored in green) are seen emerging from each ribosome. The complex of a single mRNA and multiple ribosomes is called a polysome.

## Chapter Outline



Why do we care about our genes? Let's consider this question with regard to heart disease. Researchers have identified dozens of genes that influence a person's predisposition to develop various types of heart disease. Some people carry variations in those genes that make it more likely for them to develop the disease. Such knowledge causes some people to take preventative measures, such as exercise and modifications to their diets. In addition, by identifying those genes associated with heart disease and studying the proteins specified by them, researchers may be able to develop drugs to combat various forms of the disease.

We can broadly define a gene as a unit of heredity. Geneticists view gene function at different biological levels. In this chapter, we will explore how genes work at the molecular level. You will learn how nucleotide sequences are organized to form genes and how those genes are used as a template to make RNA copies, which ultimately leads to the synthesis of functional proteins. In Chapters 15 and 16, we will examine how genes affect the traits or characteristics of individuals. The term **gene expression** can refer to gene function either at the molecular level or at the level of traits. In reality, the two phenomena are intricately woven together. The expression of genes at the molecular level affects the structure and function of cells, which in turn, determine the traits that an organism displays.

In this chapter, we will explore the steps of gene expression as they occur at the molecular level. These steps include the use of a gene as a template to make an RNA molecule, the modifications of RNA into a functional molecule (in eukaryotes), and the use of RNA to direct the formation of a protein. This chapter focuses on the expression of genes that encode polypeptides, which are called **protein-encoding genes.** The following chapter focuses on some of the interesting functions of genes that produce **non-coding RNAs (ncRNAs),** which are RNAs that do not encode polypeptides.

## 10.1 Overview of Gene Expression

### Learning Outcomes

1. Outline the general steps of gene expression at the molecular level, which together constitute the central dogma.
2. Explain how proteins are largely responsible for determining an organism's characteristics.

In this section, we will begin to examine how genes work at the molecular level. As you will learn, the expression of protein-encoding genes requires the process of transcription, in which DNA is used to make RNA, and the process of translation, in which RNA is used to make polypeptides. We will also consider how the synthesis of proteins is a key factor that determines the structure and function of cells and ultimately an organism's traits.

### Molecular Gene Expression Involves the Processes of Transcription and Translation

At the molecular level, the general steps of gene expression occur as follows:

1. **Transcription:** The first step, known as **transcription,** produces an RNA copy of a gene, also called an RNA transcript (Figure 10.1). The term transcription literally means the act of making a copy. Most genes, which are termed protein-encoding genes or structural genes,* are used to make an RNA molecule that contains the information to specify a polypeptide with a particular amino acid sequence. This type of RNA is called **messenger RNA** (abbreviated **mRNA**), because its function is to carry information from the DNA to cellular components called ribosomes.
2. **Translation**: The process of synthesizing a specific polypeptide is called **translation.** The term translation is used because a nucleotide sequence in mRNA is "translated" into an amino acid sequence of a polypeptide. The term **polypeptide** refers to a linear sequence of amino acids; it denotes structure. Most genes carry the information to make a particular polypeptide. By comparison, the term **protein** denotes function. Some proteins are composed of one polypeptide, whereas others are composed of two or more polypeptides.

Together, the transcription of DNA into mRNA and the translation of mRNA into a polypeptide constitute the **central dogma** of gene expression at the molecular level, which was first proposed in 1958 by Francis Crick (see Figure 10.1). The central dogma applies to bacteria, archaea, and eukaryotes. However, in eukaryotes, an additional step occurs between transcription and translation. During **RNA modification,** which is described later in this chapter, the RNA transcript, termed **pre-mRNA,** is modified in ways that make it a functionally active mRNA (Figure 10.1b).

Another difference between bacteria and eukaryotes is the cellular location of transcription and translation. In bacteria, both events occur in the same compartment, namely the cytoplasm. In eukaryotes, transcription occurs in the nucleus. The mRNA then

* Geneticists commonly use the term structural gene to describe all genes that encode polypeptides. Some geneticists, however, distinguish structural genes from regulatory genes—genes that encode proteins that regulate the expression of structural genes.

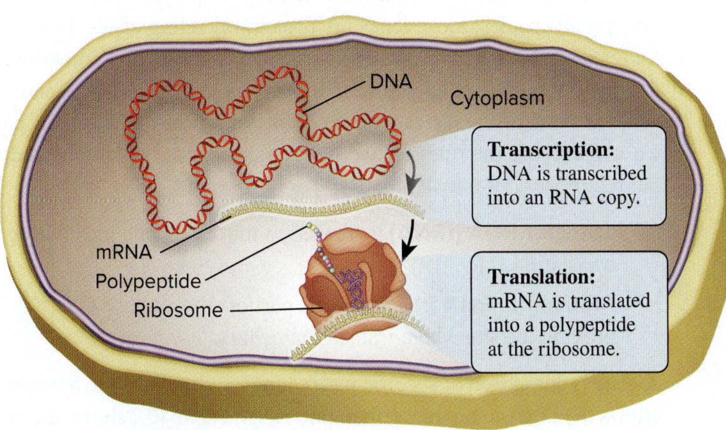


**(a) Molecular gene expression in bacteria**

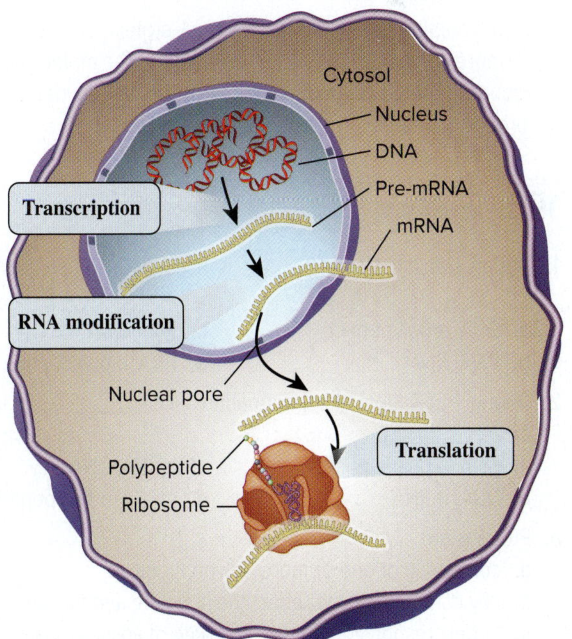


**(b) Molecular gene expression in eukaryotes**

**Figure 10.1 The central dogma of gene expression at the molecular level. (a)** In bacteria, transcription and translation occur in the cytoplasm. **(b)** In eukaryotes, transcription and RNA modification occur in the nucleus, whereas translation takes place in the cytosol.

Concept Check: *What is the direction of flow of genetic information?*

exits the nucleus through a nuclear pore, and translation occurs in the cytosol.

Though the direction of information flow, that is, from DNA to RNA to protein, is the most common pathway, exceptions do occur. For example, as we will see in Chapter 17, certain viruses called retroviruses use RNA as a template to synthesize DNA.

### The Protein Products of Genes Largely Determine an Organism's Characteristics

The genes that constitute the genetic material provide a blueprint for the characteristics of every organism. They contain the information necessary to produce an organism and allow it to favorably interact with its environment. Each protein-encoding gene stores the information for the production of a polypeptide, which then becomes a unit within a functional protein. The activities of proteins largely determine the structure and function of cells. Furthermore, the characteristics of an organism are rooted in the activities of cellular proteins. In addition, non-coding RNAs also are important in cell structure and function. This topic is discussed in Chapter 11.

One of the main purposes of the genetic material is to encode the production of proteins in the correct cell, at the proper time, and in suitable amounts. This is an intricate task, because living cells make thousands of different kinds of proteins. Genetic analyses have shown that a typical bacterium can make a few thousand different proteins, and estimates for eukaryotes range from several thousand in simpler eukaryotes to tens of thousands in more complex eukaryotes like humans.

### 10.1 Reviewing the Concepts

- At the molecular level, the central dogma states that most genes are transcribed into mRNA, and then the mRNA is translated into polypeptides. In eukaryotes, an additional step, RNA modification, occurs between transcription and translation (Figure 10.1).
- A polypeptide is a unit of structure. A protein, composed of one or more polypeptides, is a unit of function. The molecular expression of genes is an underlying factor that largely determines an organism's characteristics.

### 10.1 Testing Your Knowledge

1. During transcription, _______ is used as a template to make _______.
   a. DNA, a polypeptide
   b. DNA, RNA
   c. RNA, DNA
   d. RNA, a polypeptide
2. During translation, the information within _______ is used to make _______.
   a. DNA, a polypeptide
   b. DNA, mRNA
   c. mRNA, DNA
   d. mRNA, a polypeptide
3. Proteins are
   a. composed of one or more polypeptides.
   b. key components that affect the structure and function of cells.
   c. largely responsible for the outcome of an individual's traits.
   d. all of the above.

## 10.2 Transcription

**Learning Outcomes**

1. Give a molecular definition for the term gene.
2. Outline the three stages of transcription and the role of RNA polymerase in this process.
3. Compare and contrast transcription in bacteria and eukaryotes.

DNA is an information storage unit. For genes to be expressed, the information in them must be accessed at the molecular level. Rather than accessing the information directly, however, a working copy of the DNA, composed of RNA, is made. This occurs by the process of transcription, in which a DNA sequence is copied into an RNA sequence. Importantly, transcription does not permanently alter the structure of DNA. Therefore, the same DNA can continue to store information even after an RNA copy has been made. In this section, we will examine the steps necessary for genes to act as transcriptional units. We will also consider some differences in these steps between bacteria and eukaryotes.

### At the Molecular Level, a Gene Is Transcribed and Produces a Functional Product

What is the definition of a gene?

*At the molecular level, a* **gene** *is defined as an organized unit of nucleotide sequences that enables a segment of DNA to be transcribed into RNA and ultimately results in the formation of a functional product.*

When a protein-encoding gene is transcribed, an mRNA is made that specifies the amino acid sequence of a polypeptide. After it is made, the polypeptide becomes a functional product as a protein or part of a protein. The mRNA is an intermediary in polypeptide synthesis. Among all species, most genes are protein-encoding genes. However, for many genes, the functional product is the RNA itself. As mentioned, these RNAs are called non-coding RNAs (ncRNAs). Chapter 11 surveys some of the interesting functions of ncRNAs. In this chapter, we will explore the roles of two ncRNAs, because they are used during the process of translation.

- **Transfer RNA (tRNA)** is needed to make polypeptides during translation.
- **Ribosomal RNA (rRNA)** forms part of ribosomes, which provide the site where translation occurs.

A gene is composed of specific base sequences organized in a way that allows the DNA to be transcribed into RNA. Figure 10.2 shows the general organization of sequences in a protein-encoding gene in bacteria.

- **Promoter:** A site in the DNA where transcription begins
- **Terminator:** A site in the DNA where transcription ends
- **Regulatory sequences:** Sequences that function as sites for the binding of regulatory proteins that affect the rate of transcription. As discussed in Chapter 12, some regulatory proteins enhance the rate of transcription, whereas others inhibit it.

## Biology Principle

### The Genetic Material Provides a Blueprint for Reproduction

Transcription is the first step in accessing the information that is stored in DNA.

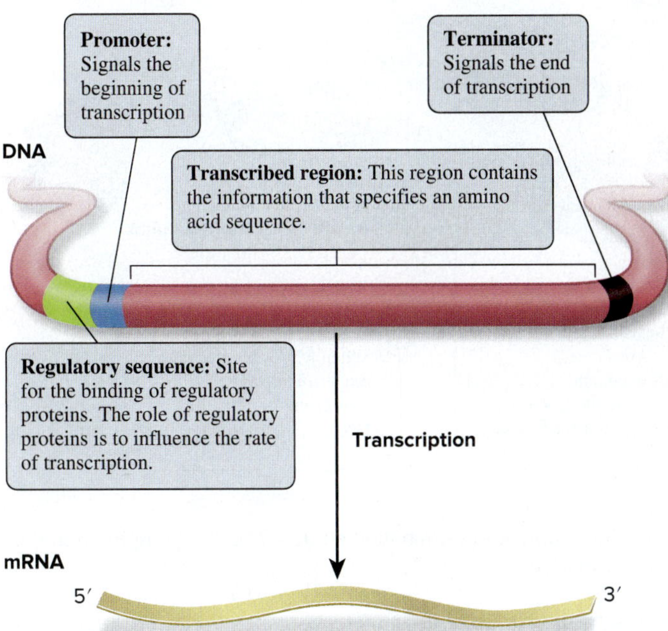


**Figure 10.2** **A bacterial protein-encoding gene as a transcriptional unit.** Note: As shown later in Figure 10.4, transcription in eukaryotes first produces a pre-mRNA that is modified before translation occurs.

**Concept Check:** *How would removing a terminator from a gene affect transcription? Where would transcription end?*

As shown in Figure 10.2, the DNA is transcribed into mRNA from the promoter to the terminator. Within this transcribed region is the information that will specify the amino acid sequence of a polypeptide when the mRNA is translated.

## During Transcription, RNA Polymerase Uses a DNA Template to Make RNA

Transcription occurs in three stages, called initiation, elongation, and termination, during which various proteins interact with DNA sequences (Figure 10.3).

1. The **initiation stage** is a recognition step. In bacteria, a protein called **sigma factor** binds to **RNA polymerase,** a type of enzyme that synthesizes strands of RNA. Sigma factor also recognizes the base sequence of a promoter and binds there. The role of sigma factor is to enable RNA polymerase to bind to the promoter. The initiation stage is completed when the DNA strands are separated near the promoter to form an **open complex** that is approximately 10–15 bp long.
2. During the **elongation stage,** RNA polymerase synthesizes the RNA transcript. For this to occur, sigma factor is released and RNA polymerase slides along the DNA in a way that maintains the open complex as it goes. The DNA strand that is used as a template for transcription is called the **template strand.** For protein-encoding genes, the opposite DNA strand is called the **coding strand.** The coding strand has the same sequence of bases as the resulting mRNA, except that it contains thymine, whereas the mRNA contains uracil. The coding strand is so named because, like mRNA, it carries the information that codes for a polypeptide.

   During the elongation stage of transcription, nucleotides bind to the template strand and are covalently connected in the 5′ to 3′ direction (see inset of step 2, Figure 10.3). The complementarity rule used in this process is similar to the AT/GC rule of DNA replication, except that in RNA, uracil (U) substitutes for thymine (T). For example, a DNA template with a sequence of 3′–TACAATGTAGCC–5′ will be transcribed into an RNA sequence reading 5′–AUGUUACAUCGG–3′. In bacteria, the rate of RNA synthesis is about 40 nucleotides per second! Behind the open complex, the DNA rewinds back into a double helix.
3. Eventually, RNA polymerase reaches a terminator, which causes it and the newly made RNA transcript to dissociate from the DNA. This event constitutes the termination stage of transcription.

## Transcription in Eukaryotes Involves More Proteins

The basic features of transcription are similar among all organisms. The genes of all species have promoters, and the transcription process occurs in the stages of initiation, elongation, and termination. However, the transcription of eukaryotic genes tends to involve a greater complexity of protein components than does the transcription of bacterial genes. For example, three forms of RNA polymerase—designated I, II, and III—are found in eukaryotes. RNA polymerase II is responsible for transcribing the mRNA from eukaryotic protein-encoding genes, whereas RNA polymerases I and III transcribe non-coding genes such as the genes that encode tRNAs and rRNAs. By comparison, bacteria have a single type of RNA polymerase that transcribes all genes, though many bacterial species have more than one type of sigma factor that can recognize different promoters.

The initiation stage of transcription is also more complex in eukaryotes. Recall that in bacteria, sigma factor recognizes the promoter of genes. By comparison, RNA polymerase II of eukaryotes always requires five **transcription factors,** proteins that influence the ability of RNA polymerase to transcribe genes, to initiate transcription. The roles of eukaryotic transcription factors are examined in Chapter 12.

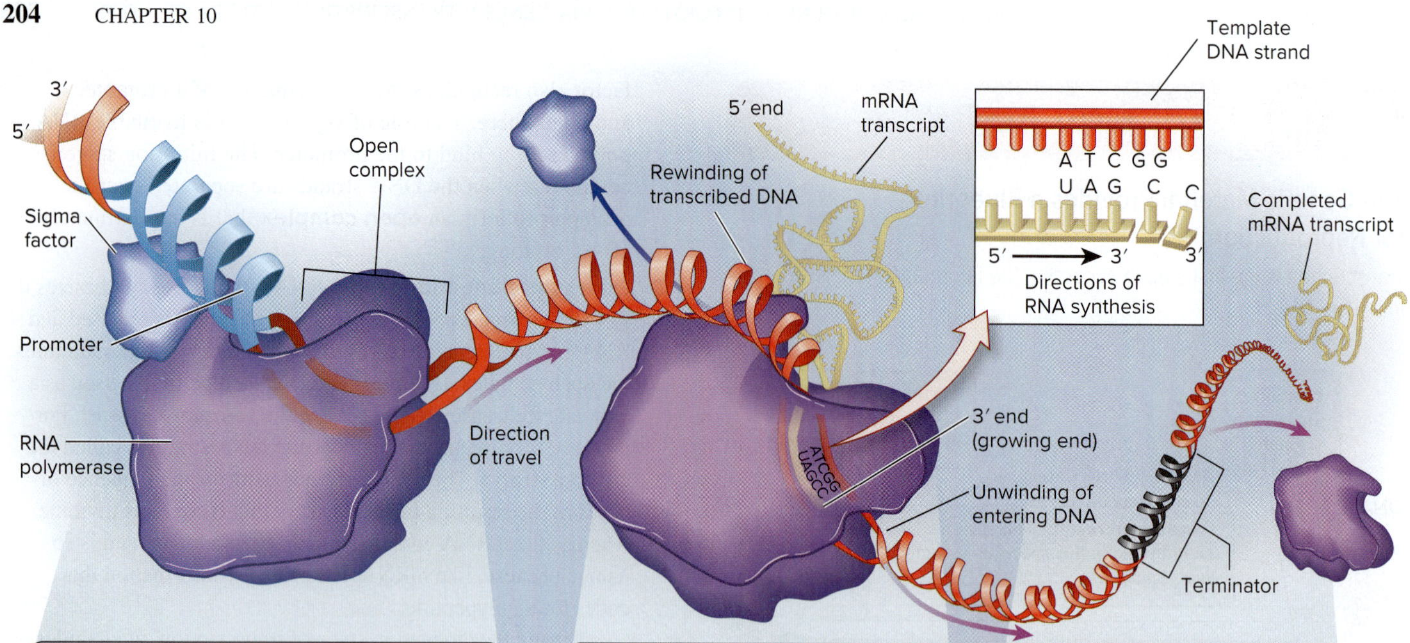

Template DNA strand
3′
5′
Sigma factor
Open complex
Promoter
RNA polymerase
Direction of travel
5′ end
mRNA transcript
Rewinding of transcribed DNA
A T C G G
U A G C C
5′ → 3′ 3′
Directions of RNA synthesis
Completed mRNA transcript
3′ end (growing end)
Unwinding of entering DNA
Terminator

1 **Initiation:**
The promoter functions as a recognition site for sigma factor. RNA polymerase is bound to sigma factor, which causes it to bind to the promoter. Following binding, the DNA is unwound to form an open complex.

2 **Elongation/synthesis of the RNA transcript:**
Sigma factor is released, and RNA polymerase slides along the DNA in an open complex to synthesize RNA.

3 **Termination:**
When RNA polymerase reaches the terminator, it and the RNA transcript dissociate from the DNA.

**Figure 10.3** **Stages of transcription.** Transcription can be divided into initiation, elongation, and termination stages. The inset emphasizes the direction of RNA synthesis and base pairing between the DNA template strand and RNA.

**BioConnections:** *Look back at DNA polymerase that is described in Figure 9.15a. What similarities and differences occur between the function of DNA polymerase and that of RNA polymerase?*

## 10.2 Reviewing the Concepts

- A site in a gene called a promoter specifies where transcription begins. A terminator specifies where transcription will end (Figure 10.2).
- In bacteria, the initiation of transcription begins when sigma factor binds to RNA polymerase and to a promoter. During elongation, an RNA transcript is synthesized due to base pairing of nucleotides to the template strand of DNA as RNA polymerase slides along the DNA. RNA polymerase and the RNA transcript dissociate from the DNA at the terminator (Figure 10.3).

## 10.2 Testing Your Knowledge

1. A region of the template strand of DNA used for transcription is 3′–AGCTTGAG–5′. What sequence of RNA would be transcribed from this region?
   a. 3′–AGCTTGAG–5′
   b. 5′–AGCTTGAG–3′
   c. 3′–UCGAACUC–5′
   d. 5′–UCGAACUC–3′
2. The stage of bacterial transcription in which sigma factor causes the binding of RNA polymerase to the promoter is
   a. initiation.
   b. elongation.
   c. termination.
   d. none of the above.

## 10.3 RNA Modifications in Eukaryotes

### Learning Outcomes

1. Describe the addition of the 5′ cap and 3′ poly A tail to eukaryotic mRNA.
2. Outline the process of splicing that produces mature eukaryotic mRNA.

As noted previously, eukaryotic transcripts, called **pre-mRNAs,** undergo certain modifications before they exit the nucleus as functionally active or **mature mRNAs.** For example, eukaryotic pre-mRNAs are modified by the addition of caps and tails to their ends (**Figure 10.4**). In addition, many eukaryotic pre-mRNAs have coding sequences that are separated by sequences that are later removed. These intervening sequences that are not translated are called **introns,** whereas sequences contained in the mature mRNA are termed **exons.** Exons are considered to be expressed regions, because they contain the coding sequence for a polypeptide. In contrast, introns are intervening regions that are not expressed, because they are removed from the pre-mRNA. To become a functional mRNA, many pre-mRNAs undergo an RNA modification known as **RNA splicing,** or simply splicing, in which introns are removed and the remaining exons are connected to each other (see Figure 10.4).

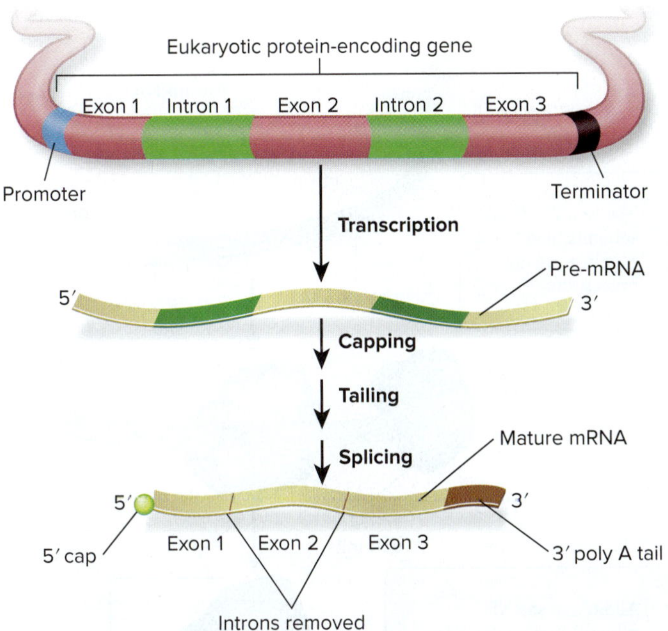


**Figure 10.4 Modifications to eukaryotic pre-mRNA that are needed to produce a mature mRNA molecule.** Note: Most RNA molecules are spliced after the pre-mRNA is completely synthesized. However, for some of them, splicing may begin before transcription of the pre-mRNA is completed.

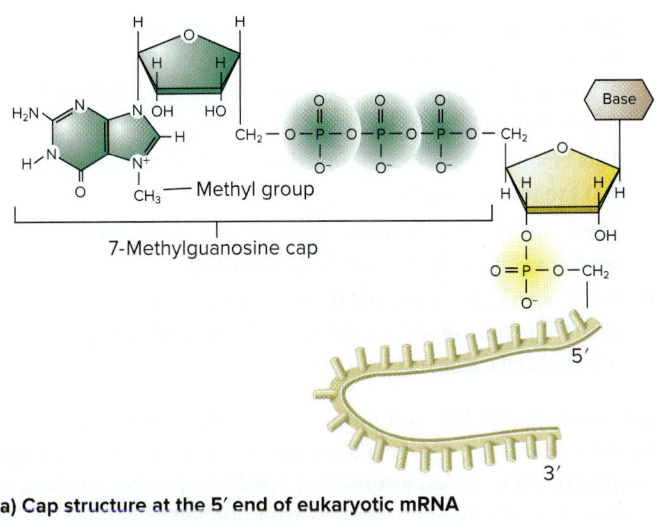


**(a) Cap structure at the 5′ end of eukaryotic mRNA**

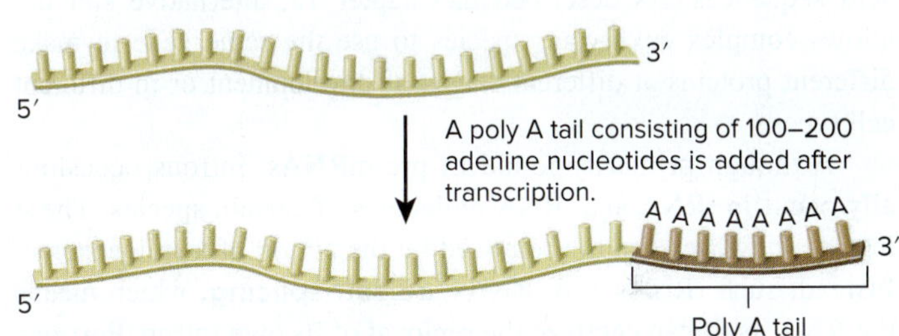


**(b) Addition of a poly A tail at the 3′ end of eukaryotic mRNA**

**Figure 10.5 Modifications that occur at the ends of mRNA in eukaryotic cells. (a)** A guanosine cap is attached to the 5′ end. This guanine base is modified by the attachment of a methyl group. The linkage between the cap and the mRNA is a 5′ to 5′ linkage. **(b)** A poly A tail is added to the 3′ end.

**Concept Check:** *Do the ends of protein-encoding genes have a poly T region that provides a template for the synthesis of a poly A tail in mRNA? Explain.*

After the modifications shown in Figure 10.4 have been completed, the mRNA leaves the nucleus and enters the cytosol, where translation occurs. In this section, we will examine the molecular mechanisms that account for RNA modifications and consider why they are functionally important.

## The Ends of Eukaryotic Pre-mRNAs Are Modified by the Addition of a 5′ Cap and a 3′ Poly A Tail

Mature mRNAs of eukaryotes have a modified form of guanine covalently attached at the 5′ end, an event known as **capping** (Figure 10.5a). Capping occurs while a pre-mRNA is being made by RNA polymerase, usually when the transcript is only 20 to 25 nucleotides in length. What are the functions of the cap? The 7-methylguanosine structure, called a **5′ cap,** is recognized by cap-binding proteins, which are needed for the proper exit of mRNAs from the nucleus. After an mRNA is in the cytosol, the cap structure helps to prevent its degradation and is recognized by other cap-binding proteins that enable the mRNA to bind to a ribosome for translation.

At the 3′ end, most mature eukaryotic mRNAs have a string of adenine nucleotides, typically 100 to 200 nucleotides in length, referred to as a **poly A tail** (Figure 10.5b). The poly A tail is not encoded in the gene sequence. Instead, the tail is added enzymatically after a pre-mRNA has been completely transcribed. A long poly A tail aids in the export of mRNAs from the nucleus. It also causes a eukaryotic mRNA to be more stable and thereby exist for a longer period of time in the cytosol.

## Splicing Involves the Removal of Introns and the Linkage of Exons

Introns are found in many but not all eukaryotic genes. Splicing is less frequent among unicellular eukaryotic species, such as yeast, but is a widespread phenomenon among more complex eukaryotes. In animals and flowering plants, most protein-encoding genes have one or more introns. For example, an average human gene has about nine introns. The sizes of introns vary from a few dozen nucleotides to over 100,000! A few bacterial genes have been found to have introns, but they are rare among bacterial and archaeal species.

Introns are precisely removed from eukaryotic pre-mRNA by a complex called a **spliceosome,** which is composed of several different subunits known as snRNPs (pronounced "snurps"). Each snRNP contains small nuclear RNA and a set of proteins. Splicing occurs in a series of steps (Figure 10.6):

1. Spliceosome subunits bind to specific sequences at three locations in the intron RNA: One is termed the branch site and two sequences at the intron-exon boundaries are called the 5′ splice site and the 3′ splice site.
2. This binding causes the intron to loop outward, which brings the two exons close together.
3. The 5′ splice site is then cut, and the 5′ end of the intron becomes covalently attached to the branch site.
4. The 3′ splice site is cut, and the two exons are covalently attached to each other. The intron, in the form of a loop, is released and eventually degraded.

In some cases, the function of the spliceosome is regulated so the splicing of exons for a given mRNA can occur in two or more ways. This phenomenon, called **alternative splicing,** allows a single gene to encode two or more polypeptides with differences in their amino acid sequences. As described in Chapter 12, alternative splicing allows complex eukaryotic species to use the same gene to make different proteins at different stages of development or in different cell types.

Although primarily found in pre-mRNAs, introns occasionally occur in rRNA and tRNA molecules of certain species. These introns, however, are not removed by the action of a spliceosome. Instead, such rRNAs and tRNAs are **self-splicing,** which means the RNA itself can catalyze the removal of its own intron. Portions of the RNA act like an enzyme to cleave the covalent bonds at the intron-exon boundaries and connect the exons together. An RNA molecule that catalyzes a chemical reaction is termed a **ribozyme.**

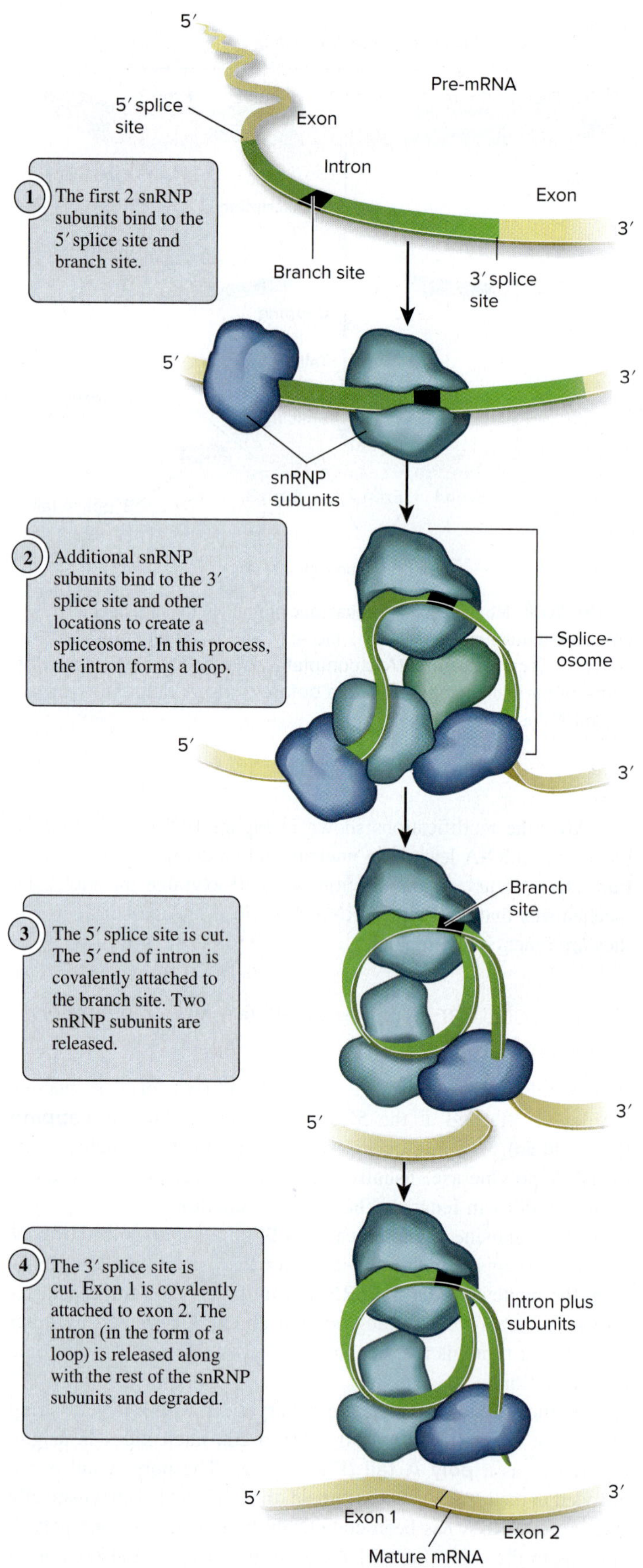


**Figure 10.6** The splicing of a eukaryotic pre-mRNA by a spliceosome.

## 10.3 Reviewing the Concepts

- In eukaryotes, transcription produces a pre-mRNA that is spliced and given a 5′ cap and a poly A tail (Figures 10.4, 10.5).
- During RNA splicing, intervening sequences called introns are removed from eukaryotic pre-mRNA by a spliceosome (Figure 10.6).

## 10.3 Testing Your Knowledge

1. The splicing of pre-mRNA in eukaryotes involves the
   a. removal of exons and the connection of introns.
   b. removal of introns and the connection of exons.
   c. removal of introns only at the 3′ end.
   d. removal of introns only at the 5′ end.
2. The function of the 5′ cap on eukaryotic mRNA is to
   a. aid in export from the nucleus.
   b. prevent degradation.
   c. aid in translation.
   d. all of the above.

## 10.4 Translation and the Genetic Code

### Learning Outcomes

1. Explain how the genetic code specifies the relationship between the sequence of codons in mRNA and the amino acid sequence of a polypeptide.
2. **SCISKILLS** ▶ Analyze the experiments of Nirenberg and Leder that led to the deciphering of the genetic code.

In the two previous sections, we considered how an RNA transcript is made from DNA and how eukaryotes process that transcript. Recall that this type of RNA is called messenger RNA (mRNA), because its function is to transmit information from DNA to ribosomes, where polypeptide synthesis occurs. In this section, we will consider the next stage—translation—which is the synthesis of polypeptides using information within mRNA. We will also explore experiments that helped to crack the code.

### The Genetic Code Specifies the Amino Acids Within a Polypeptide

The ability of mRNA to be translated into an amino acid sequence of a polypeptide relies on the **genetic code,** which specifies the relationship between the sequence of nucleotides in the mRNA and the sequence of amino acids in a polypeptide. The code is read in groups of three nucleotide bases known as **codons.** The genetic code consists of 64 different codons (Table 10.1). The sequence of three bases in most codons specifies a particular amino acid. For example, the codon CCC specifies the amino acid proline, whereas the codon GGC encodes the amino acid glycine. From the analysis of many different species, researchers have found that the genetic code is nearly universal. Only a few rare exceptions have been discovered.

Why are there 64 codons, as shown in Table 10.1? Because there are 20 types of amino acids, at least 20 different codons are needed so each amino acid can be specified by a codon. With four types of bases in mRNA (U, C, A, and G), a genetic code containing two bases in a codon would not be sufficient, because only $4^2$, or 16, different codons would be possible. A three-base system can specify $4^3$, or 64, different codons, which is more than the number of amino acids. The genetic code is said to be **degenerate,** or redundant, because more than one codon can specify the same amino acid (see Table 10.1). For example, the codons GGU, GGC, GGA, and GGG all code for the amino acid glycine. In most instances, the third base in the codon is the degenerate, or variable, base.

### During Translation, mRNA Is Used to Make a Polypeptide with a Specific Amino Acid Sequence

Let's look at the organization of a bacterial mRNA to see how translation occurs (Figure 10.7):

- **Ribosomal-binding site**: This site is located near the 5′ end of the mRNA and provides a location for the ribosome to bind to the mRNA.

**Table 10.1 The Genetic Code***

| First Position | Second Position: U | | C | | A | | G | | Third Position |
|---|---|---|---|---|---|---|---|---|---|
| | U | | C | | A | | G | | |
| U | UUU | Phe | UCU | Ser | UAU | Tyr | UGU | Cys | U |
| | UUC | | UCC | | UAC | | UGC | | C |
| | UUA | Leu | UCA | | UAA | Stop | UGA | Stop | A |
| | UUG | | UCG | | UAG | Stop | UGG | Trp | G |
| C | CUU | Leu | CCU | Pro | CAU | His | CGU | Arg | U |
| | CUC | | CCC | | CAC | | CGC | | C |
| | CUA | | CCA | | CAA | Gln | CGA | | A |
| | CUG | | CCG | | CAG | | CGG | | G |
| A | AUU | Ile | ACU | Thr | AAU | Asn | AGU | Ser | U |
| | AUC | | ACC | | AAC | | AGC | | C |
| | AUA | | ACA | | AAA | Lys | AGA | Arg | A |
| | AUG | Met/start | ACG | | AAG | | AGG | | G |
| G | GUU | Val | GCU | Ala | GAU | Asp | GGU | Gly | U |
| | GUC | | GCC | | GAC | | GGC | | C |
| | GUA | | GCA | | GAA | Glu | GGA | | A |
| | GUG | | GCG | | GAG | | GGG | | G |

*These codon sequences are in the 5′ to 3′ direction, going from left to right. Exceptions to the genetic code are rarely found among various species. For example, AUA encodes methionine in yeast and in mammalian mitochondria.

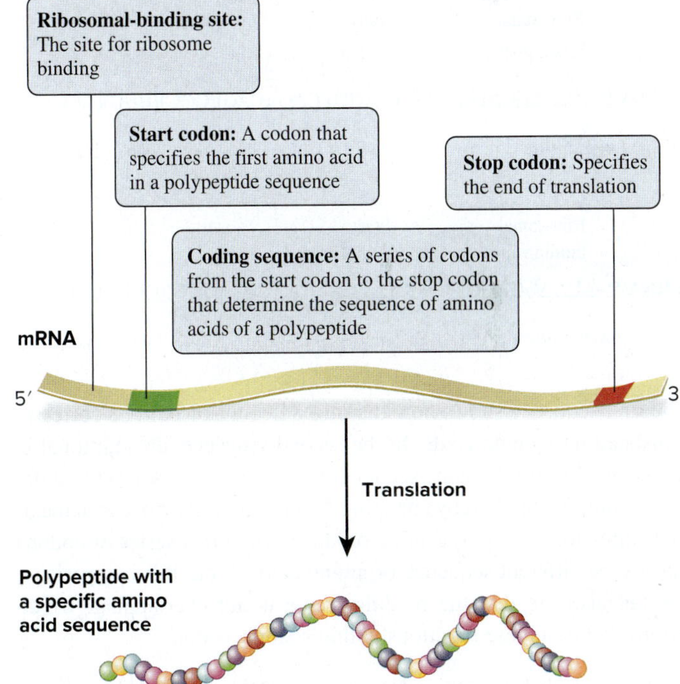


**Figure 10.7 The organization of a bacterial mRNA as a translational unit.** The string of colored balls represents a sequence of amino acids in a polypeptide. During and following translation, a sequence of amino acids folds into a more compact structure, as described in Chapter 3 (refer back to Figure 3.13).

**Concept Check:** *If a mutation eliminated the start codon from a gene, how would the mutation affect transcription, and how would it affect translation?*

- **Start codon:** The first codon that is used in translation. This codon (AUG) specifies the amino acid methionine and is only a few nucleotides from the ribosomal-binding site. The first amino acid during polypeptide synthesis is methionine, but it may be removed after the synthesis is completed.
- **Coding sequence:** A region that begins with the start codon and specifies the entire amino acid sequence of a polypeptide. A typical polypeptide is a few hundred amino acids in length. The coding sequence consists of a series of codons.
- **Stop codon:** The last codon that signals the end of translation. These codons, also known as termination codons, are UAA, UAG, and UGA. They do not code for an amino acid.

The start codon also defines the **reading frame** of an mRNA, which refers to the order in which codons are read during translation. Beginning at the start codon, each adjacent codon is read as a group of three bases, also called a **triplet,** in the 5′ to 3′ direction. For example, let's look at the following two mRNA sequences and their corresponding amino acid sequences.

Ribosomal-binding site — Start codon

**mRNA** 5′–AUAAGGAGGUUACG(AUG)(CAG)(CAG)(GGC)(UUU)(ACC)–3′

**Polypeptide** Met - Gln - Gln - Gly - Phe - Thr

Ribosomal-binding site — Start codon

**mRNA** 5′–AUAAGGAGGUUACG(AUG)(UCA)(GCA)(GGG)(CUU)(UAC)C–3′

**Polypeptide** Met - Ser - Ala - Gly - Leu - Tyr

The first sequence shows how the mRNA codons would be correctly translated into amino acids. In the second sequence, an additional U has been added to the same sequence after the start codon. This shifts the reading frame, thereby changing the codons as they occur in the 5′ to 3′ direction. The polypeptide produced from this series of codons has a very different sequence of amino acids. From this comparison, we can also see that the reading frame is not overlapping, which means that each base functions within a single codon.

## DNA Stores Information, Whereas mRNA and tRNA Access That Information to Make a Polypeptide

The relationships among the DNA sequence of a gene, the mRNA transcribed from the gene, and the polypeptide sequence are shown schematically in Figure 10.8. Recall that the template strand is used to make mRNA. The resulting mRNA strand corresponds to the coding strand of DNA, except that U in the mRNA substitutes for T in the DNA. The 5′ end of the mRNA contains an untranslated region (5′ UTR), as does the 3′ end (3′ UTR). The middle portion contains a series of codons that specify the amino acid sequence of a polypeptide.

To translate a nucleotide sequence of mRNA into an amino acid sequence, recognition occurs between mRNA and transfer RNA (tRNA) molecules. Transfer RNA, which is described in Section 10.5, functions as the "translator," or intermediary, between an mRNA codon and an amino acid. The **anticodon** is a three-base sequence in a tRNA molecule that is complementary to a codon in mRNA. Due to this complementarity, the anticodon in the tRNA and a codon in an mRNA bind to each other. Furthermore, the anticodon in a tRNA corresponds to the amino acid that it carries. For example, if the anticodon in a tRNA is 3′–AAG–5′, it is complementary to a 5′–UUC–3′ codon. According to the genetic code, a UUC codon specifies phenylalanine (Phe). Therefore, a tRNA with a 3′–AAG–5′ anticodon must carry phenylalanine.

As seen at the bottom of Figure 10.8, the direction of polypeptide synthesis parallels the 5′ to 3′ orientation of mRNA. The first amino acid is said to be at the amino end, or **N-terminus,** of the polypeptide. The term N-terminus refers to the presence of a nitrogen atom (N) at this end, whereas amino end indicates the presence of an amino group ($NH_2$). **Peptide bonds** connect the amino acids together. These covalent bonds form between the carboxyl group (COOH) of the previous amino acid and the amino group of the next amino acid. The last amino acid in a completed polypeptide does not have another amino acid attached to its carboxyl group. This last amino acid is said to be located at the carboxyl end, or **C-terminus.** A carboxyl group is always found at this end of the polypeptide. Note that at neutral pH, the amino group is positively charged ($NH_3^+$), whereas the carboxyl group is negatively charged ($COO^-$).

## Synthetic RNA Helped to Decipher the Genetic Code

Let's look at some early experiments that allowed scientists to decipher the genetic code. During the early 1960s, the genetic code was determined by the collective efforts of several researchers, including American biochemist Marshall Nirenberg, Spanish-American biochemist Severo Ochoa, and American geneticist Philip Leder. Prior to their studies, other scientists had discovered that bacterial cells can be broken open and components from the cytoplasm can synthesize polypeptides. This is termed an **in vitro translation system.** Nirenberg and Ochoa made synthetic RNA molecules using an enzyme that covalently connects nucleotides together. Using this synthetic mRNA, they then determined which amino acids were incorporated into polypeptides. For example, if an RNA molecule had only adenine-containing nucleotides (for example, 5′–AAAAAAAAAAAAAAAAAA–3′), a polypeptide was produced that contained only lysine. This result indicated that the AAA codon specifies lysine.

Another method used to decipher the genetic code involved the chemical synthesis of short RNA molecules, as described in the Feature Investigation.

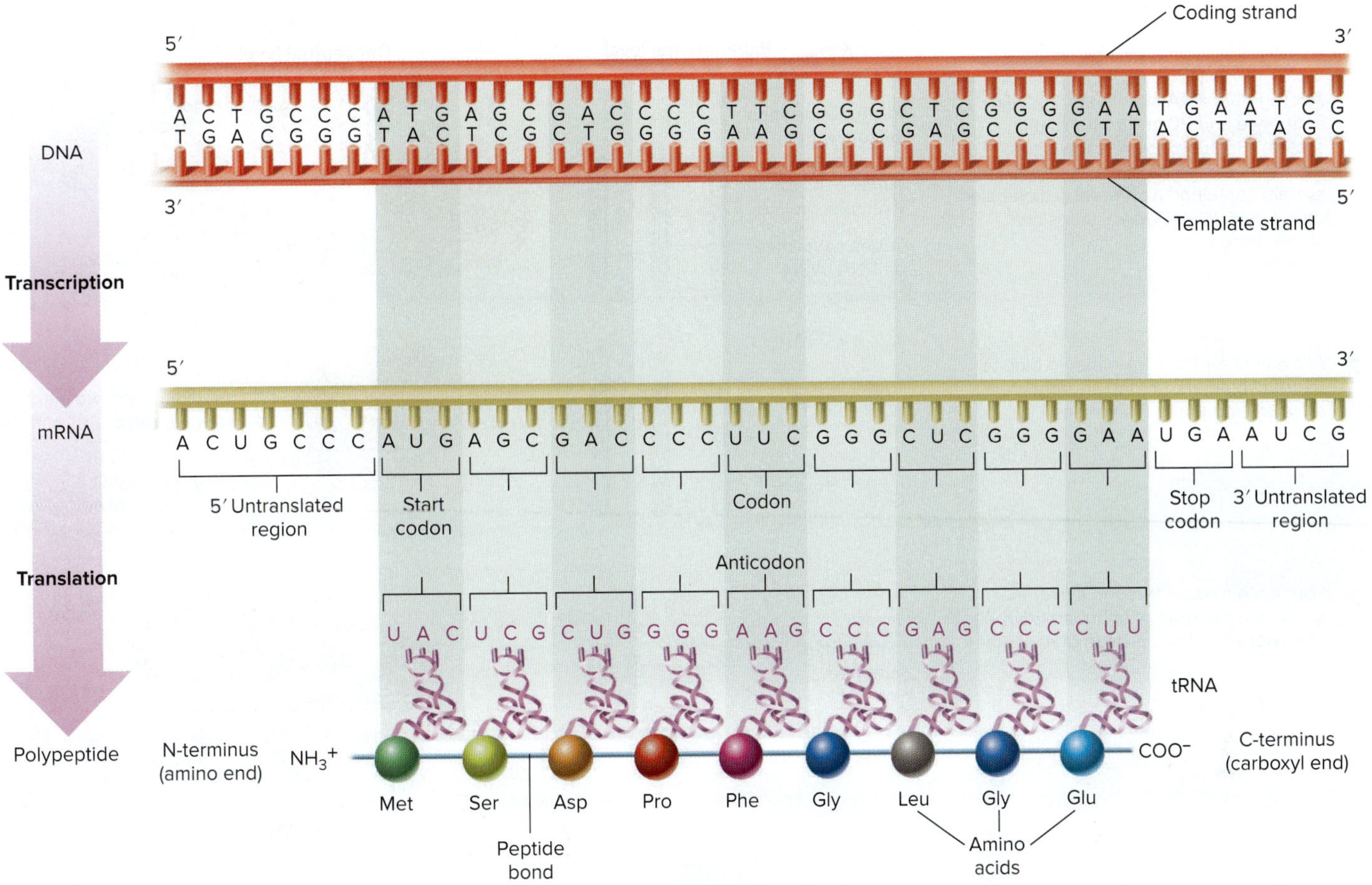

DNA
Coding strand
5′ A C T G C C C A T G A G C G A C C C C T T C G G G C T C G G G G A A T G A A T C G 3′
3′ T G A C G G G T A C T C G C T G G G G A A G C C C G A G C C C C T T A C T T A G C 5′
Template strand

Transcription

mRNA
5′ A C U G C C C A U G A G C G A C C C C U U C G G G C U C G G G G A A U G A A U C G 3′
5′ Untranslated region
Start codon
Codon
Stop codon
3′ Untranslated region

Translation

Anticodon
U A C U C G C U G G G G A A G C C C G A G C C C C U U
tRNA

Polypeptide
N-terminus (amino end) $NH_3^+$ Met Ser Asp Pro Phe Gly Leu Gly Glu $COO^-$ C-terminus (carboxyl end)
Peptide bond
Amino acids

**Figure 10.8** Relationships among the coding sequence of a gene, the coding sequence of an mRNA, the anticodons of tRNA, and the amino acid sequence of a polypeptide.

*If an anticodon in a tRNA molecule has the sequence 3′–ACC–5′, which amino acid does it carry?*

# FEATURE INVESTIGATION

## Nirenberg and Leder Found That RNA Triplets Can Promote the Binding of tRNA to Ribosomes

In 1964, Nirenberg and Leder discovered that RNA molecules containing three nucleotides (that is, a triplet) can cause a tRNA molecule to bind to a ribosome. In other words, an RNA triplet acts like a codon within an mRNA molecule. To establish the relationship between triplet sequences and specific amino acids, Nirenberg and Leder made triplets with specific base sequences (**Figure 10.9**). For example, in one experiment they studied 5′–CCC–3′ triplets. This particular triplet was added to 20 different tubes. To each tube, they next added an in vitro translation system, which contained ribosomes and tRNAs that already had amino acids attached to them. However, each translation system had only one type of radiolabeled amino acid. One translation system had only

**Figure 10.9** Nirenberg and Leder's use of triplet binding to decipher the genetic code.

**HYPOTHESIS** An RNA triplet can bind to a ribosome and promote the binding of the tRNA that carries the amino acid that the RNA triplet specifies.

**KEY MATERIALS** The researchers made 20 in vitro translation systems, which included ribosomes, tRNAs, and 20 amino acids. The 20 translation systems differed with regard to which amino acid was radiolabeled. For example, in 1 translation system, radiolabeled glycine was added, and the other 19 amino acids were unlabeled. In another system, radiolabeled proline was added, and the other 19 amino acids were unlabeled. The in vitro translation systems also contained the enzymes that attach amino acids to tRNAs.

**Experimental level** | **Conceptual level**

1 Mix together RNA triplets of a specific sequence and 20 in vitro translation systems. In the example shown here, the triplet is 5′–CCC–3′. Each translation system contained a different radiolabeled amino acid. (Note: Only 3 tubes are shown here.)

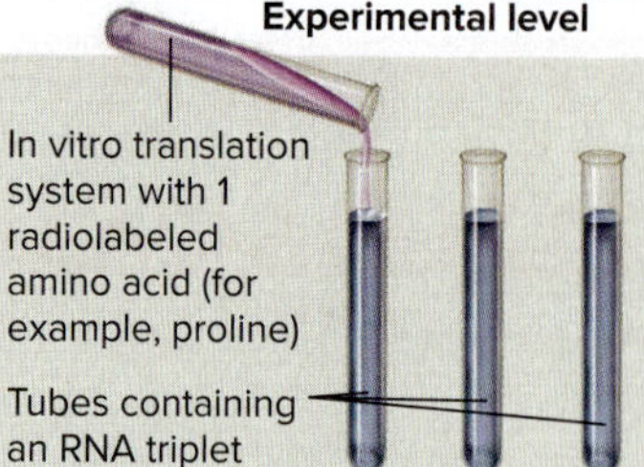


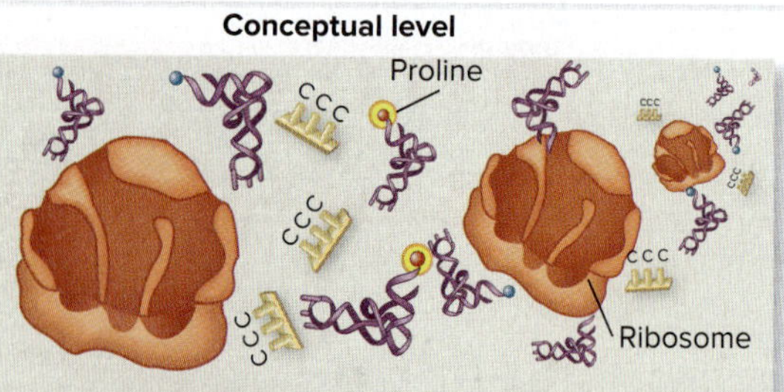


2 Allow time for the RNA triplet to bind to the ribosome and for the appropriate tRNA to bind to the RNA triplet.

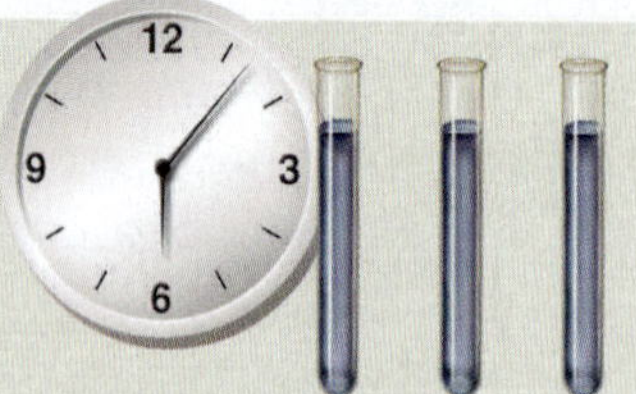


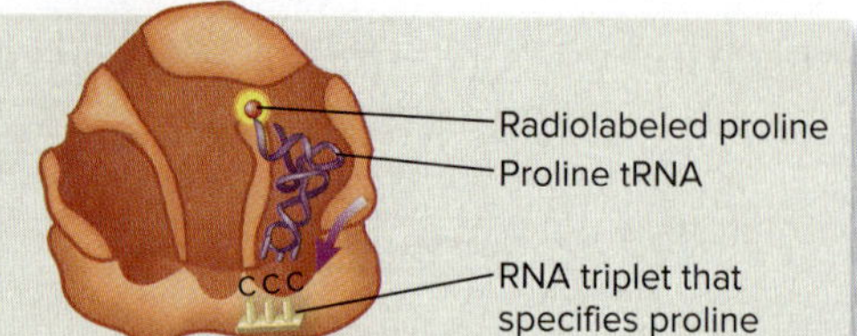


3 Pour each mixture through a filter that allows the passage of unbound tRNA but does not allow the passage of ribosomes.

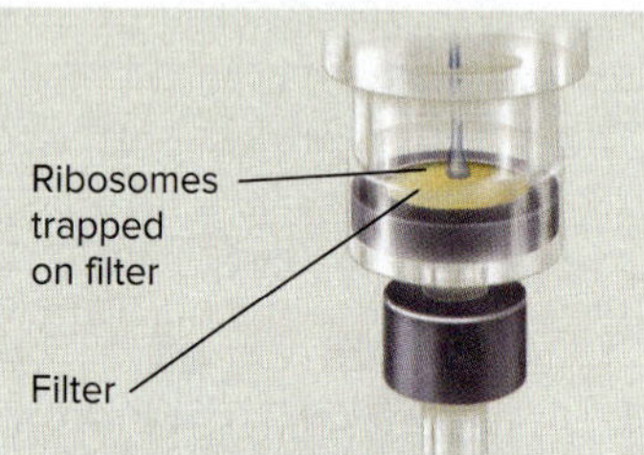


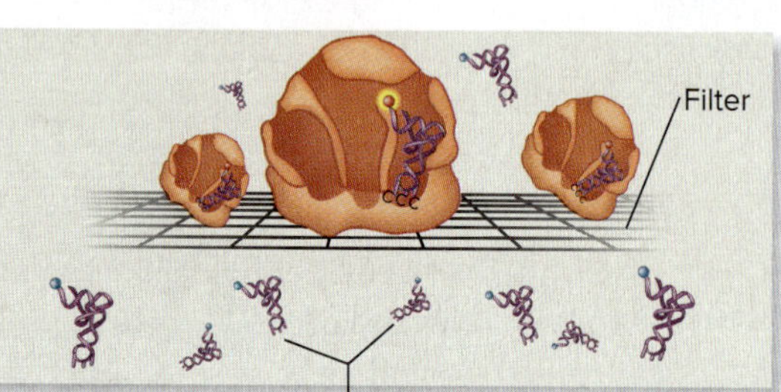


4 Count radioactivity on the filter.

5 **THE DATA**

| **Triplet** | **Radiolabeled amino acid trapped on the filter** | **Triplet** | **Radiolabeled amino acid trapped on the filter** |
|---|---|---|---|
| 5′–AAA–3′ | Lysine | 5′–GAC–3′ | Aspartic acid |
| 5′–ACA–3′ | Threonine | 5′–GCC–3′ | Alanine |
| 5′–ACC–3′ | Threonine | 5′–GGU–3′ | Glycine |
| 5′–AGA–3′ | Arginine | 5′–GGC–3′ | Glycine |
| 5′–AUA–3′ | Isoleucine | 5′–GUU–3′ | Valine |
| 5′–AUU–3′ | Isoleucine | 5′–UAU–3′ | Tyrosine |
| 5′–CCC–3′ | Proline | 5′–UGU–3′ | Cysteine |
| 5′–CGC–3′ | Arginine | 5′–UUG–3′ | Leucine |
| 5′–GAA–3′ | Glutamic acid | | |

6 **CONCLUSION** This method enabled the researchers to identify many of the codons of the genetic code.

7 **SOURCE** Leder, Philip, and Nirenberg, Marshall W. 1964. RNA Codewords and Protein Synthesis, III. On the Nucleotide Sequence of a Cysteine and a Leucine RNA Codeword. *Proceedings of the National Academy of Sciences* 52: 1521–1529.

proline that was radiolabeled, a second translation system had only radiolabeled serine, and so on.

As shown in step 2 of Figure 10.9, the triplets became bound to the ribosomes just like the binding of mRNA to a ribosome. The tRNA with an anticodon that was complementary to the added triplet would bind to the triplet, which was already bound to the ribosome. For example, when the triplet was 5′–CCC–3′, a tRNA with a 3′–GGG–5′ anticodon would bind to the triplet/ribosome complex. This tRNA carries proline.

To determine which tRNA had bound, the contents from each tube were poured through a filter that trapped the large ribosomes but did not trap tRNAs that were not bound to ribosomes (see step 3). If the tRNA carrying the radiolabeled amino acid was bound to the triplet/ribosome complex, radioactivity would be trapped on the filter. Using a scintillation counter, the researchers determined the amount of radioactivity on each filter. Because only one amino acid was radiolabeled in each in vitro translation system, they could determine which triplet corresponded to which amino acid. In the example shown here, CCC corresponds to proline. Therefore, the in vitro translation system containing radiolabeled proline showed a large amount of radioactivity on the filter. As seen in the data, by studying triplets with different sequences, Nirenberg and Leder identified many codons of the genetic code.

*Experimental Questions*

1. Briefly explain how a triplet mimics the role of an mRNA molecule. How was this observation useful in the study done by Nirenberg and Leder?
2. **SCISKILLS** ▶ What was the benefit of using radiolabeled amino acids in the Nirenberg and Leder experiment?
3. **SCISKILLS** ▶ Predict the results that Nirenberg and Leder would have found for the following triplets: AUG, UAA, UAG, and UGA.

## 10.4 Reviewing the Concepts

- Based on the genetic code, each of the 64 codons specifies a start codon (AUG, which specifies methionine), other amino acids, or a stop codon (UAA, UAG, or UGA) (Table 10.1, Figure 10.7).
- The template strand of DNA is used to make mRNA with a series of codons. Recognition between mRNA and many tRNA molecules determines the amino acid sequence of a polypeptide. A polypeptide has a directionality in which the first amino acid is at the N-terminus, or amino end, whereas the last amino acid is at the C-terminus, or carboxyl end (Figure 10.8).
- Nirenberg and Leder used the ability of RNA triplets to promote the binding of tRNA to ribosomes as a way to identify many of the codons of the genetic code (Figure 10.9).

## 10.4 Testing Your Knowledge

1. The codon 5′–GAC–3′ specifies aspartic acid. What would be the anticodon of a tRNA that recognized this aspartic acid codon?
   a. 5′–GAC–3′
   b. 3′–GAC–5′
   c. 5′–CUG–3′
   d. 3′–CUG–5′
2. The reading frame of an mRNA
   a. begins with the start codon.
   b. is read as a series of adjacent codons.
   c. is not affected if a single nucleotide is added to the coding sequence.
   d. both a and b.

## 10.5 The Machinery of Translation

**Learning Outcomes**

1. Describe the structure and function of tRNA.
2. Explain how aminoacyl-tRNA synthases attach amino acids to tRNAs.
3. Outline the structural features of bacterial and eukaryotic ribosomes.
4. **SCISKILLS** ▶ Discuss how ribosomal RNA (rRNA) is used to evaluate evolutionary relationships among different species.

In the remaining sections of this chapter, we will turn our attention to the mechanism by which living cells use the genetic code to translate mRNA into polypeptides. A cell must make many different components, including mRNAs, tRNAs, ribosomes, and translation factors, so that polypeptides can be made (**Table 10.2**).

Though the estimates vary from cell to cell and from species to species, most cells use a substantial amount of their energy to translate mRNA into polypeptides. In *E. coli*, for example, approximately 90% of the cellular energy is used for this process. This value underscores the complexity and importance of translation in living organisms. In this section, we will focus on the components of the translation machinery.

### Transfer RNAs Share Common Structural Features

To understand how tRNAs function as carriers of the correct amino acids during translation, researchers have examined their structural characteristics. The tRNAs of all species share common features.

- The two-dimensional structure of a tRNA exhibits a cloverleaf pattern. The structure has three stem-loops and a fourth stem

**Table 10.2 Components of the Translation Machinery**

| Component | Function |
|---|---|
| mRNA | Contains the information for the sequence of amino acids in a polypeptide according to the genetic code. |
| tRNA | A molecule with two functional sites: one—the anticodon—binds to a codon in mRNA; a second site—the 3′ single-stranded region—is where an appropriate amino acid is attached. |
| Ribosome | Composed of rRNA molecules and many proteins, the ribosome provides a location where mRNA and tRNA molecules can properly interact with each other. The ribosome also catalyzes the formation of covalent bonds between adjacent amino acids so that a polypeptide can be made. |
| Translation factors | Proteins needed for the three stages of translation. (1) Initiation factors are required for the assembly of mRNA, the first tRNA, and ribosomal subunits. (2) Elongation factors are needed to synthesize the polypeptide. (3) Release factors are needed to recognize the stop codon and disassemble the translation machinery. Several translation factors use GTP as an energy source to carry out their functions. |

with a 3′ single-stranded region (Figure 10.10a). The stem in a stem-loop is a region where the RNA is double-stranded due to complementary base pairing via hydrogen bonding, whereas the loop is a region without base pairing.

- The anticodon is located in the middle loop.
- The 3′ single-stranded region is the amino acid attachment site.
- The three-dimensional structure of tRNA molecules involves additional folding of the secondary structure (Figure 10.10b).

The cells of every organism make many different tRNA molecules, each encoded by a different gene. A tRNA is named according to the amino acid it carries. For example, $tRNA^{Ser}$ carries a serine. Because the genetic code contains six different serine codons, as shown in Table 10.1, a cell produces more than one type of $tRNA^{Ser}$.

## Aminoacyl-tRNA Synthetases Charge tRNAs by Attaching an Appropriate Amino Acid

To perform its role during translation, a tRNA must have the appropriate amino acid attached to its 3′ end. The enzymes that catalyze the attachment of amino acids to tRNA molecules are known as **aminoacyl-tRNA synthetases.** Cells make 20 distinct types of aminoacyl-tRNA synthetase enzymes, with each type recognizing just 1 of the 20 different amino acids. Each aminoacyl-tRNA synthetase is named for the specific amino acid it attaches to tRNA. For example, alanyl-tRNA synthetase recognizes alanine and attaches this amino acid to all tRNAs with alanine anticodons.

## Biology Principle

### Structure Determines Function

The structure of tRNA has two functional sites: a 3′ single-stranded region where an amino acid is attached and an anticodon that binds to a codon in mRNA.

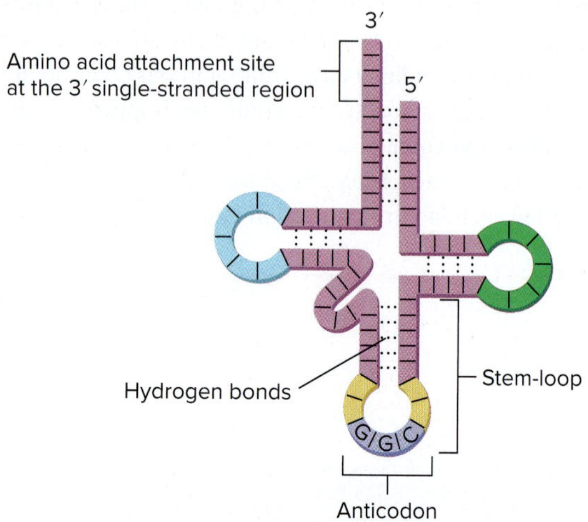


**(a) Two-dimensional structure of tRNA**

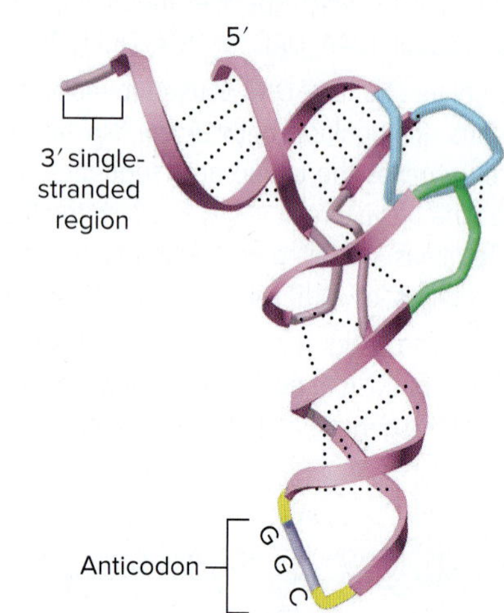


**(b) Three-dimensional structure of tRNA**

**Figure 10.10 Structure of tRNA.** **(a)** The two-dimensional or secondary structure of tRNA is that of a cloverleaf, with the anticodon within the middle loop. The 3′ single-stranded region is where an amino acid can attach. **(b)** The actual three-dimensional structure of tRNA folds in on itself.

*What is the function of the anticodon?*

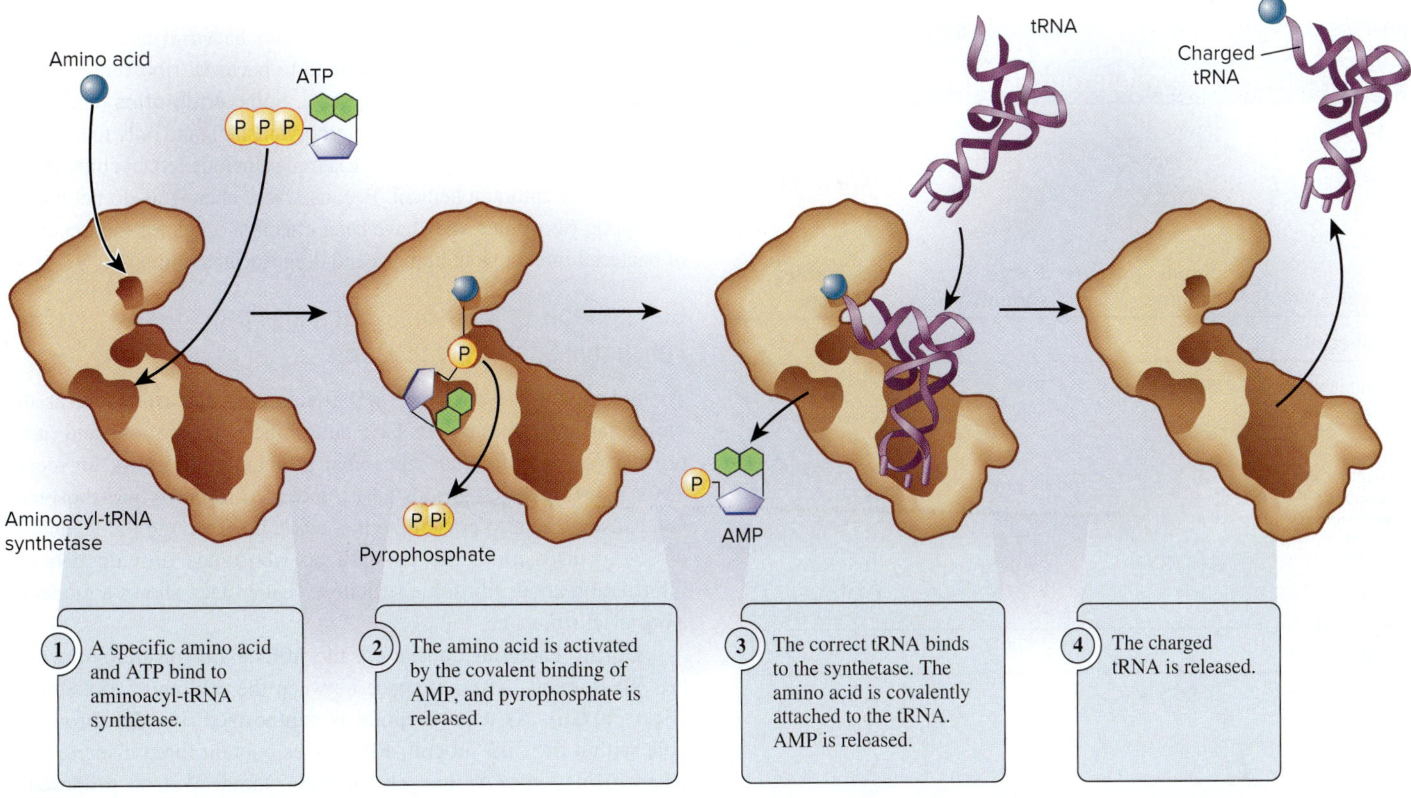


**Figure 10.11** Aminoacyl-tRNA synthetase charging a tRNA.

**BioConnections:** *Refer back to Figure 6.3, which describes the hydrolysis of ATP. Why is ATP needed to charge a tRNA?*

Aminoacyl-tRNA synthetases catalyze chemical reactions involving an amino acid, a tRNA molecule, and ATP (Figure 10.11).

1. A specific amino acid and ATP bind to the enzyme.
2. The amino acid is activated by the covalent attachment of an adenosine monophosphate (AMP) and pyrophosphate ($PP_i$) is released.
3. A specific tRNA binds to the enzyme. The activated amino acid is covalently attached to the 3′ end of the tRNA molecule, and AMP is released.
4. The tRNA with its attached amino acid, called a **charged tRNA,** or an **aminoacyl tRNA,** is released from the enzyme.

Aminoacyl-tRNA synthetases are amazingly accurate enzymes. The wrong amino acid is attached to a tRNA less than once in 100,000 times! The anticodon region of the tRNA is usually important for recognition by the correct aminoacyl-tRNA synthetase. In addition, the base sequences in other regions may facilitate binding to an aminoacyl-tRNA synthetase.

## Ribosomes Are Composed of rRNA and Proteins

Let's now turn our attention to the **ribosome,** the site where translation takes place. Bacterial cells have one type of ribosome, which translates all mRNAs in the cytoplasm. The most abundant type of eukaryotic ribosome functions in the cytosol. In addition, mitochondria have ribosomes, and plant and algal cells have ribosomes in their chloroplasts. Unless otherwise noted, the term eukaryotic ribosome refers to ribosomes in the cytosol, not to those found in mitochondria or chloroplasts.

A ribosome is a large complex composed of structures called the large and small subunits. The term subunit is perhaps misleading, because each ribosomal subunit is itself assembled from many different proteins and one or more RNA molecules. In the bacterium *E. coli*, the small ribosomal subunit is called 30S, and the large subunit is 50S. The designations 30S and 50S refer to the rate at which these subunits sediment when subjected to a centrifugal force. This rate is described as a sedimentation coefficient in Svedberg units (S) in honor of Swedish chemist Theodor Svedberg, who invented the ultracentrifuge. Together, the 30S and 50S subunits form a 70S ribosome. (Svedberg units don't add up linearly, because the sedimentation coefficient is a function of both size and shape.)

The compositions of bacterial and eukaryotic ribosomes are described in Table 10.3. In bacteria, ribosomal proteins and rRNA molecules are synthesized in the cytoplasm, and the ribosomal subunits are assembled there as well. The synthesis of eukaryotic rRNA occurs in the nucleolus, a region of the nucleus that is specialized for that purpose. The ribosomal proteins are made in the cytosol and imported into the nucleus. The rRNAs and ribosomal proteins are then assembled within the nucleolus to make the 40S and 60S subunits. The 40S and 60S subunits are exported into the cytosol, where they associate to form an 80S ribosome during translation.

**Table 10.3 Composition of Bacterial and Eukaryotic Ribosomes**

| | Small subunit | Large subunit | Assembled ribosome |
|---|---|---|---|
| **Bacterial** | | | |
| Sedimentation coefficient | 30S | 50S | 70S |
| Number of proteins | 21 | 34 | 55 |
| rRNA | 16S rRNA | 5S rRNA, 23S rRNA | 16S rRNA, 5S rRNA, 23S rRNA |
| **Eukaryotic** | | | |
| Sedimentation coefficient | 40S | 60S | 80S |
| Number of proteins | 33 | 49 | 82 |
| rRNA | 18S rRNA | 5S rRNA, 5.8S rRNA, 28S rRNA | 18S rRNA, 5S rRNA, 5.8S rRNA, 28S rRNA |

Due to structural differences between bacterial and eukaryotic ribosomes, certain chemicals may bind to bacterial ribosomes but not to eukaryotic ribosomes, and vice versa. Some **antibiotics** (chemicals that inhibit the growth of certain microorganisms) bind only to bacterial ribosomes and inhibit translation. Examples include tetracycline, erythromycin, and chloramphenicol. Because these chemicals do not inhibit eukaryotic ribosomes, they have been effective drugs for the treatment of bacterial infections in humans and domesticated animals.

## Components of Ribosomal Subunits Form Functional Sites for Translation

To understand the structure and function of the ribosome at the molecular level, researchers have determined the locations and functional roles of individual ribosomal proteins and rRNAs. In recent years, a few research groups have succeeded in purifying ribosomes and causing them to crystallize in a test tube. Through the technique of X-ray diffraction, the crystallized ribosomes provide detailed information about ribosome structure. Figure 10.12a shows a model of a bacterial ribosome.

During bacterial translation, the mRNA lies on the surface of the 30S subunit, within a space between the 30S and 50S subunits (Figure 10.12b). As a polypeptide is synthesized, it exits through a hole within the 50S subunit. Ribosomes contain three discrete sites where tRNAs may be located. In 1964, James Watson proposed a two-site model for tRNA binding to the ribosome. These sites are known as the **peptidyl site (P site)** and **aminoacyl site (A site).** In 1981, German geneticists Knud Nierhaus and Hans-Jörg Rheinberger expanded this to a three-site model (see Figure 10.12b). The third site is known as the **exit site (E site).** In Section 10.6, we will examine the roles of these sites in the synthesis of a polypeptide.

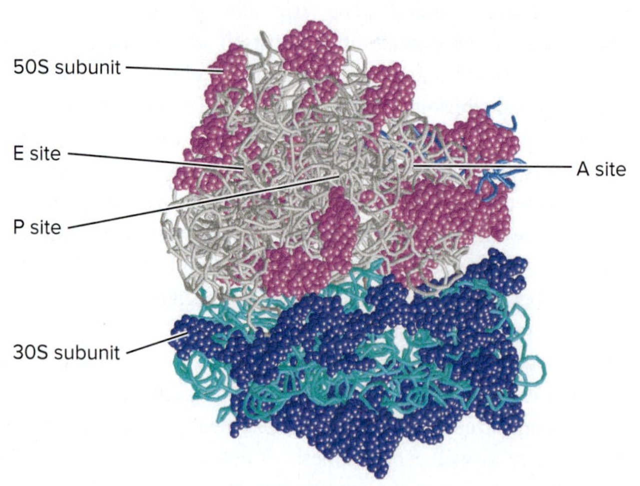


**(a) Bacterial ribosome model based on X-ray diffraction studies**

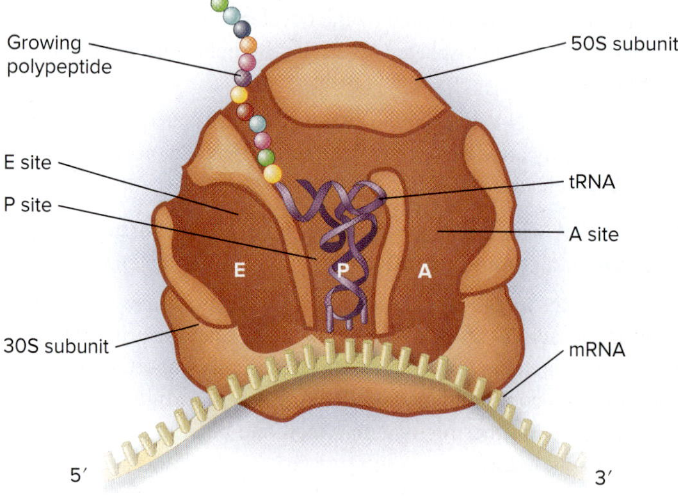


**(b) Schematic model for ribosome structure**

**Figure 10.12** **Ribosome structure. (a)** A model for the structure of a bacterial ribosome based on X-ray diffraction studies, showing the large and small subunits and the major binding sites. The rRNA is shown in gray (large subunit) and turquoise (small subunit), whereas the ribosomal proteins are magenta (large subunit) and dark blue (small subunit). **(b)** A schematic model emphasizing functional sites in the ribosome and showing bound mRNA and tRNA with an attached polypeptide. The amino acids are depicted as different-colored balls to emphasize that translation produces a polypeptide with a specific amino acid sequence.

*(a)* © Tom Pantages

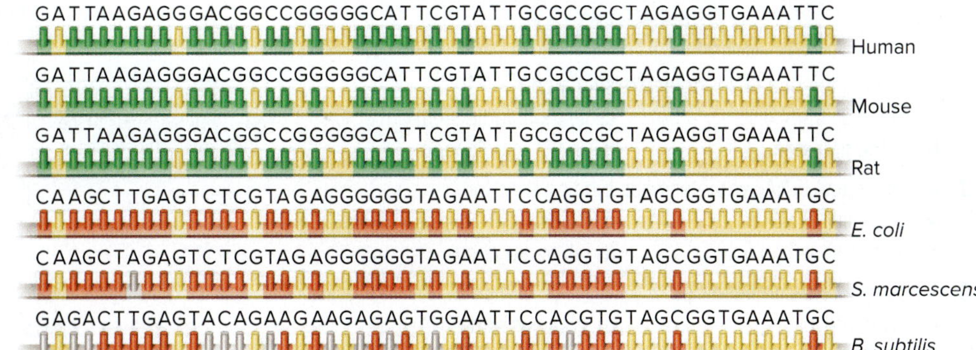


**Figure 10.13** **Comparison of small subunit rRNA gene sequences from three mammalian and three bacterial species.** Note the many similarities (yellow) and differences (green and red) among the sequences. The gray color indicates differences among the three bacterial species.

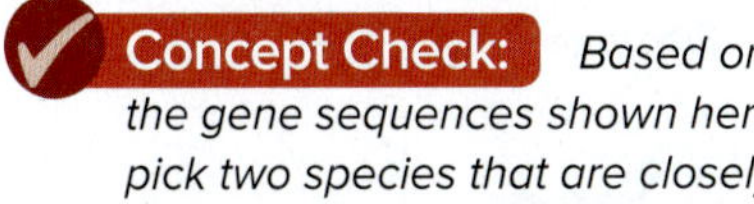

**Concept Check:** *Based on the gene sequences shown here, pick two species that are closely related evolutionarily and two that are distantly related.*

## EVOLUTIONARY CONNECTIONS

### Comparisons of Small Subunit rRNAs Among Different Species Provide a Basis for Establishing Evolutionary Relationships

Translation is a fundamental process that is vital for the existence of all living species. Research indicates that the components needed for translation arose very early in the evolution of life on our planet in ancestors that gave rise to all known living species. For this reason, all organisms have translational components that are evolutionarily related to each other. For example, the rRNA found in the small subunit of ribosomes is similar among all forms of life, though it is slightly larger in eukaryotic species (18S) than in bacterial species (16S). The gene for the small subunit rRNA (SSU rRNA) is found in all species.

How is this observation useful? One way that geneticists explore evolutionary relationships is by comparing the sequences of evolutionarily related genes. At the molecular level, gene evolution involves changes in DNA sequences. After two different species have diverged from each other during evolution, the genes of each species have an opportunity to accumulate changes, or mutations, that alter the sequences of those genes. After many generations, evolutionarily related species have genes that are similar but not identical to each other, because each species accumulates different mutations over time. In general, if a very long time has elapsed since two species diverged evolutionarily, their genes tend to be quite different. In contrast, if two species diverged relatively recently on an evolutionary timescale, their genes tend to be more similar.

**Figure 10.13** compares a portion of the sequence of the SSU rRNA gene from three mammalian and three bacterial species. The colors highlight different types of comparisons. The bases shaded in yellow are identical in five or six species. Bases that are identical in different species are said to be **evolutionarily conserved.** Presumably, these bases were found in the primordial gene that gave rise to modern species. Perhaps because these bases may have some critical function, they have not changed over evolutionary time. Those bases shaded in green are identical in all three mammals, but they differ compared with one or more bacterial species. Actually, if you scan the mammalian species, you may notice that all three sequences are identical in this region. The bases shaded in red are identical or very similar in the bacterial species, but they differ compared with the mammalian SSU rRNA genes. The sequences from *Escherichia coli* and *Serratia marcescens* are more similar to each other than the sequence from *Bacillus subtilis* is to either of them. This observation suggests that *E. coli* and *S. marcescens* are more closely related evolutionarily than either of them is to *B. subtilis*.

## 10.5 Reviewing the Concepts

- Translation requires mRNA, charged tRNAs, ribosomes, and many translation factors (Table 10.2).
- tRNA molecules have a cloverleaf structure. Two important sites are the amino acid attachment site at the 3′ single-stranded region and the anticodon, which base-pairs with a codon in mRNA (Figure 10.10).
- The enzyme aminoacyl-tRNA synthetase attaches the correct amino acid to a tRNA molecule, producing a charged tRNA (Figure 10.11).
- Ribosomes are composed of a small and a large subunit, each consisting of rRNA molecules and many proteins. Bacterial and eukaryotic ribosomes differ in their molecular composition (Table 10.3).
- Ribosomes have three sites, termed the A, P, and E sites, which are locations for the binding and release of tRNA molecules (Figure 10.12).
- The gene that encodes the small subunit rRNA (SSU rRNA) has been used to determine evolutionary relationships among different species (Figure 10.13).

## 10.5 Testing Your Knowledge

1. The anticodon of a tRNA molecule is located
   a. at the 3′ end.
   b. in the first stem.
   c. in the middle stem.
   d. in the middle loop.
2. An aminoacyl-tRNA synthetase has binding sites for
   a. a specific tRNA.
   b. a specific amino acid.
   c. ATP.
   d. all of the above.
3. Ribosomes are composed of
   a. proteins only.
   b. rRNA only.
   c. proteins and rRNA.
   d. proteins and DNA.

## 10.6 The Stages of Translation

### Learning Outcomes

1. Describe the three stages of translation.
2. Summarize the similarities and differences between translation in bacteria and eukaryotes.

Like transcription, the process of translation occurs in three stages called initiation, elongation, and termination.

1. During initiation, an mRNA, the first tRNA, and the ribosomal subunits assemble into a complex.
2. In the elongation stage, the ribosome moves in the 5′ to 3′ direction from the start codon in the mRNA toward the stop codon, synthesizing a polypeptide according to the sequence of codons in the mRNA.
3. Finally, the process is terminated when the ribosome reaches a stop codon and the complex disassembles, releasing the completed polypeptide.

In this section, we will examine the three stages of translation as they occur in living cells.

### Translation Is Initiated with the Assembly of mRNA, tRNA, and Ribosomal Subunits

During the **initiation stage,** a complex is formed between an mRNA molecule, the first tRNA, and the ribosomal subunits. The assembly of this complex requires the help of translation factors called **initiation factors** that facilitate the interactions between these components (see Table 10.2). The assembly also requires an input of energy. Guanosine triphosphate (GTP) is hydrolyzed by certain initiation factors to provide the necessary energy.

The initiation of translation in bacteria occurs as follows.

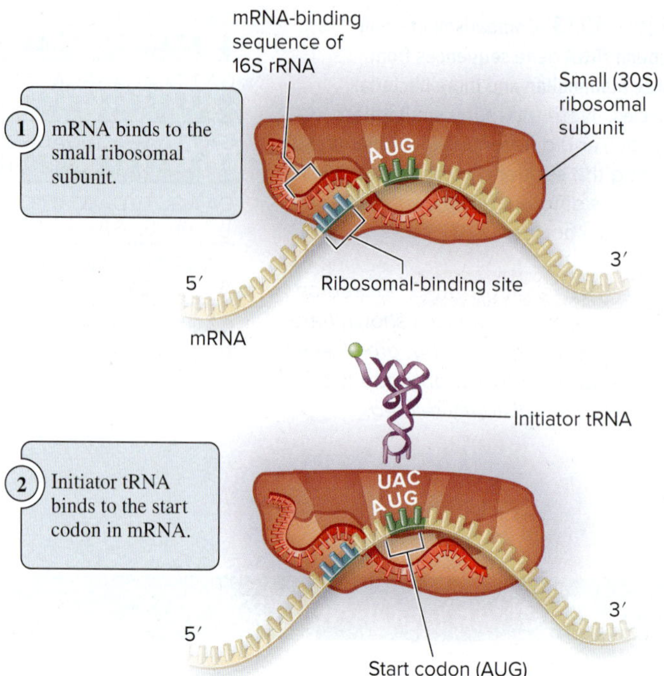


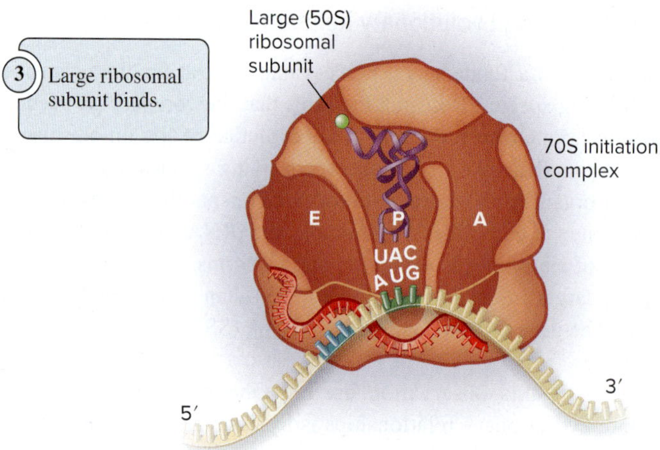


**Figure 10.14** Initiation of translation in bacteria.

**Concept Check:** *What promotes the binding between the mRNA and the small ribosomal subunit?*

1. An mRNA binds to the small ribosomal subunit (Figure 10.14). The binding of bacterial mRNA to this subunit is facilitated by a short ribosomal-binding site near the 5′ end of the mRNA. The ribosomal-binding site is a sequence of bases that is complementary to a portion of the 16S rRNA within the small ribosomal subunit. The mRNA becomes bound to the ribosome because the ribosomal-binding site and rRNA hydrogen-bond to each other by base-pairing. The start codon (AUG) is usually just a few nucleotides downstream (that is, toward the 3′ end) from the ribosomal-binding site.
2. A specific tRNA, which functions as the **initiator tRNA,** recognizes the start codon in mRNA (AUG) and binds to it. In bacteria, it carries a methionine that has been modified by the attachment of a formyl group.
3. To complete the initiation stage, the large ribosomal subunit associates with the small subunit. At the end of this stage, the initiator tRNA is located in the P site of the ribosome.

In eukaryotic species, the initiation phase of translation differs from the process in bacteria in a few ways.

- Instead of an RNA sequence that functions as a ribosomal-binding site, eukaryotic mRNAs have a guanosine cap (5′ cap) at their 5′ end. This 5′ cap is recognized by cap-binding proteins that promote the binding of the mRNA to the small ribosomal subunit.
- The first AUG codon is not always the start codon. In 1978, American biochemist Marilyn Kozak proposed that the small ribosomal subunit identifies a start codon by beginning at the 5′ end and then scanning along the mRNA in the 3′ direction. By analyzing the sequences of many eukaryotic mRNAs, Kozak and her colleagues discovered that the sequence around an AUG codon is important for it to be selected as the start codon. Aside from an AUG codon itself, a guanine just past the start codon and the sequence of six bases directly upstream from the start codon are important for start codon selection.
- In eukaryotes, the initiator tRNA carries a methionine, not formyl-methionine.

## Polypeptide Synthesis Occurs During the Elongation Stage

As its name suggests, the **elongation stage** involves the covalent bonding of amino acids to each other, one at a time, to produce a polypeptide. This process is similar in all species and involves several different components. Elongation occurs at a remarkable rate. Under normal cellular conditions, the translation machinery can elongate a polypeptide at a rate of 15 to 18 amino acids per second in bacteria and 6 amino acids per second in eukaryotes!

To elongate a polypeptide by one amino acid, a tRNA brings a new amino acid to the ribosome, where it is attached to the end of a growing polypeptide. In step 1 of Figure 10.15, translation has already proceeded to a point where a short polypeptide is attached to the tRNA located in the P site of the ribosome. This is called peptidyl tRNA. Let's consider the steps that would cause this polypeptide to become one amino acid longer.

1. **Binding of a charged tRNA:** A charged tRNA carrying a single amino acid binds to the A site. This binding occurs because the anticodon in the tRNA is complementary to the codon in the mRNA. The hydrolysis of GTP by proteins that function as **elongation factors** provides the energy for the binding of the tRNA to the A site (see Table 10.2). At this stage of translation, a peptidyl tRNA is in the P site and a charged tRNA (an aminoacyl tRNA) is in the A site. This is how the P and A sites came to be named.
2. **Peptide bond formation and peptidyl transfer:** In the second step, a peptide bond is formed between the amino acid at the A site and the growing polypeptide, thereby lengthening the polypeptide by one amino acid. As this occurs, the polypeptide is removed from the tRNA in the P site and transferred to the amino acid at the A site, an event termed a **peptidyl transfer reaction.** This reaction is catalyzed by a region of the 50S subunit known as the peptidyltransferase center, which is composed of several proteins and rRNA.

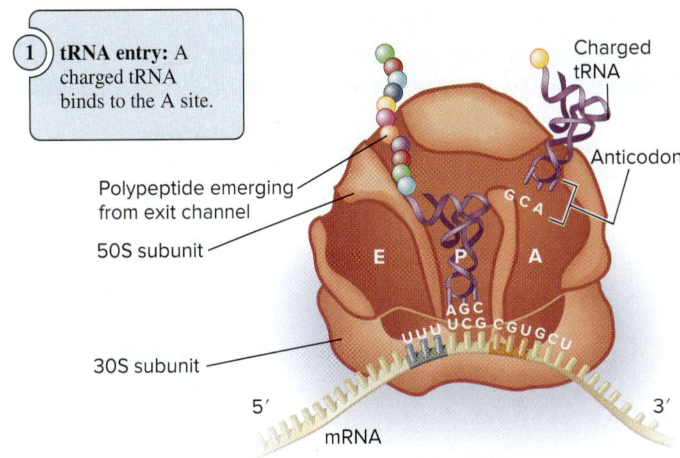


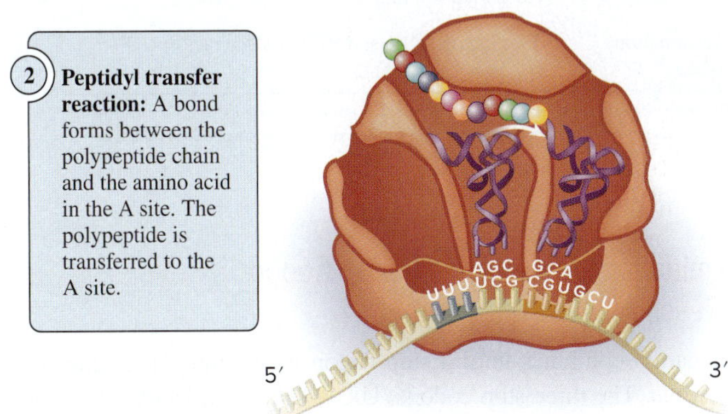

2 Peptidyl transfer reaction: A bond forms between the polypeptide chain and the amino acid in the A site. The polypeptide is transferred to the A site.
AGC GCA
UUU UCG CGU GCU
5′
3′


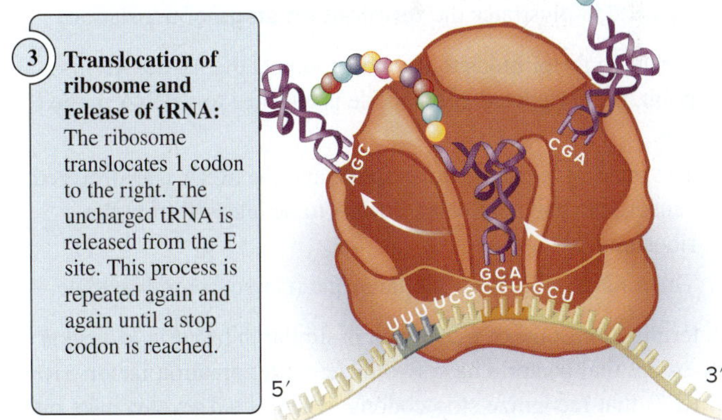

3 Translocation of ribosome and release of tRNA: The ribosome translocates 1 codon to the right. The uncharged tRNA is released from the E site. This process is repeated again and again until a stop codon is reached.
AGC
CGA
GCA
UUU UCG CGU GCU
5′
3′


**Figure 10.15** Elongation stage of translation in bacteria.

3. **Translocation and exit of uncharged tRNA:** The ribosome moves, or translocates, toward the 3′ end of the mRNA by exactly one codon. This shifts the tRNAs in the P and A sites to the E and P sites, respectively. The uncharged tRNA exits the E site.

Notice that the next codon in the mRNA (GCU in Figure 10.15) is now exposed at the unoccupied A site. At this point, a charged tRNA can enter the A site, and the same series of steps will add the next amino acid to the polypeptide.

**Table 10.4 Comparison of Bacterial and Eukaryotic Translation**

| | Bacterial | Eukaryotic |
|---|---|---|
| Cellular location | Cytoplasm | Cytosol* |
| Ribosome composition | 70S ribosomes:<br>30S subunit: 21 proteins + 1 rRNA<br>50S subunit: 34 proteins + 2 rRNAs | 80S ribosomes:<br>40S subunit: 33 proteins + 1 rRNA<br>60S subunit: 49 proteins + 3 rRNAs |
| Initiator tRNA | $tRNA^{Formyl\text{-}methionine}$ | $tRNA^{Methionine}$ |
| Initial binding of mRNA | Requires a ribosomal-binding site | Requires a 7-methylguanosine cap |
| Selection of a start codon | Just downstream from the ribosomal-binding site | According to Kozak's sequences |
| Termination factors | Two factors: RF1 and RF2 | One factor: eRF |

*The components for eukaryotic translation described in this table refer to those that are used in the cytosol. Different types of ribosomes are used for translation inside mitochondria and chloroplasts.

## Termination Occurs When a Stop Codon Is Reached in the mRNA

Elongation continues until a stop codon moves into the A site of a ribosome. The three stop codons, UAA, UAG, and UGA, are recognized by a protein known as a **release factor.** The three-dimensional structure of a release factor protein mimics the structure of tRNAs, which allows it to fit into the A site.

Figure 10.16 illustrates the **termination stage** of translation.

1. A release factor binds to the stop codon at the A site. At this point, the completed polypeptide is attached to a tRNA in the P site.
2. The bond between the polypeptide and the tRNA is hydrolyzed, causing the polypeptide and tRNA to be released from the ribosome.
3. The mRNA, ribosomal subunits, and release factor dissociate.

The termination stage of translation is similar in bacteria and eukaryotes except that bacteria have two different termination factors (RF1 and RF2) that recognize stop codons, whereas eukaryotes have only one (eRF). Table 10.4 compares some of the key differences between bacterial and eukaryotic translation.

## 10.6 Reviewing the Concepts

- Translation occurs in three stages called initiation, elongation, and termination.
- During the initiation stage, the mRNA assembles with the ribosomal subunits and the initiator tRNA molecule, which carries formyl-methionine (in bacteria) or methionine (in eukaryotes) (Figure 10.14).

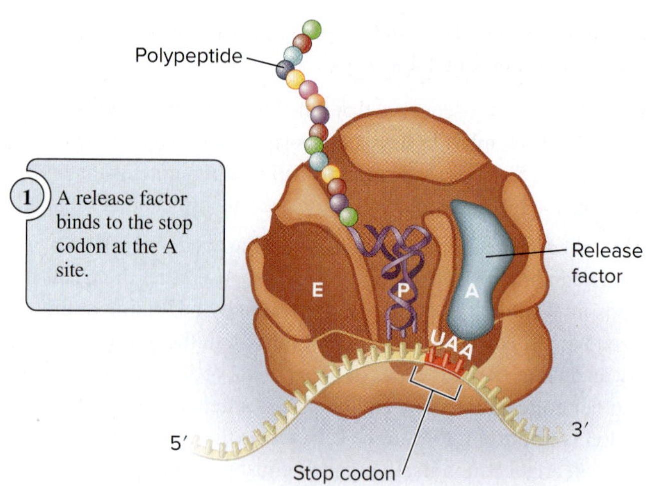


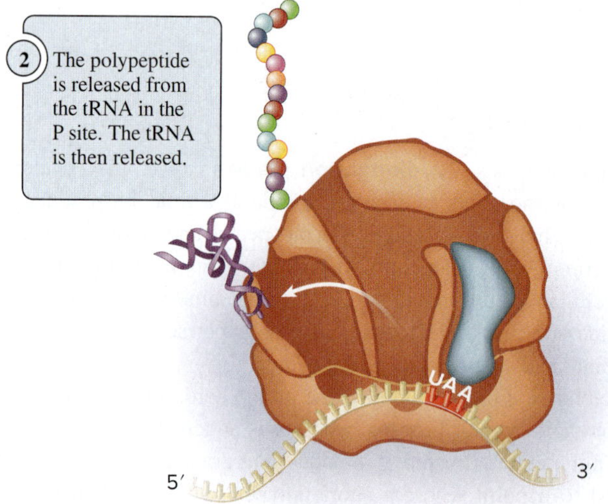


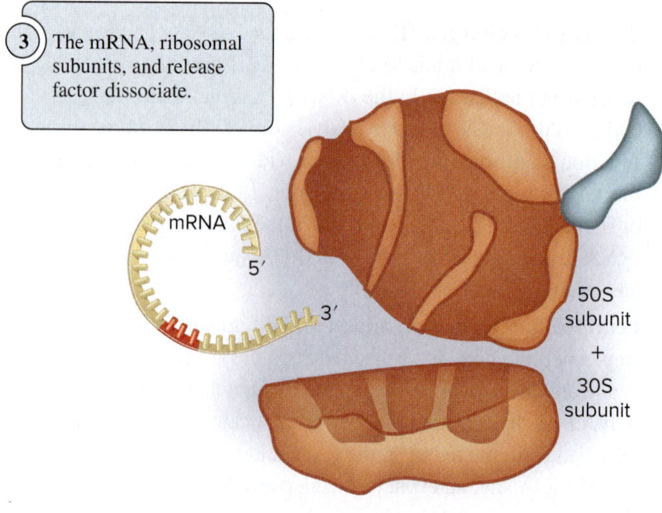


Figure 10.16 Termination of translation in bacteria.

- During the elongation stage, amino acids are added, one at a time, to a growing polypeptide (Figure 10.15).
- The termination stage occurs when the binding of a release factor to a stop codon causes the release of the completed polypeptide from the tRNA and the disassembly of the mRNA, ribosomal subunits, and the release factor (Figure 10.16).
- Though translation is strikingly similar in bacteria and eukaryotes, some key differences have been observed (Table 10.4).

## 10.6 Testing Your Knowledge

1. A key difference in the initiation phase of translation between bacteria and eukaryotes is that in eukaryotes
   - **a.** the 5′ cap promotes binding of mRNA to the small ribosomal subunit.
   - **b.** the selection of the start codon depends on the sequence around it.
   - **c.** the initiator tRNA carries methionine rather than formylmethionine.
   - **d.** all of the above.
2. A tRNA carrying a polypeptide would *never* be found in
   - **a.** the A site.
   - **b.** the P site.
   - **c.** the E site.
   - **d.** either a or c.

# Assess and Discuss

## Test Yourself

1. Which of the following best represents the central dogma of gene expression?
   - a. During transcription, DNA codes for polypeptides.
   - b. During transcription, DNA codes for mRNA, which codes for polypeptides during translation.
   - c. During translation, DNA codes for mRNA, which codes for polypeptides during transcription.
   - d. None of the above are correct.
2. A mutation prevents a gene from being transcribed into an mRNA. The mutation most likely disrupts
   - a. the promoter.
   - b. the terminator.
   - c. the start codon.
   - d. the stop codon.
   - e. both a and c.
3. The functional product of a protein-encoding gene is
   - a. tRNA.
   - b. mRNA.
   - c. rRNA.
   - d. a polypeptide.
   - e. a, b, and c.
4. Which of the following is *not* a property of the genetic code?
   - a. It specifies the amino acids within a polypeptide.
   - b. It is composed of codons, which are specific sequences of three bases.
   - c. It has a start codon, which specifies the starting point for polypeptide synthesis.
   - d. It has stop codons, which specify the end of polypeptide synthesis.
   - e. It determines the rate of transcription.
5. If a eukaryotic mRNA failed to have a cap attached to its 5′ end, what would the negative consequence(s) be?
   - a. The mRNA would not properly exit the nucleus.
   - b. The mRNA would not properly bind to a ribosome.
   - c. The mRNA would not receive a poly A tail.
   - d. The mRNA would not use the correct start codon.
   - e. Both a and b are correct.
6. The small subunit of a ribosome is composed of
   - a. a protein.
   - b. an rRNA molecule.
   - c. many proteins.
   - d. many rRNA molecules.
   - e. many proteins and one rRNA molecule.
7. The region of a tRNA that is complementary to a codon in mRNA is
   - a. the 3′ single-stranded region.
   - b. the codon.
   - c. the peptidyl site.
   - d. the anticodon.
   - e. the stem-loop.
8. During the initiation of translation, the first codon, ____, enters the ________ and associates with the initiator tRNA.
   - a. UAG, A site
   - b. AUG, A site
   - c. UAG, P site
   - d. AUG, P site
   - e. AUG, E site
9. The movement of the polypeptide from the tRNA in the P site to the tRNA in the A site is referred to as
   - a. peptide bonding.
   - b. aminoacyl binding.
   - c. translation.
   - d. the peptidyl transfer reaction.
   - e. elongation.
10. The synthesis of a polypeptide occurs during which stage of translation?
    - a. initiation
    - b. elongation
    - c. termination
    - d. splicing
    - e. both a and b

## Conceptual Questions

1. Let's suppose a gene mutation changed the sixth codon from one that specifies lysine to a stop codon. How would this mutation affect transcription? How would it affect translation?
2. What is the function of an aminoacyl-tRNA synthetase?
3. **PRINCIPLES** A principle of biology is that the genetic material provides a blueprint for reproduction. Explain how the information within the blueprint is accessed at the molecular level.

## Collaborative Questions

1. Why do you think some complexes, such as spliceosomes and ribosomes, have both protein and RNA components?
2. Discuss and make a list of the similarities and differences in the events that occur during the initiation, elongation, and termination stages of transcription and those stages in translation.

# Online Resource

# 11

# The Expression of Genetic Information via Genes II: Non-coding RNAs

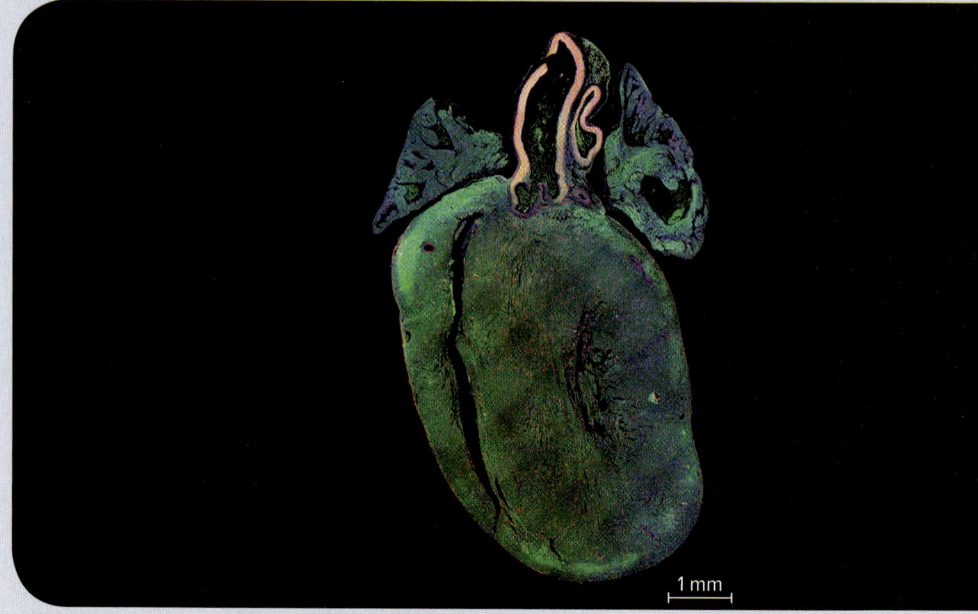


© Mauro Giacca, Ana Eulalio, Miguel Mano

**Non-coding RNAs and heart repair.** The muscles of the mammalian heart have poor regenerating abilities. Researchers have identified several non-coding RNAs that stimulate cardiac muscle regeneration. This heart is from a mouse that was treated with such a non-coding RNA, and it showed a significant increase in proliferating cells. This discovery may lead to new therapies to help heart attack victims regenerate new cardiac muscle.

## Chapter Outline



People with a rare genetic disorder called cartilage-hair hypoplasia (CHH) have short stature, underdeveloped hair, and short limbs with malformations in the cartilage. In addition, they have a higher predisposition to develop lymphomas and other cancers, and they may suffer from defective immunity. Though CHH was first identified by American geneticist Victor McKusick in 1965, the underlying genetic cause remained a mystery for 36 years. In 2001, the function of a single mutant gene was linked to this disease. The gene specifies an RNA molecule that does not encode a protein. Rather, the RNA made from this gene is the RNA component of RNase MRP, which is an RNA/protein complex involved in the modification of some ribosomal and mitochondrial RNAs. This was the first case in which geneticists determined that a human inherited disease was due to a mutation in a nuclear gene that is not a protein-encoding gene. By comparison, the first human genetic disease involving a protein-encoding gene was discovered nearly a century earlier! In 1909, Archibald Garrod proposed that patients with a disease called alkaptonuria, which is characterized by bluish-black discoloration of the cartilage and skin, is due to a mutation in a gene that encodes the protein homogentisic acid oxidase.

Why such a big time gap in our understanding of protein-encoding genes versus other types of genes? It's all about tools. The experimental tools to study the structure and function of proteins and to identify protein-encoding genes have been around for a long time. In contrast, the tools to study RNA structure and function and to identify genes that do not specify proteins are much more recent and are under rapid development. We are witnessing a revolution in molecular biology that is uncovering an unprecedented number of functions for RNA molecules.

In Chapter 10, we focused our attention on gene expression at the molecular level. The emphasis was on protein-encoding genes, which are transcribed into mRNA. During translation, the information within mRNAs is used to make polypeptides, which then assemble into functional proteins. The human genome has about 22,000 protein-encoding genes. In contrast, other genes are transcribed into **non-coding RNAs (ncRNAs),** which are RNA molecules that do not encode polypeptides. As discussed later, ncRNAs perform a very diverse set of functions. In humans, the number of genes that specify ncRNAs is still difficult to measure and a matter of controversy. Estimates range from several thousand to tens of thousands.

In the past, educators have tended to emphasize proteins and DNA in the teaching of biology at the molecular level. For example, in the cell unit (Chapters 4 through 8), we have seen many examples of how proteins affect cell structure and function, and the genetics unit (Chapters 9 through 18) largely focuses on the structure and function of DNA. Although DNA, RNA, and proteins are key molecular players in living cells, a historical bias has existed against RNA. With a few exceptions, the educational exploration of RNA has been limited to its role in making proteins (see Chapter 10).

The purpose of this chapter is to lessen the bias against RNA. New molecular tools have enabled researchers to discover that ncRNAs perform a spectacular array of cellular functions in bacteria, archaea, protists, fungi, plants, and animals. ncRNAs play important roles in a variety of processes, including DNA replication, chromatin modification, transcription, translation, and genome defense. In most cell types, ncRNAs are more abundant that mRNAs. For example, in a typical human cell, only about 20% of transcription involves the production of mRNAs, whereas 80% is associated with making ncRNAs! This observation underscores the importance of RNA in the enterprise of life, and why it deserves greater recognition and study. Furthermore, abnormalities in ncRNAs are associated with a wide range of human diseases, including CHH, cancer, neurological disorders, and cardiovascular diseases. Many ncRNAs are also critical to the growth of plants, including the crop plants that are so essential to human survival.

In this chapter, we will begin with an overview of the general properties of ncRNAs, and then examine specific examples of how they perform their functions. We will end the chapter by considering the role of ncRNAs in different human diseases and plant health.

## 11.1 Overview of Non-coding RNAs

### Learning Outcomes

1. Describe the ability of ncRNAs to bind to other molecules and macromolecules.
2. Outline the general functions of ncRNAs.
3. Define ribozyme.
4. List several examples of ncRNAs, and describe their functions.

The field of ncRNAs is rapidly expanding, and researchers speculate that many ncRNAs have yet to be discovered. Also, due to the relatively young age of this field, not all researchers agree on the names of certain ncRNAs or their primary functions. Even so, some broad themes are beginning to emerge. In this section, we will survey the general features of ncRNAs, and in later sections discuss specific examples in greater detail.

### ncRNAs Can Bind to Different Types of Molecules

The ability of ncRNAs to carry out an amazing array of functions is largely related to their ability to bind to different types of molecules. **Figure 11.1a** shows four common types of molecules that are recognized by ncRNAs. Some ncRNAs bind to DNA or another RNA due to complementary base pairing. This allows ncRNAs to affect processes such as DNA replication, transcription, and translation. In addition, ncRNAs can bind to proteins or small molecules.

As described in Chapter 10, RNA molecules, such as tRNAs, can form stem-loop structures (refer back to Figure 10.10). Similar structures in other ncRNAs may bind to pockets on the surface of proteins, or multiple stem-loops may form a binding site for a small molecule. In some cases, a single ncRNA may contain multiple binding sites. This allows an ncRNA to facilitate the formation of a large structure composed of multiple molecules, such as an ncRNA and three different proteins, as shown in **Figure 11.1b**.

### ncRNAs Can Perform a Diverse Set of Functions

In recent decades, researchers have uncovered many examples in which ncRNAs play a critical role in different biological processes. Let's first consider how ncRNAs work in a general way. The common functions of ncRNAs are the following.

**Scaffold** Some ncRNAs contain binding sites for multiple components, such as a group of different proteins. Much like the beams in a building, an ncRNA can act as a scaffold for the formation of a complex, as in Figure 11.1b.

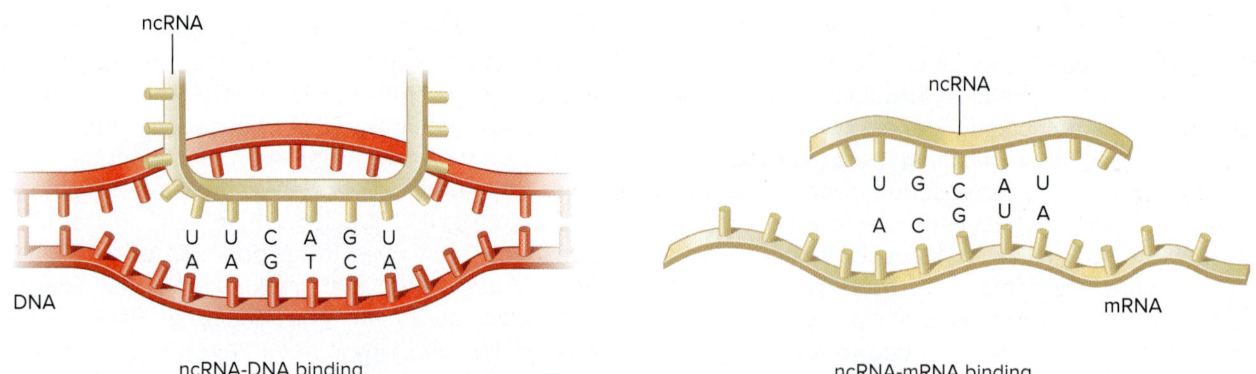


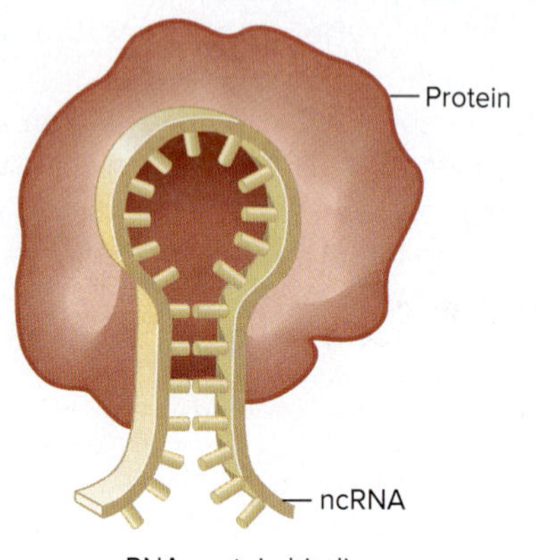


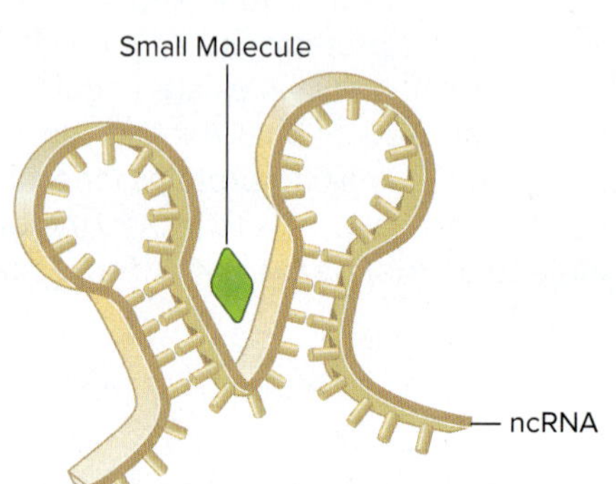


**(a) Common binding interactions between ncRNAs and other molecules**

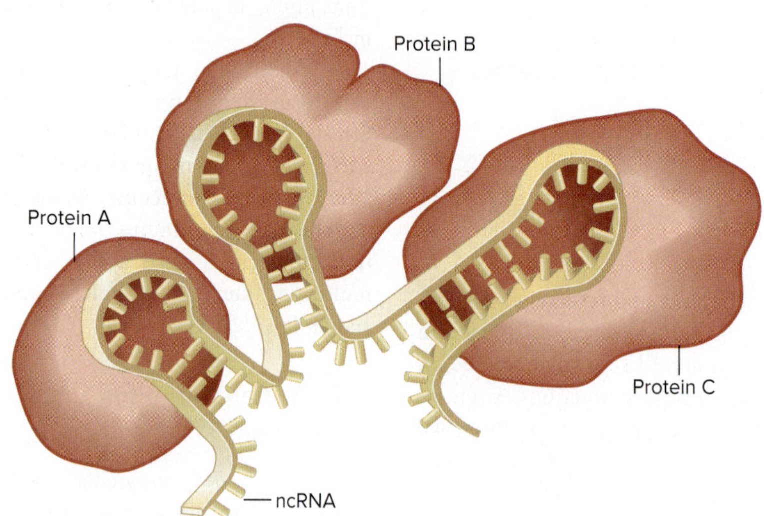


**(b) Multiple binding sites in a single ncRNA**

**Figure 11.1 Ability of ncRNAs to bind to other molecules.** **(a)** ncRNA molecules can bind to DNA, mRNA, proteins, and/or small molecules. **(b)** Some ncRNAs have multiple binding sites for different molecules, such as proteins.

*Which of these binding interactions might be inhibited by the formation of a stem-loop within the ncRNA?*

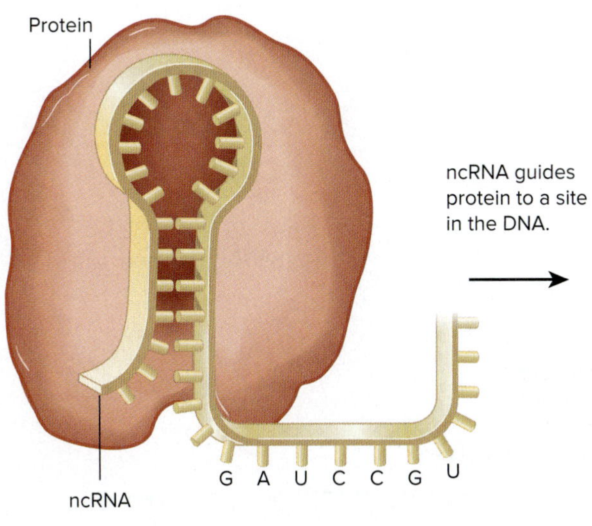


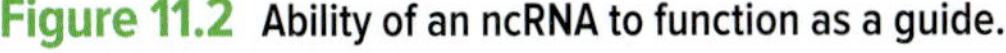

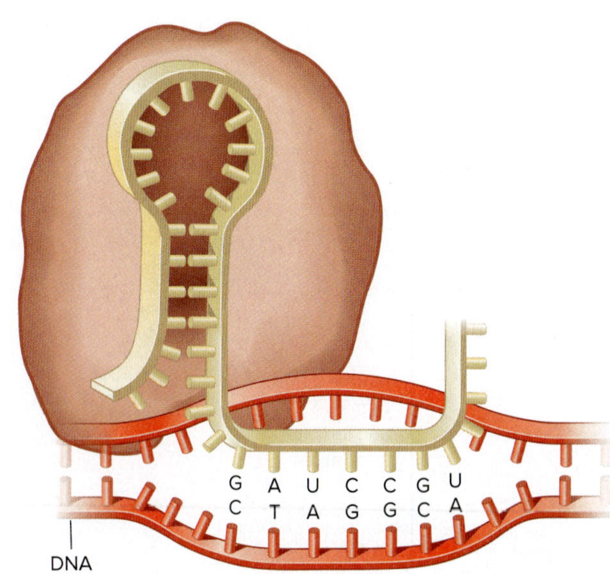


**Figure 11.2** Ability of an ncRNA to function as a guide.

**Guide** Also due to multiple binding sites, ncRNAs may guide one molecule to a specific location in a cell. For example, an ncRNA may bind to a protein and guide it to a target site in the DNA that is part of a particular gene (Figure 11.2). This occurs because the ncRNA has a binding site for the protein and another binding site for the DNA.

**Alteration of Protein Function or Stability** When it binds to a protein, an ncRNA can alter that protein's structure, which in turn can have a variety of effects. The binding of an ncRNA may affect

- the ability of the protein to act as a catalyst;
- the ability of the protein to bind to other molecules, such as proteins, DNA, or RNA; and
- the stability of the protein.

**Ribozyme** Another interesting feature of some ncRNAs is that they function as **ribozymes,** which are RNA molecules with catalytic function. For example, in Chapter 10, we considered how peptidyltransferase, which is a component of the large ribosomal subunit, catalyzes peptide bond formation during translation (refer back to Figure 10.15). It is an rRNA within peptidyltransferase that catalyzes this reaction. In other words, the ribosome is a ribozyme.

**Blocker** An ncRNA may physically prevent or block a cellular process from happening. For example, in bacteria, an antisense RNA is a type of ncRNA that is complementary to an mRNA. When an antisense RNA binds to an mRNA, it blocks the ability of the ribosome to bind to the mRNA, thereby inhibiting translation.

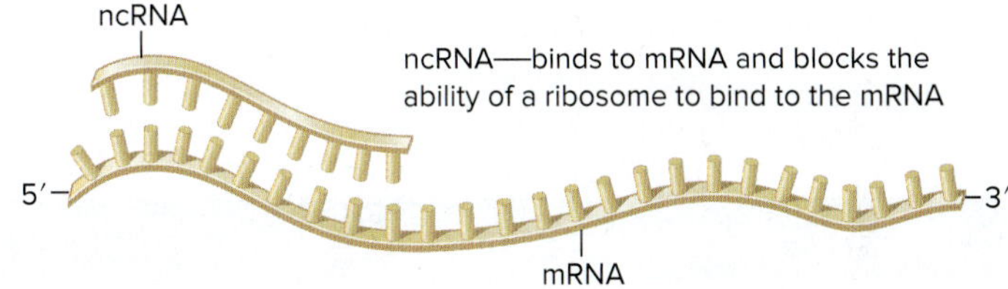


**Decoy** Some ncRNAs recognize other ncRNAs and sequester them, thereby preventing them from working. For example, a decoy ncRNA may bind to a different ncRNA called a microRNA (miRNA), which is described later in this chapter (Figure 11.3). The function of an miRNA is to inhibit the translation of a particular mRNA. However, if a decoy ncRNA binds to an miRNA, it is unable to carry out its function. The decoy ncRNA acts as a sponge by binding to the miRNA and preventing it from functioning.

## The Functions of Some ncRNAs Are Understood

Table 11.1 describes several examples of ncRNAs that have been well characterized. Some of these are described in other chapters. In the remaining sections of this chapter, we will focus on the functions of ncRNAs that are not discussed elsewhere.

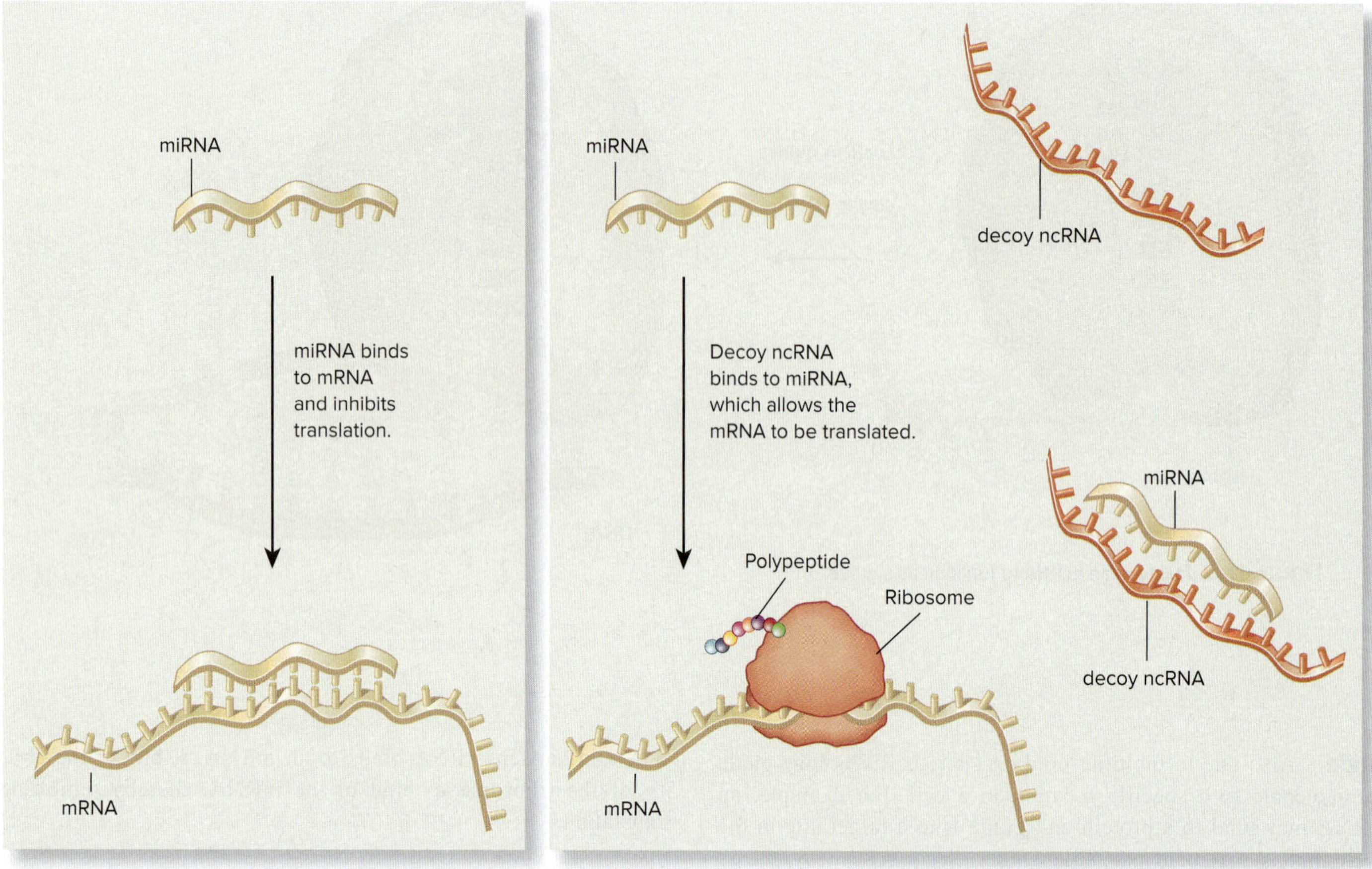


**Figure 11.3** Ability of an ncRNA to function as a decoy.

**Table 11.1 Examples of ncRNAs**

| Type of ncRNA | Plays a role in | Described in | Description |
|---|---|---|---|
| Telomerase RNA component (TERC) | DNA replication | This chapter | TERC facilitates the binding of telomerase to the telomere and acts as a template for DNA replication. |
| X inactive specific transcript (Xist RNA) | Chromatin structure, transcription | Chapter 16 | Xist RNA coats one of the X chromosomes in female mammals and plays a role in its compaction and resulting inactivation. |
| *Hox* transcript antisense intergenic RNA (HOTAIR) | Chromatin structure, transcription | This chapter | HOTAIR alters chromatin structure and thereby represses transcription by guiding histone-modifying complexes to target genes. |
| Transfer RNA (tRNA) | Translation | Chapter 10 | tRNA molecules recognize mRNA codons during translation and carry the appropriate amino acid. |
| Ribosomal RNA (rRNA) | Translation | Chapter 10 | rRNAs are components of ribosomes, which are the site of polypeptide synthesis. |
| microRNA (miRNA), small-interfering RNA (siRNA) | Translation and RNA degradation | This chapter | miRNAs and siRNAs regulate the expression and degradation of mRNAs. |
| Small nucleolar RNA (snoRNA) | RNA modification | This chapter | A snoRNA facilitates covalent modifications to rRNAs. |
| RNA component of signal recognition particle (SRP-RNA) | Protein targeting and secretion | This chapter | In bacteria, SRP directs the synthesis of some polypeptides to the plasma membrane. In eukaryotes, it directs polypeptide synthesis to the endoplasmic reticulum. |
| CRISPR RNA (crRNA) | Genome defense | This chapter | crRNA, found in bacteria and archaea, guides an endonuclease to foreign DNA, such as the DNA of a bacteriophage. |

## 11.1 Reviewing the Concepts

- Non-coding RNAs (ncRNAs) are RNA molecules that do not encode polypeptides.
- ncRNAs bind to different types of molecules, including DNA, other RNAs, proteins, and small molecules (Figure 11.1).
- An ncRNA can provide a scaffold, act as a guide, alter protein function or stability, function as a ribozyme, function as a blocker, and/or act as a decoy (Figure 11.2, 11.3).
- ncRNAs play a role in DNA replication, chromatin structure, transcription, translation, RNA degradation, RNA modification, protein secretion, and genome defense (Table 11.1).

## 11.1 Testing Your Knowledge

1. Which of the following can bind to ncRNAs?
   a. DNA
   b. RNA
   c. proteins
   d. small molecules
   e. all of the above
2. When an ncRNA functions as a decoy, it
   a. contains binding sites for many different proteins, thereby promoting the formation of a large complex.
   b. recognizes other ncRNAs and sequesters them, thereby preventing them from working.
   c. can alter its conformation, allowing it to switch between active and inactivate conformations.
   d. may physically prevent or block a cellular process from happening.
   e. does all of the above.

## 11.2 Role of Non-coding RNAs in Eukaryotic DNA Replication

### Learning Outcome

1. Explain how an ncRNA plays a role in the replication of telomeres in eukaryotes.

The topic of DNA replication is discussed in Chapter 9. A specialized form of DNA replication that requires the function of an ncRNA happens at the ends of eukaryotic chromosomes. These regions, called the **telomeres,** have a short nucleotide sequence that is repeated a few dozen to several hundred times in a row (Figure 11.4). The repeat sequence shown here, 5′–GGGTTA–3′, is the sequence found in human telomeres. Flowering plants have a similar repeat sequence, but it has an extra T (5′–GGGTTTA–3′). A telomere has a region at the 3′ end that is termed a 3′ overhang, because it does not have a complementary strand.

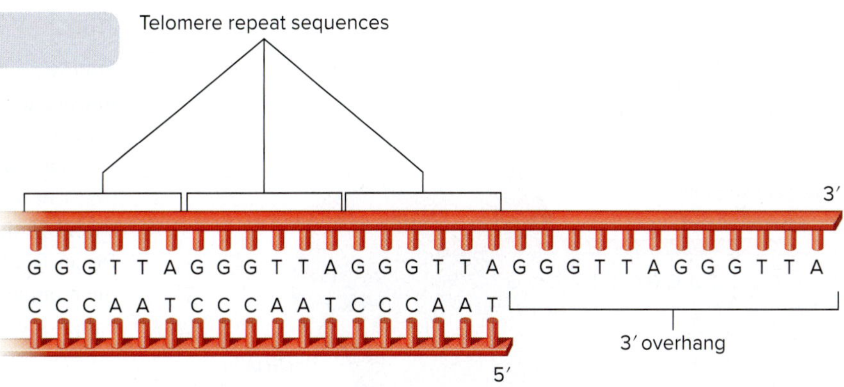


**Figure 11.4 Telomere sequences at one end of a human chromosome.** The telomere sequence shown here is found in humans and other mammals. The length of the telomere and the 3′ overhang varies among different species and cell types.

BioConnections: *Refer back to Figure 9.16. Explain why DNA polymerase cannot copy the 3′ end of a linear chromosome.*

As discussed in Chapter 9, DNA polymerase synthesizes DNA only in a 5′ to 3′ direction and requires a primer. For these reasons, DNA polymerase cannot copy the 3′ ends of double-stranded DNA. Therefore, if this replication problem were not overcome, a linear chromosome would become progressively shorter with each round of DNA replication.

In 1984, American molecular biologist Carol Greider and Australian-born American molecular biologist Elizabeth Blackburn discovered an enzyme called **telomerase** that prevents chromosome shortening by attaching many copies of a DNA repeat sequence to the 3′ ends of chromosomes (Figure 11.5). Telomerase contains both proteins and an ncRNA called telomerase RNA component (TERC). The lengthening of telomeres occurs in three steps.

### Binding of Telomerase

TERC has a sequence that is complementary to the DNA repeat sequence. In this way, TERC acts as a guide that allows telomerase to bind to the 3′ overhang region of the telomere.

### Polymerization

Following binding, TERC has a second function. This ncRNA has a sequence that functions as a template for the synthesis of a six-nucleotide sequence at the end of the DNA strand. This synthesis is called polymerization, because it is analogous to the function of DNA polymerase. Telomere lengthening is catalyzed by a protein within telomerase called telomerase reverse transcriptase (TERT). TERT's name indicates that it catalyzes the reverse of transcription; it uses an RNA template to synthesize DNA.

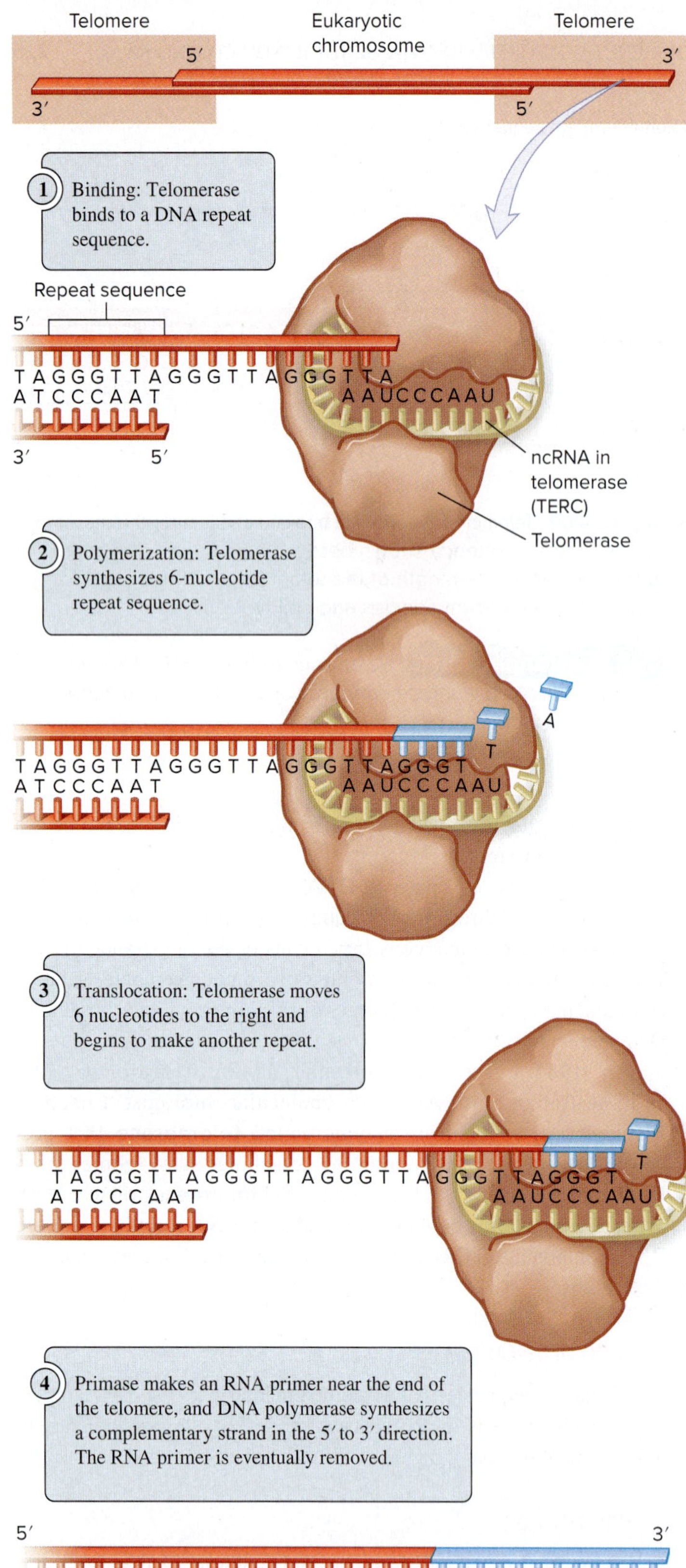


**Figure 11.5** Mechanism of DNA replication by telomerase.

**Concept Check:** *What does telomerase use as a template to make DNA?*

### Translocation

Following polymerization, telomerase then moves—a process called translocation—to the new end of the DNA strand and attaches another six nucleotides to the end.

This binding-polymerization-translocation cycle occurs many times in a row, thereby greatly lengthening the 3′ end of the DNA strand in the telomere. This lengthening provides an upstream site for an RNA primer to be made. DNA polymerase then synthesizes the complementary DNA strand. In this way, the progressive shortening of eukaryotic chromosomes is prevented.

## 11.2 Reviewing the Concepts

- The ends of linear, eukaryotic chromosomes have telomeres composed of repeat sequences. An ncRNA within telomerase called TERC guides telomerase to the telomere repeat sequence and also functions as a template for the synthesis of a six-nucleotide repeat. This happens many times in a row to lengthen one DNA strand of the telomere. This synthesis also provides a site for an RNA primer to be made, and DNA polymerase synthesizes the complementary DNA strand (Figures 11.4, 11.5).

## 11.2 Testing Your Knowledge

1. Which of the following is a function of TERC, the RNA component of telomerase?
   **a.** a guide
   **b.** a template
   **c.** a primer for DNA synthesis
   **d.** all of the above
   **e.** a and b only

# 11.3 Effects of Non-coding RNAs on Chromatin Structure and Transcription

### Learning Outcome

1. Explain how the ncRNA known as HOTAIR plays a role in gene repression.

*Hox* transcript antisense intergenic RNA, known as HOTAIR, is a recently discovered ncRNA found in humans and other mammals that alters chromatin structure and thereby represses gene transcription. The gene that encodes *HOTAIR* is located within a cluster of genes called the *HoxC* genes. (*Hox* genes, which play a role in animal development, are described in Chapter 20.) HOTAIR is so named because it is transcribed from the opposite (antisense) strand with respect to the *HoxC* genes.

Figure 11.6 shows a simplified mechanism for how HOTAIR represses gene transcription. HOTAIR acts as a scaffold for the

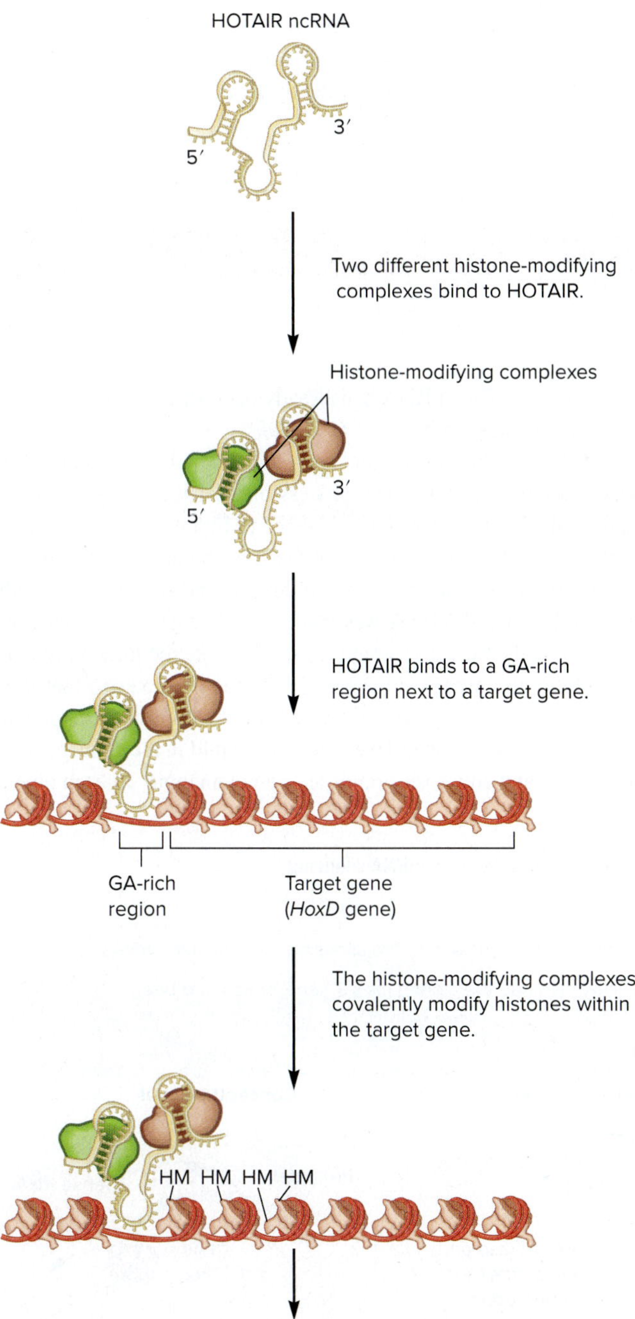


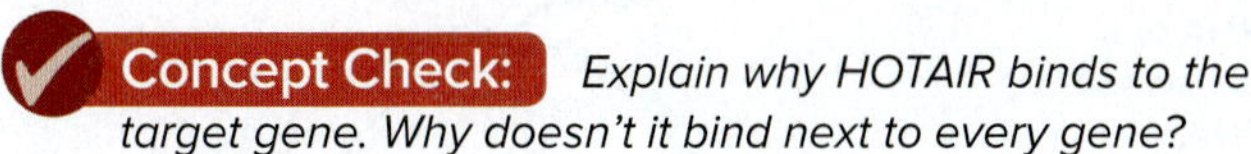


**Figure 11.6** **Simplified mechanism of HOTAIR.** This is just one proposed role of HOTAIR. This ncRNA is known to interact with other proteins as well. Note: The abbreviation HM stands for histone modification.

**Concept Check:** *Explain why HOTAIR binds to the target gene. Why doesn't it bind next to every gene?*

binding of two protein complexes that covalently modify histone proteins. One of them binds to the 5′ end of HOTAIR, and the other binds to the 3′ end. HOTAIR then guides these complexes to a target gene by binding to a region near the gene that contains many purines, which is called a GA-rich region. For example, HOTAIR binds to a GA-rich region that is next to a *HoxD* gene. A portion of HOTAIR is complementary to this GA-rich region.

The next event involves histone modifications. As described in Chapter 12, histone modifications may affect gene transcription (look ahead to Figure 12.14). The modifications that are facilitated by HOTAIR are known to inhibit transcription. This inhibition can occur in two ways:

- The modifications may directly inhibit the ability of RNA polymerase to transcribe the target gene. For example, these histone modifications may prevent RNA polymerase from forming a preinitiation complex.
- Rather than directly affecting transcription, the histone modifications may attract other chromatin-modifying enzymes to the target gene, which would lead to further changes in chromatin structure that inhibit transcription.

Great interest in the study of HOTAIR is related to its role in human disease. As discussed later in this chapter, certain types of cancer, such as breast cancer, may occur when HOTAIR is not functioning properly.

## 11.3 Reviewing the Concepts

- HOTAIR is an ncRNA found in humans and other mammals that regulates transcription by forming a scaffold that binds two protein complexes and guides them to particular genes. The protein complexes covalently modify histones, and these modifications silence the target genes (Figure 11.6).

## 11.3 Testing Your Knowledge

1. Which of the following is a function of the ncRNA known as HOTAIR?
   a. a decoy
   b. a scaffold
   c. a guide
   d. all of the above
   e. b and c only

# 11.4 Effects of Non-coding RNAs on Translation, mRNA Degradation, and RNA Modification

### Learning Outcomes

1. **SCISKILLS ▸** Analyze experimental evidence that double-stranded RNA is more potent at inhibiting mRNA than is antisense RNA.
2. Outline the steps of RNA interference.
3. Explain how snoRNAs direct covalent modifications to rRNAs.

In the previous section, we considered how an ncRNA can affect the process of transcription. In recent years, researchers have discovered that ncRNAs often exert their effects on RNA molecules that are already made. In this section, we will consider how ncRNAs can affect the ability of mRNAs to be translated or degraded, as well as the ability of rRNAs to be covalently modified.

## FEATURE INVESTIGATION

### Fire and Mello Show That Double-Stranded RNA Is More Potent Than Antisense RNA at Silencing mRNA

Specific mRNAs can be targeted for translational inhibition or degradation by a mechanism involving double-stranded RNA. This mechanism was discovered during research involving plants and the nematode worm *Caenorhabditis elegans*. The study described here involved an examination of gene expression in *C. elegans*.

American biologists Andrew Fire, Craig Mello, and colleagues used *C. elegans* as their experimental organism, because it is relatively easy to inject with RNA and the expression of many of its genes had already been established. In 1998, Fire and Mello investigated the effects of injected RNA on the expression of specific mRNAs. In the investigation described in **Figure 11.7**, we will focus on one of their experiments involving an mRNA encoded by a gene called *mex-3*. This mRNA had already been shown to be made in high amounts in early embryos of *C. elegans*.

Prior to this work, the *mex-3* gene had been identified and inserted into a plasmid. The process of inserting genes into plasmids is described in Chapter 18 (look ahead to Figure 18.2). Let's first look at the plasmid shown at the top of the conceptual level in step 1. When RNA polymerase, nucleotides, and this plasmid were mixed together in a test tube, *mex-3* mRNA was made, which is called the sense strand. In living cells, the sense strand is used to make the mex-3 protein. Fire and Mello also switched the location of the promoter so that it was at the other end of the gene and transcribed the opposite strand, which is called the antisense strand (see second plasmid in the conceptual level). The sense and antisense strands are complementary to each other.

**Figure 11.7** Injection of antisense and double-stranded RNA into *C. elegans* to compare their effects on mRNA silencing.

**GOAL** The goal was to further understand how the experimental injection of RNA was responsible for the silencing of particular mRNAs.

**KEY MATERIALS** The researchers used *C. elegans* as their model organism. They also had the cloned *mex-3* gene, which had been previously shown to be highly expressed in *C. elegans* embryos.

**Experimental level** | **Conceptual level**

1 Make sense and antisense *mex-3* RNA in vitro using cloned genes for *mex-3* with promoters on either side of the gene. RNA polymerase and nucleotides are added to synthesize the RNAs.

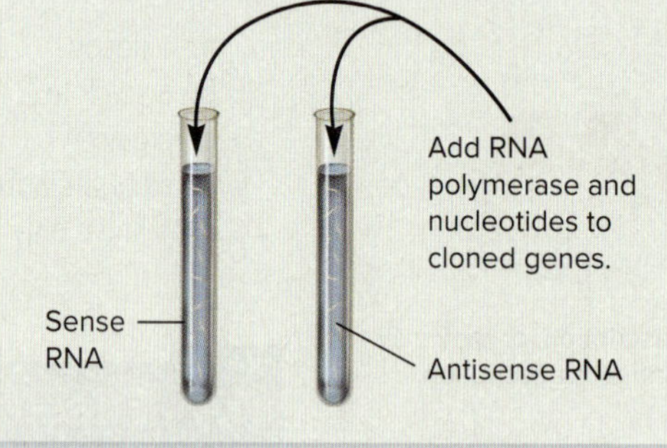


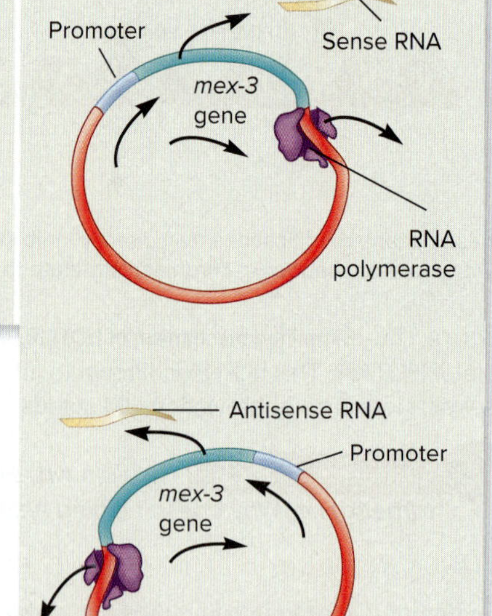


2 Inject either *mex-3* antisense RNA or a mixture of *mex-3* sense and antisense RNA into the gonads of *C. elegans*. This RNA is taken up by the eggs and early embryos. As a control, do not inject any RNA.

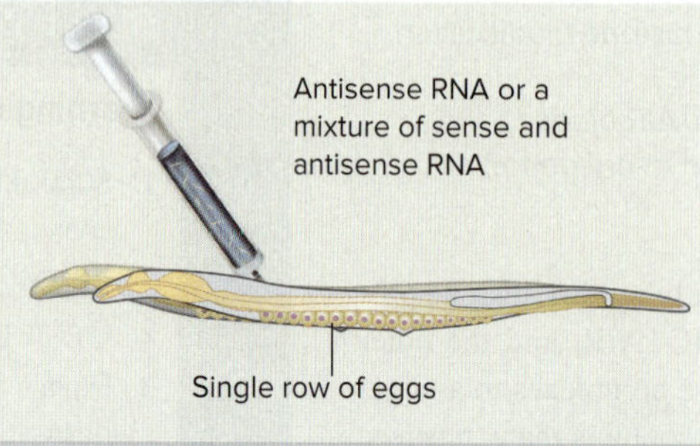

3 Incubate and then subject early embryos to in situ hybridization. In this method, a labeled probe is added that is complementary to *mex-3* mRNA. If cells express *mex-3*, the mRNA in the cells will bind to the probe and become labeled. After incubation with a labeled probe, the cells are washed to remove unbound probe.

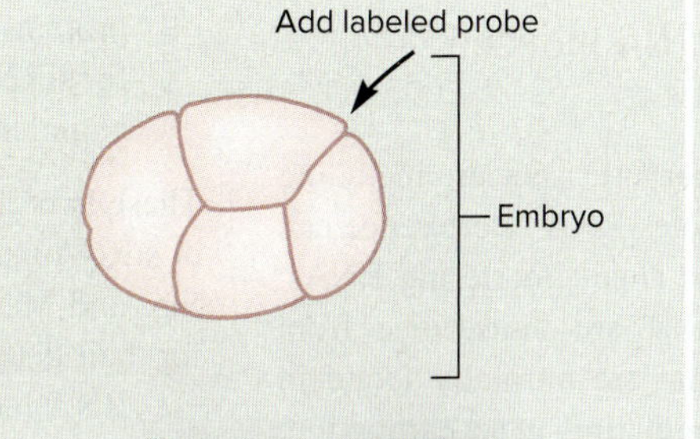


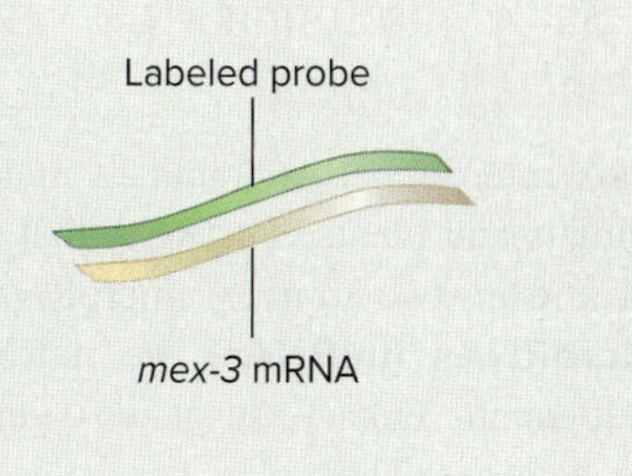


4 Observe embryos under the microscope.

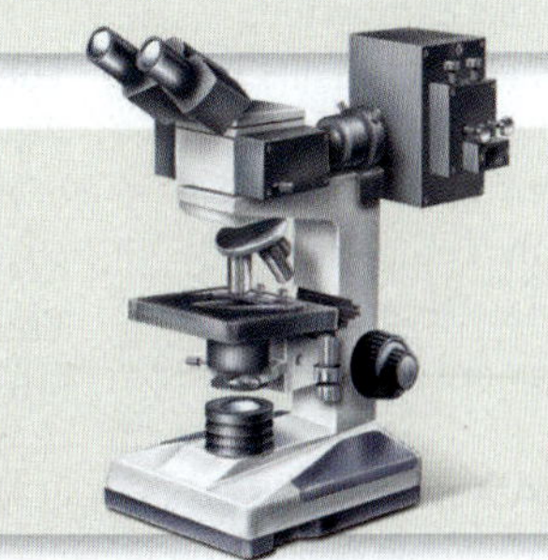

5 **THE DATA**

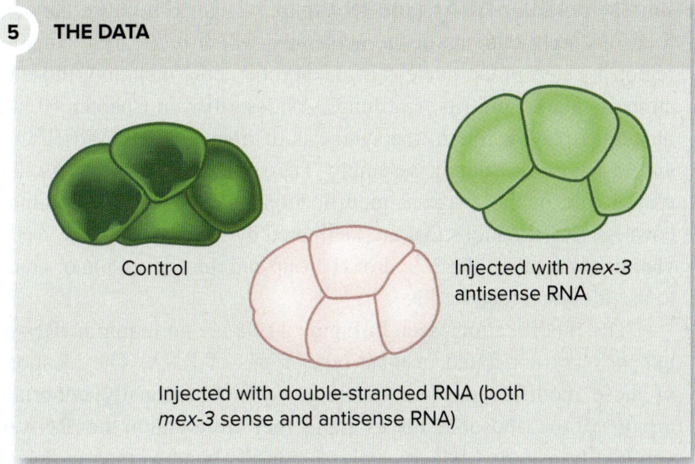


6 **CONCLUSION** Double-stranded RNA is more potent at inhibiting *mex-3* mRNA than antisense RNA alone.

7 **SOURCE** Fire, A., Xu, S., Montgomery, M.K., et al 1998. Potent and Specific Genetic Interference by Double-Stranded RNA in *Caenorhabditis elegans*. *Nature* 391: 806–811.

Next, they injected these RNAs into the gonads of *C. elegans*. Into some worms, they injected antisense RNA alone. Alternatively, they mixed sense and antisense RNA, which formed double-stranded RNA, and injected this double-stranded RNA into the gonads of other worms. They also used uninjected worms as controls. After injection, the RNA was taken up by eggs, which later developed into embryos. To determine the amount of *mex-3* mRNA present, Fire and Mello incubated the embryos with a probe that was complementary to *mex-3* mRNA. The probe was labeled so that any probe bound to *mex-3* mRNA could be observed under the microscope. After this incubation step, any probe that was not bound to this mRNA was washed away.

As seen in the schematic data of Figure 11.7, the control embryos were very darkly labeled as denoted by their green color. These results indicated that the control embryos contained a high amount of *mex-3* mRNA, which was known from previous research. In the embryos that had received antisense RNA, *mex-3* mRNA levels were decreased, but detectable, as shown by faint labeling. Remarkably, in embryos that had received double-stranded RNA, no *mex-3* mRNA was detected! These results indicated that double-stranded RNA is more potent at silencing mRNA than is antisense RNA. In this case, the double-stranded RNA caused the *mex-3* mRNA to be degraded. Fire and Mello used the term **RNA interference (RNAi)** to describe the phenomenon in which double-stranded RNA causes the silencing of mRNA. This surprising observation led researchers to investigate the underlying molecular mechanism that accounts for this phenomenon, as described next.

### *Experimental Questions*

1. In this experiment, does the *mex-3* mRNA correspond to the sense strand or antisense strand?
2. **SCISKILLS** ▶ Explain how the sense and antisense *mex-3* RNAs were made.
3. **SCISKILLS** ▶ According to the data, which material was the most effective at causing the degradation of *mex-3* mRNA?

## RNA Interference Is Mediated by MicroRNAs or Small-Interfering RNAs via the RNA-Induced Silencing Complex

RNA interference is found in most eukaryotic species, including animals and plants. The RNA that promotes RNA interference can come from two sources: microRNAs and small-interfering RNAs. **MicroRNAs (miRNAs)** are ncRNAs that are transcribed from endogenous eukaryotic genes—genes that are normally found in the genome. They play key roles in regulating gene expression, particularly during embryonic development in animals and plants. Most commonly, a single type of miRNA inhibits the translation of several different mRNAs. An miRNA and an mRNA bind to each other because they have sequences that are partially complementary. In humans, over 2,000 genes encode miRNAs. Researchers estimate that 60% of human protein-encoding genes are regulated by miRNAs.

By comparison, **small-interfering RNAs (siRNAs)** are ncRNAs that usually originate from sources that are exogenous, which means they are not normally made by cells. The siRNAs can come from viruses that infect a cell, or researchers can make siRNAs to study gene function experimentally, as in Figure 11.7. In most cases, siRNAs are a perfect match to a single type of mRNA. The functioning of siRNAs is thought to play a key role in preventing certain types of viral infections. In addition, siRNAs have become important experimental tools in molecular biology.

How do miRNAs and siRNAs cause the silencing of specific mRNAs? Figure 11.8 shows how an miRNA or an siRNA leads to RNA interference. The miRNA is first synthesized as a pri-miRNA (for primary-miRNA) in the nucleus. Due to complementary base pairing, the pri-miRNA folds into a hairpin structure (also called a stem-loop) with long, single-stranded 5′ and 3′ ends. The pri-miRNA is cleaved at both ends to form a pre-miRNA (for precursor-miRNA, not to be confused with pri-miRNA). The pre-miRNA is then exported from the nucleus.

As shown in Figure 11.8, siRNAs do not go through the processing events that occur in the nucleus. Instead, pre-siRNAs are derived from viral RNAs, or they may be made by researchers and taken up by cells. For example, in the work of Fire and Mello described in Figure 11.7, the double-stranded *mex-3* RNA is an example of a pre-siRNA. The pre-siRNA is formed from two complementary RNA molecules that base-pair with each other.

In the cytosol, both pre-miRNAs and pre-siRNAs are cut by an endonuclease called dicer (see Figure 11.8). This releases a double-stranded RNA molecule that is typically 20–25 bp long. This double-stranded RNA associates with proteins to form a complex called the **RNA-induced silencing complex (RISC).** One of the RNA strands is degraded. The remaining single-stranded miRNA or siRNA is complementary to specific mRNAs that will be silenced. The miRNA or siRNA acts as a guide that causes RISC to recognize and bind to such mRNA molecules.

After binding to an mRNA, two different things may happen.

- RISC may inhibit translation without degrading the mRNA. This is more common for miRNAs, which often are only partially complementary to their target mRNAs.
- RISC may direct the degradation of the mRNA. One of the proteins in RISC can cleave the mRNA. This usually occurs for siRNAs that typically are a perfect match (or highly complementary) to their target mRNA.

These two effects are termed RNA interference because the miRNA or siRNA interferes with the proper expression of an mRNA. In 2006, Fire and Mello received the Nobel Prize in Physiology or Medicine for their discovery of this mechanism.

RNA interference is believed to have at least two benefits.

- This mechanism represents an important form of regulation. When genes encoding pri-miRNAs are turned on, the production of miRNAs silences the expression of specific mRNAs.
- RNA interference provides a defense against viruses. This mechanism is widely used by plants to prevent viral infections.

## Small Nucleolar RNAs Direct Covalent Modifications to Ribosomal RNAs

**Small nucleolar RNAs (snoRNAs)** are so named because they are found in high amounts in the nucleolus, which is a darkly staining region located in the nucleus of eukaryotic cells, including those of animals, plants, protists, and fungi. As described in Chapter 10, the nucleolus plays a role in the synthesis of ribosomal RNAs (rRNAs) and in ribosome subunit assembly. The snoRNAs are ncRNAs that play a role in the covalent modification of rRNAs. Two common covalent modifications that are facilitated by snoRNAs are the methylation of ribose on the 2′ hydroxyl group and the conversion of uracil to pseudouracil (Figure 11.9a).

The modifications seen in Figure 11.9a are common in rRNAs, but they are not found in most other types of RNA. The locations of these modifications are concentrated in functionally important regions of the ribosome. For example, they are found in the rRNA of peptidyl transferase, which catalyzes peptide bond formation during translation (refer back to Figure 10.15). While the functional role of these covalent modifications is not entirely understood, the current view is that they fine-tune rRNA structure so it can function in an optimal manner.

To facilitate the covalent modifications described in Figure 11.9a, a snoRNA has two key functions. First, a snoRNA acts as a scaffold for the binding of several proteins to form a **snoRNA ribonucleoprotein (snoRNP).** Figure 11.9b shows the structure of a snoRNP that can methylate ribose within rRNAs. This snoRNP contains several proteins, including two proteins that catalyze the methylation reaction.

After the snoRNP has formed, the second role of a snoRNA is to act as a guide. A snoRNA has antisense sequences that are complementary to sites in rRNAs (Figure 11.9c). This enables a snoRNA to recognize and bind to rRNAs. In other words, the snoRNA guides the snoRNP to its target, which is an rRNA. Once an rRNA binds to the antisense sequence in the snoRNA, the proteins within the snoRNP (not shown in part c) catalyze the chemical modification of the rRNA. In Figure 11.9c, two different rRNAs became bound to the snoRNA and then a ribose in each rRNA was methylated.

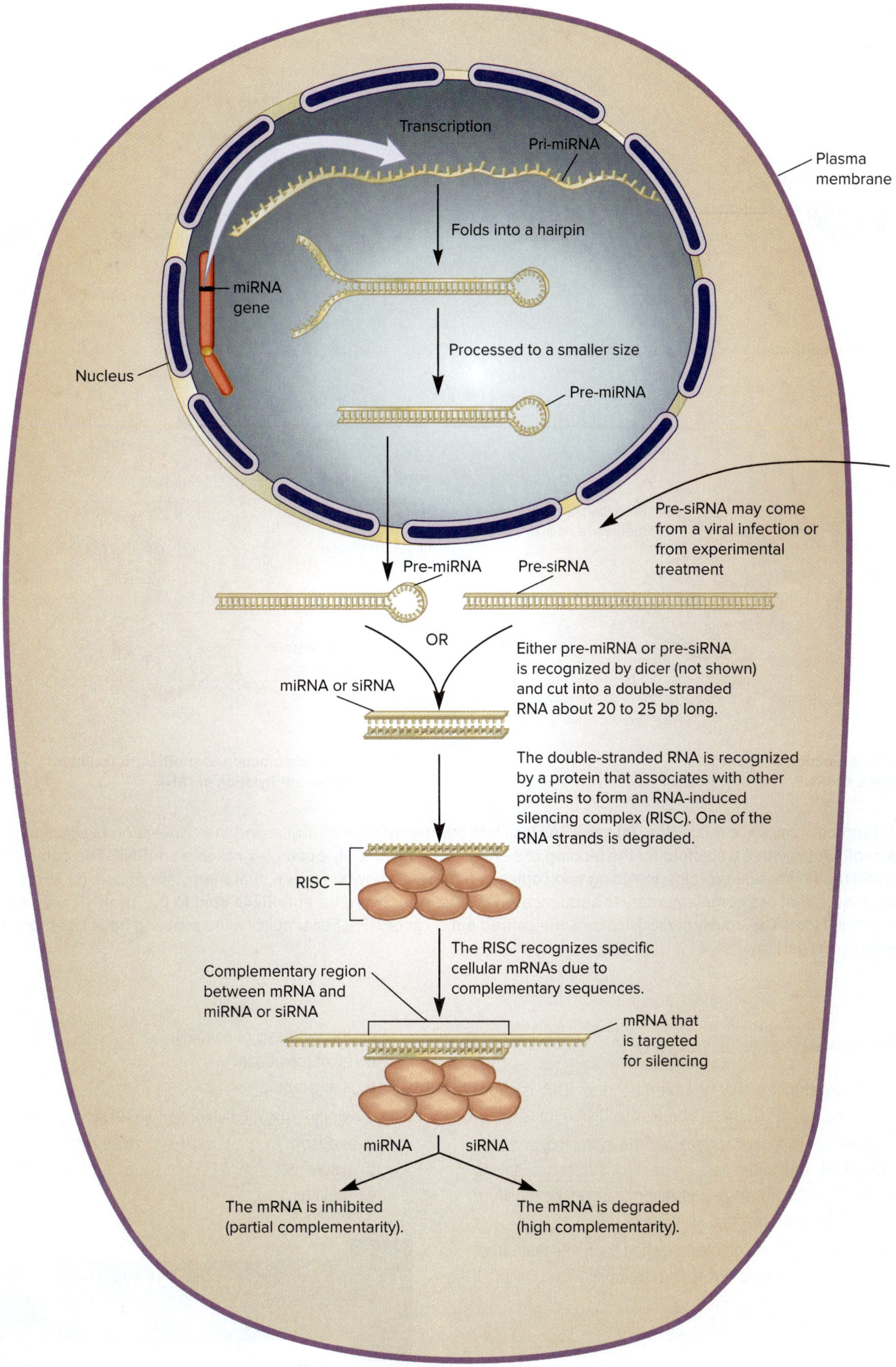

**Figure 11.8** **Mechanism of RNA interference.**

**Concept Check:** *Explain why RISC binds to a specific mRNA. What type of bonding occurs?*

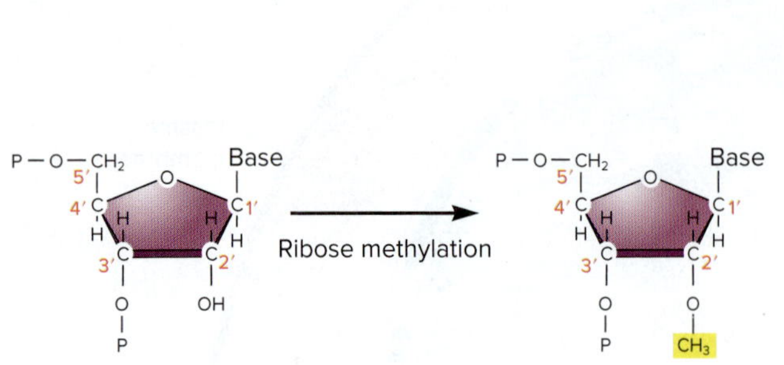


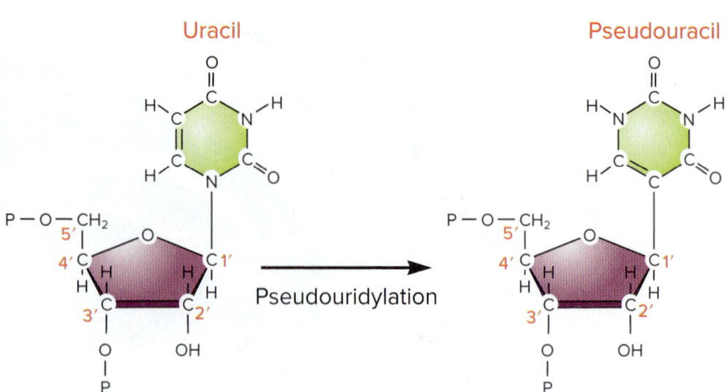


(a) Covalent modifications facilitated by snoRNAs

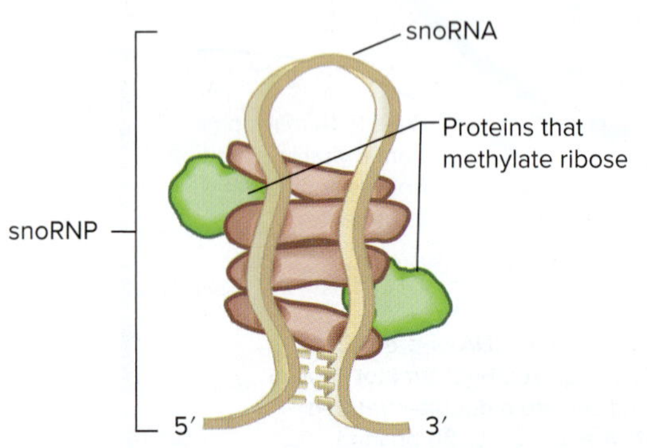


(b) Scaffold function of snoRNA to create a snoRNP

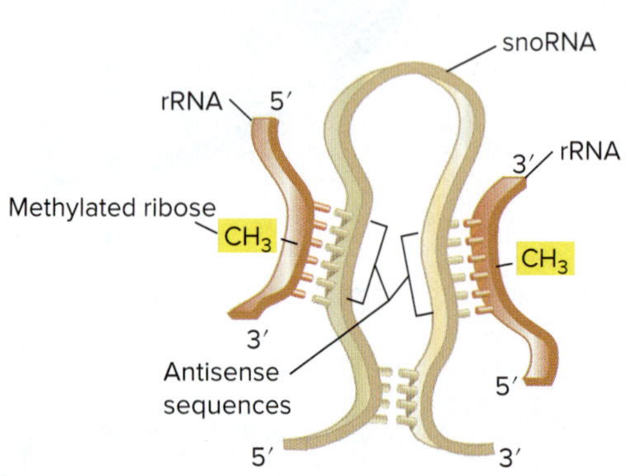


(c) Guide function of snoRNA to facilitate ribose methylation of rRNA

**Figure 11.9** **Function of small nucleolar RNAs.** **(a)** snoRNAs facilitate the methylation of ribose and the conversion of uracil to pseudouracil in rRNAs. **(b)** A snoRNA provides a scaffold for the binding of several proteins to form a complex called a snoRNP. The complex shown here contains a snoRNA and several proteins, including two copies of a protein, shown in green, that methylate ribose. **(c)** The snoRNA has antisense sequences that are complementary to sequences in rRNAs. After two different rRNAs bind to the snoRNP, they are modified by ribose methylation. Note: The covalent modifications are carried out by snoRNPs. For simplicity, the proteins have been omitted from part (c), but they are shown in part (b).

## 11.4 Reviewing the Concepts

- Fire and Mello showed that double-stranded RNA is more potent at silencing mRNA than is antisense RNA (Figure 11.7).
- RNA interference is a mechanism of RNA silencing in which miRNA or siRNA becomes part of an RNA-induced silencing complex (RISC) that inhibits the translation of a specific mRNA or causes its degradation, respectively (Figure 11.8).
- snoRNAs become part of complexes called snoRNPs that direct the covalent modification of rRNAs (Figure 11.9).

## 11.4 Testing Your Knowledge

1. The process of RNA interference may lead to
   **a.** the degradation of an mRNA.
   **b.** the inhibition of translation of an mRNA.
   **c.** the synthesis of an mRNA.
   **d.** all of the above.
   **e.** a and b only.
2. To catalyze the methylation of an rRNA, a snoRNA functions as
   **a.** a decoy.
   **b.** a scaffold.
   **c.** a guide.
   **d.** a ribozyme
   **e.** both b and c.

## 11.5 Non-coding RNAs and Protein Targeting

### Learning Outcome

1. Describe the function of SRP, and explain the roles of SRP RNA with regard to SRP function.

To carry out their functions, proteins need to be targeted to a particular location (refer back to Figure 4.30). For example, some proteins function extracellularly and need to be secreted from the cell. For such proteins to be secreted, they are first targeted to the plasma membrane in bacteria and archaea, or to the endoplasmic reticulum (ER) membrane in eukaryotic cells. This targeting process is facilitated by an RNA-protein complex called **signal recognition particle (SRP).** In bacteria, SRP is composed of one ncRNA and one protein. In eukaryotes, SRP is composed of one ncRNA and six different proteins.

Figure 11.10 shows how SRP works in eukaryotes. To be directed to the ER membrane, a polypeptide must contain a sorting signal called an **ER signal sequence,** which is a sequence of about 6–12 amino

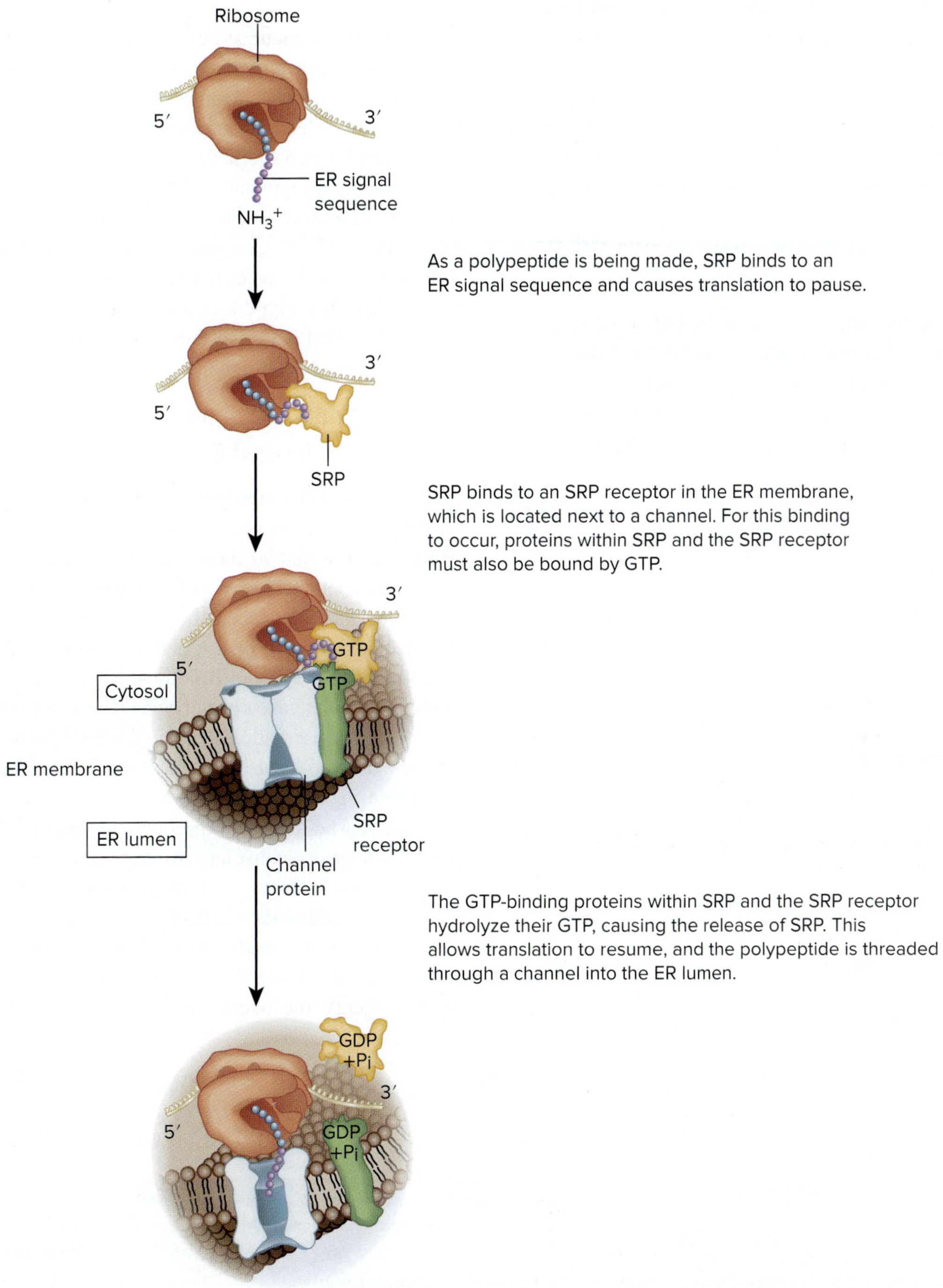

**Figure 11.10 Targeting of polypeptides to the endoplasmic reticulum membrane via SRP.** In eukaryotes, several categories of proteins are first targeted to the ER via SRP. These include proteins that are secreted from the cells as well as proteins that are destined to stay in the ER, Golgi, lysosomes, or vacuoles.

*Refer back to Figure 4.30. Which types of proteins need SRP to reach their proper location, and which do not?*

acids that are predominantly hydrophobic and usually located near the N-terminus. As the ribosome is making the polypeptide in the cytosol, the ER signal sequence emerges from the ribosome and is recognized by a protein in SRP. The binding of SRP to the polypeptide pauses translation.

SRP then binds to an SRP receptor in the ER membrane, which docks the ribosome over a channel. For this binding to occur, proteins within SRP and the SRP receptor must also be bound by GTP. Next, these GTP-binding proteins hydrolyze their GTP, which causes the release of SRP from the SRP receptor and the polypeptide. Once SRP is released, translation resumes and the growing polypeptide is threaded through a channel to cross the ER membrane. In the case of a secreted protein, the newly made polypeptide then travels through the Golgi apparatus and then to the plasma membrane, where it is released outside of the cell.

Researchers have identified at least two key roles for SRP RNA:

1. SRP RNA provides a scaffold for the binding of SRP proteins.
2. After SRP binds to the SRP receptor in the ER membrane, the SRP RNA stimulates proteins within both SRP and the SRP receptor to hydrolyze GTP. In other words, SRP RNA alters the structures of these proteins to enhance their GTPase activities. This stimulation is essential for the release of SRP.

## 11.5 Reviewing the Concepts

- Signal recognition particle (SRP), which is composed of one or more proteins and an ncRNA, plays a role in the targeting of proteins to the plasma membrane of prokaryotic cells and to the ER membrane of eukaryotic cells (Figure 11.10).

## 11.5 Testing Your Knowledge

1. Which of the following is a function of SRP?
   a. pauses translation of polypeptides that have an ER signal sequence
   b. binds to an SRP receptor in the ER membrane
   c. docks the ribosome over a channel
   d. hydrolyzes GTP to promote SRP release from the SRP receptor
   e. all of the above

## 11.6 Non-coding RNAs and Genome Defense

### Learning Outcome

1. Explain how the CRISPR-Cas system defends bacteria against bacteriophages.

Much like the immune system found in vertebrates, some species of bacteria and archaea have a system, called the **CRISPR-Cas system,** that provides defense against foreign invaders. CRISPR-Cas systems are an effective defense against bacteriophages, which are viruses that infect bacteria (discussed in Chapter 17), and transposons, which are small segments of DNA that can be inserted into the chromosomes of all species (discussed in Chapter 18). ncRNAs play a key role in the CRISPR-Cas system. About half of all bacterial species and most archaeal species have such a system. Three general types are known, designated type I, II, and III. In this section, we will focus on the type II CRISPR-Cas system and its role in providing bacteria with defense against bacteriophages.

### The CRISPR-Cas System Provides Bacteria with Defense Against Bacteriophages

In 1993, Spanish microbiologist Francisco Mojica and colleagues were the first to recognize that different species of bacteria and archaea have a site in their chromosome, now called the CRISPR locus, that contains a series of repeated sequences. In 2005, by analyzing the DNA sequences of the CRISPR locus in a variety of bacterial species, Mojica, Spanish geneticist Giles Vergnaud, and Russian microbiologist Alexander Bolotin independently proposed that it provides protection against bacteriophage infection. This hypothesis was based on the observation that the CRISPR locus contains segments that are derived from bacteriophage DNA. Their hypothesis was confirmed in 2007 by French microbiologist Philippe Horvath and colleagues, who showed experimentally that the CRISPR-Cas locus provides defense against bacterophage infection.

Figure 11.11a shows a common organization of the CRISPR-Cas system (also called the CRISPR locus), which has five genes: *tracr, Cas9, Cas1, Cas2,* and *Crispr.* A key feature of the *Crispr* gene is a group of clustered, regularly interspaced, short, palindromic repeats—hence the name CRISPR. The repeats within the *Crispr*

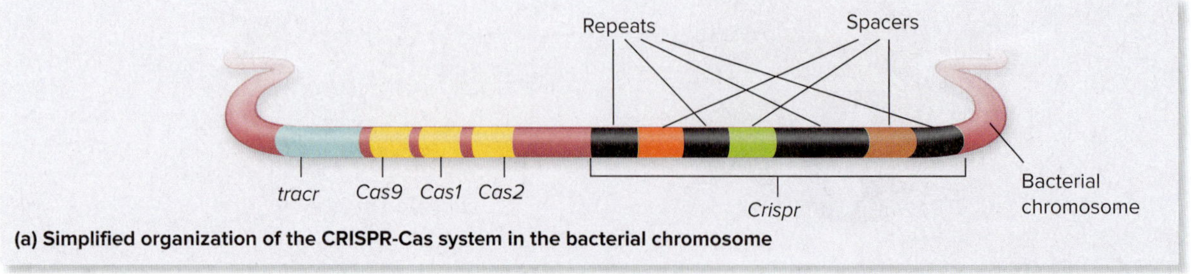


**Figure 11.11** **The CRISPR-Cas system of genome defense in bacteria.** The system shown here is a type II system, which is found in the chromosome of certain bacterial species but not in archaea. **(a)** Organization of the CRISPR-Cas system in a bacterial chromosome. This drawing shows a typical organization, but it can vary among different species. **(b–d)** A simplified mechanism of the CRISPR-Cas system. The defense occurs in three phases, called adaptation, expression, and interference.

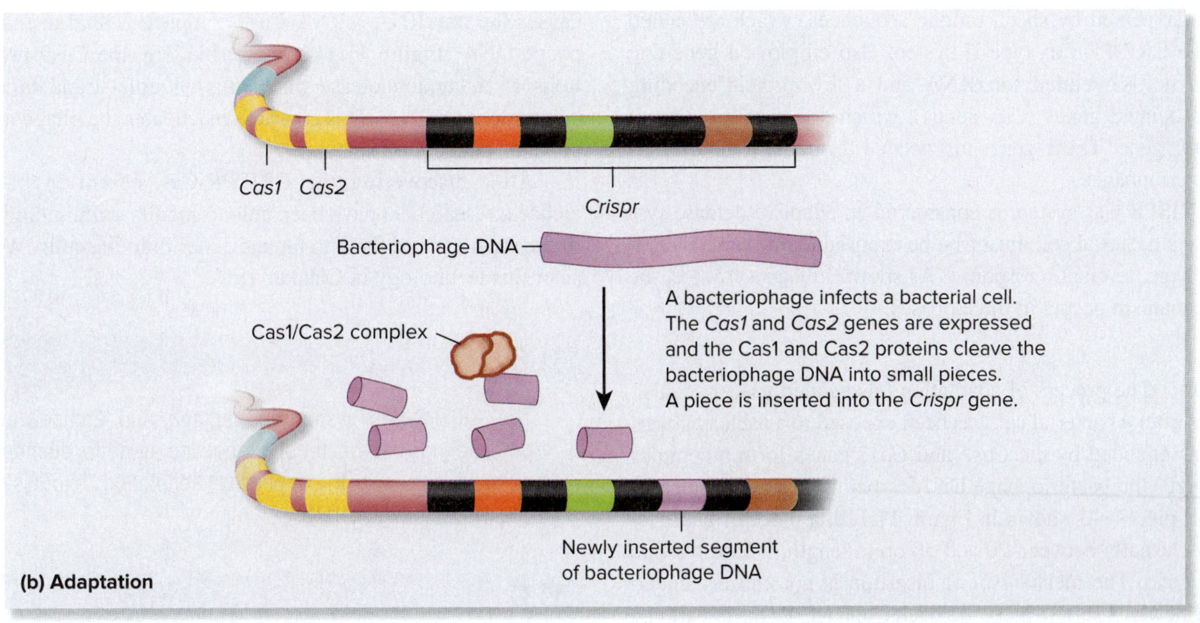


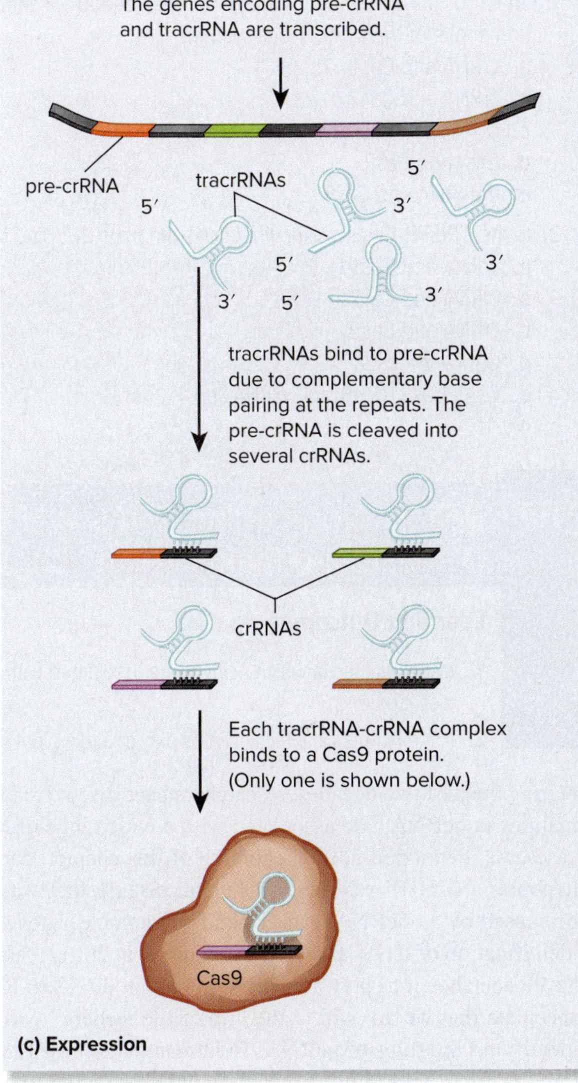

The genes encoding pre-crRNA and tracrRNA are transcribed.
pre-crRNA
tracrRNAs
5′
3′
tracrRNAs bind to pre-crRNA due to complementary base pairing at the repeats. The pre-crRNA is cleaved into several crRNAs.
crRNAs
Each tracrRNA-crRNA complex binds to a Cas9 protein. (Only one is shown below.)
Cas9
(c) Expression


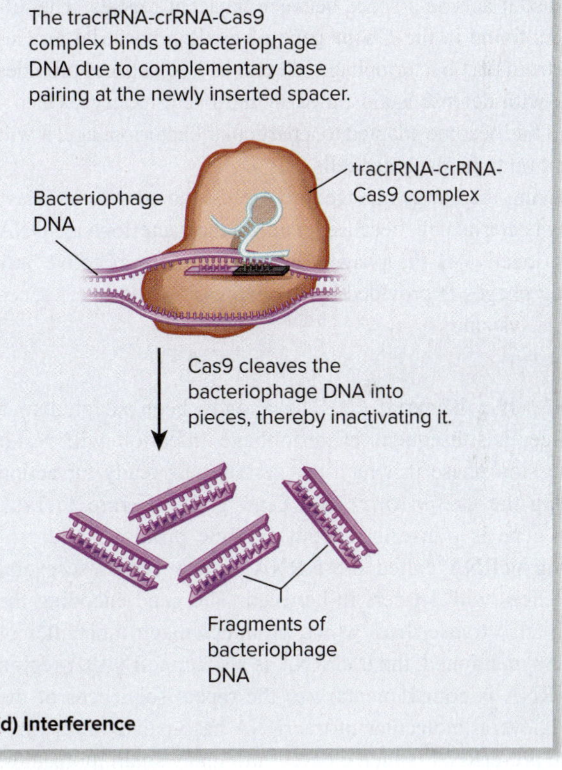


**Figure 11.11** continued

**Concept Check:** *Which component of the CRISPR-Cas system directly recognizes the bacteriophage DNA?*

gene are interspersed by short, unique sequences, which are called spacers. The CRISPR-Cas type II system also employs a gene that encodes an ncRNA called tracrRNA and a few protein-encoding CRISPR-associated genes (*Cas* genes), which are usually adjacent to the *Crispr* gene. These genes are needed to mediate the defense against bacteriophages.

The CRISPR-Cas system is considered an adaptive defense system because a bacterial cell must first be exposed to an agent, such as a bacteriophage, to elicit a response. As shown in Figure 11.11b–d, the defense mechanism occurs in three phases.

**Adaptation** The process of adaptation (also called spacer acquisition) occurs after a bacterial cell has been exposed to a bacteriophage. The proteins encoded by the *Cas1* and *Cas2* genes form a complex that recognizes the bacteriophage DNA as being foreign and cleaves it into small pieces. As shown in Figure 11.11b, a piece of bacteriophage DNA, usually between 20 and 50 bp in length, is inserted into the *Crispr* gene. The mechanism of insertion is not entirely understood. The newly inserted piece of bacteriophage DNA is called a spacer because it acts as a space between adjacent repeats. The different spacers found in the *Crispr* gene of modern bacterial species are derived from past bacteriophage infections. Each spacer provides a bacterium with defense against a particular bacteriophage. Once a bacterial cell has become adapted to a particular bacteriophage, it will pass this trait on to its daughter cells.

By cleaving the bacteriophage into pieces, the adaptation phase can protect a bacterial cell, because it cuts up the bacteriophage DNA and thereby inactivates the phage. However, a more effective way of destroying phages is provided by the expression and interference phases of this system.

**Expression** If a bacterial cell has already been adapted to a bacteriophage, a subsequent bacteriophage infection will result in the expression phase in which the system gets ready for action by expressing the *Crispr, tracr,* and *Cas9* genes (Figure 11.11c). The *Crispr* gene is transcribed from a single promoter and produces a long ncRNA called pre-crRNA, which contains several repeat sequences with spacers in-between. The gene encoding the tracrRNA is also transcribed, which produces many molecules of tracrRNA. As mentioned, the tracrRNA is also an ncRNA. A region of the tracrRNA is complementary to the repeat sequences of the pre-crRNA. Several molecules of tracrRNA base-pair with the pre-crRNA. The pre-crRNA is then cleaved into many small molecules, now called crRNA. Each crRNA is attached to a tracrRNA. A region of the tracrRNA is recognized by the Cas9 protein. The tracrRNA acts as a guide that causes the tracrRNA-crRNA complex to bind to a Cas9 protein.

**Interference** After the tracrRNA-crRNA-Cas9 complex has formed, the bacterial cell is ready to destroy the bacteriophage DNA. This phase is called interference because it resembles the process of RNA interference described earlier in this chapter (see Figure 11.8). Each spacer within a crRNA is complementary to one of the strands of a bacteriophage DNA. Therefore, the crRNA acts as a guide that causes the tracrRNA-crRNA-Cas9 complex to bind to that bacteriophage DNA (Figure 11.11d). After binding, the Cas9 protein functions as an endonuclease that makes double-strand breaks in the bacteriophage DNA. This cleavage inactivates the phage and thereby prevents phage proliferation.

After discovering the CRISPR-Cas system in bacteria and archaea, researchers have been able to modify certain components of this system and use them to mutate genes in living cells. We will consider this technology in Chapter 18.

## 11.6 Reviewing the Concepts

- The CRISPR-Cas system in bacteria and archaea provides defense against bacteriophages and genetic elements. The defense occurs in three phases: adaptation, expression, and interference (Figure 11.11).

## 11.6 Testing Your Knowledge

1. Which of the following components are needed for the adaptation phase of the CRISPR-Cas system?
   a. crRNA and Cas1
   b. crRNA and Cas2
   c. crRNA and Cas9
   d. Cas1 and Cas2
   e. tracrRNA and Cas9
2. In the CRISPR-Cas system, what does the tracrRNA bind to?
   a. crRNA and Cas1
   b. crRNA and Cas2
   c. crRNA and Cas9
   d. Cas1 and Cas2
   e. Cas1, Cas2, Cas9, and crRNA

# 11.7 Role of Non-coding RNAs in Human Disease and Plant Health

### Learning Outcomes

1. List examples in which ncRNAs are associated with human diseases.
2. List examples in which ncRNAs play a role in plant health.

During the past two decades, researchers have discovered that abnormalities in ncRNAs are associated with a wide range of human diseases. As mentioned at the beginning of this chapter, cartilage-hair hypoplasia (CHH) was the first human disease that was shown to be caused by an ncRNA transcribed from a nuclear gene. Since the identification of CHH-associated mutations in 2001, many ncRNAs have been shown to play a key role in human diseases. Researchers speculate that we are still at the "tip of the iceberg" with regard to identifying the roles of ncRNAs in human pathology. Likewise, the impact of ncRNAs on plant health is only beginning to emerge and has exciting potential in the field of agriculture. In this section, we will focus on the roles of ncRNAs in the development of cancer, neurological disorders, and cardiovascular diseases, as well as their effects on plant health.

## ncRNAs Play a Role in Many Forms of Cancer and Other Human Diseases

As we have seen throughout this chapter, ncRNAs play important roles in DNA replication, chromatin modification, gene transcription, mRNA translation, and protein function. When certain ncRNAs are expressed abnormally, that is, at too high or too low of a level, disease conditions are known to occur. Such abnormal expression levels can be caused by mutations in specific genes or by epigenetic changes, described in Chapter 16, that alter the expression of genes that encode ncRNAs. Several examples of human diseases associated with the abnormal expression of ncRNAs are listed in **Table 11.2**.

**ncRNAs and Cancer** The topic of cancer is described in Chapter 13 (look ahead to Section 13.4). The roles of ncRNAs in cancer have been most thoroughly studied with respect to miRNAs. In nearly all forms of human cancer, levels of expression of particular miRNAs differ between normal and cancer cells. In some cases, the genes behave as tumor-suppressor genes, because a lower level of expression of particular miRNAs allows tumor growth. In other cases, the genes that encode certain miRNAs act as oncogenes; their overexpression promotes cancer.

A well-studied example of the role of miRNAs in cancer involves a group of several different miRNAs called the miR-200 family. The miR-200 family plays an essential role in tumor suppression by inhibiting metastasis—the process by which cancer cells can spread through the bloodstream to other parts of the body. Low levels of expression of miR-200 members have been associated with many types of cancer, including bladder cancer, melanoma, stomach cancer, and colorectal cancer.

Though they have been less well studied, other ncRNAs are also associated with particular types of human cancers. HOTAIR, which was discussed in Section 11.3, is an ncRNA that is highly expressed in a variety of cancers, including breast cancer, lung cancer, and colorectal cancer. When overexpressed, HOTAIR behaves as an oncogene. High levels of HOTAIR expression in primary breast tumors are a significant predictor of metastasis and death. HOTAIR is known to interact with a variety of cellular components, but the mechanism by which it promotes cancer is not well understood.

**ncRNAs and Neurological Disorders** Many miRNAs are essential for the proper development and functioning of the nervous system. Approximately 70% of all miRNAs are expressed in the brain, and many of them are specific to neurons. miRNAs are involved in neuron growth and the overall development of the nervous system. Abnormal levels of expression of miRNAs have been associated with nearly all neurological disorders in which they have been investigated! Table 11.2 describes some examples in which the expression of miRNAs has been altered and associated with neurological disorders. For example, in Alzheimer disease, abnormally expressed miRNAs are thought to be involved in down-regulating the expression of the enzyme β-secretase, which leads to the overproduction of certain β-amyloid peptides—a key feature of the disease. miRNAs are also known to control the inflammatory process that leads to the development of multiple sclerosis.

**ncRNAs and Cardiovascular Diseases** Abnormalities in miRNA levels have been linked to several cardiovascular diseases. A particular miRNA, called miR-1, is associated with the development of arrhythmias—irregularities in the rate or rhythm of the heartbeat. This miRNA regulates the expression of genes that encode ion channel proteins, which are important for proper signaling between cardiac muscle cells. Other miRNAs appear to play a role in vascular disease. The formation of arterial plaques is associated with abnormal expression levels of several miRNAs, including miR10a, miR145, and miR143.

## ncRNAs Are Essential to Plant Health

In parallel to the study of human disease, plant biologists are discovering that abnormalities in ncRNAs play many essential roles that contribute to the health of plants. This realization is likely to

**Table 11.2 Examples of ncRNAs Associated with Human Diseases**

| Type of ncRNA | Disease(s)* |
|---|---|
| A group of miRNAs called the miR-200 family | Several types of cancer, including bladder cancer, melanoma, stomach cancer, and colorectal cancer |
| HOTAIR | Several types of cancer, including breast cancer, lung cancer, and colorectal cancer |
| Many miRNAs | Alzheimer disease |
| Many miRNAs | Multiple sclerosis |
| An miRNA called miR-1 | Heart arrhythmias |
| Several different miRNAs, including miR10a, miR145, and miR143 | Formation of arterial plaques |

*The diseases listed in this table show an association with abnormal levels of ncRNAs. In many cases, it is not yet clear if the disease symptoms are caused, in part, by the abnormal levels of ncRNAs or if the abnormal levels are a consequence of the disease symptoms.

**Table 11.3 Importance of ncRNAs in Plant Health**

| Type of ncRNA* | Normal role in plant structure and function |
|---|---|
| Several miRNAs, including miR156, miR157, and miR159 | Control the time of year when flowering occurs |
| An ncRNA called COOLAIR | Promotes vernalization, the process in which certain plants will only flower after being exposed to cold winter temperatures |
| Two miRNAs called miR167 and miR397 | Play a role in seed development |
| An miRNA called miR402 | Affects the rate of seed germination and seedling growth under stress conditions |
| An miRNA called miR824 | Plays a role in the development of stomata |
| An ncRNA called IPS1 | Affects the ability of plants to cope with phosphate starvation |

*Most of the examples listed are miRNAs, which are short ncRNAs. COOLAIR and IPS1 are longer ncRNAs.

have great impact in the field of agriculture as we develop methods to change the expression of ncRNAs in order to modify the characteristics of agriculturally important plants. **Table 11.3** describes several examples in which particular ncRNAs are known to play key roles in plant health.

## 11.7 Reviewing the Concepts

- Abnormalities in the expression of ncRNAs have been associated with many diseases, including cancer, neurological disorders, and cardiovascular diseases (Table 11.2).
- The proper level of expression of ncRNA is also important for plant health (Table 11.3).

## 11.7 Testing Your Knowledge

**1.** An miRNA is discovered to be overexpressed, and this overexpression promotes cancer. In this case, the miRNA is behaving as

**a.** an oncogene.
**b.** a tumor-suppressor gene.
**c.** a decoy.
**d.** both an oncogene and a tumor-suppressor gene.
**e.** both a decoy and a tumor-suppressor gene.

# Assess and Discuss

## Test Yourself

1. Which of the following types of molecules could bind to an ncRNA due to base pairing?
   a. DNA
   b. RNA
   c. protein
   d. small molecule
   e. both a and b
2. Which of the following is *not* a general function of an ncRNA?
   a. encoding a polypeptide
   b. acting as a ribozyme
   c. acting as a guide
   d. acting as a scaffold
   e. acting as a decoy
3. TERC, which is the RNA component of telomerase, plays two key roles in the synthesis of telomere repeat sequences. During this process,
   a. TERC first acts a template, then acts as a guide.
   b. TERC first acts a guide, then acts as a template.
   c. TERC first acts a template, then acts as a decoy.
   d. TERC first acts a guide, then acts as a decoy.
   e. TERC first acts as a decoy, then acts as a template.
4. HOTAIR causes certain genes to be repressed by facilitating
   a. the binding of a repressor protein.
   b. the release of an activator protein.
   c. the covalent modification of histones.
   d. the removal of nucleosomes.
   e. both a and c.
5. One of the roles of the RNA component of signal recognition particle (SRP) is to stimulate certain proteins to hydrolyze GTP. If this function of SRP RNA did not work properly, what would you expect to happen?
   a. SRP would not bind to the ER signal sequence of a polypeptide.
   b. SRP would not cause translation to pause.
   c. SRP would not bind to an SRP receptor in the ER membrane.
   d. SRP would not be released from the ER membrane.
   e. Both a and b are correct.
6. During RNA interference, what binds to an mRNA to inhibit translation?
   a. a pri-miRNA
   b. a pre-miRNA or pre-siRNA
   c. a double-stranded miRNA or double-stranded siRNA
   d. a single-stranded miRNA or single-stranded siRNA
   e. dicer
7. If the antisense sequences in a snoRNA were mutated in a way that prevented rRNA binding, this would prevent the snoRNA from acting as a
   a. guide.
   b. scaffold.
   c. decoy.
   d. ribozyme.
   e. blocker.

8. Cas1 and Cas2 proteins play a role during which of the following phases of genome defense?
   a. adaptation
   b. expression
   c. interference
   d. both adaptation and expression
   e. both expression and interference
9. Which of the following components bind to tracrRNA?
   a. crRNA and Cas1 protein
   b. crRNA and Cas2 protein
   c. crRNA and Cas9 protein
   d. crRNA only
   e. Cas1 and Cas2 proteins
10. Abnormalities in the expression of ncRNAs are associated with
   a. many forms of cancer.
   b. neurological disorders.
   c. cardiovascular diseases.
   d. all of the above.
   e. only a and b.

## Conceptual Questions

1. An ncRNA may have the following functions: scaffold, guide, alter protein function or stability, ribozyme, blocker, and/or decoy. Which of those functions are mediated by the following examples: HOTAIR, RNA of SRP, microRNA, crRNA? Note: A single ncRNA may have more than one function.
2. What is the phenomenon of RNA interference (RNAi)? Explain how the double-stranded RNA is processed during RNAi and how it leads to the silencing of a complementary mRNA.
3. **PRINCIPLES** A principle of biology is that structure determines function. Explain how the structure of HOTAIR allows it to carry out its function.

## Collaborative Questions

1. Review the concept of an RNA world described in Chapter 4, Section 4.1. Discuss which ncRNAs described in Table 11.1 may have arisen during the RNA world, and which probably arose after the modern DNA/RNA/protein came into being.
2. Go to the PubMed website and search words such as non-coding RNA and disease. Scan through the journal articles you retrieve, and make a list of the roles that ncRNAs may play in human diseases.

# 12 The Control of Genetic Information via Gene Regulation

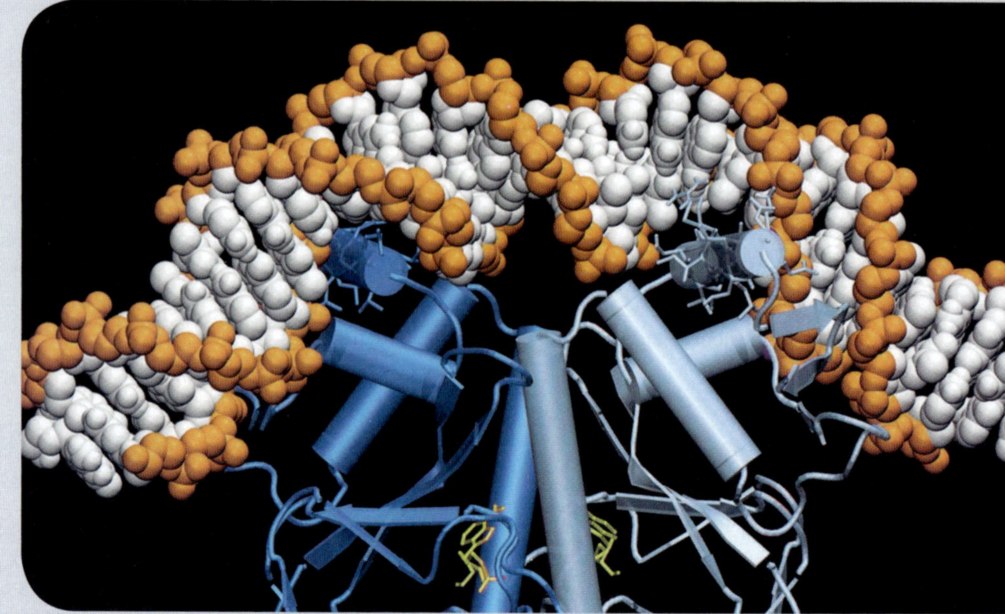

© Daniel Gage, University of Connecticut

**A model for a protein that binds to DNA and regulates genes.** The catabolite activator protein (CAP), shown in dark and light blue, is binding to the DNA double helix, shown in orange and white. The CAP, shown again in Figure 12.8, activates gene transcription.

## Chapter Outline



Consider the muscles of a professional weight lifter and those of someone who doesn't lift weights. A weight lifter's upper arm muscles (the biceps and triceps) can attain enormous sizes, whereas the same muscles in a non–weight lifter are relatively small. How do we explain such differences? Unknowingly, when people lift weights, they are "turning on," or activating, hundreds of genes in their muscle cells. For example, genes that encode actin (a cytoskeletal filament) and myosin (a motor protein) are turned on. As discussed in Chapter 35, actin and myosin are key components that promote muscle contraction. By making more of these proteins, the weight lifter increases the mass of her or his muscles and has muscles that can contract with greater force and duration.

At the molecular level, **gene expression** is the process by which the information within a gene is made into a functional product, such as an RNA molecule or a protein. The majority of genes in all species are regulated so the proteins they specify are produced at appropriate times, in the appropriate cell types, and in specific amounts. The term **gene regulation** refers to the ability of cells to control the expression of their genes. By comparison, some genes have relatively constant levels of expression in all conditions over time. These are called **constitutive genes.** Frequently, constitutive genes encode proteins that are constantly required for the survival of an organism, such as certain metabolic enzymes involved with carbohydrate breakdown.

The importance of gene regulation is underscored by the number of genes devoted to this process. For example, in *Arabidopsis thaliana,* a plant that is studied by many plant geneticists, over 5% of the genome is involved with regulating gene transcription. In humans, researchers estimate that 2,600 genes encode proteins that bind to DNA and affect gene function. With a human genome size of about 22,000 protein-encoding genes, nearly 12% of them are directly involved with gene regulation.

In this chapter, we will begin with an overview that emphasizes the benefits of gene regulation and the general mechanisms that achieve such regulation in bacteria and in eukaryotes. Later sections will describe how bacteria regulate gene expression in the face of environmental change and the more complex nature of gene regulation in eukaryotes.

## 12.1 Overview of Gene Regulation

### Learning Outcomes

1. Discuss the various ways that organisms benefit from gene regulation.
2. Identify where gene regulation can occur in the pathway of gene expression for bacteria and eukaryotes.

How do living organisms benefit from gene regulation? A key reason is that it conserves energy. RNAs and proteins that are encoded by genes are produced only when needed. In multicellular organisms, gene regulation also ensures that genes are expressed in the appropriate cell types and at the correct stage of development. In this section, we will examine a few examples that illustrate the important consequences of gene regulation. We will also survey the major points in the gene expression process at which genes are regulated in bacterial and eukaryotic cells.

### Bacteria Regulate Genes in Response to Changes in Their Environment

The bacterium *Escherichia coli* can use many types of sugars as food sources, thereby increasing its chances of survival. With regard to gene regulation, we will focus on how it uses lactose, which is a sugar found in milk. *E. coli* carries genes that code for proteins that enable it to take up lactose from the environment and metabolize it.

**Figure 12.1** illustrates the effects of lactose on the regulation of those genes. In order to utilize lactose, an *E. coli* cell requires both a transporter (lactose permease) that facilitates the uptake of lactose into the cell and an enzyme (β-galactosidase) that catalyzes the breakdown of lactose. When lactose is not present in the environment, an *E. coli* cell makes very little of these proteins. However, when lactose becomes available, the bacterium produces many more of these proteins, enabling it to readily use lactose from its environment. Eventually, all of the lactose in the environment is used up. At this point, the genes encoding these proteins will be

## Biology Principle

### Living Organisms Use Energy

Gene regulation provides a way for cells to avoid wasting energy by making proteins only when they are needed.

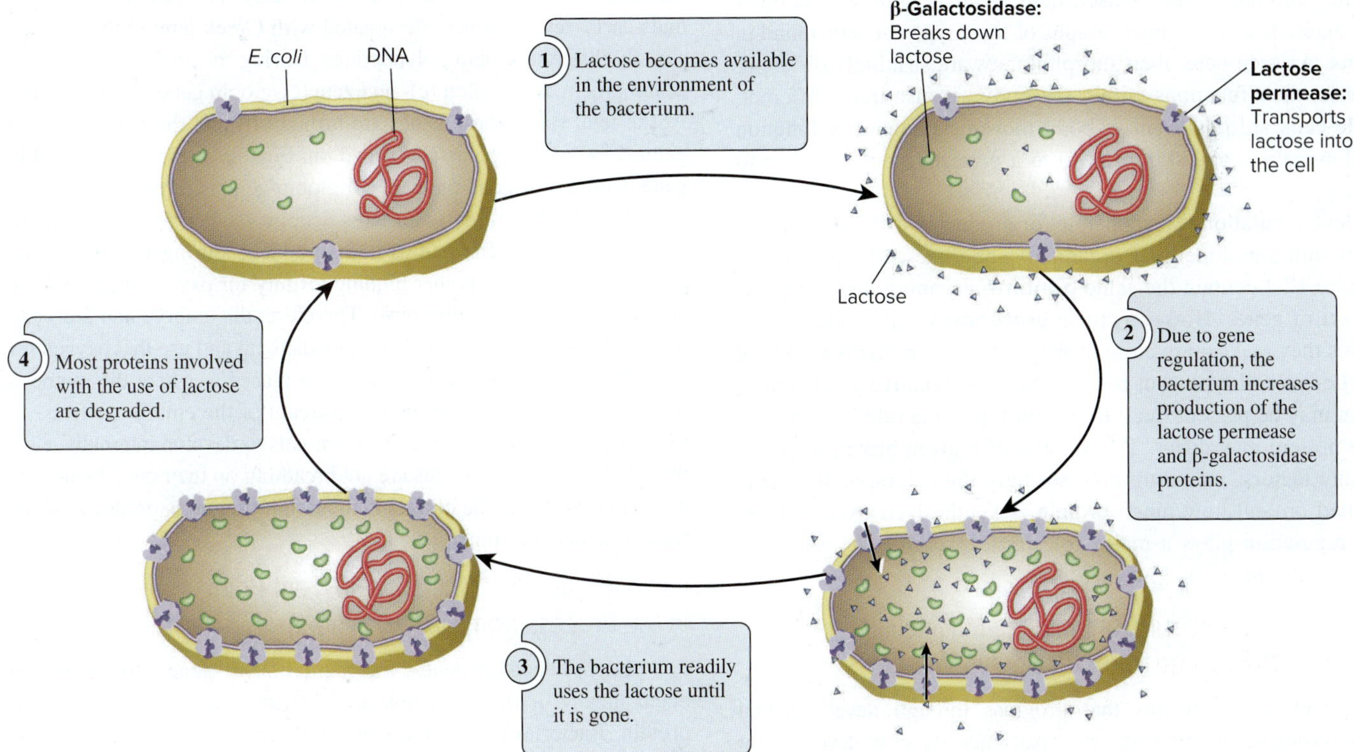


**Figure 12.1** Gene regulation of lactose utilization in *E. coli*.

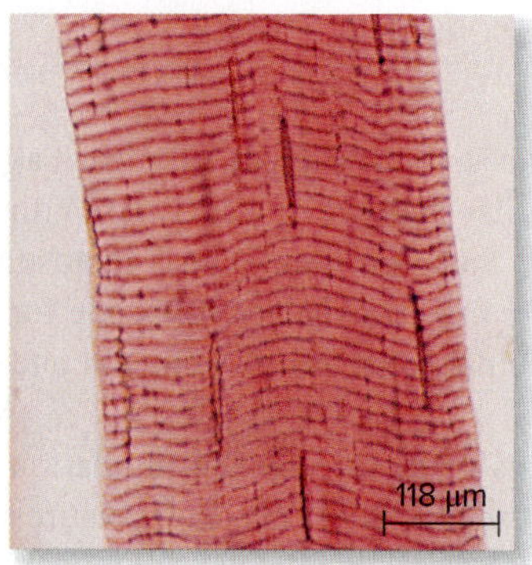


(a) Skeletal muscle cell

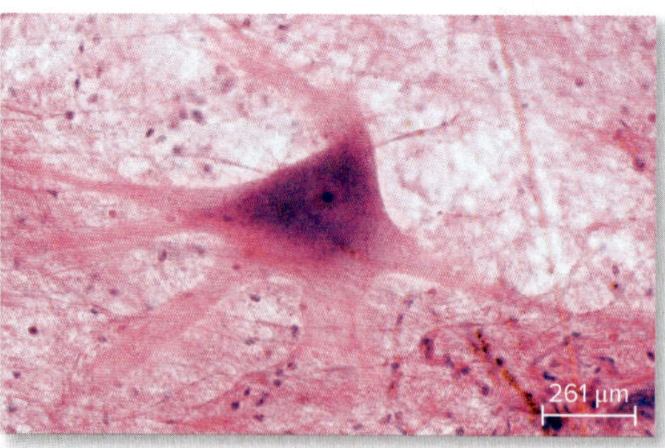


(b) Neuron

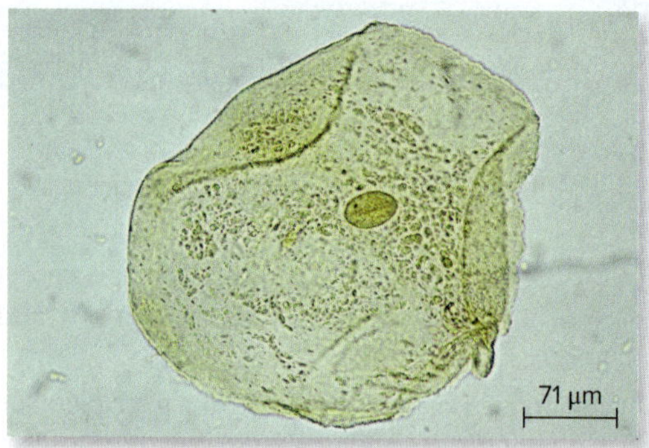


(c) Skin cell

**Figure 12.2** **Examples of different cell types in humans.** These cells have the same genetic composition. Their unique morphologies are due to differences in the proteins they make.

*(a, c)* © Ed Reschke/Getty Images *(b)* © McGraw-Hill Education/Al Telser

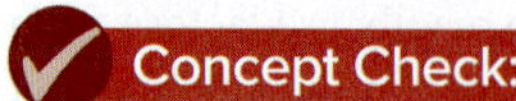

**Concept Check:** *How does gene regulation underlie the different morphologies of these cells?*

shut off, and most of the proteins will be degraded. Overall, gene regulation conserves energy because it ensures that the proteins needed for lactose utilization are made only when lactose is present in the environment.

## Eukaryotic Gene Regulation Produces Different Cell Types in a Single Organism

One of the most amazing examples of gene regulation is the phenomenon of **cell differentiation,** the process by which cells become specialized into particular types. In humans, for example, cells may differentiate into muscle cells, neurons, skin cells, or other types. Figure 12.2 shows micrographs of three types of cells found in humans. As seen here, their morphologies are strikingly different. Likewise, their functions within the body are also quite different. Muscle cells are important in body movements, neurons function in cell signaling, and skin cells form a protective outer surface to the body.

Gene regulation is responsible for producing different types of cells within a multicellular organism. The three cell types shown in Figure 12.2 contain the same **genome,** meaning they carry the same set of genes. However, their **proteomes**—the collection of proteins they make—are quite different. Certain proteins are found in particular cell types but not in others. Alternatively, whereas a protein may be present in all three cell types, the relative amounts of the protein may differ. The amount of a given protein depends on many factors, including how strongly the corresponding gene is turned on and how much protein is synthesized from mRNA. Gene regulation plays a major role in determining the proteome of each cell type.

## Eukaryotic Gene Regulation Enables Multicellular Organisms to Proceed Through Developmental Stages

In multicellular organisms that progress through developmental stages, certain genes are expressed at particular stages of development but not at others. Let's consider an example of such gene regulation in mammals. Early stages of development occur in the uterus of female mammals. In humans, the embryonic stage lasts from fertilization to 8 weeks. During this stage, major developmental changes produce the various body parts. The fetal stage occurs from 8 weeks to birth (41 weeks). This stage is characterized by a continued refinement of body parts and a large increase in size.

The oxygen demands of a rapidly growing embryo and fetus are quite different from the needs of the mother. Gene regulation plays a vital role in ensuring that an embryo and a fetus get the proper amount of oxygen. Hemoglobin is a protein that delivers oxygen to the cells of a mammal's body. A hemoglobin protein is composed of four globin polypeptides: two encoded by one globin gene and two encoded by another globin gene (Figure 12.3). The genomes of mammals carry several genes (designated with Greek letters) that encode slightly different globin polypeptides. During the embryonic stage of development, the epsilon ($\varepsilon$)- and zeta ($\zeta$)-globin genes are turned on. At the fetal stage, these genes are turned off, and the alpha ($\alpha$)- and gamma ($\gamma$)-globin genes are turned on. Finally, at birth, the $\gamma$-globin gene is turned off, and the beta ($\beta$)-globin gene is turned on.

How do the embryo and fetus acquire oxygen from their mother's bloodstream? The hemoglobin produced during the embryonic and fetal stages has a higher binding affinity for oxygen than does the hemoglobin produced after birth. Therefore, the embryo and fetus can remove oxygen from the mother's bloodstream and use that oxygen for their own needs. This occurs across the placenta, where the mother's bloodstream is adjacent to the bloodstream of the embryo or fetus. In this way, gene regulation enables mammals to develop internally, even though the embryo and fetus are not breathing on their own. Gene regulation ensures that the correct hemoglobin protein is produced at the right time in development.

## Gene Regulation Occurs at Different Points in the Process from DNA to Protein

For protein-encoding genes, the regulation of gene expression can occur at any of the steps that are needed to produce a functional protein. Bacterial gene expression can be regulated at the following stages (Figure 12.4a):

- **Transcription:** Gene regulation most commonly occurs at the level of transcription, which means that bacteria regulate

## Biology Principle

### Living Organisms Grow and Develop

Gene regulation is an important process that allows organisms to proceed through developmental stages.

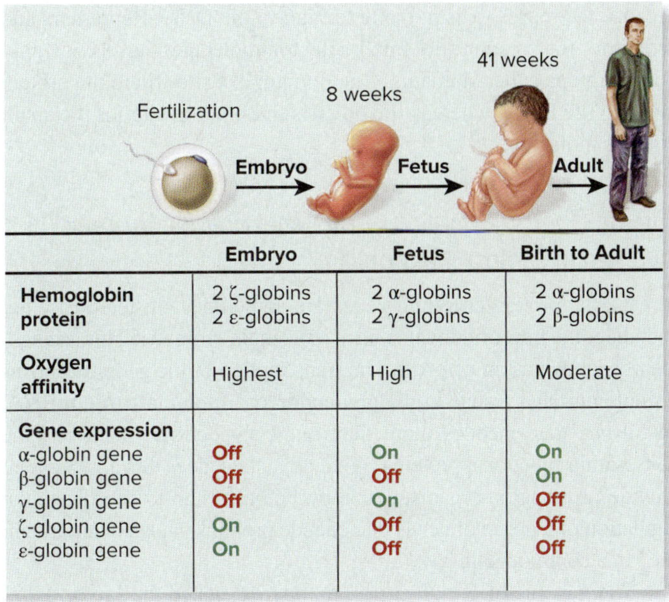


| | Embryo | Fetus | Birth to Adult |
|---|---|---|---|
| Hemoglobin protein | 2 ζ-globins<br>2 ε-globins | 2 α-globins<br>2 γ-globins | 2 α-globins<br>2 β-globins |
| Oxygen affinity | Highest | High | Moderate |
| **Gene expression** | | | |
| α-globin gene | Off | On | On |
| β-globin gene | Off | Off | On |
| γ-globin gene | Off | On | Off |
| ζ-globin gene | On | Off | Off |
| ε-globin gene | On | Off | Off |

**Figure 12.3** Regulation of human globin genes at different stages of development.

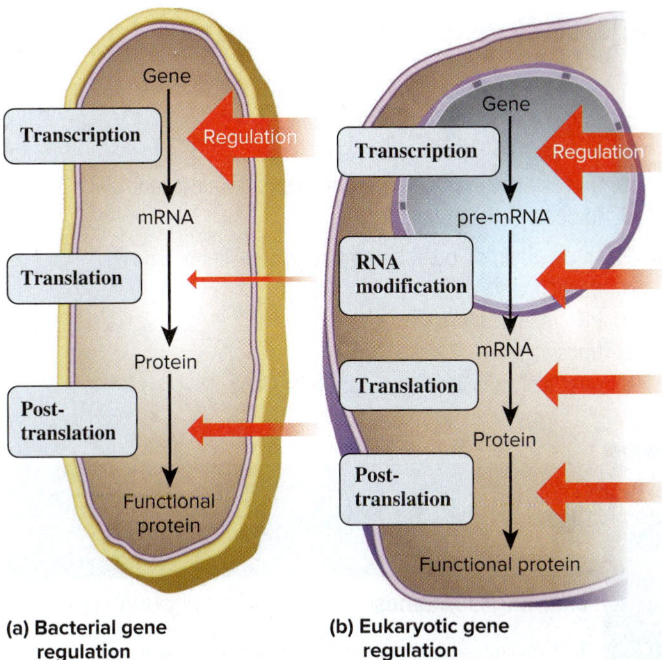


(a) Bacterial gene regulation

(b) Eukaryotic gene regulation

**Figure 12.4** **Overview of gene regulation in (a) bacteria and (b) eukaryotes.** The relative width of the red arrows indicates the prominence with which gene regulation is used to control the production of functional proteins.

how much mRNA is made from genes. Because transcription is the first step in gene expression, transcriptional regulation is a particularly efficient way to regulate genes because cells avoid wasting energy when the product of the gene is not needed.

- **Translation:** Bacteria may also control the ability of an mRNA to be translated into protein, but this mechanism is not as prominent as transcriptional regulation.
- **Post-translation:** After a protein is made, its amount or function may be regulated, which is termed post-translational regulation. This form of regulation is best understood within the context of cell biology. For example, feedback inhibition of enzymes is discussed in Chapter 6 (refer back to Figure 6.10).

In eukaryotes, gene expression can also be regulated at several stages (**Figure 12.4b**):

- **Transcription:** Transcriptional regulation is a very common form of regulation in eukaryotes.
- **RNA modification:** As discussed in Chapter 10 (refer back to Figure 10.4), eukaryotes modify their mRNA transcripts in ways that do not commonly occur in bacteria. For example, RNA splicing is a widespread phenomenon in eukaryotes. As discussed later, such events are subject to regulation.
- **Translation:** Though not as prominent as transcriptional regulation, this is a fairly common way for eukaryotes to regulate gene expression.
- **Post-translation:** This form of regulation is also prominent in eukaryotes.

## 12.1 Reviewing the Concepts

- Most genes are regulated so the level of gene expression can vary under different conditions. By comparison, constitutive genes are expressed at constant levels.
- Gene regulation ensures that gene products are made only when needed. An example is the synthesis of the gene products needed for lactose utilization in bacteria (Figure 12.1).
- In eukaryotes, gene regulation leads to the production of different cell types, such as neurons, muscle cells, and skin cells. Eukaryote gene regulation enables gene products to be produced at different developmental stages (Figures 12.2, 12.3).
- All organisms regulate gene expression at a variety of stages, including transcription, translation, and post-translation. Eukaryotes also regulate RNA processing (Figure 12.4).

## 12.1 Testing Your Knowledge

1. A reason for gene regulation is
   a. to respond to environmental changes.
   b. to make proteins in specific cell types.
   c. to make proteins at specific stages of development.
   d. to do all of the above.
2. The most common point of gene regulation in bacteria is
   a. transcription.
   b. RNA modification.
   c. translation.
   d. after a protein is made (post-translation).

## 12.2 Regulation of Transcription in Bacteria

### Learning Outcomes

1. Explain how regulatory transcription factors and small effector molecules are involved in the regulation of transcription.
2. Describe the organization of the *lac* operon and how it is regulated by negative and positive control.

As we have seen in Figure 12.1, when a bacterium is exposed to a particular nutrient in its environment, such as a sugar, the genes are expressed that encode proteins needed for the uptake and metabolism of that sugar. In addition, bacteria have enzymes that synthesize molecules such as particular amino acids. In such cases, the control of gene expression often occurs at the level of transcription. In this section, we will examine the underlying molecular mechanisms that bring about transcriptional regulation in bacteria.

### Transcriptional Regulation Involves Regulatory Transcription Factors and Small Effector Molecules

In most cases, regulation of transcription involves the actions of **regulatory transcription factors**—proteins that bind to regulatory sequences in the DNA in the vicinity of a promoter and affect the rate of transcription of one or more nearby genes. These transcription factors either decrease or increase the rate of transcription of a gene (Figure 12.5a):

- **Repressors** are regulatory transcription factors that bind to the DNA and decrease the rate of transcription. This is a form of regulation called **negative control.**
- **Activators** are regulatory transcription factors that bind to the DNA and increase the rate of transcription, a form of regulation termed **positive control.**

In conjunction with regulatory transcription factors, molecules called **small effector molecules** often play a critical role in transcriptional regulation. A small effector molecule exerts its influence by binding to a repressor or an activator and causing a conformational change in the protein. In many cases, the effect of the conformational change determines whether or not the protein can bind to the DNA. Figure 12.5b illustrates an example involving a repressor. When the small effector molecule is not present in the cytoplasm, the repressor binds to the DNA and inhibits transcription. However, when the small effector molecule is subsequently found in the cytoplasm, it will bind to the repressor and cause a conformational change that inhibits the ability of the protein to bind to the DNA. Transcription can occur because the repressor is not able to bind to the DNA. Repressors and activators that respond to small effector molecules have two functional regions called domains. One domain is a site where the protein binds to the DNA, whereas the other is the binding site for the small effector molecule.

### The *lac* Operon Contains Genes That Encode Proteins Involved in Lactose Metabolism

In bacteria, protein-encoding genes are sometimes clustered together and under the transcriptional control of a single promoter. This arrangement is known as an **operon.** The transcription of the genes occurs as a single unit and results in the production of a **polycistronic mRNA,** an mRNA that encodes more than one polypeptide, which become units within functional proteins. What advantage does this arrangement provide? An operon organization allows a bacterium to coordinate the regulation of a group of genes that encode proteins whose functions are used in a common pathway.

The genome of *E. coli* carries an operon, called the ***lac* operon,** that contains the genes for the proteins that allow it to metabolize lactose (see Figure 12.1). Figure 12.6a shows the organization of this operon as it is found in the *E. coli* chromosome, as well as the polycistronic mRNA that is transcribed from it.

- The **promoter,** *lacP*, is used to transcribe three genes: *lacZ*, *lacY*, and *lacA*.
- *LacZ* encodes β-galactosidase, which is an enzyme that breaks down lactose (Figure 12.6b). In a side reaction, β-galactosidase also converts a small percentage of lactose into allolactose, a structurally similar sugar or lactose analogue, which (as discussed later) functions as a small effector molecule.
- The *lacY* gene encodes lactose permease, which is a membrane protein required for the transport of lactose into the cytoplasm of the bacterium.
- The *lacA* gene encodes galactoside transacetylase, which covalently modifies lactose and lactose analogues by attaching an acetyl group (—$COCH_3$). The attachment of acetyl groups to nonmetabolizable lactose analogues prevents their toxic buildup in the cytoplasm.
- The **operator,** or *lacO* site, is a regulatory sequence in the DNA. The sequence of bases at the *lacO* site provides a binding site for a repressor protein.
- The **CAP site** is a regulatory sequence in the DNA that is recognized by an activator protein.

Adjacent to the *lac* operon is the *lacI* gene, which encodes the **lac repressor.** This repressor protein is important for the regulation

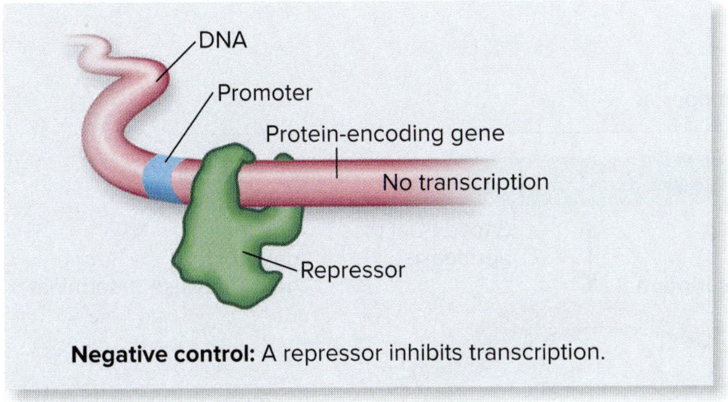


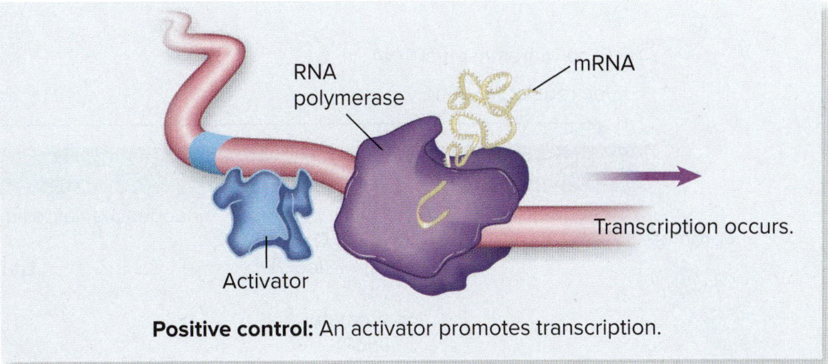


(a) Actions of regulatory transcription factors

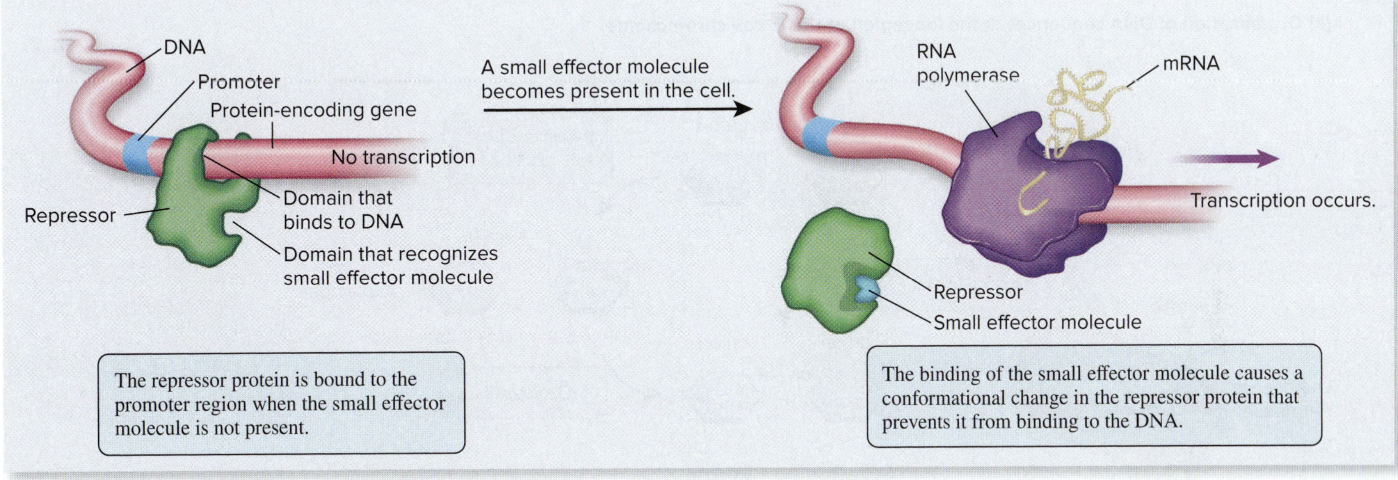


(b) Action of a small effector molecule on a repressor

**Figure 12.5** **Actions of regulatory transcription factors and small effector molecules. (a)** Regulatory transcription factors are proteins that exert negative or positive control. **(b)** One way that a small effector molecule may exert its effects is by preventing a repressor protein from binding to the DNA.

of the *lac* operon. The *lacI* gene, which is constitutively expressed at a fairly low level, has its own promoter called the *i* promoter. The *lacI* gene is not considered a part of the *lac* operon. Let's now take a look at how the *lac* operon is regulated by the lac repressor.

## The *lac* Operon Is Under Negative Control by a Repressor Protein

In the late 1950s, the first researchers to investigate gene regulation were French biologists François Jacob and Jacques Monod at the Pasteur Institute in Paris, France. Their focus on gene regulation stemmed from an interest in the phenomenon known as enzyme adaptation, which had been identified early in the 20th century. Enzyme adaptation occurs when a particular enzyme appears within a living cell only after the cell has been exposed to the substrate for that enzyme. Jacob and Monod studied lactose metabolism in *E. coli* to investigate this phenomenon. When they exposed bacteria to lactose, the levels of lactose-using enzymes in the cells increased by 1,000- to 10,000-fold. After lactose was removed, the synthesis of the enzymes abruptly stopped.

The first mechanism of regulation that Jacob and Monod discovered involved the lac repressor, which binds to the sequence of nucleotides found at the *lac* operator. Once bound, the lac repressor prevents RNA polymerase from transcribing the *lacZ*, *lacY*, and *lacA* genes (**Figure 12.7a**). RNA polymerase can bind to the promoter when the lac repressor is bound to the operator but is largely prevented from moving past the operator to transcribe the *lacZ*, *lacY*, and *lacA* genes. Therefore, if *E. coli* does not have lactose in its environment, only a small amount of lactose permease, β-galactosidase, and transacetylase are made.

Alternatively, when *E. coli* is exposed to lactose, the following events occur (**Figure 12.7b**):

1. A small amount of lactose is transported into the cytoplasm via lactose permease, and β-galactosidase converts some of it to allolactose.
2. The cytoplasmic level of allolactose gradually rises. Allolactose is an example of a small effector molecule. The lac repressor protein contains four identical subunits, each one recognizing a single allolactose molecule. When four allolactose molecules

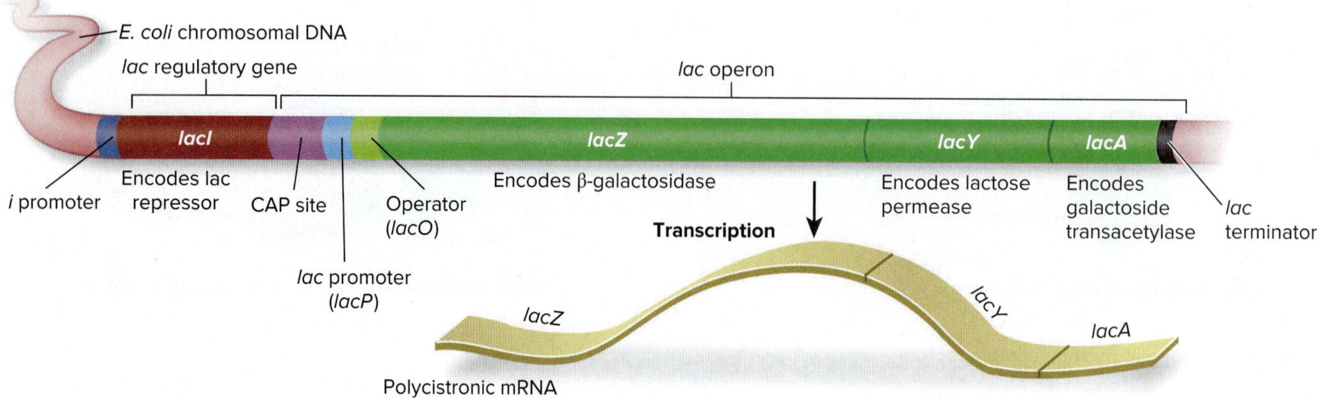


**(a) Organization of DNA sequences in the *lac* region of the *E. coli* chromosome**

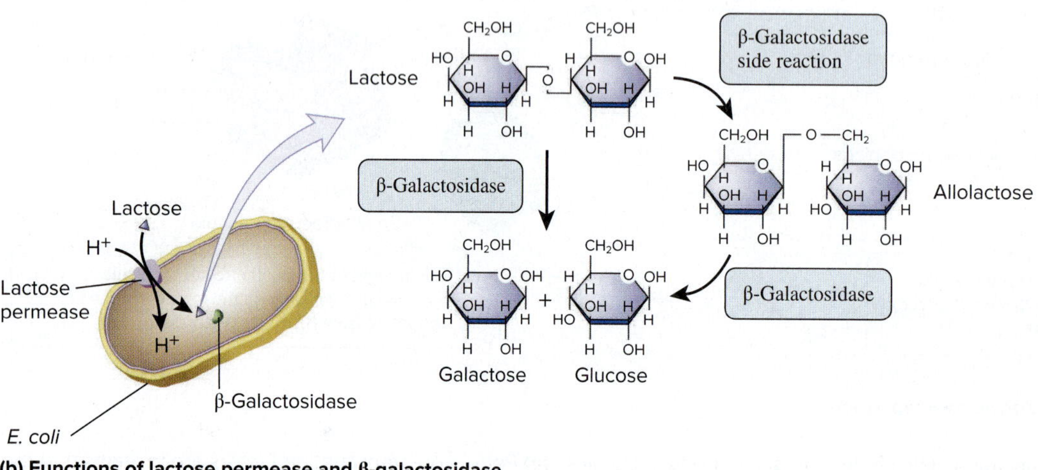


**(b) Functions of lactose permease and β-galactosidase**

**Figure 12.6** **The *lac* operon.** **(a)** This diagram depicts a region of the *E. coli* chromosome that contains the *lacI* gene and the adjacent *lac* operon, as well as the polycistronic mRNA transcribed from the operon. The mRNA is translated into three proteins: β-galactosidase, lactose permease, and galactoside transacetylase. **(b)** Lactose permease cotransports $H^+$ with lactose. Bacteria maintain an $H^+$ gradient across their cytoplasmic membrane that drives the active transport of lactose into the cytoplasm. β-galactosidase cleaves lactose into galactose and glucose. As a side reaction, it can also convert lactose into allolactose.

**Concept Check:** *Which genes are under the control of the lac promoter?*

bind to the lac repressor, a conformational change occurs that prevents the repressor from binding to the operator.

3. When the lac repressor is not bound to the operator, RNA polymerase transcribes the *lacZ*, *lacY*, and *lacA* genes at a high rate.
4. Translation of the encoded polypeptides produces an abundant amount of the proteins needed for lactose uptake and metabolism, as described previously in Figure 12.1.

The regulation of the *lac* operon enables *E. coli* to conserve energy because lactose-utilizing proteins are made in large amounts only when lactose is present in the environment. Allolactose is an **inducer,** a small effector molecule that increases the rate of transcription, and the *lac* operon is said to be an **inducible operon.**

## The *lac* Operon Is Also Under Positive Control by an Activator Protein

In addition to negative control by a repressor protein, the *lac* operon is also positively regulated by an activator called the **catabolite activator protein (CAP).** CAP is controlled by a small effector molecule, **cyclic AMP (cAMP),** which is produced from ATP via an enzyme known as adenylyl cyclase. When cAMP binds to CAP, the cAMP-CAP complex binds to the CAP site near the *lac* promoter (**Figure 12.8**). This causes a bend in the DNA that enhances the ability of RNA polymerase to bind to the promoter. However, the presence of glucose in the environment inhibits the production of cAMP, thereby preventing the binding of CAP to the DNA. In this way, glucose blocks the activation of the *lac* operon, thereby inhibiting transcription.

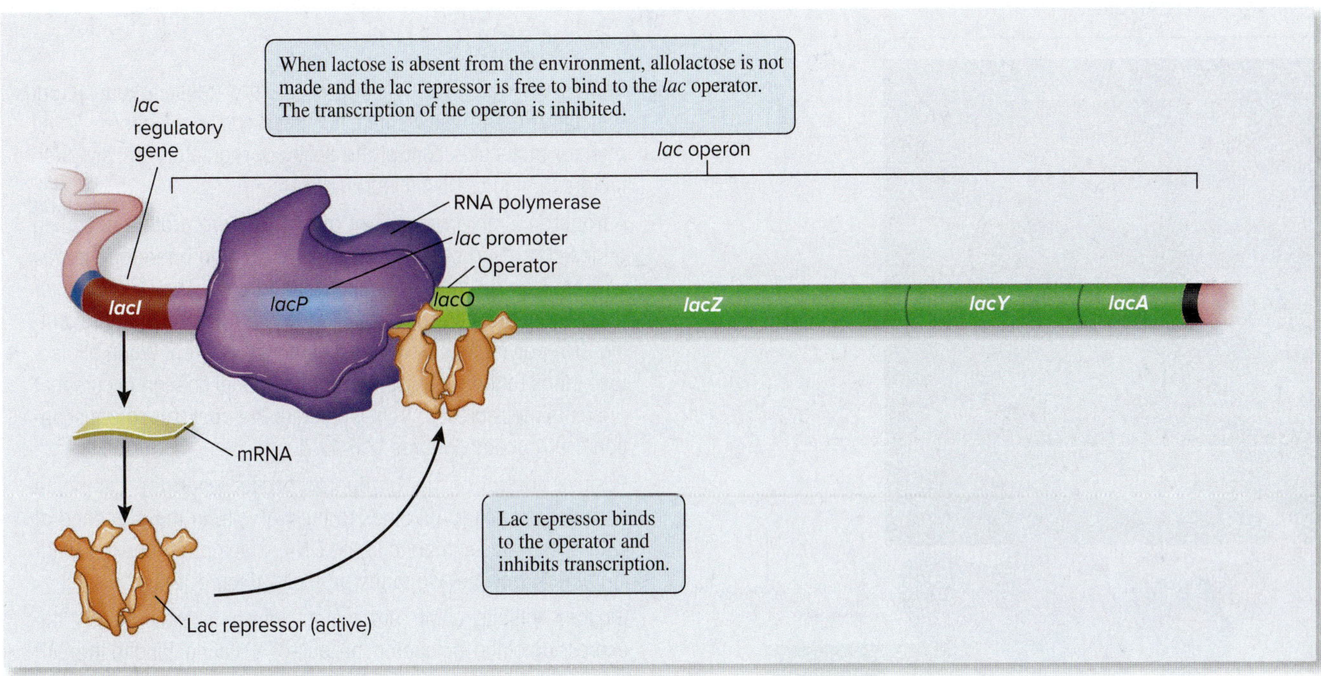


**(a) Lactose absent from the environment**

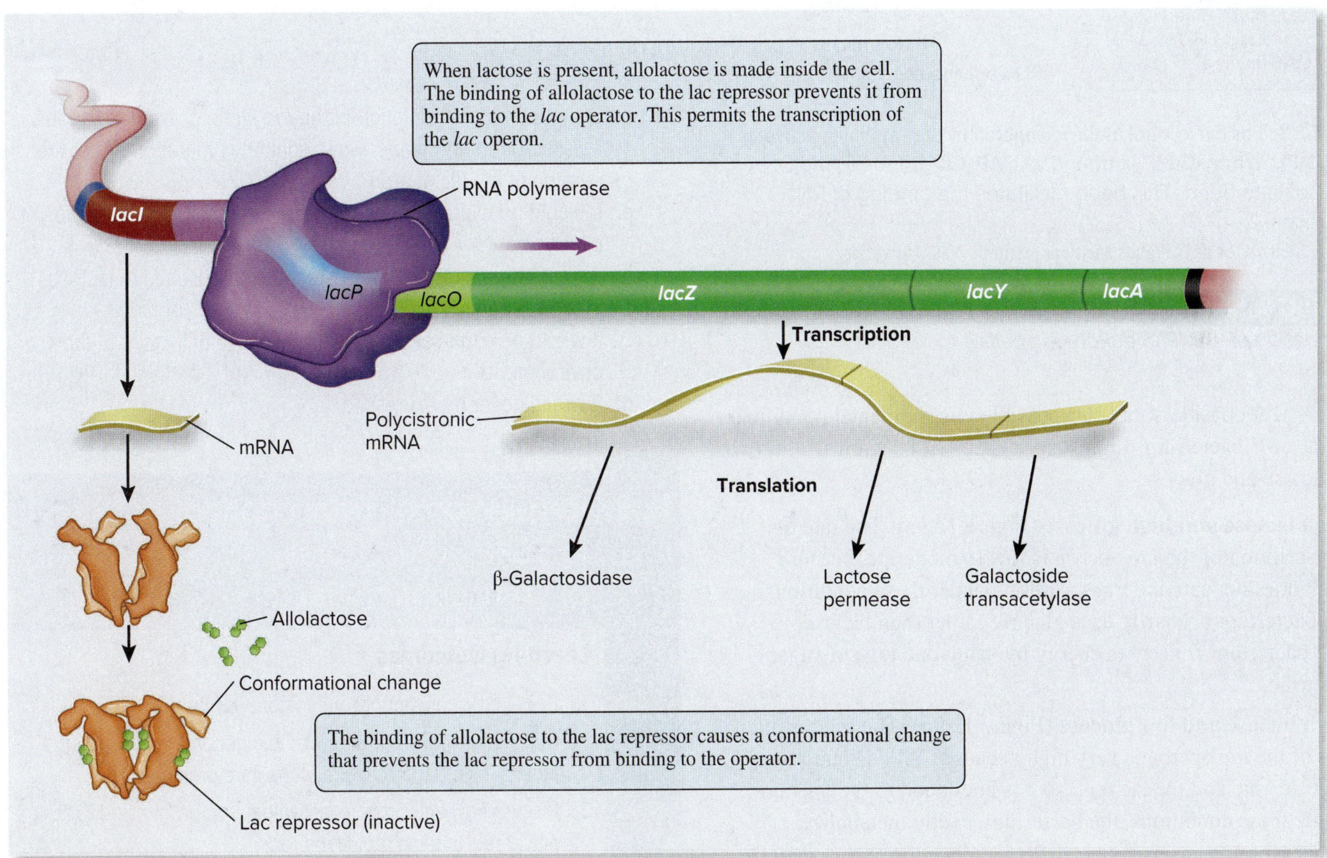


**(b) Lactose present**

**Figure 12.7** Negative control of an inducible set of genes: Function of the lac repressor in regulating the *lac* operon.

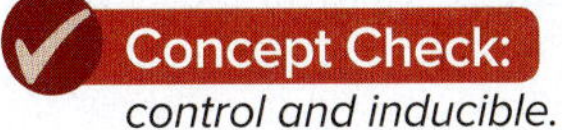

*With regard to regulatory proteins and small effector molecules, explain the meaning of the terms negative control and inducible.*

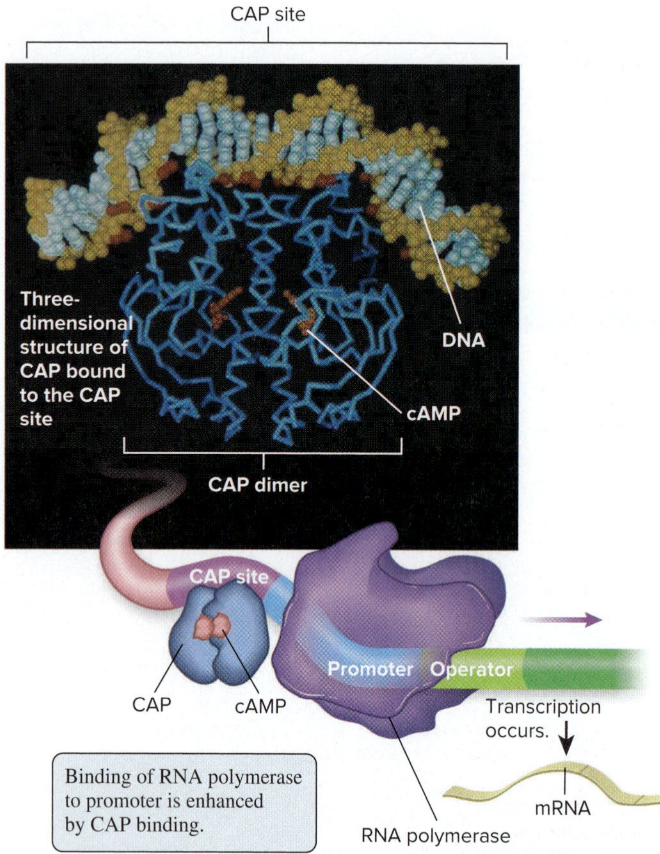


**Figure 12.8 Positive control of the *lac* operon by the catabolite activator protein (CAP).** When cAMP is bound to CAP, CAP binds to the DNA and causes it to bend. This bend facilitates the binding of RNA polymerase.



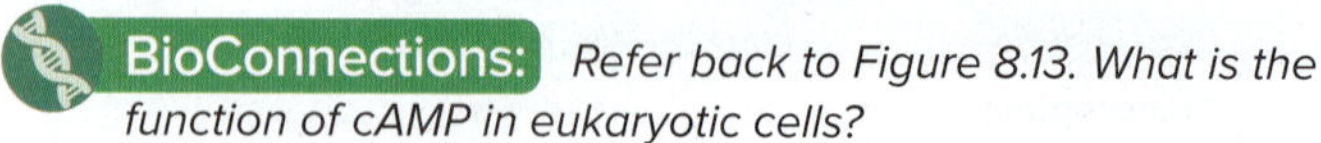

**BioConnections:** *Refer back to Figure 8.13. What is the function of cAMP in eukaryotic cells?*

Figure 12.9 considers the four possible environmental conditions that an *E. coli* bacterium might experience with regard to the two sugars lactose and glucose.

- **High lactose and high glucose** (Figure 12.9a): The rate of transcription of the *lac* operon is low to moderate, because CAP does not activate transcription. Under these conditions, the bacterium primarily uses glucose rather than lactose. The bacterium conserves energy by using one type of sugar at a time.
- **High lactose and low glucose** (Figure 12.9b): The transcription rate of the *lac* operon is very high because CAP is bound to the CAP site and the lac repressor is not bound to the operator. Under these conditions, the bacterium readily metabolizes lactose.
- **Low lactose and low or high glucose** (Figure 12.9c,d): When lactose levels are low, the lac repressor prevents transcription of the *lac* operon, whether glucose levels are high or low.

## 12.2 Reviewing the Concepts

- Repressors and activators are regulatory transcription factors that bind to DNA and affect the transcription of genes. Small effector molecules control the ability of regulatory transcription factors to bind to DNA (Figure 12.5).
- An operon is an arrangement of two or more protein-encoding genes controlled by a single promoter and an operator. The *lac* operon is an example of an inducible operon. The lac repressor exerts negative control by binding to the operator and preventing RNA polymerase from transcribing the operon. When allolactose binds to the repressor, a conformational change occurs that prevents the repressor from binding to the operator so transcription can proceed (Figures 12.6, 12.7).
- Positive control of the *lac* operon occurs when the catabolite activator protein (CAP) binds to the CAP site in the presence of cAMP. This causes a bend in the DNA, which promotes the binding of RNA polymerase to the promoter (Figure 12.8).
- Glucose inhibits cAMP production, which in turn inhibits the expression of the *lac* operon, because CAP cannot bind to the CAP site. This form of regulation provides bacteria with a more efficient utilization of their resources because the bacteria use one sugar at a time (Figure 12.9).

## 12.2 Testing Your Knowledge

1. Let's suppose that a mutation in the *lacI* gene prevented the lac repressor from being made. How would this mutation affect the expression of the *lac* operon?
   a. It would be expressed only in the presence of lactose in the environment.
   b. It would be expressed in the presence or absence of lactose in the environment.
   c. It would be expressed only in the absence of lactose in the environment.
   d. It would never be expressed.

## 12.3 Regulation of Transcription in Eukaryotes: Roles of Transcription Factors

### Learning Outcomes

1. Explain the concept of combinatorial control.
2. Describe how RNA polymerase and general transcription factors initiate transcription at the core promoter.
3. Discuss how activators, coactivators, repressors, and TFIID play a role in gene regulation.

Regulation of transcription in eukaryotes follows some of the same principles as those found in bacteria. For example, activator and repressor proteins are involved in regulating genes by influencing the

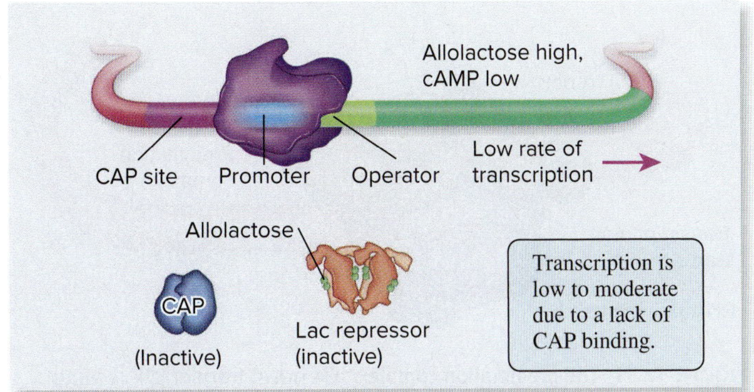


**(a) Lactose high, glucose high**

CAP site
Allolactose high,
cAMP high
Promoter
Operator
High rate of
transcription
CAP
cAMP
Allolactose
Lac repressor
(inactive)
Binding of RNA polymerase to promoter is enhanced by CAP binding. Transcription rate is high.

**(b) Lactose high, glucose low**

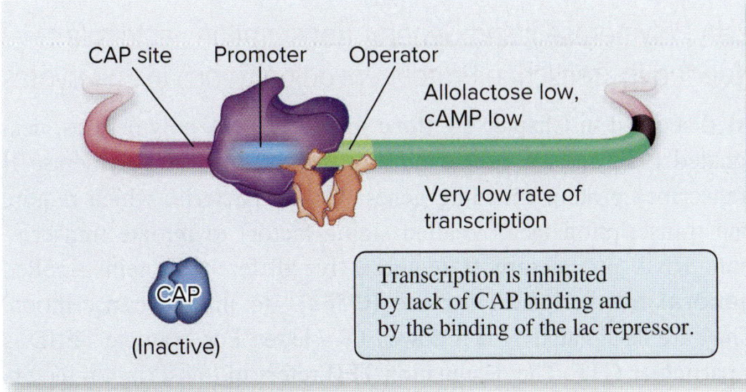


**(c) Lactose low, glucose high**

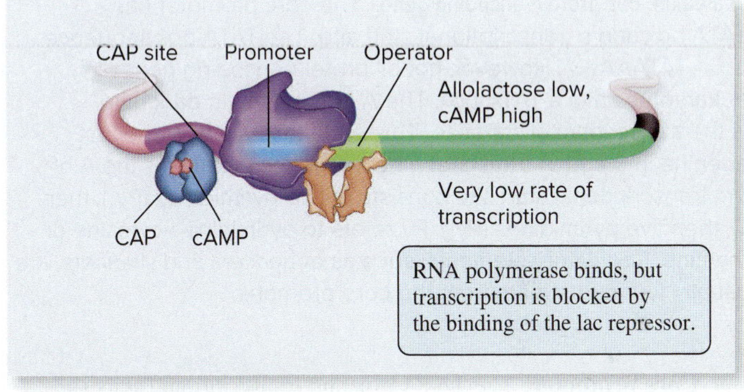


**(d) Lactose low, glucose low**

**Figure 12.9** Effects of lactose and glucose on the expression of the *lac* operon.

*What are the advantages of having both an activator and a repressor protein?*

ability of RNA polymerase to initiate transcription. In addition, many eukaryotic genes are regulated by small effector molecules. However, some important differences also occur. In eukaryotic species, genes are almost always organized individually, not in operons. In addition, eukaryotic gene regulation tends to be more intricate, because eukaryotes are faced with complexities that differ from their bacterial counterparts. For example, eukaryotes have more complicated cell structures that contain many more proteins and a variety of cell organelles.

By studying transcriptional regulation, researchers have discovered that most eukaryotic genes, particularly those found in multicellular species, are regulated by many factors. This phenomenon is called **combinatorial control** because the combination of many factors determines the expression of any given gene. At the level of transcription, common factors that contribute to combinatorial control include the following:

1. One or more activators may stimulate the ability of RNA polymerase to initiate transcription.
2. One or more repressors may inhibit the ability of RNA polymerase to initiate transcription.
3. The function of activators and repressors may be modulated in several ways, which include the binding of small effector molecules, protein-protein interactions, and covalent modifications.
4. Activators are necessary to alter chromatin structure in the region where a gene is located, thereby making it easier for the gene to be recognized and transcribed by RNA polymerase.
5. DNA methylation usually inhibits transcription, either by preventing the binding of an activator or by recruiting proteins that inhibit transcription.

All five of these factors may contribute to the regulation of a single gene, or possibly only three or four will play a role. In most cases, transcriptional regulation is aimed at controlling the initiation of transcription at the promoter. In this section and the following section, we will survey these basic types of gene regulation in eukaryotic species.

## Eukaryotic Protein-encoding Genes Have a Core Promoter and Regulatory Elements

To understand gene regulation in eukaryotes, we first need to consider the DNA sequences that are needed to initiate transcription. For eukaryotic genes that encode proteins, certain features are common among most promoters (**Figure 12.10**):

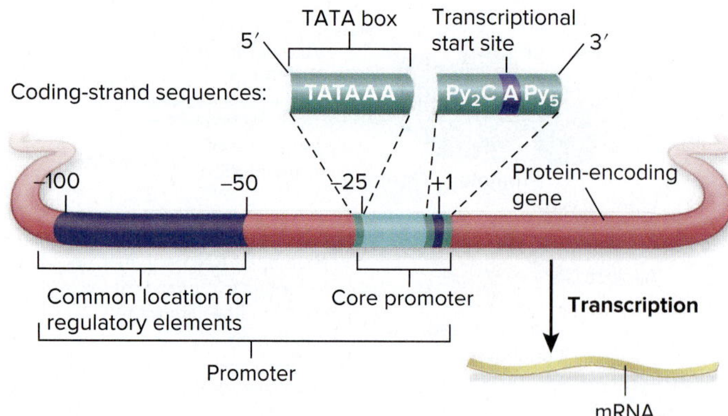


**Figure 12.10 A common organization of sequences for the promoter of a eukaryotic protein-encoding gene.** The core promoter has a TATA box and a transcriptional start site. The TATA box sequence is 5′–TATAAA–3′. However, not all protein-encoding genes in eukaryotes have a TATA box. The A highlighted in dark blue is the transcriptional start site. This A marks the site of the first adenine in the RNA transcript. The sequence that flanks the A of the transcriptional start site consists of two pyrimidines ($Py_2$), then C, then five pyrimidines ($Py_5$). Py refers to pyrimidine—cytosine or thymine. Regulatory elements, such as enhancers and silencers, are usually found upstream from the core promoter.

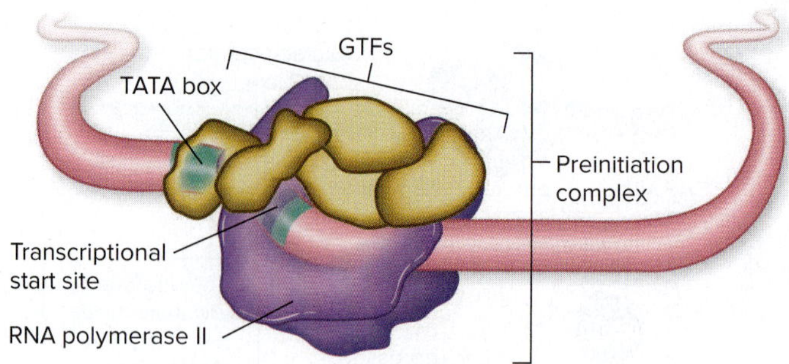


**Figure 12.11 The preinitiation complex.** General transcription factors (GTFs) and RNA polymerase II assemble into the preinitiation complex at the core promoter of eukaryotic protein-encoding genes.

- **Core promoter:** The TATA box and transcriptional start site form the **core promoter.** The **transcriptional start site** is the place in the DNA where transcription actually begins. The **TATA box,** which is a 5′–TATAAA–3′ sequence, is usually about 25 bp upstream from a transcriptional start site. The TATA box is important in determining the precise starting point for transcription. If it is missing from the core promoter, transcription may start at a variety of locations. The core promoter, by itself, results in a low level of transcription that is termed **basal transcription.**
- **Regulatory elements: Regulatory elements** (or **regulatory sequences**) are DNA segments that regulate eukaryotic genes. As described later, regulatory elements are recognized by regulatory transcription factors that control the ability of RNA polymerase to initiate transcription at the core promoter. Regulatory elements come in two general types:
    1. Some regulatory elements, known as **enhancers,** play a role in the ability of RNA polymerase to begin transcription, thereby enhancing the rate of transcription. When enhancers are not functioning, most eukaryotic genes have very low levels of basal transcription.
    2. Other regulatory elements, known as **silencers,** prevent transcription of a given gene when its expression is not needed. When these sequences function, the rate of transcription is decreased.

A common location for regulatory elements is the region that is 50–100 bp upstream from the transcriptional start site (see Figure 12.10). However, regulatory elements can be quite distant from the promoter, even 100,000 bp away, yet exert strong effects on the ability of RNA polymerase to initiate transcription at the core promoter!

## RNA Polymerase II and General Transcription Factors Are Needed to Transcribe Protein-encoding Genes in Eukaryotes

As discussed in Chapter 10, three forms of RNA polymerases, designated I, II, and III, are found in eukaryotes. RNA polymerase II transcribes protein-encoding genes. Unlike bacteria, which require one transcription factor (called sigma factor) to initiate transcription, RNA polymerase II requires five different proteins—called **general transcription factors (GTFs)**—to initiate transcription. They are designated TFII followed by a letter. For example TFIID is a particular GTF. The designation TFII refers to transcription factors that work with RNA polymerase II.

RNA polymerase II and GTFs must come together at the TATA box of the core promoter so transcription can be initiated. A series of interactions occurs between these proteins so RNA polymerase II can bind to the DNA. The completed assembly of RNA polymerase II and GTFs at the TATA box is known as the **preinitiation complex** (Figure 12.11).

## Activators and Repressors May Influence the Function of GTFs

In eukaryotes, regulatory transcription factors called activators and repressors bind to enhancers or silencers, respectively, and regulate the rate of transcription of genes. Activators and repressors commonly regulate the function of RNA polymerase II by affecting the function of GTFs. Figure 12.12 illustrates how an activator may work:

1. An activator binds to an enhancer. In this example, an activator also interacts with a **coactivator**—a protein that increases the rate of transcription but does not directly bind to the DNA itself.
2. The activator/coactivator complex improves the ability of a GTF called transcription factor IID (TFIID) to bind to the TATA box.
3. The function of TFIID is to promote the assembly of GTFs and RNA polymerase II into the preinitiation complex.

In contrast, repressors may bind to a silencer and inhibit the function of TFIID. Certain repressors exert their effects by preventing the binding of TFIID to the TATA box or by inhibiting the ability of TFIID to assemble other GTFs and RNA polymerase II at the core promoter.

Another way that regulatory transcription factors influence transcription is by recruiting proteins that affect chromatin structure in the promoter region, as described in Section 12.4.

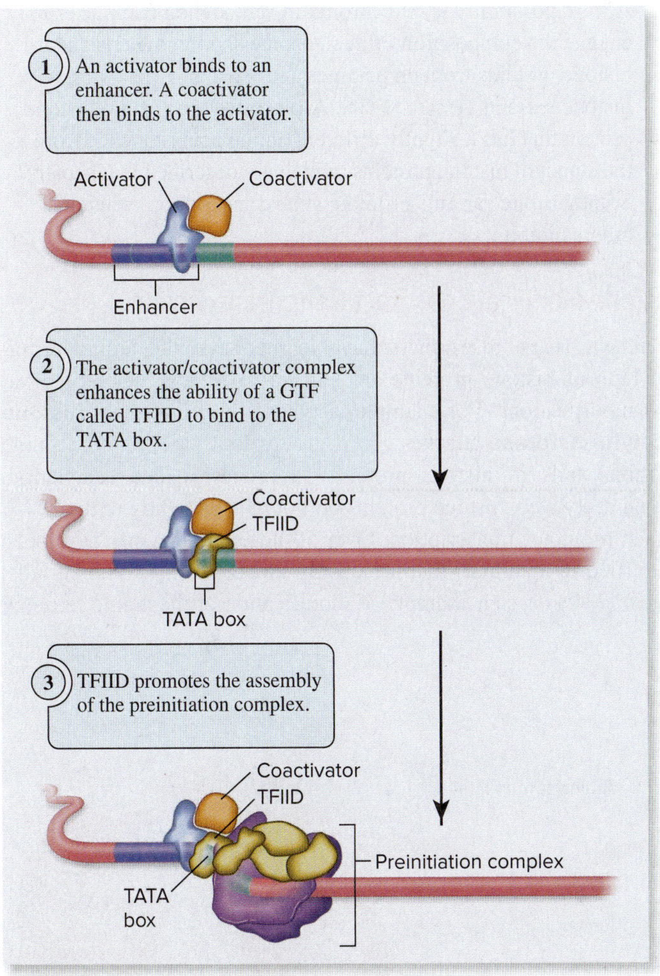


**Figure 12.12** **Effect of an activator via TFIID, a general transcription factor.**

## 12.3 Reviewing the Concepts

- Eukaryotic genes exhibit combinatorial control, meaning that many factors control the expression of a single gene.
- Eukaryotic promoters consist of a core promoter (containing a TATA box and transcriptional start site) and regulatory elements, such as enhancers or silencers, that regulate the rate of transcription (Figure 12.10).
- General transcription factors (GTFs) are needed for RNA polymerase II to bind to the core promoter, forming a preinitiation complex (Figure 12.11).
- Activators and repressors may regulate RNA polymerase II by interacting with TFIID (a GTF) (Figure 12.12).

## 12.3 Testing Your Knowledge

1. The term combinatorial control refers to the observation that
   a. GTFs must combine with each other to initiate transcription.
   b. the preinitiation complex is a combination of RNA polymerase and GTFs.
   c. a combination of different factors may regulate the expression of a given gene.
   d. eukaryotic transcription relies on a combination of a core promoter and regulatory elements.
2. The core promoter for eukaryotic protein-encoding genes usually contains
   a. a transcriptional start site.
   b. a TATA box.
   c. regulatory elements.
   d. all of the above.
   e. both a and b.

# 12.4 Regulation of Transcription in Eukaryotes: Changes in Chromatin Structure and DNA Methylation

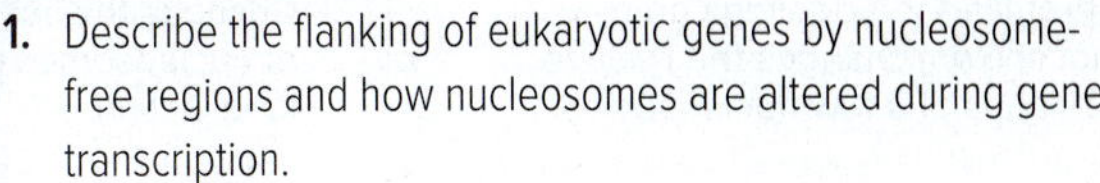

### Learning Outcomes

1. Describe the flanking of eukaryotic genes by nucleosome-free regions and how nucleosomes are altered during gene transcription.
2. Explain how DNA methylation affects transcription.

In eukaryotes, DNA is associated with proteins to form a structure called **chromatin**—the complex of DNA and proteins that make up eukaryotic chromosomes. How does the structure of chromatin affect gene transcription? Recall from Chapter 9 that nucleosomes are composed of DNA wrapped around a histone octamer (refer back to Figure 9.19). Depending on the locations and arrangements of nucleosomes, a region containing a gene may be in a **closed conformation,** and transcription may be difficult or impossible. Transcription requires changes in chromatin structure that allow transcription factors to gain access to and bind to the DNA in the promoter region. Such chromatin, said to be in an **open conformation,** is accessible to GTFs and RNA polymerase II so transcription can take place. In this section, we will examine how chromatin is converted from a closed to an open conformation. We will also explore how **DNA methylation**—the attachment of methyl groups to the base cytosine—affects chromatin conformation and gene expression.

## Transcription Is Controlled by Changes in Chromatin Structure

In recent years, geneticists have been trying to identify the steps that promote the interconversion between the closed and open conformations of chromatin. One way to change chromatin structure is through **chromatin-remodeling complexes,** which are a group of proteins that alter chromatin structure. Such complexes use energy from ATP hydrolysis to drive a change in the locations and/or compositions of nucleosomes, thereby making the DNA more or less amenable to transcription. Therefore, chromatin remodeling is important for both the activation and repression of transcription.

How do chromatin-remodeling complexes change chromatin structure? Three effects are possible:

- These complexes may bind to chromatin and change the locations of nucleosomes (**Figure 12.13a**). This may involve a shift of the relative positions of a few nucleosomes or a change in the relative spacing of nucleosomes over a long stretch of DNA.

- A second effect is that chromatin-remodeling complexes may evict a histone octamer from the DNA, thereby creating gaps where nucleosomes are not found (**Figure 12.13b**).

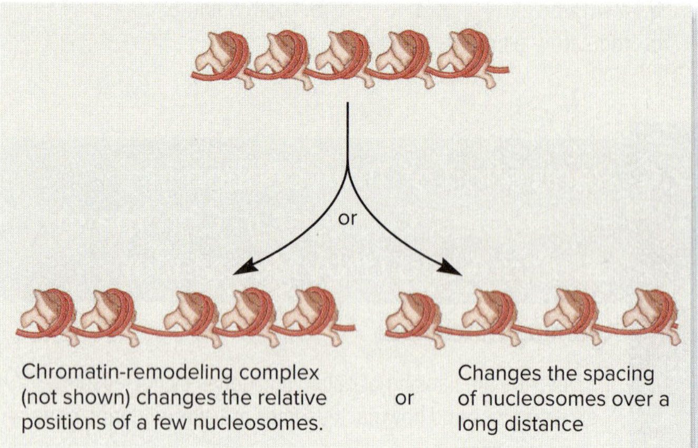

or
Chromatin-remodeling complex (not shown) changes the relative positions of a few nucleosomes.
or
Changes the spacing of nucleosomes over a long distance


**(a) Change in nucleosome position**

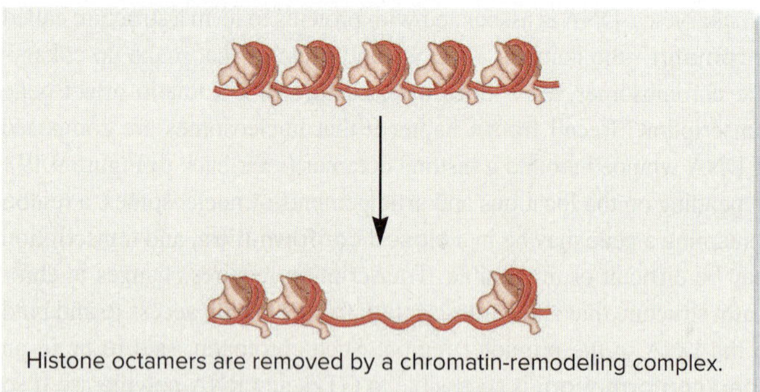

Histone octamers are removed by a chromatin-remodeling complex.


**(b) Histone eviction**

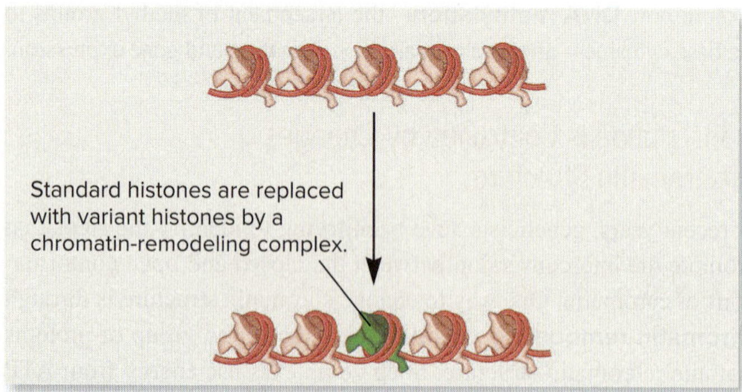

Standard histones are replaced with variant histones by a chromatin-remodeling complex.


**(c) Replacement with variant histones**

**Figure 12.13 Chromatin remodeling.** Chromatin-remodeling complexes may (**a**) change the locations of nucleosomes, (**b**) evict histones from the DNA, or (**c**) replace standard histones with variant histones. The chromatin-remodeling complex, which is composed of a group of proteins, is not shown in this figure.

- A third possibility is that chromatin-remodeling complexes may change the composition of nucleosomes by removing standard histone proteins from an octamer and replacing them with histone variants (**Figure 12.13c**). A **histone variant** is a histone protein that has a slightly different amino acid sequence from the standard histone proteins, which are described in Chapter 9. Some histone variants promote gene transcription, whereas others inhibit it.

## Histone Modifications Affect Gene Transcription

In recent years, researchers have learned that the amino terminal tails of histone proteins are subject to several types of covalent modifications. For example, a type of enzyme called **histone acetyltransferase** attaches acetyl groups (—$COCH_3$) to the amino terminal tails of histone proteins—a process called acetylation. When acetylated, histone proteins do not bind as tightly to the DNA, which promotes transcription. Over 50 different enzymes have been identified in mammals that selectively modify amino terminal tails. **Figure 12.14** shows an example of modification of the amino terminal

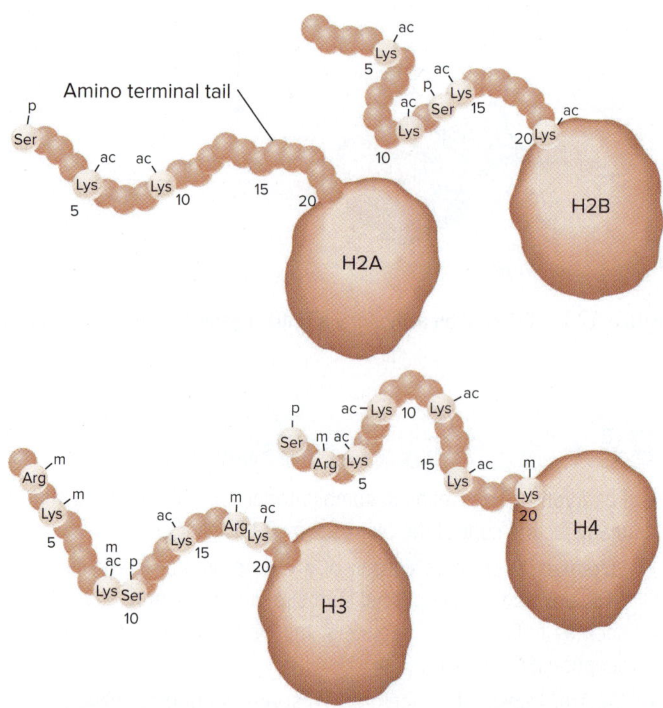


**Figure 12.14 Examples of covalent modifications that occur to the amino terminal tails of histone proteins.** The amino acids are numbered from the N-terminus, or amino end. The modifications shown here are m for methylation, p for phosphorylation, and ac for acetylation. Many more modifications can occur to the amino terminal tails. These modifications are reversible.

**Concept Check:** *What are the two opposing effects that histone modifications may have with regard to transcription?*

tails of histone proteins H2A, H2B, H3, and H4 by acetyl, methyl, and phosphate groups.

What are the effects of covalent modifications of histones?

1. Modifications may directly influence interactions between DNA and histone proteins and between adjacent nucleosomes. As mentioned, the acetylation of histones loosens their binding to DNA and aids in transcription.
2. Histone modifications provide binding sites that are recognized by other proteins. According to the **histone code hypothesis,** proposed by American biologists Brian Strahl and David Allis in 2000, the pattern of histone modification is recognized by proteins much like a language or code. One pattern of histone modification may attract proteins that inhibit transcription. Alternatively, a different combination of histone modifications may attract proteins, such as chromatin-remodeling complexes, that promote transcription. In this way, the histone code plays a key role in accessing the information within the genomes of eukaryotic species.

## Eukaryotic Genes Are Flanked by Nucleosome-Free Regions

Studies over the last 10 years or so have revealed that many eukaryotic genes show a common pattern of nucleosome organization (**Figure 12.15**). For active genes or those genes that can be activated, the core promoter is found within a **nucleosome-free region (NFR),** which is a site that is missing nucleosomes. The NFR is typically 150 bp in length. (Note: For comparison, a single nucleosome contains 146 bp of DNA.) Although the NFR may be required for transcription, it is not, by itself, sufficient for gene activation. At any given time in the life of a eukaryotic cell, many genes that contain an NFR are not being actively transcribed. The NFR is flanked by two nucleosomes that are termed the −1 and +1 nucleosomes. These nucleosomes often contain histone variants that promote transcription. The end of many eukaryotic genes is followed by another NFR at the terminator site. This arrangement at the end of genes may be important for transcriptional termination, but further research is needed.

## Transcriptional Activation Involves Changes in Nucleosome Locations, Composition, and Histone Modifications

A key role of certain activators is to recruit chromatin-remodeling complexes and histone-modifying enzymes to the promoter region of eukaryotic genes. Though the order of recruitment may differ among specific activators, this appears to be critical for transcriptional initiation and elongation. **Figure 12.16** shows a common sequence of events:

1. An activator binds to an enhancer in the NFR.
2. The activator then recruits chromatin-remodeling complexes and histone-modifying enzymes to this region. The chromatin-remodeling complex may shift nucleosomes or temporarily evict nucleosomes from the promoter region. Nucleosomes containing certain histone variants are thought to be more easily removed from the DNA than those containing the standard histones. Histone-modifying enzymes, such as histone acetyltransferase, covalently modify histone proteins and may affect nucleosome contact with the DNA.
3. The actions of chromatin-remodeling complexes and histone-modifying enzymes facilitate the binding of general transcription factors and RNA polymerase II to the core promoter, thereby allowing the formation of a preinitiation complex.
4. For transcription to occur, histones are evicted, partially displaced, or destabilized so RNA polymerase II can pass. Evicted histones are then reassembled by chaperone molecules, which aid in protein folding, and subsequently placed back on the DNA behind the moving RNA polymerase II. These histones may be deacetylated—have their acetyl groups removed—so they bind more tightly to the DNA.

## DNA Methylation Inhibits Gene Transcription

Let's now turn our attention to a mechanism that usually silences gene expression. DNA structure can be modified by the covalent attachment of methyl groups ($—CH_3$) by a type of enzyme called

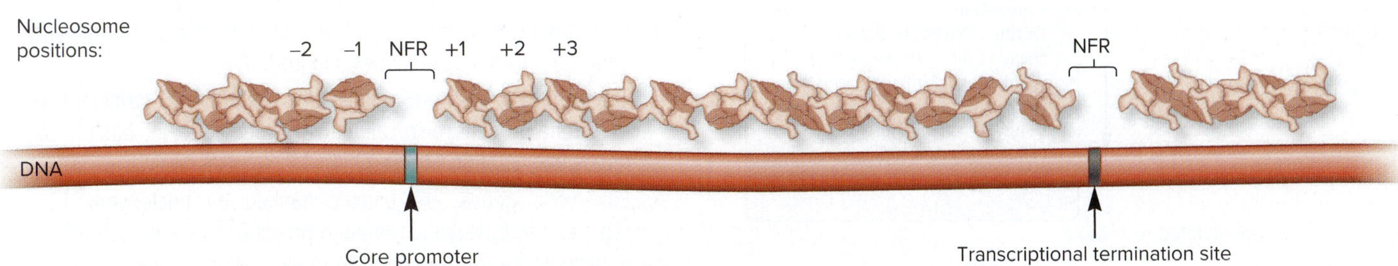


**Figure 12.15** **Nucleosome arrangements in the vicinity of a eukaryotic protein-encoding gene.**

*Refer back to Figure 9.19. What is the composition of a nucleosome?*

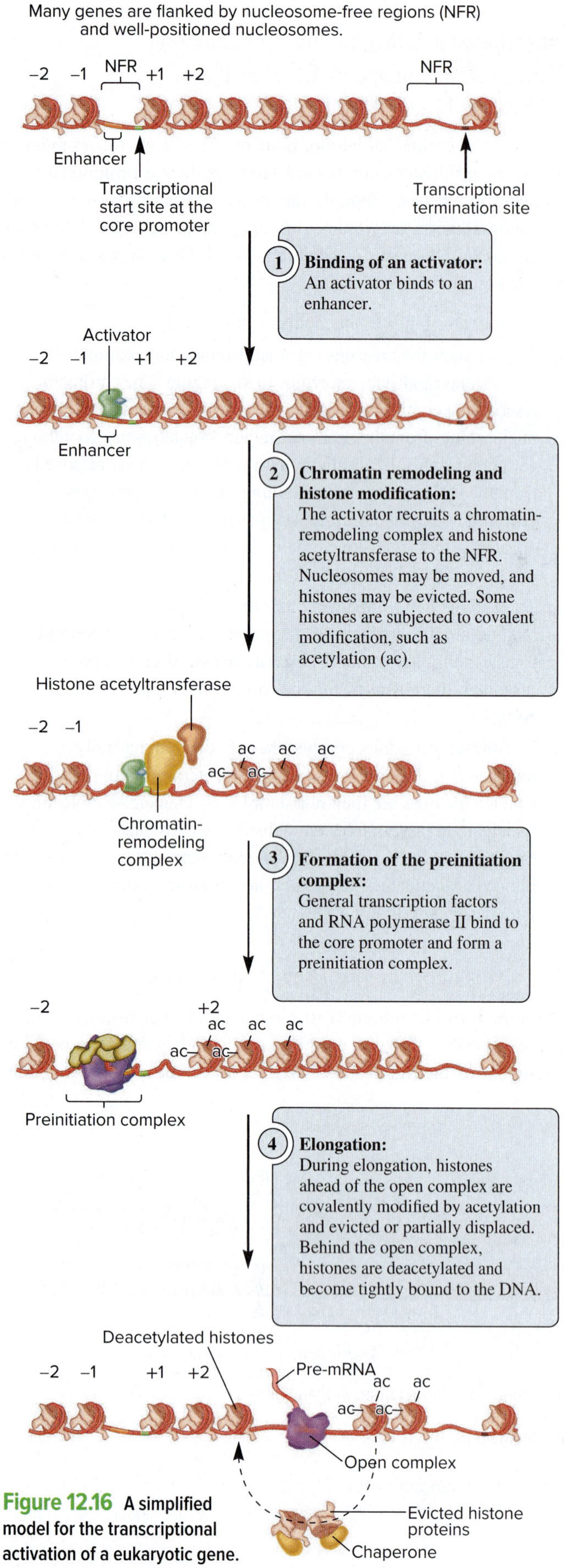

Many genes are flanked by nucleosome-free regions (NFR) and well-positioned nucleosomes.

(1) **Binding of an activator:** An activator binds to an enhancer.

(2) **Chromatin remodeling and histone modification:** The activator recruits a chromatin-remodeling complex and histone acetyltransferase to the NFR. Nucleosomes may be moved, and histones may be evicted. Some histones are subjected to covalent modification, such as acetylation (ac).

(3) **Formation of the preinitiation complex:** General transcription factors and RNA polymerase II bind to the core promoter and form a preinitiation complex.

(4) **Elongation:** During elongation, histones ahead of the open complex are covalently modified by acetylation and evicted or partially displaced. Behind the open complex, histones are deacetylated and become tightly bound to the DNA.

**Figure 12.16 A simplified model for the transcriptional activation of a eukaryotic gene.**

**DNA methyltransferase.** This modification, termed DNA methylation, is common in some eukaryotic species but not all. For example, yeast and *Drosophila* have little or no detectable methylation of their DNA, whereas DNA methylation in vertebrates and plants is relatively abundant. In mammals, approximately 5% of the DNA is methylated. Eukaryotic DNA methylation occurs on the cytosine base. The sequence that is methylated (5′–CG–3′) is shown here:

$$\begin{array}{c} CH_3 \\ | \\ 5'\text{—CG—}3' \\ 3'\text{—GC—}5' \\ | \\ CH_3 \end{array}$$

DNA methylation usually inhibits the transcription of eukaryotic genes, particularly when it occurs in the vicinity of the promoter. In vertebrates and flowering plants, many genes contain sequences called **CpG islands** near their promoters. CpG refers to the nucleotides of C and G in DNA that are connected by a phosphodiester linkage. A CpG island is a cluster of CpG sites. Unmethylated CpG islands are usually correlated with active genes, whereas repressed genes contain methylated CpG islands. In this way, DNA methylation plays an important role in the silencing of particular genes.

How does DNA methylation inhibit transcription? This can occur in two general ways:

1. Methylation of CpG islands may prevent an activator from binding to an enhancer, thus inhibiting the initiation of transcription.
2. Methylation may inhibit transcription by altering chromatin structure. Proteins known as **methyl-CpG-binding proteins** bind methylated sequences. Once bound to the DNA, the methyl-CpG-binding protein recruits other proteins to the region that inhibit transcription.

## 12.4 Reviewing the Concepts

- Chromatin-remodeling complexes change the positions and compositions of nucleosomes (Figure 12.13).
- The pattern of covalent modification of the amino terminal tails of histone proteins, also called the histone code, inhibits or promotes transcription (Figure 12.14).
- Eukaryotic genes are usually flanked by nucleosome-free regions. For eukaryotic genes, a preinitiation complex forms at a nucleosome-free region. During elongation, nucleosomes are displaced ahead of RNA polymerase and re-form after RNA polymerase has passed (Figures 12.15, 12.16).
- DNA methylation, which occurs at CpG islands near promoters, usually inhibits transcription by (1) preventing the binding of activator proteins or (2) promoting the binding of proteins that inhibit transcription.

## 12.4 Testing Your Knowledge

1. A chromatin-remodeling complex may
   a. change the locations of nucleosomes.
   b. exchange standard histones for histone variants.
   c. evict histones.
   d. do all of the above.
2. For active eukaryotic genes, the core promoter is found at a nucleosome-free region (NFR). This NFR region is needed so that
   a. DNA methylation can occur.
   b. activators can promote the formation of a preinitiation complex.
   c. histones can bind to the TATA box.
   d. the DNA can unwind.

## 12.5 Regulation of RNA Splicing and Translation in Eukaryotes

### Learning Outcomes

1. Outline the process of alternative splicing, and explain how it increases protein diversity.
2. Explain how RNA-binding proteins regulate the translation of specific mRNAs, using the regulation of iron absorption in mammals as an example.

In the preceding sections of this chapter, we focused on gene regulation at the level of transcription in bacteria and eukaryotes. Regulation of transcription in eukaryotes is an efficient form of regulation, but it takes a fair amount of time before its effects are observed at the cellular level. During transcription (1) the chromatin must be converted to an open conformation, (2) the gene must be transcribed, (3) the RNA must be processed and exported from the nucleus, and (4) the protein must be made via translation. All four steps take time, on the order of several minutes. One way to achieve faster regulation is to control steps that occur after an RNA transcript is made.

In this section, we will first examine how cells regulate pre-mRNA splicing, which can produce more than one mRNA transcript from a single gene. This allows a gene to encode two or more polypeptides, thereby increasing the complexity of the proteome. Next, we will explore how the translation of an mRNA can be regulated by the binding of a regulatory protein.

### Alternative Splicing of Pre-mRNAs Increases Protein Diversity

In eukaryotes, a pre-mRNA transcript is modified before it becomes a mature mRNA (refer back to Figure 10.4). When a pre-mRNA has multiple introns and exons, splicing may occur in more than one way, resulting in the production of two of more different polypeptides. Such **alternative splicing** is a form of gene regulation that allows an organism to use the same gene to make different proteins at different stages of development, in different cell types, and/or in response to a change in the environmental conditions. Alternative splicing is an important form of gene regulation in complex eukaryotes such as animals and plants. An advantage of alternative splicing is that two or more different polypeptides can be derived from a single gene, thereby increasing the size of the proteome while minimizing the size of the genome.

Let's consider an example of alternative splicing for a pre-mRNA that encodes a protein known as α-tropomyosin, which functions in the regulation of cell contraction in animals. It is located along the thin filaments found in smooth muscle cells, such as those in the uterus and small intestine, and in striated muscle cells that are found in cardiac and skeletal muscle. Within a multicellular organism, different types of cells must regulate their contractibility in subtly different ways. One way this may be accomplished is by the production of different forms of α-tropomyosin.

**Figure 12.17** shows the intron-exon structure of the rat α-tropomyosin pre-mRNA and two alternative ways that the pre-mRNA can be spliced. The pre-mRNA contains 14 exons, 6 of which are constitutive exons (shown in red), which are always found in the mature mRNA from all cell types. Presumably, constitutive exons encode polypeptide segments of the α-tropomyosin protein that are necessary for its general structure and function. By comparison, alternative exons (shown in green) are not always found in the mRNA after splicing has occurred. The polypeptide sequences encoded by alternative exons may subtly change the function of α-tropomyosin to meet the needs of the cell type in which it is found. For example, Figure 12.17 shows the predominant splicing products found in smooth muscle cells and striated muscle cells. Exon 2 encodes a segment of the α-tropomyosin protein that alters its function to make it suitable for smooth muscle cells. By comparison, the α-tropomyosin mRNA found in striated muscle cells does not include exon 2. Instead, this mRNA contains exon 3, which is more suitable for that cell type.

## Quantitative Analysis

### ALTERNATIVE SPLICING IS MORE PREVALENT IN COMPLEX EUKARYOTIC SPECIES

**Table 12.1** compares six species: a bacterium *(Escherichia coli),* a eukaryotic single-celled organism (yeast—*Saccharomyces cerevisiae*), a small nematode worm with about 1,000 cells *(Caenorhabditis elegans),* a fruit fly *(Drosophila melanogaster),* a flowering plant *(Arabidopsis thaliana),* and a human *(Homo sapiens).* One general trend is that less complex organisms tend to have fewer genes. For example, unicellular organisms have only a few thousand genes, whereas multicellular species have tens of thousands. However, the trend is by no means a linear one. If we compare *C. elegans* and *D. melanogaster*, the fruit fly actually has fewer genes, even though it is morphologically more complex.

A second trend you can see in Table 12.1 concerns alternative splicing. This phenomenon does not occur in bacteria and is rare in *S. cerevisiae*. The frequency of alternative splicing increases from

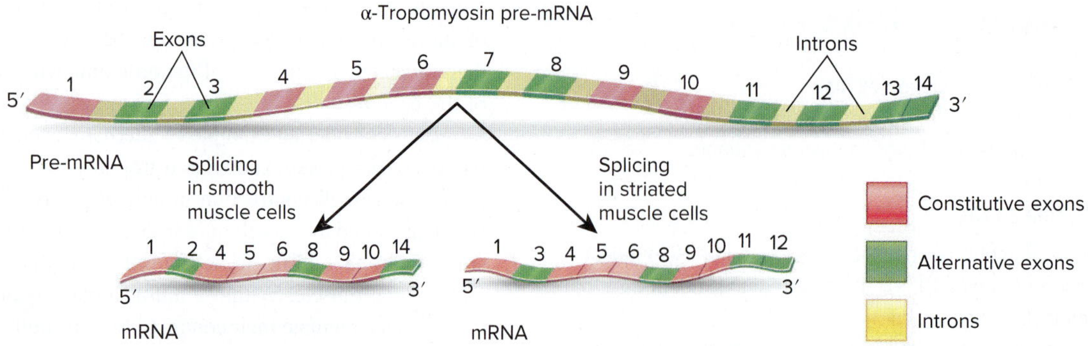


**Figure 12.17 Alternative splicing of the rat α-tropomyosin pre-mRNA.** The top part of this figure depicts the structure of the rat α-tropomyosin pre-mRNA. Exons are shown in red or green, and introns are yellow. The lower part of the figure describes the final mRNA products in smooth and striated muscle cells after alternative splicing. Note: Exon 8 is found in the final mRNA of smooth and striated muscle cells, but not in the mRNA of certain other cell types. The junction between exons 13 and 14 contains a 3′ splice site that enables exon 13 to be separated from exon 14.

*What is the biological advantage of alternative splicing?*

worms to flies to humans. For example, the level of alternative splicing is 10-fold higher in humans than in *Drosophila*. This trend can partially explain the increase in complexity among these species.

**Crunching the Numbers:** As a rough estimate, let's suppose that the pre-mRNAs from 70% of all human genes are alternatively spliced and that an average pre-mRNA that is alternatively spliced can be spliced in five different ways. By comparison, only about 7% of the pre-mRNAs from *Drosophila* are alternatively spliced; an average pre-mRNA that is alternatively spliced can be spliced in three different ways. With a human genome size of about 22,000 protein-encoding genes and a *Drosophila* genome size of 15,600 protein-encoding genes, calculate the numbers of mature mRNAs that can be made by each species. Discuss how these numbers affect the complexity of humans and *Drosophila*.

## The Prevention of Iron Toxicity in Mammals Involves the Regulation of Translation

In Chapter 11, we considered how microRNAs (miRNAs) and small-interfering RNAs (siRNAs) can silence mRNAs. Another way to regulate mRNAs involves RNA-binding proteins that directly affect the initiation of translation. The regulation of iron absorption provides a well-studied example. Although iron ($Fe^{3+}$) is a vital cofactor for many cellular enzymes, it is toxic at high levels. To prevent toxicity, mammalian cells synthesize a protein called ferritin, which forms a hollow, spherical complex that stores excess iron.

The mRNA that encodes ferritin is controlled by an RNA-binding protein known as the **iron regulatory protein (IRP).**

- **Low iron:** When iron levels in the cytosol are low and more ferritin is not needed, IRP binds to a regulatory element within the ferritin mRNA known as the **iron regulatory element (IRE).**

**Table 12.1 Genome Size and Biological Complexity**

| Species | Level of complexity | Genome size (million bp) | Approximate number of protein-encoding genes | Percentage of genes alternatively spliced |
|---|---|---|---|---|
| *Escherichia coli* | A unicellular bacterium | 4.2 | 4,300 | 0 |
| *Saccharomyces cerevisiae* | A unicellular eukaryote | 12 | 6,300 | <1 |
| *Caenorhabditis elegans* | A small worm (about 1,000 cells) | 97 | 20,500 | 2 |
| *Drosophila melanogaster* | An insect | 137 | 15,600 | 7 |
| *Arabidopsis thaliana* | A flowering plant | 142 | 27,000 | 11 |
| *Homo sapiens* | A complex mammal | 3,000 | 22,000 | 70 |

The IRE is located between the 5′ end, where the ribosome binds, and the start codon, where translation begins. Due to base pairing, it forms a stem-loop structure. The binding of IRP to the IRE inhibits translation of the ferritin mRNA (Figure 12.18a).

- **High iron:** When iron levels in the cytosol are high and ferritin is needed, the iron binds directly to IRP and causes a conformational change that prevents it from binding to the IRE. Under these conditions, the ferritin mRNA is translated to make more ferritin protein (Figure 12.18b).

Why is translational regulation of ferritin mRNA an advantage over transcriptional regulation of the ferritin gene? This mechanism of translational control allows cells to rapidly respond to changes in their environment. When cells are confronted with high levels of iron, they can quickly make more ferritin protein to prevent the toxic buildup of iron. This mechanism is faster than transcriptional regulation, which would require the activation of the ferritin gene and the transcription of ferritin mRNA prior to the synthesis of more ferritin protein.

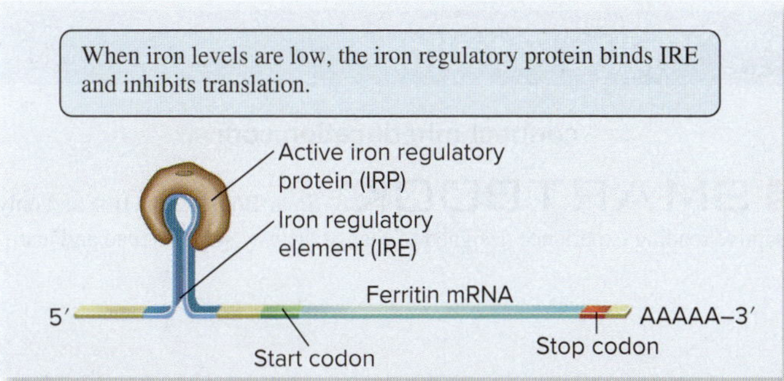


**(a) Low iron levels**

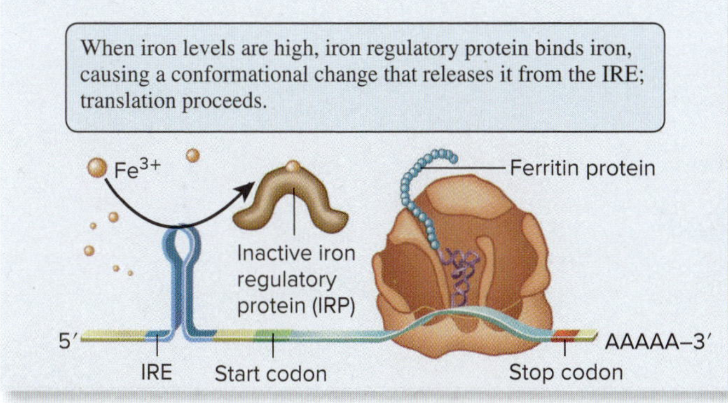


**(b) High iron levels**

**Figure 12.18 Translational regulation of ferritin mRNA by the iron regulatory protein (IRP).**

**Concept Check:** *Iron poisoning may occur when a young child finds a bottle of vitamins, such as those that taste like candy, and eats a large number of them. How does the IRP protect people from the toxic effects of too much iron?*

## 12.5 Reviewing the Concepts

- In alternative splicing, a pre-mRNA can be spliced in more than one way, producing polypeptides with somewhat different sequences. This is a common way for complex eukaryotes to increase the size of their proteomes (Figure 12.17, Table 12.1).
- RNA-binding proteins regulate the translation of specific mRNAs. An example is the regulation of iron absorption, in which the iron regulatory protein (IRP) regulates the translation of ferritin mRNA (Figure 12.18).

## 12.5 Testing Your Knowledge

1. Alternative splicing
   **a.** results in two or more mRNAs from a single type of pre-mRNA.
   **b.** increases the size of the proteome without increasing the size of the genome.
   **c.** is more prevalent in complex eukaryotes.
   **d.** is all of the above.
2. The iron regulatory protein (IRP) binds to the IRE when iron levels are ____ and ____ translation of ferritin mRNA.
   **a.** low, inhibits
   **b.** low, stimulates
   **c.** high, inhibits
   **d.** high, stimulates

# Assess and Discuss

## Test Yourself

1. Genes that are expressed at all times at relatively constant levels are known as ____ genes.
   a. inducible
   b. repressible
   c. positive
   d. constitutive
   e. protein-encoding
2. Which of the following is *not* considered a common level of gene regulation in bacteria?
   a. transcription
   b. RNA modification
   c. translation
   d. post-translation
   e. All of the above are levels at which bacteria commonly regulate gene expression.
3. Transcription factors that bind to DNA and stimulate transcription are
   a. repressors.
   b. small effector molecules.
   c. activators.
   d. promoters.
   e. operators.
4. In bacteria, the unit of DNA that contains multiple genes under the control of a single promoter is called ___. The mRNA produced from this unit is referred to as ___ mRNA.
   a. an operator, a polycistronic
   b. a template, a structural
   c. an operon, a polycistronic
   d. an operon, a monocistronic
   e. a template, a monocistronic

5. For the *lac* operon, what would be the expected effects of a mutation in the operator that prevented the binding of the repressor protein?
   a. The operon would always be turned on.
   b. The operon would always be turned off.
   c. The operon would always be turned on, except when glucose is present.
   d. The operon would be turned on only in the presence of lactose.
   e. The operon would be turned on only in the presence of lactose and the absence of glucose.
6. The presence of _____ in the environment of *E. coli* prevents CAP from binding to the DNA, resulting in _____ in transcription of the *lac* operon.
   a. lactose, an increase
   b. glucose, an increase
   c. cAMP, a decrease
   d. glucose, a decrease
   e. lactose, a decrease
7. Regulatory elements that function to increase transcription levels in eukaryotes are called
   a. promoters.
   b. silencers.
   c. enhancers.
   d. transcriptional start sites.
   e. activators.
8. DNA methylation in many eukaryotic organisms usually causes
   a. increased translation levels.
   b. decreased translation levels.
   c. increased transcription levels.
   d. decreased transcription levels.
   e. introns to be removed.
9. _____ refers to the phenomenon where a single type of pre-mRNA may give rise to multiple types of mRNAs due to different patterns of intron and exon removal.
   a. Spliceosomes
   b. Variable expression
   c. Alternative splicing
   d. Polycistronic mRNA
   e. RNA-induced silencing
10. Which of the following statements regarding alternative splicing is false?
    a. It involves different splicing patterns that alter the exons found within an mRNA.
    b. It allows cells to make different proteins at different stages of development.
    c. It allows cells to make different proteins in different cell types.
    d. It is less common in complex eukaryotes.
    e. All of the above are true.

## Conceptual Questions

1. What is the difference between inducible and repressible operons?
2. Transcriptional regulation often involves a regulatory protein that binds to a segment of DNA and a small effector molecule that binds to the regulatory protein. Do the following terms apply to a regulatory protein, a segment of DNA, or a small effector molecule? (a) repressor, (b) inducer, (c) operator, (d) corepressor, (e) activator
3. **PRINCIPLES** A principle of biology is that the genetic material provides a blueprint for reproduction. Explain how gene regulation is an important mechanism for reproduction and sustaining life.

## Collaborative Questions

1. Discuss the advantages and disadvantages of genetic regulation at the different levels described in Figure 12.4.
2. Discuss the advantages and disadvantages of combinatorial control of eukaryotic genes.

## Online Resource

**connect.mheducation.com**

SMARTBOOK® SmartBook® is the first and only adaptive reading experience designed to change the way students read and learn.

# Altering the Genetic Material: Mutation, DNA Repair, and Cancer

13

**Cigarette smoking and lung cancer.** Cigarette smoke contains chemicals that cause mutations in DNA. The accumulation of such mutations may lead to lung cancer.

## Chapter Outline



Lung cancer is diagnosed in approximately 170,000 women and men each year in the U.S., and more than 1.2 million cases are diagnosed worldwide. Nearly 90% of these cases are caused by tobacco smoking and are thus preventable (see chapter-opening photo). Unlike other cancers, for which early diagnosis is possible, lung cancer is usually detected only after it has become advanced and is difficult if not impossible to cure. The 5-year survival rate for lung cancer patients is only about 15%.

What is the underlying cause of lung cancer and other types of cancer? The answer is **mutation,** which is defined as a heritable change in the genetic material. When a mutation occurs, the order of nucleotide bases in a DNA molecule, its base sequence, is changed permanently, an alteration that can be passed from mother to daughter cells during cell division. Mutations that lead to cancer cause particular genes to be expressed in an abnormal way. For example, a mutation could affect the transcription of a gene, or it could alter the functional properties of the polypeptide that is specified by a gene. In addition to mutations, changes in chromatin structure can also affect gene expression and contribute to cancer.

Should we be afraid of mutations? Yes and no. On the positive side, mutations are essential to the long-term continuity of life. Mutations provide the foundation for evolutionary change. They supply the variation that enables species to evolve and become better adapted to their environments. On the negative side, however, new mutations are more likely to be harmful than beneficial to the individual. The genes within modern species are the products of billions of years of evolution and have evolved to work properly. Random mutations are more likely to disrupt genes rather than enhance their function. As we will see in this chapter, mutations can cause cancer. In addition, many inherited disorders, such as cystic fibrosis, are caused by gene mutations. For these and many other reasons, understanding the molecular nature of mutations is a compelling area of research.

All species have evolved several ways to repair damaged DNA. Such DNA repair systems reverse DNA damage before a permanent mutation can occur. DNA repair systems are vital to the survival of all organisms. If these systems did not exist, mutations would be so prevalent that few species, if any, would survive. In this chapter, we will examine how such DNA repair systems operate. But first, let's explore the consequences and causes of mutations.

## 13.1 Consequences of Mutations

### Learning Outcomes

1. List different ways that mutations affect the sequence of a gene.
2. Outline how mutations in a protein-encoding gene may affect the amino acid sequence of a polypeptide.
3. Explain how mutations can occur outside the coding sequence and affect gene expression.
4. Distinguish between mutations in somatic cells and germ-line cells.

How do mutations affect traits? To answer this question at the molecular level, we must understand how changes in the DNA sequence of a gene ultimately affect gene function. Most of our understanding of mutation has come from the study of experimental organisms, such as bacteria and *Drosophila*. Researchers can expose these organisms to agents that cause mutations and then study the consequences of the mutations that arise. In addition, because these organisms have a short generation time, researchers can investigate the effects of mutations when they are passed from cell to cell and from parent to offspring.

The structure and amount of genetic material can be altered in a variety of ways. For example, the structure and number of chromosomes can change. We will examine these types of genetic changes in Chapter 14. In this section, we will focus our attention on gene mutations, which are relatively small changes in the sequence of bases in a particular gene. We will also consider how the timing of new mutations during an organism's development has important consequences.

### Gene Mutations Alter the DNA Sequence of a Gene

Gene mutations cause two basic types of changes: (1) the base sequence within a gene can be changed, and (2) one or more base pairs can be added to or removed from a gene. A **point mutation** affects only a single base pair within the DNA. For example, the DNA sequence shown here has been altered by a **base substitution** in which a T (in the top strand) has been replaced by a G and the corresponding A in the bottom strand is replaced with a C:

```
5′-CCCGTAGATA-3′        5′-CCCGCGAGATA-3′
3′-GGGCATCTAT-5′   ⟶    3′-GGGCGCTCTAT-5′
```

A point mutation could also involve the addition or deletion of a single base pair to a DNA sequence. For example, in the following sequence, a single base pair (A-T) has been added to the DNA:

```
5′-GGCGCTAGATC-3′        5′-GGCAGCTAGATC-3′
3′-CCGCGATCTAG-5′   ⟶    3′-CCGTCGATCTAG-5′
```

Though point mutations may seem like small changes to a DNA sequence, they can have important consequences when genes are expressed, as we will see next.

### Gene Mutations May Affect the Amino Acid Sequence of a Polypeptide

If a mutation occurs within the coding region of a protein-encoding gene, the mutation may alter the amino acid sequence of a polypeptide in a variety of ways. **Table 13.1** considers the potential effects of point mutations.

- A **silent mutation** does not alter the amino acid sequence of the polypeptide, even though the nucleotide sequence has changed. As discussed in Chapter 10, the genetic code is degenerate, that is, more than one codon can specify the same amino acid. Silent mutations occur in the third base of many codons without changing the type of amino acid they encode.
- A **missense mutation** is a base substitution that changes a single amino acid in a polypeptide. A missense mutation may or may not alter protein function because it changes only a single

**Table 13.1 Consequences of Point Mutations Within the Coding Sequence of a Protein-encoding Gene**

| Mutation in the DNA | Effect on polypeptide | Example* |
|---|---|---|
| None | None | ATGGCCGGCCCGAAAGAGACC<br>Met-Ala-Gly-Pro-Lys-Glu-Thr |
| Base substitution | **Silent**—causes no change | ATGGCCGGCCCCAAAGAGACC<br>Met-Ala-Gly-Pro-Lys-Glu-Thr |
| Base substitution | **Missense**—changes one amino acid | ATGCCCGGCCCGAAAGAGACC<br>Met-Pro-Gly-Pro-Lys-Glu-Thr |
| Base substitution | **Nonsense**—shortens the polypeptide | ATGGCCGGCCCGTAAGAGACC<br>Met-Ala-Gly-Pro-STOP |
| Addition (or deletion) of a single nucleotide | **Frameshift**—produces a different amino acid sequence | ATGGCCGGCACCGAAAGAGACC<br>Met-Ala-Gly-Thr-Glu-Arg-Asp |

*DNA sequence in the coding strand. This sequence is the same as the mRNA sequence except that RNA contains uracil (U) instead of thymine (T).

amino acid within a polypeptide that is typically hundreds of amino acids in length. A missense mutation that substitutes an amino acid with a chemistry similar to that of the original amino acid is less likely to alter protein function. For example, a missense mutation that substitutes a glutamic acid for an aspartic acid may not alter protein function because both amino acids are negatively charged and have similar side chain structures (refer back to Figure 3.11).

- A **nonsense mutation** involves a change from a normal codon to a stop, or termination, codon. This causes translation to be terminated earlier than expected, producing a shorter polypeptide (see Table 13.1). Compared with a normal polypeptide, this shorter polypeptide is less likely to function properly.
- A **frameshift mutation** involves the addition or deletion of nucleotides that are not in multiples of three nucleotides. The example in Table 13.1 shows the addition of a single nucleotide. Because the codons are read in multiples of three, this type of insertion shifts the reading frame so a completely different amino acid sequence occurs downstream from the mutation. Such a large change in polypeptide structure is likely to inhibit proper protein function.

Changes in protein function may affect the ability of an organism to survive and to reproduce. Except for silent mutations, new mutations are more likely to produce polypeptides that have reduced rather than enhanced function. However, mutations can occasionally produce a polypeptide that has an enhanced function. Such mutations may change in frequency in a population over the course of many generations due to natural selection. This topic is discussed in Chapter 19.

## Gene Mutations That Occur Outside of Coding Sequences Can Influence Gene Expression

Thus far, we have focused our attention on mutations in the coding regions of protein-encoding genes. In Chapters 10 to 12, we explored the role of DNA sequences in gene expression. A mutation can occur within a noncoding DNA sequence and affect gene expression (**Table 13.2**).

- A mutation may alter the sequence within the promoter of a gene, thereby affecting the rate of transcription. A mutation that improves the ability of transcriptional activators to bind to the DNA may enhance transcription, whereas other mutations may block their binding, thus inhibiting transcription.
- A mutation in a regulatory element or operator site can alter the regulation of gene transcription. For example, in Chapter 12, we considered the roles of regulatory elements such as the *lac* operator in *E. coli*, which is recognized by the lac repressor protein (refer back to Figure 12.7). Mutations in the *lac* operator can disrupt the proper regulation of the *lac* operon.
- A mutation that alters splice sites may cause a pre-mRNA to be improperly spliced.
- A mutation may alter a sequence that is involved in the regulation of translation, such as the iron regulatory element (IRE) in ferritin mRNA. Such a mutation may eliminate translational regulation.
- A mutation can occur outside of a gene in an intergenic region, a DNA sequence between genes. Such mutations are less likely to have a phenotypic effect compared with gene mutations.

**Table 13.2 Effects of Mutations Outside of the Coding Sequence of a Gene**

| Sequence | Effect of mutation |
|---|---|
| Promoter | May increase or decrease the rate of transcription |
| Transcriptional regulatory element/operator site | May alter the regulation of transcription |
| Splice sites | May alter the ability of pre-mRNA to be properly spliced |
| Translational regulatory element | May alter the ability of mRNA to be translationally regulated |
| Intergenic region | Not as likely to affect gene expression |

## Mutations Can Occur in Germ-Line or Somatic Cells

Let's now consider how the timing of a mutation may have important consequences for its potential effects. Multicellular organisms typically begin their lives as a single fertilized egg cell that divides many times to produce all the cells of an adult organism. A mutation can occur in any cell of the body, either very early in life, such as in a gamete (egg or sperm) or a fertilized egg, or later in life, such as in the embryonic or adult stages. The number and location of cells with a mutation are critical both to the severity of the genetic effect and to whether the mutation can be passed on to offspring. Geneticists classify the cells of animals into two types: germ-line and somatic cells.

**Germ-Line Mutations** The term **germ line** refers to cells that give rise to gametes, such as egg and sperm cells. A germ-line mutation can occur directly in an egg or sperm cell, or it can occur in a precursor cell that produces the gametes. If a mutant human gamete participates in fertilization, all cells of the resulting offspring will contain the mutation, as indicated by the red color in **Figure 13.1a**. Likewise, when such an individual produces gametes, the mutation may be transmitted to future generations of offspring. Because humans carry two copies of most genes, a new mutation in a single gene has a 50% chance of being transmitted from parent to offspring.

**Somatic Mutations** The **somatic cells** constitute all cells of the body except for the germ line. Examples include skin cells and muscle cells. Mutations can also occur within somatic cells at early or late stages of development. What are the consequences of a mutation that happens during the embryonic stage? As shown in **Figure 13.1b**, a mutation occurred within a single embryonic cell. This single somatic cell was the precursor for many cells of the adult. Therefore, in the adult, a patch of tissue contains cells that carry the mutation. The size of any patch depends on the timing of a new mutation. In general, the earlier a mutation occurs during development, the larger the patch.

**Figure 13.2** illustrates a child who had a somatic mutation during an early stage of development. In this case, the child has a streak of white hair but the rest of his hair is black. Presumably, a single mutation happened in an embryonic cell that ultimately gave rise to the patch that produced the white hair.

## Biology Principle

### Living Organisms Grow and Develop

As a multicellular organism grows and develops, a germ-line mutation is transmitted to all cells of the body, whereas a somatic mutation is found only in a particular region.

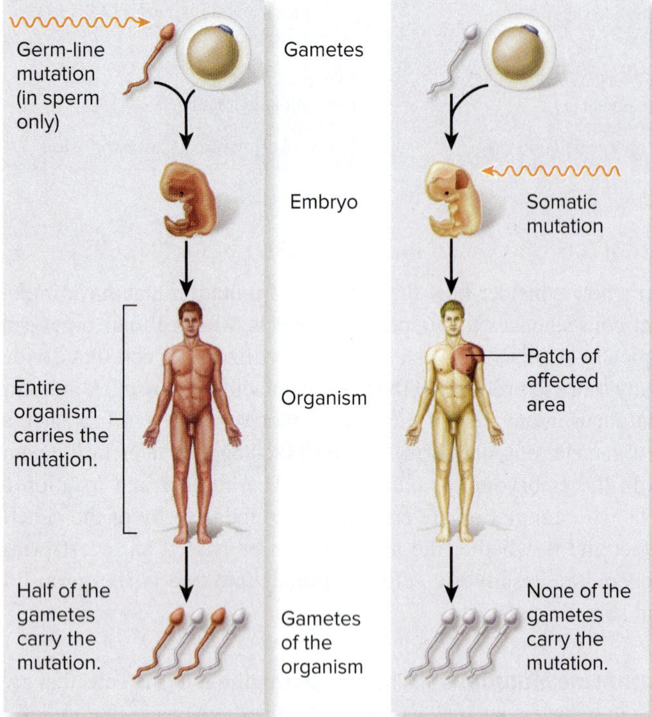


**Figure 13.1 The effects of germ-line versus somatic cell mutations.** The red color indicates which cells carry the mutation. **(a)** If a mutation is passed via gametes, such germ-line mutations occur in every cell of the body. Because humans have two copies of most genes, a germ-line mutation in one of those two copies is transmitted to only half of the gametes. **(b)** Somatic mutations affect a limited area of the body and are not transmitted to offspring. Note: This figure shows a somatic mutation in an embryo, but somatic mutations can also occur at later stages such as adulthood.

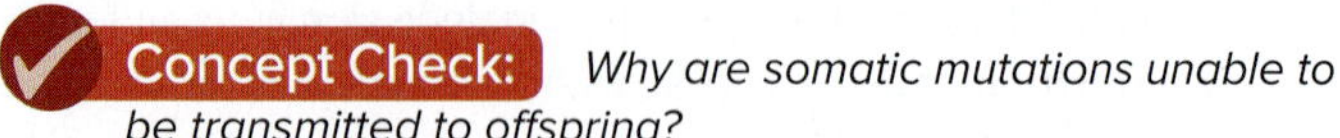

**Concept Check:** *Why are somatic mutations unable to be transmitted to offspring?*

## 13.1 Reviewing the Concepts

- A mutation is a heritable change in the genetic material; gene mutations are changes in the base sequence of a gene. Point mutations affect a single base pair and can alter the coding sequence of genes in several ways, producing silent, missense, nonsense, and frameshift mutations (Table 13.1).
- Gene mutations also alter gene function by changing DNA sequences that are not within the coding region (Table 13.2).

**Figure 13.2 Example of a somatic mutation.** This child has a streak of white hair. This is due to a somatic mutation in a single cell during embryonic development. This cell continued to divide to produce a streak of white hair.



**Concept Check:** *When he gets older, can this boy transmit this trait to his offspring?*

- Germ-line mutations affect gametes and can be passed to offspring, whereas mutations in somatic cells affect only a part of the body and cannot be passed to offspring (Figures 13.1, 13.2).

## 13.1 Testing Your Knowledge

1. A mutation changes an alanine codon into a valine codon. This is an example of a
   a. silent mutation.
   b. missense mutation.
   c. frameshift mutation.
   d. nonsense mutation.
2. A key difference between germ-line and somatic mutations is that
   a. only somatic mutations affect somatic cells.
   b. only germ-line mutations can be harmful.
   c. only somatic mutations can be passed to offspring.
   d. only germ-line mutations can be passed to offspring.

## 13.2 Causes of Mutations

### Learning Outcomes

1. **SCISKILLS ▶** Analyze the results of the Lederbergs' experiment, and explain how they were consistent with the random theory of mutation.
2. Discuss the difference between spontaneous and induced mutations, and provide examples.
3. **SCISKILLS ▶** Describe the Ames test for determining if a substance is a mutagen, and analyze data that are generated from the test.

In the previous section, we surveyed a variety of mutations with an emphasis on mutations that alter a single gene. Researchers and the general public are particularly interested in the causes of mutations because some of them may be harmful. In this section, we will explore how mutations happen.

# FEATURE INVESTIGATION

## The Lederbergs Used Replica Plating to Show That Mutations Are Random Events

In the 19th and early 20th centuries, scientists considered the following question: Are mutations that affect the traits of an individual caused by pre-existing circumstances, or are they random events that may happen in any gene of any individual?

- **Physiological adaptation hypothesis:** In the 19th century, French naturalist Jean-Baptiste Lamarck proposed that physiological events (such as use or disuse) determine whether traits are passed to offspring. For example, his hypothesis suggested that an individual who practiced and became adept at a physical activity, such as the long jump, would pass that quality to his or her offspring.
- **Random mutation hypothesis:** Geneticists in the early 20th century suggested that genetic variation occurs as a matter of chance. According to this view, those individuals whose genes happen to contain beneficial mutations are more likely to survive and pass those genes to their offspring.

These opposing views were tested in bacterial studies in the 1940s and 1950s. One such study, by American microbiologists Joshua and Esther Lederberg, focused on the occurrence of mutations in bacteria (**Figure 13.3**). First, they placed a large number of *E. coli* bacteria onto growth media and incubated them overnight, so each bacterial cell divided many times to form a bacterial colony composed of millions of cells. This is called the master plate. Using a technique known as **replica plating,** a sterile piece of velvet cloth was lightly touched to the master plate to pick up a few bacterial cells from each colony on the master plate. They then transferred this replica to two secondary plates containing an agent that selected for the growth of bacterial cells with a particular mutation.

In the example shown in Figure 13.3, the secondary plates contained T1 bacteriophages, which are viruses that infect bacteria and cause them to lyse. On these plates, only those rare cells that had acquired a mutation conferring resistance to T1, termed *ton*$^r$, could grow. All other cells were lysed by the proliferation of bacteriophages in the bacteria. Therefore, only a few colonies were observed on the secondary plates. Strikingly, these colonies occupied the same locations on each plate. How did the Lederbergs interpret these results? The data indicated that the *ton*$^r$ mutations occurred randomly while the bacterial cells were forming colonies on the nonselective master plate. The presence of T1 bacteriophages in the secondary plates did not cause the mutations to develop. Rather, the T1 bacteriophages simply selected for the growth of *ton*$^r$ mutants that were already in the population. These results supported the idea that mutations are random events, now called the random mutation theory.

### *Experimental Questions*

1. Explain the opposing views of mutation prior to the Lederbergs' study.
2. **SCISKILLS ▸** What hypothesis was being tested by the Lederbergs? What were the results of the experiment?
3. **SCISKILLS ▸** How did the results of the Lederbergs support or falsify the hypothesis?

**Figure 13.3** The experiment performed by the Lederbergs showing that mutations are random events.

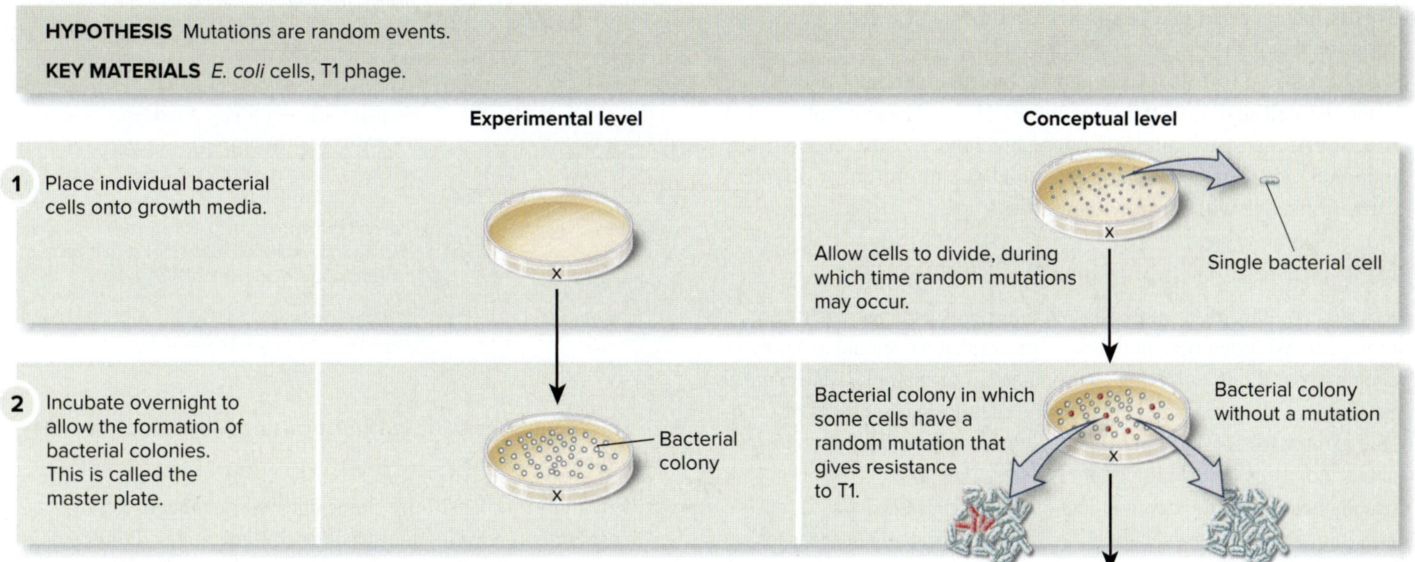

HYPOTHESIS Mutations are random events.
KEY MATERIALS E. coli cells, T1 phage.
Experimental level
Conceptual level
1 Place individual bacterial cells onto growth media.
Allow cells to divide, during which time random mutations may occur.
Single bacterial cell
2 Incubate overnight to allow the formation of bacterial colonies. This is called the master plate.
Bacterial colony
Bacterial colony in which some cells have a random mutation that gives resistance to T1.
Bacterial colony without a mutation

**3** Press a velvet cloth (wrapped over a cylinder) onto the master plate, and then lift gently to obtain a replica of each bacterial colony. Press the replica onto 2 secondary plates that contain T1 bacteriophage. Incubate overnight to allow bacterial growth.

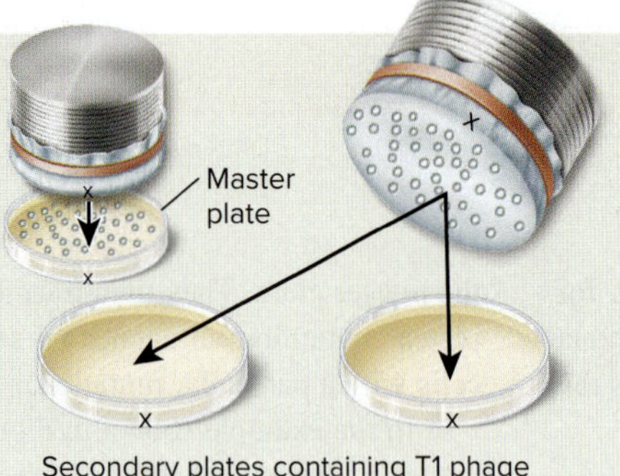


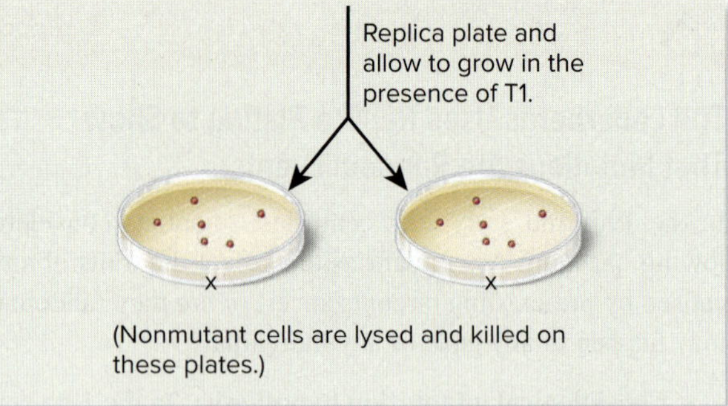


**4 THE DATA**

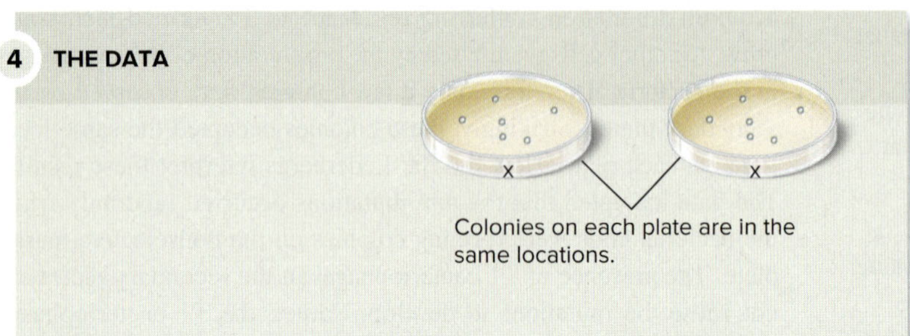


**5 CONCLUSION** Mutations are random events. In this case, the mutations occurred on the master plate prior to exposure to T1 bacteriophage.

**6 SOURCE** Lederberg, Joshua, and Lederberg, Esther M. 1952. Replica Plating and Indirect Selection of Bacterial Mutants. *Journal of Bacteriology* 63: 399–406.

## Mutations May Be Spontaneous or Induced

Biologists categorize the causes of mutation as spontaneous or induced (Table 13.3). **Spontaneous mutations** result from abnormalities in biological processes. Spontaneous mutations reflect the observation that biology isn't perfect. For example, DNA polymerase, which is discussed in Chapter 9, can make mistakes during DNA replication by putting the wrong base in a newly synthesized daughter strand, though such mistakes are very rare.

The rates of spontaneous mutations vary from species to species and from gene to gene. Larger genes are usually more likely to incur a mutation than are smaller ones. A common rate of spontaneous mutation among various species is approximately 1 mutation for every 1 million genes per cell division, which equals 1 in $10^6$, or simply $10^{-6}$. This is the expected rate of spontaneous mutation, which creates the variation that is the raw material of evolution.

**Induced mutations** are caused by environmental agents that enter the cell and alter the structure of DNA. These agents, called **mutagens,** cause the mutation rate to be higher than the spontaneous mutation rate. Mutagens can be categorized as **chemical** or **physical mutagens.** We will consider their effects next.

## Mutagens Alter DNA Structure in Different Ways

Researchers have discovered that an enormous array of agents act as mutagens. We often hear in the news media that we should avoid these agents in our foods and living environments. For example, we use products such as sunscreens to help us avoid the mutagenic effects of exposure to ultraviolet (UV) light from the Sun. The public is often concerned about mutagens for two important reasons. First, mutagenic agents are usually involved in the development of human cancers. Second, because new mutations may be deleterious, people want to avoid mutagens to prevent mutations that may have harmful effects on their future offspring.

**Table 13.3 Some Common Causes of Gene Mutations**

| Common causes of mutations | Description |
|---|---|
| **Spontaneous** | |
| Errors in DNA replication | An unrepaired mistake by DNA polymerase may cause a point mutation. |
| Toxic metabolic products | The products of normal metabolic processes may be reactive chemicals that alter the structure of DNA. |
| Changes in nucleotide structure | On rare occasions, the linkage between purines and deoxyribose can spontaneously break. Changes in base structure (isomerization) may cause mispairing during DNA replication. |
| Transposons | As discussed in Chapter 18, DNA transposons are small segments of DNA that can insert at various sites in the genome. If they insert into a gene, they may inactivate the gene. |
| **Induced** | |
| Chemical agents | Chemical substances, such as benzo(*a*)pyrene, a chemical found in cigarette smoke, may cause changes in the structure of DNA. |
| Physical agents | Physical agents such as UV (ultraviolet) light and X-rays can damage the DNA. |

How do mutagens affect DNA structure? Several effects are possible (Table 13.4):

- **Covalent modification:** Some chemical mutagens act by covalently modifying the structure of nucleotides. For example, nitrous acid ($HNO_2$) deaminates bases by replacing amino groups

**Table 13.4 Examples of Mutagens**

| Mutagen | Effect(s) on DNA structure |
|---|---|
| **Chemical** | |
| Nitrous acid | Deaminates bases |
| Nitrogen mustard | Alkylates bases |
| Ethyl methanesulfonate (EMS) | Alkylates bases |
| 5-Bromouracil | Acts as a base analogue |
| 2-Aminopurine | Acts as a base analogue |
| Benzo(*a*)pyrene | Inserts between bases in the DNA double helix and causes additions or deletions |
| **Physical** | |
| X-rays (ionizing radiation) | Causes base deletions, single nicks in DNA strands, crosslinking, and chromosomal breaks |
| UV light (nonionizing radiation) | Promotes thymine dimer formation, which involves covalent bonds between adjacent thymines |

with keto groups (—$NH_2$ to =O). This can change cytosine to uracil. When this altered DNA replicates, the modified base does not pair with the appropriate base in the newly made strand. In this case, uracil pairs with adenine (Figure 13.4).

- **Base analogues:** 5-Bromouracil and 2-aminopurine are called base analogues, because their structures are similar to those of particular bases in DNA and can substitute for them. When incorporated into DNA, they also cause errors in DNA replication.
- **Distortion of the DNA double helix:** Some chemical mutagens exert their effects by interfering with DNA structure. For example, benzo(*a*)pyrene, which is found in automobile exhaust, cigarette smoke, and charbroiled food, is metabolized to a compound (benzopyrene diol epoxide) that inserts itself between the bases of the double helix, thereby distorting the helical structure. When DNA containing such a mutagen is replicated, single-nucleotide additions and deletions may be incorporated into the newly made strands.
- **Ionizing radiation:** DNA molecules are also sensitive to physical agents such as radiation. In particular, radiation of short wavelength and high energy, known as ionizing radiation, is known to alter DNA structure. Ionizing radiation includes X-rays and gamma rays. This type of radiation can penetrate deeply into biological materials, where it creates free radicals. These molecules can alter the structure of DNA in a variety of ways. Exposure to high doses of ionizing radiation causes base deletions, breaks in one DNA strand, or even a break in both DNA strands.
- **Nonionizing radiation:** UV light is a type of nonionizing radiation. It contains less energy than ionizing radiation, and so nonionizing radiation penetrates only the surface of biological materials, such as the skin. Nevertheless, UV light is known to cause mutations. For example, UV light can cause the formation of a **thymine dimer,** which is a site where two adjacent thymine bases become covalently bonded to each other (Figure 13.5). Thymine dimers are typically repaired before or during DNA replication. However, if such repair fails to occur, a thymine dimer may cause a mutation when that DNA strand is replicated. When DNA polymerase attempts to replicate over a thymine dimer, proper base pairing does not occur between the template strand and the incoming nucleotides. This mispairing can cause gaps in the newly made strand or the incorporation of incorrect bases. Plants, in particular, must have effective ways to prevent UV damage because they are exposed to sunlight throughout the day.

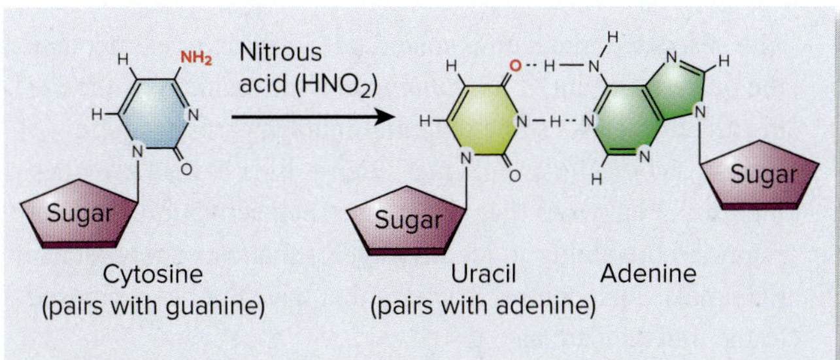


**Figure 13.4 Deamination and mispairing of modified bases by a chemical mutagen.** Nitrous acid changes cytosine to uracil by replacing $NH_2$ with an oxygen. During DNA replication, uracil pairs with adenine, thereby creating a mutation in the newly replicated strand.

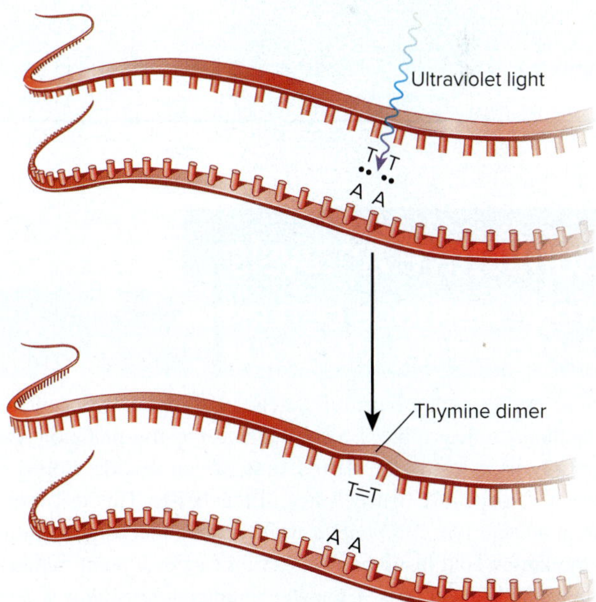


**Figure 13.5 Formation and structure of a thymine dimer, a pair of adjacent thymine bases that have covalently bonded.**

*Why is a thymine dimer harmful?*

## Biology Principle

### Biology Affects Our Society

Biologists have developed many methods, including the Ames test, for determining if a substance is a mutagen. The results of these tests have prevented the use of many different chemicals in the production of food and have resulted in warning labels on products such as cigarettes.

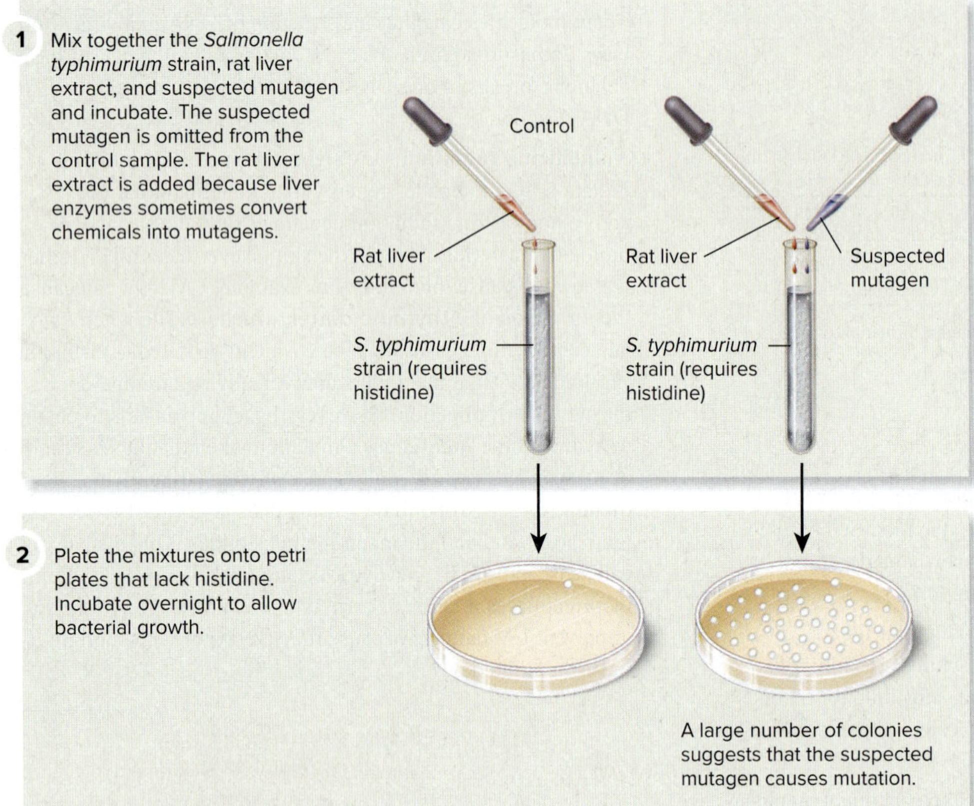

1 Mix together the *Salmonella typhimurium* strain, rat liver extract, and suspected mutagen and incubate. The suspected mutagen is omitted from the control sample. The rat liver extract is added because liver enzymes sometimes convert chemicals into mutagens.
Control
Rat liver extract
S. typhimurium strain (requires histidine)
Rat liver extract
Suspected mutagen
S. typhimurium strain (requires histidine)
2 Plate the mixtures onto petri plates that lack histidine. Incubate overnight to allow bacterial growth.
A large number of colonies suggests that the suspected mutagen causes mutation.


**Figure 13.6** **The Ames test for mutagenicity.** In this example, 2 million bacterial cells were placed on plates lacking histidine. Two colonies were observed in the control sample, whereas 44 were observed in the sample exposed to a suspected mutagen.

**Concept Check:** *Based on the results seen in this figure, what is the rate of mutation that is caused by the suspected mutagen?*

---

## Quantitative Analysis

### TESTING METHODS DETERMINE IF AN AGENT IS A MUTAGEN

Because mutagens are harmful, researchers have developed testing methods to determine if a substance can cause mutation. One commonly used test is the **Ames test,** which was developed by American biochemist Bruce Ames in the 1970s. This test uses a strain of a bacterium, *Salmonella typhimurium*, that cannot synthesize the amino acid histidine. This strain carries a point mutation within a gene that encodes an enzyme required for histidine biosynthesis. The mutation renders the enzyme inactive. The bacteria cannot grow unless histidine has been added to the growth medium. However, a second mutation may correct the first mutation, thereby restoring the ability to synthesize histidine. The Ames test monitors the rate at which this second mutation occurs and thereby indicates whether an agent increases the mutation rate above the spontaneous rate.

Figure 13.6 outlines the steps in the Ames test.

1. The suspected mutagen is mixed with a rat liver extract and the bacterial strain of *S. typhimurium* that cannot synthesize histidine. Because some potential mutagens may require activation by cellular enzymes, the rat liver extract provides a mixture of enzymes that may cause such activation. This step improves the ability to identify agents that cause mutations in mammals. As a control, bacteria that have not been exposed to the mutagen are also tested.
2. After an incubation period in which mutations may occur, a large number of bacteria are plated on a growth medium that does not contain histidine. The *S. typhimurium* strain is not expected to grow on these plates. However, if a mutation

has occurred that allows a cell to synthesize histidine, the bacterium harboring this second mutation will proliferate during an overnight incubation period to form a visible bacterial colony.

To estimate the mutation rate, the colonies that grow in the absence of histidine are counted and compared with the total number of bacterial cells that were originally placed on the plate for both the suspected-mutagen sample and the control. The control condition is a measure of the spontaneous mutation rate, whereas the other sample measures the rate of mutation in the presence of the suspected mutagen. As an example, let's suppose that 2 million bacteria were plated from both the control and the suspected-mutagen tubes. In the control experiment, 2 bacterial colonies were observed. The spontaneous mutation rate is calculated by dividing 2 (the number of mutants) by 2 million (the number of original cells). This equals 1 in 1 million, or $1 \times 10^{-6}$. By comparison, 44 colonies were observed in the suspected-mutagen sample (see Figure 13.6). In this case, the mutation rate would be 44 divided by 2 million, which equals $2.2 \times 10^{-5}$. The mutation rate in the presence of the mutagen is over 20 times higher than the spontaneous mutation rate.

How do we judge if an agent is a mutagen? Researchers compare the mutation rate in the presence and absence of the suspected mutagen. The experimental approach shown in Figure 13.6 is conducted several times. If statistics reveal that the mutation rate in the suspected-mutagen sample is significantly higher than in the control sample, they may tentatively conclude that the agent is a mutagen. Interestingly, many studies have used the Ames test to compare the urine from cigarette smokers with that from nonsmokers. This research has shown that urine from smokers contains much higher levels of mutagens.

**Crunching the Numbers:** Let's suppose a researcher studied the effects of a suspected mutagen, called mutagen X, by using the protocol described in Figure 13.6. The following data were obtained after placing 2 million cells on each plate:

| Trial | Control (number of colonies) | Plus mutagen X (number of colonies) |
|---|---|---|
| 1 | 3 | 62 |
| 2 | 2 | 77 |
| 3 | 5 | 46 |
| 4 | 2 | 55 |

Calculate the average mutation rate in the presence and absence of mutagen X. Conduct a *t* test* to determine if mutagen X is significantly affecting the mutation rate.

* A description of the *t* test and other statistical methods is provided in a statistics supplement found in Connect.

## 13.2 Reviewing the Concepts

- The Lederbergs used replica plating to show that mutations are random events (Figure 13.3).
- Spontaneous mutations are the result of errors in natural biological processes. Induced mutations are caused by mutagens—chemical or physical agents in the environment that alter DNA structure (Figures 13.4, 13.5; Tables 13.3, 13.4).
- The Ames test is a method of testing whether an agent is a mutagen (Figure 13.6).

## 13.2 Testing Your Knowledge

1. A spontaneous mutation
   - **a.** is caused by an environmental agent.
   - **b.** is caused by an abnormality in a biological process.
   - **c.** occurs very quickly and is irreversible.
   - **d.** is all of the above.
2. A chemical mutagen can cause mutation by
   - **a.** the covalent modification of bases.
   - **b.** acting as a base analogue.
   - **c.** distorting the structure of the DNA double helix.
   - **d.** doing all of the above.

## 13.3 DNA Repair

### Learning Outcomes

1. List the general features of DNA repair systems.
2. Describe the steps of nucleotide excision repair (NER).

In the previous sections, we considered the consequences and causes of mutations. As we have seen, mutations are random events that often have negative consequences. How do organisms minimize the occurrence of mutations? All living organisms have the ability to repair changes that occur in the structure of DNA. In Chapter 9, we saw how DNA polymerase has a proofreading function that helps to prevent mutations from happening during DNA replication. In addition, cells contain several DNA repair systems that can detect and repair different types of DNA alterations (Table 13.5).

Each repair system is composed of one or more proteins that play specific roles in the repair mechanism. DNA repair requires two coordinated events. In the first step, one or more proteins in the repair system detect an irregularity in DNA structure. In the second step, the abnormality is repaired. In some cases, the change in DNA structure can be directly repaired. For example, DNA may be modified by the attachment of an alkyl group, such as $—CH_2CH_3$, to a base. In **direct repair,** an enzyme removes this alkyl group, thereby restoring the structure of the original base.

More commonly, however, the altered DNA is removed, and a new segment of DNA is synthesized. As an example of how such systems operate, let's consider **nucleotide excision repair (NER).** In this type of repair, a region encompassing several nucleotides

**Table 13.5 Common Types of DNA Repair Systems***

| System | Description |
|---|---|
| Direct repair | A repair enzyme recognizes an incorrect structure in the DNA and directly converts it back to a correct structure. |
| Base excision and nucleotide excision repair | An abnormal base or nucleotide is recognized, and a portion of the strand containing the abnormality is removed. The complementary DNA strand is then used as a template for synthesizing a normal DNA strand. |
| Mismatch repair | Similar to excision repair, except that the DNA defect is a base pair mismatch in the DNA, not an abnormal nucleotide. The mismatch is recognized, and a strand of DNA in this region is removed. The complementary strand is used to synthesize a normal strand of DNA. |

*Other types of repair systems exist; these are common examples.

in the damaged strand is removed from the DNA, and the intact undamaged strand is used as a template for resynthesizing a normal complementary strand. NER can fix many different types of DNA damage, including UV-induced damage, chemically modified bases, missing bases, and thymine dimers. The system is found in all species, although its molecular mechanism is better understood in bacteria.

In *E. coli*, the NER system is composed of four key proteins: UvrA, UvrB, UvrC, and UvrD. They are named Uvr because they are involved in ultraviolet light repair of thymine dimers, although these proteins are also important in repairing chemically damaged DNA. In addition, DNA polymerase and DNA ligase are required to complete the repair process.

How does the NER system work?

1. Two UvrA proteins and one UvrB protein form a complex that tracks along the DNA (Figure 13.7). Damaged DNA has a distorted double helix, which is sensed by the UvrA-UvrB complex.
2. When the complex identifies a damaged site, the two UvrA proteins are released, and UvrC binds to UvrB at the site.
3. The UvrC protein makes incisions in one DNA strand on both sides of the damaged site.
4. After this incision process, UvrC is released. UvrD binds to UvrB. UvrD then begins to separate the DNA strands, and UvrB is released. The action of UvrD unravels the DNA, which removes a short DNA strand that contains the damaged region. UvrD is released.
5. After the damaged DNA strand is removed, a gap is left in the double helix. DNA polymerase fills in the gap using the undamaged strand as a template. Finally, DNA ligase makes the final covalent connection between the newly made DNA and the original DNA strand.

Thus far, we have considered the NER system in *E. coli*. In humans, NER systems were discovered by the analysis of genetic

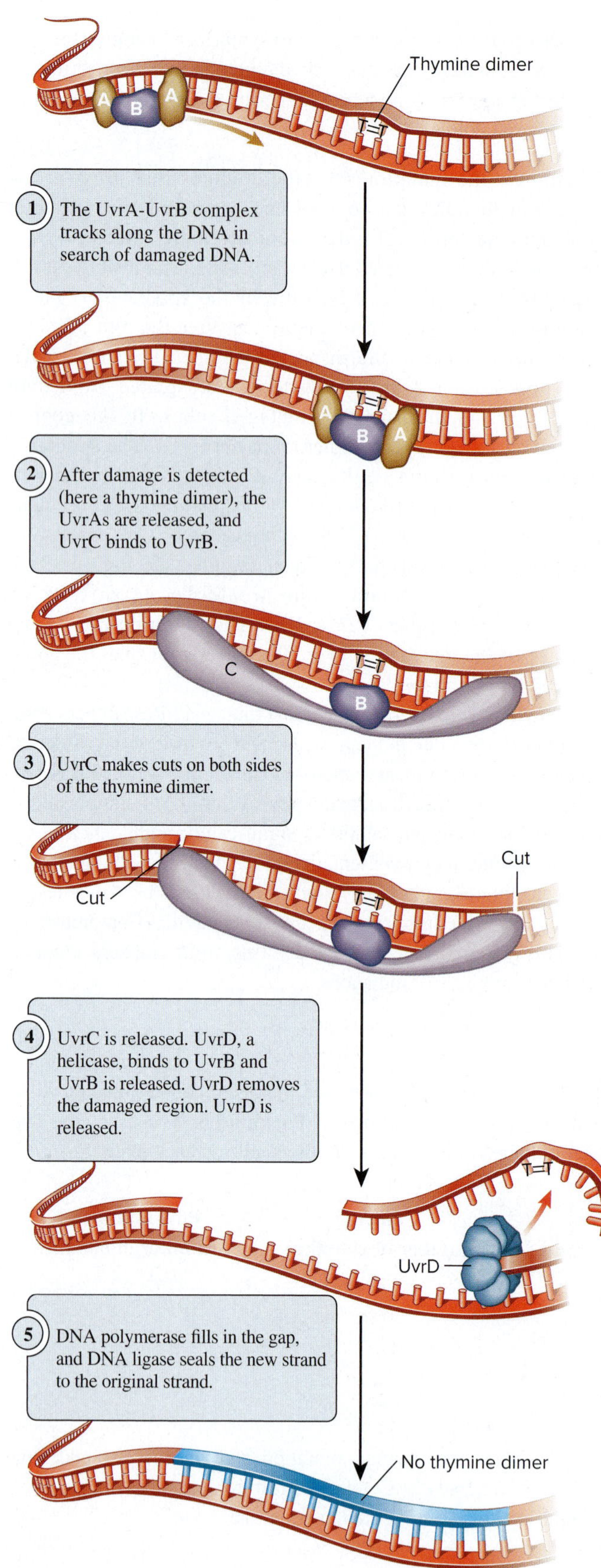


**Figure 13.7** Nucleotide excision repair (NER) in *E. coli*.

**Concept Check:** *Which components of NER are responsible for removing the damaged DNA?*

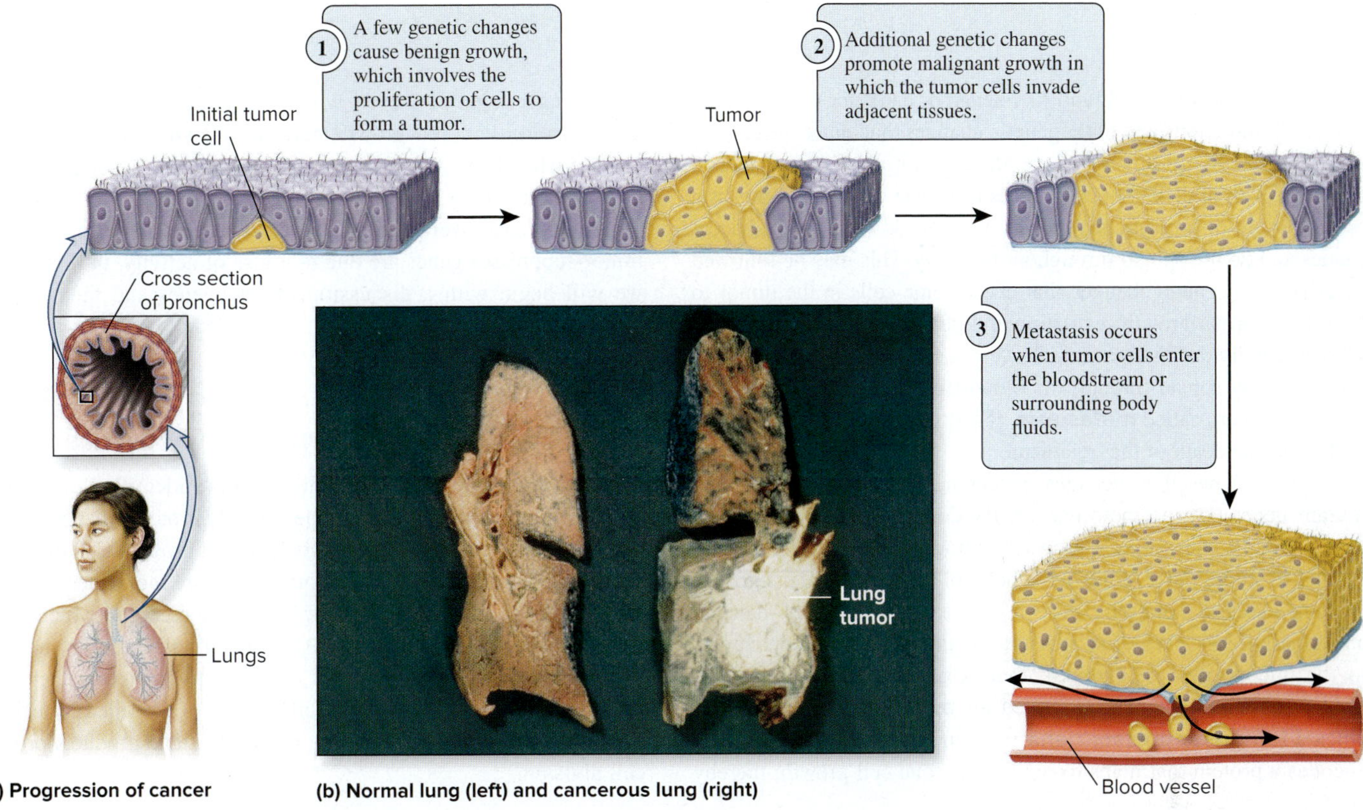

1
A few genetic changes cause benign growth, which involves the proliferation of cells to form a tumor.
2
Additional genetic changes promote malignant growth in which the tumor cells invade adjacent tissues.
3
Metastasis occurs when tumor cells enter the bloodstream or surrounding body fluids.
Initial tumor cell
Tumor
Cross section of bronchus
Lungs
Lung tumor
Blood vessel
(a) Progression of cancer
(b) Normal lung (left) and cancerous lung (right)


**Figure 13.8** Cancer: Its progression and effects. **(a)** In a healthy individual, a few mutations and other genetic changes convert a normal cell into a tumor cell. This cell divides to produce a benign tumor. Additional mutations and other genetic changes in the tumor cells may occur, leading to a malignant tumor that invades surrounding tissues. At a later stage in malignancy, some cancerous cells may metastasize by traveling through the bloodstream to other parts of the body. **(b)** On the left of the photo is a human lung that was obtained from a healthy nonsmoker. The lung shown on the right has been ravaged by lung cancer. This lung was taken from a person who was a heavy smoker.
*(b)* © St. Bartholomew's Hospital/SPL/Science Source

diseases in which human NER proteins are defective. An example is the disease called xeroderma pigmentosum (XP). Persons with this disease have pigmentation abnormalities, many precancerous lesions, and a high predisposition to developing skin cancer. These symptoms occur due to an inability to repair UV-induced lesions. Therefore, people with XP should avoid prolonged exposure to sunlight.

## 13.3 Reviewing the Concepts

- DNA repair systems consist of proteins that sense DNA damage and repair it before a mutation occurs (Table 13.5).
- In nucleotide excision repair (NER), certain Uvr proteins recognize DNA damage, such as thymine dimers. A region in the damaged strand is excised, and a new strand is synthesized, using the intact strand as a template (Figure 13.7).

## 13.3 Testing Your Knowledge

1. In NER, the function of UvrA, UvrB, UvrC, and UvrD is to
   a. detect an irregularity in DNA structure.
   b. remove a segment of the DNA strand containing the irregularity.
   c. synthesize new DNA after the irregularity is removed.
   d. do all of the above.
   e. do a and b only.

## 13.4 Cancer

### Learning Outcomes

1. Outline the general steps in the development of cancer.
2. Describe the general functions of oncogenes.
3. List the three common types of genetic changes that convert proto-oncogenes into oncogenes.
4. Identify the two general functions of proteins encoded by tumor-suppressor genes.
5. Describe three common ways that tumor-suppressor genes are inactivated.

Cancer is a disease of multicellular organisms characterized by uncontrolled cell division. Worldwide, cancer is the second leading cause of death in humans, exceeded only by heart disease. In about 10% of cancers, a higher predisposition to develop the disease is an inherited trait. Most cancers, though, do not involve genetic changes that are passed from parent to offspring. Rather, cancer is usually an acquired condition that typically occurs later in life. At least 80% of all human cancers are related to exposure to **carcinogens**—agents that increase the likelihood of developing cancer. Most carcinogens, such as UV light and certain chemicals in cigarette smoke, are mutagens that promote genetic changes in somatic cells. These genetic changes can alter gene expression in a way that ultimately affects cell division, leading to cancer.

How does cancer occur? In most cases, the development of cancer is a multistep process (**Figure 13.8**). Cancers originate from a single cell. This single cell and its lineage of daughter cells undergo a series of mutations and other genetic changes that cause the cells to grow abnormally. At an early stage, the cells form a **tumor,** which is an overgrowth of cells. For most types of cancer, a tumor begins as a precancerous, or **benign,** growth. Such tumors do not invade adjacent tissues and do not spread throughout the body. This may be followed by additional genetic changes that cause some cells in the tumor to lose their normal growth regulation and become **malignant.** At this stage, the individual has cancer. Cancerous tumors invade healthy tissues and may spread through the bloodstream or surrounding body fluids, a process called **metastasis.** If left untreated, malignant cells will cause the death of the organism.

Over the past few decades, researchers have identified many mutant genes that promote cancer. By comparing the function of each mutant gene with the corresponding nonmutant gene found in healthy cells, these genes have been placed into two categories. In some cases, a mutation causes a gene to be overactive—have an abnormally high level of expression. This overactivity contributes to the uncontrolled cell growth that is observed in cancer cells. This type of mutant gene is called an **oncogene.** Alternatively, when a **tumor-suppressor gene** is normal (that is, not mutant), it encodes a protein that helps to regulate normal cell growth, thereby preventing cancer. However, when a mutation eliminates its function, uncontrolled cell growth may occur, which is a hallmark of cancer cells. Thus, the two categories of cancer-causing genes are based on the effects of mutations. Oncogenes are the result of mutations that cause overactivity, whereas cancer-causing mutations in tumor-suppressor genes are due to a loss of activity. In this section, we will begin with a discussion of oncogenes and then consider tumor-suppressor genes.

## Oncogenes May Result from Mutations That Cause the Overactivity of Proteins Involved with Cell Division

Over the past four decades, researchers have identified many oncogenes. A large number of oncogenes encode proteins that function in signal transduction pathways involved in cell growth. Cell division is regulated, in part, by growth factors, which are molecules that regulate cell division. A growth factor binds to a receptor, which results in receptor activation (Figure 13.9). This stimulates an intracellular signal transduction pathway that activates transcription factors. In this way, the transcription of specific genes is activated in response to a growth factor. After they are made, the gene products promote cell division.

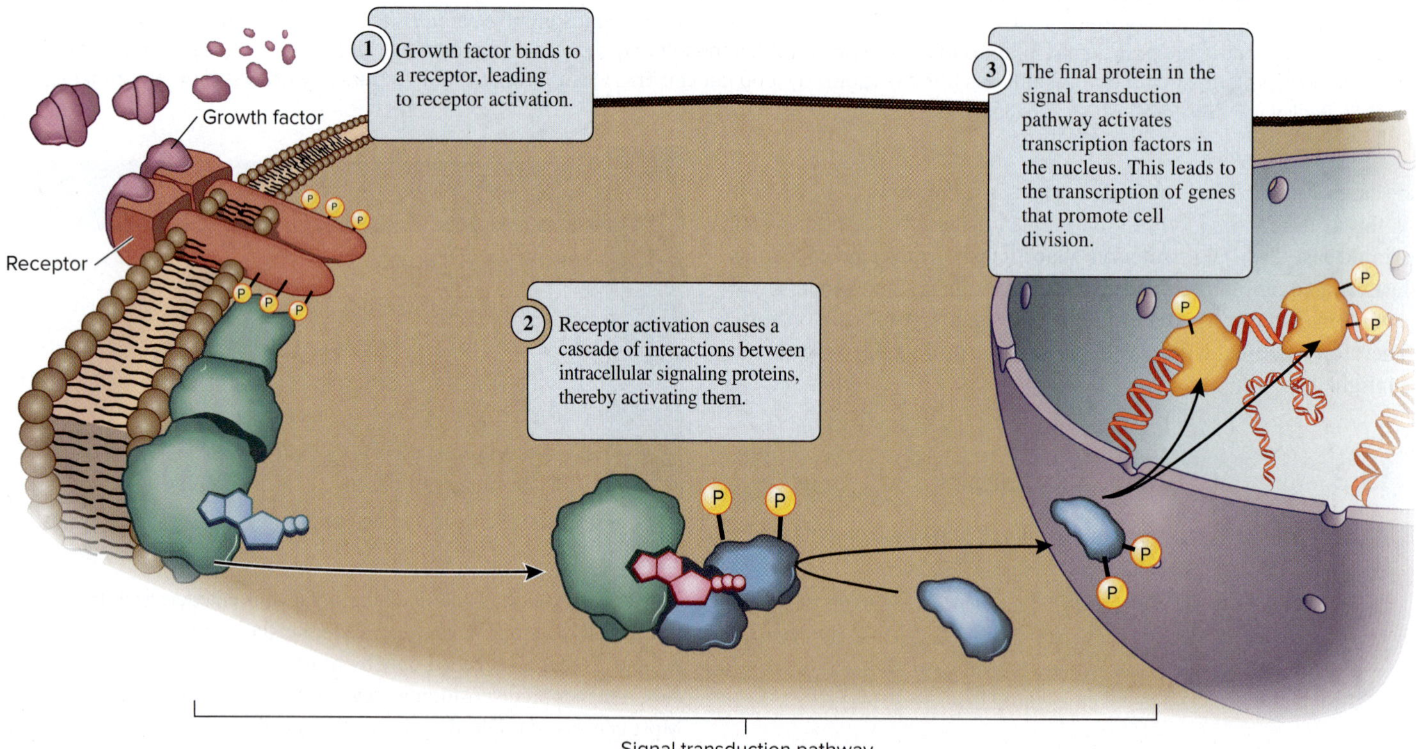


**Figure 13.9 General features of a signal transduction pathway involving a growth factor that promotes cell division.** A detailed description of this pathway is found in Chapter 8 (refer back to Figure 8.10).

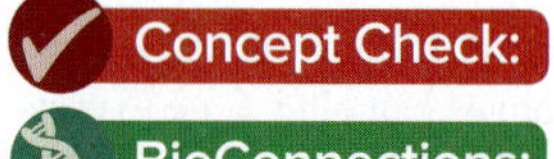
**Concept Check:** *How does the presence of a growth factor ultimately affect the function of the cell?*

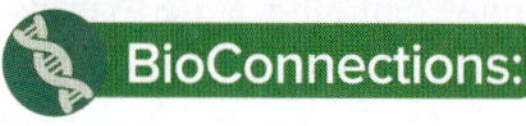
**BioConnections:** *Refer back to Figure 8.10. Could drugs that inhibit protein kinases be used to combat cancer? Explain.*

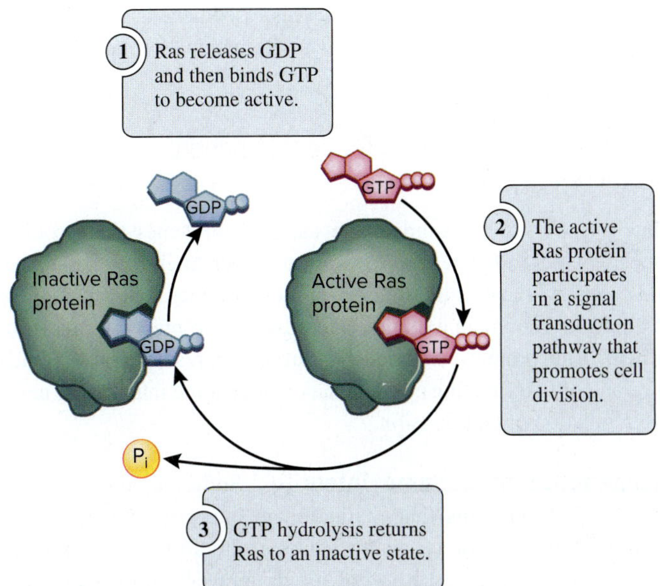

1 Ras releases GDP and then binds GTP to become active.
GDP
GTP
Inactive Ras protein
GDP
Active Ras protein
GTP
2 The active Ras protein participates in a signal transduction pathway that promotes cell division.
$P_i$
3 GTP hydrolysis returns Ras to an inactive state.


**Figure 13.10 The function of Ras, a protein that is part of signal transduction pathways.** When GTP is bound, the activated Ras protein promotes cell division. When GTP is hydrolyzed to GDP and $P_i$, Ras is inactivated and cell division is inhibited.

Eukaryotic species produce many different growth factors that play a role in cell division. Likewise, cells have several different types of signal transduction pathways (described in Chapter 8), which are composed of proteins that respond to growth factors and promote cell division. Mutations in the genes that encode these proteins can change them into oncogenes. As a specific example, let's consider how a mutation alters an intracellular signaling protein called Ras. The Ras protein is a GTPase that hydrolyzes GTP to GDP + $P_i$ (Figure 13.10). When a signal transduction pathway is activated, the Ras protein releases GDP and binds GTP. When GTP is bound, the activated Ras protein promotes cell division (also see Figure 13.9). The Ras protein returns to its inactive state by hydrolyzing its bound GTP, and cell division is no longer activated. Mutations that convert the normal *ras* gene into an oncogenic *ras* either decrease the ability of Ras protein to hydrolyze GTP or increase the rate of exchange of bound GDP for GTP. Both of these functional changes result in a greater amount of the active GTP-bound form of the Ras protein. In this way, these mutations keep the signal transduction pathway turned on when it should not be, resulting in uncontrolled cell division.

## Mutations in Proto-oncogenes Convert Them to Oncogenes

Thus far, we have examined the functions of proteins that cause cancer when they become overactive, resulting in uncontrolled cell division. Let's now consider the common types of genetic changes that create such oncogenes. A **proto-oncogene** is a normal gene that, if mutated, can become an oncogene. An oncogene is a gene that is overexpressed or expressed in the wrong cell type. Several types of genetic changes may convert a proto-oncogene into an oncogene. Figure 13.11 describes three common types: missense mutations, gene amplifications, and chromosomal translocations.

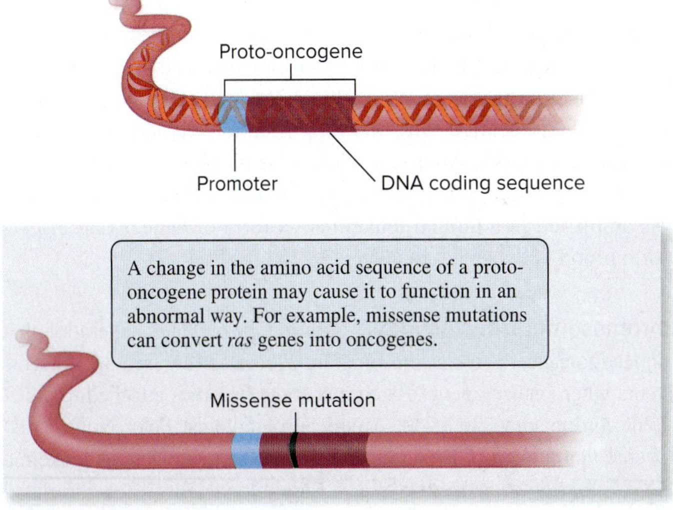

Proto-oncogene
Promoter
DNA coding sequence
A change in the amino acid sequence of a proto-oncogene protein may cause it to function in an abnormal way. For example, missense mutations can convert *ras* genes into oncogenes.
Missense mutation


**(a) Missense mutation**

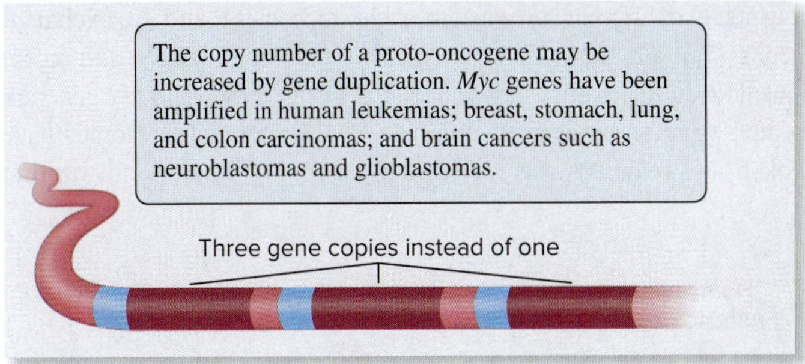

The copy number of a proto-oncogene may be increased by gene duplication. *Myc* genes have been amplified in human leukemias; breast, stomach, lung, and colon carcinomas; and brain cancers such as neuroblastomas and glioblastomas.
Three gene copies instead of one


**(b) Gene amplification**

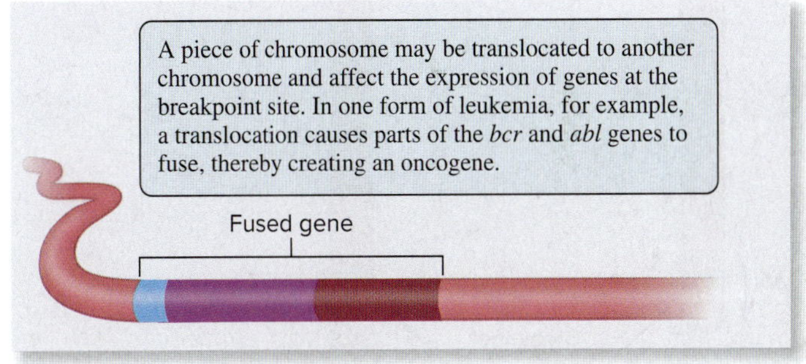

A piece of chromosome may be translocated to another chromosome and affect the expression of genes at the breakpoint site. In one form of leukemia, for example, a translocation causes parts of the *bcr* and *abl* genes to fuse, thereby creating an oncogene.
Fused gene


**(c) Chromosomal translocation**

**Figure 13.11 Common genetic changes that convert proto-oncogenes to oncogenes.**

**Missense Mutation** A missense mutation (Figure 13.11a), which changes a single amino acid in a protein, alters the function of the encoded protein in a way that promotes cancer. This type of mutation is responsible for the conversion of the *ras* gene into an oncogene. An example is a mutation in the *ras* gene that changes a specific glycine to a valine in the Ras protein. This mutation decreases the ability of the Ras protein to hydrolyze GTP, which promotes cell division (see Figure 13.10). Experimentally, chemical mutagens have been shown to cause this missense mutation, thereby leading to cancer.

**Gene Amplification** Another genetic event that occurs in some cancer cells is an increase in the number of copies of a proto-oncogene (Figure 13.11b). An abnormal increase in the number of genes results in too much of the encoded protein. Many human cancers are associated with the amplification of particular proto-oncogenes. In 1982, American molecular biologist Mark Groudine discovered that the *myc* gene, which encodes a transcription factor, was amplified in a human leukemia—a form of cancer that affects white blood cells.

**Chromosomal Translocation** A third type of genetic alteration that can lead to cancer is a chromosomal translocation (Figure 13.11c). This occurs when one segment of a chromosome becomes attached to a different chromosome. In 1960, American pathologist Peter Nowell discovered that a form of leukemia, called chronic myelogenous leukemia (CML)—a type of cancer involving white blood cells—was correlated with the presence of a shortened version of a human chromosome. This shortened chromosome is the result of a chromosome translocation in which two different chromosomes, chromosomes 9 and 22, exchange pieces (**Figure 13.12**). This activates a proto-oncogene, *abl*, in an unusual way. In healthy individuals, the *bcr* gene and the *abl* gene are located on different chromosomes. In CML, these chromosomes have broken and rejoined in a way that causes the promoter and the first part of *bcr* to fuse with part of *abl*. This fused gene acts as an oncogene that encodes a fusion protein whose functional overactivity leads to leukemia.

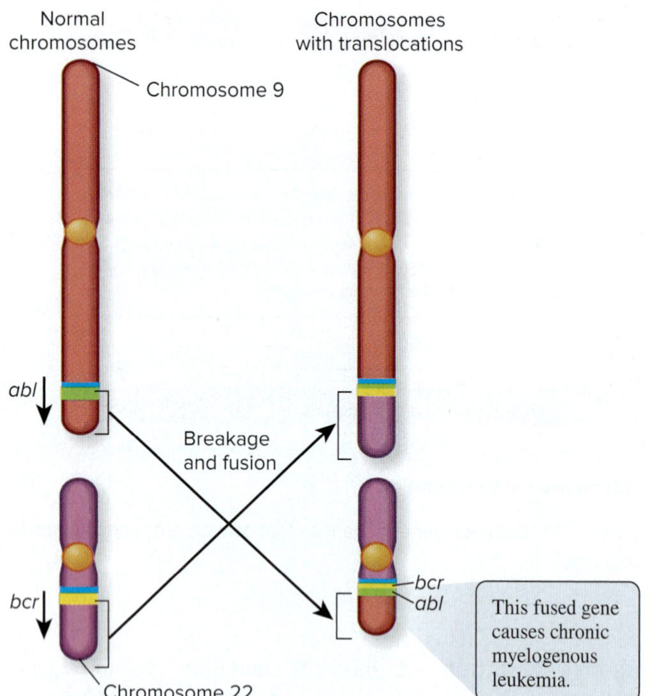


**Figure 13.12 The formation of a fused gene found in people with certain forms of leukemia.** The fusion of the *bcr* and *abl* genes creates an abnormal gene that encodes a fusion protein, leading to leukemia. The blue regions are the promoters for the *bcr* and *abl* genes.

**Concept Check:** *The bcr gene is normally expressed in white blood cells. Explain how this observation is related to the type of cancer that the translocation causes.*

## Tumor-Suppressor Genes Prevent Mutation or Cell Proliferation

Thus far, we have examined one category of genes that promote cancer, namely oncogenes. We now turn our attention to the second category, those called tumor-suppressor genes. The functioning of a normal (nonmutant) tumor-suppressor gene prevents cancerous growth. The proteins encoded by tumor-suppressor genes usually have one of two functions: maintenance of genome integrity or negative regulation of cell division.

**Maintenance of Genome Integrity** Some tumor-suppressor genes encode proteins that maintain the integrity of the genome by monitoring and/or repairing genome alterations. The proteins encoded by these genes are vital for the prevention of abnormalities such as gene mutations, DNA breaks, and improperly segregated chromosomes. Therefore, when these proteins are functioning properly, they minimize the chance that a cancer-causing mutation will occur. In some cases, the proteins encoded by tumor-suppressor genes prevent a cell from progressing through the cell cycle if an abnormality is detected. The cell cycle, which is a series of stages of cell division, is described in Chapter 14 (look ahead to Figure 14.2). Proteins that monitor the cell cycle are termed **checkpoint proteins** because their role is to check the integrity of the genome and prevent a cell from progressing past a certain point in the cell cycle. Checkpoint proteins are not always required to regulate normal, healthy cell division, but they can stop cell division if an abnormality is detected.

A specific example of a tumor-suppressor gene that encodes a checkpoint protein is *p53*, discovered in 1979 by American biologist Arnold Levine. Its name refers to the molecular mass of the p53 protein, which is 53 kDa (kilodaltons). About 50% of all human cancers, including malignant tumors of the lung, breast, esophagus, liver, bladder, and brain, as well as leukemias and lymphomas (cancer of the lymphatic system), are associated with mutations in this gene.

As shown in **Figure 13.13**, p53 is a protein that controls the ability of cells to progress from the $G_1$ stage of the cell cycle to the S phase. The expression of the *p53* gene is induced when DNA is damaged. The p53 protein functions as a regulatory transcription factor that activates several different genes, leading to the synthesis of proteins that stop the cell cycle and other proteins that repair the DNA. If the DNA is eventually repaired, a cell may later proceed through the cell cycle.

Alternatively, if the DNA damage is too severe, the p53 protein will also activate other genes that promote programmed cell death. This process, called **apoptosis,** involves cell shrinkage and DNA degradation. Enzymes known as caspases are activated during apoptosis. They function as proteases that are sometimes called the "executioners" of the cell. Caspases digest selected cellular proteins such as actin filaments, which are components of the cytoskeleton. This causes the cell to break into small vesicles that are eventually

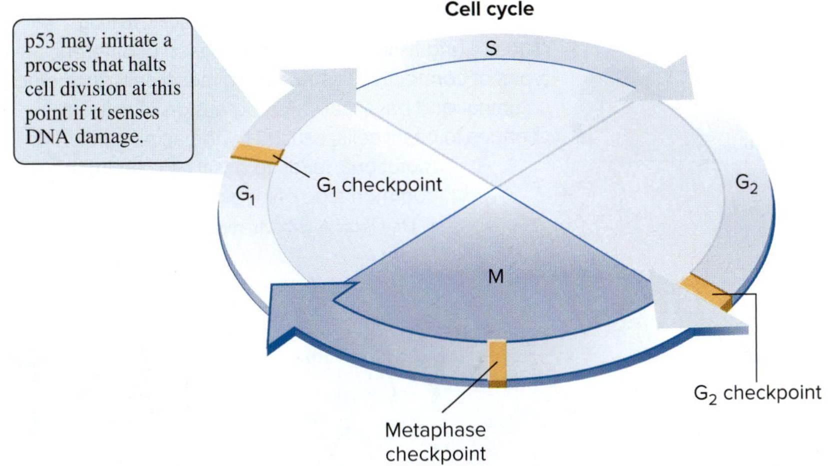


**Figure 13.13 The cell cycle and checkpoints.** As discussed in Chapter 14, eukaryotic cells progress through a cell cycle composed of $G_1$, S, $G_2$, and M phases (look ahead to Figure 14.2). The yellow bars indicate common checkpoints that stop the cell cycle if genetic abnormalities are detected. The p53 protein stops a cell at the $G_1$ checkpoint if it senses DNA damage.

**Concept Check:** *Why is it an advantage for an organism to have checkpoints that can stop the cell cycle?*

phagocytized by cells of the immune system. It is beneficial for a multicellular organism to kill an occasional cell with cancer-causing potential.

**Negative Regulation of Cell Division** A second category of tumor-suppressor genes encodes proteins that are negative regulators or inhibitors of cell division. The function of these genes is necessary to properly halt cell division. If their function is lost, cell division is abnormally accelerated.

An example of such a tumor-suppressor gene is the *Rb* gene. It was the first tumor-suppressor gene to be identified in humans by studying patients with a disease called retinoblastoma, a cancerous tumor that occurs in the retina of the eye. Some individuals have an inherited predisposition to develop this disease within the first few years of life. By comparison, the noninherited form of retinoblastoma, which is usually caused by environmental agents, is more likely to occur later in life.

Based on these differences, in 1971, American geneticist Alfred Knudson proposed a two-hit hypothesis for the occurrence of retinoblastoma. According to this hypothesis, retinoblastoma requires two mutations to occur. People have two copies of the *Rb* gene, one from each parent. Individuals with the inherited form of the disease already have received one mutant gene from one of their parents. They need only one additional mutation to develop the disease. Because the retina has more than 1 million cells, it is relatively likely that a spontaneous mutation may occur in one of these cells at an early age, leading to the disease. However, people with the noninherited form of the disease must have two *Rb* mutations in the same retinal cell to cause the disease. Because two mutations are less likely than a single mutation, the noninherited form of this disease is expected to occur much later in life, and only rarely. Since Knudson's original work, molecular studies have confirmed the two-hit hypothesis for retinoblastoma.

The Rb protein negatively controls a regulatory transcription factor called E2F that activates genes required for cell cycle advancement from $G_1$ to S phase. The binding of the Rb protein to E2F inhibits its activity and prevents cell division (Figure 13.14). When a normal cell is supposed to divide, cyclins bind to cyclin-dependent kinases (cdks). This binding activates the kinases, which catalyze

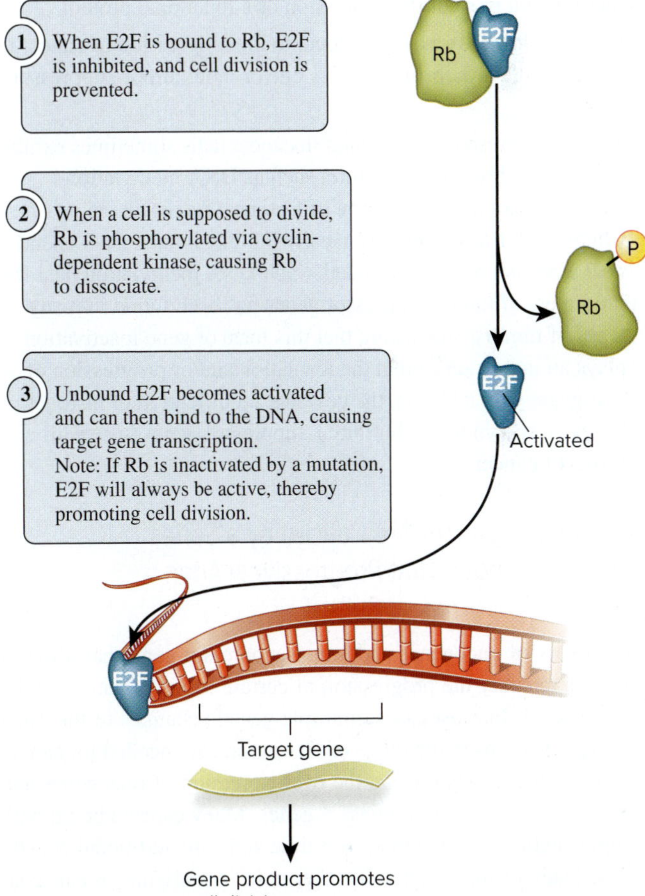

1 When E2F is bound to Rb, E2F is inhibited, and cell division is prevented.
Rb
E2F
2 When a cell is supposed to divide, Rb is phosphorylated via cyclin-dependent kinase, causing Rb to dissociate.
P
Rb
3 Unbound E2F becomes activated and can then bind to the DNA, causing target gene transcription.
Note: If Rb is inactivated by a mutation, E2F will always be active, thereby promoting cell division.
E2F
Activated
E2F
Target gene
Gene product promotes cell division.


**Figure 13.14 Function of the Rb protein.** The Rb protein inhibits the function of E2F, which turns on genes that cause a cell to divide.

**Concept Check:** *Would cancer occur if both copies of the Rb gene and both copies of the E2F gene were rendered inactive due to mutations?*

the transfer of a phosphate to the Rb protein. The phosphorylated form of the Rb protein is released from E2F, thereby allowing E2F to activate genes needed to advance through the cell cycle. When both copies of Rb are defective due to mutations, the E2F protein is always active. This explains why uncontrolled cell division occurs in retinoblastoma.

## Gene Mutations, Chromosome Loss, and Changes in Chromatin Structure Can Inhibit the Expression of Tumor-Suppressor Genes

Cancer biologists want to understand how tumor-suppressor genes are inactivated, because this knowledge may ultimately help them to prevent or combat cancer. How are tumor-suppressor genes inactivated? The function of tumor-suppressor genes is lost in three common ways:

1. A mutation can occur within a tumor-suppressor gene that inactivates its function. For example, a mutation could abolish the function of the promoter for a tumor-suppressor gene or introduce an early stop codon in its coding sequence. Either of these would prevent the expression of a functional protein.
2. Chromosome loss may contribute to the progression of cancer if the missing chromosome carries one or more tumor-suppressor genes.
3. Tumor-suppressor genes found in cancer cells sometimes exhibit changes in chromatin structure, such as DNA methylation and the covalent modification of histones. As discussed in Chapter 12, transcription is usually inhibited when CpG islands near a promoter are methylated. Such DNA methylation near the promoters of tumor-suppressor genes has been found in many types of tumors, suggesting that this form of gene inactivation plays an important role in the formation and/or progression of malignancy. In addition, the covalent modification of histones has been shown to inhibit tumor-suppressor genes in certain forms of cancer.

## Most Forms of Cancer Are Caused by a Series of Genetic Changes That Progressively Alter the Growth Properties of Cells

The discovery of oncogenes and tumor-suppressor genes has allowed researchers to study the progression of certain forms of cancer at the molecular level. In most cases, multiple genetic changes to the same cell lineage, perhaps in the range of 10 or more, are needed for cancer to occur. Such changes involve the overexpression of oncogenes and the inactivation of tumor-suppressor genes. Many cancers begin with a benign genetic alteration that, over time and with additional genetic changes, leads to malignancy. Furthermore, a malignancy can continue to accumulate genetic changes that make it even more difficult to treat because the cells divide faster or invade surrounding tissues more readily.

As an example, let's consider lung cancer, which was briefly discussed at the beginning of this chapter. Most cancers of the lung are **carcinomas**—cancers of epithelial cells (**Figure 13.15**). Epithelial cells form the lining of all internal and external body surfaces. The top

**Figure 13.15 Progression of changes leading to lung cancer.** Lung tissue is largely composed of different types of connective tissue and epithelial cells, including columnar and basal cells. A progression of cellular changes in basal cells, caused by the accumulation of mutations, leads to basal cell carcinoma, a common type of lung cancer.



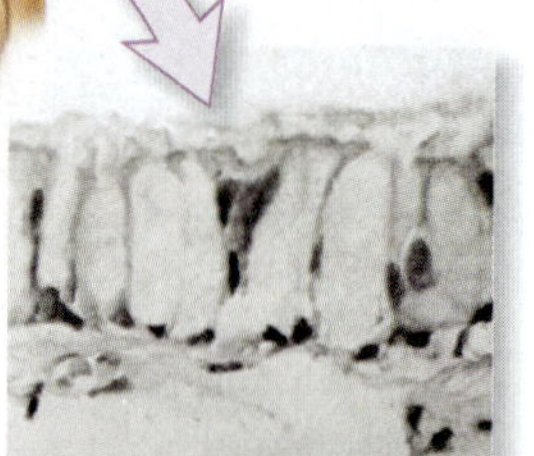

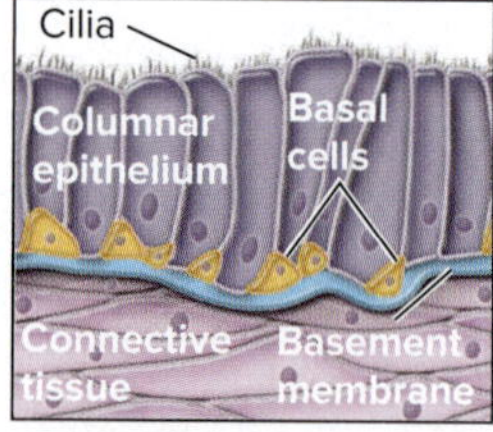


Normal lung epithelium

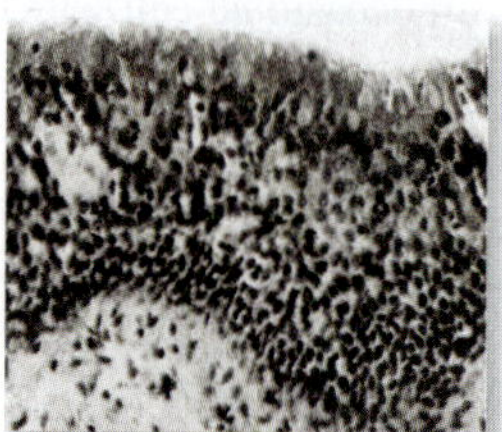

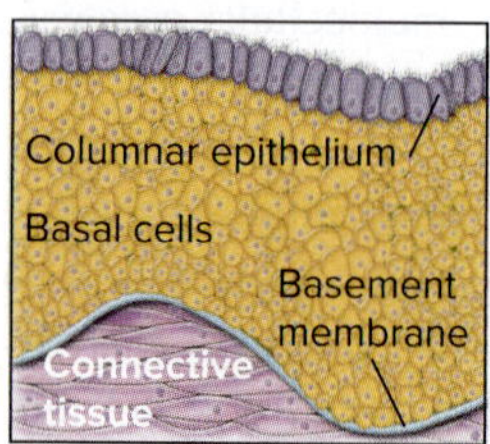


Hyperplasia

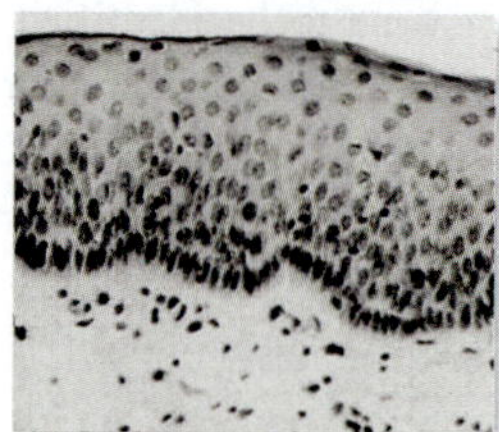

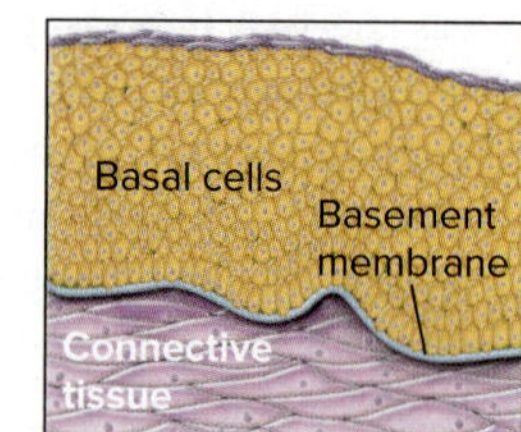


Loss of ciliated cells

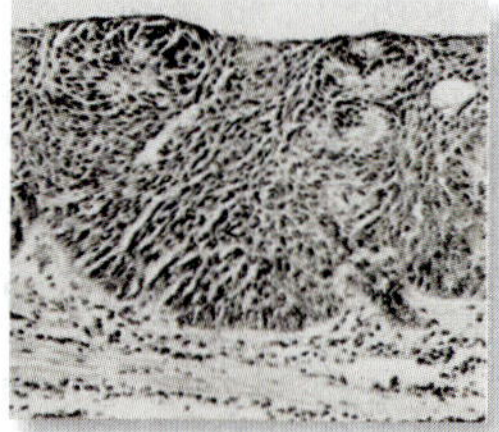

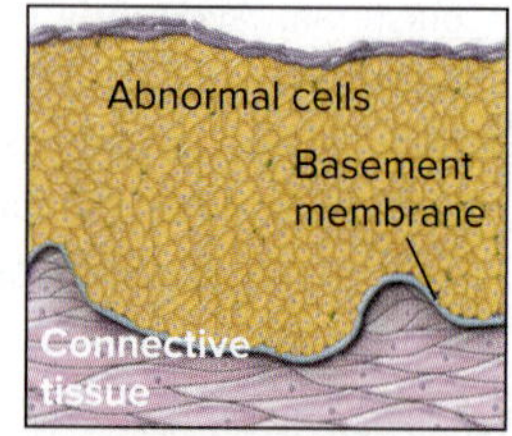


Dysplasia (initially precancerous, then cancerous)

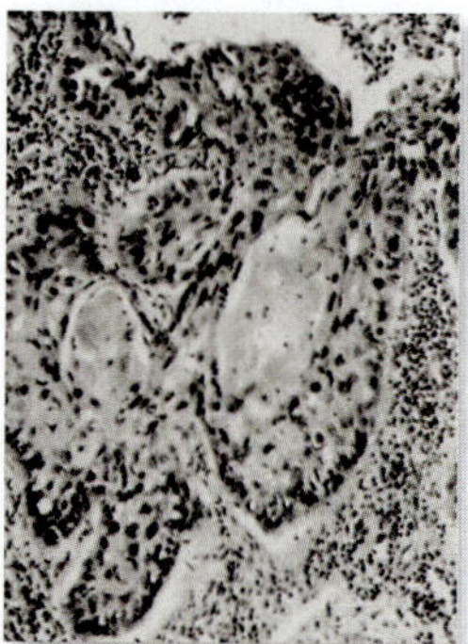

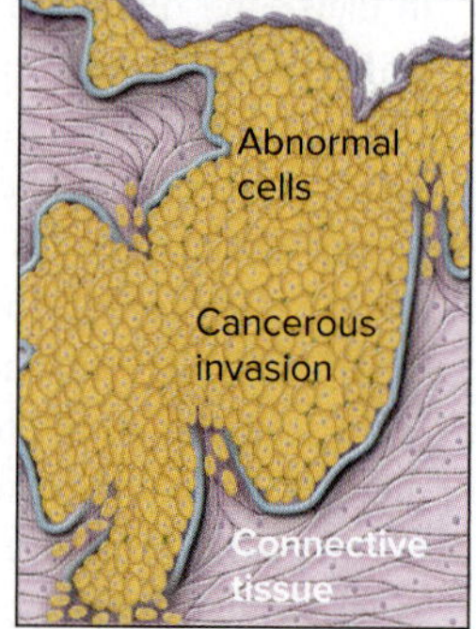


Invasive cancerous cells that can metastasize

images in this figure show the normal epithelium found in a healthy lung. The rest of the figure shows the progression of a carcinoma that is due to mutations in basal cells, a type of epithelial cell.

1. Cancer occurs due to the accumulation of mutations in a cell lineage, beginning with an initial mutant cell, which then divides multiple times to produce a population of many daughter cells. As mutations accumulate in a lineage of basal cells, their numbers increase dramatically. This causes a thickening of the epithelium, a condition called hyperplasia.
2. The proliferation of basal cells causes the loss of the ciliated, columnar epithelial cells that normally line the airways and help remove mucus and particles trapped in it from the lungs.
3. As additional mutations accumulate in this cell lineage, the basal cells develop more abnormal morphologies, a condition known as dysplasia. In the early stages of dysplasia, the abnormal basal cells are precancerous. If the source of chronic irritation (usually cigarette smoke) is eliminated, the abnormal cells are likely to disappear.
4. If smoking continues, the abnormal cells may accumulate additional genetic changes and lose the ability to stop dividing. Such cells have become cancerous—the person has basal cell carcinoma.

The basement membrane is a sheetlike layer of extracellular matrix that provides a barrier between the lung cells and the bloodstream. If the cancer cells have not yet penetrated the basement membrane, they will not have metastasized, that is, spread into the blood, bone, or other parts of the body. If the entire tumor is removed at this stage, the patient should be cured. The lower images in Figure 13.15 show a tumor that has broken through the basement membrane. The metastasis of these cells to other parts of the body will likely kill the patient, usually within a year of diagnosis.

The cellular changes that lead to lung cancer are correlated with genetic changes. These include the occurrence of mutations that create oncogenes and inhibit tumor-suppressor genes. The order of mutations is not absolute. It takes time for multiple changes to accumulate, so cancer is generally a disease of older people. Reducing your exposure to mutagens such as cigarette smoke throughout your lifetime helps minimize the risk of mutations to your genes that could promote cancer.

## 13.4 Reviewing the Concepts

- Cancer is due to the accumulation of mutations and other genetic changes in a lineage of cells that leads to uncontrolled cell growth (Figure 13.8).
- Mutations in proto-oncogenes that result in overactivity produce cancer-causing genes called oncogenes. Oncogenes often encode proteins involved in cell-signaling pathways that promote cell division (Figures 13.9, 13.10).
- Three common types of genetic changes, namely, missense mutations, gene amplifications, and chromosomal translocations, can change proto-oncogenes into oncogenes (Figures 13.11, 13.12).
- The normal function of most tumor-suppressor genes is to negatively regulate cell division or maintain genome integrity. Loss-of-function mutations in such genes promote cancer.
- Checkpoint proteins such as p53 monitor the integrity of the genome and prevent the cell from progressing through the cell cycle if abnormalities are detected (Figure 13.13).
- The Rb protein is an inhibitor of the cell cycle, because it negatively controls E2F, a transcription factor that promotes cell division (Figure 13.14).
- Tumor-suppressor genes can be inactivated by gene mutations, chromosome loss, and changes in chromatin structure.
- Most forms of cancer, such as lung cancer, involve multiple genetic changes that lead to malignancy (Figure 13.15).

## 13.4 Testing Your Knowledge

**1.** A common way to change a proto-oncogene into an oncogene is by
   **a.** a missense mutation.
   **b.** a gene amplification.
   **c.** a chromosomal translocation.
   **d.** all of the above.

**2.** A common function of a tumor-suppressor gene is
   **a.** to maintain the integrity of the genome.
   **b.** to act as a negative regulator of cell division.
   **c.** to act as a cytoskeletal protein.
   **d.** both a and b.

# Assess and Discuss

## Test Yourself

1. A mutation removes a single base pair within the coding sequence of a gene and inactivates the protein encoded by the gene. Such a mutation is
   a. a silent mutation.
   b. a missense mutation.
   c. a nonsense mutation.
   d. a frameshift mutation.
   e. both b and c.
2. Some point mutations lead to an mRNA that produces a shorter polypeptide. This type of mutation is known as a ___ mutation.
   a. neutral
   b. silent
   c. missense
   d. nonsense
   e. chromosomal
3. In which of the following locations is a mutation least likely to affect gene function?
   a. promoter
   b. coding region
   c. splice junction
   d. intergenic region
   e. regulatory site

4. Mutagens can cause mutations by
   a. chemically altering DNA nucleotides.
   b. disrupting DNA replication.
   c. altering the genetic code of an organism.
   d. all of the above.
   e. a and b only.
5. The mutagenic effect of UV light is
   a. the alteration of cytosine bases to adenine bases.
   b. the formation of adenine dimers that interfere with genetic expression.
   c. the breaking of the sugar-phosphate backbone of the DNA molecule.
   d. the formation of thymine dimers that disrupt DNA replication.
   e. the deletion of thymine bases along the DNA molecule.
6. The Ames test
   a. provides a way to determine if any type of cell has experienced a mutation.
   b. provides a way to determine if an agent is a mutagen.
   c. allows researchers to experimentally disrupt gene activity by causing a mutation in a specific gene.
   d. provides a way to repair mutations in bacterial cells.
   e. does all of the above.
7. The nucleotide excision repair (NER) system of *E. coli*
   a. recognizes a DNA abnormality and removes the strand that carries it.
   b. recognizes a DNA abnormality and directly restores an abnormal base to its original condition.
   c. prevents DNA replication from occurring over an abnormality in DNA.
   d. repairs a DNA abnormality that is initially recognized by DNA polymerase.
   e. does all of the above.
8. If a mutation eliminated the function of UvrC, which aspect of nucleotide excision repair would not work?
   a. sensing a damaged DNA site
   b. endonuclease cleavage of the damaged strand
   c. removal of the damaged strand
   d. synthesis of a new strand, using the undamaged strand as a template
   e. none of the above
9. Cancer cells are said to be metastatic when they
   a. begin to divide uncontrollably.
   b. invade healthy tissue.
   c. migrate to other parts of the body.
   d. cause mutations in other healthy cells.
   e. do all of the above.
10. Oncogenes can be produced by
    a. missense mutations.
    b. gene amplification.
    c. chromosomal translocation.
    d. all of the above.
    e. a and b only.

## Conceptual Questions

1. Is a random mutation more likely to be beneficial or harmful? Explain your answer.
2. Distinguish between spontaneous and induced mutations. Which are more harmful? Which are avoidable?
3. **PRINCIPLES** A principle of biology is that the genetic material provides a blueprint for reproduction. Explain how mutations may cause alterations to the genetic material that are detrimental to reproduction and sustaining life.

## Collaborative Questions

1. Discuss the pros and cons of mutation.
2. A large amount of research is aimed at studying mutation. However, there is not an unlimited supply of research funding. Where would you put your money for mutation research and why?
   a. testing of potential mutagens
   b. investigation of molecular effects of mutagens
   c. investigation of DNA repair mechanisms
   d. some other area (specify)

# How Eukaryotic Cells Sort and Transmit Chromosomes: Mitosis and Meiosis

# 14

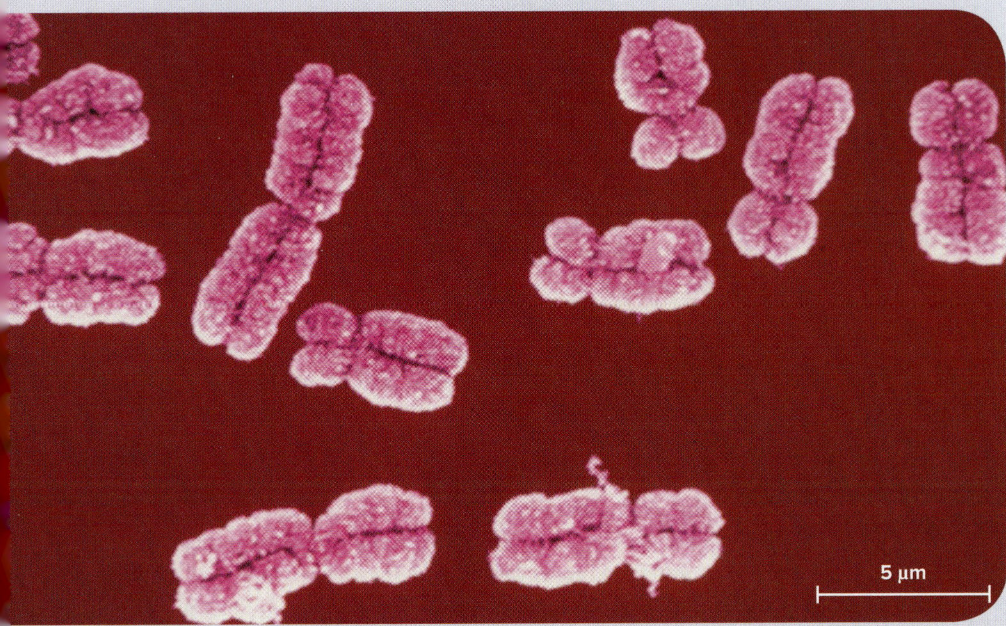




**A scanning electron micrograph of human chromosomes.** These chromosomes are highly compacted and found in a dividing cell.

## Chapter Outline



Over 10,000,000,000,000! Researchers estimate the adult human body contains somewhere between 10 trillion and 50 trillion cells. It is almost an incomprehensible number. Even more amazing is the accuracy of the process that produces these cells. After a human sperm and egg unite, the fertilized egg goes through a long series of cell divisions to produce an adult with over 10 trillion cells. Let's suppose you randomly removed a cell from your arm and compared it with a cell from your foot. If you examined the chromosomes found in both cells under the microscope, they would look identical. Likewise, the DNA sequences along those chromosomes would also be the same, barring rare mutations. Similar comparisons could be made among the trillions of cells in your body. When you consider how many cell divisions are needed to produce an adult human, the precision of cell division is truly remarkable.

What accounts for this high level of accuracy? As we will examine in this chapter, **cell division,** the reproduction of cells, is a highly regulated process that distributes and monitors the integrity of the genetic material. The eukaryotic cell cycle is a series of phases needed for cell division.

The cells of eukaryotic species follow one of two different sorting processes so that new daughter cells receive the correct number and types of chromosomes. The first sorting process we will explore, called mitosis, is needed so two daughter cells receive the same amount of genetic material as the mother cell that produced them. We will then examine another sorting process, called meiosis, which is needed for sexual reproduction. In meiosis, cells that have two sets of chromosomes produce daughter cells with a single set of chromosomes. Finally, we will explore variation in the structure and number of chromosomes. As you will see, a variety of mechanisms that alter chromosome structure and number can have important consequences for the organisms that carry them.

## 14.1 The Eukaryotic Cell Cycle

### Learning Outcomes

1. **SCISKILLS ▶** Describe the features of chromosomes and how sets of chromosomes are examined microscopically.
2. List the phases of the eukaryotic cell cycle.
3. Explain how cyclins and cdks work together to advance a cell through the eukaryotic cell cycle.

Life is a continuum in which new living cells are formed by the division of pre-existing cells. The Latin axiom *omnis cellula e cellula*, meaning every cell originates from another cell, was first proposed in 1858 by Rudolf Virchow, a German biologist. From an evolutionary perspective, cell division has a very ancient origin. All living organisms, from unicellular bacteria to multicellular plants and animals, have been produced by a series of repeated rounds of cell growth and division extending back to the beginnings of life nearly 4 billion years ago.

The **cell cycle** is a series of events that results in cell division. In all species, it is a highly regulated process so cell division occurs at the appropriate time. As discussed in Chapter 17, most bacterial cells reproduce via binary fission, which is relatively simple (look ahead to Figure 17.8). The cell cycle in eukaryotes is more complex, in part, because eukaryotic cells have sets of chromosomes that need to be sorted properly. In this section, we will examine the phases of the eukaryotic cell cycle and see how the cell cycle is controlled by proteins that carefully monitor the division process to ensure its accuracy. But first, we need to consider some general features of chromosomes in eukaryotic species.

### Chromosomes Are Inherited in Sets and Occur in Homologous Pairs

To understand the chromosomal composition of cells and the behavior of chromosomes during cell division, scientists observe cells and chromosomes with the use of microscopes. **Cytogenetics** is the field of genetics that involves the microscopic examination of chromosomes. When a cell prepares to divide, the chromosomes become tightly compacted, a process that decreases their apparent length and increases their diameter. A consequence of this compaction is that distinctive shapes and numbers of chromosomes become visible under a light microscope.

**Microscopic Examination of Chromosomes** Figure 14.1 shows the general procedure for preparing and viewing chromosomes from a eukaryotic cell. In this example, the cells were obtained from a sample of human blood. Specifically, the chromosomes within leukocytes (white blood cells) were examined.

A photographic representation of the chromosomes is shown in step 5 of Figure 14.1. Such a micrograph, called a **karyotype,** reveals the number, size, and form of chromosomes found within an actively dividing cell. The chromosomes viewed in actively dividing cells have already replicated. The two copies are still joined to each other and referred to as a pair of **sister chromatids** (see inset to Figure 14.1).

**Sets of Chromosomes** What type of information is learned from a karyotype? By studying the karyotypes of many species, scientists have discovered that eukaryotic chromosomes occur in sets. Each set is composed of several different types of chromosomes. For example, one set of human chromosomes contains 23 different types of chromosomes (see step 5 of Figure 14.1).

By convention, the chromosomes are numbered according to size, with the largest chromosomes having the smallest numbers. For example, human chromosomes 1, 2, and 3 are relatively large, whereas 21 and 22 are the two smallest. This numbering system does not apply to the **sex chromosomes,** which determine the sex of the individual. Sex chromosomes in humans are designated with the letters X and Y; females are XX and males are XY. The chromosomes that are not sex chromosomes are called **autosomes.** Humans have 22 different autosomes.

A common feature of many eukaryotic species is that most cells contain two sets of chromosomes. The karyotype shown in Figure 14.1 contains two sets of chromosomes, with 23 different chromosomes in each set. Therefore, this human cell contains a total of 46 chromosomes. Each cell has two sets because the individual inherited one set from the father and one set from the mother. When the cells of an organism carry two sets of chromosomes, that organism is said to be **diploid.** Geneticists use the letter $n$ to represent a set of chromosomes. Diploid organisms are referred to as $2n$, because they have two sets of chromosomes. For example, humans are $2n$, where $n = 23$. Most human cells are diploid. An exception involves **gametes,** the sperm and egg cells. Gametes are **haploid,** or $1n$, which means they contain one set of chromosomes.

**Homologous Pairs of Chromosomes** When an organism is diploid, the members of a pair of chromosomes are called **homologs** (see inset to Figure 14.1). The term **homology** refers to any similarity that is due to common ancestry. Pairs of homologous chromosomes are evolutionarily derived from the same chromosome. However, homologous chromosomes are not usually identical because over many generations they have accumulated some genetic changes that make them distinct.

How similar are homologous chromosomes to each other? The two chromosomes in a homologous pair are nearly identical in size and contain a very similar composition of genetic material. A particular gene found on one copy of a chromosome is also found on the homolog. However, the two homologs may vary in the way that a gene affects an organism's traits. As an example, let's consider a gene in humans called *OCA2*, which plays a major role in determining eye color. The *OCA2* gene is found on chromosome 15. One copy of chromosome 15 might carry the form of this eye color gene that confers brown eyes, whereas the gene on the homolog could confer blue eyes.

The DNA sequences on homologous chromosomes are very similar. In most cases, the sequence of bases on one homolog differs by less than 1% from the sequence on the other homolog. For example, the DNA sequence of chromosome 1 that you inherited from your mother is likely to be more than 99% identical to the DNA sequence of chromosome 1 that you inherited from your father. Nevertheless, keep in mind that the sequences are not identical. The slight differences in DNA sequence provide important variation in gene function. Again, if we use the eye color gene *OCA2* as

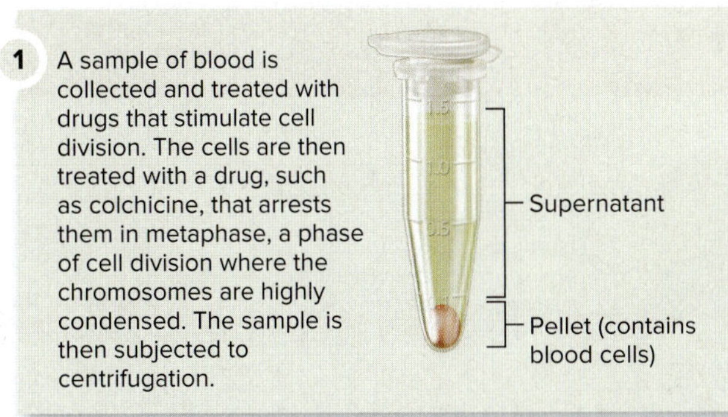

1
A sample of blood is collected and treated with drugs that stimulate cell division. The cells are then treated with a drug, such as colchicine, that arrests them in metaphase, a phase of cell division where the chromosomes are highly condensed. The sample is then subjected to centrifugation.
Supernatant
Pellet (contains blood cells)


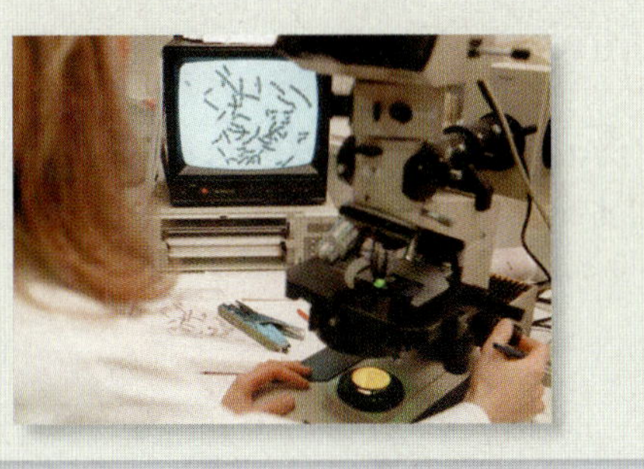

4
The slide is viewed by a light microscope equipped with a camera; the sample is seen on a computer screen. The chromosomes can be photographed and arranged electronically on the screen.


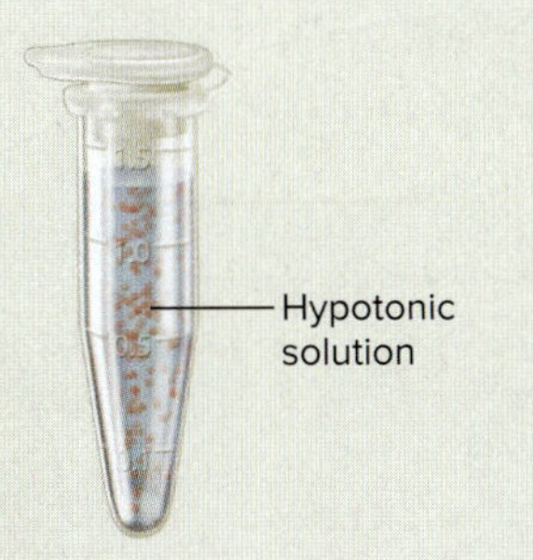

2
The supernatant is discarded, and the cell pellet is suspended in a hypotonic solution. This causes the cells to swell, which helps to spread out the chromosomes and also to burst the cells in the next step.
Hypotonic solution

3
The sample is subjected to centrifugation a second time to concentrate the cells. The cells are dropped onto a slide, where they will burst. The sample is treated with a fixative, which chemically freezes the chromosomes so they no longer move around. The chromosomes are then exposed to stains, such as Giemsa, which gives a distinctive banding pattern, called G bands.
Fix
Stain
Blood cells
5
A pair of sister chromatids
G band
Homologs
1 2 3 4 5
6 7 8 9 10 11 12
13 14 15 16 17 18
19 20 21 22 X Y
For a diploid human cell, 2 complete sets of chromosomes from a single cell constitute a karyotype of that cell.




**Figure 14.1** **The procedure for making a karyotype.** In this example, the chromosomes were treated with Giemsa stain, and the resulting dark bands are called G bands. The banding patterns make it much easier to distinguish the different chromosomes from each other.

*(4)* © Burger/SPL/Science Source; *(5)* Courtesy of the Genomic Centre for Cancer Research and Diagnosis, CancerCare Manitoba, University of Manitoba, Winnipeg, Manitoba, Canada

an example, a minor difference in DNA sequence distinguishes two forms of the gene, brown versus blue.

The striking similarity between homologous chromosomes does not apply to the sex chromosomes (for example, X and Y in humans). These chromosomes differ in size and genetic composition. Certain genes found on the X chromosome are not found on the Y chromosome, and vice versa. The X and Y chromosomes are not considered homologous chromosomes, although they do have short regions of homology.

## The Cell Cycle Is a Series of Phases That Lead to Cell Division

Eukaryotic cells that are destined to divide advance through the cell cycle, a series of changes that involves growth, replication, and division and ultimately produces new cells. Figure 14.2 provides an overview of the cell cycle. In this diagram, the mother cell has three pairs of chromosomes, for a total of six individual chromosomes. This cell is diploid ($2n$) and contains three chromosomes per set ($n = 3$). The paternal set is shown in blue, and the homologous maternal set is shown in red.

**Concept Check:** *Researchers usually treat cells with drugs that stimulate them to divide prior to the procedure for making a karyotype. Why is this useful?*

The phases of the cell cycle are

- **$G_1$ phase** (first gap),
- **S phase** (synthesis of DNA, the genetic material),
- **$G_2$ phase** (second gap), and
- **M phase** (mitosis and cytokinesis).

The $G_1$ and $G_2$ phases were originally described as gap phases to indicate the periods between DNA synthesis and mitosis. In actively

## Biology Principle

### Cells Are the Simplest Units of Life

The cells of eukaryotic species are made via cell division during the eukaryotic cell cycle.

(1) Prior to cell division, a mother cell in $G_1$ has 6 chromosomes, 2 sets of 3 each.

(2) During S phase, chromosome replication produces 6 pairs of sister chromatids.

(3) During $G_2$, the cell prepares to divide.

(4) At the beginning of mitosis, the nucleus breaks apart, and replicated chromosomes condense in preparation for mitosis.

(5) During mitosis, sister chromatids separate into two nuclei, and then 2 cells are formed during cytokinesis.

Two daughter cells form, each containing 6 chromosomes.

S
Interphase
$G_1$
$G_2$
M
Mitosis
Prophase
Prometaphase
Metaphase
Anaphase
Telophase
Cytokinesis

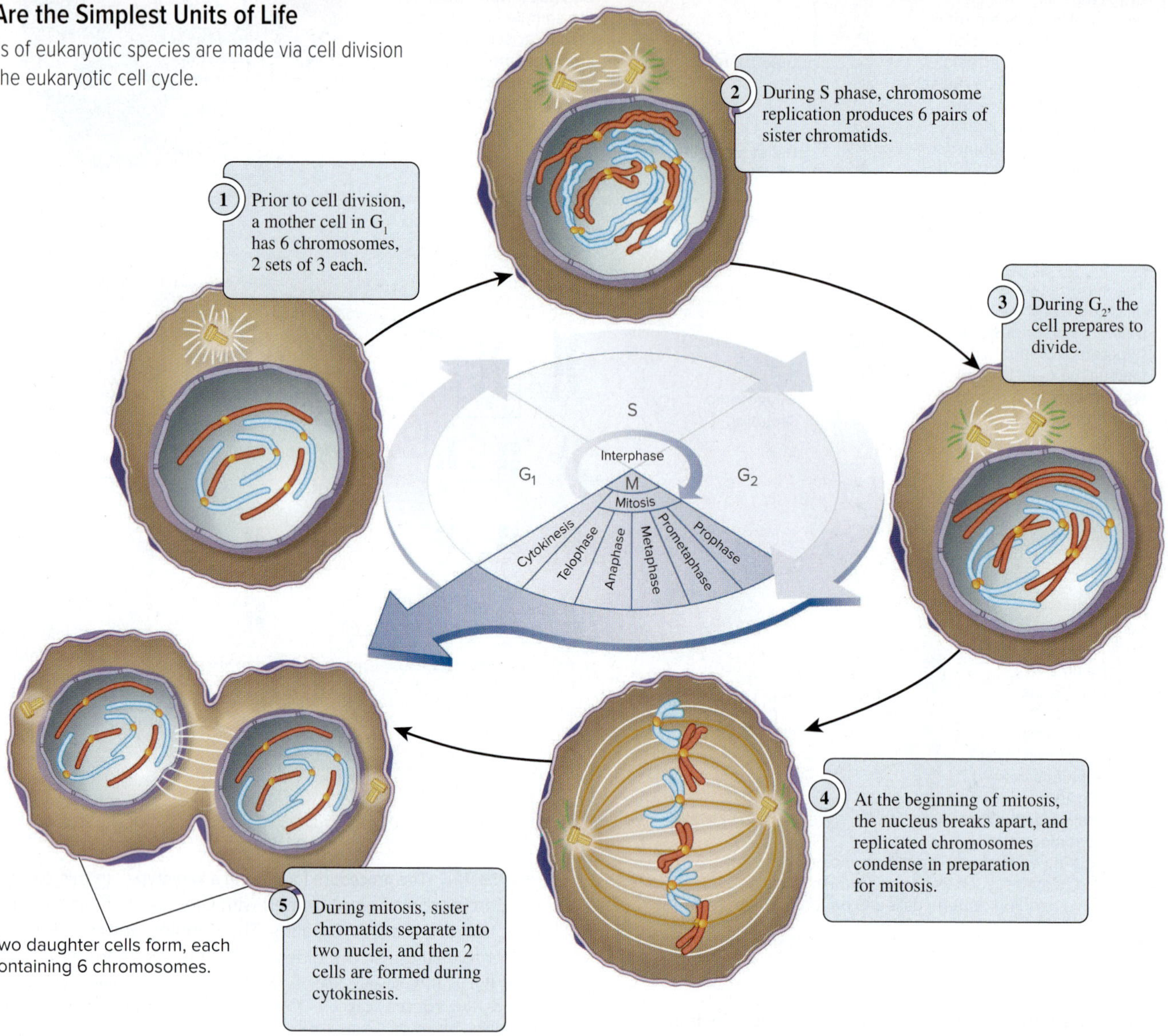

**Figure 14.2** **The eukaryotic cell cycle.** Dividing cells progress through a series of phases denoted $G_1$, S, $G_2$, and M. This diagram shows the progression of a cell through the cell cycle to produce two daughter cells. The original diploid cell had three pairs of chromosomes, for a total of six individual chromosomes. During S phase, these have replicated to yield 12 chromatids. After mitosis is completed, two daughter cells each contain six individual chromosomes. The width of the phases shown in this figure is not meant to reflect their actual length. $G_1$ is typically the longest phase of the cell cycle, whereas M phase is relatively short.

*Which phases make up interphase?*

dividing cells, the $G_1$, S, and $G_2$ phases are collectively known as **interphase.** During interphase, the cell grows and copies its chromosomes in preparation for cell division. Alternatively, cells may exit the cell cycle and remain for long periods of time in a phase called $\mathbf{G_0}$ (G zero). The $G_0$ phase is an alternative to proceeding through $G_1$. A cell in the $G_0$ phase has postponed making a decision to divide or a cell may become terminally differentiated, which means it will never divide again. $G_0$ is a nondividing phase.

**$G_1$ Phase** The $G_1$ phase is a period when a cell may become committed to divide. Depending on the environmental conditions and the presence of signaling molecules, a cell in the $G_1$ phase may accumulate

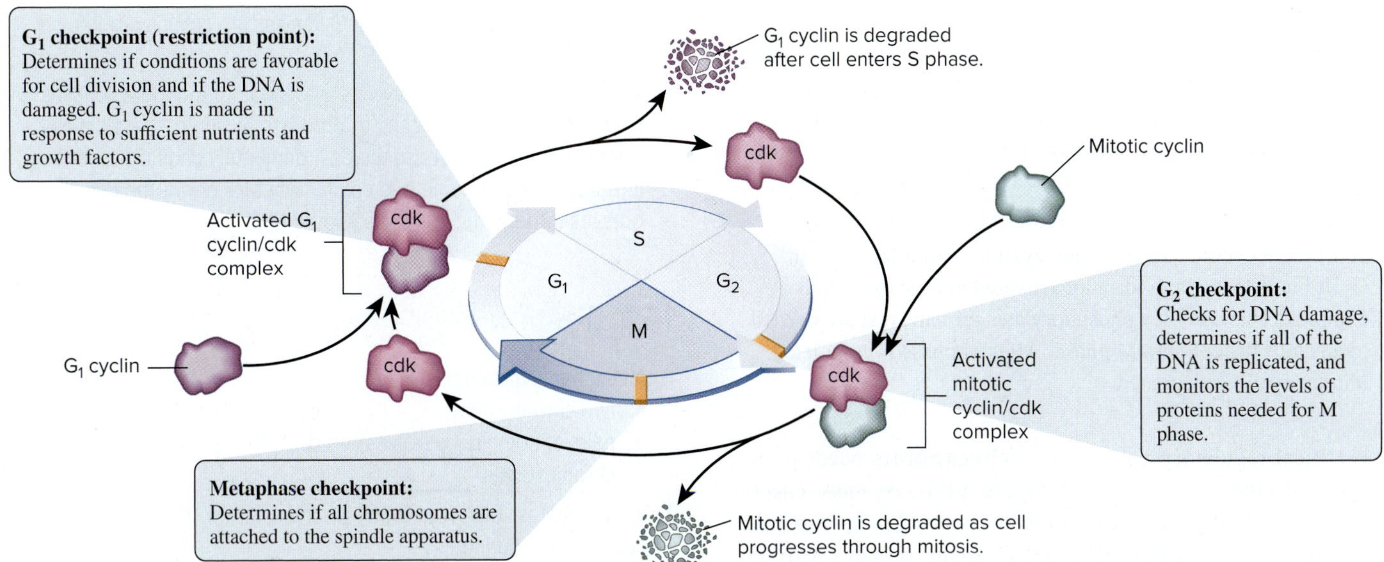


**Figure 14.3 Checkpoints in the cell cycle.** Advancement through the cell cycle requires the formation of activated cyclin/cdk complexes. Cells make different types of cyclin proteins, which are typically degraded after the cell has progressed to the next phase. The formation of activated cyclin/cdk complexes is regulated by checkpoint proteins.

*Refer back to Figure 13.13. How do checkpoint proteins prevent cancer?*

molecular changes that cause it to advance through the rest of the cell cycle. Cell growth typically occurs during the $G_1$ phase.

**S Phase** During the S phase, each chromosome is replicated to form a pair of sister chromatids (see Figure 14.1). When S phase is completed, a cell has twice as many chromatids as the number of chromosomes in the $G_1$ phase. For example, a human cell in the $G_1$ phase has 46 distinct chromosomes, whereas the same cell in $G_2$ has 46 pairs of sister chromatids, for a total of 92 chromatids.

**$G_2$ Phase** During the $G_2$ phase, a cell synthesizes the proteins necessary for chromosome sorting and cell division. Some cell growth may occur.

**M Phase** The first part of M phase is **mitosis,** in which one cell nucleus divides into two nuclei, distributing the duplicated chromosomes so each daughter cell receives the same complement of chromosomes. As noted previously, a human cell in the $G_2$ phase has 92 chromatids, which are found in 46 pairs. During mitosis, these pairs of chromatids are separated and sorted so each daughter cell receives 46 chromosomes. In most cases, mitosis is followed by **cytokinesis,** which is the division of the cytoplasm to produce two distinct daughter cells.

The length of the cell cycle varies considerably among different cell types, ranging from several minutes in quickly growing embryos to several months in slow-growing adult cells. For fast-dividing mammalian cells in adults, such as skin cells, the length of the cycle is often in the range of 10 to 24 hours. The various phases within the cell cycle also vary in length. $G_1$ is often the longest and the most variable phase, and M phase is the shortest. For a cell that divides in 24 hours, the following lengths of time for each phase are typical:

- $G_1$ phase: 11 hours
- S phase: 8 hours
- $G_2$ phase: 4 hours
- M phase: 1 hour

What factors determine whether or not a cell will divide? First, cell division is controlled by external factors, such as environmental conditions and signaling molecules. The effects of growth factors on cell division are discussed in Chapter 8 (refer back to Figure 8.10). Second, internal factors affect cell division. These include cell cycle control molecules and checkpoints, as discussed next.

## The Cell Cycle Is Controlled by Checkpoint Proteins

The advancement through the cell cycle is a process that is highly regulated to ensure that the chromosomes are intact and that the conditions are appropriate for a cell to divide. Proteins called **cyclins** and **cyclin-dependent kinases (cdks)** are responsible for advancing a cell through the phases of the cell cycle. Cyclins are so named because their amount is cyclic—their amounts rise and fall during the cell cycle. To be active, kinases controlling the cell cycle must bind to (are dependent on) a cyclin. The number of different types of cyclins and cdks varies from species to species.

Figure 14.3 shows a simplified description of how cyclins and cdks work together to advance a cell through $G_1$ and mitosis.

1. During $G_1$, the amount of a cyclin termed $G_1$ cyclin increases in response to sufficient nutrients and growth factors.

2. $G_1$ cyclin binds to a cdk to form an activated $G_1$ cyclin/cdk complex. Once activated, cdk functions as a protein kinase that phosphorylates other proteins needed to advance the cell to the next phase in the cell cycle. For example, certain proteins involved with DNA synthesis are phosphorylated and activated, thereby allowing the cell to replicate its DNA in S phase.
3. After the cell passes into the S phase, $G_1$ cyclin is degraded.
4. A different cyclin, called mitotic cyclin, accumulates late in $G_2$. It binds to a cdk to form an activated mitotic cyclin/cdk complex. This complex phosphorylates proteins that are needed to advance the cell into M phase. After M phase is completed, mitotic cyclin is degraded.

Three critical regulatory points called **checkpoints** occur in the cell cycle of eukaryotic cells (see Figure 14.3). At these checkpoints, a variety of proteins, referred to as checkpoint proteins, act as sensors to determine if a cell is in the proper condition to divide. The $G_1$ checkpoint monitors if conditions are favorable for cell division. In addition, $G_1$-checkpoint proteins can sense if the DNA has incurred damage. What happens if DNA damage is detected? The checkpoint proteins prevent the formation of active cyclin/cdk complexes, thereby stopping the advancement of the cell cycle.

A second checkpoint exists in $G_2$. This checkpoint also checks for DNA damage and ensures that all of the DNA has been replicated. In addition, the $G_2$ checkpoint monitors the levels of proteins that are needed to advance through M phase. A third checkpoint, called the metaphase checkpoint, senses the integrity of the spindle apparatus, which is involved in chromosome sorting and described later in this chapter. Metaphase is a step in mitosis during which all of the chromosomes should be attached to the spindle apparatus. If a chromosome is not correctly attached, the metaphase checkpoint will stop the cell cycle. This checkpoint prevents cells from incorrectly sorting their chromosomes during division.

Checkpoint proteins delay the cell cycle until problems are fixed or prevent cell division when problems cannot be fixed. They prevent the division of a cell that may have incurred DNA damage or that harbors abnormalities in chromosome number. As discussed in Chapter 13, when the functions of checkpoint genes are lost due to mutation, the likelihood increases that undesirable genetic changes will occur that can cause additional mutations and cancerous growth.

## 14.1 Reviewing the Concepts

- A micrograph that shows the alignment of chromosomes from a given cell is called a karyotype. Eukaryotic chromosomes are inherited in sets. A diploid cell has two sets of chromosomes. The members of each pair are called homologs (Figure 14.1).
- The eukaryotic cell cycle consists of four phases called $G_1$ (first gap), S (synthesis of DNA), $G_2$ (second gap), and M phase (mitosis and cytokinesis). The $G_1$, S, and $G_2$ phases are collectively known as interphase (Figure 14.2).
- An interaction between cyclins and cyclin-dependent kinases (cdks) is necessary for cells to progress through the cell cycle. Checkpoint proteins sense the environmental conditions and the integrity of the genome and control whether or not the cell progresses through the cell cycle (Figure 14.3).

## 14.1 Testing Your Knowledge

1. A cell that is diploid has
   a. two chromosomes.
   b. pairs of homologous chromosomes.
   c. two sets of chromosomes.
   d. both b and c.
2. In eukaryotes, DNA replication produces sister chromatids. This event occurs during
   a. $G_1$ phase.
   b. S phase.
   c. $G_2$ phase.
   d. M phase.
3. A protein that drives a cell through the cell cycle is a _______. It is directly controlled by a _______.
   a. cyclin, cyclin-dependent kinase
   b. checkpoint protein, cyclin
   c. checkpoint protein, cyclin-dependent kinase
   d. cyclin-dependent kinase, cyclin

# 14.2 Mitotic Cell Division

### Learning Outcomes

1. Explain how the replication of eukaryotic chromosomes produces sister chromatids.
2. Describe the structure and function of the mitotic spindle.
3. Outline the key events that occur during the phases of mitosis.

We now turn our attention to the process of **mitotic cell division,** during which a cell divides to produce two new cells (the daughter cells) that are genetically identical to the original cell (the mother cell). Mitotic cell division begins with mitosis—the division of one nucleus into two nuclei—and is followed by cytokinesis, in which the mother cell divides into two daughter cells.

Why is mitotic cell division important? One purpose is **asexual reproduction,** a process in which genetically identical offspring are produced from a single parent. Certain unicellular eukaryotic organisms, such as baker's yeast (*Saccharomyces cerevisiae*) and the amoeba, increase their numbers in this manner. A second important reason for mitotic cell division is multicellularity. Organisms such as plants, animals, and most fungi are derived from a single cell that subsequently undergoes repeated cellular divisions to become a multicellular organism. As discussed at the beginning of the chapter, adult humans are composed of trillions of cells.

## Biology Principle

### The Genetic Material Provides a Blueprint for Reproduction

The process of mitosis ensures that each daughter cell receives a complete copy of the genetic material.

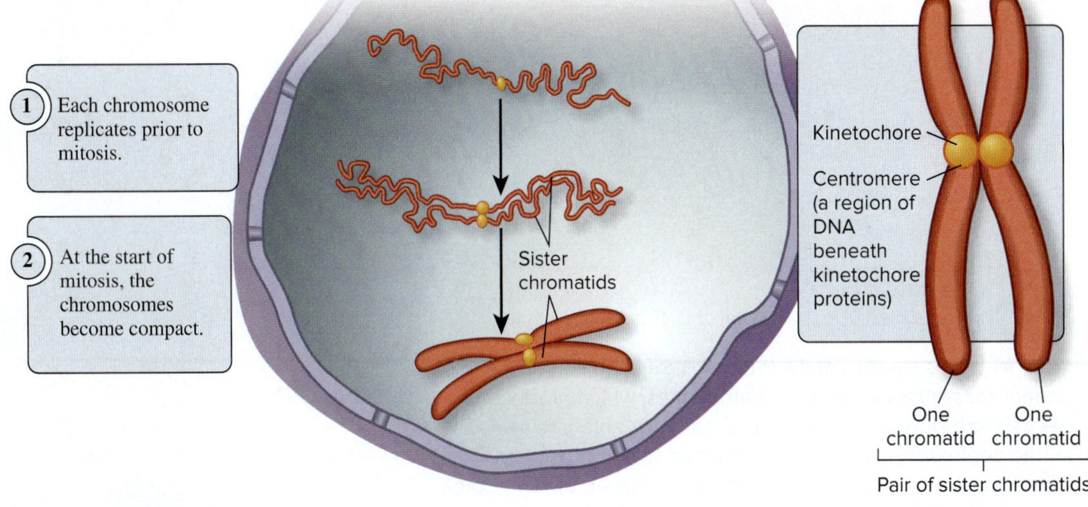


**Figure 14.4** **Replication and compaction of chromosomes into pairs of sister chromatids.** **(a)** Chromosomal replication produces a pair of sister chromatids. While the chromosomes are elongated, they are replicated during S phase to produce two copies that are connected and lie parallel to each other. This is a pair of sister chromatids. At the start of mitosis, the sister chromatids condense into more compact structures that are easily seen with a light microscope. **(b)** A schematic drawing of a metaphase chromosome. This structure has two chromatids that lie side-by-side. The two chromatids are held together by cohesin proteins (not shown in this drawing). The kinetochore is a group of proteins that are attached to the centromere and play a role during chromosome sorting.

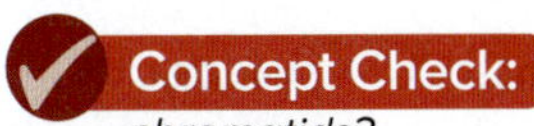
**Concept Check:** *Refer back to the karyotype in Figure 14.1. In this micrograph, is each of the 46 objects a pair of sister chromatids?*

In this section, we will explore how the process of mitotic cell division requires the replication, organization, and sorting of chromosomes. We will also examine how a single cell is separated into two distinct cells by cytokinesis.

### In Preparation for Cell Division, Eukaryotic Chromosomes Are Replicated and Compacted to Produce Pairs Called Sister Chromatids

We will now consider how chromosomes are replicated and sorted during cell division. In Chapter 9, we examined the molecular process of DNA replication. Figure 14.4 describes the process at the chromosomal level. Prior to DNA replication, the DNA of each eukaryotic chromosome consists of a linear DNA double helix that is found in the nucleus and is not highly compacted. When the DNA is replicated, two identical copies of the original double helix are produced. As discussed earlier, these copies, along with associated proteins, lie side-by-side and are termed sister chromatids. When a cell prepares to divide, the sister chromatids become highly compacted and readily visible under the microscope. As shown in Figure 14.4b, the two sister chromatids are tightly associated at a region called the **centromere.** In addition, the centromere serves as an attachment site for a group of proteins that form the **kinetochore,** a structure necessary for sorting each chromosome.

### The Mitotic Spindle Organizes and Sorts Chromosomes During Cell Division

What structure organizes and sorts the chromosomes during cell division? The answer is the **mitotic spindle** (Figure 14.5). It is composed of microtubules—long, cylindrical protein fibers that are components of the cytoskeleton (refer back to Table 4.1). In animal cells, microtubule growth and organization start at two **centrosomes,** which are also referred to as microtubule organizing centers. A single centrosome duplicates during interphase. After they separate from each other during mitosis, each centrosome defines a **pole** of the spindle apparatus, one within each of the future daughter cells. The centrosome in animal cells has a pair of **centrioles,** which are arranged perpendicular to each other. However, centrioles are not found in many other eukaryotic species, such as most plants, and are not required for spindle formation.

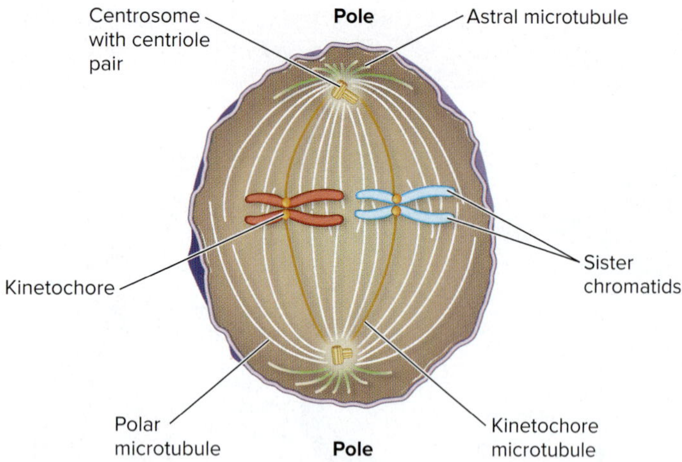


**Figure 14.5 The structure of the mitotic spindle.** The mitotic spindle in animal cells is formed by the centrosomes, which produce three types of microtubules. The astral microtubules emanate away from the centrosome and are not attached to chromosomes. The polar microtubules project into the region between the two poles. The kinetochore microtubules are attached to the kinetochores of sister chromatids. Note: For simplicity, this diagram shows only one pair of homologous chromosomes. Eukaryotic species typically have multiple chromosomes per set.

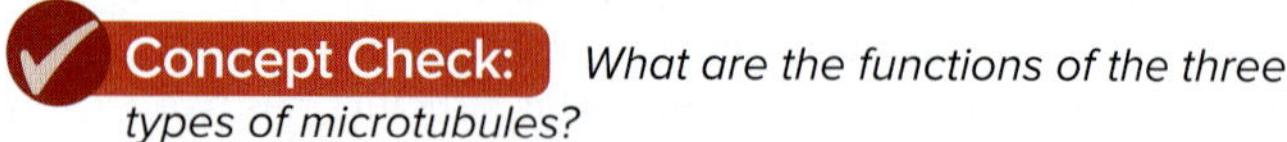


*What are the functions of the three types of microtubules?*

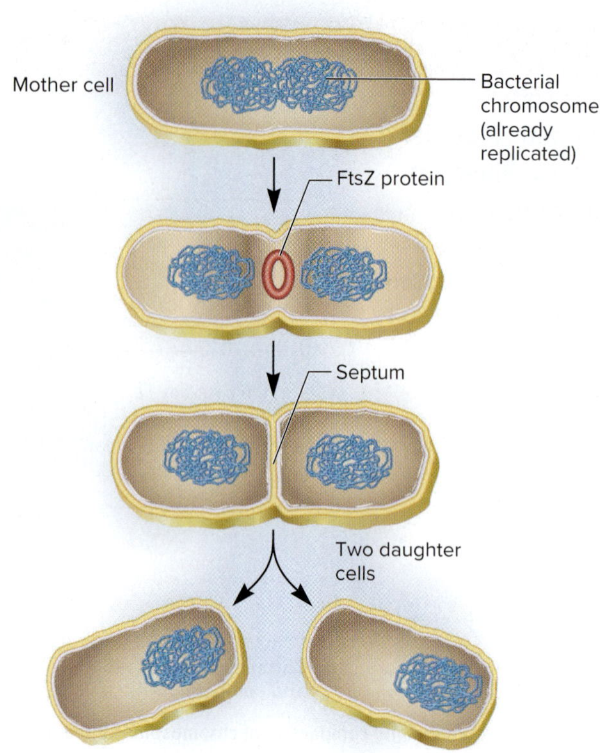


**Figure 14.6 Binary fission in bacteria.** The FtsZ protein accumulates in a ring at the future site of cell division.

Each centrosome organizes the construction of the microtubules by rapidly polymerizing tubulin proteins. The three types of spindle microtubules are termed astral, polar, and kinetochore microtubules (see Figure 14.5). The astral microtubules, which are not attached to chromosomes, are important for positioning the spindle apparatus within the cell. The polar microtubules project into the region between the two poles. Polar microtubules that overlap with each other play a role in the separation of the two poles. Finally, the kinetochore microtubules are attached to kinetochores, which are bound to the centromere of each chromosome.

## EVOLUTIONARY CONNECTIONS

### Cell Division in Bacteria Involves FtsZ, a Protein Related to Eukaryotic Tubulin

As discussed in Chapter 17 (look ahead to Figure 17.8), bacteria divide by a process called binary fission. Because bacteria usually have only one type of chromosome, the process of sorting different types of chromosomes is not necessary. Even so, events during bacterial cell division may provide insights into the manner in which mitosis evolved in eukaryotes.

Prior to cell division, bacterial cells copy, or replicate, their chromosomal DNA. This produces two identical copies of the genetic material, as shown at the top of **Figure 14.6**. During binary fission, the two daughter cells become separated from each other by the formation of a septum. Recent evidence has shown that bacterial species produce a protein called FtsZ, which is important in cell division. This protein assembles into a ring at the future site of the septum. FtsZ is thought to be the first protein to move to this division site, and it recruits other proteins that produce a new cell wall between the daughter cells.

FtsZ is evolutionarily related to the eukaryotic protein called tubulin, the main component of microtubules that are part of the mitotic spindle. In all eukaryotes, the midpoint of the mitotic spindle, which is called the metaphase plate, identifies the site for cytokinesis (look ahead to Figure 14.7d and f). This observation indicates that tubulin is also critical for cytokinesis in eukaryotic cells.

## The Transmission of Chromosomes in Eukaryotes Requires a Sorting Process Known as Mitosis

During mitosis in eukaryotes, the duplicated chromosomes are distributed so each daughter cell receives the same complement of chromosomes. **Figure 14.7** (pp. 286–287) depicts the process of mitosis in an animal cell, though the process is quite similar in a plant cell. Mitosis occurs as a continuum of phases known as prophase, prometaphase, metaphase, anaphase, and telophase. In the simplified diagrams shown along the bottom of Figure 14.7, the original cell has six chromosomes. One set of chromosomes is depicted in red, and the homologous set is blue. These different colors represent maternal and paternal chromosomes, respectively.

**Interphase** Prior to mitosis, the cells are in interphase, which consists of the $G_1$, S, and $G_2$ phases of the cell cycle. The chromosomes, which are located in the cell nucleus, have replicated in S phase and are not condensed (Figure 14.7a). The nucleolus, which is the site where the components of ribosomes assemble into ribosomal subunits, is visible during interphase.

**Prophase** At the start of mitosis, in **prophase,** the chromosomes have already replicated to produce 12 chromatids, joined as 6 pairs of sister chromatids that have condensed into highly compacted structures readily visible by light microscopy (Figure 14.7b). As prophase proceeds, the nuclear envelope begins to dissociate into small vesicles. The nucleolus is no longer visible.

**Prometaphase** During **prometaphase,** the nuclear envelope completely fragments into small vesicles, and the mitotic spindle is fully formed (Figure 14.7c). As prometaphase advances, the centrosomes move apart and demarcate the two poles. Once the nuclear envelope has dissociated, the spindle fibers can interact with the sister chromatids. How do the sister chromatids become attached to the spindle apparatus? Initially, microtubules are rapidly formed and can be seen under a microscope growing out from the two poles. As it grows, if a microtubule happens to make contact with a kinetochore, it is said to be "captured" and remains firmly attached to the kinetochore. Alternatively, if a microtubule does not collide with a kinetochore, the microtubule eventually depolymerizes and retracts to the centrosome. This random process is how sister chromatids become attached to kinetochore microtubules. As the end of prometaphase nears, the two kinetochores on each pair of sister chromatids are attached to kinetochore microtubules from opposite poles. As these events are occurring, the sister chromatids are seen under the microscope to undergo jerky movements as they are tugged, back and forth, between the two poles by the kinetochore microtubules.

**Metaphase** Eventually, the pairs of sister chromatids are aligned in a single row along the **metaphase plate,** a plane halfway between the poles. When this alignment is complete, the cell is in **metaphase** of mitosis (Figure 14.7d).

**Anaphase** During **anaphase,** the connections between the pairs of sister chromatids are broken (Figure 14.7e). Each chromatid, now an individual chromosome, is linked to only one of the two poles by one or more kinetochore microtubules. As anaphase proceeds, the kinetochore microtubules shorten, pulling the chromosomes toward the pole to which they are attached. In addition, the two poles move farther away from each other. This occurs because the overlapping polar microtubules lengthen and push against each other, thereby pushing the poles farther apart.

**Telophase** During **telophase,** the chromosomes have reached their respective poles and decondense. The nuclear envelope now re-forms to produce two separate nuclei. In Figure 14.7f, two nuclei are being produced that contain six chromosomes each.

**Cytokinesis** In most cases, mitosis is quickly followed by cytokinesis, in which the two nuclei are segregated into separate daughter cells. Whereas the phases of mitosis are similar between plant and animal cells, the process of cytokinesis is quite different. In animal cells, cytokinesis involves the formation of a **cleavage furrow,** which constricts like a drawstring to separate the cells (**Figure 14.8a**). In plants, vesicles from the Golgi apparatus move along microtubules to the center of the cell and coalesce to form a **cell plate** (**Figure 14.8b**), which then forms a cell wall between the two daughter cells.

What are the results of mitosis and cytokinesis? These processes ultimately produce two daughter cells with the same number of chromosomes as the mother cell. Barring rare mutations, the two daughter cells are genetically identical to each other and to the mother cell from which they were derived. The critical consequence of this sorting process is to ensure genetic consistency from one cell to the next. The development of multicellularity relies on the repeated process of mitosis and cytokinesis.

## 14.2 Reviewing the Concepts

- In the process of mitotic cell division, a cell divides to produce two new cells (the daughter cells) that are genetically identical to the original cell.
- During S phase, eukaryotic chromosomes are replicated to produce a pair of identical sister chromatids that remain attached to each other (Figure 14.4).
- The mitotic spindle is a network of microtubules that plays a central role in chromosome sorting during cell division (Figure 14.5).
- FtsZ in bacteria and tubulin in eukaryotes are homologous proteins that are important in identifying the site where cell division will occur (Figure 14.6).
- During mitosis, the nucleus of a mother cell divides into two genetically identical nuclei. Mitosis occurs in five phases called prophase, prometaphase, metaphase, anaphase, and telophase (Figure 14.7).
- Cytokinesis occurs by a cleavage furrow in animal cells and by the formation of a cell plate in plant cells (Figure 14.8).

## 14.2 Testing Your Knowledge

1. Let's suppose a cell is diploid, with eight chromosomes per set. How many chromatids would be found in the cell after S phase, but prior to cytokinesis?
   **a.** 8 **b.** 16 **c.** 32 **d.** 64
2. Sister chromatids become attached to spindle fibers during
   **a.** prophase. **c.** metaphase. **e.** telophase.
   **b.** prometaphase. **d.** anaphase.
3. Sister chromatids separate from each other during
   **a.** prophase. **c.** metaphase. **e.** telophase.
   **b.** prometaphase. **d.** anaphase.

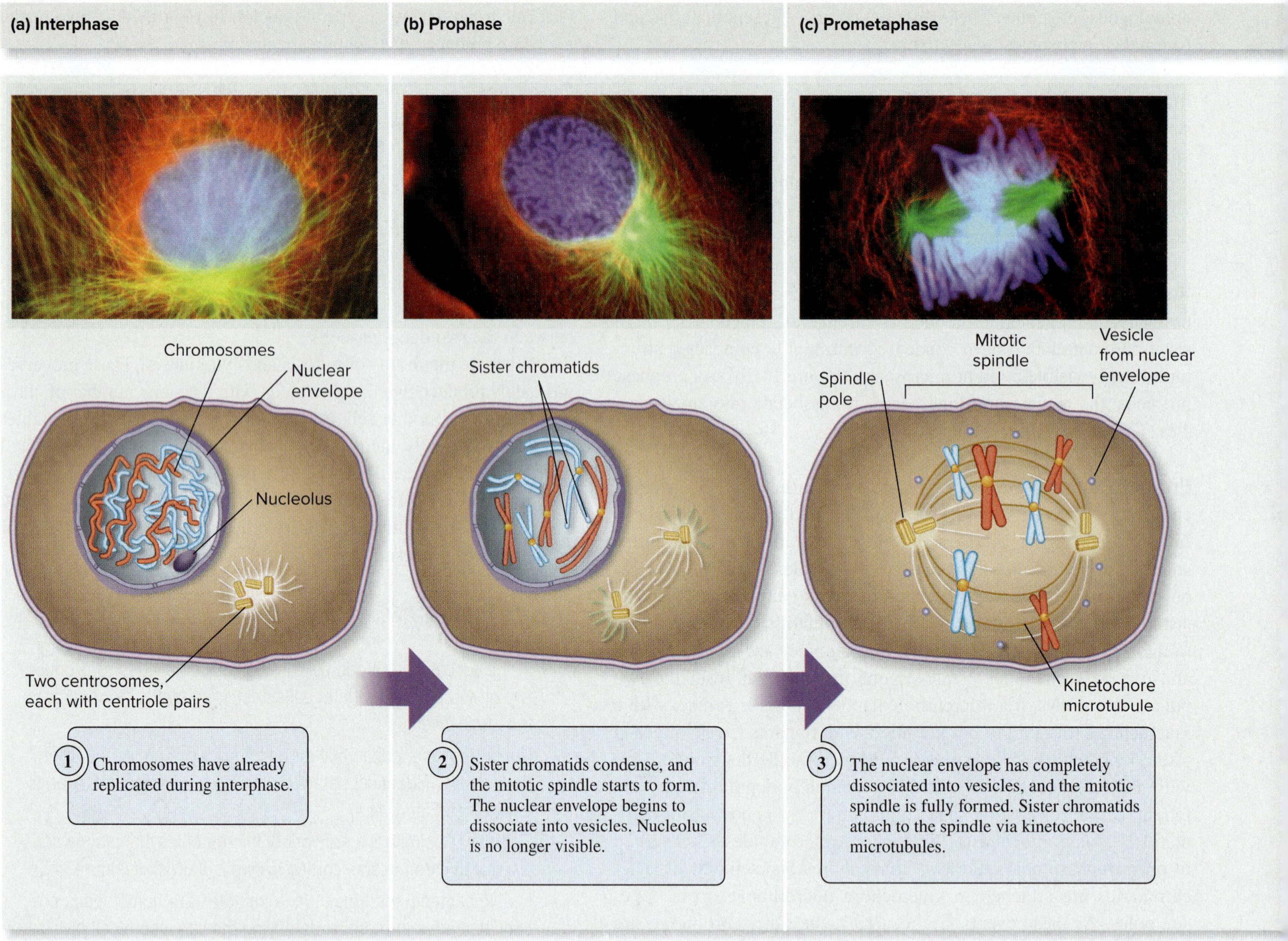

**Figure 14.7 The process of mitosis in an animal cell.** The top panels illustrate a newt cell progressing through mitosis. The bottom panels are schematic drawings that emphasize the sorting and separation of the chromosomes in which the diploid mother cell had six chromosomes (three in each set). At the start of mitosis, these have already replicated into 12 chromatids. The final result is two daughter cells, each containing six chromosomes.

Photographs by Dr. Conly L. Rieder, Wadsworth Center, Albany, New York 12201-0509

*With regard to chromosome composition, how does the mother cell compare with the two daughter cells?*

## 14.3 Meiosis and Sexual Reproduction

### Learning Outcomes

1. Describe the processes of synapsis and crossing over.
2. Outline the key events that occur during the phases of meiosis.
3. Compare and contrast mitosis and meiosis, focusing on key steps that account for the different outcomes of these two processes.
4. Distinguish between the life cycles of diploid-dominant species, haploid-dominant species, and species that exhibit an alternation of generations.

We now turn our attention to **sexual reproduction,** a process in which two haploid gametes unite to form a diploid cell called a **zygote.** For multicellular species such as animals and plants, the zygote then grows and divides by mitotic cell divisions into a multicellular organism with many diploid cells.

As discussed earlier, a diploid cell contains two homologous sets of chromosomes, whereas a haploid cell contains a single set. For example, a diploid human cell contains 46 chromosomes, but a human gamete—sperm or egg cell—is a haploid cell that contains only 23 chromosomes. **Meiosis** is the process by which haploid cells are produced from a cell that was originally diploid. The term meiosis, which means to make smaller, refers to the fewer chromosomes found in cells following this process. For this to occur, the

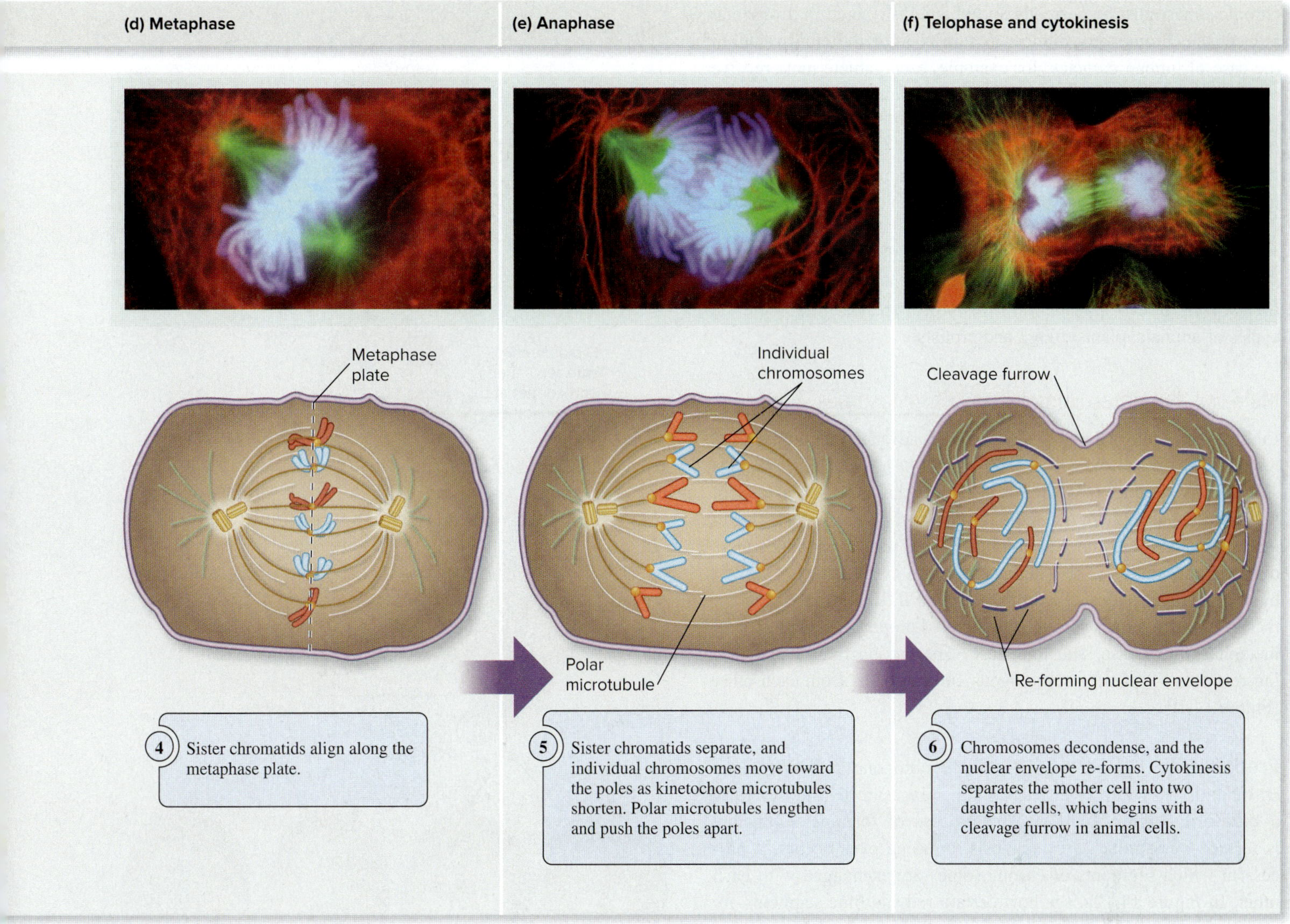

(d) Metaphase
(e) Anaphase
(f) Telophase and cytokinesis
Metaphase plate
Individual chromosomes
Cleavage furrow
Polar microtubule
Re-forming nuclear envelope
4 Sister chromatids align along the metaphase plate.
5 Sister chromatids separate, and individual chromosomes move toward the poles as kinetochore microtubules shorten. Polar microtubules lengthen and push the poles apart.
6 Chromosomes decondense, and the nuclear envelope re-forms. Cytokinesis separates the mother cell into two daughter cells, which begins with a cleavage furrow in animal cells.

Cleavage furrow
150 μm
S
$G_1$
$G_2$
Cytokinesis
Cell plate
10 μm

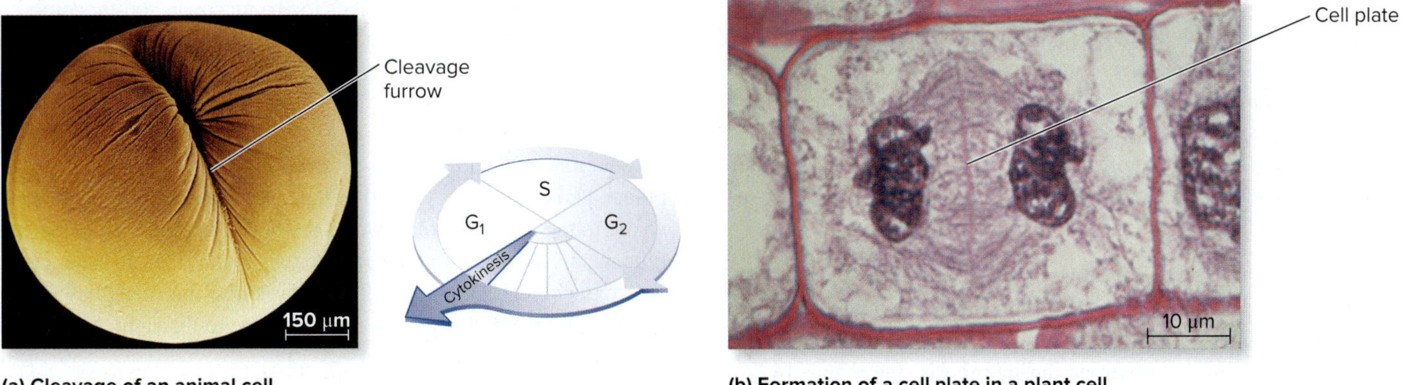

**(a) Cleavage of an animal cell**

**(b) Formation of a cell plate in a plant cell**

**Figure 14.8** Micrographs showing cytokinesis in animal and plant cells.

*(a)* © Don W. Fawcett/Science Source; *(b)* © Carolina Biological Supply Company/Phototake

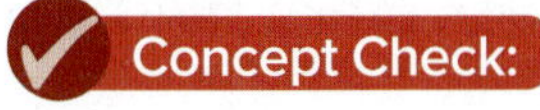

**Concept Check:** *What are the similarities and differences between cytokinesis in animal and plant cells?*

chromosomes must be correctly sorted and distributed in a way that reduces the chromosome number to half its original diploid value. In the case of human gametes, for example, each gamete must receive one chromosome from each of the 23 pairs. For this to happen, two rounds of divisions are necessary, termed meiosis I and meiosis II (Figure 14.9). When a cell begins meiosis, it has chromosomes that are found in homologous pairs. When meiosis is completed, a single diploid cell with homologous pairs of chromosomes has produced four haploid cells.

In this section, we will examine the cellular events of meiosis that reduce the chromosome number from diploid to haploid. In addition, we will briefly consider how this process plays a role in the life cycles of animals, plants, fungi, and protists.

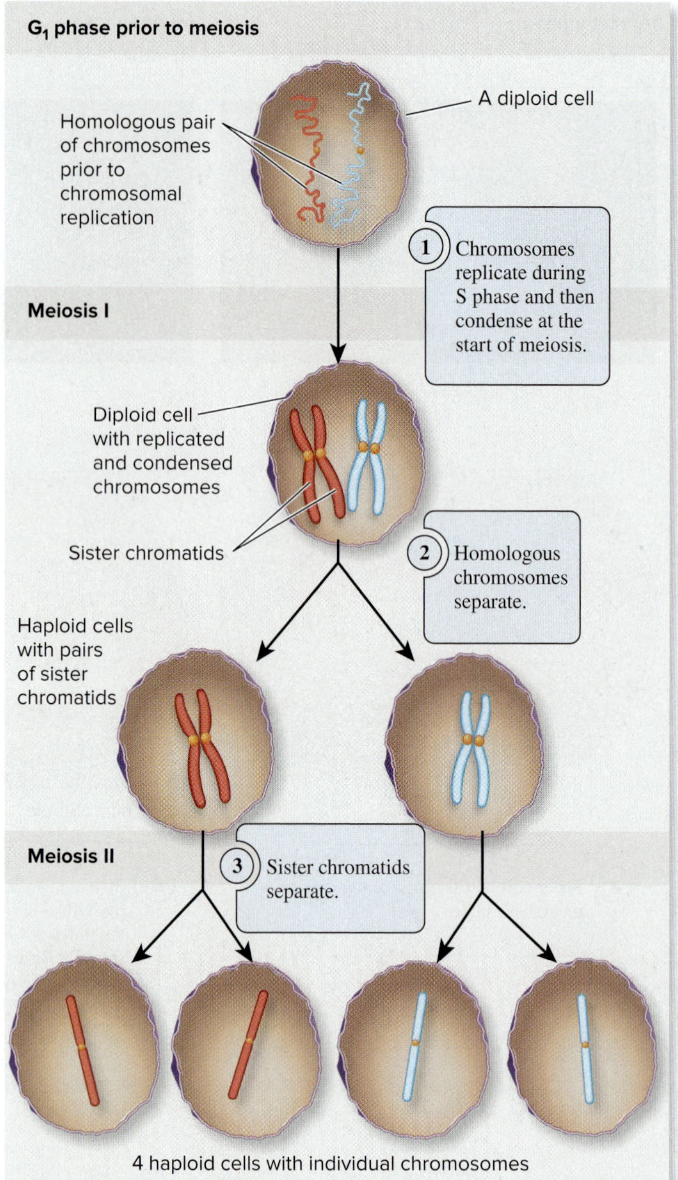


**Figure 14.9** How the process of meiosis reduces chromosome number. This simplified diagram emphasizes the reduction in chromosome number as a diploid cell divides by meiosis to produce four haploid cells.

## Meiosis I Separates Homologous Chromosomes

Like mitosis, meiosis begins after a cell has advanced through the $G_1$, S, and $G_2$ phases of the cell cycle. Figure 14.10 (pp. 290–291) depicts simplified diagrams of a diploid cell ($2n$) that contains a total of six chromosomes (as in our look at mitosis in Figure 14.7). Prior to meiosis, the chromosomes are replicated in S phase to produce pairs of sister chromatids. This single replication event is then followed by sequential divisions called meiosis I and II. Like mitosis, each of these is a continuous series of stages called prophase, prometaphase, metaphase, anaphase, and telophase. The sorting that occurs during **meiosis I** separates homologous chromosomes from each other (Figure 14.10a–e).

**Prophase I** During prophase I, homologous pairs of sister chromatids associate with each other, lying side by side to form a **bivalent** (Figure 14.10a). The process of forming a bivalent is termed **synapsis.** After synapsis, crossing over occurs, a process in which segments of chromosomes are exchanged with each other. In Figure 14.10a, see how certain red and blue segments are exchanged at a crossover site. As discussed in Chapter 16, crossing over increases the genetic variation of sexually reproducing species. During prophase I, the nuclear envelope starts to fragment into small vesicles.

**Prometaphase I** In prometaphase I, the nuclear envelope is completely broken down into vesicles, and the spindle apparatus is entirely formed. The sister chromatids become attached to kinetochore microtubules. However, a key difference occurs between mitosis and meiosis I. In mitosis, a pair of sister chromatids is attached to both poles via kinetochore microtubules (see Figure 14.7c). In meiosis I, a pair of sister chromatids is attached to just one pole (see Figure 14.10b).

**Metaphase I** At metaphase I, the bivalents are organized along the metaphase plate. Notice how this pattern of alignment is strikingly different from that observed during mitosis (see Figure 14.7d). In particular, the sister chromatids are aligned in a double row rather than a single row (as in mitosis). Furthermore, the arrangement of sister chromatids within this double row is random with regard to the (red and blue) homologs. (Remember that these different colors represent maternal and paternal chromosomes.) In Figure 14.10c, one of the red homologs is to the left of the metaphase plate, and the other two are to the right, whereas two of the blue homologs are to the left of the metaphase plate and the other one is to the right. In other cells, homologs could be arranged differently along the metaphase plate (for example, three blues to the left and none to the right, or none to the left and three to the right).

**Anaphase I** The segregation of homologs occurs during anaphase I (see Figure 14.10d). The connections between bivalents break, but not the connections that hold sister chromatids together. Each joined pair of chromatids migrates to one pole, and the homologous pair of chromatids moves to the opposite pole, both pulled by kinetochore microtubules.

**Telophase I** At telophase I (see Figure 14.10e), the sister chromatids have reached their respective poles and then decondense. The nuclear envelope now re-forms to produce two separate nuclei.

If we consider the end result of meiosis I, we see that two nuclei are produced, each with three pairs of sister chromatids; this is called a reduction division. The original diploid cell had its chromosomes in homologous pairs, whereas the two cells produced as a result of meiosis I and cytokinesis are considered haploid—they do not have pairs of homologous chromosomes.

## Meiosis II Separates Sister Chromatids

Meiosis I is followed by cytokinesis and then **meiosis II** (see Figure 14.10f–j). DNA replication does not occur between meiosis I and meiosis II. The sorting events of meiosis II are similar to those of mitosis, but the starting point is different. For a diploid cell with six chromosomes, mitosis begins with 12 chromatids that are joined as six pairs of sister chromatids (see Figure 14.7). By comparison, the two cells that begin meiosis II each have six chromatids that are joined as three pairs of sister chromatids. Otherwise, the steps that occur during prophase, prometaphase, metaphase, anaphase, and telophase of meiosis II are analogous to those in a mitotic division. Sister chromatids are separated during anaphase II, unlike anaphase I, in which bivalents are separated.

## Mitosis and Meiosis Differ in a Few Key Steps

How are the outcomes of mitosis and meiosis different from each other? Mitosis produces two diploid daughter cells that are genetically identical. In our example shown in Figure 14.7, the starting cell had six chromosomes (three homologous pairs of chromosomes), and both daughter cells received copies of the same six chromosomes. By comparison, meiosis reduces the number of sets of chromosomes. In the example shown in Figure 14.10, the starting cell also had six chromosomes, whereas the resulting four daughter cells have only three chromosomes. However, the daughter cells do not contain a random mix of three chromosomes. Each haploid daughter cell has one complete set of chromosomes, whereas the original diploid mother cell had two complete sets.

How do we explain the different outcomes of mitosis and meiosis? **Table 14.1** emphasizes the differences between certain key steps in mitosis and meiosis that account for the different outcomes of these two processes:

- DNA replication occurs prior to mitosis and meiosis I, but not between meiosis I and II.
- During prophase of meiosis I, the homologs synapse to form bivalents. This explains why crossing over occurs commonly during meiosis, but rarely during mitosis.
- During prometaphase of mitosis and meiosis II, pairs of sister chromatids are attached to both poles. In contrast, during meiosis I, each pair of sister chromatids (within a bivalent) is attached to a single pole.
- Bivalents align along the metaphase plate during metaphase of meiosis I, whereas sister chromatids align along the metaphase plate during metaphase of mitosis and meiosis II.
- At anaphase of meiosis I, the homologous chromosomes separate, but the sister chromatids remain together. In contrast, sister chromatid separation occurs during anaphase of mitosis and meiosis II.

Taken together, the steps of mitosis produce two diploid cells that are genetically identical, whereas the steps of meiosis involve two sequential cell divisions that produce four haploid cells that may not be genetically identical.

**Table 14.1 A Comparison of Mitosis, Meiosis I, and Meiosis II**

| Event | Mitosis | Meiosis I | Meiosis II |
|---|---|---|---|
| DNA replication: | Occurs prior to mitosis | Occurs prior to meiosis I | Does not occur between meiosis I and II |
| Synapsis during prophase: | No | Yes, bivalents are formed. | No |
| Crossing over during prophase: | Rarely | Commonly | Rarely |
| Attachment to poles at prometaphase: | A pair of sister chromatids is attached to kinetochore microtubules from both poles. | A pair of sister chromatids is attached to kinetochore microtubules from just one pole. | A pair of sister chromatids is attached to kinetochore microtubules from both poles. |
| Alignment along the metaphase plate: | Sister chromatids align. | Bivalents align. | Sister chromatids align. |
| Type of separation at anaphase: | Sister chromatids separate. A single chromatid, now called a chromosome, moves to each pole. | Homologous chromosomes separate. A pair of sister chromatids moves to each pole. | Sister chromatids separate. A single chromatid, now called a chromosome, moves to each pole. |
| End result when the mother cell is diploid: | Two daughter cells that are diploid | — | Four daughter cells that are haploid |

## Meiosis I

**(a) Prophase I** **(b) Prometaphase I** **(c) Metaphase I**

Bivalent
Crossovers
Sister chromatids
Spindle forming
Centrosome
Bivalent
Metaphase plate

(1) Homologous chromosomes synapse to form bivalents, and crossing over occurs. Chromosomes condense, and the nuclear envelope begins to dissociate into vesicles.

(2) The nuclear envelope completely dissociates into vesicles, and bivalents become attached to kinetochore microtubules.

(3) Bivalents randomly align along the metaphase plate. Each pair of sister chromatids is attached to one pole.

## Meiosis II

**(f) Prophase II** **(g) Prometaphase II** **(h) Metaphase II**

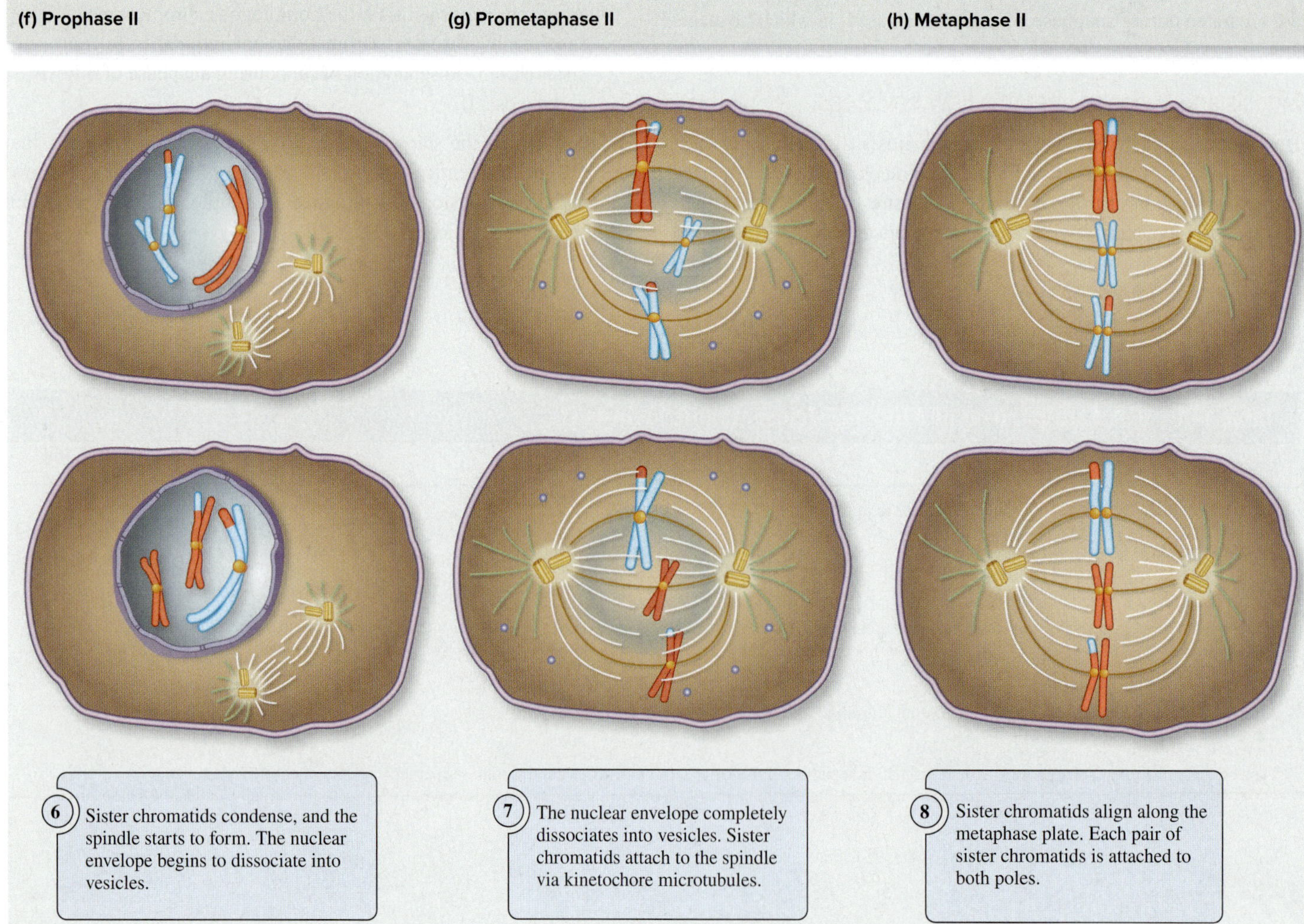

(6) Sister chromatids condense, and the spindle starts to form. The nuclear envelope begins to dissociate into vesicles.

(7) The nuclear envelope completely dissociates into vesicles. Sister chromatids attach to the spindle via kinetochore microtubules.

(8) Sister chromatids align along the metaphase plate. Each pair of sister chromatids is attached to both poles.

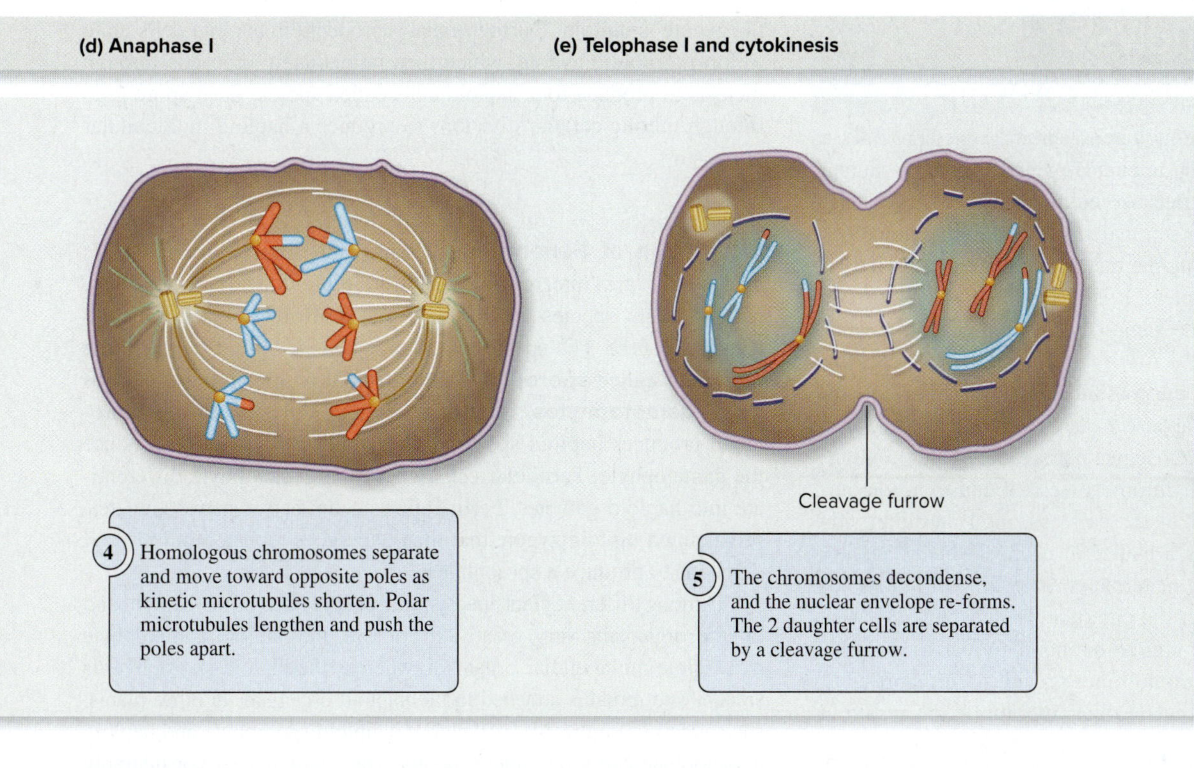

(d) Anaphase I
(e) Telophase I and cytokinesis
Cleavage furrow
4 Homologous chromosomes separate and move toward opposite poles as kinetic microtubules shorten. Polar microtubules lengthen and push the poles apart.
5 The chromosomes decondense, and the nuclear envelope re-forms. The 2 daughter cells are separated by a cleavage furrow.


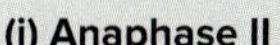

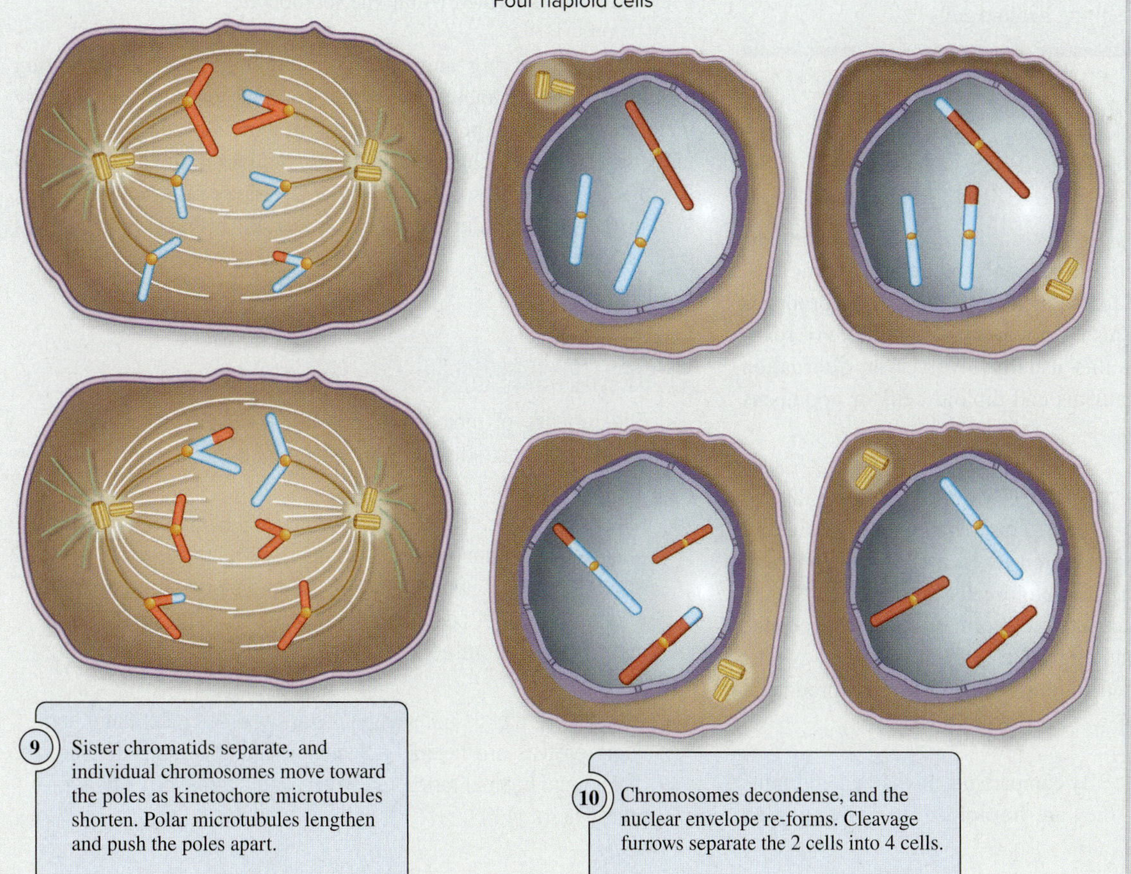

(i) Anaphase II
(j) Telophase II and cytokinesis
Four haploid cells
9 Sister chromatids separate, and individual chromosomes move toward the poles as kinetochore microtubules shorten. Polar microtubules lengthen and push the poles apart.
10 Chromosomes decondense, and the nuclear envelope re-forms. Cleavage furrows separate the 2 cells into 4 cells.


**Figure 14.10** The phases of meiosis in an animal cell.

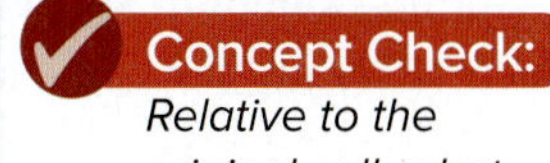


*Relative to the original cell, what is the end result of meiosis?*

## Quantitative Analysis

### MEIOSIS ENHANCES GENETIC DIVERSITY

The random alignment of homologous chromosomes provides a mechanism to promote a vast amount of genetic diversity among the resulting haploid cells. Because eukaryotic species typically have many chromosomes per set, maternal and paternal homologs can be randomly aligned along the metaphase plate in a variety of ways. When meiosis is complete, it is very unlikely that any two human gametes will have the same combination of homologous chromosomes.

For any diploid species, the possible number of different, random alignments during metaphase I of meiosis equals $2^n$, where $n$ equals the number of chromosomes per set. The random alignments equal $2^n$ because each chromosome is found in a homologous pair and each member of the pair can align on either side of the metaphase plate. It is a matter of chance which daughter cell of meiosis I will get the maternal chromosome of a homologous pair and which will get the paternal chromosome. Because the homologs are genetically similar but not identical, the random alignment of homologous chromosomes provides a mechanism to promote a vast amount of genetic diversity among the resulting haploid cells.

**Crunching the Numbers:** Humans have 23 chromosomes per set. How many possible random alignments could occur during metaphase I? How does crossing over further contribute to the genetic diversity of the resulting haploid cells?

## Sexually Reproducing Species Produce Haploid and Diploid Cells at Different Times in Their Life Cycles

Let's now turn our attention to the relationship among mitosis, meiosis, and sexual reproduction in animals, plants, fungi, and protists. For any given species, the sequence of events that produces another generation of organisms is known as a **life cycle.** For sexually reproducing organisms, this usually involves an alternation between haploid cells or organisms and diploid cells or organisms (**Figure 14.11**).

**Diploid-Dominant Species** Most species of animals are diploid ($2n$), and their haploid gametes ($1n$) are considered to be a specialized type of cell. For this reason, animals are viewed as **diploid-dominant species** (Figure 14.11a). Certain diploid cells in the testes or ovaries undergo meiosis to produce haploid sperm or eggs, respectively. During fertilization, sperm and egg unite to form a diploid zygote, which then undergoes repeated mitotic cell divisions to produce a diploid multicellular organism.

**Haploid-Dominant Species** By comparison, most fungi and some protists are just the opposite; they are **haploid-dominant species** (Figure 14.11b). In fungi, the multicellular organism is haploid; only the zygote is diploid. During sexual reproduction, haploid cells unite to form a diploid zygote, which then immediately proceeds through meiosis to produce four haploid cells called spores. Each spore goes through mitotic cellular divisions to produce a haploid multicellular organism.

**Alternation of Generations** Plants and some algae have life cycles that are intermediate between diploid or haploid dominance. Such species exhibit an **alternation of generations** (Figure 14.11c). The species alternate between diploid multicellular organisms called **sporophytes** and haploid multicellular organisms called **gametophytes.** Meiosis in certain cells within the sporophyte produces haploid spores, which divide by mitosis to produce the gametophyte. Particular cells within the gametophyte differentiate into haploid gametes. Fertilization occurs between two gametes, producing a diploid zygote that then undergoes repeated mitotic cell divisions to produce a sporophyte.

Among different plant species, the relative sizes of the haploid and diploid organisms vary greatly. In mosses, the haploid gametophyte is a visible multicellular organism, whereas the diploid sporophyte is smaller and remains attached to the haploid organism. In other plants, such as ferns (Figure 14.11c), both the diploid sporophyte and haploid gametophyte can grow independently. The sporophyte is considerably larger and is the organism we commonly think of as a fern. In seed-bearing plants, such as roses and oak trees, the diploid sporophyte is the large multicellular plant, whereas the gametophyte is composed of only a few cells and is formed within the sporophyte.

When comparing animals, plants, and fungi, it's interesting to consider how gametes are made. Animals produce gametes by meiosis. In contrast, plants and fungi produce reproductive cells by mitosis. The gametophyte of plants is a haploid multicellular organism that is created by mitotic cellular divisions of a haploid spore. Within the multicellular gametophyte, certain cells become specialized as gametes.

## 14.3 Reviewing the Concepts

- The process of meiosis begins with a diploid cell and produces four haploid cells with one set of chromosomes each (Figure 14.9).
- Meiosis consists of two divisions—meiosis I and II—each composed of prophase, prometaphase, metaphase, anaphase, and telophase. During meiosis I, the homologs are separated into two different cells, and during meiosis II, the sister chromatids are separated into four different cells (Figure 14.10; Table 14.1).
- Animals are diploid-dominant species, whereas most fungi and some protists are haploid-dominant. Plants alternate between diploid and haploid forms, which is called alternation of generations (Figure 14.11).

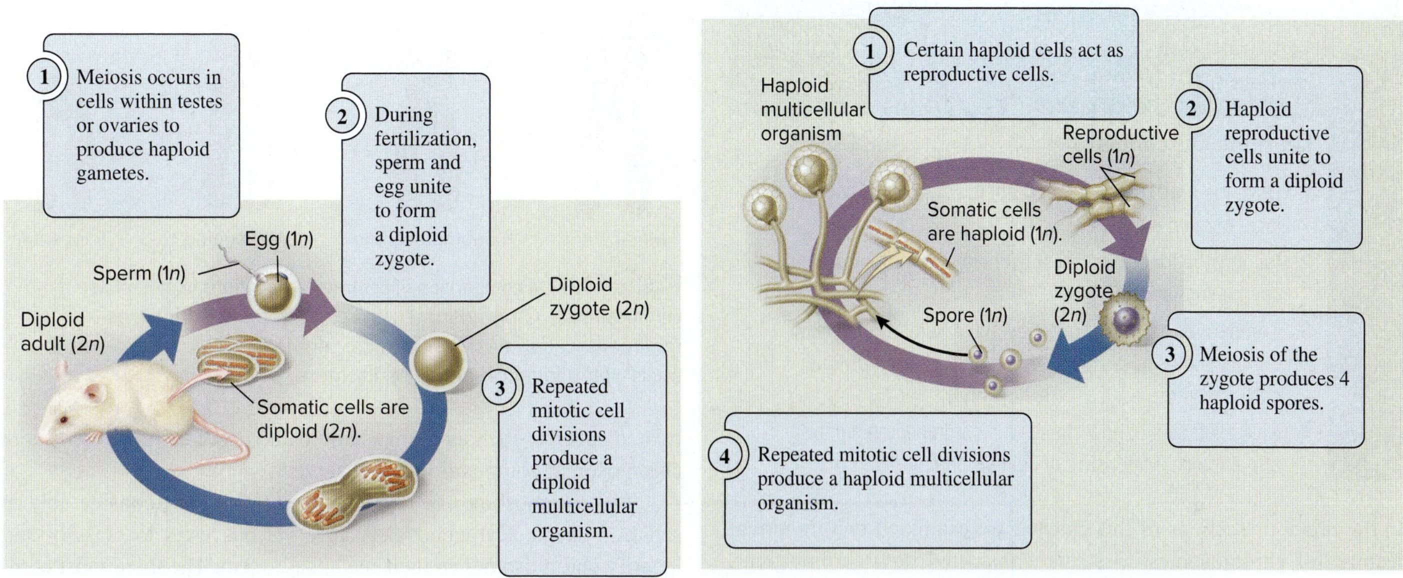


(a) Animal life cycle—diploid dominant

(b) Fungal life cycle—haploid dominant

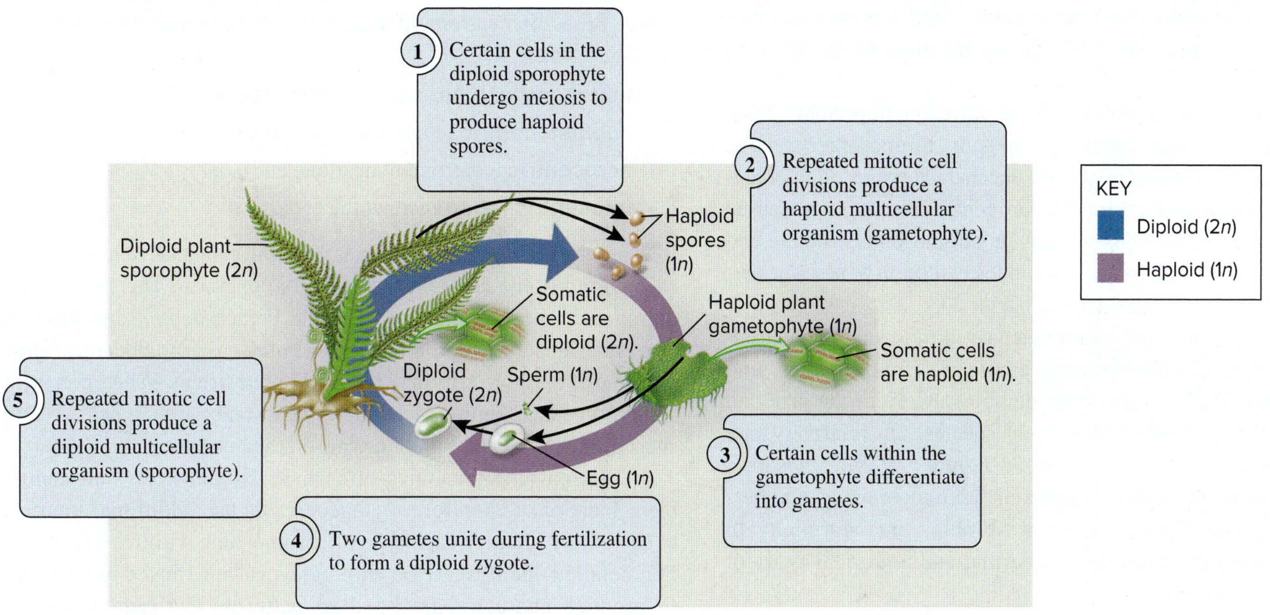

1 Certain cells in the diploid sporophyte undergo meiosis to produce haploid spores.
2 Repeated mitotic cell divisions produce a haploid multicellular organism (gametophyte).
3 Certain cells within the gametophyte differentiate into gametes.
4 Two gametes unite during fertilization to form a diploid zygote.
5 Repeated mitotic cell divisions produce a diploid multicellular organism (sporophyte).
Diploid plant sporophyte (2n)
Haploid spores (1n)
Somatic cells are diploid (2n).
Haploid plant gametophyte (1n)
Somatic cells are haploid (1n).
Diploid zygote (2n)
Sperm (1n)
Egg (1n)
KEY
Diploid (2n)
Haploid (1n)


(c) Plant life cycle—alternation of generations

**Figure 14.11** A comparison of three types of sexual life cycles.

**Concept Check:** *What is the main reason for meiosis in animals? What is the main reason for mitosis in animals?*

## 14.3 Testing Your Knowledge

1. In meiosis I, crossing over occurs during
   a. prophase.
   b. prometaphase.
   c. metaphase.
   d. anaphase.
   e. telophase.

2. The alignment of bivalents occurs during
   a. metaphase I.
   b. metaphase II.
   c. anaphase I.
   d. anaphase II.

3. A type of organism that exhibits alternation of generations is
   a. an animal.
   b. a fungus.
   c. a plant.
   d. both b and c.

## 14.4 Variation in Chromosome Structure and Number

### Learning Outcomes

1. Describe how chromosomes vary in size, centromere location, and number.
2. Identify the four ways that the structure of chromosomes can be changed.
3. Compare and contrast changes in the number of sets of chromosomes and changes in the number of individual chromosomes.
4. Give examples of how changes in chromosome number affect the characteristics of animals and plants.

In the previous sections of this chapter, we examined two important features of chromosomes. First, we considered how chromosomes occur in sets, and second, we explored two sorting processes that determine the chromosome number following cell division. In this section, we will examine how the structures and numbers of chromosomes can vary between different species and within the same species.

Why is the study of chromosomal variation important? First, geneticists have discovered that variations in chromosome structure and number can have major effects on the characteristics of an organism. We now know that several human genetic diseases are caused by such changes. In addition, changes in chromosome structure and number have been an important factor in the evolution of new species, which is a topic we will consider in Chapter 20.

Chromosome variation can be viewed in two ways. The structure and number of chromosomes among different species tend to vary greatly. There is also considerable variety in the size and shape of the chromosomes of a given species. On relatively rare occasions, however, the structure or number of chromosomes changes so that an individual is different from most other members of the same species. This is generally viewed as an abnormality. In this section, we will examine both normal and abnormal types of chromosome variation.

### Natural Variation Exists in Chromosome Structure and Number

Before we begin to examine chromosome variation, we need to have a reference point for a normal set of chromosomes. To determine what the normal chromosomes of a species look like, a cytogeneticist microscopically examines the chromosomes from several members of the species. Chromosome composition within a given species tends to remain relatively constant. In most cases, individuals of the same species have the same number and types of chromosomes. For example, as mentioned previously, the usual chromosome composition of human cells is 46 chromosomes (two sets of 23 chromosomes). Other diploid species have different numbers of chromosomes. The dog has 78 chromosomes (two sets of 39), the fruit fly has 8 chromosomes (two sets of 4), and the tomato has 24 chromosomes (two sets of 12). When comparing distantly related species, such as humans and fruit flies, major differences in chromosomal composition are observed.

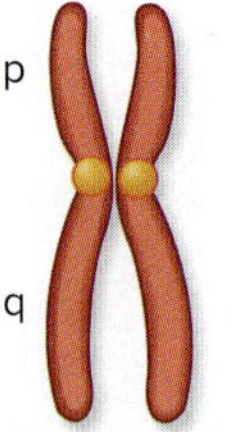


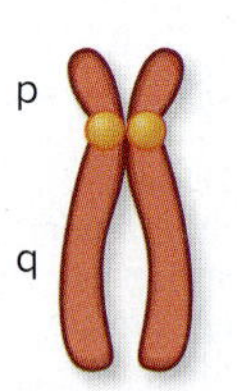


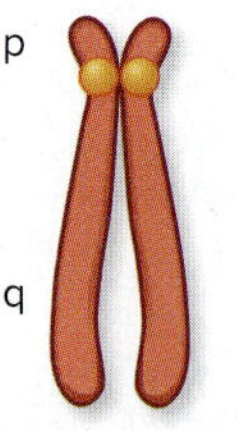


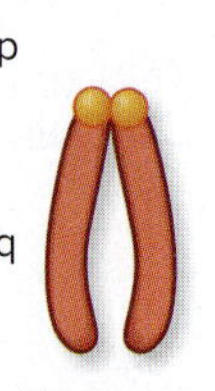


**Figure 14.12 A comparison of centromere locations in metaphase chromosomes.** Chromosomes are classified into four types—metacentric, submetacentric, acrocentric, and telocentric—on the basis of the location of the centromere.

The chromosomes of a given species also vary considerably in size and shape. Cytogeneticists have various ways to classify and identify chromosomes in their metaphase form. The three most commonly used features are size, location of the centromere, and banding patterns, which are revealed when the chromosomes are treated with stains. Based on centromere location, each chromosome is classified as (Figure 14.12)

- **metacentric** (centromere near the center),
- **submetacentric** (centromere off center),
- **acrocentric** (centromere near one end), or
- **telocentric** (centromere at the end).

Because the centromere is not exactly in the center of a chromosome, each chromosome has a short arm and a long arm. The short arm is designated with the letter "p" (for the French *petite*), and the long arm is designated with the letter "q." In the case of telocentric chromosomes, the short arm may be nearly nonexistent. When preparing a karyotype (see Figure 14.1), the chromosomes are aligned with the short arms on top and the long arms on the bottom.

Different chromosomes often have similar sizes and centromere locations. Therefore, cytogeneticists use additional methods to accurately identify each type of chromosome within a karyotype. For detailed identification, chromosomes are treated with stains to produce characteristic banding patterns. Cytogeneticists use several different staining procedures to identify specific chromosomes. An example is Giemsa stain, which produces dark G bands (see Figure 14.1). The alternating pattern of G bands is unique for each type of chromosome.

The banding pattern of eukaryotic chromosomes is useful in two ways. First, individual chromosomes can be distinguished from each other, even if they have similar sizes and centromere locations. Also, banding patterns are used to detect changes in chromosome structure that occur as a result of mutation.

### Mutations Can Alter Chromosome Structure

Let's now consider how the structures of chromosomes can be modified by a mutation—a heritable change in the genetic material. Chromosomal mutations, which involve the breaking and rejoining of chromosomes, are categorized as deletions, duplications, inversions, and translocations (Figure 14.13). Deletions and duplications

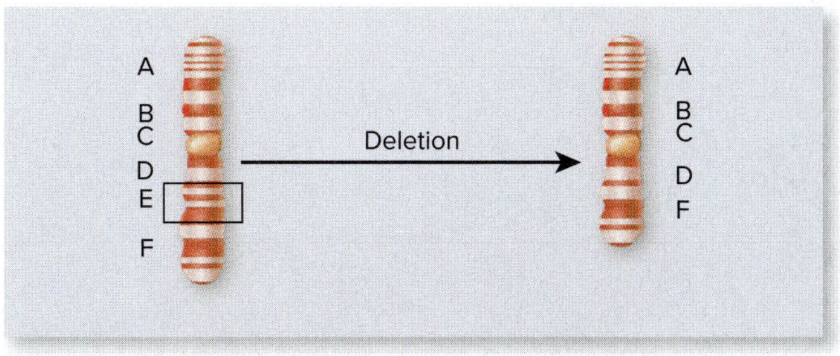


(a) Deletion: Removes a segment of chromosome

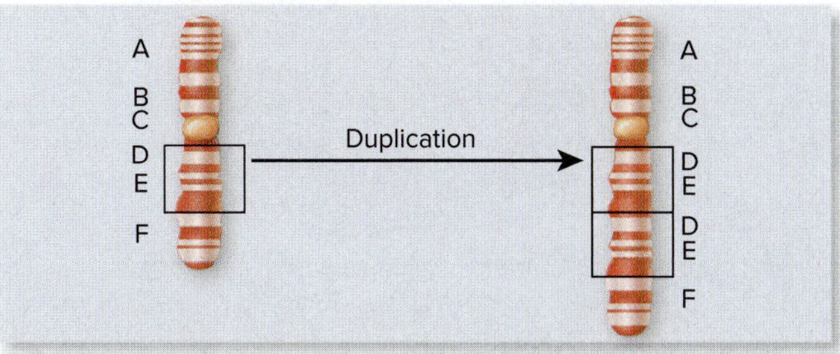


(b) Duplication: Doubles a particular region

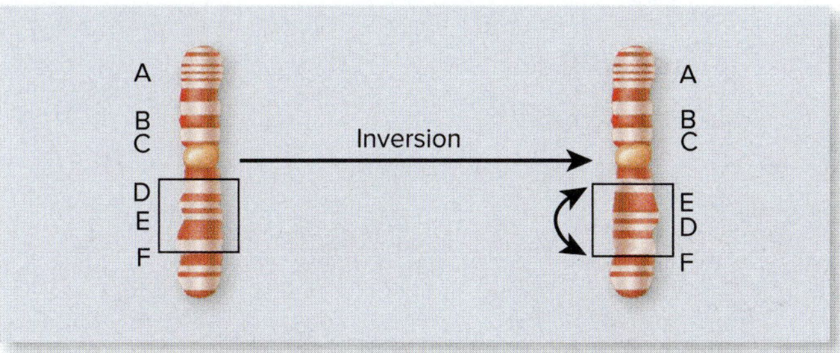


(c) Inversion: Flips a region to the opposite orientation

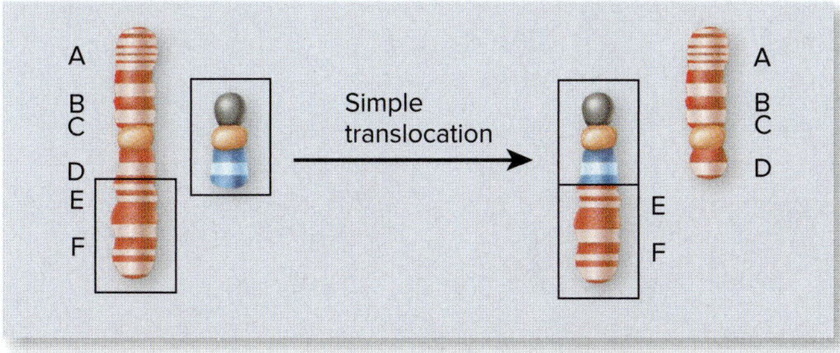


(d) Simple translocation: Moves a segment of 1 chromosome to another chromosome

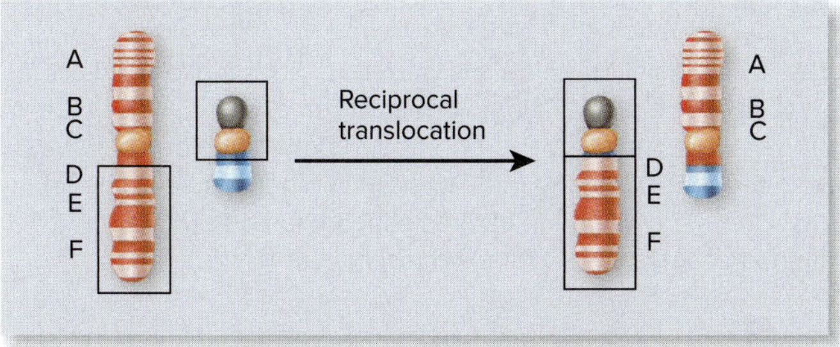


(e) Reciprocal translocation: Exchanges pieces between 2 different chromosomes

**Figure 14.13** **Types of changes in chromosome structure.** The letters alongside the chromosomes are placed there as frames of reference.

**Concept Check:** *Which types of changes shown here do not affect the total amount of genetic material?*

are changes in the total amount of genetic material in a single chromosome.

- When a **deletion** occurs, a segment of chromosomal material is removed (Figure 14.13a). The affected chromosome becomes deficient in a significant amount of genetic material.
- In a **duplication,** a section of a chromosome is repeated two or more times in a row (Figure 14.13b).

What are the consequences of a deletion or duplication? The possible effects depend on their size and whether they include genes or portions of genes that are vital to the development of the organism. When deletions or duplications have an effect, they are usually detrimental. Larger changes in the amount of genetic material tend to be more harmful because more genes are missing or duplicated.

Inversions and translocations are chromosomal rearrangements.

- An **inversion** is a change in the direction of the genetic material along a single chromosome. When a segment of one chromosome has been inverted, the order of G bands is opposite to that of a normal chromosome (Figure 14.13c).
- A **translocation** occurs when one segment of a chromosome becomes attached to a different chromosome. In a **simple translocation,** a single piece of chromosome is attached to another chromosome (Figure 14.13d). In a **reciprocal translocation,** two nonhomologous chromosomes exchange pieces by an abnormal crossing over event, thereby producing two chromosomes carrying translocations (Figure 14.13e).

## Variation Occurs in the Number of Chromosome Sets and the Number of Individual Chromosomes

Variations in chromosome number can be categorized in two ways: variation in the number of sets of chromosomes and variation in the number of particular chromosomes within a set. The suffix -ploid or -ploidy refers to a complete set of chromosomes. Organisms that are **euploid** (the prefix eu- means true) have chromosomes that occur in one or more complete sets. For example, in a species that is diploid, a euploid organism would have two sets of chromosomes in its somatic cells. In the fruit fly *(Drosophila melanogaster),* a normal individual has eight chromosomes. The species is diploid, having two sets of four chromosomes each (**Figure 14.14a**). Organisms can vary in the number of sets of chromosomes they have. For example, on rare occasions, an abnormal fruit fly can be produced with 12 chromosomes, containing three sets of 4 chromosomes each, or 16 chromosomes, containing four sets of 4 chromosomes each (**Figure 14.14b**). Organisms with three or more sets of chromosomes are called **polyploid.** A diploid organism is referred to as $2n$, a **triploid** organism as $3n$, a **tetraploid** organism as $4n$, and so forth. All such organisms are euploid because they have complete sets of chromosomes.

A second way that chromosome number can vary is a phenomenon called **aneuploidy.** This refers to an alteration in the number of particular chromosomes, so the total number of chromosomes is not an exact multiple of a set (**Figure 14.14c**). For example, an abnormal fruit fly could have nine chromosomes instead of eight because

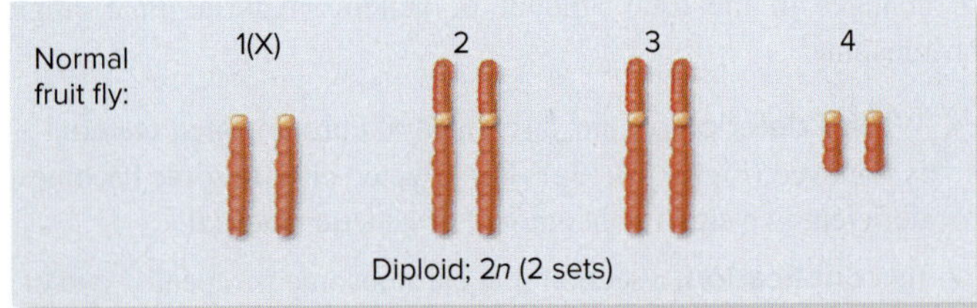


(a) Normal fruit fly chromosome composition

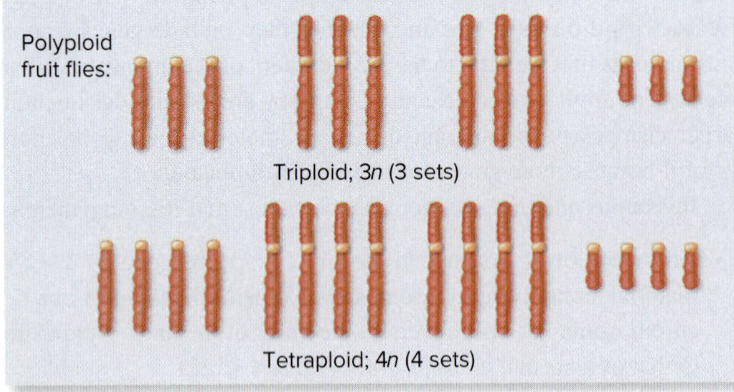


(b) Polyploidy

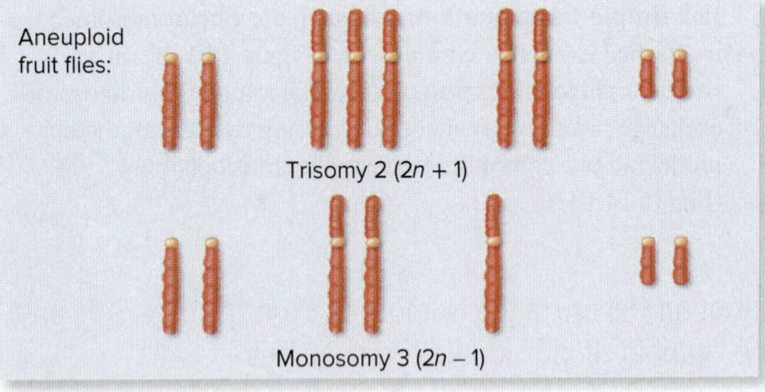


(c) Aneuploidy

**Figure 14.14** Types of variation in chromosome number in *Drosophila*. **(a)** The normal diploid number of chromosomes of a female fly. The X chromosome is also called chromosome 1. **(b)** Examples of polyploidy (triploid and tetraploid). **(c)** An example of aneuploidy (showing trisomy 2 and monosomy 3).

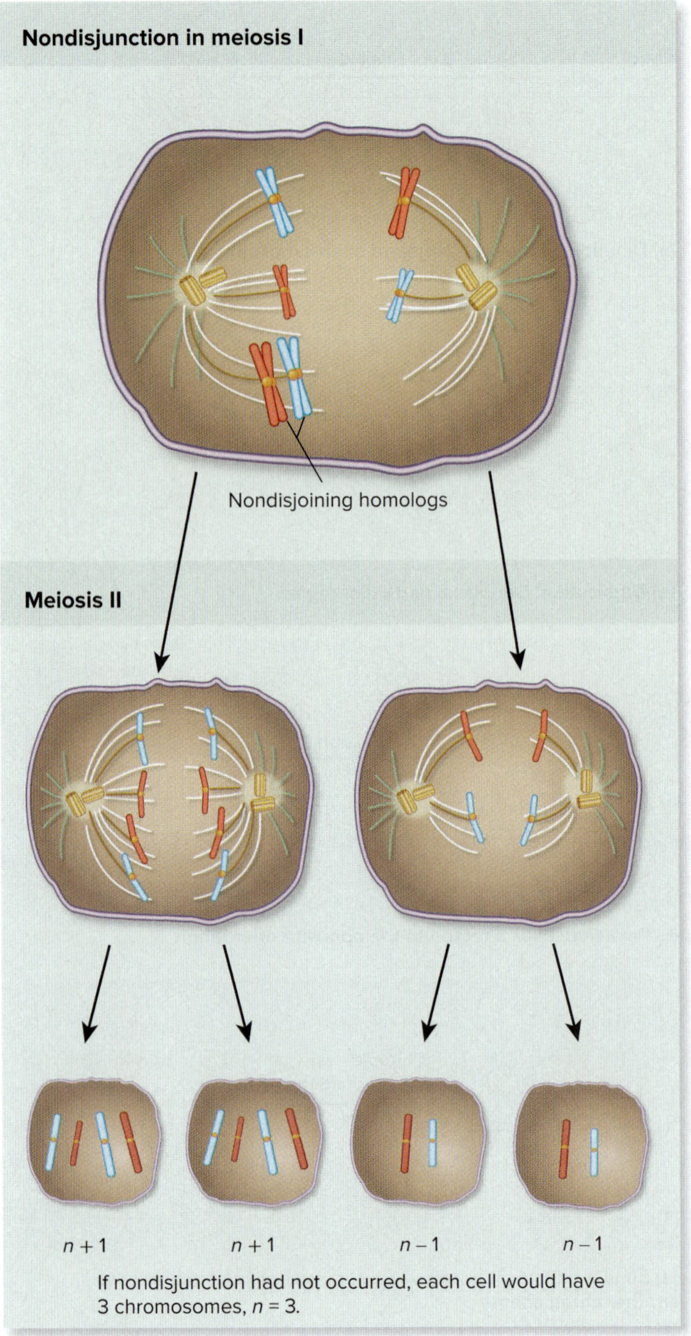


**Figure 14.15** Nondisjunction during meiosis I. For simplicity, this cell shows only three pairs of homologous chromosomes. One of the three pairs does not disjoin properly, and both homologs have moved into the cell on the left. The resulting haploid cells shown at the bottom are all aneuploid, resulting in gametes with four chromosomes and two chromosomes, instead of three.

it had three copies of chromosome 2 instead of the normal two copies. Instead of being perfectly diploid, the animal is $2n + 1$. Such a condition is called **trisomy,** and the animal is said to have trisomy 2, since chromosome 2 is affected. By comparison, a fruit fly could be lacking a single chromosome, such as chromosome 3, and have a total of seven chromosomes ($2n - 1$). This condition is called **monosomy,** and the animal has monosomy 3.

Variations in chromosome number usually have a significant effect on the characteristics of plants and animals. Therefore, researchers want to understand the mechanisms that cause these variations. In some cases, a change in chromosome number is the result of the abnormal sorting of chromosomes during cell division. The term **nondisjunction** refers to an event in which the chromosomes do not separate properly during cell division. Nondisjunction can occur during meiosis I or meiosis II and produces haploid cells that have too many or too few chromosomes. Figure 14.15 illustrates the consequences of nondisjunction during meiosis I. In this case, one pair of homologs moved into the cell on the left instead of separating from each other. This results in the production of aneuploid cells, with either too many ($n + 1$) or too few ($n - 1$) chromosomes. If such a cell becomes a gamete that fuses with another gamete during fertilization, the zygote and the resulting organism will have an abnormal number of chromosomes in all of its cells.

## Changes in Chromosome Number Have Important Consequences

How do changes in chromosome number affect the characteristics of animals and plants? Let's consider a few examples.

**Changes in Chromosome Number in Animals** In many cases, animals do not tolerate deviations from diploidy well. For example, polyploidy in mammals is generally a lethal condition. However, a few cases of naturally occurring variations from diploidy do occur in animals. Male bees, which are produced from unfertilized eggs, contain a single set of chromosomes and are therefore haploid organisms. By comparison, fertilized eggs become female bees, which are diploid. A few examples of vertebrate polyploid animals have been discovered. Interestingly, on rare occasions, animals that are morphologically very similar to each other can be found as a diploid species as well as a separate polyploid species. This situation occurs among certain amphibians and reptiles.

One important reason that geneticists are so interested in aneuploidy is its relationship to certain inherited disorders in humans. Even though most people are born with 46 chromosomes, alterations in chromosome number occur at a surprising frequency during gamete formation. About 5–10% of all fertilized human eggs result in an embryo with an abnormality in chromosome number. In most cases, these abnormal embryos do not develop properly and result in the very early miscarriage of a pregnancy, also referred to as spontaneous abortion. Approximately 50% of all miscarriages are due to alterations in chromosome number.

In some cases, an abnormality in chromosome number produces an offspring that can survive. Several human disorders are the result of abnormalities in chromosome number. The most common are trisomies of chromosome 21, 18, or 13 and abnormalities in the number of the sex chromosomes (**Table 14.2**). These syndromes are most likely due to nondisjunction. For example, Turner syndrome (X0) may occur when a gamete that is lacking a sex chromosome due to nondisjunction fuses with a normal gamete carrying an X chromosome during fertilization. By comparison, triple X syndrome (XXX) occurs when a gamete carrying two X chromosomes fuses with a gamete carrying a single X chromosome.

Most of the known trisomies involve chromosomes that are relatively small, so they carry fewer genes. Trisomies of the other human chromosomes and most monosomies are presumed to be lethal and have been found in spontaneously aborted embryos and fetuses.

Human abnormalities in chromosome number are influenced by the age of the parents. Older parents are more likely to produce children with abnormalities in chromosome number, possibly because meiotic nondisjunction is more likely to occur in older cells. **Down syndrome,** which was first described by an English physician, John Langdon Down, in 1866, provides an example. This disorder is caused by the inheritance of three copies of chromosome 21 (see Table 14.2). The incidence of Down syndrome rises with the age of either parent. In males, however, the rise occurs relatively late in life, usually past the age when most men have children. By comparison, the likelihood of having a child with Down syndrome rises dramatically during the later reproductive years in women.

**Table 14.2 Aneuploid Conditions in Humans**

| Condition | Frequency (# of live births) | Syndrome | Characteristics |
|---|---|---|---|
| **Autosomal** | | | |
| Trisomy 21 | 1/800 | Down | Mental impairment, abnormal pattern of palm creases, slanted eyes, flattened face, short stature |
| Trisomy 18 | 1/6,000 | Edward | Mental and physical impairment, facial abnormalities, extreme muscle tone, early death |
| Trisomy 13 | 1/15,000 | Patau | Mental and physical impairment, wide variety of defects in organs, large triangular nose, early death |
| **Sex chromosomal** | | | |
| XXY | 1/1,000 (males) | Klinefelter | Sexual immaturity (no sperm), breast swelling (males) |
| XYY | 1/1,000 (males) | Jacobs | Tall |
| XXX | 1/1,500 (females) | Triple X | Tall and thin, menstrual irregularity |
| X0 | 1/5,000 (females) | Turner | Short stature, webbed neck, sexually undeveloped |

**Changes in Chromosome Number in Plants** In contrast to animals, plants commonly exhibit polyploidy. Polyploidy is also important in agriculture. In many instances, polyploid species of plants display characteristics that are helpful to humans. They are often larger in size and more robust, traits that are clearly advantageous in the production of food. For example, the species of wheat that we use to make bread, *Triticum aestivum*, is a hexaploid (containing six sets of chromosomes) that arose from the union of diploid genomes from three closely related species (**Figure 14.16**). During the course of its cultivation, two diploid species must have interbred to produce a tetraploid, and then a third species interbred with the tetraploid to produce a hexaploid. Plant polyploids tend to exhibit a greater adaptability, which allows them to withstand harsher environmental conditions.

Although polyploidy is often beneficial in plants, aneuploidy in all eukaryotic species usually has detrimental consequences on the characteristics of an organism. Why is aneuploidy usually detrimental? To answer this question, we need to consider the relationship between gene expression and chromosome number. For many but not all genes, the level of gene expression is correlated with the number

**Figure 14.16 An example of a polyploid plant.** Cultivated wheat, *Triticum aestivum*, is a hexaploid. It was derived from three different diploid species of grasses that originally were found in the Middle East and were cultivated by ancient farmers in that region. Modern varieties of wheat have been produced from this hexaploid species. © irin-k/agefotostock RF

of genes per cell. Compared with a diploid cell, if a gene is carried on a chromosome that is present in three copies instead of two, approximately 150% of the normal amount of gene product is usually made. Alternatively, if only one copy of that gene is present due to a missing chromosome, only 50% of the gene product is typically made. For some genes, producing too much or too little of the gene product may not have adverse effects. However, for other genes, the over- or underexpression may interfere with the proper functioning of cells.

## 14.4 Reviewing the Concepts

- Chromosomes are classified as metacentric, submetacentric, acrocentric, and telocentric, based on their centromere location. Individual chromosomes can be distinguished from each other by their banding patterns after staining (Figure 14.12).
- Deletions, duplications, inversions, and translocations are different ways in which mutations alter chromosome structure (Figure 14.13).
- A euploid organism has chromosomes that occur in complete sets. A polyploid organism has three or more sets of chromosomes. An organism that has one too many (trisomy) or one too few (monosomy) chromosomes is termed aneuploid. Aneuploidy can be caused by nondisjunction, an event in which the chromosomes do not separate properly during cell division. Aneuploidy in humans is responsible for several types of human genetic diseases, including Down syndrome (Figures 14.14, 14.15; Table 14.2).
- Polyploid animals are relatively rare, but polyploid plants are common and tend to be larger and more robust than their diploid counterparts (Figure 14.16).

## 14.4 Testing Your Knowledge

1. A chromosome in which the telomere is at one end is called
   a. metacentric.
   b. submetacentric.
   c. acrocentric.
   d. telocentric.
2. A change in chromosome structure that does not affect the total amount of genetic material is
   a. a deletion.
   b. a duplication.
   c. an inversion.
   d. a reciprocal translocation.
   e. both c and d.
3. A person who is born with Down syndrome carries three copies of chromosome 21 in his or her cells. Which of the following terms should *not* be used to describe such an individual?
   a. trisomic
   b. trisomy 21
   c. triploid
   d. aneuploid

# Assess and Discuss

## Test Yourself

1. In which phase of the cell cycle are chromosomes replicated?
   a. $G_1$ phase
   b. S phase
   c. M phase
   d. $G_2$ phase
   e. none of the above
2. If two chromosomes are homologous, they
   a. look similar under the microscope.
   b. have very similar DNA sequences.
   c. carry the same types of genes.
   d. may carry different versions of the same gene.
   e. are all of the above.
3. Checkpoints during the cell cycle are important because they
   a. allow the organelle activity to catch up to cellular demands.
   b. ensure the integrity of the cell's DNA.
   c. allow the cell to generate sufficient ATP for cellular division.
   d. are the only time DNA replication can occur.
   e. do all of the above.
4. Which of the following is a reason for mitotic cell division?
   a. asexual reproduction
   b. gamete formation in animals
   c. multicellularity
   d. all of the above
   e. both a and c
5. A replicated chromosome is composed of
   a. two homologous chromosomes held together at the centromere.
   b. four sister chromatids held together at the centromere.
   c. two sister chromatids held together at the centromere.
   d. four homologous chromosomes held together at the centromere.
   e. one chromosome with a centromere.

6. Which of the following is *not* an event of anaphase of mitosis?
   a. The nuclear envelope breaks down.
   b. Sister chromatids separate.
   c. Kinetochore microtubules shorten, pulling the chromosomes to the pole.
   d. Polar microtubules push against each other, moving the poles farther apart.
   e. All of the above occur during anaphase.
7. A student is looking at cells under the microscope. The cells are from an organism that has a diploid number of 14. In one case, the cell has seven replicated chromosomes (sister chromatids) aligned at the metaphase plate of the cell. Which of the following statements accurately describes this particular cell?
   a. The cell is in metaphase of mitosis.
   b. The cell is in metaphase of meiosis I.
   c. The cell is in metaphase of meiosis II.
   d. All of the above are correct.
   e. Both b and c are correct.
8. Which of the following statements accurately describes a difference between mitosis and meiosis?
   a. Mitosis may produce diploid cells, whereas meiosis produces haploid cells.
   b. Homologous chromosomes synapse during meiosis but do not synapse during mitosis.
   c. Crossing over commonly occurs during meiosis, but it does not commonly occur during mitosis.
   d. All of the above are correct.
   e. Both a and c are correct.
9. During crossing over in meiosis I,
   a. homologous chromosomes are not altered.
   b. homologous chromosomes exchange genetic material.
   c. chromosomal damage occurs.
   d. genetic information is lost.
   e. cytokinesis occurs.
10. Aneuploidy may be the result of
    a. a duplication of a region of a chromosome.
    b. an inversion of a region of a chromosome.
    c. nondisjunction during meiosis.
    d. interspecies breeding.
    e. all of the above.

## Conceptual Questions

1. Distinguish between homologous chromosomes and sister chromatids.
2. The *Oca2* gene, which influences eye color in humans, is found on chromosome 15. How many copies of this gene are found in the karyotype of Figure 14.1? Is it one, two, or four?
3. **PRINCIPLES** A principle of biology is that cells are the simplest units of life. Explain how mitosis is a key process in the formation of new cells.

## Collaborative Questions

1. Why is it necessary for chromosomes to condense during mitosis and meiosis? What do you think could happen if chromosomes did not condense?
2. A diploid eukaryotic cell has 10 chromosomes (5 per set). As a group, take turns having one student draw the cell as it would look during a phase of mitosis, meiosis I, or meiosis II; then have the other students guess which phase it is.

# 15

# Transmission of Genetic Information from Parents to Offspring I: Patterns that Follow Mendel's Laws

© Tomas Rodriguez/Corbis RF

**Parents and offspring often show a striking resemblance to each other.**

## Chapter Outline



Many people are intrigued as to how the traits of children often resemble those of their parents (see chapter-opening photograph). Likewise, the offspring of other animals and plants typically are similar to those of their parents. Even so, geneticists have discovered that inheritance patterns are full of surprises. For example, if you cross a pea plant with purple flowers to a plant with white flowers, all of the offspring have purple flowers, which is the same as one of their parents. On the other hand, if you cross a four-o'clock plant with red flowers to one with white flowers, all of the offspring have pink flowers, which is not the same as either parent. A key objective of this chapter is to explain how such different outcomes can occur. The goal is to understand **inheritance**—the acquisition of traits by their transmission from parent to offspring.

Microscopic observations of chromosome transmission during mitosis and meiosis in the second half of the 19th century provided compelling evidence for **particulate inheritance**—the idea that the determinants of hereditary traits, which are now called genes, are transmitted in discrete units from one generation to the next. Remarkably, this idea was first put forward in the 1860s by an Austrian monk who knew nothing about chromosomes or DNA (Figure 15.1). Gregor Mendel used quantitative analysis of carefully designed plant breeding experiments to arrive at the concept of a gene, which is broadly defined as a unit of heredity.

**Figure 15.1** Gregor Johann Mendel.
© Science Source

**Table 15.1 Different Types of Inheritance Patterns and Their Molecular Basis**

| Type* | Description |
|---|---|
| Simple Mendelian inheritance | **Inheritance pattern:** Pattern of traits determined by a pair of alleles that display a dominant/recessive relationship and are located on an autosome. The presence of the dominant allele masks the presence of the recessive allele. |
| | **Molecular basis:** In many cases, the recessive allele is nonfunctional. Though a heterozygote may produce 50% of the functional protein compared with a dominant homozygote, this is sufficient to produce the dominant trait. |
| Incomplete dominance | **Inheritance pattern:** Pattern that occurs when the heterozygote has a phenotype intermediate to the phenotypes of the homozygotes, as when a cross between red-flowered and white-flowered plants produces pink-flowered offspring. |
| | **Molecular basis:** Fifty percent of the protein encoded by the functional (wild-type) allele is not sufficient to produce the normal trait. |
| X-linked inheritance | **Inheritance pattern:** Pattern of traits determined by genes that display a dominant/recessive relationship and are located on the X chromosome. In mammals and fruit flies, males have only one copy of the X chromosome and have one copy of X-linked genes. In these species, X-linked recessive traits occur more frequently in males than in females. |
| | **Molecular basis:** In a female with one X-linked recessive allele (a heterozygote), the protein encoded by the dominant allele is sufficient to produce the dominant trait. A male with an X-linked recessive allele (a hemizygote) does not have a dominant allele and does not make any of the functional protein. |
| Epigenetic inheritance | **Inheritance pattern:** A gene is silenced during the formation of gametes or during embryogenesis, and that silencing is maintained during the life of the individual. |
| | **Molecular basis:** Changes in chromatin structure can silence genes, and these changes are passed from cell to cell. |
| Linkage | **Inheritance pattern:** Two or more different genes tend to be transmitted as a unit. Linked genes do not assort independently. |
| | **Molecular basis:** Different genes are close together on the same chromosome. |
| Extranuclear inheritance | **Inheritance pattern:** Usually these genes are inherited from the mother. |
| | **Molecular basis:** These genes are found in the genomes of mitochondria or chloroplasts. |

*The first three patterns are described in this chapter, and the last three are discussed in Chapter 16.

In this chapter and the following chapter, we will consider inheritance patterns and how the transmission of genes is related to the transmission of chromosomes. In this chapter, we will consider the genetic patterns known as Mendelian inheritance and the relationship of these patterns to the behavior of chromosomes during meiosis. In Chapter 16, we will examine a variety of other patterns. As summarized in **Table 15.1**, our emphasis will be on two aspects of inheritance. First, we will explore how the various types of inheritance patterns produce different outcomes in a genetic cross. Second, we will examine the underlying molecular mechanisms that explain these different outcomes.

## 15.1 Mendel's Laws of Inheritance

### Learning Outcomes

1. List the advantages of using the garden pea to study inheritance.
2. Describe the difference between dominant and recessive traits.
3. Distinguish between genotype and phenotype.
4. **SCISKILLS ▶** Predict the outcome of genetic crosses using a Punnett square.
5. Define Mendel's law of segregation and law of independent assortment.

In 1856, Gregor Mendel began his historic studies on pea plants. For 8 years, he analyzed thousands of pea plants that he grew on a small plot in his monastery's garden. In 1866, he published the results of his work in a paper entitled "Experiments on Plant Hybrids." This paper was largely ignored by scientists at that time, partly because of its title and because it was published in a somewhat obscure journal *(The Proceedings of the Brünn Society of Natural History)*. Also, Mendel was clearly ahead of his time. During this period, biology had not yet become a quantitative science. In addition, the behavior of chromosomes during mitosis and meiosis, which provides a framework for understanding inheritance patterns, had yet to be studied. Prior to his death in 1884, Mendel reflected, "My scientific work has brought me a great deal of satisfaction and I am convinced it will be appreciated before long by the whole world." Sixteen years later, in 1900, Mendel's work was independently rediscovered by three biologists with an interest in plant genetics: Hugo de Vries of Holland, Carl Correns of Germany, and Erich von Tschermak of Austria. Within a few years, the influence of Mendel's landmark studies was felt around the world.

In this section, we will examine Mendel's experiments and how they led to the genetic principles known as Mendel's laws. We will see that these principles apply not only to the pea plants Mendel

## Biology Principle

### The Genetic Material Provides a Blueprint for Reproduction

The traits that Mendel studied in pea plants are governed by the genetic material of this species.

**Figure 15.2** The seven characters that Mendel studied.

**Concept Check:** *Is having blue eyes a character, a variant, or both?*

studied but also to a wide variety of sexually reproducing organisms, including humans.

## Mendel Chose the Garden Pea to Study Inheritance

Mendel chose the garden pea, *Pisum sativum*, to investigate the natural laws that govern inheritance. Why did he choose this species? Certain properties of the garden pea were particularly advantageous for studying inheritance.

**Genetic Variation** Peas are available in many varieties that differ in characteristics, such as the appearance of seeds, pods, flowers, and stems. Such general features of an organism are called **characters.** Figure 15.2 illustrates the seven characters that Mendel eventually chose to follow in his breeding experiments. Each of these characters was found in two discrete forms, called **variants.** For example, one character he followed was height, which had the variants known as tall and dwarf. A **trait** is an identifiable characteristic of an organism. The term trait usually refers to a variant for a character.* For example, seed color is a character, and green and yellow seed colors are traits.

**Self-fertilization** In **self-fertilization,** a female gamete is fertilized by a male gamete from the same plant. Like many flowering plants, peas have male and female sex organs in the same flower (Figure 15.3). Male gametes (sperm cells) are produced within pollen grains, which are formed in structures called stamens. Female gametes (egg cells) are produced in structures called ovules, which form within an organ called an ovary. For fertilization to occur, a pollen grain must land on the receptacle called a stigma, enabling a sperm to migrate to an ovule and fuse with an egg cell. In peas,

*Geneticists may also use the term trait to refer to a character.

**Figure 15.3 Flower structure in pea plants.** The pea flower produces both male and female gametes. Sperm form in the pollen produced within the stamens; egg cells form in ovules within the ovary. A modified petal called a keel encloses the stamens and stigma, encouraging self-fertilization.



the stamens and the ovaries are enclosed by a modified petal called a keel, an arrangement that greatly favors self-fertilization. Self-fertilization makes it easy to produce plants that breed true for a given trait, meaning the trait does not vary from generation to generation. For example, if a pea plant with yellow seeds breeds true for seed color, all the plants that grow from these seeds will also produce yellow seeds. A variety that continues to exhibit the same trait after several generations of self-fertilization is called a **true-breeding line,** or **strain.** Prior to conducting the studies described in this chapter, Mendel had already established that the seven characters he chose to study were true-breeding in the strains of pea plants he had obtained.

**Hybridization** A third reason for using garden peas is the ease of making crosses: The flowers are fairly large and easy to manipulate. In some cases, Mendel wanted his pea plants to self-fertilize, but in others, he wanted to cross-pollinate two plants that differed with respect to some character. When two individuals of the same species with different characteristics are bred, or crossed, to each other, the process is called **hybridization,** and the offspring are referred to as hybrids. For example, a hybridization experiment could involve breeding a purple-flowered plant to a white-flowered plant. In garden peas, hybridization requires placing pollen from one plant onto the stigma of a flower on a different plant (Figure 15.4). Mendel would pry open an immature flower and remove the stamens before they produced pollen, so the flower could not self-fertilize. He then used a paintbrush to transfer pollen from another plant to the stigma of the flower from which he had removed the stamens. In this way, Mendel was able to cross any two of his true-breeding pea plants and obtain any type of hybrid he wanted.

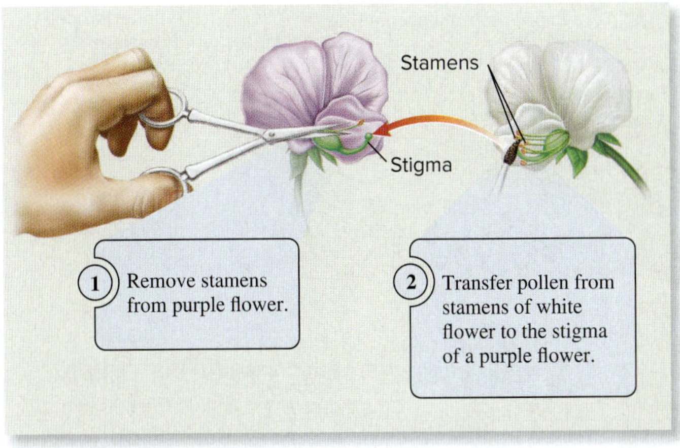


**Figure 15.4 A procedure for cross-pollinating pea plants.**

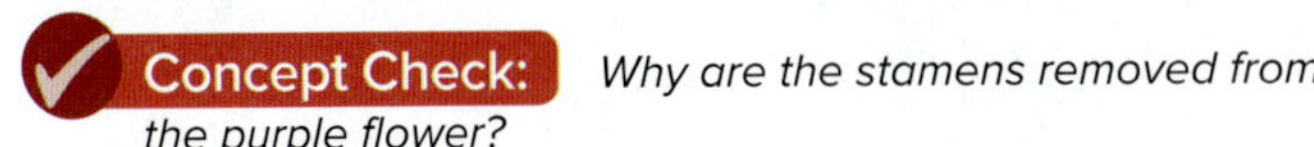


*Why are the stamens removed from the purple flower?*

## By Following the Inheritance Pattern of Single Traits, Mendel's Work Revealed the Law of Segregation

Mendel began his investigations by studying the inheritance patterns of pea plants that differed in a single character. A cross in which an experimenter follows the variants of only one character is called a **single-factor cross.** As an example, we will consider a single-factor cross in which Mendel followed the tall and dwarf variants for height (Figure 15.5). The left-hand side of Figure 15.5a shows his experimental approach. The true-breeding parents are termed the **P generation** (parental generation), and their offspring constitute the **$F_1$ generation** (first filial generation, from the Latin *filius*, meaning son). When the true-breeding parents differ in a single character, their $F_1$ offspring are called single-character hybrids, or **monohybrids.** When Mendel crossed true-breeding tall and dwarf plants, he observed that all plants of the $F_1$ generation were tall.

Next, Mendel followed the transmission of this character for a second generation. To do so, he allowed the $F_1$ monohybrids to self-fertilize, producing a generation called the **$F_2$ generation** (second filial generation). The dwarf trait reappeared in some of the $F_2$ offspring: Three-fourths of the plants were tall and one-fourth were dwarf. Mendel obtained similar results for each of the seven characters he studied, as shown in the data of Figure 15.5b. A quantitative analysis of his data allowed Mendel to postulate three important ideas about the properties and transmission of these traits from parents to offspring.

1. Traits may exist in two forms: dominant and recessive.
2. An individual carries two genes for a given character, and genes have variant forms, which are now called alleles.
3. The two alleles of a gene separate during gamete formation so that each sperm and egg receives only one allele.

**Dominant and Recessive Traits** In each of the seven cases, the $F_1$ generation displayed a trait distinctly like one of the two parents rather than an intermediate trait. Using genetic terms that Mendel originated,

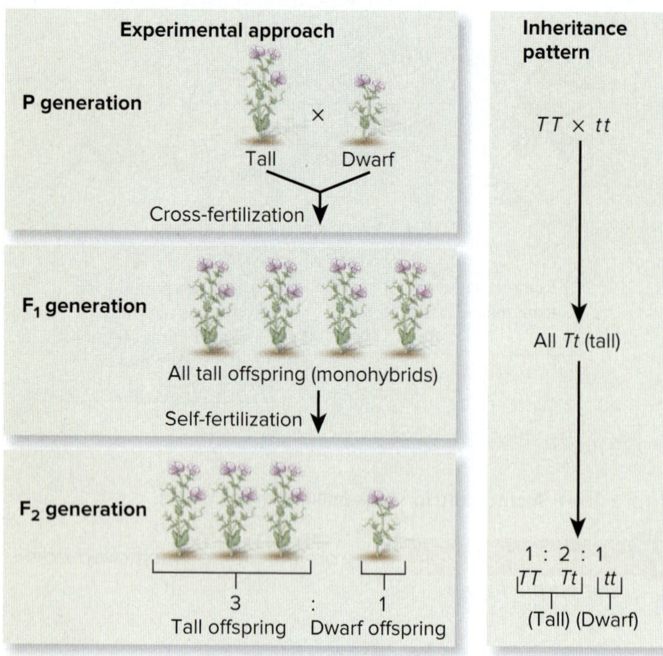


**(a) Mendel's protocol for making single-factor crosses**

**THE DATA**

| P cross | $F_1$ generation | $F_2$ generation | Ratio |
|---|---|---|---|
| Purple × white flowers | All purple | 705 purple, 224 white | 3.15:1 |
| Axial × terminal flowers | All axial | 651 axial, 207 terminal | 3.14:1 |
| Yellow × green seeds | All yellow | 6,022 yellow, 2,001 green | 3.01:1 |
| Round × wrinkled seeds | All round | 5,474 round, 1,850 wrinkled | 2.96:1 |
| Green × yellow pods | All green | 428 green, 152 yellow | 2.82:1 |
| Smooth × constricted pods | All smooth | 882 smooth, 299 constricted | 2.95:1 |
| Tall × dwarf stem | All tall | 787 tall, 277 dwarf | 2.84:1 |
| **Total** | **All dominant** | **14,949 dominant, 5,010 recessive** | **2.98:1** |

**(b) Mendel's observed data for all 7 traits**

**Figure 15.5** Mendel's analyses of single-factor crosses.

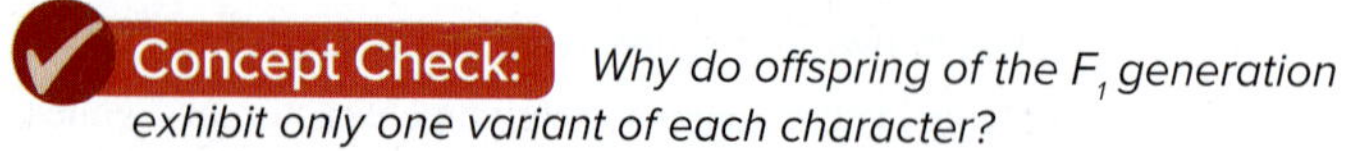

**Concept Check:** *Why do offspring of the $F_1$ generation exhibit only one variant of each character?*

we describe the alternative traits as dominant and recessive. The term **dominant** describes the displayed trait, whereas the term **recessive** describes a trait that is masked by the presence of a dominant trait. Tall stems and purple flowers are examples of dominant traits; dwarf stems and white flowers are examples of recessive traits. In this case, we say that tall is dominant to dwarf, and purple is dominant to white.

**Genes and Alleles** Mendel's results were consistent with particulate inheritance, in which the determinants of traits are inherited as unchanging, discrete units. In all seven cases, the recessive trait reappeared in the $F_2$ generation: Most $F_2$ plants displayed the dominant trait, whereas a smaller proportion showed the recessive trait. This observation led Mendel to conclude that the genetic determinants of traits are "unit factors" that are passed intact from generation to generation. These unit factors are what we now call **genes** (from the Greek *genos*, meaning birth), a term coined by Danish botanist Wilhelm Johannsen in 1909. Mendel postulated that every individual carries two genes for a given character and that the gene for each character in his pea plant exists in two variant forms, which we now call **alleles.** For example, the gene controlling height in Mendel's pea plants occurs in two variants, called the tall allele and the dwarf allele. The right-hand side of Figure 15.5a shows Mendel's conclusions, using genetic symbols (italic letters) that were adopted later. The letters *T* and *t* represent the alleles of the gene for plant height. Usually, the uppercase letter represents the dominant allele (in this case, tall), and the same letter in lowercase represents the recessive allele (dwarf).

**Segregation of Alleles** When Mendel compared the numbers of $F_2$ offspring exhibiting dominant and recessive traits, he noticed a recurring pattern. Although some experimental variation occurred, he always observed a 3:1 ratio between the dominant and the recessive trait (see Figure 15.5b). How did Mendel interpret this ratio? He concluded that each $F_1$ plant carried two versions (alleles) of a gene affecting height and that the two alleles carried by such an $F_1$ plant separate, or segregate, from each other during the process that gives rise to gametes. Therefore, each sperm or egg carried only one allele. The diagram in Figure 15.6 shows that the segregation of the $F_1$ alleles should result in equal numbers of gametes carrying the dominant allele (*T*) and the recessive allele (*t*). If these gametes combine with one another randomly at fertilization, as shown in the figure, this accounts for the 3:1 ratio of the $F_2$ generation. Note that a *Tt* individual can be produced by two different combinations—the *T* allele can come from the male gamete and the *t* allele from the female gamete, or vice versa. This accounts for the observation that *Tt* offspring are produced twice as often as either *TT* or *tt*. This work gave rise to **Mendel's law of segregation,** which can be stated as follows:

> *The two alleles of a gene separate (segregate) from each other during the process that gives rise to gametes so that every gamete receives only one allele.*

## Genotype Describes an Organism's Genetic Makeup, Whereas Phenotype Describes Its Characteristics

To continue our discussion of Mendel's results, we need to introduce a few more genetic terms.

- **Genotype** refers to the genetic composition of an individual. In the example shown in Figure 15.5a, *TT* and *tt* are the genotypes of the P generation, and *Tt* is the genotype of the $F_1$ generation.
- A **homozygote** is an individual that carries identical copies of the same allele for a given gene. In the specific cross we are

considering, the true-breeding tall plant (*TT*) is a homozygote, as is the dwarf plant (*tt*).

- A **heterozygote** carries two different alleles of a gene. Plants of the $F_1$ generation are heterozygotes with the genotype *Tt*. Every individual carries one copy of the tall allele (*T*) and one copy of the dwarf allele (*t*). The $F_2$ generation includes both homozygous individuals (homozygotes) and heterozygous individuals (heterozygotes).
- **Phenotype** refers to the characteristics of an organism that are the result of the expression of its genes. In the example in Figure 15.5a, the phenotype of one parent is tall, and the other is dwarf. Although the $F_1$ offspring are heterozygous (*Tt*), their phenotype is tall because each of them has a copy of the dominant tall allele. In contrast, the $F_2$ plants display both phenotypes in a ratio of 3:1.

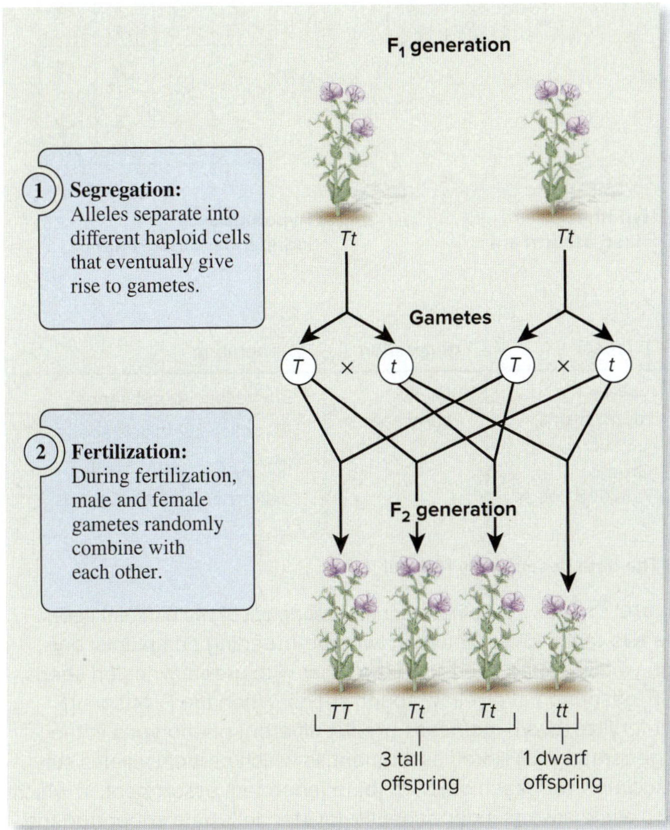


**Figure 15.6** **How the law of segregation explains Mendel's observed ratios.** The segregation of alleles in the $F_1$ generation gives rise to gametes that carry just one of the two alleles. These gametes combine randomly during fertilization, producing the allele combinations *TT*, *Tt*, and *tt* in the $F_2$ offspring. The combination *Tt* occurs twice as often as either of the other two combinations because it can be produced in two different ways. The *TT* and *Tt* offspring are tall, whereas the *tt* offspring are dwarf.

**Concept Check:** *What is the ratio of the T allele to the t allele in the $F_2$ generation? Does this ratio differ from the 3:1 phenotype ratio? If so, explain why.*

# Quantitative Analysis

## A PUNNETT SQUARE IS USED TO PREDICT THE OUTCOME OF CROSSES

A common way to predict the outcome of simple genetic crosses is to make a **Punnett square,** a method originally proposed by British geneticist Reginald Punnett. To construct a Punnett square, you must know the genotypes of the parents. What follows is a step-by-step description of the Punnett-square approach, using a cross of heterozygous tall plants.

**Step 1.** *Write down the genotypes of both parents.* In this example, a heterozygous tall plant is crossed to another heterozygous tall plant. The plant providing the pollen is considered the male parent; the plant providing the eggs, the female parent. (In self-pollination, a single individual produces both types of gametes.)

Male parent: *Tt*
Female parent: *Tt*

**Step 2.** *Write down the possible gametes that each parent can make.* Remember the law of segregation tells us that a gamete contains only one copy of each allele.

Male gametes: *T* or *t*
Female gametes: *T* or *t*

**Step 3.** *Create an empty Punnett square.* The number of columns equals the number of male gametes, and the number of rows equals the number of female gametes. Our example has two rows and two columns. Place the male gametes across the top of the Punnett square and the female gametes along the side.

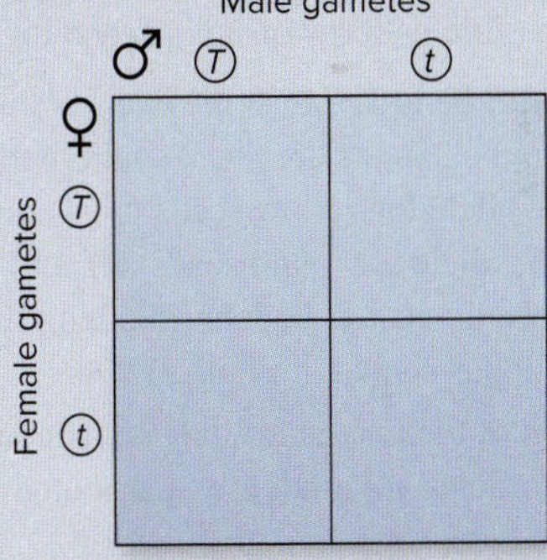


**Step 4.** *Fill in the possible genotypes of the offspring by combining the alleles of the gametes in the empty boxes.*

Male gametes

| ♀ Female gametes \ ♂ | *T* | *t* |
|---|---|---|
| *T* | ***TT*** | ***Tt*** |
| *t* | ***Tt*** | ***tt*** |

**Step 5.** *Determine the relative proportions of genotypes and phenotypes of the offspring.* The genotypes are obtained directly from the Punnett square. In this example, the genotype ratio is 1*TT* : 2*Tt* : 1*tt*. To determine the phenotypes, you must know which allele is dominant. For plant height, *T* (tall) is dominant to *t* (dwarf). The genotypes *TT* and *Tt* are tall, whereas the genotype *tt* is dwarf. Therefore, our Punnett square shows us that the phenotype ratio is expected to be 3 tall : 1 dwarf.

**Crunching the Numbers:** In pea plants, purple flowers are dominant to white. A cross is made between a heterozygous purple-flowered plant and a white plant. Set up a Punnett square, and predict the outcome of this cross.

## Analyzing the Inheritance Pattern of Two Characters Demonstrated the Law of Independent Assortment

To obtain additional insights into how genes are transmitted from parents to offspring, Mendel conducted crosses in which he simultaneously followed the inheritance of two different characters. A cross of this type is called a **two-factor cross.** We will examine a two-factor cross in which Mendel simultaneously followed the inheritance of seed color and seed shape (Figure 15.7). He began by crossing strains of pea plants that bred true for both characters. The plants of one strain had yellow, round seeds, and plants of the other strain had green, wrinkled seeds. He then allowed the $F_1$ offspring to self-fertilize and observed the phenotypes of the $F_2$ generation.

What are the possible patterns of inheritance for two characters?

1. The two genes may be linked in some way, so variants that occur together in the parents are always inherited as a unit. In our example, the allele for yellow seeds (*Y*) would always be inherited with the allele for round seeds (*R*), and the alleles for green seeds (*y*) would always be inherited with the allele for wrinkled seeds (*r*), as shown in Figure 15.7a.
2. The transmission of the two genes may be independent of one another, so their alleles are randomly distributed into gametes (Figure 15.7b).

By following the transmission pattern of two characters simultaneously, Mendel could determine whether the genes that control seed shape and seed color assort (are distributed) together as a unit or independently of each other.

What experimental results could Mendel predict for each of these two models? The two homozygous plants of the P generation can produce only two kinds of gametes, *YR* and *yr*, so in either case the $F_1$ offspring would be heterozygous for both genes; that is, they would have the genotypes *YyRr*. Because Mendel knew from his earlier experiments that yellow was dominant to green and round was dominant to wrinkled, he could predict that all $F_1$ plants would have yellow, round seeds. In contrast, as shown in Figure 15.7, the ratios he obtained in the $F_2$ generation would depend on whether the alleles of both genes assort together or independently.

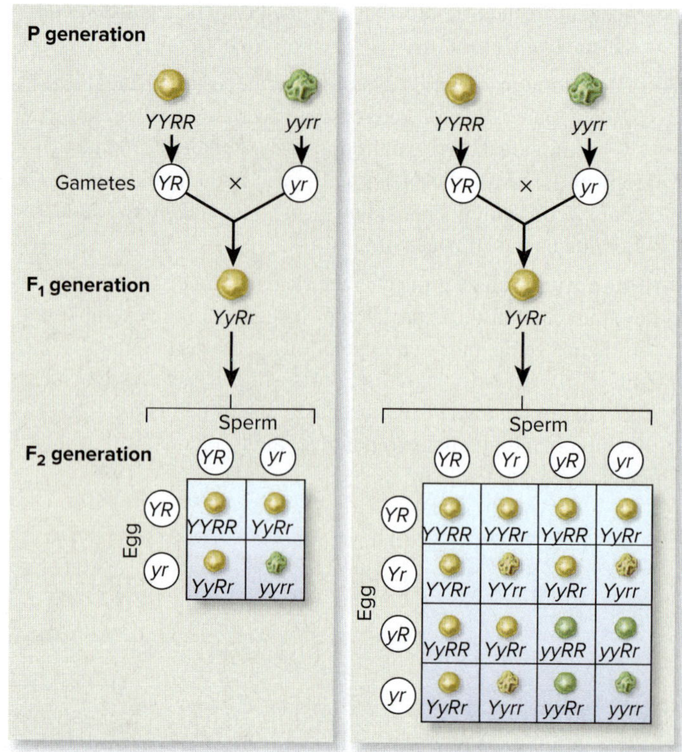


**(a) Hypothesis: linked assortment**

**(b) Hypothesis: independent assortment**

| P cross | $F_1$ generation | $F_2$ generation |
|---|---|---|
| Yellow, round seeds × Green, wrinkled seeds | Yellow, round seeds | 315 yellow, round seeds<br>101 yellow, wrinkled seeds<br>108 green, round seeds<br>32 green, wrinkled seeds |

**(c) The data observed by Mendel**

**Figure 15.7** **Two hypotheses for the assortment of two different genes.** In a two-factor cross between two true-breeding pea plants, one with yellow, round seeds and the other with green, wrinkled seeds, all $F_1$ offspring have yellow, round seeds. When the $F_1$ offspring self-fertilize, two hypotheses predict different phenotypes in the $F_2$ generation: **(a)** linked assortment, in which parental alleles stay associated with each other, or **(b)** independent assortment, in which each allele assorts independently. **(c)** Mendel's data supported the independent assortment hypothesis.

**Concept Check:** *What ratio of offspring phenotypes would have occurred if the linked hypothesis had been correct?*

If the parental genes are linked, as in Figure 15.7a, the $F_1$ plants would produce gametes that are only *YR* or *yr*. These gametes would combine to produce offspring with the genotypes *YYRR* (yellow, round), *YyRr* (yellow, round), or *yyrr* (green, wrinkled). The ratio of phenotypes would be 3 yellow, round to 1 green, wrinkled. Every $F_2$ plant would be phenotypically like one P-generation parent or the other. None would display a new combination of the parental traits. However, if the alleles assort independently, the $F_2$ generation would show a wider range of genotypes and phenotypes, as shown by the large Punnett square in Figure 15.7b. In this case, each $F_1$ parent produces four kinds of gametes—*YR, Yr, yR*, and *yr*—instead of two, so the Punnett square is constructed with four rows on each side and shows 16 possible genotypes. The $F_2$ generation includes plants with yellow, round seeds; yellow, wrinkled seeds; green, round seeds; and green, wrinkled seeds, in a ratio of 9:3:3:1.

The actual results of this two-factor cross are shown in the data of Figure 15.7c. Crossing the true-breeding parents produced **dihybrid** offspring—offspring that are hybrids with respect to both traits. These $F_1$ dihybrids all had yellow, round seeds, confirming that yellow and round are dominant traits. This result was consistent with either hypothesis. However, the data for the $F_2$ generation were consistent only with the independent assortment hypothesis. The $F_2$ offspring showed four different phenotypes in a ratio that was very close to 9:3:3:1.

In his original studies, Mendel reported that he had obtained similar results for every pair of characters he analyzed. This work gave rise to Mendel's law of independent assortment, which can be stated as follows:

> *The alleles of different genes assort independently of each other during the process that gives rise to gametes.*

Independent assortment means that a specific allele for one gene may be found in a gamete regardless of which allele for a different gene is found in the same gamete. In our example, the yellow and green alleles assort independently of the round and wrinkled alleles. The union of gametes derived from $F_1$ plants carrying these alleles produces the $F_2$ genotype and phenotype ratios shown in Figure 15.7b.

As we will see in Chapter 16, not all two-factor crosses exhibit independent assortment. In some cases, the alleles of two genes that are physically located near each other on the same chromosome do not assort independently.

## 15.1 Reviewing the Concepts

- Mendel studied seven characters found in garden peas that existed in two variants each. He allowed the peas to self-fertilize or carried out hybridization experiments (Figures 15.1, 15.2, 15.3, 15.4).
- By following the inheritance pattern of a single character (a single-factor cross) for two generations, Mendel determined the law of segregation, which states that two alleles of a gene segregate during the process that gives rise to gametes so that every gamete receives only one allele. In a single-factor cross, a 3:1 ratio is observed between the dominant and the recessive traits in the $F_2$ generation (Figures 15.5, 15.6).
- The genotype is the genetic makeup of an organism. Alleles are alternative versions of the same gene. Phenotype is a description of the traits that an organism displays.
- A Punnett square is constructed to predict the outcome of crosses.
- By conducting a two-factor cross, Mendel determined the law of independent assortment, which states that the alleles of different genes assort independently of each other during gamete formation. In a two-factor cross, this yields a 9:3:3:1 ratio in the $F_2$ generation (Figure 15.7).

## 15.1 Testing Your Knowledge

1. A cross is made between a pea plant that is *Pp* and one that is *pp*, where *P* is a dominant allele for purple flowers and *p* is the recessive allele that causes white flowers. What percentage of offspring would you expect to have purple flowers?
   a. 100%
   b. 75%
   c. 50%
   d. 25%
2. In peas, *Y* (yellow seeds) is dominant to *y* (green seeds), and *R* (round seeds) is dominant to *r* (wrinkled seeds). A plant that is heterozygous for both genes is crossed to a plant that is homozygous for both recessive alleles. What is the expected ratio of offspring?
   a. 3 yellow/round : 1 green wrinkled
   b. 9 yellow/round : 3 yellow/wrinkled : 3 green/round : 1 green wrinkled
   c. 9 yellow/wrinkled : 3 yellow/round : 3 green/round : 1 green wrinkled
   d. 1 yellow/round : 1 yellow/wrinkled : 1 green/round : 1 green wrinkled

## 15.2 Chromosome Theory of Inheritance

### Learning Outcomes

1. Outline the principles of the chromosome theory of inheritance.
2. Relate the behavior of chromosomes during meiosis to Mendel's two laws of inheritance.

Mendel's studies with pea plants eventually led to the concept of a gene, which is the foundation for our understanding of inheritance. However, at the time of Mendel's work, the physical nature and location of genes were a complete mystery. Several scientists, including German biologists Eduard Strasburger and Walther Flemming, observed dividing cells under the microscope and suggested that the chromosomes are the carriers of the genetic material. As we now know, the genetic material is the DNA within chromosomes. Other researchers noted that Mendel's laws of segregation and independent assortment showed a striking parallel with the segregation and sorting of chromosomes during meiosis (described in Chapter 14). Among these scientists were German biologist Theodor Boveri and American biologist Walter Sutton, who independently proposed the chromosome theory of inheritance. According to this theory, the inheritance patterns of traits can be explained by the transmission of chromosomes during meiosis and fertilization.

A modern view of the **chromosome theory of inheritance** consists of a few fundamental principles:

1. Chromosomes contain DNA, which is the genetic material. Genes are found within the chromosomes.
2. Chromosomes are replicated (copied) and passed from parent to offspring. They are also passed from cell to cell during the development of a multicellular organism.

3. The nucleus of a diploid cell contains two sets of chromosomes, which are found in homologous pairs. The maternal and paternal sets of homologous chromosomes are functionally equivalent; each set carries a full complement of genes.
4. At meiosis, one member of each homologous pair segregates into one daughter nucleus, and its homolog segregates into the other daughter nucleus. During the formation of haploid cells, the members of different homologous pairs segregate independently of each other.
5. Gametes are haploid cells that combine to form a diploid cell during fertilization, with each gamete transmitting one set of chromosomes to the offspring.

In this section, we will relate the chromosome theory of inheritance to Mendel's laws of inheritance.

## Mendel's Law of Segregation Is Explained by the Segregation of Homologous Chromosomes During Meiosis

The physical location of a gene on a chromosome is called the gene's **locus** (plural, loci). As shown in Figure 15.8, each member of a homologous chromosome pair carries an allele of the same gene at the same locus. The individual in this example is heterozygous (*Tt*), so each homolog has a different allele.

How can we relate the locations of genes along chromosomes to Mendel's law of segregation? Figure 15.9 shows a simplified example that follows a pair of homologous chromosomes through the events of meiosis. This example involves a pea plant, heterozygous for height, *Tt*. The top of Figure 15.9 shows the two homologous chromosomes prior to DNA replication. When a cell prepares to divide, the homologs replicate to produce pairs of sister chromatids. Each chromatid carries a copy of the allele found on the original homolog, either *T* or *t*. During meiosis I, the homologs, each consisting of two sister chromatids, pair up and then segregate into two daughter cells. One of these cells has two copies of the *T* allele, and the other has two copies of the *t* allele. The sister chromatids separate during meiosis II, which produces four haploid cells. The end result of meiosis is that each haploid cell has a copy of just one of the two original homologs. Two of the cells have a chromosome carrying the *T* allele, and the other two have a chromosome carrying the *t* allele at the same locus. If the haploid cells shown at the bottom of Figure 15.9 combine randomly during fertilization, they produce diploid offspring with the genotype and phenotype ratios shown earlier in Figure 15.6.

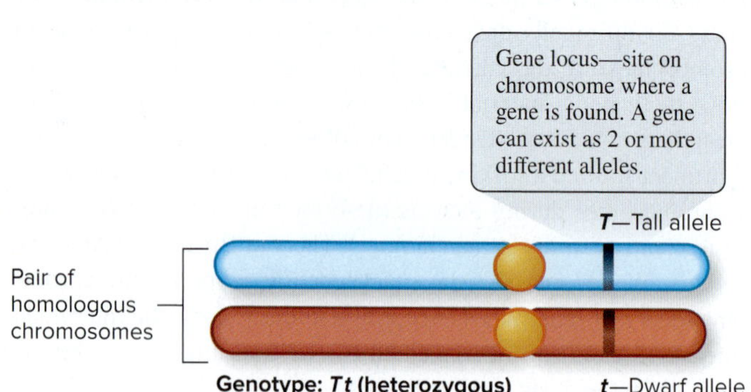


**Figure 15.8 A gene locus.** The locus (location) of a gene is the same for each member of a homologous pair, whether the individual is homozygous or heterozygous for that gene. This individual is heterozygous (*Tt*) for a gene for plant height.

**BioConnections:** *Refer back to Section 14.3 in Chapter 14. Explain the relationship between sexual reproduction and homologous chromosomes.*

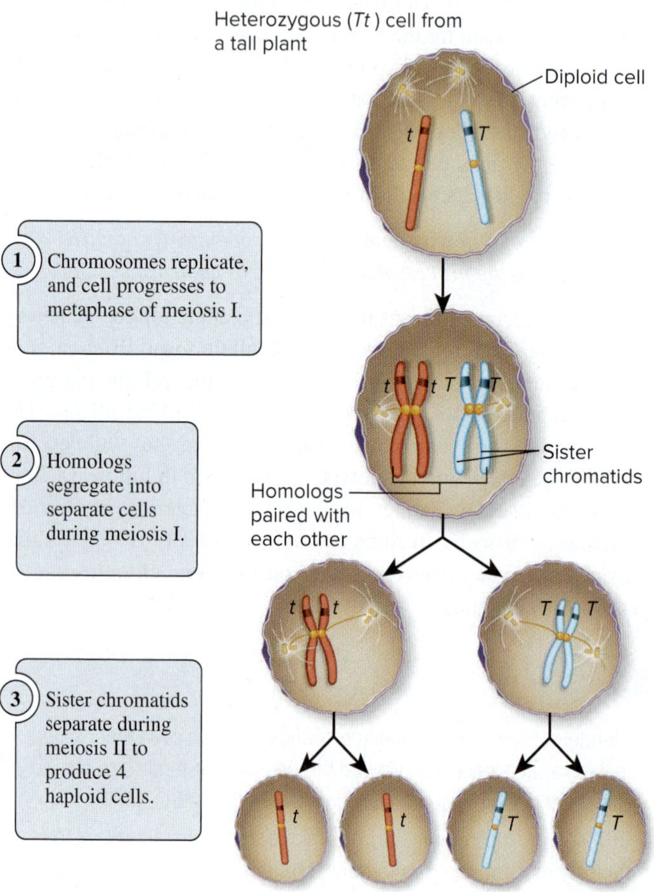


**Figure 15.9 The chromosomal basis of allele segregation.** This example shows a pair of homologous chromosomes in a cell of a pea plant. The blue chromosome was inherited from the male parent, and the red chromosome was inherited from the female parent. This individual is heterozygous (*Tt*) for a height gene. The two homologs segregate from each other during meiosis, leading to segregation of the tall allele (*T*) and the dwarf allele (*t*) into different haploid cells. Note: For simplicity, this diagram shows a single pair of homologous chromosomes, though eukaryotic cells typically have several different pairs of homologous chromosomes.

**BioConnections:** *When we say that alleles segregate, what does the word segregate mean? How is this related to meiosis, shown in Figure 14.10?*

## Mendel's Law of Independent Assortment Is Explained by the Independent Alignment of Different Chromosomes During Meiosis

How can we relate the chromosome theory of inheritance to Mendel's law of independent assortment? **Figure 15.10** shows the alignment and segregation of two pairs of chromosomes in a pea plant. One pair carries the gene for seed color: The yellow allele (*Y*) is on one chromosome, and the green allele (*y*) is on its homolog. The other pair of chromosomes carries the gene for seed shape: One member of the pair has the round allele (*R*), whereas its homolog carries the wrinkled allele (*r*). Therefore, this individual is heterozygous for both genes, with the genotype *YyRr*.

When meiosis begins, the DNA in each chromosome has already replicated, producing two sister chromatids. At metaphase I of meiosis, the two pairs of chromosomes randomly align themselves along the metaphase plate. This alignment can occur in two equally probable ways, shown on the two sides of the figure. On the left, the chromosome carrying the *y* allele is aligned on the same side of the metaphase plate as the chromosome carrying the *R* allele; *Y* is aligned with *r*. On the right, the opposite has occurred: *Y* is aligned with *R*, and *y* is with *r*. In each case, the chromosomes that aligned on the same side of the metaphase plate segregate into the same daughter cell. In this way, the random alignment of chromosome pairs during meiosis I leads to the independent assortment of alleles found on different chromosomes. For two genes found on different chromosomes, each with two variant alleles, meiosis produces four allele combinations in equal numbers (*yR, Yr, YR,* and *yr*), as seen at the bottom of the figure.

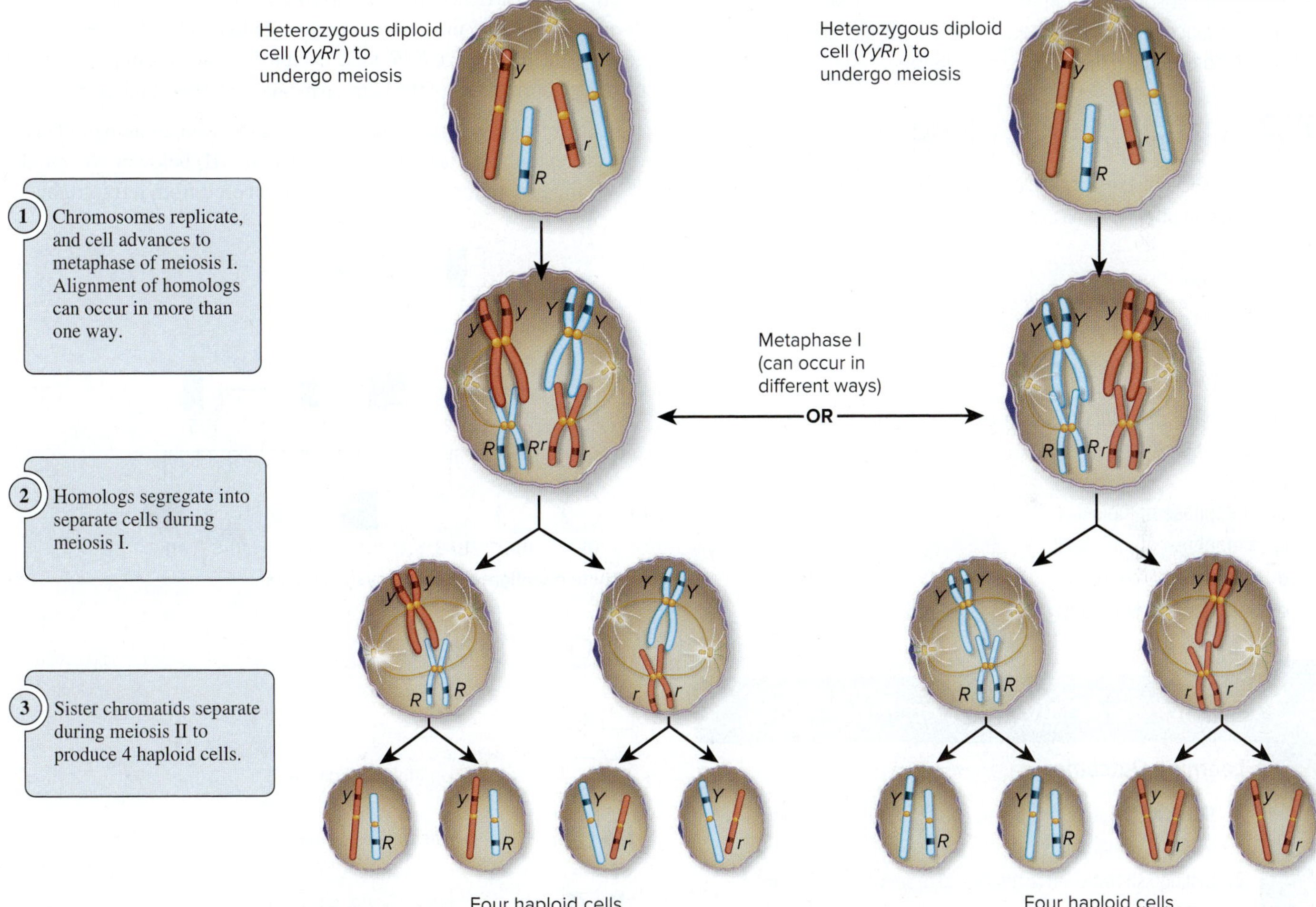


**Figure 15.10 The chromosomal basis of independent assortment.** The alleles for seed color (*Y* or *y*) and seed shape (*R* or *r*) in peas are located on different chromosomes. During metaphase of meiosis I, different arrangements of the two chromosome pairs lead to different combinations of the alleles in the resulting haploid cells. On the left, the chromosome carrying the recessive *y* allele has segregated with the chromosome carrying the dominant *R* allele. On the right, the two chromosomes carrying the dominant alleles (*Y* and *R*) have segregated together. Note: For simplicity, this diagram shows only two pairs of homologous chromosomes, though eukaryotic cells typically have several different pairs of homologous chromosomes.

**Concept Check:** *Let's suppose that a cell is heterozygous for three different genes (Aa Bb Cc) and that each gene is on a different chromosome. How many different ways can these three pairs of homologous chromosomes align themselves during metaphase I, and how many different types of gametes can be produced?*

If a *YyRr* plant undergoes self-fertilization, any two gametes can combine randomly during fertilization. Because four kinds of gametes are made, $4^2$, or 16, allele combinations (genotypes) are possible in the offspring. These genotypes, in turn, produce four phenotypes in a 9:3:3:1 ratio, as seen in Figure 15.7. This ratio is the expected outcome when a heterozygote for two genes on different chromosomes undergoes self-fertilization.

But what if two different genes are located on the same chromosome? In this case, the transmission pattern may not conform to the law of independent assortment. We will discuss this phenomenon, known as linkage, in Chapter 16.

## 15.2 Reviewing the Concepts

- The chromosome theory of inheritance explains how the behavior of chromosomes during meiosis accounts for Mendel's laws of inheritance. Each gene is located at a particular site, or locus, on a chromosome (Figures 15.8, 15.9, 15.10).

## 15.2 Testing Your Knowledge

1. During which phase of cellular division does Mendel's law of segregation physically occur?
   a. mitosis
   b. meiosis I
   c. meiosis II
   d. all of the above
   e. b and c only
2. Which phase of cellular division explains Mendel's law of independent assortment?
   a. mitosis
   b. metaphase of meiosis I
   c. metaphase of meiosis II
   d. anaphase of meiosis I
   e. anaphase of meiosis II

## 15.3 Pedigree Analysis of Human Traits

### Learning Outcomes

1. **SCISKILLS ▸** Apply pedigree analysis to deduce inheritance patterns in humans.
2. Distinguish between recessive disorders and dominant disorders.

As we have seen, Mendel conducted experiments by making selective crosses of pea plants and analyzing large numbers of offspring. Later geneticists also relied on crosses of experimental organisms, especially fruit flies *(Drosophila melanogaster)*. However, geneticists studying human traits cannot use this approach, for ethical and practical reasons. Instead, human geneticists must rely on the information contained in family trees, or pedigrees. In this approach, called **pedigree analysis,** an inherited trait is analyzed over the course of a few generations in one family.

Pedigree analysis has been used to understand the inheritance of human genetic diseases that follow simple Mendelian patterns. In human populations, genes that play a role in disease may exist in different forms:

- A wild-type allele is a common allele in a population.
- A mutant allele is rare. Some mutant alleles are known to cause inherited diseases in humans.

Pedigree analysis allows us to determine whether the mutant allele is dominant or recessive and to predict the likelihood of an individual being affected.

Let's consider a recessive condition to illustrate pedigree analysis. The pedigree in **Figure 15.11** concerns a human genetic disease known as cystic fibrosis (CF), which involves a mutation in a gene that encodes the cystic fibrosis transmembrane regulator (the *CFTR* gene; refer back to Figure 1.12). Approximately 3% of Americans of European descent are heterozygous carriers of the recessive *CFTR* allele. Carriers do not show disease symptoms; they are unaffected. Individuals who are homozygous for the recessive *CFTR* allele exhibit the disease symptoms, which include abnormalities of the lungs, pancreas, intestine, and sweat glands.

- The oldest generation (designated by the Roman numeral I) is at the top, with later generations (II and III) below it. Within the same generation, individuals are numbered from left to right.

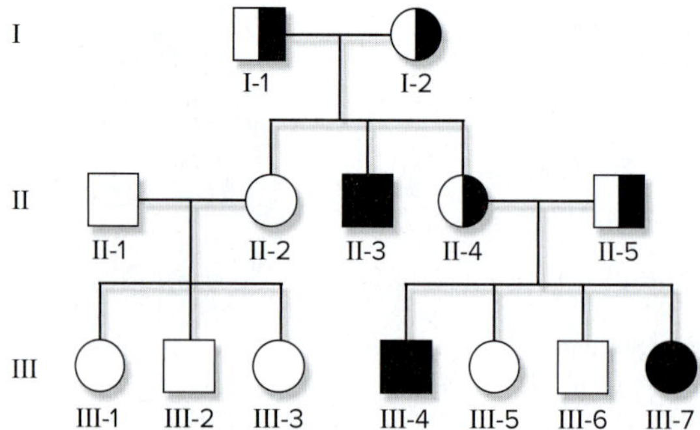


**(a) Human pedigree showing cystic fibrosis**

| | |
|---|---|
| ○ ♀ | Female |
| □ ♂ | Male |
| ○ □ | Unaffected individual |
| ● ■ | Affected individual |
| ◐ ◨ | Presumed heterozygote (carrier) |

**(b) Symbols used in a human pedigree**

**Figure 15.11 A family pedigree for a recessive trait.** Some members of the family in this pedigree are affected with cystic fibrosis. Unaffected individuals I-1, I-2, II-4, and II-5 are presumed to be heterozygotes (carriers) because they have produced affected offspring.

**Concept Check:** *Let's suppose a genetic disease is caused by a mutant allele. If two affected parents produce an unaffected offspring, can the mutant allele be recessive?*

- A male is represented by a square and a female by a circle.
- Couples who produce offspring are connected by a horizontal line.
- A vertical line connects parents with their offspring.
- Siblings (brothers and sisters) are denoted by downward projections from a single horizontal line, from left to right in the order of their birth. For example, individuals I-1 and I-2 are the parents of individuals II-2, II-3, and II-4, who are all siblings.
- Individuals affected by the disease, such as individual II-3, are depicted by filled symbols.

Why does this pedigree indicate a recessive pattern of inheritance for CF? The answer is that two unaffected individuals can produce an affected offspring. For example, II-4 and II-5 have an affected offspring. Such individuals are presumed to be heterozygotes (designated by a half-filled symbol). However, the same unaffected parents can also produce unaffected offspring (depicted by an unfilled symbol), because an individual must inherit two copies of the mutant allele to exhibit the disease. A recessive mode of inheritance is also consistent with the observation that all offspring of two affected individuals are affected themselves. However, for genetic diseases that limit survival or fertility, there may rarely, if ever, be cases where two affected individuals produce offspring.

Although many of the alleles causing human genetic diseases are recessive, some are known to be dominant. Let's consider Huntington disease, a condition that causes the degeneration of brain cells involved in emotions, intellect, and movement. If you examine the pedigree shown in Figure 15.12, you will see that every affected individual has one affected parent. This pattern is characteristic of most dominant disorders in which a heterozygote exhibits the symptoms. However, affected parents do not always produce affected offspring. For example, II-6 is a heterozygote that has passed the normal allele to his offspring, thereby producing unaffected offspring (III-3 and III-4).

Most human genes are found on the paired chromosomes known as **autosomes,** which are the same in both sexes. Mendelian inheritance patterns involving these autosomal genes are described as autosomal inheritance patterns. Huntington disease is an example of a trait with an autosomal dominant inheritance pattern, whereas cystic fibrosis displays an autosomal recessive pattern. However, some human genes are located on sex chromosomes, which are different in males and females. These genes have their own characteristic inheritance patterns, which we will consider later in this chapter.

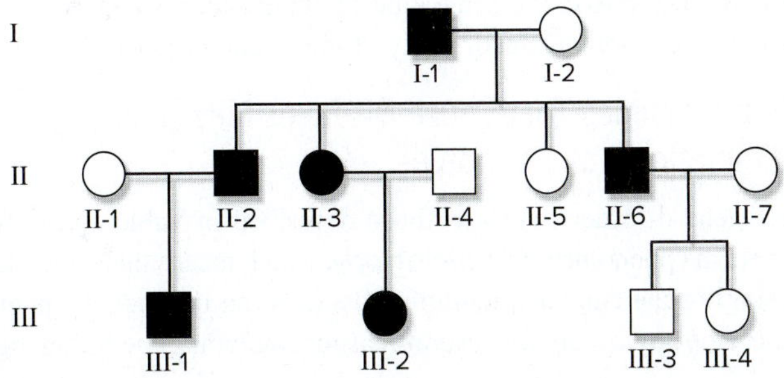


**Figure 15.12 A family pedigree for a dominant trait.** Huntington disease is caused by a dominant allele.

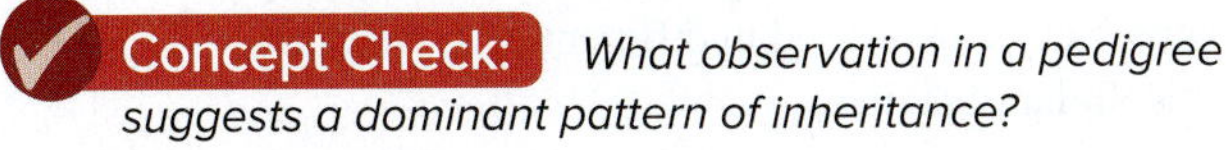


**Concept Check:** *What observation in a pedigree suggests a dominant pattern of inheritance?*

## 15.3 Reviewing the Concepts

- The inheritance patterns in humans are determined from a pedigree analysis (Figures 15.11, 15.12).

## 15.3 Testing Your Knowledge

1. Which of the following observations in a pedigree would rule out a recessive pattern of inheritance for a human disease?
   a. An affected offspring has two unaffected parents.
   b. An unaffected offspring has two affected parents.
   c. Two affected offspring are brother and sister.
   d. An affected grandfather has an affected grandson.

# 15.4 Variations in Inheritance Patterns and Their Molecular Basis

### Learning Outcomes

1. Relate dominant and recessive alleles and genotypes to protein function.
2. Describe pleiotropy, and explain why it occurs.
3. **SCISKILLS ▸** Predict the outcome of crosses that exhibit incomplete dominance.
4. Discuss how the environment plays a critical role in determining the outcome of traits.

The term **Mendelian inheritance** describes the inheritance patterns of genes that segregate and assort independently. In the first section of this chapter, we considered the inheritance pattern of traits affected by a single gene that is found in two variants, one of which is dominant to the other. This pattern is called **simple Mendelian inheritance,** because the phenotype ratios in the offspring clearly conform to Mendel's laws. In this section, we will discuss the molecular basis of dominant and recessive traits and see how the molecular expression of a gene can have widespread effects on an organism's phenotype. In addition, we will examine the inheritance patterns of genes that segregate and assort independently but do not display a simple dominant/recessive relationship. This modern knowledge also sheds light on the role of the environment in producing an organism's phenotype, which we will discuss at the end of the section.

## Protein Function Explains the Phenomenon of Dominance

As we discussed at the beginning of this chapter, Mendel studied seven characters that were found in two variants each (see Figure 15.2). The dominant variants are caused by the wild-type alleles for these traits in pea plants. A wild-type allele usually encodes a protein that is made in the proper amount and functions normally.

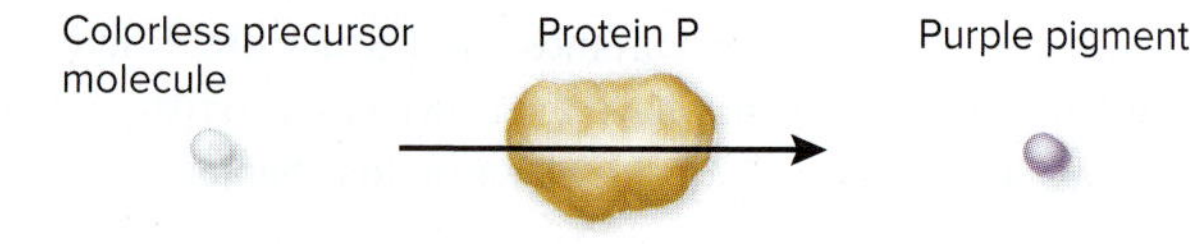


**(a) Function of protein P**

| Genotype | *PP* | *Pp* | *pp* |
|---|---|---|---|
| Amount of functional protein P produced | 100% | 50% | 0% |
| **Phenotype** | **Purple** | **Purple** | **White** |

**(b) Relationship between the amount of functional protein P and phenotype**

**Figure 15.13 How genes give rise to traits during simple Mendelian inheritance.** In many cases, the amount of protein encoded by a single dominant allele is sufficient to produce the normal phenotype. **(a)** In this example, the gene encodes an enzyme that is needed to produce a purple pigment. **(b)** A plant with one or two copies of the normal allele produces enough pigment to produce purple flowers. In a *pp* homozygote, the complete lack of the normal protein (enzyme) results in white flowers.

**Table 15.2 Examples of Recessive Human Genetic Diseases**

| Disease | Protein encoded by the gene* | Description |
|---|---|---|
| Phenylketonuria (PKU) | Phenylalanine hydroxylase | Inability to metabolize phenylalanine, an amino acid. Can lead to severe mental impairment and physical degeneration. The disease can be prevented by following a phenylalanine-free diet beginning at birth. |
| Cystic fibrosis (CF) | Cystic fibrosis transmembrane conductance regulator, which is a chloride-ion transporter | Inability to regulate ion balance in epithelial cells. Leads to a variety of abnormalities, including production of thick lung mucus and chronic lung infections. |
| Tay-Sachs disease | Hexosaminidase A | Inability to metabolize certain lipids. Leads to paralysis, blindness, and early death. |
| Hemophilia A | Coagulation factor VIII | A defect in blood clotting due to a missing clotting factor. An accident may cause excessive bleeding or internal hemorrhaging. |

*Individuals who exhibit the disease are homozygous (or hemizygous) for a recessive allele that results in a defect in the amount or function of the normal protein.

In the case of Mendel's seven characters in pea plants, the recessive alleles are due to rare mutations. By studying genes and their gene products at the molecular level, researchers have discovered that a recessive allele is often defective in its ability to express a functional protein. In other words, mutations that produce recessive alleles are likely to decrease or eliminate the synthesis or functional activity of a protein. These are called **loss-of-function alleles.**

In a simple dominant/recessive relationship, the recessive allele does not affect the phenotype of the heterozygote. A single copy of the dominant (wild-type) allele is sufficient to mask the effects of the recessive allele. How do we explain the dominant phenotype of the heterozygote? Figure 15.13 considers the example of flower color in a pea plant. The gene encodes an enzyme (protein P) that is needed to convert a colorless molecule into a purple pigment (Figure 15.13a). The *P* allele is dominant because one *P* allele encodes enough of the functional protein—50% of the amount found in a normal homozygote—to provide a purple phenotype (Figure 15.13b). Therefore, the *PP* homozygote and the *Pp* heterozygote both make enough of the purple pigment to yield purple flowers. The *pp* homozygote cannot make any of the functional protein required for pigment synthesis, so its flowers are white.

This explanation—that 50% of the functional protein is enough—is true for many dominant alleles. In such cases, the homozygote carrying the dominant allele is making much more of the protein than necessary, so if the amount is reduced to 50%, as it is in the heterozygote, the individual still has plenty of this protein to accomplish whatever cellular function it performs. In other cases, however, an allele may be dominant because the heterozygote actually produces more than 50% of the normal amount of functional protein. This increased production is due to the phenomenon of gene regulation. The dominant allele is up-regulated in the heterozygote to compensate for the lack of function of the recessive allele.

The idea that recessive alleles usually cause a substantial decrease in the expression of a functional protein is supported by analyses of many human genetic diseases. Keep in mind that many genetic diseases are caused by rare mutant alleles. Table 15.2 lists several examples of human genetic diseases in which a recessive allele fails to produce a specific cellular protein in its active form. Over 9,000 human disorders are caused by mutations in single protein-encoding genes; most of these are recessive alleles. With a human genome size of about 22,000 protein-encoding genes, roughly one-third of such genes are known to cause some kind of abnormality when mutations alter the expression or functionality of their gene product.

## Recessive Alleles That Cause Diseases May Have Multiple Effects on Phenotype

Single-gene disorders, such as those described in Table 15.2, often illustrate the phenomenon of **pleiotropy,** which means that a mutation in a single gene can have multiple effects on an individual's phenotype. Pleiotropy occurs for several reasons, including the following:

1. The expression of a single gene can affect cell function in more than one way. For example, a defect in a microtubule protein may affect cell division as well as cell movement.
2. A gene may be expressed in different cell types in a multicellular organism.
3. A gene may be expressed at different stages of development.

In this genetics unit, we tend to discuss genes as they affect a single trait. This educational approach allows us to appreciate how genes function and how they are transmitted from parents to offspring. However, this focus may also obscure how amazing genes really are. In all or nearly all cases, the expression of a gene is pleiotropic with regard to the characteristics of an organism. The expression of any given gene influences the expression of many other genes in the genome, and vice versa. Pleiotropy is revealed when researchers study the effects of gene mutations.

As an example of a pleiotropic effect, let's consider cystic fibrosis (CF), which we discussed earlier as an example of a recessive human disorder (see Figure 15.11). The normal allele encodes a protein called the cystic fibrosis transmembrane conductance regulator (CFTR), which regulates ion balance by allowing the transport of chloride ions ($Cl^-$) across epithelial cell membranes (refer back to Figure 1.12). The mutation that causes CF diminishes the function of this $Cl^-$ transporter, affecting several parts of the body in different ways.

- The most severe symptom of CF is the production of thick mucus in the lungs, which occurs because of a water imbalance.
- Similarly, thick mucus can also block the tubes that carry digestive enzymes from the pancreas to the small intestine. Without these enzymes, certain nutrients are not properly absorbed into the body. As a result, persons with CF may show poor weight gain.
- In sweat glands, the normal $Cl^-$ transporter has the function of recycling salt out of the glands and back into the skin before it can be lost to the outside world. Persons with CF have excessively salty sweat due to their inability to recycle salt back into their skin cells.

Taken together, we can see that a defect in CFTR has multiple effects throughout the body.

## Incomplete Dominance Results in an Intermediate Phenotype

We will now turn our attention to an example in which the alleles for a given gene do not show a simple dominant/recessive relationship. In some cases, a heterozygote that carries two different alleles exhibits a phenotype that is intermediate between the corresponding homozygous individuals. This phenomenon is known as **incomplete dominance.**

In 1905, Carl Correns discovered this pattern of inheritance for alleles affecting flower color in the four-o'clock plant *(Mirabilis jalapa)*. **Figure 15.14** shows a cross between two four-o'clock plants: a red-flowered homozygote and a white-flowered homozygote. The allele for red flower color is designated $C^R$, and the white allele is $C^W$. These alleles are designated with superscripts rather than upper- and lowercase letters because neither allele is dominant. The offspring of this cross have pink flowers—they are $C^RC^W$ heterozygotes with an intermediate phenotype. If these $F_1$ offspring are allowed to self-fertilize, the $F_2$ generation has 1/4 red-flowered plants, 1/2 pink-flowered plants, and 1/4 white-flowered plants. This is a 1:2:1 phenotype ratio, rather than the 3:1 ratio observed for simple Mendelian inheritance. What is the molecular explanation for this ratio? In this case, the red allele encodes a functional protein needed to produce a red pigment, whereas the white allele is a mutant allele that is nonfunctional. In the $C^RC^W$ heterozygote, 50% of the protein encoded by the $C^R$ allele is not sufficient to produce the red-flower phenotype, but it does provide enough pigment to result in pink flowers.

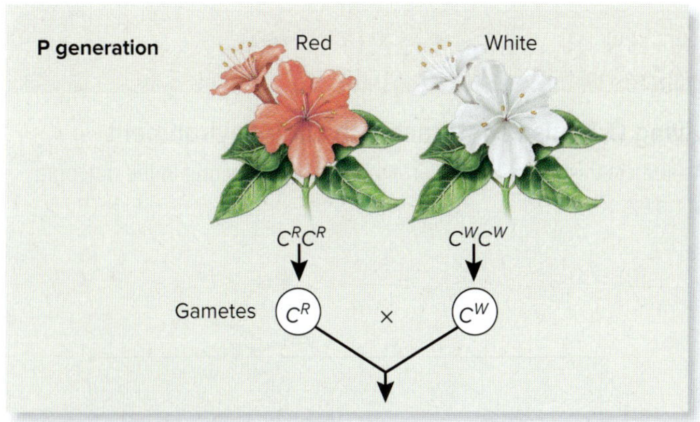

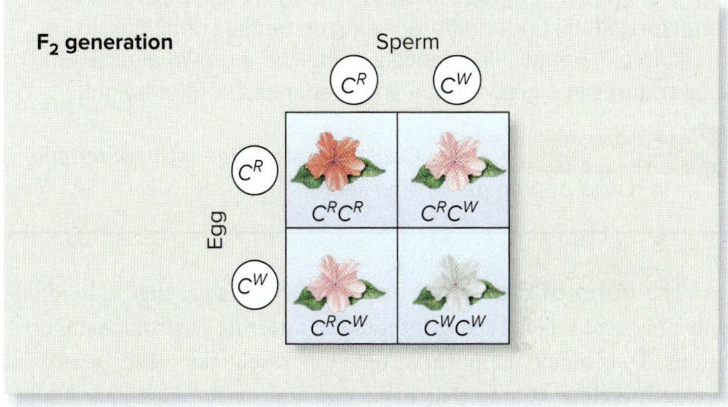


**Figure 15.14 Incomplete dominance in the four-o'clock plant.** When red-flowered and white-flowered homozygotes ($C^RC^R$ and $C^WC^W$) are crossed, the resulting heterozygote ($C^RC^W$) has an intermediate phenotype of pink flowers.

## The Environment Plays a Vital Role in the Making of a Phenotype

In this chapter, we have been mainly concerned with the effects of genes on phenotypes. In addition, phenotypes are shaped by an organism's environment. An organism cannot exist without its genes or without an environment in which to live. Both are indispensable for life. An organism's genotype provides the plan to create a phenotype, and the environment provides nutrients and energy so that plan can be executed.

## Biology Principle

### Living Organisms Interact with Their Environment

As seen in this figure, the environment plays a key role in the outcome of traits.

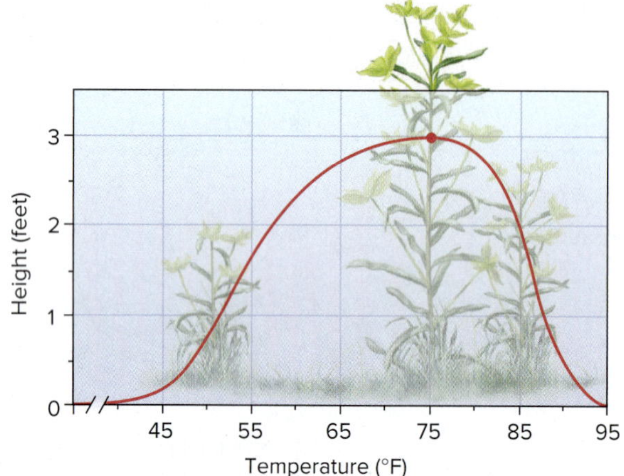


**Figure 15.15** **The norm of reaction.** The norm of reaction is the range of phenotypes that a population of organisms with a particular genotype exhibit under different environmental conditions. In this example, genetically identical plants were grown at different temperatures in a greenhouse and then measured for height.

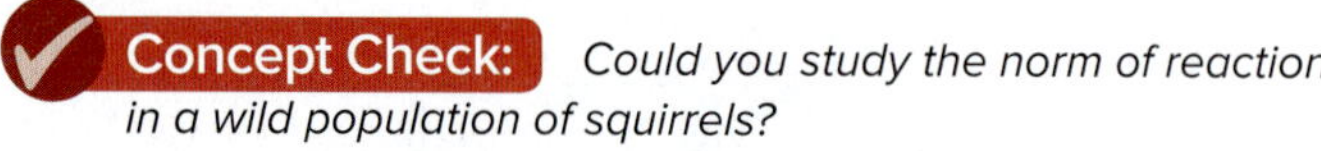

**Concept Check:** *Could you study the norm of reaction in a wild population of squirrels?*

The **norm of reaction** is the phenotype range that individuals with a particular genotype exhibit under differing environmental conditions. To evaluate the norm of reaction, researchers study members of true-breeding strains that have the same genotypes and subject them to different environmental conditions. For example, **Figure 15.15** shows the norm of reaction for genetically identical plants that are homozygous for an allele that confers maximal height (like the tall allele in pea plants). Even so, the height of the plant depends on temperature. As shown in the figure, these plants attain a maximal height when raised at 75°F. At 50°F and 85°F, the plants are substantially shorter. Growth cannot occur below 40°F or above 95°F.

## 15.4 Reviewing the Concepts

- Recessive inheritance is often due to a loss-of-function mutation. In many simple dominant/recessive relationships, the heterozygote has a dominant phenotype because 50% of the normal protein is sufficient to produce that phenotype (Figure 15.13).
- Mutant alleles, which are recessive, are responsible for many inherited diseases in humans (Table 15.2).
- In many cases, the effects of a mutant allele are pleiotropic, meaning the expression of the gene affects several different aspects of bodily structure and function.
- Incomplete dominance occurs when a heterozygote has a phenotype that is intermediate between either homozygote. This occurs because 50% of the functional protein is not enough to produce the same phenotype as a homozygote (Figure 15.14).
- Genes and the environment interact to determine an individual's phenotype. The norm of reaction describes how a phenotype may change under different environmental conditions (Figure 15.15).

## 15.4 Testing Your Knowledge

1. In pea plants, *P* is a dominant allele that confers purple flower color and *p* is a recessive allele that confers white flowers. If the function of the flower-color gene is to make purple pigment, how do you explain the white phenotype?
   a. The *P* allele is a loss-of-function allele.
   b. The *p* allele is a loss-of-function allele.
   c. The *PP* homozygote does not make any purple pigment.
   d. Both a and b are correct.
2. The effects of a gene may be pleiotropic because
   a. the expression of a gene can affect cell function in more than one way.
   b. a gene may be expressed in different cell types in a multicellular organism.
   c. a gene may be expressed at different stages of development.
   d. all of the above.

## 15.5 Sex Chromosomes and X-Linked Inheritance Patterns

### Learning Outcomes

1. Describe different systems of sex determination in animals and plants.
2. **SCISKILLS ▸** Predict the outcome of crosses when genes are located on sex chromosomes.
3. Explain why X-linked recessive traits are more likely to occur in males.

We will now turn our attention to genes located on **sex chromosomes.** As you learned in Chapter 14, this term refers to a distinctive pair of chromosomes that are different in males and females and that determine the sex of individuals. Sex chromosomes are found in many but not all species with two sexes. The study of sex chromosomes proved pivotal in confirming the chromosome theory of inheritance. The distinctive transmission patterns of genes on sex chromosomes helped early geneticists show that particular genes are located on particular chromosomes. Later, other researchers became interested in these genes because some of them were found to cause inherited diseases in humans.

In this section, we will consider a few mechanisms by which sex chromosomes in various species determine an individual's sex. We will then explore the inheritance patterns of genes on sex chromosomes and why recessive alleles are expressed more frequently in males than in females.

## In Many Species, Sex Differences Are Due to the Presence of Sex Chromosomes

Several mechanisms of sex determination have been found in different species of animals. Some examples are described in Figure 15.16. All of these mechanisms involve chromosomal differences between the sexes, and most involve a difference in a single pair of sex chromosomes.

(a) The X-Y system in mammals

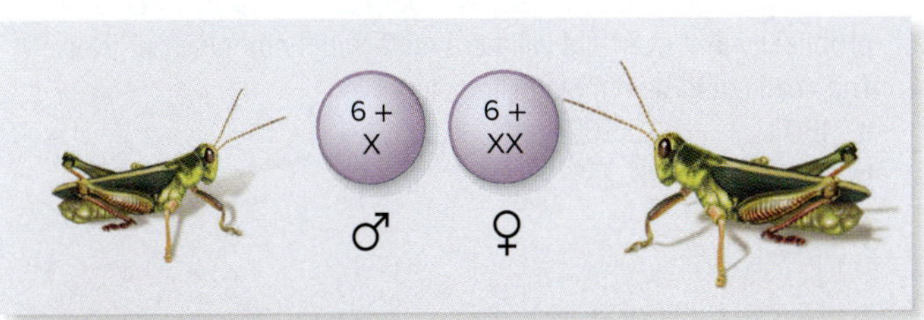


(b) The X-0 system in certain insects

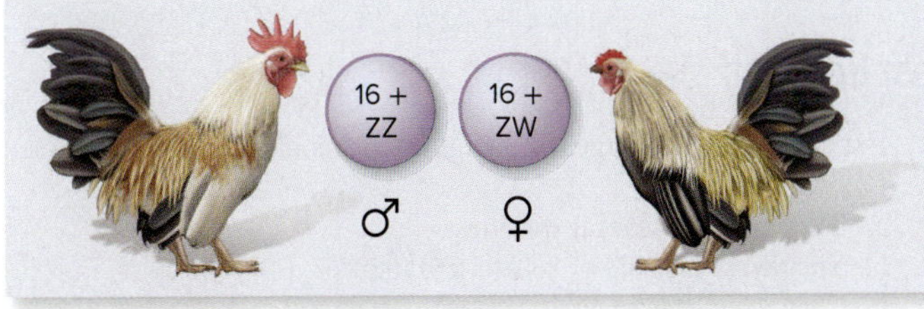


(c) The Z-W system in birds

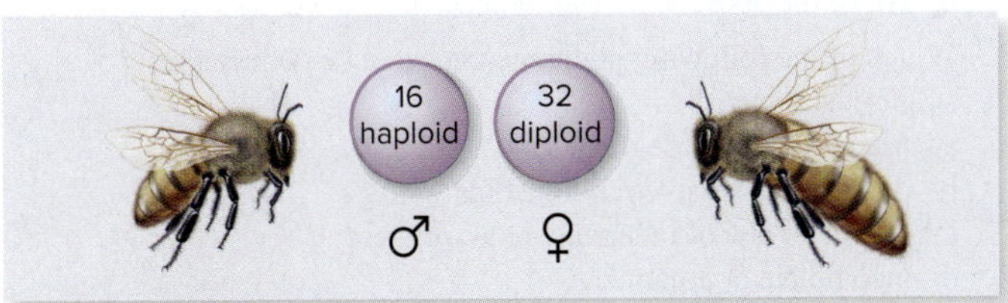


(d) The haplodiploid system in bees

**Figure 15.16 Different mechanisms of sex determination in animals.** The numbers shown in the circles indicate the numbers of autosomes.

**Concept Check:** *If a person is born with only one X chromosome and no Y chromosome, would you expect that person to be a male or a female? Explain your answer.*

**X-Y System** In the X-Y system in mammals, the somatic cells of males have one X and one Y chromosome, whereas female somatic cells contain two X chromosomes (Figure 15.16a). For example, the 46 chromosomes carried by human cells consist of 22 pairs of autosomes and 1 pair of sex chromosomes (either XY or XX). Which chromosome, the X or Y, determines sex? In mammals, the presence of the Y chromosome causes maleness. This is known from the analysis of rare individuals that carry chromosomal abnormalities. For example, mistakes that occasionally occur during meiosis may produce an individual that carries two X chromosomes and one Y chromosome. Such an individual develops into a male. A gene called the *SRY* gene located on the Y chromosome of mammals plays a key role in the developmental pathway that leads to maleness.

**X-0 System** The X-0 system operates in many insects (Figure 15.16b). Unlike the X-Y system in mammals, the presence of the Y chromosome in the X-0 system does not determine maleness. Females in this system have a pair of sex chromosomes and are designated XX. In some insect species that follow the X-0 system, the male has only one sex chromosome, the X. In other X-0 insect species, such as *Drosophila melanogaster*, the male has both an X chromosome and a Y chromosome. In all cases, an insect's sex is determined by the ratio between its X chromosomes and its sets of autosomes. If a fly has one X chromosome and is diploid for the autosomes ($2n$), this ratio is 1/2, or 0.5. This fly will become a male, whether or not it receives a Y chromosome. On the other hand, if a diploid fly receives two X chromosomes, the ratio is 2/2, or 1.0, and the fly becomes a female.

**Z-W System** In some species, such as birds and some fish, the male carries two similar chromosomes (Figure 15.16c). This is called the Z-W system to distinguish it from the X-Y system found in mammals. The male is ZZ, and the female is ZW.

**Haplodiploid System** An interesting mechanism of sex determination known as the haplodiploid system is found in bees (Figure 15.16d). The male bee, or drone, is produced from an unfertilized haploid egg. Therefore, male bees are haploid individuals. Females, both worker bees and queen bees, are produced from fertilized eggs and are diploid.

**Environmental Systems** Although sex in many species of animals is determined by chromosomes, other mechanisms are also known. In certain reptiles and fish, sex is controlled by environmental factors such as temperature. For example, in the American alligator *(Alligator mississippiensis)*, temperature controls sex development. When eggs of this alligator are incubated at 33°C, nearly all of them produce male individuals. When the eggs are incubated at a temperature significantly below 33°C, they produce nearly all females, whereas increasing percentages of females are produced above 34°C.

## In Humans, X-Linked Recessive Traits Are More Likely to Occur in Males

In humans, the X chromosome is rather large and carries over 1,000 protein-encoding genes, whereas the Y chromosome is quite small and has less than 100. Therefore, many genes are found on the X chromosome but not on the Y; these are known as **X-linked genes.** By comparison, fewer genes are known to be Y linked, meaning they are found on the Y chromosome but not on the X. **Sex-linked genes** are found on one sex chromosome but not on the other. Because fewer genes are found on the Y chromosome, the term usually refers to X-linked genes. In mammals, a male cannot be described as being homozygous or heterozygous for an X-linked gene, because these terms apply to genes that are present in two copies. Instead, the term **hemizygous** is used to describe an individual with only one copy of a particular gene. A male mammal is said to be hemizygous for an X-linked gene.

Many X-linked recessive alleles cause diseases in humans, and these diseases occur more frequently in males than in females. As an example, let's consider the X-linked recessive disorder called classical hemophilia (hemophilia A). In individuals with hemophilia, blood does not clot properly, and a minor cut may bleed for a long time. Common accidental injuries that are minor in most people pose a threat of severe internal or external bleeding for hemophiliacs. Hemophilia A is caused by an X-linked recessive allele that encodes a defective form of a clotting protein. If a mother is a heterozygous carrier of hemophilia A, each of her children has a 50% chance of inheriting the recessive allele. The following Punnett square shows a cross between an unaffected father and a heterozygous mother. $X^H$ designates an X chromosome carrying the functional allele, and $X^{h\text{-}A}$ is the X chromosome that carries the recessive allele for hemophilia A.

| Female gametes ♀ / Male gametes ♂ | $X^H$ | Y |
|---|---|---|
| $X^H$ | $X^HX^H$ Normal female | $X^HY$ Normal male |
| $X^{h\text{-}A}$ | $X^HX^{h\text{-}A}$ Carrier female | $X^{h\text{-}A}Y$ Male with hemophilia |

Although each child has a 50% chance of inheriting the hemophilia allele from the mother, only 1/2 of the sons will exhibit the disorder. Because a son inherits only one X chromosome, a son who inherits the recessive allele from his mother will have hemophilia. However, a daughter inherits an X chromosome from both her mother and her father. In this example, a daughter who inherits the recessive (hemophilia-causing) allele from her mother also inherits a functional allele from her father. This daughter will have not have hemophilia, but if she passes the recessive allele to her son, he will have hemophilia.

## 15.5 Reviewing the Concepts

- Most species of animals have separate male and female sexes. In many cases, sex chromosomes determine an individual's sex (Figure 15.16).
- X-linked genes are found on the X chromosome but not the Y chromosome. X-linked recessive alleles in humans cause diseases, such as hemophilia, which are more likely to occur in males.

## 15.5 Testing Your Knowledge

1. On rare occasions, an organism may inherit an abnormal number of sex chromosomes. If an individual was XXY, what would be the phenotype in a mammal and in an insect?
   a. male mammal and male insect
   b. male mammal and female insect
   c. female mammal and male insect
   d. female mammal and female insect
   e. none of the above
2. A woman is heterozygous for an X-linked trait, hemophilia A. If she has a child with a man without hemophilia A, what is the probability that the child will be a male with hemophilia A? (Note: The child could be a male or female.)
   a. 100% c. 50% e. 0%
   b. 75% d. 25%

# Assess and Discuss

## Test Yourself

1. Experimental advantages of using pea plants included which of the following?
   a. They came in several varieties.
   b. They were capable of self-fertilization.
   c. They were easy to cross.
   d. Their anthers could be removed prior to pollination.
   e. All of the above were advantages.
2. Which of the following phenomena cannot be deduced from a single-factor cross?
   a. Mendel's law of segregation
   b. dominant/recessive relationships
   c. Mendel's law of independent assortment
   d. incomplete dominance
   e. All of the above can be observed in a single-factor cross.
3. A *Tt* plant is crossed to a *tt* plant. What is the expected outcome for the ratio of offspring from this cross?
   a. 3 tall : 1 dwarf
   b. 1 tall : 1 dwarf
   c. 1 tall : 3 dwarf
   d. 2 tall : 1 dwarf
   e. 1 tall : 2 dwarf

4. A cross is made between a pea plant that is *RrYy* and a plant that is *rrYy*. What is the predicted outcome of the seed phenotypes?
   a. 9 round, yellow : 3 round, green : 3 wrinkled, yellow : 1 wrinkled, green
   b. 9 round, green : 3 wrinkled, green : 3 round, yellow: 1 wrinkled, yellow
   c. 3 round, yellow : 3 round, green : 1 wrinkled, yellow : 1 wrinkled, green
   d. 3 round, yellow : 1 round, green : 3 wrinkled, yellow : 1 wrinkled, green
   e. 1 round, yellow : 1 round, green : 1 wrinkled, yellow : 1 wrinkled, green
5. A pea plant has the genotype *TtRr*. The independent assortment of these two genes occurs at __________ because chromosomes with the _________ alleles line up independently of the ___________ alleles.
   a. metaphase of meiosis I, *T* and *t*, *R* and *r*
   b. metaphase of meiosis I, *T* and *R*, *t* and *r*
   c. metaphase of meiosis II, *T* and *t*, *R* and *r*
   d. metaphase of meiosis II, *T* and *R*, *t* and *r*
   e. metaphase of mitosis, *T* and *t*, *R* and *r*
6. Which of the following would not be observed in a pedigree if a genetic disorder was inherited in a recessive manner?
   a. Two unaffected parents have an affected offspring.
   b. Two unaffected parents have an unaffected offspring.
   c. Two affected parents have an unaffected offspring.
   d. One affected and one unaffected parent have an unaffected offspring.
   e. All of the above are possible for a recessive disorder.
7. A woman does not show symptoms of hemophilia A. She has a son with a man with hemophilia, and that son does have hemophilia. Their next child is a daughter. What is the probability that this daughter will have hemophilia?
   a. 100% b. 75% c. 50% d. 25% e. 0%
8. A gene that affects more than one trait is said to be
   a. dominant.
   b. wild-type.
   c. dihybrid.
   d. pleiotropic.
   e. heterozygous.
9. A hypothetical flowering plant species produces blue, light blue, and white flowers. To determine the inheritance pattern, the following crosses were conducted with the results indicated:

   blue × blue → all blue

   white × white → all white

   blue × white → all light blue

   What type of inheritance pattern does this represent?
   a. simple Mendelian
   b. X-linked
   c. codominance
   d. incomplete dominance
   e. pleiotropy
10. Genes located on one sex chromosome but not the other are said to be
   a. X-linked.
   b. dominant.
   c. hemizygous.
   d. sex-linked.
   e. autosomal.

## Conceptual Questions

1. Describe the difference between genotype and phenotype. Give three examples. Is it possible for two individuals to have the same phenotype but different genotypes?
2. When examining a human pedigree, what patterns do you look for to distinguish between X-linked recessive inheritance and autosomal recessive inheritance?
3. **PRINCIPLES** A principle of biology is that living organisms interact with their environment. Discuss how the environment plays a key role in determining the outcome of an individual's traits.

## Collaborative Questions

1. Honeybees are unusual in that male bees (drones) have only one copy of each gene but female bees have two copies of their genes. That is because drones develop from eggs that have not been fertilized by sperm cells. In bees, the trait of long wings is dominant to short wings, and the trait of black eyes is dominant to white eyes. If a drone with short wings and black eyes is mated to a queen bee that is heterozygous for both genes, what are the predicted genotypes and phenotypes of male and female offspring? What are the phenotype ratios if we assume an equal number of male and female offspring?
2. Discuss the principles of the chromosome theory of inheritance. Which principles do you think were deduced via light microscopy, and which were deduced from crosses? What modern techniques could be used to support the chromosome theory of inheritance?

## Online Resource

**connect.mheducation.com**

SMARTBOOK® SmartBook® is the first and only adaptive reading experience designed to change the way students read and learn.

# 16

# Transmission of Genetic Information from Parents to Offspring II: Epigenetics, Linkage, and Extranuclear Inheritance

Female honeybees that are fed royal jelly throughout their entire larval stage and into adulthood develop into queen bees. The larger queen bee is shown with a blue disk labeled 68. By comparison, those larvae that do not receive this diet become smaller worker bees. These differences in development are caused by epigenetic modifications.

## Chapter Outline

Female honeybees *(Apis mellifera)* are of two types: queen bees and worker bees (see chapter-opening photo). Queens are larger, live for years, and produce up to 2,000 eggs each day. By comparison, the smaller worker bees are sterile, typically live only for weeks, and engage in specialized types of work, such as constructing and cleaning comb cells, nurturing larvae, guarding the hive entrance, and foraging for pollen and nectar. The striking differences between queen and worker bees are caused by differences in their diets. Certain worker bees, called nurse bees, produce a nutritive substance called royal jelly from glands in their mouths. All female larvae are initially fed royal jelly, but those that are bathed in royal jelly throughout their larval development and feed on it into adulthood become queens. Alternatively, female larvae that are weaned at an early stage of development and switched to a diet of pollen and nectar become worker bees.

The differences between queen and worker bees are not due to differences in the alleles they carry. We cannot predict the phenotype of female bees based on their genotype. Likewise, we cannot carry out crosses and predict the ratio between queen and worker bees based on Mendel's laws. This observation does not mean Mendel was wrong, however. A large number of genes do obey his laws and follow the types of inheritance patterns that we examined in Chapter 15. However, many genes, including those that control female development in honeybees, do not.

How do we distinguish genes that follow a Mendelian pattern from genes that do not? Mendelian inheritance patterns follow three general rules:

1. Except in the case of rare mutations, genes are passed unaltered from cell to cell, and from generation to generation.
2. The genes obey Mendel's law of segregation.
3. For crosses involving two or more genes, the genes obey Mendel's law of independent assortment.

For Mendelian patterns of inheritance, we must also consider other factors, such as the dominant/recessive relationship of alleles and whether or not the gene is on the X chromosome. Once these additional factors are understood, we can predict the phenotypes of offspring from their genotypes and predict the outcome of crosses based on Mendel's laws. Most genes in eukaryotic species follow a Mendelian pattern of inheritance.

In this chapter, we will examine additional types of inheritance patterns that deviate from a Mendelian pattern because one of the three rules is broken. In Sections 16.1–16.4, we will explore epigenetics, which breaks rule number 1. In epigenetics, genes that are passed to offspring are altered, either in the gametes that produce the offspring or when the offspring are undergoing embryonic development. This alteration does not change the DNA sequence of a gene but, instead, alters the structure of chromatin and thereby affects the gene's expression. We will then examine inheritance patterns that arise because some genetic material is not located in the cell nucleus. Certain organelles, such as mitochondria and chloroplasts, contain their own genetic material. The inheritance patterns of genes within mitochondria and chloroplasts do not obey rule number 2; they do not follow the law of segregation. Finally, we will consider inheritance patterns that violate rule number 3, which is Mendel's law of independent assortment. As you will learn, some genes do not independently assort because they are located on the same chromosome.

## 16.1 Overview of Epigenetics

### Learning Outcomes

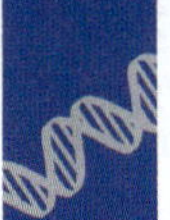

1. Define epigenetics and epigenetic inheritance.
2. Outline the types of molecular changes that underlie epigenetic effects on gene expression.

The term epigenetics was first coined by British biologist Conrad Waddington in 1941. The prefix *epi-*, which means over, suggests that some types of changes in gene expression are at a level that goes beyond changes in DNA sequences. How do geneticists distinguish an epigenetic effect from other types of gene regulation, such as those described in Chapter 12 (refer back to Section 12.4)? An epigenetic effect begins with an initial event that causes a change in gene expression. For example, DNA methylation may inhibit transcription. However, for this to be an epigenetic effect, the change must be passed from cell to cell and does not involve a change in the sequence of DNA.

Thus, a key feature of an epigenetic effect is the long-term maintenance of a change in gene expression. As an example, let's consider muscle cells in humans. Some genes in the human genome should be expressed in muscle cells, such as the genes that encode the proteins called actin and myosin, which are required for muscle contraction (see Chapter 35). In embryonic muscle cells, these genes are subjected to epigenetic changes that promote their expression through adulthood. Alternatively, in other cell types, these same genes are inhibited by epigenetic changes such as DNA methylation. As the embryo grows and eventually becomes an adult, these epigenetic changes are passed from cell to cell so that adult muscle cells express actin and myosin genes at very high levels, whereas many other cell types do not.

Some epigenetic changes, such as those involving the expression or inhibition of actin and myosin genes, are relatively permanent during the life of an individual. Alternatively, other epigenetic changes may be reversible during the life of an individual or may be reversible from one generation to the next. For example, many species of flowering plants undergo **vernalization**, which is the process in which certain species of plants require an exposure to cold temperatures in order to flower. After vernalization, plants do not necessarily initiate flowering, but they acquire the ability to do so. Most commonly, plants are vernalized by exposure to cold winter temperatures and then flower the following spring or summer. Vernalization involves epigenetic changes to genes that play a role in flowering. These epigenetic changes are induced by cold winter temperatures, and they are maintained during the flowering season. However, the epigenetic changes are reversed after the flowering season is over.

Although researchers are still debating the proper definition, one way to define epigenetics is the following.

> ***Epigenetics*** *is the study of mechanisms that lead to changes in gene expression that can be passed from cell to cell and are reversible, but do not involve a change in the sequence of DNA. This type of change may also be called an* **epimutation***—a heritable change in gene expression that does not alter the sequence of DNA.*

In multicellular species that reproduce via gametes (that is, sperm and egg cells), some types of epigenetic changes are passed from parent to offspring, a phenomenon called **epigenetic inheritance.** In Section 16.2, we will consider an example of epigenetic inheritance called genomic imprinting. However, not all epigenetic changes are inherited from parents. For example, a person may be exposed to an environmental agent in cigarette smoke that causes an epigenetic change in a lung cell that is subsequently transmitted from cell to cell and promotes lung cancer. Such a change would not be transmitted to offspring.

The molecular mechanisms of epigenetics are the subject of a large amount of recent research. The most common types of molecular changes that underlie epigenetic effects on gene expression are DNA methylation, chromatin remodeling, covalent histone modification, and the localization of histone variants (**Table 16.1**). These types of changes are also involved in transient (nonepigenetic) gene regulation that is not transmitted from cell to cell. The details

**Table 16.1 Chromatin Modifications That May Have an Epigenetic Effect on Gene Expression**

| | |
|---|---|
| DNA methylation | Methyl groups may be attached to cytosine bases in DNA. When this occurs near promoters, transcription is usually repressed. |
| Chromatin remodeling | Nucleosomes may be moved to new locations or evicted. When such changes occur in the vicinity of promoters, the level of transcription may be altered. Also, larger-scale changes in chromatin structure may occur, such as those that happen during X-chromosome inactivation in female mammals, discussed later in this chapter. |
| Covalent histone modification | Specific amino acid side chains in the amino terminal tails of histones can be covalently modified. For example, they can be acetylated or phosphorylated. Such modifications may repress or activate transcription. |
| Localization of histone variants | Histone variants may become localized to specific locations, such as near the promoters of genes, and affect transcription. |

of these mechanisms were examined in Chapter 12 (refer back to Section 12.4). In some cases, epigenetic changes repress the transcription of a given gene; in other cases, they activate gene transcription. In Sections 16.2–16.4, we will consider examples in which these types of changes have an epigenetic effect on gene expression and thereby affect an individual's phenotype.

## 16.1 Reviewing the Concepts

- Epigenetics is the study of mechanisms that lead to changes in gene expression that can be passed from cell to cell and are reversible, but do not involve a change in the sequence of DNA.
- The most common types of molecular changes that underlie epigenetic effects on gene expression are DNA methylation, chromatin remodeling, covalent histone modification, and the localization of histone variants (Table 16.1).

## 16.1 Testing Your Knowledge

1. Which of the following is an epigenetic change that alters gene expression?
   a. DNA methylation
   b. chromatin remodeling
   c. covalent histone modification
   d. localization of histone variants
   e. all of the above

## 16.2 Epigenetics: Genomic Imprinting

### Learning Outcomes

1. **SCISKILLS** ▶ Predict the outcome of crosses for imprinted genes.
2. Explain the molecular basis of genomic imprinting.

The term imprinting implies a type of marking process that has a memory. For example, newly hatched birds identify marks on their parents, which allows them to distinguish their parents from other individuals. The term **genomic imprinting** refers to an analogous situation in which a segment of DNA is marked, and that mark is retained and recognized throughout the life of the organism inheriting the marked DNA. However, depending on the gene, it may be marked by females during egg formation or by males during sperm formation, but not both. This marking process, which involves epigenetic modifications, affects whether or not the gene is expressed in the offspring. Depending on how a particular gene is marked by one of the parents, the offspring expresses either the maternal or the paternal allele, but not both. Imprinted genes do not follow a Mendelian pattern of inheritance because imprinting causes the offspring to distinguish between maternal and paternal alleles. From an epigenetic perspective, the cells in the offspring are able to "remember" an event that occurred during gamete formation in their parents.

Genomic imprinting, which was discovered in the early 1980s, occurs in numerous species, including insects, plants, and mammals. The number of human genes that are imprinted is still a matter of controversy, but most estimates are in the range of a few hundred to over 1,000. In this section, we will consider an example of genomic imprinting that occurs in mammals.

### For Imprinted Genes, the Gene from Only One Parent Is Expressed

One of the first imprinted genes to be identified is a gene called *Igf2*, found in mice and other mammals. This gene encodes a growth hormone called insulin-like growth factor 2 (Igf2), which is needed for proper growth. If a functional copy of this gene is not expressed, a mouse will be dwarf. The *Igf2* gene is known to be located on an autosome, not on a sex chromosome. Because mice are diploid, they have two copies of this gene, one from each parent.

Researchers have discovered mutations in the *Igf2* gene that block the function of the Igf2 hormone. Such an allele is designated $Igf2^-$. When a homozygous *Igf2 Igf2* mouse and a homozygous $Igf2^- Igf2^-$ mouse are crossed, the results are surprising (**Figure 16.1**). If the female parent is homozygous for the mutant allele and the male parent is homozygous for the functional allele (left side), all of the offspring grow to a normal size. By contrast, if the female is homozygous for the functional allele and the male is homozygous for the mutant allele (right side), all of the offspring are dwarf. The reason this result is so surprising is that the normal-size and dwarf offspring have the same genotype ($Igf2\ Igf2^-$) but different phenotypes! In mice, the *Igf2* gene is imprinted in such a way that only the paternal allele is expressed, which means it is transcribed into mRNA. The maternal allele is not transcribed. The newborn mice shown on the left side of the photograph of Figure 16.1 are normal size, because they express a functional paternal allele. By contrast, the mice on the right side are dwarf because they express a paternal allele that is defective and results in a nonfunctional hormone.

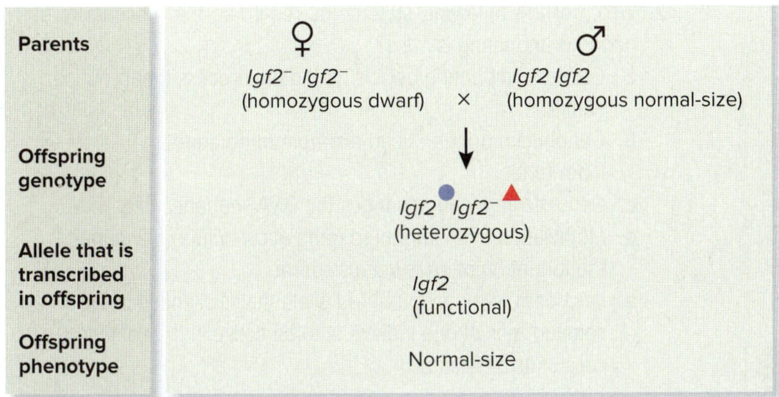


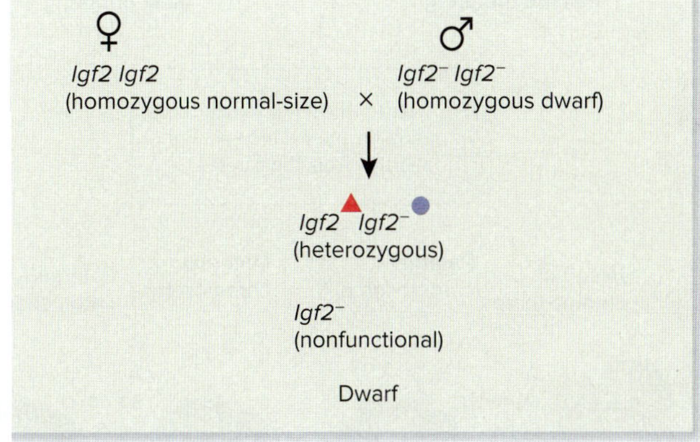

**Figure 16.1 An example of genomic imprinting in the mouse.** In the cross on the left, a homozygous female carrying the mutant allele *Igf2*$^-$ was crossed to a homozygous male with the functional *Igf2* allele. Offspring are normal size, because the paternal allele is expressed. In the cross on the right, a homozygous *Igf2 Igf2* female was crossed to a homozygous male carrying the mutant allele. In this case, offspring are dwarf because the paternal allele is defective and the maternal allele is not expressed.

**Concept Check:** *If you crossed an Igf2 Igf2$^-$ male mouse to an Igf2 Igf2$^-$ female mouse, what would be the expected results?*

## The Transcription of an Imprinted Gene Depends on Methylation

Why is the maternal gene encoding Igf2 not transcribed into mRNA? To answer this question, we need to consider the regulation of gene transcription in eukaryotes. As discussed in Chapter 12, DNA methylation, which is the attachment of methyl (—$CH_3$) groups to bases of DNA, can alter gene transcription. Researchers have discovered that DNA methylation is the marking process that occurs during the imprinting of certain genes, including the *Igf2* gene. For most genes, DNA methylation silences gene expression by inhibiting the initiation of transcription or by causing the chromatin in a region to become more compact. By contrast, for a few imprinted genes, DNA methylation may enhance gene expression by attracting activator proteins to the promoter or by preventing the binding of repressor proteins.

Figure 16.2 shows the imprinting process in which a maternal gene is methylated. The left side of the figure follows the marking process during the life of a female individual; the right side follows the same process in a male. Both individuals received a methylated gene from their mother and a nonmethylated copy of the same gene from their father. Via cell division, the zygote develops into a multicellular organism. Each time a somatic cell divides, enzymes in the cell maintain the methylation of the maternal gene, but the paternal gene remains unmethylated. If methylation inhibits transcription of this gene, only the paternal copy will be expressed in the somatic cells of both the male and female offspring.

The methylation state of an imprinted gene may be altered when individuals make gametes. First, the methylation is erased (Figure 16.2, step 2). Next, the gene may be methylated again, but that depends on whether the individual is a female or a male. In females making eggs, both copies of the gene are methylated; in males making sperm, neither copy is methylated. When we consider the effects of methylation over the course of two or more generations, we can see how this phenomenon results in an epigenetic transmission pattern. The male in Figure 16.2 has inherited a methylated gene from his mother that is transcriptionally silenced in his somatic cells. Although he does not express this gene during his lifetime, he can pass on an active, nonmethylated copy of exactly the same gene to his offspring.

Genomic imprinting is a recently discovered phenomenon that has been shown to occur for many genes in mammals. For some genes, such as *Igf2*, the maternal allele is silenced, but for other genes, the paternal allele is silenced. Although several hypotheses have been advanced, biologists are still trying to determine whether this curious marking process confers any benefit to the individual.

## 16.2 Reviewing the Concepts

- In genomic imprinting, offspring express either a maternal or paternal allele, depending on how a particular gene is marked, or imprinted (Figure 16.1).
- During gamete formation, DNA methylation of an allele from one parent is a mechanism to achieve imprinting (Figure 16.2).

## 16.2 Testing Your Knowledge

1. A male mouse that is homozygous for the functional allele of the *Igf2* gene is mated to a female that is heterozygous, carrying one functional copy and one mutant (nonfunctional) copy of the gene. What would be the expected outcome of this cross?
   a. all normal-size offspring
   b. 1/2 normal-size and 1/2 dwarf offspring
   c. all dwarf offspring
   d. 3/4 normal-size and 1/4 dwarf offspring
   e. none of the above

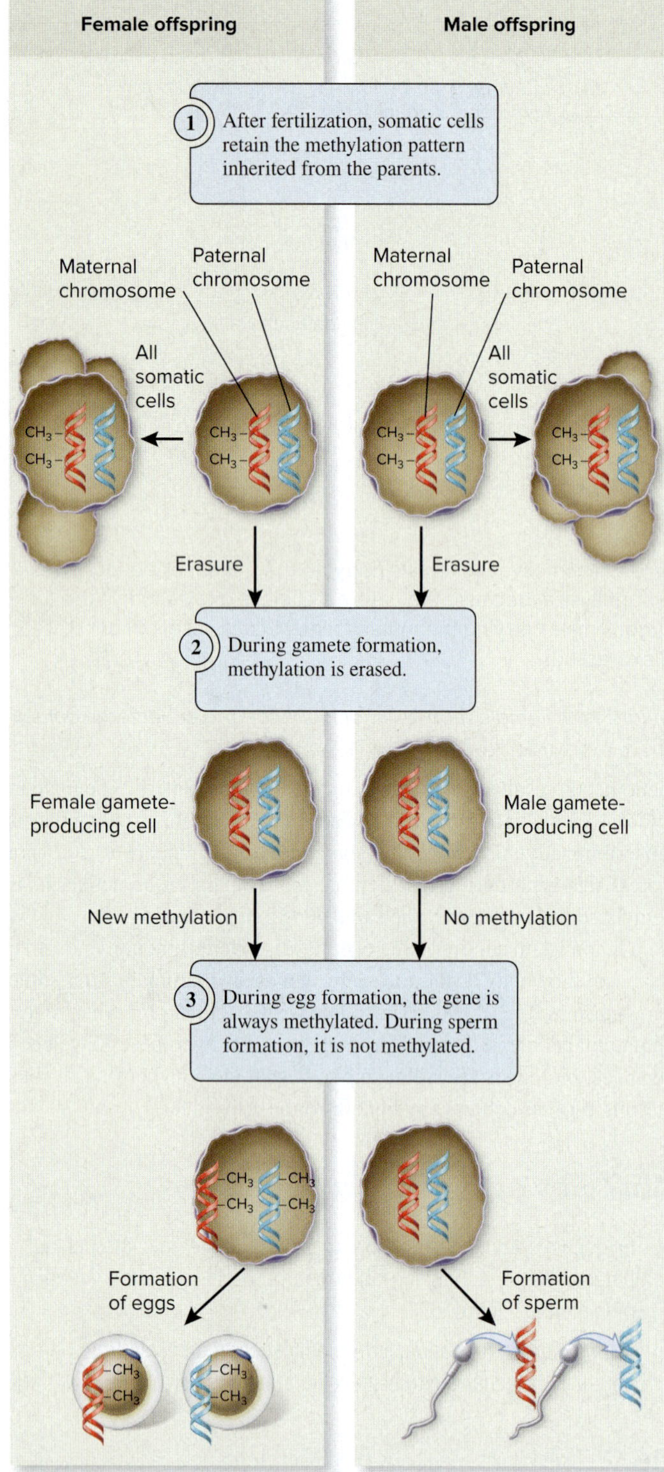


**Figure 16.2 Genomic imprinting via DNA methylation.** The cells of the female and male offspring at the top of this figure have a methylated gene inherited from the mother and a nonmethylated version of the same gene inherited from the father. This pattern of methylation is maintained in the somatic cells of both the female and male offspring. The methylation is erased during gamete formation, but in females, the gene is methylated again at a later stage in the formation of eggs. Therefore, females always transmit a methylated, transcriptionally silent copy of this gene, whereas males transmit a nonmethylated, transcriptionally active copy.

2. Which of the following statements regarding the mechanism of genomic imprinting is *false*?
   a. Genomic imprinting begins during the process that produces gametes.
   b. Genomic imprinting is an example of epigenetic inheritance.
   c. Genomic imprinting changes the DNA sequence of a gene.
   d. Methylation of an imprinted gene occurs differently during the formation of eggs versus sperm.
   e. It is possible for an imprinted allele that is silenced in the somatic cells of one individual to be active in the somatic cells of its offspring.

## 16.3 Epigenetics: X-Chromosome Inactivation

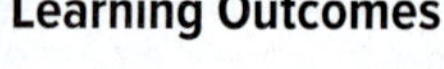

### Learning Outcomes

1. Explain how X-chromosome inactivation may affect the phenotype of female mammals.
2. Describe the process of X-chromosome inactivation at the cellular level.

As we have seen, genomic imprinting involves epigenetic changes that occur during gamete formation and are passed from parent to offspring. Other types of epigenetic changes begin later, such as during embryonic development, and then persist into adulthood. For example, as mentioned in Section 16.1, epigenetic changes may silence actin and myosin genes in some embryonic cells, and these changes are passed from cell to cell so that the same genes are silenced in nonmuscle cells of the adult.

In this section, we will consider an example of epigenetic silencing that begins during embryonic development in female mammals. As discussed in Chapter 15, female mammals carry two X chromosomes in their cells, whereas males carry one X and one Y. During embryonic development in female mammals, one of the X chromosomes undergoes an epigenetic change called **X-chromosome inactivation (XCI).** This process causes that X chromosome to become highly compacted, which silences the genes that it carries. In this section, we will examine how XCI may affect a female mammal's phenotype and how the process occurs at the cellular and molecular levels.

### In Female Mammals, One X Chromosome Is Inactivated in Each Somatic Cell

In 1961, the British geneticist Mary Lyon proposed the epigenetic phenomenon of X-chromosome inactivation, in which one of the two copies of the X chromosome in the somatic cells of female mammals is inactivated, meaning that its genes are not expressed. XCI is based on two lines of evidence.

- In 1949, Canadian physicians Murray Barr and Ewart Bertram identified a highly condensed structure in the cells of female cats that was not found in the cells of male cats. This structure was named a **Barr body** after one of its discoverers. It is now known that the Barr body is an inactivated X chromosome (Figure 16.3).

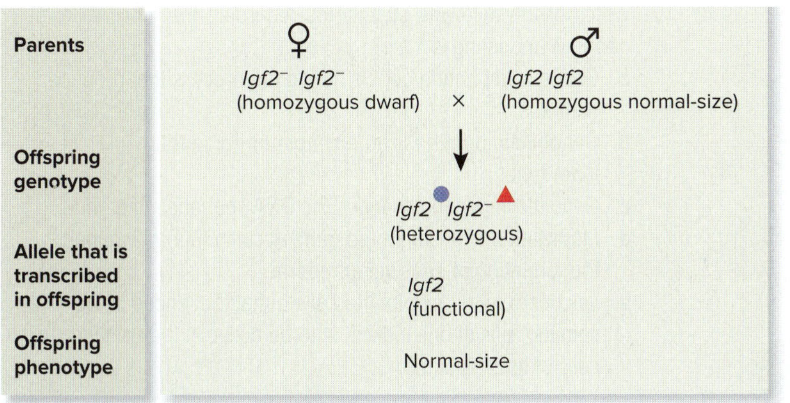


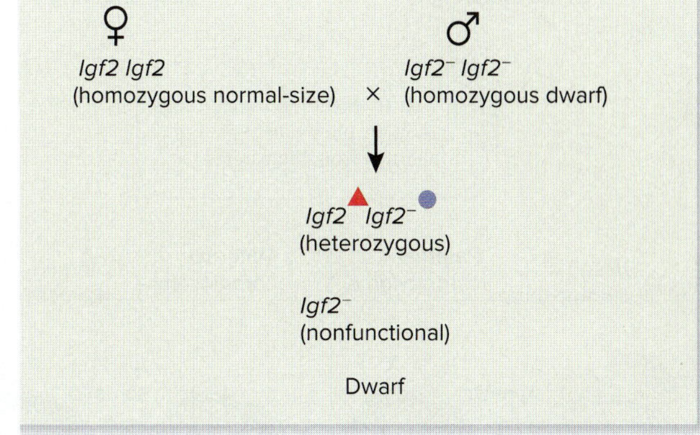


*Igf2* Functional allele
*Igf2⁻* Mutant (nonfunctional) allele
▲ Silenced allele (from female parent)
● Expressed allele (from male parent)

**Figure 16.1 An example of genomic imprinting in the mouse.** In the cross on the left, a homozygous female carrying the mutant allele *Igf2*$^-$ was crossed to a homozygous male with the functional *Igf2* allele. Offspring are normal size, because the paternal allele is expressed. In the cross on the right, a homozygous *Igf2 Igf2* female was crossed to a homozygous male carrying the mutant allele. In this case, offspring are dwarf because the paternal allele is defective and the maternal allele is not expressed.

**Concept Check:** *If you crossed an Igf2 Igf2*$^-$ *male mouse to an Igf2 Igf2*$^-$ *female mouse, what would be the expected results?*

## The Transcription of an Imprinted Gene Depends on Methylation

Why is the maternal gene encoding Igf2 not transcribed into mRNA? To answer this question, we need to consider the regulation of gene transcription in eukaryotes. As discussed in Chapter 12, DNA methylation, which is the attachment of methyl (—$CH_3$) groups to bases of DNA, can alter gene transcription. Researchers have discovered that DNA methylation is the marking process that occurs during the imprinting of certain genes, including the *Igf2* gene. For most genes, DNA methylation silences gene expression by inhibiting the initiation of transcription or by causing the chromatin in a region to become more compact. By contrast, for a few imprinted genes, DNA methylation may enhance gene expression by attracting activator proteins to the promoter or by preventing the binding of repressor proteins.

Figure 16.2 shows the imprinting process in which a maternal gene is methylated. The left side of the figure follows the marking process during the life of a female individual; the right side follows the same process in a male. Both individuals received a methylated gene from their mother and a nonmethylated copy of the same gene from their father. Via cell division, the zygote develops into a multicellular organism. Each time a somatic cell divides, enzymes in the cell maintain the methylation of the maternal gene, but the paternal gene remains unmethylated. If methylation inhibits transcription of this gene, only the paternal copy will be expressed in the somatic cells of both the male and female offspring.

The methylation state of an imprinted gene may be altered when individuals make gametes. First, the methylation is erased (Figure 16.2, step 2). Next, the gene may be methylated again, but that depends on whether the individual is a female or a male. In females making eggs, both copies of the gene are methylated; in males making sperm, neither copy is methylated. When we consider the effects of methylation over the course of two or more generations, we can see how this phenomenon results in an epigenetic transmission pattern. The male in Figure 16.2 has inherited a methylated gene from his mother that is transcriptionally silenced in his somatic cells. Although he does not express this gene during his lifetime, he can pass on an active, nonmethylated copy of exactly the same gene to his offspring.

Genomic imprinting is a recently discovered phenomenon that has been shown to occur for many genes in mammals. For some genes, such as *Igf2,* the maternal allele is silenced, but for other genes, the paternal allele is silenced. Although several hypotheses have been advanced, biologists are still trying to determine whether this curious marking process confers any benefit to the individual.

## 16.2 Reviewing the Concepts

- In genomic imprinting, offspring express either a maternal or paternal allele, depending on how a particular gene is marked, or imprinted (Figure 16.1).
- During gamete formation, DNA methylation of an allele from one parent is a mechanism to achieve imprinting (Figure 16.2).

## 16.2 Testing Your Knowledge

1. A male mouse that is homozygous for the functional allele of the *Igf2* gene is mated to a female that is heterozygous, carrying one functional copy and one mutant (nonfunctional) copy of the gene. What would be the expected outcome of this cross?
   **a.** all normal-size offspring
   **b.** 1/2 normal-size and 1/2 dwarf offspring
   **c.** all dwarf offspring
   **d.** 3/4 normal-size and 1/4 dwarf offspring
   **e.** none of the above

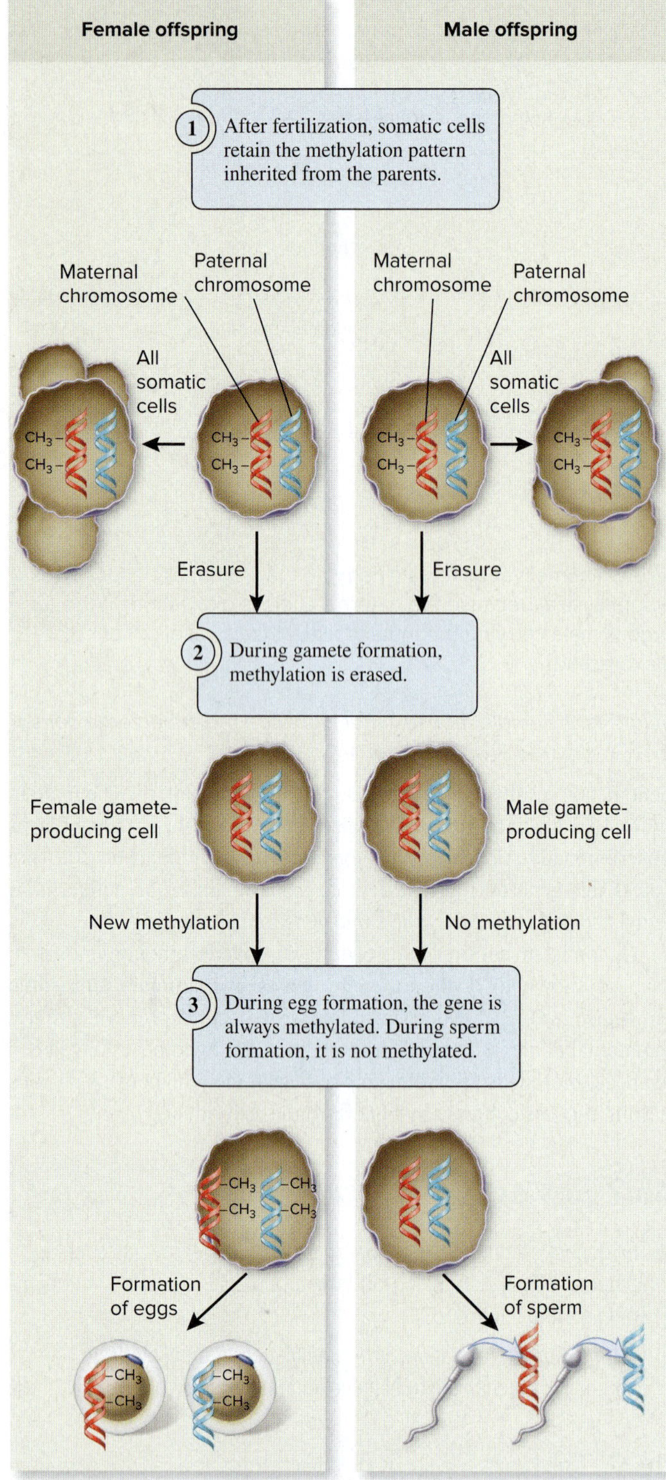


**Figure 16.2** **Genomic imprinting via DNA methylation.** The cells of the female and male offspring at the top of this figure have a methylated gene inherited from the mother and a nonmethylated version of the same gene inherited from the father. This pattern of methylation is maintained in the somatic cells of both the female and male offspring. The methylation is erased during gamete formation, but in females, the gene is methylated again at a later stage in the formation of eggs. Therefore, females always transmit a methylated, transcriptionally silent copy of this gene, whereas males transmit a nonmethylated, transcriptionally active copy.

2. Which of the following statements regarding the mechanism of genomic imprinting is *false*?
   a. Genomic imprinting begins during the process that produces gametes.
   b. Genomic imprinting is an example of epigenetic inheritance.
   c. Genomic imprinting changes the DNA sequence of a gene.
   d. Methylation of an imprinted gene occurs differently during the formation of eggs versus sperm.
   e. It is possible for an imprinted allele that is silenced in the somatic cells of one individual to be active in the somatic cells of its offspring.

## 16.3 Epigenetics: X-Chromosome Inactivation

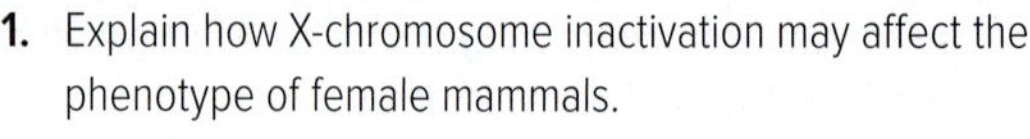

### Learning Outcomes

1. Explain how X-chromosome inactivation may affect the phenotype of female mammals.
2. Describe the process of X-chromosome inactivation at the cellular level.

As we have seen, genomic imprinting involves epigenetic changes that occur during gamete formation and are passed from parent to offspring. Other types of epigenetic changes begin later, such as during embryonic development, and then persist into adulthood. For example, as mentioned in Section 16.1, epigenetic changes may silence actin and myosin genes in some embryonic cells, and these changes are passed from cell to cell so that the same genes are silenced in nonmuscle cells of the adult.

In this section, we will consider an example of epigenetic silencing that begins during embryonic development in female mammals. As discussed in Chapter 15, female mammals carry two X chromosomes in their cells, whereas males carry one X and one Y. During embryonic development in female mammals, one of the X chromosomes undergoes an epigenetic change called **X-chromosome inactivation (XCI).** This process causes that X chromosome to become highly compacted, which silences the genes that it carries. In this section, we will examine how XCI may affect a female mammal's phenotype and how the process occurs at the cellular and molecular levels.

### In Female Mammals, One X Chromosome Is Inactivated in Each Somatic Cell

In 1961, the British geneticist Mary Lyon proposed the epigenetic phenomenon of X-chromosome inactivation, in which one of the two copies of the X chromosome in the somatic cells of female mammals is inactivated, meaning that its genes are not expressed. XCI is based on two lines of evidence.

- In 1949, Canadian physicians Murray Barr and Ewart Bertram identified a highly condensed structure in the cells of female cats that was not found in the cells of male cats. This structure was named a **Barr body** after one of its discoverers. It is now known that the Barr body is an inactivated X chromosome (Figure 16.3).

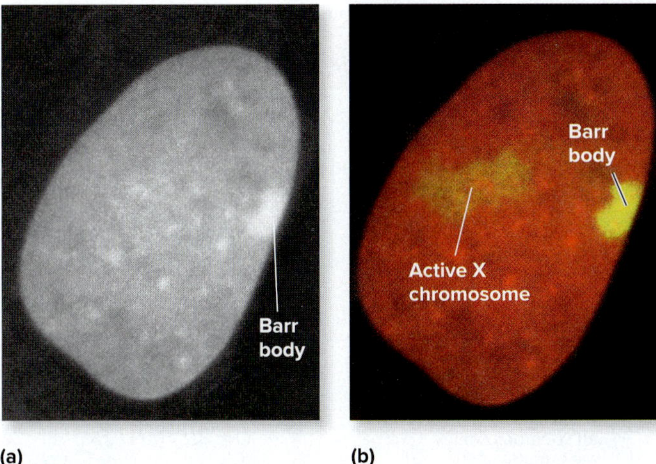


**Figure 16.3** **X-chromosome inactivation in female mammals.** **(a)** A Barr body is seen on the periphery of a human nucleus (during interphase) after staining with a DNA-specific dye. Because it is compact, the Barr body is brightly stained. **(b)** The same nucleus was labeled using a yellow fluorescent probe that recognizes the X chromosome. The Barr body has undergone epigenetic changes that make it more compact than the active X chromosome, which is to the left of the Barr body.

Courtesy of I. Solovei, University of Munich (LMU)

**Concept Check:** *How is the Barr body different from the other X chromosome in this cell?*

- Lyon's second line of evidence was the inheritance pattern of variegated coat colors in certain female mammals. A classic case is the calico cat, which has randomly distributed patches of black and orange fur (Figure 16.4a).

How do we explain this patchwork phenotype? According to Lyon's hypothesis, the calico pattern is due to the permanent inactivation of one X chromosome in each cell that forms a patch of the cat's skin, as shown in Figure 16.4b. The gene involved is an X-linked gene that occurs as an orange allele, $X^O$, and a black allele, $X^B$. A female cat heterozygous for this gene will be calico. (The cat's white underside is due to a dominant allele of a different, autosomal gene.) At an early stage of embryonic development, one of the two X chromosomes is randomly inactivated in each of the cat's somatic cells, including those that will give rise to the hair-producing skin cells. As the embryo grows and matures, the pattern of XCI is maintained during subsequent cell divisions. Skin cells derived from a single embryonic cell in which the $X^B$-carrying chromosome has been inactivated produce a patch of orange fur, because they express only the $X^O$ allele that is carried on the active chromosome. Alternatively, a group of skin cells in which the chromosome carrying $X^O$ has been inactivated express only the $X^B$ allele, producing a patch of black fur. If female mammals are heterozygous for X-linked genes, approximately half of their somatic cells express one allele, whereas the rest of their somatic cells express the other allele. The result is an animal with randomly distributed patches of black and orange fur. These heterozygotes are called **mosaics** because they are composed of two types of cells.

Why does XCI occur? Researchers have proposed that XCI achieves **dosage compensation,** a process that equalizes the expression of X-linked genes in male and female mammals. The inactivation of one X chromosome in females reduces the number of expressed copies (doses) of X-linked genes from two to one, the same as expressed in males. As a result, the expression of X-linked genes in females and males is roughly equal.

## The X Chromosome Has an X Inactivation Center That Controls Compaction into a Barr Body

After Lyon's hypothesis was confirmed, researchers became interested in the cellular and molecular mechanisms of XCI. The cells of humans and other mammals have the ability to count their X chromosomes and allow only one of them to remain active. Additional X chromosomes are converted to Barr bodies. In normal females, two X chromosomes are counted and one is inactivated. In normal males, one X chromosome is counted and none inactivated.

On rare occasions, people are born with abnormalities in the number of their sex chromosomes. In the disorders known as Turner syndrome, triple X syndrome, and Klinefelter syndrome, the cells inactivate the number of X chromosomes necessary to leave a single active chromosome (Table 16.2). For example, in triple X syndrome, in which an extra X chromosome is found in each cell, two X chromosomes are converted to Barr bodies. In spite of XCI, people with these three syndromes do exhibit some abnormalities. The symptoms associated with these disorders may be due to effects that occur prior to XCI or because not all of the genes on the Barr body are completely silenced.

Although the mechanism of XCI is not entirely understood at the molecular level, a short region on the X chromosome called the **X inactivation center (Xic)** is known to play a critical role. Finnish-born American geneticist Eeva Therman and German-born American geneticist Klaus Patau determined that XCI is accomplished by counting the number of Xics and inactivating all X chromosomes except for one. In cells with two X chromosomes, if one of them is missing its Xic due to a chromosome mutation, neither X chromosome will be inactivated, because only one Xic is counted.

**Table 16.2** **Relationship Between XCI and the Number of X Chromosomes**

| Phenotype | Chromosome composition | Number of Barr bodies | Number of active X chromosomes |
|---|---|---|---|
| Normal female | XX | 1 | 1 |
| Normal male | XY | 0 | 1 |
| Turner syndrome (female) | X0 | 0 | 1 |
| Triple X syndrome (female) | XXX | 2 | 1 |
| Klinefelter syndrome (male) | XXY | 1 | 1 |

# Biology Principle

## The Genetic Material Provides a Blueprint for Reproduction

In this example, the process of X-chromosome inactivation results in epigenetic changes and produces a variegated pattern of fur color.

(a) A calico cat

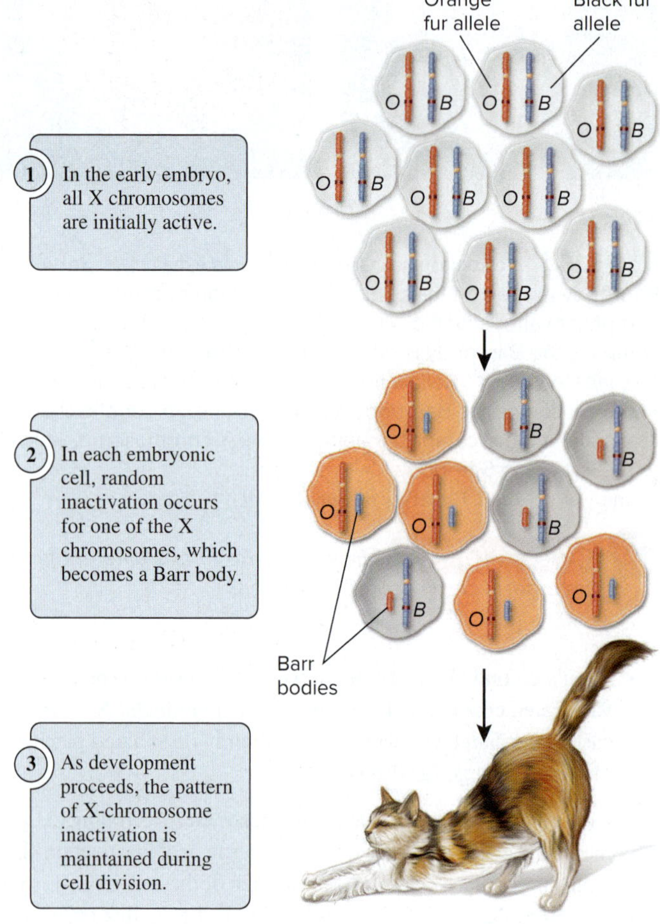

Orange fur allele
Black fur allele
O
B
1 In the early embryo, all X chromosomes are initially active.
2 In each embryonic cell, random inactivation occurs for one of the X chromosomes, which becomes a Barr body.
Barr bodies
3 As development proceeds, the pattern of X-chromosome inactivation is maintained during cell division.


(b) Process of X-chromosome inactivation

**Figure 16.4** **X-chromosome inactivation in a calico cat.** **(a)** A calico cat. **(b)** The calico pattern is due to random X-chromosome inactivation in a female that is heterozygous for an X-linked gene with black ($X^B$) and orange ($X^O$) alleles. The cells at the top of this figure represent a small mass of cells making up the very early embryo. In these cells, both X chromosomes are active. At an early stage of embryonic development, one X chromosome is randomly inactivated in each cell. The initial inactivation pattern is an epigenetic change that is maintained in the descendents of each cell as the embryo matures into an adult. The pattern of orange and black fur in the adult cat reflects the pattern of XCI in the embryo.
*(a)* © Dave King/Dorling Kindersley/Getty Images

Having two active X chromosomes is a lethal condition for a human female embryo.

The expression of a specific gene within the Xic is required for compaction of the X chromosome into a Barr body. This gene, discovered in 1991, is named *Xist* (for X inactive specific transcript). The *Xist* gene product is a long RNA molecule that does not encode a protein. Instead, the role of *Xist* RNA is to coat one of the two X chromosomes during the process of XCI. After coating, proteins associate with the *Xist* RNA and cause epigenetic changes that promote the compaction of the chromosome into a Barr body. The *Xist* gene on the Barr body continues to be expressed after other genes on this chromosome have been silenced. The expression of the *Xist* gene also maintains a chromosome as a Barr body during cell division. Whenever a somatic cell divides in a female mammal, the Barr body is replicated to produce two Barr bodies.

## 16.3 Reviewing the Concepts

- If a female is heterozygous for an X-linked gene, this can lead to a mosaic phenotype, with some of the somatic cells expressing one allele and some expressing the other (Figures 16.3, 16.4).
- XCI in female mammals occurs when one X chromosome in every somatic cell is randomly inactivated. The counting of X inactivation centers (Xics) causes somatic cells to have only one active X chromosome (Table 16.2)

## 16.3 Testing Your Knowledge

1. Calico coat pattern in cats is the result of
   a. X-chromosome inactivation.
   b. X-linked inheritance.
   c. incomplete dominance.
   d. pleiotropy.
2. Which of the following statements regarding X-chromosome inactivation is *false?*
   a. Xic is the site where X-chromosome inactivation begins.
   b. Xic is where the *Xist* gene is located.
   c. The *Xist* gene on the inactivated X chromosome (the Barr body) is not expressed.
   d. The product of the *Xist* gene coats the inactivated X chromosome.

## 16.4 Epigenetics: Effects of Environmental Agents

**Learning Outcomes**

1. Explain how chemicals in the diet may affect the individual's phenotype.
2. List examples of chemicals that cause epigenetic changes that may contribute to cancer.

One of the most active fields in genetics is the study of certain environmental agents that cause epigenetic changes and thereby affect gene expression. Two areas that have received a great deal of attention are the effects of diet and the potential effects of toxic agents, such as carcinogens—cancer-causing agents. In this section, we will consider examples in which environmental agents promote epigenetic changes that affect an individual's phenotype or cause a disease such as cancer.

### Chemicals in an Individual's Diet May Cause Epigenetic Changes That Affect Phenotype

At the beginning of this chapter, we considered how chemicals in royal jelly are responsible for producing queen bees (see chapter-opening photo). Another striking example of how chemicals in the diet can promote epigenetic changes is illustrated by studies of the *Agouti* gene (designated *A*) found in mice. This gene encodes the Agouti signaling peptide that controls the deposition of yellow pigment in developing hairs. In mice that are homozygous for a functional allele, *AA,* the expression of this gene promotes the synthesis of pheomelanin, a yellow pigment. During the growth of a hair, melanocytes (pigment-producing cells) within a hair follicle initially make eumelanin, which is black. The transient expression of the *Agouti* gene causes the melanocytes to make pheomelanin. The melanocytes then revert to making black pigment. The result is a band of yellow pigment sandwiched between layers of black pigment, which gives a brown color. The yellow pigment is not synthesized near the tip of the hair, so the hair of *AA* mice is brown with black tips.

Researchers have identified several mutations that affect the expression of the *Agouti* gene. For example, mice that are homozygous for a loss-of-function allele, *aa,* have black fur because pheomelanin is not made. Alternatively, a gain-of-function mutation that causes the *Agouti* gene to be overexpressed results in a mouse with yellow fur. One such mutation is designated $A^{vy}$. (*A* refers to Agouti, *v* refers to viable, and *y* refers to yellow. The letter *v* is used because other mutations in the *Agouti* gene are not viable.) By characterizing the $A^{vy}$ allele at the molecular level, researchers determined that it is due to the insertion of a new promoter next to the normal promoter of the *Agouti* gene. This new promoter is very active, which causes the overexpression of the *Agouti* gene.

An intriguing observation of mice carrying the $A^{vy}$ allele is that they exhibit a wide phenotypic variation, ranging from yellow to mottled to pseudo-agouti (**Figure 16.5a**). Why should mice with the same genotype show such a wide range of phenotypic variation? Although the answer is not entirely understood, researchers have speculated that the new promoter in mice carrying the $A^{vy}$ allele is very sensitive to epigenetic changes. In particular, this promoter may be more likely to be modified by DNA methylation, which would inhibit its function. Furthermore, a variety of environmental factors may cause this epigenetic change to occur. The sensitivity of this promoter to epigenetic modifications, together with variation in environmental factors, may explain the phenotypic variation seen in these mice.

A key factor that may affect epigenetic changes is diet. With regard to the $A^{vy}$ allele, the exposure of pregnant female mice to different diets affects the phenotypes of the resulting offspring. In 2003, American geneticists Robert Waterland and Randy Jirtle conducted a study in which they investigated the effects of certain dietary supplements. Their goal was to determine if nutrients that are known to inhibit DNA methylation would alter the expression of the *Agouti* gene and thereby affect coat color. A variety of dietary factors can inhibit the enzyme DNA methyltransferase, which methylates DNA. These include folic acid, vitamin $B_{12}$, betaine, and choline chloride.

Waterland and Jirtle began with female mice carrying the $A^{vy}$ allele and divided them into a control group (which was fed a standard diet) and an experimental group (which was fed a diet supplemented with folic acid, vitamin $B_{12}$, betaine, and choline chloride). Both groups were fed their respective diets before and during pregnancy and up to the stage of weaning. Offspring that inherited the $A^{vy}$ allele were then analyzed with regard to their coat color and levels of DNA methylation.

As expected, a range of coat colors was observed among the offspring (**Figure 16.5b**). However, the offspring of females that had been fed a supplemented diet tended to have darker coats. For example, over 25% of the offspring with heavily mottled coats had mothers that were fed a supplemented diet (blue bars), whereas less than 10% had mothers that were given a standard diet (red bars).

The coat colors of the offspring were correlated with the degree of methylation that occurred at the new promoter—offspring with darker coat color had greater levels of DNA methylation (**Figure 16.5c**). How do we explain these results? In the mice that are more yellow, the new promoter has undergone less DNA methylation. Therefore,

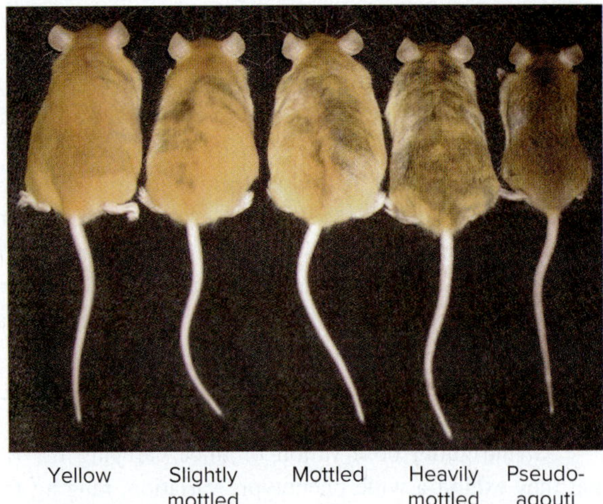


**(a) Range in coat-color phenotypes in $A^{vy}a$ mice due to epigenetic changes**

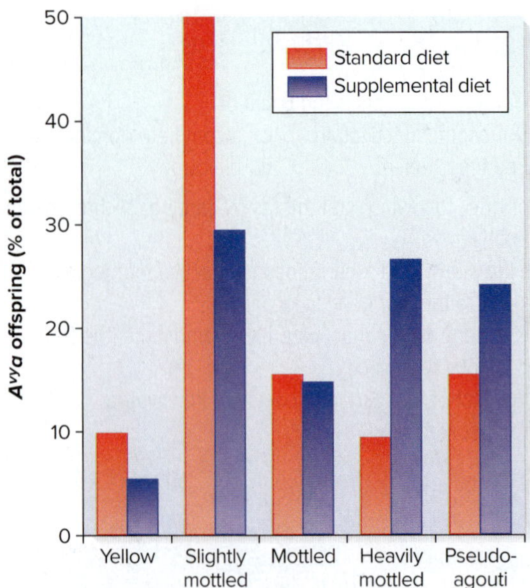


**(b) Effect of diet on coat color**

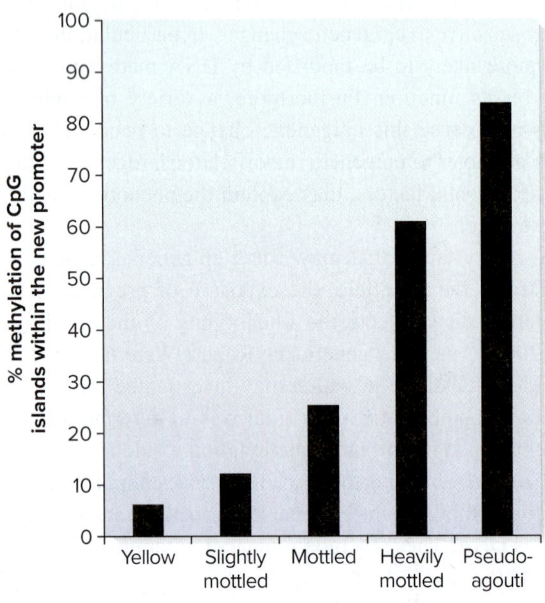


**(c) Level of DNA methylation of CpG islands within the new promoter among mice with different coat colors**

**Figure 16.5** **Dietary effects on coat color in mice.** **(a)** Mice carrying the $A^{vy}$ allele exhibit a range of phenotypes. The mice shown here are heterozygotes, $A^{vy}a$; they carry the $A^{vy}$ allele and a loss-of-function allele *(a)*. **(b)** Effects of diet supplementation on coat color. Red bars represent offspring from females given a standard diet, and blue bars represent offspring from females given a supplemented diet. **(c)** DNA methylation patterns among mice with different coat colors. Data in parts (b) and (c) are from Waterland, R. A., and Jirtle, R. L. 2003. Transposable Elements: Targets for Early Nutritional Effects on Epigenetic Gene Regulation. *Mol. Cell. Biol.* 23: 5293–5300.

Source: D.C. Dolinoy et al., "Maternal genistein alters coat color and protects $A^{vy}$ mouse offspring from obesity by modifying the fetal epigenome," *Environ Health Perspect.* 2006 Apr, 114(4): 567–572

the promoter remains active, leading to the transcription of the *Agouti* gene and the overproduction of yellow pigment. By contrast, the new promoter in the darker mice has undergone more methylation. Such methylation inhibits the overexpression of the *Agouti* gene and thereby prevents the overproduction of yellow pigment, resulting in darker fur.

## Environmental Agents May Cause Epigenetic Changes That Are Associated with Human Diseases, Such as Cancer

One of the most active fields in genetics involves the study of epigenetic changes that contribute to human diseases. We are probably at "the tip of the iceberg" in our understanding of this topic. Many medical studies have identified correlations between epigenetic changes and particular diseases. For example, a research study could compare one variable, such as the level of DNA methylation of a specific gene, to a second variable, such as the severity of a disease. If a high level of DNA methylation was associated with an increase in disease severity, this would be a positive correlation. Researchers analyze the data to decide if such a correlation is statistically significant. (The statistic called a correlation coefficient is described in a statistics primer in Connect.) When a statistically significant correlation is obtained, how do we interpret its meaning? Such a result suggests a true **association**—changes in the two variables follow a pattern. For example, in a positive correlation, when one variable increases, the other variable also increases.

**Table 16.3 Environmental Agents That Are Associated with Cancer and Are Known to Cause Epigenetic Changes**

| Environmental agent | Occurrence | Associations with particular cancers |
|---|---|---|
| Polycyclic aromatic hydrocarbons | Tobacco smoke, automobile exhaust, charbroiled food | Lung, breast, stomach, and skin cancer |
| Benzene | Tobacco smoke, automobile exhaust | Leukemia, lymphoma, multiple myeloma |
| Endocrine disruptors (such as diethylstilbestrol) | Insecticides, fungicides, herbicides, some types of plastic | Breast, prostate, and thyroid cancer |
| Cadmium | Tobacco products, production of batteries | Lung and breast cancer |
| Nickel | Occupational exposure in mining, welding, and electroplating and in the manufacturing of jewelry, stainless steel, and batteries | Lung and nasal cancer |
| Arsenic | Lead alloy, feed additive in agriculture, insecticides | Skin, bladder, kidney, and liver cancer |

However, an association does not necessarily imply a cause-and-effect relationship. When considering the role of epigenetic changes and human disease, an association can arise in three common ways:

- The epigenetic changes directly contribute to the disease symptoms. There is a cause-and-effect relationship.
- Conversely, the disease symptoms may arise first, and then they cause subsequent epigenetic changes to happen. This is also a cause-and-effect relationship, but in the opposite direction.
- The association is indirect because a third factor is involved. For example, a toxic agent in the environment may cause a disease and also cause particular types of epigenetic changes even though those epigenetic changes do not contribute to the disease.

Correlations often identify associations between two variables. We should use caution, however, because this statistic by itself cannot prove that the association is due to cause and effect. Even so, research studies that identify associations are very useful because they provide the rationale to carry out further research to determine if a cause-and-effect relationship exists.

Researchers have identified many examples in which epigenetic changes are associated with a particular disease. These include Alzheimer disease, cardiovascular diseases, diabetes, multiple sclerosis, and asthma. For these diseases, further research will be needed to determine if these epigenetic changes are directly contributing to the disease symptoms. The role of epigenetics and disease has been most extensively studied with regard to cancer. **Table 16.3** describes several examples in which an environmental factor is associated with a particular type of cancer.

In some of the examples listed in Table 16.3, scientific evidence indicates that the association is causative. For example, certain agents in tobacco smoke have been shown to cause epigenetic changes that underlie lung cancer. As described in Chapter 13, cancer-causing genes are placed into two categories: oncogenes, which are overexpressed in cancer, and tumor-suppressor genes, which are inhibited. In cases where research has shown that an environmental agent results in an epigenetic effect that contributes to cancer, it is more common for that agent to inhibit tumor-suppressor genes. However, in some cases, they may activate oncogenes. Alternatively, some of the agents listed in Table 16.3 show an association with particular cancers, but researchers are still trying to determine if the epigenetic changes caused by these agents actually promote the changes that result in cancer.

## 16.4 Reviewing the Concepts

- Dietary factors during early stages of development cause epigenetic changes that affect phenotype (Figure 16.5).
- Environmental agents have been shown to cause epigenetic changes that are associated with human diseases such as cancer (Table 16.3).

## 16.4 Testing Your Knowledge

1. When mice carrying the $A^{vy}$ allele exhibit a darker coat color, this phenotype is thought to be caused by dietary factors that result in
   a. a greater level of DNA methylation and a decrease in the expression of the *Agouti* gene.
   b. a lower level of DNA methylation and a decrease in the expression of the *Agouti* gene.
   c. a greater level of DNA methylation and the overexpression of the *Agouti* gene.
   d. a lower level of DNA methylation and the overexpression of the *Agouti* gene.
   e. both a and c.
2. When an association occurs between an environmental agent that causes an epigenetic change and a particular disease, this association can be due to which of the following factors?
   a. The epigenetic change directly contributes to the disease symptoms.
   b. The disease symptoms arise first, and they cause subsequent epigenetic changes to happen.
   c. The association is indirect because a third factor is involved.
   d. All of the above are possible.
   e. Only a and b are possible.

## 16.5 Extranuclear Inheritance: Organelle Genomes

### Learning Outcomes

1. Describe the general features of mitochondrial and chloroplast genomes.
2. Explain why chloroplast and mitochondrial genes usually exhibit maternal inheritance.
3. List human diseases associated with mutations in mitochondrial genes.

In this section, we will explore inheritance patterns that violate the law of segregation. As described in Chapter 15, the segregation of genes is explained by the pairing and segregation of homologous chromosomes during meiosis (refer back to Figure 15.9). However, some genes are not found on the chromosomes in the cell nucleus, and these genes do not segregate in the same way. The transmission of genes located outside the cell nucleus is called **extranuclear inheritance.** Two important types of extranuclear inheritance patterns involve genes found in chloroplasts and mitochondria. In this section, we will examine the transmission patterns observed for genes found in the chloroplast and mitochondrial genomes and consider how mutations in these genes may affect an individual's traits.

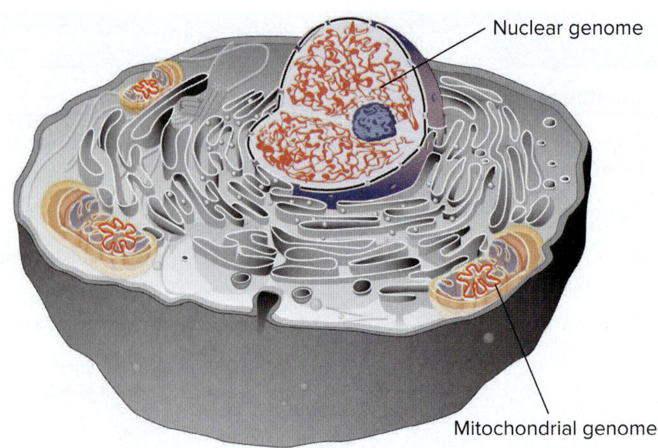


(a) An animal cell

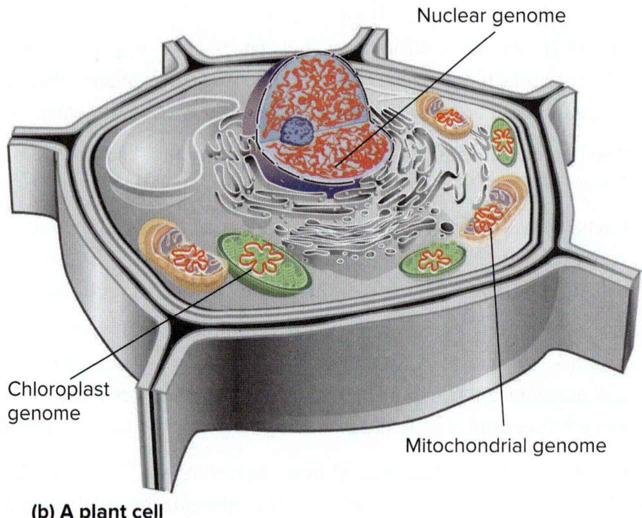


(b) A plant cell

**Figure 16.6** **The locations of genetic material in animal and plant cells.** The chromosomes in the cell nucleus are collectively known as the nuclear genome. Mitochondria and chloroplasts have small, circular chromosomes called the mitochondrial and chloroplast genomes, respectively.

**BioConnections:** *Refer back to Figure 4.29. What is the evolutionary origin of mitochondria and chloroplasts in eukaryotic cells?*

## EVOLUTIONARY CONNECTIONS

### Chloroplast and Mitochondrial Genomes Are Relatively Small but Contain Genes That Encode Important Proteins

As we discussed in Chapter 4 (refer back to Figure 4.29), mitochondria and chloroplasts are found in eukaryotic cells because of an ancient endosymbiotic relationship. Because they originated from bacterial cells, these organelles contain their own genetic material, called the mitochondrial genome and chloroplast genome, respectively (**Figure 16.6**). Mitochondrial and chloroplast genomes are composed of a single, circular DNA molecule. The mitochondrial genome of many mammalian species has been analyzed and usually contains a total of 37 genes. Twenty-four genes encode tRNAs and rRNAs, which are needed for translation inside the mitochondrion, and 13 genes encode proteins that are involved in oxidative phosphorylation. As discussed in Chapter 6, the primary function of the mitochondrion is the synthesis of ATP via oxidative phosphorylation. Among different species of flowering plants, chloroplast genomes typically contain between 100 and 200 genes. Many of these genes encode proteins that are vital to the process of photosynthesis, which is discussed in Chapter 7.

### Chloroplast Genomes Are Often Maternally Inherited

One of the first experiments showing an extranuclear inheritance pattern was carried out by German botanist Carl Correns in 1909. Correns discovered that leaf pigmentation in the four-o'clock plant (*Mirabilis jalapa*) follows a pattern of inheritance that does not obey Mendel's law of segregation. Four-o'clock leaves may be green, white, or variegated. Correns observed that the pigmentation of the offspring depended solely on the pigmentation of the female parent, a phenomenon called **maternal inheritance** (**Figure 16.7**). If the female parent had white leaves, all of the offspring had white leaves. Similarly, if

**Figure 16.7 Maternal inheritance in the four-o'clock plant.** In four-o'clocks, the egg contains all of the proplastids that are inherited by the offspring, so the phenotype of the offspring is determined by the maternal parent. The green phenotype is due to the presence of normal chloroplasts. The white phenotype in four-o'clocks is due to chloroplasts with a mutant allele that greatly reduces green pigment production. The variegated phenotype is due to a mixture of normal and mutant chloroplasts.

**Concept Check:** *In this example, where is the gene located that causes the green color of four-o'clock leaves? How is this gene transmitted from parent to offspring?*

the female was green, so were all of the offspring. The offspring of a variegated female parent could be green, white, or variegated.

What accounts for maternal inheritance? At the time, Correns did not understand that chloroplasts contain genetic material. Subsequent research identified DNA present in chloroplasts as responsible for the unusual inheritance pattern observed. We now know that the pigmentation of four-o'clock leaves can be explained by the occurrence of genetically different types of chloroplasts in the leaf cells. As discussed in Chapter 7, chloroplasts are the site of photosynthesis, and their green color is due to the presence of the pigment called chlorophyll. Certain genes required for chlorophyll synthesis are located in the chloroplast DNA. The green phenotype is due to the presence of chloroplasts that have normal genes and synthesize the usual quantity of chlorophyll. The white phenotype is caused by a mutation in a gene within the chloroplast DNA that prevents the synthesis of most of the chlorophyll. (Enough chlorophyll is made for the plant to survive.) The variegated phenotype occurs in leaves that have a mixture of the two types of chloroplasts.

Leaf pigmentation follows a maternal inheritance pattern because the chloroplasts in four-o'clocks are transmitted only through the cytoplasm of the egg (Figure 16.8). In most species of plants, the egg cell provides most of the zygote's cytoplasm, whereas the much smaller male gamete often provides little more than a nucleus. Therefore, chloroplasts are most often inherited via the egg. Recall from Chapter 4 that chloroplasts are derived from proplastids, which can later develop into chloroplasts. In four-o'clocks, the egg cell contains several proplastids that are inherited by the offspring. The sperm cell does not contribute any proplastids. For this reason, the phenotype of a four-o'clock plant reflects the types of proplastids it inherits from the maternal parent. If the maternal parent transmits only normal proplastids, all offspring will have green leaves (Figure 16.8a). Alternatively, if the maternal parent transmits only mutant proplastids, all offspring will have white leaves (Figure 16.8b). Because an egg cell contains several proplastids, an offspring from a variegated maternal parent may inherit only normal proplastids, only mutant proplastids, or a mixture of normal and mutant proplastids. Consequently, the offspring of a variegated maternal parent can be green, white, or variegated individuals (Figure 16.8c).

How do we explain the variegated phenotype at the cellular level? This phenotype is due to events that occur after fertilization. As a zygote containing both types of proplastids grows via cellular division to produce a multicellular plant, some cells may receive mostly those proplastids that develop into normal chloroplasts. Further division of these cells gives rise to a patch of green tissue. Alternatively, as a matter of chance, other cells may receive all or mostly mutant chloroplasts that are defective in chlorophyll synthesis. The result is a patch of white tissue.

In seed-bearing plants, maternal inheritance of chloroplasts is the most common transmission pattern. However, certain species exhibit a pattern called **biparental inheritance,** in which both the pollen and the egg contribute chloroplasts to the offspring. Others exhibit **paternal inheritance,** in which only the pollen contributes these organelles. For example, most species of pine trees show paternal inheritance of chloroplasts.

## Mitochondrial Genomes Are Maternally Inherited in Humans and Most Other Species

Mitochondria are found in nearly all eukaryotic species. As with the transmission of chloroplasts in plants, maternal inheritance is the

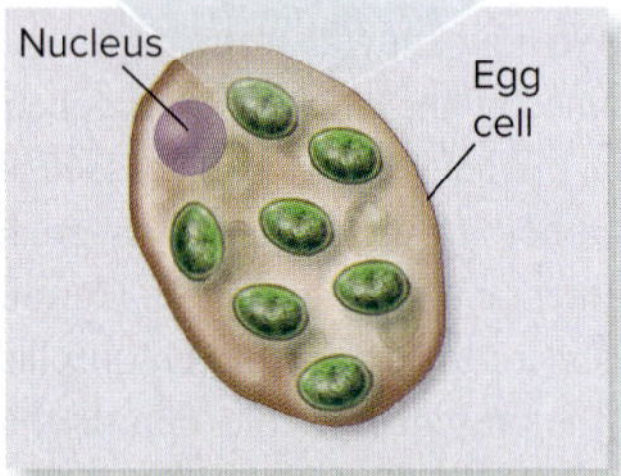

Normal proplastid will produce chloroplasts with a normal amount of green pigment.
Nucleus
Egg cell

(a) Egg cell from a maternal parent with green leaves

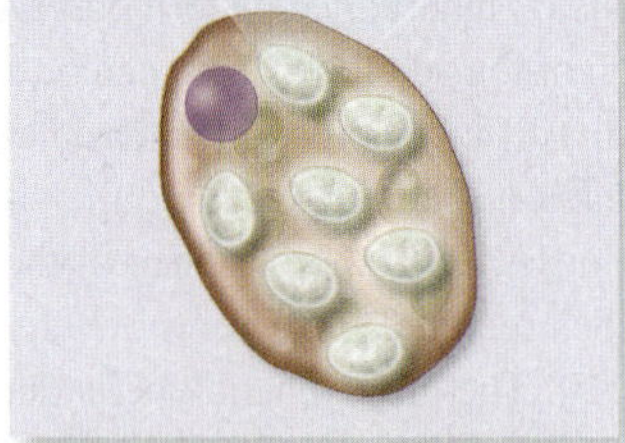

Mutant proplastid will produce chloroplasts with very little pigment.

(b) Egg cell from a maternal parent with white leaves

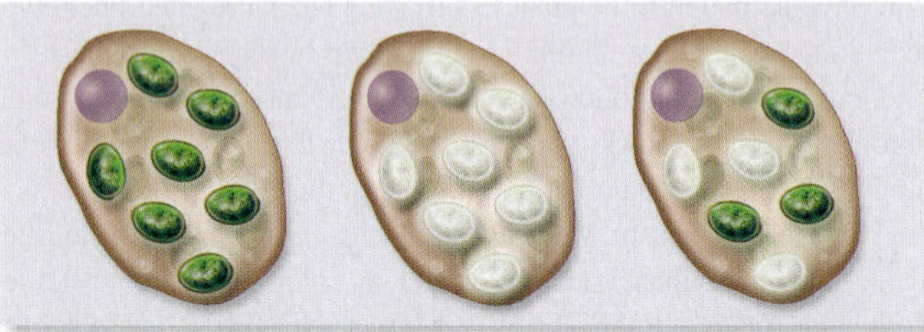

(c) Possible egg cells from a maternal parent with variegated leaves

**Figure 16.8 Plastid compositions of egg cells from green, white, and variegated four-o'clock plants.** In this drawing of four-o'clock egg cells, normal proplastids are represented as green and mutant proplastids as white. (Note: This drawing is schematic. Proplastids do not differentiate into chloroplasts in egg cells, and they are not actually green.) **(a)** A green plant produces eggs carrying normal proplastids. **(b)** A white plant produces eggs carrying mutant proplastids. **(c)** A variegated plant produces eggs that may contain either or both types of proplastids.

most common pattern of mitochondrial transmission in eukaryotes, although some species do exhibit biparental or paternal inheritance.

In humans, mitochondria are maternally inherited. Researchers have discovered that mutations in human mitochondrial genes cause a variety of rare diseases (**Table 16.4**). These are usually chronic degenerative disorders that affect organs and cells, such as the brain, eyes, heart, muscle, kidney, and endocrine glands, that require high levels of ATP. For example, Leber hereditary optic neuropathy (LHON) affects the optic nerve and leads to the progressive loss of vision in one or both eyes. LHON is caused by point mutations in several different mitochondrial genes.

## 16.5 Reviewing the Concepts

- Mitochondria and chloroplasts are derived from an ancient endosymbiotic relationship and carry a small number of genes. The inheritance of such genes is called extranuclear inheritance (Figure 16.6).
- Chloroplasts in the four-o'clock plant are transmitted via the egg, a pattern called maternal inheritance (Figures 16.7, 16.8).
- Several human diseases are known to be caused by mutations in mitochondrial genes, which follow a maternal inheritance pattern (Table 16.4).

**Table 16.4 Examples of Human Mitochondrial Diseases**

| Disease | Causes and symptoms |
|---|---|
| Leber hereditary optic neuropathy (LHON) | A mutation in one of several mitochondrial genes that encode electron transport proteins. The main symptom is loss of vision. |
| Neurogenic muscle weakness | A mutation in a mitochondrial gene that encodes a subunit of ATP synthase, which is required for ATP synthesis. Symptoms involve abnormalities in the nervous system that affect the muscles and eyes. |
| Maternal myopathy and cardiomyopathy | A mutation in a mitochondrial gene that encodes a tRNA for leucine. The primary symptoms involve muscle abnormalities, most notably in the heart. |
| Myoclonic epilepsy and ragged-red muscle fibers | A mutation in a mitochondrial gene that encodes a tRNA for lysine. Symptoms include epilepsy, dementia, blindness, deafness, and heart and kidney malfunctions. |

## 16.5 Testing Your Knowledge

1. In the four-o'clock plant (*Mirabilis jalapa*), leaf pigmentation is maternally inherited. A plant with green leaves is pollinated by a plant with variegated leaves. What would be the expected phenotype(s) of the offspring?
   a. all offspring with green leaves
   b. all offspring with variegated leaves
   c. some offspring with variegated, some with green, and some with white leaves
   d. mostly offspring with green leaves, but some with variegated leaves
   e. none of the above
2. Which of the following statements regarding human mitochondrial diseases is true?
   a. They are maternally inherited.
   b. They are due to mutations in the mitochondrial genome.
   c. They affect organs and cells that require high levels of ATP.
   d. All of the above are true.

## 16.6 Linkage of Genes on the Same Chromosome

### Learning Outcomes

1. Describe how linkage violates the law of independent assortment.
2. Explain how experimental crosses can demonstrate linkage.

In the previous section, we examined extranuclear inheritance, which violates the law of segregation. In this section, we will explore inheritance patterns that violate the law of independent assortment. A typical chromosome contains hundreds or even a few thousand different genes. When two genes are close together on the same chromosome, they tend to be transmitted as a unit, a phenomenon known as **linkage.** A group of genes that usually stay together during meiosis is called a **linkage group,** and the genes in the group are said to be linked. In a two-factor cross, linked genes that are close together on the same chromosome do not follow the law of independent assortment.

In this section, we will begin by examining the first experimental cross that demonstrated linkage. This pattern was subsequently explained by Thomas Hunt Morgan, who proposed that different genes located close to each other on the same chromosome tend to be inherited together.

## FEATURE INVESTIGATION

### Bateson and Punnett's Crosses of Sweet Peas Showed That Genes Do Not Always Assort Independently

The first study showing linkage between two different genes was a cross of sweet peas carried out by William Bateson and Reginald Punnett in 1905. A surprising result occurred when they conducted a two-factor cross involving flower color and pollen shape (**Figure 16.9**). One of the parent plants had purple flowers (*PP*) and long pollen (*LL*); the other had red flowers (*pp*) and round pollen (*ll*). As Bateson and Punnett expected, the $F_1$ plants all had purple flowers and long pollen (*PpLl*), because purple flowers and long pollen are dominant traits. The unexpected result came in the $F_2$ generation.

**Figure 16.9** **Bateson and Punnett's cross of sweet peas showing that independent assortment does not always occur.**

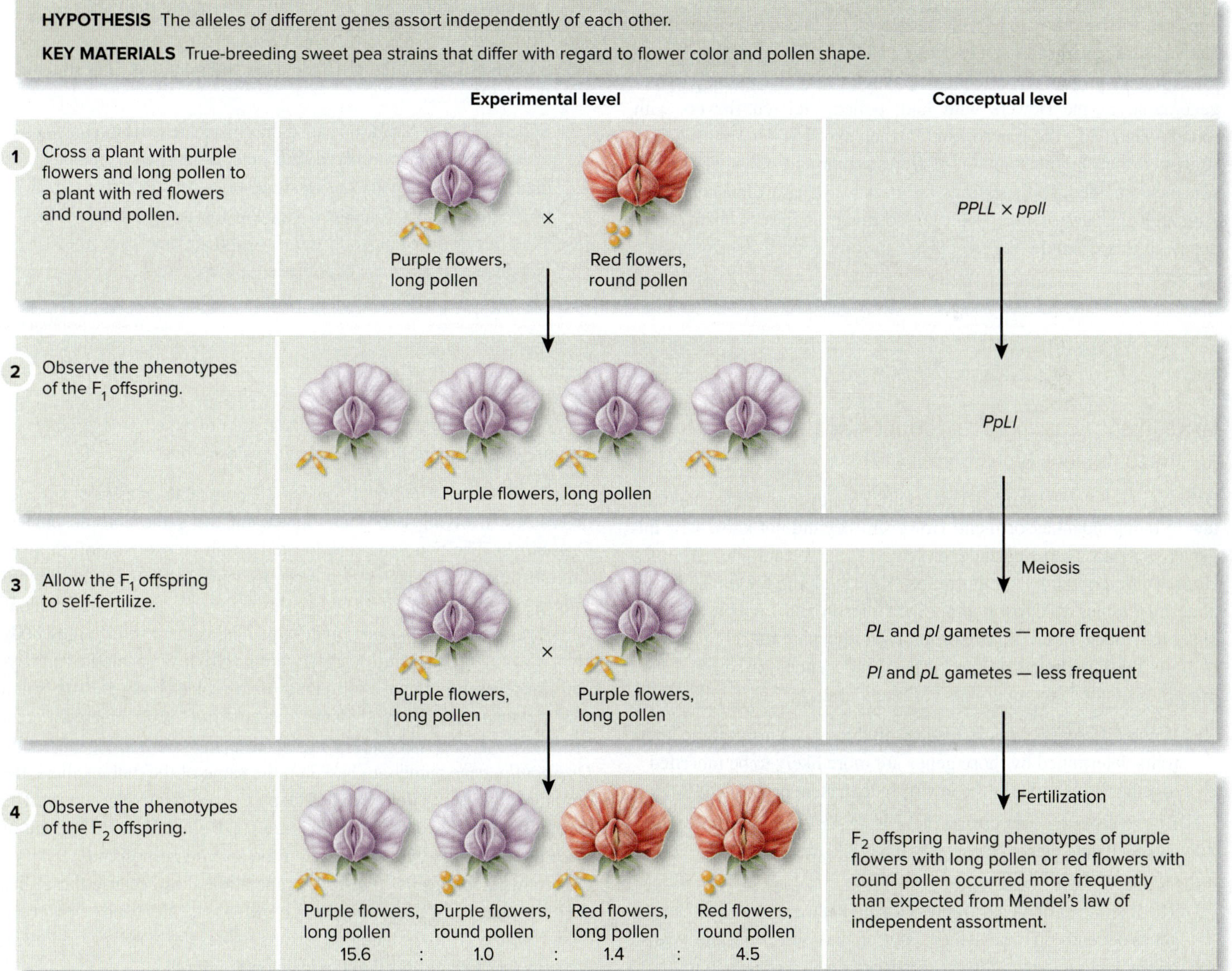

HYPOTHESIS The alleles of different genes assort independently of each other.
KEY MATERIALS True-breeding sweet pea strains that differ with regard to flower color and pollen shape.
Experimental level
Conceptual level
1 Cross a plant with purple flowers and long pollen to a plant with red flowers and round pollen.
Purple flowers, long pollen
×
Red flowers, round pollen
PPLL × ppll
2 Observe the phenotypes of the F1 offspring.
Purple flowers, long pollen
PpLl
Meiosis
3 Allow the F1 offspring to self-fertilize.
Purple flowers, long pollen
×
Purple flowers, long pollen
PL and pl gametes — more frequent
Pl and pL gametes — less frequent
Fertilization
4 Observe the phenotypes of the F2 offspring.
Purple flowers, long pollen 15.6
:
Purple flowers, round pollen 1.0
:
Red flowers, long pollen 1.4
:
Red flowers, round pollen 4.5
F2 offspring having phenotypes of purple flowers with long pollen or red flowers with round pollen occurred more frequently than expected from Mendel's law of independent assortment.

5 **THE DATA**

| Phenotypes of $F_2$ offspring | Observed number | Observed ratio | Expected number | Expected ratio |
|---|---|---|---|---|
| Purple flowers, long pollen | 296 | 15.6 | 240 | 9 |
| Purple flowers, round pollen | 19 | 1.0 | 80 | 3 |
| Red flowers, long pollen | 27 | 1.4 | 80 | 3 |
| Red flowers, round pollen | 85 | 4.5 | 27 | 1 |

6 **CONCLUSION** The data are not consistent with the law of independent assortment.

7 **SOURCE** Bateson, William, and Punnett, Reginald C. 1911. On the Inter-relations of Genetic Factors. *Proceedings of the Royal Society of London, Series B*, 84: 3–8.

Although the $F_2$ offspring displayed the four phenotypes predicted by Mendel's laws, the observed numbers of offspring did not conform to the predicted 9:3:3:1 ratio (refer back to Figure 15.7). Rather, as seen in the data in Figure 16.9, the $F_2$ generation had a much higher proportion of the two phenotypes found in the parental generation: purple flowers with long pollen, and red flowers with round pollen. How did Bateson and Punnett explain these results? They suggested that the transmission of flower color and pollen shape was somehow coupled, so these traits did not always assort independently. Although the law of independent assortment applies to many other genes, in this example, the hypothesis of independent assortment was rejected.

*Experimental Questions*

1. What hypothesis was Bateson and Punnett testing when conducting the crosses in the sweet pea?
2. **SCISKILLS** ▶ What were the expected results of Bateson and Punnett's cross?
3. **SCISKILLS** ▶ A statistic called the chi square ($\chi^2$) test can be used to evaluate whether observed data are consistent with predicted data based on a hypothesis. This statistic is described in a statistics primer in Connect. Conduct a $\chi^2$ square test using the data in this experiment. Based on your results, would you accept or reject an independent assortment hypothesis for these data?

## Linkage and Crossing Over Produce Nonrecombinant and Recombinant Types

Bateson and Punnett realized their results did not conform to Mendel's law of independent assortment. However, they did not know why the genes were not assorting independently. A few years later, Thomas Hunt Morgan obtained similar ratios while studying the transmission pattern of genes in *Drosophila*. Like Bateson and Punnett, Morgan observed many more $F_2$ offspring with the parental combination of traits than predicted on the basis of independent assortment. To explain his data, Morgan proposed three ideas:

1. When different genes are located on the same chromosome, the traits determined by those genes are more likely to be inherited together. This violates the law of independent assortment.
2. Due to crossing over during meiosis, homologous chromosomes can exchange pieces of chromosomes and create new combinations of alleles.
3. The likelihood of a crossover occurring in the region between two genes depends on the distance between the two genes. Crossovers between homologous chromosomes are much more likely to occur between two genes farther apart along a chromosome than are crossovers of two genes that are closer together.

To illustrate the first two ideas, **Figure 16.10** considers a series of crosses involving two genes linked on the same chromosome in *Drosophila*. The two genes are located on an autosome, not on a sex chromosome. The P generation cross is between flies that are homozygous for alleles that affect body color and wing shape. The female is homozygous for the dominant wild-type alleles that produce gray body color ($b^+b^+$) and straight (normal) wings ($c^+c^+$); the male is homozygous for recessive mutant alleles that produce black body color (*bb*) and curved wings (*cc*). The symbols for the genes in *Drosophila* are different from what you learned earlier in the chapter for other organisms. In *Drosophila*, gene names are based on the name of the mutant allele; the dominant wild-type allele is indicated by a superscript plus sign ($^+$). The chromosomes next to the flies in Figure 16.10 show the arrangement of these alleles. If the two genes are on the same chromosome, we know the arrangement of alleles in the P generation flies because these flies are homozygous for both genes ($b^+b^+c^+c^+$ for one parent and *bbcc* for the other parent). In the P generation female on the left, $b^+$ and $c^+$ are linked, whereas *b* and *c* are linked in the male on the right.

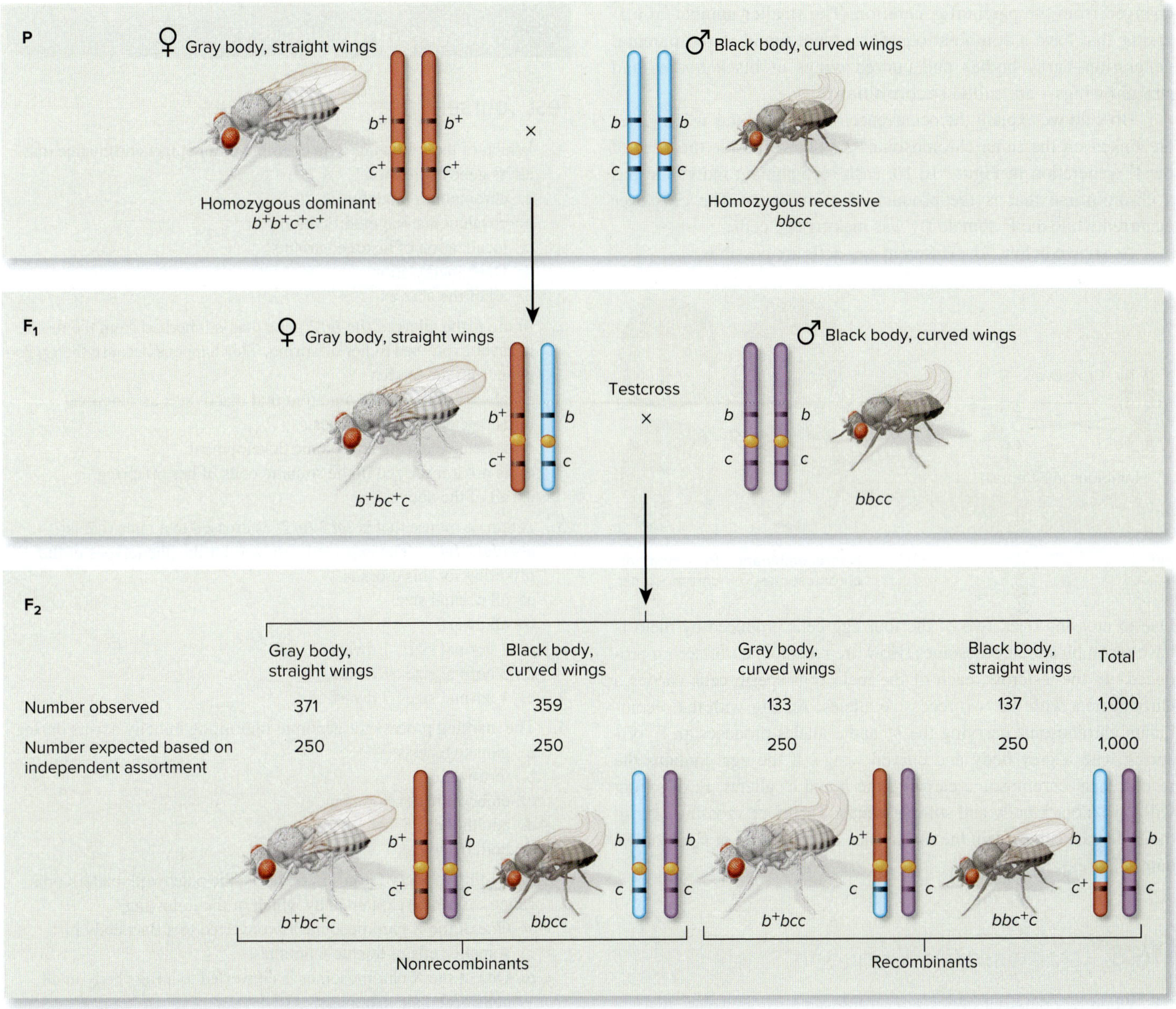

**P**

♀ Gray body, straight wings — Homozygous dominant $b^+b^+c^+c^+$ (chromosomes: $b^+$, $c^+$ / $b^+$, $c^+$)

× 

♂ Black body, curved wings — Homozygous recessive $bbcc$ (chromosomes: $b$, $c$ / $b$, $c$)

**$F_1$**

♀ Gray body, straight wings — $b^+bc^+c$ (chromosomes: $b^+$, $c^+$ / $b$, $c$)

Testcross ×

♂ Black body, curved wings — $bbcc$ (chromosomes: $b$, $c$ / $b$, $c$)

**$F_2$**

| | Gray body, straight wings | Black body, curved wings | Gray body, curved wings | Black body, straight wings | Total |
|---|---|---|---|---|---|
| Number observed | 371 | 359 | 133 | 137 | 1,000 |
| Number expected based on independent assortment | 250 | 250 | 250 | 250 | 1,000 |
| | $b^+bc^+c$ | $bbcc$ | $b^+bcc$ | $bbc^+c$ | |
| | Nonrecombinants | | Recombinants | | |

**Figure 16.10 Linkage and recombination of alleles.** An experimenter crossed $b^+b^+c^+c^+$ and $bbcc$ flies to produce $F_1$ heterozygotes. $F_1$ females were then testcrossed to $bbcc$ males. The large number of nonrecombinant phenotypes in the $F_2$ generation suggests that the two genes are linked on the same chromosome. $F_2$ recombinant phenotypes occur because the alleles can be rearranged by crossing over. Note: The $b^+$ and $c^+$ alleles are dominant, and the $b$ and $c$ alleles are recessive.

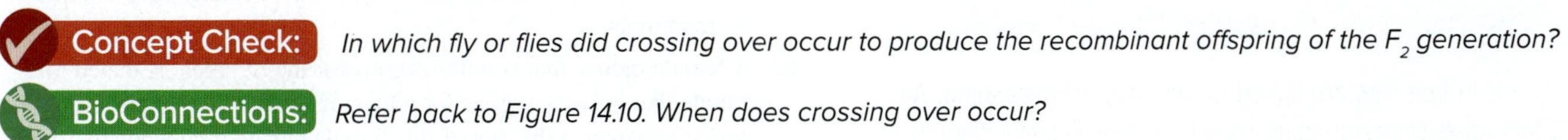


*In which fly or flies did crossing over occur to produce the recombinant offspring of the $F_2$ generation?*

*Refer back to Figure 14.10. When does crossing over occur?*

Let's now look at the outcome of the crosses in Figure 16.10. As expected, the $F_1$ offspring ($b^+bc^+c$) all had gray bodies and straight wings, confirming that these are the dominant traits. In the next cross, $F_1$ females were mated to males that were homozygous for both recessive alleles ($bbcc$). The purpose of the experiment, which is called a **testcross,** is to determine if the genes for body color and wing shape are linked. If the genes are on different chromosomes and assort independently, this cross will produce $F_2$ offspring with the four possible phenotypes in a 1:1:1:1 ratio. The observed numbers clearly conflict with this prediction. The two most abundant phenotypes are those with the combinations of characteristics in the P generation: gray bodies and straight wings or black bodies and curved wings. These offspring are termed **nonrecombinants,** because this combination of traits has not changed from the parental generation. The smaller number of offspring that have a combination of traits not found in the parental generation—gray bodies and curved wings or black bodies and straight wings—are called **recombinants.**

How do we explain the occurrence of recombinants when genes are linked on the same chromosome? As shown beside the flies of the $F_2$ generation in Figure 16.10, each recombinant individual has a chromosome that is the product of a crossover. The crossover occurred while the $F_1$ female fly was making egg cells.

As shown below, four different egg cells are possible:

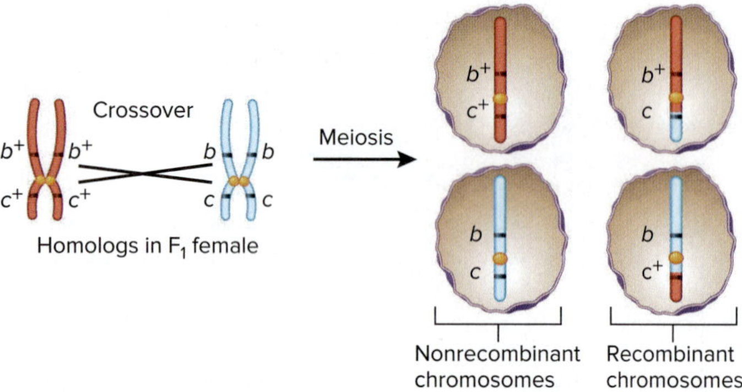


Due to crossing over, two of the four egg cells produced by meiosis have recombinant chromosomes. How are recombinant offspring produced? In the testcross, each of the male fly's sperm cells carries a chromosome with the two recessive alleles. An egg with the recombinant chromosome carrying the $b^+$ and $c$ alleles produces an $F_2$ offspring with a gray body and curved wings. If the egg contains the recombinant chromosome carrying the $b$ and $c^+$ alleles, $F_2$ offspring will have a black body and straight wings. Therefore, crossing over in the $F_1$ female can explain the occurrence of both types of $F_2$ recombinant offspring.

## 16.6 Reviewing the Concepts

- When two different genes are on the same chromosome, they are said to be linked. Linked genes tend to be inherited as a unit, unless crossing over separates them (Figures 16.9, 16.10).

## 16.6 Testing Your Knowledge

1. Two genes in fruit flies are linked on the same chromosome. An *AABB* individual is crossed to an *aabb* individual. Female offspring, which are *AaBb*, are then crossed to *aabb* males. Which offspring from this second cross do you expect to be the most common?
   **a.** *Aabb* and *aaBb*
   **b.** *AaBb* and *aaBb*
   **c.** *AaBb* and *aabb*
   **d.** *Aabb* and *aabb*
2. Genes that are linked do not conform to
   **a.** Mendel's law of segregation.
   **b.** Mendel's law of independent assortment.
   **c.** both of Mendel's laws.
   **d.** the chromosome theory of inheritance.

# Assess and Discuss

## Test Yourself

1. Which of the following is an example of an epigenetic change that alters gene expression?
   a. chromatin remodeling
   b. covalent histone modification
   c. localization of histone variants
   d. DNA methylation
   e. all of the above
2. In mice, the allele of the *Igf2* gene that is inherited from the mother is never expressed in her offspring. This happens because the *Igf2* gene from the mother
   a. always undergoes a mutation that inactivates its function.
   b. is deleted during oogenesis.
   c. is deleted during embryonic development.
   d. is not transcribed in the somatic cells of her offspring.
   e. all of the above.
3. A female mouse that is *Igf2 Igf2*$^-$ is crossed to a male that is also *Igf2 Igf2*$^-$. The expected outcome for the phenotypes of the offspring for this cross is
   a. all normal size.
   b. all dwarf.
   c. 1 normal size : 1 dwarf
   d. 3 normal size : 1 dwarf.
   e. 1 normal size : 3 dwarf.
4. The marking process for genomic imprinting initially occurs during
   a. gametogenesis.
   b. fertilization.
   c. embryonic development.
   d. adulthood.
   e. both b and c.
5. According to Mary Lyon's hypothesis, the patchwork pattern on a calico cat can be explained by which of the following?
   a. One of the X chromosomes is converted to a Barr body in somatic cells of female mammals.
   b. One of the X chromosomes is converted to a Barr body in all cells of female mammals.
   c. Both of the X chromosomes are converted to Barr bodies in somatic cells of female mammals.
   d. Both of the X chromosomes are converted to Barr bodies in all cells of female mammals.
   e. One of the X chromosomes is lost in the somatic cells of female mammals.
6. A female mouse that is homozygous for the $A^{vy}$ allele is mated to a male that is homozygous for a loss-of-function *(a)* allele. How would you expect the diet of this female during pregnancy to affect her offspring?
   a. Dietary agents that promote a greater level of DNA methylation would produce offspring with more yellow fur.
   b. Dietary agents that promote a lower level of DNA methylation would produce offspring with more yellow fur.
   c. Dietary agents that promote a greater level of DNA methylation would produce offspring with darker fur.
   d. Dietary agents that promote a lower level of DNA methylation would produce offspring with darker fur.
   e. Both b and c are correct.

7. Environmental agents may cause epigenetic changes that alter the expression of specific genes. Which of the following epigenetic changes would be expected to promote cancer?
   a. a change that resulted in the overexpression of a tumor-suppressor gene
   b. a change that resulted in the inhibition of a tumor-suppressor gene
   c. a change that resulted in the overexpression of an oncogene
   d. a change that resulted in the inhibition of an oncogene
   e. both b and c
8. In many organisms, organelles, such as the mitochondria, are transmitted only by the egg. This phenomenon is known as
   a. biparental inheritance.
   b. paternal inheritance.
   c. X-linked inheritance.
   d. maternal inheritance.
   e. both c and d.
9. Based on the ideas proposed by Morgan, which of the following statements concerning linkage is *false?*
   a. Traits determined by genes located close together on the same chromosome are likely to be inherited together.
   b. Crossing over between homologous chromosomes can create new allele combinations.
   c. A crossover is more likely to occur in a region between two genes that are close together than in a region between two genes that are farther apart.
   d. The probability of crossing over depends on the distance between the genes.
   e. Genes that tend to be transmitted together are physically located on the same chromosome.
10. Extranuclear inheritance occurs because
    a. certain genes are found on the X chromosome.
    b. chromosomes in the nucleus may be transferred to the cytoplasm.
    c. some cellular organelles contain DNA.
    d. the nuclear membrane breaks down during cell division.
    e. both a and c.

## Conceptual Questions

1. Define epigenetics. Are all epigenetic changes passed from parent to offspring? Explain.
2. What is a Barr body? How is its structure different from that of other chromosomes in the cell? How does the structure of a Barr body affect the level of X-linked gene expression?
3. **PRINCIPLES** A principle of biology is that the genetic material provides a blueprint for reproduction. Explain how epigenetics affects that blueprint.

## Collaborative Questions

1. Go to the PubMed website and search words such as epigenetic and cancer. Scan through the journal articles you retrieve, and make a list of environmental agents that may cause epigenetic changes that contribute to cancer.
2. Mendel studied seven traits in pea plants, and the garden pea happens to have seven different chromosomes. It has been pointed out that Mendel was very lucky not to have conducted crosses involving two traits governed by genes that are closely linked on the same chromosome, because the results would have confounded his theory of independent assortment. It has even been suggested that Mendel may not have published data involving traits that were linked! An article by Stig Blixt ("Why Didn't Gregor Mendel Find Linkage?" *Nature* 256: 206, 1975) considers this issue. Look up this article, and discuss why Mendel did not find linkage.

## Online Resource

# 17 The Simpler Genetic Systems of Viruses and Bacteria

**A tobacco plant infected with tobacco mosaic virus (TMV).** Note the mottled lesions on the surface of the leaves.

## Chapter Outline



We often view viruses as agents that infect people and cause illness. Many people don't realize that viruses infect all types of organisms, including bacteria and plants. The first virus to be discovered, the tobacco mosaic virus (TMV), is a virus that infects tobacco and many other plant species, causing mosaic-like patterns in which normal-colored patches are interspersed with light green or yellowish patches on the leaves (see chapter-opening photo). TMV damages leaves, flowers, and fruit but almost never kills the plant.

In 1883, German chemist Adolf Mayer determined that this disease could be spread by spraying the sap from one plant onto another. By subjecting this sap to filtration, Russian scientist Dmitri Ivanovski demonstrated that the disease-causing agent was too small to be a bacterium. At first, some researchers suggested the agent was a chemical toxin. However, Dutch botanist Martinus Beijerinck ruled out this possibility by showing that sap continued to transmit the disease after many plant generations. A toxin would have been diluted after many generations, but Beijerinck's results indicated the disease agent was multiplying in the plant.

In this chapter, we first turn our attention to the genetic analyses of viruses. Once a cell is infected, the genetic material of a virus orchestrates a series of events that ultimately leads to the production of new virus particles. We will consider the biological complexity of viruses and explore viral reproductive cycles.

In the remaining sections of this chapter, we will examine the bacterial genome and the methods used in its investigation. Like their eukaryotic counterparts, bacteria have genetic differences that affect their cellular traits, and techniques of modern microbiology make many of these differences, such as sensitivity to antibiotics and differences in nutritional requirements, easy to study. Although bacteria reproduce asexually by cell division, their genetic diversity is enhanced by the phenomenon known as gene transfer, in which genes are passed from one bacterial cell to another. Like sexual reproduction in eukaryotes, gene transfer enhances the genetic diversity observed among bacterial species. In this chapter, we will explore three interesting ways that bacteria transfer genetic material.

## 17.1 Genetic Properties of Viruses

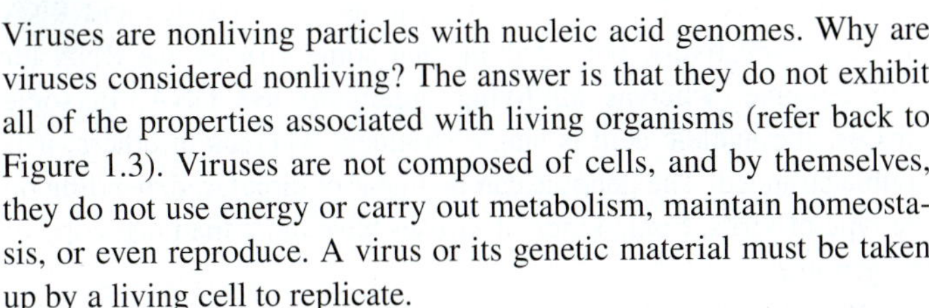

### Learning Outcomes

1. Compare and contrast how viruses differ with regard to their host range, structure, and genomes.
2. List the six steps in viral reproductive cycles and distinguish between the lysogenic and lytic cycles.
3. Describe how emerging viruses such as HIV arise and spread through populations.
4. Explain how an understanding of virus structure and reproduction can aid in the development of drugs to combat viruses.

Viruses are nonliving particles with nucleic acid genomes. Why are viruses considered nonliving? The answer is that they do not exhibit all of the properties associated with living organisms (refer back to Figure 1.3). Viruses are not composed of cells, and by themselves, they do not use energy or carry out metabolism, maintain homeostasis, or even reproduce. A virus or its genetic material must be taken up by a living cell to replicate.

Because of the ability to cause disease, microbiologists, geneticists, and molecular biologists have taken a great interest in the structure, genetic composition, and replication of viruses. In this section, we will discuss the structure of viruses and examine viral reproductive cycles in detail, paying particular attention to human immunodeficiency virus (HIV), the virus that causes acquired immunodeficiency syndrome (AIDS) in humans.

### Viruses Are Remarkably Varied, Despite Their Simple Structure

A **virus** is a small, infectious particle that consists of nucleic acid enclosed in a protein coat. Researchers have identified and studied over 4,000 different types of viruses. Although all viruses share some similarities, such as small size and the reliance on a living cell for replication, they vary greatly in their characteristics, including their host range, structure, and genome composition. Some of the major differences are described next, and characteristics of selected viruses are shown in Table 17.1.

**Differences in Host Range** A cell that is infected by a virus is called a **host cell,** and a species that can be infected by a specific virus is called a host species for that virus. Viruses differ greatly in their **host range**—the number of species and cell types they can infect. Table 17.1 lists a few examples of viruses with widely different ranges of host species. For example, TMV has a broad host range. It is known to infect over 150 different species of plants.

**Table 17.1 Hosts and Characteristics of Selected Viruses**

| Virus or group of viruses | Host | Effect on host | Nucleic acid* | Genome size (kb)† | Number of genes† |
|---|---|---|---|---|---|
| Phage λ | *E. coli* | Can exist harmlessly in the host cell or cause lysis | dsDNA | 48.5 | 36 |
| Phage T4 | *E. coli* | Causes lysis | dsDNA | 169 | 288 |
| Tobacco mosaic virus (TMV) | Many plants | Causes mottling and necrosis of leaves and other plant parts | ssRNA | 6.4 | 6 |
| Baculoviruses | Insects | Most baculoviruses are species specific; they usually kill the insect | dsDNA | 133.9 | 154 |
| Influenza virus | Mammals | Causes classical “flu,” with fever, cough, sore throat, and headache | ssRNA | 13.5 | 11 |
| Epstein-Barr virus | Humans | Causes mononucleosis, with fever, sore throat, and fatigue | dsDNA | 172 | 80 |
| Adenovirus | Humans | Causes respiratory symptoms and diarrhea | dsDNA | 34 | 35 |
| Herpes simplex type II | Humans | Causes blistering sores around the genital region | dsDNA | 158.4 | 77 |
| HIV (type I) | Humans | Causes AIDS, an immunodeficiency syndrome eventually leading to death | ssRNA | 9.7 | 9 |

*The abbreviations ss and ds refer to single stranded and double stranded, respectively.

†Several of the viruses listed in this table are found in different strains that vary with regard to genome size and number of genes. The numbers reported in this table are typical values. The abbreviation kb refers to kilobase, which equals 1,000 bases.

**Table 17.2 Examples of Human Viruses**

| Area of body infected | Virus | Name of disease |
|---|---|---|
| Brain and central nervous system | Flavivirus | Yellow fever |
| | Rhabdovirus | Rabies |
| Skin | Herpes simplex I | Cold sores |
| | Variola virus | Smallpox |
| Respiratory tract | Influenza virus | Flu |
| | Rhinovirus | Common cold |
| Immune system | Human immunodeficiency virus (HIV) | Acquired immunodeficiency syndrome (AIDS) |
| | Epstein-Barr virus | Mononucleosis |
| Digestive system | Hepatitis B virus | Viral hepatitis |
| | Norwalk virus | Viral gastroenteritis |
| Reproductive system | Herpes simplex II | Genital herpes |
| | Papillomavirus | Genital warts, cervical cancer |
| Blood | Ebola virus | Hemorrhagic fever |
| | Hantavirus | Hemorrhagic fever with renal syndrome |

By comparison, other viruses have a narrow host range, with some infecting only a single species. Furthermore, a virus may infect only a specific cell type in a host species. **Table 17.2** describes some viruses that infect human cells in certain parts of the body and thereby cause disease.

**Structural Differences** Viruses are so small that they cannot be resolved by even the best light microscope. Although the existence of viruses was postulated in the 1890s, viruses were not observed until the 1930s, when the electron microscope was invented. Viruses range in size from about 20 to 400 nm in diameter (1 nm [nanometer] = $10^{-9}$ m). For comparison, a typical bacterium is 1,000 nm in diameter, and the diameter of most eukaryotic cells is 10 to 1,000 times that of a bacterium. Adenoviruses, which cause infections of the respiratory and gastrointestinal tracts, have an average diameter of 75 nm. Over 50 million adenoviruses could fit into an average-sized human cell.

What are the common structural features of all viruses? As shown in **Figure 17.1**, all viruses have a protein coat called a **capsid,** which encloses a genome consisting of one or more molecules of nucleic acid. Capsids are composed of one or several different protein subunits called capsomers. Capsids have a variety of shapes, including helical and polyhedral. Figure 17.1a shows the structure of TMV, which has a helical capsid made of identical capsomers. Figure 17.1b shows an adenovirus, which has a polyhedral capsid. Protein fibers with a terminal knob are located at the corners of the polyhedral capsid. Many viruses that infect animal cells, such as influenza virus, shown in Figure 17.1c, have a **viral envelope** enclosing the capsid. The envelope consists of a lipid bilayer that is derived from a cellular membrane of the host cell and is embedded with virally encoded spike glycoproteins.

In addition to encasing and protecting the genetic material, the capsid and envelope enable viruses to infect their hosts. In many viruses, protein fibers with a knob (Figure 17.1b) or spike glycoproteins (Figure 17.1c) help them bind to the surface of a host cell. Viruses that infect bacteria, called **bacteriophages,** or simply **phages,** may have more complex protein coats, with accessory structures used for anchoring the virus to a host cell and injecting the viral nucleic acid (Figure 17.1d). For example, the tail fibers of such bacteriophages are needed to attach the virus to the bacterial cell wall.

**Genome Differences** The genetic material in a virus is called a **viral genome.** The composition of viral genomes varies markedly among different types of viruses, as indicated by the examples in Table 17.1. The nucleic acid of some viruses is DNA, whereas in others it is RNA. These are referred to as DNA viruses and RNA viruses, respectively. It is striking that some viruses use RNA for their genome, whereas all living organisms use DNA. In some viruses, the nucleic acid is single stranded, whereas in others, it is double stranded. The genome can be linear or circular, depending on the type of virus. Some kinds of viruses have more than one copy of the genome.

Viral genomes also vary considerably in size, ranging from a few thousand to more than a hundred thousand nucleotides in length (see Table 17.1). For example, the genome of TMV is only 6,400 base pairs in length and contains only six genes. Other viruses, particularly those with a complex structure, such as phage T4, contain many more genes. These extra genes encode many different proteins that are involved in the formation of the elaborate structure shown in Figure 17.1d.

## Viral Reproductive Cycles Consist of a Few Basic Steps

When a virus infects a host cell, the expression of viral genes leads to a series of steps, called a **viral reproductive cycle,** which results in the production of new viruses. The details of the steps differ among various types of viruses, and even the same virus may have the capacity to follow alternative cycles. Even so, by studying the reproductive cycles of hundreds of different viruses, researchers have determined that the viral reproductive cycle consists of five or six basic steps.

To illustrate the general features of viral reproductive cycles, **Figure 17.2** considers these steps for two types of viruses. Figure 17.2a shows the cycle of phage λ (lambda), a bacteriophage with double-stranded DNA as its genome, and Figure 17.2b depicts the cycle of HIV, an enveloped animal virus containing single-stranded RNA. The descriptions that follow compare the reproductive cycles of these two very different viruses.

**Step 1: Attachment** In the first step of a viral reproductive cycle, the virus attaches to the surface of a host cell. This attachment is usually specific for one or just a few types of cells because proteins in the virus recognize and bind to specific molecules on the cell surface. In

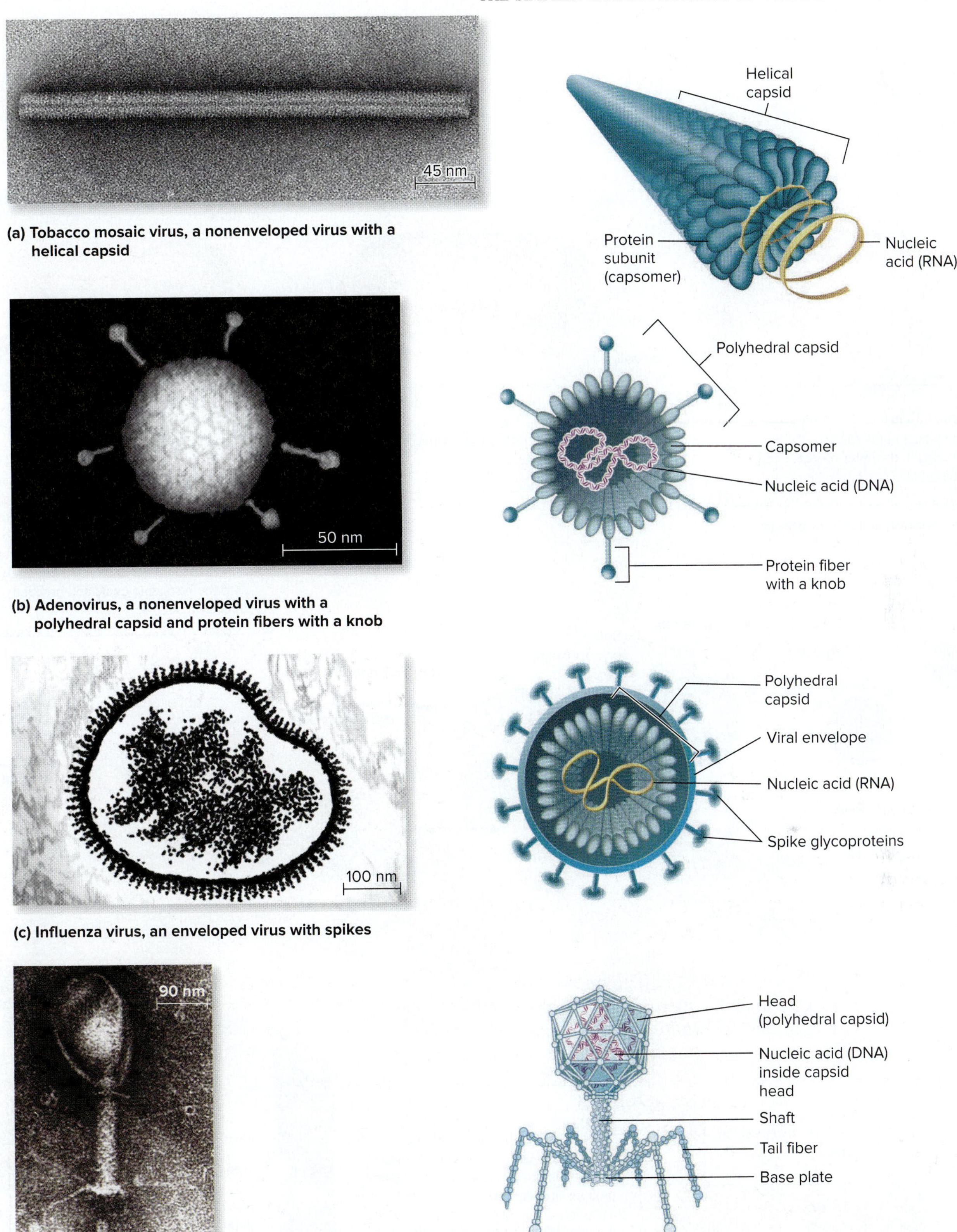


**Figure 17.1 Variations in the structure of viruses, as shown by transmission electron microscopy (TEM).** All viruses contain nucleic acid (DNA or RNA) surrounded by a protein capsid. They may or may not have an outer envelope surrounding the capsid. **(a)** Tobacco mosaic virus (TMV) has a capsid made of 2,130 identical protein subunits, helically arranged around a strand of RNA. **(b)** Adenoviruses have polyhedral capsids containing protein fibers with a knob. **(c)** Many viruses that infect animals, including the influenza virus, have an envelope composed of a lipid bilayer and spike glycoproteins. The envelope is obtained from the host cell when the virus buds from the plasma membrane. **(d)** Some bacteriophages, such as T4, have protein coats with accessory structures such as tail fibers that facilitate invasion of a bacterial cell.
*(a)* © Science Source; *(b)* © Dr. Linda M. Stannard, University of Cape Town/SPL/Science Source; *(c)* © Chris Bjornberg/Science Source; *(d)* © Omikron/Science Source

**Concept Check:** *What features vary among different types of viruses?*

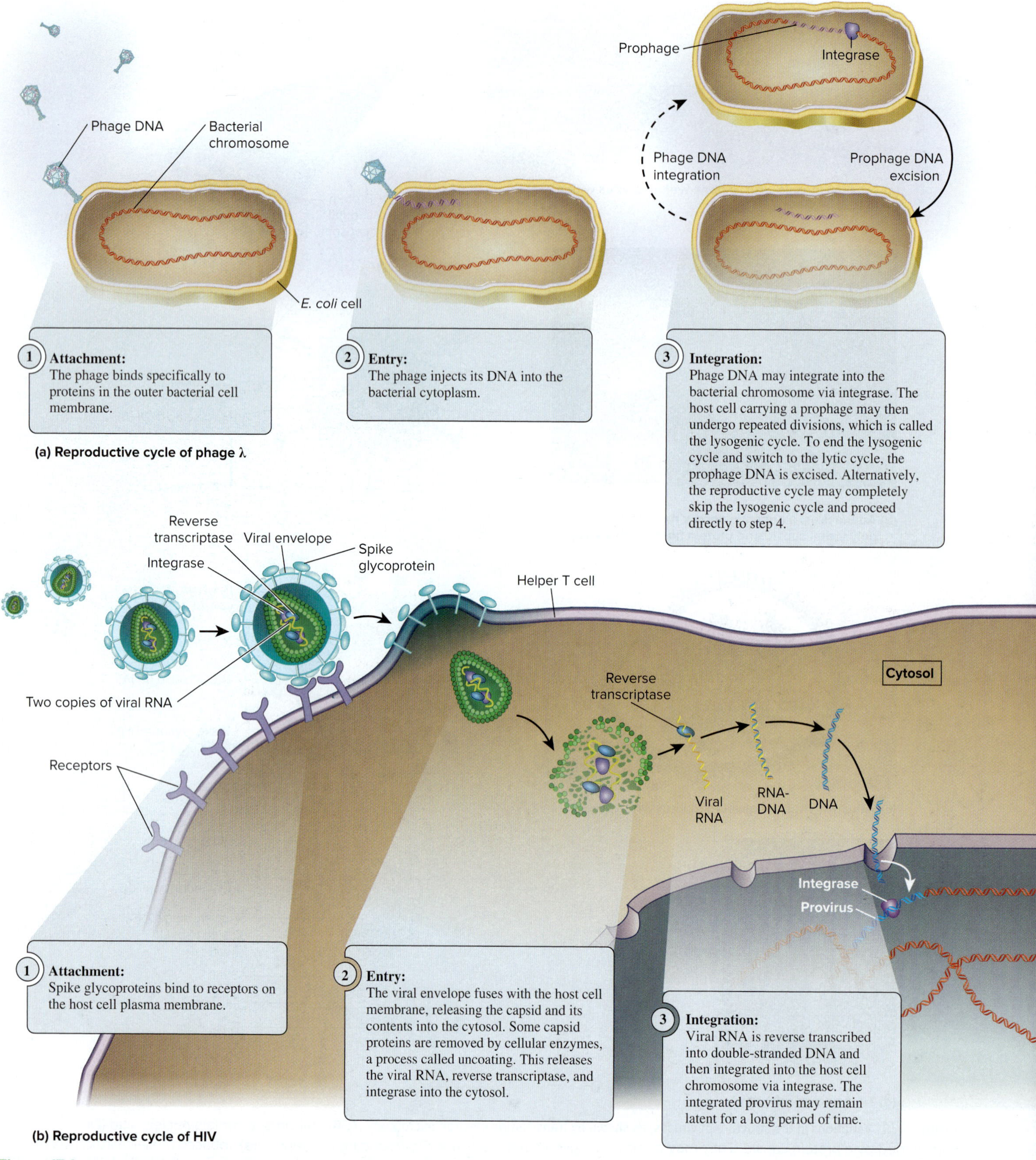

Phage DNA
Bacterial chromosome
E. coli cell
Prophage
Integrase
Phage DNA integration
Prophage DNA excision
1 Attachment:
The phage binds specifically to proteins in the outer bacterial cell membrane.
2 Entry:
The phage injects its DNA into the bacterial cytoplasm.
3 Integration:
Phage DNA may integrate into the bacterial chromosome via integrase. The host cell carrying a prophage may then undergo repeated divisions, which is called the lysogenic cycle. To end the lysogenic cycle and switch to the lytic cycle, the prophage DNA is excised. Alternatively, the reproductive cycle may completely skip the lysogenic cycle and proceed directly to step 4.
(a) Reproductive cycle of phage λ
Reverse transcriptase
Viral envelope
Spike glycoprotein
Integrase
Helper T cell
Two copies of viral RNA
Cytosol
Reverse transcriptase
Receptors
Viral RNA
RNA-DNA
DNA
Integrase
Provirus
1 Attachment:
Spike glycoproteins bind to receptors on the host cell plasma membrane.
2 Entry:
The viral envelope fuses with the host cell membrane, releasing the capsid and its contents into the cytosol. Some capsid proteins are removed by cellular enzymes, a process called uncoating. This releases the viral RNA, reverse transcriptase, and integrase into the cytosol.
3 Integration:
Viral RNA is reverse transcribed into double-stranded DNA and then integrated into the host cell chromosome via integrase. The integrated provirus may remain latent for a long period of time.
(b) Reproductive cycle of HIV


**Figure 17.2** **Comparison of the steps of two viral reproductive cycles.** **(a)** The reproductive cycle of phage λ, a bacteriophage with a double-stranded DNA genome. **(b)** The reproductive cycle of HIV, an enveloped animal virus with a single-stranded RNA genome.

**BioConnections:** *Refer back to Figure 5.20. How does the release of HIV resemble exocytosis?*

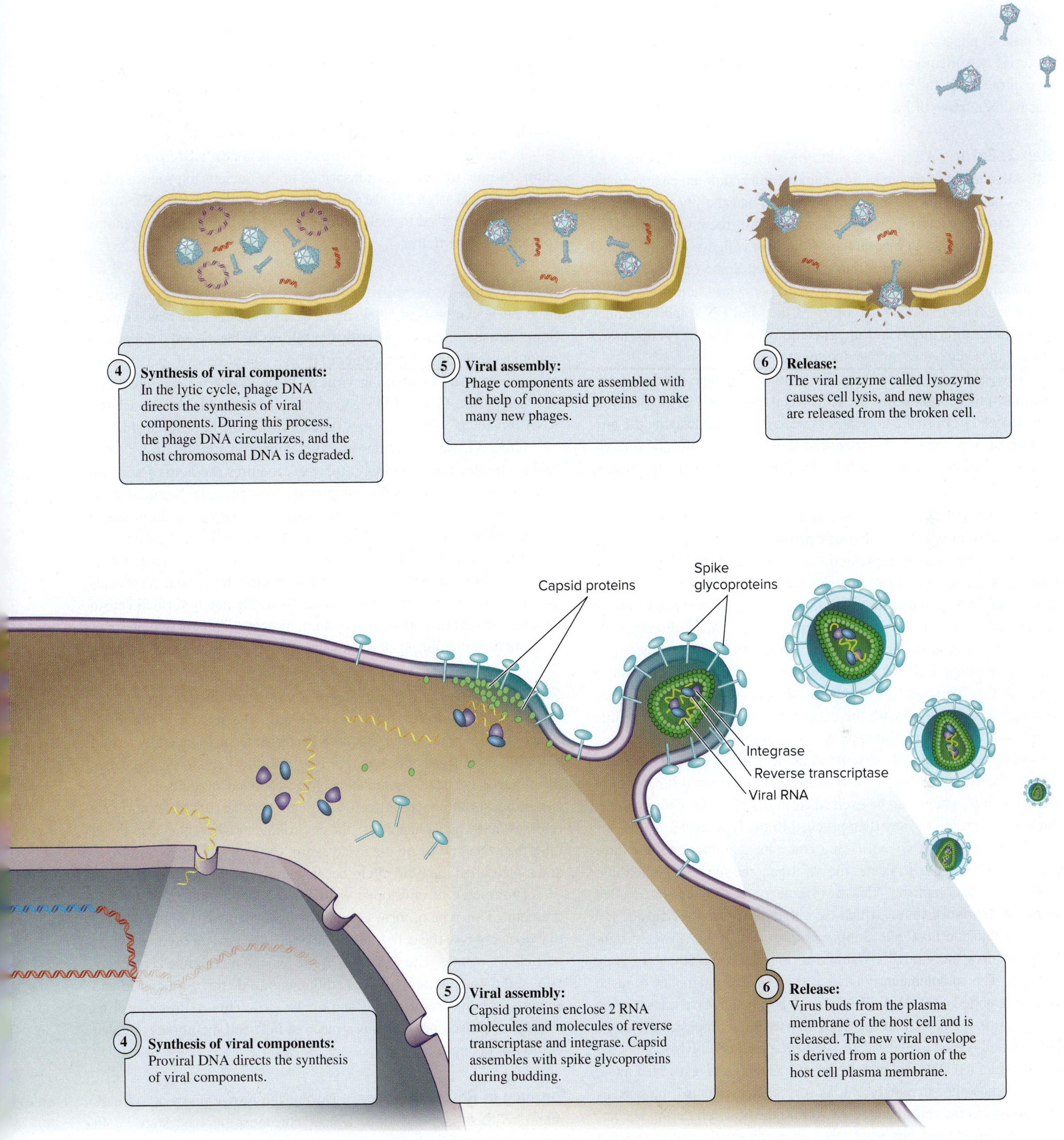

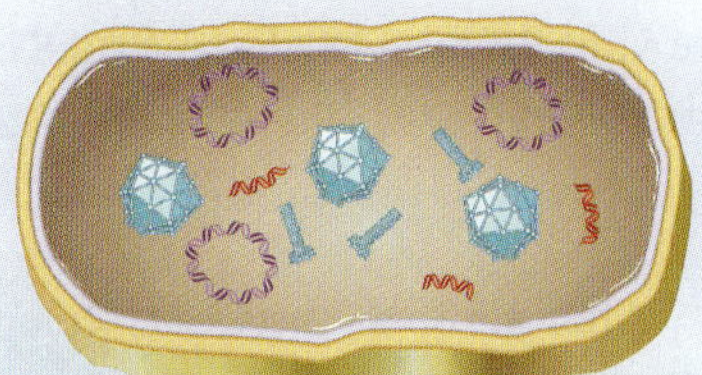

**4 Synthesis of viral components:**
In the lytic cycle, phage DNA directs the synthesis of viral components. During this process, the phage DNA circularizes, and the host chromosomal DNA is degraded.

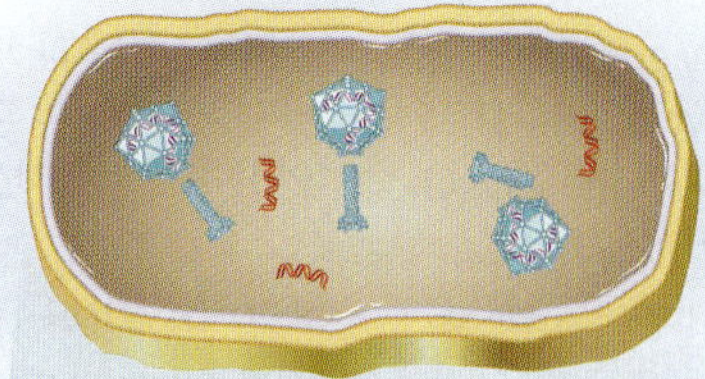

**5 Viral assembly:**
Phage components are assembled with the help of noncapsid proteins to make many new phages.

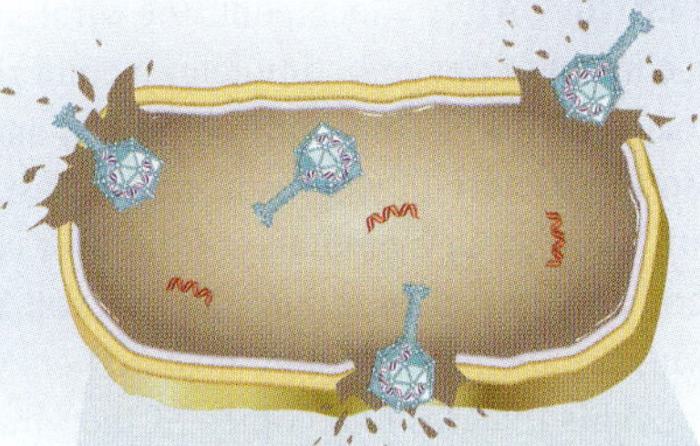

**6 Release:**
The viral enzyme called lysozyme causes cell lysis, and new phages are released from the broken cell.

Capsid proteins

Spike glycoproteins

Integrase

Reverse transcriptase

Viral RNA

**4 Synthesis of viral components:**
Proviral DNA directs the synthesis of viral components.

**5 Viral assembly:**
Capsid proteins enclose 2 RNA molecules and molecules of reverse transcriptase and integrase. Capsid assembles with spike glycoproteins during budding.

**6 Release:**
Virus buds from the plasma membrane of the host cell and is released. The new viral envelope is derived from a portion of the host cell plasma membrane.

the case of phage λ, the phage tail fibers bind to proteins in the outer bacterial cell membrane of *E. coli* cells. In the case of HIV, spike glycoproteins in the viral envelope bind to protein receptors in the plasma membrane of human blood cells called helper T cells.

**Step 2: Entry** After attachment, the viral genome enters the host cell. Attachment of phage λ stimulates a conformational change in the phage coat proteins, so the shaft contracts, and the phage injects its DNA into the bacterial cytoplasm. In contrast, the envelope of HIV fuses with the plasma membrane of the host cell, so both the capsid and its contents are released into the cytosol. Some of the HIV capsid proteins are then removed by host cell enzymes, a process called uncoating. This releases two copies of the viral RNA and molecules of reverse transcriptase and integrase into the cytosol. As discussed shortly, these enzymes are needed for step 3.

The genome of some viruses, including both phage λ and HIV, can integrate into a chromosome of the host cell. For such viruses, the cycle may proceed from step 2 to step 3, as described next, delaying the production of new viruses. Alternatively, the cycle may proceed directly from step 2 to step 4 and quickly lead to the production of new viruses.

**Step 3: Integration** Viruses capable of integration carry a gene that encodes an enzyme called **integrase.** For integration of phage λ to occur, this gene is expressed soon after entry and the integrase protein is made. Integrase cuts the host's chromosomal DNA and inserts the viral genome into the chromosome. In the case of phage λ, the double-stranded DNA that entered the cell can be directly integrated into the double-stranded DNA of the chromosome. Once integrated, the phage DNA in a bacterium is called a **prophage.** When a bacterial cell divides, the prophage DNA is copied and transmitted to daughter cells along with the bacterial chromosomal DNA. While it exists as a prophage, this type of viral reproductive cycle is called the **lysogenic cycle.** As discussed later, new phages are not made during the lysogenic cycle, and the host cell is not destroyed. On occasion, a prophage can be excised from the bacterial chromosome. An excised prophage may then proceed to the lytic cycle.

How can an RNA virus integrate its genome into the host cell's DNA? For this to occur, the viral genome must be copied into DNA. HIV accomplishes this by means of a viral enzyme called **reverse transcriptase,** which is carried within the capsid and released into the host cell along with the viral RNA. Reverse transcriptase uses the viral RNA strand to make a complementary copy of DNA. The complementary DNA is then used as a template to make double-stranded viral DNA. This process is called reverse transcription because it is the reverse of the usual transcription process, in which a DNA strand is used to make a complementary strand of RNA. The viral double-stranded DNA enters the host cell nucleus and is inserted into a host chromosome via integrase. Like reverse transcriptase, integrase is carried within the capsid and released into the host cell during uncoating. Once integrated, the viral DNA in a eukaryotic cell is called a **provirus.** Viruses that follow this mechanism are called **retroviruses.**

**Step 4: Synthesis of Viral Components** The production of new viruses by a host cell involves the synthesis of new copies of the viral genome and the viral proteins that make up the protein coat. A prophage must be excised as described in step 3 before the synthesis of new viral components can occur. Following excision, host cell enzymes make many copies of the phage DNA and transcribe the genes within these copies into mRNA. Host cell ribosomes translate this viral mRNA into viral proteins. The expression of phage genes also leads to the degradation of the host chromosomal DNA.

In the case of HIV, the provirus DNA is not excised from the host chromosome. Instead, it is transcribed in the nucleus to produce many copies of viral RNA. These viral RNA molecules enter the cytosol, where they are used to make viral proteins and serve as the genome for new viral particles.

**Step 5: Viral Assembly** After all of the necessary components have been synthesized, they must be assembled into new viruses. Some viruses with a simple structure self-assemble—viral components spontaneously bind to each other to form a complete virus particle. An example of a self-assembling virus is TMV, which we examined earlier (see Figure 17.1a). TMV capsid proteins assemble around a TMV RNA molecule, which becomes trapped inside the hollow capsid.

Other viruses, including the two shown in Figure 17.2, do not self-assemble. The assembly of phage λ requires the help of noncapsid proteins not found in the completed phage particle. Some of these noncapsid proteins function as enzymes that modify capsid proteins, whereas others serve as scaffolding for the assembly of the capsid.

The assembly of HIV occurs in two stages. First, capsid proteins assemble around two molecules of viral RNA and molecules of reverse transcriptase and integrase. Next, the newly formed capsid acquires its outer envelope in a budding process. This second phase of assembly occurs during step 6, as the virus is released from the cell.

**Step 6: Release** The last step of a viral reproductive cycle is the release of new viruses from the host cell. The release of bacteriophages is a dramatic event. Because bacteria are surrounded by a rigid cell wall, the phages must burst, or lyse, their host cell to escape. After the phages have been assembled, a phage-encoded enzyme called lysozyme digests the bacterial cell wall, causing the cell to burst. Lysis releases many new phages into the environment, where they can infect other bacteria and begin the cycle again. Collectively, steps 1, 2, 4, 5, and 6 are called the **lytic cycle** because they lead to cell lysis.

The release of enveloped viruses from an animal cell is far less dramatic. This type of virus escapes by a mechanism, called budding, that does not lyse the cell. In the case of HIV, a newly assembled virus particle associates with a portion of the plasma membrane containing HIV spike glycoproteins. The membrane enfolds the viral capsid and eventually buds from the surface of the cell. This is how the virus acquires its envelope, which is a piece of host cell membrane studded with viral glycoproteins.

**Latency in Bacteriophages** During the lysogenic cycle, viruses integrate their genomes into a host chromosome (see step 3). In some cases, the resulting prophage or provirus may remain inactive, or **latent,** for a long time. Most of the viral genes are silent during latency, and the viral reproductive cycle does not progress to step 4. When a lysogenic bacterium prepares to divide, it copies the prophage DNA along with its own DNA, so each daughter cell inherits a copy of the prophage. A prophage can be replicated repeatedly in this way without killing the host cell or producing new phage particles.

Environmental conditions influence whether or not viral DNA is integrated into a host chromosome and how long the virus remains in the lysogenic cycle. If nutrients are readily available, phage λ usually proceeds directly to the lytic cycle after its DNA enters the cell. Alternatively, if nutrients are in short supply, the lysogenic cycle is favored because sufficient material may not be available to make new viruses. If more nutrients become available later, the prophage may become activated. At this point, the viral reproductive cycle switches to the lytic cycle, and new viruses are made and released.

**Latency in Human Viruses** Latency among human viruses occurs in two different ways. For HIV, latency occurs because the virus has integrated into the host genome and may remain dormant for a long time. In addition, the genomes of other viruses can exist as an **episome**—a genetic element that replicates independently of the chromosomal DNA but also can occasionally integrate into chromosomal DNA. Examples of viral genomes that exist as episomes include different types of herpesviruses that cause cold sores (usually herpes simplex type I), genital herpes (usually herpes simplex type II), and chickenpox (varicella-zoster). A person infected with a given type of herpesvirus may have periodic outbreaks of disease symptoms when the virus switches from the latent, episomal form to the active form that produces new virus particles.

As an example, let's consider the herpesvirus called varicella-zoster. The initial infection by this virus causes chickenpox, after which the virus may remain latent for many years as an episome. If the varicella-zoster virus becomes active and starts making new virus particles, it can cause a disease called shingles. Shingles begins as a painful rash that eventually erupts into blisters. The blisters follow the path of the neurons that carry the latent varicella-zoster virus. The blisters often form a ring across the back of the patient's body. The term shingles is derived from a Latin word meaning girdle, referring to the observation that the blisters girdle a part of the body.

## Emerging Viruses, Such as HIV and Zika Virus, Have Arisen Recently and May Rapidly Spread Through a Population

A key reason researchers are interested in viral reproductive cycles is the ability of many viruses to cause diseases in humans and other hosts. **Emerging viruses** are ones that have arisen recently or are likely to have a greater probability of causing infection. They typically arise via mutations in pre-existing viruses. Such emerging viruses, which may lead to a significant loss of human life, often cause public alarm. New strains of influenza virus arise fairly regularly due to mutations. One example is the strain H1N1, also called swine flu. In the U.S., despite attempts to minimize deaths by vaccination, over 30,000 people die annually from influenza.

Another example of an emerging virus is Zika virus, a type of flavivirus. It is an enveloped virus with a genome composed of single-stranded RNA. Its name comes from the Zika Forest of Uganda, where the virus was first isolated in 1947. Zika virus is primarily spread by mosquitoes of the genus *Aedes*. The most common symptoms of a Zika infection are fever, rash, joint pain, and conjunctivitis. In most people, the illness is usually mild, but in rare cases, Zika infections in adults can cause a more serious illness called Guillain-Barré syndrome. In addition, Zika virus infection during pregnancy can cause a serious birth defect called microcephaly, as well as other severe fetal brain defects. The Zika virus has spread globally from Africa into Asia, South America, and North America. Though predictions for infection rates vary, some epidemiologists expect that millions of people will become infected with the virus in the coming years.

During the past few decades, the most devastating example of an emerging virus is **human immunodeficiency virus (HIV),** the causative agent of **acquired immunodeficiency syndrome (AIDS).** HIV is primarily spread by sexual contact between infected and uninfected individuals, but it can also be spread by the sharing of needles among intravenous drug users, and from infected mother to unborn child. Since AIDS was first recognized in 1981, the total number of AIDS deaths is over 30 million, making it one of the most deadly diseases in human history. More than 0.5 million of these deaths have occurred in the U.S. In 2014, about 34 million people were living with HIV; approximately 2.3 million of them were infected that year. In that same year, 2 million died from AIDS. Worldwide, nearly 1 in every 100 adults between ages 15 and 49 is infected. In the U.S., about 55,000 new HIV infections occur each year, 70% of which are in men and 30% in women.

The devastating effects of AIDS result from viral destruction of helper T cells, a type of white blood cell that plays an essential role in the immune system of mammals, which is discussed in Chapter 40. Figure 17.3 shows HIV virus particles invading a helper T cell, which interacts with other cells of the immune system to facilitate the production of antibodies and other molecules that target and kill foreign invaders of the body. When large numbers of helper T cells are destroyed by HIV, the function of the immune system is seriously compromised, and the individual becomes susceptible to diseases, called opportunistic infections, that would not normally occur in a healthy person. For example, *Pneumocystis jiroveci*, a fungus that causes a type of pneumonia, is easily destroyed by a healthy immune system. However, in people with AIDS, *P. jiroveci* pneumonia can be fatal.

An insidious feature of HIV replication, described in Figure 17.2b, is that reverse transcriptase, the enzyme that copies the RNA genome into DNA, lacks a proofreading function. In Chapter 9, we learned that DNA polymerase can identify and remove mismatched nucleotides in newly synthesized DNA. Because reverse transcriptase lacks this function, it makes more errors and thereby

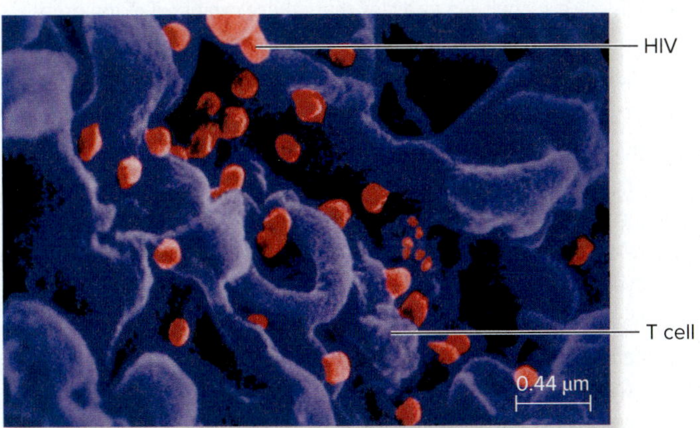


**Figure 17.3 Micrograph of HIV invading a human helper T cell.** This is a colorized scanning electron micrograph (SEM). The surface of the T cell is purple, and HIV particles are red.
Source: Cynthia Goldsmith/CDC

tends to create mutant strains of HIV. This high rate of mutation undermines the ability of the body to combat HIV because mutant strains may be resistant to the body's defenses. In addition, mutant strains of HIV may be resistant to antiviral drugs, as described next.

### Drugs Have Been Developed to Combat the Proliferation of HIV

A compelling reason to understand the reproductive cycle of HIV and other disease-causing viruses is that such knowledge may be used to develop drugs that stop viral proliferation. For example, in the U.S., the highest rate of AIDS-related deaths was approximately 17 per year per 100,000 people in 1994 and 1995. The current rate is about 4 to 5 deaths per year per 100,000 people, owing in part to the use of new antiviral drugs. These drugs inhibit viral proliferation, though they cannot eliminate the virus from the body.

One approach to the design of antiviral treatments is to develop drugs that specifically bind to proteins encoded by the viral genome. For example, azidothymidine (AZT) mimics the structure of a normal nucleotide and binds to the enzyme reverse transcriptase. In this way, AZT inhibits reverse transcription, thereby inhibiting viral replication. Another way to combat HIV involves the use of antiviral drugs that inhibit proteases, enzymes that are needed during the assembly of the HIV capsid. An HIV protease cuts capsid proteins, which makes them smaller and able to assemble into a capsid structure. If the protease does not function, the capsid will not assemble, and new HIV particles will not be made. Several drugs known as protease inhibitors have been developed that bind to HIV protease and inhibits its function.

A major challenge in AIDS research is to discover drugs that inhibit viral proteins without also binding to host cell proteins and inhibiting normal cellular functions. A second challenge is to develop drugs to which mutant strains will not become resistant. As mentioned, HIV readily accumulates mutations during viral replication. A current strategy is to treat HIV patients with a "cocktail" of three or more HIV drugs, making it less likely that any mutant strain will overcome all of the inhibitory effects.

## 17.1 Reviewing the Concepts

- Viruses are nonliving particles that do not exhibit all of the properties associated with living organisms. A virus or its genetic material must be taken up by a living cell to replicate.
- Viruses vary with regard to their host range, structure, and genome composition (Tables 17.1, 17.2; Figure 17.1).
- The viral reproductive cycle consists of a series of basic steps, including attachment, entry, integration, synthesis, assembly, and release. Some bacteriophages follow two different reproductive cycles: the lytic cycle and the lysogenic cycle (Figure 17.2).
- Emerging viruses, such as human immunodeficiency virus (HIV), cause significant loss of human life. HIV, the causative agent of the disease AIDS, is a retrovirus whose reproductive cycle involves the integration of the viral genome into a chromosome in the host cell. HIV accomplishes this by means of a viral enzyme called reverse transcriptase. Drugs to combat HIV have been developed that specifically inhibit viral proteins (Figure 17.3).

## 17.1 Testing Your Knowledge

1. Viruses may differ with regard to their
   a. host range.
   b. structure.
   c. use of DNA or RNA as the genetic material.
   d. all of the above.
2. The step in the viral cycle that follows attachment is
   a. entry.
   b. integration into the host genome.
   c. synthesis of new viral components.
   d. release.

## 17.2 Genetic Properties of Bacteria

**Learning Outcomes**

1. Outline the key features of a bacterial chromosome.
2. Describe the two processes that compact the bacterial chromosome.
3. Outline the structure and functions of plasmids.
4. Diagram the process of cell division in bacteria.

Bacterial cells may exist as single units or remain associated with each other after cell division, forming pairs, chains, or clumps. Bacteria are widespread on Earth, and numerous species are known to cause various types of infectious diseases. We begin this section by exploring the structure and replication of the bacterial genome and the organization of DNA sequences along a bacterial chromosome. We will then examine how the bacterial chromosome is compacted to fit inside a bacterium and how it is transmitted during asexual reproduction.

### Bacteria Typically Have Circular Chromosomes That Carry a Few Thousand Genes

The genes of bacteria are found within bacterial chromosomes.

- Although a bacterial cell usually has a single type of chromosome, it may have more than one copy of that chromosome. The number of copies depends on the species and on growth conditions, but a bacterium typically has one to four identical chromosomes.
- Each bacterial chromosome is tightly packed within a distinct **nucleoid** of the cell (**Figure 17.4**). Unlike the eukaryotic nucleus, the nucleoid is not a separate cellular compartment bounded by a membrane. The DNA in the nucleoid is in direct contact with the cytoplasm of the cell.
- Bacterial chromosomes contain molecules of double-stranded DNA along with many different proteins. They are usually circular and, typically only a few million base pairs (bp) long. For example, the chromosome of *Escherichia coli* has approximately 4.6 million bp, and the *Haemophilus influenzae* chromosome has roughly 1.8 million bp.
- A typical bacterial chromosome contains a few thousand unique genes that are found throughout the chromosome. Gene sequences, primarily those that encode proteins, account for the largest part of bacterial DNA.

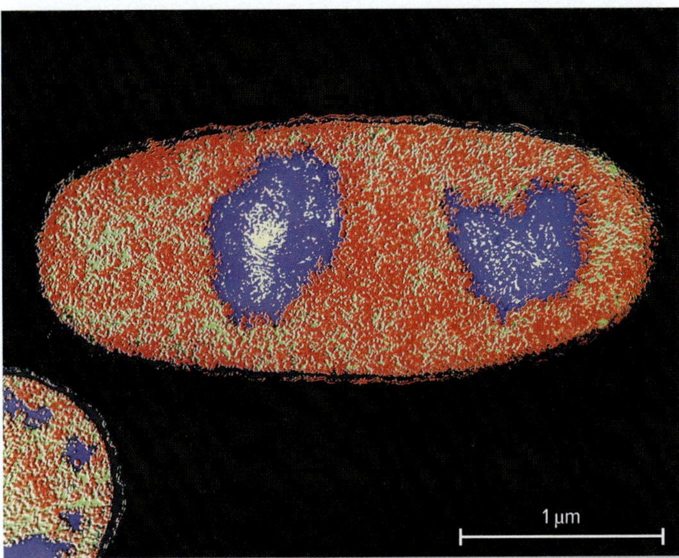


**Figure 17.4 Nucleoids within the bacterium *Bacillus subtilis.*** In the light micrograph shown here, the nucleoids are fluorescently labeled and seen as purple oval-shaped areas within the bacterial cytoplasm. Two or more nucleoids are usually found within each cell.


**Concept Check:** *How many nucleoids are in this bacterial cell?*

**BioConnections:** *Refer back to Figures 4.8 and 4.9. How is a nucleoid different from a nucleus found in a eukaryotic cell?*

- Other nucleotide sequences in the chromosome play a role in DNA replication, gene expression, and chromosome structure. One of these sequences is the origin of replication, which is a few hundred base pairs long. Bacterial chromosomes have a single origin of replication that functions as an initiation site for the assembly of several proteins that are required for DNA replication.

## The Formation of Chromosomal Loops and DNA Supercoiling Make the Bacterial Chromosome Compact

Bacterial cells are much smaller than most eukaryotic cells (refer back to Figure 4.5). *E. coli* cells, for example, are approximately 1 μm wide and 2 μm long. To fit within a bacterial cell, the DNA of a typical bacterial chromosome must be compacted about 1,000-fold. How does this occur? The compaction of a bacterial chromosome, shown in **Figure 17.5**, occurs by two processes: the formation of loops and DNA supercoiling.

Unlike eukaryotic DNA, bacterial DNA is not wound around histone proteins to form nucleosomes. However, the binding of proteins to bacterial DNA is important in the formation of **loop domains**—chromosomal segments that are folded into loops. As seen in Figure 17.5, DNA-binding proteins anchor the bases of the loops in place. The number of loops varies according to the size of a bacterial chromosome and the species. The *E. coli* chromosome has 50 to 100 loop domains, each with about 40,000 to 80,000 bp. This looping compacts the circular chromosome about 10-fold.

**DNA supercoiling** is a second important way to compact the bacterial chromosome. Because DNA is a long, thin molecule, twisting can dramatically change its conformation. This compaction is similar to what happens to a rubber band if you twist it in one direction. Because the two strands of DNA already coil around each other, the formation of additional coils due to twisting is referred to as supercoiling. Bacterial enzymes called topoisomerases twist the DNA and control the degree of DNA supercoiling.

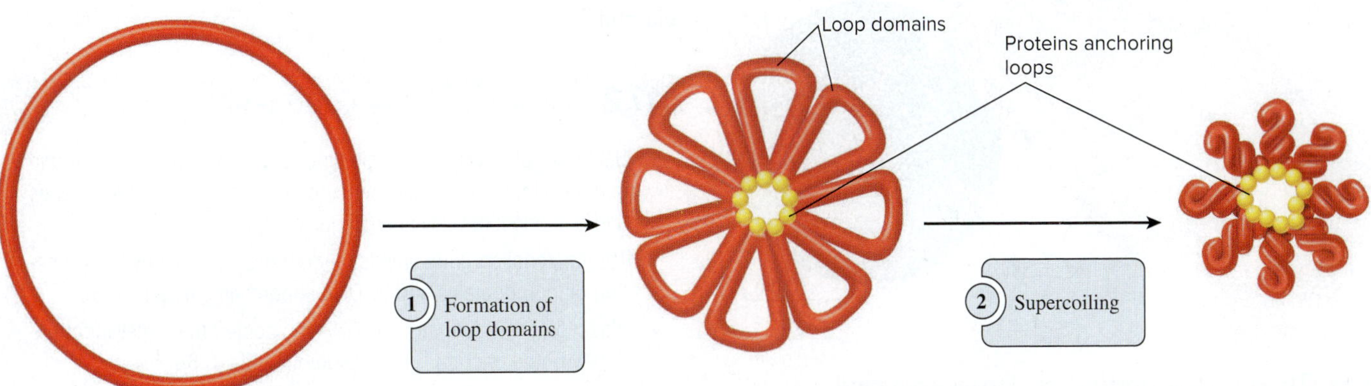


**Figure 17.5 The compaction of a bacterial chromosome.** As a way to compact the large, circular chromosome, segments are organized into smaller loop domains by binding to proteins at the bottoms of the loops. These loops are made more compact by DNA supercoiling. Note: This is a simplified drawing; bacterial chromosomes typically have 50 to 100 loop domains.

**Concept Check:** *Describe how the loop domains are held in place.*

## Plasmids Are Small, Circular Pieces of Extrachromosomal DNA

In addition to chromosomal DNA, bacterial cells commonly contain **plasmids**, small, circular pieces of DNA that exist separately from the bacterial chromosome (Figure 17.6). Plasmids occur naturally in many strains of bacteria and some eukaryotic cells, such as yeast. The smallest plasmids consist of just a few thousand base pairs and carry only a gene or two. The largest are in the range of 100,000 to 500,000 bp and carry several dozen or even hundreds of genes. A plasmid has its own origin of replication that allows it to be replicated independently of the bacterial chromosome. The DNA sequence of the origin of replication influences how many copies of the plasmid are found within a cell. Some origins are said to be very strong because they result in many copies of the plasmid, perhaps as many as 100 per cell. Other origins of replication have sequences that are much weaker, so the number of copies is relatively low, such as one or two per cell.

Why do bacteria have plasmids? Certain genes within a plasmid usually provide some type of growth advantage to the cell or may aid in survival under certain conditions. By studying plasmids in many different species, researchers have discovered that most plasmids fall into five categories:

1. Resistance plasmids, also known as R factors, contain genes that confer resistance against antibiotics and other types of toxins.
2. Degradative plasmids carry genes that enable the bacterium to digest and utilize an unusual substance. For example, a degradative plasmid may carry genes that allow a bacterium to digest an organic solvent such as toluene.
3. Col-plasmids contain genes that encode colicins, which are proteins that kill other bacteria.
4. Virulence plasmids carry genes that turn a bacterium into a pathogenic strain.
5. Fertility plasmids, also known as F factors, allow bacteria to transfer genes to each other, a topic described later in this chapter.

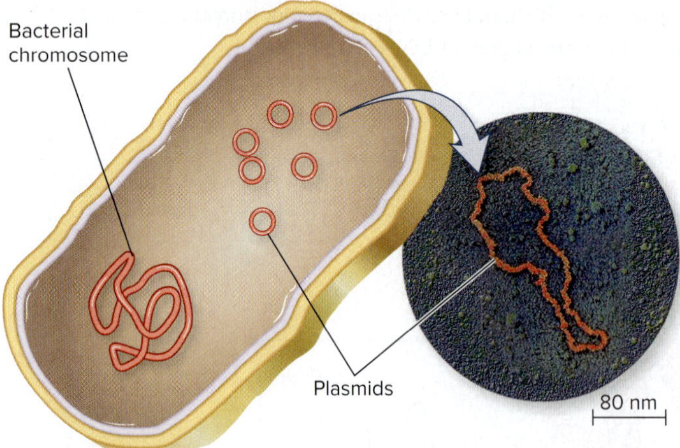


**Figure 17.6 Plasmids in a bacterial cell.** Plasmids are small, circular DNA molecules that exist independently of the bacterial chromosome.



**Concept Check:** *Describe the similarities and differences between a bacterial chromosome and a plasmid.*

On occasion, a plasmid may integrate into the bacterial chromosome. Such plasmids, which can integrate or remain independent of the chromosome, are termed episomes.

## Most Bacteria Reproduce by Binary Fission

Thus far, we have considered the genetic material of bacteria and how the bacterial chromosome is compacted to fit inside the cell. Let's now turn our attention to the process of bacterial cell division. The capacity of bacteria to divide is quite astounding. The cells of some species, such as *E. coli*, can divide every 20–30 minutes. When placed on a solid growth medium in a petri dish, an *E. coli* cell and its daughter cells undergo repeated cellular divisions and form a group of genetically identical cells called a **bacterial colony** (Figure 17.7). Starting with a single cell that is invisible to the naked eye, a visible bacterial colony containing 10 to 100 million cells forms in less than a day!

Cell division of most bacterial species occurs by a process called **binary fission,** during which a cell divides into two daughter cells. Figure 17.8 shows this process for a cell with a single chromosome. Before it divides, the cell replicates its DNA. This produces two identical copies of the chromosome. Next, the cell's plasma membrane is drawn inward and deposits new cell-wall material, separating the two daughter cells. Each daughter cell receives one of the copies of the original chromosome. Therefore, except when a mutation occurs, each daughter cell contains an identical copy of the mother cell's genetic material.

Plasmids replicate independently of the bacterial chromosome. During binary fission, the plasmids are distributed to daughter cells, so each daughter cell usually receives one or more copies of the plasmid.

## 17.2 Reviewing the Concepts

- Bacteria typically have a single type of circular chromosome found in the nucleoid of the cell. The chromosome contains many genes and one origin of replication (Figure 17.4).
- The bacterial chromosome is made more compact by the formation of loop domains and by DNA supercoiling (Figure 17.5).
- Plasmids are small, circular DNA molecules that exist independently of the bacterial chromosome (Figure 17.6).
- When placed on solid growth media, a single bacterial cell will divide many times to produce a colony composed of many cells (Figure 17.7).
- Bacteria reproduce by asexual reproduction in a process called binary fission, during which a cell divides to form two daughter cells (Figure 17.8).

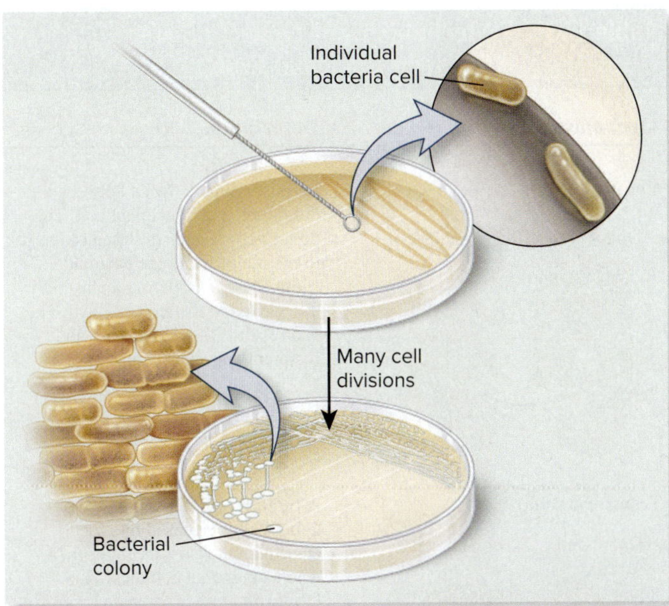


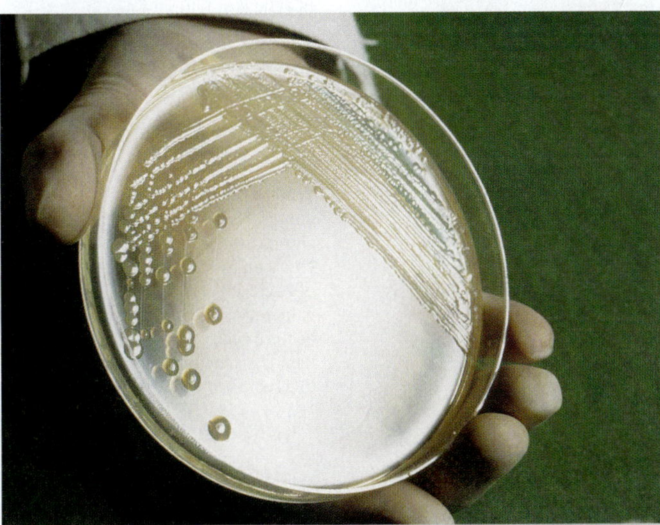

**Figure 17.7 Growth of a bacterial colony.** Through successive cell divisions, a single bacterial cell of *E. coli* forms a genetically identical group of cells called a bacterial colony.
© Dr. Jeremy Burgess/SPL/Science Source

**Concept Check:** *Let's suppose a bacterial strain divides every 30 minutes. If a single cell is placed on a plate, how many cells will be in the colony after 16 hours?*

## 17.2 Testing Your Knowledge

1. Which of the following would *not* be found in a typical bacterial chromosome?
   a. a few thousand genes
   b. one origin of replication
   c. a centromere
   d. All of the above would be found in a bacterial chromosome.

## Biology Principle

### Living Organisms Grow and Develop

After cell division, a bacterial cell increases in size.

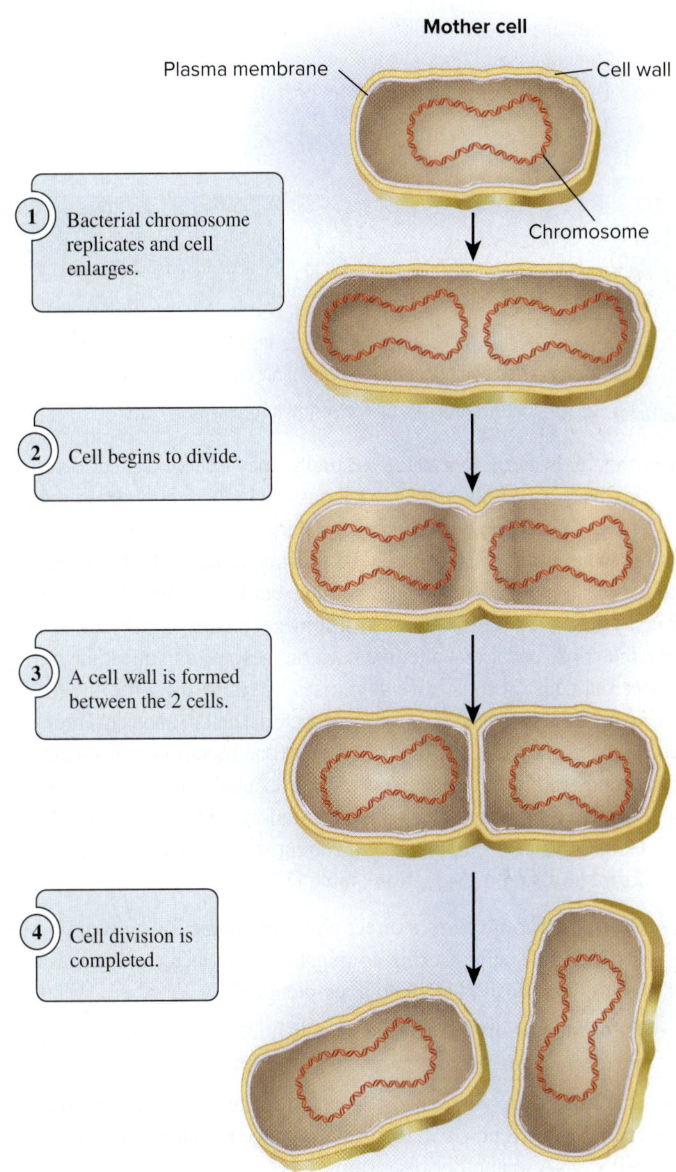


**Figure 17.8 Bacterial cell division.** Bacteria reproduce by binary fission. Before a bacterium divides, the bacterial chromosome is replicated to produce two identical copies. These two copies segregate from each other during cell division, with one copy going to each daughter cell.

2. Bacterial chromosomes are made more compact by
   a. attaching to the plasma membrane.
   b. the formation of loop domains.
   c. DNA supercoiling.
   d. both b and c.
3. Which of the following is a characteristic of plasmids?
   a. They have an origin of replication.
   b. They are usually circular.
   c. They may carry genes, such as antibiotic-resistance genes.
   d. They exist independently of the bacterial chromosome.
   e. All of the above are correct.

**Table 17.3 Mechanisms of Gene Transfer Between Bacterial Cells**

| Mechanism | Description |
|---|---|
| **Conjugation:**<br>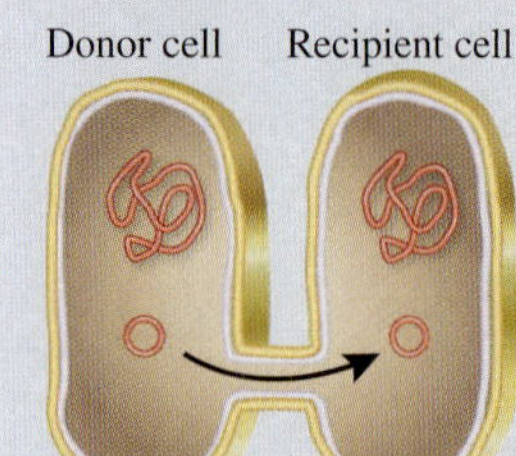<br> | Requires direct contact between a donor cell and a recipient cell. The donor cell transfers a strand of DNA to the recipient. In the example shown here, DNA from a plasmid is transferred to the recipient cell. The end result is that both donor and recipient cells have a plasmid. |
| **Transformation:**<br>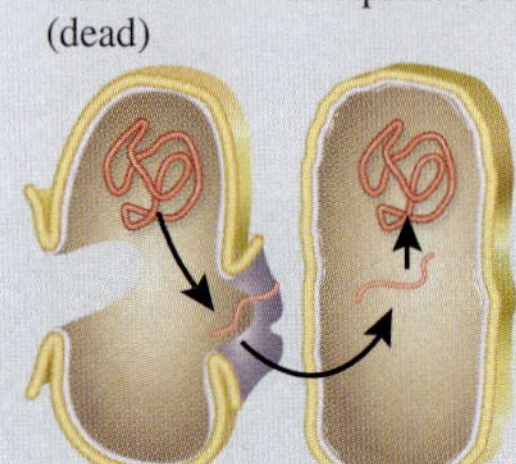<br> | A fragment of DNA from a donor cell is released into the environment. This may happen when a bacterial cell dies. This DNA fragment is taken up by a recipient cell, which incorporates the DNA into its chromosome. |
| **Transduction:**<br>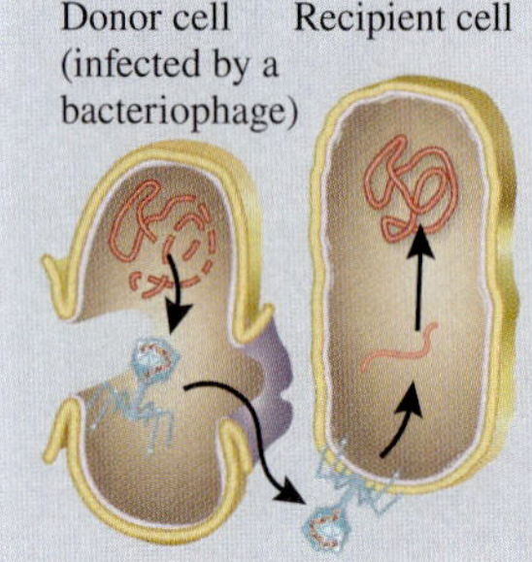<br> | When a bacteriophage infects a donor cell, it causes the bacterial chromosome of the donor cell to break up into fragments. A fragment of bacterial chromosomal DNA is incorporated into a newly made bacteriophage. The bacteriophage then transfers this fragment of DNA to a recipient cell. |

## 17.3 Gene Transfer Between Bacteria

### Learning Outcomes

1. Compare and contrast the three forms of gene transfer between bacteria: conjugation, transformation, and transduction.
2. Define the process of horizontal gene transfer.

Even though bacteria reproduce asexually, they exhibit a great deal of genetic diversity. Within a given bacterial species, the term **strain** refers to a lineage that has genetic differences compared with another strain. For example, one strain of *E. coli* may be resistant to an antibiotic, whereas another strain may be sensitive to the same antibiotic. How does genetic diversity arise in an asexual species? It comes primarily from two sources. First, mutations occur that alter the bacterial genome and affect the traits of bacterial cells. Second, diversity arises by **gene transfer,** in which genetic material is transferred from one bacterial cell to another. Through gene transfer, genetic variation that arises in one bacterium can spread to other strains and even to other species. For example, an antibiotic-resistance gene may be transferred from a resistant strain to a sensitive strain.

Gene transfer occurs in three different ways, termed conjugation, transformation, and transduction (Table 17.3).

- **Conjugation** involves a direct physical interaction between two bacterial cells. During conjugation, one bacterium acts as a donor and transfers DNA to a recipient cell.
- **Transformation** is a process in which DNA is released into the environment and taken up by another bacterial cell.
- **Transduction** occurs when a bacteriophage infects a bacterial cell and then a newly made bacteriophage transfers some of that cell's DNA to another bacterium.

These three types of gene transfer have been extensively investigated in research laboratories, and their molecular pathways continue to be studied with great interest. In this section, we will examine these mechanisms in greater detail.

### During Conjugation, DNA Is Transferred from a Donor Cell to a Recipient Cell

In the early 1950s, American microbiologists Joshua and Esther Lederberg, Irish physician William Hayes, and Italian geneticist Luca Cavalli-Sforza independently discovered that certain bacterial strains can donate genetic material during conjugation. For example, about 5% of *E. coli* strains found in nature act as donor strains. Further research showed that a strain that cannot act as a donor may acquire this ability after being mixed with a donor strain. Hayes correctly proposed that donor strains contain a type of plasmid called a fertility factor, or **F factor,** that can be transferred to recipient strains. Also, other donor *E. coli* strains were later identified that transfer portions of the bacterial chromosome at high frequencies. After a segment of the chromosome is transferred, it then inserts, or recombines, into the chromosome of the recipient cell. Such donor strains were named Hfr (for High frequency of recombination). In our discussion, we will focus on donor strains that carry F factors.

The micrograph in Figure 17.9a shows two conjugating *E. coli* cells. The cell on the left is designated F$^+$, meaning that it has an F factor. The F$^+$ cell, also called the donor cell, makes **sex pili** that bind specifically to recipient cells, which are designated F$^-$ cells. The F factor found in F$^+$ cells carries several genes that are required for conjugation, such as those required for sex pili formation. Also, some

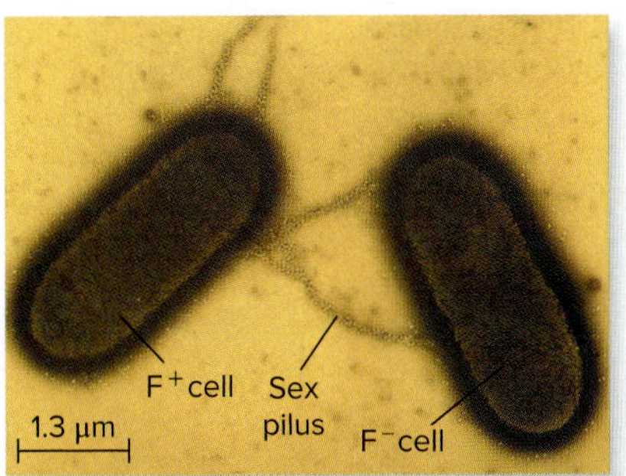


(a) Micrograph of conjugating cells

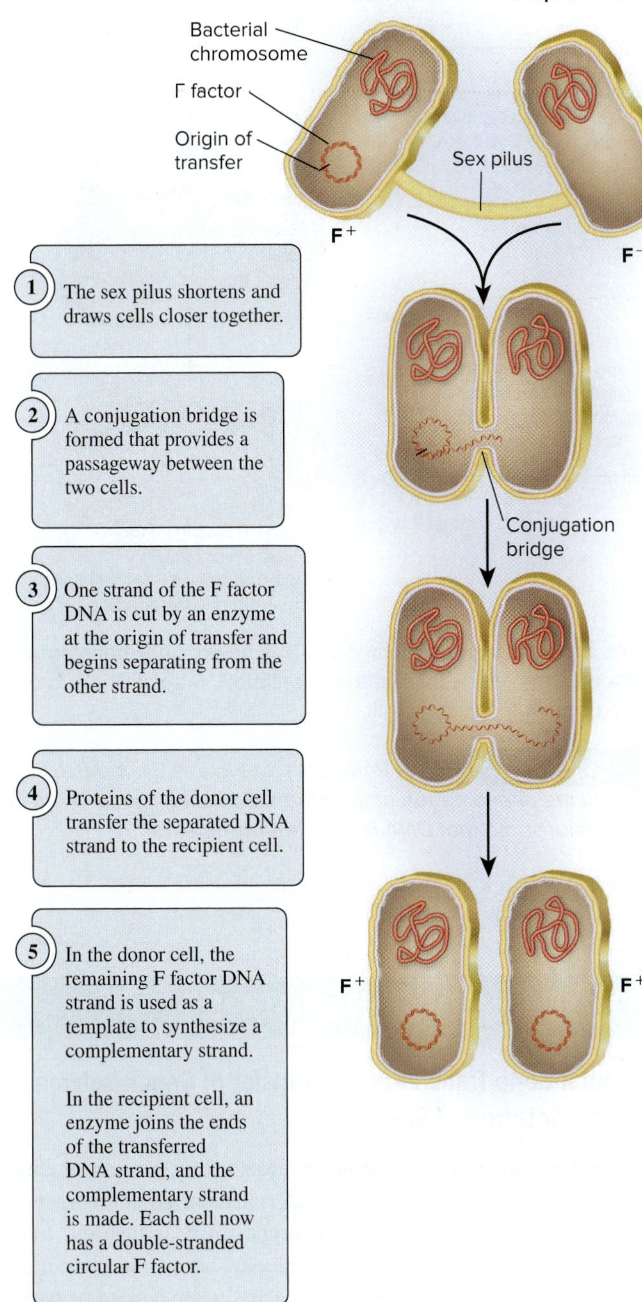

Donor cell
Recipient cell
Bacterial chromosome
F factor
Origin of transfer
Sex pilus
F⁺
F⁻
1 The sex pilus shortens and draws cells closer together.
2 A conjugation bridge is formed that provides a passageway between the two cells.
Conjugation bridge
3 One strand of the F factor DNA is cut by an enzyme at the origin of transfer and begins separating from the other strand.
4 Proteins of the donor cell transfer the separated DNA strand to the recipient cell.
5 In the donor cell, the remaining F factor DNA strand is used as a template to synthesize a complementary strand.
In the recipient cell, an enzyme joins the ends of the transferred DNA strand, and the complementary strand is made. Each cell now has a double-stranded circular F factor.
F⁺
F⁺


(b) Transfer of an F factor during conjugation

F factors, called F′ factors (F prime factors) carry additional bacterial genes that are not needed for conjugation.

Figure 17.9b describes the events that occur during conjugation in *E. coli*. The process is similar in other bacteria that are capable of conjugating, although the details vary somewhat from one species to another.

1. In *E. coli* and some other species, $F^+$ cells make very long pili that attempt to make contact with nearby $F^-$ cells. Once contact is made, the pili shorten, drawing the donor and recipient cells closer together.
2. A conjugation bridge is formed that provides a direct passageway for DNA transfer.
3. One strand of F factor DNA is cut at the origin of transfer and separates from the other strand.
4. Proteins of the donor cell transfer the separated DNA strand through the conjugation bridge into the recipient cell. The other strand remains in the donor cell.
5. In the donor cell, the complementary strand is synthesized, thereby restoring the F factor DNA to its original double-stranded condition. In the recipient cell, the two ends of the newly acquired F factor DNA strand are joined to form a circular molecule, and its complementary strand is synthesized to produce a double-stranded F factor.

If conjugation is successful, the end result is that the recipient cell has acquired an F factor, converting it from an $F^-$ to an $F^+$ cell. The genetic composition of the donor strain has not been changed.

## In Transformation, Bacteria Take Up DNA from the Environment

In contrast to conjugation, the process of gene transfer known as bacterial transformation does not require direct contact between bacterial cells. Frederick Griffith first discovered this process in 1928 while working with strains of *Streptococcus pneumoniae* (refer back to Figure 9.1).

How does a bacterial cell become transformed? First, it imports a strand of DNA from the environment. This DNA strand, which is typically derived from a dead bacterial cell, may then insert or recombine into the bacterial chromosome. The live bacterium then carries genes from the dead bacterium—the live bacterium has been transformed.

Not all bacterial strains have the ability to take up DNA. Those that do have this ability are described as naturally **competent,** and they have genes that encode proteins called competence factors. Competence factors facilitate the binding of DNA fragments to the bacterial cell

**Figure 17.9 Bacterial conjugation. (a)** A micrograph of two *E. coli* cells that are conjugating. The cell on the left, designated $F^+$, is the donor; the cell on the right, designated $F^-$, is the recipient. The two cells make contact via sex pili made by the $F^+$ cell. **(b)** The transfer of an F factor during conjugation. At the end of conjugation, both the donor cell and the recipient cell are $F^+$.

*(a)* © Dr. L. Caro/SPL/Science Source

**Concept Check:** *If a donor cell has only one F factor, explain how the donor and recipient cell both contain one F factor following the transfer of an F factor during conjugation.*

surface, the uptake of DNA into the cytoplasm, and the incorporation of the imported DNA into the bacterial chromosome. Temperature, ionic conditions, and the availability of nutrients affect whether or not a bacterium will be competent to take up genetic material.

In recent years, biologists have unraveled some of the steps that occur when competent bacterial cells are transformed by taking up genetic material from the environment. In the example shown in Figure 17.10, the DNA released from a dead bacterium carries a gene, $tet^R$, that confers resistance to the antibiotic tetracycline.

1. A large fragment of the DNA binds to a cell surface receptor on the outside of a bacterial cell that is sensitive to tetracycline.
2. Enzymes secreted by the bacterium cut this large fragment into fragments of DNA small enough to enter the cell.
3. One of the two strands of a fragment of DNA containing the $tet^R$ gene is degraded. The other strand enters the bacterial cytoplasm via a DNA uptake system that transports the DNA across the plasma membrane.
4. Finally, the imported single-stranded DNA strand is incorporated into the bacterial chromosome, and the complementary strand is synthesized.

Following transformation, the recipient cell has been transformed from a tetracycline-sensitive cell to a tetracycline-resistant cell.

## In Transduction, Bacteriophages Transfer Genetic Material from One Bacterium to Another

A third mechanism of gene transfer is transduction, in which bacteriophages transfer bacterial genes from one bacterium to another. As discussed earlier in this chapter, a bacteriophage (phage) is a virus that uses the cellular machinery of a bacterium for its own replication. The new viral particles made in this way usually contain only viral genes. On rare occasions, however, a phage may pick up a piece of DNA from the bacterial chromosome. When a phage carrying a segment of bacterial DNA infects another bacterium, it transfers this segment into the chromosome of its new bacterial host.

Transduction is actually an error in a phage lytic cycle, as shown in Figure 17.11.

1. In this example, a phage called P1 infects an *E. coli* cell that has a gene ($his^+$) for histidine synthesis.
2. Phage P1 causes the host cell chromosome to be broken into small pieces.
3. New phage DNA and proteins are synthesized. When new phages are assembled, coat proteins may accidentally enclose a piece of host DNA, such as the $his^+$ gene. This produces a transducing phage carrying bacterial chromosomal DNA in its capsid.
4. In the example shown in Figure 17.11, this transducing phage is released and binds to an *E. coli* cell that lacks the $his^+$ gene. It inserts the bacterial DNA fragment into the recipient cell.
5. This fragment carrying the $his^+$ gene is recombined into the host chromosome.

In this case, gene transfer by transduction converts a $his^-$ strain of *E. coli* to a $his^+$ strain.

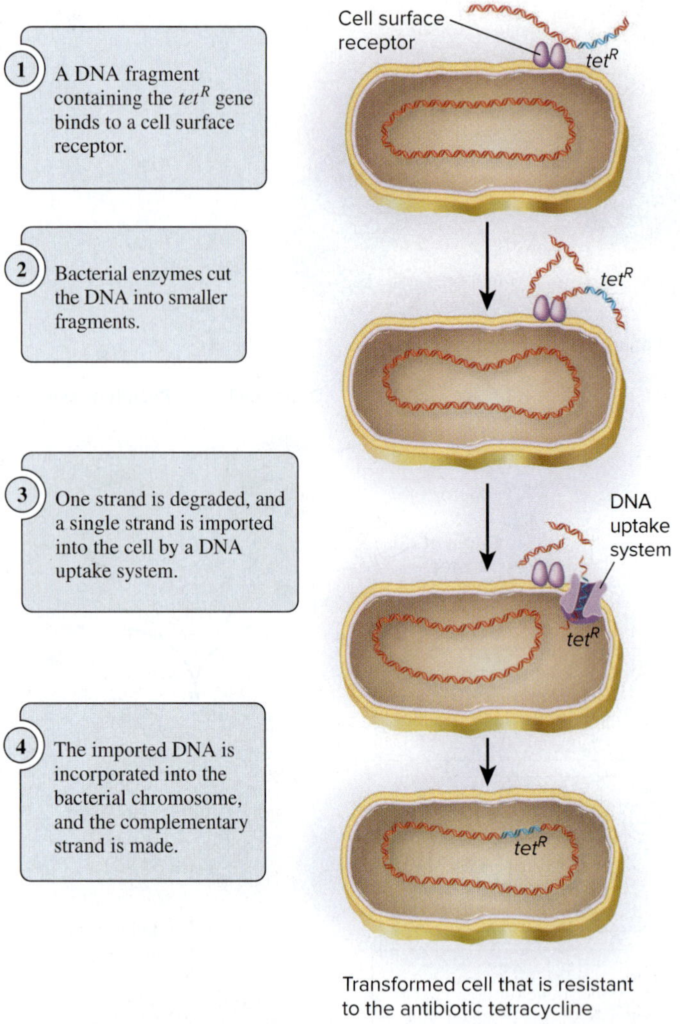

1 A DNA fragment containing the $tet^R$ gene binds to a cell surface receptor.
Cell surface receptor
$tet^R$
2 Bacterial enzymes cut the DNA into smaller fragments.
$tet^R$
3 One strand is degraded, and a single strand is imported into the cell by a DNA uptake system.
DNA uptake system
$tet^R$
4 The imported DNA is incorporated into the bacterial chromosome, and the complementary strand is made.
$tet^R$
Transformed cell that is resistant to the antibiotic tetracycline

**Figure 17.10 Bacterial transformation.** This process has transformed a bacterium that was sensitive to the antibiotic tetracycline into one that is resistant to this antibiotic.

**BioConnections:** *Refer back to Figures 9.1 and 9.2. How did the phenomenon of transformation allow researchers to demonstrate that DNA is the genetic material?*

# EVOLUTIONARY CONNECTIONS

## Horizontal Gene Transfer Is the Transfer of Genes Between the Same or Different Species

The term **horizontal gene transfer** refers to any process in which an organism incorporates genetic material from another organism without being the offspring of that organism. Conjugation, transformation, and transduction are examples of horizontal gene transfer. In contrast, vertical gene transfer occurs when genes are passed from one generation to the next—from parents to offspring and from mother cells to daughter cells.

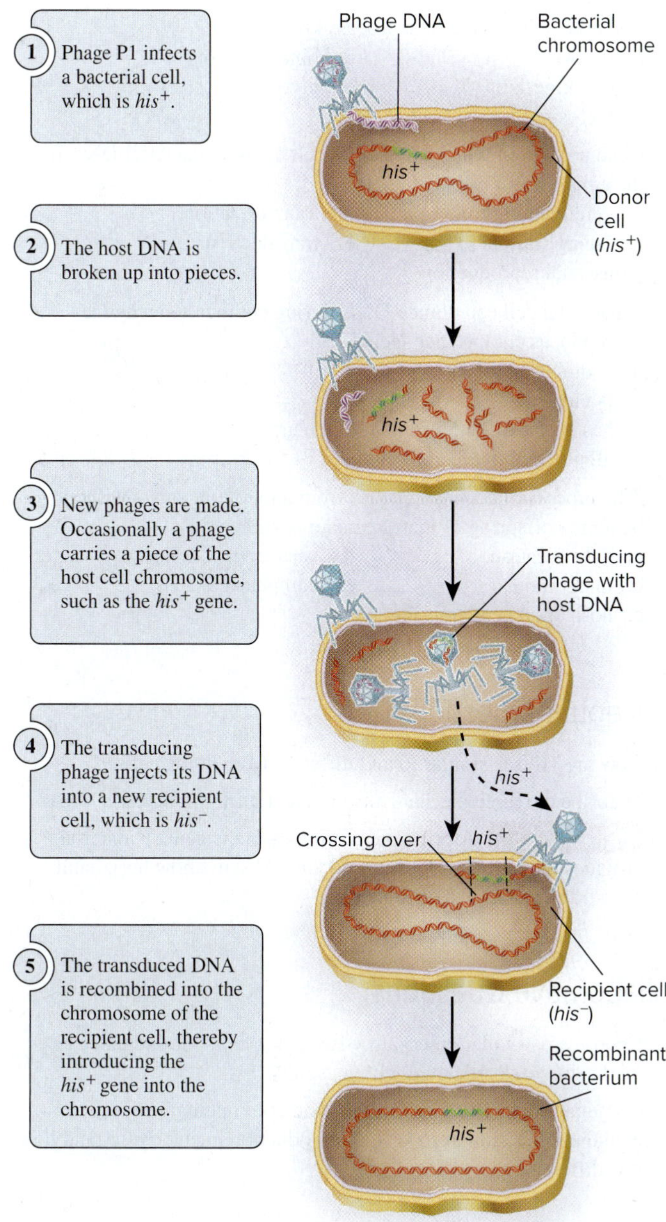

1 Phage P1 infects a bacterial cell, which is *his*+.
Phage DNA
Bacterial chromosome
his+
Donor cell (his+)
2 The host DNA is broken up into pieces.
his+
3 New phages are made. Occasionally a phage carries a piece of the host cell chromosome, such as the *his*+ gene.
Transducing phage with host DNA
4 The transducing phage injects its DNA into a new recipient cell, which is *his*−.
his+
Crossing over
his+
5 The transduced DNA is recombined into the chromosome of the recipient cell, thereby introducing the *his*+ gene into the chromosome.
Recipient cell (his−)
Recombinant bacterium
his+
The recombinant bacterium has a genotype (his+) that is different from the original recipient bacterial cell (his−).


**Figure 17.11** **Bacterial transduction by phage P1.**

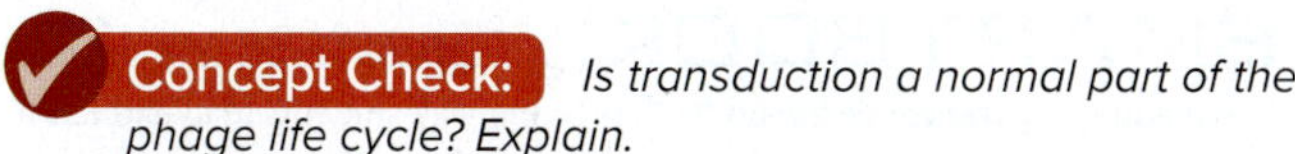

**Concept Check:** *Is transduction a normal part of the phage life cycle? Explain.*

Conjugation, transformation, and transduction occasionally occur between cells of different bacterial species. In recent years, analyses of bacterial genomes have shown that a sizeable fraction of bacterial genes are derived from horizontal gene transfer. For example, roughly 17% of the genes of *E. coli* and of *Salmonella typhimurium* have been acquired from other species by horizontal gene transfer during the past 100 million years. Many of these acquired genes affect traits that give cells a selective advantage, including genes that confer antibiotic resistance, the ability to degrade toxic compounds, and the ability to withstand extreme environments. Some horizontally transferred genes confer pathogenicity, turning a harmless bacterial strain into one that can cause disease. Geneticists have suggested that horizontal gene transfer has played a major role in the evolution of different bacterial species. In many cases, the transferred genes may encode proteins that allow a bacterium to survive in a new type of environment and can eventually lead to the formation of a new species.

Horizontal gene transfer is also important for its medical relevance. Let's consider the topic of antibiotic resistance. Antibiotics are widely prescribed to treat bacterial infections in humans. The term **acquired antibiotic resistance** refers to the common phenomenon of a previously susceptible strain becoming resistant to a specific antibiotic. This change may result from genetic alterations in the bacteria's own genome, but it is often due to the horizontal transfer of resistance genes from a resistant strain. As an example, some *Staphylococcus aureus* strains have developed resistance to methicillin and all penicillins. Evidence suggests that these so-called methicillin-resistant strains of *Staphylococcus aureus* (MRSA) acquired the methicillin-resistance gene by horizontal gene transfer, possibly from a strain of *Enterococcus faecalis*. MRSA strains cause skin infections that are more difficult to treat than infections caused by nonresistant strains of *S. aureus*.

## 17.3 Reviewing the Concepts

- Three common modes of gene transfer between bacteria are conjugation, transformation, and transduction (Table 17.3).
- During conjugation, a strand of DNA is transferred from a donor to a recipient cell by physical contact (Figure 17.9).
- Transformation is the process in which a segment of DNA from the environment is taken up by a competent cell and incorporated into the bacterial chromosome (Figure 17.10).
- During transduction, a bacteriophage transfers a segment of bacterial chromosomal DNA from one bacterium to another (Figure 17.11).
- Horizontal gene transfer is a process in which an organism incorporates genetic material from another organism without being the offspring of that organism.

## 17.3 Testing Your Knowledge

1. During conjugation, an $F^+$ cell transfers genetic material to an $F^-$ cell. After conjugation is completed, what is the result?
   a. The donor cell is $F^-$ and the recipient cell is $F^+$.
   b. The donor cell is $F^+$ and the recipient cell is $F^+$.
   c. The donor cell is $F^-$ and the recipient cell is $F^-$.
   d. The donor cell is $F^+$ and the recipient cell is $F^-$.
2. During transformation, where does the transforming DNA come from?
   a. the environment
   b. a living bacterial cell
   c. a virus
   d. all of the above

# Assess and Discuss

## Test Yourself

1. The __________ is the protein coat of a virus that surrounds the genetic material.
   a. host
   b. prion
   c. capsid
   d. viroid
   e. cell wall
2. Which of the following does *not* describe a typical characteristic of viral genomes?
   a. The genetic material may be DNA or RNA.
   b. The nucleic acid may be single stranded or double stranded.
   c. The genome may carry just a few genes or several dozen.
   d. The number of copies of the genome may vary.
   e. All of the above describe typical variation in viral genomes.
3. During viral infection, attachment is usually specific to a particular cell type because
   a. the virus is attracted to the appropriate host cells by proteins secreted into the extracellular fluid.
   b. the virus recognizes and binds to specific molecules in the cytoplasm of the host cell.
   c. the virus recognizes and binds to specific molecules on the surface of the host cell.
   d. the host cell produces channel proteins that provide passageways for viruses to enter the cytoplasm.
   e. the virus releases specific proteins that make holes in the membrane large enough for the virus to enter.
4. HIV, a retrovirus, has a high mutation rate because
   a. the HIV genome is less stable than other viral genomes.
   b. reverse transcriptase lacks a proofreading function.
   c. the viral genome is altered every time it is incorporated into the host genome.
   d. antibodies produced by the host cell mutate the viral genome when infection occurs.
   e. all of the above are correct.
5. Which of the following is *not* a correct statement regarding an emerging virus?
   a. Emerging viruses have arisen recently.
   b. Emerging viruses do not need host cells to replicate.
   c. Emerging viruses have a greater capacity to cause disease.
   d. HIV is an example of an emerging virus.
   e. All of the above are correct statements.
6. Genetic diversity is *not* maintained in bacterial populations by
   a. binary fission.
   b. mutation.
   c. transformation.
   d. transduction.
   e. conjugation.
7. Bacterial cells divide by a process known as
   a. mitosis.
   b. cytokinesis.
   c. meiosis.
   d. binary fission.
   e. glycolysis.
8. Gene transfer in which a bacterial cell takes up bacterial DNA from the environment is called
   a. conjugation.
   b. binary fission.
   c. asexual reproduction.
   d. transformation.
   e. transduction.
9. A bacterial cell can donate DNA during conjugation when it
   a. produces competence factors.
   b. contains an F factor.
   c. is virulent.
   d. has been infected by a bacteriophage.
   e. all of the above.
10. The process of acquiring genes from another organism without being the offspring of that organism is known as
   a. hybridization.
   b. integration.
   c. horizontal gene transfer.
   d. vertical gene transfer.
   e. competence.

## Conceptual Questions

1. How are viruses similar to and different from living cells?
2. Describe the three mechanisms of gene transfer in bacteria.
3. PRINCIPLES A principle of biology is the genetic material provides the blueprint for reproduction. Discuss how horizontal gene transfer alters the blueprint.

## Collaborative Questions

1. Choose a particular emerging virus and discuss the societal and medical issues associated with its spread.
2. Conjugation is sometimes called "bacterial mating." Discuss how conjugation is similar to sexual reproduction in eukaryotes and how it is different.

# Online Resource

# Genetic Technologies: How Biologists Study Genes and Genomes

# 18

Japanese researchers used an exciting new genetic approach, called CRISPR-Cas technology, to silence a gene in mice that encodes tyrosinase, an enzyme needed for pigment production. In the seven albino newborn mice on the right, both copies of this gene have been silenced. In the ones with a mottled appearance, partial silencing was achieved. The gene was not silenced in the black newborn on the left.

## Chapter Outline



Jon was born with hemophilia A, which is a blood clotting disorder. As discussed in Chapter 15, hemophilia is inherited as an X-linked recessive trait. In Jon's case, his blood is unable to clot properly because he is missing a protein, called factor VIII, that is needed in a pathway required for normal blood clotting. Fortunately, Jon can take injections of purified factor VIII and thereby minimize the harmful effects of hemophilia. You might be surprised to learn that purified factor VIII is not obtained from human blood. Instead, it's made by cells grown in the laboratory. These cells have been genetically modified to synthesize factor VIII in large amounts. This is just one example of how researchers have been able to apply **recombinant DNA technology**—the use of laboratory techniques to bring together fragments of DNA from multiple sources—to benefit humans.

In the early 1970s, the first successes in making recombinant DNA molecules were accomplished independently by two groups at Stanford University: David Jackson, Robert Symons, and Paul Berg; and Peter Lobban and A. Dale Kaiser. Both groups were able to isolate and purify pieces of DNA in a test tube and then covalently link two or more DNA fragments. Once inside a host cell, the recombinant molecules were replicated to produce many identical copies. The process of making multiple copies of a particular gene is known as **gene cloning.** In the first section of the chapter, we will explore recombinant DNA technology and gene cloning, techniques that have enabled geneticists to probe the relationships between gene sequences and phenotypic consequences.

In the second section, we will consider the topic of **genomics**—the molecular analysis of the entire genome of a species. In recent years, molecular techniques have progressed to the point where researchers can study the structure and function of many genes as large, integrated networks and can make changes to the genome in order to silence particular genes (see chapter-opening photo) or alter their DNA sequence in a specific way.

In the remaining sections of this chapter, we will consider the genome characteristics of bacteria, archaea, and eukaryotes. In many species, such as multicellular plants and animals, the genome may have a high abundance of repetitive sequences—sequences that are repeated multiple times in a genome. We will examine how such repetitive sequences increase in number.

## 18.1 Gene Cloning

### Learning Outcomes

1. Outline the steps of gene cloning using vectors.
2. Distinguish between a genomic library and a cDNA library.
3. Explain how gel electrophoresis is used to separate DNA fragments.
4. Describe the steps of gene cloning using polymerase chain reaction (PCR).

As mentioned, the term gene cloning refers to the process of making many copies of a particular gene. Why is gene cloning useful? **Figure 18.1** provides an overview of the steps and goals of gene cloning. The process is usually done with one of two goals in mind. One goal is that a researcher or clinician wants many copies of the gene, perhaps to study the DNA directly or to use the DNA as a tool. For example, geneticists may want to determine the sequence of a gene from a person with a disease to see if the gene carries a mutation. Alternatively, the goal may be to obtain a large amount of the gene product, such as a specific protein. Along these lines, biochemists use gene cloning to obtain large amounts of proteins to study their structure and function. In recent years, gene cloning has provided the foundation for critical technical advances in a variety of disciplines, including molecular biology, genetics, cell biology, biochemistry, and medicine. In this section, we will examine the procedures that are used in gene cloning.

### Step 1: Vector DNA and Chromosomal DNA Are the Starting Materials for Gene Cloning

One way to carry out gene cloning uses a type of DNA known as a **vector** (from the Latin, for "one who carries") (see Figure 18.1). Vector DNA acts as a carrier of the DNA segment that is to be cloned. In cloning experiments, a vector may carry a small segment of chromosomal DNA, perhaps only a single gene. By comparison, a chromosome carries up to a few thousand genes. When a vector is introduced into a living cell, it can replicate, and so the DNA that it carries is also replicated. This produces many identical copies of the inserted gene.

The vectors commonly used in gene cloning experiments were originally derived from two natural sources: plasmids or viruses.

- **Plasmids** are small, circular pieces of DNA that are found naturally in many strains of bacteria and exist independently of the bacterial chromosome. Commercially available plasmids have been genetically engineered for effective use in cloning experiments. They contain unique sites into which geneticists can easily insert pieces of chromosomal DNA.
- **Viral vectors** are derived from viruses, which can infect living cells and propagate themselves by taking control of the host cell's metabolic machinery. When a chromosomal gene is inserted into a viral vector, the gene is replicated whenever the viral DNA is replicated. Therefore, viruses can be used as vectors to carry other pieces of DNA.

The second material necessary for cloning a gene is the gene itself, which we will call the gene of interest. The source of the gene is the chromosomal DNA that carries the gene. The preparation of chromosomal DNA involves breaking open cells and extracting and purifying the DNA using biochemical separation techniques such as chromatography and centrifugation.

### Step 2: Cutting Chromosomal and Vector DNA into Pieces and Linking Them Together Produces Recombinant Vectors

The second step in a gene cloning experiment is the insertion of the gene of interest into the vector (see Figure 18.1 step 2). How is this accomplished? DNA molecules are cut and pasted into vectors to produce recombinant vectors. To cut DNA, researchers use enzymes known as **restriction enzymes.** These enzymes, which were discovered by Swiss geneticist Werner Arber and American microbiologists Hamilton Smith and Daniel Nathans in the 1960s and 1970s, are made naturally by many different species of bacteria. Restriction enzymes protect bacterial cells from invasion by viruses by degrading the viral DNA into small fragments. Several hundred different restriction enzymes from various bacterial species have been identified and are commercially available to molecular biologists.

The restriction enzymes used in cloning experiments bind to a specific base sequence and then cleave the DNA backbone at two defined locations, one in each strand. Most restriction enzymes recognize sequences that are palindromic, which means the sequence is identical when read in the opposite direction in the complementary strand (**Table 18.1**). For example, the sequence recognized by the restriction enzyme *Eco*RI is 5′–GAATTC–3′ in the top strand. Read in the opposite direction in the bottom strand, this sequence is also 5′–GAATTC–3′. Certain restriction enzymes are useful in cloning because they digest DNA into fragments with single-stranded ends (termed "sticky" ends) that hydrogen-bond to other DNA fragments that are cut with the same enzyme and thus have complementary sequences.

## Biology Principle

### Biology Is an Experimental Science

The technique of gene cloning allows researchers to study genes and gene products in greater detail.

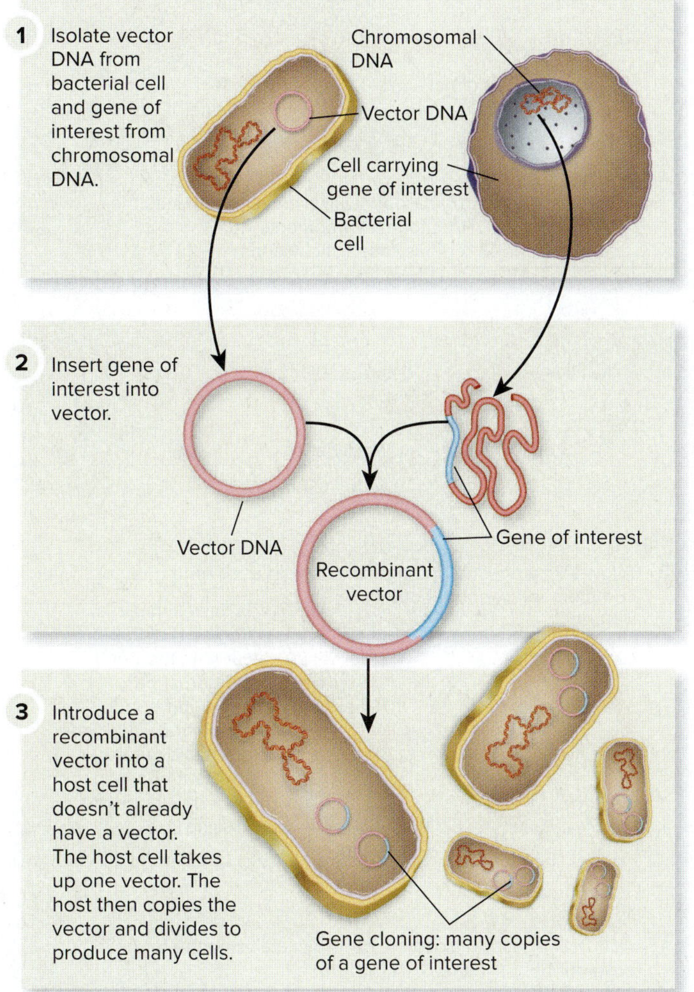

1 Isolate vector DNA from bacterial cell and gene of interest from chromosomal DNA.
Chromosomal DNA
Vector DNA
Cell carrying gene of interest
Bacterial cell
2 Insert gene of interest into vector.
Vector DNA
Gene of interest
Recombinant vector
3 Introduce a recombinant vector into a host cell that doesn't already have a vector. The host cell takes up one vector. The host then copies the vector and divides to produce many cells.
Gene cloning: many copies of a gene of interest


**Gene cloning is done to achieve one of two main goals:**

| Producing large amounts of DNA of a specific gene | Expressing the cloned gene to produce the encoded protein |
|---|---|
| *Examples*<br>• Cloned genes provide enough DNA for DNA sequencing. The sequence of a gene can help us understand how a gene works and identify mutations that cause diseases.<br>• Cloned DNA can be used as a probe to identify the same gene or similar genes in other organisms. | *Examples*<br>• Large amounts of the protein can be purified to study its structure and function.<br>• Cloned genes can be introduced into bacteria or livestock to make pharmaceutical products such as insulin.<br>• Cloned genes can be introduced into plants and animals to alter their traits.<br>• Cloned genes can be used to treat diseases—a clinical approach called gene therapy. |

**Figure 18.1 Overview of gene cloning.** The process of gene cloning is used to produce large amounts of a gene or its protein product.

**Figure 18.2** shows the action of a restriction enzyme (*Eco*RI) and the insertion of a gene into a vector. This vector, which is a plasmid, carries the $amp^R$ and *lacZ* genes, whose useful functions will be discussed later. *Eco*RI binds to specific sequences in both the vector and chromosomal DNA. It then cleaves the DNA backbones, producing DNA fragments with sticky ends (Figure 18.2, step 1). The sticky ends of a piece of chromosomal DNA and the vector DNA can hydrogen-bond with each other (Figure 18.2, step 2), a process called annealing. However, this interaction is not stable, because it involves only a few hydrogen bonds between complementary bases. To establish a permanent connection between two DNA fragments, the sugar-phosphate backbones of the DNA strands must be covalently linked or ligated together. This linkage is catalyzed by DNA ligase (Figure 18.2, step 3). Recall from Chapter 9 that DNA ligase is an enzyme that catalyzes the formation of a covalent bond between two adjacent DNA fragments.

In some cases, the two ends of the vector simply ligate back together, restoring it to its original circular structure; this forms a recircularized vector. In other cases, a fragment of chromosomal DNA may become ligated to both ends of the vector. When this happens, a segment of chromosomal DNA has been inserted into the vector. The result is a vector containing a piece of chromosomal DNA, which is called a **recombinant vector.** Such a vector is ready to be cloned. A recombinant vector may contain the gene of interest, or it may contain a different piece of chromosomal DNA.

**Table 18.1 Examples of Restriction Enzymes Used in Gene Cloning**

| Restriction enzyme* | Bacterial source | Sequence recognized† |
|---|---|---|
| *Eco*RI | *Escherichia coli* (strain RY13) | ↓<br>5′–GAATTC–3′<br>3′–CTTAAG–5′<br>↑ |
| *Sac*I | *Streptomyces achromogenes* | ↓<br>5′–GAGCTC–3′<br>3′–CTCGAG–5′<br>↑ |

*Restriction enzymes are named according to the species in which they are found. The first three letters are italicized because they indicate the genus and species names. Because a species may produce more than one restriction enzyme, the enzymes are designated I, II, III, and so on, to indicate the order in which they were discovered in a given species.

†The arrows show the locations in the upper and lower DNA strands where the restriction enzymes cleave the DNA backbone.

1 Cut vector and chromosomal DNA with *Eco*RI, a restriction enzyme that recognizes the sequence **GAATTC** and cuts at the arrows. **CTTAAG**

*lacZ*
*Eco*RI site
*amp*$^R$
Origin of replication
GAATTC
CTTAAG
*Eco*RI sites
Gene of interest

Vector DNA has one *Eco*RI site.

Chromosomal DNA has many *Eco*RI sites.

The restriction enzyme opens up the vector and cuts the chromosomal DNA into many fragments with short single-stranded regions called sticky ends.

Sticky ends
G
CTTAA
AATTC
G
Sticky end

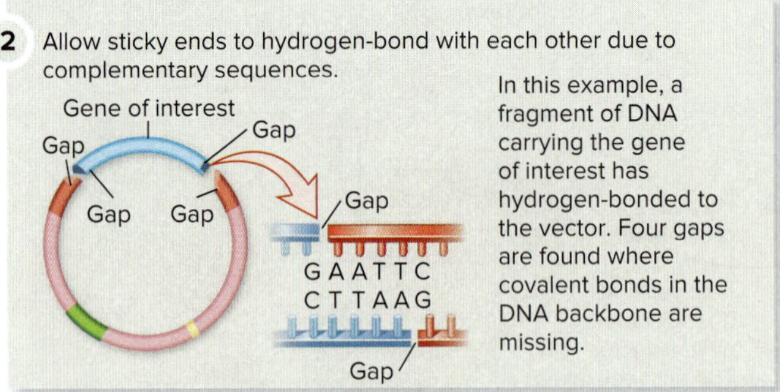

2 Allow sticky ends to hydrogen-bond with each other due to complementary sequences.
Gene of interest
Gap
Gap
Gap
Gap
Gap
GAATTC
CTTAAG
Gap
In this example, a fragment of DNA carrying the gene of interest has hydrogen-bonded to the vector. Four gaps are found where covalent bonds in the DNA backbone are missing.


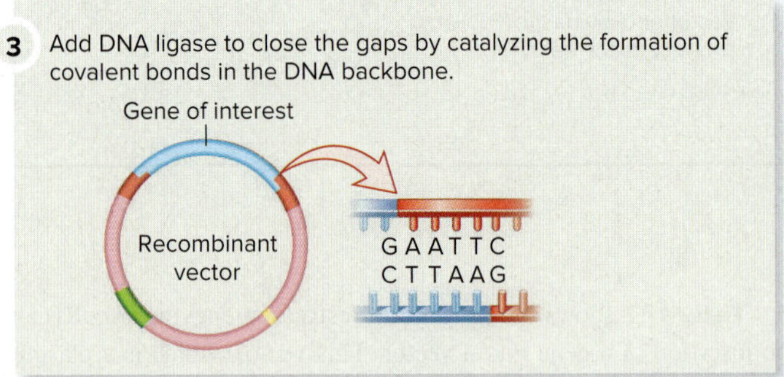


**Figure 18.2** **Step 2 of gene cloning: The actions of a restriction enzyme and DNA ligase to produce a recombinant vector.**

**Concept Check:** *In the procedure shown in this figure, has the gene of interest been cloned?*

**BioConnections:** *Refer back to Figure 17.6. What are the general properties of plasmids?*

## Step 3: Putting Recombinant Vectors into Host Cells and Allowing Those Cells to Propagate Achieves Gene Cloning

The third step in gene cloning is the actual cloning of the gene of interest. In this step, the goal is for the recombinant vector carrying the desired gene to be taken up by bacterial cells that have been treated with agents that render them permeable to DNA molecules (**Figure 18.3**). The process, called transformation, is described in Chapter 17 (refer back to Figure 17.10). Bacterial cells with the ability to take up DNA are called **competent.** Some bacterial cells take up a single plasmid, whereas most cells fail to take up a plasmid. The bacteria are then plated on petri plates containing a bacterial growth medium and ampicillin.

In the experiment shown in Figure 18.3, the bacterial cells were originally sensitive to ampicillin. The plasmid already carries an antibiotic-resistance gene, called the *amp*$^R$ gene. What is the purpose of this gene in a cloning experiment? Such a gene is called a **selectable marker** because the presence of the antibiotic selects for the growth of cells expressing the *amp*$^R$ gene. The *amp*$^R$ gene encodes an enzyme known as β-lactamase that degrades the antibiotic ampicillin, which normally kills bacteria. Bacteria that have not taken up a plasmid are killed by the antibiotic. In contrast, any bacterium that has taken up a plasmid carrying the *amp*$^R$ gene grows and divides many times to form a bacterial colony containing tens of millions of cells. Because each cell in a single colony is derived from the same original cell that took up a single plasmid, all cells within a colony contain the same type of plasmid DNA.

The experimenter also needs a way to distinguish bacterial colonies that contain cells with a recombinant vector from those containing cells with a recircularized vector. In a recombinant vector, a piece of chromosomal DNA has been inserted into a region of the vector that contains the *lacZ* gene, which encodes the enzyme β-galactosidase. The insertion of chromosomal DNA into the vector disrupts the *lacZ* gene, thereby preventing the synthesis of β-galactosidase. By comparison, a recircularized vector has a functional *lacZ* gene. The functionality of *lacZ* can be determined by adding to the growth medium a colorless compound, X-Gal, which is cleaved by β-galactosidase into a blue dye. Bacteria grown in the presence of X-Gal form blue colonies if they produce a

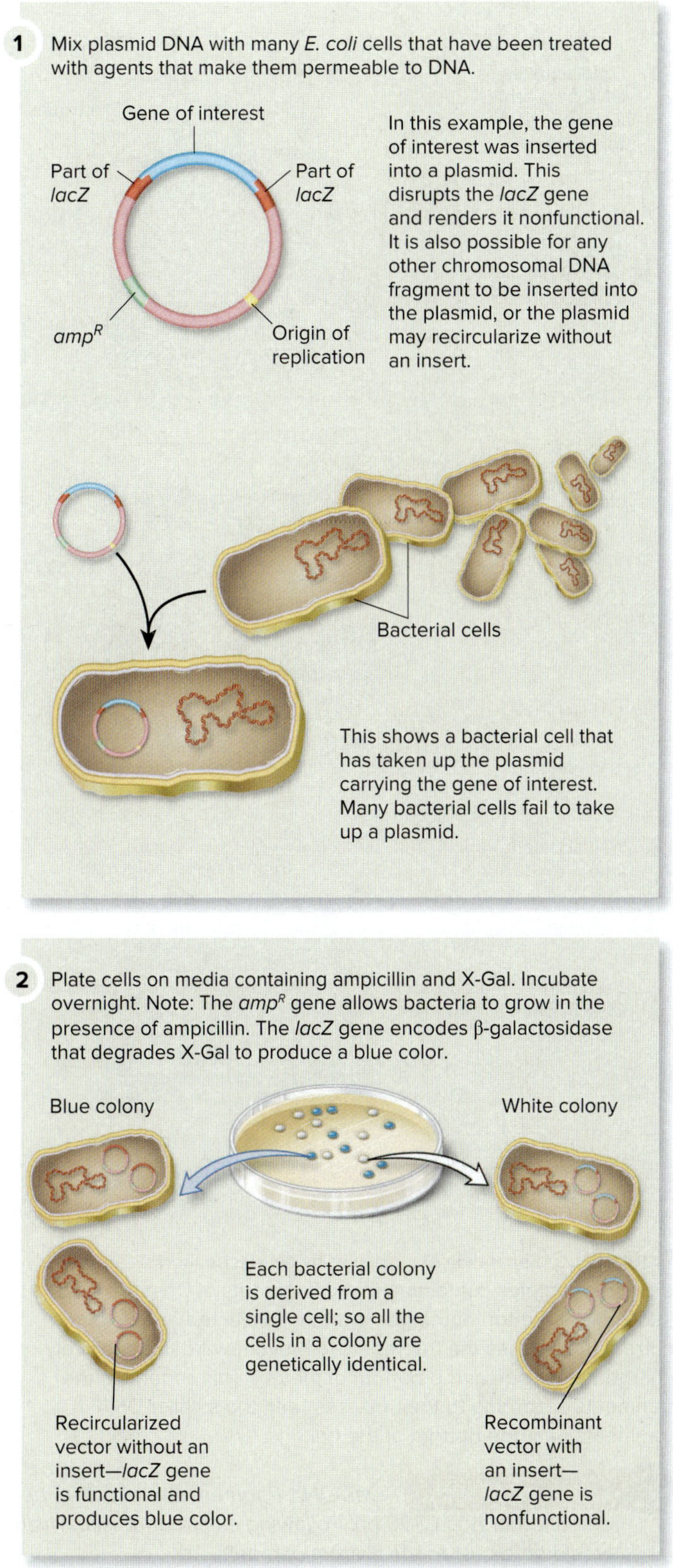

**Figure 18.3 Step 3 of gene cloning: Introduction of a recombinant vector into a host cell.** For cloning to occur, a recombinant vector is introduced into a host cell, which copies the vector and divides to produce many cells. This produces many copies of the gene of interest.

**Concept Check:** *In this cloning experiment, what is the purpose of having the lacZ gene in the vector?*

functional β-galactosidase enzyme, and white colonies if they do not. In this experiment, therefore, bacterial colonies containing recircularized vectors form blue colonies, whereas colonies containing recombinant vectors carrying a segment of chromosomal DNA are white.

After a bacterial cell has taken up a recombinant vector, two subsequent events lead to the production of many copies of that vector. First, when the vector has a highly active origin of replication, the bacterial host cell produces many copies of the recombinant vector per cell. Second, the bacterial cells divide approximately every 20 minutes. Following overnight growth, a population of many millions of bacteria is obtained from a single cell. For example, a bacterial colony may comprise 10 million cells, with each cell containing 50 copies of the recombinant vector. Therefore, this bacterial colony has 500 million copies of the cloned gene!

## Quantitative Analysis

### A DNA LIBRARY IS A COLLECTION OF MANY DIFFERENT DNA FRAGMENTS CLONED INTO VECTORS

In a typical cloning experiment, such as the one described in Figures 18.2 and 18.3, the treatment of chromosomal DNA with restriction enzymes actually yields many different DNA fragments. Therefore, after the DNA fragments are ligated individually to vectors, a researcher has a collection of many recombinant vectors, with each vector containing a particular fragment of chromosomal DNA. A collection of recombinant vectors containing DNA fragments of a given organism is known as a **DNA library** (Figure 18.4). Researchers make DNA libraries using the methods shown in Figures 18.2 and 18.3 and then use those libraries to obtain clones that carry a gene of interest.

Two types of DNA libraries are commonly made. When the inserts are derived from chromosomal DNA, the library is called a **genomic library.** Alternatively, researchers can isolate mRNA and use the enzyme reverse transcriptase, which is described in Chapter 17, to make DNA molecules using mRNA as a starting material. Such DNA is called **complementary DNA,** or **cDNA.** A **cDNA library** is a collection of recombinant vectors that have cDNA inserts. From a research perspective, an important advantage of cDNA is that it lacks introns—intervening sequences that are not translated into proteins. Because introns can be quite large, it is much simpler for researchers to insert cDNAs into vectors rather than work with chromosomal DNA if they want to focus their attention on the coding sequence of a gene. In addition, because bacteria do not splice out introns, cDNAs are an advantage if researchers want to express their gene of interest in bacteria.

**Crunching the Numbers:** The restriction enzyme *Eco*RI recognizes a six-base-pair sequence (see Table 18.1). Assuming that this sequence is found randomly in the human genome, how often (on average) will this restriction enzyme cut the chromosomal DNA? With a human genome size of approximately 3 billion base pairs, how many DNA fragments will be generated by cutting with *Eco*RI?

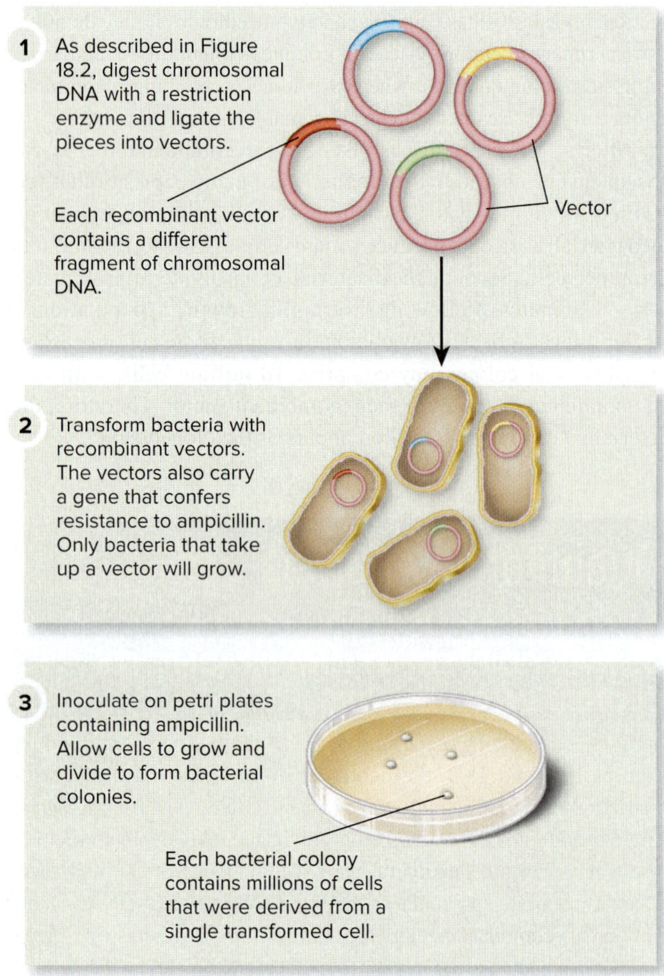

1
As described in Figure 18.2, digest chromosomal DNA with a restriction enzyme and ligate the pieces into vectors.
Each recombinant vector contains a different fragment of chromosomal DNA.
Vector
2
Transform bacteria with recombinant vectors. The vectors also carry a gene that confers resistance to ampicillin. Only bacteria that take up a vector will grow.
3
Inoculate on petri plates containing ampicillin. Allow cells to grow and divide to form bacterial colonies.
Each bacterial colony contains millions of cells that were derived from a single transformed cell.


**Figure 18.4** **A DNA library.** Each colony in a DNA library contains a vector with a different piece of chromosomal DNA.

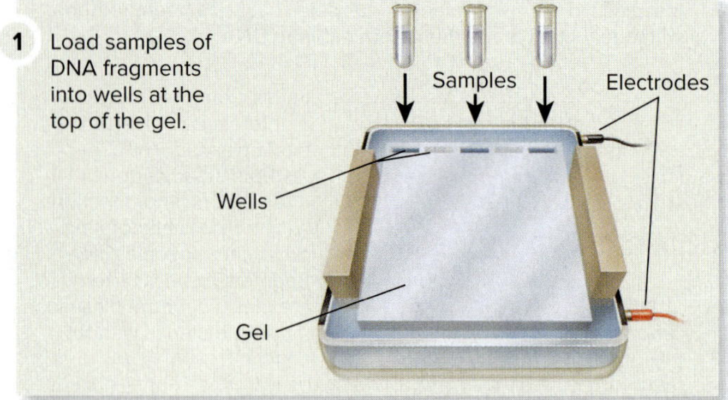


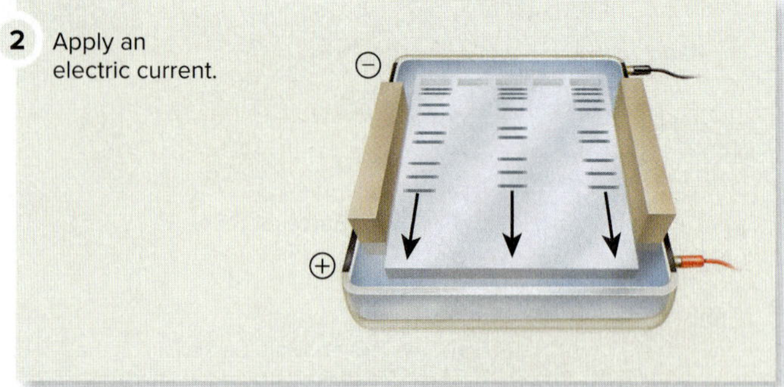


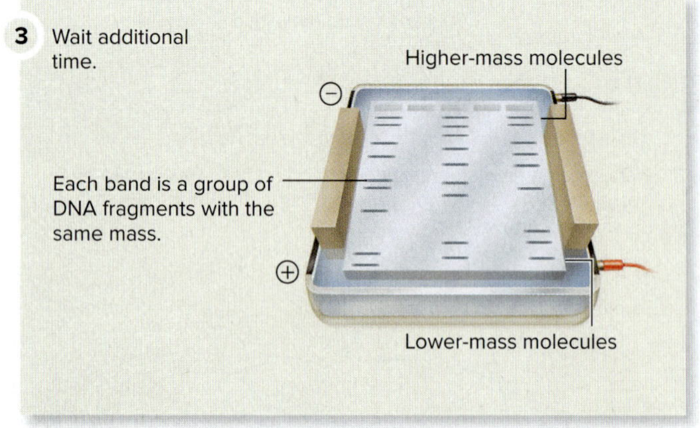


**Figure 18.5** **Separation of molecules by gel electrophoresis.** In this example, samples containing many fragments of DNA are loaded into wells at the top of the gel and then subjected to an electric current that causes the fragments to move toward the positively charged electrode at the bottom of the gel. This separates the fragments according to their masses, with the smaller DNA fragments near the bottom of the gel.

**Concept Check:** *One DNA fragment contains 600 bp and another has 1,300 bp. Following electrophoresis, which would be closer to the bottom of a gel?*

## Gel Electrophoresis Separates Macromolecules, Such as DNA Fragments

**Gel electrophoresis** is a technique for separating macromolecules, such as DNA and proteins, as they migrate through a gel. This method is often used to evaluate the results of a cloning experiment. For example, gel electrophoresis is used to determine the sizes of DNA fragments that have been inserted into recombinant vectors.

Gel electrophoresis can separate molecules based on their charge, size/length, and mass. In the example shown in Figure 18.5, gel electrophoresis is used to separate different fragments of chromosomal DNA based on their masses. The flat slab of semisolid gel has depressions at the top called wells, where samples are added. Electrodes are found at each end of the gel. An electric current is applied to the gel, which causes charged molecules, either proteins or nucleic acids, to migrate from the top of the gel toward the bottom—a process called electrophoresis. DNA is negatively charged and moves toward the positively charged electrode, which is at the bottom in this figure. As the gel runs, the DNA fragments are separated into distinct bands within the gel. Smaller DNA fragments move more quickly through the gel than larger ones in a given amount of time and therefore are located closer to the bottom of the gel than the larger ones. The fragments in each band can then be stained with a dye for identification.

## Polymerase Chain Reaction (PCR) Is Also Used to Make Many Copies of DNA

As we have seen, one method of cloning involves an approach in which the gene of interest is inserted into a vector, introduced into a host cell, and then propagated. Another cloning technique, in which DNA is copied without the aid of vectors and host cells, is a process called **polymerase chain reaction (PCR),** which was developed by American biochemist Kary Mullis in 1985 (**Figure 18.6**). The goal of PCR is to make many copies of DNA in a defined region, perhaps encompassing a gene or part of a gene. Several reagents are required for the synthesis of DNA. First, two different primers are needed that are complementary to sequences at each end of the DNA region to be amplified. These primers are usually about 20 nucleotides long. One primer is called the forward primer, and the other is the reverse primer. PCR also requires all four deoxynucleoside triphosphates (dNTPs) and a heat-stable form of DNA polymerase called *Taq* polymerase. *Taq* polymerase is isolated from the bacterium *Thermus aquaticus*, which lives in hot springs and can tolerate temperatures up to 95°C. A heat-stable form of DNA polymerase is necessary because PCR is conducted at high temperatures that would inactivate DNA polymerase from most other bacteria.

To make copies, the following three steps occur:

1. A sample of chromosomal DNA, called the template DNA, is heated to separate (denature) the DNA into single-stranded molecules.
2. The primers bind to the DNA as the temperature is lowered (see Figure 18.6). The binding of the primers to the specific sites in the template DNA is called primer annealing.
3. After the primers have annealed, the temperature is slightly raised and *Taq* polymerase uses dNTPs to catalyze the synthesis of complementary DNA strands, thereby doubling the amount of DNA in the region that is flanked by the primers. This step is called primer extension because the length of the primers is extended by the synthesis of DNA.

The sequential process of denaturation—primer annealing—primer extension is repeated many times in a row. This method is called a chain reaction because the products of each previous step are used as reactants in subsequent steps. A device that controls

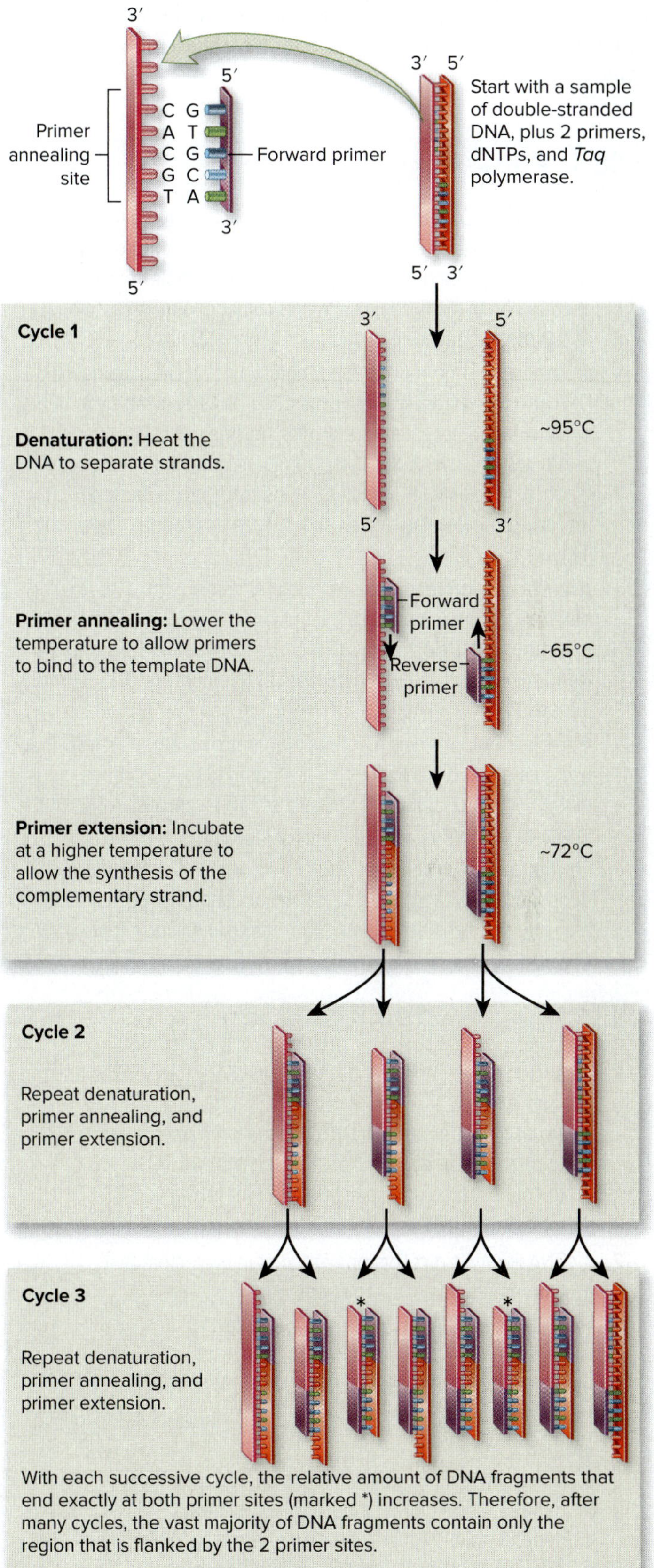

3′
5′
Primer annealing site
C G
A T
C G
G C
T A
Forward primer
3′
5′
3′ 5′
Start with a sample of double-stranded DNA, plus 2 primers, dNTPs, and Taq polymerase.
5′ 3′
Cycle 1
3′
5′
Denaturation: Heat the DNA to separate strands.
~95°C
5′
3′
Primer annealing: Lower the temperature to allow primers to bind to the template DNA.
Forward primer
Reverse primer
~65°C
Primer extension: Incubate at a higher temperature to allow the synthesis of the complementary strand.
~72°C
Cycle 2
Repeat denaturation, primer annealing, and primer extension.
Cycle 3
Repeat denaturation, primer annealing, and primer extension.
*
*
With each successive cycle, the relative amount of DNA fragments that end exactly at both primer sites (marked *) increases. Therefore, after many cycles, the vast majority of DNA fragments contain only the region that is flanked by the 2 primer sites.


**Figure 18.6 Polymerase chain reaction (PCR).** During each PCR cycle, the steps of denaturation, primer annealing, and primer extension take place. The net result of PCR is the synthesis of many copies of DNA in the region that is flanked by the two primers. To conduct this type of PCR experiment, researchers must have prior knowledge about the base sequence of the template DNA so they can make primers with base sequences that are complementary to the ends of the template DNA. Note: The temperatures shown in this figure are approximate and may vary depending on the primer sequence and length.

**Concept Check:** *Why do the PCR primers bind specifically to the primer-annealing sites?*

**BioConnections:** *Refer back to Figure 9.16. Why are primers needed in a PCR experiment?*

the temperature and automates the timing of each step, known as a thermocycler, is used to carry out PCR. The PCR technique can amplify the sample of DNA by a staggering amount. After 30 cycles of denaturation, primer annealing, and primer extension, a DNA sample will have increased by $2^{30}$, approximately a billion-fold, in a few hours!

## 18.1 Reviewing the Concepts

- Recombinant DNA technology is the use of laboratory techniques to bring together DNA fragments from two or more sources. Gene cloning, the process of making multiple copies of a gene, is used to obtain many copies of a particular gene or large amounts of the protein encoded by the gene (Figure 18.1).
- In one method of gene cloning, both a vector and chromosomal DNA are cut with restriction enzymes. The DNA fragments hydrogen-bond to each other at their sticky ends, and the pieces are covalently linked together via DNA ligase, producing recombinant vectors. When a recombinant vector is introduced into a bacterial cell, the cell replicates the vector and divides to produce many cells (Figures 18.2, 18.3).
- A collection of recombinant vectors, each with a particular piece of chromosomal DNA, is introduced into bacterial cells to produce a DNA library. If the DNA molecules are derived from mRNA, this creates a cDNA (complementary DNA) library (Figure 18.4).
- Gel electrophoresis is used to separate macromolecules by using an electric current that causes them to move through a gel matrix. Gel electrophoresis typically separates molecules according to their charge, size, and mass (Figure 18.5).
- Polymerase chain reaction (PCR) is an alternative technique for gene cloning without the use of vectors and host cells. Primers are used that flank the region of DNA to be amplified (Figure 18.6).

## 18.1 Testing Your Knowledge

1. In a cloning experiment using plasmid vectors, a restriction enzyme is used
   - a. to cut the chromosomal DNA into many fragments.
   - b. to open up the vector at a single location.
   - c. to connect a chromosomal DNA fragment to a vector.
   - d. for both a and b.
2. A cDNA library is a collection of recombinant vectors that
   - a. carry fragments of chromosomal DNA.
   - b. carry DNA segments derived from RNA.
   - c. only carry a gene of interest.
   - d. carry DNA segments from many different species.
3. The three steps in a PCR cycle occur in which of the following orders?
   - a. primer annealing—primer extension—denaturation
   - b. denaturation—primer annealing—primer extension
   - c. denaturation—primer extension—primer annealing
   - d. primer extension—primer annealing—denaturation

## 18.2 Genomics: Techniques for Studying and Altering Genomes

### Learning Outcomes

1. Distinguish between genomics and functional genomics.
2. Outline the steps of DNA sequencing using the dideoxy chain-termination method.
3. Explain what a DNA microarray is and how it is used to identify the genes expressed by a sample of cells.
4. Describe how the CRISPR-Cas technology can be used to mutate genes.

As discussed throughout Unit III, the genome is the complete genetic composition of a cell, an organism, or a species. As genetic technology has progressed over the past few decades, researchers have gained an increasing ability to analyze the composition of genomes as a whole unit. The term genomics refers to the use of techniques to study a genome. Segments of chromosomes are cloned and analyzed in progressively smaller pieces, the locations of which are known on the intact chromosomes. This is the mapping phase of genomics. The mapping of a genome ultimately progresses to the determination of the complete DNA sequence, which provides the most detailed description available of an organism's genome at the molecular level. By comparison, **functional genomics** studies the expression of a genome. For example, functional genomics can be used to analyze which genes are turned on or off in normal versus cancer cells. In this section, we will consider a few of the methods that are used in genomics and functional genomics.

### The Dideoxy Chain-Termination Method Is Used to Determine the Base Sequence of DNA

The term **DNA sequencing** refers to a procedure that is aimed at determining the base sequence of DNA. Scientists can learn a great deal of information about the function of a gene if its nucleotide sequence is known. For example, the investigation of genetic sequences has been vital in our understanding of the genetic basis of human diseases.

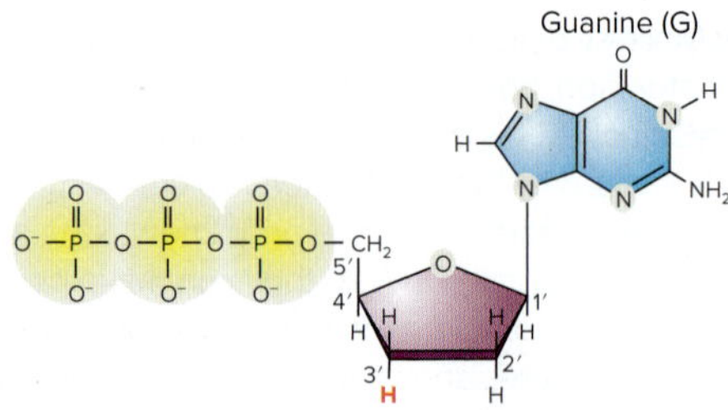


**Figure 18.7 Structure of a dideoxynucleotide triphosphate.** This figure shows the structure of dideoxyguanosine triphosphate (ddGTP). It has a hydrogen, shown in red, instead of a hydroxyl group at the 3′ position. The prefix, dideoxy-, means it has two (di) missing (de) oxygens (oxy) compared with ribose, which has —OH groups at both the 2′ and 3′ positions.

One type of DNA sequencing, developed in 1977 by English biochemist Frederick Sanger and colleagues, is known as the **dideoxy chain-termination method,** or more simply, **dideoxy sequencing.** Dideoxy sequencing is based on our knowledge of DNA replication. As described in Chapter 9, DNA polymerase connects adjacent deoxynucleoside triphosphates (dNTPs) by catalyzing a covalent linkage between the phosphate group at the 5′ position on an incoming nucleotide and the —OH group at the 3′ position on the growing strand. Chemists, however, can synthesize nucleotides, called dideoxynucleoside triphosphates (ddNTPs), that are missing the —OH group at the 3′ position (**Figure 18.7** on previous page). What happens if a ddNTP is incorporated during DNA replication? If a ddNTP is added to a growing DNA strand, the strand can no longer grow because the 3′ —OH group, the site of attachment for the next nucleotide, is missing. This ending of DNA synthesis is called chain termination.

Before describing the steps of this DNA sequencing procedure, let's first consider the DNA segment that is analyzed in a sequencing experiment. The segment of DNA to be sequenced, the target DNA, is obtained in large amounts by using the gene cloning techniques that were described earlier in this chapter. In **Figure 18.8a**, the target DNA was inserted into a vector next to a primer-annealing site, the place where a primer will bind. The target DNA is initially double stranded, but Figure 18.8a shows the DNA after it has been denatured into a single strand by heat treatment.

Let's now examine the steps involved in DNA sequencing.

1. Many copies of single-stranded template DNA are placed into a tube and mixed with primers that bind to the primer-annealing site. DNA polymerase and all four types of regular dNTPs are also added. In addition, a low concentration of the four possible dideoxynucleoside triphosphates—ddGTP, ddATP, ddTTP, and ddCTP—is added. Each type of ddNTP is tagged with a different colored fluorescent molecule; typically, ddA is green, ddT is red, ddG is yellow, and ddC is blue.
2. The tube is then incubated to allow DNA polymerase to synthesize strands that are complementary to the target DNA sequence. However, addition of a ddNTP causes DNA synthesis to terminate early, preventing further elongation of the strand. For example, let's consider ddTTP. Synthesis of new DNA strands stops at the sixth or thirteenth position after the annealing site if a ddTTP, instead of a dTTP, is incorporated into the growing DNA strand (Figure 18.8a). This means the target DNA has a complementary A at the sixth and thirteenth positions. Eventually, a set of fluorescently tagged strands results, with the color of the ddNTP representing the last nucleotide added to the strand.
3. After the samples have been incubated for several minutes, the newly made DNA strands are separated according to their lengths

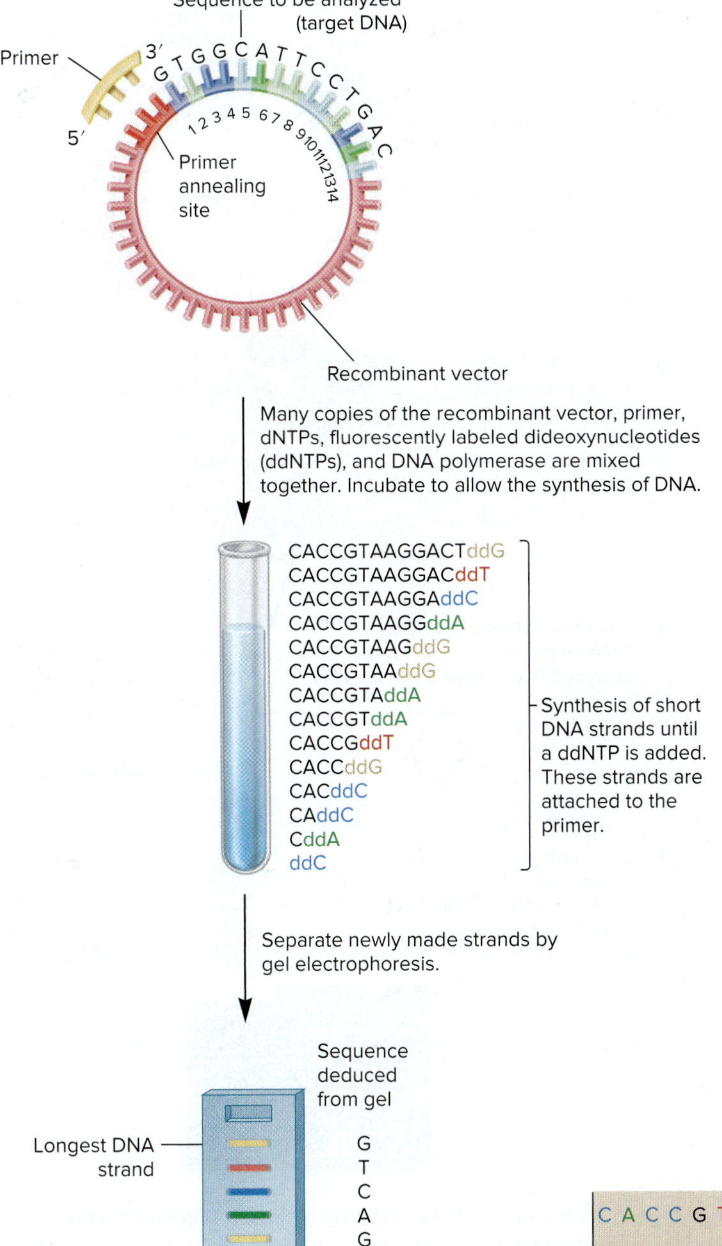


**(a) Dideoxy DNA sequencing method**

C A C C G T A A G G A C T G

**(b) Output from automated sequencing**

**Figure 18.8 Dideoxy sequencing of DNA.** **(a)** The procedure of dideoxy sequencing uses fluorescently labeled ddNTPs. **(b)** This method uses a fluorescence detector that measures the four kinds of ddNTPs as they emerge from the gel.

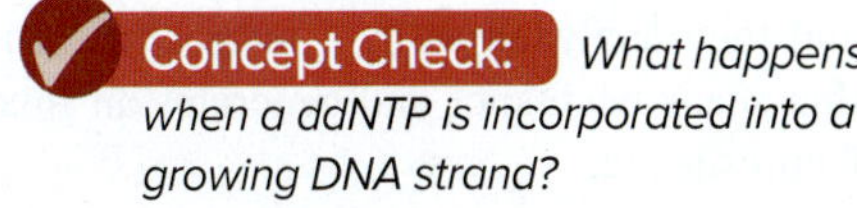


*What happens when a ddNTP is incorporated into a growing DNA strand?*

by subjecting them to gel electrophoresis. This can be done using a slab gel, as shown in Figure 18.8a, or more commonly by running them through a gel-filled capillary tube. The shorter strands move to the bottom of the gel more quickly than the longer ones. Electrophoresis is continued until each band emerges from the bottom of the gel, where a laser excites the fluorescent dye. A fluorescence detector records the amount of fluorescence emission at four wavelengths, corresponding to the four dyes.

An example of a printout from a fluorescence detector is shown in Figure 18.8b. The peaks of fluorescence correspond to the DNA sequence that is complementary to the target DNA. The heights of the fluorescent peaks are not always the same, because ddNTPs get incorporated at some sites more readily than at others. Note that although ddG is usually labeled with a yellow dye, it is converted to black ink on the printout for ease of reading. Though improvements in automated sequencing continue to be made, a typical sequencing run can provide a DNA sequence that is approximately 700 to 900 bases long, and perhaps even longer.

Researchers are also developing alternative methods to the dideoxy chain-termination method in order to sequence DNA. For example, pyrosequencing is a newer method of DNA sequencing that is based on the detection of released pyrophosphate ($PP_i$) during DNA synthesis.

## A Microarray Can Identify Which Genes Are Transcribed by a Cell

Let's now turn our attention to functional genomics. Researchers have developed an exciting new technology, called a **DNA microarray,** that is used to monitor the expression of thousands of genes simultaneously. A DNA microarray is a small silica, glass, or plastic slide that is dotted with many different sequences of single-stranded DNA, each corresponding to a short sequence within a known gene. Each spot contains multiple copies of a known DNA sequence. For example, one spot in a microarray may correspond to a sequence within the β-globin gene; another might correspond to a different gene, such as a gene that encodes a glucose transporter. A single slide contains tens of thousands of different spots in an area the size of a postage stamp. These microarrays are typically produced using a technology that "prints" spots of DNA sequences onto a slide, similar to the way that an inkjet printer deposits ink on paper.

What is the purpose of using a DNA microarray? In the experiment shown in Figure 18.9, the goal is to determine which genes are transcribed into mRNA from a particular sample of cells. In other words, which genes in the genome are expressed? To conduct this experiment, the mRNA was isolated from the cells and then used to make fluorescently labeled cDNAs. The labeled cDNAs were then incubated with a DNA microarray. The single-stranded DNA in the microarray corresponds to the coding strand—the strand that has a sequence that is similar to mRNA. Those cDNAs that are complementary to the DNAs in the microarray hybridize, thereby remaining bound to the microarray. The array is washed and then analyzed using a microscope equipped with a computer that scans each spot and generates an image of the spots' relative fluorescence.

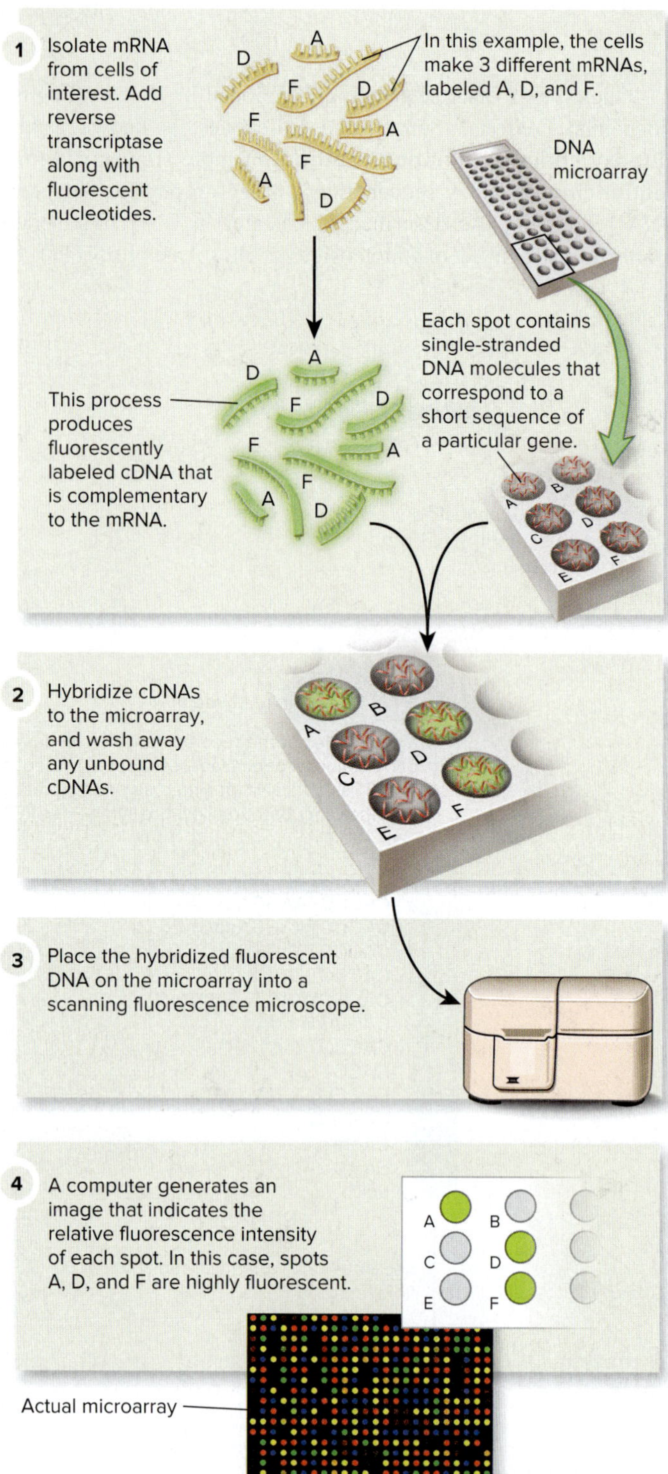

1 Isolate mRNA from cells of interest. Add reverse transcriptase along with fluorescent nucleotides.
In this example, the cells make 3 different mRNAs, labeled A, D, and F.
DNA microarray
Each spot contains single-stranded DNA molecules that correspond to a short sequence of a particular gene.
This process produces fluorescently labeled cDNA that is complementary to the mRNA.
2 Hybridize cDNAs to the microarray, and wash away any unbound cDNAs.
3 Place the hybridized fluorescent DNA on the microarray into a scanning fluorescence microscope.
4 A computer generates an image that indicates the relative fluorescence intensity of each spot. In this case, spots A, D, and F are highly fluorescent.
Actual microarray


**Figure 18.9** **Identifying transcribed genes within a DNA microarray.** In this simplified example, only three cDNAs specifically hybridize to spots on the microarray. Those genes were expressed in the cells from which the mRNA was isolated. In an actual experiment, the array typically contains hundreds or thousands of different cDNAs and tens of thousands of different spots.



**Concept Check:** *If a fluorescent spot appears on a microarray, what information does this provide regarding gene expression?*

If the fluorescence intensity in a spot is high, a large amount of cDNA was in the sample that hybridized to the DNA at this location. For example, if the β-globin gene was expressed in the cells being tested, a large amount of cDNA for this gene would be made, and the fluorescence intensity for that spot would be high. Because the DNA sequence of each spot is already known, a fluorescent spot identifies cDNAs that are complementary to those DNA sequences. Furthermore, because the cDNA was generated from mRNA, this technique identifies genes that have been transcribed in a particular cell type under a given set of conditions. However, the amount of protein encoded by an mRNA may not always correlate with the amount of mRNA, due to variation in the rates of mRNA translation and protein degradation.

Thus far, the most common use of DNA microarrays is to study gene expression patterns. In addition, the technology of DNA microarrays has found several other important uses, as described in Table 18.2.

## Genes in Living Cells Can Be Altered Using CRISPR-Cas Technology

In Chapter 11, we considered how the CRISPR-Cas system provides bacteria with a defense against bacteriophages (refer back to Figure 11.9). Researchers realized that the components of this system can be used to mutate genes in living cells. This approach is called **CRISPR-Cas technology.** Researchers have made a modification to the natural system to make it efficient for gene mutagenesis. They create a single RNA in which the tracrRNA and crRNA are linked to each other (Figure 18.10a). This is called the single guide RNA (sgRNA). The spacer region of the sgRNA is designed to be complementary to one of the DNA strands of a target gene that a researcher wants to mutate. The sgRNA binds to Cas9 and guides it to the target gene. Cas9 then makes a double-strand break in this gene.

Following this break, two different DNA repair events are possible. If the break is repaired by a process called end joining, the gene may incur a small deletion (see left side of Figure 18.10b). This may inactivate the gene, particularly if it causes a frameshift mutation in the coding sequence. Alternatively, a researcher can also add a double-stranded segment of DNA, called the donor DNA, that is homologous to the region where the break occurs (see right side of Figure 18.10b). This homologous DNA is made synthetically and is designed to carry a particular mutation, such as a point mutation, that the researcher wants to make. A different DNA repair system swaps in the donor DNA by a double crossover. In this way, researchers can introduce a specific mutation into a gene.

The experiment described in Figure 18.10b can be performed on different cell types and even on whole organisms. By studying how a gene mutation affects cell structure and function or by examining the phenotypic effects of a mutation in a whole organism (see chapter-opening photo), researchers often gain a better understanding of how genes function.

**Table 18.2 Applications of DNA Microarrays**

| Application | Description |
|---|---|
| Cell-specific gene expression | A comparison of microarray data using cDNAs derived from mRNA of different cell types can identify genes that are expressed in a cell-specific manner. |
| Gene regulation | Because environmental conditions play an important role in gene regulation, a comparison of microarray data using cDNA derived from mRNA from cells exposed to two different environmental conditions may reveal genes that are induced under one set of conditions and repressed under another set. |
| Elucidation of metabolic pathways | Genes that encode proteins that participate in a common metabolic pathway are often expressed together and can be revealed from a microarray analysis. |
| Tumor profiling | Different types of cancer cells exhibit striking differences in their gene expression profiles, which can be revealed by a DNA microarray analysis. This approach is gaining use as a tool to classify tumors that are sometimes morphologically indistinguishable. |
| Genetic variation | A mutant allele may not hybridize to a spot on a microarray as well as a wild-type allele. Therefore, microarrays are gaining use as a tool for detecting genetic variation. This application has been used to identify disease-causing alleles in humans and to identify mutations that contribute to quantitative traits in plants and other species. |
| Microbial strain identification | Microarrays can distinguish between closely related bacterial species and subspecies. |

## 18.2 Reviewing the Concepts

- Genomics is the study of genomes as whole units; functional genomics studies the expression of a genome.
- Dideoxy sequencing uses ddNTPs to determine the base sequence of a segment of DNA (Figures 18.7, 18.8).
- A DNA microarray is a small silica, glass, or plastic slide dotted with different sequences of single-stranded DNA, each corresponding to a short sequence within a known gene. It is used to study gene expression patterns (Figure 18.9; Table 18.2).
- CRISPR-Cas technology can be used to mutate genes (Figure 18.10).

## 18.2 Testing Your Knowledge

1. In dideoxy sequencing, what is a key property of a dideoxy nucleotide?
   a. It stops the further growth of a DNA strand.
   b. It is fluorescently labeled.
   c. It cannot be incorporated into a growing DNA strand.
   d. All of the above are correct.
   e. Both a and b are correct.
2. A microarray allows researchers to
   a. clone genes.
   b. make recombinant DNA molecules.
   c. study the expression of many genes simultaneously.
   d. make mutations within genes.
   e. do all of the above.

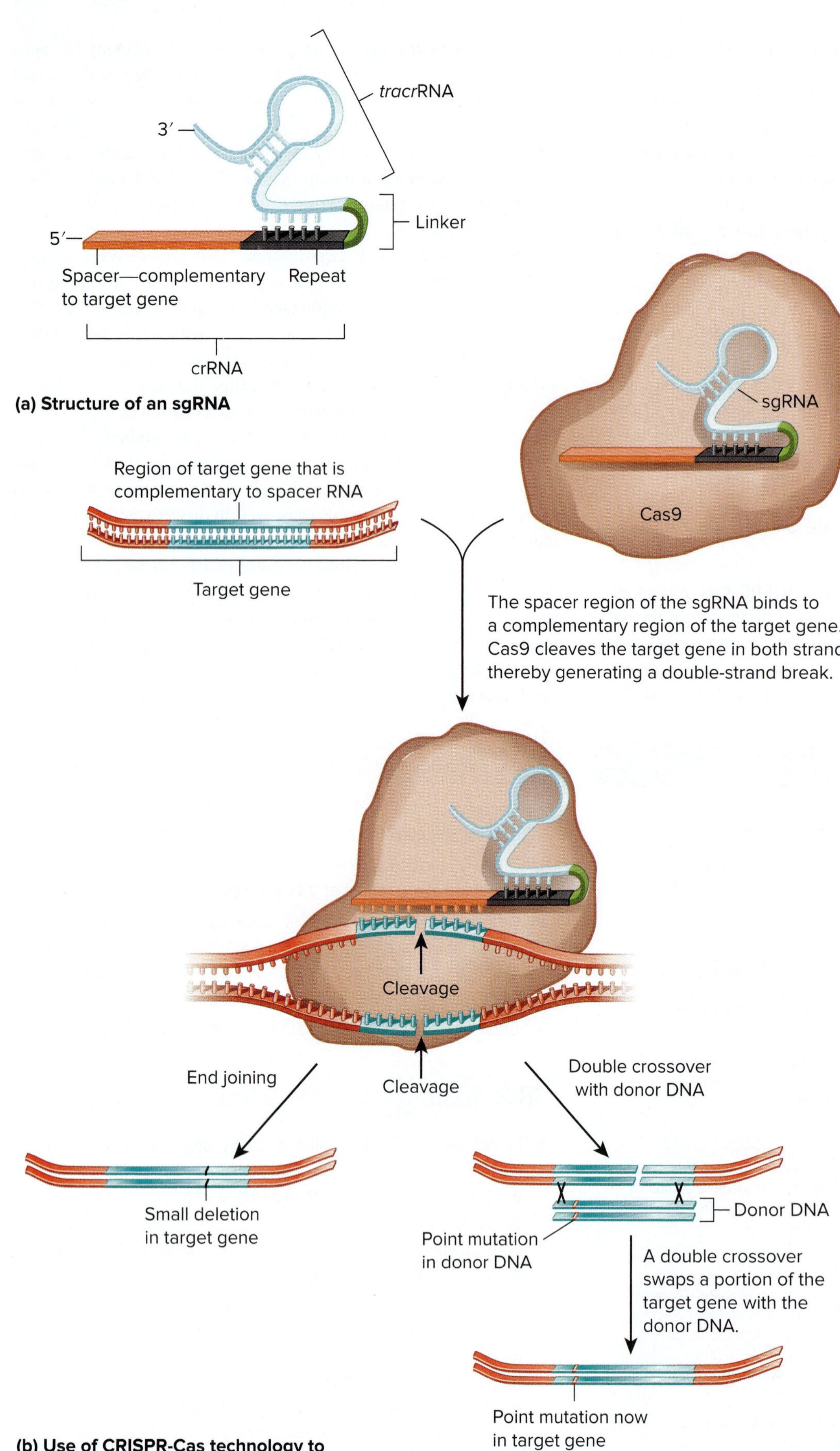


**Figure 18.10 The use of CRISPR-Cas technology to cause a gene mutation.** **(a)** Structure of an sgRNA, which is composed of a crRNA that is connected to a tracrRNA via a linker. **(b)** Use of CRISPR-Cas technology. On the left side, the gene has been repaired by end joining and suffers a small deletion. This may result in gene inactivation. On the right side, a double crossover has occurred that swaps in the donor DNA. The target gene now carries a point mutation.

**Concept Check:** *How is the sgRNA different from certain components of the bacterial defense system described in Chapter 11 (refer back to Figure 11.11)?*

## 18.3 Bacterial and Archaeal Genomes

**Learning Outcome**

1. List the key characteristics of bacterial and archaeal genomes.

The past decade has seen remarkable advances in our overall understanding of the entire genome of many species. As genetic technology has progressed, researchers have gained an increasing ability to analyze the composition of genomes as a whole unit. The complete DNA sequence is now known for many species, which provides the most detailed description available of an organism's genome at the molecular level. In this section, we will survey the sizes and composition of genomes in selected species of bacteria and archaea.

### Studying the Genomes of Bacteria and Archaea Has Important Applications

Why are researchers interested in the genomes of bacteria and archaea?

1. Bacteria cause many different diseases that affect humans as well as other animals and plants. Studying the genomes of bacteria reveals important clues about the process of infection, which may also help us find ways to combat those infections.
2. The knowledge that is obtained by studying bacterial and archaeal genomes often applies to larger and more complex organisms.
3. A third reason is evolution. The origin of the first eukaryotic cell probably involved a union between an archaeal cell and a bacterial cell, as we will explore in Chapter 22. The study of bacterial and archaeal genomes helps us understand how all living species evolved.
4. Bacteria are often used as tools in research, which was discussed in Section 18.1. A better understanding of their genomes can make them more effective tools.

### The Genomes of Bacteria and Archaea Typically Consist of a Circular Chromosome with a Few Thousand Genes

Geneticists have made great progress in the study of bacterial and archaeal genomes. The genomes of thousands of bacterial and archaeal species have been sequenced and analyzed. The chromosomes of bacteria and archaea are usually a few million base pairs in length. Genomic researchers refer to 1 million base pairs as 1 megabase pair, abbreviated 1 Mb. Most bacteria and archaea contain a single type of chromosome, though multiple copies may be present in a single cell. However, some bacteria are known to have more than one type of chromosome. For example, *Vibrio cholerae*, the bacterium that causes the diarrheal disease cholera, has two different chromosomes in each cell, one 2.9 Mb and the other 1.1 Mb.

Bacterial and archaeal chromosomes are usually circular. For example, the two chromosomes in *V. cholerae* are circular, as is the single type of chromosome found in *E. coli*. However, linear chromosomes are found in some species, such as *Borrelia burgdorferi*, the bacterium that causes Lyme disease, the most common tick-borne disease in the United States. Certain bacterial species may even contain both linear and circular chromosomes. *Agrobacterium tumefaciens*, which infects plants and causes crown gall tumors, has one linear chromosome (2.1 Mb) and one circular chromosome (3.0 Mb).

Table 18.3 compares the sequenced genomes from several bacterial and archaeal species. They range in size from 1.7 to 5.2 Mb. The total number of genes is correlated with the total genome size. Roughly 1,000 genes are found for every megabase pair of DNA. Compared with eukaryotic genomes, bacterial and archaeal genomes are less complex. Their chromosomes lack centromeres and telomeres and have a single origin of replication. Also, chromosomes of bacteria and archaea have relatively little repetitive DNA, whereas repetitive sequences, which are discussed later in this chapter, are often abundant in eukaryotic genomes.

In addition to one or more chromosomes, bacteria may have plasmids, circular pieces of DNA that exist independently of the bacterial chromosome (refer back to Figure 17.6). Plasmids are typically small, in the range of a few thousand to tens of thousands of base pairs in length, though some can be quite large, even hundreds of thousands of base pairs. The various functions of plasmids were described in Chapter 17, and their use as vectors in gene cloning was discussed in Section 18.1.

**Table 18.3 Examples of Bacterial and Archaeal Genomes That Have Been Sequenced***

| Species | Genome size (Mb)[†] | Number of genes[‡] | Description |
|---|---|---|---|
| *Methanobacterium thermoautotrophicum* | 1.7 | 1,921 | An archaeal species that produces methane |
| *Haemophilus influenzae* | 1.8 | 1,753 | One of several different bacterial species that cause respiratory illness and meningitis |
| *Sulfolobus solfataricus* | 3.0 | 3,032 | An archaeal species that metabolizes sulfur-containing compounds |
| *Lactobacillus plantarum* | 3.3 | 3,052 | A type of lactic acid–producing bacterium used in the production of cheese and yogurt |
| *Mycobacterium tuberculosis* | 4.4 | 4,294 | The bacterium that causes the respiratory disease tuberculosis |
| *Escherichia coli* | 4.6 | 4,377 | A naturally occurring intestinal bacterium; certain strains cause human illness |
| *Bacillus anthracis* | 5.2 | 5,439 | The bacterium that causes the disease anthrax |

*Bacterial and archaeal species often exist in different strains that may differ slightly in their genome size and number of genes. The data are from common strains of the indicated species. The species shown in this table have only one type of chromosome.

†Mb equals 1 million base pairs, or a megabase pair.

‡The number of genes is an estimate based on the analysis of genome sequences.

## 18.3 Reviewing the Concepts

- The genome is the complete genetic makeup of a cell, an organism, or a species.
- Bacterial and archaeal genomes are typically a single circular chromosome with a few million base pairs of DNA. Such genomes usually have a few thousand different genes. Bacteria often have additional plasmids (Table 18.3).

## 18.3 Testing Your Knowledge

1. Which of the following is *not* a reason to study bacterial and archaeal genomes?
   a. to better understand how bacteria and archaea cause disease
   b. to better understand the biology of more complex organisms
   c. to better understand evolution
   d. to make them more effective tools in research and biotechnology
   e. All of the above are good reasons.

## 18.4 Eukaryotic Genomes

### Learning Outcomes

1. Describe the key features of eukaryotic genomes.
2. Explain how gene duplications occur and lead to the formation of gene families.
3. List the goals and results of the Human Genome Project.

In the previous section, we examined bacterial and archaeal genomes. We now turn to eukaryotes, which include protists, fungi, animals, and plants. As you will learn, their genomes are larger and more complex than those of their bacterial and archaeal counterparts. We will also examine how the duplication of genes can lead to families of related genes and will survey the goals of the Human Genome Project, which was aimed at mapping and sequencing the human genome.

### Studying the Genomes of Eukaryotes Has Important Applications

Why are researchers interested in the genomes of eukaryotes? Motivation to sequence eukaryotic genomes comes from four main sources.

1. The availability of genome sequences makes it easier for researchers to identify and characterize the genes of model organisms. This was the impetus for genome projects involving baker's yeast (*Saccharomyces cerevisiae*), the fruit fly (*Drosophila melanogaster*), a nematode worm (*Caenorhabditis elegans*), the flowering plant called thale cress (*Arabidopsis thaliana*), and the mouse (*Mus musculus*).
2. Studying eukaryotic genomes enables researchers to gather more information to identify and treat human diseases. Researchers hope that knowing the DNA sequence of the human genome will help to identify genes in which mutation plays a role in disease.
3. By sequencing the genomes of agriculturally important species, new strains of livestock and plant species with improved traits can be developed.
4. Biologists are increasingly relying on genome sequences as a way to establish evolutionary relationships.

### The Nuclear Genomes of Eukaryotes Are Sets of Linear Chromosomes That Vary Greatly in Size and Composition Among Different Species

As discussed in Chapter 14 (refer back to Figure 14.1), the genome located in the nucleus of eukaryotic species is usually found in sets of linear chromosomes. In humans, for example, one set contains 23 linear chromosomes—22 autosomes and 1 sex chromosome, X or Y. In addition, certain organelles in eukaryotic cells contain a small amount of DNA, including the mitochondrion, which plays a role in ATP synthesis, and the chloroplast (found in plants and algae), which carries out photosynthesis. The genetic material in these organelles is referred to as the mitochondrial or the chloroplast genome to distinguish it from the nuclear genome, which is located in the cell nucleus. In this chapter, we will focus on the nuclear genome of eukaryotes.

**Sizes of Nuclear Genomes** In the past decade or so, the DNA sequence of entire nuclear genomes has been determined for hundreds of eukaryotic species, including several dozen mammalian genomes. Examples are shown in **Table 18.4**.

**Relationship Between Genome Sizes and Repetitive Sequences** Eukaryotic genomes are generally larger than bacterial and archaeal genomes, in terms of both the number of genes and genome size. The genomes of simpler eukaryotes, such as yeast, carry several thousand different genes, whereas the genomes of more complex eukaryotes contain tens of thousands of genes (see Table 18.4). Note that the number of genes is not the same as genome size. When we speak of genome size, we mean the total amount of DNA, often measured in megabase pairs. The relative sizes of nuclear genomes vary dramatically among different eukaryotic species (**Figure 18.11a**). In general, increases in the amount of DNA are correlated with increases in cell complexity and body complexity. For example, yeast have smaller genomes than animals.

However, major variations in genome sizes are often observed among species that are similar in form and function. For example, the total amount of DNA found within different species of amphibians varies over 100-fold. As another example, let's consider two closely related species of the plant called the globe thistle, *Echinops bannaticus* and *Echinops nanus* (**Figure 18.11b,c**). These species have similar numbers of chromosomes, but *E. bannaticus* has nearly double the amount of DNA in *E. nanus*. What is the explanation for the larger genome of *E. bannaticus*? The genome of *E. bannaticus* is not likely to contain twice as many genes. Rather, its genome composition includes many repetitive sequences, which are short DNA sequences that are present in many copies throughout the genome. Repetitive sequences are often abundant in eukaryotic species. We will examine the characteristics of such repetitive sequences in Section 18.5.

**Table 18.4 Examples of Eukaryotic Nuclear Genomes That Have Been Sequenced**

| Species | Nuclear genome size (Mb) | Number of protein-encoding genes | Description |
|---|---|---|---|
| *Saccharomyces cerevisiae* (baker's yeast) | 12.1 | ~6,600 | One of the simplest eukaryotic species; it has been extensively studied by researchers to understand eukaryotic molecular biology |
| *Caenorhabditis elegans* (a nematode worm) | 100 | ~20,000 | A model organism used to study animal development |
| *Drosophila melanogaster* (fruit fly) | 180 | ~15,000 | A model organism used to study many genetic phenomena, including development |
| *Arabidopsis thaliana* (thale cress) | 120 | ~26,000 | A model organism studied by plant biologists |
| *Oryza sativa* (rice) | 440 | ~40,000 | A cereal grain with a relatively small genome; it is very important worldwide as a food crop |
| *Mus musculus* (mouse) | 2,500 | ~21,000 | A model mammalian organism used to study genetics, cell biology, and development |
| *Homo sapiens* (humans) | 3,200 | ~22,000 | Our own genome, the sequencing of which will help to elucidate our understanding of inherited traits and aid in the identification and treatment of diseases |

Note: The genome size refers to the number of megabase (Mb) pairs in one set of chromosomes. For species with sex chromosomes, it would include both sex chromosomes.

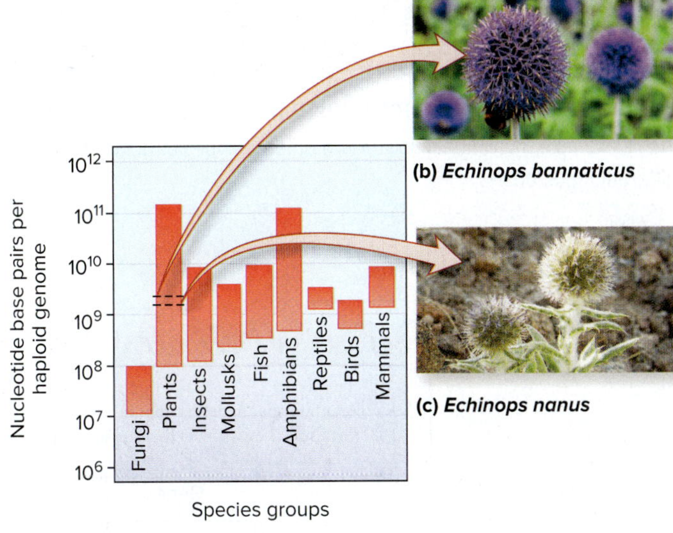


**Figure 18.11 Genome sizes among selected groups of eukaryotes.** **(a)** Genome sizes among various groups of eukaryotes are shown on a log scale. As an example for comparison, two closely related species of globe thistle are pictured. These species have similar characteristics, but *Echinops bannaticus* **(b)** has nearly double the amount of DNA as does *E. nanus* **(c)** due to the accumulation of repetitive DNA sequences.

*(b)* © The Picture Store/SPL/Science Source; *(c)* © Photo by Michael Beckmann, Institute of Geobotany and Botanical Garden, Halle, Germany

**Concept Check:** *What are two reasons the groups of species shown in part (a) vary in their total amount of DNA?*

## EVOLUTIONARY CONNECTIONS

### Gene Duplications Provide Additional Material for Genome Evolution, Sometimes Leading to the Formation of Gene Families

Let's now turn our attention to gene duplications, a way that the number of genes in a genome can increase. Gene duplications are important because they provide raw material for the addition of more genes into a species' genome. Such duplications produce **homologous genes,** two or more genes that are derived from the same ancestral gene (**Figure 18.12a**). Over the course of many generations, each version of the gene accumulates different mutations, resulting in genes with similar but not identical DNA sequences.

How do gene duplications occur? One mechanism that produces gene duplications is a misaligned crossover (**Figure 18.12b**). In this example, two homologous chromosomes have paired with each other during meiosis, but the homologs are misaligned. A crossover produces one chromosome with a gene duplication, one with a gene deletion, and two normal chromosomes. Each of these chromosomes is segregated into different haploid cells. If a haploid cell carrying the chromosome with the gene duplication participates in fertilization with another gamete, an offspring with a gene duplication is produced. In this way, gene duplications can form and be transmitted to future generations.

During evolution, gene duplications can occur several times. Two or more homologous genes within a single species are also called paralogous genes, or **paralogs.** Multiple gene duplications followed by the accumulation of mutations in each paralog result in a **gene family**—a group of paralogs that carry out related functions. A well-studied example is the globin gene family found in animals. The globin genes encode polypeptides that are subunits of proteins that function in oxygen binding. Hemoglobin, which is made in red blood cells, carries oxygen throughout the body. In humans, the globin gene family is composed of 14 paralogs that were originally derived from a single ancestral globin gene (**Figure 18.13**). According to an evolutionary analysis, the ancestral globin gene duplicated between 500 and 600 mya. Since that time, additional duplication events and

## Biology Principle

### Populations of Organisms Evolve from One Generation to the Next

In this example, evolution involves two different types of genetic changes. First, a gene duplication occurs; second, each copy of the gene accumulates different mutations.

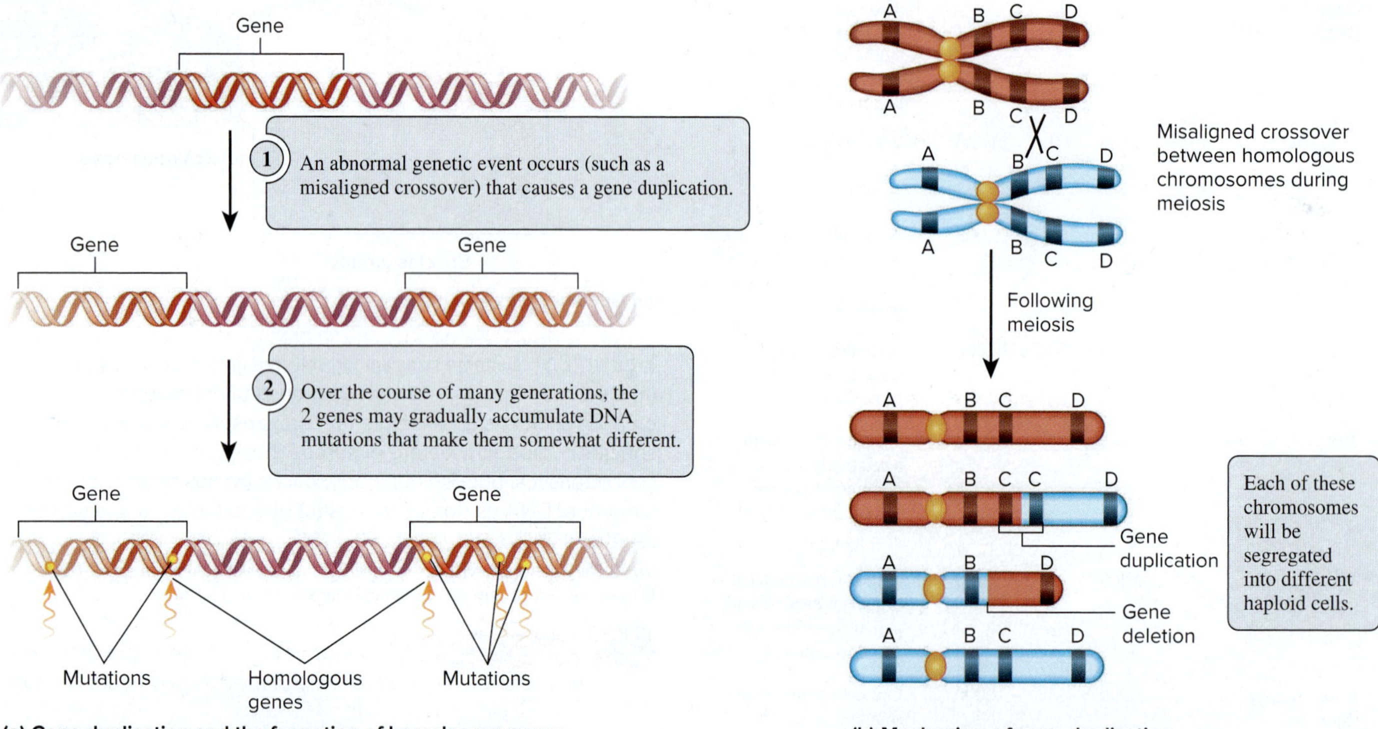


**Figure 18.12 Gene duplication and the evolution of homologous genes. (a)** A gene duplication produces two copies of the same gene. Over time, these copies accumulate different random mutations, which results in homologous genes with similar but not identical DNA sequences. **(b)** A mechanism of gene duplication. If two homologous chromosomes misalign during meiosis, a crossover may produce a chromosome with a gene duplication.

chromosomal rearrangements have occurred to produce the current number of 14 genes on three different human chromosomes. Four of these are pseudogenes—genes that have been produced by gene duplication but have accumulated mutations that make them nonfunctional, so they are not transcribed into RNA.

The accumulation of different mutations in the various family members has produced globins that are specialized in their function. For example, myoglobin binds and stores oxygen in muscle cells, whereas the hemoglobins bind and transport oxygen via red blood cells. Also, different globin genes are expressed during different stages of development. The zeta ($\zeta$)-globin and epsilon ($\varepsilon$)-globin genes are expressed very early in embryonic life. During the second trimester of gestation, the alpha ($\alpha$)-globin and gamma ($\gamma$)-globin genes are turned on. Following birth, the $\gamma$-globin genes are turned off, and the $\beta$-globin gene is turned on. These differences in the expression of the globin genes reflect the differences in the oxygen transport needs of humans during the embryonic, fetal, and postpartum stages of life (refer back to Figure 12.3).

## The Human Genome Project Has Stimulated Genomic Research

Before ending our discussion of genomes, let's consider the **Human Genome Project,** a research effort to identify and map all human genes. Scientists had been discussing how to undertake this project since the mid-1980s. In 1988, the National Institutes of Health (NIH) in Bethesda, Maryland, established an Office of Human Genome Research, with James Watson as its first director. The Human Genome Project officially began on October 1, 1990, and was largely finished by the end of 2003. It was an international consortium that included research institutions in the U.S., U.K., France, Germany, Japan, and China. From its outset, the Human Genome Project had the following goals:

- *To obtain the DNA sequence of the entire human genome.* The first draft of a nearly completed DNA sequence was published in February 2001, and a second draft was published in 2003. The entire genome is approximately 3.2 billion base pairs in length.
- *To identify all human genes.* This involved mapping the locations of genes throughout the entire genome.

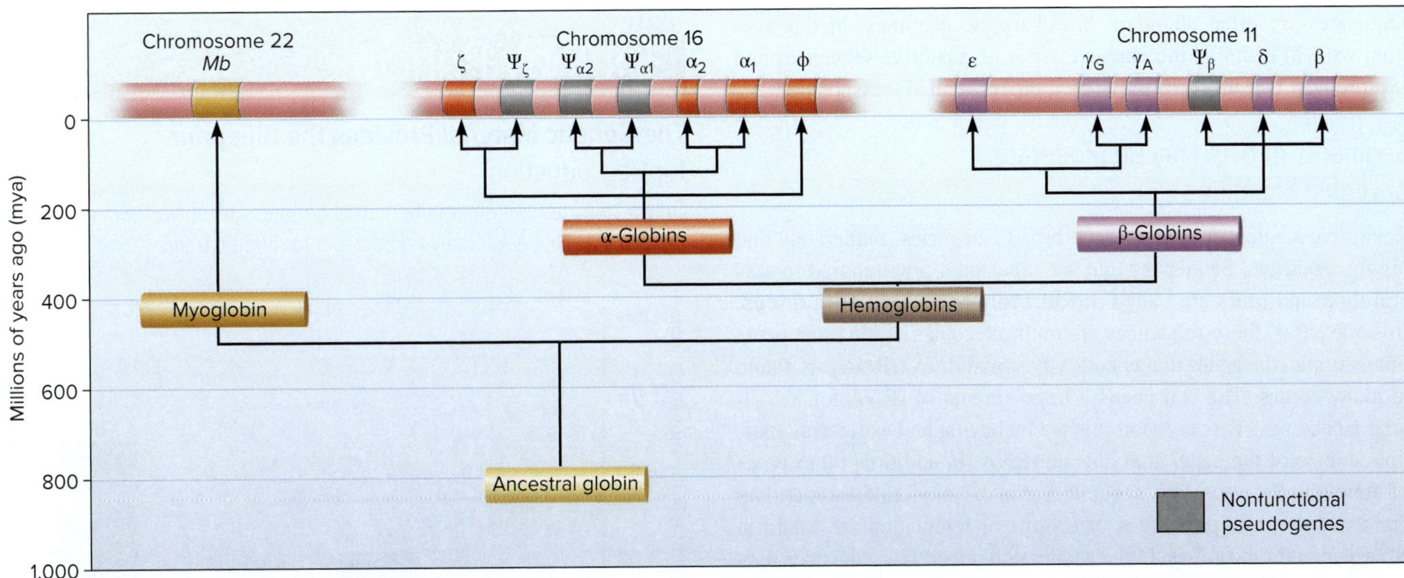


**Figure 18.13** **The evolution of the globin gene family in humans.** The globin gene family evolved from a single ancestral globin gene.

**BioConnections:** *Refer back to Figure 12.3. How do different gene family members vary in their affinity for oxygen?*

- *To develop technology for the generation and management of human genome information.* Some of the efforts of the Human Genome Project have involved improvements in molecular genetic technology, such as gene cloning, DNA sequencing, and so forth. The Human Genome Project has also developed computer tools that allow scientists to easily access up-to-date information from the project and analytical tools to interpret genomic information.
- *To analyze the genomes of model organisms.* These include *E. coli*, *S. cerevisiae*, *D. melanogaster*, *C. elegans*, *A. thaliana*, and *M. musculus*.
- *To develop programs focused on understanding and addressing the ethical, legal, and social implications of the results obtained from the Human Genome Project.* The Human Genome Project raised many ethical issues regarding genetic information and genetic engineering. Who should have access to genetic information? Should employers, insurance companies, law enforcement agencies, and schools have access to our genetic makeup?

Some current and potential applications of the Human Genome Project include the improved diagnosis and treatment of genetic diseases such as cystic fibrosis, Huntington disease, and Duchenne muscular dystrophy. The project may also enable researchers to identify the genetic basis of common disorders such as cancer, diabetes, and heart disease, which involve alterations in several genes.

## 18.4 Reviewing the Concepts

- The nuclear genomes of eukaryotic species are composed of sets of linear chromosomes with a total length of several million to billions of base pairs. They typically contain several thousand to tens of thousands of genes. Genome sizes vary greatly among eukaryotic species (Figure 18.11; Table 18.4).
- One way that the number of genes in a genome can increase is via gene duplication. Gene duplication may occur by a misaligned crossover during meiosis. This is one mechanism that can produce a gene family, two or more homologous genes in a species that have related functions (Figures 18.12, 18.13).
- The Human Genome Project, an international effort to map and sequence the entire human genome, was completed by an international consortium in 2003.

## 18.4 Testing Your Knowledge

1. The sizes of eukaryotic genomes vary because
   a. more complex eukaryotes tend to have more genes.
   b. the amount of repetitive sequences can vary greatly among different species.
   c. the diameter of the cell nucleus may be different.
   d. both a and b.
2. The members of a gene family
   a. are called paralogs.
   b. have similar but not identical DNA sequences.
   c. usually carry out similar functions.
   d. all of the above.

## 18.5 Repetitive Sequences and Transposable Elements

### Learning Outcomes

1. Describe the characteristics of repetitive sequences.
2. Name the two major types of transposable elements and explain how they move about the genome.

**Repetitive sequences** are segments of DNA that are repeated multiple times within a genome. They are found in bacterial, archaeal, and eukaryotic genomes. As mentioned in Section 18.4, repetitive sequences are often abundant in eukaryotic genomes. In this section, we will examine the characteristics of repetitive sequences and explore how certain types move by a process called transposition.

## Segments of DNA May Be Moderately or Highly Repetitive

Repetitive sequences fall into two broad categories, moderately and highly repetitive. Sequences that are repeated a few hundred to several thousand times are called **moderately repetitive sequences.** In some cases, these sequences are multiple copies of the same gene. For example, the genes that encode ribosomal RNA (rRNA) are found in many copies. The cell needs a large amount of rRNA for its cellular ribosomes. This is accomplished by having and expressing multiple copies of the genes that encode rRNA. In addition, other types of functionally important sequences can be moderately repetitive. For example, multiple copies of origins of replication are found in eukaryotic chromosomes. Other moderately repetitive sequences may play a role in the regulation of gene transcription and translation.

**Highly repetitive sequences** are those that are repeated tens of thousands to millions of times throughout the genome. Each copy of a highly repetitive sequence is relatively short, ranging from a few to several hundred nucleotides in length.

Some highly repetitive sequences are clustered together in a tandem array, in which a very short nucleotide sequence is repeated many times in a row. In *Drosophila*, for example, 19% of the chromosomal DNA is highly repetitive DNA found in tandem arrays. An example is shown here:

AATATAATATATAATATAATATAATATAT
TTATATTATATATTATATTATATTATATA

In this particular tandem array, two related sequences, AATAT and AATATAT (in the top strand), are repeated multiple times. Highly repetitive sequences, which contain tandem arrays of short sequences, can be quite long, sometimes more than 1 million bp in length!

Other highly repetitive sequences are interspersed throughout the genome. A widely studied example is the *Alu* family of sequences found in humans and other primates. The *Alu* sequence is approximately 300 bp long. This sequence derives its name from the observation that it contains a site for cleavage by a restriction enzyme known as *Alu*I. It represents about 10% of the total human DNA and occurs (on average) approximately every 5,000–6,000 bases. Evolutionary studies suggest that the *Alu* sequence arose 65 million years ago (mya) from a section of a single ancestral gene known as the 7SL RNA gene. Remarkably, over the course of 65 million years, the *Alu* sequence has been copied and inserted into the human genome so often that it now occurs more than 1 million times! The mechanism for the proliferation of *Alu* sequences will be described later.

Figure 18.14 shows the composition of the relative classes of DNA sequences found in the nuclear genome of humans. Surprisingly, exons, the coding regions of protein-encoding genes, and the genes that give rise to rRNA and tRNA make up only about 2% of our genome! The other 98% is composed of noncoding sequences. Though we often think of genomes as being the repository of sequences that code for proteins, most eukaryotic genomes are largely composed of other types of sequences. Intron DNA comprises about 24% of

## Biology Principle

### The Genetic Material Provides the Blueprint for Reproduction

Only a small percentage of the human genome is involved with encoding the proteins that are largely responsible for human traits.

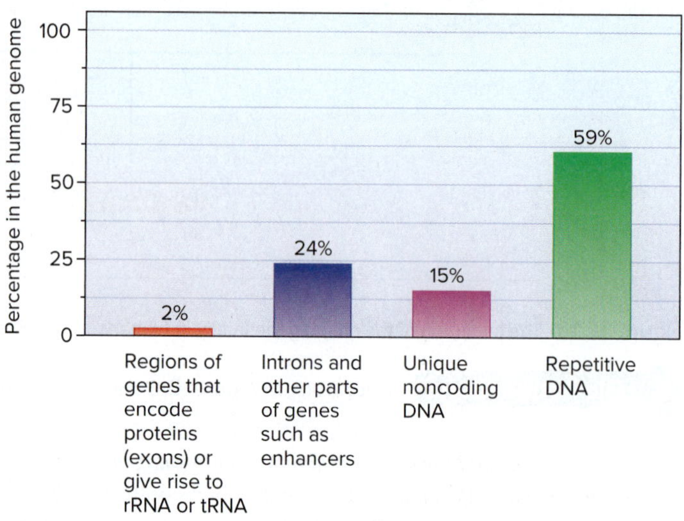


**Figure 18.14 The composition of DNA sequences in the nuclear genome of humans.** Only about 2% of our genome codes for proteins. Most of our genome is made up of repetitive sequences.

the human genome, and unique noncoding DNA constitutes 15%. Repetitive DNA makes up 59% of the DNA in the genome.

Scientists once described noncoding sequences as "junk DNA" because it was believed to have no biological function. However, in 2003, the Encyclopedia of DNA Elements (ENCODE) Project began, which has involved more than 440 researchers in 32 laboratories. The overall goal of the ENCODE Project is to identify functional sequences in the human genome. In 2012, the researchers announced that they were able to assign function to approximately 80% of the human genome. Much of the previously described "junk DNA" appears to play a role in a complex network of gene regulation, a topic discussed in Chapters 11 and 12.

## Transposable Elements Move from One Chromosomal Location to Another

As we have seen, repetitive sequences can be moderately or highly repetitive, and they can be found in tandem arrays or interspersed throughout a genome. Some types of repetitive sequences, which are interspersed in genomes, are **transposable elements (TEs)**—DNA segments that can move throughout the genome. The *Alu* sequence discussed earlier is an example of a TE. The process in which a TE moves to a new site in a genome is called **transposition.** TEs range from a few hundred to several thousand base pairs in length. They have sometimes been referred to as "jumping genes," because they are inherently mobile.

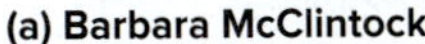

(a) Barbara McClintock

(b) Speckled corn kernels caused by transposable elements

**Figure 18.15** **Barbara McClintock, who discovered transposable elements (TEs).** As shown in part **(b)**, when a TE is found within a pigment gene in corn, its frequent movement disrupts the gene, causing the kernel color to be speckled.

*(a)* © Topham/The Image Works; *(b)* © Kenneth Keifer/Getty Images RF

American cytogeneticist Barbara McClintock first identified TEs in the late 1940s from her studies with corn plants (Figure 18.15). She identified a segment of DNA that could move into and out of a gene that affected the color of corn kernels, producing a speckled appearance. Since that time, biologists have discovered many different types of TEs in nearly all species examined.

Though McClintock identified TEs in corn in the late 1940s, her work was met with great skepticism because many researchers had trouble believing that DNA segments could be mobile. The advent of molecular technology in the 1960s and 1970s allowed scientists to understand more about the characteristics of TEs that enable their movement. Most notably, research involving bacterial TEs eventually progressed to a molecular understanding of the transposition process. In 1983, more than 30 years after her initial discovery, McClintock was awarded the Nobel Prize in Physiology or Medicine.

Researchers have studied TEs from many species, including bacteria, archaea, and eukaryotes. They have discovered that TEs fall into two groups—DNA transposons and retrotransposons—based on different mechanisms of movement.

**DNA Transposons** Transposable elements that move via a DNA molecule are called **DNA transposons.** Both ends of DNA transposons have inverted repeats (IRs)—DNA sequences that are identical (or very similar) but run in opposite directions (Figure 18.16a), such as the following:

| | | |
|---|---|---|
| 5′-CTGACTCTT-3′<br>3′-GACTGAGAA-5′ | and | 5′-AAGAGTCAG-3′<br>3′-TTCTCAGTC-5′ |

Depending on the particular TE, inverted repeats range from 9 to 40 bp in length. In addition, DNA transposons may contain a central region that encodes **transposase,** an enzyme that facilitates transposition.

As shown in Figure 18.16b, transposition of DNA transposons occurs by a cut-and-paste mechanism.

1. Transposase first recognizes the inverted repeats (IR) in the transposon.

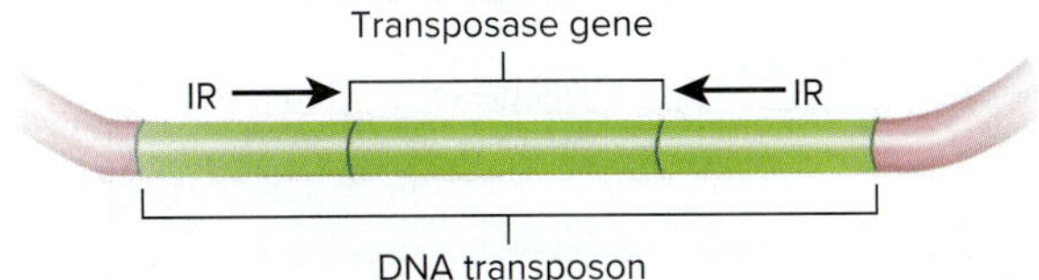


(a) Organization of a DNA transposon

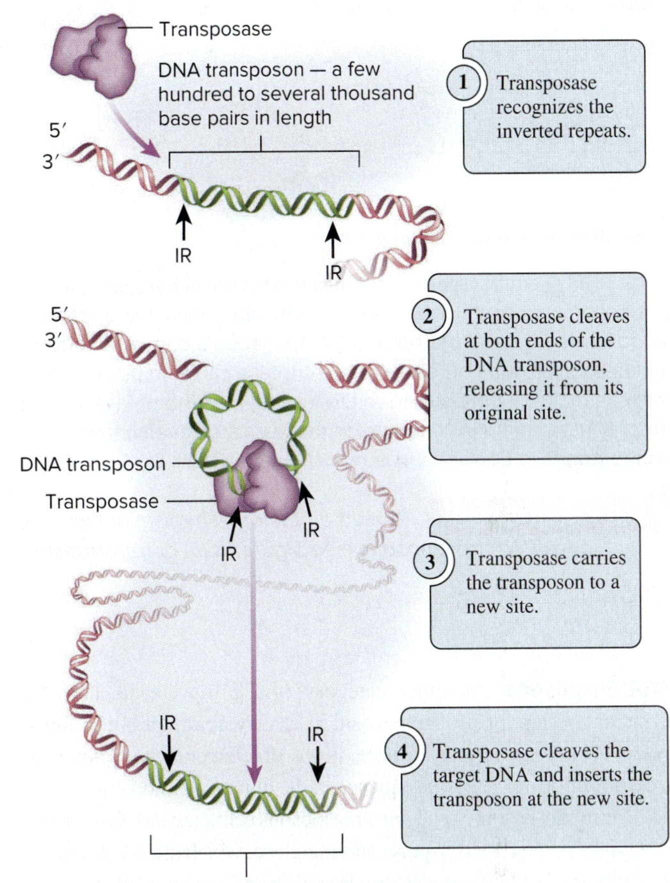


(b) Cut-and-paste mechanism of transposition

**Figure 18.16** **DNA transposons and their mechanism of transposition.** **(a)** DNA transposons contain inverted repeat (IR) sequences at each end and may contain a gene that encodes transposase in the middle. **(b)** Transposition occurs by a cut-and-paste mechanism.

**Concept Check:** *What is the role of the inverted repeats in the mechanism of transposition?*

2. It then cleaves both ends of the DNA transposon and removes it from its original site.
3. Next, the transposase/transposon complex moves to a new location, where transposase cleaves the target DNA and inserts the transposon into the site.

Transposition may occur when a cell is in the process of DNA replication. If a TE is removed from a site that has already replicated and is inserted into a chromosomal site that has not yet replicated, the transposon will increase in number after DNA replication is complete. This is one way for transposons to become more prevalent in a genome.

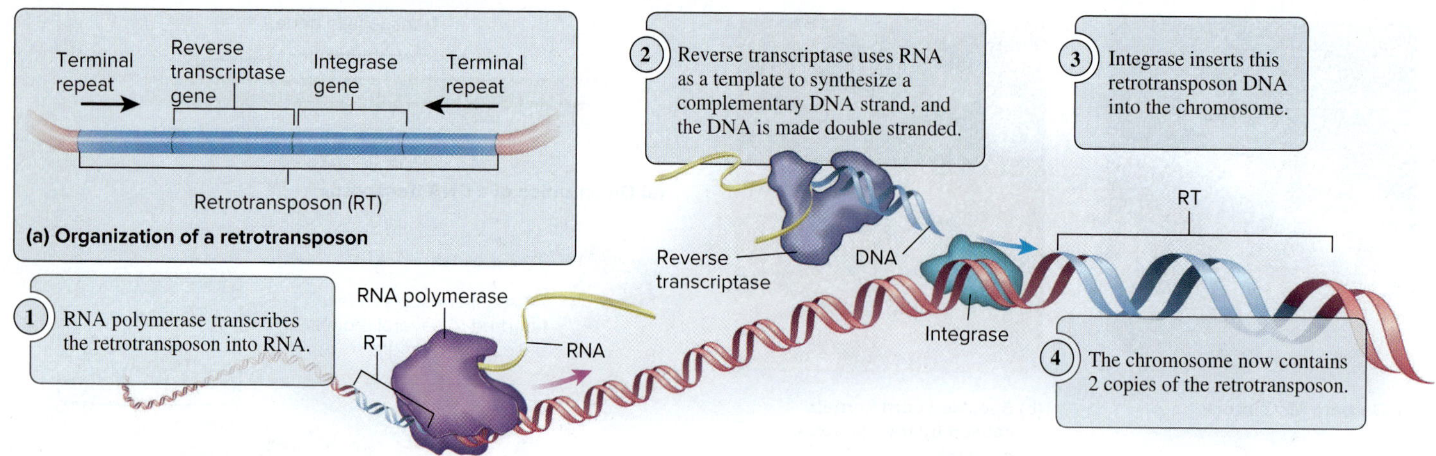


**Figure 18.17** **Retrotransposons and their mechanism of transposition.** Retrotransposons are found only in eukaryotic species. **(a)** Some retrotransposons contain terminal repeats and genes that encode the enzymes reverse transcriptase and integrase, which are needed in the transposition process. **(b)** The process that adds a copy of a retrotransposon into a host chromosome. Note: In addition to using RNA as a template, some forms of reverse transcriptase can also use a DNA template to make a complementary DNA strand. As depicted here, these forms can make double-stranded DNA using a strand of RNA as a starting material. However, some forms of reverse transcriptase can make DNA only using an RNA template. In those cases, reverse transcriptase makes a complementary DNA strand from the RNA template and then the opposite DNA strand is made via a host-cell DNA polymerase.

**Concept Check:** *Based on their mechanism of movement, which type of TE do you think would proliferate more rapidly in a genome, DNA transposons (see Figure 18.16) or retrotransposons?*

**Retrotransposons** Another category of TE moves via an RNA intermediate. This form of transposition is very common but is found only in eukaryotic species. These types of elements are known as **retrotransposons.** The *Alu* sequence in the human genome is an example of a retrotransposon. Some retrotransposons contain genes that encode the enzymes reverse transcriptase and integrase, which are needed in the transposition process (Figure 18.17a). Recall from Chapter 17 that reverse transcriptase uses RNA as a template to synthesize a complementary copy of DNA. Retrotransposons may also contain repeated sequences called terminal repeats at each end that facilitate their recognition.

The mechanism of movement for some types of retrotransposons is shown in Figure 18.17b.

1. First, the enzyme RNA polymerase transcribes the retrotransposon into RNA.
2. Reverse transcriptase uses this RNA as a template to synthesize a double-stranded DNA molecule.
3. The ends of the double-stranded DNA are then recognized by integrase, which catalyzes the insertion of the retrotransposon DNA into the host chromosomal DNA.
4. The host chromosome now contains two copies of the retrotransposon.

The integration of retrotransposons can occur at many locations within the genome. Furthermore, because a single retrotransposon can be copied into many RNA transcripts, retrotransposons may accumulate rapidly within a genome. This explains how the *Alu* sequence in the human genome was able to proliferate and constitute 10% of the human genome.

## 18.5 Reviewing the Concepts

- Repetitive DNA is found in multiple copies in the genome, which can be moderately or highly repetitive. Repetitive DNA comes from different sources, including multiple copies of the same gene, tandem arrays, and transposable elements (TEs). It is often abundant in eukaryotic genomes, such as the human genome (Figure 18.14).
- Transposable elements are segments of DNA that can move from one site to another through a process called transposition (Figure 18.15).
- Transposable elements fall into two groups that move by two different molecular mechanisms. DNA transposons move by a cut-and-paste mechanism facilitated by the enzyme transposase. Retrotransposons move to new sites in the genome via RNA intermediates (Figures 18.16, 18.17).

## 18.5 Testing Your Knowledge

1. Repetitive sequences may be
   a. moderately or highly repetitive.
   b. tandem arrays or interspersed.
   c. transposable elements.
   d. all of the above.
2. A segment of DNA that moves via an RNA intermediate is *not* a
   a. repetitive sequence.
   b. transposable element.
   c. DNA transposon.
   d. retrotransposon.

# Assess and Discuss

## Test Yourself

1. Restriction enzymes used in most cloning experiments
   a. are used to cut DNA into pieces for gene cloning.
   b. are produced by bacteria cells to prevent viral infection.
   c. produce sticky ends on DNA fragments.
   d. all of the above
   e. only a and c
2. DNA ligase is needed in a cloning experiment
   a. to promote hydrogen bonding between sticky ends.
   b. to covalently link the backbone of DNA strands.
   c. to digest the chromosomal DNA into small pieces.
   d. only a and b
   e. answers a, b, and c
3. Let's suppose you followed the protocols described in Figures 18.2 and 18.3. Which experiment(s) would you conduct to confirm that a white colony really contained a recombinant vector with an insert?
   a. Pick a white bacterial colony and restreak on plates containing X-Gal to confirm that the cells really form white colonies.
   b. Pick a white bacterial colony, isolate plasmid DNA, digest the plasmid DNA with a restriction enzyme, and then perform gel electrophoresis with the DNA.
   c. Pick a white bacterial colony and test it to see if β-galactosidase is functional within the bacterial cells.
   d. Pick a white bacterial colony and retest it on ampicillin-containing plates to double-check that the cells are really ampicillin resistant.
   e. Both c and d should be conducted.
4. Why is *Taq* polymerase used in PCR rather than other DNA polymerases?
   a. *Taq* polymerase is a synthetic enzyme that produces DNA strands at a faster rate than natural polymerases.
   b. *Taq* polymerase is a heat-stable form of DNA polymerase that can function after exposure to the high temperatures necessary for PCR.
   c. *Taq* polymerase is easier to isolate than other DNA polymerases.
   d. *Taq* polymerase is the DNA polymerase commonly produced by most eukaryotic cells.
   e. All of the above are correct.
5. Let's suppose you want to clone a gene that has never been analyzed before by DNA sequencing. Which of the following statements do you agree with the most?
   a. Do PCR to clone the gene because it is much faster.
   b. Do PCR to clone the gene because it is very specific and gives a high yield.
   c. You can't do PCR because you can't make forward and reverse primers.
   d. Do cloning using a vector because it will give you a higher yield.
   e. Do cloning by insertion into a vector because it is easier than PCR.
6. In the CRISPR-Cas technology used for mutating genes, what is(are) the function(s) of the sgRNA?
   a. to bind to the target gene
   b. to bind to Cas9
   c. to cause a double-strand break in the target gene
   d. all of the above
   e. both a and b
7. Important reasons for studying the genomes of bacteria and archaea include all of the following *except*
   a. it may provide information that helps us understand how bacteria infect other organisms.
   b. it may provide a basic understanding of cellular processes that allows us to determine eukaryotic cellular function.
   c. it may provide the means of understanding evolutionary processes.
   d. it will reveal the approximate number of genes that an organism has in its genome.
   e. All of the above are important reasons.
8. The enzyme that allows short segments of DNA to move from one chromosomal location to another is
   a. transposase.
   b. DNA polymerase.
   c. protease.
   d. restriction endonuclease.
   e. DNA ligase.
9. A gene family includes
   a. one specific gene found in several different species.
   b. all of the genes on the same chromosome.
   c. two or more homologous genes found within a single species.
   d. genes that code for structural proteins.
   e. both a and c.
10. Which of the following was *not* a goal of the Human Genome Project?
    a. identify all human genes
    b. sequence the entire human genome
    c. address the legal and ethical implications resulting from the project
    d. develop programs to manage the information gathered from the project
    e. be able to clone a human

## Conceptual Questions

1. Draw the structure of a dideoxyribonucleotide triphosphate and explain how it causes chain termination.
2. Briefly describe whether or not each of the following can be appropriately described as a genome.
   a. the *E. coli* chromosome
   b. human chromosome 11
   c. a complete set of 10 chromosomes in corn
   d. a copy of the single-stranded RNA packaged into human immunodeficiency virus (HIV)
3. **PRINCIPLES** A principle of biology is that the genetic material provides a blueprint for reproduction. Explain the roles of the genome with regard to this principle.

## Collaborative Questions

1. Identify and discuss three important advances that have resulted from gene cloning.
2. Compare and contrast the characteristics of the genomes of bacteria, archaea, and eukaryotes.

# Online Resource

**connect.mheducation.com**

# UNIT IV
# EVOLUTION

**Evolution** is a change in one or more heritable characteristics of a population from one generation to the next. This process not only alters the characteristics of populations but also leads to the formation of new species.

We will begin this unit by considering the fundamental concepts of evolution and examining observations of evolutionary change, which include (1) the fossil record, (2) a comparison of the characteristics of modern species, and (3) an analysis of molecular data. We will explore how evolution occurs at the molecular level and focus on how changes in allele and genotype frequencies from one generation to the next are driven by a variety of factors. By comparison, Chapter 20 shifts the emphasis of evolution to the level of species. We will examine how species are identified and discuss the mechanisms by which new species arise via evolution. In Chapter 21, we will examine how biologists determine the evolutionary relationships among different species and produce branching diagrams, called evolutionary trees, to describe those relationships. Finally, in Chapter 22, we will explore a time line for the evolution of species from 4 billion years ago to the present, which includes a description of how humans evolved from primate ancestors.

***The following biology principles will be emphasized in this unit:***

- ***Populations of organisms evolve from one generation to the next.*** *This concept will be emphasized throughout the entire unit.*
- ***Living organisms interact with their environment.*** *As discussed in Chapters 19, 20, and 21, natural selection is a process in which certain individuals have greater reproductive success. This success is often due to their ability to survive in a given environment.*
- ***Structure determines function.*** *Chapters 19 and 20 will also consider how structural features change during the evolution of new species. Such changes are related to changes in function.*
- ***All species (past and present) are related by an evolutionary history.*** *Chapter 21 is devoted to examining how biologists determine evolutionary relationships among different species.*
- ***Biology is an experimental science.*** *Most chapters have a Feature Investigation describing a pivotal experiment that provided insights into our understanding of evolution.*

19

© Mark Dadswell/Getty Images

© Nigel Pavitt/AWL Images/Getty Images

© Photos M. Kuntner

© George Bernard/SPL/Science Source

# Evolution of Life I: How Populations Change from Generation to Generation

# 19

© Mark Dadswell/Getty Images

Selective breeding. The horses in this race have been bred for a particular trait—in this case, speed. Such a practice, called selective breeding, can dramatically change the traits of organisms over several generations.

## Chapter Outline

Organic life beneath the shoreless waves
Was born and nurs'd in Ocean's pearly caves
First forms minute, unseen by spheric glass,
Move on the mud, or pierce the watery mass;
These, as successive generations bloom,
New powers acquire, and larger limbs assume;
Whence countless groups of vegetation spring,
And breathing realms of fin, and feet, and wing.

From *The Temple of Nature* by Erasmus Darwin, grandfather of Charles Darwin. Published posthumously in 1803.

The term **evolution** is used to describe a change in one or more heritable characteristics of a population from one generation to the next. It is a process of change over time. In the first part of this chapter, we will examine the development of evolutionary thought and some of the basic tenets of evolution, particularly those proposed by the British naturalist Charles Darwin in the mid-1800s. Our understanding of evolution has been refined over the past 150 years or so, but the fundamental principle of evolution remains unchanged and has provided a cornerstone for our understanding of biology.

We will then survey the extensive data that illustrate the processes by which evolution occurs. These data not only support the idea that populations change over time, or evolve, but also allow us to understand the interrelatedness of different species, whose similarities are often due to descent from a common ancestor. Much of the early evidence supporting evolution came from direct observations and comparisons of living and extinct species. More recently, advances in molecular genetics, particularly those related to DNA sequencing and genomics, have revolutionized the study of evolution. Scientists now have information that allows us to understand how evolution involves changes in the DNA sequences of a given species that affect a species' genes and the proteins they encode.

The remainder of the chapter is concerned with **population genetics,** the study of genes and genotypes in a population. The central issue in population genetics is genetic variation—its extent within populations, why it exists, how it is maintained, and how it changes over the course of many generations. We will examine the various evolutionary mechanisms that promote genetic change in a population, including mutation, natural selection, genetic drift, migration, and nonrandom mating.

## 19.1 Overview of Evolution

### Learning Outcomes

1. Describe the process of evolution.
2. List the factors that led Darwin to propose the theory of descent with modification through variation and natural selection.
3. Explain how natural selection works.

Undoubtedly, the question "Where did we come from?" has been asked and debated by people for thousands of years. Many of the early ideas regarding the existence of living organisms were strongly influenced by religion and philosophy. Some of these ideas suggested that all forms of life have remained the same since their creation. In the 17th century, however, scholars in Europe began a revolution that created the basis of empirical thought. **Empirical thought** relies on observation to form an idea or a hypothesis rather than trying to understand life from a nonphysical or spiritual point of view. As described in this section, the shift toward empirical thought encouraged scholars to look for the basic rationale behind a given process or phenomenon.

### The Work of Several Scientists Set the Stage for Darwin's Ideas

Late in the 1700s, a small number of European scientists began to quietly challenge the belief that life-forms are fixed and unchanging. A French zoologist, Georges Buffon, actually proposed that populations of living things change through time. However, Buffon was careful to hide his views in a 44-volume series of books on natural history. Around the same time, a French naturalist named Jean-Baptiste Lamarck suggested an intimate relationship between variation and evolution. By examining fossils, he realized that some species had remained the same over the millennia and others had changed. Lamarck hypothesized that species change over the course of many generations by adapting to new environments. According to Lamarck, organisms altered their behavior in response to environmental change. He thought that behavioral changes modified traits and hypothesized that these traits were inherited by offspring. He called this idea the **inheritance of acquired characteristics.** For example, according to Lamarck's hypothesis, giraffes developed their elongated necks and front legs by feeding on the leaves at the top of trees. The exercise of stretching up to the leaves altered the neck and legs, and Lamarck presumed that these acquired characteristics were transmitted to offspring. However, further research has rejected Lamarck's idea that acquired traits can be inherited. Even so, Lamarck's work was important in promoting the idea of evolutionary change.

Interestingly, Erasmus Darwin, the grandfather of Charles Darwin, was a contemporary of Buffon and Lamarck and an early advocate of evolutionary change. He was a physician, a plant biologist, and a poet (see poem at the beginning of the chapter). He was aware that modern species were different from related types of fossilized organisms and noted how plant and animal breeders used breeding practices to change the traits of domesticated species such as horses (see chapter-opening photo). He knew that offspring inherited features from their parents and went so far as to say that life on Earth could have descended from a common ancestor.

### Darwin Suggested That Existing Species Are Derived from Pre-existing Species

Charles Darwin played a central role in developing the theory that existing species have evolved from pre-existing ones. Darwin's unique perspective and ability to formulate evolutionary principles were shaped by several different fields of study, including ideas of his time about geology and population growth, as well as his own biological observations.

Two main hypotheses about geological processes predominated in the early 19th century.

1. Catastrophism was first proposed by French zoologist and paleontologist Georges Cuvier to explain the age of the Earth. Cuvier suggested that the Earth was just 6,000 years old and that only catastrophic events had changed its geological structure. This idea fit well with certain religious teachings.
2. Uniformitarianism, proposed by Scottish geologist James Hutton and popularized by another Scottish geologist, Charles Lyell, suggested that changes in the Earth are directly caused by recurring events. For example, they suggested that geological processes such as erosion existed in the past and happened at the same gradual and uniform rate as they do in the present. For such slow geological processes to eventually lead to substantial changes in the Earth's characteristics, a great deal of time was required. Hutton and Lyell were the first to propose that the age of the Earth is well beyond 6,000 years. The ideas of Hutton and Lyell helped to shape Darwin's view of the world.

Darwin's thinking was also influenced by a paper published in 1798 called *Essay on the Principle of Population* by Thomas Malthus, an English economist. Malthus asserted that the population size of humans can, at best, increase linearly due to increased land usage and improvements in agriculture, whereas our reproductive potential is exponential (for example, doubling with each generation). He argued that famine, war, and disease, especially among the poor, keep population growth within existing resources. The relevant message from Malthus's work was that not all members of any population will survive and reproduce.

Darwin's ideas, however, were most influenced by his own experiences and observations. His work as a young man aboard the HMS *Beagle*, a survey ship, lasted from 1831 to 1836 and involved a careful examination of many different species (**Figure 19.1**). The main mission of the *Beagle* was to map the coastline of southern South America and take oceanographic measurements. As the ship's naturalist, Darwin's job was to record information about the weather, geological features, plants, animals, fossils, rocks, minerals, and indigenous people.

Though Darwin made many interesting observations on his journey, he was particularly struck by the distinctive traits of island species. For example, Darwin observed several species of finches found on the Galápagos Islands, a group of volcanic islands 600 miles off the coast of Ecuador. Though it is often assumed that

**Figure 19.1** **Charles Darwin and the voyage of the *Beagle*, 1831–1836. (a)** A portrait of Charles Darwin (1809–1882) at age 31. **(b)** Darwin's voyage on the *Beagle*, which took almost 5 years to circumnavigate the world.

*(a)* © PoodlesRock/Corbis

Darwin's personal observations of these finches directly inspired his ideas regarding evolution, this is not the case. Initially, Darwin thought the birds were various species of blackbirds, grosbeaks, and finches. Later, however, the bird specimens from the islands were given to the British ornithologist John Gould, who identified them as several new finch species. Gould's observations helped Darwin in the later formulation of his theory.

As seen in **Table 19.1,** the finches differed widely in the size and shape of their beaks and in their feeding habits. Darwin clearly saw the similarities among these species, yet he noted the differences that provided them with specialized feeding strategies. It is now known these finches all evolved from a single species similar to the dull-colored grassquit finch (*Tiaris obscura*), commonly found along the Pacific coast of South America. Once they arrived on the Galápagos Islands, the finches' ability to obtain particular types of food in their new habitat depended, in part, on the relative sizes and shapes of their beaks, which, in turn, influenced their abilities to survive and reproduce.

In 1856, Darwin began to write a long book to explain his ideas regarding evolution. In 1858, however, Alfred Wallace, a British naturalist working in the East Indies, sent Darwin an unpublished manuscript to read prior to its publication. In it, Wallace proposed the same ideas concerning evolution. In response to this, Darwin decided to publish some of his own writings on this subject, and two papers, one by Darwin and one by Wallace, appeared in the *Proceedings of the Linnean Society of London*. These papers were not widely recognized. A year later, however, Darwin finished his book *On the Origin of Species* (1859), which described his ideas in greater detail and included observational support. This book, which received high praise from many scientists but scorn from others, started a great debate concerning evolution. Although some of his ideas were incomplete because the genetic basis of traits was not understood at that time, Darwin's work remains a foundation of our understanding of biology.

## Natural Selection Changes Populations from Generation to Generation

Darwin proposed that existing life-forms on our planet result from the modification of pre-existing life-forms. He expressed this concept of evolution as "the theory of descent with modification through variation and natural selection." The term evolution refers to the process of change. What factors bring about evolutionary change? According to Darwin's ideas, evolution occurs from generation to generation due to two interacting factors: genetic variation and natural selection.

1. Variation in traits may occur among individuals of a given species. Those traits that are heritable are then passed from parents to offspring. The genetic basis for variation within a species was not understood during Darwin's time. We now know that such variation is due to different types of genetic changes such as random mutations in genes. Even though Darwin did not fully appreciate the genetic basis of variation, he and many other people before him observed that offspring resemble their parents more than they do unrelated individuals. Therefore, he assumed that some traits are passed from parent to offspring.
2. In each generation, many more offspring are usually produced than will survive and reproduce. Often, resources in the

**Table 19.1 A Comparison of Beak Type and Diet Among the Galápagos Finches Darwin Studied**

| Type of finch species | | Type of beak/diet |
|---|---|---|
| **Ground finches** | | Crushing |
| Large ground finch (*Geospiza magnirostris*) | | Seeds—ground finches have crushing beaks to crush various sizes of seeds; large beaks can crush large seeds, whereas smaller beaks are better for crushing small seeds |
| Medium ground finch (*G. fortis*) | | |
| Small ground finch (*G. fuliginosa*) | 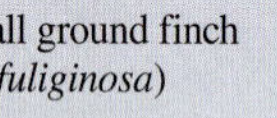  | |
| Sharp-billed ground finch (*G. difficilis*) |  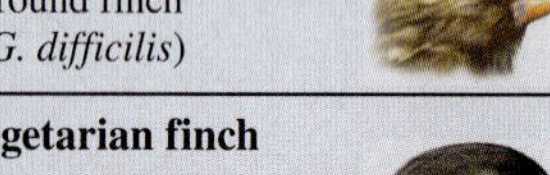 | |
| **Vegetarian finch** | | Crushing |
| Vegetarian finch (*Platyspiza crassirostris*) |  | Buds—vegetarian finches have crushing beaks to pull buds from branches |
| **Tree finches** | | Grasping |
| Large tree finch (*Camarhynchus psittacula*) |  | Insects—tree finches have grasping beaks to pick insects from trees. Those with heavier beaks can also break apart wood in search of insects. |
| Medium tree finch (*Camarhynchus pauper*) | 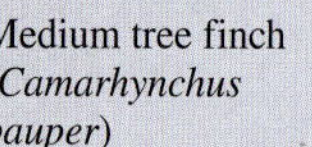 | |
| Small tree finch (*Camarhynchus parvulus*) | 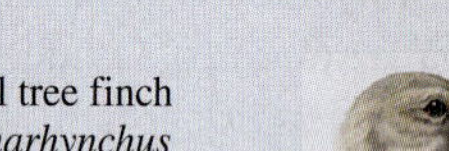  | |
| **Tree and warbler finches** | | Probing |
| Mangrove finch (*Cactospiza heliobates*) | | Insects—these finches have probing beaks to search for insects in crevices and then to pick them up. The woodpecker finch can also use a cactus spine for probing. |
| Woodpecker finch (*Camarhynchus pallidus*) | | |
| Warbler finch (*Certhidea olivacea*) | 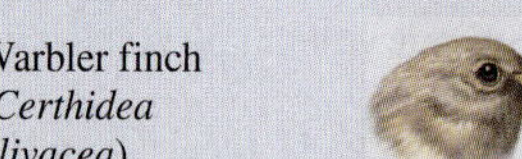  | |
| **Cactus finches** | | Probing |
| Large cactus finch (*G. conirostris*) |  | Seeds—cactus finches have probing beaks to open cactus fruits and take out seeds |
| Cactus finch (*G. scandens*) |  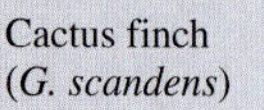 | |

environment are limiting so that some members of a population do not survive. During the process of **natural selection,** individuals with heritable traits that make them better suited to their native environment tend to flourish and reproduce, whereas other individuals are less likely to survive and reproduce. As a result of natural selection, certain traits that favor reproductive success become more prevalent in a population over time.

As an example, let's consider a population of finches that migrates from the South American mainland to a distant island (Figure 19.2). Beak sizes vary among the migrating birds. If the seeds produced on the distant island were larger than those produced on the mainland, those birds with larger beaks would be better able to feed on these larger seeds and therefore would be more likely to survive and pass that trait to their offspring. What are the consequences of this selection process? In succeeding generations, the population tends to have a greater proportion of finches with larger beaks.

Alternatively, if a trait happens to be detrimental to an individual's ability to survive and reproduce, natural selection is likely to eliminate this type of variation. For example, if a finch in the same environment had a small beak, this bird would be less likely to acquire enough food, which would decrease its ability to survive and pass this trait to its offspring. Natural selection may ultimately result in a new species with a combination of multiple traits that are quite different from those of the original species, such as finches with larger beaks and changes in coloration. In other words, the newer species has evolved from a pre-existing one.

## 19.1 Reviewing the Concepts

- Evolution is the process of change in one or more heritable characteristics of a population from one generation to the next.
- Charles Darwin proposed the theory of descent with modification through variation and natural selection based on his understanding of geology and population growth and his observations of species in their natural settings. His voyage on the *Beagle*, during which he studied many species, including finches on the Galápagos Islands, was particularly influential in the development of his ideas (Figure 19.1; Table 19.1).
- Darwin proposed that evolution occurs from generation to generation through two interacting factors: genetic variation and natural selection. As a result of natural selection, certain traits that favor reproductive success become more prevalent in a population over time (Figure 19.2).

## 19.1 Testing Your Knowledge

1. What core concept underlies the process of evolution?
   **a.** inheritance of acquired characteristics
   **b.** genetic variation
   **c.** natural selection
   **d.** all of the above
   **e.** both b and c
2. What may change as a result of natural selection?
   **a.** an individual's characteristics
   **b.** the genetic composition of populations from one generation to the next
   **c.** the characteristics found in populations from one generation to the next
   **d.** all of the above
   **e.** both b and c

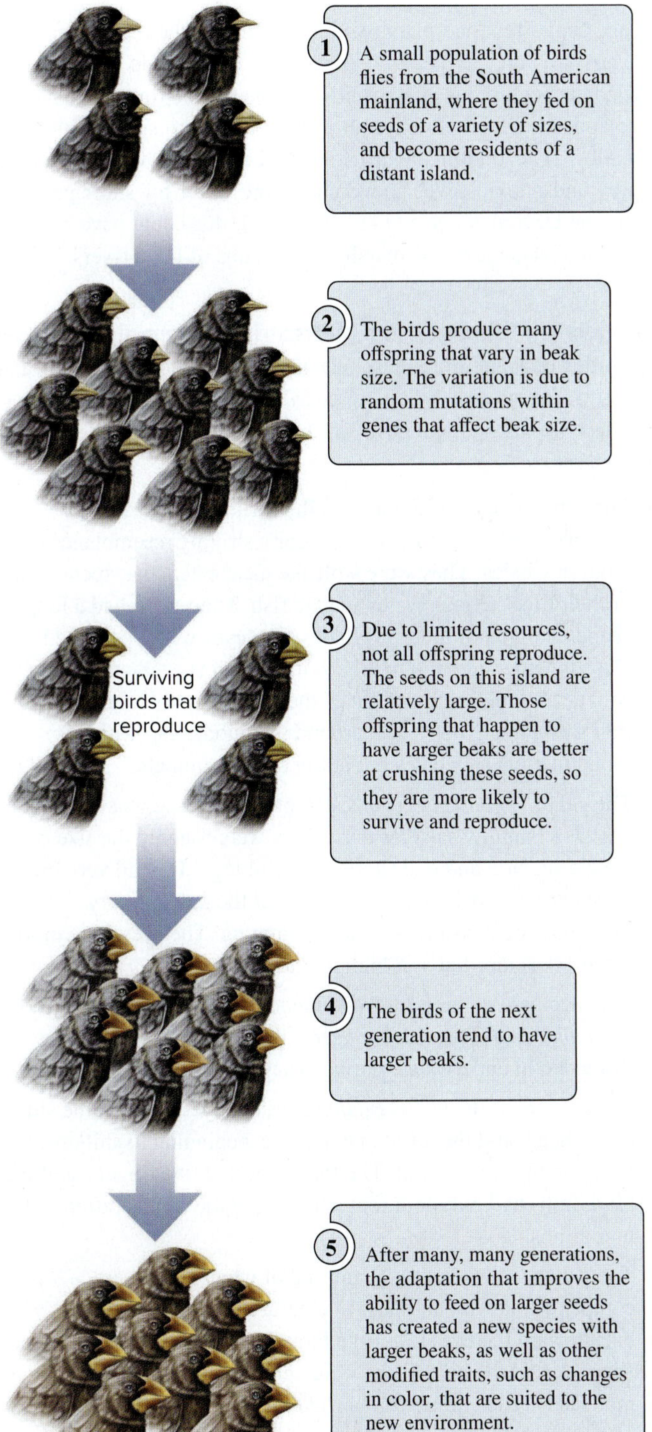

1 A small population of birds flies from the South American mainland, where they fed on seeds of a variety of sizes, and become residents of a distant island.
2 The birds produce many offspring that vary in beak size. The variation is due to random mutations within genes that affect beak size.
Surviving birds that reproduce
3 Due to limited resources, not all offspring reproduce. The seeds on this island are relatively large. Those offspring that happen to have larger beaks are better at crushing these seeds, so they are more likely to survive and reproduce.
4 The birds of the next generation tend to have larger beaks.
5 After many, many generations, the adaptation that improves the ability to feed on larger seeds has created a new species with larger beaks, as well as other modified traits, such as changes in color, that are suited to the new environment.


**Figure 19.2 Evolutionary adaptation to a new environment via natural selection.** The example shown here involves a species of finch adapting to a new environment on a distant island. According to Darwin, the process of adaptation can lead to the formation of a new species with traits that are better suited to the new environment.

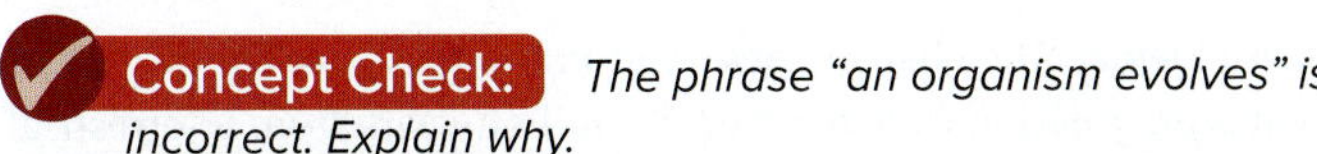

**Concept Check:** *The phrase "an organism evolves" is incorrect. Explain why.*

**BioConnections:** *Refer back to Figure 4.4. How is natural selection similar to chemical selection? How are they different?*

## 19.2 Evidence of Evolutionary Change

### Learning Outcomes

1. Summarize different types of evidence for evolutionary change, including the fossil record, biogeography, convergent traits, selective breeding, and homologies.
2. Provide examples of three types of homologies.

Evidence that reflects the process of evolution has been obtained from many sources (**Table 19.2**). Historically, the first evidence of biological evolution came from studies of the characteristics of populations over time, studies of the fossil record, the distribution of species on Earth, the comparison of similar anatomical features in different species, and selective breeding experiments. More recently, additional evidence that illustrates the process of evolution has been found at the molecular level. By comparing DNA sequences from many different species, biologists have gained great insight into the relationship

**Table 19.2 Evidence of Biological Evolution**

| Type of observation | Description |
|---|---|
| Studies of natural selection | By following the characteristics of populations over time, researchers have observed how natural selection alters populations in response to environmental changes. |
| Fossil record | When fossils are compared according to their age, from oldest to youngest, successive evolutionary change becomes apparent. |
| Biogeography | Unique species found on islands and other remote areas have arisen because the species in these locations have evolved in isolation from the rest of the world. |
| Convergent evolution | Two different species from different lineages sometimes become anatomically similar because they occupy similar environments. This indicates that natural selection results in adaptation to a given environment. |
| Selective breeding | The traits in domesticated species have been profoundly modified by selective breeding (also called artificial selection) in which breeders choose the parents that have desirable traits. |
| **Homologies** | |
| Anatomical | Homologous structures are structures that are anatomically similar to each other because they evolved from a structure in a common ancestor. In some cases, such structures have lost their original function and become vestigial. |
| Developmental | An analysis of embryonic development often reveals similar features that point to past evolutionary relationships. |
| Molecular | At the molecular level, certain characteristics are found in all living cells, suggesting that all modern species are derived from an interrelated group of common ancestors. In addition, the genes of species that are closely related evolutionarily have DNA sequences that are more similar to each other than they are to the DNA sequences of distantly related organisms. |

between the evolution of species and the associated changes in the genetic material. In this section, we will survey a variety of evidence that shows the process of evolutionary change.

## Fossils Show Successive Evolutionary Change

As discussed in Chapter 22, fossils are the preserved remains of past life. The fossil record has provided biologists with evidence of the history of life on Earth. Let's consider a couple of examples in which paleontologists have studied fossils that illustrate evolutionary change.

**Evolution of Terrestrial Tetrapods from Fish** In 2005, fossils of *Tiktaalik roseae*, nicknamed fishapod, were discovered by paleontologists Ted Daeschler, Neil Shubin, and Farish Jenkins. The discovery of fishapod illuminates one of several steps that led to the evolution of tetrapods, which are animals with four legs. *T. roseae* is called a **transitional form** because it displays an intermediate state between an ancestral form and the form of its descendants (Figure 19.3). In this case, the fishapod is a transitional form between fishes, which have fins for locomotion, and tetrapods, which are four-limbed animals. Unlike a true fish, *T. roseae* had a broad skull, a flexible neck, and eyes mounted on the top of its head, like a crocodile. Its interlocking rib cage suggests it had primitive lungs. Perhaps the most surprising discovery was that its pectoral fins (those on the side of the body) revealed the beginnings of a primitive wrist and five finger-like bones. These appendages would have allowed *T. roseae* to support its body on shallow river bottoms and lift its head above the water to search for prey and perhaps even move out of the water for short periods. During the Devonian period (417–354 mya), this could have been an important advantage in the marshy floodplains of large rivers.

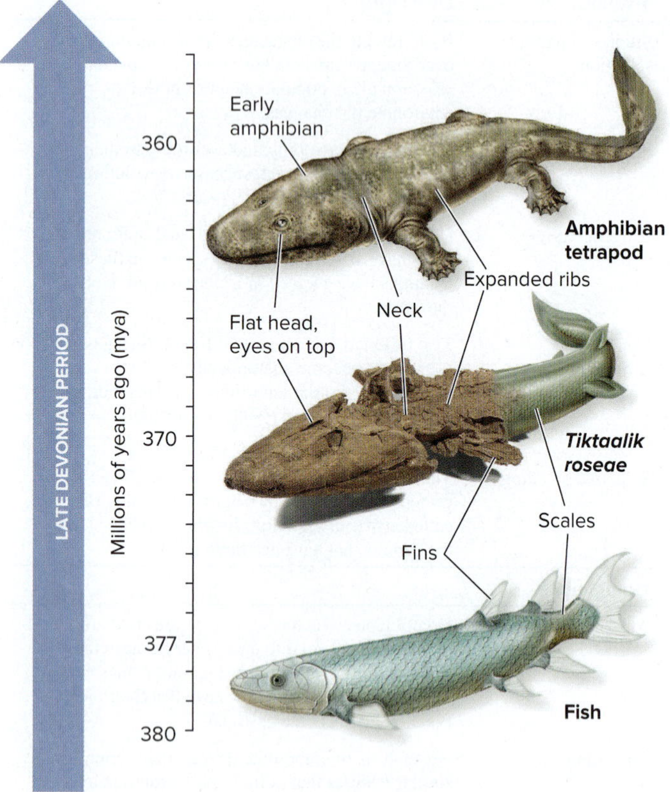


**Figure 19.3** **A transitional form in the tetrapod lineage.** This figure shows two early tetrapod ancestors, a Devonian fish and the transitional form *Tiktaalik roseae*, as well as a descendant, an early amphibian. An analysis of the fossils shows that *T. roseae*, also known as a fishapod, had both fish and amphibian characteristics, so it was likely able to survive brief periods out of the water.
*(middle)* © T. Daeschler/VIREO

**Evolution of Whales from Terrestrial Mammals** Cetacea is an order of aquatic animals that includes whales, dolphins, and porpoises. The closest living relatives of cetaceans are the hippos. Figure 19.4 presents a hypothesis for the evolution of whales based on the fossil record.

- The genus *Pakicetus* comprised the earliest known whales. However, these animals did not bear a striking resemblance to modern whales. They were wolflike meat eaters that spent some of their time in fresh water and ate fish. Their skull had a long shape, like that of modern whales. The eyes were positioned close together and high on the skull, which is characteristic of aquatic animals that peer out of the water. A particularly striking trait was a thick, bony wall around the middle ear, which is found in modern whales but not in other mammals.
- The genus *Ambulocetus* consisted of semiaquatic whales of brackish (slightly salty) waters. They were roughly the size of a male sea lion and had short, powerful legs. The tail vertebrae were particularly large, suggesting that the tail was very muscular and possibly used for swimming. The eyes were more toward the sides but still high on the skull.
- The members of the genus *Remingtonocetus* were similar to those of *Ambulocetus* but with a longer snout and a fat pad in the jaw that aided in underwater hearing. They lived in saltwater habitats.
- In members of the genus *Rodhocetus*, the eyes were on the side of the head, and the nasal opening was beginning to shift away from the tip of the snout. The forelimbs had five fingers, and the hind limbs had only four toes, suggesting the degeneration of the hind limbs.
- The genus *Dorudon* was composed of whales that were completely aquatic animals. The nasal opening was shifted back toward the eyes to form a blowhole. The forelimbs became flippers, and the hind limbs were very tiny. The tail was modified at the end to form a fluke.
- Odontoceti and Mysticeti are suborders of the order Cetacea, which includes many extinct species as well as all modern species of whales, dolphins, and porpoises. These animals show a complete loss of the hind limbs in the adult. The nasal opening is the blowhole seen in modern whales. In odontocetes, echolocation is used for hunting. In mysticetes, baleen is used for filtering food.

Taken together, the changes observed in the fossil record of whales reveal a progression over the past 50 million years from a terrestrial tetrapod to aquatic animals that lack hindlimbs and have many adaptations that are beneficial in an aquatic environment.

## Biology Principle

### All Species (Past and Present) Are Related by an Evolutionary History

This diagram shows the morphological changes that occurred in the evolution of whales.

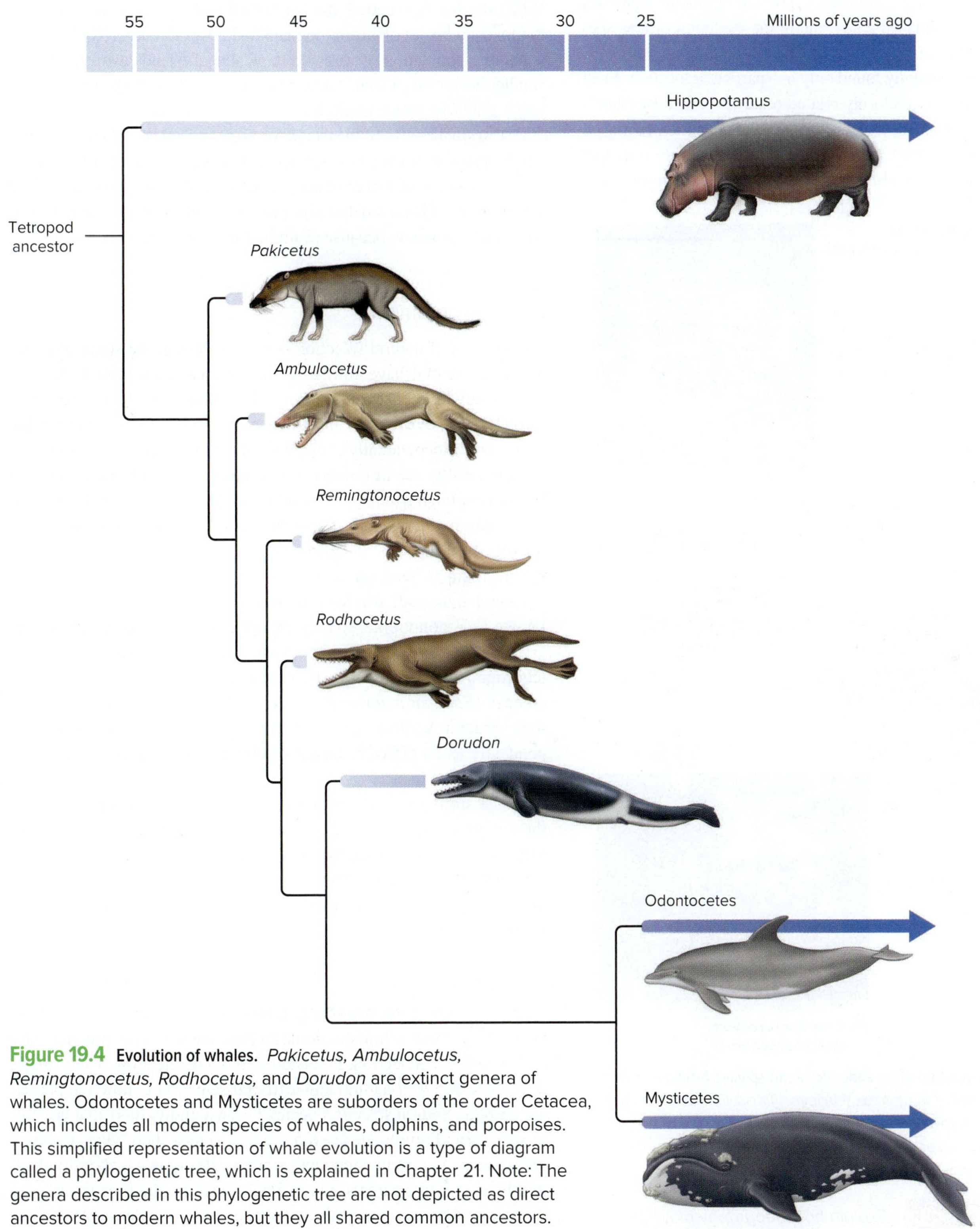


**Figure 19.4 Evolution of whales.** *Pakicetus, Ambulocetus, Remingtonocetus, Rodhocetus,* and *Dorudon* are extinct genera of whales. Odontocetes and Mysticetes are suborders of the order Cetacea, which includes all modern species of whales, dolphins, and porpoises. This simplified representation of whale evolution is a type of diagram called a phylogenetic tree, which is explained in Chapter 21. Note: The genera described in this phylogenetic tree are not depicted as direct ancestors to modern whales, but they all shared common ancestors.

## Biogeography Indicates That Species in a Given Area Have Evolved from Pre-existing Species

**Biogeography** is the study of the geographic distribution of extinct and living species. Patterns of past evolution are often found in the natural geographic distribution of related species. For example, islands, which are isolated from other large landmasses, provide numerous examples in which geography has played a key role in the evolution of new species. Islands often have many species of **endemic** plants and animals, which means they are naturally found only in a particular location. Most endemic island species have closely related relatives on nearby islands or the mainland. As an example, let's consider the island fox (*Urocyon littoralis*), which lives on the Channel Islands located off the coast near Santa Barbara in southern California (**Figure 19.5**). This type of fox is found nowhere else in the world. It weighs about 3–6 pounds and feeds largely on insects, mice, and fruits. The island fox evolved from the mainland gray fox (*Urocyon cinereoargenteus*), which is much larger, usually 7–11 pounds. During the last Ice Age, about 16,000–18,000 years ago, the Santa Barbara channel was frozen and narrow enough for ancestors of the mainland gray fox to cross over to the Channel Islands. When the Ice Age ended, the ice melted and sea levels rose, causing the foxes to be cut off from the mainland. Over the last 16,000–18,000 years, the population of foxes on the Channel Islands evolved into the smaller island fox, which is now considered a different species from the larger gray fox, which is still found on the mainland.

The smaller size of the island fox is an example of island dwarfing, a phenomenon in which the size of large animals on an isolated island shrinks dramatically over many generations. It is the result of natural selection in which a smaller size provides a survival and reproductive advantage, probably because of limited food and other resources.

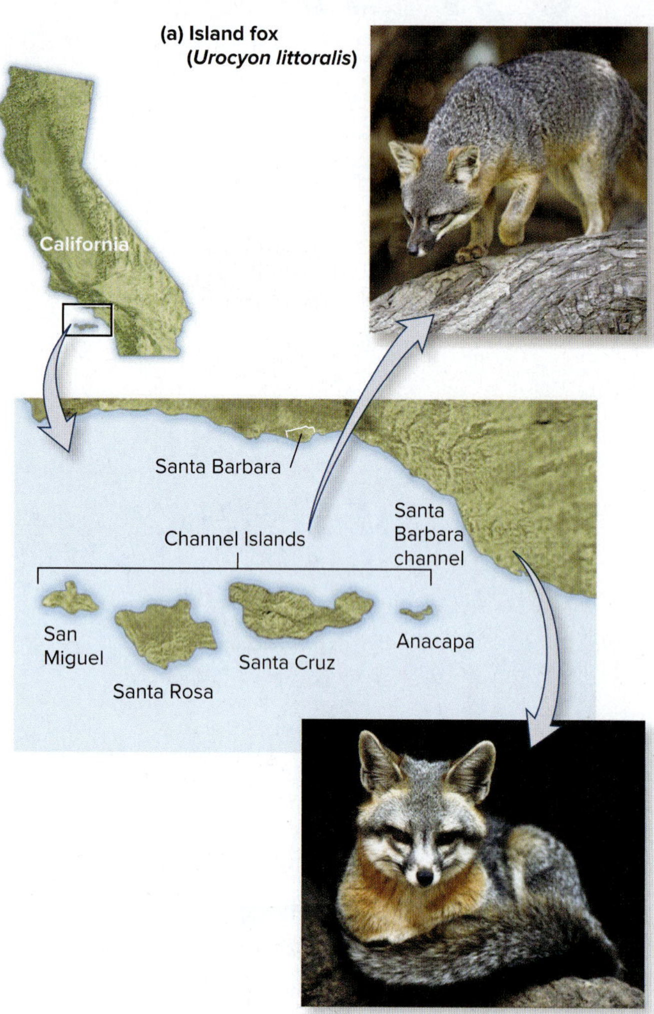


**Figure 19.5** **The evolution of an endemic island species from a mainland species.** **(a)** The smaller island fox (*Urocyon littoralis*) found on the Channel Islands is hypothesized to have evolved from **(b)** the gray fox (*Urocyon cinereoargenteus*) found on the California mainland.
*(a)* © Kevin Schafer/Getty Images; *(b)* © Prisma Bildagentur AG/Alamy

**Concept Check:** *Explain how geography played a key role in the evolution of the island fox.*

## Convergent Evolution Suggests Adaptation to the Environment

The process of natural selection is also evident in the study of plants and animals that have similar characteristics, even though they are not closely related evolutionarily. This similarity is the result of **convergent evolution,** in which two species from different lineages have independently evolved similar characteristics because they occupy similar environments. For example, both the giant anteater (*Myrmecophaga tridactyla*), found in South America, and the echidna (*Tachyglossus aculeatus*), found in Australia, have a long snout and tongue. Both species independently evolved these adaptations that enable them to feed on ants (**Figure 19.6a**). The giant anteater is a placental mammal, whereas the echidna is an egg-laying mammal known as a monotreme, so they are not closely related evolutionarily.

Another example of convergent evolution involves aerial rootlets found in vines such as English ivy (*Hedera helix*) and wintercreeper (*Euonymus fortunei*) (**Figure 19.6b**). Based on differences in their structures, these aerial rootlets appear to have developed independently as an effective means of clinging to the support on which a vine attaches itself.

The similar characteristics shown in Figure 19.6—for example, the snouts of the anteater and the echidna—are called **analogous structures** or convergent traits. They represent cases in which characteristics have arisen independently, two or more times, because different species have occupied similar types of environments on the Earth.

## Selective Breeding Is a Human-Driven Form of Selection

The term **selective breeding** refers to programs and procedures designed to modify traits in domesticated species. This practice, also called artificial selection, is related to natural selection. The primary difference between natural and artificial selection is how the parents are chosen. Natural selection occurs because individuals that are able to survive and reproduce are more likely to pass their genes to future generations. Environmental factors often determine which individuals will be successful parents. In artificial selection, the breeder chooses as parents those individuals with traits that are desirable from a human perspective.

(a) The long snouts and tongues of the giant anteater (left) and the echidna (right) allow them to feed on ants.

(b) The aerial rootlets of English ivy (left) and wintercreeper (right) enable them to climb up supports.

**Figure 19.6 Examples of convergent evolution.** Both pairs of species shown in this figure are not closely related evolutionarily but occupy similar environments, suggesting that natural selection results in similar adaptations to a particular environment.

*(a left)* © Peter Schoen/Getty Images RF; *(a right)* © Tom McHugh/Science Source; *(b left)* © mm88/Getty Images RF; *(b right)* © 2003 Steve Baskauf/bioimages .vanderbilt.edu

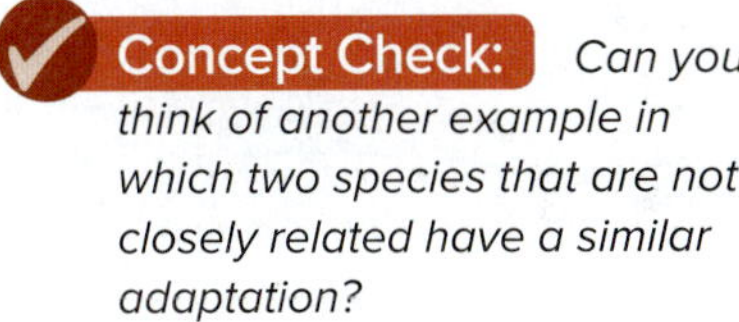


*Can you think of another example in which two species that are not closely related have a similar adaptation?*

(a) Bulldog

(b) Greyhound

(c) Dachshund

**Figure 19.7 Common breeds of dogs that have been obtained by selective breeding.** By selecting individuals carrying the alleles that influence traits desirable to humans, dog breeders have produced breeds with distinctive features. The dogs in this figure carry the same kinds of genes (for example, genes that affect their size, shape, and fur color). However, the alleles for many of these genes are different among these dogs, thereby allowing breeders to select for or against them and produce breeds with strikingly different phenotypes.

*(a)* © WilleeCole/Getty Images RF; *(b)* © Martin Rugner/agefotostock; *(c)* © Juniors Bildarchiv/agefotostock

The underlying phenomenon that makes selective breeding possible is genetic variation. Within a population, variation may exist in a trait of interest. The underlying cause of the phenotypic variation is usually related to differences in **alleles,** variant forms of a particular gene, that determine the trait. The breeder chooses parents with desirable phenotypic characteristics. For centuries, humans have employed selective breeding to obtain domesticated species with interesting or agriculturally useful characteristics. The various breeds of dog are the result of selective breeding strategies (**Figure 19.7**).

All dogs are members of the same species, *Canis lupus*, subspecies *familiaris*, so they can interbreed to produce offspring. Selective breeding can dramatically modify the traits in a species. When you compare certain breeds of dogs (for example, a greyhound and a dachshund), they hardly look like members of the same species! Recent work in 2007 by American geneticist Nathan Sutter and colleagues indicates that the size of dogs may be determined by alleles in the *Igf1* gene that encodes a growth hormone called insulin-like growth factor 1. A particular allele of this gene was found to be common to all small breeds of dogs and nearly absent from very large breeds, suggesting that this allele is a major contributor to body size in small breeds of dogs.

## A Comparison of Homologies Shows Evolution of Related Species from a Common Ancestor

Let's now consider other widespread observations of the process of evolution among different species. In biology, the term **homology** refers to a similarity that occurs due to descent from a common ancestor. Two species may have a similar trait because the trait was originally found in a common ancestor. As described next, such homologies may involve anatomical, developmental, or molecular features.

**Anatomical Homologies** An examination of the limbs of modern vertebrate species reveals similarities that indicate the same set of bones has undergone evolutionary changes, becoming modified to perform different functions in different species. As seen in Figure 19.8, the forelimbs of vertebrates have a strikingly similar pattern of bone arrangements. These are termed **homologous structures**—structures that are similar to each other because they evolved from a common ancestor. The forearm has developed different functions among various vertebrates, including grasping, walking, flying, swimming, and climbing. The process of evolution explains how these animals have descended from a common ancestor and how natural selection has resulted in modifications to the structure of the original set of bones in ways that ultimately allowed them to be used for several different functions.

Another result of evolution is the phenomenon of **vestigial structures,** anatomical features that have no current function but resemble structures of their presumed ancestors (Table 19.3). An interesting case is found in humans. People have a complete set of muscles for moving their ears, even though most people are unable to do so. By comparison, many modern mammals can move their ears to determine the direction of sounds, and presumably this was an important trait in a distant human ancestor. Why would organisms have structures that are no longer useful? Within the context of evolutionary theory, vestigial structures are evolutionary relics. Natural selection maintains functional structures in a population of individuals. However, if a species changes its lifestyle, so the structure loses its purpose, the selection that would normally keep the structure in a functional condition is no longer present. When this occurs, the structure may degenerate over the course of many generations due to the accumulation of mutations that limit its size and shape. Natural selection may eventually eliminate such traits due to the inefficiency and cost of producing unused structures.

**Developmental Homologies** Species that differ substantially at the adult stage often bear striking similarities during early stages of embryonic development. These temporary similarities are called developmental homologies. In addition, evolutionary history is revealed during development in certain organisms, such as vertebrates. For example, if we consider human development, several features are seen in the embryo that are not present at birth. Human embryos have rudimentary gill ridges, like a fish embryo, even though human embryos receive oxygen via the umbilical cord. The presence

## Biology Principle

### Structure Determines Function

These homologous sets of bones have evolved into somewhat different structures, which result in differences in their functions in humans, turtles, bats, and whales.

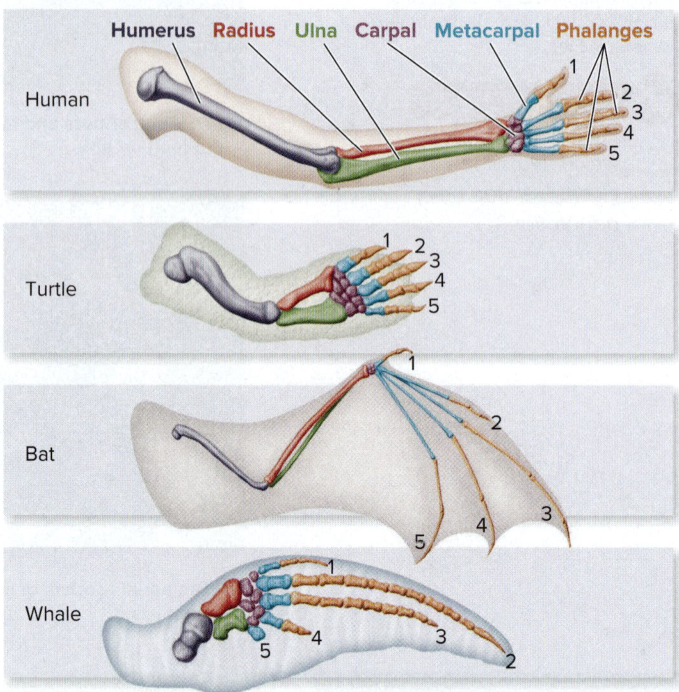


**Figure 19.8 An example of anatomical homology: Homologous structures found in vertebrates.** The same set of bones is found in the human arm, turtle arm, bat wing, and whale flipper, although their relative sizes and shapes differ significantly. This homology suggests that all of these animals evolved from a common ancestor.

**Table 19.3 Examples of Vestigial Structures in Animals**

| Organism | Vestigial structure(s) |
|---|---|
| Human | Bony tail in embryo and muscles to wiggle ears in adult |
| Boa constrictor | Skeletal remnants of hip and hind leg bones |
| Whale | Skeletal remnants of a pelvis |
| Manatee | Fingernails on the flippers |

| | Short amino acid sequence within the p53 protein | Percentages of amino acids in the whole p53 protein that are identical to human p53 |
|---|---|---|
| Human (*Homo sapiens*) | Val Pro Ser Gln Lys Thr Tyr Gln Gly Ser Tyr Gly Phe Arg Leu Gly Phe Leu His Ser Gly Thr | 100 |
| Rhesus monkey (*Macaca mulatta*) | Val Pro Ser Gln Lys Thr Tyr His Gly Ser Tyr Gly Phe Arg Leu Gly Phe Leu His Ser Gly Thr | 95 |
| Green monkey (*Cercopithecus aethiops*) | Val Pro Ser Gln Lys Thr Tyr His Gly Ser Tyr Gly Phe Arg Leu Gly Phe Leu His Ser Gly Thr | 95 |
| Rabbit (*Oryctolagus cuniculus*) | Val Pro Ser Gln Lys Thr Tyr His Gly Asn Tyr Gly Phe Arg Leu Gly Phe Leu His Ser Gly Thr | 86 |
| Dog (*Canis lupus familiaris*) | Val Pro Ser Pro Lys Thr Tyr Pro Gly Thr Tyr Gly Phe Arg Leu Gly Phe Leu His Ser Gly Thr | 80 |
| Chicken (*Gallus gallus*) | Val Pro Ser Thr Glu Asp Tyr Gly Gly Asp Phe Asp Phe Arg Val Gly Phe Val Glu Ala Gly Thr | 53 |
| Channel catfish (*Ictalurus punctatus*) | Val Pro Val Thr Ser Asp Tyr Pro Gly Leu Leu Asn Phe Thr Leu His Phe Gln Glu Ser Ser Gly | 48 |
| European flounder (*Platichthys flesus*) | Val Pro Val Val Thr Asp Tyr Pro Gly Glu Tyr Gly Phe Gln Leu Arg Phe Gln Lys Ser Gly Thr | 46 |
| Congo puffer fish (*Tetraodon miurus*) | Val Pro Val Thr Thr Asp Tyr Pro Gly Glu Tyr Gly Phe Lys Leu Arg Phe Gln Lys Ser Gly Thr | 41 |

**Figure 19.9 An example of genetic homology: A comparison of a short amino acid sequence within the p53 protein from nine different animals.** This figure compares a short region of the p53 protein, a checkpoint protein that plays a role in preventing cancer. Amino acids are represented by three-letter abbreviations. The orange-colored amino acids in the sequences are identical to those in the human sequence. The numbers in the right column indicate the percentage of amino acids within the whole p53 protein that is identical with the human p53 protein, which is 393 amino acids in length. For example, 95% of the amino acids, or 373 of 393, are identical between the p53 sequence found in humans and the one in rhesus monkeys.

**Concept Check:** *In the sequence shown in this figure, how many amino acid differences occur between the following pairs: rhesus and green monkeys, Congo puffer fish and European flounder, and rhesus monkey and Congo puffer fish? What do these differences tell you about the evolutionary relationships among these four species?*

of gill ridges indicates that humans evolved from an aquatic species that had gill slits. A second observation is that human embryos have a bony tail. It is difficult to see the advantage of such a structure in utero, but easier to understand its presence assuming that an ancestor of the human lineage possessed a tail. These observations, and many others, illustrate that closely related species share similar developmental pathways.

**Molecular Homologies** A similarity between organisms at the molecular level due to descent from a common ancestor is called a **molecular homology.** Compelling molecular evidence indicating that modern life-forms are derived from an interrelated group of common ancestors is revealed by analyzing genetic sequences. The same type of gene is often found in diverse organisms. Furthermore, the degree of similarity between genetic sequences from different species reflects the evolutionary relatedness of those species. As an example, let's consider the *p53* gene, which encodes the p53 protein—a checkpoint protein of the cell cycle (see Chapter 14, Figure 14.3). Figure 19.9 shows a short amino acid sequence that makes up part of the p53 protein from a variety of species, including five mammals, one bird, and three fish. The top sequence is the human p53 sequence, and the right column gives the percentages of amino acids within each animal's entire sequence that are identical to those in the entire human sequence. Amino acids in the other species that are identical to those in humans are highlighted in orange. The sequences from the two monkeys are the most similar to those in humans, followed by the other two mammalian species (rabbit and dog). The three fish sequences are the least similar to the human sequence, but the fish sequences tend to be similar to each other.

Taken together, the data shown in Figure 19.9 illustrate two critical points about evolution at the gene level. First, specific genes are found in a diverse array of species such as mammals, birds, and fishes. Second, the sequences of closely related species tend to be more similar to each other than they are to distantly related species.

## 19.2 Reviewing the Concepts

- Evidence of evolutionary change is found in studies of the fossil record, biogeography, convergent evolution, selective breeding, and homologies (Table 19.2).
- Fossils provide evidence of evolutionary change in a series of related organisms. The fossil record often reveals transitional forms that link past ancestors to modern species (Figures 19.3, 19.4).
- Biogeography provides information on the geographic distribution of related species. When populations become isolated on islands or continents, they often evolve into new species (Figure 19.5).
- In convergent evolution, independent adaptations result in similar characteristics, or analogous structures, because different species occupy similar environments (Figure 19.6).
- Selective breeding, the selecting and breeding of individual organisms having desired traits, is a human-driven form of selection (Figure 19.7).

- Homologies are similarities that occur due to descent from a common ancestor. Homologous structures, such as the set of bones in the forearms of vertebrates, and vestigial structures, structures that were functional in an ancestor but no longer have a useful function in modern species, are evidence of evolutionary change. Homologies can also be seen during embryonic development and at the molecular level (Figures 19.8, 19.9; Table 19.3).

## 19.2 Testing Your Knowledge

1. Which of the following provides evidence of evolutionary change?
   a. an analysis of fossils of the whale lineage
   b. the identification of transitional species in the fossil record
   c. the identification of island species that are closely related to mainland species
   d. the similarity of homologous genes
   e. all of the above
2. Homologous traits show similarities because the species exhibiting those traits
   a. occupy similar environments.
   b. evolved from a common ancestor.
   c. evolved in a similar environment.
   d. are geographically close to each other.

## 19.3 Genes in Populations

### Learning Outcomes

1. Define gene pool.
2. Distinguish between allele and genotype frequency.
3. **SCISKILLS ▶** Use the Hardy-Weinberg equation to calculate allele and genotype frequencies of a given population.
4. List the conditions that must be met for a population to be in Hardy-Weinberg equilibrium.
5. Describe the factors that cause microevolution to happen.

Population genetics is an extension of our understanding of Darwin's theory of natural selection, Mendel's laws of inheritance, and newer studies in molecular genetics. All of the alleles for every gene in a given population make up the **gene pool.** Each member of the population receives its genes from its parents, which, in turn, are members of the gene pool. Individuals that reproduce contribute to the gene pool of the next generation. Population geneticists study the genetic variation within a gene pool and how such variation changes from one generation to the next. The emphasis is often on understanding the variation in alleles among members of a population. In this section, we will examine some of the general features of populations and gene pools.

### Populations Are Dynamic Units

A population is a group of individuals of the same species that occupy the same environment at the same time and (for sexually reproducing organisms) can interbreed with one another. Certain species occupy a wide geographic range and are divided into discrete populations due to geographic isolation. For example, distinct populations of a given species may be located on different sides of a physical barrier, such as a mountain range.

Populations may change in size and geographic location from one generation to the next. As the size and location of a population change, their genetic composition generally changes as well. Some of the genetic changes involve adaptation, in which a population becomes better suited to its environment, consisting of individuals that are more likely to survive and successfully reproduce. For example, a population of mammals may move from a warmer to a colder geographic location. Natural selection may favor those individuals in the population that have thicker fur, which over time will produce a population of individuals that are better insulated against the colder temperatures.

## EVOLUTIONARY CONNECTIONS

### Genes Are Usually Polymorphic

The term **polymorphism** (from the Greek, meaning many forms) refers to the presence of two or more variants or traits for a given character within a population. **Figure 19.10** illustrates a striking example of polymorphism in the elder-flowered orchid (*Dactylorhiza sambucina*). Throughout the range of this species in Europe, both yellow- and red-flowered individuals are prevalent.

Polymorphism in a character is usually due to two or more alleles of a gene that influences the character. Geneticists also use the term polymorphism to describe the variation in the DNA sequence of genes. A gene that commonly exists as two or more alleles in a population is a **polymorphic gene.** To be considered polymorphic, a gene must exist in at least two different forms or alleles, and each allele must occur at a frequency that is greater than 1%. By comparison, a **monomorphic gene** exists predominantly as a single allele in a population. When 99% or more of the alleles of a given gene are identical in a population, the gene is considered to be monomorphic.

**Figure 19.10** An example of polymorphism: The two color variations found in the orchid *Dactylorhiza sambucina*.

What types of molecular changes cause genes to be polymorphic? A polymorphism may involve various types of changes, such as a deletion of a significant region of the gene, a duplication of a region, or a change in a single nucleotide. This last type of variation is called a **single-nucleotide polymorphism (SNP).** SNPs (pronounced "snips") are the smallest type of genetic variation that can occur within a given gene and the most common. For example, the allele that causes the recessive disorder called sickle cell disease involves a single-nucleotide change in the human β-globin gene, which encodes a subunit of the oxygen-carrying protein called hemoglobin. The non-disease-causing allele and sickle cell allele represent a SNP of the β-globin gene:

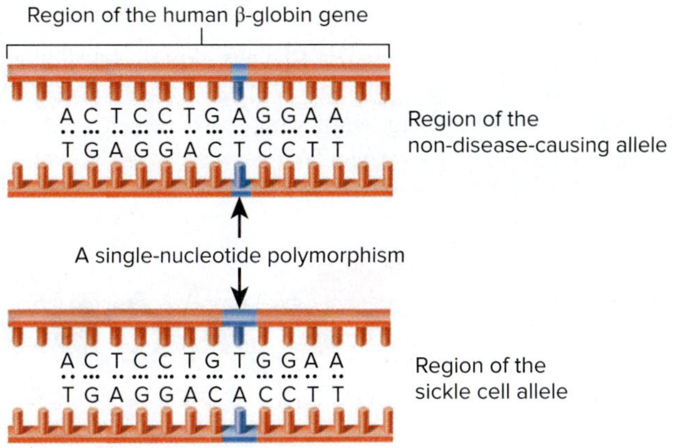

Region of the human β-globin gene

ACTCCTGAGGAA
TGAGGACTCCTT
Region of the non-disease-causing allele

A single-nucleotide polymorphism

ACTCCTGTGGAA
TGAGGACACCTT
Region of the sickle cell allele

Relative to the non-disease-causing allele, this is a single-nucleotide substitution of an A (in the top strand) to a T (in the sickle cell allele).

SNPs represent 99% of all variation in human DNA sequences that occurs among different people. In human populations, a gene that is 2,000–3,000 bp in length, on average, contains 10 different SNPs. Likewise, SNPs with a frequency of 1% or more are found very frequently among genes of nearly all species. Polymorphism is the norm for relatively large, healthy populations of nearly all species, as evidenced by the occurrence of SNPs within most genes.

## Population Genetics Is Concerned with Allele and Genotype Frequencies

One approach to analyzing genetic variation in populations is to consider the frequency of specific alleles and genotypes in a quantitative way. Two fundamental calculations are central to population genetics: **allele frequency** and **genotype frequency.** Allele and genotype frequency are defined as follows:

$$\text{Allele frequency} = \frac{\text{Number of copies of a specific allele in a population}}{\text{Total number of all alleles for that gene in the population}}$$

$$\text{Genotype frequency} = \frac{\text{Number of individuals with a particular genotype in a population}}{\text{Total number of individuals in the population}}$$

Although allele and genotype frequencies are related, make sure you clearly distinguish between them. As an example, let's consider a population of 100 four-o'clock plants (*Mirabilis jalapa*) with the following genotypes:

49 red-flowered plants with the genotype $C^RC^R$

42 pink-flowered plants with the genotype $C^RC^W$

9 white-flowered plants with the genotype $C^WC^W$

When calculating an allele frequency for a diploid species, remember that homozygous individuals have two copies of a given allele, whereas heterozygotes have only one. For example, in tallying the $C^W$ allele, each of the 42 heterozygotes has one copy of the $C^W$ allele, and each white-flowered plant has two copies. Therefore, the allele frequency for $C^W$ (the white-color allele) equals

$$\text{Frequency of } C^W = \frac{(C^RC^W) + 2(C^WC^W)}{2(C^RC^R) + 2(C^RC^W) + 2(C^WC^W)}$$

$$\text{Frequency of } C^W = \frac{42 + (2)(9)}{(2)(49) + (2)(42) + (2)(9)}$$

$$= \frac{60}{200} = 0.3\text{, or } 30\%$$

This result tells us that the allele frequency of $C^W$ is 0.3. In other words, 30% of the alleles for this gene in the population are the white-color ($C^W$) allele.

Let's now calculate the genotype frequency of $C^WC^W$ homozygotes (white-flowered plants).

$$\text{Frequency of } C^WC^W = \frac{9}{49 + 42 + 9}$$

$$= \frac{9}{100} = 0.09\text{, or } 9\%$$

We see that 9% of the individuals in this population have the homozygous white-flower genotype.

## Quantitative Analysis

### THE HARDY-WEINBERG EQUATION RELATES ALLELE AND GENOTYPE FREQUENCIES IN A POPULATION

In 1908, Godfrey Harold Hardy, an English mathematician, and Wilhelm Weinberg, a German physician, independently derived a simple mathematical expression, now called the Hardy-Weinberg equation, that describes the relationship between allele and genotype frequencies when a population is not evolving. Let's examine the Hardy-Weinberg equation using the population of four-o'clock plants that we just considered. If the allele frequency of $C^R$ is denoted by the symbol $p$ and the allele frequency of $C^W$ by $q$, then

$$p + q = 1$$

For example, if $p = 0.7$, then $q$ must be 0.3. In other words, if the allele frequency of $C^R$ equals 70%, the remaining 30% of alleles must be $C^W$, because together they equal 100%.

For a gene that exists in two alleles, the **Hardy-Weinberg equation** states that

$$p^2 + 2pq + q^2 = 1$$

If we apply this equation to our flower color gene, then

$p^2$ = the genotype frequency of $C^RC^R$ homozygotes

$2pq$ = the genotype frequency of $C^RC^W$ heterozygotes

$q^2$ = the genotype frequency of $C^WC^W$ homozygotes

If $p = 0.7$ and $q = 0.3$, then

Frequency of $C^RC^R = p^2 = (0.7)^2 = 0.49$

Frequency of $C^RC^W = 2pq = 2(0.7)(0.3) = 0.42$

Frequency of $C^WC^W = q^2 = (0.3)^2 = 0.09$

In other words, if the allele frequency of $C^R$ is 70% and the allele frequency of $C^W$ is 30%, the expected genotype frequency of $C^RC^R$ is 49%, $C^RC^W$ is 42%, and $C^WC^W$ is 9%.

Figure 19.11 uses a Punnett square to illustrate the relationship between allele frequencies and the way that gametes combine to produce genotypes. To be valid, the Hardy-Weinberg equation carries the assumption that two gametes combine randomly with each other to produce offspring. In a population, the frequency of a gamete carrying a particular allele is equal to the allele frequency in that population. For example, if the allele frequency of $C^R$ equals 0.7, the frequency of a gamete carrying the $C^R$ allele also equals 0.7. The probability of producing a $C^RC^R$ homozygote with red flowers is $0.7 \times 0.7 = 0.49$, or 49%. The probability of inheriting both $C^W$ alleles, which produces white flowers, is $0.3 \times 0.3 = 0.09$, or 9%. Two different gamete combinations produce heterozygotes with pink flowers. An offspring could inherit the $C^R$ allele from the pollen and $C^W$ from the egg, or $C^R$ from the egg and $C^W$ from the pollen. Therefore, the frequency of heterozygotes is $pq + pq$, which equals $2pq$. In our example, this is $2(0.7)(0.3) = 0.42$, or 42%. Note that the frequencies for all three genotypes total 100%.

The Hardy-Weinberg equation predicts that allele and genotype frequencies will remain the same, generation after generation, provided that a population is in equilibrium. To be in equilibrium, evolutionary mechanisms that can change allele and genotype frequencies are not acting on a population. For this to occur, the following conditions must be met:

- No new mutations occur to alter allele frequencies.
- No natural selection occurs; that is, no survival or reproductive advantage exists for any of the genotypes.
- The population is so large that allele frequencies do not change due to chance.
- No migration occurs between different populations, altering the allele frequencies.
- Random mating occurs; that is, the members of the population mate with each other without regard to their genotypes.

Why is Hardy-Weinberg equilibrium a useful concept? An equilibrium is a null hypothesis, which suggests that evolutionary change is not occurring. In reality, however, populations rarely achieve equilibrium, though in large natural populations with little migration and negligible natural selection, Hardy-Weinberg equilibrium may be nearly approximated for certain genes. Sometimes, when researchers experimentally examine allele and genotype frequencies for one or more genes in a given species, they discover that the frequencies are not in Hardy-Weinberg equilibrium. In such cases, they assume that one or more of the conditions are being violated—in other words, mechanisms of evolutionary change are affecting the population. Conservation biologists and wildlife managers may wish to determine why such disequilibrium has occurred because it may affect the future survival of the species.

**Crunching the Numbers:** A disease-causing allele is present at a frequency of 0.01 (that is, 1%) in a population. Assuming Hardy-Weinberg equilibrium, what percentage of individuals are heterozygous carriers?

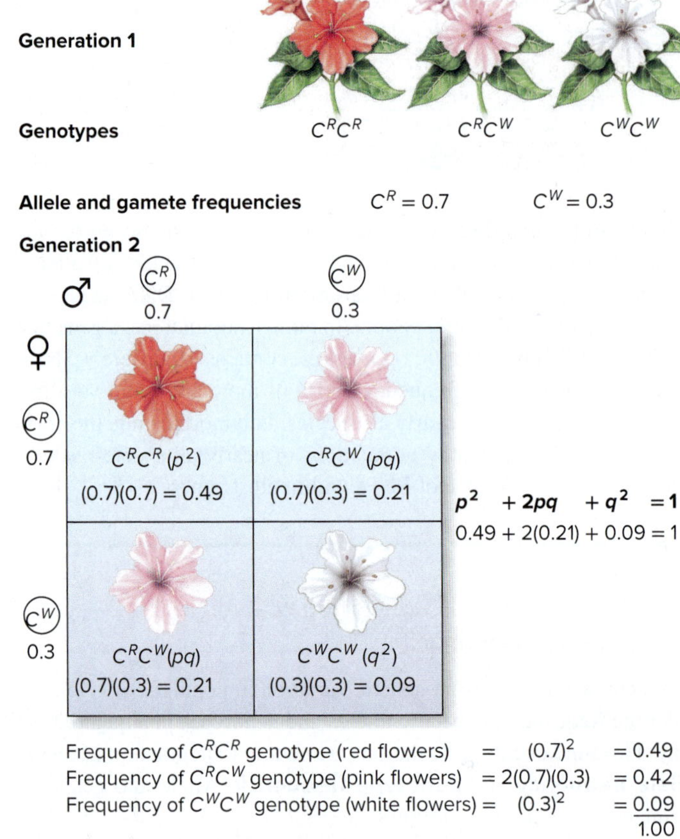


**Figure 19.11 Calculating allele and genotype frequencies with the Hardy-Weinberg equation.** A population of four-o'clock plants has allele and gamete frequencies of 0.7 for the $C^R$ allele and 0.3 for the $C^W$ allele. Knowing the allele frequencies allows us to calculate the genotype frequencies in the population.

**Concept Check:** *What is the frequency of pink flowers in a population in which the allele frequency of $C^R$ is 0.4 and the population is in Hardy-Weinberg equilibrium? Assume that $C^R$ and $C^W$ are the only two alleles.*

## Microevolution Involves Changes in Allele Frequencies from One Generation to the Next

The term **microevolution** is used to describe changes in a population's gene pool, such as changes in allele frequencies, from generation to generation. What causes microevolution to happen? Such change is rooted in two related phenomena (Table 19.4). First, the introduction of new genetic variation into a population is one essential aspect of microevolution. New alleles of pre-existing genes arise by random mutation and, as discussed in Chapters 18 and 21, new genes can be introduced into a population by gene duplication and horizontal gene transfer. Such changes, albeit rare, provide a continuous source of new variation to populations. In 1926, the Russian geneticist Sergei Chetverikov was the first to suggest that random mutations are the raw material for evolution.

**Table 19.4 Factors That Govern Microevolution**

| | |
|---|---|
| **Sources of new genetic variation*** | |
| New mutations within genes that produce new alleles | Random mutations within pre-existing genes introduce new alleles into populations, but at a very low rate. New mutations may be neutral, deleterious, or beneficial. Because mutations are rare, the change from one generation to the next is very small. For alleles to rise to a significant percentage in a population, evolutionary mechanisms, such as natural selection, genetic drift, and migration, must operate on them. |
| Gene duplication | Abnormal crossover events and transposable elements may increase the number of copies of a gene. Over time, the additional copies accumulate random mutations and constitute a gene family. |
| Horizontal gene transfer | Genes from one species may be introduced into another species. The transferred gene may be acted on by evolutionary mechanisms. |
| **Evolutionary mechanisms that alter the frequencies of existing genetic variation** | |
| Natural selection | Natural selection is the process by which individuals that possess certain traits are more likely to survive and reproduce than individuals without those traits. Over the course of many generations, beneficial traits that are heritable become more common and detrimental traits become less common. |
| Genetic drift | Genetic drift is a change in genetic variation from generation to generation due to random chance. Allele frequencies may change as a matter of chance from one generation to the next. Genetic drift has a greater influence in a small population. |
| Migration | Migration can occur between two populations that have different allele frequencies. The introduction of migrants into a recipient population may change the allele frequencies of that population. |
| Nonrandom mating | Nonrandom mating is the phenomenon in which individuals select mates based on their phenotypes or genetic lineage. This alters the relative proportion of homozygotes and heterozygotes that is predicted by the Hardy-Weinberg equation, but it does not change allele frequencies. |

* These are examples that affect single genes. Other events, such as crossing over, independent assortment, and changes in chromosome structure and number, may alter the genetic variation among many genes.

However, due to their low rate of occurrence, mutations by themselves do not have a major effect in changing allele frequencies in a population over time. They do not significantly disrupt Hardy-Weinberg equilibrium.

The second phenomenon required for evolution to occur is one or more mechanisms that alter the prevalence of a given allele or genotype in a population. These mechanisms are natural selection, genetic drift, migration, and nonrandom mating (see Table 19.4). Over the course of many generations, these mechanisms may promote widespread genetic changes in a population. In the remainder of this chapter, we will examine how natural selection, genetic drift, migration, and nonrandom mating affect genetic variation from one generation to the next.

## 19.3 Reviewing the Concepts

- Population genetics is the study of genes and genotypes in a population. A population is a group of individuals of the same species that occupy the same environment and can interbreed. All of the alleles for every gene in a population constitute a gene pool.
- Polymorphism, which is very common in nearly all populations, refers to two or more variants of a character in a population. A monomorphic gene exists as a single allele (>99%) in a population (Figure 19.10).
- Allele frequency is the number of copies of a specific allele divided by the total number of all alleles in a population. Genotype frequency is the number of individuals with a given genotype divided by the total number of individuals in a population.
- The Hardy-Weinberg equation ($p^2 + 2pq + q^2 = 1$) predicts that allele and genotype frequencies will remain in equilibrium if no new mutations are formed, no natural selection occurs, the population size is very large, migration does not occur, and mating is random (Figure 19.11).
- Sources of new genetic variation include random gene mutations, gene duplications, and horizontal gene transfer. Natural selection, genetic drift, migration, and nonrandom mating may alter allele and genotype frequencies and cause a population to evolve (Table 19.4).

## 19.3 Testing Your Knowledge

1. The frequency of individuals in a population that exhibit a rare recessive disorder is 0.0025. The value of 0.0025 is an example of
   a. an allele frequency.
   b. a genotype frequency.
   c. a phenotype frequency.
   d. all of the above.
   e. both b and c.
2. Hardy-Weinberg equilibrium assumes
   a. a very large population with no random fluctuations (no genetic drift).
   b. no natural selection.
   c. random mating.
   d. no new mutations.
   e. all of the above.

## 19.4 Natural Selection

### Learning Outcomes

1. Explain how natural selection results in a population that is better adapted to its environment and more successful at reproduction.
2. **SCISKILLS ▶** Calculate the fitness values of given genotypes.
3. Compare and contrast four different types of natural selection.
4. Define sexual selection and distinguish between intrasexual and intersexual selection.

As discussed earlier in this chapter, natural selection is the process by which individuals with certain heritable traits tend to survive and reproduce at higher rates than those without those traits. As a result, favorable heritable traits become more common, while detrimental heritable traits become less common in a population over many generations. Keep in mind that natural selection itself is not evolution. Rather, it is a key mechanism that causes evolution to happen. Over time, natural selection results in **adaptations**—changes in populations of living organisms that increase their ability to survive and reproduce in a particular environment. In this section, we will examine various ways that natural selection produces such adaptations.

### Natural Selection Favors Individuals with Greater Reproductive Success

**Reproductive success** is the likelihood of an individual contributing fertile offspring to the next generation. Natural selection occurs because some individuals in a population have greater reproductive success than other individuals. Those individuals having heritable traits that favor reproductive success are more likely to pass such traits to their offspring. Reproductive success is commonly attributed to two categories of traits:

1. Certain characteristics make organisms better adapted to their environment and therefore more likely to survive to reproductive age. Therefore, natural selection favors individuals with characteristics that provide a survival advantage.
2. Reproductive success may involve traits that are directly associated with reproduction, such as the abilities to find a mate and produce viable gametes and offspring. Traits that enhance the ability of individuals to reproduce, such as brightly colored plumage in male birds, are often subject to natural selection.

A modern description of the principles of natural selection can relate our knowledge of molecular genetics to the process of evolution:

1. Within a population, allelic variation arises from random mutations that cause differences in DNA sequences. A mutation that creates a new allele may alter the amino acid sequence of the encoded protein. This, in turn, may alter the function of the protein.
2. Some alleles encode proteins that enhance an individual's survival or reproductive capability over other members of the population. For example, an allele may produce a protein that is more efficient at a higher temperature, conferring on the individual a greater probability of survival in a hot climate.
3. Individuals with beneficial alleles are more likely to survive and reproduce and thereby contribute their alleles to the gene pool of the next generation.
4. Over the course of many generations, allele frequencies of many different genes may change through natural selection, thereby significantly altering the characteristics of a population. The net result of natural selection is a population that is better adapted to its environment and more successful at reproduction.

### Fitness Is a Quantitative Measure of Reproductive Success

To continue our discussion of natural selection, we need to consider the concept of **fitness,** which is the relative likelihood that one genotype will contribute to the gene pool of the next generation compared with other genotypes. Although this property often correlates with physical fitness, the two ideas should not be confused. Fitness is a measure of reproductive success. An extremely fertile individual may have a higher fitness than a less fertile individual that appears more physically fit.

To examine fitness, let's consider an example of a hypothetical gene existing in *A* and *a* alleles. We assign fitness values to each of the three possible genotypes according to their relative reproductive success. For example, let's suppose the average reproductive successes of the three genotypes are

*AA* produces 5 offspring

*Aa* produces 4 offspring

*aa* produces 1 offspring

By convention, the genotype with the highest reproductive success is given a fitness value of 1.0. Fitness values are denoted by the variable *w*. The fitness values of the other genotypes are assigned values relative to this 1.0 value.

Fitness of *AA*: $w_{AA} = 1.0$

Fitness of *Aa*: $w_{Aa} = 4/5 = 0.8$

Fitness of *aa*: $w_{aa} = 1/5 = 0.2$

Variation in fitness occurs because certain genotypes result in individuals that have greater reproductive success than other genotypes.

### Natural Selection Follows Different Patterns

By studying species in their native environments, population geneticists have discovered that natural selection can occur in several ways. In most of the examples described next, natural selection leads to adaptations that make certain members of a species more likely to survive to reproductive age.

**Directional Selection** During **directional selection,** individuals at one extreme of a phenotypic range have greater reproductive success in a particular environment. Different phenomena may initiate the process of directional selection. A common reason for directional selection is that a population may be exposed to a prolonged change in its living environment. Under the new environmental conditions, the relative fitness values may change to favor one genotype, which will promote the elimination of other genotypes. As an example, let's suppose a population of finches on a mainland already has genetic variation that affects beak size (see Figure 19.2). A small number of birds migrate to an island where the seeds are generally larger than on the mainland. In this new environment, birds with larger beaks have a higher fitness because they are better able to crack open the larger seeds and thereby survive to reproduce. Over the course of many generations, directional selection produces a population of birds carrying alleles that promote larger beak size.

Another way that directional selection may arise is that a new allele may be introduced into a population by mutation, and the new allele may confer a higher fitness in individuals that carry it (Figure 19.12). What are the long-term effects of directional selection? If the homozygote carrying the favored allele has the highest fitness value, directional selection may cause this favored allele to eventually predominate in the population, perhaps even leading to a monomorphic gene.

**Stabilizing Selection** A type of natural selection called **stabilizing selection** favors the survival of individuals with intermediate phenotypes and selects against those with extreme phenotypes. Stabilizing selection tends to decrease genetic diversity, because it eliminates alleles that cause extreme phenotypes. An example of stabilizing selection involves clutch size (number of eggs laid) in birds, which was first studied by British biologist David Lack in 1947. Under stabilizing selection, birds that lay too many or too few eggs per nest have lower fitness values than do those that lay an intermediate number. When a bird lays too many eggs, many offspring die due to inadequate parental care and food. In addition, the strain on the parents themselves may decrease their likelihood of survival and consequently their ability to produce more offspring. Having too few offspring, however, does not contribute many individuals to the next generation. Therefore, the most successful parents are those that produce an intermediate clutch size. In the 1980s, Swedish evolutionary biologist Lars Gustafsson and his colleagues examined the phenomenon of stabilizing selection in the collared flycatcher (*Ficedula albicollis*) on the Swedish island of Gotland. They discovered that Lack's hypothesis concerning an optimal clutch size appears to be true for this species (Figure 19.13).

**Diversifying Selection** **Diversifying selection** (also known as disruptive selection) favors the survival of two or more different genotypes that produce different phenotypes. In diversifying selection,

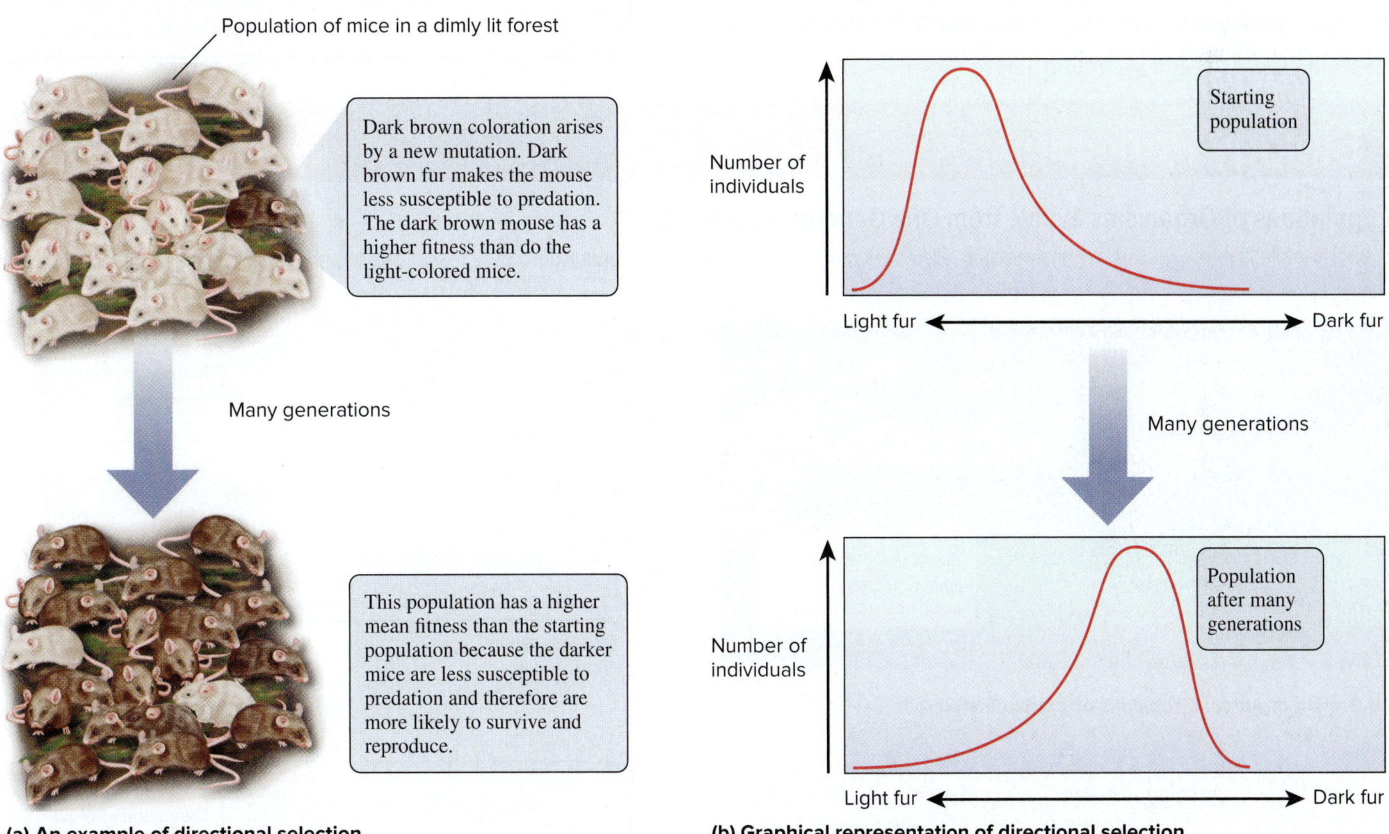

Population of mice in a dimly lit forest
Dark brown coloration arises by a new mutation. Dark brown fur makes the mouse less susceptible to predation. The dark brown mouse has a higher fitness than do the light-colored mice.
Many generations
This population has a higher mean fitness than the starting population because the darker mice are less susceptible to predation and therefore are more likely to survive and reproduce.
(a) An example of directional selection
Number of individuals
Starting population
Light fur
Dark fur
Many generations
Number of individuals
Population after many generations
Light fur
Dark fur
(b) Graphical representation of directional selection


**Figure 19.12** **Directional selection.** This pattern of natural selection selects for one extreme of a phenotype that confers the highest fitness in the population's environment. **(a)** In this example, a mutation causing darker fur arises in a population of mice. This new genotype confers higher fitness, because mice with darker fur can better evade predators and are more likely to survive and reproduce. Over many generations, directional selection favors the prevalence of individuals with darker fur. **(b)** These graphs show the change in fur color phenotypes before and after directional selection.

**Concept Check:** *Let's suppose the climate on an island abruptly changed such that the average temperature was 10°C higher. The climate change is permanent. How would directional selection affect the genetic diversity in a population of mice on the island (1) over the short run and (2) over the long run?*

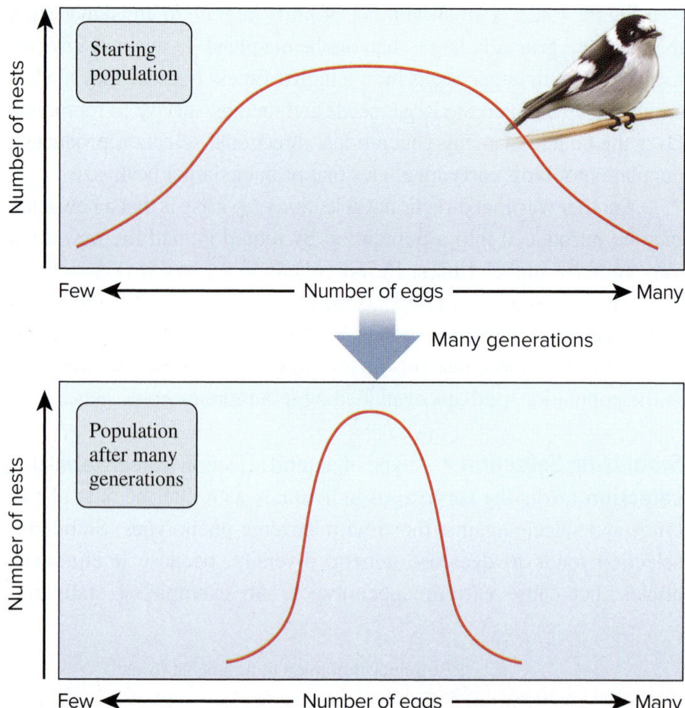


**Figure 19.13 Stabilizing selection.** In this pattern of natural selection, the extremes of a phenotypic distribution are selected against. Those individuals with intermediate traits have the highest fitness. These graphs show the results of stabilizing selection on clutch size in a population of collared flycatchers (*Ficedula albicollis*). This process results in a population with less diversity and more uniform traits.

**Concept Check:** *Why does stabilizing selection decrease genetic diversity?*

the fitness values of a particular genotype are higher in one environment and lower in a different one, whereas the fitness values of the second genotype vary in an opposite manner. Diversifying selection is likely to occur in populations that occupy heterogeneous environments, so some members of the species are more likely to survive in each type of environmental condition.

An example of diversifying selection involves colonial bentgrass (*Agrostis capillaris*) (**Figure 19.14**). In certain locations where this grass is found, such as near abandoned mines in Wales, places occur where the soil is contaminated with high levels of heavy metals due to mining. The relatively recent metal contamination has selected for the proliferation of mutant strains of *A. capillaris* that are tolerant of the

# Biology Principle

## Populations of Organisms Evolve from One Generation to the Next

In this example, the frequencies of metal-resistant alleles become more prevalent when populations of *Agrostis capillaris* are exposed to toxic metals in the soil.

**(a) Growth of *Agrostis capillaris* on contaminated soil**

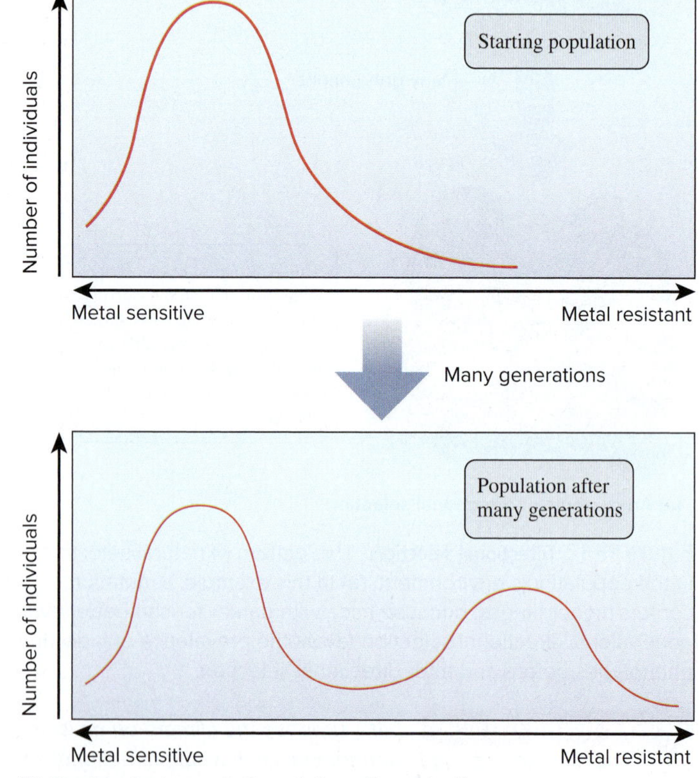


**(b) Graphical representation of disruptive selection**

**Figure 19.14 Diversifying selection.** This pattern of natural selection selects for two different phenotypes, each of which is most fit in a particular environment. **(a)** In this example, random mutations have resulted in metal-resistant alleles in colonial bentgrass (*Agrostis capillaris*) that allow it to grow on contaminated soil. In uncontaminated soils, the grass does not show metal tolerance. The existence of both metal-resistant and metal-sensitive alleles in the population is an example of diversifying selection due to heterogeneous environments. **(b)** Graphs showing the change in phenotypes in this bentgrass population before and after diversifying selection.

*(a)* Courtesy Mark MacNair, University of Exeter

heavy metals (see Figure 19.14a). Such genetic changes enable these mutant strains to grow on contaminated soil but tend to inhibit their growth on normal, noncontaminated soil. These metal-resistant plants often grow on contaminated sites that are close to plants that grow on uncontaminated land and do not show metal tolerance.

**Balancing Selection** Contrary to a popular misconception, natural selection does not always cause the elimination of "weaker," or less-fit, alleles. **Balancing selection** is a type of natural selection that maintains genetic diversity in a population. Over many generations, balancing selection results in a **balanced polymorphism,** in which two or more alleles are kept in balance and therefore are maintained in a population over many generations.

How does balancing selection maintain a polymorphism? Population geneticists have identified two common ways that balancing selection occurs. First, for genetic variation involving a single gene, balancing selection can favor the heterozygote over either corresponding homozygote. This situation is called **heterozygote advantage.** Heterozygote advantage sometimes explains the persistence of alleles that are deleterious in a homozygous condition.

A classic example of heterozygote advantage involves the $H^S$ allele of the human β-globin gene. A homozygous $H^SH^S$ individual has sickle cell disease, which causes the red blood cells to have a deformed, sickle shape. Sickle-shaped cells deliver less oxygen to the body's tissues and can block the flow of blood through the vessels. The $H^SH^S$ homozygote has a lower fitness than a homozygote with two copies of the more common β-globin allele, $H^AH^A$. Heterozygotes, $H^AH^S$, do not typically show symptoms of the disease, but they have an increased resistance to malaria. Compared with $H^AH^A$ homozygotes, heterozygotes have the highest fitness because they have a 10–15% better chance of surviving if infected by the malarial parasite *Plasmodium falciparum*. Therefore, the $H^S$ allele is maintained in populations where malaria is prevalent even though the allele is detrimental in the homozygous state (**Figure 19.15**).

**Negative frequency-dependent selection** is a second way that natural selection produces a balanced polymorphism. In this pattern of natural selection, the fitness of a genotype decreases when its frequency becomes higher. In other words, common individuals have a lower fitness, and rare individuals have a higher fitness. Therefore, common individuals are less likely to reproduce, whereas rare individuals are more likely to reproduce, thereby producing a balanced polymorphism in which no genotype becomes too rare or too common.

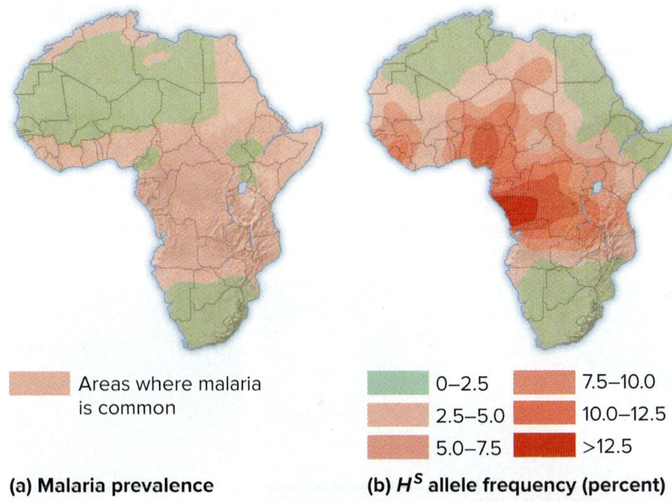


**Figure 19.15 Balancing selection and heterozygote advantage. (a)** The geographic prevalence of malaria in Africa. **(b)** The frequency of the $H^S$ allele of the β-globin gene in the same area. In the homozygous condition, the $H^S$ allele causes sickle cell disease. This allele is maintained in human populations in areas where malaria is prevalent, because the heterozygote ($H^AH^S$) has a higher fitness than either of the corresponding homozygotes ($H^AH^A$ or $H^SH^S$).

**Concept Check:** *If malaria was eradicated, what would you expect to happen to the frequencies of the $H^A$ and $H^S$ alleles over the long run?*

Negative frequency-dependent selection is thought to maintain polymorphisms among species that are preyed upon by predators. Research has shown that certain predators form a mental "search image" for their prey, which is usually based on the common type of prey in an area. A prey that exhibits a rare polymorphism that affects its appearance is less likely to be recognized by the predator. For example, a prey that is a different color from most other members of its species may not be readily recognized by the predator. Such relatively rare organisms are subject to a lower rate of predation. This type of selection maintains polymorphism among certain prey populations.

## FEATURE INVESTIGATION

### The Grants Observed Natural Selection in Galápagos Finches

Let's now turn to an example that shows natural selection occurring as a recent event. Since 1973, British evolutionary biologists Peter Grant, Rosemary Grant, and their colleagues have studied natural selection in finches found on the Galápagos Islands. For over 40 years, the Grants have focused much of their work on one of the Galápagos Islands known as Daphne Major (**Figure 19.16a**). This small island (0.34 km$^2$) has a moderate degree of isolation (it is 8 km from the nearest island), an undisturbed habitat, and a resident population of *Geospiza fortis*, the medium ground finch (**Figure 19.16b**).

To study natural selection, the Grants observed various traits in finches over the course of many years. One trait they observed was beak size. The medium ground finch has a relatively small crushing beak, allowing it to more easily feed on small, tender seeds (see Table 19.1). The Grants quantified beak size among the medium ground finches of Daphne Major by carefully measuring beak

## Biology Principle

### Populations of Organisms Evolve from One Generation to the Next

On this island, the Grants analyzed how beak size may change from one generation to the next.

(a) Daphne Major

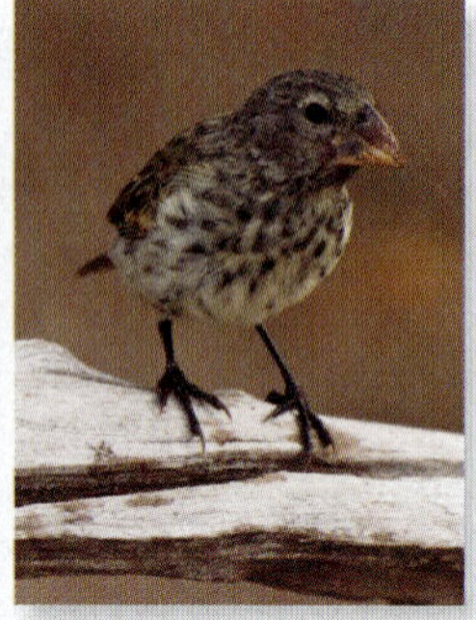

(b) Medium ground finch

**Figure 19.16** **Daphne Major and a resident finch.** **(a)** Daphne Major, one of the Galápagos Islands. **(b)** One of the medium ground finches (*Geospiza fortis*) that populate this island.

*(a)* © Worldwide Picture Library/Alamy; *(b)* © Ralph Lee Hopkins/Getty Images RF

depth—a measurement of the beak from top to bottom (**Figure 19.17**). The small size of the island made it possible for them to measure a large percentage of birds and their offspring. During the course of their studies, they compared the beak depths of parents and offspring by examining many broods over several years and found that the depth of the beak was transmitted from parents to offspring, indicating that differences in beak depths are due to genetic differences in the population. In other words, they found that beak depth is a heritable trait.

By measuring many birds every year, the Grants were able to assemble a detailed portrait of natural selection in action. In the study shown in Figure 19.17, they measured beak depth from 1976 to 1978. In the wet year of 1976, the plants of Daphne Major produced an abundance of the small, tender seeds that these finches could easily eat. However, a severe drought occurred in 1977. During this year, the plants on Daphne Major tended to produce few of the smaller seeds, which the finches rapidly consumed. Therefore, the finches resorted to eating larger, drier seeds, which are harder to crush. As a result, birds with larger beaks were more likely to survive and reproduce because they were better at breaking open the large seeds. As shown in the data, the average beak depth of birds in the population increased substantially, from 8.8 mm in predrought offspring to 9.8 mm in postdrought offspring.

How do we explain these results? Birds with larger beaks were more likely to survive and pass this trait to their offspring. Overall, these results illustrate the power of natural selection to alter the features of a trait—in this case, beak depth—in a given population over time. This is an example of directional selection.

*Experimental Questions*

1. What features of Daphne Major made it a suitable field site for studying the effects of natural selection?
2. **SCISKILLS ▸** Why is beak depth in finches a good trait for a study of natural selection? What environmental conditions were important in allowing the Grants to collect information concerning natural selection?
3. **SCISKILLS ▸** Analyze the results of the Grants' study, and explain what they mean with regard to natural selection and evolution.

**Figure 19.17** **The Grants' investigation of natural selection in the medium ground finch.**

**HYPOTHESIS** Dry conditions produce larger seeds and may result in larger beaks in succeeding generations of *Geospiza fortis* due to natural selection.

**KEY MATERIALS** A population of *G. fortis* on the Galápagos Island called Daphne Major.

| | Experimental level | | Conceptual level |
|---|---|---|---|
| **1** In 1976, measure beak depth in parents and offspring of the species *G. fortis*. | Capture birds and measure beak depth. |  | This is a way to measure a trait that may be subject to natural selection. |
| **2** Repeat the procedure on offspring that were born in 1978 and had reached mature size. A drought had occurred in 1977 that caused plants on the island to produce mostly large dry seeds and relatively few small seeds. | Capture birds and measure beak depth. | 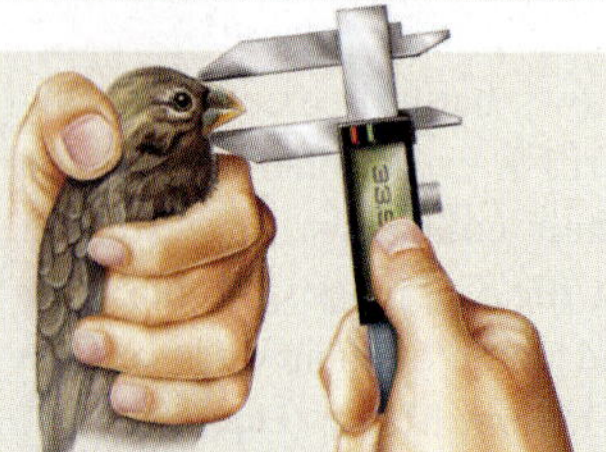 | This is a way to measure a trait that may be subject to natural selection. |

**3 THE DATA**

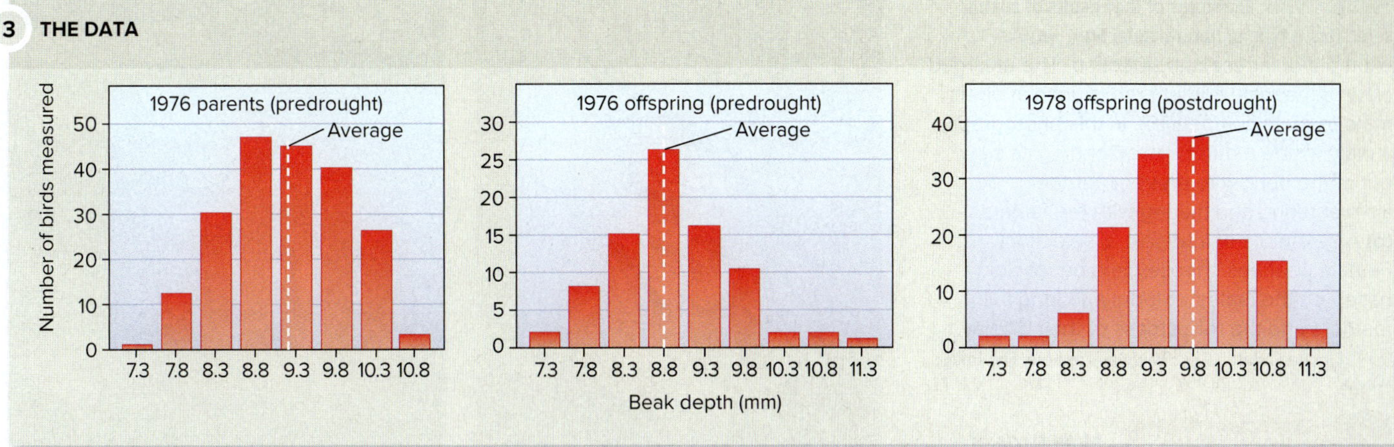


**4 CONCLUSION** Because a drought produced larger seeds, birds with larger beaks were more likely to survive and reproduce. The process of natural selection produced postdrought offspring that had larger beaks compared to predrought offspring.

**5 SOURCE** Grant, B. Rosemary, and Grant, Peter R. 2003. What Darwin's Finches Can Teach Us About the Evolutionary Origin and Regulation of Biodiversity. *Bioscience* 53: 965–975.

## Sexual Selection Is a Type of Natural Selection Pertaining to Traits That Are Directly Involved with Reproduction

Thus far, we have largely focused on examples of natural selection that produce adaptations for survival in particular environments. Now let's turn our attention to a form of natural selection, called **sexual selection,** in which individuals with certain traits are more likely to engage in successful reproduction than other individuals. Darwin originally described sexual selection as "the advantage that certain individuals have over others of the same sex and species solely with respect to reproduction."

In many species of animals, sexual selection affects the characteristics of males more intensely than those of females. Unlike females, which tend to be fairly uniform in their reproductive success, male success tends to be more variable, with some males mating with many females and others not mating at all. Sexual selection results in the prevalence of traits, called secondary sex characteristics, that favor reproductive success. The process can result in **sexual dimorphism**—a significant difference between the appearances of the two sexes within a species.

Sexual selection operates in one of two ways.

- In **intrasexual selection,** members of one sex, usually males, directly compete with each other for the opportunity to mate with individuals of the opposite sex. Examples of traits that result from intrasexual selection in animals include horns in male sheep, antlers in male moose, and the enlarged claw of male fiddler crabs (Figure 19.18a). In fiddler crabs (*Uca paradussumieri*), males enter the burrows of females that are ready to mate. If another male attempts to enter the burrow, the male already inside stands in the burrow shaft and blocks the entrance with his enlarged claw. Males with the largest claws are more likely to be successful at driving off their rivals and being able to mate; therefore, they are more likely to pass on their genes to future generations.
- In **intersexual selection,** also called mate choice, members of one sex, usually females, choose their mates from individuals of the other sex on the basis of certain desirable characteristics. This type of sexual selection often results in showy characteristics in males. Figure 19.18b shows a classic example that involves the Indian peafowl (*Pavo cristatus*). Male peacocks have long and brightly colored tail feathers, which they fan out as a mating behavior. Female peahens select among males based on feather color and pattern, as well as the physical prowess of the display.

Many animals have secondary sexual characteristics, and evolutionary biologists generally agree that sexual selection is responsible for such traits. But why should males compete, and why should females be choosy? Researchers have proposed various hypotheses to explain the underlying mechanisms. One possible reason is related to the different roles that males and females play in the nurturing of offspring. In some animal species, the female is the primary caregiver, whereas the male plays a minor role. In such species, mating behavior may influence the fitness of both males and females. Males increase their fitness by mating with multiple females. This increases their likelihood of passing their genes on to the next generation. By comparison, females may produce relatively fewer offspring, and their reproductive success may not be limited by the number of available males. In these circumstances, females will have higher fitness if they choose males that are good defenders of their territory and have alleles that confer a survival advantage to their offspring.

(a) Intrasexual selection

(b) Intersexual selection

**Figure 19.18 Examples of the results of sexual selection, a type of natural selection. (a)** An example of intrasexual selection. The enlarged claw of the male fiddler crab is used in direct male-to-male competition. In this photograph, a male inside a burrow is extending its claw out of the burrow to prevent another male from entering and mating with the female. **(b)** An example of intersexual selection. Female peahens choose male peacocks based on the males' colorful and long tail feathers and the robustness of their display.

*(a)* © Gerald Cubitt; *(b)* © Topham/TopFoto/The Image Works

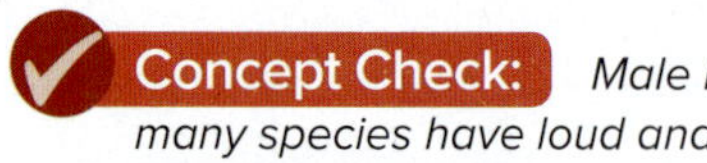

*Male birds of many species have loud and elaborate courtship songs. Is this likely to be the result of intersexual or intrasexual selection? Explain.*

## 19.4 Reviewing the Concepts

- Natural selection is the process by which individuals with certain heritable traits that favor survival and reproduction tend to become more prevalent in a population. Fitness, the relative likelihood that a genotype will contribute to the gene pool of the next generation, is a measure of reproductive success.
- Directional selection is the process by which one extreme of a phenotypic distribution is favored (Figure 19.12).
- Stabilizing selection is the process by which an intermediate phenotype is favored (Figure 19.13).
- Diversifying selection is the process by which two or more phenotypes are favored. An example is a population that occupies a diverse environment (Figure 19.14).
- Balancing selection maintains genetic polymorphism in a population. Examples include heterozygote advantage and negative frequency-dependent selection (Figure 19.15).
- A study by the Grants showed natural selection occurring in a population of finches (Figures 19.16, 19.17).
- Sexual selection is a form of natural selection by which individuals with certain traits are more likely than others to engage in successful mating.
- In intrasexual selection, members of one sex compete with each other for the opportunity to mate with individuals of the opposite sex. In intersexual selection, members of one sex choose their mates on the basis of certain desirable characteristics (Figure 19.18).

## 19.4 Testing Your Knowledge

1. In a population of fish, body coloration varies from a light shade, almost white, to a very dark shade of green. Changes in the environment that result in decreased predation of individuals with the lightest coloration is an example of __________ selection.
   a. diversifying
   b. stabilizing
   c. directional
   d. balancing
2. Considering the same population of fish described in question 1, if the stream environment included several areas of sandy, light-colored bottom areas and a lot of dark-colored vegetation, both the light- and dark-colored fish would have selective advantage and increased survival in certain places. This type of scenario could explain the occurrence of which type of natural selection?
   a. diversifying
   b. stabilizing
   c. directional
   d. balancing
3. The enlarged claw of male fiddler crabs is an example of
   a. intrasexual selection.
   b. intersexual selection.
   c. both intrasexual and intersexual selection.
   d. neither intrasexual or intersexual selection.

## 19.5 Genetic Drift

### Learning Outcomes

1. Define genetic drift, and explain its effects on allele frequencies over time.
2. Compare and contrast the bottleneck and founder effects.
3. Explain how neutral mutations can spread through a population.

Thus far, we have focused on natural selection as a mechanism that promotes widespread genetic changes in a population. Let's now turn our attention to a second important way the gene pool of a population can change. In the 1930s, Sewall Wright played a large role in developing the concept of **genetic drift** (also called random genetic drift), which refers to changes in allele frequencies due to random chance. The term genetic drift is derived from the observation that allele frequencies may "drift" randomly from generation to generation as a matter of chance.

Changes in allele frequencies due to genetic drift happen regardless of the fitness of individuals that carry those alleles. For example, an individual with a high fitness value may, by chance, not encounter a member of the opposite sex. Likewise, random chance can influence which alleles happen to be found in the gametes that fuse with each other in a successful fertilization. In this section, we will examine how genetic drift alters allele frequencies in populations.

## Genetic Drift Has a Greater Effect in Small Populations

What are the effects of genetic drift? Over the long run, genetic drift favors either the elimination or the fixation of an allele—that is, when its frequency reaches 0% or 100% in a population, respectively. However, the number of generations it takes for an allele to be lost or fixed greatly depends on the population size. **Figure 19.19** illustrates the potential consequences of genetic drift in one large ($N = 1{,}000$) and two small ($N = 10$) populations. This simulation involves the frequency of hypothetical *B* and *b* alleles of a gene for fur color in a population of mice—*B* is the black allele, and *b* is the white allele.

At the beginning of this hypothetical simulation, which runs for 50 generations, all three populations had identical allele frequencies: $B = 0.5$ and $b = 0.5$. In the small populations, the allele frequencies fluctuated substantially from generation to generation. Eventually, in one population, the *b* allele was eliminated; in another, it was fixed at 100%. These small populations then consisted of only black mice or white mice, respectively. At this point, the gene became monomorphic, and allele frequencies cannot fluctuate due to genetic drift.

By comparison, the frequencies of *B* and *b* in the large population fluctuated much less. The relative effect of random chance, also termed random sampling error, is much smaller when the sample size is large. Nevertheless, genetic drift can eventually lead to allele elimination or fixation even in large populations, but this will take many more generations to occur than it does in small populations.

In nature, genetic drift may rapidly alter allele frequencies when the size of a population dramatically decreases. Two examples of this phenomenon are the bottleneck effect and the founder effect, which are described next.

**Bottleneck Effect** A population can be dramatically reduced in size by events such as earthquakes, floods, drought, and human destruction of habitat. These occurrences may eliminate most members of the population without regard to their genetic composition. The population is said to have passed through a bottleneck. The change in allele frequencies of the resulting population due to genetic drift is called the **bottleneck effect.** Some alleles may be overrepresented, whereas others may even be eliminated. Such changes may happen for two reasons. First, the surviving members may have allele frequencies that differ from those of the original population that was much larger. Second, as we saw in Figure 19.19, genetic drift acts more quickly to reduce genetic variation when the population size is small. Eventually, a population that has gone through a bottleneck may regain its original size. However, the new population is likely to have less genetic variation than the original one.

A hypothetical example of the bottleneck effect is shown with a population of frogs in **Figure 19.20.** In this example, a starting population of frogs is found in three phenotypes: yellow, dark green, and striped. Due to a bottleneck caused by a drought, the dark green variety is lost from the population.

As a real-life example, the northern elephant seal (*Mirounga angustirostris*) has lost much of its genetic variation. This was caused by a bottleneck effect in which the population decreased

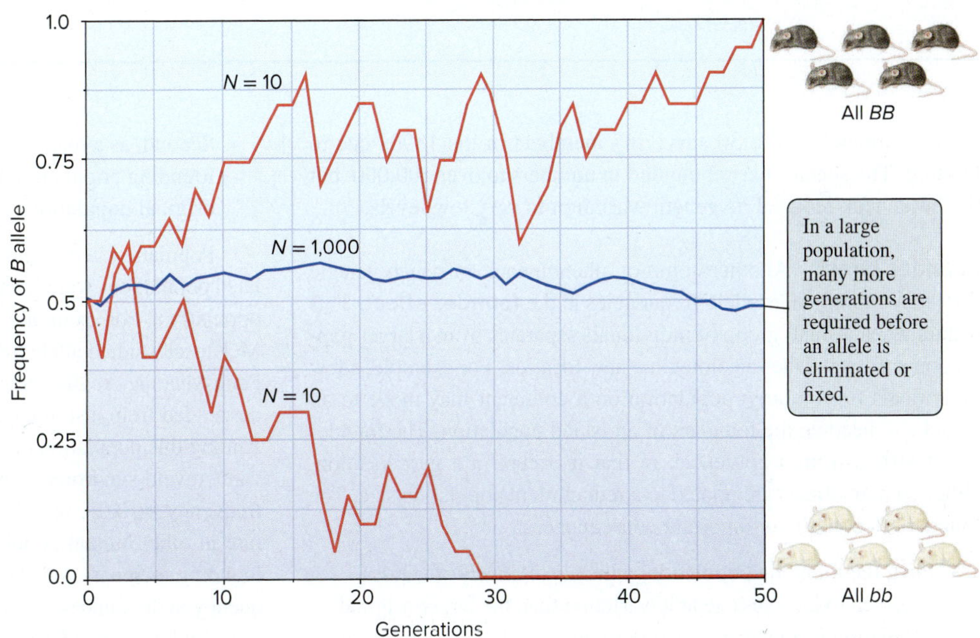


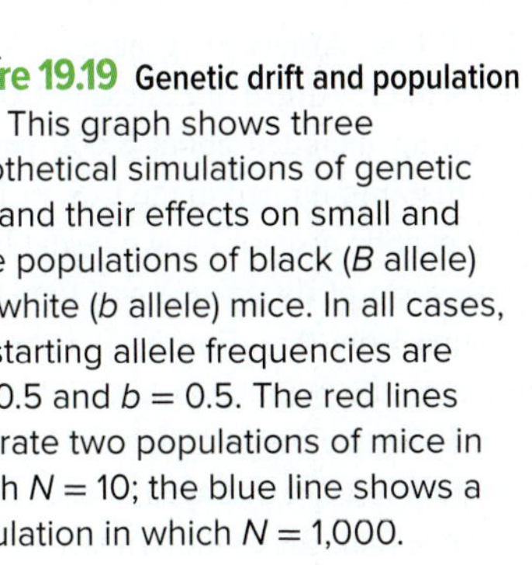

**Figure 19.19 Genetic drift and population size.** This graph shows three hypothetical simulations of genetic drift and their effects on small and large populations of black (*B* allele) and white (*b* allele) mice. In all cases, the starting allele frequencies are $B = 0.5$ and $b = 0.5$. The red lines illustrate two populations of mice in which $N = 10$; the blue line shows a population in which $N = 1{,}000$.

## Biology Principle

### Populations of Organisms Evolve from One Generation to the Next

Genetic drift randomly changes allele frequencies and (in the long run) leads to a loss or fixation of an allele.

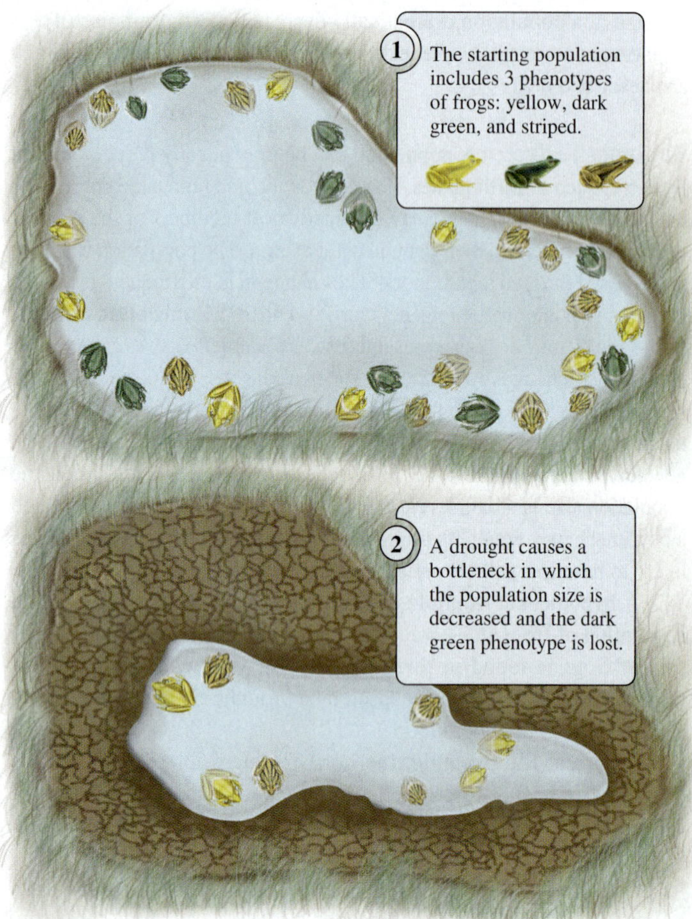

1 The starting population includes 3 phenotypes of frogs: yellow, dark green, and striped.
2 A drought causes a bottleneck in which the population size is decreased and the dark green phenotype is lost.


**Figure 19.20** **A hypothetical example of the bottleneck effect.** This example involves a population of frogs in which a drought dramatically reduced population size, resulting in a bottleneck. The bottleneck reduced the genetic diversity in the population.

**Concept Check:** *How does the bottleneck effect undermine the efforts of conservation biologists who are trying to save species nearing extinction?*

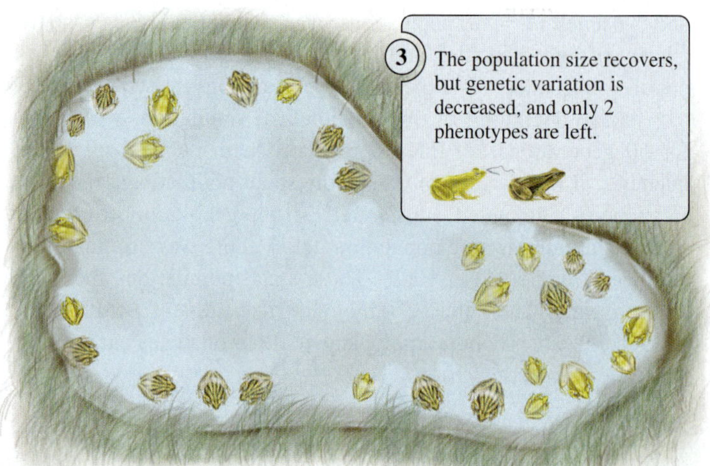


to approximately 20 to 30 surviving members in the 1890s due to hunting. The species has rebounded in numbers to over 100,000, but the bottleneck reduced its genetic variation to very low levels.

**Founder Effect** Another common phenomenon in which genetic drift may rapidly alter allele frequencies is the **founder effect.** This occurs when a small group of individuals separates from a larger population and establishes a colony in a new location. For example, a few individuals from a large population on a continent may move to an island and become the founders of an island population. The founder effect differs from a bottleneck in that it occurs in a new location, although both effects are related to a reduction in population size. The founder effect has two important consequences.

- First, the founding population, which is relatively small, is expected to have less genetic variation than the larger original population from which it was derived.
- Second, as a matter of chance, the allele frequencies in the founding population may differ markedly from those of the original population.

Population geneticists have studied many examples in which isolated populations were founded via colonization by members of another population. For example, in the 1960s, American geneticist Victor McKusick studied allele frequencies in the Amish of Lancaster County, Pennsylvania. At that time, this group included about 8,000 people, descended from just three couples that immigrated to the U.S. in 1770. Among this population of 8,000, a genetic disease known as the Ellis–van Creveld syndrome (a recessive form of dwarfism) was found at a frequency of 0.07, or 7%. By comparison, this disorder is extremely rare in other human populations, even the population from which the founding members had originated. Evidence suggests that the high frequency in the Lancaster County population can be traced to one couple, one of whom carried the mutated gene that causes the syndrome.

## Genetic Drift Plays an Important Role in Promoting Genetic Change

In 1968, Japanese evolutionary biologist Motoo Kimura proposed that much of the DNA sequence variation seen in genes in natural populations is the result of genetic drift rather than natural selection. Genetic drift is a random process that does not preferentially select for any particular allele—it can alter the frequencies of both beneficial and deleterious alleles. Much of the time, genetic drift promotes **neutral variation**—changes in genes and proteins that do not have an effect on reproductive success.

According to Kimura, most variation in DNA sequences is due to the accumulation of neutral mutations that have attained high frequencies in a population via genetic drift. This is called the **neutral theory of evolution.** For example, a new mutation within a gene that changes a glycine codon from GGG to GGC would not affect the amino acid sequence of the encoded protein. Both genotypes may be equal in fitness. However, such new mutations can spread throughout a population due to genetic drift (Figure 19.21). Kimura agreed with Darwin that natural selection is responsible for adaptive changes in a species during evolution. The long snout of an anteater is the result of natural selection. His main idea is that much of the variation in DNA sequences is explained by neutral variation rather than adaptive variation.

The sequencing of genomes from many species is consistent with Kimura's theory. When researchers examine changes of the coding sequence within protein-encoding genes, nucleotide substitutions are more prevalent in the third base of a codon than in the first or second base. Mutations in the third base are often neutral; that is, they do not change the amino acid sequence of the protein (refer back to Table 10.1). In contrast, random mutations at the first or second base are more likely to be harmful than beneficial and tend to be eliminated from a population.

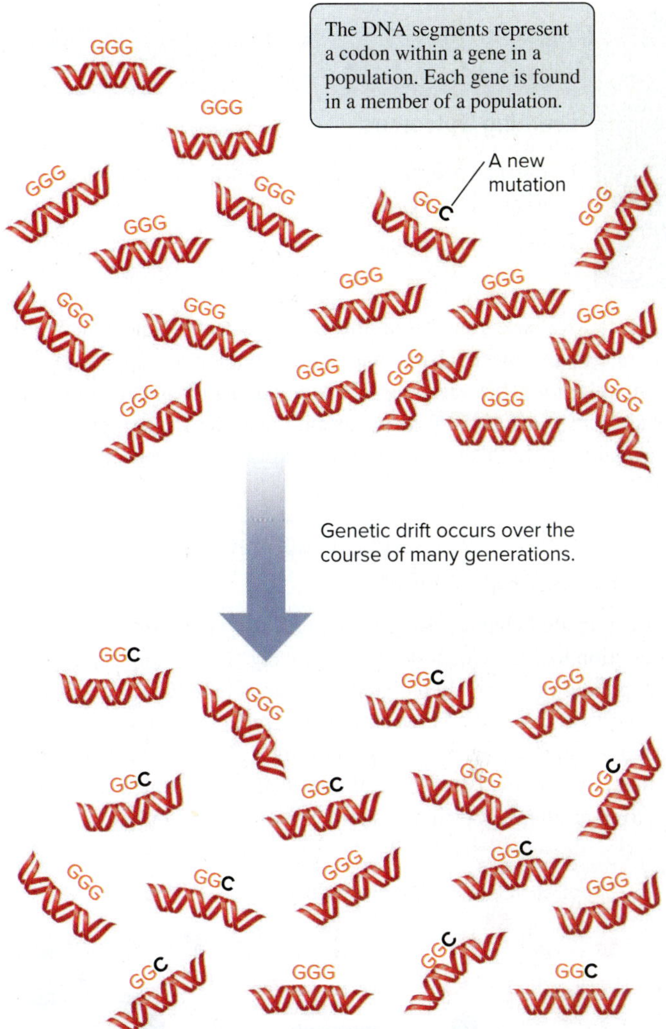

The DNA segments represent a codon within a gene in a population. Each gene is found in a member of a population.
A new mutation
GGG
GGC
Genetic drift occurs over the course of many generations.


**Figure 19.21 Neutral evolution in a population.** In this example, a mutation within a gene changes a glycine codon from GGG to GGC, which does not affect the amino acid sequence of the encoded protein. Each gene shown represents a copy of the gene in a member of a population. Over the course of many generations, genetic drift may cause this neutral allele to become prevalent in the population, perhaps even monomorphic.

**BioConnections:** *Refer back to the genetic code described in Table 10.1. Describe three different genetic changes that you would expect to be neutral.*

## 19.5 Reviewing the Concepts

- Genetic drift involves changes in allele frequencies over time due to chance events. It occurs more rapidly in small populations and leads to either the elimination or the fixation of alleles (Figure 19.19).
- In the bottleneck effect, an environmental event dramatically reduces a population size and the allele frequencies of the resulting population change due to genetic drift (Figure 19.20).
- The founder effect occurs when a small population moves to a new geographic location and genetic drift alters the genetic composition of that population.
- Kimura proposed the neutral theory of evolution, in which genetic drift promotes the accumulation of neutral genetic changes that do not affect reproductive success. Much of the variation in DNA sequences in populations appears to be the result of genetic drift rather than natural selection (Figure 19.21).

## 19.5 Testing Your Knowledge

1. A hurricane kills 98% of a population of cranes. The population eventually rebounds to its original size, but the amount of genetic variation in the rebounded population is significantly less. This is an example of
   a. the founder effect.
   b. the bottleneck effect.
   c. diversifying selection.
   d. both a and b.
2. Kimura's proposal regarding neutral mutations differs from Darwinian evolution in that
   a. natural selection does not exist.
   b. most of the genetic variation in a population is due to neutral mutations, which do not affect reproductive success.
   c. neutral variation alters survival and reproductive success.
   d. neutral mutations are not affected by population size.
   e. both b and c.

## 19.6 Migration and Nonrandom Mating

### Learning Outcomes

1. Describe how gene flow affects genetic variation in neighboring populations.
2. Define inbreeding, and explain how it may have detrimental consequences.

Thus far, we have considered how natural selection and genetic drift operate as key mechanisms that cause evolution to happen. In addition, migration between neighboring populations and nonrandom mating may influence genetic variation and the relative proportions of genotypes. In this section, we will explore how these mechanisms work.

### Migration Between Two Populations Tends to Increase Genetic Variation

Earlier in this chapter, we considered the founder effect, in which migration to a new location by a relatively small group can result in a population with an altered genetic composition due to genetic drift. In addition, migration between two different established populations can alter allele frequencies. As an example, let's consider two populations of a particular species of deer that are separated by a mountain range running north and south (Figure 19.22). On rare occasions, a few deer from the western population may travel through a narrow pass between the mountains and become members of the eastern population. If the two populations are different with regard to genetic variation, this migration will alter the frequencies of certain alleles in the eastern population. Of course, this migration could occur in the opposite direction as well and would then affect the western population. This transfer of alleles into or out of a population, called **gene flow,** occurs whenever individuals move between populations having different allele frequencies.

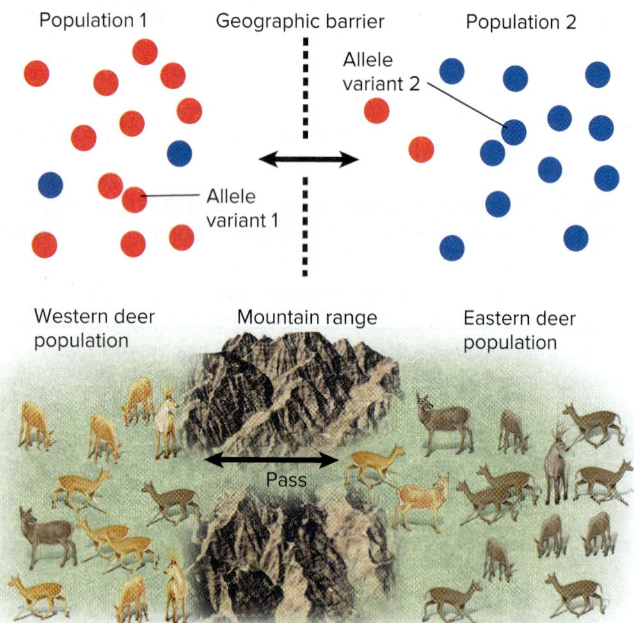


**Figure 19.22 Migration and gene flow.** In this example, two populations of a deer species are separated by a mountain range. On rare occasions, a few deer from one population travel through a narrow pass and become members of the other population. If the two populations differ in regard to genetic variation, this migration will alter the frequencies of alleles in the populations.

**Concept Check:** *How does migration affect the genetic compositions of populations?*

What are the consequences of migration? First, migration tends to reduce differences in allele frequencies between neighboring populations. Population geneticists can evaluate the extent of migration between two populations by analyzing the similarities and differences between their allele frequencies. Populations that frequently mix their gene pools via migration tend to have similar allele frequencies, whereas the allele frequencies of isolated populations tend to differ due to the effects of natural selection and genetic drift. Second, migration generally increases genetic diversity within populations. As discussed earlier in this chapter, new mutations are relatively rare events. Therefore, a new mutation may arise in only one population, and migration may then introduce this new allele into a neighboring population.

### Nonrandom Mating Affects the Relative Proportion of Homozygotes and Heterozygotes in a Population

As mentioned earlier, one of the conditions required to establish Hardy-Weinberg equilibrium is random mating, which means that members of a population choose their mates irrespective of their genotypes or phenotypes. In many species, including human populations, this condition is violated. Such **nonrandom mating** takes different forms. Assortative mating occurs when individuals with similar phenotypes are more likely to mate. If the similar phenotypes are due to similar genotypes, assortative mating tends to increase the proportion of homozygotes and decrease the proportion of heterozygotes in the population. The opposite situation, where dissimilar phenotypes mate preferentially, causes heterozygosity to increase.

Another form of nonrandom mating involves the choice of mates based on their genetic history rather than their phenotypes. Individuals may choose a mate that is part of the same genetic lineage. The mating of two genetically related individuals, such as cousins, is called **inbreeding.** This sometimes occurs in human societies and is more likely to take place in nature when population size becomes very small.

In the absence of other evolutionary factors, nonrandom mating does not affect allele frequencies in a population. However, it will alter the balance of genotypes predicted by the Hardy-Weinberg equation. As an example, let's consider a human pedigree involving a mating between cousins (Figure 19.23). Individuals III-2 and III-3 are cousins and have produced the daughter labeled IV-1. She is said to be inbred, because her parents are genetically related. The parents of an inbred individual have one or more common ancestors. In the pedigree of Figure 19.23, I-2 is the grandfather of both III-2 and III-3.

Inbreeding increases the relative proportions of homozygotes and decreases the likelihood of heterozygotes in a population.

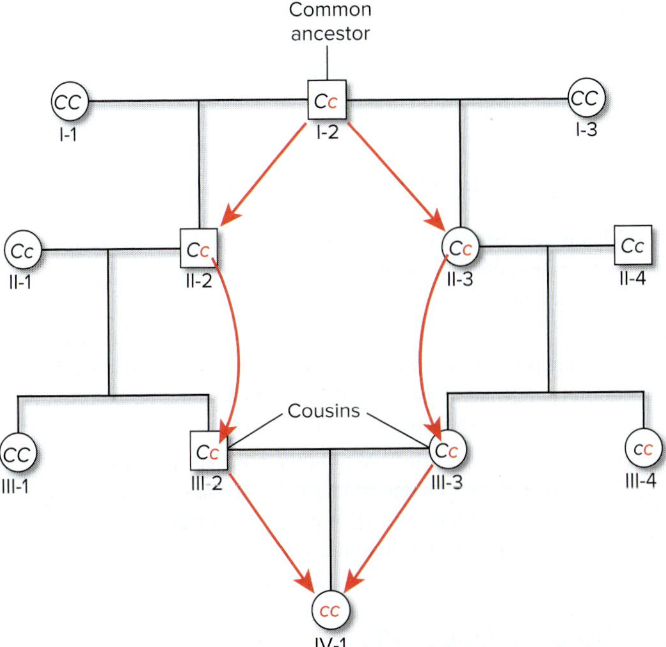


**Figure 19.23 A human pedigree containing inbreeding.** The parents of individual IV-1 are genetically related (cousins), and therefore individual IV-1 is inbred. Inbreeding increases the likelihood that an individual will be homozygous for any given gene. The red arrows show how IV-1 could become homozygous by inheriting the same allele (*c*) from the common ancestor (I-2) to both of her parents.

**BioConnections:** *Many inherited human diseases show a recessive pattern of inheritance (refer back to Table 15.2). Explain whether inbreeding would increase or decrease the likelihood of such diseases.*

Why does this happen? An inbred individual has a higher chance of being homozygous for any given allele because the same allele for that gene could be inherited twice from a common ancestor. For example, individual I-2 is a heterozygote, *Cc*. The *c* allele could pass from I-2 to II-2 to III-2 and finally to IV-1 (see red lines in Figure 19.23). Likewise, the *c* allele could pass from I-2 to II-3 to III-3 and then to IV-1. Therefore, IV-1 has a chance of being homozygous because she inherited both copies of the *c* allele from a common ancestor to both of her parents. Inbreeding does not favor any particular allele—it does not favor *c* over *C*—but it does increase the likelihood that an individual will be homozygous for any given gene.

Although inbreeding by itself does not affect allele frequencies, it may have negative consequences with regard to recessive alleles. Rare recessive alleles that are harmful in the homozygous condition are found in nearly all natural populations. Such alleles do not usually pose a problem because heterozygotes carrying a rare recessive allele are also rare, making it very unlikely that two such heterozygotes will mate with each other. However, related individuals share some of their genes, including recessive alleles. Therefore, if inbreeding occurs, homozygous offspring are more likely to be produced. For example, rare recessive diseases in humans are more frequent when inbreeding occurs.

## 19.6 Reviewing the Concepts

- Gene flow occurs when individuals migrate between populations with different allele frequencies. It reduces differences in allele frequencies between populations and enhances genetic diversity (Figure 19.22).
- Inbreeding, a form of nonrandom mating in which genetically related individuals have offspring with each other, tends to increase the proportion of homozygotes relative to heterozygotes (Figure 19.23).

## 19.6 Testing Your Knowledge

**1.** If the members of one population frequently migrate to a neighboring population and vice versa, which of the following would you expect to be correct?
   **a.** The two populations would tend to have similar allele frequencies.
   **b.** The two populations would tend to be more genetically diverse than if no migration had occurred.
   **c.** The two populations would be more susceptible to genetic drift.
   **d.** All of the above are correct.
   **e.** Only a and b are correct.

**2.** Populations that experience inbreeding may also experience
   **a.** a decrease in fitness due to an increased frequency of recessive genetic diseases.
   **b.** an increase in fitness due to increases in heterozygosity.
   **c.** very little genetic drift.
   **d.** no apparent change.
   **e.** increased mutation rates.

# Assess and Discuss

## Test Yourself

1. A change in one or more heritable characteristics of a population that occurs from one generation to the next is called
   a. natural selection.
   b. sexual selection.
   c. population genetics.
   d. evolution.
   e. inheritance of acquired characteristics.
2. An evolutionary change in which a population of organisms changes its characteristics over many generations in ways that make it better suited to its environment is
   a. natural selection.
   b. an adaptation.
   c. an acquired characteristic.
   d. evolution.
   e. both a and c.
3. Homology occurs because
   a. different species occupy similar environments.
   b. different species evolved from a common ancestor.
   c. new mutations modify existing structures.
   d. populations evolve from one generation to the next.
   e. both a and b.

4. Vestigial structures are anatomical structures that
   a. have more than one function.
   b. were functional in an ancestor but no longer have a useful function.
   c. look similar in different species but have different functions.
   d. have the same function in different species but have very different appearances.
   e. all of the above.
5. Which of the following is an example of a developmental homology seen in human embryonic development and other vertebrate species that are not mammals?
   a. gill ridges
   b. umbilical cord
   c. tail
   d. both a and c
   e. all of the above
6. Population geneticists are interested in the genetic variation in populations. The most common type of genetic change that causes polymorphism in a population is
   a. a deletion of a gene sequence.
   b. a duplication of a region of a gene.
   c. a rearrangement of a gene sequence.
   d. a change in a single nucleotide.
   e. an inversion of a segment of a chromosome.
7. In the Hardy-Weinberg equation, what portion of the equation would be used to calculate the frequency of individuals that do *not* exhibit a recessive disease but are carriers of a recessive allele?
   a. $q$
   b. $p^2$
   c. $2pq$
   d. $q^2$
   e. both b and d
8. By itself, which of the following is *not* likely to have a major influence on allele frequencies?
   a. natural selection
   b. genetic drift
   c. mutation
   d. inbreeding
   e. both c and d
9. Which of the following statements regarding mutations is *true?*
   a. Mutations are not important in evolution.
   b. Mutations provide the source for genetic variation, but other evolutionary factors are more important in determining allele frequencies in a population.
   c. Mutations occur at such a high rate that they promote major changes in the gene pool from one generation to the next.
   d. Mutations are of greater importance in smaller populations than in larger ones.
   e. Mutations are of greater importance in larger populations than in smaller ones.
10. The microevolutionary factor most sensitive to population size is
   a. mutation.
   b. migration.
   c. natural selection.
   d. genetic drift.
   e. all of the above.

## Conceptual Questions

1. What is convergent evolution? How do the pairs of organisms shown in Figure 19.6 support the theory of descent with modification?
2. The percentage of individuals exhibiting a recessive disease in a population is 0.04, which is 4%. Based on Hardy-Weinberg equilibrium, what percentage of individuals would be expected to be heterozygous carriers?
3. **PRINCIPLES** A principle of biology is that populations of organisms evolve from one generation to the next. Explain how the homologous forelimbs of vertebrates indicate that populations evolve from one generation to the next.

## Collaborative Questions

1. The term natural selection is sometimes confused with the term evolution. Discuss the meanings of these two terms. Explain how the processes are different and how they are related to each other.
2. Antibiotics are commonly used to combat bacterial and fungal infections. During the past several decades, however, antibiotic-resistant strains of microorganisms have become alarmingly prevalent. This has undermined the ability of physicians to treat many types of infectious disease. Discuss how the following processes that alter allele frequencies may have contributed to the emergence of antibiotic-resistant strains:
   a. random mutation
   b. genetic drift
   c. natural selection

# Evolution of Life II: The Emergence of New Species

# 20

**Two different species of zebras.** Grevy's zebra (*Equus grevyi*) is shown on the left, and Grant's zebra (*E. quagga boehmi*), which has fewer and thicker stripes, is shown on the right.

## Chapter Outline

The origin of living organisms has been described by philosophers as the great "mystery of mysteries." Perhaps that is why so many different views have been put forth to explain the existence of living species. At the time of Aristotle (4th century B.C.E.), most people believed that some living organisms came into being by spontaneous generation—the idea that nonliving materials could give rise to living organisms. For example, it was commonly believed that worms and frogs could arise from mud, and mice could come from grain. By comparison, many religious teachings contend that species were divinely made and have remained the same since their creation. In contrast to these ideas, Charles Darwin proposed the scientific theory of evolution by descent with modification.

This chapter provides an exciting way to build on the information that we have considered in previous chapters. In Chapter 4, we examined how the first primitive cells may have arisen in an RNA world. In Chapter 19, we surveyed the tenets on which the theory of evolution is built and also explored **microevolution—**evolution on a small scale as it relates to allele frequencies in a population. In this chapter, we will consider evolution on a larger scale. **Macroevolution** refers to evolutionary changes that produce new species and groups of species.

To biologists, the concept of a **species** refers to a group of related organisms that share a distinctive set of attributes in nature. Members of the same species share an evolutionary history, which makes them more genetically similar to each other than they are to members of a different species. You may already have an intuitive sense of this concept. It is obvious that zebras and mice are different species. However, as discussed in the first section of this chapter, the distinction between different, closely related species is often blurred in natural environments. Two closely related species may look very similar, as the chapter-opening photo illustrates.

In this chapter, we will also focus on the mechanisms that promote the formation of new species, a phenomenon called **speciation.** Such macroevolution typically occurs by the accumulation of microevolutionary changes, those that occur in single genes. We will also consider how variations in the genes that control development play a role in the evolution of new species.

## 20.1 Identification of Species

### Learning Outcomes

1. Outline the characteristics that biologists use to distinguish different species.
2. Describe different species concepts.

How many different species are on Earth? The number is astounding. A study done by American biologist E. O. Wilson and colleagues in 1990 estimated the known number of species at approximately 1.4 million. Currently, about 1.8 million species have been identified and catalogued. However, a vast number of species have yet to be classified. This is particularly true among bacteria and archaea, which are difficult to categorize into distinct species. Also, new invertebrate and even vertebrate species are still being found in the far reaches of pristine habitats. Common estimates of the total number of species range from 5 to 50 million!

When studying natural populations, evolutionary biologists are often confronted with situations in which some differences between two populations are apparent, but it is difficult to decide whether the two populations truly represent separate species. When two or more geographically restricted groups of the same species display one or more traits that are somewhat different but not enough to warrant their placement into different species, biologists sometimes classify such groups as **subspecies**. Similarly, many bacterial species are subdivided into **ecotypes**. Each ecotype is a genetically distinct population adapted to its local environment. In this section, we will consider the characteristics that biologists examine when deciding if two groups of organisms constitute different species.

### Each Species Is Established Using Characteristics and Histories That Distinguish It from Other Species

As mentioned, a species is a group of organisms that share a distinctive set of attributes in nature. In the case of sexually reproducing species, members of one species usually cannot successfully interbreed with members of other species. Members of the same species share an evolutionary history that is distinct from other species. Although this may seem like a reasonable way to characterize a given species, biologists would agree that distinguishing between species is a more difficult undertaking. What criteria do we use to distinguish species? How many differences must exist between two populations to classify them as different species? Such questions are often difficult to answer.

The characteristics that a biologist uses to identify a species depend, in large part, on the species in question. For example, the traits used to distinguish insect species are quite different from those used to identify different bacterial species. The relatively high level of horizontal gene transfer among bacteria presents special challenges in the grouping of bacterial species. Among bacteria, it is sometimes very difficult and perhaps arbitrary to divide closely related organisms into separate species.

The most commonly used characteristics for identifying species are morphological traits, the ability to interbreed, molecular features, ecological factors, and evolutionary relationships. A comparison of these concepts will help you appreciate the various approaches that biologists use to identify the bewildering array of species on our planet.

**Morphological Traits** One way to establish that a population constitutes a unique species is based on their physical characteristics. Organisms are classified as the same species if their anatomical traits appear to be very similar. Likewise, microorganisms can be classified according to morphological traits at the cellular level. By comparing many different morphological traits, biologists may be able to decide that certain populations constitute a unique species.

Although an analysis of morphological traits is a common way for biologists to establish that a particular group constitutes a species, this approach has drawbacks. First, researchers may have difficulty deciding how many traits to consider. In addition, quantitative traits, such as size and weight, that vary in a continuous way among members of the same species are not easy to analyze. Another drawback is that the degree of dissimilarity that distinguishes different species may not show a simple relationship. The members of the same species sometimes look very different, and conversely, members of different species sometimes look remarkably similar to each other. For example, Figure 20.1a shows two different frogs of the species *Dendrobates tinctorius*, commonly called the dyeing poison frog. This species exists in many different-colored morphs, which are individuals of the same species that have noticeably dissimilar appearances. In contrast, Figure 20.1b shows two different species of frogs, the northern leopard frog (*Rana pipiens*) and the southern leopard frog (*Rana utricularia*), which look fairly similar.

**Reproductive Isolation** Why would biologists describe two species, such as the northern leopard frog and southern leopard frog, as being different if they are morphologically similar? One reason is that biologists have discovered that the two species of frogs are unable to breed with each other in nature. Therefore, a second way of identifying a species is by its ability to interbreed. In the late 1920s, geneticist Theodosius Dobzhansky proposed that each species is reproductively isolated from other species. Such **reproductive isolation** prevents one species from successfully interbreeding with other species. In 1942, German evolutionary biologist Ernst Mayr expanded on the ideas of Dobzhansky to provide a definition of a species. According to Mayr, a key feature of sexually reproducing species is that, in nature, the members of one species have the potential to interbreed with one another to produce viable, fertile offspring but cannot successfully interbreed with members of other species.

Reproductive isolation has been used to distinguish many plant and animal species, especially those that look alike but do not interbreed. Even so, this criterion suffers from four main problems.

- In nature, it may be difficult to determine if two populations are reproductively isolated, particularly if the populations have nonoverlapping geographic ranges.
- Biologists have noted many cases in which two different species can interbreed in nature yet consistently maintain themselves as separate species. For example, different species of yucca plants, such as *Yucca pallida* and *Yucca constricta*, do interbreed

**(a) Frogs of the same species**

**(b) Frogs of different species**

**Figure 20.1 Difficulties of using morphological traits to identify species.** In some cases, members of the same species appear quite different. In other cases, members of different species look very similar. **(a)** Two frogs of the same species, the dyeing poison frog (*Dendrobates tinctorius*). **(b)** Two different species of frog, the northern leopard frog (*Rana pipiens*, left) and the southern leopard frog (*Rana utricularia,* right).

*(a) (left)* © Mark Smith/Science Source; *(a) (right)* © Pascal Goetgheluck/ardea.com; *(b) (left)* © Ron Erwin/Getty Images; *(b) (right)* © Robin Chittenden/Alamy

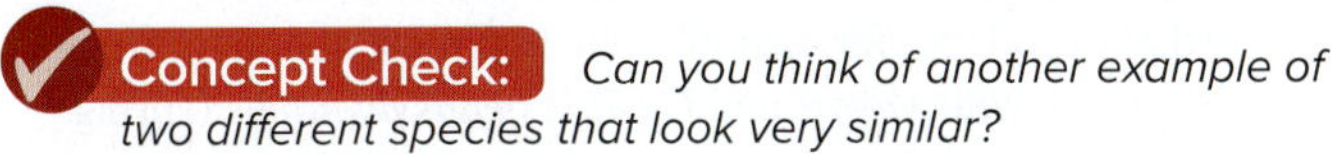

*Can you think of another example of two different species that look very similar?*

in nature yet typically maintain populations with distinct characteristics. For this reason, they are viewed as distinct species.

- Another drawback of reproductive isolation is that it does not apply to asexual species such as bacteria and some species of plants and fungi.
- A fourth drawback is that reproductive isolation cannot be applied to extinct species.

For these reasons, reproductive isolation has been primarily used to distinguish closely related species of modern animals and plants that reproduce sexually.

**Molecular Features** Molecular features are now commonly used to determine if two different populations are different species. Evolutionary biologists often compare DNA sequences within genes, gene order along chromosomes, chromosome structure, and chromosome number in order to identify similarities and differences among different populations. For example, researchers may compare the DNA sequence of the *16S rRNA* gene between different bacterial populations as a way of determining if two populations represent different species. When the sequences are very similar, such populations would probably be judged as the same species. However, it may be difficult to draw the line when separating groups into different species. How much difference must be present for species to be considered separate? Is a 2% difference in their genome sequences sufficient to warrant placement into two different species, or do we need a 5% difference?

**Ecological Factors** A variety of factors related to an organism's habitat may be used to distinguish one species from another. For example, certain species of warblers are distinguished by the habitat in which they forage for food. Some species search the ground for food, others forage in bushes or small trees, and some primarily forage in tall trees. Such habitat differences are used to distinguish different species that look morphologically similar.

Many bacterial species have been categorized as distinct based on ecological factors. Bacterial cells of the same species are likely to use the same types of resources (such as sugars and vitamins) and grow under the same types of conditions (such as temperature and pH). However, a drawback of this approach is that different groups of bacteria sometimes display very similar growth characteristics, and even the same species may show great variation in the growth conditions it will tolerate.

**Evolutionary Relationships** In Chapter 21, we will examine the methods used to produce "evolutionary trees," diagrams that describe the relationships between ancestral species and modern species. In some cases, such relationships are based on an analysis of the fossil record. For example, we will consider how the fossil record was used to construct a tree that shows the ancestors that led to modern horse species. Alternatively, another way of establishing evolutionary relationships is by the analysis of DNA sequences. Researchers obtain samples of cells from different individuals and compare the genes within those cells to see how similar or different they are.

## Species Concepts Emphasize Particular Features to Define and Distinguish Species

A **species concept** is a way of defining the concept of a species and/or of providing an approach to distinguish one species from another. However, even Darwin realized the difficulty in defining a species. In 1859, he said, "No one definition [of species] has as yet satisfied all naturalists; yet every naturalist knows vaguely what he means when he speaks of a species." Since 1942, over 20 different species concepts have been proposed by a variety of evolutionary biologists. Let's consider a few of the more common ones.

- Ernst Mayr proposed one of the first species concepts, called the **biological species concept,** which defines a species as a group of individuals whose members have the potential to interbreed with one another in nature to produce viable, fertile offspring but cannot successfully interbreed with members of other species.
- According to the **evolutionary lineage concept** proposed by American paleontologist George Gaylord Simpson in 1961, species should be defined based on the separate evolution of lineages. A **lineage** is a series of species that forms a line of

descent, with each new species the direct result of speciation from an immediate ancestral species.

- The **ecological species concept,** described by American evolutionary biologist Leigh Van Valen in 1976, defines each species based on an ecological niche, which is the unique set of habitat resources that a species requires, as well as its influence on the environment and other species.
- In 1998, American zoologist Kevin de Queiroz suggested that there is only a single general species concept, which concurs with Simpson's evolutionary lineage concept and includes all previous concepts. According to de Queiroz's **general lineage concept,** each species is a population of an independently evolving lineage. Each species has evolved from a specific series of ancestors and, as a consequence, forms a group of organisms with a particular set of characteristics. Multiple criteria are used to determine if a population is part of an independent evolutionary lineage, and thus a species, which is distinct from others. Because of its generality, the general lineage concept has received significant support.

## 20.1 Reviewing the Concepts

- A species is a group of related organisms that share a distinctive set of attributes in nature. Speciation is the process by which new species are formed. Macroevolution refers to evolutionary changes that produce new species and groups of species.
- Different characteristics, including morphological traits, reproductive isolation, molecular features, ecological factors, and evolutionary relationships, are used to identify species (Figure 20.1).

## 20.1 Testing Your Knowledge

1. According to Mayr's biological species concept, a species is defined by
   a. ecological factors.
   b. morphological factors.
   c. the lineage to which it belongs.
   d. reproductive isolation.
   e. all of the above.
2. According to de Queiroz's general lineage concept, each species
   a. is a population of an independently evolving lineage.
   b. occupies its own niche.
   c. is reproductively isolated.
   d. all of the above.
   e. both a and b.

## 20.2 Reproductive Isolation

### Learning Outcome

1. Compare and contrast prezygotic and postzygotic isolating mechanisms.

Thus far, we have considered various ways of differentiating species. From the discussion, you may have realized that the identification of a species is not always a simple matter. The phenomenon of reproductive isolation has played a major role in the way biologists study plant and animal species, partly because it identifies a possible mechanism for the process of forming new species. For this reason, much research has been done to try to understand **reproductive isolating mechanisms,** the mechanisms that prevent interbreeding between different species. Why do reproductive isolating mechanisms occur? Populations do not intentionally erect these reproductive barriers. Rather, reproductive isolation is a consequence of genetic changes that usually occur because a species becomes adapted to its own particular environment. The view of evolutionary biologists is that reproductive isolation typically evolves as a by-product of genetic divergence. Over time, as a species evolves its own unique characteristics, some of those traits are likely to prevent breeding with other species.

Reproductive isolating mechanisms fall into two categories: **prezygotic isolating mechanisms,** which prevent the formation of a zygote, and **postzygotic isolating mechanisms,** which block the development of a viable and fertile individual after fertilization has taken place. Figure 20.2 summarizes some of the more common ways that reproductive isolating mechanisms prevent reproduction between different species. When two species do produce offspring, such an offspring is called an **interspecies hybrid.**

**Prezygotic Isolating Mechanisms** We will consider five types of prezygotic isolating mechanisms.

1. **Habitat isolation:** One obvious way to prevent interbreeding is for members of different species to never come in contact with each other. This phenomenon, called habitat isolation, may involve a geographic barrier to interbreeding.
2. **Temporal isolation:** In temporal isolation, species happen to reproduce at different times of the day or year. In the northeastern U.S., for example, the two most abundant field crickets, *Gryllus veletis* and *Gryllus pennsylvanicus* (spring and fall field crickets, respectively), do not differ in song or habitat and are morphologically very similar (Figure 20.3). *G. veletis* matures in the spring, whereas *G. pennsylvanicus* matures in the fall, which minimizes interbreeding between the two species.
3. **Behavioral isolation:** In the case of animals, mating behavior and anatomy often play key roles in promoting reproductive isolation. An example is found between the western meadowlark (*Sturnella neglecta*) and eastern meadowlark (*Sturnella magna*) (Figure 20.4). In the zone of overlap, very little interspecies mating takes place, largely due to differences in their songs, which enable meadowlarks to recognize potential mates as members of their own species.
4. **Mechanical isolation:** Morphological features such as size or incompatible genitalia prevent two species from interbreeding. For example, male dragonflies use a pair of special appendages to grasp females during copulation. When a male tries to mate with a female of a different species, his grasping appendages do not fit her body shape.

**Prezygotic isolating mechanisms**

**Habitat isolation:** Species occupy different habitats, so they never come in contact with each other.

**Temporal isolation:** Species have different mating or flowering seasons or times of day or become sexually mature at different times of the year.

**Behavioral isolation:** Sexual attraction between males and females of different animal species is limited due to differences in behavior or physiology.

Attempted mating

**Mechanical isolation:** Morphological features such as size and incompatible genitalia prevent 2 members of different species from interbreeding.

**Gametic isolation:** Gametic transfer takes place, but the gametes fail to unite with each other. This can occur because the male and female gametes fail to attract, because they are unable to fuse, or because the male gametes are inviable in the female reproductive tract of another species. In plants, the pollen of one species usually cannot generate a pollen tube to fertilize the egg cells of another species.

Fertilization

**Postzygotic isolating mechanisms**

**Hybrid inviability:** The egg of one species is fertilized by the sperm from another species, but the fertilized egg fails to develop past the early embryonic stages.

**Hybrid sterility:** An interspecies hybrid survives, but it is sterile. For example, the mule, which is sterile, is produced from a cross between a male donkey (*Equus asinus*) and a female horse (*Equus caballus*).

**Hybrid breakdown:** The $F_1$ interspecies hybrid is viable and fertile, but succeeding generations ($F_2$, and so on) become increasingly inviable. This is usually due to the formation of less-fit genotypes by genetic recombination.

Interspecies hybrid

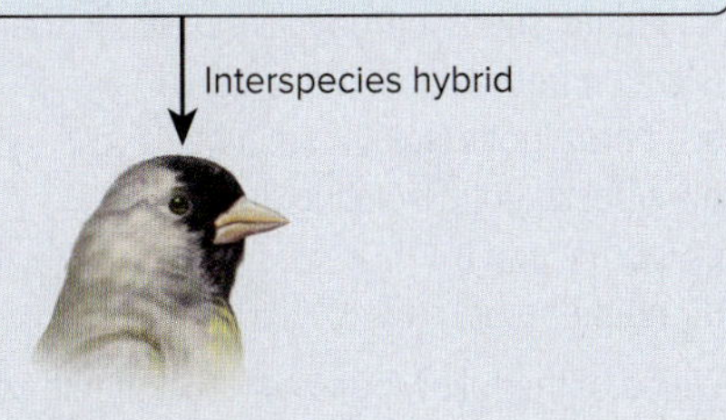

**Figure 20.2 Reproductive isolating mechanisms.** These mechanisms prevent successful breeding between different species. They can occur prior to fertilization (prezygotic) or after fertilization (postzygotic).

**BioConnections:** *Refer back to Figure 19.18b. Is female choice an example of a prezygotic or postzygotic isolating mechanism?*

**(a) Spring field cricket (*Gryllus veletis*)**

**(b) Fall field cricket (*Gryllus pennsylvanicus*)**

**Figure 20.3 Temporal isolation.** Interbreeding between these two species of crickets does not usually occur because *Gryllus veletis* matures in the spring, whereas *Gryllus pennsylvanicus* matures in the fall.

*(a)* © C. Allan Morgan/Getty Images; *(b)* © Bryan E. Reynolds

**Concept Check:** *Is this an example of a prezygotic or a postzygotic isolating mechanism?*

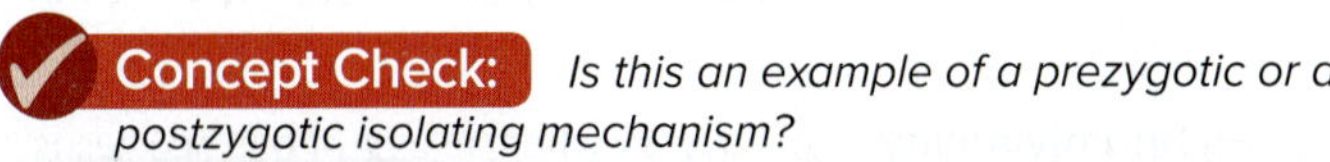

## Biology Principle

### Populations of Organisms Evolve from One Generation to the Next

One of the evolutionary changes that took place in these two species of meadowlarks is that their mating songs became different.

**(a) Western meadowlark (*Sturnella neglecta*)**

**(b) Eastern meadowlark (*Sturnella magna*)**

**Figure 20.4 Behavioral isolation. (a)** The western meadowlark (*Sturnella neglecta*) and **(b)** eastern meadowlark (*Sturnella magna*) are very similar in appearance. The red region in this map shows where the two species' ranges overlap. However, very little interspecies mating takes place due to differences in their songs.

*(a)* © Rod Planck/Science Source; *(b)* © Ron Austing/Science Source

5. **Gametic isolation:** Two species may attempt to interbreed, but the gametes fail to unite in a successful fertilization event. In flowering plants, gametic isolation is commonly associated with pollination. Plant fertilization is initiated when a pollen grain lands on the stigma of a flower and sprouts a pollen tube that ultimately reaches an egg cell. When pollen is released from a plant, it could be transferred to the stigma of many different plant species. In most cases, when a pollen grain lands on the stigma of a different species, either it fails to generate a pollen tube or the tube does not grow properly and cannot reach the egg cell.

**Postzygotic Isolating Mechanisms** Let's now turn to postzygotic mechanisms of reproductive isolation, of which there are three common types.

1. **Hybrid inviability:** This occurs when an egg of one species is fertilized by a sperm from another species, but the fertilized egg cannot develop past early embryonic stages.
2. **Hybrid sterility:** An interspecies hybrid may be viable but sterile. A classic example of hybrid sterility is the mule, which is produced by a mating between a male donkey (*Equus asinus*) and a female horse (*Equus ferus caballus*) (Figure 20.5). All male mules and most female mules are sterile. Why are mules usually sterile? Two reasons explain the sterility. Because the horse has 32 chromosomes per set and a donkey has 31, a mule inherits 63 chromosomes (32 + 31). Due to the uneven number, not all of the chromosomes can pair evenly. Also, the chromosomes of the horse and donkey have structural differences, which either prevent them from pairing correctly or lead to chromosomal abnormalities if crossing over occurs during meiosis. For these reasons, mules usually produce inviable gametes.
3. **Hybrid breakdown:** Interspecies hybrids may be viable and fertile, but the subsequent generation(s) may harbor genetic abnormalities that are detrimental. This mechanism, called hybrid breakdown, can be caused by changes in chromosome structure. The chromosomes of closely related species may have structural differences from each other, such as inversions. In hybrids, a crossover may occur in the region that is inverted in one species but not the other. This will produce gametes with too little or too much genetic material. Such hybrids often have offspring with developmental abnormalities.

Male donkey (*Equus asinus*)

×

Female horse (*Equus ferus caballus*)

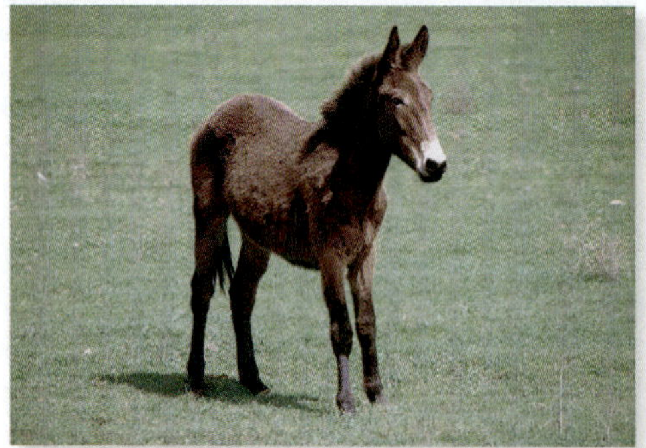

Mule

**Figure 20.5** **Hybrid sterility.** When a male donkey (*Equus asinus*) mates with a female horse (*Equus ferus caballus*), their offspring is a mule, which is usually sterile.

*(top left)* © Mark Boulton/Science Source; *(top right)* © MyLoupe/UIG/Getty Images; *(bottom)* © Stephen L. Saks/Science Source

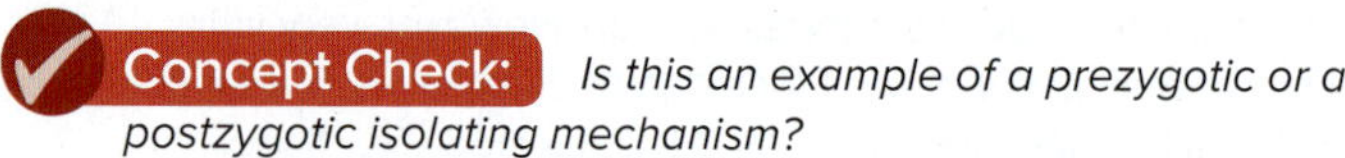

**Concept Check:** *Is this an example of a prezygotic or a postzygotic isolating mechanism?*

Postzygotic isolating mechanisms tend to be uncommon in nature compared with prezygotic mechanisms. Why are postzygotic mechanisms rare? One explanation is they are more costly in terms of energy and resources used. For example, a female mammal would use a large amount of energy to produce an offspring that is sterile. Evolutionary biologists hypothesize that natural selection has favored prezygotic isolating mechanisms because they do not waste a lot of energy.

## 20.2 Reviewing the Concepts

- Reproductive isolating mechanisms prevent two different species from breeding with each other (Figure 20.2).
- Prezygotic isolating mechanisms include habitat isolation, temporal isolation, behavioral isolation, mechanical isolation, and gametic isolation (Figures 20.3, 20.4).
- Postzygotic isolating mechanisms include hybrid inviability, hybrid sterility, and hybrid breakdown (Figure 20.5).

## 20.2 Testing Your Knowledge

1. A mechanism of reproductive isolation that does *not* prevent fertilization from happening is called
   a. a prezygotic isolating mechanism.
   b. a postzygotic isolating mechanism.
   c. both a and b.
   d. neither a or b.
2. Which of the following is *not* an example of a prezygotic isolating mechanism?
   a. Two species breed at different times of the year.
   b. An egg from one species is fertilized by the sperm of another, but the embryo dies at an early stage.
   c. Two different bird species have different mating calls.
   d. Two different species are geographically isolated from each other.
   e. All of the above are prezygotic isolating mechanisms.

## 20.3 Mechanisms of Speciation

### Learning Outcomes

1. Describe how allopatric speciation can occur and how it can lead to adaptive radiation.
2. Outline two different mechanisms of sympatric speciation.

Speciation, the formation of a new species, is caused by genetic changes in a particular group that make it different from the species from which it was derived. As discussed in Chapter 19, mutations in genes can be acted on by natural selection and other evolutionary mechanisms to alter the genetic composition of a population. New species commonly evolve in this manner. In addition, interspecies matings, changes in chromosome number, and horizontal gene transfer may also contribute to the formation of new species. In all of these cases, the underlying cause of speciation is the accumulation of genetic changes that ultimately promote enough differences that a population can be recognized as a unique species.

Even though genetic changes account for the phenotypic differences observed among living organisms, such changes do not fully explain the existence of many distinct species on our planet. Why does life often diversify into the more or less discrete populations that we recognize as species? Two main explanations have been proposed:

1. In some cases, speciation may occur due to abrupt events, such as changes in chromosome number, that cause reproductive isolation.
2. More commonly, species arise as a consequence of adaptation to different ecological niches. For sexually reproducing organisms, reproductive isolation is typically a by-product of that adaptation.

Depending on the species involved, one or both factors may play a dominant role in the formation of new species. In this section, we will consider how reproductive isolating mechanisms and adaptation to particular environments are critical aspects of the speciation process.

### Geographic and Habitat Isolation Can Promote Allopatric Speciation

**Cladogenesis** is the splitting or diverging of a population into two or more species. In the case of sexually reproducing organisms, the process of cladogenesis requires that gene flow becomes interrupted between two or more populations, limiting or eliminating reproduction between members of different populations. **Allopatric speciation** (from the Greek *allos*, meaning other, and the Latin *patria*, meaning homeland) is the most prevalent way for cladogenesis to occur. This form of speciation occurs when a population becomes isolated from other populations and evolves into one or more new species. Typically, this isolation may involve a geographic barrier such as a large area of land or body of water.

In some cases, geographic separation may be caused by slow geological events that eventually produce large geographic barriers. For example, a mountain range may emerge and split one species that occupies the lowland regions, or a creeping glacier may divide a population. Figure 20.6 shows an interesting example in which geological separation promoted speciation. A fish called the Panamic porkfish (*Anisotremus taeniatus*) is found in the Pacific Ocean, whereas the porkfish (*Anisotremus virginicus*) is found in the Caribbean Sea. These two species were derived from an ancestral species whose population was split by the formation of the Isthmus of Panama about 3.5 mya. Before that event, the waters of the Pacific Ocean and Caribbean Sea mixed freely. Since the formation of the isthmus, the two populations have been geographically isolated and have evolved into distinct species.

Allopatric speciation can also occur when a small population moves to a new location that is geographically isolated from the main population. The Hawaiian Islands are a showcase of this type of allopatric speciation. The islands' extreme isolation coupled with their phenomenal array of ecological niches has enabled a small number of founding species to evolve into a vast assortment of different species. Biologists have investigated several examples of **adaptive radiation,** in which a single ancestral species has evolved into a wide array of descendant species that differ in their habitat, form, or behavior.

As shown in Figure 20.7, an example of adaptive radiation is seen with a family of birds called honeycreepers (Drepanidinae). Researchers

## Biology Principle

### All Species (Past and Present) Are Related by an Evolutionary History

These two species of fish look similar because they share a common ancestor that existed in the fairly recent past.

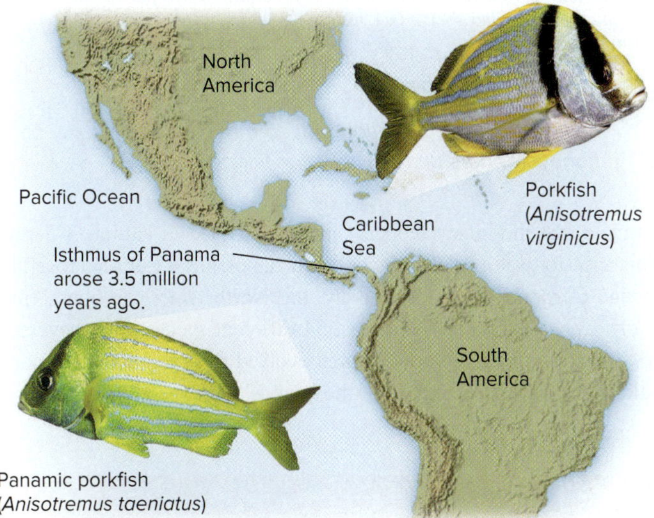


**Figure 20.6 Allopatric speciation.** An ancestral fish population was split into two by the formation of the Isthmus of Panama about 3.5 mya. Since that time, different genetic changes occurred in the two populations. These changes eventually led to the formation of different species: the Panamic porkfish (*Anisotremus taeniatus*) is found in the Pacific Ocean, and the porkfish (*Anisotremus virginicus*) is found in the Caribbean Sea.

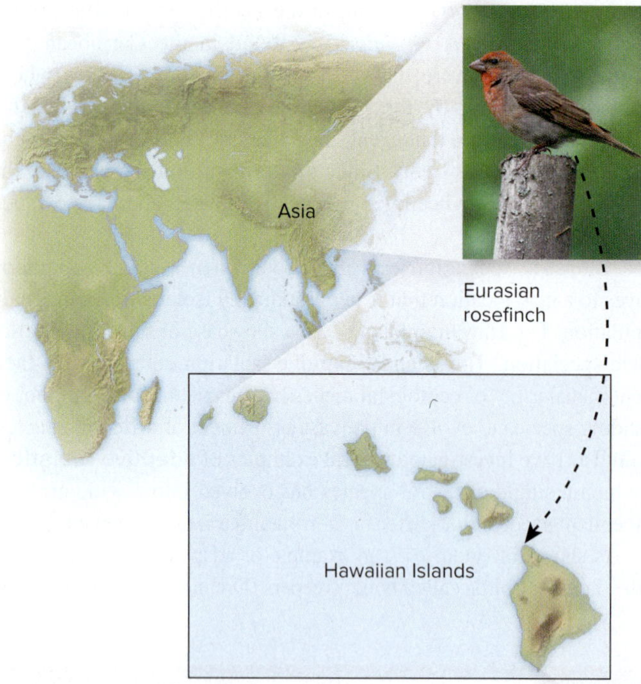


**(a) Migration of ancestor to the Hawaiian Islands**



**(b) Examples of Hawaiian honeycreepers**

**Figure 20.7 Adaptive radiation.** **(a)** The honeycreepers' ancestor is believed to be related to a Eurasian rosefinch that arrived on the Hawaiian Islands approximately 3–7 mya. Since that time, at least 54 different species of honeycreepers (Drepanidinae) have evolved on the islands. **(b)** Adaptations to feeding have produced honeycreeper species with notable differences in beak morphology.

*(a) (Eurasian rosefinch)* © FLPA/Alamy; *(b) (Palila), (Nihoa finch), (Maui Alauahio),* and *(Akohekohe)* © Jack Jeffrey Photography; *(Akikiki)* and *(I'iwi)* © Jim Denny

**BioConnections:** *Refer back to Figure 19.14. Discuss how diversifying selection played a role in the diversity of honeycreepers on the Hawaiian Islands.*

estimate that the honeycreepers' ancestor arrived in Hawaii 3–7 mya. This ancestor was a single species of finch, possibly a Eurasian rosefinch (genus *Carpodacus*) or, less likely, the North American house finch (*Carpodacus mexicanus*). At least 54 different species of honeycreepers, many of which are now extinct, evolved from this founding population to fill available niches in the islands' habitats. Natural selection resulted in the formation of many species with different feeding strategies. Seed eaters have stouter, stronger bills capable of cracking tough husks. Insect-eating honeycreepers have thin, warbler-like bills adapted for picking insects from foliage or strong, hooked bills to root out wood-boring insects. The curved bills of nectar-feeding honeycreepers enable them to extract nectar from the flowers of Hawaii's endemic plants.

# FEATURE INVESTIGATION

## Podos Found That an Adaptation for Feeding May Have Promoted Reproductive Isolation in Finches

In 2001, American evolutionary biologist Jeffrey Podos analyzed the songs of Darwin's finches on the Galápagos Islands to determine how environmental adaptation may contribute to reproductive isolation. As in honeycreepers, the differences in beak sizes and shapes among the various species of finches are adaptations to different feeding strategies. Podos hypothesized that changes in beak morphology could also affect the songs that the birds produce, thereby having the potential to affect mate choice. The components of the vocal tract of birds, including the trachea, larynx, and beak, work collectively to produce a bird's song. Birds actively modify the shape of their vocal tracts during singing, and beak movements are normally very rapid and precise.

Podos focused on two aspects of a bird's song. The first feature is the frequency range, which is a measure of the minimum and maximum frequencies in a bird's song, measured in kilohertz (kHz). The second feature is the trill rate. A trill is a series of notes or group of notes repeated in succession. Figure 20.8 shows a graphical depiction of the songs of Darwin's finches. As you can see, the song patterns of these finches are quite different from each other.

To quantitatively study the relationship between beak size and song, Podos first captured male finches on Santa Cruz, one of the Galápagos Islands, and measured their beak sizes (Figure 20.9). The birds were banded and then released into the wild. The banding provided a way of identifying the birds whose beaks had already been measured. After release, the songs of the banded birds were recorded on a tape recorder, and their range of frequencies and trill rate were analyzed. Podos then compared the data for the Galápagos finches to a large body of data that had been collected on many other

**Figure 20.8** **Differences in the songs of Galápagos finches.** These spectrograms depict the frequency of each bird's song over time, measured in kilohertz (kHz). The songs are produced in a series of trills that have a particular pattern and occur at regular intervals. Notice the differences in frequency and trill rate between different species of birds.

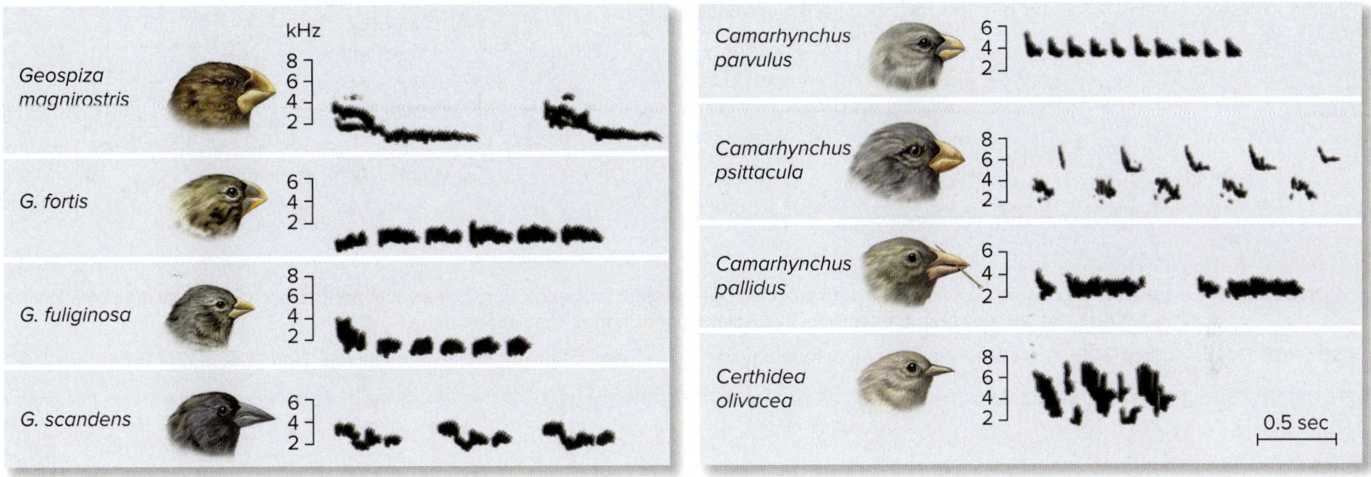


**Figure 20.9** **Study by Podos investigating the effects of beak depth on song among different species of Galápagos finches.**

**HYPOTHESIS** Changes in beak morphology that are an adaptation for feeding may also affect the songs of Galápagos finches and thereby lead to reproductive isolation between species.

**KEY MATERIALS** This study was conducted on finch populations of the Galápagos Island of Santa Cruz.

4 Analyze the songs with regard to frequency range and trill rate.

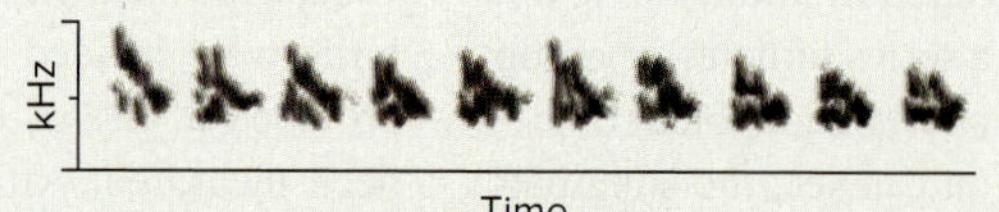


The frequency range is the value between high and low frequencies. The trill rate is the number of repeats per unit time.

5 **THE DATA**

The data for the Galápagos finches were compared to a large body of data that had been collected on many other bird species. The relative constraint on vocal performance is higher if a bird has a narrower frequency range and/or a slower trill rate. These constraints were analyzed with regard to each bird's beak depth.

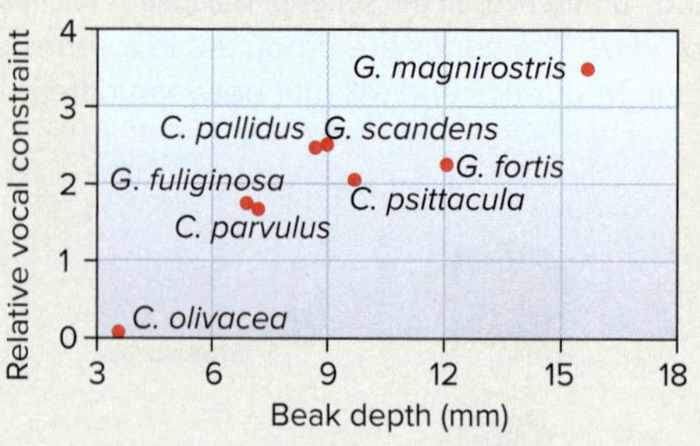


6 **CONCLUSION** Larger beak size, which is an adaptation to cracking open large, hard seeds, constrains vocal performance. This may affect mating song patterns and thereby promote reproductive isolation and, in turn, speciation.

7 **SOURCE** Podos, Jeffrey. 2001. Correlated Evolution of Morphology and Vocal Signal Structure in Darwin's Finches. *Nature* 409: 185–188.

bird species. This comparison was used to evaluate whether beak size, in this case, beak depth—the measurement of the beak from top to bottom, at its base—constrained the frequency range and/or the trill rate of the finches.

The results of this comparison are shown in the data of Figure 20.9. As seen here, the relative constraint on vocal performance became higher as the beak depth became larger. This means that birds with larger beaks had a narrower frequency range and/or a slower trill rate. Podos proposed that as jaws and beaks became adapted for strength to crack open larger, harder seeds, they became less able to perform the rapid movements associated with certain types of songs. In contrast, the finches with smaller beaks adapted to probe for insects or eat smaller seeds had less constraint on their vocal performance. From the perspective of speciation, the changes observed in song patterns for the Galápagos finches could have played an important role in promoting reproductive isolation, because song pattern is an important factor in mate selection in birds. Therefore, a by-product of beak adaptation for feeding is that it also appears to have affected song pattern, possibly promoting reproductive isolation and eventually the formation of distinct species.

### *Experimental Questions*

1. What did Podos hypothesize regarding the effects of beak size on a bird's song? How could changes in beak size and shape lead to reproductive isolation among the finches?
2. **SCISKILLS ▸** How did Podos test the hypothesis that beak morphology caused changes in the birds' songs?
3. **SCISKILLS ▸** Analyze the results of Podos's study, and explain whether they support his original hypothesis. What is meant by by-product of adaptation, and how does it apply to this particular study?

## Sympatric Speciation Occurs When Populations Are in Direct Contact

**Sympatric speciation** (from the Greek *sym*, meaning together) occurs when members of a species that are within the same range diverge into two or more different species even though there are no physical barriers to interbreeding. Although sympatric speciation is believed to be less common than allopatric speciation, particularly in animals, evolutionary biologists have discovered several ways in which it can occur. These include polyploidy and adaptation to local environments.

**Polyploidy** A type of genetic change that can cause immediate reproductive isolation is **polyploidy,** in which an organism has more than two sets of chromosomes. Plants tend to be more tolerant of changes in chromosome number than animals. For example, many crops and decorative species of plants are polyploid. How does polyploidy occur? One mechanism is complete nondisjunction of chromosomes, which increases the number of chromosome sets in a given species (autopolyploidy). Such changes can result in an abrupt sympatric speciation. For example, nondisjunction could produce a tetraploid plant with four sets of chromosomes from a species that was diploid with two sets. A cross between a tetraploid and a diploid produces a triploid offspring with three sets of chromosomes. Triploid offspring are usually sterile because an odd number of chromosomes cannot be evenly segregated during meiosis. This hybrid sterility causes reproductive isolation between the tetraploid and diploid species.

Another mechanism that leads to polyploidy is interspecies breeding. For example, interbreeding between two different species may produce an **allodiploid,** an organism that has one set of chromosomes from two different species. This term refers to the occurrence of chromosome sets (ploidy) from the genomes of different (allo-) species. Diploid species that are close evolutionary relatives can sometimes interbreed and produce allodiploid offspring. An organism containing two sets of chromosomes from two different species, for a total of four sets, is called an **allotetraploid.** An allotetraploid can occur as a result of nondisjunction in an allodiploid organism. This can also abruptly lead to reproductive isolation, because the allotetraploid would be reproductively isolated from related diploid species.

Polyploidy is so frequent in plants that it is a major mechanism of their speciation. In ferns and flowering plants, about 40–70% of the species are polyploid. By comparison, polyploidy can occur in animals, but it is much less common. For example, less than 1% of reptiles and amphibians are polyploids derived from diploid ancestors. The reason polyploidy is not usually tolerated in animals is not understood.

**Adaptation to Local Environments** In some cases, populations occupying different local environments that are continuous with each other may diverge into different species. An early example of this type of sympatric speciation was described by American biologists Jeffrey Feder, Guy Bush, and colleagues. They studied the North American apple maggot fly (*Rhagoletis pomenella*). This fly originally fed on native hawthorn trees. However, the introduction of apple trees approximately 200 years ago provided a new local environment for this species. The apple-feeding populations of this species develop more rapidly because apples mature more quickly than hawthorn fruit. The result is partial temporal isolation, which is an example of a prezygotic isolating mechanism. Although the two populations—those that feed on apple trees and those that feed on hawthorn trees—are considered subspecies, evolutionary biologists speculate they may eventually become distinct species due to reproductive isolation and the accumulation of independent mutations.

American entomologist Sara Via and colleagues have studied the beginnings of sympatric speciation in pea aphids (*Acyrthosiphon pisum*), a small, plant-eating insect. Pea aphids in the same geographic area can be found on both alfalfa (*Medicago sativa*) and red clover (*Trifolium pratenae*) (Figure 20.10). Although pea aphids on these two host plants look identical, they show significant genetic differences and are highly ecologically specialized. Pea aphids that are found on alfalfa exhibit a lower fitness when transferred to red clover, whereas pea aphids found on red clover exhibit a lower fitness when transferred to alfalfa. The same traits involved in this host specialization cause these two groups of pea aphids to be substantially reproductively isolated. Taken together, the observations of the North American apple maggot fly, pea aphids, and other insect species suggest that diversifying selection (described in Chapter 19) occurs because some members within the same range evolve to feed on a different host. This may be an important mechanism of sympatric speciation among insects.

## Biology Principle

### Populations of Organisms Evolve from One Generation to the Next

Populations of pea aphids are evolving based on preference for different food sources—alfalfa or red clover. The populations may eventually evolve into separate species.

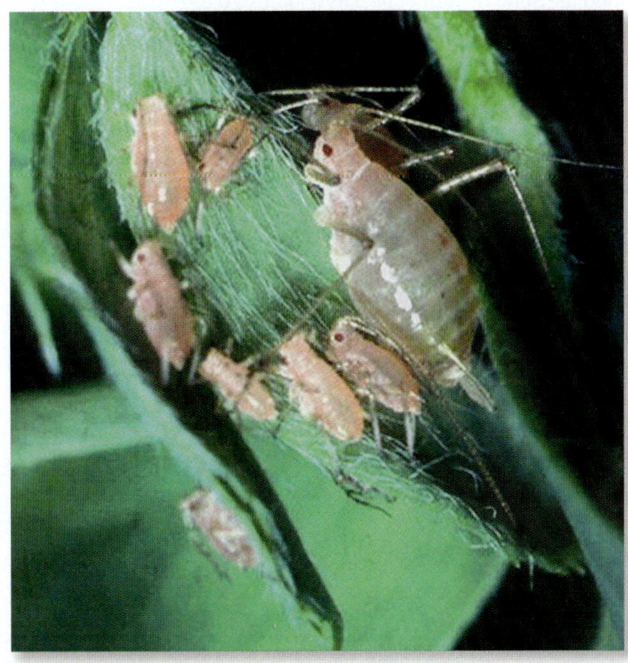

**Figure 20.10 Pea aphids, a possible example of sympatric speciation in progress.** Some pea aphids prefer alfalfa, whereas others prefer red clover. These two populations may be in the process of sympatric speciation.



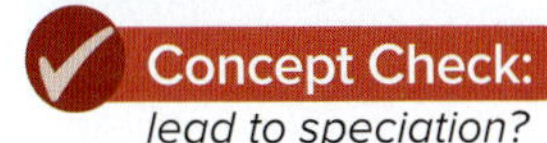


*How may host preference eventually lead to speciation?*

## 20.3 Reviewing the Concepts

- Allopatric speciation occurs when a population becomes isolated from other populations and evolves into one or more new species. When speciation from a single ancestral species occurs multiple times, the process is called adaptive radiation (Figures 20.6, 20.7).
- Podos hypothesized that changes in beak depth, associated with adaptation for feeding, promoted reproductive isolation by altering the song pattern of finches (Figures 20.8, 20.9).
- Sympatric speciation involves the formation of different species in populations that are not geographically isolated from one another. Polyploidy and adaptation to local environments are mechanisms that promote sympatric speciation (Figure 20.10).

## 20.3 Testing Your Knowledge

1. Which of the following is an example of allopatric speciation?
   a. A few birds fly to a distant island, and after many generations a new bird species is formed.
   b. A diploid species becomes a new tetraploid species due to nondisjunction.
   c. A species diverges into two different species due to adaptation to local environments.
   d. All of the above are examples of allopatric speciation.
2. Which of the following would be the most likely to promote an abrupt speciation event?
   a. formation of a new species on an island
   b. adaptation of a species to local environments
   c. formation of a polyploid species from a single diploid species
   d. formation of an allotetraploid from two diploid species
   e. both c and d

## 20.4 Evo-Devo: Evolutionary Developmental Biology

### Learning Outcomes

1. Describe how the spatial expression of genes, such as *BMP4* and *Gremlin,* affects pattern formation.
2. Explain the relationship between the number of *Hox* genes and the body pattern of an animal species.
3. Outline how differences in the growth rates of body parts can change the characteristics of species.

In biology, **development** refers to a series of changes in the state of a cell, a tissue, an organ, or an organism. Development is the process that gives rise to structures and functions of living organisms. In recent years, many evolutionary biologists have begun to investigate how genetic variation produces species and groups of species with novel shapes and forms. The underlying reasons for such changes are often rooted in the developmental pathways that control an organism's morphology.

**Evolutionary developmental biology** (referred to as **evo-devo**) is an exciting and relatively new field of biology that compares the development of different organisms in an attempt to understand ancestral relationships between organisms and the mechanisms that bring about evolutionary change. During the past few decades, developmental geneticists have gained a better understanding of biological development at the molecular level. Much of this work has involved the discovery of genes that control development in model organisms. As the genomes of more organisms have been analyzed, researchers have become interested in the similarities and differences that occur between closely related and distantly related species. The field of evolutionary developmental biology has arisen out of this interest. In this section, we will see that proteins that control developmental changes, such as cell-signaling proteins and transcription factors, often play a key role in promoting the morphological changes that occur during evolution.

### The Spatial Expression of Genes That Affect Development Has a Dramatic Effect on Phenotype

Genes that play a role in development influence cell division, cell migration, cell differentiation, and cell death. The interplay among these four processes produces an organism with a specific body pattern, a process called **pattern formation.** As you might imagine, developmental genes are very important to the phenotypes of individuals. They affect traits such as the shape of a bird's beak, the length of a giraffe's neck, and the size of a plant's flower. In recent years, the study of development has revealed that developmental genes are key players in the evolution of many types of traits. Changes in such genes affect traits that can be acted on by natural selection. Furthermore, variation in the expression of these genes may be commonly involved in the acquisition of new traits that promote speciation.

As an example, let's compare pattern formation of a chicken's foot with that of a duck's foot. Two different patterns occur: a nonwebbed pattern, in which the digits are not interconnected, and a webbed pattern, in which the digits are connected by sheets of skin. Developmental biologists have discovered that the morphological differences between a nonwebbed and a webbed pattern are due to the differential expression of two different cell-signaling proteins called bone morphogenetic protein 4 (BMP4) and gremlin. The *BMP4* gene is expressed throughout the developing limb of both the chicken and duck; this is shown in **Figure 20.11a**, in which the BMP4 protein is stained purple. The BMP4 protein causes cells to undergo programmed cell death. The gremlin protein, which is stained brown in **Figure 20.11b**, inhibits the function of BMP4, thereby allowing cells to survive. In the developing chicken limb, the *Gremlin* gene is expressed throughout the limb, except in the regions between each digit. Therefore, in these regions, the cells die, and a chicken develops a nonwebbed foot (**Figure 20.11c**). By comparison, in the duck, *Gremlin* is expressed throughout the entire limb, including the interdigit regions, and the duck develops a webbed foot. Interestingly, researchers have been able to introduce gremlin protein into the interdigit regions of developing chicken limbs. This produces a chicken with webbed feet!

How are these observations related to evolution? During the evolution of birds, genetic variation arose such that some individuals

Chicken Duck

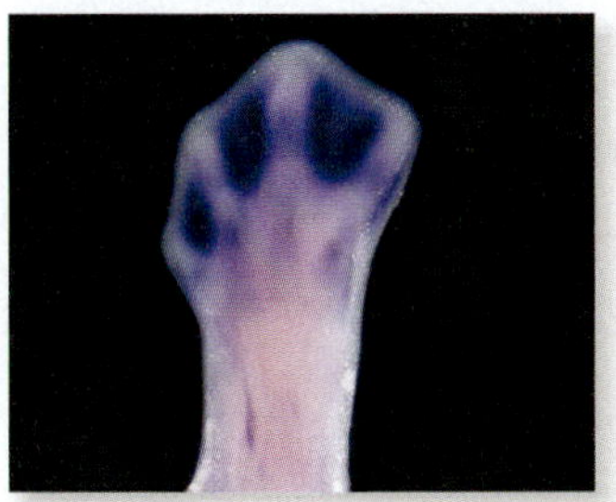

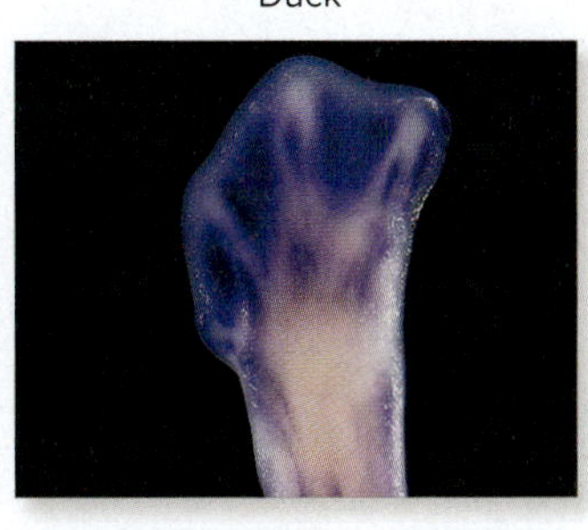

(a) BMP4 protein levels–similar expression in chicken and duck

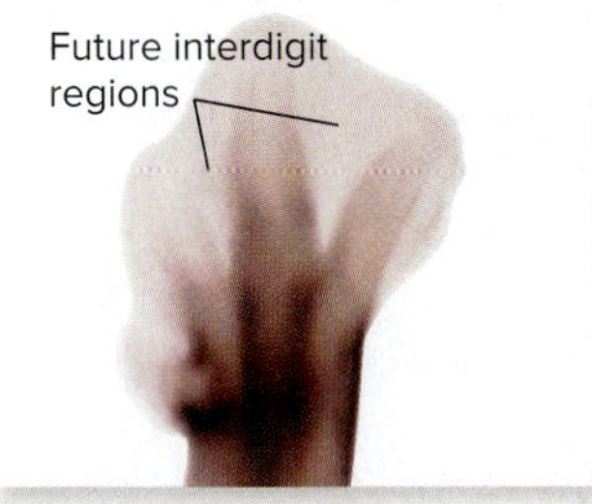

(b) Gremlin protein levels–not expressed in interdigit region in chicken

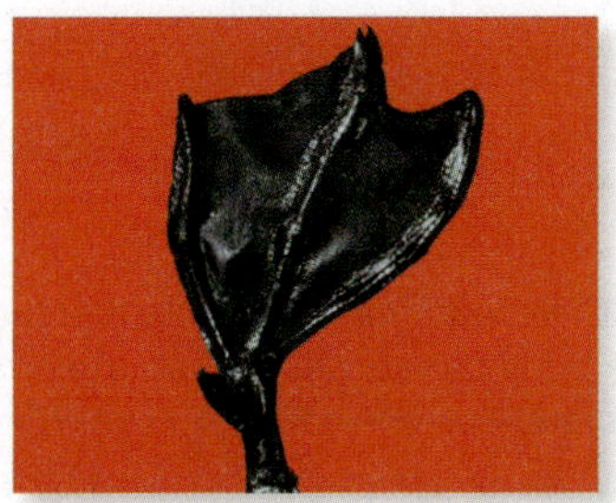

(c) Comparison of a chicken foot and a duck foot

**Figure 20.11** **The role of cell-signaling proteins in the pattern of birds' feet.** This figure shows how changes in developmental gene expression can affect webbing between the toes. (a) Expression of the *BMP4* gene in the developing limbs. BMP4 protein is stained purple here and is expressed throughout the limb. (b) Expression of the *Gremlin* gene in the developing limbs. Gremlin protein is stained brown here. Note that *Gremlin* is not expressed in the interdigit regions of the chicken but is expressed in these regions of the duck. Gremlin inhibits BMP4, which causes programmed cell death. (c) Because BMP4 is not inhibited in the interdigit regions in the chicken, the cells in this region die, and the foot is not webbed. By comparison, inhibition of BMP4 in the interdigit regions in the duck results in a webbed foot.

*(a) (left) (right)* Courtesy Ed Laufer; *(b–c) (left) (right)* Courtesy of Dr. J.M. Hurle. Originally published in *Development*. 1999 Dec. 126(23): 5515–5522

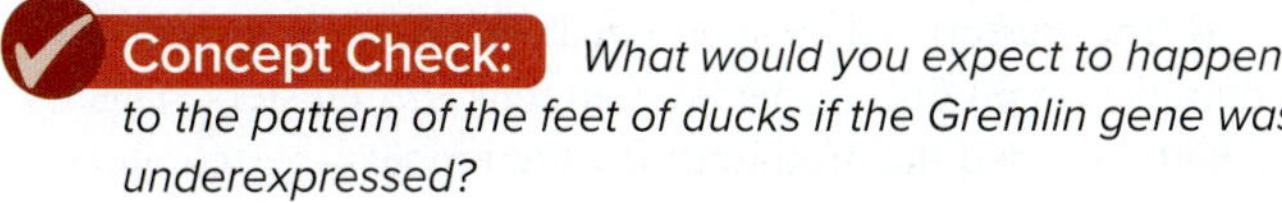

**Concept Check:** *What would you expect to happen to the pattern of the feet of ducks if the Gremlin gene was underexpressed?*

expressed the *Gremlin* gene in the regions between each digit, but others did not. This variation determined whether or not a bird's feet were webbed. In terrestrial settings, having nonwebbed feet is an advantage because it enables the individual to hold onto perches, run along the ground, and snatch prey. Therefore, natural selection would favor nonwebbed feet in terrestrial environments. This process explains the occurrence of nonwebbed feet in chickens, hawks, crows, and many other terrestrial birds. In aquatic environments, however, webbed feet are an advantage because they act as paddles for swimming, so genetic variation that produced webbed feet in aquatic birds would have been acted on by natural selection. Over time, this gave rise to the webbed feet now found in a wide variety of aquatic birds, including ducks, geese, and penguins.

## EVOLUTIONARY CONNECTIONS

### The *Hox* Genes Have Been Important in the Evolution of a Variety of Body Patterns

The study of developmental genes has revealed interesting trends among large groups of species. *Hox* genes are found in all animals, indicating they have originated very early in animal evolution. They specify the fate of a particular segment or region of the body. Developmental biologists have hypothesized that variation in the *Hox* genes has spawned the formation of many new body patterns. As shown in **Figure 20.12**, the number and arrangement of *Hox* genes vary considerably among different types of animals. Sponges, the simplest of animals, have at least one gene that is homologous to *Hox* genes. Insects typically have nine or more *Hox* genes. In most cases, multiple *Hox* genes occur in a cluster in which the genes are close to each other along a chromosome. In mammals, *Hox* gene clusters have been duplicated twice during the course of evolution to form four clusters, all slightly different, containing a total of 39 genes.

Researchers propose that increases in the number of *Hox* genes have been instrumental in the evolution of many animal species with greater complexity in body structure. To understand how, let's first consider *Hox* gene function. All *Hox* genes encode transcription factors that act as master control proteins for directing the formation of particular regions of the body. Each *Hox* gene controls a hierarchy of many regulatory genes that regulate the expression of genes encoding proteins that ultimately affect the morphology of the organism. The evolution of complex body patterns is associated with an increase not only in the number of regulatory genes—as evidenced by the increase in *Hox* gene complexity during evolution—but also in genes that encode proteins that directly affect an organism's form and function.

How would an increase in *Hox* genes enable more complex body patterns to evolve? Part of the answer lies in the spatial expression of the *Hox* genes. Different *Hox* genes are expressed in different regions of the body along the body axis called the anteroposterior axis, which runs from the head toward the abdomen (**Figure 20.13**). Therefore, an increase in the number of *Hox* genes allows each of these master control genes to become more specialized in the region that it controls. In fruit flies, one segment in the middle of the body can be controlled by a particular *Hox* gene and forms wings and legs, whereas a segment in the head region can be controlled by a different *Hox* gene and develops antennae. Therefore, research suggests that one way for new, more complex body patterns to evolve is by increasing the number of

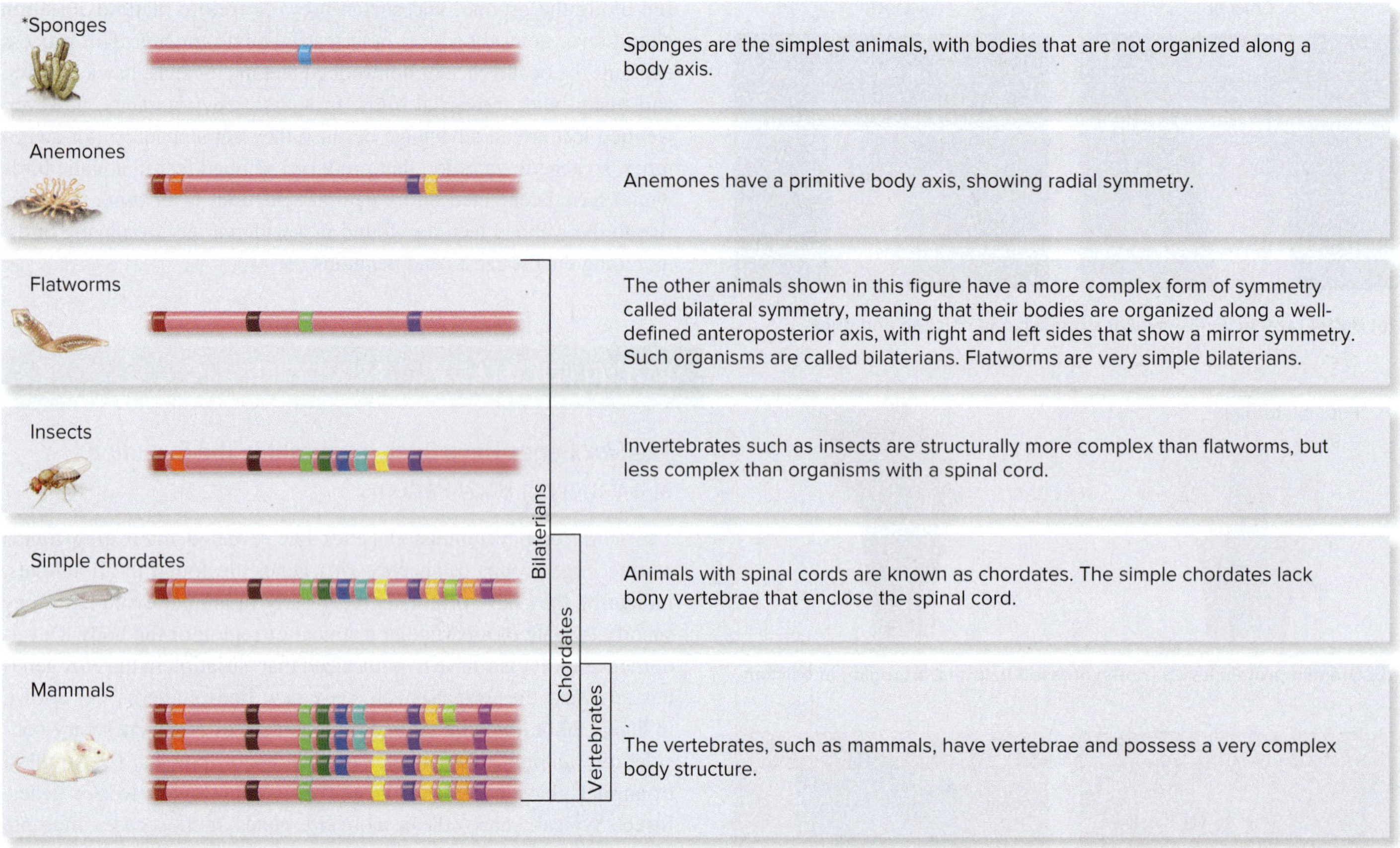

*Sponges

Sponges are the simplest animals, with bodies that are not organized along a body axis.

Anemones

Anemones have a primitive body axis, showing radial symmetry.

Flatworms

The other animals shown in this figure have a more complex form of symmetry called bilateral symmetry, meaning that their bodies are organized along a well-defined anteroposterior axis, with right and left sides that show a mirror symmetry. Such organisms are called bilaterians. Flatworms are very simple bilaterians.

Insects

Invertebrates such as insects are structurally more complex than flatworms, but less complex than organisms with a spinal cord.

Simple chordates

Animals with spinal cords are known as chordates. The simple chordates lack bony vertebrae that enclose the spinal cord.

Mammals

The vertebrates, such as mammals, have vertebrae and possess a very complex body structure.

Bilaterians

Chordates

Vertebrates

**Figure 20.12** ***Hox*** **gene number and body complexity in different types of animals.** Researchers speculate that the duplication of *Hox* genes and *Hox* gene clusters played a key role in the evolution of more complex body patterns in animals. A correlation is observed between increasing numbers of *Hox* genes and increasing complexity of body structure. The different colors of *Hox* genes correspond to genes that are most closely related in different animal groups.

*Note: Sponges, which are the simplest animals, have no true tissues. They do not have true *Hox* genes, though they have an evolutionarily related gene called an *NK-like* gene. Some species of sponges have more than one copy of this gene.

*What is the relationship between the total number of Hox genes in an animal species and its morphological complexity?*

*Hox* genes, thereby making it possible to form many specialized parts of the body that are organized along a body axis.

Three lines of evidence support the idea that increases in *Hox* gene number have been instrumental in the evolution and speciation of animals with more complex body patterns.

- *Hox* genes are known to control the fate of regions along the anteroposterior axis.
- As described in Figure 20.12, a general trend is observed in which animals with a more complex body structure tend to have more *Hox* genes and *Hox* clusters in their genomes than do simpler animals.
- A comparison of *Hox* gene evolution and animal evolution reveals striking parallels. Researchers have analyzed *Hox* gene sequences among modern species and made estimates regarding the timing of past events. Though the date is difficult to precisely pinpoint, the first *Hox* gene arose well over 600 mya. In addition, gene duplications of this primordial gene produced clusters of *Hox* genes in other species. Clusters such as those found in modern insects were likely to be present approximately 600 mya. A duplication of that cluster is estimated to have occurred around 520 mya.

Estimates of *Hox* gene origins correlate with major diversification events in the history of animals. As described in Chapter 22 the Cambrian period, which occurred from 543 to 490 mya, saw a great diversification of animal species. This diversification occurred after the *Hox* cluster was formed and was possibly undergoing its first duplication to produce two *Hox* clusters. Also, approximately 420 mya, a second duplication produced species with four *Hox* clusters. This event preceded the proliferation of tetrapods—vertebrates with four limbs—that occurred during the Devonian period, approximately 417–354 mya. Modern tetrapods have four *Hox* clusters. This second duplication may have been a critical event that led to the evolution of complex terrestrial vertebrates with four limbs, such as amphibians, reptiles, and mammals.

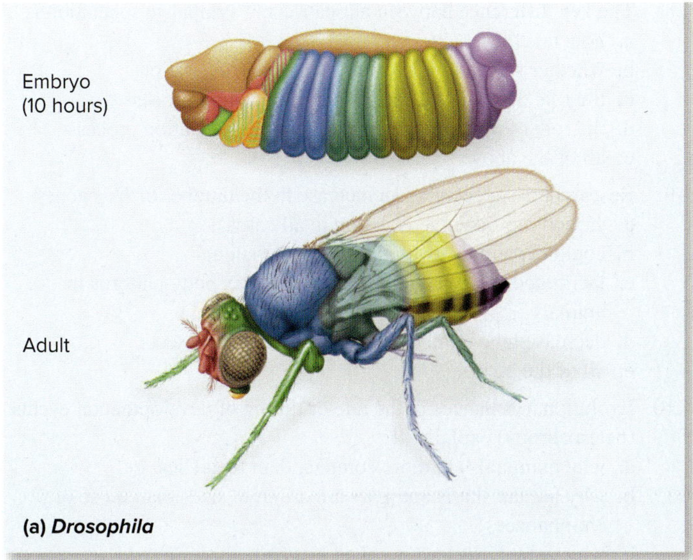


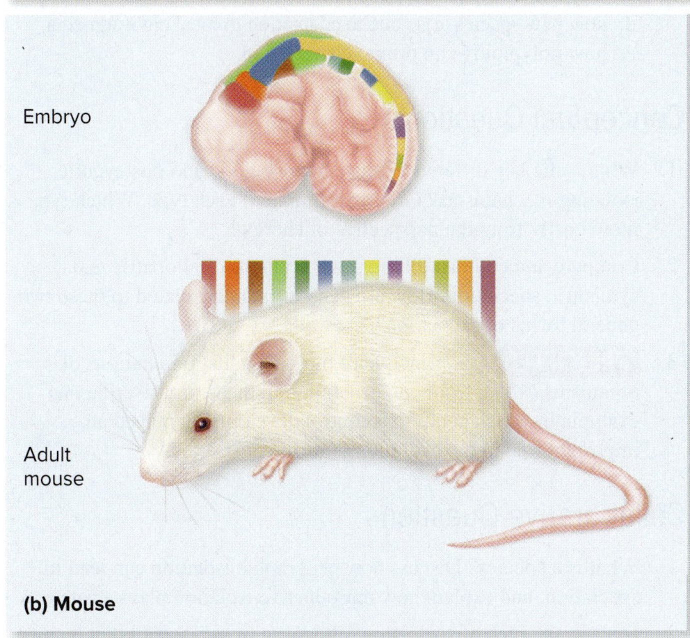


**Figure 20.13 Expression of *Hox* genes along the anteroposterior axis in animals.** The colors shown in this figure correspond to the gene colors in Figure 20.12, from left to right.

## Variation in Growth Rates Can Have a Dramatic Effect on Morphology

Another way that developmental genes influence morphology is by controlling the relative growth rates of different parts of the body during development. The term **heterochrony** refers to evolutionary changes in the rate or timing of developmental events. The speeding up or slowing down of growth appears to be a common occurrence in evolution and leads to different species with striking morphological differences. With regard to the pace of evolution, such changes may rapidly lead to the formation of new species.

As an example, Figure 20.14 compares the progressive growth of human and chimpanzee skulls. At the fetal stage, the size and shape of the skulls look fairly similar. However, after this stage, the relative growth rates of certain regions become markedly different, thereby affecting the shape and size of the adult skull. In the chimpanzee, the jaw region grows faster, giving the adult chimpanzee a much larger and longer jaw. In the human, the jaw grows more slowly, and the region of the skull that surrounds the brain—the cranium—grows faster. The result is that adult humans have a smaller jaw but a larger cranium.

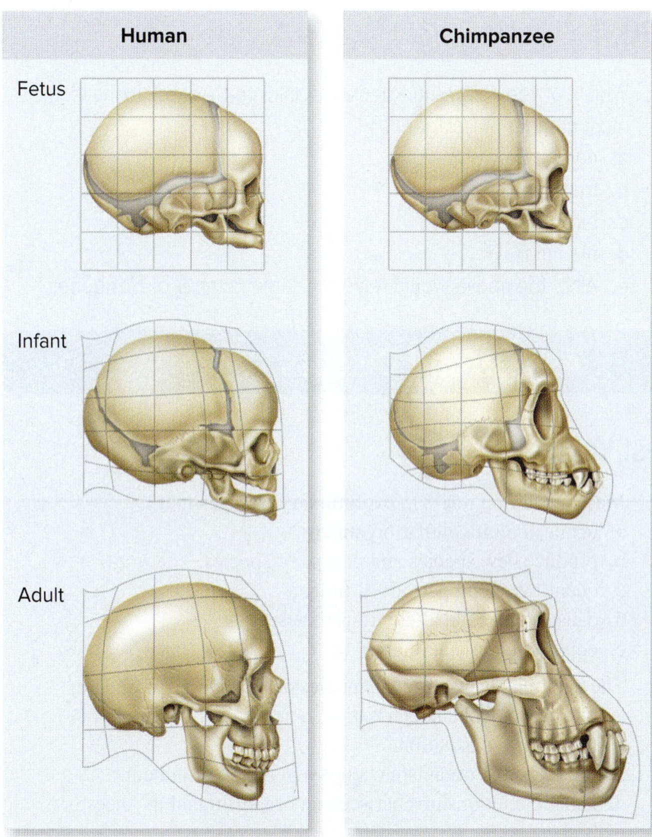


**Figure 20.14 Heterochrony.** Heterochrony is the phenomenon in which one region of the body grows faster than another among different species. The phenomenon explains why the skulls of adult chimpanzees and humans have different shapes even though their fetal skull shapes are quite similar.

## 20.4 Reviewing the Concepts

- Evolutionary developmental biology (evo-devo) compares the development of different species in order to understand ancestral relationships and the mechanisms that bring about evolutionary change. These changes often involve variation in the expression of cell-signaling proteins and transcription factors.
- The spatial expression of genes that affect development can affect phenotypes dramatically, as shown by the expression of the *BMP4* and *Gremlin* genes in birds with nonwebbed or webbed feet (Figure 20.11).
- An increase in the number of *Hox* genes played an important role in the evolution of more complex body patterns in animals (Figures 20.12, 20.13).
- A difference in the relative growth rates of body parts among different species, called heterochrony, can have a major effect on morphology (Figures 20.14).

## 20.4 Testing Your Knowledge

1. Which of the following species would you expect to have the most number of *Hox* genes?
   a. nematode worm
   b. fruit fly
   c. bird
   d. mammal
   e. All of the above would have the same number of *Hox* genes.

# Assess and Discuss

## Test Yourself

1. Macroevolution refers to evolutionary changes that
   a. occur in multicellular organisms.
   b. produce new species and groups of species.
   c. occur over long periods of time.
   d. cause changes in allele frequencies.
   e. occur in large mammals.
2. The ecological species concept classifies a species based on
   a. morphological characteristics.
   b. reproductive isolation.
   c. the niche the organism occupies in the environment.
   d. genetic relationships between an organism and its ancestors.
   e. both a and b.
3. Which of the following is an example of a postzygotic isolating mechanism?
   a. incompatible genitalia
   b. different mating seasons
   c. incompatible gametes
   d. a mountain range separating two populations
   e. the failure of fertilized eggs to develop normally
4. Hybrid breakdown occurs when interspecies hybrids
   a. do not develop past the early embryonic stages.
   b. have a reduced life span.
   c. are infertile.
   d. are fertile but produce offspring with reduced viability and fertility.
   e. produce offspring that express the traits of only one of the original species.
5. The evolution of one species into two or more species is called
   a. gradualism.
   b. punctuated equilibrium.
   c. cladogenesis.
   d. horizontal gene transfer.
   e. microevolution.
6. A large number of honeycreeper species on the Hawaiian Islands is an example of
   a. adaptive radiation.
   b. genetic drift.
   c. stabilizing selection.
   d. horizontal gene transfer.
   e. microevolution.
7. A major mechanism of speciation in plants but not in animals is
   a. adaptation to new environments.
   b. polyploidy.
   c. hybrid breakdown.
   d. genetic changes that alter the organism's niche.
   e. both a and d.
8. The key difference between allopatric and sympatric speciation is
   a. how fast they occur.
   b. whether speciation involves geographic separation.
   c. the role of sexual selection in the speciation process.
   d. the role of reproductive isolation in the speciation process.
   e. all of the above.
9. Researchers suggest that an increase in the number of *Hox* genes
   a. caused reproductive isolation in all cases.
   b. could explain the evolution of color vision.
   c. facilitated the evolution of more complex body patterns in animals.
   d. decreased the number of body segments in insects.
   e. all of the above.
10. Evolutionary changes in the rate or timing of developmental events (heterochrony) explain
    a. why mammals are more complex than fruit flies.
    b. why human skulls and jaws are different sizes than those of chimpanzees.
    c. why some birds have webbed feet and others do not.
    d. how new species arise due to adaptation to local environments.
    e. how polyploidy can promote speciation.

## Conceptual Questions

1. What is the key difference between prezygotic and postzygotic isolating mechanisms? Give an example of each type. Which type is more costly from the perspective of energy?
2. Compare and contrast different mechanisms of allopatric and sympatric speciation. How are genetic changes related to these two general forms of speciation?
3. **PRINCIPLES** A principle of biology is that populations of organisms evolve from one generation to the next. Describe one example in which genes that control development played an important role in the evolution of different species.

## Collaborative Questions

1. What is a species? Discuss how geographic isolation can lead to speciation, and explain how reproductive isolation plays a role.
2. Discuss the type of speciation (allopatric or sympatric) that is most likely to occur under each of the following conditions:
   a. A pregnant female rat is transported by an ocean liner to a new continent.
   b. A meadow containing several species of grasses is exposed to a pesticide that promotes nondisjunction.
   c. In a very large lake containing several species of fishes, the water level gradually falls over the course of several years. Eventually, the large lake becomes subdivided into smaller lakes, some of which are connected by narrow streams.

# Online Resource

**connect.mheducation.com**

# How Biologists Classify Species and Study Their Evolutionary Relationships

# 21

© Photos M. Kuntner

**Darwin's bark spider, *Caerostris darwini* (left side), which was discovered in 2010.** The right side shows an example of the enormous web that a female spider of this species can weave.

## Chapter Outline



New species are discovered on a regular basis. In 2010, researchers reported the discovery of a new species of spider in Madagascar, which they named Darwin's bark spider, *Caerostris darwini*. As seen in the chapter-opening photo, this species builds the largest orb-style webs that are known. Such webs have been found spanning rivers, streams, and lakes, reaching up to 25 m in length. The silk spun by these spiders is thought to be the toughest biological material ever studied. It is over 10 times stronger than Kevlar—the synthetic fiber used in bullet-proof vests, bicycle tires, and racing sails!

The rules for the classification of newly described species, such as Darwin's bark spider, are governed by the discipline of taxonomy (from the Greek *taxis*, meaning order, and *nomos*, meaning law). **Taxonomy** is the science of describing, naming, and classifying **extant** species, those that still exist today, as well as **extinct** species, those that have died out. Taxonomy results in the ordered division of species into groups based on similarities and dissimilarities in their characteristics. This task has been ongoing for over 300 years. In the mid- to late 1600s, naturalist John Ray made the first attempt to broadly classify all known forms of life. Ray's ideas were later extended by naturalist Carolus Linnaeus in the mid-1700s, which is considered by some to be the official birth of taxonomy.

**Systematics** is the study of biological diversity and the evolutionary relationships among organisms, both extant and extinct. In the 1950s, German entomologist Willi Hennig proposed that evolutionary relationships should be inferred from features shared by descendants of a common ancestor. Researchers now try to place new species into taxonomic groups based on evolutionary relationships with other species. In addition, previously established taxonomic groups are revised as new data shed light on evolutionary relationships.

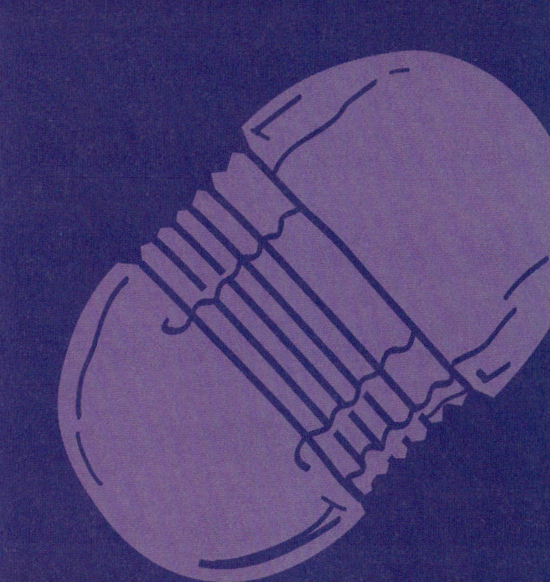

In this chapter, we will begin with a discussion of taxonomy and the concept of taxonomic groups. We will then examine how biologists use systematics to determine evolutionary relationships among organisms, looking in particular at how those relationships are portrayed in diagrams called phylogenetic trees. We will then explore how analyses of morphological data and molecular genetic data are used to understand the evolutionary history of life on Earth.

## 21.1 Taxonomy

### Learning Outcomes

1. Identify the three domains of life.
2. Outline the hierarchy of groupings in taxonomy.
3. Explain how species are named using binomial nomenclature.

A hierarchy is a system of organization that classifies entities into successive levels. In biological taxonomy, every species is placed into several different nested groups within a hierarchy. For example, a leopard and a fruit fly are both classified into a large group known as animals, though they differ in many traits. By comparison, leopards and lions are placed together into a group with a smaller number of species called felines (more formally named Felidae), which are predatory cats. The felines are a subset of the animal group, which has species that share many similar traits. The species that are placed together into small taxonomic groups are likely to share many of the same characteristics. In this section, we will consider how biologists use a hierarchy to group similar species.

### Living Species Are Subdivided into Three Domains of Life

Modern taxonomy places species into progressively smaller hierarchical groups. Each group at any level is called a **taxon** (plural, taxa). In the late 1970s, based on information in the sequences of genes, American biologist Carl Woese proposed the idea of creating a category called a **domain.** Under this system, all forms of life are grouped within three domains: **Bacteria, Archaea,** and **Eukarya** (Figure 21.1). The terms Bacteria and Archaea are capitalized when referring to the domains, but they are not capitalized when referring to individual species. A single bacterial cell is called a bacterium, and a single archaeal cell is an archaeon.

The domain Eukarya formerly consisted of four kingdoms called Protista, Fungi, Plantae, and Animalia. Researchers later discovered that protists do not constitute a separate kingdom but, instead, are a very broad collection of species. Taxonomists now subdivide the eukaryotic domain into seven groups called supergroups. In the taxonomy of eukaryotes, a **supergroup** lies between a domain and a kingdom (Figure 21.1). As discussed in Chapter 23, all seven supergroups contain a distinctive group of protists. In addition, kingdoms Fungi and Animalia are within the supergroup Opisthokonta, because they are closely related to other protists in this supergroup. Kingdom

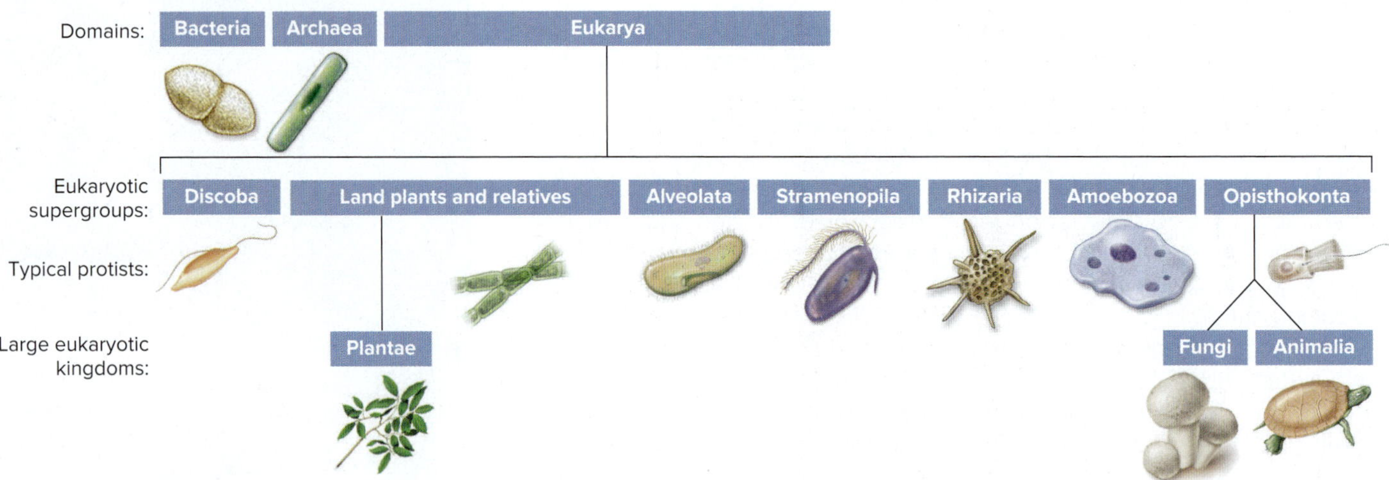


**Figure 21.1 A classification system for living and extinct organisms.** All organisms are grouped into three domains: Bacteria, Archaea, and Eukarya. Eukaryotes are divided into seven supergroups. These seven supergroups encompass most eukaryotes. However, several small eukaryotic lineages are classified outside of the seven supergroups.

**BioConnections:** *Refer back to Figure 4.8. Which of the three domains contains organisms with prokaryotic cells?*

**Table 21.1 Distinguishing Cellular and Molecular Features of Domains Bacteria, Archaea, and Eukarya***

| Characteristic | Bacteria | Archaea | Eukarya |
|---|---|---|---|
| Chromosomes | Usually circular | Circular | Usually linear |
| Nucleosome structure | No | No | Yes |
| Chromosome segregation | Binary fission | Binary fission | Mitosis/ meiosis |
| Introns in genes | Rarely | Rarely | Commonly |
| Ribosomes | 70S | 70S | 80S |
| Initiator tRNA | Formyl-methionine | Methionine | Methionine |
| Operons | Yes | Yes | No |
| Capping of mRNA | No | No | Yes |
| RNA polymerases | One | Several | Three |
| Promoters of structural genes | −35 and −10 sequences | TATA box | TATA box |
| Cell compartmentalization | No | No | Yes |
| Membrane lipids | Ester-linked | Ether-linked | Ester-linked |

*The descriptions in this table represent the general features of most species in each domain. Some exceptions are observed. For example, certain bacterial species have linear chromosomes, and operons occasionally are found in eukaryotes, such as the nematode worm *Caenorhabditis elegans*.

Plantae is within the supergroup called land plants and relatives. Green plants are closely related to green algae, which are protists in this supergroup. **Table 21.1** compares a variety of molecular and cellular characteristics among the domains Bacteria, Archaea, and Eukarya.

## EVOLUTIONARY CONNECTIONS

### Every Species Is Placed into a Taxonomic Hierarchy

Why is it useful to categorize species into groups? The three domains of life contain millions of different species. Subdividing them into progressively smaller taxonomic groups makes it easier for biologists to appreciate the relationships among such a large number of species. The order of hierarchy is

**domain** > **supergroup** > **kingdom** > **phylum** (plural, phyla) > **class** > **order** > **family** > **genus** (plural, genera) > **species**

Each of these taxa contains progressively fewer species that are more similar to each other than they are to the members of the taxa above them in the hierarchy. For example, the taxon Animalia, which is at the kingdom level, has a larger number of fairly diverse species than does the class Mammalia, which contains fewer species that are relatively similar to each other.

To further understand taxonomy, let's consider the classification of a species such as the gray wolf (*Canis lupus*) (**Figure 21.2**). The classification of this species is as follows:

- Domain Eukarya (several million species): The gray wolf is a eukaryote. See Table 21.1 for a description of eukaryotes.
- Supergroup Opisthokonta (over 1 million species): As described in Chapter 23, this supergroup includes the animal and fungal kingdoms and related protists.
- Kingdom Animalia (over 1 million species): Animals are multicellular and heterotrophic, which means they must consume other organisms or parts of other organisms to survive. Their cells lack a cell wall. Except for sponges, animal bodies contain tissues that have specialized cell types. Examples are muscle and nervous tissue.
- Phylum Chordata (~50,000 species): Animals in this phylum all have four common features at some stage of their development. These are a notochord (a cartilaginous rod that runs along the back), a tubular nerve or spinal cord located above the notochord, gill slits or arches, and a postanal tail.
- Class Mammalia (5,513 species): Two distinguishing features are hair, which helps the body maintain a constant body temperature, and mammary glands, which produce milk to nourish the young.
- Order Carnivora (282 species): Carnivores typically have sharp claws, prominent canine teeth, and binocular vision. They usually follow a diet that primarily consists of meat.
- Family Canidae (34 species): All species in the family Canidae are doglike animals. This family is composed of different species of wolves, jackals, foxes, wild dogs, and the coyote.
- Genus *Canis* (7 species): This genus is composed of closely related doglike species, which includes four species of jackals, the coyote, and two types of wolves. The species *Canis lupus* encompasses several subspecies, including the domestic dog (*Canis lupus familiaris*).

### Binomial Nomenclature Is Used to Name Species

As originally advocated by Linnaeus, **binomial nomenclature** is the standard method for naming species. The scientific name of every species has two names, its genus name and its unique, specific epithet. An example is the gray wolf, *Canis lupus*. The genus name is always capitalized, but the specific epithet is not. Both names are italicized. After the first mention, the genus name is often abbreviated to a single letter. For example, we write that *Canis lupus* is the gray wolf, and in subsequent sentences, the species is referred to as *C. lupus*.

When naming a new species, genus names are always nouns or treated as nouns, whereas species epithets may be either nouns or adjectives. The names often have a Latin or Greek origin and refer to characteristics of the species or to features of its habitat. For example, the genus name of the gray wolf, *Canis,* is from the Latin, *canis*,

# Biology Principle

## All Species (Past and Present) Are Related by an Evolutionary History

A goal of taxonomy is to relate the diversity of species according to their evolutionary relationships.

| Taxonomic group | Gray wolf found in | Number of current species |
|---|---|---|
| Domain | Eukarya | ~4–10 million |
| Supergroup | Opisthokonta | >1 million |
| Kingdom | Animalia | >1 million |
| Phylum | Chordata | ~50,000 |
| Class | Mammalia | 5,513 |
| Order | Carnivora | 282 |
| Family | Canidae | 34 |
| Genus | *Canis* | 7 |
| Species | *lupus* | 1 |

**Figure 21.2** **A taxonomic classification of the gray wolf (*Canis lupus*).**

*Which group is broader, a phylum or a family?*

meaning dog, and the species name, *lupus*, is from the Latin, *lupus*, meaning wolf. However, the choice of name for a species can be lighthearted. For example, there is a beetle named *Agra vation*.

## 21.1 Reviewing the Concepts

- Taxonomy is the field of biology concerned with describing, naming, and classifying living and extinct organisms. Systematics is the study and classification of evolutionary relationships among organisms through time.
- Taxonomy places all living organisms into progressively smaller hierarchical groups called taxa (singular, taxon). The broadest groups are the three domains, called Bacteria, Archaea, and Eukarya, followed by supergroups, kingdoms, phyla, classes, orders, families, genera, and species (Figures 21.1, 21.2; Table 21.1).
- Binomial nomenclature is a naming convention that provides each species with two names: its genus name and species epithet.

## 21.1 Testing Your Knowledge

1. The three domains of life are
   a. Animals, Bacteria, and Archaea.
   b. Eukarya, Prokaryotes, and Viruses.
   c. Eukarya, Bacteria, and Archaea.
   d. Protists, Animals, and Fungi.
2. Which of the following is a correct order of taxa?
   a. supergroup > domain > kingdom > phylum > class > order > family > genus > species
   b. domain > supergroup > kingdom > phylum > class > order > family > genus > species
   c. domain > supergroup > kingdom > class > phylum > order > family > genus > species
   d. supergoup > domain > kingdom > phylum > class > family > order > genus > species

## 21.2 Phylogenetic Trees

### Learning Outcomes

1. Explain the concept of phylogeny and its basis for the construction of phylogenetic trees.
2. Explain the process of cladogenesis, the primary way that new species arise.
3. Describe how morphological and genetic homologies are used to construct phylogenetic trees.

As mentioned, systematics is the study of biological diversity and evolutionary relationships. By studying the similarities and differences among species, biologists can construct a **phylogeny—**the evolutionary history of a species or group of species. To propose a phylogeny, biologists use the tools of systematics. For example, the classification of the gray wolf in Figure 21.2 is based on systematics. Therefore, one use of systematics is to place species into taxa and to understand the evolutionary relationships among different taxa.

In this section, we will consider the features of diagrams, or "trees," that describe the evolutionary relationships among various species, both extant and extinct. As you will learn, such trees are usually based on morphological or genetic data.

### A Phylogenetic Tree Depicts Evolutionary Relationships Among Species

A **phylogenetic tree** is a diagram that describes the evolutionary relationships among various species, based on the information available to and gathered by systematists. A phylogenetic tree should be viewed as a hypothesis that is proposed, tested, and later refined as additional data become available. Let's look at what information a phylogenetic tree contains and the form in which it is presented. Figure 21.3 shows a hypothetical phylogenetic tree of the relationships among various flowering plant species, in which the species are labeled A through K. The vertical axis represents time, with the oldest species at the bottom.

New species can be formed by **anagenesis,** in which a single species evolves into a different species. However, the primary way that new species arise is by **cladogenesis,** in which a species diverges into two or more species. Some evolutionary biologists focus only on cladogenesis in the construction of their trees. The branch points in a phylogenetic tree, also called **nodes,** indicate times when cladogenesis has occurred. For example, approximately 12 mya, species A diverged into species A and species B. Figure 21.3 also shows anagenesis in which species C evolved into species G. The tips of branches may represent species that became extinct in the past, such

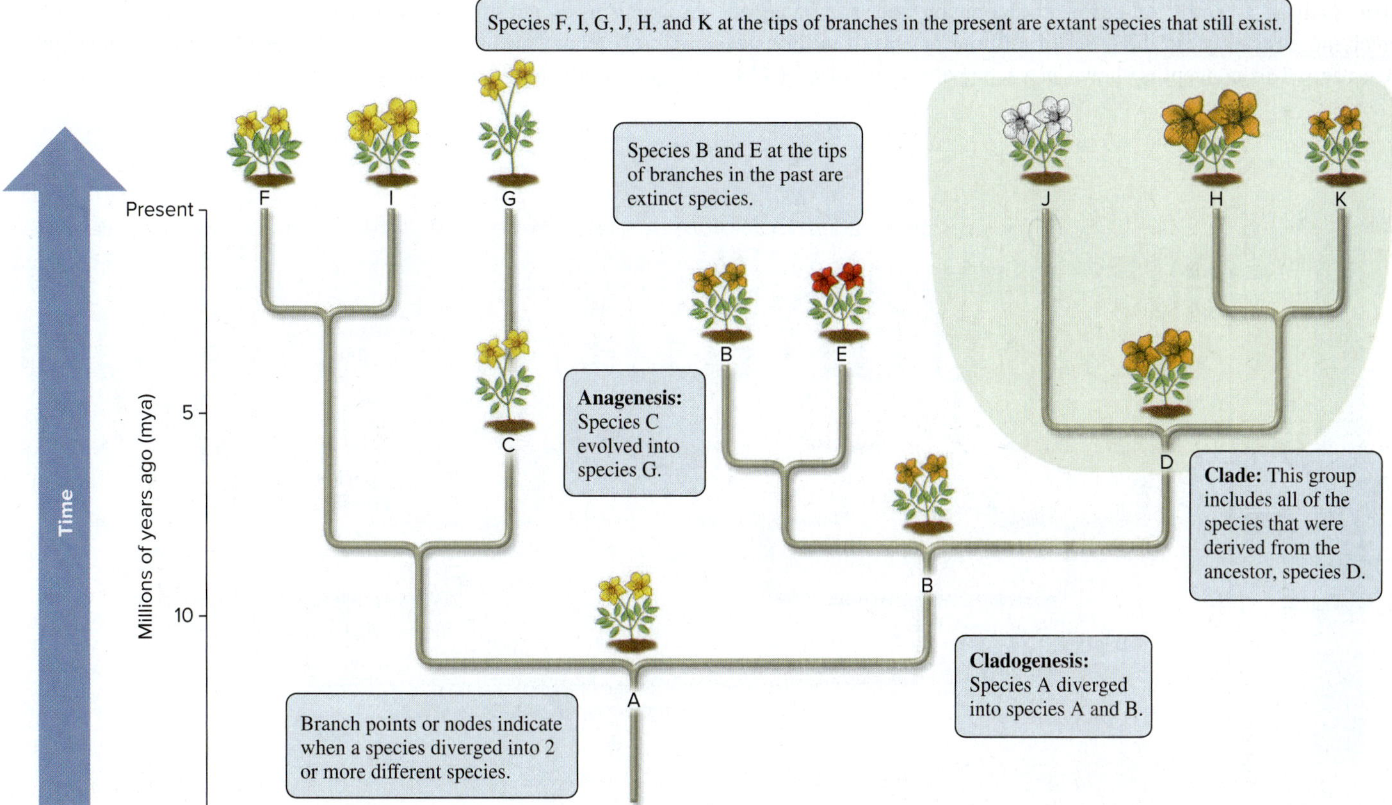


**Figure 21.3** **How to read a phylogenetic tree.** This hypothetical tree shows the proposed relationships between various plant species. Species are placed into clades, groups of organisms containing an ancestral organism and all of its descendants. Note: Anagenesis is a possible way for a new species to arise, but cladogenesis is the primary mechanism.

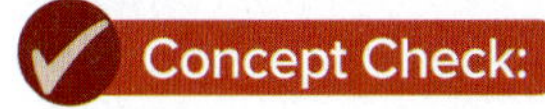

Concept Check: *Can two different species have more than one common ancestor?*

as species B and E, or living species, such as F, I, G, J, H, and K, which are at the top of the tree. Species A and D are also extinct but gave rise to species that are still in existence.

A **clade** consists of a common ancestral species and all of its descendant species. For example, the group highlighted in light green in Figure 21.3 is a clade derived from the common ancestral species labeled D. Likewise, the entire tree forms a clade, with species A as a common ancestor. Therefore, smaller and more recent clades are nested within larger clades that have older common ancestors.

## Systematics Constructs Taxonomic Groups Based on Evolutionary Relationships

A key goal of modern systematics is to create taxonomic groups that reflect evolutionary relationships. Systematics attempts to organize species into clades, which means that each group includes an ancestral species and all of its descendants. A **monophyletic group** is a taxon that is a clade. Ideally, every taxon, whether it is a domain, supergroup, kingdom, phylum, class, order, family, or genus, should be a monophyletic group.

What is the relationship between a phylogenetic tree and taxonomy? The relationship depends on how far back we go to identify a common ancestor. For broader taxa, such as a kingdom or phylum, the common ancestor existed a very long time ago, on the order of hundreds of millions or even billions of years ago. For smaller taxa, such as a family or genus, the common ancestor occurred much more recently, on the order of millions or tens of millions of years ago. This concept is shown in a schematic way in **Figure 21.4.** This small, hypothetical kingdom is a clade that contains 64 living species. (Actual kingdoms are obviously larger and exceedingly more complex.) The diagram emphasizes the taxa that contain the species designated number 43. The common ancestor that gave rise to this kingdom existed approximately 1 bya. Over time, more recent species arose that subsequently became the common ancestors to the phylum, class, order, family, and genus that contain species number 43.

## The Study of Systematics Is Usually Based on Morphological or Genetic Homology

As discussed in Chapter 19, the term **homology** refers to a similarity that occurs due to descent from a common ancestor. Such features are said to be homologous. For example, the arm of a human, the wing of a bat, and the flipper of a whale are homologous structures (refer back to Figure 19.8). Similarly, genes found in different species are homologous if they have been derived from the same ancestral gene (refer back to Figure 19.9).

In systematics, researchers identify homologous features that are shared by some species but not by others, which allows them to group species based on shared similarities. Researchers usually study homology by examining morphological features—those related to the structure of an organism—or they analyze genetic data. In addition, the data they gather are viewed in light of geographic data. Many organisms do not migrate extremely long distances. Species that are closely related evolutionarily are relatively likely to inhabit neighboring or overlapping geographic regions, though many exceptions occur.

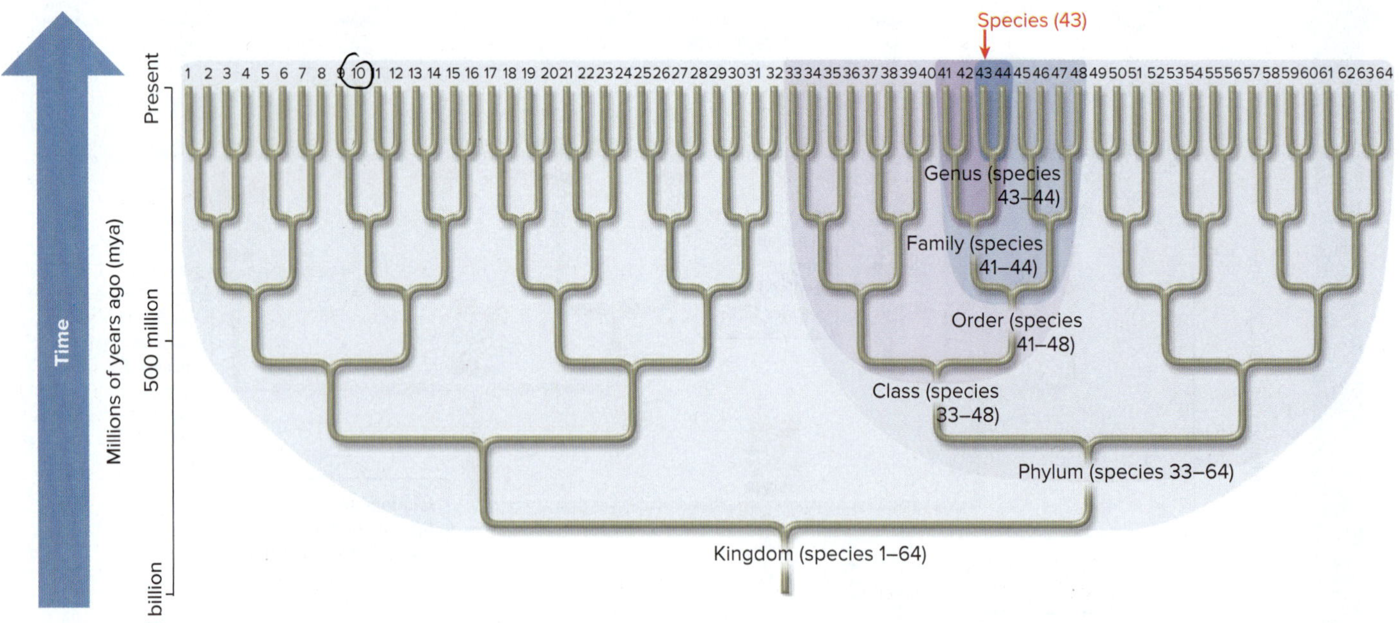


**Figure 21.4** **Schematic relationship between a phylogenetic tree and taxonomy, when taxonomy is correctly based on evolutionary relationships.** The shaded areas highlight the kingdom, phylum, class, order, family, and genus for species number 43. All of the taxa are clades. Broader taxa, such as phyla and classes, are derived from more ancient common ancestors. Smaller taxa, such as families and genera, are derived from more recent common ancestors. These smaller taxa are subsets of the broader taxa.

*Which taxon would have a more recent common ancestor, a phylum or an order?*

**Analysis of Morphological Features** The first studies in systematics focused on morphological features (also called anatomical features) of extinct and living species. Morphological traits continue to be widely used in systematic studies, particularly in those studies pertaining to extinct species and those involving groups that have not been extensively studied at the molecular level. To establish evolutionary relationships based on morphological features, many traits have to be analyzed to identify similarities and differences.

By studying morphological features of extinct species in the fossil record, paleontologists can propose phylogenetic trees that chart the evolutionary lineages of species, including those that still exist. In this approach, the trees are based on morphological features that change over the course of many generations. As an example, Figure 21.5 depicts a current hypothesis of the evolutionary changes that led to the development of the modern horse. This figure shows representative species from various genera. Many morphological features were used to create this tree. Because hard parts of the body are more commonly preserved in the fossil record, this tree is largely based on the analysis of skeletal changes in foot structure, lengths and shapes of various leg bones, skull shape and size, and jaw and teeth. Over an evolutionary timescale, the accumulation of many genetic changes has had a dramatic effect on the species' characteristics. In the genera depicted in this figure, a variety of morphological changes occurred, such as an increase in size, a reduction in the number of toes,

## Biology Principle

### Structure Determines Function

The changes in structural features during horse evolution are related to changes in their functional needs. During this time, horse populations shifted from feeding on leaves in forested regions to feeding on abrasive grasses in more wide-open spaces.

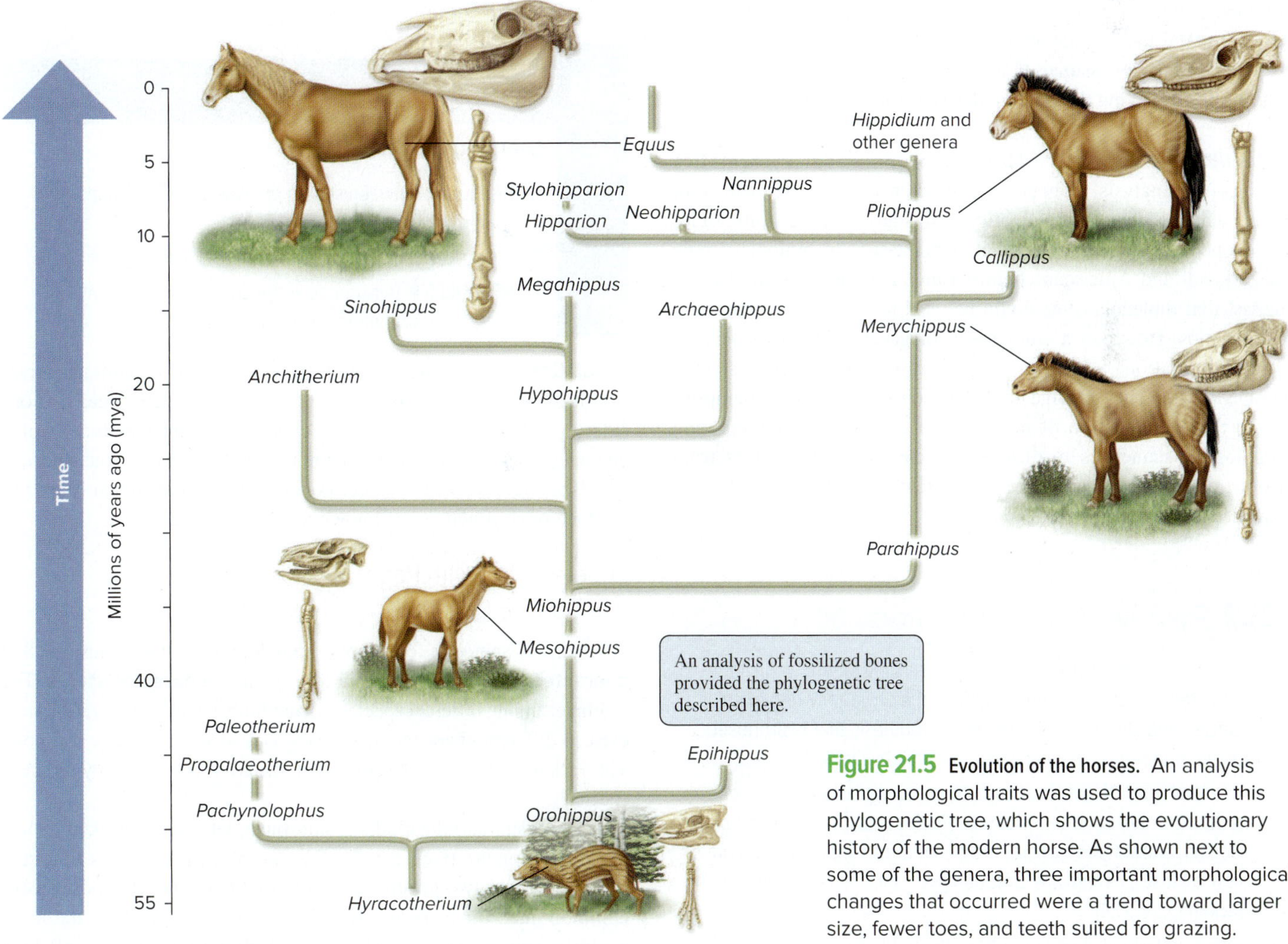


**Figure 21.5** **Evolution of the horses.** An analysis of morphological traits was used to produce this phylogenetic tree, which shows the evolutionary history of the modern horse. As shown next to some of the genera, three important morphological changes that occurred were a trend toward larger size, fewer toes, and teeth suited for grazing.

and modifications in the jaw and teeth consistent with a dietary shift from tender leaves to fibrous grasses.

How do evolutionary biologists explain these changes in horses' traits? The changes can be attributed to natural selection, which acted on existing variation and resulted in adaptations to changes in climate. Over North America, where much of horse evolution occurred, changes in climate caused large areas of dense forests to be replaced with grasslands. The increase in size and changes in foot structure enabled horses to escape predators more easily and travel greater distances in search of food. The changes seen in horses' teeth are consistent with a shift from eating the tender leaves of bushes and trees to eating grasses and other, more abrasive types of vegetation that require more chewing.

A potential drawback of using morphological features for constructing phylogenetic trees is that similar morphological features among different species may be due to convergent evolution. As described in Chapter 19, convergent evolution results in analogous structures—characteristics that arose independently in different lineages because the species evolved in similar environments. For example, the giant anteater and the echidna have long snouts and tongues that enable these animals to feed on ants (refer back to Figure 19.6). These traits are not derived from a common ancestor. Rather, they arose independently during evolution due to adaptation to similar environments. The use of analogous structures in systematics can cause errors if a researcher assumes that a particular trait arose only once and that all species having the trait are derived from a common ancestor.

**Molecular Systematics** The field of **molecular systematics** involves the analysis of genetic data, such as DNA sequences or amino acid sequences, to identify and study genetic homologies and propose phylogenetic trees. In 1963, Austrian biologist Emile Zuckerkandl and American chemist Linus Pauling were the first to suggest that molecular data could be used to establish evolutionary relationships. How can a comparison of genetic sequences help to establish evolutionary relationships? As discussed later in this chapter, DNA sequences change over the course of many generations due to the accumulation of mutations. Therefore, when comparing homologous sequences in different species, DNA sequences from closely related species are more similar to each other than they are to sequences from distantly related ones.

## 21.2 Reviewing the Concepts

- The evolutionary history of a species is its phylogeny. A phylogenetic tree is a diagram that describes the phylogeny of particular species and should be viewed as a hypothesis that is proposed, tested, and refined as more data become available (Figure 21.3).
- A central goal of systematics is to construct taxa and phylogenetic trees based on evolutionary relationships. Smaller taxa, such as families and genera, are derived from more recent common ancestors than are broader taxa such as kingdoms and phyla (Figure 21.4).
- Both morphological and genetic data are used to propose phylogenetic trees (Figure 21.5).

## 21.2 Testing Your Knowledge

1. A phylogenetic tree is
   a. a hypothesis about the evolutionary relationships among different species.
   b. a diagram that depicts the evolutionary relationships among different species.
   c. a diagram that should be organized according to the grouping of species into clades.
   d. all of the above.
2. Earlier in this chapter, we considered the taxonomy of the gray wolf. Which of the following taxa would have a common ancestor that is the oldest?
   a. kingdom Animalia
   b. phylum Chordata
   c. order Carnivora
   d. genus *Canis*
3. To construct a phylogenetic tree, a researcher compares the DNA sequence of a gene that encodes the cytoskeletal protein called tubulin among many different species. This is an example of using
   a. morphological homology.
   b. genetic homology.
   c. the fossil record.
   d. both a and b.

# 21.3 Cladistics

### Learning Outcomes

1. Distinguish between shared primitive characters and shared derived characters.
2. Explain the use of cladistics to construct a phylogenetic tree.
3. **SCISKILLS ▸** Describe how the principle of parsimony is used to choose among phylogenetic trees.

**Cladistics** is the classification of species based on evolutionary relationships. A cladistic approach produces phylogenetic trees by considering the possible pathways of evolutionary changes that involve characteristics that are shared or not shared among various species. In this section, we will consider how the cladistic approach is used to produce phylogenetic trees known as **cladograms.**

## Species Differ with Regard to Primitive and Derived Characters

A cladistic approach compares homologous features, also called **characters,** which may exist in two or more **character states.** For example, among different species, a front limb, which is a character, may exist in different character states such as a wing, an arm, or a flipper. The various character states are either shared or not shared by different species.

To understand the cladistic approach, let's take a look at a simplified phylogeny (**Figure 21.6**). We can place the species that currently exist into two groups: D and E, and F and G. The most recent common ancestor to D and E is B, whereas species C is the most recent common ancestor to F and G. With these ideas in mind, let's focus on the front limbs (flippers versus legs) and eyes.

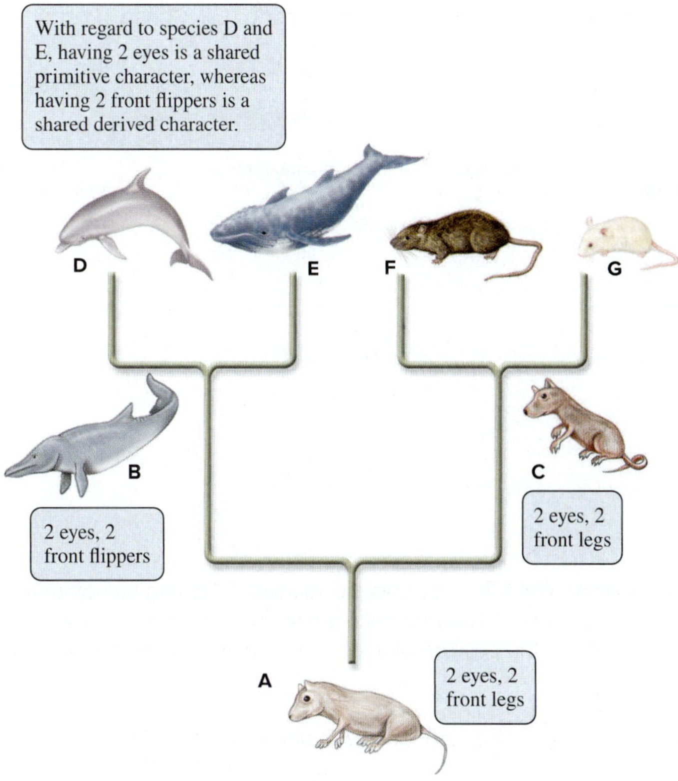


**Figure 21.6** A comparison of shared primitive characters and shared derived characters.

| | Lancelet | Lamprey | Salmon | Lizard | Rabbit |
|---|---|---|---|---|---|
| Notochord | Yes | Yes | Yes | Yes | Yes |
| Vertebrae | No | Yes | Yes | Yes | Yes |
| Hinged jaw | No | No | Yes | Yes | Yes |
| Tetrapod | No | No | No | Yes | Yes |
| Mammary glands | No | No | No | No | Yes |

**(a) Characteristics among species**

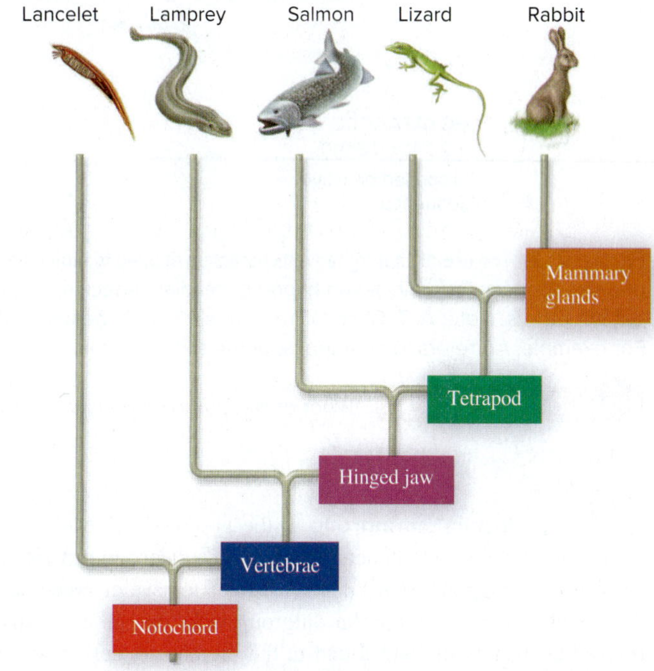


**(b) Cladogram based on morphological traits**

**Figure 21.7** **Using shared primitive characters and shared derived characters to propose a phylogenetic tree. (a)** A comparison of characteristics among these species. **(b)** This phylogenetic tree illustrates both shared primitive and shared derived characters in a cladogram of five animal species.

A character that is shared by two or more different taxa and inherited from ancestors older than their last common ancestor is called a **shared primitive character.** Such characters are viewed as being older—ones that occurred earlier in evolution. With regard to species D, E, F, and G, having two eyes is a shared primitive character. It originated prior to species B and C.

By comparison, a **shared derived character** is a character that is shared by two or more species or taxa and has originated in their most recent common ancestor. With regard to species D and E, having two front flippers is a shared derived character that originated in species B, their most recent common ancestor (see Figure 21.6). Shared derived characters are more recent traits on an evolutionary timescale than are shared primitive characters. For example, among mammals, only some species have flippers, such as whales and dolphins. In this case, flippers were derived from the two front limbs of an ancestral species. The word derived indicates that evolution involves the modification of traits in pre-existing species. In other words, populations of organisms with new traits are derived from changes in pre-existing populations. The basis of the cladistic approach is to analyze many shared derived characters among groups of species to deduce the pathway that gave rise to those species.

The terms primitive and derived do not indicate the complexity of a character. For example, the flippers of a dolphin do not appear more complex than the front limbs of ancestral species A (see Figure 21.6), which were limbs with individual toes. Derived characters can be similar in complexity, less complex, or more complex than primitive characters.

**Concept Check:** *What shared derived character is common to the salmon, lizard, and rabbit, but not the lamprey?*

## A Cladistic Approach Produces a Cladogram Based on Shared Derived Characters

To illustrate how shared derived characters are used to propose a cladogram, **Figure 21.7a** compares several traits among five species of animals. The cladogram shown in **Figure 21.7b** is consistent with the distribution of shared derived characters among these species. A branch point is where two species differ in a character. The oldest common ancestor, which is now extinct, had a notochord and was an ancestor to all five species. Vertebrae are a shared derived character of the lamprey, salmon, lizard, and rabbit, but not the lancelet, which is an invertebrate. By comparison, a hinged jaw is a shared derived character of the salmon, lizard, and rabbit, but not of the lamprey or lancelet.

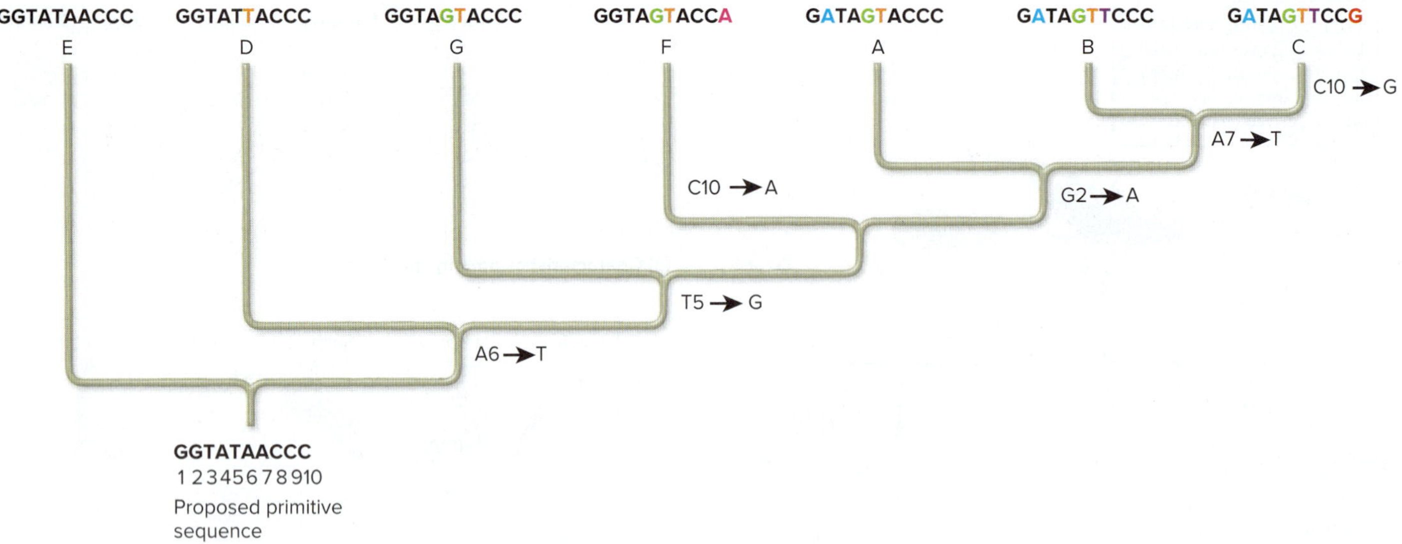


**Figure 21.8** **The use of shared derived characters applied to molecular data.** This phylogenetic tree illustrates a cladogram involving homologous gene sequences found in seven hypothetical plant species. Mutations that alter a primitive DNA sequence are shared among certain species but not others. Note: A, T, G, and C refer to nucleotide bases, and the numbers refer to the position of the base in the nucleotide sequences. For example, A6 refers to an adenine at the sixth position.

*What nucleotide change is a shared derived character for species A, B, and C, but not for species G?*

In a cladogram, an **ingroup** is the group whose evolutionary relationships we wish to understand. By comparison, an **outgroup** is a species or group of species that is assumed to have diverged before the species in the ingroup. An outgroup lacks one or more shared derived characters that are found in the ingroup. A designated outgroup can be closely related or more distantly related to the ingroup. In the tree shown in Figure 21.7, if the salmon, lizard, and rabbit are an ingroup, the lamprey is an outgroup. The lamprey has a notochord and vertebrae but lacks a character shared by the ingroup, namely a hinged jaw. Thus, for the ingroup, the notochord and vertebrae are shared primitive characters, whereas the hinged jaw is a shared derived character not found in the outgroup.

Likewise, the concept of shared derived characters can apply to molecular data, such as a gene sequence. Let's consider an example to illustrate this idea. Our example involves molecular data obtained from seven different hypothetical plant species called A–G. In these species, a homologous region of DNA was sequenced as shown here:

```
   12345678910
A: GATAGTACCC
B: GATAGTTCCC
C: GATAGTTCCG
D: GGTATTACCC
E: GGTATAACCC
F: GGTAGTACCA
G: GGTAGTACCC
```

The cladogram of **Figure 21.8** is a hypothesis of how these DNA sequences arose. In this case, a mutation that changes the sequence of nucleotides is comparable to a modification of a character. For example, let's designate plant species D as an outgroup and species A, B, C, F, and G as the ingroup. In this case, a G (guanine) at the fifth position is a shared derived character. The genetic sequence carrying this G is derived from an older primitive sequence.

In this textbook, most phylogenetic trees are rooted, which means that a single node at the bottom of the tree represents a common ancestor of all species or groups of species in the tree. A method for rooting trees is the use of a noncontroversial outgroup. Such an outgroup typically shares enough morphological traits and/or DNA sequence similarities with the members of the ingroup to allow a comparison between the ingroup and outgroup. Even so, the outgroup must be noncontroversial in that it has enough distinctive differences with the ingroup to be considered a clear outgroup. For example, if the ingroup was a group of mammalian species, an outgroup could be a reptile.

## Quantitative Analysis

### THE PRINCIPLE OF PARSIMONY IS USED TO CHOOSE FROM AMONG POSSIBLE CLADOGRAMS

Although different methods may be used for choosing among possible cladograms, one commonly used approach is to assume that the best hypothesis is the one that requires the fewest number of evolutionary changes to explain the differences among observed characters. This concept, called the **principle of parsimony,** states that the preferred hypothesis is the one that is the simplest for all the characters and their states. For example, if two species possess a tail, we would initially assume that a tail arose once during evolution and that both species have descended from a common ancestor with a tail. Such a hypothesis is simpler, and more likely to be correct, than assuming that tails arose twice during evolution and that the tails in the two species are not due to descent from a common ancestor.

The principle of parsimony can also be applied to genetic data, in which case the most likely hypothesis is the one requiring the fewest base changes. Let's consider a hypothetical example involving a DNA sequence from four taxa (A–D), where A is presumed to be the outgroup.

| | 12345 |
|---|---|
| A: | GTACA (outgroup) |
| B: | GACAG |
| C: | GTCAA |
| D: | GACCG |

Given that B, C and D are the ingroup, three hypotheses for phylogenetic trees are shown in **Figure 21.9**, although more are possible. Tree 1 requires seven mutations, and tree 2 requires six, whereas tree 3 requires only five. Therefore, tree 3 requires the smallest number of mutations and is considered the most parsimonious. Based on the principle of parsimony, the tree with the fewest number of base changes is the hypothesis that is the most likely to accurately reflect the evolutionary history of the ingroup in question. In practice, when researchers have multiple sequences that are longer than the ones shown here, computer programs are used to find the most parsimonious tree.

**Crunching the Numbers:** According to the principle of parsimony, which of these trees would you consider to be the least likely? Explain why.

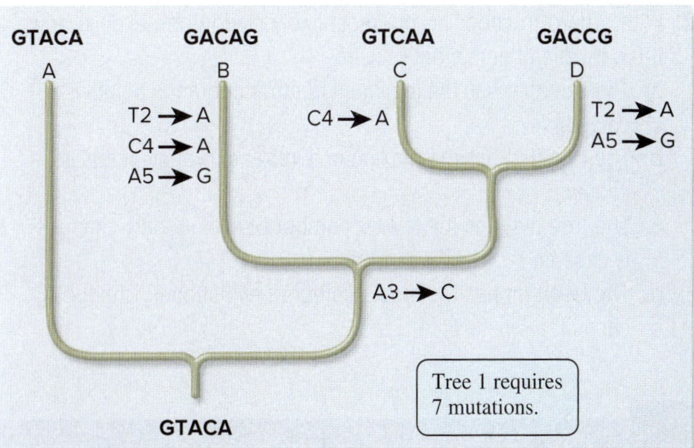


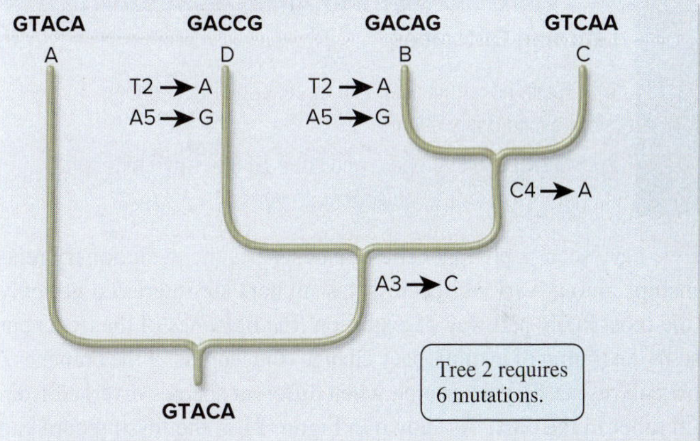


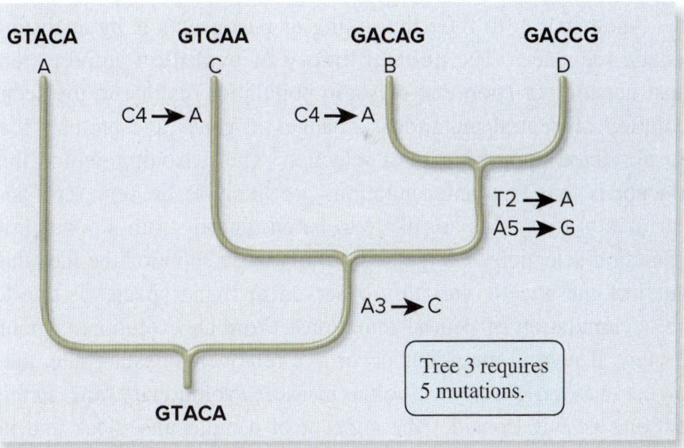


**Figure 21.9 Using the principle of parsimony and molecular genetic data to choose a phylogenetic tree.** Shown are three possible phylogenetic trees for the evolution of a short DNA sequence (although many more are possible). Changes in nucleotide sequence are indicated for each tree. For example, T2 → A means that the second base, a T, was changed to an A. According to the principle of parsimony, tree number 3 is the most likely choice because it requires only five mutations.

## 21.3 Reviewing the Concepts

- In the cladistic approach to creating a phylogenetic tree, also called a cladogram, species are grouped together according to shared derived characters, characters shared by two or more species or taxa and originating in their most recent common ancestor (Figure 21.6).
- An ingroup is the group of interest, whereas an outgroup is a species or group of species that lacks one or more shared derived characters found in the ingroup. A comparison of the ingroup and outgroup is used to determine which character states are derived or primitive (Figures 21.7, 21.8).
- The cladistic approach produces many possible cladograms. The most likely phylogenetic tree may be chosen by the principle of parsimony (Figure 21.9).

## 21.3 Testing Your Knowledge

1. In a cladogram, the branch points, or nodes, are determined by differences in

   a. species complexity.
   b. shared derived characters.
   c. genetic homology but not molecular homology.
   d. molecular homology but not genetic homology.

2. With regard to choosing the most likely cladogram, explain how the principle of parsimony is used.
   a. The tree that has the minimum number of branch points is chosen.
   b. The tree that has the maximum number of branch points is chosen.
   c. The tree that has the fewest number of evolutionary changes is chosen.
   d. The tree that has the most number of evolutionary changes is chosen.

## 21.4 Molecular Clocks

### Learning Outcomes

1. Explain how molecular clocks are used in the dating of evolutionary events.
2. **SCISKILLS ▶** Compare and contrast the use of different genes to produce phylogenetic trees.

As we have seen, a phylogenetic tree describes the evolutionary relationships among various species. Researchers are interested not only in the most likely pathway of evolution (the branches of the trees) but also in the timing of evolutionary change (the lengths of the branches). How can researchers determine when different species diverged from each other in the past? As shown in Figure 21.5, the fossil record can sometimes help researchers apply a timescale to a phylogeny.

Another way to infer the timing of past events is by analyzing genetic sequences. The **neutral theory of evolution** proposes that most genetic variation that exists in populations is due to the accumulation of neutral mutations—changes in genes and proteins that are not acted upon by natural selection. The reasoning behind this concept is that favorable mutations are likely to be very rare, and detrimental mutations are likely to be eliminated from a population by natural selection. A large body of evidence supports the idea that much of the genetic variation observed in living species is due to the accumulation of neutral mutations. From an evolutionary point of view, if neutral mutations occur at a relatively constant rate, they can serve as a **molecular clock** to measure evolutionary time. In this section, we will consider the concept of a molecular clock and its application in phylogenetic trees.

### The Timing of Evolutionary Change May Be Inferred from Molecular Clock Data

**Figure 21.10** illustrates the concept of a molecular clock. The graph's vertical axis is the number of base-pair differences in a homologous gene between different pairs of species. The horizontal axis plots the amount of time that has elapsed since each pair of species shared a common ancestor.

As an example, let's suppose a researcher compared a gene sequence that was 500 bp long. Between species A and species B, this sequence might differ at 10 places and be identical at 490 places. By comparison, the 500-bp sequence might differ at 20 places between species A and species C and be the same at 480 places. Such a result is consistent with the idea that species A and species B share a more recent common ancestor than do species A and species C. The explanation for this phenomenon is that the gene sequences of various species accumulate independent mutations after they have diverged from each other. A longer period of time since their divergence allows for a greater accumulation of mutations, which makes their sequences of base pairs more different.

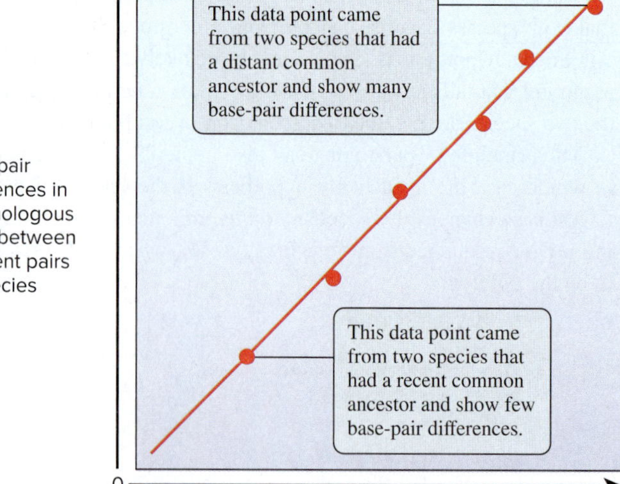


**Figure 21.10** **A molecular clock.** According to the concept of a molecular clock, neutral mutations accumulate at a relatively constant rate over evolutionary time. When comparing homologous genes between species, those species that diverged more recently tend to have fewer differences than do those whose common ancestor occurred in the distant past.

**BioConnections:** *Look back at Table 10.1, which shows the genetic code. Propose a mutation that would change the sequence of a codon and also be neutral.*

Figure 21.10 suggests a linear relationship between the number of base-pair changes and the time of divergence. For example, a linear relationship predicts that a pair of species that has, say, 20 differences in a given gene sequence would have a common ancestor that is roughly twice as old as that of a pair showing 10 differences. Although actual data sometimes show a relatively linear relationship over a defined time period, evolutionary biologists do not think that molecular clocks are perfectly linear over very long periods of time. Several factors can contribute to nonlinearity of molecular clocks. These include differences in the generation times of the species being analyzed and variation in the mutation rates of genes between different species.

To obtain reliable data, researchers must calibrate their molecular clocks. How much time does it take to accumulate a certain percentage of base-pair changes? To perform such a calibration, researchers must have information regarding the date when two species diverged from a common ancestor. Such information could come from the fossil record, for instance. The genetic differences between those species are then divided by the amount of time since their last common ancestor to calculate a rate of evolutionary change.

As an example of clock calibration, let's consider primates. The fossil evidence suggests that humans and chimpanzees diverged from a common ancestor approximately 6 mya. The percentage of base-pair differences between the mitochondrial DNA of humans and chimpanzees is 12%. From these data, the molecular clock for changes in mitochondrial DNA sequences of primates is calibrated at roughly 2% base-pair changes per million years.

## Different Genes Are Analyzed to Study Phylogeny and Evaluate the Timing of Evolutionary Change

For evolutionary comparisons, the DNA sequences of many genes have been obtained from a wide range of sources. Many different genes have been used to propose phylogenetic trees and evaluate the timing of past events. For example, the gene that encodes an RNA found in the small ribosomal subunit (SSU rRNA) has been commonly used in evolutionary studies. As noted in Chapter 10, the gene for SSU rRNA is found in the genomes of all living organisms. Therefore, its function must have been established at an early stage in the evolution of life on this planet, and its sequence has changed fairly slowly. Furthermore, SSU rRNA is a rather large molecule, so it contains a large amount of sequence information. This gene has been sequenced from thousands of different species (refer back to Figure 10.13). Slowly changing genes such as the gene that encodes SSU rRNA are useful for evaluating distant evolutionary relationships, such as comparing higher taxa. For example, SSU rRNA data can be used to place eukaryotic species into their proper phyla or orders.

Other genes have changed more rapidly during evolution because of a greater tolerance of neutral mutations. For example, the mitochondrial genome and DNA sequences within introns can more easily incur neutral mutations (compared to the coding sequences within exons), and so their sequences change frequently during evolution. More rapidly changing DNA sequences have been used to study recent evolutionary relationships, particularly among eukaryotic species such as large animals that have longer generation times and tend to evolve more slowly. In these cases, slowly evolving genes may not be very useful for establishing evolutionary relationships because two closely related species are likely to have identical or nearly identical DNA sequences for such genes.

**Figure 21.11** shows a simplified phylogeny of closely related species of primates. This tree was proposed by comparing DNA sequence

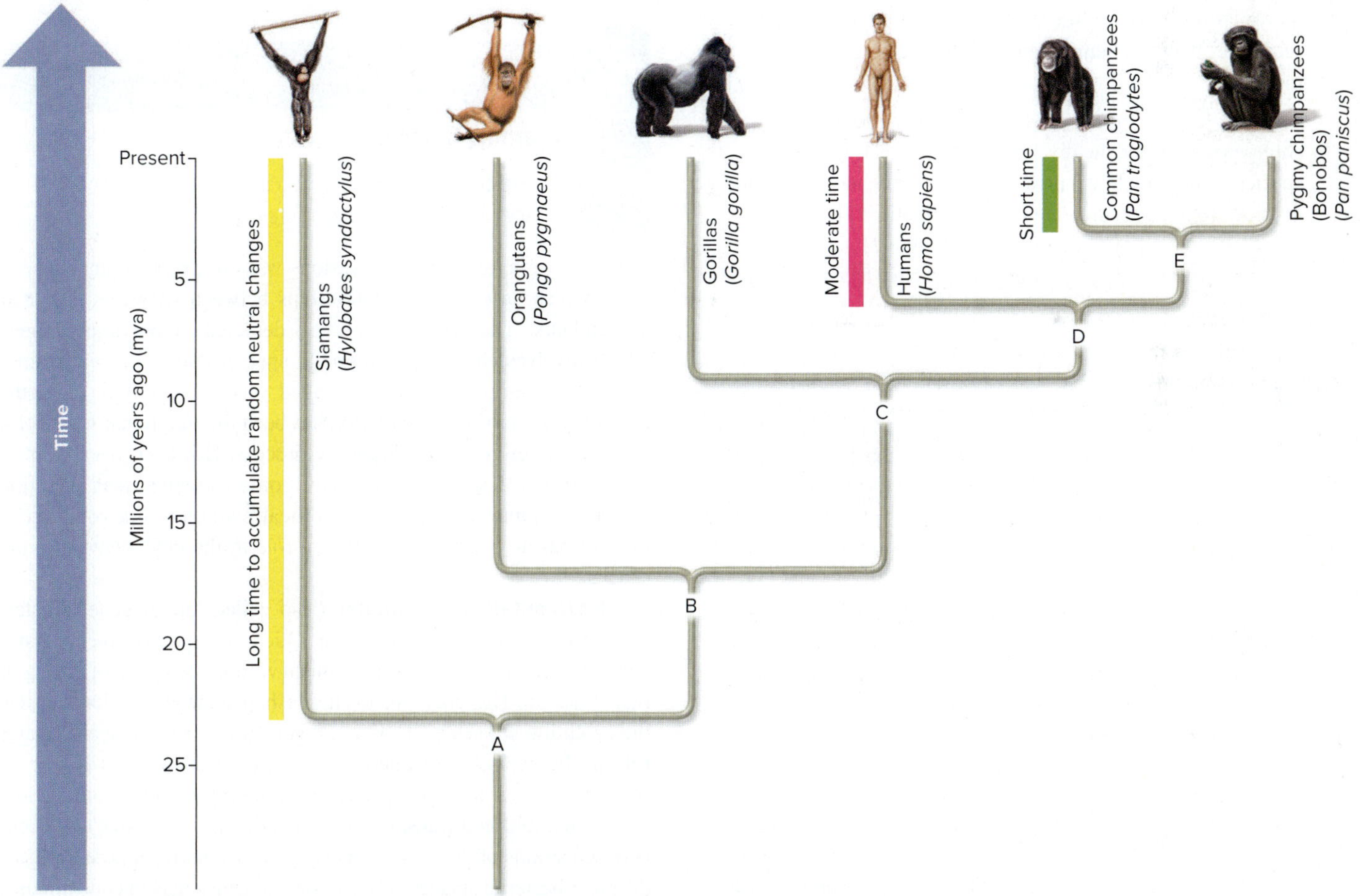

**Figure 21.11 The use of DNA sequence changes to study primate evolution.** This phylogenetic tree, which shows relationships among closely related species of primates, is based on a comparison of mitochondrial gene sequences encoding the protein cytochrome oxidase subunit II.

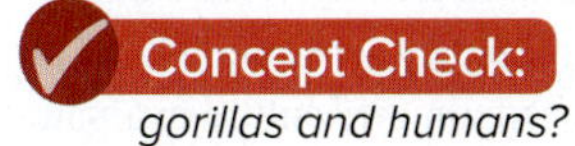

*Which pair of species would be expected to have fewer genetic differences: orangutans and gorillas or gorillas and humans?*

changes in the gene for cytochrome oxidase subunit II, one of several subunits of cytochrome oxidase, a protein in the mitochondrial inner membrane that is involved in cellular respiration. This gene tends to change fairly rapidly on an evolutionary timescale. The vertical scale on Figure 21.11 represents time, and the branch points that are labeled with letters represent common ancestors. Let's take a look at three branch points (labeled A, D, and E) and relate them to the accumulation of neutral mutations.

- *Ancestor A:* This ancestor diverged into two species that ultimately gave rise to siamangs and the other five species. Since this divergence, there has been a long time (approximately 23 million years) for the siamang genome to accumulate a relatively high number of random neutral changes that would be different from the random changes that have occurred in the genomes of the other five species (see the yellow bar in Figure 21.11). Therefore, the gene in the siamangs is fairly different from the genes in the other five species.
- *Ancestor D:* This ancestor diverged into two species that eventually gave rise to humans and chimpanzees. This divergence occurred a moderate time ago, approximately 6 mya, as illustrated by the red bar. The differences in gene sequences between humans and chimpanzees are relatively moderate.
- *Ancestor E:* This ancestor diverged into two species of chimpanzees. Since the divergence of species E into two species, approximately 3 mya, the time for the molecular clock to "tick" (that is, accumulate random mutations) is relatively short, as depicted by the green bar in Figure 21.11. Therefore, the two existing species of chimpanzees have fewer differences in their gene sequences than between other primates.

## 21.4 Reviewing the Concepts

- The neutral theory of evolution proposes that most genetic variation in a population is due to neutral mutations. Assuming that neutral mutations occur at a relatively constant rate, genetic data can serve as a molecular clock to measure evolutionary time (Figure 21.10).
- Slowly changing genes are useful for analyzing distant evolutionary relationships, whereas rapidly changing genes are used to analyze more recent evolutionary relationships, particularly among eukaryotes that have long generation times and evolve more slowly (Figure 21.11).

## 21.4 Testing Your Knowledge

1. When the same homologous gene is compared between closely related species (for example, a lion and a tiger) versus distantly related species (for example, a lion and a turtle),
   a. the genes between *closely* related species have *fewer* differences because their common ancestor occurred more recently.
   b. the genes between *distantly* related species have *fewer* differences because their common ancestor occurred more recently.
   c. the genes between *closely* related species have *more* differences because their common ancestor occurred more recently.
   d. the genes between *distantly* related species have *more* differences because their common ancestor occurred more recently.
2. An analysis of the number of neutral genetic changes among different species can be used to estimate the dates of past evolutionary events, such as the divergence of one species into two species. For this to be valid,
   a. researchers must have a way to calibrate their clock, such as using information from the fossil record.
   b. the rate of neutral mutations must be relatively constant over a given time period.
   c. the gene sequences that are analyzed must be relatively short.
   d. both a and b.

## 21.5 Horizontal Gene Transfer

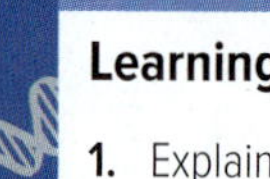

### Learning Outcome

1. Explain how horizontal gene transfer affects evolution and the relationships among different taxa.

Thus far, we have considered various ways to propose phylogenetic trees, which describe the relationships between ancestors and their descendants. The type of evolution depicted in previous figures, which involves changes in groups of species due to descent from a common ancestor, is sometimes called vertical evolution. Since the time of Darwin, vertical evolution has been the traditional way biologists have viewed the evolutionary process. However, over the past couple of decades, researchers have come to realize that evolution is not so simple. In addition to vertical evolution, horizontal gene transfer has also played a significant role in the phylogeny of living species.

**Horizontal gene transfer** (also called lateral gene transfer) is used to describe any process in which an organism incorporates genetic material from another organism without being the offspring of that organism. This phenomenon has reshaped the way biologists view the evolution of species. Horizontal gene transfer has played a major role in the evolution of many species. As discussed in Chapter 17 (refer back to Table 17.3), bacteria can transfer genes via conjugation, transformation, and transduction. Bacterial gene transfer can occur between strains of the same species or, occasionally, between cells of different bacterial species. The transferred genes may encode proteins that provide a survival advantage, such as resistance to antibiotics or the ability to metabolize an organic molecule in the environment. Horizontal gene transfer is also fairly common among certain unicellular eukaryotes. However, its relative frequency and importance in the evolution of multicellular eukaryotes remain difficult to evaluate.

Scientists have debated the role of horizontal gene transfer in the earliest stages of evolution, prior to the divergence of the bacterial and archaeal domains. The traditional viewpoint was that the three domains of life—Bacteria, Archaea, and Eukarya—arose from a single type of prokaryotic (or pre-prokaryotic) cell called the universal ancestor. However, genomic research has suggested that horizontal gene transfer may have been particularly common during the early stages of evolution on Earth, when all species were unicellular. Horizontal gene transfer may have been so prevalent that the universal ancestor may have actually been an ancestral community of cell lineages that evolved as a whole. If that were the case, the tree of life cannot be traced back to a single prokaryotic ancestor.

**Figure 21.12** illustrates a schematic scenario for the evolution of life on Earth that includes the roles of both vertical evolution and horizontal gene transfer. This has been described as a "web of life" rather than a "tree of life." In this scenario, instead of a universal ancestor, a community of primitive cells frequently transferred genetic material in a horizontal fashion. Horizontal gene transfer was also prevalent during the early evolution of bacteria and archaea, and when eukaryotes first emerged as unicellular species. In living bacteria and archaea, it remains a prominent way to foster evolutionary change. By comparison, the region of the diagram that contains eukaryotic species has a more treelike structure. Researchers have speculated that multicellularity and sexual reproduction have presented barriers to horizontal gene transfer in most eukaryotes. For a gene to be transmitted to eukaryotic offspring, it would have to be transferred into a eukaryotic cell that is a gamete or a cell that gives rise to gametes. Horizontal gene transfer has become less common in eukaryotes, though it does occur occasionally.

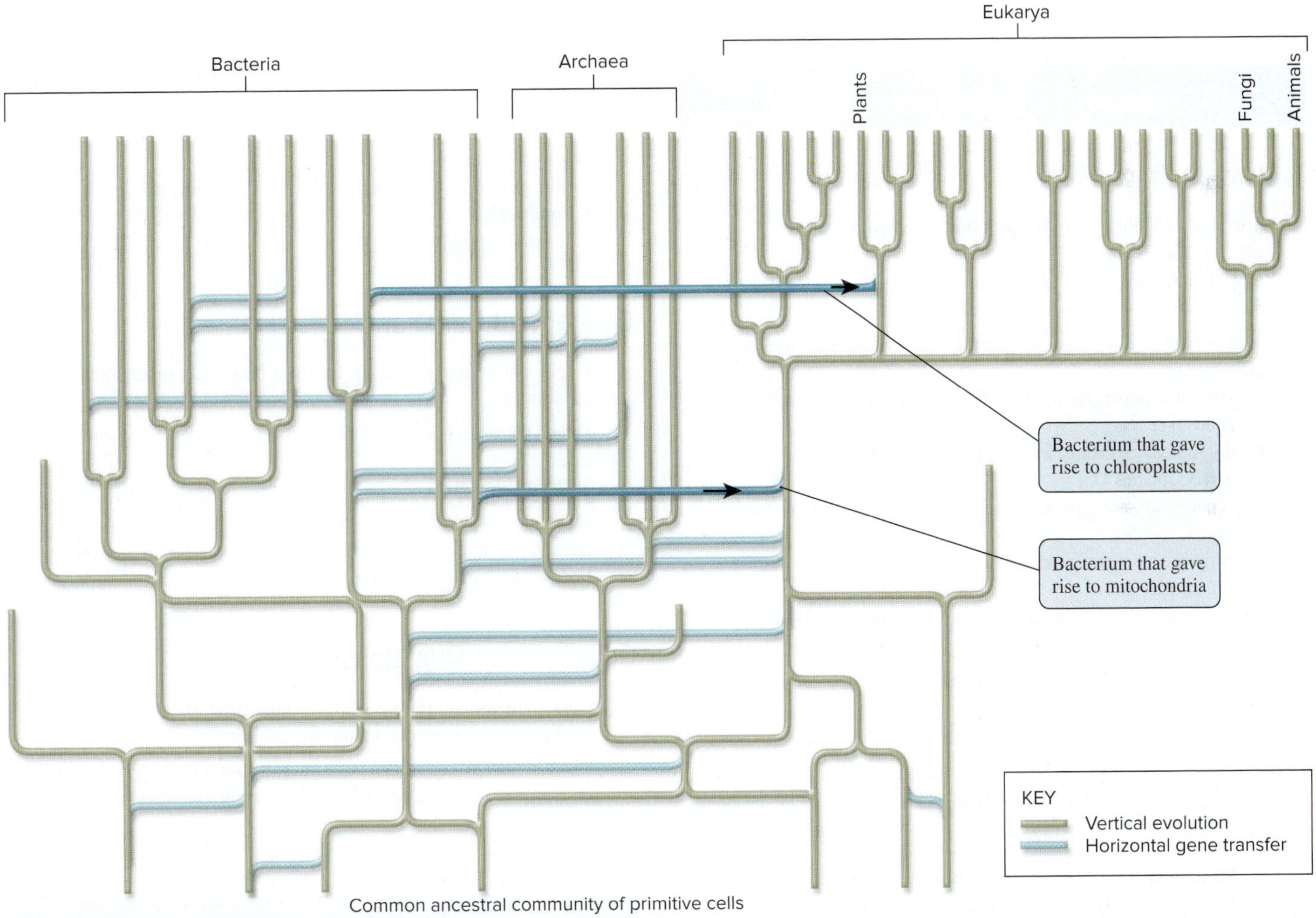

Eukarya
Bacteria
Archaea
Plants
Fungi
Animals
Bacterium that gave rise to chloroplasts
Bacterium that gave rise to mitochondria
KEY
Vertical evolution
Horizontal gene transfer
Common ancestral community of primitive cells

**Figure 21.12 A web of life.** This phylogenetic tree shows not only the vertical evolution of life on Earth but also the contribution of horizontal gene transfer. In this scenario, horizontal gene transfer was prevalent during the early stages of evolution, when all organisms were unicellular, and continues to be a prominent factor in the speciation of Bacteria and Archaea. Note: This tree is schematic. Also, while the introduction of chloroplasts into the eukaryotic domain is shown as a single event, such events have occurred multiple times and by different mechanisms.

*How does the phenomenon of horizontal gene transfer muddle the concept of monophyletic groups?*

## 21.5 Reviewing the Concepts

- Horizontal gene transfer is the phenomenon in which an organism incorporates genetic material from another organism without being the offspring of that organism. Due to the prevalence of horizontal gene transfer, the tree of life may more accurately be described as a web of life (Figure 21.12).

## 21.5 Testing Your Knowledge

1. Which of the following is an example of horizontal gene transfer?
   a. a mating between two different species, such as a horse and a donkey, to produce a mule
   b. the passage of genes from a mother cell to two daughter cells during cell division
   c. the transfer of an antibiotic-resistance gene from one bacterial species to a different species
   d. both a and c

# Assess and Discuss

## Test Yourself

1. The study of biological diversity based on evolutionary relationships is
   a. paleontology. b. evolution. c. systematics. d. natural selection. e. both a and b.
2. Which of the following is the correct order of the taxa used to classify organisms?
   a. kingdom, supergroup, domain, phylum, class, order, family, genus, species
   b. domain, supergroup, kingdom, class, phylum, order, family, genus, species
   c. supergroup, domain, kingdom, phylum, class, family, order, genus, species
   d. domain, supergroup, kingdom, phylum, class, order, family, genus, species
   e. kingdom, supergroup, domain, phylum, order, class, family, species, genus
3. Which type of taxon consists of organisms with the greatest similarity?
   a. kingdom b. class c. order d. family e. genus
4. Which of the following characteristics is *not* shared by bacteria, archaea, and eukaryotes?
   a. DNA is the genetic material.
   b. Messenger RNA encodes the information to produce proteins.
   c. All cells are surrounded by a plasma membrane.
   d. The cytoplasm is compartmentalized into organelles.
   e. Both a and d are correct.
5. Which of the following occurs at branch points, or nodes, in a phylogenetic tree?
   a. anagenesis b. cladogenesis c. horizontal gene transfer d. both a and b e. both b and c
6. The evolutionary history of a species is its
   a. domain. b. taxonomy. c. evolution. d. phylogeny. e. embryology.
7. A taxon composed of all species derived from a common ancestor is referred to as
   a. a phylum.
   b. a monophyletic group or clade.
   c. a genus.
   d. an outgroup.
   e. all of the above.
8. A goal of modern taxonomy is to
   a. classify all organisms based on morphological similarities.
   b. classify all organisms into monophyletic groups.
   c. classify all organisms based solely on genetic similarities.
   d. determine the evolutionary relationships only between similar species.
   e. none of the above.
9. The concept that the preferred hypothesis is the one that is the simplest is termed
   a. homology. b. cladistics. c. the principle of parsimony. d. a molecular clock. e. both b and d.
10. Research indicates that horizontal gene transfer is less prevalent in eukaryotes because of
    a. the presence of organelles. b. multicellularity. c. sexual reproduction. d. all of the above. e. b and c only.

## Conceptual Questions

1. Explain how species' names follow a binomial nomenclature. Give an example.
2. What is a molecular clock? How is it used in depicting phylogenetic trees?
3. **PRINCIPLES** A principle of biology is that populations of organisms evolve from one generation to the next. What are some advantages and potential pitfalls of using changes in morphological features to construct phylogenetic trees?

## Collaborative Questions

1. Discuss how taxonomy is useful. Make a list of some practical applications that are derived from taxonomy.
2. Discuss how systematics is used to propose a phylogenetic tree. Discuss the rationale behind using the principle of parsimony.

# Online Resource

# The History of Life on Earth and Human Evolution

# 22

© George Bernard/SPL/Science Source

**A fossil fish.** This 50-million-year-old fossil of a unicorn fish (*Naso rectifrons*) is an example of the many different kinds of organisms that have existed during the history of life on Earth.

## Chapter Outline



The amazing origin of the universe is difficult to comprehend. Astronomers propose that the universe began with a cosmic explosion called the Big Bang about 13.8 billion years ago (bya), when the first clouds of the elements hydrogen and helium were formed. Over a long time period, gravitational forces collapsed these clouds to create stars that converted hydrogen and helium into heavier elements, including carbon, nitrogen, and oxygen, which are the atomic building blocks of life on Earth. These elements were returned to interstellar space by exploding stars called supernovas, which created clouds in which simple molecules such as water, carbon monoxide, and hydrocarbons formed. The clouds then collapsed to make a new generation of stars and solar systems.

Our solar system began about 4.6 bya after one or more local supernova explosions. According to one widely accepted scenario, hundreds of planetesimals (small celestial bodies, such as asteroids) occupied the region where Venus, Earth, and Mars are now found. The Earth, which is estimated to be 4.55 billion years old, grew from the aggregation of such planetesimals over a period of 100 to 200 million years. For the first half billion years or so after its formation, the Earth was too hot to allow liquid water to accumulate on its surface. By 4 bya, the Earth had cooled enough for the outer layers of the planet to solidify and for oceans to begin to form.

The period between 4.0 and 3.5 bya marked the emergence of life on our planet. The first forms of life that we know about produced microscopic fossils, the preserved remains of organisms that existed in the past. These fossils, estimated to be about 3.5 billion years old, resemble modern cyanobacteria, which are photosynthetic bacteria (Figure 22.1). Researchers cannot travel back through time and observe how the first life-forms came into being. However, plausible hypotheses regarding how life first arose have emerged from our understanding of modern life.

(a) Fossil prokaryote

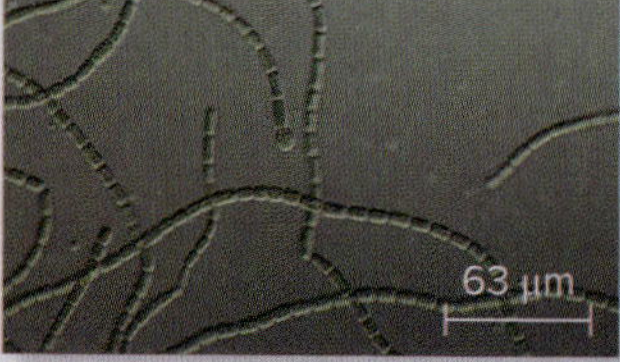


(b) Modern cyanobacteria

**Figure 22.1** **Earliest fossils and living cyanobacteria.** (**a**) A fossilized prokaryote about 3.5 billion years old that is thought to be an early cyanobacterium. (**b**) A modern cyanobacterium, which has a similar morphology. Cyanobacterial cells connect to each other to form chains, as shown here.

*(a)* © J.W. Schopf; *(b)* © Michael Abbey/Science Source

This chapter emphasizes when particular forms of life arose. We will begin by considering how researchers analyze and date fossils, which are the remains of past life-forms. We will then consider how fossils, such as the one shown in the chapter-opening photo, have provided biologists with evidence of the history of life on Earth from its earliest beginnings to the present day. The last section emphasizes one of the most interesting stories of evolution, which is the lineage that gave rise to modern humans.

## 22.1 The Fossil Record

### Learning Outcomes

1. Describe how fossils are formed.
2. **SCISKILLS ▸** Explain how radiometric dating is used to estimate the age of a fossil.
3. List several factors that affect the completeness of the fossil record.

Let's begin by turning our attention to a process that has given us a window into the history of life over the past 3.5 billion years. **Fossils** are the preserved remains of past life on Earth. They can take many forms, including bones, shells, and leaves, and the impression of cells or other evidence, such as footprints or burrows. Scientists who study fossils are called **paleontologists** (from the Greek *palaios,* meaning ancient). Because our understanding of the history of life is derived primarily from the fossil record, it is important to appreciate how fossils are formed and dated, and to understand why the fossil record cannot be viewed as complete.

### Fossils Are Formed Within Sedimentary Rock

How are fossils usually formed? Most fossils are found in sedimentary rocks that were formed from particles of older rocks broken apart by water or wind. These particles—in the form of gravel, sand, or mud—settle and bury living and dead organisms at the bottoms of rivers, lakes, and oceans. Over time, more particles pile up, and sediments at the bottom of the pile eventually become compressed into rock. Most fossils are formed when organisms are buried quickly, and then during the process of sedimentary rock formation, their hard parts are gradually replaced over millions of years by minerals, producing a recognizable representation of the original organism (see, for example, the chapter-opening photo).

The relative ages of fossils can sometimes be revealed by their locations in sedimentary rock formations. Because sedimentary rocks are formed by small particles settling in layers, the layers are piled one on top of the other. In a sequence of layered rock, the lower layers are usually older than the upper layers. Paleontologists often study changes in life-forms over time by studying the fossils in layers from bottom to top (**Figure 22.2**). The more ancient life-forms are found in the lower layers, and newer species are found in the upper ones. However, such an assumption can occasionally be misleading when geological processes, such as folding, have flipped the layers.

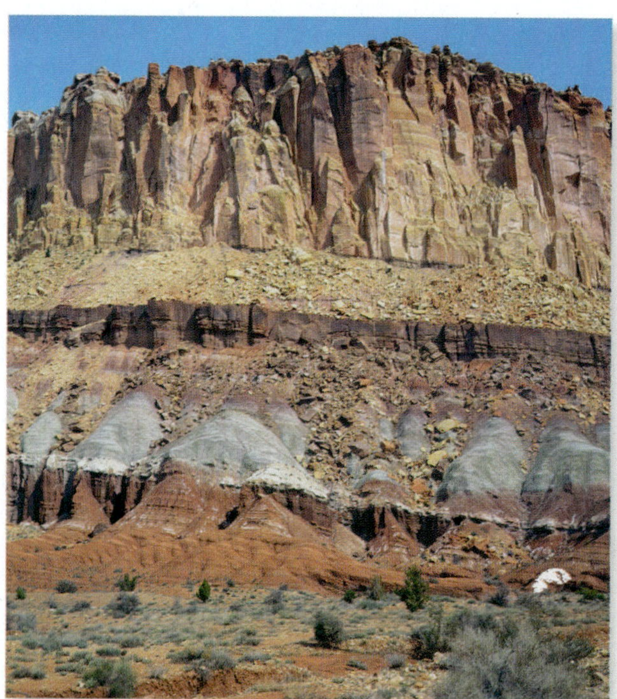

**Figure 22.2** **An example of layers of sedimentary rock that contain fossils.**

© Simon Fraser/SPL/Science Source

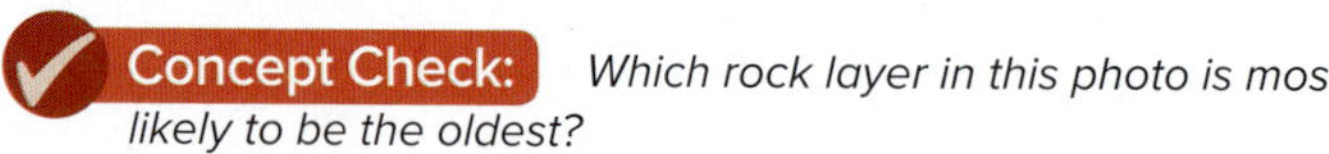

**Concept Check:** *Which rock layer in this photo is most likely to be the oldest?*

## Quantitative Analysis

### RADIOISOTOPES PROVIDE A WAY TO DATE FOSSILS

A common way to estimate the age of a fossil is by analyzing the decay of radioisotopes within the accompanying rock, a process called **radiometric dating.** The atoms of certain elements can exist in multiple forms, called isotopes, that differ in the number of neutrons they contain. A radioisotope is an unstable isotope of an element that decays spontaneously, releasing radiation at a constant rate. The **half-life** is the length of time required for a radioisotope to decay to exactly one-half of its initial quantity. Each radioisotope has its own unique half-life (**Figure 22.3a**). Within a sample of rock, scientists can measure the amount of a given radioisotope as well as the amount of the decay product—the isotope that is produced when the original isotope decays. For dating geological materials, several types of isotope decay patterns are particularly useful, including carbon-14 to nitrogen-14 and uranium-235 to lead-207 (**Figure 22.3b**).

To determine the age of a rock using radiometric dating, paleontologists need to have a way to set the clock—extrapolate back to a starting point in which a rock did not have any amount of the decay product. Except for fossils less than 50,000 years old, in which carbon-14 ($^{14}C$) dating can be employed, fossil dating is not usually conducted on the fossil itself or on the sedimentary rock in which the fossil is found. Most commonly, igneous rock—rock formed through the cooling and solidification of lava—in the vicinity of the sedimentary rock is dated. Why is igneous rock chosen? One reason is that igneous rock derived from an ancient lava flow initially contains uranium-235 ($^{235}U$) but no lead-207 ($^{207}Pb$). The ultimate decay product of $^{235}U$ is $^{207}Pb$. By comparing the relative proportions of $^{235}U$ and $^{207}Pb$ in a sample, the age of igneous rock can be accurately determined.

The relationship between a radioisotope and the process of decay is represented by the following equation:

$$N = N_0 e^{-(0.693t/T_{1/2})}$$

where

$N$ is the number of atoms of a radioisotope after a certain time period.
$N_0$ is the number of atoms of that radioisotope that were originally present prior to any decay.
$e$ is the natural logarithm.
$t$ is the time period during which decay has occurred.
$T_{1/2}$ is the half-life of the radioisotope.

**Crunching the Numbers:** A paleontologist discovered a fossil of a previously unidentified reptile. A sample of nearby igneous rock that was 1 kg in weight contained 0.11 mg of uranium-235 and 0.035 mg of lead-207. Estimate the age of this new fossil. (Note: Uranium and lead have different molecular masses, so you first need to calculate the number of uranium and lead atoms in this sample.)

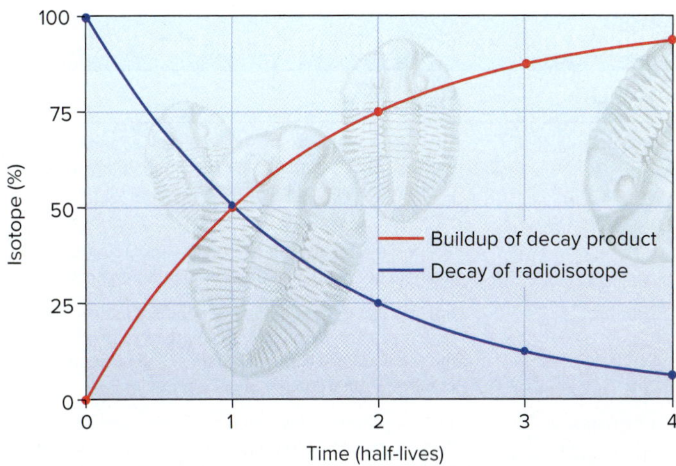


**(a) Decay of a radioisotope**

| Radioisotope | Decay product | Half-life (years) | Useful dating range (years) |
|---|---|---|---|
| Carbon-14 | Nitrogen-14 | 5,730 | 100–50,000 |
| Potassium-40 | Argon-40 | 1.3 billion | 100,000–4.5 billion |
| Rubidium-87 | Strontium-87 | 47 billion | 10 million–4.5 billion |
| Uranium-235 | Lead-207 | 710 million | 10 million–4.5 billion |
| Uranium-238 | Lead-206 | 4.5 billion | 10 million–4.5 billion |

**(b) Radioisotopes that are useful for geological dating**

**Figure 22.3** **Radiometric dating of fossils.** **(a)** A rock can be dated by measuring the relative amounts of a radioisotope and its decay product within the rock. **(b)** These five radioisotopes, each of which has a useful dating range, are commonly analyzed for the dating of fossils.

**Concept Check:** *If you suspected a fossil is 50 million years old, which pair of radioisotopes would you choose to analyze?*

### Several Factors Affect the Completeness of the Fossil Record

The fossil record should not be viewed as a complete and balanced representation of the species that existed in the past. Several factors affect the likelihood that extinct organisms have been preserved as fossils and will be identified by paleontologists (**Table 22.1**).

- Certain organisms are more likely than others to become fossilized. Organisms with hard shells or bones tend to be overrepresented.
- Factors such as anatomy, size, and numbers of the species, and the environment and time in which they lived play important roles in determining the likelihood that organisms will be preserved in the fossil record.
- Geological processes may favor the fossilization of certain types of organisms.

**Table 22.1 Factors That Affect the Fossil Record**

| Factor | Description |
|---|---|
| Anatomy | Organisms with hard body parts, such as animals with a skeleton or thick shell, are more likely to be preserved than are organisms composed of only soft tissues. |
| Size | The fossil remains of larger organisms are more likely to be found than those of smaller organisms. |
| Number | Species that existed in greater numbers or over a larger area are more likely to be preserved within the fossil record than those that existed in smaller numbers or in a smaller area. |
| Environment | Inland species are less likely to become fossilized than are those that lived in a marine environment or near the edge of water because sedimentary rock is more likely to be formed in or near water. |
| Time | Species that lived relatively recently or existed for a long time are more likely to be found as fossils than species that lived very long ago or for a relatively short time. |
| Geological processes | Due to the chemistry of fossilization, certain organisms are more likely to be preserved than are other organisms. |
| Paleontology | Certain types of fossils may be more interesting to paleontologists. In addition, a significant bias exists with regard to the locations where paleontologists search for fossils. For example, they tend to search in regions where other fossils have already been found. |

- Unintentional biases arise that are related to the efforts of paleontologists. For example, scientific interests may favor searching for and analyzing certain species over others. Many paleontologists have been greatly interested in finding the remains of dinosaurs.

Although the fossil record is incomplete, it has provided a wealth of information regarding the history of the types of life that existed on Earth. The rest of this chapter will survey the emergence of life-forms from 3.5 bya to the present.

## 22.1 Reviewing the Concepts

- Fossils, which are preserved remnants of past life-forms, are formed in sedimentary rock. Our understanding of the history of life on Earth is derived primarily from the fossil record (Figure 22.2).
- Radiometric dating is one way of estimating the age of a fossil (Figure 22.3).
- Several factors affect the likelihood that organisms have been preserved in the fossil record (Table 22.1).

## 22.1 Testing Your Knowledge

1. Fossils are typically formed
   a. within igneous rock.
   b. within sedimentary rock.
   c. when dead organisms are exposed to intense sunlight.
   d. when dead organisms are left undisturbed for long periods of time.
   e. when dead organisms are quickly frozen.
2. Which of the following may account for biases in the fossil record?
   a. Certain organisms are more likely than others to become fossilized.
   b. Various factors, such as anatomy and number, affect the likelihood that organisms will be preserved in the fossil record.
   c. Geological processes may favor the fossilization of certain types of organisms.
   d. Unintentional biases arise that are related to the efforts of paleontologists.
   e. All of the above may cause biases.

## 22.2 History of Life on Earth

### Learning Outcomes

1. List the types of environmental changes that have affected the history of life on Earth.
2. Describe the cell structure and energy utilization of the first living organisms that arose during the Archaean eon.
3. Explain how the origin of eukaryotic cells involved a union between bacterial and archaeal cells.
4. Describe the key features of multicellular organisms, which arose during the Proterozoic eon.
5. Outline the major events and changes in species diversity during the Paleozoic, Mesozoic, and Cenozoic eras.

The oldest fossils that have been discovered thus far were from single-celled organisms, which were preserved approximately 3.5 bya. In this section, we will begin with a brief description of the geological changes on Earth that have affected the emergence of new forms of life and then examine some of the major changes in life that have occurred since it began.

### Many Environmental Changes Have Occurred Since the Origin of the Earth

The **geological timescale** is a timeline of the Earth's history and major events from its origin approximately 4.55 bya to the present (Figure 22.4). This timeline is subdivided into four eons—the Hadean, Archaean, Proterozoic, and Phanerozoic—and then further subdivided into eras. The first three eons are collectively known as the Precambrian because they preceded the Cambrian era, a geological era that saw a rapid increase in the diversity of life. We will examine these time periods later in this section.

The changes that occurred in species over the past 4 billion years are the result of two interactive processes. First, as discussed in Chapters 19–21, genetic changes in organisms can affect their characteristics. Such changes can influence the ability of organisms to survive and reproduce in their native environment. Second, the

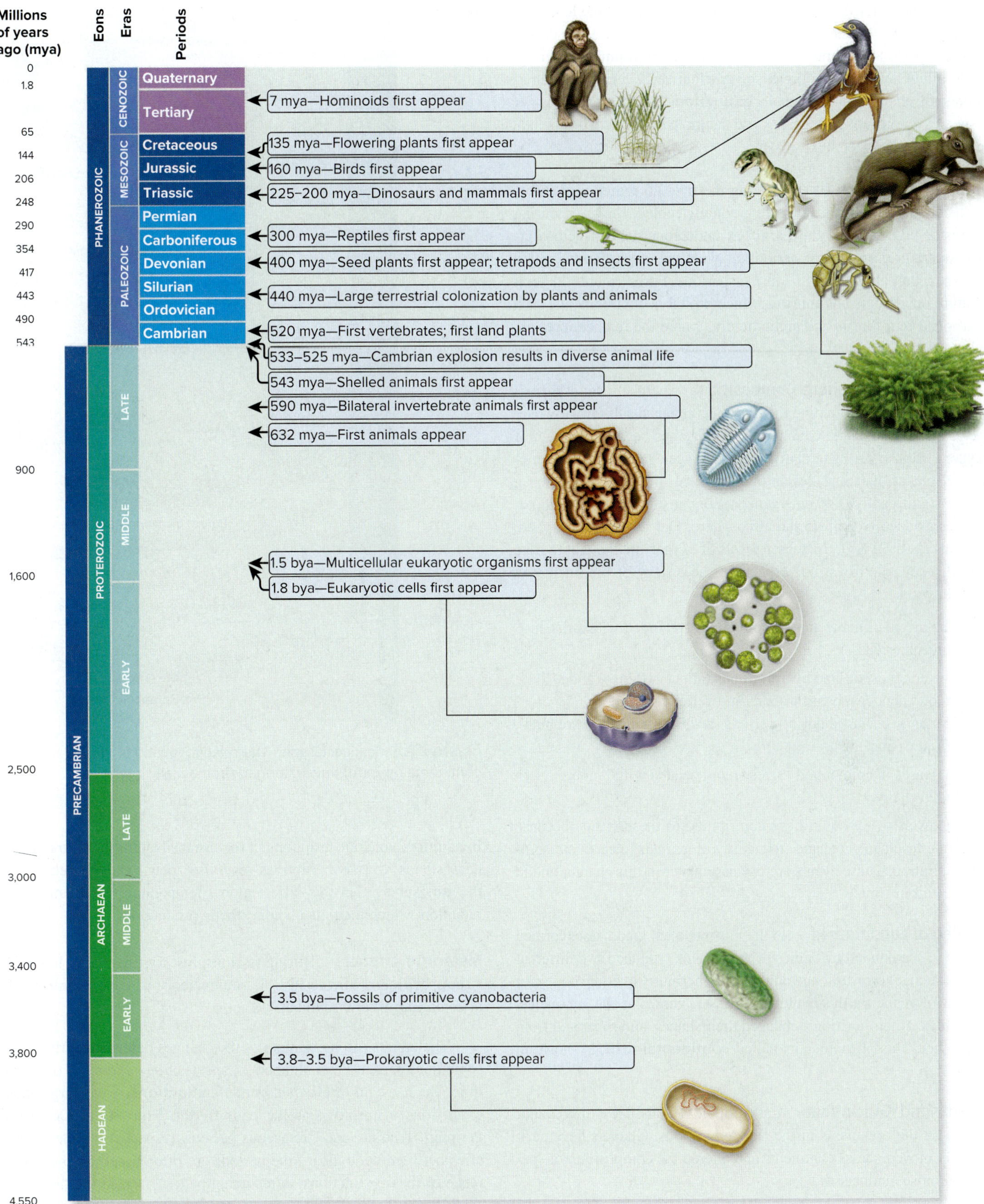
Millions of years ago (mya)
Eons
Eras
Periods
0
1.8
65
144
206
248
290
354
417
443
490
543
900
1,600
2,500
3,000
3,400
3,800
4,550
PHANEROZOIC
CENOZOIC
MESOZOIC
PALEOZOIC
Quaternary
Tertiary
Cretaceous
Jurassic
Triassic
Permian
Carboniferous
Devonian
Silurian
Ordovician
Cambrian
PRECAMBRIAN
PROTEROZOIC
LATE
MIDDLE
EARLY
ARCHAEAN
LATE
MIDDLE
EARLY
HADEAN
7 mya—Hominoids first appear
135 mya—Flowering plants first appear
160 mya—Birds first appear
225–200 mya—Dinosaurs and mammals first appear
300 mya—Reptiles first appear
400 mya—Seed plants first appear; tetrapods and insects first appear
440 mya—Large terrestrial colonization by plants and animals
520 mya—First vertebrates; first land plants
533–525 mya—Cambrian explosion results in diverse animal life
543 mya—Shelled animals first appear
590 mya—Bilateral invertebrate animals first appear
632 mya—First animals appear
1.5 bya—Multicellular eukaryotic organisms first appear
1.8 bya—Eukaryotic cells first appear
3.5 bya—Fossils of primitive cyanobacteria
3.8–3.5 bya—Prokaryotic cells first appear

**Figure 22.4** The geological timescale and an overview of the history of life on Earth.

environment on Earth has undergone dramatic changes that have profoundly influenced the types of organisms that have existed during different periods of time. In some cases, an environmental change might allow new types of organisms to flourish. Alternatively, environmental changes can result in **extinction**—the complete loss of a species or group of species. The major types of environmental changes are described next.

**Changes in Temperature** During the first 2.5 billion years of its existence, the surface of the Earth gradually cooled. However, during the last 2 billion years, the Earth has undergone major fluctuations in temperature, producing periods of widespread glaciation, called Ice Ages, that alternate with warmer periods. Furthermore, the temperature on Earth is not uniform, which produces environments, such as tropical rain forests and arctic tundra, where the temperatures are quite different.

**Changes in Atmosphere Composition** The chemical composition of the gases surrounding the Earth has changed substantially over the past 4 billion years. One notable change has been the amount of oxygen. Prior to 2.4 bya, relatively little oxygen gas was in the atmosphere, but at that time, levels of oxygen in the form of $O_2$ began to rise significantly. The emergence of organisms capable of photosynthesis added oxygen to the atmosphere. Our current atmosphere contains about 21% $O_2$. Increased levels of oxygen are thought to have played a key role in various aspects of the history of life, including the following:

- The origin of many animal body plans coincided with a rise in atmospheric $O_2$.
- The conquest of land by arthropods (about 410 mya) and a second conquest by arthropods and vertebrates (about 350 mya) occurred during periods in which $O_2$ levels were high or increasing.
- Increases in animal body sizes are associated with higher $O_2$ levels.

Higher levels of $O_2$ could have contributed to these events because higher $O_2$ levels may enhance the ability of animals to carry out aerobic respiration. These events are discussed later in this chapter and in more detail in Unit V.

**Shifting of Landmasses** As the Earth cooled, landmasses formed that were surrounded by bodies of water. This produced two different environments: terrestrial and aquatic. Furthermore, over the course of billions of years, Earth's major landmasses, known as the continents, have shifted their positions, changed their shapes, and separated from each other. This phenomenon, called **continental drift,** is shown in Figure 22.5.

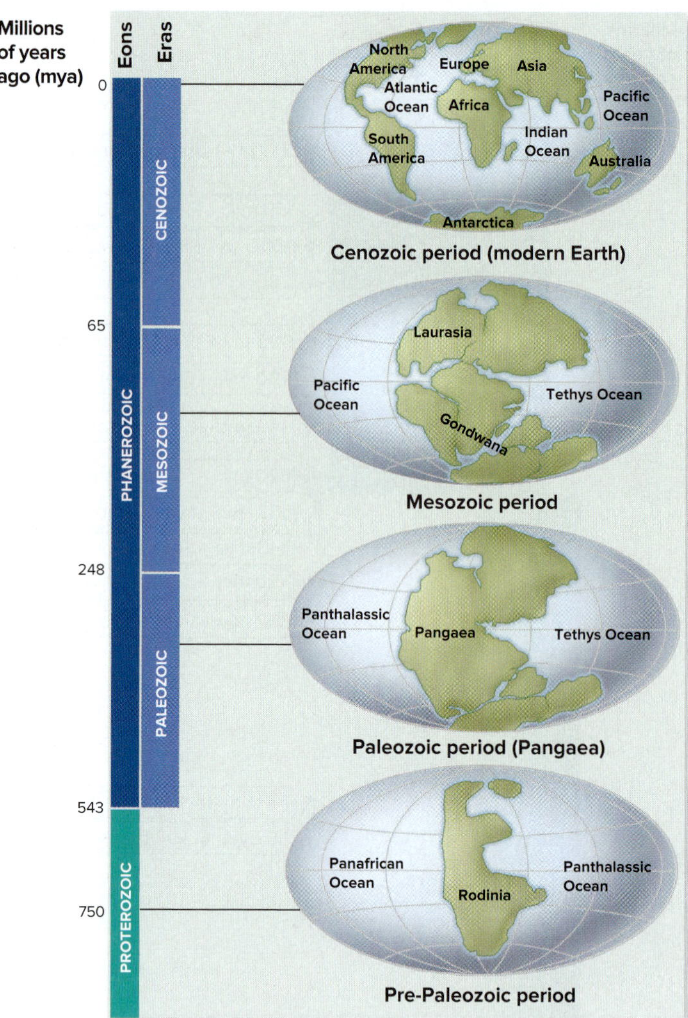


**Figure 22.5** **Continental drift.** The relative locations of the continents on Earth have changed dramatically over time.

**Floods and Glaciations** Catastrophic floods have had major effects on the organisms in the flooded regions. Glaciers have periodically moved across continents and altered the composition of species on those landmasses.

**Volcanic Eruptions** The eruptions of volcanoes harm organisms in the vicinity of an eruption, sometimes causing extinctions. In addition, volcanic eruptions in the oceans lead to the formation of new islands. Massive eruptions may also spew so much debris into the atmosphere that they affect global temperatures and limit solar radiation, which restricts photosynthetic production.

**Meteorite Impacts** During its long history, the Earth has been struck by many meteorites. Large meteorites have significantly affected Earth's environment.

The effects of one or more of the changes described earlier have sometimes caused large numbers of species to go extinct at the same time. Such events are called **mass extinctions.** Five large mass extinctions occurred near the ends of the Ordovician, Devonian, Permian, Triassic, and Cretaceous periods, respectively. The boundaries between geological time periods are often based on the occurrence of mass extinctions. A recurring pattern seen in the history of life is the extinction of some species and the emergence of new ones. The rapid extinction of many modern species due to human activities is sometimes referred to as the sixth mass extinction.

## Prokaryotic Cells Arose During the Archaean Eon

The Archaean (from the Greek, meaning ancient) was an eon when diverse microbial life flourished in the primordial oceans. As mentioned previously, the oldest known fossils of living cells were preserved in rocks that are about 3.5 billion years old (see Figure 22.1), though scientists postulate that cells arose many millions of years prior to this time. Based on the morphology of their fossilized remains, these first cells were prokaryotic. During the more than 1 billion years of the Archaean eon, all life-forms were prokaryotic. Because Earth's atmosphere had very little free oxygen, the single-celled microorganisms of this eon almost certainly used only anaerobic respiration, which occurs in the absence of oxygen.

Organisms with prokaryotic cells are divided into two groups: bacteria and archaea. Bacteria are more prevalent on modern Earth, though many species of archaea have also been identified. Archaea are found in many different environments, with some found in extreme environments such as hot springs. Bacteria and archaea share fundamental similarities, indicating that they are derived from a common ancestor. Even so, certain differences suggest that these two types of prokaryotes diverged from each other quite early in the history of life. In particular, bacteria and archaea show some interesting differences in cellular and molecular features (refer back to Chapter 21, Table 21.1).

## Biologists Are Undecided About Whether Heterotrophs or Autotrophs Came First

An important factor that greatly influenced the emergence of new species is the availability of energy. As we learned in Unit II, all organisms require energy to survive and reproduce. Organisms follow two different strategies to obtain energy.

- **Heterotrophs** derive their energy from the chemical bonds within the organic molecules they consume. Because the most common sources of organic molecules today are other organisms, heterotrophs typically consume other organisms or materials from other organisms.
- **Autotrophs** directly harness energy from either light or inorganic molecules. Among modern species, plants are an important example of autotrophs. Plants can directly absorb light energy and use it (via photosynthesis) to synthesize organic molecules such as glucose. On modern Earth, heterotrophs ultimately rely on autotrophs for the production of food.

Were the first forms of life heterotrophs or autotrophs? The answer is not resolved. Some biologists have speculated that autotrophs, such as those living near deep-sea vents, may have arisen first. These organisms would have used chemicals that were made near the vents as an energy source to make organic molecules. Alternatively, many scientists have hypothesized that the first living cells were heterotrophs. They reason that it would have been simpler for the first primitive cells to use the organic molecules in the prebiotic soup as a source of energy.

If heterotrophs came first, why were cyanobacteria, which are autotrophs, preserved in the earliest fossils rather than heterotrophs? One possible reason is related to their manner of growth.

**(a) Fossil stromatolite**

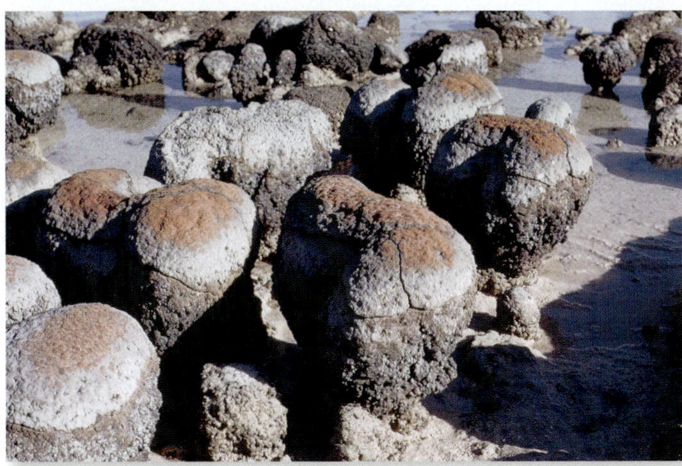

**(b) Modern stromatolites**

**Figure 22.6 Fossil and modern stromatolites: Evidence of early autotrophs.** Each stromatolite is a rocklike structure, typically 1 meter in diameter. **(a)** Section of a fossilized stromatolite. These layers are mats of mineralized cyanobacteria, one layer on top of the other. The existence of fossil stromatolites provides evidence of early autotrophic organisms. **(b)** Modern stromatolites that have formed in western Australia.

*(a)* © Dirk Wiersma/SPL/Science Source; *(b)* © Roger Garwood & Trish Ainslie/Corbis

Certain cyanobacteria promote the formation of a layered structure called a **stromatolite** (Figure 22.6a). The aquatic environment where these cyanobacteria survive is rich in minerals such as calcium. The cyanobacteria grow in large mats that form layers. As they grow, they deplete the carbon dioxide ($CO_2$) in the surrounding water. This causes calcium carbonate in the water to gradually precipitate over the bacterial cells, calcifying the older cells in the lower layers into rock and trapping grains of sediment. Newer cells produce a layer on top. Over time, many layers of calcified cells and sediment are formed, thereby producing a stromatolite. This process still occurs today in places such as Shark Bay in western Australia, which is renowned for the stromatolites along its beaches (Figure 22.6b).

The emergence and proliferation of ancient cyanobacteria had two critical consequences. First, the autotrophic nature of these bacteria enabled them to produce organic molecules from $CO_2$. This prevented the depletion of organic foodstuffs that would have been exhausted if only heterotrophs existed. Second, cyanobacteria produce $O_2$ as a waste product of photosynthesis. During the Archaean and Proterozoic eons, the activity of cyanobacteria led to the gradual rise in atmospheric $O_2$ discussed earlier. The increase in $O_2$ spelled doom for many anaerobic species, which became restricted to a few anoxic (without oxygen) environments, such as deep within the ground. However, $O_2$ enabled the emergence of new bacterial and archaeal species that used aerobic (with oxygen) respiration (see Chapter 6). In addition, aerobic respiration is likely to have played a key role in the emergence and eventual explosion of eukaryotic life-forms, which typically have high energy demands. These eukaryotic life-forms are described next.

# EVOLUTIONARY CONNECTIONS

## The Origin of Eukaryotic Cells Is Hypothesized to Involve a Union Between Bacterial and Archaeal Cells

Eukaryotic cells arose during the Proterozoic eon, which began 2.5 bya and ended 543 mya (see Figure 22.4). The manner in which the first eukaryotic cells originated is not entirely understood. In modern eukaryotic cells, genetic material is found in three distinct organelles. All eukaryotic cells contain DNA in the nucleus and mitochondria, and plant and algal cells also have DNA in their chloroplasts. To address the issue of the origin of eukaryotic species, scientists have examined the DNA sequences found in these three organelles. They have concluded that the nuclear, mitochondrial, and chloroplast genomes appear to be derived from once-separate cells that came together.

**Nuclear Genome** Both bacteria and archaea contributed substantially to the nuclear genome of eukaryotic cells. Eukaryotic nuclear genes encoding proteins involved in metabolic pathways and lipid biosynthesis appear to be derived from ancient bacteria, whereas genes involved with transcription and translation appear to be derived from an archaeal ancestor.

To explain the origin of the nuclear genome, several hypotheses have been proposed. The most widely accepted hypothesis involves a symbiotic relationship between ancient bacteria and archaea. **Endosymbiosis** describes a relationship in which a smaller organism (the endosymbiont) lives inside a larger organism (the host).

Researchers have suggested that an archaeal species evolved the ability to invaginate its plasma membrane, which could have two results (**Figure 22.7**). First, it could eventually lead to the formation of an extensive internal membrane system and enclose the genetic material in a nuclear envelope. Second, the ability to invaginate the plasma membrane provides a mechanism for taking up materials from the environment via endocytosis, which is described in Chapter 5. In the scenario described in Figure 22.7, an ancient archeon (an archaeal cell) engulfed a bacterium via endocytosis, maintaining the bacterium

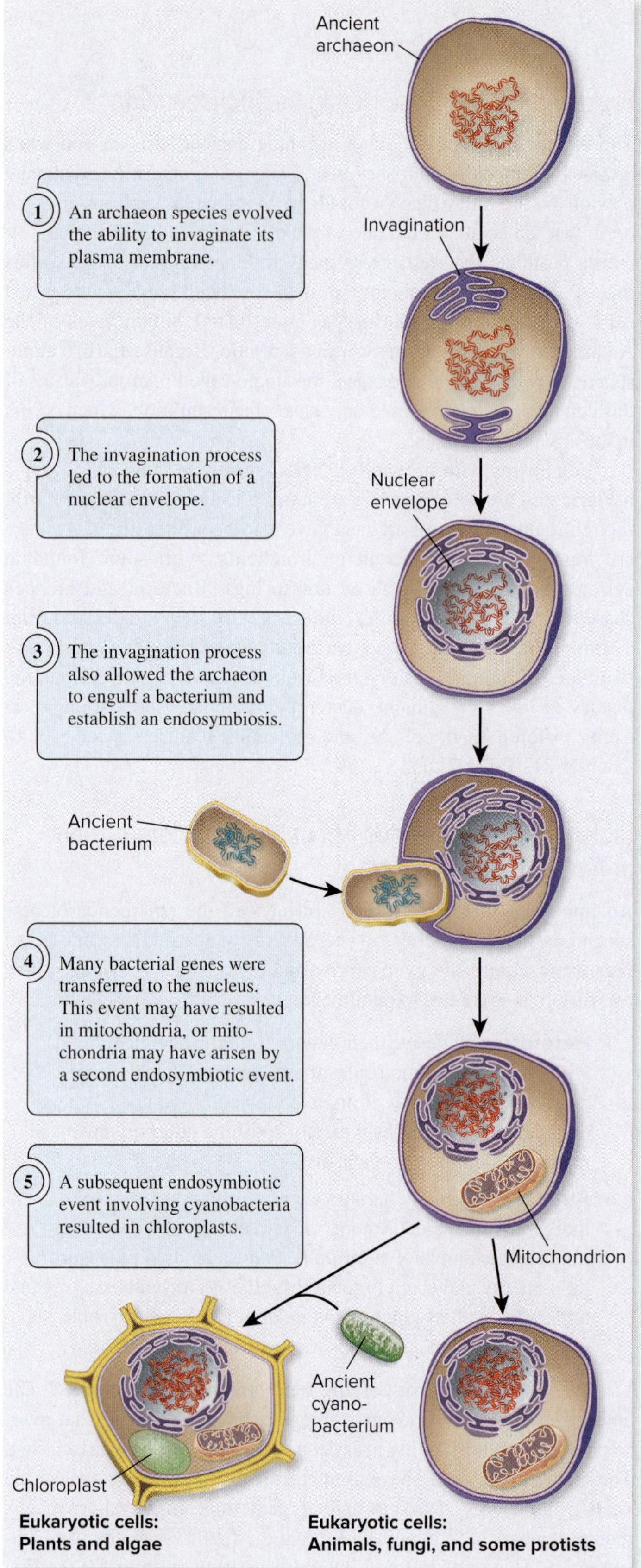

Ancient archaeon
1 An archaeon species evolved the ability to invaginate its plasma membrane.
Invagination
2 The invagination process led to the formation of a nuclear envelope.
Nuclear envelope
3 The invagination process also allowed the archaeon to engulf a bacterium and establish an endosymbiosis.
Ancient bacterium
4 Many bacterial genes were transferred to the nucleus. This event may have resulted in mitochondria, or mitochondria may have arisen by a second endosymbiotic event.
5 A subsequent endosymbiotic event involving cyanobacteria resulted in chloroplasts.
Mitochondrion
Ancient cyano-bacterium
Chloroplast
Eukaryotic cells: Plants and algae
Eukaryotic cells: Animals, fungi, and some protists


**Figure 22.7** Possible endosymbiotic relationships that gave rise to the first eukaryotic cells.

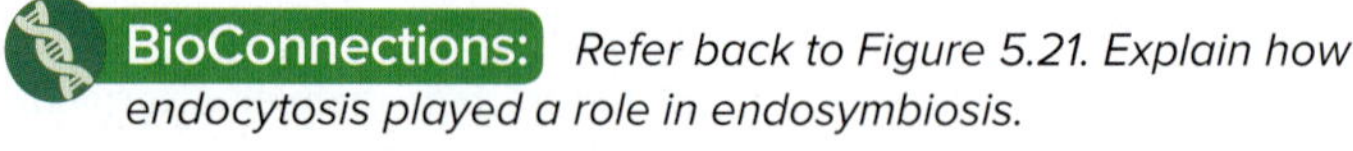
**BioConnections:** *Refer back to Figure 5.21. Explain how endocytosis played a role in endosymbiosis.*

in its cytoplasm as an endosymbiont. Over time, many genes from the bacterium were transferred to the archaeal host cell, and the resulting genetic material eventually became the nuclear genome.

**Mitochondrial and Chloroplast Genomes** Mitochondria found in eukaryotic cells are likely derived from a bacterial species that resembles modern alphaproteobacteria, a diverse group of bacteria that carry out oxidative phosphorylation to make ATP. One possibility is that an endosymbiotic event involving an ancestor of this bacterial species produced the first eukaryotic cell and that the mitochondrion is a remnant of that event. Alternatively, endosymbiosis may have produced the first eukaryotic cell, and then a subsequent endosymbiosis resulted in mitochondria (see Figure 22.7). DNA-sequencing data indicate that chloroplasts were derived from a separate endosymbiotic relationship between a primitive eukaryotic cell and a cyanobacterium. As discussed in Chapter 23, plastids, such as chloroplasts, have arisen on several independent occasions via primary, secondary, and tertiary endosymbiosis (look ahead to Figure 23.23).

## Multicellular Eukaryotes and the Earliest Animals Arose During the Proterozoic Eon

The first multicellular eukaryotes are thought to have emerged about 1.5 bya, in the middle of the Proterozoic eon. The oldest fossil evidence for multicellular eukaryotes was an organism that resembled modern red algae; this fossil was dated at approximately 1.2 billion years old.

Simple multicellular organisms are believed to have originated in one of two ways. One mechanism is that many individual cells can find each other and aggregate to form a colony. Cellular slime molds are examples of modern organisms in which groups of single-celled organisms can come together to form a small multicellular organism. According to the fossil record, such organisms have remained very simple for hundreds of millions of years.

Alternatively, another way that multicellularity can occur is when a single cell divides and the resulting cells stick together. This pattern occurs in many simple multicellular organisms, such as algae and fungi, as well as in species with more complex body plans, such as plants and animals. Biologists cannot be certain whether the first multicellular organisms arose by an aggregation process or by cell division and adhesion. However, the development of complex, multicellular organisms now occurs by cell division and adhesion.

An interesting example showing changes in the level of complexity from unicellular organisms to more complex multicellular organisms is found among evolutionarily related species of volvocine green algae. These algae exist as unicellular species, as small clumps of cells of the same cell type, or as larger groups of cells with two distinct cell types. **Figure 22.8** compares four species of volvocine algae. *Chlamydomonas reinhardtii* is a unicellular alga (Figure 22.8a). It is called a biflagellate because each cell has two flagella. *Gonium pectorale* is a multicellular organism composed of eight cells (Figure 22.8b). This simple multicellular organism is formed from a single cell by cell division and adhesion. All of the cells in this species are biflagellate. Other volvocine algae have evolved into larger and more complex organisms. *Pleodorina californica* has 64–128 cells

## Biology Principle

### New Properties Emerge from Complex Interactions

The formation of different cell types is an emergent property of multicellularity.

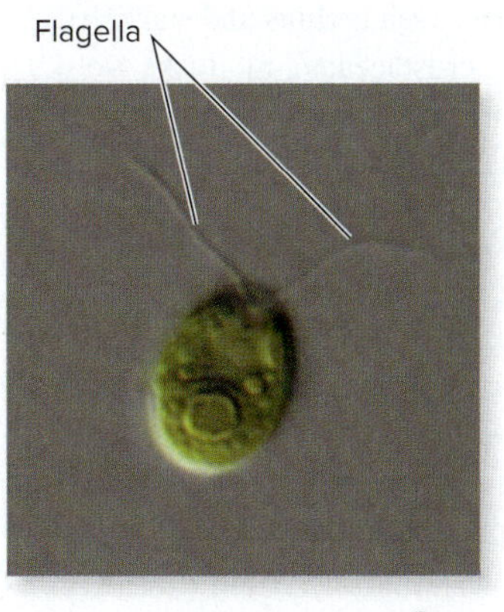


**(a)** ***Chlamydomonas reinhardtii*, a unicellular alga**

**(b)** ***Gonium pectorale*, composed of 16 identical cells**

**(c)** ***Pleodorina californica*, composed of 64 to 128 cells, has 2 cell types, somatic and reproductive**

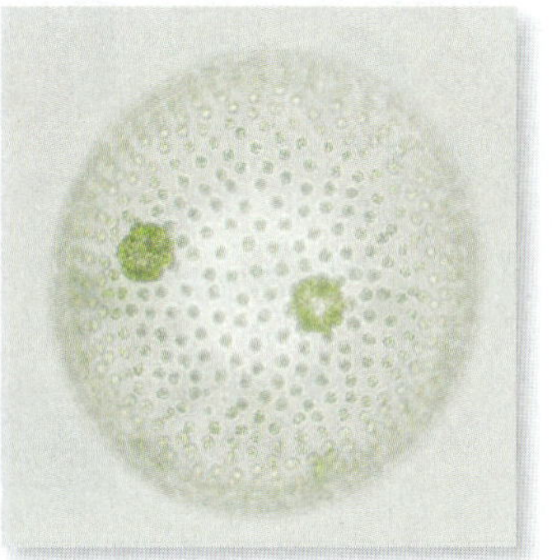


**(d)** ***Volvox aureus*, composed of about 1,000 to 2,000 cells, has 2 cell types, somatic and reproductive**

**Figure 22.8** Variation in the level of multicellularity among volvocine algae.

*(a)* Courtesy of Dr. Barbara Surek, Culture Collection of Algae at the University of Cologne (CCAC); *(b)* © Bill Bourland/micro*scope; *(c, d)* © Dr. Cristian A. Solari, Department of Ecology and Evolutionary Biology, University of Arizona

(Figure 22.8c), and *Volvox aureus* has about 1,000–2,000 cells (Figure 22.8d). A feature of these more complex organisms is they have two cell types: somatic and reproductive cells. The somatic cells are biflagellate cells, but the reproductive cells are not. *V. aureus*, which is larger, has a higher percentage of somatic cells than *P. californica*.

Overall, an analysis of these four species of algae illustrates three important principles found among complex multicellular species:

1. Multicellular organisms arise from a single cell that divides to produce daughter cells that adhere to one another.
2. The daughter cells can follow different fates, thereby producing multicellular organisms with different cell types.
3. As organisms get larger, a greater percentage of their cells tend to be somatic cells. The somatic cells carry out the activities required for the survival of the multicellular organism, whereas the reproductive cells are specialized for the sole purpose of producing offspring.

Toward the end of the Proterozoic eon, the first multicellular animals emerged. The first animals were invertebrates—animals without a backbone. Most animals, except for organisms such as sponges and jellyfish, exhibit bilateral symmetry—a two-sided body plan with a right and left side that are mirror images. Because each side of the body has appendages such as legs, one advantage of bilateral symmetry is that it facilitates locomotion. Bilateral animals also have anterior and posterior ends, with the mouth at the anterior end. In 2004, Chinese paleontologist Jun-Yuan Chen, American paleobiologist David Bottjer, and their colleagues discovered a fossil in southern China, which they described as the earliest known ancestor of animals with bilateral symmetry. This small animal, with a shape like a flattened helmet, is barely visible to the naked eye (**Figure 22.9**).

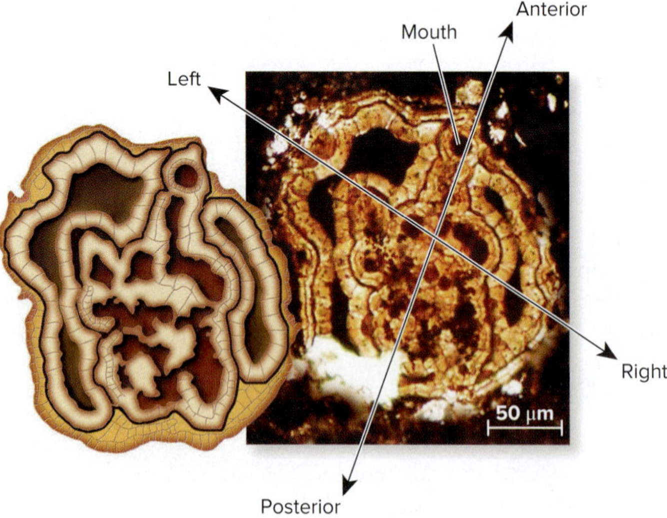


**Figure 22.9 Fossil of an early invertebrate animal showing bilateral symmetry.** This fossil of an early animal, *Vernanimalcula guizhouena*, dates from 580 to 600 mya.

*(right)* We thank Prof. Jun-Yuan Chen for permission to use this image

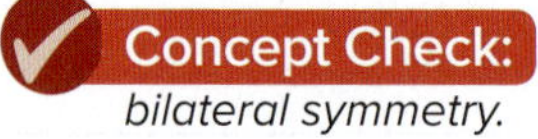

*Name three other species that exhibit bilateral symmetry.*

The fossil, called *Vernanimalcula guizhouena,* is approximately 580–600 million years old. However, the interpretation that *V. guizhouena* was the earliest ancestor of animals with bilateral symmetry remains controversial and is under active investigation.

## Phanerozoic Eon: The Paleozoic Era Saw the Diversification of Invertebrates and the Colonization of Land by Plants and Animals

The proliferation of multicellular eukaryotic life has been extensive during the Phanerozoic eon, which started 543 mya and extends to the present day. Phanerozoic means well-displayed life, referring to the abundance of fossils of plants and animals that have been identified from this eon. As described in Figure 22.4, the Phanerozoic eon is subdivided into three eras: Paleozoic, Mesozoic, and Cenozoic. Because they are relatively recent and researchers have identified many fossils from these eras, each of them is further subdivided into periods. We will consider each era with its associated conditions and prevalent forms of life separately.

The term Paleozoic means ancient animal life. The Paleozoic era covers approximately 300 million years, from 543 to 248 mya, and is subdivided into six periods: Cambrian, Ordovician, Silurian, Devonian, Carboniferous, and Permian. Periods are usually named after regions where rocks and fossils of that age were first discovered.

**Cambrian Period (543–490 mya)** The climate in the Cambrian period was generally warm and wet, with no evidence of ice at the poles. During this time, the diversity of animal species increased rapidly, an event called the **Cambrian explosion.** However, recent evidence suggests that many types of animal groups present during the Cambrian period actually arose prior to this time.

- By the middle of the Cambrian period, all of the existing major types of invertebrates were present, plus many others that no longer exist.
- Examples of animals that first appeared in the Cambrian period and still exist include echinoderms (sea urchins and starfish), arthropods (insects, spiders, and crustaceans), mollusks (clams and snails), chordates (organisms with a dorsal nerve chord), and vertebrates (animals with backbones).

The cause of the Cambrian explosion is not understood. Because it occurred shortly after marine animals evolved shells, some scientists have speculated that the changes observed in animal species may have allowed them to exploit new environments. Alternatively, others have suggested that the increase in diversity may be related to atmospheric oxygen levels. During this period, oxygen levels were increasing, and perhaps more complex body plans became possible only after the atmospheric oxygen surpassed a certain threshold. In addition, as atmospheric oxygen reached its present levels, an ozone ($O_3$) layer was produced that screens out harmful ultraviolet radiation, thereby allowing complex life to live in shallow water and eventually on land. Another possible contributor to the Cambrian explosion was an "evolutionary arms race" between interacting species. The ability of predators to capture prey and the ability of prey to avoid predators may have been a major factor that resulted in a diversification of animals into many different species.

**Ordovician Period (490–443 mya)** As in the Cambrian period, the climate of the early and middle parts of the Ordovician period was warm, and the atmosphere was moist.

- A diverse group of hard-shelled marine invertebrates, including trilobites and brachiopods, appeared in the fossil record (Figure 22.10). Marine communities consisted of invertebrates, algae, early jawless fishes (a type of early vertebrate), mollusks, and corals.
- Early land plants and arthropods may have first invaded the land during this period.

Toward the end of the Ordovician period, the climate changed rather dramatically. Large glaciers formed, which drained the relatively shallow oceans, causing the water levels to drop. This resulted in a mass extinction in which as much as 60% of the existing marine invertebrates became extinct.

**Silurian Period (443–417 mya)** In contrast to the dramatic climate changes observed during the Ordovician period, the climate during the Silurian period was relatively stable. The glaciers largely melted, which caused the ocean levels to rise.

- No new major types of invertebrate animals appeared during this period, but significant changes were observed among existing vertebrate and plant species.
- Many new types of fishes appeared in the fossil record.
- Coral reefs made their first appearance during this period.
- The Silurian marked a major colonization of land by terrestrial plants and animals. For this to occur, certain species evolved adaptations that prevented them from drying out, such as a hard cuticle.
- Ancestral relatives of spiders and centipedes became widespread.
- The earliest vascular plants, which have tissues that are specialized for the transport of water, sugar, and salts throughout the plant body, arose during this period.

(a) Trilobite

(b) Brachiopod

**Figure 22.10** **Shelled invertebrate fossils of the Ordovician period.** Trilobites existed for millions of years before becoming extinct about 250 mya. Many species of brachiopods exist today.
*(a)* © Francois Gohier/Science Source; *(b)* © DK Limited/Corbis

**Devonian Period (417–354 mya)** In the Devonian period, generally dry conditions occurred across much of the northern landmasses. However, the southern landmasses were mostly covered by cool, temperate oceans.

- The Devonian saw a major increase in the number of terrestrial plant species. At first, the vegetation consisted primarily of small plants, only a meter tall or less. Later, ferns, horsetails, and seed plants, such as gymnosperms, also emerged. By the end of the Devonian, the first trees and forests were formed.
- A major expansion of terrestrial animals also occurred. Insects first appeared in the fossil record, and other invertebrates became plentiful. In addition, the first tetrapods—vertebrates with four legs—are believed to have arisen in the Devonian. Early tetrapods included amphibians, which lived on land but required water in which to lay their eggs.
- In the oceans, many types of invertebrates flourished, including brachiopods, echinoderms, and corals. This period is sometimes called the Age of Fishes, as many new types of fishes emerged.

During a period of approximately 20 million years near the end of the Devonian period, a prolonged series of extinctions eliminated many marine species. The cause of this mass extinction is not well understood.

**Carboniferous Period (354–290 mya)** The term Carboniferous refers to the deposits of coal, a sedimentary rock primarily composed of carbon, that were formed during this period. The Carboniferous period had the ideal conditions for the subsequent formation of coal. It was a cooler period, and much of the land was covered by forest swamps. Coal was formed over many millions of years from compressed layers of rotting vegetation.

- Very large plants and trees became prevalent. For example, tree ferns such as *Psaronius* grew to a height of 15 meters or more (Figure 22.11).

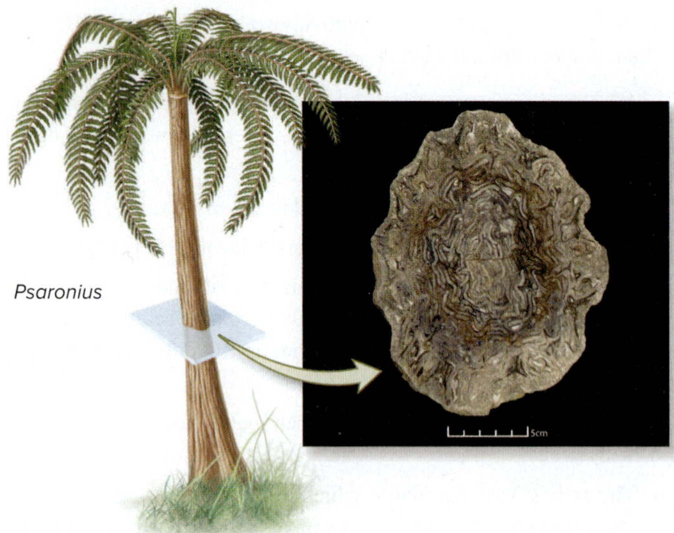


**Figure 22.11** **A giant tree fern, *Psaronius*, from the Carboniferous period.** This genus became extinct during the Permian. The illustration is a re-creation based on fossil evidence. The inset shows a fossilized section of the trunk.



- The first flying insects emerged. Giant dragonflies with a wingspan of over 2 feet inhabited the forest swamps.
- Terrestrial vertebrates also became more diverse.
- Amphibians were very widespread.
- The amniotic egg was one of the most important evolutionary innovations of the Carboniferous period. This innovation was critical for the emergence of reptiles, tetrapods that had the ability to lay eggs on land. In reptiles, the amniotic egg was covered with a leathery or hard shell, which prevented the desiccation of the embryo inside.

**Permian Period (290–248 mya)** At the beginning of the Permian, continental drift had brought much of the total land together into a supercontinent known as Pangaea (see Figure 22.5). The interior regions of Pangaea were dry, with large seasonal fluctuations.

- The forests of fernlike plants were replaced with gymnosperms, seed-bearing land plants. Species resembling modern conifers first appeared in the fossil record.
- Amphibians were prevalent, but reptiles became the dominant vertebrate species.

At the end of the Permian period, the largest known mass extinction in the history of life on Earth occurred; 90–95% of marine species and a large proportion of terrestrial species were eliminated. The cause of the Permian extinction is the subject of much research and controversy. One possibility is that glaciation destroyed the habitats of terrestrial species and lowered ocean levels, which would have caused greater competition among marine species. Another hypothesis is that enormous volcanic eruptions in Siberia produced large ash clouds that abruptly changed the climate on Earth.

## Phanerozoic Eon: The Mesozoic Era Saw the Rise and Fall of the Dinosaurs

The Permian extinction marks the division between the Paleozoic and Mesozoic eras. Mesozoic means middle animals. It was a time that saw great changes in animal and plant species. This era is sometimes called the Age of Dinosaurs, because those animals flourished during this time. The climate during the Mesozoic era was consistently hot, and terrestrial environments were relatively dry. Little if any ice was found at either pole. The Mesozoic is divided into three periods: Triassic, Jurassic, and Cretaceous.

### Triassic Period (248–206 mya)

- Reptiles were plentiful in this period, including new groups such as crocodiles and turtles.
- The first dinosaurs, which were a diverse group of reptiles that typically had an erect posture and sometimes attained a very large size, emerged during the middle of the Triassic. The first mammals, such as the small *Megazostrodon* also appeared (**Figure 22.12**).
- Gymnosperms were the dominant land plant.

Volcanic eruptions near the end of the Triassic are thought to have caused global warming, resulting in mass extinctions that eliminated many marine and terrestrial species.

### Jurassic Period (206–144 mya)

- Gymnosperms, such as conifers, continued to be the dominant vegetation.
- Mammals were not prevalent.

**Figure 22.12** ***Megazostrodon*, the first known mammal of the Triassic period.** The illustration is a re-creation based on fossilized skeletons. The *Megazostrodon* was 10 to 12 cm long.

**BioConnections:** *Look ahead to Table 27.1. What are the common characteristics of mammals?*

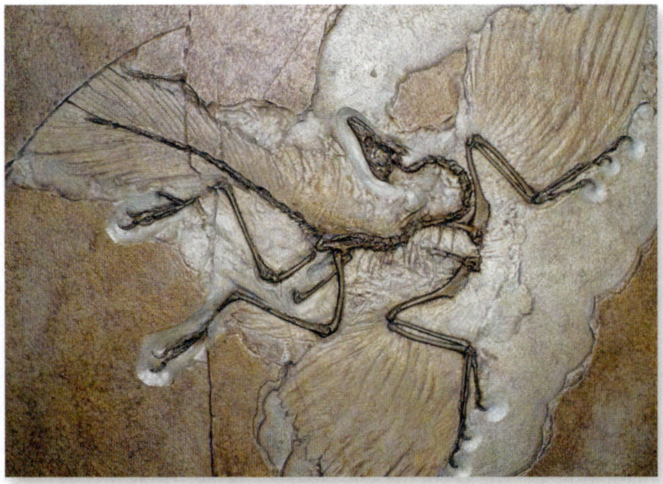

**Figure 22.13** A fossil of an early birdlike animal, *Archaeopteryx lithographica*, which emerged in the Jurassic period.
© Jason Edwards/Getty Images RF

- Dinosaurs and other reptiles continued to be the dominant land vertebrates. Modern birds are descendants of a dinosaur lineage called theropod (meaning beast-footed) dinosaurs. *Tyrannosaurus rex* is one of the best known theropod dinosaurs. An early birdlike animal, *Archaeopteryx lithographica* (Figure 22.13), emerged in the Jurassic period. However, paleontologists are debating whether or not *Archaeopteryx* is a true ancestor of modern birds.

#### Cretaceous Period (144–65 mya)

- On land, dinosaurs continued to be the dominant animals.
- The earliest flowering plants, called angiosperms, which form seeds within a protective chamber, emerged and began to diversify.

The end of the Cretaceous period witnessed another mass extinction, which brought an end to many previously successful groups of organisms. Except for the lineage that gave rise to birds, dinosaurs abruptly died out, as did many other species. As with the Permian extinction, the cause or causes of this mass extinction are still debated. One plausible hypothesis suggests that a large meteorite hit the region that is now the Yucatan Peninsula of Mexico, lifting massive amounts of debris into the air and thereby blocking the sunlight from reaching the Earth's surface. Such a dense haze could have cooled the Earth's surface by 11–15°C (20–30°F). Evidence also points to huge volcanic eruptions as a contributing factor to this mass extinction.

### Phanerozoic Eon: Mammals and Flowering Plants Diversified During the Cenozoic Era

The Cenozoic era spans the most recent 65 million years. It is divided into two periods: Tertiary and Quaternary. In many parts of the world, tropical conditions were replaced by a colder, drier climate. During this time, mammals became the largest terrestrial animals, which is why the Cenozoic is sometimes called the Age of Mammals. However, the Cenozoic era also saw an amazing diversification of many types of organisms, including birds, fishes, insects, and flowering plants.

#### Tertiary Period (65–1.8 mya)

- On land, the mammals that survived from the Cretaceous period began to diversify rapidly during the early part of the Tertiary period.
- Angiosperms became the dominant land plant, and insects became important for their pollination.
- Fishes also diversified, and sharks became abundant.
- Toward the end of the Tertiary period, about 7 mya, hominoids came into existence. **Hominoids** include humans, chimpanzees, gorillas, orangutans, and gibbons, plus all of their recent ancestors. The subset of hominoids called **hominins** includes modern humans, extinct human species (for example, of the *Homo* genus), and our immediate ancestors.

**Quaternary Period (1.8 mya–present)** Periodic Ice Ages have been prevalent during the last 1.8 million years, covering much of Europe and North America.

- This period has witnessed the widespread extinction of many species of mammals, particularly larger ones.
- Certain species of hominins became increasingly more like modern humans. *Homo sapiens*—modern humans—first appeared about 200,000 years ago. The evolution of hominins is discussed in more detail in Section 22.3.

## 22.2 Reviewing the Concepts

- The geological timescale, which is divided into four eons and many eras and periods, charts the major events that occurred during the history of life on Earth (Figure 22.4).
- Both the emergence of new species and mass extinctions are correlated with changes in temperature, amount of $O_2$ in the atmosphere, landmass locations, floods and glaciation, volcanic eruptions, and meteorite impacts (Figure 22.5).
- During the Archaean eon, bacteria and archaea arose. The proliferation of cyanobacteria led to a gradual rise in $O_2$ levels (Figure 22.6).
- Eukaryotic cells arose during the Proterozoic eon. This origin involved a union between bacterial and archaeal cells that is hypothesized to have been endosymbiotic. The origin of mitochondria and chloroplasts was also the result of endosymbiosis (Figure 22.7).
- Multicellular eukaryotes arose about 1.5 bya during the Proterozoic eon. Multicellularity now occurs via cell division and the adherence of the resulting cells to each other. A multicellular organism can have multiple cell types. The first animal showing bilateral symmetry emerged toward the end of the Proterozoic eon (Figures 22.8, 22.9).
- The Phanerozoic eon is subdivided into the Paleozoic, Mesozoic, and Cenozoic eras. During the Paleozoic era, invertebrates greatly diversified, particularly during the Cambrian explosion, and the land became colonized by plants and animals. Terrestrial vertebrates, including tetrapods, became more diverse (Figures 22.10, 22.11).
- Mammals, dinosaurs, and birds emerged during the Mesozoic era. Dinosaurs were particularly prevalent during the Jurassic period (Figures 22.12, 22.13).
- During the Cenozoic era, mammals diversified, and flowering plants became the dominant plant species. The first hominoids

emerged approximately 7 mya. *Homo sapiens*, our species, first appeared about 200,000 years ago.

## 22.2 Testing Your Knowledge

1. A heterotroph is an organism that
   a. only consumes autotrophs.
   b. only consumes other heterotrophs.
   c. derives its energy from the chemical bonds within organic molecules.
   d. directly harnesses energy from inorganic sources.
2. Eukaryotic cells were originally derived from
   a. an endosymbiotic relationship between archaeal and bacterial cells.
   b. an endosymbiotic relationship between different archaeal cells.
   c. an endosymbiotic relationship between different bacterial cells.
   d. a single archaeal species.
   e. a single bacterial species.
3. Multicellularity in most modern species occurs because
   a. individual cells find each other and aggregate.
   b. a single cell undergoes multiple divisions and the cells adhere to one another.
   c. both a and b.
   d. neither a or b.

## 22.3 Human Evolution

### Learning Outcomes

1. List the common characteristics of primates and describe their evolutionary relationships.
2. Explain how human species evolved from other primate species and describe how they spread across the Earth.
3. Provide examples of how populations of *Homo sapiens* are still evolving.
4. Compare and contrast modern human variation at the phenotype and genotype levels.

Hardly a topic in biology has evoked more interest or public debate than human evolution. The question of "where did we come from" has been considered by people for thousands of years. In this section, we will tackle this question from an evolutionary perspective.

We begin with an overview of primate evolution, and explore how humans are evolutionarily related to their closest nonhuman relatives. We then take a closer look at the evolutionary events that gave rise to modern humans. As you will see, many extinct species of humans have existed, including the Denisovans, which were discovered as recently as 2010! Finally, we will turn our attention to the level of existing genetic variation among modern humans, and consider whether humans are still evolving.

### Primates Evolved from a Tree-Dwelling Species and Exhibit a Distinctive Set of Characteristics

Primates are primarily tree-dwelling species that evolved from a group of small, arboreal, insect-eating mammals about 85 mya, before dinosaurs went extinct. Primates have several defining characteristics, mostly relating to their tree-dwelling nature:

- **Binocular vision.** Primates have forward-facing eyes that are positioned close together on a flattened face. Jumping from branch to branch requires accurate judgment of distances. This is facilitated by binocular vision in which the field of vision for both eyes overlaps, producing a single image.
- **At least some digits with flat nails instead of claws.** This feature is believed to aid in the manipulation of objects.
- **Grasping hands.** All primates have grasping hands, a characteristic that enables them to hold onto branches. The monkeys, gibbons, orangutans, gorillas, chimpanzees, and humans also possess an opposable thumb, a thumb that can be placed opposite the fingers of the same hand, which gives them a precision grip and enables the manipulation of small objects. These same primates except humans also have an opposable big toe.
- **Large brain.** Acute vision and other senses enhancing the ability to move quickly through the trees require the efficient processing of large amounts of information. As a result, primate brains are relatively large for their body sizes and they are well developed.
- **Complex social behavior and well-developed parental care.** Compared to other mammals, primates have a tendency toward complex social behavior and relatively long parental care.

Some of these characteristics are found in other animals. For example, binocular vision occurs in owls and some other birds, grasping hands are found in raccoons, and relatively large brains occur in marine mammals. Primates are defined by possessing the whole suite of these characteristics together.

Primates are classified in multiple ways. Taxonomists divide them into two broad groups: the strepsirrhini and the haplorrhini (Figure 22.14). The **strepsirrhini** are smaller species, such as bush babies, lemurs, and pottos. These are generally nocturnal and smaller-brained primates with eyes positioned a little more toward the side of their heads (Figure 22.15a). The strepsirrhini are named for their wet noses with no fur at the tip. The **haplorrhini** have dry noses and fully forward-facing eyes. This group consists of the tarsiers and the larger-brained and diurnal **anthropoidea:** monkeys and the **hominoidea** (gibbons, orangutans, gorillas, chimpanzees, and humans) (Figure 22.15b and c). A key feature of anthropoidea is an opposable thumb, which makes it easier to grasp and handle objects.

What differentiates monkeys from hominoids? Most monkeys have tails, but hominoids do not. In addition, hominoids have more mobile shoulder joints, broader rib cages, and a shorter spine. These features aid in brachiation, a swinging movement in trees. Hominoids also possess relatively long limbs and short legs and, with the exception of gibbons, are much larger than monkeys. The 20 species of hominoids are split into two groups: the lesser apes (family Hylobatidae), or the gibbons; and the greater apes (family Hominidae), or the orangutans, gorillas, chimpanzees, and humans (Figure 22.16). The lesser apes are strictly arboreal, whereas the greater apes often descend to the ground to feed.

Although humans are closely related to chimpanzees and gorillas, they did not evolve directly from them. Rather, all hominoid species

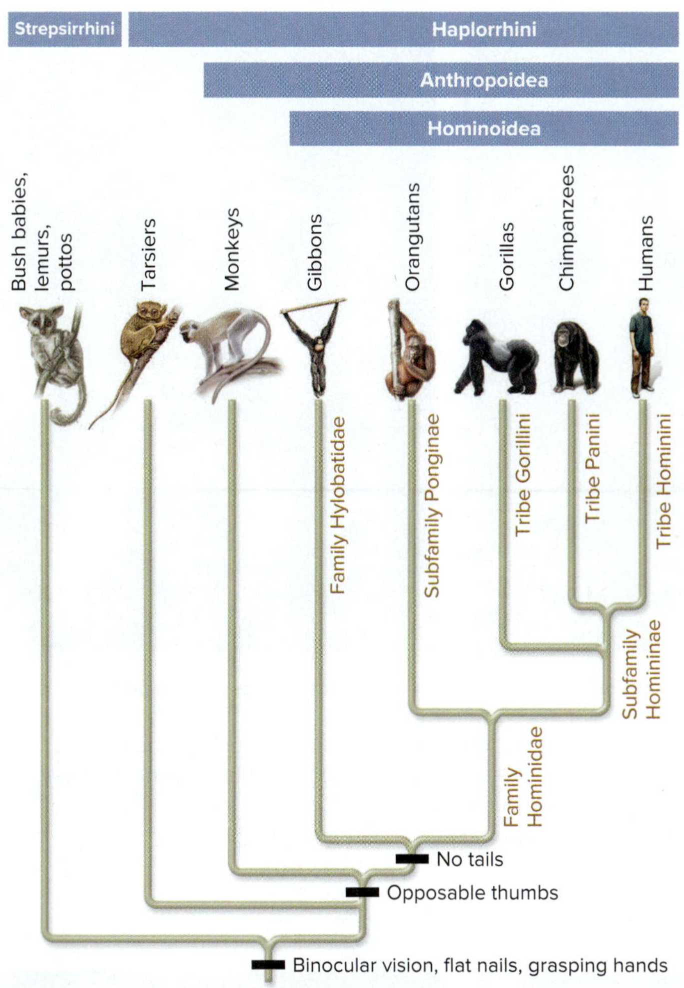


**Figure 22.14** Evolutionary tree of the primates.

shared a common ancestor. Recent molecular studies show that gorillas, chimpanzees, and humans are more closely related to one another than they are to gibbons and orangutans, so scientists have split the family Hominidae into groups, including the subfamily Ponginae (orangutans) and the subfamily Homininae (gorillas, chimpanzees, and humans and their ancestors) (see Figure 22.14). In turn, the Homininae are split into three tribes: the Gorillini (gorillas), the Panini (chimpanzees), and the Hominini (humans and their ancestors).

## EVOLUTIONARY CONNECTIONS

### Comparing the Genomes of Humans and Chimpanzees

A male chimp called Clint, which lived at a primate research center in Atlanta provided the DNA used to sequence the chimp genome. In 2005, the Chimpanzee Sequencing and Analysis Consortium published an initial sequence of the chimpanzee genome. The draft sequence followed the 2003 publication of the human genome (see Chapter 18) and allowed scientists to make detailed comparisons between the two species. These comparisons revealed that the sequences of base pairs of both species' genomes differ by only 1.23%, which is 10 times less than the difference between the mouse and rat genomes. Comparisons of human and chimpanzee proteomes also showed that 29% of all proteins are identical, with most others differing by one or two amino acid substitutions.

Many of the genetic differences between chimps and humans result from chromosome inversions and duplications. Over 1,500 inversions occur between the chimp and human genomes. Although many inversions are in the noncoding regions of the genome, the DNA in these regions may regulate the expression of the genes in the coding regions. Duplications and deletions are also common. For example, humans have lost a gene called *caspase-12,* which in other primates may protect against Alzheimer's disease.

Some interesting genetic differences were apparent between chimps and humans even before their entire genomes were sequenced. In 2002, Swedish molecular geneticist Svante Pääbo discovered differences between humans and chimps in a gene called *FOXP2,* which plays a role in speech development. Proteins encoded by this gene differ in just two amino acids of a 715-amino-acid sequence. Researchers propose that the mutations in this gene have been crucial for the development of human speech.

In 2006, a team led by American geneticist David Reich discovered that the human X chromosome diverged from the chimpanzee X chromosome about 1.2 million years more recently than the other chromosomes. This indicated to the researchers that the human and chimp lineages split apart, then began interbreeding before diverging again. This explains why many fossils appear to exhibit traits of both humans and chimps, because they may actually have been human-chimpanzee hybrids.

## Bipedalism Is a Distinguishing Feature of Humans

About 7 mya in Africa, a lineage that led to modern humans diverged from other primate lineages. The evolution of humans should not be viewed as a neat, stepwise progression from one species to another. Rather, human evolution, like the evolution of all species, can be visualized more like a tree, with one or two **hominin** species—members of the Hominini tribe—likely coexisting at the same point in time, with some eventually going extinct and some giving rise to other species (**Figure 22.17**).

The key characteristic differentiating hominins from other apes is that hominins walk on two feet; that is, they are **bipedal.** At about the time when hominins diverged from other ape lineages, the Earth's climate had cooled, and the forests of Africa gave way to grassy savannas. Bipedal locomotion and an upright stance may have been advantageous in allowing hominins to peer over the tall grasses of the savanna to see predators or prey.

Bipedalism is correlated with important anatomical changes in hominins. First, the opening of the skull where the spinal cord enters shifted forward, causing the spine to be more directly underneath the head. Second, the hominin pelvis became broader to support the additional weight. And third, the lower limbs, used for walking, became relatively larger than those in other apes. These are the types of anatomical changes paleontologists look for in the fossil record to determine whether fossil remains are hominin.

**(a) Strepsirrhini (lesser bush baby)**

**(b) Anthropoidea (capuchin monkey)**

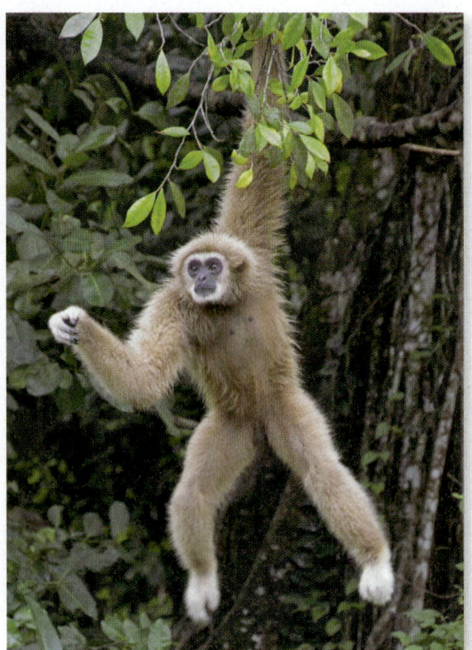

**(c) Hominoidea (white-handed gibbon)**

**Figure 22.15 Primate classification.** The primates are divided into two groups: **(a)** the strepsirrhini (smaller, nocturnal species such as this bush baby), and the haplorrhini (larger, diurnal species). Haplorrhini comprise **(b)** the monkeys and tarsiers, such as this capuchin monkey (*Cebus capucinus*), and **(c)** the hominoids, species such as this white-handed gibbon (*Hylobates lar*).
*(a)* © David Haring/DUPC/Getty Images; *(b)* © Brand X Pictures/PunchStock RF; *(c)* © Keren Su/Getty Images

Concept Check: *What are the defining features of primates?*

**(a) Gorilla (*Gorilla gorilla*)**

**(b) Chimpanzee (*Pan troglodytes*)**

**(c) Human (*Homo sapiens*)**

**Figure 22.16 Members of the family Hominidae. (a)** Gorillas, the largest of the living primates, are ground-dwelling herbivores that inhabit the forests of Africa. **(b)** Chimpanzees are smaller, omnivorous primates that also live in Africa. The chimpanzees are close living relatives of modern humans. **(c)** Humans are also members of the family Hominidae. The orangutan is also a member of this group.
*(a)* Source: Richard Ruggiero, USFWS/Flickr/CC BY 2.0; *(b)* © DLILLC/Masterfile RF; *(c)* © Tetra Images/Getty Images RF

The earliest known hominin, *Sahelanthropus tchadensis*, was discovered in Central Africa in 2002. Fossils of this species are dated at 7 million years old. The evolutionary relationship between *S. tchadensis* and later hominin species is unclear. Another early group of hominins included several species of the genus *Australopithecus* which first emerged in Africa about 4 mya. As shown in Figure 22.17, *Australopithecus afarensis* is generally regarded as the direct ancestor of most hominin species, but it could be a close relative of an unknown species that was the direct ancestor. From there, the evolution of different human species is still debated. It is generally agreed that two genera evolved from *Australopithecus:* the more robust *Paranthropus* and the genus *Homo*.

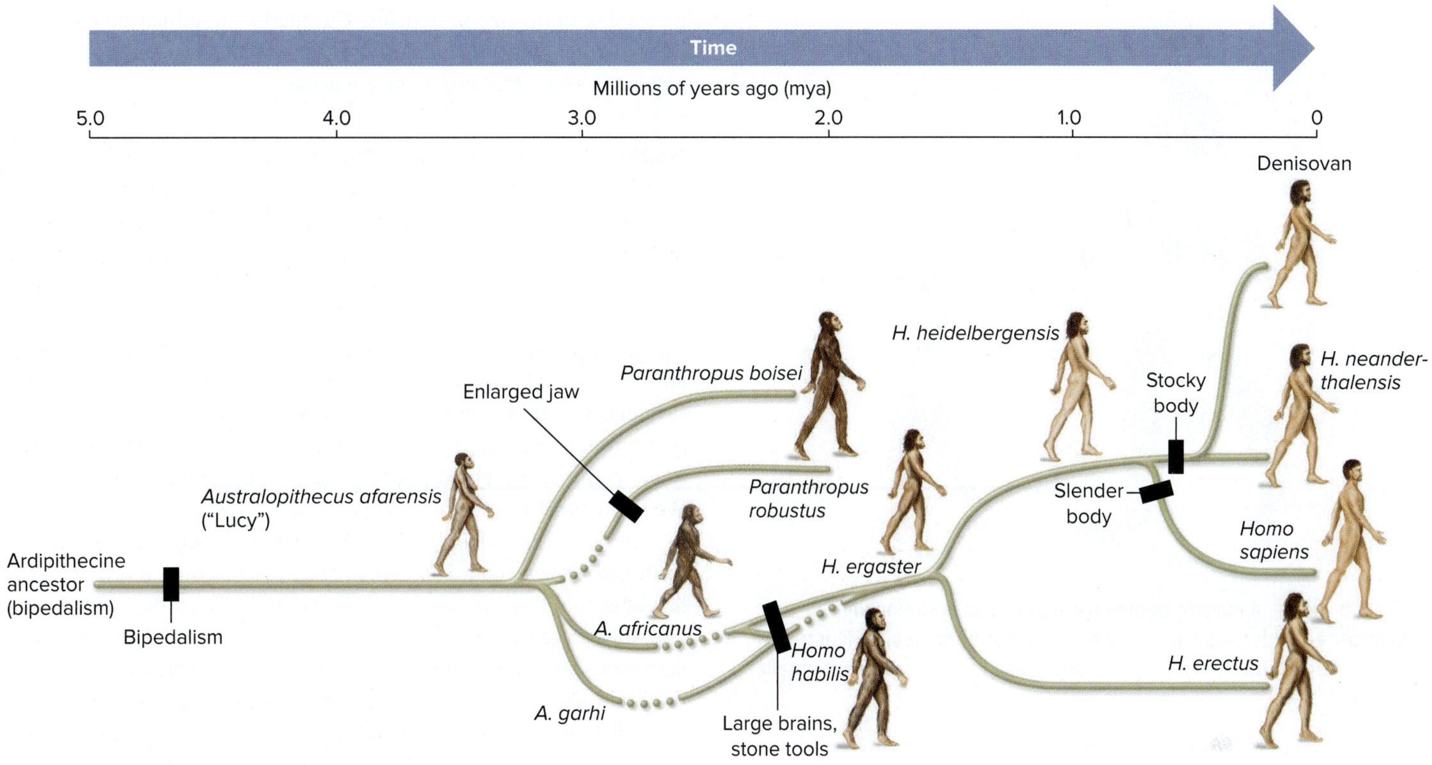


**Figure 22.17 A possible scenario for human evolution.** In this human family tree (based on the work of Donald Johanson and others), several hominin species lived during the same time period as others, but only one lineage gave rise to modern humans (*Homo sapiens*).

The early stages of the evolution of *Homo* species, and their differentiation from at least two possible *Australopithecus* species, have not yet been determined with great certainty. However, the later divergence of various *Homo* species is better understood.

## *Australopithecus* and *Paranthropus* Are Early Human Genera

Let's consider some of the general features of the two early human genera, *Australopithecus* and *Paranthropus*. In 1924, the first fossil was found in South Africa for a member of the genus *Australopithecus* (from the Latin *austral,* meaning southern, and the Greek *pithecus*, meaning ape). Since then, hundreds of fossils of this group have been unearthed all over southern and eastern Africa, the areas where fossil deposits are best exposed to paleontologists. This group was widespread, with at least six species. In 1974, American paleontologist Donald Johanson discovered the skeleton of a female *A. afarensis* in the Afar region of Ethiopia and dubbed her Lucy. (The Beatles' song "Lucy in the Sky with Diamonds" was playing in the camp the night when Johanson was sorting the unearthed bones.) Over 40% of the skeleton had been preserved, enough to provide a good idea of the physical appearance of australopithecines. Compared with modern humans, they were relatively small, about 1–1.5 m in height and approximately 18 kg in weight (**Figure 22.18**). Females were much smaller than males, a characteristic known as sexual dimorphism. Examination of the bones revealed that *A. afarensis* walked on two legs. They had a facial structure and brain size (about 500 cubic centimeters [$cm^3$]) similar to those of a chimp.

In the 1930s, the remains of bigger-boned hominins were found. Two of the larger species now considered to be a separate genus, *Paranthropus,* weighed about 40 kg and lived during the same time period as australopithecines and early *Homo* species. *Paranthropus* were vegetarians with enormous jaws used for grinding tough roots and tubers. Both *Paranthropus* species died out rather suddenly about 1.5–2.0 mya.

Although *Australopithecus africanus* was thought to have evolved slightly later than *A. afarensis,* its bones had been found much earlier than those of *A. afarensis.* In the 1920s, Australian anthropologist Raymond Dart described *A. africanus* from infant bones discovered in a cave in Taung, South Africa. The well-preserved skull was small but was well rounded, unlike the skulls of chimpanzees and gorillas. Also, the positioning of the head on the vertebral column indicated bipedalism. These observations suggested to Dart that he had found a transitional form between apes and humans. However, it would take another 20 years and the discovery of more fossils to convince the scientific world to support Dart's view. In 1996, remains of another species, *Australopithecus garhi,* were also found in the Afar region. They were a surprise in that the teeth suggested similarities with *Paranthropus boisei.* Garhi means surprise in the local Afar

**Figure 22.18 A modern woman compared to an australopithecine.** Compared with modern humans, Australopithecines, as illustrated by this reconstruction of the famous fossil Lucy, were much smaller and lighter.

language. The proposal that both *A. garhi* and *A. africanus* are ancestors of modern humans has been the subject of much debate. They have been viewed as dead-end cousins or the ancestors of the first members of the genus *Homo.*

## The Genus *Homo* Includes Modern Humans and Their Most Recent Relatives

Seven different species in the *Homo* genus are shown in Figure 22.17. Let's consider how they are evolutionarily related to each other.

**Homo habilis** In the 1960s, British paleontologist Louis Leakey found hominin fossils estimated to be about 2 million years old in Olduvai Gorge, Tanzania. Two particularly interesting observations about these fossils stand out. First, reconstruction of the skull showed a brain size of about 680 $cm^3$, larger than that of *Australopithecus.* Second, the fossils were found with a wealth of stone tools. As a result, Leakey assigned the fossils to a new species, *Homo habilis,* from the Latin, meaning handy man. The discovery of several more *Homo* fossils followed, but there have been no extensive finds, as there were with Lucy. This makes it difficult to determine which *Australopithecus* lineage gave rise to the *Homo* lineage (see Figure 22.17), and paleontologists remain divided on this point.

*Homo habilis* lived alongside *Paranthropus* in East Africa but had much smaller jaws and teeth, indicating that it probably ate large quantities of meat. The smaller jaw provided more space in the skull for brain development (look ahead to Figure 35.6). *Homo habilis* probably scavenged most of its meat from the kills of large predators. A meatier diet is easier to digest and is rich in nutrients and calories. The human brain uses a lot of energy, 20% of the body's total energy production. The meat-eating habit thus helped propel the evolution of increasing brain size in humans. Cut marks on animal bones of the period reveal that early humans used stone tools to smash open bones and extract the protein-rich bone marrow.

**Homo ergaster** Although the evidence is not entirely clear, *H. habilis* probably gave rise to one of the most important species of *Homo, Homo ergaster. Homo ergaster* was a hominin that arose in Africa about 2 million years ago; it had a human-looking face and skull, with downward-facing nostrils. *Homo ergaster* was also a tool user, and the tools, such as hand axes, were larger and more sophisticated. *Homo ergaster* evolved in a period of global cooling and drying that reduced tropical forests even more and promoted savanna conditions. Hairlessness and the regulation of body temperature through sweating may also have evolved at this time as adaptations to the sunny environment. A leaner body shape was evident. We know this from so-called Turkana boy, a fossil teenage boy found in Kenya in 1984. Though only 13 years old, scientists predict he would have been about 185 cm (6 ft 1 in.) when adult, much the same height as the Masai tribesman that inhabit the area today. A dark skin probably protected *H. ergaster* from the Sun's rays. The pelvis had narrowed, promoting efficiencies in walking upright, and the size of the brain and hence the skull increased, which may have produced more difficulty in childbirth. Mothers had to push increasingly large-brained infants through a narrowed pelvis. Researchers think that as a result, the human gestation period was shortened. Earlier birth leads to prolonged care of human infants compared with that in other apes. Prolonged child care required well-nourished mothers, who would have benefited from the support of their male partner and other members of a social group. Some anthropologists have suggested this was the beginning of the family.

*H. ergaster* is thought to have given rise to many species, including *Homo erectus, Homo heidelbergensis, Homo neanderthalensis,* and *Homo sapiens.* A possible time line and geographic locations for these species are shown in Figure 22.19. *Homo ergaster* probably was the first type of human to leave Africa, as similar bones have been found in the Eurasian country of Georgia. *H. ergaster* is believed to be a direct ancestor of modern humans, with *Homo heidelbergensis* viewed as an intermediary step. Living at the same time as *H. heidelbergensis* was another descendent of *H. ergaster, H. erectus,* though some researchers consider *H. ergaster* and *H. erectus* as the same species. We will treat them as separate species here.

**Homo erectus** *H. erectus* was a large hominin, as large as a modern human but with heavier bones and a smaller brain capacity of between 750 and 1,225 $cm^3$ (modern brain size is about 1,350 $cm^3$). Fossil evidence shows that *H. erectus* was a social species that used tools, hunted animals, and cooked over fires. The meat-eating habit may have sparked the migration of *H. erectus.* Carnivores tend to have larger ranges than similar-sized herbivores, because the food sources of carnivores (prey) are usually scarcer per unit area. *H. erectus* spread out of Africa soon after the species appeared, over a million years ago, and fossils have been found as far away as China and Indonesia. The first fossil was found by Dutch physician Eugene Dubois in 1891 on the Indonesian island of Java. Stone tools are rarely found in these Asian sites, suggesting *H. erectus* based their

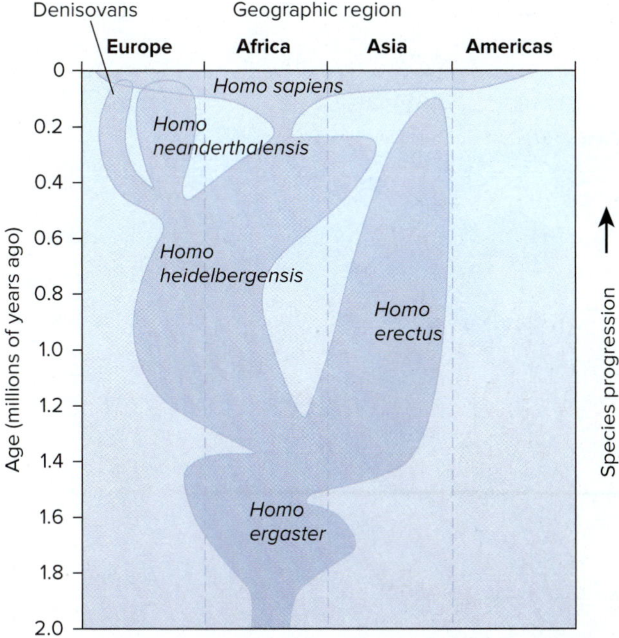


**Figure 22.19 One view of the temporal and geographic evolution of hominid populations.** Though not shown in this figure, the range of the Denisovans and *Homo neanderthalensis* also extended into Asia.

technology on other materials, such as bamboo, which was abundant at that time. Bamboo is strong yet lightweight and could have been used to make spears. These humans may even have used rafts to take to the seas. *H. erectus* went extinct about 100,000 years ago, for reasons that are unclear but may be related to the spread of *H. sapiens* into its range.

***Homo heidelbergensis*** *H. heidelbergensis* was similar in body form to modern humans. Large caches of this species' bones were found in Spain, at the bottom of a 14-m (45-ft) shaft known as La Sima de Los Huesos (the pit of bones). Similar remains were also found at Boxgrove in England. Shinbones recovered from Boxgrove suggest males stood around 180 cm (6 ft) and weighed 88 kg (196 pounds). Skulls were large, with brain volumes from 1,100 to 1,400 cm$^3$, similar to modern humans. Animal bones from these sites showed cut marks from stone blades beneath tooth marks from carnivores. This indicated that humans were killing large prey before scavengers arrived. Horses, giant deer, and rhinoceroses were common prey and would have required much skill and cooperation to hunt. *H. heidelbergensis* gave rise to three species, *Homo neanderthalensis,* the Denisovans, and *Homo sapiens*. However, some scientists view these as three subspecies. In the discussion that follows, we will consider them to be three different species.

***Homo neanderthalensis*** *H. neanderthalensis* was named for the Neander Valley of Germany, where the first fossils of its type were found. In the Pleistocene epoch, glaciers were locked in a cycle of advance and retreat, and the European landscape was often covered with snow. Over the course of many generations, the more slender body form of *H. heidelbergensis* evolved into a shorter, stockier build that was better equipped to conserve heat; we now call this type of human Neanderthal. Neanderthals also possessed a more massive skull and larger brain size than modern humans, about 1,450 cm$^3$, perhaps associated with their bulk. Males were about 168 cm (5 ft 6 in.) in height and would have been very strong by modern standards. They had a large face with a prominent bridge over the eyebrows, a large nose, and no chin (**Figure 22.20a**). They lived predominantly in Europe, with a range extending to the Middle East and Asia (**Figure 22.20b**). Their stocky and muscular physique was well suited to the rigors of cold climates and hunting prey. Paleontologists have found a high rate of head and neck injuries in Neanderthal bones, similar to that seen in present-day rodeo riders. This suggests that close encounters with large prey often resulted in blows that knocked the hunters off their feet. The hyoid bone, which holds the larynx (voice box) in place, was well developed, suggesting speech was used. The Neanderthals went extinct about 40,000 to 30,000 years ago.

**Denisovans** In 2010, scientists sequenced DNA from a fossilized pinky finger of a young female found in Denisova Cave in Siberia, Russia, and found it to be genetically distinct from the DNA of *H. neanderthalensis* and *H. sapiens*. Because it is such a recent discovery, its species name is yet to be agreed upon. The common name given to this type of human is a Denisovan. Using carbon radioisotope dating, the fossil was estimated to be about 400,000 years old. Scientists now hypothesize that Denisovans were a sister group to the Neanderthals, diverging from the Neanderthals about 400,000 years ago. It is difficult to reconstruct the physical traits of Denisovans, because only a finger bone and a few molar teeth have been discovered thus far. Even so, the finger bone was unusually broad and robust, suggesting the Denisovans were similar in build to the Neanderthals. Like the Neanderthals, the Denisovans are thought to have gone extinct about 40,000 to 30,000 years ago.

***Homo sapiens*** *H. sapiens,* (from the Latin, meaning wise man), is our own species. *H. sapiens* is a slender, lighter-weight species with a slightly smaller brain capacity than that of the Neanderthals. Researchers hypothesize a variety of reasons why *H. sapiens* thrived while the Neanderthals disappeared, including possessing a more efficient body type with lower energy needs, increased longevity, and differences in social structure and cultural adaptations.

## Evidence Suggests That *Homo sapiens* Arose in Africa and Then Migrated to Other Parts of the World

Two models have been proposed to explain where the species of modern humans, *Homo sapiens,* arose. The first model, the out-of-Africa hypothesis, suggests that *H. sapiens* evolved in Africa from *H. heidelbergensis*. Some members of *H. sapiens* later migrated to other parts of the world, and gradually replaced species such as *H. erectus* and *H. neanderthalensis.* An alternative hypothesis, called the multiregional hypothesis, which is not widely accepted, suggests that human groups have evolved from *H. ergaster* populations in a number of different parts of the world. According to this hypothesis, gene flow

(a) An adult male Neanderthal

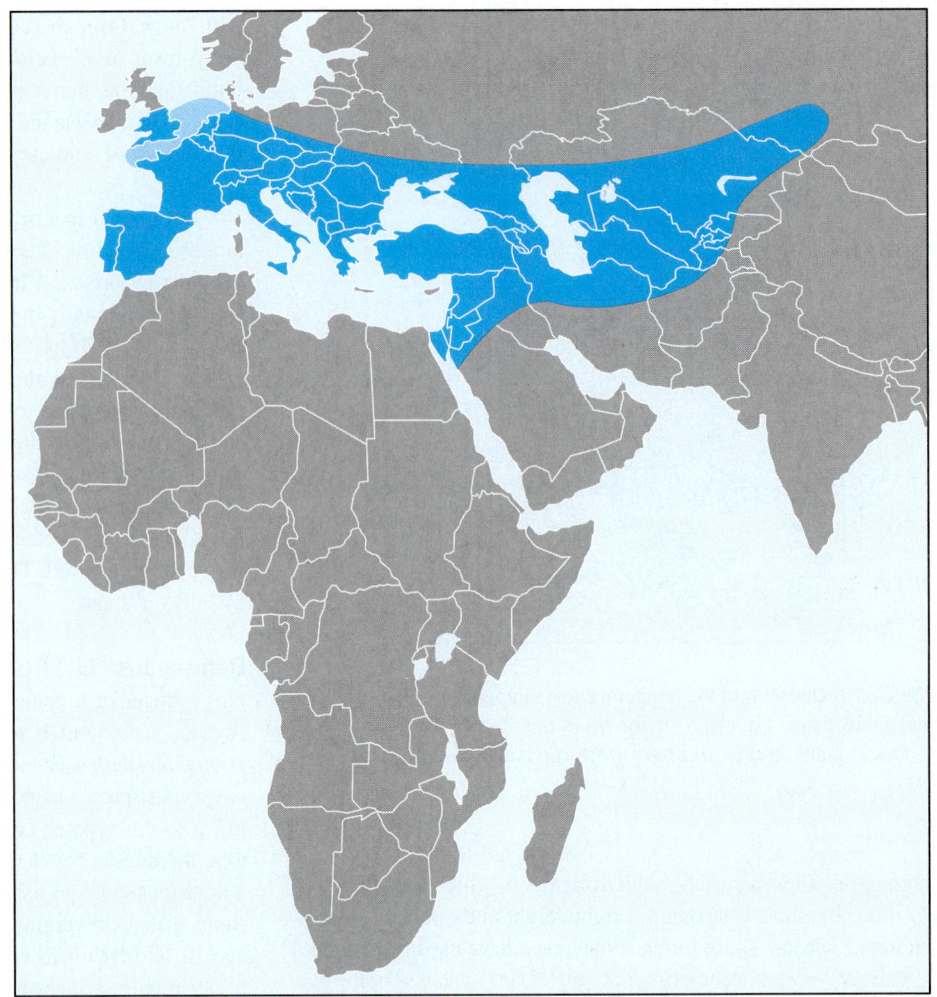

(b) Geographic range of Neanderthals

**Figure 22.20** **Neanderthals.** **(a)** Artist's rendition of a Neanderthal human. Neanderthals were shorter and stockier than modern humans, with larger elbow and ankle joints, shorter forearms, and a larger, broader rib cage. **(b)** The range of Neanderthals was confined to Europe and western Asia, with a northerly limit that corresponds to about 50° north, the southern limit of glaciation. Total population size may only have been 70,000 at its peak.

between neighboring populations prevented the formation of several different *H. sapiens* species.

In 1987, American evolutionary biologists Rebecca Cann, Allan Wilson, and colleagues analyzed the sequences of mitochondrial DNA that was collected from many different people from around the world. By comparing these sequences to each other, they were able to conclude that *H. sapiens* arose in Africa about 200,000 years ago. These results and those of subsequent studies support the out-of Africa hypothesis. According to this hypothesis, *H. ergaster* arose in Africa and spread to Asia and Europe. Later, *H. erectus* diverged from *H. ergaster* in Africa and spread into Asia, and *H. neanderthalensis* diverged from *H. heidelbergensis* in Europe. Both of these species and the Denisovans became extinct as *H. sapiens* migrated across the world. Some researchers have suggested that the extinction of these other human species may have occurred because they were outcompeted by *H. sapiens*. However, further research is needed to confirm or refute that idea.

The analysis of DNA sequences from modern humans across the world can also be used to construct a map that describes the migration of humans out of Africa. Figure 22.21 describes such a map. However, the time periods should be considered as approximate. As we gather more data from people in various regions of the world, this map is under frequent revision, and is more complicated than the simplified map shown in Figure 22.21. Modern humans spread first into the Middle East and Asia, then later into Europe and Australia, finally crossing the Bering Strait to the Americas.

Much remains to be resolved in our understanding of human evolution, and new data provide paleontologists with information to revise their hypotheses. For example, in 2004, the remains of a small human on the Indonesian island of Flores were discovered and were given the name *Homo florensiensis,* nicknamed hobbits by the media. Many species—for example, deer and elephants—develop into small forms in insular situations, so hobbit humans seemed plausible. Since then, many researchers have suggested these people were modern humans who were suffering from a genetic disorder. Even modern humans on Flores are pygmies. Pathological dwarfism would have made these people even smaller. Only *H. sapiens* tools have been found at the area where the bones occur, suggesting these individuals were indeed dwarf forms of modern humans.

A 2014 study showed that the small brain size of one of the *H. floresiensis* fossils was in the range predicted for an individual with Down syndrome. The authors went on to argue that *H. florensiensis* is an invalid species.

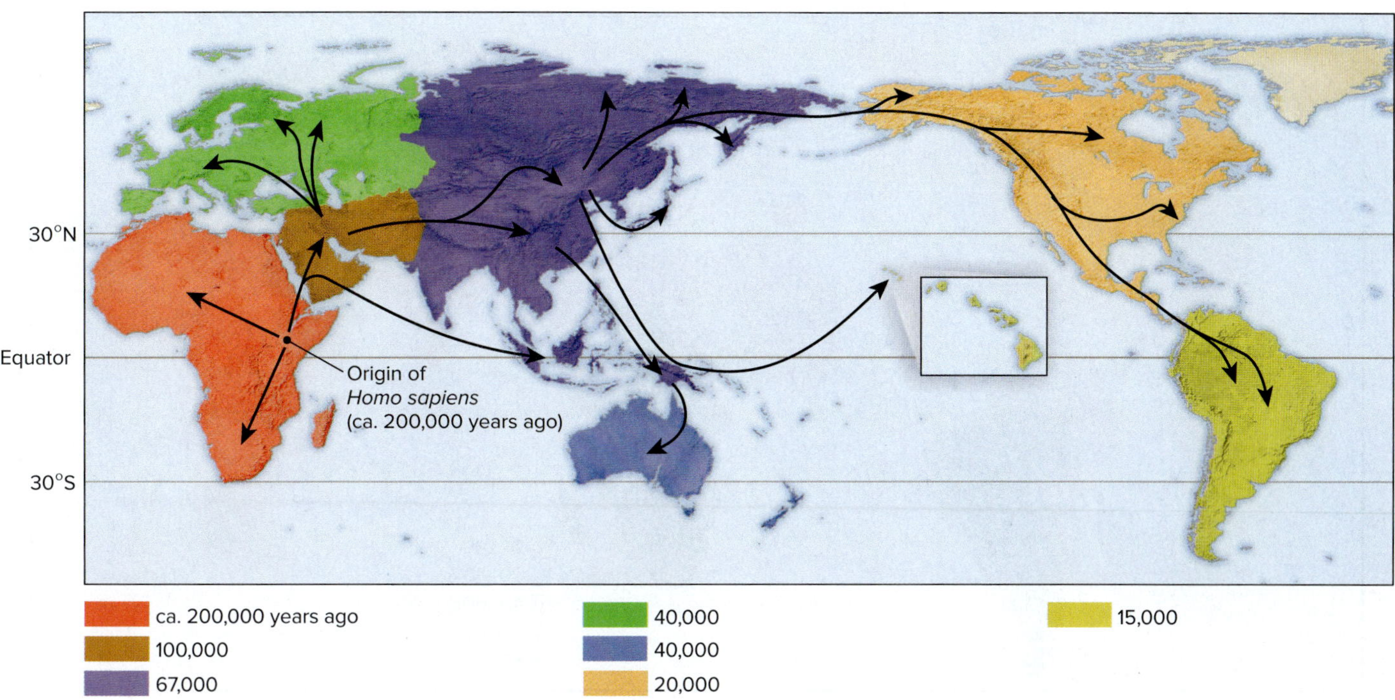


**Figure 22.21 A simplified model for the origin and spread of *Homo sapiens* throughout the world.** This map, based on differences of mtDNA throughout current members of the world's population, suggests *Homo sapiens* originated in East Africa. About 100,000 years ago, the species spread into the Middle East and from there to Europe, Asia, Australia, and the Americas.

## Human Evolution Has Involved Interbreeding Among Closely Related Species

Because the remains of Neanderthals and Denisovans are less than 50,000 years old, researchers have been able to extract DNA from their fossils and compare their sequences to each other and to *H. sapiens*. Detailed comparisons of Neanderthal, Denisovan, and modern human genomes have revealed interbreeding among the three species as *H. sapiens* spread into Europe and Asia. The genomes of modern humans of African descent contain little or no DNA that is derived from Neanderthals or Denisovans. However, 1% to 4% of the DNA from a person of European descent is derived from Neanderthals, and 4% to 6% of the DNA from a person of Southeast Asia descent is derived from Denisovans. These results indicate that *H. sapiens* ancestors interbred with Neanderthals and Denisovans as *H. sapiens* spread across Europe and Asia. Interestingly, some people of African descent carry very small amounts of Neanderthal DNA. How is this possible? By analyzing the DNA sequences of certain African people, researchers speculate that some *H. sapiens* from Europe may have migrated back to Africa about 3,000 years ago and interbred with a few isolated African populations of *H. sapiens*.

When analyzing the human genome on a population level, at least 20% of the Neanderthal genome is found in the genome of modern humans of European descent. No one individual has all 20%; rather, any given person of European descent has about 1% to 4%. Researchers have speculated that certain Neanderthal genes may have provided a survival advantage, and that may explain why they have been retained in particular human populations. For example, a group of related proteins called keratins form filaments that play a role in the formation of human skin, hair, and nails. Such filaments may differ among populations that have evolved in a warm climate versus a cold one. Some alleles of keratin genes encode keratins that provide better insulation, a trait that is advantageous in cold climates. In people of European descent, genes that encode keratins are often of Neanderthal origin. This observation is consistent with the idea that some Neanderthal alleles may have helped European *H. sapiens* adapt to colder European environments.

On the downside, scientists also speculate that Neanderthals carried an allele of a gene involved with fatty acid uptake that helped them store fat better than *H. sapiens*. Such a gene was an advantage for a Neanderthal lifestyle in which hunter-gatherers gorged on prey and then went for days without eating. However, in modern humans who overeat, fat storage can put them at risk for developing various diseases, such as type 2 diabetes. In certain human populations, particularly native Americans and those of Latin American descent, this Neanderthal allele is fairly common. Some people are heterozygous, carrying one allele that is derived from the Neanderthal genome and one from the *H. sapiens* genome. Such individuals are 20% more likely to develop type 2 diabetes compared to people who are homozygous for the *H. sapiens* allele. Furthermore, people who are homozygous for the Neanderthal allele are 40% more likely to develop diabetes.

## Are Human Populations Still Evolving?

As discussed in Chapter 19, natural selection is the process in which individuals with greater reproductive success are more likely to pass their genes to future generations. Natural selection results in evolution. (Other processes, such as genetic drift, can also promote

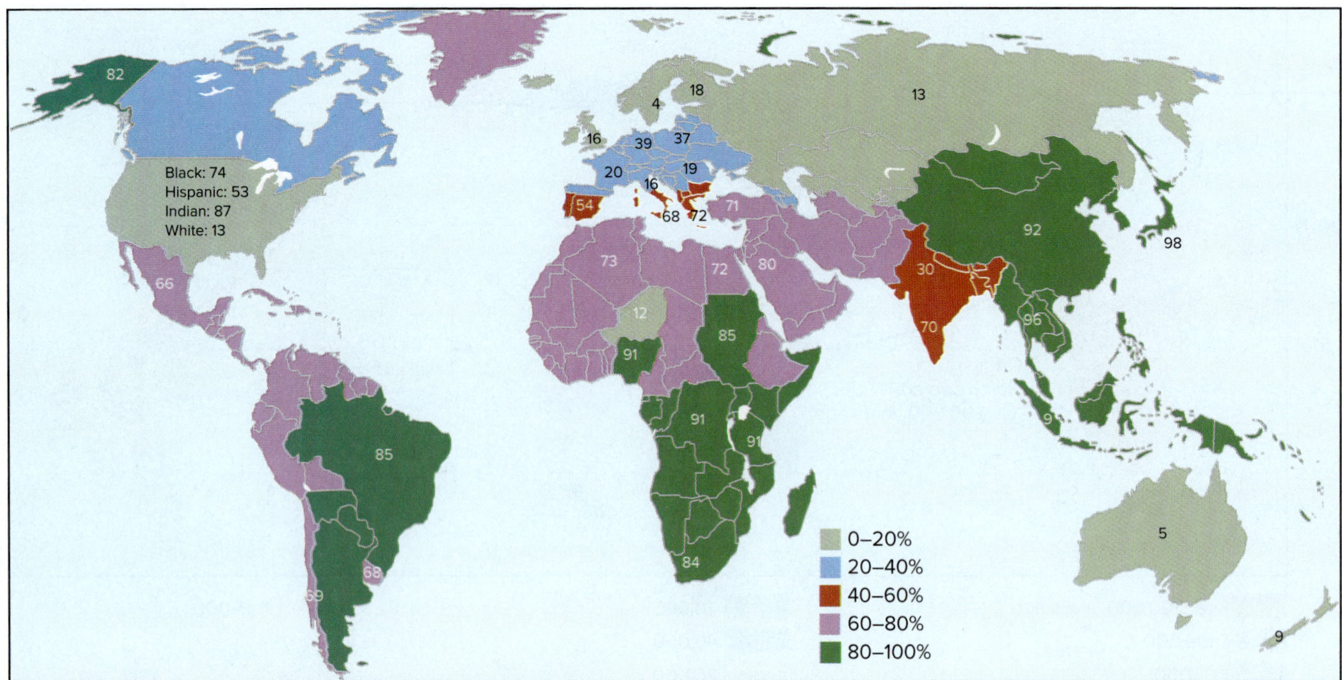


**Figure 22.22 The frequency of lactose intolerance in human populations.** This map emphasizes lactose intolerance, which is the inability to digest lactose after weaning. In contrast, most northern Europeans can digest the sugar lactose found in dairy products; they are lactose tolerant.

evolutionary change.) One factor that often plays a role in natural selection is the environment. Various types of environmental factors such as temperature, predators, and food sources can affect reproductive success and thereby cause a population to evolve in a particular direction. For example, a prolonged decrease in temperature may favor the survival of mammals with thicker fur, which will increase in frequency over the course of many generations.

Modern humans have great control over their environment. They live in dwellings where they can control the temperature, they can largely avoid predators, and they usually don't rely entirely on locally grown food for their survival. For such reasons, the impact of the natural environment on the evolution of human populations may have lessened compared to humans that lived long ago. Even so, human evolution via natural selection is still occurring. For example, we continue to evolve genetic resistance to infectious diseases. The bubonic plague of the 14th century killed about one-third of the Asian and European populations, yet many people survived and passed on alleles that confer greater resistance to this terrible disease. Similarly, alleles that confer resistance to malaria are becoming more common in certain African populations.

A classic example of recent human evolution is the ability to digest lactose, a sugar found in milk. In most human populations, the ability to readily digest lactose is lost after the age when babies are weaned from their mother's milk. After weaning, lactose becomes indigestible to most people and they suffer bloating, abdominal cramps, flatulence, diarrhea, nausea, or vomiting if they eat or drink dairy products. However, the ability to consume dairy products may have provided a survival advantage as people began to domesticate cows, sheep, and goats. In human populations where such domestication took place, the ability to digest lactose, called lactose tolerance, is expected to increase in frequency due to natural selection if it provided people with greater reproductive success.

Recent studies suggest that mutations that confer lactose tolerance arose several thousand years ago in a few different places. The frequency of this mutation increased dramatically over the past few thousand years in populations where dairy products are commonly consumed. The genetic mutation for digesting lactose after weaning is now carried by most northern Europeans. Lactose intolerance is less common in this geographical region (**Figure 22.22**). For this population, the trait is linked to a single mutation that affects the expression of the lactase gene, which encodes an enzyme that is needed to digest lactose. The mutation prevents the gene from being turned off after weaning. Lactose tolerance has also been found in other populations, such as Africa and the Middle East, but these mutations occurred independently of the one that is usually found in lactose-tolerant people of European descent.

As another example, eye color in certain human populations has also changed in recent times. Over 10,000 years ago, all or nearly all humans had brown eyes. About 10,000 years ago, someone who lived near the Baltic Sea inherited a mutation that resulted in blue eyes in the homozygous state. In 2008, researchers discovered that this mutation decreases the expression of the *Oca2* gene, which encodes a protein that is needed for the production of melanin pigment in the iris and other parts of the body. The frequency of this allele increased, especially in northern Europe (**Figure 22.23**), and by about 3,000 years ago, blue eyes had spread across Europe.

Why did the frequency of blue eyes increase in such a rapid fashion in human populations? The dramatic rise in blue eye color suggests that natural selection was playing a role, but researchers have

(a) Blue eye color

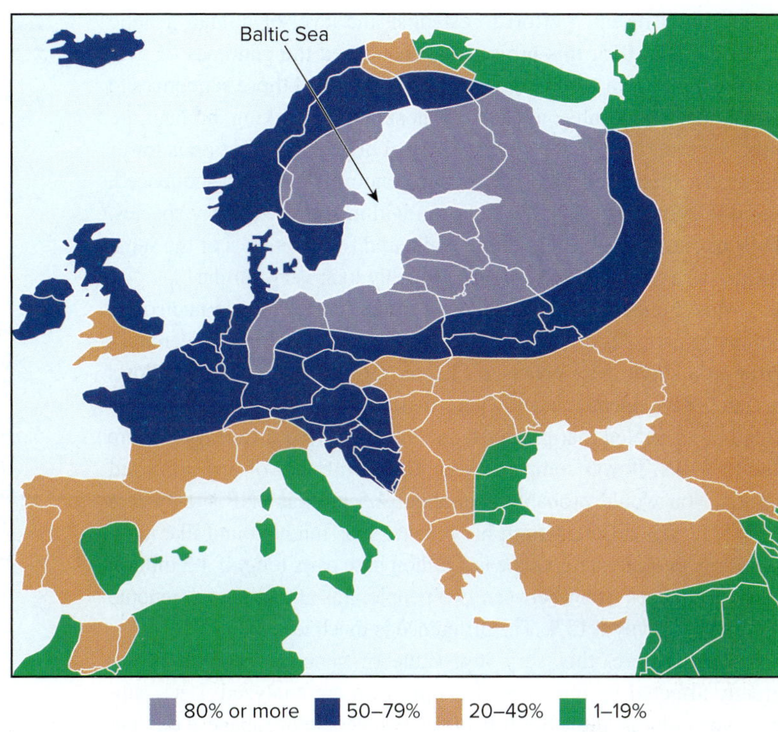


(b) Percentage of people with blue eye color in Europe and surrounding regions

**Figure 22.23** **Blue eye color and its spread throughout Europe.** **(a)** In humans, blue eye color is due to a lack of melanin in the iris. This mutation probably appeared only about 10,000 years ago. **(b)** Many people in western and northern Europe have blue eyes due to the spread of this single mutation that originated in someone living near the Baltic Sea. Note: The indicated percentages include blue eye color and other light colors of eyes, such as green.

*(a)* © harpazo_hope/Getty Images RF

yet to come up with a definitive answer regarding its selective advantage. One possible explanation has to do with vitamin D deficiency. Vitamin D is an important human vitamin that requires skin exposure to the UV rays in sunlight in order to be made. People living in northern latitudes are exposed to much less sunlight compared to those living nearer the equator, putting them at greater risk for vitamin D deficiency. A decrease in melanin synthesis not only affects eye color (brown to blue) but also results in lighter skin, which more easily absorbs UV rays. Therefore, one hypothesis for the spread of blue eyes (and lighter skin) through the human population is that it may have enabled humans to avoid the harmful health effects of vitamin D deficiency, which includes weakness and bone abnormalities. In this scenario, natural selection acted on skin color, and the eye color phenotype increased due to its association with a lighter skin color.

## Modern Humans Show Relatively Small Genetic Variation Yet Exhibit a Significant Amount of Phenotypic Variation

Looks can be deceiving. If you take a plane ride around the world and make stops in Japan, Nigeria, Norway, Brazil, and Australia, you might get the impression that the human species is genetically diverse (Figure 22.24). The physical differences among humans in many parts of the globe are striking. To assess the level of genetic variation, the **1000 Genomes Project** was launched in January 2008, which is an

(a) Japan (b) Nigeria (c) Norway (d) Brazil (e) Australia

**Figure 22.24** **Examples of human phenotypic variation in different parts of the world.** The women seen here are from **(a)** Japan, **(b)** Nigeria, **(c)** Norway, **(d)** Brazil, and **(e)** Australia. They are genetically very similar even though they look somewhat different.

*(a)* © William Perugini/123RF.com; *(b)* © Filipe Frazao/Shutterstock RF; *(c)* © Andrea Magugliani/Alamy; *(d)* © Daniel Ernst/Alamy RF; *(e)* © David Freund/Getty Images RF

international research effort to establish the level of human genetic variation. By 2015, this project had sequenced the genomes of over 2,500 people from around the world and compared those sequences to each other. The results indicate that, genetically speaking, humans are very, very similar to each other. Our level of genetic variation is lower than most species of mammals whose genomes have been sequenced, and it is even lower than the variation within certain fruit fly species. This may seem surprising; if you compared two fruit flies of the same species to each other, they would probably look very similar!

Most human genetic variation is in the form of single nucleotide polymorphisms (SNPs), which are places where two humans' genomes differ at a single base pair (see Chapter 19). DNA sequencing allows researchers to identify sites where two people's genomes differ. The data have revealed that greater than 99.9% of those differing sites are due to SNPs. If you compared your DNA with that of any unrelated person, you would probably find about 4.5 million SNP sites where your SNP was different from his or hers. This might sound like a lot, but consider that your genome is 3 billion base pairs long. If 4.5 million SNP differences occur between two people, this amounts to a genome difference of only 0.15%. The difference is much less than 1%!

Furthermore, this very low value of genetic variation is not greatly affected by the pairs of people who are analyzed. Let's suppose you compared the SNPs between two people of Japanese descent and compared the SNPs between a person of Japanese descent and a person of northern European descent. Both pairs would show a low level of SNP variation, which would be around 0.15%. The value for the first pair (Japanese-Japanese) would probably be a little lower (say, 0.14%) and the other pair (Japanese–northern European) might be a little higher (say, 0.16%), but both pairs would be remarkably similar. Genetic variation in human populations is very low and most of it occurs within all populations. Relatively little additional genetic variation is observed when comparing individuals from populations that are geographically separated.

Why is the genetic diversity of our species low? Various factors are involved. First, *Homo sapiens* has not been around that long. Even though 200,000 years might seem like a long time, humans have a long generation time, and evolution occurs over the course of generations. Second, until recently, human populations have been relatively small. About 10,000 years ago, the human population size is estimated to have been about 5 million. Small populations tend to be less genetically diverse than larger ones. Although the human population has grown enormously over the past few centuries, this expansion is very recent on an evolutionary timescale.

How do we explain the disparity between a low level of genetic variation and a seemingly higher level of phenotypic diversity? The answer is not entirely understood, but it may be related to the traits that influence our perception of diversity. Some of the traits that are most visually obvious, such as eye and skin color, may be dramatically influenced by natural selection even though they involve changes in a relatively small number of genes. For example, we have already considered how blue eye color and light skin color spread rapidly throughout Europe and this change involved a mutation in a single gene. This observation underscores how our visual perception of diversity may be biased by traits that may not be rooted in major differences in genetic diversity.

## 22.3 Reviewing the Concepts

- Many defining characteristics of primates relate to their tree-dwelling nature and include grasping hands, large brain, nails instead of claws, and binocular vision (Figures 22.14–22.16).
- About 7 mya in Africa, a lineage that led to humans began to separate from other primate lineages. A key characteristic of hominins (extinct and modern humans) is bipedalism. Human evolution can be visualized like a tree, with a few hominin species coexisting at the same point in time, some eventually going extinct, and some giving rise to other species (Figures 22.17–22.20).
- Data from the sequencing of human mitochondrial DNA suggest that *H. sapiens* originated in East Africa. From there, *H. sapiens* spread to Asia and then to all other parts of the globe (Figure 22.21).
- Human evolution has involved the interbreeding between closely related species, such as *H. sapiens* to the Neanderthals and to the Denisovans.
- Evidence that human populations are still evolving include traits such as lactose tolerance and blue eye color (Figures 22.22, 22.23).
- Modern humans show relatively small genetic variation yet exhibit a significant amount of phenotypic variation (Figure 22.24).

## 22.3 Testing Your Knowledge

**1.** Which of the following is *not* a trait that is found in some or all primates?
   **a.** binocular vision
   **b.** opposable thumb
   **c.** flat nails
   **d.** large brain
   **e.** All of the above are found in some or all primates.

**2.** Some modern humans have a small amount of Neanderthal DNA. This observation suggests that
   **a.** *H. sapiens* and *N. neanderthalensis* interbred with each other.
   **b.** *H. sapiens* and *N. neanderthalensis* were not derived from a common ancestor.
   **c.** *H. sapiens* incurred new mutations that make them more like *N. neanderthalensis*.
   **d.** All of the above
   **e.** Only a and b are correct.

# Assess and Discuss

## Test Yourself

1. The movement of landmasses that have changed their positions, shapes, and association with other landmasses is called
   a. glaciation.
   b. Pangaea.
   c. continental drift.
   d. biogeography.
   e. geological scale.

2. Paleontologists estimate the dates of fossils by
   a. the layer of rock in which the fossils are found.
   b. analysis of radioisotopes found in nearby igneous rock.
   c. the complexity of the body pattern of the organism.
   d. all of the above.
   e. a and b only.
3. The fossil record does not give us a complete picture of the history of life because
   a. not all past organisms have become fossilized.
   b. only organisms with hard skeletons can become fossilized.
   c. fossils of very small organisms have not been found.
   d. fossils of early organisms are located too deep in the crust of the Earth to be found.
   e. all of the above.
4. According to endosymbiosis theory, eukaryotic cells came into existence by
   a. endocytosis of an archaeal cell by a bacterial cell.
   b. endocytosis of a bacterial cell by an archaeal cell.
   c. fusion of an archaeal cell and a bacterial cell.
   d. fusion of a bacterial cell and an archaeal cell.
   e. a and c only.
5. Which of the following explanations of multicellularity in eukaryotes is seen in the development of complex, multicellular organisms today?
   a. endosymbiosis
   b. aggregation of cells to form a colony
   c. division of cells, with the resulting cells adhering together
   d. multiple cell types aggregating to form a complex organism
   e. none of the above
6. Which of the following evolutionary innovations was advantageous for survival in a terrestrial environment?
   a. the amniotic egg in animals
   b. the seed in plants
   c. the shell in marine invertebrates
   d. all of the above
   e. both a and b
7. The first mammal arose during the ___________ period.
   a. Triassic c. Cretaceous e. Quaternary
   b. Jurassic d. Tertiary
8. The earliest fossils of vascular plants were formed during the ___________ period.
   a. Ordovician c. Devonian e. Jurassic
   b. Silurian d. Triassic
9. The appearance of the first hominoids dates to the ___________ period.
   a. Triassic c. Cretaceous e. Quaternary
   b. Jurassic d. Tertiary
10. Which of the following statements regarding modern humans, *H. sapiens,* is *false?*
   a. Some modern humans have a small amount of DNA that is derived from Neanderthals.
   b. Some modern humans have a small amount of DNA that is derived from Denisovans.
   c. *H. sapiens* probably arose in Africa.
   d. Modern humans are very genetically diverse compared to most other species.
   e. Modern humans are still evolving.

## Conceptual Questions

1. How are the ages of fossils determined? In your answer, you should discuss which types of rocks are analyzed and explain the concepts of radiometric dating and half-life.
2. How was the phenomenon of endosymbiosis important in the evolution of the first eukaryotic cells?
3. **PRINCIPLES** Two principles of biology are (1) living organisms interact with their environment and (2) populations of organisms evolve from one generation to the next. Describe two examples in which changes in the global climate affected the evolution of species.

## Collaborative Questions

1. Discuss the factors that have contributed to the dramatic changes in life-forms since the origin of life on Earth about 3.5 to 4 billion years ago.
2. Discuss how the human body has changed since the human lineage diverged from other primates about 7 million years ago.

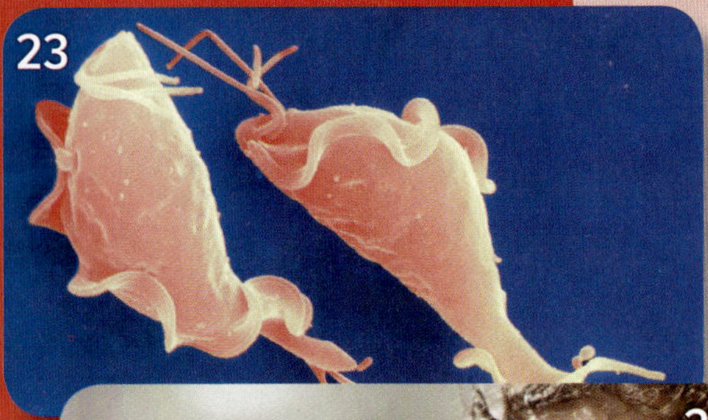

23

24

25

26

27

(23) © David M. Phillips/Science Source; (24) © KidStock/Getty Images RF; (25) © Linda Graham; (26) © NHPA/SuperStock; (27) © The Africa Image Library/Alamy

# UNIT V
# DIVERSITY

Biological diversity encompasses the variety of living things that exist on Earth now, as well as all the life-forms that lived in the past. Knowing about the many different kinds of modern organisms helps us to understand how life-forms are structured in ways that allow them to function differently in nature (described in Units VI and VII) and how species interact with each other and with their environments (described in Unit VIII—Ecology).

Unit V begins with microorganisms, which include bacteria, archaea, diverse lineages of protists, and fungi, Earth's oldest, simplest, and most numerous life-forms; Chapter 23 reveals the diversity of microorganisms that influence our lives. In Chapter 24, we learn how microbes occur in complex systems known as microbiomes, which influence our environments and health. Chapter 25 explores the diversification of plants, a process that changed features of planet Earth, culminating in seed plants that are vital sources of human food, fiber, and medicine. The simplest animals, the invertebrates, are covered in Chapter 26. More complex animals, including mammals, are the focus of Chapter 27.

***The following biology principles will be emphasized in this unit:***

- ***Cells are the simplest units of life.*** *Most present and past microbial species have bodies consisting of only one or a few cells, whereas some microbes, plants, animals, and many fungi display more complex bodies composed of many cells.*
- ***Living organisms interact with their environment.*** *Chapters 23–25 explain how ancient microorganisms and plants dramatically changed the composition of Earth's atmosphere, and how their modern descendants continue to influence today's atmosphere and climate.*
- ***All species (past and present) are related by an evolutionary history.*** *This concept, emphasized throughout the unit, helps us to comprehend how humans are related to other living things on Earth.*
- ***Structure determines function.*** *The chapters in this unit provide many examples of ways in which evolutionary change in molecular, cellular, or body structure has allowed organisms to achieve new and diverse functions.*

# Diversity of Microbial Life: Archaea, Bacteria, Protists, and Fungi

# 23

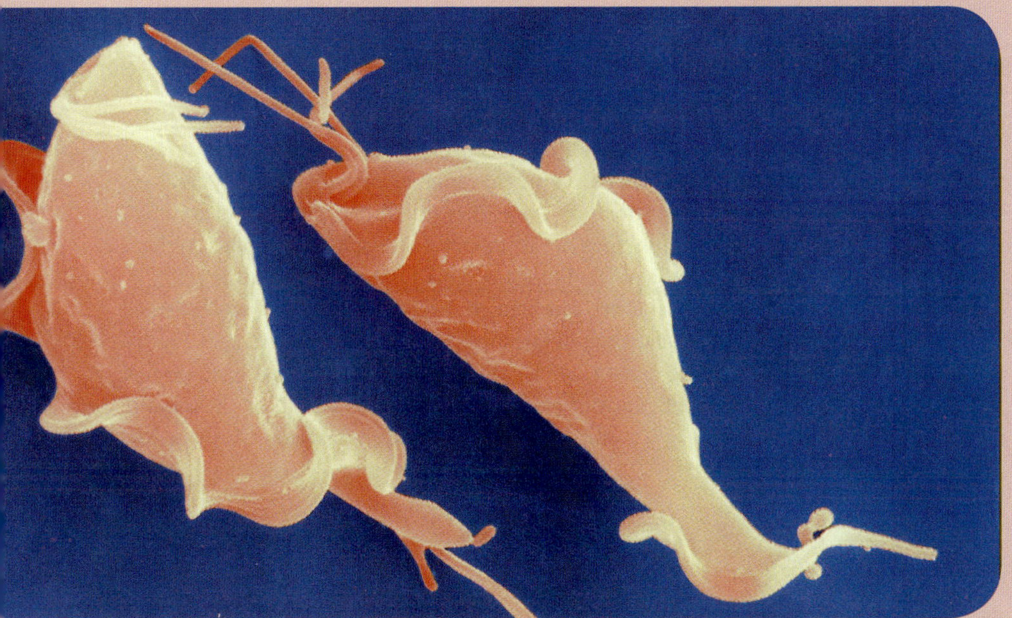

© David M. Phillips/Science Source

The protist *Trichomonas vaginalis* is one type of microorganism that causes infection of the human urogenital tract.

## Chapter Outline



Millions of people around the world suffer from infections with urogenital tract microorganisms, which include prokaryotic bacteria as well as eukaryotic protists and fungi, such as yeasts. *Trichomonas vaginalis*, for example, is a sexually transmitted protist that causes disease symptoms when it invades the human genitourinary tract. In this location *T. vaginalis* consumes bacteria and host epithelial and red blood cells, as well as carbohydrates and proteins released from damaged host cells. An estimated 170 million cases of infection with this organism occur each year around the globe, and infections can predispose humans to other diseases. *T. vaginalis* has an undulating membrane and flagella that allow it to move over mucus-coated skin (see chapter-opening photo). *Giardia intestinalis,* which infects the human intestinal tract, is another example of a protist whose flagella are critical to the ability to cause disease. Many types of disease-causing bacteria and fungi likewise may inhabit human bodies.

These examples illustrate how knowledge of microbial diversity and structure is essential in medicine, but such knowledge is also important in comprehending the many ecological roles that Earth's microorganisms play. By influencing global environment, many microbes that do not cause disease nevertheless influence human health and sustainability. In this chapter, we will survey microbial diversity, including the prokaryotic domains Archaea and Bacteria, as well as diverse eukaryotic protists and fungi.

## 23.1 Introduction to Microorganisms

### Learning Outcomes

1. Define the terms microorganism, archaea, bacteria, protists, and fungi.
2. Draw a diagram that illustrates horizontal gene transfer, and explain how it benefits the recipient organism.

Microorganisms, also known as microbes, are organisms that are typically so small in size that they can be seen only with the use of a microscope. Although a few animals are microscopic in size, most animals grow large enough to see without a microscope. In this chapter, we focus on groups that are primarily microscopic:

- The prokaryotic domain Archaea (informally known as archaea)
- The prokaryotic domain Bacteria (informally, bacteria)
- Diverse phyla of protists and fungi having eukaryotic cells, which are classified in the domain Eukarya, together with animals and plants

Recall that prokaryotic cells lack many of the features that occur in eukaryotic cells. Examples of distinctive eukaryotic cell features include nuclei having an envelope with complex pores, mitochondria, and other membrane-bound organelles. Comparative study of modern microbes helps to reveal how eukaryotic cells first arose from prokaryotic ancestors.

### Archaea and Bacteria Are Ancient Lineages That Continue to Be Abundant Today

The domains Archaea and Bacteria are both monophyletic groups, meaning they each have a single common ancestor. Although archaea and bacteria together are sometimes referred to as the "prokaryotes," this combination does not represent a monophyletic group; Archaea are more closely related to the domain Eukarya (Figure 23.1).

Archaea and bacteria include Earth's smallest known cells and are the most abundant organisms in the world. About half of Earth's total biomass consists of an estimated $10^{30}$ individual bacteria and archaea. Just a pinch of garden soil can contain 2 billion prokaryotic cells, and about a million occur in 1 mL of seawater. Archaea and

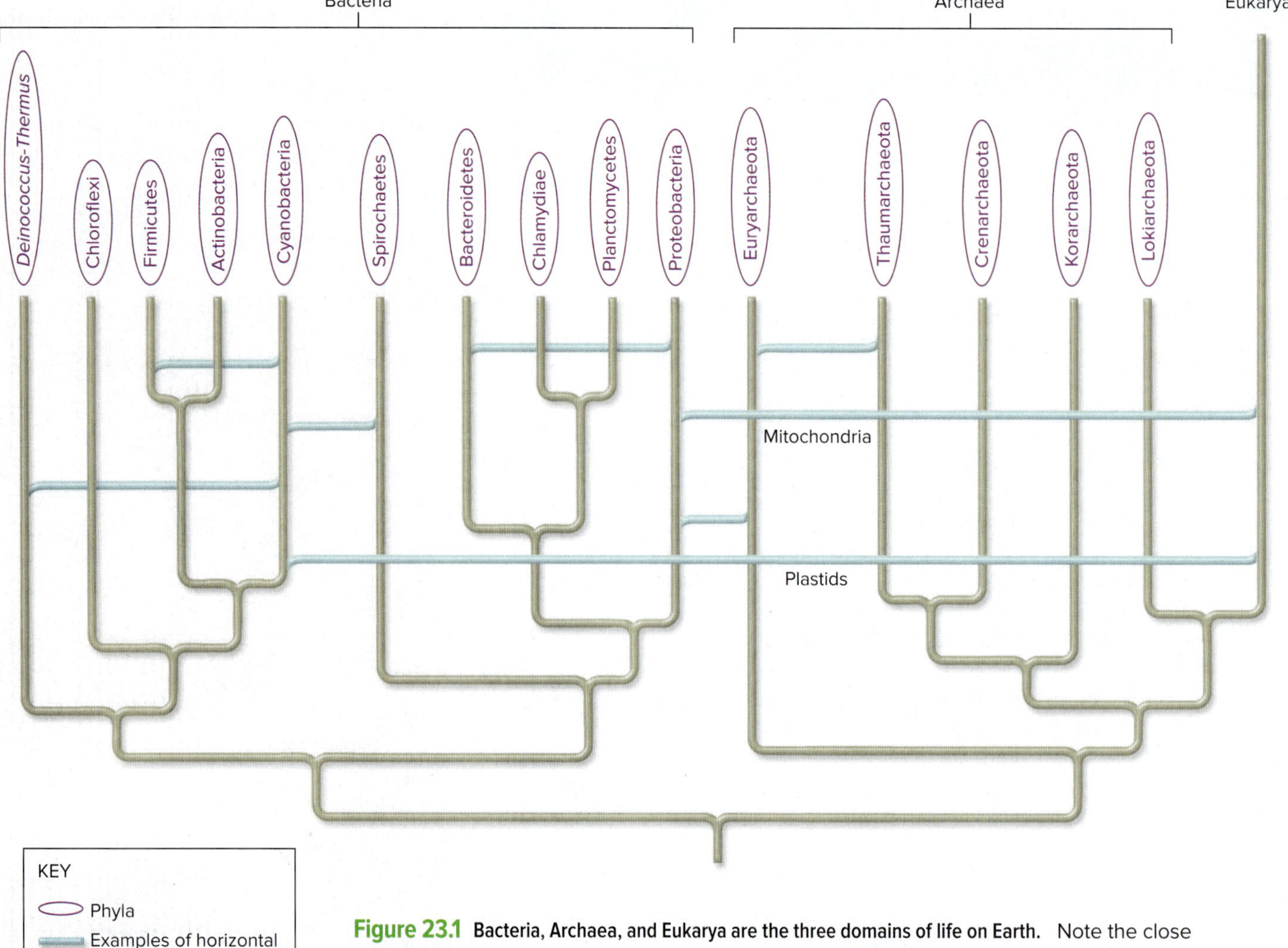


**Figure 23.1 Bacteria, Archaea, and Eukarya are the three domains of life on Earth.** Note the close relationship between Archaea and Eukarya. Horizontal gene transfer among organisms belonging to the same or different domains is a key evolutionary process.

## Biology Principle

### Cells Are the Simplest Units of Life

In the case of unicellular bacteria and archaea, a single cell represents an entire organism, but some bacteria have multicellular bodies.

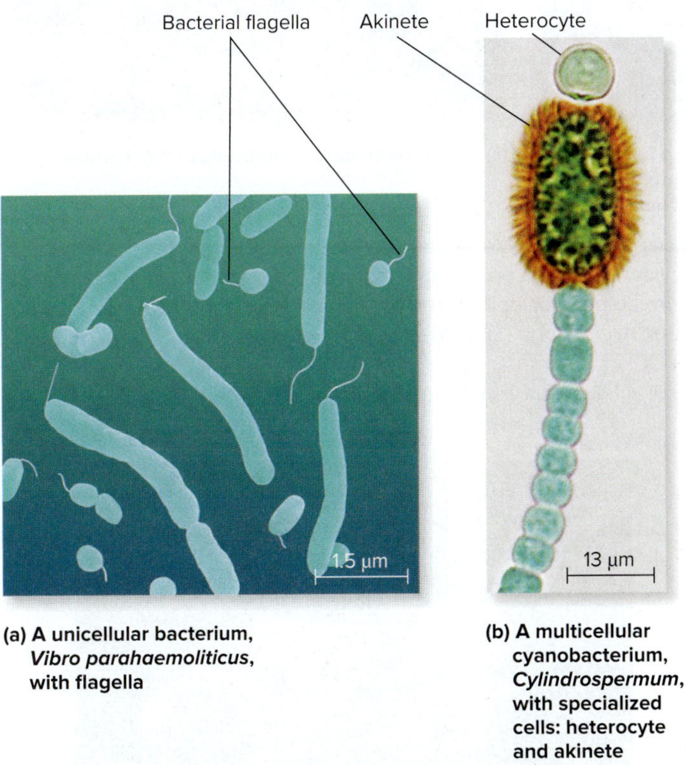


**(a) A unicellular bacterium, *Vibro parahaemoliticus*, with flagella**

**(b) A multicellular cyanobacterium, *Cylindrospermum*, with specialized cells: heterocyte and akinete**

**Figure 23.2** Bacterial cells. **(a)** *Vibrio parahaemoliticus,* a unicellular bacterium that causes seafood poisoning, artificially colored. **(b)** *Cylindrospermum,* a multicellular blue-green bacterium with specialized cells.

*(a)* © Dennis Kunkel Microscopy, Inc./Phototake; *(b)* © Lee W. Wilcox

bacteria live in nearly every conceivable habitat, including extremely hot or salty waters that support no other life, and they are Earth's most ancient organisms, having originated more than 3.5 bya. Their great age and varied habitats have resulted in extraordinarily high metabolic diversity. Such high metabolic diversity explains why modern archaea and bacteria play diverse ecological roles and can be used in many technological applications. Prokaryotic organisms are often single-celled (Figure 23.2a), though many achieve larger body size by aggregating cells into colonies or rows of cells known as **filaments.** Some bacteria display the defining features of true multicellularity, such as specialized cells (Figure 23.2b).

## Protists Are Early-Diverging Eukaryotes Displaying Diverse Structural Types

The term protist comes from the Greek word *protos,* meaning first, reflecting the observation that protists were Earth's first eukaryotes. **Protists** are diverse eukaryotic organisms that cannot be classified into the kingdoms Fungi, Plantae, or Animalia. Protists that lack chloroplasts are informally termed protozoa (Figure 23.3a) or fungus-like protists (Figure 23.3b), whereas protists having one or more chloroplasts are informally termed algae (Figure 23.3c). Although most protists are microscopic in size (see Figures 23.3a,b), many can be observed without using a microscope, examples being the multicellular marine photosynthetic protists commonly known as seaweeds (see Figure 23.3c).

## Fungi Form a Kingdom of Diverse Plastidless Eukaryotes

Fungi are eukaryotic organisms that lack plastids so are not photosynthetic, and they live by absorbing organic food. Some fungi are unicellular, and others occur as microscopic threads known as hyphae (Figure 23.4). Although most of the body of most fungi is microscopic in size, the hyphae of many species can merge to form structures visible with the naked eye—mushrooms are examples. Another distinctive feature of fungi is that most species produce cell walls containing the polysaccharide chitin.

## Horizontal Gene Transfer Fosters Microbial Evolution

Archaea, bacteria, protists, and fungi often live in communities that occupy the same types of habitats. Living closely together in communities fosters the exchange of genetic material between microbial species, an important evolutionary process known as horizontal gene transfer. **Horizontal gene transfer (HGT)** is the process in which an organism receives genetic material from another organism without being the offspring of that organism. This process contrasts with vertical gene transfer, which occurs from parent to progeny. Horizontal gene transfer is increasingly recognized as an important evolutionary mechanism. Recipient organisms benefit from this process by acquiring new metabolic capabilities.

Horizontal gene transfer is particularly common among archaea and bacteria (see Figure 23.1). For example, at least 17% of the genes present in the common human gut bacterium *Escherichia coli* came from other bacteria. The study of nearly 200 genomes has revealed that about 80% of bacterial and archaeal genes have been involved in horizontal transfer at some point in their history. Genes also move among the bacterial, archaeal, and eukaryotic domains. For example, about a third of the genes present in the archaeal species *Methanosarcina mazei* originally came from bacteria. Bacterial genes have also been transferred to many types of eukaryotic protists, fungi, and plants. Horizontal gene transfer also occurs among different fungal species, as well as between fungi and protists, animals, and plants. Such gene transfer increases the genetic diversity of microorganisms and other living things, influencing their evolution.

Horizontal gene transfer can occur between different prokaryotic species via both transduction and transformation, as discussed in Chapter 17. Horizontal gene transfer also occurs when eukaryotes feed or when microbes invade the cells of other organisms, known as hosts.

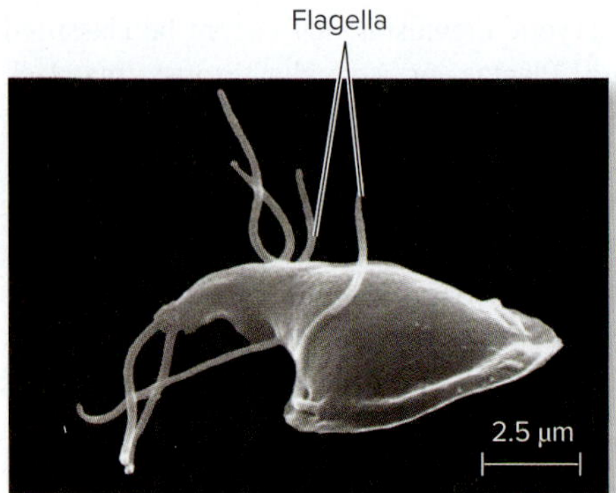


(a) A single-celled protist, the flagellate protozoan, *Giardia*

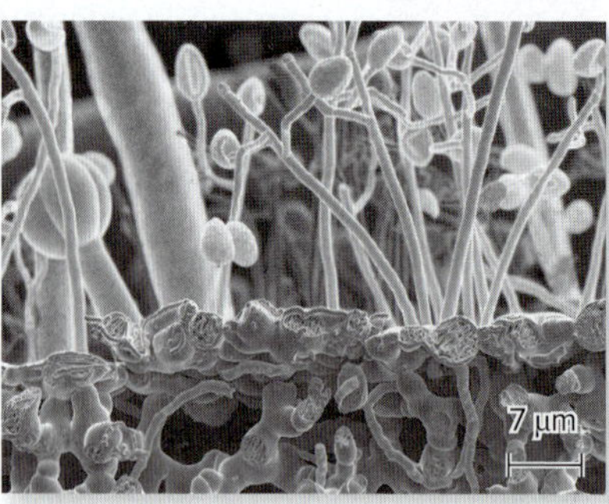


(b) *Phytophthora infestans*, a fungus-like protist whose filaments grow within and emerge from a plant leaf

(c) *Chondrus crispus*, a multicellular algal protist

**Figure 23.3** **Protist bodies take diverse forms that can be unicellular or multicellular.** **(a)** Scanning electron microscopic (SEM) view of the single-celled flagellate protozoan, *Giardia*, which infects the human gut. **(b)** SEM view of microscopic filaments produced by the fungus-like protist *Phytophthora infestans*, which causes disease of wild and crop plants. **(c)** An edible multicellular red alga that is conspicuous to the unaided eye, *Chondrus crispus*. Note: SEM images are made with a scanning electron microscope (SEM) that employs electrons rather than visible light, with the result that cellular structures do not normally appear in color.

*(a)* Source: Dr. Stan Erlandsen, Dr. Dennis Feely/CDC; *(b)* © Andrew Syred/Science Source; *(c)* © Andrew J. Martinez/Science Source

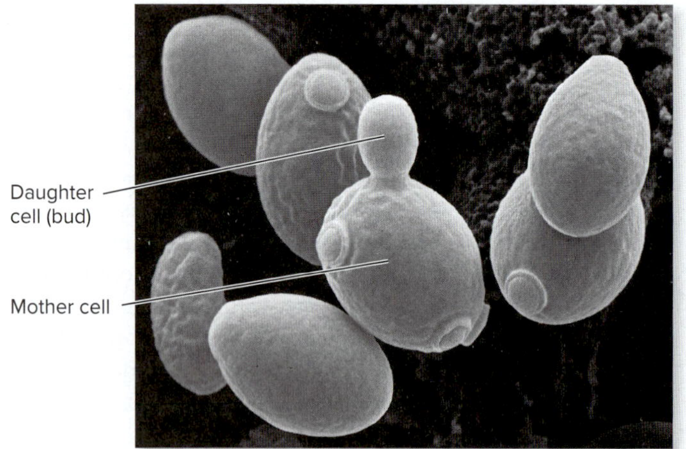


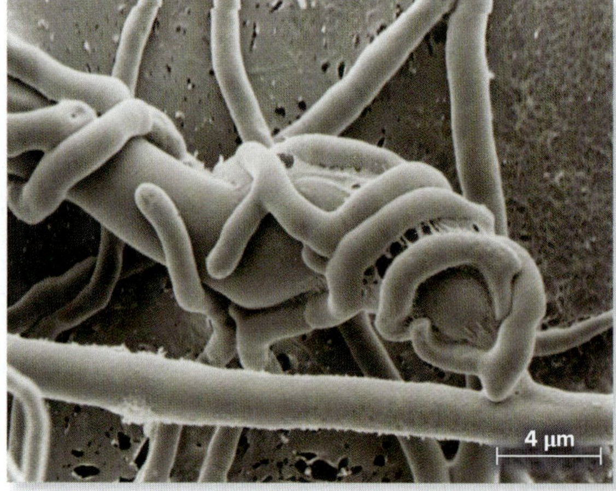


**Figure 23.4** **Fungal bodies can be unicellular or composed of microscopic filaments known as hyphae.** **(a)** Scanning electron microscopic (SEM) view of the single-celled yeast fungus *Saccharomyces cerevisiae*, important in the brewing and baking industries. **(b)** SEM of *Trichoderma* hyphae attacking the wider hyphae of another fungus.

*(a)* © Mediscan/Alamy; *(b)* © Biophoto Associates/Science Source

## 23.1 Reviewing the Concepts

- Prokaryotic Bacteria (informally bacteria) and Archaea (informally archaea) form two of the three domains of life. Prokaryotic species are often single-celled, though cellular aggregations are common, and examples of multicellularity occur (Figures 23.1, 23.2).
- Protists are classified into the domain Eukarya and occur as single eukaryotic cells or more complex bodies; multicellularity is common and some protists, such as photosynthetic seaweeds, are quite large (Figure 23.3).
- Fungi form a eukaryotic kingdom of heterotrophic unicellular or filamentous organisms having chitin in their cell walls (Figure 23.4).
- Horizontal gene transfer (HGT) commonly occurs among prokaryotic species and between domains of life, fostering evolutionary diversification.

## 23.1 Testing Your Knowledge

1. Imagine that you have found a new type of organism that is composed of several cells but can only be observed by using a microscope. Is it possible that the cells of this organism have prokaryotic organization?
   a. Yes; although many prokaryotes are single-celled, some prokaryotes are multicellular.
   b. No; so far as is known, prokaryotes occur only as single cells.

2. What process can serve as a mechanism of horizontal gene transfer?
   a. viral transduction
   b. transformation
   c. feeding by eukaryotes
   d. invasion of host cells by microbes
   e. all of the above

## 23.2 Diversity and Ecological Importance of Archaea

### Learning Outcomes

1. Describe the evolutionary and ecological importance of archaea.
2. List the features of archaea that enable them to grow in extreme habitats.

From an evolutionary perspective, archaea are important because they share a number of features with eukaryotes, suggesting common ancestry. For example, histone proteins are typically associated with the DNA of both archaea and eukaryotes, but such proteins are absent from most bacteria. Archaea and eukaryotes also share more than 30 ribosomal proteins that are not present in bacteria, and archaeal RNA polymerases are closely related to their eukaryotic counterparts. These similarities help evolutionary biologists learn how the information systems of eukaryotes evolved.

Archaea are also ecologically important. Though many archaea occur in soils and surface ocean waters of moderate conditions, diverse archaea occupy habitats with very high salt content, acidity, methane levels, or temperatures that would kill most bacteria and eukaryotes. Organisms that occur primarily in extreme habitats are known as **extremophiles** (meaning lover of extreme conditions). One example is *Methanopyrus,* which grows best at deep-sea thermal vent sites where the temperature is 98°C and is thus known as a hyperthermophile (meaning lover of high temperatures). At this temperature, the proteins of most organisms would denature, but those of *Methanopyrus* are resistant to such damage. Archaea help biologists to better understand the origin of life, the origin of eukaryotes, how life on Earth has evolved in extreme environments, and what kinds of extraterrestrial life might exist.

The domain Archaea includes the phyla Lokiarchaeota, Korarchaeota, Thaumarchaeota, Crenarchaeota, and Euryarchaeota (see Figure 23.1). Lokiarchaeota are known for their particularly close evolutionary relationship to Eukarya. Thaumarchaeota species that oxidize ammonia are important in global nitrogen cycling. The early-diverging Euryarchaeota includes some hyperthermophiles, diverse methane producers, and extreme halophiles—species able to grow in higher than usual salt concentrations.

Archaea possess distinctive membrane lipids, which are formed with ether bonds; in contrast, ester bonds characterize the membrane lipids of bacteria and eukaryotes (Figure 23.5). Ether-bonded membranes are resistant to damage by heat and other extreme conditions, which helps explain why many archaea are able to grow in extremely harsh environments. Also note that archaea use isoprene chains instead of fatty acid chains to build membranes. Although some archaea lack cell walls, most possess a wall composed of protein (which differs from the chemical composition of most bacterial cell walls described in Section 23.4).

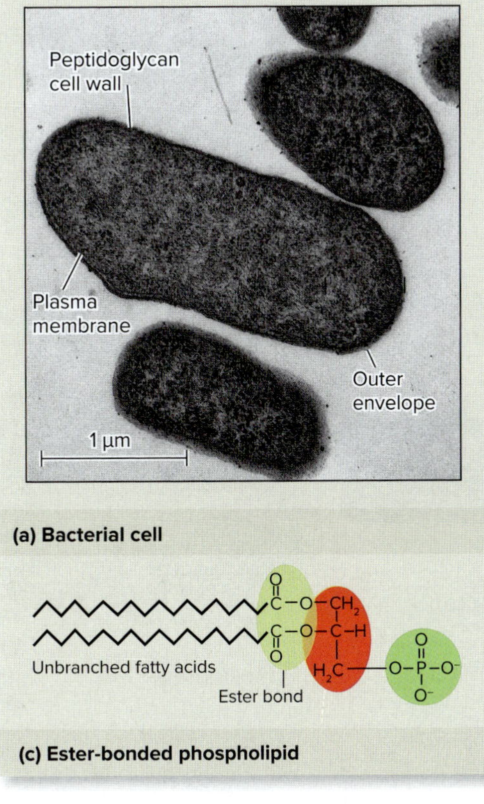


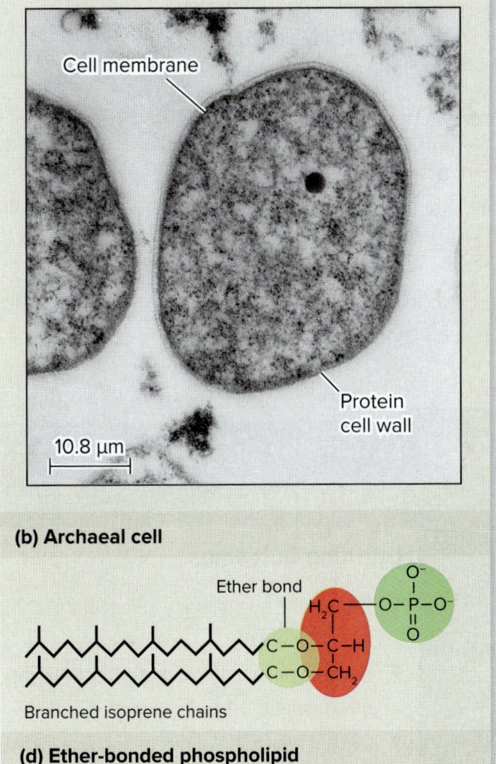


**Figure 23.5 Structure of bacteria and archaea.** **(a)** Bacteria and **(b)** archaea both have prokaryotic cell structure, but **(c)** bacterial membrane lipids are formed with ester linkages, whereas **(d)** archaeal membrane lipids feature ether linkages, which are thought to be more stable under extreme environmental conditions. As shown in transmission electron microscopic (TEM) images (a) and (b), most bacteria feature cell walls made of a material known as peptidoglycan, which is often enclosed by an outer envelope, whereas archaea lack these features. Most archaea have outer coverings made of protein.

*(a)* © Linda Graham; *(b)* © Eye of Science/Science Source

Archaea are also ecologically important as producers of methane ($CH_4$)—the major component of natural gas. Methane is a greenhouse gas that increases global warming over 20 times more per molecule than does $CO_2$. In recent years, the level of $CH_4$ has been increasing in Earth's atmosphere as the result of human activities. Several groups of anaerobic archaea convert $CO_2$, methyl groups, or acetate to $CH_4$ and release $CH_4$ from their cells into the atmosphere. Methane-producing archaea live in swampy wetlands, the bottoms of lakes, deep-sea habitats, and the digestive systems of animals such as cattle and humans.

## 23.2 Reviewing the Concepts

- Domain Archaea, particularly the phylum Lokiarchaeota, is more closely related to Domain Eukarya than either is to Domain Bacteria (see Figure 23.1).
- Many representatives of Domain Archaea occur in extremely hot, salty, or acidic habitats. Ether-linked membrane lipids are among the features of archaea that enable their survival in extreme habitats (Figure 23.5).
- Several groups of archaea that occupy watery habitats generate the powerful greenhouse gas methane ($CH_4$).

## 23.2 Testing Your Knowledge

1. What features are characteristic of archaea but not bacteria?
   a. Archaea possess membrane lipids linked by ether bonds—not ester bonds, as in bacteria.
   b. Archaea typically have cell walls composed of protein—not peptidoglycan, as in the case of most bacteria.
   c. The DNA of archaea is bound to histone proteins, as in eukaryotes but not bacteria.
   d. All of the above are correct.
   e. None of the above are correct.
2. How does methane production by certain archaea contribute directly to global climate change?
   a. Methane has a cooling effect on Earth's climate.
   b. Methane in Earth's atmosphere increases rainfall.
   c. Methane is a powerful greenhouse gas that helps to warm Earth's climate.
   d. Methane in Earth's atmosphere increases the chance of fire.
   e. None of the above.

## 23.3 Diversity and Ecological Importance of Bacteria

### Learning Outcomes

1. Explain the evolutionary and ecological importance of cyanobacteria.
2. List some examples of proteobacteria important in the environment.
3. Explain how populations of bacteria increase and how some bacteria survive under stressful conditions.

Molecular studies suggest the existence of 50 or more bacterial phyla, though many are poorly known. Though some members of Domain Bacteria live in extreme environments, most favor moderate conditions. The characteristics of 10 prominent bacterial phyla (see Figure 23.1) are briefly summarized in **Table 23.1.** Among these, the Cyanobacteria and the Proteobacteria are particularly diverse and relevant to eukaryotic cell evolution, global ecology, and human affairs.

**Table 23.1 Representative Bacterial Phyla**

| Phyla | Characteristics |
|---|---|
| *Deinococcus-Thermus* | Extremophiles. The genus *Deinococcus* is known for high resistance to ionizing radiation, and the genus *Thermus* inhabits hot springs. *Thermus aquaticus* has been used in commercial production of Taq polymerase enzyme used in the polymerase chain reaction (PCR), an important procedure in molecular biology laboratories. |
| Chloroflexi | Known as the green nonsulfur bacteria; conduct photosynthesis without releasing oxygen (anoxygenic photosynthesis). |
| Firmicutes | Diverse Gram-positive bacteria, some of which produce endospores. The disease-causing *Clostridium difficile* is an example. |
| Actinobacteria | Gram-positive bacteria producing branched filaments; many form spores. *Mycobacterium tuberculosis,* the agent of tuberculosis in humans, is an example. Actinobacteria are notable antibiotic producers; over 500 different antibiotics are known from this group. |
| Cyanobacteria | The oxygen-producing photosynthetic bacteria. Photosynthetic pigments include chlorophyll *a* and blue-green or red accessory pigments. Occur as unicells, colonies, unbranched filaments, and branched filaments. Many of the filamentous species produce specialized cells: dormant akinetes and heterocytes in which nitrogen fixation occurs. In waters having excess nutrients, cyanobacteria produce blooms and may release toxins harmful to the health of humans and wild and domesticated animals. |
| Spirochaetes | Motile bacteria having distinctive corkscrew shapes, with flagella held close to the body. They include the pathogens *Treponema pallidum,* the agent of syphilis, and *Borrelia burgdorferi,* which causes Lyme disease. |
| Bacteroidetes | Includes representatives of diverse metabolism types; some are common in the human intestinal tract, and others are primarily aquatic. |
| Chlamydiae | Notably tiny, obligate intracellular parasites. Some cause eye disease in newborns or sexually transmitted diseases. |
| Planctomycetes | Reproduce by budding rather than binary fission; cell walls lack peptidoglycan; cytoplasm may contain nucleus-like bodies as in the case of *Gemmata obscuriglobus.* |
| Proteobacteria | A very large group of Gram-negative bacteria, collectively having high metabolic diversity. Includes many species important in medicine, agriculture, and industry. |

## Cyanobacteria Are Photosynthetic Bacteria That Produce Oxygen and Play Other Important Ecological Roles

The phylum Cyanobacteria contains photosynthetic bacteria that are abundant in fresh waters, oceans, and wetlands and on the surfaces of arid soils. Cyanobacteria are named for blue-green (cyan) coloration conferred by photosynthetic pigments that help chlorophyll absorb light energy (see Figure 23.2b). Cyanobacteria are the only bacteria known to generate oxygen as a product of photosynthesis. Ancient cyanobacteria produced Earth's first oxygen-rich atmosphere, which allowed the eventual rise of eukaryotes. The chloroplasts of eukaryotic algae and plants evolved from cyanobacteria, a process that involved massive horizontal transfer of bacterial genes into eukaryotic nuclear genomes.

In phosphorus-rich bodies of water, some species of cyanobacteria grow rapidly into large, visible populations (blooms) that color the water blue-green, or cyan. The individual cells release small amounts of toxins that help to keep small aquatic animals from eating them, but when blooms occur, toxins can rise to levels that harm humans, pets, livestock, and wildlife. Consequently, public health authorities often warn people not to swim in waters with visible blue-green blooms or allow pets and livestock to drink such water. People can prevent the formation of cyanobacterial blooms by reducing the input of phosphorus-rich fertilizers, manure, and sewage into bodies of water.

Despite the harmful effects of some species, cyanobacteria provide important benefits to humans and other organisms, such as producing atmospheric oxygen. Many cyanobacteria also have the ability to convert abundant but inert atmospheric nitrogen gas into ammonia, which algae and plants can use to synthesize amino acids and proteins. This process, known as **nitrogen fixation,** enriches nutrient-poor soils, particularly wet paddy fields where rice is grown in many regions of the world, thereby helping to provide food for billions of people. Cyanobacteria illustrate the wide range of body diversity found among bacterial phyla:

- Unicells (**Figure 23.6a**) are single-celled bodies.
- Colonies are groups of cells held together by a thick, gluey substance called mucilage (**Figure 23.6b**).
- Filaments are cells that are attached end-to-end (**Figure 23.6c**), some of which are branched (**Figure 23.6d**). Some filamentous cyanobacteria display the hallmarks of multicellularity that also mark multicellular plants and animals: cell-to-cell attachment, specialized cells, intercellular chemical communication, and programmed cell death.

## Proteobacteria Do Not Produce Oxygen but Play Other Ecologically Important Roles

Although some proteobacteria are photosynthetic, these species do not produce oxygen as do the cyanobacteria. Even so, certain proteobacteria play an important role by using oxygen to consume methane, thereby reducing the impact of this greenhouse gas. Methane-destroying proteobacteria often live in close association with photosynthetic plants and algae, which provide the oxygen the bacteria need to oxidize methane. Though proteobacteria share molecular and cell-wall features, this phylum displays amazing diversity of form and metabolism in five major classes:

- Alphaproteobacteria, which are closely related to the ancestry of mitochondria, include nitrogen-fixing species that associate with plants, thereby fostering plant growth. Some species destroy methane, thereby reducing potential environmental impacts of this gas.
- Betaproteobacteria, such as the soil inhabitant *Nitrosomonas,* are important in the global nitrogen cycle.
- Gammaproteobacteria include human pathogens such as *Neisseria gonorrhoeae,* the agent of the sexually transmitted disease gonorrhea; *Vibrio cholerae,* which causes cholera epidemics when drinking water becomes contaminated with

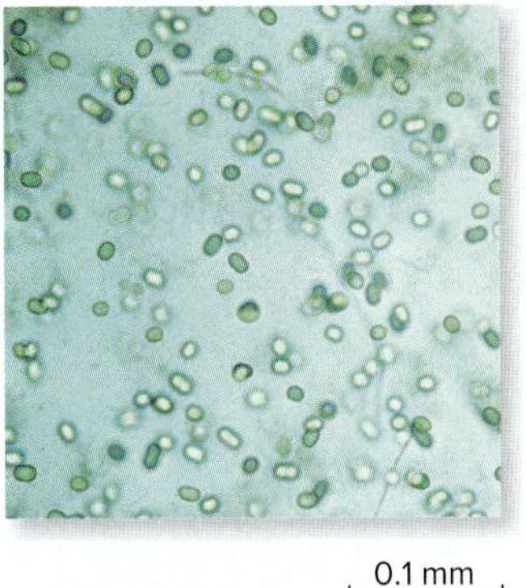


(a) Unicells

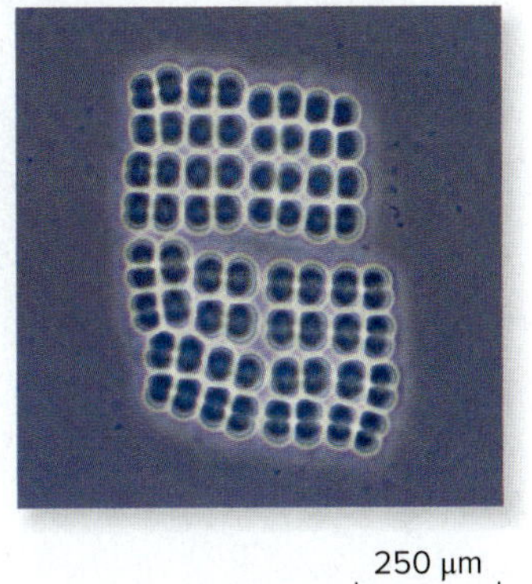


(b) Colony of cells

(c) Unbranched filaments

(d) Branched filaments

**Figure 23.6 Major body types found in the phylum Cyanobacteria. (a)** The genus *Chroococcus* occurs as unicells. **(b)** The genus *Merismopedia* is a flat colony of cells held together by mucilage. **(c)** The genus *Oscillatoria* is an unbranched filament. **(d)** The genus *Stigonema* is a branched filament having a mucilage sheath; sunscreen compounds that protect the cells from damage by ultraviolet (UV) radiation cause the brown color of the sheath.

*(a)* © Linda Graham; *(b)* © Michael Abbey/Science Source; *(c)* © Sinclair Stammers/Science Source; *(d)* © Lee W. Wilcox

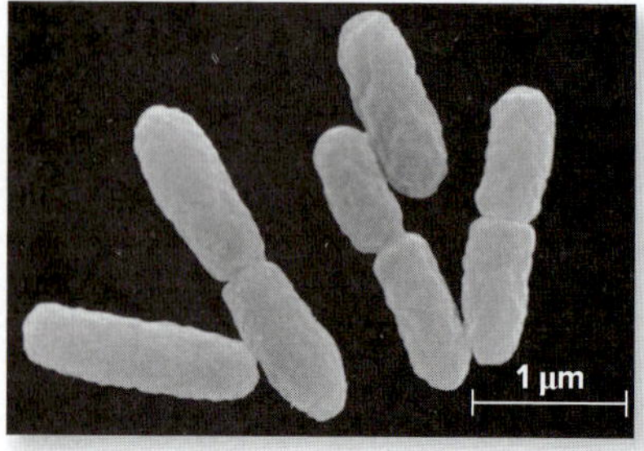


(a) Bacterium undergoing binary fission

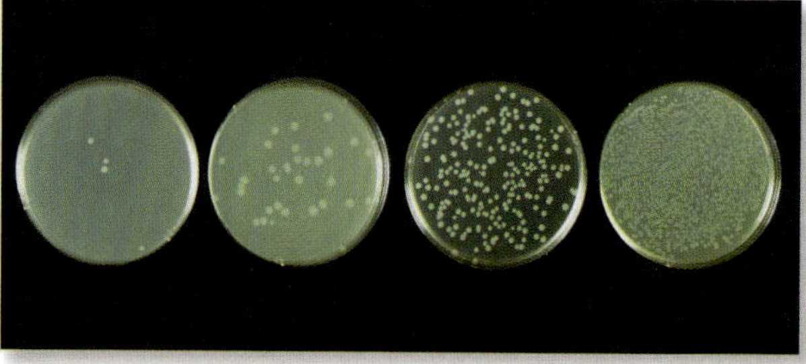

(b) Colonies developed from single cells

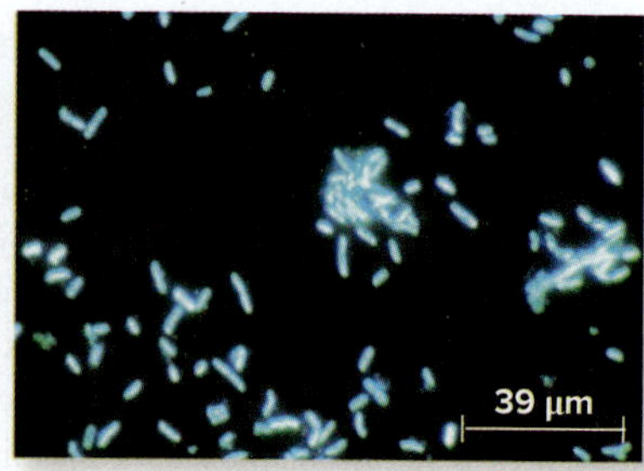


(c) Bacteria stained with fluorescent DNA-binding dye

**Figure 23.7** Bacteria can be visualized and counted by means of colonies formed by repeated binary fission or by using fluorescent dye.
*(a)* © SCIMAT/Science Source; *(b)* © Linda Graham; *(c)* © Lee W. Wilcox

animal waste during floods and other natural disasters; and *Salmonella enterica* and *Escherichia coli* strain O157:H7, which can contaminate food and water, causing intestinal infections. Some species destroy the greenhouse gas methane.

- Deltaproteobacteria include colony-forming myxobacteria and predatory bdellovibrios, which drill through the cell walls of other bacteria in order to consume them.
- Epsilonproteobacteria are represented by *Helicobacter pylori,* a risk factor for peptic ulcers and stomach cancer. Other species play important roles in the global sulfur cycle.

## Bacterial Reproduction Helps to Explain Ecological Function

Bacteria lack eukaryote-type sexual reproduction involving specialized gametes, gamete fusion (syngamy), and meiosis, though they can exchange some genes by conjugation, transformation, and transduction (described in Chapter 17). Some bacteria reproduce asexually by budding, producing small progeny cells from larger parental cells. However, most bacteria reproduce asexually by division of a single cell into two equal progeny cells, a process known as **binary fission** (Figure 23.7a).

Binary fission is the basis of a widely used method for detecting and counting bacteria in food, water samples, or patient fluids. Microbiologists who study the spread of disease need to quantify bacterial cells in samples taken from the environment. Medical technicians often need to count bacteria in body fluid samples to assess the level of infection. However, because bacterial cells are small and often unpigmented, they are difficult to view and count directly. One way that microbiologists count bacteria is to place a measured volume of sample into plastic dishes filled with a semisolid nutrient medium. Bacteria in the sample undergo repeated binary fission to form colonies of cells visible to the unaided eye. Because each colony is assumed to represent a single cell that was present in the original sample, the number of colonies in the dish approximates the number of living bacteria in the original sample (Figure 23.7b). Bacteria that cannot be cultivated in the laboratory can, however, be observed and counted with the use of a fluorescent dye that binds bacterial DNA (Figure 23.7c).

Some bacteria produce thick-walled cells that are able to survive unfavorable conditions in a dormant state. These specialized cells develop when bacteria have experienced stress, such as low nutrients or unfavorable temperatures, and are able to germinate into metabolically active cells when conditions improve. For example, when winter approaches, aquatic filamentous cyanobacteria often produce **akinetes**—large, thick-walled, food-filled cells (see Figure 23.2b). Akinetes are able to survive winter at the bottoms of lakes, and they produce new filaments in spring when they are carried by water currents to the brightly lit surface. Persistence of such akinetes, followed by their growth into filaments by means of binary fission, explains how harmful cyanobacterial blooms can develop year after year in overly fertile lakes.

**Endospores** (Figure 23.8) are produced inside the cells of some bacterial species. DNA and other materials become enclosed within a tough coat and then are released when the enclosing cell dies and breaks down. Bacterial endospores can remain alive, though in a dormant state, for long periods, then reactivate when conditions are suitable. The ability to produce endospores allows some Gram-positive

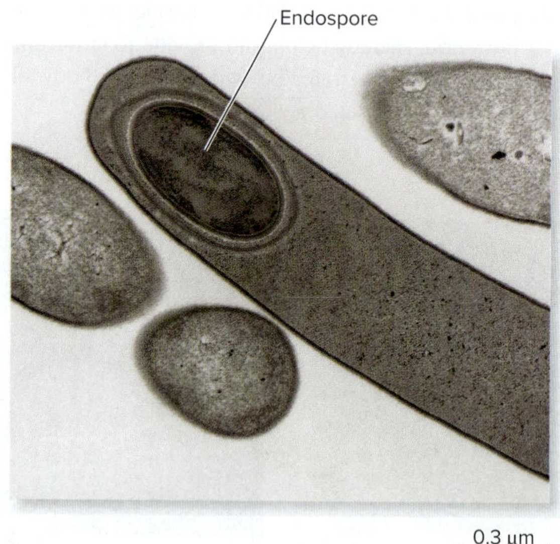


**Figure 23.8** Specialized bacterial cells capable of dormancy. An endospore with a resistant wall develops within the cytoplasm of the pathogen *Clostridium difficile.*
© Dr. Kari Lounatmaa/Science Source

**Concept Check:** *How do endospores influence the ability of some bacteria to cause disease?*

Firmicutes bacteria to cause serious diseases. For example, *Bacillus anthracis* causes the disease anthrax, a potential agent in bioterrorism and germ warfare. Most cases of human anthrax result when endospores of *B. anthracis* enter breaks in the skin, causing skin infections that are relatively easily cured by antibiotic treatment. But sometimes the endospores are inhaled or consumed in undercooked, contaminated meat, potentially causing more serious illness or death. *Clostridium botulinum* can contaminate improperly canned food that has not been heated to temperatures high enough to destroy its tough endospores. When the endospores germinate and bacterial cells grow in the food, they produce a deadly toxin, as well as $NH_3$ and $CO_2$ gas, which causes can lids to bulge. If humans consume the food, the toxin causes botulism, a severe type of food poisoning that can lead to respiratory and muscular paralysis. The toxin is so potent that only 400 g of *C. botulinum* toxin would be sufficient to kill every human on Earth. That being the case, it is interesting that people use a commercial preparation of botulinum toxin in the form of Botox, which is injected into the skin to paralyze facial muscles, thereby reducing the appearance of wrinkles.

*Clostridium tetani* produces a nerve toxin that causes lockjaw, also known as tetanus, when bacterial cells or endospores from soil enter wounds. The ability of the genera *Bacillus* and *Clostridium* to produce resistant endospores helps to explain their widespread presence in nature and their effect on humans.

## 23.3 Reviewing the Concepts

- Domain Bacteria includes 50 or more phyla, including Cyanobacteria and Proteobacteria (Figure 23.1; Table 23.1).
- Cyanobacteria were ancestral to all modern chloroplasts, are the only prokaryotic organisms that produce oxygen as a result of photosynthesis, sometimes form harmful bloom populations in overly fertile waters, and illustrate prokaryotic body types (Figure 23.6).
- Proteobacteria were ancestral to mitochondria, do not produce oxygen, and include environmentally important species.
- Bacteria reproduce by binary fission or budding and some produce akinetes or endospores that allow them to survive under unfavorable conditions (Figures 23.7, 23.8).

## 23.3 Testing Your Knowledge

1. What features of cyanobacteria can be harmful to humans and other animals?
   a. oxygen production
   b. toxin production
   c. production of blue-green photosynthetic pigments
   d. carbon fixation
   e. nitrogen fixation
2. What essential role did early alphaproteobacteria play in the origin of eukaryotic cells?
   a. They were precursors of plastids.
   b. They were precursors of nuclei.
   c. Some evolved into mitochondria.
   d. Some evolved into the endomembrane system.
   e. None of the above.

## 23.4 Diversity in Bacterial Cell Structure and Metabolism

### Learning Outcomes

1. Discuss structural adaptations that have increased the complexity of prokaryotic cells.
2. Describe the structural differences between Gram-positive and Gram-negative bacterial cells.
3. Explain how mucilage influences bacterial behavior.
4. List the different means by which prokaryotic cells can move.

Even though bacteria have a much simpler cellular organization than do eukaryotes, many bacteria display structural adaptations that increase their complexity. Cyanobacteria and other photosynthetic bacteria, for example, are able to use light energy to produce organic compounds because their cells contain large numbers of thylakoids—flattened, tubular membranes that grow inward from the plasma membrane (Figure 23.9a). The extensive membrane surface of the thylakoids bears large amounts of chlorophyll and other components required for photosynthesis. This explains why thylakoids are also abundant in plant chloroplasts, which descended from cyanobacterial ancestors. Thylakoids enable photosynthetic bacteria and chloroplasts to take maximum advantage of light energy in their environments.

In other bacteria, plasma membrane ingrowth has generated additional intriguing adaptations—magnetosomes and nucleus-like bodies—that are sometimes described as bacterial organelles. Magnetosomes are tiny crystals of an iron mineral known as magnetite, each surrounded by a membrane. These structures occur in the bacterium *Magnetospirillum* and related genera (Figure 23.9b). In each cell, about 15 to 20 magnetosomes occur in a row, together acting as a compass needle that responds to the Earth's magnetic field. Magnetosomes help the bacteria to orient themselves in space and thereby locate the submerged, low-oxygen habitats they prefer. Plasma membrane ingrowths may also surround cellular DNA, producing nucleus-like structures within bacteria of the phylum Planctomycetes (Figure 23.9c). Common in aquatic habitats, such bacteria are important to evolutionary biologists who are interested in the evolutionary origin of the eukaryotic nucleus.

### Prokaryotic Cells Vary in Shape

Bacterial cells occur in five common shapes (Figure 23.10):

- Spheres, known as **cocci,** maximize surface area/volume ratio, thereby enhancing exchange of materials with the environment.
- Elongate rods, called **bacilli,** are able to store more nutrients than some other shapes.
- Comma-shaped cells are called **vibrios.**
- Spiral-shaped cells that are flexible are known as **spirochaetes.**
- Spiral-shaped cells that are rigid are termed **spirilli.**

Cytoskeletal proteins similar to those present in eukaryotic cells control these cell shapes. For example, helical strands of an actin-like protein are responsible for the rod shape of bacilli; if this protein is not produced, bacilli become spherical in shape.

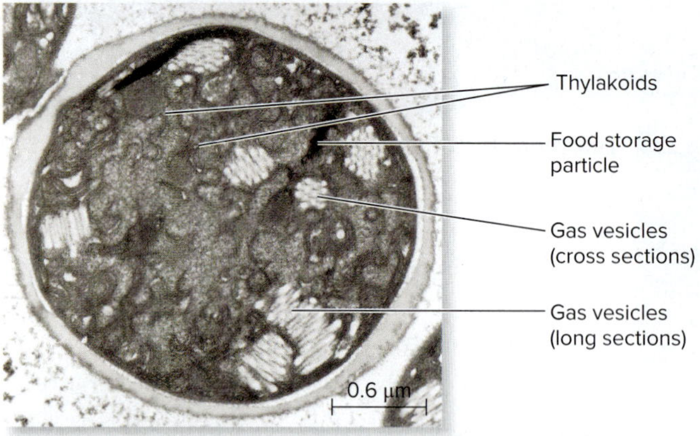


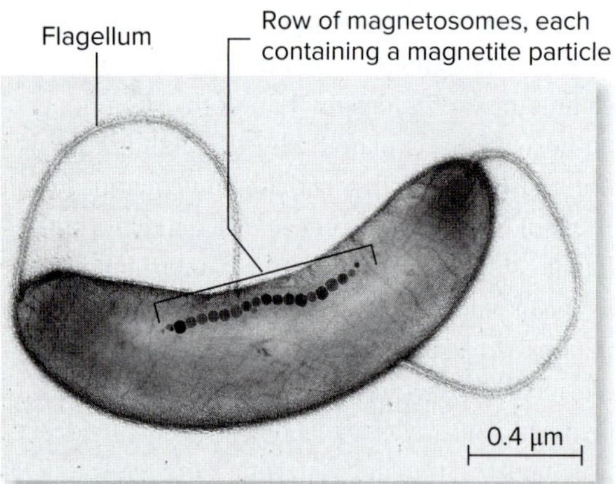


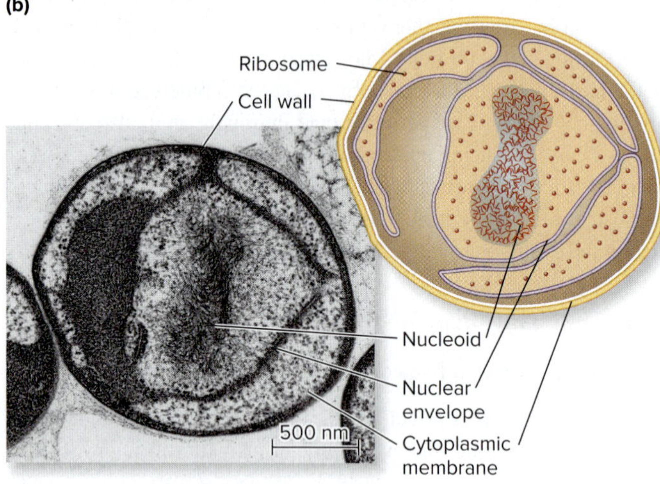


**Figure 23.9 Bacterial internal cell structures (a)** Photosynthetic thylakoid membranes and numerous gas vesicles found in a cell of the aquatic cyanobacterial genus *Microcystis.* **(b)** Magnetosomes found in the spirillum *Magnetospirillum magnetotacticum*. An internal row of iron-rich magnetite crystals, each enclosed by a membrane derived from the plasma membrane. **(c)** A nucleus-like structure in *Gemmata obscuriglobus,* a genus of planctomycete bacteria. Although the nuclear envelope-like structure lacks pores, it encloses the bacterial cell's DNA, known as a nucleoid.

*(a)* © Norma Lang; *(b)* © Dr. Richard P. Blakemore, University of New Hampshire; *(c)* © Richard Webb and John Fuerst, The University of Queensland, Australia/Centre for Microscopy and Microanalysis

## Bacterial Cells Vary in Cell-Wall Structure

Most prokaryotic cells possess a rigid cell wall outside the plasma membrane. Cell walls maintain cell shape and help protect against attack by viruses or predatory bacteria. Cell walls also help bacteria and other microbes avoid lysing in hypotonic conditions, when the solute concentration is higher inside the cell than outside. The structure and composition of bacterial cell walls are medically important.

A polymer known as **peptidoglycan** is an important component of most bacterial cell walls. Peptidoglycan is composed of carbohydrates that are crosslinked by peptides. Bacterial cell walls occur in two major forms that differ in peptidoglycan thickness, presence or absence of a membrane occurring outside the peptidoglycan, staining properties, and response to antibiotics. Bacteria having these chemically different walls are called Gram-positive or Gram-negative bacteria, after the staining process used to distinguish them (**Figure 23.11**). The stain is named for its inventor, Danish scientist Hans Christian Gram.

Gram-positive bacteria classified in the phyla Firmicutes and Actinobacteria have walls with a relatively thick peptidoglycan layer (**Figure 23.12a**). By contrast, the cell walls of cyanobacteria, proteobacteria, and other Gram-negative species have a thinner peptidoglycan layer and are enclosed by a thin, outer envelope whose outer leaflet is rich in **lipopolysaccharides** (**Figure 23.12b**). This outer envelope of Gram-negative bacteria is a lipid bilayer but is distinct from the plasma membrane. Peptidoglycan and lipopolysaccharides can affect disease symptoms, the composition of vaccines, and bacterial responses to antibiotics.

The lipopolysaccharide-rich outer membrane of Gram-negative bacteria helps them to resist the entry of some antibiotics and can contain proteins that help disease-causing bacteria to attach to target cells. However, this outer envelope also impedes the secretion of proteins from bacterial cells into the environment, a process that normally allows cells to communicate with each other. Gram-negative bacteria have adapted to the presence of an outer membrane by evolving several types of protein systems that function in secretion. In some disease-causing bacteria, these secretion systems have evolved into weapons used to attack plant or animal cells.

Distinguishing Gram-positive from Gram-negative bacteria is an important factor in choosing the best antibiotics for treating infectious diseases. For example, Gram-positive bacteria are typically more susceptible than Gram-negative bacteria to penicillin and related antibiotics because these antibiotics interfere with synthesis of peptidoglycan, which Gram-positive bacteria require in larger amounts. For this reason, penicillin or related antibiotics such as methicillin are widely used to treat infections caused by Gram-positive bacteria. However, it is of societal concern that some strains of Gram-positive bacteria have become resistant to some antibiotics, an example being methicillin-resistant *Staphylococcus aureus,* or MRSA. The evolution of antibiotic resistance in bacteria can often be traced to a horizontal gene transfer event.

## Slimy Mucilage Often Coats Cellular Surfaces

Many bacteria exude a coat of slimy mucilage, also called a capsule, glycocalyx, or extracellular polymeric substance. Mucilage, which varies in consistency and thickness, is largely composed of hydrated

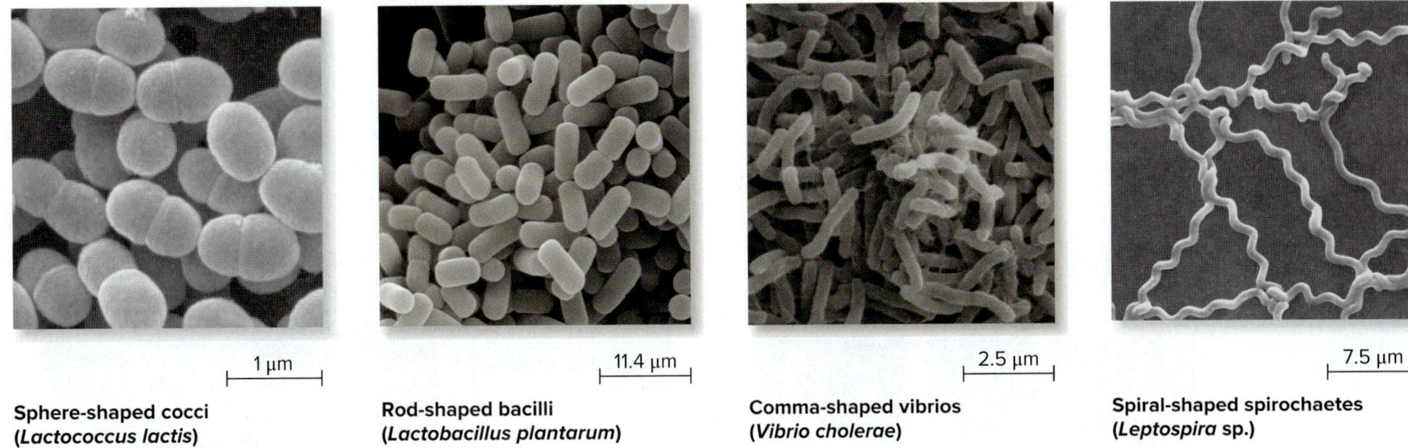


**Sphere-shaped cocci (*Lactococcus lactis*)**

**Rod-shaped bacilli (*Lactobacillus plantarum*)**

**Comma-shaped vibrios (*Vibrio cholerae*)**

**Spiral-shaped spirochaetes (*Leptospira* sp.)**

**Figure 23.10 Major types of prokaryotic cell shapes.** Scanning electron microscopic (SEM) views, which naturally lack color.

*(1)* © SCIMAT/Science Source; *(2)* © Dennis Kunkel Microscopy, Inc./Phototake; *(3)* © Media for Medical/UIG/Getty Images; *(4)* © Dennis Kunkel Microscopy, Inc./Phototake

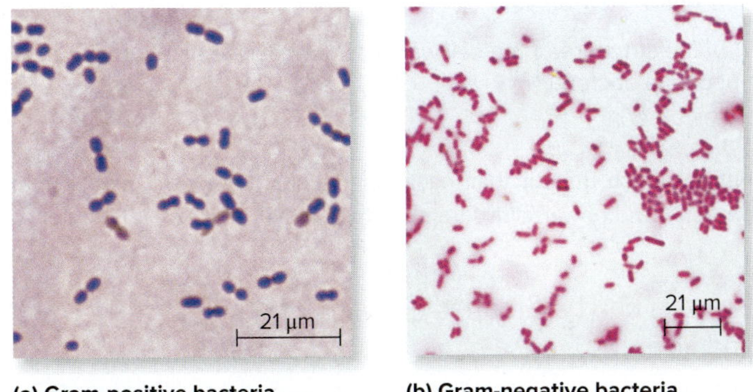


**(a) Gram-positive bacteria**

**(b) Gram-negative bacteria**

**Figure 23.11 Gram-positive and Gram-negative bacteria.**
**(a)** *Streptococcus pneumoniae,* a member of the phylum Firmicutes, stains positive (purple) with the Gram stain. **(b)** *Escherichia coli,* a member of the Proteobacteria, stains negative (pink) when the Gram stain procedure is applied.

*(a)* © CNRI/Science Source; *(b)* © Lee W. Wilcox

polysaccharides and protein. A capsule helps some pathogenic bacteria evade the defense system of their host. You may recall that Frederick Griffith discovered the transfer of genetic material while experimenting with capsule-producing pathogenic strains and capsule-less, nonpathogenic strains of the bacterium *Streptococcus pneumoniae*. The immune system cells of mice are able to destroy this bacterium only if it lacks a capsule.

Mucilage plays many additional roles: holding cells together closely enough for chemical communication and DNA exchange to occur, helping aquatic species to float in water, binding mineral nutrients, helping cells to stick to surfaces where they may form coatings known as **biofilms** (see Chapter 24), and repelling attack. Pigmented slime sheaths (see Figure 23.6d) coat some bacterial filaments, where they help to prevent UV damage.

## Bacteria Move by a Variety of Cellular Adaptations

Many bacteria have structures at the cell surface or within cells that enable them to change position in their environment, a process known as motility. Motility allows bacteria and other microbes to respond to chemical signals emitted from other cells and to move to favorable conditions within gradients of light, gases, or nutrients.

- Internal gas vesicles (see Figure 23.9a) help cyanobacteria to float into well-illuminated waters that allow photosynthesis to occur. Cyanobacteria filled with gas vesicles can collect at the water surface, forming noticeable scums.
- Threadlike cell surface structures known as pili (singular, pilus) allow some species to twitch or glide across surfaces and may aid reproduction.
- Bacterial flagella (see Figure 23.2a) allow cells to move by twitching, gliding, or swimming in liquids at rates of more than 150 μm per second and thus can foster the spread of infection through an animal or plant body.

Bacterial flagella (singular, flagellum) differ from eukaryotic flagella in several ways. Although bacterial flagella are largely built of about 30 types of proteins, they lack a plasma membrane covering, an internal cytoskeleton of microtubules, and motor proteins—all features that characterize eukaryotic flagella. Unlike eukaryotic flagella, prokaryotic flagella do not repeatedly bend and straighten. Instead, prokaryotic flagella spin, propelled by molecular machines composed of a filament, hook, and motor that work together, somewhat like a boat's outboard motor and propeller (**Figure 23.13**). Lying outside the cell, the long, stiff, curved filament acts as a propeller. The hook links the filament with the motor that contains a set of protein rings at the cell surface. Hydrogen ions (protons), which have been pumped out of the cytoplasm, usually via the electron transport system, diffuse back into the cell through channel proteins within the motor. This proton flow powers the turning of the hook and filament at rates of hundreds of revolutions per second. (Archaea may have flagella that also rotate

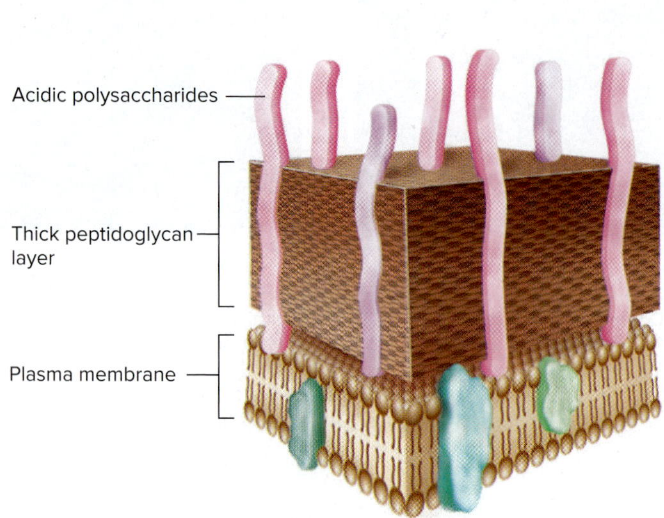


(a) Gram-positive: thick peptidoglycan cell-wall layer, no outer envelope

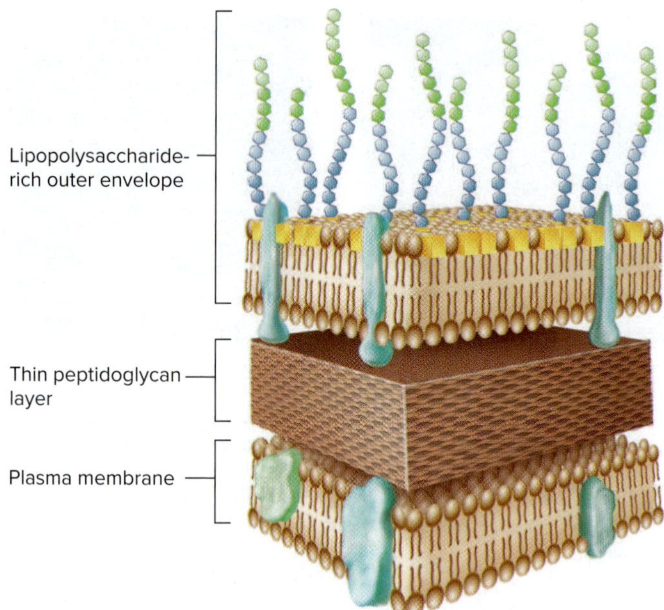


(b) Gram-negative: thinner peptidoglycan cell-wall layer, with outer envelope

**Figure 23.12 Cell-wall structures of Gram-positive and Gram-negative bacteria. (a)** The structure of the cell wall of Gram-positive bacteria. **(b)** The structure of the cell wall and lipopolysaccharide envelope typical of Gram-negative bacteria.

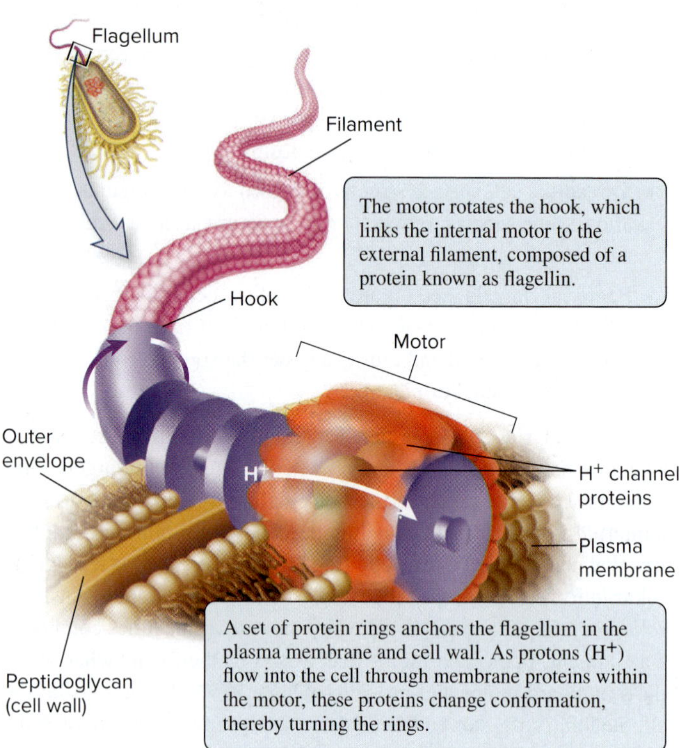


**Figure 23.13 Diagram of a prokaryotic flagellum, showing a filament, hook, and motor.**

**Concept Check:** *Does the filament move more like the arms of a human swimmer or the shaft of a boat propeller?*

but are much thinner than bacterial flagella, composed of different proteins, and powered differently—by the hydrolysis of ATP.)

## Bacteria Can be Grouped by Type of Nutrition and Response to Oxygen

Microbes can be classified according to their energy source, carbon source, response to oxygen, and presence of specialized metabolic processes such as nitrogen fixation.

- **Autotrophs** (from the Greek, meaning self-feeders) are organisms that are able to produce all or most of their own organic compounds from inorganic sources. Autotrophs fall into two categories: photoautotrophs and chemolithoautotrophs.
- **Photoautotrophs** such as cyanobacteria (like most species of eukaryotic algae) use light as a source of energy for the synthesis of organic compounds from $CO_2$ and $H_2O$ or from $H_2S$.
- **Chemolithoautotrophs** use energy obtained by chemical modifications of inorganic compounds to synthesize organic compounds. Such chemical modifications include nitrification (the conversion of ammonia to nitrate) and the oxidation of sulfur, iron, or hydrogen.
- **Heterotrophs** (from the Greek, meaning other feeders) are organisms that require at least one organic compound, and often more, from their environment.
- **Photoheterotrophs** are able to use light energy to generate ATP, but they must take in organic compounds from their environment as a source of carbon.

- **Chemoheterotrophs** must obtain organic molecules both for energy and as a carbon source. Fungi, archaea, and many bacteria are chemoheterotrophs. Among the many types of bacterial chemoheterotrophs is the Gram-positive species *Propionibacterium acnes,* which contributes to the skin condition acne, affecting up to 80% of adolescents in the U.S. The genome sequence of *P. acnes* has revealed numerous genes that allow it to break down skin cells and consume the products.

Bacterial species also differ in their need for and responses to oxygen. These variations are important to the ecological or medical roles of microbes.

- **Obligate aerobes** require $O_2$ in order to survive.
- **Obligate anaerobes,** such as the bacterial Firmicutes genus *Clostridium,* are poisoned by $O_2$. People suffering from gas gangrene caused by *Clostridium perfringens* and related species are usually treated by placement in a chamber with a high oxygen content (called a hyperbaric chamber), which kills the organisms and deactivates the toxins.
- **Aerotolerant anaerobes** do not use $O_2$, but they are not poisoned by it, either. These organisms obtain their energy by fermentation or anaerobic respiration, which uses electron acceptors other than oxygen in electron transport processes. Anaerobic metabolic processes include denitrification (the conversion of nitrate into $N_2$ gas) and the reduction of manganese, iron, and sulfate, which are all important in the Earth's cycling of minerals.
- **Facultative anaerobes** can use $O_2$ via aerobic respiration, obtain energy via anaerobic fermentation, or use inorganic chemical reactions to obtain energy, shifting between modes depending on environmental conditions.

Cyanobacteria and some other prokaryotic species are able to convert the abundant atmospheric nitrogen gas into a reduced form that algae and plants can use to produce amino acids and proteins. Oxygen poisons the principal enzyme involved in the nitrogen fixation process, so many cyanobacteria accomplish nitrogen fixation within specialized cells known as heterocytes, where oxygen concentrations are low (see Figure 23.2b). Some cyanobacteria that lack heterocytes accomplish nitrogen fixation at night, when photosynthetic oxygen production does not occur.

Cyanobacteria and other autotrophic bacteria, together with plants and many eukaryotic algae, are important **producers,** organisms that use photosynthesis to synthesize the organic compounds used by other organisms for food. **Decomposers,** also known as saprobes, include heterotrophic microorganisms (as well as animals). Bacteria, together with fungi, play important ecological roles by breaking down dead organisms and organic matter, releasing minerals for uptake by living things.

## 23.4 Reviewing the Concepts

- Intracellular structures such as thylakoids, magnetosomes, and nucleus-like organelles are examples of prokaryotic cell structure complexity (Figure 23.9).
- Major bacterial cell shape types are spherical cocci, rod-shaped bacilli, comma-shaped vibrios, and coiled spirochaetes and spirilli (Figure 23.10).
- Most bacterial cell walls contain peptidoglycan, which is composed of carbohydrates crosslinked by peptides. Gram-positive bacterial cells have thick peptidoglycan walls, whereas Gram-negative cells have less peptidoglycan in their walls and are enclosed by an outer lipopolysaccharide envelope (Figures 23.11, 23.12).
- Motility enables bacteria to change positions within their environment, which aids in locating favorable conditions for growth. Buoyancy vesicles, pili, and flagella are structures that enable motility (Figure 23.13).
- Bacteria display diverse types of metabolism, ways in which they obtain energy and carbon, and interact with oxygen.

## 23.4 Testing Your Knowledge

1. Which structure is associated with motility in one or more types of bacteria?
   a. flagellum
   b. peptidoglycan
   c. lipopolysaccharide envelope
   d. mucilage
   e. all of the above

## 23.5 Diversity and Ecological Importance of Protists

### Learning Outcomes

1. List three features that define protists.
2. Distinguish among algae, protozoa, and fungus-like protists.
3. Define the terms phytoplankton, periphyton, flagellate, ciliate, and amoeba.
4. Describe a distinctive structural characteristic for each of seven eukaryotic supergroups.

The term protist comes from the Greek word *protos*, meaning first, reflecting the observation that protists were Earth's first eukaryotes. Protists are eukaryotes that are not classified in the plant, animal, or fungal kingdoms. Protists display two additional common characteristics: They are most abundant in moist habitats, and most of them are microscopic in size. Despite their small size, protists have a greater influence on global ecology and human affairs than most people realize. For example, the photosynthetic protists known as algae generate at least half of the oxygen in the Earth's atmosphere and produce organic compounds that feed marine and freshwater animals. The oil that fuels our cars and industry is derived from pressure-cooked algae that accumulated on the ocean floor over millions of years. Because fossil oil deposits are becoming depleted, algae are being engineered into systems for producing renewable biofuels that simultaneously clean pollutants from water and air.

Protists also include some parasites that cause serious human illnesses, as illustrated by the chapter-opening photo. Another example is the waterborne protist *Cryptosporidium parvum,* which in 1993 sickened 400,000 people in Milwaukee, Wisconsin, costing $96 million in medical expenses and lost work time. Species of the related protist *Plasmodium,* which is carried by mosquitoes in many warm regions of the world, cause the disease malaria. Every year, nearly 500 million people become ill with malaria, and more than 2 million die of this disease.

## Protists Can Be Informally Labeled According to Their Ecological Roles

Protists are often labeled according to their ecological roles, which occur in three major types: algae, protozoa, and fungus-like protists (see Figure 23.3). All of these protists occur widely in aquatic habitats.

- **Algae** (singular, alga, from the Latin, meaning seaweeds) are protists that are generally photoautotrophic, meaning that most possess chlorophyll and other photosynthetic pigments and can produce organic compounds from inorganic sources by means of photosynthesis (**Figure 23.14**). In addition to organic compounds, photosynthetic algae produce oxygen. Despite the common occurrence of photosynthesis, algae do not form a monophyletic group descended from a single common ancestor.
- **Protozoa** (from the Greek, meaning first life) are diverse types of heterotrophic protists that cannot produce their own organic food and must obtain it by feeding. Protozoa feed by absorbing small organic molecules or by ingesting prey. For example, the protozoa known as ciliates consume smaller cells such as the single-celled photosynthetic algae known as diatoms (see Figure 23.14). Like the algae, the protozoa do not form a monophyletic group.
- **Fungus-like protists** are a nonmonophyletic assemblage of organisms having bodies, nutrition, or reproduction mechanisms similar to those of the true fungi (see Section 23.6). However, fungus-like protists are not closely related to fungi; their similar features represent cases of convergent evolution, in which species from different lineages have independently evolved similar characteristics. Water molds, some of which cause diseases of fish, and *Phytophthora infestans*, which causes diseases of many wild and crop plants, are examples of fungus-like protists (see Figure 23.3b).

## Biology Principle

### Living Organisms Use Energy

The organic components of the diatoms are digested by the ciliate and used as food, whereas the indigestible diatom walls are excreted.

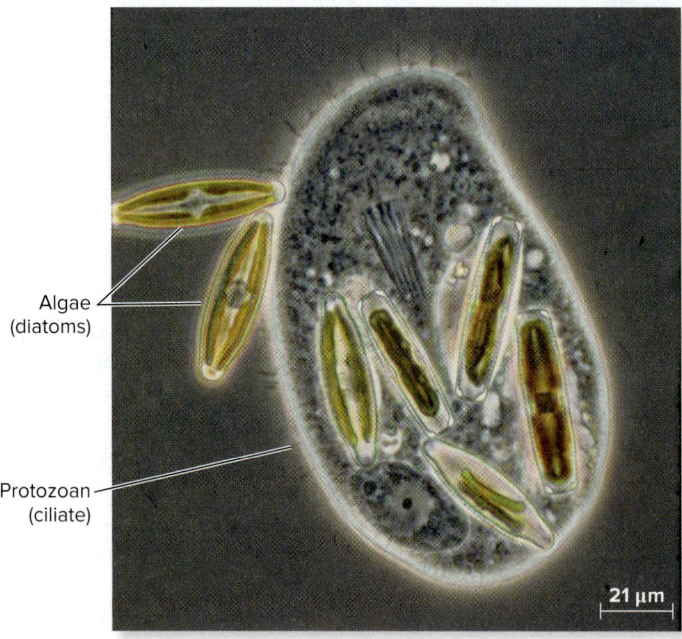


**Figure 23.14 A heterotrophic protozoan feeding on photosynthetic algae.** The ciliate shown here has consumed several oil-rich, golden-pigmented, silica-walled algal cells known as diatoms. Diatom cells that have avoided capture glide nearby.


## Protists Can Be Informally Labeled According to Their Type of Motility

Microscopic protists have evolved diverse ways to propel themselves in moist environments of diverse types. Swimming by means of flagella, cilia, and amoeboid movement are major types of protist movements.

- **Flagellates** are the many types of photosynthetic and heterotrophic protists that are able to swim because they produce one or a few eukaryotic flagella (**Figure 23.15a**). Recall that eukaryotic flagella are cellular extensions whose movement is based on interactions between microtubules and the motor protein dynein. Eukaryotic flagella rapidly bend and straighten, thereby pulling or pushing cells through the water. Flagellates are typically composed of one or only a few cells and are small—usually from 2 to 20 µm long—because flagellar motion is not powerful enough to keep larger bodies from sinking. Some flagellate protists are sedentary, living attached to underwater surfaces. These protists use flagella to collect bacteria and other small particles for food. Macroalgae and other immobile protists often produce small, flagellate reproductive cells that allow these protists to mate and disperse to new habitats.
- **Ciliates** are protists that use many tiny, hairlike surface extensions, known as **cilia,** to move. Cilia are structurally similar to eukaryotic flagella but are shorter and more abundant on cells (**Figure 23.15b**). Having many cilia allows ciliates to achieve larger sizes than flagellates yet still remain buoyant in water.
- **Amoebae** are protists that move by extending cytoplasm into lobes, known as pseudopodia (from the Greek, meaning false feet) (**Figure 23.15c**). Once these pseudopodia move toward a food source or other stimulus, the rest of the cytoplasm flows after them, thereby changing the shape of the entire organism as it creeps along.

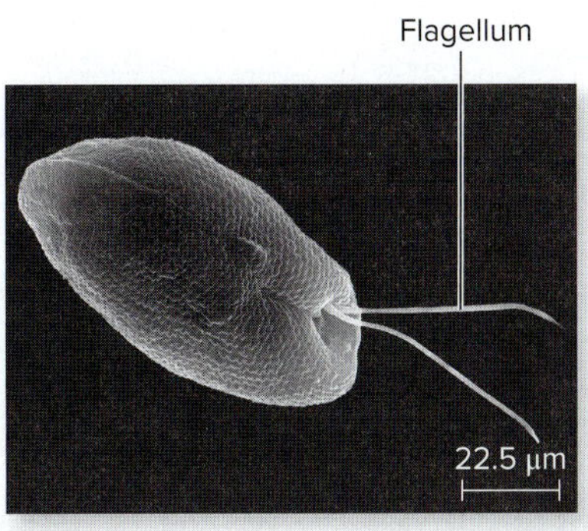


(a) A cryptomonad with eukaryotic flagella

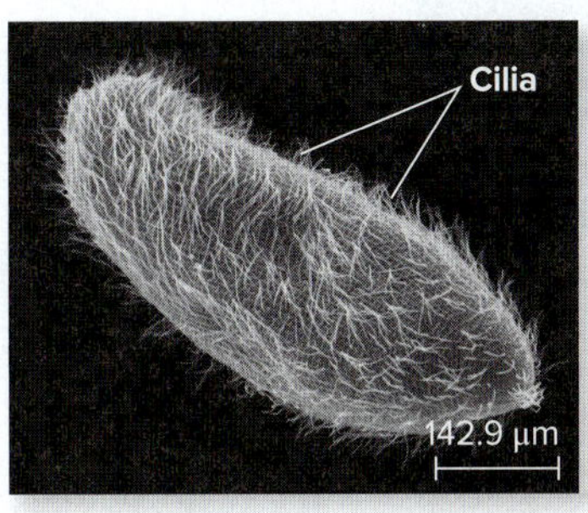


(b) A protzoan ciliate covered by cilia

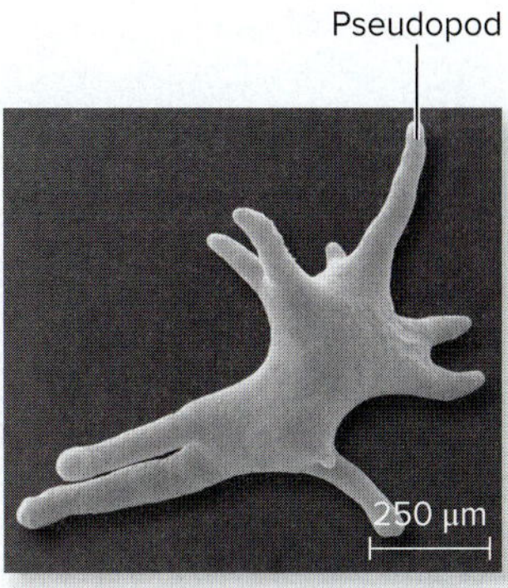


(c) An amoeba extending many pseudopodia

**Figure 23.15** **SEMs of three protists: A flagellate, a ciliate, and an amoeba.** **(a)** This cryptomonad, with two flagella, is an example of a flagellate. **(b)** The ciliate *Paramecium*, showing numerous cilia on the cell surface. **(c)** An amoeba of the genus *Pelomyxa,* showing pseudopodia.



## Protists Can Be Informally Labeled According to Their Habitats

Although protists occupy nearly every type of moist habitat, they are particularly common and diverse in oceans, lakes, wetlands, and rivers. Even extreme aquatic environments such as Arctic and Antarctic ice and acidic hot springs serve as habitats for some protists. In such places, protists may swim or float in open water or live attached to surfaces such as rocks or beach sand. As noted earlier, these different habitats influence protist structure and size. When humans pollute natural waters, harmful or nuisance growths of protists may result, but it is also the case that protists provide important ecological services.

- Planktonic protists swim or float in fresh or salt water, together with planktonic bacteria, viruses, and small animals. The photosynthetic protists in plankton are known as **phytoplankton** (plantlike plankton). Planktonic protists are necessarily quite small in size; otherwise, they would readily sink to the bottom. Staying afloat is a particularly important characteristic of phytoplankton, which need light for photosynthesis, so planktonic protists are microscopic in size. Aquatic environments that have become polluted with too many nutrients such as phosphorus may foster the growth of red tides or other conspicuously large populations of harmful protists (**Figure 23.16a**).
- Periphytic protists attach themselves to underwater surfaces such as rocks, sand, and plants, where they occur in communities known as **periphyton** that also harbor attached prokaryotic species and fungi. Because sinking is not a problem for attached protists, their bodies can be larger than those of the plankton. In water that is polluted with too many nutrients such as phosphorus, growths of periphytic protists may become conspicuous (**Figure 23.16b**). Bushy periphyton growths can be considered a nuisance, but also provide ecological benefits such as habitat for beneficial microbes and small animals.
- Seaweeds, also known as macroalgae, are photosynthetic protists that are large enough to see with the unaided eye. They usually grow attached to underwater surfaces such as rocks, sand, docks, ship hulls, or offshore oil platforms. Seaweeds require sunlight and carbon dioxide for photosynthesis and growth, so most of them grow along coastal shorelines fairly near the water's surface. Kelp forests are communities dominated by tree-sized seaweeds that serve as refuges for aquatic animals such as sea otters, generate large amounts of organic carbon that enter aquatic food chains, and play additional important ecological roles (**Figure 23.16c**).

## Eukaryotic Supergroups Include Protists and Related Plant, Animal, and Fungal Kingdoms

At one time, protists were classified into a single kingdom. However, modern phylogenetic analyses based on comparative analysis of DNA sequences and cellular features reveal that protists do not form a monophyletic group (**Figure 23.17**). That is because some protists are more closely related to plants, animals, or fungi than to other protists. The relationships of some protists are uncertain or disputed, and new protist species are continuously being discovered. As a result, concepts of protist evolution and relationships have been changing as new information becomes available.

Even so, molecular and cellular data reveal that many protist phyla can be classified within seven eukaryotic **supergroups** that each display distinctive features (see Figure 23.17). All of the eukaryotic supergroups include phyla of protists; some, in fact, contain only protist phyla. The supergroup Opisthokonta includes the multicellular animal and fungal kingdoms and related protists, whereas another supergroup includes the multicellular plant kingdom and the protists most closely related to it. The study of protists helps to reveal how multicellularity originated in animals, fungi, and plants. The following survey of eukaryotic supergroups reveals the ecological importance of modern protists.

(a) Coastal red tide of phytoplankton

(b) Periphyton in shallow, nearshore water

(c) Offshore kelp forest

**Figure 23.16** **Protists can be classified according to habitats.** **(a)** Swimming and floating protists occur in the plankton and photosynthetic plankton known as phytoplankton, sometimes forming large populations known as blooms, or red tides, that can be harmful. **(b)** Many types of protists occur as part of the periphyton, a diverse community of organisms that attach to substrata in shallow, nearshore waters. The green periphytic growths shown here are dominated by the common, abundant green alga *Cladophora,* which is associated with many other protists as well as bacteria, fungi, and small animals. **(c)** Kelp forests are extensive growths of tree-sized brown algae that grow in deeper offshore waters and harbor many other organisms.

*(a)* © Purestock/Alamy RF; *(b)* © Linda Graham; *(c)* © Jeff Rotman/Science Source

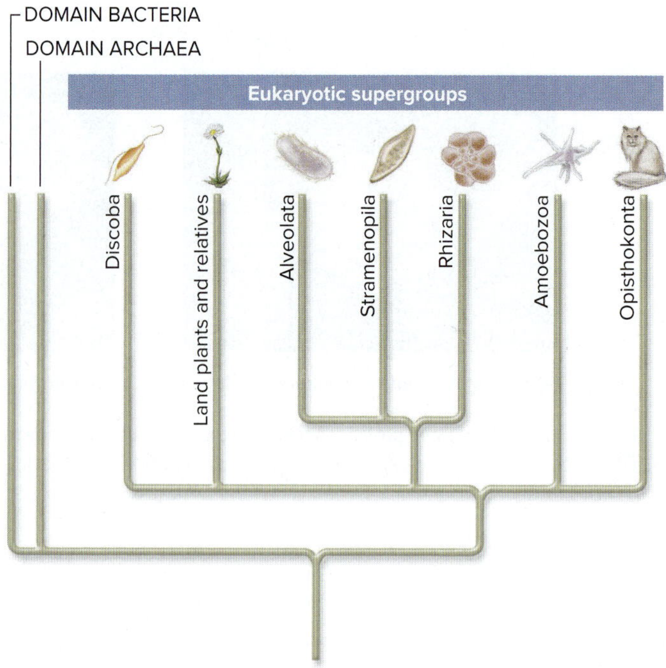


**Figure 23.17** **Phylogenetic diagram showing relationships among seven major eukaryotic supergroups.** All eukaryotic supergroups include protists, and some supergroups include only protists.

## A Feeding Groove Characterizes Many Protists Classified in the Discoba

The protist supergroup known as the Discoba (named for their disk-shaped mitochondrial cristae) originated very early among eukaryotes, so this supergroup is important in understanding the early evolution of eukaryotes. Many Discoba, such as the genus *Jakoba,* and other protists display a feeding groove (**Figure 23.18**). The feeding groove is an important adaptation that allows these single-celled organisms to ingest small particles of food in their aquatic habitats. Once food particles are collected within the feeding groove, they are then taken into cells by a type of endocytosis known as **phagocytosis** (from the Greek, meaning cellular eating) (see Figure 23.18). During phagocytosis, a vesicle of plasma membrane surrounds each food particle and pinches off within the cytoplasm. Enzymes within these food vesicles break the food particles down into small molecules that, upon their release into the cytoplasm, can be used for energy.

Phagocytosis is also the basis for an important evolutionary process known as **endosymbiosis,** a symbiotic association in which a smaller species known as the **endosymbiont** lives within the body of a larger host species. Phagocytosis provides a way for protist cells that function as hosts to take in diverse types of prokaryotic or eukaryotic cells. If not digested, ingested cells may become endosymbionts that confer valuable traits. Examples include, endosymbiotic proteobacteria that evolved into mitochondria and cyanobacteria that evolved into chloroplasts. Invasion of cells by microbes that become endosymbionts is a major way in which many new genes can be horizontally transferred into protist cells.

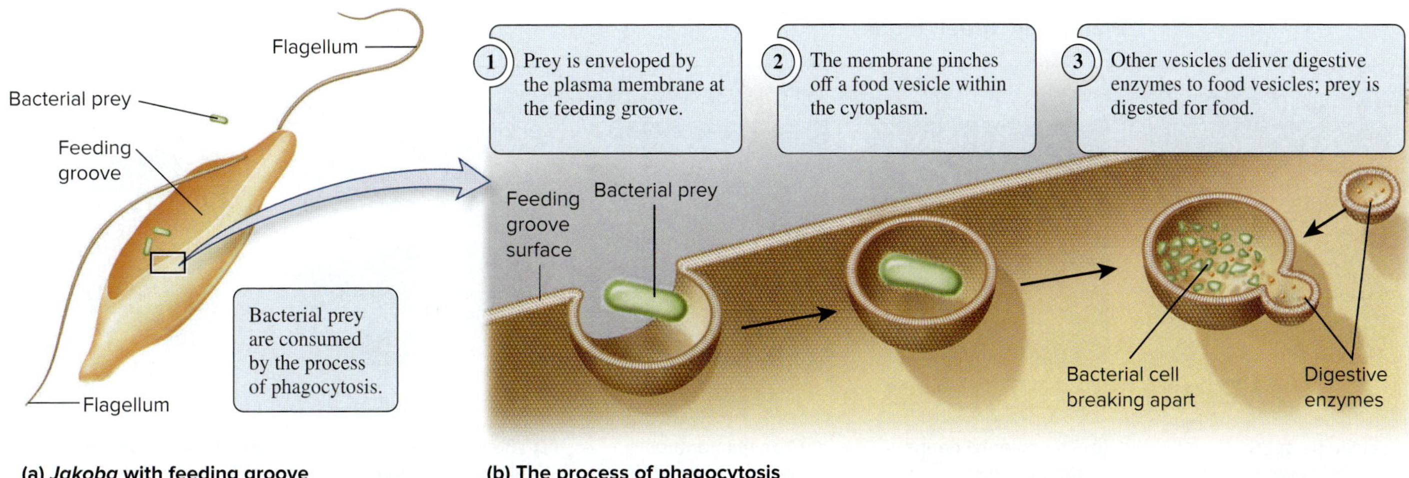


**Figure 23.18** **Discoba. (a)** The flagellate *Jakoba intestinalis* is an example of the supergroup Discoba. **(b)** Many protists display a characteristic feeding groove that functions in phagocytosis.

The supergroup Discoba includes euglenoid flagellates (**Figure 23.19**). Many euglenoids lack plastids, but others contain green plastids acquired by endosymbiosis. Photosynthetic euglenoids display conspicuous particles of carbohydrate and light-sensing systems that include red eyespots, which enable the cells to swim to favorable light environments.

Some plastidless members of Discoba have become parasitic within animals, including human hosts. In addition to feeding by phagocytosis, parasitic species attack host cells and absorb food molecules released from them. *Trichomonas vaginalis* (see chapter-opening photo) is one example. *Giardia intestinalis* (previously known as *G. lamblia*) contains two active nuclei and produces eight flagella (see Figure 23.3a). *G. intestinalis* causes giardiasis, an intestinal infection that can result from drinking untreated water or from unsanitary conditions in day-care centers. Nearly 300 million human infections occur every year, and the disease also harms young farm animals, dogs, and cats, as well as wild animals. In the animal body, flagellate cells cause disease and produce tough infectious stages known as cysts that are transmitted in feces and can survive several weeks outside a host. When an animal ingests as few as 10 of these cysts, within 15 minutes stomach acids induce the flagellate stage to develop and adhere to cells of the small intestine. *T. vaginalis* and *G. intestinalis* were once thought to lack mitochondria, but they are now known to possess simpler structures that are highly modified mitochondria.

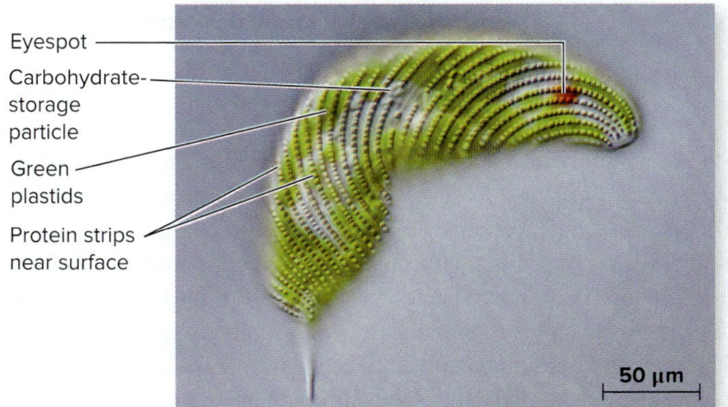


**Figure 23.19** ***Euglena*, a common Discoba protist that contains plastids.**
© Gerd Guenther/SPL/Science Source

## Land Plants and Related Algae Share Similar Genetic Features

Genetic analyses indicate that the supergroup that includes land plants also encompasses several protist phyla, many of which are photosynthetic. Land plants and some closely related green algae together form one lineage (see Chapter 25), the phylum Chlorophyta includes most other green algae, a third phylum (Rhodophyta) is otherwise known as the red algae, and a fourth phylum (Cryptophyta) includes flagellates known as cryptomonads (see Figure 23.15a).

**Green Algae** Diverse structural types of green algae (**Figure 23.20**) occur in fresh water, in the ocean, and on land. Some green algae are planktonic, some occur as attached periphyton, and others are more conspicuous seaweeds. Most of the green algae are photosynthetic, and their cells contain the same types of plastids and photosynthetic pigments that are present in land plants (Kingdom Plantae). The green algae that are most closely related to land plants are known as streptophyte algae (see Chapter 25). Some green algae are responsible for harmful algal growths, but others are useful as food for aquatic animals and as model organisms.

**Red Algae** Most species of red algae are multicellular marine seaweeds (see Figure 23.3c). The red appearance of these algae is caused by the presence of distinctive photosynthetic pigments that are

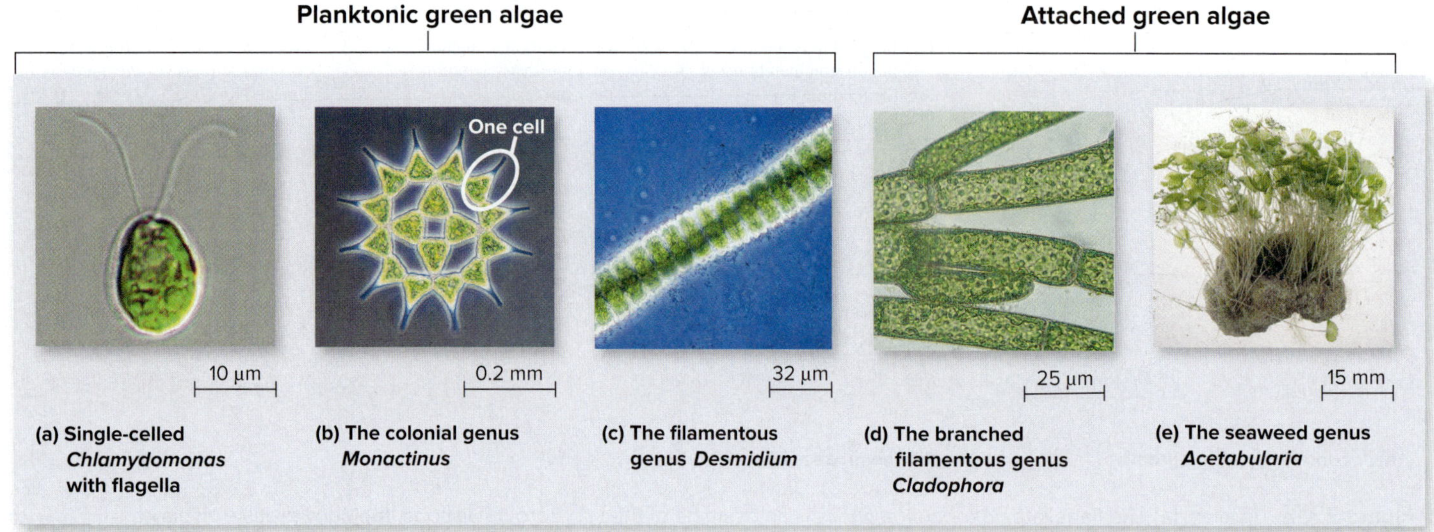


**Figure 23.20** **Green algal body diversity fosters survival in aquatic habitats.** **(a)** The single-celled, swimming genus *Chlamydomonas* occurs in open water plankton, as do many other types of small, suspended microorganisms. **(b)** The genus *Monactinus* is composed of several cells that are associated to form a colony, whose lacy star shape helps to keep this photosynthetic protist afloat in brightly illuminated surface waters. Colonies can also avoid being eaten by predators that consume single-celled prey. **(c)** The filamentous genus *Desmidium* occurs as a twisted row of cells, another way to achieve larger body size while remaining afloat. **(d)** The large-celled, branched, filamentous, attached body of *Cladophora* is too large for most aquatic animals to consume. **(e)** The seaweed genus *Acetabularia* represents many amazing species of tropical green algae whose bodies are the size of a dandelion plant or larger yet are composed of only one extremely large cell.
*(a)* © Brian P. Piasecki; *(b)* © Roland Birke/Phototake; *(c–e)* © Linda Graham

**Figure 23.21** **Alveolata.** **(a)** This single-celled dinoflagellate is a representative of the Alveolata. In this image made with a fluorescence microscope, cellulose wall pieces that occur within alveoli fluoresce blue, chlorophyll-containing plastids fluoresce red, and the cell nucleus fluoresces white. **(b)** Surface sacs called alveoli characterize the alveolate protists and are illustrated in this TEM of a dinoflagellate.
*(a)* Courtesy Joseph Wong and Alvin Kwok; *(b)* © Lee W. Wilcox

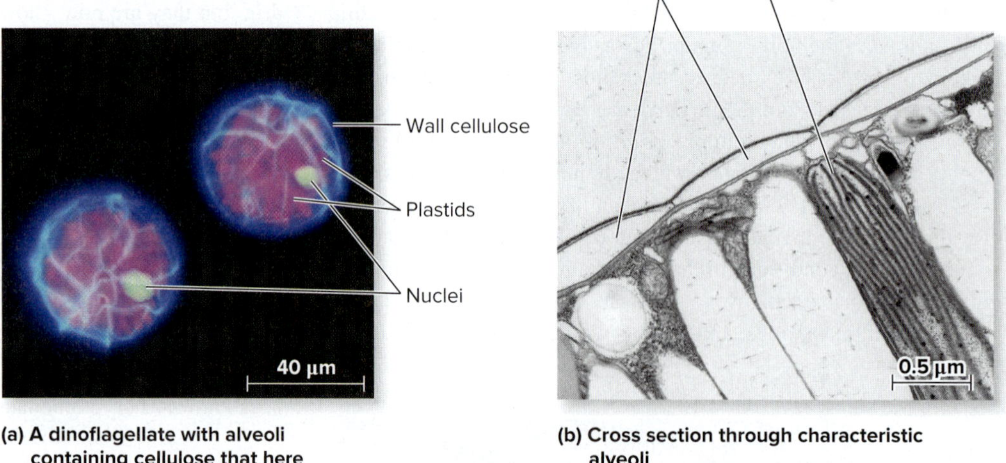


**(a) A dinoflagellate with alveoli containing cellulose that here appears blue**

**(b) Cross section through characteristic alveoli**

absent from green algae or land plants. Red algae characteristically lack flagella—a feature that has strongly influenced the evolution of this group, resulting in unusually complex life cycles.

## Membrane Sacs Lie at the Cell Periphery of Alveolata

The three supergroups Alveolata, Stramenopila, and Rhizaria have been shown to be closely related in recent phylogenetic studies (see Figure 23.17). Alveolata includes three important phyla: (1) the Ciliophora, or ciliates, which feed on bacteria and small algae (see Figures 23.14, 23.15b); (2) the Dinozoa, informally known as dinoflagellates (**Figure 23.21a**), recognized for their symbiotic relationships with reef-building corals and harmful red tide blooms that some species produce (see Figure 23.16a); and (3) the Apicomplexa, a medically important group of parasites. The Alveolata is named for saclike, membranous vesicles known as alveoli that are present at the cell periphery in all of these phyla (**Figure 23.21b**).

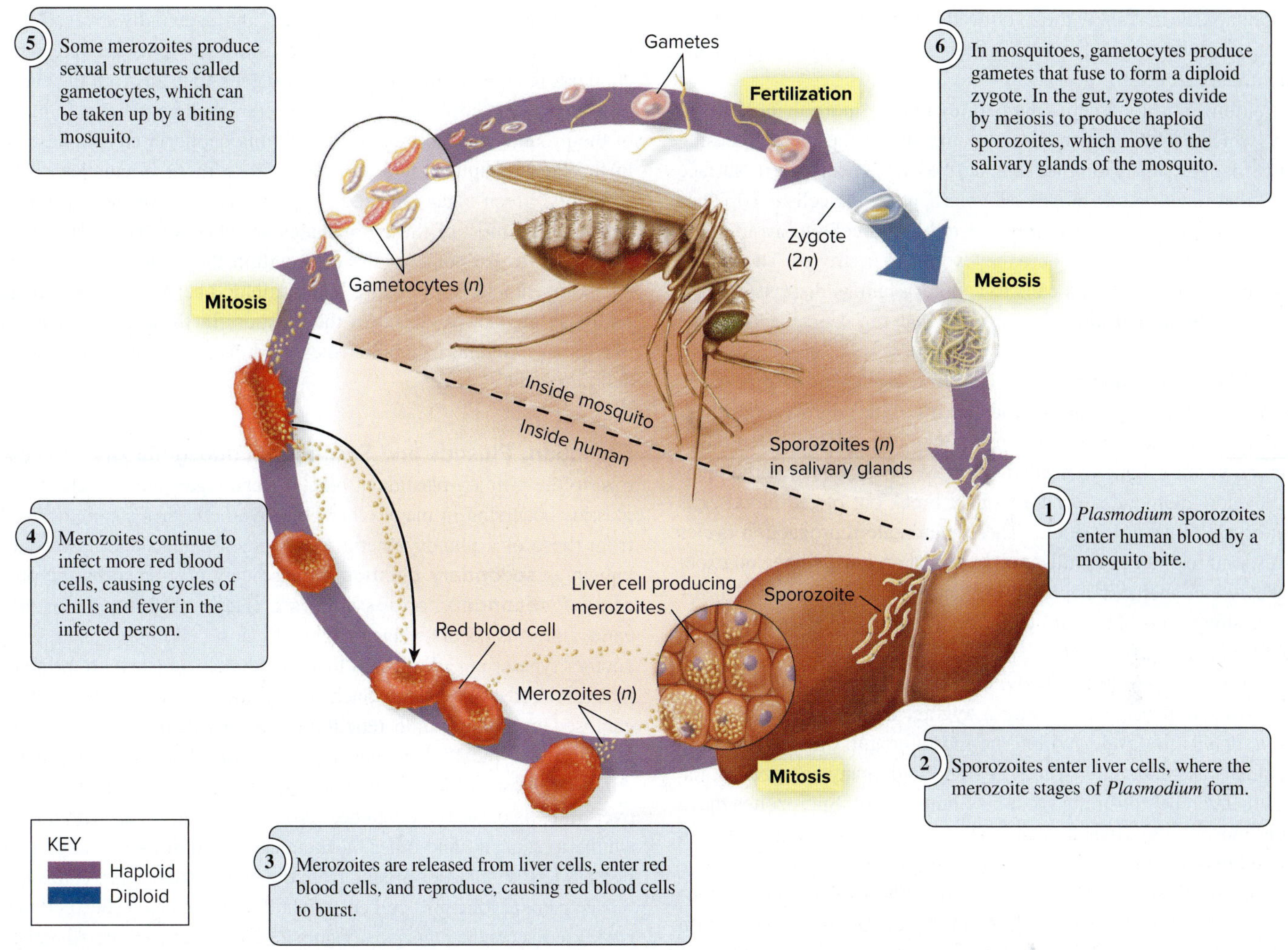

(1) *Plasmodium* sporozoites enter human blood by a mosquito bite.

(2) Sporozoites enter liver cells, where the merozoite stages of *Plasmodium* form.

(3) Merozoites are released from liver cells, enter red blood cells, and reproduce, causing red blood cells to burst.

(4) Merozoites continue to infect more red blood cells, causing cycles of chills and fever in the infected person.

(5) Some merozoites produce sexual structures called gametocytes, which can be taken up by a biting mosquito.

(6) In mosquitoes, gametocytes produce gametes that fuse to form a diploid zygote. In the gut, zygotes divide by meiosis to produce haploid sporozoites, which move to the salivary glands of the mosquito.

Gametes
Fertilization
Zygote (2*n*)
Meiosis
Gametocytes (*n*)
Mitosis
Inside mosquito
Inside human
Sporozoites (*n*) in salivary glands
Sporozoite
Liver cell producing merozoites
Red blood cell
Merozoites (*n*)
Mitosis

KEY
Haploid
Diploid

**Figure 23.22 Diagram of the life cycle of *Plasmodium falciparum,* an alveolate species that causes malaria.** This life cycle requires two alternate hosts, humans (or great apes) and *Anopheles* mosquitoes.

*In which of the hosts does sexual mating of P. falciparum gametes occur?*

About half of dinoflagellate species are heterotrophic, and half possess photosynthetic plastids of diverse types that originated by secondary or tertiary endosymbiosis. Tertiary endosymbiosis is the acquisition by hosts of plastids from cells that possessed secondary plastids (see Figure 23.23c). Species having tertiary plastids resulting from tertiary endosymbiosis have received genes by horizontal transfer from diverse endosymbiont genomes.

Apicomplexa include the protist genus that causes malaria, *Plasmodium*. About 40% of humans live in tropical regions of the world where malaria occurs, and millions of infections and human deaths result each year. Malaria is particularly deadly for young children. In addition to humans, the malarial parasite's alternate host is the mosquito classified in the genus *Anopheles,* which can also transmit malaria to the great apes. Though insecticides can be used to control mosquito populations and though antimalarial drugs exist, malarial parasites can develop drug resistance. Experts are concerned that cases may double in the next 20 years.

When a mosquito bites a human or a great ape, *Plasmodium* enters the bloodstream as an asexual life stage known as a sporozoite (**Figure 23.22**). Upon reaching a victim's liver, sporozoites enter liver cells, where they divide to form an asexual life stage known as merozoites. Hundreds of merozoites are produced within liver cells, which then release into the bloodstream packages of merozoites enclosed by a host-derived cell membrane. This membrane protects merozoites from destruction by host immune cells, which would otherwise engulf merozoites by phagocytosis and then destroy the invaders. In the bloodstream, the protective host membranes disintegrate, releasing merozoites. The merozoites have protein complexes at their front ends, or apices, that allow them to invade human red blood cells. (The presence of these apical complexes gives rise to the phylum name Apicomplexa.)

Within red blood cells, merozoites release more than 200 proteins, which enable the parasites to commandeer these cells, causing many changes. For example, infected red blood cells form surface knobs that function like molecular Velcro, attaching cells to capillary linings. This process allows infected red blood cells to avoid being transported to the spleen, where they would be destroyed. The attachment of infected red blood cells to capillary linings disrupts circulation in the brain and kidney, a process that can cause death of the animal host.

While living within red blood cells, merozoites form rings, which can be visualized by staining and the use of a microscope, allowing diagnosis. The merozoites consume the hemoglobin in red blood cells, providing resources needed to reproduce asexually. Large numbers of new merozoites synchronously break out of red blood cells at intervals of 48 or 72 hours. These merozoite reproduction cycles correspond to cycles of chills and fever that an infected person experiences. Some merozoites produce sexual structures—gametocytes—which, along with blood, are transmitted to a female mosquito after she bites an infected person.

Within the mosquito's body, the gametocytes produce gametes and fertilization occurs, yielding a zygote, the only diploid cell in *Plasmodium*'s life cycle. Within the mosquito gut, the zygote undergoes meiosis, generating structures filled with many sporozoites, the stage that can be transmitted to a new human host. Sporozoites move to the mosquito's salivary glands, where they remain until they are injected into a human host when the mosquito feeds.

*Plasmodium falciparum* and some other apicomplexan protists possess plastids that are descended from algal ancestors with photosynthetic secondary plastids. About 550 (some 10%) of *Plasmodium*'s nuclear-encoded proteins are imported into a nonphotosynthetic plastid known as an apicoplast, where they are needed for fatty acid metabolism and other processes. Because plastids are not present in mammalian cells, enzymes in apicoplast pathways are possible targets for development of drugs that will kill the parasite without harming the host. Mammals also lack calcium-dependent protein kinases (CDPKs), enzymes that are essential to merozoite release from red blood cells and the parasite's sexual development, offering another potential way to develop new antimalarial medicines.

## EVOLUTIONARY CONNECTIONS

### Primary Plastids and Primary Endosymbiosis

The plastids of red algae, green algae, and land plants have an enclosing envelope composed of two membranes. Such plastids, known as **primary plastids,** are thought to have originated via a process known as **primary endosymbiosis** (**Figure 23.23a**). During primary endosymbiosis, heterotrophic host cells captured cyanobacterial cells via phagocytosis but did not digest them. These endosymbiotic cyanobacteria provided host cells with photosynthetic capability and other useful biochemical pathways and eventually evolved into primary plastids. Endosymbiotic acquisitions of plastids and mitochondria resulted in massive horizontal gene transfer from the endosymbiont to the host nucleus. As a result of such gene transfer, many of the proteins needed by plastids and mitochondria are synthesized in the host cytoplasm and then targeted to these organelles. Cells of plants, green algae, and red algae contain one or more primary plastids, and most of these organisms are photosynthetic. However, some species (or some of the cells within the multicellular bodies of photosynthetic species) are heterotrophic because photosynthetic pigments are not produced in the plastids. In these cases, plastids play other essential metabolic roles, such as producing amino acids and fatty acids.

**Secondary Plastids and Secondary Endosymbiosis** In contrast to the primary plastids of plants, green algae, and red algae, the plastids occurring in many other photosynthetic protists are derived from those of a photosynthetic red or green alga. Such plastids are known as **secondary plastids** because they originated by the process of **secondary endosymbiosis** (**Figure 23.23b**). Secondary endosymbiosis occurs when a eukaryotic host cell ingests and retains another type of eukaryotic cell that already has one or more primary plastids, a red or green alga. Such eukaryotic endosymbionts are often enclosed by endoplasmic reticulum (ER), explaining why secondary plastids typically have envelopes of more than two membranes. Although most of the endosymbiont's cellular components were lost over time, its plastids are retained, providing the host cell with photosynthetic capacity and other biochemical capabilities. Some protists possess tertiary plastids arising from the process of tertiary endosymbiosis; a eukaryotic host cell engulfs and retains a eukaryotic cell that possessed secondary plastids (**Figure 23.23c**). Many of the Alveolata have secondary or tertiary plastids obtained from a eukaryotic alga. In these amazing cases, the alveolate protists contain genes derived from diverse microbial organisms.

## Flagellar Hairs Distinguish Stramenopila

The supergroup Stramenopila (informally known as the stramenopiles) encompasses a wide range of algae, protozoa, and fungus-like protists that usually produce flagellate reproductive cells at some point in their lives. The Stramenopila (from the Greek *stramen*, meaning straw, and *pila*, meaning hair) is named for distinctive, strawlike hairs that occur on the surfaces of flagella (**Figure 23.24**). These flagellar hairs function something like oars to greatly increase swimming efficiency.

Heterotrophic stramenopiles include the fungus-like protist *Phytophthora infestans*, which causes the serious potato disease known as late blight that results in the loss of billions of dollars of crops every year (see Figure 23.3b). Photosynthetic stramenopiles include diatoms (Bacillariophyceae) (see Figure 23.14) and the brown algae known as giant kelps, whose ecological importance has previously been described (see Figure 23.16c).

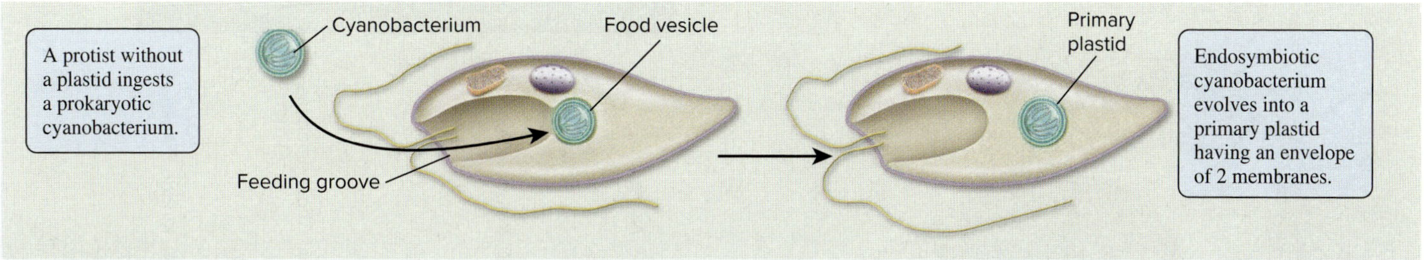


**(a) Primary endosymbiosis**

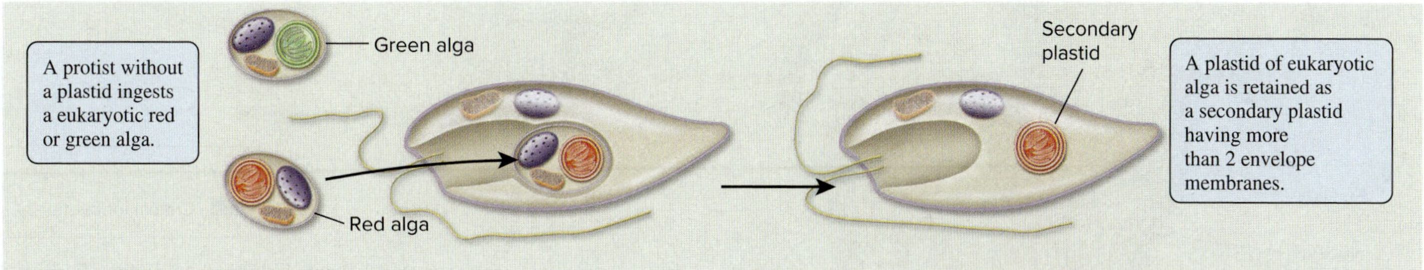


**(b) Secondary endosymbiosis**

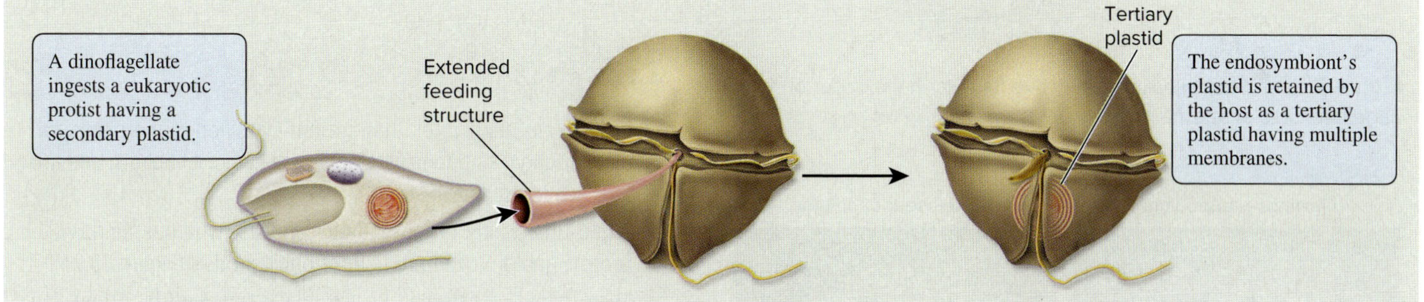


**(c) Tertiary endosymbiosis**

**Figure 23.23** **Primary, secondary, and tertiary endosymbiosis.** **(a)** Primary endosymbiosis involves the acquisition of a cyanobacterial endosymbiont by a host cell without a plastid. During the evolution of a primary plastid, the bacterial cell wall is lost, and most endosymbiont genes are transferred to the host nucleus. **(b)** Secondary endosymbiosis involves the acquisition by a host cell of a eukaryotic endosymbiont that contains one or more primary plastids, a red or green alga. During the evolution of a secondary plastid, most components of the endosymbiont cell are lost, but a plastid is often retained within an envelope of endoplasmic reticulum. **(c)** Tertiary endosymbiosis involves the acquisition by a host cell of a eukaryotic endosymbiont that possesses secondary plastids.

## Spiky Cytoplasmic Extensions Are Present on the Cells of Many Protists Classified in Rhizaria

Several groups of flagellates and amoebae that have thin, hairlike extensions of their cytoplasm—known as filose pseudopodia—are classified into the supergroup Rhizaria (from the Greek *rhiza*, meaning root) (**Figure 23.25**). Rhizaria includes the Radiolaria (Figure 23.25a) and Foraminifera (Figure 23.25b)—two phyla of ocean plankton that produce exquisite mineral shells. Radiolaria and Foraminifera commonly shelter symbiotic algal cells that provide benefits to their nonphotosynthetic hosts.

## Amoebozoa Includes Many Types of Amoebae with Pseudopodia

The supergroup Amoebozoa includes many types of amoebae that move by extension of pseudopodia (see Figure 23.15c). Several types of protists known as slime molds are classified in this supergroup. One example, *Dictyostelium discoideum*, is widely used as a model organism for understanding movement, communication among cells, and development. During reproduction, in response to starvation, single *Dictyostelium* amoebae aggregate into a multicellular slug that produces a cellulose-stalked structure containing many single-celled, asexual spores. In favorable conditions, these spores produce new amoebae, which feed on bacteria.

## A Single Flagellum Occurs on Swimming Cells of Opisthokonta

The supergroup Opisthokonta includes the animal and fungal kingdoms and related protists (**Figure 23.26**). This supergroup is named for the presence of a single posterior flagellum on swimming cells. The more than 125 species of opisthokont protists known as

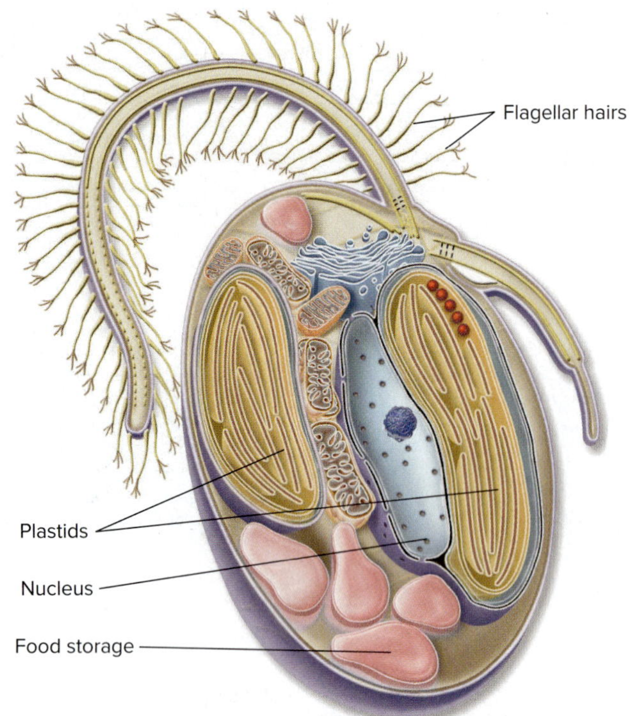


**Figure 23.24** **Stramenopila.** Distinctive flagellar hairs occur on the flagellate cells of stramenopile protists.

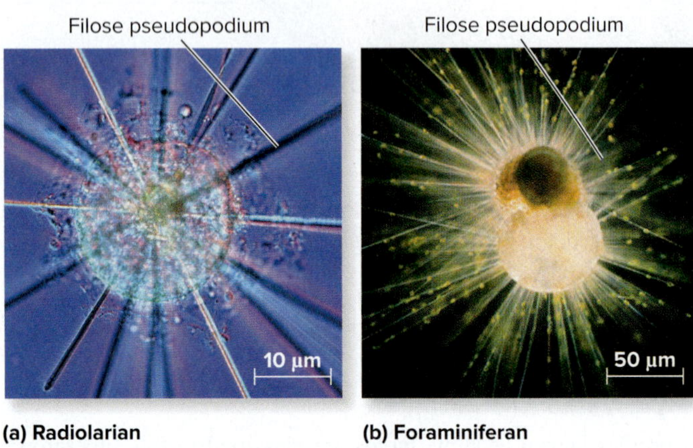


**(a) Radiolarian** **(b) Foraminiferan**

**Figure 23.25** **Rhizaria. (a)** The radiolarian, *Acanthoplegma* spp. **(b)** A foraminiferan showing long, filose pseudopodia.

*(a)* © Nuridsany et Perennou/Science Source; *(b)* © O. Roger Anderson, Columbia University, Lamont-Doherty Earth Observatory

choanoflagellates (formally the Choanomonada) are single-celled or colonial protists featuring a distinctive collar surrounding the single flagellum (**Figure 23.27a**). The collar is made of cytoplasmic extensions that filter bacterial food from water currents generated by flagellar motion. Among protists, choanoflagellates are believed to represent the closest living relatives of animals. *Nuclearia* is an opisthokont genus that is very closely related to the kingdom Fungi (**Figure 23.27b**). The preceding survey of protists, summarized in **Table 23.2,** represents just some of the enormous diversity of protists on Earth.

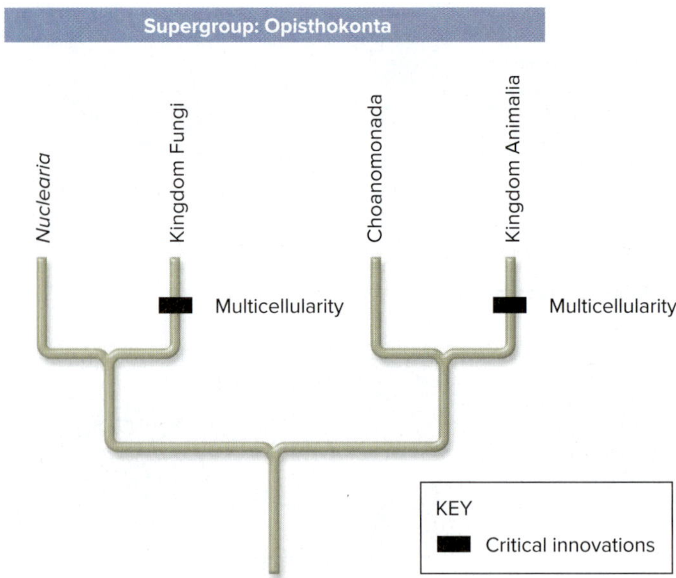


**Figure 23.26** **Supergroup Opisthokonta includes the kingdoms Fungi and Animalia and closely related protists.**

## 23.5 Reviewing the Concepts

- Protists are eukaryotes that are not classified in the plant, animal, or fungal kingdoms; are abundant in moist habitats; and are mostly microscopic in size.
- Algae are mostly photosynthetic protists, whereas fungus-like protists and protozoa are heterotrophic protists (Figure 23.24).
- Microscopic protists can be informally classified by propulsion method: flagella (flagellates), cilia (ciliates), or pseudopodia (amoebae) (Figure 23.15).
- Protists can also be labeled according to size and habitat (Figure 23.16).
- Modern phylogenetic analysis has revealed that many protists can be formally classified into one of seven major eukaryotic supergroups, each displaying one or more distinctive features (Figure 23.17).
- Primary, secondary, and tertiary endosymbiosis, all based on the cellular process of phagocytosis, have increased eukaryotic diversity, giving rise to photosynthetic protists. Protists are members of the supergroups Discoba, land plants and relatives, alveolates, stramenopiles, Rhizaria, Amoebozoa, and opisthokonts (Figures 23.18, 23.19, 23.20, 23.21, 23.22, 23.23, 23.24, 23.25, 23.26).

## 23.5 Testing Your Knowledge

**1–4.** Match the four eukaryotic supergroups listed below with characteristic structural features.

| | |
|---|---|
| **1.** Alveolata | **a.** primary plastids |
| **2.** Land plants and relatives | **b.** flagella occur singly |
| **3.** Opisthokonts | **c.** peripheral membrane-bound sacs |
| **4.** Stramenopiles | **d.** distinctive hairs on flagella |

## Biology Principle

### All Species (Past and Present) Are Related by an Evolutionary History

Many more eukaryotic branches exist than are shown in the streamlined diagram of Figure 23.17 or are listed in Table 23.2.

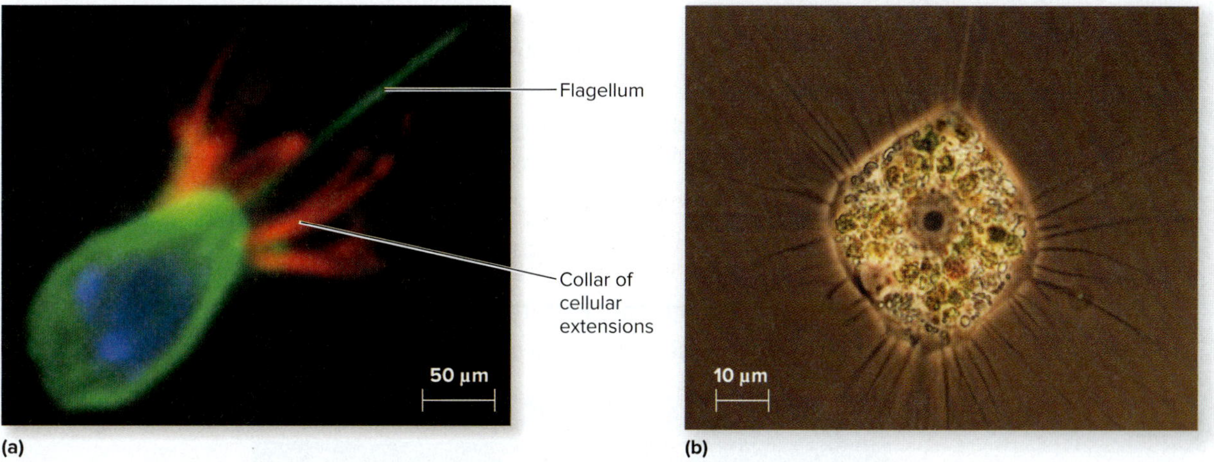


**Figure 23.27** **Opisthokonta, illustrated by protists related to the animal and fungal kingdoms.** **(a)** A fluorescence image of a choanoflagellate, representing modern protists most closely related to the animal kingdom. **(b)** The amoeba *Nuclearia,* representing modern protists most closely related to the fungal kingdom. Study of such protists helps to reveal how multicellularity evolved.

*(a)* © Stephen Fairclough, King Lab, University of California at Berkeley; *(b)* © David J. Patterson, Maple Ferryman Pty Ltd.

**Table 23.2 Eukaryotic Supergroups and Examples of Constituent Kingdoms, Phyla, Classes, or Species**

| Supergroup | KINGDOMS, Phyla, classes, or *species* | Distinguishing features |
|---|---|---|
| Discoba | *Jakoba* spp.<br>*Giardia intestinalis*<br>*Trichomonas vaginalis*<br>*Euglena* | Unicellular flagellates, often with feeding groove. The euglenoids (many of which are photosynthetic) and diverse heterotrophic protists (some of which are pathogens) are grouped with Discoba. |
| Land Plants and Relatives | **Rhodophyta** (red algae)<br>**Chlorophyta** (most green algae)<br>KINGDOM PLANTAE and close green algal relatives | Land plants, green algae, and red algae have primary plastids derived from cyanobacteria; such plastids have two envelope membranes. Cryptomonads (most of which possess secondary red plastids) and plastidless relatives have recently been linked to this supergroup. |
| Alveolata | **Ciliophora** (ciliates)<br>**Apicomplexa** (apicomplexans)<br>*Plasmodium falciparum*<br>*Cryptosporidium parvum*<br>**Dinozoa** (dinoflagellates) | Peripheral membrane sacs (alveoli)<br>Apicomplexa often have nonphotosynthetic secondary plastids; some Dinozoa have secondary plastids derived from red algae, some have secondary plastids derived from green algae, and some have tertiary plastids. |
| Stramenopila | Bacillariophyceae (diatoms)<br>Phaeophyceae (brown algae)<br>*Phytophthora infestans* (fungus-like) | Strawlike flagellar hairs; secondary plastids (when present) derived from red algae. |
| Rhizaria | **Radiolaria**<br>**Foraminifera** | Thin, cytoplasmic projections from plastidless cells having mineral shells; symbiotic relationships with diverse types of photosynthetic algal cells are common. |
| Amoebozoa | **Dictyostelia** (a slime mold phylum)<br>*Dictyostelium discoideum* | Amoeboid movement by pseudopodia |
| Opisthokonta | *Nuclearia* spp.<br>KINGDOM FUNGI<br>**Choanomonada** (choanoflagellates)<br>KINGDOM ANIMALIA | Swimming cells possess a single posterior flagellum. |

## 23.6 Diversity and Ecological Importance of Fungi

### Learning Outcomes

1. List three features that define fungi.
2. List three groups of fungi that lack flagellate cells.
3. Explain how most fungi feed.

The microbial eukaryotes known as fungi are so distinct from other organisms that they are placed in their own kingdom, the kingdom Fungi, also known as the true fungi (**Figure 23.28**). Fungi diverged from other opisthokonts more than a billion years ago, during the middle of the Proterozoic era.

Fungi form a monophyletic group of more than 100,000 species. Like their close protist relative *Nuclearia,* the early-diverging fungi (Cryptomycota, Blastocladiomycota, and Chytridiomycota) are simple in body structure and primarily occupy aquatic environments. By contrast, later-diverging fungal groups (AM fungi and relatives, Ascomycetes, and Basidiomycetes) are adapted to life on land (**Table 23.3**).

- **Cryptomycota.** The earliest-diverging fungi are classified as Cryptomycota, which occur in diverse genetic types in soil and

## Biology Principle

### All Species (Past and Present) Are Related by an Evolutionary History

More than 15 fungal phyla occur, but their relationships and names are still being determined.

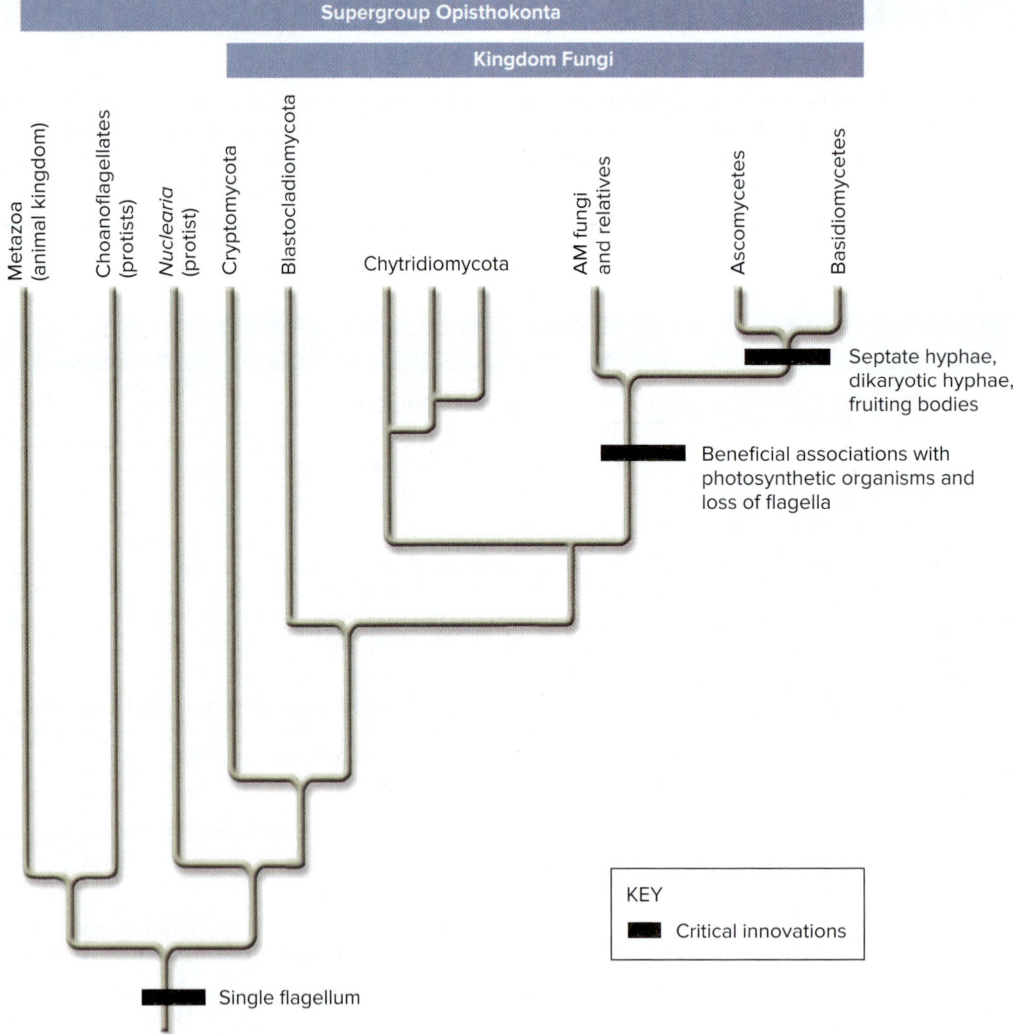


**Figure 23.28** **Evolutionary relationships of the fungi.** The kingdom Fungi arose from a unicellular protist ancestor similar to the modern amoeba *Nuclearia*. Fungal groups described in this chapter include early-diverging cryptomycota and chytrids and later-diverging AM fungi, ascomycetes, and basidiomycetes.

**Table 23.3 Major Fungal Groups**

| Informal name | Habitat | Ecological role | Reproduction |
|---|---|---|---|
| Cryptomycota | Water and soil | Some (microsporidians) are parasites | Flagellate cells or nonflagellate spores in the case of microsporidia |
| Chytrids | Water and soil | Mostly decomposers; some parasites | Flagellate spores or gametes |
| AM Fungi | Terrestrial | Form mutually beneficial associations with plants | Distinctively large, nonflagellate, multinucleate asexual spores |
| Ascomycetes | Mostly terrestrial | Decomposers; some are pathogens; many form lichen partnerships with cyanobacteria or green algae; some partner with plant roots | Asexual conidia; nonflagellate sexual spores (ascospores) in sacs (asci) on fruiting bodies (ascocarps) |
| Basidiomycetes | Terrestrial | Decomposers; many form beneficial associations with plants; some form lichen partnerships | Several types of asexual spores; nonflagellate sexual spores (basidiospores) on club-shaped basidia on fruiting bodies (basidiocarps) |

water. Some members of the Cryptomycota produce cells with flagella. Microsporidia are tiny cells that parasitize the cells of diverse animals, including humans. Some microsporidian species are linked to the decline of honeybee populations (Figure 23.29a). Microsporidian genomes contain diverse genes acquired from other organisms by horizontal gene transfer.

- **Chytrids.** Several aquatic lineages of microscopic species, informally known as chytrids, display cell walls that contain a tough, nitrogen-containing carbohydrate known as **chitin** (Figure 23.29b), as do most fungi. Presence of a chitin-containing wall enables fungi to resist high osmotic pressure that results when they feed by absorbing small organic molecules, a process known as osmotrophy. Chytrids produce flagellate reproductive cells, which are absent from most other fungi. The loss of flagella is linked to an ecological transition from aquatic habitats to land, where the chitin-containing cell wall is also advantageous.
- **AM Fungi.** The arbuscular mycorrhizal fungi, abbreviated AM fungi, are well known for their close associations with plant roots (Figure 23.29c). In these associations, the fungus provides the plant with minerals and the plant provides organic food to the fungus. Ancient fossils similar to modern AM fungi and molecular evidence suggest that early plants may likewise have depended on associations with fungi. AM fungi are closely related to diverse terrestrial fungi that produce tough sexual structures known as zygospores. The black mold of bread known as *Rhizopus* is an example of a zygospore-producing fungus.
- **Ascomycetes.** The name ascomycetes derives from unique reproductive structures known as asci (singular, ascus) from the Greek *asco*, meaning bags or sacs. Ascomycetes are ecologically important as decomposers and disease-causing organisms. Edible truffles and morels are the reproductive structures of particular ascomycetes (Figure 23.29d).
- **Basidiomycetes.** The name given to the basidiomycetes derives from basidia, club-shaped reproductive cells. An estimated 30,000 modern basidiomycete species are known and are very important as decomposers and in associations with plants. Reproductive structures occur as mushrooms, puffballs, stinkhorns, shelf fungi, rusts, and smuts (Figure 23.29e). Ascomycetes and basidomycetes are the two most recently diverged fungal lineages (see Figure 23.28).

## Fungi Have Absorptive Nutrition and a Unique Body Form

Because fungi are closely related to the animal kingdom, fungi and animals display some common features. For example, both animals and fungi are heterotrophic, meaning that they cannot produce their own food but must obtain it from the environment. Fungi use an amazing array of organic compounds as food, which is termed their substrate. The substrate could be the soil, a rotting log, a piece of bread, a living tissue, or a wide array of other materials. Fungi are also like animals in having **absorptive nutrition.** Both fungi and the cells of animal digestive systems secrete enzymes that break down complex organic materials and absorb the resulting small organic food molecules. In addition, both fungi and animals store surplus food as the carbohydrate glycogen in their cells. Despite these nutritional commonalities, fungal body structure, growth, and reproduction differ from those in animals.

Most fungi have a distinctive type of body known as a **mycelium** (plural, mycelia), which is composed of individual microscopic, branched filaments known as **hyphae** (singular, hypha) (see Figures 23.4b and 23.29b,c). Hyphae and mycelia evolved even before fungi made the transition from aquatic to terrestrial habitats. The hyphae of ascomycetes and basidiomycetes are septate, that is, divided into smaller cellular units by cross walls known as septa. Other fungal hyphae are not septate.

A fungal mycelium may be very extensive, but is often inconspicuous because the component hyphae are so tiny and spread out in the substrate. The diffuse form of the fungal mycelium makes sense because most hyphae function to absorb organic food from the substrate. By spreading out, hyphae can absorb food from a large volume of substrate. The absorbed food is used for mycelial growth and for reproduction by means of **fruiting bodies,** such as mushrooms. Fruiting bodies are often the more conspicuous parts of fungal bodies because they typically emerge from substrata (see Figure 23.29d,e). A fungal mycelium that has not yet produced fruiting bodies represents a complete fungal body but can be quite inconspicuous.

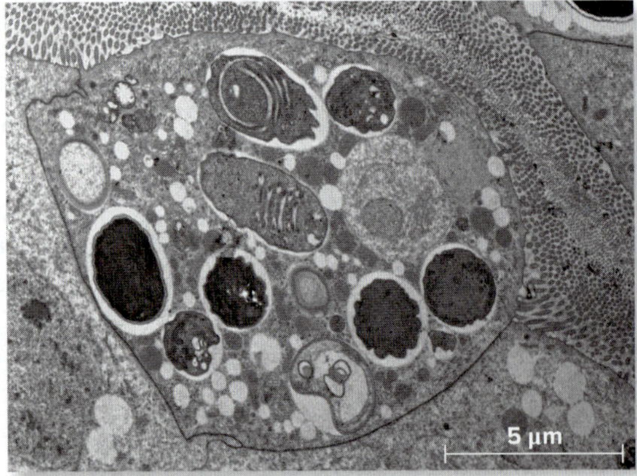


**(a) Representative of cryptomycota**

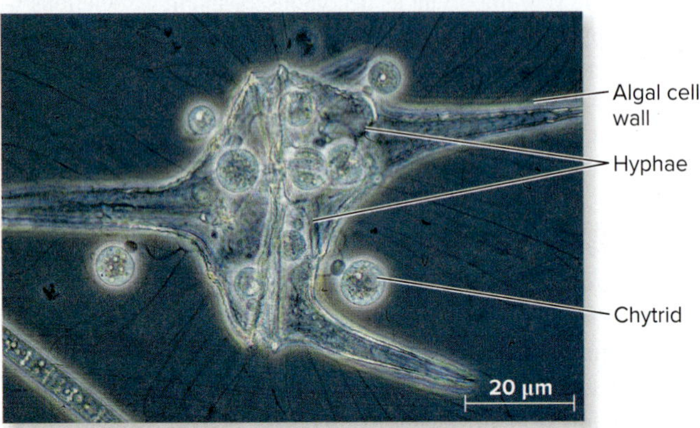


**(b) Chytrids attacking algal cell**

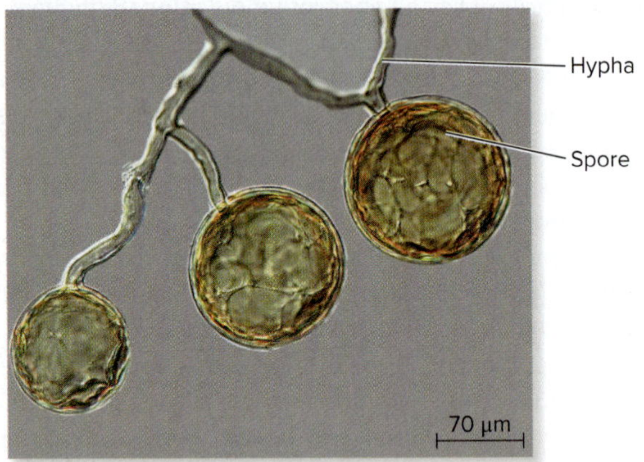


**(c) An AM fungus**

**(d) An ascomycete**

**(e) A basidiomycete**

**Figure 23.29** **Representatives of five fungal lineages.**
**(a)** Cryptomycota representative. Oblong cells of the microsporidian fungus *Nosema ceranae* are parasitic within honeybee cells. **(b)** The colorless chytrids produce hyphae that penetrate the cellulose cell walls of the aquatic dinoflagellate *Ceratium hirundinella,* absorbing organic materials. Chytrids use these materials to produce spherical, flagellate reproductive cells called spores that swim away to attack other algal cells. **(c)** The genus *Glomus,* an example of an AM fungus. These fungi are found in and near the roots of many types of plants, aiding them in acquiring water and nutrients. AM fungi produce distinctive large, multinucleate reproductive spores. **(d)** Fruiting bodies of the ascomycetes commonly known as cup fungi. **(e)** Shelf fungi, such as this *Laetiporus sulphureus,* are the fruiting bodies of basidiomycete fungi that have infected trees.



Fruiting bodies produced by ascomycetes and basidiomycetes are composed of densely packed hyphae that have undergone a sexual mating process (Figure 23.30). During mating, hyphae of different but compatible individuals are attracted to each other and fuse. The resulting mated hyphae differ genetically and biochemically from unmated hyphae. When environmental conditions are right, mated hyphae pack together and enlarge to produce mature fruiting bodies that emerge from the substrate. Sexual reproduction generates new allele combinations that may allow fungi to colonize new types of habitats.

Amazingly diverse in form, color, and odor, mature fruiting bodies are specialized to produce and disperse chitin-walled reproductive cells known as **spores.** Produced by the processes of meiosis or mitosis and protected by tough walls, spores reflect a major adaptation to the terrestrial habitat. Fungal spores are generally adapted for transport by wind or by animals that move around above the substrate. When fungal spores settle in places where conditions are favorable for growth, they produce new mycelia. When the new mycelia undergo sexual reproduction, they produce

## Biology Principle

### Living Organisms Grow and Develop

In ascomycetes and basidiomycetes, after a mating process occurs, mated hyphae produce fruiting bodies whose form fosters spore production and dispersal. In suitable sites, spores may germinate, producing new mycelia.

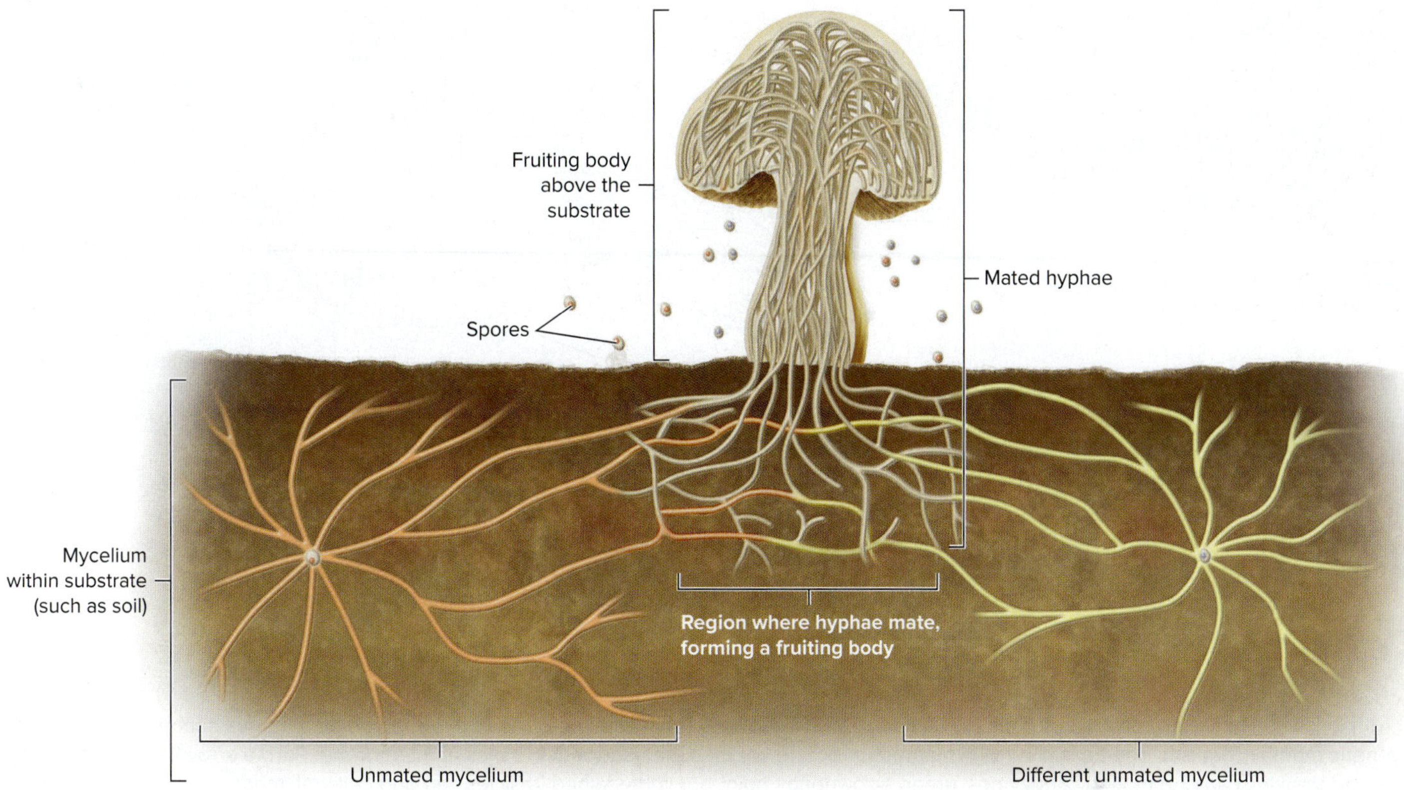


**Figure 23.30 Fungal body structure and life cycle.** A young fungus consists of haploid food-gathering hyphae that grow and branch from a central point to form a diffuse mycelium within a food substrate, such as soil. When compatible mates are present, a sexual reproduction process generates a mycelium composed of mated hyphae that can produce a conspicuous fruiting body (such as a mushroom). Cells at the surface of the fruiting body contain diploid zygote nuclei that undergo meiosis, thereby producing haploid spores of diverse genetic types. Spore dispersal allows fungi to colonize new habitats.

new fruiting bodies and spores, completing the fungal life cycle (Figure 23.30).

## Fungi Have Distinctive Growth Processes

If you have ever watched bread or ripe fruit become increasingly moldy over the course of several days, you have observed fungal growth. When a food source is plentiful, fungal mycelia can grow rapidly at their tips, adding as much as a kilometer of new hyphae per day! The mycelia grow at their edges as the fungal hyphae extend their tips through the undigested substrate. The narrow dimensions and extensive branching of hyphae provide a very high surface area for absorption of organic molecules, water, and minerals.

**Hyphal Tip Growth** Cytoplasmic streaming and osmosis are important cellular processes in hyphal growth. Recall that osmosis (see Chapter 5) is the diffusion of water through a membrane, from a solution with a lower solute concentration into a solution with a higher solute concentration. Water enters fungal hyphae by means of osmosis because their cytoplasm is rich in sugars, ions, and other solutes. Water entry swells the hyphal tip, producing the force necessary for tip extension. Masses of tiny vesicles carrying enzymes and cell-wall materials made in the Golgi apparatus collect in the hyphal tip (Figure 23.31). The vesicles then fuse with the plasma membrane. Some vesicles release enzymes that break down complex materials in the environment into small organic molecules that are taken up as food. Other vesicles deliver cell-wall materials to the hyphal tip, allowing it to extend.

**Variations in Mycelium Growth Form** Fungal hyphae grow rapidly through a substrate from areas where the food has become depleted to food-rich areas. In nature, mycelia may take an irregular shape,

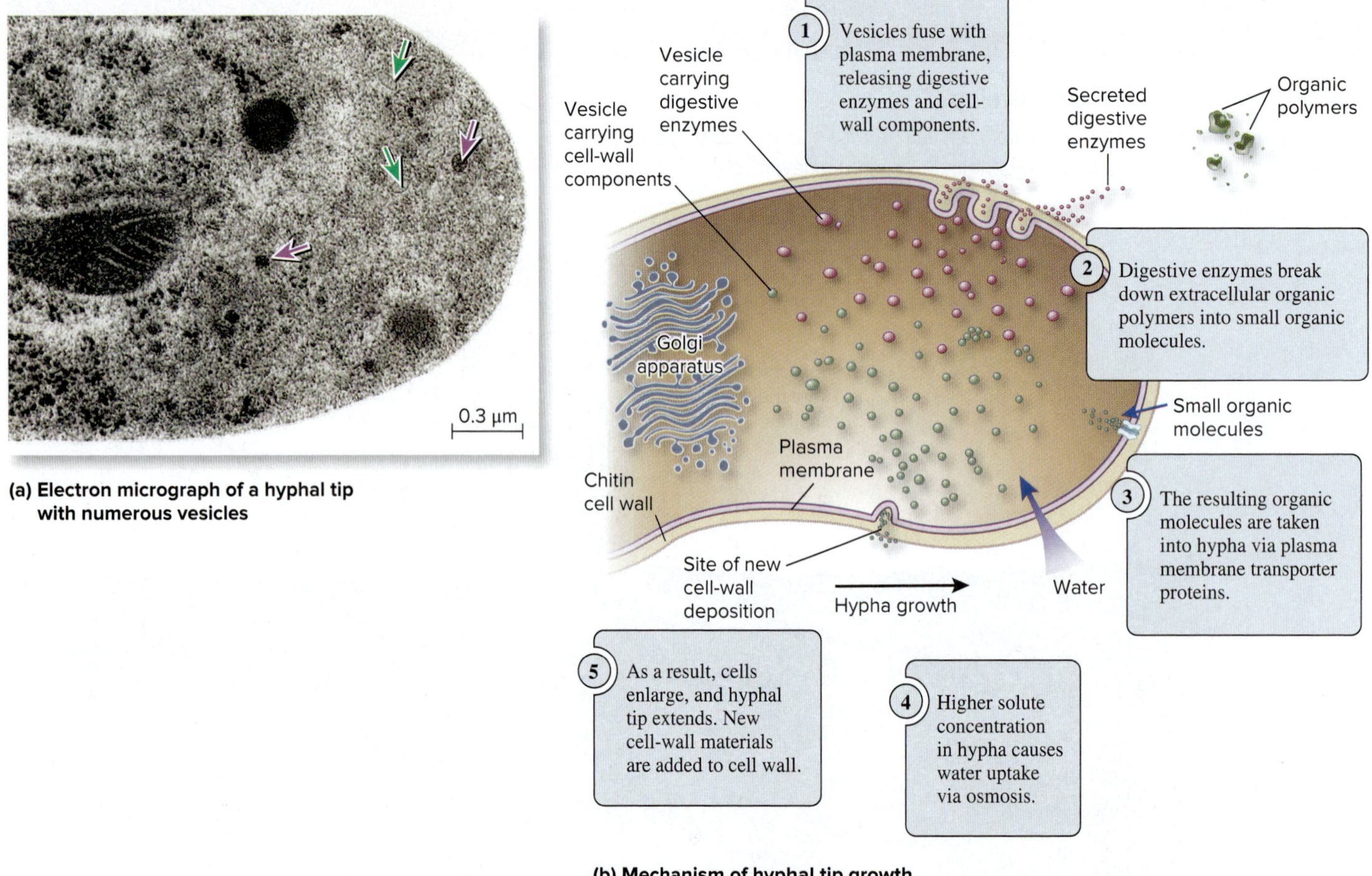


(a) Electron micrograph of a hyphal tip with numerous vesicles

Vesicle carrying cell-wall components

**1** Vesicles fuse with plasma membrane, releasing digestive enzymes and cell-wall components.

Secreted digestive enzymes

Organic polymers

Golgi apparatus

**2** Digestive enzymes break down extracellular organic polymers into small organic molecules.

Small organic molecules

Chitin cell wall

Plasma membrane

**3** The resulting organic molecules are taken into hypha via plasma membrane transporter proteins.

Site of new cell-wall deposition

Hypha growth

Water

**5** As a result, cells enlarge, and hyphal tip extends. New cell-wall materials are added to cell wall.

**4** Higher solute concentration in hypha causes water uptake via osmosis.

(b) Mechanism of hyphal tip growth

**Figure 23.31 Hyphal tip growth and absorptive nutrition. (a)** TEM showing the hyphal tip of *Aspergillus nidulans,* a fungus commonly used as a genetic model organism. The tip is filled with membrane-bound vesicles that fuse with the plasma membrane. Purple arrowheads show dark-stained vesicles carrying digestive enzymes; green arrowheads point out light-stained vesicles carrying cell-wall materials. **(b)** Diagram of a hyphal tip, with vesicles of the same two types, showing the steps of hyphal tip growth.

*(a)* S.G.W. Kaminskyj, and I.B. Heath (1996), "Studies on *Saprolegnia ferax* suggest the general importance of the cytoplasm in determining hyphal morphology," *Mycologia,* 88: 20-37, Fig. 16. Mycological Society of America. Allen Press, Lawrence Kansas

**Concept Check:** *What do you think would happen to fungal hyphae that begin to grow into a substrate with a higher solute concentration? How might your answer be related to food preservation techniques such as drying or salting?*

depending on the distribution of the food substrate. A fungal mycelium may extend into food-rich areas for great distances. In liquid laboratory media, fungi will grow as a spherical mycelium that resembles a cotton ball floating in water (Figure 23.32a). Grown in flat laboratory dishes, the mycelium assumes a more two-dimensional growth form (Figure 23.32b). Sometimes cells or hyphae of different fungal species fuse, forming hybrid mycelia whose traits may differ from those of the parents.

## Fungi Reproduce Asexually by Dispersing Specialized Cells

Asexual reproduction is particularly important to fungi, allowing them to spread rapidly. To reproduce asexually, fungi do not need to find compatible mates or expend resources on fruiting-body formation and meiosis. More than 17,000 fungal species reproduce primarily or exclusively by asexual means. DNA-sequencing studies have revealed that many types of modern fungi that reproduce only asexually have evolved from ancestors that had both sexual and asexual reproduction.

Many fungi produce asexual spores known as **conidia** (from the Greek *konis*, meaning dust) at the tips of hyphae (Figure 23.33). When they land on a favorable substrate, conidia germinate into a new mycelium that produces many more conidia. The green molds that form on citrus fruits are familiar examples of conidial fungi. A single fungus can produce as many as 40 million conidia per hour over a period of 2 days.

Because they can spread so rapidly, asexual fungi are responsible for costly fungal food spoilage, allergies, and diseases. Medically important fungi that reproduce primarily by asexual means include the athlete's foot fungus (*Epidermophyton floccosum*) and the infectious yeast (*Candida albicans*). **Yeasts** are unicellular fungi of various lineages. Asexual reproduction in some yeasts occurs by budding, the production of a new cell on the surface of a larger parental cell (see Figure 23.4a).

**(a) Mycelium growing in liquid medium**

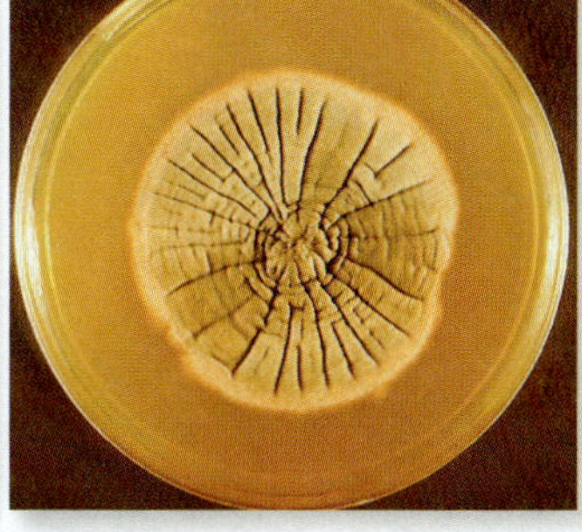

**(b) Mycelium growing on flat, solid medium**

**Figure 23.32 Fungal shape shifting. (a)** When a mycelium, such as that of this *Rhizoctonia solani,* is surrounded by food substrate in a liquid medium, it will grow into a spherical form. **(b)** When the food supply is limited to a two-dimensional supply, as shown by *Neotestudina rosatii* in a laboratory dish, the mycelium will form a disc. Likewise, distribution of the food substrate determines the mycelium shape in nature.

*(a)* © Agriculture and Agri-Food Canada, Southern Crop Protection and Food Research Centre, London ON; *(b)* Source: CDC

**Figure 23.33 Asexual reproductive cells of fungi.** SEM of the asexual spores (conidia) of *Aspergillus versicolor,* which causes skin infections in burn victims and lung infections in AIDS patients. Each of these small cells is able to detach and grow into an individual that is genetically identical to the parent fungus and so is able to grow in similar conditions.

© Dennis Kunkel Microscopy, Inc./Phototake

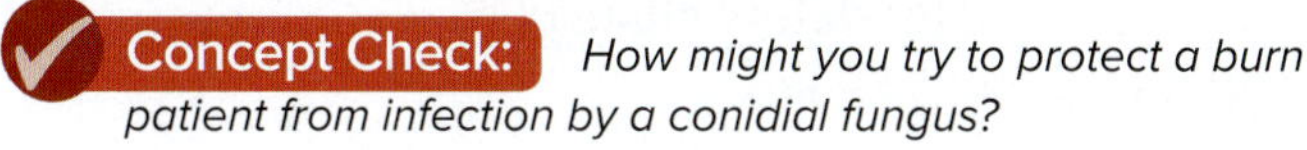

**Concept Check:** *How might you try to protect a burn patient from infection by a conidial fungus?*

## Fungi Have Distinctive Sexual Reproductive Processes

As is typical for eukaryotes, the fungal sexual reproductive cycle involves the union of gametes, the formation of zygotes, and the process of meiosis. In contrast to plants, whose life cycle is an alternation of haploid and diploid generations, and diploid-dominant animals, the fungal life cycle is mostly haploid-dominant, as is the life cycle for many protists.

**Fungal Gametes and Mating** Early-diverging fungi that live in the water produce flagellate sperm that swim to nonmotile eggs, as do animals and many protists and plants. By contrast, the gametes of terrestrial fungi are cells of hyphal branches rather than distinguishable male and female gametes. Fungal mycelia occur in multiple mating types that differ biochemically, controlled by particular genes. During fungal sexual reproduction, hyphal branches of different but compatible mycelia are attracted to each other by secreted peptides, and when hyphae have grown sufficiently close, they fuse. The mating of hyphal branches represents an adaptation to terrestrial life.

**Fruiting Bodies** As we have seen, under appropriate environmental conditions, such as seasonal change, a mated mycelium may produce a fruiting body such as a mushroom (see Figure 23.30). Fruiting bodies typically disperse haploid spores that can grow into haploid mycelia. If a haploid mycelium encounters hyphae of an appropriate mating type, hyphal branches will fuse and start the sexual cycle over again. The structures of fruiting bodies vary in ways that reflect different adaptations that foster spore dispersal by wind, rain, or animals.

- Mature puffballs have delicate surfaces upon which just a slight pressure causes the spores to puff out into wind currents (**Figure 23.34a**). Birds' nest fungi form characteristic, egg-shaped spore clusters. Raindrops splash on these clusters and disperse the spores.
- The fruiting bodies of stinkhorn fungi smell and look like rotting meat (**Figure 23.34b**), which attracts carrion flies. The flies land on the fungi to investigate the potential meal and then fly away, in the process dispersing spores that stick to their bodies.
- The fruiting bodies of fungal truffles (**Figure 23.34c**) are unusual in being produced underground. Truffles have evolved a spore dispersal process that depends on animals that eat fungi. Mature truffles emit an odor that attracts wild pigs and dogs, which break up the fruiting bodies while digging for them, thereby dispersing the spores. Collectors use trained leashed pigs or dogs to locate valuable truffles from forests for the market.

Many fungal fruiting bodies such as truffles and morels are edible, and several species of edible fungi are cultivated for human consumption. However, the bodies of many other fungi produce toxic substances that may deter animals from consuming them. For example, several fungi that attack stored grains, fruits, and spices produce aflatoxins that cause liver cancer and are a major health concern worldwide. The forest mushroom *Amanita virosa* (**Figure 23.35a**), known as the "destroying angel," contains a powerful toxin that can cause liver failure so severe that death may ensue unless a liver transplant is performed. Each year, many people in North America are poisoned when they consume similarly toxic mushrooms gathered in the wild. There is no reliable way for nonexperts to distinguish poisonous from nontoxic fungi; it is essential to receive instruction from an expert before foraging for mushrooms in the woods. Therefore,

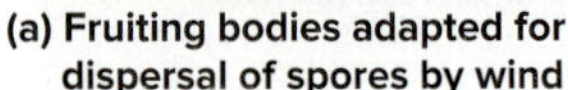

(a) Fruiting bodies adapted for dispersal of spores by wind

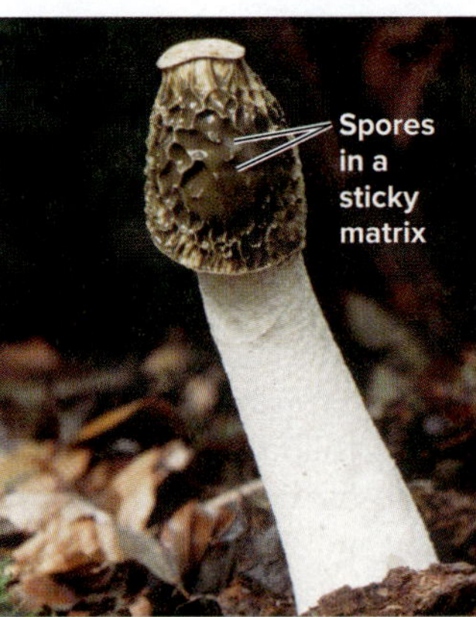


(b) Fruiting body adapted for dispersal of spores by insects

(c) The black truffle *Tuber melanosporum*, an ascomycete fungus, highly valued for its taste.

**Figure 23.34** **Fruiting body adaptations that foster spore dispersal.** **(a)** When disturbed by wind gusts or animal movements, spores puff from fruiting bodies of the puffball fungus (*Lycoperdon perlatum*). **(b)** The fruiting bodies of stinkhorn fungi, such as this *Phallus impudicus,* smell and look like dung or rotting meat. This attracts flies, which come into contact with the sticky fungal spores, thereby dispersing them. **(c)** The black truffle *Tuber melanosporum,* an ascomycete fungus, produces fruiting bodies underground. The smell of mature truffle fruiting bodies attracts animals that dig them up, thereby dispersing spores.

*(a)* © Chris Howes/Wild Places Photography/Alamy; *(b)* © Bob Gibbons/ardea.com; *(c)* © Nacivet/Getty Images

(a) *Amanita*

(b) *Claviceps*

**Figure 23.35** **Toxic or hallucinogenic fruiting bodies.** **(a)** Common in conifer forests, *Amanita muscaria* is both toxic and hallucinogenic. Ancient people used this fungus to induce spiritual visions and to reduce fear during raids. This fungus produces a toxin, amanitin, which specifically inhibits RNA polymerase II of eukaryotes. **(b)** The fungus *Claviceps purpurea* infects rye and other grasses, producing hard masses of mycelia known as ergots in place of some of the grains (fruits). Ergots such as the one illustrated produce alkaloids related to LSD and thus cause psychotic delusions in humans and animals that consume products made with infected rye.

*(a)* © MyLoupe/UIG CALC/Universal Images Group/Getty Images; *(b)* © imageBROKER/Superstock RF

**BioConnections:** *Refer back to Figure 12.11, which illustrates the cellular role of RNA polymerase II in eukaryotes. What effect would the amanitin toxin have on human cells?*

many authorities recommend that it is better to search for mushrooms in the grocery store than in the wild.

Several types of fungal fruiting bodies produce hallucinogenic or psychoactive substances. As in the case of fungal toxins, fungal hallucinogens may have evolved as herbivore deterrents, but humans have inadvertently experienced their effects. For example, *Claviceps purpurea*, which causes a disease of rye crops and other grasses known as ergot, produces a psychogenic compound related to LSD (lysergic acid diethylamide) (**Figure 23.35b**). Some experts speculate that cases of hysteria, convulsions, infertility, and a burning sensation of the skin that occurred in Europe during the Middle Ages and that were attributed to witchcraft resulted from ergot-contaminated rye used in foods. Another example of a hallucinogenic fungus is the "magic mushroom" *Psilocybe*, which is used in traditional rituals in some cultures. Like ergot, the magic mushroom produces a compound similar to LSD. Consuming hallucinogenic fungi is risky because the amount used to achieve psychoactive effects is dangerously close to a poisonous dose.

## Decomposer and Predatory Fungi Play Important Ecological Roles

Decomposer fungi are essential components of the Earth's ecosystems. Together with bacteria, they decompose dead organisms and wastes, preventing the buildup of organic debris in ecosystems. For example, only certain bacteria and fungi can break down cellulose and lignin, the main components of wood. Decomposer fungi and bacteria are Earth's recycling engineers. They release $CO_2$ into the air and other minerals into the soil and water, making these essential nutrients available to plants and algae.

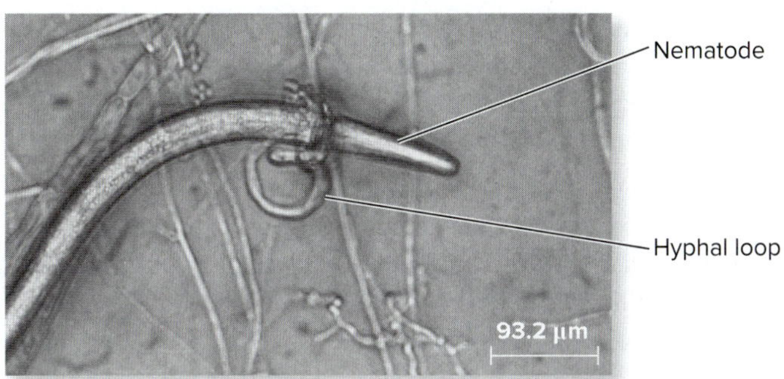


**Figure 23.36** **A predatory fungus.** The fungus *Arthrobotrys anchonia* traps nematode worms in hyphal loops that suddenly swell in response to the animal's presence. Fungal hyphae then grow into the worm's body and digest it.
© Science Source

More than 200 species of predatory soil fungi use special adhesive or nooselike hyphae to trap tiny soil animals, such as nematodes, and absorb nutrients from their bodies (Figure 23.36). Such fungi help to control populations of nematodes, some of which attack plant roots. Other fungi obtain nutrients by attacking insects, and certain of these species have been used as biological control agents to kill black field crickets, red-legged earth mites, and other pests.

## Pathogenic Fungi Cause Plant and Animal Diseases

One of the most important ways in which fungi affect humans is by causing diseases of wild and crop plants and animals. Five thousand fungal species are known as plant pathogens because they cause serious crop diseases. Pathogenic fungi use the organic compounds absorbed from plants to grow, attack additional plant cells, and produce reproductive spores capable of infecting more plants. Wheat rust is an example of a common crop disease caused by fungi (Figure 23.37). Rusts are named for the reddish spores that emerge from the surfaces of infected plants. Agricultural customs inspectors closely monitor the entry of plants, soil, foods, and other materials that might harbor plant pathogenic fungi.

Athlete's foot and ringworm are common human skin diseases caused by several types of fungi that are known as dermatophytes because they colonize the human epidermis. *Pneumocystis jiroveci* and *Cryptococcus neoformans* are fungal pathogens that infect individuals with weakened immune systems, such as those with AIDS, sometimes causing death. *Blastomyces dermatitidis*, which causes the disease blastomycosis; *Coccidioides immitis*, the cause of coccidioidomycosis; and *Histoplasma capsulatum*, the agent of histoplasmosis, are fungal pathogens that affect the lungs and may spread to other parts of the body, causing severe illness. Though fungal diseases that attack humans and domesticated animals are of societal concern, in nature, fungal pathogens play an important ecological role in controlling population growth of other organisms.

## Fungi Form Beneficial Associations with Other Species

Symbioses are close associations of two or more species, and mutualism is a symbiotic interaction in which all partners in the association benefit. Fungi form several types of mutualisms with animals, plants, algae, bacteria, and even viruses. In the next chapter we will see how fungi occur in microbial systems such as the associations with plant roots known as mycorrhizae, and with bacteria and algae to form lichens.

## Biology Principle

### Biology Affects Our Society

Recent analyses indicate that environmental changes linked to human activities correlate with increases in the incidence of new fungal pathogen infestations that threaten human health and agricultural sustainability.

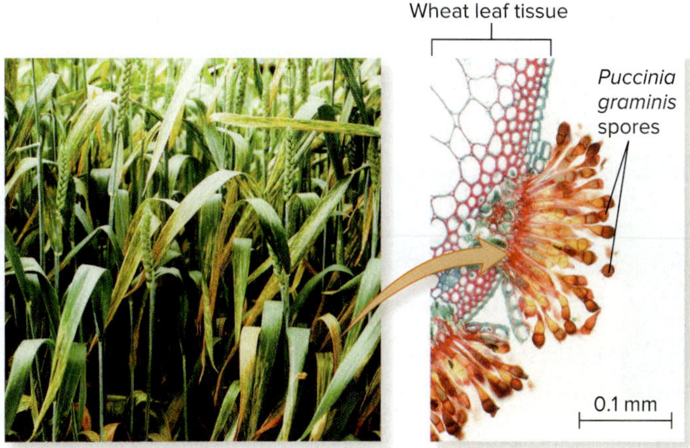


**Figure 23.37** **Wheat rust.** The plant pathogenic fungus *Puccinia graminis* grows within the tissues of wheat plants, using plant nutrients to produce rusty streaks of red spores that erupt at the stem and leaf surface where spores can be dispersed. Red spore production is but one stage of a complex life cycle involving several types of spores. Rusts infect many other crops in addition to wheat, causing immense economic damage.
© Nigel Cattlin/Science Source; *(Inset)* © Herve Conge/ISM/Phototake

## 23.6 Reviewing the Concepts

- Fungi form a monophyletic kingdom that is part of the eukaryotic supergroup Opisthokonta (Figure 23.28).
- Major fungal lineages include early-diverging cryptomycota (including microsporidia) and chytrids, and later-diverging AM fungi, ascomycetes, and basidiomycetes (Figure 23.29).
- Most fungi produce filamentous hyphae that have absorptive nutrition and form bodies known as mycelia that can take different shapes (Figure 23.30, 23.31, 23.32).
- Fungi spread rapidly in the environment by means of asexual conidia and sexual spores produced in fruiting bodies of diverse types (Figure 23.33, 23.34, 23.35).
- Fungi play important roles in nature as decomposers, predators, and pathogens, and by forming beneficial associations with other organisms (Figures 23.36, 23.37).

## 23.6 Testing Your Knowledge

1. How does asexual reproduction benefit fungi?
   a. Asexual reproduction allows fungi to spread rapidly in a favorable environment without expending resources required for mating and the formation of fruiting bodies.
   b. Asexual reproduction fosters genetic diversity.
   c. Asexual reproduction allows fungi to produce fruiting bodies in which meiosis occurs.
   d. Asexual reproduction does not benefit fungi.
   e. None of the above is correct.

## 23.7 Technological Applications of Microorganisms

### Learning Outcome

1. List some examples of ways in which bacteria, protists, and fungi are used in human technology.

Bacteria, protists, and fungi are used in diverse technological applications.

- The food industry uses bacteria to produce chemical changes in food that improve consistency or flavor. Cheese makers add pure cultures of certain bacteria to milk. The bacteria consume milk sugar (lactose) and produce lactic acid, which aids in curdling the milk.
- The chemical industry produces enzymes, vinegar, amino acids, vitamins, insulin, vaccines, antibiotics, and other useful pharmaceuticals by growing particular bacteria in giant vats. For example, the hot springs bacterial species *Thermus aquaticus* is a source of a form of DNA polymerase widely used in biology laboratories to amplify DNA in polymerase chain reaction (PCR). Industrially grown bacteria produce the antibiotics streptomycin, tetracycline, kanamycin, gentamycin, bacitracin, polymyxin-B, and neomycin.
- Vast accumulations of the silica-rich walls of ancient diatoms, known as diatomite or diatomaceous earth, are mined for use in reflective paint and other industrial products.
- Diverse, photosynthetic brown algae (Phaeophyceae) are sources of polysaccharides known as alginates that are used to make many industrial products.
- Many red algae are harvested for extraction of food industry emulsifiers known as carageenan and for production of agar and agarose that are widely used in biology laboratories.
- Bacterial and algal cells can be used as chemical factories. By genetically modifying microbial genomes, microorganisms can be made to produce particular useful compounds, including pharmaceuticals and renewable biofuels.
- The ability of some microorganisms to break down organic compounds, precipitate metals from mine waste, or absorb water pollutants such as phosphorus makes them very useful in treating wastewater, industrial discharges, and oil spills. This process, known as bioremediation, is used to reduce levels of harmful materials in the environment.
- A variety of industrial processes use fungi to convert inexpensive organic compounds into valuable materials such as citric acid used in the food industry; glycerol; antibiotics such as penicillin; and cyclosporine, a drug widely used to prevent rejection of organ transplants.
- Enzymes extracted from fungi are used to break down plant materials for renewable bioenergy production.
- In the food industry, fungi are used to produce the distinctive flavors of blue cheese and other cheeses.
- Other fungi secrete enzymes that are used in the manufacture of protein-rich tempeh and other food products from soybeans.
- The brewing and winemaking industries find yeasts essential, and the baking industry depends on the yeast *Saccharomyces cerevisiae* for bread production.
- *S. cerevisiae* is also widely used as a model organism for fundamental biological studies. Yeasts are useful in the laboratory because they have short life cycles, they are easy and safe for lab workers to maintain, and their genomes are similar to those of animals.

## 23.7 Reviewing the Concepts

- Many bacteria, protists, and fungi are useful in industrial and other applications to make food products, medicines, or renewable biofuels, or to clean up polluted environments.

## 23.7 Testing Your Knowledge

1–4. Match the microbes listed below with an appropriate technological application.

1. the stramenopiles known as diatoms
2. *Bacillus thuringiensis*
3. *Thermus aquaticus*
4. red algae

a. source of genes for Bt toxin that kills insect pests
b. source of commercial DNA polymerase
c. ingredient in reflective paint
d. agar is extracted for use in microbiology labs

# Assess and Discuss

## Test Yourself

1. The bacterial phylum that typically produces oxygen gas as the result of photosynthesis is
   a. the proteobacteria.
   b. the cyanobacteria.
   c. the Gram-positive bacteria.
   d. all of the above.
   e. none of the above.

2. The Gram stain is a procedure that microbiologists use to
   a. determine if a bacterial strain is a pathogen.
   b. determine if a bacterial sample can break down oil.
   c. infer the structure of a bacterial cell wall and bacterial response to antibiotics.
   d. count bacteria in medical or environmental samples.
   e. do all of the above.
3. The structures that enable some Gram-positive bacteria to remain dormant for extremely long periods of time are known as
   a. akinetes.
   b. endospores.
   c. biofilms.
   d. lipopolysaccharide envelopes.
   e. pili.
4. What features do protists have in common:
   a. Protists primarily live in wet or moist places.
   b. Protists are eukaryotes that are not classified as fungi, plants, or animals
   c. Protists include algae, protozoa, and species that resemble fungi but are not closely related to them (fungus-like protists)
   d. All of the above.
   e. None of the above.
5. Protists that move around by means of cilia are the
   a. cryptomonad algae.
   b. euglenoids.
   c. ciliate protozoa.
   d. red algae
   e. green algae
6. The protists most closely related to the fungal, plant, and animal kingdoms are
   a. *Nuclearia,* green algae, and choanoflagellates.
   b. *Plasmodium,* diatoms, and Discoba.
   c. *Jakoba,* red algae, and dinoflagellates.
   d. *Phytophthora,* euglenoids, and ciliates.
   e. None of the above.
7. Fungal cells differ from animal cells in that fungal cells
   a. lack ribosomes, though these are present in animal cells.
   b. lack mitochondria, though these occur in animal cells.
   c. have chitin cell walls, whereas animal cells lack rigid walls.
   d. lack cell walls, whereas animal cells possess walls.
   e. none of the above.
8. Protists that lack flagella or cilia are the
   a. cryptomonad algae.
   b. euglenoids.
   c. ciliate protozoa.
   d. red algae
   e. green algae
9. People use bacteria, protists, or fungi to:
   a. make beer and wine.
   b. generate renewable biofuels.
   c. produce antibiotics and other pharmaceuticals.
   d. extract useful enzymes such as DNA polymerase.
   e. All of the above.

## Conceptual Questions

1. Explain why many microbial populations grow rapidly, thereby influencing the rate of food spoilage, infection, and the formation of harmful blooms in aquatic habitats.
2. Explain how a bacterial cell entering through a wound could rapidly spread throughout the body of a plant or an animal.
3. **PRINCIPLES** A principle of biology is that biology affects our society. Why are the endospores produced by certain bacteria important in human life?

## Collaborative Questions

1. When most people think of infectious diseases they may think of disease-causing bacteria or viruses. Make a list of serious human diseases that are caused by protists or fungi. How might your list influence how you would choose antibiotics to treat these diseases?
2. Imagine that you are searching for new antibiotic compounds. Where in nature might be a good place to look?

# 24 Microbiomes: Microbial Systems on and Around Us

Nutrition can affect the microbiomes of infants and young children, which in turn affect proper growth.

## Chapter Outline

Ideally, all of the world's children would have good food sufficient to start healthy lives. Unfortunately, malnutrition has left nearly 180 million children around the world stunted in their growth. Public health experts have tried to use dietary supplements to restore normal growth, a strategy that does not always work. New research on microbes living in the bodies of these children has revealed why. Stunted children retain an infantile set of gut microbes, in contrast to children whose more mature collection of gut microbes stimulates growth hormones. This understanding may help people to devise new ways to improve the health of millions of children.

Chapter 23 introduced the diversity of microorganisms—archaea, bacteria, protists, and fungi—that influence our health and environments in many ways. This chapter builds on that foundation, describing how diverse microbes occur together in microbiomes, complex systems associated with particular environments, such as the human body. Medical aspects of the human microbiome are often in the news. Oceans, ice, fresh waters, soils, and the bodies of other organisms likewise have distinctive microbiomes that affect human life.

Because microbes are generally small, microbiomes are often inconspicuous. New molecular approaches described in this chapter have enabled biologists to study microbiome composition and functions. Using these new methods, medical scientists are identifying new ways in which people can improve their health. Agricultural scientists are discovering new strategies for engineering crop microbiomes to promote plant health. Exploring the microbiomes of other organisms and environments has revealed previously unrecognized global ecological effects. In this chapter, we will examine the properties of microbiomes, which has become an important and fast-growing area of modern biology.

## 24.1 Microbiomes: Diversity of Microbes and Functions

### Learning Outcomes

1. Define microbiome.
2. List some reasons why microbiomes are considered to be complex systems.
3. Discuss how biologists use ribosomal RNA gene sequences and whole metagenomic sequencing to catalog diverse types of microbes in a microbiome.
4. Explain how biologists identify microbiome functions.

Microbes include millions of species of archaea, bacteria, protists, and fungi that play important ecological roles worldwide (see Chapter 23). A **microbiome** is a particular assemblage of microbes and genes that occurs in a defined environment. A biome is defined as a major type of habitat characterized by distinctive life forms; in the case of a microbiome, those life-forms are microscopic organisms (**Figure 24.1**).

### Microbiomes Are Complex Biological Systems

Microbiomes are complex systems, in part, because they often contain thousands of different microbial species that are challenging to identify. Microbiome studies often reveal new types of microbes that have not been studied in the lab and, so, have not even been formally named. Diverse species of bacteria or archaea dominate some microbiomes (as in Figure 24.1), whereas others include many types of eukaryotic protists and fungi, in addition to prokaryotic species (**Figure 24.2**). Some microbiomes also contain a surprising number of microscopic invertebrate animals; viruses may also be common.

These diverse life forms carry out many types of metabolism that influence the environments and other members of the same microbiome. Microbes also communicate with each other by means of chemical or electrical signals that biologists are just beginning to understand. In these ways, microbiomes resemble culturally diverse human groups whose social networks and responses to outside influences are complex. Identifying what species occur together, how microbes affect each other and their environments, and the effects of environmental change are major goals of microbiome research.

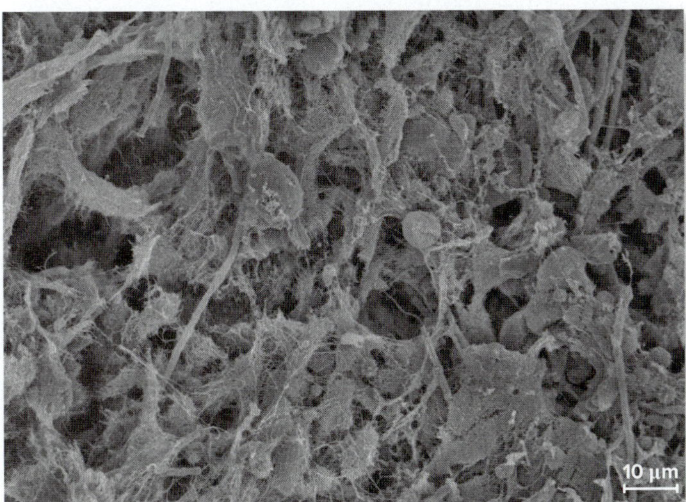


**Figure 24.1 A microbiome dominated by diverse bacterial types.** This microbiome, viewed by SEM, occurs on the surface of a photosynthetic organism. Photosynthesis provides organic molecules and oxygen to members of the microbiome.
© Linda Graham

**Figure 24.2 A complex microbiome that includes diverse prokaryotic and eukaryotic species.** The green alga *Cladophora* functions as the host, providing high surface area, organic molecules, and oxygen to the members of the microbiome. The microbiome forms a biofilm on the surface of this alga (look ahead to Figure 24.9). In addition to tiny bacteria too small to distinguish by light microscopy, this microbiome includes numerous eukaryotic microbes such as diatoms whose brown chloroplasts are visible here.
© Lee W. Wilcox

Determining the species composition, functions, and responses of microbiomes are challenging because microbes are so small and difficult to distinguish. For example, different bacterial species often have similar body structure such as single round or rod-shaped cells about 1 µm in diameter (see Figure 24.1). Likewise, the bodies of millions of fungal species are composed of thin hyphae that often look alike, even with the use of a microscope (see Chapter 23).

For these reasons, biologists commonly use genetic differences to distinguish and identify microbial species and genes present in a complex microbiome, as well as to infer their functions. In addition to cataloging the types of microbes and genes present in a microbiome, biologists examine proteins and metabolites.

### The Complexity of Microbiomes May Be Evaluated by Analyzing Genes That Encode Ribosomal RNA

We will first consider how biologists catalog organisms in a particular microbiome by using gene sequences that encode ribosomal RNA. All of Earth's living things, including microbes, produce proteins by means of ribosomes, which contain ribosomal RNA (rRNA). Ribosomal RNA is so important to living things that rRNA structure tends to be conserved in evolution. Likewise, the sequence of bases of rRNAs within a given species or closely related species is highly conserved. As described in Chapter 10, changes in the sequences of rRNAs can be used to evaluate evolutionary relationships (refer back to Figure 10.13). The genes that encode rRNAs are called rDNAs. Differences in rDNA sequences are used to identify and classify the microbes present in a microbiome, even if it includes thousands of microbial species.

1 Obtain a sample to analyze.

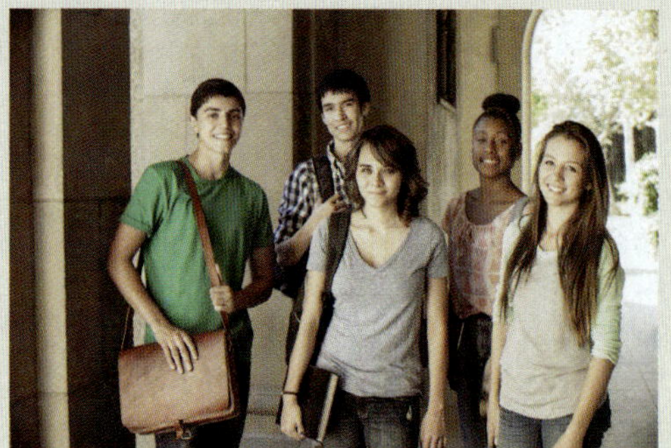

Sample of saliva or fecal sample

Sample of seawater

2 Extract the DNA from the sample. This yields a collection of many DNA fragments from many different microbial species.

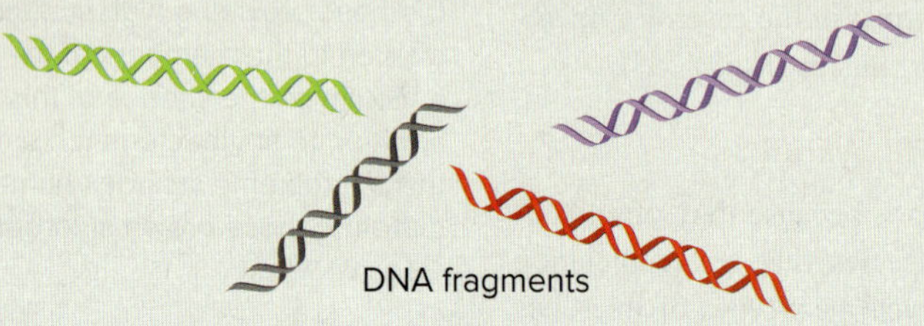


3 **3a Amplicon Analysis:** Use primers that recognize a region of an rDNA gene and make many copies of that region using polymerase chain reaction (PCR). Subject the amplified segments of the *rDNA* gene to DNA sequencing.

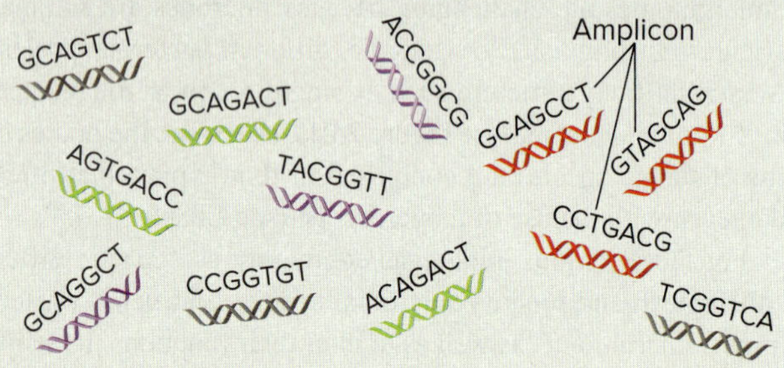


Sequences of rDNA genes that were in the sample.

**3b Whole metagenome sequencing:** Subject all of the DNA in the sample to shotgun DNA sequencing.

GTAGCATCCGGTGTGGTGAG
TCGGTCACCTGACGGTAGCA
GCAGTCTACCGGCTTCGGTC
CCGGTGTGGTGACTTACGGT
GCAGACTTCGGTCACCTGAT
ACCGGCTGCAGACTTCGGTC
TACGGTTGCAGCCTGCAGAA
ACTGACGGCAGGCTGTAGCA

Sequences of all of the DNA fragments that were in the sample.

4 Compare the DNA sequences obtained in step 3 to DNA sequences in a database. The database sequences are already known to come from particular microbial species. This allows researchers to match the sequences obtained in step 3 to known sequences. For example, if a DNA sequence obtained in step 3a matches an rDNA sequence from the bacterium *E. coli*, this result indicates that *E. coli* was in the microbiome of that sample.

**Figure 24.3** Characterization of microbiomes via amplicon analysis or whole metagenome sequencing.

Biologists usually begin a microbiome study by obtaining a sample and then extracting the DNA (Figure 24.3). As seen in step 3a, polymerase chain reaction (PCR) can be used to copy a particular region of an rDNA gene. This process yields many copies of that region from many different species that are in the sample. These copied rDNA regions are known as **amplicons.** Amplifying the DNA is generally needed to generate sufficient DNA for sequencing (described in Chapter 18).

The amplicons are then subjected to DNA sequencing, which yields the base sequence of each rDNA type that was amplified in the original sample. Biologists use computers to compare the DNA sequences of each amplicon to reference sequences in databases. These reference sequences come from microbes, the names and metabolic functions of which are already known. This relatively inexpensive way to examine the microbial diversity in a microbiome is called **amplicon analysis.**

In amplicon analysis, researchers often use PCR primers that amplify a region of the *16S rRNA* gene, which is a particular rDNA gene. 16S rRNA is found in the small ribosomal subunit of bacteria and archaea. 16S rRNA sequences are commonly used to classify prokaryotic species because such genes are distinctive for prokaryotic organisms. Figure 24.4 shows the evolutionary relationships among the members of a microbiome found on the surface of a photosynthetic organism, the same microbiome illustrated by SEM in Figure 24.1. As you can see, the microbiome composition is complex and contains many members that are found within distinct microbial groups.

In eukaryotes, the small ribosomal subunit rRNA is longer and designated 18S rRNA. This 18S rRNA is used to identify and classify eukaryotic microbiome components: protists, fungi, or small animals. Other types of nucleic acid sequences are employed to detect and identify viruses in environmental samples. Taken together, these identification methods reveal the diversity of particular types of microbes present in a microbiome.

## The Complexity of Microbiomes May Also Be Analyzed by Whole Metagenomic Sequencing

An alternative method for characterizing microbial diversity is to obtain base sequences of all the DNA present in a sample, a process known as **whole metagenomic sequencing** (**WMS**) (see Figure 24.3, step 3b). A **metagenome** is defined as the genomes of all the organisms present in a sample. WMS is carried out using an approach known as **shotgun sequencing,** because the process generates many tiny pieces of DNA, reminding people of the tiny metal pellets of shot sprayed by a shotgun blast. This approach does not have to focus on the sequencing of one particular gene, such as the *16S rRNA* gene. Instead, many fragments of DNA are randomly sequenced, and then biologists use computers to identify places where the ends of DNA fragments have the same DNA sequences (Figure 24.5). These overlapping regions allow researchers to align the DNA fragments into longer sequences known as contiguous sequences, or **contigs.** What is the advantage of a contig? One advantage is that longer contigs provide greater amounts of information needed for more detailed classification.

If metagenomic sequences are relatively long, or if a microbiome contains relatively few microbes, it may be possible to use computer methods to assemble contigs into whole microbial genomes (see Chapter 18). A number of microbial species were discovered in this way and even today are known only from their genomic sequence. Some experts consider that one goal of WMS should be to assemble entire microbe genome sequences, a process known as **genome-centric metagenomics.**

If sufficient DNA has been analyzed, WMS and computer methods can be used to identify both prokaryotic and eukaryotic species in a microbiome at the same time. By contrast, amplicon analyses typically focus on the amplification of a particular gene from a selected group of species. For example, researchers may focus on *16S rRNA* gene amplicons, which will only identify many or most of the prokaryotic species in a given sample. For this reason, many experts consider that the term microbiome should be limited to microbial communities characterized by WMS. The term **microbiota** is commonly used to describe collections of microbial life catalogued by limited amplicon analysis, such as when only *16S rDNA* sequence data are used.

## Certain Functions Within Microbiomes Are Determined by Identifying Protein-Encoding Genes

When analyzing microbiomes by WMS, another goal is to find and classify protein-encoding genes, providing a deeper view of microbial function. To gain more information about microbiome function, biologists may look for particular protein-coding genes that indicate specialized microbial functions. Three examples are described next.

**Nitrogen Fixation** One important microbial function is nitrogen fixation, the process in which atmospheric nitrogen gas is reduced to form ammonia, which is useful as crop fertilizer. Only certain prokaryotic species have the natural ability to accomplish nitrogen fixation (see Chapter 23). Crop plants and other photosynthetic organisms require ammonia or another source of fixed nitrogen to make amino acids, chlorophyll, and other essential molecules. To obtain evidence for bacterial species within a microbiome that are able to fix nitrogen, crop scientists may identify gene sequences known to encode enzymes essential for nitrogen fixation. One such gene is *nifH*, an indicator of nitrogen fixation. Such genes are known as **marker genes,** because they "mark" the occurrence of a particular function—in this case, nitrogen fixation.

**Methane Oxidation** Additional marker genes encode subunits of the enzyme methane monooxygenase (MMO). This enzyme uses oxygen gas to oxidize the greenhouse gas methane, important in global carbon cycles and climate warming. A microbiome that contains these genes indicates that a function of the microbiome is to oxidize methane.

**Metabolites** Some microbes produce very specific compounds that are not produced by most species of microbes. These compounds are called metabolites, because they are the products of metabolic pathways. Examples include certain vitamins and toxins. In many cases, previous research has identified the enzymes that are needed to produce a particular metabolite. When a WMS identifies the genes that encode these enzymes, this result indicates that one function of the microbiome is to produce that metabolite. One example of a metabolite is vitamin $B_{12}$, which some bacteria produce and some other organisms require. The presence of genes that encode the enzymes needed to synthesize vitamin $B_{12}$ indicates that a microbiome may provide this service.

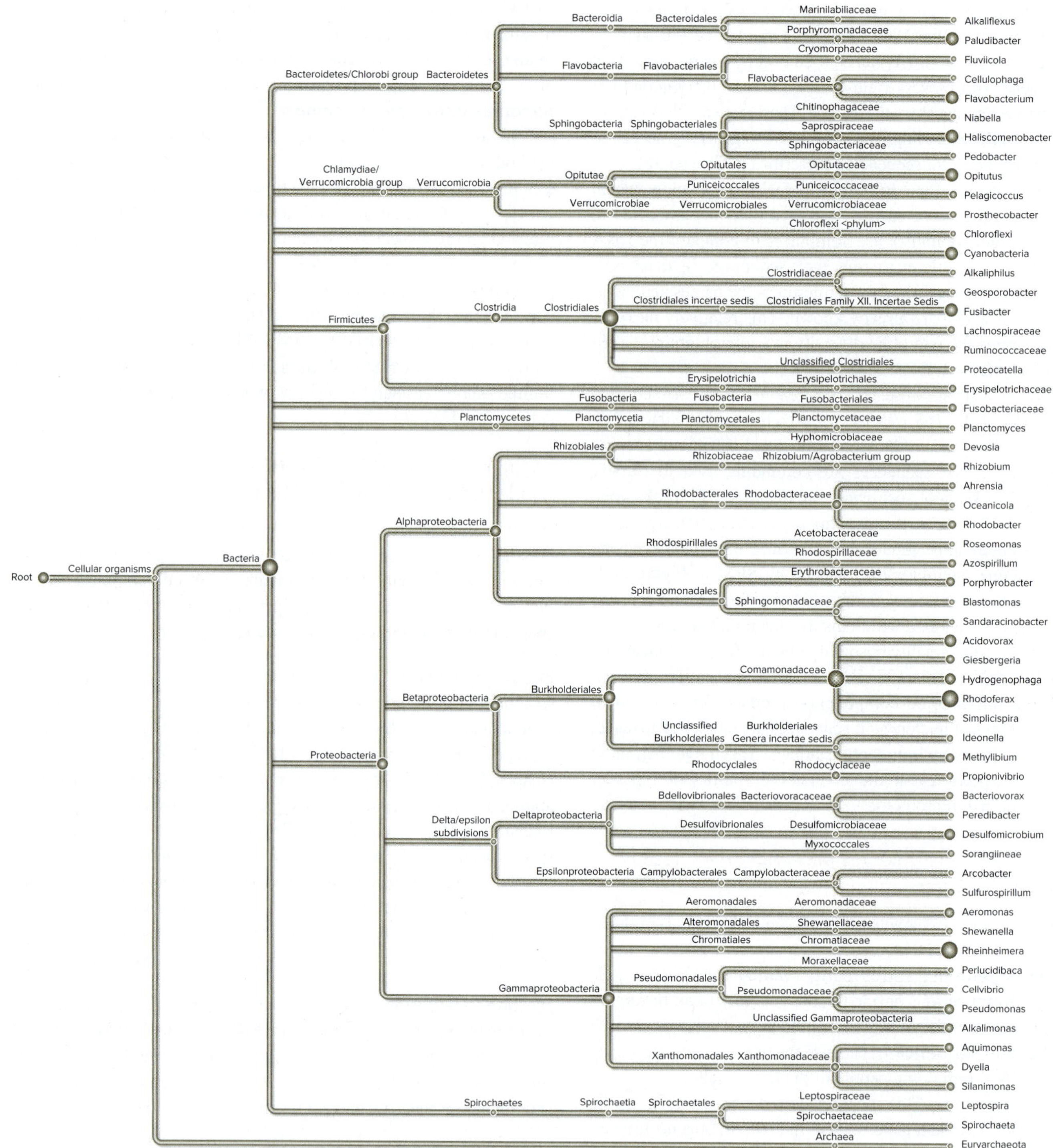


**Figure 24.4 Example of using bacterial *16S ribosomal RNA* gene sequences from the microbiome of the photosynthetic organism shown in Figure 24.1 to produce a phylogenetic tree.** Circle sizes are relative sequence abundances, indicators of the relative abundance of bacterial and archaeal genera.

Source: Graham, L. E., Knack, J. J., Piotrowski, M. J., Wilcox, L. W., Cook, M. E., Wellman, C. H., Taylor, W., Lewis, L. A., Arancibia-Avila, P. (2014), Lacustrine *Nostoc* (Nostocales) and associated microbiome generate a new type of modern clotted microbialite. *Journal of Phycology 50:* 280–291. doi: 10.1111/jpy.12152 

Overlapping region

TTACGGTACCAGTTACAAATTCCAGACCTAGTACC
AATGCCATGGTCAATGTTTAAGGTCTGGATCATGG

GACCTAGTACCGGACTTATTCGATCCCCAATTTTGCAT
CTGGATCATGGCCTGAATAAGCTAGGGGTTAAAACGTA

**Figure 24.5 A comparison of two DNA fragments that contain an overlapping region.** A contig is a series of DNA fragments that contain overlapping regions.

## The Analysis of mRNAs, Proteins, and Metabolites Provides Additional Information About Microbiome Function

Catalogs of microbial species and genes obtained by amplicon analysis or WMS don't reveal which genes were actually being transcribed and which transcripts were being translated at the time a sample was collected.

**Metatranscriptome** To learn which mRNAs were present in a microbiome at the time of sampling, biologists analyze transcriptomes. A **transcriptome** is a collection of all the mRNA sequences produced by a single organism under defined conditions. A **metatranscriptome** is a collection of all the mRNAs present in an environmental sample, that is, all of the mRNAs produced by all of the organisms sampled at a particular time.

**Metaproteome** Biologists sometimes use the number of mRNA sequences of a particular type to infer abundance or activity level of a translated protein. However, mRNA abundance is influenced by the extent to which microbes were actively growing when they were collected, the lifetime of a particular transcript, and how often that transcript is translated. A proteome analysis can provide more direct information about what proteins are present in a particular microbiome. A proteome analysis, accomplished by chromatographic and spectroscopy methods, reveals which proteins are present in a particular sample. The **metaproteome** is all the proteins produced by all the members of a microbiome.

**Meta-metabolome** Because proteins, even if present, might not be functionally active, researchers may analyze the products of metabolism. **Metabolomes** are collections of information about the types and abundances of molecules produced by metabolism, such as sugars or fatty acids, by a single organism. A **meta-metabolome** provides similar information for an entire microbiome.

### 24.1 Reviewing the Concepts

- Microbiomes are complex biological systems that are difficult to characterize using a microscope alone (Figures 24.1, 24.2).
- The members of microbiomes are usually identified using genetic techniques such as amplicon analysis or whole metagenomic sequencing (WMS)(Figures 24.3–24.5).
- The analysis of protein-encoding genes can reveal important functions of microbiomes.
- Additional information regarding microbiome function can be obtained by characterizing the metatranscriptome, metaproteome, and meta-metabolome, which are all the types of RNAs, proteins, and metabolic products, respectively, that are present in a microbiome.

### 24.1 Testing Your Knowledge

1. A key difference between an amplicon analysis and whole metagenome sequencing (WMS) is that
   a. only an amplicon analysis involves DNA sequencing.
   b. only WMS involves DNA sequencing.
   c. an amplicon analysis focuses on specific genes, such as an rDNA gene, in a microbiome, whereas WMS involves all genes.
   d. a WMS focuses on specific genes, such as an rDNA gene, in a microbiome, whereas an amplicon analysis involves all genes.
   e. both a and c.
2. All of the proteins produced by a microbiome is called a
   a. transcriptome.
   b. metatranscriptome.
   c. metaproteome.
   d. meta-metabolome.
   e. both b and c.

## 24.2 Microbiomes of Physical Systems

### Learning Outcomes

1. List some types of microbiomes that are found within physical systems.

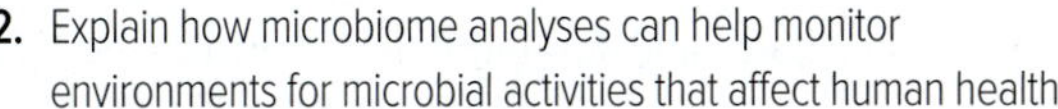

2. Explain how microbiome analyses can help monitor environments for microbial activities that affect human health.

Thus far, our discussion of microbiomes as complex systems, and our survey of the methods that biologists use to study microbiome composition and function, provides the foundation for surveying the diversity of Earth's microbiomes. Some microbiomes are found within **physical systems,** which include oceans, ice, fresh waters, and soils. For example, biologists may study the microbiome in a sample of seawater or in a block of ice. Alternatively, other microbiomes are associated with living organisms known as hosts. In this section, we will consider microbiomes within physical systems, and then Section 24.3 will explore their occurrence within or on the surfaces of hosts.

## Microbiomes Are Abundant in the World's Oceans and in Ice

Although you might imagine that animals such as fish and whales dominate Earth's oceans, in fact, microbes are far more numerous. Oceans occupy 71% of Earth's surface and a volume of 1.37 billion cubic kilometers, with a concentration of $10^4$–$10^6$ microbial cells per mL; the number of microbes in 1liter of seawater reaches into the billions. That's a lot of microbes!

Together, ocean microbes represent an immense amount of genetic and functional diversity that influences the entire planet. For example, floating or swimming cyanobacteria and algae (see Chapter 23) are so abundant in Earth's oceans that these photosynthetic microbes produce about half of Earth's annual organic carbon and oxygen. Other ocean microbes play essential roles in degrading organic molecules and recycling dissolved minerals, a process essential to ocean productivity. The cyanobacterial genus *Synechococcus* and its phages, together with stramenopile, alveolate, and rhizarian protists (see Chapter 23), are key to the movement of organic molecules to deep-ocean waters. Retention of carbon in the deep ocean for long periods affects global climate and is the mechanism by which extensive ocean oil and methane (fossil-fuel) deposits form.

Biologists have recently used genetic techniques (described in Section 24.1) to catalog viruses, prokaryotic species, and small eukaryotes from 68 ocean locations worldwide. By also monitoring the physical features of these places, they have discovered that water temperature is a major factor influencing the composition of ocean microbiomes, raising questions about the impact of global climate change.

Earth's icy environments—collectively known as the **cryosphere**—likewise contain a surprisingly large number of microbes, an estimated $10^{25}$–$10^{28}$ cells. The sampling of microbes from sea ice and glaciers is challenging, so adventurous biologists use ice-breaking ships, helicopters, planes, tractors, drilling rigs, and remotely operated vehicles to access polar oceans, sea ice sheets, snow, and glaciers (Figure 24.6a). Biologists are intrigued by the possibility that Earth's cold microbiomes might be similar to life on icy planetary bodies such as Mars, Jupiter's moon Europa, and Saturn's moon Enceladus.

Microbiome studies of cold habitats have revealed surprisingly diverse types of microbiota that colorize their otherwise white environments. Beneath floating sheets of sea ice live conspicuous brown diatom growths that dangle into the cold ocean. These photosynthetic algae supply organic molecules and oxygen to heterotrophic bacteria; ciliate, flagellate, and foraminiferan protists (see Chapter 23); and small animals. Algal cells on the surfaces of glaciers can be so abundant that they color the ice green, red, yellow, purple, or gray, in this way influencing the amount of energy that glaciers reflect into space (Figure 24.6b). Blood Falls in Antarctica gets its dramatic red color from dissolved iron released from subsurface minerals by iron-metabolizing bacteria living in the cold darkness (Figure 24.6c).

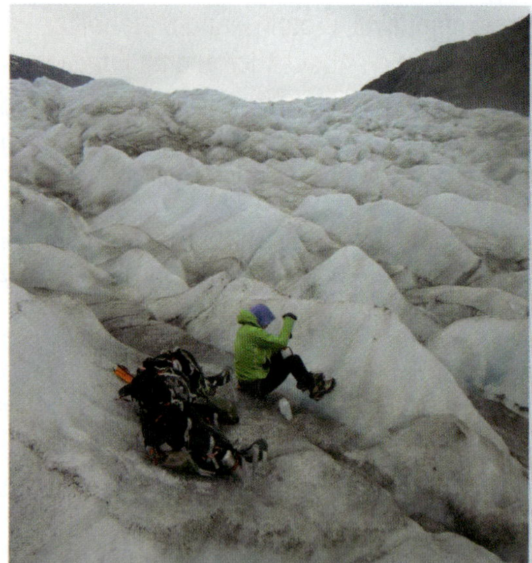

**(a) Researchers sampling the cryosphere**

**(b) A glacier colored by algae in ice microbiomes**

**(c) Blood Falls in Antarctica colored red by iron released by bacteria**

**Figure 24.6 Microbiomes within and on ice.** **(a)** Researchers obtaining samples of ice for microbiome analysis. **(b)** The color of the ice seen here is due to the presence of algae. **(c)** Blood Falls in Antarctica. The red color is from iron that is released by bacteria, which are part of a microbiome within the ice.

*(a)* Courtesy Cody S. Sheik; *(b)* © Jason Edwards/National Geographic Creative *(c)* Source: Peter Rejcek, NSF

## Microbiomes Are Found in Fresh Water and Within the Soil

Freshwater and soil microbiomes are important to diverse human concerns, including drinking water safety and agricultural production. *16S rDNA* and other genes are commonly analyzed to detect infectious or toxic organisms in the water used for drinking and recreation. Experts are particularly concerned about the effects of global climate warming on freshwater and soil cyanobacteria, because these organisms grow abundantly in warmer temperatures, particularly where humans have polluted environments with excess minerals.

Some abundant cyanobacteria produce persistent and potent toxins that harm people and wildlife. For example, the cyanobacterial genus *Microcystis* (Figure 24.7) produces more than 100 different chemical forms of the toxin **microcystin,** which binds to and inhibits eukaryotic phosphatases, enzymes that cells use to remove phosphate from organic compounds. The inability to remove phosphate interferes with many cellular processes, including cell signaling (see Chapter 8). By this mechanism, microcystin can cause severe liver damage and hemorrhage in mammals, and chronic exposure to microcystin is associated with high rates of human liver cancer. Humans and other animals become exposed to microcystin when large numbers of cyanobacteria occur in water used for drinking or swimming, a growing public health concern. Ribosomal DNA amplicon analyses are increasingly used to monitor aquatic microbiomes for cyanobacterial species likely to produce toxins. Because the genes encoding all enzymes required for the biosynthesis of microcystin and other cyanobacterial toxins are known, metagenomic data can be analyzed for the presence of these genes and thereby detect the potential for toxin production.

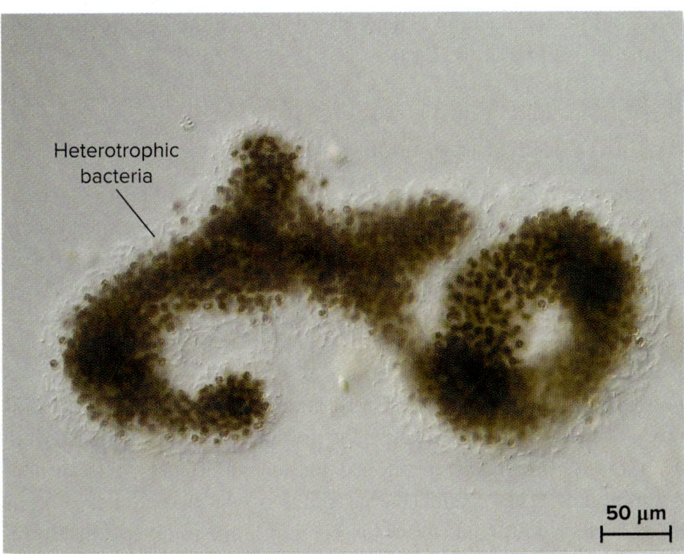


**Figure 24.7 The colony-forming cyanobacterial genus *Microcystis* with a colorless microbiome containing heterotrophic bacteria on its surface.** *Microcystis* cells look dark because they contain many light-refracting gas vesicles that aid in floatation. *Microcystis* and its microbiome produce toxins that may occur in drinking water and are harmful to human health.
© Lee W. Wilcox

Soil microbiomes are key to our ability to grow crops, because various soil microbes foster or decrease plant health (see Chapter 23). A single gram of soil contains as many as 50,000 bacterial species and diverse fungi, many of which affect plants in some way. Soil microbiomes are important for understanding how terrestrial plants acquire microbiomes, described in Section 24.3.

## 24.2 Reviewing the Concepts

- Seawater serves as a physical system for microbiomes, which include important photosynthetic microbes such as cyanobacteria and algae.
- Biologists also explore the microbiomes found within ice. These microbiomes can affect the color of ice and its ability to absorb light (Figures 24.5, 24.6).
- Microbiomes are also found in freshwater habitats and may elicit human health concerns, such as toxins produced by *Microcystis* (Figure 24.7).
- The soil provides a physical system for microbiomes, which are known to have a great impact on the growth of crops.

## 24.2 Testing Your Knowledge

1. Which of the following is *not* a microbiome of a physical system?
   a. a microbiome in a block of ice
   b. a microbiome in the human gut
   c. a microbiome in a soil sample
   d. a microbiome in a sample of seawater
   e. All of the above are examples of microbiomes of physical systems.

# 24.3 Host-Associated Microbiomes

### Learning Outcomes

1. Define the terms holobiont and hologenome.
2. Describe a few examples in which animal and plant hosts acquire microbiomes.
3. Make a list of the benefits that are derived from the associations between microbiomes and their hosts.
4. **SCISKILLS ▶** Analyze the evidence that host microbiomes are influenced by both host genetics and environment and play a role in human health.

Many people have heard news media stories about the human microbiome, which includes thousands of different microbial species that live in various locations within and on the surface of the human body. Consequently, humans are said to be microbiome **hosts.** The human body serves as an environment for such microbes, some of which provide us with nutritional or protective benefits. Our microbes, in turn, receive benefits from us, such as organic molecules that serve as microbial food. Likewise, the bodies of other animals, plants, fungi, protists, and even prokaryotes support complex microbiomes that promote both host and microbial survival. In this section, we will examine the composition and functional properties of host-associated microbiomes.

## Microbiomes Extend Host Capabilities and Are Acquired by Their Hosts in Different Ways

The combination of a host organism and its microbiome is known as a **holobiont.** The host and microbiome genomes together are called the **hologenome.** Microbiomes contribute many more genes to the hologenome than their hosts. For example, the human genome contains about 22,000 protein-encoding genes, whereas the human microbiome is estimated to have a few million! In this way, microbiomes extend host metabolic capabilities and influence host evolution. Chemical signals produced by the host act on particular microbes that serve as information hubs, transmitting information to the broader microbial community. Microbiomes thereby function as complex biological networks.

Having a functionally useful microbiome aids the survival of the young and thereby increases fitness. The acquisition of a microbiome by its host can occur in a variety of ways.

- Certain insects coat the casings of their eggs with bacteria; when the young hatch, they become inoculated with beneficial microbes.
- Newborn bees get their microbiomes from sibling worker bees.
- Mammals, including humans, transmit important microbes as the young transit the birth canal.
- Termites use specific behaviors to transfer among themselves microbes they need to break down plant materials into food.
- Plant seedlings acquire their microbiomes from the surrounding soil and air but use inherited mechanisms, often secretion of particular organic compounds, to attract beneficial microbes.

Both host genetics and the environment are important for the types of microbiomes that are acquired by the host. For example, closely related tropical plants have microbiomes more similar to each other than to the microbiomes of more distantly related plant species, providing evidence that host genetics is important for sustaining microbiomes. This phenomenon is also illustrated by studies of the gut microbiomes of certain primates, including African apes (chimpanzees, bonobos, and gorillas) and their close relative, humans (see Chapter 22). The gut microbiomes of African apes resemble human microbiomes in particular ways (**Figure 24.8**). This observation is consistent with the idea that African apes and humans possess common genes and have some common dietary factors that support the growth of certain microbial species within their gut microbiomes.

However, after humans diverged from African apes, the African-ape diet remained plant-rich, whereas the human diet became increasingly animal-rich. These dietary differences, in turn, affected the environment of gut microbiomes, which were exposed to different types of nutrients. Compared to African apes, some human-associated microbes flourished under an animal-rich diet, whereas others were lost. For example, humans have more gut microbes from the Bacteroidetes phylum, such as the genus *Bacteroides*. This feature is associated with diets rich in animal fats and proteins. In contrast to apes, humans have lower numbers of the archaeon genus *Methanobrevibacter*, which degrades complex plant polysaccharides to methane. Humans also have lower amounts of *Fibrobacter*, a bacterial genus common in ape microbiomes, where it helps to break down the plant foods that apes consume. Overall, these results indicate that microbiome composition is determined by both the genetic and environmental factors of the host.

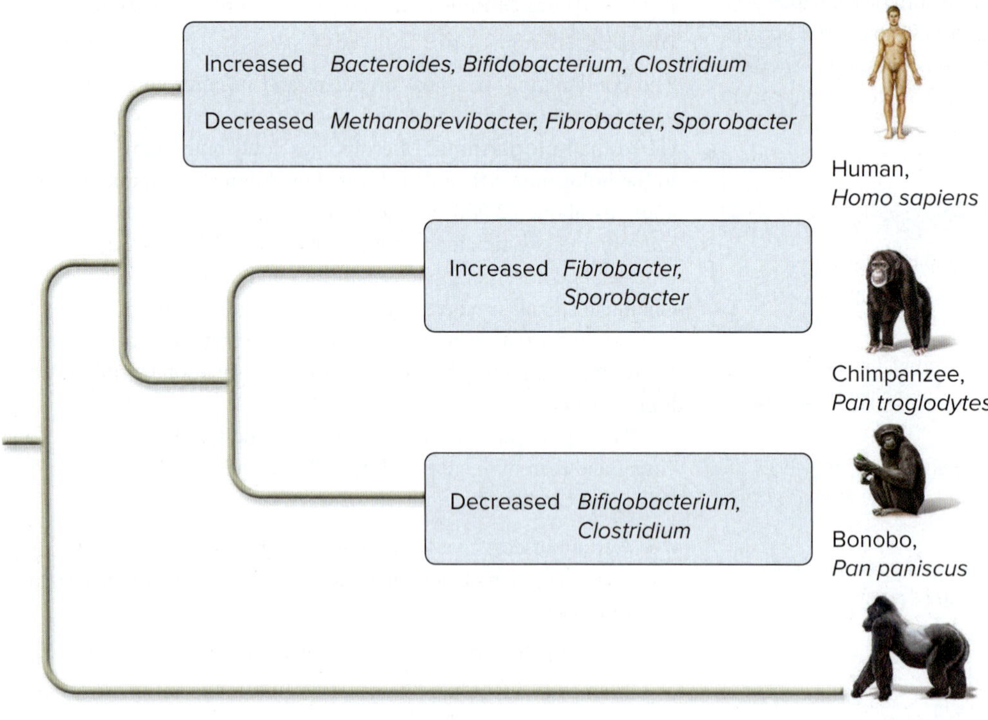


**Figure 24.8 Changes in gut microbiomes during the evolution of African apes and humans.** The changes seen in humans, chimpanzees, and bonobos are relative to the gut microbiome of gorillas.
Source: Proc Natl Acad Sci U S A. 2014 Nov 18; 111(46): 16431–16435

## Bacterial, Protist, and Fungal Microbiomes Perform Important Functions

Thus far, we have considered some general properties of host-associated microbiomes, using examples from animals and plants. It might seem surprising that microbes also can serve as microbiome hosts. These microbiomes carry out important functions for the growth and survival of their hosts.

**Bacterial Microbiomes** Certain species of bacteria, such as cyanobacteria, can produce relatively large bodies that host microbiomes of ecological significance. One example is the cyanobacterium *Microcystis*, previously mentioned for its importance in freshwater microbiomes. This cyanobacterial genus occurs as colonies of cells held together with mucilage, which is also home to diverse heterotrophic bacteria (see Figure 24.7). These bacteria benefit from the oxygen and organic molecules produced by the photosynthetic host. WMS shows that the microbiome bacteria synthesize vitamin $B_{12}$, which the cyanobacterial host requires but cannot produce. The microbiome bacteria also produce toxins different from those generated by the host.

**Protist Microbiomes** Microbiomes associated with protist hosts are likewise ecologically important. The green alga *Cladophora*, which grows along freshwater and marine shorelines worldwide (see Figure 24.2), provides an example. WMS reveals that this highly branched green protist operates something like a large hotel by hosting hundreds of species of bacterial, protist, and fungal microbes, and at least a hundred species of microscopic animals. Bacterial species attach themselves to the algal cell wall by secreting mucilage to form a **biofilm,** which is a group of microbes that stick together (**Figure 24.9**). In addition to a congenial living space, the photosynthetic host provides room service in the form of oxygen and organic molecules. Bacterial guests include diverse and numerous methane-oxidizers, which reduce the amount of the greenhouse gas methane that enters Earth's atmosphere from aquatic environments. Some bacterial guests pay their hotel bill by producing vitamin $B_{12}$, which the algal host requires, as do the methane-oxidizers. These and many other microbiome functions are important on a global scale, because the algal host has such a large surface area and is so globally widespread and abundant.

**Fungal Microbiomes** Fungi also act as hosts for microbiomes, most conspicuously as the partnerships known as lichens. About 20% of

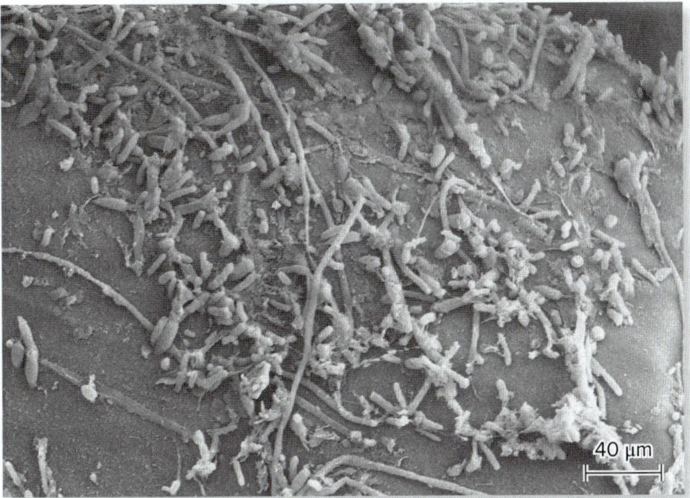


**Figure 24.9 A biofilm composed of numerous microbes attached to a surface by mucilage.** This is an SEM of a biofilm that is on the cell wall of the green protist shown Figure 24.2. Biofilms occur on many other biological and nonliving surfaces.
© Linda Graham

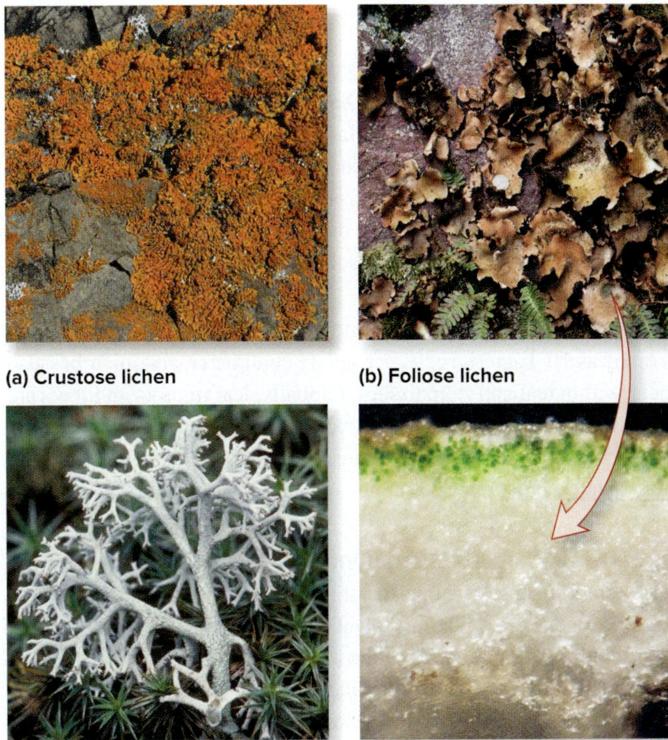


**Figure 24.10 Lichen structure. (a)** An orange-colored crustose lichen grows tightly pressed to the substrate. **(b)** The flattened, leaf-shaped genus *Umbilicaria* is a common foliose lichen. **(c)** The highly branched genus *Cladonia* is a common fruticose lichen. **(d)** A handmade thin slice of *Umbilicaria* viewed with a light microscope reveals that the photosynthetic algae occur in a thin upper layer. Fungal hyphae make up the rest of the lichen.
*(a)* © Perry Mastrovito/Corbis RF; *(b, d)* © Lee W. Wilcox; *(c)* © Ed Reschke/Getty Images

fungal species occur in 18,500 kinds of lichen. In the past, lichens were regarded as relatively simple associations between one fungal species and a single green algal or cyanobacterial species (sometimes both). According to this hypothesis, the fungus obtained essential organic carbon and fixed nitrogen from its algal or cyanobacterial partner. The algae or cyanobacteria could receive inorganic nutrients and water from the spongy fungal body, as well as benefit from fungal compounds that protect against intense sunlight and other organisms.

However, WMS studies reveal that lichens are actually microbiomes that include hundreds of bacterial species and multiple fungal species. Because fungal hyphae generally make up most of a lichen body, fungi are considered to function as the host. As noted earlier, cyanobacteria and green algae independently host microbiomes and, so, may function in lichens as co-hosts.

Lichen bodies exhibit different morphological forms and attach to surfaces in different ways:

- Crustose lichens are flat and adhere tightly to an underlying surface (Figure 24.10a).
- Foliose lichens are flattened and leaflike (Figure 24.10b).
- Fruticose lichens grow upright (Figure 24.10c) or hang down from tree branches.

In a lichen, photosynthetic green algae or cyanobacteria typically occur in a distinct layer close to the lichen's surface, with fungal hyphae making up the rest (Figure 24.10d). Lichen structure differs dramatically from that of the fungal component grown separately, demonstrating that the photosynthetic components influence lichen form.

Lichens often grow on rocks, buildings, tombstones, tree bark, soil, or other surfaces that easily become dry. When water is not available, the lichens lie dormant until moisture returns. Thus, lichens may spend much of their time in an inactive state, and for this reason, they often grow very slowly. However, because they can persist for long periods, lichens can be very old; some are estimated to be more than 4,500 years old. Lichens occur in diverse types of habitats, and a number grow in some of the most extreme, forbidding terrestrial sites on Earth—deserts, mountaintops, and the Arctic and Antarctic—places where most plants cannot survive. In these locations, lichens serve as a food source for reindeer and other hardy organisms. Though unpalatable, lichens are not toxic to humans and have also served as survival foods for indigenous peoples in times of shortages.

Soil building is another important lichen function. Lichen acids help to break up the surfaces of rocks, beginning the process of soil development. Lichens having nitrogen-fixing cyanobacterial partners (see Chapter 23) can also increase soil fertility. One study showed that such lichens released 20% of the nitrogen they fixed into the environment, where it is available for uptake by plants.

## Microbiomes Are Found on the Surfaces of Plants and Within Certain Tissues

Comparative WMS studies of the microbiomes of modern bryophytes, which model the earliest land plants, and related green algae that model plant ancestors (see Chapter 25) indicate that plants have hosted beneficial microbiomes throughout their evolutionary history. Modern plants host microbiomes on both their aboveground leaf and stem surfaces and their subterranean roots.

**Aboveground Plant Microbiomes** Leaf surfaces commonly host as many as $10^7$ bacterial cells per square centimeter, and microbes also occur within leaf and stem tissues. Globally, leaf microbe numbers are estimated at $10^{26}$. Leaves in tropical forest trees may host more than 400 types of bacteria. Among these are methanol-producing alphaproteobacteria that generate about $10^{24}$ grams of methanol every year.

**Subterranean Root Microbiomes** Legumes and some other plants are known to form partnerships with soil bacteria that provide fixed nitrogen, greatly aiding plant growth. Certain fungi have long been important components of most plant microbiomes, because fungal hyphae efficiently absorb minerals from the soil and transport these essential nutrients to plant partners. Such partnerships, known as mycorrhizae, are described more fully later in this section. Plants acquire these beneficial bacterial and fungal partners from a more diverse assortment of microbes in the soil close to plant root surfaces, a region called the **rhizosphere.**

Plant microbiomes change with age, and microbiomes in turn influence plant hormones, molecules that control plant growth and development. By secreting organic compounds that serve as signals, young plants increase the microbial transcription of genes related to nitrogen cycling. By contrast, older plants induce increased transcription of microbial genes involved in the biosynthesis of organic compounds that enhance plant growth. Such age-related changes in plant microbiomes parallel microbiome changes that occur in young humans, which likewise influence growth (see chapter introduction).

**Mycorrhizae** Associations between the hyphae of certain fungi and the roots of most seed plants are known as **mycorrhizae** (from the Greek, meaning fungus roots). Similar associations also occur between fungi and bryophytes, which lack roots, suggesting that fungi have been key to plant success on land from the beginning. Modern fungus-root associations are very important in nature and agriculture; more than 80% of terrestrial plants form mycorrhizae. Plants that have mycorrhizal partners receive an increased supply of water and mineral nutrients, primarily phosphate, copper, and zinc. They do so because an extensive fungal mycelium is able to absorb minerals from a much larger volume of soil than can roots alone (Figure 24.11). Added together, the branches of a fungal mycelium in 1 $m^3$ of soil can reach 20,000 km in total length. Experiments have shown that mycorrhizae greatly enhance plant growth compared with plants lacking fungal partners. In return, plants provide fungi with organic food molecules, sometimes contributing as much as 20% of their photosynthetic products.

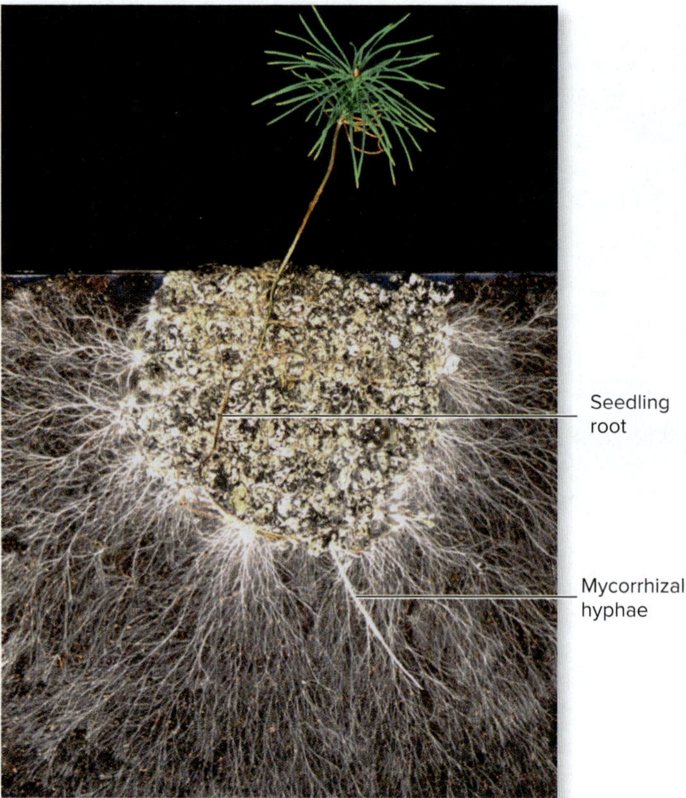


**Figure 24.11 Tree seedling with mycorrhizal fungi.** Hyphae of a mycorrhizal fungus extend farther into the soil than do the seedling's roots, helping the plant to obtain water and mineral nutrients.



The two most common types of mycorrhizae are endomycorrhizae, which occur within root tissue, and ectomycorrhizae, which coat roots. **Endomycorrhizae** (from the Greek *endo*, meaning inside) are partnerships between plants and fungi in which the fungal hyphae penetrate the spaces between root cell walls and plasma membranes and grow along the outer surface of the plasma membrane. In such spaces, endomycorrhizal fungi often form highly branched, bushy arbuscules (from the word arbor, referring to its tree-like shape). As the arbuscules develop, the root plasma membrane also expands. Consequently, the arbuscules and the root plasma membranes surrounding them have a very high surface area that facilitates rapid and efficient exchange of materials: Minerals flow from fungal hyphae to root cells, while organic food molecules move from root cells to hyphae. These fungus-root associations are known as **arbuscular mycorrhizae,** abbreviated **AM** (Figure 24.12). AM fungi are associated with apple and peach trees, coffee shrubs, and many herbaceous plants, including legumes, grasses, tomatoes, and strawberries.

**Ectomycorrhizae** (from the Greek *ecto*, meaning outside) are partnerships between temperate forest trees and soil fungi, particularly basidiomycetes. The fungi that engage in such associations are known as ectomycorrhizal fungi (Figure 24.13a). The hyphae of ectomycorrhizal fungi coat tree-root surfaces (Figure 24.13b)

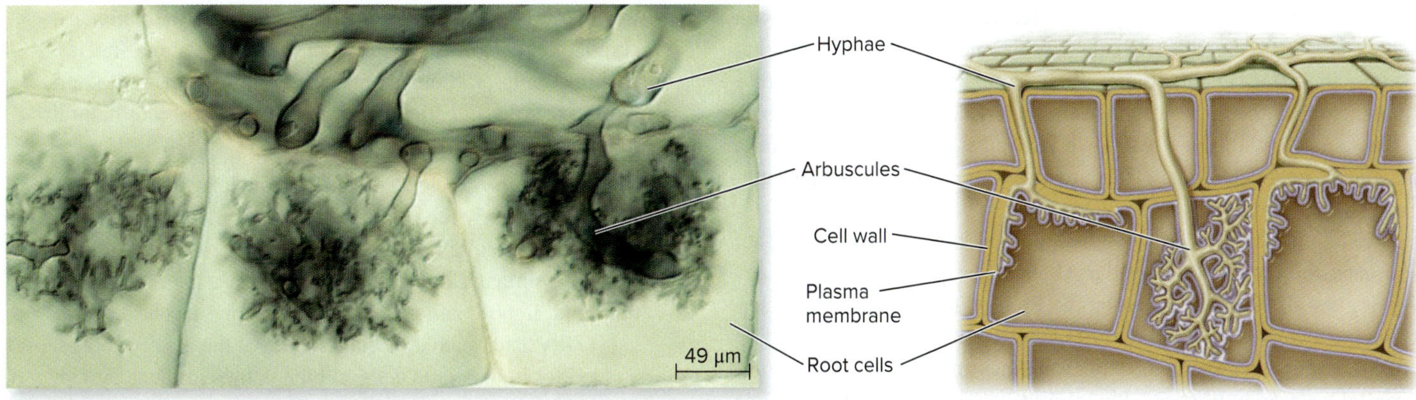


**(a) Micrograph of arbuscular mycorrhizae**

**(b) Hyphae growing between cell walls and plasma membranes**

**Figure 24.12** **Endomycorrhizae.** **(a)** Light micrograph showing black-stained AM fungi within the roots of the forest herb *Asarum canadensis*. Endomycorrhizal fungal hyphae penetrate plant root cell walls, then branch into the space between root cell walls and plasma membranes. **(b)** Diagram showing the position of highly branched arbuscules. Hyphal branches or arbuscules are found on the surface of the plasma membrane, which becomes highly invaginated. The result is that both hyphae and plant membranes have very high surface areas.


and grow into the spaces between root cells but do not penetrate the cell membrane (**Figure 24.13c**). Some species of oak, beech, pine, and spruce trees will not grow unless their ectomycorrhizal partners are also present. Mycorrhizae are thus essential to the success of commercial nursery tree production and reforestation projects.

## Animal Microbiomes Serve Many Useful Functions

Animal microbiomes contain viruses, archaea, bacteria, fungi, protists, and microscopic animals. These species affect animal health and may play important environmental roles or have medical applications. Termites, for example, are globally important in recycling plant biomass, a function enhanced by hundreds of gut bacterial species. Tunicates, an early-diverging lineage of chordates (see Chapter 26), occur worldwide as colonies of animals that filter microbial food from seawater. The guts of tunicates contain complex microbiomes, including beneficial cyanobacteria that synthesize sterols useful to the animal and nonphotosynthetic bacteria that produce defensive molecules. Such microbiomes are potential sources of new antibiotics that control disease-causing microbes without harming the animal host.

Microbiomes are likewise important to mammals, such as the brown bear, *Ursus arctos*. Brown bears are notable for their seasonal lifestyle, which involves fat accumulation in summer, with reliance

**(a) Ectomycorrhizal fruiting body**

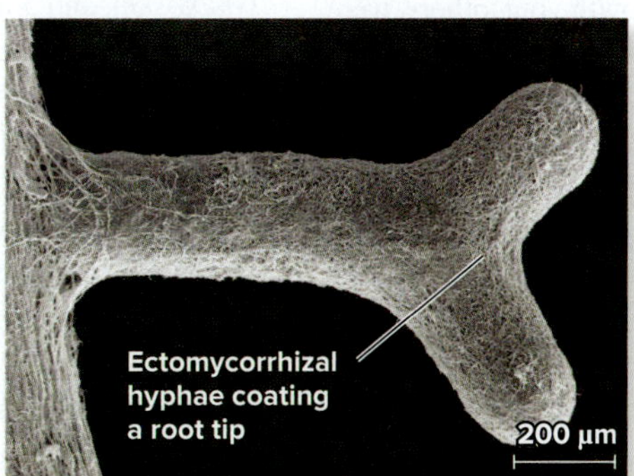


**(b) SEM of ectomycorrhizal hyphae**

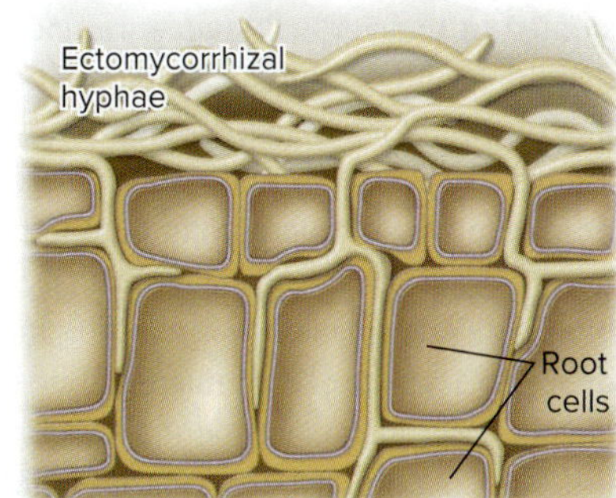


**(c) Hyphae invading intercellular spaces**

**Figure 24.13** **Ectomycorrhizae.** **(a)** The fruiting structure of the common forest fungus *Laccaria bicolor*. This is an ectomycorrhizal fungus that is associated with tree roots. **(b)** SEM showing ectomycorrhizal fungal hyphae of *L. bicolor* covering the surfaces of young *Pinus resinosa* root tips. **(c)** Diagram showing that the hyphae of ectomycorrhizal fungi do not penetrate root cell walls but grow within intercellular spaces. In this location, fungal hyphae are able to obtain organic food molecules produced by plant photosynthesis.

*What benefits do plants obtain from the association with fungi?*

on that fat during hibernation in winter. Ribosomal DNA amplicon analysis (see Figure 24.3) reveals that the microbiome in a bear's gut differs in winter versus summer. Winter microbiomes are less diverse and include higher levels of the bacterial phylum Bacteroidetes, which is associated with the breakdown of lipids and proteins. Bacteria linked to diets rich in plant fiber are also lower in winter than in summer, when bear diets include plants.

The most intensively studied animal microbiome is that of humans, the subject of several large scientific projects involving many biologists. The Human Microbiome Project, for example, characterized the microbes of 18 body sites on 300 healthy U.S. adults, finding that humans have distinctive microbiomes in the gut, vagina, urogenital tract, mouth, nose, skin, and teeth. The Human Microbiome Project and other studies have shown that human bodies host 100 trillion microbes that make up 1–3% of our body weight! The microbiomes associated with humans affect our health in many ways.

- The microbiomes of teeth form biofilms, known as plaque, that are detrimental to dental health.
- Up to 240 bacterial genera can be associated with human skin alone, performing beneficial functions. Some species within skin microbiomes break down dead skin, and others help to prevent infections or transform skin oil into natural moisturizer.
- Microbes of the digestive system, particularly the gut, are very important in early life. *Bifidobacterium longum* subspecies *infantis*, for example, is the most prevalent microbe in the gut of healthy infants. This bacterium has genes encoding proteins that bind, import, and metabolize milk polysaccharides into short fatty acids such as acetate. Some of these fatty acids serve as food for infant colon cells, aid the baby's immune system, and reduce gut pH, which deters some disease microbes. The best foods for *B. longum* are polysaccharides that are abundant in human milk but rare or absent in that of other animals. Breastfeeding thereby fosters the growth of these beneficial microbes.

## FEATURE INVESTIGATION

### Blanton, Gordon, and Associates Found That Gut Microbiomes Affect the Growth of Malnourished Children

As noted at the beginning of this chapter, information about the human microbiome has the potential to improve child growth. In a study published in 2016, Laura Blanton, Jeffrey Gordon, and colleagues described the effects of differences in gut microbiomes on the growth of malnourished children. These researchers began their work by analyzing *16S rDNA* to determine how microbiome bacteria change during the first 32 months of life in malnourished children from the same locale (**Figure 24.14**). Some of these malnourished children appeared healthy based on their growth, but others were stunted in their growth and were underweight. The results revealed that the healthy children had microbiomes that changed over time, so that older children had a different, more mature microbiome than did infants. By contrast, children who were stunted in growth retained an immature type of microbiome (see left side of data of Figure 24.14).

In a next phase of their work (see step 3 of Figure 24.14), these biologists transplanted the microbiomes of healthy or stunted children into separate sets of microbe-free ("germ-free") mice, so that the effects on growth could be monitored in ways that would have been difficult to accomplish with children. After this fecal transplantation, the mice were fed germ-free food of the same type eaten by the children. This experiment revealed that mice having microbiomes transplanted from the stunted children gained less weight than mice that received microbiome transplants from healthy children, even though the mice consumed the same amount of food (see right side of data of Figure 24.14). Though not shown in Figure 24.14, these researchers also used magnetic resonance imaging technology to determine lean and fat body mass, as well as micro-CT (computerized tomography) to evaluate the femur bone structure of the mice. The results of these imaging studies also showed that the gut microbiome played an important role in proper growth, but the basis of the effect remained unclear.

In further work that is not shown in Figure 24.14, the researchers took a closer look at the composition of the microbiomes to identify potential microbes that were responsible for better growth. By performing an amplicon analysis, they determined that the microbiomes of mice that had received fecal transplants from healthy donors tended to be dominated by a few bacterial species—*Ruminococcus gnavus*, *Clostridium symbiosum*, and a few others. To determine if any of these species were responsible for greater growth, researchers grew cultures of these bacterial species in the laboratory and then introduced them into mice that had received fecal transplants from stunted children. When these bacterial species were transplanted into the smaller mice, they grew larger! The researchers then characterized the gut microbiomes from these mice and determined that they were dominated by *Ruminococcus gnavus* and *Clostridium symbiosum*. Metabolomic studies revealed that these two bacterial species had the effect of decreasing amino acid oxidation and increasing amino acid incorporation into proteins that contributed to increases in body mass. In this way, the microbes were inferred to help malnourished children better use amino acids for growth. Together, these experiments revealed a causal relationship between microbiome composition and mouse growth, rather than merely a correlation.

#### *Experimental Questions*

1. Why did investigators feed experimental mice germ-free food?
2. **SCISKILLS ▸** Explain how the microbiomes from healthy or stunted children affected the growth of mice.

**Figure 24.14** Impact of the gut microbiome on growth.

**HYPOTHESIS** Microbiomes from children who were stunted in their growth due to malnourishment will impair the growth of mice.

**KEY MATERIALS** Fecal samples from healthy and stunted children, germ-free mice which are mice that have been raised in an aseptic environment and do not have any microbes in or on their bodies.

| | Experimental level | Conceptual level |
|---|---|---|
| 1 Obtain fecal samples from many healthy and stunted children of the same locale from birth to 32 months of age. | 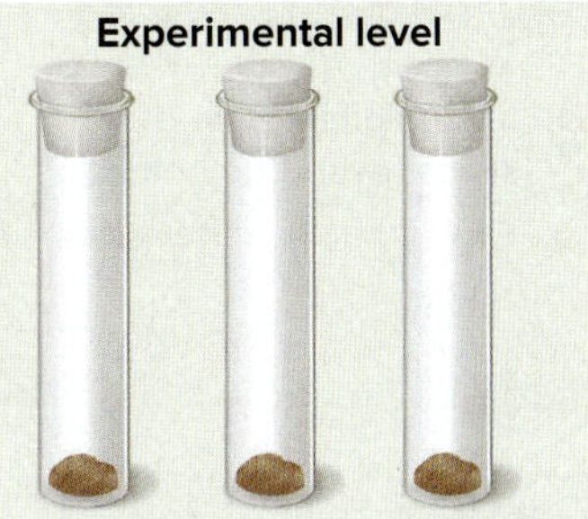 | 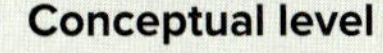 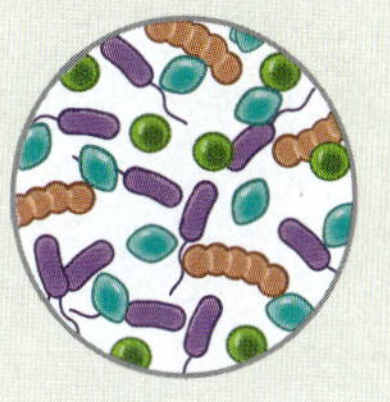 |
| 2 Perform an amplicon analysis on the fecal samples from birth to 32 months old. | See Figure 24.3. | This method reveals which prokaryote species are in the gut microbiome and how they change over the course of 32 months. |
| 3 Using a tube, introduce fecal samples from healthy and stunted children into the gut of 5-week-old germ-free mice. | 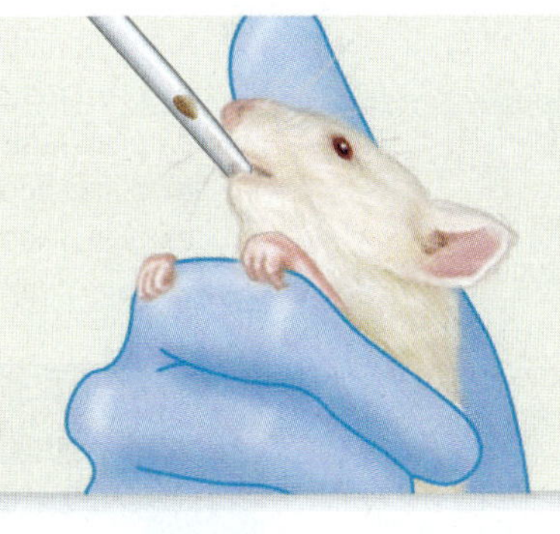 | 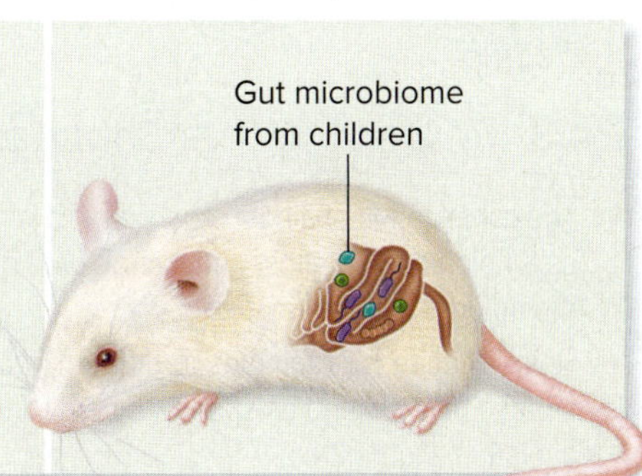  |
| 4 Give the mice a diet that corresponds to a diet that the children eat. |  | The two different types of gut microbiomes may affect the ability of the mice to gain weight. |
| 5 Monitor the weight of the mice over a 5-week period. | 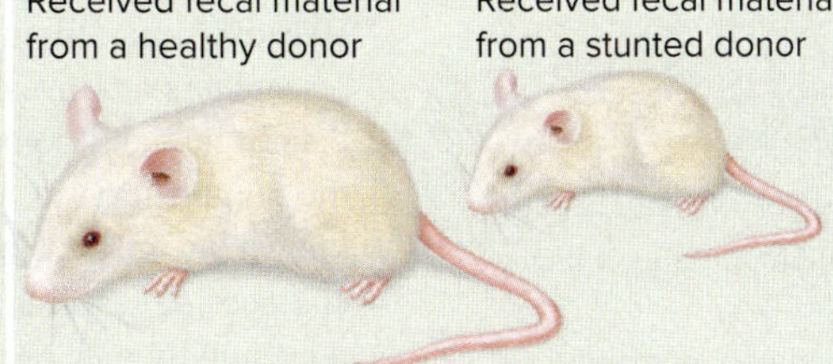  | Weight gain is a measure of how the gut microbiomes affect growth. |

6 **THE DATA**

**From step 2:**

Healthy children: From birth to 32 months → Microbiomes changed over 32 months

Stunted children: From birth to 32 months → Microbiomes stayed similar over 32 months

**From step 6:**

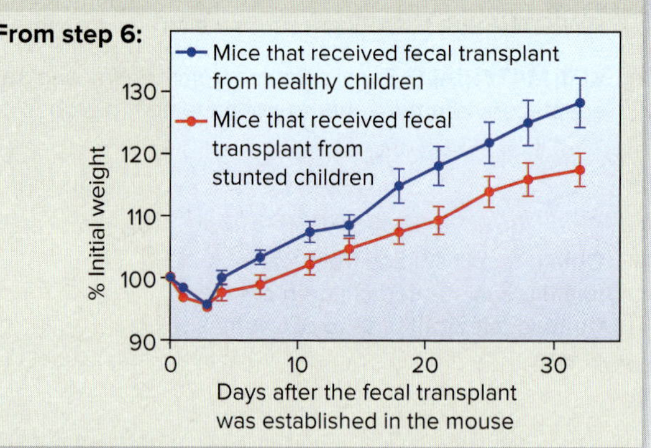


7 **CONCLUSION** The gut microbiomes from stunted donors impaired the growth of mice.

8 **SOURCE** Blanton, Laura B., et al. 2016. Gut Bacteria That Prevent Growth Impairments Transmitted by Microbiota from Malnourished Children. *Science* 351: 830.

## 24.3 Reviewing the Concepts

- The combination of a host organism and its microbiome is known as a holobiont. The host and microbiome genomes together are the hologenome.
- Hosts acquire microbiomes in different ways, and these microbiomes may change as species diverge from each other during evolution (Figure 24.8).
- A variety of species, including other microbes, plants, and animals, can act as hosts for microbiomes. The associations between microbiomes and their hosts result in an astonishing array of benefits for both the hosts and their microbial partners (Figures 24.9–24.13).
- Blanton, Gordon, and associates found that gut microbiomes affect the growth of children, and this observation was confirmed by studies in mice (Figure 24.14).

## 24.3 Testing Your Knowledge

1. Which of the following statements regarding genome size is true?
   a. The human genome is larger than the metagenome of the human microbiome.
   b. The human genome is smaller than the metagenome of the human microbiome.
   c. The human genome is about the same size as the metagenome of the human microbiome.
2. Mycorrhizae are associations between
   a. plant roots and protists.
   b. plant roots and fungi.
   c. plant roots and bacteria.
   d. plant shoots and protists.
   e. plant shoots and fungi.

## 24.4 Engineering Animal and Plant Microbiomes

### Learning Outcomes

1. Explain why people want to engineer animal and plant microbiomes.
2. Discuss how biologists produce synthetic microbiomes.
3. Describe artificial selection methods for identifying the most beneficial microbiomes for hosts.

Because microbiomes affect the health of animals and plants, researchers are investigating the potential of altering microbiomes to benefit humans, domesticated animals, and crop plants. Manipulating the composition of a microbiome to improve host characteristics is known as **microbiome engineering.** Much animal microbiome engineering research is done on mice or other laboratory animals, but it eventually may be applied to humans.

**Transplantion of Gut Microbiomes** As we saw in Figure 24.14, the gut microbiome can have an important impact on health. For this reason, fecal transplants are performed to introduce the entire gut microbiome of a healthy animal host into the gut of an unhealthy host. However, because thousands of microbial species and many complex interactions may be involved, the biological basis for success or failure of this treatment process can be difficult to determine. Alternatively, biologists perform experiments to determine more precisely which particular microbiome members are consistently associated with host health. As shown by the studies of Blanton, Gordon, and colleagues, described earlier, comparing the microbiomes of healthy hosts with those of unhealthy hosts helps to reveal these key beneficial species.

Additional information about healthy human microbiomes is gained by studying humans in nonindustrial societies. For example, the microbiomes of isolated Amerindians having no previous contact with Westerners have the highest known diversity of gut and skin bacteria and microbiome functions. By contrast to humans living in industrialized societies, isolated Amerindians have lower diversity of bacterial species from the phylum Bacteroidetes and higher diversity of other important microbiome species. Such differences indicate that humans living in industrialized countries have undergone shifts in their microbiomes that are implicated in gastrointestinal disorders, obesity, and autoimmune disease. For example, by comparison with isolated populations, people in industrialized countries don't consume enough dietary fiber, composed of complex plant carbohydrates. Studies of mice show that the chronic lack of dietary fiber reduces gut microbiome diversity.

Researchers use such information to assemble **synthetic microbiomes** by mixing cultures of beneficial microbial species. The effects on host health are determined by implanting these synthetic microbiomes into germ-free hosts, also known as axenic or gnotobiotic hosts, such as mice, rats, or Guinea pigs. The microbe combinations associated with the most positive effects on host health may be considered for use as treatments in humans.

**Probiotic Treatment** A **probiotic treatment** involves the introduction of a single microbial strain or a few strains into the microbiome of a host. For example, the bacterial genus *Lactobacillus* has been used to treat vaginal conditions in humans by replacing beneficial microbiome components that have become lost. Plants can also benefit from probiotic treatments. For example, strawberry plants that grew in a particular soil for many years without being attacked by harmful fungi were found to benefit from the presence in the soil of a fungus that produces an antifungal antibiotic. This observation suggests that the inoculation of soil with this fungus or the antibiotic could be beneficial to strawberry growers.

**Artificial Selection** Yet another engineering approach relies on artificial selection, which is a human-controlled form of natural selection (see Chapter 19). In this process, biologists engineer communities by artificial selection of favorable microbiomes. As a plant example, the process of engineering by artificial selection begins by planting sterilized soil with seedlings bearing different natural communities of microbes. At the end of a period of growth, plants are assessed for an agriculturally important trait such as biomass or time of flowering. Those soils that resulted in hosts with the most desirable traits are presumed to have acquired microbes that produce the most beneficial microbiomes. The microbiomes from the soils of the plants with favorable phenotypes are selected and used to inoculate soils for the next generation of plants. In this way, scientists use plant characteristics to select the most beneficial microbiomes (**Figure 24.15**). Amplicon analysis or WMS is then used to determine the composition of the winning microbiome.

Similar artificial selection strategies have been used to engineer the microbiomes of animals, such as bees (**Figure 24.16**). Bees having different gut microbiomes can be evaluated for health by examining particular phenotypic traits. Newborn bees lacking microbiomes can then be exposed to the microbiomes of the bees judged to have the best health. By repeating this process, the healthiest microbiomes can be inferred from indicators of bee health. Bees aid plant reproduction in natural and agricultural systems. Because bees around the world suffer from microbial infections (see Chapter 23), microbiome engineering might be a valuable way to help improve bee survival.

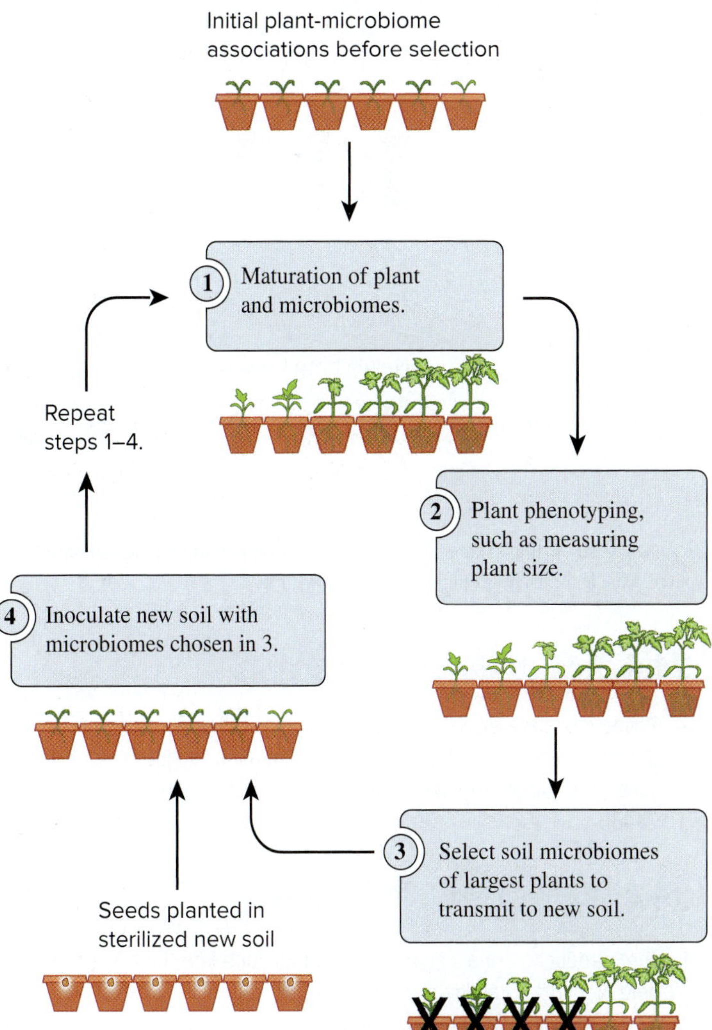


**Figure 24.15** **Artificial selection as a tool to engineer plant microbiomes.**
Source: Mueller & Sachs. Trends in Microbiology, Fig. 1

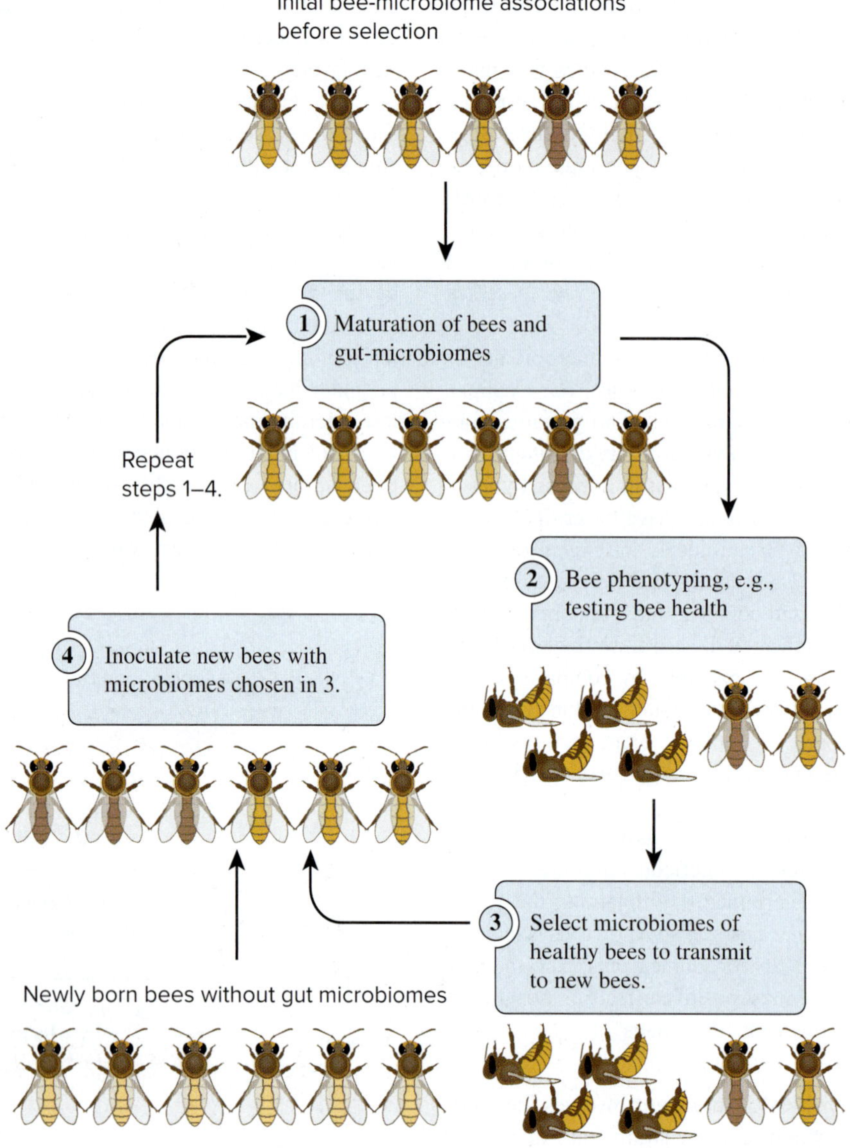


**Figure 24.16** **Artificial selection as a tool to engineer animal microbiomes.** Engineering the bee microbiome provides an example.
Source: Mueller & Sachs. Trends in Microbiology, Fig. 1

## 24.4 Reviewing the Concepts

- Researchers are exploring a variety of approaches for engineering microbiomes, including the transplantation of gut microbiomes, the use of probiotics, and artificial selection (Figures 24.15, 24.16).

## 24.4 Testing Your Knowledge

1. The introduction of a single species of microbe into the microbiome of a host is called
   a. gut microbiome transplantation.
   b. probiotics.
   c. artificial selection.
   d. all of the above.

# Assess and Discuss

## Test Yourself

1. A microbiome is
   a. an interaction between two different species of microorganisms.
   b. an environment that is microscopic.
   c. a particular assemblage of microbes and genes that occurs in a defined environment.
   d. the entire genetic makeup of a particular microorganism.
   e. both a and d.
2. An example of a microbiome function is
   a. nitrogen fixation.
   b. methane oxidation.
   c. the production of particular metabolites, such as vitamin $B_{12}$.

d. all of the above.
e. b and c only.

3. What would be the order of the four DNA fragments below if they were aligned into a contig? With the exception of black, DNA regions with the same sequence are shown in the same color.

Fragment 1:
Fragment 2:
Fragment 3:
Fragment 4:

a. 1, 2, 3, 4
b. 2, 3, 4, 1
c. 1, 3, 2, 4
d. 3, 2, 1, 4
e. b or d

4. Which of the following is a microbiome of a physical system?
a. a microbiome on the surface of a leaf
b. a microbiome in the human gut
c. a microbiome in a soil sample
d. a microbiome in a sample of human saliva
e. all of the above

5. The combination of a host organism and its microbiome is known as
a. a microbiome.
b. a holobiont.
c. a metagenome.
d. a metabolome.
e. both a and c.

6. Which of the following categories of organisms can function as a host for a microbiome?
a. cyanobacteria
b. algae
c. fungi
d. plants and animals
e. all of the above

7. Which of the following is a way in which an animal can acquire a microbiome?
a. Certain insects coat the casings of their eggs with bacteria.
b. Newborn bees get microbes from sibling worker bees.
c. Mammals, including humans, transmit important microbes as the young transit the birth canal.
d. Termites use specific behaviors to transfer among themselves microbes they need to break down plant materials into food.
e. All of the above are ways that animals can acquire microbiomes.

8. What is a biofilm?
a. a microbiome within the gut of an animal
b. a group of microbes that secrete mucilage and stick together
c. a microbiome that forms an opaque film on ice
d. a group of microbes that perform a metabolic function the host cannot perform
e. a microbiome that floats on the surface of seawater

9. The goal of microbiome engineering is to
a. eliminate an unwanted microbiome from the host.
b. manipulate the composition of a microbiome to cause its self-destruction.
c. manipulate the composition of a microbiome to benefit the host.
d. identify all of the microbial species within a microbiome.
e. alter the genome of the host so that it can support a different microbiome.

10. Which of the following is *not* an approach for microbiome engineering?
a. transplantation of gut microbiomes
b. probiotics
c. artificial selection
d. All of the above are approaches for microbiome engineering.
e. Only a and c are not approaches for microbiome engineering.

## Conceptual Questions

1. Describe two examples of microbiomes that are found within a physical environment. List examples of how such microbiomes can affect Earth's environment on a global scale.
2. Pick one host-associated microbiome. Describe how the host benefits from the microbiome and how the microbes within the microbiome benefit from their association with the host.
3. **PRINCIPLES** A principle of biology is that living organisms interact with their environment. Give one example of how a microbiome of a plant can influence its ability to interact with the environment.

## Collaborative Questions

1. Microbiomes can occur in places that you may have not imagined. Starting with the term microbiome or microbiota, search the literature using scientific search engines such as Pubmed, Google Scholar, or BioOne, and identify microbiomes that you never knew existed. Discuss your findings with fellow students.
2. Your textbook describes how the microbiomes of African apes and humans have changed since their evolutionary divergence. Pick your own example of a small group of closely related species and hypothesize how the microbiomes of these species may have changed since they diverged from each other. Explain your reasoning.

# 25 Plant Evolution: How Plant Diversification Changed Planet Earth

© Linda Graham

**The drought-resistant water fern *Marsilea drummondii* in the Australian outback.**

## Chapter Outline



During an expedition to the desert outback of Australia in which the goal was to learn more about ancient events in the evolutionary history of land plants, a team of plant biologists was amazed to find extensive growths of the so-called water fern *Marsilea drummondii* (chapter-opening photo). They were surprised because most ferns live in relatively moist environments, including the humid interiors of buildings where ferns are often used as ornamentals. As indicated by their common name, most water ferns live in very moist places indeed. But some fern species, including *M. drummondii*, are so resistant to drying that they can live in arid places. This is possible because this fern's leaves are fuzzy with surface hairs that act as a blanket to retard drying and excessive heating. The leaves are also able to fold themselves, thereby reducing surface area from which water can evaporate. The plant can become so dry that it crunches underfoot; it is not dead but is instead living in a state of reduced metabolism until sporadic rains stimulate active photosynthesis and growth. With these adaptations, *M. drummondii* is sufficiently abundant that aboriginal Australians, who call this plant "nardoo," have long gathered its small, brown, food-rich reproductive structures, using grinding stones to make flour for cakes.

Nardoo is just one of the hundreds of thousands of plants that have become adapted in diverse ways to life on land, even very arid places. As is illustrated by aboriginal Australians' use of nardoo and developed societies' reliance on agricultural crops, humans depend on this plant diversity for food and many other materials. In this chapter, we will see how, in addition to their modern importance, ferns and many other types of plants have played dramatic roles in the Earth's past. Throughout their evolutionary history, diverse plants have influenced Earth's atmospheric chemistry, climate, and soils, as well as the evolution of many other groups of organisms.

## 25.1 Ancestry and Diversity of Land Plants

### Learning Outcomes

1. Name several characteristics unique to land plants.
2. Compare and contrast the features of vascular and nonvascular plants.
3. **SCISKILLS ▶** Explain how cuticles, epidermal stomata, and internal conduction systems work together to maintain stable water content in vascular plants.

Several hundred thousand modern species are formally classified into the kingdom Plantae, informally known as the land plants or simply plants (Figure 25.1). Although some species live in watery places, plants can be defined as multicellular, mostly photosynthetic eukaryotes that are adapted in many ways for life on land. Molecular and other evidence indicates that the plant kingdom evolved from green algal ancestors similar to particular modern algae. Together, plants and their closest green algal relatives are known as **streptophytes** (from Latin words meaning twisted and plant) (Figure 25.2). Land plants are distinguished from closely related algae by the presence of traits that foster survival in terrestrial conditions, which are drier, sunnier, hotter, colder, and less physically supportive than aquatic habitats. Likewise, the first land animals had to acquire similar structural and reproductive adaptations.

**Figure 25.1 A temperate rain forest containing diverse plant phyla in Olympic National Park in Washington State.**
© Craig Tuttle/Corbis

### Distinctive Features of the Land Plants

Land plants display distinctive features that represent early adaptations to the land habitat:

- The bodies of all land plants are primarily composed of **tissues,** defined as close associations of cells. Bodies composed of tissues have lower surface area/volume ratios than simpler green algal bodies and, thus, less readily lose water in dry air by evaporation.
- Land plant tissues arise from one or more actively dividing cells that occur at growing tips, forming **apical meristems.** The tissue-producing apical meristems of land plants produce relatively thick, robust bodies able to withstand drought and mechanical stress and produce tissues and organs with specialized functions. (Related green algae have simpler growth and body structure.)
- The life cycle of land plants, known as **alternation of generations** (or sporic life cycle), involves two types of multicellular bodies. A haploid generation known as the gametophyte produces eggs and/or sperm; a diploid generation known as the sporophyte produces spores by meiosis. (Related green algae possess a simpler zygotic life cycle in which a single-celled zygote is the only diploid stage.)
- Land plant embryos are young, multicellular sporophytes that are produced by mitosis following fertilization of an egg cell and depend on maternal tissues for food during early development. Embryos are not present in related green algae but are such a basic feature of land plants that the latter are also known as embryophytes.
- Tough-walled plant **spores** allow land plants to disperse offspring through dry air. Plant spores are tough because their walls contain an impervious material known as **sporopollenin.** (Although related green algae incorporate sporopollenin into zygote walls, and algal zygotes undergo meiosis to produce spores, such algal spores don't contain sporopollenin in their walls and thus cannot survive long in air.)
- Specialized structures called **gametangia** generate, protect, and disperse land plant gametes; tough **sporangia** likewise produce, protect, and disperse the spores of land plants.

### Modern Land Plants Can Be Classified into Nine Phyla

Plant systematists use molecular and structural information to classify plants into phyla and organize phyla into an evolutionary sequence (see Figure 25.2). The modern plant phyla are listed as follows, together with a brief description of their evolutionary and ecological importance.

- Plants informally known as liverworts and mosses were the earliest to appear and thus serve as useful models of the earliest land plants.
- Mosses play a particularly important role in stabilizing Earth's climate now and have likely been doing so for the past hundreds of millions of years.
- Hornworts are evolutionarily important because they are closely related to the lycophytes and other land plants that have a well-developed system for internal conduction of water, minerals, and organic compounds. As a consequence of their phylogenetic position, hornworts are useful in learning how plant conduction systems originated.

# Biology Principle

## All Species (Past and Present) Are Related by an Evolutionary History

This figure shows evolutionary divergence times indicated by molecular clock and some fossil evidence, suggesting when plant groups may first have arisen.

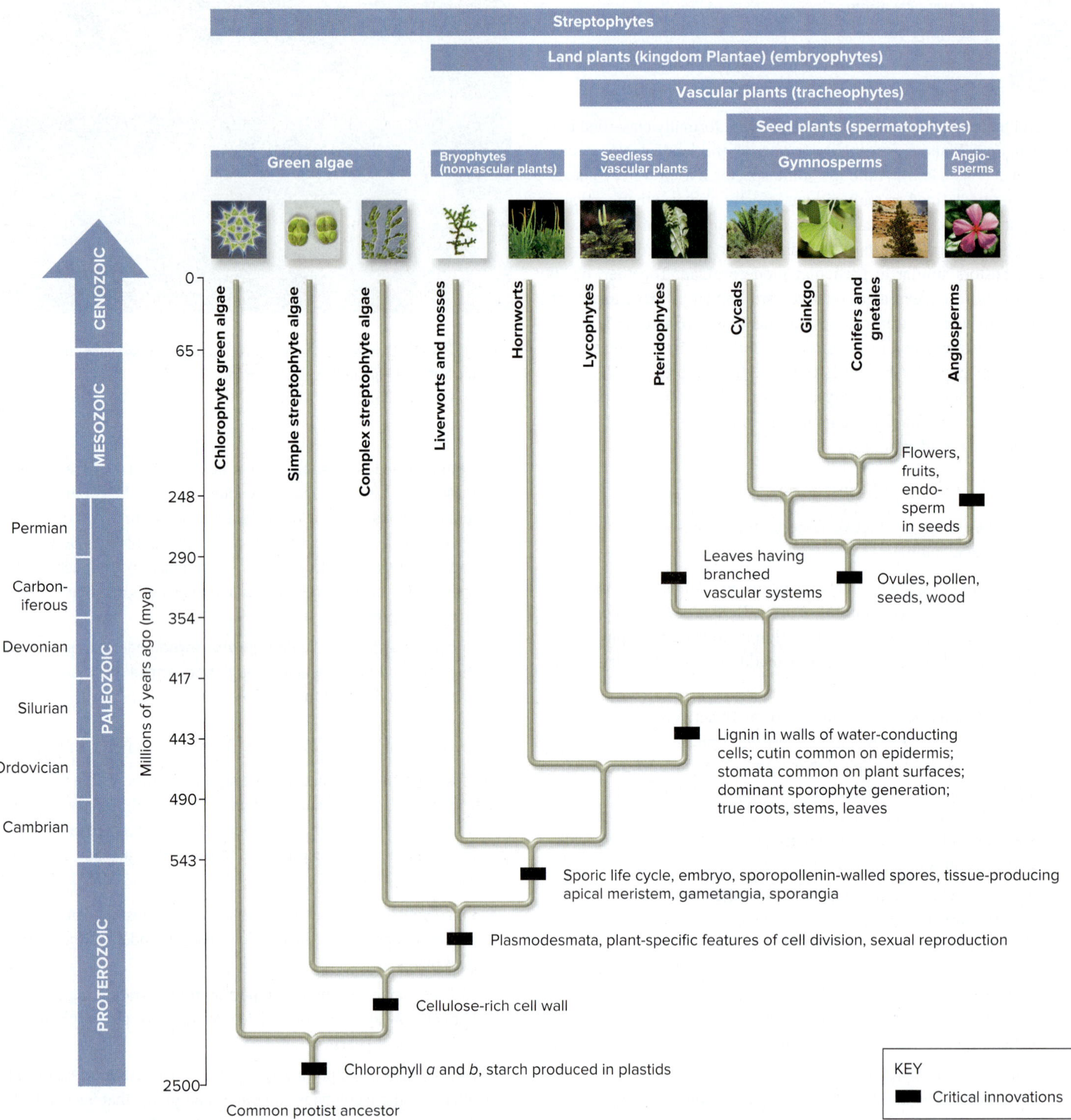


**Figure 25.2 Evolutionary relationships of the modern plant phyla.** Land plants gradually acquired diverse structural, biochemical, and reproductive adaptations to land that are known as critical innovations.

*(1)* © Roland Birke/Phototake; *(2)* © the CAUP image database, http://botany.natur.cuni.cz/algo/database; *(3–5)* © Lee W. Wilcox; *(6)* © Ed Reschke/Getty Images; *(7)* © Patrick Johns/Corbis; *(8)* © Philippe Psaila/SPL/Science Source; *(9)* © Fancy Photography/Veer RF; *(10)* © imageBROKER/Alamy RF; *(11)* © Gallo Images/Corbis

- Lycophytes are the earliest-divergent group of vascular land plants—those having conduction systems that also provide skeletal strength. In the ancient past, lycophytes were more abundant, much larger in size, and more diverse than are modern representatives.
- Pteridophytes include many modern species of ferns that are particularly common in tropical habitats, and are valued for horticultural plantings. Lycophytes and pteridophytes do not produce seeds but display several adaptations that foster stable tissue-water content, a feature critical to the ability to survive through dry periods.
- Cycads are an early-diverging lineage of seed plants and, so, are important in understanding how seeds first evolved.
- Ginkgos are seed plants that were more diverse in the past than today. Modern ginkgos are valued as hardy and attractive street trees.
- Conifers, cone-bearing seed plants, are particularly important to humans as sources of wood and other commercial materials.
- Flowering plants, also known as angiosperms, are the plants upon which humans and many other animals most depend.

## Liverworts, Mosses, and Hornworts Are the Simplest Land Plants

Liverworts, mosses, and hornworts (Figures 25.3, 25.4) are Earth's simplest land plants. There are about 6,500 species of modern liverworts, 12,000 or more species of mosses, and about 100 species of hornworts. Collectively, liverworts, mosses, and hornworts are known informally as the **bryophytes** (from the Greek *bryon,* meaning moss, and *phyton*, meaning plant). The bryophytes do not form a clade, but the term bryophyte is useful for expressing common structural, reproductive, and ecological features: Bryophytes are all relatively small in stature and are most common and diverse in moist habitats. Bryophytes lack traits allowing them to grow tall or reproduce in dry places. Although many bryophytes possess a type of conducting tissue, bryophytes are unable to grow tall because such conducting tissues are not tough enough to also provide structural support. (In contrast, the tougher conducting tissues of vascular plants also provide support.)

Bryophytes require moist habitats in order to accomplish sexual reproduction because they produce flagellate sperm that must swim through water films to fertilize eggs. Recently, Todd Rosenstiel, Sarah Eppley, and associates discovered that at least some bryophytes emit perfume-like chemical attractants that lure small animals such as mites and springtails, which then help to disperse bryophyte sperm. This was a surprising discovery because animal-aided sexual reproduction was previously regarded as a seed plant trait.

Bryophytes are also distinctive in having a dominant (larger and longer-lived) gametophyte generation; the bryophyte sporophyte is a relatively small and short-lived life phase that does not branch

**Figure 25.3 Liverwort, moss, and hornwort representatives. (a)** The liverwort *Marchantia polymorpha* has a flat body that produces umbrella-shaped structures bearing egg-shaped sporophytes on the undersides. When sporophytes mature, they release tough-walled spores into the air. **(b)** The moss genus *Mnium* has a conspicuous upright, leafy, green gametophyte that generates eggs and sperm and a less noticeable attached, unbranched sporophyte that bears a spore-producing sporangium at its tip. Inset: This SEM shows that the tips of moss sporangia often produce a number of tooth-shaped structures separated by spaces. Such tips function something like a salt shaker to sprinkle spores into the wind over a period of time, rather than releasing spores all at once. **(c)** Hornwort gametophytes grow close to the ground and produce upright sporophytes that release spores from tips into the air.

*(a–c)* © Lee W. Wilcox; *(b inset)* © Eye of Science/Meckes/Ottawa/Science Source

*Why do you think liverworts, mosses, and hornworts produce their reproductive spores on raised structures?*

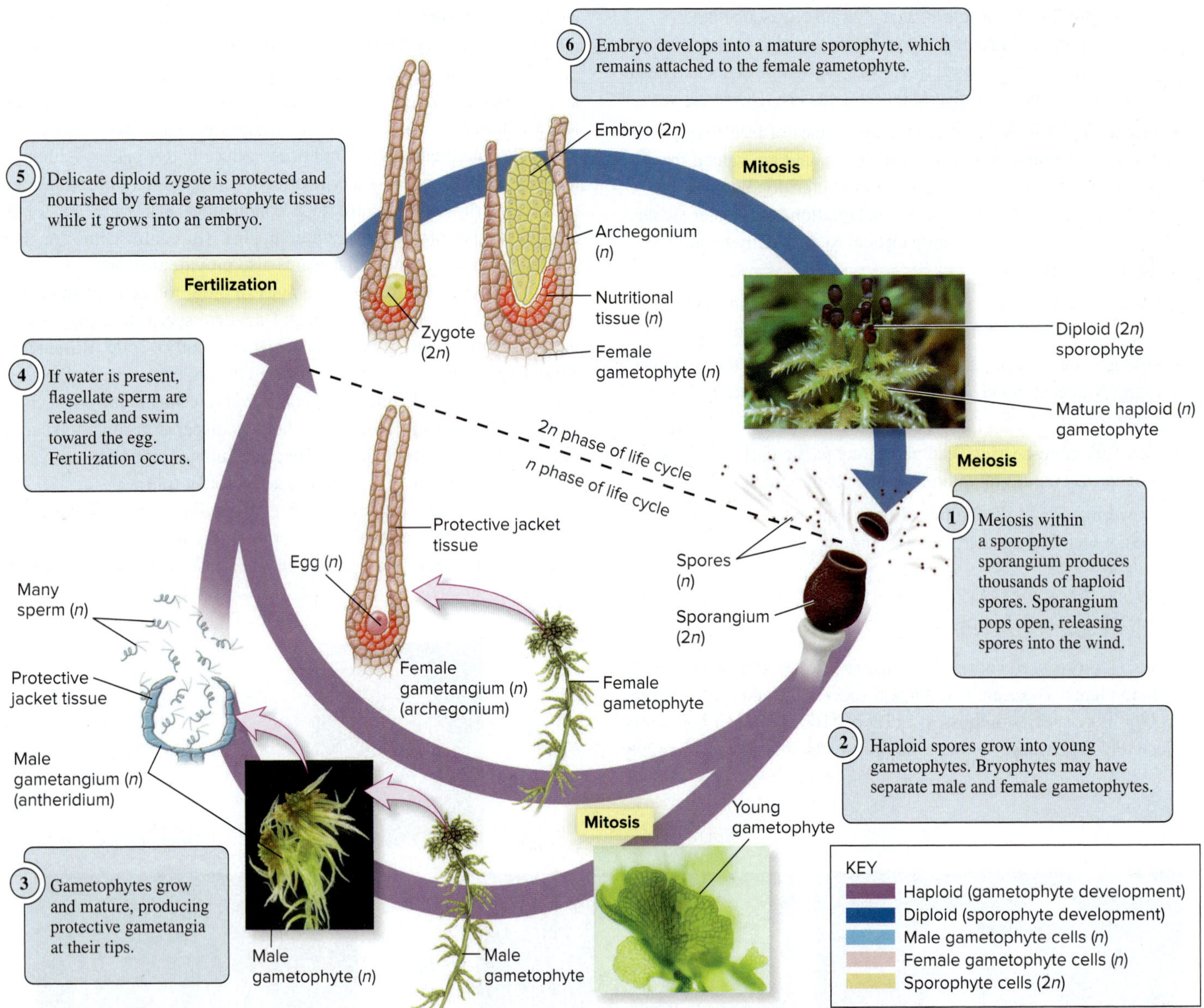


**Figure 25.4 The life cycle of the early-diverging peat moss genus *Sphagnum*.** The life cycle of this common peat moss illustrates reproductive adaptations commonly found in bryophytes that were also present in early land plants.

*(top inset)* © Larry West/Science Source; *(bottom insets)* © Linda Graham

(see Figure 25.3b, Figure 25.4). By contrast, as described next, vascular land plants possess a dominant sporophyte generation that branches, giving rise to a larger body, whereas the gametophyte generation is relatively small and inconspicuous.

## Lycophytes and Pteridophytes Are Vascular Plants That Do Not Produce Seeds

Vascular plants, which include all groups of modern plants except bryophytes, possess a conduction system that also provides structural support. Vascular plants have been important to Earth's ecology for several hundreds of millions of years. Fossils indicate that the first vascular plants appeared later than the earliest bryophytes and that several early vascular plant lineages once existed but became extinct. Molecular data demonstrate that the lycophytes are the oldest phylum of living vascular plants and that pteridophytes are the next oldest living plant phylum (see Figure 25.2). In the past, lycophytes were very diverse and included tall trees, but now only about 1,000 relatively small species exist (**Figure 25.5a**). Pteridophytes diversified more recently, and there are about 12,000 modern species, including horsetails, whisk ferns, and other ferns (**Figure 25.5b–d**). Because the lycophytes and pteridophytes diverged prior to the origin of seeds, they are informally known as seedless vascular plants. Gymnosperms and angiosperms are the seed-producing vascular plants.

Together, lycophytes, pteridophytes, and seed-producing vascular plants are also known as the **tracheophytes,** named for **tracheids,** a type of specialized conducting cell. Tracheids occur in specialized tissues known as **xylem** that conduct water and minerals throughout

**Figure 25.5 Representative seedless vascular plants.** **(a)** The lycophyte *Lycopodium obscurum.* The sporophyte stems bear many tiny leaves, and leaves bearing sporangia generally occur in club-shaped clusters. For this reason, lycophytes are informally known as club mosses or spike mosses, though they are not true mosses. The gametophytes of lycophytes are small structures that often occur underground, where they are better protected from drying. **(b)** The leafless, rootless, green stems of the whisk fern (*Psilotum nudum*) branch by forking and bear many clusters of yellow sporangia that disperse spores via wind. The gametophyte of this plant is a tiny, pale structure that lives underground in a symbiotic partnership with fungi. **(c)** Another pteridophyte is the giant horsetail (*Equisetum telmateia*), which displays branches in whorls around the green stems. The leaves of this plant are tiny, light brown structures that encircle stems and branches at intervals. This plant produces spores in cone-shaped structures, and the wind-dispersed spores grow into small, green gametophytes. **(d)** The early-diverging fern *Botrychium lunaria,* showing a green, photosynthetic leaf with leaflets and a modified leaf that bears many round sporangia.

*(a–b)* © Lee W. Wilcox; *(c)* © S. Solum/PhotoLink/Getty Images RF; *(d)* © Patrick Johns/Corbis

the plant body. The xylem also provides structural support, allowing vascular plants to grow taller than nonvascular plants. This support function arises from the presence of a compression and decay-resistant waterproofing material known as **lignin,** which occurs in the cell walls of tracheids and some other types of plant cells. A second type of vascular tissue known as **phloem** conducts a watery sap containing organic molecules such as sugar throughout the plant body. Xylem and phloem, together known as **vascular tissues,** occur in the major plant organs: stems, roots, and leaves.

Tracheophyte **stems** are branching organs that bear reproductive sporangia and leaves. Most vascular plants also produce **roots—**branching organs specialized for uptake of water and minerals from the soil—and **leaves** that generally have a photosynthetic function. Like stems, roots and leaves contain both xylem- and phloem-conducting tissues. Together, the branching vascular tissues of the stem, roots, and leaves are known as the vascular system. The vascular system of a plant functions much like the vascular system of an animal; both systems carry nutrients and other materials throughout the body.

In relatively dry habitats, lycophytes, pteridophytes, and other vascular plants are able to grow to larger sizes and remain metabolically active for longer periods than can bryophytes. Vascular plants have this advantage because they are better able to maintain stable internal water content by means of several adaptations, including a vascular system, a waxy cuticle, and stomata. A vascular system allows water and minerals taken up by roots to move through stems and leaves and allows sugar produced in leaves (and sometimes stems) to move into locations where the sugar is metabolized to produce

ATP. A protective waxy **cuticle** present on most plant surfaces contains a polyester polymer known as cutin that deters pathogen attack and wax that helps to prevent drying (Figure 25.6a). The surface tissue of vascular plant stems and leaves contains many **stomata** (singular, stoma or stomate)—pores that can open and close (Figure 25.6b). Stomata allow plants to take in the carbon dioxide needed for photosynthesis and release oxygen to the air, while conserving water. When the environment is moist, the pores open, allowing photosynthetic gas exchange to occur. When the environment is very dry, the pores close, thereby reducing water loss by evaporation into the air. In vascular plants, a conducting system, a waxy cuticle, and stomata function together to maintain moisture homeostasis, allowing tracheophytes to exploit a wide spectrum of land habitats. Additional information about the function of angiosperm vascular systems and stomata can be found in Chapter 29.

Lycophytes and pteridophytes have a life cycle dominated by the diploid sporophyte, which is large enough to produce many spores. Clusters of sporangia are often visible on the undersides of fern leaves (Figure 25.7). The fern gametophyte is small but can be seen with the unaided eye and, so, is widely used in biology classes to demonstrate the plant life cycle. By contrast, seed plants produce both spores and seeds, and seed plant gametophytes are microscopic in size.

## Gymnosperms and Angiosperms Are the Modern Seed Plants

Among the vascular plants, the seed plant phyla known as gymnosperms and angiosperms dominate most modern landscapes. Because both groups produce seeds, gymnosperms and angiosperms are together known as the **spermatophytes. Seeds** are complex structures having specialized tissues that protectively enclose embryos and contain stores of carbohydrate, lipid, and protein that enable embryos to grow and develop. Seed plants produce **pollen,** small air- or animal-borne spores that contain and protect microscopic male gametophytes. The ability to produce seeds and pollen helps to free seed plants from reproductive limitations experienced by the seedless plants, revealing why seed plants are the dominant plants on Earth today.

The modern seed plant phyla commonly known as cycads, ginkgos, and conifers (including a group known as the gnetophytes) are collectively known as gymnosperms (Figure 25.8a–c). The term **gymnosperm** comes from the Greek, meaning naked seeds, reflecting the observation that gymnosperm seeds are not enclosed within fruits. Though gymnosperms produce pollen and seeds, they lack flowers, fruits, and endosperm.

The **angiosperms** (Figure 25.8d) produce seeds and pollen but are distinguished from gymnosperms by the presence of flowers, fruits, and a specialized seed tissue known as endosperm. A **flower** is a short stem bearing reproductive organs that are specialized in ways that enhance seed production. **Fruits** are structures that develop from flowers, enclose seeds, and foster seed dispersal in the environment. The term angiosperm comes from the Greek, meaning enclosed seeds, reflecting the observation that the flowering plants produce seeds within fruits. **Endosperm** is a nutritive seed tissue that increases the efficiency with which food is stored in the seeds of flowering plants, and it contributes a large proportion of human food such as cereal crops. Flowers, fruits, and endosperm are defining features of the angiosperms, and they are integral components of the diet of many animals.

## Biology Principle

### Living Organisms Maintain Homeostasis

The structures illustrated in this figure explain how vascular land plants maintain homeostasis in water content.

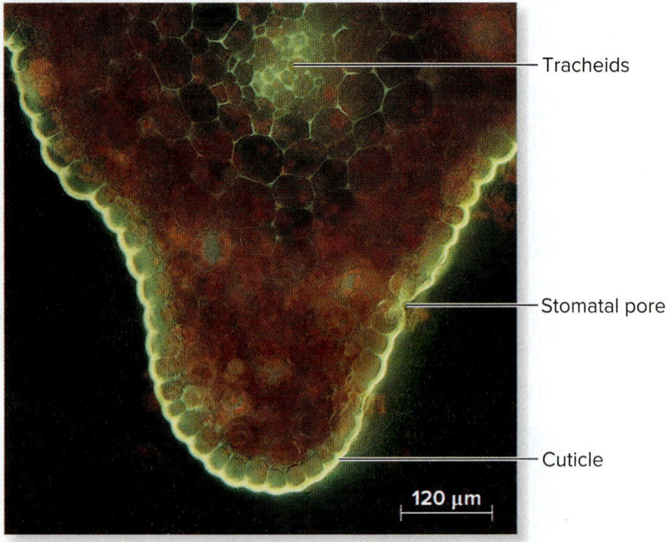


**(a) Stem showing tracheophyte adaptations**

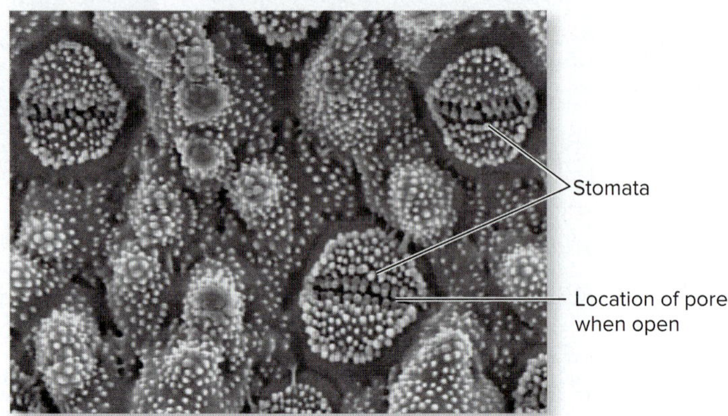


**(b) Close-up of stomata**

**Figure 25.6 Tracheophyte adaptations for transporting and conserving water. (a)** A cross section through a stem of the whisk fern *Psilotum nudum.* When viewed with fluorescence microscopy and illuminated with violet light, an internal core of xylem tracheids glows yellow, as does the surface cuticle. **(b)** Surface pores associated with specialized cells—the complexes known as stomata—allow for gas exchange between plant and atmosphere. This photo naturally lacks color because it was made with a scanning electron microscope (SEM), which uses electrons rather than visible light to form magnified images.

*(a)* © Linda Graham; *(b)* © Martha Cook

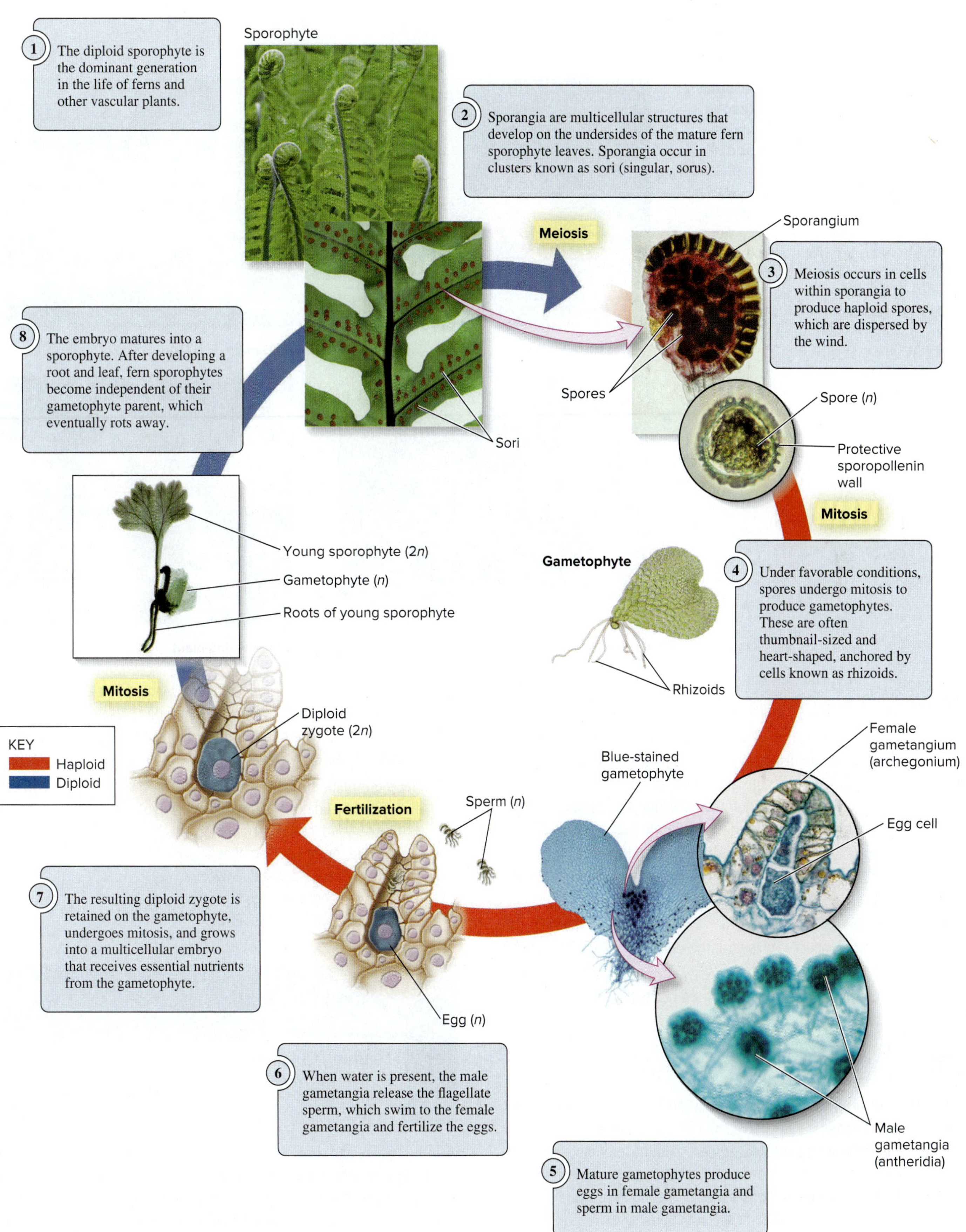

1 The diploid sporophyte is the dominant generation in the life of ferns and other vascular plants.
Sporophyte
2 Sporangia are multicellular structures that develop on the undersides of the mature fern sporophyte leaves. Sporangia occur in clusters known as sori (singular, sorus).
Meiosis
Sporangium
3 Meiosis occurs in cells within sporangia to produce haploid spores, which are dispersed by the wind.
Spores
Sori
Spore (n)
Protective sporopollenin wall
Mitosis
8 The embryo matures into a sporophyte. After developing a root and leaf, fern sporophytes become independent of their gametophyte parent, which eventually rots away.
Young sporophyte (2n)
Gametophyte (n)
Roots of young sporophyte
Gametophyte
4 Under favorable conditions, spores undergo mitosis to produce gametophytes. These are often thumbnail-sized and heart-shaped, anchored by cells known as rhizoids.
Rhizoids
Mitosis
Diploid zygote (2n)
KEY
Haploid
Diploid
Female gametangium (archegonium)
Blue-stained gametophyte
Egg cell
Fertilization
Sperm (n)
7 The resulting diploid zygote is retained on the gametophyte, undergoes mitosis, and grows into a multicellular embryo that receives essential nutrients from the gametophyte.
Egg (n)
6 When water is present, the male gametangia release the flagellate sperm, which swim to the female gametangia and fertilize the eggs.
Male gametangia (antheridia)
5 Mature gametophytes produce eggs in female gametangia and sperm in male gametangia.


**Figure 25.7 The life cycle of a typical fern.** The fern life cycle is often used to illustrate plant alternation of generations because both sporophyte and gametophyte are large enough for people to see with the unaided eye.

*(Sporophyte)* © Ernst Kucklich/Getty Images; *(sori)* © Carolina Biological Supply Company/Phototake; *(Sporangium, Spore)* © Linda Graham; *(Gametophyte, Female and Male gametangium, Blue-stained gametophyte)* © Lee W. Wilcox; *(Young sporophyte)* © Mike Peres RBP SPAS/CMSP Biology/Newscom

**Figure 25.8** **Representative seed plants.** **(a)** Palmlike foliage and conspicuous seed-producing cones are features of most cycads. **(b)** *Ginkgo biloba* has leaves with forked veins and seeds with fleshy, foul-smelling seed coats. **(c)** An example of a conifer, the pine (*Pinus*). **(d)** Citrus is an example of a flowering plant, showing the distinctive features of flowers and fruits.

*(a)* © Ed Reschke/Getty Images; *(b)* © Topic Photo Agency Inc./agefotostock; *(c)* © Lee W. Wilcox; *(d)* © Bill Ross/Corbis

**(a) Cycad**

**(b) *Ginkgo***

**(c) Conifer**

**(d) Flowering plant**

## 25.1 Reviewing the Concepts

- Plants are multicellular eukaryotic organisms that are mostly photosynthetic and display many adaptations to life on land. The modern plant kingdom consists of nine phyla, informally known as the liverworts, mosses, hornworts, lycophytes, pteridophytes, cycads, ginkgos, conifers, and angiosperms (Figures 25.1, 25.2).
- The monophyletic liverwort, moss, and hornwort phyla are together known informally as the bryophytes (Figures 25.3, 25.4).
- Lycophytes, pteridophytes, and other vascular plants (tracheophytes) generally possess stems, roots, and leaves having vascular systems composed of conductive phloem and xylem (Figures 25.5, 25.6, 25.7).
- Cycads, ginkgos, and conifers (the gymnosperms) and flowering plants (the angiosperms) are the seed plants (Figure 25.8).

## 25.1 Testing Your Knowledge

1. Which of the following descriptions most accurately represents how plant gametophytes have changed through time?
   a. The first plant gametophytes were, like those of modern bryophytes, larger than the earliest plant sporophytes.
   b. The first plant gametophytes were, like modern bryophytes, the same size as the earliest plant sporophytes.
   c. The first plant gametophytes were, like those of modern bryophytes, smaller than the earliest plant sporophytes.
   d. The earliest type of plant life cycle, represented by modern bryophytes, was similar to that of related green algae; the only diploid cells produced by early land plants were zygotes, so a multicellular sporophyte generation was lacking.
   e. The earliest type of plant life cycle involved the dispersal of small gametophytes within pollen, as in modern seed plants.
2. Place the plant structures in correct order, starting with earliest evolved and ending with most recently evolved.
   a. flower, seed, embryo, tracheid
   b. tracheid, embryo, seed, flower
   c. embryo, tracheid, seed, flower
   d. embryo, seed, tracheid, flower
   e. seed, embryo, tracheid, flower

## 25.2 An Evolutionary History of Land Plants

### Learning Outcome

1. Describe two major events in plant history and how they affected other life on Earth.

A billion years ago, Earth's terrestrial surface was comparatively devoid of life. Green or brown crusts of cyanobacteria most likely grew in moist places, but there would have been very little soil, no plants, and no animal life. The origin of the first land plants was a pivotal event in the history of life on Earth because the first plants enabled development of the first substantial soils and the evolution of modern plant communities. Had early land plants not played these key roles, animals would not have found sufficient food and shelter to allow land colonization.

How can we know about events such as the origin and diversification of land plants? One line of information comes from comparing molecular and other features of modern plants. For example, the genome sequence of the moss *Physcomitrella patens*, first reported in 2007, reveals the presence of genes that aid heat and drought tolerance, which are especially useful in the terrestrial habitat. Plant fossils, the preserved remains of plants that lived in earlier times, provide another line of information. Tough plant materials such as sporopollenin, cutin, and lignin do not readily decay and therefore foster the fossilization of plant parts that contain these tough materials (**Figure 25.9**). The study of fossils and the molecular, structural, and functional features of modern plants have revealed an amazing story—how plants gradually acquired adaptations, allowing them to conquer the land.

## Past and Present Seedless Plants Have Transformed Earth's Ecology

Past and present bryophytes have played important ecological roles by storing $CO_2$ as decay-resistant organic compounds. The abundant modern moss genus *Sphagnum* (see Figure 25.4) is widely harvested from natural wetlands for use as a garden soil conditioner because this moss does not easily decay and, so, adds valuable organic material that improves plant growth. *Sphagnum* moss contains so much decay-resistant mass that in many places, dead moss has accumulated over thousands of years, forming deep peat deposits. By storing very large amounts of organic carbon for long periods, *Sphagnum* helps to keep Earth's climate steady. Under cooler than normal conditions, *Sphagnum* grows more slowly and thus absorbs less $CO_2$, allowing atmospheric $CO_2$ to rise a bit, warming the climate a little. As the climate warms, *Sphagnum* grows faster and sponges up more $CO_2$, storing it in peat deposits. Such a reduction in atmospheric $CO_2$ returns the climate to slightly cooler conditions. In this way, ancient peat mosses, for which fossils more than 450 million years old are known, and modern peat mosses have long helped to keep the world's climate from changing dramatically. Today, experts are concerned that large regions currently dominated by peat mosses are being affected by land use alterations, overharvesting of moss peat, and global environmental changes that could reduce the ability of peat moss to moderate Earth's climate.

Fossils tell us that extensive forests dominated by tree-sized lycophytes, ferns, and early seed plants occurred in widespread swampy regions during the warm, moist Carboniferous period (354–290 mya) (**Figure 25.10**). Plants that died and fell into oxygen-poor sediments didn't completely decay and were eventually transformed into coal.

**Figure 25.9** **Fossil of *Pseudosalix handleyi*, an angiosperm.**

© Photo by Steven R. Manchester, University of Florida. Courtesy Botanical Society of America, St. Louis, MO, www.botany.org

**Concept Check:** *What biochemical components of plants favor the formation of fossils?*

**Figure 25.10** **Reconstruction of a Carboniferous (Coal Age) forest.** This ancient swampy forest was dominated by tree-sized lycophytes and pteridophytes, which later contributed to the formation of large coal deposits.

© Lee W. Wilcox

**Concept Check:** *Why did giant dragonflies occur during the Carboniferous, but not now?*

Much of today's coal is derived from the abundant remains of ancient plants, explaining why the Carboniferous is commonly known as the Coal Age. During this period, plants and coal stored so much carbon dioxide as organic carbon that Earth's atmospheric $CO_2$ fell dramatically and oxygen rose to the highest known level. This large increase in the oxygen content of Earth's atmosphere fostered the evolution of giant insects (see Figure 25.10). In today's lower atmospheric oxygen levels, these insects' modern relatives have a harder time obtaining sufficient oxygen and, so, are smaller.

### An Ancient Cataclysm Marked the Rise of Angiosperms

Diverse phyla of gymnosperms dominated Earth's vegetation through the Mesozoic era (248–65 mya), which is sometimes called the Age of Dinosaurs. In addition, fossils provide evidence that early mammals and flowering plants existed in the Mesozoic. Gymnosperms and early angiosperms were probably sources of food for early mammals as well as for herbivorous dinosaurs.

One fateful day about 65 mya, disaster struck from the sky, causing a dramatic change in the types of plants and animals that dominated terrestrial ecosystems. That day, at least one large meteorite crashed into the Earth near the present-day Yucatán Peninsula in Mexico. This episode is known as the **Cretaceous-Paleogene event** (also sometimes referred to as the K/T event). The impact, together with substantial volcanic activity that also occurred at this time, is thought to have produced large amounts of ash, smoke, and haze that dimmed the Sun's light long enough to kill many of the world's plants. Many types of plants became extinct, though some survived and their descendants persist to the present time. With a severely reduced food supply, most dinosaurs were also doomed, the exceptions being their descendants, the birds. The demise of the dinosaurs left room for birds and mammals to expand into many kinds of terrestrial habitats formerly inhabited by dinosaurs.

After the Cretaceous-Paleogene event, ferns dominated long enough to leave large numbers of fossil spores, and then surviving groups of flowering plants began to diversify into the space left by the extinction of previous plants. The rise of angiosperms fostered the diversification of beetles and other types of insects that associate with modern plants. Because most mammals, including humans, rely directly or indirectly upon food that angiosperms produce, the rise of angiosperms was also key to the evolutionary diversification of humans and other mammals. Diversification of angiosperms was also critical to the origin of agriculture, upon which most modern humans depend for food, shelter, and many other materials.

### 25.2 Reviewing the Concepts

- Paleobiologists and plant evolutionary biologists infer the history of land plants by analyzing the molecular features of modern plants and by comparing the structural features of fossil and modern plants (Figure 25.9).
- Ancient and modern plants transformed Earth's ecology by altering atmospheric chemistry and climate, and by influencing the evolutionary diversification of animals (Figure 25.10).

### 25.2 Testing Your Knowledge

1. Plants lack bones, but they have left many fossils. What decay-resistant plant materials are largely responsible for the occurrence of plant fossils?
   a. lignin in the xylem of the plant conducting system
   b. cutin produced on plant surfaces
   c. sporopollenin produced on spore surfaces
   d. all of the above
   e. none of the above
2. How have plants influenced the evolution of animals?
   a. The earliest land plants generated substantial soils, thereby fostering the growth of later plants that provided earliest land animals with food and shelter.
   b. Plant evolutionary diversification generated many different types of habitat, thereby influencing the evolutionary diversification of animals.
   c. The photosynthetic activity and fossilization of abundant Coal Age plants decreased the amount of $CO_2$ and increased the amount of $O_2$ in Earth's atmosphere, thereby affecting insect size and causing global climate change.
   d. The diversification of angiosperms following the Cretaceous-Paleogene event fostered the diversification of modern mammals and other animal lineages.
   e. All of the above are correct.

## 25.3 Diversity of Modern Gymnosperms

**Learning Outcome**

1. List three gymnosperm phyla and describe their importance to humans.

Figure 25.11 shows our current understanding of the relationships among modern seed plants, the flowering plants (angiosperms), and three phyla of gymnosperms. Gymnosperms appear to have originated from ancestral plants that had the capacity to produce wood, a tough material composed of xylem tissue, allowing such plants to grow to tree height. Tall height allows trees to gain an advantage in capturing light energy for photosynthesis. Consequently, wood is a common trait in gymnosperms. Following the early diversification of gymnosperms, an unknown, extinct gymnosperm lineage gave rise to the angiosperms—the flowering plants. Angiosperms inherited many traits from gymnosperm ancestors, including the capacity to produce seeds and wood. Table 25.1 provides a summary of the characteristics of modern seed plants.

### Cycads Are Endangered in the Wild but Are Widely Used as Ornamentals

Cycads (see Figure 25.8a) are regarded as the earliest-diverging modern gymnosperm phylum, originating more than 300 mya. Although cycads are an ancient group, molecular clock analyses indicate that modern cycad species are about 12 million years old. Nearly 300 cycad species occur today, primarily in tropical and subtropical regions. However, many species of cycads are rare, and their

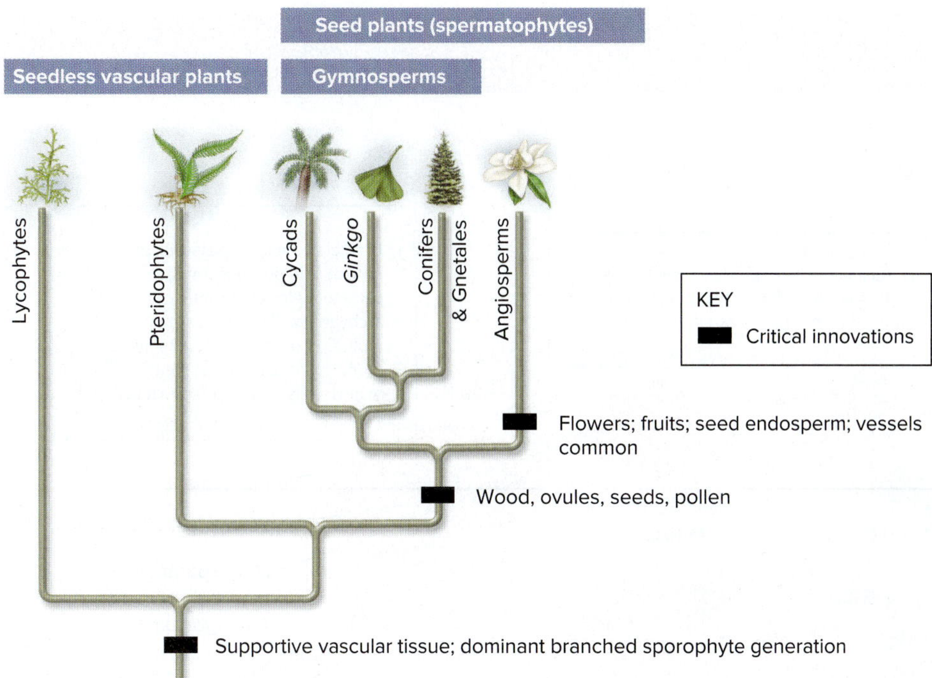


**Figure 25.11** **A phylogeny of modern seedless and seed plants.**

Cycads display spreading, palmlike leaves (*cycad* comes from a Greek word, meaning palm). Mature leaves of the African cycad *Encephalartos laurentianus* can reach an astounding 8.8 m in length! Individual cycad plants produce conspicuous cone-like structures that produce either pollen or **ovules,** structures that contain female gametophyte tissues, including egg cells (see Figure 25.8a). When mature, both types of reproductive structures emit odors that attract beetles. These insects carry pollen to ovules, where the pollen produces tubes that deliver sperm to eggs. Fertilization of an egg to form a zygote triggers development of the ovule into a seed.

tropical forest homes are increasingly threatened by human activities. Many cycads are listed as endangered, and commercial trade in cycads is regulated by CITES (Convention on International Trade in Endangered Species of Wild Fauna and Flora), a voluntary international agreement between governments to protect such species.

The structure of cycads is so interesting and attractive that many species are cultivated for use in outdoor plantings or as houseplants.

## *Ginkgo biloba* Is the Last Survivor of a Once Diverse Group

The beautiful tree *Ginkgo biloba* (see Figure 25.8b) is the single remaining species of a phylum that was much more diverse during the Age of Dinosaurs. Today, *G. biloba* may be nearly extinct in the wild; widely cultivated modern *Ginkgo* trees are descended from seeds produced by a tree found in a remote Japanese temple garden and taken to Europe by 17th-century explorers.

*G. biloba* trees are widely planted along city streets because they are ornamental and they tolerate cold, heat, and pollution better than many other trees. In addition, these trees are long-lived—individuals can live for more than a thousand years and grow to 30 m in height.

**Table 25.1 Distinguishing Features of Modern Streptophyte Algae and Land Plants**

Streptophyte Algae

Primarily aquatic habitat; zygotic life cycle; embryos and tough-walled spores absent

LAND PLANTS (EMBRYOPHYTES)

Primarily terrestrial habitat; alternation of two multicellular generations—diploid sporophyte and haploid gametophyte; multicellular embryos are nutritionally dependent on maternal gametophyte for at least some time during development; spore-producing sporangia; gamete-producing gametangia; tough-walled spores

Nonvascular plants (Bryophytes) (**liverworts, mosses, hornworts**)

Dominant gametophyte generation; supportive, lignin-containing vascular tissue absent; true roots, stems, leaves absent; sporophytes unbranched and unable to grow independently of gametophytes

VASCULAR PLANTS (TRACHEOPHYTES) (lycophytes, pteridophytes, spermatophytes)

Dominant sporophyte generation; lignin-walled water-conducting tissue—xylem; specialized organic food-conducting tissue—phloem; sporophytes branched and eventually become able to grow independently of gametophytes; all have stems; most also have leaves and roots

**Lycophytes** Leaves generally small with a single, unbranched vein (lycophylls); sporangia borne on sides of stems

PTERIDOPHYTES + SEED PLANTS

**Pteridophytes** Leaves generally large with extensively branched vein system; sporangia borne on leaves; seeds absent

SEED PLANTS (SPERMATOPHYTES)

Seeds present; leaves have branched vein systems

GYMNOSPERMS (**cycads, ginkgos, conifers**)

Flowers and fruits absent; seed food stored in female gametophyte, endosperm absent

**Angiosperms (flowering plants)**

Flowers and fruits present; seed food stored after fertilization in endosperm formed by double fertilization

Key: **Phyla;** LARGER MONOPHYLETIC CLADES (synonyms). All other classification terms are not clades.

## Biology Principle

### Living Organisms Grow and Develop

This diagram illustrates the entire seed-to-seed growth and development cycle of conifers.

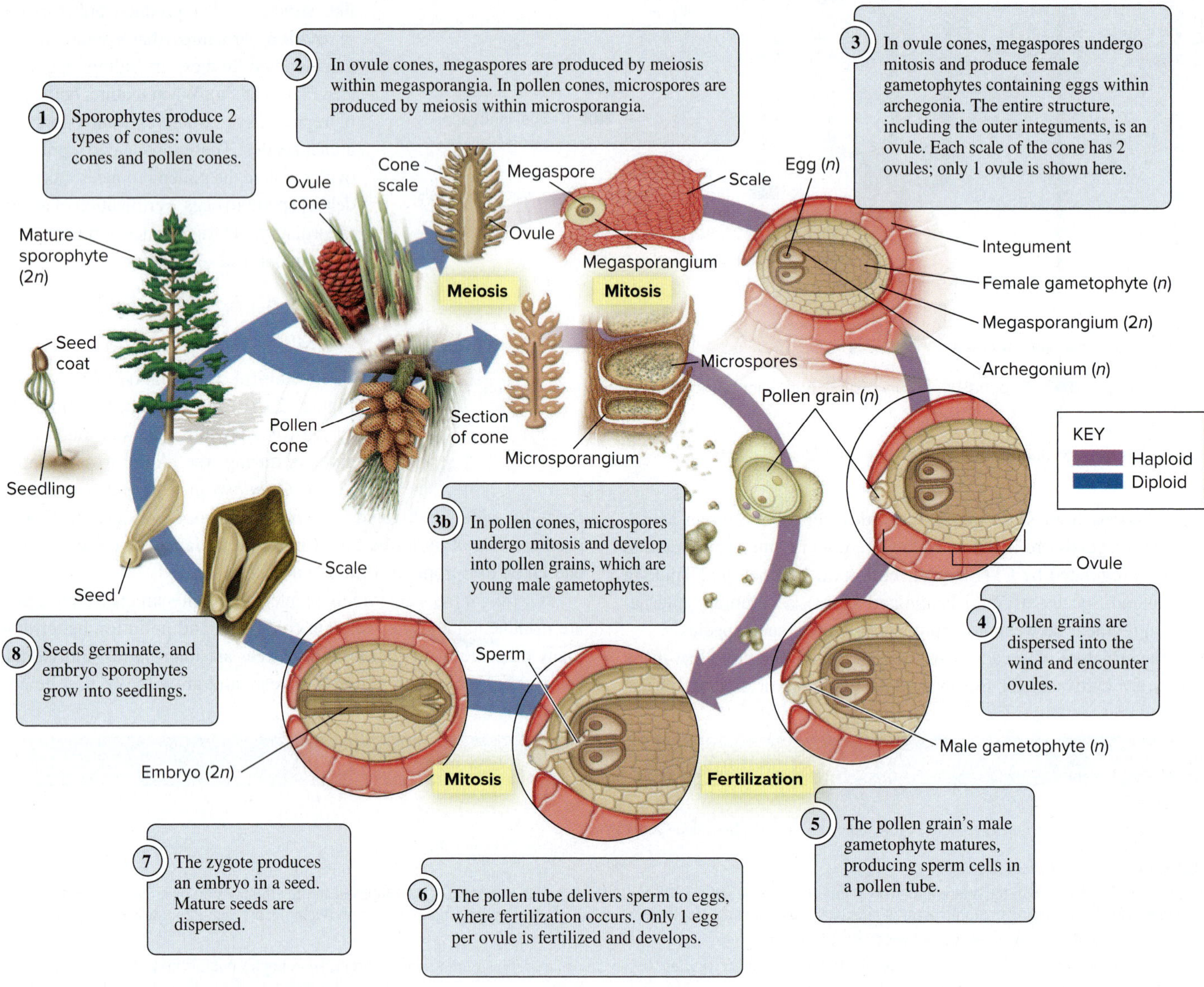


**Figure 25.12** The life cycle of the genus *Pinus*.

Individual trees have sex chromosomes much like those of humans, so they produce either ovules and seeds or pollen. Seed embryos are protected by a fleshy, bad-smelling outer seed coat and a hard, inner seed coat. For street-side or garden plantings, people usually select the pollen-producing trees to avoid the stinky seeds.

### Conifers Are the Most Diverse Modern Gymnosperm Lineage

The conifers (see Figure 25.8c) are a lineage of trees named for their seed cones, of which pinecones are familiar examples. Modern conifer families include more than 50 genera. Conifers are particularly common in mountain and high-latitude forests and are important sources of wood and paper pulp.

Conifers produce simple pollen cones and more complex ovule-bearing cones (Figure 25.12). The pollen cones of conifers bear many leaflike structures, each bearing a sporangium in which meiosis occurs and pollen grains develop. This pollen-producing sporangium is called a microsporangium because the pollen grains develop from small spores known as microspores. The ovule cones are composed of many short branch systems that bear ovules, within which female gametophyte tissues produce eggs.

When conifer pollen is mature, it is released into the wind, which transports pollen to ovules. When released from pollen tubes, sperm fuse with eggs, generating zygotes that grow into the embryos within seeds. Altogether, it takes nearly 2 years for pine (the genus *Pinus*) to complete the processes of male and female gamete development, fertilization, and seed development. The seeds of pine and some other conifers develop wings that aid in wind dispersal. Other conifers, such as yew and juniper, produce seeds or cones with bright-colored, fleshy coatings that are attractive to birds, which help to disperse the seeds.

Many conifers occur in cold climates and thus display numerous adaptations to such environments. Their conical shapes and flexible branches help conifer trees shed snow, preventing heavy snow accumulations from breaking branches. People who use conifers in landscape plantings also value these traits. Conifer leaf shape and structure are adapted to resist damage from drought that occurs in both summer and winter, when liquid water is scarce. Conifer leaves are often scalelike or needle-shaped; these shapes reduce the area of leaf surface from which water can evaporate. A thick, waxy cuticle coats conifer leaf surfaces, retarding water loss and preventing attack by disease organisms. Many conifers are evergreen; that is, their leaves live for more than 1 year before being shed and are not all shed during the same season. Retaining leaves through winter helps conifers start up photosynthesis earlier than deciduous trees, which in spring must replace leaves lost during the previous autumn. Evergreen leaves thus provide an advantage in the short growth season of alpine or high-latitude environments.

The conifer phylum also includes the Gnetales, an order of three genera, *Gnetum, Ephedra,* and *Welwitschia,* that feature distinctive adaptations:

- *Gnetum* is unusual among modern gymnosperms in having broad leaves similar to those of many tropical plants (**Figure 25.13a**). Such leaves foster light capture in the dim forest habitat. More than 30 species of the genus *Gnetum* occur as vines, shrubs, or trees in tropical Africa or Asia.
- *Ephedra*, native to arid regions of the southwestern U.S., has tiny, brown, scalelike leaves and green, photosynthetic stems (**Figure 25.13b**). These adaptations help to conserve water by preventing water loss that would otherwise occur from the surfaces of larger leaves. Like all plants, *Ephedra* produces **secondary metabolites,** compounds that are not essential for basic cellular structure and growth but that aid plant structure, protection, or reproduction. Plant secondary compounds influence humans in many ways (see Section 25.4), and *Ephedra*'s secondary compounds have long been known to affect humans. For example, early settlers of the western U.S. used *Ephedra* to treat colds and other medical conditions. The modern decongestant drug pseudoephedrine is based on the chemical structure of ephedrine, which was named for and originally obtained from *Ephedra*. Pseudoephedrine sales are now restricted in many places because this compound can be used as a starting point for the synthesis of illegal drugs. Ephedrine has also been used to enhance sports performance, a practice that has elicited medical concern.
- *Welwitschia* has only one living representative species. *Welwitschia mirabilis* is a strange-looking plant that grows in the coastal Namib Desert of southwestern Africa, one of the driest places on Earth (**Figure 25.13c**). A long taproot anchors a stubby stem that barely emerges from the ground. Two very long leaves grow from the stem but are rapidly shredded by the wind into many strips. The plant is thought to obtain most of its water from coastal fog that accumulates on the leaves, explaining how *W. mirabilis* can grow and reproduce in such a dry place.

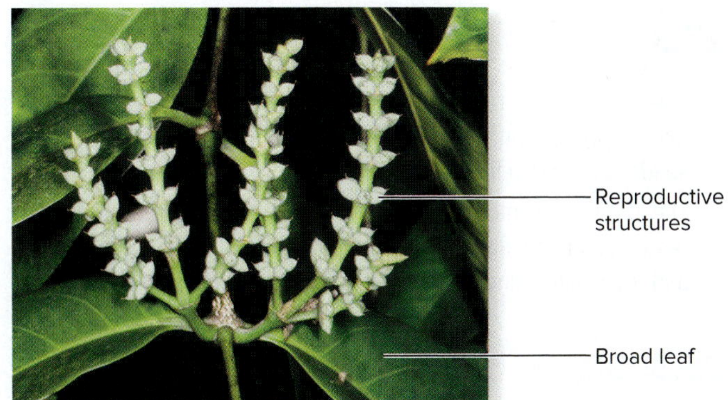


**(a) Genus *Gnetum***

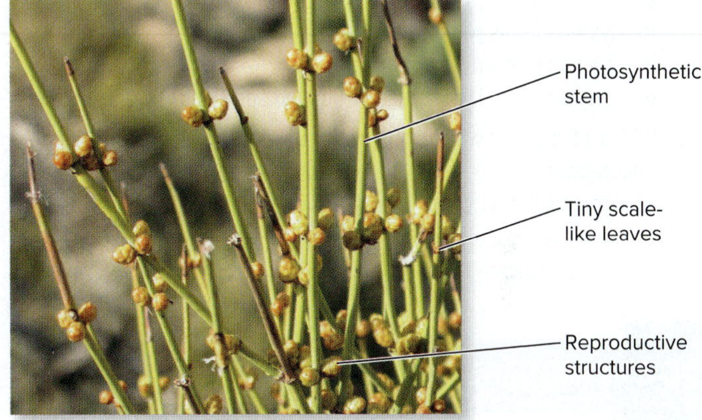


**(b) *Ephedra californica***

**(c) *Welwitschia mirabilis***

**Figure 25.13** **Gnetales. (a)** A tropical plant of the genus *Gnetum,* displaying broad leaves and reproductive structures. **(b)** *Ephedra californica* growing in deserts of North America, showing minuscule, brown leaves on green, photosynthetic stems and reproductive structures. **(c)** *Welwitschia mirabilis* growing in the Namib Desert of southwestern Africa, showing long, wind-shredded leaves and reproductive structures.

## 25.3 Reviewing the Concepts

- The seed plants (also called spermatophytes) consist of the gymnosperms, with exposed seeds, and the angiosperms, with seeds enclosed in fruits (Figures 25.11, 25.12; Table 25.1).
- The diversity of modern gymnosperms includes three modern phyla that are all important to human life: cycads, *Ginkgo biloba,* and the conifers (including Gnetales) (Figure 25.13).

## 25.3 Testing Your Knowledge

1. In what way are gymnosperms different from angiosperms?
   a. Gymnosperms possess embryos.
   b. Gymnosperms possess lignified xylem.
   c. Gymnosperms possess seeds.
   d. Gymnosperms possess wood.
   e. Gymnosperms lack flowers, fruit, and seed endosperm.
2. Which of the following plant traits represent examples of plant adaptations that are useful to humans?
   a. lignin-walled xylem
   b. stems, leaves, and roots
   c. seed endosperm
   d. secondary metabolites
   e. all of the above

# 25.4 Diversity of Modern Angiosperms

### Learning Outcomes

1. List four flower organs and their functions, and explain how each flower part may have first evolved.
2. Describe how diversification of flowers and fruits enhances seed production and dispersal.
3. Name three major types of angiosperm secondary metabolites and how these affect animals.

The flowering plants, including hundreds of thousands of species, are the most diverse group of modern land plants. The evolutionary success of angiosperms is based upon traits that are absent or rare in other plants—flowers, fruits, and endosperm. Recall that a flower is a short stem bearing reproductive organs, fruits develop from flowers, and endosperm is a nutritive seed tissue. Flowers foster seed production, fruits favor seed dispersal, and endosperm food helps embryos within seeds grow into seedlings. Angiosperms also possess more efficient water-conducting systems than do other plants. In addition to tracheids, most angiosperms also possess water-conducting **vessels,** which are wider than tracheids and therefore increase the efficiency of water flow through plants. Though similar structures occur in some seedless plants and gymnosperms, these are thought to have evolved independently. Angiosperms also possess vascular systems that branch more densely than do those of other plants, a trait that allows flowering plants to more effectively conduct materials through the body.

More than 124 mya, an extinct group of gymnosperms gave rise to the angiosperms—the flowering plants. Charles Darwin famously referred to the origin of the flowering plants as "an abominable mystery," one that has not been fully solved even today. Modern plant evolutionary biologists continue to study the evolutionary origin of distinctive angiosperm traits, such as the flower, that help to explain why flowering plants are so diverse.

## Flowers Include Stamens and/or Carpels and Often Petals and Sepals

Flowers are produced at stem tips and may contain four types of organs: sepals, petals, pollen-producing stamens, and ovule-producing carpels (**Figure 25.14**). These flower organs are supported by tissue located at the tip of a flower stem. The functioning of several genes that control flower organ development explains why carpels are the most central flower organs, why stamens surround carpels, and why petals and sepals are the outermost flower organs.

Many flowers produce attractive **petals** that play a role in **pollination,** the transfer of pollen among flowers of similar type. **Sepals** of many flowers are green and form the outer layer of flower buds. The sepals of some flowers, such as tulip and lily, look similar to petals, in which case both sepals and petals are known as tepals. All of a flower's sepals and petals are collectively known as the perianth. Most flowers produce one or more **stamens,** the structures that produce and disperse pollen. Most flowers also contain a single or multiple **carpels,** structures that produce ovules. Some flowers produce only a single carpel, others display several separate carpels, and many possess several carpels that are fused together into a compound structure. Both a single carpel and fused carpels are referred to as a **pistil** (from the Latin *pistillum,* meaning pestle) because it resembles the device people use to grind materials to powder in a mortar (see Figure 25.14). Pistil structure can be divided into three regions having distinct functions. The topmost portion of the pistil, known as the stigma, receives and recognizes pollen of the appropriate species or genotype. The elongate middle portion of the pistil is called the style. The lowermost portion of the pistil is the ovary, which encloses and protects ovules.

During the flowering plant life cycle (**Figure 25.15**), the stigma allows pollen of appropriate genetic type to germinate, producing a

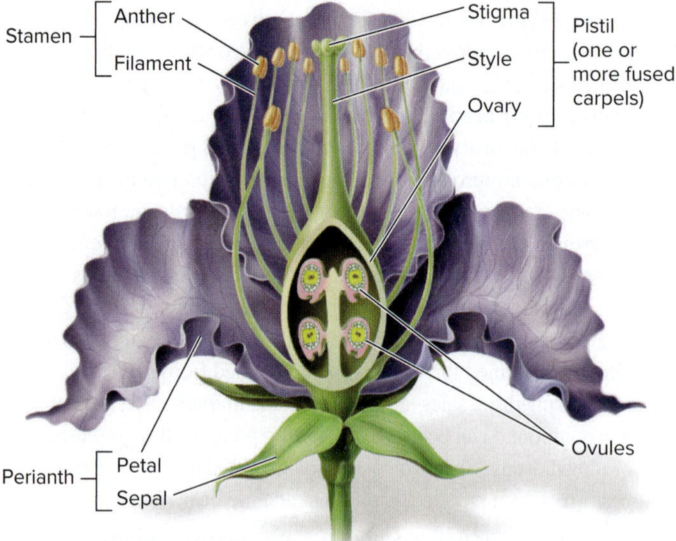


**Figure 25.14** **Generalized flower structure.** Although flowers are diverse in size, shape, and color, they commonly have the parts illustrated here.

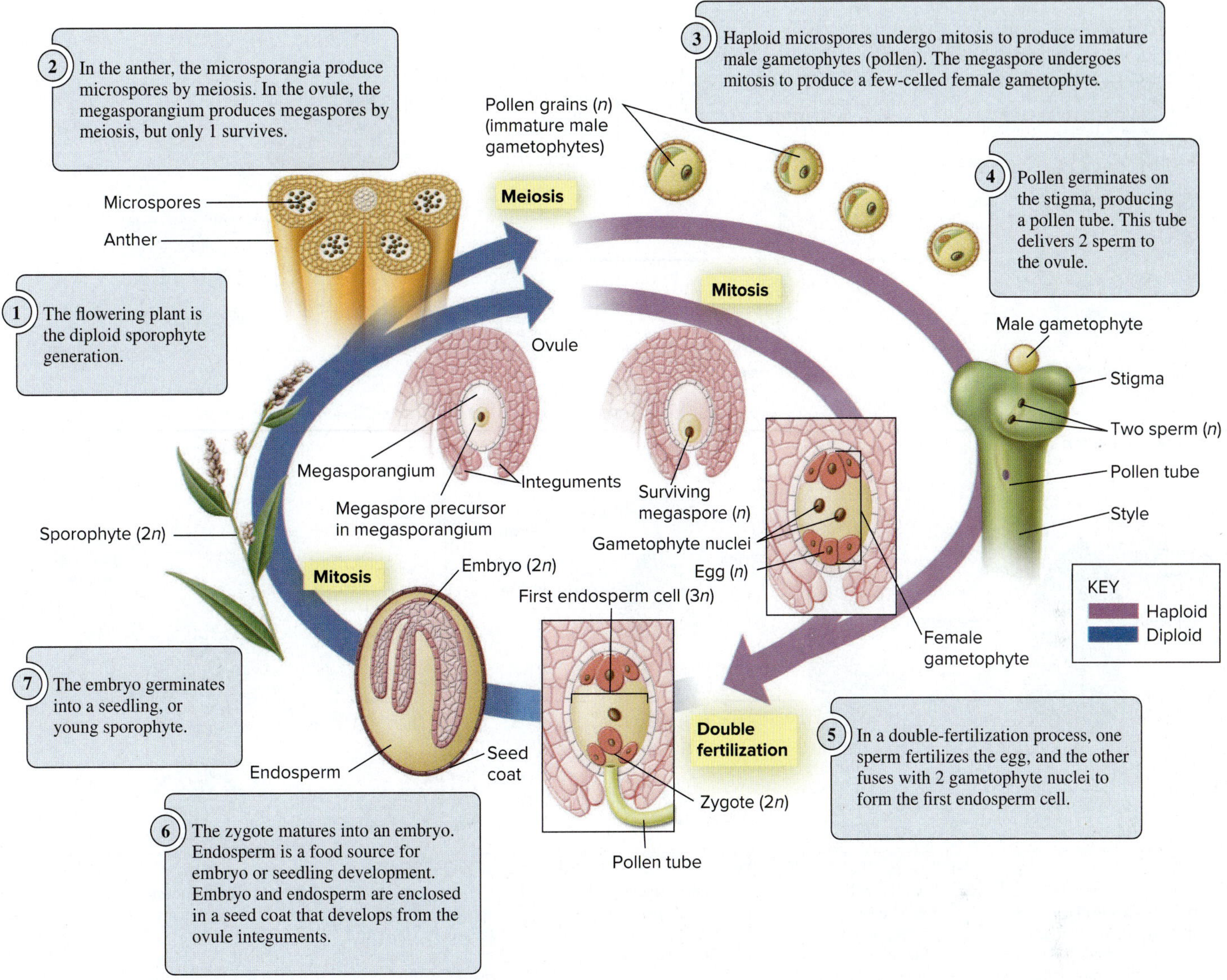


**Figure 25.15 The life cycle of a flowering plant, illustrated by the genus *Polygonum*.** Flowering plant life cycles differ in the length of the cycle and the number of cells and nuclei in the female gametophyte, the seven cells and eight nuclei of *Polygonum* being common.

long pollen tube that grows through the style. The pollen tube delivers two sperm cells to ovules. In the distinctive angiosperm process known as **double fertilization,** one sperm nucleus fuses with the egg nucleus to form a zygote, and the other sperm nucleus fuses with the nuclei of other haploid cells of the female gametophyte. The latter is the first step in the development of a characteristic angiosperm nutritive tissue known as endosperm. Fed by the endosperm, the zygote develops into an embryo, and the ovule develops into a seed. Ovaries (and sometimes additional flower parts) develop into fruits.

## EVOLUTIONARY CONNECTIONS

### Flower Organs Evolved from Leaflike Structures

Petals and sepals are often similar to leaves in shape and structure of their vascular system, features that suggest the evolution of petals and sepals from leaves. Structural comparison and molecular data indicate that stamens and pistils likewise originated from leaflike structures. Angiosperm stamens resemble leaflike structures of gymnosperms that produce the microspores that develop into pollen. Plant evolutionary biologists hypothesize that over time, the microspore-producing tissues became restricted to the stamen tips, generating anthers on a supporting filament (**Figure 25.16a**). Plant evolutionary biologists likewise hypothesize that carpels are homologous to leaflike structures of gymnosperms that bear ovules on their surfaces. The carpels of some early-diverging modern plants are leaflike structures that fold over ovules, with the carpel edges stuck together by secretions (**Figure 25.16b**). Most modern flowers produce carpels whose edges have fused together into a tube whose lower portion (ovary) encloses and protects ovules, improving plant fitness. The first flowers arose when early stamens, carpels, and perianth parts aggregated into a single structure.

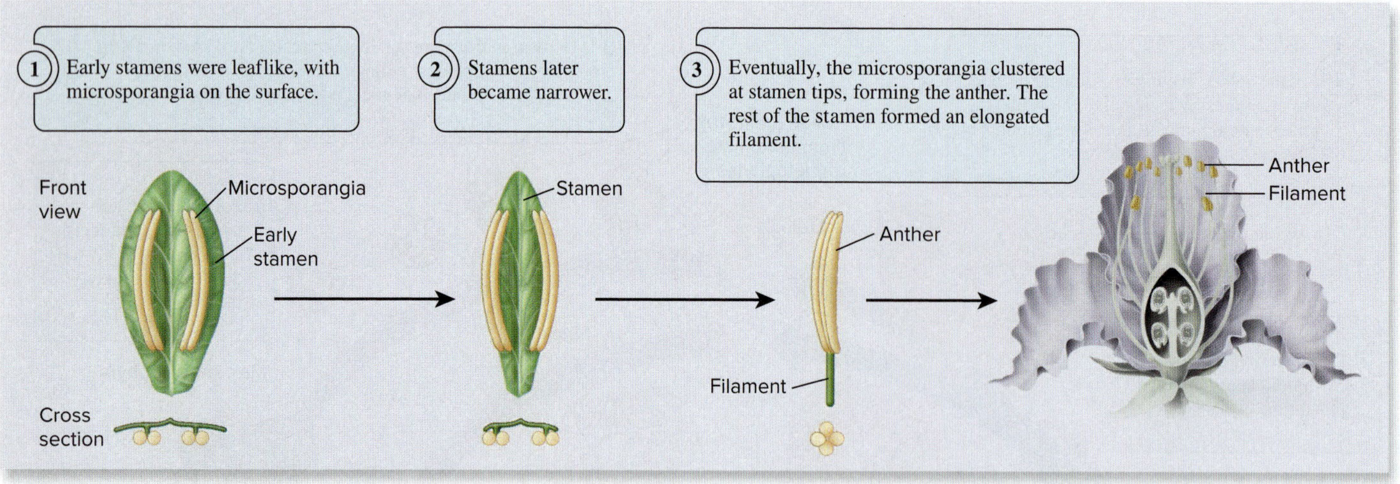


**(a) Stamen evolution**

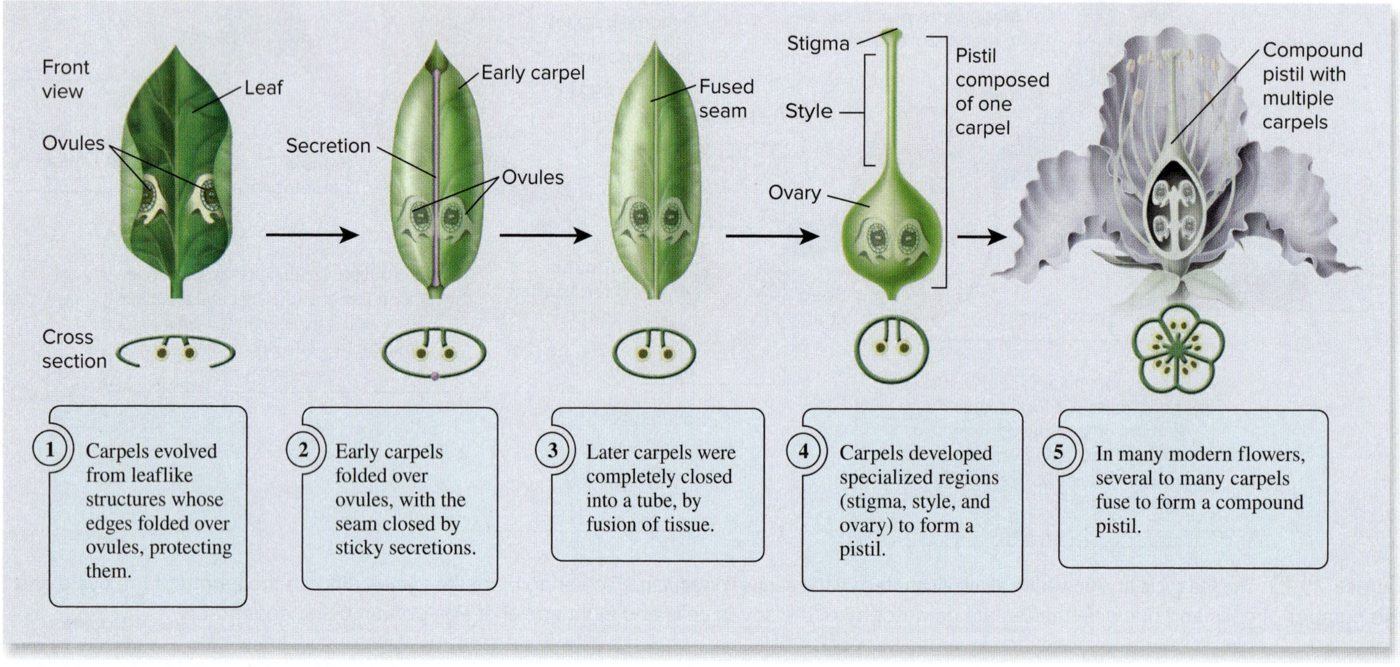


**(b) Carpel evolution**

**Figure 25.16 Hypothetical evolutionary origin of stamens, carpels, and pistils.** Plant biologists test these models by searching for new fossils or generating additional molecular data.

## Flowering Plants Diversified into Several Lineages, Including Monocots and Eudicots

Figure 25.17 presents our current understanding of the relationships among modern angiosperm groups. According to gene-sequencing studies, the earliest-diverging modern angiosperms are represented by a single species called *Amborella trichopoda*, a shrub that lives in forests on the South Pacific island of New Caledonia. Later-diverging groups of angiosperms include water lilies, the star anise plant, and other close relatives (see Figure 25.17). Magnoliids, represented by the genus *Magnolia*, are the next-diverging group. Magnoliids are closely related to two very large and diverse angiosperm lineages: the **monocots** and the **eudicots.**

Monocots and eudicots are named for differences in the number of embryonic leaves called cotyledons. Monocot embryos possess one cotyledon, whereas eudicots possess two cotyledons. Also, monocots typically have petals, stamens, or other floral parts numbering three or some multiple of three, whereas eudicot flower parts tend to occur in fours, fives, or a multiple of four or five. Tulips, daffodils, and irises are examples of monocots, whereas roses, snapdragons, and daisies are examples of eudicots.

**Figure 25.17** **A phylogeny showing the major modern angiosperm lineages.**

## Flower Diversification Has Fostered Efficient Seed Production

During the diversification of flowering plants, flower evolution has involved several types of changes that foster the transfer of pollen from one plant to another. Effective pollination is essential to efficient seed production because it minimizes the amount of energy plants must expend to accomplish sexual reproduction. The fusion of flower organs, clustering of flowers into groups, and reduction of the perianth are some examples of changes leading to effective pollination.

- Many flowers have fused petals that form floral tubes. Such tubes tend to accumulate sugar-rich nectar that provides a reward for **pollinators,** animals that transfer pollen among plants. The diameters of floral tubes vary among flowers and are evolutionarily tuned to the feeding structures of diverse animals, which range from the narrow tongues of butterflies to the wider bills of nectar-feeding birds (Figure 25.18). Nectar-feeding bats stick their heads into even larger, tubular flowers to lap up nectar with their tongues.
- Many plants produce **inflorescences,** groups of flowers tightly clustered together, which occur in several types. The zinnia features a type of inflorescence in which many small flowers are clustered into a head (Figure 25.18a). The flowers at the center of a sunflower head function in reproduction and lack showy petals, but flowers at the rim have showy petals that attract pollinators. Flower heads allow pollinators to transfer pollen among a large number of flowers at the same time.
- The grass family features flowers with few or no perianths, which explains why grass flowers are not showy. This adaptation fosters pollination by wind, since petals would get in the way of such pollen transfer. Since grasses often grow in dense populations, wind pollination can be very effective.

## Diverse Types of Fruits Function in Seed Dispersal

Fruits are structures that develop from ovary walls in diverse ways that aid the dispersal of the enclosed seeds. Seed dispersal helps to prevent seedlings from competing with their larger parents for scarce

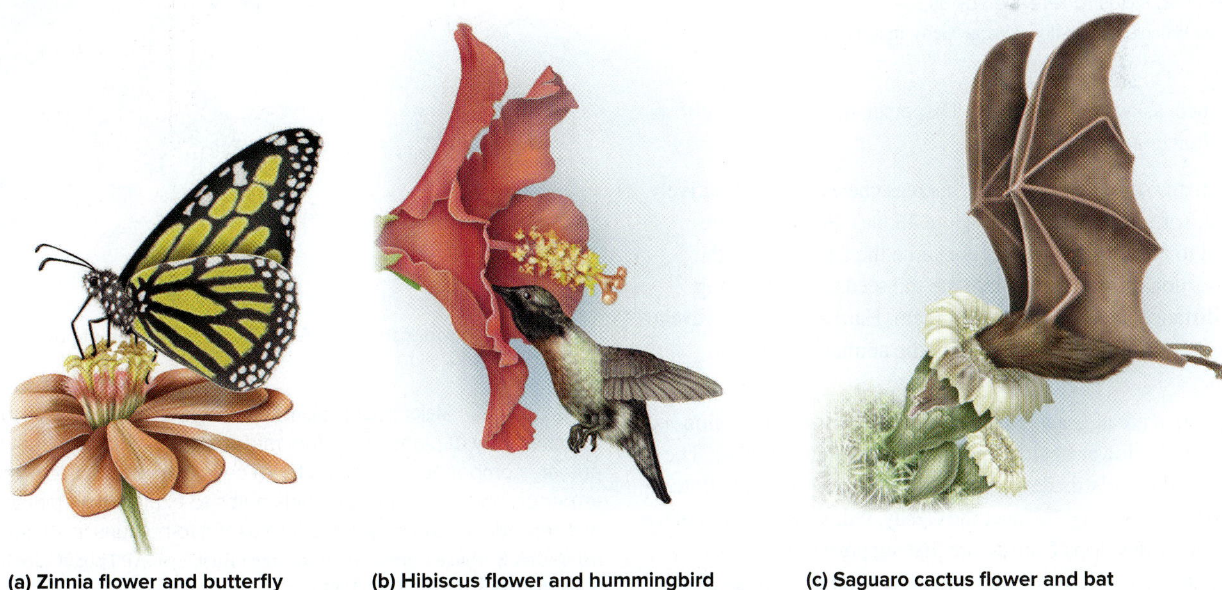

**(a) Zinnia flower and butterfly** **(b) Hibiscus flower and hummingbird** **(c) Saguaro cactus flower and bat**

**Figure 25.18** **Flowers whose perianths form nectar-containing floral tubes of different widths that accommodate different pollinators.** **(a)** This zinnia is composed of an outer rim of showy flowers and a central disc of narrow, tubular flowers that produce nectar. Butterflies, but not other pollinators, are able to reach the nectar by means of narrow tongues. **(b)** The hibiscus flower produces nectar in a floral tube whose diameter corresponds to the dimensions of a hummingbird bill. **(c)** The saguaro cactus (*Carnegiea gigantea*) flower forms a floral tube that is wide enough for nectar-feeding bats to get their heads inside. The cactus flower has been drawn here as if it were transparent, to illustrate bat pollination.

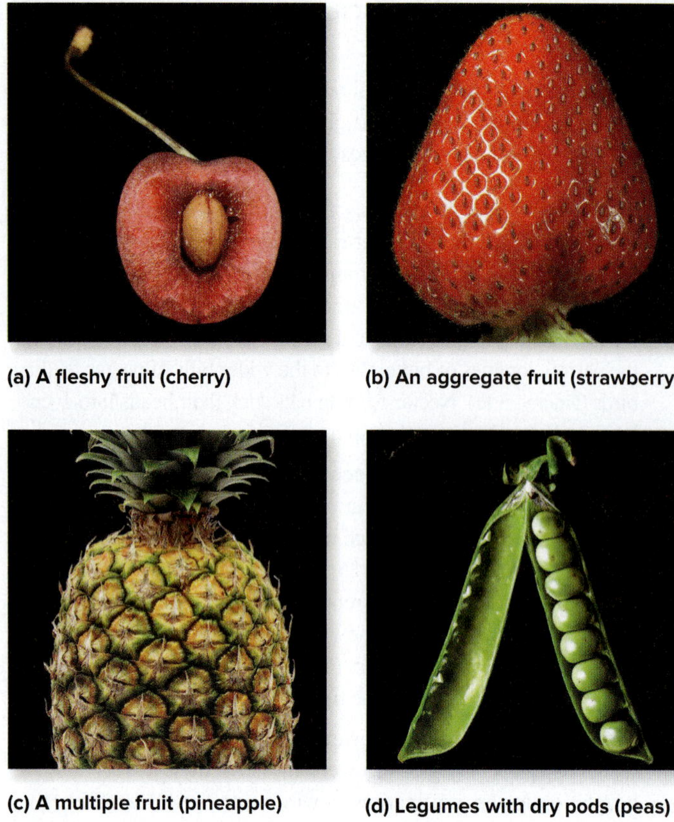


**Figure 25.19** **Representative fruit types.** **(a)** The cherry is an example of a fleshy fruit that is adapted to attract animals that consume the fruits and excrete the seeds. **(b)** Strawberry is an aggregate fruit, consisting of many tiny, single-seeded fruits produced by a single flower. The fruits are embedded in the surface of tissue that originally supported the flower. **(c)** Pineapple is a large, multiple fruit formed by the aggregation of smaller fruits, each produced by one of the flowers in an inflorescence. **(d)** Peas produce legumes, fruits that open on two sides to release seeds.
*(a–c)* © Lee W. Wilcox; *(d)* © Gloomerique/Getty Images RF

resources such as water and light. Dispersal of seeds also allows plants to colonize new habitats.

- Many mature angiosperm fruits, such as cherries, are attractively colored, soft, juicy, and tasty (Figure 25.19a). Such fruits are adapted to attract animals that consume the fruits, digest the outer portion as food, and eliminate the seeds after traveling some distance, thereby dispersing them. Hard seed coats prevent such seeds from being destroyed by the animal's digestive system.
- Strawberries are aggregate fruits, many fruits that all develop from a single flower having multiple pistils (Figure 25.19b). The ovaries of these pistils develop into tiny, single-seeded, yellow fruits on a strawberry surface; the fleshy, red, sweet portion of a strawberry develops from tissue that supported the flower. Aggregate fruits allow a single animal consumer, such as a bird, to disperse many seeds at the same time.
- Pineapples (Figure 25.19c) are juicy, multiple fruits that develop when many ovaries of an inflorescence fuse together. Such multiple fruits are larger and attract relatively large animals that have the ability to disperse seeds over long distances.

**(a) Natural rubber produced by *Hevea brasiliensis* is an example of a complex terpene.**

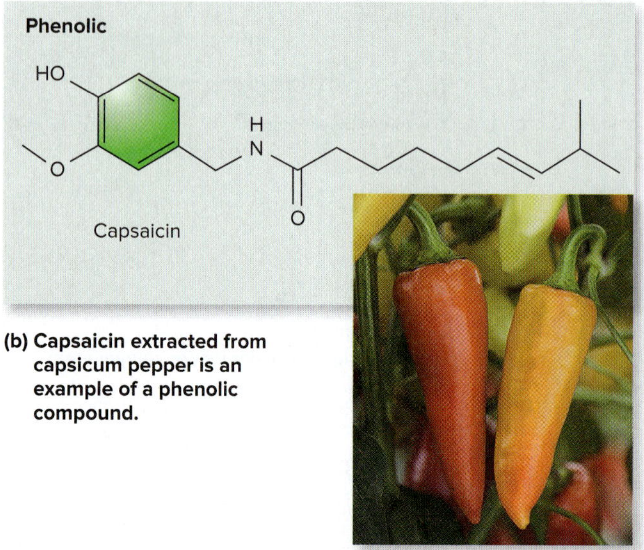


**(b) Capsaicin extracted from capsicum pepper is an example of a phenolic compound.**

**(c) Caffeine produced by *Coffea arabica* is an example of an alkaloid.**

**Figure 25.20** **Major types of plant secondary metabolites.** Note that the chemistry of plant secondary metabolites differs from that of the primary compounds produced by all cells. The production by plants of terpenes, phenolics, and alkaloids helps to explain how plants survive and reproduce, and why plants are useful to humans in so many ways.
*(a)* © Eddi Boehnke/Corbis; *(b)* © Jonathan Buckley/GAP Photo/Getty Images; *(c)* © Science Photo Library/Alamy

- The plant family informally known as legumes is named for its distinctive fruits, dry pods that split open down both sides when seeds are mature, thereby releasing them (Figure 25.19d).

- Nuts and grains are additional examples of dry fruits. Grains are the characteristic single-seeded fruits of cereal grasses such as rice, corn (maize), barley, and wheat. Other plants produce dry fruits with surface burrs that attach to animal fur.

## Angiosperms Produce Diverse Secondary Metabolites That Play Important Roles in Structure, Reproduction, and Protection

As noted in the discussion of the gymnosperm *Ephedra* (see Section 25.3), secondary metabolites are organic compounds that are not essential for basic cell structure and growth but aid survival, structure, and reproduction. Certain prokaryotes, protists, and fungi, as well as some animals and all plants, produce secondary metabolites, but these compounds are most diverse in the angiosperms. Most of the 100,000 or so known types of secondary metabolites are produced by flowering plants. The diversification of these compounds has strongly influenced flowering plant evolution, as well as everyday human life. Three major classes of plant secondary metabolites occur: (1) terpenes and terpenoids, (2) phenolics, and (3) alkaloids (Figure 25.20).

- About 25,000 types of plant terpenes and terpenoids are derived from different arrangements of the simple hydrocarbon gas isoprene. Taxol, used in the treatment of cancer, is a terpene, as are citronella and a variety of other compounds that repel insects. Rubber, turpentine, rosin, and amber are complex terpenoids that likewise serve important roles in plant biology, as well as having useful human applications.
- Phenolic compounds are responsible for some flower and fruit colors as well as the distinctive flavors of cinnamon, nutmeg, ginger, cloves, chilies, and vanilla. Phenolics absorb ultraviolet radiation, thereby preventing damage to cellular DNA. They also help to defend plants against insects and disease microbes. Some phenolic compounds found in tea, red wine, grape juice, and blueberries are antioxidants that detoxify free radicals, highly reactive by-products of normal metabolism. Such phenolics thereby prevent cellular damage in the plant, as well as in animals that consume them.
- Alkaloids are nitrogen-containing secondary metabolites that often have potent effects on the animal nervous system. Plants produce at least 12,000 types of alkaloids, and certain species produce many alkaloids. Caffeine, nicotine, morphine, ephedrine, cocaine, and codeine are examples of alkaloids that influence the physiology and behavior of humans and, so, their use and abuse are of societal concern. Like flowers and fruit structures, secondary metabolites are useful in distinguishing among Earth's hundreds of thousands of flowering plant species.

# FEATURE INVESTIGATION

## Hillig and Mahlberg Analyzed Secondary Metabolites to Explore Species Diversification in the Genus *Cannabis*

The genus *Cannabis* has long been a source of hemp fiber used for ropes and fabric. People have also used *Cannabis* (also known as marijuana) in traditional medicine and as a hallucinogenic drug. *Cannabis* produces THC (tetrahydrocannabinol), a type of alkaloid called a cannabinoid. THC and other cannabinoids are produced in glandular hairs that cover most of the *Cannabis* plant's surface but are particularly rich in leaves located near the flowers. THC mimics compounds known as endocannabinoids, which are naturally produced and act in the animal brain and elsewhere in the body. THC affects humans by binding to receptor proteins in plasma membranes in the same way as natural endocannabinoids. Cancer and AIDS patients sometimes choose to use cannabis to reduce nausea and stimulate their appetite, counteracting side effects of cancer treatment and the wasting that accompanies both diseases.

Because humans have subjected cultivated *Cannabis* plants to artificial selection for so long, plant biologists have been uncertain how cultivated *Cannabis* species are related to those in the wild. In the past, plants cultivated for drug production were often identified as *Cannabis indica*, whereas those grown for hemp were typically known as *Cannabis sativa*. However, these species are difficult to distinguish on the basis of structural features, and the relevance of these names to wild cannabis was unknown. At the same time, species identification has become important for biodiversity studies, agriculture, and law enforcement. For these reasons, plant biologists Karl Hillig and Paul Mahlberg hypothesized that ratios of THC to another cannabinoid known as CBD (cannabidiol) might aid in defining *Cannabis* species and identifying plant samples at the species level, as shown in Figure 25.21. Like THC, CBD has important applications in modern medicine.

To test their hypothesis, the investigators began by collecting *Cannabis* fruits from nearly a hundred diverse locations around the world and then growing these plants from seed under uniform conditions in a greenhouse. The investigators next extracted cannabinoids, analyzed them by means of gas chromatography (a laboratory technique used to identify components of a mixture), and determined the ratios of THC to CBD. The results, published in 2004, suggested that the wild and cultivated *Cannabis* samples evaluated in this study could be classified into two species: *C. sativa*, displaying relatively low THC levels, and *C. indica*, having relatively high THC levels. As a result of this work, ecologists, agricultural scientists, and forensic scientists can reliably use ratios of THC to CBD to classify samples. Similar studies of plant secondary metabolites offer the benefit of uncovering potential new medicinal compounds or other applications of significance to humans.

### *Experimental Questions*

1. **SCISKILLS ▸** Designing an experiment requires a plan to achieve an adequate number of samples in order to allow statistical analysis. Hillig and Mahlberg obtained nearly a hundred *Cannabis* fruit samples from around the world. Why were so many samples needed?
2. Why did Hillig and Mahlberg collect samples from the leaves growing nearest the flowers?

**Figure 25.21** Hillig and Mahlberg's analysis of secondary metabolites in the genus *Cannabis*.
(*1*) © Phil Schermeister/Getty Images; (*2*) © Matthew Kellett/Alamy

**GOAL** To determine if cannabinoids aid in distinguishing *Cannabis* species.

**KEY MATERIALS** *Cannabis* fruits obtained from nearly 100 different worldwide sources.

| | Experimental level | Conceptual level |
|---|---|---|
| 1 | Grow multiple *Cannabis* plants from seeds under standard conditions in a greenhouse. | Eliminates differential environmental effects on cannabinoid content. |
| 2 | Extract cannabinoids from leaves surrounding flowers. | Extracts were made from tissues richest in cannabinoids; this reduces the chance that cannabinoids present in lower levels would be missed. |
| 3 | Analyze cannabinoids by gas chromatography. Determine ratios of THC (tetrahydrocannabinol) to CBD (cannabidiol) in about 200 *Cannabis* plants. | Previous data suggested that ratios of THC to CBD might be different in separate species. |

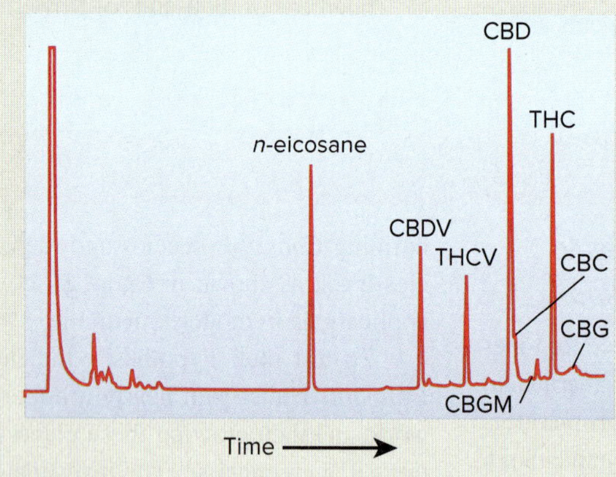


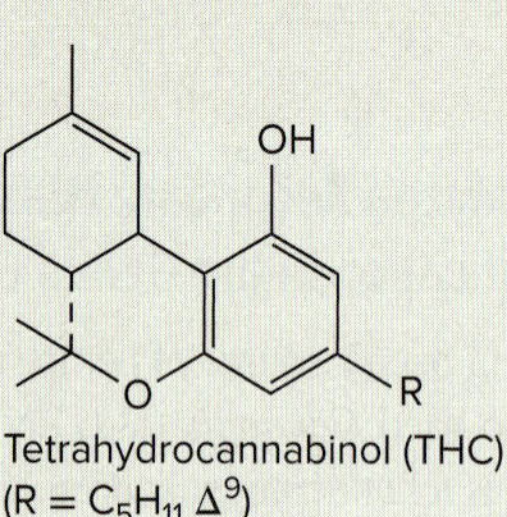

Tetrahydrocannabinol (THC)
($R = C_5H_{11}\ \Delta^9$)

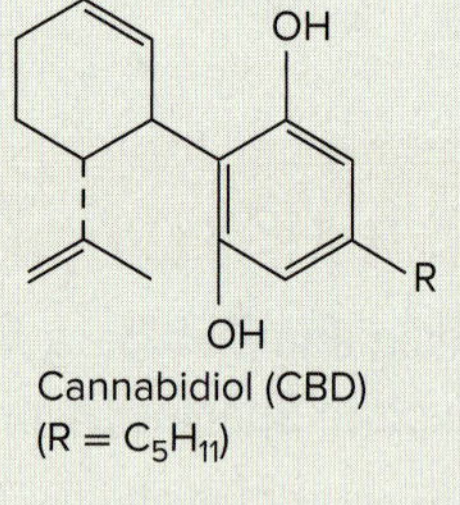

Cannabidiol (CBD)
($R = C_5H_{11}$)

**4 THE DATA**

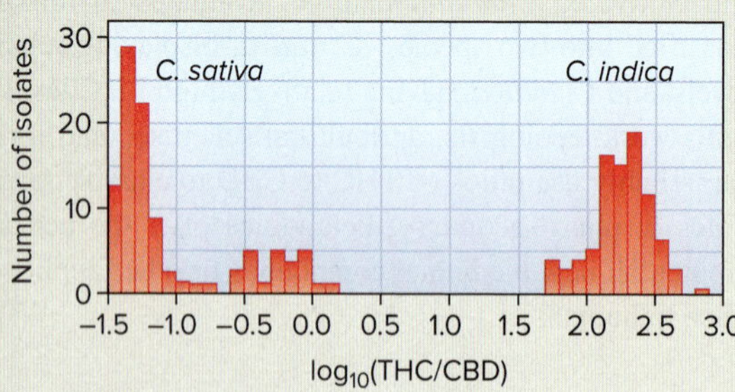


*Cannabis* plants isolated from diverse sources worldwide formed 2 groups—those having relatively high THC to CBD ratios and those having lower THC to CBD ratios.

Plants having low THC to CBD ratios, often used as hemp fiber sources, corresponded to the species *C. sativa*.

Plants having high THC to CBD ratios, often used as drug sources, corresponded to the species *C. indica*.

**5 CONCLUSION** Differing cannabinoid ratios support a concept of 2 *Cannabis* species.

**6 SOURCE** Hillig, K. W., and Mahlberg, P. G. 2004. A Chemotaxonomic Analysis of Cannabinoid Variation in *Cannabis* (Cannabaceae). *American Journal of Botany* 91: 966–975.

## 25.4 Reviewing the Concepts

- Flowers foster seed production and are adapted in various ways that aid pollination. Flower parts likely evolved from leaflike structures (Figures 25.14, 25.15, 25.16).
- The two largest and most diverse lineages of flowering plants are the monocots and eudicots (Figure 25.17).
- Flower diversification involved evolutionary changes such as the fusion of petals, clustering of flowers into inflorescences, and a reduced perianth, which improve pollination effectiveness and seed production (Figure 25.18).
- Fruits are structures that enclose seeds and aid in their dispersal (Figure 25.19).
- Angiosperms produce three main groups of secondary metabolites—(1) terpenes and terpenoids, (2) phenolics, and (3) alkaloids—which play essential roles in plant structure, reproduction, and defense, respectively (Figures 25.20, 25.21).

## 25.4 Testing Your Knowledge

1. A plant produces very small, petal-less flowers that are clustered into a dangling inflorescence. Among the choices given, which pollination agent is most likely?
   **a.** butterfly **c.** bat **e.** water
   **b.** hummingbird **d.** wind
2. A plant produces juicy, sweet, red fruits having relatively small seeds. Among the choices given, which seed dispersal agent is most likely in nature?
   **a.** butterfly **c.** bear **e.** water
   **b.** bird **d.** wind

## 25.5 Human Influences on Angiosperm Diversification

### Learning Outcome

1. Explain how humans created domesticated wheat, corn, and rice grain crops widely planted around the world today.

By means of the process known as **domestication,** which involves artificial selection for traits desirable to humans, ancient humans transformed wild plant species into new crop species. Cultivated bread wheat (*Triticum aestivum*) was probably among the earliest food crops, having originated more than 8,000 years ago in what is now southeastern Turkey and northern Syria. Bread wheat originated by a series of steps from wild ancestors (*Triticum boeoticum* and *Triticum dicoccoides*). Wheat, corn, and rice produce grains (a type of fruit) at the tip of a stem called the ear. Among the earliest changes that occurred during wheat domestication was the loss of shattering, the process by which ears of wild grain crops break apart and disperse their grains. A mutation probably caused the ears of some wheat plants to remain intact, a trait that is disadvantageous in nature but beneficial to humans. Nonshattering ears would have been easier for humans to harvest than normal ears. Early farmers probably selected seed stock from plants having nonshattering ears and other favorable traits such as larger grains. These ancient artificial selection processes, together with modern breeding efforts, explain why cultivated wheat differs from its wild relatives in shattering and other properties. The accumulation of these trait differences explains why cultivated and wild wheat plants are classified as different species.

About 9,000 years ago, people living in what is now Mexico domesticated a native grass known as teosinte (of the genus *Zea*), producing a new species, *Zea mays*, known as corn or maize. The evidence for this pivotal event includes ancient ears that were larger than wild ones and distinctive fossil pollen. Modern ears of corn are much larger than those of teosinte, with many more rows and larger and softer corn grains, and modern corn ears do not shatter, as do those of ancestral teosinte (Figure 25.22).

Molecular analyses indicate that domesticated rice (*Oryza sativa*) originated from ancestral wild species of grasses (*Oryza nivara* and/or *Oryza rufipogon*). As in the cases of wheat and corn, domestication of rice involved loss of ear shattering, in this case resulting from a key amino acid substitution. Ancient humans might have unknowingly selected for this mutation while gathering rice from wild populations, because the mutants would not so easily have shed grains during the harvesting process. Eventually, the nonshattering mutant became a

## Biology Principle

### Biology Affects Our Society

The domestication of corn from a wild grass to one of the world's largest production crops is an amazing feat of artificial selection.

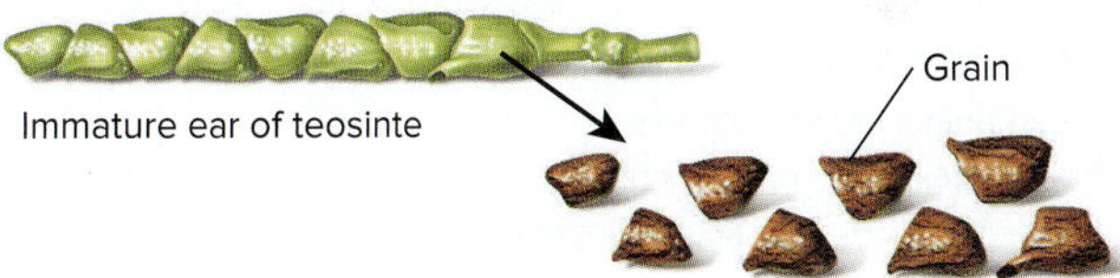

**Figure 25.22** **Ears and grains of modern corn and its ancestor, teosinte.** This illustration shows that domesticated corn ears are much larger than those of the ancestral grass teosinte. In addition, corn fruits are softer and more edible than the grains of teosinte, which are enclosed in a hard casing.

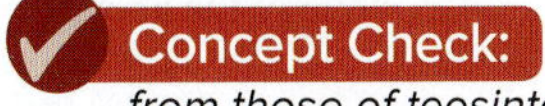

**Concept Check:** *In what other way do corn ears differ from those of teosinte?*

widely planted crop throughout Asia, and today it is the food staple for millions of people. Although humans generated these and other new plant species, in modern times humans have caused the extinction of plant species as the result of habitat destruction and other threats to species. Protecting biodiversity will continue to challenge humans as populations and demands on the Earth's resources increase.

## 25.5 Reviewing the Concepts

- Humans have produced new crop species by domesticating wild plants. The process of domestication involved artificial selection for traits such as nonshattering ears of wheat, corn, and rice (Figure 25.22).

## 25.5 Testing Your Knowledge

**1.** What feature of wild food did humans alter during crop domestication?
**a.** fruit size
**b.** fruit softness
**c.** seed dispersal
**d.** all of the above

# Assess and Discuss

## Test Yourself

1. The simplest and most ancient of modern land plants are probably
a. the ferns.
b. the cycads.
c. the liverworts and mosses.
d. the angiosperms.
e. none of the above.

2. Plants possess a life cycle that involves alternation of two multicellular generations: the gametophyte and
a. the lycophyte.
b. the bryophyte.
c. the spermatophyte.
d. the lignophyte.
e. the sporophyte.

3. The seed plants are also known as
a. bryophytes.
b. ferns.
c. lycophytes.
d. spermatophytes
e. none of the above.

4. A waxy cuticle is an adaptation that
a. helps to prevent water loss from tracheophytes.
b. helps to prevent water loss from algae.
c. helps to increase water loss from tracheophytes.
d. aids in water transport within the bodies of vascular plants.
e. does all of the above.

5. Vascular plant photosynthesis transformed a very large amount of carbon dioxide into organic compounds that were incompletely degraded and transformed into coal, thereby causing a dramatic decrease in atmospheric carbon dioxide levels during the geological period known as the
a. Cambrian.
b. Ordovician.
c. Carboniferous.
d. Permian.
e. Pleistocene.

6. Which phylum among the plants listed has vascular tissue?
a. liverworts
b. hornworts
c. mosses
d. ferns
e. none of the above

7. Which sequence of critical adaptations reflects the order of their appearance in time?
a. embryos, vascular tissue, seeds, flowers
b. vascular tissue, embryos, flowers, seeds
c. vascular tissue, seeds, embryos, flowers
d. seeds, embryos, flowers, vascular tissue
e. seeds, vascular tissue, embryos, flowers

8. The primary function of a fruit is to
a. provide food for the developing seed.
b. provide food for the developing seedling.
c. foster pollen dispersal.
d. foster seed dispersal.
e. none of the above.

9. How did human domestication affect the transformation of wild grain plants into modern crops?
a. Domestication reduced ear shattering.
b. Domestication increased ear size.
c. Domestication increased grain edibility.
d. All of the above.
e. None of the above.

10. Examples of plant groups whose biodiversity is much reduced from past levels, or currently restricted to limited geographical regions include
a. lycophytes.
b. whisk ferns.
c. Ginkgo trees.
d. cycads.
e. all of the above.

## Conceptual Questions

1. How have flowering plants been able to diversify into so many species?
2. Why are some fruits useful for human food?
3. **PRINCIPLES** A principle of biology is that living organisms maintain homeostasis. Explain how several structural features help vascular plants to maintain stable internal water content.

## Collaborative Questions

1. Discuss at least one difference in environmental conditions experienced by Coal Age plants as compared to modern plants.
2. List as many plant adaptations to land as you can.

# Online Resource

**connect.mheducation.com**

# Invertebrates: The Vast Array of Animal Life Without a Backbone

# 26

© NHPA/SuperStock

**Scarab beetles have been revered in Egyptian culture for millennia.** Invertebrate animals have tremendous impacts on human societies.

## Chapter Outline



There are more beetles (about 400,000 species) than any other type of organism on Earth. When the British biologist J. B. S. Haldane was asked by a group of theologians what one could conclude about the nature of the Creator, he is said to have replied, "An inordinate fondness for beetles."

The dung beetle, or scarab beetle, shown in the chapter-opening photo, is found in the area surrounding the Mediterranean Sea. Many scarab beetles feed exclusively on dung. In some species, males form the dung into a ball and roll it to a place where they bury it. Males can roll balls up to 50 times their own body weight. The female lays an egg in the dung ball, and when the egg hatches, the larva feeds on the dung surrounding it. After a period of weeks to months, the larva completely changes its appearance from a grub to a fully formed beetle before it emerges. Scarabs featured prominently in ancient Egyptian culture. The Egyptians saw the rolling of balls of dung across the ground as a symbol of the forces moving the Sun across the sky and considered the scarab beetle to be sacred. The apparent self-creation of the beetles from a dung ball buried underground and the mummy-like appearance of the pupa conveyed the ideas of transformation and resurrection from an underground tomb.

Animals constitute the most species-rich kingdom. Many different kinds of animals and their products are part of our diet. In addition, humans enjoy many animal species as companions and depend on other species to test lifesaving drugs. However, other animals threaten our food supply and transmit deadly diseases.

Since the time of Carolus Linnaeus in the 1700s, scientists have classified animals based on their morphology, that is, on their physical structure. In the 1990s, animal classifications based on similarities in DNA and rRNA sequences became more common. Quite often, traditional classifications based on morphology and those based on molecular data were similar, but some important differences arose. In this chapter, we will begin by defining the key characteristics of animals and then take a look at the major types of invertebrates. In Chapter 27, we will look specifically at the vertebrates, animals more familiar to us, including fish, amphibians, reptiles, and mammals. Throughout Chapters 26 and 27, we will explore how new molecular data have enabled scientists to revise and refine the animal phylogenetic tree.

## 26.1 Characteristics of Animals

### Learning Outcome

1. List the key characteristics of animals that distinguish them from others organisms.

The Earth contains a dazzling diversity of animal species, living in environments from the deep sea to the desert and exhibiting an amazing array of characteristics. Most animals move and eat multicellular prey, and therefore they are loosely differentiated from species in other kingdoms. However, coming up with a firm definition of an animal can be tricky because animals are so diverse that biologists can find exceptions to nearly any given characteristic. Even so, the following features can help us broadly characterize the group we call animals.

### Cell Structure

- Animals are multicellular. Animal cells lack cell walls and are flexible. This flexibility facilitates movement.
- Animal cells gain structural support from an extensive extracellular matrix (ECM) that forms strong fibers outside the cell (refer back to Figure 4.31).
- A group of unique cell junctions—anchoring and tight junctions—play an important additional role in holding animal cells in place and allowing communication between cells (refer back to Section 5.7).

### Mode of Nutrition

- Animals are heterotrophs; that is, they ingest other organisms or their products to sustain life. Many different modes of feeding exist among animals, including suspension feeding—filtering food out of the surrounding water; bulk feeding—eating large food pieces; and fluid feeding—sucking plant sap or animal body fluids (Figure 26.1).

### Movement

- Muscle tissue is unique to animals, and most animals are capable of some type of locomotion in order to acquire food or escape predators.
- Nervous tissue is also unique to animals, and a nervous system coordinates movement.

### Genomes

- Most animals possess *Hox* genes, which function in patterning the body axis (refer back to Figures 20.12 and 20.13).
- Animals have very similar genes that encode small ribosomal subunit (SSU) rRNA (refer back to Figure 10.13).

### Reproduction and Development

- Nearly all members of the animal kingdom reproduce sexually, whereby a small, mobile sperm unites with a much larger egg to form a fertilized egg, or zygote. Fertilization can occur internally, which is common in terrestrial species, or externally, which is more common in aquatic species.
- Similarly, embryos develop inside the mother or outside in the mother's environment. A particularly unusual developmental phenomenon is the occurrence of **metamorphosis,** by which an organism changes from a juvenile to an adult form—for example, when a caterpillar changes into a butterfly or when a scarab larva changes into an adult scarab beetle.

## 26.1 Reviewing the Concepts

- Animals constitute a very species-rich kingdom, with a number of characteristics that most of them have in common, including multicellularity, an extracellular matrix (ECM), and unique cell junctions, in addition to heterotrophic feeding, internal digestion, and the possession of nervous and muscle tissues.
- Many different feeding modes exist among animals, including suspension feeding, bulk feeding, and fluid feeding (Figure 26.1).

## 26.1 Testing Your Knowledge

1. Which of the following is *not* a distinguishing characteristic of animals?
   a. the capacity to move at some point in their life cycle
   b. possession of cell walls
   c. multicellularity
   d. heterotrophy
   e. All of the above are characteristics of animals.

(a)

(b)

(c)

**Figure 26.1 Modes of animal nutrition. (a)** Suspension feeders, such as this tube worm, filter food particles from the water column. **(b)** Bears and other bulk feeders tear off large pieces of their food and chew it or swallow it whole. **(c)** Fluid feeders, such as these aphids, suck fluid from their food source.

## 26.2 Animal Classification

### Learning Outcomes

1. Discuss why choanoflagellates are believed to be the closest living relatives of animals.
2. Describe how the presence of symmetry, germ layers, and embryonic development form the basis of animal classification.
3. Describe how differences in body cavities and segmentation may also be used to distinguish groups of animals.
4. **SCISKILLS ▶** Draw an animal phylogeny based solely on body plans, and contrast it with the animal phylogeny based on body plans and molecular data.

Although animals constitute an extremely diverse kingdom, most biologists agree that the kingdom is monophyletic, meaning that all taxa have evolved from a single common ancestor. Today, scientists recognize about 35 animal phyla. At first glance, many of these phyla seem so distantly related to one another (for example, chordates and jellyfish) that making sense of this diversity with a classification scheme seems nearly impossible. However, over the course of centuries, scientists have come to some basic conclusions about the evolutionary relationships among animals. In this section, we explore the major features of animal body plans that form the basis of animal phylogeny (Figure 26.2).

### Animals Evolved from a Choanoflagellate-Like Ancestor

Animal life on Earth has evolved over hundreds of millions of years. Some scientists suggest that changing environmental conditions, such as a buildup of dissolved oxygen and minerals in the ocean or an increase in atmospheric oxygen, eventually permitted higher metabolic rates and increased the activity of a wide range of animals. Others suggest that with the development of sophisticated locomotor skills, a wide range of predators and prey evolved, leading to an evolutionary

## Biology Principle

**All Species (Past and Present) Are Related by an Evolutionary History**

All animals are believed to be derived from a choanoflagellate-like ancestor.

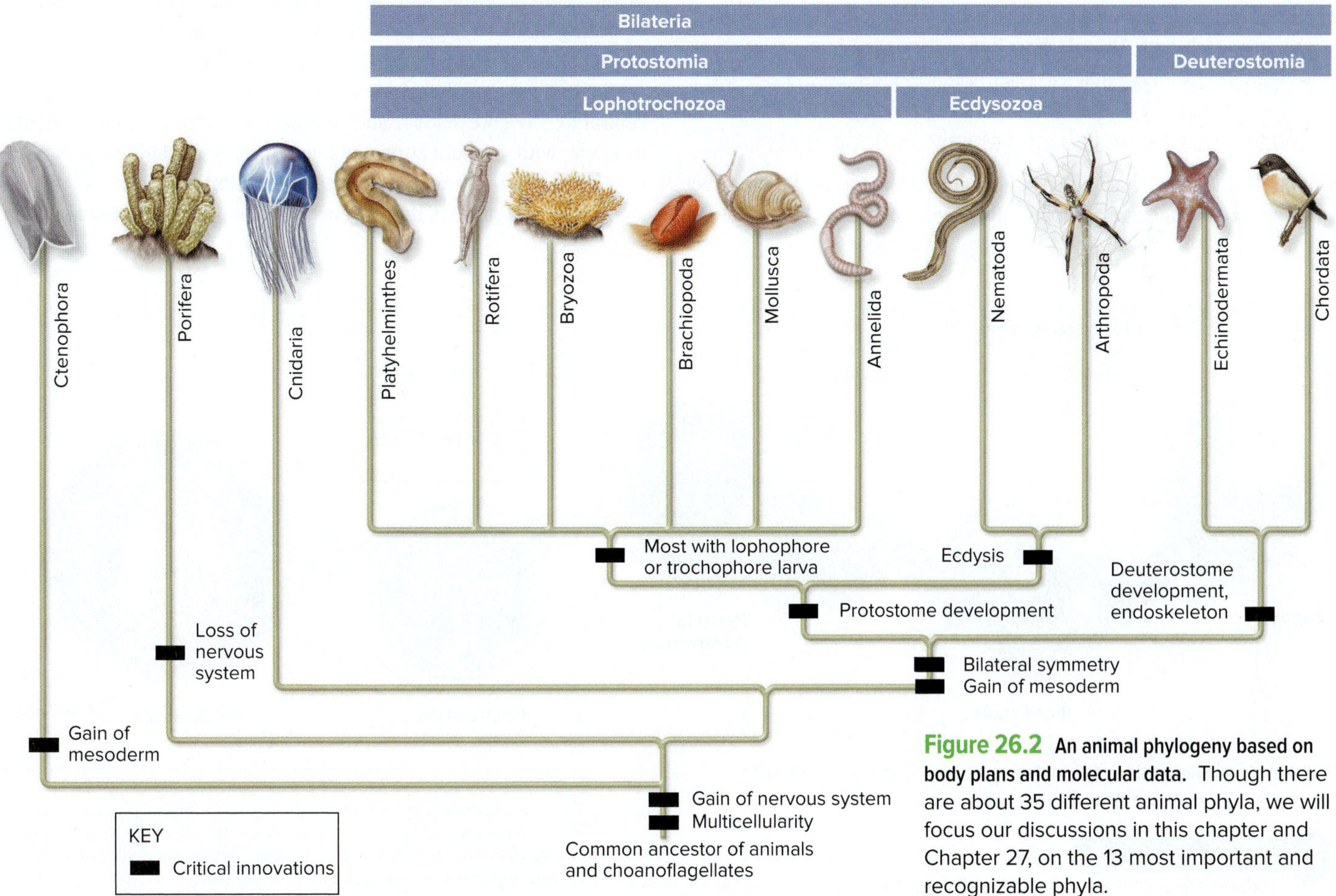


**Figure 26.2 An animal phylogeny based on body plans and molecular data.** Though there are about 35 different animal phyla, we will focus our discussions in this chapter and Chapter 27, on the 13 most important and recognizable phyla.

(a) Porifera: no symmetry

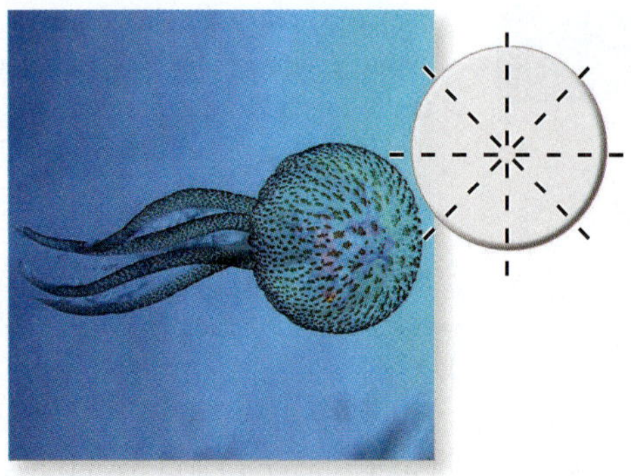

(b) Cnidaria: radial symmetry

(c) Bilateria: bilateral symmetry

**Figure 26.3** **Early divisions in the animal phylogeny.** Animals can be categorized based on symmetry: **(a)** the absence of symmetry (Porifera; the sponges), **(b)** radial symmetry (the cnidarians), or **(c)** bilateral symmetry (Bilateria; all other animals).
*(a)* © E. Teister/agefotostock; *(b)* © Gavin Parsons/Getty Images; *(c)* © Jens Kuhfs/Getty Images

"arms race" in which predators evolved powerful weapons and prey evolved more powerful defenses against them. Such adaptations would have led to a proliferation of different lifestyles and taxa. Finally, the evolution of *Hox* genes may have resulted in an increase in diversity.

With the monophyletic nature of the animal kingdom in mind, scientists have attempted to characterize the organism from which animals most likely evolved. According to research, the closest living relative of animals is believed to be a flagellated protist known as a choanoflagellate. Choanoflagellates are tiny, single-celled or colonial organisms, each with a single flagellum surrounded by a collar composed of cytoplasmic tentacles (refer back to Figure 23.27).

## Animal Classification Is Based Mainly on Body Plans

Biologists traditionally classified animal diversity in terms of these three main morphological and developmental features of animal body plans:

1. Type of body symmetry
2. Number of germ layers
3. Specific features of embryonic development

We will discuss each of these major features of animal body plans next.

**Symmetry** Animals may be categorized according to their type of symmetry. Symmetry refers to the existence of balanced proportions of the body on either side of a median plane. Some of the earliest diverging animals, such as sponges, were asymmetric, meaning they had no plane of symmetry (Figure 26.3a). Radially symmetric animals can be divided equally by any longitudinal plane passing through the central axis (Figure 26.3b). Such animals are often circular or tubular in shape, with a mouth at one end, and include cnidarians (jellyfish).

Bilaterally symmetric animals, the **Bilateria,** can be divided along a vertical plane at the midline to create two halves (Figure 26.3c).

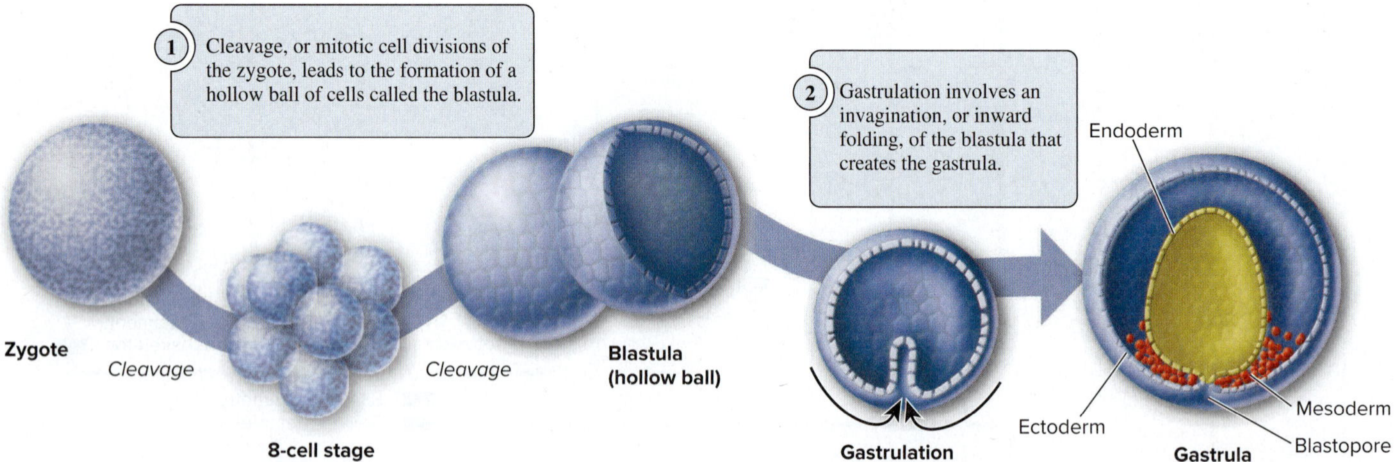

1 Cleavage, or mitotic cell divisions of the zygote, leads to the formation of a hollow ball of cells called the blastula.
Zygote
Cleavage
8-cell stage
Cleavage
Blastula (hollow ball)
2 Gastrulation involves an invagination, or inward folding, of the blastula that creates the gastrula.
Gastrulation
Endoderm
Ectoderm
Gastrula
Mesoderm
Blastopore
3 In the gastrula, the layer of cells lining the primitive digestive tract becomes the endoderm. The cells on the outside of the gastrula form the ectoderm. In the Bilateria, a middle layer termed the mesoderm develops between the ectoderm and endoderm.

**Figure 26.4** **Formation of germ layers.** Note: Radially symmetric animals (Radiata) do not form mesoderm.

**BioConnections:** *Refer back to Figure 21.7. Is the existence of three layers in the Bilateria a shared primitive character or a shared derived character?*

Thus, a bilateral animal has a left side and a right side, which are mirror images, as well as a **dorsal** (upper) and a **ventral** (lower) side, which are not identical, and an **anterior** (head) and a **posterior** (tail) end. Bilateral symmetry is strongly correlated with both the ability to move through the environment and **cephalization**—the localization of sensory structures at the anterior end of the body. Such abilities allow animals to encounter their environment initially with their head, which is best equipped to detect and consume prey and to detect and respond to predators and other dangers. Most animals are bilaterally symmetric.

**Germ Layers** Fertilization of an egg by a sperm creates a diploid zygote. The zygote then undergoes **cleavage**—a succession of rapid cell divisions with no significant growth that produces a hollow sphere of cells called a **blastula.** In all animals except the sponges, the growing zygote develops different layers of cells, called **germ layers,** during a process known as **gastrulation** (Figure 26.4). In gastrulation, an area in the blastula folds inward, or invaginates, creating in the process a structure called a **gastrula.** The inner layer of cells becomes the **endoderm,** which lines the primitive digestive tract. The outer layer, or **ectoderm,** covers the surface of the embryo and differentiates into the epidermis and nervous system.

A key difference between Bilateria and most other animals is that the Bilateria develop a third layer of cells, termed the **mesoderm,** between the ectoderm and endoderm. Mesoderm forms the muscles and most other organs between the digestive tract and the ectoderm (look ahead to Figure 40.17). Because the Bilateria have these three distinct germ layers, they are often referred to as **triploblastic,** whereas the cnidarians, which have only ectoderm and endoderm, are termed **diploblastic.** Interestingly, the mesoderm of the earliest-diverging animals, the ctenophores, probably originated independently of the mesoderm found in bilaterians.

**Specific Features of Embryonic Development** The most fundamental feature of embryonic development is the formation of a mouth and an anus (Figure 26.5a). In gastrulation, the endoderm forms an indentation, the **blastopore,** which is the opening of the digestive

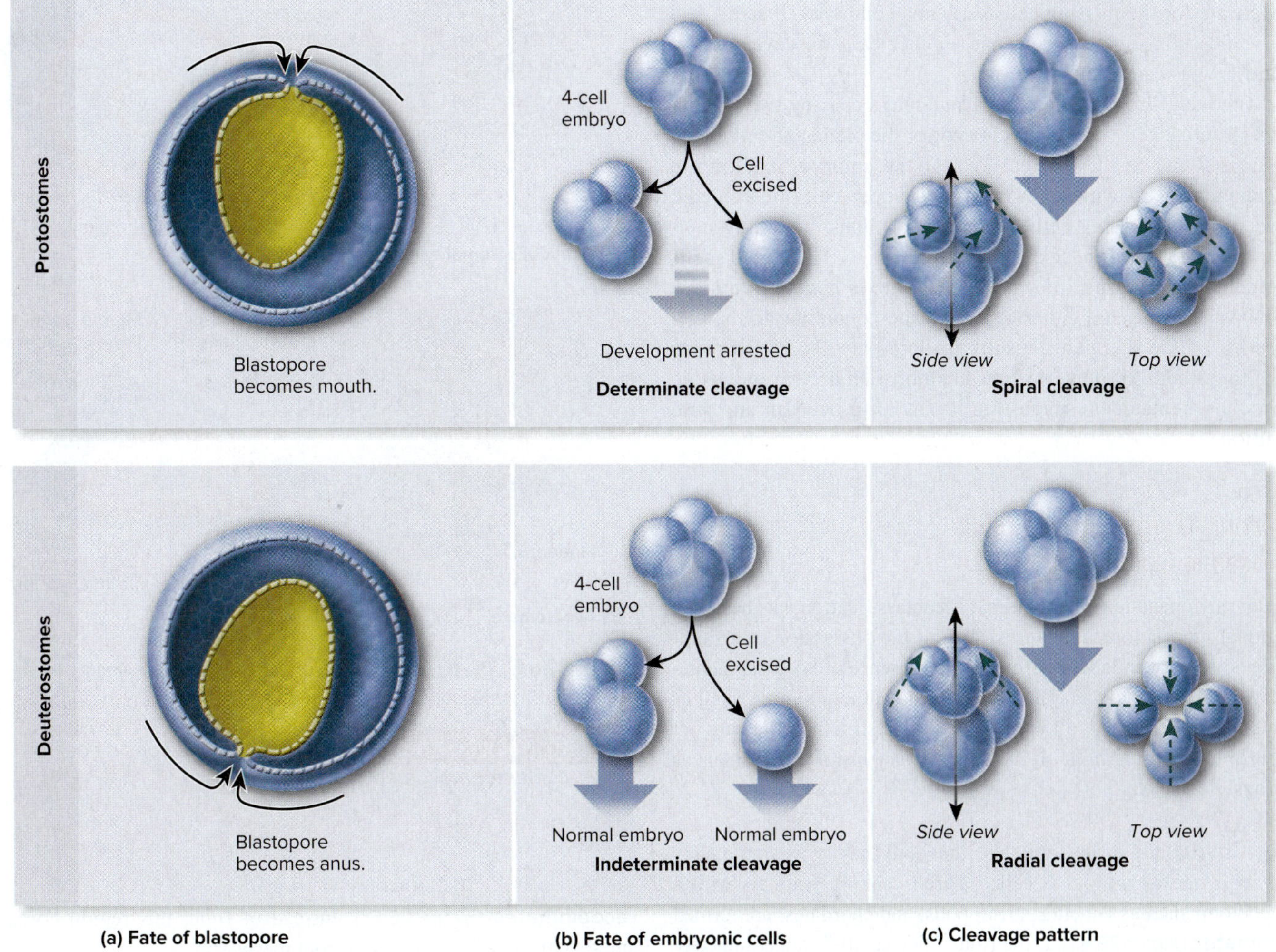


**Figure 26.5** **Differences in embryonic development between protostomes and deuterostomes.** See Figure 26.2 for a comparison of protostome and deuterostome groups. **(a)** In protostomes, the blastopore becomes the mouth. In deuterostomes, the blastopore becomes the anus. **(b)** Protostomes have determinate cleavage, whereas deuterostomes have indeterminate cleavage. **(c)** Many protostomes have spiral cleavage, and all deuterostomes have radial cleavage. The dashed arrows indicate the direction of cleavage.

tract to the outside. In **protostomes** (from the Greek *protos*, meaning first, and *stoma*, meaning mouth), the blastopore becomes the mouth. If an anus is formed in a protostome, it develops from a secondary opening. In contrast, in **deuterostomes** (from the Greek *deuteros*, meaning second), the blastopore becomes the anus, and the mouth develops from a secondary opening.

In addition, protostomes and deuterostomes differ in some other embryonic features. In the early stages of embryonic development, repeated cell divisions occur without cell growth, a process known as cleavage. Protostome development is generally characterized by so-called **determinate cleavage,** in which the fate of each embryonic cell is determined very early (Figure 26.5b). If one of the cells is removed from a four-cell protostome embryo, neither the single cell nor the remaining three-cell mass can form viable embryos, and development is halted. In contrast, most deuterostome development is characterized by **indeterminate cleavage,** in which each cell produced by early cleavage retains the ability to develop into a complete embryo. For example, when one cell is excised from a four-cell sea urchin embryo, both the single cell and the remaining three can go on to form viable embryos. Other embryonic cells compensate for the missing cells. In human embryos, if individual embryonic cells separate from one another early in development, identical twins can result.

In the developing zygote, cleavage may occur by two mechanisms (Figure 26.5c). In **spiral cleavage,** the planes of cell cleavage are oblique to the vertical axis of the embryo, resulting in an arrangement in which newly formed upper cells lie centered between the underlying cells. Many protostomes, including mollusks and annelid worms, exhibit spiral cleavage. The coiled shells of some mollusks result from spiral cleavage. In **radial cleavage,** the cleavage planes are either parallel or perpendicular to the vertical axis of the egg. This results in tiers of cells, one directly above the other. All deuterostomes exhibit radial cleavage, as do insects and nematodes, suggesting it may have been an ancestral condition.

## Additional Morphological Criteria Are Used to Classify Animals

In older phylogenetic trees of animal life, classification was based on additional morphological features, such as the possession of a fluid-filled body cavity called a **coelom** or the presence of body segmentation. More recent molecular data suggest that although these features are helpful in describing differences in animal structure, they are not as useful in shedding light on the evolutionary history of animals as previously believed.

**Body Cavity** In many animals, the body cavity is completely lined with mesoderm and is called a true coelom. Animals with a true coelom, such as annelids, arthropods, and chordates, are termed **coelomates** (Figure 26.6a). If the fluid-filled cavity is not completely lined by tissue derived from mesoderm, it is known as a pseudocoelom (Figure 26.6b). Animals with a pseudocoelom, including rotifers and nematodes, are termed **pseudocoelomates.** Some animals, such as flatworms, lack a fluid-filled body cavity and are termed **acoelomates** (Figure 26.6c). Instead of fluid, this region contains mesenchyme, a tissue derived from mesoderm.

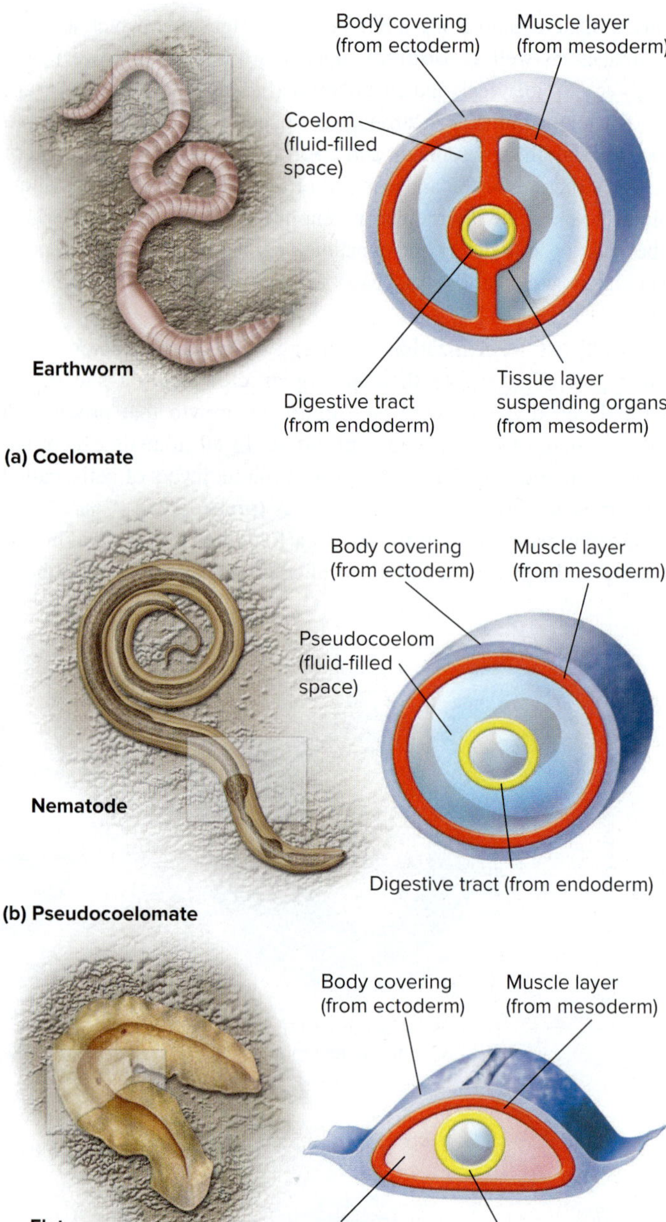


**Figure 26.6** **The three basic body cavities of bilaterally symmetric animals.** Cross sections of each animal are shown on the right.

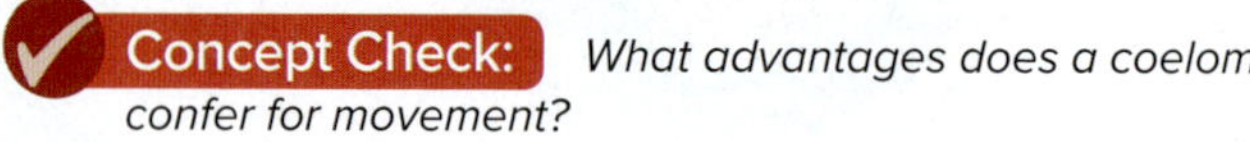

*What advantages does a coelom confer for movement?*

In some soft-bodied invertebrates, such as earthworms, the coelom functions as a **hydrostatic skeleton**—a fluid-filled body cavity surrounded by muscles that gives support and shape to the body. Muscle contractions at one part of the body push this fluid

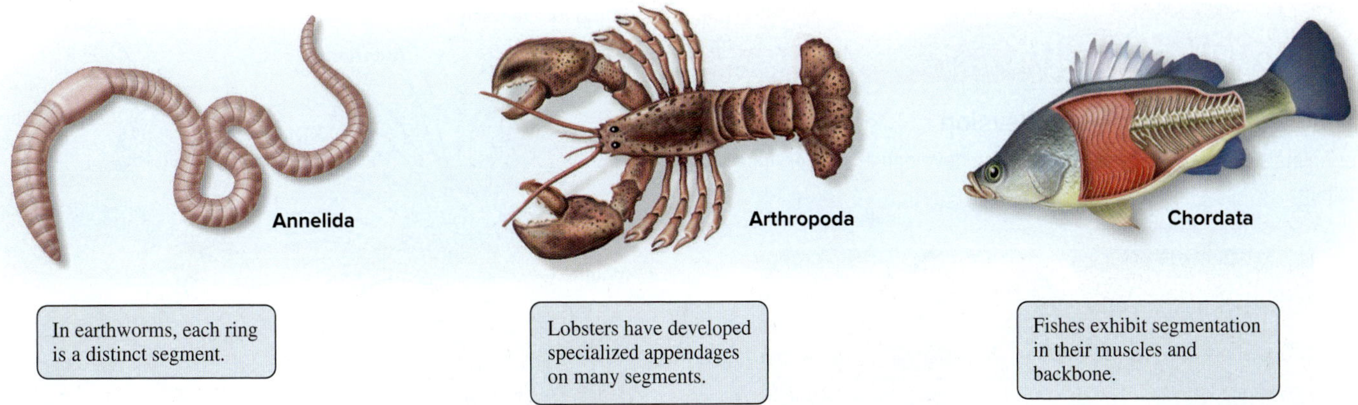

Annelida
Arthropoda
Chordata
In earthworms, each ring is a distinct segment.
Lobsters have developed specialized appendages on many segments.
Fishes exhibit segmentation in their muscles and backbone.


**Figure 26.7** **Segmentation.** Annelids, arthropods, and chordates all exhibit segmentation.

toward another part of the body. This type of movement can best be observed in an earthworm. In some organisms, such as insects and mollusks, the fluid in the body cavity also acts as a simple circulatory system. Finally, the fluid is relatively incompressible and therefore cushions internal organs such as the heart and intestinal tract in all coelomate animals, helping to prevent injury from external forces.

**Segmentation** Another well-known feature of the animal body plan is the presence or absence of segmentation. In segmentation, the body is divided into regions called segments. Segmentation is most obvious in the annelids, or segmented worms, but it is also evident in arthropods and chordates (Figure 26.7). In annelids, most segments are similar in structure and contain the same set of blood vessels, nerves, and muscles. However, some segments may differ, such as those containing the sex organs. In chordates, we can see segmentation in the backbone and muscles. The advantage of segmentation is that it allows specialization of body regions and the appendages used for movement. Many insects have wings and three pairs of legs, whereas centipedes have no wings and many legs. Crabs, lobsters, and shrimp have highly specialized thoracic appendages that aid in movement and feeding (see Figure 26.30).

## EVOLUTIONARY CONNECTIONS

### The Protostomes Consist of Two Major Clades—the Ecdysozoa and the Lophotrochozoa

In 1997, American molecular biologists Anna Marie Aguinaldo, James Lake, and colleagues analyzed the relationships between protostome clades by sequencing the complete gene that encodes SSU (small subunit) rRNA from a variety of representative taxa. Total genomic DNA was isolated using standard techniques and amplified by the polymerase chain reaction (PCR; refer back to Figure 18.6). PCR fragments were then subjected to DNA sequencing, a technique also described in Chapter 18, and the evolutionary relationships among 50 species were examined. The resulting data indicated the existence of a new clade of molting animals, the **Ecdysozoa** (pronounced ek-die-so-zo-ah), consisting of the nematodes and arthropods. This study represented a scientific breakthrough because it underscored the value of molecular phylogenies. According to molecular evidence, the other major protostome clade is the **Lophotrochozoa** (pronounced low-fo-tro-ko-zo-ah), which encompasses the mollusks, annelids, and several other phyla (see Figure 26.2). When some morphologists reviewed their data in light of this new information, they found there was also morphological support for these new groupings.

The Ecdysozoa is so named because all of its members secrete a nonliving cuticle, an external skeleton (exoskeleton); think of the hard shell of a beetle or that of a crab. As these animals grow, the exoskeleton becomes too small, and the animal molts, or breaks out of its old exoskeleton, and secretes a newer, larger one (Figure 26.8) This molting process is called **ecdysis,** hence the name Ecdysozoa. Although this group was named for this morphological characteristic, it was first strongly supported as a separate clade by molecular evidence such as similarities in DNA.

Although the Lophotrochozoa clade was organized primarily through analysis of molecular data, its name also stems from two morphological features seen in many organisms of this clade. The lopho part is derived from the **lophophore,** a horseshoe-shaped crown of tentacles used for feeding that is present on some phyla in this clade, such as the rotifers, bryozoans, and brachiopods (Figure 26.9a). The trocho part refers to the **trochophore larva,** a distinct larval stage characterized by a band of cilia around its middle that is used for swimming (Figure 26.9b). Trochophore larvae are found in several Lophotrochozoa phyla, such as annelid worms and mollusks, indicating their similar ancestry. Other members of the clade, such as the platyhelminthes, have neither of these morphological features and are classified as lophotrochozoans based strictly on molecular data.

## 26.2 Reviewing the Concepts

- Animals can be categorized according to their type of symmetry, whether asymmetric (the sponges), radial (the cnidarians and ctenophores), or bilateral (Bilateria, all other animals) (Figure 26.3).

## Biology Principle

### Living Organisms Grow and Develop

For animals with exoskeletons, growth and development necessitate molting.

**Figure 26.8** **Ecdysis.** This dragonfly, emerging from a discarded exoskeleton, is a member of the Ecdysozoa—a clade of animals exhibiting ecdysis, the periodic shedding (molting) and re-formation of the exoskeleton.



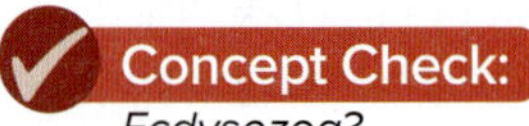


*What are the main members of the Ecdysozoa?*

- While some animals have two embryonic cell layers (germ layers): the endoderm and the ectoderm, the Bilateria have a third germ layer termed the mesoderm, which develops between the endoderm and the ectoderm (Figure 26.4).
- Animals are also classified according to patterns of embryonic development. In protostomes, the blastopore becomes the mouth; in deuterostomes, the blastopore becomes the anus (Figure 26.5). Most protostomes have spiral cleavage and all deuterostomes have radial cleavage.
- Recent molecular studies propose a division of the protostomes into two major clades: the Ecdysozoa and the Lophotrochozoa. Members of the Ecdysozoa secrete and periodically shed a nonliving cuticle, typically an exoskeleton, or external skeleton

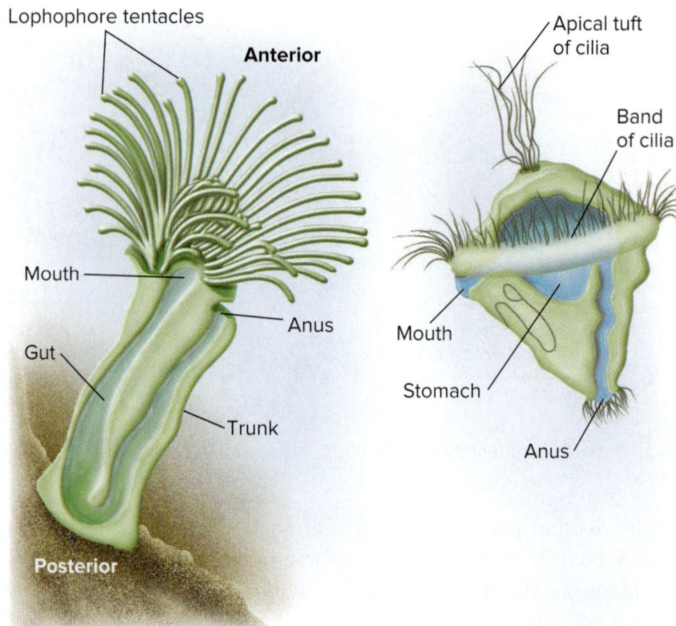


**Figure 26.9** **Characteristics of the Lophotrochozoa.** **(a)** A lophophore, a crown of ciliated tentacles, generates a current to bring food particles into the mouth. **(b)** The trochophore larval form is found in several animal lineages.

(Figure 26.8). Most members of the Lophotrochozoa are distinguished by two morphological features: the lophophore, a crown of tentacles used for feeding, and the trochophore larva, a distinct larval stage (Figure 26.9).

## 26.2 Testing Your Knowledge

1. In protostomes
   - **a.** the blastopore becomes the mouth.
   - **b.** the blastopore becomes the anus.
   - **c.** development is characterized by indeterminate cleavage.
   - **d.** both a and c.
   - **e.** both b and c.
2. In triploblastic animals, the inner lining of the digestive tract is derived from the
   - **a.** ectoderm.
   - **b.** mesoderm.
   - **c.** endoderm.
   - **d.** pseudocoelom.
   - **e.** coelom.
3. Pseudocoelomates
   - **a.** lack a fluid-filled cavity.
   - **b.** have a fluid-filled cavity that is completely lined with mesoderm.
   - **c.** have a fluid-filled cavity that is partially lined with mesoderm.
   - **d.** have a fluid-filled cavity that is not lined with mesoderm.
   - **e.** have an air-filled cavity that is partially lined with mesoderm.

## 26.3 Ctenophores: The Earliest Animals

### Learning Outcome

1. Outline the unique features of ctenophores.

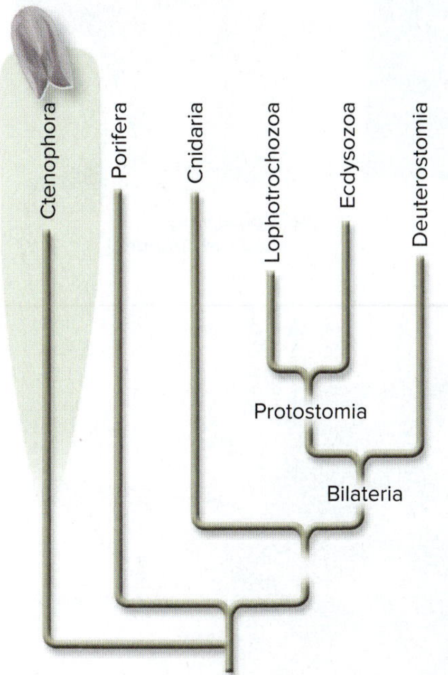


Ctenophores, also known as comb jellies, constitute the earliest-diverging animal lineage. Jellies are a small phylum of about 200 species, all of which are marine and look very much like jellyfish, with which they were improperly classified for a long time (**Figure 26.10**). They have eight rows of cilia on their surfaces that resemble combs.

The name Ctenophora comes from the Greek *ktenos,* meaning comb, and *phora,* meaning bearing (pronounced teen-o-for'-ah). The coordinated beating of the cilia propels the ctenophores. Averaging about 1–10 cm in length, comb jellies are probably the largest animals to use cilia for locomotion.

Comb jellies possess two long tentacles but lack stinging cells. Instead, they have cells on the tentacles that secrete a sticky substance onto which small prey adhere. The tentacles are then drawn over the mouth. Digestion occurs in a body cavity called a **gastrovascular cavity,** but waste and water are eliminated through two anal pores. Prey are generally small and may include tiny crustaceans called copepods and small fishes. Comb jellies are often transported around the world in ships' ballast water. *Mnemiopsis leidyi*, a ctenophore species native to the Atlantic coast of North and South America, was accidentally introduced into the Black and Caspian Seas in the 1980s. With a plentiful food supply and a lack of predators, *Mnemiopsis* underwent a population explosion and ultimately devastated the local fishing industries.

All ctenophores are hermaphroditic, possessing both ovaries and testes, and gametes are shed into the water to eventually form a free-swimming larva that is very similar in form to the adult. Nearly all ctenophores exhibit **bioluminescence,** a phenomenon that results from chemical reactions that give off light rather than heat. Individuals are particularly evident at night, and ctenophores that wash up on shore can make the sand or mud appear luminescent.

Like jellyfish, ctenophores contain both muscle and nerve cells organized as a diffuse net centralized at an elementary brain. However, the nervous system uses different neurotransmitters than those in bilaterians and jellyfish and has different types of synapses. The presence of muscle cells originating from mesoderm suggests that ctenophores share a three-germ-layer architecture with bilaterians. However, recent analysis of the genome of the ctenophore *Mnemiopsis leidyi* suggests that ctenophores lack many of the genes involved in specifying bilaterian mesoderm. In addition, ctenophores lack *Hox* genes and possess a ctenophore-specific cleavage program. Finally, many bilaterian neuron-specific genes are absent or not expressed. These findings argue against a linear march of evolutionary forms from more simple animals such as sponges to complex bilaterians. Instead, it appears more likely that ctenophores were the earliest animals to diverge from a choanoflagellate-like ancestor that was multicellular and had a simple nervous system. After diverging from the rest of the animal kingdom, ctenophores evolved their own unique way of forming mesoderm, which is different from other animal groups.

**Figure 26.10** **A ctenophore.** Ctenophores are called comb jellies because the eight rows of cilia on their surfaces resemble combs.


### 26.3 Reviewing the Concepts

- Ctenophores are the earliest-diverging animals (Figure 25.10). They possess a unique nervous system and mesoderm germ layer. They are predatory, possess a complete gut, and use cilia to propel themselves.

### 26.3 Testing Your Knowledge

1. Ctenophores evolved
   a. after sponges.
   b. after cnidarians.
   c. after bilaterians.
   d. before sponges.
   e. before choanoflagellates.

## 26.4 Porifera: The Sponges

### Learning Outcomes

1. Outline the body plan and unique characteristics of sponges.
2. Describe how sponges defend themselves against predators.

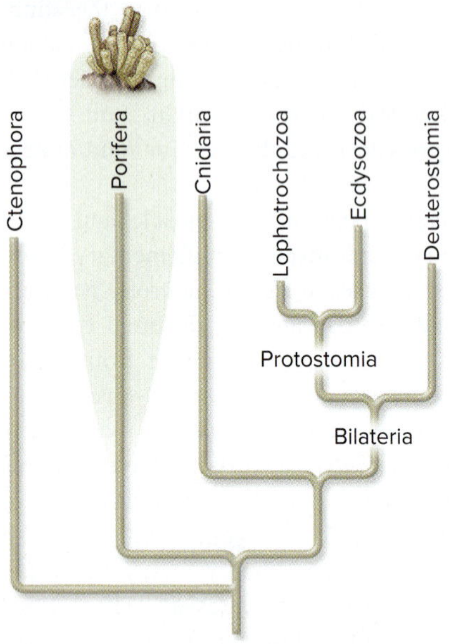


Members of the phylum Porifera (from the Latin, meaning pore bearers), are commonly referred to as sponges. Sponges lack true tissues—groups of cells that have a similar structure and function. However, sponges are multicellular and possess several types of cells that perform different functions. Even though sponges lack a nervous system, they have the necessary genetic machinery for a functioning nervous system but have lost these cell types. Biologists have identified approximately 8,000 species of sponges, the vast majority of which are marine. Sponges range in size from only a few millimeters across to more than 2 m in diameter. Smaller sponges may be radially symmetric, but most have no apparent symmetry. Some sponges have a low, encrusting growth form, whereas others grow tall and erect (Figure 26.11a). Although adult sponges are sessile—that is, anchored in place—the larvae are free-swimming.

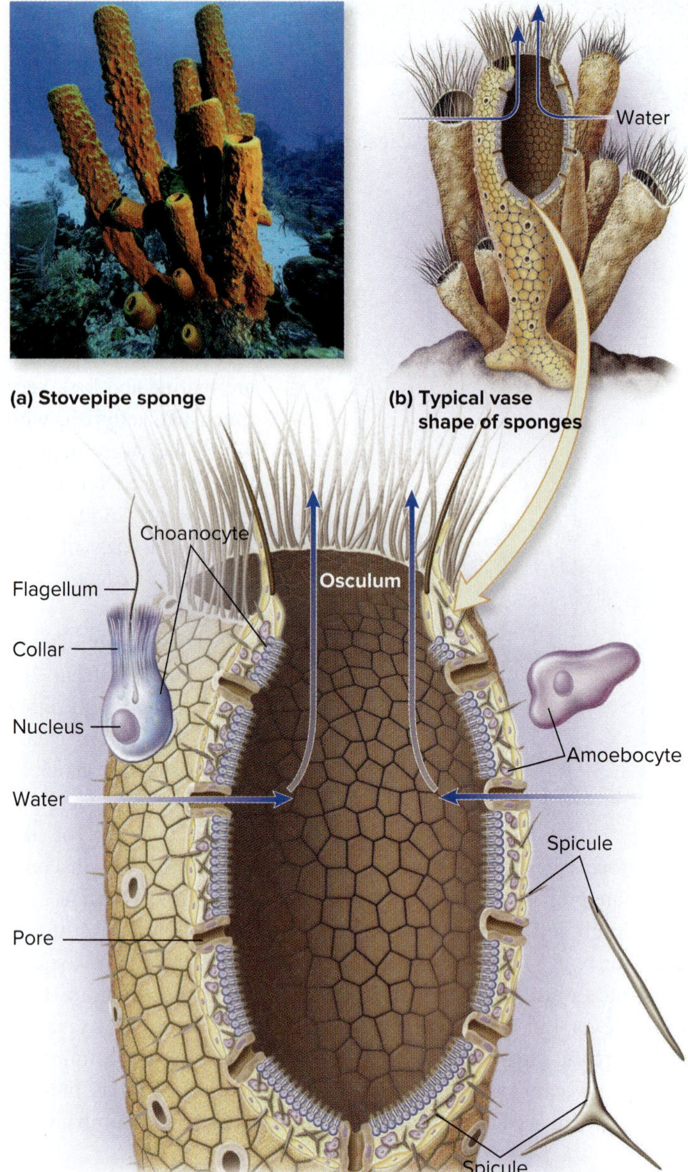


(a) Stovepipe sponge

(b) Typical vase shape of sponges

(c) Cross section of sponge morphology

**Figure 26.11** **(a)** The stovepipe sponge (*Aplysina archeri*) is common on Caribbean reefs. **(b)** Many sponges have a vaselike shape. **(c)** A cross section reveals that sponges are truly multicellular animals, having various cell types but no distinct tissues.

*(a)* © Norbert Probst/agefotostock

### Choanocytes Help Circulate Water

The body of most sponges looks similar to a vase pierced with small holes, or pores (Figure 26.11b). Water is drawn through these pores into a central cavity and flows out through the large opening at the top, called the osculum. The water enters the pores by the beating action of the flagella of the **choanocytes,** or collar cells, that line the central cavity (Figure 26.11c). In the process, the choanocytes trap and eat small particulate matter and tiny plankton. Also lining the central cavity are mobile cells called amoebocytes that absorb food from choanocytes, digest it, and carry the nutrients to other cells. Sponges are unique among the major animal phyla in using intracellular digestion, the uptake of food particles by cells, as a mode of feeding.

**Concept Check:** *Since sponges are soft and sessile, why aren't they eaten by other organisms?*

### Sponges Have Mechanical and Chemical Defenses Against Predators

Some amoebocytes can also form tough skeletal fibers that support the body. In many sponges, this skeleton consists of sharp **spicules** formed of protein, calcium carbonate, or silica. For example, some deep-ocean species, called glass sponges, are distinguished by needle-like silica spicules that form elaborate, lattice-like skeletons. The presence of such tough spicules may help explain why there is not much predation of sponges. Other sponges have fibers of a tough protein called **spongin** that lend skeletal support. Spongin skeletons are still commercially harvested and sold as bath sponges. Many species produce toxic defensive chemicals, some of which are thought to have possible antibiotic and anti-inflammatory effects in humans.

## Sponges Reproduce Sexually and Asexually

Sponges reproduce through both sexual and asexual means. Most sponges are **hermaphrodites** (from the Greek, for the Greek god Hermes and the goddess Aphrodite), individuals that can produce both sperm and eggs. Gametes are derived from amoebocytes or choanocytes. Sperm are released into the water and carried by water currents to fertilize the eggs of neighboring sponges. Zygotes develop into flagellated swimming larvae that eventually settle on a suitable substrate to become sessile adults. In asexual reproduction, a small fragment, or bud, may detach and form a new sponge.

## 26.4 Reviewing the Concepts

- The phylum Porifera, or sponges, lack true tissues but are multicellular animals possessing several types of cells (Figure 26.11).
- They are asymmetric marine filter feeders.

## 26.4 Testing Your Knowledge

1. How do sponges protect themselves from predators?
   a. They are protected by silica spicules.
   b. They are protected by toxic defensive chemicals.
   c. They are eaten, and the leftover cells reaggregate into new, smaller sponges.
   d. Both a and b are correct.
   e. All of the above are correct.
2. Choanocytes are
   a. a group of protists believed to have given rise to animals.
   b. specialized cells of sponges that trap and eat small particles.
   c. cells that make up the gelatinous layer in sponges.
   d. cells of sponges that transfer nutrients to other cells.
   e. cells that form spicules in sponges.

# 26.5 Cnidaria: Jellyfish and Other Radially Symmetric Animals

### Learning Outcomes

1. Compare and contrast the two body forms of cnidarians.
2. Describe how cnidarians defend themselves.

The members of the phylum Cnidaria (from the Greek *knide*, meaning nettle, and *aria*, meaning related to; pronounced nid-air'-e-ah) are mostly found in marine environments, although a few are freshwater species. Cnidaria includes hydra, jellyfish, box jellies, sea anemones, and corals. The cnidarians have only two embryonic germ layers: the ectoderm and the endoderm. A gelatinous substance called the **mesoglea** connects the two layers. In jellyfish, the mesoglea is enlarged and forms a transparent jelly, whereas in hydra and corals, the mesoglea is very thin (Figure 26.12). Most cnidarians have tentacles around the mouth that aid in food detection and capture.

## The Cnidarians Exist in Two Different Body Forms

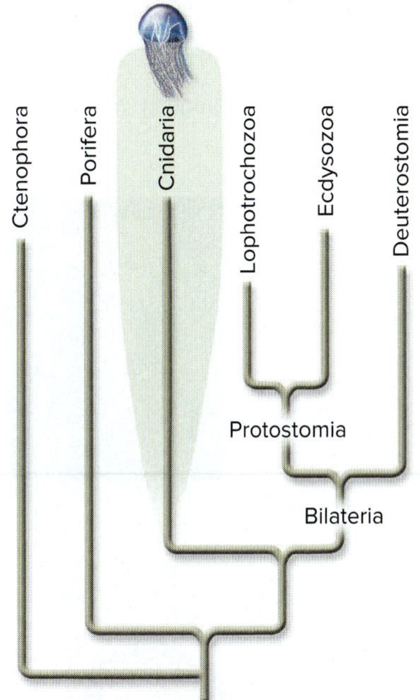


Cnidarians may exist as two different body forms and associated lifestyles: the sessile **polyp** and the motile **medusa** (Figure 26.12b). For example, corals exhibit only the polyp form, and jellyfish exist predominantly in the medusa form.

The polyp form has a tubular body with an opening at the oral (top) end that is surrounded by tentacles and functions as both mouth and anus (Figure 26.12a). The aboral (bottom) end is attached to the substrate. Polyps exist colonially, as they do in corals, or alone, as in sea anemones. Corals take dissolved calcium and carbonate ions from seawater and precipitate them as limestone underneath their bodies. In some species, this leads to a buildup of limestone deposits. As each successive generation of polyps dies, the limestone remains in place, and new polyps grow on top. Thus, huge underwater limestone deposits called coral reefs are formed (look ahead to Figure 45.22b). The largest of these is Australia's Great Barrier Reef, which stretches over 2,300 km. Many other extensive coral reefs are known, including the reef system along the Florida Keys, all of which occur in warm water, generally between 20°C and 30°C.

The free-swimming medusa form has an umbrella-shaped body with an opening that serves as both mouth and anus on the concave underside, which is surrounded by tentacles (see Figure 26.12b). Medusae possess simple sense organs near the bell margin, including organs of equilibrium called statocysts and photosensitive organs known as ocelli. When one side of the bell tips upward, the statocysts on that side are stimulated, and muscle contraction is initiated to reorient the medusa so that the tentacles are pointing downward. The ocelli allow medusae to position themselves in particular light levels.

The phylum Cnidaria consists of four classes: Hydrozoa (including the Portuguese man-of-war), Scyphozoa (jellyfish), Anthozoa (sea anemones and corals), and Cubozoa (box jellies). The distinguishing characteristics of these classes are shown in Table 26.1.

## Cnidarians Have Specialized Stinging Cells

One of the unique and characteristic features of the cnidarians is the existence of stinging cells called **cnidocytes,** which function in defense or the capture of prey (Figure 26.13a). Cnidocytes contain **nematocysts,** powerful capsules with an inverted coiled and barbed thread. Each cnidocyte has a hairlike trigger called a **cnidocil** on its surface. When the cnidocil is touched or a chemical stimulus is detected, the nematocyst is discharged; its filament penetrates the prey and injects a small amount of toxin. Small prey are immobilized and passed into the mouth by the tentacles. After

## Biology Principle

### Structure Determines Function

The inverted, umbrella-shaped medusae are free-swimming, whereas the tubular, polyp forms are sedentary.

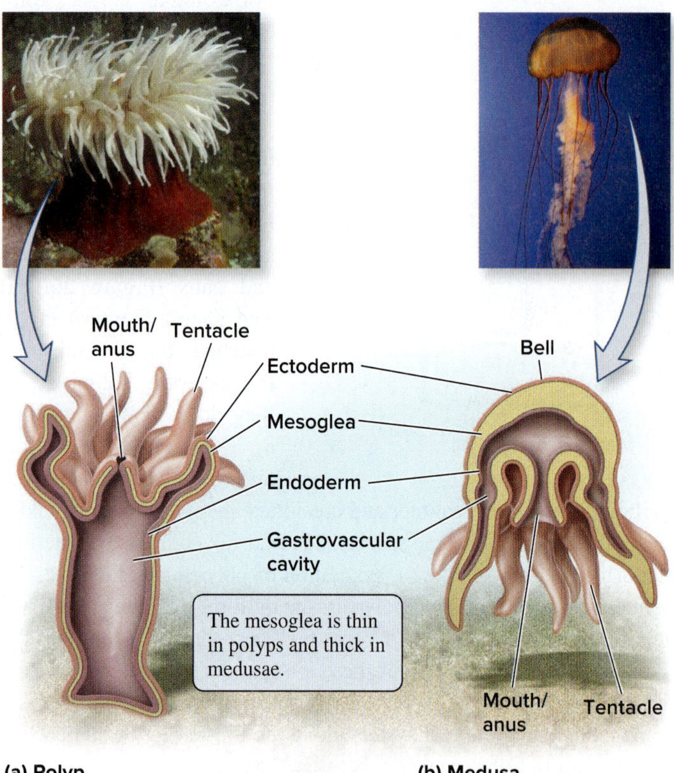


**Figure 26.12** **Polyp and medusa forms of cnidarians.** Both **(a)** polyp and **(b)** medusa forms have two layers of cells: an outer (ectoderm) and an inner (endoderm). In between is a layer of mesoglea, which is thin in polyps and thick in medusae.

*(a)* Source: Linda Snook, NOAA/CBNMS; *(b)* © Kick Images/Getty Images RF

**Concept Check:** *What are the dominant life stages of the following types of cnidarians: jellyfish, sea anemone, and Portuguese man-of-war?*

**Table 26.1 Main Classes and Characteristics of the Cnidaria**

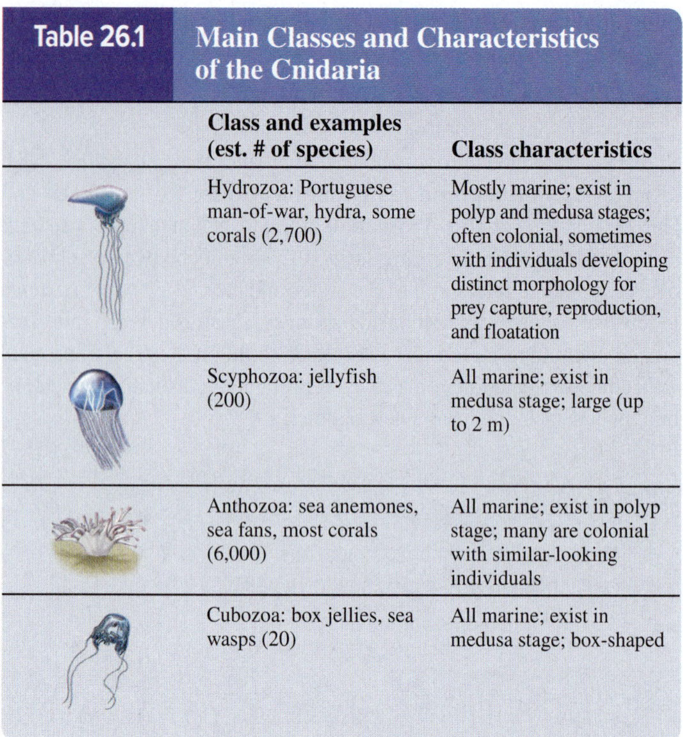

| | Class and examples (est. # of species) | Class characteristics |
|---|---|---|
| | Hydrozoa: Portuguese man-of-war, hydra, some corals (2,700) | Mostly marine; exist in polyp and medusa stages; often colonial, sometimes with individuals developing distinct morphology for prey capture, reproduction, and floatation |
| | Scyphozoa: jellyfish (200) | All marine; exist in medusa stage; large (up to 2 m) |
| | Anthozoa: sea anemones, sea fans, most corals (6,000) | All marine; exist in polyp stage; many are colonial with similar-looking individuals |
| | Cubozoa: box jellies, sea wasps (20) | All marine; exist in medusa stage; box-shaped |

discharge, the cnidocyte is absorbed, and a new one grows to replace it. The nematocysts of most cnidarians are not harmful to humans, but those on the tentacles of the larger jellyfish and the Portuguese man-of-war (Figure 26.13b) can be extremely painful and even fatal.

## 26.5 Reviewing the Concepts

- Cnidaria includes jellyfish, box jellies, sea anemones, corals, and the Portuguese man-of-war (Table 26.1). Cnidarians have only two embryonic germ layers: the ectoderm and endoderm, with a gelatinous substance (mesoglea) connecting the two layers.
- Cnidarians exist in two forms: polyp and medusa. A characteristic feature of cnidarians is their stinging cells, or cnidocytes, which function in defense and prey capture (Figures 26.12, 26.13).

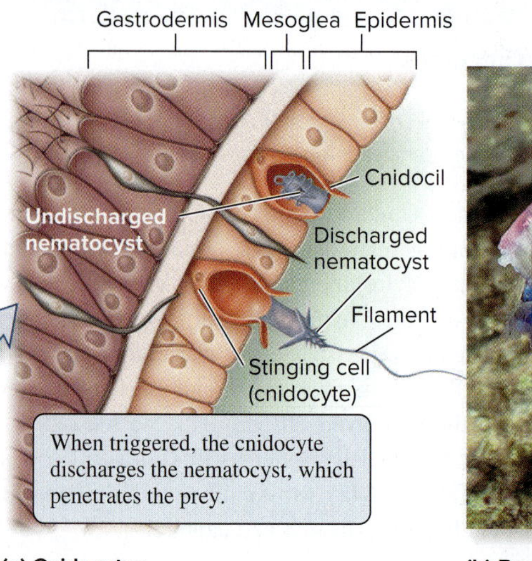

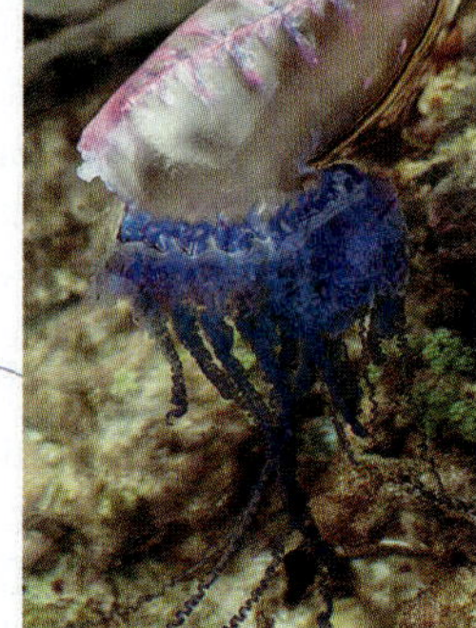


**Figure 26.13** **Specialized stinging cells of cnidarians, called cnidocytes.** **(a)** Cnidocytes, which contain stinging capsules called nematocysts, are situated in the tentacles. **(b)** The Portuguese man-of-war (*Physalia physalis*) employs cnidocytes that can be lethal to humans.

*(b)* © Wild Horizon/Getty Images

## 26.5 Testing Your Knowledge

1. Cnidarians defend themselves using
   a. a nerve net.
   b. bioluminescence.
   c. nematocysts.
   d. medusae.
   e. cnidocils.

## 26.6 Lophotrochozoa: The Flatworms, Rotifers, Bryozoans, Brachiopods, Mollusks, and Annelids

### Learning Outcomes

1. Describe the unique features of platyhelminthes, rotifers, bryozoans, and brachiopods.
2. Outline the main biological features and list the main classes of the mollusks.
3. List the advantages of segmentation in the annelids.

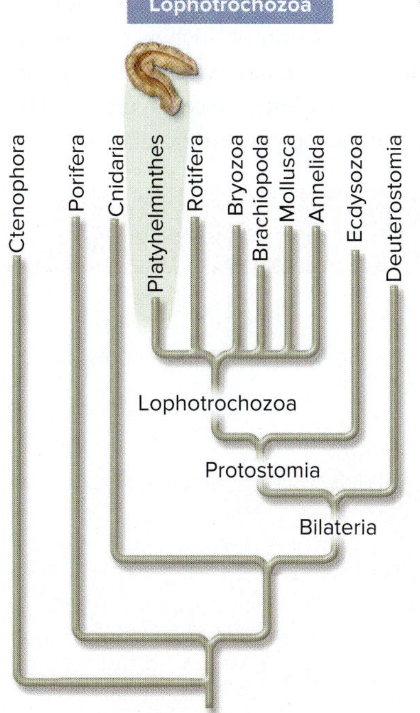


As noted earlier, morphological and molecular data are consistent with three clades of bilateral animals: the Lophotrochozoa and the Ecdysozoa (collectively known as the Protostomia, or protostomes) and the Deuterostomia (deuterostomes). In this section, we explore the distinguishing characteristics of the Lophotrochozoa, a diverse group that includes taxa that possess either a lophophore (a crown of ciliated tentacles, Bryozoa and Brachiopoda) or a distinct larval stage called a trochophore (Mollusca and Annelida). Also included in this clade are the Platyhelminthes (some of which have trochophore-like larvae) and the Rotifera (which have a lophophore-like feeding device), both of which share molecular similarities with the other members of the Lophotrochozoa.

### The Phylum Platyhelminthes Consists of Flatworms with No Coelom

Platyhelminthes (from the Greek *platy*, meaning flat, and *helminth*, meaning worm), or flatworms, were among the first animals to develop an active predatory lifestyle. Platyhelminthes, and indeed most animals, possess mesoderm and are bilaterally symmetric, with a head bearing sensory appendages, a feature called cephalization (**Figure 26.14**).

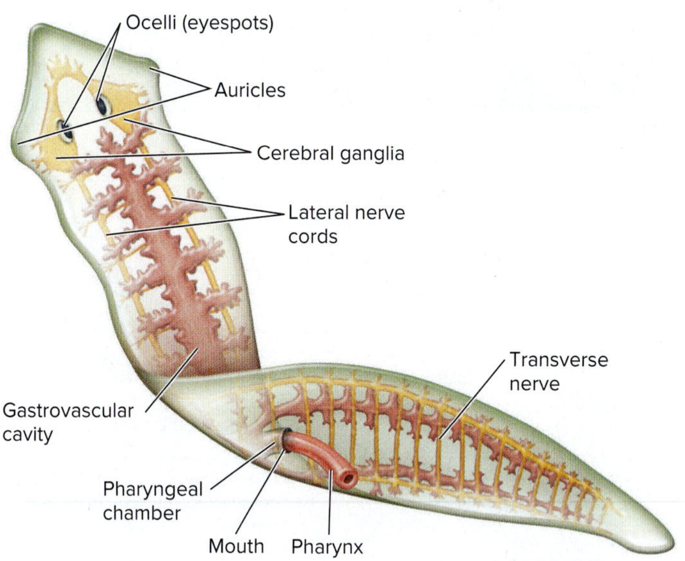


**Figure 26.14 Body plan of a flatworm.** Flatworm morphology is represented by a planarian, a member of the class Turbellaria.

Concept Check: *How do flatworms breathe?*

**Development of Mesoderm** The flatworms are believed to be the first bilaterian animals to develop three distinctive embryonic germ layers—ectoderm, endoderm, and mesoderm—with mesoderm replacing the simpler gelatinous mesoglea of cnidarians. As such, they are said to be triploblastic. The muscles in flatworms, which are derived from mesoderm, are well developed. The development of mesoderm was therefore a critical evolutionary innovation in bilaterian animals, leading to complex movement and the development of more sophisticated organs. Flatworms lack a fluid-filled body cavity in which the gut is suspended and, instead, mesoderm fills the body spaces around the gastrovascular cavity; hence, they are described as acoelomates (see Figure 26.6c). Flatworms also lack a specialized respiratory or circulatory system and must respire by diffusion. Thus, no cell can be too far from the surface, making a flattened shape necessary. The digestive system of flatworms is incomplete, with only one opening, which serves as both mouth and anus, as in cnidarians. Most flatworms possess a muscular pharynx that may be extended through the mouth. The pharynx opens to a gastrovascular cavity, where food is digested. In large flatworms, the gastrovascular cavity is highly branched to distribute nutrients to all parts of the body. The incomplete digestive system of flatworms prevents continuous feeding. Some flatworms are predators, but many species invade other animals as parasites.

**Cephalization** At the anterior end of some free-living flatworms are light-sensitive eyespots, called ocelli, as well as chemoreceptive and sensory cells that are concentrated in organs called auricles. A pair of **cerebral ganglia,** which are clusters of nerve cell bodies, receives input from photoreceptors in eyespots and sensory cells. From the ganglia, a pair of lateral nerve cords running the length of the body allows rapid movement of information from anterior to posterior. In addition, transverse nerves form a nerve net on the ventral surface, similar to that of cnidarians. Thus, flatworms retain the cnidarian-style nervous system, while possessing the beginnings of the more centralized type of nervous system seen throughout much of the rest of the animal kingdom.

**Table 26.2 Main Classes and Characteristics of Platyhelminthes**

| | Class and examples (est. # of species) | Class characteristics |
|---|---|---|
| | Turbellaria: planarian (3,000) | Mostly marine; free-living flatworms; predators or scavengers |
| | Monogenea: flukes (1,000) | Marine and freshwater; usually external parasites of fish; simple life cycle (no intermediate host) |
| | Trematoda: flukes (11,000) | Internal parasites of vertebrates; complex life cycle with several intermediate hosts |
| | Cestoda: tapeworms (5,000) | Internal parasites of vertebrates; complex life cycle, usually with one intermediate host; no digestive system; nutrients absorbed across epidermis |

In Platyhelminthes, reproduction is either sexual or asexual. Most species are hermaphroditic but do not fertilize their own eggs. Flatworms can also reproduce asexually by splitting into two parts, with each half regenerating the missing fragment.

**Flatworm Diversity** The four classes of flatworms are the Turbellaria, Monogenea, Trematoda (flukes), and Cestoda (tapeworms) (**Table 26.2**).

- Turbellarians are the only free-living class of flatworms and are widespread in lakes, ponds, and marine environments (**Figure 26.15a**).
- Monogeneans are relatively simple external parasites with just one host species (a fish).
- Trematodes and cestodes are internally parasitic in humans and therefore are of great medical and veterinary importance. They possess a variety of attachment organs, such as hooks and suckers, that enable them to remain embedded within their hosts. For example, cestodes attach to their host by means of an organ at the head end called a scolex (**Figure 26.15b**). They have no mouth or gastrovascular cavity and absorb nutrients across the body surface.

Flatworm life cycles can be complex. Cestodes often require two separate vertebrate host species: one host, such as pigs or cattle, to begin their life cycle and another host, a primate such as a human, to complete their development. Behind the scolex in cestodes is a long ribbon of identical segments called proglottids (see Figure 26.15b). These are segments of sex organs that produce thousands of eggs.

## Biology Principle

### Biology Affects Our Society

About 1% of U.S. cattle are infected by beef tapeworms. Consuming beef that is not sufficiently well cooked can lead to infection by these parasites.

(a)

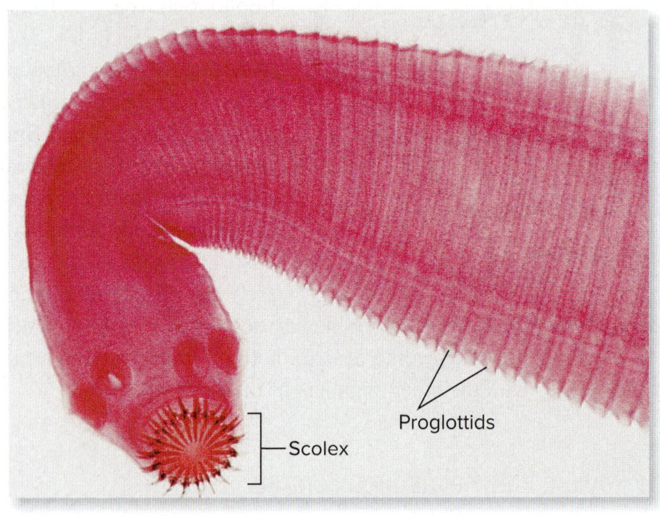

(b)

**Figure 26.15 Flatworms. (a)** Many free-living marine turbellarians are brightly colored such as this racing stripe flatworm, *Pseudoceros bifurcus,* from Bali, Indonesia. **(b)** The tapeworm *Taenia pisiformis* is a member of the class Cestoda. Note the tiny hooks and suckers that make up the scolex. Each segment is a proglottid filled with eggs.

*(a)* © Wolfgang Poelzer/Wa./agefotostock; *(b)* © Biophoto Associates/Science Source

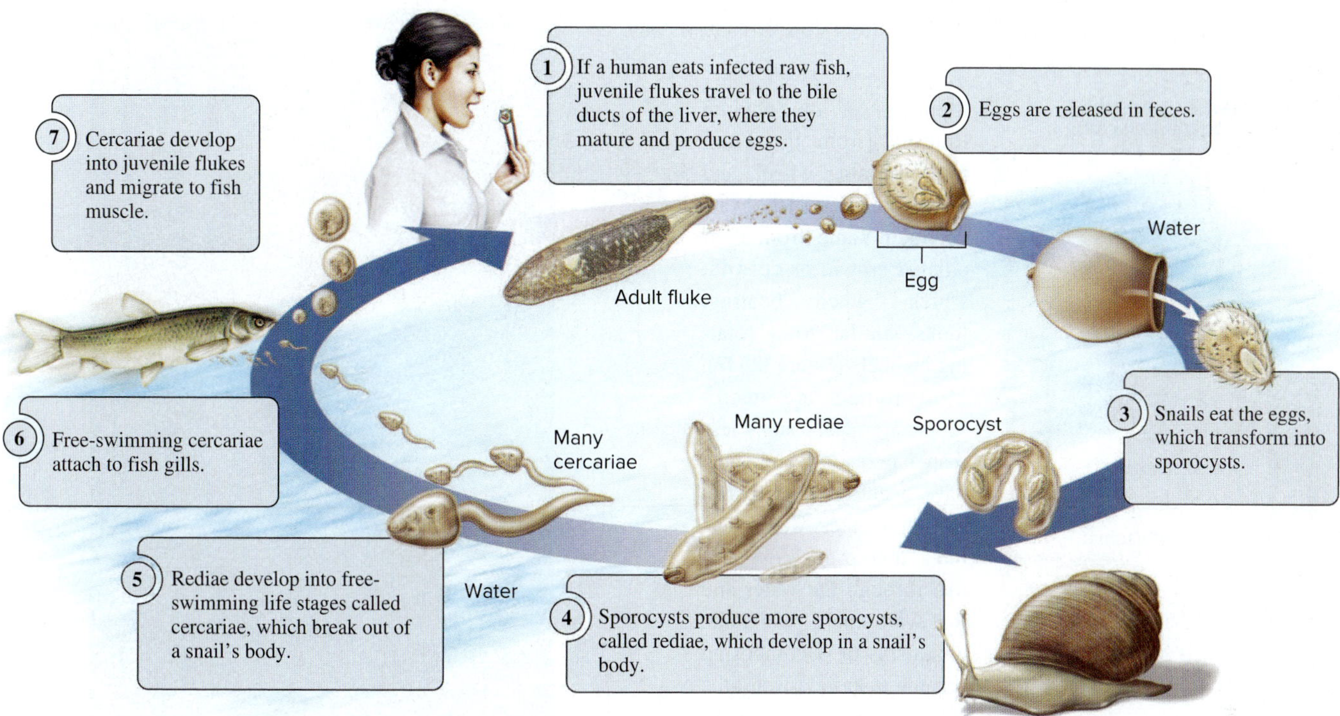


**Figure 26.16 The complete life cycle of a trematode.** This figure shows the life cycle of the Chinese liver fluke (*Clonorchis sinensis*).

The proglottids are continually shed in the host's feces. Human feces passed out into the ground are eaten with grass by pigs or cattle. Many tapeworms are then ingested by humans when they consume undercooked, infected meat—hence the value of thoroughly cooking meat.

The life history of trematodes is even more complex than that of cestodes, involving multiple hosts. The first host, called the intermediate host, is usually a mollusk, and the final host, or definitive host, is usually a vertebrate, but often a second or even a third intermediate host is involved (Figure 26.16). Because the flukes multiply in each host, prodigious numbers of offspring can be produced.

Blood flukes are the most common parasitic trematodes infecting humans; they cause the disease known as schistosomiasis. Over 200 million people worldwide, primarily in tropical Asia, Africa, and South America, are infected with this disease. The inch-long adult flukes can live for years in human hosts, and the release of eggs may cause chronic inflammation and blockage in many organs. Untreated schistosomiasis can lead to severe damage to the liver, intestines, and lungs and can eventually lead to death. Sewage treatment and access to clean water can greatly reduce infection rates.

## Quantitative Analysis

### HOW MANY FLUKES?

Many parasites face the challenge of reproducing and having their offspring find a new host. How can offspring relocate to a new host when their parent is locked up inside the host's body? As noted previously, in tapeworms each proglottid contains thousands of eggs. This ensures that vast numbers of eggs pass out of the host's body in the hope of finding a new host. Flukes, such as Chinese liver fluke (*Clonorchis sinensis*), have multiple host species. Eggs, on release from the human host, pass into water, where snails eat them (see Figure 26.16, steps 2 and 3). Eggs develop into sporocysts inside the snail, and each sporocyst produces more sporocysts, called rediae. The rediae produce cercariae, which break out of the snail's body to attach to fish. The cercariae eventually lodge in fish muscles, which are then consumed by humans. The chances of infection at each of these stages are very low, so flukes produce huge numbers of offspring to compensate. We can appreciate the vast reproductive effort involved by calculating the number of offspring at each stage:

- Average number of eggs produced per fluke per day = 28,000
- Average length of life of adult fluke = 1,800 days (about 5 years)
- Number of sporocysts produced per lifetime = 28,000 × 1,800 = about 50 million
- Average number of cercariae produced per sporocyst = 200
- Total number of cercariae produced per lifetime = 200 × 50 million = 10 billion

**Crunching the Numbers:** Let's suppose a tapeworm sheds about 5 proglottids per day. Each proglottid contains about 50,000 eggs, and most tapeworms live in colonies of about 10 per host. How many eggs are shed by each host per day?

## Members of the Phylum Rotifera Have a Pseudocoelom and a Ciliated Crown

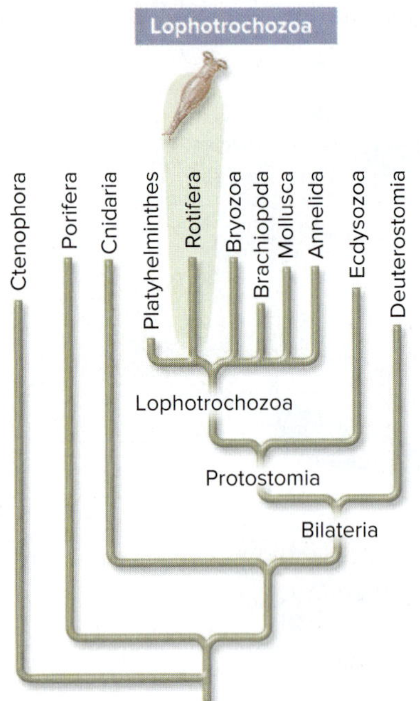


Members of the phylum Rotifera (from the Latin *rota*, meaning wheel, and *fera*, meaning to bear) get their name from their ciliated crown, or **corona,** which, when beating, looks similar to a rotating wheel (**Figure 26.17**). Most rotifers are microscopic animals, usually less than 1 mm long, and some have beautiful colors. The corona creates water currents that propel the animal through the water and that waft small, planktonic organisms or decomposing organic material toward the mouth. Rotifers have an alimentary canal, a digestive tract with a separate mouth and anus, which means they can feed continuously. There are about 2,000 species of rotifers, most of which inhabit fresh water, with a few marine and terrestrial species. Most often, they are bottom-dwelling organisms, living on the pond floor or along lakeside vegetation.

**Figure 26.17** ***Philodina roseola*, a common rotifer, showing the crown of cilia (corona).**



## Bryozoa and Brachiopoda Are Closely Related Phyla

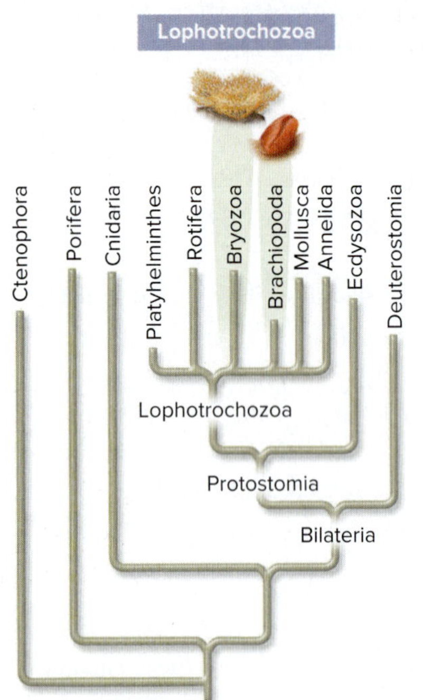


The Bryozoa and the Brachiopoda both possess a lophophore, a ciliary feeding device (see Figure 26.9a), and a true coelom (see Figure 26.6a). The lophophore is a circular fold of the body wall bearing tentacles that draw water toward the mouth. Because a thin extension of the coelom penetrates each tentacle, the tentacles also serve as a respiratory organ. Gases diffuse across the tentacles and into or out of the coelomic fluid and are carried throughout the body.

The spacey bryozoans (from the Greek *bryon*, meaning moss, and *zoon*, meaning animal) are small, colonial animals, most of which are less than 0.5 mm long, found encrusted on rocks in shallow aquatic environments (**Figure 26.18a**). For this reason, bryozoans have been important reef-builders. Bryozoans date back to the Paleozoic era, and thousands of fossil forms have been discovered and identified.

Brachiopods (from the Greek *brachio*, meaning arm, and *podos*, meaning foot) are small, 1–8 cm bottom-dwelling marine organisms with two shell halves, much like clams (**Figure 26.18b**). Although they are now a relatively small group, with about 300 living species, brachiopods flourished in the Paleozoic and Mesozoic eras—about 30,000 fossil species have been identified.

## Mollusca Is a Large Phylum Containing Snails, Slugs, Clams, Oysters, Octopuses, and Squids

Mollusks (from the Latin *mollis*, meaning soft) constitute a very large phylum, with over 100,000 living species, including organisms as diverse as snails, clams and oysters, octopuses and squid, and chitons. They are an ancient group, as evidenced by the classification of about 35,000 fossil species. Mollusks have a considerable economic, aesthetic, and ecological importance to humans. Many, including scallops, oysters, clams, and squids, serve as sources of food. A significant industry involves the farming of oysters to produce cultured pearls, and rare and beautiful mollusk shells are extremely valuable to collectors. Snails and slugs can damage vegetables and ornamental plants, and boring mollusks can penetrate wooden ships and wharfs. Mollusks are intermediate hosts to many parasites, and several invasive species have become serious pests. For example, populations of

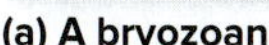
(a) A bryozoan

(b) A brachiopod, the northern lamp shell

**Figure 26.18** **Bryozoans and brachiopods.** **(a)** Bryozoans are colonial animals that reside inside a nonliving case. **(b)** Brachiopods, such as this northern lamp shell (*Terebratulina septentrionalis*), in Nova Scotia, Canada, have dorsal and ventral shells.

*(a)* © Hecker/Sauer/agefotostock; *(b)* © Gordon MacSkimming/agefotostock

*What are the two main functions of the lophophore?*

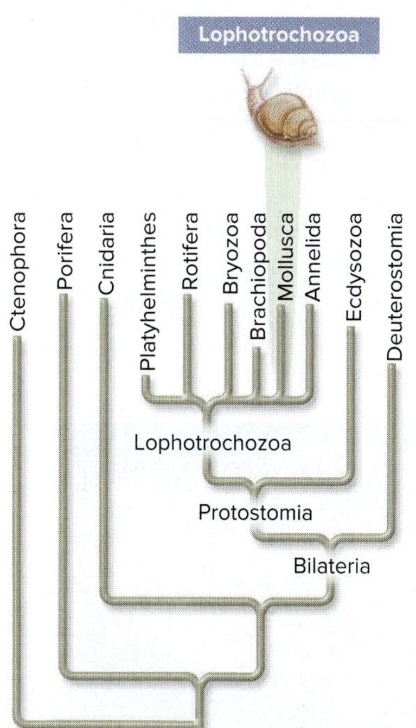


the zebra mussel (*Dreissena polymorpha*) have been introduced into North America from Asia via ballast water from transoceanic ships. Since their introduction, they have spread rapidly throughout the Great Lakes and an increasing number of inland waterways, adversely affecting native organisms and clogging water intake valves to municipal water-treatment plants around the lakes.

### The Mollusk Body Plan

One common feature of the mollusks is their soft body, which in many species exists under a protective external shell. Most mollusks are marine, although some have colonized fresh water. Many snails and slugs have moved onto land but survive only in humid areas and where the calcium necessary for shell formation is abundant in the soil. The ability to colonize freshwater and terrestrial habitats has led to a diversification of mollusk body plans. Thus, in the amazing diversity of mollusks we see how organismal diversity is related to environmental diversity.

Although great variation in morphology occurs between classes, mollusks have a basic body plan consisting of three parts (Figure 26.19).

- A muscular **foot** is usually used for movement, and a **visceral mass** containing the internal organs rests atop the foot.
- The **mantle,** a fold of skin draped over the visceral mass, secretes a shell in those species that form shells.
- The **mantle cavity** houses delicate **gills,** filamentous organs that are specialized for gas exchange. A continuous current of water, often induced by cilia present on the gills or by muscular pumping, flushes out the wastes from the mantle cavity and brings in new, oxygen-rich water.

Mollusks have an **open circulatory system** with a heart that pumps body fluid, called hemolymph, through vessels and into sinuses. Sinuses are the open, fluid-filled cavities between the internal organs. The organs and tissues are therefore continually bathed in hemolymph. The sinuses coalesce to form an open cavity known as the hemocoel (blood cavity). From these sinuses, the hemolymph drains into vessels that take it to the gills and then back to the heart.

The mollusk's mouth may contain a **radula,** a unique, protrusible, tonguelike organ that has many teeth and is used to eat plants, scrape food particles off rocks, or, if the mollusk is predatory, bore into shells of other species and tear flesh (see inset Figure 26.19). Other mollusks, particularly bivalves, have lost their radula and are filter feeders that strain water brought in by ciliary currents.

**Mollusk Reproduction** Most mollusks have separate sexes, although some are hermaphroditic. Gametes are usually released into the water, where they mix and fertilization occurs. In some snails, however, fertilization is internal, with the male inserting sperm directly into the female. Internal fertilization was a key evolutionary development, enabling some snails to colonize land, and can be considered a critical innovation that fostered extensive adaptive radiation.

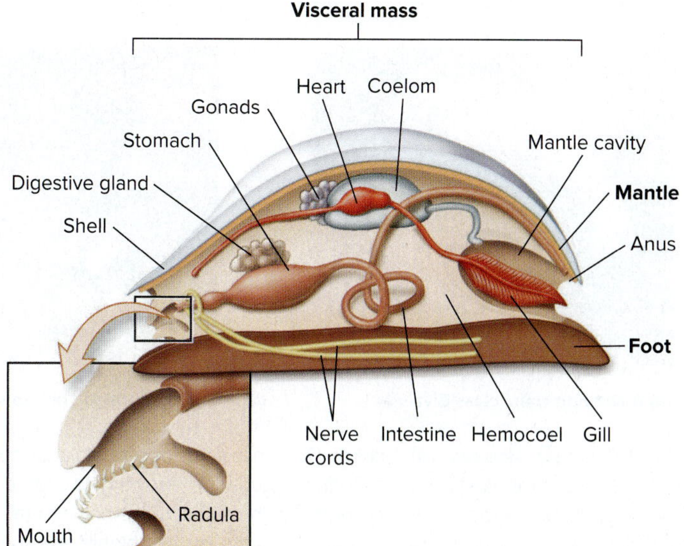


**Figure 26.19** **The mollusk body plan.** The generalized body plan of a mollusk includes the characteristic foot, mantle, and visceral mass.

**Table 26.3 Main Classes and Characteristics of Mollusks**

| Class and examples (est. # of species) | Class characteristics |
|---|---|
| Polyplacophora: chitons (860) | Marine; eight-plated shell |
| Gastropoda: snails, slugs, nudibranchs (75,000) | Marine, freshwater, or terrestrial; most with coiled shell, but shell absent in slugs and nudibranchs; radula present |
| Bivalvia: clams, mussels, oysters (30,000) | Marine or freshwater; shell with two halves or valves; primarily filter feeders with siphons |
| Cephalopoda: octopuses, squids, nautiluses (780) | Marine; predatory, with tentacles around mouth, often with suckers; shell often absent or reduced; closed circulatory system; jet propulsion via siphon |

In many species, reproduction involves the production of a trochophore larva that develops into a free-swimming larva with a rudimentary foot, shell, and mantle.

**The Major Molluscan Classes** The four most common molluscan classes are the Polyplacophora (chitons), Gastropoda (snails and slugs), Bivalvia (clams and mussels), and Cephalopoda (octopuses, squids, and nautiluses) (**Table 26.3**). Chitons are marine mollusks with a shell composed of eight separate plates (**Figure 26.20a**). Chitons are common in the intertidal zone, an area above water at low tide and under water at high tide. The class Gastropoda (from the Greek *gaster*, meaning stomach, and *podos*, meaning foot) is the largest group of mollusks and encompasses about 75,000 living species, including snails, periwinkles, limpets, and other shelled members (**Figure 26.20b**). The class also includes species such as slugs and nudibranchs, whose shells have been greatly reduced or completely lost during their evolution (**Figure 26.20c**). Most gastropods are marine or freshwater species, but some species, including snails and slugs, have also colonized land. Most gastropods are slow-moving animals that are weighed down by their shell. Unlike bivalves, gastropods have a one-piece shell, into which the animal can withdraw

**(a) A chiton, class Polyplacophora**

**(b) A snail, class Gastropoda**

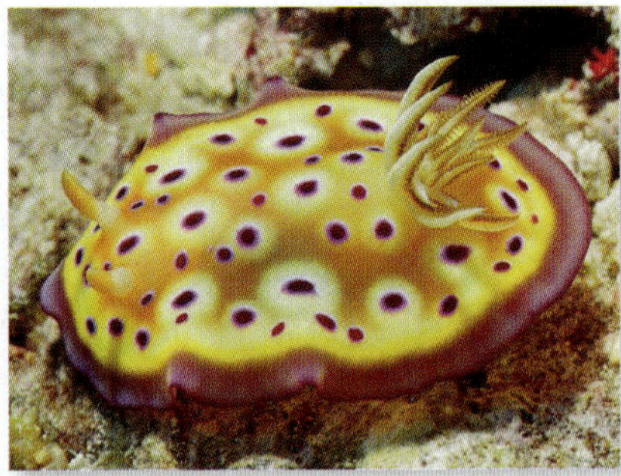

**(c) A sea slug, class Gastropoda**

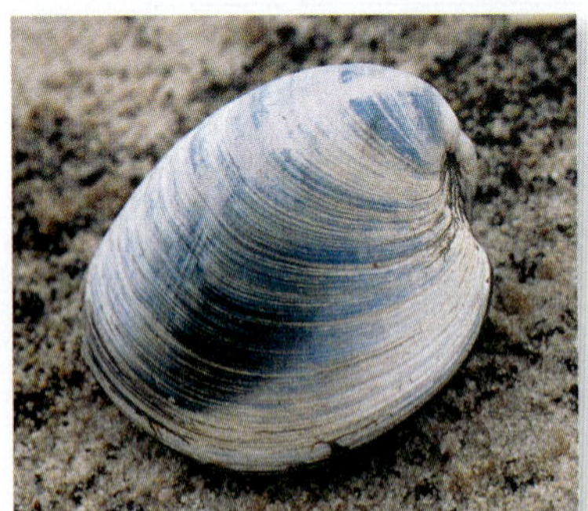

**(d) A Quahog clam, class Bivalvia**

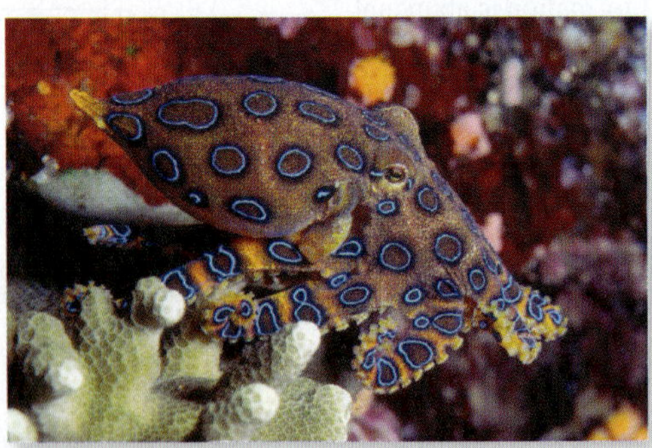

**(e) A blue-ringed octopus, class Cephalopoda**

**Figure 26.20** **Mollusks.** **(a)** A chiton (*Tonicella lineata*), a polyplacophoran with a shell made up of eight separate plates. **(b)** A gastropod tree snail (*Liguus fasciatus*) from the Florida Everglades, showing its characteristic coiled shell. **(c)** A nudibranch (*Chromodoris kuniei*). The nudibranchs are a gastropod subclass whose members have lost their shell altogether. **(d)** A bivalve shell, class Bivalvia, with growth rings. Quahog clams (*Mercenaria mercenaria*) can live over 20 years. **(e)** The highly poisonous blue-ringed octopus (*Hapalochlaena lunulata*) is a cephalopod.



**BioConnections:** *In what period did mollusks arise? Refer back to Figure 22.4.*

to escape predators. Bivalves are marine or freshwater mollusks with two halves to their shell. Most are filter feeders, using their gills to strain food particles from the water (Figure 26.20d).

The 780 species of Cephalopoda (from the Greek *kephalo*, meaning head, and *podos*, meaning foot) are the most morphologically complex of the mollusks and indeed among the most complex of all invertebrates. Most are fast-swimming marine predators that range from organisms just a few centimeters in size to the colossal squid (*Mesonychoteuthis hamiltoni*), which is known to reach over 13 m in length and 495 kg (1,091 lb) in weight. A cephalopod's mouth is surrounded by many long arms commonly armed with suckers. All cephalopods have a beaklike jaw that allows them to bite their prey, and some, such as the blue-ringed octopus (*Hapalochlaena lunulata*), deliver a deadly poison through their saliva (Figure 26.20e).

The foot of some cephalopods has become modified into a muscular siphon. Water drawn into the mantle cavity is quickly expelled through the siphon, propelling the organisms forward or backward in a kind of jet propulsion. Such vigorous movement requires powerful muscles and a very efficient circulatory system to deliver oxygen and nutrients to the muscles. Cephalopods are the only mollusks with a **closed circulatory system,** in which blood flows throughout an animal entirely within a series of vessels. One of the advantages of this type of system is that the heart can pump blood through the tissues rapidly, making oxygen more readily available. The blood of cephalopods contains the copper-rich protein hemocyanin for transporting oxygen. Less efficient than the iron-rich hemoglobin of vertebrates, hemocyanin gives the blood a blue color.

## The Phylum Annelida Consists of the Segmented Worms

Annelids are a large phylum with about 15,000 described species. Its members include free-ranging marine worms, tube worms, the familiar earthworm, and leeches. They range in size from less than 1 mm to enormous Australian earthworms that can reach a size of 3 m.

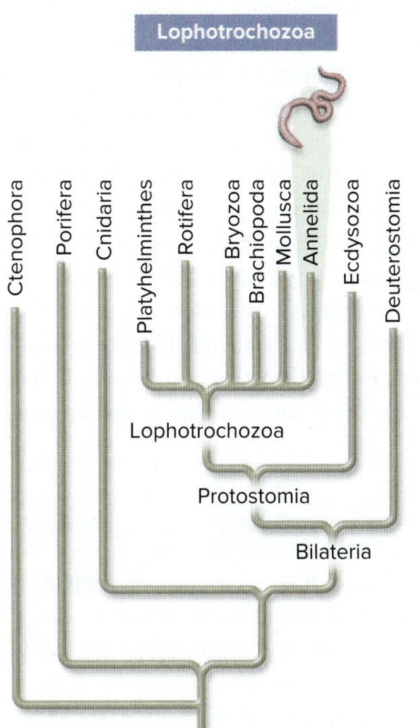


### The Annelid Body Plan

If you look at an earthworm, you will see little rings all down its body. Indeed, the phylum name Annelida is derived from the Latin *annulus*, meaning little ring. Each ring is a distinct segment of the annelid's body, with each segment separated from the one in front and the one behind by septa (Figure 26.21). All annelids except the leeches have chitinous bristles, called **setae,** on each segment. In some, these are situated on fleshy, footlike **parapodia** (from the Greek, meaning almost feet) that are pushed into the substrate to provide traction during movement. In others, the setae are held closer to the body. Many annelid species burrow into soil or into muddy marine sediments and extract nutrients from ingested soil or mud. Some annelids also feed on dead or living vegetation, whereas others are predatory or parasitic.

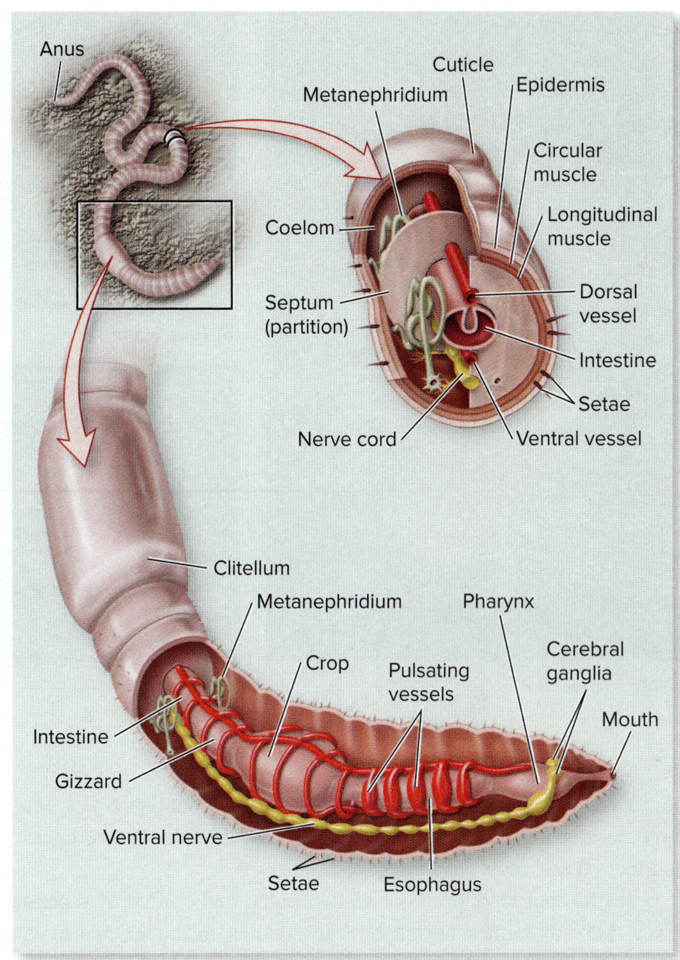


**Figure 26.21 The segmented body plan of an annelid, as illustrated by an earthworm.** The segmented nature of the worm is apparent internally as well as externally. Individual segments are separated by septa.

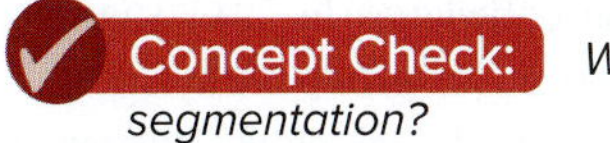

*What are some of the advantages of segmentation?*

Annelids have a double transport system. Both the circulatory system and the coelomic fluid carry nutrients, wastes, and respiratory gases, to some degree. Annelids have a closed circulatory system, with dorsal and ventral vessels connected by pairs of pulsating vessels (look ahead to Figure 36.1b). The blood of most annelid species contains the respiratory pigment hemoglobin. Respiration occurs directly through the permeable skin surface, which restricts annelids to moist environments. The digestive system is complete and unsegmented, with many specialized regions: mouth, pharynx, esophagus, crop, gizzard, intestine, and anus. Annelids have a relatively sophisticated nervous system involving a pair of cerebral ganglia that connect to a large ventral nerve cord running down the entire length of the

(a) Marine worm

(b) Tube worm

(c) Earthworm

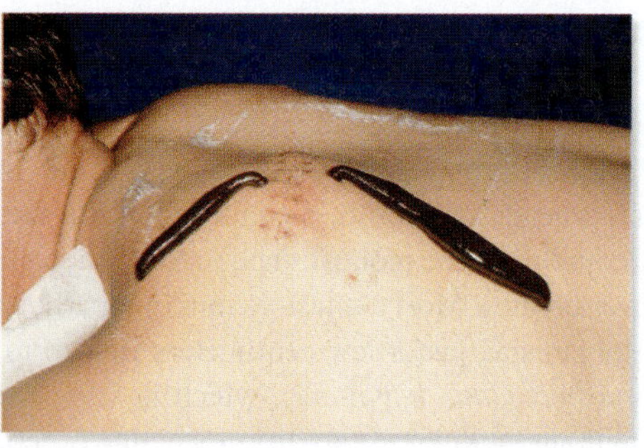

(d) Leech

**Figure 26.22** **Annelids.** **(a)** This free-ranging marine worm from Indonesia is a member of the group Errantia. Members of the group Sedentaria include **(b)** tube worms, **(c)** earthworms, and **(d)** leeches. This species, *Hirudo medicinalis,* is sucking blood from a hematoma, a swelling of blood that can occur after surgery.



body. The ventral nerve is unusual because it contains a few very large nerve cells called **giant axons** that facilitate high-speed nerve conduction and rapid responses to stimuli.

Sexual reproduction involves two individuals, often of separate sexes, but sometimes hermaphrodites, which exchange sperm via internal fertilization. In some species, asexual reproduction by fission occurs, in which the posterior part of the body breaks off and forms a new individual.

**The Major Annelid Groups** Evidence published by German evolutionary biologist Torsten Struck and colleagues in 2011 suggests that the phylum Annelida contains two major groups: the Errantia, active, free-ranging worms, and the Sedentaria, which are burrowers or parasites.

Members of the Errantia have many long setae bristling out of their body and are supported on footlike parapodia (Figure 26.22a). Most of them are free-ranging predators with well-developed eyes and powerful jaws. Many are brightly colored. In turn, most species are important prey for fishes and crustaceans.

In the Sedentaria, setae are in close proximity to the body wall, which facilitates better anchorage in tubes and burrows. Their more sedentary lifestyle is associated with reductions in head appendages. Within the Sedentaria, three types of lifestyles are apparent. Tube worms are marine sedentarians that exhibit beautiful tentacle crowns for filtering food items, such as plankton, from the water column (Figure 26.22b). The bulk of these worms remain hidden in a tube deep in the mud or sand.

Earthworms ingest soil and leaf tissue to extract nutrients and in the process create burrows in the Earth (Figure 26.22c). As plant material and soil pass through the earthworm's digestive system, they are finely ground in the gizzard into smaller fragments. The biologist Charles Darwin, who penned the first detailed study of earthworm ecology, wrote, "All the fertile areas of this planet have at least once passed through the bodies of earthworms."

Leeches are primarily found in freshwater environments, but there are also some marine species as well as terrestrial species that inhabit warm, moist areas such as tropical forests. Leeches have a fixed number of segments, usually 34, though in most species the septa have disappeared. Most leeches feed as blood-sucking parasites of vertebrates. They have powerful suckers at both ends of the body, and the anterior sucker is equipped with razor-sharp jaws that can bore or slice into the host's tissues. The salivary secretion of leeches (hirudin) acts as an anticoagulant to stop the prey's blood from clotting and an anesthetic to numb the pain. Leeches can suck up to several times their own weight in blood. They were once used in the medical field in the practice of bloodletting, the withdrawal of often considerable quantities of blood from a patient in the erroneous belief that this would prevent or cure illness and disease. Even today, leeches may be used after surgeries (Figure 26.22d). In these cases, the blood vessels are not fully reconnected and excess blood accumulates, causing a swelling called a hematoma. The accumulated blood blocks the delivery of new blood and stops the formation of new vessels. The leeches remove the accumulated blood, and new capillaries are more likely to form.

## 26.6 Reviewing the Concepts

- Most Lophotrochozoa include taxa that possess either a lophophore or trochophore larva. Platyhelminthes, or flatworms, are regarded as the first animals to have the organ-system level of organization (Figure 26.14; Table 26.2). Flukes and tapeworms are internally parasitic, with complex life cycles (Figures 26.15, 26.16).

- The bryozoa and brachiopods both possess a lophophore, a ciliary feeding structure (Figure 26.18).
- The mollusks, which constitute a large phylum with over 100,000 diverse living species, have a basic body plan with three parts—a foot, a visceral mass, and a mantle—and an open circulatory system (Figures 26.19, 26.20). The four most common mollusk classes are the polyplacophora (chitons), gastropoda (snails and slugs), bivalvia (clams and mussels), and cephalopoda (octopuses, squids, and nautiluses) (Table 26.3).
- Annelids are a large phylum with two main groups: Errantia, which includes free-ranging marine worms, and Sedentaria, which includes tube worms, earthworms, and leeches (Figure 26.22). Segmentation, in which the body is divided into compartments, is a critical evolutionary innovation in the annelids.

## 26.6 Testing Your Knowledge

1. Platyhelminthes possess
   **a.** a lophophore.
   **b.** a coelom.
   **c.** mesoderm.
   **d.** a muscular foot.
   **e.** giant axons.
2. All mollusks possess
   **a.** a radula.
   **b.** an open circulatory system.
   **c.** a soft body.
   **d.** an external shell.
   **e.** segmentation.
3. What is an advantage of segmentation in annelids?
   **a.** It allows specialization of body regions and appendages.
   **b.** It allows sexual reproduction.
   **c.** It allows for radial symmetry.
   **d.** It is correlated with cephalization.
   **e.** It permits the development of a complete gut.

## 26.7 Ecdysozoa: The Nematodes and Arthropods

### Learning Outcomes

1. List the distinguishing characteristics of nematodes.
2. Describe the arthropod body plan and its major features.
3. Give examples of the arthropod subphyla Chelicerata, Myriapoda, Hexapoda, and Crustacea.
4. List the features that help account for the diversity of insect species.

The Ecdysozoa is the sister group to the Lophotrochozoa. Although the separation is supported by molecular evidence, Ecdysozoa is named for a morphological characteristic, the physical phenomenon of ecdysis, or the periodic molting of the exoskeleton (see Figure 26.8). All ecdysozoans possess a **cuticle,** a nonliving cover, such as an exoskeleton, that both supports and protects the animal. Once formed, however, the cuticle typically cannot increase in size, which restricts the growth of the animal inside. The solution for growth is the formation of a new, softer cuticle under the old one. The old one then splits open and is sloughed off, allowing the new, soft cuticle to expand to a bigger size before it hardens. Where the cuticle is thick, as in arthropods, it impedes the diffusion of oxygen across the skin. Such species acquire oxygen by lungs, gills, or a set of branching, air-filled tubes called tracheae. A variety of appendages specialized for locomotion evolved in many species, including legs for walking or swimming and wings for flying. The ability to shed the cuticle opened up developmental options for the ecdysozoans. For example, many species undergo a complete metamorphosis, changing from a wormlike larva into a winged adult. Another significant adaptation is the development of internal fertilization, which permitted species to live in dry environments.

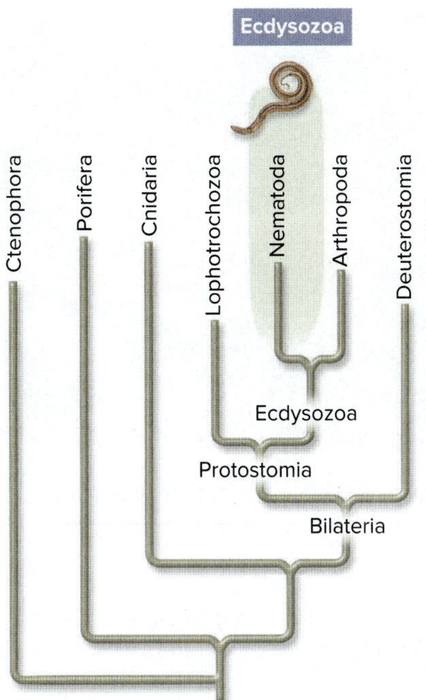


Because of these innovations, ecdysozoans are an incredibly successful group. Here, we will consider the two most common ecdysozoan phyla: the nematodes and arthropods. The grouping of nematodes and arthropods is a relatively new concept supported by molecular data and implies that the process of molting arose only once in animal evolution. In support of this, certain hormones that stimulate molting have been discovered to exist only in both nematodes and arthropods.

### The Phylum Nematoda Consists of Small Pseudocoelomate Worms Covered by a Tough Cuticle

The nematodes (from the Greek *nematos*, meaning thread), also called roundworms, are small, thin worms that range from less than 1 mm to about 5 cm (**Figure 26.23**), although some parasitic species measuring 1 m or more have been found in the placenta of sperm whales. Nematodes are ubiquitous organisms that exist in nearly all habitats, from the poles to the tropics. They are found in the soil, in both freshwater and marine environments, and inside plants and animals as parasites. A shovelful of soil may contain a million nematodes. Over 20,000 species are known, but there are probably at least five times as many undiscovered species.

**The Nematode Body Plan** Nematodes have several distinguishing characteristics. A tough cuticle covers the body. The cuticle is secreted by the epidermis and is made primarily of **collagen,** a structural protein also present in vertebrates (see Figure 4.31). The cuticle is shed periodically as the nematode grows. The pseudocoelom functions as both a fluid-filled skeleton and a circulatory system.

**Figure 26.23** **Scanning electron micrograph of a nematode within a plant leaf.**


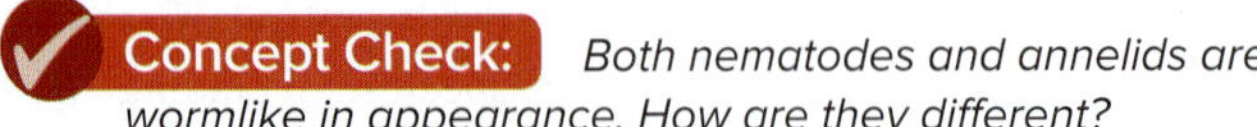

**Concept Check:** *Both nematodes and annelids are wormlike in appearance. How are they different?*

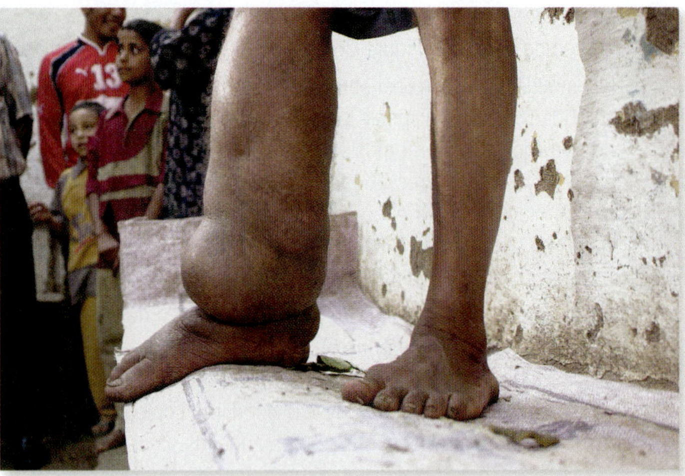

**Figure 26.24** **Elephantiasis in a human leg.** The disease is caused by the nematode parasite *Wuchereria bancrofti,* which lives in the lymphatic system and blocks the flow of lymph.


Diffusion of gases occurs through the cuticle. Roundworms have a complete digestive tract composed of a mouth, pharynx, intestine, and anus. The mouth often contains sharp, piercing organs called **stylets,** and the muscular pharynx functions to suck in food.

**Nematode Reproduction** Nematode reproduction is usually sexual, with separate males and females, and fertilization takes place internally. Females can produce prodigious numbers of eggs—in some cases, over 100,000 per day. Development in some nematodes is easily observed because the organism is transparent and the generation time is short. For these reasons, the small, free-living nematode *Caenorhabditis elegans* has become a model organism for researchers to study. This nematode has 1,090 somatic cells, but 131 die, leaving exactly 959 cells. The cells die via genetically controlled programmed cell death, or apoptosis. Many diseases in humans, including acquired immunodeficiency syndrome (AIDS), cause extensive apoptosis, whereas others, such as cancer and autoimmune diseases, reduce apoptosis so that cells that should die do not. Researchers are studying the process of apoptosis in *C. elegans* in the hope of finding treatments for these and other human diseases.

**Parasitic Nematodes** A large number of nematodes are parasitic in humans and other vertebrates. The large roundworm *Ascaris lumbricoides* is a parasite of the small intestine that can reach up to 30 cm in length. Over a billion people worldwide carry this parasite, although infections are most prevalent in tropical or developing countries. Eggs pass out in feces and can remain viable in the soil for years, although they require ingestion before hatching into an infective stage. Hookworms (*Necator americanus*), so named because their anterior end curves dorsally like a hook, are also parasites of the human intestine. The eggs pass out in feces, and recently hatched hookworms can penetrate the skin of a host's foot to establish a new infection. In areas with modern plumbing, these diseases are uncommon.

Pinworms (*Enterobius vermicularis*), although a nuisance, have relatively benign effects on their hosts. The rate of infection in the U.S., however, is staggering: 30% of children and 16% of adults are believed to be hosts. Adult pinworms live in the large intestine and migrate to the anal region at night to lay their eggs, which causes intense itching. The resultant scratching can spread the eggs from the hand to the mouth. In the tropics, some 250 million people are infected with *Wuchereria bancrofti*, a fairly large (100 mm) worm that lives in the lymphatic system, blocking the flow of lymph and, in rare cases, causing elephantiasis, an extreme swelling of the legs and other body parts (Figure 26.24). Females release tiny, live young called microfilariae, which are transmitted to new hosts via mosquitoes.

## The Phylum Arthropoda Contains the Spiders, Millipedes and Centipedes, Insects, and Crustaceans—Species with Jointed Appendages

The arthropods (from the Greek *arthron*, meaning joint, and *podos*, meaning foot) constitute perhaps the most successful phylum on Earth. About three-quarters of all described living species present on Earth are arthropods, and scientists have estimated they are also numerically common, with an estimated $10^{18}$ (a billion billion) individual organisms. The huge success of the arthropods, in terms of their sheer numbers and diversity, is related to features that permit these animals to live in all the major biomes on Earth, from the poles to the tropics and from marine and freshwater habitats to dry land. Such features include an exoskeleton, segmentation, and jointed appendages.

**The Arthropod Body Plan** The body of a typical arthropod is covered by a hard cuticle, an **exoskeleton** (external skeleton), made of layers of chitin and protein. The exoskeleton provides protection and a point of attachment for muscles, all of which are internal. It is

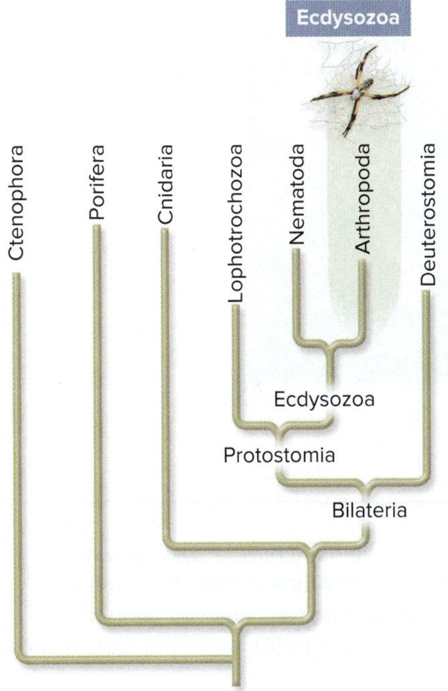


also relatively impermeable to water, a feature that may have enabled many arthropods to conserve water and colonize land. From this point of view, the development of a hard cuticle was a critical innovation. It also reminds us that the ability to adapt to diverse environmental conditions can lead to increased organismal diversity.

Arthropods are segmented, and many of the segments bear jointed appendages, which are used for such functions as walking, swimming, sensing, breathing, food handling, and reproduction. In many orders, the body segments have become fused into functional units, or **tagmata,** such as the head, thorax, and abdomen of an insect (Figure 26.25). Cephalization is extensive, and arthropods have well-developed sensory organs, including organs of sight, touch, smell, hearing, and balance.

Like most mollusks, arthropods have an open circulatory system (look ahead to Figure 36.1a), in which hemolymph is pumped from a tubelike heart into the aorta or short arteries and then into open sinuses from where the gases and nutrients diffuse into tissues. The hemolymph flows back into the heart via pores, called ostia, that are equipped with valves.

Because the cuticle impedes the diffusion of gases through the body surface, arthropods possess special organs that permit gas exchange. Aquatic arthropods have feathery gills, and terrestrial species have a highly developed **tracheal system** (look ahead to Figure 36.4). On the body surface, pores called **spiracles** provide openings to a series of finely branched air tubes within the body called trachea. The tracheal system delivers oxygen directly to tissues and cells.

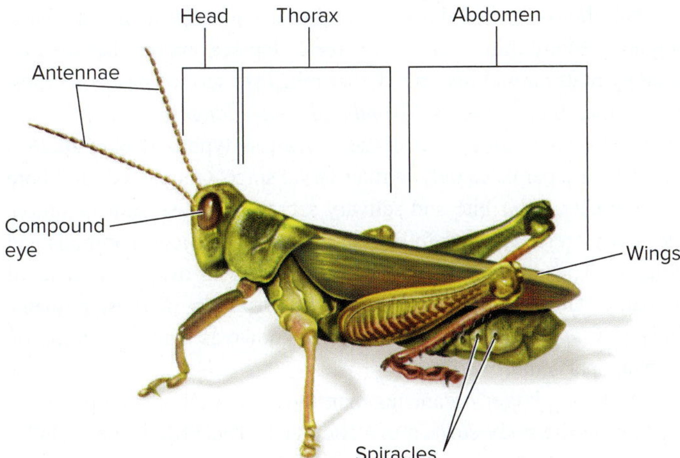


**Figure 26.25** Body plan of an arthropod, as represented by a grasshopper.

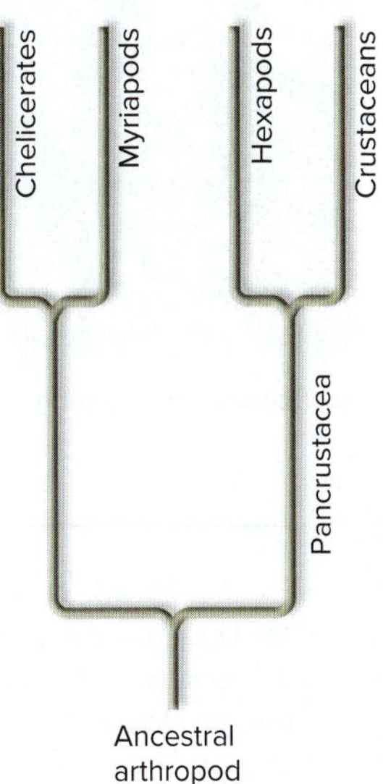


**Arthropod Diversity** Arthropod classification is complex and ongoing. Although many classifications have been proposed, a 1995 study of the mitochondrial DNA of arthropod species by American geneticist Jeffrey Boore and colleagues suggests a phylogeny with five main subphyla: one now-extinct subphylum, Trilobita (trilobites), and four living subphyla—Chelicerata (spiders and relatives), Myriapoda (millipedes and centipedes), Hexapoda (insects and relatives), and Crustacea (crabs and relatives) (Table 26.4). Molecular evidence thus suggests insects are more closely related to crustaceans than they are to spiders or millipedes and centipedes.

**Table 26.4 Main Subphyla and Characteristics of Arthropods**

| | Subphyla and examples (est. # of species) | Class characteristics |
|---|---|---|
| | Chelicerata: spiders, scorpions, mites, ticks, horseshoe crabs, sea spiders (74,000) | Body usually with cephalothorax and abdomen only; six pairs of appendages, including four pairs of legs, one pair of fangs, and one pair of pedipalps; terrestrial; predatory or parasitic |
| | Myriapoda: millipedes, centipedes (13,000) | Body with head and highly segmented trunk. In millipedes, each segment with two pairs of walking legs; terrestrial; herbivorous. In centipedes, each segment with one pair of walking legs; terrestrial; predatory, poison jaws |
| | Hexapoda: insects such as beetles, butterflies, flies, fleas, grasshoppers, ants, bees, wasps, termites, springtails (>1 million) | Body with head, thorax, and abdomen; mouthparts modified for biting, chewing, sucking, or lapping; usually with two pairs of wings and three pairs of legs; mostly terrestrial, some freshwater; herbivorous, parasitic, or predatory |
| | Crustacea: crabs, lobsters, shrimp (45,000) | Body of two to three parts; three or more pairs of legs; chewing mouthparts; usually marine |

(a) Black widow spider

(b) Scorpion with young

(c) Chigger mite

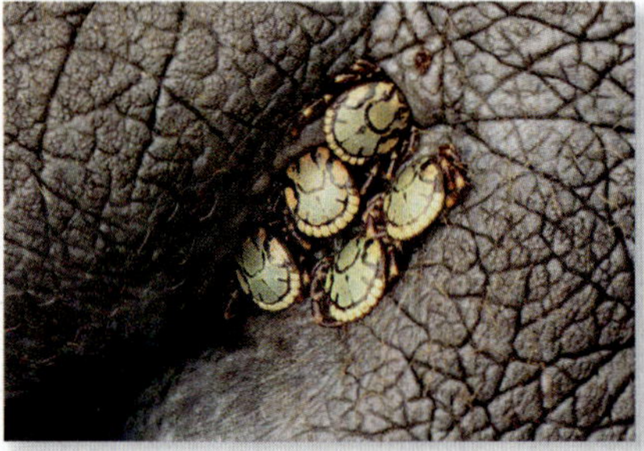

(d) Bont ticks

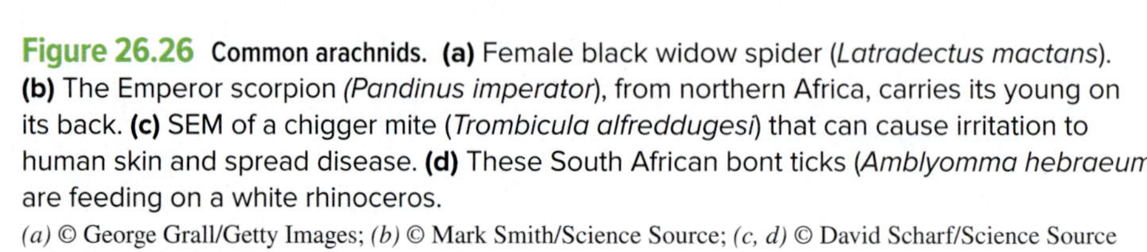

**Figure 26.26 Common arachnids.** **(a)** Female black widow spider (*Latradectus mactans*). **(b)** The Emperor scorpion (*Pandinus imperator*), from northern Africa, carries its young on its back. **(c)** SEM of a chigger mite (*Trombicula alfreddugesi*) that can cause irritation to human skin and spread disease. **(d)** These South African bont ticks (*Amblyomma hebraeum*) are feeding on a white rhinoceros.

*(a)* © George Grall/Getty Images; *(b)* © Mark Smith/Science Source; *(c, d)* © David Scharf/Science Source

Concept Check: *What is one of the main characteristics distinguishing arachnids from insects?*

**Subphylum Chelicerata: Spiders and Relatives** Chelicerata consists mainly of the class Arachnida, which contains predatory spiders and scorpions as well as the ticks and mites, some of which are blood-sucking parasites that feed on vertebrates. All species have a body consisting of two tagmata: a fused head and thorax (called a cephalothorax) and an abdomen. All species also possess six pairs of appendages: the chelicerae, or fangs; a pair of pedipalps, which have various sensory, predatory, or reproductive functions; and four pairs of walking legs.

Spider fangs are supplied with venom from poison glands. Most spider bites are harmless to humans, although they are very effective in immobilizing and/or killing their insect prey. Venom from some species, including the black widow (*Latrodectus mactans*) (Figure 26.26a) and the brown recluse (*Loxosceles reclusa*), are potentially, although rarely, fatal to humans. The toxin of the black widow is a neurotoxin, which interferes with the functioning of the nervous system, whereas that of the brown recluse is hemolytic, meaning it destroys red blood cells around the bite. After the spider has subdued its prey, it pumps digestive fluid into the tissues via the fangs and sucks out the partially digested meal.

Spiders have abdominal silk glands, called spinnerets, and many spin webs to catch prey. The silk is a protein that stiffens after extrusion from the body because the mechanical shearing causes a change in the organization of the amino acids. Silk is stronger than steel of the same diameter and more elastic than Kevlar, the material used in bulletproof vests. Each spider family constructs a characteristic size and style of web and can do it perfectly on its first attempt, indicating that web spinning is an innate (instinctual) behavior (see Chapter 42).

Scorpions (order Scorpionida) are generally tropical or subtropical animals that feed primarily on insects, though they may eat spiders and other arthropods as well as smaller reptiles and mice. Their pedipalps are modified into large claws, and their abdomen tapers into a stinger, which is used to inject venom. Although the venom of most North American species is generally not fatal to humans, that of the *Centruroides* genus from deserts in the U.S. Southwest and Mexico can be deadly. Fatal species are also found in India, Africa, and other countries. Unlike spiders, which lay eggs, scorpions bear live young, which the mother then carries around on her back until they have their first molt (Figure 26.26b).

In mites and ticks (order Acari), the two main body segments (cephalothorax and abdomen) are fused and appear as one large segment. Many mite species are free-living scavengers that feed on dead plant or animal material. Other mites are serious pests on crops, and some, like chiggers (*Trombicula alfreddugesi*), are parasites of humans that can spread diseases such as typhus (Figure 26.26c). Chiggers are parasites only in their larval stage. Chiggers do not bore into the skin; their bite and salivary secretions cause skin irritation. *Demodex brevis* is a hair-follicle mite that is common in animals and humans. The mite is estimated to be present in over 90% of adult humans. Although the mite causes no irritation in most humans, *Demodex canis* causes the skin disease known as mange in domestic animals, particularly dogs.

Ticks are larger organisms than mites, and all are ectoparasitic, feeding on the body surface of vertebrates. Their life cycle includes attachment to a host, sucking blood until they are replete and dropping off the host to molt (Figure 26.26d). Ticks can carry a huge variety of viral and bacterial diseases, including Lyme disease, a global

**(a) Two millipedes**

**(b) A centipede**

**Figure 26.27 Millipedes and centipedes. (a)** Millipedes have two pairs of legs per segment. **(b)** The venom of the giant centipede (*Scolopendra heros*) is known to produce significant swelling and pain in humans.

*(a)* © David Aubrey/Corbis; *(b)* © Larry Miller/Science Source

bacterial disease so named because it was first found in the town of Lyme, Connecticut, in the 1970s. Infection rates in the U.S. are highest in the Northeast, mid-Atlantic, and upper Midwest regions.

### Subphylum Myriapoda: The Millipedes and Centipedes

Myriapods have one pair of antennae on the head and three pairs of appendages that are modified as mouth parts, including mandibles that act as jaws. The millipedes and centipedes, both wormlike arthropods with legs, are among the earliest terrestrial animal phyla known. Millipedes (class Diplopoda) have two pairs of legs per segment, as their class name denotes (from the Latin *diplo*, meaning two, and *podos*, meaning feet)—not 1,000 legs, as their common name suggests (Figure 26.27a). They are slow-moving, herbivorous creatures that eat decaying leaves and other plant material. When threatened, the millipede's response is to roll up into a protective coil. Many millipede species also have glands on their underside that can eject a variety of toxic, repellent secretions. Some millipedes are brightly colored, warning potential predators that they can protect themselves.

Class Chilopoda (from the Latin *chilo*, meaning lip, and *podos*, meaning feet), or centipedes, are fast-moving carnivores that have one pair of walking legs per segment (Figure 26.27b). The head of centipedes has powerful claws connected to poison glands. The toxin from venom of some of the larger species, such as *Scolopendra heros*, is powerful enough to cause pain in humans. Most species do not have a waxy waterproofing layer on their cuticle and, so, are restricted to moist environments under leaf litter or in decaying logs, usually coming out at night to actively hunt their prey.

### Subphylum Hexapoda: Insects and Relatives

Hexapods are six-legged arthropods. Most are insects, but there are a few earlier-diverging noninsect hexapods, including soil-dwelling groups such as collembolans, that molecular studies have shown represent a separate but related lineage. Insects are in a class by themselves (Insecta), literally and figuratively. There are more species of insects than all other species of animal life combined. One million species of insects have been described, and, according to best estimates, 2–5 million more species await description. At least 90,000 species of insects have been identified in the U.S. and Canada alone.

Insects are the subject of an entire field of scientific study, **entomology.** Insects live in all terrestrial habitats, and virtually all species of plants are fed upon by many insect species. Because approximately one-quarter of the world's crops are lost annually to insects, scientists are constantly trying to find ways to reduce pest densities. Insect pest reduction often involves chemical control (the use of pesticides) or biological control (the use of living organisms). Many species of insects are also important pests or parasites of humans and livestock, both by their own actions and as vectors of diseases such as malaria and sleeping sickness.

In contrast, insects also provide us with many types of essential biological services. We depend on insects such as honeybees and butterflies to pollinate our crops. Bees also produce honey, and silkworms are the source of silk fiber. Despite the revulsion they provoke in us, fly larvae (maggots) are important in the decomposition process of dead plants and animals. In addition, we use insects in the biological control of insect pests of crops.

#### The Insect Body Plan

Of paramount importance to the success of insects was the evolution of wings, a feature possessed by no other arthropod and indeed no other living animal except birds and bats. Unlike vertebrate wings, however, insect wings are actually outgrowths of the body wall cuticle and are not true segmental appendages. This means that insects still have all their walking legs. In contrast, birds and bats have one pair of appendages (arms) modified for flight, which leaves them considerably less agile on the ground.

Insects in different orders have also evolved a variety of mouthparts in which the constituent parts are modified for different functions.

- Grasshoppers, beetles, dragonflies, and many others have mouthparts adapted for chewing.
- Mosquitoes and many plant pests have mouthparts adapted for piercing and sucking.
- Butterflies and moths have a coiled tongue (proboscis) that can be uncoiled, enabling them to drink nectar from flowers.
- Some flies have lapping, spongelike mouthparts that sop up liquid food.

Their varied mouthparts are adaptations that allow insects to specialize their feeding on virtually anything: plant matter, decaying organic matter, and other living animals. The biological diversity of insects is therefore related to environmental diversity—in this case, the variety of foods that insects eat.

#### Insect Diversity

The great diversity of insects is illustrated by the fact that there are 35 different orders, some of which have over 100,000 species. The most common of the orders are discussed in Table 26.5. Although all insects have six legs, different orders have slightly different wing structures, and many of the orders are based on wing type (their names often include the root pter-, from the Greek *pteron*, meaning wing).

- In beetles (Coleoptera), only the back pair of wings is functional, as the front wings have been hardened into protective, shell-like coverings under which the back pair folds when not in use.

**Table 26.5 Main Orders and Characteristics of Insects**

| Order and examples (est. # of species) | Order characteristics |
|---|---|
| Coleoptera: beetles, weevils (500,000) | Two pairs of wings (front pair thick and leathery, acting as wing cases, back pair membranous); armored exoskeleton; biting and chewing mouthparts; complete metamorphosis; largest order of insects |
| Hymenoptera: ants, bees, wasps (190,000) | Two pairs of membranous wings; chewing or sucking mouthparts; many have posterior stinging organ on females; complete metamorphosis; many species social; important pollinators |
| Diptera: flies, mosquitoes (190,000) | One pair of wings with hind wings modified into halteres (balancing organs); sucking, piercing, or lapping mouthparts; complete metamorphosis; larvae are grublike maggots in various food sources; some adults are disease vectors |
| Lepidoptera: butterflies, moths (180,000) | Two pairs of colorful wings covered with tiny scales; long tubelike tongue for sucking; complete metamorphosis; larvae are plant-feeding caterpillars; adults are important pollinators |
| Hemiptera: true bugs; bedbug, chinch bug, cicada (100,000) | Two pairs of membranous wings; piercing or sucking mouthparts; incomplete metamorphosis; many plant feeders; some predatory or blood feeders; vectors of plant diseases |
| Orthoptera: crickets, grasshoppers (30,000) | Two pairs of wings (front pair leathery, back pair membranous); chewing mouthparts; mostly herbivorous; incomplete metamorphosis; powerful hind legs for jumping |
| Odonata: damselflies, dragonflies (6,500) | Two pairs of long, membranous wings; chewing mouthparts; large eyes; predatory on other insects; incomplete metamorphosis; nymphs aquatic; considered early-diverging insects |
| Siphonaptera: fleas (2,600) | Wingless, laterally flattened; piercing and sucking mouthparts; adults are bloodsuckers on birds and mammals; jumping legs; complete metamorphosis; vectors of plague |
| Phthiraptera: sucking lice (2,400) | Wingless ectoparasites; sucking mouthparts; flattened body; reduced eyes; legs with claws for clinging to skin; incomplete metamorphosis; very host-specific; vectors of typhus |
| Isoptera: termites (2,000) | Two pairs of membranous wings when present; some stages wingless; chewing mouthparts; social species; incomplete metamorphosis |

- Wasps and bees (Hymenoptera) have two pairs of wings hooked together that move as one wing.
- Flies (Diptera) possess only one pair of wings (the front pair); the back pair has been modified into a small pair of balancing organs, called halteres, that act as miniature gyroscopes.
- Butterflies (Lepidoptera) have wings that are covered in scales (from the Greek *lepido*, meaning scale); other insects generally have clear, membranous wings.
- In ants and termites, the young queens and drones (males) usually have wings, whereas female individuals called workers do not have wings. Other species, such as fleas and lice, are completely wingless.

**Insect Reproduction and Development** All insects have separate sexes, and fertilization is internal. During development, the majority (approximately 85%) of insects undergo a change in body form known

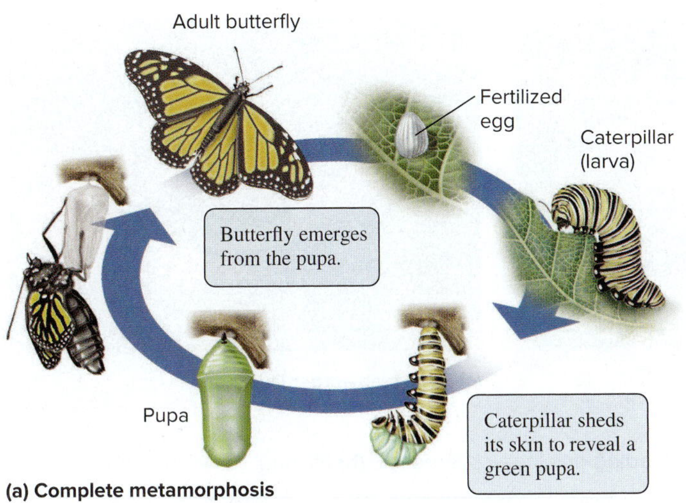


**(a) Complete metamorphosis**

Adult grasshopper
Fertilized egg
Nymphs look like miniature adults.
Nymph stages

**(b) Incomplete metamorphosis**

**Figure 26.28 Metamorphosis. (a)** Complete metamorphosis, as illustrated by the life cycle of a monarch butterfly. The adult butterfly has a completely different appearance than the larval caterpillar. **(b)** Incomplete metamorphosis, as illustrated by the life cycle of a grasshopper. The eggs hatch into nymphs, essentially miniature versions of the adult.

Concept Check: *Name two key insect innovations.*

as **complete metamorphosis** (from the Greek *meta*, meaning change, and *morph*, meaning form) (Figure 26.28a). Animals that undergo complete metamorphosis have four stages: egg, larva, pupa, and adult. The dramatic body transformation from larva to adult occurs in the pupa stage. The larval stage is often spent in an entirely different habitat from that of the adult, and larval and adult forms use different food sources. Metamorphosis permits existence of a feeding stage, often a saclike caterpillar or larva, and a reproductive stage, usually a winged adult.

The remaining insects undergo **incomplete metamorphosis,** in which change is more gradual (Figure 26.28b). Incomplete metamorphosis has only three stages: egg, nymph, and adult. Young insects, called nymphs, look like miniature adults when they hatch from their eggs but usually don't have wings. As they grow and feed, they shed their exoskeleton and replace it with a larger one several times, each time entering a new stage of growth. When the insects reach their adult size, they have also grown wings.

Some insects, such as bees, wasps, ants, and termites, have developed complex social behavior and live cooperatively in underground or aboveground nests. Such colonies exhibit a division of labor, in that some individuals forage for food and care for the brood (workers), others protect the nest (soldiers), and some only reproduce (the queen and drones).

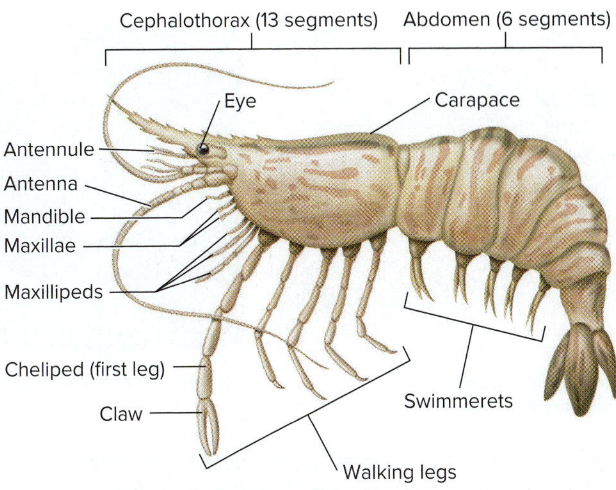


**Figure 26.29 Body plan of a crustacean, as represented by a shrimp.**

BioConnections: *Look ahead to Figure 34.8. Where are a crustacean's organs of balance located?*

**Subphylum Crustacea: Crabs, Lobsters, Barnacles, and Shrimp** The crustaceans are common inhabitants of marine environments, although some species live in fresh water and a few are terrestrial. Many species, including crabs, lobsters, crayfish, and shrimp, are economically important food items for humans; smaller species are important food sources for other predators.

**The Crustacean Body Plan** The crustaceans are unique among the arthropods in that they possess two pairs of antennae at the anterior end of the body—the antennule (first pair) and antenna (second pair) (Figure 26.29). In addition, they have multiple sensory and feeding appendages that are modified mouthparts. These are followed by walking legs and, often, additional abdominal appendages, called swimmerets, and a powerful tail. In some orders, the first pair of walking legs, or chelipeds, is modified to form powerful claws. The head and thorax are often fused together, forming the cephalothorax. In many species, such as crabs, the cuticle covering the head extends over most of the cephalothorax, forming a hard, protective fold called the **carapace.** For growth to occur, a crustacean must molt, or shed the entire exoskeleton.

Many crustaceans are predators, but others are scavengers, and some, such as barnacles, are filter feeders. Gas exchange typically

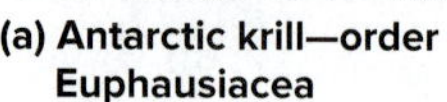

**(a) Antarctic krill—order Euphausiacea**

**(b) Pill bug—order Isopoda**

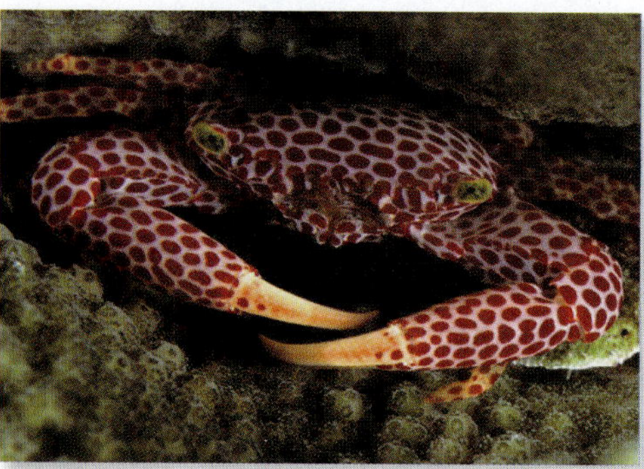

**(c) Coral crab—order Decapoda**

**Figure 26.30 Common crustaceans.** **(a)** Antarctic krill (*Euphausia superba*) feeding on algae-covered ice. **(b)** Pill bug, or wood louse (*Armadillidium vulgare*). **(c)** Coral crab (*Carpilius maculatus*).

*(a)* © Dmytro Pylypenko/123RF.com; *(b)* © Miyuki Satake/iStock/Getty Images RF; *(c)* © Franco Banfi/AGF/UIG/Getty Images

occurs via gills, and crustaceans, like other arthropods, have an open circulatory system. Reproduction usually involves separate sexes, and fertilization is internal.

**Crustacean Diversity** There are many crustacean clades, but most are small and obscure, although many smaller orders contain important prey items for other marine organisms. For example, copepods are tiny and abundant planktonic crustaceans, which are a food source for filter-feeding organisms and small fish.

The order Euphausiacea are shrimplike krill that grow to about 3 cm and provide a large part of the diet of many whales, seals, penguins, fish, and squid (Figure 26.30a). The order Isopoda contains many small species that are parasitic on marine fishes. There are also terrestrial isopods, better known as pill bugs, or wood lice, that retain a strong connection to water and need to live in moist environments such as leaf litter or decaying logs (Figure 26.30b). When threatened, they curl up into a tight ball, making it difficult for predators to get a grip on them.

The most famous crustacean order is the Decapoda, which includes the crabs and lobsters, the largest crustacean species (Figure 26.30c). As their name suggests, these decapods have 10 legs (five pairs), but one pair, called the chelipeds, have large claws and are not used for walking. Most decapods are marine, but many are freshwater species, such as crayfish, and in hot, moist tropical areas, even some terrestrial species called land crabs.

## 26.7 Reviewing the Concepts

- The ecdysozoans are so named for their ability to shed their cuticle, a nonliving cover providing support and protection. The two most common ecdysozoan phyla are the nematodes and the arthropods.
- Nematodes, which exist in nearly all habitats, have a cuticle made of collagen, a structural protein. Many nematodes are parasitic in humans (Figures 26.23, 26.24).
- Arthropods are perhaps the most successful phylum on Earth. The arthropod body is covered by a cuticle made of layers of chitin and protein, and it is segmented, with segments fused into functional units called tagmata (Figure 26.25).
- The five main subphyla of arthropods are Trilobita (trilobites; now extinct), Chelicerata (spiders, scorpions, and relatives), Myriapoda (millipedes and centipedes), Hexapoda (insects), and Crustacea (crabs and relatives) (Table 26.4; Figures 26.26, 26.27).

## 26.7 Testing Your Knowledge

1. Nematodes are distinguished from members of the Lophotrochozoa in that they possess
   a. tracheae.
   b. a cuticle.
   c. a coelom.
   d. incomplete metamorphosis.
   e. stylets.
2. Which is *not* a feature of the arthropod body plan?
   a. an exoskeleton
   b. segmentation
   c. tagmata
   d. a closed circulatory system
   e. metamorphosis
3. Characteristics of the class Arachnida include
   a. two tagmata.
   b. six walking legs.
   c. an aquatic lifestyle.
   d. a lobed body.
   e. both b and d.

## 26.8 Deuterostomia: The Echinoderms and Chordates

### Learning Outcomes

1. Identify the distinguishing characteristics of echinoderms.
2. List the four critical innovations in the body plan of chordates.
3. Describe the two invertebrate subphyla of the phylum Chordata and their relationship to the vertebrates.

As mentioned at the beginning of the chapter, the deuterostomes are grouped together because they share similarities in patterns of development (see Figure 26.5). Molecular evidence also supports a deuterostome

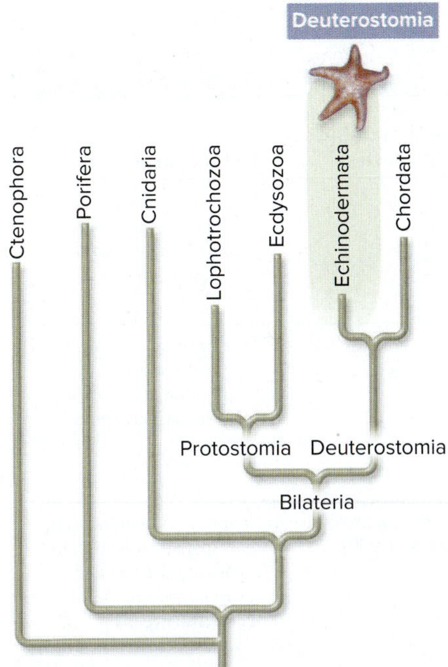


clade. All animals in the phylum Chordata (from the Greek *chorde*, meaning string, referring to the spinal cord), which includes the vertebrates, are deuterostomes. Interestingly, there is one invertebrate deuterostome group, the phylum Echinodermata, which includes the sea stars, sea urchins, and sea cucumbers. Although there are far fewer phyla and species of deuterostomes than protostomes, the deuterostomes are generally much more familiar to us. After all, humans are deuterostomes. We will conclude our discussion of invertebrate biology by turning our attention to the invertebrate deuterostomes. In this section, we will explore the phylum Echinodermata and then introduce the phylum Chordata, looking in particular at its distinguishing characteristics and at its two invertebrate subphyla: the cephalochordates, commonly referred to as the lancelets, and the urochordates, also known as the tunicates. We will discuss the subphylum Vertebrata in Chapter 27.

## The Phylum Echinodermata Includes Sea Stars and Sea Urchins—Species with a Water Vascular System

The phylum Echinodermata (from the Greek *echinos*, meaning spiny, and *derma*, meaning skin) consists of a unique grouping of deuterostomes. A striking feature of all echinoderms is their modified radial symmetry (see Figure 26.3b). The body of most species can be divided into five parts pointing out from the center. As a consequence, cephalization is absent in most classes. Only a simple nervous system is present, consisting of a central nerve ring from which arise radial branches to each limb. The radial symmetry of echinoderms is secondary, present only in adults. The free-swimming larvae have bilateral symmetry and metamorphose into the radially symmetric adult form.

### The Echinoderm Body Plan

Most echinoderms have an **endoskeleton,** an internal hard skeleton composed of calcareous plates overlaid by a thin skin (**Figure 26.31**).

Echinoderms possess a true coelom, and a portion of the coelom has been adapted to serve as a unique **water vascular system,** a network of canals that branch into tiny **tube feet** that function in movement, gas exchange, feeding, and excretion (see inset to Figure 26.31). The water vascular system uses hydraulic power (water pressure generated by the contraction of muscles), which enables the tube feet to extend and contract, allowing echinoderms to move only very slowly.

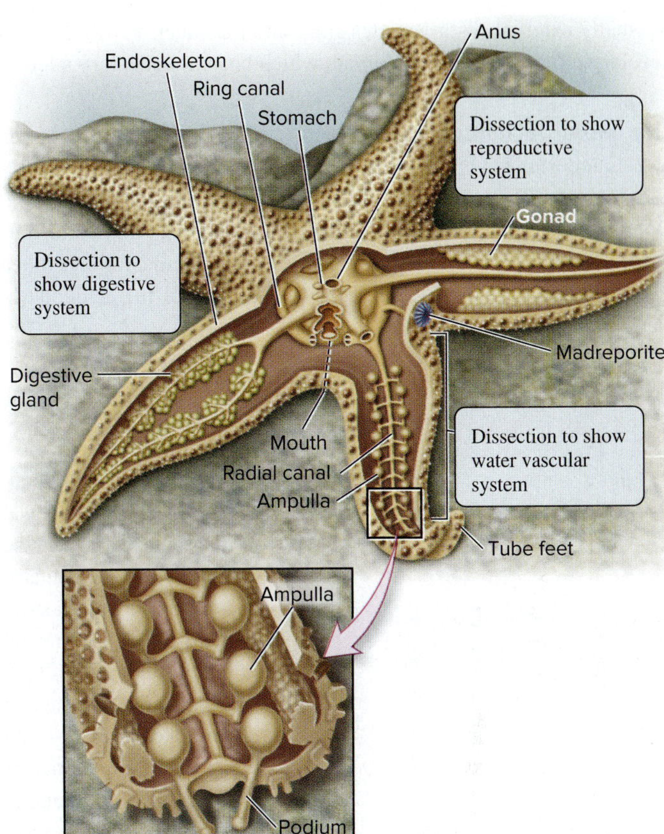


**Figure 26.31** **Body plan of an echinoderm, as represented by a sea star.** The arms of this sea star have been dissected to different degrees to show the echinoderm's various organs. The inset shows a close-up view of the tube feet, part of the water vascular system characteristic of echinoderms.

**Concept Check:** *Echinoderms and chordates are both deuterostomes. What are three defining features of deuterostomes?*

Water enters the water vascular system through the **madreporite,** a sievelike plate on the animal's surface. From there it flows into a **ring canal** in the central disc, into five radial canals, and into the tube feet. At the base of each tube foot is a muscular sac called an **ampulla,** which stores water. Contractions of the ampullae force water into the tube feet, causing them to straighten and extend. When the foot contacts a solid surface, muscles in the foot contract, forcing water back into the ampullae. Sea stars also use their tube feet in feeding, where they can exert a constant and strong pressure on their preferred prey, bivalves, whose adductor muscles open and close the shell. The adductor muscles eventually tire, allowing the shell to open slightly. At this stage, the sea star everts its stomach and inserts it into the shell's opening. It then digests its prey, using juices secreted from extensive digestive glands. Sea stars also feed on sea urchins, brittle stars, and sand dollars, prey that cannot easily escape them.

Most echinoderms exhibit **autotomy,** the ability to intentionally detach a body part, such as a limb, that will later regenerate. In some species, a broken limb can even regenerate into a whole animal. Most echinoderms reproduce sexually and have separate sexes. Fertilization is usually external, with gametes shed into the water. Fertilized eggs develop into free-swimming larvae, which become sedentary adults.

**The Major Echinoderm Classes** Although over 20 classes of echinoderms have been described from the fossil record, only 5 main classes of echinoderms exist today: the Asteroidea (sea stars), Ophiuroidea (brittle stars), Echinoidea (sea urchins and sand dollars), Crinoidea (sea lilies and feather stars), and Holothuroidea (sea cucumbers) (Figure 26.32). The key features of the echinoderm classes are listed in Table 26.6.

**Table 26.6 Main Classes and Characteristics of Echinoderms**

| Class and examples (est. # of species) | Class characteristics |
|---|---|
| Asteroidea: sea stars (1,600) | Five arms; tube feet; predatory on bivalves and other echinoderms; eversible stomach |
| Ophiuroidea: brittle stars (2,000) | Five long, slender arms; tube feet not used for locomotion; browse on sea bottom or filter feed |
| Echinoidea: sea urchins, sand dollars (1,900) | Spherical (sea urchins) or disc-shaped (sand dollars); no arms; tube feet and movable spines; many feed on seaweeds |
| Crinoidea: sea lilies, feather stars (700) | Cup-shaped; often attached to substrate via stalk; arms feathery and used in filter feeding; very abundant in fossil record |
| Holothuroidea: sea cucumbers (1,200) | Cucumber-shaped; no arms; spines absent; endoskeleton reduced; tube feet; browse on sea bottom |

## The Phylum Chordata Includes All the Vertebrates and Some Invertebrates

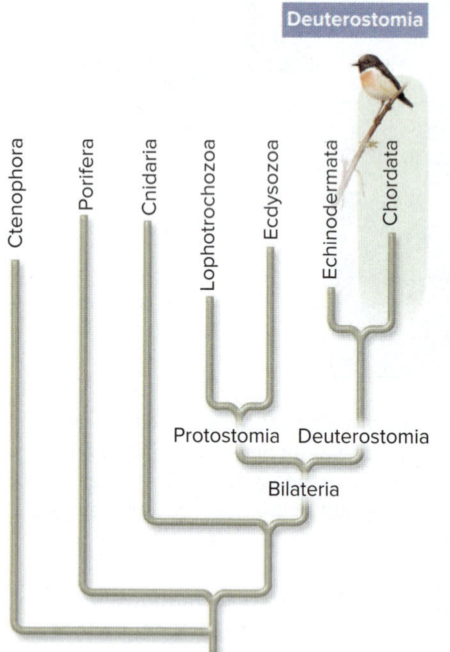


The deuterostomes consist of two major phyla: the echinoderms and the chordates. As deuterostomes, both phyla share similar developmental traits (see Figure 26.5). In addition, both have an endoskeleton, consisting in the echinoderms of calcareous plates and in chordates, for the most part, of bone. However, the echinoderm endoskeleton functions in much the same way as the arthropod exoskeleton, in that an important function is providing protection. The chordate endoskeleton serves a very different purpose. In early-diverging chordates, the endoskeleton is composed of a single flexible rod situated dorsally, deep inside the body. Muscles move this rod, and their contractions cause the back and tail end to move from side to side, permitting a swimming motion in water. The endoskeleton becomes more complex in different lineages that develop limbs, as we will see in Chapter 27, but it is always internal, with muscles attached. This arrangement permits the possibility of complex movements, including the ability to move on land.

Let's take a look at the four critical innovations in the body plan of chordates that distinguish them from all other animal life (Figure 26.33):

1. **Notochord.** Chordates are named for the **notochord,** a single flexible rod that lies between the digestive tract and the nerve cord. In most chordates, such as vertebrates, a more complex jointed backbone usually replaces the notochord; its remnants exist only as the soft material within the discs between each vertebra.
2. **Dorsal hollow nerve cord.** Many animals have a long nerve cord, but in nonchordate invertebrates, it is a solid tube that lies ventral to the alimentary canal. In contrast, the nerve cord in chordates is a hollow tube that develops dorsal to the alimentary canal. In vertebrates, the **dorsal hollow nerve cord** develops into the brain and spinal cord.
3. **Pharyngeal slits.** Chordates, like many animals, have a complete gut, with two openings—a mouth and an anus. However, in chordates, slits develop in the pharyngeal region, close to the mouth, that open to the outside. This permits water to enter through the mouth and exit via the slits, without having to go through the digestive tract. In early-diverging chordates, **pharyngeal slits** function as a filter-feeding device, whereas in later-diverging chordates, they develop into gills for gas exchange. In terrestrial chordates, the slits do not fully form, and they become modified for other purposes.
4. **Postanal tail.** In most nonchordate phyla, the anus is at the end of the body. Chordates possess a **postanal tail** of variable length that extends posterior to the anal opening. In aquatic chordates such as fishes, the tail is used in locomotion. In terrestrial chordates, the tail may be used in a variety of functions or may be present only during an early developmental stage.

Although few chordates apart from fishes possess all of these characteristics in their adult life, they all exhibit them at some time during development. For example, in adult humans, the notochord becomes the spinal column, and the dorsal hollow nerve cord becomes

(a) Necklace sea star, *Fromia monilis*, Baa Atoll, Maldives

(b) Brittle star, *Ophiarachna* spp., Gulf of Mexico

(c) Sea urchin, *Heterocentrotus trigonarius*, Hawaii

(d) Sea lily, *Proisocrinus ruberrimus*, Indonesia

(e) Bronze-spot sea cucumber, *Holothuria argus*

**Figure 26.32** Echinoderms. (a) Sea star. (b) Brittle star. (c) Sea urchin. (d) Sea lily. (e) Sea cucumber.

*(a)* © ullstein bild/Getty Images; *(b)* Source: NOAA Okeanos Explorer Program, Gulf of Mexico 2012 Expedition; *(c)* Source: David Burdick/NOAA; *(d)* Source: NOAA Okeanos Explorer Program, INDEX-SATAL 2010; *(e)* © Datacraft Co Ltd/Getty Images RF

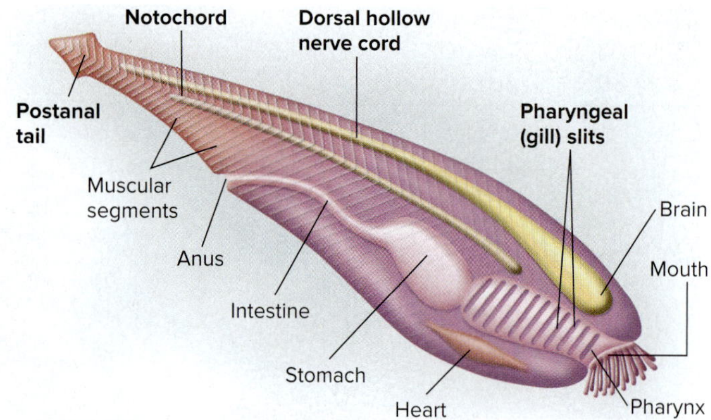


**Figure 26.33 Chordate characteristics.** The generalized chordate body plan has four main features: notochord, dorsal hollow nerve cord, pharyngeal slits, and postanal tail.

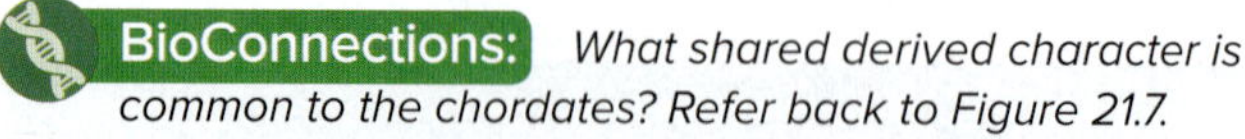

*What shared derived character is common to the chordates? Refer back to Figure 21.7.*

**(a) Lancelet with its anterior end protruding into the water**

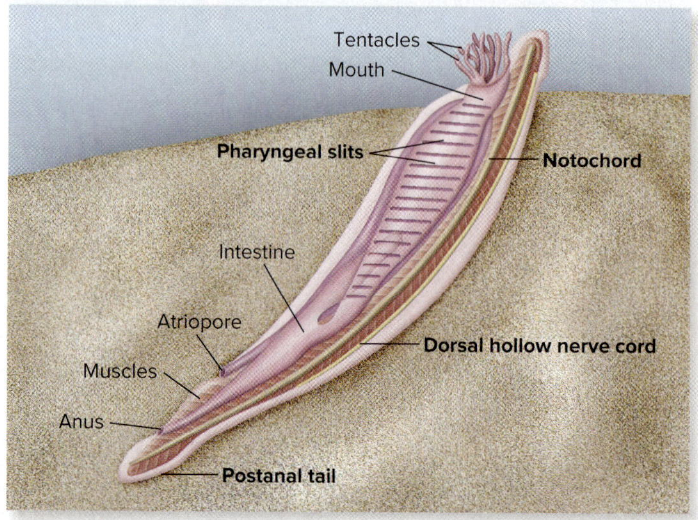


**(b) Body plan of the lancelet**

**Figure 26.34 Lancelets. (a)** A bladelike lancelet. **(b)** The body plan of the lancelet clearly displays the four characteristic chordate features.

*(a)* © Natural Visions/Alamy

the central nervous system. However, humans exhibit pharyngeal slits and a postanal tail only during early embryonic development. All the pharyngeal slits, except one, which forms the auditory (Eustachian) tube in the ear, are eventually lost, and the postanal tail regresses to form the tailbone (the coccyx).

The phylum Chordata consists of the invertebrate chordates—the subphylum Cephalochordata (lancelets) and the subphylum Urochordata (tunicates)—along with the subphylum Vertebrata (vertebrates). Although the Vertebrata is by far the largest of these subphyla, biologists have focused on the cephalochordates and urochordates for clues to better understand the evolution of the chordate phylum.

**Subphylum Cephalochordata: The Lancelets** The cephalochordates (from the Greek *kephalo*, meaning head) are commonly referred to as lancelets, in reference to their bladelike shape and size, about 5–7 cm in length (Figure 26.34a). Lancelets are a small subphylum of 26 species, all marine filter feeders, with 4 species occurring in North American waters.

The lancelets live mostly buried in sand, with only the anterior end protruding into the water. Lancelets have the four distinguishing chordate characteristics: a clearly discernible notochord (extending well into the head), a dorsal hollow nerve cord, pharyngeal slits, and a postanal tail (Figure 26.34b). They are filter feeders, drawing water through the mouth and into the pharynx, where it is filtered through the pharyngeal slits. A mucous net across the pharyngeal slits traps food particles, and ciliary action takes the food into the intestine, while water exits via the opening to the exterior called the atriopore. Gas exchange generally takes place across the body surface. Although the lancelet is usually sessile, it can leave its sandy burrow and swim to a new spot, using a sequence of serially arranged muscles that appear like chevrons (<<<<) along its sides. These muscles reflect the segmented nature of the lancelet body and permit a fishlike swimming motion.

**Subphylum Urochordata: The Tunicates** The urochordates (from the Greek *oura*, meaning tail) are a group of 3,000 marine species also known as tunicates. Looking at an adult tunicate, you might never guess that it is a relative of modern vertebrates. The only one of the four distinguishing chordate characteristics that it possesses is pharyngeal slits (Figure 26.35a). The larval tunicate, in contrast, looks like a tadpole and exhibits all four chordate hallmarks (Figure 26.35b). The larval tadpole swims for only a few days, usually without feeding. Larval tunicates settle on and attach to a rock surface, where they metamorphose into adults and, in the process, lose most of their chordate characteristics.

Adult tunicates are marine animals, some colonial and others solitary, that superficially resemble sponges or cnidarians. Tunicates are filter feeders that draw water into the mouth through an **incurrent siphon,** using a ciliated pharynx, filter it through extensive pharyngeal slits and expel it through an **excurrent siphon.** The food is trapped on a mucous sheet; passes via ciliary action to the stomach and intestine, and exits through the anus. The whole animal is enclosed in a nonliving **tunic** made of a protein and a cellulose-like material called tunicin. Tunicates are also known as sea squirts for their ability to squirt out water from the excurrent siphon when

## Biology Principle

### The Genetic Material Provides a Blueprint for Sustaining Life

Genetic similarities between the tunicates and the vertebrates suggest they are indeed our closest invertebrate relatives.

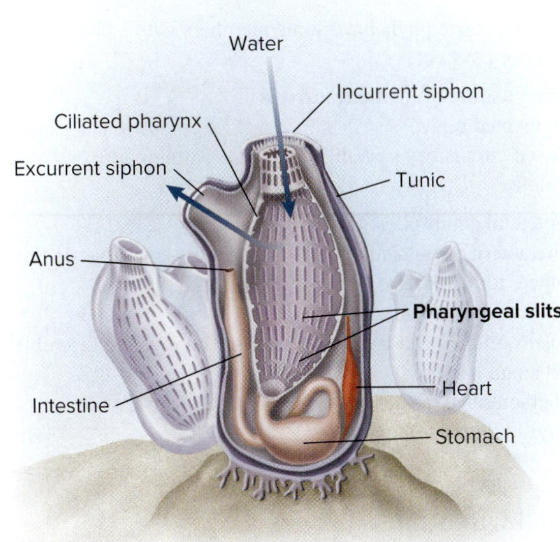


**(a) Adult tunicate**

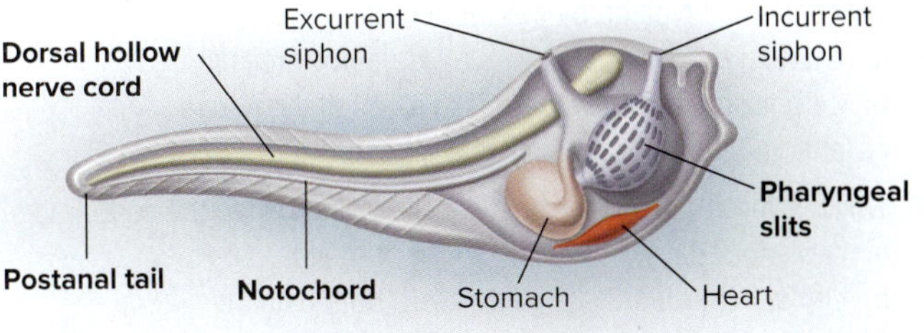


**(b) The larval form of the tunicate**

**(c) Typical tunicate**

**Figure 26.35** **Tunicates.** **(a)** Body plan of the sessile, filter-feeding adult tunicate. **(b)** The larval form of the tunicate, which shows the four characteristic chordate features. **(c)** The blue tunicate, *Rhopalaea crassa.*

*(c)* © Franco Banfi/Getty Images

disturbed. They have a rudimentary circulatory system with a heart and a simple nervous system of relatively few nerves connected to sensory tentacles around the incurrent siphon. The animals are mostly hermaphroditic.

## 26.8 Reviewing the Concepts

- Deuterostomia includes the phyla Echinodermata and Chordata. A striking feature of the echinoderms is their radial symmetry, which is present only in adults; the free-swimming larvae are bilaterally symmetric. Echinoderms possess a unique water vascular system (Figure 26.31).
- Five main classes of echinoderms exist today: the Asteroidea (sea stars), Ophiuroidea (brittle stars), Echinoidea (sea urchins and sand dollars), Crinoidea (sea lilies and feather stars), and Holothuroidea (sea cucumbers) (Table 26.6; Figure 26.32)
- The phylum Chordata is distinguished by four critical innovations: the notochord, dorsal hollow nerve chord, pharyngeal slits, and postanal tail (Figure 26.33).
- The subphylum Cephalochordata (lancelets) and subphylum Urochordata (tunicates) are invertebrate chordates. Genetic studies have shown that tunicates are the closest invertebrate relatives of the vertebrate chordates (subphylum Vertebrata) (Figures 26.34, 26.35).

## 26.8 Testing Your Knowledge

1. Echinodermata are a member of which clade?
   a. protostomia
   b. radiata
   c. eumetazoa
   d. lophotrochozoa
   e. ecdysozoa
2. Chordates are typified by possession of
   a. a notochord.
   b. a dorsal hollow nerve cord.
   c. pharyngeal slits.
   d. a postanal tail.
   e. all of the above.
3. Which invertebrate phylum is most closely related to the vertebrates?
   a. cephalopoda
   b. cephalochordata
   c. urochordata
   d. radiata
   e. both b and c

# Assess and Discuss

## Test Yourself

1. Which is the correct hierarchy of divisions in the animal kingdom, from most inclusive to least inclusive?
   a. Deuterostomia, Protostomia, Ecdysozoa
   b. Bilateria, Lophotrochozoa, Deuterostomia
   c. Bilateria, Ecdysozoa, Protostomia
   d. Deuterostomia, Ecdysozoa, Bilateria
   e. Bilateria, Protostomia, Ecdysozoa
2. Bilateral symmetry is strongly correlated with
   a. the ability to move through the environment.
   b. cephalization.
   c. the ability to detect prey.
   d. both a and b.
   e. all of the above.
3. Protostomes and deuterostomes can be classified based on
   a. cleavage pattern.
   b. destiny of the blastopore.
   c. whether the fate of the embryonic cells is fixed early during development.
   d. all of the above.
4. Indeterminate cleavage is found in
   a. annelids.
   b. mollusks.
   c. nematodes.
   d. vertebrates.
   e. all of the above.
5. Naturally occurring identical twins are possible only in animals that
   a. have spiral cleavage.
   b. have determinate cleavage.
   c. are protostomes.
   d. have indeterminate cleavage.
   e. a, b, and c.
6. Which of the following taxa are members of the Lophotrochozoa?
   a. Cnidaria
   b. Nematoda
   c. Platyhelminthes
   d. Arthropoda
   e. Echinodermata
7. What organisms can survive without a mouth, a digestive system, or an anus?
   a. cnidarians
   b. rotifers
   c. echinoderms
   d. cestodes
   e. nematodes
8. Which phylum does *not* have at least some members with a closed circulatory system?
   a. Lophotrochozoa
   b. Arthopoda
   c. Annelida
   d. Mollusca
   e. All of the above phyla have some members with a closed circulatory system.
9. A defining feature of the Ecdysozoa is
   a. a segmented body.
   b. a closed circulatory system.
   c. a cuticle.
   d. a complete gut.
   e. a lophophore.
10. Incomplete metamorphosis
    a. is characterized by distinct larval and adult stages that do not compete for resources.
    b. is typically seen in arachnids.
    c. involves gradual changes in life stages where young resemble the adult stage.
    d. is characteristic of the majority of insects.
    e. always includes a pupal stage.

## Conceptual Questions

1. Early morphological phylogenies were based on what three features of animal body plans?
2. Compare and contrast the five main feeding types (not modes of nutrition) discussed in the chapter.
3. **PRINCIPLES** A principle of biology is that all species (past and present) are related by an evolutionary history. Annelids, arthropods, and chordates all exhibit segmentation. Are these monophyletic, paraphyletic, or polyphyletic groups? Explain your answer.

## Collaborative Questions

1. Discuss the many ways that animals can affect humans, both positively and negatively.
2. Why are there more species of insects than any other taxa?

# Online Resource

# Vertebrates: Fishes, Amphibians, Reptiles, and Mammals

# 27

**Labordi's chameleon (*Furcifer labordi*).** This species has the shortest life cycle of any known land vertebrate, with individuals living about 4 months after hatching.

## Chapter Outline



Most terrestrial vertebrates, such as amphibians, reptiles, birds, and mammals, typically live between 2 to 10 years, with some having much longer or much shorter lives. Humans and turtles may live to 100 years, while some lizards die after a year. Labordi's chameleon (*Furcifer labordi*), discovered only this century in Madagascar, lives only 1 year. Even more unusual is that the chameleon spends two-thirds of that time as an egg. After hatching, the chameleons live only about 4 months, growing 2–3 mm each day. They reach sexual maturity and reproduce, after which the entire adult population dies off, leaving only the developing eggs. When the next generation of hatchlings emerges, adult chameleons are not present. With such a precarious life cycle, one bad year in which the eggs fail to hatch would severely affect the species' survival. Scientists are studying *F. labordi*'s unusual life cycle in the hope that it may prove valuable in discovering genes and other factors that affect aging and longevity.

*F. labordi* is a **vertebrate** (from the Latin *vertebratus*, meaning joint of the spine), an animal with a backbone. Vertebrates range in size from tiny fishes weighing 0.1 g to huge whales of over 100,000 kg. They occupy nearly all of Earth's habitats, from the deepest depths of the oceans to mountaintops and the sky beyond. Throughout history, humans have depended on many vertebrate species for their welfare—domesticating species such as horses, cattle, pigs, sheep, and chickens; using skin and fur for clothes; and keeping countless species, including cats and dogs, as pets. Many vertebrate species are the subjects of conservation efforts, as we will see in Chapter 46.

In Chapter 26, we discussed two chordate subphyla: the cephalochordates (lancelets) and urochordates (tunicates). The third subphylum of chordates, the Vertebrata, with about 53,000 species, is by far the largest and most dominant group of chordates. In this chapter, we will explore the characteristics of vertebrates and the evolutionary development of the major vertebrate classes, including fishes, amphibians, reptiles, and mammals.

## 27.1 Vertebrates: Chordates with a Backbone

### Learning Outcomes

1. List the main distinguishing characteristics of vertebrates.
2. Identify the two classes of existing jawless vertebrates.

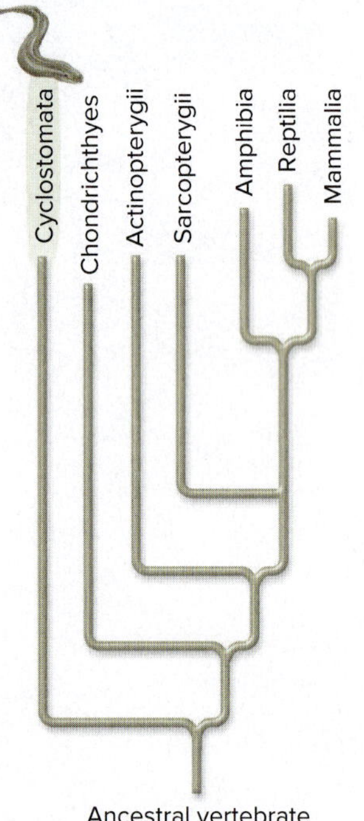


Our current understanding of the relationships between vertebrate groups is shown in **Figure 27.1.** Nested within the vertebrates are various clades based on genetic and morphological data. The vertebrates retain all chordate characteristics we outlined in Chapter 26, as well as possessing several additional traits, including the following:

1. **Vertebral column.** In vertebrates, the notochord is replaced by a bony or cartilaginous column of interlocking **vertebrae** that provides support and protects the nerve cord, which lies within its tubelike structure.
2. **Cranium.** The anterior end of the nerve cord is expanded to form a more developed brain that is encased in a protective bony or cartilaginous housing called the **cranium.**
3. **Endoskeleton of cartilage or bone.** The cranium and vertebral column are parts of the endoskeleton, the living skeleton of vertebrates that forms within the animal's body. The endoskeleton contains living cells that secrete the skeleton, which grows with the animal, unlike the nonliving exoskeleton of arthropods.

Although these are the main distinguishing characteristics of vertebrates, there are others. For example, vertebrates have multiple clusters of *Hox* genes, compared with the single cluster of *Hox* genes in tunicates and lancelets. These additional gene clusters are believed to have permitted increasingly complex morphologies beyond those possessed by invertebrate chordates. Vertebrates possess a great diversity of internal organs, including a liver, kidneys, endocrine glands, and a heart with at least two chambers. The liver is unique to vertebrates, and the vertebrate kidneys, endocrine system, and heart are more complex than are analogous structures in invertebrate taxa.

Although these features are exhibited in all vertebrate classes, some classes evolved critical innovations that helped them succeed in specific environments such as on land or in the air. For example, birds developed feathers and wings, structures that enable most species to fly. In fact, each of the vertebrate classes is distinctly different from one another, as outlined in **Table 27.1**. One of the earliest innovations was the development of jaws. All vertebrates except some early-diverging fishes possessed jaws. Today, the only jawless vertebrates are hagfish and lampreys, together known as the Cyclostomata or cyclostomes (circle mouths), eel-like animals that do not possess jaws. Sequencing of RNA libraries in 2010, together with genomic surveys, yields strong support that the Cyclostomata is monophyletic.

### The Hagfish Are the Simplest Living Cyclostomes

The hagfish are entirely marine cyclostomes that lack eyes, jaws, fins, and even vertebrae (**Figure 27.2**). The hagfish skeleton consists largely of a notochord and a cartilaginous skull that encloses the brain. The lack of a vertebral column leads to extensive flexibility. How, then, can hagfish be vertebrates without a vertebral column? Strong molecular support for a cyclostome clade suggests hagfish possessed the trait but that it was secondarily lost during evolution. Only the cranium and diversity of organs provide evidence of the organism's vertebrate ancestry.

Hagfish live in the cold waters of northern oceans, close to the muddy bottom, feeding on marine worms and other invertebrates. Essentially blind, hagfish have a very keen sense of smell and are attracted to dead and dying fish, which they attach themselves to via toothed plates on the mouth. The powerful tongue then scrapes off pieces of tissue. Though the hagfish cannot see approaching predators, they have special glands that produce copious amounts of slime. When provoked, the hagfish's slime production increases dramatically, enough to potentially distract predators or coat their gills and interfere with breathing. Hagfish can sneeze to clear their nostrils of their own slime.

**Figure 27.2** A hagfish, *Myxine glutinosa.*
© Pat Morris/ardea.com

**Figure 27.1 The major clades of vertebrates.** Critical innovations of each group are shown. The nested labels across the top of the figure refer to groupings within the vertebrates.

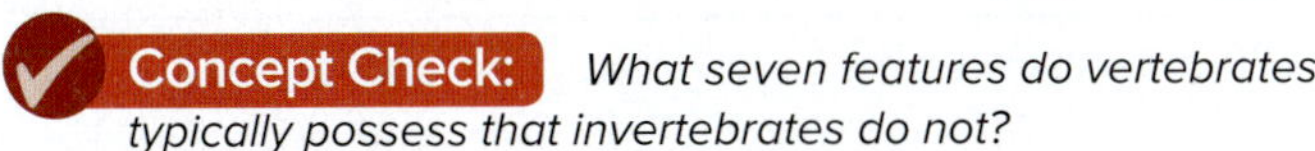

**Concept Check:** *What seven features do vertebrates typically possess that invertebrates do not?*

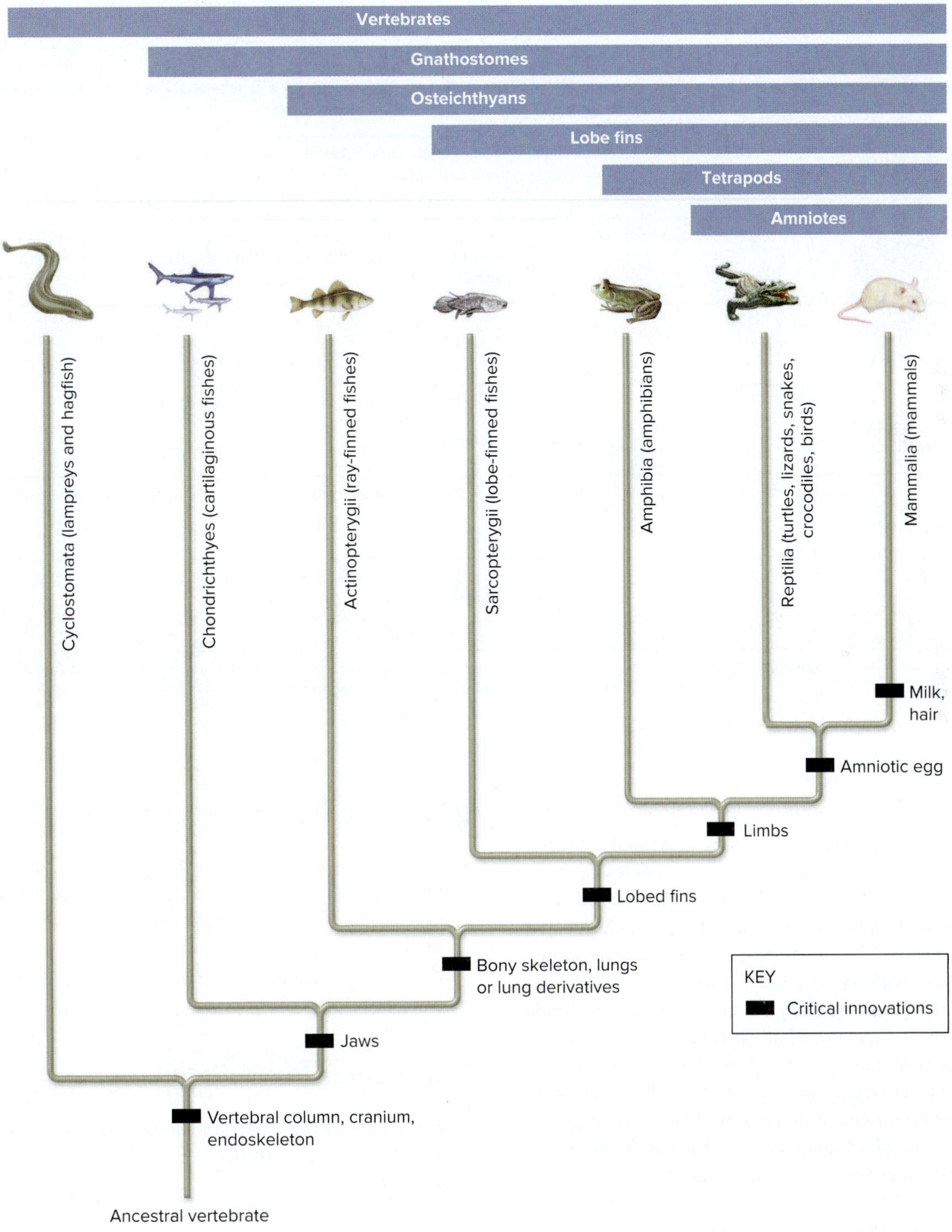

Vertebrates
Gnathostomes
Osteichthyans
Lobe fins
Tetrapods
Amniotes

Cyclostomata (lampreys and hagfish)
Chondrichthyes (cartilaginous fishes)
Actinopterygii (ray-finned fishes)
Sarcopterygii (lobe-finned fishes)
Amphibia (amphibians)
Reptilia (turtles, lizards, snakes, crocodiles, birds)
Mammalia (mammals)

Milk, hair
Amniotic egg
Limbs
Lobed fins
Bony skeleton, lungs or lung derivatives
Jaws
Vertebral column, cranium, endoskeleton
Ancestral vertebrate

KEY
Critical innovations

**Table 27.1 The Main Clades and Characteristics of Living Vertebrates**

| Clade | Examples (approx. # of species) | Main characteristics |
|---|---|---|
| Cyclostomata | Lampreys, hagfish (100) | Jawless fishes; no appendages |
| Chondrichthyes | Sharks, skates, rays (970) | Fishes with cartilaginous skeleton; teeth not fused to jaw; no swim bladder; well-developed fins; internal fertilization; single blood circulation |
| Actinopterygii | Ray-finned fishes, most bony fish (27,000) | Fishes with ossified skeleton; gill opening covered by operculum; fins supported by rays, fin muscles within body; swim bladder often present; mucous glands in skin |
| Sarcopterygii | Lobe-finned fishes, of which coelacanths (2) and lungfishes (6) are the only living members | Fishes with ossified skeleton; bony extensions, together with muscles, project into pectoral and pelvic fins |
| Amphibia | Frogs, toads, salamanders (6,300) | Adults able to live on land; fresh water needed for reproduction; development usually involving metamorphosis from tadpoles; adults with lungs and double blood circulation; moist skin; shell-less eggs |
| Testudines | Turtles (310) | Body encased in hard shell; no teeth; head and neck retractable into shell; eggs laid on land |
| Squamata | Lizards, snakes (7,900) | Lower jaw not attached to skull; skin covered in scales |
| Crocodilia | Crocodiles, alligators (23) | Four-chambered heart; large aquatic predators; parental care of young |
| Aves | Birds (10,000) | Feathers; hollow bones; air sacs; reduced internal organs; endothermic; four-chambered heart |
| Mammalia | Mammals (5,500) | Mammary glands; hair; specialized teeth; enlarged skull; external ears; endothermic; four-chambered heart; highly developed brains; diversity of body forms |

## The Lampreys Are Eel-like Animals That Lack Jaws

Lampreys are similar to hagfish in that they lack both a hinged jaw and true appendages. However, lampreys do possess a notochord surrounded by a cartilaginous rod that represents a rudimentary vertebral column. Lampreys can be found in both marine and freshwater environments. Marine lampreys are parasitic as adults. They grasp other fish with their circular mouth (**Figure 27.3a**) and rasp a hole in the fish's side, sucking out blood, tissue, and fluids (**Figure 27.3b**). Males and females of all species spawn in freshwater streams, and the resultant larval lampreys bury into the sand or mud, much like lancelets (refer back to Figure 26.34a), emerging to feed on small invertebrates or detritus at night. This stage can last for 3 to 7 years, at which time the larvae metamorphose into adults. In most freshwater species, the adults do not feed at all but quickly mate and die. Marine species migrate from fresh water back to the ocean, until they return to fresh water to spawn and then die.

## 27.1 Reviewing the Concepts

- Vertebrates have several characteristic features, including a vertebral column, a cranium, an endoskeleton of cartilage or bone, multiple clusters of *Hox* genes, and complex internal organs (Figure 27.1; Table 27.1).
- Early-diverging vertebrates lacked jaws. Today the only jawless vertebrates are the hagfish and lampreys (Figures 27.2, 27.3).

## 27.1 Testing Your Knowledge

1. Which of the following is *not* a defining characteristic of vertebrates?
   **a.** great diversity of organs
   **b.** hinged jaw
   **c.** multiple clusters of *Hox* genes
   **d.** vertebral column
   **e.** endoskeleton
2. Which of the following organisms are considered cyclostomes?
   **a.** tunicates
   **b.** hagfish
   **c.** lampreys
   **d.** both a and b
   **e.** both b and c

(a) Jawless mouth of a sea lamprey

(b) A sea lamprey feeding

**Figure 27.3** **The lamprey, a modern jawless fish.** **(a)** The sea lamprey (*Petromyzon marinus*) has a circular, jawless mouth. **(b)** A sea lamprey feeding on a fish.

*(a)* © R. Duran/Getty Images RF; *(b)* © Jacana/Science Source

## 27.2 Gnathostomes: Jawed Vertebrates

### Learning Outcomes

1. Describe how jaws evolved.
2. Discuss the distinguishing features of sharks.
3. List the three features that distinguish bony fishes from cartilaginous fishes.
4. Outline the differences between the ray-finned fishes and the lobe-finned fishes.

All vertebrate species that possess jaws are called **gnathostomes** (from the Greek, meaning jaw mouth) (see Figure 27.1). Gnathostomes are a diverse clade of vertebrates that include fishes, amphibians, reptiles, and mammals. The earliest-diverging gnathostomes were fishes. Biologists have identified about 28,000 species of living fishes, more than all other species of vertebrates combined. Most are aquatic, gill-breathing species that usually possess fins and a scaly skin. The three separate clades of jawed fishes—the Chondrichthyes (cartilaginous fishes), Actinopterygii (ray-finned fishes), and Sarcopterygii (lobe-finned fishes)—each have distinguishing characteristics (see Table 27.1).

The jawed mouth was a significant evolutionary development. It enabled an animal to grip its prey more firmly, which may have increased its rate of capture, and to attack larger prey species, thus increasing its potential food supply. Accompanying the jawed mouth was the development of more sophisticated head and body structures, including appendages such as fins. Gnathostomes also possess two additional *Hox* gene clusters beyond those in the cyclostomes (bringing their total to four), which likely led to increased morphological complexity.

The hinged jaw developed from the gill arches, cartilaginous or bony rods that help to support gills. Similarities between cells that make up jaws and gill arches support this view. Primitive jawless fishes had nine gill arches surrounding the eight gill slits (**Figure 27.4a**). During the late Silurian period (about 417 mya), some of these gill arches became modified. The first and second gill arches were lost, and the third and fourth pairs evolved to form the jaws (**Figure 27.4b,c**). This is how evolution typically works; body features do not appear de novo but, instead, existing features become modified to serve other functions.

### Chondrichthyans Are Fishes with Cartilaginous Skeletons

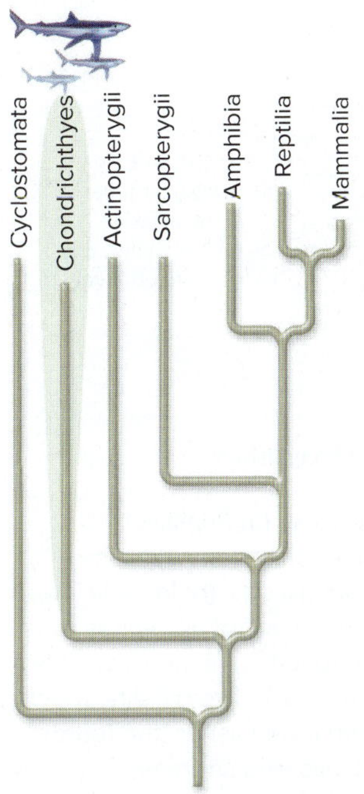


Members of the clade Chondrichthyes (the **chondrichthyans**)—sharks, skates, and rays—are called cartilaginous fishes because their skeleton is composed of flexible cartilage rather than bone. The cartilaginous skeleton is considered a derived character, meaning that the ancestors of the chondrichthyans had bony skeletons, but that members of this class subsequently lost this feature. This hypothesis is reinforced by the observation that during development, the skeleton of most vertebrates is cartilaginous and then it becomes bony (ossified) as a hard calcium-phosphate matrix replaces the softer cartilage. A change in the developmental sequence of the cartilaginous fishes is believed to prevent the ossification process.

**The Chondrichthyan Body Plan** All chondrichthyans are denser than water, which means that they would sink if they stopped swimming. Many sharks never stop swimming and maintain buoyancy via the use of their fins and a large, oil-filled liver. Perhaps the most important fin for propulsion in sharks is the large and powerful caudal fin, or tail fin, which, when swept from side to side, thrusts the fish forward at great speed (**Figure 27.5a**). The paired pelvic fins

## Biology Principle

### Structure Determines Function

The development of a jaw increased the predatory capabilities of gnathostomes.

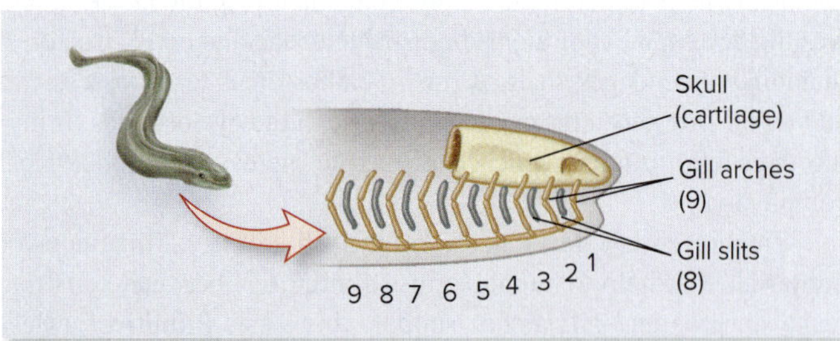


**(a) Primitive jawless fishes**

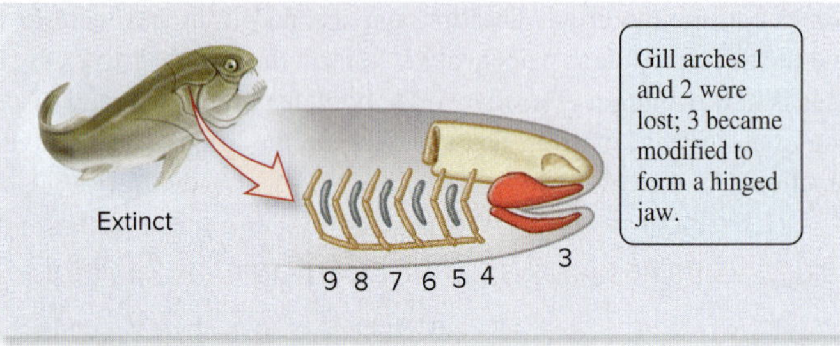


**(b) Early jawed fishes (placoderms)**

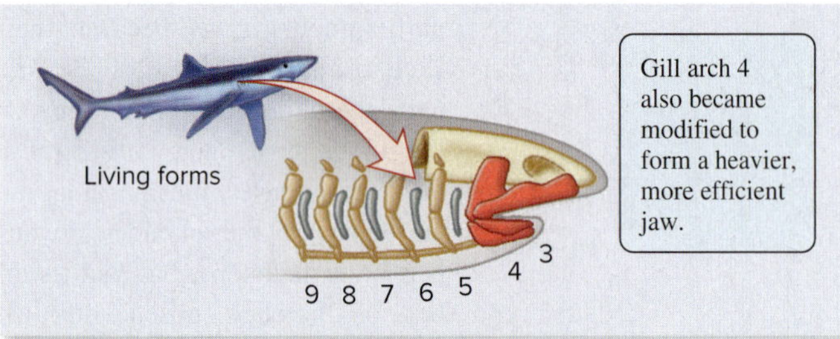


**(c) Modern jawed fishes (cartilaginous and bony fishes)**

**Figure 27.4** **The evolution of the vertebrate jaw.** **(a)** Primitive fishes and modern jawless fishes, such as lampreys, have nine cartilaginous gill arches that support eight gill slits. **(b)** In early jawed fishes such as the placoderms, the first two pairs of gill arches were lost, and the third pair became modified to form a hinged jaw. This left six gill arches (4–9) to support the remaining five gill slits, which were still used in breathing. **(c)** In modern jawed fishes, the fourth gill arch also contributes to jaw support, allowing stronger, more powerful bites to be delivered.

(at the back) and pectoral fins (at the front) act like flaps on airplane wings, allowing the shark to dive deeper or rise to the surface. They also aid in steering. In addition, the dorsal fin (on the shark's back) acts as a stabilizer to prevent the shark from rolling in the water as the tail fin pushes it forward. Skates and rays are essentially flattened sharks that cruise along the ocean floor by using hugely expanded pectoral fins (Figure 27.5b).

During swimming, water continually enters the shark's mouth and is forced over the gills, allowing the shark to extract oxygen and breathe. How, then, do skates and rays breathe when they rest on the ocean floor? These species, and a few sharks such as the nurse shark, use a muscular pharynx and jaw muscles to pump water over the gills. In these and all species of fishes, the heart consists of two chambers, an atrium and a ventricle, that contract in sequence. They employ what is known as a single circulation, in which blood is pumped from the heart to capillaries in the gills to collect oxygen, and then it flows through arteries to the tissues of the body before returning to the heart (look ahead to Figure 36.2a).

Sharks were among the earliest fishes to develop teeth, which evolved from rough scales on the skin. Although shark's teeth are very sharp and hard, they are not set into the jaw, as are human teeth, so they break off easily and are continually replaced, row by row (Figure 27.5c). Sharks have a powerful sense of smell, facilitated by sense organs in the nostrils (sharks and other fishes do not use nostrils for breathing). They can see well but cannot distinguish colors. All jawed fishes have a row of microscopic organs in the skin, arranged in a line that runs laterally down each side of the body, which can sense movements in the surrounding water. This system of sense organs, known as the **lateral line,** senses pressure waves and sends nerve signals to the brain.

**The Chondrichthyan Life Cycle** Fertilization is internal in chondrichthyans, with the male transferring sperm to the female via a pair of claspers, modifications of the pelvic fins. Some shark species exhibit **oviparity;** that is, they lay eggs, often inside a protective pouch called a mermaid's purse (Figure 27.5d). In species with **ovoviparity** the eggs are retained within the female's body, but there is no placenta to nourish the young. A few species exhibit **viviparity;** that is, the eggs develop within the uterus, receiving nourishment from the mother via a placenta. Both ovoviparous and viviparous sharks give birth to live young.

## Osteichthyans Are Fishes with Bony Skeletons

Unlike the cartilaginous fishes, all other gnathostomes have a bony skeleton and belong to the clade known as Osteichthyes. This term means bony fish and was originally proposed for just that group. With the advent of modern phylogenetic systematics, however, the term **osteichthyans** has expanded to include all vertebrates with a bony skeleton, including tetrapods (see Figure 27.1).

**Bony Fish Body Plan** Bony fishes are the most numerous of all types of fishes, with more species (about 27,000) than any other. The skin of bony fishes, unlike the rough skin of sharks, is covered by a thin epidermal layer containing glands that produce mucus, an adaptation that reduces drag during swimming. Just as in the cartilaginous fishes, water is drawn over the gills for breathing, but in bony fishes, a protective flap called an **operculum** covers the gills (look ahead to Figure 36.4). Muscle contractions around the gills and operculum

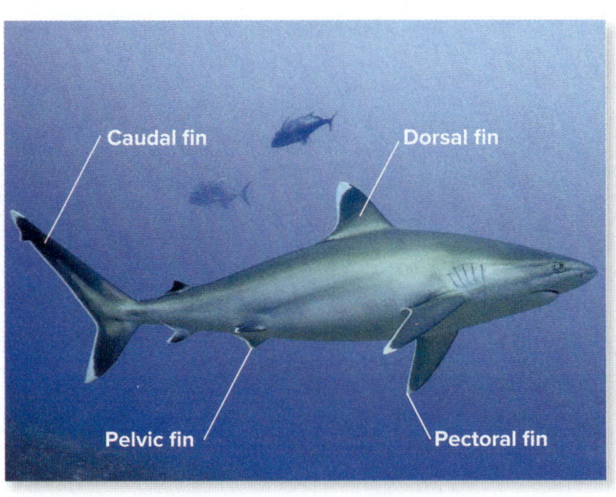


(a) Silvertip shark

(b) Stingray

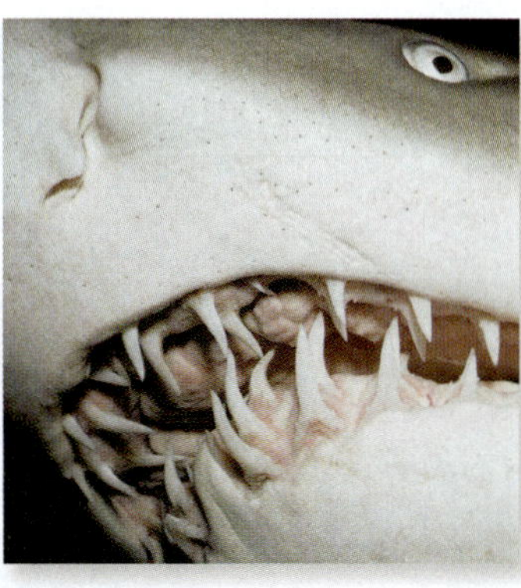

(c) Rows of shark teeth

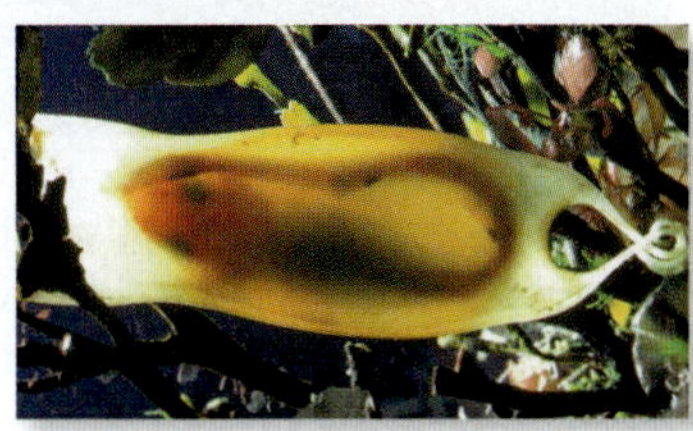

(d) Shark egg pouch

**Figure 27.5** **Cartilaginous fishes.** **(a)** The silvertip shark (*Carcharhinus albimarginatus*) is one of the ocean's most powerful predators. **(b)** Stingrays are essentially flattened sharks with very large pectoral fins. **(c)** Close-up of the mouth of a sand tiger shark (*Carcharias taurus*), showing rows of teeth. **(d)** This mermaid's purse (egg pouch) of a dogfish shark (*Scyliorhinus canicula*) is entwined in vegetation to keep it stationary.

*(a)* © Valerie & Ron Taylor/ardea.com; *(b)* © Bill Curtsinger/National Geographic/Getty Images; *(c)* © Jeff Rotman/naturepl.com; *(d)* © Oxford Scientific/Getty Images

**BioConnections:** *Look ahead to Figure 32.5. What types of cells are responsible for forming cartilage?*

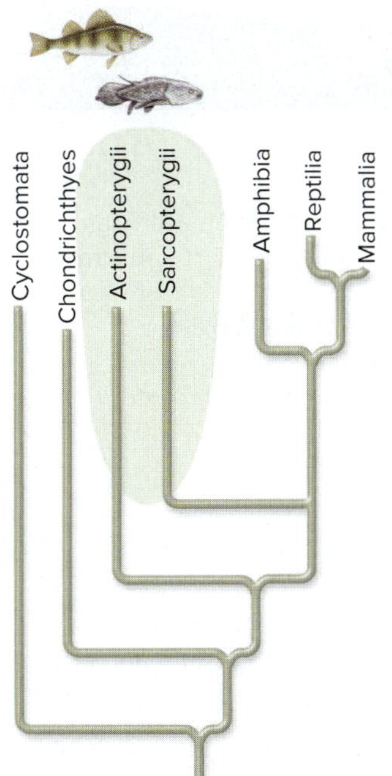


draw water across the gills so that bony fishes do not need to swim continuously in order to breathe.

Some early bony fishes lived in shallow, oxygen-poor waters and developed lungs as an embryological offshoot of the pharynx. These fish could rise to the water surface and gulp air. As we will see, modern lungfishes operate in much the same fashion. Many other fishes, known as bimodal breathers, can breathe through their gills and by gulping air, absorbing oxygen through their digestive tracts. For example, Siamese fighting fish (*Betta splendens*, known as betta), a popular freshwater aquarium fish, is a bimodal breather that is relatively easy to care for, since it can survive without an air pump in its aquarium. In most bony fishes, the lungs evolved into a **swim bladder,** a gas-filled, balloon-like structure that helps the fish remain buoyant in the water even when it is completely stationary. These three features—bony skeleton, operculum, and swim bladder—distinguish bony fishes from cartilaginous fishes.

**Bony Fish Reproduction and Diversity** Reproductive strategies of bony fishes vary tremendously, but most species reproduce via external fertilization, with the female shedding her eggs and the male depositing sperm on top of them. Although adult bony fishes can maintain their buoyancy, their eggs tend to sink. This helps explain why many species spawn in shallow, more oxygen- and food-rich waters and why coastal areas are important fish nurseries.

Bony fishes have colonized nearly all aquatic habitats. Following the cooling of the newly formed planet Earth, water condensed into rain and over a vast period of time filled what are now the oceans. Later, as water evaporated from the oceans and sodium, potassium, and calcium were added via runoff from the land, the oceans became salty. Therefore, most fishes probably evolved in freshwater habitats and secondarily became adapted to marine environments. This, of course, required the development of physiological adaptations to the different osmotic problems seawater presents compared with fresh water.

Most authorities now recognize two living clades of bony fishes: the Actinopterygii (ray-finned fishes) and the Sacropterygii (lobe-finned fishes).

**Actinopterygii, the Ray-Finned Fish** The most species-rich clade of bony fishes is the Actinopterygii, or **ray-finned fishes.** In Actinopterygii, the fins are supported by thin, bony, flexible rays and are moved by muscles on the interior of the body (Figure 27.5). The class has a diversity of forms, from large predatory moray eels to delicate sea dragons. Whole fisheries are built around the harvest of such ray-finned species as cod, anchovies, and salmon.

**Figure 27.6** **An Actinopterygian fish (ray-finned fish).** Black tip grouper (*Epinephelus fasciatus*), Maldives, Indian Ocean.
© Ian Cartwright/Getty Images RF

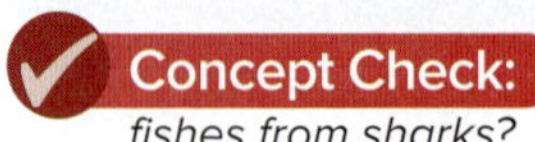
**Concept Check:** *What features distinguish ray-finned fishes from sharks?*

**(a) A coelacanth**

**(b) Australian lungfish**

**Figure 27.7** **The Sarcopterygii (lobe-finned fish).** **(a)** An actinisian, the coelacanth (*Latimeria chalumnae*). **(b)** A Dipnoi, the Australian lungfish (*Neoceratodus forsteri*).
*(a)* © Peter Scoones/Science Source; *(b)* © Oceans-Image/Photoshot

**Concept Check:** *How are lungfishes similar to coelacanths?*

**Sarcopterygii, the Lobe-Finned Fish** In the Sarcopterygii, or **lobe-finned fishes,** the fins are supported by skeletal extensions of the pectoral and pelvic areas that are moved by muscles within the fins. The Actinistia (coelacanths) and Dipnoi (lungfishes) are both considered Sarcopterygii. The name Sarcopterygii used to refer solely to the lobe-finned fishes, but since it has become clear that terrestrial vertebrates (tetrapods) evolved from such fishes, the definition of the group has been expanded to include both lobe-finned fishes and tetrapods (see Figure 27.1).

The Actinistia, or coelacanths, were a very successful group in the Devonian period, but all fishes of the group were once believed to have died off at the end of the Mesozoic era (some 65 mya). You can therefore imagine the scientific excitement when in 1938, a modern coelacanth was discovered as part of the catch of a boat fishing off the coast of South Africa (Figure 27.7a). Another species was found more recently in Indonesian waters.

The Dipnoi, or **lungfishes,** like the coelacanths, are also not currently a very species-rich group, having just three genera and six species (Figure 27.7b). Lungfishes live in oxygen-poor freshwater swamps and ponds. They have both gills and lungs, the latter of which enable them to come to the surface and gulp air. In fact, lungfish will drown if they are unable to breathe air. Because they also have muscular lobe fins, they are often able to successfully traverse long distances across land if their ponds dry out.

The morphological features of coelacanths, lungfishes, and primitive terrestrial vertebrates, together with the similarity of their nuclear genes, suggest to many scientists that lobe-finned ancestors gave rise to three lineages: the coelacanths, the lungfishes, and the tetrapods. In the next section, we will examine the biology of tetrapods in more detail.

## 27.2 Reviewing the Concepts

- A critical innovation in vertebrate evolution is the hinged jaw, which first developed in fishes. Gnathostomes are vertebrate species that possess a hinged jaw (Figure 27.4).
- The chondrichthyans (sharks, skates, and rays) have a skeleton composed of flexible cartilage and powerful appendages called fins. They are active predators with acute senses and were among the earliest fishes to develop teeth (Figure 27.5).
- Bony fishes consist of the Actinopterygii (ray-finned fishes, the most species-rich clade) and the lobe-finned fishes, which include the Actinistia (coelacanths) and the Dipnoi (lungfishes). In Actinopterygii, the fins are supported by thin, flexible rays and moved by muscles inside the body (Figure 27.6).

- The lobe fins comprise the lobe-finned fishes (Actinistia and Dipnoi) and the tetrapods. In the lobe-finned fishes, the fins are supported by extensions of the pectoral and pelvic areas and are moved by their own muscles (Figure 27.7).

## 27.2 Testing Your Knowledge

1. In modern jawed fishes, which gill arches have been modified to form the jaws?
   **a.** 1 and 2 **b.** 2 and 3 **c.** 3 and 4 **d.** 4 and 5 **e.** 1 and 9
2. What is *not* a morphological feature of sharks?
   **a.** paired pelvic fins
   **b.** paired pectoral fins
   **c.** cartilaginous skeleton
   **d.** swim bladder
   **e.** multiple rows of teeth
3. Which type of fish is a lobe fin?
   **a.** moray eel
   **b.** sea dragon
   **c.** lamprey
   **d.** stingray
   **e.** lungfish

## 27.3 Tetrapods: Gnathostomes with Four Limbs

### Learning Outcomes

1. List adaptations that the transition to life on land required.
2. Describe the different amphibian orders and what differentiates them.
3. **SCISKILLS ▶** Explain how relatively simple experiments with mice showed that mutations in just two genes can cause large changes in limb development.

During the Devonian period (from about 417 to 354 mya, refer back to Figure 22.4), a diversity of plants and animals colonized the land. The presence of plants served as both a source of oxygen and a potential food source for animals that ventured out of the aquatic environment. Terrestrial arthropods appeared during the Devonian, as did the first terrestrial vertebrates.

The transition to life on land involved a large number of adaptations. Paramount among these were adaptations that prevented desiccation, or drying out, and made locomotion and reproduction on land possible. We have seen that some fish evolved the ability to breathe air. In this section, we begin by outlining the development of early **tetrapods,** vertebrate animals having four legs or leglike appendages, and the characteristics of their immediate descendants, the amphibians.

### The Origin of Tetrapods Involved the Development of Four Limbs

Over the Devonian period, the fossil record demonstrates the evolution of sturdy lobe-finned fishes to fishes with four limbs. The abundance of light and nutrients in shallow waters encouraged a profusion of plant life and the invertebrates that fed on them. The development of lungs enabled lungfishes to colonize these productive yet often oxygen-poor waters. Here, the ability to move in shallow water clogged with plants and debris was more vital than the ability to swim swiftly through open water and may have favored the progressive development of sturdy limbs. As an animal's weight began to be borne more by the limbs, the adaptations of a strengthened vertebral column and hip bones and shoulder bones that were braced against the backbone for added support were favored. Such modifications are the result of changes in the expression of genes, especially *Hox* genes (see the Feature Investigation next).

## FEATURE INVESTIGATION

### Davis and Colleagues Provide a Genetic-Developmental Explanation for Limb Length in Tetrapods

The development of limbs in tetrapods was a vital step that allowed animals to colonize land. The diversity of vertebrate limb types is amazing, from fins in fish and marine mammals, to different wing types in bats and birds, to legs and arms in primates. Early in vertebrate evolution, an ancestral gene complex was duplicated twice to give rise to four groups of genes, called *Hox A, B, C,* and *D,* which control limb development. In 1995, Allen Davis, Mario Capecchi, and colleagues analyzed the effects of mutations in specific *Hox* genes that are responsible for determining limb formation in mice. The vertebrate forelimb is divided into three zones: humerus (upper arm); radius and ulna (forearm); and carpals, metacarpals, and phalanges (digits). The authors had no specific hypothesis in mind; their goal was to understand the role of *Hox* genes in limb formation. As described in **Figure 27.8**, they began with strains of mice carrying loss-of-function mutations in *HoxA-11* or *HoxD-11* that, on their own, did not cause dramatic changes in limb formation. They bred the mice and obtained offspring carrying one, two, three, or four loss-of-function mutations. Double mutants (*aadd*) exhibited dramatically different phenotypes not seen in mice homozygous for the individual mutations. The radius and ulna were almost entirely eliminated.

As seen in the data, the mutations affected the formation of limbs. For example, the wrist contains seven bones: three proximal carpals—called navicular lunate (NL), triangular (T), and pisiform (P)—and four distal carpals (d1–d4). In mice with the genotypes *aaDD* and *AAdd*, the proximal carpal bones are usually fused together. Individuals having one heterozygous recessive

**Figure 27.8** Relatively simple changes in *Hox* genes control limb formation in tetrapods.

**GOAL** To determine the role of *Hox* genes in limb development in mice.

**KEY MATERIALS** Mice with individual mutations in *HoxA-11* and *HoxD-11* genes.

**Experimental level** | **Conceptual level**

**1** Breed mice with individual mutations in *HoxA-11* and *HoxD-11* genes. (The *A* and *D* refer to wild-type alleles; *a* and *d* are mutant alleles.)

The mice bred were heterozygous for both genes (*AaDd*).

Based on previous studies, researchers expect mutant mice to produce viable offspring, perhaps with altered limb morphologies.

**2** Using molecular techniques described in Chapter 18, obtain DNA from the tail and determine the genotypes of offspring.

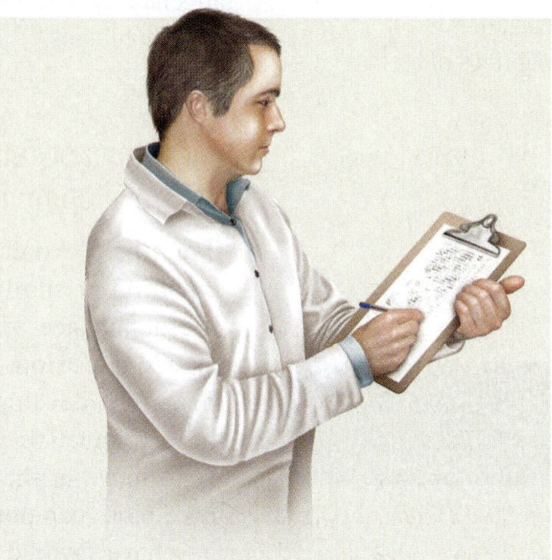

The resulting genotypes occur in Mendelian ratios, generating mice with different combinations of wild-type and mutant alleles.

| | *AD* | *Ad* | *aD* | *ad* |
|---|---|---|---|---|
| *AD* | *AADD* | *AADd* | *AaDD* | *AaDd* |
| *Ad* | *AADd* | *AAdd* | *AaDd* | *Aadd* |
| *aD* | *AaDD* | *AaDd* | *aaDD* | *aaDd* |
| *ad* | *AaDd* | *Aadd* | *aaDd* | *aadd* ← Double mutant |

**3** Stain the skeletons and compare the limb characteristics of the wild-type mice (*AADD*) to those of strains carrying mutant alleles in one or both genes.

Mutant mice may have altered bone morphologies.

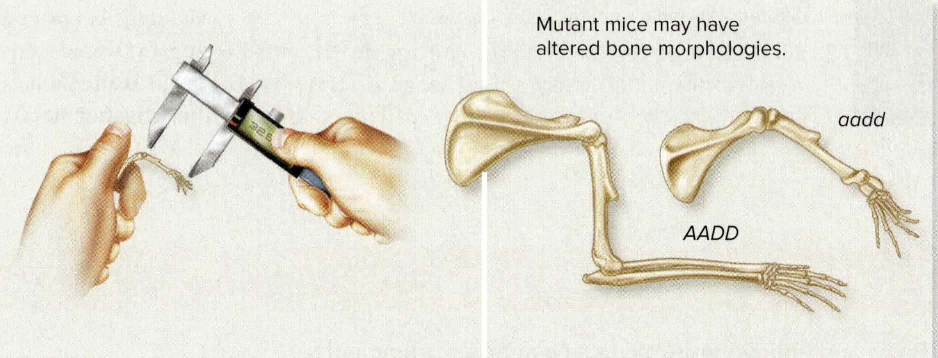


**4 THE DATA**

| Genotype | Carpal bone fusions (% of mice showing the fusion) | | | |
|---|---|---|---|---|
| | Normal (none fused) | NL fused to T | T fused to P | NL fused to T and P |
| *AADD* | 100 | 0 | 0 | 0 |
| *AaDD* | 100 | 0 | 0 | 0 |
| *aaDD* | 33 | 17 | 50 | 0 |
| *AADd* | 100 | 0 | 0 | 0 |
| *AAdd* | 0 | 17 | 17 | 67 |
| *AaDd* | 17 | 17 | 33 | 33 |

5 **CONCLUSION** Relatively simple mutations involving two genes can cause large changes in limb development.

6 **SOURCE** Davis, A.P., et al. 1995. Absence of Radius and Ulna in Mice Lacking *HoxA-11* and *HoxD-11. Nature* 375: 791–795.

allele (*AADd* and *AaDD*) do not show this defect, but individuals heterozygous for two recessive alleles (*AaDd*) often do. Therefore, any two mutant alleles (either from both *HoxA-11* and *HoxD-11* or one from each locus) cause carpal fusions. Deformities became even more severe with three mutant alleles (*Aadd* or *aaDd*) or four mutant alleles (*aadd*) (data not shown in the figure). Thus, scientists have shown that relatively simple mutations can control relatively large changes in limb development.

*Experimental Questions*

1. What was the purpose of the study conducted by Davis and colleagues?
2. How were the researchers able to study the effects of individual genes?
3. **SCISKILLS** ▶ Explain the results of the experiment and how this relates to limb development in vertebrates.

## Amphibian Lungs and Limbs Are Adaptations to a Semiterrestrial Lifestyle

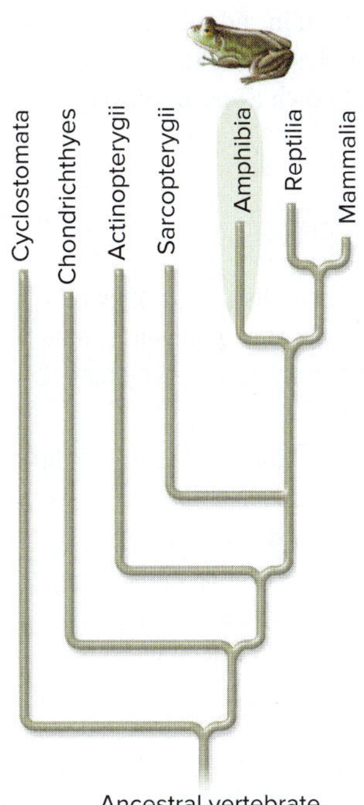


**Amphibians** (from the Greek, *amphibios* meaning both ways of life) live in two worlds: They have successfully invaded the land, but most must return to the water to reproduce. By the middle of the Carboniferous period (about 320 mya), species similar to modern amphibians had become common in the terrestrial environment. With a bonanza of terrestrial arthropods to feast on, the amphibians became numerous and species-rich, and often large in body size. The mid-Permian period (some 260 mya) is sometimes known as the Age of Amphibians. However, most of the large amphibians became extinct at the end of the Permian period. This was the largest known mass extinction in Earth's history, with the extinction of 90–95% of marine species and a large proportion of terrestrial species. Most surviving amphibians were smaller organisms resembling modern species.

One of the first challenges terrestrial animals had to overcome was breathing air when on land. Amphibians breathe in several different ways. They can use the same technique as lungfishes, opening their mouths to let in air. Alternatively, they may take in air through their nostrils. They then close their nostrils and close and raise the floor of the mouth, creating a positive pressure that pumps air into the lungs. This method of breathing is called **buccal pumping.** In addition, the skin of amphibians is much thinner than that of fishes, and amphibians absorb oxygen from the air directly through their outer moist skin or through the skin lining of the inside of the mouth.

Because the skin of amphibians is so thin, the animals face the problem of desiccation on land. As a consequence, even amphibian adults are more abundant in damp habitats, such as swamps or rain forests, than in dry areas. Also, most amphibians cannot venture too far from water because their larval stages are still aquatic. In frogs and toads, fertilization is generally external, with males shedding sperm over the gelatinous egg masses laid by the females in water (**Figure 27.9a**). The fertilized eggs lack a shell and would quickly dry out if exposed to the air for long. They soon hatch into tadpoles (**Figure 27.9b**)—small, fishlike animals that lack limbs and breathe through gills. As the tadpole nears the adult stage, the tail and gills are resorbed, and limbs and lungs appear (**Figure 27.9c**). Such a dramatic change from a juvenile to an adult body form is known as **metamorphosis.** A few species of amphibians do not require water to reproduce. These species are ovoviparous or viviparous—retaining the eggs in the reproductive tract and giving birth to live young.

## Modern Amphibians Include a Variety of Frogs, Toads, Salamanders, and Caecilians

Approximately 6,300 living amphibian species are known, and the vast majority of these, some 5,600 species, are frogs and toads of the order Anura (from the Greek, meaning tail-less ones) (**Figure 27.10a**). The other two orders are the Apoda (from the Greek, meaning legless ones), the wormlike caecilians; and the Urodela (from the Latin, meaning tailed ones), the salamanders.

Adult anurans are carnivores, eating a variety of invertebrates by catching them on a long, sticky tongue. In contrast, their aquatic larvae (tadpoles) are primarily herbivores. In addition to secreting mucus, which keeps their skin moist, some frogs can also secrete poisonous chemicals that deter would-be predators. Some amphibians advertise the poisonous nature of their skin with warning coloration (look ahead to Figure 43.12b). Global warming is currently threatening many anurans with extinction.

(a) Gelatinous mass of amphibian eggs

(b) Tadpole

(c) Tadpole undergoing metamorphosis

**Figure 27.9** **Amphibian development in the European common frog (*Rana temporaria*).** **(a)** Amphibian eggs are laid in gelatinous masses in water. **(b)** The eggs develop into tadpoles, aquatic herbivores with a fishlike tail that breathe through gills. **(c)** During metamorphosis, the tadpole loses its gills and tail and develops limbs and lungs.

*(a)* © Don Vail/Alamy; *(b)* © Nature Picture Library/Alamy; *(c)* © NHPA/Photoshot

*What were some advantages to animals of moving onto land?*

(a) Tree frog

(b) A caecilian

(c) Mud salamander

**Figure 27.10** **Amphibians.** **(a)** Most amphibians are frogs and toads of the order Anura, including this red-eyed tree frog (*Agalychnis callidryas*). **(b)** The order Apoda includes wormlike caecilians such as this species from Equador, *Siphonops annulatus*. **(c)** The order Urodela includes species such as this mud salamander (*Pseudotriton montanus*).

*(a)* © Gregory G. Dimijian/Science Source; *(b)* © Danita Delimont/Alamy; *(c)* © Gary Meszaros/Science Source

*Do all amphibians produce tadpoles?*

Caecilians are a small order of about 170 species of legless, nearly blind amphibians (**Figure 27.10b**). Most are tropical and burrow in forest soils, but a few live in ponds and streams.

The more than 550 species of salamanders possess legs and a long tail and have a more elongate body than anurans (**Figure 27.10c**). Like frogs, salamanders often have colorful skin patterns that advertise their distastefulness to predators. A few species, such as Cope's giant salamander (*Dicamptodon copei*), retain the gills and tail fins characteristic of the larval stage into adulthood, and they mature sexually in the larval stage, a phenomenon known as paedomorphosis.

## 27.3 Reviewing the Concepts

- Fossils record the evolution of lobe-finned fishes to fishes with four limbs. Recent research has shown that relatively simple mutations can control large changes in limb development (Figure 27.8).
- Amphibians live on land but return to the water to reproduce. In frogs and toads, the larval stage undergoes metamorphosis, losing gills and tail and gaining lungs and limbs (Figure 27.9).
- The majority of amphibians belong to the order Anura (frogs and toads). Other orders are the Apoda (caecilians) and Urodela (salamanders) (Figure 27.10).

## 27.3 Testing Your Knowledge

1. The transition of tetrapods from water to life on land required
   a. the prevention of desiccation.
   b. internal fertilization.
   c. the ability to breathe air.
   d. both a and c.
   e. all of the above.
2. Which of the following amphibian orders have legs?
   a. salamanders
   b. frogs and toads
   c. caecilians
   d. both a and b
   e. all amphibian orders

## 27.4 Amniotes: Tetrapods with a Desiccation-Resistant Egg

### Learning Outcomes

1. Diagram the structure of the amniotic egg.
2. Identify the critical innovations of the amniotes.
3. Describe the distinguishing features of the major amniote clades.
4. List the features that enable birds to fly.

Although amphibians live successfully in a terrestrial environment, they must lay their eggs in water or in a very moist place so that their shell-less eggs do not dry out on exposure to air. Thus, a critical innovation in animal evolution was the development of a shelled egg that sheltered the embryo from desiccation on land. A shelled egg containing fluids was like a personal enclosed pond for each developing individual. Such an egg evolved in the common ancestor of turtles, lizards, snakes, crocodiles, birds, and mammals—a group of tetrapods collectively known as the **amniotes.** The amniotic egg permitted animals to lay their eggs in a dry place, so that reproduction was no longer tied to water. It was truly a critical innovation, untethering animals from water in much the same way as the development of seeds liberated plants from water (see Chapter 25).

Over time, the amniotes became very diverse in species and morphology. Mammals are considered amniotes, too, because even though most of them do not lay eggs, they retain other features of amniotic reproduction. In this section, we begin by discussing the morphology of the amniotic egg and other adaptations that permitted animal species to become fully terrestrial. We then examine the biology of the reptiles, the first group of vertebrates to fully exploit land.

### The Amniotic Egg and Other Innovations Permitted Life on Land

The **amniotic egg** (Figure 27.11) contains the developing embryo and the four separate extraembryonic membranes that it produces:

1. The innermost membrane is the **amnion,** which protects the developing embryo in a fluid-filled sac called the amniotic cavity.
2. The **yolk sac** encloses a stockpile of nutrients, in the form of yolk, for the developing embryo.
3. The **allantois** functions as a disposal sac for metabolic wastes.
4. The **chorion,** along with the allantois, provides gas exchange between the embryo and the surrounding air.

Surrounding the chorion is the albumin, or egg white, which also stores nutrients. The **shell** provides a tough, protective covering that is not very permeable to water and prevents the embryo from drying out. However, the shell remains permeable to oxygen and carbon dioxide, so the embryo can breathe. In birds, this shell is hard and calcareous, whereas in reptiles and early-diverging mammals such as the platypus and echidna, it is soft and leathery. In most mammals, however, the embryos embed into the wall of the uterus and receive their nutrients directly from the mother.

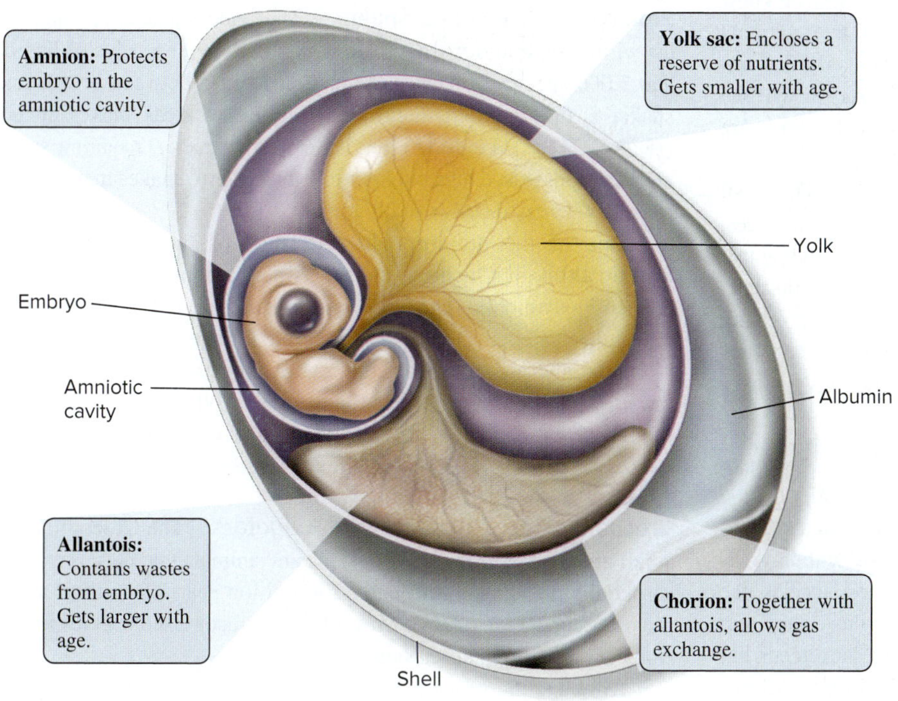

**Amnion:** Protects embryo in the amniotic cavity.
**Yolk sac:** Encloses a reserve of nutrients. Gets smaller with age.
Yolk
Embryo
Amniotic cavity
Albumin
**Allantois:** Contains wastes from embryo. Gets larger with age.
**Chorion:** Together with allantois, allows gas exchange.
Shell

**Figure 27.11** The amniotic egg.

Concept Check: *What are the other critical innovations of amniotes?*

Along with the amniotic egg, other critical innovations that enabled the conquest of land include the following:

- **Desiccation-resistant skin.** Whereas the skin of amphibians is moist and aids in respiration, the skin of amniotes is thicker and water-resistant and contains keratin, a tough protein. As a result, most gas exchange takes place through the lungs.
- **Thoracic breathing.** Amphibians use buccal pumping to breathe, contracting the mouth to force air into the lungs. In contrast, amniotes use **thoracic breathing,** in which coordinated contractions of muscles expand the rib cage, creating a negative pressure to suck air in and then forcing it out later. This results in a greater volume of air being displaced with each breath than with buccal pumping.
- **Water-conserving kidneys.** The ability to concentrate waste prior to elimination and thus conserve water is an important role of the amniotic kidneys.

(a) Green turtle

(b) Common collared lizard

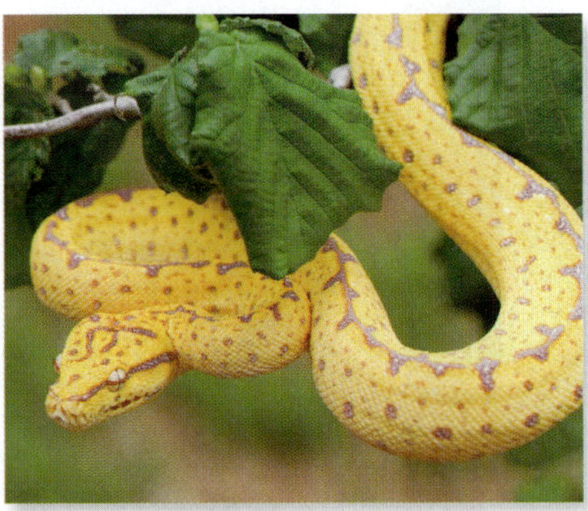

(c) Green tree python

**Figure 27.12** **A variety of reptiles.** **(a)** A green turtle (*Chelonia mydas*) laying eggs in the sand in Malaysia. **(b)** Common collared lizard (*Crotaphytus collarus*). **(c)** Juvenile tree python (*Morelia viridis*).

*(a)* © Pat Morris/ardea.com; *(b)* © Royalty-Free/Corbis; *(c)* © Mark Kostich/Getty Images RF

*If snakes are limbless, how can they be considered tetrapods?*

- **Internal fertilization.** Because sperm cannot penetrate a shelled egg, fertilization occurs internally, within the female's body before the shell is secreted. In this process, the male of the species often uses a copulatory organ (penis) to transfer sperm into the female reproductive tract.

## Reptiles Include Turtles, Lizards, Snakes, Crocodilians, and Birds

Early amniote ancestors gave rise to all the modern amniotes we know today, from lizards and snakes to birds and mammals. The traditional view of amniotes involved three living classes: the reptiles (turtles, lizards, snakes, and crocodilians), birds, and mammals. As we will see later in this chapter, modern systematists now consider that birds are part of the reptilian lineage. This is the classification scheme that we will follow in this chapter. The fossil record includes other reptilian clades, all of which are extinct, including two groups of dinosaurs (ornithischian and saurischian dinosaurs), flying reptiles (pterosaurs), and two groups of ancient aquatic reptiles (icthyosaurs and plesiosaurs).

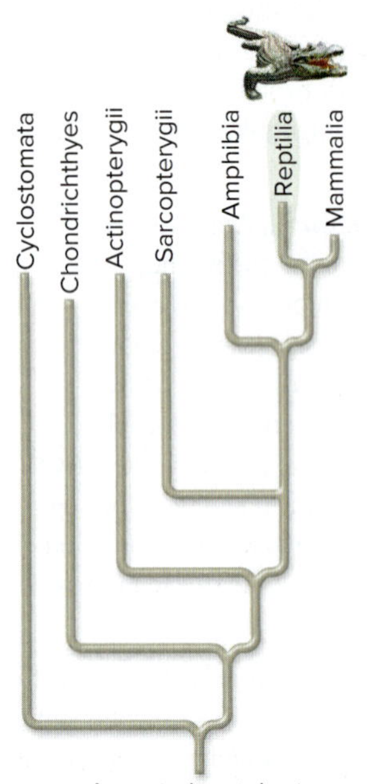


**Testudines: The Turtles** The major distinguishing characteristic of the turtles is a hard, protective shell into which the animal can withdraw its head and limbs. In most species, the vertebrae and ribs are fused to form this shell. All turtles lack teeth but have sharp beaks for biting. All turtles, even the aquatic species, lay their eggs on land, usually in soft sand. The gender of hatchlings is dependent on the temperature of the nest, with high temperatures producing more females. Marine species often make long migrations to sandy beaches to lay their eggs (Figure 27.12a). Even though turtles are very long-lived species, often surviving for 120 years or more, they do not appear to show reproductive senescence or aging and reproduce continually throughout their lifetime. Most organs, such as the liver, lungs, and kidneys, of a centenarian turtle function as effectively as do organs in young individuals, prompting genetic researchers to examine the turtle genome for longevity genes.

**Squamata: Lizards and Snakes** The clade Squamata is the largest within the traditional reptiles, with about 4,900 species of lizards and 3,000 species of snakes. A main difference between lizards and snakes is that lizards generally have limbs, whereas snakes do not (Figures 27.12b,c). Leglessness is a derived character, meaning snake ancestors possessed legs but later lost them. Also, snakes may be venomous, whereas lizards usually are not.

One of the defining characteristics of the Squamata is a **kinetic skull,** in which the joints between various parts of the skull are extremely mobile. The lower jaw does not join directly to the skull but, rather, is connected by a multijointed hinge, and the upper jaw is hinged and movable from the rest of the head. This allows their jaws to open relatively wider than those of other vertebrates, with the result that lizards, and especially snakes, can swallow large prey (Figure 27.13). Nearly all species are carnivores.

**Crocodilia: The Crocodiles and Alligators** The Crocodilia is a small group of large, carnivorous, aquatic animals that have remained essentially unchanged for nearly 200 million years since the time of the dinosaurs (Figure 27.14). Most of the 23 recognized species live in tropical or subtropical regions.

Although the group is small, it is evolutionarily very important. Crocodiles have a four-chambered heart, a feature they share with birds and mammals (look ahead to Figure 36.2b). In this regard, crocodiles are more closely related to birds than to any other living reptiles. Their teeth are set in sockets, a feature typical of the earliest

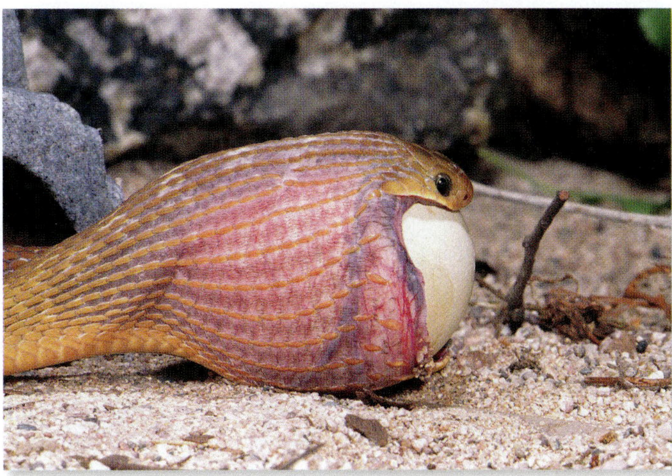

**Figure 27.13** **The kinetic skull.** In snakes and lizards, both the top and bottom of the jaw are hinged on the skull, thereby allowing them to swallow large prey. African egg-eating snake (*Dasypeltis scabra*) eating an egg.
© Danita Delimont/Getty Images

**Figure 27.14** **Crocodilians.** Crocodilia is an ancient class that has existed unchanged for millions of years. The Australian saltwater crocodile (*Crocodylus porous*), at up to 6.2 m (20 ft), is the largest species of all.
© DEA/C. Dani-I. Jeske/Getty Images

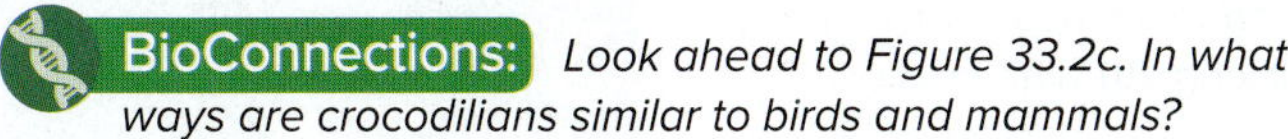

**BioConnections:** *Look ahead to Figure 33.2c. In what ways are crocodilians similar to birds and mammals?*

birds. Similarly, crocodiles care for their young, another trait they have in common with birds. These and other features suggest that crocodiles and birds are closely related. As with turtles, the gender of the offspring is dependent on nest temperature.

**Aves: The Birds** The defining characteristics of birds are that they have feathers and that nearly all species can fly. Flight has shaped nearly every feature of the bird body. The other vertebrates that have evolved the ability to fly—the bats and the now-extinct pterosaurs—used skin stretched tightly over elongated limbs and digits to form wings. Such a surface can be irreparably damaged, however, though some holes may heal remarkably quickly. In contrast, birds use feathers, epidermal outgrowths that can be replaced if damaged. Recent research shows that feathers evolved in dinosaurs before the appearance of birds.

One of the first known fossils exhibiting the faint impression of feathers was *Archaeopteryx lithographica* (from the Greek, meaning ancient wings and stone picture), found in a limestone quarry in Germany in 1861. The fossil was dated at 150 million years old, which places it during the Jurassic period. Except for the presence of feathers, *Archaeopteryx* appears to have had features similar to those of many other dinosaurs (**Figure 27.15**).

## Feathers, a Lightweight Skeleton, Air Sacs, and Reduced Organs Are Bird Adaptations That Aid in Flying

Modern birds possess many characteristics, including scales on their feet and legs and shelled eggs, that reveal their reptilian ancestry. In addition, however, among living animals birds have four unique features, all of which are associated with flight.

1. **Feathers.** Feathers are modified scales that keep birds warm and enable flight. Soft, downy feathers, which are close to the body, maintain heat, whereas stiffer contour feathers, supported on a modified forelimb, give the wing the airfoil shape it needs to generate lift.

**Figure 27.15** ***Archaeopteryx lithographica*** **was a Jurassic animal with dinosaur-like features as well as wings and feathers.** This supports the idea that birds and dinosaurs are closely related.
© Jason Edwards/Getty Images RF

**Concept Check:** *What adaptations in birds help reduce their body weight to enable flight?*

2. **Air sacs.** Flight requires a great deal of energy generated from an active metabolism, which requires abundant oxygen. Birds have nine air sacs—large, hollow sacs that may extend into the bones—that expand and contract when a bird inhales and exhales, while the lungs remain stationary. Air is therefore being constantly moved across the lungs during inhalation and exhalation. Although making bird breathing very efficient, this process also makes birds especially susceptible to airborne toxins (hence, the utility of the canary in the coal mine; the bird's death signaled the presence of harmful carbon dioxide or methane gas that was otherwise unnoticed by miners).
3. **Reduction of organs.** Some organs are reduced in size or are lacking altogether in birds, which reduces the total mass that the bird carries. For example, birds have only one ovary and can carry relatively few eggs. As a result, they lay fewer eggs than most reptile species. In fact, the gonads of both males and females are reduced, except during the breeding season, when they increase in size. Most birds also lack a urinary bladder. In addition, their lack of teeth has reduced weight at the head end.
4. **Lightweight bones.** Most bird bones are thin and hollow and are crisscrossed internally by tiny pieces of bone to give them a strong but lightweight honeycomb structure.

Because birds need a rapid metabolism and the quick production of adenosine triphosphate (ATP) to fuel flight, body temperatures are generally 40–42°C, considerably warmer than the human body's average of 37°C. Birds also have a double circulation and a four-chambered heart that ensures rapid blood circulation. Flight requires good vision, and bird vision is the best in the vertebrate world. Birds generally have high energy needs, and most birds are carnivores, eating insects or other invertebrates. However, some birds, such as parrots, eat nutrient-rich fruits and seeds. Bird eggs also need to be kept warm for successful development, which entails sitting on the nest until the eggs have hatched, a process called brooding. Often, the males and females take turns brooding so that one parent can feed and maintain its strength. Picking successful partners is therefore an important breeding task, and birds often engage in complex courtship rituals (look ahead to Figure 42.17).

With about 10,000 species, birds are the most species-rich group of terrestrial vertebrates. Despite this diversity, birds lack the variety of body shapes that exists in the other class of vertebrates, the mammals, some of which can swim, some of which can fly, others walk on four legs, and yet others walk only on two legs. Most birds fly, and therefore most have the same general body shape. However, birds feed in many different ways, which is reflected in the variety of their beak shapes (Figure 27.16).

## Biology Principle

### Structure Determines Function

Each of these beak shapes permits a different method of feeding.

(a) Cracking beak

(d) Probing beak

(b) Scooping beak

(e) Nectar-feeding beak

(c) Tearing beak

(f) Sieving beak

**Figure 27.16** **A variety of bird beaks.** Birds have evolved a variety of beak shapes used in different types of food gathering. **(a)** Hyacinth macaw (*Anodorhynchus hyacinthinus*)—cracking. **(b)** White pelican (*Pelecanus onocrotalus*)—scooping. **(c)** Verreaux's eagle (*Aquila verreauxii*)—tearing. **(d)** American avocet (*Recurvirostra americana*)—probing. **(e)** Lucifer hummingbird (*Calypte anna*)—nectar feeding. **(f)** Roseate spoonbill (*Platalea ajaja*)—sieving.

## 27.4 Reviewing the Concepts

- The amniotic egg permitted animals to become fully terrestrial. Other critical innovations included desiccation-resistant skin, thoracic breathing, water-conserving kidneys, and internal fertilization (Figure 27.11).

- Living reptilian clades include the Testudines (turtles), Squamata (lizards and snakes), Crocodilia (crocodiles), and Aves (birds) (Figures 27.12–27.14).
- The fossil bird *Archaeopteryx lithographica* helps trace a lineage from dinosaurs to birds (Figure 27.15).
- The four key characteristics of birds are feathers, air sacs, reduced organs, and a lightweight skeleton. Birds are the most species-rich clade of terrestrial vertebrates. The diversity of bird beaks reflects the varied methods they use for feeding (Figure 27.16).

## 27.4 Testing Your Knowledge

1. The membrane of the amniotic egg that serves as a site for waste storage is the
   a. amnion.
   b. yolk sac.
   c. allantois.
   d. chorion.
2. Which is a critical innovation of amniotes?
   a. buccal pumping
   b. moist skin that aids in respiration
   c. external fertilization
   d. water-conserving kidneys
   e. bony skeleton
3. Possession of a four-chambered heart suggests crocodiles are most closely related to which other living vertebrates?
   a. lizards
   b. turtles
   c. birds
   d. dinosaurs
   e. mammals
4. Which of the following features is associated with the ability of birds to fly?
   a. scaly legs
   b. feathers
   c. air sacs
   d. double circulation
   e. both b and c

## 27.5 Mammals: Milk-Producing Amniotes

### Learning Outcome

1. Identify the four features that separate mammals from other vertebrate clades.

Mammals evolved from amniote ancestors earlier than birds. About 225 mya, the first mammals appeared in the mid-Triassic period (refer back to Figure 22.4). The extinction of the dinosaurs in the Cretaceous period, some 65 mya, paved the way for mammals to increase in diversity of size and species. Today, biologists have identified about 5,500 species of mammals, from fishlike dolphins to birdlike bats, and from small insectivores such as shrews to large herbivores such as giraffes and elephants. The range of sizes and body forms of mammals is unmatched by any other living vertebrate group, and mammals are prime illustrations of the concept that organismal diversity is related to environmental diversity. In this section, we will outline the features that distinguish mammals from other taxa. Chapter 22 also considers the evolution of humans and other primates in greater detail (refer back to Section 22.3).

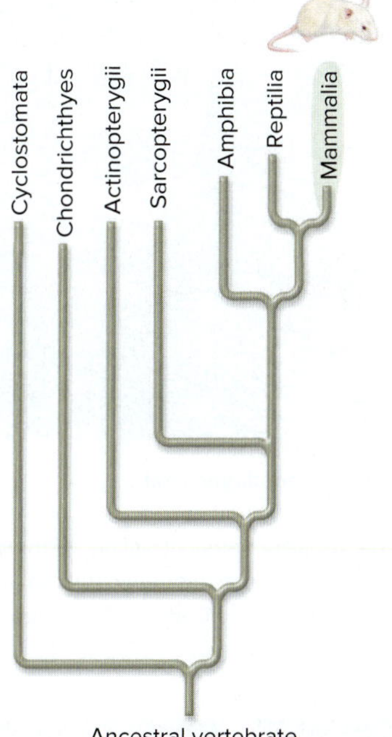


### Mammals Have Mammary Glands, Hair, Specialized Teeth, and an Enlarged Skull

Four characteristics distinguish mammals: the possession of mammary glands, hair, specialized teeth, and an enlarged skull.

- **Mammary glands.** Mammals, or the clade Mammalia (from the Latin *mamma*, meaning breast), are named after the female's distinctive **mammary glands,** which secrete milk. Newborn mammals suckle this fluid, which helps promote rapid growth.
- **Hair.** All mammals have hair, although some have more than others. Whales have hair in utero, but adults are hairless or retain only a few hairs on their snout. Mammals are endothermic, and their fur is an efficient insulator. Hair can also take on functions other than insulation. Many mammals, including cats, dogs, walruses, and whales, have sensory hairs called vibrissae (**Figure 27.17a**). Hair can be of many colors, to allow the mammals to blend into their background (**Figure 27.17b**). In some cases, as in porcupines and hedgehogs, the hairs become long, stiffened, and sharp (quills) and serve as a defense mechanism (**Figure 27.17c**).
- **Specialized teeth.** Mammals are the only vertebrates with highly differentiated teeth—incisors, canines, premolars, and molars—that are adapted for different types of diets (**Figure 27.18**).
- **Enlarged skull.** The mammalian skull differs from other amniote skulls in several ways. First, the brain is enlarged and is contained within a relatively large skull. Second, mammals have a single lower jawbone, unlike reptiles, whose lower jaw is composed of multiple bones. Third, mammals have three bones in the middle ear, as opposed to reptiles, which have one bone in the middle ear.

(a) Sensory hairs

(b) Camouflaged coat

(c) Defensive quills

**Figure 27.17** **Mammalian hair.** **(a)** The sensory hairs (vibrissae) of the walrus (*Odobenus rosmarus*). **(b)** The camouflaged coat of a bobcat (*Lynx rufus*). **(c)** The defensive quills of the crested porcupine (*Hystrix africaeaustralis*).
*(a)* © Mark Carwardine/Getty Images; *(b)* © Charles Krebs/Corbis; *(c)* © Anthony Bannister/Science Source

(a) Biting teeth

(b) Grinding teeth

(c) Gnawing teeth

(d) Tusks

(e) Grasping teeth

**Figure 27.18** **Mammalian teeth.** Mammals have different types of teeth, according to their diet. **(a)** The wolf has long canine teeth that bite and tear its prey. **(b)** The deer has a long row of molars that grind plant material. **(c)** The beaver, a rodent, has long, continually growing incisors used to gnaw wood. **(d)** The elephant's incisors are modified into tusks for defense and combat. **(e)** Dolphins and other fish or plankton feeders have numerous small teeth used to grasp prey.
*(a)* © Image Source/Corbis RF; *(b)* © Sam Camp/Getty Images RF; *(c)* © Mauritius Images GmbH/Alamy; *(d)* © DLILLC/Corbis RF; *(e)* © Jim Watt/Getty Images

## Mammals Are Morphologically Diverse

Modern mammals are incredibly diverse in size and lifestyle (Table 27.2). They vary in size from tiny insect-eating bats, weighing in at only 2 g, to leviathans such as the blue whale, the largest animal ever known, which tips the scales at 100 tonnes (over 200,000 lb). Mammalian orders are divided into two distinct subclasses. The subclass Prototheria contains only the order Monotremata, or **monotremes,** which are found in Australia and New Guinea. There are only five species of monotremes: the duck-billed platypus (Figure 27.19a) and four species of echidna, a spiny animal resembling a hedgehog. Monotremes are early-diverging mammals that lay eggs rather than bear live young, lack a placenta, and have mammary glands with poorly developed nipples. The mothers incubate the eggs, and upon hatching, the young simply lap up the milk as it oozes onto the fur.

The subclass Theria, which contains all remaining mammals, is divided into two clades, the Metatheria and the Eutheria. The clade Metatheria, or the **marsupials,** is a group of seven orders, with about

**Table 27.2 The Main Orders of Mammals, in Order of Species Richness**

| Order | Examples (approx. # of species) | Main characteristics |
|---|---|---|
| Rodentia | Mice, rats, squirrels, beavers, porcupines (2,337) | Plant eaters; gnawing habit, with two pairs of continually growing incisor teeth |
| Chiroptera | Bats (1,171) | Insect or fruit eaters; small; have ability to fly; navigate by sonar; nocturnal |
| Eulipotyphla | Shrews, moles, hedgehogs (452) | Insect eaters; primitive placental mammals |
| Primates | Monkeys, apes, humans (404) | Opposable thumb; binocular vision; large brains |
| Carnivora | Cats, dogs, weasels, bears, seals, sea lions (286) | Flesh-eating mammals; canine teeth |
| Artiodactyla | Deer, antelopes, cattle, sheep, goats, camels, pigs (240) | Herbivorous hoofed mammals, usually with two toes, hippopotamus and others with four toes; many with horns or antlers |
| Diprotodontia | Kangaroos, koalas, opossums, wombats (143) | Marsupials mainly found in Australia |
| Lagomorpha | Rabbits, hares (92) | Powerful hind legs; rodent-like teeth |
| Cetacea | Whales, dolphins (84) | Marine fishes or plankton feeders; front limbs modified into flippers; no hind limbs; little hair except on snout |
| Perissodactyla | Horses, zebras, tapirs, rhinoceroses (18) | Hoofed herbivorous mammals with odd-numbered toes, one (horses) or three (rhinoceroses) |
| Monotremata | Duck-billed platypuses, echidna (5) | Egg-laying mammals found only in Australia and New Guinea |
| Proboscidea | Elephants (3) | Long trunk; large, upper incisors modified as tusks |

280 species, including the rock wallaby pictured in Figure 27.19b. Once widespread, members of this order are now largely confined to Australia, although some marsupials exist in South America, and one species—the opossum—is found in North America. Fertilization is internal, and reproduction is viviparous in marsupials. Marsupials have a relatively simple, short-lived placenta that nourishes the embryo. Unlike other mammals, however, marsupials are extremely small when they are born (often only 1–2 cm) and make their way to a ventral pouch called a marsupium for further development.

All the other mammalian orders are members of the clade Eutheria and are considered **eutherians,** or placental mammals, such as the orangutans shown in Figure 27.19c. Eutherians have a long-lived and complex placenta, compared with that of marsupials. In eutherians, fertilization is internal, and reproduction is viviparous, but the developmental period, or gestation, of the fetus is prolonged.

## 27.5 Reviewing the Concepts

- The distinguishing characteristics of mammals are mammary glands, hair, specialized teeth, and an enlarged skull (Figures 27.17, 27.18).
- Two subclasses of mammals exist: the Prototheria (monotremes) and the Theria (the live-bearing mammals). The live-bearing mammals consist of the Metatheria (marsupials) and Eutheria (placental mammals) (Table 27.2; Figure 27.19).

**Figure 27.19 Diversity among mammals.** **(a)** Prototherians, such as this duck-billed platypus (*Ornithorhynchus anatinus*), lay eggs, lack a placenta, and possess mammary glands with poorly developed nipples. **(b)** Metatherians, or marsupials, such as this rock wallaby (*Petrogale assimilis*), feed and carry their developing young, or "joeys," in a ventral pouch. **(c)** Gestation lasts longer in eutherians, and their young are more developed at birth, as illustrated by this gray langur (*Presbytis entellus*).

*(a)* © Dave Watts/naturepl.com; *(b)* © Nigel Pavitt/Getty Images; *(c)* © Education Images/Getty Images

**(a) Prototherian (duck-billed platypus)**

**(b) Metatherian (rock wallaby)**

**(c) Eutherian (gray langur)**

## 27.5 Testing Your Knowledge

1. Mammals can be distinguished from other vertebrate classes because of the possession of
   - a. amniotic eggs.
   - b. mammary glands.
   - c. warmbloodedness.
   - d. reduction of organs.
   - e. thoracic breathing.

# Assess and Discuss

## Test Yourself

1. To which of the following clades does a jawed fish belong?
   - a. gnathostomes
   - b. lobe fins
   - c. tetrapods
   - d. amniotes
   - e. cyclostomes
2. The presence of a bony skeleton, an operculum, and a swim bladder are all defining characteristics of
   - a. class Myxini.
   - b. lampreys.
   - c. class Chondrichthyes.
   - d. bony fishes.
   - e. amphibians.
3. Organisms that lay eggs are said to be
   - a. oviparous.
   - b. ovoviparous.
   - c. viviparous.
   - d. placental.
   - e. none of the above.
4. Which clade does *not* include frogs?
   - a. vertebrates
   - b. gnathostomes
   - c. tetrapods
   - d. amniotes
   - e. lobe fins
5. The earliest-diverging tetrapods were
   - a. coelacanths.
   - b. lungfish.
   - c. amphibians.
   - d. reptiles.
   - e. mammals.

6. The outermost membrane of the amniotic egg is the
   a. amnion.
   b. yolk sac.
   c. allantois.
   d. chorion.
   e. albumin.
7. Which characteristic qualifies lizards as gnathostomes?
   a. a cranium
   b. a skeleton of bone or cartilage
   c. a hinged jaw
   d. the possession of limbs
   e. amniotic eggs
8. Which clade is *not* considered part of the reptiles?
   a. turtles
   b. crocodiles
   c. birds
   d. mammals
   e. both c and d
9. Which of the following is *not* a distinguishing characteristic of birds?
   a. amniotic egg
   b. feathers
   c. air sacs
   d. lack of certain organs
   e. lightweight skeleton
10. The mammalian subclass Prototheria contains which order(s)?
   a. Diprodantia
   b. Perissodactyla
   c. Monotremata
   d. a and b
   e. b and c

## Conceptual Questions

1. How is vertebrate movement similar to arthropod movement, and how is it different?
2. Why aren't all reptiles endothermic if both birds and mammals are?
3. **PRINCIPLES** A principle of biology is that all species (past and present) are related by an evolutionary history. Are birds living dinosaurs?

## Collaborative Questions

1. By what means can vertebrates move?
2. Why are amphibians considered good indicator species, which are species whose status provides information on the overall health of an ecosystem?

# UNIT VI
# FLOWERING PLANTS

**Flowering plants,** also known as the angiosperms, are essential to the lives of humans and most other organisms on Earth. Flowering plants provide most of our food, either directly in the form of vegetables and fruits (which include grains), or indirectly as animal fodder. Cotton, linen, and other fibers that we use for clothing and wood that we use for construction and fuel—as well as powerful cancer drugs and many other medicines—come from the flowering plants. Unit VI reveals molecular, biochemical, structural, evolutionary, and ecological features of the hundreds of thousands of flowering plants that support Earth's life.

Unit VI begins with Chapter 28, which provides an overview of flowering plant structure and function, focusing on the seed-to-seed life cycle. By comparing plant bodies with those of animals, you will learn how plants are constructed and how they grow. Building on this background, Chapter 29 explains the genetic and physiological bases of plant responses to external stimuli, such as day length, that are mediated by internally produced hormones. In this chapter, you will learn that plants, like animals, have evolved sophisticated sensory systems that monitor environmental conditions and allow plants to respond in predictable ways. Chapter 30 explains the nutritional requirements of plants and how plants transport materials, a deep understanding of which is critical to human agriculture and our ability to feed increasing populations of humans. Chapter 31 focuses on the molecular and cellular bases of plant reproduction, the process that generates seeds and fruits. This chapter ties together key concepts presented throughout Unit VI: the seed-to-seed life cycle of plants and their distinctively structured bodies, plant development and growth in response to environmental and hormonal influences, and plant acquisition and transport of materials that support growth and reproduction.

***The following biology principles will be emphasized in this unit:***

- ***Living organisms use energy.*** *Though sunlight is the major source of energy for photosynthetic plants, hundreds of flowering plant species as well as diverse tissues occurring in all flowering plants are heterotrophic and thus require a supply of organic food as a source of energy.*
- ***Living organisms interact with their environment.*** *Some of the ways in which plants detect and respond to light and other environmental factors are similar to those operating in microbes or animals, but others are distinctive.*
- ***Living organisms maintain homeostasis.*** *As is the case for animals, it is essential for flowering plants to maintain body water content and energy balance within tolerance limits, or they will die.*
- ***Living organisms grow and develop.*** *Some of the cellular and molecular bases of plant growth and development are also features of microbes and animals, but plants display some unique modes of growth and development.*
- ***Structure determines function.*** *Differences in the structures of animal and plant bodies—and structural differences among plants—correspond with differences in function.*

28

© John Swithinbank/agefotostock

© Linda Graham

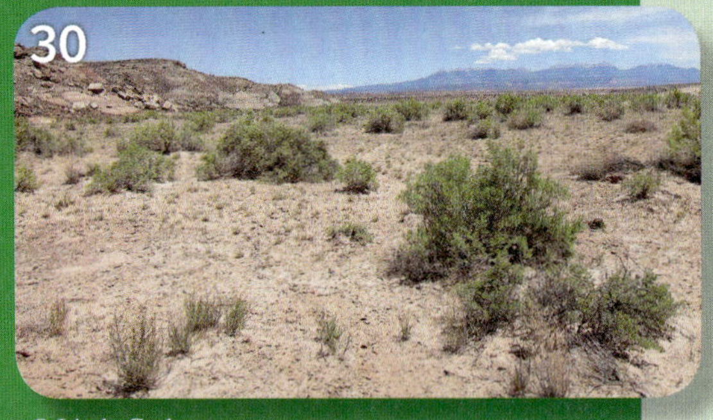


© Linda Graham

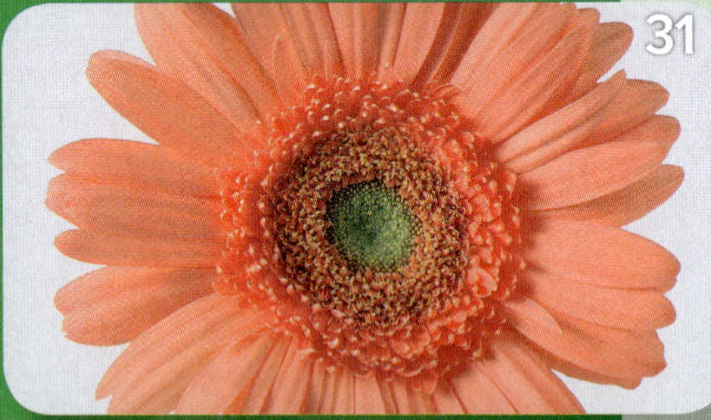


© Burke/Triolo Productions/Getty Images RF

# 28 An Introduction to Flowering Plant Form and Function

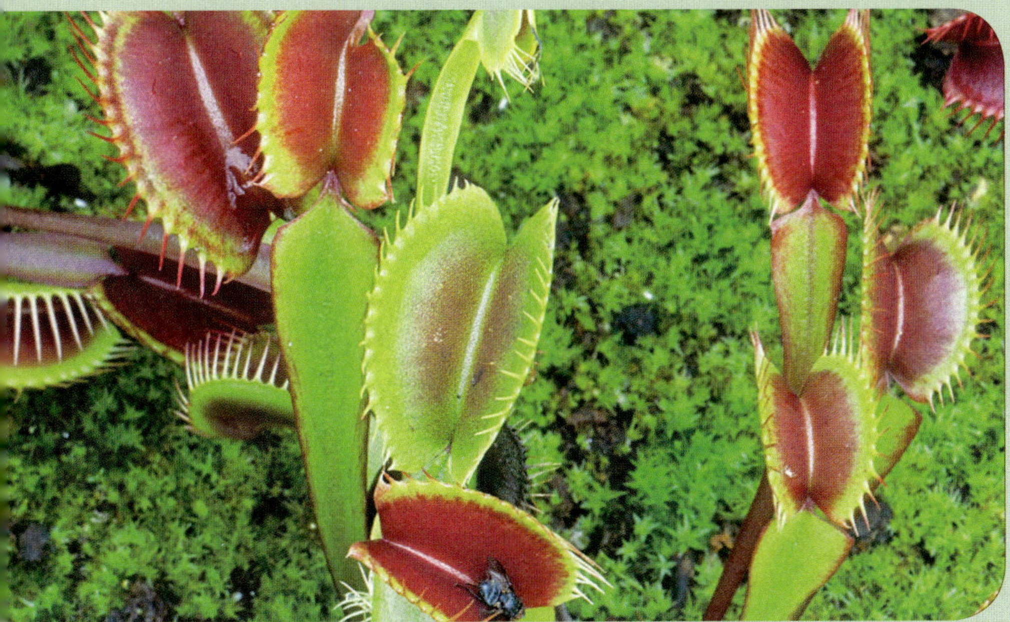

© John Swithinbank/agefotostock

**Venus flytrap.** This carnivorous plant has distinctive leaves that can snap together, trapping flies and absorbing essential nutrients from their decomposing bodies.

## Chapter Outline



Many types of fascinating carnivorous (meat-eating) flowering plants occur abundantly in wetlands around the world, even though the soils in these places are often infertile. Using modified leaves that snap shut to form jail cells (see the chapter-opening photo) or that have other unusual shapes, carnivorous plants trap insects and other small animals. As the trapped animals die, the plant leaves absorb essential nutrients released from the animal bodies, allowing plant growth and reproduction by flowers and seeds. The example of carnivorous plants reveals that plant leaves are not just photosynthetic organs but can have additional functions. Many other plants produce leaves, stems, or roots whose distinctive structures likewise foster survival under challenging circumstances.

Chapter 25 provided an overview of modern plant evolution and diversity, which includes bryophytes, lycophytes, ferns, gymnosperms, and angiosperms—the flowering plants. The next four chapters focus on the flowering plants, the most species-rich modern plant group and the one most important in human agriculture. This chapter provides an overview of the principles of plant growth, development, and reproduction. These principles are important to plant biologists because they underlie the ability of humans to produce agricultural crops and understand how plants function in nature. We begin by considering the life of a flowering plant, from seed germination to seedling growth into mature plants that are capable of producing the next generation.

## 28.1 From Seed to Seed: The Life of a Flowering Plant

### Learning Outcomes

1. List ways in which flowering plant reproduction and growth differ from those of animals.
2. Describe how the two major types of flowering plants—monocots and eudicots—differ in form.
3. Distinguish among annual, biennial, and perennial plants.

Several major events punctuate the lives of flowering plants, also known as the **angiosperms.**

- When seeds germinate, dormant embryos wake to metabolic activity and begin the process of seedling development.
- Seedlings grow and develop into mature sporophyte bodies capable of reproduction. Growth is an increase in weight or size, and development is a series of changes in the state of a cell, tissue, organ, or body.
- As they develop, the mature sporophytes of all flowering plants display some architectural features in common, though plant species vary widely in organ and body structure and life span.
- Flowers produce gametophytes that engage in sexual reproduction, and fruits disperse the next generation of seeds.

### Seedlings Develop from Embryos in Seeds

**Seeds** are reproductive structures produced by flowering plants and other seed plants, usually as the result of sexual reproduction. Seeds allow plants to survive periods unsuitable for growth, remaining in a dormant condition until environmental conditions improve. Seeds contain embryos that develop into young plants—seedlings—when the seeds germinate. Embryos and seedlings are young **sporophytes,** the larger, spore-producing generation of flowering plants (**Figure 28.1**). By contrast, a microscopic alternating life phase known as the **gametophyte,** which occurs in flowers, has the function of sexual reproduction.

In addition to an embryo, plant seeds also contain a supply of stored food. The embryo and stored food are enclosed by a tough, protective seed coat (**Figure 28.2a**). The seed coat protects the delicate plant embryo during the dispersal of seeds from parent plants into the environment. Dispersed seeds may remain dormant in the soil—sometimes for long periods—but germinate when temperature, moisture, and light conditions are favorable. Such conditions activate embryo metabolism, in which stored food is respired for energy needed for growth and development. Enlarging plant embryos break the seed coat and grow into seedlings (**Figure 28.2b**). If sufficient resources such as water and minerals are available, and if they are able to avoid being killed by disease organisms or being eaten, seedlings develop into mature plants (**Figure 28.2c**). The seedling stage is often the most vulnerable phase of a plant's life.

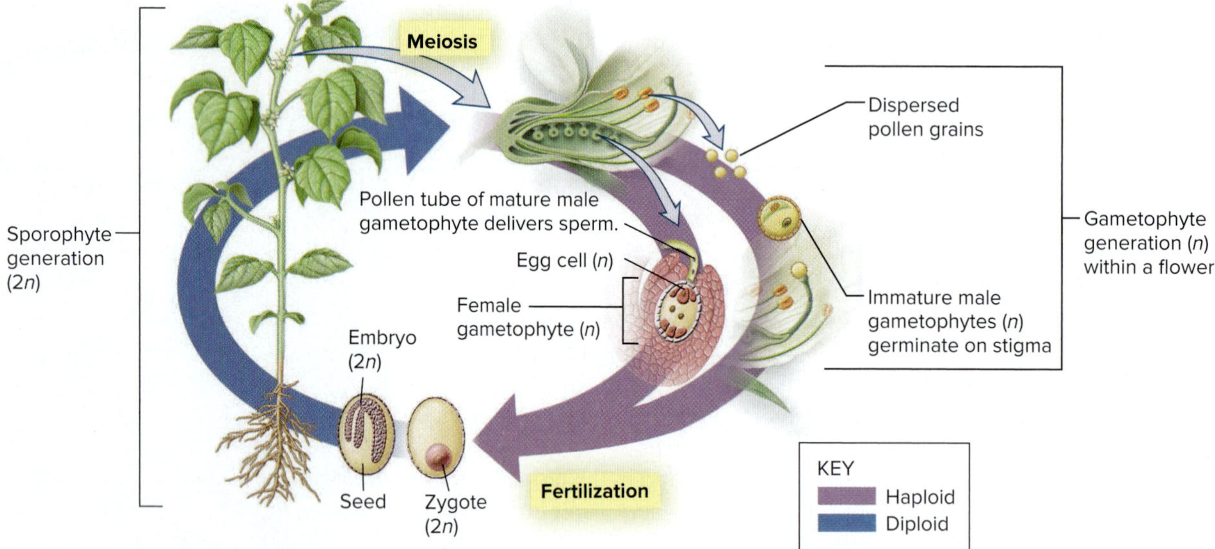


**Figure 28.1 The plant sexual cycle.** The sexual cycle of all flowering plants involves alternation of sporophyte and gametophyte generations. In flowering plants, the sporophyte is the dominant, conspicuous generation, whereas the tiny gametophytes are mostly hidden within flowers. The bean plant shown here produces both male and female gametophytes. Embryos form when eggs produced by female gametophytes are fertilized by sperm produced by male gametophytes. Embryos, which are the young sporophytes of the next generation, are formed within seeds and dispersed from plants.

**BioConnections:** *Refer back to Figure 25.4 to see the life cycle of a typical bryophyte, Figure 25.7 to see the life cycle of a typical fern, Figure 25.12 to see the life cycle of a typical gymnosperm, and Figure 25.15 to see an angiosperm life cycle. How does the life cycle of the flowering plant differ from that of the moss and the fern?*

## Biology Principle

### Living Organisms Grow and Develop

The plant *Arabidopsis thaliana,* shown here, is widely used as a model organism for understanding the genetics of plant structure and development.

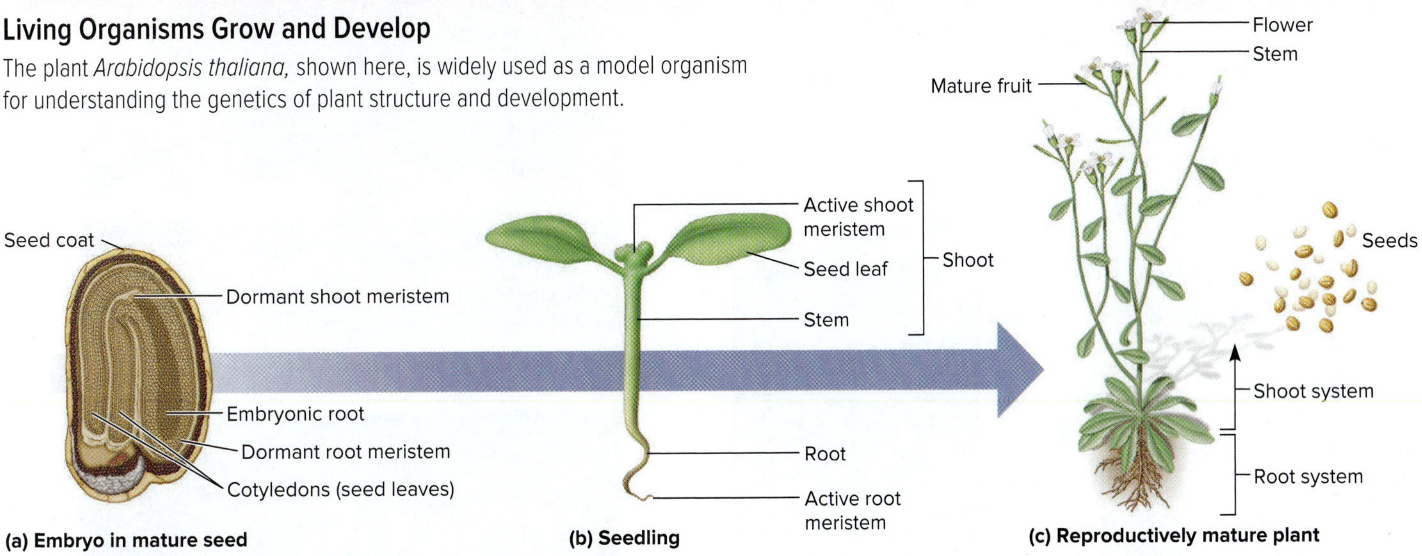

**Figure 28.2** **The seed-to-seed life of flowering plants.** **(a)** Seed embryos possess embryonic leaves, known as cotyledons; a dormant shoot meristem; an embryonic root; and a dormant root meristem. **(b)** When seeds germinate, the shoot and root meristems become active. Meristem activity allows the embryonic root to produce the seedling root and the young shoot of the seedling to grow and produce leaves. **(c)** Reproductively mature plants have branched shoot and root systems and bear flowers and fruits that disperse seeds.

The angiosperm plant body is composed of only three types of organs: stems, leaves, and roots. **Stems** produce leaves and branches and bear the reproductive structures of mature plants. **Leaves** are flattened structures that emerge from stems and are often specialized in ways that enable photosynthesis. Stems and leaves together make up the plant **shoot** (see Figure 28.2b). Mature plants often possess multiple stems bearing many leaves, which together form the shoot system. **Roots** provide anchorage in the soil and foster efficient uptake of water and minerals. The aggregate of a plant's roots make up the root system (see Figure 28.2c).

Plant seedlings and mature plants produce new tissues in specific areas called meristems. A **meristem** (from the Greek *merizein,* meaning to divide) is a region of undifferentiated cells that produces new tissues by cell division. A dormant meristem occurs at the shoot and root tips of seed embryos, and these meristems become active in seedlings (see Figure 28.2a,b). In mature plants, active meristems occur at each stem and root tip. Such meristems are known as shoot and root apical meristems because shoot and root tips are called apices. The term shoot apical meristem is commonly known by the acronym SAM, and the root apical meristem is the RAM.

## Mature Sporophytes Develop from Seedlings

As seedlings develop into mature sporophytes, the aboveground shoot typically becomes green and photosynthetic and thus able to produce organic food. Photosynthesis powers the transformation of seedlings into mature plants. The development of mature plants encompasses both vegetative growth, a process that increases the size of the shoot and root systems, and reproductive development. Vegetative growth and reproductive development involve **organ systems,** structures that are composed of more than one organ. Branches, buds, flowers, seeds, and fruits are organ systems, analogous to organ systems such as the circulatory, skeletal, and reproductive systems in the animal body. The hierarchy of structure in a mature plant, ranging from specialized cells, tissues, organs, and organ systems to root and shoot systems, is shown in **Figure 28.3**.

**Vegetative Growth** **Vegetative growth** is the growth of nonreproductive parts of the plant body. During their growth, plant shoots produce **buds**—miniature shoots—each having a dormant shoot apical meristem. A bud is an example of an organ system because it contains more than one organ. Scaly modified leaves protect the bud contents. Under favorable conditions, the bud scales fall off, and the buds open. Newly opened vegetative buds display young leaves on a short shoot. The shoot apical meristem then becomes active, producing new stem tissue and leaves. In this way, buds generate leafy branches that contribute to vegetative growth. By contrast, during reproduction, floral buds are produced; these open to produce flowers, modified shoots that do not contribute to vegetative growth.

Vegetative shoots often display **indeterminate growth,** meaning that apical meristems continuously produce new stem tissues and leaves as long as conditions remain favorable. This process explains how very large plants, such as trees, can develop from seedlings (**Figure 28.4a**). However, plant size is also under genetic control, so some plants remain small even when they are mature. The tiny, floating plants of *Lemna* species, commonly known as duckweeds, which sometimes cover the surfaces of ponds in summer, are examples of plants whose

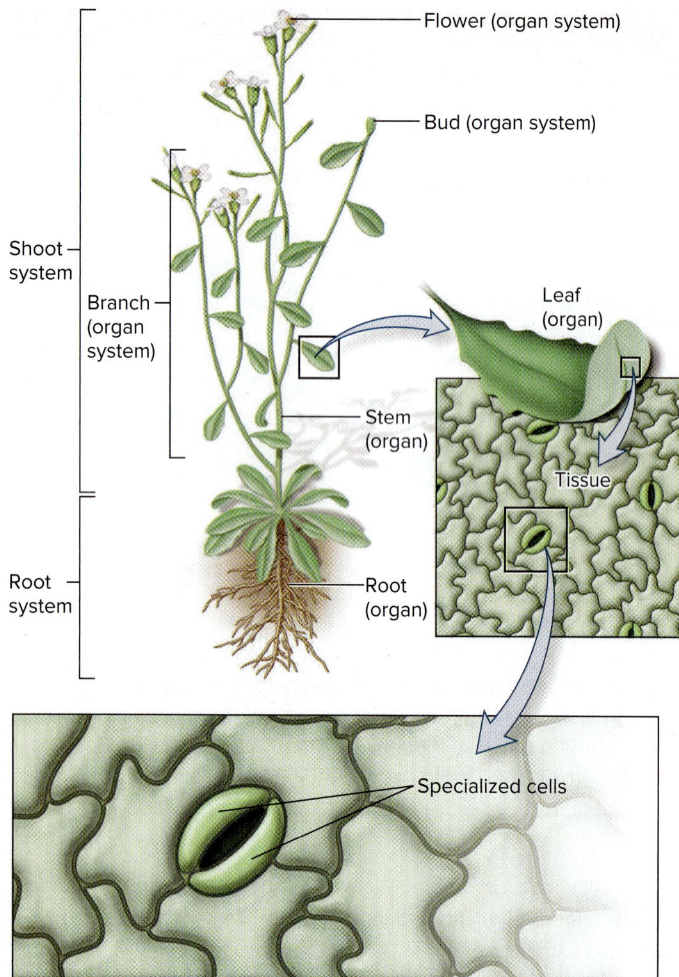


**Figure 28.3** **Levels of biological organization in a plant.** Flowering plant sporophyte bodies consist of a root system and a shoot system. Shoot systems produce organ systems such as buds, flowers, fruits, and seeds, which are composed of organs, tissues, and specialized cells. Root systems are composed of organs, tissues, and specialized cells.

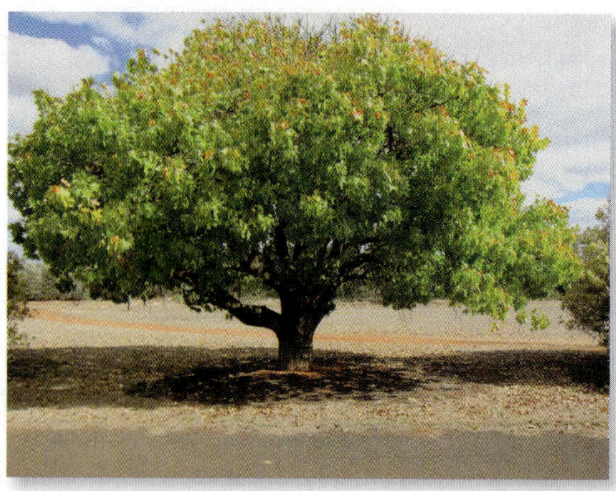

**(a)** ***Brachychiton*****, a tree native to Australia**

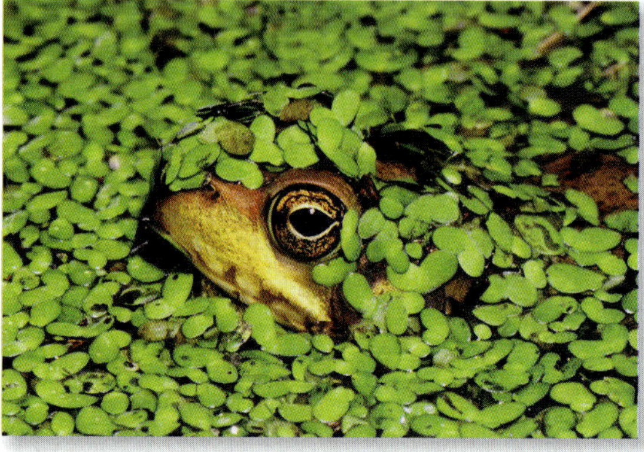

**(b)** ***Lemna*** **(common duckweed)**

**Figure 28.4** **Plants display indeterminate growth but vary in size.** **(a)** A large, woody angiosperm has the same indeterminate growth pattern as **(b)** tiny, floating duckweed plants, each smaller than the eye of the frog hiding in the water beneath.
*(a)* © Linda Graham; *(b)* © David Davis/Science Source

small size is genetically determined (Figure 28.4b). Indeterminate growth allows plants to adapt their vegetative body structure to environmental conditions. By contrast, animal bodies and flowers display **determinate growth,** which is growth of limited duration.

**Reproductive Development** Under favorable conditions, mature plants produce reproductive structures: flowers, seeds, and fruits. **Flowers** and floral buds are reproductive shoots that develop when shoot apical meristems produce flower parts instead of new stem tissues and leaves. Flower development occurs under the control of several genes whose roles are well understood. The gene *LEAFY*, for example, acts in the SAM to flip the molecular switch from vegetative to floral development, and interactions of *APETALA*, *PISTILLATA*, *AGAMOUS*, and *SEPALLATA* genes control the development of flower parts.

In contrast to shoots, which often show indeterminate growth, flowers are produced by determinate growth. A floral shoot no longer produces new stem growth or leaves. Therefore, vegetative growth and reproductive development are alternative processes. In order to flower, a plant must give up some of its potential to continue vegetative growth.

Flower tissues enclose and protect tiny male and female gametophytes during their growth and development (see Figure 28.1). Female gametophytes produce eggs within structures known as ovules, produced in the ovary of a flower pistil (refer back to Figure 25.15). Male gametophytes begin their development within pollen grains produced in the anthers of a flower stamen. Pollen is dispersed to the flower pistil, whereupon pollen grains may germinate, producing a tube that delivers sperm to eggs. Fertilization generates zygotes, which develop into embryos, and triggers the process by which ovules develop into seeds and flower parts develop into fruits. **Fruits** thus enclose seeds and function in seed dispersal. Flowers and fruits are features that distinguish flowering plants from other plant phyla (see Chapter 25).

## Flowering Plants Vary in Structure and Life Span

With some exceptions, flowering plants occur in two groups, informally known as the **eudicots** and the **monocots.** These groups take their names from the number of seed leaves (cotyledons) that are present on seed embryos. For example, bean plants and relatives, which possess two (*di*) seed leaves, are examples of eudicots. Most woody trees, shrubs, and vines are also eudicots. Corn, which has only one (*mono*) seed leaf, is an example of a monocot, as is tiny duckweed (see Figure 28.4b). Eudicots and monocots also vary in the structure of other organs and organ systems. For example, eudicot flowers typically have petals and other parts numbering four, five, or a multiple of those numbers, whereas monocot flower parts usually occur in threes or a multiple of three. Stems, roots, leaves, and pollen of eudicots and monocots also vary in distinctive ways, as shown in **Table 28.1.**

The life span of a flowering plant can vary from a few weeks to many years.

- Plants that die after producing seed during their first year of life are known as **annuals.** Corn and the common bean are examples of annual crops whose nutrient-rich seeds are harvested within a few months after planting and must be replanted at the beginning of each new growing season.
- Plants that do not reproduce during the first year of life but may reproduce within the following year are known as **biennials.** Such plants often store food in fleshy roots during the first year of growth, and this food fuels reproduction during the second or later year of life. Humans use some of these fleshy roots for food, including carrots, parsnips, and sugar beets.
- Trees are examples of **perennials,** plants that live for more than 2 years, often producing seed each year after they reach reproductive maturity.

**Table 28.1 Distinguishing Features of Eudicots and Monocots, Two Major Groups of Flowering Plants**

| Feature | Eudicots | Monocots |
|---|---|---|
| Number of seed leaves (cotyledons) | Two | One |
| Number of flower parts | Usually four, five, or multiples of these | Usually three or multiple of three |
| Stem vascular bundles | Arranged in a ring | Scattered |
| Root system | Branched taproot | Fibrous; adventitious |
| Leaf venation | Netted or branched | Often parallel |
| Pollen | Three pores or slits | One pore or slit |

## 28.1 Reviewing the Concepts

- Seed embryos, seedlings, and mature plants are components of the sporophyte generation in the flowering plant sexual cycle; tiny gametophytes develop and grow within flowers (Figure 28.1).
- Plant organs are composed of tissues that contain specialized cells. The basic plant organs are stems, leaves, and roots. Shoot systems include stems and stem branches, and stems produce leaves, buds, flowers, and fruits. Root systems include one or more main roots with branches. Buds, flowers, fruits, and seeds are organ systems, composed of more than one organ (Figures 28.2, 28.3).
- The two major groups of flowering plants—eudicots and monocots—differ in the structure of their seed embryos, flowers, stems, roots, leaves, and pollen. Plants can be categorized according to their life span as annuals, biennials, and perennials (Figure 28.4; Table 28.1).

## 28.1 Testing Your Knowledge

1. Which of the following statements represents plant vegetative growth rather than plant reproduction?
   a. A plant produces new root tissue from a root apical meristem.
   b. A plant produces new shoot tissue from a shoot apical meristem.
   c. A vegetative bud opens to generate a new leafy shoot.
   d. All of the above are correct.
   e. None of the above are correct.
2. How do annual plants differ from biennials and perennials?
   a. An annual plant has a seed-to-seed life span of a year or less, whereas the life spans of biennials and perennials are longer.
   b. An annual plant reproduces in the second year of its life span, whereas biennials and perennials reproduce in the first year of life.
   c. An annual plant reproduces every year of its life span, which can be several years, whereas biennials and perennials reproduce only in the second year of life.

d. An annual plant reproduces only in the final year of a multiyear life span, whereas biennials and perennials reproduce throughout their multiyear life spans.
e. None of the above are correct.

## 28.2 Plant Growth and Development

### Learning Outcomes

1. Explain how plant development is similar to and different from animal development.
2. Discuss how branching of the plant shoot system differs from branching in the root system.

As plants grow from seedlings, their development depends on four processes that are also essential to animal growth and development: cell division, growth, cell specialization, and programmed cell death (a process similar to apoptosis in animals). Additional and distinctive aspects of plant growth and development include

1. development and maintenance of a plant-specific architecture throughout life,
2. increase in length by the activity of apical meristems,
3. maintenance of a population of youthful stem cells in meristems, and
4. expansion of cells by water uptake and in controlled directions.

### Plants Display a Distinctive Architecture

In plant biology, the term apical has two distinct meanings. As we have seen, apical refers to the tips or apices of shoots and roots, as in shoot apical meristems or root apical meristems. A second meaning for apical is the part of a plant that typically projects upward, which is the top of the shoot. By contrast, the bottom of a root is termed the basal region. So the shoot apical meristem occurs at the apical pole, and the root apical meristem occurs at the basal pole. This property, known as **apical-basal polarity,** explains why plants produce shoots at their tops and roots at their lower regions.

Apical-basal polarity originates during embryo development. As seedlings and maturing plants grow in length by the activity of shoot and root meristems, apical-basal polarity is maintained (**Figure 28.5a**). Animals likewise have anterior and posterior ends, whose development is influenced by *Hox* genes. In contrast, plant apical-basal polarity is under the control of genes such as *GNOM*; mutations in such genes result in plant embryos that are cone-shaped or spherical and thus lack normal apical-basal architecture (**Figure 28.5b**).

A second architectural feature of the typical plant body is **radial symmetry.** Plant embryos normally display a cylindrical shape, also known as an axis, which is retained in the stems and roots of seedlings and mature plants. A thin slice or cross section of an embryo, a stem, or a root is typically circular in shape. Most plants produce new leaves or flower parts in circular whorls, or spirals, around shoot tips (**Figure 28.6**). Buds and branches likewise emerge from stems in radial patterns, as do lateral roots from a central root axis. Together, apical-basal polarity and radial symmetry explain why diverse plant species have a fundamentally similar architecture.

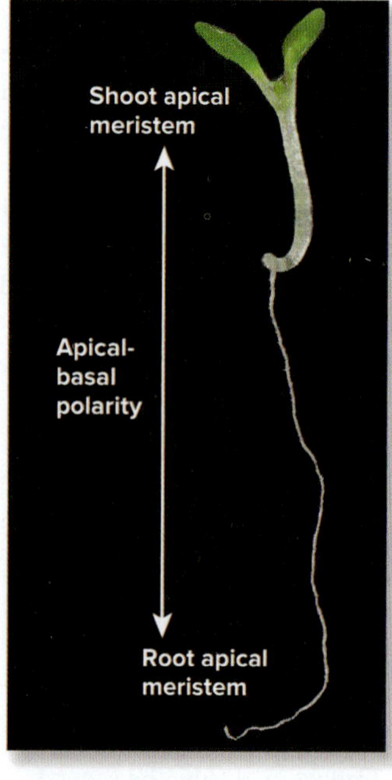


(a) Normal seedling

(b) Abnormal *GNOM* mutants

**Figure 28.5 Plant apical-basal polarity.** **(a)** Normal plants exhibit apical-basal polarity, as shown by this seedling. Such growth occurs at two meristems, one at the shoot (SAM) and one at the root (RAM). **(b)** *GNOM* mutants of *Arabidopsis thaliana* lack apical-basal polarity and thus produce abnormal embryos and seedlings.

*(a)* © James Mann/ABRC/CAPS; *(b)* © Prof. Dr. Gerd Jürgens/Universität Tübingen. Image Courtesy Hanno Wolters

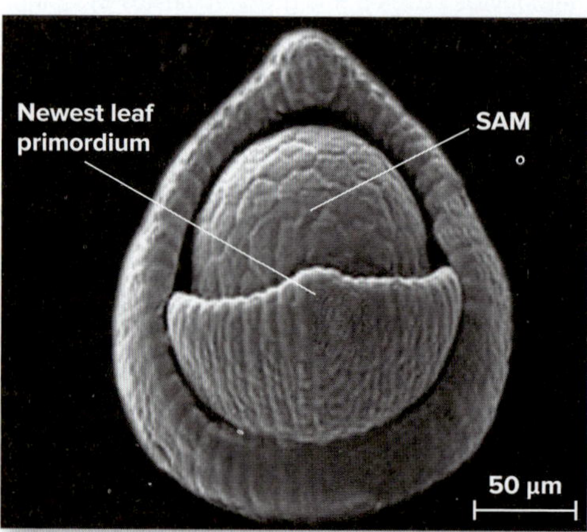


**Figure 28.6 Plant radial symmetry.** This top-down view of a shoot apical meristem reveals the radial symmetry of the shoot, illustrated by its cylindrical shape. Leaf primordia are produced in circles or spirals around the shoot axis.

Figure adapted from D. Jackson, and S. Hake (1999), "Control of Phylotaxy in Maize by the ABPHYL1 Gene," *Development,* 126: 315–323, Fig. 3A © The Company of Biologists Limited

Concept Check: *How could you determine that roots have radial symmetry?*

## Primary Meristems Increase Plant Length and Produce Plant Organs

Plant embryos grow into seedlings, and seedlings grow into mature plants by adding new cells from only two growth points: the shoot apical meristem (SAM) and the root apical meristem (RAM). During plant development, the embryonic SAM and RAM give rise to many more apical meristems that are located in the buds of shoots and at the tips of roots.

As a SAM grows, it generates additional **primary meristem** tissues that increase plant length and produce new organs. In the process known as **primary growth,** primary meristems ultimately produce **primary tissues** made of specialized cells, and these tissues occur in organs of diverse types (Table 28.2). The primary meristems of woody plants also give rise to secondary or lateral meristems. In a process known as **secondary growth** described in Section 28.3, the **secondary meristems** increase the girth of woody plant stems and roots by producing specialized secondary tissues (see Table 28.2). Plant cell specialization and tissue development depend on chemical influences such as hormones, transcription factor proteins, and microRNAs that move through plants.

**Table 28.2 Examples of Tissues and Specialized Cells Found in Flowering Plants***

| Primary Growth | |
|---|---|
| **Simple primary tissues (composed of one or two cell types)** | **Plant cell types found in those tissues** |
| Parenchyma | Parenchyma cells |
| Collenchyma | Collenchyma cells |
| Sclerenchyma | Fibers and sclereids |
| Root endodermis | Endodermal cells |
| Root pericycle | Pericycle cells |
| **Complex primary tissues (composed of at least two cell types)** | **Plant cell types found in those tissues** |
| Leaf or stem epidermis | Flattened epidermal cells, trichomes, stomatal guard cells |
| Root epidermis | Flattened epidermal cells, root hairs |
| Leaf mesophyll | Spongy parenchyma cells, palisade parenchyma cells |
| Leaf, stem, or root xylem | Tracheids, vessel elements, fibers, parenchyma cells |
| Leaf, stem, or root phloem | Sieve-tube elements, companion cells, fibers, parenchyma cells |
| **Secondary Growth** | |
| **Simple and complex secondary tissues** | **Plant cell types found in those tissues** |
| Secondary xylem (wood) | Tracheids, vessel elements, fibers, parenchyma |
| Secondary phloem (inner bark) | Sieve-tube elements, companion cells, fibers, parenchyma |
| Outer bark | Cork cells |

*This list does not include all of the tissues and cell types found in flowering plants.

**Primary Stem Structure and Development** New primary stem tissues arise by the cell division activities of SAMs.

- A layer of outermost tissue known as the **epidermis** develops at the stem surface. The epidermis produces a waxy surface coating known as the **cuticle,** which reduces water loss from the plant surface and protects plants from damage by ultraviolet (UV) light, animals, and disease microorganisms.
- Beneath the epidermis lies the stem **cortex,** which is largely composed of **parenchyma tissue** (Figure 28.7a). This tissue is composed of only one cell type, thin-walled cells known as **parenchyma cells.** These cells often store starch in plastids and therefore serve as an organic food reserve. Stem parenchyma also has the ability to undergo cell division, which aids wound healing when stems are damaged. Stems also contain **collenchyma tissue** (Figure 28.7b), composed of flexible cells, and rigid **sclerenchyma tissue** (Figure 28.7c), composed of tough-walled cells that provide strength and protection to the plant stem.
- New water- and food-conducting **primary vascular tissues** develop inside young shoots and roots of all vascular plants.

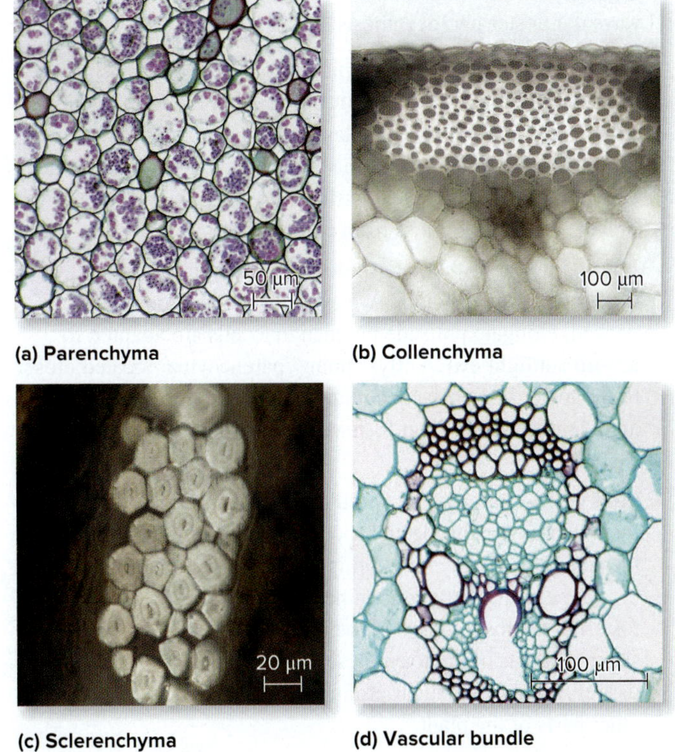


**Figure 28.7** **Examples of tissues produced by primary shoot meristems.** **(a)** Parenchyma, **(b)** collenchyma, **(c)** sclerenchyma, and **(d)** a vascular bundle composed of complex xylem and phloem tissues.
*(a–d)* © Lee W. Wilcox

Plant vascular tissues occur in two forms—**xylem** and **phloem—**which are each composed of several types of specialized tissues and cells. In plants, materials are transported directly from cell to cell, contrasting with extracellular transport, which occurs in animal circulatory systems. Newly formed stem xylem and phloem connect with older conducting tissues that extend throughout the stem and root systems. In contrast to the circulatory system of most animals, the plant conduction system is not closed but, instead, is open to the environment (described more completely in Chapter 30).

- Primary vascular tissues are typically arranged in elongate clusters known as **vascular bundles** that appear round or oval when cross-cut (Figure 28.7d). In the primary stems of beans and other eudicots, the vascular bundles are arranged in a ring, which is easily seen in thin slices made across a stem (see Table 28.1). By contrast, in the stems of corn and other monocots, the vascular bundles are scattered.

**Leaf Structure and Development** Young leaves are produced at the sides of a SAM as small bumps known as **leaf primordia** (see Figure 28.6). As young leaves develop, they acquire vascular tissue that is connected to the stem xylem and phloem (Figure 28.8a) and become flattened, a process that expands the area of leaf surface available for light collection during photosynthesis. Leaf thinness is an adaptation that helps to shed excess heat. Leaves also become bilaterally symmetric, meaning that they can be divided into two equal halves in only one direction, from the leaf tip to its base (Figure 28.8b).

Upper, lower, and internal leaf tissues develop differently in several ways that foster photosynthesis (Figure 28.8c).

- The more shaded lower leaf epidermis usually displays larger numbers of pores, known as **stomata** (from the Greek word *stoma*, meaning mouth; singular, stoma), than the sunnier upper leaf surface. When open, stomata allow $CO_2$ to enter and water vapor and $O_2$ to escape leaf tissues. Closure of stomata helps to prevent excess water loss from plant surfaces.
- Internal photosynthetic tissues of the leaf, known as **leaf mesophyll,** include two types of tissue. Upper, closely packed, elongate palisade parenchyma cells are adapted to absorb sunlight efficiently. Spongy parenchyma, located closer to the lower leaf surface, contains rounder cells separated by abundant air spaces that foster $CO_2$ absorption and $O_2$ release by leaves.
- **Leaf veins** composed of vascular tissue commonly occur at the junction of palisade and spongy parenchyma, or within the spongy parenchyma (see Figure 28.8c). The xylem tissues of veins conduct water and minerals throughout leaf tissues, fostering photosynthesis. Phloem tissues of leaf veins carry the sugar products of photosynthesis from leaf cells to stem vascular tissues. In this way, sugar produced in leaves can be exported to other parts of the plant.

**Root System Structure and Development** In beans and most other eudicots, a main root develops from the embryonic root and then produces branch roots, also known as lateral roots. The root system of eudicots, known as a **taproot system,** has one main root with many branch roots (see Table 28.1). In contrast, the embryonic root of most monocots dies soon after seed germination, and it is replaced by a **fibrous root system** consisting of multiple roots that grow from the stem base (see Table 28.1). Fibrous roots are examples of adventitious roots, structures that are produced on the surfaces of stems (and sometimes leaves) of both monocots and eudicots. Roots that develop at the bases of stem cuttings are also adventitious.

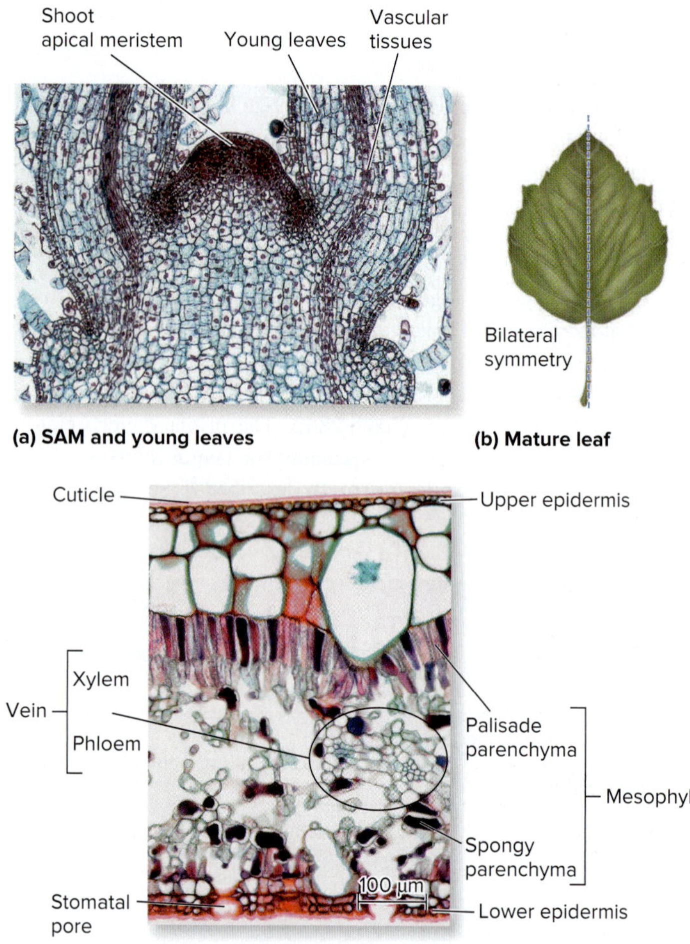


**Figure 28.8** **Leaf development and structure.** **(a)** Young leaves develop at the sides of SAMs, as shown in this thinly sliced, stained shoot tip. Note that the darkly stained vascular tissues of young leaves are connected to those of the stem. **(b)** Mature leaves are typically thin and flat and show bilateral symmetry. **(c)** An internal view of a thinly sliced and stained leaf reveals upper to lower surface tissue differentiation. A layer of palisade parenchyma lies just beneath the upper epidermis capped with a waxy cuticle. Veins of conducting tissue (xylem and phloem) are embedded in the photosynthetic mesophyll. Spongy parenchyma lies above the lower epidermis, which displays stomata. Stomatal pores can open or close. These structural features of mature foliage leaves facilitate photosynthesis.
*(a, c)* © Lee W. Wilcox

**Concept Check:** *What advantage do plant leaves obtain by having stomata on the lower epidermal surface?*

The tips of roots and their branches each possess an apical meristem that adds new cells. Expansion of these new cells allows roots to grow into the soil. As they lengthen, roots produce branches but not from buds, as is the case for stems. Instead, branch roots develop from meristematic tissues located within the root (see Section 28.4). The root system both anchors plants in the soil and plays an essential role in absorbing water and mineral nutrients. Root tissues are usually not green and photosynthetic. They must rely on organic compounds transported from the shoot. The plant root system and shoot system are therefore interdependent.

## Plant Meristems Contain Youthful Stem Cells

Plant meristems include undifferentiated cells referred to as **stem cells,** which function similarly to the stem cells of animals. When plant stem cells divide, they produce two cells: one that remains young and unspecialized plus another cell. This second cell may differentiate into various types of specialized cells, but it often retains the ability to divide. As a result of these properties, stem cell numbers influence the size of a meristem, which in turn affects plant growth. Normal function of plant stem cells depends on a plant gene that is closely related to the animal tumor-suppressor gene *Rb*. The protein encoded by this gene functions to regulate the cell cycle in both animals and plants. In animals, mutations in *Rb* cause retinoblastoma, a cancerous tumor of the eye; in plants, mutations cause meristematic cells to fail to produce differentiated cells and organs.

## Plant Cells Expand in a Controlled Way by Absorbing Water

Meristem production of new cells is essential to plant growth, but plant growth also requires cell expansion, a less important part of animal growth. Recall that plant cells typically possess a relatively large vacuole. Cell extension occurs when water enters the central vacuole by osmosis (Figure 28.9). As the central vacuole expands, the cell wall also expands and increases the cell's volume. By taking up water, plant cells can enlarge quickly, allowing rapid plant growth. Bamboo, for instance, can grow taller by 2 m within a week and can grow up to 30 m in less than 3 months! The importance of water uptake in cell expansion helps to explain why plant growth is so dependent on water supply.

Plant cell walls contain cellulose microfibrils that are held together by crosslinking polysaccharides. When plant cells and their vacuoles absorb water, pressure builds on cell walls. In response to this pressure and under acidic conditions, proteins unique to plants—known as expansins—are produced. **Expansins** unzip crosslinking cell-wall polysaccharides from cellulose microfibrils so that the cell wall can stretch (Figure 28.10). As a result, cells enlarge, often by elongating in a particular direction, which is important to plant form. Some plant cells are able to elongate up to 20 times their original length.

The direction in which a plant cell expands depends on the arrangement of cellulose microfibrils in its cell wall, which is in turn determined by the orientation of cytoplasmic microtubules. These microtubules are thought to influence the positions of cellulose-synthesizing protein complexes located in the plant plasma membrane. The protein complexes connect sugars to form cellulose

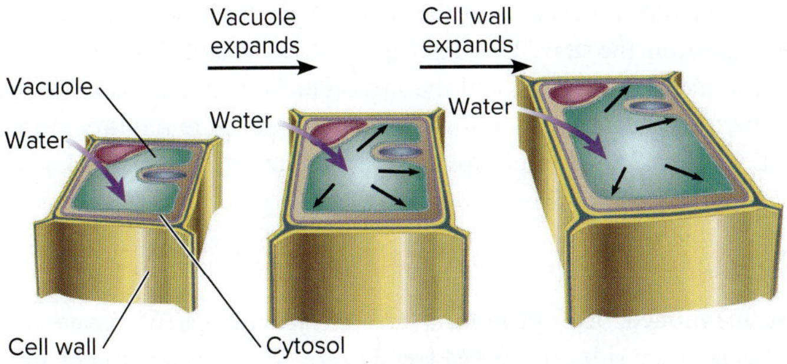

Vacuole expands → Cell wall expands →
Vacuole
Water
Water
Water
Cell wall
Cytosol

**Figure 28.9** **Plant cells expand by taking up water into their vacuoles.**

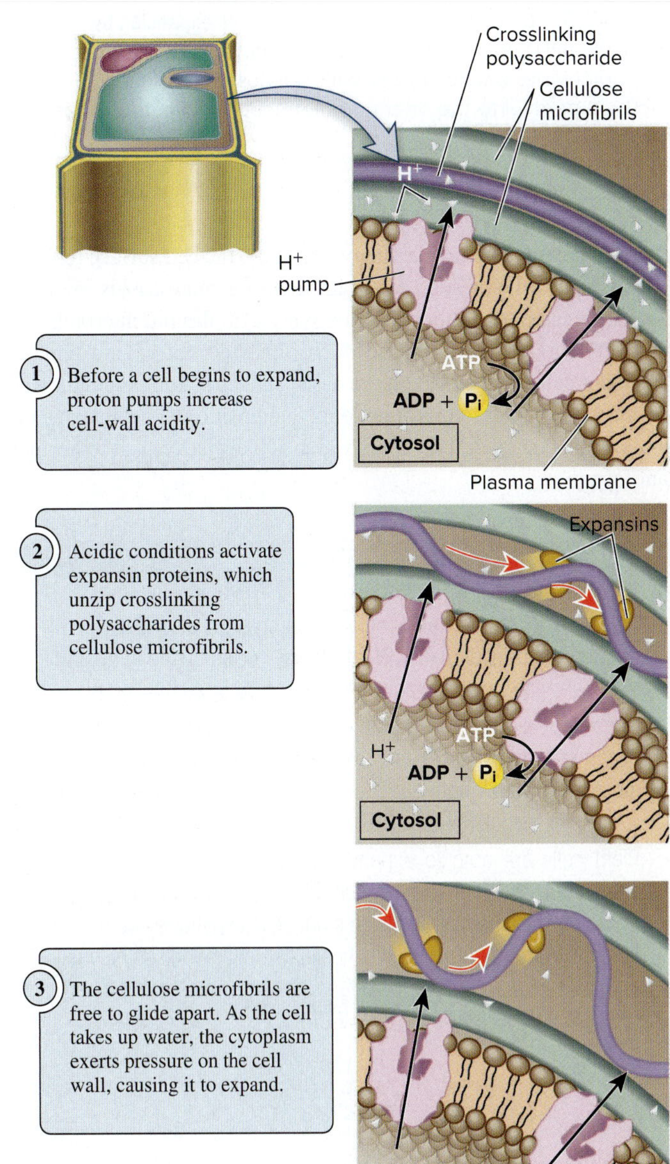


**Figure 28.10** **A hypothetical model of the process of cell-wall expansion.**

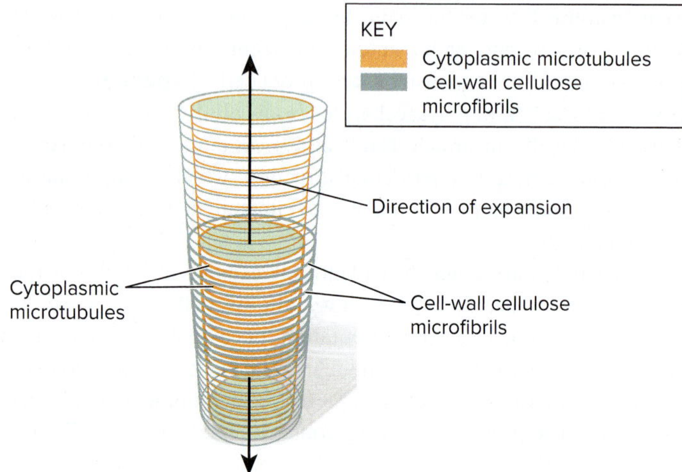


**Figure 28.11** **Control of the direction of plant cell expansion by microfibrils and microtubules.** Plant cells enlarge in the direction perpendicular to encircling cell-wall cellulose microfibrils, which run parallel to the orientation of underlying cytoplasmic microtubules.

polymers, spinning cellulose microfibrils onto the cell surface to form the cell wall. As a result, cell-wall cellulose microfibrils encircle cells in the same orientation as underlying cytoplasmic microtubules (**Figure 28.11**). Because cellulose microfibrils do not extend lengthwise, plant cell walls expand more easily in a direction perpendicular to them. Microtubules control not only the direction of cell expansion but also the plane of cell division, which is likewise critical to normal plant form.

## 28.2 Reviewing the Concepts

- Apical-basal polarity and radial symmetry are distinctive architectural features of plants. Plant growth occurs by the addition of new cells and by cell expansion (Figures 28.5, 28.6).
- Shoot apical meristems contain undifferentiated stem cells and produce primary meristems that generate new cells, thereby increasing plant length and producing tissues and organs: stems, roots, and leaves (Figures 28.7, 28.8); (Table 28.2).
- Plant cells are able to expand by water uptake into vacuoles and under conditions that result in loosening of cell-wall components. The direction in which plant cells expand is determined by the arrangement of cell-wall cellulose microfibrils, which is in turn influenced by the orientation of microtubules in the nearby cytoplasm (Figures 28.9, 28.10, 28.11).

## 28.2 Testing Your Knowledge

1. Which feature distinguishes plant growth and development from those of animals?
   a. an apical-basal polarity that is maintained throughout life
   b. increase in length by the activity of apical meristems
   c. maintenance of a population of undifferentiated stem cells within meristems
   d. expansion of cells by water uptake and in controlled directions
   e. all of the above are correct
2. How does a gene that is closely related to the animal *Rb* (*Retinoblastoma*) gene function in plants?
   a. The plant *Rb* gene controls the ability to detect and respond to light.
   b. The plant *Rb* gene is essential to normal cell cycle function of plant stem cells.
   c. Mutations in the plant *Rb* gene cause a form of plant cancer.
   d. All of the above are correct.
   e. None of the above are correct.

# 28.3 The Shoot System: Stem and Leaf Adaptations

### Learning Outcomes

1. Discuss why plant shoots are said to have a modular structure.
2. Explain why leaves having different shapes and vein patterns exist in nature.
3. Compare the structure and function of the conducting tissues xylem and phloem.
4. Create a drawing showing how bark and wood are produced in woody plants.

As we have seen, the shoot system includes all of a plant's stems, branches, leaves, and buds. It also produces flowers and fruits when the plant has reached reproductive maturity. Thus, the shoot system is essential to plant growth, photosynthesis, and reproduction.

## Shoot Systems Have a Modular Structure

More than 200 years ago, German writer, politician, and scientist Johann Wolfgang von Goethe realized that plants are modular organisms, composed of repeated units. Shoots are notably modular (**Figure 28.12**). Each shoot module consists of four parts: a stem node, an internode, a leaf, and an axillary bud. A **node** is the stem region from which one or more leaves emerge. An **internode** is the region of stem between adjacent nodes. Differences in numbers and lengths of internodes help to explain why plants differ in height (see Figure 28.4). Each time a young leaf is produced at the SAM, a new meristem develops in the upper angle formed where the leaf emerges from the stem. This angle is known as an axil (from the Greek *axilla*, meaning armpit), and the meristem formed there is called an axillary meristem. Such axillary meristems generate **axillary buds,** which bear a SAM at their tips and can produce leafy branches known as lateral shoots, or flowers.

## Hormones and MicroRNAs Influence Leaf Development

As mentioned, leaf primordia are surface bumps of tissue that develop at the sides of a SAM (see Figure 28.6). Production of such leaf primordia is under the control of a hormone known as **auxin.**

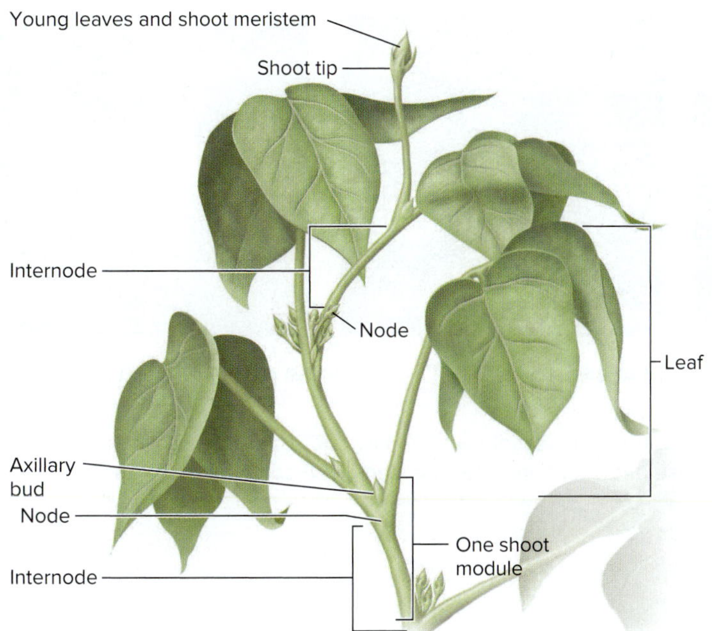


**(a) Modular structure of herbaceous shoot**

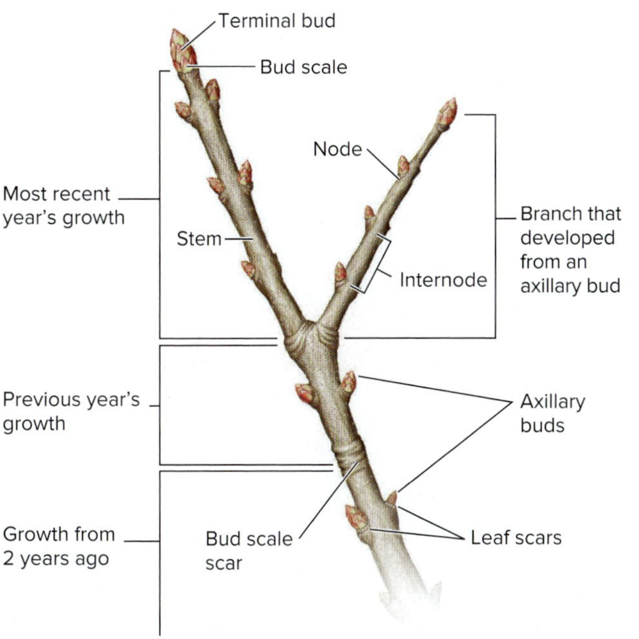


**(b) Modular structure of woody shoot in winter**

**Figure 28.12** **The modular organization of plant shoots. (a)** The top end of a herbaceous stem showing the shoot modules. Each module consists of a node with its associated leaf and axillary bud and an internode. **(b)** The modular organization shown by a woody stem as it appears during winter. Axillary buds lie above the scars left by leaf fall. Regions between successive sets of bud scale scars mark each year's growth.

**Concept Check:** *If a twig has five sets of bud scale scars, how old is the twig likely to be?*

In general, **hormones** are signaling molecules that are produced at one site and exert their effects on distant target cells (hormones are described in more detail in Chapter 29). The outermost epidermal layer of cells at shoot tips produces auxin, which moves from cell to cell by means of specific membrane transport proteins. Auxin accumulates in particular locations because cells of the shoot apex differ in their ability to import and export auxin. When auxin accumulates in a particular apical region, the hormone causes expansin gene expression to increase. When expansin loosens their cell walls (see Figure 28.10), cells expand by taking up water, thereby forming a tissue bulge—a leaf primordium. The development of leaf primordia depletes auxin from nearby tissue, with the result that the next leaf primordium will develop in a different place on the shoot apex, where the auxin level is higher. Such changes in auxin concentration on the surface of the shoot explain why leaf (and flower) primordia develop in spiral or whorled patterns around the shoot tip. The youngest leaf primordia occur closest to the shoot tip, and successively older leaf primordia occur on the sides of the shoot tip (see Figure 28.6).

Hormones also influence the transformation of primordia into leaves. The cells of leaf primordia do not produce a protein known as KNOX, which is produced by other shoot meristematic cells. KNOX is a transcription factor, a protein that regulates gene transcription. The absence of KNOX proteins induces leaf primordia to produce a plant hormone known as **gibberellic acid.** This hormone stimulates both cell division and cell enlargement, causing young leaves to grow.

Other molecules produced by a SAM direct leaf flattening and differentiation of the upper and lower leaf tissues (see Figure 28.8c). A particular set of genes expressed in young primordia causes uppermost leaf tissue layers to develop, and a different set of genes specifies development of lower leaf tissue layers. This differential gene expression is caused by differences in the location of a type of microRNA (miRNA), a small RNA molecule that silences the expression of preexisting mRNAs. The miRNA accumulates in the lower regions of a primordium. In this location, the miRNA interrupts the expression of genes that would otherwise stimulate development of upper leaf tissues while allowing expression of the genes specifying lower leaf tissue development. Other types of miRNA molecules influence leaf shape, which helps to explain the diversity of leaf shapes produced by different plant species.

## Leaf Shape and Surface Features Reflect Adaptation to Environmental Stress

As previously noted, leaf flatness facilitates solar energy collection, and thinness helps leaves to avoid overheating. Leaf shape and surface features also reflect adaptation to stressful environmental conditions.

**Leaf Shape** The flattened portion of a leaf is known as the leaf **blade.** In beans and most other eudicots, blades are attached to the stem by means of a stalk known as a **petiole,** and an axillary bud occurs at the junction of the stem and petiole (Figure 28.13a). In contrast, corn and other monocots have leaf blades that grow directly from the stem, encircling it to form a leaf sheath (Figure 28.13b).

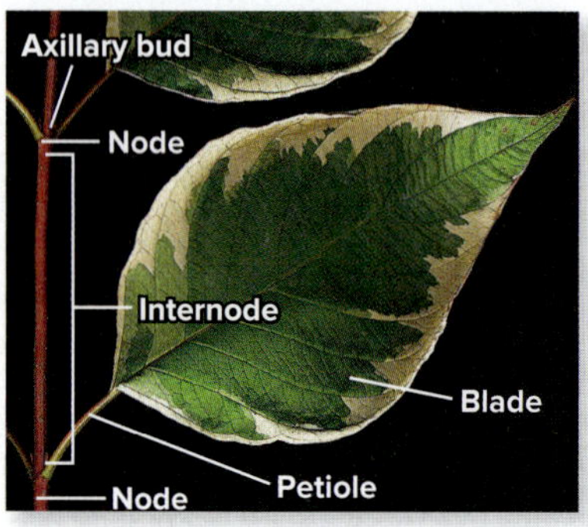


(a) Eudicot stem with simple leaf and pinnate venation

(b) Monocot stem and leaf

(c) A compound leaf

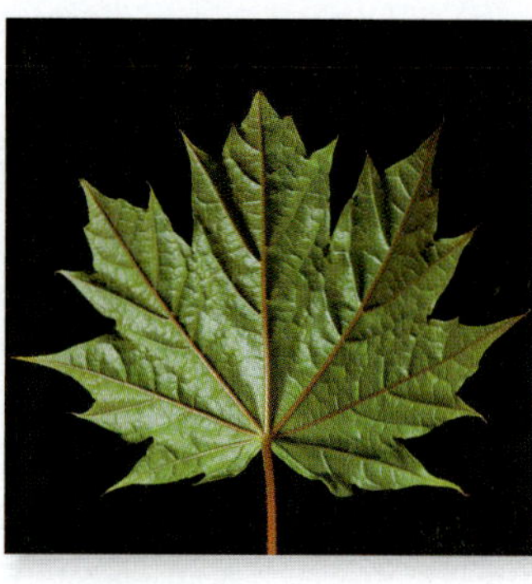

(d) A simple leaf and palmate venation

**Figure 28.13** **Examples of variation in leaf form.** **(a)** A simple eudicot leaf, showing blade, petiole, and axillary bud. This leaf has a pinnate venation pattern. **(b)** The leaf of a monocot, showing parallel veins. The base of the leaf encircles the stem. **(c)** A compound leaf divided into leaflets. **(d)** A simple leaf having palmate venation.
*(a–d)* © Lee W. Wilcox

Leaf shape can be simple or compound, each having particular advantages. Simple leaves have only one blade, though the edges may be smooth, toothed, or lobed. Simple leaves are advantageous in shady environments because they provide maximal light absorption surface, but they can overheat in sunny environments. As an evolutionary response to heating stress, the blades of some leaves have become highly dissected into leaflets. Such leaves are known as compound leaves (Figure 28.13c). Leaflets can be distinguished from leaves because leaflets lack axillary buds at their bases. Compound leaves are common in hot environments because leaflets foster heat dissipation.

**Leaf Vein Patterns** Leaf vein patterns are known as venation. Eudicot leaves occur in two major venation forms. They may have a single main vein from which smaller lateral veins diverge in a feather-like pattern known as **pinnate** venation (see Figure 28.13a). Alternatively, several main veins may spread from a common point on the petiole, like the fingers of your hand, a pattern known as **palmate** venation (Figure 28.13d). In eudicot leaves, small veins connect in a netted pattern, but most monocot leaves have a distinctive parallel venation (see Figure 28.13b and Table 28.1).

## FEATURE INVESTIGATION

### Lawren Sack and Colleagues Showed That Palmate Venation Confers Tolerance of Leaf Vein Breakage

In 2008, Lawren Sack and associates studied the adaptive value of leaf venation patterns by comparing water conduction after injury in pinnately and palmately veined leaves (Figure 28.14). Leaves with palmate venation have several main veins, whereas leaves with pinnate venation have a single main vein. The investigators hypothesized that the multiple main veins of palmate leaves could confer greater tolerance of vein breakage of the type that would occur during mechanical injury or insect damage; if one main vein were damaged, water flow could continue through the other main veins. To test the hypothesis, the investigators experimentally cut a main vein in the leaves of several plants belonging to seven different plant species: four having pinnately veined leaves and three having palmately veined leaves. They conducted the experiments in vivo—that is, in the plant's natural forest environment. After the wounds had healed, the investigators measured the extent of water flow within the leaves at two or three places on each leaf. They found that across all species examined, palmately veined leaves tolerated the disruption in water flow better than pinnately veined leaves. Although palmate venation provides redundancy in case a main vein becomes damaged, it is more costly in terms of materials needed to construct the additional main veins. Hence, leaves with pinnate venation are less costly to produce and work well when the potential for vein damage is low.

*Experimental Questions*

1. Why did Sack and associates conduct their studies of palmate venation on plants growing in a forest rather than in a greenhouse?
2. **SCISKILLS ▸** Why did Sack and colleagues use splints on uncut control leaves as well as for experimentally cut leaves?
3. **SCISKILLS ▸** Why did Sack and associates measure leaf water conduction at two or more places on each leaf?

**Figure 28.14** Sack and colleagues investigated the function of palmate venation.

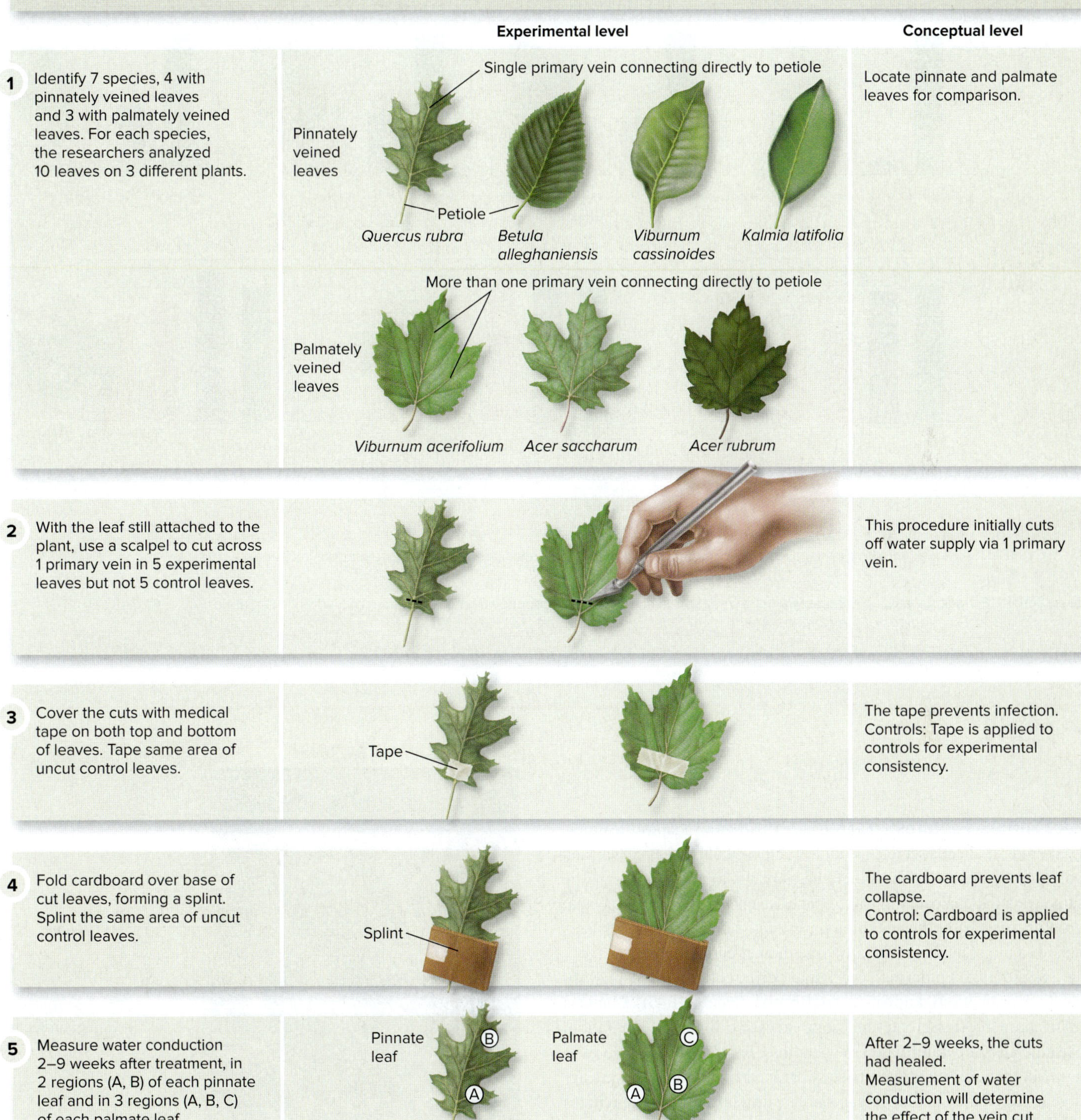

HYPOTHESIS Palmate venation provides vascular redundancy, which allows leaves to tolerate vein breakage.
KEY MATERIALS Seven species of trees or shrubs at Harvard Forest, Petersham, MA.
Experimental level
Conceptual level
1 Identify 7 species, 4 with pinnately veined leaves and 3 with palmately veined leaves. For each species, the researchers analyzed 10 leaves on 3 different plants.
Single primary vein connecting directly to petiole
Pinnately veined leaves
Petiole
Quercus rubra
Betula alleghaniensis
Viburnum cassinoides
Kalmia latifolia
More than one primary vein connecting directly to petiole
Palmately veined leaves
Viburnum acerifolium
Acer saccharum
Acer rubrum
Locate pinnate and palmate leaves for comparison.
2 With the leaf still attached to the plant, use a scalpel to cut across 1 primary vein in 5 experimental leaves but not 5 control leaves.
This procedure initially cuts off water supply via 1 primary vein.
3 Cover the cuts with medical tape on both top and bottom of leaves. Tape same area of uncut control leaves.
Tape
The tape prevents infection. Controls: Tape is applied to controls for experimental consistency.
4 Fold cardboard over base of cut leaves, forming a splint. Splint the same area of uncut control leaves.
Splint
The cardboard prevents leaf collapse.
Control: Cardboard is applied to controls for experimental consistency.
5 Measure water conduction 2–9 weeks after treatment, in 2 regions (A, B) of each pinnate leaf and in 3 regions (A, B, C) of each palmate leaf.
Pinnate leaf
B
A
Palmate leaf
C
A
B
After 2–9 weeks, the cuts had healed.
Measurement of water conduction will determine the effect of the vein cut.

6 **THE DATA**

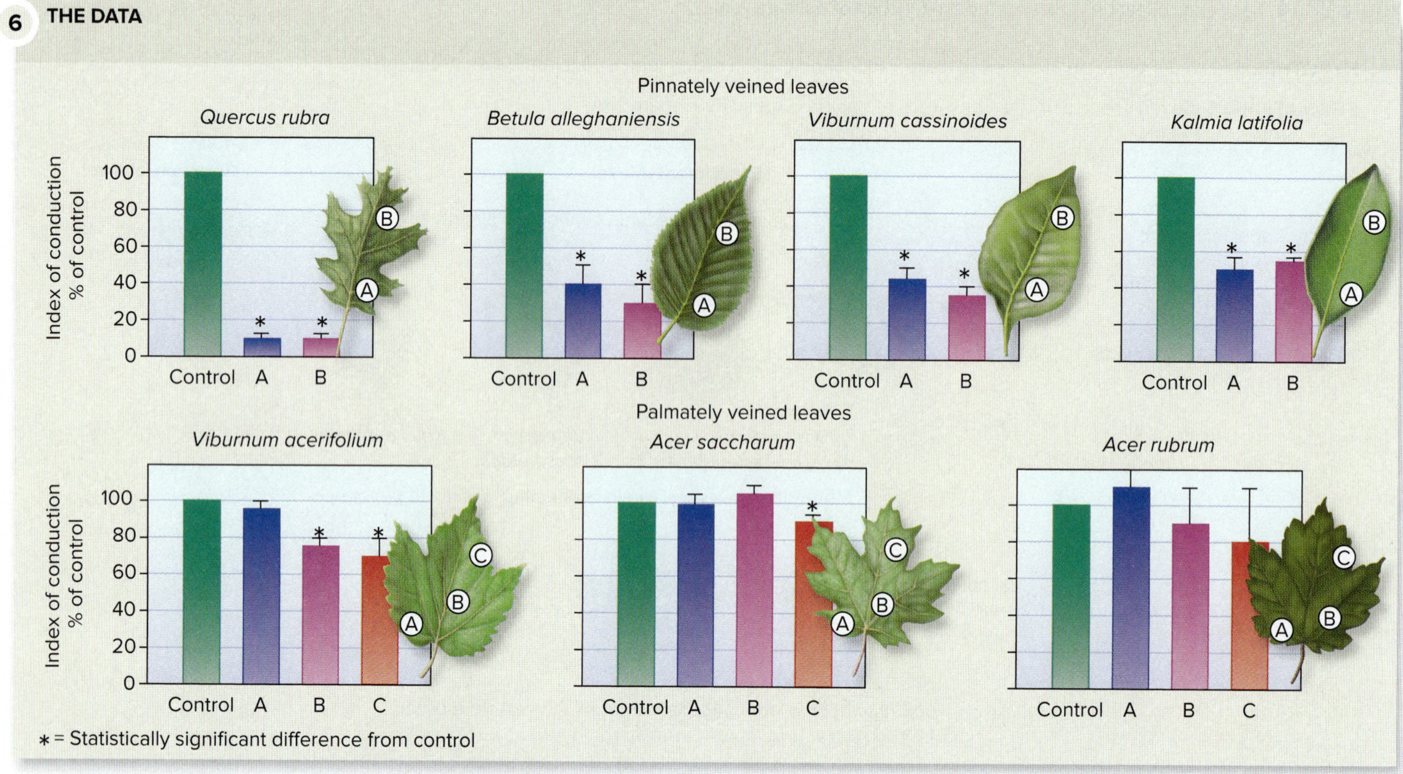


7 **CONCLUSION** Palmately veined leaves did not suffer as much conduction loss from a primary vein cut as did pinnately veined leaves.

8 **SOURCE** Sack, L., et al. 2008. Leaf Palmate Venation and Vascular Redundancy Confer Tolerance of Hydraulic Disruption. *Proceedings of the National Academy of Sciences of the U.S.* 105: 1567–1572.

**Leaf Surface Features** Leaf surfaces also show adaptive features. Leaf epidermal cells secrete a cuticle composed of protective wax and polyester compounds. The cuticle helps plants to avoid drying in the same way that enclosure in waxed paper keeps food moist. Plants that grow in very arid climates often have thick cuticles, whereas plants native to moist habitats typically have thinner cuticles.

Some leaf epidermal cells may differentiate into spiky or hairlike projections known as **trichomes** (Figure 28.15). Blankets of trichomes offer protection from excessive light, UV radiation, extreme air temperature, excess water loss, or attack by herbivores—animals that consume plant tissues. Broken trichomes of the stinging nettle, for example, release a caustic substance that irritates animals' skin, causing them to avoid these plants. Leaf epidermal cells include pairs of specialized guard cells located on either side of stomata (see Figure 28.15). These **guard cells** allow stomata to be open during moist conditions and to close when conditions are dry, thereby preventing plants from losing too much water. The opening and closing of stomata are described in more detail in Chapter 30.

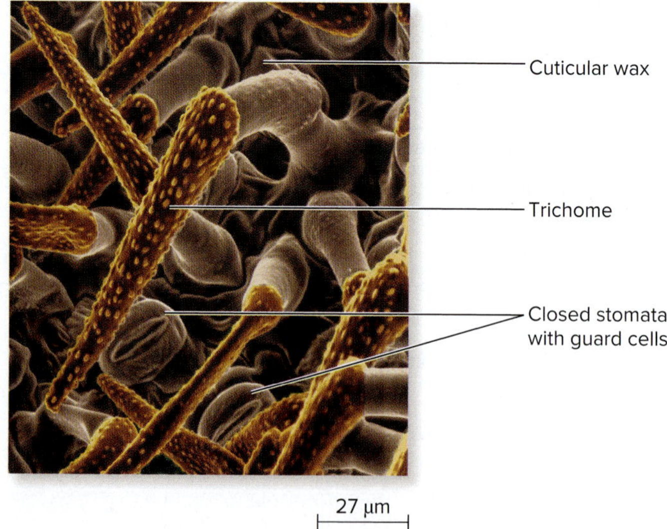


**Figure 28.15** **Leaf surface features.** These features, viewed by SEM, include cuticular wax, trichomes, and stomatal guard cells.

(a) Tendrils

(b) Bud scales

(c) Bracts

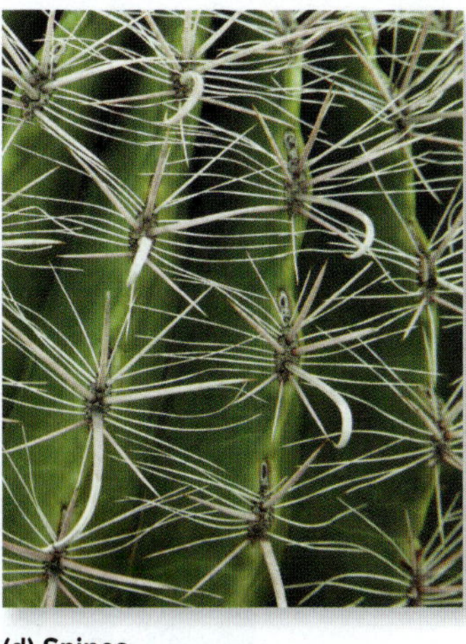

(d) Spines

**Figure 28.16 Examples of modified leaves.** **(a)** The tendrils of an American vetch plant are modified leaves that help the plant attach to a trellis. **(b)** Bud scales, such as those on this sycamore bud, are modified leaves that protect buds from winter damage. **(c)** The attractive red bracts of poinsettia are modified leaves that function like flower petals to attract pollinator insects to the small flowers. **(d)** Cactus spines, such as these on this giant saguaro, are modified leaves that function in defense.

*(a)* © Ed Reschke/Getty Images; *(b)* © John Farmar; *(c)* © Steve Terrill/Corbis; *(d)* © Don Paulson Photography/Purestock/SuperStock RF

**Concept Check:** *Because cactus leaves are so highly modified for defense that they cannot effectively accomplish photosynthesis, how do cacti obtain organic compounds?*

## Modified Leaves Perform Diverse Functions

Though most leaves function primarily as photosynthetic organs, some plants produce leaves that are modified in ways that allow them to play other roles. For example, threadlike tendrils that help some plants attach to a supporting structure are modified leaves or leaflets (Figure 28.16a). The tough scales that protect buds on plants, such as the sycamore, from winter damage are modified leaves (Figure 28.16b). Poinsettia "petals" are actually modified leaves known as bracts, which are larger and more brightly colored than the flowers they surround and help attract pollinators (Figure 28.16c). Cactus spines are modified leaves that have taken on a defensive role, leaving photosynthesis to the cactus stem (Figure 28.16d).

## Primary and Secondary Vascular Tissues

Stems, leaves, roots, buds, flowers, and fruits all contain vascular tissues, in the form of xylem and phloem, which conduct water, minerals, and organic compounds. **Herbaceous plants** such as corn and bean produce mostly primary vascular tissues. In contrast, **woody plants** produce both primary and secondary vascular tissues. A comparison of primary and secondary vascular tissues will aid in understanding their roles.

### Primary Vascular Tissues

**Primary vascular tissues** occur in primary xylem and phloem, described in more detail in Chapter 30. Primary xylem is a complex tissue containing several cell types (see Table 28.2): unspecialized parenchyma cells, stiff fibers that provide structural support, and two types of cells that facilitate water transport. Water transport cells of angiosperms include narrower tracheids and wider vessel elements. Arranged in pipeline-like arrays, tracheids and vessel elements conduct water, along with dissolved minerals and certain organic compounds (Figure 28.17). Mature tracheids and vessel elements are no longer living cells, and the absence of cytoplasm facilitates water flow. During development, these cells

## Biology Principle

### Structure Determines Function

Because lignin strengthens these cells, they do not change shape as large volumes of water move through them.

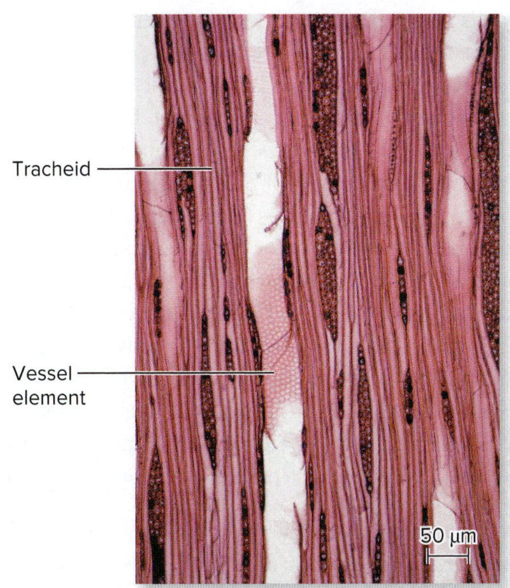


**Figure 28.17 Water-conducting cells of the xylem.** In this thinly sliced portion of a stem, the red-stained, lignin-impregnated walls of narrow tracheids and wider vessel elements can be distinguished.

© Lee W. Wilcox

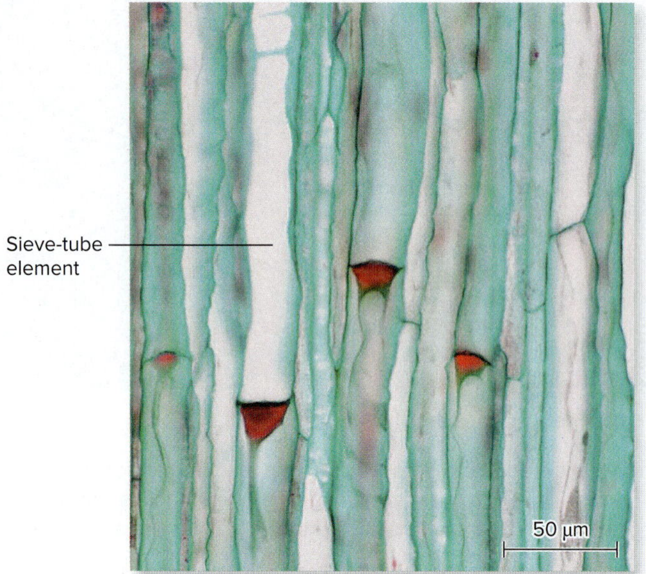


**Figure 28.18 Food-conducting cells of the phloem.** This thinly sliced portion of a stem shows green-stained, thin-walled sieve-tube elements that conduct watery solutions of certain minerals and organic compounds such as sugar. The red-stained material is a protein that seals wounded phloem cells, thereby preventing loss of fluids, much like blood clotting in humans.



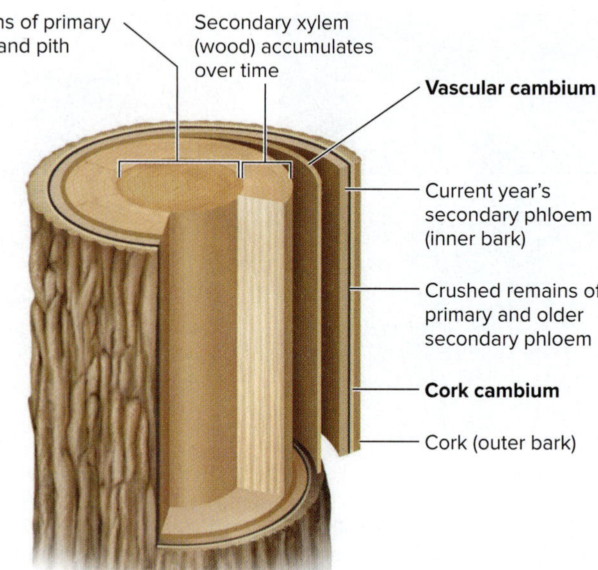


**Figure 28.19 Formation of wood and bark by secondary (lateral) meristems.** The vascular cambium is a thin cylinder of tissue that produces a thick cylinder of wood (secondary xylem) toward the inside of the stem and a thinner cylinder of inner bark (secondary phloem) toward the outside of the stem. The cork cambium forms an outer coating of protective cork (outer bark).

lose their cytoplasm by the process of programmed cell death, a process that resembles apoptosis in animals. However, the cell walls of tracheids and vessel elements don't easily break down or collapse because they are impregnated with a tough polymer known as lignin (see Chapter 25). The rigid cell walls of tracheids and vessel elements not only foster water conduction but also help support the plant body. Like the walls of plastic plumbing pipes, lignin provides the hydrophobic surface needed for water movement, as well as the strength to support trees weighing more than 2,000 metric tons.

Primary phloem, which is composed of living cells, transports organic compounds such as sugars and certain minerals in a watery solution. Phloem tissue has sieve-tube elements, thin-walled cells that are arranged end to end to form pipelines (Figure 28.18). Pores in the end walls of sieve-tube elements allow solutions to move from one cell to another. Phloem tissue also includes companion cells that aid sieve-tube element metabolism, supportive fibers, and parenchyma cells (see Table 28.2). Phloem fibers are tough-walled sclerenchyma cells that are surprisingly long (20–50 mm) and valued for their high strength. The phloem fibers of hemp (*Cannabis sativa*), flax (*Linum usitatissimum*), jute (*Corchorus capsularis*), kenaf (*Hibiscus cannabinus*), and ramie (*Boehmeria nivea*) are commercially important in the production of rope, textiles, and paper.

**Secondary Vascular Tissues** Woody plants begin life as herbaceous seedlings that possess only primary vascular tissues. But as these plants mature, they produce secondary vascular tissues and bark. Secondary vascular tissues are composed of secondary xylem and secondary phloem. **Secondary xylem** is also known as **wood,** a component of plants that plays many important roles in human life. Wood is composed of about 25% lignin, 45% cellulose, and 25% other polysaccharides that are together known as hemicelluloses.

**Secondary phloem** is the **inner bark. Outer bark** is composed of protective layers of mostly dead cork cells that cover the outside of woody stems and roots. Therefore, bark includes both inner bark (secondary phloem) and outer bark (cork). Woody plants produce secondary vascular tissues by means of **secondary meristems,** also known as lateral meristems, which form rings of actively dividing cells that encircle the stem. The two types of secondary meristems are vascular cambium and cork cambium, which are derived from primary meristems.

The secondary meristem, known as **vascular cambium,** is a ring of dividing cells that produces secondary xylem to its interior and secondary phloem to its exterior (Figure 28.19). Secondary xylem conducts most of a woody plant's water and minerals. Cell divisions that occur in secondary meristems increase the girth of woody stems. During each new growing season, the vascular cambium produces new cylinders of secondary xylem and secondary phloem. In temperate trees, each year's addition of new secondary xylem forms growth rings that can be observed on the cut stem surfaces (Figure 28.20). The growth rings of secondary xylem surround the remains of the primary xylem and a central cylinder of parenchyma cells known as the pith. If environmental conditions favor plant growth, the growth rings formed at that time will be wider than those formed during stressful conditions. Climatologists use growth ring widths in samples of old wood to deduce past climatic conditions, and archeologists use growth ring data to determine the age of wood constructions and artifacts left by ancient cultures.

Secondary xylem may transport water for several years, but usually only the current year's production of secondary phloem is active in food transport. This is because thin-walled sieve elements typically live for only a year. Thus, only a thin layer of phloem, the inner bark, is responsible for most of the sugar transport in a large tree. Deep abrasion of tree bark may damage this thin phloem layer, disrupting a tree's food transport.

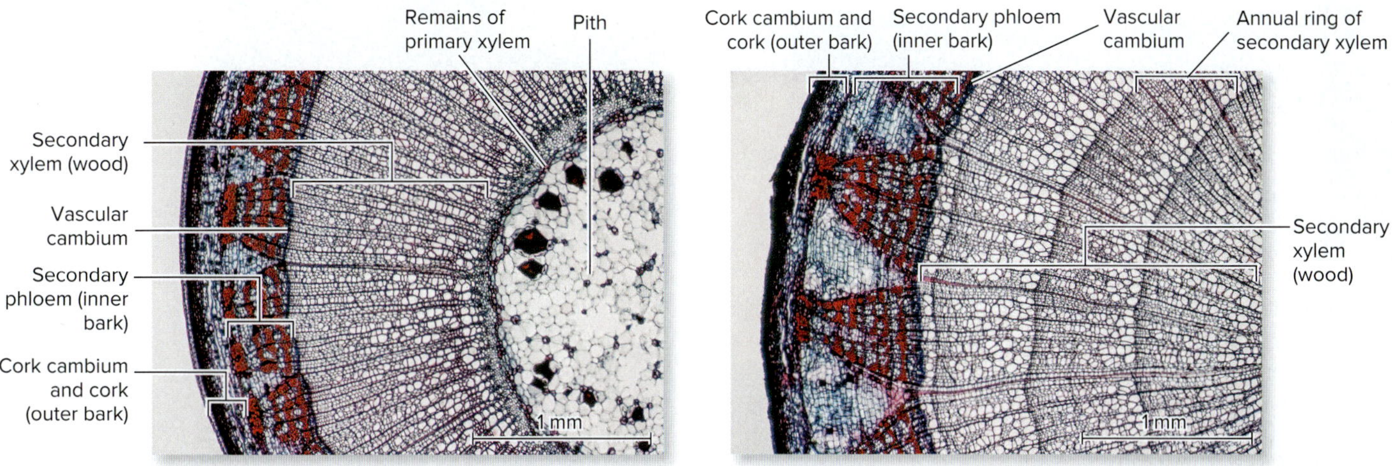


**Figure 28.20 The anatomy of a tree trunk.** Each year, a new cylinder of wood is produced; this yearly wood production appears as annual rings on the cut surface of a woody stem. The stem partial cross section at left is from a 1-year-old stem having a single secondary xylem growth ring. The stem partial cross section at right is from a stem that is more than 3 years old and thus has more wood growth rings.

*Why do tree trunks have a thicker layer of wood (secondary xylem) than of inner bark (secondary phloem)?*

As a young woody stem begins to increase in diameter, its thin epidermis eventually ruptures and is replaced by outer bark, which is composed of protective cork tissues. Cork is produced by a secondary meristem called the **cork cambium,** another ring of actively dividing cells. The cork cambium surrounds the secondary phloem (see Figures 28.19 and 28.20). Together, the cork cambium, layers of cork tissue produced by the cambium, and associated parenchyma cells are known as a **periderm.** The outer bark becomes thicker as woody stems accumulate multiple periderm layers. The outer bark surface is often interrupted by passages known as **lenticels** that allow inner stem tissues to accomplish gas exchange.

Cork cells are dead when mature, and their walls are layered with suberin, a material that helps to prevent both attack by microbial pathogens and water loss from the stem surface. Cork tissues also produce tannins, compounds that protect against pathogens by inactivating their proteins. The cracked surfaces of tree trunks are dead cork tissues of the outer bark. Commercial cork is sustainably harvested from the cork oak tree *(Quercus suber)* for production of flooring material, bottle stoppers, and other items.

## Modified Stems Display Diverse Forms and Functions

Stems mostly grow upright because light is required for photosynthesis. But some stems, known as rhizomes, occur underground and grow horizontally. For example, potato tubers are the swollen, food-storing tips of rhizomes. Grass stems also grow horizontally, as either rhizomes just beneath the soil surface or stolons, which grow along the soil surface. The leaves and reproductive shoots of grasses grow upward from the point where they are attached to these horizontal stems. Grass blades continue to elongate from their bases even if you cut their tips off, explaining why lawns must be mowed repeatedly during the growing season. The horizontal stems of grasses are adaptations that help to protect vulnerable shoot apical meristems against natural hazards such as fire and grazing animals.

## 28.3 Reviewing the Concepts

- Shoots are modular systems; each module includes a node, an internode, a leaf, and an axillary bud. Axillary buds develop from axillary meristems and may grow into new branches (Figure 28.12).
- Variations in leaf structure reflect adaptations that aid photosynthesis or protect against stress (Figures 28.13, 28.14, 28.15).
- Herbaceous plants are those whose stems produce little or no wood and are mostly composed of primary vascular tissues in the form of primary xylem and primary phloem (Figures 28.16, 28.17, 28.18).
- In addition to primary tissues, woody plants—trees, shrubs, and woody vines—possess secondary meristems that produce wood and bark. The vascular cambium produces secondary xylem (wood) and secondary phloem (inner bark). The cork cambium produces cork tissues that form outer bark (Figures 28.19, 28.20).

## 28.3 Testing Your Knowledge

1. Imagine that you are asked to draw a typical plant stem. Where would you correctly draw buds?
   - **a.** anywhere along the stem
   - **b.** anywhere on the surfaces of the internodes
   - **c.** on the surfaces of leaves
   - **d.** at any junction of leaf and stem
   - **e.** only at the junction of the topmost leaf and the stem
2. What function might a leaf serve?
   - **a.** photosynthesis
   - **b.** attachment of the plant to a supporting structure
   - **c.** protective bud scales
   - **d.** defensive spines
   - **e.** all of the above

## 28.4 Root System Adaptations

### Learning Outcomes

1. List ways that root structure has been modified in different plants in response to different habitats.
2. Create a drawing that shows how root systems and branch roots develop.

Roots play the essential roles of absorbing water and minerals, anchoring plants in soil, and storing nutrients. The external form of roots varies among flowering plants, reflecting adaptation to particular life spans or habitats. In contrast, root internal structure is more uniform.

### Modified Roots Display Diverse External Forms and Functions

In addition to the underground taproot systems of eudicots and the fibrous root systems of monocots (see Table 28.1), several root modifications provide adaptive advantages in particular habitats. For example, corn and many other plants produce supportive prop roots from the lower portions of their stems. Many tropical trees grow in such thin soils that the trees are vulnerable to being blown down in windstorms. Such trees often produce dramatic aboveground buttress roots that help keep trees upright (**Figure 28.21a**). Many mangrove trees that grow along tropical coasts produce pneumatophores (Greek, meaning breath bearers), roots that grow upward into the air (**Figure 28.21b**). Functioning as snorkels, pneumatophores absorb oxygen-rich air, which diffuses to submerged roots growing in oxygen-poor sediments. This is necessary because, like animals, all roots require a supply of oxygen in order to produce ATP. Roots use this ATP to power root growth and the uptake of mineral nutrients (see Chapter 30).

**(a) Buttress roots**

**(b) Pneumatophores**

**Figure 28.21 Modified aboveground roots. (a)** Buttress roots help to keep tropical trees such as this *Pterocarpus hayesii* from toppling in windstorms. **(b)** Pneumatophores produced by mangroves are roots that extend upward into the air. These roots take up air and then transmit it to underwater roots that grow in oxygen-poor sediments.
*(a)* © Konrad Wothe/imagebroker.net/SuperStock; *(b)* © Peter E. Smith, Natural Sciences Image Library

### Root Internal Growth and Tissue Specialization Occur in Distinct Zones

Studies of gene expression in roots of the model plant *Arabidopsis thaliana* reveal remarkably complex internal structure, with at least 15 distinct cell types. For our purposes, a simpler microscopic examination of root internal structure reveals three major zones: a root apical meristem (RAM) protected by a root cap, a zone of root elongation, and a zone of maturation in which specialized cells can be observed (**Figure 28.22**).

- The root apical meristem (RAM) contains stem cells that surround a tiny region of cells that rarely divide, known as the quiescent center. Signals emanating from the quiescent center keep nearby stem cells in an undifferentiated state. Root stem cells farther away from the quiescent center produce new cells in multiple directions. (1) Toward the root tip, stem cells produce columella cells that sense gravity and touch, which helps roots extend downward into the soil and around obstacles such as rocks. (2) At the sides of the quiescent center, stem cells produce the protective root cap and epidermal cells. Root tip epidermal cells secrete a sticky substance called mucigel that lubricates root growth through the soil. (3) Toward the shoot, stem cells generate cells that become internal root tissues such as xylem, phloem, endodermis, and cortex. Root tip damage reduces the ability of root systems to grow.
- Above the RAM lies the **zone of elongation,** in which cells extend by water uptake, thereby dramatically increasing root length.
- The **zone of maturation,** located above and overlapping with the root zone of elongation, is where most root cell differentiation and tissue specialization occur. Specialized root tissues include (1) mature vascular tissues at the root core, (2) an enclosing cylinder of cells known as the pericycle that produces branch roots (**Figure 28.23**), (3) another cell cylinder called the endodermis (meaning inside skin) that helps roots avoid conducting toxic materials, (4) relatively unspecialized parenchyma cells forming a starch-storing cortex, and (5) epidermal cells at the root surface. The zone of maturation can be identified by the presence of numerous microscopic hairs that emerge from the root epidermis.
- **Root hairs** are specialized epidermal cells that can be as long as 1.3 cm, about the width of your little finger, but are only 10 μm in diameter. Their small diameter allows root hairs to obtain water and minerals from soil pores that are too narrow for even the smallest roots to enter. Root hair plasma membranes are rich in transport proteins that use ATP to selectively absorb materials from the soil. Damage to root hairs reduces a plant's ability to take up essential resources.

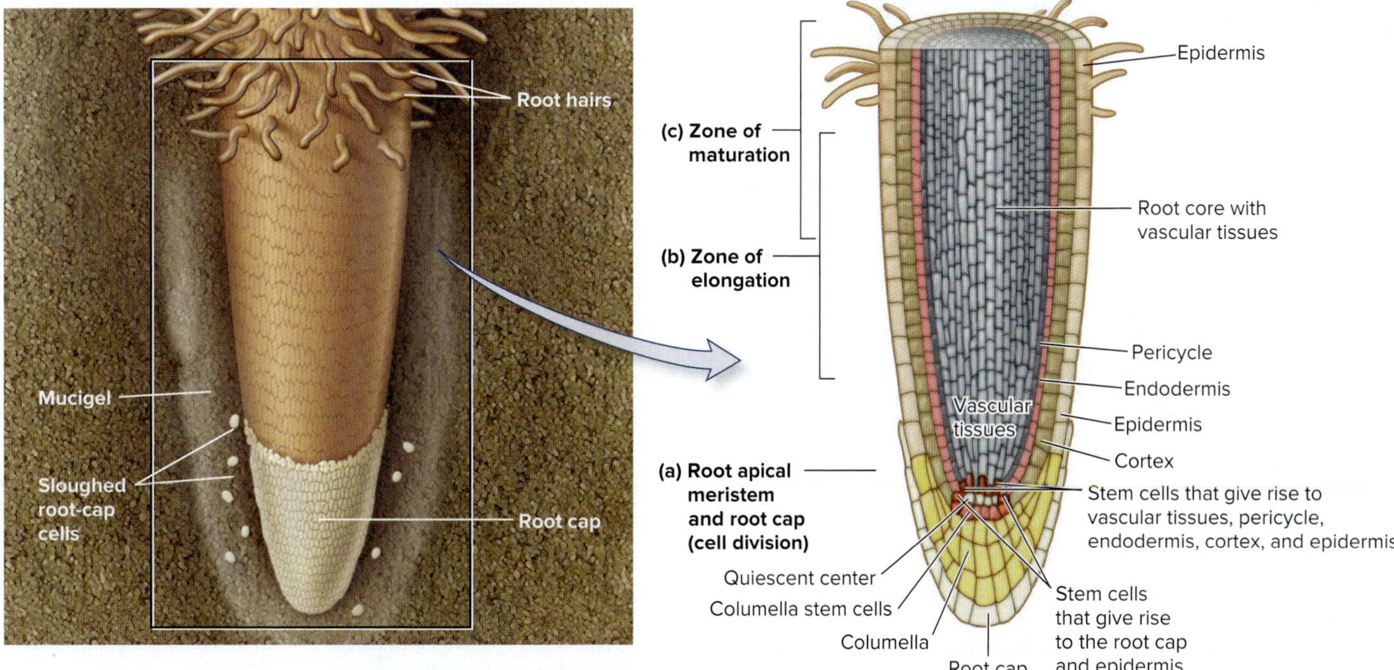


**Figure 28.22** **Three zones of root growth.** A longitudinal view of a typical root reveals three major zones: **(a)** a root apical meristem region that includes stem cells, a quiescent zone, columella, and a root cap; **(b)** a zone of elongation; and **(c)** a zone of maturation, characterized by specialized cells and tissues, including epidermal root hairs, cylinders of endodermis and pericycle tissue, and a core of vascular tissue.

## 28.4 Reviewing the Concepts

- Roots occur in multiple forms that reflect adaptation to environmental conditions (Figure 28.21).
- The internal organization of roots is comparatively uniform, and three major zones can be recognized on microscopic examination: the root apical meristem and root cap, a zone of cell and root elongation, and a zone of tissue maturation. Features of the mature root include an inner core of vascular tissue, a pericycle that produces lateral (branch roots), an endodermis that functions in mineral selection, and epidermal root hairs that aid nutrient uptake (Figures 28.22, 28.23).

## 28.4 Testing Your Knowledge

1. Which of the following structures is a type of root?
   a. endodermis
   b. epidermis
   c. cortex
   d. pneumatophore
   e. none of the above
2. Which of the following series of words most accurately describes the external structure of a root, starting at the tip and moving toward the shoot?
   a. zone bearing root hairs, zone lacking root hairs
   b. quiescent center, columella, stem cells
   c. root cap, zone of elongation, zone of maturation
   d. conducting tissues, endodermis, pericycle
   e. none of the above

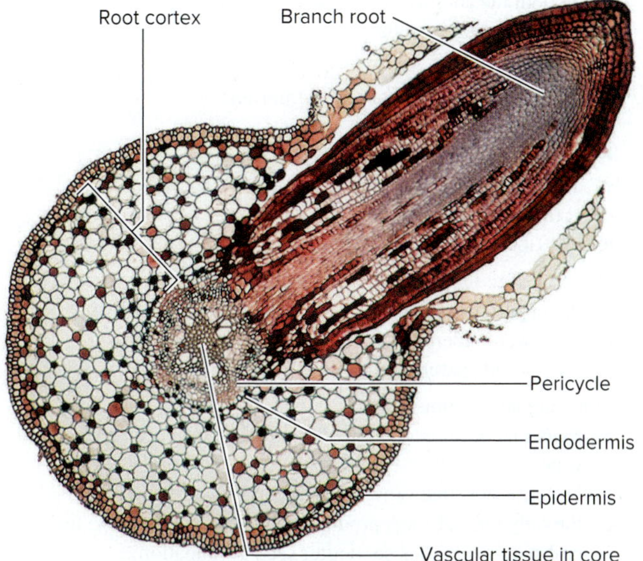


**Figure 28.23** **Cross section of a mature root.** This stained light micrograph shows the epidermis and cortex of a root surrounding a central core of vascular tissue. An inner cortex layer is the endodermis, which surrounds a cylinder of meristematic pericycle tissue. The pericycle has produced a young branch root that has grown through the cortex and the epidermis.


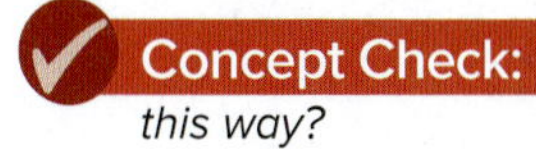

*Why are lateral roots produced in this way?*

# Assess and Discuss

## Test Yourself

1. Where would you look to find the gametophyte generation of a flowering plant?
   a. at the shoot apical meristem
   b. at the root apical meristem
   c. in seeds
   d. in flower parts
   e. Flowering plants lack a gametophyte generation.
2. Which type of plant is most likely to have food-rich roots that are useful as human food?
   a. an annual
   b. a biennial
   c. a perennial
   d. a centennial
   e. a plant that grows along coastal shorelines
3. Which of the following terms best describes the distinctive architecture of plants?
   a. radial symmetry and apical-basal polarity
   b. bilateral symmetry and apical-basal polarity
   c. radial symmetry and absence of apical-basal polarity
   d. bilateral symmetry and absence of apical-basal polarity
   e. absence of symmetry and absence of apical-basal polarity
4. Which is the most accurate description of how plants grow?
   a. by the addition of new cells at meristems that include stem cells
   b. by cell enlargement as the result of water uptake
   c. by both the addition of new cells and cell expansion
   d. by addition of fat cells
   e. all of the above
5. Where would you look for leaf primordia?
   a. at a vegetative shoot tip
   b. at the root apical meristem
   c. at the vascular cambium
   d. at the cork cambium
   e. in a floral bud
6. Which leaf tissues display the greatest amount of air space?
   a. the upper epidermis
   b. the lower epidermis
   c. the palisade parenchyma
   d. the spongy parenchyma
   e. the vascular tissues
7. What are adventitious roots?
   a. roots that develop on plant cuttings that have been placed in water
   b. buttress roots that grow from tree trunks
   c. the only kind of roots produced by monocots, because their embryonic root dies soon after seed germination
   d. any roots that are produced by stem (or sometimes leaf) tissue, rather than developing directly from the embryonic root
   e. all of the above
8. During its development, a water-conducting tracheid elongates in a direction parallel to the shoot or root axis. Based on this information, what can you say about the orientation of cellulose cell-wall microfibrils and cytoplasmic microtubules in this developing tracheid?
   a. The microfibrils will be oriented perpendicularly (at right angles) to the long axis of the developing tracheid, encircling it, but the cytoplasmic microtubules will be oriented parallel to the direction in which the tracheid is elongating.
   b. Microfibrils and microtubules will both be oriented perpendicularly (at right angles) to the elongating axis of the tracheid.
   c. Microfibrils and microtubules will both be oriented parallel to the direction of tracheid elongation.
   d. Microfibrils will be oriented parallel to the direction of tracheid elongation, but microtubules will be perpendicular (at right angles) to both the microfibrils and the elongating tracheid.
   e. None of the above are correct.

## Conceptual Questions

1. What would be the consequences if overall plant architecture were bilaterally symmetric?
2. What would be the consequences if leaves were radially symmetric (shaped like spheres or cylinders)?
3. **PRINCIPLES** A principle of biology is that structure determines function. Why are most tall plants woody, rather than herbaceous?

## Collaborative Questions

1. Find a tree stump or a large limb that has recently been cut from a tree (or imagine doing so). Which of the following features could you locate with the unaided eye: the outer bark, the inner bark, the secondary xylem, the vascular cambium, annual rings?
2. Which physical factors would you expect to influence shoot growth most strongly? Which physical factors would you expect to influence underground root growth most strongly?

# Online Resource

# How Flowering Plants Sense and Interact with Their Environments

# 29

© Linda Graham

**Behavior of the sensitive plant, *Mimosa pudica*.** When touched, the leaves of the sensitive plant quickly fold, and only slowly unfold.

## Chapter Outline



If you touch a leaf of a sensitive plant *(Mimosa pudica)*, in less than a second the leaf will fold, only slowly returning to its original flat position (see chapter-opening photo). Such rapid leaf folding is an example of plant behavior, in this case a response to touch. Experts are not completely sure why sensitive plants fold their leaves when touched. One possibility is that leaf folding in response to the touch of wind reduces the amount of leaf surface exposed to evaporation, thereby helping the plant to retain water. Sensitive plant leaf folding is one of the fastest of plant behavioral responses, but there are many more. Plants also respond to changes in Sun position, the period of daily illumination, gravity, attack by animals and disease-causing microorganisms, and other environmental stimuli. This chapter provides examples of ways in which understanding plant behavior has been key to increasing agricultural productivity and protecting natural ecosystems for human benefit.

We begin with a survey of the diverse types of internal and external stimuli that induce plant behavior and review how cells perceive and respond to stimuli by means of signal transduction pathways. Next, we focus on plant hormones, which are major types of internal mobile molecules that influence plant behavior. Finally, we consider how responses to light, gravity, touch, and attack foster plant survival and reproduction.

## 29.1 Overview of Plant Behavioral Responses

### Learning Outcomes

1. List examples of plant responses to biological and physical stimuli.
2. Describe the three stages of cell signaling.

You might not have noticed that plants display behavior, possibly because much of it occurs on a slower timescale than humans easily observe. But plants, like all living things, respond to various types of stimuli, which is a definition of behavior. Examples of plant behavior include plant movements, some types of which were described in 1880 by Charles Darwin and his son Francis in their book *The Power of Movement in Plants*. Modern time-lapse photography, which makes plant behavior more obvious to humans, reveals that most plants are constantly in motion, bending, twisting, or rotating in dancelike movements known as nutation (**Figure 29.1**). Many other intriguing examples of plant behavior exist:

- Plant shoots typically grow toward light and against the pull of gravity, and most roots grow toward water and in the same direction as the gravitational force.
- Seeds may germinate when they detect the presence of sufficient light and moisture for successful seedling growth.
- Plants produce flowers, fruit, and seeds only at the season most favorable for reproductive success.
- Plants take protective actions when they sense attack by disease microbes or hungry animals, thereby preventing excessive damage to their own bodies or those of neighboring plants.

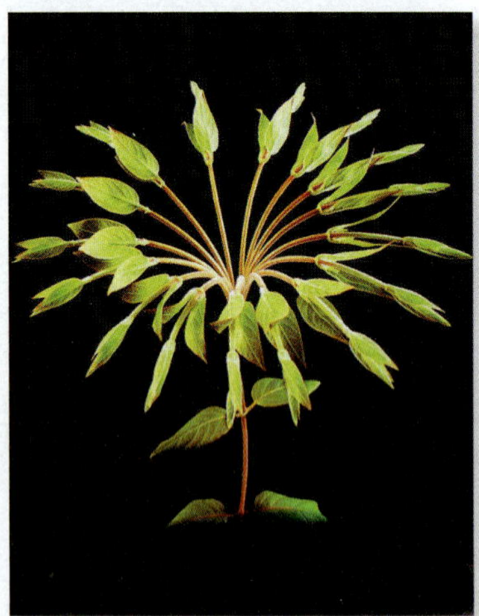

**Figure 29.1 Example of plant movement.** Sixteen superimposed photographs of a shoot of the honeysuckle vine *Lonicera japonica,* taken over a period of 2 hours, reveal the circular movement known as nutation.



### Plant Behavior Involves Responses to Biological and Physical Stimuli

Most people are aware that both biological and physical stimuli influence animal behavior. Bird nesting behavior in spring, for example, involves hormonal changes triggered by seasonal conditions. Plants likewise respond to biological and physical stimuli (**Figure 29.2**).

**Biological Stimuli** Biological stimuli can originate from inside or outside the plant body. Plants respond to two types of internal stimuli: internal biological clocks and mobile chemical signals. Internal biological clocks, based on **circadian rhythms,** occur not only in plants but also in animals and other organisms. The word circadian comes

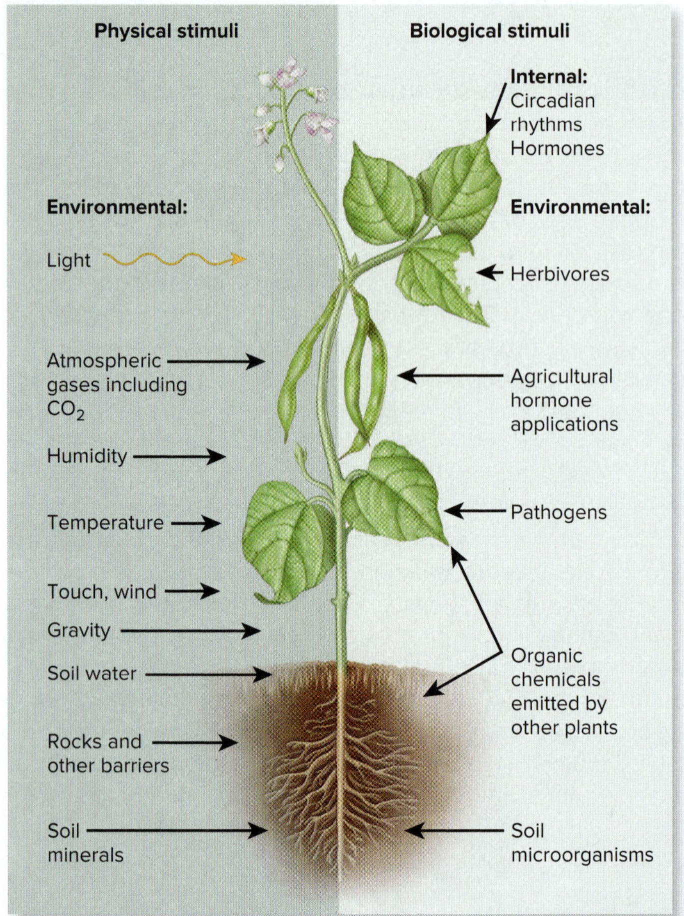


**Figure 29.2 Types of plant stimuli.** Plants respond to both physical and biological stimuli. Stimuli may be internal to the plant or come from the environment.

from the Latin words meaning about and day. Circadian rhythms evolved under the influence of Earth's rotation, which causes the regular alternation of night and day. Leaf movements, flower opening, fragrance emission, and many other behaviors result from the operation of circadian rhythms.

Like animals, plants also respond to mobile chemical signals that are produced within the body and move from one location to another, acting at very low concentrations. In plants, these chemical signals may move short distances from cell to cell, or long distances within the vascular system. Mobile plant chemical signals include transcription factor proteins and microRNAs as well as chemically different compounds known as **hormones.** Some plant hormones are similar to particular animal hormones. One example is the plant hormone jasmonic acid, which is chemically similar to prostaglandins produced by animals. The gaseous hormone nitric oxide (NO) functions in both animals and plants. Other hormones are distinctive to plants. As in the case of animal hormones, plant hormones often interact with each other and with external biological and physical stimuli to maintain a stable internal environment (homeostasis) and enable progress through the life stages described in Chapters 28 and 31.

Other biological stimuli arise from outside the plant body; these include

- herbivores, animals that consume plant parts;
- airborne or soil pathogens, disease-causing microbes;
- beneficial microorganisms; and
- chemicals emitted from neighboring plants.

**Physical Stimuli** Plants sense and respond to many types of physical stimuli (see Figure 29.2). Physical stimuli in natural plant environments include light, atmospheric gases such as $CO_2$ and water vapor, temperature, touch, wind, gravity, soil water, rocks and other barriers to root growth, and soil minerals. Crop plants also respond to applications of agricultural chemicals. That plants have evolved such a broad array of sensory capacity is not surprising, because all of the listed environmental influences affect plant survival and reproduction.

**Sensing and Responding to Environmental Stimuli** Though plants lack the specialized sense organs typical of animals, receptor molecules located in plant cells sense stimuli and cause responses. When many cells of a tissue receive and respond to the same biological or physical stimuli, entire organs or plant bodies display behavior. For example, plants tend to grow toward a light source. This process, known as positive **phototropism,** involves both a cellular perception of light and a growth response of stem tissue to an internal chemical signal. In general, a tropism is a growth response that depends on a stimulus that occurs in a particular direction. In the case of phototropism, cells in plant shoot tissues that face a light source sense the light and cause a plant hormone known as auxin to move across the stem into cells not facing the light. In stem tissues not facing the light, the inward movement of auxin influences gene expression, causing cells to expand by taking up water (refer back to Figure 28.9). Because cells not facing the light expand more than the cells facing the light, the stem bends toward light. Bending toward light enables the plant to acquire more light energy for photosynthesis than if bending did not occur. Phototropism is only one of many examples of ways in which plants perceive their physical environments and respond to them.

We will next review how plant cells more generally receive signals and transmit them intracellularly. This process, known as cell communication (also called cell signaling), occurs in all cells.

## Plant Cell Communication Involves Receptors, Messengers, and Effectors

**Cell communication** is the process in which a cell perceives a physical or chemical signal, thereby switching on an intracellular pathway that leads to a cellular response. Cell communication often involves three types of molecules and processes:

1. Receptors that may become activated in the process known as receptor activation
2. Second messengers that transmit signals in the process known as signal transduction
3. Effector molecules that cause a cellular response

**Receptors** are sensor proteins that become activated when they receive a specific type of signal (Figure 29.3). Receptors occur in diverse cellular locations. Some plant defense receptors are located within the plasma membrane, light receptors may occur in the cytosol, and auxin receptors are found in the nucleus. Some activated receptors directly generate a response, such as an increased flow of ions across a membrane. In contrast, many activated receptors first bind to molecules that initiate an intracellular communication pathway. For example, the binding of certain defense hormones to plasma membrane receptors results in the intracellular production of second-messenger molecules that transmit the signal.

**Second messengers** transmit messages from many types of activated receptors. Cyclic AMP (cAMP), inositol trisphosphate ($IP_3$), and calcium ions ($Ca^{2+}$) are major types of second messengers in animal and plant cells. $Ca^{2+}$ is a particularly common messenger in plant cells. Touch and various other stimuli cause $Ca^{2+}$ to flow from storage sites in the endoplasmic reticulum (ER) lumen into the cytosol. Calmodulin or other calcium-binding proteins then bind the $Ca^{2+}$. Calcium binding alters the structure of the proteins, causing them to interact with other cell proteins or to alter enzymatic function.

**Effectors** are molecules that directly influence cellular responses. In plants, calcium-dependent protein kinases (CDPKs) are particularly important effector molecules. Cell communication ends when an effector causes a cellular response, such as opening or closing an ion channel or switching the transcription of particular genes on or off. A single activated receptor can dispatch many second-messenger molecules, which in turn can activate scores of effectors, leading to many molecular responses within a single cell.

# Biology Principle

## Cells Are the Simplest Units of Life

Internal and environmental stimuli are received and elicit responses at the cellular level.

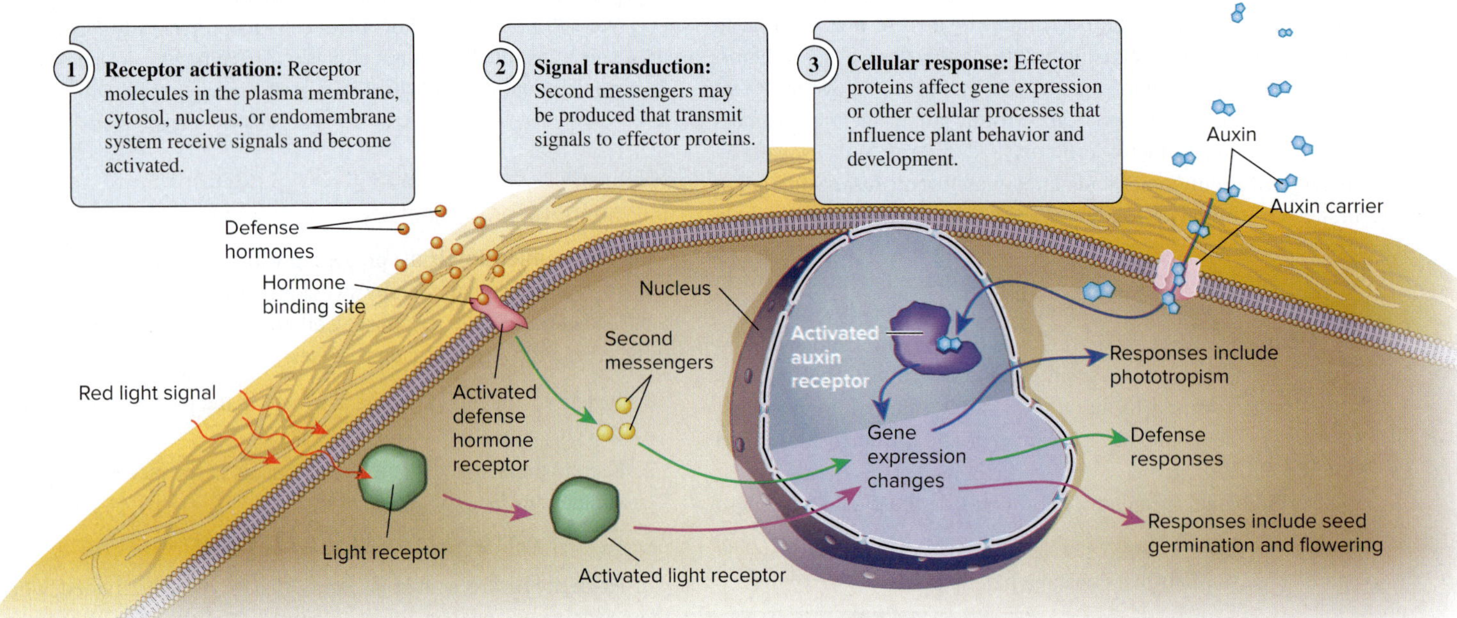


**Figure 29.3** **An overview of plant cell communication.** Plant cells respond to hormonal signals produced within the plant body, as well as to environmental stimuli. Three different signal transduction processes are shown here: one initiated when a compound senses light, one initiated when a defense hormone binds to a plasma membrane receptor, and one initiated when the plant hormone auxin binds a receptor located in the nucleus.

**BioConnections:** *Refer back to Figure 8.4, which shows an example of a cell-communication process in an animal cell. Which of the plant processes illustrated above is most similar?*

## 29.1 Reviewing the Concepts

- Plants sense and respond to diverse biological and physical signals and thus display behavior (Figures 29.1, 29.2).
- During the process of cell communication, cellular receptors respond to environmental physical and biological stimuli as well as to internal hormonal signals. The communication process involves receptor activation, signal transduction by messengers, and a cellular response (Figure 29.3).
- Cellular responses include changes in gene expression and ion channels that influence plant growth, development, reproduction, defense, and movements such as leaf folding.

## 29.1 Testing Your Knowledge

1. Which of the following processes most accurately reflects cell communication during the plant behavior known as phototropism?

   a. In a stem, plant cells that face a light source sense this physical signal, causing the movement of the hormone auxin across the stem to cells not facing the light; cells not facing the light respond to auxin entry by expanding, causing the stem to bend toward the light.

   b. In a stem, plant cells not facing a light source cause auxin to move across the stem to cells facing the light; in response to auxin, cells facing the light contract, causing the stem to bend toward the light.

   c. In a stem, plant cells facing a light source cause auxin to move across the stem to cells not facing the light; the contraction of cells not facing the light causes the stem to bend away from the light.

   d. In a stem, plant cells not facing a light source cause auxin to move across the stem to cells facing the light; cells facing the light expand, causing the stem to bend away from the light.

   e. None of the above are correct.

## 29.2 Plant Hormones

### Learning Outcomes

1. Explain what plant hormones are.
2. Discuss the general mechanism by which plant hormones affect gene expression.
3. List some ways in which plant hormones are relevant to agriculture.

Plant hormones are chemical signals transported within the plant body; when taken up by target cells, the signals elicit responses. Plant hormones are small molecules synthesized in metabolic pathways that also make amino acids, nucleotides, sterols, or secondary metabolites. Here, we focus on the major plant hormones known as auxins, cytokinins, gibberellins, ethylene, abscisic acid, and brassinosteroids (**Table 29.1**).

Individual plant hormones often have multiple effects, and different concentrations or combinations of hormones can produce distinct growth or developmental responses. Several plant hormones are

**Table 29.1 The Major Types of Plant Hormones**

| Type of plant hormone | Chemical structure of an example | Functions (note: this is a partial list) |
|---|---|---|
| **Auxins** | Indoleacetic acid (IAA) 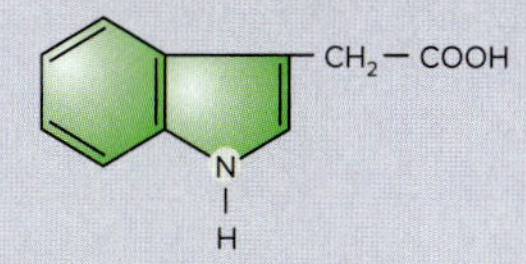 | Establish apical-basal polarity, induce vascular tissue development, mediate phototropism, promote formation of roots, inhibit leaf and fruit drop, and stimulate fruit development |
| **Cytokinins** | Zeatin 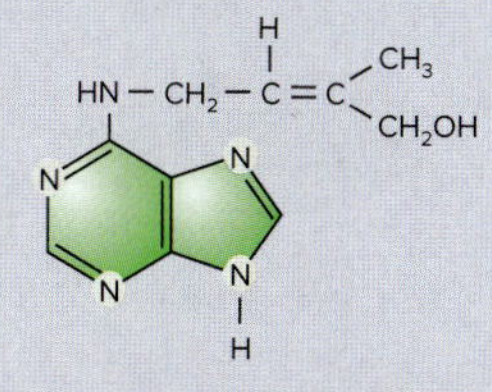 | Promote cell division, influence cell specialization and plant aging, activate secondary meristem development, promote adventitious root growth, and promote shoot development in plant tissue culture |
| **Gibberellins** | Gibberellic acid 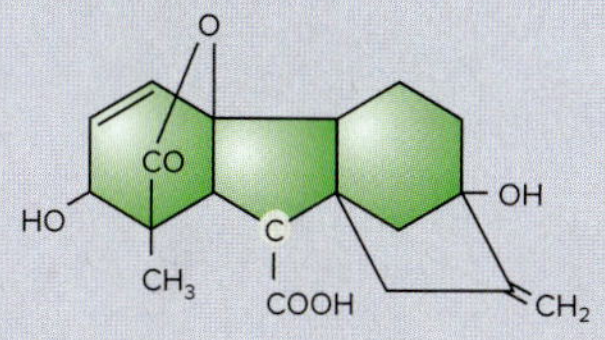 | Stimulate cell division and cell elongation, stimulate stem elongation and flowering, and promote seed germination |
| **Ethylene** | Ethylene $H_2C=CH_2$ | Promotes seedling growth, induces fruit ripening, plays a role in leaf and petal aging and drop, and coordinates defenses against osmotic stress and pathogen attack |
| **Abscisic acid** | Abscisic acid 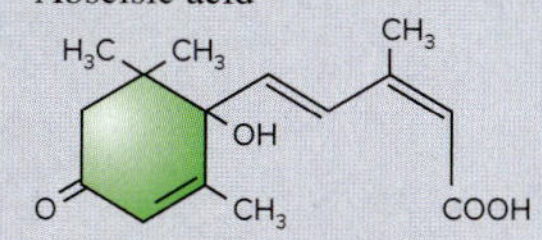 | Slows or stops metabolism during environmental stress, induces bud and seed dormancy, prevents seed germination in unfavorable conditions, and promotes stomatal closing |
| **Brassinosteroids** | Brassinolide 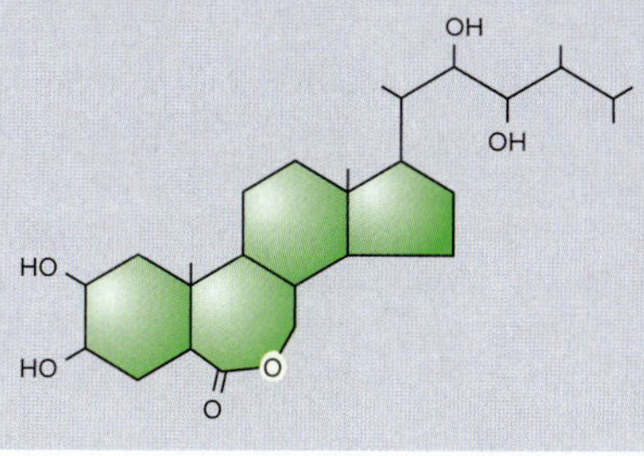 | Promote cell expansion, stimulate shoot elongation, retard leaf drop, stimulate xylem development, and promote stress responses |

known to act by removing gene repressors, thereby allowing gene expression to occur. This general mechanism, which is akin to releasing the brake on a car, allows hormones to cause relatively rapid responses. A closer look at the major types of plant hormones reveals their multifaceted roles.

## Auxins Are the Master Plant Hormones

Plants produce several types of **auxins,** which are often considered to be the master plant hormones because they, often working with other hormones, influence plant structure, development, and behavior in many ways. Indoleacetic acid (IAA) is one plant auxin (see Table 29.1), but other natural and artificial compounds have similar structures and effects. In this section, we will refer to this family of related compounds simply as auxin.

Auxin exerts so many effects because it promotes the expression of diverse genes, together known as **auxin-responsive genes.** Under low auxin conditions, proteins called Aux/IAA repressors prevent plant cells from expressing these genes. The repressor proteins prevent gene expression by binding to activator proteins at gene promoters. When the auxin concentration is high enough, auxin molecules glue repressors onto a protein complex called TIR1, which causes the breakdown of the repressors. Free of the repressors, the activator proteins enhance the expression of auxin-responsive genes.

**Auxin Transport** The way in which auxin is transported into and out of cells is integral to its effects in plants. Auxin is produced in apical shoot tips and young leaves, and it is directionally transported from one living parenchyma cell to another. Auxin in an uncharged form (IAAH) may enter cells from intercellular spaces by means of diffusion; however, the negatively charged form ($IAA^-$) requires the aid of a plasma membrane protein known as the **auxin influx carrier (AUX1)**. Several types of proteins, called PIN proteins, transport auxin out of cells. They are named for the pin-shaped shoot apices of plants having mutations in *PIN* genes. Because they transport auxin out of cells, PIN proteins are called **auxin efflux carriers**. They are necessary because in the cytoplasm, auxin occurs as a charged ion that does not readily diffuse out of cells.

In shoots, AUX1 is located at the apical ends of cells, whereas PIN proteins often occur at the basal ends (Figure 29.4a). This polar distribution of auxin carriers explains polar auxin movement (also called **polar transport**) in which auxin primarily flows downward in shoots and into roots (Figure 29.4b). However, the locations of auxin

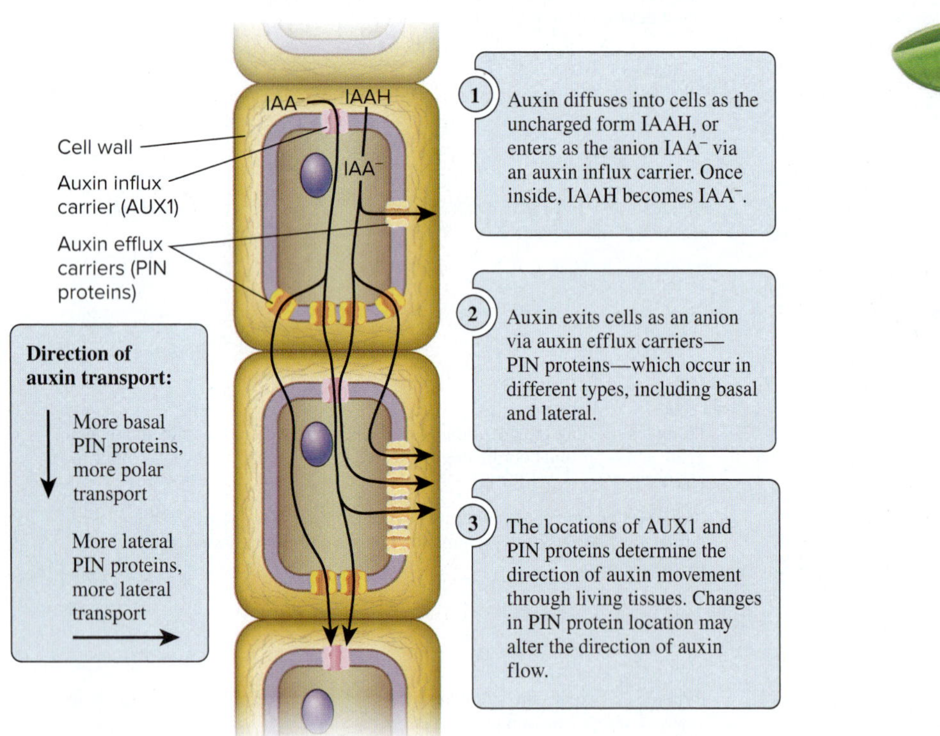


**Figure 29.4 Auxin transport.** **(a)** Polar and lateral auxin movement is controlled by the distribution of auxin efflux carriers located in the plasma membrane. When efflux carriers primarily occur at the basal ends of cells, auxin will flow downward. Auxin may flow laterally when auxin efflux carriers occur at the sides of cells. **(b)** In a whole plant, auxin primarily flows downward from shoot tips to root tips, where it then flows upward for a short distance.

*How could auxin carriers be organized to allow auxin to move upward in roots?*

carriers can also change within cell plasma membranes, allowing lateral (see Figure 29.4b) or upward transport of auxin. Differences in the presence and positions of auxin carrier proteins explain variations in auxin concentration within plants. By measuring the local auxin concentration, plant cells determine their position within the plant body and respond by dividing, expanding, or specializing.

**Auxin Effects** In nature, auxin influences plants throughout their lifetimes. In addition to mediating phototropism, auxin establishes the apical-basal polarity of seed embryos, induces vascular tissue to differentiate, promotes formation of roots, and stimulates fruit development. Many of auxin's effects are of practical importance to humans:

- Auxin treatment is used to produce some types of seedless fruit.
- Growers may use auxin sprays to retard premature fruit drop in orchards.
- To foster growth of whole plants from stem cuttings, people often dip cut stems into powdered auxin to stimulate root development.
- To make many clones of a valuable plant, in a process called plant tissue culture, growers first culture cells taken from the plant, grow many cultures from these cells, then induce mature plants to develop from the cell cultures by using a hormone treatment that includes auxin. This process can be used to generate many plants for market or to help preserve rare species.

## Cytokinins Stimulate Cell Division

Like auxins, the plant hormones known as **cytokinins** play varied and important roles throughout the lives of plants. One example of a cytokinin is the compound zeatin, named for the corn genus *Zea* from which zeatin can be obtained (see Table 29.1). The name cytokinin reflects these hormones' major effect—an increase in the rate of plant cytokinesis, or cell division. Root tips are major sites of cytokinin production, but shoots and seeds also make this plant hormone. Transported in the xylem to meristems and other plants parts, cytokinins bind to receptors thought to be located in the plasma membrane. At shoot and root tips, cytokinins influence meristem size, stem cell activity, and vascular tissue development. Cytokinins are also involved in root and shoot growth and branching, the production of flowers and seeds, and leaf aging. Plant tissue culture depends on hormone treatment with cytokinin, in addition to auxin and gibberellins.

## Gibberellins Stimulate Cell Division and Elongation

The **gibberellins** (also known as gibberellic acids) (see Table 29.1) are another family of plant hormones. Gibberellins are produced in apical buds, roots, young leaves, and seed embryos. In addition to promoting shoot development in plant tissue culture, gibberellins interact with light and other hormones to foster seed germination and enhance stem elongation and flowering in nature. Gibberellins also retard leaf and fruit aging. These multiple effects largely arise from gibberellin's stimulatory effects on cell division and elongation.

More than a hundred different forms of gibberellin have been found. Many kinds of dwarf plants are short because they produce less gibberellin than taller varieties of the same species. The dwarf strain of pea plants Mendel used in some of his breeding experiments is an example. When dwarf varieties of plants are experimentally sprayed with gibberellin, their stems grow to normal heights. However, dwarf wheat and rice crops are valued in agriculture because they can be more productive and less vulnerable to storm damage than taller varieties. Since the discovery of gibberellin, plant scientists have determined how this plant hormone works at the molecular level and how gibberellin regulation of plant growth evolved.

## EVOLUTIONARY CONNECTIONS

### Plant Gibberellin Responses Evolved in a Stepwise Manner

In flowering plants, gibberellin works by helping to liberate repressed transcription factors, allowing gene transcription to proceed. When gibberellin is absent, certain proteins (known as DELLA proteins) bind to particular transcription factor proteins needed for the expression of gibberellin-responsive genes. In this way, DELLAs function as brakes that restrain cell division and expansion. When sufficient gibberellin is present, it releases the brakes by binding to a protein known as GID1 protein. Gibberellin binding to GID1 causes DELLAs to release transcription factor proteins and fosters the destruction of DELLAs. In the absence of DELLA proteins, transcription factor proteins are able to bind the promoter regions of gibberellin-responsive genes, allowing their expression. As a result, cell division and expansion occur, leading to growth.

In 2007, Yuki Yasumura, Nicholas Harberd, and their colleagues reported that the components of the gibberellin-DELLA mechanism in flowering plants arose from features present in more ancient lycophyte and bryophyte lineages (described in Chapter 25). Modern bryophytes and lycophytes possess GID1 as well as DELLA and the transcription factor proteins it binds, but these substances do not affect plant growth. This observation suggests that DELLA-mediated repression of plant growth evolved after the divergence of lycophytes but prior to the appearance of the first gymnosperms and angiosperms (**Figure 29.5**). Though the necessary components were present earlier, they did not assemble into a growth regulation system until relatively late in plant evolutionary history.

## Ethylene's Effects Include Cell Expansion

The plant hormone **ethylene** is particularly important in coordinating plant developmental and stress responses. Ethylene is a simple hydrocarbon gas produced during seedling growth, flower development, and fruit ripening (see Table 29.1). In the root tip, ethylene determines how many stem cells remain inactive and how many cells undergo divisions. This hormone also plays important roles in defense against osmotic stress and pathogen attack, as well as

## Biology Principle

### All Species (Past and Present) Are Related by an Evolutionary History

The gibberellin-DELLA system that controls growth of flowering plants evolved in a step-by-step fashion.

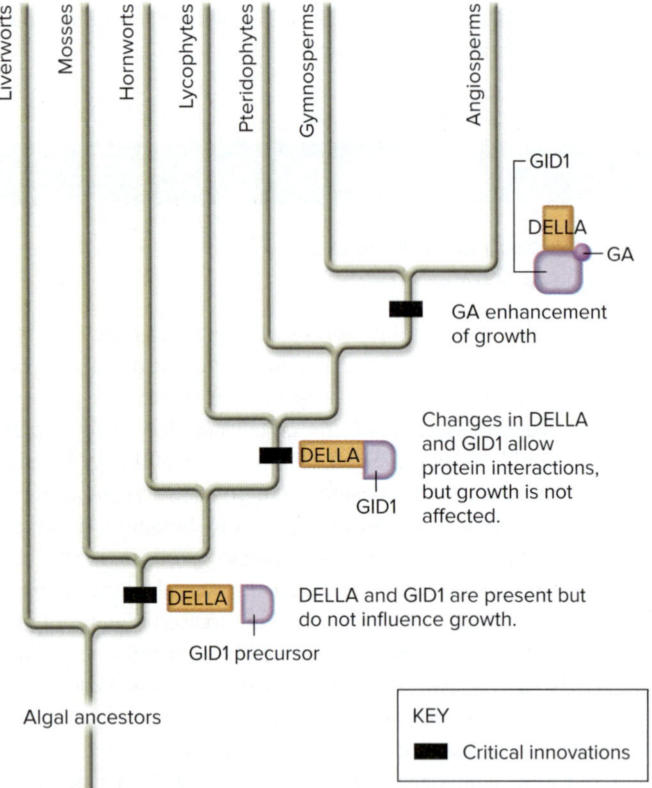


**Figure 29.5 Evolution of the gibberellin-DELLA system.** Although DELLA, and GID1 proteins are present in a bryophyte, they do not interact. In lycophytes, DELLA binds to the GID1 protein, a process that also occurs in seed plants, but the lycophyte DELLA system does not influence growth. Gymnosperms and angiosperms have a fully functional system in which gibberellin binds to GID1, thereby also binding to DELLA proteins. After this gibberellin-responsive gene expression occurs. This pattern suggests that the process by which gibberellin operates in flowering plants evolved from components present in earlier-diverging plant lineages.

leaf and petal aging and drop. As a gas, ethylene is able to diffuse through the plasma membrane and cytosol to bind to ethylene receptors localized in the ER. When activated by ethylene binding, these receptors inactivate a particular protein, enabling the transcription of various genes.

**Figure 29.6 Seedling growth showing the triple response to ethylene.**
© Nigel Cattlin/Alamy

**Concept Check:** *What adaptive advantage does this seedling behavior provide?*

One effect of ethylene is to increase the disorder of microtubules within cells, thereby causing random orientation of cell-wall microfibrils. As a result, cells exposed to ethylene tend to expand in all directions rather than elongating (refer back to Figure 28.11). Ethylene's effects on cell expansion explain this hormone's important role in dicot seedling growth. The tender apical meristems of seedlings could be easily damaged during their growth through crusty soil. Ethylene helps seedlings avoid such damage by inducing what is known as the triple response (Figure 29.6). First, ethylene prevents the seedling stem and root from elongating. Second, the hormone induces the stem and root to swell radially, thereby increasing in thickness. Together, these responses strengthen the seedling stem and root. Third, the seedling stem bends so that embryonic leaves and the delicate meristem grow horizontally rather than vertically. The bent portion of the stem, known as a hook, then pushes up through the soil. The hook forms as the result of an imbalance of auxin across the stem axis, which causes cells on one side of the stem to elongate faster than cells on the other side. Ethylene drives this auxin imbalance.

Knowledge of the effects of ethylene on fruit has been very useful commercially. Ripe fruit can be easily damaged during transit, but tomatoes and apples can be picked before they ripen for transport with minimal damage. At their destination, such fruit can be ripened by treatment with ethylene. However, fruit that becomes overripe may exude ethylene, which hastens ripening in nearby, unripe fruit. For this reason, fruit that must be stored for extended periods is kept in ethylene-free environments.

## Several Hormones Help Plants Cope with Environmental Stresses

Several plant hormones share the property of helping plants respond to environmental stresses such as flooding, drought, high salinity, cold, heat, and attack by disease microorganisms and animal herbivores. These protective hormones include the major plant hormones known as **abscisic acid** and **brassinosteroids** (see Table 29.1). Additional protective hormones are salicylic acid (SA), whose chemical structure is similar to that of aspirin, and a peptide known as systemin. The fragrant compound jasmonic acid and the gas nitric oxide (NO) are also protective plant hormones.

**Abscisic Acid** Abscisic acid (ABA) was named at a time when plant biologists thought that it played a role in leaf or fruit drop, also known as abscission. Later, they discovered that ethylene actually causes leaf and fruit abscission, whereas abscisic acid slows or stops plant metabolism when growing conditions are poor. Other important effects of ABA include

- **Induction of bud and seed dormancy.** In preparation for winter, ABA stimulates the formation of tough, protective scales around the dormant buds of perennial plants. Seed coats of apple, cherry, and other plants also accumulate ABA, which prevents seeds from germinating unless temperature and moisture conditions are favorable for seedling growth. Dormant buds and seeds resume growth only when specific environmental signals reveal the onset of conditions suitable for survival.
- **Stomatal closure.** Water-stressed roots produce ABA, which is then transported to shoots, where (together with ABA produced by water-stressed leaf mesophyll) it helps to prevent evaporation of water from leaf surfaces by inducing leaf pores (stomata) to close.

**Brassinosteroids** Named after the cruciferous plant genus *Brassica* (which includes cabbage and broccoli), from which they were first identified, brassinosteroids occur in seeds, fruit, shoots, leaves, and flower buds of all types of plants. Brassinolide is an example of a brassinosteroid (see Table 29.1). Brassinosteroids are chemically related to animal steroid hormones, such as the sex hormones testosterone and estrogen. In plants, brassinosteroids induce vacuole water uptake and influence enzymes that alter cell-wall carbohydrates, thereby fostering cell expansion. Mutations that affect brassinosteroid synthesis cause plants to exhibit dwarfism. Such plants have small, dark green cells because their tissues are unable to expand. Brassinosteroids also impede leaf drop, help grass leaves to unroll, and stimulate xylem development. They can be applied to crops to help protect plants from heat, cold, high salinity, and herbicide injury.

## 29.2 Reviewing the Concepts

- Plant hormones interact with environmental stimuli to control plant development, growth, and behavior (Table 29.1).
- Auxin plays an important role in many aspects of plant behavior, including phototropism. The position of auxin influx and efflux carriers determines the direction of auxin transport (Figure 29.4).
- The major effect of cytokinins is to increase the rate of cell division. Auxin and cytokinin are used in the process of plant tissue culture.
- Gibberellin function illustrates the general principle that plant hormones often act to release cellular brakes on gene expression. The flowering plant gibberellin system evolved from components present in earlier-diverging plant lineages (Figure 29.5).
- The gaseous hormone ethylene plays an important role in seedling development (Figure 29.6).
- Abscisic acid (ABA) and brassinosteroids help plants to respond to environmental stress.

## 29.2 Testing Your Knowledge

**1.** If you were asked to draw a picture showing how several plant hormones affect gene expression, which of the following auto parts would be most useful as a component of your drawing?
**a.** crankshaft **c.** axle **e.** steering wheel
**b.** tires **d.** brakes

**2–6.** Match the major plant hormones with a major function.

| | |
|---|---|
| **2.** auxins | **a.** promote cell division and shoot development in tissue cultured plants |
| **3.** cytokinins | **b.** mediate phototropism |
| **4.** gibberellin | **c.** promote seed germination |
| **5.** ethylene | **d.** promote stomatal closing |
| **6.** abscisic acid | **e.** induce fruit ripening |

## 29.3 Plant Responses to Light

### Learning Outcome

1. Describe how phytochrome allows plants to respond to light, including day length.

How do buried seeds know if enough light is available to support seedling photosynthesis? How do plants know to flower at times of the year that are most beneficial for achieving pollination or seed dispersal? The answer in both cases is that plants possess cellular systems for measuring light and determining the season of the year. Using these systems, plant seeds germinate only when there is sufficient light for seedling growth, and flowering occurs during the most advantageous season.

## Plants Detect Light and Measure Day Length

A plant's ability to measure and respond to day length, a process called **photoperiodism,** is based on the presence of several types of light receptors within cells. Such light sensors are known as photoreceptors; here we focus on the particular type known as phytochrome.

**Phytochrome** is a red- and far-red-light receptor that influences many plant processes. This light sensor operates much like a light switch, flipping back and forth between two conformations: $P_r$, which absorbs red light, and $P_{fr}$, which absorbs only far-red light (light at the far-red end of the visible spectrum) (Figure 29.7). When red light is abundant, as in full sunlight, $P_r$ absorbs red light and changes to $P_{fr}$, which activates cellular responses such as seed germination. When left in the dark for a long period, $P_{fr}$ slowly transforms into $P_r$. The phytochrome light switch controls (1) seed germination, (2) seasonal flowering, and (3) plant response to shading.

1. The role of phytochrome as a plant "light switch" has been shown experimentally in studies of lettuce-seed germination. Researchers have found that water-soaked lettuce seeds will not germinate in darkness, but they will germinate if exposed to as little as 1 minute of red light (Figure 29.8). This amount of light exposure is sufficient to transform a critical amount of $P_r$ to the active $P_{fr}$ conformation, which stimulates germination. However, if this brief red-light treatment is followed by a few minutes of treatment with far-red light, the lettuce seeds will not germinate. This short period of far-red illumination is enough to convert seed $P_{fr}$ back to the inactive $P_r$ conformation.

   The most recent light exposure determines whether the phytochrome occurs in the active or inactive conformation. In nature, if seeds are close enough to the surface that their phytochrome is switched on by red light, the seeds will germinate. But if seeds are buried too deeply for red light to penetrate, they will not germinate. In this way, seeds can sense if they are close enough to the surface to begin the germination process.

2. Phytochromes also play a critical role in photoperiodism. Flowering plants can be classified as long-day, short-day, or day-neutral plants. When scientists named these groups, they thought that plants measured the amount of daylight; researchers later discovered that plants actually measure night length.
   - Lettuce, spinach, radish, beet, clover, gladiolus, and iris are examples of **long-day plants** because they flower in spring or early summer, when the night period is shorter (and thus the day length is longer) than a defined period (Figure 29.9).

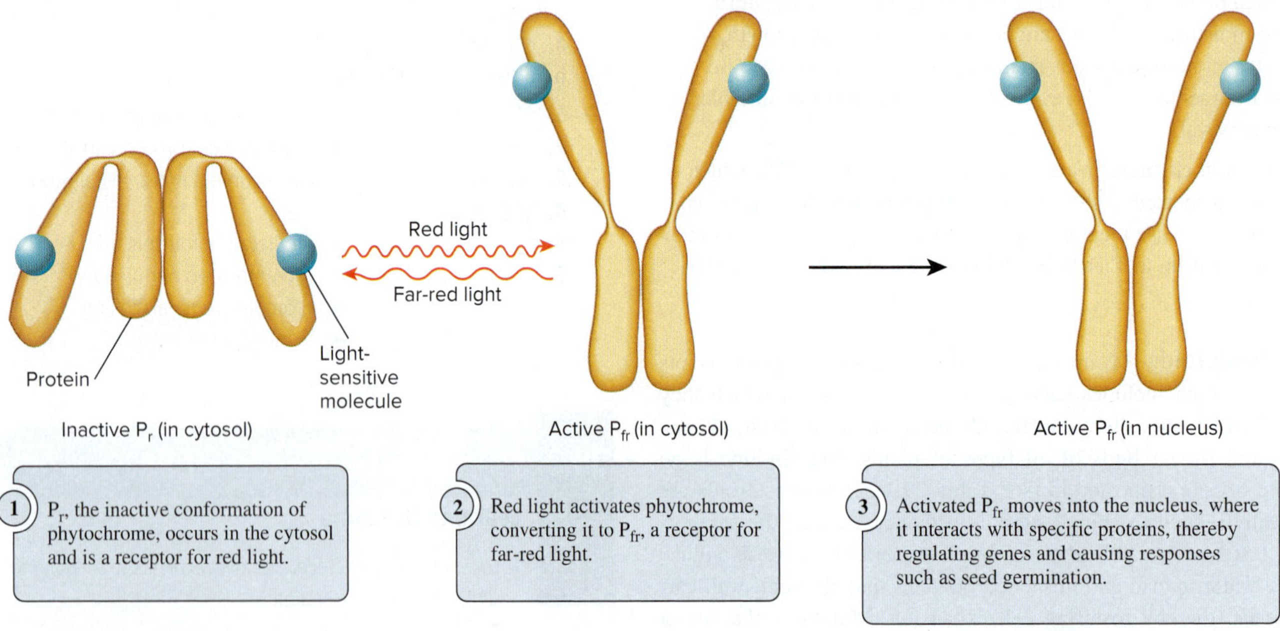

Red light
Far-red light
Protein
Light-sensitive molecule
Inactive $P_r$ (in cytosol)
Active $P_{fr}$ (in cytosol)
Active $P_{fr}$ (in nucleus)
1 $P_r$, the inactive conformation of phytochrome, occurs in the cytosol and is a receptor for red light.
2 Red light activates phytochrome, converting it to $P_{fr}$, a receptor for far-red light.
3 Activated $P_{fr}$ moves into the nucleus, where it interacts with specific proteins, thereby regulating genes and causing responses such as seed germination.


**Figure 29.7** How phytochrome acts as a molecular light switch.

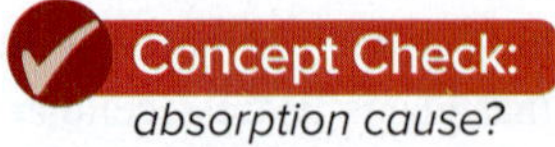


*What kind of light does the active conformation of phytochrome absorb, and what kind of change does such absorption cause?*

**DARKNESS**

In darkness, seeds do not germinate because phytochrome remains in the inactive $P_r$ conformation.

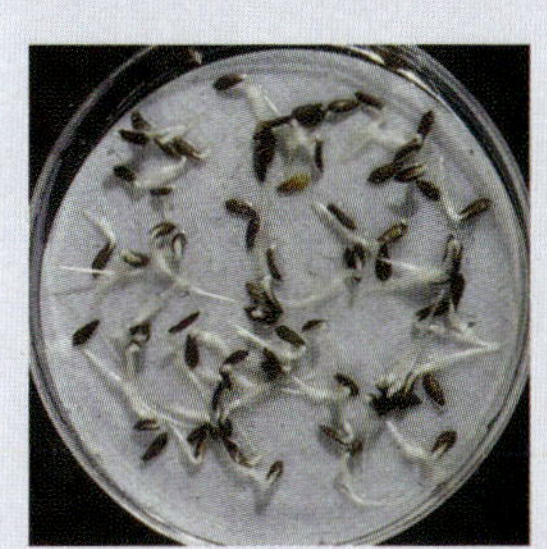

Red

Even a brief exposure to red light generates the active $P_{fr}$ conformation of phytochrome, allowing seeds to germinate.

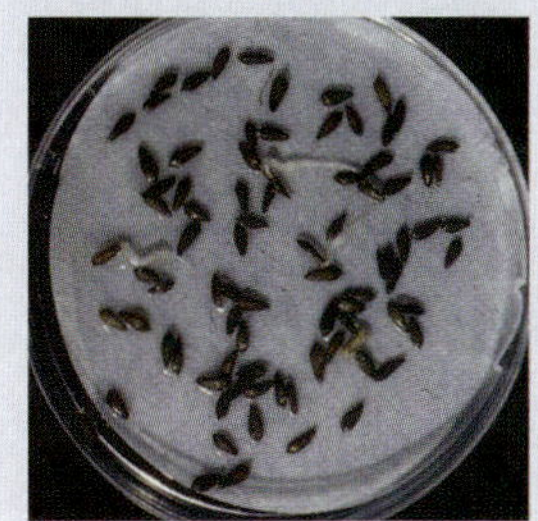

Red | Far red

Exposure to far-red light after red-light exposure converts active $P_{fr}$ to inactive $P_r$, so seeds do not germinate.

Red | Far red | Red

Exposure to red light after far-red light switches phytochrome back to the active $P_{fr}$ conformation, so seeds germinate.

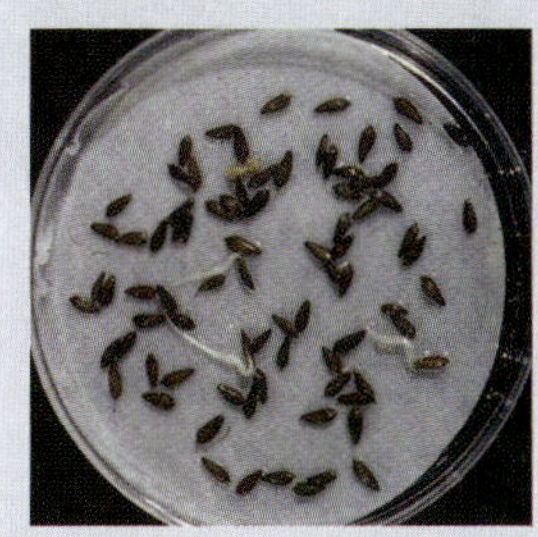

Red | Far red | Red | Far red

The most recent light exposure determines whether phytochrome occurs in the active $P_{fr}$ or in the inactive $P_r$ conformation. If in the latter, most seeds do not germinate.

**Figure 29.8** How phytochrome influences seed germination.
*(1–5)* © Prof. and Mrs. M.B. Wilkins/University of Glasgow

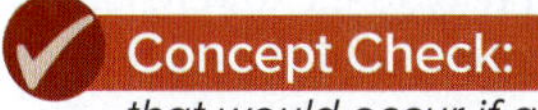

**Concept Check:** *Describe the change in phytochrome that would occur if a deeply buried seed were uncovered enough to receive sunlight.*

- Asters, strawberries, dahlias, poinsettias, potatoes, soybeans, and goldenrods are examples of **short-day plants** because they flower only when the night length is longer than a defined period. Such night lengths occur in late summer, fall, or winter, when days are short (see Figure 29.9).
- Roses, snapdragons, cotton, carnations, dandelions, sunflowers, tomatoes, and cucumbers flower regardless of the night length, as long as day length meets the minimal requirements for plant growth, and are thus known as **day-neutral plants.**

Ornamental plant growers manipulate night length to produce flowers for market during seasons when they are not naturally available. For example, chrysanthemums are short-day plants that usually flower in the fall, but growers use light-blocking shades to increase night length in order to produce flowering chrysanthemums at any season.

3. Phytochrome also mediates plant responses to shading. These responses include the extension of leaves from shady portions of a dense tree canopy into the light, and growth that allows plants to avoid being shaded by neighboring plants. These growth responses occur by the elongation of shoot internodes. Leaves detect shade as an increased proportion of far-red light to red light. This means that more of the phytochrome in shaded leaves is in the inactive ($P_r$) state than is the case for leaves in sunlight. Activated phytochrome ($P_{fr}$) inhibits the growth of shoot internodes, but phytochrome in the inactivated state does not, so branches bearing shaded leaves extend toward sunlight.

## 29.3 Reviewing the Concepts

- Light-sensitive pigments such as phytochrome allow plants to respond to light stimuli and influence seed germination, control of seasonal flowering, and response to shading (Figures 29.7, 29.8, 29.9).

## 29.3 Testing Your Knowledge

**1–4.** Imagine a garden that has been planted with irises, dahlias, carnations and in which goldenrods also occur as weeds. Match each with the seasonal period during which blooming will occur.

**1.** iris (a long-day plant)
**2.** dahlia (a short-day plant)
**3.** carnation (a day-neutral plant)
**4.** goldenrod

**a.** spring or early summer when nights are relatively short
**b.** anytime that light amounts meet plant growth needs
**c.** late summer, fall, or winter when days are short
**d.** none of the above

**Figure 29.9 Flowering and photoperiodism.** Iris is a long-day plant that flowers in response to the short nights of late spring and early summer, whereas goldenrod is a short-day plant that flowers in response to the longer nights of autumn. The length of night is the critical factor, as shown by the effects of light flashes.

*(top left) (bottom left)* © Ray Bulson/Newscom; *(top right) (bottom right)* © Lee W. Wilcox; *(middle left)* © Garden Picture Library/Getty Images; *(middle right)* © Comstock Images/PictureQuest RF

*What would happen if you gave these plants a brief exposure to darkness in the middle of the daytime?*

## 29.4 Plant Responses to Gravity and Touch

### Learning Outcome

1. Discuss how plant roots respond to gravity and touch.

Have you ever wondered what causes plant stems to generally grow upward and roots downward? The upward growth of shoots and the downward growth of roots are behaviors known as **gravitropism,** growth in response to the force of gravity. Shoots are said to be negatively gravitropic because they often grow in the direction opposite to gravitational force. If a potted plant is turned over on its side, the shoot will eventually bend and begin to grow vertically again (Figure 29.10). Most roots are said to be positively gravitropic because they grow in the same direction as the gravitational force.

Both roots and shoots detect gravity by means of starch-heavy plastids known as **statoliths,** which are located in specialized gravity-sensing cells called statocytes. In shoots, statocytes are located in a tissue known as the endodermis, which forms a sheath around vascular tissues. Although roots also possess an endodermis, in roots, gravity-sensing cells primarily occur in the center of the root cap (refer back to Figure 28.22).

Gravity causes the relatively heavy statoliths to sink, which causes changes in calcium ion messengers that affect the direction of auxin transport. This process induces changes in the direction of shoot or root growth. For example, in a root that becomes oriented horizontally, the statoliths are pulled by gravity to the lower sides of statocytes (Figure 29.11). The change in statolith position causes auxin to move to cells on the lower sides of roots. In roots, auxin inhibits

**Figure 29.10 Negative gravitropism in a shoot.** This tomato shoot system has resumed upward growth after being placed on its side. Upward growth started about 4 hours after the plant was turned sideways; this photo was taken 20 hours later. Shoots sense gravity by means of starchy statoliths present in endodermal tissue.

© Lee W. Wilcox

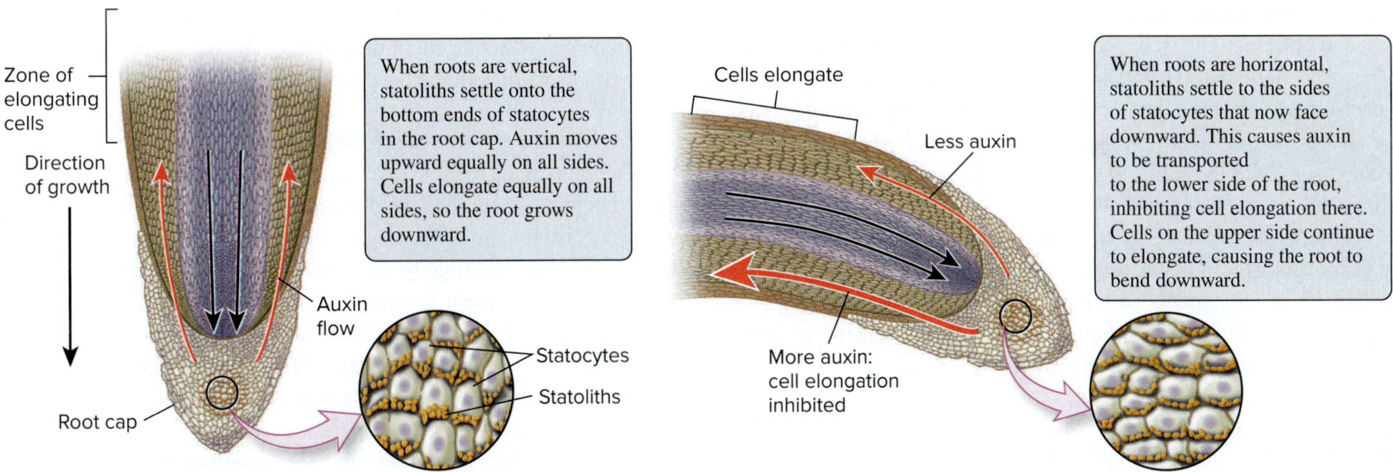


**Figure 29.11 Positive gravitropism in a root.** Root-tip cells sense gravity by means of starchy statoliths present in statocytes at the center of the root cap.

*Is there any other environmental signal that roots could use to achieve downward growth?*

cell elongation (in contrast to its action in shoots). Therefore, root growth slows on the lower side, while cell elongation continues normally on the upper side. This process causes the root to bend, so that it eventually grows downward again.

Recent studies suggest that gravity responses are related to touch responses. For example, when roots encounter rocks or other barriers to their downward growth in the soil, they display a touch response that temporarily supersedes their response to gravity. Such roots grow horizontally until they get around the barrier, whereupon downward growth in response to gravity resumes.

Plant shoots also respond to touch—for example, vines with tendrils that wind around or clasp supporting structures. Wind also induces touch responses. In very windy places, trees tend to be shorter than normal, giving them the advantage of being less likely to blow over than are taller trees. In the laboratory, plant scientists have simulated natural touch responses by rubbing plant stems and found that this treatment can result in shorter plants. Touch causes the release of calcium ion messengers that influence gene expression.

More rapid responses to touch, such as leaf folding by the sensitive plant (see chapter-opening photo), are based on changes in the water content of cells within a structure known as a **pulvinus** (plural, pulvini), a swelling located at the base of attachment of each pair of leaflets in complex leaves. A pulvinus consists of a thick layer of parenchyma cells that surrounds a core of vascular tissue (Figure 29.12). When the leaflet of a sensitive plant is touched, an electrical signal known as an action potential opens ion channels in parenchyma cells

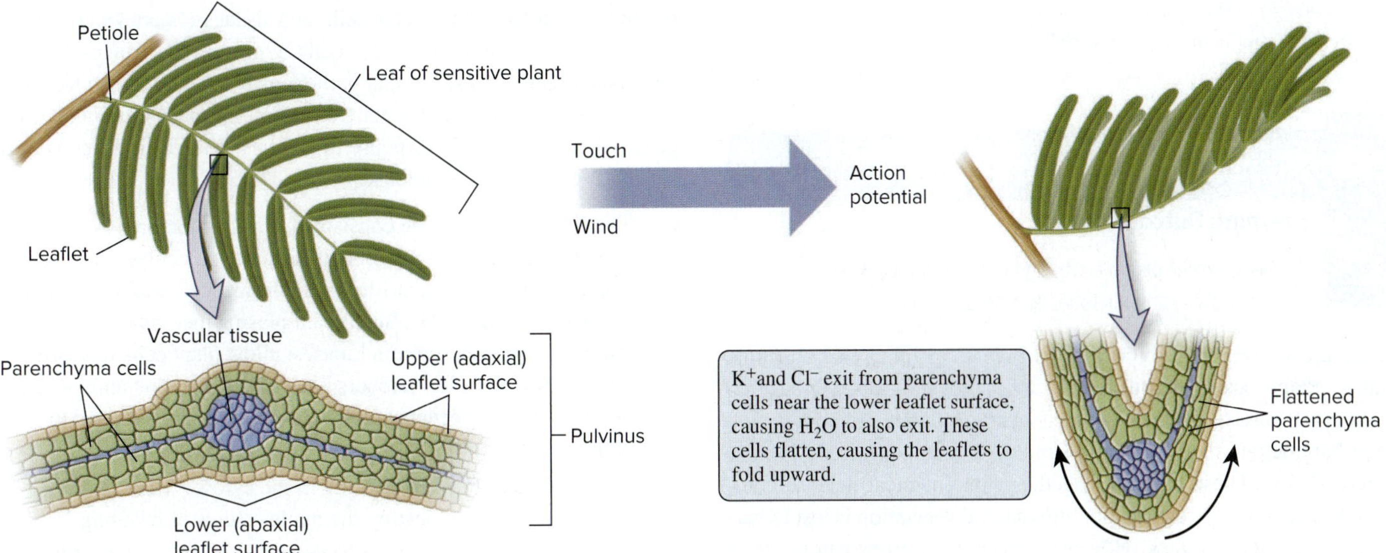


**Figure 29.12 Leaf folding in the sensitive plant: How pulvini change the positions of leaflets.** The electrical signals known as action potentials result from the rapid flow of ions through membrane ion channels. Action potentials spread from one plant cell to another through plasmodesmata. Swelling of cells on one side of the pulvinus and loss of water from cells on the other side cause bending.

near the lower surfaces of the pulvini. These cells expel potassium ($K^+$) and chloride ($Cl^-$) ions, causing water to flow out and the cells to become flattened. This bends the leaflets together, starting the leaflet-folding process. The action potential generated at the touch site also flows through the leaflet, causing many or all of the leaflets to also bend, with the result that the entire leaf folds. Reversal of this process allows the leaf to unfold.

Some plant movements that are unrelated to touch, including sleep movements and Sun tracking, also result from the action of pulvini. Sleep movements are changes in leaf position that occur in response to day-night cycles. Sun tracking is the movement of leaves or flowers in response to the Sun's position.

## 29.4 Reviewing the Concepts

- Plant shoots and roots respond to gravity (in the process of gravitropism) by means of starch-heavy statoliths located within statocytes. Rapid touch responses, such as leaf folding in sensitive plants, depend on changes in the water content of cells in structures known as pulvini (Figures 29.10, 29.11, 29.12).

## 29.4 Testing Your Knowledge

1. If you turn an evenly illuminated potted tomato plant onto its side, after some period of time the shoot will turn upward and the roots will turn downward. Which description of this plant growth response is most accurate?
   a. The shoot displays phototropism, and the root displays a touch response.
   b. The shoot displays a touch response, and the root displays negative phototropism.
   c. The shoot displays negative gravitropism, and the root displays positive gravitropism.
   d. The shoot displays positive gravitropism, and the root displays negative gravitropism.
   e. None of the above are correct.

## 29.5 Plant Responses to Attack

### Learning Outcome

1. Discuss how plants sense and respond to attack by disease-causing microbes and herbivorous animals.

Plants are vulnerable to attack by herbivores, animals that consume plants. Plants are also vulnerable to pathogens—disease-causing microorganisms. Structural barriers such as cuticles, trichomes, and outer bark help to reduce infection and herbivore attack (refer back to Figure 28.15). These barriers, together with chemical defense compounds, explain why remarkably little natural vegetation is lost to herbivore or pathogen attack. However, agricultural crops can be more vulnerable to attack than their wild counterparts. This is because some protective adaptations have been lost during crop domestication as the result of genetic changes that increase edibility. For this reason, crop scientists are particularly interested in understanding plant defense, with the goal of being able to breed or genetically engineer crop plants that are better protected from pests.

**Plants Defend Themselves Against Herbivores** Herbivory defense compounds include diverse secondary metabolites—terpenes and terpenoids, phenolics, and alkaloids (refer back to Figure 25.20)—but also poisonous hydrogen cyanide and other molecules. Some of these substances act directly on herbivores, making plants taste bad so that herbivores learn to avoid them. Other chemical compounds function indirectly, on the principle that "the enemy of my enemy is my friend." For example, when attacked by insect caterpillars, cruciferous plants release terpenoids that attract the bodyguard wasp (*Cotesia rubecula*), which attacks the caterpillars.

Protective plant hormones are often involved in plant defenses (**Figure 29.13**). For example, when insects wound tomato plants, damaged cells release the peptide hormone systemin. This hormone induces undamaged cells to produce the hormone jasmonic acid (JA), which functions as an alarm system. JA travels in the phloem or through the air to undamaged parts of the plant, inducing widespread, preemptive production of defensive compounds (see Figure 29.13, step 2a). Airborne compounds released from herbivore-damaged plants may also attract the enemies of insect attackers. Nearby plants may detect these airborne signals and respond by similarly arming themselves against attack (see Figure 29.13, step 2b).

**Plants Defend Themselves Against Pathogen Attack** Every year, about 15% of worldwide crop production is lost to disease caused by certain bacteria, fungi, protists, or viruses. During their evolution, plants have acquired mechanisms for detecting and responding to disease microbes, but microbes can rapidly evolve new disease processes. Agricultural scientists aim to understand how pathogenic microbes attack plants and how plants are able to prevent disease, in order to develop disease-resistant crop varieties.

Plant pathogens produce compounds known as **elicitors,** which promote the infection of plant cells and tissues (**Figure 29.14**). Some bacteria inject elicitors into plant cells by means of syringe-like systems (see Figure 29.14). Fungal pathogens often deliver elicitors into plant cells and absorb nutrients from them by means of specialized penetration structures. In response to elicitors, plant cells have evolved first and second lines of defense:

1. The first line of defense consists of plasma membrane receptors that specifically bind microbe-associated molecules, such as bacterial lipopolysaccharides or the fungal cell-wall compound chitin (see Chapter 23). Such plant plasma membrane receptors, which are typically protein kinases, allow plant cells to sense disease microbes. Researchers have discovered that animal cells use homologous receptor molecules to sense and respond to pathogens.
2. A second line of defense occurs in the cytosol. Here, microRNAs help to destroy the nucleic acids of invading viruses, in processes similar to those present in animal cells. In addition, receptors that occur in the cytosol allow plant cells to recognize injected elicitors. The binding of receptors with elicitors triggers the production of chemical defense

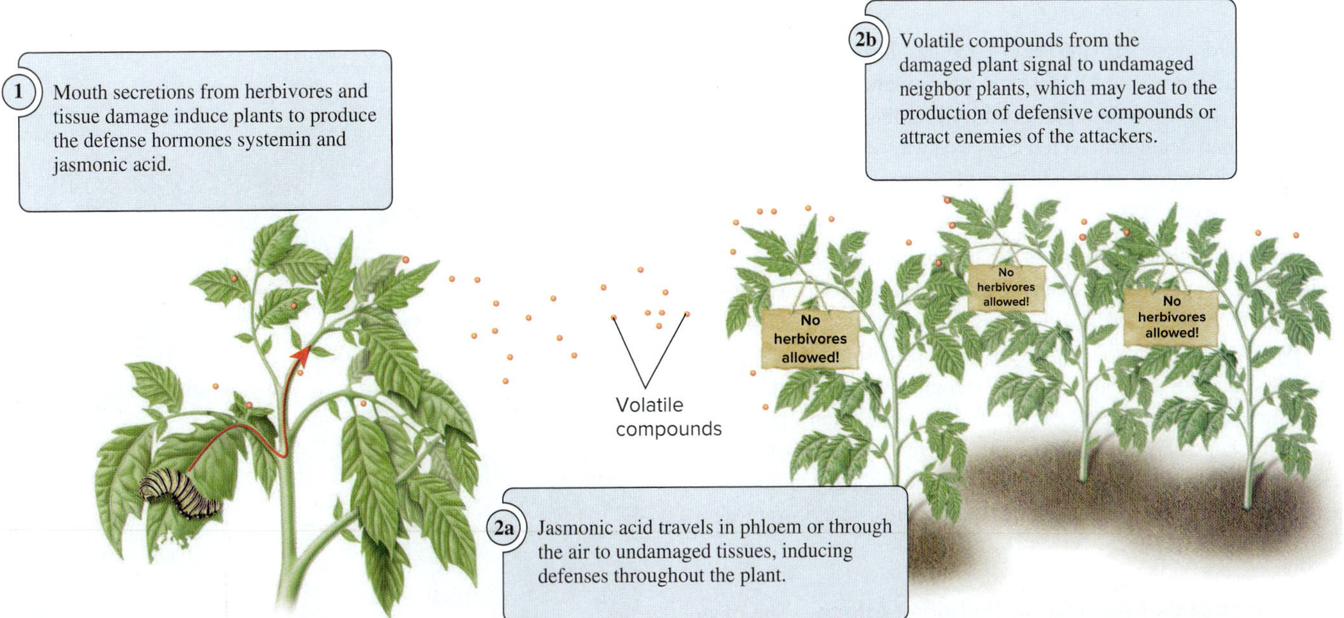

1 Mouth secretions from herbivores and tissue damage induce plants to produce the defense hormones systemin and jasmonic acid.
2b Volatile compounds from the damaged plant signal to undamaged neighbor plants, which may lead to the production of defensive compounds or attract enemies of the attackers.
No herbivores allowed!
No herbivores allowed!
No herbivores allowed!
Volatile compounds
2a Jasmonic acid travels in phloem or through the air to undamaged tissues, inducing defenses throughout the plant.


**Figure 29.13** **Plant responses to herbivore attack.** **(1)** Leaf damage induces local defensive responses, including production of compounds that travel through the plant and cause defense responses elsewhere. **(2a)** In addition, attacked plants may release volatile compounds, such as jasmonic acid. These volatile compounds foster the development of defenses in undamaged parts of the same plant. **(2b)** Such volatile compounds may also induce defenses in neighboring plants, which then become less vulnerable to herbivores. Volatile compounds may also attract predators that feed on the herbivore attacker.

**Concept Check:** *What advantage would a predator obtain by responding to the volatile signals emitted from herbivore-damaged plants? What adaptive advantage could a plant obtain by attracting the enemies of its attackers?*

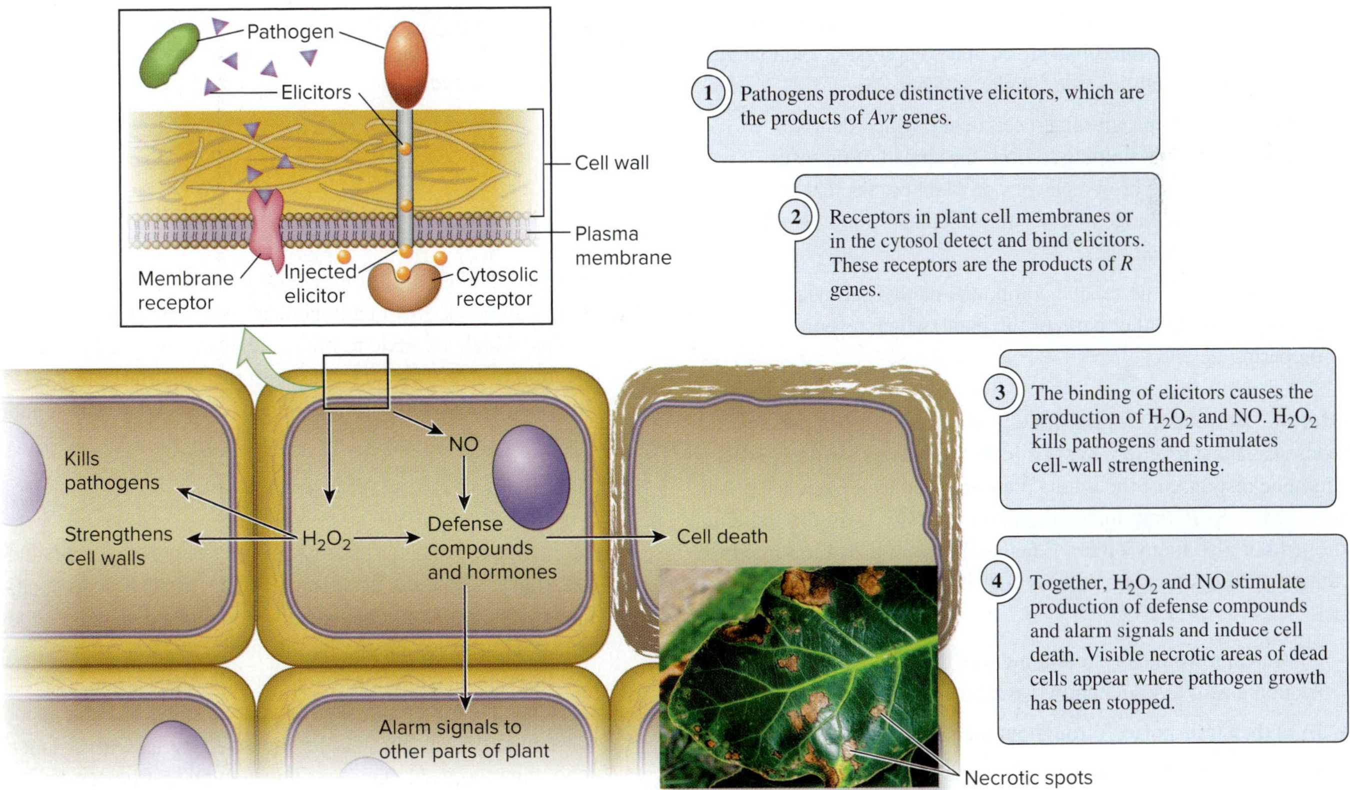

Pathogen
Elicitors
Cell wall
Plasma membrane
Membrane receptor
Injected elicitor
Cytosolic receptor
1 Pathogens produce distinctive elicitors, which are the products of *Avr* genes.
2 Receptors in plant cell membranes or in the cytosol detect and bind elicitors. These receptors are the products of *R* genes.
3 The binding of elicitors causes the production of $H_2O_2$ and NO. $H_2O_2$ kills pathogens and stimulates cell-wall strengthening.
NO
Kills pathogens
Strengthens cell walls
$H_2O_2$
Defense compounds and hormones
Cell death
4 Together, $H_2O_2$ and NO stimulate production of defense compounds and alarm signals and induce cell death. Visible necrotic areas of dead cells appear where pathogen growth has been stopped.
Alarm signals to other parts of plant
Necrotic spots


**Figure 29.14** **Pathogen/plant interactions and the hypersensitive response to pathogen attack.**

*(inset)* © G.R. "Dick" Roberts/Natural Sciences Image Library

signals: hydrogen peroxide ($H_2O_2$), the gas nitrous oxide (NO), and mobile hormones. Defense responses include deposition of polymers that strengthen plant cell walls at the site of attack, programmed death of infected cells as a way of limiting the spread of infection, and production of compounds that kill pathogens (see Figure 29.14).

Many types of **resistance genes (*R* genes)** encode plant disease receptor proteins. If a plant is genetically unable to produce a receptor that can recognize a pathogen's presence or elicitor, disease may result. By contrast, plants successfully resist disease when the product of a dominant *R* gene (a receptor) recognizes a pathogen's presence. Diverse alleles of *R* genes occur in plant populations, providing plants with a large capacity to cope with different types of pathogens. These receptors, together with plant responses to pathogen attack—the hypersensitive response and systemic acquired resistance—represent the plant immune system.

**The Hypersensitive Response to Pathogen Attack** The plant **hypersensitive response** is a local reaction to pathogen attack that limits the progression of disease (see Figure 29.14, steps 3 and 4). There are several components to the hypersensitive response:

- Plant-produced hydrogen peroxide ($H_2O_2$) can kill pathogens and helps strengthen the cell wall by deposition of cross-binding polymers.
- The gaseous hormone NO works with $H_2O_2$ to stimulate the synthesis of defensive secondary metabolites, the hormone salicylic acid (SA), and tough lignin in the cell walls of nearby tissues. Recall that lignin is a plant cell-wall polymer that helps cell walls to resist microbial attack and compression and that has waterproofing qualities (see Chapters 25 and 28).
- NO also induces programmed cell death, which deprives pathogens of food, helping to limit the spread of disease. Necrotic spots are brown patches on plant organs that reveal where disease pathogens have attacked plants and where plant tissues have battled back.
- Salicylic acid or the related compound methyl salicylate sends alarm signals to other parts of the plant, which respond by preparing defenses.

**Systemic Acquired Resistance** Whereas the hypersensitive response is a local reaction, **systemic acquired resistance (SAR)** is an immune response of the whole plant induced by a pathogen attack (**Figure 29.15**). SAR immunizes the plant not only from the inducing pathogen but also many others. The SAR system uses the same type of long-distance signaling process that is associated with herbivore attack (see Figure 29.13):

- At or near a wound, systemin induces the production of jasmonic acid (JA) in nearby vascular tissues.
- JA is then transported throughout the plant.
- In addition, the hormone salicylic acid can be converted to volatile methyl salicylate, which diffuses into the air surrounding a plant, inducing resistance in noninfected tissues. These signals involve changes in transcription of up to 10% of plant genes, including plant homologs of the *BRCA* genes associated with human breast cancer.
- Responding plant tissues may produce defensive enzymes, which can break down pathogen cell walls, or they may generate tannins, which are toxic to microorganisms. Experimental infection of *Arabidopsis* leaves with the bacterial pathogen *Pseudomonas syringae* led to induction of SAR within 6 hours.

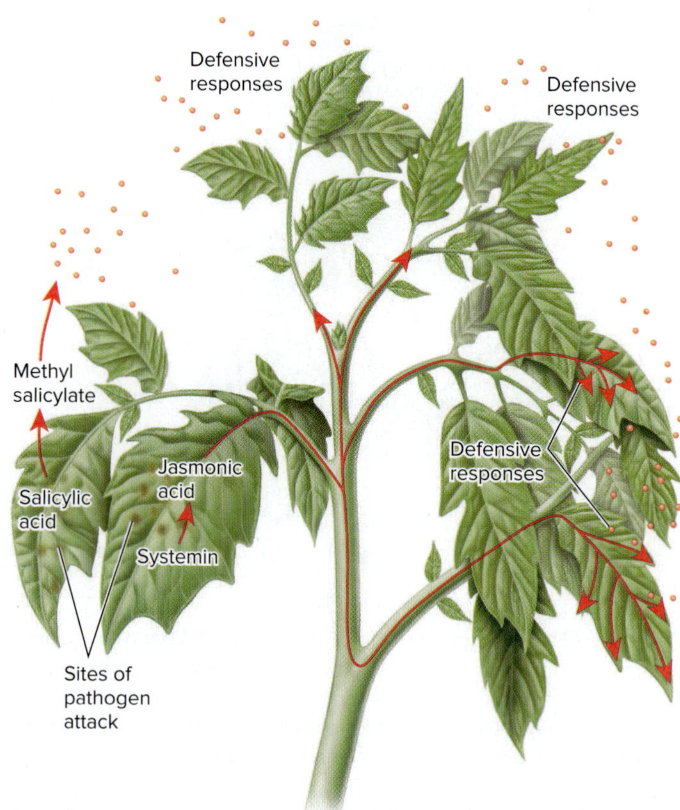


**Figure 29.15** Systemic acquired resistance to pathogen attack.

**Concept Check:** *In what way is systemic acquired resistance to pathogens similar to plant responses to herbivore attack?*

## 29.5 Reviewing the Concepts

- Plants cope with biological stresses such as herbivore and pathogen attack by means of structural and chemical adaptations. Injured plant parts produce volatile hormones that signal other parts of the same plant and nearby plants to produce defensive responses. Plant pathogens begin their attack by producing elicitor compounds. If an attacked plant can produce a compound that interferes with that elicitor, encoded by an *R* (resistance) gene, then the plant can resist infection by that pathogen. The hypersensitive response (HR) is a local defensive response to pathogen attack, whereas systemic acquired resistance (SAR) is a whole-plant immune response (Figures 29.13, 29.14, 29.15).

## 29.5 Testing Your Knowledge

1. Which of the following materials is involved in plant defensive responses to pathogen attack?
   a. systemin and jasmonic acid
   b. protein kinase receptor molecules
   c. microRNAs
   d. hydrogen peroxide
   e. all of the above

# Assess and Discuss

## Test Yourself

1. The major types of plant hormones include
   a. cyclic AMP, $IP_3$, and calcium ions.
   b. calcium, CDPKs, and DELLA proteins.
   c. auxin, cytokinin, and gibberellin.
   d. chlorophyll, carotenoids, and phytochrome.
   e. statoliths, pulvini, and parenchyma.
2. Phototropism is
   a. the production of flowers in response to a particular day length.
   b. the production of flowers in response to a particular night length.
   c. the growth response of a plant, an organ system, or an organ to directional light.
   d. the growth response of a plant, an organ system, or an organ to gravity.
   e. the growth response of a plant, an organ system, or an organ to touch.
3. What is the most accurate order of events during signal transduction?
   a. first, receptor activation; then, messenger signaling; and last, an effector response
   b. first, an effector response; then, messenger signaling; and last, receptor activation
   c. first, messenger signaling; then, receptor activation; and last, an effector response
   d. first, an effector response; then, receptor activation; and last, messenger signaling
   e. none of the above
4. Which of the plant hormones is known as the "master hormone," and why?
   a. cytokinin, because many plant functions require cell division
   b. gibberellins, because growth is essential to many plant responses
   c. abscisic acid, because it is necessary for leaf and fruit drop
   d. brassinosteroids, because water uptake is so fundamental to plant growth
   e. auxin, because there are many different auxin-responsive genes
5. Gaseous hormones are able to enter cells without requiring special membrane transporter systems. Which of the major plant hormones is a diffusible gas?
   a. auxin
   b. gibberellin
   c. cytokinin
   d. ethylene
   e. abscisic acid
6. Photoreceptor molecules allow plant cells to detect light of particular wavelengths. To which wavelengths of light is phytochrome responsive?
   a. violet and ultraviolet
   b. blue and green
   c. red and far-red
   d. all of the above
   e. none of the above
7. Gravitropism is a plant response to
   a. light.
   b. cold.
   c. touch.
   d. gravity.
   e. drought.
8. How do plants defend themselves against pathogens?
   a. Plants produce resistance molecules (usually proteins) that bind pathogen elicitors, thereby preventing disease.
   b. Plants display a hypersensitive response that limits the ability of pathogens to survive and spread.
   c. Plants display systemic acquired resistance, whereby an infection induces immunity to diverse pathogens in other parts of a plant.
   d. All of the above are correct.
   e. None of the above are correct.

## Conceptual Questions

1. Why can plants be said to display behavior?
2. Why do plants produce so many types of resistance (*R*) genes?
3. **PRINCIPLES** A principle of biology is that living organisms interact with their environment. Because diverse plants exude volatile compounds in response to herbivore or pathogen attack, some experts have written about "talking trees." Is there any such thing?

## Collaborative Questions

1. Why are most wild plants distasteful, and some even poisonous, to people?
2. How could you increase the resistance of a particular crop plant species to particular types of herbivores?

# Online Resource

# 30

# How Flowering Plants Obtain and Transport Nutrients

© Linda Graham

**Arid land vegetation in the Mojave Desert of the United States.**

## Chapter Outline



Millions of people live in arid lands. Arid lands are places where natural and crop plant growth is limited by water availability, as shown in the chapter-opening photo. Past indigenous societies invented clever agricultural practices that effectively supported relatively low human populations. But more recently, increasing human populations and global climate change have challenged the ability of people to survive in dry lands. Famines can arise when droughts decrease food production. To aid food production everywhere, and particularly in dry lands, biologists work closely with climate and soil scientists to better understand how plants interact with water and other materials in the soil and atmosphere.

One amazing discovery resulting from such multidisciplinary work is that plants actually increase local rainfall by moving huge amounts of water from the soil to the atmosphere. Plants take up $CO_2$ from the air and absorb water and dissolved mineral nutrients from the soil for use in photosynthesis. Organic food produced via photosynthesis supports plant growth and reproduction, and herbivorous animals also depend on it. But many people don't realize that the process that plants use to obtain and transport minerals and food within their bodies results in the release of large amounts of water into the atmosphere. For every $CO_2$ molecule that plants convert to organic food during the process of photosynthesis, plants transport 500 water molecules from the soil through their bodies, releasing the water to the air as vapor! Because rain is condensed water vapor, plants actually generate much of the precipitation they and humans need, particularly in arid regions. In addition, by exuding water vapor, plants cool themselves and the surrounding air. Consequently, when vegetation is lost as the result of livestock overgrazing and other effects of burgeoning human populations, local rainfall decreases and temperatures increase. Agriculture and human life then become even more challenging, particularly in arid regions.

Dry-land agricultural concerns illustrate why scientists interested in helping people to achieve more healthful and sustainable lives need a basic understanding of how plants obtain nutrition and transport materials. These topics, covered in this chapter, are both related to plant interactions with water.

## 30.1 Plant Nutritional Requirements

### Learning Outcomes

1. List the major nutritional resources that green plants need for healthy growth.
2. Distinguish between plant macronutrients and micronutrients.

Plants, like animals, have requirements for **nutrients—**substances that are metabolized by or incorporated into an organism. Photosynthetic plants require water ($H_2O$), carbon dioxide ($CO_2$), and more than a dozen elements to produce organic food by means of photosynthesis. $CO_2$ is primarily absorbed from air, whereas $H_2O$ and elements in the form of mineral ions—including potassium (as $K^+$), nitrogen (as $NH_4^+$ or $NO_3^-$), and calcium (as $Ca^{2+}$)—are primarily taken up from soil (**Figure 30.1**). Although this chapter focuses on flowering plants, the principles described here apply more broadly to all plants (refer back to Chapter 25 for a survey of plant diversity: bryophytes, lycophytes, ferns, gymnosperms, and flowering plants).

The chemical elements required by plants play many roles in plant metabolism, often functioning as enzyme cofactors (**Table 30.1**). Elements that are generally required in amounts of at least 1 g/kg of plant dry matter are known as **macronutrients.** In contrast, elements that are needed in amounts at or less than 0.1 g/kg of plant dry mass are known as **micronutrients,** or trace elements. As is also the case for animals, an insufficient amount of essential nutrients can cause deficiency symptoms. Chlorosis, a yellowing of plant tissue resulting from an insufficiency of one or more minerals needed to make chlorophyll, is an example of a plant nutrient deficiency symptom (**Figure 30.2**). Because insufficient amounts of water, light, carbon

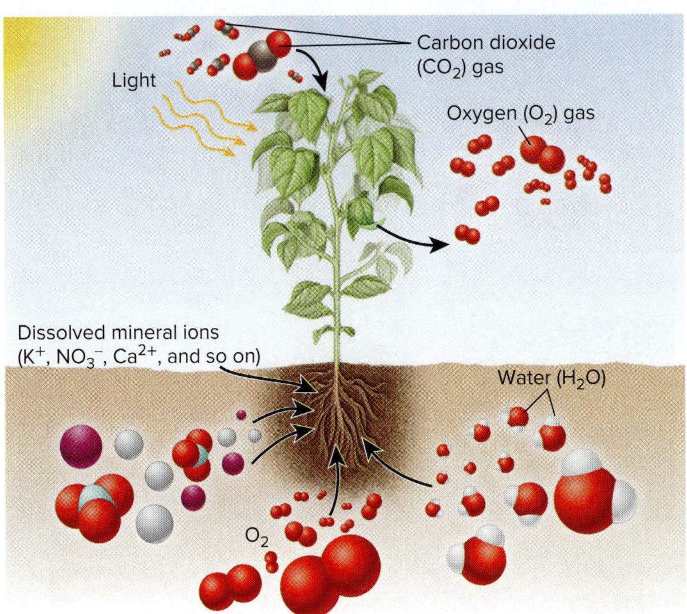


**Figure 30.1** The major types of plant nutrients and their sources.

**Table 30.1 Plant Essential Mineral Nutrients**

| Element (chemical symbol) | Percent of plant dry mass | Major source | Form taken up by plants | Function(s) |
|---|---|---|---|---|
| **Macronutrients** | | | | |
| Carbon (C) | 45 | Air | $CO_2$ | Component of all organic molecules |
| Oxygen (O) | 45 | Air, soil, water | $CO_2$, $O_2$, $H_2O$ | Component of all organic molecules |
| Hydrogen (H) | 6 | Water | $H_2O$ | Component of all organic molecules; protons used in chemiosmosis and cotransport |
| Nitrogen (N) | 1.5 | Soil | $NO_3^-$, $NH_4^+$ | Component of proteins, nucleic acids, chlorophyll, coenzymes, and alkaloids |
| Potassium (K) | 1.0 | Soil | $K^+$ | Essential role in cell ionic balance |
| Calcium (Ca) | 0.5 | Soil | $Ca^{2+}$ | Component of cell walls; messenger in signal transduction |
| Magnesium (Mg) | 0.2 | Soil | $Mg^{2+}$ | Component of chlorophyll; activates some enzymes |
| Phosphorus (P) | 0.2 | Soil | $HPO_4^{2-}$ | Component of nucleic acids, ATP, phospholipids, and some coenzymes |
| Sulfur (S) | 0.1 | Soil | $SO_4^{2-}$ | Component of proteins, some coenzymes, and defense compounds |
| **Micronutrients** | | | | |
| Chlorine (Cl) | 0.01 | Soil | $Cl^-$ | Required for water splitting in photosystem; cell ion balance |
| Iron (Fe) | 0.01 | Soil | $Fe^{3+}$, $Fe^{2+}$ | Enzyme cofactor; component of cytochromes |
| Manganese (Mn) | 0.005 | Soil | $Mn^{2+}$ | Enzyme cofactor |
| Boron (B) | 0.002 | Soil | $B(OH)_3$ | Enzyme cofactor; component of cell walls |
| Zinc (Zn) | 0.002 | Soil | $Zn^{2+}$ | Enzyme cofactor |
| Sodium (Na) | 0.001 | Soil | $Na^+$ | Required to generate PEP in $C_4$ and CAM plants |
| Copper (Cu) | 0.0006 | Soil | $Cu^+$, $Cu^{2+}$ | Enzyme cofactor |
| Molybdenum (Mo) | 0.00001 | Soil | $MoO_4^{2-}$ | Enzyme cofactor |
| Nickel (Ni) | 0.000005 | Soil | $Ni^{2+}$ | Enzyme cofactor |

**Figure 30.2** **Chlorosis as a symptom of mineral deficiency.** This camellia plant is suffering from an iron deficiency, as revealed by its yellow leaves, a symptom known as chlorosis. Other mineral deficiencies can also cause chlorosis.
© Geoff Kidd/SPL/Science Source

dioxide, and other mineral nutrients can limit the extent of green plant growth, these resources are known as limiting factors.

**Water** Though plants vary in water content, water typically makes up about 90% of the weight of living plants. Although many types of plants living in arid lands display adaptations that allow them to survive for extended periods in nearly dry conditions, all plants require an adequate supply of water for active metabolism and growth. Water is essential to plants for several reasons:

- Water may act as a nutrient by supplying most of the hydrogen atoms for incorporation into organic compounds. Hydrogen from water is incorporated into plants when $CO_2$ is reduced during the process of photosynthesis.
- Water is the solvent for other mineral nutrients and is the main transport medium in plants, allowing movement of minerals and other solutes throughout the plant body via the vascular tissues.
- Evaporative loss of water from plant surfaces helps to cool plants and their environments.
- Cytoplasmic and vacuolar water help to support plants by maintaining hydrostatic pressure on the cell wall.

**Light** Photosynthetic plants require light energy for the formation of the covalent bonds of organic compounds that make up the plant's body. Green plants' use of light energy as an essential resource parallels animals' nutritional requirements for organic food as a source of chemical energy.

Remarkably, several hundred species of flowering plants have lost their photosynthetic capacity; for these heterotrophic plants, essential nutrients include absorbed organic compounds that replace light as a source of energy. An example is the ghostly white Indian pipe (*Monotropa uniflora*), which grows in forests (Figure 30.3). Studies of early-diverging bryophyte plants and green algal relatives performed by evolutionary biologist Linda Graham and others reveal that heterotrophic capacity is an ancient plant trait; modern angiosperms inherited heterotrophic capacity from their ancestors. This evolutionary insight explains why heterotrophic nutrition occurs in diverse groups of modern flowering plants that have lost photosynthetic capacity.

For photosynthetic plants, the amount of light available may be less than or greater than optimal, and different plant species are adapted to cope with varying amounts of light and shade:

- In forests, light availability limits the growth of tree seedlings and other small plants that are shaded by the

## Biology Principle

### Living Organisms Use Energy

*Monotropa uniflora* is one of many types of nongreen heterotrophic plants that lack photosynthetic capacity and therefore must absorb organic compounds for use as an energy source. In contrast, nearby autotrophic green plants use sunlight as a source of energy.

**Figure 30.3** **The heterotrophic flowering plant, Indian pipe (*Monotropa uniflora*).**
© Robert Ziemba/agefotostock

leafy tree canopy overhead. Forest seedlings often rely on organic food stored within the seed or supplied by underground networks of mycorrhizal fungi (refer back to Figure 24.12), which provide organic food tapped from the roots of mature trees. Forest saplings may not be able to grow maximally until a light gap appears in the leafy canopy.

- Conversely, plants growing in deserts or on mountains often experience light so intense that it can damage their photosynthetic components. Plants have evolved light-absorbing carotenoid pigments and other adaptations that help them cope with too much light.
- Plants can adjust the thickness of leaves in response to variations in light availability. Some plants produce sun-leaves in sunnier parts and shade-leaves in more shaded parts. Sun-leaves have a thicker layer of chlorophyll-rich mesophyll and are able to harvest more of the bright sunlight that penetrates deeply into the leaf. In contrast, shade-leaves have a thinner mesophyll layer, lacking enough light to supply a thicker layer of photosynthetic tissue (**Figure 30.4**).

## Biology Principle

### Structure Determines Function

These scanning electron micrographs of cut leaves reveal the thinner mesophyll and greater amount of air spaces in **(a)** shade-leaves compared with **(b)** sun-leaves.

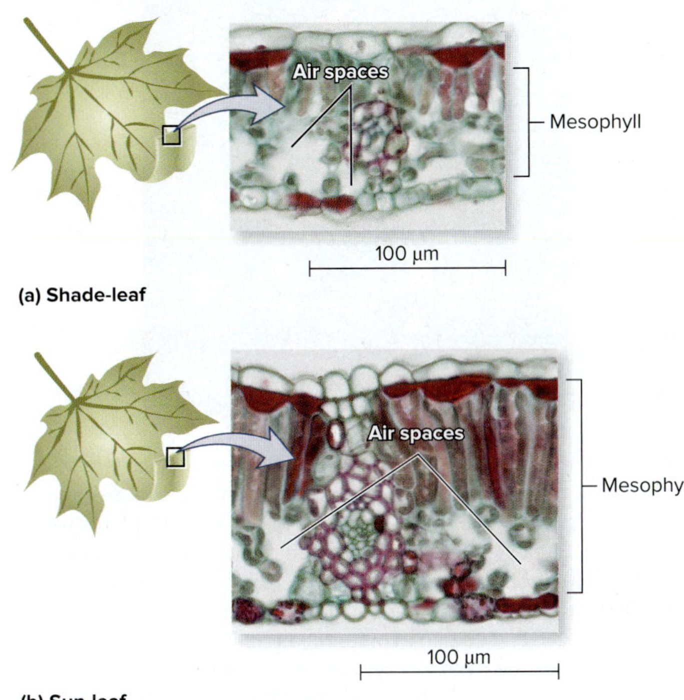


**Figure 30.4** Shade- and sun-leaves.
*(a–b)* © Lee W. Wilcox

**Carbon Dioxide** Light provides the energy for plant photosynthesis, but most plant dry mass originates from carbon dioxide ($CO_2$). Most plants obtain $CO_2$ gas from the atmosphere by absorption through **stomata,** pores that occur in the plant epidermis, which plants can close to retain water or open to allow the entry of $CO_2$ needed for photosynthesis. Under experimental conditions, plant photosynthetic rates increase with $CO_2$ concentration until the Calvin cycle enzyme rubisco has become fully supplied—that is, saturated with $CO_2$. In nature, plants are often unable to obtain enough $CO_2$ for maximal photosynthesis. This is partly because the modern atmospheric concentration of $CO_2$ is only 390 $\mu$L/L (a small fraction of atmospheric content), whereas considerably more $CO_2$ would be required to saturate photosynthesis in most plants.

Climate plays an important role in the ability of plants to acquire $CO_2$. Plants that experience hot, dry conditions are particularly vulnerable to low $CO_2$ levels. When their stomata close to conserve water, the plants cannot absorb $CO_2$, which limits photosynthesis. About 30,000 species of $C_4$ and CAM plants, common in arid lands, possess various types of structural and biochemical adaptations that help them cope with $CO_2$ limitation by improving their $CO_2$ absorption (see Chapter 7). You may recall that $C_4$ photosynthesis typically involves two types of leaf cells; mesophyll cells use a special enzyme (PEP carboxylase) to trap $CO_2$ into a four-carbon compound that moves into bundle-sheath cells that surround leaf veins (vascular bundles). Within the bundle-sheath cells, $CO_2$ is released and utilized by the Calvin cycle to produce organic compounds. This system allows $C_4$ plants to keep the Calvin cycle supplied with $CO_2$ even when stomata must be partially closed to conserve water. *Bienertia cycloptera*, which grows in Iranian deserts, is remarkable because $C_4$ photosynthesis occurs within single cells. Plastids at the cell periphery use PEP carboxylase to trap $CO_2$, thereby functioning like mesophyll cells of most $C_4$ plants, and plastids clustered near the nucleus function like the bundle-sheath cells of most $C_4$ plants (**Figure 30.5**). Plant biologists study diverse $C_4$ plants to gain insight into ways that crop plants might be engineered as a way to increase food production.

When grown in greenhouses and supplied with sufficient light, moisture, and air that has been enriched with $CO_2$, tomatoes, cucumbers, leafy vegetables, and some other crops can double their growth rate. This observation has stimulated plant biologists to perform experiments designed to help them predict how wild and crop plants might respond to expected increases in Earth's atmospheric $CO_2$ content.

In addition to light, $CO_2$, and water, plants require additional elements that occur naturally in soil or that can be added in the

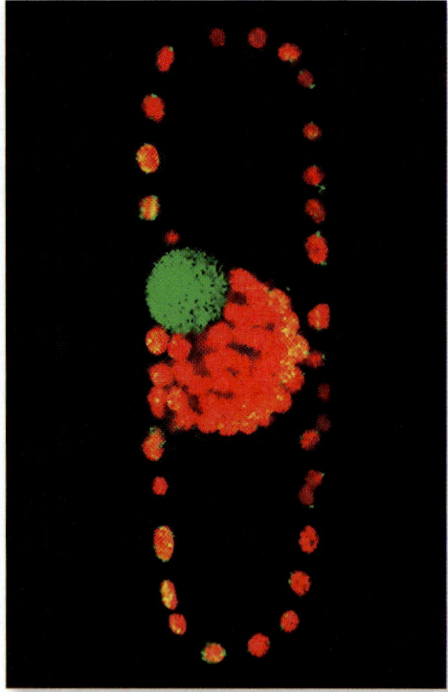

**Figure 30.5 A $CO_2$-acquisition adaptation.** *Bienertia cycloptera,* native to desert regions, is able to conduct $C_4$ photosynthesis within the confines of a single cell (by contrast, most $C_4$ plants use two different types of cells to accomplish photosynthesis). In this fluorescence photograph of a *B. cycloptera* cell, the red plastids at the cell periphery function like the mesophyll cells of most $C_4$ plants, which take up $CO_2$ and use it to produce four-carbon organic acids that eventually move into the cytoplasm. In contrast, the red plastids clustered near the green-stained nucleus function much like the bundle-sheath cells of most $C_4$ plants, in which $CO_2$ (released from organic acids absorbed from the cytoplasm) is used in the Calvin cycle to make sugar. The plastids appear red because chlorophyll emits red light when it fluoresces.

Confocal fluorescence micrograph courtesy Simon D.X. Chuong

form of fertilizers (see Table 30.1). Section 30.2 focuses on the roles of soil in plant nutrition.

## 30.1 Reviewing the Concepts

- The nutritional requirements of terrestrial green plants are $H_2O$, light energy, $CO_2$, and mineral ions such $K^+$, $NO_3^-$, and $Ca^{2+}$ absorbed from soil. Deficiency symptoms such as chlorosis signal the lack of sufficient nutrients (Figures 30.1, 30.2; Table 30.1).
- Like animals and fungi, diverse heterotrophic plants obtain chemical energy by metabolizing organic compounds absorbed from their environment (Figure 30.3).
- Plants display adaptations that allow them to cope with resource levels that are not optimal. Sun- and shade-leaves are adaptations to variation in light availability; $C_4$ and CAM photosynthesis represent adaptation to low $CO_2$ (Figures 30.4, 30.5).

## 30.1 Testing Your Knowledge

1. Animals require $O_2$ and food in the form of organic compounds that supply energy and provide essential minerals, such as calcium. What essential resources do green plants require?
   a. water as a source of H and O
   b. light as a source of energy
   c. $CO_2$ as a source of carbon for making organic compounds
   d. elements in the form of mineral ions such as $K^+$, $NH_4^+$ or $NO^{3-}$, and $Ca^{2+}$
   e. all of the above
2. Which environment is likely to display more types of plants that possess $C_4$ photosynthesis?
   a. cool, moist environment
   b. hot, dry environment
   c. cool, dry environment
   d. hot, moist environment
   e. all of the above

## 30.2 The Roles of Soil in Plant Nutrition

### Learning Outcomes

1. Discuss the advantages of using soils high in organic matter for plant cultivation.
2. Explain the advantages of loam soils for plant cultivation.
3. List reasons that plants require a source of fixed nitrogen.

In addition to light, $CO_2$, and water, plants require additional elements that occur naturally in soil or that can be added in the form of fertilizers (see Table 30.1). Soil is an essential resource for most wild and cultivated plants, providing water and other essential nutrients. For this reason, extensive loss of soil by wind and water erosion is of wide concern. Soils vary greatly in fertility—their ability to support plant growth—because they vary in degree of aeration, water-holding capacity, and mineral content. These properties arise from differences in the physical structure and chemical features of soils.

### Soils Have Both a Physical and Chemical Structure

Natural soils display layers, the topsoil being the uppermost layer and the one most important to plant growth (**Figure 30.6a**). The remains of plants that have recently died and other organisms form a layer of organic litter above the topsoil. Many of the inorganic minerals and organic materials that enrich high-quality topsoil arise from the activities of microorganisms that decompose the litter. In this way, the minerals contained in living plants are eventually recycled to subsequent generations. Beneath the topsoil lie layers called subsoil and soil base, which are largely composed of mineral materials. Bedrock is the bottom layer that supports the soil layers. Plant roots play an important role in conveying deep-lying minerals to the surface, thereby helping to enrich the topsoil.

Soil layers vary in composition and thickness, depending on various factors, including climate, vegetation, bedrock type, and human influences. For example, natural grasslands produce deep topsoil rich in both organic matter and minerals useful to plants, with the result

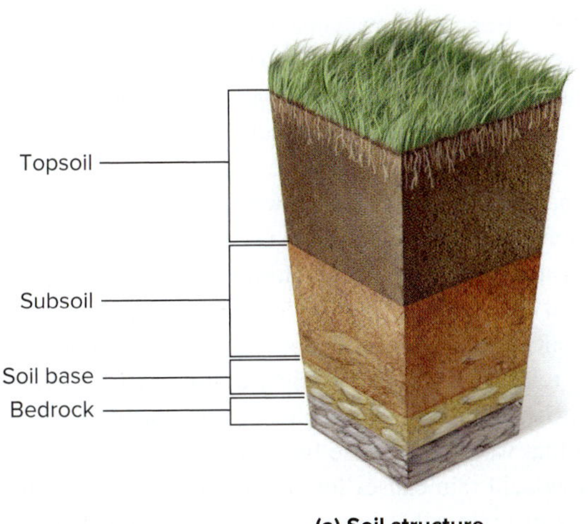


(a) Soil structure

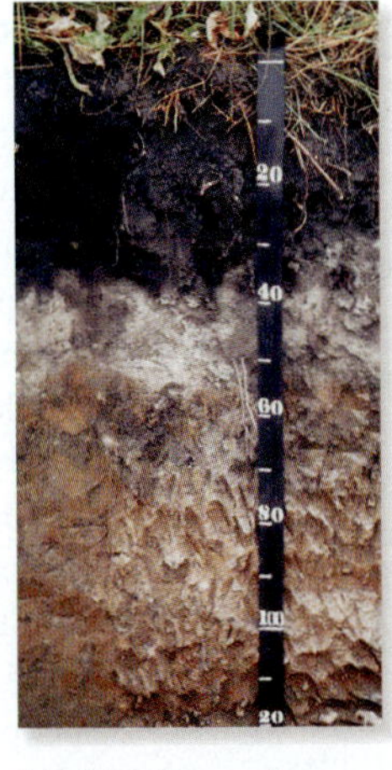

(b) Thick layer of topsoil in cropland

(c) Thin layer of topsoil in rain forest

**Figure 30.6** **The structural layers of soil.** **(a)** Diagram showing general soil structure. **(b)** A vertical view of an agricultural soil, showing a relatively deep layer of dark topsoil. **(c)** A vertical view of a tropical rain forest soil, showing a thin layer of dark topsoil.

*(b)* Courtesy of C.A. Stiles, USDA-NRCS; *(c)* © Ruddy Gold/agefotostock

that grassland soils have been transformed into croplands in many regions of the world (Figure 30.6b). In dramatic contrast, tropical rain forests often have only thin layers of topsoil; their low-fertility soils are composed mostly of inorganic materials of types that are not useful to plants (Figure 30.6c). Understanding the roles of soil organic matter and inorganic soil components is essential to maintaining and improving soil fertility.

**Soil Organic Matter** Soil organic matter, also known as humus, is largely derived from plant detritus, the dead and decaying remains of plants, although animal wastes and decayed animal bodies also contribute to the organic content of soils. Maintaining or improving soil organic matter provides many benefits:

- Organic-rich soils, containing 8% or so organic matter, are less likely to erode—wash or blow away with water or wind.
- Soil organic matter binds mineral nutrients, thereby fostering soil fertility.
- Soil organic matter softens the soil's consistency, thereby fostering plant root growth and farmers' ability to cultivate the soil.
- Gardeners often produce their own organic-rich compost by layering vegetable waste from the kitchen and yard waste with soil, and watering and turning the pile occasionally to introduce the water and oxygen necessary for decomposition. Mature compost is then mixed with garden soil to improve the soil's fertility.
- After harvesting a crop such as corn, farmers often leave the plant stubble to help hold soil in place and then plough the rotting remains back into the soil to increase its organic content.

**Inorganic Soil Components** Inorganic materials in soil are derived from the physical and chemical breakdown of rock by (1) cycles of freezing and thawing and (2) organic acids exuded from plant roots or lichens. Inorganic materials needed by plants, such as $K^+$, $NH_4^+$ or $NO_3^-$, and $Ca^{2+}$ (see Table 30.1), can be lost from soil by **leaching,** the dissolution and removal of minerals as water percolates through materials. Heavy rainfall can reduce the fertility of soils by leaching large amounts of nutrients from them. For example, high annual rainfall in tropical regions causes extensive leaching of minerals, explaining why tropical soils can be low in nutrients. In arid lands, irrigation of crops with ion-laden water, followed by evaporation, can increase soil saltiness to a degree that inhibits plant growth.

**Soil Types** Soils can be classified according to their relative content of coarse and fine inorganic materials, classified as sand, silt, or clay (Figure 30.7):

- Sand grains range from 20 to 2,000 μm (micrometers) in diameter.
- Particles of silt range from 2 to 20 μm in diameter.
- Clay particles are smaller than 2 μm in diameter.

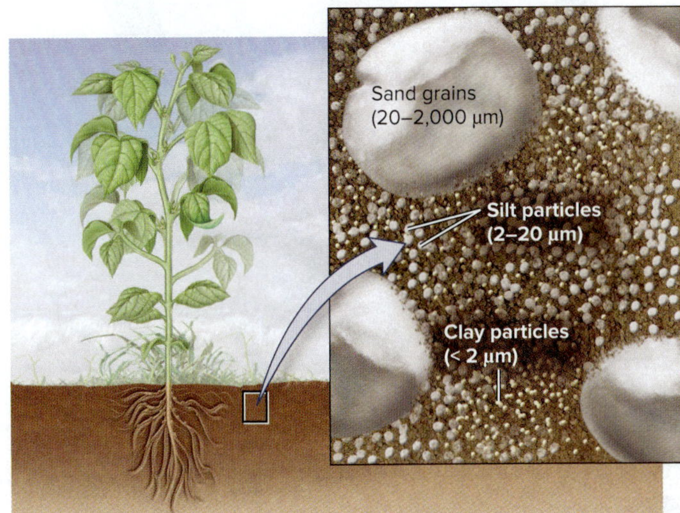


**Figure 30.7** **The relative sizes of inorganic soil components.**

Because of their size differences, sand, silt, and clay particles confer different properties on soils, allowing them to be classified as sandy, silty-clay, or loam soils.

- Soils that contain 45% or more sand and 35% or less clay are classified as sandy soils. The relatively large size and irregular shapes of sand particles allow air and water to move rapidly through sandy soils, which are said to be porous (Figure 30.8). Sandy soils are well aerated, which is favorable to the growth of plant roots, because they require oxygen for cellular respiration. However, sandy soils hold less water than the same volume of clay, and rapid percolation of water through sandy soils both reduces the amount of water available to the roots and leaches minerals from the soil.
- Silty-clay soils contain more than 35% of silt and clay particles. Silt and clay particles fit closely together, so such soils are less porous than sandy soils. Water percolates less easily through silty-clay soils, which therefore retain more minerals than do sandy soils. In addition, clay particles have negative charges on their surfaces that electrostatically bind positively charged ions (cations) such as $K^+$, $NH_4^+$, and $Ca^{2+}$. Because these and some other soil cations are plant nutrients, their retention benefits plant growth.
- Loam soils contain a balanced mixture of sand, silt, and clay and thus combine the aeration benefits of sandy soils with the water- and cation-retention properties of silty-clay soils. Hence, loam soil is considered ideal for the cultivation of many plants.

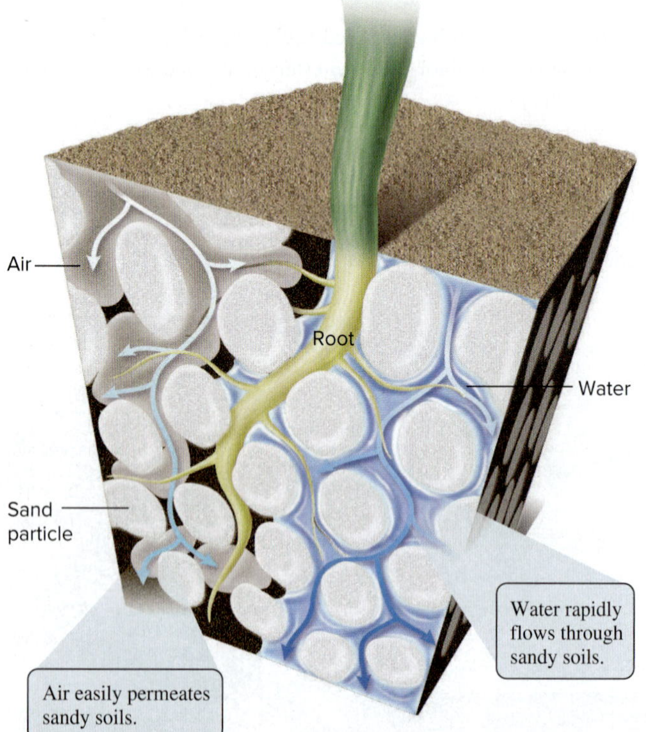


**Figure 30.8** The movement of air and water in sandy soil.

**Concept Check:** *In which type of soil is nutrient leaching a more common problem: sandy or clay soil?*

**Cation Exchange** To become available to plants, cations must be detached from clay particles. Plant roots often exude organic acids that help to release cations. Recall that an acid is a molecule that releases hydrogen ions ($H^+$, protons) into solutions. Hydrogen ions released into soil water are able to replace mineral cations on the surfaces of clay particles in a process known as **cation exchange** (Figure 30.9). Cation exchange releases cations into soil water, making them available for uptake by plant roots. However, free ions are also more easily washed out of soil.

If the $H^+$, concentration becomes too high, large numbers of mineral ions are released and can be leached from soil by heavy rainfall. Such leached minerals may include heavy metals such as aluminum ($Al^{3+}$) that would otherwise be bound in soil. These processes explain how acid rain causes loss of soil fertility and pollution of streams with toxic substances such as aluminum that can harm human health. Humans foster acid rain by burning fossil fuels and setting fire to woodlands, processes that add acidic contaminants to the atmosphere.

**The Role of Fertilizers** Soil additions known as fertilizers enhance plant growth by providing essential elements that are otherwise absent or present in less than optimal amounts. Fertilizers occur in organic and inorganic forms:

- Organic fertilizers are those in which most of the minerals are bound to organic molecules and are thus released relatively slowly. Manure and compost are examples of organic fertilizers. They play an important role in organic farming, the production of crops without the use of inorganic fertilizers, growth substances, and pesticides.
- Inorganic fertilizers consist largely of more water-soluble inorganic minerals, which are immediately useful to plants but can more easily leach from soils by heavy rainfall. Nitrogen (N), phosphorus (P), and potassium (K) are the mineral nutrients that most frequently limit crop growth and are common components of commercial inorganic fertilizers. Such fertilizers are available in different ratios of minerals that are optimal for different types of plants.

Even though fertilizers can improve the growth of crops, excessive application of fertilizers to fields and lawns is undesirable because (1) fertilizers are costly, and (2) minerals not taken up by plant roots are easily washed away by rain into waterways. In aquatic ecosystems, excess nutrients can fuel large growths of algae and aquatic plants that can harm other aquatic life. For example, large areas of the Gulf of Mexico and other coastal regions are now known as "dead zones" because microbial decomposition of large algal populations is a process that depletes oxygen from the water, suffocating the animal life. These large populations of algae are fostered by fertilizers that wash from farm soils into rivers, such as the Mississippi River, and then drain into coastal oceans. By the overuse of fertilizers, humans are directly responsible for the occurrence of coastal dead zones. More careful application of fertilizers and planting stream and river edges with vegetation that helps to absorb mineral nutrients can help reduce or prevent dead zones.

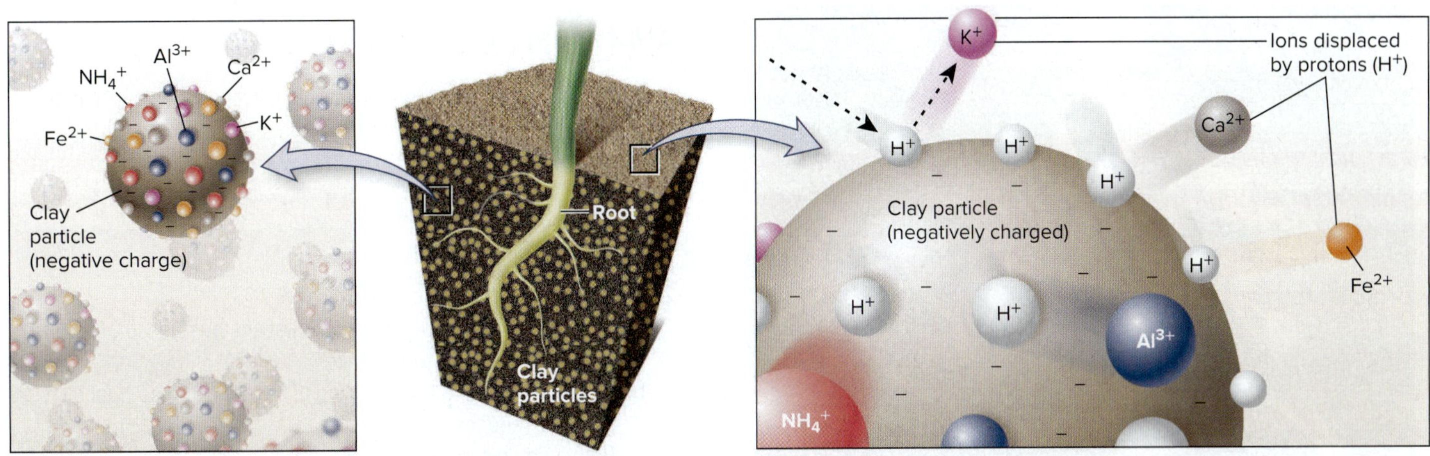


**Figure 30.9** **Cation binding and exchange.** **(a)** Clay and organic particles in the soil display negative electrostatic surface charges that bind cations. Bound cations include not only plant mineral nutrients such as ammonium ($NH_4^+$) but also cations that are not plant nutrients, such as aluminum ion ($Al^{3+}$). **(b)** Cation exchange occurs when protons ($H^+$) displace other cations, releasing them to soil water. This process makes cations more available for uptake by plant roots, but it also increases the potential for cations to leach away during rains or floods.

## Plants Require Fixed Nitrogen

As we have seen, plants require a variety of mineral nutrients obtained from the soil (see Table 30.1). Here, we focus on nitrogen as an important example. Nitrogen availability frequently limits plant growth in nature and in crop fields, because large amounts of it are required for plant synthesis of amino acids, proteins, nucleotides, chlorophyll, and alkaloids, among many other cellular materials.

Plants take up nitrogen in a chemically combined form, ammonium ($NH_4^+$) or nitrate ($NO_3^-$), and such combined forms of nitrogen are known as **fixed nitrogen.** New fixed nitrogen can be added to soils by the action of lightning, fire, and air pollution, as well as biological and industrial nitrogen fixation. **Nitrogen fixation** is the process by which atmospheric $N_2$ is combined with hydrogen (that is, reduced) to produce ammonium ($NH_4^+$). Most of the fixed nitrogen in soils is produced by biological nitrogen fixation, which is performed in nature only by certain prokaryotic species. These species possess an enzyme known as nitrogenase that reduces atmospheric $N_2$ at normal temperatures and pressures. Nitrogen fertilizers applied to crops are produced by industrial nitrogen fixation, a process that requires very high temperatures and pressures and thus consumes large amounts of energy.

**Biological Nitrogen Fixation by Bacteria** Nitrogen-fixing bacteria include many types of cyanobacteria, which are photosynthetic organisms that occur in oceans, lakes, and other aquatic systems (see Chapter 23). Certain nitrogen-fixing cyanobacteria live symbiotically within the leaves of water ferns that naturally occur in wet rice paddies. Fixed nitrogen produced by these cyanobacteria fertilizes not only the water ferns but also the rice plants, thereby helping to feed millions of people who depend on rice as their staple food. Nitrogen-fixing cyanobacteria also occur in surface soil crusts that are common in arid lands; fixed nitrogen produced by soil crusts enhances soil fertility (Figure 30.10).

Various types of nonphotosynthetic nitrogen-fixing bacteria also enrich water and soil. Nitrogen-fixing bacterial symbionts are taken up and fostered within special nodules on the roots of legume plants such as beans, peas, and peanuts (Figure 30.11). This close association allows the bacteria to transfer fixed nitrogen directly to plant cells and explains why food and forage legumes are valued for their high protein content. Biologists have been trying to incorporate nitrogen-fixing symbionts or genes into cereal crops, which lack such effective nitrogen-fixation capacity.

**Figure 30.10** **A biological soil surface crust that includes nitrogen-fixing, soil-enriching cyanobacteria.** Such crusts are widespread in grasslands and other arid regions where they are essential to maintaining the fertility and stability of soils. Biological crusts help to prevent arid soils from blowing away and forming destructive dust storms.
Source: Photo courtesy of Jayne Belnap, U.S. Geological Survey

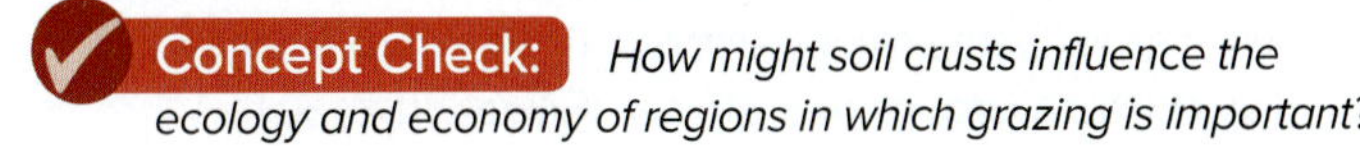

*How might soil crusts influence the ecology and economy of regions in which grazing is important?*

**Figure 30.11** **Legume root nodules.** The cells of nodules on the roots of this soybean plant (*Glycine max*) and other legumes contain heterotrophic, nitrogen-fixing bacteria known as rhizobia. The rhizobia provide fixed nitrogen to the legume, and the plant provides the bacteria with essential organic molecules.


**Industrial Nitrogen Fixation** Worldwide, farmers apply more than 80 million metric tons of nitrogen fertilizer per year. The fixed nitrogen found in fertilizer is produced industrially from $N_2$ by means of a procedure invented by German chemists Fritz Haber and Carl Bosch in 1909. The reduction of $N_2$ gas to $NH_3$ is energetically favorable at room temperature, but the activation energy is very high, so the reaction occurs extremely slowly. Using an iron catalyst, temperatures of 400–650°C (752–1,202°F), and high pressures (150–400 atm), the Haber-Bosch process generates $NH_3$ rapidly. However, because of its high energy requirement, industrial nitrogen fixation can be costly from the perspective of many of the world's farmers.

## 30.2 Reviewing the Concepts

- Soils are composed of organic material and inorganic minerals. Soil organic matter is largely derived from plant detritus, animal wastes, and decayed animal bodies (Figure 30.6).
- Inorganic soil components occur as particles that can be categorized according to their size as sand, silt, or clay. Cation exchange releases cations to soil water, making them available for uptake by plant roots, but also more vulnerable to loss by leaching (Figures 30.7, 30.8, 30.9).
- Biological or industrial processes can convert atmospheric nitrogen gas into fixed nitrogen that plants can utilize. Nitrogen-fixing bacteria living in soil or symbiotically within the tissues of legumes and some other plants provide them with a continuous and reliable source of fixed nitrogen (Figures 30.10, 30.11).

## 30.2 Testing Your Knowledge

1. Which soil type is considered best for cultivation of many plants?
   a. sandy soil
   b. silty-clay soil
   c. loam soil
   d. all of the above
   e. none of the above
2. Which process involves cation exchange?
   a. Plant roots release organic acids that aid the release of bound minerals for root uptake.
   b. Acid rain, resulting from burning of forests and fossil fuel by humans, reduces soil fertility.
   c. Both a and b are correct.
   d. Neither a nor b is correct.

## 30.3 Transport at the Cellular Level

### Learning Outcomes

1. Describe the processes of osmosis and passive and active transport.
2. Explain the water potential equation.

As discussed in Chapter 28, plant root-hair cells absorb water and dissolved minerals from the soil, and plants take in $CO_2$ via stomatal pores on the epidermal surfaces of leaves and stems (**Figure 30.12**). Photosynthetic cells use these materials to produce sugar and other organic compounds needed for overall plant growth and reproduction. Nonphotosynthetic plant cells—such as those of roots, flowers, fruits, and seeds—depend on organic food produced by photosynthetic cells. Many plants can grow to considerable heights, the tallest trees being over 100 m tall. Therefore, plants must transport water, food, and minerals for short distances within and among tissues, as well as long distances. As a basis for learning how plants accomplish short- and long-distance transport, we start by reviewing the processes by which materials are taken up and move at the cellular level.

**Review of Osmosis and Membrane Transport** Chapter 5 described how all cells use both passive and active processes to import or export materials across membranes. These cellular transport processes are briefly reviewed here in the context of plant biology.

- Water, gases, and certain small, uncharged compounds can passively (that is, without the expenditure of ATP) diffuse across membranes in the direction of their concentration gradients. Osmosis is the diffusion of water across a selectively permeable membrane in response to differences in solute concentrations. Osmosis explains how water enters root cells having a higher solute concentration than is present in soil water. After first dissolving in water present within leaf intercellular spaces, the gas $CO_2$ enters photosynthetic mesophyll cells by diffusing across their plasma membranes.

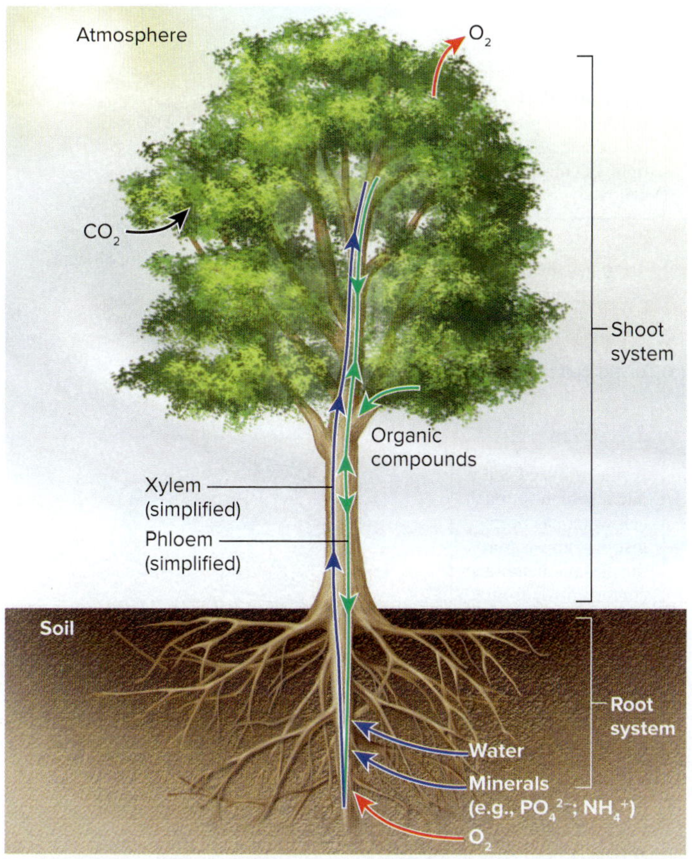


**Figure 30.12** Overview of material uptake and long-distance transport processes in plants.

- Facilitated diffusion is the passive transport of molecules across membranes down a concentration gradient with the aid of membrane transport proteins: channels and transporters (**Figure 30.13a**). Because diffusion often occurs too slowly to meet the high water needs of plants, plant cells often produce aquaporins, channel proteins that facilitate water flow across the plasma membrane.
- Transporter proteins move molecules by binding them on one side of a membrane and then changing conformation so that the molecule is released to the other side of the membrane (refer back to Figure 5.14). Transporters also increase the rate at which specific mineral ions and organic molecules are able to enter or leave plant cells and vacuoles. The presence of plasma membrane transporters of specific types allows plant roots to take up from the soil essential ions that are required without absorbing unneeded or harmful materials.
- If a substance must be transported across a plasma membrane against its concentration gradient, work must be performed in the process known as active transport. During active transport, membrane transporter proteins use energy to move substances against their concentration gradients. An example is the $H^+$-ATPase proton pump found in the plasma membranes of plant cells, which uses ATP to pump protons ($H^+$) against a gradient (**Figure 30.13b**).

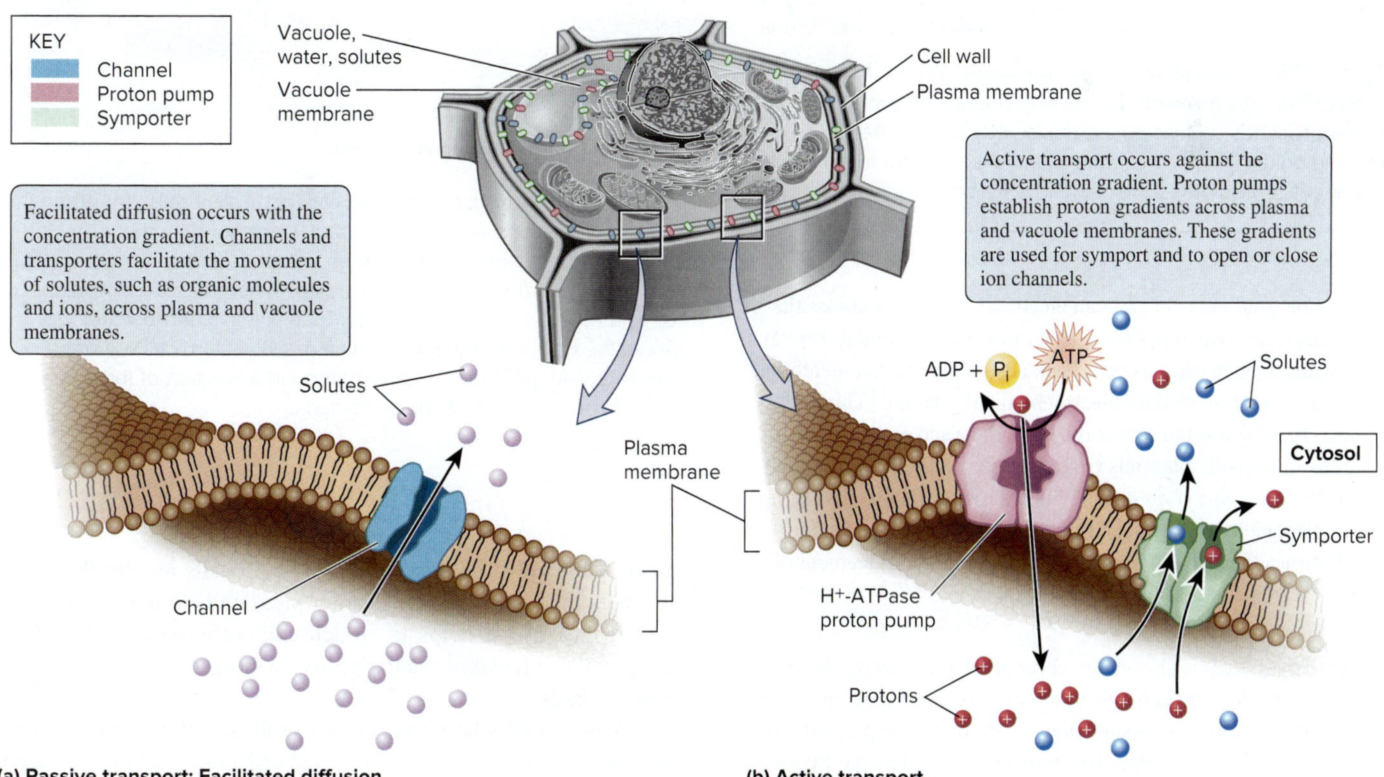


**Figure 30.13** Facilitated diffusion (a type of passive transport) and active transport.

## Biology Principle

### Living Organisms Use Energy

When soil mineral ion concentrations are lower than those within cells, root-hair plasma membranes take up nutrient ions by active transport, which requires energy in the form of ATP.

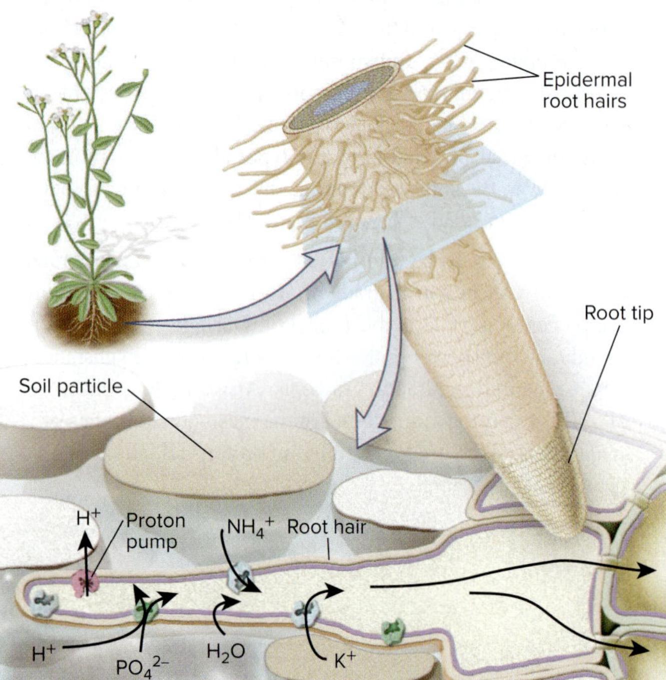


**Figure 30.14 Ion uptake at root-hair membranes.** The $H^+$-ATPase proton pump establishes an electrochemical gradient that drives the active uptake of solutes such as potassium ion ($K^+$), phosphate ($PO_4^{2-}$), and ammonium ($NH_4^+$). The resulting increase in intracellular solute concentration also drives the osmotic diffusion of water into the cell.

- Proton gradients generate an electrical difference across the membrane, which is known as a membrane potential. Energy is released when protons pass back across the plasma membrane, in the direction of their electrochemical gradient. This energy can then be used to power the secondary active transport of ions or organic materials (refer back to Figure 5.16). Secondary active transport (Figure 30.14) allows root-hair cells to take up and concentrate dissolved mineral nutrients to more than 75 times their abundance in soil, fostering the movement of soil water into root hairs by osmosis.

**Water Potential** The water content of plant cells depends on osmosis, which is influenced by solute concentration and a property known as turgor pressure that arises from the presence of the rigid plant-cell wall. **Turgor pressure** is defined as the hydrostatic pressure required to stop the net flow of water across a plasma membrane due to osmosis. Turgor pressure increases as water

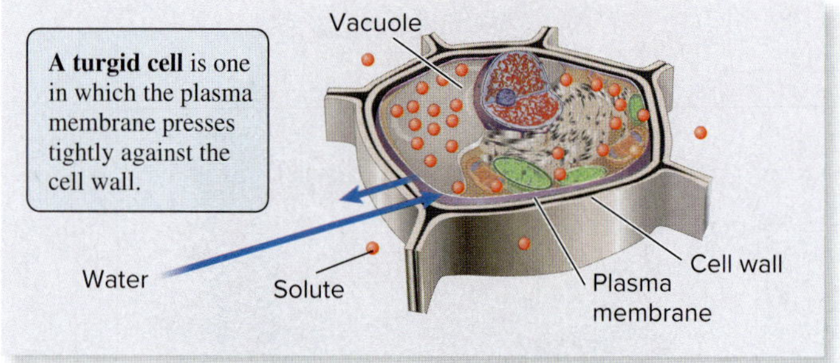


**(a) Turgid cell in a hypotonic solution**

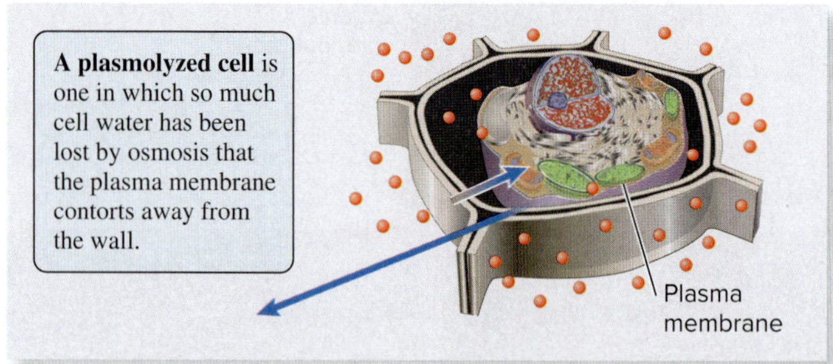


**(b) Plasmolyzed cell in a hypertonic solution**

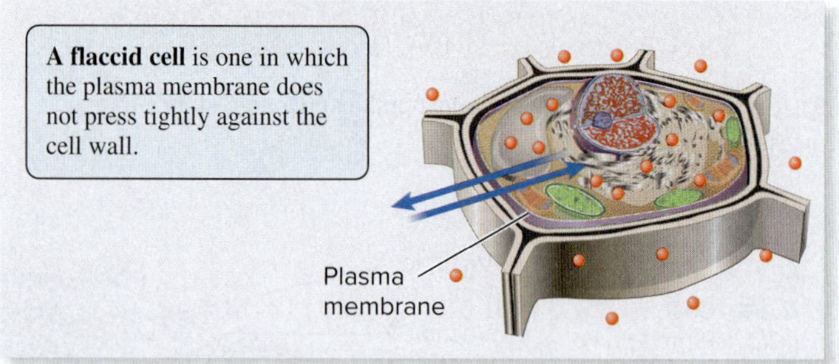


**(c) Flaccid cell in an isotonic solution**

**Figure 30.15 Turgid, plasmolyzed, and flaccid plant cells. (a)** When the concentration of solutes inside a cell is greater than that outside, more water may enter the cell than will leave it. As a result, the plant cell may become swollen, or turgid. **(b)** When the concentration of solutes outside a cell is greater than within it, more water will leave the cell than will enter it. As a result, a plant cell will become plasmolyzed. **(c)** When a cell is bathed in a solution of the same solute concentration, it will be flaccid.

enters plant cells, because their rigid cell walls restrict the extent to which the cells can swell. Turgor pressure is important for plant growth because it helps to stretch cell walls (see Figure 30.9). Turgor also increases plant rigidity; wilting occurs when turgor pressure falls.

A plant cell whose cytosol is so full of water that the plasma membrane presses right up against the cell wall is said to be **turgid** (Figure 30.15a). The pressure relationship between the cytosol and cell wall is similar to the way that a soccer ball's leather skin presses inward upon the air within, while at the same time the internal air presses on the ball's cover. If you add more air to a limp ball, the ball will stiffen. In the same way, if a nonturgid cell absorbs more water, it will become more rigid as the water exerts pressure on the cell wall. By contrast, a plasmolyzed cell is one that has lost so much water by osmosis that turgor pressure has also been lost. **Plasmolysis** is the condition in which the plasma membrane no longer presses on the cell wall (**Figure 30.15b**). The concentration of solutes is the same outside and inside a **flaccid** cell, which will have a water content higher than that of a plasmolyzed cell but lower than that in a turgid cell (**Figure 30.15c**).

Together, solute concentration and presence of a cell wall influence an important plant cell property known as **water potential ($\psi_W$)**, the potential energy of water. Using specialized equipment, water potential is measured in pressure units known as megapascals (MPa) (a pascal is equal to 1 newton per square meter). One megapascal is equal to 10 times the average air pressure at sea level, about the same pressure that occurs within an inflated bicycle tire. As another reference point, 1 MPa is several times the pressure in typical home plumbing pipes, which you experience when turning on a water faucet.

A water potential equation can be used to predict the direction of cellular water movement, given information about the solute concentrations inside and outside plant cells, and a measure of pressure at the cell-wall–membrane interface. In this equation, cellular water potential is symbolized by the Greek letter psi ($\psi$) with the subscript W for water: $\psi_W$. In its simplest form, total $\psi_W$ is calculated as

$$\psi_W = \psi_S + \psi_P$$

where $\psi_S$ is solute potential and $\psi_P$ is pressure potential.

**Solute Potential** **Solute potential ($\psi_S$)** is the component of water potential due to the presence of solute molecules. As you might expect, solute potential is proportional to the concentration of solutes in a solution. The solute potential of pure water open to the air, at sea level and room temperature, is defined as zero. When solutes are added, they occupy space, thereby diluting the water. As a result, fewer free water molecules are present, which reduces the potential energy of water. Thus, in the absence of a pressure potential, water that contains solutes always has a negative solute potential. The higher the concentration of dissolved solutes, the lower (more negative) the solute potential.

**Pressure Potential** **Pressure potential ($\psi_P$)** is the component of water potential due to hydrostatic pressure. In plant cells, the hydrostatic pressure is determined in part by the resistance provided by the cell wall. Because of this resistance, the value for pressure potential can be either positive or negative. For example, a turgid cell has a positive pressure potential, which typically measures about 1 MPa. This high pressure inside turgid plant cells is a testimony to the strength of their cellulose-rich plant cell walls. In contrast to turgid cells, both flaccid and plasmolyzed cells have a pressure potential of zero (**Figure 30.16**).

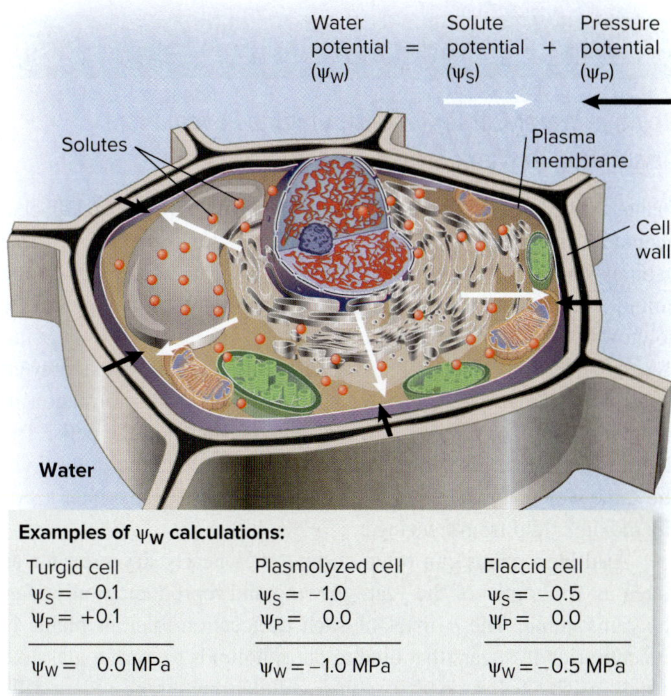


**Examples of $\psi_W$ calculations:**

| Turgid cell | Plasmolyzed cell | Flaccid cell |
|---|---|---|
| $\psi_S = -0.1$ | $\psi_S = -1.0$ | $\psi_S = -0.5$ |
| $\psi_P = +0.1$ | $\psi_P = 0.0$ | $\psi_P = 0.0$ |
| $\psi_W =$ 0.0 MPa | $\psi_W = -1.0$ MPa | $\psi_W = -0.5$ MPa |

**Figure 30.16** **Plant-cell water potential.** The water potential of a plant cell, which predicts the direction of water movement into or out of a cell, can be simplified as the sum of the solute potential and the pressure potential resulting from pressure exerted by the cell-wall–plasma membrane complex. Examples of water potential calculations are shown for a turgid cell, a plasmolyzed cell, and a flaccid cell.

**SCISKILLS ▶** *Make a drawing that shows the direction of water movement when a cell of each of these types is placed into a solution of pure water (whose water potential, $\psi_W$, is defined as 0 MPa).*

## Quantitative Analysis

### THE WATER POTENTIAL EQUATION CAN BE USED TO PREDICT CELLULAR WATER STATUS

How can researchers use water potential concepts to predict the water status of plant cells?

**Crunching the Numbers:** Imagine that you are able to use specialized equipment to measure the solute potential ($\psi_S$) of a plant cell as −0.15 MPa and measure the pressure potential ($\psi_P$) as +0.15 MPa. Use the simple form of the water potential equation, $\psi_W = \psi_S + \psi_P$, to calculate this cell's water potential. From that number, is this cell turgid, plasmolyzed, or flaccid?

## EVOLUTIONARY CONNECTIONS

### Relative Water Content Measurements Reveal Plant Adaptation to Water Stress

Plants native to cold, dry, or saline environments have evolved diverse adaptations that allow them to cope with low water availability in these habitats. Plant cells may increase their cytosolic concentration of the amino acid proline, sugars, or sugar alcohols such as mannitol. This solute concentration increase draws water into cells. By increasing the concentration of solutes inside cells, cold-resistant plants prevent water from moving out of their cells when ice crystal formation in intercellular spaces lowers the water potential outside the cells. The additional solutes also lower the freezing point of the cytosol, in the same way that adding antifreeze to a car's radiator in winter keeps the radiator fluid from freezing.

Arid-land plants can often survive in a nearly dry state for as much as 10 months of the year, growing and reproducing only after the rains come. The cytosol of such desiccation-tolerant plants is typically rich in sugars that bind to phospholipids to form a glasslike structure. This helps to stabilize their cellular membranes. Plant cells under water stress may also increase the number of plasma membrane aquaporins, allowing cells to quickly absorb water when it becomes available.

Researchers measure a property known as **relative water content (RWC)** in studies of natural plant adaptation to cold, drought, or salt stress and in agricultural research for developing drought-tolerant crops. RWC can be used to predict a plant's ability to recover from the wilted condition. An RWC of less than 50% spells death for most plants. However, some plants, particularly those of arid habitats, can tolerate lower water content for substantial time periods, an evolutionary adaptation to water stress. A standard method for determining RWC involves three simple weight measurements:

- Sample tissue taken from a plant under a given set of conditions is first weighed to obtain the fresh weight.
- Then the sample is completely hydrated in water within an enclosed, lighted chamber until constant turgid weight is achieved.
- Finally, the sample is dried to a constant dry weight.
- RWC is calculated as 100 × [(fresh weight minus dry weight) divided by (turgid weight minus dry weight)].

Comparisons of the RWC of plants sampled from or grown in environments varying in water availability, followed by assessment of the plants' ability to recover from water stress, reveal that some plants have evolved in ways that allow them to survive for longer periods in arid conditions.

## 30.3 Reviewing the Concepts

- Plant root hairs take up minerals via plasma membrane transporters, a process that involves ATP hydrolysis, and water flows into solute-rich root cells by osmosis. Short- and long-distance transport allows these materials to move from roots to stems and leaves, where they and carbon dioxide absorbed from the air are used in photosynthesis. Organic compounds produced by photosynthetic tissues move to nonphotosynthetic tissues by short- and long-distance transport (Figures 30.12, 30.13, 30.14).
- Turgor pressure increases as water enters plant cells because cell walls restrict the extent to which cells can swell. A cell that is so full of water that the plasma membrane presses closely against the cell wall is turgid. A cell that contains so little water that the plasma membrane pulls away from the cell wall is plasmolyzed. A flaccid cell has a water content between these extremes (Figure 30.15).
- Solute potential ($\psi_S$) and pressure potential ($\psi_P$) arising from the presence of a cell wall are major factors affecting cellular water potential ($\psi_W$). Water moves from a region of higher water potential to a region of lower water potential (Figure 30.16).
- The measurement known as relative water content (RWC) can be used to compare the drought recovery capacities of different plants.

## 30.3 Testing Your Knowledge

1. Assuming equally thick cell walls of the same composition, which plant cell would have the highest turgor pressure?
   a. a plant cell that displays a lower concentration of solutes than its environment
   b. a plant cell that displays the same concentration of solutes as its environment
   c. a plant cell that displays a higher concentration of solutes than its environment
   d. All of these plant cells would have the same turgor pressure.
2. A plant is able to recover from a relative water content (RWC) of 40%. To what type of environment is this plant most likely adapted?
   a. an area of tropical rain forest that receives plentiful rainfall year round
   b. an area of temperate forest that receives plentiful rainfall during the growing season
   c. an area of grassland that receives relatively low levels of rainfall that is evenly distributed throughout the growing season
   d. an area of arid desert in which water is not available during much of the year
   e. all of the above

## 30.4 Plant Transport at the Tissue Level

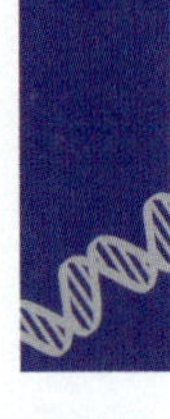

### Learning Outcomes

1. Distinguish among transmembrane transport, symplastic transport, and apoplastic transport.
2. Explain how the root endodermis functions as a diffusion barrier.

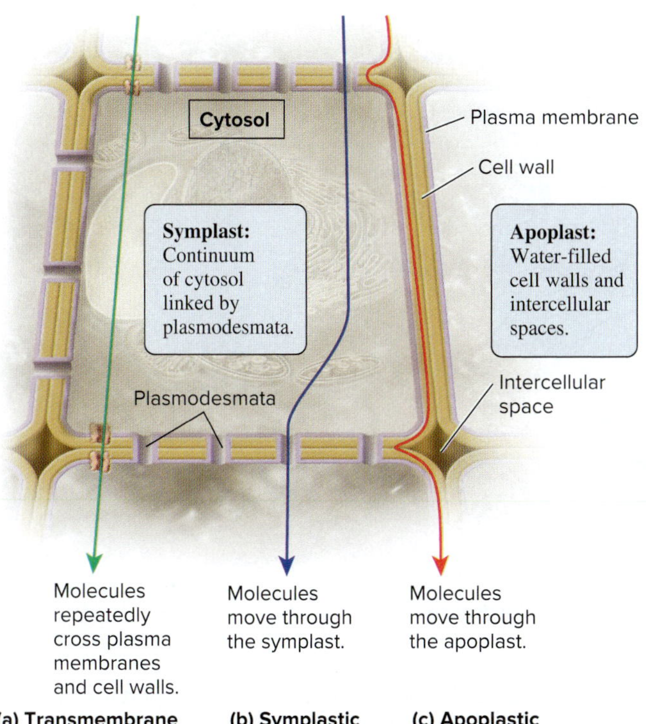


**Figure 30.17** Three routes of tissue-level transport in plants: (a) transmembrane, (b) symplastic, and (c) apoplastic.

Now that we have reviewed how water, dissolved minerals, and organic compounds enter or leave plant cells, we are prepared to consider short-distance transport within and among nearby plant tissues. Understanding tissue-level transport is an essential step toward comprehending how plants accomplish long-distance transport. Tissue-level transport occurs in three forms: transmembrane transport, symplastic transport, and apoplastic transport (**Figure 30.17**).

**Transmembrane transport** involves the export of a material from one cell via membrane proteins, followed by import of the same substance by an adjacent cell (Figure 30.17a). One prominent example of transmembrane transport is the movement of the plant hormone auxin downward in shoots. Auxin travels from one parenchyma cell to another with the aid of carrier proteins (refer back to Figure 29.4a). This process explains how auxin produced in one part of the plant body can influence more distant tissues.

**Symplastic transport** is the movement of a substance from the cytosol of one cell to the cytosol of an adjacent cell via membrane-lined channels called plasmodesmata (Figure 30.17b). Plasmodesmata are large enough in diameter to allow transport of proteins and nucleic acids, as well as smaller molecules. Together, all of a plant's protoplasts (the cell contents without the cell walls) and plasmodesmata form the **symplast.** Symplastic transport has the potential to move molecules widely among the cells and tissues of the plant body.

**Apoplastic transport** is the movement of water and solutes along cell walls and the spaces between cells. Both symplastic and apoplastic transport play important roles in the transport of mineral nutrients through the outer tissues of roots (**Figure 30.18**).

**The Key Roles of the Root Endodermis** Water and dissolved minerals can move apoplastically into root epidermal and cortex tissues. However, apoplastic movement of water and minerals stops at the **endodermis,** a term meaning inside skin. In roots, the endodermis is a thin cylinder of tissue whose close-fitting cells and specialized cell walls form a barrier to diffusion between the cortex and the central core of vascular tissue. Root endodermal cells are able to function as diffusion barriers because their walls possess ribbon-like strips of waterproof suberin, composed of wax and phenolic polymers. Suberin is also abundant in cork, the protective outer bark of trees and shrubs (see Chapter 28). Known as **Casparian strips,** these suberin ribbons wrap all the way around each endodermal cell, forming a tight seal (**Figure 30.19**).

The plant endodermis is functionally analogous to animal epithelial tissues whose cellular tight junctions likewise form diffusion barriers.

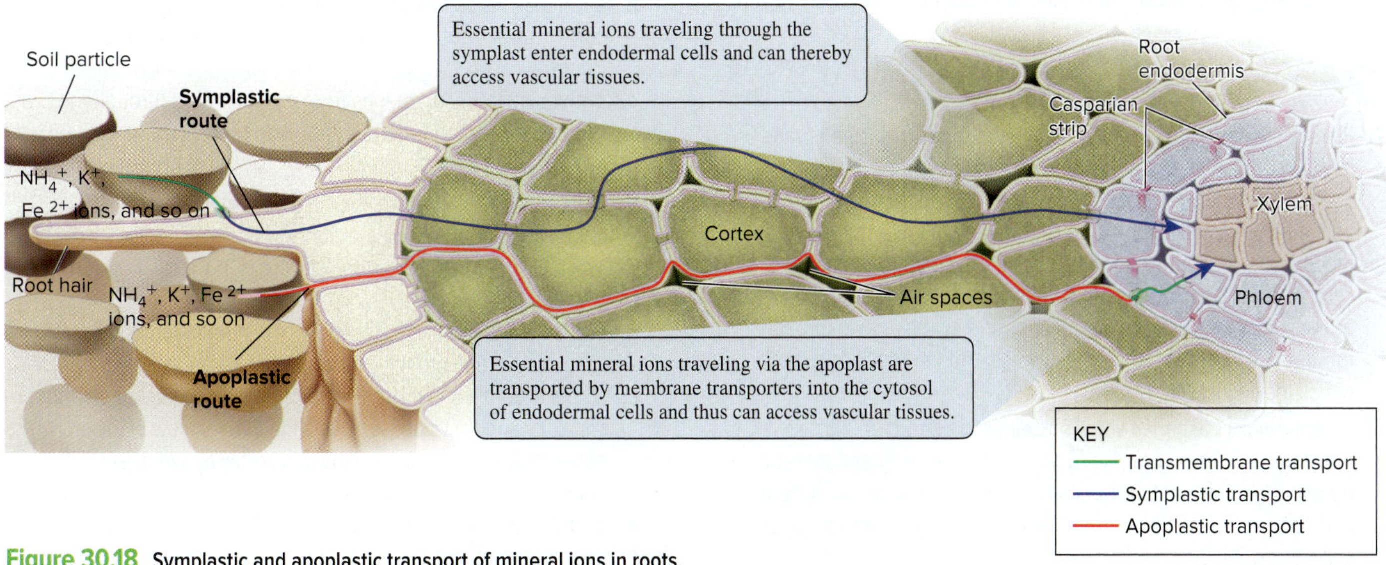


**Figure 30.18** Symplastic and apoplastic transport of mineral ions in roots.

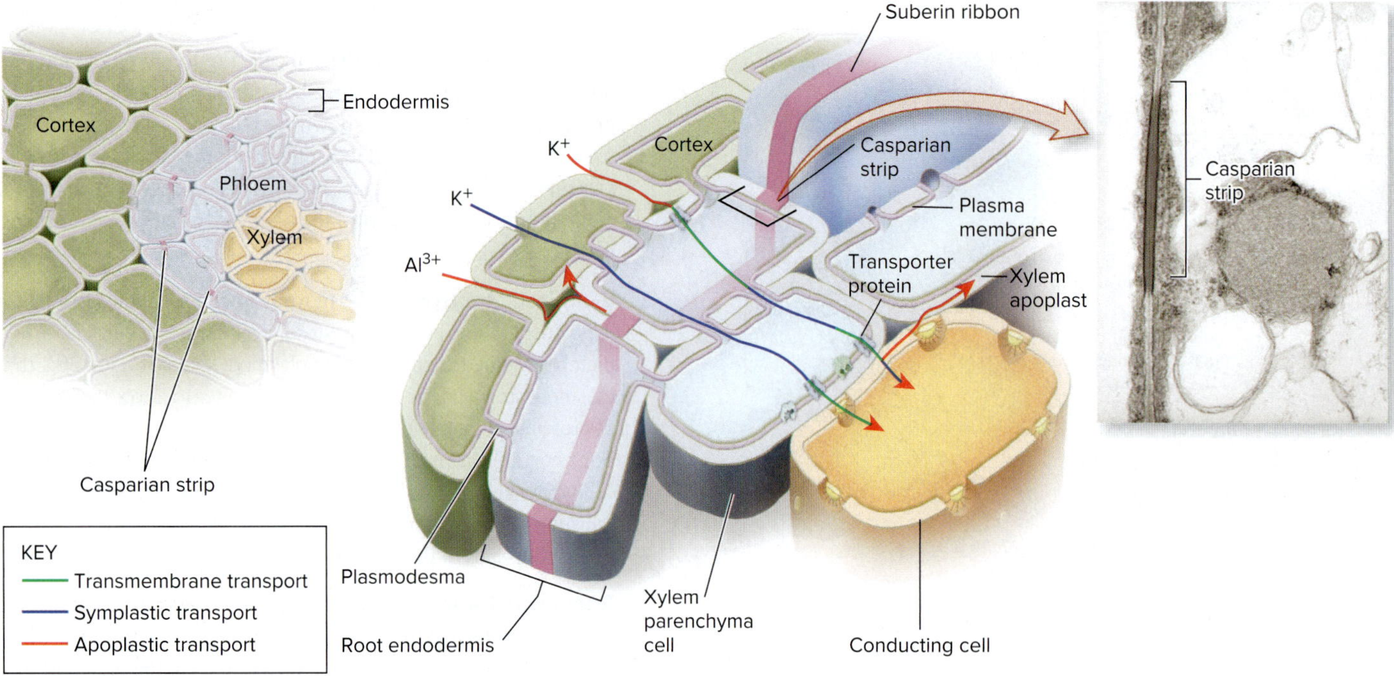


**Figure 30.19** **Ion transport pathways across the root endodermis.** Casparian strips in endodermal cell walls prevent apoplastic transport across the root endodermis, limiting entry of harmful soil minerals such as aluminum ion ($Al^{3+}$) and exit of useful minerals. Mineral nutrients that are transported into the cytosol of endodermal cells are able to pass through the endodermal barrier to xylem parenchyma cells via plasmodesmata. Once past the endodermis, nutrient ions such as $K^+$ are moved across plasma membranes to the apoplast of the vascular tissue and are thus able to enter xylem. Inset shows a transmission electron micrograph (TEM) of a Casparian strip in the wall of an endodermal cell.
*(right)* © James S. Busse

**BioConnection:** *Refer back to Figure 5.23 to see a diagram of tight junctions between adjacent cells of animal intestinal epithelium. How is the root endodermis similar in structure and function?*

Because of the Casparian strips, materials in the root apoplast cannot penetrate farther into the root unless endodermal cells transport them into their cytosol, a process that requires specific transporter proteins. In this way, the root endodermis prevents harmful solutes such as toxic metal ions from moving through the apoplast to vascular tissues and being transported to the shoot. In addition, the endodermis prevents xylem solutes such as $K^+$, $NO_3^-$, or $Ca^{2+}$ from flowing back into root cortex or soil.

## 30.4 Reviewing the Concepts

- Plant tissue-level transport occurs in three forms. Transmembrane transport involves the movement of materials from one cell to another from intercellular spaces, across plasma membranes, and into the cytosol. Symplastic transport allows materials to move from one cell to another through the symplast, the cells' cytosol and plasmodesmata, without crossing plasma membranes. In apoplastic transport, water and solutes move through the apoplast, the water-filled cell walls and intercellular spaces of tissues (Figures 30.17, 30.18).
- In roots, waxy Casparian strips on cell walls of endodermal tissue function as diffusion barriers to apoplastic transport and thereby reduce the movement of toxic ions into the plant vascular system and reduce ion backflow from the xylem into the root cortex or soil (Figure 30.19).

## 30.4 Testing Your Knowledge

1. A magnesium ion ($Mg^{2+}$) travels from the soil to a plant leaf, where the ion is used to make chlorophyll. Which of the following processes is likely involved?
   a. active transport of ions from soil into root hairs
   b. apoplastic transport of ions from the soil through the root cortex, then through endodermal transporters to the root xylem
   c. symplastic transport of ions from root-hair cells through the root cortex and endodermis, then into root xylem
   d. all of the above
   e. none of the above
2. What functions does the root endodermis serve?
   a. prevents the apoplastic entry into the root vascular system of harmful minerals
   b. prevents the apoplastic backflow of minerals from the root vascular system into the root cortex and soil
   c. allows absorption of mineral nutrients that have entered the root cortex by apoplastic transport, and the through-flow of mineral nutrients that have traversed the root cortex by symplastic transport
   d. all of the above
   e. none of the above

## 30.5 Long-Distance Transport in Plants

### Learning Outcomes

1. Draw a diagram that illustrates the cohesion-tension concept of long-distance water movement in plants.
2. Define the term transpiration and explain its importance to local and global climate.
3. Explain how stomatal behavior and plant-controlled leaf abscission can reduce transpirational water loss from the plant body.
4. Contrast cohesion-tension in xylem with the pressure-flow concept for long-distance movement of photosynthetic products in phloem.

Plants rely on long-distance transport to move water and dissolved materials from roots to shoots and among organs. Tall trees are able to transport water and minerals to astounding heights, more than 110 m in some cases. This is possible because plant bodies possess an extensive, branched, long-distance vascular system composed of xylem and phloem tissues. Recall that tracheids and vessels of mature xylem lack cytoplasmic components (see Chapter 28); because such obstructions are absent, **bulk flow** can occur, which is the mass movement of liquid molecules by pressure, gravity, or both.

Bulk flow occurs much faster than diffusion, allowing water with dissolved minerals to move rapidly through soil toward plant roots and, once inside the plant, through xylem conducting tissues from the root system to the top of the shoot. Bulk flow also promotes the movement of organic solutes from leaves to other parts of the plant body through the phloem. Bulk flow in xylem occurs by means of negative pressure, as explained by the cohesion-tension theory, whereas bulk flow in phloem occurs via positive pressure, described by the pressure-flow concept.

### Transport in the Xylem Is Explained by the Cohesion-Tension Theory

Water and dissolved materials move by bulk flow through xylem tissues as the result of differences in root and shoot water pressure. When soil water is abundantly available to roots, water pressure will be higher in root xylem than in shoot xylem, causing bulk flow into the shoot. Water pressure is higher in shoot xylem located closer to roots than in xylem located in the upper parts of plant shoots, where water diffuses from the plant through stomata into the atmosphere (Figure 30.20). The extent to which plants lose water to the atmosphere depends on atmospheric humidity; water will diffuse outward more readily when the humidity is lower than when it is higher. This evaporation process is known as **transpiration** (from the French *transpirer*, meaning to perspire) (Figure 30.21a). Transpiration is capable of pulling water by bulk flow up to the tops of the tallest trees and is the primary way in which water is transported for long distances in plants. Plants expend no energy to transport water and minerals by transpiration. Rather, the Sun's energy indirectly powers this process by generating a water pressure difference between moist soil and drier air.

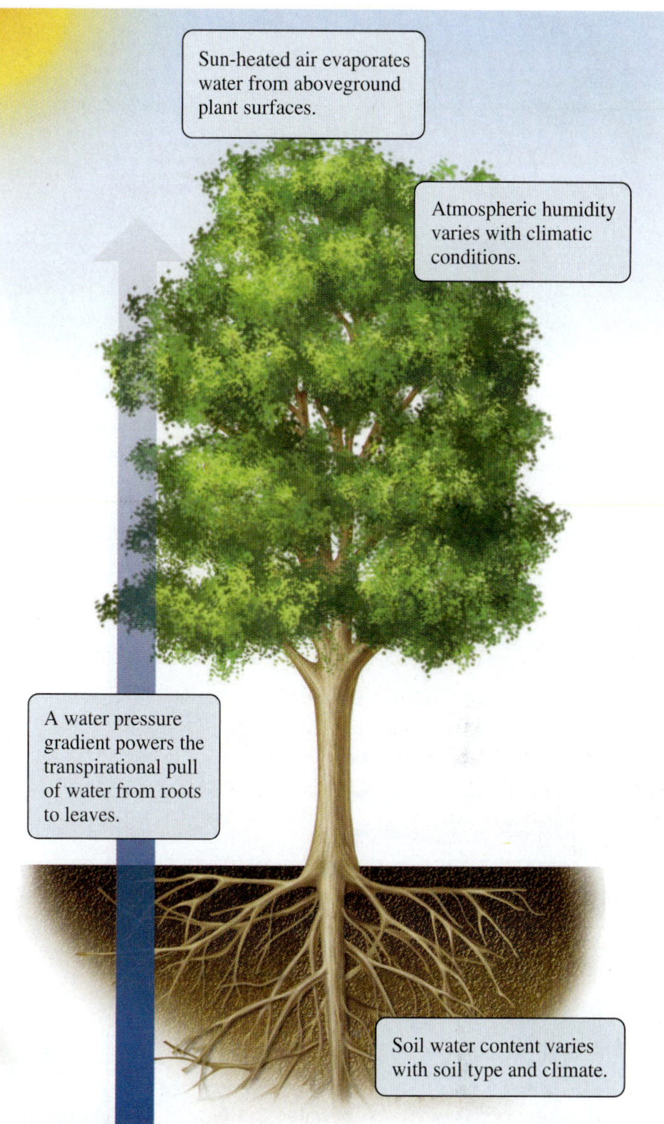

Sun-heated air evaporates water from aboveground plant surfaces.
Atmospheric humidity varies with climatic conditions.
A water pressure gradient powers the transpirational pull of water from roots to leaves.
Soil water content varies with soil type and climate.


**Figure 30.20** **Upward transport of water in xylem.** Water pressure differences between moist soil and drier air drive the upward movement of water in plants. In drier air, water evaporates from plant surfaces, pulling a stream of liquid water through the plant body.

The evaporative loss of water at shoot surfaces generates a negative pressure that pulls a stream of highly H-bonded, cohesive water molecules up the plant, much as you would suck liquids up a straw (Figure 30.21b,c). The movement of water through plants, from soil to root to shoot to atmosphere, is also analogous to sucking up liquids into a syringe by pulling on the plunger, which exerts a tension. Because the cohesion of water molecules and the tension exerted by atmospheric evaporation are both key elements of xylem transport, the concept used to explain how water moves long distances through plants is known as the **cohesion-tension theory.**

During tracheid and vessel element development, the cytoplasm is lost by controlled cell death, and cell walls become coated with the plastic-like polymer lignin, which confers strength and waterproofing. Consequently, tracheids and vessels function as water-conducting

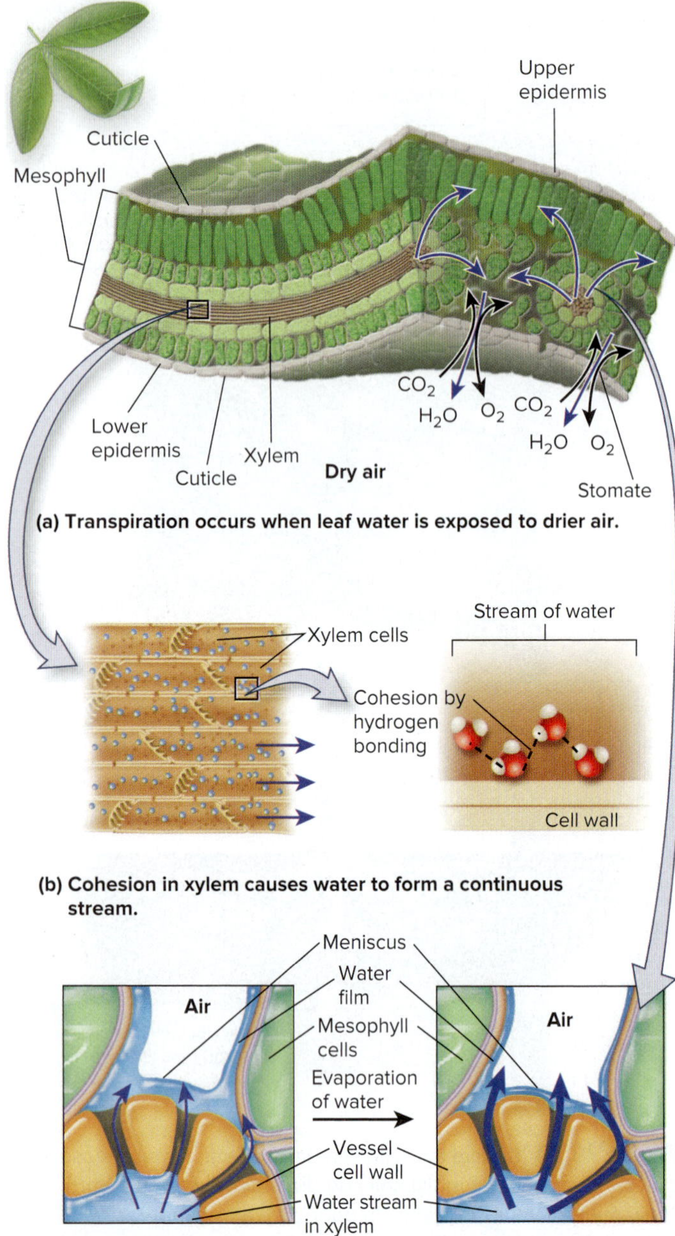


**Figure 30.21 The roles of transpiration, cohesion, and tension in long-distance water transport. (a)** Transpiration is the loss of water by evaporation to drier air that occurs at plant surfaces, primarily through open stomata. **(b)** Cohesion is the tendency of water molecules to stick together, allowing evaporation at the plant surface to pull a column of water through leaves, stem, and roots. **(c)** Tension is the force exerted on the water column as water evaporates at plant surfaces. Adhesion of water molecules to the surfaces of xylem tracheids and vessels also plays a role in water transport.

tubes that do not easily collapse, even under high negative pressures. Tracheids are arranged end-to-end to form long arrays, and vessel elements are lined up to form tubular vessels—arrangements that facilitate long-distance water movement. The tubular structure and tough walls of xylem water-conducting components function much like plastic plumbing pipes in buildings.

## Plant Transpiration Has Global Climate Effects

Plant transpiration results in the movement of huge amounts of water from the soil to the atmosphere. About 97% of the water that enters plants via roots is generally lost as water vapor during transpiration. Each crop season, a single corn plant (*Zea mays*) loses more than 200 L of water, which is more than 100 times the corn plant's mass. A typical tree loses 400 L of water per day! These examples illustrate why water availability is so important to agriculture and natural vegetation, particularly in arid regions.

On a regional and global basis, plant transpiration has enormous climate effects. An estimated one-half to three-quarters of rainfall received by the Amazon tropical rain forest originates from plant-transpired water vapor, often visible as mist (Figure 30.22). About half of the solar heat received by Amazonian plants is dispersed to the atmosphere during transpiration. This heat dispersal has a cooling effect on regional ground temperature, which would be much higher in the absence of plant transpiration. Such cooling effects result from water's unusually high heat of vaporization, the amount of heat needed to isolate water molecules from the liquid phase and move them to the vapor phase. Water has one of the highest heat of vaporization of any liquid, 44 kJ/mol. This energy is needed to break the

**Figure 30.22 Plant-transpired water vapor mist rising from a tropical rain forest.** This mist visually illustrates the enormous amount of water that is transpired from the surfaces of plants into the atmosphere. Water vapor derived from plant transpiration is an important source of rainfall, and the process of evaporation cools plant surfaces and affects the local and global climate.



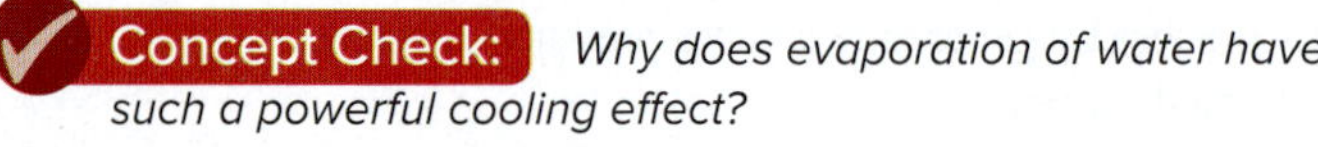

*Why does evaporation of water have such a powerful cooling effect?*

large numbers of hydrogen bonds that occur in liquid water, allowing water vapor to form. The evaporation of large amounts of water from plant surfaces effectively dissipates heat, explaining how plants cool themselves and their environments. Transpiration helps to keep plant temperature within tolerable ranges and fosters local and global climate stability.

Although evaporation of water from plant surfaces plays an essential role in bulk flow through xylem, if plants lose too much water, they will die. Plant surfaces are typically coated with a cuticle, a wax-containing layer that retards evaporative water loss. More than 90% of the water that evaporates from plants is lost through their stomata. When the stomata are open, water vapor also exits that plant, particularly when atmospheric humidity is relatively low. Plants face a constant dilemma: whether to open their stomata to take in $CO_2$ and suffer the effect of reduced water content or to close stomata to retain water, thereby preventing $CO_2$ uptake.

## Plant Adaptations Help to Reduce Transpirational Water Loss

Water stress is common for plants of the world's arid regions, and their growth is often limited by water availability (see chapter-opening photo). But even plants in moist, forested regions of the world can experience water stress during drier or colder seasons or under windy conditions. Control of stomatal opening and closing and leaf drop are two ways in which plants can reduce transpirational water loss.

**Stomatal Opening and Closing** Plant stomata are pores that can close to conserve water under conditions of water stress and open when the stress has been relieved. Stomata are bordered by a pair of **guard cells,** which are sausage-shaped, chloroplast-containing cells attached at their ends (Figure 30.23a). The distinctive structural features of guard cells explain how they are able to open and close a pore. As guard cells become fully turgid, their volume expands by 40–100%. This expansion does not occur evenly, however, because the innermost cell walls are thicker and less extensible than are other parts of the guard cell walls. In addition, the cells expand primarily in the lengthwise direction because bands of radially oriented cellulose microfibrils prevent the guard cells from expanding laterally (Figure 30.23b). Thus, when guard cells are turgid, a stomatal pore opens between them, allowing air exchange with the leaf's spongy mesophyll. Conversely, when the guard cells lose their turgor, their volume decreases, and the stomatal pore closes.

What causes the change in guard cell turgor? In flowering plants, stomata often open early in the morning, in response to sunlight. This response makes sense, given that light, water, and carbon dioxide are all required for photosynthesis. Blue light stimulates $H^+$-ATPase proton pumps, leading to guard cell uptake of ions, especially potassium ($K^+$), and other solutes. As a result of increases in solute concentrations inside guard cells, osmotic water uptake occurs via plasma membrane aquaporins, resulting in cell expansion and stomatal opening (Figure 30.24a).

At night, the reverse process closes flowering plant stomata. Potassium ($K^+$) and other solutes are pumped out of guard cells, causing water to exit and deflate guard cells, resulting in pore closure. Flowering plants also close stomata during the daytime under conditions of water stress, a process mediated by the stress hormone abscisic acid (ABA). Water stress causes a 50-fold increase in ABA, which is transported in the xylem sap to guard cells. There, ABA binds to a membrane receptor, which elicits a $Ca^{2+}$ second messenger, causing the guard cells to lose solutes and deflate (Figure 30.24b).

**Leaf Abscission** Angiosperm trees and shrubs of seasonally cold habitats experience water stress every winter, when evaporation from plant surfaces occurs but soil water is frozen and therefore unavailable for uptake by roots. Under these conditions, air bubbles can form in the xylem, preventing bulk water transport, much as the introduction of air into an animal's circulatory system can block conduction. Desert plants also experience water stress conditions, but at less predictable times and for much of the year. Many types of plants that live in cold climates or deserts are adapted to cope with water stress by dropping their leaves, a process known as leaf

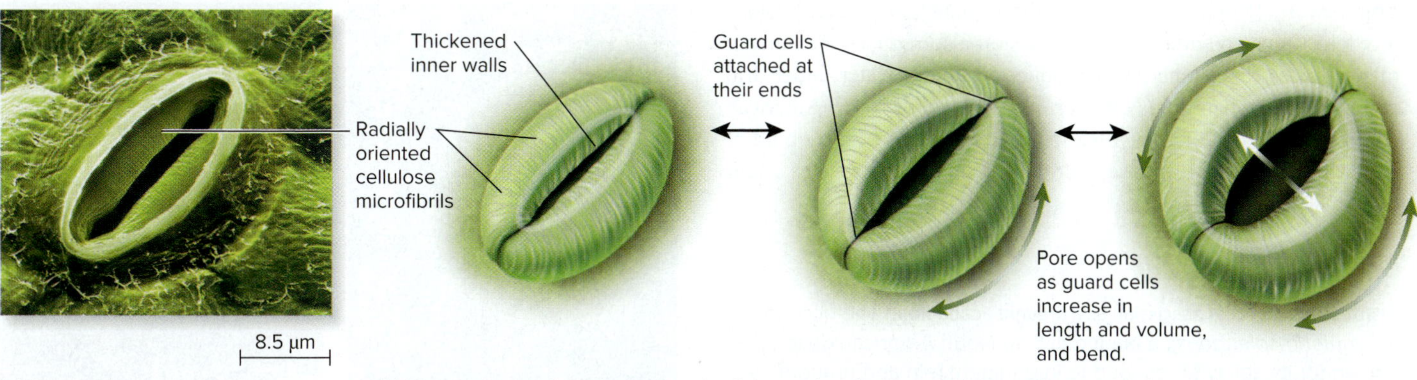


**(a) Stomatal guard cells**

**(b) The roles of radial orientation of cellulose microfibrils and thickened inner walls in opening or closing guard cells**

**Figure 30.23 The structure of guard cells.** When flaccid, guard cells close stomata. Turgid guard cells produce a stomatal opening. **(a)** An SEM of a stomate in a rose leaf, showing the two guard cells bordering a partly open pore. **(b)** Thickened inner cell walls and radial orientation of cellulose microfibrils in the guard cell walls explain why they separate when turgid, forming a pore.

*(a)* © Andrew Syred/Science Source

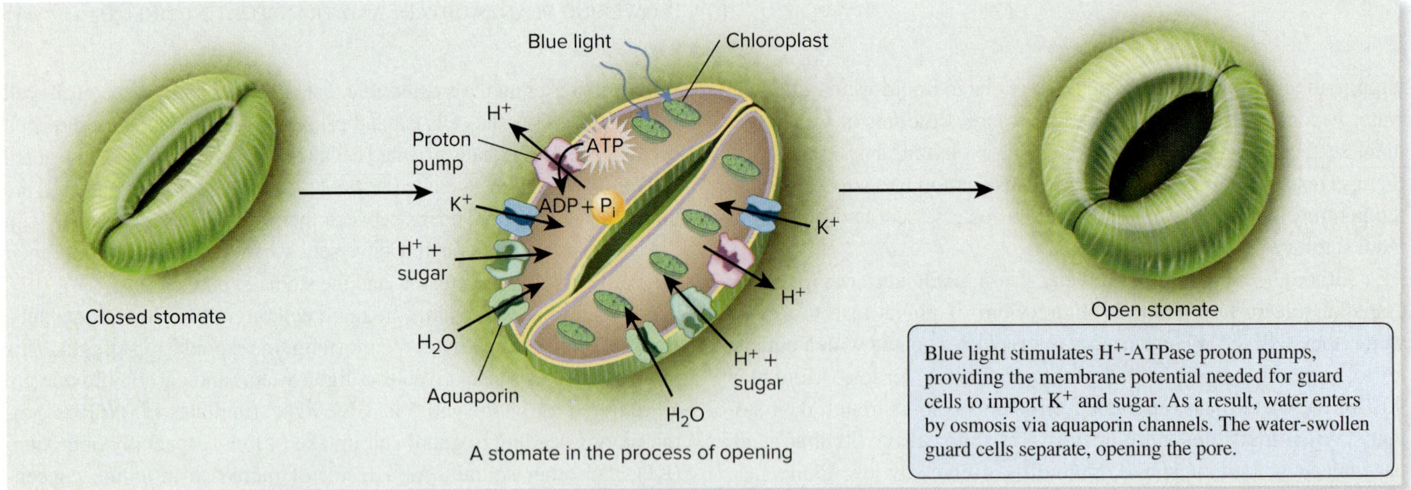


**(a) The process of stomate opening**

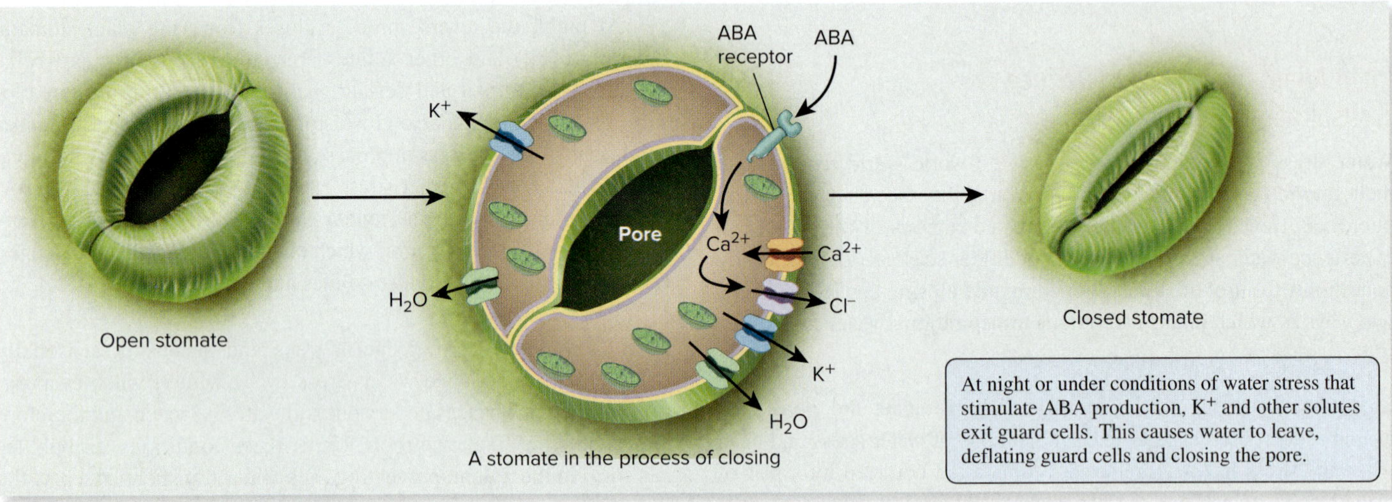


**(b) The process of stomate closing**

**Figure 30.24** **How stomata of flowering plants open and close.** **(a)** Angiosperm stomata usually open in response to sunlight. **(b)** Angiosperm stomata usually close in response to lack of sunlight. They can also close during the day under conditions of water stress, which induces plants to produce more of the hormone abscisic acid (ABA). Guard cell plasma membranes possess ABA receptors, which receive the drought signal.

**abscission.** Dropping leaves lets these plants avoid dangerously low leaf water potentials. Leaf abscission also reduces the amount of root mass that plants must produce to obtain water under arid conditions. The ocotillo (*Fouquieria splendens*) of North American deserts can produce leaves after sporadic rains and then drop all of its leaves as a direct response to drought as many as six times a year (**Figure 30.25**).

**(a) Ocotillo with leaves**

**(b) Ocotillo without leaves**

**Figure 30.25** **Leaf abscission as a drought adaptation.** The ocotillo (*Fouquieria splendens*), a plant native to North American deserts, is known for its ability to respond to intermittent rain and drought by producing and dropping leaves multiple times within a year.
*(a)* © Dan Suzio/Science Source; *(b)* © Matthew Heinrichs/Alamy RF

**Concept Check:** *Why doesn't the ocotillo drop its leaves at a single predictable time each year, as do temperate angiosperm trees and shrubs?*

## Biology Principle

### Living Organisms Maintain Homeostasis

This is illustrated by leaf drop in response to, or in anticipation of, arid conditions or cold temperatures.

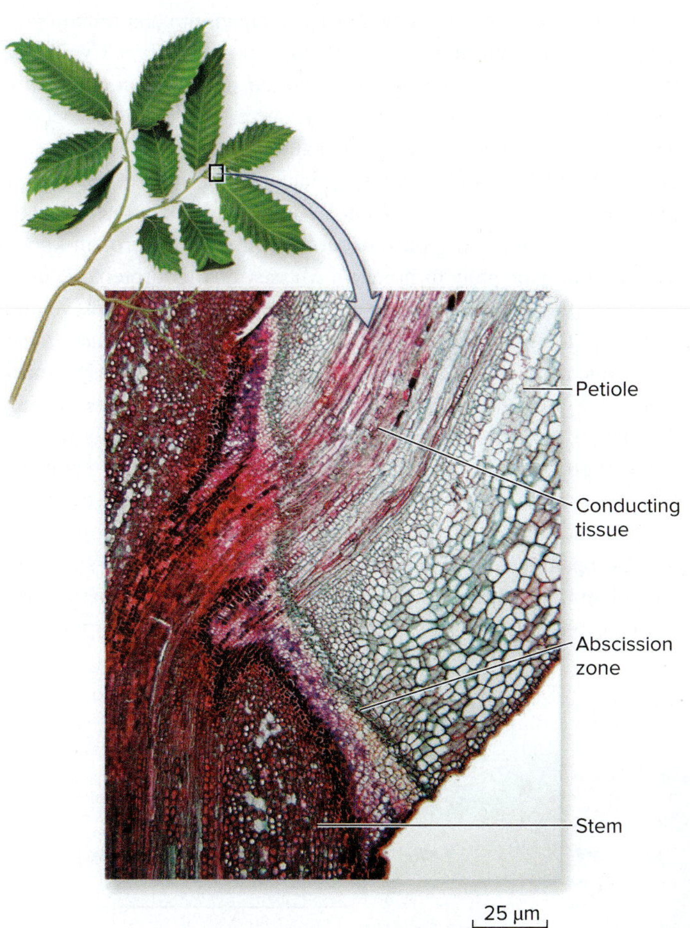


**(a) LM of leaf abscission zone at the junction of a petiole and stem, stained with dyes**

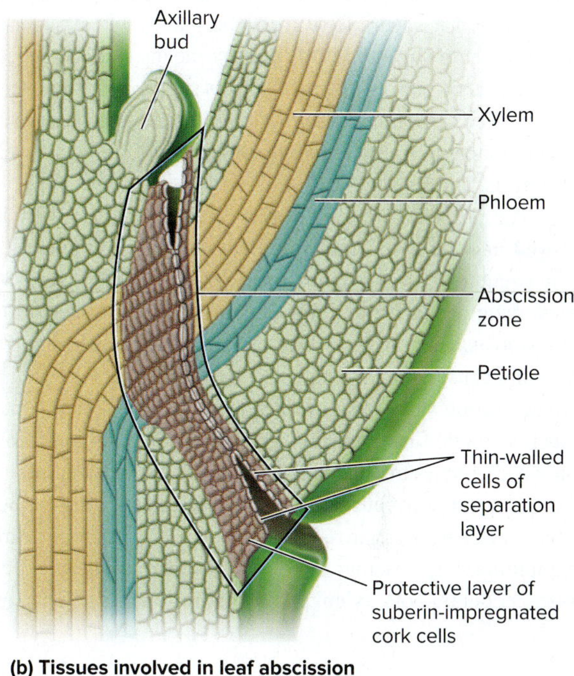


**(b) Tissues involved in leaf abscission**

**Figure 30.26** **Leaf abscission.** Leaf abscission helps maintain plant hydration.
*(a)* © Lee W. Wilcox

The sugar maple (*Acer saccharum*) is an example of the many types of trees or shrubs of temperate forests that drop their leaves each autumn and are known as deciduous plants. During their evolution, temperate-zone angiosperm trees and shrubs have acquired the genetic capacity to use seasonal changes to predict the onset of cold, dry winter conditions and respond with preemptive leaf abscission. Leaf abscission reduces the potential for evaporative loss of water from a perennial plant during the period when soil water is frozen and thus cannot be taken up by roots.

Leaf abscission is a highly coordinated developmental process. The hormone ethylene stimulates an abscission zone to develop at the bases of leaf petioles (Figure 30.26a). The abscission zone contains two types of tissues: a separation layer of short, thin-walled cells and an underlying protective layer of suberin-impregnated cork cells (Figure 30.26b). As noted in the discussion of endodermal Casparian strips, suberin contains both waterproofing wax and phenolic polymers that retard microbial attack. As the abscission layer develops across the vein linking the petiole with the stem, it eventually cuts off the water supply to the leaf. Chlorophyll in the leaves degrades, revealing colorful orange and yellow carotenoid and xanthophyll pigments that were hidden beneath. In addition, some plants synthesize red and reddish blue pigments in response to changing environmental conditions. The presence of these pigments explains colorful autumn vegetation in temperate zones. Enzymes eventually break down the cell-wall components of the separation layer, causing the petiole to break from the stem. The underlying protective layer forms a leaf scar that seals the wound, helping to protect the plant stem from water loss and pathogen attack.

## Transport in the Phloem Occurs by Pressure Flow

As we have seen, transport of water and dissolved minerals occurs within dead, empty xylem tracheids and vessels by suction, explained by the cohesion-tension hypothesis. Movement of materials in the watery sap of living phloem tissues involves a contrasting process that is explained by the **pressure-flow hypothesis.** This hypothesis states that sugars flow from the places where sugars are made to sites where sugars are utilized, by means of differences in turgor pressure. Leaf mesophyll tissue makes sugars and is known as a sugar source. Cells that receive and utilize sugar, referred to as sinks, occur in the nonphotosynthetic tissues of roots and stems, as well as flowers, fruits, and seeds.

The process by which sugars made in leaf mesophyll cells are moved into the living cells of phloem is known as sugar loading and often requires energy input in the form of ATP (**Figure 30.27**). The movement of sugar into living phloem cells causes water to also enter by osmosis, with the result that the pressure of phloem sap rises. This pressure causes phloem contents to flow from regions of higher solute concentration (leaf veins) to regions of lower solute concentrations (such as roots, flowers, fruits, and seeds). Phloem bulk flow is based on positive pressure, much like you would exert when squirting water from a syringe by pushing on the plunger (see Figure 30.27). The water that enters sugar-rich leaf phloem comes from nearby xylem, explaining why vascular bundles that make up the plant conduction system contain both xylem and phloem tissues (see Chapter 28).

The living food-conducting components of phloem are known as sieve-tube elements. Joined end-to-end, sieve-tube elements form long sieve tubes whose structure fosters the flow of sap for long distances through the plant body. During their development, sieve tubes arise by the mitotic division of precursor cells that also produce narrow cells known as companion cells (see Figure 28.18). Sieve-tube elements and neighboring companion cells remain connected by plasmodesmata, channels that run through cell walls. As they mature, sieve-tube elements lose their nucleus, but companion cells retain a nucleus. Consequently, sieve-tube elements become dependent upon a neighboring companion cell for essential messenger RNA (mRNA) and proteins, which are supplied via plasmodesmata. For example, when a plant's conducting system is damaged, a short-term wound response occurs that involves a protein known as P protein (for phloem protein). Masses of this protein accumulate along the end walls of damaged sieve-tube elements, preventing loss of phloem sap (refer back to Figure 28.18). This mass functions much like a clot that helps reduce blood loss from wounded animals. P protein also binds to the cell walls of pathogens, thereby helping to prevent infection at wounds. Companion cells provide either P protein mRNA or the protein itself to sieve-tube elements. In a longer-term response, plants deposit the carbohydrate callose as a sealant at the wound site.

Companion cells also play an essential role in loading sugar into the phloem. Although glucose and some other monosaccharides can occur in phloem, the more stable disaccharide sucrose is the main

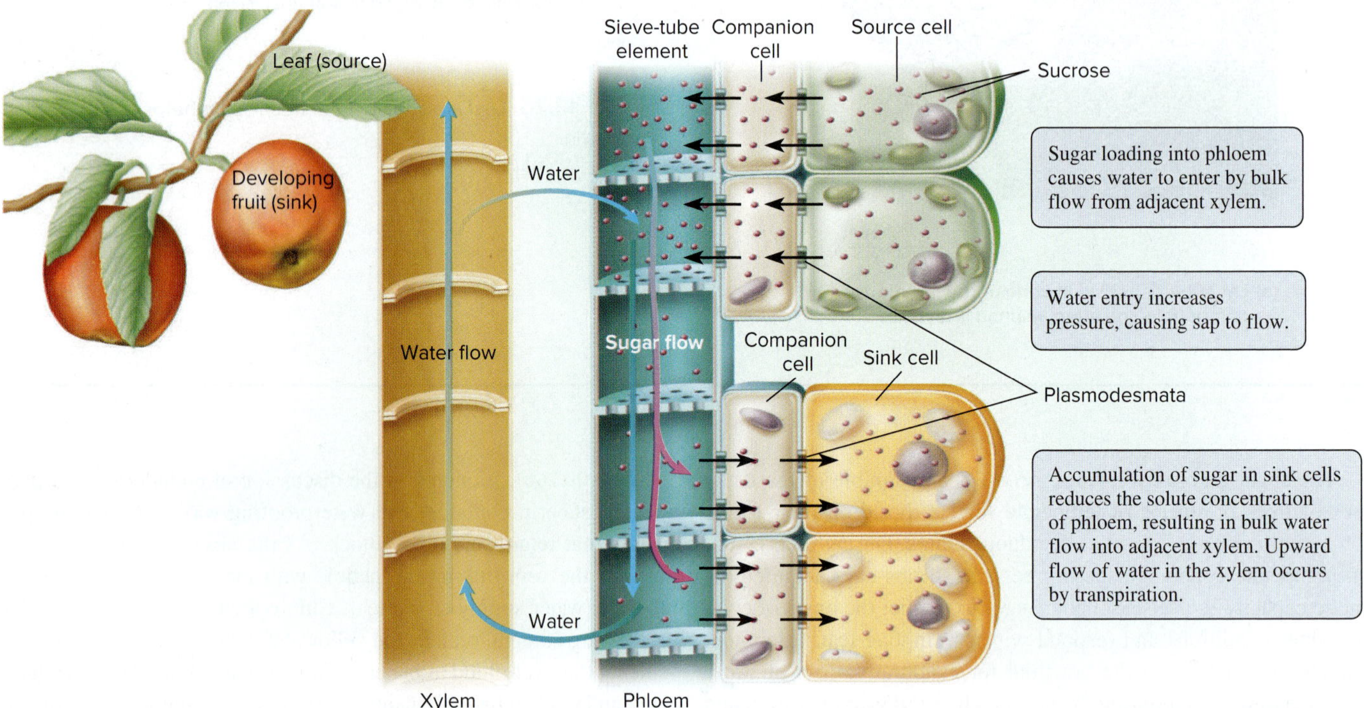


**Figure 30.27 Pressure-flow hypothesis for phloem transport.** Water flows by osmosis into solute-rich source phloem tissues from nearby xylem. The increased water pressure pushes solute-rich phloem sap toward sink tissues, which are lower in solute concentration. This process explains why phloem transport from sources to sinks is referred to as pressure-flow. At sink tissues, sugars are unloaded from the phloem, thereby decreasing phloem solute concentration, and these processes allow water to flow back into xylem. The role of water in the phloem pressure-flow process explains why phloem and xylem tissues occur close together in vascular bundles that form the plant conduction system.

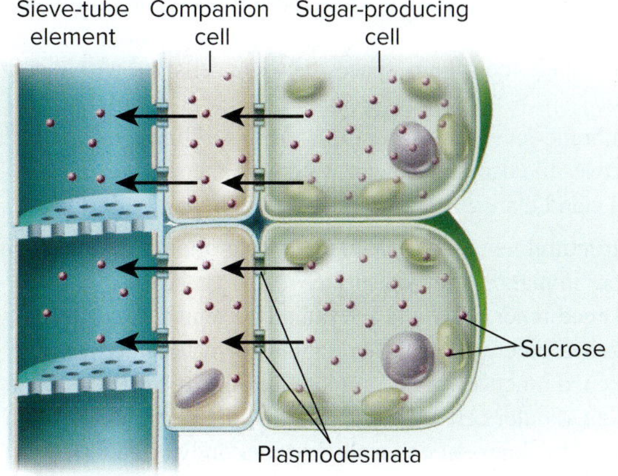


(a) Symplastic phloem loading

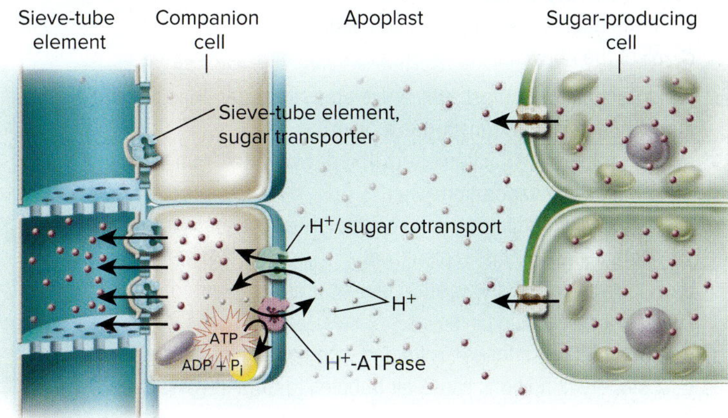


(b) Partly apoplastic phloem loading

**Figure 30.28** **Symplastic and partly apoplastic phloem loading.** Variations in sugar loading that occur among plants illustrate important transport concepts and are important in crop production.

form in which most plants transport sugar over long distances. Two types of phloem loading occur: symplastic and partly apoplastic. Many woody plants transport sucrose from sugar-producing cells of the leaf mesophyll to companion cells and then to sieve-tube elements via plasmodesmata, a process known as symplastic phloem loading (**Figure 30.28a**). The advantage of symplastic loading is that it does not require ATP; by moving through plasmodesmata, sugar does not have to cross plasma membranes.

In contrast, most herbaceous plants, including important crop plants and the model plant *Arabidopsis*, load sugar into sieve-tube elements or companion cells from intercellular spaces, often against a concentration gradient. ATP must be used to move the sugar across a plasma membrane into a companion cell or sieve-tube element (**Figure 30.28b**). Therefore, this second type of phloem loading is partly apoplastic and partly a transmembrane process. Understanding how sugar-loading processes differ among plants is essential to improving crop production. Variations in plant sugar-loading processes also illustrate the importance of osmosis, turgor pressure, active transport, symplastic transport, and apoplastic transport, concepts described earlier in this chapter.

## 30.5 Reviewing the Concepts

- Water and solutes move for long distances by bulk flow within the xylem and phloem. Plant tracheids, vessels, and sieve-tube elements are structured in ways that foster bulk flow. Bulk flow of water upward in xylem is powered by the water pressure difference between moist soil and drier air, the latter resulting from solar heating (Figures 30.20, 30.21).
- The cohesion-tension theory proposes that as the result of water's cohesion and the tension exerted on water at the plant's surface by evaporation, a continuous stream of water can be pulled up through the plant body from the soil, into roots, through stems, and into leaves. Transpiration, the evaporative loss of water from plant surfaces, strongly influences local and global climate (Figure 30.22).
- Regulation of stomata helps plants prevent excessive water loss by transpiration. Expansion of guard cells causes stomata to open, allowing $CO_2$ intake. Guard cell deflation causes pores to close, limiting water loss (Figures 30.23, 30.24).
- Plants under existing or predicted water stress often drop their leaves in a process known as abscission, an adaptive response that lets plants avoid very low water potentials (Figures 30.25, 30.26).
- The pressure-flow hypothesis helps to explain movement of phloem sap as a process driven by differences in turgor pressure that occur between a sugar source (for example, leaves) and a sugar sink (for example, developing fruit). Phloem loading, the movement of sugars into sieve-tube elements for long-distance transport, occurs by symplastic or partly apoplastic transport (Figures 30.27, 30.28).

## 30.5 Testing Your Knowledge

1. What structural feature of xylem tracheids and vessel elements fosters bulk flow of water?
   a. absence of cytoplasmic contents, resulting from controlled cell death
   b. arrangement of cells end-to-end to form long arrays
   c. cell walls coated with lignin
   d. all of the above
   e. none of the above

2. Under what environmental condition are plant stomata most likely to be open?
   a. at night
   b. on a hot, dry, windy day
   c. in the daytime under humid conditions
   d. all of the above
3. Which scenario is most closely related to sugar loading and long-distance sugar transport in a plant?
   a. Dry air causes a suction effect that draws water from plant surfaces by evaporation.
   b. The uptake of mineral ions from moist soil causes water to osmotically flow into a plant root.
   c. A long period of dry weather causes a desert plant to respond by losing its leaves.
   d. A tree with lush foliage is loaded with developing fruit.
   e. Cold, dry conditions cause air bubbles to form in vessels.

## Assess and Discuss

### Test Yourself

1. Which of the following can limit plant growth in nature?
   a. sunlight
   b. water
   c. carbon dioxide
   d. fixed nitrogen
   e. all of the above
2. In what form do plants take up most soil minerals?
   a. as ions dissolved in water
   b. as neutral salts
   c. as mineral-clay complexes
   d. linked to particles of organic carbon
   e. none of the above
3. Soil organic matter provides the benefit of
   a. allowing water to percolate rapidly through soil.
   b. making soil softer in consistency.
   c. increasing the aluminum content of soil.
   d. causing minerals to be leached more rapidly from soil.
   e. none of the above.
4. What are ways in which plants accomplish tissue-level transport?
   a. transmembrane transport of solutes from one cell to another
   b. symplastic transport of materials from one cell to another via plasmodesmata
   c. apoplastic transport of water and dissolved solutes through cell walls and intercellular spaces
   d. all of the above
   e. none of the above
5. A root endodermis is
   a. an innermost layer of cortex cells that each display characteristic Casparian strips.
   b. a layer of cells just inside the epidermis of a root.
   c. a layer of cells just outside the epidermis of a root.
   d. a group of specialized cells within the root epidermis.
   e. none of the above.
6. Which of the following statements explains how water enters root cells from the soil?
   a. Roots accumulate sugars from shoots, thereby increasing root cell ability to absorb water by osmosis.
   b. Roots actively pump water from the soil using the chemical energy of ATP.
   c. Water enters the spongy spaces between cells and within cell walls.
   d. Membrane-level transport of ions from soil into root-hair cells increases the osmotic flow of water into them.
   e. Both c and d are correct.
7. What structural features of guard cells foster their ability to form an open pore in plant epidermal surfaces?
   a. thickened inner cell walls and radially oriented microfibrils
   b. thickened outer cell walls and radially oriented microfibrils
   c. thickened inner cell walls and longitudinal microfibrils
   d. thickened outer cell walls and longitudinal microfibrils
   e. uniform thickness of cell walls and randomly arranged microfibrils
8. What substances plug wounded sieve-tube elements, thereby preventing the leakage of phloem sap?
   a. X protein and callose
   b. C protein and callose
   c. P protein and callose
   d. P protein and sucrose
   e. none of the above

### Conceptual Questions

1. Why is it a bad idea to overfertilize your houseplants? If the amount recommended on the package is good, wouldn't more be better?
2. Why should a farmer try to avoid overfertilizing farm fields that lie near clean trout streams?
3. **PRINCIPLES** A principle of biology is that biology affects our society. Why should farmers (those barely able to grow enough crops to feed themselves) in arid lands prevent livestock from grazing natural vegetation to the point that it disappears?

### Collaborative Questions

1. Imagine that you are part of a team assigned to determine what environmental conditions best suit a new crop so that the crop can be recommended to farmers in appropriate climate regions. What features of the crop plants might you investigate?
2. Take a look outside or imagine a forest or grassland. What can you deduce about the availability of soil water from the types of plants that occur?

## Online Resource

**connect.mheducation.com**

SMARTBOOK® SmartBook® is the first and only adaptive reading experience designed to change the way students read and learn.

# How Flowering Plants Reproduce and Develop

# 31

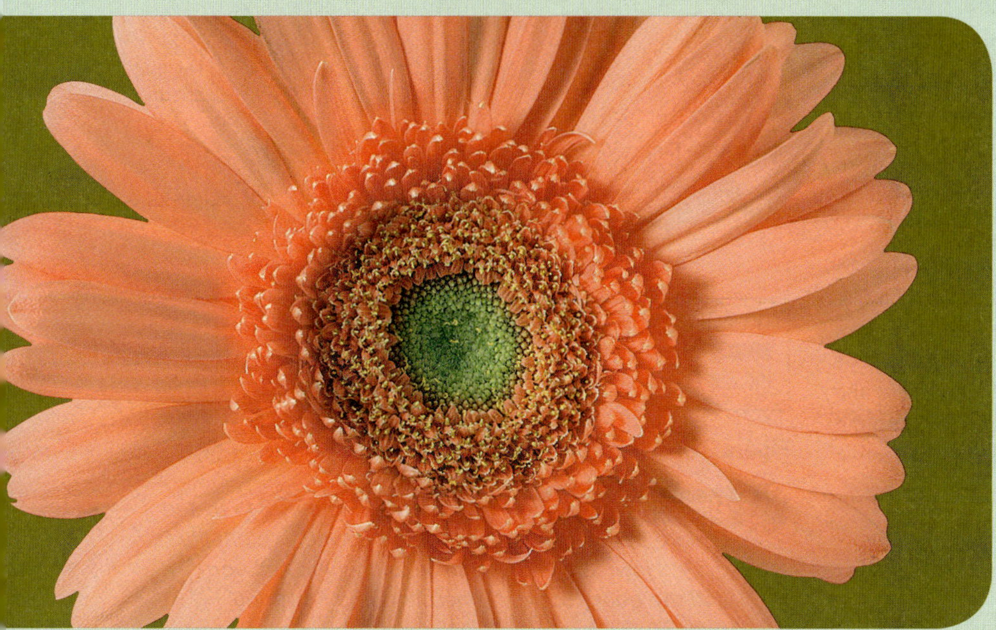



The Gerbera daisy, *Gerbera hybrida.*

## Chapter Outline



A type of daisy known as the *Gerbera* (chapter-opening photo) is one of the most popular plants you might find at the florist or garden shop. As is the case for other daisies and sunflowers, the attractive "flowers" of a *Gerbera* are actually inflorescences, hundreds of small flowers clustered together. The centermost disk flowers of a *Gerbera* produce sexual organs, pollen, and seeds, but the outermost ray flowers do not. Instead, their petals are fused together to form large, colorful flags that attract bees, butterflies, or birds. These animals accomplish cross-pollination, the exchange of pollen between *Gerbera* flowers of different genetic types. As a result of cross-pollination, the next generation of seeds will be genetically diverse, fostering species survival and evolution. The inner disk and outer ray flowers of a Gerbera daisy both play essential roles in the plant's reproductive success.

The Gerbera daisy is also a popular model organism, which molecular biologists use to explore the genetic basis for flower traits. Molecular biologists have discovered how *Gerbera* produces both disk and ray flowers in the same inflorescence. During development, a particular transcription factor is expressed less strongly at the inflorescence center than at the margins, where the protein starts a developmental process that shapes the beautiful ray flowers and cancels their production of sexual organs. Less affected by the protein, the central disk flowers produce seeds within fruits that aid seed dispersal.

Flowers, seeds, and fruits are defining reproductive features of the diverse types of flowering plants that are found in nature and that humans rely upon for food and many other products. This chapter begins with an overview of the sexual reproductive cycle of flowering plants. This background allows a closer look at flower structure and development and some of the genes that control flower production and appearance. We will also see how dandelions and some other interesting plants reproduce without using the sexual process.

# 31.1 An Overview of Flowering Plant Reproduction

## Learning Outcomes

1. Explain how the sexual life cycle of flowering plants differs from that of earlier-evolved mosses.
2. List the four organs found in many flowers and the functions of each organ.
3. Describe the phenomenon of double fertilization.
4. List the parts of an angiosperm seed and explain how each part is produced and functions.

Most flowering plants display **sexual reproduction,** the process by which two gametes fuse to produce offspring that have unique combinations of genes. Flowering plants, also known as angiosperms, inherited their sexual life cycle, known as **alternation of generations** (Figure 31.1), from ancestors extending back to the earliest land plants (see Chapter 25). Though all plants share the same basic life cycle, flowering plants display unique reproductive features.

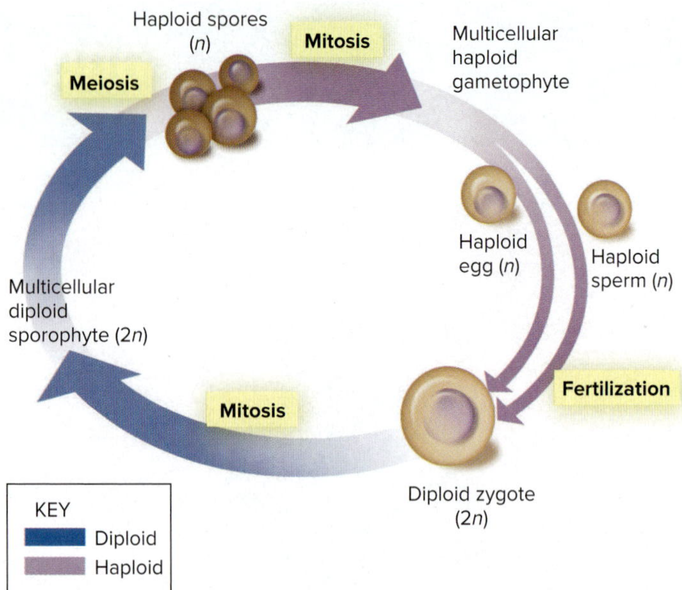


**Figure 31.1** Alternation of generations, the plant life cycle.

*BioConnections: Refer back to Figure 25.15 to see a more detailed illustration of the flowering plant life cycle. How can you recognize and where can you find the gametophyte generation of a flowering plant?*

## Alternation of Generations Is the Sexual Cycle of Flowering Plants

All groups of land plants produce two alternating, multicellular life cycle stages, in essence, two distinct types of plant body. These two life cycle stages are the diploid, spore-producing **sporophyte** and the haploid, gamete-producing **gametophyte.** In all groups of plants, haploid spores are typically produced by diploid sporophytes as the result of meiosis. These spores undergo mitotic cell divisions to produce multicellular gametophytes. Certain cells within the gametophytes differentiate into gametes. Thus, meiosis does not directly generate the gametes of plants, in contrast with animals. The processes of meiosis and fertilization form the transitions between the plant sporophyte and gametophyte life stages and link them in a cycle (see Figures 31.1 and 25.15). The life cycles of bryophytes (see Figure 25.4), ferns (see Figure 25.7), and gymnosperms (see Figure 25.12) also illustrate alternation of generations.

During the evolutionary diversification of land plants, the sporophyte generation has become larger and more complex, whereas the gametophyte generation has become smaller and less complex. To illustrate this, we can compare the life cycle stages of mosses to those of angiosperms. Mosses diverged early in the history of land plants, whereas angiosperms appeared much later. During the intervening time, the relative sizes and dependence of the sporophyte and gametophyte generations changed dramatically. Moss sporophytes are small structures that always grow attached to larger, photosynthetic gametophytes, because moss sporophytes are incapable of independent life and depend on gametophytes to supply them with essential nutrients (Figure 31.2a).

In contrast, flowering plant sporophytes are notably larger and more complex than gametophytes. A tall oak tree, for example, is a single sporophyte. However, oak gametophytes are few-celled, microscopic structures that develop and grow within flowers (Figure 31.2b). In addition, photosynthetic oak seedlings and trees grow independently, but nonphotosynthetic oak gametophytes depend completely on the sporophyte for their nutrition. A closer look at flower structure will help us to gain a more complete understanding of angiosperm gametophyte structure and function.

## Flowers Produce and Nurture Male and Female Gametophytes

A **flower** is defined as a reproductive shoot, a stem that produces reproductive organs instead of leaves. Flowers are organ systems because several different organs typically occur within a flower (see Chapter 25). Flower organs are produced by shoot apical meristems much like those that generate leaves and are thought to have evolved from leaflike structures by descent with modification (refer back to Figure 25.16).

Most flowers contain four types of organs: sepals, petals, stamens, and carpels (Figure 31.3). **Sepals** often protect the unopened flower bud. **Petals** usually attract insects or other animals for pollen transport. **Stamens** and **carpels** each produce distinctive types of spores by the process of meiosis. From these spores, tiny, multicellular gametophytes develop, and certain cells of these gametophytes become specialized gametes.

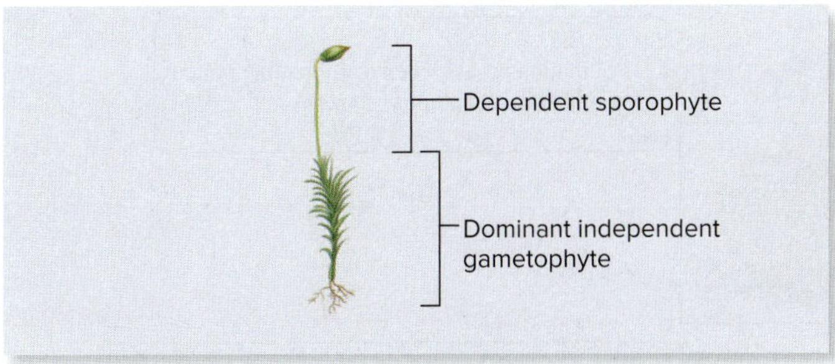


**(a) Gametophyte-dominant bryophyte (moss)**

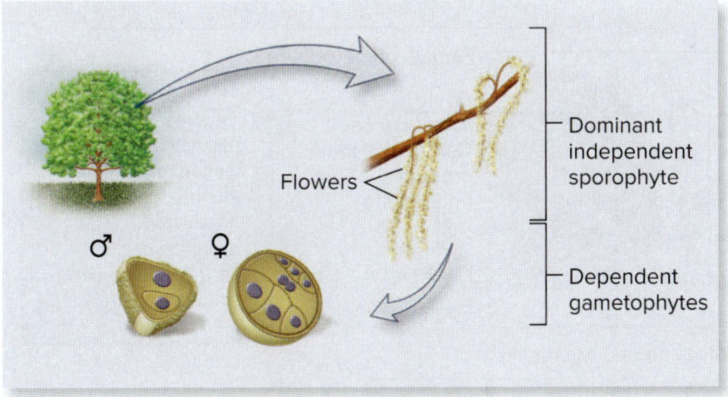


**(b) Sporophyte-dominant flowering plant (oak)**

**Figure 31.2 Evolutionary shift in plant life cycle stage dominance. (a)** In mosses, the gametophyte is the dominant life cycle stage, and the sporophyte depends on the gametophyte for resources. **(b)** In flowering plants such as oak trees, the sporophyte life cycle stage is dominant. Microscopic flowering plant gametophytes develop and grow within flower tissues and depend completely on the sporophyte generation.

**Concept Check:** *What advantages do flowering plants obtain by having such small and dependent gametophytes?*

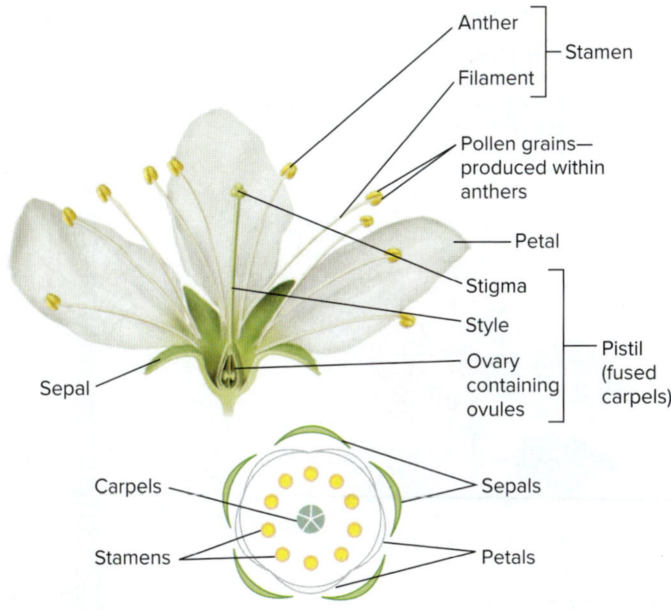


**(a) Flower parts**

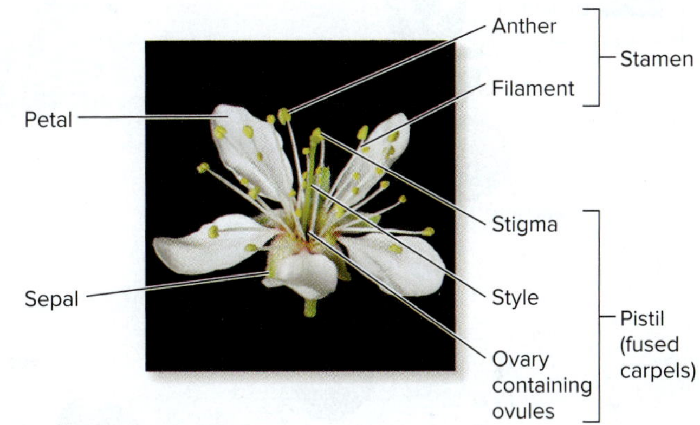


**(b) *Prunus americana* (plum)**

**Figure 31.3 The structure of a typical flower.**
*(b)* © Lee W. Wilcox

**Concept Check:** *Do all flowers have all of the structures illustrated here?*

**Stamens** Stamens produce **male gametophytes** and foster their early development. Most stamens display an elongate stalk, known as a **filament,** which is topped by an anther. Filaments contain vascular tissue that delivers nutrients from the parental sporophyte to the anthers. Each **anther** is a group of four **sporangia—**structures in which spores are produced. Within the anther's sporangia, many diploid cells undergo meiosis, each producing four tiny, haploid spores. Because they are so small, the spores produced within anthers are known as **microspores.** Immature male gametophytes, known as **pollen grains,** develop from microspores. The term pollen comes from a Latin word meaning fine flour, reflecting the small size of pollen grains. Pollen grains are eventually dispersed through pores or slits in the anthers. At the time of dispersal, the pollen grain is a two- or three-celled, immature male gametophyte produced by mitotic division. During a later phase of development, a mature male gametophyte produces **sperm** cells.

**Carpels** Carpels are vase-shaped structures that produce, enclose, and nurture **female gametophytes.** Carpels contain veins of vascular tissue that deliver nutrients from the parent sporophyte to the developing gametophytes. The term **pistil** (named for its resemblance to the pestle used to grind materials to a powder) refers to a single carpel or several fused carpels (see Figure 31.3). The topmost portion of a pistil, known as a stigma (Greek, meaning mark), receives pollen grains. The style is the middle portion of the pistil, and an ovary is at the bottom of the pistil. The **ovary** produces and nourishes one or more ovules. An **ovule** consists of a spore-producing sporangium and enclosing tissues known as integument. Within an ovule, a diploid cell produces four haploid **megaspores** by meiosis; megaspores

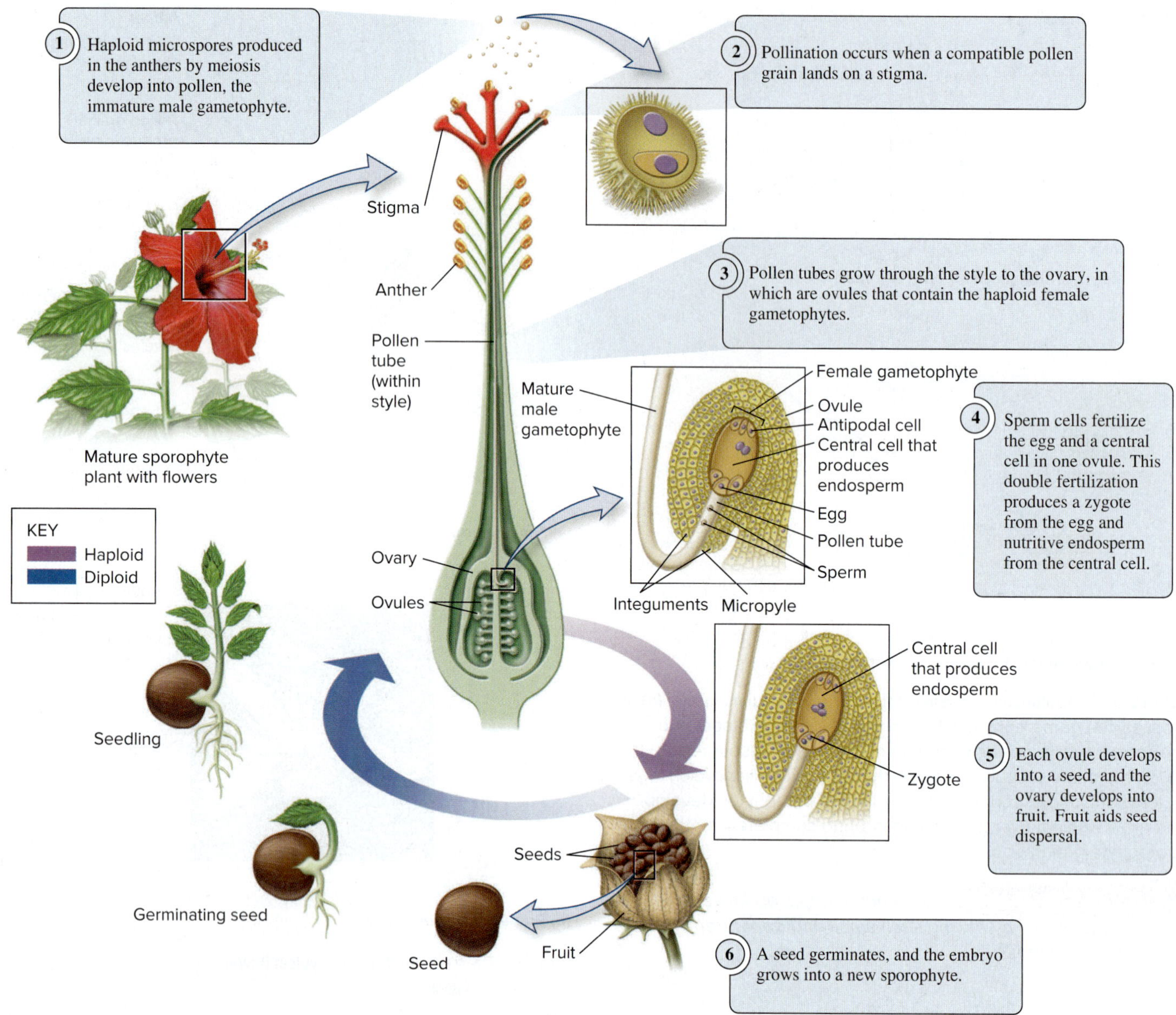

1 Haploid microspores produced in the anthers by meiosis develop into pollen, the immature male gametophyte.
2 Pollination occurs when a compatible pollen grain lands on a stigma.
Stigma
Anther
Pollen tube (within style)
Mature male gametophyte
3 Pollen tubes grow through the style to the ovary, in which are ovules that contain the haploid female gametophytes.
Female gametophyte
Ovule
Antipodal cell
Central cell that produces endosperm
Egg
Pollen tube
Sperm
Integuments
Micropyle
4 Sperm cells fertilize the egg and a central cell in one ovule. This double fertilization produces a zygote from the egg and nutritive endosperm from the central cell.
Mature sporophyte plant with flowers
KEY
Haploid
Diploid
Ovary
Ovules
Seedling
Central cell that produces endosperm
Zygote
5 Each ovule develops into a seed, and the ovary develops into fruit. Fruit aids seed dispersal.
Seeds
Germinating seed
Seed
Fruit
6 A seed germinates, and the embryo grows into a new sporophyte.


**Figure 31.4** **The life cycle of a flowering plant.** The plant reproductive cycle is illustrated here by hibiscus.

**Concept Check:** *What advantage does the hibiscus flower gain by clustering its stamens around the pistil?*

are larger than the microspores produced in anthers, and megaspores have a different fate. Within each group of four megaspores, three die, which allows the surviving one to gain more nutrients. This survivor generates a female gametophyte by mitosis. The female gametophytes of flowering plants typically consist of seven cells, one of which is the female gamete, the **egg** cell. The female gametophytes of flowering plants also include a large central cell that typically has two nuclei, as well as antipodal cells, whose function is not well understood. This basic information about male and female gametophytes will help us to understand how they function to produce a young sporophyte within a seed.

## Fertilization Triggers the Development of Embryonic Sporophytes, Seeds, and Fruits

In flowering plants, fertilization leads to the production of a young sporophyte that lies within a seed, completing the life cycle (Figure 31.4). Prior to fertilization, pollen grains released from anthers first find their way to the stigma of a compatible flower, the process known as pollination. **Self-pollination** occurs when pollen from the anthers of a flower is transferred to the stigma of the same flower or between flowers of the same plant. **Cross-pollination** occurs when a stigma receives pollen from a different plant of the same species. Flowers are adapted for effective pollination by diverse pollination mechanisms. Many flowers, including the Gerbera daisy, are attractive to insects or other animals that transport pollen, whereas oak flowers and those of some other angiosperms are adapted for pollen transport by wind. A few plants move pollen by means of water currents.

**Pollen Germination** When pollen grains land on the stigma, the stigma may function as a "gatekeeper," allowing only pollen of appropriate genotype to germinate. During germination, a pollen grain produces a long, thin **pollen tube** that contains two sperm cells. The pollen tube grows through the style toward the ovary. Upon reaching the ovules, the pollen tube grows through an opening in the ovule and delivers sperm to the female gametophyte (see Figure 31.4). These sperm unite with haploid cells of the female gametophyte in the process of **fertilization.** Note that pollination and fertilization are distinct processes in flowering plants.

**Double Fertilization** Angiosperms display a phenomenon known as **double fertilization.** In this process, two different fertilization events occur. One of the two sperm cells delivered by a pollen tube fertilizes the egg cell, thereby forming a diploid **zygote.** This zygote may develop by mitotic division into a young sporophyte, known as an **embryo.** Fertilization thus begins a new cycle of alternation between sporophyte and gametophyte generations. The other sperm delivered by the same pollen tube fuses with the two nuclei in the central cell of the female gametophyte (see Figure 31.4). The cell formed by this second fertilization undergoes mitosis, eventually producing a nutritive tissue known as the **endosperm.** The embryo and the endosperm are essential parts of maturing seeds. Fertilization not only starts the development of zygotes into embryos but also triggers the transformation of ovules into seeds and ovaries into fruits. Embryo, seed, and fruit development occur at the same time.

**Embryos and Seeds** An embryo is a young, multicellular, diploid sporophyte that develops from a single-celled zygote by mitosis. Because they are not yet capable of photosynthesis, embryos depend on organic food and other materials supplied by sporophytes. Therefore, embryo development occurs within developing seeds located in a flower ovary. Seeds develop from fertilized ovules. Each developing seed contains an embryo and nutritive endosperm tissue, enclosed and protected by a **seed coat** that develops from the ovule integuments. When embryos and the seed coat have fully matured, they undergo drying, and the seed enters a phase of metabolic slowdown known as dormancy. Fully mature, dormant seeds are ready to be dispersed.

**Fruit and Seed Dispersal** A **fruit** is a structure that encloses and helps to disperse seeds (see Figure 31.4). Seed dispersal benefits plants by reducing competition for resources among seedlings and parental plants, and it allows plants to colonize new sites.

Fruits develop from the flower's ovary and sometimes include other flower parts. Young fruits bearing immature seeds are typically small and green. While embryos and seeds are developing, the fruit also matures. The ovary wall changes into a fruit wall known as a **pericarp** (from the Greek, meaning surrounding the fruit). Mature fruits vary greatly among plant species in size, shape, color, and water content. These variations represent adaptations for seed dispersal in different ways. Fruit variation is also extremely important to wild animals and in human agriculture. For example, most fruit crops are juicy and sweet, with relatively small seeds. In nature, these features foster dispersal by birds and other animals that feed on such fruits.

**Seed Germination and Seedlings** If a dispersed seed encounters favorable conditions, including sufficient sunlight and water, it will undergo seed **germination.** During seed germination, the embryo absorbs water, becomes metabolically active, and grows out of the seed coat, producing a seedling. If the seedling obtains sufficient nutrients from the environment, it grows into a mature sporophyte capable of producing flowers. In the next section, we focus on flower production, structure, and development.

## 31.1 Reviewing the Concepts

- Flowering plants display a sexual life cycle known as alternation of generations, a characteristic of all land plants. The gamete-producing male and female gametophytes of flowering plants are very small and depend entirely on nurturing sporophytic tissues. Plant gametes arise by the process of mitosis (Figures 31.1, 31.2).
- Flowers are reproductive shoots that develop from a shoot apical meristem. The role of flowers is to promote seed production. A flower shoot generally produces four types of organs: sepals, petals, stamens, and carpels. Stamens produce pollen grains, which are immature male gametophytes. Carpels produce, enclose, and nurture female gametophytes in ovules (Figure 31.3).
- A mature male gametophyte, or pollen tube, produces two sperm and delivers them to ovules within the ovary. Flowering plants display double fertilization: One of the two sperm released from a pollen tube combines with an egg cell to form a zygote, while the other fuses with two nuclei located in a central cell of the female gametophyte, producing the first cell of endosperm tissue. Seed endosperm is a nutritive tissue that supports development of an embryonic sporophyte (Figure 31.4).
- Seeds are reproductive structures containing a dormant embryo enclosed by a protective seed coat that develops from ovule integuments. Fruits are structures that contain seeds and foster seed dispersal. Like flowers and endosperm, fruits are unique features of flowering plants.

## 31.1 Testing Your Knowledge

1. How does the sexual life cycle of a flowering plant differ from the life cycle of an animal?
   a. Two multicellular bodies alternate in the sexual life cycles of flowering plants, but not in the sexual cycles of animals.
   b. Meiosis generates spores in plants, whereas meiosis generates gametes in animals.
   c. The gamete-producing plant gametophyte is haploid, whereas a diploid animal produces gametes.
   d. All of the above are correct.
   e. None of the above are correct.

2. What is a unique combination of reproductive features found in flowering plants?
   a. flowers, fruits, and endosperm
   b. stems, roots, and leaves
   c. alternation of generations and sporophytes
   d. gametophytes, eggs, sporangia, and spores
   e. all of the above

## 31.2 Flower Production, Structure, and Development

### Learning Outcomes

1. Give an example of how gene expression affects flowering time or flower appearance.
2. List examples of complete, incomplete, perfect, imperfect, radially symmetric, and bilaterally symmetric flowers and inflorescences.
3. **SCISKILLS ▶** Explain recent research findings on how flowers bloom.

Flowers are essential sources of food for many animal pollinators. As the result of coevolutionary relationships with such animals, flowers occur in a spectacular array of colors and forms that attract particular animal pollinators. Flowers also attract humans because we possess sensory systems much like those of animal pollinators. We give bouquets to show love and appreciation; decorate homes, workplaces, and objects with flowers; display flower arrangements on ceremonial occasions; and make perfume from flowers. Consequently, many types of flowers are grown for the florist and perfume industries. Flowers are necessary for the production of grain and other fruit crops. For these and other reasons, biologists investigate how flower development is controlled by environmental signals and changes in gene expression.

### Environmental Signals Interact with Genes to Control Flower Production

You've probably noticed that different plants flower at particular times of the year. How do plants know when to flower? Flowering time is controlled by the integration of environmental information such as temperature and day length (photoperiod) with hormonal influences and circadian rhythms (see Chapter 29). These stimuli are perceived and integrated by leaves, which then use a mobile protein known as FT (flowering time) to signal shoot meristems to produce flowers. FT protein interacts with other proteins in the shoot meristem, initiating the process by which a leaf-producing apical meristem transforms into a reproductive meristem.

Some plants—such as winter wheat—are planted and sprout in fall, are dormant in winter, and then flower the following spring. In these plants a special transcription factor represses flowering genes, thereby preventing flowers from appearing too soon. Exposure to cold winter conditions causes the production of a small RNA molecule that silences the transcription factor, allowing these plants to flower when the appropriate day length occurs in spring. The process by which cold exposure allows plants to flower in spring is known as vernalization.

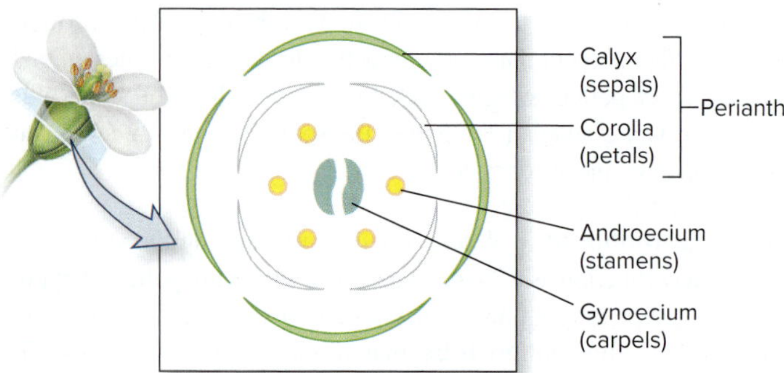


**Figure 31.5** The occurrence of flower parts in concentric whorls.

### Developmental Genes Control Flower Structure

Organ identity genes specify the four basic flower organs: sepals, petals, stamens, and carpels. Other genes determine flower shape, color, odor, or grouping into bunches known as inflorescences, such as those of the *Gerbera*.

**The Genetic Basis of Flower Organ Identity** Sepals, petals, stamens, and carpels occur in four concentric rings known as whorls. Sepals (collectively known as the calyx) form the outermost whorl, and petals (together known as the corolla) form an adjacent whorl type. Stamens (together, the androecium) create a third type of whorl, and carpels (the gynoecium) form the innermost whorl (Figure 31.5). The **perianth** consists of the calyx plus the corolla.

Gene classes referred to as A, B, C, and E encode transcription factors that control the production and arrangement of these flower organ whorls. The expression of A genes is required for the formation of sepals and petals, B gene expression is needed for petal and stamen development, and C genes specify stamen and carpel production. Consequently, the effects of the A, B, and C genes in determining flower organs are together known as the ABC model of floral organ identity. More recently, a fourth class of genes, the E genes, have been recognized as essential for the formation of petals, stamens, and carpels. Gymnosperms possess genes similar to the class B and C genes of flowering plants, suggesting an early stage in the evolution of the ABCE system.

**Variation in Number of Whorls** Flowers that possess all four types of flower whorls—calyx, corolla, androecium (stamens), and gynoecium (one or more carpels)—are known as **complete flowers.** In contrast, flowers that lack one or more flower whorls are described as **incomplete flowers.** Flowers having both stamens and carpels are said to be **perfect flowers,** whereas flowers lacking stamens or carpels are described as **imperfect flowers.** An imperfect flower that produces only carpels is known as a carpellate flower (or pistillate

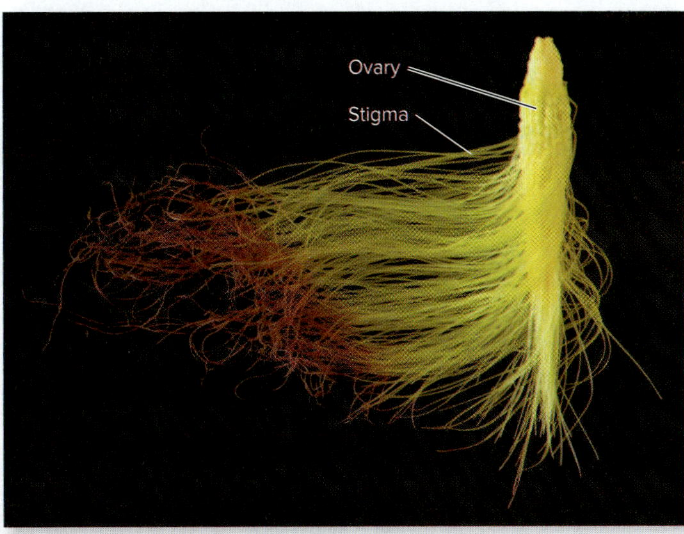


**(a) Staminate flowers of *Zea mays* (corn)** **(b) Carpellate flowers of *Zea mays* (corn)**

**Figure 31.6 Imperfect flowers of corn, a monoecious plant. (a)** Staminate flowers lack carpels, and **(b)** carpellate flowers lack stamens, but both types of flowers occur on a single corn plant. In contrast, dioecious plants produce staminate and carpellate flowers on separate plants. These features foster cross-pollination.

*(a)* © Scott Sinklier/agefotostock; *(b)* © Robert and Jean Pollock

**Concept Check:** *What inference can you draw from the observation that flowers of corn lack showy petals?*

flower). Imperfect flowers that produce only stamens are described as staminate flowers.

Corn produces both imperfect staminate and carpellate flowers on an individual plant (Figure 31.6). The flowers of corn start to develop as perfect flowers, but in carpellate flowers, the stamens stop developing. After pollination and fertilization, each carpellate flower produces one of the kernels on a cob of corn. In contrast, staminate flowers of corn, which are found in corn tassels, produce the pollen. Corn produces staminate and carpellate flowers on the same plant. Holly and willow are examples of the many types of plants that produce staminate and carpellate flowers, though on separate plants.

**Variation in Flower Organ Number** In addition to variation in whorls, flowers vary in number of organs. You may recall that flowering plants occur in two major groups differing in flower structure and other features: the eudicots and the monocots (refer back to Table 28.1). Eudicot flower organs usually occur in fours or fives or a multiple of these numbers. By contrast, monocot flower organs typically occur in threes or a multiple of three. Other flowers possess relatively few organs. For example, the minuscule flowers of the tiny aquatic flowering plant *Lemna gibba* (commonly known as duckweed) each produce no perianths and only two stamens.

Many plants sold for use in gardens have been bred so that the flowers produce multiple organs. Garden roses, for example, typically have many more whorls of petals than do wild roses. This change results from a mutation that causes organs that would have become stamens to develop instead into additional petals.

**Variation in Flower Color** Different flower parts may vary in color. The calyx and corolla of monocot flowers such as tulip and lily are often similar in appearance and attractive function. In contrast, eudicot flowers tend to have green, leaflike sepals (see Figure 31.3) that are quite distinct from petals, which are often colorful and fragrant. Color variations arise from differences in gene action that influence pigment biosynthesis pathways.

**Variation in Flower Fragrance** The fragrances of flowers result from secondary metabolites that diffuse into the air from petals and other flower organs. One example is the terpene known as geraniol, which gives roses their distinctive fragrance. Diverse fragrances function to attract particular types of animal pollinators. Because humans possess sensory systems similar to those of many pollinators, we are also attracted to many of these fragrances. Genetic variation in the synthesis of different types of secondary metabolites is responsible for the many types of flower scents.

**Flower Shape Variation Resulting from Organ Fusion** During their development, many flowers undergo genetically controlled fusion between whorls or fusion of the organs within a whorl. For example, pistils are often composed of two or more fused carpels. In addition, stamen filaments may partially fuse with petals, or with the carpel, or form a tube surrounding the pistil, a feature displayed by hibiscus flowers (see Figure 31.4). As mentioned, *Gerbera* ray flowers arise from a fusion process. Some flowers have petals that are fused together to form a tube, which holds nectar consumed by animal pollinators (refer back to Figure 25.18).

## Biology Principle

(a) Normal snapdragon flower

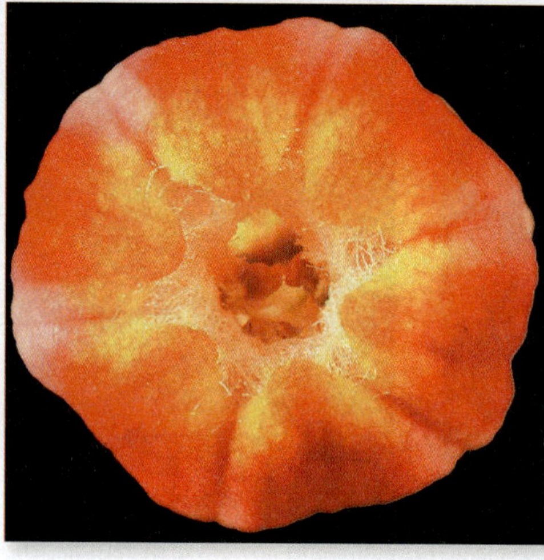

(b) Snapdragon flower with *CYCLOIDEA* mutation

**The Genetic Material Provides a Blueprint for Reproduction**

This principle is illustrated by a comparison of normal and mutant snapdragon flowers, which shows alternate phenotypes conferred by distinct gene sequences.

**Figure 31.7** **Genetic control of flower symmetry.** **(a)** Normal snapdragon flowers, with functioning *CYCLOIDEA* genes, are bilaterally symmetric. **(b)** Snapdragon plants carrying mutations in the *CYCLOIDEA* gene produce flowers that have radial symmetry.
*(a–b)* Courtesy of John Innes Centre

**Variations in Flower Symmetry** Flower shape variation can also result from changes in symmetry. Flowers having radial symmetry can be divided into two equal parts by more than one longitudinal plane inserted through the center of the flower. Other flowers display bilateral symmetry. Flowers having bilateral symmetry can be divided into two equal parts by only a single plane inserted through the center.

Symmetry, like other flower features, is under genetic control. The production of flowers having bilateral symmetry is controlled by transcription factors such as those encoded by the *CYCLOIDEA* gene. For example, snapdragon (*Antirrhinum majus*) flowers are normally bilaterally symmetric, but a loss-of-function mutation in the *CYCLOIDEA* gene causes these flowers to display radial symmetry (Figure 31.7).

## FEATURE INVESTIGATION

### Liang and Mahadevan Used Time-Lapse Video and Mathematical Modeling to Explain How Flowers Bloom

Like the curled limbs of a human embryo within the womb, concave and overlapping petals occupy minimal space within a flower bud. But when flowers bloom, petals rapidly become convex, spreading outward from each other. Various hypotheses had been proposed to explain petal growth during the opening of flowers. One was that growth started at the midrib running along the center of each petal; another proposed that the blooming process was driven by differential growth rates of top and bottom petal surfaces. In 2011, Haiyi Liang and Lakshminarayanan Mahadevan reported a new explanation for blooming based on their studies of the common lily, *Lilium casablanca* (Figure 31.8).

The investigators placed cut stems bearing young, green buds into water, in an environment with a constant humidity, temperature, and light. To evaluate growth during blooming, they first painted small black dots 1 cm apart along the edges and midribs of many sepals and petals. By tracking the movements of these dots on many sepals and petals, the researchers repeated, or replicated, the experiment. Replication is key to the experimental process because it helps researchers to avoid making erroneous conclusions that might be reached on the basis of just a few observations. The investigators then used time-lapse video to track changes in the positions of the black dots during the opening of young, green buds, until the flowers had fully bloomed $4^1/_2$ days later. This process allowed investigators to measure and graph growth at the petal edge versus the center. These measurements also enabled them to mathematically model changes in petal shape, starting with equations that describe changes in the shapes of thin sheets (see Figure 31.8).

The biologists observed that by the end of the fourth day, each bud had absorbed one-fifth of a liter of water, increased length by 10% and width by 20%, and turned white. The inner three petals had become wrinkled, especially at the edges, a key clue to the mechanism underlying the blooming process. The wrinkling reflected greater growth at the edges than in the petal center. This difference in growth generated physical stress values large enough to cause the flower to bloom rapidly as individual petals and sepals simultaneously reversed curvature and bent outward, a process predicted by mathematical modeling. The study revealed that flower blooming is based upon a different mechanism than previously thought and illustrates the value of applying mathematical models to biological phenomena.

### *Experimental Questions*

1. Why do you think Liang and Mahadevan used *Lilium casablanca* for their analysis of flower blooming?
2. How did time-lapse video improve data gathering in this study?
3. What is the value of using mathematical models to predict the blooming process?

**Figure 31.8** **Liang and Mahadevan used time-lapse video and mathematical modeling to explain how flowers bloom.**
*(6)* H. Liang and L. Mahadevana (2011), "Growth, geometry, and mechanics of a blooming lily," *PNAS,* 108(14): 5516–5521, Fig. 2c

**GOAL** To better understand flower blooming.

**KEY MATERIALS** Asiatic lily, *Lilium casablanca*, cut stems with flower buds.

| | Experimental level | Conceptual level |
|---|---|---|
| **1** Paint small black dots at 1 cm intervals along edges and centers of perianth parts on replicate flower buds. |  Painted dots | Compare rate of growth at center and edges of petals/sepals. |
| **2** Maintain cut stems at constant temperature, humidity, and light conditions for 4.5 days, until flowers open. |  | Avoid confounding the experiment with environmental changes. |
| **3** Set up an automatic time-lapse video system that records blooming at intervals of 1 minute. |  | Generates a record of changes in the positions of dots on petals/sepals. |
| **4** Use equations for growth of thin films to model petal/sepal shape changes. | | Compare model to video-based observations. |

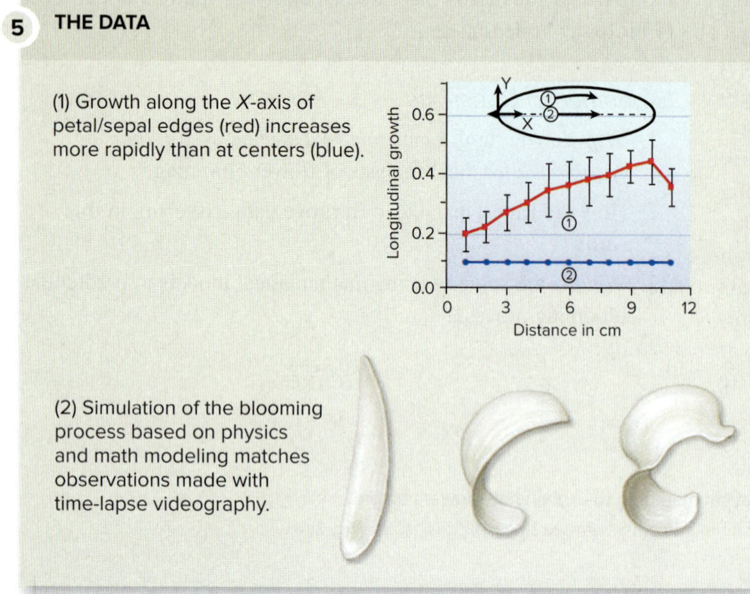

5 THE DATA
(1) Growth along the X-axis of petal/sepal edges (red) increases more rapidly than at centers (blue).
Longitudinal growth
0.0
0.2
0.4
0.6
0
3
6
9
12
Distance in cm
Y
X
(2) Simulation of the blooming process based on physics and math modeling matches observations made with time-lapse videography.


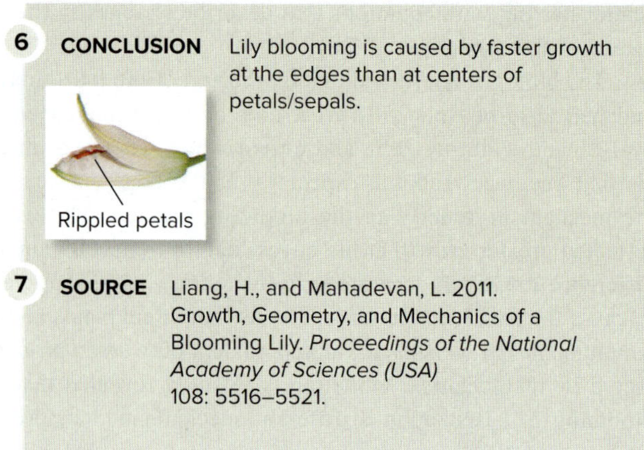

6 CONCLUSION Lily blooming is caused by faster growth at the edges than at centers of petals/sepals.
Rippled petals
7 SOURCE Liang, H., and Mahadevan, L. 2011. Growth, Geometry, and Mechanics of a Blooming Lily. Proceedings of the National Academy of Sciences (USA) 108: 5516–5521.


## 31.2 Reviewing the Concepts

- Plants flower in response to environmental stimuli, such as temperature and day length, by the conversion of a leaf-producing apical meristem into a reproductive meristem.
- Flowers vary in the type of whorls present, the number of flower organs, color, fragrance, shape, organ fusion, symmetry, and whether single or in inflorescences. These variations are related to pollination mechanisms and are genetically controlled (Figures 31.5, 31.6, 31.7).
- Flower blooming involves dramatic changes in petal shape caused by faster growth at petal edges than at their centers (Figure 31.8).

## 31.2 Testing Your Knowledge

1. Put the following events in the development and blooming of a flower in the correct order:
   - **I.** Environmental stimuli are perceived by leaves, which then produce FT protein.
   - **II.** As they develop, flowers may undergo genetically controlled fusion between whorls or organs within a whorl, and may undergo genetically controlled changes in symmetry.
   - **III.** Flowering gene classes A, B, C, and E become active, generating transcription factors that specify calyx, corolla, androecium, and gynoecium, or at least some of these whorls.
   - **IV.** During blooming, flowers open from buds by absorbing water and by greater growth at petal and sepal edges than at the center (midrib).
   - **V.** FT protein moves to shoot tips, causing leaf-producing meristems to develop into reproductive meristems.

   **a.** I, II, III, IV, V **b.** I, III, II, V, IV **c.** I, V, III, II, IV **d.** I, III, II, IV, V **e.** V, I, II, III, IV

## 31.3 Male and Female Gametophytes and Double Fertilization

### Learning Outcomes

1. Explain how plant sperm production differs from sperm production in animals.
2. Outline the different fates of the two sperm cells transmitted by each pollen tube.

In flowering plants, immature male gametophytes are called pollen grains; mature male gametophytes are pollen tubes that deliver sperm cells, the male gametes. Mature female gametophytes are located within ovules and produce egg cells, the female gametes.

### Pollen Grains Are Immature Male Gametophytes

In animals, the cells that will produce gametes—known as the germ line—are set aside from other body cells during early development, but this is not the case for plants. Instead, plants make gamete-producing cells from gametophytic body cells each time that stamens form, on an "as needed" basis.

Pollen grains develop within sporangia located in the anthers of stamens (see Figure 31.4, step 1). Sporangia are structures produced

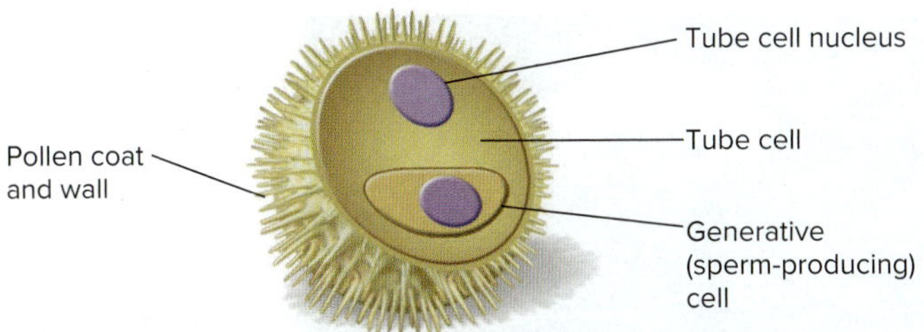


**(a) A cut pollen grain showing immature male gametophyte**

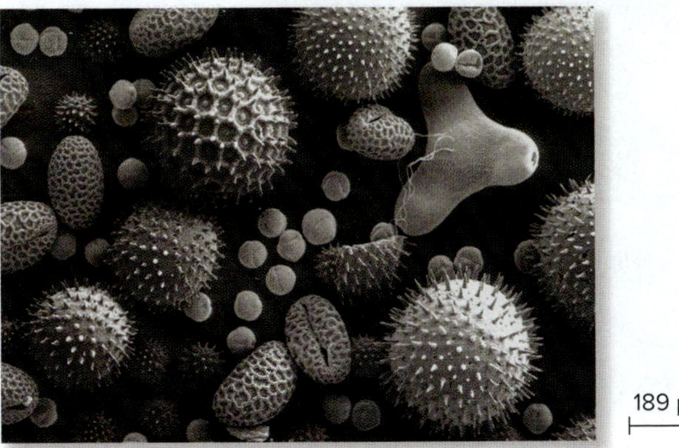


**(b) SEM of whole pollen grains showing distinctive pollen wall ornamentation**

**Figure 31.9** **Pollen grains. (a)** Diagram of cut pollen grain. **(b)** SEM of whole pollen grains of different species.
*(b)* © Phanie/Alamy

**Concept Check:** *What is the maximum number of cells in a mature male gametophyte of a flowering plant?*

by all plants, within which meiosis generates spores. (Note that in plants, meiosis generates haploid spores, not haploid gametes as in animals.) Diploid cells that are located inside protective sporangia undergo meiosis to produce a cluster of four haploid microspores, each having a thin cellulose cell wall. The development of microspores into pollen grains involves two processes that occur at the same time: (1) microspore division to produce the youngest type of male gametophyte and (2) development of a tough pollen wall that protects the male gametophyte during pollen transport. Both of these processes are completed before anthers release pollen.

Each microspore nucleus undergoes one or two mitotic divisions to form an immature male gametophyte. The first division gives rise to two specialized cells: a tube cell and a generative cell suspended within the tube cell (Figure 31.9a). The **generative cell** divides to produce two sperm cells, either before or (more commonly) after pollination. The **tube cell** produces the pollen tube, which delivers sperm to the female gametophyte.

A mature pollen grain has a tough wall, and each plant species produces pollen whose wall has a distinctive sculptural shape (Figure 31.9b). The pollen wall, which surrounds the plasma membrane of the tube cell, is composed largely of a tough polymer known as sporopollenin. Named for its presence on the surfaces of mature spores and pollen, sporopollenin protects spores and pollen from damage. As pollen grains mature, anther cells secrete a pollen coat, a layer of material that covers the sporopollenin-rich pollen wall. Coat materials include additional sporopollenin and pigments that give pollen its typical yellow, orange, or brown coloration, as well as lipids and proteins that aid in pollen attachment to carpels. Certain of these pollen coat compounds are responsible for allergic reactions in people exposed to particular types of airborne pollen. About 10% of flowering plants are wind-pollinated, and such plants produce copious amounts of pollen. For example, ragweed plants (genus *Ambrosia*), which are commonly associated with allergies, each produce an estimated 1 billion pollen grains during a growing season.

## After Pollination, the Pistil Controls Pollen Germination

Recall that pollination is the process by which pollen is delivered to surfaces of the stigma, the uppermost part of the flower pistil. However, even if a pollen grain reaches the stigma of the right flower species, it may not be able to germinate and produce a sperm-delivery tube. The stigma and the style determine whether or not pollen grains germinate and pollen tubes grow toward ovules. How is this accomplished?

About half of plant species can serve as both mother and father to their progeny, because pollen produced by those plants is able to germinate on pistils of the same plants. Such self-pollinating plants are also termed **self-compatible (SC).** By contrast, plants that prevent the germination of pollen that is too genetically similar to the pistil are **self-incompatible (SI).** As in the case of human cultural practices that prevent mating between close relatives, plant SI helps to decrease the likelihood of recessive disorders in offspring. In many plants, SI involves the *S* gene locus, which encodes S proteins. Each locus contains genetic sequences that determine pollen compatibility traits and pistil compatibility traits. Multiple *S* alleles for both genes occur in plant populations.

## A Female Gametophyte Develops Within Each Ovule

Each ovule produces a single female gametophyte, which often consists of seven cells and eight nuclei (Figure 31.10). One of these cells is an egg cell, which lies wedged between two supporting cells that attract pollen tubes. As a pollen tube approaches, synergids use molecular communication processes to detect the tube and help prepare the egg cell to be fertilized. The other four cells of the angiosperm female gametophyte consist of three cells (antipodals) whose functions are not well understood and a large **central cell** that contains two nuclei. This central cell and the egg cell are involved in the process of double fertilization.

## Pollen Tubes Deliver Sperm Cells That Accomplish Double Fertilization

When pollen grains germinate successfully, they take up water, and the tube cell produces a long pollen tube. Within the tube, the generative cell divides by mitosis to produce two sperm cells. To deliver sperm to female gametophytes, the tube must grow at its tip from the stigma, through the style, to reach the ovule (Figure 31.11).

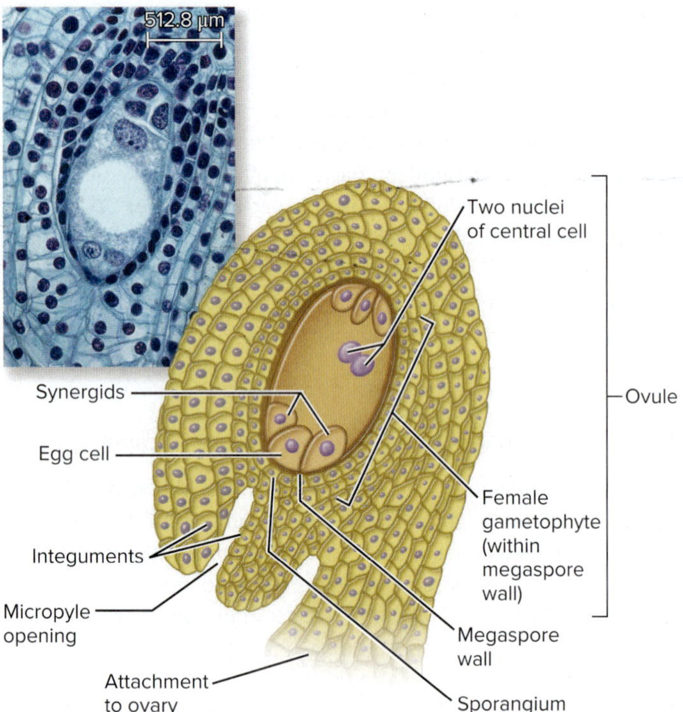


**Figure 31.10** **Angiosperm female gametophyte within an ovule.**
*(top left)* © Ed Reschke/Getty Images

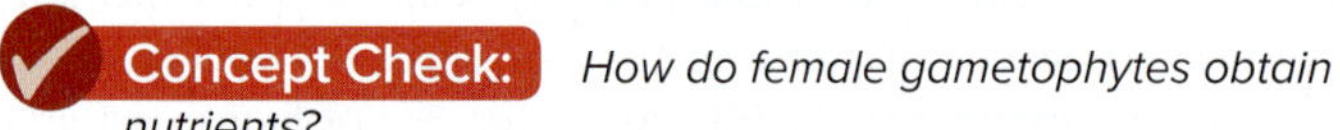

*How do female gametophytes obtain nutrients?*

**Figure 31.11** **Pollen tubes delivering sperm to ovules.** This fluorescence microscopic view shows pollen grains (the pale objects) germinating on the stigma surface (SG) and pollen tubes (PT) growing through the style (ST) toward ovules.
Courtesy J.M. Escobar-Restrepo and A.J. Johnston, University of Zurich, Institute of Plant Biology. From: Bernasconi et al. (2004), "Evolutionary ecology of the prezygotic stage," *Science,* 303(5660): 971–975, Fig. 2

**BioConnections:** *Look forward to Figure 38.5, which illustrates the male sexual organ of humans, an adaptation that allows internal fertilization. How is the mature male gametophyte of plants like a human penis?*

Tip growth is controlled by the tube cell nucleus. During tip growth, new cytoplasm and cell-wall material are added to the tip of an elongating cell. Pollen tubes have been observed to grow toward ovules at about 0.5 mm per hour, commonly taking from 1 hour to 2 days to reach their destination. When a pollen tube encounters the opening (known as a micropyle) of an ovule, the tube stops growing and its thin tip wall bursts, releasing the sperm. The bursting process propels the two sperm toward the egg and the central cell. One sperm nucleus fuses with the egg cell to produce a zygote, the first cell of a new sporophyte generation (see step 4 in Figure 31.4). The other sperm fuses with the two nuclei of the central cell to form the first endosperm cell (see step 5 in Figure 31.4). The term double fertilization arises from these two processes.

The zygote and fertilized central cell have dramatically different fates. The endosperm absorbs protein, lipid, carbohydrates, vitamins, and minerals from the mother plant and stores these nutrients. The stored nutrients provide material and energy needed by the zygote to develop into an embryo and then into a seedling. The nutritive role of endosperm explains why a large percentage of human and animal food comes from seed endosperm of grain crops; corn, wheat, rice, and other grain crops generate more than 380 billion pounds of endosperm per year in the U.S. alone.

You might wonder why the embryo and endosperm resulting from double fertilization have such different fates despite similar genetic composition. Part of the answer is that during its development, plant endosperm undergoes genomic imprinting by DNA methylation, a process also known to occur in mammals. The imprinting process causes gene expression to occur differently in plant endosperm than it does in the embryo.

## 31.3 Reviewing the Concepts

- Pollen grains are immature male gametophytes protected by a tough sporopollenin wall. Female gametophyte development occurs within an ovule. Mature female gametophytes include an egg and two supporting cells, a central cell with two nuclei, and three additional cells (Figures 31.9, 31.10).
- After pollination, interactions between proteins of pistil cells and those of pollen determine pollen germination.
- Germinated pollen delivers two sperm to female gametophytes by means of a long pollen tube. One sperm nucleus fuses with the egg to produce a zygote, the first cell of a new sporophyte generation; the other sperm nucleus fuses with the two nuclei of the central cell, generating the first cell of the nutritive endosperm tissue (Figure 31.11).

## 31.3 Testing Your Knowledge

1. Put the events in the life of a pollen grain in the correct order:
   **I.** Microspore cytoplasm divides by mitosis to produce a two-celled male gametophyte, protected by a tough pollen wall.
   **II.** Mature pollen is transported, most commonly by animal pollinators or wind, to the stigma of a flower pistil.
   **III.** Diploid cells of sporangia located within flower anthers undergo meiosis to produce haploid microspores.
   **IV.** Compatible pollen grain takes up water and germinates: The tube cell produces the pollen tube and the generative cell has divided to produce two sperm cells.
   **V.** Double fertilization occurs, leading to the formation of a zygote and the first endosperm cell.
   **VI.** At the ovule, the pollen tube bursts, propelling sperm toward the egg and central cell.
   **a.** I, II, III, IV, V, VI
   **b.** III, I, II, IV, VI, V
   **c.** I, II, III, VI, V, IV
   **d.** V, IV, III, II, I, VI
   **e.** none of the above

## 31.4 Embryo, Seed, Fruit, and Seedling Development

### Learning Outcomes

1. Discuss the major stages in angiosperm embryo development.
2. List examples of environmental and internal factors that influence seed germination.

Seeds and fruits are major components of plant reproduction and are essential to the nutrition of animals, including humans. Seeds contain dormant plant embryos that may, under favorable conditions, develop into seedlings. As noted earlier, fruits aid seed dispersal, which allows plants to colonize new sites. Embryos, seeds, and fruits mature simultaneously, and their development is coordinated by hormonal signals, which were introduced in Chapters 28 and 29. Seedling development is also hormonally regulated.

### Embryos Develop from Zygotes

Fueled by endosperm nutrients, angiosperm embryos undergo development in a series of stages known as **embryogenesis** (Figure 31.12). Sometime within a period of days to several weeks following fertilization, a zygote begins to divide. At this point, a zygote is blanketed with a layer of callose, which helps to seal it off from the environment, thereby fostering embryo-specific gene expression. The zygote's first cell division is unequal, producing a smaller cell and a larger cell (Figure 31.12, step 1). This unequal division helps to establish the apical-basal (top-bottom) polarity of the embryo, which persists through the life of the plant. The smaller cell develops into the embryo, whose radial symmetry is established at this point and continues in adult *Arabidopsis* plants. The larger cell develops into a short chain of cells, called a suspensor, that is anchored near the micropyle at the ovule entrance (Figure 31.12, step 2). These cells channel nutrients and hormones into the young embryo, which absorbs them at its surfaces.

Young eudicot embryos are spherical, but they soon become heart-shaped as the seedling leaves, called **cotyledons,** start to develop (Figure 31.12, step 3). At this point, auxin and a mobile protein transcription factor are involved in establishing the young

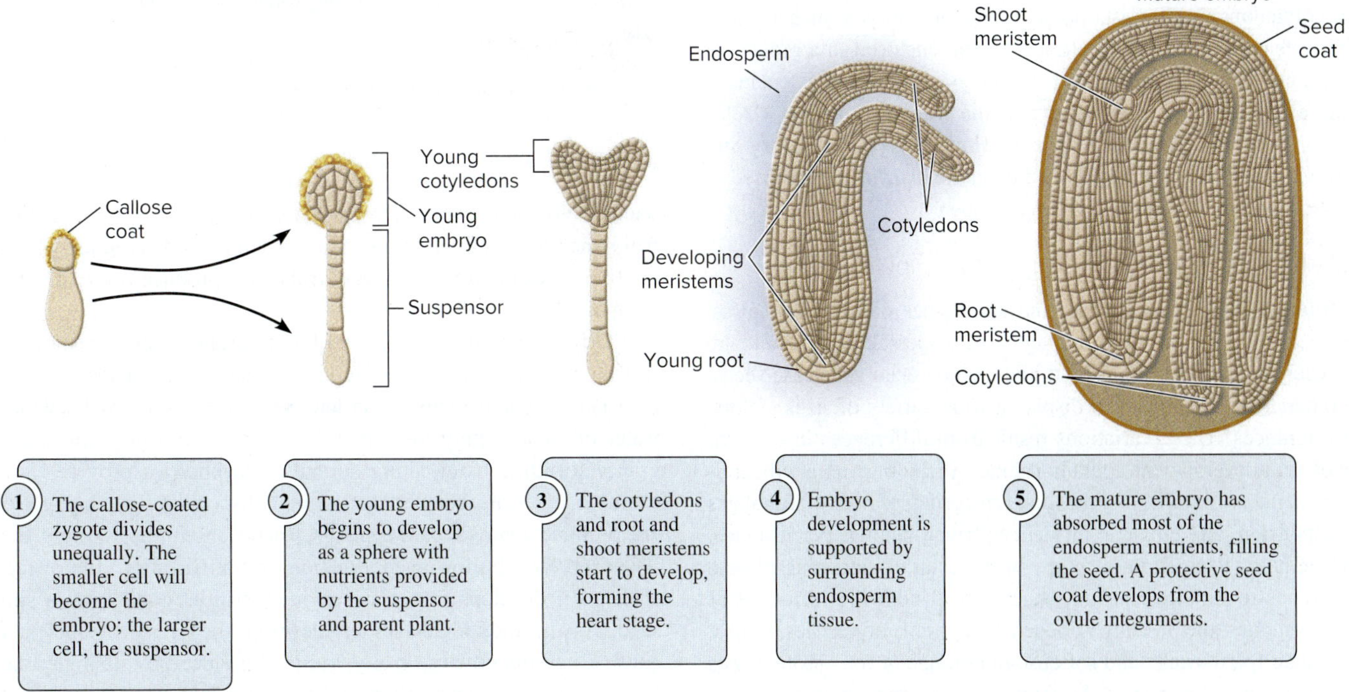

Callose coat
Young cotyledons
Young embryo
Suspensor
Endosperm
Developing meristems
Young root
Cotyledons
Mature embryo
Shoot meristem
Seed coat
Root meristem
Cotyledons
1 The callose-coated zygote divides unequally. The smaller cell will become the embryo; the larger cell, the suspensor.
2 The young embryo begins to develop as a sphere with nutrients provided by the suspensor and parent plant.
3 The cotyledons and root and shoot meristems start to develop, forming the heart stage.
4 Embryo development is supported by surrounding endosperm tissue.
5 The mature embryo has absorbed most of the endosperm nutrients, filling the seed. A protective seed coat develops from the ovule integuments.


**Figure 31.12** Embryogenesis in the eudicot *Arabidopsis*.

shoot and root at the apical and basal poles, respectively. Eudicot embryos such as *Arabidopsis* then become torpedo-shaped, and as the cotyledons grow, they often curl to fit within the developing seed (Figure 31.12, step 4). In contrast, mature monocot embryos are cylindrical, with a single cotyledon and a side notch where the apical meristem forms. Mature embryos become dormant as seeds mature.

## Mature Seeds Contain Dormant Embryos

As they mature, seeds undergo changes leading to dormancy, an adaptation that prevents them from germinating when environmental conditions are not suitable for seedling growth. During this process, embryos become dry and thus able to survive in the absence of water. Seed maturation includes transformation of the ovule's integuments into a tough seed coat (Figure 31.12, step 5). The seed coat restrains seedlings from growing and prevents the entry of water and oxygen, which maintains low seed metabolism. In addition, the coats of some seeds are darkly colored with pigments that may help to prevent damage by UV radiation or microbial attack. Another change leading to seed dormancy is gradual, controlled loss of water from the embryo and other seed tissues. As a result, the water content of dispersed seeds is only 5–15%. Abscisic acid (ABA) is a hormone that induces the activity of genes that help embryo tissues to survive the drying process.

The structure of mature monocot and eudicot seeds differs. Within eudicot seeds such as beans and peas, mature embryos often display an epicotyl. An **epicotyl** is the portion of an embryonic stem having a first bud with two tiny leaves; this bud is located above the point of attachment of the cotyledons (Figure 31.13a). The **hypocotyl** is the portion of an embryonic stem located below the point of attachment of the cotyledons. An embryonic root, the radicle, extends from the hypocotyl. Much of the endosperm has been absorbed into the large cotyledons. In contrast, mature monocot embryos, such as those of corn, feature an epicotyl with a first bud enclosed in a protective sheath known as the coleoptile. The young monocot root is enclosed within a protective envelope known as the coleorhiza (Figure 31.13b). When the embryo-containing seeds of flowering plants are dry and ready for dispersal, they are released from the plant while enclosed in a fruit or released when the fruit breaks open.

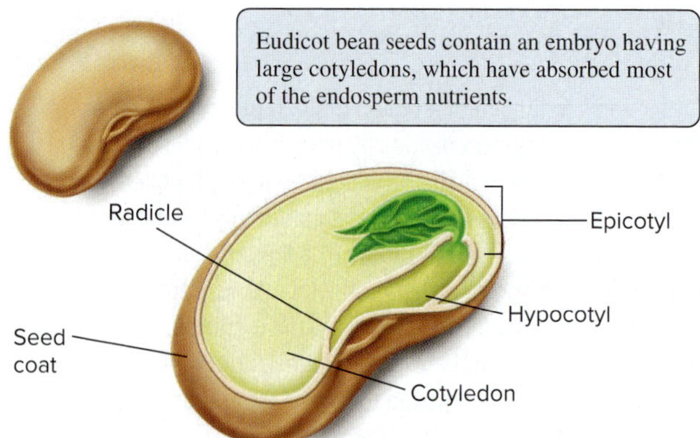


**(a) Eudicot bean seed, showing embryo with epicotyl, hypocotyl, and radicle**

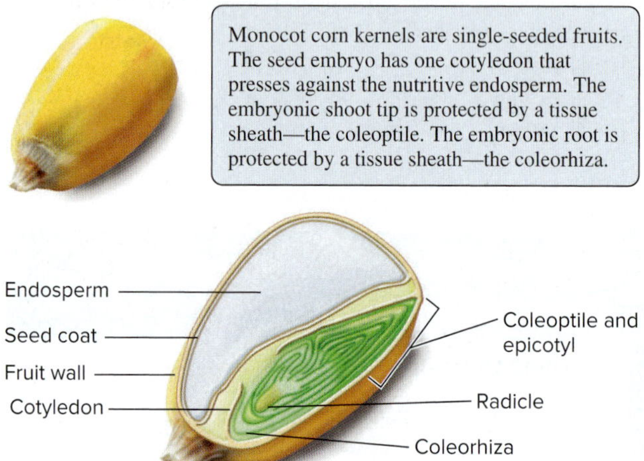

Monocot corn kernels are single-seeded fruits. The seed embryo has one cotyledon that presses against the nutritive endosperm. The embryonic shoot tip is protected by a tissue sheath—the coleoptile. The embryonic root is protected by a tissue sheath—the coleorhiza.
Endosperm
Seed coat
Fruit wall
Cotyledon
Coleoptile and epicotyl
Radicle
Coleorhiza


**(b) Monocot corn seed, showing an embryo protected by coleoptile and coleorhiza**

**Figure 31.13** Structure of mature angiosperm seeds and embryos.

**Concept Check:** *Why do mature seeds of eudicots lack extensive amounts of endosperm?*

## Fruits Develop from Ovaries and Other Flower Parts

All fruits develop from ovaries and sometimes other flower parts. They occur in diverse forms that aid seed dispersal. Some fruits are dry, whereas others are moist and juicy; some open to release seeds, and others do not. Fruits also display a wide variety of sizes, colors, and fragrances. These variations result from differences in the process of fruit development. Plant hormones, including auxin, gibberellic acid, and cytokinins, control this transformation. ABA stimulates cell expansion, and ethylene influences fruit ripening. For instance, ethylene helps to ripen nuts, a type of dry fruit, by inducing plasma membranes to rupture, causing water loss. Under the influence of plant hormones, the pericarp (ripened ovary wall) of peaches, plums, and related fruits swells and softens, and orange or red chromoplasts replace green chloroplasts. As fruits mature, the outer protective cuticle often becomes very thick, contributing to peel toughness, which helps to prevent microbe attack. In addition, many maturing fruits increase their sugar and acid content, which produces the distinctive tastes of ripe fruit. Many fruits also produce fragrant volatile compounds.

Differences in the shape, color, fragrance, and moisture content of wild fruits reflect evolutionary adaptation for effective seed dispersal. Though many fruits and seeds are dispersed by wind or water or by attaching to animal fur, others use fruit color and fragrance to attract fruit-eating animals. Blackberries provide a good example of fruits adapted for animal dispersal. Blackberry flowers produce many separate pistils, each containing a single ovule (Figure 31.14a). Following pollination and fertilization, the ovary of each pistil develops into a sweet, juicy fruitlet containing a single seed. As the individual fruitlets develop, they fuse together at the sides. Consequently, the many fruitlets produced by a single blackberry flower are dispersed together (Figure 31.14b). Attracted by the color, birds consume the whole aggregate and excrete the seeds,

# Biology Principle

## Living Organisms Interact with Their Environment

This principle is illustrated by the structure of blackberry fruits, which evolved as an effective way to achieve dispersal by birds.

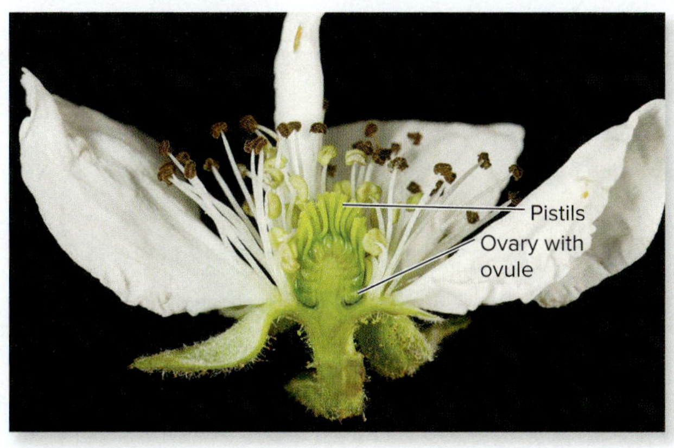


(a) ***Rubus allegheniensis*** **(common blackberry) flower**

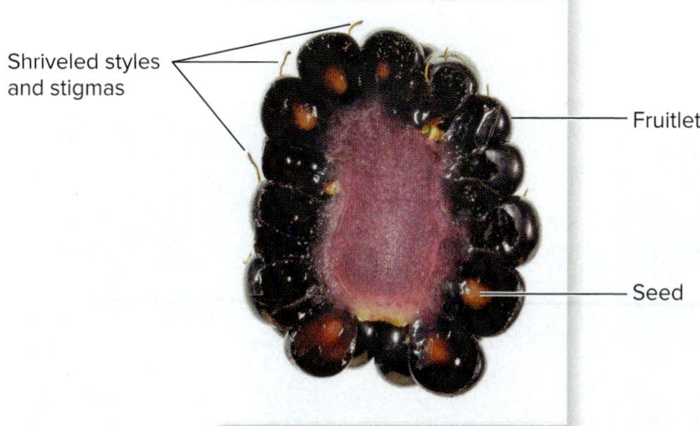


**(b) Blackberry fruit**

**Figure 31.14** **Blackberry flower and fruit. (a)** Each of the many separate pistils in a blackberry flower is able to produce a single one-seed fruit (called a fruitlet) if fertilization occurs. **(b)** Together, the individual fruitlets of the blackberry compose an aggregate fruit that allows many seeds to be efficiently dispersed at the same time by the same animal agent.
*(a,b)* © Lee W. Wilcox

thereby dispersing many at a time. Many other types of fruits occur, and these likewise represent adaptations that foster seed dispersal (refer back to Figure 25.19). Although a fruit is usually defined as a mature ovary containing seeds, commercial seedless fruits such as watermelon are produced by genetic modification or treatment with artificial auxin.

## Environmental and Internal Factors Influence Seed Germination

Seeds vary greatly in their ability to germinate after dispersal. Small seeds such as those of dandelions and lettuces germinate quickly if light is available. Other seeds require a period of dormancy before germination occurs. Some seeds can remain dormant for amazingly long time periods. For example, a lotus (*Nelumbo nucifera*) seed collected from a lakebed in China germinated at the age of 1,300 years, as determined by radiocarbon dating. In 2005, plant scientists germinated a 2,000-year-old date seed found in Israel.

Water is generally required to rehydrate seeds so that embryos can resume their metabolic activity. Water absorption also swells seeds, helping to break the seed coat and allowing embryonic organs to emerge. In some cases, rainfall of sufficient duration to leach germination-inhibiting compounds out of seeds is required. The optimal temperature for germination of most seeds lies between 25°C and 31.25°C (77°F and 86.25°F). This explains why gardeners wait until the soil is warm before planting seeds outdoors in spring. However, some seeds need a period of cold treatment or seed coat abrasion before they will germinate. Such physical stimuli induce the activity of more than 2,000 genes associated with seed germination.

When grass seeds rehydrate, the young shoot secretes the hormone gibberellic acid from the seed cotyledon into the outermost endosperm layer, known as the aleurone. In response, the aleurone secretes digestive enzymes into the central endosperm, releasing sugars from stored starch (**Figure 31.15**). The seedling uses these sugars for growth. This highly coordinated process allows grass seeds to quickly germinate when it rains, an advantage in arid grassland habitats. Humans also use this basic process of germination to make beer. In the process known as malting, beer brewers apply gibberellic acid to barley seeds to induce them to germinate simultaneously. The barley seeds are then baked at a high temperature to stop germination, a process that produces malt. Brewers then treat malt with water and heat, add the dried flowers of the hop plant (the genus *Humulus*), and add yeasts to ferment the plant sugars to alcohol.

Once seeds have germinated, plants vary in the process by which the embryonic shoot emerges. When bean and onion seeds germinate, the hypocotyl forms a hook that first breaches the soil surface and then straightens, thereby pulling the rest of the seedling and cotyledons aboveground (**Figure 31.16a,b**). In contrast, when pea seeds germinate, the epicotyl forms a hook that pulls the shoot tip out of the ground, leaving the cotyledons beneath the soil surface (**Figure 31.16c**). The tough hooks bear the brunt of passage through hard surface soil crusts, thereby protecting the delicate shoot tips. The shoot tips of corn seedlings and those of other grasses are

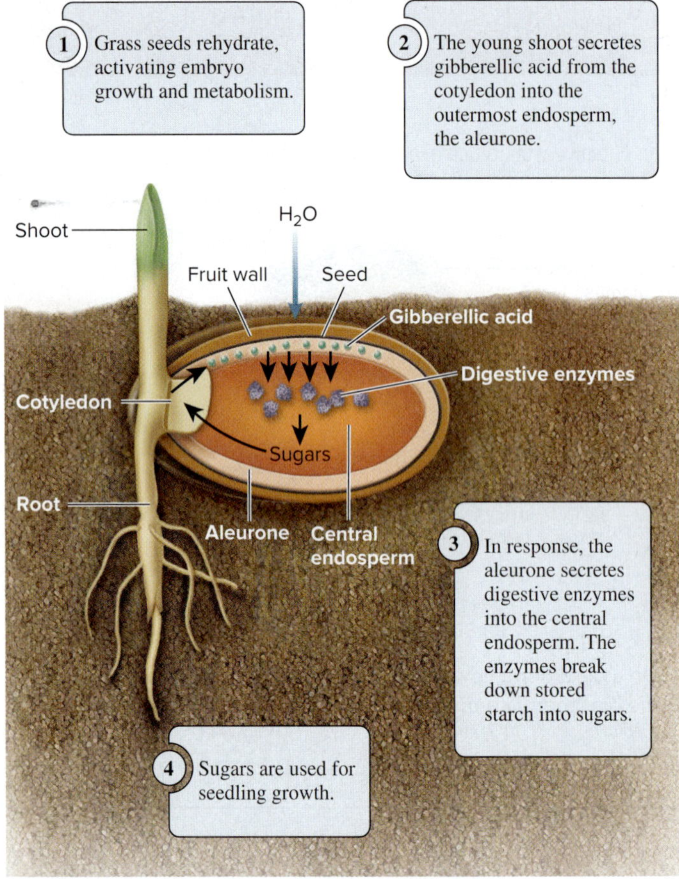

1 Grass seeds rehydrate, activating embryo growth and metabolism.
2 The young shoot secretes gibberellic acid from the cotyledon into the outermost endosperm, the aleurone.
Shoot
$H_2O$
Fruit wall
Seed
Gibberellic acid
Digestive enzymes
Cotyledon
Sugars
Root
Aleurone
Central endosperm
3 In response, the aleurone secretes digestive enzymes into the central endosperm. The enzymes break down stored starch into sugars.
4 Sugars are used for seedling growth.


**Figure 31.15** Germination of grass seeds.

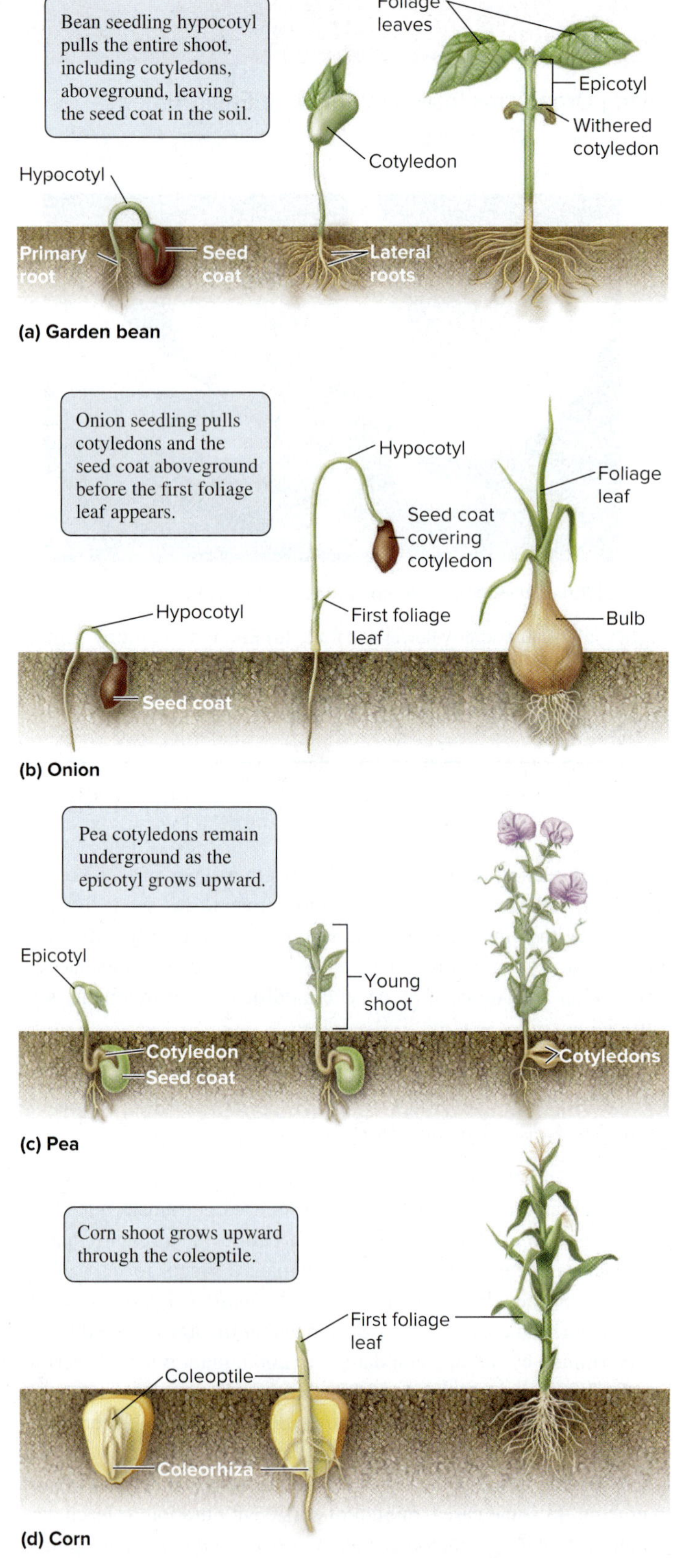


**Figure 31.16** Variations in seed germination and seedling growth patterns.

protected by the **coleoptile,** a protective tube that encloses the first foliage leaves (**Figure 31.16d**).

## 31.4 Reviewing the Concepts

- Unequal division of a zygote leads to development of an embryo and nutritive suspensor. (Figures 31.12, 31.13).
- Mature seeds contain embryos that become dry and are protected by desiccation-resistance proteins and a tough seed coat. These adaptations enable seeds to withstand long periods of dormancy, germinating only when conditions are favorable for seedling survival. Mature fruits develop from ovaries and aid in seed dispersal (Figure 31.14).
- Seed germination is influenced by environmental and internal factors (Figures 31.15, 31.16).

## 31.4 Testing Your Knowledge

1. Put the following events in the life of an embryo flowering plant in the correct order:
   I. The zygote uses endosperm nutrients to undergo mitotic division to form a smaller cell that develops into the embryo and a larger cell that becomes the nutrient-channeling suspensor.

**II.** In maturing seeds, embryos dry and become dormant.
**III.** Young embryos take up nutrients and divide mitotically to form cotyledons, young shoot, and young root.
**IV.** Under favorable light and moisture conditions, seeds germinate and embryos grow into seedlings.
**V.** Mature fruits, often with the aid of wind or animals, disperse mature seeds containing dry embryos into the environment.

**a.** I, II, III, IV, V
**b.** I, III, II, V, IV
**c.** I, II, III, V, IV
**d.** V, IV, II, III, I
**e.** none of the above

## 31.5 Asexual Reproduction in Flowering Plants

### Learning Outcomes

1. Explain the benefits of asexual reproduction in flowering plants.
2. List several examples of ways in which plant asexual reproduction is important to agriculture.

Many plants rely on sexual reproduction. However, a wide variety of angiosperms reproduce primarily by asexual means, and other plants commonly utilize both sexual and asexual reproduction. **Asexual reproduction** is the production of new individuals from a single parent without the occurrence of fertilization. Although sexual reproduction provides beneficial genetic variation, asexual reproduction can be advantageous in other ways. For example, asexual reproduction maintains favorable gene combinations that allow faster population growth in stable environments. Asexual reproduction is also advantageous in stressful habitats where pollinators or mates can be rare, because it allows a single individual to start a new population. Finally, asexual reproduction allows some plants to persist for very long periods of time. Among the oldest known plants are creosote bushes that are asexual clones of a parent that grew from a seed about 12,000 years ago!

The artificial propagation of plants from cuttings is a form of asexual reproduction that is widely used commercially and by home gardeners. In this section, we will explore the three main mechanisms of plant asexual reproduction: production of plant clones from nonreproductive plant organs, somatic embryogenesis, and apomixis.

### Asexual Reproduction Generates Plant Clones from Organs

Roots, stems, and leaves are plant organs that can function as asexual reproductive structures. For example, root sprouts, such as those produced by aspens, can generate entire groves of genetically identical trees. Sucker shoots, such as those appearing at the bases of banana plants and date palms, and the pieces of tuber-bearing "eyes" (which are buds) that develop into potato plants are examples of vegetative plant organs that have agricultural importance. Attractive horticultural varieties of African violets and other plants can be propagated from leaf cuttings. Such asexual offspring grow into adult plants that have the same valued properties as their parent plant. In contrast, these crops grown from seed would produce diverse progeny, not all of which would have economically prized properties.

**Figure 31.17 Asexual reproduction via vegetative structures.** The leaves of this *Kalanchoë* plant bear small plantlets around the edges. When mature, these plantlets drop off and, under the right conditions, grow into new plants.
© Lee W. Wilcox

### Somatic Embryogenesis Is the Production of Plant Embryos from Body (Somatic) Cells

Somatic embryos can develop from many types of plant somatic cells. Unlike the embryo described earlier in Figure 31.12, somatic embryos do not have endosperm and do not develop within a seed. Even so, somatic embryos produce root and shoot systems and can develop into mature plants. Somatic embryogenesis occurs naturally in citrus, mango, onion, and tobacco plants, but agricultural scientists have made use of it as well.

In the 1950s, British plant biologist F. C. Steward and associates were the first researchers to successfully clone a complex organism, carrot plants, by means of somatic embryogenesis. They used differentiated cells from carrot roots, grew the cells in conditions that caused some cells to lose their specialized properties and develop into embryos, and then cultivated each embryo in conditions that favored development into a mature carrot plant. Many types of plants are now cloned by somatic embryogenesis, which allows commercial growers to produce large numbers of genetically identical individuals.

The leaves of the common houseplant *Kalanchoë daigremontiana* produce many tiny plantlets at their edges via somatic embryogenesis. When these detach, they are able to take root and grow into new individuals (Figure 31.17). Recent molecular studies have revealed that evolutionary changes in embryogenesis genes are involved in this process.

## EVOLUTIONARY CONNECTIONS

### The Evolution of Plantlet Production in *Kalanchoë*

*Kalanchoë daigremontiana* is informally known as Mother of Thousands because it produces many plantlets at the edges of leaves. Helena Garces, Neelima Sinha, and their colleagues investigated the evolution of this type of asexual reproduction by studying genes that are involved in the development of organs and embryos in four species of *Kalanchoë*. In addition to *K. daigremontiana,* they investigated *K. marmorata,* which does not produce plantlets; *K. pinnata,* which produces plantlets only under certain stressful conditions; and *K. gastonis-bonnieri,* which produces plantlets both normally and when stressed. In leaves of these species, the biologists looked for expression of the gene *STM,* which encodes a key regulator of leaf production at the shoot meristem. They found that *STM* is expressed in cells at the leaf margins of all *Kalanchoë* species that produce leaf plantlets, but not the species that do not produce plantlets. Then the investigators checked for the expression in leaves of two genes (*LEC1* and *FUS3*) that are involved in embryo development. They discovered that these embryo-linked genes were expressed only in the leaf margins of species that normally form plantlets, not the species in which plantlet formation is induced by stress.

In a survey of a larger number of species, these investigators also discovered that *Kalanchoë* species that normally produce plantlets have an altered LEC1 protein. The normal form of LEC1 protein is essential to the process by which embryos become dry and thus tolerant of arid conditions. LEC1 is therefore necessary for the production of viable seeds, those able to germinate. *Kalanchoë* species that produce plantlets only under stressful conditions and those that do not produce plantlets at all were able to produce viable seeds. By contrast, the seeds of plantlet-producing species were not viable.

Together, these data allowed the investigators to infer that the evolution of plantlet formation began when certain leaf cells of some species gained the ability to function like a shoot meristem, thereby producing structures resembling small shoots. In some of the descendants of these species, normal LEC1 function in seeds was lost, as was the ability to produce viable seeds. Some species adapted by expressing the embryo-development process in leaf margin cells, a process that allowed them to produce plantlets. Such plantlets are not affected by loss of LEC1 function because, unlike seeds, they do not undergo a drying process during development.

### Apomixis Is Seed Production Without Fertilization

**Apomixis** (from the Greek, meaning away from mixing—that is, genetic mixing) is a natural asexual reproductive process in which fruits and seeds are produced in the absence of fertilization. By contrast, somatic embryogenesis does not involve fruit and seed production. More than 300 species of flowering plants, including hawkweeds, dandelions, and some types of citrus, are able to reproduce asexually by apomixis. Dandelions and some other apomictic plants require pollination to stimulate seed development, but others do not. Agricultural scientists are interested in apomixis as a potential method for producing genetically uniform seeds, propagating hybrids, and removing the need for fertilization in crop plants.

## 31.5 Reviewing the Concepts

- Asexual reproduction is the production of new individuals from a single parent without the occurrence of fertilization. Plant clones can develop asexually from roots, stems, and leaves.
- Somatic embryogenesis is the production of embryos from individual body cells.
- Apomixis is a mechanism by which some plants produce seeds and fruits from flowers without the occurrence of fertilization (Figure 31.17).

## 31.5 Testing Your Knowledge

**1.** What is a reasonable way to try to produce many clones (copies) of a particularly valuable plant?
   **a.** You could try planting cuttings from the valuable plant.
   **b.** You could try to grow cells from the valuable plant under conditions that allow somatic embryogenesis in carrot, as in F. C. Steward's cloning experiments.
   **c.** You could try to genetically alter the plant so that apomictic seeds and fruits are produced, as occurs naturally in hawkweeds and dandelions.
   **d.** All of the above are reasonable.
   **e.** Plant cloning has not yet been achieved.

# Assess and Discuss

## Test Yourself

1. Where do the pollen grains of flowering plants develop?
   a. in the anthers of a flower
   b. in the carpels of a flower
   c. while being dispersed by wind, water, or animals
   d. within ovules
   e. within pistils
2. Where do mature male gametophytes of flowering plants primarily develop?
   a. in the anthers of a flower
   b. in the carpels of a flower
   c. while being dispersed by wind, water, or animals
   d. within ovules
   e. on the surfaces of leaves
3. Where would you find female gametophytes of a flowering plant?
   a. in the anthers of a flower
   b. at the stigma of a pistil
   c. in the style
   d. within ovules in a flower's ovary
   e. in structures that are dispersed by wind, water, or animals
4. How does double fertilization occur in flowering plants?
   a. The two sperm in a pollen tube fertilize the two egg cells present in each female gametophyte.
   b. One of the two sperm in a pollen tube fertilizes the single egg in a female gametophyte, and the other fuses with the two nuclei present in the central cell.
   c. Two sperm, one contributed by each of two different pollen tubes, fertilize the two egg cells in a single female gametophyte.
   d. Two sperm contributed by separate pollen tubes enter a single female gametophyte; one of the sperm fertilizes the egg cell; the other fertilizes the central cell.
   e. None of the above are correct.

5. A seed is
   a. an embryo produced by the fertilization of an egg, which is protected by a seed coat.
   b. a structure that germinates to form a seedling under the right conditions.
   c. an embryo produced by parthenogenesis that is enclosed by a seed coat.
   d. all of the above.
   e. none of the above.
6. What is the likely chemical composition of the chemical stimulus of flowering that is produced by leaves and transported to the shoot apical meristem?
   a. the hormone auxin
   b. the protein STM
   c. the carbohydrate callose
   d. the mineral ion $K^+$
   e. none of the above
7. How many whorls of organs occur in complete flowers?
   a. 2
   b. 4
   c. 6
   d. 8
   e. 10
8. If an ovary contains eight ovules, how many seeds can result if pollen tubes reach all eight ovules?
   a. 1
   b. 4
   c. 8
   d. more than 20
   e. none of the above
9. From what structure does the fleshy, outer part of a juicy fruit primarily develop?
   a. the style
   b. a stamen filament
   c. the ovary wall
   d. a group of fused sepals
   e. the stigma

## Conceptual Questions

1. Why are pollen grain walls composed of sporopollenin?
2. Why are seed coats often tough?
3. **PRINCIPLES** A principle of biology is that all species (past and present) are related by an evolutionary history. Modern flowering plants likely possessed a single common ancestor. Why, then, do flowers occur in such a diversity of shapes and colors?

## Collaborative Questions

1. Observe or view images of orchid flowers. Are these flowers bilaterally symmetric or radially symmetric? Are these flowers more likely to be pollinated by wind or by animals? What gene might be involved in the production of orchid flower shape?

## Online Resource

**connect.mheducation.com**

32

33

34

35

36

37

38

39

40

41

(32) © John Rowley/Getty Images RF; (33) Courtesy Ann-Shyn Chiang, Tsing Hua Chair Professor/Brain Research Center & Institute of Biotechnology/National Tsing Hua University; (34) © Gustavo Mazzarollo/Getty Images; (35) © Don W. Fawcett/Science Source; (36) © SPL/Science Source; (37) © Richard Hutchings/PhotoEdit; (38) © Sovereign/ISM/Phototake; (39) © David Liebman/Pink Guppy; (40) © SPL/Science Source; (41) © BSIP SA/Alamy

# UNIT VII
# ANIMALS

In Chapters 26 and 27, we considered the amazing diversity of animal life. Despite the obvious differences between animals, they are nevertheless linked by fundamental similarities, which we will explore in this unit. The basic features of animal bodies, the relationship between structure and function, and the ability of animals to maintain homeostasis will be introduced in Chapter 32. Chapters 33 and 34 will describe major principles of nervous systems in animals. The ability of animals to move through their environments will be covered in Chapter 35. Circulatory, respiratory, digestive, excretory, and endocrine systems are then covered in Chapters 36–38. Chapters 39 and 40 cover animal reproduction, development, and immune systems. In the last chapter in this unit (41), a synthesis of a whole-body response to a major homeostatic challenge that involves all the organ systems is explored.

Biologists study the various organ systems in animals at two levels. Anatomy is the study of the structure (or form) of an animal's body, and physiology is the study of how body structures function. Collectively, anatomy and physiology form the foundation of animal biology. We begin the unit, therefore, with an examination of the structure and organization of animal bodies.

***The following biology principles will be emphasized in this unit:***

- ***Living organisms maintain homeostasis.*** *This key principle of biology is related to all aspects of animal biology and is covered throughout the unit.*
- ***Living organisms use energy.*** *We will see how many of the processes that achieve and maintain homeostasis require energy; for example, it takes energy to obtain oxygen and to maintain blood flow (Chapter 36), to regulate body temperature (Chapter 32), and to maintain stable concentrations of sodium and other ions in body fluids (Chapter 37).*
- ***Living organisms grow and develop.*** *In Chapter 39, we will explore the factors that regulate how an embryo develops into a mature animal.*
- ***Living organisms interact with their environment.*** *How animals sense changes in their external environment, such as light and dark, smells, and vibrations, will be covered in Chapter 34.*
- ***Structure determines function.*** *The relationship between structure and function in animals will be evident at multiple levels: cells, tissues, organs, and whole bodies.*
- ***All species (past and present) are related by an evolutionary history.*** *This principle will be explored in the Evolutionary Connections feature throughout the unit. As one example, the evolution of a four-chambered heart is described in Chapter 36.*
- ***Biology affects our society.*** *At the ends of Chapters 33 to 40 are sections entitled Impact on Public Health. These sections relate what you've learned in a chapter to important facets of human disease, including their costs to society.*
- ***Biology is an experimental science.*** *Several chapters include a Feature Investigation that describes an experiment notable for its innovation and importance, and for moving the study of animal biology forward.*

# General Features of Animal Bodies, and Homeostasis as a Key Principle of Animal Biology

32

© John Rowley/Getty Images RF

Perspiring and drinking water are both mechanisms that help achieve homeostasis—a stable internal body environment.

## Chapter Outline



In 2002, a healthy young woman died shortly after completing the Boston Marathon. The cause of death was traced to an imbalance in the concentrations of ions such as sodium ($Na^+$) in her body fluids. This occurred because the woman drank an extreme excess of water before and during the race, causing the concentrations of $Na^+$ and other ions in her body fluids to decrease. When this happens, water moves by osmosis into cells, causing them to swell. Whereas some body cells can withstand considerable swelling like this, the cells of the brain cannot. Swelling of brain cells increases pressure in the brain, which in turn can squeeze and compress local blood vessels. This decreases blood flow in the brain, depriving it of the oxygen it needs to function. Brain damage, coma, and death can result. This tragic event underscores the importance of matching fluid and ion (salt) intake with the body's requirements, since too much or too little of either can have serious consequences.

The process by which the different aspects of an animal's internal environment are maintained within normal limits, even in the face of changes in the external surroundings, is known as **homeostasis** (from the Greek *homoios*, meaning similar, and *stasis*, meaning to stand still). Before we can fully appreciate the importance of homeostasis to animals and how it is achieved, however, we first need to understand some basic features of animal bodies. We begin this chapter with a discussion of how cells are organized into tissues, and tissues into organs, and how fluids exist in different body compartments. We will look at tissues and organs from the perspective of the whole animal, examining how the properties of life arise from the complex interactions of its components. Next, we will discuss a principle of biology that also helps us understand homeostasis, namely that structure determines function. We then link these principles together with a detailed look at what homeostasis means for different animals, how it may be challenged, and how it is restored or maintained.

## 32.1 Organization of Animal Bodies

### Learning Outcomes

1. List the different types of animal tissue, providing general functions and specific examples of each.
2. Name the various organ systems found in many animals, list major components of each, and describe their general functions.
3. Describe how the bodies of animals are organized into fluid compartments.
4. **SCISKILLS ▶** Predict how water will move between body fluid compartments if solute concentrations differ between the compartments.

All animals share similarities in the ways in which they exchange materials with their surroundings, obtain energy from organic nutrients, synthesize complex molecules, detect and respond to signals in their immediate environment, and reproduce. Animals typically begin life as a single cell—most commonly a fertilized egg—which divides to create two cells, each of which divides in turn, resulting in four cells, and so on. If cell multiplication were the only event occurring, the end result would be a spherical mass of identical cells. As we will see in Chapter 39, however, cells become specialized during development to perform a particular function (that is, they differentiate). Examples of differentiated cells are muscle and blood cells. As shown in **Figure 32.1**, the cells of an animal's body are organized into progressively more complex structures, including tissues, organs, and organ systems, as shown for the urinary system in mammals. In this section, we will explore this organization in greater detail.

### Specialized Cells Are Organized into Tissues

A **tissue** is an association of many cells having a similar structure and function. The tissues in a typical animal's body can be classified into four types, according to the cell types that constitute them and the functions they perform: muscle, nervous, epithelial, and connective tissues (see Figure 32.1). Within each of these functional categories, subtypes of tissues perform variations of that function, as illustrated by the three types of muscle tissue.

## Biology Principle

### New Properties Emerge from Complex Interactions

By themselves, none of the four tissues that constitute a bladder or kidney could perform the functions of those organs, but when combined in precise ways, the result is a functional organ system capable of removing soluble wastes from the fluids of an animal's body.

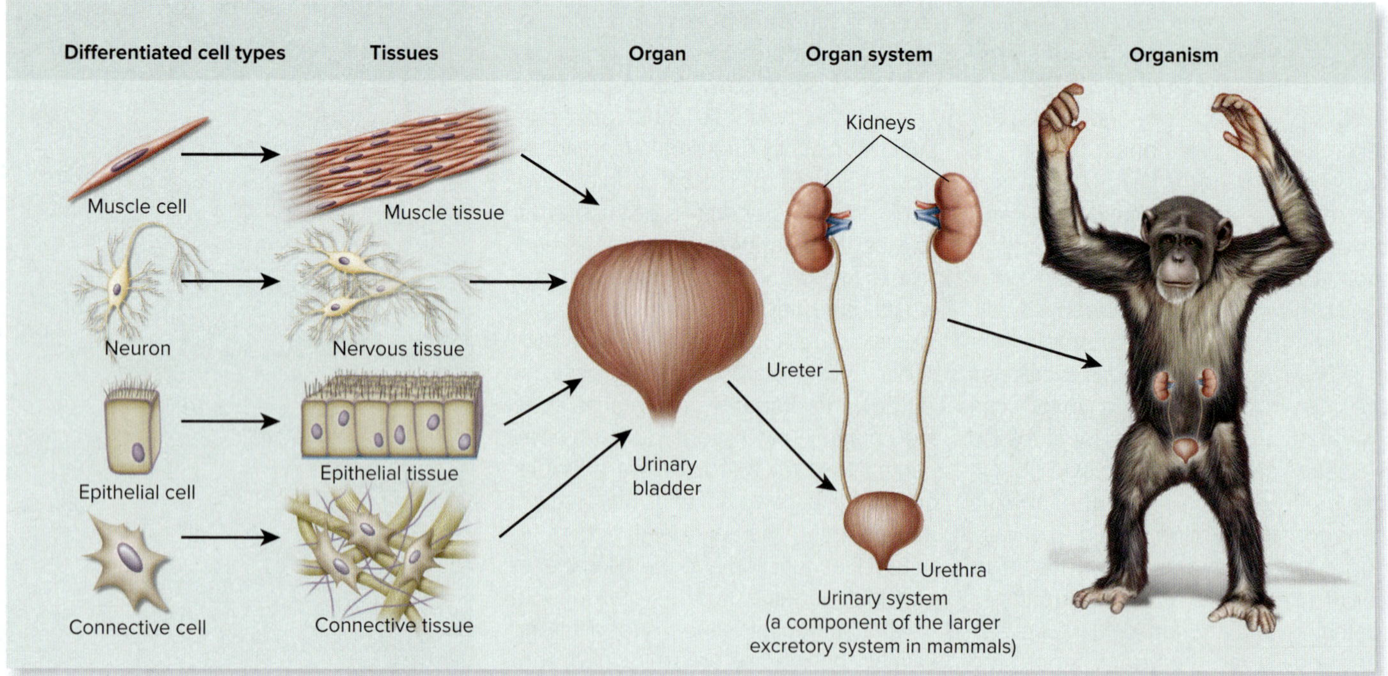


**Figure 32.1 The internal organization of cells, tissues, organs, and organ systems in a mammal.** Most animals share the same four tissue types.

**Muscle Tissues** **Muscle tissues** consist of cells specialized to shorten, or contract, generating the mechanical forces that may:

- produce body movements,
- decrease the diameter of a tube, or
- exert pressure on a fluid-filled cavity.

Three types of muscle tissue may be found in animals (Figure 32.2). **Skeletal muscles** are generally linked to bones in vertebrates via bundles of collagen fibers called tendons, and to the exoskeleton of invertebrates. When skeletal muscles are stimulated by signals from the nervous system, they generate force that leads to the contraction of the muscle (see Chapter 35). Contraction of these muscles may be under voluntary control and can produce the types of movements required for locomotion, such as extending limbs or flapping wings. **Smooth muscles** surround hollow tubes and cavities inside the body's organs, such that their contraction or relaxation can control the movement of the contents of those organs. For example, the contraction or relaxation of smooth muscle in the lung tubes called bronchioles regulates the rate at which air can flow through those tubes. Contraction of all smooth muscle is involuntary—that is, it occurs without conscious control. In the third type, **cardiac muscle,** physical and electrical connections between individual cells enable many cells to contract almost in unison. It is found only in the heart, where it provides the force that generates pressure sufficient to propel blood through an animal's body. As with smooth muscle, contraction of cardiac muscle is involuntary.

**Nervous Tissues** **Nervous tissues** are complex networks of cells called **neurons** that are specialized to initiate and conduct electrical signals from one part of an animal's body to another part (Figure 32.3). An electrical signal produced in one neuron may

## Biology Principle

### Living Organisms Interact with Their Environment

Animals interact with their environment in many ways; these interactions largely depend on the use of nervous tissue, which allows animals to sense and respond to changes in the environment.

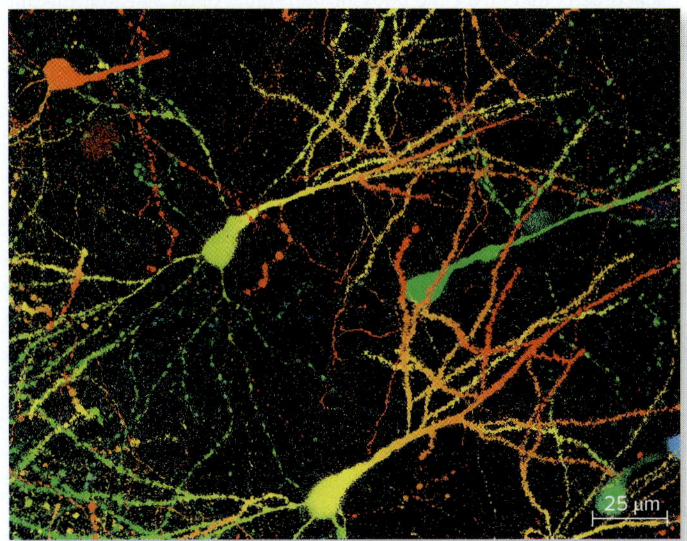


**Figure 32.3 Nervous tissue in the brain of a vertebrate.** Nervous tissue consists of neurons with extensive cell-cell contacts, as shown in this confocal micrograph of a section from a human brain. The cells are labeled with fluorescent markers; different colors signify different depths in the section.

© Dr. Gopal Murti/Science Source

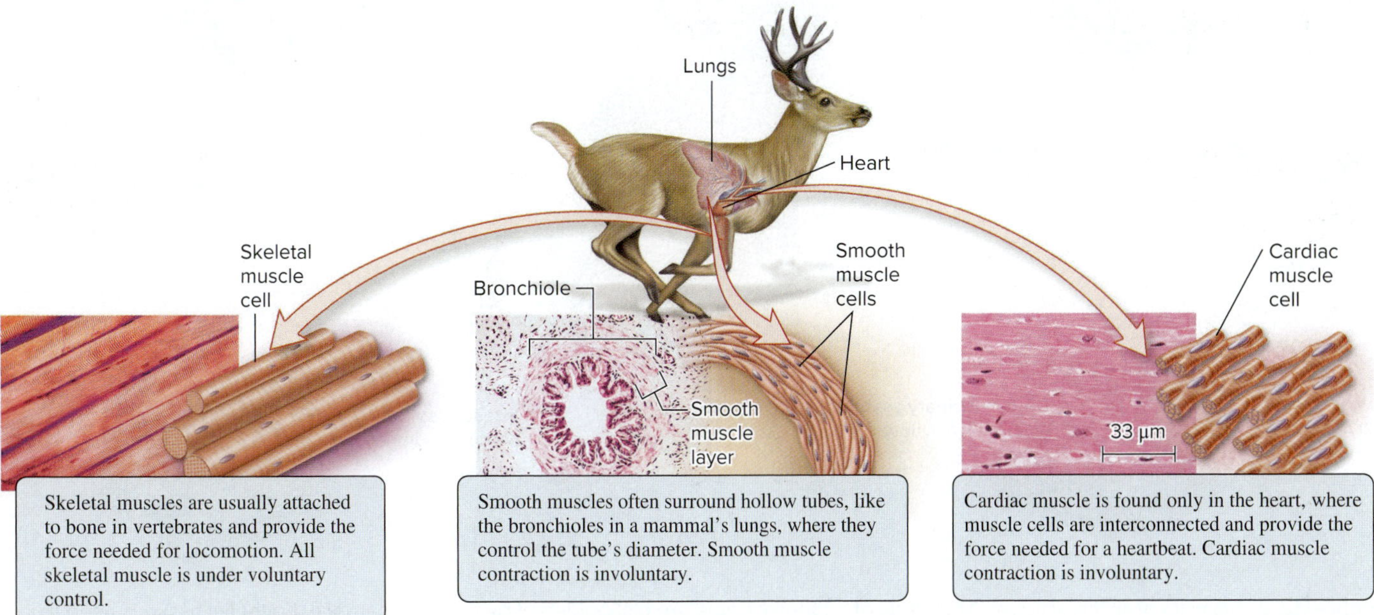


**Figure 32.2 Three types of muscle tissue: skeletal, smooth, and cardiac.** All three types produce force, but they differ in their appearance and in their locations within animals' bodies.

*(left)* © Michael Abbey/Science Source; *(middle)* © Sinclair Stammers/Science Source; *(right)* © Innerspace Imaging/Science Source

stimulate or inhibit other neurons to initiate new electrical signals, stimulate muscle cells to contract, or stimulate glandular cells to secrete substances into the animal's body fluids. Thus, nervous tissue provides a means of controlling many diverse activities in an animal's body.

**Epithelial Tissues** **Epithelial tissues** are sheets of densely packed cells that cover the body or individual organs or that line the interiors of various cavities inside an animal's body. Epithelial cells are specialized to protect structures and to secrete and transport ions and organic molecules. For example, epithelial tissue can invaginate (fold inward) to form sweat glands that secrete water and ions onto the surface of an animal's skin.

Epithelial cells come in a variety of shapes, including cuboidal (cube-shaped), squamous (flattened), and columnar (elongated). These three types of epithelial cells are typically arranged in epithelial tissues as simple epithelia (one layer), stratified epithelia (multiple layers), pseudostratified epithelia (one layer, but with nuclei located in such a way that it appears stratified), or transitional epithelia (multiple layers with the ability to expand and contract) (Figure 32.4).

Regardless of their shape, organization into tissues, or location, all epithelial cells are asymmetric, or polarized. This means that one side of such a cell is anchored to an extracellular matrix (ECM) called the basal lamina, or basement membrane; this side of the cell is called the basal or basolateral membrane. The other side, called the apical membrane, faces the internal or external environment of the animal. Within epithelial tissue, the cells generally align such that their membranes face the same way.

**Connective Tissues** As their name implies, **connective tissues** connect, surround, anchor, and support the structures of an animal's body. In many cases, connective tissues are composed of a relatively sparse number of cells within a noncellular matrix of some type. Connective tissues include adipose (fat-storing), bone, cartilage, loose connective tissue, dense connective tissue, and blood, a type of fluid connective tissue (Figure 32.5). Many of the types and functions of proteins that make up connective tissue were covered in Chapter 4.

## Different Tissue Types Combine to Form Organs and Organ Systems

An **organ** is composed of two or more kinds of tissues arranged in various proportions and patterns, such as sheets, tubes, layers, bundles, or strips. The different tissues interact with each other through

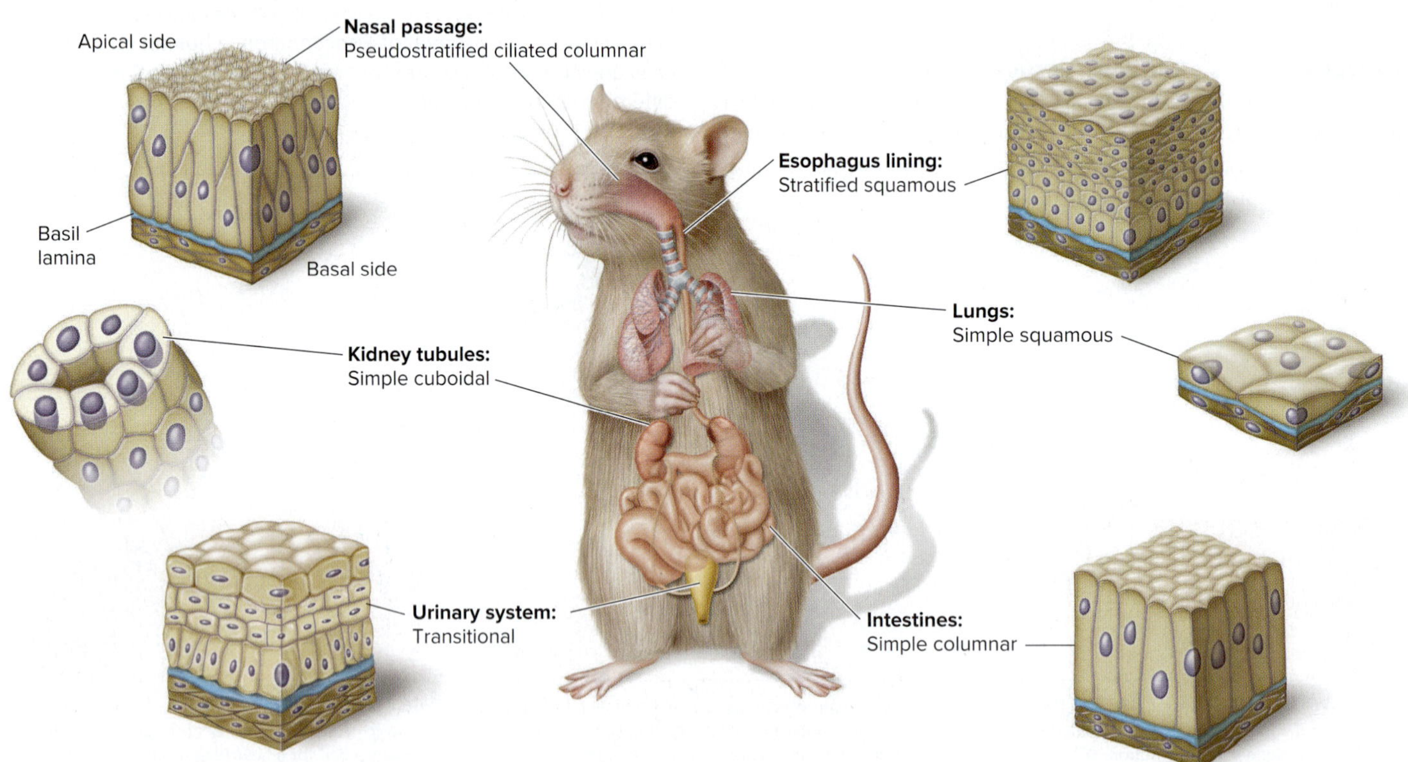


**Figure 32.4** **Examples of epithelial tissue.** The types of epithelial tissue are distinguished by their appearance. Epithelial tissue is used to construct body coverings and the protective sheets that line and cover hollow tubes and cavities. The epithelial cells that make up epithelial tissues have apical and basal (or basolateral) membranes; the apical side typically faces the exterior of the body or the lumen of a structure such as the intestines. The shapes and sizes of these cells determine their properties. For example, ciliated epithelial cells in the nasal passage help trap and filter out airborne particles that could damage the lungs.

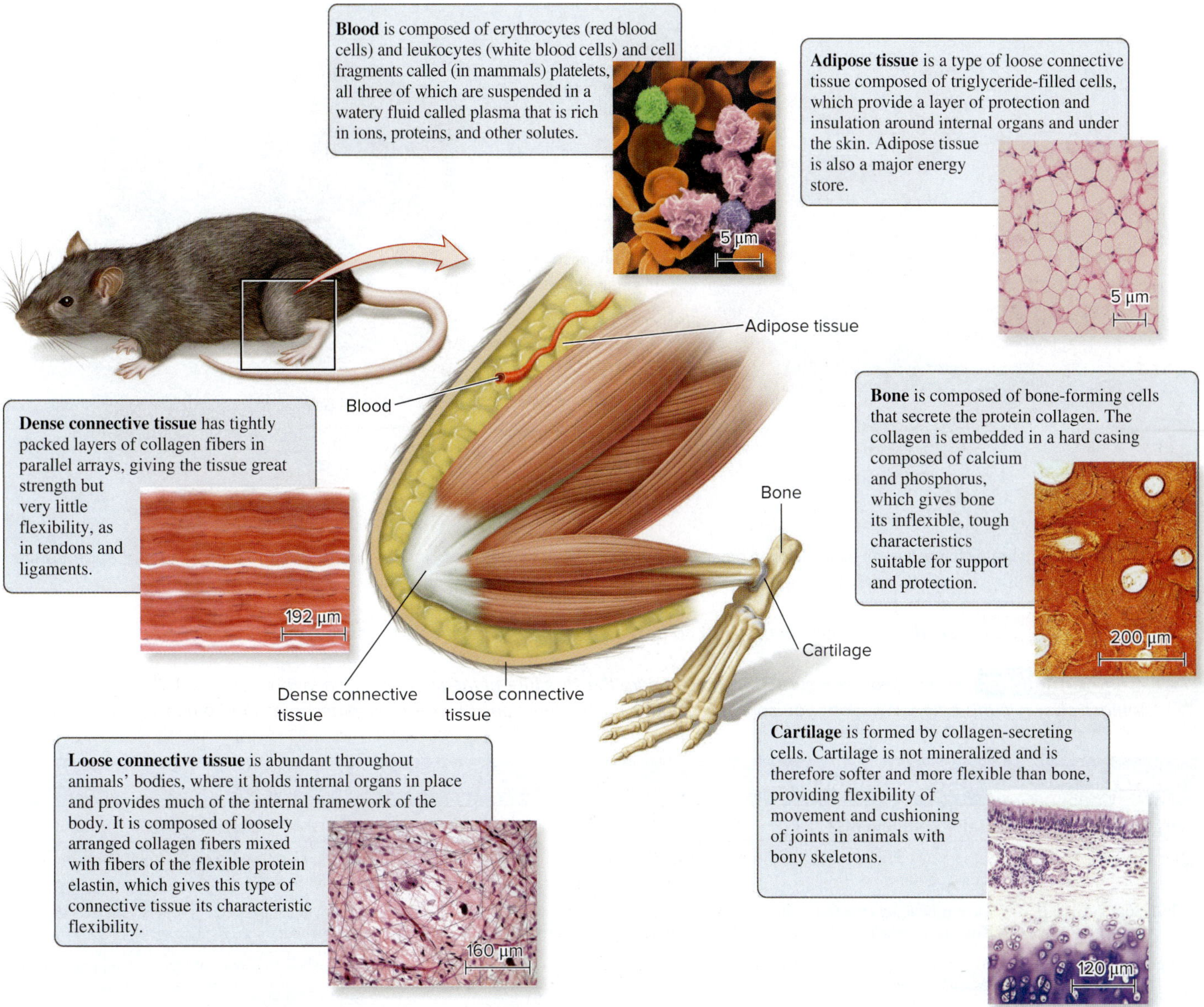

**Blood** is composed of erythrocytes (red blood cells) and leukocytes (white blood cells) and cell fragments called (in mammals) platelets, all three of which are suspended in a watery fluid called plasma that is rich in ions, proteins, and other solutes.
5 μm
**Adipose tissue** is a type of loose connective tissue composed of triglyceride-filled cells, which provide a layer of protection and insulation around internal organs and under the skin. Adipose tissue is also a major energy store.
5 μm
Adipose tissue
Blood
**Dense connective tissue** has tightly packed layers of collagen fibers in parallel arrays, giving the tissue great strength but very little flexibility, as in tendons and ligaments.
192 μm
**Bone** is composed of bone-forming cells that secrete the protein collagen. The collagen is embedded in a hard casing composed of calcium and phosphorus, which gives bone its inflexible, tough characteristics suitable for support and protection.
Bone
200 μm
Cartilage
Dense connective tissue
Loose connective tissue
**Loose connective tissue** is abundant throughout animals' bodies, where it holds internal organs in place and provides much of the internal framework of the body. It is composed of loosely arranged collagen fibers mixed with fibers of the flexible protein elastin, which gives this type of connective tissue its characteristic flexibility.
160 μm
**Cartilage** is formed by collagen-secreting cells. Cartilage is not mineralized and is therefore softer and more flexible than bone, providing flexibility of movement and cushioning of joints in animals with bony skeletons.
120 μm


**Figure 32.5** **Examples of connective tissue in mammals.** Connective tissue connects, surrounds, anchors, and supports other tissues and may exist as a suspension of cells in a fluid (blood), clumps of cells (fat), or tough, rigid material (bone and cartilage). The samples have been stained or the micrographs have been computer-colorized to reveal connective tissue.



direct contact and secretions. An example is the vertebrate stomach (Figure 32.6), which consists of

- an outer and inner layer of epithelial tissue,
- connective tissue that covers and cements the organ together,
- smooth muscle tissue that helps propel the stomach contents into the small intestine, and
- nervous tissue that regulates smooth muscle activity.

In an **organ system,** different organs function together to perform an overall function. In the example just described, the stomach is part of the digestive system, along with other structures, such as the mouth, esophagus, small and large intestines, and anus. In another example, the urinary system (which is part of the larger excretory system) in mammals was shown in Figure 32.1. The organ systems commonly found in animals are listed in Table 32.1.

Organ systems should not be considered as functioning in isolation. Instead, they influence and depend on each other in many ways. For example, signals from the nervous, endocrine, and circulatory systems all strongly influence how much water the vertebrate urinary system retains in the body as it forms urine, an adaptation that can be lifesaving under certain circumstances.

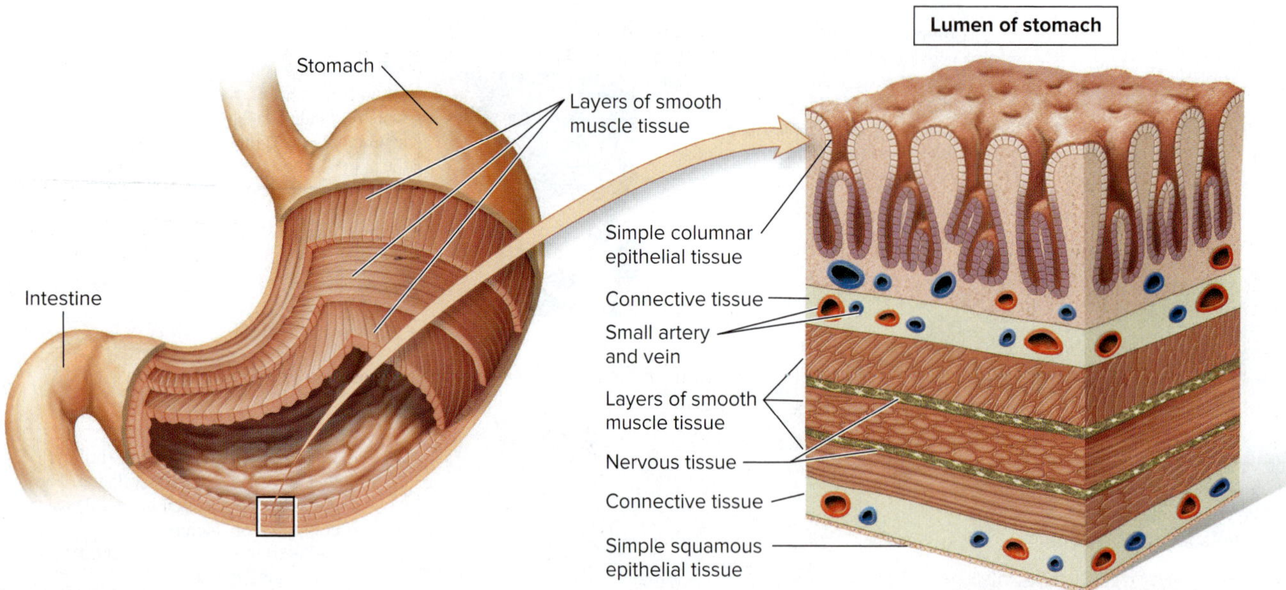


**Figure 32.6** **The vertebrate stomach as an example of an organ composed of all four tissue types.** In this illustration, the thickness and appearance of the layers of nervous tissue have been considerably exaggerated for visual clarity.

**Concept Check:** *Do animal organs necessarily contain equal proportions of all four tissue types?*

**BioConnections:** *Look ahead to Figure 37.6 and compare the structure of the small intestine in that figure with the structure of the stomach shown in this figure. What can you conclude about common functions of the stomach and intestine from this comparison?*

**Table 32.1 Organ Systems Found in Animals**

| Organ system | Major components* | Major functions |
|---|---|---|
| Nervous | Sensory and processing organs (brain); structures for signal delivery (spinal cord, nerves, ganglia, sense organs) | Regulates and coordinates movement, sensation, organ functions, behavior, and learning |
| Musculoskeletal | Force-producing structures (muscles); support structures (bones, cartilage, exoskeleton); connective structures (tendons, ligaments) | Produces locomotion; generates force; propels materials through body organs; supports body |
| Circulatory | Contractile element (heart or vessel); distribution network (blood vessels); blood | Distributes solutes such as nutrients and gases to all parts of an animal's body |
| Respiratory | Gas-exchange structures (gills, skin, tracheae, lungs) | Exchanges oxygen and carbon dioxide with environment; regulates blood pH |
| Digestive | Ingestion structures (mouth, mouthparts); storage structures (crop, stomach); digestive and absorptive structures (stomach, intestines); egestion structures (rectum, anus); accessory structures (pancreas, gallbladder, liver, salivary glands) | Breaks complex molecules from food into absorbable units; absorbs organic and inorganic nutrients and water; eliminates solid wastes |
| Excretory | All organs, including respiratory structures (such as gills and lungs), kidneys, and related structures that are involved in removing soluble wastes from the body; the vertebrate urinary system is a part of the excretory system and includes the kidneys, ureters, bladder, and urethra | Eliminates soluble metabolic wastes; regulates body fluid volume and solute concentrations |
| Endocrine | All glands, organs, tissues, or cells that secrete hormones | Regulates and coordinates growth, development, metabolism, mineral balance, water balance, blood pressure, behavior, and reproduction |
| Reproductive | Gonads and associated structures | Produces gametes (sperm and egg); in some animals, provides nutritive environment for embryo and fetus |
| Immune and lymphatic | Circulating leukocytes; lymphatic vessels and lymph nodes | Defends against pathogens, cancer cells, and toxins |
| Integumentary | Body surfaces (skin) | Protects from dehydration and injury; provides a barrier against pathogens; in some animals, plays a role in regulation of body temperature |

*Selected examples only; these do not necessarily pertain to all animals.

The spatial arrangement of organs into organ systems is part of the overall body plan of animals. Organ systems develop at specific times and locations within the body and along the anteroposterior body axis (the formation of the "head" and "tail" of the body), as do other structures, such as limbs, tentacles, antennae, and other animal appendages. Scientists have long questioned how the layout of animal bodies is determined during the period when an embryo is developing. Remarkably, organ development in animals appears to be under the control of a highly conserved family of body-plan genes with homologs in all animals, as described next.

## EVOLUTIONARY CONNECTIONS

### Organ Development and Function Are Controlled by *Hox* Genes

*Hox* genes are a family of ancient, highly conserved genes that are found in all animals. The evolution of *Hox* genes was a major event that led to the precise timing and spatial patterning of the anteroposterior body axis during animal development. For example, these genes determine the number and position of legs and wings in *Drosophila* and other insects. *Hox* genes play a similar role in determining the spatial patterning of the vertebrate body and appendages (refer back to Figures 20.12 and 20.13). Recently, however, scientists have begun exploring the role of *Hox* genes in the development and spatial patterning of the organs that make up animals' organ systems.

By generating mutant mice that fail to express one or more *Hox* genes (the genes are said to have been knocked out), researchers have discovered the important role these genes play in determining where within the vertebrate body particular organs form. For example, one member of the *Hox* gene family in mice has been found to be important for development of anterior parts of the body, including the neck. When this gene is knocked out, mouse embryos show defects in neck structure, such as abnormal blood vessels. Also, the organs within the neck—including the thymus, thyroid, and other glands—do not develop normally or even fail to form at all. Experimental deletion of *Hox* genes associated with other body segments does not affect the development and function of neck glands and organs. Likewise, investigators have uncovered vital roles of different *Hox* genes in lung development within the thorax and the proper positioning and development of the vertebrate kidneys in the abdomen.

Of particular interest is the discovery that *Hox* genes are important not only for spatial patterning of organs but also for their growth, development, and function. For instance, certain *Hox* genes help determine the branching patterns of the airways of the lungs, the size of the lungs, and the ability of the lungs to produce secretions that are important for breathing air after birth. Other *Hox* genes control cell proliferation, shape changes, apoptosis, cell migration, and cell-to-cell adhesion within various organs. This is also true in invertebrates, such as the leech, where *Hox* genes are first expressed during organ formation, and in *Drosophila,* where the final shape and size of the heart are partly controlled by *Hox* genes. In all animals, therefore, this evolutionarily ancient gene family plays different but broadly related roles in how an animal's body and organs form.

## 32.1 Reviewing the Concepts

- Groups of cells with similar properties form tissues, which combine with other types of tissues to form organs, which are functionally linked to form organ systems (Figure 32.1). The four types of tissues are muscle, nervous, epithelial, and connective tissues (Figures 32.2, 32.3, 32.4, 32.5). An organ is composed of two or more kinds of tissues. In an organ system, different organs work together to perform an overall function (Figure 32.6; Table 32.1).
- The development, spatial positioning, and function of many body organs are under the control of highly conserved *Hox* genes in animals.

## 32.1 Testing Your Knowledge

1. Which statement is *false?*
   a. All organs consist of at least two different tissue types.
   b. Epithelial tissue is specialized to initiate and conduct electrical signals.
   c. Cardiac muscle is found only in the heart.
   d. Blood is a type of fluid connective tissue.
   e. The final location of organs in an animal's body is under genetic control.
2. Match the representative functions with the organ system.

| | |
|---|---|
| Supports the body and produces locomotion | Digestive system |
| Absorbs nutrients into the body | Circulatory system |
| Regulates growth, metabolism, reproduction, mineral balance | Integumentary system |
| Distributes solutes throughout the body | Endocrine system |
| Provides a barrier against pathogens entering the body | Musculoskeletal system |

## 32.2 The Relationship Between Form and Function

### Learning Outcomes

1. Provide an example of how the structure of an animal's tissues or organs can help predict their function.
2. **SCISKILLS ▶** Explain the quantitative relationship between surface area of an object and its volume, and explain the importance of this to animal form and function.

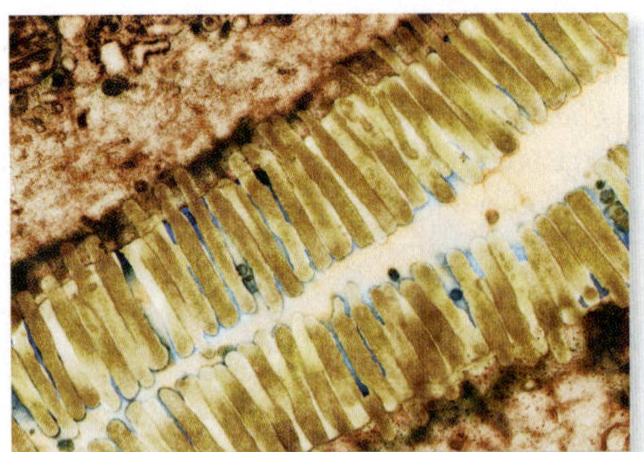

(a) Human intestine (longitudinal section)

(b) Frog skin

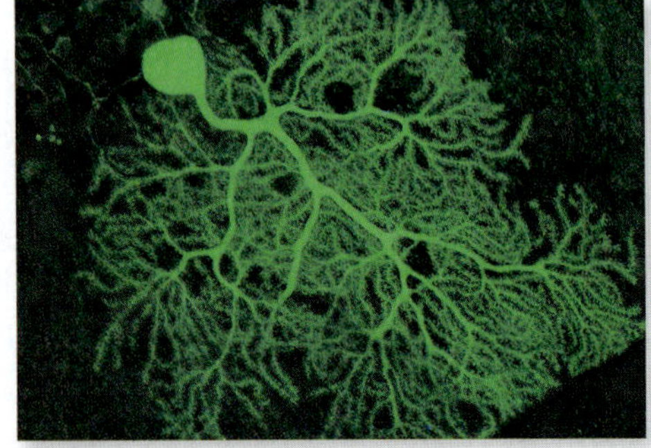

(c) Mouse brain neuron stained with a fluorescent marker

(d) Moth antennae

**Figure 32.7 Examples of structures in which extensive surface area is important for function.** A large surface area allows **(a)** high rates of transport of nutrients across the intestine of a human, **(b)** increased diffusion of oxygen across the folds of skin of a frog living at high altitude in Lake Titicaca, where $O_2$ is less available, **(c)** extensive communication between neurons in a mouse's brain, and **(d)** detection of airborne chemicals by moth antennae.



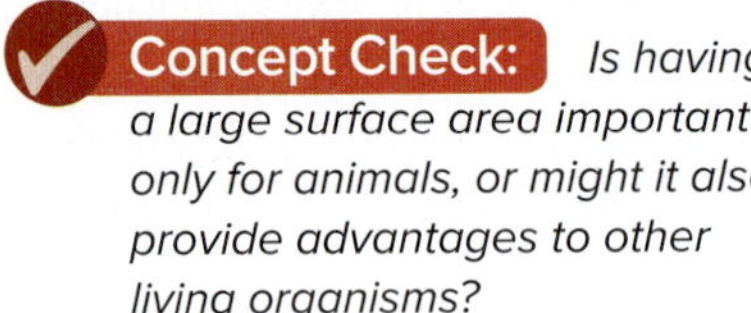

**Concept Check:** *Is having a large surface area important only for animals, or might it also provide advantages to other living organisms?*

A key principle of biology throughout this unit is that form (structure) determines function. Consider, for instance, the similarities in the finger-like projections of the small intestine of a human, the skin folds of some high-altitude frogs, the cellular extensions of the neurons of a mouse, and the feathery antennae of a moth (**Figure 32.7**). What do these structures have in common? They are all adaptations that increase surface area, which often predicts a function related to transport, absorption, or detection of environmental stimuli. In the examples shown in Figure 32.7, these features maximize the intestine's ability to absorb nutrients, the skin's ability to obtain oxygen from the environment, a neuron's ability to form contacts with many other cells, and an antenna's ability to detect airborne chemicals. A large increase in surface area of a structure is an important adaptation, but it comes at the expense of greatly increasing a structure's volume unless the shape of the structure is changed. The reason is that as an object enlarges, its volume grows relatively more than its surface area; surface area increases by the power of 2, whereas volume increases by the power of 3 (refer back to Figure 4.12).

The challenge of packaging an extensive surface area into a confined space is overcome by changes in shape (see, for example, the way the inner surface of the intestines in Figure 32.7a folds inward to form finger-like extensions). The relationship between a structure's surface area and its volume is called the **surface area/volume (SA/V) ratio.** A high SA/V ratio is ideal for exchange of heat, solutes, and water across a surface without contributing greatly to the mass or volume of a body part. This concept will appear throughout this unit as we explore the ways in which animals obtain energy, regulate their metabolism and body temperature, obtain oxygen, and eliminate wastes.

When all of an animal's organ systems function properly and body fluid levels and solute composition are maintained within normal limits, an animal is generally considered to be in a healthy condition. In the rest of this chapter, we will examine the process of achieving and maintaining this condition, which is known as homeostasis.

## 32.2 Reviewing the Concepts

- The structure of an animal's tissues and organs is related to the function of those structures. A large surface area maximizes the ability of a tissue or an organ to absorb nutrients, exchange oxygen and carbon dioxide with the environment, communicate with other cells, and receive sensory information from the environment (Figure 32.7).

## 32.2 Testing Your Knowledge

1. Refer to Figure 32.7a. What would happen if the dimensions of the finger-like projections shown in that figure were to double in size?
   a. Their surface area would double.
   b. Their surface area would increase eightfold.
   c. Their volume would double.
   d. Their volume would increase eightfold.
   e. Their ratio of surface area/volume would increase.

## 32.3 Homeostasis

### Learning Outcomes

1. Discuss the concept of homeostasis as it applies to the internal environment of animals.
2. Distinguish between conforming and regulating as mechanisms for maintaining homeostasis.
3. List several variables that are regulated within a homeostatic range in vertebrate animals.
4. Name the components of a homeostatic control system, and describe the importance of each to the regulation of an animal's internal environment.
5. **SCISKILLS ▶** Contrast negative feedback, positive feedback, and feedforward regulation, and explain how they do or do not contribute to the maintenance of homeostasis in animals.

The environmental conditions in which organisms—including all animals—live are rarely, if ever, constant. For example, animals may be exposed to fluctuations in air and water temperatures, nutrient and water availability, pH, and oxygen availability. Any one of these environmental changes can be harmful or even fatal if an organism is unable to respond appropriately. However, as you might expect from the incredible diversity of environments in which they exist, animals can adjust in many ways to their surroundings and thrive. As noted earlier, the process of maintaining a relatively stable internal environment despite changes in the external surroundings is known as homeostasis, which is an adaptation that is essential for an animal's health and survival. In this section, we will examine several of the most important features of homeostasis, including how it is achieved and maintained.

### Some Animals Conform to External Environments; Others Regulate Their Internal Environments

Some animals conform to their environments so that some feature of their internal environment matches that of their external surroundings. A marine crab, for example, has about the same solute concentration in its body fluids as is found in seawater. Likewise, the body temperatures of many fishes and aquatic invertebrates match the temperatures of the surrounding waters. Energetically, conforming is a cheap strategy for survival. A small fish would have to expend a great deal of energy to maintain its body temperature at, say, 37°C when swimming in the waters near Antarctica. However, because they do not actively adjust their internal body composition, conformers are often restricted to living in environments that are relatively unchanging.

Other animals homeostatically regulate their body temperature and the composition of their fluids and solutes, allowing them to differ from the external environment. Regulating the internal environment requires considerable energy. For this reason, homeostasis for regulators comes with a high energy price tag. However, the price paid by regulators makes it possible for them to inhabit a wide range of environments even if those environments fluctuate significantly, something that is difficult for conformers.

### Vertebrates Maintain Most Physiological Variables Within a Narrow Range

In vertebrates, the common physiological variables are usually maintained within a certain range, despite fluctuating external environmental conditions. Examples of such variables include the concentrations of blood-borne solutes such as minerals, glucose, and oxygen; the pH of the body fluids; blood pressure; and energy stores. At first glance, homeostasis may appear to be a state of stable balance of physiological variables. However, this simple idea does not capture the scope of homeostasis. For example, no physiological function is constant for very long, which is why we call them variables. Some variables may fluctuate around an average value during the course of a single day yet still be considered in balance. Homeostasis is a dynamic process, not a static one.

Consider an example in your own body. Normally, blood sugar (glucose) remains at fairly steady and predictable concentrations in any healthy individual. After a meal, however, the concentration of glucose in your blood can increase quickly, especially if you have just eaten something sweet. Conversely, if you skip a few meals, your blood glucose concentration may decrease slightly. Such fluctuations above and below the normal value might suggest that blood glucose concentrations are not homeostatic, but this is incorrect. Once the blood glucose concentration increases or decreases, homeostatic mechanisms restore it back toward normal. In the case of glucose, the endocrine and nervous systems are primarily responsible for this adjustment, but in other examples, a wide variety of control systems may be initiated. In later chapters, we will see how every organ and tissue of an animal's body contribute to homeostasis, sometimes in multiple ways and usually in concert with each other.

Homeostasis, then, does not imply that a given physiological function or variable is rigidly constant. Instead, homeostasis means that a variable fluctuates within a certain normal range and that if disturbed from that range, compensatory mechanisms restore the variable toward normal.

### Homeostatic Control Systems Maintain the Internal Environment

The activities of cells, tissues, and organs must be regulated and coordinated with each other so that any change in the extracellular fluid—a part of the body's internal environment—initiates a response to correct the change. These compensating regulatory responses are performed by homeostatic control systems. A **homeostatic control system** must have several components, including

- a **set point**—the normal value for a controlled variable,
- a **sensor,** which monitors the level or activity of a particular variable,
- an **integrator,** which compares signals from the sensor with the set point, and
- an **effector,** which compensates for any deviation between the actual value and the set point.

Figure 32.8 shows an example of a homeostatic control system that regulates body temperature in mammals. This system is somewhat

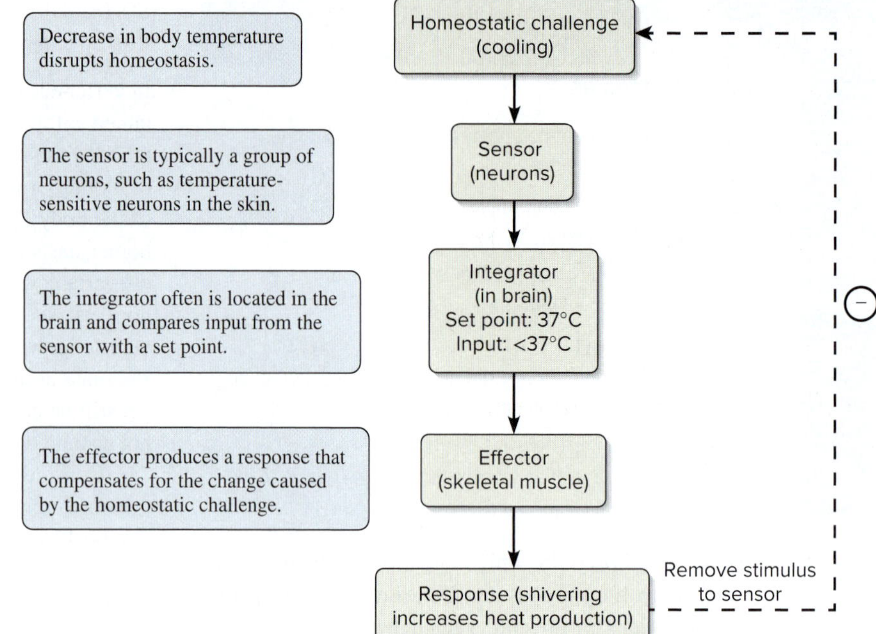


**Figure 32.8 An example of a homeostatic control system.** The sensor, integrator, and effector for responding to a decrease in body temperature are shown. Negative feedback (dashed line and minus symbol) prevents the response from continuing above normal. Different homeostatic control systems have different sensors and effectors.

**Concept Check:** *Would you expect all animals to have similar set points for a given homeostatic variable such as body temperature?*

**BioConnections:** *Refer back to Figure 6.10. Can negative feedback occur at a molecular level and contribute to homeostasis?*

analogous to the heating system of a home. In that case, a sensor and integrator within the thermostat compare the actual room temperature with the set point temperature that was determined by setting the thermostat to a given temperature. If the room temperature becomes cooler than the thermostat setting, the effector (furnace) is activated and adds heat to the room. In a mammal, the sensors are temperature-sensitive neurons in the skin and a part of the brain, whereas the integrator is within another brain structure. Signals from the brain are sent along nerves to the effectors, which include skeletal muscles. If body temperature decreases, the muscles contract vigorously in response to these signals, resulting in shivering—a major way in which mammals generate heat. We will discuss other heat-conserving and heat-generating mechanisms that contribute to this important homeostatic control system in Section 32.4. BIO TIPS ONLINE

## Negative Feedback Is a Key Feature of Homeostasis

A fundamental feature of homeostasis is that disturbances to a physiological variable, when they occur, are minimized and corrected. The temperature-regulation system just described is an example of a **negative feedback loop,** in which a change in the variable being regulated brings about responses that move the variable in the opposite direction. Thus, a decrease in body temperature leads to responses that increase body temperature—that is, move it back toward its original value. This removes the input to the sensor in the control system, and the effectors (in this example, skeletal muscles that cause shivering) are no longer stimulated.

Negative feedback prevents homeostatic responses from overcompensating. Whereas it is beneficial to initiate warming mechanisms when an animal's body temperature decreases, it would not be beneficial to continue to generate heat beyond the normal, healthy range.

Not all forms of feedback contribute to homeostasis. In some cases, a **positive feedback loop** may accelerate a process (think of an avalanche that begins with a small snowball rolling down a steep hill). Once begun, a positive feedback loop has no obvious means of stopping, since a change in a variable or process leads to events that amplify that change. Look again at Figure 32.8. Imagine what would happen if shivering not only generated heat but also stimulated the sensor neurons to send additional stimulatory inputs to the effectors. Heat production would continue in such a circumstance without any means of stopping, until a potentially dangerous or even life-threatening body temperature was reached. This is contrary to the principle of homeostasis, in which large fluctuations in a variable are minimized and reversed. Not surprisingly, therefore, positive feedback is far less common than negative feedback in animals. However, it does have important roles in certain processes, such as ovulation and birth in mammals, blood clotting, and the generation of action potentials in neurons, all of which will be described in later chapters.

## Feedforward Regulation Prepares for an Upcoming Challenge to Homeostasis

In animals with well-developed nervous systems, homeostasis is aided by **feedforward regulation,** in which an animal's body prepares for a possible challenge to homeostasis before it even begins. For example, consider the increase in heart rate that occurs prior to an increase in an animal's activity (think of a racehorse at the starting gate). Another classic example is that demonstrated by Russian physiologist Ivan Pavlov. When a hungry dog (or human!) sees or smells food, it begins to salivate and its stomach starts churning and producing acid. Salivation and activity of the stomach are important components of the digestive process, yet at this stage, the animal has not actually eaten any food. Instead, its digestive system is already preparing for the arrival of food in order to maximize digestive efficiency, speed the flow of nutrients into the blood, and minimize the time required for active cells to replenish energy stores. Therefore, feedforward regulation helps prepare an animal for an ensuing challenge such as the requirement for increased activity or the assimilation of nutrients.

## 32.3 Reviewing the Concepts

- Homeostasis is the process of maintaining a relatively stable internal environment despite changes in the external environment. Some animals conform to their environment, but many others homeostatically regulate internal processes in response to their environment.
- Homeostatic control systems coordinate homeostasis. Negative feedback loops minimize changes in a variable and prevent homeostatic responses from overcompensating. Feedforward regulation prepares the body for an upcoming challenge to homeostasis (Figure 32.8).

## 32.3 Testing Your Knowledge

1. The normal value for a controlled variable such as body temperature is called the
   **a.** set point. **c.** effector. **e.** sensor.
   **b.** integrator. **d.** feedback.
2. A person consumes a very sugary meal. After a short time, her blood glucose concentration increases from a normal resting value of 100 mg/dL to 210 mg/dL. A little while later, it is down to 50 mg/dL. Which of the following is correct?
   **a.** She demonstrates normal glucose homeostasis.
   **b.** She is unable to produce compensating mechanisms for the sugar that enters her blood.
   **c.** She has some compensating mechanisms, but they are not functioning homeostatically.

# 32.4 Homeostatic Regulation of Body Temperature

### Learning Outcomes

1. Provide examples of how changes in temperature affect chemical reactions, protein function, and membrane structure.
2. Define metabolic rate, and describe how it is related to heat production and body temperature.
3. List and define the four terms used to categorize organisms based on their source of heat and ability to maintain body temperature.
4. Identify the four main mechanisms animals use to exchange heat with the environment.
5. Define several mechanisms by which animals can alter their rate of heat gain or loss.

As we have seen, the control of body temperature is an example of homeostasis. Look at the image of the young woman in the chapter-opening photo. What is happening at that moment? It is a hot day, and the woman is perspiring and thirsty; it appears that she has just been exercising outdoors. Under such conditions, her body must work to prevent her body temperature from increasing, because high body temperature can destroy enzymes and damage cellular membranes. One way some animals, including humans, eliminate heat from the body is to perspire; recall from Chapter 2 that water has a high heat of vaporization, and therefore the evaporation of water from the body surface helps eliminate body heat. In this section, we begin by discussing why body temperature homeostasis is important for the health and survival of animals. We next examine how metabolic rate is related to body temperature. We then consider the homeostatic mechanisms by which animals gain or lose heat, and how these mechanisms may be regulated.

## Temperature Affects Chemical Reactions, Protein Function, and Membrane Structure

Most animals can survive only in a relatively narrow range of body temperatures. Mammals, for example, typically have a resting body temperature of 35–38°C (95–100°F). In humans, a body temperature of 41°C (106°F) may result in organ damage (notably the brain), and a body temperature of 42–43°C (107–109°F) is fatal. Temperature affects animals' bodies in three main ways: chemical reactions (and the enzymes that catalyze them), protein function, and membrane structure. First, chemical reactions depend on temperature. Heat accelerates the motion of molecules, so as an animal's body temperature increases, the rates at which the molecules in its body move and encounter each other also increase. Consequently, the rate of most chemical reactions in animals doubles or even triples for every 10°C increase in body temperature. Low temperatures slow down chemical reactions, making it harder for an animal to remain active and carry out functions such as digestion, reproduction, and immunity. An effect on an animal's immunity is particularly important, as many vertebrates become susceptible to disease when their body temperatures are decreased for long periods.

A second effect of temperature is that very high temperature can cause proteins to become denatured (refer back to Figure 6.7); that is, they lose the normal three-dimensional structure that is crucial to their ability to function properly. This occurs because the bonds that form tertiary and quaternary protein structures result from weak interactions, such as hydrogen bonds, and can be disrupted by heat. Denaturation of enzymes, in particular, is especially serious because of the major role they have in the metabolism of nutrients for the generation of ATP.

A third effect of temperature is on the structures of plasma membranes and the membranes of intracellular organelles. At low temperatures, membranes become less fluid and more rigid, partly due to changes in membrane phospholipids. Rigid membranes are less able to perform biological functions, such as transporting ions and nutrients. Alternatively, if the temperature becomes too high, membranes can become leaky.

## Heat Production Is Related to Metabolic Rate

The amount of energy an organism uses in a given period of time to power its activities is called its **metabolic rate.** The greater the metabolic rate, the more heat an animal generates. This heat arises from chemical reactions and, in turn, speeds up the rate of other chemical reactions.

The breakdown of organic molecules liberates energy in their chemical bonds that can be transferred to and stored in the bonds in ATP (see Chapter 6). This is the energy that cells use to perform various biological activities such as muscle contraction, active transport, and molecular synthesis. We refer to these activities as work. Not all of the energy is used to do work, however. Some of it appears as heat, which contributes to an animal's body temperature or is dissipated to the environment. Biologists have historically quantified the energy of metabolism in calories. A **calorie** is approximately the amount of heat required to raise the temperature of 1 gram of water by 1 degree Celsius. Most biological activities, however, require much greater amounts of energy than a calorie, and consequently the more common unit of measurement is the kilocalorie (1,000 calories, or 1 **kcal**). (In food labeling, 1 Calorie with a capital C is the same as 1 kcal.) Biologists often measure and compare the metabolic rates of different animals to learn, for example, how some animals are capable of hibernating, how an animal's body temperature influences its metabolic rate, and how hormones and other factors alter an animal's metabolism.

The most common measure used to compare the metabolic rates of different species is the **basal metabolic rate (BMR).** BMR can be thought of as the metabolic cost of living in a resting animal under controlled conditions. In mammals, most of it can be attributed to the routine functions of the heart, liver, kidneys, and brain. To compare the BMRs of different animals, however, a distinction must be made between animals that generate their own internal heat and those that do not (or do so less well). **Endotherms** are animals that generate their own internal heat through their metabolism. By contrast, **ectotherms** are animals that rely on heat from the external environment to warm themselves (**Figure 32.9**).

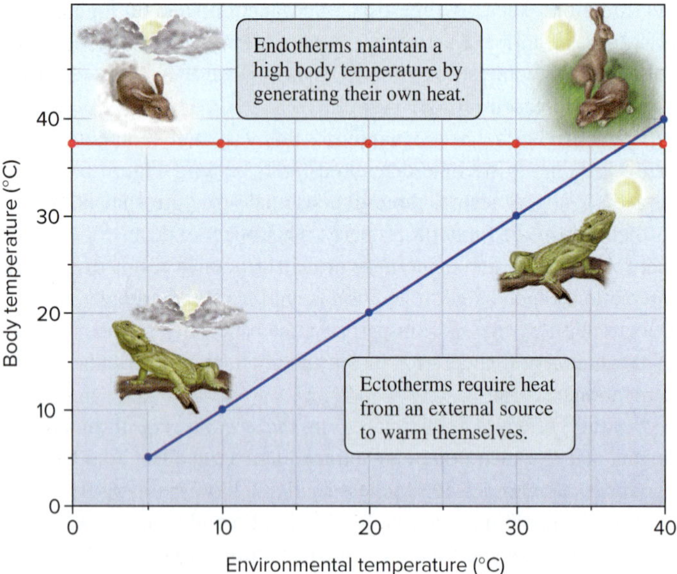


**Figure 32.9 Body temperature and environmental temperature in endotherms and ectotherms.** In this example, the rabbit is also a homeotherm because its body temperature doesn't change; the lizard is also a heterotherm because its body temperature changes considerably.

**Concept Check:** *Into which thermoregulatory categories do humans fit?*

In addition, biologists define **homeotherms** as animals capable of maintaining their body temperature within a narrow range, whereas **heterotherms** have body temperatures that may undergo considerable variation (see Figure 32.9). Most animals can be categorized as either endotherms or ectotherms and as either homeotherms or heterotherms. Generally, birds and mammals are endothermic and homeothermic. Other vertebrates and most invertebrates are ectothermic and heterothermic.

Not all animals, however, can be neatly classified into two categories at all times. Hibernating mammals, for example, are endotherms and homeotherms, except during the transition from fall to winter and again from winter to spring, when they are heterotherms. Generally, though, birds and mammals can quickly adjust the body's mechanisms for retaining or releasing heat such that body temperature remains within a narrow homeostatic range. This provides the advantage that the body's chemical reactions occur at optimal rates even when the environment imposes extreme challenges. The metabolic rate of a resting mammal, for example, is roughly six times greater than that of a comparably sized reptile. A suddenly awakened mammal is instantly capable of intense activity even on a winter's day, but an icy-cold reptile could be at the mercy of a predator because of the time required to warm itself in order to flee.

Endothermy does have three major disadvantages, however.

- **Requirement for larger amounts of energy.** First, to produce sufficient heat by metabolic processes, endotherms must consume larger amounts of food to provide the nutrients used by cells in the synthesis of ATP, during which heat is generated. Endotherms with high BMRs, such as shrews, must eat almost continuously and may die if deprived of food for as little as a day. By contrast, many ectotherms, such as snakes, can go for weeks without eating.
- **Risk of overheating.** Second, endotherms have a greater risk of overheating during periods of intense activity, even in cold weather.
- **Loss of body fluids.** Third, as described shortly, one mechanism for preventing overheating is the evaporation of body fluids by, for example, perspiration; this can result in an imbalance in water levels in an animal's body.

## Animals Exchange Heat with the Environment in Four Ways

Body temperature homeostasis is aided through the exchange of heat between an animal's body and its environment. The surface of an animal's body can lose or gain heat from the external environment via four mechanisms: radiation, evaporation, conduction, and convection.

- **Radiation** is the emission of electromagnetic waves by the surfaces of objects. The rate of emission is determined by the temperature of the radiating surface. Thus, if the body surface is warmer than the environment, the body loses heat at a rate that depends on the temperature difference. If the outside temperature is warmer than body temperature, the body gains heat by radiation—for instance, from sunlight. We can observe radiated heat from an animal's body with imaging devices that detect infrared light, the wavelength at which thermal energy is radiated from animal bodies (**Figure 32.10**).

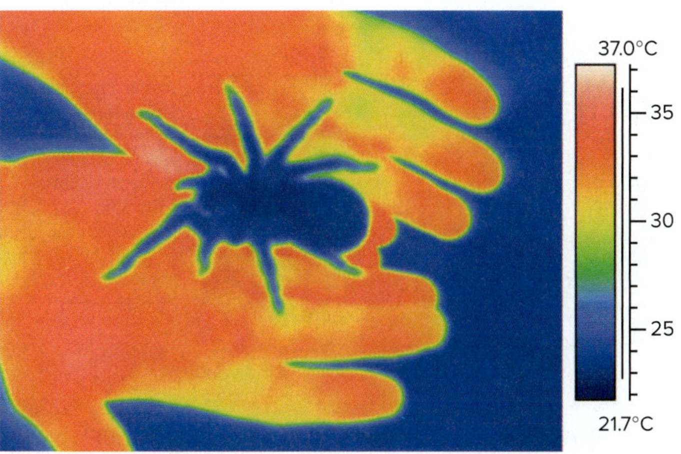


**Figure 32.10 Visualization of heat exchange in an ectotherm and an endotherm.** Thermal-imaging cameras can detect heat radiated from an animal's body. Note the warm skin of the endotherm (the human) and the cold skin of the ectotherm (the tarantula), even though both animals are at the same environmental temperature of about 25°C.



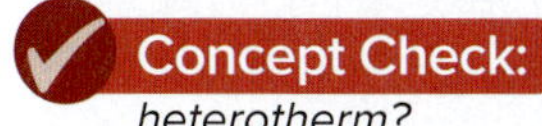

*Which animal in this figure is a heterotherm?*

- **Evaporation** is the conversion of water from the liquid to the gaseous state. Animals can lose body heat through evaporation of water from the skin and membranes lining the respiratory tract, including the surface of the tongue. A large amount of energy in the form of heat is required to transform water from liquid to gas. Whenever water vaporizes from the body's surface, the heat required to drive the process is conducted from the surface, thereby cooling the animal.
- **Conduction** is a process whereby the body surface loses or gains heat through direct contact with cooler or warmer objects. The greater the temperature difference, the greater is the rate of heat transfer. Different materials have different abilities to absorb heat, however. Water has a higher specific heat (see Chapter 2) than air, meaning that at any temperature, water will retain greater amounts of heat than will air. Consequently, aquatic animals in water that is 10°C lose considerably more heat in a given time than terrestrial animals lose in air that is 10°C. On a hot day, terrestrial animals can efficiently lose heat by immersing themselves in water.
- **Convection** is the transfer of heat by the movement of air or fluid next to the body. For example, the air close to an endotherm's body is heated by conduction. Because warm air is less dense than cold air, the warm air near the body is forced upward and carries away heat by convection. Convection is aided by creating currents of air around an animal's body. Humans do this by sitting near fans, but other animals can create cooling air currents themselves, as when an elephant waves its ears.

The four processes of heat transfer just described can be homeostatically regulated in animals, such that heat is retained within the body at some times and lost at other times, as we see next.

## Several Mechanisms Can Alter Rates of Heat Gain or Loss in Endotherms

For purposes of temperature control, think of an endotherm's body as a central core surrounded by a shell consisting of skin and subcutaneous (just below the skin) tissue. Depending on the species, the temperature of the central core of endotherms is regulated at approximately 37–40°C (97–104°F), but the temperature of the outer surface of the skin varies considerably. If the skin were a perfect insulator, the body would never lose or gain heat by conduction. The skin of animals does not insulate completely, however, so the temperature of its outer surface generally is somewhere between that of the external environment and the core. Only in animals that store large amounts of subcutaneous fat (blubber) does the body surface provide considerable insulation. In endotherms without blubber, the main form of insulation is a covering of hair, fur, or feathers, which traps heat from the body in a layer of warm air near the skin, reducing heat loss by conduction. Let's take a look at four mechanisms that different endotherms use to regulate how much heat is gained or lost between the environment and their bodies.

**Changes in Skin Blood Flow** Rather than acting as an insulator, as in some animals, in most endotherms the skin functions as a variable heat exchanger that can be adjusted to increase or decrease heat loss from the body (Figure 32.11). In such animals, blood vessels just beneath the surface of the skin dilate (widen) on hot days. This increases the flow of warm blood to the skin and helps to dissipate heat to the environment across the skin surface. These same vessels constrict (get narrower) on cold days to retain body heat. Signals from the nervous system regulate the dilation and constriction of these blood vessels. Diving birds and diving mammals are good examples of animals that use this mechanism. Ducks, seals, and walruses greatly reduce the amount of blood flowing to the skin when they dive in cold waters. This allows them to retain body heat that would otherwise be conducted into the water.

**Countercurrent Exchange** Many endotherms and ectotherms regulate heat loss to the environment through **countercurrent exchange,** in which heat is transferred between fluids flowing in opposite directions. Countercurrent exchange regulates heat loss to the environment by returning heat to the body's core and keeping the core much warmer than the extremities. In endotherms, countercurrent exchange occurs primarily in the extremities—the flippers of dolphins, for example, or the legs of birds and certain other terrestrial animals (Figure 32.12a). As warm blood travels from the body core through arteries down a bird's leg, heat moves by conduction from the artery to adjacent veins carrying cooler blood in the opposite direction (Figure 32.12b). By the time the arterial blood reaches the tip of the leg, its temperature has decreased considerably, reducing the amount of heat lost to the environment and returning the heat to the body's core.

**Evaporative Heat Loss** Recall that some animals can lose body heat through evaporation of water from the skin and membranes

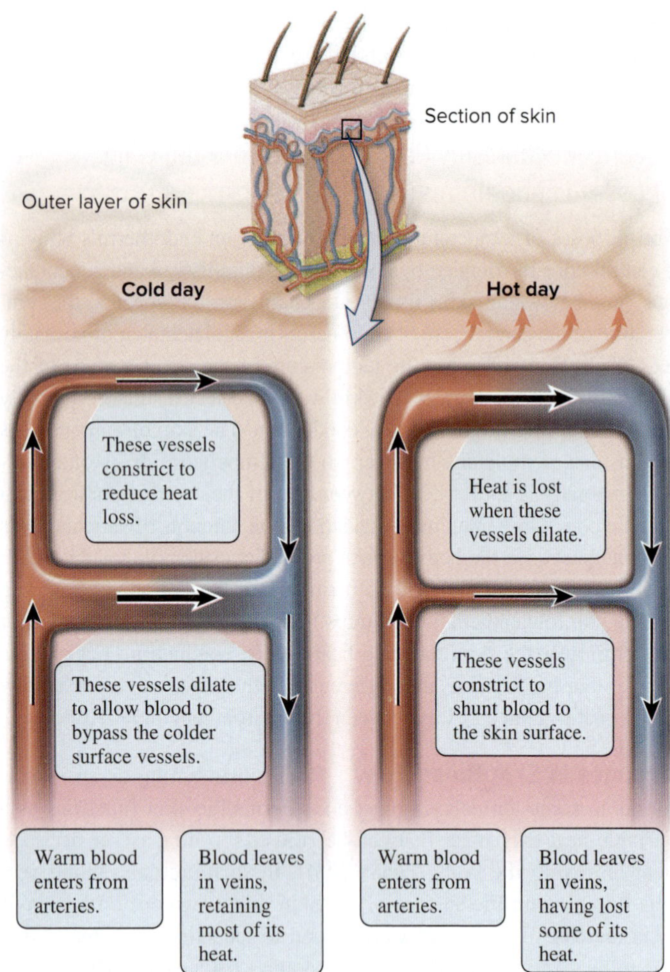


**Figure 32.11** **Regulation of heat exchange in the skin.** As shown in this schematic illustration, the skin functions as a variable heat exchanger. The arrows in the blood vessels indicate direction of blood flow.

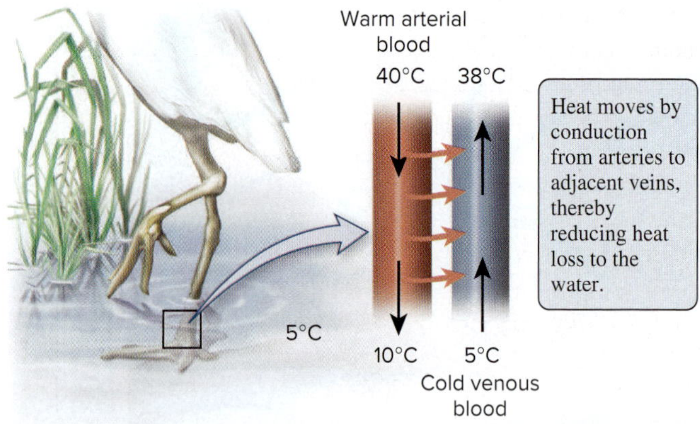


**(a) Countercurrent exchange of heat in the leg of an endotherm**

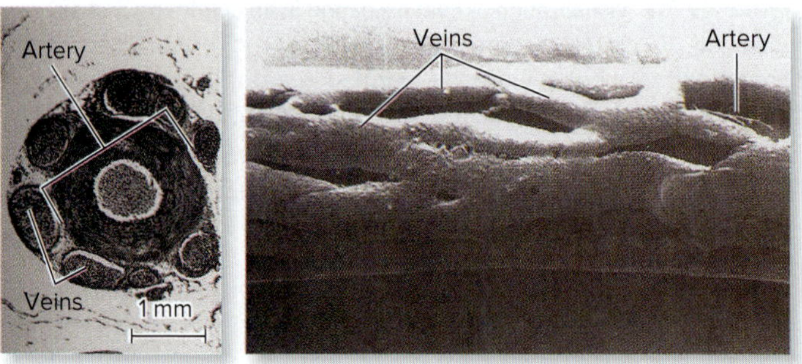


**(b) Cross section and surface view of veins covering an artery**

**Figure 32.12** **Countercurrent exchange. (a)** Countercurrent exchange retains heat in the leg of an endotherm such as this bird. Black arrows in vessels indicate direction of blood flow. **(b)** An SEM of the arrangement of veins surrounding an artery in a wading bird's leg. The artery is almost completely covered by overlying veins, allowing efficient heat exchange between the vessels.

*(b)* Courtesy of Uffe Midtgård, University of Copenhagen

**Concept Check:** *In which direction does heat transfer in this example, from artery to veins or from veins to artery?*

lining the respiratory tract. Heat exchange in some mammals can be regulated by changing the rate of water evaporation through perspiration (as was shown in the chapter-opening photo). Nerves to the sweat glands stimulate the production of sweat, a dilute solution containing sodium chloride.

In endotherms that lack sweat glands, such as birds, or those that have very few glands, such as dogs and cats, panting (short, rapid breaths with the mouth open) promotes evaporation of water from the tongue surface. Panting has advantages over sweating, because only water but no salt is lost, and panting provides the air current that promotes heat exchange by convection. However, the surface area of the mouth and tongue is relatively small, which limits the rate at which heat can be eliminated.

**Behavioral Adaptations** Behavioral mechanisms can also alter heat loss by radiation, conduction, and evaporation. For example, on hot days birds may ruffle their feathers and raise their wings, whereas many mammals will reduce their activity and spread their limbs; these postural changes increase the surface area available for heat transfer. Terrestrial endotherms seek shade, partially immerse themselves in water, or burrow into the ground when the sun is high. Animals that neither sweat nor pant can still benefit from evaporative heat loss. The evaporation of fluids deposited on the body surface by licking the skin or splashing the skin with water also draws heat from the body.

Similarly, animals respond to cold temperatures with numerous behavioral adaptations. Huddling in groups, curling up into a ball, hunching the shoulders, burying the head and feet in feathers, and similar maneuvers reduce the surface area exposed to a cold environment and decrease heat loss by radiation and conduction.

## Muscle Activity and Brown Adipose Tissue Metabolism Increase Heat Production

We have discussed how heat is gained or lost to the environment and how heat can be retained by reducing blood flow to the skin on a cold day. Body temperature, however, is a balance between these factors and heat production. Changes in muscle activity constitute a major mechanism for the control of heat production in endotherms.

When an endotherm is exposed to low environmental temperatures, core body temperature begins to decrease. The primary response to decreasing temperatures is to decrease the flow of blood to regions that permit conduction of heat. If this does not adequately reduce heat loss, skeletal muscle contraction is increased. This may lead to shivering, which consists of rapid skeletal muscle contractions without any locomotion. Virtually all of the energy liberated by the chemical reactions that power the contracting muscles appears as internal heat, a process known as **shivering thermogenesis.** Many birds that remain in cold climates during the winter shiver almost continuously.

In many mammals, chronic cold exposure also induces **nonshivering thermogenesis,** an increase in the metabolic rate and therefore heat production that is not due to increased muscle activity. Nonshivering thermogenesis occurs primarily in **brown adipose tissue** (also called brown fat), a specialized tissue that occurs in small or young mammals in which the shivering mechanism is not present or not yet fully developed. Brown adipose tissue is responsive to hormones and signals from the nervous system, which are activated when body temperature decreases. Unlike ordinary white adipose tissue, which stores energy in the form of fat, brown adipose tissue metabolizes fat and generates heat as a by-product.

## 32.4 Reviewing the Concepts

- The amount of energy an organism expends to power its activities is called its metabolic rate. In the resting state, this is called the basal metabolic rate (BMR). The greater the BMR of an animal, the more heat the animal generates.
- Ectotherms depend on external heat sources to warm their bodies, whereas endotherms use their own metabolically generated heat to warm themselves. Homeotherms maintain their body temperature within a narrow range, but the body temperature of heterotherms varies with environmental conditions (Figure 32.9).
- The surface of an animal's body can lose or gain heat from the environment via four mechanisms: radiation, evaporation, convection, and conduction (Figure 32.10).
- The skin can function as a variable heat exchanger; blood vessels near the skin surface dilate to dissipate heat or constrict to retain it. Heat exchange can also be controlled by countercurrent exchange, perspiration, and behavioral mechanisms (Figures 32.11, 32.12).
- Muscle activity (shivering thermogenesis) and brown adipose tissue metabolism (nonshivering thermogenesis) increase the production of heat.

## 32.4 Testing Your Knowledge

1. Animals that have body temperatures that are maintained within a narrow range are
   **a.** endotherms. **c.** homeotherms. **e.** both b and d.
   **b.** ectotherms. **d.** heterotherms.
2. The rate of heat loss in a mammal is regulated by
   **a.** the degree of blood flow at the surface of the skin.
   **b.** the amount of perspiration.
   **c.** behavioral adaptations.
   **d.** air currents near the animal's body.
   **e.** all of the above.

# 32.5 Principles of Homeostasis of Internal Fluids

### Learning Outcomes

1. **SCISKILLS ▶** Predict outcomes of imbalances of water and ions on animals' function and survival.
2. Describe what osmolarity is and how it relates to movement of water and cell shape.
3. Describe the factors that result in obligatory exchanges of water and ions between an animal's body and its environment.

Look again at the chapter-opening photo of a young woman who has been exercising on a hot day. As she perspires, she loses heat, but this comes with a cost because of the water and ions she loses. Animal bodies are comprised in large part of water, which is contained in body fluid compartments. Within the fluids are many salts, such as NaCl and KCl. Salts dissociate in solution into ions. (Because ions from dissolved salts are electrically charged, they are sometimes referred to as electrolytes.) In this section, we will first examine how water is distributed in the body and why water and ion homeostasis is so important. We then explore some of the challenges that must be overcome to maintain that homeostasis.

## Body Fluids Are Distributed into Fluid Compartments

Up to now, we have examined organ systems and their component structures without any reference to the other major feature of animal bodies, namely water. All animal bodies are composed primarily of water. This water is distributed within fluid compartments (**Figure 32.13**). Most of the water in an animal's body is contained inside its cells; this fluid compartment is called **intracellular fluid** (from the Latin *intra*, meaning inside of). The rest of the water in its body exists outside of the cells and is therefore called **extracellular fluid** (from the Latin *extra*, meaning outside of). In vertebrates, extracellular fluid includes the fluid part of blood, called **plasma,** and the fluid-filled regions that surround cells, called **interstitial fluid** (from the Latin *inter*, meaning between). Water can move between body fluid compartments by the process of osmosis, as described in Chapter 5. This has important consequences for a variety of processes that occur in animals, such as the control of the volume of blood in an animal's body. For example, if the solute concentration of an animal's blood were to increase (perhaps due to ingestion of a salty meal), water would move by osmosis from the interstitial fluid into the plasma. This would enlarge the blood volume and could cause an increase in blood pressure.

The solute composition of the extracellular fluid is very different from that of the intracellular fluid. Maintaining differences in solute composition across the plasma membrane is an important

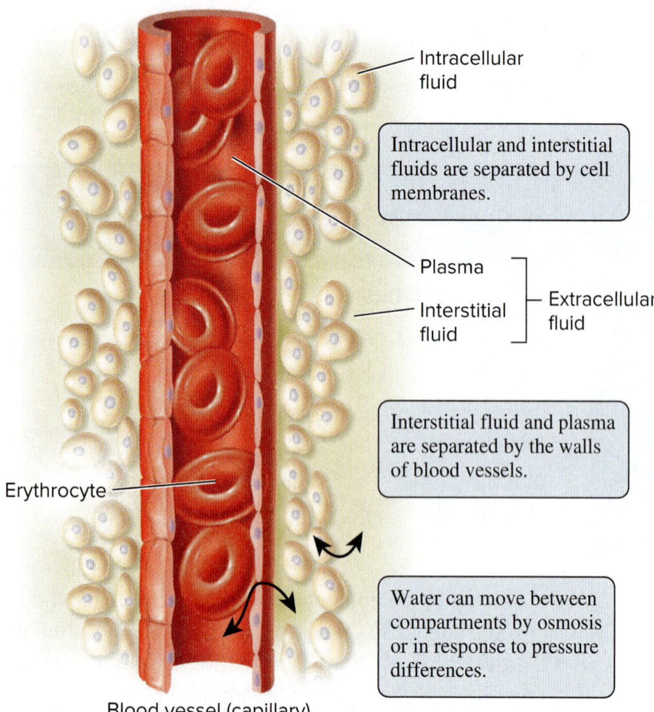


**Figure 32.13 Fluid compartments in a typical vertebrate.** Most of the fluid within an animal's body exists within cells (intracellular fluid). Extracellular fluid is that portion of the body's fluid that lies outside cells (interstitial fluid) or within blood vessels (plasma), such as the capillary shown here. Arrows indicate directions of water movement between adjacent compartments.

**Concept Check:** *What would happen to the distribution of water in the fluid compartments of an animal's body if a blood vessel was damaged and leaked its contents internally?*

way in which animal cells regulate their own activity. For example, differences in ion concentrations on either side of the plasma membrane of a cell create the potential for moving electrical charges across the membrane; this is a key mechanism by which cells of the nervous system communicate with each other. The mechanisms by which solutes move across plasma membranes were also described in Chapter 5.

## A Balance of Water and Ions Is Critical for Survival

Maintenance of normal body water levels is of great importance for all animals for several reasons (refer back to Figure 2.13). Water is the solvent that permits dissolved solutes to participate in chemical reactions. It also directly participates in hydrolysis reactions, and it is the transport vehicle that brings oxygen and nutrients to cells and removes wastes generated by metabolism.

When an animal's water volume is decreased below the normal range, we say the animal is dehydrated. In terrestrial animals, dehydration may occur if sufficient drinking water is not available or when water is lost by evaporation (perspiring or panting). Dehydration can be a serious, potentially life-threatening condition. For example, because blood is roughly 50% water (plasma), blood volume tends to decrease in dehydrated animals. Decreased blood volume compromises the ability of the circulatory system to move nutrients and wastes throughout the body and, in endotherms, to assist in the regulation of body temperature on hot days.

Ion balance is also very important for animals. A change of only a few percent in the extracellular concentration of $K^+$, for example, can trigger changes in nerve, heart, and skeletal muscle function by altering membrane potentials (see Chapter 33). Other ions, such as $Ca^{2+}$, $Mg^{2+}$, $PO_4^{3-}$, and $SO_4^{2-}$, also participate in various biological activities. These functions include serving as cofactors for enzyme activation, participating in bone formation, forming part of the extracellular matrices around cells, and activating cellular events such as exocytosis and muscle contraction. An imbalance in any of these ions can seriously disrupt cellular activities; dehydration is one way in which such an imbalance could arise. That is because when the amount of water in a body fluid compartment decreases, the concentrations of solutes in the remaining water increases.

Water moves between adjacent body compartments by osmosis (refer back to Figure 5.10). Changes in the ion concentration in one compartment will lead to changes in fluid distribution between the compartments. These changes can cause cells to shrink or swell. If, for example, the ion concentration of the extracellular fluid increases, water moves by osmosis from inside cells to the extracellular fluid, causing the cells to shrink. Shrinking or swelling of cells in the brain, heart, and other vital organs can rupture plasma membranes, leading to cell death.

The solute concentration of a solution of water is known as the solution's **osmolarity,** expressed as milliosmoles/liter (mOsm/L). The number of dissolved solute particles determines a solution's osmolarity. For example, a 150 mM NaCl solution has an osmolarity of 300 mOsm/L, because each NaCl molecule dissociates into two ions, one $Na^+$ and one $Cl^-$ ($2 \times 150 = 300$). For reference, the value 300 mOsm/L is well within the range of typical osmolarities of animal body fluids.

## Many Exchanges of Ions and Water with the Environment Are Obligatory

Many vital processes—eliminating nitrogenous wastes, obtaining oxygen and eliminating carbon dioxide ($CO_2$), consuming and metabolizing food, and regulating body temperature—have the potential to disturb ion and water balance (Figure 32.14). Therefore, these processes generally require additional energy expenditure to minimize or reverse the disturbance. Exchanges of ions and water with the environment that occur as a consequence of such vital processes are called obligatory exchanges (because the animal is "obliged" to make them).

**Exchanges Due to Elimination of Nitrogenous Wastes** When carbohydrates and fats are metabolized by animal cells, the major waste product is $CO_2$, which is exhaled or diffuses across the body surface. By contrast, proteins and nucleic acids contain nitrogen; when these molecules are broken down and metabolized, nitrogenous wastes are generated. **Nitrogenous wastes** are molecules that include nitrogen from amino groups ($—NH_2$). These wastes are toxic

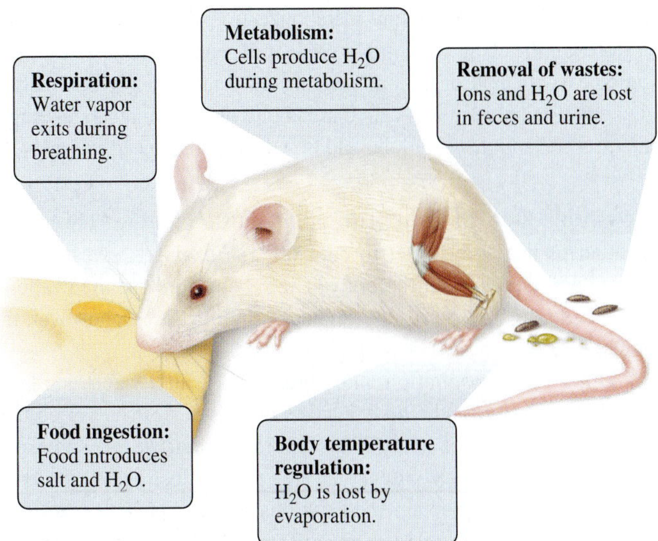


**Respiration:** Water vapor exits during breathing.

**Metabolism:** Cells produce $H_2O$ during metabolism.

**Removal of wastes:** Ions and $H_2O$ are lost in feces and urine.

**Food ingestion:** Food introduces salt and $H_2O$.

**Body temperature regulation:** $H_2O$ is lost by evaporation.

**Figure 32.14** **Types of obligatory ion and water exchanges in a terrestrial animal.** Obligatory exchanges with the environment occur as the result of necessary life processes.

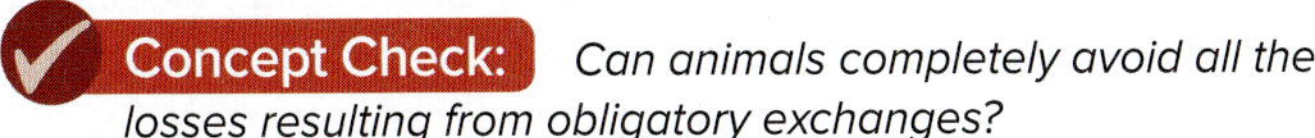

**Concept Check:** *Can animals completely avoid all the losses resulting from obligatory exchanges?*

at high concentrations and must be eliminated from the body, but, unlike carbon dioxide gas, they cannot be eliminated by exhaling. As we will see in Chapter 37, the excretion of nitrogenous wastes is carried out by excretory organs and may require body water.

**Exchanges Due to Respiration** The requirements for both respiration and water and ion balance present different challenges to air- and water-breathing animals. To ventilate its lungs, an air-breathing animal moves air into and out of its airways. Water vapor in the mouth, nasal cavity, and upper airways exits the body with each exhalation. As an animal becomes more active, breathing becomes deeper and more rapid, which, in turn, increases the rate of water loss from the body. Therefore, respiration in animals with lungs is associated with significant water loss, as you can observe in cold weather when you can "see your breath."

In water-breathing animals, the challenge of water and ion balance is more complex, because such animals move water, not air, over their respiratory organs (gills). Gills, like all respiratory organs, are thin structures with large amounts of surface area and an extensive network of blood vessels. Although these features make gills suitable for gas exchange by diffusion between the blood and the surrounding water, they also make them suitable for ion and water movement by diffusion and osmosis, respectively. This can create imbalances in the volume and osmolarity of fluids in an animal's body.

Fishes or other water-breathing animals that live in fresh water and those that live in salt water face opposite challenges in maintaining this balance (Figure 32.15). The internal fluid osmolarity of most fishes is usually around 300 mOsm/L, similar to that of most other vertebrates. Because freshwater lakes and rivers have very little salt content (usually <50 mOsm/L), they are hypo-osmotic relative to an animal's body fluids. Thus, an osmotic gradient favors the movement of water across the gills into the body fluids of a freshwater fish. Likewise, there is a large concentration gradient for

## Biology Principle

### Living Organisms Use Energy, and Living Organisms Maintain Homeostasis

Energy from the hydrolysis of ATP is required for active transport processes associated with the balance of ions in body fluids. This energy is used to maintain a homeostatic body fluid osmolarity despite the osmotic challenges imposed by very different environments.

Does not drink.

$H_2O$ enters gill capillaries by osmosis.

**Osmolarity of body fluids: ~300 mOsm/L**

Some ions and $H_2O$ from food

Active uptake of $Na^+$ and $Cl^-$ across gill epithelia into capillaries

Very dilute, copious urine containing very few ions

Osmolarity of fresh water: 0–50 mOsm/L

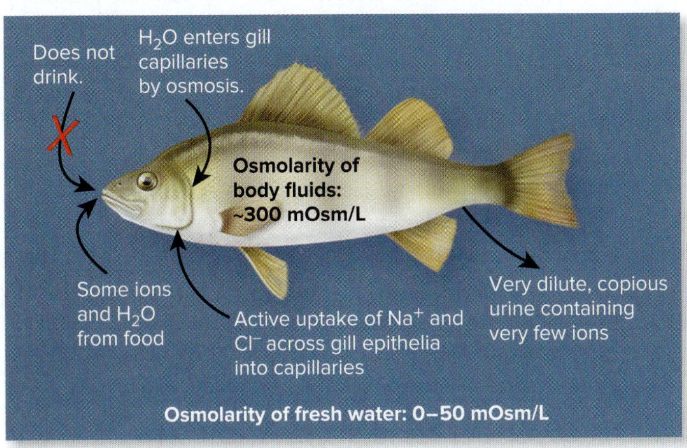


**(a) Freshwater fish**

Drinks seawater.

$H_2O$ leaves gill capillaries and enters seawater by osmosis.

**Osmolarity of body fluids: ~300 mOsm/L**

Some ions and $H_2O$ from food

Active excretion of $Na^+$ and $Cl^-$ across gill epithelia

Small amount of concentrated urine containing large amounts of ions

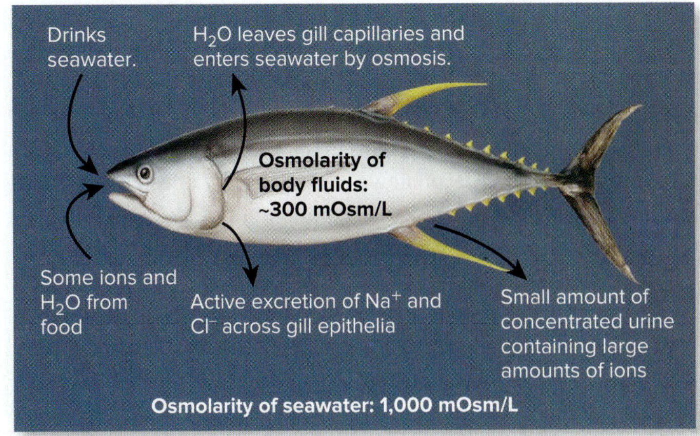


**(b) Saltwater fish**

**Figure 32.15** **Water and ion balance in fish.** Water-breathing creates osmoregulatory challenges due to diffusion of ions and osmosis of water across gills. These challenges differ between **(a)** freshwater and **(b)** saltwater fishes and are met by behaviors (whether or not they drink water), by active transport of ions across the gills, and by alterations in urine output.

ions that promotes a loss of ions from a fish's body into the fresh water. Freshwater fishes, therefore, gain water and lose ions when ventilating their gills (Figure 32.15a). Without a means to reverse this, a dangerous decrease in ion concentrations in the blood could arise. This doesn't happen, however, because such fish excrete into the urine most of the water that enters across the gills; they also rarely, if ever, drink water. Cells in the gills can also actively transport ions into the body.

Marine fishes have the opposite situation. They tend to gain ions and lose water across their gills, because seawater is hyperosmotic (about 1,000 mOsm/L) relative to their body fluids (Figure 32.15b). These changes are partly offset by producing only a very small volume of urine so that water can be retained in the body; the urine is also highly concentrated with ions. In addition, marine fishes drink seawater. The ingested salt is eliminated across the gills by active transport.

**Exchanges Due to Feeding** Because foods contain salts and water, eating also involves an obligatory exchange of these substances. Some plant products are over 95% water by weight, and other foods may contain high amounts of $Na^+$ or other ions. Therefore, the type of diet an animal consumes determines how much salt and water it ingests.

When food molecules are metabolized to provide energy that will be stored in the chemical bonds of ATP, oxygen captures electrons and combines with hydrogen ions, thereby making water (refer back to Figure 6.18). This water is called metabolic water to indicate its origin and, in some animals, is all the additional water they need to survive.

**Exchanges Due to Evaporation of Water** As we have seen, endotherms use body water to cool off. For example, sweating and panting are used to cool the body. These behaviors use the evaporation of water to draw heat out of the body. In the process, however, the animal loses water and, in sweat, some ions.

Other than perspiration and panting, very little water is gained or lost directly across the body surface of most terrestrial vertebrates (with the exception of amphibians), because their skin is impermeable to water. In invertebrates, the rate of water loss across the body surface depends on whether the animal is soft-bodied, like worms, or covered in a waxy, water-impermeable cuticle, like most insects.

The significance of obligatory exchanges and their effects on homeostasis was dramatically illustrated by a long-term investigation by a research team at the University of Florida, as described next. Their discovery would lead to a revolution in our understanding of exercise physiology and ion and water homeostasis in humans.

## FEATURE INVESTIGATION

### Cade and Colleagues Discovered Why Athletes' Performances Wane on Hot Days

On a typically hot summer day in the mid-1960s in Gainesville, Florida, the University of Florida football team was practicing in full equipment. The players were becoming dehydrated and, unbeknownst to them, the osmolarity of their body fluids was increasing as their bodies produced copious amounts of sweat in an effort to maintain body temperature. The athletes became aware of two things. First, they discovered that they did not need to urinate for long periods after a strenuous practice session and, second, their performance on the field suffered as they became increasingly fatigued and more susceptible to severe muscle cramps. Occasionally, players required medical treatment for their symptoms. In extreme cases, athletes exercising in these conditions have been known to occasionally develop seizures—uncontrolled activity of neurons in the brain. This situation did not escape the notice of the team physicians and, notably, university faculty member and kidney specialist Robert Cade.

Many of the symptoms experienced by the players could be readily explained. The fatigue was directly related to loss of water from the body, which put a strain on the circulatory system and reduced the amount of blood flow to muscles and other organs. It was worsened by a decrease in blood glucose concentration that was not compensated for during the long periods of strenuous activity. The muscle cramps and even occasional seizures arose from an imbalance in extracellular ions—notably $Na^+$ and $K^+$—which are secreted outside the body by sweat glands in the process of perspiration. Sweat is a hypo-osmotic solution relative to body fluids, and consequently sweating results in an imbalance in extracellular ion concentrations. In the athletes, this caused a change in the electrical potential across muscle and neuronal cell membranes, which triggered the symptoms. Finally, the decreased urine production is one of the body's mechanisms for reducing fluid loss when body water is decreasing.

Cade and his colleagues rejected what was then the prevailing view that drinking any fluids during heavy exercise somehow contributed to cramps and other problems. Instead, they hypothesized that the best way to maintain ion and water homeostasis in a profusely sweating person is to restore to the body exactly what was lost; that is, the person should drink a solution that resembles sweat!

The first thing Cade did was analyze how much $Na^+$, $K^+$, and other ions are present in sweat. Fortunately, he had an abundance of human sweat at his disposal to analyze. Once the players left the field, their jerseys were wrung out, and the composition of the collected sweat was determined with an ion analyzer such as the flame spectrophotometer shown in **Figure 32.16**. The concentrations could then be compared with known values of ion concentrations in human blood. Today, we know that the composition of human sweat can change under certain conditions and can vary among people, but Cade's results were typical. The athletes' sweat contained mostly $Na^+$, $K^+$, and $Cl^-$ at concentrations that indicated the solution was dilute compared with blood. Once Cade completed this analysis, he simply prepared an artificial solution of a composition similar to human sweat. The next step was to have the players drink the solution before and during the practice sessions and games. Improving its taste—adding some lemon flavoring and sugar—removed any inhibitions the players may have had about drinking it, while also providing an energy boost.

For the first trial, Cade gave the solution to the freshman players during an intrasquad scrimmage against the more experienced varsity B-team, whose members received only pure water to drink as a control. At first, the freshman team appeared overmatched by the B-team, as might be expected. In the second half of the scrimmage, however, the freshman team vastly outperformed the more experienced players and did not suffer the characteristic late-game fatigue the B-team experienced. Based on this test, the varsity A-team was given a similar solution to drink the next day during a game against a heavily favored opponent, whom they beat handily on a hot 39°C (102°F) day.

In subsequent years, Cade and other researchers conducted carefully controlled experiments with humans and laboratory animals to confirm that a balanced solution of ions similar to that present in sweat effectively improves exercise performance and decreases the possibility of dehydration and its consequences. Because the solution was envisioned as "aid" for the team known as the University of Florida "Gators," the drink eventually came to be called Gatorade.

The story of Gatorade is one of a practical application based on solid scientific principles of osmolarity and ion and water homeostasis. You can now understand why drinking a dilute salt solution (in any of numerous sports drinks) during strenuous exercise is better than drinking water. Although drinking pure water prevents dehydration, if drunk in excess, it will actually decrease plasma ion concentrations below normal. In other words, it will replace one type of imbalance with another as happened to the marathoner in the chapter introduction.

### *Experimental Questions*

1. What symptoms are sometimes seen in athletes after prolonged, strenuous exercise, particularly in hot weather? How are these symptoms related to water loss during exercise, and what did Cade and his colleagues hypothesize about this?
2. **SCISKILLS** ▶ How did the researchers test their hypothesis?
3. **SCISKILLS** ▶ Propose a hypothesis: Would drinking a sports drink like Gatorade be beneficial for someone at rest or someone exercising only lightly?

**Figure 32.16** **Cade and colleagues discovered a way to improve athletic performance and prevent ion and water imbalance during strenuous exercise.**

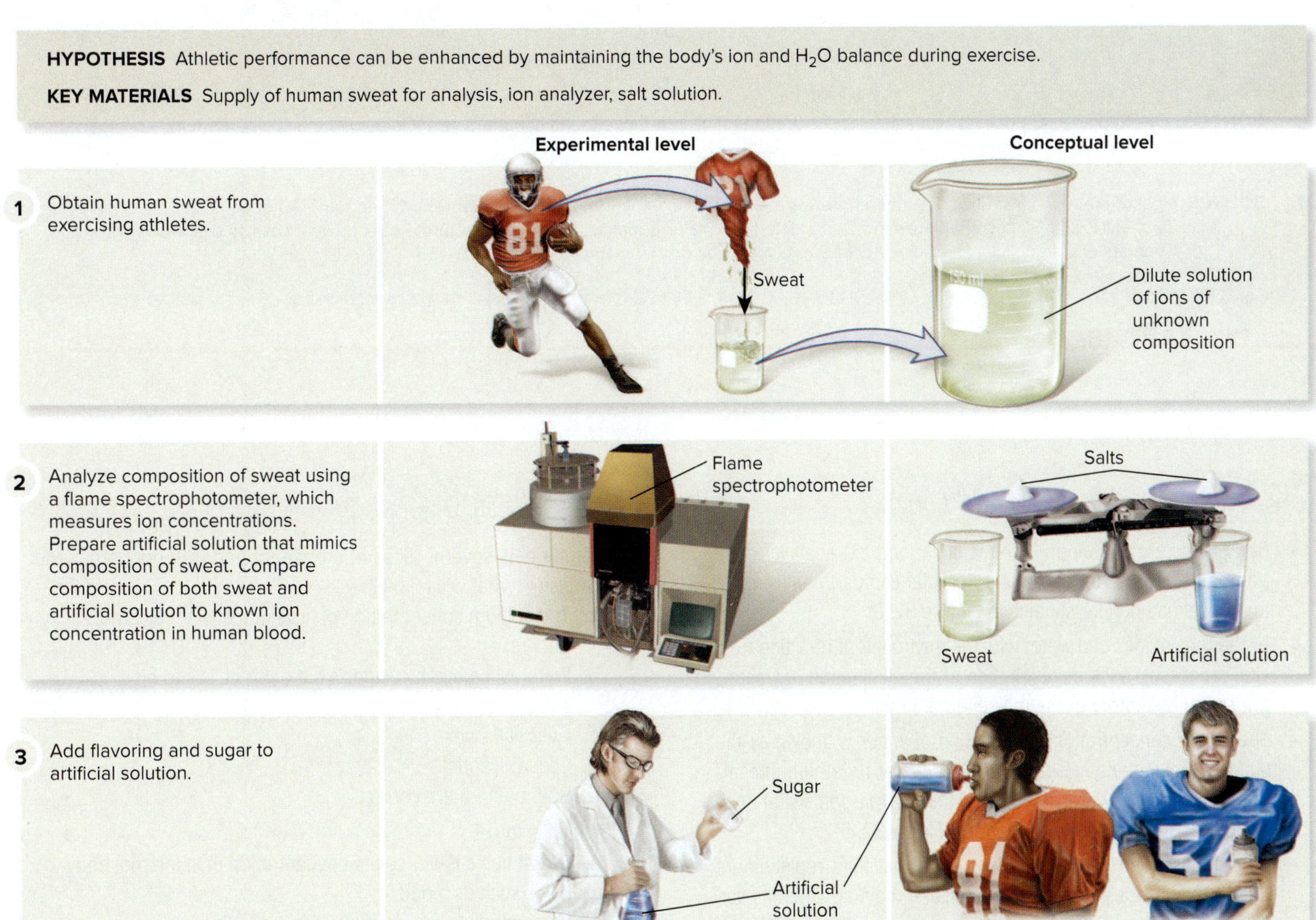

HYPOTHESIS Athletic performance can be enhanced by maintaining the body's ion and $H_2O$ balance during exercise.
KEY MATERIALS Supply of human sweat for analysis, ion analyzer, salt solution.
Experimental level
Conceptual level
1 Obtain human sweat from exercising athletes.
Sweat
Dilute solution of ions of unknown composition
2 Analyze composition of sweat using a flame spectrophotometer, which measures ion concentrations. Prepare artificial solution that mimics composition of sweat. Compare composition of both sweat and artificial solution to known ion concentration in human blood.
Flame spectrophotometer
Salts
Sweat
Artificial solution
3 Add flavoring and sugar to artificial solution.
Sugar
Artificial solution
Sugar improves flavor and provides energy.

4 Provide freshman team with the artificial solution and varsity B-team with water. Hold scrimmage.

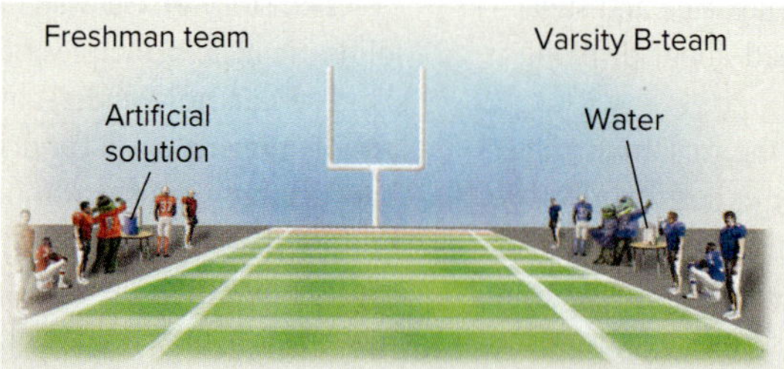

5 **THE DATA**

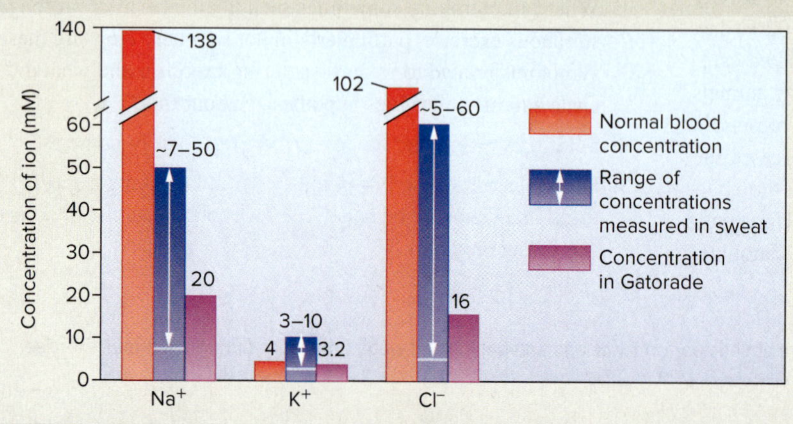


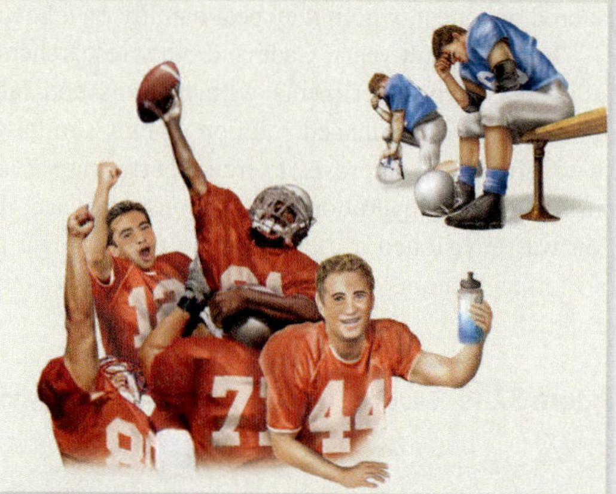

6 **CONCLUSION** Replacement of fluid with solute concentrations similar to those found in human sweat improves athletic performance compared with water replacement alone.

7 **SOURCE** Most of the original studies described here were published in expanded form in a later report. See Cade, R., et al. 1972. Effect of Fluid, Electrolyte, and Glucose Replacement During Exercise on Performance, Body Temperature, Rate of Sweat Loss, and Compositional Changes of Extracellular Fluid. *Journal of Sports Medicine and Physical Fitness* 12: 150–156.

## 32.5 Reviewing the Concepts

- Body fluids are contained in intracellular and extracellular compartments. In vertebrates, the extracellular compartment is made up of plasma and interstitial fluid (Figure 32.13).
- Exchanges of ions and water with the environment resulting from vital processes, such as respiration or the elimination of wastes, are called obligatory exchanges (Figure 32.14).
- The solute concentration of a solution of water is known as the solution's osmolarity. Fishes and other water-breathing animals that live in fresh water and those that live in salt water face opposite osmoregulatory challenges (Figure 32.15).
- Robert Cade and coworkers discovered that fluid replacement during exercise is particularly beneficial if the fluid contains solutes at concentrations similar to those in sweat (Figure 32.16).

## 32.5 Testing Your Knowledge

1. A man eats several slices of a pizza topped with pepperoni and sausage. What will follow?
   a. The sodium concentration of his extracellular fluids will decrease.
   b. Water will move by osmosis out of his cells into his extracellular fluid.
   c. Cells will swell.
   d. Cells will shrink.
   e. Both b and d are correct.
2. Which is *false?*
   a. Sweat has a lower solute concentration than internal body fluids such as blood.
   b. Some animals can survive without drinking water.

**c.** A solution of 1,000 mOsm/L is hyperosmotic relative to the body fluids of a fish.

**d.** Marine fishes, but not freshwater fishes, drink the water they swim in.

**e.** A 300 mOsm/L solution of NaCl contains 300 mM $Na^+$ and 300 mM $Cl^-$.

## Assess and Discuss

### Test Yourself

1. Tissue that is specialized to conduct electrical signals from one structure in the body to another structure is __________ tissue.
   a. epithelial  c. nervous
   b. connective  d. muscle

2. Structures composed of two or more tissue types arranged in various proportions and patterns are
   a. cells.  c. organs.  e. tissue systems.
   b. tissues.  d. organ systems.

3. Which is *not* a type of connective tissue?
   a. adipose tissue
   b. cartilage
   c. blood
   d. neurons
   e. collagen fibers found in tendons and ligaments

4. When examining the structure of many animal organs,
   a. it is apparent that an organ's surface area increases more than its volume as an organ enlarges.
   b. a general function can sometimes be predicted based on the structural adaptations.
   c. different tissues do not come into contact with each other within an organ.
   d. all four tissue types are equally represented in all organs.
   e. none of the above are correct.

5. Most of the water in an animal's body
   a. lacks any type of ions or other solutes.
   b. is found in the spaces between cells.
   c. is contained inside the cells.
   d. is located in the extracellular fluid.
   e. is unable to move between body compartments.

6. The folds, convolutions, or extensions in many animal structures results in
   a. decreased level of activity in those structures.
   b. interruption in the normal functioning of the structures.
   c. increased surface area for absorption, communication, or exchange.
   d. higher susceptibility to infection.
   e. all of the above.

7. Adapting to changes in the external environment and maintaining internal variables within normal ranges is/are called
   a. equilibrium.  d. homeostasis.
   b. negative feedback.  e. both c and d.
   c. positive feedback.

8. Which of the following statements regarding negative feedback is/are *false?*
   a. It helps regulate variables such as body temperature.
   b. It explains why an animal may salivate when it sees food.
   c. It is a major feature of homeostatic control systems.
   d. It prevents homeostatic responses from overcompensating.
   e. Both b and d are false.

9. The ability of an animal's body to prepare for a change in some variable before the change occurs is called
   a. positive feedback.  d. feedforward regulation.
   b. negative feedback.  e. autoregulation.
   c. steady state.

10. Frogs have skin that is relatively permeable to water. When submerged in a freshwater pond, what do you predict will happen?
   a. The frog will lose body water to the pond.
   b. There will be no net gain or loss of water.
   c. The frog will gain water across its skin.
   d. The solute concentration of the frog's body fluids will increase.
   e. Both a and d will occur.

### Conceptual Questions

1. Describe the relationship between structure and function in animals, and how surface area and volume are related.
2. Define homeostasis, and give examples of homeostatic variables in animals. Include in your discussion the differences between ectotherms and endotherms and between heterotherms and homeotherms.
3. **PRINCIPLES** Two principles of biology are that living organisms use energy and living organisms maintain homeostasis. How are energy use and the maintenance of homeostasis related?

### Collaborative Questions

1. Define the terms negative feedback loop and positive feedback loop, and describe how they differ.
2. Discuss the organization of animal bodies from the cellular to the organ system level.

## Online Resource

**connect.mheducation.com**

SMARTBOOK® SmartBook® is the first and only adaptive reading experience designed to change the way students read and learn.

33

# Neuroscience I: The Structure, Function, and Evolution of Nervous Systems

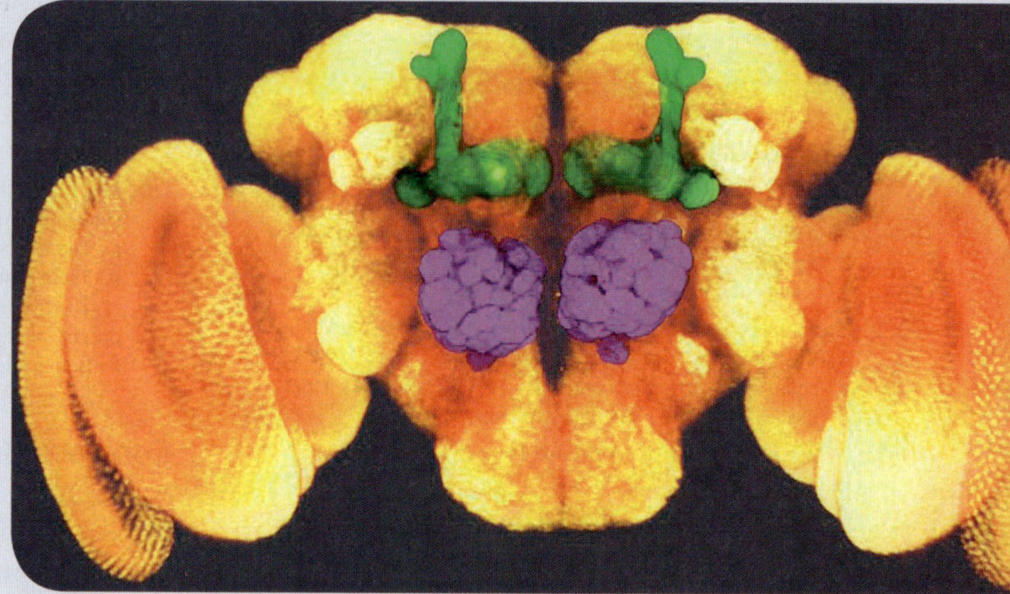

Courtesy Ann-Shyn Chiang, Tsing Hua Chair Professor/Brain Research Center & Institute of Biotechnology/National Tsing Hua University

**Three-dimensional image of the brain of a fruit fly.** The brains of animals are organized into anatomical structures with specialized functions. For example, the green structures in this image are important for learning and memory, the purple structures are important for olfaction, and the large, orange structures on either side process visual information.

## Chapter Outline



After learning that she was going to lose her job, a 42-year-old woman began losing interest in her daily activities. A mother of two, she found it increasingly difficult to find the energy or motivation to care for them, whereas in the past she had been an attentive, involved, and nurturing parent. Her relationship with her husband deteriorated as she found herself spending more and more time isolated and unwilling even to engage in ordinary conversation. Eventually, simply getting out of bed in the morning became overwhelming. As her condition worsened, she finally agreed to see a physician, who diagnosed her with major depressive disorder, commonly called depression. This disorder affects how brain cells signal each other through the release of chemical messengers called **neurotransmitters.** In depression, certain neurotransmitters that are important for maintaining a positive mood may be deficient. She was treated, therefore, with a drug that increases the brain concentration of one of these neurotransmitters. After several months, she began to feel "like her old self" again and eventually the medication was withdrawn. Her case illustrates just how crucial it is that a proper balance of signaling molecules be present in the human brain.

The **nervous system** is composed of coordinated groups of cells that sense internal and environmental changes and transmit signals that enable animals to respond in an appropriate way. Nervous systems provide animals with advantages that promote survival. For example, they help animals avoid predation and other environmental dangers; form social bonds that enhance the chances of survival for the individual and the group; and even perform the complex tasks of thinking, learning, remembering, and, as illustrated by the woman with depression, maintaining a stable mood.

Nervous systems are composed of circuits of **neurons**—highly specialized cells that communicate with each other and with other types of cells by electrical or chemical signals. In many animals, neurons become organized into a central-processing area of the nervous system called a **brain** (see the chapter-opening photo). The brain sends commands to parts of the body through **nerves**—bundles of neuronal cell extensions encased in connective tissue that project to and from various tissues and organs.

We begin by investigating the features of neurons that make them suited for rapid communication. We then examine neurotransmitters, how nervous systems are organized, how they evolved, and how the human brain is affected by disease.

## 33.1 Cellular Components of Nervous Systems

### Learning Outcomes

1. Explain the difference between the central nervous system (CNS) and the peripheral nervous system (PNS).
2. List the cellular components of the nervous system, and describe the function of each cell type.
3. **SCISKILLS ▶** Identify the different parts of a typical neuron, and explain how the structure of each relates to its function.
4. Describe the functional relationships among sensory neurons, motor neurons, and interneurons.
5. **SCISKILLS ▶** Explain what a reflex is, and propose reasons for why reflexes are important in animals.

The organization of the nervous system permits extremely rapid responses to changes in an animal's external or internal environment. In many animals, the **central nervous system (CNS)** consists of a brain and a nerve cord, which in vertebrates extends from the brain through the vertebral column and is called the **spinal cord.** The **peripheral nervous system (PNS)** consists of all neurons and projections of their plasma membranes that are outside of but connect with the CNS, such as projections that end on muscle and gland cells. In certain invertebrates with simpler nervous systems, the distinction between the CNS and PNS is less clear or not present.

In this section, we will survey the general properties of the cells that are common to all nervous systems.

### Cells of the Nervous System Are Specialized to Receive and Send Signals

The major signaling cells of nervous systems are neurons. All animals except sponges have neurons, which comprise the following parts (Figure 33.1a,b):

- A **cell body** contains the cell nucleus and other organelles.
- **Dendrites** (from the Greek word *dendron*, meaning tree) are projections of the cell body with numerous, branching extensions that provide a large surface area for contacts with other neurons. Electrical and chemical messages from other neurons are received by the dendrites and from there move toward the cell body (Figure 33.1c). Some messages also arrive directly at the cell body.
- An **axon** is an extension of the cell body that transmits signals along its length and eventually to neighboring cells. A typical neuron has a single axon.
- An **axon hillock** is the part of the axon closest to the cell body and is important in the initial generation of the electrical signals that travel along an axon.
- **Axon terminals** convey electrical or chemical messages to other cells, such as other neurons or muscle cells.

Within an animal's body, axons from many neurons are arranged in parallel bundles to form nerves, which are covered by a protective layer of connective tissue (Figure 33.1d). Nerves enter and leave the CNS and transmit signals in both directions between the PNS and the CNS.

Nervous systems also contain **glia,** which are non-neuronal cells that surround, support, and protect the neurons. In vertebrates, specialized glial cells wrap around certain axons of both the CNS and the PNS at regular intervals to form an insulating layer called a **myelin sheath** (see Figure 33.1). The sheath is periodically interrupted by noninsulated gaps called nodes of Ranvier. As we will see later, myelin and the nodes of Ranvier increase the speed with which electrical signals pass along the axon.

### Neurons Are Specialized for Different Functions

Neurons can be categorized into three main types: sensory neurons, motor neurons, and interneurons. The structures of each type reflect their specialized functions.

**Sensory Neurons** As their name suggests, **sensory neurons** detect or sense information from the outside world, such as light, odors, touch, or heat. In addition, they detect internal body conditions such as blood pressure or body temperature. Sensory neurons are also called afferent (from the Latin, meaning to bring toward) neurons because they transmit information from the periphery to the CNS.

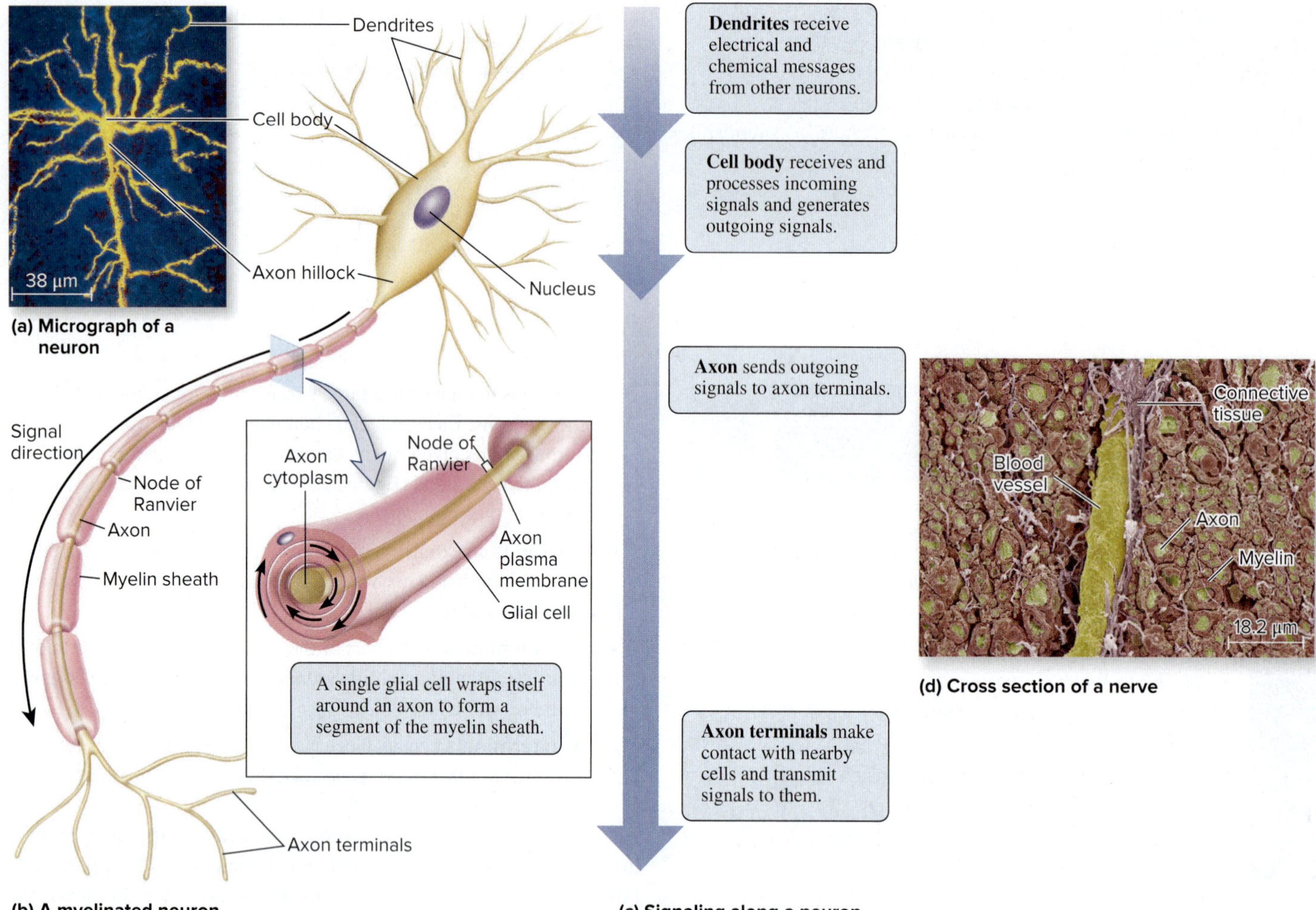


**Figure 33.1** **Structure and basic function of a typical vertebrate neuron and associated glial cells.** **(a)** A stained neuron seen at high magnification (confocal fluorescence microscopy). **(b)** A diagrammatic representation of a peripheral neuron with associated glial cells. The glial cells wrap their membranes around the axon at regular intervals, creating a myelin sheath that is interrupted by bare patches called nodes of Ranvier. **(c)** The structures involved in signaling by a neuron. Signals flow in the direction shown. **(d)** Cross section of a nerve, as seen in a false-color scanning electron micrograph. Axons are in green, surrounded by myelin sheaths (red). Connective tissue fibers (purple) and a portion of a blood vessel (yellow) can also be seen.



Many sensory neurons have a long axon that branches into two main processes, with the cell body in between (Figure 33.2a).

**Motor Neurons** **Motor neurons** transmit signals away from the CNS and elicit some type of response that depends on the cells receiving the signals (Figure 33.2b). They are so named because one type of response they cause is movement. In addition, motor neurons may cause other effects such as the secretion of hormones from endocrine glands. Because they transmit signals away from the CNS, motor neurons are also called efferent (from the Latin, meaning to carry from) neurons. Like sensory neurons, motor neurons tend to have long axons but do not branch into two main processes.

**Interneurons** A third type of neuron, called the **interneuron,** forms connections between other neurons in the CNS. The signals sent between interneurons are critical in the interpretation of information that the CNS receives, as well as the response that it may elicit. Interneurons tend to have many dendrites, and their axons are typically short and highly branched (Figure 33.2c). This arrangement allows interneurons to form complex connections with many other cells.

**Reflex Arcs** As a way to understand the interplay between sensory neurons, interneurons, and motor neurons, let's consider a simple example in which these types of neurons form interconnections with each other. Neurons transmit signals to each other through a series

## Biology Principle

### Cells Are the Simplest Units of Life

Cells are the simplest units of an organ system, including the nervous system. However, the variety and complexity of cells within an organ system can be great, as shown here.

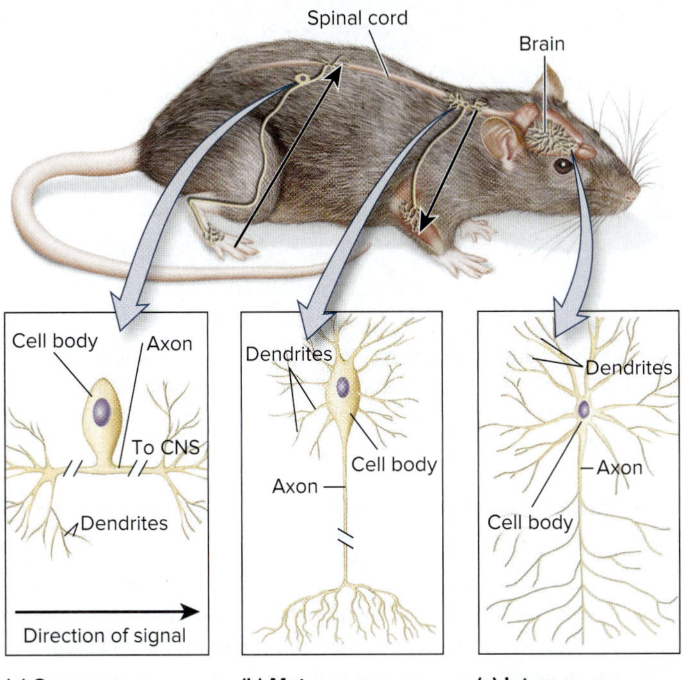


**Figure 33.2** Types of neurons. **(a)** Vertebrate sensory neurons have an axon that bypasses the cell body and projects to the CNS. **(b)** Motor neurons transmit signals away from the CNS and usually have long axons that enable them to act on distant cells. In parts **a** and **b,** the hatch marks indicate that the axons are not to scale, but are actually much longer than shown. **(c)** Interneurons are usually short neurons that connect two or more other neurons within the CNS. Although short, the axons and dendrites may have extensive branches, allowing them to receive many inputs and transmit signals to many neurons.

of connections that form a circuit. An example of a simple circuit is a **reflex arc,** which allows an organism to respond rapidly to inputs from sensory neurons and consists of only a few neurons (Figure 33.3). For example, the stimulus from sensory neurons that sense a tap to the patellar tendon is sent to the CNS, but there is little or no interpretation of the signal. Typically, very few, if any, interneurons are involved, as in the example shown in Figure 33.3. A signal is then transmitted to motor neurons, the axons of which leave the CNS and elicit a response, such as a knee jerk. Such a response is very quick and automatic.

Reflexes are among the evolutionarily oldest and most important features of nervous systems, because they allow animals to respond quickly to potentially dangerous events. For instance, many vertebrates will immediately cringe, jump, leap, or take flight in response to a loud noise, which could represent sudden danger. Other reflexes, such as the patellar tendon reflex just described, are important for postural changes and locomotion. There are a great many examples of reflexes and their importance is evident from the observation that they exist in abundance in nearly all animals.

## 33.1 Reviewing the Concepts

- The central nervous system (CNS) is composed of a brain and a nerve cord. The peripheral nervous system (PNS) consists of all neurons and their projections that are outside of and connect with the CNS. Nerves transmit signals between the PNS and CNS.
- In neurons, signals flow from dendrites to the cell body and then to the axon and axon terminals. Types of neurons include sensory neurons, motor neurons, and interneurons. A neuron's structure is a reflection of its function (Figures 33.1, 33.2). Glia are non-neuronal cells that surround neurons and perform numerous supportive functions.
- The most basic neuronal circuit is a reflex arc, which occurs rapidly in response to inputs from sensory neurons and consists of one or a few sensory and motor neurons (Figure 33.3).

## 33.1 Testing Your Knowledge

1. Which of the following is *false?*
   a. The CNS of vertebrates comprises the brain and spinal cord.
   b. Certain glial cells produce a myelin sheath that wraps around axons.
   c. A typical reflex arc consists of only a small number of both sensory and motor cells.
   d. Motor neurons are afferent neurons that convey information into the CNS.
   e. Interneurons are found in the CNS and are specialized for communication with many other cells.
2. Signals arrive at a neuron at the
   a. axon.
   b. axon terminal.
   c. dendrites and cell body.
   d. axon hillock.
   e. myelin sheath.

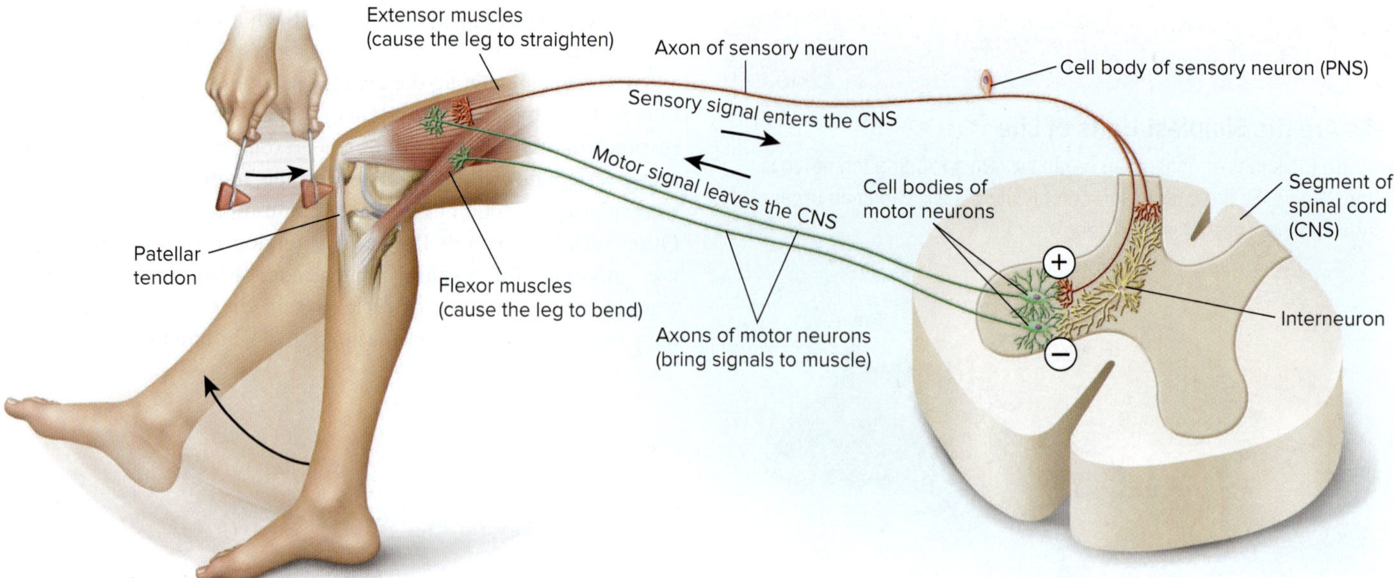


**Figure 33.3** **A reflex arc.** The knee-jerk response is an example of a reflex arc. A tap below the kneecap (also known as the patella) stretches the patellar tendon, which acts as a stimulus for a sensory neuron. This initiates a reflex arc that activates (+) a motor neuron that causes the extensor muscle on top of the thigh to contract, which extends the leg. At the same time, a different interneuron inhibits (−) the motor neuron of the flexor muscle, causing it to relax so that it does not bend the leg.

**Concept Check:** *Animals have many types of reflexes. Once initiated, must all reflexes occur to completion, or do you think that in some cases they may be overridden or partially suppressed?*

## 33.2 Electrical Properties of Neurons and the Resting Membrane Potential

### Learning Outcomes

1. Explain what a membrane potential is.
2. Outline how the resting membrane potential is established and maintained.
3. Describe what is meant by an electrochemical gradient.
4. **SCISKILLS ▶** State the Nernst equation, and use it to predict the direction in which different ions move across a plasma membrane.

Neurons use electrical signals called action potentials (described in Section 33.3) to communicate with other neurons, muscle cells, or gland cells. These signals involve changes in the amount of electric charge across a neuron's plasma membrane. In this section, we first examine the electrical and chemical (concentration) gradients across the plasma membrane of neurons. In subsequent sections, we explore how these gradients provide a way for neurons to generate and conduct action potentials.

### Neurons Establish Differences in Ion Concentration and Electric Charge Across Their Membranes

Like all cell membranes, the plasma membrane of a neuron acts as a hydrophobic barrier that separates charges. The charges are positively and negatively charged ions. Ion concentrations differ between the interior and exterior of the cell. Such differences create a charge imbalance across the plasma membrane; this produces an electrical potential measurable in units called millivolts (mV). Analogous to a battery, neurons have negative and positive poles, in this case the inside and outside surfaces, respectively, of the plasma membrane. For this reason, a neuron is said to be electrically polarized. The difference between the electric charges along the inside and outside surfaces of a cell membrane is called a membrane potential. The **resting membrane potential** refers to the membrane potential of a cell that is not producing action potentials; the cell is considered "at rest" and is not at that moment sending signals to other cells.

Let's examine how the resting membrane potential is established and maintained. When investigators first measured the resting membrane potential of a typical neuron, they registered a voltage of about −70 millivolts (mV) inside the cell with respect to the outside. This means that the interior of the resting cell had a more negative charge than the exterior (**Figure 33.4a**). Negative ions within the cell and positive ions outside the cell attract each other and line up along the membrane. Although there are more positive charges along the outside surface of a neuron and more negative charges inside, the actual number of ions that contribute to the resting membrane potential is extremely small compared with the total number of ions inside and outside the cell. The ions that are most critical for establishing the resting potential are $Na^+$ and $K^+$.

The following are major factors responsible for the resting membrane potential (**Figure 33.4b**).

1. **Establishment of ion concentration gradients:** The $Na^+$/$K^+$-ATPase pump within the plasma membrane continually transports $Na^+$ out of the cell and $K^+$ into the cell (refer back to

# Biology Principle

## Living Organisms Use Energy

The $Na^+/K^+$-ATPase pump uses the energy stored in ATP to pump ions across the neuronal plasma membrane. Consequently, establishing and maintaining resting membrane potentials accounts for a significant fraction of the energy expended per day in an animal.

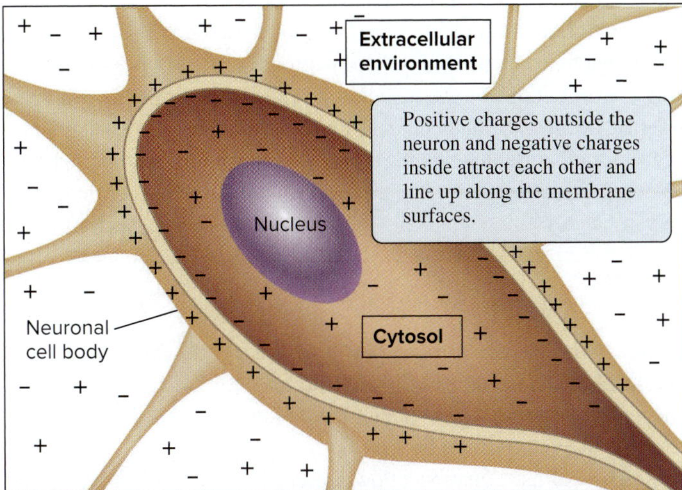


**(a) Distribution of charges across the neuronal plasma membrane**

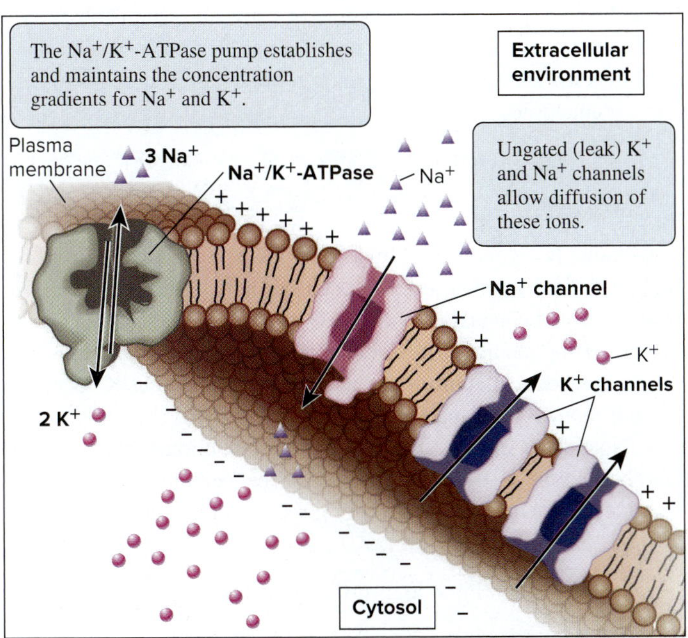


**(b) Two major factors that influence the resting membrane potential**

**Figure 33.4 The resting membrane potential.** The slight excess of negative charges inside and positive charges outside the cell membrane is shown in part **(a)**. In part **(b)**, the major factors that contribute to this charge distribution are shown: the $Na^+/K^+$-ATPase pump that establishes ion concentration gradients, and leak ion channels that permit diffusion of $Na^+$ but especially $K^+$ across the membrane.

Figure 5.17). The $Na^+/K^+$-ATPase pump transports three $Na^+$ out of the cell for every two $K^+$ it moves into the cell. Consequently, the pump contributes modestly to a charge difference across the plasma membrane. More importantly, however, it establishes concentration gradients for $Na^+$ and $K^+$ by actively transporting them across the membrane in opposite directions. These gradients will contribute to movement of these ions across the membrane, which in turn will create the membrane potential.

2. **Membrane permeability to different ions:** The plasma membrane contains ion channels that affect the permeability of $Na^+$ and $K^+$ through the membrane. These channels are relatively specific for $Na^+$ or $K^+$, and they influence the resting membrane potential by allowing the diffusion of these ions across the plasma membrane. They are referred to as ungated, or leak, channels that are open in a resting, unstimulated cell. In most neurons, there are anywhere from 10 to 100 times more leak $K^+$ channels than leak $Na^+$ channels, so at rest, the membrane is more permeable to $K^+$ than to $Na^+$. As you will learn shortly, when the cell is at rest, $Na^+$ tends to diffuse into the cell and $K^+$ diffuses out of the cell through these channels. The greater number of leak channels for $K^+$ means that diffusion of $K^+$ will make a more significant contribution to establishing a membrane potential than will diffusion of $Na^+$. The movement of positively charged $K^+$ out of the cell will make the interior of the cell negatively charged with respect to the outside. This is the resting membrane potential. Let's now see how this happens.

## An Electrochemical Gradient Affects the Diffusion of Ions Across a Membrane

Both the membrane potential and the concentration of ions influence the direction of ion diffusion across a membrane. The direction that an ion moves depends on the **electrochemical gradient** for that ion, which is the combined effect of both an electrical and concentration (chemical) gradient (refer back to Figure 5.8). **Figure 33.5** considers the concept of an electrochemical gradient for $K^+$. This hypothetical situation illustrates two compartments containing solutions separated by a membrane that permits only the diffusion of $K^+$. Figure 33.5a illustrates an electrical gradient. In this case, the concentration of $K^+$ is equal across the membrane, but the concentrations of other ions such as $Na^+$ and $Cl^-$ are unequal on opposite sides of the membrane and thereby produce an electrical gradient. Because $K^+$ is positively charged, it is attracted to the side of the membrane with more negative charge. Figure 33.5b shows a concentration gradient in which $K^+$ concentration is higher on one side than the other, but there is no electrical gradient. In this scenario, $K^+$ diffuses from a region of high to one of low concentration. Finally, Figure 33.5c shows an electrochemical gradient. The electrical gradient would favor the movement of $K^+$ from left to right, but the concentration gradient would favor movement from right to left. If perfectly balanced, these opposing forces can create an electrochemical equilibrium in which there is no net diffusion of $K^+$ in either direction. The membrane potential at which this occurs for a particular ion at a given concentration gradient is referred to as that ion's **equilibrium potential.** The equilibrium potential for a given ion can be calculated under a variety of starting conditions, as described next.

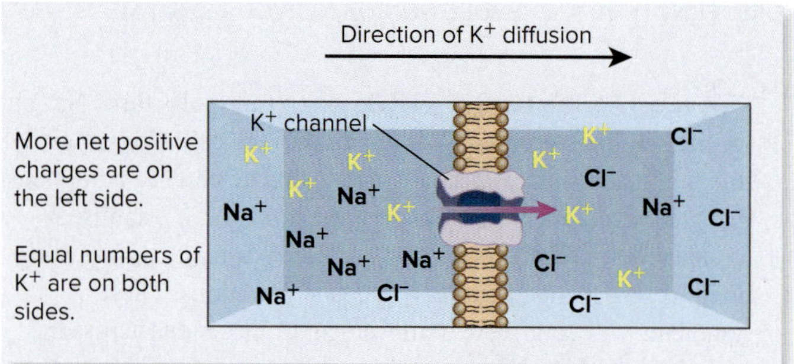


**(a) Electrical gradient, no concentration gradient for $K^+$**

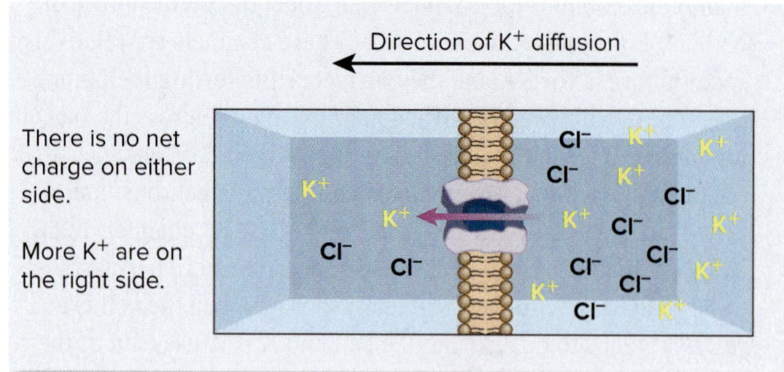


**(b) Concentration gradient for $K^+$, no electrical gradient**

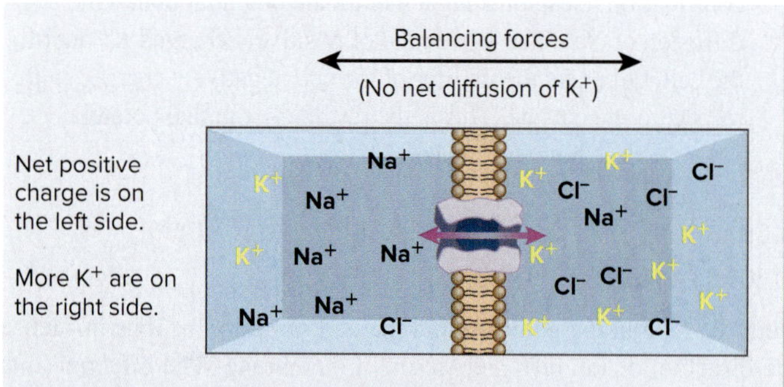


**(c) Electrochemical gradient, $K^+$ equilibrium**

**Figure 33.5** **Electrical and concentration gradients.** This hypothetical example depicts two chambers separated by a membrane that is permeable only to $K^+$. **(a)** In this example, the compartments initially contain equal concentrations of $K^+$, but an electrical gradient exists due to an unequal distribution of $Na^+$ and $Cl^-$. Potassium ions are attracted to the greater amount of negative charge on the right side of the membrane. **(b)** In this case, there is initially no electrical gradient across the membrane, and the left compartment contains a lower concentration of $K^+$ than in the right compartment. Under these conditions, $K^+$ diffuses down its concentration gradient from right to left. **(c)** This example illustrates opposing electrical and chemical (concentration) gradients. The right compartment contains a higher concentration of $K^+$, and the left side has a higher net amount of positive charge. These gradients balance each other, so no net movement of $K^+$ occurs and a state of electrochemical equilibrium is reached.

**Concept Check:** *In part (b), does the diffusion of $K^+$ down its concentration gradient result in an electrical gradient? What will eventually stop the net diffusion of $K^+$?*

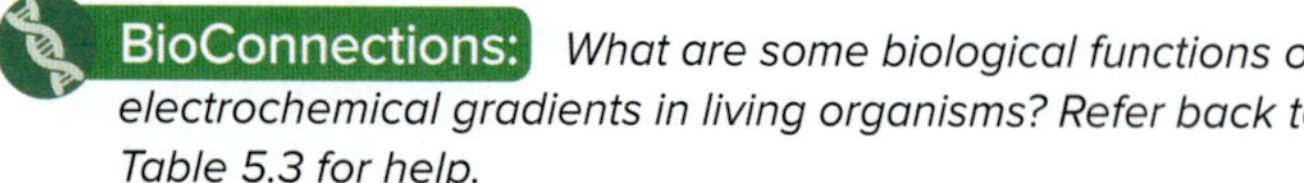

**BioConnections:** *What are some biological functions of electrochemical gradients in living organisms? Refer back to Table 5.3 for help.*

# Quantitative Analysis

## AN ION'S EQUILIBRIUM POTENTIAL DEPENDS ON ITS CONCENTRATION GRADIENT

Given the charge difference across a resting neuron's membrane, is it possible to predict the direction that an ion will diffuse across the membrane at any given concentration gradient for that ion? In other words, can we compare the relative strengths of the electrical and concentration gradients and predict their net effect? By measuring the membrane potential of isolated neurons in the presence of changing concentrations of extracellular ions, German researcher Walther Nernst deduced a mathematical formula that relates electrical and concentration gradients to each other. This formula, now called the **Nernst equation,** gives the equilibrium potential for an ion at any given concentration gradient. For monovalent cations such as $Na^+$ and $K^+$ at 37°C, the Nernst equation can be expressed as

$$E = 60 \text{ mV} \log_{10} ([X_{\text{extracellular}}]/[X_{\text{intracellular}}])$$

where $E$ is the equilibrium potential; $[X]$ is the extracellular or intracellular concentration of an ion; and 60 mV is a value that depends on temperature, valence of the ion, and other factors. (For anions, the value would be −60 mV.)

The Nernst equation allows researchers to predict when an ion is or is not at electrochemical equilibrium. To understand the usefulness of this equation, consider two examples. First, let's suppose the resting membrane potential of a given neuron is observed to be −88.6 mV and the $K^+$ concentration is 5 mM outside and 150 mM inside the cell. If we enter these concentrations into the Nernst equation,

$$E = 60 \text{ mV} \log_{10} (5/150) = 60 \text{ mV} (-1.48)$$
$$= -88.6 \text{ mV}$$

This value is a good approximation of the true equilibrium potential for $K^+$ in many neurons. Under these conditions, the $K^+$ equilibrium potential calculated by the Nernst equation exactly equals the actual resting membrane potential in our example. Therefore, $K^+$ is in electrochemical equilibrium, and no net diffusion of $K^+$ occurs, even if many $K^+$ channels are open.

Now let's suppose that the resting membrane potential of another neuron is −70 mV, and the $Na^+$ concentration is 100 mM outside and 10 mM inside the cell. If we enter these concentrations into the Nernst equation, we get

$$E = 60 \text{ mV} \log_{10} (100/10) = 60 \text{ mV} (1)$$
$$= 60 \text{ mV}$$

This turns out to be a typical value for the equilibrium potential for $Na^+$ in many neurons. At a resting membrane potential of −70 mV, as in this example, the value of +60 mV tells us that $Na^+$ is not in electrochemical equilibrium. When the equilibrium potential for a given ion—calculated by the Nernst equation—and the resting membrane potential do not match, there is a driving force for that ion to diffuse across the membrane. In this example, $Na^+$ will diffuse into the cell, because both the electrical and concentration gradients favor an inward flow. However, if the actual membrane potential was +60 mV instead of −70 mV, $Na^+$ would be in equilibrium at those concentrations, and therefore there would be no net diffusion of $Na^+$ in either direction.

Typically, the resting membrane potential of neurons is not at the equilibrium potentials for either $K^+$ or $Na^+$ but is somewhere in between. Because neuronal membranes generally have so many more leak $K^+$ channels than $Na^+$ ones, the actual resting membrane potential is typically much closer to the equilibrium potential for $K^+$ than to that for $Na^+$. Therefore, these ions will move across the plasma membrane if there are open ion channels. The diffusion of a charged ion across the plasma membrane down its electrochemical gradient results in an electric current. This small current provides the necessary electrical signal that neurons use to communicate with one another, as we will see in the next section.

**Crunching the Numbers:** A certain neuron has an intracellular concentration of $K^+$ of 145 mM and an extracellular concentration of 6.5 mM. The resting membrane potential is −70 mV. What is the equilibrium potential for $K^+$ under these conditions? Is $K^+$ at its equilibrium potential in this neuron? If not, in which way will $K^+$ diffuse if the membrane has open $K^+$ channels?

## 33.2 Reviewing the Concepts

- The resting membrane potential of a neuron is determined by the distribution and relative permeabilities of ions across the plasma membrane (Figure 33.4).
- Diffusion of ions through ion channels occurs as a result of an electrochemical gradient across the membrane (Figure 33.5).
- The Nernst equation gives the equilibrium potential for an ion at any given concentration gradient. Differences between this value and the resting membrane potential determine whether and in which direction an ion will diffuse across a membrane.

## 33.2 Testing Your Knowledge

1. A neuron has a resting membrane potential of −80 mV. The equilibrium potential for $Na^+$ in that cell is calculated to be +58 mV. In which direction will $Na^+$ diffuse if the membrane is permeable to it?
   a. It will move into the cell.
   b. It will move out of the cell.
   c. It will not move in either net direction.
2. The resting membrane potential occurs because
   a. the membrane is leaky to $Na^+$ and $K^+$.
   b. the concentrations of $Na^+$ and $K^+$ are unequal across the membrane.
   c. the $Na^+/K^+$-ATPase pump moves two $K^+$ into the cell for every three $Na^+$ it moves out of the cell.
   d. a and b are correct.
   e. a, b, and c are all correct.

## 33.3 Generation and Transmission of Electrical Signals Along Neurons

### Learning Outcomes

1. Distinguish between ligand-gated and voltage-gated ion channels.
2. Compare and contrast graded potentials and action potentials.
3. **SCISKILLS ▶** Describe the steps involved in the generation of an action potential, beginning with the spread of excitation to the axon hillock.
4. Discuss how myelination influences the propagation of an action potential.

Communication between neurons begins when one cell receives a stimulus and in turn sends an electrical signal along its axon via currents generated by ion movements. This signal then influences the next neuron in a pathway. In this section, we will survey the amazing ability of neurons to generate the electrical changes that produce action potentials, and how those potentials are transmitted along an axon.

### Signaling by a Neuron Occurs Through Changes in the Membrane Potential

Recall that a neuron is polarized because of the separation of charges across its plasma membrane. Changes in the membrane potential, therefore, are changes in the degree of polarization. **Depolarization** occurs when the cell membrane becomes less polarized—that is, when it becomes less negative inside the cell relative to the outside. When a neuron is stimulated by an electrical or chemical stimulus, one or more types of ion channels open; this may, for example, result in the diffusion of $Na^+$ into the cell down its electrochemical gradient, thereby bringing positive charge into the cell. This makes the new membrane potential somewhat less negative than the original resting membrane potential. In contrast, **hyperpolarization** occurs when the cell membrane becomes more polarized than at rest. For example, an increase in the diffusion of $K^+$ out of the cell would make the charge along the inside of the cell membrane more negative than it normally is when at the resting membrane potential.

Whereas all cells in an animal's body have a membrane potential, neurons are called excitable cells because they can change their resting membrane potential and generate electrical signals. This is accomplished by another class of ion channels called gated ion channels, because they open and close in a manner analogous to a gate in a fence (Figure 33.6). **Voltage-gated ion channels** open and close in response to changes in voltage across the membrane. **Ligand-gated ion channels** open or close when ligands, such as neurotransmitters, bind to them. The opening and closing of voltage-gated and ligand-gated ion channels are responsible for two types of changes in a neuron's membrane potential: graded potentials and action potentials.

### Graded Potentials Vary in Size Depending on the Strength of a Stimulus

A **graded potential** is any depolarization or hyperpolarization that varies with the strength of the stimulus. A large change in membrane potential occurs when a strong stimulus opens many ion channels,

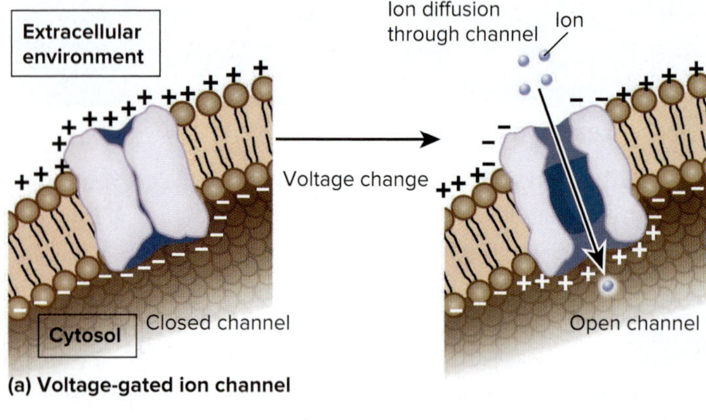

(a) Voltage-gated ion channel

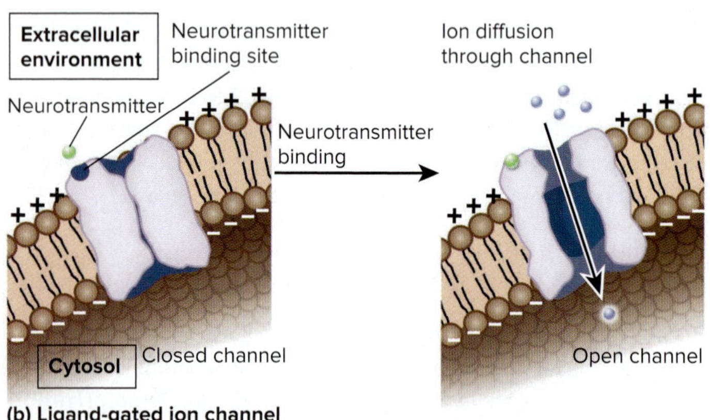

(b) Ligand-gated ion channel

**Figure 33.6** **Examples of gated ion channels.** **(a)** A voltage-gated ion channel allows ions to diffuse into the cell. These channels open or close depending on changes in charge (voltage) across the membrane. **(b)** A ligand-gated ion channel opens or closes in response to ligand binding. In the example here, the binding of a neurotransmitter opens the channel.

**BioConnections:** *Are ions the only substances that can move through channels in plasma membranes? Refer back to Figure 5.13 for an important molecule that diffuses through membranes via channels.*

whereas a weak stimulus causes a small change because only a small number of ion channels are opened (Figure 33.7).

Graded potentials occur locally where an electrical or chemical stimulus opens gated ion channels on a region of the plasma membrane of dendrites or the cell body. From this initial site of stimulation, a graded potential spreads a small distance across nearby regions of the plasma membrane. When the stimulus stops, the gated ion channels close, and ion pumps restore the ion concentration gradients. This allows the resting membrane potential to return to normal as $Na^+$ and $K^+$ diffuse through their leak channels.

Graded potentials occur on all neurons and are particularly important for the function of sensory neurons, which must distinguish between strong and weak stimuli reaching the organism from the environment. In addition, however, graded potentials can act as triggers for the second type of electrical signal, the action potential.

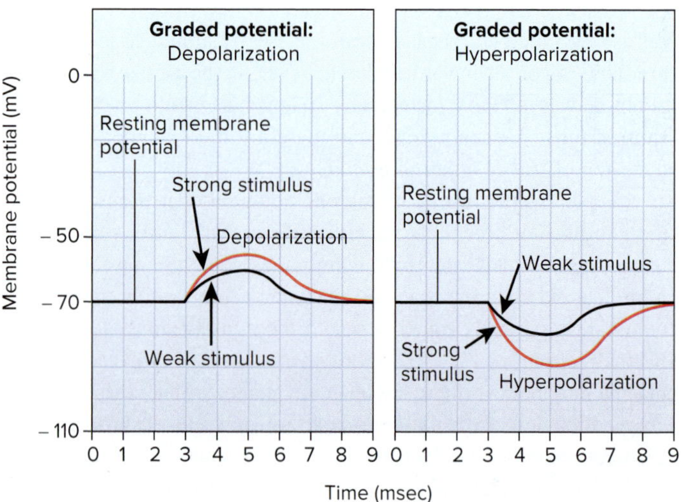


**Figure 33.7** **Graded membrane potentials.** Within a limited range, the change in the membrane potential occurs in proportion to the intensity of the stimulus. If the membrane potential becomes less negative following a stimulus, the resulting change is called a depolarization. If large enough, a depolarization can elicit an action potential. If the membrane potential becomes more negative, it is called hyperpolarization. A cell that is hyperpolarized is less likely to elicit an action potential.

## Action Potentials Transmit Electrical Signals Along the Length of Axons

An **action potential** is the electrical event that transmits a signal along an axon. In contrast to a graded potential, an action potential is always a large depolarization. Once an action potential has been triggered, it occurs in an all-or-none fashion. In other words, under normal circumstances it cannot be graded; a cell cannot have a partial action potential. Unlike a graded potential, an action potential is conducted, or propagated, along the entire axon, regenerating itself as it travels. Action potentials travel rapidly down the axon to the axon terminals, where they initiate a response at a junction with another cell.

Figure 33.8 shows the electrical changes that happen in a localized region of an axon when an action potential is occurring. Voltage-gated $Na^+$ and $K^+$ channels are both present in very high numbers from the axon hillock to the axon terminals. An action potential begins when a cell receives a strong stimulus that results in a graded potential that is large enough to spread from the cell body to the axon hillock and depolarize the membrane there. This opens some voltage-gated $Na^+$ channels, depolarizing the membrane further to a value called the threshold potential (step 2 in Figure 33.8). The **threshold potential** is the membrane potential, typically around −55 to −50 mV, at which sufficient voltage-gated $Na^+$ channels are open that the all-or-none phase of the action potential begins.

The opening of voltage-gated $Na^+$ channels involves a change in the conformation of the membrane-spanning region of the channel (see Figure 33.8). When the channel opens, $Na^+$ rapidly diffuses into the cell down its electrochemical gradient (recall that the equilibrium potential for $Na^+$ is about +60 mV). This influx of charges further depolarizes the cell, causing even more voltage-gated $Na^+$ channels to open, resulting in the spike in membrane potential that characterizes the action potential. This positive feedback process is so rapid that the membrane potential reaches its peak positive value in less than 1 millisecond!

As the membrane potential becomes more and more depolarized, a second, delayed conformational change in the voltage-gated $Na^+$ channel blocks the continued flow of $Na^+$ into the cell. This conformational change involves the **inactivation gate** (step 3 in Figure 33.8), a string of amino acids that juts out from the channel protein into the cytosol. The inactivation gate swings into the opening of the channel, thereby preventing any further movement of $Na^+$ through the channel into the cell. Thus, the inactivation gate terminates the depolarization phase of an action potential; this gate does not swing out of the channel until the membrane potential returns to a value near the resting potential.

Voltage-gated $K^+$ channels are also opened by the change in voltage to the threshold potential, but they open more slowly, about the time the $Na^+$ channels become inactivated. When $K^+$ channels open, some $K^+$ diffuses out of the cell down its electrochemical gradient, and the membrane potential becomes repolarized. The membrane has a brief period of hyperpolarization while the $K^+$ channels remain open and the membrane potential approaches the equilibrium potential for $K^+$, which is more negative than the resting potential (step 4 in Figure 33.8). The repolarization then closes the voltage-gated $K^+$ channels. The membrane returns to the resting potential as the $Na^+$/$K^+$-ATPase pump restores the original ion concentration gradients (step 5 in Figure 33.8) and $Na^+$ and $K^+$ diffuse through their leak channels. At this stage, both the $Na^+$ and $K^+$ voltage-gated channels have the ability to reopen if the membrane potential is again depolarized to the threshold potential by a new stimulus.

During the time when the voltage-gated $Na^+$ channels are open, and again while they are inactivated, the membrane at that region is in its **absolute refractory period** (see Figure 33.8 step 3), during which time that portion of membrane is unresponsive to another stimulus. As the cell repolarizes, there is a brief period when the inactivation gate is released and the voltage-gated $Na^+$ channels revert to the closed state, but the voltage-gated $K^+$ channels are still open. During this time, the membrane is in its **relative refractory period** and becomes even more polarized than the resting membrane potential (see Figure 33.8 step 4). A new action potential may be generated at this time, but only in response to an unusually large stimulus. The reason is that the continued $K^+$ efflux through the open voltage-gated $K^+$ channels would oppose some of the $Na^+$ entry that normally

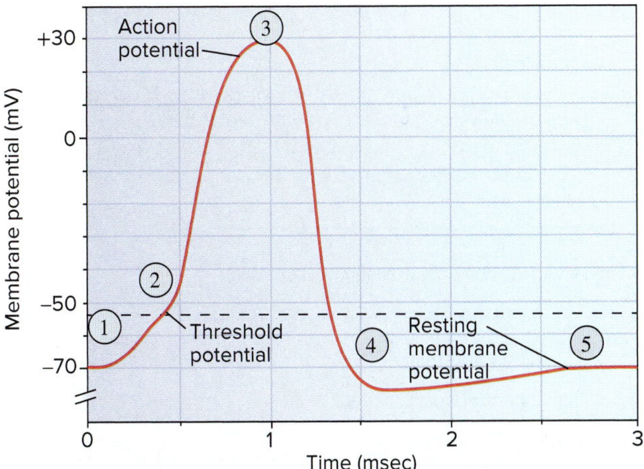


(1) **Resting membrane potential:** The membrane is at the resting potential. Leak channels are open but voltage-gated channels are closed.

Closed — Closed — Cytosol — Voltage-gated $Na^+$ channel — Voltage-gated $K^+$ channel

(2) **Depolarization to threshold:** An action potential is triggered if the cell is sufficiently stimulated such that the threshold potential of about −55 to −50 mV is reached. Voltage-gated $Na^+$ channels open. $Na^+$ diffuses into the cell and depolarizes the membrane.

Open — Closed — $Na^+$

(3) **Peak of action potential:** At about +30 mV, voltage-gated $Na^+$ channels are inactivated, and voltage-gated $K^+$ channels open. $K^+$ exits the cell and repolarizes the membrane. At this time, the membrane is in its absolute refractory period.

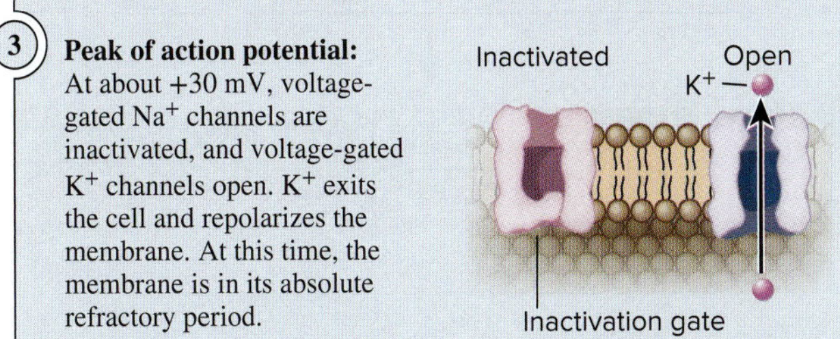


(4) **Repolarization:** Voltage-gated $Na^+$ channels change from inactivated to closed. Voltage-gated $K^+$ channels remain open, causing a hyperpolarization of the membrane. The membrane is now in its relative refractory period.

Closed — Open

(5) **Restoration of resting membrane potential:** Voltage-gated $K^+$ channels close. The resting potential of the membrane is restored by the $Na^+$/$K^+$-ATPase pump and the leak channels.

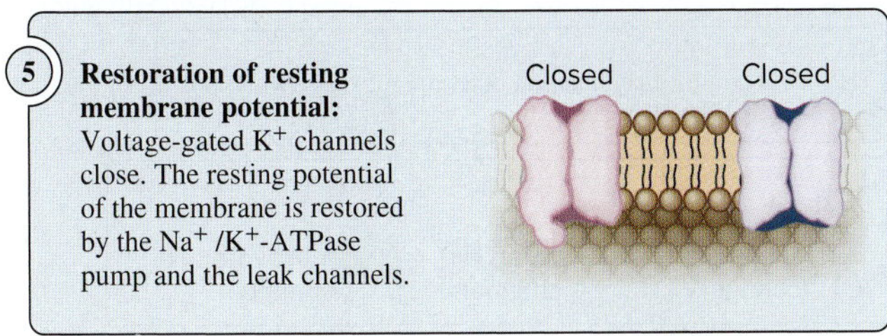


**Figure 33.8** **Changes that occur during an action potential.** The movement of ions across the plasma membrane of an axon first depolarizes and then repolarizes the cell. The values given for membrane potential are representative and not necessarily the same in all neurons. These changes in the membrane potential are caused by the opening and closing of voltage-gated $Na^+$ and $K^+$ channels.

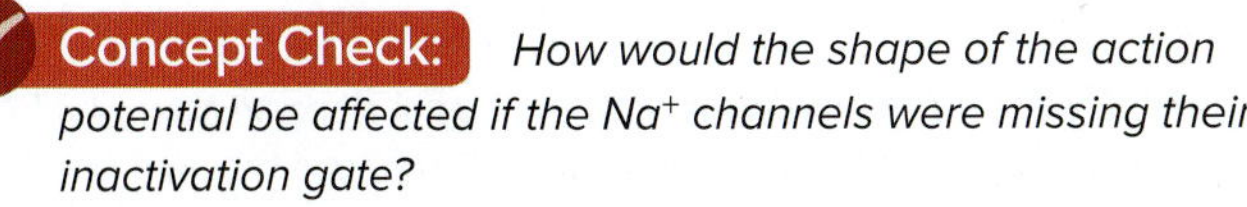

**Concept Check:** *How would the shape of the action potential be affected if the $Na^+$ channels were missing their inactivation gate?*

occurs during the beginning of a new action potential. Therefore, the refractory periods place a limit on how many action potentials can be generated in a given time; they also allow the brain to distinguish signals arriving from weak (few action potentials/sec) and strong (many action potentials/sec) stimuli; only the latter can "break through" the relative refractory period. BIO TIPS ONLINE

## Action Potentials Are Propagated Along the Axon to the Axon Terminals

Let's now consider how action potentials are propagated along an axon from the axon hillock to the axon terminals (**Figure 33.9**). Each action potential in a membrane results in $Na^+$ entering the cell and diffusing along the membrane. The absolute number of ions that cross the membrane in a single action potential is actually extremely small and does not alter the concentration gradient in a measurable way. Nonetheless, the charges carried with the $Na^+$ are sufficient to open nearby voltage-gated $Na^+$ channels in an adjacent strip of membrane along the axon toward the axon terminals. ($Na^+$ that diffuses in the other direction can be ignored, as that portion of the membrane is in its absolute refractory period.) A new action potential is then generated, which again depolarizes the next strip of membrane, and so on. In this way, the sequential opening of $Na^+$ channels along the axon membrane conducts a wave of depolarization from the axon hillock to the axon terminals.

An action potential can be conducted down the axon as fast as 100 m/sec or as slow as 1 cm/sec. The speed is determined by two factors: the axon diameter and the presence or absence of a myelin sheath. The axon diameter influences the rate at which incoming ions can spread along the inner surface of the plasma membrane. The flow of ions meets less resistance in a wide axon than it does in a narrow axon. Therefore, in a wider axon, the action potentials propagate faster. The large axons of squids and lobsters, for example, conduct action potentials very rapidly, allowing the animals to move quickly when threatened.

Myelination also increases the speed at which action potentials travel along an axon. An insulating myelin sheath reduces the leakage of charge across the membrane of an axon (see Figure 33.1). The exposed nodes of Ranvier are the only areas of myelinated axons that have sufficient numbers of voltage-gated $Na^+$ channels to generate an action potential (**Figure 33.10**). When $Na^+$ ions diffuse into the cell at one node, the charge spreads through the cytosol, causing the $Na^+$ channels at the next node to open, where an action potential is regenerated. This type of conduction is called **saltatory conduction** (from the Latin *saltare*, meaning to leap) because the action potential seems to "jump" from one node to the next. In reality, action potentials do not jump from place to place. Saltatory conduction speeds up the conduction process because it takes less time for charge to travel from node to node, regenerating new action potentials only at the nodes, than it would if each tiny strip of membrane generated action potentials all along the length of the axon. Saltatory conduction is an adaptation that permits fast propagation of signals without a

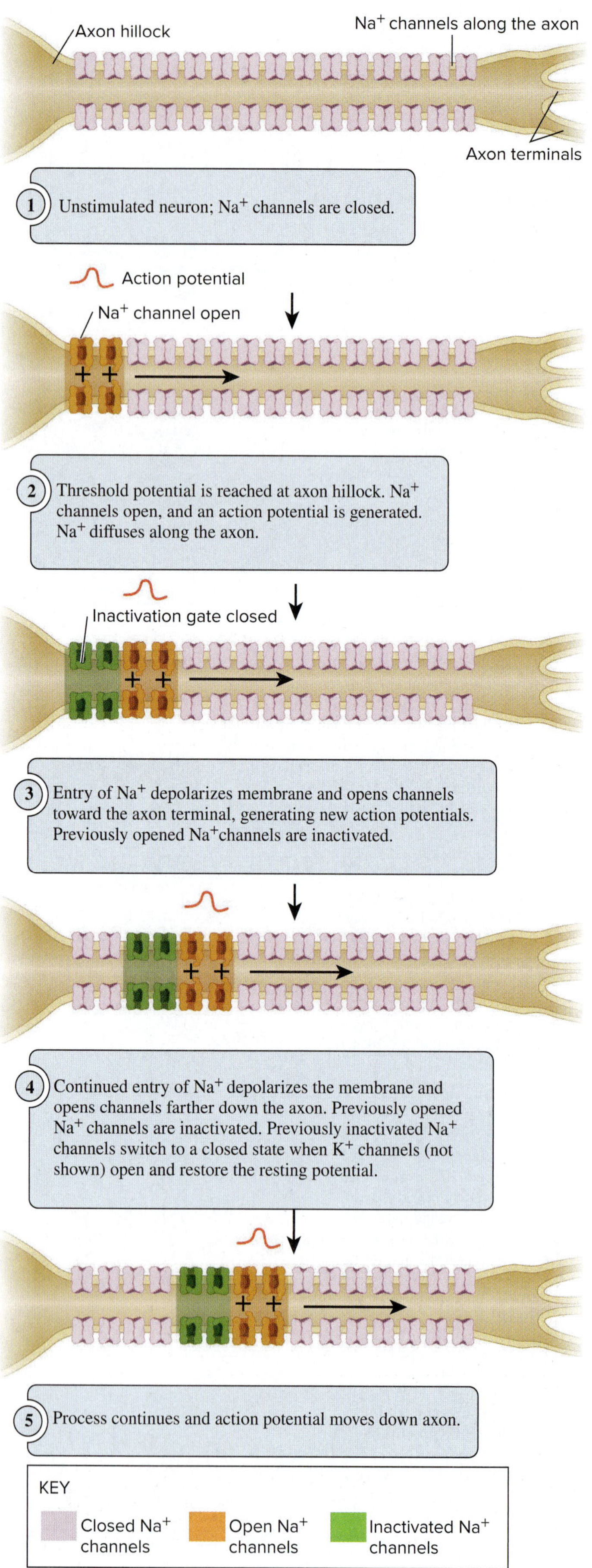

Axon hillock
Na+ channels along the axon
Axon terminals
1 Unstimulated neuron; Na+ channels are closed.
Action potential
Na+ channel open
2 Threshold potential is reached at axon hillock. Na+ channels open, and an action potential is generated. Na+ diffuses along the axon.
Inactivation gate closed
3 Entry of Na+ depolarizes membrane and opens channels toward the axon terminal, generating new action potentials. Previously opened Na+ channels are inactivated.
4 Continued entry of Na+ depolarizes the membrane and opens channels farther down the axon. Previously opened Na+ channels are inactivated. Previously inactivated Na+ channels switch to a closed state when K+ channels (not shown) open and restore the resting potential.
5 Process continues and action potential moves down axon.
KEY
Closed Na+ channels
Open Na+ channels
Inactivated Na+ channels


**Figure 33.9** **Propagation of the action potential along an axon.** All $Na^+$ channels shown refer to voltage-gated channels.

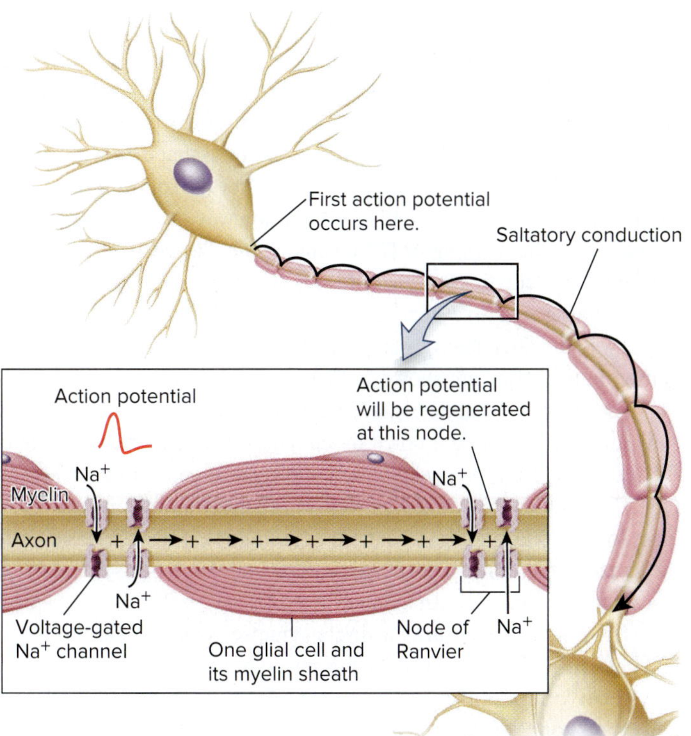


**Figure 33.10** **Saltatory conduction along a myelinated axon.** Action potentials are generated only at the nodes of Ranvier, which lack a surrounding sheath of myelin.

**Concept Check:** *What is the main advantage of saltatory conduction?*

requirement for increasing the diameter of axons. It occurs primarily in long axons in the PNS and CNS; very short neurons such as interneurons are typically not myelinated.

## 33.3 Reviewing the Concepts

- Gated ion channels enable a cell to rapidly change its membrane potential and form graded potentials (Figures 33.6, 33.7).
- Graded potentials that reach the threshold potential will trigger an action potential, an event that carries an electrical signal along an axon. Axon diameter and myelination influence the speed at which an action potential propagates (Figures 33.8, 33.9, 33.10).

## 33.3 Testing Your Knowledge

1. Which is false?
   a. Graded potentials may die out or trigger an action potential.
   b. Action potentials arise at the axon hillock and move toward the axon terminals.
   c. The diameter of an unmyelinated axon is directly related to the speed at which action potentials travel along its length.
   d. Saltatory conduction occurs in all axons.
   e. Action potentials are all-or-none, but graded potentials can have different sizes.
2. Action potentials
   a. require a threshold potential around −55 to −50 mV in order to occur.
   b. have many different shapes and sizes, depending on how a cell was stimulated.
   c. occur only in myelinated neurons.
   d. occur whenever a neuron is sufficiently hyperpolarized.
   e. are limited in their frequency by inactivation gates of $K^+$ channels.

## 33.4 Communication at Synapses

### Learning Outcomes

1. Describe the structural features of a synapse.
2. **SCISKILLS ▶** Predict the effects of excitatory postsynaptic potentials (EPSPs) and inhibitory postsynaptic potentials (IPSPs) on the ability of neurons to generate an action potential.
3. List the classes of neurotransmitters, and provide brief descriptions of their generalized functions.
4. Describe the two general types of postsynaptic membrane receptors.

Neurons communicate with other cells at a **synapse,** which is a junction where an axon terminal meets another neuron or other cell. In this section, we will examine the structure and function of different types of synapses, including the role that neurotransmitters have in neuronal communication.

### Synapses May Be Electrical or Chemical

A synapse includes an axon terminal of the neuron that is sending the signal, the nearby plasma membrane of the receiving cell, and often a synaptic cleft, or extracellular space between the two cells. By definition, the presynaptic cell sends the signal, and the postsynaptic cell receives it. A given cell may be presynaptic to one cell, and postsynaptic to another (**Figure 33.11**).

Two types of synapses occur in animals: electrical and chemical. In the first type, the **electrical synapse,** electric current passes directly from the presynaptic to the postsynaptic cell via gap junctions (refer back to Figure 5.18). The most well-studied examples occur in certain aquatic invertebrates. In those animals, electrical synapses occur where a group of neurons must fire extremely quickly and synchronously, such as when a crayfish must coordinate a number of muscles to dart away from a predator. An advantage of electrical synapses is their great speed of transmission.

The second type of synapse is a **chemical synapse,** in which a neurotransmitter is released from an axon terminal into the synaptic cleft. The neurotransmitter diffuses to the membrane of the postsynaptic cell, where it interacts with receptor proteins and triggers a change in the membrane potential of the postsynaptic cell. As a result of the time required for these events, chemical synapses are slower than electrical ones. However, unlike electrical synapses, chemical

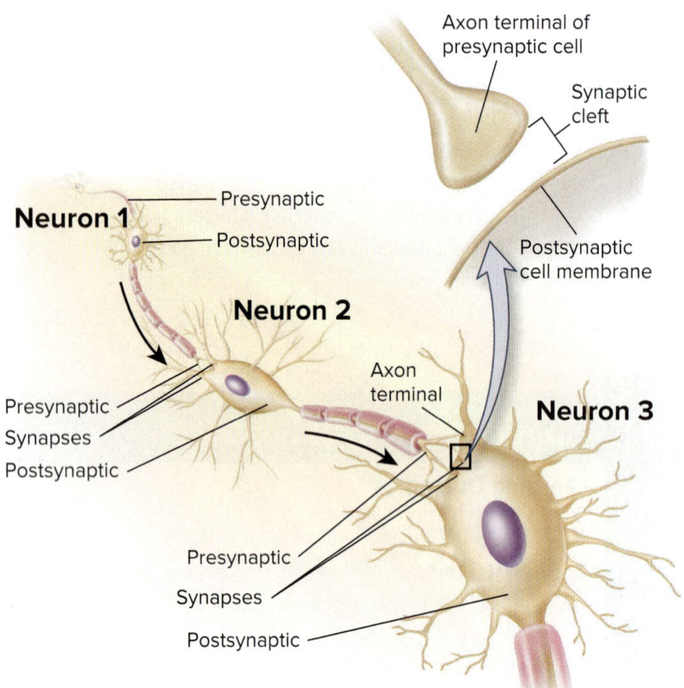


**Figure 33.11 Presynaptic and postsynaptic cells.** The arrows show the direction of signal transmission from one neuron to the next. Note that neuron 2 is postsynaptic with respect to neuron 1 and presynaptic with respect to neuron 3.

synapses allow for complex modulation of the response of the postsynaptic cell, as we see next.

## In Chemical Synapses, Neurotransmitters Trigger Changes in Membrane Potential of Postsynaptic Cells

**Figure 33.12** describes the steps that occur when two neurons communicate via a chemical synapse. The cytosol of an axon terminal of a presynaptic cell contains vesicles—small, membrane-enclosed packets, each containing thousands of molecules of neurotransmitters. The membranes of axon terminals contain voltage-gated $Ca^{2+}$ channels. When an action potential arrives at an axon terminal, these channels open, allowing $Ca^{2+}$ ions to diffuse down their electrochemical gradient into the cell. Calcium ions bind to a protein associated with the vesicle membrane. This triggers exocytosis, in which the vesicle fuses with the presynaptic axon terminal membrane, thereby releasing its neurotransmitter molecules into the synaptic cleft. The neurotransmitter molecules then diffuse across the cleft and bind to receptor proteins in the postsynaptic cell membrane.

The binding of neurotransmitter molecules to their receptors opens or closes ion channels, thereby changing the membrane potential of the postsynaptic cell. Some neurotransmitters are called excitatory, because they depolarize the postsynaptic membrane; the response is an **excitatory postsynaptic potential (EPSP).** An EPSP is a graded potential that brings the membrane potential closer to the threshold potential that would trigger an action potential. Two ways in which an EPSP can be generated are by the opening of $Na^+$ channels in a postsynaptic membrane or by the closing of $K^+$ channels. In both cases, positive charges accumulate inside the cell. Conversely, an inhibitory neurotransmitter usually hyperpolarizes the postsynaptic membrane, producing an **inhibitory postsynaptic potential (IPSP).** An IPSP is a graded potential that brings the postsynaptic cell's membrane potential farther away from the threshold for producing an action potential. A common way in which an IPSP is formed is by the opening of $Cl^-$ channels; the electrochemical gradient for $Cl^-$ is such that it usually favors diffusion of $Cl^-$ into neurons, thus bringing negative charge into the cell. Hyperpolarization is a major way in which neurons are inhibited; see, for example, the motor neuron in the reflex arc shown in Figure 33.3.

To end the synaptic signal, neurotransmitter molecules in the synaptic cleft are broken down by enzymes or transported back into the terminal of the presynaptic cell and repackaged into vesicles for reuse. The latter event is called reuptake and is an efficient mechanism for recapturing unused neurotransmitters that were released into the synaptic cleft. These mechanisms also prevent a postsynaptic cell from being chronically stimulated.

The effect of activating a single synapse is usually far too weak to elicit an action potential in the postsynaptic neuron. A neuron may receive inputs from many synapses on its dendrites and cell body, and some of these synapses may release their neurotransmitter onto the neuron at the same or nearly the same time. When two or more EPSPs are generated at one time in a postsynaptic cell, their depolarizations sum together. The resulting larger depolarization may bring the membrane potential at the axon hillock to the threshold potential, initiating an action potential. Conversely, if EPSPs and IPSPs arrive together at a postsynaptic cell, the two types of signals may cancel each other out, and no action potential is elicited. This is a major advantage of chemical synapses; responses of postsynaptic cells can be modified to reflect the sum of many different inputs arriving at once or in close succession. It is also a way in which an animal's brain can determine the relative importance of competing stimuli. Imagine a hungry fish confronted with a worm and a predator. IPSPs sent to hunger-sensitive neurons can suppress those cells, while other neurons that respond to danger cues may receive only EPSPs, and the fish ignores the worm and seeks safety.

## Several Classes of Neurotransmitters Elicit Responses in Postsynaptic Cells

Neuroscientists have identified more than 100 different neurotransmitters in animals. Generally, neurotransmitters are categorized by structure. The changing balance between excitatory and inhibitory neurotransmission controls the state of nervous system circuits at any one time. To understand how neurotransmitters work, imagine driving a car with one foot on the gas pedal (which you can think of here as an EPSP) and one foot on the brake (an IPSP). To speed up, you could press down on the gas pedal, ease up on the brake, or both, whereas to slow down, you could do the opposite. All nervous systems operate in this way, with combined excitatory and inhibitory actions of neurotransmitters, creating opportunities to fine-tune signaling between neurons.

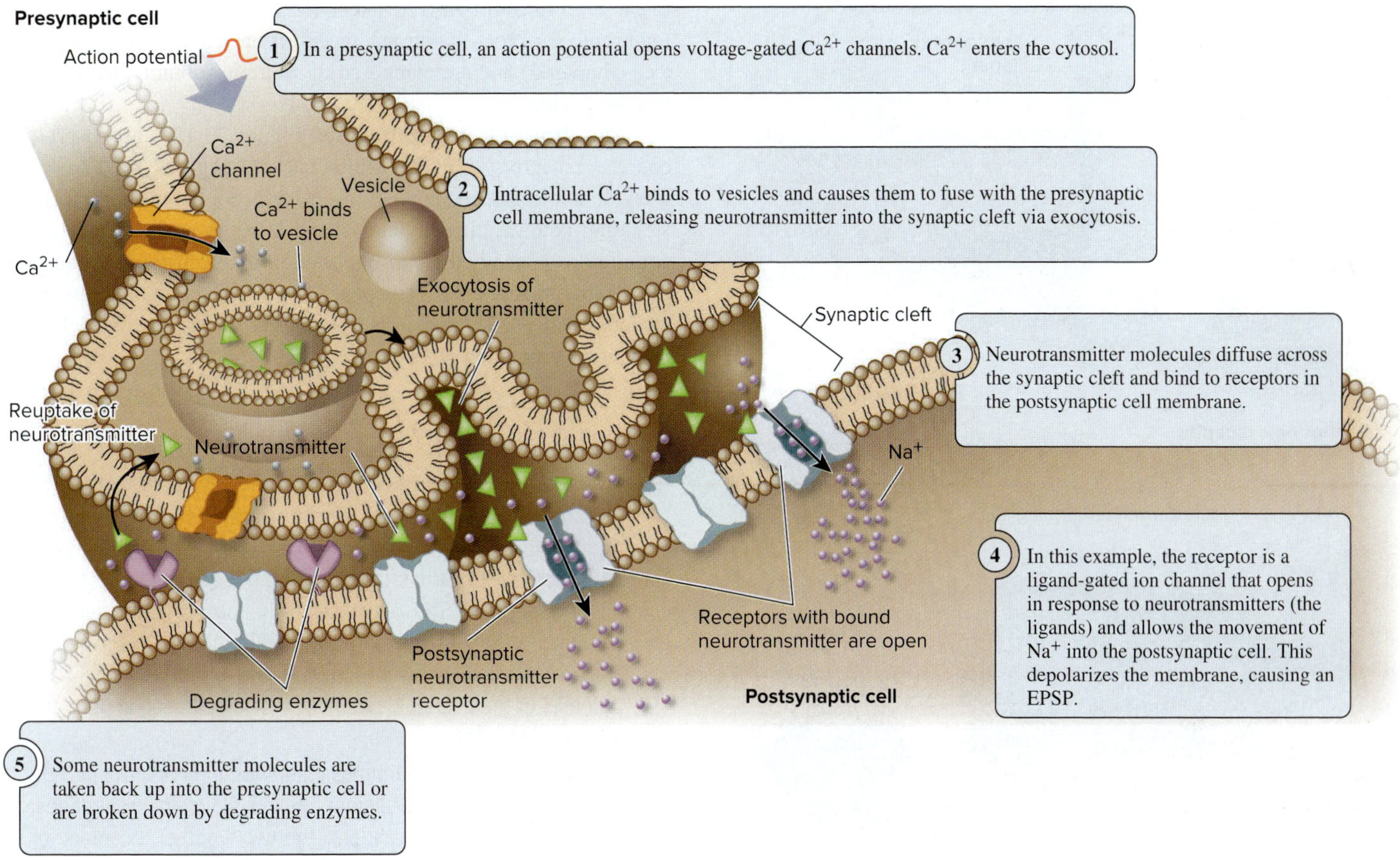

Presynaptic cell

Action potential

(1) In a presynaptic cell, an action potential opens voltage-gated $Ca^{2+}$ channels. $Ca^{2+}$ enters the cytosol.

$Ca^{2+}$ channel

Vesicle

$Ca^{2+}$ binds to vesicle

$Ca^{2+}$

(2) Intracellular $Ca^{2+}$ binds to vesicles and causes them to fuse with the presynaptic cell membrane, releasing neurotransmitter into the synaptic cleft via exocytosis.

Exocytosis of neurotransmitter

Synaptic cleft

(3) Neurotransmitter molecules diffuse across the synaptic cleft and bind to receptors in the postsynaptic cell membrane.

Reuptake of neurotransmitter

Neurotransmitter

$Na^+$

(4) In this example, the receptor is a ligand-gated ion channel that opens in response to neurotransmitters (the ligands) and allows the movement of $Na^+$ into the postsynaptic cell. This depolarizes the membrane, causing an EPSP.

Receptors with bound neurotransmitter are open

Postsynaptic neurotransmitter receptor

Degrading enzymes

Postsynaptic cell

(5) Some neurotransmitter molecules are taken back up into the presynaptic cell or are broken down by degrading enzymes.

**Figure 33.12 Structure and function of a chemical synapse.** In response to an action potential, $Ca^{2+}$ enters the presynaptic neuron axon terminal. This results in vesicle fusion with the plasma membrane, releasing neurotransmitter molecules into the synaptic cleft. The neurotransmitter molecules then bind to receptors in the plasma membrane of the postsynaptic cell. This causes ion channels to open or close, which in turn changes the membrane potential of the postsynaptic cell. In this example, $Na^+$ channels open, which would create an EPSP in the postsynaptic cell.

*What is the benefit of having degrading enzymes in the synaptic cleft break down neurotransmitter molecules?*

The different types of neurotransmitters found in animals fall into the following classes:

- **Acetylcholine.** Acetylcholine is released at synapses where a neuron contacts skeletal or cardiac muscle. It is also released at synapses within the brain and on many gland cells. Depending on where it acts, it may be excitatory or inhibitory. For example, it stimulates contraction of skeletal muscle but inhibits cardiac muscle activity.
- **Biogenic amines.** The biogenic amines are compounds containing amine groups that are derived from amino acids. They include the catecholamines—dopamine, norepinephrine, and epinephrine—and serotonin and histamine. The catecholamines are formed from the amino acid tyrosine, serotonin is formed from tryptophan, and histamine from histidine. They have widespread physiological effects such as control of blood pressure, metabolism, mood, attention, behavior, and learning.
- **Amino acids.** The amino acids glutamate, aspartate, glycine, and $\gamma$-aminobutyric acid (GABA) function as neurotransmitters. Glutamate is the most widespread excitatory neurotransmitter found in animal nervous systems, whereas GABA is the most common inhibitory neurotransmitter.
- **Neuropeptides.** Neuropeptides are short polypeptides that can be excitatory or inhibitory. Neuropeptides are often called neuromodulators, because they can modulate the response of a postsynaptic neuron to other neurotransmitters. For example, a neuropeptide released from a presynaptic cell may stimulate an increase in the number of receptors for another neurotransmitter on the postsynaptic cell, which makes the cell more responsive to that neurotransmitter.
- **Gases.** Certain gases such as nitric oxide (NO) are produced locally in many tissues, including some neurons where they function as neurotransmitters. Unlike other neurotransmitters, they are not sequestered into vesicles but are released from terminals by diffusion. They also differ by not acting on postsynaptic receptors but, rather, by diffusing into the postsynaptic cell and acting on intracellular enzymes. These

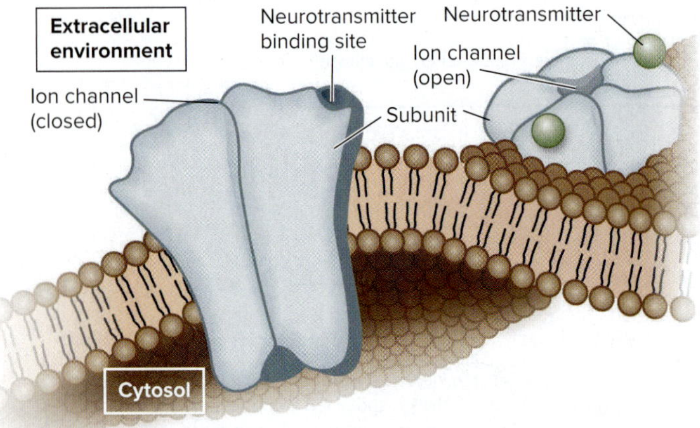


**(a) Ionotropic receptor**

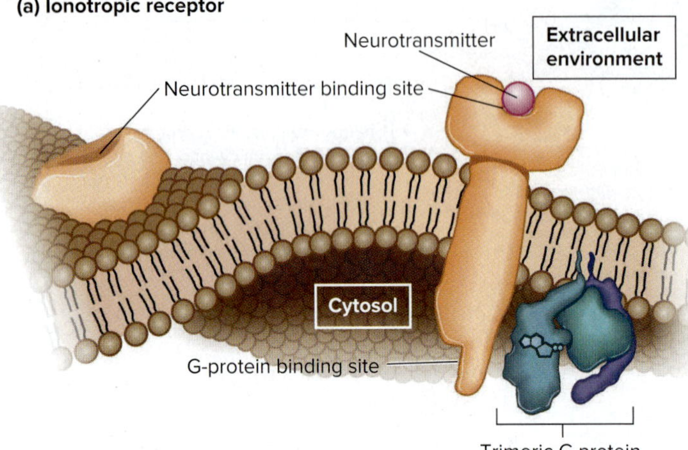


**(b) Metabotropic receptor**

**Figure 33.13** **The two major categories of postsynaptic receptors.** **(a)** Ionotropic receptors are ligand-gated channels with several subunits. Neurotransmitters bind to ionotropic receptors and directly open ion channels in the membrane. **(b)** Metabotropic receptors are G-protein-coupled receptors. Neurotransmitters bind to metabotropic receptors and initiate a signaling pathway that typically opens or closes nearby ion channels (refer back to Figures 8.7 and 8.13).

enzymes may alter membrane potential in a postsynaptic neuron, or, in non-neuronal cells, they may affect such actions as smooth muscle contraction.

## Postsynaptic Membrane Receptors Determine the Type of Response to Neurotransmitters

As we have just learned, in some cases the same neurotransmitter can have both excitatory and inhibitory effects depending on the cell on which it acts. The response of the postsynaptic cell depends on the type of receptor present in the postsynaptic membrane. The two major types of postsynaptic membrane receptors are ionotropic and metabotropic, and many neurotransmitters act on both (**Figure 33.13**).

**Ionotropic receptors** are ligand-gated ion channels that open in response to neurotransmitter binding (see Figure 33.13a). When neurotransmitter molecules bind to these receptors, the conformation of the receptors changes, allowing ions to flow through the channels to cause an EPSP or an IPSP. Acetylcholine and amino acids bind to ionotropic receptors. Ionotropic receptors have a quaternary structure (see Chapter 3) and are composed of multiple subunits that form the receptor's channel.

**Metabotropic receptors** (see Figure 33.13b) are G-protein-coupled receptors (GPCRs) (refer back to Figures 8.7 and 8.13). They do not form a channel but instead are coupled to an intracellular signaling pathway that initiates changes in the postsynaptic cell. A common response is the phosphorylation of specific ion channels in the plasma membrane. In Chapter 34, we will see that metabotropic receptors are particularly important in activating sensory cells that respond to visual and other stimuli.

## 33.4 Reviewing the Concepts

- In an electrical synapse, an electric current is conducted from one cell to another via gap junctions. In a chemical synapse, a neurotransmitter carries the signal from the presynaptic to the postsynaptic cell. The response in the postsynaptic cell may be inhibitory (IPSP) or excitatory (EPSP). Many EPSPs generated at one time can sum together and bring the membrane potential to the threshold potential, initiating an action potential (Figures 33.11, 33.12).
- Chemical classes of neurotransmitters found in animals include acetylcholine, biogenic amines, amino acids, neuropeptides, and gaseous neurotransmitters. The two types of receptors for neurotransmitters are ionotropic and metabotropic receptors (Figure 33.13).

## 33.4 Testing Your Knowledge

1. When two or more EPSPs arrive at a postsynaptic cell at the same location in quick succession, what will happen?
   - a. Their depolarizations will sum and this might produce an action potential.
   - b. Each EPSP will create a graded potential, some of which will depolarize the cell and some of which will hyperpolarize the cell.
   - c. Each EPSP will cause a separate action potential.
   - d. They will cancel each other out, leading to no action potentials.
   - e. Their depolarizations will sum, but this cannot produce an action potential.
2. Which is true?
   - a. An EPSP may be caused by opening $K^+$ channels in a postsynaptic membrane.
   - b. An IPSP may be caused by closing $K^+$ channels in a postsynaptic membrane.
   - c. It is possible for certain neurotransmitters to be either excitatory or inhibitory in different cells.
   - d. Metabotropic receptors are also ligand-gated ion channels.
   - e. One way for a neurotransmitter to depolarize a postsynaptic cell would be to activate ionotropic receptors that are linked to $Cl^-$ channels.

## 33.5 The Evolution and Development of Nervous Systems

### Learning Outcomes

1. List some different types of nervous systems found in animals.
2. Describe the three basic divisions of the vertebrate brain.
3. Discuss anatomical changes that accompanied the evolution of brain complexity.

Studying the evolution and development of animal nervous systems helps us understand how particular nervous systems are adapted to different functions. The characteristics of an animal's nervous system determine the behaviors that an animal displays. In this section, we will survey the nervous systems of animals and examine the brains of vertebrates in greater detail.

### The Evolution of Nervous Systems Enabled Animals to Sense and Respond to Changes in the Environment

Precisely when nervous systems first arose and whether or not the nervous systems of most or all animals can be traced back to a common ancestor are questions of active investigation. For example, recent genetic studies have uncovered remarkable similarities in the genes that regulate neuronal development in bilaterally symmetric animals. Those studies suggest that the patterning of nervous system development in bilateral animals may be traced to a common ancestor that lived more than 500 mya!

All multicellular animals except sponges have a nervous system. Let's begin with a survey of the variety of nervous systems found in animals. **Nerve nets** are found in the radially symmetric cnidarians (Figure 33.14a). The neurons are arranged in a network of connections between the inner and outer body layers of the animals. Activation of neurons in any one region leads to excitation that spreads in all directions at once, activating contractile cells to contract. This causes the animal to move or swim, enabling it to capture food or evade a predator. Recent research has identified regions of specialized function in the nerve nets of some cnidarians, such as local sensory neurons in the outer body wall. These findings push the origins of specialized nervous system structures and functions further back in evolutionary time than previously thought.

Sea stars and other echinoderms have a slightly more complex nervous system, with a nerve ring around the mouth that is connected to radial nerves extending into the limbs (Figure 33.14b). This arrangement allows the mouth and limbs to operate independently.

During the evolution of animals, more complex body types have been associated with a concentration of sense organs at the anterior end

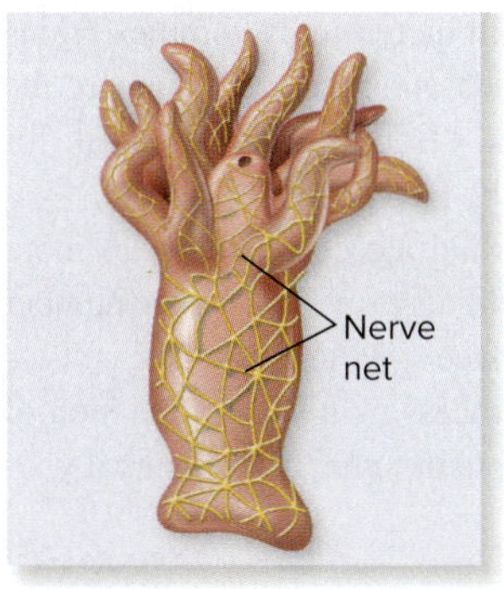


(a) Cnidarian

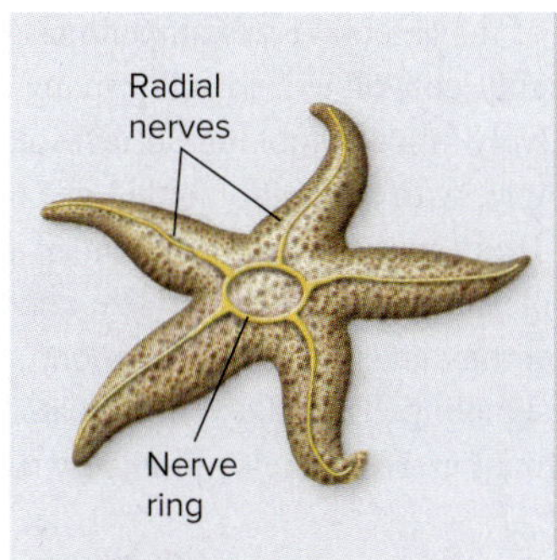


(b) Echinoderm

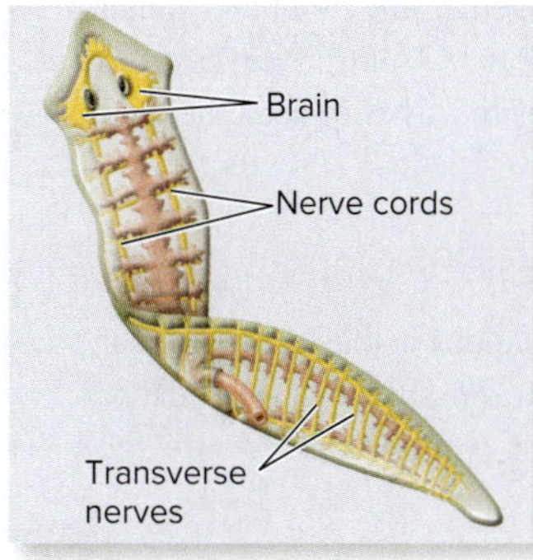


(c) Platyhelminth

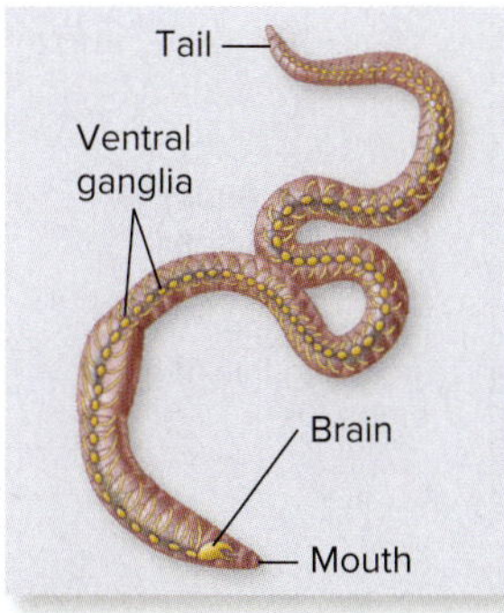


(d) Annelid

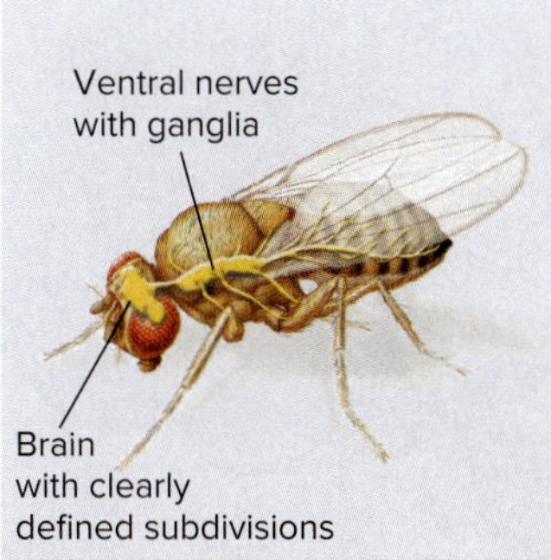


(e) Arthropod

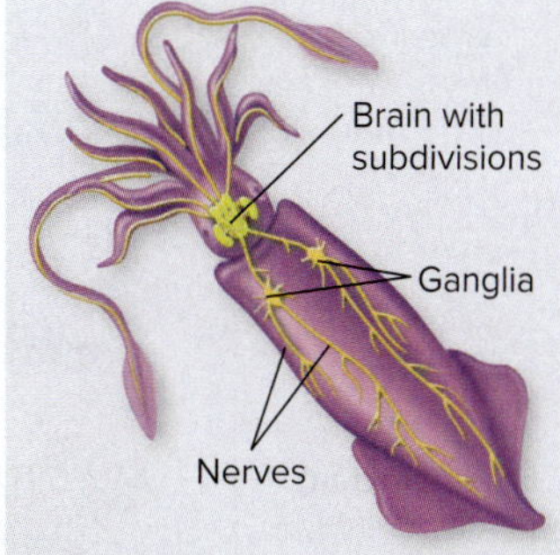


(f) Mollusk

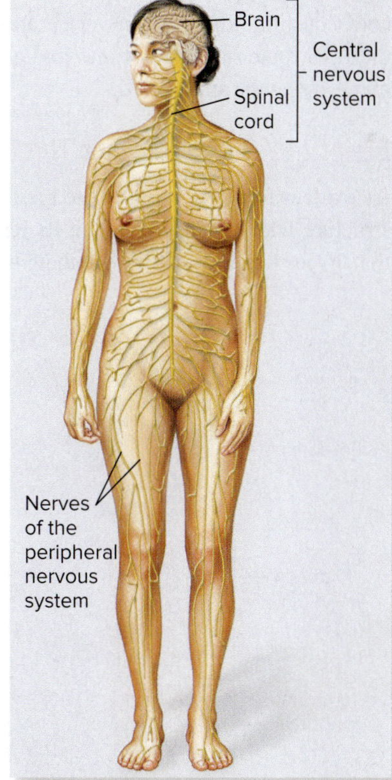


(g) Chordate

**Figure 33.14** Representative nervous systems throughout the animal kingdom.

**BioConnections:** *Nervous systems are one of the defining features of animals. In Chapter 26, you learned that several other features define animals and distinguish them from other living organisms. What were some of these features?*

of the body, along with an increasingly complex brain that controls sensory and motor functions of the entire body. Within the brain, neuronal pathways provide the integrative functions necessary for an animal to make more sophisticated responses to its environment. Brains are found in all vertebrates and most invertebrates, and they are composed of more than one anatomical and functional region with considerable synaptic complexity. The first animals to have evolved a brain as defined by regionalization and synaptic complexity are the platyhelminthes such as *Planaria*. In these animals, different regions of the brain appear to integrate inputs from sense organs, such as the eyes, and control motor outputs such as those involved in swimming (Figure 33.14c). Two **nerve cords** extend along the ventral surface of the animal and are connected to each other by short nerves. In annelids, ganglia and nerves are present at each body segment, where they coordinate local sensory and motor activities (Figure 33.14d). **Ganglia** are collections of neuron cell bodies with limited processing ability, limited synapses, and little or no subdivisions as in a brain; they are present in most animals and often serve to coordinate local signaling in a body part.

In *Drosophila* (Figure 33.14e and see the chapter-opening photo), the brain has several subdivisions with distinct, well-defined functions, such as a region devoted to learning and memory. Some mollusks, such as the squid and octopus, have brains with well-developed subdivisions that allow these animals to coordinate the complex visual and motor behaviors necessary for their predatory lifestyle (Figure 33.14f). In chordates, a dorsally located nerve cord is present that, in vertebrates, forms the brain and spinal cord. Together, the brain and spinal cord constitute the CNS in vertebrates (Figure 33.14g). Nerves from the PNS route information into and out of the CNS at separate segments along the spinal cord.

## Brains of Vertebrates Have Three Basic Divisions

As evident from Figure 33.15, the brain develops as basically a linear structure with expansions along its length. Fossils of jawless fishes that lived 400 mya show that their brains were organized into the three basic divisions that have been retained in all modern vertebrates: the **hindbrain, midbrain,** and **forebrain.**

The vertebrate hindbrain coordinates many basic reflexes, motor functions, and other bodily functions, such as breathing, that maintain the normal homeostatic processes of the animal. The midbrain processes several types of sensory inputs, including vision, olfaction (smell), and audition (hearing), and has a role in alertness. The forebrain initiates motor functions and processes sensory inputs. Many forebrain functions are attributed to a region called the **cerebrum,** the surface layer of which is especially developed in mammals and is called the **cerebral cortex.** As we will learn later, in humans this region is critical for thought, reasoning, and learning, among other functions. Other structures of the forebrain lie beneath the cerebral cortex and will be discussed later in this chapter.

## Evolution of Increased Brain Complexity Involved a Larger, Highly Folded Cerebrum

With the increased complexity of the brain, the size of the cerebrum also increased, making up a greater proportion of the brain. In mammals, many of the important functions of the cerebrum are carried out by the cerebral cortex. Therefore, the increased complexity of the brain in these animals is also correlated with an increased surface area of the cerebral cortex. During the evolution of mammals, this increase in surface area occurred more rapidly than an expansion in the size of the skull. How could this occur? The answer is that the external surface of the cerebral cortex in animals with increasingly complex brains is highly convoluted, forming many folds and grooves. Compare the relatively smooth-looking surface of the cerebral cortex of a rat, for example, with the highly folded one of the human in Figure 33.16.

Brain mass and the amount of folding are correlated with more complex behaviors. Greater size and folding provide a larger number of neurons and synapses and more surface area available for cell-to-cell signaling. This allows for greater processing and interpretation of information. Even so, it would be wrong to assume that an animal with

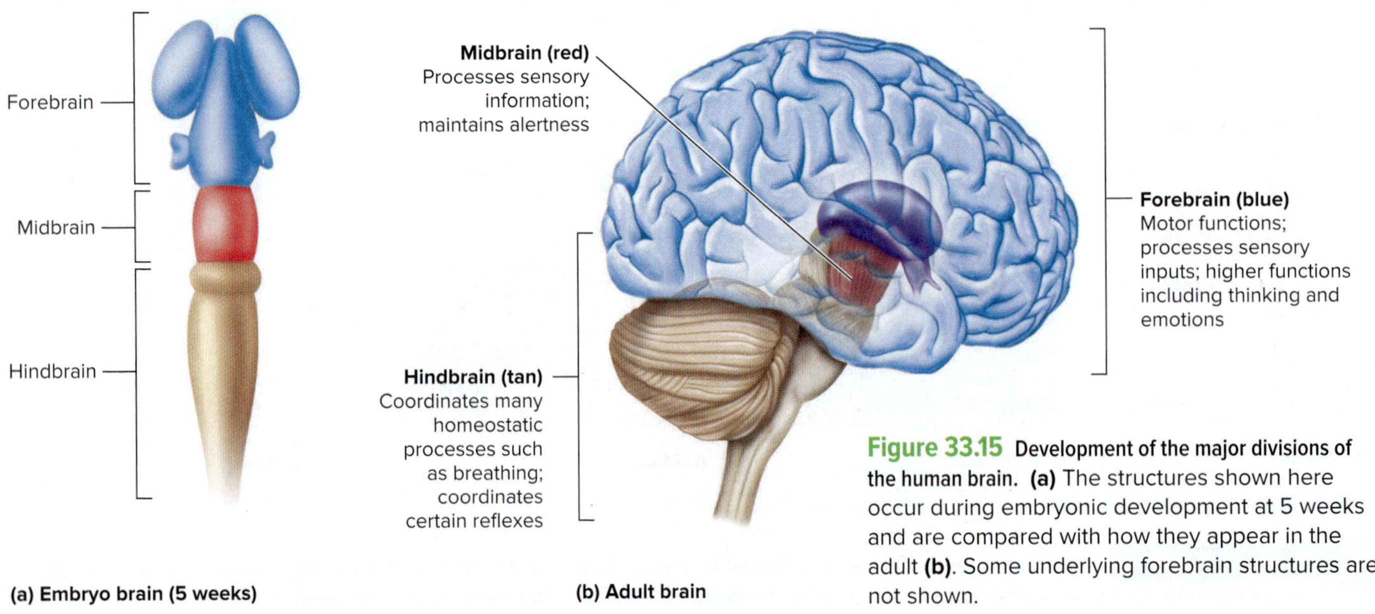


**Figure 33.15 Development of the major divisions of the human brain. (a)** The structures shown here occur during embryonic development at 5 weeks and are compared with how they appear in the adult **(b)**. Some underlying forebrain structures are not shown.

## Biology Principle

### Structure Determines Function

The increased folding of the cerebral cortex of certain mammals increases the surface area of this part of the brain, allowing for more neuronal connections and complex behaviors, including conscious thought.

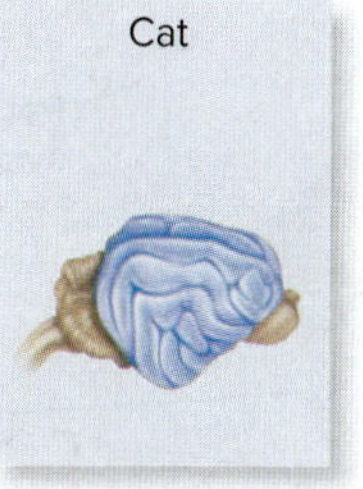

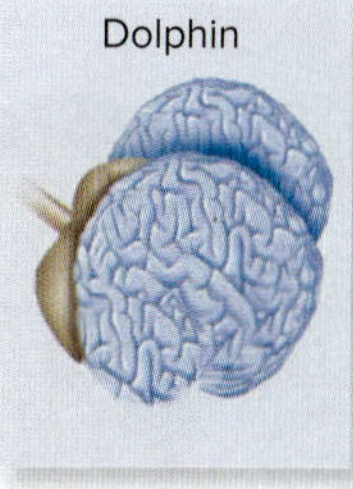

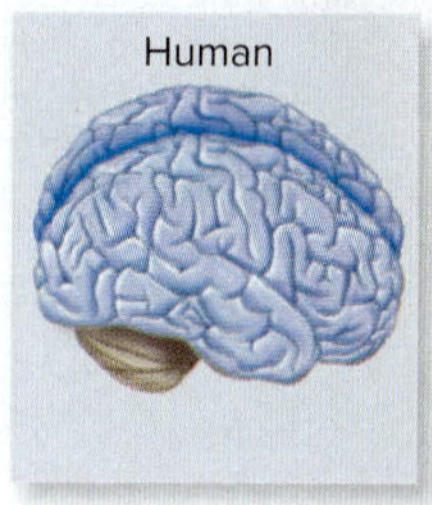


**Figure 33.16** **The degree of cerebral cortex folding in different mammalian species.** The brains are not shown to scale.

a small brain is profoundly limited in its behavioral repertoire. A bat with a 0.9-g brain and an elephant with a 2,500-g brain can both perform an astounding variety of complex behaviors, such as navigating across great distances and maintaining sophisticated social hierarchies with other members of their species.

## 33.5 Reviewing the Concepts

- Nervous systems in animals range from nerve nets in cnidarians to increasingly complex systems with nerve cords and ganglia and a brain (Figure 33.14).
- In all vertebrates, the three major divisions of the brain are the hindbrain, midbrain, and forebrain (Figure 33.15).
- Increased mass and folding of the brain's cerebral cortex allow for expansion of regions associated with conscious thought, reasoning, and learning (Figure 33.16).

## 33.5 Testing Your Knowledge

1. Match the nervous system feature with the animal.

| | |
|---|---|
| nerve net | sea star (echinoderm) |
| two ventral nerve cords | cnidarian |
| nerve ring and radial nerves | *Planaria* |
| brain with several well-defined subdivisions | *Drosophila* |

2. Scientists have determined that the brain mass of *Homo neanderthalensis* (Neanderthal man) was likely greater than that of our own. Does this mean that Neanderthals were more intelligent and capable of more complex behaviors than we are?
   a. Yes. Increased brain mass suggests an animal with a more complex set of behaviors.
   b. No. As brain mass increases, complex behavior and intelligence decrease.
   c. No, because a larger brain mass does not necessarily mean that the brain has more surface area and more neuronal complexity.

## 33.6 Structure and Function of the Nervous Systems of Humans and Other Vertebrates

### Learning Outcomes

1. Outline the anatomical organization of the vertebrate nervous system.
2. Describe the organization of the peripheral nervous system.
3. Distinguish between the somatic and autonomic branches of nervous systems.
4. Briefly describe the major structures and functions of the human brain.

The nervous systems of vertebrates are amazingly complex—in humans, the brain alone has over 100 billion neurons and an even greater number of glial cells. Moreover, complexity is defined by more than just numbers of cells. Within the brain, for example, are enormous numbers of connections between neurons—a single neuron in the cerebellum may have as many as 100,000 synapses with other cells! In this section, we will examine the structure and function of the vertebrate nervous system, with special attention to the human nervous system.

### Information in and Between the CNS and PNS Is Conveyed by Nerves and Tracts

In vertebrates, the CNS and PNS are connected anatomically and functionally by nerves (Figure 33.17). Nerves that enter and exit the brain are called **cranial nerves;** those that enter and exit through the spinal cord are called **spinal nerves.**

Information may flow in both directions within nerves between the CNS and PNS. For example, suppose you accidentally lean against a newly painted fence. Neuronal endings in your skin, which are part of the PNS, transmit tactile (touch) information directly into the spinal cord via spinal nerves. From there, the information travels to your brain, where the sensation is analyzed and identified as

# Biology Principle

## All Species (Past and Present) Are Related by an Evolutionary History

The similarities between the nervous systems of these two very different-looking animals point to a common ancestor.

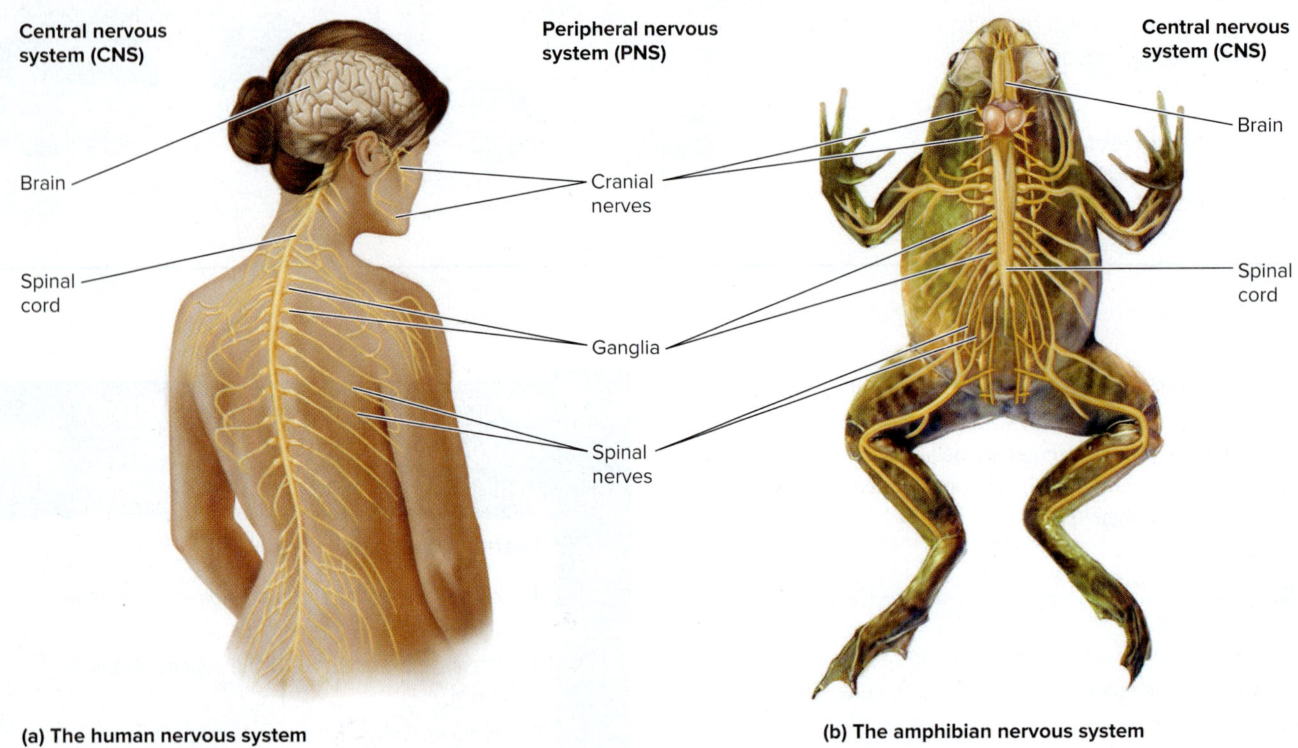


**Figure 33.17** **Organization of the vertebrate nervous system.** The CNS consists of the brain and spinal cord, both of which are encased in bone (not shown). The PNS includes cranial nerves, ganglia, and spinal nerves, which carry information to and from the CNS, and many other neurons throughout the body. Note the similarities between two widely divergent vertebrates, **(a)** humans and **(b)** frogs.

something sticky. Signals are sent from your brain, down your spinal cord, and out through motor neurons in spinal nerves to your muscles, causing you to move away.

Information is also conveyed from region to region within the CNS. In that case, axons travel in bundles known as **tracts.** One of the most obvious characteristics of the CNS is that some parts look white, and others appear gray, leading to the terms white matter and gray matter (Figure 33.18). The white matter gets its appearance from myelin; it consists of myelinated axons bundled together in large numbers to form tracts. The gray matter is darker in appearance and consists of neuronal cell bodies, dendrites, and unmyelinated axons. In the spinal cord, for example, the gray matter is located in the center and forms two dorsal horns and two ventral horns (see Figure 33.18). Each dorsal horn connects to a dorsal root, which is part of a nerve. Dorsal roots receive incoming information from sensory neurons of the PNS. The ventral horn connects to the ventral root, which transmits outgoing information through motor neurons.

Unlike the PNS, the CNS is encased in protective structures, including bone (the skull and vertebrae) and sheathlike membranes called meninges. The meninges are protective structures encasing a watery solution called cerebrospinal fluid, which absorbs physical shocks to the CNS that result from sudden movements, injuries, or blows to the head.

## The PNS Consists of the Somatic and Autonomic Nervous Systems

The PNS of vertebrates consists of two major functional and anatomical divisions: the somatic nervous system and the autonomic nervous system.

**Somatic Nervous System** The primary function of the **somatic nervous system** is to sense the external environment and activate skeletal muscles. The afferent sensory neurons of the somatic nervous system receive stimuli, such as heat, light, odors, chemicals (in food), sounds, and touch, and transmit signals to the CNS. Efferent motor neurons of the somatic nervous system activate skeletal muscles. The cell bodies of these motor neurons are located within the CNS.

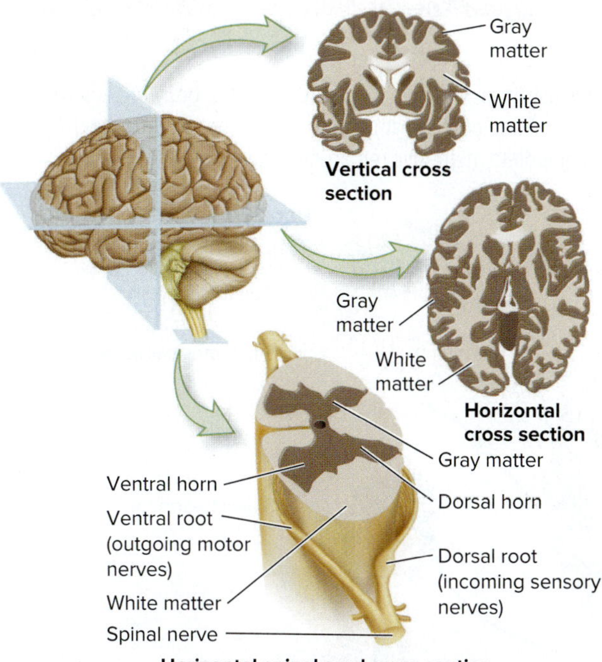


**Figure 33.18** **Gray matter and white matter in the CNS.** The gray matter is composed of groups of cell bodies, dendrites, and unmyelinated axons. The white matter consists of tracks of myelinated axons.

The axons from these cells leave the spinal cord and project directly onto skeletal muscle cells, where they release acetylcholine at the synapse. Acetylcholine stimulates the muscle cells to contract.

**Autonomic Nervous System** The **autonomic nervous system** regulates homeostasis and organ function. For example, it is involved in regulating heart rate, blood pressure, glucose homeostasis, and digestion. For the most part, the autonomic nervous system is not subject to voluntary control. We usually cannot consciously change our blood glucose concentration, for example.

The efferent pathways of the autonomic nervous system involve two neurons. The cell body of the first neuron is within the CNS and synapses on a second neuron in ganglia outside the spinal cord; these ganglia, therefore, are part of the PNS. This second neuron sends its axon to smooth muscles, cardiac muscle, and glands.

The motor nerves of the autonomic system are subdivided into the sympathetic and parasympathetic divisions (Figure 33.19). Both divisions typically act on the same organs and usually have opposing actions. The **sympathetic division** is responsible for rapidly activating systems that prepare the body for danger or stress. Imagine, for example, the physiological responses that would occur if you were hiking and came upon a dangerous animal. You would react instinctively to prepare yourself to confront (fight) or avoid (flight) a potentially life-threatening situation. This is the **fight-or-flight response,** and it is characterized by increased heart rate, stronger pumping action of the heart, relaxed (dilated) airways for taking fast, deep breaths, inhibition of digestive activity, increased blood flow to skeletal muscles, and increased secretion by the liver of glucose into the blood. All of these actions are mediated by the release of norepinephrine from axon terminals of neurons of the sympathetic nervous system.

The **parasympathetic division** of the autonomic nervous system is involved in maintaining and restoring body functions. It is particularly active during restful periods or after a meal, which is why it is sometimes said to mediate the **rest-or-digest response.** Neurons of the parasympathetic division promote digestion and absorption of food from the intestines, slow the heart rate, and decrease the amount of fuel supplied to the blood from the liver and adipose tissue. These effects are all mediated by acetylcholine, the neurotransmitter of the parasympathetic division.

## The Human Hindbrain Is Important for Homeostasis and Essential Bodily Functions

Let's turn our attention to a more detailed analysis of the structure and function of the human brain (Figure 33.20). We begin with the three major structures of the hindbrain, which is the evolutionarily oldest part of the brain and contains structures that control the basic processes that sustain life.

**Medulla Oblongata** The **medulla oblongata** regulates numerous homeostatic processes and, so, is essential for life. It is involved in the control of heart rate, blood pressure, breathing, digestion, glucose balance, and swallowing, among many other functions.

**Cerebellum** The **cerebellum** sits dorsal to the medulla oblongata and receives sensory inputs from the cerebral cortex. It also receives inputs from the spinal cord and inner ears that convey information about the position of the limbs and head, thereby helping to maintain balance and coordinate hand-eye movements. In addition, the cerebellum synchronizes fine motor activities such as texting or touching the fingers to the tip of the nose with eyes closed. When the cerebellum is damaged or injured, such as in an accident, a person finds it difficult to maintain balance and coordination.

**Pons** The **pons** sits anterior to the medulla oblongata and ventral to the cerebellum. Large tracts pass through the pons into and out of the cerebellum, which allows the pons to serve as a relay between the cerebellum and other areas of the brain. In addition to this integrative motor function, the pons has a very important role in regulating the unconscious control of breathing.

## The Midbrain Processes Sensory Inputs and Maintains Alertness

The human midbrain lies anterior to the pons. It receives sensory inputs from visual, olfactory, and auditory pathways. It supplies tracts that pass this information to other parts of the brain for further processing and interpretation. As one example, the midbrain is responsible for activating neural reflexes that change the diameter of the pupil of the eye in response to a change in the amount of ambient light.

Together the midbrain, medulla oblongata, and pons constitute the **brainstem.** All three parts of the brainstem contain nuclei (collections of many neurons in one location that have a similar function) that form a network of neuronal connections called the

# Biology Principle

## Living Organisms Maintain Homeostasis

The autonomic nervous system is an excellent illustration of a major way in which homeostasis is maintained. Many organs in the body are controlled in opposite ways by the two divisions of this branch of the nervous system. Therefore, the function of these structures can be modulated in two directions; for example, heart rate can be increased or decreased to match an animal's immediate metabolic requirements.

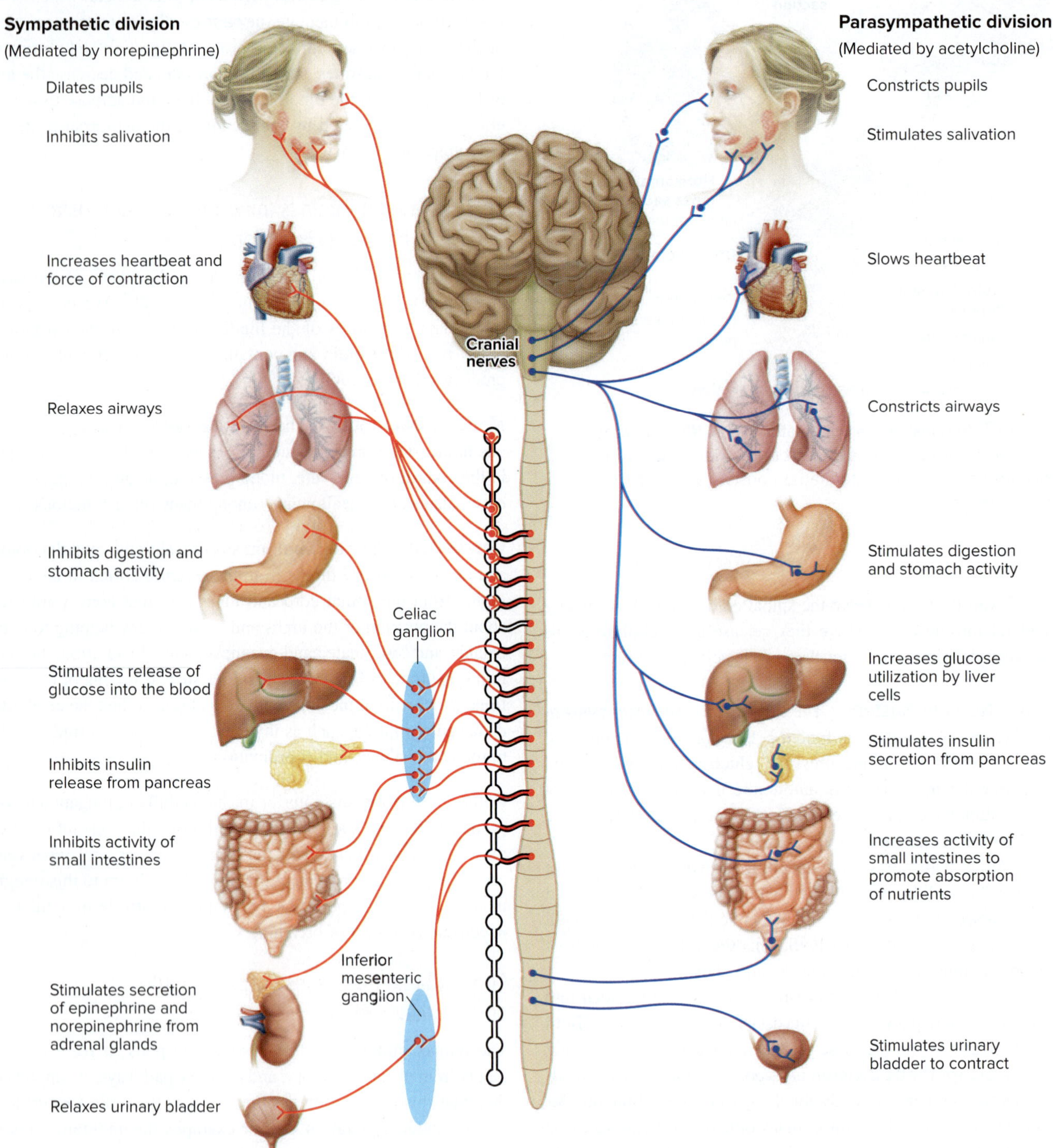


**Figure 33.19** **Sympathetic and parasympathetic divisions of the autonomic nervous system.** For simplicity, only some of the major functions of each division are shown in this figure. The sympathetic and parasympathetic systems tend to have opposing effects, and most parts of the body receive inputs from both divisions. Neurons from the sympathetic division make connections with a chain of ganglia, many of which are alongside the spinal cord. Neurons of the parasympathetic division make connections in ganglia near or in their targets (for clarity, the ganglia are shown with only one representative synapse).

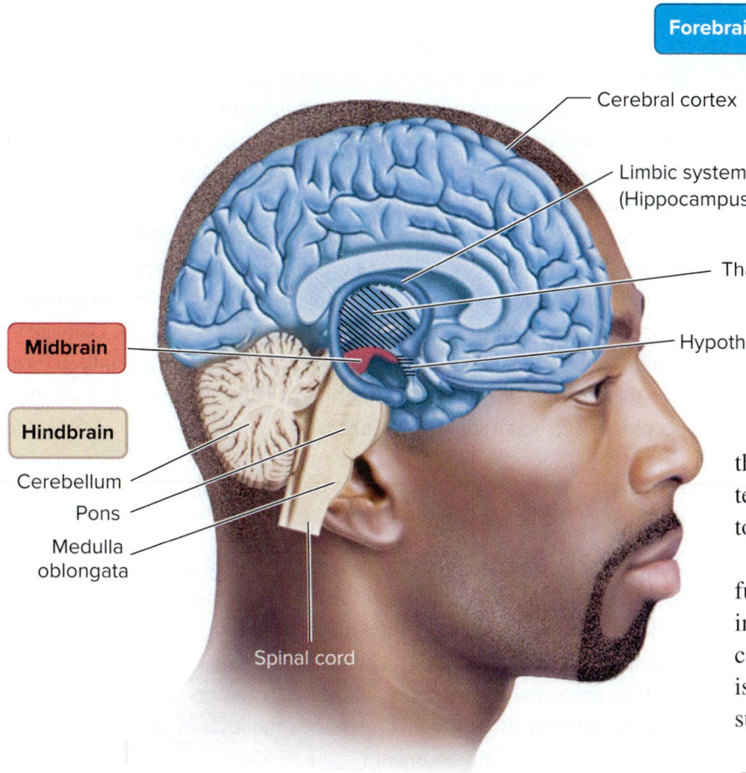


**Figure 33.20** **Major structures of the human brain.** An overview of the human brain, showing several internal structures. The epithalamus is obscured by the limbic system in this illustration. Only a portion of the limbic system is shown. The basal nuclei are not shown for clarity; they are located beneath the cerebral cortex in the region of the limbic system and thalamus. The midbrain, pons, and medulla oblongata collectively constitute the brainstem.

**Concept Check:** *Are the three major divisions of hindbrain, midbrain, and forebrain unique to humans? To mammals in general?*

**reticular formation.** Axons from the brainstem reticular formation nuclei travel in tracts that extend throughout the CNS. The reticular formation controls consciousness, alertness, sleep, arousal, and essential functions such as regulation of the heart. Because of the importance of the brainstem's functions, damage to it is catastrophic and may result in coma or death.

## The Human Forebrain Is Responsible for Motor, Sensory, and Higher Functions

The human forebrain comprises the **diencephalon—**which in turn encompasses the thalamus, hypothalamus, and epithalamus—and the cerebrum. The cerebrum includes the cerebral cortex, basal nuclei, and limbic system (Figure 33.21; also see Figure 33.20).

### Diencephalon (Thalamus, Hypothalamus and Epithalamus)

The **thalamus** has a major role in relaying sensory information to appropriate parts of the cerebrum. It receives input from all sensory systems except olfaction. One type of processing performed by neurons in the thalamus is that of filtering out sensory information in a way that allows us to pay attention to important cues while temporarily ignoring less important ones. This filtering mechanism begins in the reticular formation, which relays information about sensory inputs to the thalamus. Together, these brain regions permit selective attention to certain stimuli. A good example of this is a new parent's ability to sleep through a thunderstorm but awaken immediately to the cry of a baby in the next room.

The **hypothalamus,** located below the thalamus at the bottom of the forebrain, controls functions of the digestive and reproductive systems, body temperature, circadian rhythms, and certain drives related to homeostasis such as appetite and thirst.

The **epithalamus** is a collection of structures that have several functions in the interpretation of olfactory and visceral inputs. It also includes the pineal gland, an endocrine gland that secretes a hormone called melatonin into the blood. The function of melatonin in humans is still debated, but it has been suggested to function in daily cycles such as our sleep/wake rhythm.

### Cerebrum (Cerebral Cortex, Basal Nuclei, and Limbic System)

The cerebrum is divided into two halves (hemispheres) connected to each other by a large tract called the corpus callosum.

The cerebral cortex is the surface layer that covers the cerebrum and comprises four lobes in each hemisphere: the frontal, parietal, occipital, and temporal lobes (Figure 33.22). A few general functions of each lobe are listed here:

- The **frontal lobe** is important for the conscious motor commands that initiate movement. It is also chiefly responsible for such higher functions as making decisions, controlling impulses, planning for the future, exhibiting judgment, and some aspects of short-term memory, as well as for conscious thought and social awareness.

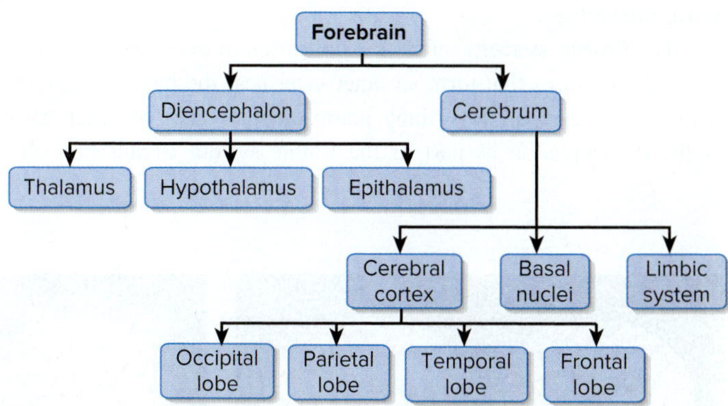


**Figure 33.21** **Relationships among major structures that make up the human forebrain.**

**Concept Check:** *Is the occipital lobe part of the diencephalon or cerebrum?*

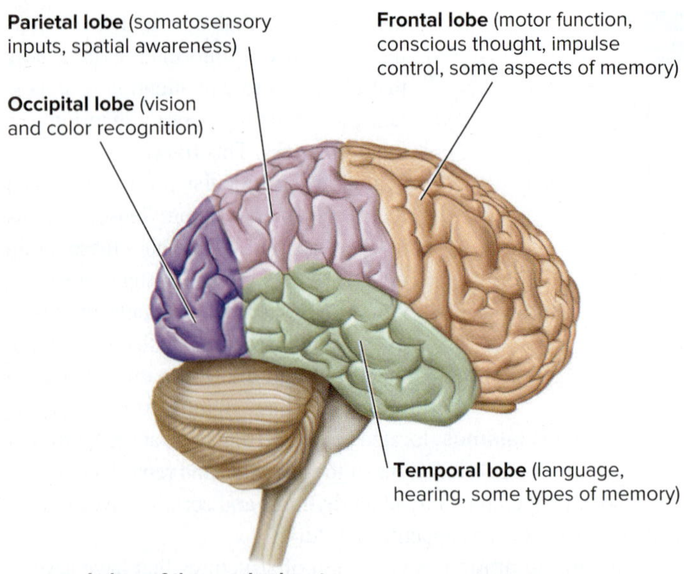


**Figure 33.22** The four lobes of the cerebral cortex as seen on the right hemisphere, and some major functions they control.

- The **parietal lobe** receives and interprets sensory input from somatic pathways, including touch from the surface of the body. This lobe also has an important role in our ability to use visual cues to orient ourselves in space.
- The **occipital lobe** receives visual information and controls many aspects of visual perception and color recognition.
- The **temporal lobe** is necessary for language, hearing, and some types of memory.

The **basal nuclei** are a group of forebrain nuclei that surround the thalamus and lie beneath the cerebral cortex. The basal nuclei are involved in planning, learning, and fine-tuning movements. They also function via a complex circuitry to smoothly initiate or inhibit movements. Parkinson's disease is a relatively common neurological disorder that affects the basal nuclei and results in numerous and severe motor difficulties.

The **limbic system** refers to a collection of evolutionarily older cerebral structures that form an inner layer near the base of the forebrain (see Figure 33.20). Many neuroscientists also consider parts of the diencephalon as part of the limbic system because of very extensive connections between these regions. The limbic system is primarily involved in the formation and expression of emotions, and it has a role in learning, memory, and the perception and recognition of smells. The expression of emotions occurs early in childhood before the more advanced functions of the cerebrum are evident. Thus, even very young babies can express fear, distress, and anger as well as bond emotionally with their parents.

A major part of the limbic system is the **hippocampus,** which includes several layers of cells that are connected together in a complex circuit (see Figure 33.20). Its main function appears to be establishing memories for spatial locations, facts, and sequences of events. Damage to certain parts of the hippocampus in humans results in an inability to form new memories, a devastating condition that prevents recognition of other people or even an awareness of daily events.

## Imaging Studies Help Scientists Understand the Functions of Brain Regions

A key way in which scientists have learned about the functions of the parts of the brain is through the use of imaging techniques. These include a technique called **magnetic resonance imaging (MRI),** which can detect the size and shape of individual brain structures, as well as abnormal structures such as brain tumors. A related tool is **functional MRI (fMRI),** which measures the level of activity of different brain regions at any moment (Figure 33.23).

The use of fMRI has revealed many fascinating aspects of the activities of different brain regions, notably in people who have suffered brain damage or loss of sensory inputs. For example, individuals who are blind from birth might be predicted to have occipital lobes that are less functional or active than are those in sighted persons (recall that the occipital lobes have the major role in visual perception). However, the work of American researcher Harold Burton and others has revealed with fMRI that the occipital lobes of blind persons are active but have become adapted to other sensory functions such as tactile signals from the fingers, including those arising from using Braille, a tactile reading system. Amazingly, this reassignment of occipital function occurs to some extent even in individuals who lose their vision later in life. Most likely, this does not represent a new function of the occipital lobes but, rather, an expanded ability of an existing function that remains relatively minor in sighted persons.

fMRI is also revealing differences in brain structure and function in individuals due to the types of activities in which they regularly engage, as described next.

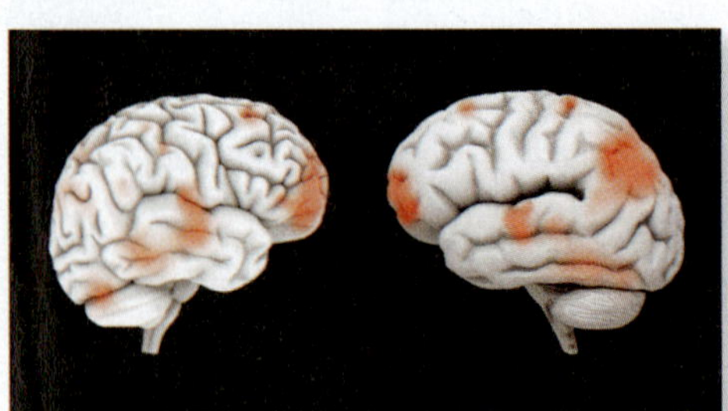

(a) Brain activity of a person thinking about a task that requires finger movements

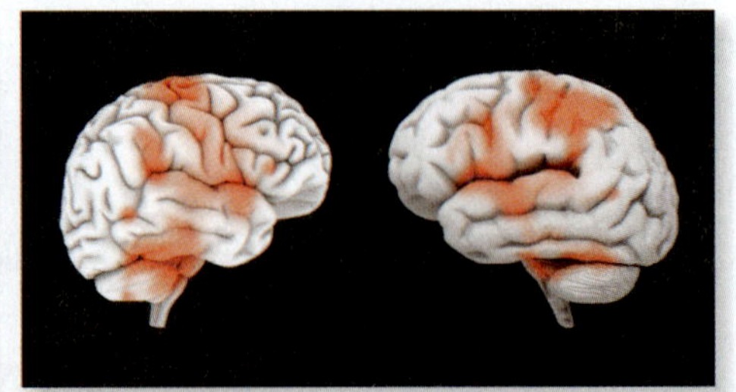

(b) Brain activity when the same person is performing this task

**Figure 33.23** **Exploring the functional activity of brain regions using an fMRI scan.** Red color indicates higher $O_2$ use; both hemispheres are shown.

*(a & b: brain on left)* © Science Photo Library/Alamy RF; *(a & b: brain on right)* © Simone Brandt/Getty Images RF

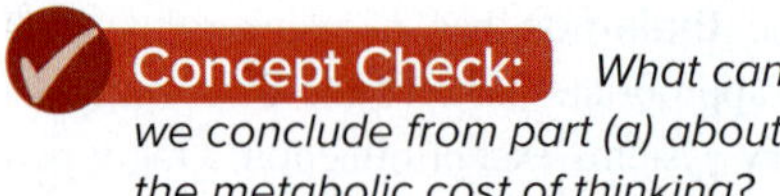


*What can we conclude from part (a) about the metabolic cost of thinking?*

# FEATURE INVESTIGATION

## Gaser and Schlaug Showed That the Sizes of Certain Brain Structures Differ Between Musicians and Nonmusicians

MRI and fMRI have been extremely useful in revealing which brain areas are involved in a particular function. Such research has also shown that the human brain is surprisingly adaptable. A number of studies have been carried out on musicians, because they practice extensively throughout their lives, enabling researchers to study the effects of repeated use on brain function.

In 2003, American neuroscientist Christian Gaser and German neuroscientist Gottfried Schlaug used MRI to examine the sizes of brain structures in three groups of people—professional musicians, amateur musicians, and nonmusicians (**Figure 33.24**). Individuals were assigned to each group based on their reported history of musical training: (1) professional musicians with over 2 hours of concentrated musical practice time each day, (2) amateur musicians who played a musical instrument regularly but not professionally (practicing about 1 hour/day), and (3) those who had never played a musical instrument. The researchers hypothesized that repeated exposure to musical training would increase the size of brain areas associated with visual, motor, and auditory skills, because all of these activities are used to read, hear, make, and interpret music. As seen in the data in part 4, brain areas involved in hearing, moving the fingers, and coordinating movements with vision and hearing were larger in professionals than in amateur musicians, and larger in amateurs than in nonmusicians. The region of the brain that controls finger movements was particularly well developed in the professional musicians, an interesting finding because all of the musicians in this study played keyboard instruments such as the piano.

In subsequent studies, Schlaug and colleagues examined the brains of people of different ages who either did or did not have musical training beginning in childhood. The researchers were able to correlate the degree of training with the sizes of different brain regions. They identified one particular region within the temporal lobe that was highly correlated with the extent of musical training in the course of one's life. This region, called the planum temporale, is believed to be especially important for recognizing and interpreting sound, identifying its source, and translating auditory signals (such as music) into motor processes (such as playing an instrument or humming a melody).

In a related study, American researchers Vincent Schmithorst and Scott Holland used fMRI to determine if musicians' brains were activated differently than nonmusicians' brains when they heard music. They found that one area of the cerebral cortex was selectively activated by melodies only in musicians. This study differed from that of Gaser and Schlaug, because Schmithorst and Holland examined the activity of brain areas as well as their sizes. Their results showed that listening to music activates certain neurons and pathways in the brains of musicians but not in nonmusicians.

The human studies of Gaser, Schlaug, Schmithorst, and Holland have not determined the underlying reason(s) for increased brain size. One possibility is that people with increased brain size in these regions are more likely to become musicians. Alternatively, musical training may actually cause certain regions of the brain to grow larger and alter their neuronal connections. In other research involving experimental animals, groups of animals have been randomly separated into those learning a task versus controls (that do not learn the task). Such experiments have shown increases in the size of brain regions that are associated with learning and memory. The increased size may result from formation of new synapses, growth of blood vessels to the region, and/or production of more glial cells.

### *Experimental Questions*

1. What was the hypothesis proposed by Gaser and Schlaug? How did Gaser and Schlaug test this hypothesis? What were the results of their experiment?
2. **SCISKILLS ▸** How did the research of Schmithorst and Holland differ from that of Gaser and Schlaug? Were their results generally supportive of Gaser and Schlaug's hypothesis?
3. **SCISKILLS ▸** Based on the results described here and what you have learned in this chapter, predict general differences that might be found between the brain of a professional athlete, such as a tennis player, and that of a nonathlete. What about the brain of a person who has been deaf from birth compared to that of a person with normal hearing?

**Figure 33.24** Gaser and Schlaug's study of the size of visual, motor, and auditory nuclei in the brains of musicians and nonmusicians.

**HYPOTHESIS** Musical training is associated with structural differences in the brain.

**KEY MATERIALS** Volunteer subjects with different degrees of musical training.

| | Experimental level | Conceptual level |
|---|---|---|
| 1 Establish 3 groups of subjects with different musical backgrounds. | Interview subjects for musical history and assign to 1 of 3 groups.  | **Controls (nonmusicians)** No musical training. **Amateur musicians** Play an instrument about 1 hr/day and are not employed as musicians. **Professional musicians** Employed as musicians and practice their instrument > 2 hr/day. |

2 Control for possible factors that might affect results.

Make group assignments in a way that matches subjects' age and other characteristics.

| | Controls | Amateurs | Professionals |
|---|---|---|---|
| Age (average) | 23 | 26 | 27 |
| Type of instrument | None | Keyboard | Keyboard |
| Mental IQ | 119 | 123 | 118 |
| Sex | Male | Male | Male |

3 Perform MRI and calculate volume of gray matter in different brain regions.

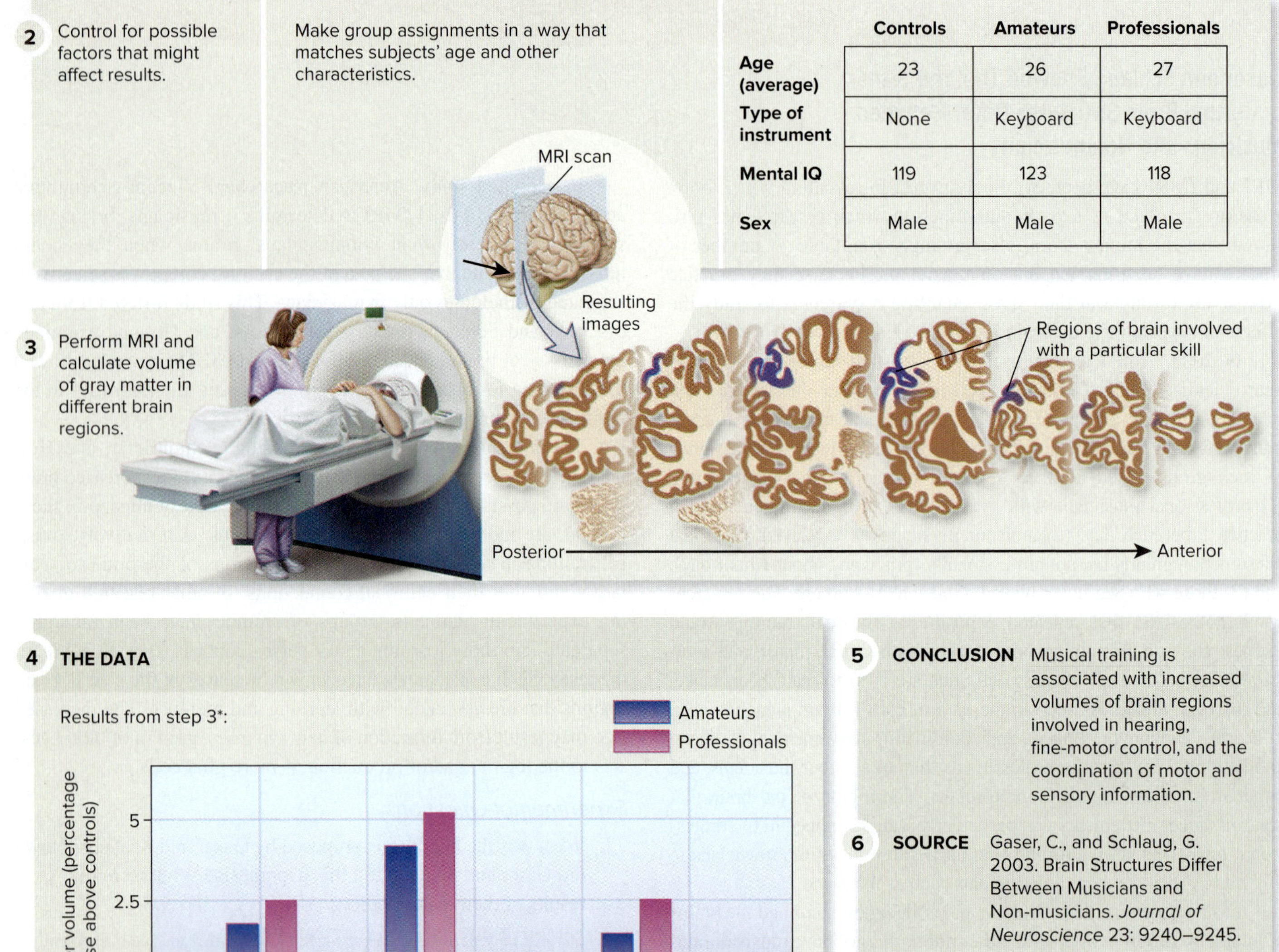


4 **THE DATA**

Results from step 3*:

*Controls are not shown separately because the data are expressed relative to controls.

5 **CONCLUSION** Musical training is associated with increased volumes of brain regions involved in hearing, fine-motor control, and the coordination of motor and sensory information.

6 **SOURCE** Gaser, C., and Schlaug, G. 2003. Brain Structures Differ Between Musicians and Non-musicians. *Journal of Neuroscience* 23: 9240–9245.

## 33.6 Reviewing the Concepts

- In humans and other vertebrates, the brain and spinal cord make up the central nervous system (CNS). The neurons and all axons outside the CNS, including the cranial and spinal nerves, constitute the peripheral nervous system (PNS). The CNS receives sensory input from the PNS, and the PNS receives commands from the CNS (Figure 33.17).
- The CNS is composed of gray and white matter and is protected by meninges and cushioned by cerebrospinal fluid (Figure 33.18).
- The PNS consists of the somatic nervous system, which senses external environmental conditions and controls skeletal muscles, and the autonomic nervous system which controls homeostatic functions (Figure 33.19).
- The human brain consists of the hindbrain, which controls the basic processes that sustain life; the midbrain, which processes several types of sensory inputs, including vision, olfaction, and audition; and the forebrain, which is made of the diencephalon (thalamus, hypothalamus, and epithalamus) and the cerebrum (cerebral cortex, limbic system, and basal nuclei) (Figures 33.20, 33.21, 33.22).
- Imaging techniques such as MRI and fMRI allow researchers to examine the structure and activity of the brain (Figures 33.23, 33.24).

## 33.6 Testing Your Knowledge

1. The division of the nervous system that controls voluntary muscle movement is
   a. the autonomic nervous system.
   b. the sensory division.
   c. the somatic nervous system.
   d. the parasympathetic division.
   e. the sympathetic division.
2. Match the human brain region with its representative functions.

| | |
|---|---|
| thalamus | making decisions and exercising judgment |
| hypothalamus | controlling breathing/relay between cerebellum and rest of brain |
| basal nuclei | filtering incoming sensory information |
| cerebral cortex | fine-tuning, starting, or inhibiting movement |
| pons | controlling body temperature and circadian rhythms |

**Table 33.1 Diseases or Disorders Affecting the Human Central Nervous System**

| Category | Examples |
|---|---|
| Infectious | Meningitis (inflammation of the meninges), encephalitis (inflammation of the brain) |
| Neurodegenerative | Alzheimer's disease |
| Movement | Parkinson's disease (death of neurons in the midbrain and basal nuclei) |
| Seizure | Epilepsy |
| Sleep | Sleep apnea (in one form, the brain fails to regulate breathing during sleep) |
| Tumors | Glioma (a tumor arising from glial cells) |
| Headache | Migraine (severe, recurring headache) |
| Mood | Major depressive disorder, bipolar disorder |
| Demyelinating | Multiple sclerosis (damage to the myelin sheath around neurons in the central nervous system) |
| Injury-related | Brain and spinal cord injuries due to accidents |

## 33.7 Impact on Public Health

### Learning Outcomes

1. Describe how disorders of neurotransmission can affect human health.
2. Discuss the cellular changes that occur in Alzheimer's disease and its impact on human health.
3. List several recreational and illicit drugs and their effects on the nervous system.

When neurons fail to develop properly or their function becomes impaired, the consequences can be devastating—affecting mood, behavior, and even the ability to think or move. Over 100 neurological disorders have been identified in humans. Most neurological disorders can be classified into several broad groups (Table 33.1). These disorders collectively affect hundreds of millions of people around the world. In addition, all so-called recreational drugs, including alcohol and tobacco, as well as illicit drugs such as marijuana, cocaine, heroin, LSD, and methamphetamine, exert their effects by altering neurotransmission, as summarized in Table 33.2. The use of these drugs, therefore, can result in symptoms similar to those of neurological disorders. Certain therapeutic drugs also exert their effects by altering neurotransmission—for example, to correct a neurotransmitter deficiency.

### Disorders of Neurotransmission Can Impact Mood

Several neurological disorders result from disrupted neurotransmission between cells. Genetic processes involved in the synthesis of neurotransmitters or malfunction of synaptic events can increase or decrease activity at synapses, which in turn affects emotions and behavior. The most common mood disorder is **major depressive disorder,** or depression. This illness results in prolonged periods of sadness, despair, and most commonly a profound lack of interest in daily activities. Depression affects 5–12% of men and 10–25% of women at some time during their lives. This condition is thought to result from decreased activity of synapses that release biogenic amines such as serotonin, which changes neuronal activity within specific areas in the brain that are involved in processing mood and emotion. Drugs used to treat depression include the **selective serotonin reuptake inhibitors (SSRIs),** such as Prozac, Zoloft, and Lexapro, which reduce the reuptake of serotonin into the presynaptic terminal after it is released. As a result, serotonin accumulates in the synaptic cleft, counteracting the presumed deficit that causes the alteration in mood. Patients taking SSRIs often report a greater interest in daily activities and a gradual return to normal, as was the case for the woman in the chapter-opening story.

### Alzheimer's Disease Impairs Memory, Thought, and Language

**Alzheimer's disease (AD)** is the leading cause of dementia worldwide. It is characterized by a loss of memory and cognitive function. AD is a progressive, incurable disease that begins with small memory lapses, leading in later stages to problems with language and abstract thinking and finally to loss of normal motor control and eventual death. The disease typically appears after age 65, though some inherited forms can strike people in their 30s and 40s.

Although psychological testing can help to diagnose AD, a definitive diagnosis is possible only after death when the brain is examined microscopically. Brains of AD patients show two noticeable changes: the existence of plaques and neurofibrillary tangles. Plaques are extracellular deposits of a misfolded protein, β-amyloid, that forms large, sticky aggregates. These plaques were first noted in 1906 by German physician Alois Alzheimer, after whom the disease is named. Neurofibrillary tangles are intracellular, twisted accumulations of cytoskeletal fibers. Scientists are unsure how these changes influence intellectual function

**Table 33.2 Representative Effects of Drugs on Neurotransmitter Action, Mood, and Behavior**

| Name of drug* | Actions on neurotransmission | Effects on mood or behavior | Effects of abuse or overdose |
|---|---|---|---|
| ***Illicit or Recreational***[†] | | | |
| Alcohol (ethanol) | Enhances inhibitory GABA transmission; increases dopamine transmission; inhibits glutamate transmission | Relaxation; euphoria; sleepiness | Liver damage; brain damage |
| Amphetamines: "uppers," "crystal meth," "speed" | Stimulate the release of dopamine and norepinephrine | Euphoria; increased activity | High blood pressure; psychosis |
| Cocaine | Blocks norepinephrine and dopamine reuptake | Intense euphoria followed by depression | Convulsions; hallucinations; death from overdose |
| LSD (lysergic acid diethylamide) | Binds to serotonin receptors | Hallucinations; sensory distortions | Unpredictable and irrational behavior |
| Marijuana (tetrahydrocannabinol) | Binds to receptors for natural cannabinoids | Increased sense of well-being; decreased short-term memory; decreased goal-directed behavior; increased appetite | Delusions; paranoia; confusion |
| Narcotics: heroin, morphine, meperidine (Demerol), codeine | Bind to opiate receptors | Pain relief; euphoria; sedation | Slowed breathing; death from overdose |
| Nicotine | Initially stimulates but then depresses activity in adrenal medulla and neurons in the peripheral nervous system; increases dopamine in brain | Increased attention; decreased irritability | Heart disease and lung disease |
| PCP (phencyclidine)—"angel dust" | Blocks channel for excitatory amino acid neurotransmitters; increases dopamine activity | Violent behavior; feelings of power; numbness; disorganized thoughts | Psychosis; convulsions; coma; death |
| ***Therapeutic*** | | | |
| Antidepressants: tricyclic antidepressants (for example, Elavil, Anafranil) | Block the reuptake of norepinephrine from synapses | Relieve depression and obsessive-compulsive disorder | Drowsiness; confusion |
| Selective serotonin reuptake inhibitors (SSRIs) (for example, Prozac, Zoloft, Paxil, Lexapro) | Block the reuptake of serotonin from synapses | Relieve depression and obsessive-compulsive disorder | Insomnia; anxiety; headache; sexual dysfunction |
| Monoamine oxidase inhibitors (for example, Parnate, Nardil) | Block the breakdown of biogenic amine neurotransmitters | Relieve depression | Liver damage; hyperexcitability |
| Antianxiety drugs: benzodiazepines [for example, Xanax, Valium, Librium, Rohypnol ("date rape drug," "roofies")] | Bind to GABA receptors and increase inhibitory neurotransmission | Relieve anxiety; cause sleepiness and in some cases amnesia | Drowsiness; memory loss in some cases |
| Antipsychotic drugs: phenothiazines (for example, Thorazine, Mellaril, Stelazine), and atypical antipsychotics (Abilify, Risperdal) | Block dopamine receptors | Ease schizophrenic symptoms | Decreased control of movement |

*The capitalized drug names represent only some of the brand name varieties available in each category. In some cases, generic versions are also available.
[†]Note: Some of these drugs have some therapeutic value under certain conditions.

and memory. AD is also associated with the degeneration and death of neurons, particularly in the hippocampus and parietal lobes, which is why it is considered a neurodegenerative disease.

Researchers have identified variation in a few genes whose products are associated with the likelihood of developing AD later in life, but the underlying changes that result in the expression of these and other possible AD-related genes are still the subject of considerable research. Although genetics undoubtedly has a role in AD, it is not the only possible cause. For example, when one identical twin develops AD, the other appears to be at increased risk but does not always develop the disease, even if he or she survives to a very old age. Moreover, evidence suggests that severe head injuries, metabolic diseases such as diabetes, and heart and blood vessel disease may increase a person's risk to develop AD in later life.

Currently, AD cannot be prevented or cured. However, three major clinical approaches are currently being tested to slow down its progression. These are designed to (1) induce a person's immune system to destroy β-amyloid as soon as it is formed, (2) prevent the formation of β-amyloid with drugs that block its synthesis, or (3) prevent the accumulation of β-amyloid into large aggregates using antiaggregation drugs. These approaches hold promise but remain unproven.

Until a cure for AD is found, its impact on public health remains enormous. Currently, about 4–5 million Americans have AD, a number that is expected to grow to nearly 16 million by 2050. The prevalence of the disease is about 3% for people between the ages of 65 and 74, and 25–50% for people older than 85. Estimated costs associated with providing health care and housing for AD patients (30% of whom live in nursing homes), as well as lost productivity in the workplace, total a staggering $100 billion in the U.S. per year. This number will rise substantially now that the oldest members of the population spike known as the baby boom generation have passed the age of 70.

## Many Illicit Drugs Disrupt Normal Neurotransmission

Many illicit drugs work at the synapse to either enhance or interfere with the normal mechanisms of neurotransmission (see Table 33.2). In the presynaptic terminal, such drugs can decrease neurotransmitter release by decreasing $Ca^{2+}$ entry into the cell and preventing the exocytosis of vesicles containing stored neurotransmitters. In the synaptic cleft, drugs can slow the rate at which the neurotransmitter is broken down into an inactive form or taken back up into the presynaptic neuron, thereby prolonging the action of the neurotransmitter in the synaptic cleft. Some substances act on the postsynaptic membrane by either preventing the neurotransmitter from binding to its receptor or by acting as a substitute for the neurotransmitter by stimulating its receptor.

In effect, these drugs produce changes or imbalances in neurotransmission similar to those observed in some neurological disorders. These substances can increase activity, alter mood, and produce hallucinations. They can also have life-threatening effects and be highly addictive.

Some drugs, such as cocaine, block the removal of the amine neurotransmitters dopamine and norepinephrine from the synaptic cleft by preventing their reuptake into the presynaptic terminal. Morphine and marijuana mimic the actions of biological substances already in the brain, binding to receptors on the postsynaptic membrane. With these drugs, the resulting effects are much stronger than are the effects of natural neurotransmitters. It is no surprise that many of these drugs are termed mind-altering. They do, after all, change the ways in which neurons communicate with each other.

## 33.7 Reviewing the Concepts

- Disorders of the human central nervous system can be placed into several broad categories. Most neurological conditions can be classified as disorders of either neurotransmission or conduction. Mood disorders caused by disrupted neurotransmission include major depressive disorder (Table 33.1).
- Alzheimer's disease is a progressive disorder characterized by the formation of plaques and neurofibrillary tangles in brain tissue.
- Drugs used in the treatment of neurological disorders and many recreational and illicit drugs usually alter neurotransmission (Table 33.2).

## 33.7 Testing Your Knowledge

1. __________ is a progressive disease that causes a loss of memory and intellectual and emotional function.
   a. Major depressive disorder
   b. Parkinson's disease
   c. Amnesia
   d. Alzheimer's disease
   e. Stroke
2. Therapeutic drugs that reduce symptoms of depression may act
   a. by decreasing exocytosis of neurotransmitters in the brain.
   b. by increasing serotonin levels in brain synapses.
   c. by decreasing formation of plaques and neurofibrillary tangles.
   d. by blocking calcium entry into presynaptic neurons.
   e. by decreasing acetylcholine levels in brain synapses.

# Assess and Discuss

## Test Yourself

1. In vertebrates, the brain and spinal cord constitute
   a. the peripheral nervous system.
   b. the efferent division of the nervous system.
   c. the central nervous system.
   d. the autonomic nervous system.
   e. the central and peripheral nervous systems.
2. Which part of a neuron contains voltage-gated $Ca^{2+}$ channels and stored neurotransmitters?
   a. the myelin sheath
   b. the axon hillock
   c. the dendrites
   d. the axon terminals
   e. the cell body
3. The myelin sheath
   a. is produced by neurons in the peripheral nervous system.
   b. is only formed around neurons in the brain.
   c. is present on all neurons.
   d. is generally present around long axons in either the CNS or the PNS.
   e. significantly slows transmission along neurons.
4. Neurons that function mainly in connecting other neurons in the central nervous system are
   a. sensory neurons.
   b. efferent neurons.
   c. motor neurons.
   d. afferent neurons.
   e. interneurons.
5. The difference in charges across the plasma membrane of an unstimulated neuron is called
   a. an EPSP.
   b. the resting membrane potential.
   c. an IPSP.
   d. the graded potential.
   e. the action potential.

6. Which of the following contributes to the resting membrane potential?
   a. the relative leakiness of the membrane to $Na^+$ and $K^+$
   b. active transport of ions across the membrane
   c. concentrations of $Na^+$ and $K^+$ inside and outside the cell
   d. all of the above
   e. b and c only
7. A neuron has reached a threshold potential when it has depolarized to the point where
   a. the voltage-gated $Na^+$ channels become inactivated.
   b. sufficient numbers of voltage-gated $Na^+$ channels open to initiate a positive feedback cycle, resulting in an action potential.
   c. voltage-gated $K^+$ channels close.
   d. voltage-gated $Na^+$ channels close.
   e. both b and c occur.
8. The speed of transmission of an action potential along an axon is greatest in
   a. wide, unmyelinated axons.
   b. narrow, unmyelinated axons.
   c. wide, myelinated axons.
   d. narrow, myelinated axons.
9. Which of the following is *not* a response to activation of the sympathetic division of the autonomic nervous system?
   a. decreased heart rate
   b. increased breathing rate
   c. increased blood flow to the skeletal muscles
   d. increased blood glucose levels
   e. All of the above are characteristic responses to activation of the sympathetic division of the autonomic nervous system.
10. The _______ is a portion of the limbic system that is important for memory formation.
   a. medulla oblongata
   b. hippocampus
   c. pons
   d. epithalamus
   e. midbrain

## Conceptual Questions

1. Describe the differences between graded and action potentials.
2. In certain diseases, such as kidney failure, the $Na^+$ concentration in the body's extracellular fluid can increase. What effect might a high extracellular $Na^+$ concentration have on neurons?
3. **PRINCIPLES** A principle of biology is that new properties emerge from complex interactions. How is this principle evident by the structure and function of animal nervous systems?

## Collaborative Questions

1. Discuss how nervous systems are organized into central and peripheral nervous systems in many animals and how these two parts of the nervous system interact with each other.
2. An important feature of all nervous systems is the reflex. Try to think of as many reflexes as you can that occur in animals, including humans. Why are reflexes beneficial?

# Neuroscience II: How Sensory Systems Allow Animals to Interact with the Environment

© Gustavo Mazzarollo/Getty Images

Compound eyes of a robber fly (family Asilidae).

## Chapter Outline



Two young parents were alarmed to see their 15-month-old child's mouth profusely bleeding one morning. The child wasn't crying, nor did he seem frightened or in pain; however, the parents could see that a small piece of the boy's tongue had been chewed off. After extensive testing by the pediatrician and a neurologist, it was determined that the child suffered from a rare genetic disorder in which individuals cannot sense pain (called congenital insensitivity to pain). This is a potentially catastrophic disorder, because very young children cannot tell their parents when something hurts; this could be a twisted or broken ankle, a stomachache, a cut, or a bee sting. Such children must be watched almost constantly. Because they do not feel internal pain associated with illness, they typically must have their temperature taken every day. Small cuts may go completely unnoticed, increasing the likelihood of infections, and sprains and broken bones don't heal properly because the injury may go undetected for long periods. What must it be like to experience life without an ability to perceive pain? What do scientists even mean when they use words like sense, sensation, and perception?

In neuroscience, a **sense** is broadly defined as a system that consists of specialized cells that detect a specific type of chemical or physical stimulus and send signals to the central nervous system (CNS); there, the signals are interpreted (that is, they are perceived). Senses are the windows through which animals experience the world around them. Common examples include sight, smell, taste, touch, hearing, balance, and the ability to sense heat, cold, and pain.

In this chapter, we will examine how nervous systems detect incoming sensory information and how membrane potentials of specialized neurons change in response to sensory inputs. We will then learn how other structures of the nervous system interpret this neural activity. Finally, we will discuss how problems with sensory systems—in particular, vision and hearing deficits—can affect human health.

## 34.1 Introduction to Sensation

### Learning Outcomes

1. Describe the relationship between sensory transduction and perception.
2. Explain how sensory receptors communicate information about the intensity of a stimulus.

Sensory systems convert chemical or physical stimuli from an animal's body or the external environment into a signal that causes a change in the membrane potential of sensory neurons. **Sensory transduction** is the process by which incoming stimuli are converted into neural (electrical) signals. Sensory transduction involves cellular changes, such as opening of ion channels, which cause either graded potentials or action potentials in neurons.

**Perception** is an awareness of the sensations that are experienced. For instance, touching a hot object generates a thermal sensation, which initiates a neuronal response in the brain, giving us the perception that this stimulus is hot. Not all sensations are consciously perceived by an organism. Most of the time, for example, we are not aware of the touch of our clothing.

We begin our study of sensory systems by examining the specialized cells that receive sensory inputs. A **sensory receptor** is a specialized neuron or epithelial cell that recognizes an internal or external (environmental) stimulus and initiates sensory transduction by creating graded potentials (described in Chapter 33) in itself or an adjacent cell (Figure 34.1). If a response is strong enough, sensory receptors initiate electrical responses to stimuli, such as chemicals, light, and heat, which result in action potentials that are sent to the CNS.

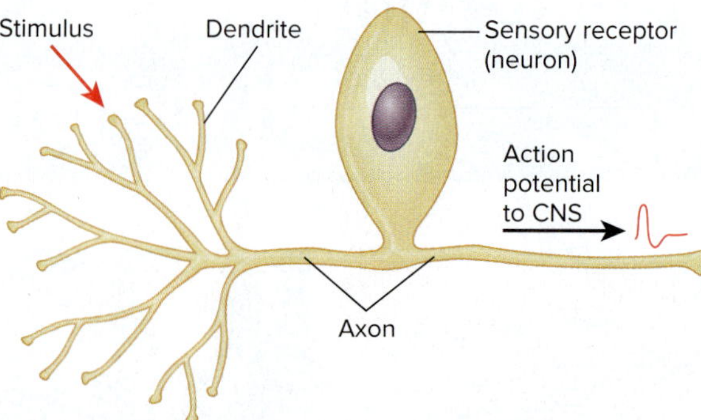


**(a) A neuron as a sensory receptor**

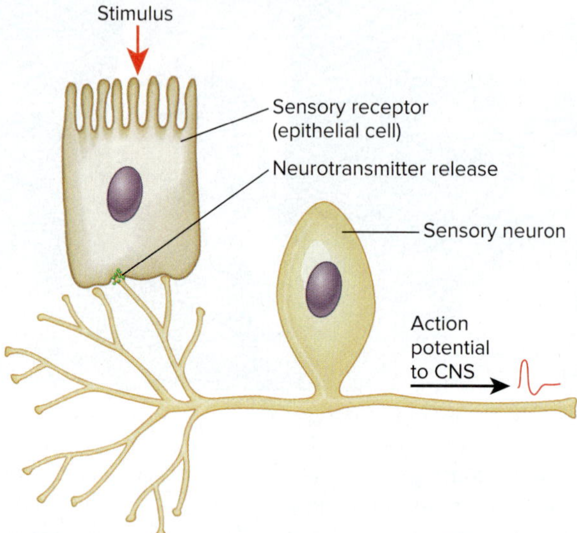


**(b) A specialized epithelial cell as a sensory receptor**

**Figure 34.1 Sensory receptors.** **(a)** Many sensory receptors are neurons that directly sense stimuli that arrive at the neuron's dendrites. **(b)** Other sensory receptors are specialized epithelial cells that sense stimuli and secrete neurotransmitters that stimulate nearby sensory neurons. In both cases, when stimulated, the neurons send action potentials to the CNS, where the signals are interpreted.

**BioConnections:** *The term receptor is used in more than one way in biology. What is the difference between the sensory receptor described in this chapter and the cell surface receptors described in Chapter 8?*

### A Strong Stimulus Generates More Frequent Action Potentials

How do sensory receptors pass along the intensity of a stimulus? Let's consider an example involving weak and strong stimuli to the sense of touch (Figure 34.2). Sensory transduction begins when the specialized endings of a sensory receptor respond to a stimulus. Such a stimulus—in this case, the touch of a glass rod—opens ion channels that allow sodium ions to diffuse down their electrochemical gradient into the cell, depolarizing the sensory receptor. The amount of depolarization is directly related to the intensity of the stimulus, because a stronger stimulus opens more ion channels.

The first response of a sensory neuron is usually a graded change in the membrane potential of the cell body that is proportional to the intensity of the stimulus (see Figure 34.2). The membrane potential, known as a **receptor potential** in these cells, becomes less and less negative as the strength of the stimulus increases. If a stimulus is strong enough, it depolarizes the membrane to the threshold potential at the axon hillock and produces an action potential in a sensory neuron (refer back to Figure 33.8).

Recall from Chapter 33 that action potentials proceed in an all-or-none fashion, regardless of the nature or strength of the stimulus that elicits them. How, then, can action potentials provide information about the intensity of a stimulus? The answer is that the strength of the stimulus is encoded by the frequency of action potentials generated. A particularly strong stimulus generates many action potentials in a short period of time because it can overcome the membrane's refractory period (refer back to Figure 33.8). As a result, the frequency of action potentials is higher when the stimulus is strong than when it is weak. The action potentials are transmitted into the CNS and carried to the brain for interpretation. The brain interprets a higher frequency of action potentials as a more intense stimulus.

### The CNS Processes Each Sense Within Its Own Pathway

Different stimuli produce different sensations because they activate specific neural pathways that are dedicated to processing only that type of stimulus. We know that we are seeing light because the signals generated by visual sensory receptors in the eye are transmitted along a

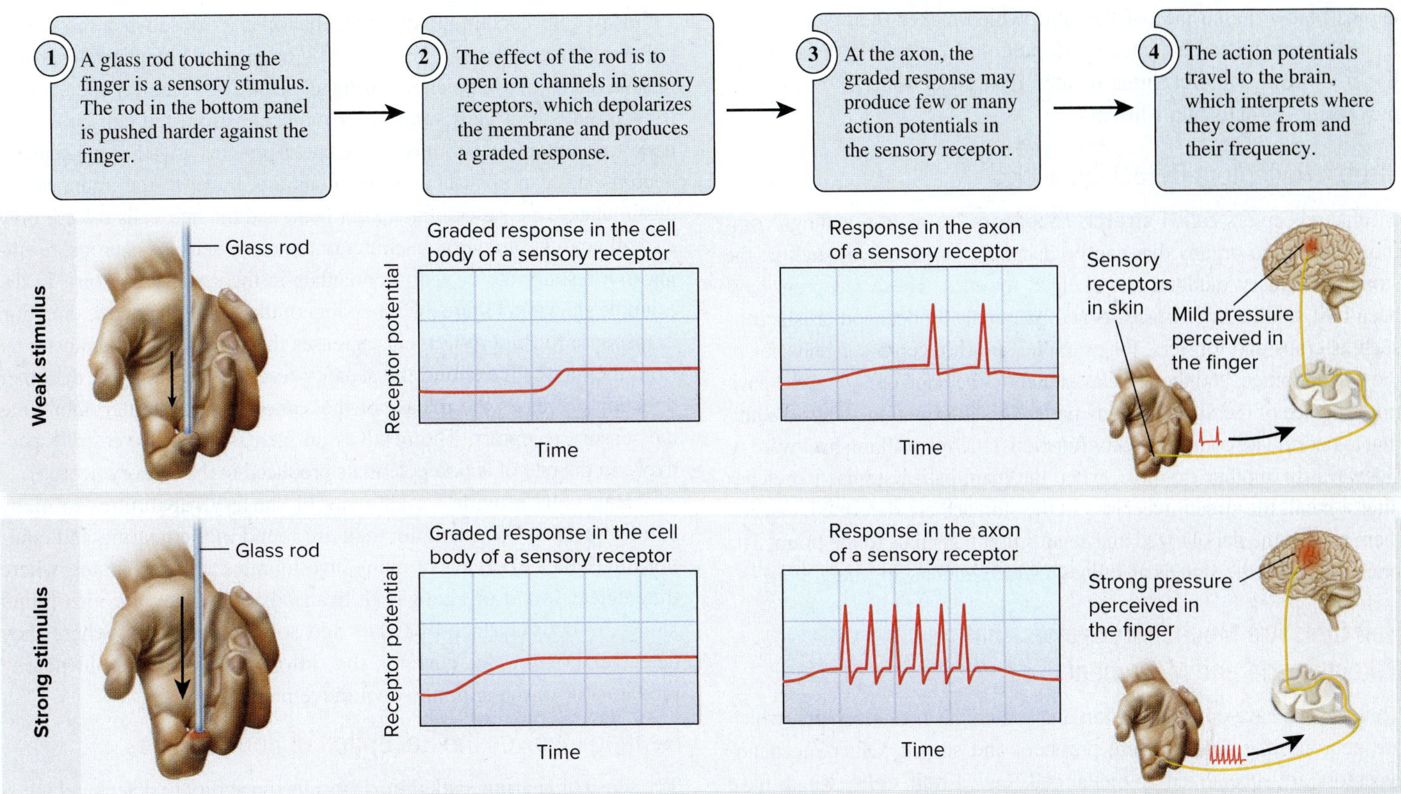


**Figure 34.2 Transduction of a sensory stimulus of two different intensities.** In this example, the sensory receptor is a neuron. Note the faster and larger graded response following the stronger stimulus. The perception of stimulus strength occurs in the brain.

neural pathway that sends action potentials into areas of the brain that are devoted to processing vision. For this reason, the brain interprets such signals as visual stimuli. The brain can separate and identify each sense because each sense uses its own dedicated pathway.

Sensory receptors can be divided into general types based on the type of stimulus to which they respond. Each type uses a different mechanism to detect stimuli and to transmit the information to different regions of the CNS. In the following sections of this chapter, we will examine the different types of stimuli and the structures and functions of different sensory receptors.

## 34.1 Reviewing the Concepts

- Sensory transduction is the process by which incoming stimuli are converted to neural signals. Perception is an awareness of sensations. Sensory receptors are neurons or specialized epithelial cells that respond to stimuli and begin the process of sensory transduction (Figures 34.1, 34.2).

## 34.1 Testing Your Knowledge

1. The process by which incoming stimuli are converted into neural signals is called
   a. perception.
   b. sensory transduction.
   c. sensory reception.
   d. sensory perception.
   e. perceptual transduction.

## 34.2 Mechanoreception

### Learning Outcomes

1. Describe what mechanoreceptors are and the various sensory abilities they impart to animals.
2. Outline the structure of the mammalian ear, and explain how mechanical forces move through it.
3. Describe how body position and movement are detected by sense organs.

**Mechanoreceptors** are cells that detect physical stimuli such as touch, pressure, stretch, sound, and movement. Physically touching or deforming a mechanoreceptor cell opens ion channels in its plasma membrane (see Figure 34.2). Some mechanoreceptors are neurons that send action potentials to the CNS in response to physical stimuli. Other mechanoreceptors are specialized epithelial cells that contain hairlike structures that bend in response to mechanical forces.

### Skin Receptors Detect Touch and Pressure

Several types of receptors in the skin of many animals detect touch, deep pressure, or the bending of hairs on the skin. Some of these specialized receptors consist of neuronal dendrites covered in dense connective tissue. When these structures become deformed (for example, by pressure on the skin), the neurons generate action potentials that travel to the brain. In mammals, these receptors are located at different depths below the surface of the skin, which makes them suitable for responding to different types of stimuli, such as soft or hard touch. Other skin mechanoreceptors located in the hair follicles respond to movements of hairs and whiskers.

## Stretch Receptors Detect Expansion

Mechanoreceptors called **stretch receptors** are neuron endings commonly found in organs that can be distended or stretched, such as the stomach, urinary bladder, and skeletal muscles. These receptors have been best studied in crustaceans and mammals. In decapod crustaceans such as crabs and lobsters, for example, stretch receptors in muscles of the tail, abdomen, and thorax relay signals to the brain regarding the position in space of the different body parts. This information allows the animal to coordinate complex motor functions, such as walking backward or sideways. In another example, when the mammalian stomach stretches after a meal, the stretch receptors in the stomach are deformed, causing them to become depolarized and send action potentials to the brain. The brain interprets the signals as fullness, which inhibits appetite.

## Hair Cells Are Mechanoreceptors That Detect Sound and Movement

Thus far, we have considered skin and stretch mechanoreceptors, which are neurons that detect touch, pressure, and stretch. Other mechanoreceptors are specialized epithelial cells called **hair cells,** which have deformable projections termed **stereocilia** that detect movement. The stereocilia are different from true cilia (refer back to Figure 4.17) because they do not contain motor proteins in their structure. They are displaced by movements of fluid or other physical stimuli (Figure 34.3). Stereocilia allow hair cells to be responsive to movements made by an animal.

Hair cells contain ion channels that open or close when the stereocilia bend, thereby changing the cell's membrane potential. When the plasma membrane depolarizes, voltage-gated $Ca^{2+}$ channels open, which triggers release of neurotransmitter molecules from the hair cells. The neurotransmitter then binds to receptor proteins in adjacent sensory neurons and can elicit action potentials that travel to the brain. When unstimulated, the stereocilia are not bent, and the hair cells release only a small amount of neurotransmitter onto nearby sensory neurons, resulting in a resting rate of action potentials in the sensory neurons. In the example shown in Figure 34.3, bending of the stereocilia in one direction in response to fluid movement increases the release of neurotransmitter from the hair cell, exciting the sensory receptors. Bending in the other direction decreases the release of the same neurotransmitter, inhibiting the sensory receptors. The result is an increase or a decrease, respectively, in the rate of action potentials produced in the sensory neurons.

Hair cells provide a rich array of sensory capabilities in many animal species. For example, they are found in the hearing and equilibrium (balance) organs of many invertebrates and vertebrates, where they detect sound or changes in head position. They are also found along the body surface of fishes and some amphibians, where they detect external water currents that inform an animal about nearby predators or of the animal's own movements.

## Hearing Involves the Reception of Sound Waves

The sense of hearing, called **audition,** is the ability to detect and interpret sound. This sense is critical for the survival and reproduction of many types of animals. For example, many animals locate their young by hearing its calls, or they use sound to attract a mate. Hearing is also important for detecting the approach of danger—a predator, a thunderstorm, an automobile—and locating its source.

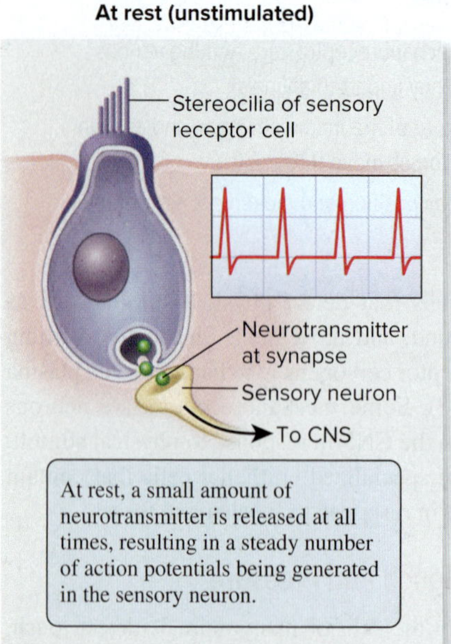


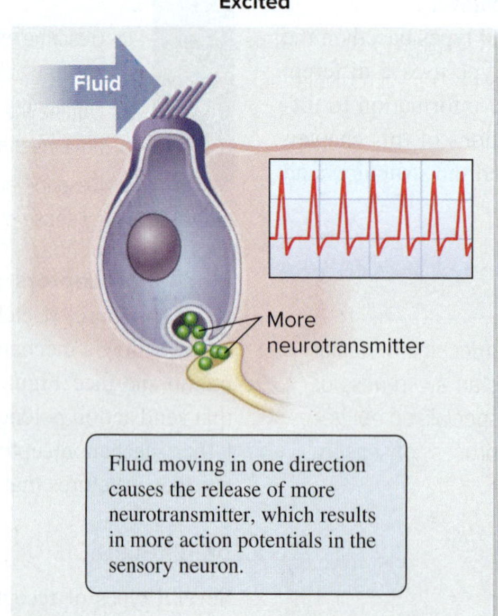

Excited
Fluid
More neurotransmitter
Fluid moving in one direction causes the release of more neurotransmitter, which results in more action potentials in the sensory neuron.


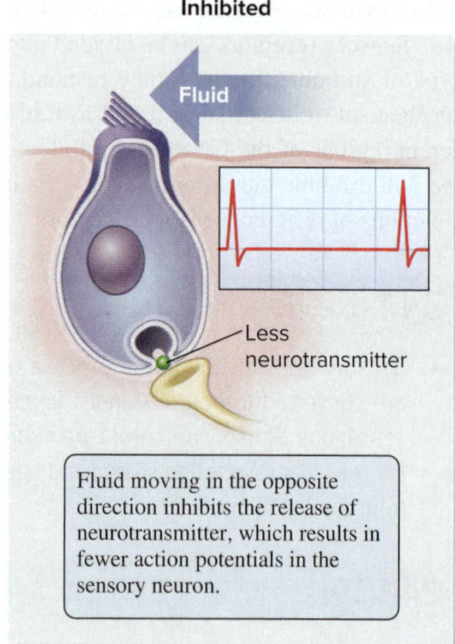

Inhibited
Fluid
Less neurotransmitter
Fluid moving in the opposite direction inhibits the release of neurotransmitter, which results in fewer action potentials in the sensory neuron.


**Figure 34.3 The response of hair cells to mechanical stimulation.** The stereocilia of these hair cells are hairlike projections of the plasma membrane that bend in response to fluid movement outside the cell.

*How are stereocilia different from cilia, which are described in Chapter 4?*

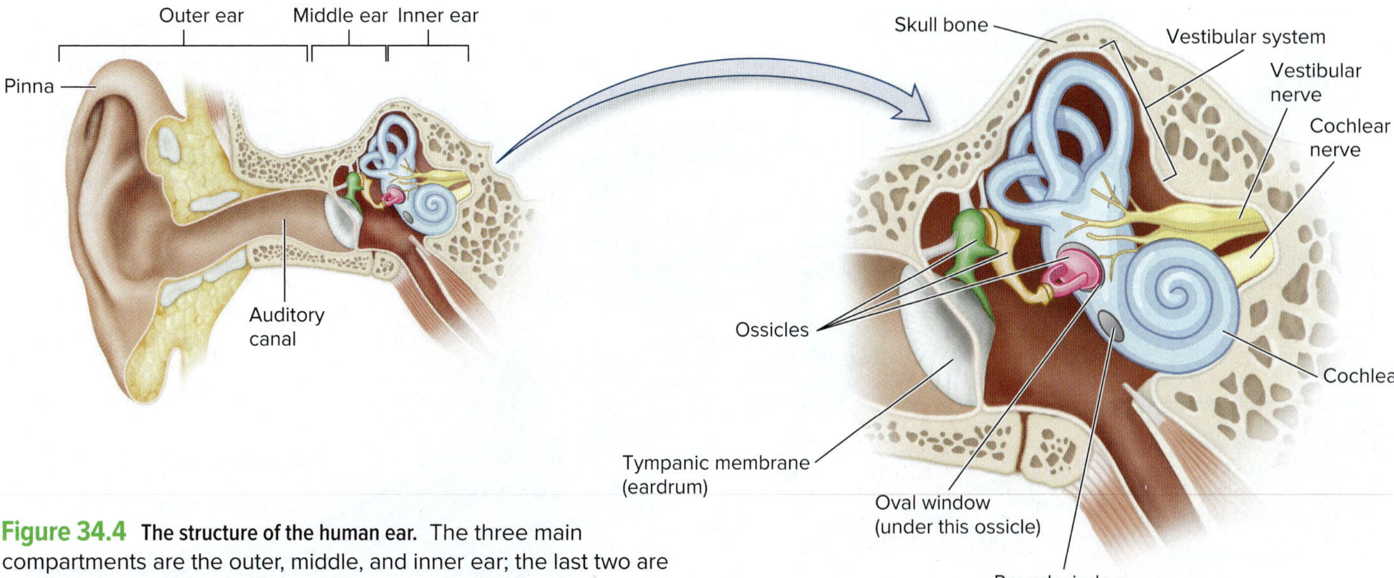


**Figure 34.4** **The structure of the human ear.** The three main compartments are the outer, middle, and inner ear; the last two are shown in more detail in the inset.

Sound consists of oscillating pressure waves; when we hear a sound, we are actually detecting waves of air pressure. The distance from the peak of one wave to the peak of the next is a wavelength. The number of complete wavelengths that occur in 1 sec is called the frequency of the sound (1 wave/sec = 1 Hertz, or Hz). The wavelength and frequency of sound waves impart certain characteristics to the stimulus. Short wavelengths have high frequencies that are perceived as a high pitch or tone, and long wavelengths have lower frequencies and a lower pitch. The human hearing range is 20–20,000 Hz.

The sense of hearing is present in vertebrates and to a limited extent in arthropods. Hearing, however, is especially well developed in birds and mammals; we turn now to a detailed discussion of the mammalian ear and the mechanism by which it detects sound, including the importance of hair cells in this mechanism.

**Structure of the Mammalian Ear** The mammalian ear has three main compartments (Figure 34.4). The first is the outer ear, which consists of the external ear, or pinna (plural, pinnae), and the auditory canal. The middle ear is separated from the outer ear by the tympanic membrane (eardrum). It contains three small bones called ossicles that link movements of the eardrum with another membrane, called the oval window. The oval window separates the middle ear from the inner ear. Within the inner ear is the **cochlea**—a coiled chamber of bone containing hair cells and the membrane-like round window—and the **vestibular system,** which functions in balance. It is the structures in the inner ear that generate the signals that travel via a nerve called the cochlear nerve to auditory centers of the brain.

**Generation of Electrical Signals in the Mammalian Ear** To understand how mammals hear, let's first consider how mechanical forces are transmitted through the ear. The outer ear collects sound waves and funnels them to the tympanic membrane, causing it to vibrate back and forth (Figure 34.5). The ossicles transfer the vibration of the tympanic membrane to the oval window, which vibrates against the cochlea. This sends pressure waves through two fluid-filled passages called the vestibular and tympanic canals. The waves eventually strike the round window, where they dissipate. Along the way, however, the waves cause the vibration of a membrane called the **basilar membrane,** which is formed from elastic fibers tensed across a duct called the cochlear duct. The basilar membrane is stiff near the oval window and gets progressively less stiff toward its end. The fluid pressure waves interact with a portion or all of the basilar membrane, depending on the frequency of the sound that initiated them; for example, only the lowest-frequency sound waves push fluid along the entire route shown in Figure 34.5.

Within the cochlea, mechanical vibrations are changed, or transduced, into electrical signals. This happens in a structure called the **organ of Corti,** which rests on top of the basilar membrane along its length. To understand how this works, look at a cross section through the cochlea (Figure 34.6). The organ of Corti contains supporting cells and rows of hair cells. The stereocilia of the hair cells are embedded in a gelatinous membrane called the tectorial membrane. The back-and-forth vibration of the basilar membrane bends the stereocilia in one direction and then the other. When bent in one direction, the hair cells depolarize and release neurotransmitter onto sensory neurons, which then send action potentials to the brain. When bent in the other direction, the hair cells hyperpolarize and stop releasing neurotransmitter. In this way, the frequency of action potentials generated by the sensory neurons is determined by the up-and-down vibration of the basilar membrane.

## The Sense of Balance Is Mediated by Statocysts in Invertebrates and the Vestibular System in Vertebrates

Let's now turn our attention to another form of mechanoreception, the sense of balance, also called equilibrium. This includes an animal's ability to sense the position, orientation, and movement of its body. Being able to sense body position is vital for the survival of animals. This is how a lobster, for example, rights itself when flipped over by a predator or how a bird maintains its balance while flying.

**Statocysts** Many aquatic invertebrates have sensory organs called **statocysts** that send information to the brain about the position of the animal in space (Figure 34.7). Statocysts are small, round structures

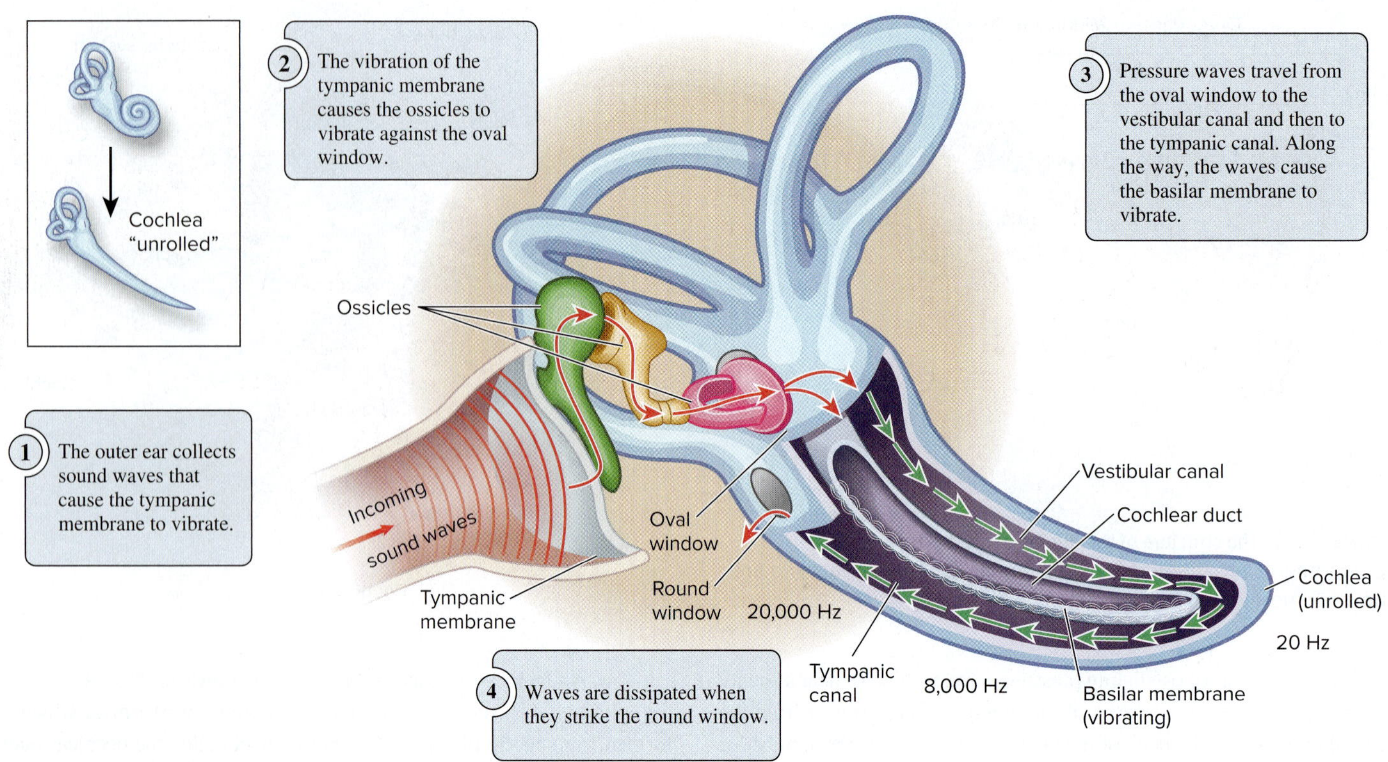


**Figure 34.5** **Movement of sound waves through the human ear.**

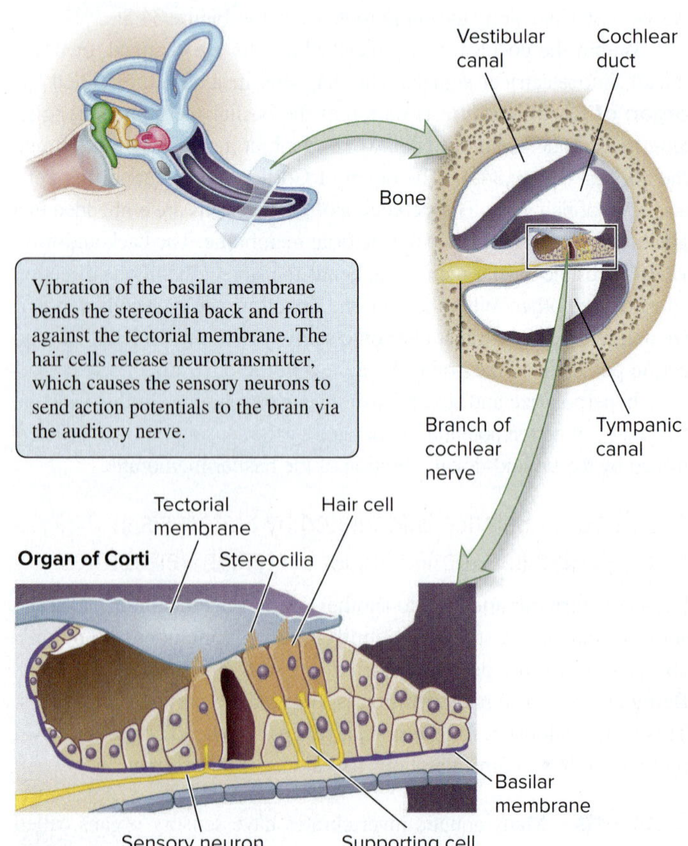


**Figure 34.6** **Transduction of mechanical vibrations to action potentials in the organ of Corti.**

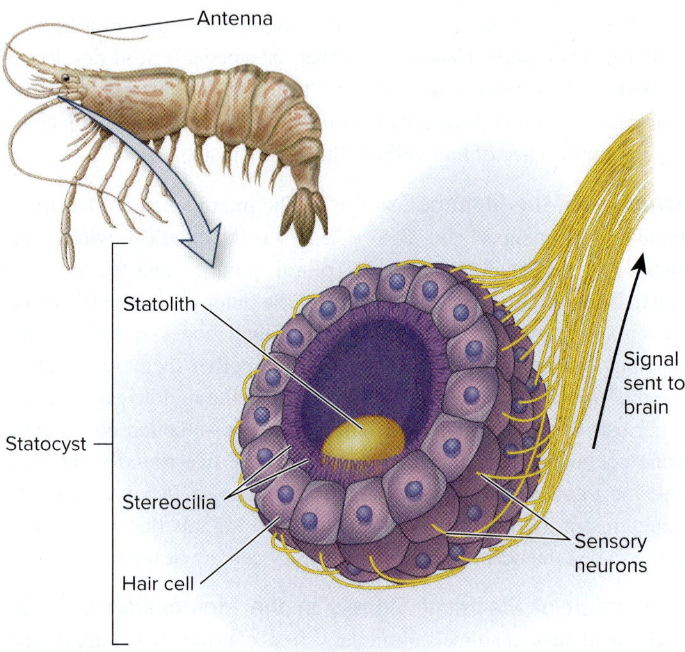


**Figure 34.7** **Sensing of balance in an invertebrate.** Statocysts located near the antennae consist of a sphere of sensory hair cells surrounding one or more statoliths. When the animal moves, gravity shifts the statolith and stimulates the hair cells beneath it.

**BioConnections:** *Is the use of statoliths unique to animals? Refer back to Figure 29.10 for a hint.*

consisting of an outer sphere of hair cells and one or more **statoliths,** tiny granules of sand or other dense objects that normally sit at the bottom of the statocyst when the animal is at rest. When the animal moves, gravity alters the statoliths' position and they move across hair cells, bending the stereocilia in that region. If the animal is turned over on its side, for example, the movement of statoliths stimulates a new set of hair cells to release neurotransmitter. This generates action potentials in sensory neurons that inform the brain of the change in body position and trigger a reflex that allows the animal to right itself.

**The Vestibular System** The organs of balance in vertebrates, known as the vestibular system, are located in the inner ear next to the cochlea (Figure 34.8). The vestibular system is composed of a series of fluid-filled sacs and tubules that provide information about linear or rotational movements. The utricle and saccule—the two sacs nearest the cochlea—detect linear movements of the head (see Figure 34.8a), such as those that occur when an animal runs, jumps, or changes its posture. The hair cells within these structures are embedded in a gelatinous substance that contains granules of calcium carbonate called **otoliths** (from the Latin, meaning ear stones), which are analogous to statoliths. When the head moves forward, the heavy otoliths are temporarily "left behind" as they are dragged forward more slowly, and the weight of the otoliths bends the stereocilia of the hair cells in the direction opposite that of the linear movement. This changes the membrane potential of the hair cells and alters the electrical responses of nearby sensory neurons. These signals are sent to the brain, which uses them to interpret how the head has moved.

Three **semicircular canals** connect to the utricle at bulbous regions called ampullae (singular, ampulla). The function of the semicircular canals is to detect rotational motions of the head (see Figure 34.8b). The hair cells in the semicircular canals are embedded in the ampullae in a gelatinous cone called the cupula. When the head moves, the fluid in the canal shifts in the opposite direction. This movement of fluid pushes on the cupula and bends the stereocilia of the hair cells in the direction of the fluid flow, which is opposite that of the motion of the head. The three canals are oriented at right angles to each other, and each canal is maximally sensitive to motion in its own plane. For example, the canal that is oriented horizontally would respond most to rotations such as shaking the head "no," whereas the other canals respond to "yes" motions or to tipping the ear to the shoulder. Overall, by comparing the signals from the three canals, the brain can interpret the motion of the head in three dimensions.

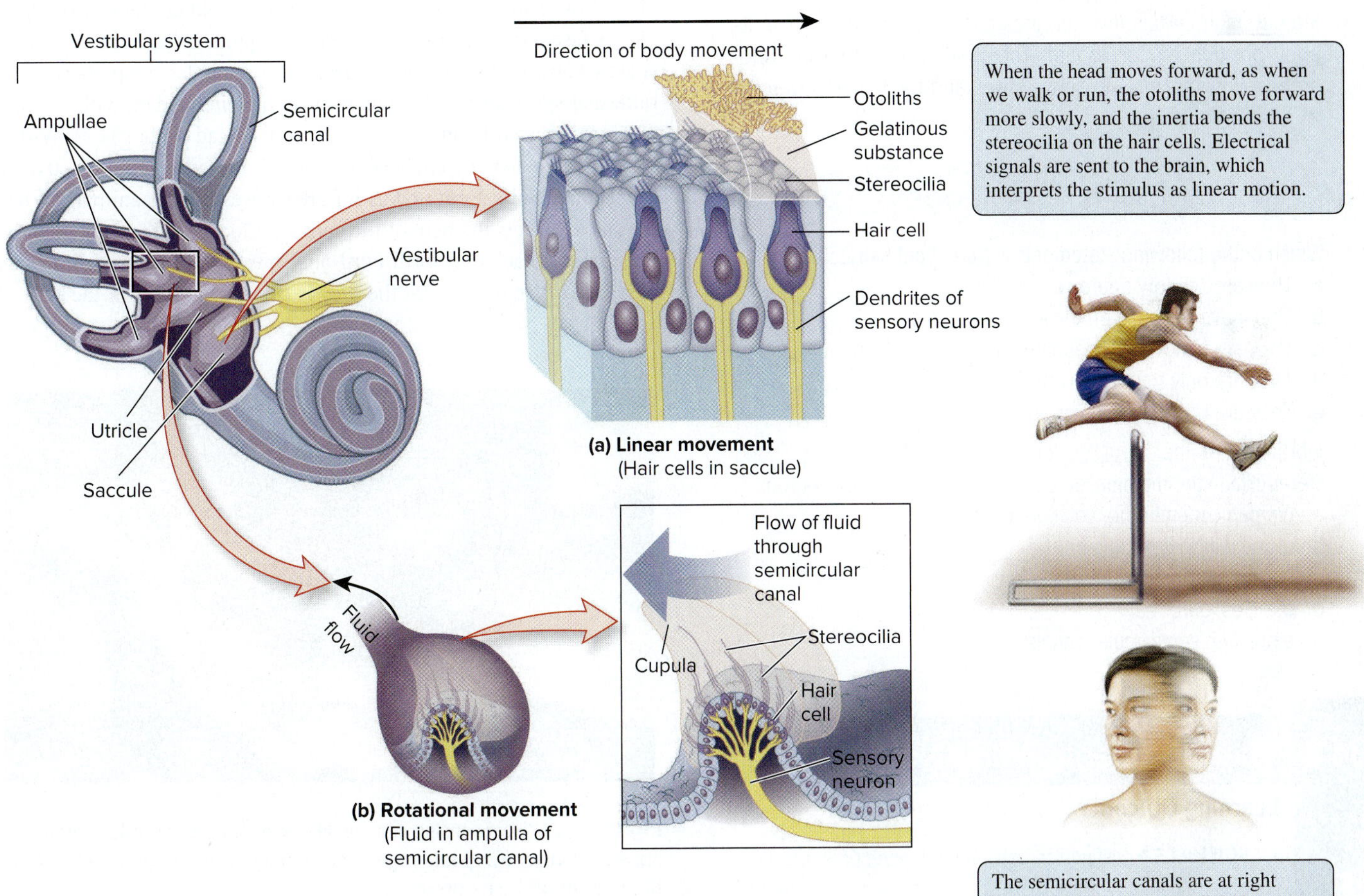


**Figure 34.8** The vertebrate vestibular system.

**Concept Check:** *Note the orientation of the three semicircular canals with respect to each other. Why are the canals oriented in three different planes?*

The vestibular system of vertebrates provides conscious information about body position and movement. It also supplies unconscious information for reflexes that maintain normal posture, control head and eye movements, and assist in locomotion. Researchers have discovered clear correlations between the types of locomotion an animal engages in and the size of its semicircular canals relative to an animal's body mass. Within primates, for instance, agile animals with jerky forms of locomotion, such as leaping tarsiers, have much larger canals than lorises, which are quadripedal and move slowly.

## 34.2 Reviewing the Concepts

- Mechanoreceptors respond to physical stimuli such as touch, pressure, stretch, movement, and sound.
- Hair cells have stereocilia that respond to movements of fluid or other stimuli; these cells release neurotransmitters that can trigger action potentials in adjacent sensory neurons (Figure 34.3).
- The mammalian ear has three main compartments: the outer, middle, and inner ear. Sound (pressure) waves are transmitted through the outer and middle ear to the cochlea of the inner ear, where the hair cells are located and signals are generated that travel to the brain for interpretation (Figures 34.4, 34.5, 34.6).
- Statocysts in certain invertebrates allow an animal to sense its body position or equilibrium. The vestibular system in vertebrates allows animals to sense linear and rotational movement (Figures 34.7, 34.8).

## 34.2 Testing Your Knowledge

1. Which of the following statements is *true* about hair cells?
   a. They are sensory neurons.
   b. They have true cilia on their surface.
   c. They are important for audition and balance in many animals.
   d. They are only found in invertebrates.
   e. They are only found in vertebrates.
2. Vibrations of the ______, located within the ______, bend stereocilia in the mammalian ear and activate sensory neurons.
   a. tympanic membrane, cochlear duct
   b. round window, middle ear
   c. basilar membrane, organ of Corti
   d. ossicles, outer ear
   e. ampullae, semicircular canals

## 34.3 Thermoreception and Nociception

### Learning Outcome

1. **SCISKILLS ▶** Distinguish between thermoreception and nociception, and propose reasons why they are vital to the safety and survival of an animal.

The perception of temperature and pain enables animals to respond effectively to potentially dangerous changes in their environments. These sensory stimuli are related to each other in that their receptors are located in some of the same areas, share similar physical features, and under certain conditions result in similar perceptions.

### Thermoreceptors Detect Temperature

An ability to sense heat is important for animals because their body temperature is affected by the external temperature. This is particularly true for ectotherms, animals that rely on heat from the environment to warm themselves. As you learned in Chapter 32, most animals can survive only if their body temperature remains within certain limits, because cell membranes and proteins function optimally only within a particular temperature range. Thermoreceptors respond to cold or hot temperatures by activating or inhibiting enzymes within their plasma membranes, which alters membrane channels. There are two types of thermoreceptors: those that respond to hot and those that respond to cold. Both of these types of sensory receptors are neurons. Thermoreceptors are often linked with reflexive behaviors, as when an animal steps on a hot surface and pulls its foot away.

In addition to skin receptors that sense the outside temperature, endotherms have thermoreceptors in the brain that detect changes in core body temperature. Activation of skin or brain thermoreceptors triggers physiological and behavioral adjustments that help maintain temperature homeostasis. These adjustments include changes in skin blood flow, shivering, and behaviors such as seeking shade or sunlight.

A fascinating example of thermoreceptors is found in venomous snakes known as pit vipers (a group that includes copperheads and rattlesnakes). These snakes can localize prey in the dark with thermoreceptors located in pits on each side of the head (**Figure 34.9**). Within the pit, a thin, nerve-rich, heat-sensitive membrane becomes activated in response to heat emitted as infrared waves by live animals. When the snake detects the heat of the animal, it localizes the prey by moving its head back and forth until both pits detect the same intensity of heat. This indicates that the prey is centered in front of the snake.

**Figure 34.9 Heat sensing in a reptile.** Sensory pits enable a white-lipped pit viper (*Cryptelytrops albolabris*) to detect and localize the heat given off by its prey.

**Concept Check:** *What advantage is provided by the presence of sensory pits on both sides of the pit viper's head, rather than just one side?*

## Nociceptors Warn of Pain

**Nociceptors** are neurons in the skin and internal organs. They respond to tissue damage or to stimuli that are about to cause tissue damage. Nociceptors are unusual because they can respond not only to external stimuli, such as extreme temperatures, but also to internal stimuli, such as molecules released into the extracellular space from injured cells. Damaged cells release a number of substances, including acids and small signaling molecules called prostaglandins, that cause inflammation and make nociceptors more sensitive to painful stimuli. Anti-inflammatory drugs such as aspirin and ibuprofen reduce pain by preventing the production of prostaglandins.

Signals arising from nociceptors travel to the CNS and reach the cerebrum, where the type or cause of the pain is interpreted. The signals are also sent to the limbic system, which holds memories and emotions associated with pain, and to the brainstem reticular formation, which increases alertness and arousal—an important response to a painful stimulus (see Chapter 33 for a discussion of brain regions and their functions).

The sense of pain—nociception—is one of the most important of all the senses, as was described in the chapter-opening story about a young child with congenital insensitivity to pain. It informs an animal whether it has been injured and triggers behavioral responses that protect it from further danger. Although in many cases we cannot know whether or how animals perceive pain, nociceptors have been identified in all classes of vertebrates and in many invertebrates.

## 34.3 Reviewing the Concepts

- Thermoreceptors in the skin and brain allow an animal to sense external and internal temperatures, respectively. Nociceptors in the skin and internal organs sense pain, an important survival adaptation in animals (Figure 34.9).

## 34.3 Testing Your Knowledge

1. Which of the following is false?
   a. Nociceptors and thermoreceptors are sensory neurons.
   b. Thermoreceptors are found in the skin and the brain of some animals and are of two types: hot- and cold-sensing thermoreceptors.
   c. Signals from painful stimuli are not sent to the CNS but are interpreted and acted upon entirely by neurons within the PNS.
   d. Thermoreceptors are often linked with reflexive behaviors.
   e. Nociceptors may respond not only to pain but also to certain other things such as extreme temperature.

# 34.4 Photoreception

### Learning Outcomes

1. Describe the structures of the various types of visual organs in animals.
2. Compare and contrast the structure and function of rods and cones, and explain how they detect and respond to light.

A type of sensory reception called electromagnetic reception is widespread in animals. It includes the ability of different animals to detect magnetic or electrical fields, such as the ability of some sharks and rays to detect the tiny electrical fields generated by the beating of the hearts of their prey. The most common type of electromagnetic sensing, however, is photoreception. Visual systems employ specialized neurons called **photoreceptors** that detect light. Some of the properties of light are described in Chapter 7. In this section, we will examine the variety of animal visual organs that detect light and send signals to the brain, where the image is interpreted. We then take a detailed look at the structure and function of the vertebrate eye.

## Compound Eyes Are Found in Many Invertebrates

The ability to sense light is an ancient adaptation found even in many unicellular organisms, where it may provide a selection advantage for photosynthesis or protection. Typically, such organisms may have a single eyespot that contains a small number of light-sensitive molecules but lack the ability to discern the direction of a light source or to interpret a visual image.

In contrast, arthropods and some annelids have image-forming **compound eyes** (Figure 34.10a), which consist of hundreds to as many as 10,000 light detectors called ommatidia (singular, ommatidium) (Figure 34.10b). Each ommatidium makes up one facet of the eye. Within the ommatidium, a two-part **lens**—composed of an outer cornea and an inner crystalline cone—focuses light onto a long, central structure called a rhabdom (Figure 34.10c). The rhabdom is a column of light-sensitive microvilli that project from the cell membranes of the photoreceptor cells of the ommatidium (Figure 34.10d). The light-sensitive molecules required for vision are located in the microvilli; the extensive surface area imparted by the microvilli provides the eye with increased sensitivity to light. Pigment cells surrounding the photoreceptor cells absorb excess light, thereby isolating each ommatidium from its neighbors.

Each ommatidium senses the intensity and color of light. Individually, each ommatidium receives light from only a very narrow field, but collectively they provide animals with a wide viewing area. In addition, animals with compound eyes are very sensitive to the movement of objects as they travel across the viewing areas of successive ommatidia. Combining the different inputs from neighboring ommatidia, the compound eye is believed to form a mosaic-type image that is interpreted by the brain. Animals such as bees and fruit flies, with large numbers of ommatidia, presumably have sharper vision than those with fewer sensory cells, such as grasshoppers. Behavioral studies have shown, however, that the resolving power of even the best compound eye is considerably less than that of the single-lens eye, which we will consider next.

## Vertebrates and Some Invertebrates Have a Single-Lens Eye

**Single-lens eyes** are found in vertebrates and in some mollusks, such as the squid and octopus, and in some annelids. In such eyes, different patterns of light emitted from or reflected off objects in the animal's field of view are transmitted through a small opening, or pupil, through a lens, to a sheetlike layer of photoreceptors called the **retina** at the back of the eye (Figure 34.11a). The light inputs form a visual image of the environment on the retina. The activation of these photoreceptors triggers electrical changes in neurons that pass out of

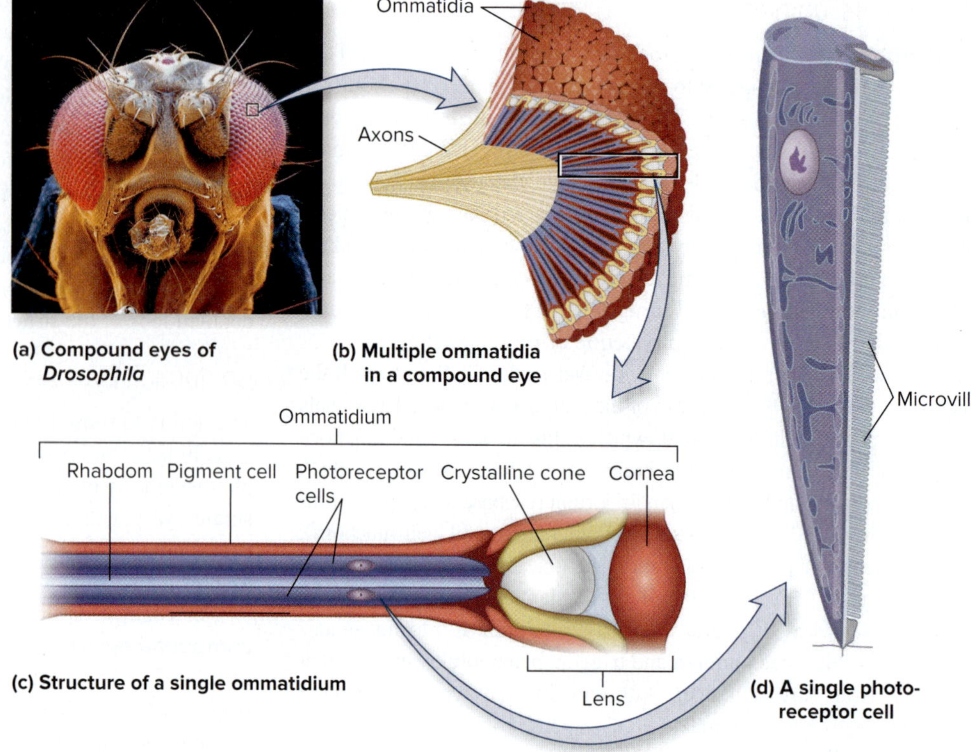


**Figure 34.10** **The compound eye of insects.** **(a)** Close-up of the eyes of a fruit fly (*Drosophila melanogaster*). **(b)** Each eye has approximately 1,000 ommatidia, which form a sheet on the surface of the eye. **(c)** Each ommatidium has a lens that directs light to the photosensitive rhabdom. **(d)** Extending from each photoreceptor cell and forming the rhabdom are many light-sensitive microvilli.

*(a)* © Eye of Science/Science Source

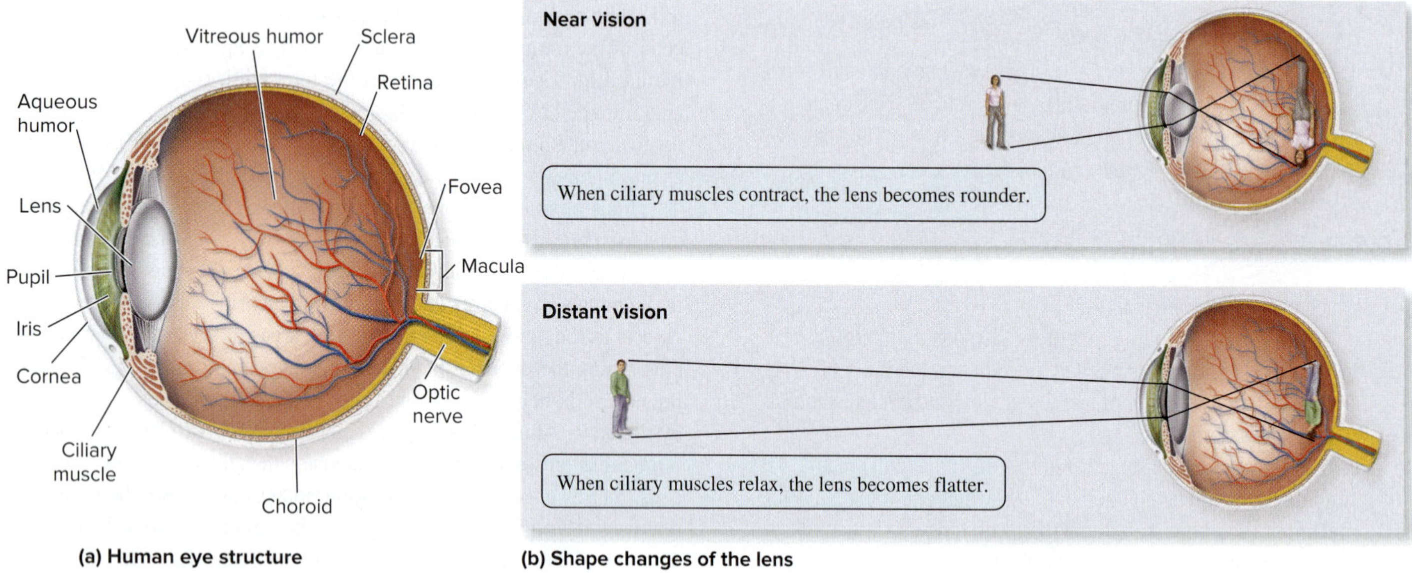


**Figure 34.11** **The vertebrate single-lens eye.** **(a)** Major structural features of the human eye. **(b)** Changes in lens shape help the eye focus at different distances. When an object is near, the ciliary muscles contract and the lens becomes rounder, causing light to bend more. When the object is far away, the ciliary muscles relax and the lens flattens.

the eye through the optic nerve, carrying the signals to the brain. The brain then interprets the visual image that was transmitted.

As illustrated in Figure 34.11a, the vertebrate eye has an outer sheath called the sclera (the white of the eye). At the front of the eye, the sclera is continuous with a thin, clear layer known as the **cornea.** The iris is a circle of pigmented smooth muscle responsible for eye color, with the pupil in its center. The size of the pupil changes when the muscles of the iris reflexively relax or contract to allow more or less light, respectively, to enter the eye. Between the sclera and the retina is a layer of blood vessels called the choroid. Within the eye are two chambers containing fluids—the aqueous and vitreous humors—that help maintain eye shape.

Because light radiates in all directions from a light source, light must be bent (refracted) inward toward the photoreceptors at the back of the eye. This is accomplished by the cornea and the lens. Whenever light passes from one medium to another medium of a different density, light waves will bend (try looking at a pencil in a glass partly filled with water). The cornea initially refracts the light. The light then passes through the lens, where it is refracted again and focused onto the layer of photoreceptors, the retina, at the back of the eye. The sharpest images are those that form directly at the back, at the macula; this region includes the fovea, which contains a particularly high density of photoreceptors. The bending of the incoming light results in an upside-down and laterally inverted image on the retina, but the brain adjusts for this, and the image is perceived correctly (Figure 34.11b).

The lens is adjusted to focus light that comes from different distances. In the mammalian eye, the lens changes shape to become more or less round depending on whether the object being looked at is near or far. Contraction and relaxation of the ciliary muscles adjust the lens according to the angle at which light enters the eye. When the lens is stretched, it flattens, and light passing through it bends less than when it is round (see Figure 34.11b).

## Rods and Cones Are Photoreceptor Cells

We turn now to the structure and function of the photoreceptors. Many animals, including humans, have two types of photoreceptors with names that are derived from their shapes: **rods** and **cones** (Figure 34.12). Rods and cones are cells with three functional parts. The outer segment of the cell contains folds of membranes that form stacks of discs. These discs contain the molecules that absorb light. The inner segment of the cell contains the cell nucleus and other cytoplasmic organelles. Rods and cones do not have axons but have synaptic terminals with vesicles that contain neurotransmitter. These terminals synapse with neurons within the retina.

Rods are extremely sensitive to low-intensity light, but they do not discriminate different colors. Rods are useful mostly at night, and they send signals to the brain that generate a black-and-white visual image. Unlike rods, cones are sensitive to wavelengths of light that allow animals to perceive color. Cones function best in daylight and are used by diurnal vertebrate species and by some insects such as the honeybee, which can detect the yellow color of pollen. Compared with rods, the human retina has fewer cones, but they are very densely clustered in and around the fovea; their density provides for sharp vision when we look directly at an object. Cones are less sensitive to light than rods, but this is less critical in daylight because the amount of light reaching the eyes at this time far exceeds what is needed to stimulate any photoreceptor cell.

## Rods and Cones Contain Visual Pigments That Absorb Light

**Visual pigments** are the molecules that absorb light; they are embedded in the disc membranes of the outer segment of rods and cones (Figure 34.13). These pigments consist of two components bonded together. The first is **retinal,** a derivative of vitamin A that is capable of absorbing energy from light. The second component of visual pigments is a protein called **opsin,** of which there are several types. Each type binds with retinal in different ways, which alters the wavelength of light that retinal absorbs.

Rods have only one visual pigment, called **rhodopsin** (see inset, Figure 34.13). Cones may contain any one of several types of visual pigments called cone pigments. The structural differences of the opsins in different cone pigments determine the wavelengths of light that the retinal in a given cone can absorb; in humans, these correspond to red, green, or blue light.

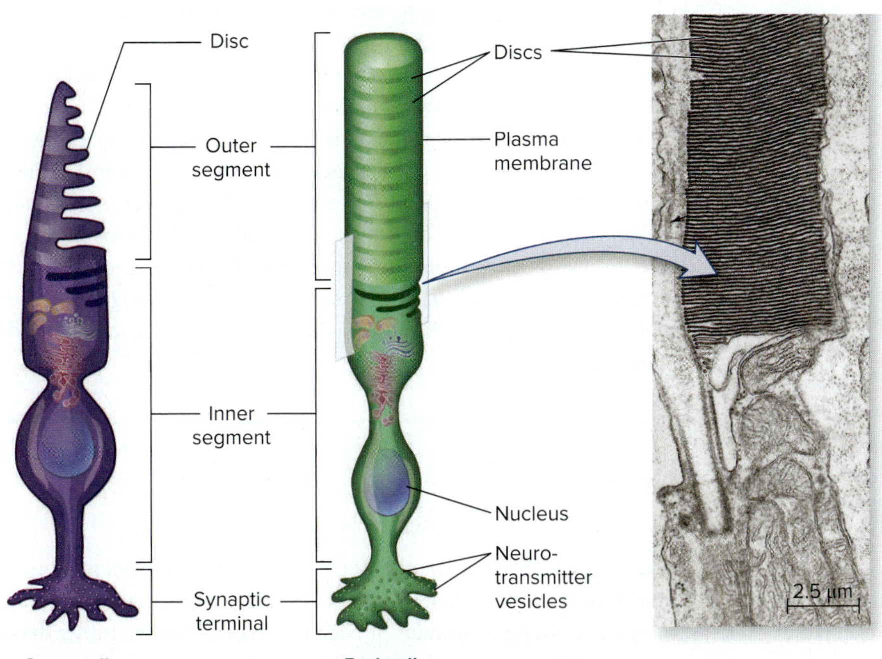


**Figure 34.12** **Structure of cones and rods.** The illustration shows the structure of a cone and rod photoreceptor, and a section of a rod in a transmission electron micrograph. Note the multiple stacks of membranous discs in the outer segment of the cell.

*(right)* © Don W. Fawcett/Science Source

## Biology Principle

### Structure Determines Function

The stacks of discs in the photoreceptor outer segment are an excellent example of a structural adaptation that increases surface area without significantly increasing volume, thereby allowing for greater light-capturing ability and improving the function of the cell as a sensory cell.

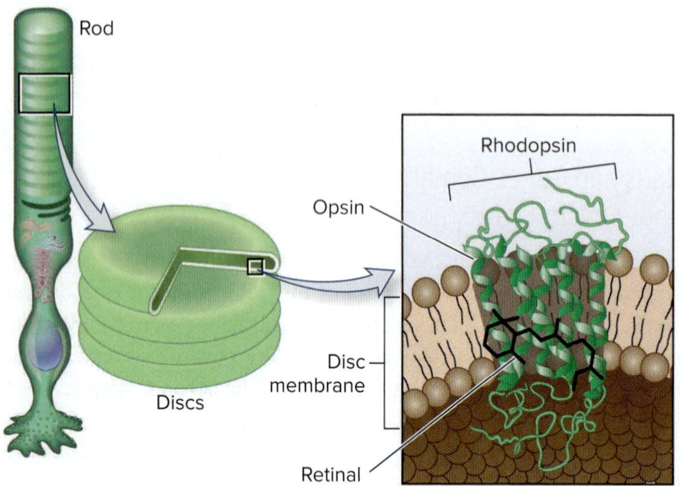


**Figure 34.13** **A visual pigment.** The visual pigment rhodopsin spans the membrane of the rod photoreceptor discs. It is composed of a membrane protein, opsin, that is bound to a molecule of retinal, a derivative of vitamin A that is capable of absorbing energy from light.

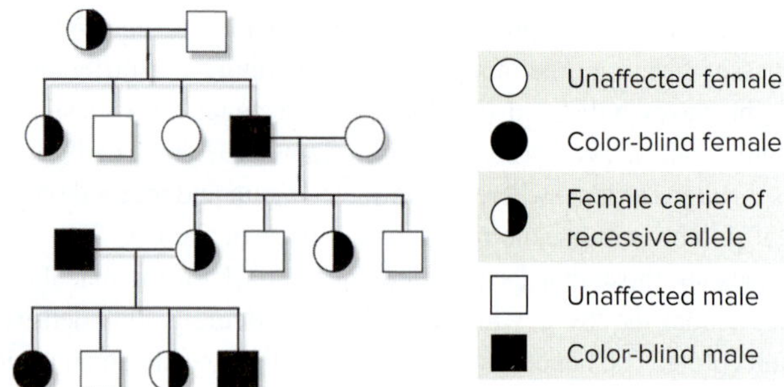


**Figure 34.14** **Color blindness.** An example of a pedigree for red-green color blindness showing all possible offspring.

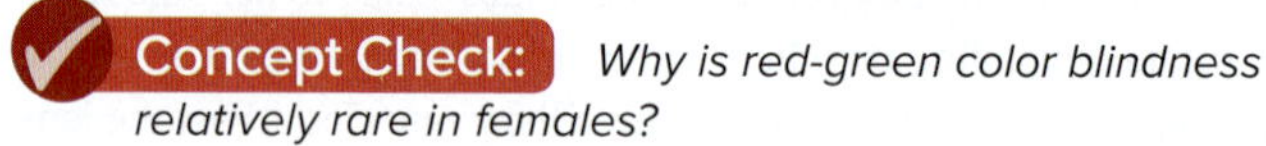
**Concept Check:** *Why is red-green color blindness relatively rare in females?*

## EVOLUTIONARY CONNECTIONS

### Color Vision Is an Ancient Adaptation in Animals

Color vision requires the presence of at least two types of opsins with optimal sensitivities to light of different wavelengths. Genome analyses and behavioral and neurophysiological testing of many animals have confirmed that color vision is widespread across invertebrate and vertebrate taxa, and it arose early in animal evolution. It is debatable why color vision provided a selective advantage to animals. Some investigators believe that it produced more acute vision due to an overall improved contrast and was not originally related to such commonly observed phenomena today as colorful plumage displays or brightly colored flowers. Coloration in plants may have evolved in response to the color sensitivity in animals, not vice versa.

It appears that the primordial color sensitivity was to short wavelengths in the ultraviolet/blue region of visible light (refer back to Figure 7.4), and then additional opsins evolved with sensitivities to medium and longer wavelengths. Most vertebrates that have been studied have four types of color-sensitive photoreceptors. By contrast, some insects, including the butterfly *Papilio xuthus,* appear to have up to six types of light receptors! It is impossible for us to understand how the world must appear to such an animal, which presumably sees shades of colors we cannot.

An intriguing exception to vertebrates are most mammals and, in particular, primates. The first mammals were most likely nocturnal animals, and therefore there would have been no selection pressure to retain sensitive color vision; recall that cones are less sensitive to low levels of light and color vision is limited at night. Most mammals today are dichromatic; that is, they have only two types of opsins and thus see a more limited color palette than do other vertebrates. Primates, however, are an exception, because a third opsin reappeared in the course of primate evolution, one that was sensitive to middle (green) wavelengths. It has been postulated that this may have imparted an advantage that allowed fruit- and leaf-eating primates to better discern orange, red, and yellow fruits against a background of green leaves.

Most humans have normal trichromatic color vision. However, problems in color vision may result from defects in the cone pigments arising from mutations in the opsin genes. The most common is red-green color blindness, which occurs predominantly in men (1 in 12 males compared with 1 in 200 females). Individuals with red-green color blindness either lack the red or green cone pigments entirely or, more commonly, have one or both of them in an abnormal form. In one form, for example, an abnormal green pigment responds to red light as well as green, making it difficult to discriminate between the two colors.

Researchers have determined that color blindness results from a recessive mutation in one or more genes encoding the opsins. Genes encoding the red and green opsins are located on the X chromosome, but the gene encoding the blue opsin is located on a different chromosome. In males, the presence of only one X chromosome means that a single recessive allele from the mother results in red-green color blindness, even though the mother herself may not be color blind (**Figure 34.14**).

### Photons Change Photoreceptor Activity by Altering the Conformation of Visual Pigments

Photoreceptors differ from other sensory receptor cells because at rest in the dark their membrane potential is slightly depolarized, whereas in response to a light stimulus, it becomes hyperpolarized rather than depolarized (**Figure 34.15**). In the dark, the cell membranes of the outer

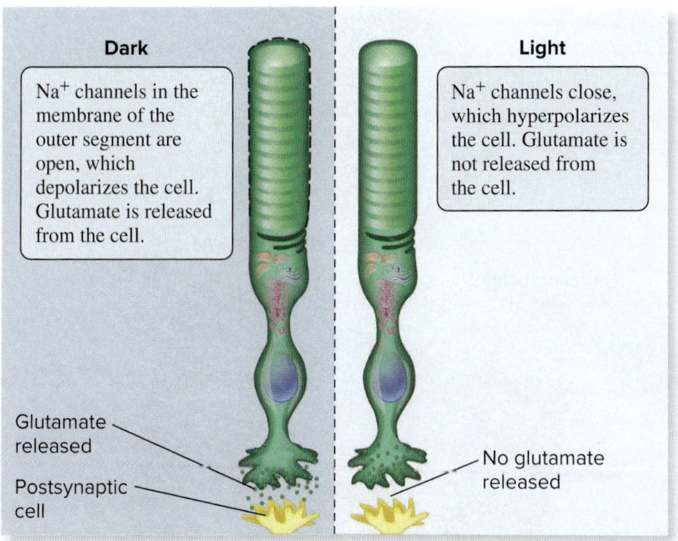

Dark
$Na^+$ channels in the membrane of the outer segment are open, which depolarizes the cell. Glutamate is released from the cell.
Light
$Na^+$ channels close, which hyperpolarizes the cell. Glutamate is not released from the cell.
Glutamate released
Postsynaptic cell
No glutamate released


**Figure 34.15** **Membrane potential response of photoreceptors to dark and light.**

segments of resting cells are highly permeable to $Na^+$. These ions diffuse into the cytosol of the cell through open $Na^+$ channels in the outer segment membrane. The $Na^+$ channels are opened by intracellular cyclic guanosine monophosphate (cGMP). In the dark, cytosolic concentrations of cGMP are high, keeping $Na^+$ channels open and depolarizing the cell. This depolarization results in a continu release of the neurotransmitter glutamate from the synaptic termi of the photoreceptor. The photoreceptor synapses with a postsyna tic cell that is the next neuron in the visual pathway. This initiate a series of events within the retina that is interpreted by the brain as an absence of light. In contrast, when exposed to light, the $Na^+$ channels in the outer segment membranes of the photoreceptor close. The resulting decrease in $Na^+$ concentration leads to a hyperpolarization of the cell. In response, the release of glutamate is stopped. This results in a series of cellular activations within the retina and brain that is interpreted as a visual image.

Let's take a more detailed look at the steps in the signal transduction pathway that allow a photoreceptor to respond to light (**Figure 34.16**).

1. When the photoreceptor is exposed to light, the retinal within the visual pigment absorbs a photon, the smallest unit of electromagnetic energy, including light (see Chapter 7). The energy of the photon alters the form of retinal from *cis*-retinal to *trans*-retinal, an isomer with a slightly different conformation due to a rotation at one of the molecule's double bonds (see Figure 34.16).
2. This change results in retinal briefly dissociating from the opsin protein. This causes a conformational change in the opsin that, in turn, activates a G protein called transducin, located in the

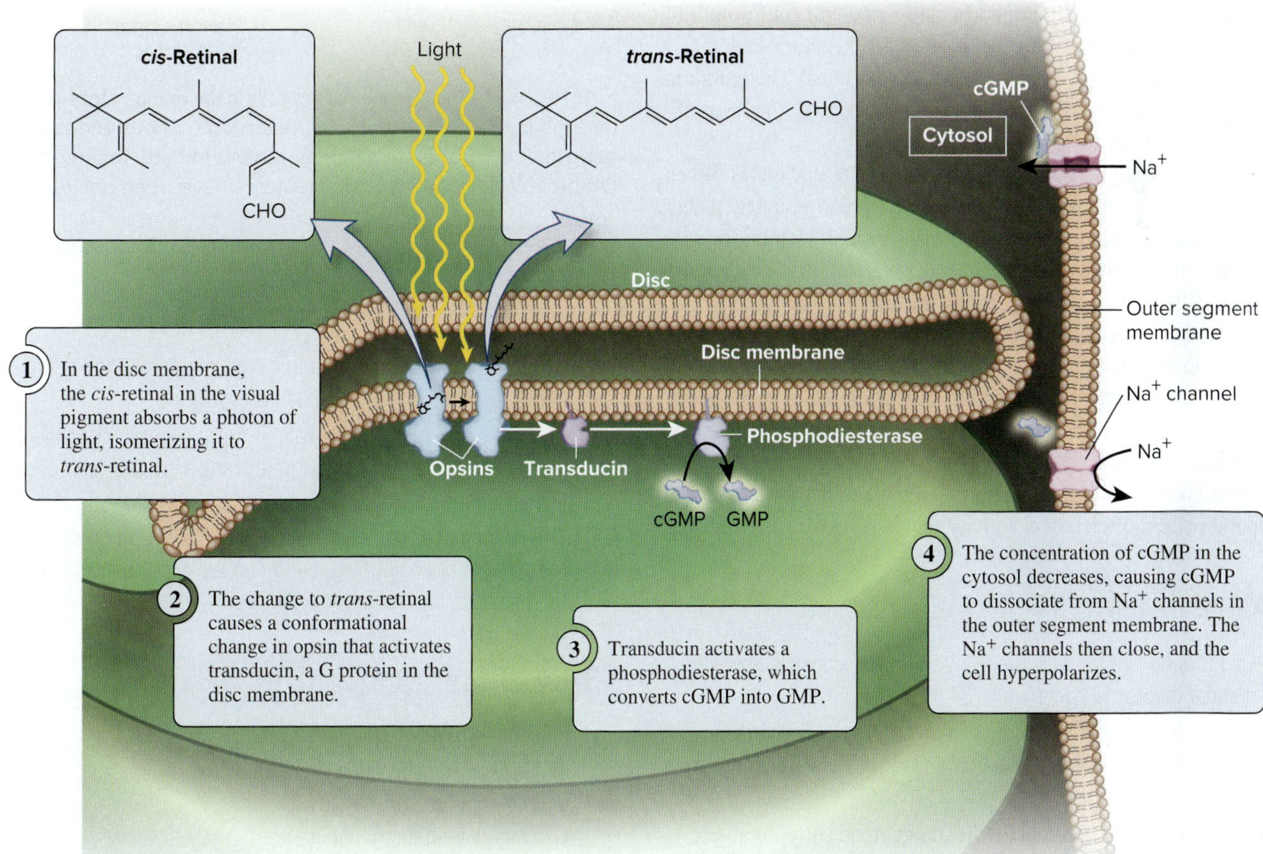

*cis*-Retinal
CHO
Light
*trans*-Retinal
CHO
cGMP
Cytosol
$Na^+$
Disc
Outer segment membrane
Disc membrane
$Na^+$ channel
Phosphodiesterase
$Na^+$
Opsins
Transducin
cGMP
GMP
1 In the disc membrane, the *cis*-retinal in the visual pigment absorbs a photon of light, isomerizing it to *trans*-retinal.
2 The change to *trans*-retinal causes a conformational change in opsin that activates transducin, a G protein in the disc membrane.
3 Transducin activates a phosphodiesterase, which converts cGMP into GMP.
4 The concentration of cGMP in the cytosol decreases, causing cGMP to dissociate from $Na^+$ channels in the outer segment membrane. The $Na^+$ channels then close, and the cell hyperpolarizes.


**Figure 34.16** **Signal transduction pathway in photoreceptor (rod) cells in response to light.**

disc membrane. (Thus, opsins belong to the class of proteins called G-protein-coupled receptors.)

3. The activated transducin activates another disc protein, a phosphodiesterase enzyme that is specific for cGMP. When phosphodiesterase is activated, it decreases the concentration of cytosolic cGMP by converting it to GMP.
4. The action of phosphodiesterase results in the closure of $Na^+$ channels in the outer segment membrane. Remember that in the dark, these channels are kept open by intracellular cGMP. However, as cGMP decreases, the $Na^+$ channels of the outer segment membrane close, and there is no longer any net diffusion of $Na^+$ into the cell. The membrane potential of the cell becomes less positive than it was in the dark. Therefore, the response of the cell is a hyperpolarization that is proportional to the intensity of the light.

As a result of hyperpolarization, there is a decrease in glutamate release from the photoreceptor (see Figure 34.15), ultimately leading to a visual image. The sequential activation of enzymes following activation of a single photoreceptor results in an amplification of the original signal (refer back to Figure 8.15). Because of this property, many animals are able to detect extremely low levels of light.

## The Visual Image Is Refined in the Retina

We now turn our attention to the neural pathway through which the visual signal travels to reach the brain. The vertebrate retina has several layers of cells (**Figure 34.17**). The photoreceptors (rods and cones) are positioned at the back of the retina. The next layer of the retina contains bipolar cells, so named because one end (or "pole") of the cell synapses with the photoreceptors, and the other end relays responses to the front layer of cells, the ganglion cells. The ganglion cells send their axons into the optic nerves, which carry the information to visual centers of the brain.

The pathway for light reception begins at the rods and cones. When the photoreceptors absorb photons, the cells hyperpolarize and no longer release glutamate onto bipolar cells. Glutamate normally inhibits bipolar cells, and thus its removal leads to their depolarization. The depolarized bipolar cells release their neurotransmitters onto ganglion cells, stimulating action potentials in those cells. These signals travel along pathways that include the thalamus, brainstem, cerebellum, and cerebral cortex. The cerebral cortex responds to such characteristics of the visual scene as whether something is moving, how far away it is, how one color compares with another, and the nature of the image (for example, a face). The cortex does not form a picture in the brain, but forms a spatial and temporal pattern of electrical activity that is perceived as an image.

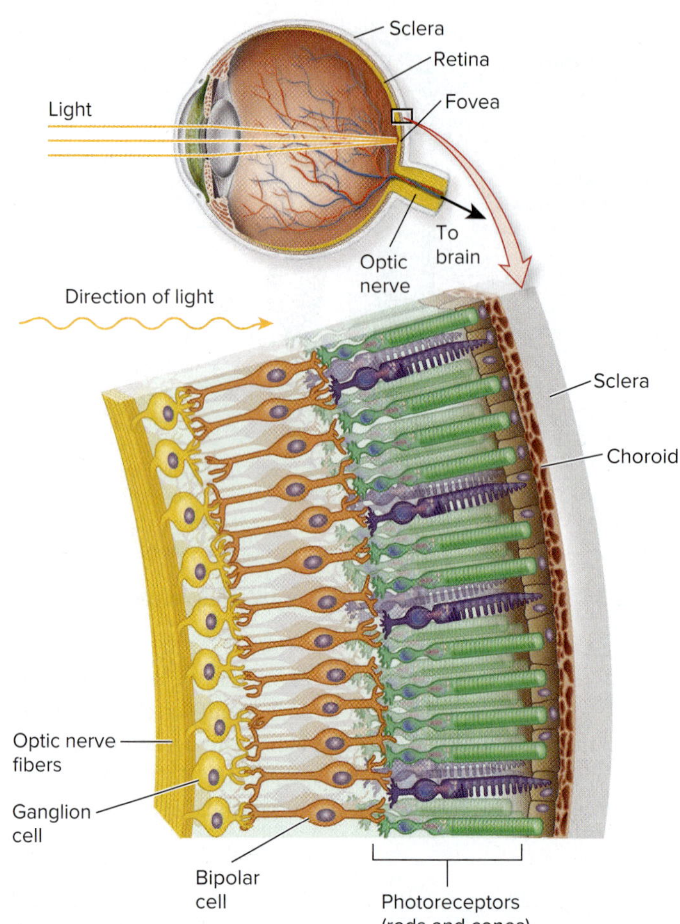


**Figure 34.17** **The arrangement of cells in the retina.** Light passes through layers of cells before it reaches the photoreceptors. The ganglion cells generate action potentials to carry the information to the brain. Additional structural features have been omitted for clarity.

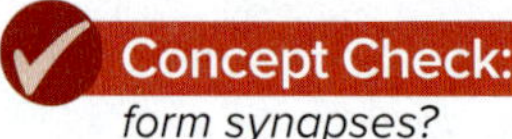


*With what cells do photoreceptors form synapses?*

## 34.4 Reviewing the Concepts

- The compound eye found in many invertebrates consists of many ommatidia that focus light (Figure 34.10).
- The single-lens eye is found in vertebrates and certain invertebrates. These eyes form a visual image of the environment on the retina (Figure 34.11).
- Rods and cones are photoreceptors found in the vertebrate eye. Only cones detect color. The visual pigment in rods and cones consists of retinal and a protein called opsin. Red-green color blindness is due to a defect in a type of opsin found in cone pigments (Figures 34.12, 34.13, 34.14).
- Absorption of photons by retinal triggers a signal transduction pathway that causes hyperpolarization of the photoreceptor (Figures 34.15, 34.16).
- The retina is composed of layers of cells, including photoreceptors, that receive light input and convey it to the visual centers of the brain via the optic nerve (Figure 34.17).

## 34.4 Testing Your Knowledge

1. Which is *true* of compound eyes?
   a. They do not contain a lens or lenslike structure.
   b. They cannot sense color.
   c. They have one ommatidium.
   d. They are found in insects and many vertebrates.
   e. They probably have less resolving power than single-lens eyes.

2. In the vertebrate eye, the pigment molecules that absorb light
   a. are found in the inner segment of all photoreceptors.
   b. are only found in cones.
   c. are found in the outer segment of both rods and cones.
   d. are composed of retinal or an opsin, but not both.
   e. reflect photons.

## 34.5 Chemoreception

### Learning Outcomes

1. **SCISKILLS ▸** Describe olfaction and gustation in animals, and propose reasons for why these senses are advantageous.
2. Explain how olfactory receptors respond to the binding of odor molecules.
3. Outline how receptor cells within taste buds respond to the binding of food molecules.

Chemoreception includes the senses of smell **(olfaction)** and taste **(gustation),** both of which involve detecting chemicals in the environment or food. These chemicals bind to sensory receptors called **chemoreceptors,** which in turn initiate electrical responses that travel to the brain. Airborne molecules that bind to olfactory receptors must be small enough to be carried in the air and into the nose. Taste molecules can be larger because they are conveyed in food and liquid.

Taste and smell are important mechanisms by which many species sense and learn about their environment. The two senses are closely related. In fact, the distinction is largely meaningless for aquatic animals, because for them all chemoreception comes through the water. Even in terrestrial animals, about 80% of the perception of taste is actually due to activation of olfactory receptors. This is why food loses its flavor when the sense of smell is impaired, such as when you have a cold. In this section, we explore chemoreception in mammals.

### Mammalian Olfactory Receptors Respond to the Binding of Odor Molecules

The olfactory sensitivity of mammals varies widely depending on their supply of olfactory sensory receptor cells, which ranges from 5 or 6 million in humans to 100 million in rabbits and 220 million in dogs. Olfactory sensory receptors are neurons located in the epithelial tissue at the upper part of the nasal cavity (Figure 34.18a). These cells are surrounded by two additional cell types: supporting cells and basal cells. Supporting cells provide physical support for the sensory receptors. In humans, basal cells differentiate into new sensory receptors every 30–60 days, replacing those that have died.

Olfactory sensory receptors have dendrites from which long, thin cilia extend into a mucous layer that covers the epithelium. These cells do not function like the mechanoreceptor hair cells of the auditory and vestibular systems; the cilia are different from hair cell stereocilia, which bend. Instead, olfactory sensory receptor cells have receptor proteins located in the plasma membranes of the cilia (Figure 34.18b). Airborne molecules dissolve in the mucous layer and bind to these receptor proteins. When an odor molecule binds to its receptor protein, it initiates a signal transduction pathway that opens $Na^+$ channels in the plasma membrane; this permits $Na^+$ to diffuse into the cell due to the ion's electrochemical gradient. The subsequent depolarization results in

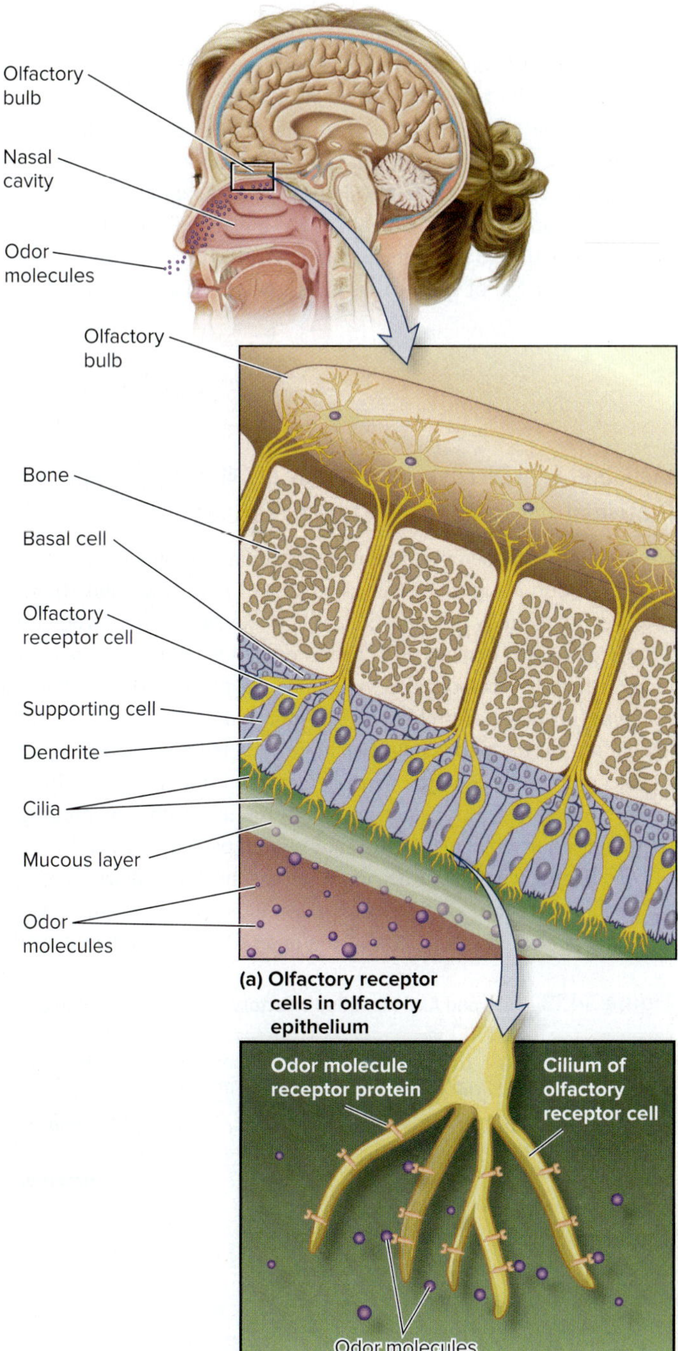


**Figure 34.18** **Olfactory structures in the human nose.** Odor molecules dissolve in a layer of mucus that coats the olfactory receptor cells. The molecules bind to receptor proteins in the membrane of cilia that extend from the olfactory receptor cells. Action potentials in the olfactory receptor cells are conducted to cells in the olfactory bulb, and from there to the brain for interpretation. Basal cells periodically differentiate into new olfactory receptor cells, replacing dead or damaged cells.

action potentials being transmitted to the next series of cells, which are located in the olfactory bulbs of the brain. The olfactory bulbs, which are part of the limbic system, are a collection of neurons that act as an initial processing center of olfactory information and relay it to the cerebral cortex for further processing and interpretation.

The relative size of the olfactory bulbs indicates the importance of olfaction to an animal. In humans, the olfactory bulbs make up only about 5% of the weight of the brain, whereas in nocturnal animals, like rats and mice, they can constitute as much as 20%. Even with their relatively limited olfactory sensitivity, however, humans have the capacity to detect up to 10,000 different odors, and other mammals are thought to respond to many more. Recent research suggests the possibility that this number could theoretically be as great as 1 trillion! Exactly how it is possible to detect such a large number of odors remained a mystery until 1991, when two scientists uncovered the molecular basis of olfaction.

## FEATURE INVESTIGATION

### Buck and Axel Discovered a Family of Olfactory Receptor Proteins That Bind Specific Odor Molecules

How does the olfactory system discriminate among thousands of different odors? American neuroscientists Linda Buck and Richard Axel set out to study this question. When they began, two hypotheses were proposed to explain this phenomenon. One possibility was that many different types of odor molecules might bind to one or just a few types of receptor proteins, with the brain responding differently depending on the number or distribution of the activated receptors. The second hypothesis was that olfactory sensory receptor cells can make many different types of receptor proteins, each type binding a particular odor molecule or group of related odor molecules.

To begin their study, Buck and Axel assumed that olfactory receptor proteins would be highly expressed in the olfactory sensory receptor cells, but not in other parts of the body. Based on previous work, they also postulated that the receptor proteins would be members of the large family of G-protein-coupled receptors (GPCRs). As shown in **Figure 34.19**, they isolated olfactory sensory receptor cells from rats and then used a homogenizer to break open the cells to release their mRNA. The mRNA was purified and used to make complementary DNA (cDNA) using the enzyme reverse transcriptase. This generated a large pool of cDNAs, representing all of the genes that were expressed in the receptor cells at the time of mRNA collection. To determine if any of these cDNAs encoded GPCRs, they used primers that recognized conserved regions within previously identified genes that encoded GPCRs. A conserved region is a DNA sequence that rarely changes among different members of a gene family (refer back to Chapter 18 for a description of gene families). The primers were used

**Figure 34.19** Buck and Axel identified olfactory receptor proteins in olfactory receptor cells.

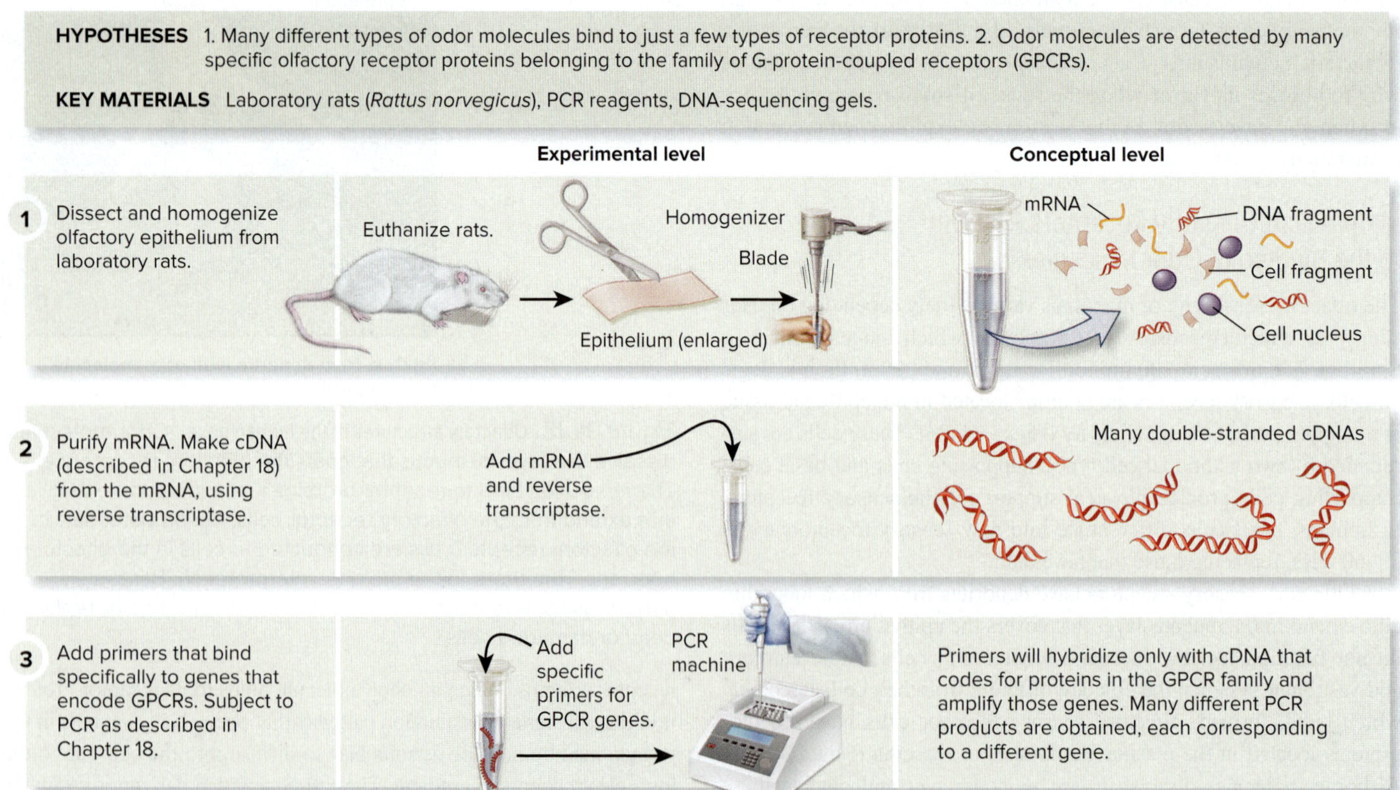

HYPOTHESES 1. Many different types of odor molecules bind to just a few types of receptor proteins. 2. Odor molecules are detected by many specific olfactory receptor proteins belonging to the family of G-protein-coupled receptors (GPCRs).
KEY MATERIALS Laboratory rats (Rattus norvegicus), PCR reagents, DNA-sequencing gels.
Experimental level
Conceptual level
1 Dissect and homogenize olfactory epithelium from laboratory rats.
Euthanize rats.
Homogenizer
Blade
Epithelium (enlarged)
mRNA
DNA fragment
Cell fragment
Cell nucleus
2 Purify mRNA. Make cDNA (described in Chapter 18) from the mRNA, using reverse transcriptase.
Add mRNA and reverse transcriptase.
Many double-stranded cDNAs
3 Add primers that bind specifically to genes that encode GPCRs. Subject to PCR as described in Chapter 18.
Add specific primers for GPCR genes.
PCR machine
Primers will hybridize only with cDNA that codes for proteins in the GPCR family and amplify those genes. Many different PCR products are obtained, each corresponding to a different gene.

**4** Subject each PCR product to DNA sequencing, also described in Chapter 18.

Output from automated sequencing (example)

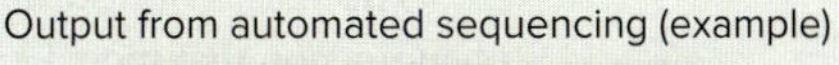


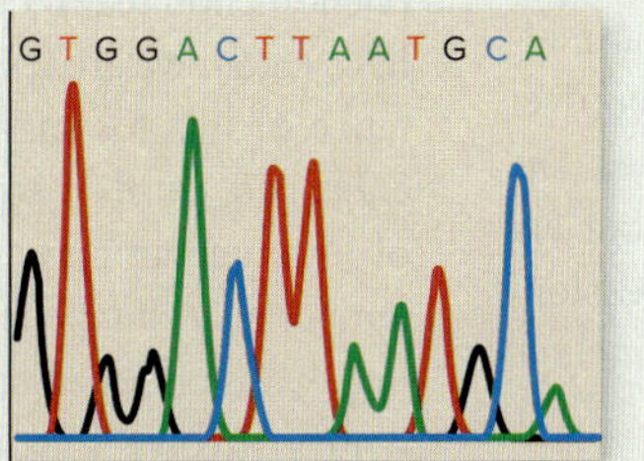

Different GPCRs will have slightly different DNA sequences.

**5 THE DATA**

At least 100 different GPCRs were uniquely expressed in olfactory sensory receptor cells. Analysis of their predicted amino acid sequences revealed significant variability in the putative ligand-binding regions (transmembrane domains 3–5).

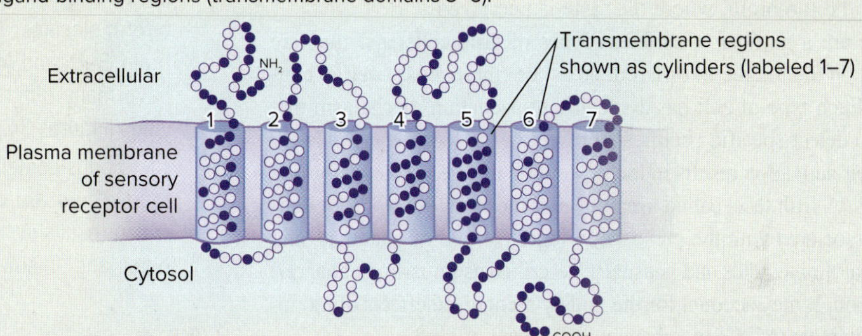


**6 CONCLUSION** Olfactory receptor cells express many different receptor proteins that account for an animal's ability to detect a wide variety of odors.

**7 SOURCE** Buck, L., and Axel, R. 1991. A Novel Multigene Family May Encode Odorant Receptors: A Molecular Basis for Odor Recognition. *Cell* 65: 175–187.

in the polymerase chain reaction (PCR) technique to amplify cDNAs that encoded GPCRs. This produced many PCR products that were then subjected to DNA sequencing. Buck and Axel identified at least 100 different genes, each encoding a GPCR with a slightly different amino acid sequence as predicted from the DNA sequence. Significant variability in the predicted amino acid sequences between the olfactory GPCRs was found in the putative ligand-binding region of the molecules (believed to be transmembrane domains 3–5). Further research demonstrated that these genes were expressed in olfactory cells, but not in other parts of the body. These results were consistent with the second hypothesis—namely, that organisms produce a large number of distinct olfactory receptor proteins, each type binding a particular odor molecule or a group of related odor molecules.

Since these studies, researchers have determined that this family of olfactory genes in mammals is surprisingly large. In humans, roughly 400 genes encode olfactory receptor proteins. Each olfactory receptor cell is thought to express only one type of GPCR that recognizes its own specific odor molecule or group of closely related molecules. Most odors an animal encounters, however, are due to multiple chemicals that activate many different types of odor receptors at the same time. We perceive odors based on the combination of receptors that become activated. This ability to combine different inputs allows animals such as humans to perceive thousands of different odors despite having only a few hundred different olfactory receptor proteins.

The research of Buck and Axel explained, in part, how animals detect a myriad of odors. In 2004, they received the Nobel Prize in Physiology or Medicine for this pioneering work.

### *Experimental Questions*

1. What were the two major hypotheses to explain how animals discriminate between different odors? How did Buck and Axel test the hypothesis of multiple olfactory receptor proteins?
2. **SCISKILLS ▶** Of the two hypotheses proposing how animals discriminate between different odors (see question 1), which one was supported by the results of this experiment? With the evidence presented by Buck and Axel, what is the current hypothesis regarding the discrimination of odors in animals?
3. **SCISKILLS ▶** Predict the survival or reproductive advantages that may be provided to animals by the ability to sense thousands or more different types of odors.

## Taste Buds Detect Food Molecules

**Taste buds** are structures containing cells that detect particular molecules in food. Humans have about 9,000 taste buds, whereas other mammals, such as pigs and rabbits, can have double that number. The bumps that you see on your tongue are not the taste buds but are papillae, elevated structures that collect food molecules in depressions called taste pores. These pores contain many taste buds that include the sensory receptor cells along with supporting cells (Figure 34.20). The tips of the sensory receptor cells have microvilli that extend into the taste pore. Here, molecules in food that have dissolved in saliva bind to receptor proteins. This triggers intracellular signals that alter ion permeability and membrane potentials. The sensory receptor cells then release neurotransmitters onto underlying sensory neurons. Action potentials travel from these neurons to the thalamus and other regions of the forebrain, where the taste is perceived.

There are a number of different types of taste cells, and they are distributed on the tongue in different areas, but these areas overlap considerably. Each type of cell has a specific transduction mechanism that allows it to detect specific chemicals present in the foods and fluids we ingest. Their activation results in the perception of sweet, sour, salty, and bitter tastes. A fifth taste called umami (after the Japanese word for delicious) is associated with the presence in ingested food of glutamate and other similar amino acids and is usually described as imparting enhanced flavor to food. It may account for the widely recognized effects of monosodium glutamate (MSG) in enhancing the flavor of food.

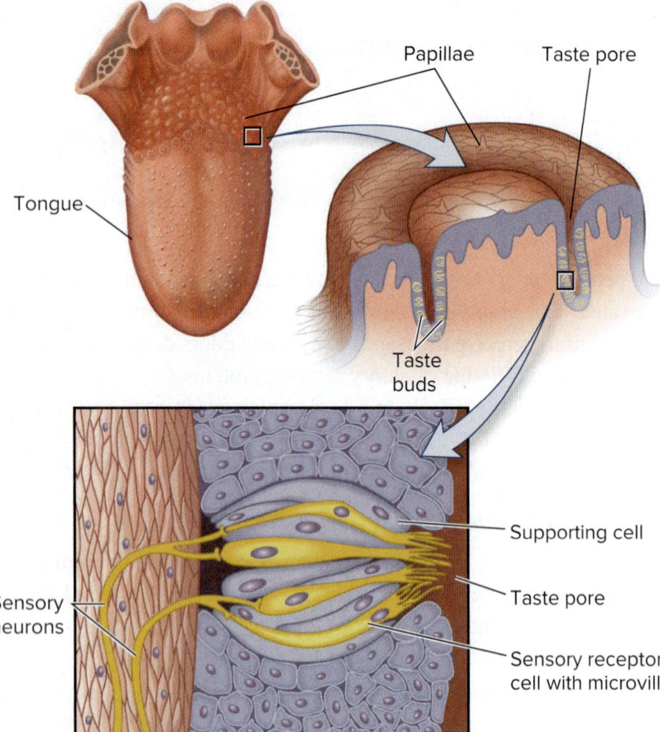


**Figure 34.20 Structures involved in the sense of taste.** This sense occurs in taste buds, which contain the sensory receptor cells that respond to dissolved food molecules.

**Concept Check:** *Where are the receptor proteins that bind to food molecules located?*

The senses of taste and smell are enhanced when we are hungry, a phenomenon that most likely occurs in other animals as well. Once we have eaten, we are less aware of the smell and taste of food. The importance of this for survival is clear: A hungry animal needs to eat. An enhanced sense of smell aids in locating food, and a heightened sense of taste encourages an animal to eat. Afterward, these senses become temporarily dampened so as not to distract the animal from its other needs. The mechanism by which these changes occur is uncertain, but it may involve a temporary alteration in the number of smell and taste receptors, or in their ability to bind ligands. BIO TIPS ONLINE

## 34.5 Reviewing the Concepts

- In mammals, olfactory sensory receptors are neurons in the nasal cavity. They have cilia with receptor proteins that bind specific odor molecules. The binding initiates a signal that is sent to the brain. Buck and Axel discovered that olfactory sensory receptors have many different types of receptor proteins for odor molecules (Figures 34.18, 34.19).
- Taste buds contain chemosensory cells that detect molecules in food. There are different types of taste cells, the activation of which results in the perception of five tastes: sweet, sour, salty, bitter, and umami (Figure 34.20).

## 34.5 Testing Your Knowledge

1. Which is *not* true of chemoreception?
   a. It is not unique to mammals.
   b. It includes gustation and olfaction.
   c. The sensory cells are activated by bending of stereocilia.
   d. Some chemoreceptors are located within taste buds.
   e. It requires receptor proteins.
2. Olfactory sensory cells in humans
   a. are much greater in number than in many other mammals, including dogs.
   b. have a relatively short lifetime and cannot be replaced.
   c. allow humans to sense as many as 10,000 or more different odors.
   d. can all be classified into one of just a few types.
   e. are modified epithelial cells, not neurons, and cannot produce action potentials.

## 34.6 Impact on Public Health

### Learning Outcome

1. Identify common types of visual and hearing deficits in humans and the causes of each.

Sensory disorders are among the most common neurological problems found in humans and range from mild (needing eyeglasses or a hearing aid) to severe (blindness or deafness). In this section, we present an overview of a few representative sensory disorders that have a major impact on human health.

**4** Subject each PCR product to DNA sequencing, also described in Chapter 18.

Output from automated sequencing (example)

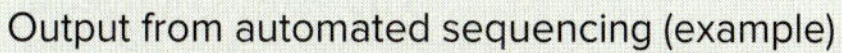


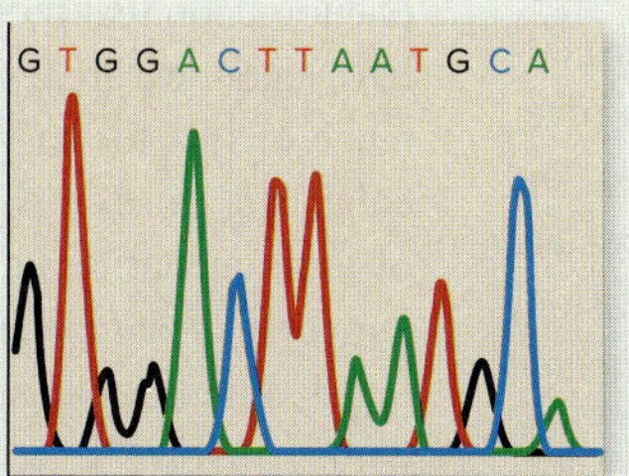

Different GPCRs will have slightly different DNA sequences.

**5 THE DATA**

At least 100 different GPCRs were uniquely expressed in olfactory sensory receptor cells. Analysis of their predicted amino acid sequences revealed significant variability in the putative ligand-binding regions (transmembrane domains 3–5).

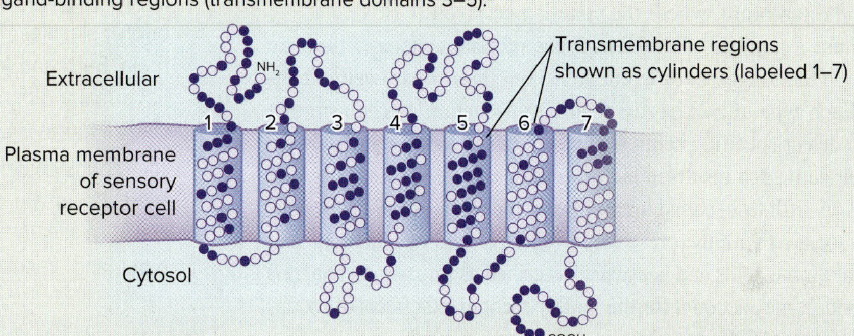


**6 CONCLUSION** Olfactory receptor cells express many different receptor proteins that account for an animal's ability to detect a wide variety of odors.

**7 SOURCE** Buck, L., and Axel, R. 1991. A Novel Multigene Family May Encode Odorant Receptors: A Molecular Basis for Odor Recognition. *Cell* 65: 175–187.

in the polymerase chain reaction (PCR) technique to amplify cDNAs that encoded GPCRs. This produced many PCR products that were then subjected to DNA sequencing. Buck and Axel identified at least 100 different genes, each encoding a GPCR with a slightly different amino acid sequence as predicted from the DNA sequence. Significant variability in the predicted amino acid sequences between the olfactory GPCRs was found in the putative ligand-binding region of the molecules (believed to be transmembrane domains 3–5). Further research demonstrated that these genes were expressed in olfactory cells, but not in other parts of the body. These results were consistent with the second hypothesis—namely, that organisms produce a large number of distinct olfactory receptor proteins, each type binding a particular odor molecule or a group of related odor molecules.

Since these studies, researchers have determined that this family of olfactory genes in mammals is surprisingly large. In humans, roughly 400 genes encode olfactory receptor proteins. Each olfactory receptor cell is thought to express only one type of GPCR that recognizes its own specific odor molecule or group of closely related molecules. Most odors an animal encounters, however, are due to multiple chemicals that activate many different types of odor receptors at the same time. We perceive odors based on the combination of receptors that become activated. This ability to combine different inputs allows animals such as humans to perceive thousands of different odors despite having only a few hundred different olfactory receptor proteins.

The research of Buck and Axel explained, in part, how animals detect a myriad of odors. In 2004, they received the Nobel Prize in Physiology or Medicine for this pioneering work.

### *Experimental Questions*

1. What were the two major hypotheses to explain how animals discriminate between different odors? How did Buck and Axel test the hypothesis of multiple olfactory receptor proteins?
2. **SCISKILLS ▶** Of the two hypotheses proposing how animals discriminate between different odors (see question 1), which one was supported by the results of this experiment? With the evidence presented by Buck and Axel, what is the current hypothesis regarding the discrimination of odors in animals?
3. **SCISKILLS ▶** Predict the survival or reproductive advantages that may be provided to animals by the ability to sense thousands or more different types of odors.

## Taste Buds Detect Food Molecules

**Taste buds** are structures containing cells that detect particular molecules in food. Humans have about 9,000 taste buds, whereas other mammals, such as pigs and rabbits, can have double that number. The bumps that you see on your tongue are not the taste buds but are papillae, elevated structures that collect food molecules in depressions called taste pores. These pores contain many taste buds that include the sensory receptor cells along with supporting cells (**Figure 34.20**). The tips of the sensory receptor cells have microvilli that extend into the taste pore. Here, molecules in food that have dissolved in saliva bind to receptor proteins. This triggers intracellular signals that alter ion permeability and membrane potentials. The sensory receptor cells then release neurotransmitters onto underlying sensory neurons. Action potentials travel from these neurons to the thalamus and other regions of the forebrain, where the taste is perceived.

There are a number of different types of taste cells, and they are distributed on the tongue in different areas, but these areas overlap considerably. Each type of cell has a specific transduction mechanism that allows it to detect specific chemicals present in the foods and fluids we ingest. Their activation results in the perception of sweet, sour, salty, and bitter tastes. A fifth taste called umami (after the Japanese word for delicious) is associated with the presence in ingested food of glutamate and other similar amino acids and is usually described as imparting enhanced flavor to food. It may account for the widely recognized effects of monosodium glutamate (MSG) in enhancing the flavor of food.

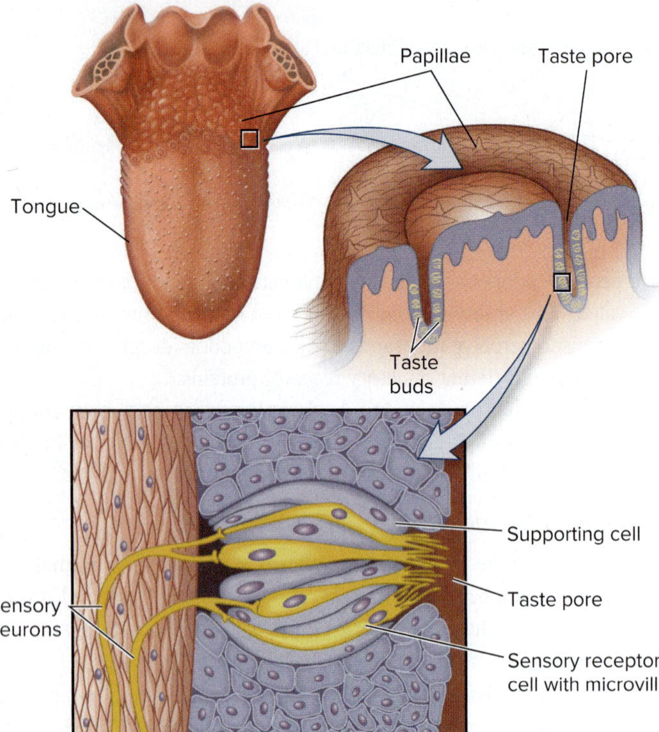


**Figure 34.20 Structures involved in the sense of taste.** This sense occurs in taste buds, which contain the sensory receptor cells that respond to dissolved food molecules.

**Concept Check:** *Where are the receptor proteins that bind to food molecules located?*

The senses of taste and smell are enhanced when we are hungry, a phenomenon that most likely occurs in other animals as well. Once we have eaten, we are less aware of the smell and taste of food. The importance of this for survival is clear: A hungry animal needs to eat. An enhanced sense of smell aids in locating food, and a heightened sense of taste encourages an animal to eat. Afterward, these senses become temporarily dampened so as not to distract the animal from its other needs. The mechanism by which these changes occur is uncertain, but it may involve a temporary alteration in the number of smell and taste receptors, or in their ability to bind ligands. BIO TIPS ONLINE

## 34.5 Reviewing the Concepts

- In mammals, olfactory sensory receptors are neurons in the nasal cavity. They have cilia with receptor proteins that bind specific odor molecules. The binding initiates a signal that is sent to the brain. Buck and Axel discovered that olfactory sensory receptors have many different types of receptor proteins for odor molecules (Figures 34.18, 34.19).
- Taste buds contain chemosensory cells that detect molecules in food. There are different types of taste cells, the activation of which results in the perception of five tastes: sweet, sour, salty, bitter, and umami (Figure 34.20).

## 34.5 Testing Your Knowledge

1. Which is *not* true of chemoreception?
   a. It is not unique to mammals.
   b. It includes gustation and olfaction.
   c. The sensory cells are activated by bending of stereocilia.
   d. Some chemoreceptors are located within taste buds.
   e. It requires receptor proteins.
2. Olfactory sensory cells in humans
   a. are much greater in number than in many other mammals, including dogs.
   b. have a relatively short lifetime and cannot be replaced.
   c. allow humans to sense as many as 10,000 or more different odors.
   d. can all be classified into one of just a few types.
   e. are modified epithelial cells, not neurons, and cannot produce action potentials.

## 34.6 Impact on Public Health

### Learning Outcome

1. Identify common types of visual and hearing deficits in humans and the causes of each.

Sensory disorders are among the most common neurological problems found in humans and range from mild (needing eyeglasses or a hearing aid) to severe (blindness or deafness). In this section, we present an overview of a few representative sensory disorders that have a major impact on human health.

## Visual Disorders Include Glaucoma, Macular Degeneration, and Cataracts

Visual disorders affect an enormous number of people worldwide. In the U.S. alone, more than 10 million people have severe eye problems that cannot be corrected by eyeglasses, and more than 1 million people are blind (42 million worldwide). The costs to the U.S. government associated with blindness and severe visual loss amount to nearly $4 billion yearly. Although vision loss has many causes, three disorders account for over half of all cases: glaucoma, macular degeneration, and cataracts.

**Glaucoma** Normally, the fluid that makes up the aqueous humor in the eye is produced and reabsorbed (drained) in a circulation that keeps the fluid level constant. In **glaucoma,** drainage of aqueous humor becomes blocked, and the pressure inside the eye increases as the fluid accumulates. If untreated, this eventually damages cells in the retina and leads to irreversible loss of vision (Figure 34.21a). The cause of glaucoma is not always known, but in some cases it is due to trauma to or severe infection of the eye, chronic use of certain medicines, diseases such as diabetes, disease of the blood vessels of the eye, or genetics. Glaucoma can be treated with eyedrops that contain a drug that decreases fluid production in the eye or with laser surgery to reshape the drainage area in the eye. The American Foundation for the Blind estimates that 400,000 new cases of glaucoma are diagnosed each year in the U.S., and up to 10 million individuals have elevated eye pressure. Glaucoma accounts for roughly 10% of all cases of blindness in the U.S.

**Macular Degeneration** In **macular degeneration,** photoreceptor cells in and around the macula are lost. Because this is the region where cones are most densely packed, this condition is associated with loss of sharpness and color vision, particularly in the central field because of the location of the macula (Figure 34.21b and see Figure 34.11). It usually does not occur before age 60, but can occur at any age. It is also associated with smoking, and in some people may have a genetic component. Macular degeneration is the leading cause of blindness in the U.S., accounting for roughly 25% of all cases, but its causes remain obscure.

**Cataracts** **Cataracts,** the accumulation of protein in the lens, cloud the lens and cause blurring, poor night vision, and difficulty focusing on nearby objects (Figure 34.21c). By age 65, as many as 50% of individuals have one or more cataracts in either eye, a number that jumps to 70% by age 75. Many cataracts are small enough not to affect vision. In fact, many people do not even realize they have cataracts until they undergo an eye exam. The causes of cataracts are not all known but include trauma, certain medications such as adrenal steroids, diabetes, and heredity. Cataracts are also associated with chronic alcohol abuse, excessive exposure to UV radiation from sunlight, hypertension, and smoking. The treatment, when needed, is usually to have the affected lens surgically removed and replaced with an artificial lens. Nearly 1.4 million such surgeries are performed each year in the U.S.

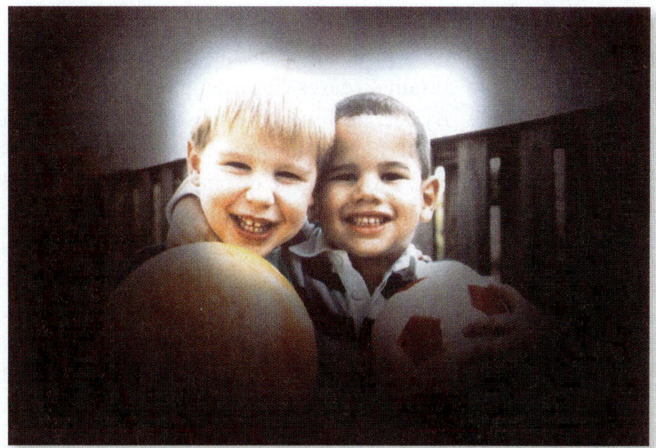

**(a) How the world is seen by a person with glaucoma**

**(b) How the world is seen by a person with macular degeneration**

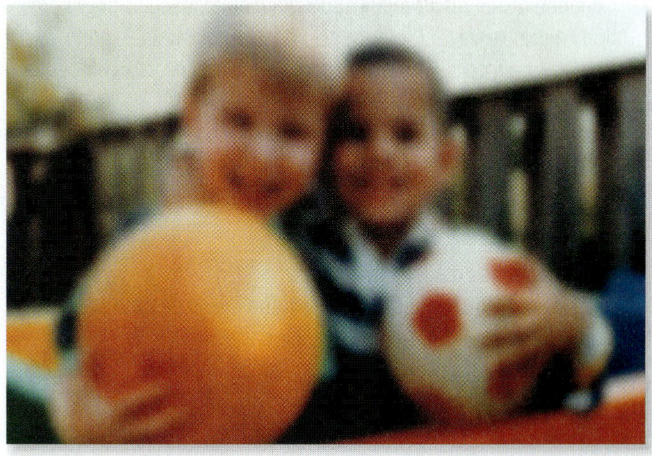

**(c) How the world is seen by a person with cataracts**

**Figure 34.21** Approximate appearance of the visual field in a person with (a) glaucoma, (b) macular degeneration, and (c) cataracts.

*(a–c)* Source: National Eye Institute, National Institutes of Health

## Damaged Hair Cells Within the Cochlea Can Cause Deafness

**Deafness** (partial to complete hearing loss) is usually caused by damage to the hair cells within the cochlea, although some cases result from functional problems in brain areas that process sound or in the nerves that carry information from the hair cells to the brain. When the hair cells are damaged, sounds have to be louder to be detected, and an affected person may require the use of hearing aids to amplify incoming sounds.

Hearing loss may be mild or severe and has many causes, including injury to the ear or head, hereditary defects of the inner ear, and exposure to certain diseases (for example, rubella) or toxins during fetal life. By far, however, the most significant cause of hearing loss is repeated, long-term exposure to loud noise. The amplitude of a sound wave (that is, the distance between the peak and the trough of the wave) determines its loudness; as the amplitude of a sound wave gets larger, the perceived loudness of the sound increases. The loudness of sound is measured in decibels (technically, decibels measure the intensity of sound pressure, which we perceive as loudness). Normal conversation is about 60 dB, a chainsaw is 108 dB, and a jet plane taking off can reach 150 dB. Long-term exposure to sounds of greater than 100 dB can damage the sensory apparatus of the ear. Noise-induced hearing loss usually results from job-related activities and is estimated to account for as many as 20% of all cases of hearing loss worldwide; this makes it the leading occupational disorder in the U.S. It is, therefore, one of the most significant disabilities in the U.S. in terms of numbers of people affected and costs to society. Scientists estimate that nearly 40 million people are exposed daily to dangerous noise levels in the U.S. as a result of their occupation.

If a cochlea is severely damaged, it can be surgically replaced with artificial cochlear implants. In response to sound waves, these devices generate electrical signals that can stimulate the nerves from the ear, which communicate with the brain. Cochlear implants cannot restore hearing to normal, but they can make it possible to hear conversations.

## 34.6 Reviewing the Concepts

- Common visual disorders include glaucoma, macular degeneration, and cataracts. Prolonged exposure to intense sound can damage hair cells within the cochlea and lead to deafness, a partial to complete loss of hearing (Figure 34.21).

## 34.6 Testing Your Knowledge

1. Which describes glaucoma?
   a. The amount of aqueous humor in the eye is decreased.
   b. The pressure inside the eye is increased above normal.
   c. There is a loss of photoreceptor cells around the macula.
   d. It is the leading cause of blindness in the U.S.
   e. It results from proteins accumulating in the lens.
2. Which is *false?*
   a. Cataracts result in blurred vision and poor night vision.
   b. Hearing loss may result from head injury.
   c. The most significant cause of hearing loss is hereditary defects of the inner ear.
   d. Macular degeneration is associated with loss of visual sharpness and color vision.
   e. The development of cataracts is linked to diabetes, heredity, and certain medications.

# Assess and Discuss

## Test Yourself

1. In which process or processes are hair cells involved?
   a. balance in vertebrates and invertebrates
   b. hearing in mammals
   c. vision in animals with compound eyes
   d. heat sensing in pit vipers
   e. both a and b
2. ______ sense pain, __________ sense heat or cold, and __________ sense touch.
   a. Mechanoreceptors, thermoreceptors, nociceptors
   b. Nociceptors, thermoreceptors, mechanoreceptors
   c. Nociceptors, thermoreceptors, stretch receptors
   d. Mechanoreceptors, nociceptors, stretch receptors
   e. Nociceptors, photoreceptors, mechanoreceptors
3. The sensory receptors for audition (hearing) are located in
   a. the organ of Corti.
   b. the middle ear.
   c. the utricle and saccule.
   d. the tympanic membrane.
   e. the outer ear.
4. In an experiment to test the function of statocysts, researchers replaced the statoliths of a crayfish with iron filings and then held a magnet directly above the animal's head. Normally, statoliths sit at the bottom of a statocyst unless an animal moves or turns over. What do you think happened in response to the magnet?
   a. The animal turned in circles.
   b. The animal flipped itself over.
   c. The animal began moving backwards.
   d. The animal fell over onto its side.
5. Cone pigments detect different wavelengths of light due to
   a. their location in the retina.
   b. the amount of light they absorb.
   c. the type of retinal they have.
   d. the type of opsin protein they have.
   e. interactions with bipolar cells.
6. In the mammalian eye, light from near or far objects is focused on the retina when
   a. the lens moves forward or backward.
   b. the lens flattens or rounds up.
   c. the eyeball changes shape.
   d. the cornea changes shape.
   e. the retina moves toward or away from the front of the eye.
7. Depolarization of rods and cones occurs
   a. in response to the release of neurotransmitters from bipolar cells.
   b. in the light.
   c. in the dark.
   d. when they are stimulated by signals from the brain.
   e. both a and b are correct.
8. The amount of glutamate released from photoreceptors of the vertebrate eye is highest when
   a. an animal is in full sunlight.
   b. an animal is in a completely dark area.
   c. an animal is in a dimly lit area.
   d. $Na^+$ channels of the photoreceptor are closed.
   e. both a and d.

9. Which is *true?*
   a. A loss of taste buds would reduce a person's sense of smell.
   b. Olfactory sensory receptors have a limited lifetime and can't be replaced.
   c. Chemoreception only occurs in terrestrial animals.
   d. The relative size of the olfactory bulbs is a good predictor of a mammal's ability to sense odors.
   e. There are a very small number of types of odor receptor proteins in animals.
10. The stimulation for olfaction involves odorant molecules
   a. bending the cilia of olfactory sensory receptor cells.
   b. binding to protein receptors of olfactory sensory receptor cells.
   c. entering the cytoplasm of olfactory sensory receptor cells.
   d. opening $K^+$ channels of olfactory sensory receptor cells.
   e. binding to cells of the olfactory bulb.

## Conceptual Questions

1. Distinguish between sensory transduction and perception.
2. Explain how the structures of the mammalian ear allow an organism to hear sounds.
3. **PRINCIPLES** A principle of biology is that living organisms interact with their environment. This is certainly true of the sensory systems of animals. Of the ways in which you use your senses to interact with the environment, which do you think is the least important?

## Collaborative Questions

1. Discuss the types of sensory stimuli that animals can detect and how sensory receptors are adapted for each of these stimuli.
2. Despite the differences in their appearances, try to find structural and functional similarities between the compound and single-lens eyes found in animals.

# 35 How Muscles and Skeletons Are Adaptations for Movement, Support, and Protection

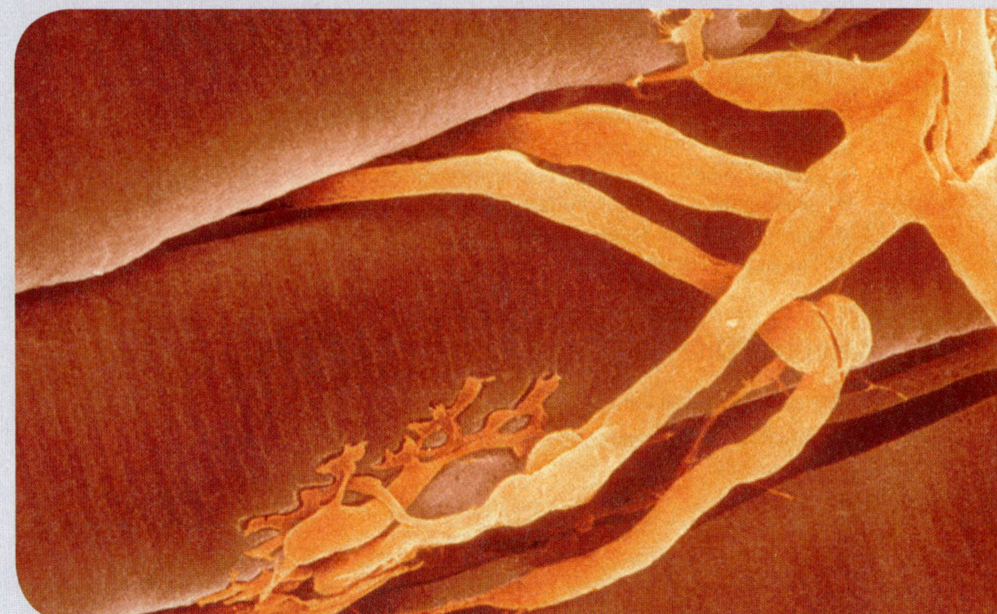

False-color scanning electron micrograph of a neuromuscular junction in a mammal (neuronal structures in yellow; muscle cells in red).



## Chapter Outline



An active 24-year-old woman began experiencing fatigue after even simple activities and found that she needed to curtail her regular gym workouts. Around the same time, she noticed that she had difficulty swallowing, and one of her eyelids began to droop. Alarmed by these symptoms, she visited her physician. After a series of tests, it was determined that she had developed the disease called myasthenia gravis. This disease ranges in severity, but at its worst, it can make chewing, swallowing, talking, and even breathing difficult. It affects an estimated 10,000–30,000 Americans, usually not becoming apparent until adolescence or later. It cannot be cured, but a variety of medical treatments are available that can decrease its severity.

Myasthenia gravis is a disease of skeletal muscle, the structure and function of which are subjects of this chapter. In this disease, the body's immune system inactivates receptor proteins on skeletal muscle cells. (This type of disease is called an autoimmune disease, which you will learn more about in Chapter 40.) These receptor proteins bind the neurotransmitter acetylcholine (ACh), which is released by motor neurons onto skeletal muscle cells, stimulating the contraction of those cells. Consequently, myasthenia gravis results in a degree of paralysis that depends on the severity of the disease. For this reason, the woman in this example was given a medication that increases the concentration of ACh in the synapses on her skeletal muscle cells. She will have to take this medication for the rest of her life.

All muscles are composed of highly specialized cells that have the ability to contract. The three types of muscle—skeletal, smooth, and cardiac—were introduced in Chapter 32. In this chapter, you will learn about the structure and function of skeletal muscle and how skeletal muscle controls movements such as lifting an object, breathing, and locomotion—the movement of an animal from place to place. For skeletal muscles to produce their effects, however, they usually must exert a force on an animal's skeleton. We begin the chapter, therefore, with an overview of animal skeletons. Then we examine the structure and function of skeletal muscle and how the interaction of two muscle proteins—actin and myosin—produces muscle contraction. We conclude with a consideration of important bone and muscle diseases in humans.

## 35.1 Types of Animal Skeletons

### Learning Outcomes

1. Distinguish between exoskeletons and endoskeletons.
2. List the advantages and disadvantages of an exoskeleton.
3. List the major functions of the vertebrate skeleton.
4. Describe the composition of vertebrate bone.

A broad definition of a **skeleton** is a body structure that serves one or more functions related to support, protection, and locomotion. In this section, we examine the two major types of skeletons found in animals: exoskeletons (found in many invertebrates) and endoskeletons (found in all vertebrates and a few invertebrates).

### Exoskeletons Are on the Outside of Animal Bodies

Many invertebrates have a hard outer casing or shell that covers all or part of their body surface. Bivalve mollusks, for example, are covered by a tough shell made of calcium carbonate. The body surface of arthropods is covered by layers of the polysaccharide chitin and minerals, including calcium. The term **exoskeleton** generally refers to any tough exterior body covering of these types in an invertebrate (Figure 35.1a). An exoskeleton surrounds most or all of the body surface and provides one or more of the following: support for the body, protection from the environment (for example, acting as a barrier to dessication in terrestrial species), protection from predators, and protection for internal organs.

Exoskeletons are often tough and durable and, in arthropods, are segmented to allow for flexibility and movement. They provide a support mechanism for the attachment of muscles that participate in movement or locomotion. In arthropods, the exoskeleton must be periodically shed, regrown, and strengthened again to allow the animal to grow, a process called ecdysis, or molting (see Figure 35.1a). One disadvantage of such exoskeletons is that when an animal is molting, its new exoskeleton is temporarily soft, making the animal more vulnerable to predators and the environment.

Exoskeletons vary enormously in their complexity, thickness, and durability. The differences in exoskeletons are usually adaptations that enhance an animal's survival. Think, for example, of the difference between the body surfaces of a butterfly and a lobster. A butterfly's exoskeleton must be light enough for the animal to fly, whereas the thick, tough exoskeleton of the lobster provides a very effective defense against predators in the sea. Arthropods are a good example of the benefits of exoskeletons. These animals are among the most successful of all animal phyla living today, having survived for hundreds of millions of years and inhabiting nearly every possible ecological niche on the planet. To some degree, the amazing success of arthropods can be attributed to the exoskeleton.

### Endoskeletons Are Internal Support Structures

An **endoskeleton** is an internal, mineralized structure that provides support and protection, including in some cases to internal organs such as those in the thorax of vertebrates. Unlike exoskeletons,

## Biology Principle

### Living Organisms Grow and Develop

Part of the growth and development of animals involves their skeletons, as shown in this figure. For animals with exoskeletons, however, this growth may create a brief period when the animal is particularly vulnerable to predation.

**(a) An arthropod next to its recently shed exoskeleton**

**(b) Echinoderm (sea star) and (right) SEM of its endoskeleton**

**Figure 35.1 Types of skeletons.** **(a)** Exoskeleton. An arthropod's skeleton covers and protects its body, but it must be periodically shed and replaced to allow growth, as shown in this photo. **(b)** Endoskeleton. Echinoderms such as the sea star have endoskeletons of bony plates made of calcium carbonate (inset). Vertebrate endoskeletons will be described later.

*(a)* © Tom McHugh/Science Source; *(b left)* © Georgette Douwma/Science Source; *(b right)* © The Trustees of the Natural History Museum, London/Alamy

**Concept Check:** *Unlike exoskeletons, endoskeletons do not provide protection for the body surface of animals. How can the lack of an external skeleton also be an advantage for animals?*

endoskeletons do not protect the outer surface of the body and do not need to be shed as an animal grows.

Various types of endoskeletons are found in some species of sponges and all echinoderms and vertebrates. Minerals such as calcium, magnesium, phosphate, and carbonate supply the hardening material that gives the skeleton its firm structure. The endoskeletons of invertebrates are primarily support structures that consist of spiky networks of proteins and minerals in sponges (refer back to Figure 26.11) or mineralized, platelike structures in echinoderms. Beneath the body surface of echinoderms, for example, arrays of mineralized plates made largely of calcium carbonate extend into the spines and arms that radiate from the main body (**Figure 35.1b**). Vertebrate endoskeletons, by contrast, are composed of cartilage, bone, or both and have important functions beyond providing support for an animal's body. Next, we consider the structure and function of the vertebrate skeleton in greater detail.

## The Vertebrate Skeleton Performs Several Important Functions

Vertebrate skeletons are composed of either cartilage—as in cartilaginous fishes (sharks, rays, and skates)—or both cartilage and bone (refer back to Figure 32.5), as in bony fishes, amphibians, reptiles, birds, and mammals. Let's consider the vertebrate bony skeleton as a whole before examining bones in greater detail.

In the vertebrate skeleton, bones are connected in ways that allow for support, protection, and movement. The vertebrate skeleton is often considered in two parts: the axial and appendicular skeletons (**Figure 35.2**). The axial skeleton is composed of the bones that form the main longitudinal axis of an animal's body, including the skull, vertebrae, sternum, and ribs. The appendicular skeleton consists of the limb bones and the bones that connect them to the axial skeleton. A joint is formed where two or more bones come together. Some joints permit free movement (for example, the shoulders). Other joints allow no movement (fused joints, like those interlocking the skull bones) or allow only limited movement (such as those of the vertebral column).

The skeleton of vertebrates serves several functions in addition to support, protection, and movement. For example, blood cells and platelets, the latter of which help blood to clot (see Chapter 36), are formed within the marrow of certain bones including the ilia, the vertebrae, and the ends of the femurs. In addition, homeostasis of the concentrations of calcium and phosphate ions in body fluids is achieved in large part through exchanges of these ions between bone and blood.

## Bone Consists of a Mixture of Organic and Mineral Components

**Bone** is a living, dynamic tissue with both organic and mineral components. Organic materials include cells that form bone—called osteoblasts and osteocytes—and cells that break it down—called osteoclasts. The organic part of bone is secreted by osteoblasts and osteocytes and consists of the protein collagen. This protein has a unique triple helical structure that gives bone both strength and flexibility (refer back to Figure 4.31). The mineral component is composed of a crystalline mixture of primarily $Ca^{2+}$ and $PO_4^{3-}$ plus a few other ions that provide the rigidity of bone. These ions must be obtained in an animal's diet, absorbed into the blood, and deposited in bone.

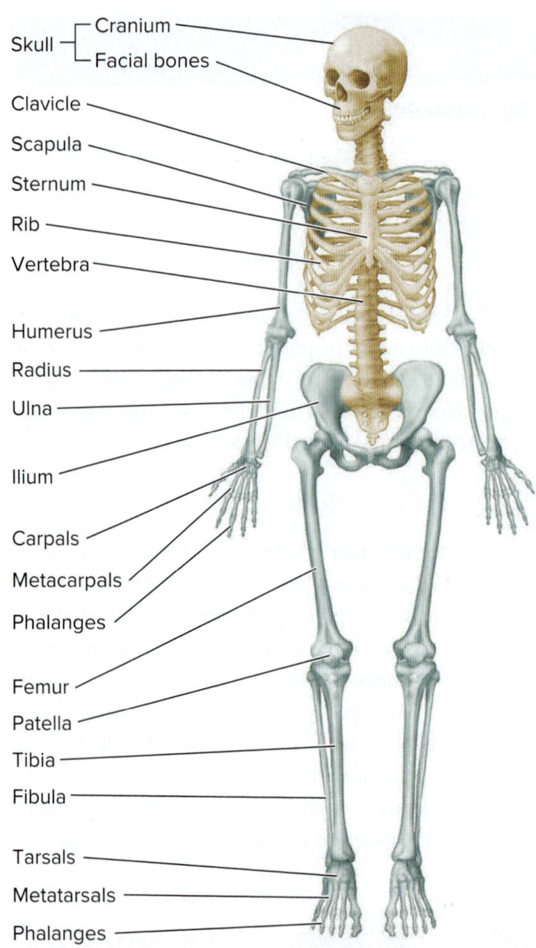


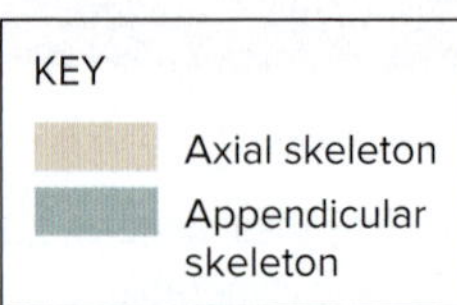


**Figure 35.2** **The adult human skeleton.** This diagram shows the axial (beige) and appendicular (green) parts of the skeleton in an adult human, an animal with an endoskeleton. The adult human skeleton consists of 206 separate bones.

Bones cannot move by themselves but, instead, provide the scaffold upon which skeletal muscles act to cause body movement. We turn now to a discussion of skeletal muscle and the mechanism by which it generates force.

## 35.1 Reviewing the Concepts

- Skeletons are structures that provide support and protection and function in locomotion.
- Two major types of skeletons are found in animals. Exoskeletons are protective external structures that typically must be shed and regrown to accommodate an animal's growth. Endoskeletons are internal structures that grow with the animal but do not protect its body surface (Figure 35.1). Vertebrate endoskeletons are made up of the axial and appendicular skeletons (Figure 35.2).
- In addition to the functions of support, protection, and locomotion, the vertebrate skeleton produces blood cells and constitutes a reservoir for ions crucial to homeostasis.

## 35.1 Testing Your Knowledge

1. A disadvantage of exoskeletons is that
   a. they cannot protect an animal's internal organs.
   b. they must be periodically shed, leaving the animal in a vulnerable state.
   c. they do not provide any flexibility for an animal to move easily.
   d. they are easily damaged, soft structures.
   e. they cannot protect the outside of the body surface.
2. Endoskeletons
   a. are found in all arthropods.
   b. do not protect the outer surface of an animal's body.
   c. must be periodically shed and regrown.
   d. are found in certain invertebrates but not in vertebrates.
   e. are only found in vertebrates.

## 35.2 Skeletal Muscle Structure and the Mechanism of Force Generation

### Learning Outcomes

1. List the three types of muscle found in vertebrates and where they are found in the body.
2. Identify the structural components of a muscle down to the level of the sarcomere.
3. Outline the sliding-filament mechanism of muscle contraction.
4. Explain how tropomyosin and troponin help regulate muscle contraction.
5. Describe how electrical excitation and muscle contraction are linked by events at the neuromuscular junction.

Vertebrates have three types of muscle, which are classified according to their structure, function, and control mechanisms, as first introduced in Chapter 32:

- **Cardiac muscle** is found only in the heart and provides the force required for the heart to pump blood. Cardiac muscle will be discussed in Chapter 36.
- **Smooth muscle** surrounds and forms part of the lining of hollow organs and tubes, including those of the digestive tract, urinary bladder, uterus, blood vessels, and airways. Smooth muscle cell contraction is not under voluntary control. Instead, as described in Chapter 33, it is largely controlled by the autonomic nervous system.
- **Skeletal muscle** is found throughout the body and is directly involved in movements and locomotion. Let's take a detailed look at the structure and function of skeletal muscle and its cells.

### A Skeletal Muscle Is a Contractile Organ That Supports and Moves Bones

A skeletal muscle is a grouping of cells—called **muscle fibers**—bound together in bundles by a succession of connective tissue layers (Figure 35.3). Skeletal muscles are usually linked to bones by collections of collagen fibers known as tendons. The transmission of force from contracting muscle to bone can be likened to a number of people pulling on a rope attached to a heavy object. Each person corresponds to a single muscle fiber, the rope corresponds to the tendons, and the bone is the heavy object.

When the force generated by a contracting muscle is great enough, the bone to which it is connected moves as the muscle shortens. A contracting muscle exerts only a pulling force, so as the muscle shortens, the attached bones are pulled toward or away from each other. Muscles that bend a limb at a joint (that is, reduce the angle between two bones)

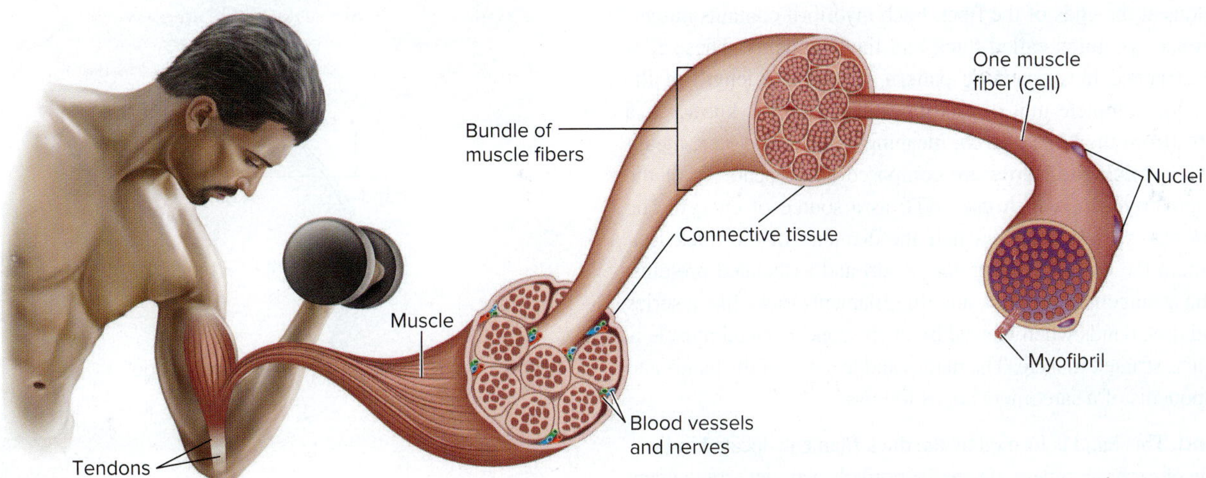


**Figure 35.3 Skeletal muscle structure.** Skeletal muscles attach to bone by tendons, which are bundles of collagen fibers. Each muscle consists of bundles of muscle fibers (muscle cells) bound together by connective tissue. Myofibrils are the contractile elements of muscle fibers.

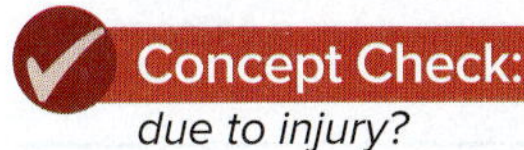

**Concept Check:** *What would happen to the ability of a muscle to move a bone if its tendon were torn—for example, due to injury?*

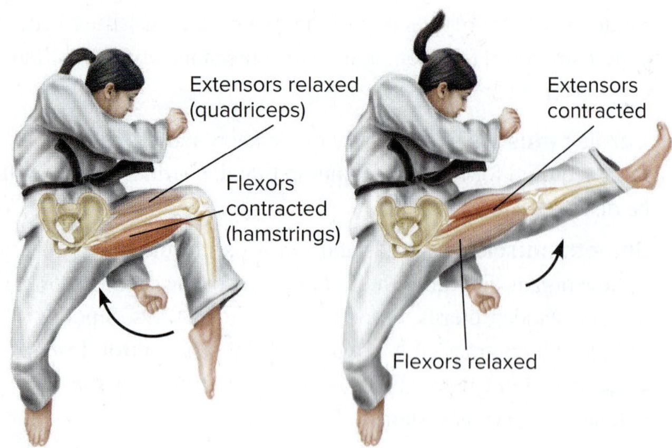


**Figure 35.4 Actions of flexors and extensors.** The figure shows how skeletal muscles cause flexion or extension of a limb. When the flexor muscle contracts, the extensor relaxes, and vice versa.

are called **flexors,** whereas muscles that straighten a limb (increase the angle between two bones) are called **extensors.** Groups of muscles that produce oppositely directed movements at a joint are known as **antagonists.** For example, in Figure 35.4, we can see that contraction of the hamstrings (along with relaxation of the quadriceps) flexes the leg at the knee joint, whereas contraction of the quadriceps (with relaxation of hamstrings) causes the leg to extend. Both antagonistic muscles exert only a pulling force when they contract; where their tendons connect to the shin bones determines whether the leg is flexed or extended.

## Muscle Fibers Contain Myofibrils Composed of Arrays of Filaments

Each skeletal muscle fiber arises from several cells that fuse to form a single mature cell with multiple nuclei (see Figure 35.3). Each fiber contains numerous cylindrical bundles known as **myofibrils** (Figure 35.5). Myofibrils extend from one end of the fiber to the other and are linked to the tendons at the ends of the fiber. Each myofibril contains numerous functional structures called thick and thin filaments. These filaments are arranged in a repeating pattern running the length of the myofibril. One complete unit of this repeating pattern is known as a **sarcomere** (from the Greek *sarco*, meaning muscle, and *mer*, meaning part). The **thick filaments** are composed almost entirely of the motor protein **myosin,** which uses ATP as a source of energy. The **thin filaments,** which are about half the diameter of the thick filaments, contain the cytoskeletal protein **actin** and associated proteins. Because the arrangement of thick and thin filaments looks like a series of light and dark bands when viewed by microscope, skeletal muscle is also known as striated muscle. The names and features of the bands and other components of a sarcomere are as follows:

- **A band.** This band is formed by the thick filaments located in the middle of each sarcomere, where their orderly, parallel arrangement produces a wide, dark band. A portion of the thin filaments overlap the thick filaments in this band (see Figure 35.5).
- **Z line.** This is a network of proteins to which thin filaments are attached. Two successive Z lines define the boundaries of one sarcomere.

## Biology Principle

### Cells Are the Simplest Units of Life

A muscle consists of many cells called muscle fibers, each containing a highly ordered arrangement of intracellular filaments. As the cells change their shape by shortening, the shape of the entire muscle also changes.

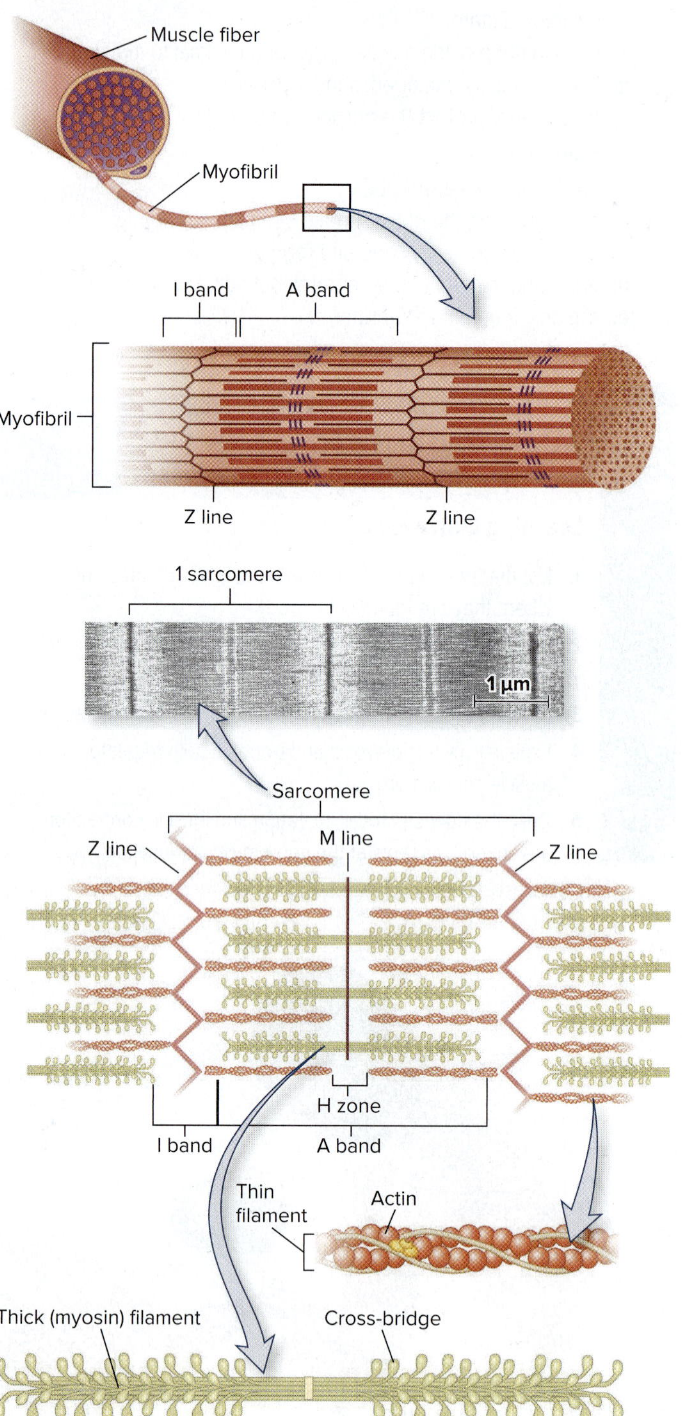


**Figure 35.5 Myofibril structure.** Each muscle fiber consists of numerous myofibrils containing thick and thin filaments. Their arrangement produces a striated banding pattern seen with microscopy (inset). Each repeating unit of the arrayed filaments constitutes a sarcomere.
*(middle)* © Dr. H.E. Huxley

- **I band.** This band lies between the A bands of two adjacent sarcomeres. Each I band contains those portions of the thin filaments that do not overlap the thick filaments, and each I band is bisected by a Z line.
- **H zone.** This is a narrow region in the center of the A band. It corresponds to the space between the two sets of thin filaments in each sarcomere.
- **M line.** This band is in the center of the H zone, and it corresponds to proteins that link the central regions of adjacent thick filaments.

The spaces between overlapping thick and thin filaments are bridged by projections known as **cross-bridges,** which are regions of myosin molecules that extend from the surface of the thick filaments toward the thin filaments (see Figure 35.5).

The protein myosin was first discovered in extracts of frog leg skeletal muscle in the 19th century by the German physiologist Wilhelm Kühne. We now know that eukaryotic genomes encode a family of related myosin proteins and that at least one of them may have made a significant contribution to human evolution, as described next.

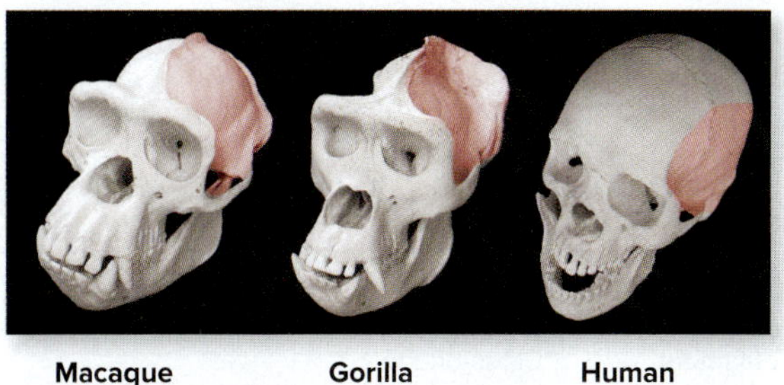


**Figure 35.6** Comparison of jaw size and area of muscle attachment (shown in red) in modern primates.


## EVOLUTIONARY CONNECTIONS

### Did an Ancient Mutation in Myosin Play a Role in the Development of the Human Brain?

Myosins are among the most ancient of eukaryotic proteins and are found throughout the animal kingdom. Nonetheless, small differences in the sequences of genes that arose from a single primordial myosin gene have led to the tissue-specific expression of various myosin proteins, each with a characteristic ability to bind actin and hydrolyze ATP. One member of this gene family, called *MYH16*, codes for a subunit of a myosin molecule that is expressed mainly in the muscles of the jaw in primates. In 2004, American researcher Hansell Stedman and colleagues discovered that this gene, although present in humans, did not code for a functional protein because of a mutation that deleted two base pairs from it. This mutation was present in 100% of people tested worldwide but was not found in the eight nonhuman primate species tested, which included chimpanzees.

Based on genetic comparisons and estimates of genetic divergence among species, the researchers estimated that the mutation occurred approximately 2.4 mya. Significantly, the genus *Homo*, with its smaller jaw and larger braincase, is also thought to have first appeared about 2.4 mya, leading Stedman to suggest that the loss of the MYH16 protein led to the smaller, less muscular jaws characteristic of modern humans (**Figure 35.6**). Because jaw muscles are attached to skull bones, massive jaw muscles may have placed a mechanical constraint on skull growth in early hominins (as well as in modern nonhuman primates). Smaller jaw-closing muscles would place less constraint on bone formation on the outside of the skull (such as the illustrated anchor points obscuring the sutures or growth plates). This in turn would prevent the early cessation of brain growth seen in the nonhuman primates. This may have eliminated a major constraint on the evolution of the larger brains seen in modern humans.

Stedman's hypothesis is compelling, given that the myosin mutation affects the jaw-closing muscles and is present only in humans. Another research group analyzed gene sequences made available through the Chimpanzee Sequencing and Analysis Consortium—an effort to determine the complete DNA sequence of the chimpanzee genome—and concluded that the mutation may have arisen earlier, perhaps 4.0 to 5.4 mya, before small-jawed species of *Homo* appear in the fossil record. Although scientists are certain that a human-specific mutation in *MYH16* arose at some point in the evolution of hominins, whether the mutation facilitated expansion of the cranium, which in turn may have permitted enlargement of the human brain, remains an intriguing question.

### Skeletal Muscle Shortens When Thin Filaments Slide over Thick Filaments

Movement requires shortening of muscles to pull against attached tendons and bones. However, a muscle can generate force, or contract, without producing movement. Holding a heavy weight at a constant position, for example, requires muscle contraction, but not muscle shortening. Therefore, as used in muscle physiology, the term contraction refers to activation of the cross-bridges within muscle fibers, which initiates the generation of force. When these mechanisms are turned off, contractions end, allowing the muscle fiber to relax.

Let's look at how the structure of the thin and thick filaments allows them to move (**Figure 35.7**). In thin filaments, actin molecules form polymers that are arranged into two intertwined, helical chains. These chains are closely associated with two proteins, called tropomyosin and troponin, that have important functions in regulating contraction.

Myosin proteins have a three-domain structure composed of two intertwined tails, two hinges, and two heads (see Figure 35.7a). The hinges are flexible regions that connect the heads to the tails. Each filament is made of many myosin proteins associated in a parallel array, with the hinges and heads extending out to the sides, forming cross-bridges. Each head contains two binding sites: one for actin and one for ATP. The hydrolysis of ATP by the ATPase activity of myosin provides the energy for the cross-bridge to move via a bending motion at the hinge.

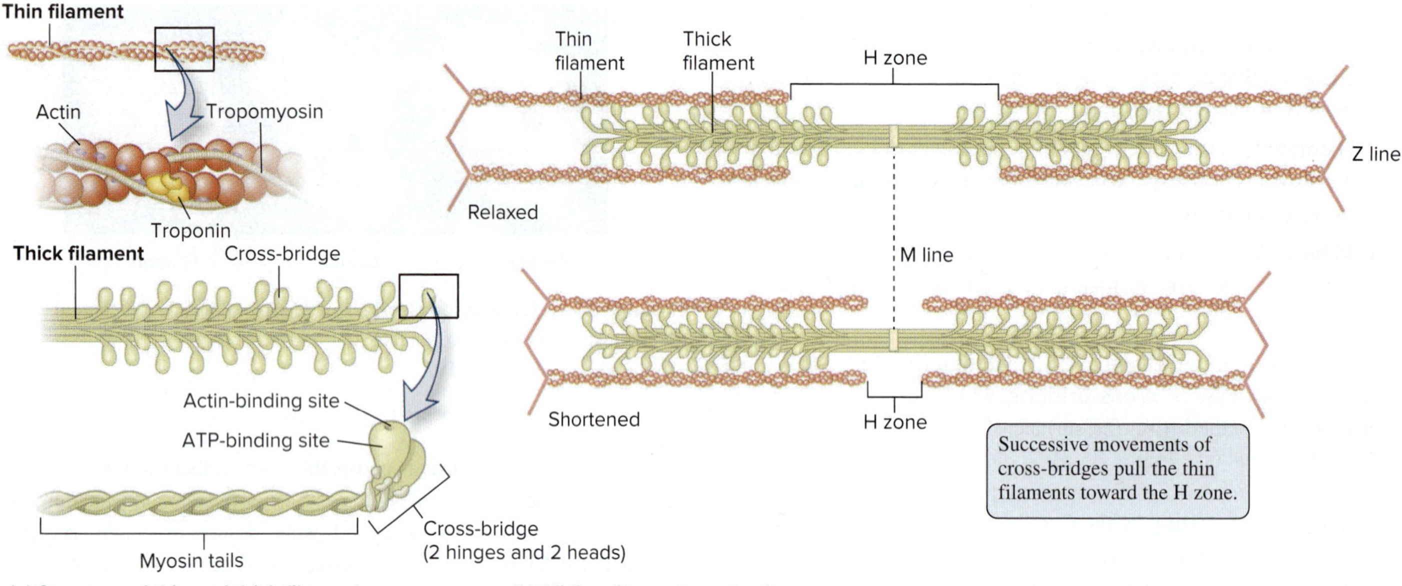


**Figure 35.7** **Structure and function of the filaments and the sliding-filament mechanism.** **(a)** Thin filaments are composed of two intertwined actin molecules and associated proteins, tropomyosin and troponin. Thick filaments are made of the protein myosin, which has two intertwined tails, two hinges, two heads, an actin-binding site, and an ATP-binding site. The end of a myosin molecule that contains the actin-binding sites is bent at an angle to form the cross-bridges. **(b)** When the cross-bridges on myosin molecules bind to actin, the thin filaments are pulled toward the H zone, shortening the sarcomere. The sliding of thin filaments past the overlapping thick filaments shortens the sarcomere but does not change the lengths of the filaments themselves.

How a muscle fiber actually shortens is known as the **sliding-filament mechanism** of muscle contraction (see Figure 35.7b). In this mechanism, the sarcomeres shorten, but neither the thick nor the thin filaments change in length. Instead, the thick filaments remain stationary while the thin filaments slide, pulling on the Z lines and shortening the sarcomere. The sliding-filament movement is propelled by the myosin cross-bridges. During shortening, each cross-bridge attaches to an actin molecule in a thin filament and moves in a motion much like your fist bending at the wrist. Because of the opposing orientation of the thick filaments, the movement of a cross-bridge forces the thin filaments toward the center of the sarcomere (the M line in the H zone), thereby narrowing the H zone and shortening the sarcomere. One stroke of a cross-bridge produces only a very small movement of a thin filament relative to a thick filament. As long as a muscle fiber continues to be stimulated to contract, however, each cross-bridge repeats its motion many times, resulting in continued sliding of the thin filaments. Thus, the ability of a muscle fiber to generate force and movement depends on the amount of interaction between actin and myosin.

## The Cross-Bridge Cycle Requires ATP and $Ca^{2+}$

Let's now turn our attention to how actin and myosin interact to promote muscle contraction and shortening. The sequence of events that occurs between the time when a cross-bridge binds to a thin filament and when it is set to repeat the process is known as a **cross-bridge cycle.**

Cross-bridge cycling occurs when the cytosolic $Ca^{2+}$ concentration exceeds a critical threshold. This usually occurs when neural input stimulates processes that result in the release of $Ca^{2+}$ from intracellular storage sites (described in detail shortly). In other words, the contraction of skeletal muscle fibers is under nervous control.

Figure 35.8 illustrates the events that occur during the four steps of the cross-bridge cycle: (1) cross-bridge binding, (2) power stroke (moving), (3) detaching, and (4) resetting. As the cycle begins, the myosin cross-bridges are in an energized state, which is produced by the hydrolysis of their bound ATP to ADP and inorganic phosphate ($P_i$). The ADP and $P_i$ remain for the moment bound to the cross-bridge. The sequence of storage and release of energy by myosin is analogous to the operation of a mousetrap: Energy is first stored in the trap by cocking the spring (ATP hydrolysis) and is then released by the springing of the trap (head binds to actin and moves in power stroke). Let's look at each step of the process more closely.

**Step 1: $Ca^{2+}$ concentration increases, triggering the cross-bridge to bind to actin.** When the $Ca^{2+}$ concentration in the cytosol increases, an energized myosin cross-bridge along with its associated ADP and $P_i$ binds to an actin molecule on a thin filament.

**Step 2: Release of phosphate ($P_i$) fuels the power stroke: The cross-bridge and thin filaments move.** The binding of an energized myosin cross-bridge to actin in step 1 triggers the release of $P_i$. This release causes a conformational change in the hinge domain of the myosin molecule. This causes the cross-bridge to rotate toward the M line in the H zone at the center of the sarcomere as myosin returns to a lower energy conformation. This process is known as the **power stroke,** which moves the actin filament. At this time, ADP is released from the cross-bridge.

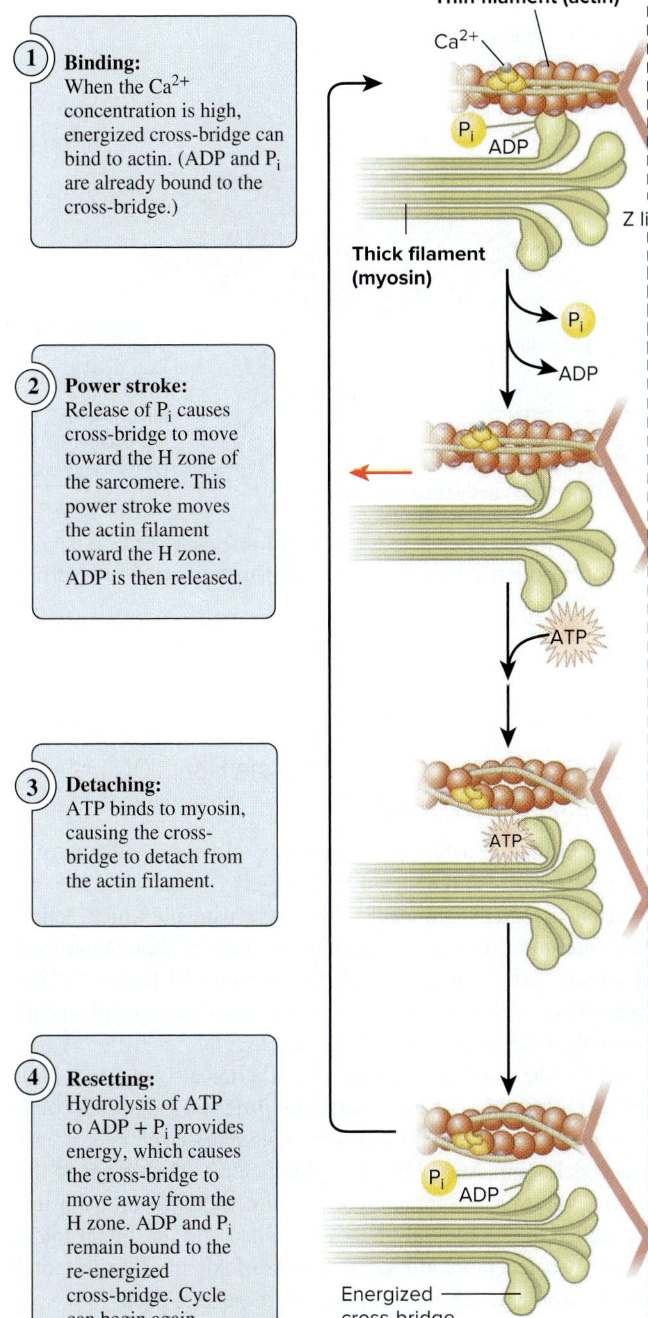


**Figure 35.8** The four stages of cross-bridge cycling in skeletal muscle.

**Concept Check:** *Rigor mortis is the condition that occurs shortly after death, during which the skeletal muscles contract. Why would the muscle remain contracted for a while after death, and why does such contraction eventually stop? (Hint: Think about the concentrations of ATP in the dying muscle cell.)*

**Step 3: ATP binds to myosin, causing the cross-bridge to detach.** During the power stroke, myosin is bound very firmly to actin; this linkage must be broken to allow the cross-bridge to be reenergized and repeat the cycle. The binding of a new molecule of ATP to the myosin cross-bridge breaks the link between actin and myosin. The dissociation of actin and myosin by ATP is an example of allosteric regulation of protein activity. When ATP binds at one site on myosin (called an allosteric site; refer back to Figure 6.6c), myosin's affinity for actin decreases. ATP is not hydrolyzed in this step. Instead, ATP functions at this stage as an allosteric modulator of the myosin head, weakening the binding of myosin to actin and leading to their dissociation.

**Step 4: ATP hydrolysis reenergizes and resets the cross-bridge.** After actin and myosin dissociate, the ATP bound to myosin is hydrolyzed. This hydrolysis reenergizes myosin, causing it to reset to the conformation that allows actin binding. If $Ca^{2+}$ is still available at this time, the cross-bridge can reattach to a new actin molecule in the thin filament, and the cross-bridge cycle repeats, causing the muscle fiber to shorten further.

As we have seen, ATP performs two different roles in the cross-bridge cycle. First, the energy released from ATP hydrolysis provides the energy for cross-bridge movement. Second, ATP binding breaks the link formed between actin and myosin during the cycle, causing the cross-bridge to detach, allowing a new cycle to start. Although the precise mechanisms may differ slightly between vertebrates and invertebrates, biologists think that all skeletal muscle functions according to steps similar to those just described. As mentioned, skeletal muscle function requires the actions of $Ca^{2+}$. Next, we examine the mechanism by which $Ca^{2+}$ regulates skeletal muscle contraction.

## The Regulation of Muscle Contraction by $Ca^{2+}$ Is Mediated by Tropomyosin and Troponin

How does the presence of $Ca^{2+}$ in the cytosol of muscle cells regulate the cycling of cross-bridges? The answer requires a closer look at the two additional thin filament proteins mentioned earlier—tropomyosin and troponin.

**Tropomyosin** proteins are arranged end to end along the actin thin filament (Figure 35.9a). In the absence of $Ca^{2+}$, they partially cover the myosin-binding site on each actin molecule, thereby preventing cross-bridges from making contact with actin. Each tropomyosin is held in this blocking position by **troponin,** a smaller, globular-shaped protein with three subunits that is bound to both tropomyosin and actin. In this way, troponin and tropomyosin block access to myosin-binding sites on actin molecules in the relaxed muscle fiber.

One of the three subunits of troponin is capable of binding $Ca^{2+}$. When $Ca^{2+}$ enters the cytosol, it binds to troponin. The binding of $Ca^{2+}$ produces a change in the shape of troponin, which—through troponin's linkage to tropomyosin—causes tropomyosin to move away from the myosin-binding site on each actin molecule (Figure 35.9b). This movement allows cross-bridge cycling to occur. Conversely, release of $Ca^{2+}$ from troponin reverses the process, blocking the myosin-binding site and stopping contractile activity.

Thus, the concentration of cytosolic $Ca^{2+}$ determines whether or not $Ca^{2+}$ is bound to troponin molecules, which in turn determines the number of actin sites available for cross-bridge binding. The cytosolic concentration of $Ca^{2+}$, however, is very low in resting muscle. Let's see how the $Ca^{2+}$ concentration is increased so that contraction can occur.

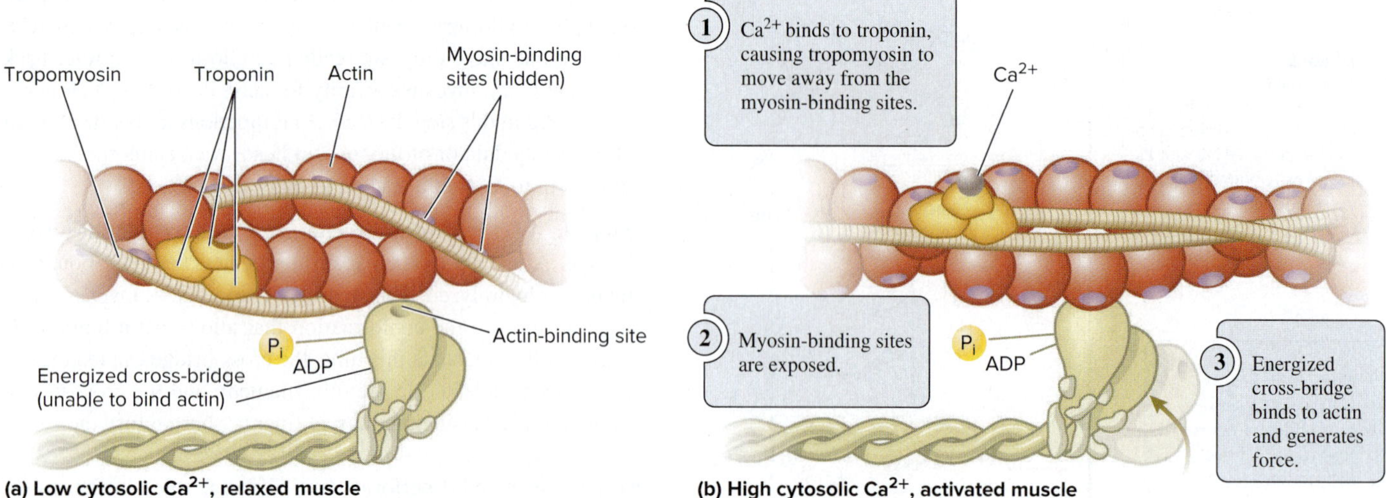


**Figure 35.9** **Function of $Ca^{2+}$, tropomyosin, and troponin in cross-bridge cycling.** **(a)** When cytosolic $Ca^{2+}$ is low, the myosin-binding sites on actin are blocked by tropomyosin. **(b)** When cytosolic $Ca^{2+}$ increases, $Ca^{2+}$ binds to troponin, which in turn causes tropomyosin to move away from the myosin-binding sites on actin.

## $Ca^{2+}$ Concentration and Contraction of Skeletal Muscle Are Coupled with Electrical Excitation

Like neurons, vertebrate skeletal muscle cells generate and propagate action potentials in response to an appropriate stimulus. This causes an increase in the concentration of cytosolic $Ca^{2+}$, which triggers contraction of a muscle fiber. The sequence of events by which an action potential in the plasma membrane of a muscle fiber leads to cross-bridge activity by the mechanisms just described is called **excitation-contraction coupling.** The electrical activity in the plasma membrane does not act directly on the contractile proteins but, instead, acts as a stimulus to increase the cytosolic $Ca^{2+}$ concentration. The increased $Ca^{2+}$ concentration continues to activate the contractile apparatus long after electrical activity in the membrane has ceased.

The source of the increased cytosolic $Ca^{2+}$ that occurs following a muscle action potential is the fiber's sarcoplasmic reticulum, which acts as a $Ca^{2+}$ reservoir. The **sarcoplasmic reticulum,** which is a specialized form of the endoplasmic reticulum found in other cells, is composed of interconnected, sleevelike compartments and sacs around each myofibril (Figure 35.10). Separate tubular structures, the **transverse tubules (T-tubules),** are invaginations of the plasma membrane that lie close to the sarcoplasmic reticulum. The T-tubules course around each myofibril and conduct action potentials from the outer surface of the muscle fiber inside to the myofibrils. An action potential causes the opening of $Ca^{2+}$ channels in the sarcoplasmic reticulum, which allows $Ca^{2+}$ to diffuse into the cytosol and bind to troponin, initiating cross-bridge cycling.

A contraction continues until $Ca^{2+}$ is removed from troponin and the cytosol. This is achieved by ATP-dependent $Ca^{2+}$ transporters in the sarcoplasmic reticulum that restore the $Ca^{2+}$ concentration in the cytosol back to its resting concentration.

## Electrical Stimulation of Skeletal Muscle Fibers Occurs at the Neuromuscular Junction

We have seen that an action potential in the plasma membrane of a skeletal muscle fiber is the signal that triggers contraction. How are these action potentials initiated? The mechanism by which action potentials are initiated in a skeletal muscle involves stimulation by a motor neuron. These are neurons that have their cell bodies located in the central nervous system (CNS), with axons that transmit signals away from the CNS to control skeletal muscle cells.

The site where a motor neuron's axon terminals synapse with a muscle fiber is known as a **neuromuscular junction** (Figure 35.11 and see chapter-opening photo). These terminals release the neurotransmitter acetylcholine (ACh) into the synaptic cleft. The region of the muscle fiber plasma membrane that lies directly under an axon terminal is called the motor end plate; it is folded into junctional folds, where the ACh receptors are located. These folds increase the total surface area available for the membrane to respond to ACh.

When an action potential in a motor neuron arrives at an axon terminal, it triggers the release of stored ACh, which diffuses across the synaptic cleft and binds to receptor proteins in the junctional folds of the muscle fiber. The ACh receptor is a ligand-gated ion channel (refer back to Chapters 5 and 33). The binding of ACh opens the channel and causes an influx of $Na^+$ into the muscle fiber, which causes the muscle fiber to depolarize; this results in an action potential that spreads along the membrane of the muscle fiber and through the T-tubules.

Overstimulation of a muscle fiber is prevented by the action of the enzyme **acetylcholinesterase.** Acetylcholinesterase breaks down excess ACh in the synapse to inactive molecules, ending neurotransmission (see Figure 35.11).

Now that we have learned about the structural characteristics of skeletal muscle and the events that initiate and produce force

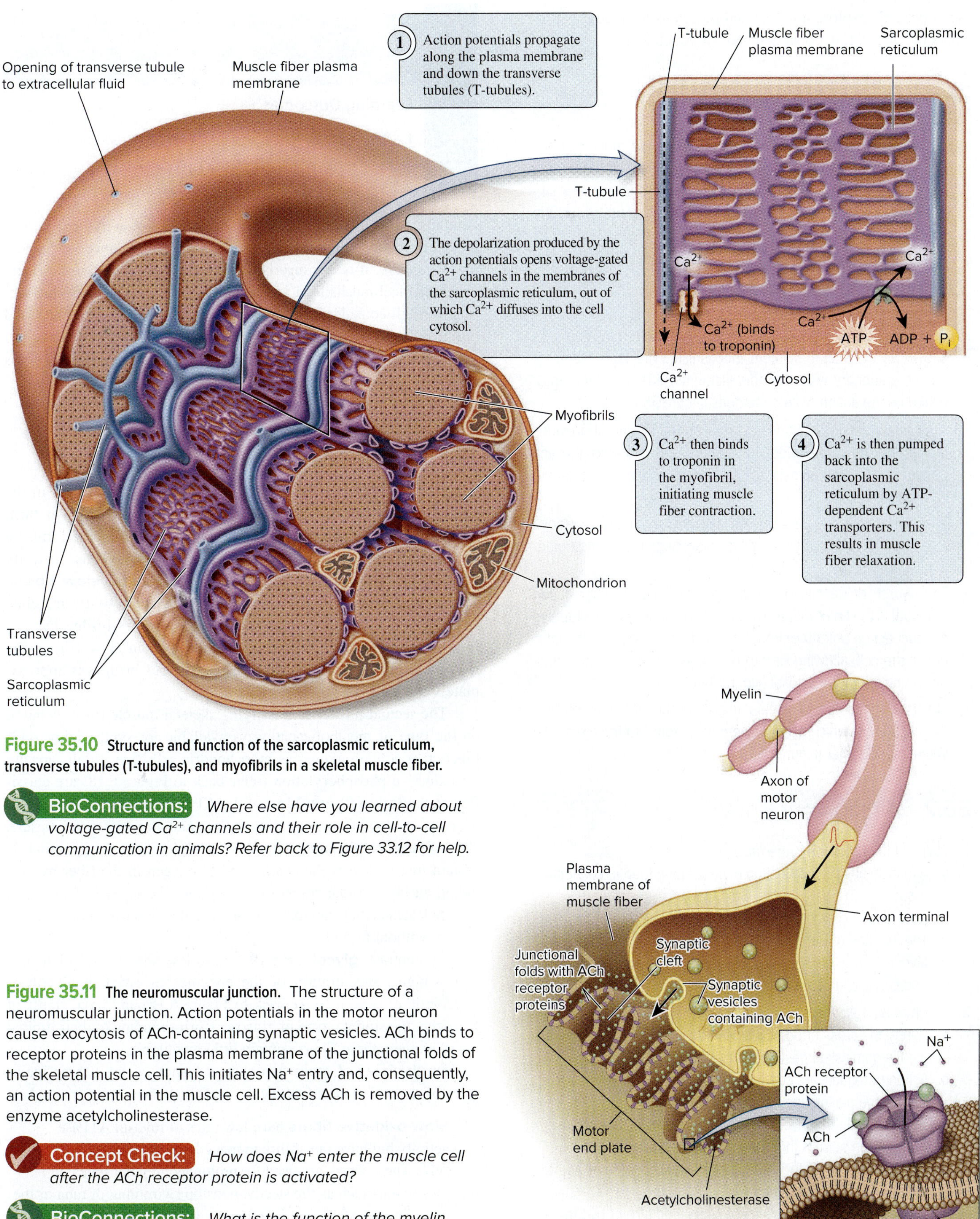


**Figure 35.10** **Structure and function of the sarcoplasmic reticulum, transverse tubules (T-tubules), and myofibrils in a skeletal muscle fiber.**

**BioConnections:** *Where else have you learned about voltage-gated $Ca^{2+}$ channels and their role in cell-to-cell communication in animals? Refer back to Figure 33.12 for help.*

**Figure 35.11** **The neuromuscular junction.** The structure of a neuromuscular junction. Action potentials in the motor neuron cause exocytosis of ACh-containing synaptic vesicles. ACh binds to receptor proteins in the plasma membrane of the junctional folds of the skeletal muscle cell. This initiates $Na^+$ entry and, consequently, an action potential in the muscle cell. Excess ACh is removed by the enzyme acetylcholinesterase.

**Concept Check:** *How does $Na^+$ enter the muscle cell after the ACh receptor protein is activated?*

**BioConnections:** *What is the function of the myelin indicated in this figure? Refer back to Figure 33.10 for help.*

generation, let's explore how skeletal muscle is adapted to meet the varied functional demands of vertebrates.

## 35.2 Reviewing the Concepts

- A skeletal muscle is a grouping of cells, called muscle fibers, bound together by connective tissue layers. Within each fiber are myofibrils, which contain thick and thin filaments arranged in sarcomeres. Regions of the thick filaments that extend toward the thin filaments are called cross-bridges (Figures 35.3, 35.4, 35.5).
- Mutations in a myosin gene expressed in jaw muscles of primates may have contributed to the evolution of larger brain size in humans (Figure 35.6).
- During muscle contraction, the sarcomeres shorten by a process known as the sliding-filament mechanism. The thick filaments remain stationary while the thin filaments slide past them, propelled by the action of cross-bridges (Figure 35.7).
- ATP has two major functions in cross-bridge cycling; it provides the energy for cross-bridge movement and it breaks the link between actin and myosin so that another cycle of shortening can occur (Figure 35.8).
- The binding of $Ca^{2+}$ to troponin allows tropomyosin to move away from the myosin-binding sites, initiating cross-bridge binding (Figure 35.9).
- The source of $Ca^{2+}$ involved in a skeletal muscle fiber action potential is the fiber's sarcoplasmic reticulum. Transverse tubules conduct action potentials from the plasma membrane to the interior of the cell, allowing $Ca^{2+}$ to be released from the sarcoplasmic reticulum to the cytosol (Figure 35.10).
- Electrical stimulation of skeletal muscle occurs at a neuromuscular junction, in which a motor neuron's axon terminals synapse with a muscle fiber (Figure 35.11).

## 35.2 Testing Your Knowledge

1. Which of the following is/are *false?*
   a. A muscle fiber is a collection of cells embedded in connective tissue.
   b. A sarcomere contains both actin and myosin arranged in parallel.
   c. The function of $Ca^{2+}$ in contraction is to bind to tropomyosin.
   d. Myofibrils are individual muscle cells.
   e. Choices a, c, and d are false.
2. Which of the following is correct?
   a. During the cross-bridge cycle, sarcomeres are lengthened as the filaments slide past each other.
   b. Action potentials are conducted from the interior of muscle cells to the plasma membrane via T-tubules.
   c. Troponin and tropomyosin are bound to myosin; their removal from myosin permits the filaments to slide past each other.
   d. Acetylcholine receptors are located in junctional folds at the motor end plate of muscle cells.
   e. The actions of acetylcholine are stimulated further by the enzyme acetylcholinesterase.

## 35.3 Types of Skeletal Muscle Fibers and Their Functions

**Learning Outcomes**

1. Outline the general characteristics of the three types of skeletal muscle fibers.
2. Explain how muscles adapt to exercise.

Animals use skeletal muscle for a wide variety of activities, such as locomotion, stretching, chewing, breathing, and balance, to name a few. Therefore, it is not surprising that not all skeletal muscle fibers share identical mechanical and metabolic characteristics. In this section, we will consider how different types of fibers can be classified on the basis of their rates of shortening (as either fast or slow) and the way in which they produce the ATP required for contraction (as oxidative or glycolytic).

### Skeletal Muscle Fibers Are Adapted for Different Types of Movement

Different muscle fibers contain forms of myosin that differ in the maximal rates at which they can hydrolyze ATP. This, in turn, determines the maximal rates of cross-bridge cycling and muscle shortening. Fibers containing myosin with low ATPase activity (that is, slow rates of hydrolyzing ATP) are called **slow fibers.** Those containing myosin with higher ATPase activity are classified as **fast fibers.** Although the rate of cross-bridge cycling is about four times faster in fast fibers than in slow fibers, the maximal force produced by both types of cross-bridges is approximately the same.

The second means of classifying skeletal muscle fibers is based on the type of metabolic pathways available for synthesizing ATP. Fibers that contain numerous mitochondria and have a high capacity for oxidative phosphorylation (refer back to Figure 6.18) are classified as **oxidative fibers.** Most of the ATP production by such fibers depends on blood flow delivering oxygen and nutrients to the muscle. These fibers contain large amounts of myoglobin, an oxygen-binding protein that increases the availability of oxygen in the fiber by providing an intracellular reservoir of oxygen. The benefit of oxidative muscle fibers is they can maintain sustained action over a long period of time without fatigue.

By contrast, **glycolytic fibers** have few mitochondria but possess a high concentration of the enzymes involved in glycolysis (glycolytic enzymes) (refer back to Figure 6.12) and large stores of glycogen, the storage form of glucose. Corresponding to their limited use of oxygen, these fibers contain little myoglobin.

On the basis of these two characteristics, at least three major types of skeletal muscle fibers can be distinguished:

1. **Slow-oxidative fibers** have low rates of myosin ATPase activity but have the ability to make large amounts of ATP. These fibers are used for prolonged, regular types of movement, such as the steady, nonstop swimming of tuna or the maintenance of posture in a human. Long-distance runners have a high proportion of these fibers in their leg muscles. These

types of activities require muscles that do not fatigue easily and are not needed for intense bursts of activity.

2. **Fast-oxidative fibers** have high myosin ATPase activity and can make large amounts of ATP. Like slow-oxidative fibers, these fibers do not fatigue quickly and can be used for long-term activities. They are also particularly suited for rapid actions, however, such as the rapid trilling sounds made by the throat muscles in songbirds or the clicking sounds generated by the shaking of a rattlesnake's tail.
3. **Fast-glycolytic fibers** have high myosin ATPase activity but cannot make as much ATP as oxidative fibers, because their source of ATP is glycolysis; thus, they fatigue much more quickly than oxidative fibers. They also tend to be larger cells with greater amounts of myosin and actin. These fibers are best suited for rapid, short-lasting but powerful actions. An example is the sudden burst of flight of chickens or turkeys when eluding a predator; these relatively large birds can quickly take flight but cannot sustain it for long.

Different muscle groups within an animal's body have different proportions of each fiber type interspersed with one another; many activities require the action of all three types of fibers at once. This is important when you consider the wide range of animal activities related to locomotion alone, including walking, climbing, running, swimming, flying, crawling, and jumping. Depending on the needs of an animal at any given moment, the motor nerve inputs can be adjusted to stimulate different ratios of fiber types. For example, when a crab uses its pincers to grab prey, fast-glycolytic muscles quickly snap the claws closed, and slow-oxidative fibers maintain a tight grip for as long as required.

### Muscles Adapt to Exercise

The regularity with which a muscle is used, as well as the duration and intensity of the activity and whether it includes resistance, affects the properties of the muscle. For example, increased amounts of resistance exercise—such as weightlifting—result in hypertrophy (increase in size) of muscle fibers. Normally, the number of fibers in a muscle remains essentially constant throughout adult life. The increases in muscle size that occur with resistance exercise result primarily from increases in the size of each fiber. These fibers undergo an increase in fiber diameter due to the increased synthesis of actin and myosin filaments, which form more myofibrils.

Exercise of relatively low intensity but long duration—popularly called aerobic exercise, including running and swimming—increases the number of mitochondria in the fibers that are required in this type of activity. In addition, the number of blood vessels around these fibers increases to supply the greater energy demands of active muscle. All of these changes increase muscle endurance.

By contrast, short-duration, high-intensity exercise, such as weightlifting, primarily affects fast-glycolytic fibers, which are used during strong contractions. In addition, this type of exercise stimulates the synthesis of glycolytic enzymes, thereby increasing the glycolytic capacity of the muscle. The results of such high-intensity exercise are the increased strength and enlarged muscles of a conditioned weight lifter. Such muscles, although very powerful, have little capacity for endurance and therefore fatigue rapidly.

## 35.3 Reviewing the Concepts

- Slow-oxidative muscle fibers have low rates of myosin ATP hydrolysis and are used for prolonged, regular activities. Fast-oxidative fibers have high rates of myosin ATP hydrolysis and are suited for rapid, long-term actions. Fast-glycolytic fibers have high myosin ATP hydrolysis but cannot make as much ATP as oxidative fibers; they are best suited for rapid, intense, short-term actions.
- Increased amounts of exercise affect the property of muscle and can result in hypertrophy.

## 35.3 Testing Your Knowledge

1. Muscles that contain large amounts of fast-glycolytic fibers
   a. are suited for prolonged, steady activity such as running a marathon.
   b. are ideal for the flight muscles of birds during long migrations.
   c. are suited for relatively brief, intense activity such as that of a predator chasing a prey for a few seconds.
   d. have low amounts of myosin ATPase activity.
   e. can make more ATP than can muscles containing oxidative fibers.

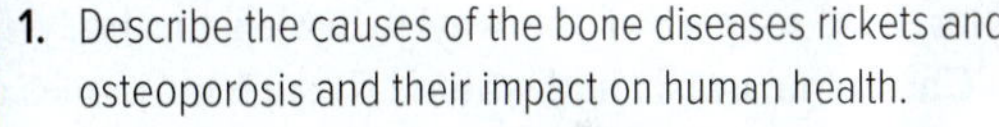

## 35.4 Impact on Public Health

**Learning Outcomes**

1. Describe the causes of the bone diseases rickets and osteoporosis and their impact on human health.
2. Describe the general features of muscular dystrophy.

In this chapter, we have seen how skeletal muscles and, in vertebrates, bones work together to enable animals to move around in their environments. A number of diseases affect bone structure and function in humans. Bone disease may involve defects in either the mineral or organic components of bone. Poor bone formation and structure may result from inadequate nutrition, hormonal imbalances, aging, or skeletal muscle atrophy, to name just a few of the common causes.

In addition, many diseases or disorders directly affect the contraction of skeletal muscle. Some of them are temporary and not serious, such as muscle cramps, whereas others, such as the disease muscular dystrophy, are chronic and severe. Also, some muscle diseases result from defects in parts of the nervous system that control contraction of the muscle fibers rather than from defects that originate in the fibers themselves. One example is amyotrophic lateral sclerosis (also known as Lou Gehrig disease), in which motor neurons die, leading to the atrophy of the muscles those neurons normally stimulate. Still other diseases that affect skeletal muscle function result when normal processes go awry. For example, the system that normally protects the body from foreign pathogens may turn on itself, as occurred in the woman with myasthenia gravis described in the chapter-opening story. This section looks at a few of these conditions in more detail.

## Rickets and Osteoporosis Affect the Bones of Millions of People

Bone diseases are fairly common, particularly among individuals older than age 50. Two major abnormalities can occur in bone. The first is improper mineral deposition in bone, usually due to inadequate dietary $Ca^{2+}$ intake or inadequate absorption of $Ca^{2+}$ from the small intestine. Without adequate minerals, bone becomes soft and easily deformed, as occurs in the weight-bearing bones of the legs of children with **rickets** (or **osteomalacia,** as it is called in adults) (Figure 35.12a). These disorders are prevented or treated with vitamin D supplements, because a hormone derived from this vitamin is the most important factor in promoting absorption of ingested $Ca^{2+}$ from the small intestine.

The second major abnormality is a more common disease called **osteoporosis,** in which both the mineral and organic portions of bone are decreased (Figure 35.12b). This disease, which affects four times as many women as men, occurs when the normal balance between bone formation and bone breakdown is disrupted, resulting in decreased bone density. Osteoporosis is the most prevalent bone disease in the U.S., affecting an estimated 15–30 million individuals. It results in annual national expenditures of approximately $15–20 billion in hospital and other medical costs.

Most commonly, osteoporosis results from hormonal imbalances. Some hormones—for example, estrogen—are required to maintain bone mass. When estrogen concentrations decrease after menopause (the period of time when a woman's reproductive cycles cease), bone density may decline, increasing the risk of bone fractures. By contrast, some hormones—such as parathyroid hormone (see Chapter 38)—act to demineralize bone, releasing $Ca^{2+}$ into the extracellular fluid as part of the way in which the body normally maintains $Ca^{2+}$ homeostasis in the blood. If such hormones are present in excess, however, they can cause sufficient demineralization of bone to result in osteoporosis. This may happen in rare cases when the glands that make these hormones malfunction and overproduce the hormones.

Another cause of osteoporosis is prolonged disuse of muscles. In ways that are not completely clear, the force produced by active skeletal muscle contractions helps maintain bone mass. When muscles are not or cannot be used—due to paralysis or long-term immobilizing illnesses or even during prolonged space flight in which the decreased gravity reduces the need for muscles to work hard—bone mass declines.

Osteoporosis can be minimized with adequate $Ca^{2+}$ and vitamin D intake and weight-bearing exercise programs. In some cases, postmenopausal women may be given estrogen to replace what their bodies are no longer producing, but this therapy is controversial due to the potential adverse effects of estrogen on cardiovascular health and possible increased risk of developing breast cancer. Drugs that inhibit bone-resorbing osteoclasts are also commonly used to minimize bone damage due to this disease.

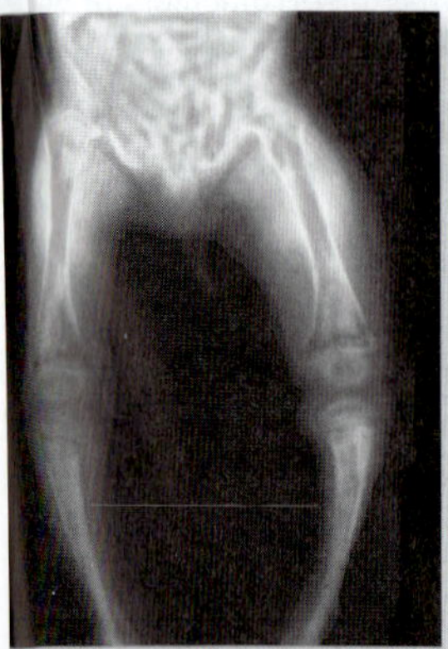

(a) X-ray image of leg bones of a child with rickets

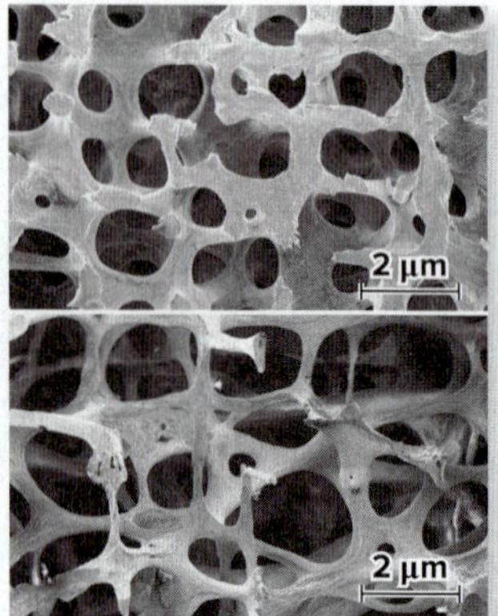


(b) Histologic appearance of normal bone (top) and bone from a person with osteoporosis (bottom)

Figure 35.12 Human bone diseases.

(a) © Dr. LR/Science Source; (b) © Tim Arnett, University College London

## Muscular Dystrophy Is a Rare Genetic Disease That Causes Muscle Degeneration

The group of diseases collectively called muscular dystrophy affects 1 of every 3,500 American males; it is much less common in females. **Muscular dystrophy** is associated with the progressive degeneration of skeletal and cardiac muscle fibers, weakening the muscles and leading ultimately to death from heart failure and other causes. The signs and symptoms become evident at about 2 to 6 years of age, and most affected individuals do not survive beyond the age of 30.

The most common form of muscular dystrophy, called **Duchenne muscular dystrophy,** is an X-linked recessive disorder resulting from a defective gene on the X chromosome. Since females have two X chromosomes, and males only one, a heterozygote female with one abnormal and one normal allele will not generally develop Duchenne muscular dystrophy. If she passes the abnormal allele to a son, however, he will have the disease. The affected gene normally encodes a protein known as dystrophin, which is absent in patients with the disease. Dystrophin is a large protein that links cytoskeletal proteins to the plasma membrane. Scientists think it is involved in maintaining the structural integrity of the plasma membrane in muscle fibers. In the absence of dystrophin, the plasma membrane of muscle fibers is disrupted, causing extracellular fluid to enter the cell. Eventually, the cell ruptures and dies.

One area of interest in the treatment of muscular dystrophy involves a protein called myostatin, which is one of a large number of proteins that control muscle growth in mammals. Myostatin functions to prevent overgrowth of skeletal muscle by inhibiting stem cell maturation into new muscle fibers, and preventing excessive growth of mature fibers. Researchers have identified animals with mutations in the myostatin gene that result in the production of an inactive protein; such animals show astonishing muscle development (Figure 35.13). Whether or not targeting the myostatin protein will prove successful in reversing muscle loss in muscular dystrophy and other diseases remains to be determined.

types of activities require muscles that do not fatigue easily and are not needed for intense bursts of activity.

2. **Fast-oxidative fibers** have high myosin ATPase activity and can make large amounts of ATP. Like slow-oxidative fibers, these fibers do not fatigue quickly and can be used for long-term activities. They are also particularly suited for rapid actions, however, such as the rapid trilling sounds made by the throat muscles in songbirds or the clicking sounds generated by the shaking of a rattlesnake's tail.
3. **Fast-glycolytic fibers** have high myosin ATPase activity but cannot make as much ATP as oxidative fibers, because their source of ATP is glycolysis; thus, they fatigue much more quickly than oxidative fibers. They also tend to be larger cells with greater amounts of myosin and actin. These fibers are best suited for rapid, short-lasting but powerful actions. An example is the sudden burst of flight of chickens or turkeys when eluding a predator; these relatively large birds can quickly take flight but cannot sustain it for long.

Different muscle groups within an animal's body have different proportions of each fiber type interspersed with one another; many activities require the action of all three types of fibers at once. This is important when you consider the wide range of animal activities related to locomotion alone, including walking, climbing, running, swimming, flying, crawling, and jumping. Depending on the needs of an animal at any given moment, the motor nerve inputs can be adjusted to stimulate different ratios of fiber types. For example, when a crab uses its pincers to grab prey, fast-glycolytic muscles quickly snap the claws closed, and slow-oxidative fibers maintain a tight grip for as long as required.

### Muscles Adapt to Exercise

The regularity with which a muscle is used, as well as the duration and intensity of the activity and whether it includes resistance, affects the properties of the muscle. For example, increased amounts of resistance exercise—such as weightlifting—result in hypertrophy (increase in size) of muscle fibers. Normally, the number of fibers in a muscle remains essentially constant throughout adult life. The increases in muscle size that occur with resistance exercise result primarily from increases in the size of each fiber. These fibers undergo an increase in fiber diameter due to the increased synthesis of actin and myosin filaments, which form more myofibrils.

Exercise of relatively low intensity but long duration—popularly called aerobic exercise, including running and swimming—increases the number of mitochondria in the fibers that are required in this type of activity. In addition, the number of blood vessels around these fibers increases to supply the greater energy demands of active muscle. All of these changes increase muscle endurance.

By contrast, short-duration, high-intensity exercise, such as weightlifting, primarily affects fast-glycolytic fibers, which are used during strong contractions. In addition, this type of exercise stimulates the synthesis of glycolytic enzymes, thereby increasing the glycolytic capacity of the muscle. The results of such high-intensity exercise are the increased strength and enlarged muscles of a conditioned weight lifter. Such muscles, although very powerful, have little capacity for endurance and therefore fatigue rapidly.

## 35.3 Reviewing the Concepts

- Slow-oxidative muscle fibers have low rates of myosin ATP hydrolysis and are used for prolonged, regular activities. Fast-oxidative fibers have high rates of myosin ATP hydrolysis and are suited for rapid, long-term actions. Fast-glycolytic fibers have high myosin ATP hydrolysis but cannot make as much ATP as oxidative fibers; they are best suited for rapid, intense, short-term actions.
- Increased amounts of exercise affect the property of muscle and can result in hypertrophy.

## 35.3 Testing Your Knowledge

1. Muscles that contain large amounts of fast-glycolytic fibers
   a. are suited for prolonged, steady activity such as running a marathon.
   b. are ideal for the flight muscles of birds during long migrations.
   c. are suited for relatively brief, intense activity such as that of a predator chasing a prey for a few seconds.
   d. have low amounts of myosin ATPase activity.
   e. can make more ATP than can muscles containing oxidative fibers.

## 35.4 Impact on Public Health

**Learning Outcomes**

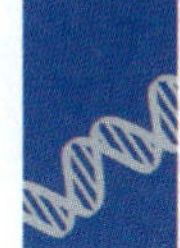

1. Describe the causes of the bone diseases rickets and osteoporosis and their impact on human health.
2. Describe the general features of muscular dystrophy.

In this chapter, we have seen how skeletal muscles and, in vertebrates, bones work together to enable animals to move around in their environments. A number of diseases affect bone structure and function in humans. Bone disease may involve defects in either the mineral or organic components of bone. Poor bone formation and structure may result from inadequate nutrition, hormonal imbalances, aging, or skeletal muscle atrophy, to name just a few of the common causes.

In addition, many diseases or disorders directly affect the contraction of skeletal muscle. Some of them are temporary and not serious, such as muscle cramps, whereas others, such as the disease muscular dystrophy, are chronic and severe. Also, some muscle diseases result from defects in parts of the nervous system that control contraction of the muscle fibers rather than from defects that originate in the fibers themselves. One example is amyotrophic lateral sclerosis (also known as Lou Gehrig disease), in which motor neurons die, leading to the atrophy of the muscles those neurons normally stimulate. Still other diseases that affect skeletal muscle function result when normal processes go awry. For example, the system that normally protects the body from foreign pathogens may turn on itself, as occurred in the woman with myasthenia gravis described in the chapter-opening story. This section looks at a few of these conditions in more detail.

## Rickets and Osteoporosis Affect the Bones of Millions of People

Bone diseases are fairly common, particularly among individuals older than age 50. Two major abnormalities can occur in bone. The first is improper mineral deposition in bone, usually due to inadequate dietary $Ca^{2+}$ intake or inadequate absorption of $Ca^{2+}$ from the small intestine. Without adequate minerals, bone becomes soft and easily deformed, as occurs in the weight-bearing bones of the legs of children with **rickets** (or **osteomalacia,** as it is called in adults) (Figure 35.12a). These disorders are prevented or treated with vitamin D supplements, because a hormone derived from this vitamin is the most important factor in promoting absorption of ingested $Ca^{2+}$ from the small intestine.

The second major abnormality is a more common disease called **osteoporosis,** in which both the mineral and organic portions of bone are decreased (Figure 35.12b). This disease, which affects four times as many women as men, occurs when the normal balance between bone formation and bone breakdown is disrupted, resulting in decreased bone density. Osteoporosis is the most prevalent bone disease in the U.S., affecting an estimated 15–30 million individuals. It results in annual national expenditures of approximately $15–20 billion in hospital and other medical costs.

Most commonly, osteoporosis results from hormonal imbalances. Some hormones—for example, estrogen—are required to maintain bone mass. When estrogen concentrations decrease after menopause (the period of time when a woman's reproductive cycles cease), bone density may decline, increasing the risk of bone fractures. By contrast, some hormones—such as parathyroid hormone (see Chapter 38)—act to demineralize bone, releasing $Ca^{2+}$ into the extracellular fluid as part of the way in which the body normally maintains $Ca^{2+}$ homeostasis in the blood. If such hormones are present in excess, however, they can cause sufficient demineralization of bone to result in osteoporosis. This may happen in rare cases when the glands that make these hormones malfunction and overproduce the hormones.

Another cause of osteoporosis is prolonged disuse of muscles. In ways that are not completely clear, the force produced by active skeletal muscle contractions helps maintain bone mass. When muscles are not or cannot be used—due to paralysis or long-term immobilizing illnesses or even during prolonged space flight in which the decreased gravity reduces the need for muscles to work hard—bone mass declines.

Osteoporosis can be minimized with adequate $Ca^{2+}$ and vitamin D intake and weight-bearing exercise programs. In some cases, postmenopausal women may be given estrogen to replace what their bodies are no longer producing, but this therapy is controversial due to the potential adverse effects of estrogen on cardiovascular health and possible increased risk of developing breast cancer. Drugs that inhibit bone-resorbing osteoclasts are also commonly used to minimize bone damage due to this disease.

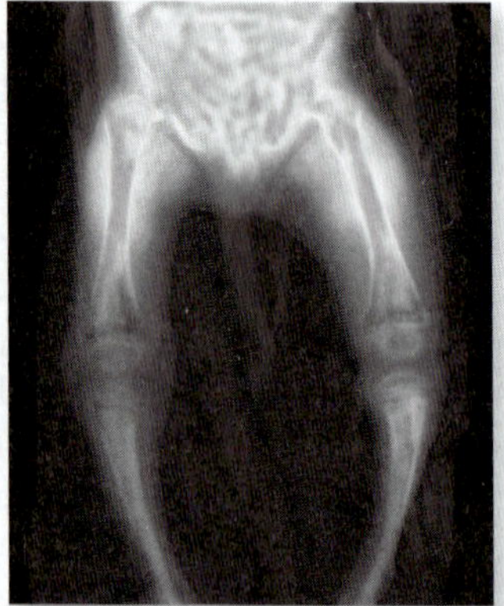

(a) X-ray image of leg bones of a child with rickets

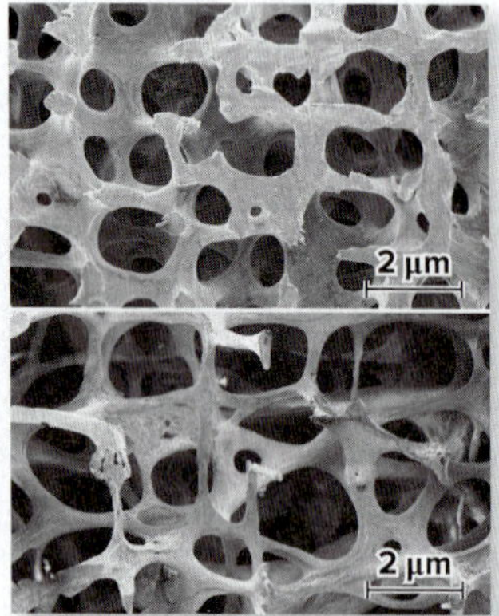


(b) Histologic appearance of normal bone (top) and bone from a person with osteoporosis (bottom)

**Figure 35.12** Human bone diseases.

*(a)* © Dr. LR/Science Source; *(b)* © Tim Arnett, University College London

## Muscular Dystrophy Is a Rare Genetic Disease That Causes Muscle Degeneration

The group of diseases collectively called muscular dystrophy affects 1 of every 3,500 American males; it is much less common in females. **Muscular dystrophy** is associated with the progressive degeneration of skeletal and cardiac muscle fibers, weakening the muscles and leading ultimately to death from heart failure and other causes. The signs and symptoms become evident at about 2 to 6 years of age, and most affected individuals do not survive beyond the age of 30.

The most common form of muscular dystrophy, called **Duchenne muscular dystrophy,** is an X-linked recessive disorder resulting from a defective gene on the X chromosome. Since females have two X chromosomes, and males only one, a heterozygote female with one abnormal and one normal allele will not generally develop Duchenne muscular dystrophy. If she passes the abnormal allele to a son, however, he will have the disease. The affected gene normally encodes a protein known as dystrophin, which is absent in patients with the disease. Dystrophin is a large protein that links cytoskeletal proteins to the plasma membrane. Scientists think it is involved in maintaining the structural integrity of the plasma membrane in muscle fibers. In the absence of dystrophin, the plasma membrane of muscle fibers is disrupted, causing extracellular fluid to enter the cell. Eventually, the cell ruptures and dies.

One area of interest in the treatment of muscular dystrophy involves a protein called myostatin, which is one of a large number of proteins that control muscle growth in mammals. Myostatin functions to prevent overgrowth of skeletal muscle by inhibiting stem cell maturation into new muscle fibers, and preventing excessive growth of mature fibers. Researchers have identified animals with mutations in the myostatin gene that result in the production of an inactive protein; such animals show astonishing muscle development (Figure 35.13). Whether or not targeting the myostatin protein will prove successful in reversing muscle loss in muscular dystrophy and other diseases remains to be determined.

**Figure 35.13** **An example of an animal with a mutation in the gene that encodes the protein myostatin, an inhibitor of skeletal muscle growth.**
© Bruce Stotesbury/ZUMAPRESS.com/Newscom.com

## 35.4 Reviewing the Concepts

- Rickets in children (osteomalacia in adults) is characterized by soft, deformed bones, usually resulting from insufficient dietary intake or absorption of $Ca^{2+}$. In osteoporosis, bone density is reduced when bone formation fails to keep pace with normal bone breakdown (Figure 35.12).
- Muscular dystrophy is an ultimately fatal genetic disease associated with the progressive degeneration of skeletal and cardiac muscle fibers (Figure 35.13).

## 35.4 Testing Your Knowledge

1. Which of the following statements is/are *false?*
   a. Rickets is a disease of the organic component of bone in children.
   b. Osteoporosis is characterized by thickened bones in adults.
   c. In osteomalacia, bone hardens and becomes filled with excess mineral deposits.
   d. Osteoporosis is typically caused by excessive intake of $Ca^{2+}$ and vitamin D.
   e. All of the above are false.
2. Duchenne muscular dystrophy is primarily characterized by
   a. bone loss due to $Ca^{2+}$ depletion.
   b. an X-linked disorder, making it more common in females.
   c. disruption of plasma membranes of muscle fibers.
   d. increased amounts of the protein dystrophin.
   e. degeneration of smooth muscle.

# Assess and Discuss

## Test Yourself

1. Which two minerals are the major components of bone?
   a. $Na^+$ and $K^+$
   b. $Na^+$ and $Ca^{2+}$
   c. $Ca^{2+}$ and $K^+$
   d. $Ca^{2+}$ and $PO_4^{2-}$
   e. $PO_4^{2-}$ and $Mg^{2+}$
2. Which, if any, of the following is *not* a function of the vertebrate skeleton?
   a. structural support
   b. protection of internal organs
   c. calcium reserve
   d. blood cell production
   e. All of the above are functions of the vertebrate skeleton.
3. The protein that provides strength and flexibility to bone is
   a. actin.
   b. myosin.
   c. myoglobin.
   d. collagen.
   e. elastin.
4. In a sarcomere, the ________ contain(s) thin filaments and no thick filaments.
   a. A band
   b. M line
   c. I band
   d. H zone
   e. both a and d
5. The function of ATP during muscle contraction is to
   a. cause an allosteric change in myosin so that it detaches from actin.
   b. provide the energy necessary for the movement of the cross-bridge.
   c. expose the myosin-binding sites on the thin filaments.
   d. do all of the above.
   e. do a and b only.
6. During cross-bridge cycling, $Ca^{2+}$ binds to ________, which causes _________.
   a. tropomyosin, detachment of the cross-bridge from actin
   b. myosin, binding of the cross-bridge to actin
   c. actin, binding of the cross-bridge to myosin
   d. troponin, exposure of the myosin-binding sites on actin
   e. troponin, exposure of the actin-binding sites on tropomyosin
7. Stimulation of a muscle fiber by a motor neuron occurs at
   a. the neuromuscular junction.
   b. the transverse tubules.
   c. the myofibril.
   d. the sarcoplasmic reticulum.
   e. none of the above.
8. Muscle fibers that have a high number of mitochondria, contain large amounts of myoglobin, and exhibit low rates of ATP hydrolysis are called __________ fibers.
   a. slow-glycolytic
   b. fast-glycolytic
   c. intermediate
   d. fast-oxidative
   e. slow-oxidative

9. _____ is released from axon terminals of motor neurons and directly causes _______.
   a. $Ca^{2+}$, an action potential in the muscle fiber
   b. Acetylcholine, $Ca^{2+}$ diffusion out of the muscle fiber
   c. Acetylcholine, $Na^{+}$ diffusion into the muscle fiber
   d. $Na^{+}$, $Ca^{2+}$ diffusion out of the muscle fiber
   e. Acetylcholine, $Ca^{2+}$ diffusion into the muscle fiber
10. Which of the following is *false?*
   a. Flexors bend a limb.
   b. Antagonist muscles have opposite effects on limb movement.
   c. The increase in muscle mass observed with weight training is primarily due to an increase in the number of myofibrils inside each muscle cell.
   d. Glycolytic muscle fibers synthesize less ATP and have fewer mitochondria than oxidative fibers.
   e. Fast fibers have lower myosin ATPase activity than slow fibers.

## Conceptual Questions

1. Explain the link between electrical excitation and skeletal muscle contraction. Which of these two events occurs first, and why?
2. List and briefly describe the steps in the cross-bridge cycle.
3. **PRINCIPLES** A principle of biology is that living organisms use energy. How is that principle illustrated in this chapter, particularly with regard to the mechanisms by which skeletal muscle cells contract?

## Collaborative Questions

1. Discuss the structure and function of the two major types of skeletons found in animals, giving examples of animals in which such skeletons are found.
2. Discuss the three types of muscle tissues found in vertebrates, and identify distinguishing functional features of each.

**Figure 35.13 An example of an animal with a mutation in the gene that encodes the protein myostatin, an inhibitor of skeletal muscle growth.**
© Bruce Stotesbury/ZUMAPRESS.com/Newscom.com

## 35.4 Reviewing the Concepts

- Rickets in children (osteomalacia in adults) is characterized by soft, deformed bones, usually resulting from insufficient dietary intake or absorption of $Ca^{2+}$. In osteoporosis, bone density is reduced when bone formation fails to keep pace with normal bone breakdown (Figure 35.12).
- Muscular dystrophy is an ultimately fatal genetic disease associated with the progressive degeneration of skeletal and cardiac muscle fibers (Figure 35.13).

## 35.4 Testing Your Knowledge

1. Which of the following statements is/are *false?*
   a. Rickets is a disease of the organic component of bone in children.
   b. Osteoporosis is characterized by thickened bones in adults.
   c. In osteomalacia, bone hardens and becomes filled with excess mineral deposits.
   d. Osteoporosis is typically caused by excessive intake of $Ca^{2+}$ and vitamin D.
   e. All of the above are false.
2. Duchenne muscular dystrophy is primarily characterized by
   a. bone loss due to $Ca^{2+}$ depletion.
   b. an X-linked disorder, making it more common in females.
   c. disruption of plasma membranes of muscle fibers.
   d. increased amounts of the protein dystrophin.
   e. degeneration of smooth muscle.

# Assess and Discuss

## Test Yourself

1. Which two minerals are the major components of bone?
   a. $Na^+$ and $K^+$
   b. $Na^+$ and $Ca^{2+}$
   c. $Ca^{2+}$ and $K^+$
   d. $Ca^{2+}$ and $PO_4^{2-}$
   e. $PO_4^{2-}$ and $Mg^{2+}$
2. Which, if any, of the following is *not* a function of the vertebrate skeleton?
   a. structural support
   b. protection of internal organs
   c. calcium reserve
   d. blood cell production
   e. All of the above are functions of the vertebrate skeleton.
3. The protein that provides strength and flexibility to bone is
   a. actin.
   b. myosin.
   c. myoglobin.
   d. collagen.
   e. elastin.
4. In a sarcomere, the ________ contain(s) thin filaments and no thick filaments.
   a. A band
   b. M line
   c. I band
   d. H zone
   e. both a and d
5. The function of ATP during muscle contraction is to
   a. cause an allosteric change in myosin so that it detaches from actin.
   b. provide the energy necessary for the movement of the cross-bridge.
   c. expose the myosin-binding sites on the thin filaments.
   d. do all of the above.
   e. do a and b only.
6. During cross-bridge cycling, $Ca^{2+}$ binds to ________, which causes ________.
   a. tropomyosin, detachment of the cross-bridge from actin
   b. myosin, binding of the cross-bridge to actin
   c. actin, binding of the cross-bridge to myosin
   d. troponin, exposure of the myosin-binding sites on actin
   e. troponin, exposure of the actin-binding sites on tropomyosin
7. Stimulation of a muscle fiber by a motor neuron occurs at
   a. the neuromuscular junction.
   b. the transverse tubules.
   c. the myofibril.
   d. the sarcoplasmic reticulum.
   e. none of the above.
8. Muscle fibers that have a high number of mitochondria, contain large amounts of myoglobin, and exhibit low rates of ATP hydrolysis are called ________ fibers.
   a. slow-glycolytic
   b. fast-glycolytic
   c. intermediate
   d. fast-oxidative
   e. slow-oxidative

9. _____ is released from axon terminals of motor neurons and directly causes _______.
   a. $Ca^{2+}$, an action potential in the muscle fiber
   b. Acetylcholine, $Ca^{2+}$ diffusion out of the muscle fiber
   c. Acetylcholine, $Na^+$ diffusion into the muscle fiber
   d. $Na^+$, $Ca^{2+}$ diffusion out of the muscle fiber
   e. Acetylcholine, $Ca^{2+}$ diffusion into the muscle fiber
10. Which of the following is *false?*
   a. Flexors bend a limb.
   b. Antagonist muscles have opposite effects on limb movement.
   c. The increase in muscle mass observed with weight training is primarily due to an increase in the number of myofibrils inside each muscle cell.
   d. Glycolytic muscle fibers synthesize less ATP and have fewer mitochondria than oxidative fibers.
   e. Fast fibers have lower myosin ATPase activity than slow fibers.

## Conceptual Questions

1. Explain the link between electrical excitation and skeletal muscle contraction. Which of these two events occurs first, and why?
2. List and briefly describe the steps in the cross-bridge cycle.
3. **PRINCIPLES** A principle of biology is that living organisms use energy. How is that principle illustrated in this chapter, particularly with regard to the mechanisms by which skeletal muscle cells contract?

## Collaborative Questions

1. Discuss the structure and function of the two major types of skeletons found in animals, giving examples of animals in which such skeletons are found.
2. Discuss the three types of muscle tissues found in vertebrates, and identify distinguishing functional features of each.

## Online Resource

# Circulatory and Respiratory Systems: Transporting Solutes and Exchanging Gases

36

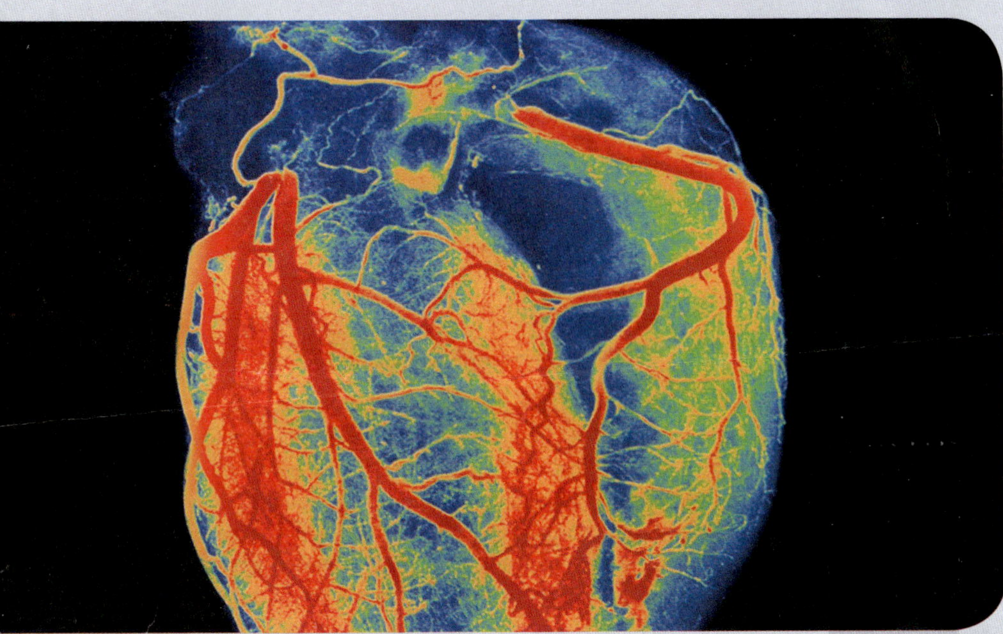

© SPL/Science Source

Image from cardiac angiography of a human heart (dye contrast injection to make the coronary blood vessels [in red] easier to visualize).

## Chapter Outline



An overweight 62-year-old woman decided to get herself "back in shape" after years of a sedentary lifestyle. Rather than gradually increasing her activity level, however, she embarked on a rigorous, demanding exercise program at a local gym. While running at 7 mph on a 10% inclined treadmill, she felt an acute, crushing sensation in her chest. She became extremely anxious and stopped exercising, but the pain did not go away. She felt light-headed, short of breath and nauseated, and called for help. An ambulance arrived, and the paramedic determined from her vital signs that the woman was experiencing a heart attack. She was transported to the emergency department of a local hospital and remained there for further treatment. Eventually, she was released and placed on several medications to control her blood pressure, prevent blood clots from forming, and improve the ability of the healthy parts of her heart to function. She also received counseling from a nutritionist and a physical therapist, who provided her with a sensible plan for improving her overall health.

Cardiovascular disease—disease of the heart or blood vessels—is the leading cause of death in the U.S. and in much of the rest of the developed world. In the U.S. alone, an estimated 80 million or more people have one or more diseases that affect the heart and blood vessels. About 1–1.5 million of those individuals will have a heart attack in the next year, and between 400,000 and 600,000 of them will die as a result.

The heart is a muscular pump that requires considerable nutrients and oxygen to sustain its unceasing muscular effort. The chapter-opening photo shows the extensive network of blood vessels coursing through a typical mammalian heart. Should any of those vessels become diseased, the regions of the heart supplied by them can die. That is what happens during a heart attack. When this occurs, the damaged heart may not be able to pump blood forcefully enough to generate the pressure required for sufficient blood to reach all the cells of the body.

Circulatory systems transport necessary materials to all the cells of an animal's body and transport waste products away from the cells so that they can be released into the environment. Two of the most important solutes carried in the blood are oxygen ($O_2$) and carbon dioxide ($CO_2$). Oxygen enters an animal's body via a respiratory organ such as gills or lungs. It then dissolves into body fluids and is distributed to cells. As described in Chapter 6, $O_2$ is used by mitochondria in the formation of ATP. One of the waste products of those reactions is $CO_2$, which diffuses out of cells. In vertebrates, $CO_2$ is returned by the circulatory system to the respiratory organ and is then released into the environment. The process of moving $O_2$ and $CO_2$ in opposite directions between the environment, body fluids, and cells is called **gas exchange,** or **respiration.** (This differs from cellular respiration, in which cells generate energy from metabolism and produce ATP, as described in Chapter 6.) A **respiratory system** includes all of an animal's structures that contribute to gas exchange.

This chapter deals with the mechanisms that control the circulation of body fluids, the means by which animals obtain and transport $O_2$, rid themselves of $CO_2$, and cope with the challenges imposed by changing metabolic demands.

## 36.1 Types of Circulatory Systems

### Learning Outcomes

1. Compare and contrast open and closed circulatory systems.
2. Describe the differences between single and double circulations in vertebrates.

Apart from the simple type of circulation that exists in certain invertebrates that have a gastrovascular cavity (look ahead to Figure 37.2), two basic types of animal circulatory systems exist: open systems and closed systems. Here, we compare and contrast some of the key features of the two major types of circulatory systems found in animals.

### In Open Circulatory Systems, Hemolymph Enters the Body Cavity

As illustrated in **Figure 36.1a**, an **open circulatory system** is characterized by a fluid that is pumped by one or more contractile hearts into the body cavity (hemocoel) of an animal. Therefore, the fluid in the blood vessels and the interstitial fluid that surrounds cells mingle in one large, mixed compartment, rather than being located in separate body compartments. The mixed fluid is called **hemolymph.** Nutrients and wastes are exchanged by diffusion between the hemolymph and body cells, and the hemolymph is eventually returned to the heart through vessels or through small openings called ostia. Oxygen and $CO_2$ are not transported in hemolymph; in animals with open circulatory systems, these gases are exchanged with the environment by a different mechanism, which we will examine later.

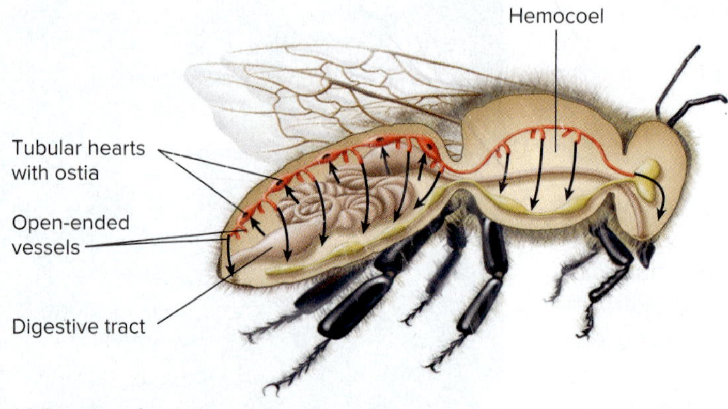


**(a) An open circulatory system**

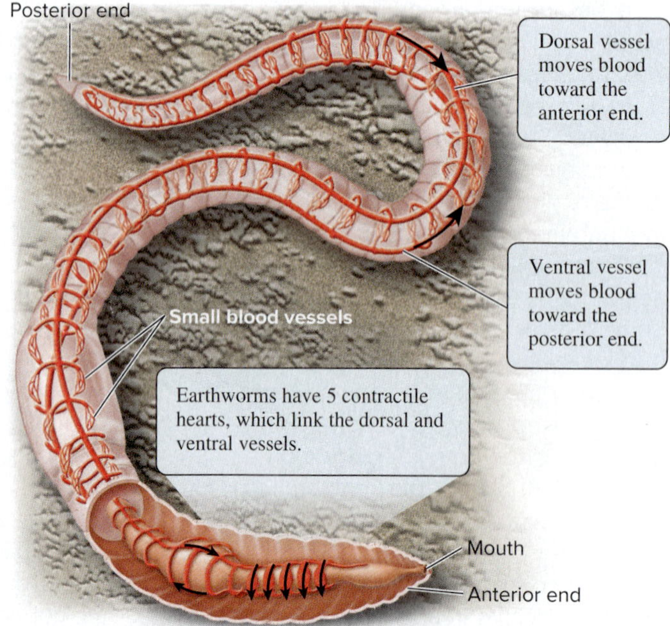


**(b) A closed circulatory system**

**Figure 36.1** **Types of circulatory systems. (a)** In open circulatory systems, one or more muscular, tubular hearts pump hemolymph through open-ended vessels, where it percolates through the body. In arthropods such as this honeybee, hemolymph reenters the heart through ostia. The arrows show the movement of hemolymph. **(b)** In closed circulatory systems, such as that of this earthworm, blood remains within vessels and hearts, and it recirculates without emptying into the body cavity. Arrows indicate direction of blood flow.

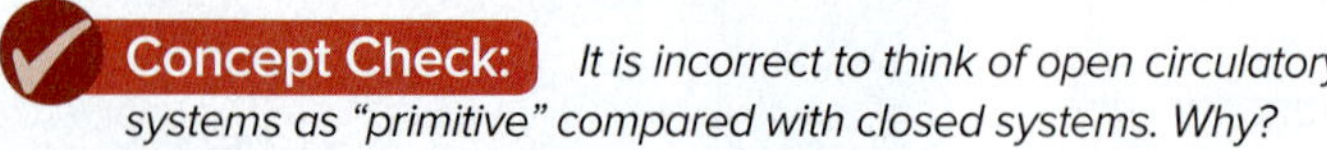

**Concept Check:** *It is incorrect to think of open circulatory systems as "primitive" compared with closed systems. Why?*

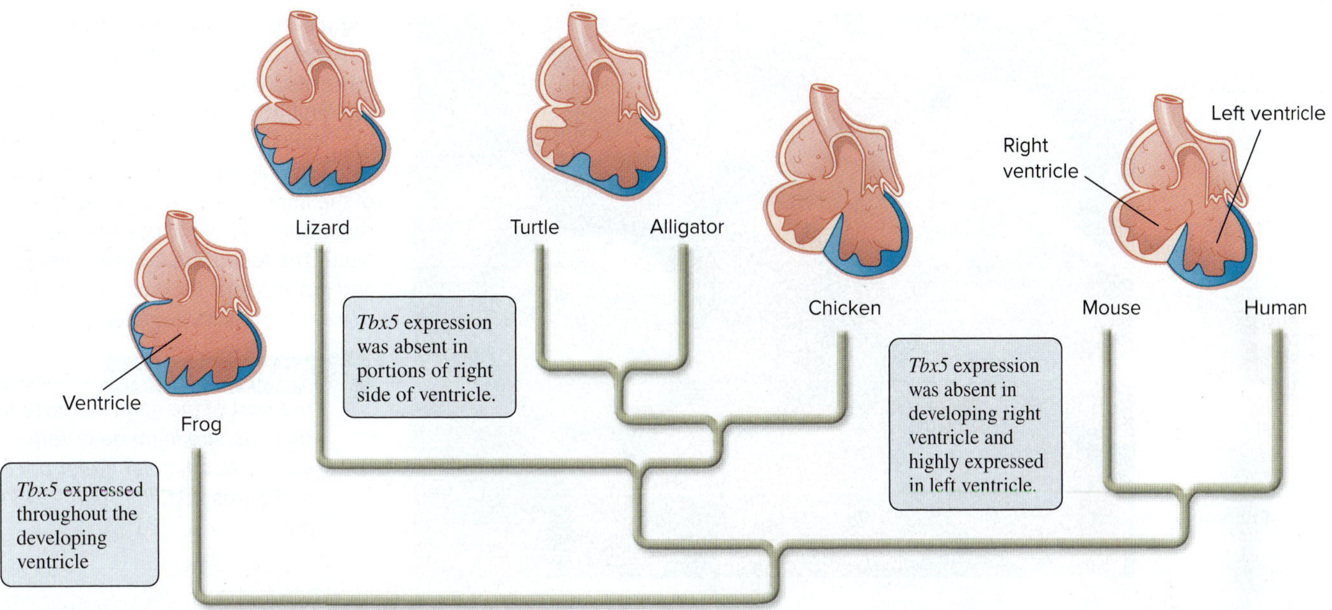


**Figure 36.3** **Expression of the *Tbx5* gene in embryonic vertebrate hearts.** The blue color indicates where in the developing heart muscle the *Tbx5* gene was expressed. Note: Structural details have been omitted for simplicity.

Source: Kazuko Koshiba-Takeuchi et al., Reptilian heart development and the molecular basis of cardiac chamber evolution. Nature 461, 95–98, 2009.

expression was experimentally induced in the heart of an embryonic mouse in a pattern that mimicked that seen in lizards, the mouse heart developed only three clearly defined chambers—there was no distinction between left and right ventricles!

From these and other studies, it appears that the expression of a core set of genes is vital for the formation of a heart or heartlike structure, and a subset of genes, like *Tbx5,* is critical for the division of the heart into chambers. The most common congenital organ defects in humans are those associated with the heart. The relatively high incidence of heart-related birth defects makes an understanding of the genetic control of the growth and development of the heart a great concern in medicine today.

## 36.1 Reviewing the Concepts

- Circulatory systems transport necessary materials to all cells of an animal's body and transport waste products away from cells. The two major types of circulatory systems are open and closed circulatory systems.
- An open circulatory system does not transport $O_2$ and $CO_2$. It contains three components: hemolymph, vessels, and one or more hearts. The vessels open into the animal's body cavity. In a closed circulatory system, blood and interstitial fluid are physically separated by blood vessels (Figure 36.1).
- The closed circulatory systems of vertebrates transport $O_2$ and $CO_2$ and may be single or double circulations. In a double circulation, blood circulates at two different pressures through the systemic circulation and pulmonary circulation (Figures 36.2, 36.3).

## 36.1 Testing Your Knowledge

1. Which statement is *true*?
   - **a.** Open circulatory systems cannot adapt to changes in an animal's activity.
   - **b.** Open circulatory systems are only found in certain invertebrates and a small number of vertebrates.
   - **c.** Closed circulatory systems may be either single or double circulations.
   - **d.** An advantage of open circulatory systems is the ability to distribute blood to body parts in proportion to each part's metabolic requirements at a given moment.
   - **e.** Animals with single circulations generally have higher blood pressures than those with double circulations.
2. Fishes have ______ atria/atrium and ______ ventricle(s).
   **a.** one, one **b.** one, two **c.** two, one **d.** two, two

## 36.2 The Composition of Blood

### Learning Outcomes

1. List the four components of blood and the composition and functions of each.
2. Describe the process of blood clotting in mammals.

Blood is the transport medium of animals with closed circulatory systems. It moves necessary materials—including nutrients and gases such as $O_2$—to all cells and takes away waste products, including $CO_2$ and other breakdown products of metabolism. What are the components of blood that allow it to perform its functions?

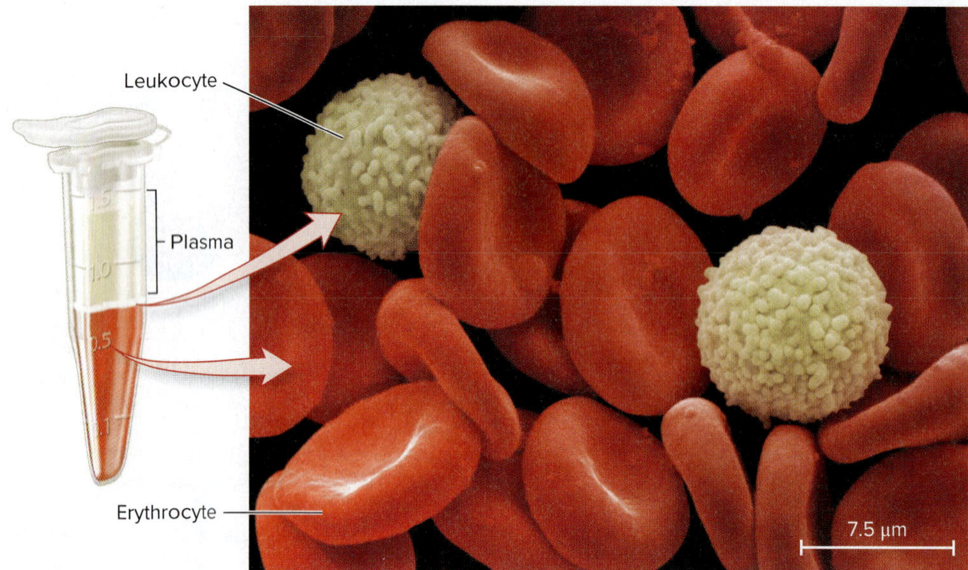


**Figure 36.4 Components of blood.** When a blood sample is centrifuged, it forms three visible layers: plasma, leukocytes, and erythrocytes. Leukocytes, shown in the scanning electron micrograph, are the white blood cells that make up part of the immune system. Erythrocytes are red blood cells, which carry oxygen. Note: The leukocyte layer is enlarged and not to scale, for illustrative purposes.
*(right)* © Power and Syred/Science Source

**BioConnections:** *Leukocytes are part of the immune system of animals. Are immune defenses unique to animals? (Refer back to Figures 29.13 through 29.15 for a hint.)*

## Blood Is Composed of Cells Suspended in Plasma

Blood consists of cells and, in mammals, membrane-enclosed fragments of cytosol, suspended in a solution containing dissolved nutrients, proteins, gases, and other molecules. If we collect a blood sample and spin it in a centrifuge, the blood separates into three visible layers (Figure 36.4). Let's take a look at each of these.

**Plasma** The top layer of the centrifuged blood sample is a yellowish solution called **plasma** (see Figure 36.4). Plasma typically makes up about half of the total volume of blood in most vertebrates. It is made up of water and dissolved nutrients, oxygen, waste products of metabolism, and many other molecules released by cells, such as hormones and proteins. These proteins serve numerous functions, such as helping in the formation of blood clots, which seal off wounds to blood vessels.

**Leukocytes** Beneath the plasma in the centrifuged blood sample is a narrow white layer of **leukocytes,** also known as white blood cells (see Figure 36.4). Leukocytes develop from a specialized connective tissue (the marrow) of certain bones in vertebrates. Although there are several types—which we describe further in Chapter 40—all leukocytes perform vital functions that defend the body against infection and disease and thus are key components of an animal's immune system.

**Erythrocytes** The bottom visible layer of the centrifuged blood sample consists of **erythrocytes,** also called red blood cells because of their color (see Figure 36.4). The term **hematocrit** refers to the volume of blood (expressed as percentage) that is composed of erythrocytes, usually between about 35% and 65% among vertebrates. Erythrocytes serve the critical function of transporting oxygen throughout the body. There are approximately a thousand times more erythrocytes than leukocytes in the circulation. Like leukocytes, erythrocytes are derived from cells in the bone marrow. In most vertebrates, mature erythrocytes retain their nuclei and other cellular organelles, but in all mammals (and a few species of fishes and amphibians), the nuclei are lost upon maturation. The lack of a nucleus and many other organelles in the mammalian erythrocyte increases the cell's oxygen-carrying capacity and contributes to its characteristic biconcave shape (see Figure 36.4). The biconcave shape of the mammalian erythrocyte increases its surface area relative to the flattened disc or oval shape seen in most other vertebrates. This is believed to increase the efficiency of gas exchange between the erythrocyte and the body fluids.

Oxygen is poorly soluble in plasma. Consequently, the amount of oxygen that dissolves in plasma usually cannot support a vertebrate's basal metabolic rate, let alone more strenuous activity. Within the cytosol of erythrocytes, however, are large amounts of the protein **hemoglobin,** which can reversibly bind to oxygen. In this way, hemoglobin greatly increases the reservoir of oxygen in the blood and enables animals to be more active. In a later section, we will consider the mechanisms by which hemoglobin binds and releases oxygen.

**Platelets** Vertebrate blood has a fourth component not visible in a centrifuged sample of blood called **platelets** in mammals and thrombocytes in other vertebrates. Platelets, which are derived from bone marrow cells, are cell fragments that lack a nucleus; they serve a crucial role in the formation of blood clots that limit blood loss after injury. The formation of a blood clot requires several steps, two of which include platelets (Figure 36.5). First, when a blood vessel is injured, platelets secrete substances that cause them to clump together and bind to collagen fibers in the surrounding connective tissue at the wound site; this forms a plug that prevents continued blood loss. Second, other platelet secretions interact with plasma proteins to cause the precipitation from solution of the fibrous protein fibrin. Fibrin forms a meshwork of threadlike fibers that wrap around and between platelets and erythrocytes, enlarging and thickening the plug

## Biology Principle

### Living Organisms Maintain Homeostasis

Because the blood carries with it the nutrients, hormones, proteins, and waste products that must circulate throughout an animal's body, a loss of blood due to injury can be catastrophic unless the wound is healed and the blood loss is stopped. In this way, blood clotting is a key feature that contributes to the ability of animals to maintain homeostasis.

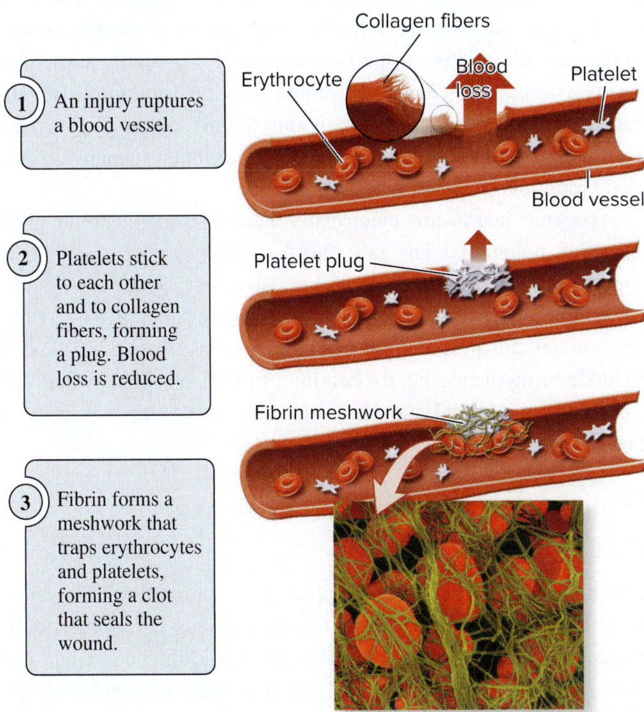

Collagen fibers
Erythrocyte
Blood loss
Platelet
1 An injury ruptures a blood vessel.
Blood vessel
2 Platelets stick to each other and to collagen fibers, forming a plug. Blood loss is reduced.
Platelet plug
Fibrin meshwork
3 Fibrin forms a meshwork that traps erythrocytes and platelets, forming a clot that seals the wound.


**Figure 36.5** **Platelets and the process of blood clot formation.** A blood clot forms in two major steps: A platelet plug reduces initial blood loss, and a fibrin clot then seals the wound. An example of a fibrin clot is shown in the scanning electron micrograph.

*(bottom)* © Dennis Kunkel Microscopy, Inc./Phototake

to form a clot. Blood clotting begins within seconds and helps prevent injured animals from bleeding to death. Eventually, the body absorbs the clot as the injured vessel heals.

## 36.2 Reviewing the Concepts

- Blood is a fluid connective tissue consisting of cells suspended in a solution containing dissolved nutrients, proteins, gases, and other molecules. Blood has four components: plasma, leukocytes, erythrocytes, and (in vertebrates) platelets or thrombocytes (Figures 36.4, 36.5).
- Platelets contribute to clot formation in two ways: by binding together to form part of the clot itself and by secreting substances that lead to precipitation of the protein fibrin.

## 36.2 Testing Your Knowledge

1. In an animal whose blood is 53% plasma by volume,
   a. the hematocrit is 53%.
   b. the hematocrit is 47%.
   c. the erythrocytes make up about 47% of the volume.
   d. the leukocytes make up about 47% of the volume.
   e. both b and c are correct.
2. Which is *false?*
   a. Platelets are nucleated cells that participate in clot formation.
   b. In mammals, erythrocytes are non-nucleated cells that contain hemoglobin.
   c. Erythrocytes contain a reservoir of oxygen carried in the blood.
   d. Leukocytes are key components of the immune system of some animals.
   e. Platelets form clots both by clumping together with other platelets and by inducing the precipitation of fibrin.

## 36.3 The Vertebrate Heart and Its Function

### Learning Outcomes

1. Describe the structure of the vertebrate heart.
2. **SCISKILLS ▶** Describe the sequence of events of the cardiac cycle, and draw conclusions regarding the functional relationship between electrical and contractile events.
3. **SCISKILLS ▶** Interpret the meaning of the tracings on an electrocardiogram.

As described earlier, in animals with a double circulation, such as mammals, the heart has a left atrium and ventricle and a right atrium and ventricle (Figure 36.6). The two sides of the heart are physically separated by a muscular and fibrous wall called a septum. Blood enters the atria from veins from the systemic or pulmonary circulation; the superior and inferior vena cavae return blood from the body, and the left and right pulmonary veins return blood from the lungs. The left ventricle ejects blood into the **aorta,** which branches to other arteries that distribute the blood. The right ventricle ejects blood into the pulmonary trunk, which divides into **pulmonary arteries** that lead to the right and left lungs.

Between the atria and ventricles are one-way valves called **atrioventricular (AV) valves** that control the movement of blood between the chambers. Blood normally can flow in only one direction through a valve. Blood moves down a pressure gradient through the AV valves from the atria into the ventricles. One-way valves are also found between each ventricle and the artery that a ventricle sends blood to (either the aorta or the pulmonary trunk); these are known as **semilunar valves.** The right ventricle pumps deoxygenated blood to the lungs through one semilunar valve, and the left ventricle pumps oxygenated blood to the rest of the body through another semilunar valve.

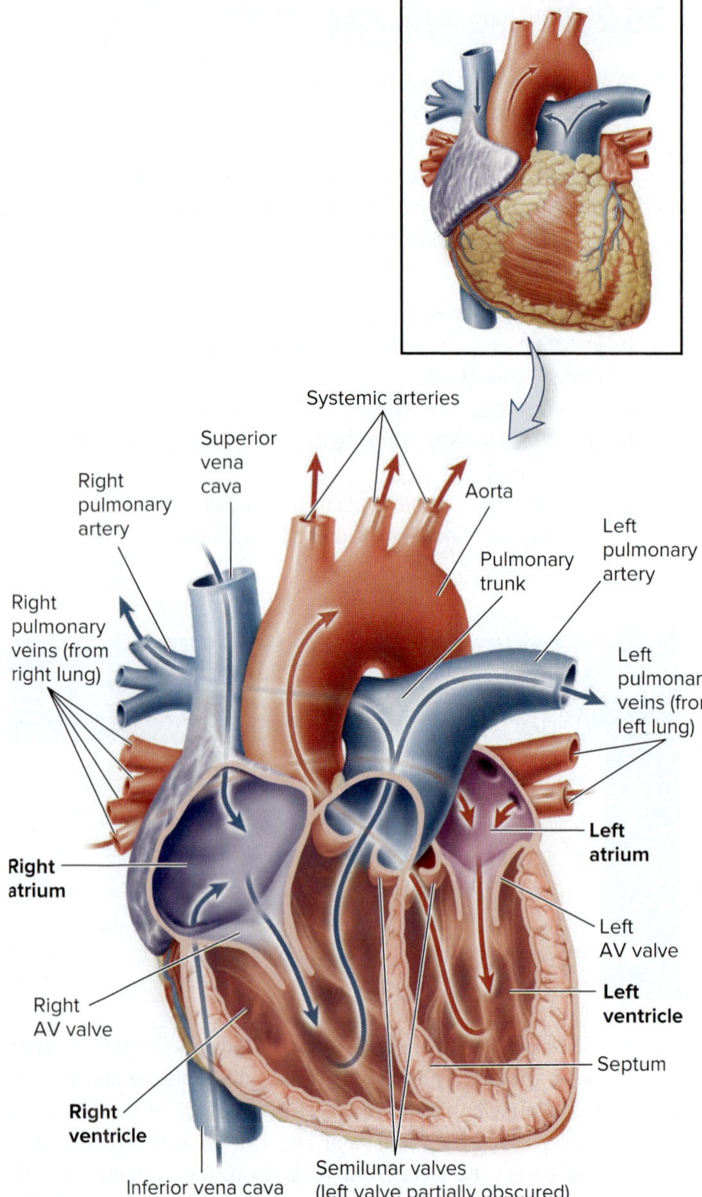


**Figure 36.6 The mammalian heart and circulation.** This cutaway illustration of a human heart shows the major blood vessels entering and leaving the heart, and the relationships of the four chambers (see inset for intact heart). Regions that contain oxygenated blood are shown in red, deoxygenated in blue. Note that the pulmonary veins carry oxygenated blood because they return blood from the lungs, whereas veins from the systemic circulation carry deoxygenated blood.

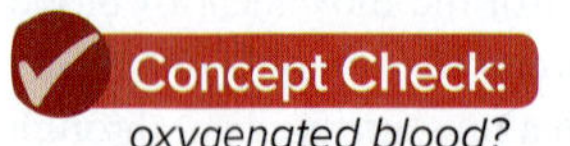

*Which arteries in this illustration carry oxygenated blood?*

## Vertebrates Have a Myogenic Heart

What causes the heart to beat so steadily? Animals cannot consciously initiate heart contractions. The beating of the heart is initiated either by nerves or by intrinsic activity of the heart muscle cells themselves. Many arthropods, for example, have a neurogenic heart that will not beat unless it receives regular electrical impulses from the nervous system. All vertebrates, however, have a myogenic heart, in which the signaling mechanism that initiates contraction resides within the cardiac muscle itself.

In myogenic hearts, cardiac muscle is distinguished by the interconnectedness between individual cardiac muscle cells, or cardiac myocytes. Each myocyte has membrane extensions that form interlocking networks with other myocytes. Within these networks, or intercalated discs, as they are called, are many gap junctions (refer back to Figure 5.18) that electrically couple the myocytes. The large number of gap junctions permits the rapid spread of electric current from cell to cell, so that all parts of the heart are rapidly stimulated virtually as one. This electric current is the signal that increases the intracellular concentration of $Ca^{2+}$ within the myocytes. The increase in $Ca^{2+}$ triggers contraction in a manner that is similar in many ways to skeletal muscle (see Chapter 35).

Myogenic hearts are electrically excitable and generate their own action potentials. The rate and forcefulness of the beating of myogenic hearts can, however, be regulated by the nervous system. For example, the autonomic nervous system can increase heart rate in vertebrates during the fight-or-flight response (see Chapter 33). Nonetheless, myogenic hearts continue to beat on their own if dissected out of an animal and placed in a nutrient bath, even with no nerves present.

## Excitation of the Heart Begins in the Atria and Spreads to the Ventricles

As with skeletal muscle (see Chapter 35), contraction of the vertebrate heart cannot occur until the muscle has been electrically excited. The electrical excitation of the vertebrate heart has two phases: atrial and ventricular. In atrial excitation, electrical signals are generated within the wall of the right atrium at the **sinoatrial (SA) node,** or **pacemaker** (Figure 36.7). The SA node is a collection of modified myocytes that have an inherently unstable resting membrane potential. Ion channels in the membranes of these cells are opened spontaneously and allow the influx of positively charged ions into the cytosol, thereby depolarizing the cell. These depolarizations produce action potentials in the SA node cells with a range of frequencies that are characteristic for a given species; the frequency determines an animal's heart rate.

Once the SA node cells generate action potentials, the potentials quickly spread across one or both atria through the gap junctions described earlier. The action potentials trigger an influx of $Ca^{2+}$ into the muscle cell cytosol, which activates contraction. Because the impulses spread very rapidly across the atria, both atria contract together almost as if they were one large muscle cell. Atrial contraction assists in providing a final push of blood through the open AV valves into the ventricles.

To begin ventricular excitation, action potentials initiated in the SA node first reach another node of specialized myocytes, the atrioventricular node. The **atrioventricular (AV) node** is located at the junction between the atria and ventricles and conducts the electrical signals from the atria to the ventricles. This is important, because the atria and ventricles are separated by a connective tissue septum

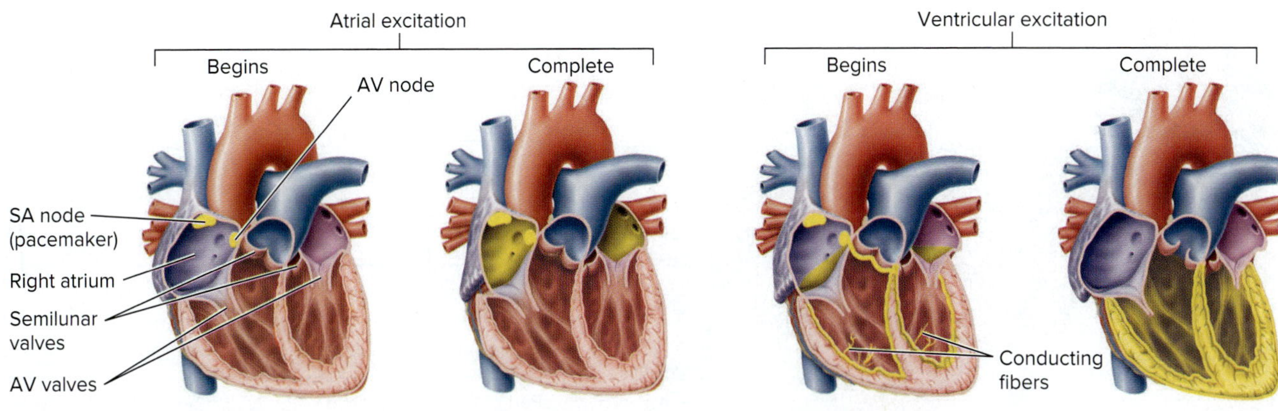


| | | | | |
|---|---|---|---|---|
| **AV valves** | OPEN | OPEN | Beginning to close | CLOSED |
| **Semilunar valves** | CLOSED | CLOSED | CLOSED | OPEN |
| **Phase of cardiac cycle** | Diastole | Diastole | Systole beginning | Systole |
| **Atria** | Relaxed | Contracted | Beginning to relax | Relaxed |
| **Ventricles** | Relaxed | Relaxed | Beginning to contract | Contracted |
| **Chambers with highest pressure** | All are low pressure | Atria (slightly) | Ventricles | Ventricles |

**Figure 36.7 Electrical activity in the mammalian heart and the cardiac cycle.** The spread of electrical activity, which is depicted in yellow, begins in the sinoatrial (SA) node and quickly spreads through the atria to the atrioventricular (AV) node. Branches from the AV node transmit electrical activity throughout the ventricles via conducting fibers. The atria and ventricles do not contract until they have become electrically excited. Changes in pressure gradients between the atria and ventricles, and between the ventricles and the aorta or pulmonary trunk, are the forces that open or close the four heart valves. Note: The left semilunar valve is only partly visible in this orientation.

that does not contain gap junctions. Like the SA node, the AV node is electrically excitable, but its cells require a longer time to become excited than do the cells of the SA node. This electrical delay allows time for the atria to contract before the ventricles do. Fibers branching from the AV node spread electrical impulses along a conducting system of cardiac muscle cells that branch throughout the muscular walls of the ventricles. This ensures that all of the ventricular muscle gets depolarized quickly and nearly simultaneously (see Figure 36.7). In this way, the entire mass of both ventricles contracts in a coordinated way in response to depolarization.

## The Cardiac Cycle Has Two Phases: Diastole and Systole

Each beat of the vertebrate heart requires the coordinated activities of the atria and ventricles. The contraction and relaxation events that produce a single heartbeat are known as the **cardiac cycle,** which can be divided into two phases (see Figure 36.7). In the first phase, **diastole,** the ventricles are relaxed and fill with blood coming from the atria through the open AV valves. At the end of diastole, the atria are electrically stimulated, and therefore they contract, which provides a boost of additional blood to the ventricles. In the second phase, **systole,** the ventricles contract and eject the blood through the open semilunar valves.

The valves open and close (**Figure 36.8**) as a result of pressure gradients established between the atria and ventricles, and between the ventricles and arteries. During diastole, when the ventricles are in a relaxed state, pressure in the ventricles is lower than in the atria and the arteries; therefore, the AV valves are open, but the semilunar valves are closed. Once the ventricles are electrically excited, they begin contracting, and the pressure within the ventricles rapidly increases. This marks the beginning of systole. When the pressure in the ventricles exceeds that in the atria, the AV valves are forced closed. Because the ventricles are thicker and stronger than the atria, ventricular contractions generate much greater pressure. The pressure exerted by the ventricles squeezes the blood upward, causing the AV valves to shut; this prevents blood from flowing backward into the atria. Closure of the AV valves makes the "lub" sound of the familiar "lub-dub" heard through a stethoscope. Ventricular pressure continues to increase as the ventricles contract. When the pressure in the ventricles exceeds the pressure in the arteries, the semilunar valves are forced open (see Figure 36.8b). Blood is then ejected through the open semilunar valves into the arteries.

The pressure in the arteries increases during systole because they are expanding with the blood that was pumped into them. At this time, the ventricles begin to return to their resting, unexcited state, and the pressure in the ventricles decreases below that in the arteries. The high pressure in the arteries causes backflow of blood toward the ventricles, which closes the semilunar valves and thereby prevents blood from flowing back into the heart. Closure of the semilunar valves creates the second heart sound, "dub," heard through a stethoscope. Throughout systole, meanwhile, the atria continued to fill with

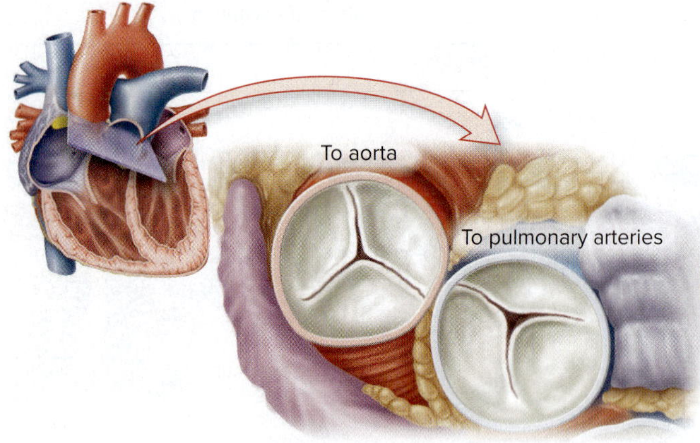


**(a) Semilunar valves in nearly closed position**

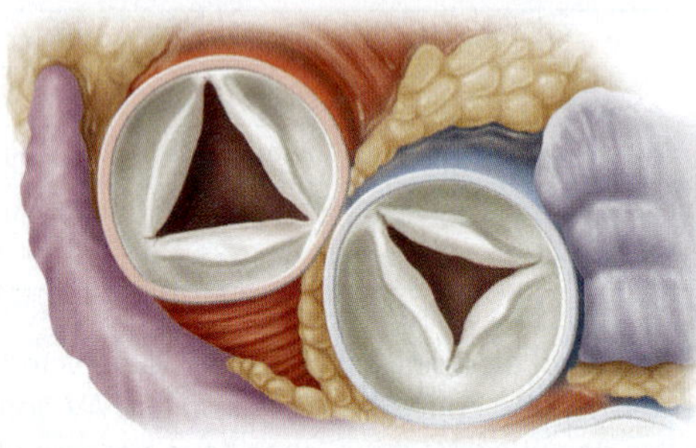

**(b) Semilunar valves in opened position**

**Figure 36.8 Appearance and function of heart valves.** The heart valves consist of two or three flaps of tissue that open in one direction. Shown in this illustration are the semilunar valves of a human heart.

**Concept Check:** *Which heart chambers pump blood through the semilunar valves?*



**Figure 36.9 An electrocardiogram (ECG).** Electrodes placed on the skin detect electrical impulses occurring in the heart, and an ECG is a recording of them. The resultant waveform is a useful indicator of cardiac health. Note that the ventricular wave (QRS complex) is taller than the atrial wave (P wave). This is because the ventricles are larger than the atria and generate more electrical activity.

**Concept Check:** *Why is it possible to detect electrical changes in the heart using electrodes placed on the surface of the skin? (Hint: Think about the composition of body fluids and their ability to conduct electricity.)*

blood, which raised the pressure in the atria. Soon the pressure in the atria exceeds that in the relaxing ventricles, and the AV valves open again, bringing a new volume of blood into the ventricles and starting a new diastole.

**Blood pressure,** which is defined as the force exerted by blood on the walls of blood vessels, is highest in the arteries during systole and lowest during diastole. For this reason, blood pressure is measured with two numbers, the systolic and diastolic pressures. For historical reasons, the units are usually given in millimeters of mercury (mmHg). Of vertebrates, blood pressures are highest in mammals and birds, and lowest in fishes. For example, a typical healthy blood pressure in humans is around 120/80 mmHg (systolic/diastolic).

## An ECG Tracks Electrical Events During the Cardiac Cycle

An **electrocardiogram** (**ECG** or **EKG,** the latter reflecting the original German spelling) is a medical test used to investigate the function of the heart. An ECG is a record of the summed electrical potentials between various points on the body. These potentials arise from the electrical signals generated by myocytes during the cardiac cycle. Sensitive electrodes are placed on the surface of the body to monitor the wave of electricity initiated by the SA node, which spreads in sequence through the atria, AV node, and ventricles. This procedure works because the body fluids that surround the heart conduct electricity, even the very weak impulses generated by a beating heart.

The tracing of an ECG reveals several waves of electrical excitation (**Figure 36.9**):

- The P wave begins when the SA node generates action potentials and ends when the two atria are completely depolarized. In the cardiac cycle, the P wave is followed by atrial contraction, while the ventricles are in diastole and filling with blood through the open AV valves.
- The QRS complex begins when the branches from the AV node excite the ventricles and ends when both ventricles depolarize completely. In the cardiac cycle, the QRS complex is followed by ventricular contraction during systole.
- The T wave results from the repolarization of the ventricles back to their resting state. In the cardiac cycle, this causes ventricular relaxation as diastole begins again. (No wave is visible for atrial repolarization because it occurs simultaneously with the large QRS complex.)

The ECG record displays both the amplitude (strength) of the electrical signal and the direction that the signal is moving in the chest. From this information, physicians can determine whether a person's heart signals have a normal frequency, strength, duration, and pattern.

BIO TIPS ONLINE

## 36.3 Reviewing the Concepts

- Blood enters the atria from veins of the systemic or pulmonary circulation, flows down a pressure gradient through the AV valves into the ventricles, and is pumped out through the semilunar valves into the arteries of the systemic and pulmonary circulations (Figure 36.6).
- Excitation of the atria begins at the sinoatrial (SA) node; excitation of the ventricles begins at the atrioventricular (AV) node. The electrical impulses spread through the atria and ventricles and stimulate the heart muscle to contract. The cardiac cycle has two phases: diastole and systole (Figures 36.7, 36.8).
- The electrical activity of the heart can be monitored using an electrocardiogram (Figure 36.9).

## 36.3 Testing Your Knowledge

1. The AV valves of the vertebrate heart
   a. are open during systole when the ventricles contract.
   b. conduct the electrical current from the atria to the ventricles.
   c. open so that blood moves from ventricles to atria.
   d. create the second ("dub") heart sound when they close.
   e. are open during diastole.
2. The pace of the heart rate in vertebrates is set by
   a. the AV node.
   b. the SA node.
   c. the rate of electrical conduction along fibers that branch from the AV node.
   d. the QRS complex.
   e. the opening of the AV valves.

## 36.4 Blood Vessels

### Learning Outcome

1. Explain the structural and functional distinctions among arteries, arterioles, capillaries, venules, and veins.

Now that we have discussed the vertebrate heart, let's take a more detailed look at the vessels that transport blood throughout the body. **Figure 36.10a** illustrates the route that blood follows in a closed circulatory system. Blood is pumped by the heart to large arteries and then flows to branching small arteries and eventually to the smallest arteries, which are called arterioles. Arterioles bring blood to the smallest vessels, the capillaries. Gas and nutrient exchange occurs between the blood in the capillaries and the cells surrounding the capillaries. From capillaries, the blood then flows back to the heart through the smallest veins, called venules, to small veins and finally to large veins.

### Arteries Distribute Blood to Organs and Tissues

Arteries are thick-walled vessels that consist of layers of smooth muscle and connective tissue wrapped around a single-celled inner layer of specialized epithelial cells called an endothelium. The endothelium forms a smooth lining in contact with the blood (**Figure 36.10b**). Because thick layers of tissue surround the endothelium, most dissolved substances cannot diffuse across arteries. Instead, arteries act as conducting tubes that distribute blood leaving the heart to all the organs and tissues of an animal's body.

In vertebrates, the walls of the largest arteries, such as the aorta, also contain one or more layers of elastin, a protein with elastic properties (refer back to Figure 4.32). As the aorta stretches to accommodate blood arriving from the heart, the elastin layers also stretch. The thick layers of tissue in the aorta and other large arteries prevent them from stretching more than a small amount. When the heart relaxes as it readies for another beat, the elastin layers in the aorta and largest arteries recoil to their original state, something like releasing a stretched rubber band. The recoiling vessels generate a force on the blood within them; this helps prevent blood pressure from decreasing too much while the heart is refilling during diastole.

### Arterioles Distribute Blood to Capillaries

As arteries carry blood away from the heart, they branch repeatedly and become smaller in diameter. Eventually, the vessels are little more than a single-celled layer of endothelium surrounded by one or two layers of smooth muscle and connective tissue (see Figure 36.10b). These are the arterioles, which deliver blood to regions of the body in proportion to metabolic demands. This is accomplished by changing the diameter of arterioles. They widen, or dilate, in areas of high metabolic activity and narrow, or constrict, in inactive regions. Arterioles dilate when their smooth muscle cells relax and constrict when these cells contract.

### Capillaries Are the Site of Gas and Nutrient Exchange

Arterioles branch into the tiny, thin-walled capillaries. Capillaries are tubes composed of a single-celled layer of endothelium resting on a layer of extracellular matrix called a basal lamina (refer back to Figure 32.4). Capillaries are the narrowest blood vessels in the body; essentially, every cell in an animal's body is near one. The diameter of a capillary is about the same as the width of erythrocytes, which move through the capillaries in single file (**Figure 36.11**).

Solutes diffuse readily between the plasma in a capillary and the surrounding interstitial fluid and intracellular fluid. Nutrients, hormones, and other solutes diffuse from the plasma and into cells; waste products generated by cells diffuse in the opposite direction. Gases are exchanged across capillaries between the blood and the environment via a respiratory organ, such as the lungs, and between the blood and tissues (**Figure 36.12**).

Blood enters capillaries under pressure that is created by the beating of the heart. This pressure forces some of the fluid in blood out through tiny openings in capillary walls into the interstitial fluid. These openings are wide enough to permit water with its solutes, including $O_2$, to leave the capillary, but not cells and most proteins. The movement of fluid between capillaries and interstitial fluid is important for maintaining a normal distribution of water between the body fluid compartments.

If the fluid that leaves a capillary were to remain in the interstitial fluid, the volume of plasma in the blood would decrease, and the interstitial fluid would swell. Most of the fluid that leaves at the beginning of a capillary, however, is recaptured at the capillary's end.

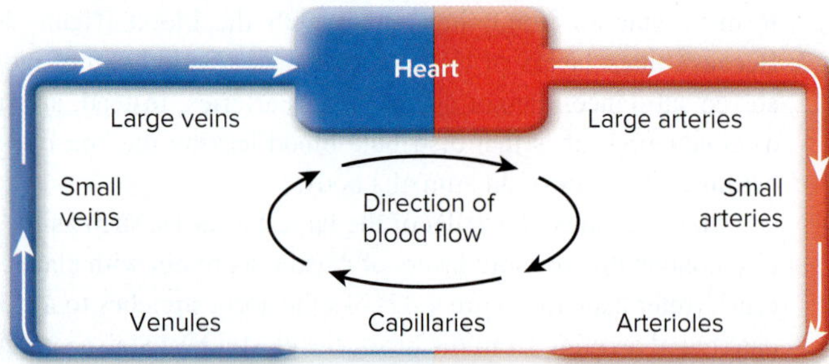


(a) The sequence of blood flow in a closed circulatory system

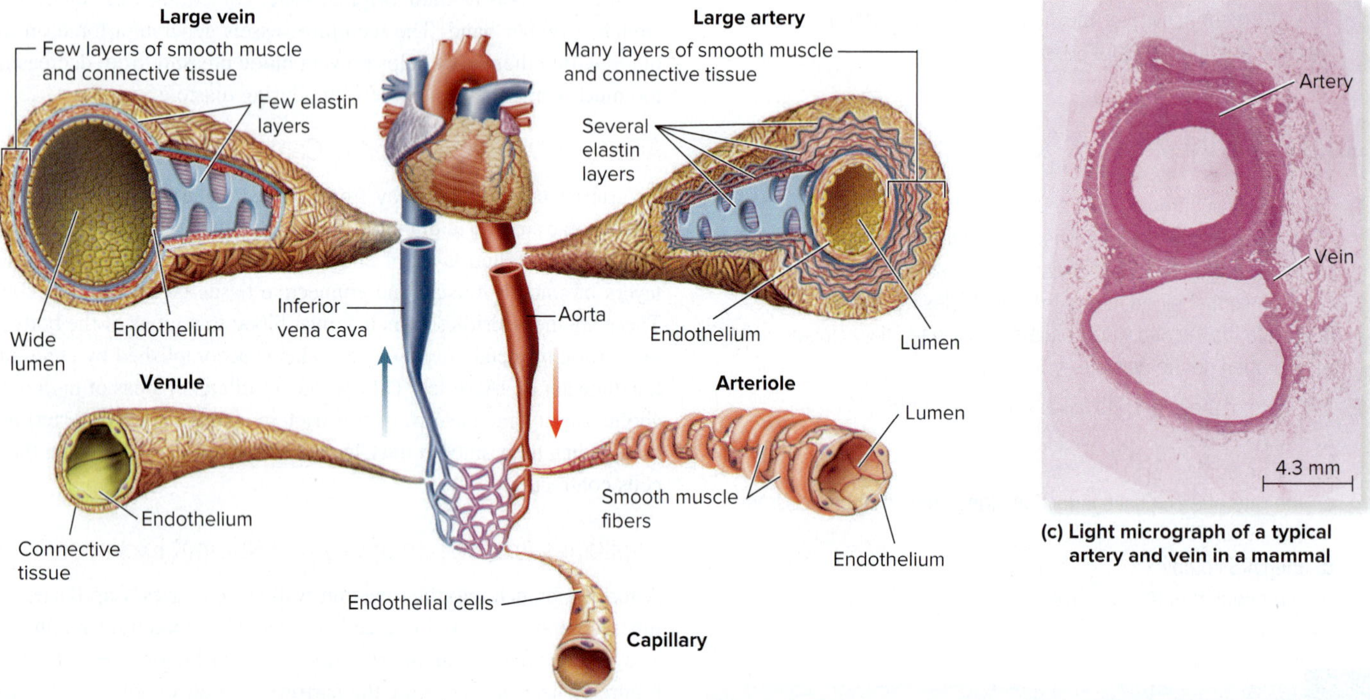


(b) The structure of blood vessels in a closed circulatory system

(c) Light micrograph of a typical artery and vein in a mammal

**Figure 36.10** **Comparative features of blood vessels.** **(a)** Overview of blood flow through vessels in a closed circulatory system. Regions of gas exchange with the environment (such as the lungs), have been omitted for simplicity. **(b)** Blood vessel structure. Sizes are not drawn to scale. **(c)** Light micrograph of a medium-sized artery near a vein. Note the difference between the two vessels in thickness and lumen diameter.



Any excess fluid is picked up by lymphatic vessels and later returned to veins.

## Venules and Veins Return Blood to the Heart

Once blood travels through capillaries, picking up any substances secreted from the cells of the body, it enters the venules, which are small, thin-walled extensions of capillaries. The venules empty into veins that return blood to the heart. The walls of veins are much thinner, less muscular, and more easily distended, or stretched, than those of arteries (see Figure 36.10b,c).

By the time blood has reached the veins, the blood pressure there is very low. Veins can fill with considerable volumes of blood, particularly veins in the lower parts of an animal's body, such as the legs, where gravity tends to cause blood to pool. However, several factors assist the blood on its way toward the heart when flowing against gravity:

- Neurotransmitters released by neurons of the sympathetic nervous system stimulate contraction of smooth muscles of leg veins. This compresses the veins and helps force blood back to the heart.
- Activity of skeletal muscles in the limbs assists the return of venous blood to the heart, by squeezing the veins that travel through them (Figure 36.13).
- One-way valves inside veins ensure that blood returning from below the heart moves in only one direction, toward the heart

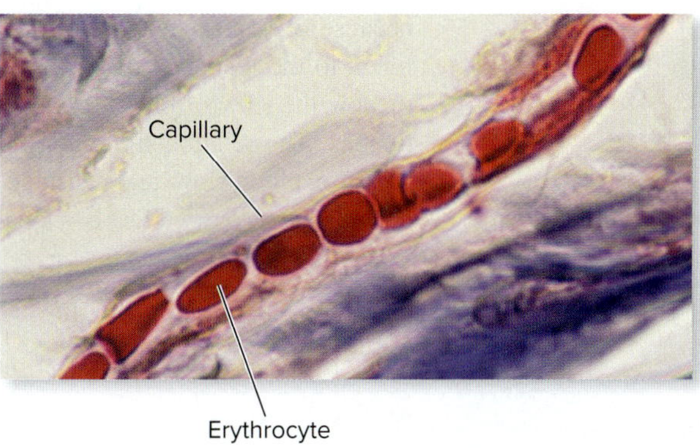


**Figure 36.11 Erythrocytes moving through a capillary.** Note the cells move along in single file, as seen through a light microscope. The diameter of the capillary is approximately 7 μm.


**Concept Check:** *Do erythrocytes enter and exit through pores in capillaries?*

(see Figure 36.13). The valves can open toward the heart, but not in the other direction. When a leg vein, for example, is compressed by the mechanisms just mentioned, the blood forces open the valves and blood moves in one direction. When the vein is relaxed, the valves close again, preventing blood from flowing backward, away from the heart. (By contrast, veins located above the heart, like those in the necks of bipeds and some quadrupeds, lack valves because gravity is sufficient to return blood toward the heart.)

Many people can easily observe the effects of gravity on venous blood flow. When the arms are held down by their side, the veins are visible on the backs of their hands. When the arms are raised above the head, the bulging veins quickly lose blood and become less visible. When blood from the veins returns to the heart, it must travel against gravity when your arms are at your side and with gravity when your arms are elevated. Blood drains from veins much more effectively when gravity works in its favor.

## 36.4 Reviewing the Concepts

- Arteries carry blood away from the heart. Arterioles distribute blood to capillaries, which are the site of gas and nutrient exchange. Blood from the capillaries flows into venules and then into veins, which return it to the heart (Figures 36.10, 36.11, 36.12).
- Muscle activity and one-way valves help move venous blood toward the heart against gravity (Figure 36.13).

## 36.4 Testing Your Knowledge

1. The vessels that mediate fluid and solute exchange are the ______, which then empty into ______ .
   a. veins, the heart
   b. venules, capillaries
   c. capillaries, venules
   d. capillaries, arterioles
   e. arterioles, capillaries

## Biology Principle

### Living Organisms Maintain Homeostasis

Keep this principle in mind as you read the rest of this chapter. The exchange of $O_2$ and $CO_2$ illustrated here is absolutely essential for all homeostatic processes in animals, in part because oxygen is required for the production of much of an animal's ATP. The ATP, in turn, provides the energy required to sustain processes that contribute to homeostasis.

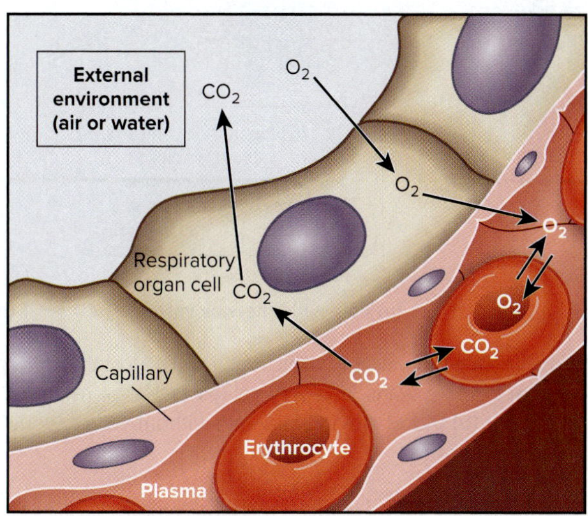


**(a) Gas exchange between the environment and a respiratory organ**

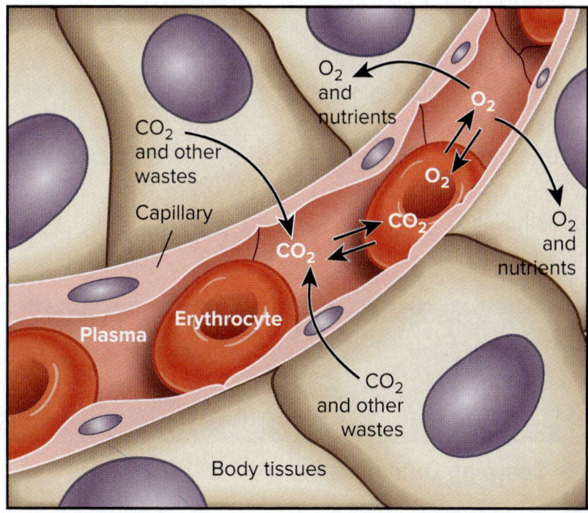


**(b) Gas, nutrient, and waste exchange between the blood and tissues**

**Figure 36.12 Overview of gas and other solute exchange between capillaries and cells. (a)** In vertebrates, oxygen ($O_2$) diffuses from the environment across the cells of a respiratory organ into the blood (plasma and erythrocytes); carbon dioxide ($CO_2$) diffuses in the opposite direction. **(b)** From the blood, $O_2$ and nutrients diffuse into tissue cells, where they are used during the formation of ATP and other activities. Cells generate $CO_2$ and other waste products, which diffuse out of cells and into blood. Note: Interstitial fluid is omitted for simplicity.

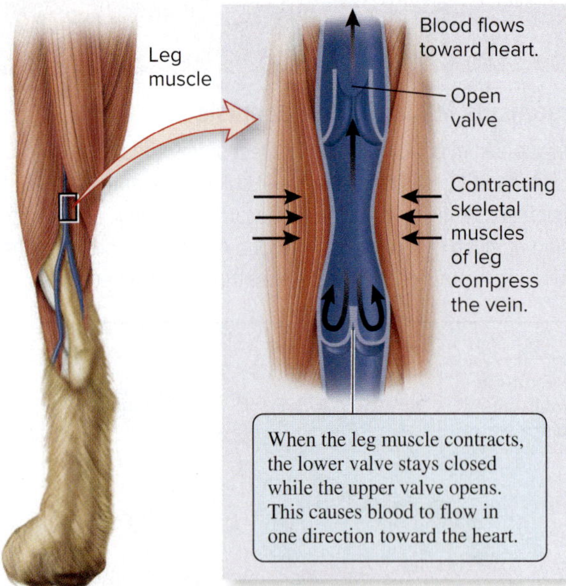


(a) Vein with one-way valves

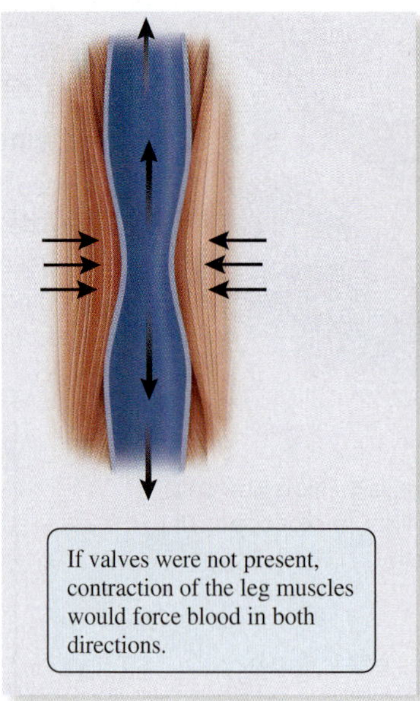

If valves were not present, contraction of the leg muscles would force blood in both directions.


(b) Vein without valves

**Figure 36.13** **One-way valves in veins.** Valves are typically present in the limbs, as shown in this dog's leg. **(a)** In some veins, one-way valves assist the return of blood to the heart against the force of gravity. **(b)** Blood moving in a vein without valves would flow in both directions if the vein were compressed.

**Concept Check:** *Unlike most mammals, giraffes have one-way valves in the veins of their long necks. Which way do you think the valves open, toward the heart or toward the head, and why? (Hint: Search online for a picture of a giraffe drinking from a body of water.)*

2. Which characteristic matches with arterioles?
   a. They have elastic properties.
   b. They consist of an endothelium plus one or two layers of smooth muscle.
   c. They affect local blood flow by dilating or constricting.
   d. They empty into venules.
   e. Both b and c match.

## 36.5 Relationship Among Blood Pressure, Blood Flow, and Resistance

### Learning Outcomes

1. **SCISKILLS ▶** Be able to relate blood pressure, blood flow, and resistance in a mathematical way.
2. Distinguish between the effects of vasodilation and vasoconstriction on blood flow.
3. **SCISKILLS ▶** Predict the effects of changes in resistance and cardiac output on blood pressure.

Blood pressure—the force exerted by blood on vessel walls—is responsible for blood flow, the movement of blood through the vessels. It is not the same in all regions of an animal's body, however, because of resistance. **Resistance (R)** refers to the tendency of blood vessels to slow the flow of blood through their lumens. The relationship among blood pressure, blood flow, and resistance can be considered on two levels—local and systemic—as we see next.

### Blood Pressure, Blood Flow, and Resistance Are Mathematically Related

The relationship among blood pressure, blood flow, and resistance is stated by Poiseuille's law, which was derived in the 1840s by Jean Marie Louis Poiseuille, a French physician and physiologist. His law is simplified here:

$$\text{Flow } (F) = \Delta\text{Pressure } (P)/\text{Resistance } (R)$$

Stated mathematically, blood flow through a blood vessel is directly proportional to the difference ($\Delta$) in pressure of the blood between the beginning and end of the vessel, and inversely proportional to the resistance created by that vessel. The equation can be rearranged as $\Delta P = F \times R$, which demonstrates that blood pressure depends on both blood flow and resistance. Poiseuille's law applies to blood flow through a single vessel, an organ, or the entire body.

### Resistance to Flow Depends on the Radius of Arterioles

Changes in arteriolar resistance are the major mechanism for increasing or decreasing blood flow to a region. The relationship between arteriolar radius and resistance is not linear. Resistance is inversely proportional to the radius of the vessel raised to the fourth power: $R \propto 1/r^4$, where $\propto$ means "proportional to" and $r$ is the radius of the arteriolar lumen. Let's consider an arteriole with a radius that increases by a factor of 2. This would occur if the smooth muscles of the arteriole relaxed sufficiently to allow the vessel to dilate and double its original radius. Because resistance is inversely proportional to the fourth power of vessel radius, an increase in radius by a factor of 2 will result in a decrease in resistance of $2^4$, or 16-fold.

**Vasodilation** refers to an increase in blood vessel radius, and **vasoconstriction** refers to a decrease in blood vessel radius. The signals that control arteriolar radius come from three sources:

- **Local factors.** Locally, metabolic by-products such as carbon dioxide, lactic acid, and other substances secreted by metabolically active tissues cause nearby arterioles to vasodilate, thereby decreasing their resistance to blood flow. This permits more blood flow to the active region, facilitating oxygen and nutrient delivery and waste removal.

- **Hormones.** Hormones secreted by glands throughout the body can also regulate arteriolar radius. For example, during the fight-or-flight response, some hormones cause the arterioles that deliver blood to the small intestine to vasoconstrict, routing blood away from the intestine and to more vital areas.
- **Nervous system.** Signals from the autonomic nervous system cause contraction or relaxation of arteriolar smooth muscle; this effect is similar to that described for hormones.

## Quantitative Analysis

### CARDIAC OUTPUT AND RESISTANCE DETERMINE BLOOD PRESSURE

Because blood vessels provide resistance to blood flow, the heart must beat forcefully enough to overcome that resistance throughout the whole body. **Cardiac output (CO)** is the amount of blood the heart pumps per unit of time, usually expressed in units of liters per minute (L/min). Poiseuille's law can be adapted to the whole body. In this case, pressure refers to arterial blood pressure (BP) recorded using a blood pressure cuff, flow refers to CO, and resistance refers to the total resistance provided by all the arterial vessels of the systemic circulation (**total peripheral resistance,** or **TPR**). Thus, we have BP = CO × TPR. The CO and TPR determine the pressure the blood exerts in the arteries of a closed circulatory system.

Cardiac output depends on the size of an animal's heart, how often it beats each minute, and how much blood it ejects with each beat. Each beat, or stroke, of the heart ejects an amount of blood known as the **stroke volume (SV),** which is roughly proportional to the size of the heart (Table 36.1). Thus, if we know the stroke volume of a heart and can measure the heart rate (HR, the number of beats per minute, bpm), we can determine the CO. Simply put, CO = SV × HR. As one example from Table 36.1, the CO of a blue whale would be

$$100 \text{ L/beat} \times 10 \text{ beats/min} = 1{,}000 \text{ L/min}$$

To put that number in perspective, 1,000 L/min is roughly equivalent to 250 gal/min, or 25 ordinary fish aquaria!

Of course, the CO of a blue whale is far greater than that of a human, a dog, or a shrew. The heart of a typical shrew, for example, is the size of a small pea, whereas the heart of a blue whale is as large as a cow. Typically, heart size varies in proportion to body mass within a given class of vertebrates, with birds and mammals having larger hearts than similarly sized fishes, amphibians, or reptiles. Note in Table 36.1 that heart rate decreases as mammals get larger, but heart mass and stroke volume increase roughly in proportion with body mass. Smaller animals have smaller hearts and, therefore, smaller stroke volumes. However, small animals have faster heart rates than do large animals. A shrew's resting heart rate may be more than 800 bpm, whereas a blue whale's heart may beat only 10 times per minute (although the volume of blood ejected with each of those beats is enormous!). The faster heart rates of small animals give them a greater cardiac output than predicted from the size of their hearts, which helps them meet the extraordinary oxygen and nutrient demands of their high metabolisms.

**Table 36.1 Comparative Features of Representative Mammalian Hearts***

| Animal | Body mass (kg) | Heart mass (kg) | Stroke volume (L) | Heart rate (bpm)† |
|---|---|---|---|---|
| Shrew‡ | 0.0024 (2.4 g) | 0.000035 (35 mg) | 0.000008 (8 μl) | 835 |
| Rat | 0.20 | 0.001 | 0.0002 | 360 |
| Rabbit | 2 | 0.012 | 0.0013 | 189 |
| Small dog | 5 | 0.030 | 0.007 | 120 |
| Large dog | 30 | 0.180 | 0.040 | 88 |
| Human | 75 | 0.380 | 0.075 | 70 |
| Horse | 450 | 3.50 | 0.90 | 38 |
| Elephant | 4,000 | 25 | 4.0 | 25 |
| Blue whale | 100,000 | 600 | 100 | 10 |

*Values are based on average body masses and resting conditions. In some cases, stroke volumes are estimates based on heart size.

†bpm = beats per minute.

‡The shrew reported here is the Etruscan shrew, one of the smallest known mammals. Its heart is somewhat larger than would be predicted for its body mass. Note its heart rate; at 835 bpm, the heart beats 14 times per second!

The greater the cardiac output and resistance to blood flow, the higher the blood pressure will be. Imagine that the circulatory system is like a faucet (the heart) connected to a garden hose (the arteries and arterioles) (Figure 36.14). If the faucet is fully open (analogous to maximal cardiac output) and the hose is not blocked (analogous to low resistance), the amount of water rushing into the hose will be high, and so will the water pressure, representing blood pressure. If the faucet is only partially open, the water pressure will be lower. However, now imagine that the faucet is partially open but a region of the hose is kinked, or constricted, representing a region of high resistance. In that case, the pressure of the water in the hose will increase between the faucet and the point where the hose is kinked, and will decrease beyond the constriction, as will the flow of water.

Arterial blood pressure, therefore, is a function of how hard the heart is working and how constricted or dilated the various arterioles are. Blood pressure must be high enough for blood to reach all body tissues even at the farthest extremities, but not high enough to damage blood vessels or force excess plasma out of capillaries.

**Crunching the Numbers:** Using the data in Table 36.1 and a graphing program (such as Excel) on your computer, create a plot of body mass (*x*-axis) vs. heart mass (*y*-axis). Now create another

(a) Maximal cardiac output with low resistance

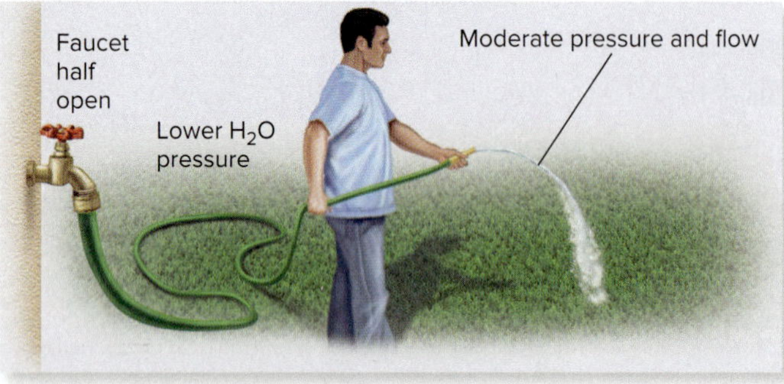


(b) Moderate cardiac output with low resistance

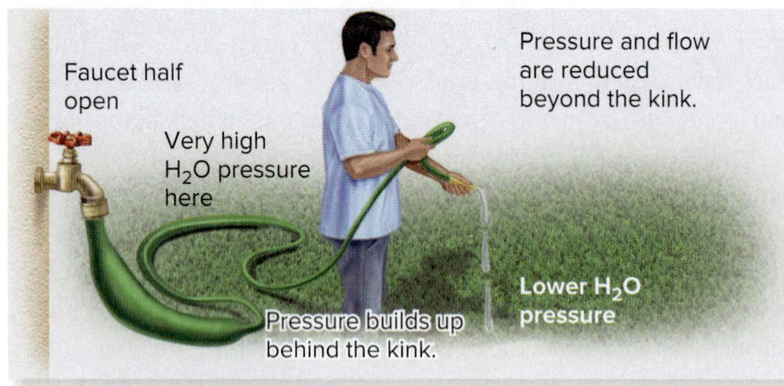


(c) Moderate cardiac output with high resistance

**Figure 36.14 The relationship among cardiac output, resistance, and blood pressure.** A hose analogy shows the way in which cardiac output (the faucet) and resistance (constriction of the hose) affect blood pressure.

of body mass versus heart rate. How do the graphs differ? Is one more linear than the other? Based on these relationships, can you determine which variable—heart mass or heart rate—makes the most significant contribution to the relatively greater cardiac output in smaller animals?

## 36.5 Reviewing the Concepts

- Blood pressure, which is the force exerted by blood on the walls of blood vessels, is responsible for moving blood through the vessels. Resistance refers to the tendency of blood vessels to slow the flow of blood through their lumens.
- Local blood flow through a vessel is directly proportional to the pressure of the blood entering the vessel and inversely proportional to the resistance created by that vessel. Arteriole radius is the major factor that regulates resistance.
- Cardiac output overcomes resistance to generate systemic blood pressure and depends on the size of an animal's heart, how often it beats each minute, and how strongly it contracts with each beat (Table 36.1).
- Cardiac output (CO) and total peripheral resistance (TPR) determine blood pressure (BP): BP = CO × TPR (Figure 36.14).

## 36.5 Testing Your Knowledge

1. The radius of an arteriole decreases 50%. What will happen to the flow of blood through that arteriole?
   a. It will be decreased by 50%.
   b. It will decrease to 1/4 of normal.
   c. It will decrease to 1/16 of normal.
   d. It will be decreased by 25%.
   e. It will be decreased by a factor of 8.
2. What would happen to an animal's blood pressure if its cardiac output increased by a factor of 2 and its total peripheral resistance was decreased 50%?
   a. It would decrease 16-fold.
   b. It would decrease 8-fold.
   c. It would increase 1.5-fold.
   d. It would increase 24-fold.
   e. It would not change.

## 36.6 Physical Properties of Gases

### Learning Outcomes

1. List the relative amounts of each gas that make up the majority of the air we breathe.
2. Define partial pressure, and explain how it is calculated.
3. List three factors that determine how much gas dissolves in a solution.

We will now turn our attention to the process of gas exchange between the environment and the fluids of an animal's body. In many animal species, gas exchange involves an interplay between the circulatory system and the respiratory system. The exchange of $O_2$ and $CO_2$ depends on the solubility of the gas in water and the rate of diffusion of the gas. Air is composed of about 21% $O_2$, 78% nitrogen ($N_2$), and roughly 1% $CO_2$ and other gases. From a respiratory standpoint, $N_2$ can usually be ignored because it plays no role in energy production nor is it created as a waste product of metabolism. In this section, we begin by examining some of the basic properties of the two major gases that are important in respiration.

### Gases Exert Pressure, Which Depends on Altitude

The gases in air exert pressure on the body surfaces of animals, although the pressure is not perceptible (unless it changes suddenly, as when your ears "pop" in a descending airplane). This pressure is called **atmospheric pressure.** It can be measured by noting how high a column of mercury is forced upward by the air pressure in a device called a mercury manometer. The traditional unit of gas pressure, therefore, is millimeters of mercury (mmHg)—the same as for blood pressure.

At sea level, atmospheric pressure is 760 mmHg. It decreases as we ascend to higher elevations because the gravitational pull of the Earth decreases; consequently, gases expand and fewer gas molecules are in a given volume of air at high altitude. To visualize why gas pressure decreases at higher altitudes, think about how hydrostatic pressure decreases from the ocean floor to the ocean surface. Now, imagine that you are standing at the bottom of an "ocean" of air. The closer you get to the "ocean's" surface (the top of the atmosphere), the lower the pressure exerted on your body.

Atmospheric pressure is the sum of the pressures exerted by each gas in air, in exact proportion to their amounts. The individual pressure of each gas is its **partial pressure,** symbolized by a capital P and a subscript depending on the gas. At sea level, the partial pressure of oxygen ($PO_2$) in the air we breathe is 21% of the atmospheric pressure at sea level, or 160 mmHg (0.21 × 760 mmHg). The percentage of oxygen and other gases in air remains the same regardless of altitude, but the lower the atmospheric pressure, the lower the partial pressure of oxygen in air. In Denver, for example, where the atmospheric pressure is only about 640 mmHg, the $PO_2$ would be 21% of 640 mmHg, or 134.4 mmHg.

The partial pressure of oxygen in the environment provides the driving force for its diffusion from air or water across an animal's respiratory surface and into its blood. All gases diffuse from regions of higher pressure to regions of lower pressure. Consequently, the rate of oxygen diffusion into the blood of a terrestrial animal decreases when the animal moves from sea level to a higher altitude, where the $PO_2$ is lower. This explains why humans who climb to high altitudes often need to breathe from a tank containing gas that is enriched in oxygen to compensate for the lower $PO_2$.

## Pressure, Temperature, and Other Solutes Influence the Solubility of Gases

Gases dissolve in water, including fresh water, seawater, and all body fluids. Gases such as $O_2$ exert their biological effects while in solution. However, most gases dissolve rather poorly in water. There is less $O_2$ in a given volume of water than in air, for example, which places constraints on how active many aquatic organisms can be. Among the factors that influence the solubility of a gas in water, the following are particularly important.

- **Pressure of the gas.** As the pressure of a gas that comes into contact with water increases, more of that gas will dissolve, up to a limit that is specific for each gas at a given temperature. The partial pressure of gas in water is given in the same units as atmospheric pressure (mmHg). For example, if water is in contact with air that contains a $PO_2$ of 160 mmHg, the $PO_2$ of the water is also 160 mmHg.
- **Temperature of water.** More gas dissolves in a given volume of cold water than in warm water. At higher temperatures, gases in solution have more thermal energy and are therefore more likely to escape from the liquid. This means that animals inhabiting warm waters generally have less $O_2$ available to them than do animals in cold waters.
- **Presence of other solutes.** Ions and other solutes decrease the amount of gas that dissolves in water. Thus, blood—which contains many solutes—dissolves less $O_2$ than does pure water. Likewise, less $O_2$ dissolves in seawater than fresh water at any given temperature and pressure. Animals living in cold freshwater lakes, therefore, have more $O_2$ available to them than animals living in warm, salty seas.

## 36.6 Reviewing the Concepts

- The partial pressure of oxygen ($PO_2$) in the environment provides the driving force for its diffusion. Atmospheric pressure decreases at higher elevations.
- Three factors—the pressure of the gas, temperature of the water, and presence of any other solutes—affect the solubility of a gas in water.

## 36.6 Testing Your Knowledge

1. In which solution would you expect the most oxygen to be dissolved?
   a. cold seawater
   b. a cold freshwater lake
   c. warm seawater
   d. a very warm freshwater pond

# 36.7 Types of Respiratory Systems

### Learning Outcomes

1. Describe the different ways in which animals obtain oxygen from the environment.
2. **SCISKILLS ▶** Diagram the process of countercurrent flow in gill ventilation.
3. Describe Boyle's law and how it relates to ventilation in air-breathing vertebrates.

Animals obtain oxygen from their surroundings with one of four major types of gas-exchange organs: the body surface, gills, tracheae, or lungs. **Ventilation** is the process of bringing oxygenated water or air into contact with a gas-exchange (respiratory) organ. In this section, we will examine the mechanisms of ventilation used by animals with different respiratory systems.

## Some Animals Exchange Gases Across the Body Surface

In those invertebrates that are only a few cell layers thick, such as cnidarians and platyhelminthes, $O_2$ and $CO_2$ can diffuse directly across the body surface (refer back to Figures 26.12 and 26.14). In this way, $O_2$ reaches all the interior cells, in some cases without any specialized circulatory system.

Even in some vertebrates, the body surface may be permeable to gases. Amphibians have unusually moist, permeable skin. On land, amphibians rely primarily on their lungs, but when they are under water and thus cannot breathe air, $O_2$ and $CO_2$ can diffuse across the skin. In some species, skin folds increase surface area and thereby increase the rates of diffusion of gases (refer back to Figure 32.7b). The ability to exchange gases across the skin is an adaptation that permits amphibians to spend prolonged times under water.

## Water-Breathing Animals Use Gills for Gas Exchange

Water-breathing animals use specialized respiratory structures called **gills,** which can be either external or internal. External gills are uncovered extensions from the body surface, found in many invertebrates and the larval forms of some amphibians. Internal gills, which occur in fishes, are enclosed in a protective cavity.

**External Gills** External gills vary widely in appearance, but all have a large surface area, often in the form of elaborate projections (Figure 36.15). In many cases, external gills are ventilated by being moved back and forth through the water. The ability to move external gills is particularly important for sessile invertebrates, which must otherwise rely on sporadic local water currents or muscular efforts of their bodies to create local currents for ventilation.

Despite the success of marine invertebrates, external gills have several limitations. First, they are unprotected and therefore are susceptible to damage from the environment. Second, because water is much denser than air, considerable energy is required to continually wave the gills back and forth through the water (think of the difference between waving your hands through air and waving them through water). Finally, their appearance and motion may draw the attention of predators.

**Biology Principle**

**Structure Determines Function**

Note how the gills of this nudibranch have extensive surface modifications to increase the total area available for gas exchange. This is not unique to external gills; all gas-exchange surfaces have structural modifications that maximize their function.

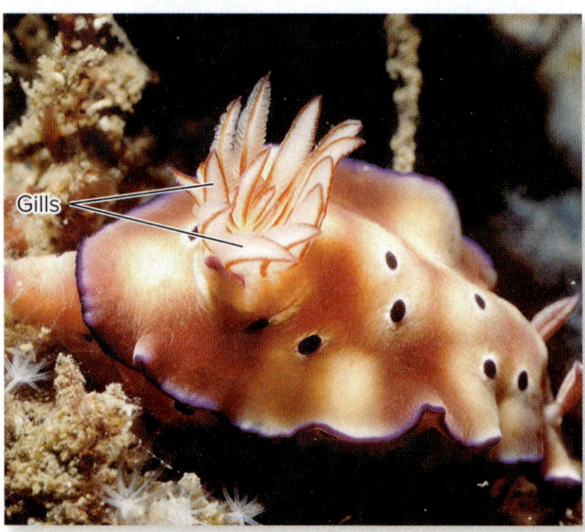


**Figure 36.15** **Example of an animal with external gills.** The gills have extensive projections to increase surface area, which facilitates oxygen diffusion from the surrounding water.
© Hal Beral/V&W/The Image Works

**Internal Gills** By contrast, fishes have internal gills, which are covered by a bony plate called an operculum (Figure 36.16a). Fish gills are confined within the opercular cavity, the space beneath the operculum, which protects the gills and helps streamline the body.

The main support structures of gills are the gill arches, from which project gill filaments composed of numerous platelike structures called **lamellae** (Figure 36.16b). Blood vessels run the length of the filaments. Oxygen-poor blood travels through a vessel called the afferent vessel along one side of the filament, and oxygen-rich blood travels through another vessel called the efferent vessel along the other side. Within the lamellae are numerous capillaries, all oriented with blood flowing from the oxygen-poor vessel to the oxygen-rich one.

Water enters a fish's mouth and flows between the lamellae in the opposite direction from blood flowing through the lamellar capillaries. This arrangement of water and blood flow is an example of **countercurrent exchange** (refer back to Figure 32.12). As oxygenated water encounters the lamellae, it comes into close proximity to blood in the gill capillaries. Recall that $O_2$ diffuses along a pressure gradient from a region of higher $O_2$ pressure to one of lower $O_2$ pressure. Thus, $O_2$ diffuses from the water into the capillaries of the lamellae. As water continues to flow across the lamellar surface, it encounters regions of capillaries that have not yet picked up $O_2$—in other words, even as $O_2$ begins to diffuse from the water into the gill capillaries, a sufficient pressure gradient remains along the lamellae to permit diffusion of more of the remaining $O_2$ from the water. This is an extremely efficient way to remove as much $O_2$ from the water as possible before the water passes out of the operculum.

Different fishes can ventilate their gills in three possible ways:

- Actively drawing water in through the mouth and out the operculum
- Swimming with the mouth open so that water continually moves across the gills
- Facing into a current of water while resting, but keeping the mouth open

Each of these ventilation mechanisms is a flow-through system—water moves only in one direction so that the gills are constantly in contact with fresh, oxygenated water. This improves gas exchange.

## Insects Use Tracheal Systems to Exchange Gases with the Air

The delicate nature of gill lamellae generally makes them unsuitable for gas exchange in air. Air-breathing probably evolved as an adaptation in aquatic animals inhabiting regions that were subject to periodic drought.

One of the major mechanisms that animals evolved to breathe air is the **tracheal system** found in insects. Along the surface of an insect's body are tiny openings to the outside called **spiracles.** Arising from the spiracles are sturdy tubes called **tracheae**

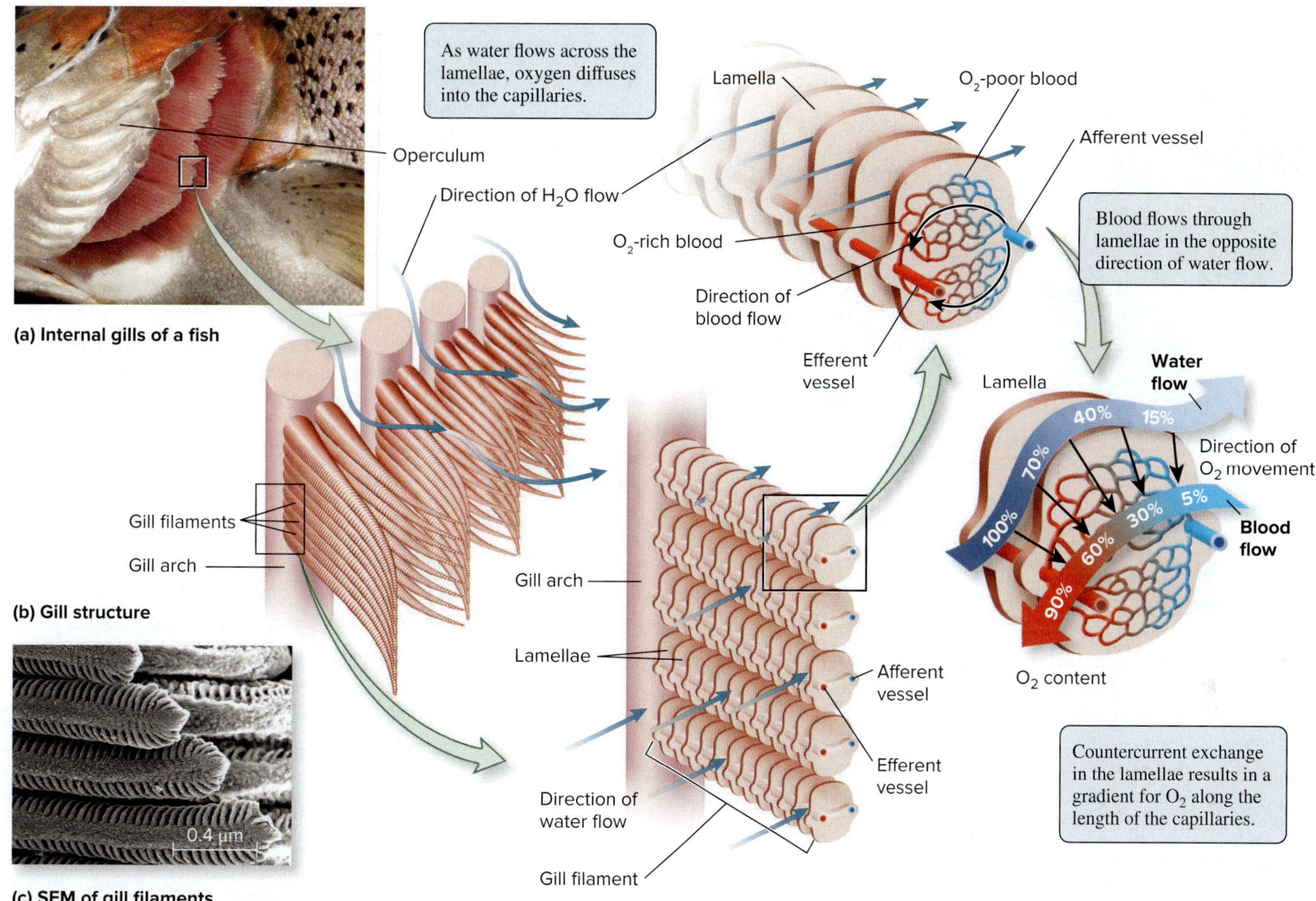


**Figure 36.16** **Structure of fish gills.** **(a)** The operculum, which has been lifted up in this photo, protects the gills underneath. **(b)** The gills are composed of gill arches, from which numerous pairs of filaments arise. Thin, platelike lamellae are arrayed along the filaments. Blood flows through the capillaries of the gill filaments in the opposite direction of water flowing between lamellae, a process called countercurrent exchange. **(c)** Several filaments with their lamellae, as revealed in a scanning electron micrograph (SEM).

*(a)* © Sarah Ahrens/Alamy RF; *(c)* © Electron Microscopy Unit, Royal Holloway, University of London

**Concept Check:** *Why do fishes die when out of water? (Hint: Consider what happens if you hold two thin, wet paper towels together.)*

**BioConnections:** *You have learned about countercurrent exchange already in another context (refer back to Figure 32.12). How does it relate to heat exchange in vertebrates?*

(singular, trachea) (**Figure 36.17**). Tracheae branch extensively into ever-smaller tubes called tracheoles, which eventually become small enough that their tips contact virtually every cell in the body. At their tips, tracheoles are filled with a small amount of fluid. Air flowing into the tracheoles comes into contact with this fluid. Oxygen from the air dissolves in the fluid, and from there it diffuses across the tracheoles and into nearby cells. Carbon dioxide diffuses in the opposite direction, from cells into the tracheoles, and from there to the environment.

When an insect's oxygen demands increase due to increased activity, muscular movements of its abdomen and thorax draw air into and out of the tracheae a little like a bellows. An insect's muscles and tracheal system match ventilation with the animal's exercise intensity and $O_2$ requirements. This is particularly important in flying insects, which have very great metabolic demands.

As discussed earlier, the open circulatory system of insects does not participate in gas exchange. Oxygen diffuses directly from air to trachea to tracheoles and finally to body cells. This mechanism of ventilation and $O_2$ delivery is very effective. The relative metabolic rate of insect flight muscles is among the highest known of any tissue in any animal, and the tracheal system supplies enough $O_2$ to meet those enormous demands.

## Air-Breathing Vertebrates Use Lungs to Exchange Gases

Except for some amphibians such as lungless salamanders, all air-breathing terrestrial vertebrates use lungs to bring $O_2$ into the circulatory system and remove $CO_2$. **Lungs** are internal, paired structures that arise from the pharynx during embryonic life. All lungs receive deoxygenated blood from the heart and return oxygenated blood to the heart.

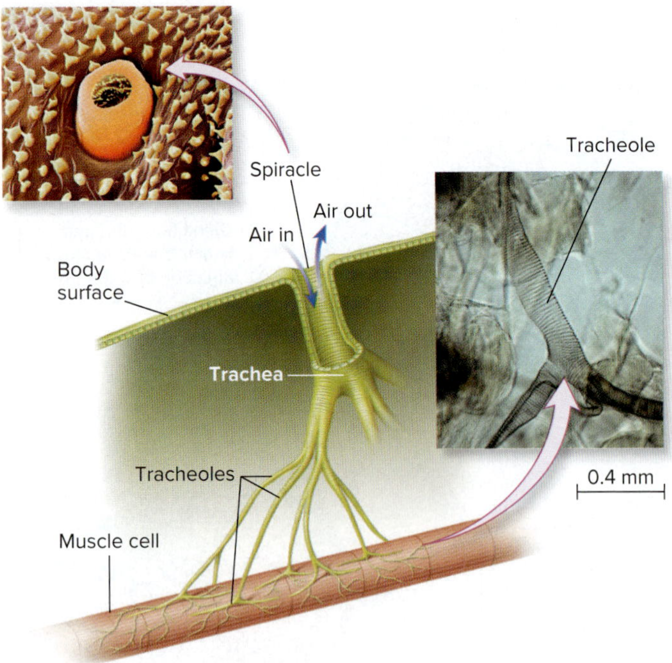


**Figure 36.17** **The tracheal system of insects.** Air enters holes on the body surface called spiracles. Oxygen diffuses directly from the fluid-filled tracheole tips to cells that come into contact with the tips. The circulatory system plays no role in gas exchange. The micrographs illustrate a single spiracle and a branching tracheole.

*(left)* © Microfield Scientific Ltd/Science Source; *(right)* © Ed Reschke/Getty Images

**Concept Check:** *How might the structure of the respiratory system of insects be related to why insects tend to be small?*

Most vertebrates ventilate their lungs by a process called **negative pressure filling,** in which the pressure of air in the lungs is decreased below that of the environment. Boyle's law states that the pressure and volume of a gas are inversely related (Figure 36.18). For example, when the volume in which a gas is contained increases, the pressure of the gas in that container decreases. By expanding its lungs (inhaling), an animal creates a pressure gradient for air to move from the atmosphere (higher pressure) into its lungs (lower pressure). When an animal exhales, the lungs become compressed, increasing the pressure of the air inside them; this causes air to leave the lungs.

To understand the mechanisms that achieve negative pressure filling, we will next examine the structure and function of the mammalian respiratory system as a model.

## 36.7 Reviewing the Concepts

- Ventilation is the process of bringing oxygenated water or air into contact with a respiratory organ.
- The body surface is permeable to gases in some invertebrates and in amphibians. Water-breathing animals use external or internal gills for gas exchange (Figures 36.15, 36.16).
- In insects, air in the tracheoles comes into contact with fluid at the tracheole tips. Oxygen from the air dissolves in this fluid and diffuses across the tracheole wall and into nearby cells (Figure 36.17).
- Air-breathing vertebrates generally use lungs to obtain $O_2$ and eliminate $CO_2$. All lungs receive deoxygenated blood from the heart and return oxygenated blood to the heart. Most vertebrates fill their lungs by negative pressure filling, which follows the principles of Boyle's law (Figure 36.18).

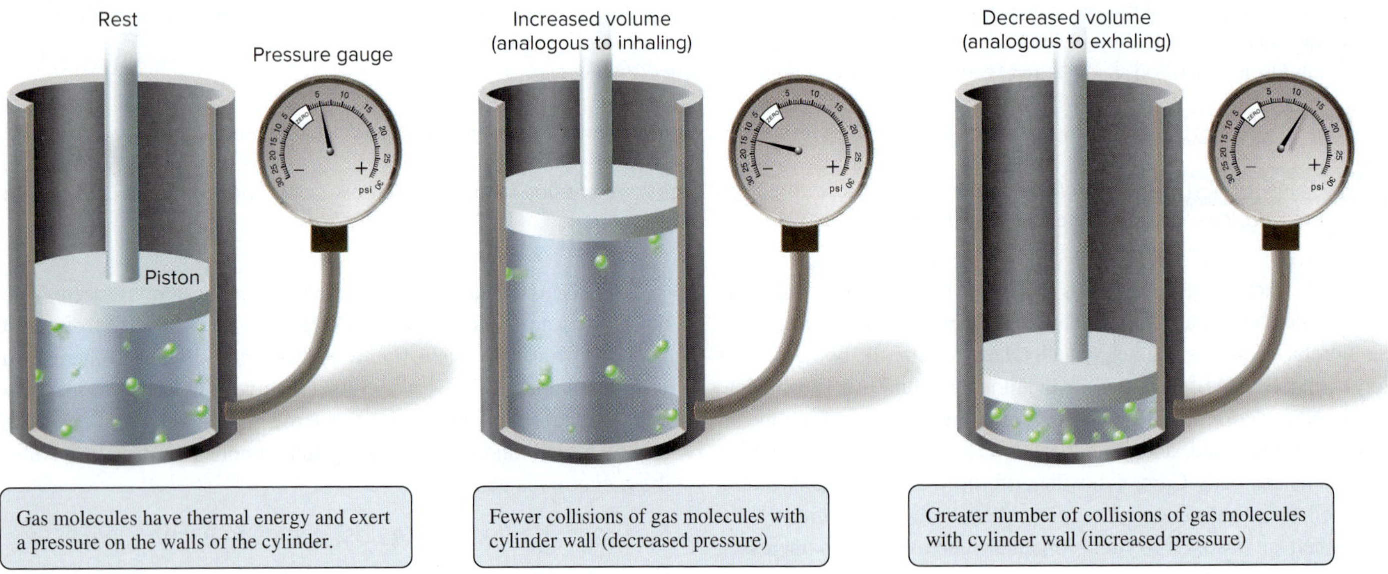

Rest
Pressure gauge
Piston
Increased volume
(analogous to inhaling)
Decreased volume
(analogous to exhaling)
Gas molecules have thermal energy and exert a pressure on the walls of the cylinder.
Fewer collisions of gas molecules with cylinder wall (decreased pressure)
Greater number of collisions of gas molecules with cylinder wall (increased pressure)


**Figure 36.18** **Boyle's law.** The volume and pressure of a gas are inversely related; that is, when the volume is increased, the pressure is decreased and vice versa. This relationship creates the gas pressure gradients that ventilate the vertebrate lungs.

**Concept Check:** *As lungs expand during inhalation, what happens to the pressure of air inside them?*

## 36.7 Testing Your Knowledge

1. Boyle's law states that
   a. the pressure and volume of a gas are directly related.
   b. the pressure exerted by a gas will increase when its volume decreases.
   c. increasing the volume of a gas increases its pressure.
   d. decreasing the volume of a gas decreases its pressure.
   e. both c and d are correct.
2. Which of the following statements is *false*?
   a. The insect respiratory system can adapt to increased metabolic demands of the animal.
   b. The tracheal system of insects is not sufficient to support the oxygen demands of very active tissues.
   c. Exchanging gases across the body surface is a feature found in certain invertebrates and some vertebrates.
   d. External gills are more susceptible to damage than are internal gills.
   e. The flow of blood and water through the gill lamellae follows a countercurrent arrangement.

## 36.8 Structure and Function of the Mammalian Respiratory System

### Learning Outcomes

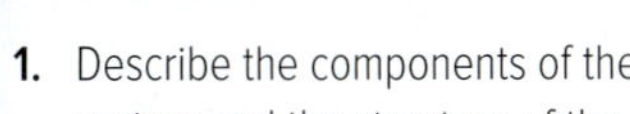

1. Describe the components of the mammalian respiratory system and the structure of the mammalian lung.
2. Outline the process of ventilation in mammalian lungs.

In this section, we will examine in detail the structures of the mammalian respiratory system and the mechanisms by which mammals ventilate their lungs.

### During Ventilation, Air Flows Through a Series of Branching Tubes

In mammals, the respiratory system includes the nose, mouth, pharynx, larynx, trachea, branching tubes, lungs, and muscles and connective tissues that encase these structures within the thoracic (chest) cavity (Figure 36.19a). When humans and other mammals breathe, air first enters the nose and mouth, where it is warmed and humidified. These processes protect the lungs from drying out. While in the nose, the air is partially purified as it flows over a coating of sticky mucus in the nasal cavity. The mucus and hairs in the nasal cavity trap some of the larger dust and other particles that are inhaled with air. These are then removed by the body's immune cells or swallowed.

The inhaled air from the mouth and nose converges at the back of the throat, or **pharynx,** a common passageway for air and food. From there, air passes through the **larynx,** which contains the vocal cords. Air flows from the larynx into the **trachea,** a tube that leads to the lungs.

The trachea is partially ringed by cartilage that provides rigidity and ensures that the trachea always remains open. Inhaled air flows down the trachea as it branches into two smaller tubes, called **bronchi** (singular, bronchus), which lead to each lung. The bronchi branch repeatedly into smaller and smaller tubes, eventually becoming thin-walled **bronchioles** surrounded by circular rings of smooth muscle (Figure 36.19b). Bronchioles can dilate or constrict in a manner analogous to that of arterioles, the small blood vessels that deliver blood to capillaries (see Figure 36.10).

The **alveoli** (singular, alveolus) are the saclike regions of the lungs where gas exchange occurs (Figure 36.19c). The alveoli are highly adapted for gas exchange and consist of two major types of cells. Gases diffuse across type I cells, whereas type II cells are secretory cells (described later). The alveoli are only one cell thick and resemble extremely thin sacs, appearing like bunches of grapes on a stem. Deoxygenated blood pumped from the right ventricle of the heart flows to the many capillaries surrounding the alveoli. Oxygen diffuses from the lumen of each alveolus across the type I alveolar cells, through the interstitial space outside the cells, and into the capillaries (see Figure 36.19b). Carbon dioxide diffuses in the opposite direction. The oxygenated blood from the lungs then flows to the left atrium of the heart and from there enters the left ventricle, where it is pumped out through the aorta to the rest of the body.

### The Pleural Sacs Protect the Lungs

The lungs are soft, delicate tissues that could easily be damaged by the surrounding bone, muscle, and connective tissue of the thorax if not protected. Each lung is encased in a **pleural sac,** a double layer of thin, moist connective tissue. Between the two layers is a microscopically thin layer of water that acts as a lubricant and makes the two tissue layers adhere to each other.

In addition to protecting the lungs, the inner pleural sac adheres to its lung, and the outer pleural sac adheres to the chest wall. In this way, movements of the chest wall result in similar movements of the lungs. This is important because the lungs are not muscular and, so, cannot inflate themselves. Instead, as we will see, the lungs are inflated by the expansion of the thoracic cavity, which results from the contraction of muscles in the thorax.

### The Lungs Expand by Negative Pressure Filling

The way in which you inflate a balloon, by forcing air from your mouth into the balloon, is called positive pressure filling. Negative pressure filling, by contrast, is the mechanism by which mammals and many other vertebrates ventilate their lungs. In this process, the volume of the lungs expands, creating a decreased pressure that draws air into the lungs (see Figure 36.18). The process differs in some ways among classes of vertebrates, but in mammals, the work is provided by the intercostal muscles, which surround and connect the ribs in the chest, and a large muscle called the **diaphragm** (see Figure 36.19a), which divides the thoracic cavity from the abdomen.

Let's follow the process when a mammal ventilates its lungs (Figure 36.20). At the start of a breath, the diaphragm contracts, pulling downward and enlarging the thoracic cavity. Simultaneously, the intercostal muscles contract, moving the chest upward and outward, which also helps to enlarge the thoracic cavity. Recall that the pleural sacs adhere the lungs to the chest wall, so as the chest expands, the lungs expand with it. According to Boyle's law, as the volume of the lungs

Nasal cavity
Nostril
Mouth
Larynx
Pharynx
Trachea
Left lung
Right lung
Rib
Right bronchus
Left bronchus
Intercostal muscles
Diaphragm

**(a) Human respiratory system**

Blood flow
Branch of pulmonary vein
Bronchiole
Branch of pulmonary artery
Smooth muscle
Alveoli
Capillaries

**(b) Structure of a bronchiole and alveoli**

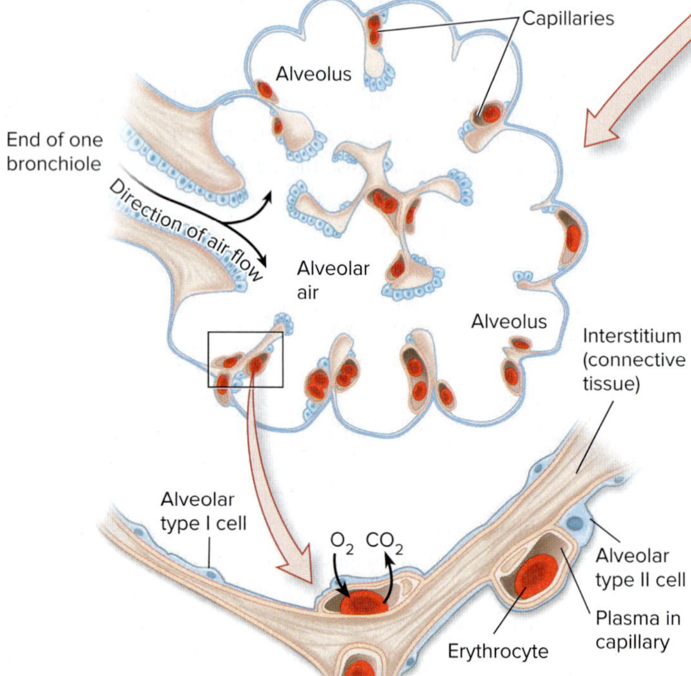


**(c) Cross section of an alveolar cluster, with enlarged region**

**Figure 36.19 The mammalian respiratory system.** **(a)** The lungs and major airways. The thoracic cavity is bounded by the ribs and intercostal muscles and the muscular diaphragm. For simplicity, the ribs have been removed in front. **(b)** The bronchioles deliver air to clusters of alveoli. Smooth muscle cells around the bronchioles can cause the bronchioles to constrict (narrow) or dilate (widen). Capillaries surround the alveoli. Red represents oxygenated blood; blue represents partly deoxygenated blood. **(c)** Cross section through a cluster of alveoli. Note the single layer of alveoli cells and their close proximity to adjacent capillaries.

increases, the pressures of the gases within them must decrease. In other words, the pressure in the lungs becomes negative with respect to the outside air. Air, therefore, flows down its pressure gradient from outside the mouth and nose, into the lungs.

Once the lungs are inflated with air, the chest muscles and diaphragm relax and recoil back to their original positions as an animal exhales. This compresses the lungs and forces air out of the airways. Whereas inhaling requires the expenditure of significant amounts of energy, exhaling is mostly passive and normally does not require much energy. This is possible because the lungs and chest have large numbers of elastin fibers, which, as mentioned earlier in the case of arteries, have elastic properties.

## Mammals Breathe by Tidal Ventilation

When mammals exhale, air leaves via the same route that it entered during inhalation, and no new oxygen is delivered to the airways at that time. This type of breathing is called **tidal ventilation** (think of

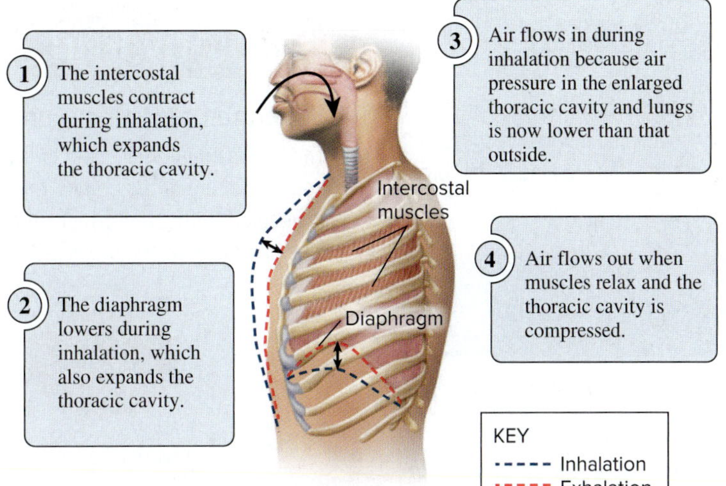

1 The intercostal muscles contract during inhalation, which expands the thoracic cavity.
2 The diaphragm lowers during inhalation, which also expands the thoracic cavity.
3 Air flows in during inhalation because air pressure in the enlarged thoracic cavity and lungs is now lower than that outside.
4 Air flows out when muscles relax and the thoracic cavity is compressed.
Intercostal muscles
Diaphragm
KEY
Inhalation
Exhalation


(a) Action of muscles during ventilation

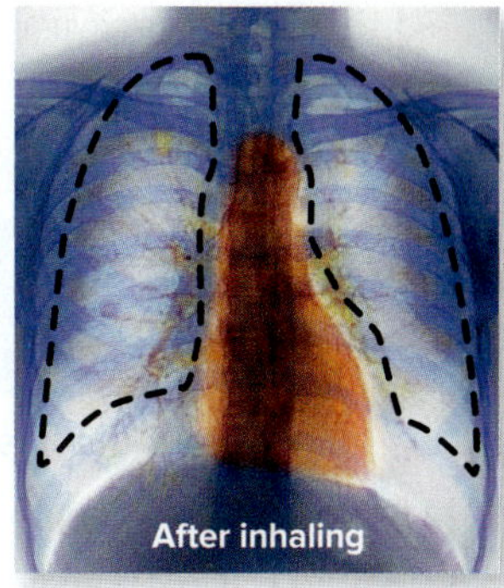


(b) Change in lung volume during ventilation

**Figure 36.20 Ventilation of the mammalian lung by negative pressure filling.** **(a)** The intercostal muscles contract, which expands the chest cavity by moving the ribs up and out. The diaphragm also contracts, causing it to pull downward, further expanding the cavity. The muscular efforts of inhalation require energy, whereas the return to the resting state by exhaling occurs primarily by recoil. **(b)** X-ray image of the chest of an adult man after inhaling. The volume of the lungs after exhaling is superimposed using dashed lines to illustrate the relative change in lung volume.

*(b)* © Pr. M. Brauner/Science Source

air flowing into and out of the lungs, like the ebb and flow of ocean tides). Tidal ventilation is less efficient than the unidirectional, flow-through system of fishes in which gills are always exposed to oxygenated water during all phases of the respiratory cycle.

As you likely know from experience, the lungs are neither fully inflated nor deflated at rest. For example, you could easily take a larger breath than normal if you wished, or exhale more than the usual amount of air. The volume of air that is normally breathed in and out at rest is the tidal volume, about 0.5 L in an average-sized human. Lung size and tidal volume are proportional to body size, both among humans and between species. A 6-foot-tall adult human, for example, has a larger tidal volume than a 4-foot-tall child because the adult has larger lungs. Similarly, horses have larger tidal volumes than humans, and humans have larger tidal volumes than dogs.

During exertion, the lungs can be inflated further than the resting tidal volume to provide additional oxygen. Likewise, the lungs can be deflated beyond their normal limits at rest, by exerting a strong effort during exhalation. The lungs never fully deflate, however, partly because they are held open by their adherence to the chest wall. This is important for a simple reason. Think again of our analogy of a balloon. It is much easier to fill a balloon that is already partly inflated than it is to inflate a completely empty balloon. The same is true of the lungs. The most difficult breath is the very first one that a newborn mammal takes—the only time its lungs are ever completely empty of air.

## Surfactant Facilitates Lung Inflation by Decreasing Surface Tension

Like all cells, those that make up the lining of the alveoli are surrounded by extracellular fluid. This fluid layer is where gases dissolve. Unlike other internal body cells, however, alveolar cells come into contact with air, creating an air/liquid interface along the inner surface of the alveoli. This results in surface tension within the alveoli. Surface tension results from the attractive forces between water molecules at the air/liquid interface and partly explains why droplets of water form beads. This tension produces a force that tends to collapse alveoli as water molecules lining its surfaces are attracted to each other. If many or all of the alveoli collapsed, however, the amount of surface area available for gas exchange in the lungs would be greatly reduced. What prevents them from collapsing? The type II cells of the alveoli produce **surfactant,** a mixture of proteins and amphipathic lipids (that is, lipids with both polar and nonpolar regions), and secrete it into the alveolar lumen. Surfactant forms a barrier between the air and the fluid layer inside the alveoli. This barrier reduces surface tension in the alveolar walls, allowing them to remain open.

## 36.8 Reviewing the Concepts

- The mammalian respiratory system includes the nose, mouth, airways, lungs, muscles, and connective tissues that encase these structures within the thoracic cavity (Figure 36.19).
- Mammals ventilate their lungs by negative pressure filling. The work is provided by the intercostal muscles and diaphragm. Surfactant forms a barrier between air and the fluid layer of the alveoli, helping keep them open (Figure 36.20).

## 36.8 Testing Your Knowledge

1. The _______ separate(s) the thoracic cavity from the abdomen and _______ during exhalation.
   a. pleural sacs, relax
   b. pleural sacs, contract
   c. diaphragm, relaxes
   d. diaphragm, contracts
   e. intercostal muscles, contract

2. The tidal volume
   a. decreases with age from childhood to adulthood.
   b. is about 5 L in a typical adult.
   c. is the maximum volume of air that can be inhaled in one breath.
   d. is the minimum volume of air that can be inhaled in one breath.
   e. is the volume of air normally inhaled and exhaled in one breath.

## 36.9 Mechanisms of Gas Transport in Blood

### Learning Outcomes

1. Discuss the characteristics of respiratory pigments.
2. **SCISKILLS ▶** Analyze the quantitative relationship between oxygen and hemoglobin, and explain how certain factors can modify the shape of the dissociation curve.
3. Write the reversible reaction between hemoglobin and $O_2$.
4. Describe the ways in which $CO_2$ is carried in blood.

The amount of $O_2$ that can be dissolved in the body fluids is not sufficient to sustain life in most animals. In nearly all animals, therefore, the amount of $O_2$ in the body fluids must be increased above that which can be physically dissolved in water. This is made possible because of the widespread occurrence of oxygen-binding proteins, which increase the total reservoir of $O_2$ available to cells. In this section, we examine the structure, function, and evolution of these molecules and examine how $CO_2$ is transported.

### Oxygen Binds to Respiratory Pigments

The oxygen-binding proteins that have evolved in animals are called **respiratory pigments** because they have a color (blue or red). In vertebrates, the pigments are contained within erythrocytes, whereas many invertebrates have these pigments in their hemolymph. Respiratory pigments are proteins containing one or more metal atoms that bind to oxygen. In vertebrates and many marine invertebrates, the metal is typically iron ($Fe^{2+}$). As stated earlier, hemoglobin is the major iron-containing pigment and gives blood its red color. In decapod crustaceans, arachnids, and many mollusks, the metal is copper ($Cu^{2+}$). The copper-containing pigment **hemocyanin** gives the blood or hemolymph a bluish tint.

Hemoglobin gets its name because it is a globular protein—which refers to the shape and water solubility of a protein—and because it contains a chemical group called a heme in its core. An atom of iron is bound within the heme group. In vertebrates, each molecule of hemoglobin consists of four polypeptide subunits, each with its own iron atom to which a molecule of $O_2$ can bind. Thus, a single hemoglobin molecule can bind up to four molecules of oxygen (Figure 36.21).

Respiratory pigments share certain characteristics that make them ideal for transporting $O_2$. First, they all have a high affinity for binding $O_2$. Second, the binding between the pigment and $O_2$ is noncovalent and reversible. Clearly, it would be of no benefit for a protein to bind $O_2$ if it could not later unload the $O_2$ to cells that require it. The reaction that describes the reversible binding of $O_2$ to hemoglobin (Hb) is

$$Hb + O_2 \rightleftharpoons HbO_2$$

## Biology Principle

### Structure Determines Function

The precise quaternary structure of hemoglobin permits its association with heme groups, which contain the iron atoms that bind oxygen molecules. Without a correct structure, the ability of hemoglobin to bind oxygen would be compromised.

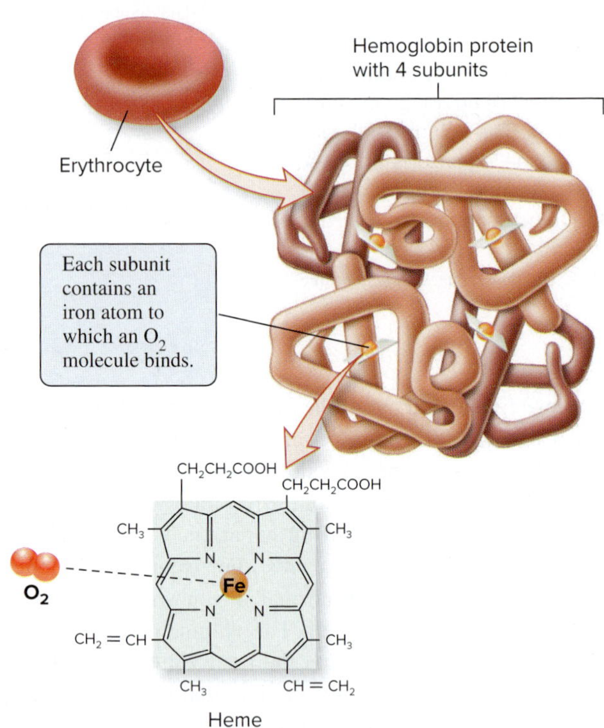


**Figure 36.21** **Hemoglobin.** Erythrocytes contain large amounts of the protein hemoglobin. Oxygen binds reversibly to iron atoms in the heme portion of each subunit of hemoglobin.

where Hb is hemoglobin, $HbO_2$ is oxyhemoglobin (hemoglobin with bound $O_2$), and the double arrows ($\rightleftharpoons$) indicate that the reaction is reversible.

The amount of pigment present in the blood is sufficient to provide enough $O_2$ to meet most oxygen demands, except for those that occur under the most strenuous exertion. In humans, for example, the presence of hemoglobin gives blood about 45 times more $O_2$-carrying capacity than plasma alone.

### The Amount of Oxygen Bound to Hemoglobin Depends on the $Po_2$ of Blood

The partial pressure of $O_2$ ($Po_2$) in blood is a measure of its dissolved concentration. When $Po_2$ is high, more $O_2$ binds to hemoglobin, whereas fewer $O_2$ molecules will be bound when $Po_2$ is low. Figure 36.22 shows the relationship between $O_2$ binding and $Po_2$, known as an **oxygen-hemoglobin dissociation curve,** for

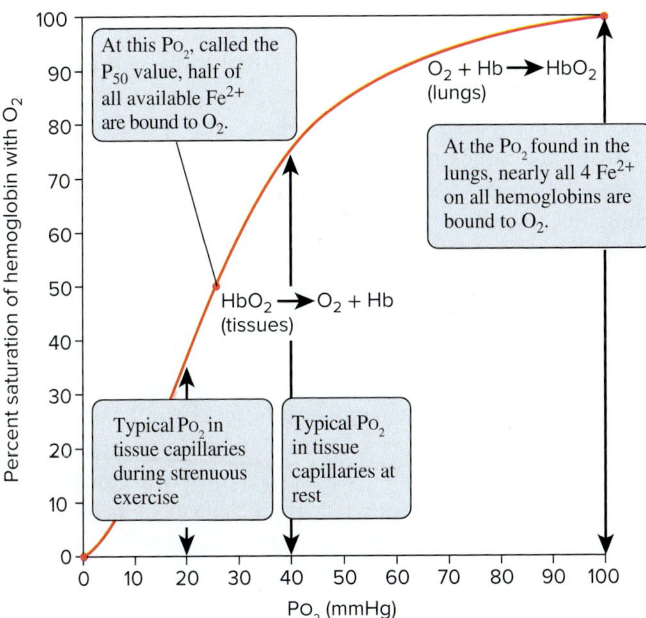


**Figure 36.22** **The human oxygen-hemoglobin dissociation curve.** Depending on the partial pressure of oxygen ($Po_2$), oxygen is either loaded onto hemoglobin, as in the lungs, or unloaded from hemoglobin, as in the rest of the body tissues. When $Po_2$ is high, more $Fe^{2+}$ atoms are bound to $O_2$, and therefore the hemoglobin is more saturated with $O_2$.

humans. At a $Po_2$ of 100 mmHg, which is typical of the oxygenated blood leaving the lungs, nearly every hemoglobin molecule is bound to four $O_2$ molecules (far right of graph). It is essentially 100% saturated with $O_2$ at that $Po_2$. The $Po_2$ of blood leaving the tissue capillaries of other parts of the body is lower and depends on metabolic activity and exercise. At rest, the average $Po_2$ of blood capillaries in these other parts of the body is typically around 40 mmHg. At this $Po_2$, hemoglobin releases some $O_2$ molecules, decreasing to about 75% saturation with $O_2$. During strenuous exercise, $Po_2$ in the capillaries drops even further (as low as 20 mmHg); consequently, during exercise hemoglobin releases even more $O_2$ and becomes less saturated. In this way, hemoglobin performs its role of $O_2$ delivery. In the lungs, it binds $O_2$, and elsewhere it releases $O_2$ as required.

The curve in Figure 36.22 is not linear but S-shaped (sigmoidal). This is because the subunits of hemoglobin cooperate with each other in binding $O_2$. Once a molecule of $O_2$ binds to one subunit's iron atom, the shape of the entire hemoglobin protein changes, making it easier for a second $O_2$ to bind to the next subunit, and so on. Thus, the relationship between $Po_2$ and the amount of $O_2$ bound to hemoglobin becomes very steep in the midrange of the curve, which represents the pressures that occur in the tissue capillaries throughout the body. This steepness allows $O_2$ release from hemoglobin to be very sensitive to even small decreases in $Po_2$ generated by the diffusion of $O_2$ from capillaries to cells. The curve levels off at high oxygen pressures as 100% saturation is approached.

## Quantitative Analysis

### THE ABILITY OF HEMOGLOBIN TO BIND OXYGEN IS DECREASED BY FACTORS SUCH AS TEMPERATURE, $CO_2$, AND pH

One of the remarkable features of the oxygen-hemoglobin binding relationship is that it can be influenced by metabolic waste products such as $CO_2$ and $H^+$ and by heat (temperature). Figure 36.23a shows three curves, one obtained under normal resting conditions and the others in the presence of low or high levels of $CO_2$. Carbon dioxide binds to amino acids in the hemoglobin protein (not to the iron, like $O_2$), and when it does, it decreases the affinity of hemoglobin for $O_2$. Note how an increase in $CO_2$ shifts the curve to the right, such that at any $Po_2$, less $O_2$ is bound to hemoglobin. Another way of saying this is that at any $Po_2$, more $O_2$ has been released from hemoglobin, thus becoming available to cells. A similar shift in the curve occurs with an increase in acidity (decreased pH), because $H^+$ can also bind to hemoglobin and alter its oxygen-carrying capacity; the effect of $CO_2$ and $H^+$ on the oxygen-hemoglobin dissociation curve is known as the **Bohr effect.** Elevated temperature also reduces the affinity of hemoglobin for $O_2$, resulting in a right-shifted curve.

Cells generate each of these products—$CO_2$, $H^+$, and heat—when they are actively metabolizing nutrients such as glucose. The metabolic products enter the surrounding capillaries and diffuse into erythrocytes, where they alter the shape of hemoglobin, causing it to release more of its $O_2$ than would normally occur at that $Po_2$. This phenomenon is a way in which individual body tissues obtain more $O_2$ from the blood to match their higher metabolic demands. Thus, when an animal increases its physical activity, the skeletal muscles generate more $CO_2$, $H^+$, and heat than do some other tissues. Therefore, the hemoglobin in muscle capillaries releases more $O_2$ to muscle cells compared to the hemoglobin in the capillaries of these other tissues.

The shift in the oxygen-hemoglobin dissociation curve occurs in all classes of vertebrates (although not in all species), but it has different magnitudes in different species. Not surprisingly, perhaps, animals with relatively high metabolic rates, such as mice, show a greater Bohr effect than do animals with lower relative metabolic rates. Recall that these same waste products of metabolism also cause local vasodilation of arterioles. The more metabolically active a tissue is, therefore, the more blood flow it receives, which means more oxygen-bound hemoglobin. Moreover, the oxygen is unloaded from hemoglobin more readily due to the shift in the curve. This is an excellent example of how adaptive changes in circulatory and respiratory functions often are complementary.

Regardless of their sensitivity to $CO_2$ and other factors, the hemoglobins of metabolically active animals also have a lower affinity for oxygen (Figure 36.23b) even in the absence of a greater than normal amount of $CO_2$, $H^+$, and heat. In small, active animals, the curves are displaced to the right relative to the human curve (in other words, their hemoglobin $P_{50}$ value is higher than it is in larger animals with lower relative metabolic rates).

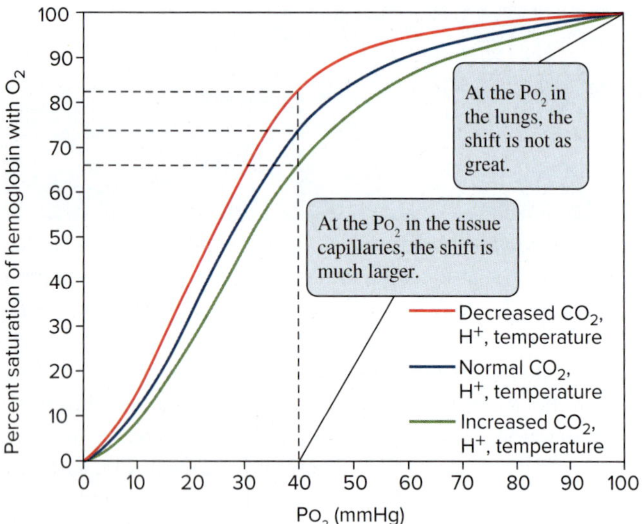


**(a) Shifts in the oxygen-hemoglobin dissociation curve**

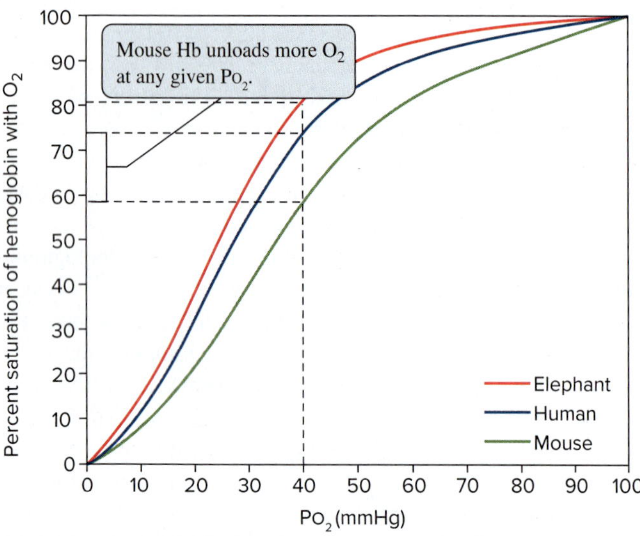


**(b) Oxygen-hemoglobin dissociation curves of different animals**

**Figure 36.23 Changes in oxygen-hemoglobin dissociation curves under different conditions and among different species. (a)** Increasing or decreasing the amounts of $CO_2$ or $H^+$ (pH), or the temperature of the blood, shifts the oxygen-hemoglobin dissociation curve. Metabolically active tissues generate more of these products. The change in affinity of hemoglobin for oxygen ($O_2$) allows different tissues to obtain $O_2$ in proportion to their metabolic requirements. **(b)** Oxygen-hemoglobin dissociation curves for three mammals with low (elephant), moderate (human), or high (mouse) relative metabolic rates. For any $Po_2$, such as the one shown (40 mmHg), which is typical of the $Po_2$ of tissue capillaries, less $O_2$ is bound to mouse hemoglobin than to human or elephant hemoglobin, and less $O_2$ is bound to human hemoglobin than to elephant hemoglobin. Therefore, $O_2$ is unloaded from hemoglobin more readily in smaller animals.

**Concept Check:** *What would happen to the position of the middle curve in part (a) following infusion of an alkaline compound such as bicarbonate ions ($HCO_3^-$) into the blood of a resting, healthy individual?*

In contrast, larger animals with slower relative metabolic rates, such as the elephant, have curves shifted to the left compared with that of humans. At high oxygen pressures, such as those that occur in the lungs, these shifts have little relevance because nearly all the hemoglobin is bound to oxygen at those pressures. At lower $Po_2$, however, such as those that would occur in the capillaries of metabolically active tissues, the difference in the curves becomes significant. For example, look at the three curves at a $Po_2$ of 40 mmHg, a typical value found in tissues that are using oxygen at a resting rate. The mouse hemoglobin has less oxygen bound (its hemoglobin is less saturated) at that pressure than does the human hemoglobin, which in turn has less $O_2$ bound than does the elephant hemoglobin. In other words, the mouse hemoglobin has released more of its $O_2$ to the active tissues than have the other animal hemoglobins. The different properties of hemoglobin among species result from changes in the sequence of hemoglobin, the result of numerous evolutionary changes in the genes encoding hemoglobin subunits in vertebrates.

**Crunching the Numbers:** Refer to Figure 36.23a. In which case would hemoglobin be more saturated, at a $Po_2$ of 35 mmHg and with increased $CO_2$, or at a $Po_2$ of 30 mmHg but with decreased $CO_2$? Now refer to Figure 36.23b. What percent saturation of hemoglobin would you estimate for a small mammal such as a rabbit at a $Po_2$ of 30 mmHg?

## Carbon Dioxide Is Transported in the Blood in Three Forms

As with oxygen, only a limited amount of $CO_2$ physically dissolves in blood. An additional amount of $CO_2$ is carried bound to hemoglobin, as noted earlier. The majority of $CO_2$, however, is converted into highly soluble bicarbonate ions ($HCO_3^-$). This occurs according to the following reaction, where $H_2CO_3$ is a short-lived compound called carbonic acid that immediately dissociates to an $H^+$ and an $HCO_3^-$:

$$CO_2 + H_2O \rightleftharpoons H_2CO_3 \rightleftharpoons H^+ + HCO_3^-$$

Note that one $H^+$ is formed for every $CO_2$ that enters this reaction. The resulting pH change is what makes $CO_2$ a dangerous waste product, because the activities of most enzymes in an animal's body are very sensitive to changes in pH.

These reactions are readily reversible. The first step is catalyzed in both directions by the enzyme carbonic anhydrase, which is present in high amounts in erythrocytes. As you learned in Chapter 2, the concentrations of reactants and products affect the direction of a chemical reaction. For example, as tissues release $CO_2$ into capillaries, the forward reaction (from left to right, as illustrated here) would be favored, and the pH of the blood would decrease because the concentration of $H^+$ would increase. Conversely, when $CO_2$ diffuses out of lung capillaries and is exhaled, the reactions would proceed from right to left, and the pH of the blood would increase as $H^+$ combined with $HCO_3^-$.

## 36.9 Reviewing the Concepts

- A limited amount of any gas can dissolve in water; such limits are overcome in part by respiratory pigments that provide additional oxygen-carrying capacity. Hemoglobin is the major iron-containing pigment in vertebrates and many invertebrates (Figure 36.21).
- The amount of oxygen bound to hemoglobin depends on the partial pressure of $O_2$ ($PO_2$) in the blood. Metabolic products such as $CO_2$ and $H^+$ and heat can influence the oxygen-hemoglobin binding relationship (Figures 36.22, 36.23).
- $CO_2$ is transported dissolved in solution, bound to hemoglobin, and in the form of $HCO_3^-$.

## 36.9 Testing Your Knowledge

1. Hemoglobin
   a. is identical in all mammals.
   b. has decreased affinity for oxygen in blood that is more acidic (higher $H^+$ concentration) than normal.
   c. has increased affinity for oxygen in the presence of increased $CO_2$.
   d. contains two subunits, each of which binds two molecules of oxygen.
   e. forms nonreversible bonds with oxygen.
2. Compared to humans, the typical affinity of hemoglobin for oxygen in small mammals is ______, and the Bohr effect is ______.
   a. lower, less
   b. lower, greater
   c. higher, less
   d. higher, greater

## 36.10 Control of Ventilation

### Learning Outcome

1. Describe the role of respiratory centers and chemoreceptors in the regulation of ventilation.

In the previous sections, you saw that different species ventilate their respiratory organs in different ways. Now let's look at how the mechanisms of breathing are controlled in mammals. Lungs are neither muscles nor electrically excitable tissue. Therefore, they cannot initiate or regulate their own expansion. Nonetheless, lungs require a mechanism to rhythmically expand and recoil because animals cannot always consciously control breathing, such as during sleep. In this section, we examine the ways in which the nervous system and chemoreceptors control ventilation in mammals.

### The Nervous System Contains the Control Center for Ventilation

The control center that initiates rhythmic expansion of the lungs is a collection of nuclei in the central nervous system. In mammals, these **respiratory centers** are located in the pons and medulla oblongata of the brainstem (**Figure 36.24**). Neurons within these regions rhythmically

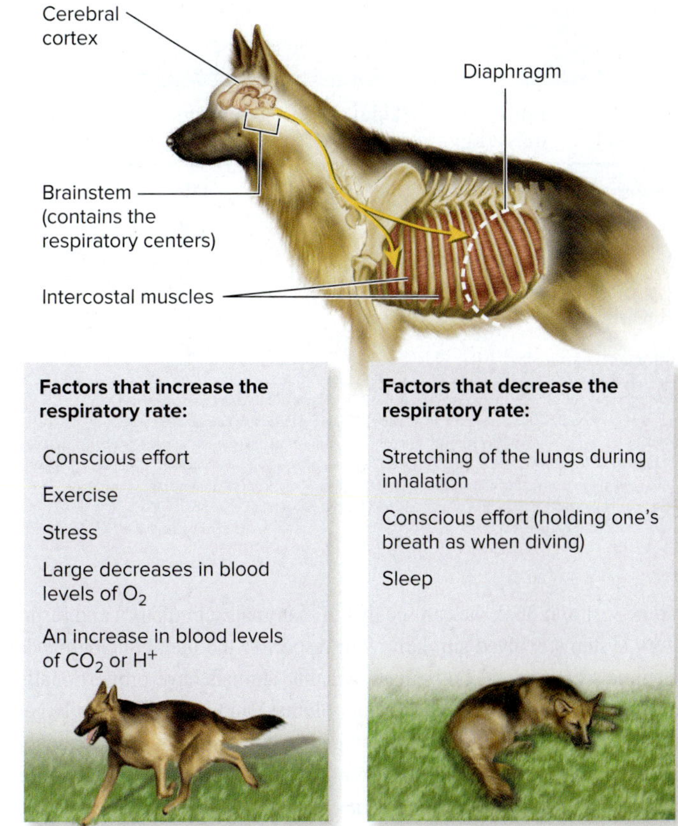

Cerebral cortex
Diaphragm
Brainstem (contains the respiratory centers)
Intercostal muscles
Factors that increase the respiratory rate:
Conscious effort
Exercise
Stress
Large decreases in blood levels of $O_2$
An increase in blood levels of $CO_2$ or $H^+$
Factors that decrease the respiratory rate:
Stretching of the lungs during inhalation
Conscious effort (holding one's breath as when diving)
Sleep


**Figure 36.24** **The control of breathing via respiratory centers in the mammalian brain.** Neurons in the brainstem send action potentials along neurons in nerves that stimulate the intercostal muscles and diaphragm. The factors listed here can modulate the rate of action potential generation and therefore the respiratory rate.

**BioConnections:** *The brainstem of vertebrates was described in Chapter 33. What parts of the brain are included in the brainstem?*

generate action potentials. These electrical impulses travel from the brainstem through two sets of nerves. The first set stimulates the intercostal muscles, and the second set stimulates the diaphragm. When the lungs expand in response to the contraction of these muscles, stretch-sensitive neurons in the lungs and chest send signals to the respiratory centers, informing them that the lungs are inflated. This temporarily turns off the stimulating signal until the animal exhales, whereupon a new signal is sent to the breathing muscles.

Although the brainstem automatically generates a steady rhythm of breathing, it can be modified or overridden. For example, animals that dive underwater—including humans when we swim—can temporarily hold their breath. The respiratory centers decrease their activity during sleep and increase it during stress. Normally, however, increased breathing occurs in response to physical activity. At such times, a variety of factors converge on the respiratory centers to increase the rate and strength of signals to the breathing muscles, resulting in faster and deeper breaths.

As with cardiovascular parameters, respiratory activity varies with body mass, as seen for mammals in **Table 36.2**. Comparing

**Table 36.2 Respiratory Characteristics of Different Mammals**

| Animal | Body mass (kg) | Tidal volume (L) | Breaths/ min | $P_{50}$ value |
|---|---|---|---|---|
| Shrew | 0.0024 | 0.00003 | 700 | 37 |
| Rat | 0.20 | 0.0016 | 85 | 35 |
| Dog | 25 | 0.27 | 20 | 29 |
| Human | 75 | 0.50 | 12 | 26 |
| Horse | 450 | 6.50 | 9 | 24 |

Note: All values are averages from resting animals. Tidal volume is the volume of air breathed in with each breath. Note that tidal volume increases as an animal's mass (and therefore lung size) increases. By contrast, breathing rate is higher in smaller animals. Similar relationships occur in other vertebrates, notably birds. The $P_{50}$ value is the oxygen pressure at which an animal's hemoglobin is 50% saturated with $O_2$. Higher $P_{50}$ values correspond to lower affinities of hemoglobin for $O_2$ (in other words, animals with high $P_{50}$ values unload $O_2$ from hemoglobin more readily than do animals with low $P_{50}$ values; see Figure 36.23).

Tables 36.1 and 36.2, we can see that in mammals, circulatory and respiratory systems evolved similarly with respect to the metabolic demands of animals of different body mass. Smaller animals have proportionally smaller lungs (as evidenced by tidal volumes) and hearts than do larger animals, but they have faster breathing rates and heart rates. These adaptations make it possible to deliver oxygen to tissues at a rate sufficient for the relatively high metabolic demands of very small species.

## Chemoreceptors Modulate the Activity of the Respiratory Centers

The respiratory centers are influenced by the partial pressures of oxygen and carbon dioxide in the arteries, as well as the concentration of hydrogen ions (in other words, the pH of the blood). Recall from Chapter 34 that chemoreceptors recognize specific chemicals in the air, water, body fluids, or food. Chemoreceptors located in the aorta, carotid arteries, and the brainstem detect the circulating levels of $O_2$, $CO_2$, and $H^+$ and relay that information through nerves or interneurons to the respiratory centers.

If the arterial $Po_2$ decreases well below normal, as might occur at high altitude or in certain respiratory diseases, the chemoreceptors signal the respiratory centers to increase the rate and depth of breathing to increase ventilation of the lungs. This brings in more oxygen. Similarly, a buildup of $CO_2$ in the blood, which would occur if an animal's ventilation were lower than normal (again, often the result of respiratory disease), signals the respiratory centers to stimulate breathing. The increased ventilation not only brings in more $O_2$, but also helps eliminate more $CO_2$. Finally, an increased concentration of $H^+$ in the blood (such as, during exercise) activates chemoreceptors that signal the brain that the blood is too acidic. This leads to an increase in the rate of breathing, which eliminates more $CO_2$, thereby allowing more $H^+$ to bind to $HCO_3^-$.

BIO TIPS ONLINE

## 36.10 Reviewing the Concepts

- In mammals, respiratory centers in the brainstem initiate the rhythmic expansion of the lungs (Figure 36.24).
- Chemoreceptors detect arterial blood levels of $H^+$, $CO_2$, and $O_2$. The chemoreceptors relay this information to the respiratory centers, which in turn affect the breathing rate.

## 36.10 Testing Your Knowledge

1. According to the following reaction, what would happen if a person at rest were to hold his breath for a minute or so?

$$CO_2 + H_2O \rightleftharpoons H_2CO_3 \rightleftharpoons H^+ + HCO_3^-$$

   a. The arterial levels of $CO_2$ would decrease.
   b. The arterial levels of $CO_2$ would decrease and those of $H^+$ would increase.
   c. The blood would become more alkaline.
   d. The blood would become more acidic, and the arterial levels of $CO_2$ would increase.
   e. The arterial levels of $H^+$ and $HCO_3^-$ would both decrease.

## 36.11 Impact on Public Health

**Learning Outcomes**

1. Define hypertension and atherosclerosis, and explain their impact on human health.
2. List the causes, symptoms, and current medical treatments for myocardial infarction (MI), or heart attack.
3. Outline the causes, symptoms, and current treatments of asthma.
4. Discuss the impact of smoking tobacco on respiratory health.

Diseases of the heart and blood vessels account for more deaths each year in the U.S. than any other cause. Why is cardiovascular disease so devastating? One reason is that damage to structures of the circulatory system often occurs slowly, over many years, and without warning symptoms until the disease has reached late stages. Cardiovascular disease not only has a dramatic impact on the health of many Americans but also has a staggering impact on expenditures for health-care services, medications, and lost productivity.

Respiratory diseases of all kinds (including lung cancer) afflict as many as 10% of the U.S. population and result in an estimated 300,000–400,000 deaths per year, making lung disease among the top three causes of death in the U.S. The economic impact of respiratory diseases on the U.S. economy is immense, with recent estimates of up to $150 billion per year in health-related costs and lost productivity. Many of these diseases are chronic—once they appear, they last for the rest of a person's life. Lifestyle factors, such as smoking tobacco and exposure to air pollution, cause some respiratory disorders or make existing conditions worse. In this section, we examine the nature and some common causes of cardiovascular and respiratory disease in humans.

## Hypertension and Atherosclerosis Contribute to Heart and Blood Vessel Disease

**Hypertension,** or high blood pressure, refers to an arterial blood pressure that is chronically above normal. The normal range of blood

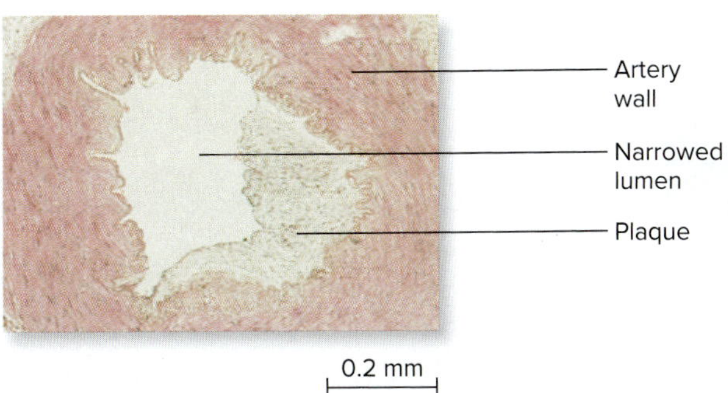


**Figure 36.25** **An atherosclerotic plaque in an artery.** Compare this with the normal artery shown in the inset in Figure 36.10.
© Biophoto Associates/Science Source

pressure in humans varies from about 90/60 to 120/80 mmHg. Values between 120/80 and 140/90 mmHg are considered borderline, and a resting blood pressure above 140/90 mmHg defines hypertension. Hypertension can have many causes, including obesity, smoking, aging, kidney disease, excess male hormones, and genetic factors, although in many cases, the cause is unknown. It can often be treated with diet and exercise and with drugs that cause vasodilation, thereby reducing total peripheral resistance.

Hypertension rarely has any noticeable symptoms, which is why it is often referred to as a "silent killer." For this reason, it is important to have blood pressure checked regularly. Without treatment, hypertension can damage arteries, contributing to the formation of **plaques**—accumulations of lipids, fibrous tissue, and smooth muscle cells—inside arterial walls. Plaques may lead to **atherosclerosis,** in which plaque causes the arteries to narrow and harden (Figure 36.25). Large plaques may occlude (block) the lumen of an artery entirely. Arterial plaques are known to arise from a variety of factors in addition to hypertension, including calcium and fat deposits, and are also correlated with obesity, high blood cholesterol concentrations, and smoking.

If plaques form in an artery, the regions of the body supplied with blood by that artery receive less oxygen and nutrients. Although atherosclerosis is dangerous anywhere, it is especially significant if it affects the coronary arteries (see the chapter-opening image), which carry oxygen and nutrients to the heart muscle. **Coronary artery disease** occurs when plaques form in the coronary vessels—a condition that can be life-threatening. One warning sign of coronary artery disease is **angina pectoris,** chest pain during exertion due to the heart being deprived of oxygen.

## Myocardial Infarction Results in Death of Cardiac Muscle Cells

If a portion of heart muscle is deprived of its normal blood flow for an extended time, the result may be a **myocardial infarction (MI),** or heart attack. This is usually caused by atherosclerosis in one of the coronary arteries. Some heart attacks are relatively minor; in fact, the discomfort of a small heart attack may not even alarm someone enough to seek medical attention. A heart attack with no symptoms is called a silent heart attack. More serious heart attacks can lead to significant damage or death to a portion of the heart. Dead cardiac muscle tissue does not regenerate; therefore, the heart's ability to pump is permanently decreased. Reduced pumping activity of the heart can result in congestive heart failure, in which the heart can't pump enough blood to meet the body's needs. As noted in the chapter introduction, each year in the U.S., between 1 and 1.5 million people suffer a heart attack, many of which are fatal.

Preventing a heart attack from occurring at all is the best way to increase survival rates. Procedures are available that allow physicians to monitor the status of the coronary vessels in people thought to have heart disease. In the procedure called **cardiac angiography,** the coronary arteries can be visualized by injecting a dye into a person's veins and then taking an X-ray image of the chest. The resulting image allows a physician to determine if the vessels are narrowed by disease.

If a blockage is found, several common treatments can restore blood flow through a coronary artery. One is **balloon angioplasty,** in which a thin tube with a tiny, inflatable balloon at its tip is threaded through the artery to the diseased area. Inflating the balloon compresses the plaque against the arterial wall, widening the lumen. In most cases, a wire-mesh device called a stent is inserted into the artery after angioplasty has expanded it, providing a sort of lattice to hold the artery open (Figure 36.26). A treatment for more

## Biology Principle

### Biology Affects Our Society

Biologists investigating the mechanisms of blood flow, heart function, and blood pressure in animals can use that knowledge to develop treatments for heart diseases in humans.

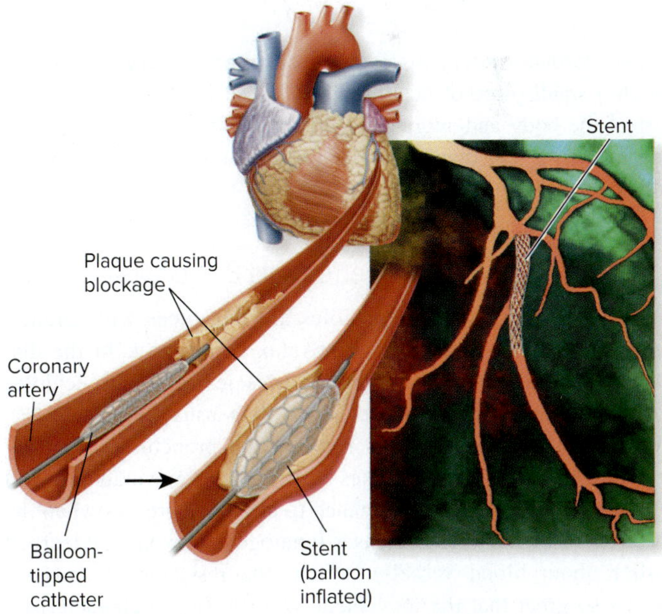


**Figure 36.26** **A treatment for blocked blood vessels.** Balloon angioplasty can widen diseased arteries, followed by insertion of a stent. The inset shows a stent placed in a coronary artery of a human patient.
*(inset)* © Sovereign/ISM/Phototake

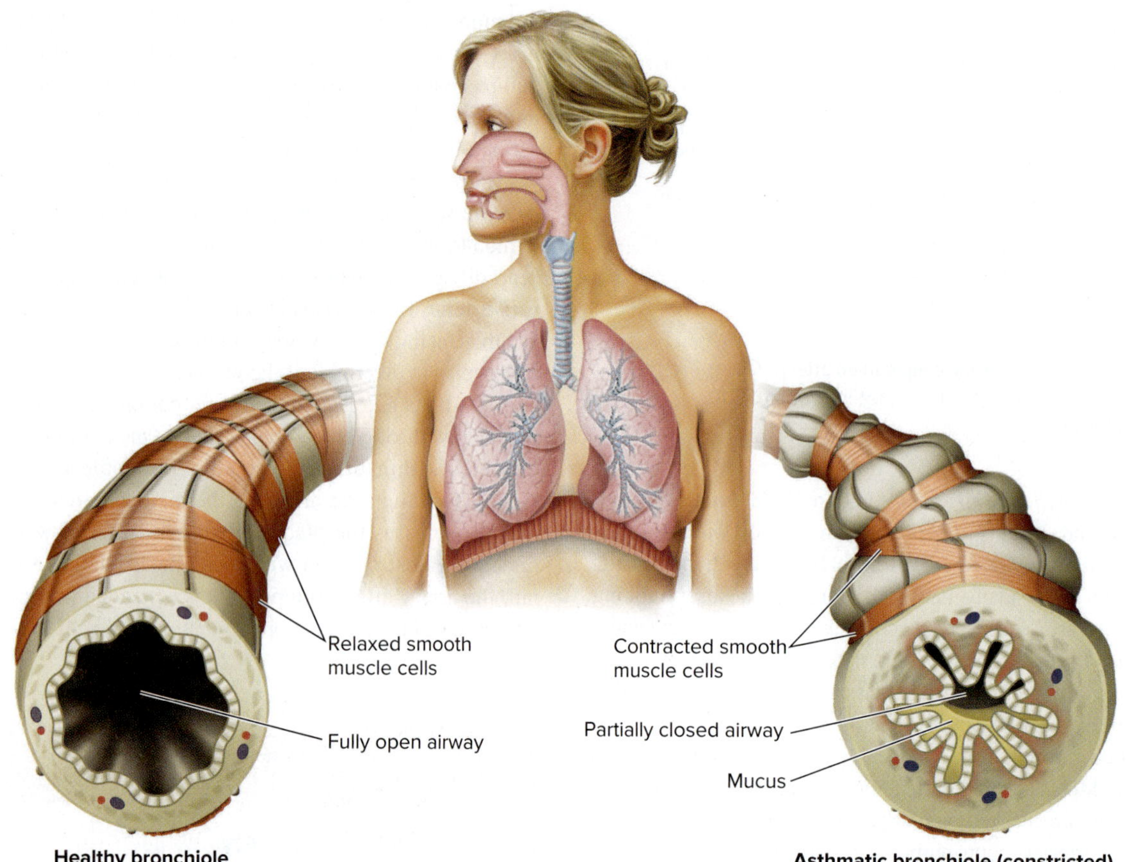


**Figure 36.27** Comparison of healthy and asthmatic bronchioles in a human.

serious coronary artery disease is a **coronary artery bypass,** in which a small piece of healthy blood vessel is removed from one part of the body and surgically grafted onto the coronary circulation in such a way that blood bypasses the diseased region of the unhealthy artery.

## Asthma Is a Disease of Hyperreactive Bronchioles

We saw previously that bronchioles are thin tubes with circular rings of smooth muscles that can relax or contract. In the disease **asthma,** however, the smooth muscles of the bronchioles are hyperreactive and contract more than usual (**Figure 36.27**). Contraction of these muscles narrows the bronchioles, causing bronchoconstriction. This makes it difficult to move air into and out of the lungs, because resistance to airflow increases when the diameter of the airways decreases (analogous to what you learned earlier about blood vessels). Often, the resistance to airflow can be so great that the movement of air creates a characteristic wheezing sound.

Asthma tends to run in families and therefore has a genetic basis. Several known triggers can elicit bronchoconstriction, including exercise, cold air, and allergic reactions. The last of these—allergic reactions—is of interest because asthma is believed to be partly the result of an imbalance in the immune system, which controls inflammation and other allergic responses. During flare-ups of asthma, inflammation results in the secretion of a viscous, mucus-like fluid; this inhibits airflow and worsens symptoms.

The symptoms of asthma can be alleviated by inhaling an aerosol mist containing **bronchodilators,** compounds that bind to receptors located on the plasma membranes of smooth muscle cells that make up the outer part of bronchioles. These compounds, which are related to the neurotransmitter norepinephrine, cause bronchiolar smooth muscle cells to relax. This, in turn, allows the bronchioles to dilate (widen). To help reduce the inflammation of the lungs, the medication may contain a steroid hormone with anti-inflammatory actions. Currently, there is no cure for asthma, but with regular treatment and the avoidance of known triggers, most people with this disease can lead normal lives with few restrictions.

## Emphysema Causes Permanent Lung Damage

Unlike asthma, in which the major problems are inflamed airways and hyperreactive bronchioles, **emphysema** is a progressive

## Biology Principle

### Biology Affects Society

Lung disease affects millions of people. Biologists study the causes, mechanisms, and possible treatments of human diseases such as emphysema.

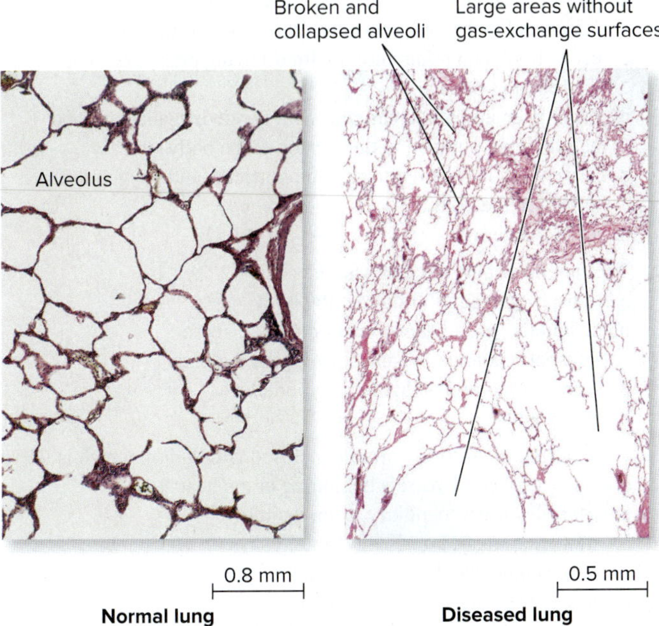


**Figure 36.28** **The effects of emphysema.** These light micrographs compare a section of a normal lung (left) with that of a lung from a person who died of emphysema (right). The collapse and destruction of alveoli caused by this disease reduces the surface area for gas exchange in the lungs.



disease that involves extensive lung damage (**Figure 36.28**). The disease decreases the total surface area of the alveoli, which greatly decreases the rate of oxygen diffusion from the lungs into the circulation. Consequently, one sign of emphysema is a $Po_2$ in the arteries that is lower than normal. It is also physically harder for those with emphysema to exhale because the recoil ability of the lungs is decreased due to the loss of elastin fibers, and therefore, arterial $CO_2$ concentrations increase. Finally, the terminal ends of the bronchioles are often damaged, which increases resistance to airflow and creates one of the disease's main symptoms, shortness of breath.

Reduced blood $O_2$ and poor lung function limit the patient's ability to function, and in its late stages, emphysema results in a person being essentially bedridden. Oxygen therapy, in which the person breathes a mixture of air and pure $O_2$ from a portable gas tank, can provide some help. The extra oxygen increases the pressure gradient for $O_2$ from the alveoli to the lung capillaries, promoting $O_2$ diffusion into the blood.

In some cases, emphysema results from an enzyme deficiency in the lungs that destroys the elastic protein that provides the recoil during exhalation or from chronic exposure to air pollution. However, the overwhelming majority of cases, 85%, are due to smoking. Toxins in cigarettes and other tobacco products damage the lungs by stimulating leukocytes to release proteolytic enzymes that degrade lung tissue. The likelihood of developing emphysema is strongly correlated with the quantity of cigarettes smoked during a person's lifetime.

Estimating how many people have emphysema is difficult because the symptoms appear gradually, but more than 3 million people have severe cases of the disease in the U.S., and at least 15,000 people die from it each year. Emphysema is a progressive disease, worsening with time. Although medical care can sometimes slow the rate of emphysema's progress, the disease is not curable and does not get better.

## Tobacco Smoke Causes Respiratory Health Problems and Cancer

Smoking tobacco products is one of the leading global causes of death, contributing to about 430,000 deaths each year in the U.S. alone and over 5 million annually worldwide. According to the Centers for Disease Control and Prevention and the American Lung Association, people who smoke up to one pack of cigarettes each day live on average 7 years less than nonsmokers, and heavy smokers live on average 15–25 years less. Pregnant women who smoke run a high risk of their babies being born underweight, a potentially serious condition that may affect the newborn's long-term health.

Up to 85% of all new cases of lung cancer diagnosed each year are attributable to smoking, making lung cancer the leading cause of preventable death. Equally important, however, is that smoking is estimated to be responsible for nearly 30% of all cancers, including cancer of the mouth and throat, esophagus, bladder, pancreas, and ovaries. Smoking is also a leading cause of cardiovascular disease, high blood pressure, and stroke. Smoking only a few cigarettes per day increases the risk of heart disease.

Because tobacco smoke is inhaled directly into the lungs, the chemicals in the smoke can do considerable damage to lung tissue. Even adolescents who have only recently started smoking have increased mucus (phlegm) production in their airways, shortness of breath, and reduced lung growth. Thousands of chemicals, including over 40 known cancer-causing compounds, have been identified in cigarette smoke. Some of these chemicals—such as formaldehyde—are toxic to all cells. Others, like the odorless gas carbon monoxide (CO), have harmful effects on lung function in particular. CO competes with $O_2$ for binding sites in hemoglobin, thereby reducing hemoglobin saturation. Heavy smokers who smoke more than a pack of cigarettes each day may have as much as 15% less $O_2$-carrying capacity in their blood.

## 36.11 Reviewing the Concepts

- Cardiovascular disease accounts for more deaths each year in the U.S. than any other cause. Cardiovascular disorders include hypertension, atherosclerosis, coronary artery disease, angina pectoris, and myocardial infarction (MI, or heart attack) (Figure 36.25).
- Cardiovascular diagnostic techniques and treatments include cardiac angiography, balloon angioplasty, and coronary artery bypass (Figure 36.26).
- In asthma, the muscles of the bronchioles contract more than usual, increasing resistance to airflow (Figure 36.27).
- Smoking tobacco products is one of the leading global causes of death. Smoking is strongly linked to cancer, cardiovascular disease, stroke, and emphysema, in which lung tissues are severely damaged (Figure 36.28).

## 36.11 Testing Your Knowledge

1. A person with chronically elevated blood pressure has an increased risk of
   a. coronary artery disease.
   b. atherosclerosis.
   c. plaque formation in arteries.
   d. heart attack.
   e. all of the above.
2. Asthma
   a. is primarily a disease of decreased resistance in the airways.
   b. is treatable with drugs that cause bronchoconstriction.
   c. is associated with decreased diameter of bronchioles in response to certain triggers.
   d. can lead to other lung diseases such as emphysema.
   e. rarely runs in families and thus probably has no genetic cause.
3. One similarity between asthma and emphysema is that in both cases
   a. there is increased resistance to airflow into the lungs.
   b. there is significant loss of elastic tissue in the lungs.
   c. they result from smoking.
   d. there is extensive lung damage.
   e. the total surface area of the alveoli is decreased.

# Assess and Discuss

## Test Yourself

1. Hemolymph differs from blood in that it
   a. does not contain blood cells.
   b. is a mixture of fluid in blood vessels and the hemocoel.
   c. circulates through closed circulatory systems only.
   d. functions only in defense of the body and not transport.
   e. does not pass through a heart.
2. A typical hematocrit for a human is around 42%. This means that
   a. the typical fluid portion of blood is about 42% of the total volume.
   b. the leukocytes make up 42% of the blood volume.
   c. the erythrocytes make up 42% of the blood volume.
   d. the leukocytes and erythrocytes together make up 58% of the blood volume.
   e. the erythrocytes alone make up 58% of the blood volume.
3. A major advantage of a double circulation is that
   a. blood can be pumped to the upper portions of the body by one circuit and to the lower portions of the body by the other circuit.
   b. each circuit can pump blood with differing pressures to optimize the function of each.
   c. the oxygenated blood can mix with the deoxygenated blood before being pumped to the tissues of the body.
   d. less energy is required to provide nutrients and oxygen to the tissues of the body.
   e. all of the above.
4. The function of erythrocytes is to
   a. transport oxygen throughout the body.
   b. defend the body against infection and disease.
   c. transport chemical signals throughout the body.
   d. secrete the proteins that form blood clots.
   e. both a and d.
5. Considering blood flow through a closed circulation, which is the correct sequence of vessels beginning at the heart?
   a. arteriole, artery, capillary, vein, venule
   b. artery, capillary, arteriole, venule, vein
   c. vein, venule, capillary, arteriole, artery
   d. artery, arteriole, capillary, venule, vein
   e. artery, arteriole, capillary, vein, venule
6. Carbon dioxide is considered a harmful waste product of cellular respiration because it
   a. lowers the pH of the blood.
   b. lowers the $H^+$ concentration in the blood.
   c. competes with oxygen for transport in the blood.
   d. does all of the above.
   e. does a and b only.
7. The countercurrent exchange mechanism in fish gills
   a. maximizes oxygen diffusion into the bloodstream.
   b. is a less efficient mechanism for gas exchange than that in mammalian lungs.
   c. occurs because the flow of blood is in the same direction as water flowing across the gills.
   d. facilitates diffusion of carbon dioxide into the blood of the fish.
   e. facilitates diffusion of oxygen to the environment.
8. The tracheal system of insects
   a. consists of several tracheae that connect to multiple lungs within the different segments of the body.
   b. consists of extensively branching tubes that are in close contact with all the cells of the body.
   c. allows oxygen to diffuse directly across the thin exoskeleton of the insect to the bloodstream.
   d. cannot function without constant movement of the wings to move air into and out of the body.
   e. provides oxygen that is carried through the animal's body in hemolymph.

9. Which of the following factors does *not* increase the rate of breathing by influencing the chemoreceptors?
   a. an increase in $P_{CO_2}$ in the arterial blood
   b. an increase in $P_{O_2}$ in the arterial blood
   c. a decrease in the arterial blood pH
   d. a decrease in $P_{O_2}$ in the arterial blood
   e. an increase in $H^+$ concentration in the arterial blood
10. The majority of oxygen is transported in the blood of vertebrates
   a. by binding to plasma proteins.
   b. by binding to hemoglobin in erythrocytes.
   c. as dissolved gas in the plasma.
   d. as dissolved gas in the cytoplasm in the erythrocytes.
   e. by binding to hemoglobin in the plasma.

## Conceptual Questions

1. Discuss the difference between closed and open circulatory systems. What advantages does a closed circulatory system provide?
2. Explain the advantage of $CO_2$, $H^+$, and heat being major factors in decreasing the affinity of hemoglobin for oxygen.
3. **PRINCIPLES** Two principles of biology are that new properties emerge from complex interactions and structure determines function. How are these principles related to what you have learned in this chapter about hemoglobin?

## Collaborative Questions

1. Describe the cardiac cycle, and explain why heart valves must open in only one direction.
2. List the components of the mammalian respiratory system, and describe the major functions of each.

## Online Resource

# 37

# Digestive and Excretory Systems Help Maintain Nutrient, Water, and Energy Balance and Remove Waste Products from Animal Bodies

© Richard Hutchings/PhotoEdit

**A balanced meal containing many nutrients, including carbohydrates, lipids, proteins, vitamins, minerals, and water.**

## Chapter Outline



A 37-year-old man visited his physician because of chronic heartburn. The man was obese, weighing 247 pounds and standing 5 feet, 8 inches tall. He was in considerable pain, particularly after eating. His physician determined that his heartburn was directly related to his obesity. The distension of his abdomen and his practice of eating large meals increased the likelihood of stomach acid being regurgitated into the esophagus—the tube that carries food from the pharynx to the stomach. The reflux of acid into the esophagus creates the symptoms of heartburn and can lead to ulcers. The man was counseled by a nutritionist regarding a sensible weight loss program and was treated with a medication that decreases acid production in the stomach. Fortunately, weight loss can reverse most of the man's symptoms. Unfortunately, however, although this person initially lost about ten percent of his body weight within a short time, he regained it all within 2 years, at which time his condition worsened.

This case illustrates several topics covered in this chapter, including the importance of sufficient and sensible consumption of nutrients and the control of digestive processes such as acid production. In animals, a **nutrient** is any organic or inorganic substance that is consumed and is required for survival, growth, development, tissue repair, or reproduction; the process of consuming and using nutrients is called **nutrition.**

In this chapter, we will begin by looking at the types of nutrients that animals require. Next, we will discuss the diverse ways in which animals obtain, digest, and absorb those nutrients and how these processes are regulated. We then consider how energy homeostasis is maintained while an animal is fasting. We then examine how different excretory organs participate in removing waste products from an animal's body. We conclude by highlighting some major features of the vertebrate and mammalian kidney and examine how the kidneys eliminate soluble wastes and regulate homeostasis.

## 37.1 Overview of Animal Nutrition

### Learning Outcomes

1. List the major categories of nutrients consumed by animals and some of their general functions.
2. Identify four groups of essential nutrients, listing several examples of each.
3. **SCISKILLS ▶** Predict some of the consequences of a deficiency of a given nutrient.
4. Describe some examples of the diversity of animal feeding habits.

Food processing in animals occurs in four phases: ingestion, digestion, absorption, and egestion (Figure 37.1). **Ingestion** is the act of taking food into the body via a structure such as a mouth. From there, the food moves into a digestive cavity of some sort or, as shown for the vertebrate in Figure 37.1, a specialized tube called an **alimentary canal.** If the nutrients in the food are in a form that cannot be directly used by cells, they must be broken down into smaller molecules—a process known as **digestion.** This is followed by the process of **absorption,** in which ions, water, and small molecules diffuse or are transported from the digestive cavity into an animal's circulatory system or body fluids. **Egestion** (or defecation) is the process by which animals pass undigested material out of the body.

Animals require both organic and inorganic nutrients. Organic nutrients fall into five categories: carbohydrates, proteins, lipids, nucleic acids, and vitamins. These provide energy and the building blocks of new molecules or serve as coenzymes in many enzymatic reactions. Inorganic nutrients include minerals such as calcium, copper, and iron. Minerals serve many functions, including acting as cofactors in enzymatic reactions and other processes.

In this section, we will examine the various nutrients consumed by animals and some of the major functions of these molecules. We will then see that despite the enormous number of animal species and the highly varied environments in which they live, the ways in which animals obtain and ingest food can be classified into just a few categories.

### Animals Require Nutrients for Energy and the Synthesis of New Molecules

Although different species consume a wide variety of foods, all animals require the same fundamental organic molecules (**Table 37.1**; see also Chapter 3 for a discussion of the chemical nature of organic molecules).

Ingested organic molecules are used for two general purposes: to provide energy and to make new molecules. As described in Chapter 6, the bonds of organic molecules may be enzymatically broken down to release energy; this energy can be used in the synthesis of ATP, a primary energy source for all cells. Indirectly, therefore, organic molecules provide the energy required for most of the chemical reactions that occur in animals' bodies. In addition, organic molecules serve as building blocks for synthesizing new cellular molecules. For example, the muscles of animals use amino acids obtained from food to make the specialized proteins that allow their muscle fibers to contract.

### Essential Nutrients Must Be Obtained from the Diet

Animal cells can synthesize many organic molecules, but certain compounds cannot be synthesized from any ingested or stored precursor molecule. These **essential nutrients** must be obtained in the diet in their complete form. The essential nutrients can be classified into four groups:

- **Essential amino acids** are required in the diet of many but not all animals. In humans, they include isoleucine, leucine, lysine, methionine, phenylalanine, histidine, threonine, tryptophan, and valine (refer back to Figure 3.11 for the structures of the 20 amino acids in animals). Without a recurring supply of essential amino acids, protein synthesis in each cell in an animal's body would slow down or stop. Because all 20 amino acids are

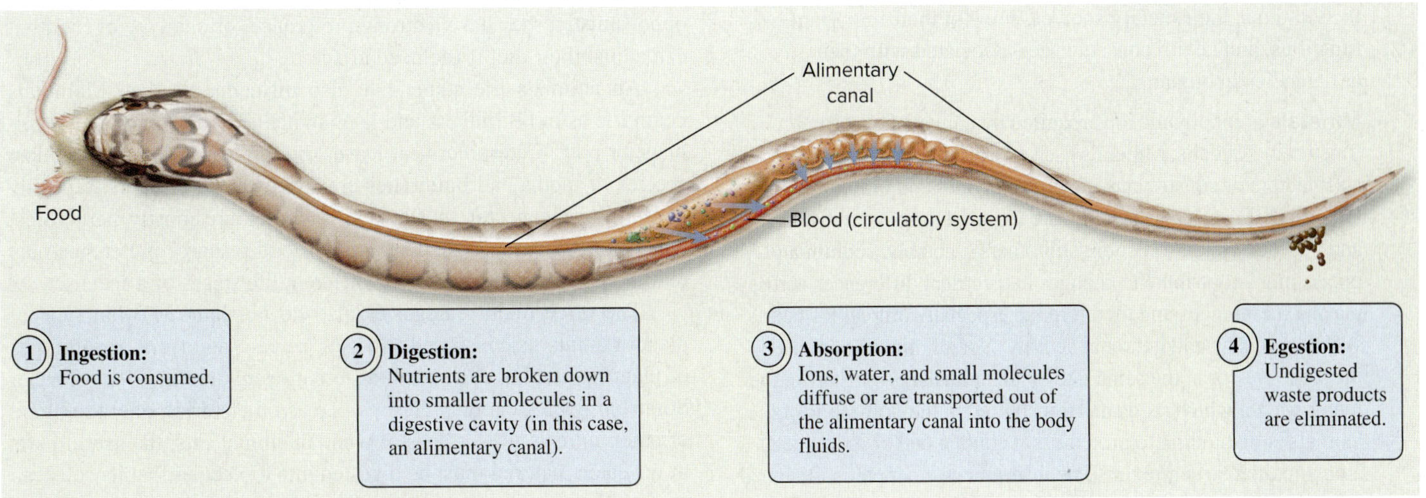

Alimentary canal
Food
Blood (circulatory system)
1 Ingestion:
Food is consumed.
2 Digestion:
Nutrients are broken down into smaller molecules in a digestive cavity (in this case, an alimentary canal).
3 Absorption:
Ions, water, and small molecules diffuse or are transported out of the alimentary canal into the body fluids.
4 Egestion:
Undigested waste products are eliminated.


**Figure 37.1** An overview of the four phases of food processing in animals, as shown in a snake.

**Table 37.1 Major Nutrients, Their Dietary Sources, and Some of Their Functions and Symptoms of Deficiency**

| Class of nutrient | Dietary sources | Functions in vertebrates | Symptoms of deficiency in humans |
|---|---|---|---|
| Carbohydrates | All food sources, especially starchy plants | Energy source; component of some proteins; source of carbon | Muscle weakness; weight loss |
| Proteins | All food sources, especially meat, legumes, cereals, and roots | Provide amino acids to make new proteins; build muscle; some amino acids used as energy source | Weight loss; muscle loss; weakness; weakened immune system; increased likelihood of infections |
| Lipids | All food sources, especially fatty meats, dairy products, and plant oils | Major component of cell membranes; energy source; thermal insulator; building blocks of some hormones | Hair loss; dry skin; weight loss; hormonal and reproductive disorders |
| Nucleic acids | All food sources | Provide sugars, bases, and phosphates that can be used to make DNA, RNA, and ATP | None; components of nucleic acids can be synthesized by cells from amino acids and sugars |

readily available in all meats but not as abundant in plants, some vegetation-eaters, such as cows, have evolved the capacity to synthesize the essential amino acids.

- **Essential fatty acids** are certain unsaturated fatty acids, such as linoleic acid (refer back to Figure 3.9), that cannot be synthesized by animal cells. Linoleic acid is vital to an animal's health because it is converted in cells to another fatty acid—arachidonic acid—which is the precursor required for production of several compounds important in many aspects of animal physiology. Such compounds include the prostaglandins, which function in pain, blood clotting, and smooth muscle contraction. Unsaturated fatty acids are abundant in plants, which provide a dietary source of essential fatty acids for many animals. Animals that never consume plants, however, may obtain their essential fatty acids from fishes or from the adipose (fat) tissue of birds and mammals.
- **Vitamins** are important organic nutrients that serve as coenzymes for many metabolic and biosynthetic reactions. Water-soluble vitamins, such as vitamin C, are not stored in the body and must be regularly ingested. Fat-soluble vitamins, such as vitamin A, are stored in adipose tissue. **Table 37.2** summarizes the vitamins, their dietary sources, some of their important functions, and health consequences associated with their deficiencies in humans.
- **Minerals** are inorganic ions required by animals for normal functioning of cells. Minerals such as iron and zinc are required as cofactors or constituents of some enzymes and other proteins. Other minerals such as calcium are required for bone, muscle, and nervous system function; still others—notably, sodium and potassium—contribute to changes in electrical differences across plasma membranes and therefore are especially critical for heart, skeletal muscle, and neuronal activity. Not all minerals are used the same way or at the same rate by all animals. For instance, the major way in which oxygen is transported in the body fluids of animals is by binding to proteins that contain one or more metal atoms in their structure. In some invertebrates, copper is the metal atom that binds oxygen, whereas in all vertebrates and in most other invertebrates, iron serves this function.

## Animals Have Very Diverse Feeding Habits

Animals share many of the same nutritional requirements, but they obtain and process those nutrients in diverse ways. Several factors determine how and when an animal obtains food, including its energetic demands, whether or not it is capable of locomotion, its local environment, and the structure and function of its digestive system. The digestive systems of **herbivores,** animals that normally eat only vegetation, contain microorganisms that assist in the digestion of cellulose, the main macromolecule of plant cell walls. By contrast, **carnivores** are primarily adapted to consume animal flesh or fluids, and **omnivores,** such as humans, eat both plant and animal products.

Although broadly useful, the three dietary categories just described are limited when considering the diversity of animal feeding habits. For example, some animals eat protists, algae, and/or fungi as a major source of food. Other animals are almost strictly carnivores at one time of year but herbivores at other times. Many nonmigratory birds, for example, feed on insects and worms during the summer but switch to eating whatever vegetation, buds, or seeds they can find during the winter. Animals like these are said to be opportunistic; they have a strong preference for one type of food but can adjust their diet if the need arises.

An animal's life stage may also influence its diet. Mammals begin life as milk-drinkers and later switch to consuming plants, animals, or both. Caterpillars eat leaves, but after metamorphosis, most species of moths and butterflies are strictly fluid-drinkers, typically consuming the nectar in flowers. Tadpoles are mostly herbivores, whereas frogs eat insects. This type of food resource preference may reduce competition between the different life stages of a species.

The three major categories also do not indicate what type of plant or animal is consumed. Grasses, cereals, and fruits are all types of plant matter, but each has a different energy and nutrient content. Some carnivores eat only flesh, whereas others drink only the blood of other animals. Regardless of what an animal eats, the useful parts of the eaten material must be digested into molecules that its cells can absorb. Next, we turn to a discussion of how animals digest their food and absorb the nutrients.

**Table 37.2 Vitamins Required by Animals***

| Class of nutrient | Dietary sources | Functions in vertebrates | Symptoms of deficiency in humans |
|---|---|---|---|
| **Water-soluble vitamins** | | | |
| Biotin | Liver; legumes; soybeans; eggs; nuts; mushrooms; some green vegetables | Coenzyme for gluconeogenesis and fatty acid and amino acid metabolism | Skin rash; nausea; loss of appetite; mental disorders (depression or hallucinations) |
| Folic acid | Green vegetables; nuts; legumes; whole grains; organ meats (especially liver, kidney, heart) | Coenzyme required for synthesis of nucleic acids | Anemia (a lower than normal number of erythrocytes in the blood); depression; birth defects |
| Niacin | Legumes; nuts; milk; eggs; meat | Involved in many oxidation-reduction reactions | Skin rashes; diarrhea; mental confusion; memory loss |
| Pantothenic acid | Nearly all foods | Part of coenzyme A, which is involved in numerous synthetic reactions, including formation of cholesterol | Burning sensation in hands and feet; GI symptoms; depression |
| Vitamin $B_1$ (thiamine) | Meats; legumes; whole grains | Coenzyme involved in metabolism of sugars and some amino acids | Beriberi (muscular weakness, anemia, heart problems, loss of weight) |
| Vitamin $B_2$ (riboflavin) | Dairy foods; meats; organ meats; cereals; some vegetables | Respiratory coenzyme; required for metabolism of fats, carbohydrates, proteins | Seborrhea (excessive oil secretion from skin glands resulting in skin lesions) |
| Vitamin $B_6$ (pyridoxine) | Meats; liver; fish; nuts; whole grains; legumes | Coenzyme for over 100 enzymes that participate in amino acid metabolism, lipid metabolism, and heme synthesis | Seborrhea; nerve disorders; depression; confusion; muscle spasms |
| Vitamin $B_{12}$ | Meats; liver; eggs; some shellfish; dairy foods | Required for red blood cell formation, DNA synthesis, and nervous system function | Anemia; nervous system disorders leading to sensory problems; balance and gait problems; loss of bladder and bowel control |
| Vitamin C (ascorbic acid) | Citrus fruits; green vegetables; tomatoes; potatoes | Antioxidant and free radical scavenger; aids in iron absorption; helps maintain healthy connective tissue and gums | Scurvy (connective tissue disease associated with skin lesions, weakness, poor wound healing, tooth decay); bleeding gums |
| **Fat-soluble vitamins** | | | |
| Vitamin A (retinol) | Liver; green and yellow vegetables; some fruits in small amounts | Component of visual pigments; regulatory molecule affecting transcription; important for reproduction and immunity | Night blindness due to loss of visual ability; skin lesions; impaired immunity |
| Vitamin D | Fish oils; fish; egg yolk; liver; synthesized in skin via sunlight | Required for calcium and phosphorus absorption from intestines; bone growth | Rickets (weakened, deformed bones) in children; osteomalacia (weak bones) in adults |
| Vitamin E | Meats; vegetable oils; grains; nuts; seeds; small amounts in some fruits and vegetables | Antioxidant; inhibits prostaglandin synthesis | Unknown, possibly skeletal muscle atrophy; peripheral nerve disorders |
| Vitamin K | Legumes; green vegetables; some fruits; some vegetable oils (olive oil, soybean oil); liver; synthesized by bacteria in the large intestine | Component of blood clotting mechanism | Reduced blood clotting ability |

*Not all animals require each of these vitamins. For example, most vertebrates with the exception of fishes and a few mammals (including humans) can synthesize vitamin C.

## 37.1 Reviewing the Concepts

- The four phases of animal nutrition are ingestion, digestion, absorption, and egestion. Animals require organic nutrients—carbohydrates, proteins, lipids, nucleic acids, and vitamins—and inorganic nutrients (water and minerals) (Figure 37.1).
- Essential nutrients consist of vitamins, minerals, and those amino acids and fatty acids that cannot be synthesized by animals (Tables 37.1, 37.2).
- Herbivores eat only vegetation, carnivores consume animal flesh or fluids, and omnivores eat both plant and animal products.

## 37.1 Testing Your Knowledge

1. Which of the following is *not* an organic nutrient?
   a. lipid  c. mineral  e. protein
   b. vitamin  d. carbohydrate
2. Which statement is *false*?
   a. Water is an inorganic nutrient.
   b. All nutrients can be synthesized in animal cells.
   c. Copper and potassium are examples of minerals consumed by animals.
   d. In addition to other functions, carbohydrates, lipids, and proteins can all supply energy.
   e. An animal's diet depends in some cases on its life stage.

## 37.2 General Principles of Digestion and Absorption of Nutrients

### Learning Outcomes

1. Describe the general structure of an alimentary canal.
2. **SCISKILLS** ▶ Make predictions about the function of parts of the alimentary canal based on their structural adaptations.
3. Distinguish between passive and active absorption of food.

Once food has been ingested, some of the nutrients from the food must be broken down (digested) so that they can be absorbed by the cells of the digestive system. In this section, we will examine some of the major principles of digestion and absorption in animals, beginning with where digestion takes place.

### Digestion Usually Occurs Extracellularly

Food is digested either inside cells (intracellularly) or outside cells (extracellularly). Intracellular digestion occurs only in some very simple invertebrates such as sponges and single-celled organisms and to a limited extent in cnidarians. It involves using phagocytosis to bring food particles directly into a cell, where the food is segregated from the rest of the cytoplasm in food vacuoles. Once inside these vacuoles, hydrolytic enzymes digest the macromolecules in food into monomers (the building blocks of polymers), which then are moved out of the vacuole to be used directly by that cell. Intracellular digestion cannot support the metabolic demands of an active animal for long, because only tiny bits of food can be engulfed and digested at one time. It also does not provide a mechanism for storing large quantities of food so that an animal can digest it slowly while going about its other activities.

Most animals digest food via extracellular digestion in a cavity of some sort. Extracellular digestion protects the interior of the cells from the actions of hydrolytic enzymes and allows animals to consume large prey or vegetation. Food enters the digestive cavity, where it is stored, slowly digested, and absorbed gradually over long periods of time, ranging from hours (for example, after a human eats a pizza) to weeks (after a python eats a gazelle).

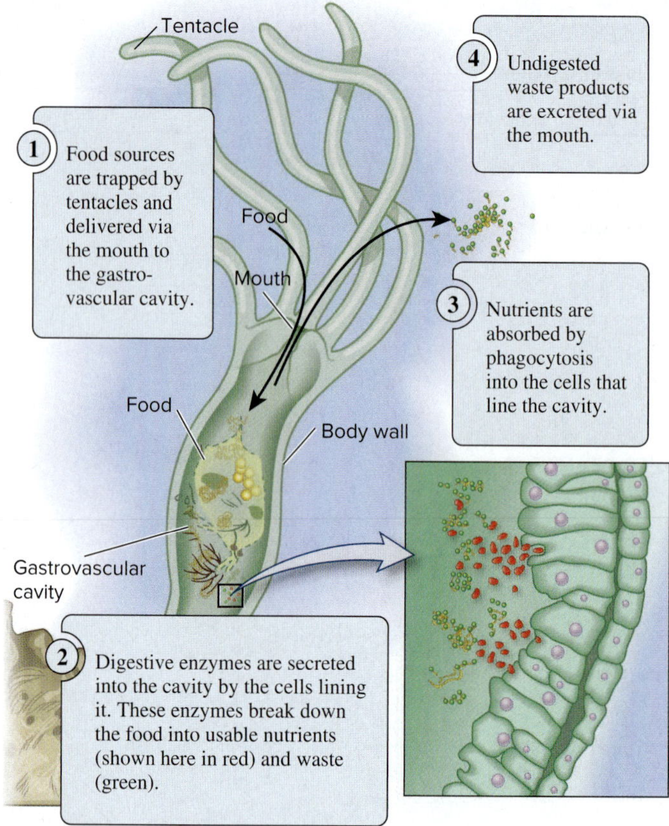


**Figure 37.2 Extracellular digestion in a gastrovascular cavity.** In animals with gastrovascular cavities, such as the cnidarian illustrated here, most digestion occurs extracellularly within the cavity.

*What is an advantage of extracellular digestion?*

In the simplest form of extracellular digestion—seen in invertebrates such as cnidarians and flatworms—the digestive cavity has one opening that serves as both an entry and an exit port (Figure 37.2). The digestive cavity of these animals is called a **gastrovascular cavity** because not only does digestion occur within it but fluid movements in the cavity also serve as a circulatory—or vascular—system that distributes digested nutrients throughout the animal's body. Food within a gastrovascular cavity is partially digested by enzymes that are secreted into the cavity by the cells lining the cavity. As the food particles become small enough, they are engulfed by the lining cells and further digested intracellularly. Undigested material that remains in the gastrovascular cavity is expelled.

### Most Digestive Cavities Are Tubes with Specialized Regions and Openings at Opposite Ends

In contrast to the gastrovascular cavities of simple invertebrates, all other animals possess digestive systems that consist of a single elongated tube, usually called an alimentary canal, with an opening at each end, through which food passes from one end to the other (see Figure 37.1). Along its length, the tube contains smooth muscle as part of its structure. When the muscle contracts, it helps churn up the ingested food so that it is mechanically broken into smaller fragments. The canal is lined on its interior surface by a layer of epithelial cells. These cells synthesize digestive enzymes and secrete other factors into the hollow cavity, or lumen, of the alimentary canal; they also secrete hormones into the blood that help regulate digestive processes. The cells are also involved with transporting digested material out of the canal.

The alimentary canal has several specialized regions along its length that vary according to species. Because of these specializations, digestive processes requiring low pH can be segregated from those requiring higher pH, and undigested food can be stored in one region while digestion continues in another area. The ability of some animals to store food in a stomach, for example, allows them to eat less frequently, leaving time for other activities.

### Absorption of Food May Be Passive or Active

Once food is digested, the nutrients must be absorbed across the epithelium of specialized portions of the alimentary canal. This occurs in different ways, either by simple or facilitated diffusion or by active transport. Small, hydrophobic molecules such as fatty acids diffuse down concentration gradients across the epithelium. By contrast, minerals are hydrophilic; as you learned in Chapter 5, hydrophilic substances do not readily cross the lipid-rich plasma membranes of cells. Instead, minerals are usually actively transported across the membranes of the epithelial cells of the canal by ATP-dependent ion transporters. In other cases, small, hydrophilic organic nutrients are transported by facilitated diffusion or secondary active transport, usually with $Na^+$ (refer back to Figure 5.16b for a description of secondary active transport).

After nutrients enter the epithelial cells of the alimentary canal, the cells use some of the nutrients for their own requirements. Most of the nutrients, however, are transported across the basolateral surface of the epithelial cells (the side opposite the lumen of the canal), where they can diffuse into nearby blood or lymphatic vessels and circulate to the other cells of the body. Thus, nutrients enter the alimentary canal in food, are digested within the canal into smaller molecules that can be transported into epithelial cells, and from there are released into the blood, where they can reach all of the body's cells. In the special case of water, osmotic gradients established by the transport of ions and other nutrients out of the epithelial cells draw water by osmosis from the canal, across the epithelium, and from there into the blood.

The mechanisms that activate and control the digestive and absorptive functions of the alimentary canal have been extensively studied in vertebrates and have great importance for human health. In the rest of this chapter, we will explore the structure and function of the digestive systems of vertebrates.

## 37.2 Reviewing the Concepts

- Intracellular digestion occurs in single-celled organisms and some simple invertebrates. Most animals digest food via extracellular digestion. Flatworms and cnidarians have a gastrovascular cavity with one opening that serves as both an entry and exit port (Figure 37.2).
- Nearly all other animals have an alimentary canal, open at both ends and segregated into specialized regions, through which food passes from one end to the other.
- Once food has been digested, the nutrients must be absorbed via diffusion, facilitated diffusion, or primary or secondary active transport.

## 37.2 Testing Your Knowledge

1. Which is correct?
   a. Intracellular digestion commonly occurs in most vertebrates.
   b. Absorption of all nutrients requires active transport.
   c. Alimentary canals have two openings, whereas gastrovascular cavities have only one.
   d. Extracellular, but not intracellular, digestion requires enzymes.
   e. Most minerals are absorbed by simple diffusion.

## 37.3 Vertebrate Digestive Systems

**Learning Outcomes**

1. Describe the structure of vertebrate digestive systems.
2. Describe how food moves through regions of the alimentary canal and how different parts contribute to the processes of digestion and absorption.
3. **SCISKILLS ▸** Predict problems that might arise if saliva or acid cannot be secreted into the mouth or stomach, respectively.
4. Outline how microorganisms can help digest cellulose in ruminants and other herbivores.
5. **SCISKILLS ▸** Propose a hypothesis explaining why most human adults are lactose-intolerant to some degree.
6. Describe the mechanisms of digestion and absorption of carbohydrates, proteins, and fats in vertebrates and the importance of enzymes in these processes.
7. Explain why some nutrients do not require digestion prior to being absorbed.
8. Name three major hormones important for the regulation of digestion in vertebrates, and describe the function of each.

The vertebrate **digestive system** consists of the alimentary canal—also known as the **gastrointestinal (or GI) tract**—plus several accessory glands and organs (Figure 37.3). The human GI tract consists of the mouth, pharynx, esophagus, stomach, small and large intestines, and anus. The accessory structures, not all of which are found in all vertebrates, are the tongue, teeth, salivary glands, liver, gallbladder, and pancreas. The differences between the digestive systems of various vertebrates reveal much about their respective feeding behaviors and mechanisms. In this section, we will look at some of the most important differences as we discuss the form and function of each part of the vertebrate digestive system.

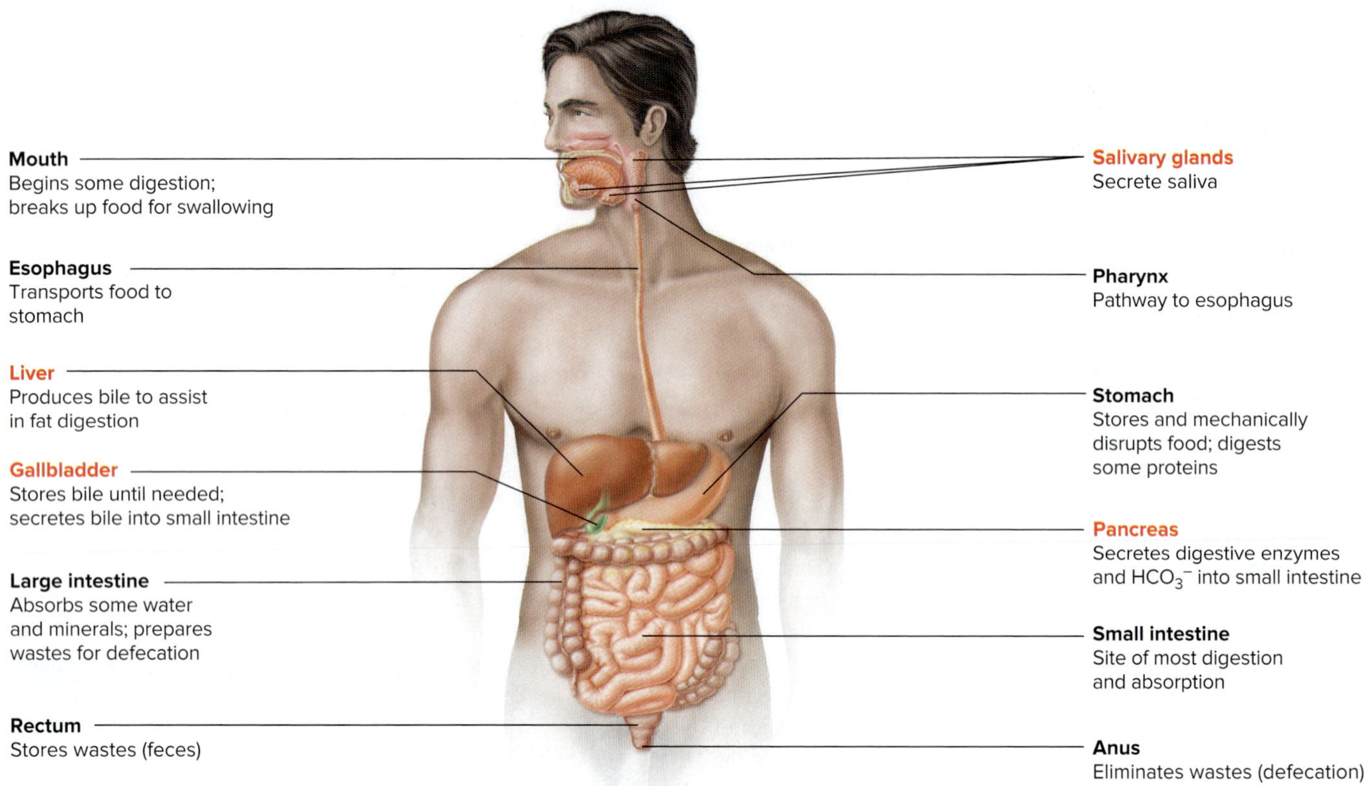


**Figure 37.3** **A vertebrate digestive system, as shown in the human.** This figure shows the organs of the gastrointestinal tract (labeled in black) and the major accessory structures (labeled in red). Not all vertebrates share identical features of the digestive system; for example, many birds and some mammals lack a gallbladder.

## The Vertebrate Alimentary Canal Is Divided into Functional Regions

The alimentary canal is one continuous tube that changes in structure and function along its length. The first section, at the anterior end, functions primarily in the ingestion of food. It contains the mouth, salivary glands, pharynx (throat), and esophagus. The middle section, which functions in the storage and initial digestion of food, contains one or more food-storage or digestive organs, including the crop, gizzard, and stomach, depending on the species. This section also contains the upper part of the small intestine—where most of the digestion and absorption of food take place—and accessory structures that connect with the intestine, including the pancreas, liver, and gallbladder. The final, posterior part of the canal functions in some absorption and the elimination of nondigestible waste. It consists of the remainder of the small intestine and, in most vertebrates other than fishes, a large intestine. Undigested material is defecated through an opening called an **anus,** or in many amphibians, reptiles, and birds, a cloaca (a common opening for the digestive and urogenital tracts).

From the middle of the esophagus to the anus or cloaca, the GI tract has the same general structure, with a lumen that is lined by a layer of epithelial cells. Included in the epithelial layer are secretory cells that release a protective coating of mucus into the lumen of the tract, and for this reason the epithelial layer together with some underlying tissue is also referred to as the **mucosa.** Other cells in the mucosa release hormones into the blood in response to the presence of food; these hormones control certain aspects of the digestive process. Passing through the mucosa are ducts from secretory glands that release acid, enzymes, water, and ions into the lumen. From the stomach onward, the epithelial cells are linked along the edges of their apical surfaces by tight junctions (refer back to Figure 5.23) that prevent digestive enzymes and undigested food from moving between the cells and out of the alimentary canal. In some places, such as the small intestine, the apical surface of the mucosa is highly convoluted, a feature that increases the surface area available for digestion and absorption.

The epithelial cell layer is surrounded by layers of tissue made up of smooth muscles, neurons, connective tissue, and blood vessels. The neurons are activated by signals coming from regions of the central nervous system that respond to the sight and smell of food. The neurons are also activated directly by the presence of food in the gastrointestinal tract. Contraction of the smooth muscles is controlled by these neurons and results in mechanical mixing of the contents within the stomach and intestine. This helps speed up digestion and brings digested foods into contact with the epithelium to facilitate absorption.

### Food Processing and Carbohydrate Digestion Begin in the Mouth

Ingestion and the start of digestion begin once food enters the mouth. In terrestrial vertebrates, the presence of food stimulates salivary glands in and around the mouth, cheeks, tongue, and throat to produce a flow of saliva—a watery fluid containing proteins, mucus, and antibacterial substances. Fishes, which lack true salivary glands, secrete mucus from specialized cells in their mouth and pharynx. Saliva has several functions, not all of which pertain to all vertebrates:

- It moistens and lubricates food to facilitate swallowing.
- It dissolves food particles to facilitate the ability of sensory cells in taste buds to taste food.
- It kills ingested bacteria with a variety of antibacterial compounds, including antibodies.
- It initiates digestion of carbohydrates through the action of a secreted enzyme called salivary amylase.

Digestion is the least important of these functions in most vertebrates, only a few of which produce salivary amylase. In humans and other primates, salivary amylase is present and accounts for a very small percent of total carbohydrate digestion. Depending on how much salivary amylase you produce, you may be able to detect the action of amylase in your own saliva with a simple test. If you chew on a starchy soda cracker and leave it in your mouth for a while, you may notice it begins to taste sweet. That is because some of the polysaccharides (polymers of glucose) in the cracker have been digested to sweet-tasting disaccharides.

The other functions of saliva, however, are very important. For example, imagine trying to chew and swallow food with a completely dry mouth. Also, the antibiotic properties of saliva help cleanse the mouth. In people who have had cancerous salivary glands removed, the teeth and gums often become so diseased that tooth loss occurs.

### Peristalsis Moves Swallowed Food Through the Esophagus to a Storage Organ

As food is swallowed, it moves into the next segments of the alimentary canal, the **pharynx** (throat) and **esophagus.** These structures, although not contributing to digestion or absorption, serve as a pathway to storage organs such as the stomach.

In the pharynx, swallowing begins as a voluntary action but continues in the esophagus by the process of **peristalsis**—rhythmic, spontaneous waves of smooth muscle contraction that begin near the mouth and end at the stomach. When most vertebrates eat, the mouth and stomach are roughly horizontal with respect to each other. In fact, the head of a terrestrial grazing animal is at times lower than its stomach when swallowing. The wavelike action of peristalsis ensures that swallowed food is moved toward the stomach and does not remain in the esophagus or even move backward into the mouth if the head is lowered.

In some animals, food does not move directly from the esophagus to the stomach but first enters a storage organ called the **crop,** which is a dilation of the lower esophagus (Figure 37.4). Crops are found in most birds (and many invertebrates, including insects and some worms). Food is stored and softened by watery secretions in the crop, but little or no digestion occurs there. Because they process large amounts of tough food, birds that eat primarily grains and seeds have larger crops than do birds that eat insects and worms. The material that birds regurgitate to their young comes from the crop.

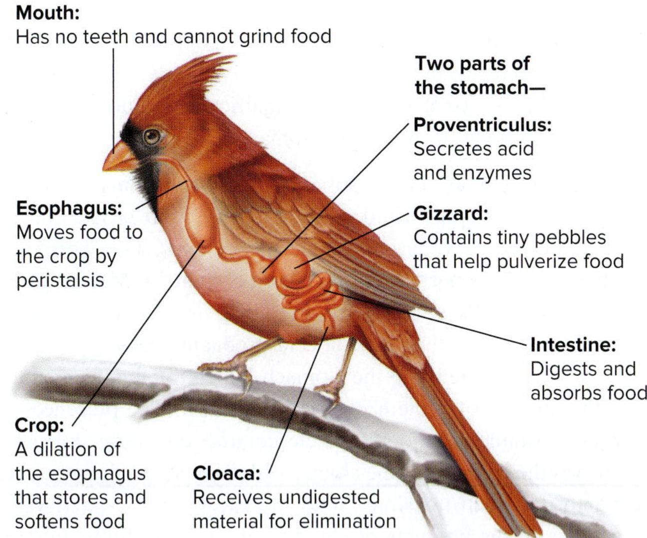


**Figure 37.4 The alimentary canal of birds.** The avian alimentary canal contains specialized regions for storing and softening food (the crop) and pulverizing food (the gizzard). The gizzard and proventriculus constitute the stomach. Undigested material is eliminated through the cloaca.

### Food Processing Continues and Protein Digestion Begins in the Stomach

Once food passes through the esophagus (and crop, if present), it reaches the stomach. A **stomach** is a muscular, saclike organ that most likely evolved as a means of storing food. In addition to its storage function, the muscular nature of the stomach helps mechanically break up large chunks of food into smaller, more easily digestible fragments. The stomach also partially digests some of the macromolecules in food and regulates the rate at which the contents empty into the small intestine.

Glands within the stomach wall known as gastric glands secrete hydrochloric acid (HCl) and an inactive molecule called pepsinogen into the stomach lumen. The acid serves four major functions in the stomach lumen:

- It converts pepsinogen into an active enzyme called **pepsin,** which begins the digestion of protein. (Stomach gland cells do not make pepsin itself because, if they did, it would digest the cells' own proteins.)
- It kills many of the microorganisms that may have been ingested with food.
- It helps dissolve the particulate matter in food.

- It alters the ionization of polar molecules, especially proteins. This disrupts the structural framework of the tissues in food and makes the proteins more accessible to the digestive actions of pepsin. (In contrast to proteins, no significant digestion of other organic molecules occurs in the stomach.)

After a period of time, the pepsin and HCl in the stomach reduce food particles to a pulpy fluid called **chyme.** Acid production is controlled by the nervous and endocrine systems. One hormone that is particularly important in this regard is **gastrin.** Molecules in chyme stimulate cells in the stomach epithelium to release gastrin into the blood. Gastrin then acts on acid-producing cells to increase the rate of HCl secreted into the stomach lumen roughly in proportion to the amount of protein ingested in a meal. The presence of gastrin also stimulates smooth muscle contraction in the stomach, which mixes the contents of the chyme with the acid and helps move chyme into the small intestine. With the exception of water, very little or none of the products of chyme can cross the epithelium of the stomach; therefore, little or no absorption of nutrients occurs in the stomach.

In birds, the stomach is divided into two parts: the proventriculus and the gizzard (see Figure 37.4). The **proventriculus** is the glandular portion of the stomach that secretes acid and pepsinogen. Partially digested and acidified food then moves to the **gizzard,** a muscular structure with a rough inner lining that grinds food into smaller fragments. The gizzard contains sand or tiny stones swallowed by the bird. The gritty sand and stones take the place of teeth and help mash and grind ingested food. Grain-eating birds, particularly chickens and other fowl, many passerines (or perching birds), and pigeons and doves, generally have more muscular gizzards than do insectivorous birds, because of the difficulty in breaking down plant matter.

## The Stomach of Some Herbivores Is Part of a Complex Structure That Allows the Animals to Process Cellulose

Cellulose, the main macromolecule of plant cell walls (refer back to Figures 3.6, 4.34, 28.10, and 28.11), is a very important part of the diet of herbivores and many omnivores. In human nutrition, cellulose is also called fiber or roughage and is useful in eliminating solid wastes. The human digestive system is not equipped with the enzymes to digest cellulose, and it therefore passes through the system intact. **Ruminants** (sheep, goats, llamas, and cows) also lack the enzymes, but they are able to digest cellulose with the help of microorganisms living within their GI tract. The microorganisms break down the cellulose into monosaccharides that can be absorbed along with other by-products of microbial digestion, such as fatty acids and some vitamins. In this way, bacteria and protists predigest the food, and the animal absorbs the breakdown products.

Ruminants have a system of several chambers, beginning with three outpouchings of the lower esophagus composed of the rumen, reticulum, and omasum, followed by a true stomach called an abomasum (Figure 37.5). The rumen and reticulum contain the microorganisms that digest cellulose, and the omasum absorbs some of the water and ions released from the chewed and partially digested food. The tough, partially digested food (called the cud) is occasionally regurgitated, rechewed, and swallowed again. Eventually, the partially digested food, the microorganisms, and the useful by-products of microbial digestion reach the abomasum, which secretes the acid and proteases typical of the vertebrate stomach. From the abomasum, the material passes to the intestines, where digestion and absorption are completed. The microorganisms are killed by the stomach acid, digested, or eliminated. Some microorganisms remain in the rumen and quickly multiply to replenish their populations, ensuring that a well-balanced mutualistic relationship is maintained.

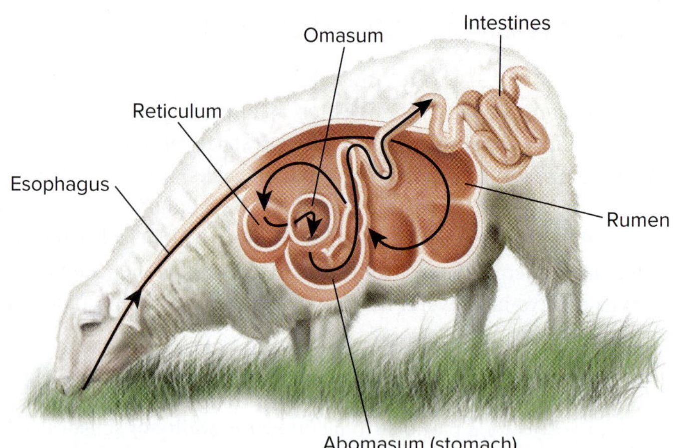


**Figure 37.5 Digestive system of a ruminant.** Ruminants have a complex arrangement of three modified pouches arising from the esophagus: the rumen, reticulum, and omasum. The rumen and reticulum act as storage and processing sites (in large ruminants, the rumen may store up to 95 L of undigested food); the omasum absorbs some water and ions. Digestion by acid and pepsin takes place in the abomasum, which connects with the intestines.

## Most Digestion and Absorption Occur in the Small Intestine

Most digestion of food and absorption of nutrients occurs in the **small intestine**—the portion of the alimentary canal that leads from the stomach to the large intestine (or, in animals without a large intestine, directly to the anus or cloaca). Hydrolytic enzymes break down macromolecules of organic nutrients into monomers. Some of these hydrolytic enzymes are located in the apical membrane of the intestinal epithelial cells where they face the intestinal lumen; others are secreted by the pancreas and enter the lumen through a duct. The products of digestion are absorbed across the epithelial cells and enter the blood. Vitamins and minerals, which do not require enzymatic digestion, are also absorbed in the small intestine. Water is absorbed by osmosis from the small intestine in response to the movement of nutrients across the intestinal epithelium.

The ability of the small intestine to carry out the bulk of digestion and absorption is aided by mucosal infoldings and specializations along its length. Finger-like projections known as **villi** (singular, villus) extend into the lumen of the vertebrate small intestine (Figure 37.6). The surface of each villus is covered with a layer of epithelial cells with plasma membranes that have many small projections called **microvilli,** known collectively as a **brush border.**

## Biology Principle

### Structure Determines Function

Note the arrangement of muscle layers in the small intestine. Some are oriented in circles around the lumen, whereas others are lengthwise. This structure allows the intestine to be squeezed and shortened, not necessarily at the same time, maximizing the mixing and churning of the lumen contents. Also, the structure of the villi and microvilli provide extensive surface area in a compact volume, increasing the efficiency of digestion and absorption.

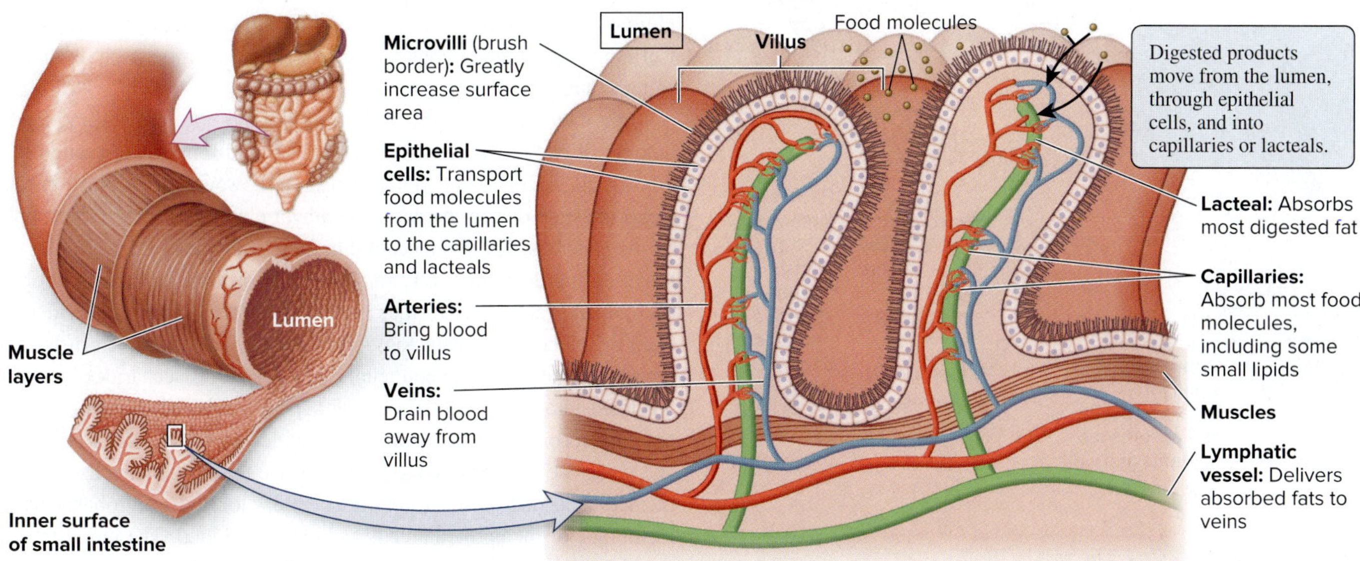


**Figure 37.6 The specialized arrangement of tissues in the small intestine.** The small intestine is folded into numerous villi, which increase the surface area for digestion and absorption. Within each villus are capillaries and lymphatic vessels (lacteals), into which absorbed nutrients are transported. The epithelial cells of the villi have extensions from their surface called microvilli. The microvilli constitute the brush border of the intestine and greatly add to the total surface area.

The combination of folded mucosa, villi, and microvilli increases the small intestine's surface area enormously. This means that the likelihood of an ingested food particle encountering a digestive enzyme and being absorbed across the epithelium is very high, so digestion and absorption proceed more rapidly.

The center of each intestinal villus is occupied by capillaries, the smallest blood vessels in the body, and by a type of vessel called a lacteal, which is part of the lymphatic system (see Figure 37.6). Most nutrients are absorbed directly into the capillaries and from there into veins. Most of the fat absorbed in the small intestine, however, exists as bulky, protein-coated particles that are too large to enter capillaries. Consequently, absorbed fat enters the larger, wider lacteals. Material absorbed by the lacteals eventually empties into large veins in the circulatory system.

### The Pancreas Secretes Substances into the Small Intestine That Aid Digestion

As the chyme moves through the small intestine, the **pancreas** secretes substances that flow via ducts into the first portion of the intestine, the **duodenum** (Figure 37.7). The pancreas has numerous functions, but here we will focus on those that are directly involved in digestion.

The arrival of chyme into the duodenum triggers the release of two hormones from intestinal epithelial cells. The first, cholecystokinin (CCK), enters the blood and stimulates the pancreas to secrete digestive enzymes into the intestine, including those that break down large molecules–proteins, lipids, carbohydrates, and nucleic acids–into smaller units. It also relaxes a sphincter, a circular ring of smooth muscle, that synchronizes the entry of material from the pancreatic duct into the intestine with the presence of chyme in the intestine. The second hormone, secretin, stimulates other cells of the pancreas to secrete bicarbonate ions ($HCO_3^-$) into the small intestine. The $HCO_3^-$ neutralizes the acidity of chyme, which would otherwise inactivate the pancreatic enzymes in the small intestine and could damage the intestinal epithelium. Thus, through the actions of CCK and secretin, the secretion of digestive enzymes and $HCO_3^-$ into the intestine is precisely coordinated with the arrival of chyme.

The digested nutrients, along with water, are absorbed across the plasma membranes of the brush border cells. Peristalsis slowly propels the remaining contents through the posterior portions of the small intestine, called the jejunum and the ileum, where further absorption occurs.

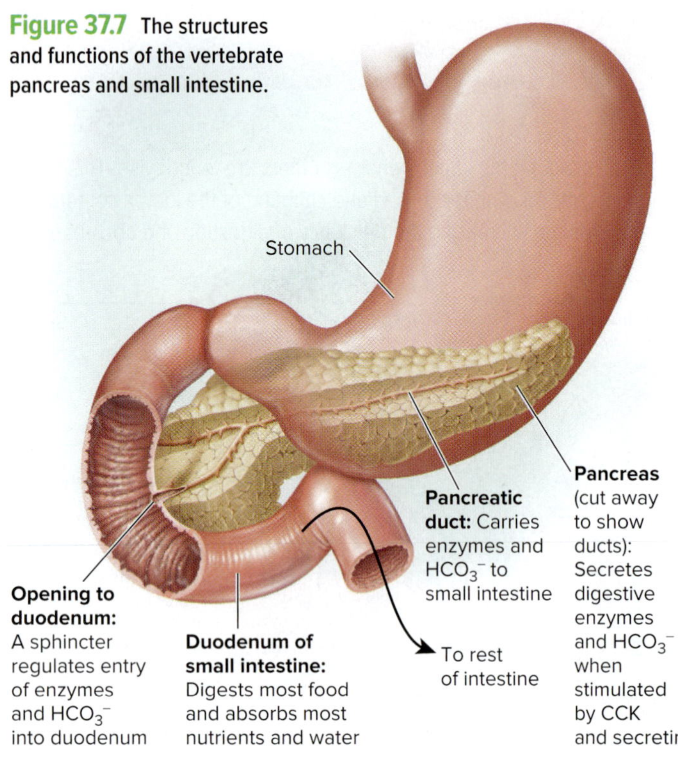


**Figure 37.7** The structures and functions of the vertebrate pancreas and small intestine.

## Carbohydrates Are Digested by Amylase and Brush Border Enzymes

In omnivores such as humans, most of the ingested carbohydrates are the polysaccharides starch and cellulose from plants and glycogen from animals (refer back to Figures 3.4–3.6 for the structures of carbohydrates). The remainder consists of simple carbohydrates, such as the monosaccharides fructose and glucose in fruit, and disaccharides (combinations of two monosaccharides), such as lactose in milk. Humans also add the disaccharide sucrose (table sugar) to their food. Certain other animals consume sucrose from sources such as maple sap and sugarcane.

As mentioned earlier, depending on the species, a very small amount of carbohydrate is digested in the mouth by salivary amylase. Almost all glycogen and starch digestion, however, takes place in the small intestine due to the action of amylase secreted into the intestine by the pancreas. The products of digestion by amylase are molecules of the disaccharide maltose (Figure 37.8). Maltose, along with any ingested sucrose and lactose, is further broken down into monosaccharides—fructose, glucose, and galactose—by enzymes located on the brush border of the small intestine epithelial cells. The monosaccharides are then absorbed into the epithelial cells either by facilitated diffusion (fructose) or secondary active transport coupled to sodium ions (glucose and galactose). Monosaccharides then leave the epithelial cells by way of facilitated diffusion through transporters located in the basolateral membrane of the epithelial cells and enter the blood. The bloodstream distributes the monosaccharides and other absorbed nutrients to the cells of the body.

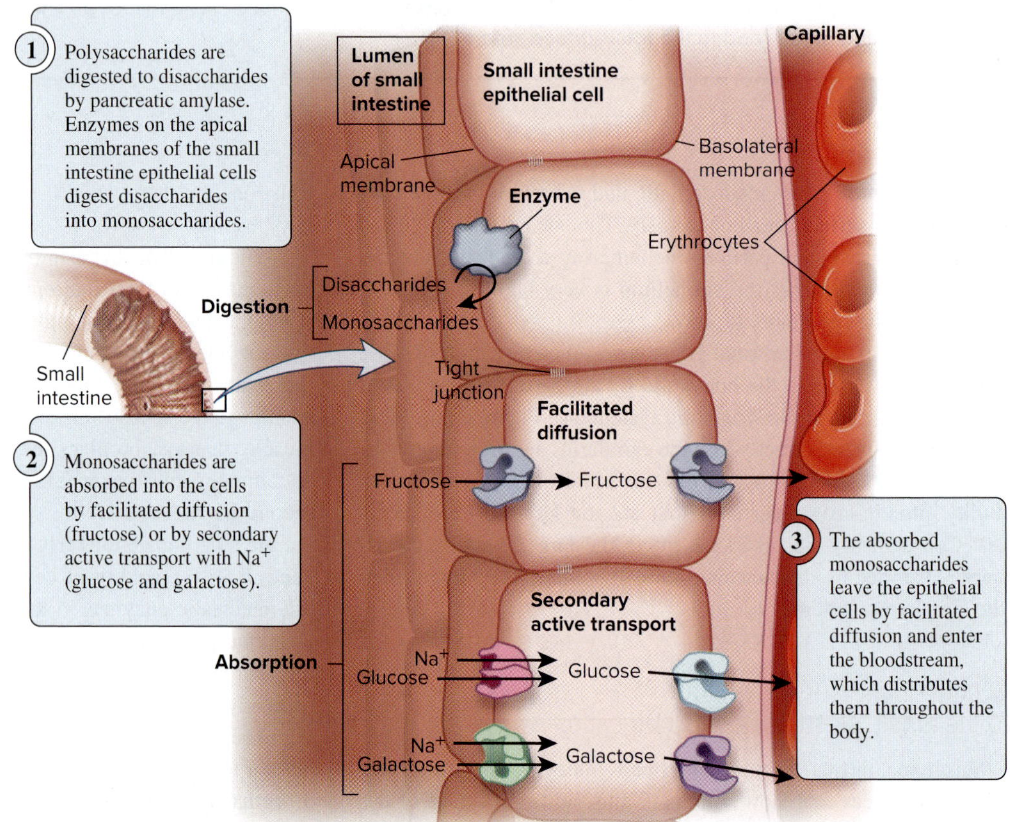


**Figure 37.8** Digestion and absorption of carbohydrates in the small intestine. In this and subsequent figures, the microvilli (brush border) are omitted for simplicity.

**Concept Check:** *The absorption of many nutrients requires active transport. What can we conclude from that about the energetic cost of absorbing the nutrients from a meal?*

**BioConnections:** *Are the transport processes shown here, including facilitated diffusion and active transport, unique to animal cells? Refer back to Figure 30.13 for help.*

Do you feel ill after drinking milk or consuming certain dairy products? If so, you may be among the vast majority of people who cannot adequately digest lactose, the chief disaccharide in milk. A small percentage of humans, however, retain the ability to fully digest lactose throughout life. Next, we'll examine this phenomenon along with its genetic and evolutionary basis.

## EVOLUTIONARY CONNECTIONS

### Genetics Explains Lactose Intolerance

One of the defining features of mammals is that they produce and consume milk, a nutrient-rich solution containing all the nutrients required by their young offspring. With rare exceptions, the milk of all mammals contains the dissacharide lactose. Lactose is digested by the brush border enzyme lactase, which cleaves lactose into glucose and galactose. Humans with lactose intolerance cannot adequately digest lactose because their lactase is inactive, absent, or present in small amounts.

Milk is the sole food of most mammals shortly after birth and the primary food for various lengths of time thereafter until weaning, the transition from consuming mother's milk to eating a diet of solid foods. Once weaned, mammals never again drink milk, except, of course, for humans.

The only dietary source for lactose is milk, and therefore it is not surprising that older mammals lose the ability to digest this disaccharide. This occurs because the gene that encodes the enzyme lactase is shut off around the age of weaning or shortly thereafter. The developmental mechanisms that turn off lactase production and activity are not firmly understood, but they are known to involve decreased transcription of the lactase gene.

If an adult mammal were to drink milk, the undigested lactose would remain in the GI tract. As a result, water that normally would be absorbed by osmosis with the digested monosaccharides formed from lactose would also remain in the GI tract. Farther along the alimentary canal, microorganisms in the large intestine digest some of the lactose for their own use and, in the process, release by-products such as hydrogen and other gases. The combination of water retention and bacterial action results in GI symptoms such as diarrhea, gas, and cramps.

About 90% of the human population cannot fully digest lactose after early childhood and experience these symptoms to some degree (refer back to Figure 22.22). In other words, lactose intolerance is not a disorder; it is a common condition that evolved in all mammals, including humans. Why, then, are some adults able to consume dairy products without getting ill? Researchers have discovered that the ability of human adults to digest lactose is clearly linked to genetic background. Lactose intolerance in humans is an example of a polymorphism, a genetic trait that varies among people. Other than individuals who trace their ancestry to northern Europe and a few isolated regions in Africa, nearly every population of humans shows a considerable degree of lactose intolerance, reaching nearly 100% in most of south and east Asia. One hypothesis for this phenomenon appears to be a behavioral and cultural change that occurred in Neolithic times, when certain populations domesticated cattle and added cow's milk to their diet. Adults who were able to digest lactose—due to a mutation affecting the expression of the lactase gene—enjoyed a selective advantage and tended to thrive and pass on their genes more frequently than those whose digestive systems could not handle milk.

Recently, Finnish investigators have uncovered two single-nucleotide changes located in regulatory sites that control the expression of the lactase gene in humans. These changes are associated with lactase expression that continues long after weaning. People carrying these mutations are lactose-tolerant. By comparison, adults who lack these mutations can consume dairy products only in small amounts or not at all because they do not express the lactase enzyme. These observations suggest that natural selection may have resulted in certain populations carrying mutations that caused the lactase gene to continue to be expressed long after weaning. Eventually, a substantial percentage of people in those regions of the world retained sufficient intestinal lactase to use milk as a staple food.

### Most Protein Digestion Occurs in the Small Intestine by the Actions of Proteases

As mentioned earlier, protein digestion begins in the stomach. The partially digested protein in the chyme coming from the stomach is further digested in the small intestine by several proteases, including **trypsin** (Figure 37.9). The pancreas secretes this enzyme as an inactive precursor called trypsinogen, which prevents the active enzyme from digesting the pancreas itself. Once trypsinogen enters the small intestine, it is cleaved into the active molecule trypsin by another enzyme, called enterokinase (also called enteropeptidase), located on the apical membranes of the intestinal cells.

The polypeptide fragments produced by trypsin and other proteases are further digested into individual amino acids by the actions of proteases located on the membranes of the brush border in the small intestine. These proteases cleave off one amino acid at a time from the N-terminus and C-terminus of polypeptides. Individual amino acids then enter the epithelial cells by secondary active transport coupled to sodium ions. Amino acids leave these cells and enter the blood by facilitated diffusion across the basolateral membrane. As with carbohydrates, protein digestion and absorption are largely completed in the duodenum.

### Lipid Digestion and Absorption in the Small Intestine Are Facilitated by Bile

Most ingested lipid is in the form of triglycerides, which are too large to cross the intestinal epithelium and must be broken down. This process occurs almost entirely in the small intestine. The major digestive enzyme in this process is **pancreatic lipase,** secreted with the other digestive enzymes by the pancreas into the small intestine (see Figure 37.7). The lipase reaction catalyzes the splitting of bonds in triglycerides, producing two free fatty acids and a monoglyceride:

$$\text{1 Triglyceride} \longrightarrow \text{2 Free fatty acids} + \text{1 Monoglyceride}$$

Triglycerides, however, are poorly soluble in water and aggregate into large lipid droplets, as you can see if you shake a salad dressing

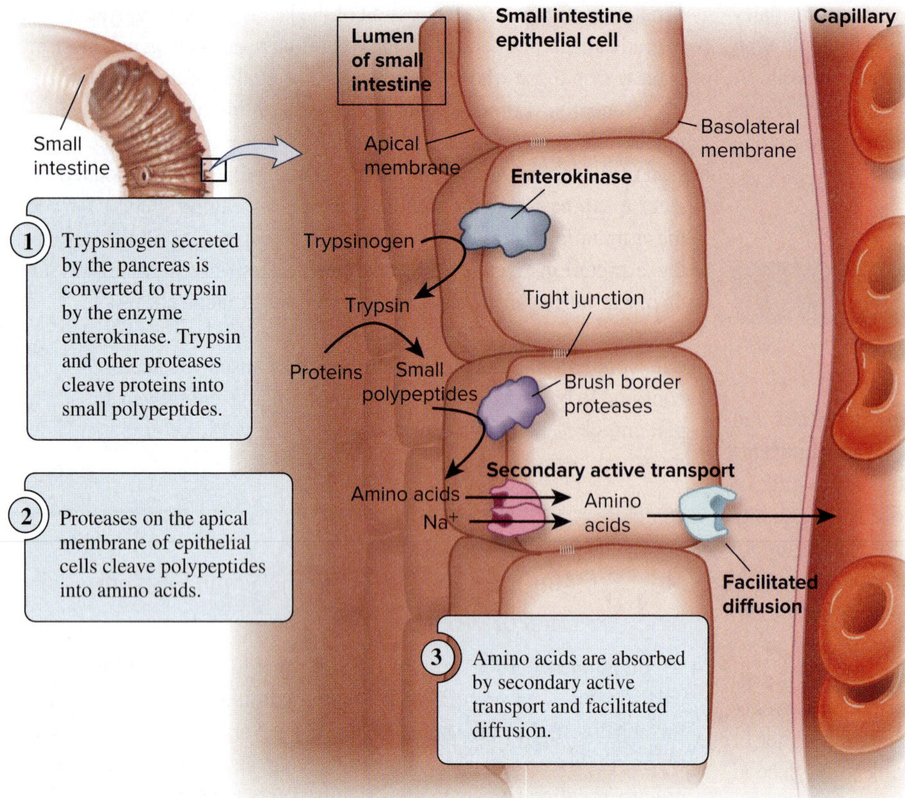


**Figure 37.9** Digestion of proteins and absorption of amino acids in the small intestine.

made of oil and vinegar. Because lipase is a water-soluble enzyme, its digestive action in the small intestine can take place only at the surface layer of a lipid droplet. The rate of digestion is substantially increased by the process of emulsification, which disrupts the large lipid droplets into many tiny droplets, increasing their total surface area and exposure to the lipase. This occurs partly by the muscular contractions of the stomach and small intestine, which mechanically break apart the fat droplets.

To prevent the smaller droplets from coalescing, however, something else is needed. The liver secretes a mixture of substances called bile (Figure 37.10). **Bile** contains $HCO_3^-$, cholesterol, phospholipids, a number of organic wastes, and cholesterol-derived substances called bile salts. The $HCO_3^-$, like that from the pancreas, helps neutralize acid ($H^+$) from the stomach, and the phospholipids and bile salts coat the small lipid droplets, thereby preventing them from coalescing into large droplets. They can do this because they are amphipathic molecules; that is, bile salts and phospholipids have a nonpolar region that dissolves in the lipid droplets and polar regions that face the watery environment of the intestinal lumen. The polar regions are charged and repel each other on nearby droplets.

The liver secretes bile into small ducts that join to form one large duct that delivers the bile to the **gallbladder,** a small sac underneath the liver found in many vertebrates, including humans. During a meal, the smooth muscles in the gallbladder contract, injecting the bile solution into another duct that empties into the small intestine through the same opening as the pancreatic duct (see Figure 37.10). The gallbladder, therefore, is a storage organ that allows the release of large amounts of bile to be precisely timed to the consumption of fats. This timing is under the control of the same hormone, CCK, that we learned earlier is linked with the arrival of chyme into the small intestine. In addition to its actions on the pancreas, CCK also stimulates the gallbladder to contract and relaxes the sphincter that controls the flow of bile into the intestine.

Although bile aids the digestion of fats, absorption of the poorly soluble products of the lipase reaction is facilitated by a second action of phospholipids and bile salts, the formation of **micelles** (Figure 37.11a). Micelles consist of bile salts, phospholipids, fatty acids, and monoglycerides clustered together. Micelles continuously break apart and re-form near the epithelium of the intestine. When a micelle breaks down, the small lipid molecules are released into the solution and diffuse across the intestinal epithelium. As the lipids diffuse into epithelial cells, micelles release more lipids into the aqueous phase. Thus, the micelles keep most of the insoluble fat digestion products in small, soluble aggregates while gradually releasing very small quantities of lipids that diffuse into the intestinal epithelium. Note that it is not the micelle that is absorbed but, rather, the individual lipid molecules that are released from the micelle.

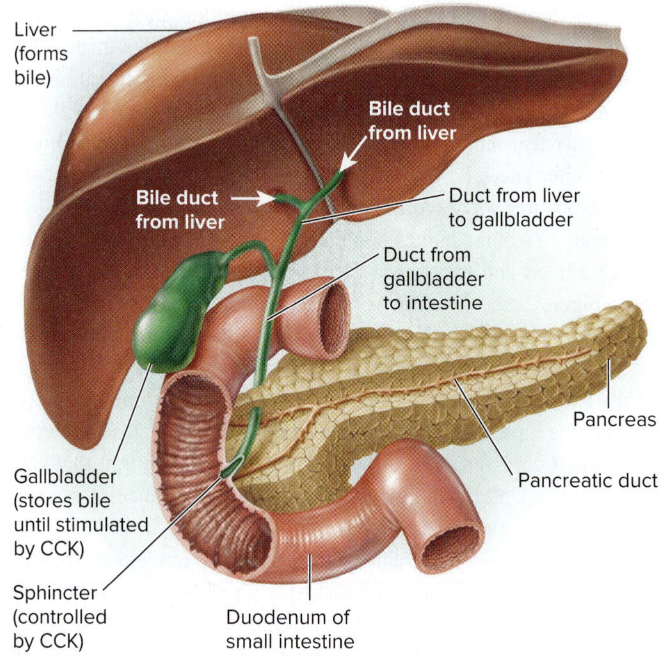


**Figure 37.10** Sites of production, storage, and secretion of bile.

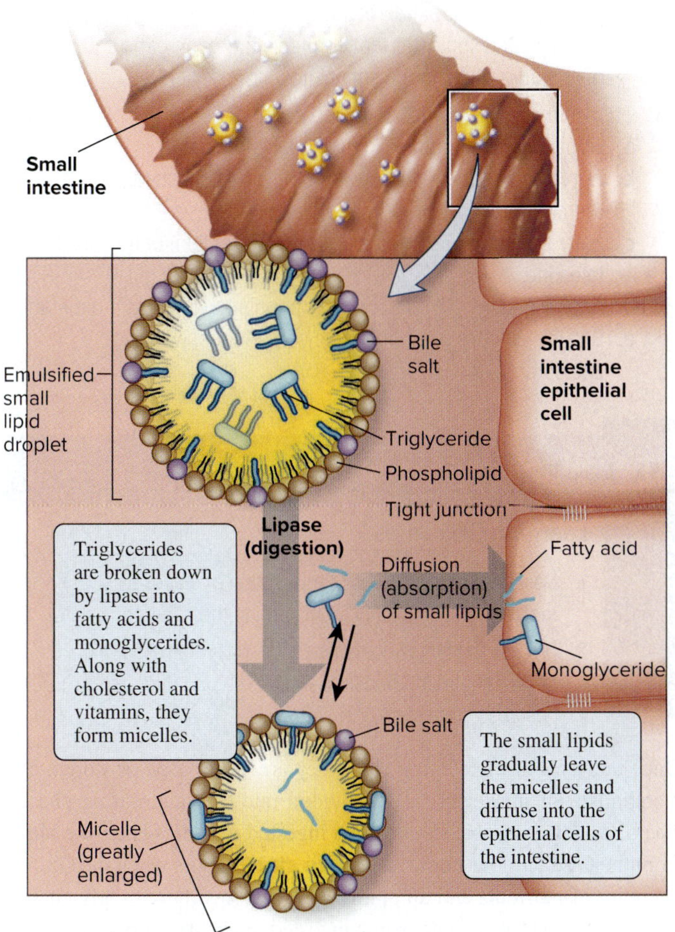


**(a) Digestion of emulsified fats into micelles, and absorption into intestinal cells**

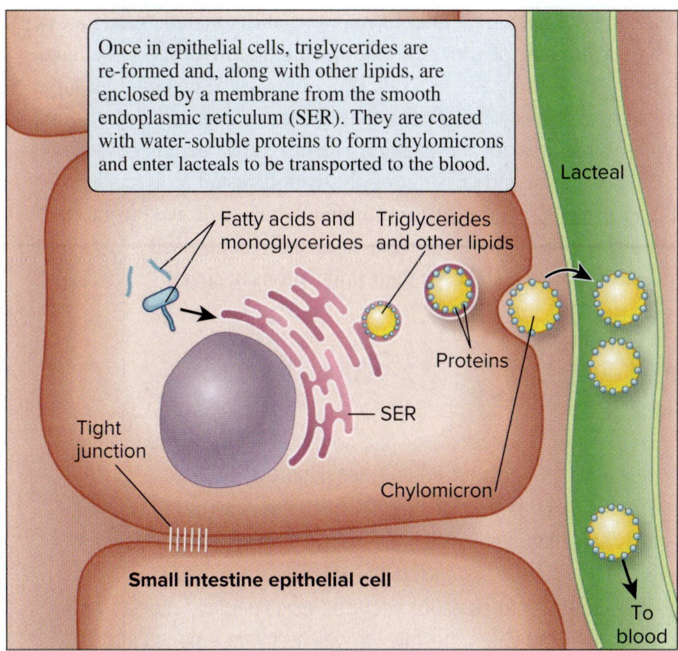


**(b) Absorption of triglycerides via the formation and release of chylomicrons**

**Figure 37.11 Formation of micelles and absorption of lipids in the small intestine.** Note: Droplets, triglycerides, and other molecules are not drawn to scale.

**Concept Check:** *Why are absorbed fatty acids and monoglycerides resynthesized into triglycerides in the epithelial cells?*

During their passage through the epithelial cells, fatty acids and monoglycerides are resynthesized into triglycerides in the smooth endoplasmic reticulum (SER). This process decreases the concentration of cytosolic free fatty acids and monoglycerides in the epithelial cells and, so, maintains a diffusion gradient for these molecules from the lumen into the cell. The resynthesized triglycerides aggregate into large droplets called **chylomicrons.** The chylomicrons are coated with proteins that perform an emulsifying function similar to that of bile salts (Figure 37.11b), and they keep the absorbed fats in a soluble form in the blood. In addition to triglycerides, chylomicrons contain any absorbed phospholipids, cholesterol, and fat-soluble vitamins.

Chylomicrons are released by exocytosis from the epithelial cells and enter lacteals. The fluid from the lacteals eventually empties into a large vein and from there into the general blood circulation.

## The Large Intestine Concentrates Undigested Material, Which Is Then Egested via the Anus

Once food has been digested and the nutrients have been absorbed, the remnants of chyme pass to the large intestine. The size of the large intestine varies greatly among different vertebrates, and the organ is vestigial or even absent in many animals, notably fishes. Its first portion, the cecum, forms a small pouch from which extends the appendix, a finger-like projection having no certain essential function in humans but that may contribute to the body's immune defense mechanisms. The next part of the large intestine in humans and other mammals is called the **colon.** Much of the colon in humans is S-shaped, forming the sigmoid colon, which empties into the **rectum,** a short segment of the large intestine that ends at the anus. The characteristic appearance of the mammalian colon is not widely found in other vertebrates, most of whom have a simple, straight large intestine.

The primary functions of the large intestine are to store and concentrate fecal material before defecation and to absorb some of the remaining ions and water that were not absorbed in the small intestine. Because most substances are absorbed in the small intestine, only a small volume of water and ions, along with undigested material (feces), is passed on to the large intestine. **Defecation** occurs when contractions of the rectum and relaxation of associated sphincter muscles expel the feces through the final portion of the canal, the anus (see Figure 37.3).

Located within the large intestine are large populations of various bacteria, some of which provide benefits to animals. For example, some bacteria release by-products such as certain vitamins into the lumen of the large intestine, where these substances can be absorbed across the intestinal epithelium. Although this source of vitamins generally provides only a small part of the normal daily requirement, it may make a significant contribution when dietary vitamin intake is low. People may even develop a vitamin deficiency if treated with antibiotics that kill these species of bacteria. Other bacterial products include gas (flatus), which is a mixture of nitrogen, carbon dioxide, hydrogen, methane, and hydrogen sulfide. Some foods (for example, beans) contain large amounts of certain carbohydrates that cannot be digested by intestinal enzymes in humans but are readily metabolized by bacteria in the large intestine. Bacterial processing of these undigested polysaccharides produces these gases, except for nitrogen, which is derived from swallowed air. Much of the fecal mass is actually comprised of these bacteria, which keeps their populations in check.

BIO TIPS ONLINE

## 37.3 Reviewing the Concepts

- The vertebrate digestive system consists of the alimentary canal plus associated glands and organs (Figure 37.3).
- Saliva facilitates swallowing, dissolves food particles, and initiates digestion. Peristalsis moves food through the pharynx and esophagus to the stomach (Figure 37.4).
- The stomach stores food, secretes acid, partially digests proteins, and regulates the rate at which chyme empties into the small intestine. Ruminants have a stomach that is associated with several chambers that facilitate cellulose digestion (Figure 37.5).
- Nearly all digestion and absorption occur in the small intestine. Villi and microvilli increase the small intestine's surface area and maximize the efficiency of digestion and absorption. The pancreas secretes digestive enzymes and a bicarbonate-rich fluid that neutralizes the acidic chyme (Figures 37.6, 37.7).
- In the small intestine, carbohydrates are digested to monomers by amylase and brush border enzymes and are then transported into the blood. Proteins are broken down to smaller fragments in the stomach by pepsin and further broken down in the small intestine into amino acids by proteases. Lipid digestion and absorption occur entirely in the small intestine by the action of pancreatic lipase (Figures 37.8, 37.9, 37.10, 37.11).
- The large intestine concentrates undigested material, which is then egested through the anus or cloaca.

## 37.3 Testing Your Knowledge

1. Which is *not* part of the vertebrate alimentary canal?
   a. salivary glands
   b. pharynx
   c. esophagus
   d. stomach
   e. rectum
2. CCK does all except which of the following?
   a. It stimulates the pancreas to secrete digestive enzymes into the small intestine.
   b. It stimulates the production of bile by the liver.
   c. It stimulates contraction of the gallbladder.
   d. It relaxes the sphincter that connects the pancreatic and gallbladder ducts to the small intestine.
3. Choose the *incorrect* statement.
   a. Trypsin digests proteins in the small intestine.
   b. Enzymes on the brush border of the small intestine complete the process of carbohydrate digestion.
   c. Fatty acids are absorbed across the epithelium of the small intestine.
   d. Approximately 10% of the world's population is lactose-intolerant.
   e. Bile is a mixture of amphipathic compounds, wastes, cholesterol, and $HCO_3^-$.

## 37.4 Nutrient Use and Storage

### Learning Outcomes

1. Describe the processes of absorption of glucose, triglycerides, and amino acids during the absorptive state.
2. Discuss the two major ways that vertebrates can maintain the concentration of glucose in their blood during the postabsorptive state.

Once nutrients have been ingested, they are either used or stored. The metabolism of nutrients can be divided into two alternating phases. The **absorptive state** occurs when ingested nutrients enter the blood from the GI tract. The **postabsorptive state** occurs when the GI tract is empty of nutrients and an animal's own stores must supply energy. During the absorptive state, some ingested nutrients supply the immediate energy requirements of the body. The rest are added to the body's energy stores to be called upon during the next postabsorptive state. Total body energy stores are adequate for the average human to withstand a fast of several weeks. By contrast, some animals can barely survive a single missed meal—particularly if they have low energy reserves—because their relative metabolic requirements are much greater than our own. In this section, we will focus on nutrient absorption, use, and storage in vertebrates, with a closer look at mammals.

### In the Absorptive State, Nutrients Are Absorbed and Used for Energy or Stored

The categories of nutrients that are absorbed during the absorptive state include carbohydrates, lipids, proteins, nucleic acids, vitamins, minerals, and water. We will look in depth at the first three of these; **Figure 37.12** gives an overview of what happens to these nutrients, as summarized here:

- **Absorbed carbohydrates.** The chief carbohydrate monomer absorbed from the GI tract of vertebrates is glucose. Much of the absorbed glucose enters all cells and is enzymatically broken down, releasing energy that can be used in the synthesis of ATP. If more glucose is absorbed than is required for immediate energy, a portion of the excess is incorporated into glycogen, primarily in the liver and skeletal muscle, and the remainder is broken down to provide sources of carbon for building triglycerides in adipose cells.

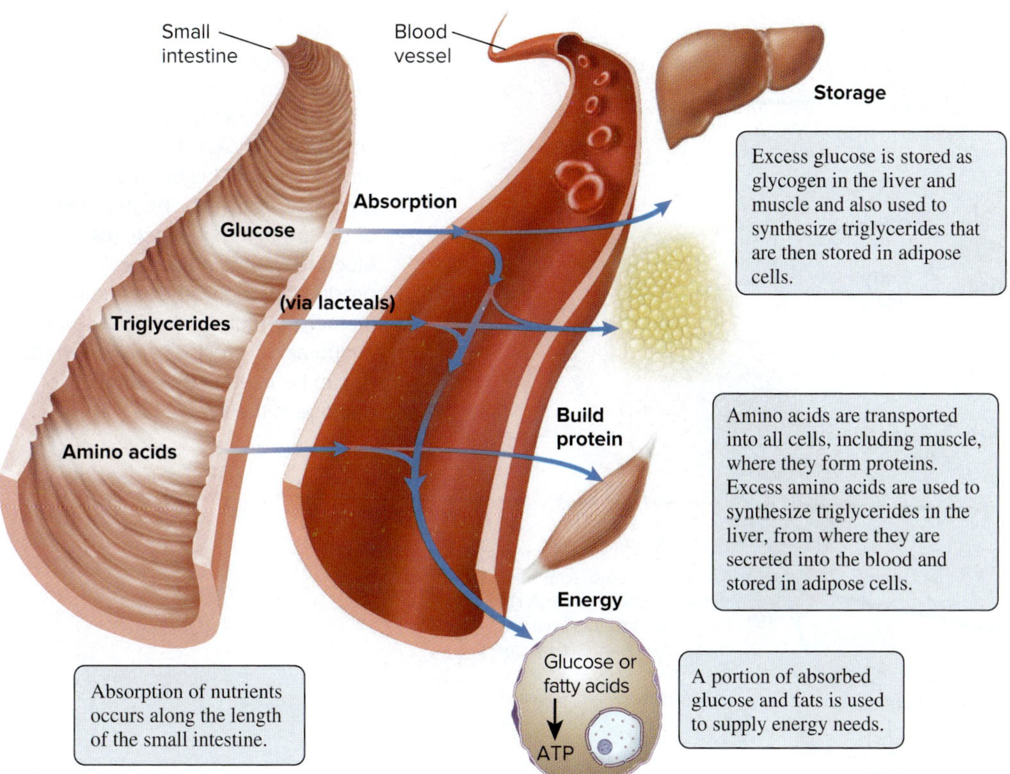


**Figure 37.12 Events of the absorptive state.** The products of digestion are absorbed into the blood along the length of the small intestine. These nutrients are broken down for immediate energy demands or they are incorporated into macromolecules and used for energy storage, such as glycogen, or cell function, such as proteins.

- **Absorbed triglycerides.** Triglycerides are packaged into chylomicrons, which enter lymph and from there the blood circulation (see Figure 37.11). As blood moves through adipose tissue, an enzyme there releases the fatty acids from the triglycerides in the chylomicrons. The fatty acids then diffuse into adipose cells and combine with glycerol to re-form triglycerides. These triglycerides are stored in adipose cells until the body requires additional energy.
- **Absorbed amino acids.** Amino acids are taken up from the blood by all cells of an animal's body, where they are used to synthesize proteins. In vertebrates, excess amino acids are enzymatically broken down in liver cells and the products are used to synthesize fatty acids, which then are incorporated into triglycerides. The triglycerides are then released into the blood and taken up and stored in adipose cells. Thus, an excess in the diet of any of these nutrients—carbohydrates, triglycerides, or proteins—will result in fat deposition in adipose cells.

## In the Postabsorptive State, Stored Nutrients Are Released and Used

As the postabsorptive state begins, synthesis of glycogen and triglycerides slows and the breakdown of these substances begins. During this state, macromolecules formed during the absorptive state are broken down to supply monomers that can be used for energy. No glucose is available to be absorbed from the intestines during this time, yet the blood glucose concentration must be maintained because the cells of the central nervous system (CNS) normally can use only glucose for energy. A large decrease in blood glucose concentration can disrupt CNS functions, ranging from subtle impairment of mental function to seizures, coma, or even death.

The events that maintain the blood glucose concentration fall into two categories: (1) reactions that provide glucose to the blood and (2) cellular use of fatty acids for energy.

**Glucose from Glycogen and Other Sources** Vertebrates can increase the blood glucose concentration during the postabsorptive state in two major ways: (1) by breaking down glycogen into glucose and (2) by making new glucose from noncarbohydrate sources. In the first, the glycogen that was formed during the absorptive state can be broken back down into molecules of glucose by hydrolysis, a process known as **glycogenolysis** (Figure 37.13a). This occurs primarily in the liver, after which the glucose is released into the blood, where it can be made available to all cells. Muscle glycogen is also broken down into glucose by glycogenolysis, but this glucose is used exclusively by muscle cells and not secreted into the blood.

The amount of liver glycogen available to provide glucose during the postabsorptive state varies among animals, but it is generally sufficient to maintain the blood glucose concentration for only a brief time, such as an overnight fast in a human. Therefore, a second mechanism for maintaining the blood glucose is required if the postabsorptive state continues. In the process of **gluconeogenesis** (literally, creation of new glucose), enzymes in the liver synthesize glucose from noncarbohydrate precursors; the glucose is then secreted into the blood (Figure 37.13b). This process occurs in all vertebrates but appears to be especially important in mammals.

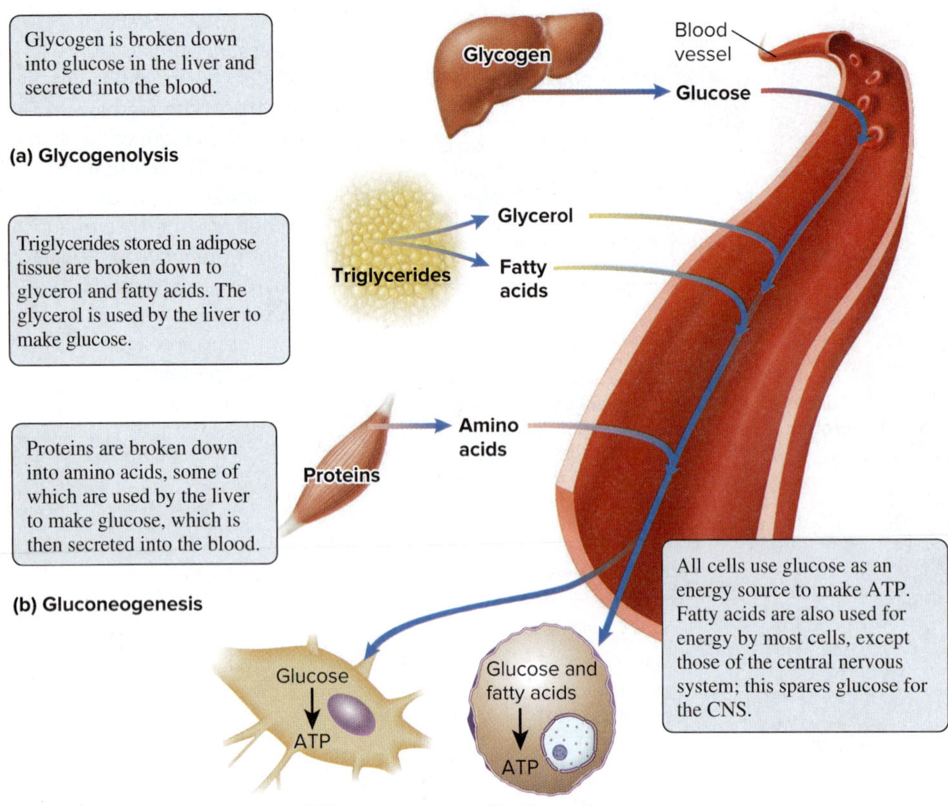

Glycogen is broken down into glucose in the liver and secreted into the blood.
(a) Glycogenolysis
Glycogen
Blood vessel
Glucose
Triglycerides stored in adipose tissue are broken down to glycerol and fatty acids. The glycerol is used by the liver to make glucose.
Triglycerides
Glycerol
Fatty acids
Proteins are broken down into amino acids, some of which are used by the liver to make glucose, which is then secreted into the blood.
Proteins
Amino acids
(b) Gluconeogenesis
All cells use glucose as an energy source to make ATP. Fatty acids are also used for energy by most cells, except those of the central nervous system; this spares glucose for the CNS.
Glucose
ATP
CNS
Glucose and fatty acids
ATP
All other cells


**Figure 37.13 Events of the postabsorptive state.** Macromolecules formed during the absorptive state are broken down to supply smaller molecules that can be used for energy. **(a)** This begins with glycogenolysis—the breakdown of glycogen into glucose. **(b)** In gluconeogenesis, the breakdown products of triglycerides and proteins (namely, glycerol, fatty acids, and amino acids) can be used for energy or can be used in the synthesis of glucose in the liver.

A major precursor for gluconeogenesis is glycerol, which is released from triglycerides in adipose tissue by the breakdown process called **lipolysis.** In lipolysis, lipase enzymes within adipose cells hydrolyze triglycerides into fatty acids and glycerol, both of which enter the blood. The fatty acids diffuse into cells, where they are used as an alternative energy source to glucose (except for the CNS, which continues to require glucose). The glycerol is taken up by the liver, where enzymes use it to synthesize glucose by gluconeogenesis; the glucose is then released back into the blood.

When the postabsorptive state continues for an extended period of time—as when an animal fails to find food—protein becomes an important source of gluconeogenesis. Large quantities of protein in muscle and other tissues can be broken down by hydrolysis to amino acids without serious tissue damage or loss of function. Certain of the amino acids that enter the blood are taken up by the liver. In the liver, the amino group is removed from each amino acid, and the remaining molecule is used to synthesize glucose by a stepwise series of enzyme-catalyzed reactions. This process, however, has limits. Continued protein loss can result in the death of cells throughout the body because they depend on proteins for such vital processes as plasma membrane function, enzymatic activity, and formation of organelles.

The combined effects of glycogenolysis, gluconeogenesis, and the provision and use of fatty acids as an alternative energy source are so effective that after several days of fasting, the concentration of glucose in the blood of humans decreases by only a small percentage.

## 37.4 Reviewing the Concepts

- In the absorptive state, ingested nutrients are entering the blood from the alimentary canal. In the postabsorptive state, the canal is empty of nutrients and the body's own stores must supply energy.
- Much of the absorbed glucose immediately enters cells and is enzymatically broken down, providing energy required to synthesize ATP. Most absorbed triglycerides are stored in adipose cells until the body requires additional energy. Amino acids are taken up by all body cells and used to synthesize proteins (Figure 37.12).
- The events that maintain the blood glucose concentration in the postabsorptive state fall into two categories: (1) glycogenolysis and gluconeogenesis, which provide glucose to the blood, and (2) cellular use of fatty acids for energy, which spares glucose for use by the nervous system (Figure 37.13).

## 37.4 Testing Your Knowledge

1. During the absorptive state, an animal is
   a. fasting.
   b. relying entirely on stored molecules for energy.
   c. absorbing nutrients from a recently ingested meal.
   d. metabolizing lipids stored in adipose tissue to supply ATP to its cells.
   e. both a and b.

2. Gluconeogenesis
   a. occurs when the liver synthesizes glucose from noncarbohydrate precursors.
   b. is the process by which glycogen is broken down to glucose.
   c. occurs primarily when an animal is in the absorptive state.
   d. occurs when triglycerides are being formed and stored in adipose cells.
   e. occurs primarily in skeletal muscle.

## 37.5 Regulation of the Absorptive and Postabsorptive States

### Learning Outcomes

1. **SCISKILLS ▶** Predict how the blood concentration of insulin will change as the blood concentration of glucose changes, and vice versa.
2. Describe how the use and storage of glucose are regulated.
3. Describe the effect of exercise on energy demands and the mechanisms that are activated to meet those demands.

Tight control mechanisms are required to maintain homeostatic levels of energy-providing molecules in the blood. These controls come in two forms: endocrine and nervous. Cells of the endocrine system produce long-distance signaling molecules called hormones. Several hormones work together with signals arising from the nervous system. Here, we will see that one common function of the endocrine and nervous systems is to regulate the processes of glycogenolysis and gluconeogenesis so that glucose is made available to cells at all times.

### Insulin Is a Key Regulator of Metabolism

The blood concentration of **insulin,** a polypeptide hormone made by the pancreas, increases during the absorptive state and decreases during the postabsorptive state. Insulin regulates metabolism primarily by regulating the blood glucose concentration. It does this by promoting the transport of glucose from extracellular fluid into cells, where it can be used for metabolism. Glucose is a polar molecule that cannot cross plasma membranes without the aid of a transporter protein. Insulin increases glucose uptake by binding to a cell surface receptor and stimulating an intracellular signaling pathway. This pathway increases the availability of transport proteins called glucose transporters (GLUTs) in the plasma membrane (Figure 37.14). Prior to insulin exposure, most GLUTs are located within pre-formed vesicles stored in the cytosol of cells. When a cell is stimulated by insulin, the vesicles (along with their GLUTs) fuse with the plasma membrane, making more GLUTs available to transport glucose into the cell. Consequently, insulin functions to lower the blood glucose concentration because it increases uptake of glucose into cells.

Insulin exerts its effects mainly on skeletal muscle cells and adipose cells, because these cells have insulin receptors in their plasma membranes. Cells of the brain contain GLUTs that are always present in the plasma membrane, even in the absence of insulin. This ensures that brain cells will obtain glucose at all times (recall that neurons of the CNS cannot normally use any alternative energy source).

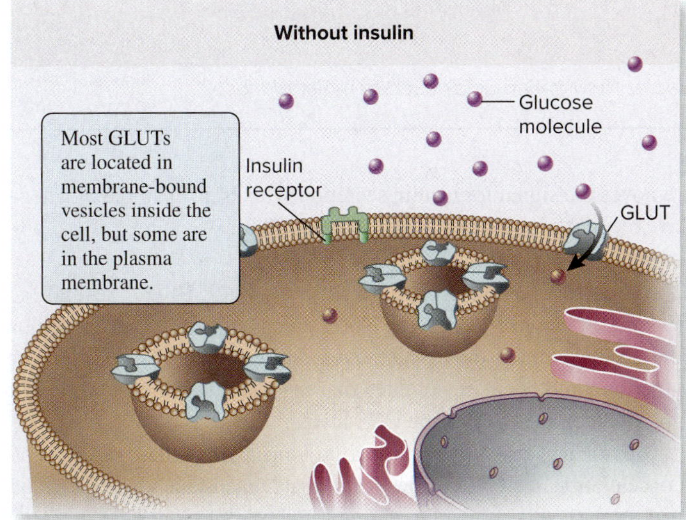


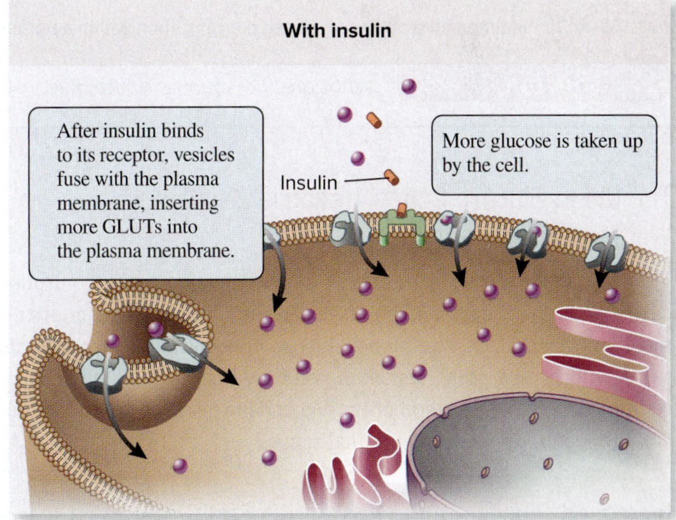


**Figure 37.14 The effect of insulin on glucose transport (GLUT) proteins.** In skeletal muscle and adipose cells, insulin stimulates the fusion of intracellular vesicles containing GLUT proteins with the plasma membrane, where they facilitate glucose uptake into the cell.

**Concept Check:** *What benefit is there to having GLUT proteins premanufactured and stored in intracellular vesicles?*

**BioConnections:** *In what other contexts have you learned about the fusion of intracellular vesicles with a cell's plasma membrane? (Refer back to Figure 5.21 for help.)*

## Biology Principle

### Living Organisms Maintain Homeostasis

Most animals cannot feed constantly, and thus the ability to tap into stored reserves of energy allows them to maintain the concentration of glucose in their body fluids within a homeostatic range. This ensures that homeostasis can be maintained because cells are supplied with energy even when an animal has not recently ingested nutrients.

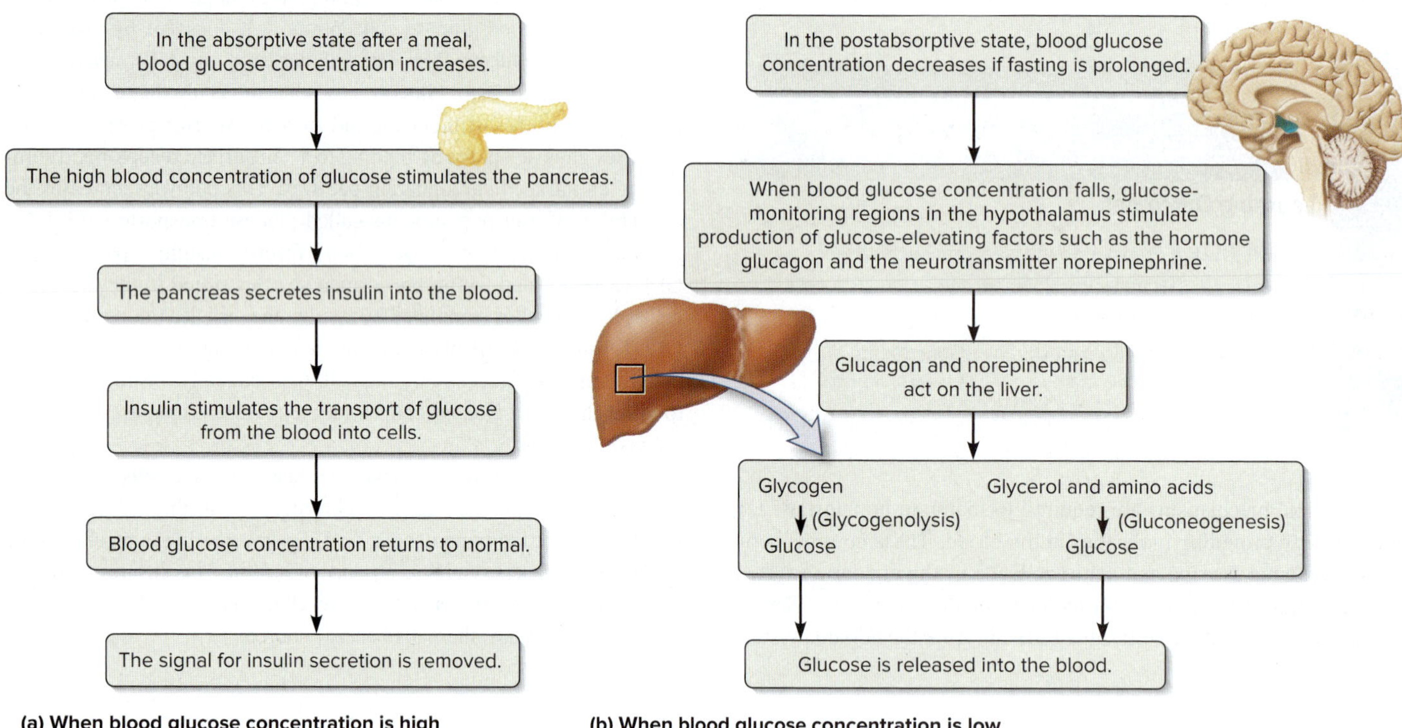


**Figure 37.15** Maintenance of blood glucose concentration within a normal range.

*What are the sources of the glycerol, fatty acids, and amino acids used to make glucose?*

## The Blood Glucose Concentration Is Maintained Within a Normal Range

To maintain homeostasis the blood glucose concentration is controlled by a system of checks and balances. In the absorptive state, after a meal has been eaten, the blood glucose concentration increases; in the postabsorptive state, depending on how long it lasts, the concentration decreases. What homeostatic mechanisms keep the blood glucose concentration within a normal range?

**Increased Glucose in the Absorptive State** The primary factor controlling the secretion of insulin from the pancreas is the blood glucose concentration itself (Figure 37.15a). An increase in an animal's blood glucose concentration directly stimulates the cells of the pancreas to secrete insulin in proportion to the amount of glucose in the blood. Later, after insulin promotes glucose uptake into its target cells, the resulting decrease in the blood glucose concentration removes the signal for insulin secretion. This is an example of a negative feedback loop (refer back to Figure 32.8).

**Decreased Glucose in the Postabsorptive State** Several factors act in concert to prevent blood glucose from decreasing below the normal homeostatic range, even during a short fast. Otherwise, glucose could decrease so much that there would not be enough glucose to keep the cells of the CNS alive.

If for any reason—such as a prolonged fast—the blood glucose concentration decreases below the normal homeostatic range for an animal, neurons that respond to changes in the extracellular concentration of glucose are activated within the hypothalamus (Figure 37.15b). Signals from the hypothalamus then stimulate the production of glucose-elevating factors. These include numerous hormones, notably **glucagon,** another polypeptide hormone secreted by the pancreas. Glucagon stimulates the processes of glycogenolysis and gluconeogenesis in the liver. In addition, the neurotransmitter norepinephrine, released from neurons of the autonomic nervous system, stimulates adipose tissue to release fatty acids into the blood. The fatty acids diffuse across plasma membranes and provide another source of energy for the synthesis of ATP. The overall effect is to increase the blood concentrations of glucose and fatty acids during the postabsorptive state or during a prolonged fast.

## More Energy Is Required During Exercise

We think of exercise as something humans do for fun or fitness, but in its broadest sense, **exercise** can be defined as any physical activity that increases an animal's metabolic rate (refer back to Chapter 32). Generally, an animal becomes physically active to seek something, such as food, shelter, or a mate, or to elude something, such as a predator or a storm. The types of activities animals engage in can be quite varied. When a cheetah sprints after a small antelope, for example, the activity of both predator and prey is brief and intense, perhaps lasting only a few seconds. By contrast, a migrating bird may fly a hundred miles a day or more over a span of weeks.

For any type of activity or exercise, including those that result from an animal's behavioral responses to stress, nutrients must be available to provide the energy required for such things as skeletal muscle contraction, increased heart and lung activity, and increased activity of the nervous system. These energy-providing nutrients include glucose and fatty acids as well as the muscle's own glycogen. The liver supplies the blood with the glucose used during exercise by glycogenolysis and gluconeogenesis. This occurs even in the absorptive state, and thus the blood glucose concentration increases above normal when an animal is active at such times. In addition, an increase in adipose tissue lipolysis releases fatty acids into the blood, which provides an additional source of energy for the exercising muscle.

These events are mediated by the same hormones and nerves responsible for the regulation of the postabsorptive state. For example, inputs from the autonomic nervous system to the pancreas inhibit insulin secretion during exercise or acutely stressful situations. Consequently, glucose transport into muscle and adipose cells is decreased, which tends to increase blood glucose concentrations (this is part of the fight-or-flight response described in Chapter 33). Because the brain does not depend on insulin for glucose transport across neuronal membranes, as described earlier, more of the body's supply of glucose is available to the brain at such times, whereas other cells can use fatty acids for energy. Therefore, the body uses all available forms of energy in response to fasting, exercise, and stress.

## 37.5 Reviewing the Concepts

- Hormones and the nervous system maintain homeostatic levels of energy sources in the blood. Insulin acts in muscle and adipose tissue to facilitate diffusion of glucose from blood into the cell cytosol via glucose transporters (GLUTs) (Figure 37.14).
- In vertebrates, an increase in blood glucose concentration in the absorptive state stimulates the pancreas to secrete insulin; a decrease in glucose concentration removes the signal for secretion. Should the blood glucose concentration decrease, glucose-monitoring regions in the hypothalamus stimulate production of glucose-elevating factors such as glucagon and norepinephrine (Figure 37.15).
- Exercise is any type of physical activity that increases an animal's metabolic rate. Exercise activates pathways similar to those of the postabsorptive state to meet an animal's increased metabolic demands.

## 37.5 Testing Your Knowledge

1. Insulin primarily regulates blood glucose concentrations by
   a. stimulating the recruitment of glucose transporter proteins from the cytosol to the plasma membrane for transport of glucose from extracellular to intracellular fluid.
   b. stimulating gluconeogenesis.
   c. suppressing glucose uptake by muscle tissue.
   d. stimulating the release of glucose from glycogen reserves in the liver.
   e. inhibiting the synthesis of new GLUT proteins.
2. Which of the following increase the concentration of glucose in the blood during exercise?
   a. insulin and glucagon
   b. glucagon and norepinephrine
   c. norepinephrine and GLUTs
   d. glucagon and GLUTs
   e. insulin and norepinephrine

# 37.6 Excretory Systems in Different Animal Groups

### Learning Outcomes

1. Describe the forms of nitrogenous wastes generated by animals.
2. Describe the general processes of filtration, reabsorption, secretion, and excretion.
3. Compare and contrast different invertebrate osmoregulatory organs.
4. List the general features of kidneys that are common to all vertebrates.

We turn now to an examination of the mechanisms by which non-useful products of metabolism are removed from an animal's body. An **excretory system** includes all of the tissues and organs (such as gills, lungs, skin, and kidneys) of an animal that function to remove soluble wastes such as $CO_2$ and nitrogenous wastes from the body, and to regulate ion and water balance. Respiratory structures and their ability to remove $CO_2$–the major waste product of metabolism–were covered in Chapter 36. Here, we focus on excretory organs that remove nitrogenous and other soluble wastes and, in the process, help regulate ion and water balance. Most such excretory organs contain tubular structures lined with epithelial cells that have the capacity to actively transport ions and other substances across their membranes. Wastes are excreted out of the body by means of these tubes.

In some cases, animals may have considerable ability to regulate the rate at which waste is excreted and how much water is lost in

## Biology Principle

### Living Organisms Use Energy

Energy from ATP is required to metabolize nitrogenous wastes into urea and uric acid. Many animals, therefore, must expend energy each day to rid themselves of nitrogenous wastes.

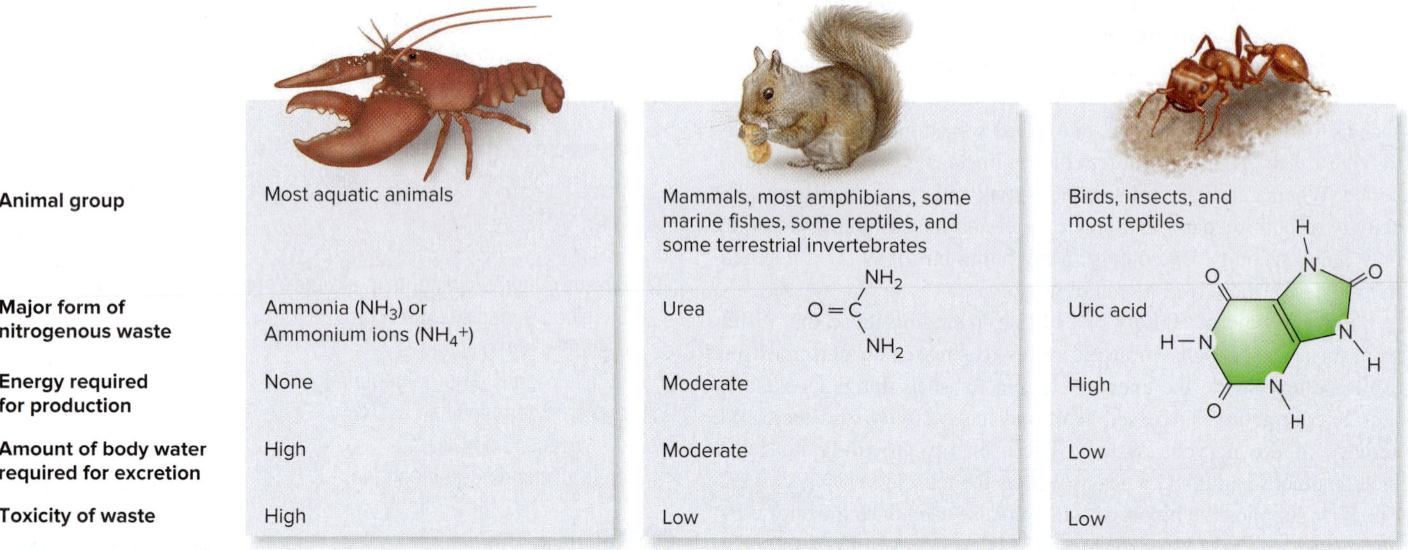

| | | | |
|---|---|---|---|
| **Animal group** | Most aquatic animals | Mammals, most amphibians, some marine fishes, some reptiles, and some terrestrial invertebrates | Birds, insects, and most reptiles |
| **Major form of nitrogenous waste** | Ammonia ($NH_3$) or Ammonium ions ($NH_4^+$) | Urea | Uric acid |
| **Energy required for production** | None | Moderate | High |
| **Amount of body water required for excretion** | High | Moderate | Low |
| **Toxicity of waste** | High | Low | Low |

**Figure 37.16** **Nitrogenous wastes produced by different animal groups.** The three forms of nitrogenous wastes, which are derived from the breakdown of proteins or nucleic acids, have different properties.

the process. For example, even though a thirsty mammal on a hot, sunny day must continue to rid its body of soluble waste products generated by its metabolism, it must also conserve water. In this section, we consider the structure and function of invertebrate and vertebrate excretory organs. First, however, we examine the major forms of nitrogenous wastes generated in animals due to the metabolism of proteins and nucleic acids.

## Nitrogenous Wastes Are Found in Three Major Forms

Nitrogenous wastes are usually found in three forms—ammonia (or ammonium ions), urea, or uric acid (Figure 37.16), depending on the species and the environment in which they live.

- **Ammonia ($NH_3$)** and **ammonium ions ($NH_4^+$)** are the most toxic of the nitrogenous wastes because they disrupt pH, ion electrochemical gradients, and many chemical reactions that involve oxidations and reductions. With some exceptions, animals that excrete wastes in this form typically live in water, where the ammonia is rapidly diluted. These animals generally excrete the ammonia as soon as it is formed, via the skin, gills, or kidneys. The chief advantage of excreting nitrogenous wastes as $NH_3$ and $NH_4^+$ is that energy is not required for their conversion to a less toxic product. A disadvantage is that it requires considerable body water to keep these wastes at a safe, dilute concentration in the body.
- **Urea** is produced by all mammals, most amphibians, some marine fishes, some reptiles, and some terrestrial invertebrates. One advantage of producing urea is that it is less toxic than $NH_3$, and thus animals can tolerate some accumulation of urea in their blood, tissues, and storage organs such as the urinary bladder. This also means that less body water is required for its excretion. One drawback of producing urea is that the metabolic conversion of $NH_3$ into urea requires a moderate expenditure of ATP and thus consumes part of an animal's total daily energy budget.
- **Uric acid** is the major form of nitrogenous waste produced in birds, insects, and most reptiles. Like urea, it is less toxic than ammonia but is even more energetically costly to synthesize. However, because it is poorly soluble in water, uric acid is not excreted in a watery urine but instead is packaged with other waste products and excess salts into a semisolid, partly dried precipitate that is excreted. The energy investment required to produce uric acid, therefore, is balanced against the water conserved by excreting nitrogenous wastes in this form.

## Excretory Systems Use Four Processes to Remove Nitrogenous and Other Soluble Wastes from the Body

Most excretory systems function by using one or more of four processes, as shown in Figure 37.17.

1. **Filtration**. In filtration, an organ acts as a filter or sieve, removing some of the water and its small solutes from the blood, interstitial fluid, or hemolymph, while excluding blood cells and large solutes such as proteins. A typical filtration system is that seen

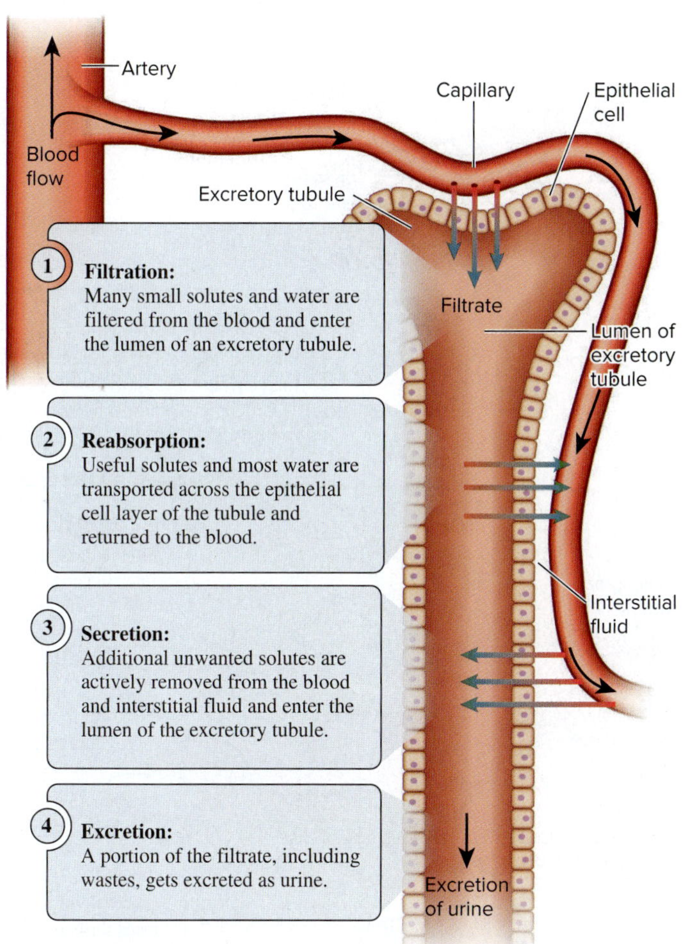

Artery
Blood flow
Capillary
Epithelial cell
Excretory tubule
1 Filtration:
Many small solutes and water are filtered from the blood and enter the lumen of an excretory tubule.
Filtrate
Lumen of excretory tubule
2 Reabsorption:
Useful solutes and most water are transported across the epithelial cell layer of the tubule and returned to the blood.
Interstitial fluid
3 Secretion:
Additional unwanted solutes are actively removed from the blood and interstitial fluid and enter the lumen of the excretory tubule.
4 Excretion:
A portion of the filtrate, including wastes, gets excreted as urine.
Excretion of urine


**Figure 37.17** **Basic features of the function of many excretory systems.**

**Concept Check:** *What is the benefit of secreting substances into the tubule?*

in the mammalian kidney, in which a portion of the plasma component of the blood is forced out under pressure through leaky capillaries and into the kidney tubules. The material that passes through the filter and enters the excretory organ for further processing or excretion is called a **filtrate.**

2. **Reabsorption.** Some of the material in the filtrate can be recaptured and returned to the blood in the process called reabsorption. This is an important feature of many excretory organs, because the formation of a filtrate is not selective, apart from the exclusion of proteins and blood cells. In order to filter the blood and remove soluble wastes, necessary molecules such as ions, sugars, amino acids, and water also get filtered in the process. Reabsorbing some or all of the useful solutes requires active transport pumps or other transport systems, and osmosis in the case of water.
3. **Secretion.** In some cases, solutes may get excreted from the body in quantities greater than those found in the filtrate. How is this possible? Some solutes are actively transported from the extracellular fluid surrounding the epithelial cells of the tubules into the tubule lumens. This process, called secretion, supplements the amount of a solute that can be removed by filtration alone. This is often a way in which excretory organs eliminate particularly toxic compounds from an animal's body.
4. **Excretion.** The process of expelling soluble waste or harmful materials from the body is called excretion. In animals that form a filtrate, the part of the filtrate that remains after reabsorption has been completed and that gets excreted is called **urine.**

## Many Invertebrates Remove Soluble Wastes from Their Body Fluids by Filtration or Secretion

Invertebrates use a variety of mechanisms to remove wastes from their body fluids. As one example, annelids have a series of tubular filtration structures called **metanephridia** (Figure 37.18a). Pairs of metanephridia (singular, metanephridium) are located in each body segment and consist of a tubular network that begins with a funnel-like structure called a nephrostome. The nephrostomes collect coelomic fluid, which contains nitrogenous wastes, through tiny pores that exclude large solutes. $Na^+$, $Cl^-$, and other solutes are reabsorbed by active transport along the length of the tubules that extend from the nephrostomes and from there diffuse into nearby capillaries. The nitrogenous wastes remain behind in the tubules and are excreted through pores in the body wall called nephridiopores. Many annelids live in watery environments and thus excrete a hypo-osmotic urine, which helps them eliminate excess water that has entered their bodies from the environment.

By contrast, the insect excretory system is quite different from that of other invertebrates, because it involves secretion rather than filtration of body fluids. In insects, a series of narrow, extensive tubes called **Malpighian tubules** arises from the midgut and extends into the surrounding hemolymph (Figure 37.18b). The cells lining the tubules actively transport ions and uric acid from the hemolymph into the tubule lumen. This secretion process creates an osmotic gradient that draws water into the tubules. The fluid moves from the tubules into the intestine and rectum, where much of the useful solutes and water are reabsorbed. The nitrogenous wastes, any excess ions, and other waste compounds are excreted together with the feces through the anus.

Unlike other invertebrates, most terrestrial insects excrete urine that is either iso-osmotic or hyperosmotic to body fluids. This reflects the general principle that life in terrestrial environments is associated with a risk of dehydration and a need to conserve water.

## The Kidney Is the Major Filtration Organ in Vertebrates

The major filtration organ that eliminates soluble wastes found in all vertebrates is the **kidney.** The kidneys of all vertebrates have many features in common. They all function by filtration, with the exception of purely secretory kidneys found in some marine fishes. They usually contain specialized tubules composed of epithelial cells that participate in both solute and water homeostasis by promoting active transport of $Na^+$, $K^+$, and other ions across their membranes. Finally, their activity can be adjusted to meet an animal's changing ion and water requirements.

In Section 37.7, we will take a detailed look at the structure and function of the mammalian kidney. Note that in the following sections, we often use the adjective renal, which means of or pertaining to the kidneys.

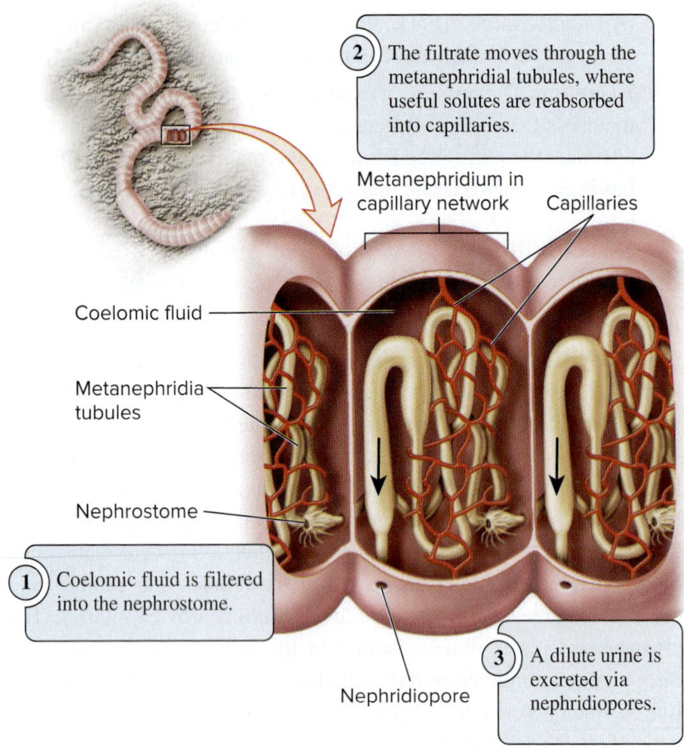

(2) The filtrate moves through the metanephridial tubules, where useful solutes are reabsorbed into capillaries.
Metanephridium in capillary network
Capillaries
Coelomic fluid
Metanephridia tubules
Nephrostome
(1) Coelomic fluid is filtered into the nephrostome.
Nephridiopore
(3) A dilute urine is excreted via nephridiopores.


**(a) Filtration: Metanephridial system in an annelid**

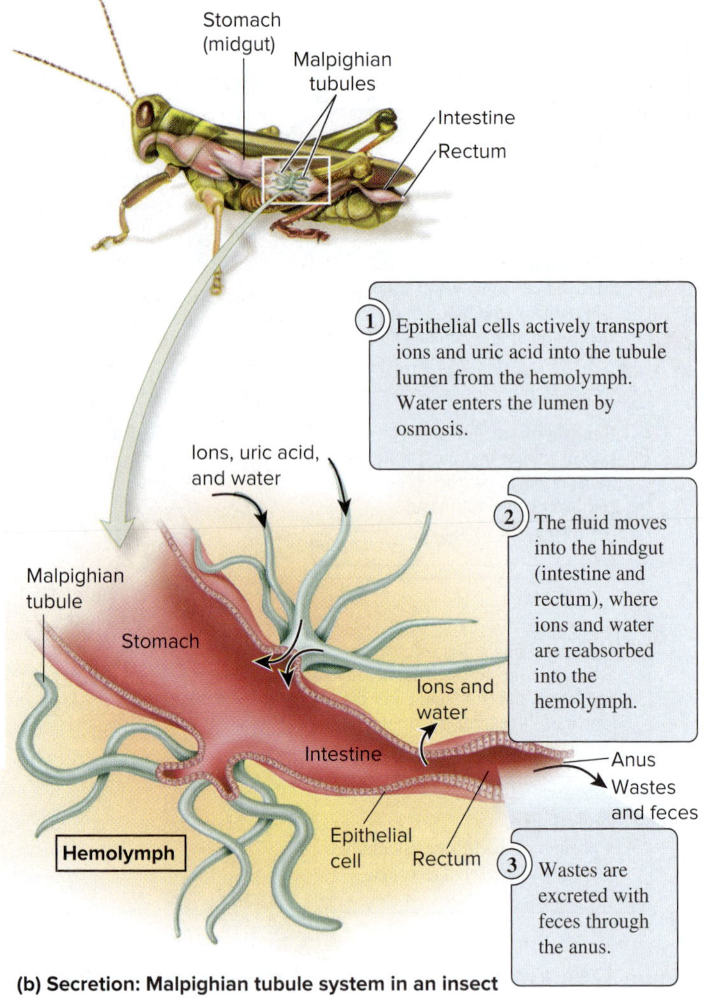


**(b) Secretion: Malpighian tubule system in an insect**

**Figure 37.18 Representative waste-removal systems in invertebrates.** Most internal body structures have been omitted for clarity. **(a)** The metanephridia of annelids. Only one of the two metanephridia in each segment is depicted; the other would be located behind the one shown. **(b)** Malpighian tubules of an insect. The tubules, which are longer and more convoluted than shown in this simplified illustration, extend into the body cavity, where they are surrounded by hemolymph.

**Concept Check:** *Do insects remove soluble wastes by filtration, secretion, or both?*

## 37.6 Reviewing the Concepts

- Nitrogenous wastes, molecules that include nitrogen from amino groups ($NH_2$), are formed from the breakdown of proteins and nucleic acids. These include ammonia ($NH_3$) and ammonium ions ($NH_4^+$), urea, and uric acid (Figure 37.16).
- Most excretory organs function by one or more of the following processes: (1) filtration, the removal of water and small solutes from the body fluids; (2) reabsorption, the return of useful filtered solutes to body fluids via transport systems; (3) secretion, the transport of unnecessary solutes from blood or hemolymph into the excretory tubule; and (4) excretion, the passing of soluble waste out of the body (Figure 37.17).
- In annelids, metanephridia filter interstitial fluid, and dilute urine is excreted via nephridiopores. In insects, Malpighian tubules secrete ions and uric acid from hemolymph into the tubule lumen; water follows by osmosis. After ions and water are reabsorbed into the hemolymph, wastes are excreted from the body (Figure 37.18).
- The kidney is the major filtration organ in vertebrates and is important for homeostatic control of solute and water balance.

## 37.6 Testing Your Knowledge

1. Which is *not* an example of a filtration mechanism for removing soluble wastes from body fluids?
   a. Malpighian tubules
   b. the vertebrate kidney
   c. the excretory system of annelids
   d. metanephridia
2. The active transport of solutes into the lumen of an excretory organ is called
   a. filtration. b. reabsorption. c. secretion. d. excretion.

## 37.7 Structure and Function of the Mammalian Kidneys

### Learning Outcomes

1. Name the primary components of the urinary system in mammals and the major anatomical features of kidneys.
2. Describe the main parts of a nephron, and outline the steps in which each contributes to the formation of urine.
3. Explain how the actions of two hormones, aldosterone and antidiuretic hormone (ADH), mediate the final composition of urine.
4. **SCISKILLS ▶** Predict the effects of changes in the blood concentrations of aldosterone and antidiuretic hormone on the osmolarity of blood.

In mammals, the two kidneys lie in the abdominal cavity (Figure 37.19a). The urine formed in each kidney flows through tubes called ureters into the urinary bladder. Urine is eliminated via the urethra. Collectively, the kidneys, ureters, urinary bladder, and urethra constitute the part of the excretory system called the **urinary system** in mammals.

Each kidney has an outer portion called the renal cortex and an inner portion called the renal medulla (Figure 37.19b). The cortex is the primary site of blood filtration. In the medulla, the filtrate becomes concentrated by the reabsorption of water back into the blood. When it reaches the renal pelvis, the filtrate is called urine, which moves from there to the urinary bladder. In this section, we will examine the structural features of kidneys that allow them to remove soluble wastes from the blood and to regulate ion and water balance.

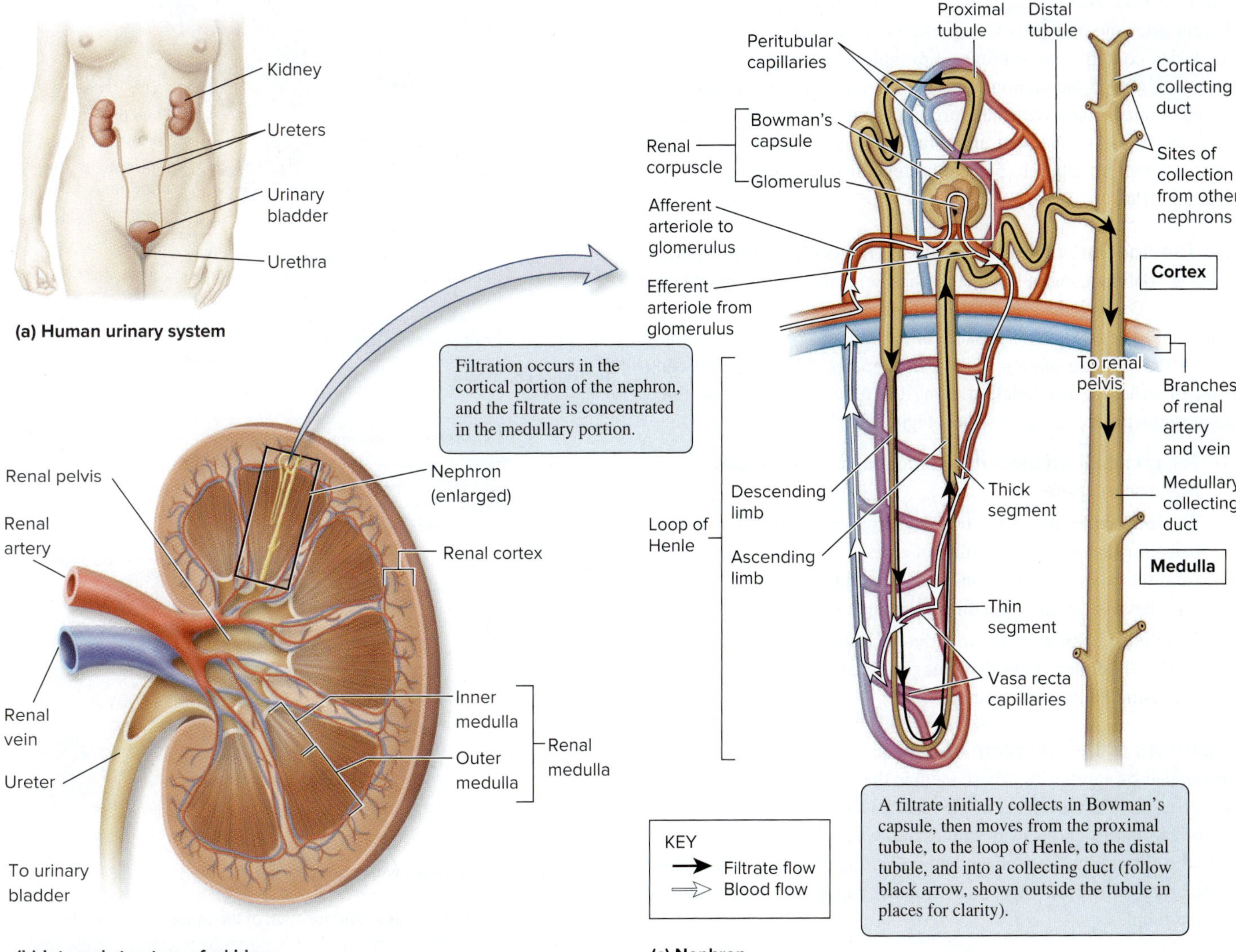


**Figure 37.19** **The mammalian urinary system, including its basic functional unit—the nephron. (a)** The organs of the human urinary system. **(b)** View of a section through a kidney, showing the locations of the major internal structures and a single nephron (enlarged; at the scale of this illustration, nephrons would be microscopic). **(c)** Structure of a nephron. The nephron begins at Bowman's capsule and empties into a collecting duct. Many nephrons empty into a given collecting duct. Surrounding the nephron are capillaries, called peritubular capillaries in the cortex and vasa recta in the medulla.

## The Functional Units of Kidneys Are Called Nephrons

Depending on its size, a mammalian kidney contains as many as several million functional units called **nephrons.** As shown in Figure 37.19c, the filtering process begins at a region of the nephron called the **renal corpuscle,** where a filtrate that is free of cells and proteins is formed from blood. This filtrate then leaves the renal corpuscle and passes through the different regions of the nephron. As the filtrate moves through the nephron, substances are reabsorbed from it or secreted into it. Ultimately the filtrate remaining at the end of each nephron combines in a collecting duct, which is functionally divided into a cortical and a medullary region. Let's look more closely at each part of the nephron and its associated structures.

**The Renal Corpuscle** Each renal corpuscle consists of a cluster of capillaries called the **glomerulus** (plural, glomeruli) that is surrounded by a membranous structure called Bowman's capsule (Figure 37.20a). Blood enters the glomerulus under pressure via an afferent arteriole and exits via an efferent arteriole. The glomerular capillaries contain fenestrations—tiny holes that permit rapid flow of plasma out of the capillaries. This filtration is further modified by cells called podocytes that surround and support the capillaries and form filtration slits that allow the passage of small solutes, but which are believed to help prevent proteins from entering the filtrate (Figure 37.20b).

**The Tubular Part of the Nephron** The tubule of the nephron, which is continuous with Bowman's capsule, is made of a single layer of epithelial cells joined by tight junctions and resting on a basement membrane. The basolateral surfaces of the epithelial cells have numerous $Na^+/K^+$-ATPase pumps in their membranes. The epithelial cells differ in structure and function along the tubule's length, and three major segments are recognized (see Figure 37.19c):

- The **proximal tubule** is the nephron segment that drains Bowman's capsule.
- The **loop of Henle** arises from the proximal tubule; it is a long, hairpin-shaped loop consisting of a descending limb coming from the proximal tubule and an ascending limb. The ascending limb has two segments: a thin and a thick segment.
- The **distal tubule** arises from the thick segment of the loop of Henle; fluid then flows from the distal tubule into one of the collecting ducts in the kidney.

**Capillaries of the Nephron** All along its length, each tubule is surrounded by capillaries. These include the peritubular capillaries in the cortex and the vasa recta in the medulla (see Figure 37.19c). Both sets of capillaries carry away solutes and water that were reabsorbed from the filtrate in the nephron and collecting duct and return them to the blood. However, only the peritubular capillaries secrete solutes into the tubules.

## A Filtrate Is Produced by Hydrostatic Pressure in Leaky Capillaries

Filtration begins as blood flows through the glomerulus and a portion of the plasma leaves the glomerular capillaries and filters into

## Biology Principle

### Structure Determines Function

The structure of the capillaries in the glomerulus is suited to their function. Fenestrations (tiny holes) permit the passage of large volumes of plasma but not blood cells out of the capillaries. The structure of the podocytes is also suited to their function, in that filtration slits may help prevent the passage of plasma proteins into the filtrate. Damage to the structure of the podocytes, for example, causes protein to leak into the filtrate and be lost in the urine.

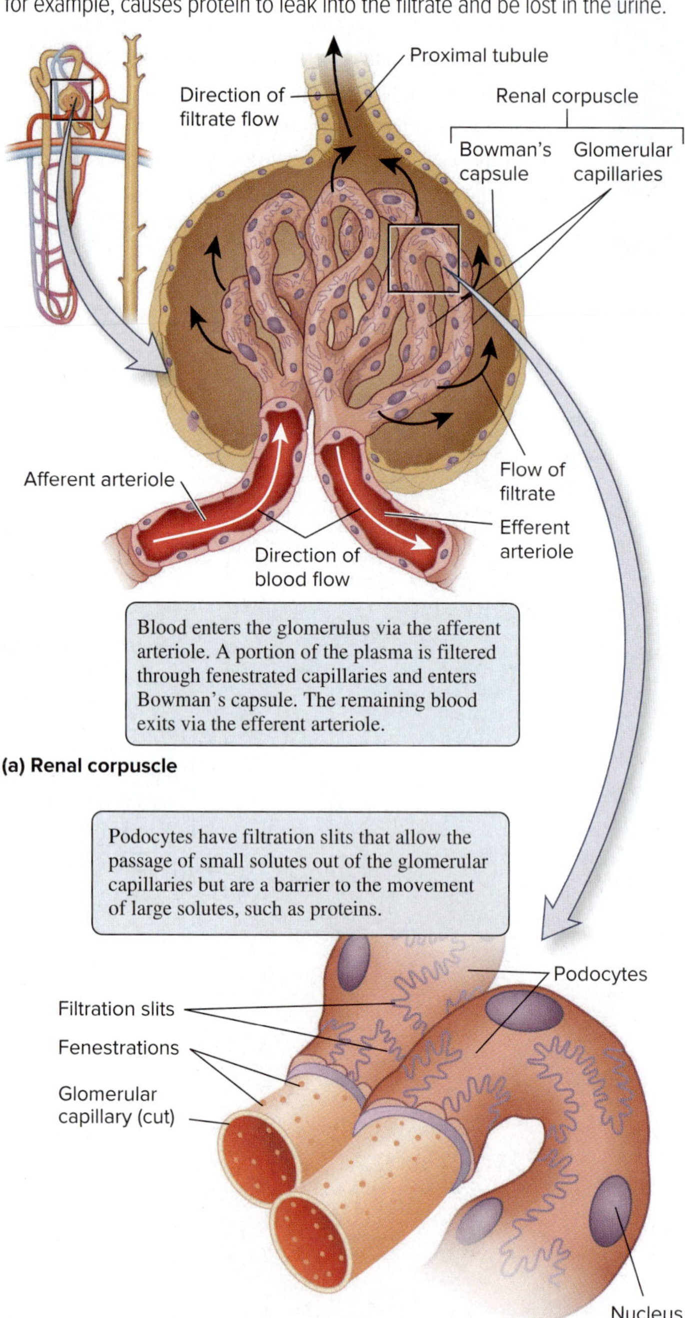


**Figure 37.20 The structure and function of the renal corpuscle. (a)** A renal corpuscle comprises Bowman's capsule and the capillaries that make up the glomerulus. It is here that the filtrate is first formed. **(b)** The capillaries are completely encased in podocytes, specialized cells that support the glomerulus and are believed to act in part as a filtration barrier.

Bowman's capsule. Blood cells do not enter the filtrate. About 15–20% of the plasma is filtered from the blood as it circulates through a glomerulus. The remainder of the plasma exits the glomerulus by the efferent arteriole (see Figure 37.20a). Most proteins are prevented from leaving the glomerular capillaries because of the small diameter of the fenestrations and the filtration slits.

The rate at which the filtrate enters Bowman's capsule is called the **glomerular filtration rate (GFR).** GFR can be increased by dilation (widening) of the afferent arteriole. When the afferent arteriole dilates, more blood enters the glomerulus, increasing the hydrostatic pressure in those capillaries and forcing more plasma through the fenestrations in the glomerular capillaries and into Bowman's capsule. This might happen, for example, when excess water in the body must be excreted in the urine.

By contrast, constriction (narrowing) of the afferent arteriole decreases the amount of blood entering the glomerular capillaries and therefore decreases the GFR. This might occur in response to dehydration or a loss of blood due to a severe injury. In such situations, decreasing the GFR results in less urine production, which in turn minimizes how much water is lost from the body and, in the case of injury, helps compensate for the blood that was lost due to bleeding.

## Useful Solutes Are Reabsorbed from the Filtrate in the Proximal Tubule

The filtrate flows from the renal corpuscle to the proximal tubule (**Figure 37.21**). Anywhere from two-thirds to all of a particular useful solute is reabsorbed from the filtrate in the proximal tubule. Useful solutes include $Ca^{2+}$, $Na^+$, $K^+$, $Cl^-$, $HCO_3^-$ , and organic molecules such as glucose, vitamins, and amino acids. Some ions diffuse through channels in the membranes of the epithelial cells that form the proximal tubule. Others are actively transported across the epithelium. Organic molecules generally are reabsorbed by being coupled to transport of ions such as $Na^+$.

Most of the water in the filtrate is reabsorbed by osmosis as the ions and organic molecules are transported from the lumen of the proximal tubule to the interstitial fluid. From the interstitial fluid, the solutes and water enter peritubular capillaries to return to the blood. By contrast, some solutes that are not required by an animal or that are potentially toxic at high concentrations are removed from the extracellular fluid and then secreted by active transport mechanisms into the proximal tubule. Examples of such substances include drugs (for example, penicillin), naturally occurring toxins, nucleoside metabolites, and ions such as $K^+$ and $H^+$ if they should ever be in excess. These solutes are excreted in the final urine.

By the time the filtrate leaves the proximal tubule, its volume and composition have changed considerably. The amount of solutes and water it contains is much reduced, and the organic molecules have usually all been removed and returned to the blood.

## Additional Water and Ions Are Reabsorbed from the Filtrate Along the Loop of Henle

In the loop of Henle, the filtrate flows down the descending limb of the loop, makes a U-turn at the bottom, and then moves back up the ascending limb of the loop. The permeabilities and transport

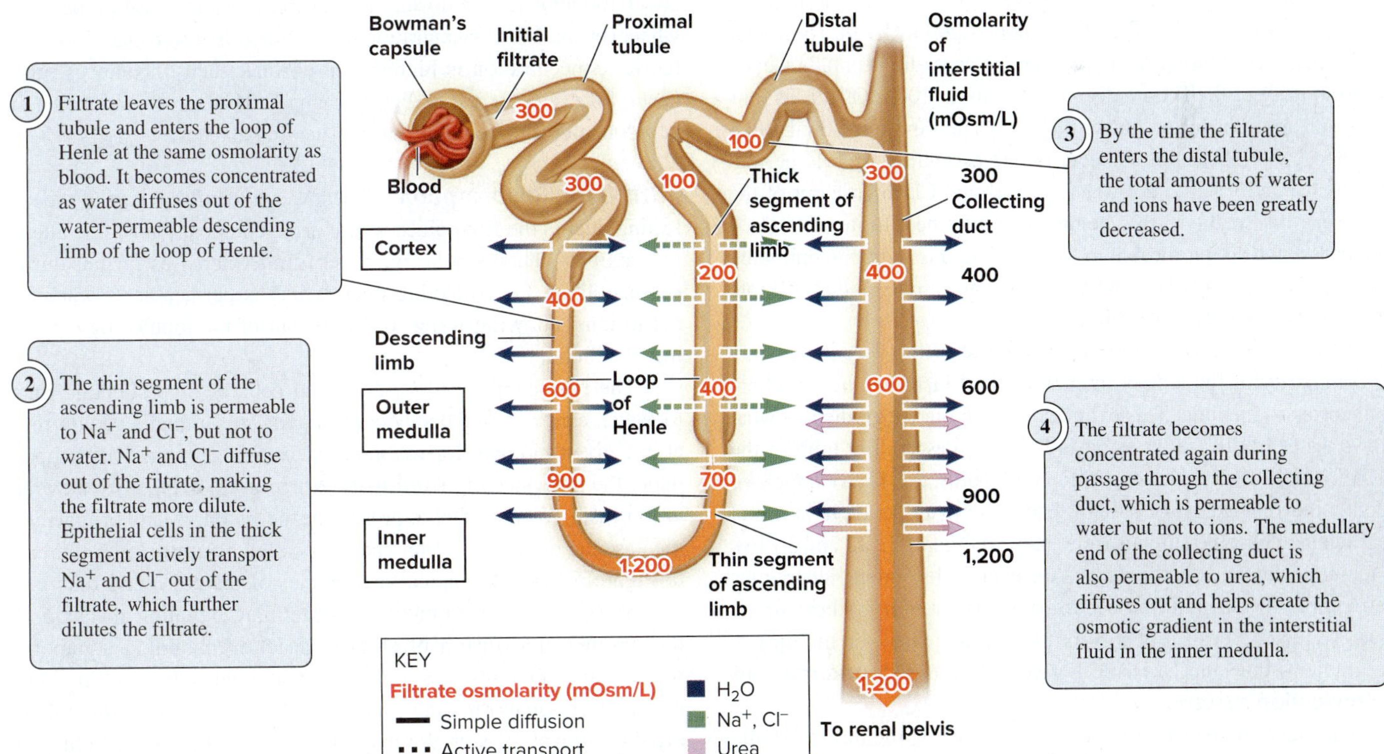


**Figure 37.21** Tubule permeabilities and osmolarities of the filtrate in the nephron and collecting duct.

characteristics of the epithelial cells lining the loop change over its length as it descends from the cortex into the medulla and then ascends to the cortex again (see Figure 37.21).

The filtrate that leaves the proximal tubule has had solutes and water reabsorbed from it in about equal proportions, and its osmolarity is therefore still about 300 mOsm/L. This filtrate enters the descending limb of the loop of Henle, which is very permeable to water but not to $Na^+$ and $Cl^-$. Water leaves the filtrate by osmosis in this region because the surrounding interstitial fluid is hyperosmotic relative to the tubule contents. The hyperosmolarity of the interstitial fluid originates from three sources:

1. The initial upturn of the thin segment of the ascending limb of the loop of Henle is very permeable to $Na^+$ and $Cl^-$, but not to water. Therefore, these ions diffuse at high rates out of this part of the loop and into the interstitial fluid, significantly increasing the osmolarity of the inner medulla.
2. The epithelial cells of the thick segment of the ascending limb of the loop of Henle actively transport some of the remaining $Na^+$ and $Cl^-$ out of the filtrate and into the interstitial fluid of the outer medulla.
3. Although urea is a waste product, some of the urea that is present in the filtrate does not get excreted in the urine but, instead, diffuses out of the lower ends of the collecting ducts and into the interstitial fluid of the medulla. Collectively, urea and ions create the osmotic force that draws water out of the filtrate in the descending limb. The water then enters the vasa recta capillaries and rejoins the blood circulation.

As water diffuses out of the filtrate in the descending limb of the loop of Henle, the filtrate becomes more and more concentrated and its osmolarity increases from 300 to about 1,200 mOsm/L (or higher in some species). During its passage up the ascending limb, however, the osmolarity of the filtrate decreases to about 200 mOsm/L as ions diffuse or are transported out of the tubule into the interstitial fluid. As noted earlier, the thin segment of the ascending limb is permeable to $Na^+$ and $Cl^-$, but not water, so $Na^+$ and $Cl^-$ diffuse out of the filtrate in this segment; this begins to dilute the filtrate. The thick segment is also not permeable to water, but in this segment, epithelial cells actively transport $Na^+$ and $Cl^-$ out of the filtrate, and the filtrate becomes increasingly more dilute.

As a consequence of ion movement out of the ascending limb, the osmolarity of the kidney interstitial fluid increases in a gradient from cortex to medulla. This extracellular osmolarity gradient is what allows water to diffuse by osmosis from the descending limb of the loop of Henle all along its length. In other words, this is an example of a countercurrent exchange system, like those that operate in heat and gas exchange in some animals (refer back to Figures 32.12 and 36.16). A major difference, however, is that energy-dependent ion pumps are required to maintain the necessary concentration gradient in the loop of Henle. Because energy is used to increase—or multiply—the gradient, the loop of Henle is referred to as a **countercurrent multiplication system.**

The chief advantage provided to an animal by the loop of Henle is that the final volume of urine produced has been reduced and its contents concentrated by the reabsorption of water along the tubule. This is especially important in animals in which total body water stores are regularly in danger of being depleted. For example, desert mammals such as the kangaroo rat tend to have longer loops of Henle than other mammals. The extra length of the loop provides for a very large osmotic gradient in the medulla and, therefore, more water-reabsorbing capacity.

## The Concentrations of Ions in Urine Are Fine-Tuned in the Distal Tubule and Cortical Collecting Duct

By the time the filtrate reaches the distal tubule and cortical collecting duct, most of the reabsorbed ions and water have already been restored to the blood, and the filtrate has been diluted to an osmolarity of about 100 mOsm/L (see Figure 37.21). However, the remaining concentrations of a few ions in the filtrate, including $Na^+$ and $K^+$, can still be fine-tuned to precisely match an animal's requirements for retaining or eliminating water and ions. This process is mediated by the actions of two hormones called aldosterone and antidiuretic hormone (ADH).

**Aldosterone: $Na^+$ Reabsorption and $K^+$ Secretion** **Aldosterone** is a steroid hormone made by the adrenal glands. It acts on epithelial cells of the distal tubule and cortical collecting duct, stimulating the $Na^+/K^+$-ATPase pump activity in the basolateral membranes. $Na^+$ is transported out of the filtrate into the blood (reabsorption), and $K^+$ is transported in the opposite direction, into the filtrate (secretion) (Figure 37.22). Because this pump moves three $Na^+$ for every two $K^+$, it creates an osmotic gradient that causes water to be reabsorbed. This makes the filtrate that moves to the medullary part of the collecting duct more concentrated than it was before, with an osmolarity of about 400 mOsm/L. Aldosterone concentrations increase in the blood whenever the $Na^+$ concentration of the blood is lower than normal or the $K^+$ concentration is higher than normal; such imbalances might occur, for example, due to dietary changes. Through its actions on the nephron, aldosterone helps correct such imbalances.

**ADH: Water Reabsorption** The lower (medullary) part of the collecting duct is the final place where urine composition can be altered in mammals. The osmolarity of the filtrate increases further during passage along the collecting duct, which is permeable to water but not to ions, allowing water to diffuse out of the tubules by osmosis into the blood.

The permeability of the epithelial cells of the collecting ducts to water can be regulated, depending on an animal's requirement at any given moment for retaining or excreting water. This happens under the influence of **antidiuretic hormone (ADH),** also known as vasopressin. When ADH is present in the blood, it acts primarily to increase the number of water channels called **aquaporins** (refer back to Figure 5.13) in the apical membranes of the collecting duct cells (Figure 37.23). ADH stimulates intracellular signaling mechanisms that promote the fusion of storage vesicles containing aquaporins with the apical membranes of the duct epithelial cells, resulting in the insertion of aquaporins into those membranes. Because the collecting ducts travel through the hyperosmotic medulla of the kidney, an osmotic gradient draws water from the filtrate in the duct into the collecting duct epithelial cells. The water exits the cells on the other side

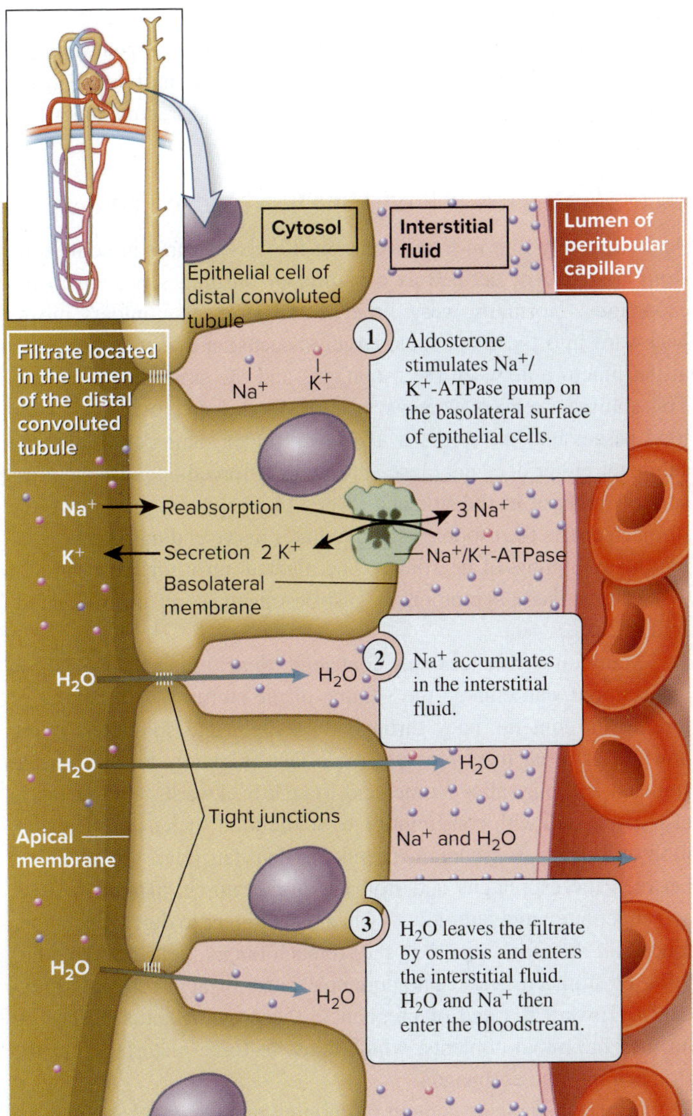


**Figure 37.22** **Action of aldosterone on distal tubule and cortical collecting duct epithelial cells.** Aldosterone stimulates the $Na^+/K^+$-ATPase pumps in cells of the distal tubule and cortical collecting duct; these ion pumps are localized only on the side of the cells facing the interstitial fluid (the basolateral surface). This activity creates an osmotic gradient, as three $Na^+$ are reabsorbed from the filtrate for every two $K^+$ secreted into it. Water then leaves the tubule by osmosis. The pump creates a diffusion gradient for $Na^+$ entry and $K^+$ exit on the apical side. The net effect is reabsorption of $Na^+$ and water into the blood, and secretion of $K^+$, which gets excreted in the urine.

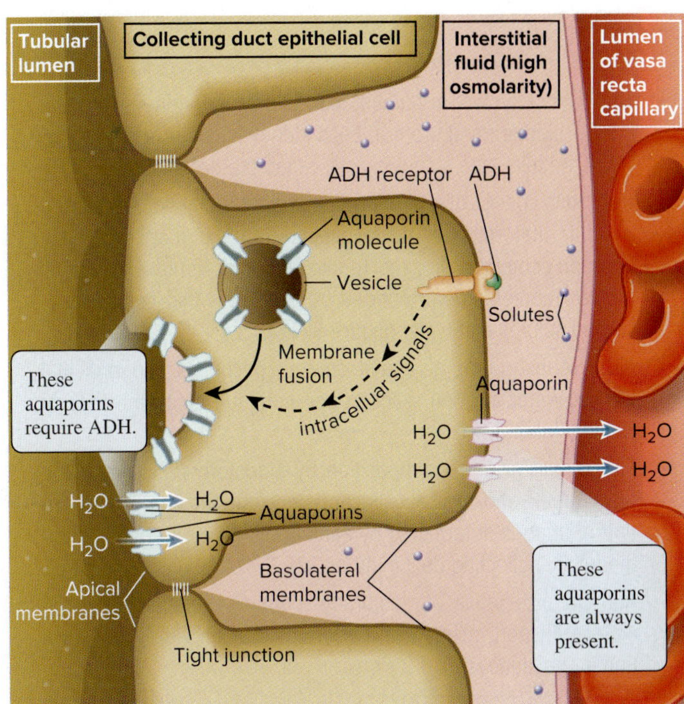


**Figure 37.23** **The effect of antidiuretic hormone (ADH) on water reabsorption in the collecting ducts of the kidney.** $H_2O$ molecules move through membranes through channel proteins called aquaporins. The epithelial cells of the collecting duct express two types of aquaporins. One type is always present on the basolateral side of the cells; the other inserts into the apical membrane only in the presence of ADH. ADH activates cell-signaling mechanisms that stimulate the fusion of intracellular storage vesicles containing aquaporins with the apical membrane. Water moves by osmosis across the cell and into the interstitial fluid, and from there enters the vasa recta to be transported into the circulation. When ADH concentrations in the blood decrease (for example, when an animal is fully hydrated), the aquaporins on the luminal membrane are returned to the intracellular storage vesicles by endocytosis.

**BioConnections:** *Do you recall a similar type of situation in which transporters are present in cytosolic vesicles and recruited to the plasma membrane due to the action of a hormone? What was that mechanism? See Figure 37.14 for help.*

(the basolateral surface), where another set of aquaporins is present. These latter aquaporins are always present in the basolateral membrane. Unlike those on the apical side of the cells, they do not require the presence of ADH for their insertion into the membrane. Once the water moves from the cells into the interstitial fluid, it enters the vasa recta capillaries. In this way, as the filtrate travels through the collecting ducts, it becomes greatly concentrated (hyperosmotic) relative to blood—as much as four to five times more concentrated in humans, for example.

ADH concentrations increase in the blood during situations where it is important to conserve water, such as when an animal is dehydrated. The ability of the kidneys to produce hyperosmotic urine is a major determinant of an animal's ability to survive in conditions where water availability is limited. By contrast, when the body's stores of water are plentiful, the concentration of ADH in the blood decreases. This results in endocytosis of portions of the apical membranes, along with their aquaporins. This, in turn, decreases the water permeability of the collecting ducts. In such a case, urine increases in volume and becomes more dilute because less water diffuses out of the collecting ducts.

BIO TIPS ONLINE

## 37.7 Reviewing the Concepts

- The urinary system in humans consists of the kidneys, ureters, urinary bladder, and urethra. Each kidney is composed of a cortex, a medulla, and a central area called the pelvis (Figure 37.19).
- Nephrons consist of a filtering component (renal corpuscle) and a tubule that empties into a collecting duct. In mammals, each nephron contains a proximal tubule, a loop of Henle, and a distal tubule. Tubules are surrounded by capillaries that collect reabsorbed solutes and water (Figure 37.20).
- Most reabsorption of useful solutes occurs in the proximal tubule. Water and ion reabsorption continue along the loop of Henle (Figure 37.21).
- Fine-tuning of urine composition by aldosterone occurs in the distal tubule and cortical collecting duct, and the final concentration of urine is determined by antidiuretic hormone (ADH) in the medullary collecting duct (Figures 37.22, 37.23).
- Aquaporins, a family of membrane-spanning proteins involved in water transport, mediate the effect of ADH on water reabsorption in the kidneys.

## 37.7 Testing Your Knowledge

1. If the afferent arterioles in the renal glomeruli were to dilate, which would follow?
   a. The GFR would increase.
   b. The GFR would decrease.
   c. The rate of urine production would increase.
   d. The volume of fluid in the body would decrease.
   e. a, c and d would all follow.
2. A deficiency of aldosterone would be expected to result in
   a. increased plasma $Na^+$ and decreased plasma $K^+$ concentrations.
   b. decreased plasma $Na^+$ and decreased plasma $K^+$ concentrations.
   c. decreased plasma $Na^+$ and increased plasma $K^+$ concentrations.
   d. increased plasma $Na^+$ and increased plasma $K^+$ concentrations.

## 37.8 Impact on Public Health

### Learning Outcomes

1. Describe the symptom called heartburn and some of its causes.
2. Identify the primary causes of ulcers.
3. Describe several common health issues related to diseases of the kidneys.
4. **SCISKILLS ▶** Diagram the process of hemodialysis, and predict its strengths and limitations.

As we have seen, the functioning of the vertebrate digestive system is extraordinarily fine-tuned by the brain, nerves, and hormones. When people are well nourished, therefore, you might assume that digestive problems would be rare. However, each year in the U.S. alone, gastrointestinal complaints account for approximately 31 million visits to doctors and 10 million visits to hospitals and emergency rooms. Here, we revisit the problem experienced by the individual in the chapter-opening case and examine the causes and treatments of kidney disease in humans.

### Excess Stomach Acid Production Can Lead to Heartburn

Approximately one in four people in the U.S. suffers at some time from **heartburn,** defined as a painful or burning sensation in the esophagus. Normally, very little of the stomach contents moves backward into the esophagus, largely because a muscular sphincter at the juncture between the esophagus and the stomach prevents it. This sphincter relaxes when we swallow food, then closes again once food passes into the stomach. However, under some conditions, the sphincter either does not close entirely or is forced open by the pressure of material in the stomach. When this happens, the acid in the stomach enters the esophagus and irritates nerve endings there, causing the symptoms of heartburn. The medical term for the condition that causes these symptoms is gastroesophageal reflux, or simply acid reflux.

Many circumstances may contribute to acid reflux. Overeating, for example, can enlarge the volume of the stomach to the point of forcing its contents back through the esophageal sphincter. Lying down after a big meal removes the effect of gravity on food in the stomach and may allow some acid to leak backward. Acid reflux is also associated with consumption of acidic food such as citrus fruits; food and drink that contain chemicals that can relax the sphincter between the esophagus and stomach (for example, alcohol, coffee, and chocolate); and fatty foods (which take longer to digest than other foods and therefore remain in the stomach longer, delaying the emptying of stomach acid). One common cause of acid reflux is pregnancy. Toward the end of pregnancy, the growing fetus pushes up on the abdominal contents, which tends to force material from the stomach into the esophagus.

Common antacids contain calcium carbonate, which buffers the acid in the stomach and esophagus. In severe cases, acid reflux can sufficiently damage the walls of the esophagus to cause a chronic cough and pain, or even perforate the esophagus. Antacids may not be sufficient to treat these patients, who are instead given drugs that inhibit the stomach's ability to produce acid. These include common brand name drugs such as Zantac, Prilosec, and Nexium, some of which are among the most widely prescribed medications in the U.S. Left untreated, acid reflux disease can lead to cancer of the esophagus.

### Erosion of the Walls of the Alimentary Canal Causes an Ulcer

Erosion of any portion of the alimentary canal due to any cause is called an **ulcer.** Most ulcers occur in the stomach (gastric ulcers) and the duodenum but occasionally also in the lower esophagus, because these sites have the greatest exposure to acid (Figure 37.24). If left untreated, contents of the lumen may leak into the surrounding body cavity, where enzymes and acids from the stomach can do considerable damage. As many as 20 million Americans have an ulcer. Each year, over 30,000 patients require surgery to repair tissue damaged by ulcers, and around 6,000 die due to complications from the disease.

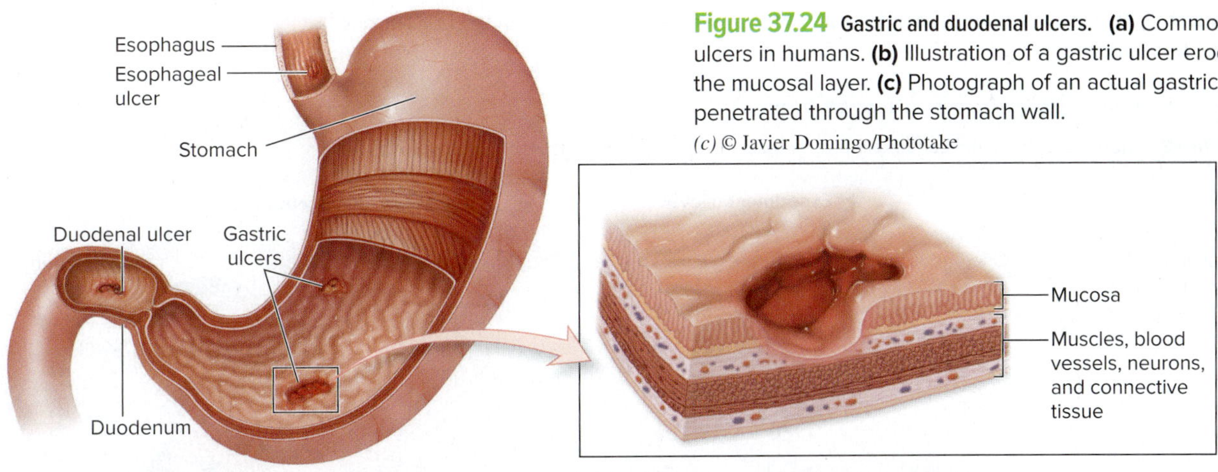


**(a) Common locations of ulcers**

**(b) Gastric ulcer penetrating mucosal layer**

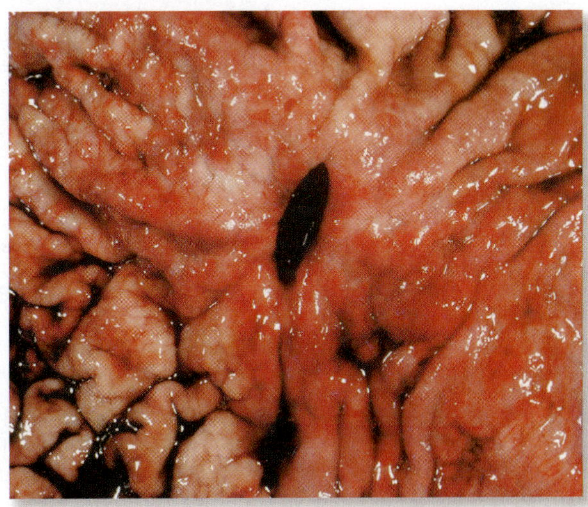

**(c) Human gastric ulcer**

**Figure 37.24** **Gastric and duodenal ulcers.** **(a)** Common locations of ulcers in humans. **(b)** Illustration of a gastric ulcer eroding through the mucosal layer. **(c)** Photograph of an actual gastric ulcer that has penetrated through the stomach wall.

*(c)* © Javier Domingo/Phototake

Acid is essential for ulcer formation, but we now know that it is not usually the primary cause. Many patients with ulcers have perfectly normal rates of acid production. Contrary to popular belief, stress or spicy food is probably not a major cause of ulcers, either. So what is the main cause of ulcers? In the 1980s, a team of Australian scientists demonstrated that many ulcers actually arise from a bacterial infection. Let's take a look at their revolutionary discovery, for which they earned the Nobel Prize in Physiology and Medicine in 2005.

## FEATURE INVESTIGATION

### Barry Marshall, Robin Warren, and Coworkers Demonstrated a Link Between Bacterial Infection and Ulcers

For many years, the conventional wisdom regarding ulcers was that they occurred as a consequence of overproduction of stomach acid, and that this in turn was due to factors such as stress, spicy foods, and smoking. In the early 1980s, Barry Marshall and Robin Warren had obtained evidence that some individuals with gastritis (inflammation of the stomach lining) or ulcers had active colonies of a bacterium (initially thought to be *Campylobacter pylori* but later identified as *Heliobacter pylori*) in their stomachs. This was a startling observation, because it had been assumed that no organism could survive the harsh pH and enzyme environment of the stomach. In preliminary studies, they further observed that treatment of such individuals with compounds that kill bacteria could reduce colonies of *H. pylori* and provide relief from gastritis. Based on these observations, they hypothesized that symptoms of gastritis and ulcers in certain individuals were caused by *H. pylori* infection and that eradication of the infection would cure the disease (**Figure 37.25**).

To test their hypothesis, the investigators recruited individuals who were being treated for ulcers at a local clinic. First, the presence of *H. pylori* and an ulcer were confirmed with visual inspection, biopsies, enzymatic tests, and bacterial cultures in vitro. On the basis of these exams and tests, 100 patients were chosen for the study. They were assigned to one of four groups based on the treatments they were to be given: an antacid plus placebo; antacid plus an antibiotic; bismuth plus placebo; or bismuth plus the antibiotic. Bismuth

**Figure 37.25** Marshall, Warren, and coworkers demonstrate that *H. pylori* infection is a cause of ulcers in humans.
*(inset)* © Juergen Berger/Science Source

**HYPOTHESIS** *H. pylori* infection is a cause of ulcers in humans.

**KEY MATERIALS** Endoscope, bacterial culture plates and medium, histologic stains, biopsy instruments.

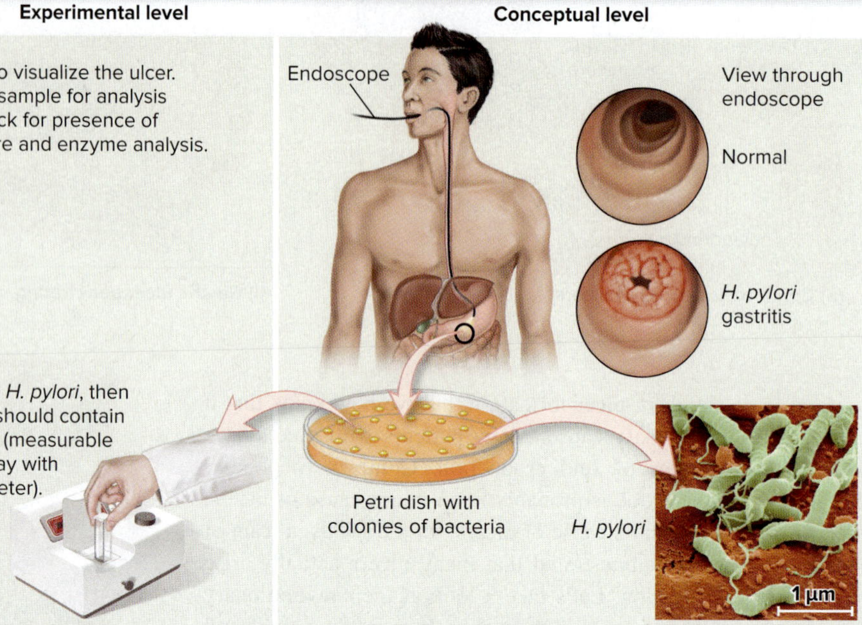

| | Experimental level | Conceptual level |
|---|---|---|
| 1 Confirm presence of ulcer and *H. pylori* in human subjects. | Use endoscopy to visualize the ulcer. Remove a tissue sample for analysis (biopsy), and check for presence of *H. pylori* by culture and enzyme analysis.<br><br>If the cultures are *H. pylori*, then extracts of them should contain specific enzymes (measurable in a sensitive assay with a spectrophotometer). | Endoscope<br>View through endoscope<br>Normal<br>*H. pylori* gastritis<br>Petri dish with colonies of bacteria<br>*H. pylori*<br>1 µm |
| 2 Randomly assign subjects to 4 treatment groups. | Treatment Group 1: Antacid + Placebo<br>Treatment Group 2: Antacid + Antibiotic<br>Treatment Group 3: Bismuth + Placebo<br>Treatment Group 4: Bismuth + Antibiotic<br>↓ 10 weeks<br>Re-evaluate for ulcers and *H. pylori*<br>↓ 1 year<br>Follow-up exams | Group 1 is a control, because antacids help the symptoms of ulcers, but do not cure them. A placebo is an ineffective treatment with no function. An antibiotic is a compound that kills bacteria. Bismuth is an element that also has bacteria-killing properties. |

**3 THE DATA**

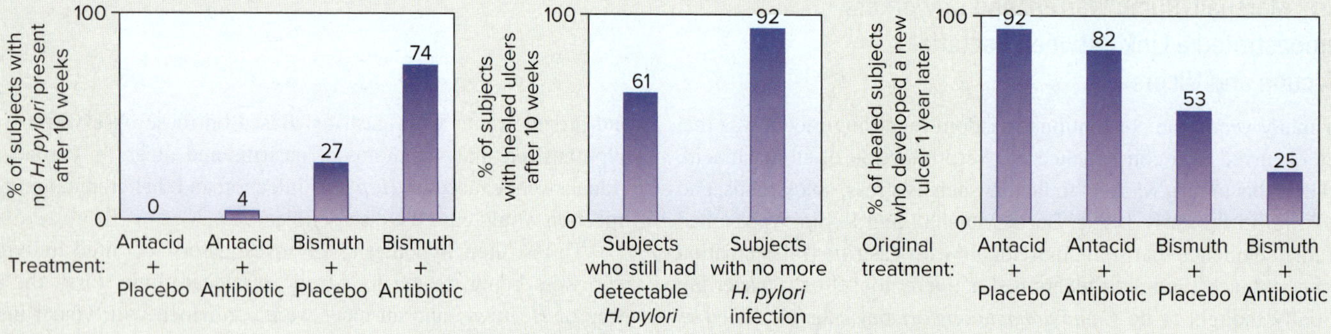

**4 CONCLUSION** Infection with *H. pylori* is a significant cause of ulcers in humans.

**5 SOURCE** Marshall, B. J., et al. 1988. Prospective Double-Blind Trial of Duodenal Ulcer Relapse After Eradication of *Campylobacter pylori*. *The Lancet* 332: 1437–1442.

is known to have antibacterial properties, but its mechanism of action is unclear. The antacid used in the study is a drug that temporarily relieves the symptoms of an ulcer by neutralizing acid; it does not cure ulcers, however, and therefore served as a control.

After 10 weeks, the patients were examined again. The results were striking: The antacid by itself or with the antibiotic had little effect on the elimination of *H. pylori* infection. Bismuth cleared the infection from 27% of patients, and bismuth plus the antibiotic cleared it from 74%. Interestingly, the use of antibiotic alone led to antibiotic-resistant strains of *H. pylori*, accounting for the failure of the antibiotic on its own to eliminate the infection. The presence of bismuth somehow prevented this resistance from occurring.

A trend was observed in which patients with cured ulcers were more likely to no longer have *H. pylori* infections. As seen in step 3, 92% of those individuals with no more evidence of infection had healed ulcers, compared to only 61% of those who still had signs of infection. Finally, a 1-year follow-up of those subjects whose ulcers had fully healed after 10 weeks (regardless of treatment) was performed. A combined bismuth/antibiotic treatment 1 year previously had reduced the rate of relapses (new ulcer formation) to only 25% compared with a high of 92% in the antacid/placebo group, a statistically highly significant result.

The investigators concluded that *H. pylori* infection is a significant cause of ulcers in humans. The mechanism seems to involve toxic by-products released by the bacterium and the body's inflammatory reaction to the microbe. These responses erode the mucous layer in the stomach and cause acid-producing cells to become overactive. Today, patients with ulcers can be easily tested for the presence of colonies of this bacterium in their stomach, and combined treatment with bismuth and one or two different antibiotics is remarkably effective at curing the disease.

### *Experimental Questions*

1. What hypothesis was tested in this experiment, and what background information led to that hypothesis?
2. Did the results support the hypothesis? Do the results indicate that *H. pylori* infection is the only cause of ulcers in people?
3. What treatment therapy would you recommend for someone with confirmed presence of *H. pylori* and an ulcer? From the data, can you conclude that some ulcers heal on their own?

## Kidney Disease Disrupts Homeostasis

Diseases and disorders of the kidneys are a major cause of illness in the human population. Approximately 20 million people in the U.S. suffer from kidney disease, with about 15,000–20,000 individuals receiving kidney transplants each year. Many diseases affect the kidneys. Diabetes, bacterial infections, allergies, congenital defects, kidney stones (accumulation of mineral deposits in nephron tubules), tumors, and toxic chemicals are some sources of kidney damage or disease. A buildup of pressure due to obstruction of the urethra or a ureter may damage one or both kidneys and increase the likelihood of a bacterial infection and eventual renal failure. The symptoms of renal failure include fatigue, weakness, swelling of the abdomen and other regions, changes in urine amount and composition, anemia, ion imbalances, nausea, and muscle disturbances. These symptoms are similar regardless of the cause of the disease, and all stem from the condition known as **uremia,** the retention of urea and other waste products in the blood.

Assuming that a person with poorly functioning kidneys continues to ingest a normal diet containing the usual quantities of nutrients and salts, what problems might arise? Potentially toxic waste products that would normally enter the nephron tubules by filtration instead build up in the blood, because kidney damage significantly reduces the number of functioning nephrons. In addition, the usual excretion of $K^+$ in the urine is impaired because too few nephrons remain capable of normal tubular secretion of this ion. Increased $K^+$ in the blood is an extremely serious condition, because of the importance of stable extracellular concentrations of $K^+$ in the control of heart and neuron function.

The kidneys are still able to perform their homeostatic functions to a degree as long as at least 20% or so of the nephrons are functioning normally. The remaining nephrons undergo alterations in function—filtration, reabsorption, and secretion—to partially compensate for the missing nephrons. For example, each remaining nephron increases its rate of $K^+$ secretion so that the total amount of $K^+$ excreted in the urine can be maintained at or near normal levels. The kidneys' regulatory abilities are limited, however. If, for example, someone with severe renal disease were to eat a diet high in $K^+$, the remaining nephrons would not likely be able to excrete enough $K^+$ to prevent its concentrations from increasing in the extracellular fluid.

## Kidney Disease May Be Treated with Hemodialysis or Transplantation

Diseased kidneys may eventually reach a point where they can no longer excrete and reabsorb water and solutes at rates that maintain homeostasis or excrete waste products as fast as they are produced. In such cases, doctors must use various procedures that artificially perform the kidneys' filtering functions.

The most important of these procedures is **hemodialysis.** The general term dialysis means to separate substances in solution using a porous membrane. In hemodialysis, blood from an artery is purified by redirecting it through a dialysis machine, called a dialyzer (Figure 37.26). Within the dialyzer, blood flows through cellophane tubing that is surrounded by a special dialysis fluid. The tubing is highly permeable to most solutes but relatively impermeable to protein and completely impermeable to blood cells. These characteristics are designed to be quite similar to those of the kidneys' own glomerular capillaries. The dialysis fluid has ion concentrations similar to those in plasma but contains no urea or other substances that are to be completely removed from the blood.

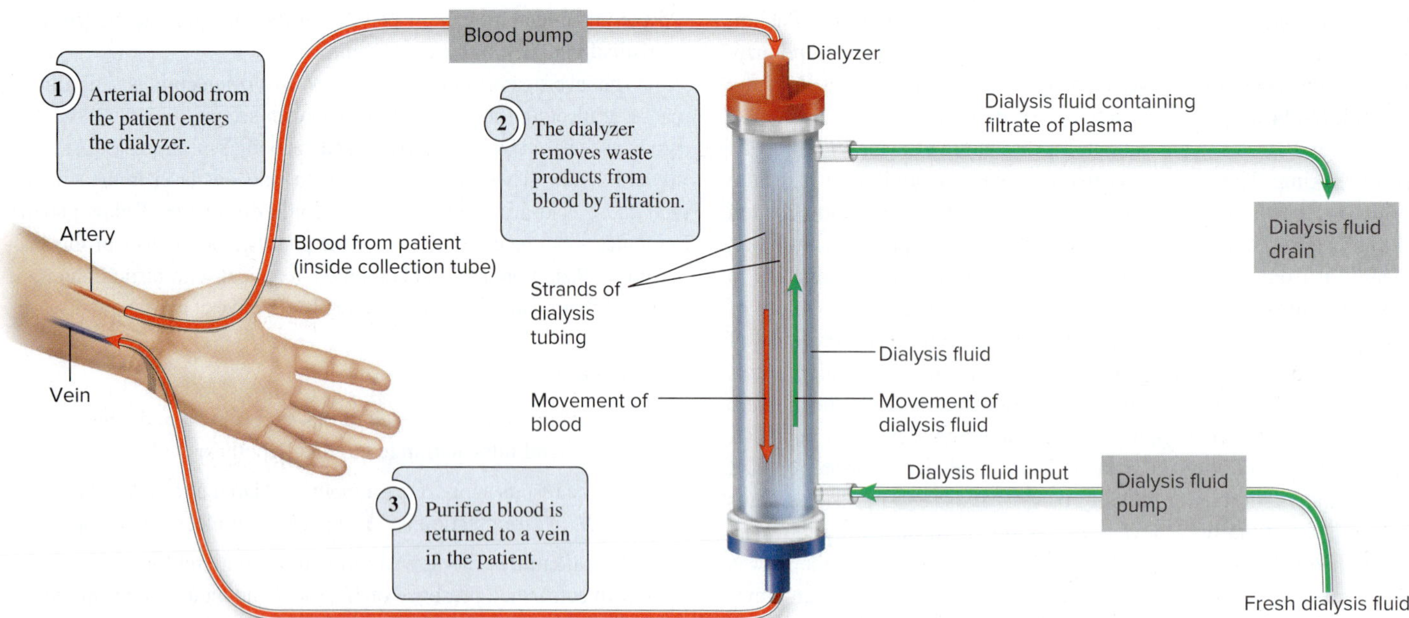


**Figure 37.26 Simplified diagram of hemodialysis.** The dialyzer is composed of many strands of very thin, sievelike tubing. In the dialyzer, blood within the dialysis tubing and the dialysis fluid bathing the tubing move in opposite directions (a countercurrent), which maximizes diffusion of substances out of the blood. The dialyzer provides a large surface area for diffusion of waste products out of the blood and into the dialysis fluid.

**BioConnections:** *Note that modern medical technology is using a feature found in many animals—countercurrent exchange. What other examples of countercurrent exchange have you learned about? Refer back to Figures 32.12 and 36.16 for help.*

As blood flows through the tubing in the dialyzer, small solutes diffuse out into the dialysis fluid until an equilibrium is reached. If, for example, the patient's plasma $K^+$ concentration is above normal, $K^+$ diffuses out of the blood across the cellophane tubing and into the dialysis fluid. Similarly, waste products and excess amounts of other substances also diffuse into the dialysis fluid and thus are eliminated from the body. Note in Figure 37.26 that blood and dialysis fluid move in opposite directions through the dialyzer. This establishes an artificial countercurrent exchange system that increases the efficiency with which solutes are removed from the blood. The dialyzed blood is then returned to one of the patient's veins through another tube that leaves the dialyzer. Each year nearly 400,000 Americans undergo some type of hemodialysis.

The treatment of choice for most patients with permanent renal disease is kidney transplantation. Rejection of the transplanted kidney by the recipient's body is a potential problem, but great strides have been made in reducing the frequency of rejection. Many people who might benefit from a transplant, however, do not receive one, because the number of people needing a transplant far exceeds the number of donors. Currently, the major source of kidneys for transplanting is recently deceased persons. However, donation from a living, related donor has become more common, particularly with improved methods for preventing rejection. As noted earlier, the mammalian kidney can perform its functions with only a fraction of its nephrons intact. Due to this large safety margin, a person who donates one of his or her kidneys can function quite normally with only one kidney.

## 37.8 Reviewing the Concepts

- Heartburn (acid reflux) and ulcers are common disorders affecting millions of people, involving several regions of the GI tract. Many cases of ulcer are due to an infection with the *H. pylori* bacterium (Figures 37.24, 37.25).
- The symptoms of renal malfunction from any cause are similar and stem from uremia—the retention of urea and other waste products in the blood.
- Treatments for kidney disease include hemodialysis, in which a machine called a dialyzer removes wastes from the blood and returns the purified blood to the body, and kidney transplantation (Figure 37.26).

## 37.8 Testing Your Knowledge

1. Which is *false?*
   a. Gastroesophageal reflux can result from eating a high-fat meal.
   b. Gastroesophageal reflux is common in late pregnancy.
   c. Gastroesophageal reflux is made worse by lying down after a large meal.
   d. The primary cause of ulcers is overproduction of acid due to stress and spicy foods.
   e. Ulcers may occur in regions of the alimentary canal other than the stomach.

**2.** Hemodialysis
   **a.** is the treatment of choice for people with advanced renal disease.
   **b.** directly mixes a person's blood with the dialysis fluid.
   **c.** can lead to a loss of blood cells.
   **d.** normalizes concentrations of small solutes in the blood by a diffusion mechanism.
   **e.** is only useful for removing nitrogenous wastes from the blood.

## Assess and Discuss

### Test Yourself

1. ________ is the process of enzymatically breaking down large molecules into smaller molecules that can be used by cells.
   a. Absorption
   b. Secretion
   c. Ingestion
   d. Digestion
2. Which of the following statements is correct?
   a. Hydrolytic enzymes are released from the gallbladder into the small intestine.
   b. Pancreatic enzyme release is stimulated by an increase in CCK in the blood.
   c. The pancreas secretes CCK through a sphincter into the small intestine.
   d. All vertebrates have a gallbladder.
   e. Secretin stimulates the release of bile from the liver.
3. Which of the following statements regarding the vertebrate stomach is *false*?
   a. Its cells secrete the protease pepsin in its active form.
   b. It is a saclike organ that may have evolved to store food.
   c. Its cells secrete hydrochloric acid.
   d. It is the initial site of protein digestion.
   e. Little or no absorption of nutrients occurs there.
4. Bile is produced by the __________ and stored in the ______________
   a. liver, small intestine.
   b. gallbladder, liver.
   c. pancreas, small intestine.
   d. small intestine, gallbladder.
   e. liver, gallbladder.
5. Which of the following is *true* of the large intestine?
   a. It contains microorganisms that may produce useful by-products that can be absorbed.
   b. It stores and concentrates fecal material.
   c. Its cells absorb ions and water that remain in chyme after it leaves the small intestine.
   d. It varies considerably in size and may even be absent in some vertebrates.
   e. All of the above are true of the large intestine.
6. ADH-dependent aquaporins are
   a. receptors for ADH.
   b. ion channels.
   c. small pores in glomerular capillaries.
   d. synthesized each time they are required.
   e. pre-formed and delivered to the apical membranes of collecting duct cells when required.
7. The excretory systems found in insects are
   a. nephrostomes.
   b. metanephridia.
   c. filtration kidneys.
   d. Malpighian tubules.
   e. secretory kidneys.
8. If a person drinks an excess of water, what would happen as a result?
   a. The GFR would decrease.
   b. ADH concentrations in the blood would decrease.
   c. ADH concentrations in the blood would increase.
   d. GFR would increase.
   e. Both b and d would occur.
9. In the mammalian kidney, filtration is driven by the
   a. solute concentration in the proximal tubule.
   b. ion concentration in the blood.
   c. osmolarity of the blood.
   d. osmolarity of the filtrate in the distal tubule.
   e. hydrostatic pressure in the blood vessels of the glomerulus.
10. Which of the following cause(s) an increase in $Na^+$ reabsorption in the distal tubule and cortical collecting duct?
    a. an increase in aldosterone concentration
    b. an increase in antidiuretic hormone concentration
    c. a decrease in aldosterone concentration
    d. a decrease in antidiuretic hormone concentration
    e. both a and b

### Conceptual Questions

1. Explain the functions of the crop and gizzard in birds. Why don't humans have a crop or gizzard? Smooth, polished stones have been found in the stomach region of fossilized skeletons of ancient sauropod dinosaurs. What does this suggest about the alimentary canal of such animals?
2. List and define the major processes involved in urine production. Does each of these processes occur for every substance that enters an excretory organ such as the kidney?
3. **PRINCIPLES** A principle of biology is that living organisms maintain homeostasis. The mammalian kidneys are an excellent example of this principle. Briefly, how many homeostatic processes can you describe in which the kidneys play a role?

### Collaborative Questions

1. Discuss the major ways in which glucose homeostasis is maintained even when a vertebrate is fasting.
2. Briefly discuss the parts and functions of the nephron in the mammalian kidney. Which functions occur in multiple regions of the nephron and which occur in only one portion?

## Online Resource

# 38 How Endocrine Systems Influence the Activities of All Other Organ Systems

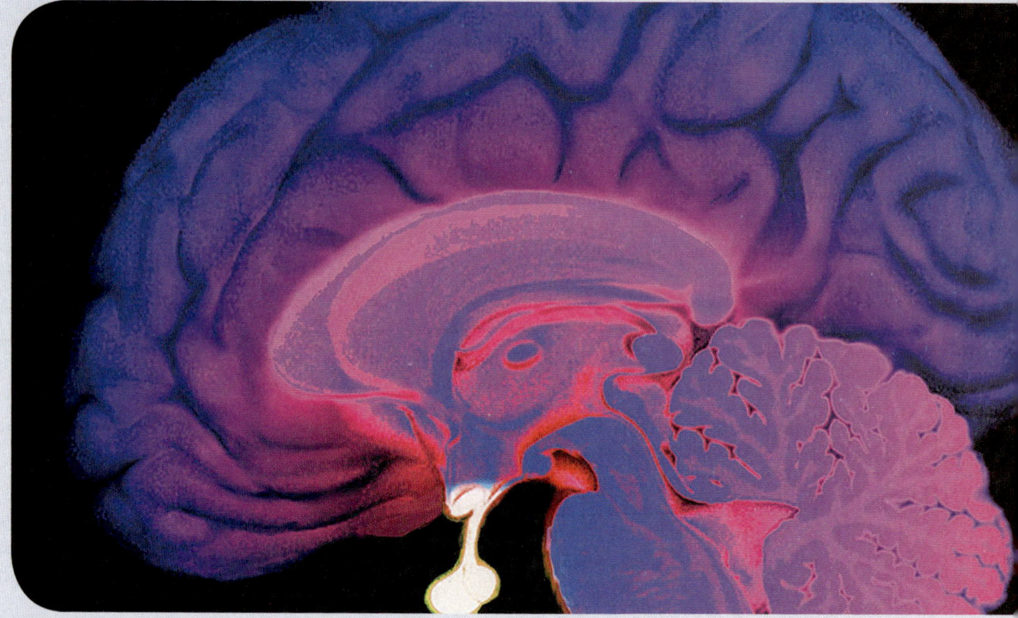


**A section through a human brain, highlighting the pituitary gland and its connection to the hypothalamus (white).** Both of these structures secrete numerous hormones. (Image is a three-dimensional MRI.)

## Chapter Outline



A fit, 23-year-old woman became alarmed when her friends repeatedly commented on her recent weight loss and voracious appetite. She had attributed these changes to her regular exercise program, a daily regimen of rigorous gym workouts. However, despite her appetite she continued losing weight until she reached a body weight of 102 pounds, down from her normal weight of about 122 pounds. At this point, she visited her doctor, who ordered a blood test for thyroid hormone.

In animals, a **hormone** is a signaling molecule produced by cells and secreted into the blood, from where it reaches one or more distant target tissues to alter their functions. Thyroid hormone is made by the thyroid gland. One of its important functions is the stimulation of metabolic rate. In this woman's case, the concentration of thyroid hormone in her blood was discovered to be greatly elevated, which caused her weight loss; her appetite increased to compensate for her increased metabolism. It was later determined that she had a disease that was caused by cells of her thyroid gland overproducing thyroid hormone. She was treated with a procedure that destroys the gland, and as a result, she will have to take a daily thyroid hormone pill for the rest of her life to normalize her hormone levels. As you will learn in this chapter, there are many examples of hormone disease, and many situations in which hormones are used to treat disease.

A chief function of hormones is to maintain homeostasis. Not surprisingly, therefore, hormones affect a wide range of animal body functions, including digestion, blood pressure, fluid and mineral balance, reproduction, and metabolism.

Although hormones are made by cells in nearly all the organs of an animal's body, hormone-producing cells are often found in specialized glands, called **endocrine glands,** the primary function of which is hormone synthesis and secretion. Examples include the pituitary gland, highlighted in the chapter-opening photo, and the thyroid gland. Collectively, all the endocrine glands and other hormone-producing structures constitute an animal's **endocrine system.**

In this chapter, we will first learn about the chemical nature of hormones and their mechanisms of action, how the endocrine and nervous systems interact, and the ways in which hormones influence homeostasis and other functions. We conclude with a discussion of how hormones are used therapeutically in humans as well as misused in certain situations.

## 38.1 Types of Hormones and Their Mechanisms of Action

### Learning Outcomes

1. List and give examples of the three different structural classes of hormones.
2. Identify the cellular location of receptors for lipid-soluble hormones and for water-soluble hormones.
3. Describe the factors that regulate the concentration of a hormone in the blood.

**Figure 38.1** is an overview of the major vertebrate endocrine structures and their hormones. Take a moment and review this figure as preparation for the rest of the chapter. In this section, we will examine some of the general characteristics of hormones, how they act, and how their concentrations in the blood are maintained.

### Hormones Are Classified by Their Structures

Hormones fall into three broad classes:

- Amine hormones are synthesized from an amino acid—either tyrosine or tryptophan—and include epinephrine, norepinephrine, dopamine, thyroid hormone, and melatonin.
- Polypeptide hormones are the most abundant class of hormones. They are synthesized like any other polypeptide—by gene transcription and translation of mRNA. With the exception of thyroid hormone, amine hormones and polypeptide hormones are packaged into secretory vesicles that provide a reservoir of stored hormone that is available for immediate release by exocytosis when required.
- Steroid hormones are enzymatically synthesized from cholesterol, as depicted in **Figure 38.2**. Unlike amines and the polypeptide hormones, steroid hormones are not packaged into secretory vesicles because the hydrophobic steroid can diffuse across lipid membranes. Instead, steroid hormones are made on demand and immediately diffuse out of the cell.

### Hormones Act Through Plasma Membrane or Intracellular Receptors

The amine and polypeptide hormones are generally water-soluble, and the steroid hormones are lipid-soluble. An exception is the thyroid hormone. Although derived from an amino acid, it is sufficiently modified that it is mostly lipid-soluble. Water-soluble hormones cannot cross plasma membranes, and therefore they reversibly bind to a receptor protein on the surface of a target cell. Steroid hormones and thyroid hormones, however, diffuse across a target cell's plasma membrane and bind to a receptor protein located in either the cytosol or nucleus. Hormones that act by binding to plasma membrane receptors tend to elicit fast responses, whereas those that act via intracellular receptors generally take longer.

The general mechanisms by which chemical messengers such as hormones act were covered in Chapter 8. Water-soluble hormones activate intracellular signaling pathways such as the cAMP pathway, which rapidly influences the activity of numerous proteins in a given target cell (refer back to Figure 8.13). By contrast, steroid and thyroid hormones when bound to their receptors act as transcription factors to regulate gene expression (refer back to Figure 8.9). Steroid and thyroid hormones can influence several genes within a single cell or different cells. In these ways, one hormone can exert a variety of actions throughout the body.

### Hormone Concentrations Depend on Rates of Synthesis and Removal

When hormones are produced by endocrine cells, they are released into the interstitial fluid surrounding the cells; from there, the hormones diffuse into the blood. Although most hormones are present at all times in the blood, their concentrations can increase when necessary. This can be accomplished by changing the rate of hormone secretion or by changing the rates at which hormones are inactivated or removed from the blood. The latter is important because although hormones carry out vital functions, excessive stimulation of cells by hormones often produces detrimental effects. Therefore, once a hormone has performed its functions, it is usually prevented from exerting its effects indefinitely. This is accomplished by negative feedback (see Chapter 32), removal of the hormone from the circulation by the liver or kidneys, or by receptor-mediated endocytosis of a hormone bound to a membrane receptor.

Generally, these processes ensure that hormone concentrations in the blood remain within a normal range but have the capacity to be increased beyond that range if required. One of the ways in which changes in hormone concentrations are initiated is through sensory input to an animal's brain. As we will see next, the nervous and endocrine systems are functionally linked in many animals, including all vertebrates.

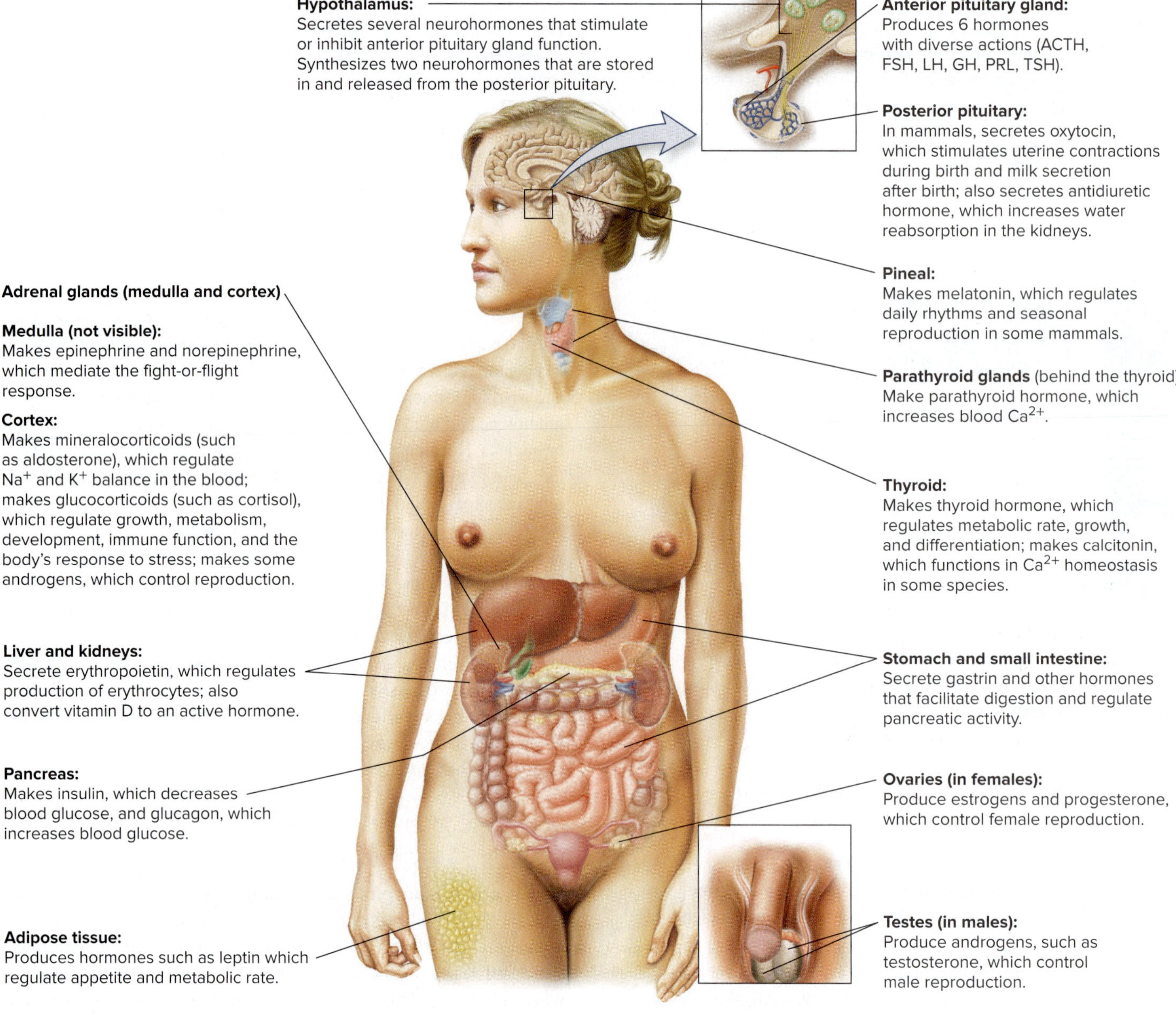

Hypothalamus:
Secretes several neurohormones that stimulate or inhibit anterior pituitary gland function. Synthesizes two neurohormones that are stored in and released from the posterior pituitary.
Anterior pituitary gland:
Produces 6 hormones with diverse actions (ACTH, FSH, LH, GH, PRL, TSH).
Posterior pituitary:
In mammals, secretes oxytocin, which stimulates uterine contractions during birth and milk secretion after birth; also secretes antidiuretic hormone, which increases water reabsorption in the kidneys.
Pineal:
Makes melatonin, which regulates daily rhythms and seasonal reproduction in some mammals.
Parathyroid glands (behind the thyroid):
Make parathyroid hormone, which increases blood $Ca^{2+}$.
Thyroid:
Makes thyroid hormone, which regulates metabolic rate, growth, and differentiation; makes calcitonin, which functions in $Ca^{2+}$ homeostasis in some species.
Stomach and small intestine:
Secrete gastrin and other hormones that facilitate digestion and regulate pancreatic activity.
Ovaries (in females):
Produce estrogens and progesterone, which control female reproduction.
Testes (in males):
Produce androgens, such as testosterone, which control male reproduction.
Adrenal glands (medulla and cortex)
Medulla (not visible):
Makes epinephrine and norepinephrine, which mediate the fight-or-flight response.
Cortex:
Makes mineralocorticoids (such as aldosterone), which regulate $Na^+$ and $K^+$ balance in the blood; makes glucocorticoids (such as cortisol), which regulate growth, metabolism, development, immune function, and the body's response to stress; makes some androgens, which control reproduction.
Liver and kidneys:
Secrete erythropoietin, which regulates production of erythrocytes; also convert vitamin D to an active hormone.
Pancreas:
Makes insulin, which decreases blood glucose, and glucagon, which increases blood glucose.
Adipose tissue:
Produces hormones such as leptin which regulate appetite and metabolic rate.


**Figure 38.1 Overview of the endocrine system in vertebrates, as seen in a human.** This figure shows the major endocrine glands and selected other structures that constitute the human endocrine system, along with some major functions of the hormones produced by those glands.

## 38.1 Reviewing the Concepts

- The endocrine glands and other organs with hormone-secreting cells constitute the endocrine system. Hormones include amines, polypeptides, and steroids. Water-soluble hormones act on receptor molecules located in the plasma membrane, whereas lipid-soluble hormones act on intracellular receptors (Figures 38.1, 38.2).
- Concentrations of hormones can be increased or decreased by changing the rate of hormone secretion and the rates at which hormones are inactivated or removed from the blood.

## 38.1 Testing Your Knowledge

1. ______ hormones are hydrophilic, whereas __________ hormones diffuse through plasma membranes.
   a. Steroid, thyroid
   b. Steroid, all amine
   c. Polypeptide, steroid and thyroid
   d. Thyroid, polypeptide
   e. Polypeptide, all amine
2. The synthesis of which class of hormone requires transcription and translation?
   a. amine hormones
   b. polypeptide hormones
   c. steroid hormones
   d. a and b only
   e. a, b, and c

# Biology Principle

## Structure Determines Function

Notice how slight changes in the arrangements of functional groups attached to the four-ring backbone (and in one case an open ring) create products with completely different functions. Keep in mind that the three-dimensional shape of a molecule may be changed much more by such chemical modifications than is visible in a two-dimensional depiction.

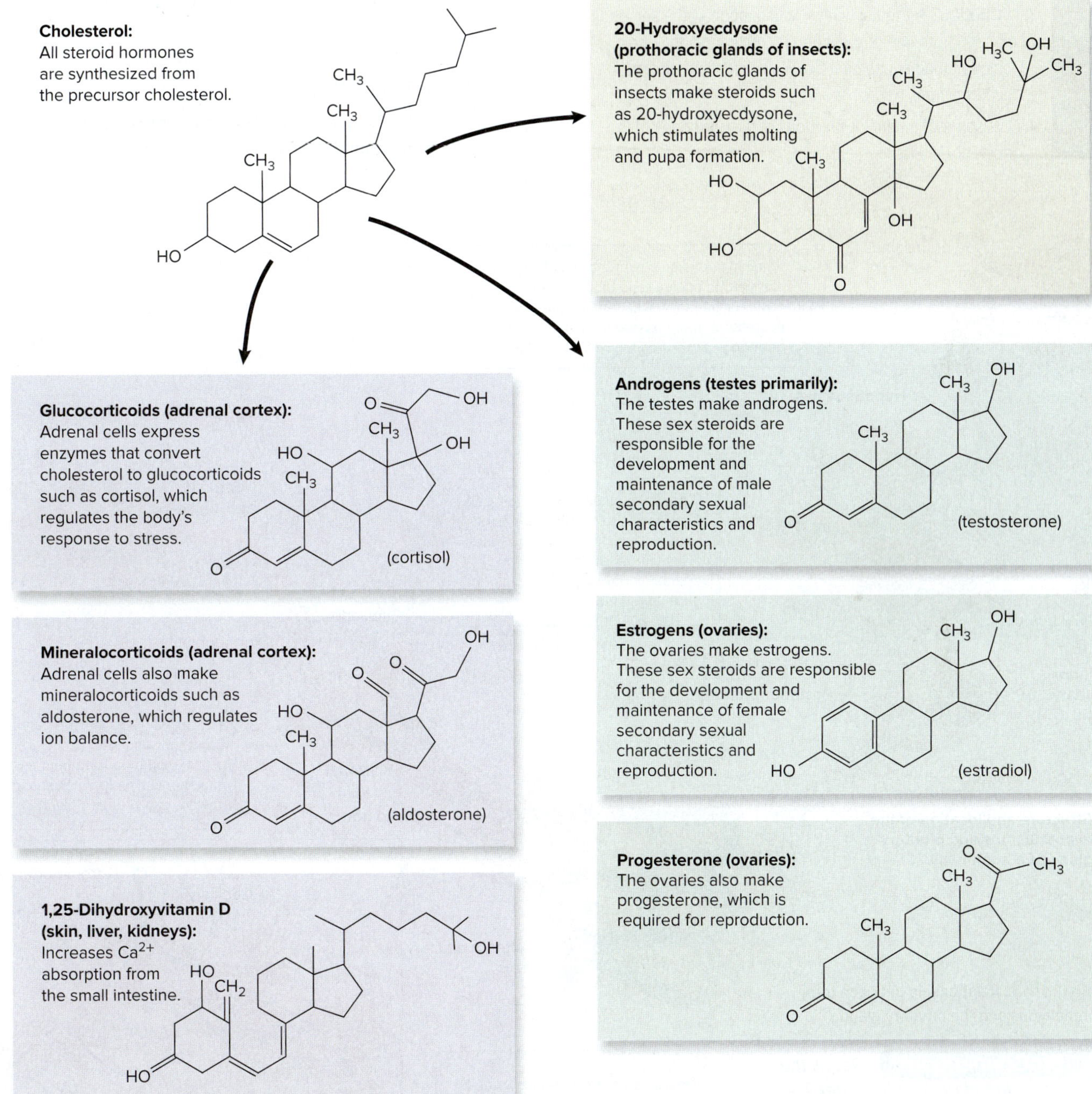


**Figure 38.2** Synthesis and functions of the major steroid hormones in animals.

## 38.2 Links Between the Endocrine and Nervous Systems

### Learning Outcomes

1. Describe the anatomical connections between the hypothalamus and pituitary gland.
2. Discuss the role of the hypothalamus and anterior pituitary gland in regulation of endocrine function.
3. **SCISKILLS ▶** Predict what would happen to the secretion of each hormone of the anterior pituitary gland if the unique blood connection between that gland and the hypothalamus was severed.
4. List some of the major functions of the hormones of the posterior pituitary.

A key feature of the endocrine system in most animals is that the concentrations of many hormones in the extracellular fluid rise and fall in response to changes in an animal's environment. For this to happen, the cells of the endocrine system must receive signals that indicate environmental changes. A sensory signal must be detected by a sensory receptor, and this information must then be converted into an endocrine response. Electrical signals are transmitted from the sensory receptors and are eventually conveyed to the hypothalamus. Here, we explore how the hypothalamus and pituitary gland function in linking the nervous and endocrine systems of vertebrates.

### The Hypothalamus and Pituitary Gland Are Physically Connected

As described in Chapter 33, the **hypothalamus** is a collection of nuclei located at the ventral surface of the vertebrate brain (**Figure 38.3a**).

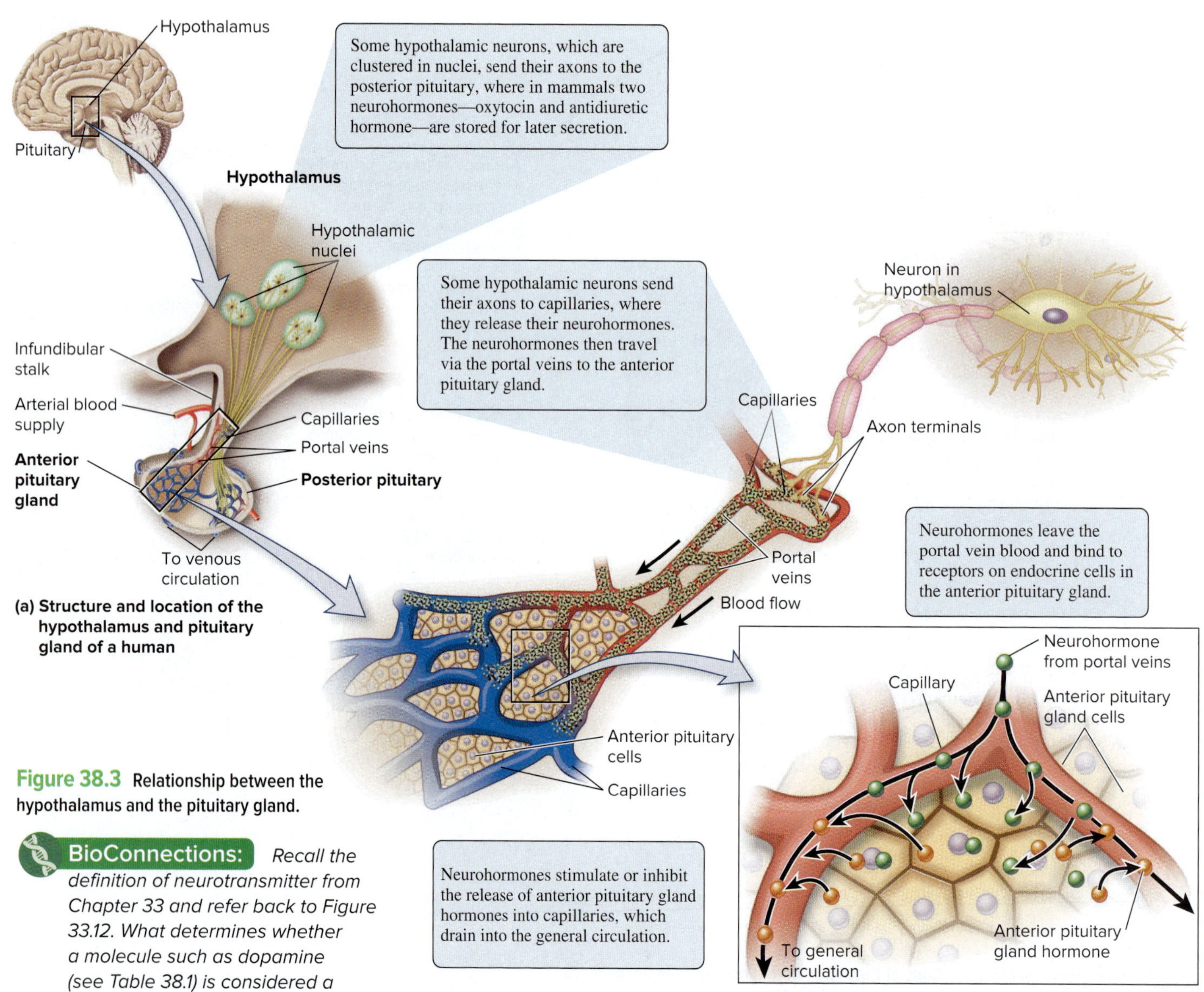


(a) Structure and location of the hypothalamus and pituitary gland of a human

(b) Stimulation of the anterior pituitary gland by the hypothalamus

**Figure 38.3** Relationship between the hypothalamus and the pituitary gland.

**BioConnections:** *Recall the definition of neurotransmitter from Chapter 33 and refer back to Figure 33.12. What determines whether a molecule such as dopamine (see Table 38.1) is considered a neurotransmitter or a neurohormone?*

**Table 38.1 Hormones of the Anterior Pituitary Gland and Hypothalamus**

| Anterior pituitary gland hormone | Stimulatory neurohormone from hypothalamus | Inhibitory neurohormone from hypothalamus | Major functions |
|---|---|---|---|
| Adrenocorticotropic hormone (ACTH) | Corticotropin-releasing hormone (CRH) | None known | Stimulates adrenal cortex to make glucocorticoids |
| Follicle-stimulating hormone (FSH) | Gonadotropin-releasing hormone (GnRH) | None known | Stimulates germ cell development and sex steroid production in gonads |
| Luteinizing hormone (LH) | GnRH | None known | Stimulates release of eggs in females; stimulates sex steroid production from gonads |
| Growth hormone (GH) | Growth hormone-releasing hormone (GHRH) | Somatostatin | Promotes growth; regulates glucose and fatty acid balance in blood |
| Thyroid-stimulating hormone (TSH) | Thyrotropin-releasing hormone (TRH) | None known | Stimulates thyroid gland to make thyroid hormones |
| Prolactin (PRL) | TRH and other factors have been suggested as stimulators of PRL. | Dopamine | Stimulates milk formation in mammals; participates in mineral balance in other vertebrates |

Neurons in these nuclei are connected to an endocrine gland sitting directly below the hypothalamus, called the **pituitary gland** (see the white structure in the chapter-opening photo). The pituitary gland in humans is made up of two lobes, the anterior and posterior lobes. Some vertebrate species have an intermediate lobe that is believed to be vestigial in others, including primates.

The hypothalamus and pituitary are connected by a thin piece of tissue called the **infundibular stalk** and by a system of blood vessels called portal veins. **Portal veins** differ from ordinary veins because not only do they collect blood from capillaries—as all veins do—but they also empty into another set of capillaries, as opposed to returning the blood directly to the heart. The second set of capillaries drain into veins that return blood to the heart. This arrangement of blood vessels bypasses the general circulation and allows the hypothalamus to communicate directly with the anterior lobe of the pituitary gland (often simply called the anterior pituitary gland).

As described in Chapter 33, the hypothalamic nuclei are vital for regulating such diverse functions as reproduction, body temperature homeostasis, circadian rhythms, appetite, thirst, metabolism, and responses to stress. The hypothalamus has such wide-ranging effects in part because it acts as a master control, signaling the pituitary gland when to produce and secrete one or more of its several hormones. However, the hypothalamus communicates differently with the anterior and posterior lobes of the pituitary gland. Let's look at the interaction between the hypothalamus and anterior pituitary gland first.

## The Hypothalamus and Anterior Pituitary Gland Have Integrated Functions

Within the different nuclei of the hypothalamus are neurons that synthesize a class of hormones called neurohormones. A **neurohormone** is any hormone that is made in and secreted by neurons. All neurohormones are either amines or polypeptides. The axon terminals from the hypothalamus end next to capillaries, into which they secrete their neurohormones (Figure 38.3b). The neurohormones are delivered directly from the hypothalamus to the cells of the anterior pituitary gland via the portal veins.

In response to these hypothalamic neurohormones, cells of the anterior pituitary gland synthesize several hormones. Table 38.1 lists the six major anterior pituitary hormones that have well-defined functions in vertebrates and the neurohormones that either stimulate or inhibit them. The stimulatory action of several of the neurohormones has historically led to them also being known as hypothalamic-releasing hormones, because they cause the release, or secretion, of other hormones from the anterior pituitary. The six major hormones of the anterior pituitary gland are secreted into the general blood circulation, where they act on other endocrine glands or structures.

## The Posterior Pituitary Contains Axon Terminals from Hypothalamic Neurons That Store and Secrete Oxytocin and Antidiuretic Hormone

The posterior pituitary of mammals has a blood supply, but, in contrast to the anterior pituitary gland, it is not connected to the hypothalamus by portal veins and does not receive neurohormones from the hypothalamus. The posterior pituitary is not actually a gland but is instead an extension of the hypothalamus that lies in close contact with the anterior pituitary gland (see Figure 38.3a). Axons from hypothalamic neurons terminate in the posterior pituitary. In mammals, the axon terminals in the posterior pituitary store one of two polypeptide hormones—oxytocin or antidiuretic hormone. When the hypothalamus receives information that these hormones are required, they are released directly from the neuron axon terminals into the blood.

**Oxytocin** increases in the blood of pregnant mammals just prior to birth. It stimulates contractions of the smooth muscles in the uterus, which facilitates the birth process and shortly afterward helps expel the placenta. Oxytocin also stimulates the secretion of milk from the mammary glands of lactating females. When the mother's nipples are stimulated by the suckling of a newborn, sensory neurons transmit a signal from there to the mother's hypothalamus, which stimulates the cells that produce and secrete oxytocin. Oxytocin stimulates smooth muscle cells surrounding the secretory components of the mammary glands. This stimulates the release of milk. Recent evidence also supports a role of oxytocin in maternal/infant bonding and prosocial behavior in some species.

**Antidiuretic hormone (ADH)** gets its name because it acts on cells of the kidneys to decrease urine production—a process known as antidiuresis. (Diuresis is an increased loss of water in the urine, as happens, for example, when you drink large amounts of fluids). If the fluid content of the body is low—for example, during dehydration or after a significant loss of blood—more ADH is secreted from the posterior pituitary. It acts to increase the number of water channels called aquaporins (refer back to Figure 37.23) present in the apical membranes of kidney tubule cells; the aquaporins permit reabsorption of water from the forming urine back into the blood. Minimizing the volume of water used to form urine is an adaptation that conserves body water when necessary.

At high concentrations, ADH also increases blood pressure by stimulating vasoconstriction of blood vessels, a function that accounts for the other common name of ADH—vasopressin. Both of the major functions of ADH are related in that they contribute to maintaining body fluid levels and blood pressure.

## 38.2 Reviewing the Concepts

- Sensory input from an animal's nervous system modulates the activity of certain endocrine glands and influences blood concentrations of many hormones.
- The hypothalamus is connected to the pituitary gland, which consists of the anterior and posterior lobes (Figure 38.3).
- Neurons of the hypothalamus synthesize neurohormones that stimulate or inhibit the anterior pituitary gland, which can secrete six different hormones (Table 38.1).
- In mammals, the axon terminals in the posterior pituitary secrete oxytocin or antidiuretic hormone (ADH).

## 38.2 Testing Your Knowledge

1. Choose the correct sequence to depict a portal vascular network.
   a. arteries→capillaries→portal veins→capillaries →veins→heart
   b. arteries→portal veins→capillaries→veins→capillaries →veins→heart
   c. veins→capillaries→arteries→capillaries→portal veins→heart
   d. capillaries→veins→capillaries→arteries→portal veins→heart
   e. arteries→capillaries→portal veins→heart
2. In a mammal, activation of the hypothalamic cells that have axon terminals that end in the posterior pituitary would be expected to cause
   a. decreased urine volume.
   b. decreased blood pressure.
   c. a diuresis (increased loss of water in the urine).
   d. inhibition of milk secretion in a lactating mammal.
   e. both a and d.

# 38.3 Hormonal Control of Metabolism and Energy Balance

### Learning Outcomes

1. Describe the important anatomical features of the thyroid gland and major functions of thyroid hormone in adult animals.
2. List the hormones produced by the pancreas and adrenal glands, and provide a brief description of the major metabolic functions of each.
3. **SCISKILLS ▶** Think of one or more ways in which type 1 and type 2 diabetes mellitus could be distinguished in an individual with the symptoms of diabetes.

An important function of the endocrine system is to regulate metabolic rate and energy balance. Hormones do this by modulating appetite, digestion, absorption of nutrients, and the availability to cells of energy sources such as glucose. Although many hormones are involved in these processes, those from the thyroid gland, the adrenal glands, and the pancreas have particularly important roles, as described in this section.

## Thyroid Hormone Requires Iodine for Its Function

The thyroid gland lies within the neck of vertebrates, straddling the trachea just below the larynx in mammals (**Figure 38.4a**). Thyroid hormone is produced when the neurohormone thyrotropin-releasing hormone (TRH) (see Table 38.1) stimulates the anterior pituitary gland to secrete thyroid-stimulating hormone (TSH) into the blood.

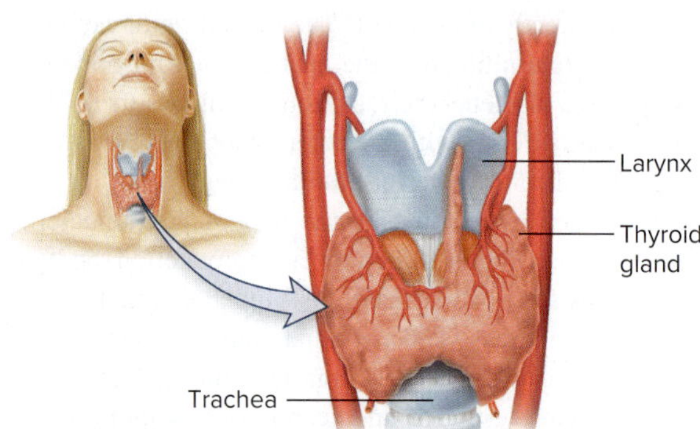


**(a) Location and structure of the thyroid gland**

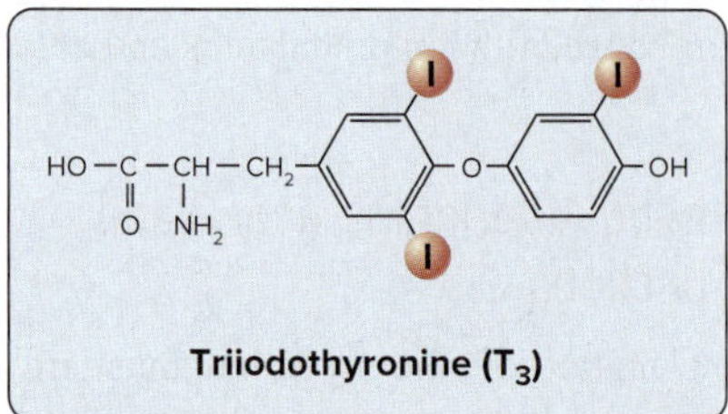


**(b) Structure of triiodothyronine (thyroid hormone)**

**Figure 38.4 The thyroid gland and synthesis of thyroid hormones.** **(a)** Location and structure of the thyroid gland in a human. **(b)** The structure of triiodothyronine , or $T_3$.

TSH stimulates certain cells of the thyroid gland to begin the process of synthesizing thyroid hormone. Thyroid hormone exerts a negative feedback effect on the hypothalamus and anterior pituitary gland, preventing them from overproducing TRH and TSH, thereby keeping the circulating concentration of thyroid hormone in check.

Thyroid hormone is unique among hormones in that it contains three or four iodine molecules. The most active form of thyroid hormone has three iodines and is called **triiodothyronine ($T_3$)** (Figure 38.4b).

## Thyroid Hormone Stimulates Metabolism

A major action of thyroid hormone in adult animals is to stimulate energy utilization by many different cell types. This occurs in large part by increasing the number and activity of the $Na^+/K^+$-ATPase pumps in plasma membranes. As these pumps hydrolyze ATP, the cellular concentration of ATP decreases. This decrease is compensated for by increasing the cell's metabolism of glucose, which provides the energy to produce more ATP. Whenever metabolism is increased, heat production is increased because heat is a by-product of the enzymatic breakdown of glucose and other energy sources. Consequently, a person with a hyperactive thyroid gland (**hyperthyroidism**) generally feels warm, whereas the opposite condition (**hypothyroidism**) results in a sensation of coldness. The increased metabolism associated with hyperthyroidism also stimulates a person's appetite. Despite that, however, such individuals generally lose weight as the body continues to use energy at very high rates (see the chapter-opening story). By contrast, hypothyroidism is usually associated with a sluggish metabolism and weight gain, despite reduced appetite.

The observation that thyroid hormone cannot be made without iodine presents some interesting and unique features of this hormone. The left side of Figure 38.5a shows the pattern of $T_3$ production from a healthy thyroid gland when iodine ingestion is adequate. The right side of the figure shows the consequences when thyroid hormone is not produced in normal amounts, as occurs due to a lack of iodine in the diet. In such a case, the decreased $T_3$ concentration provides less negative feedback on the hypothalamus and anterior pituitary gland, resulting in elevated TRH and, consequently, elevated TSH concentrations. The thyroid gland responds to the increased TSH by increasing the cellular machinery needed to produce more and more $T_3$, even though, in the absence of iodine, no additional $T_3$ can be synthesized. What results is an overgrown thyroid gland that still lacks the

## Biology Principle

### Living Organisms Maintain Homeostasis

Recall from Chapter 32 that a key way in which homeostasis is maintained is via negative feedback. Notice in this figure how a change in dietary intake of iodine results in decreased negative feedback and subsequent overgrowth of the thyroid gland.

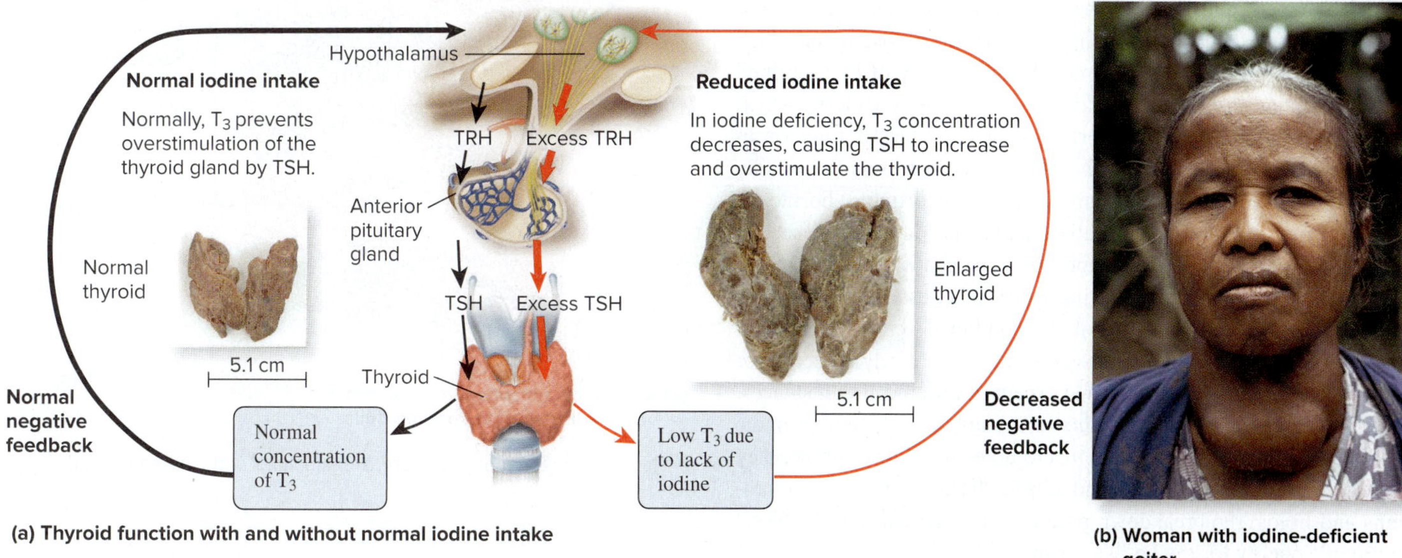


**(a) Thyroid function with and without normal iodine intake**

**(b) Woman with iodine-deficient goiter**

**Figure 38.5** **Consequences of normal and inadequate iodine in the diet. (a)** With normal iodine intake, as shown on the left, $T_3$ inhibits TSH secretion by negative feedback, thereby keeping the concentrations of the hormones in check. Without enough iodine, as shown on the right, less $T_3$ is synthesized, and the TSH concentration increases as feedback is removed. This leads to the enlargement of the thyroid gland. **(b)** An extreme example of an enlarged thyroid gland, or goiter, due to iodine deficiency.

*(a left and right inset)* © S. Goodwin & Dr. Max Hincke, Division of Clinical and Functional Anatomy, University of Ottawa; *(b)* © Bruce Coleman Inc./Alamy

resources to make thyroid hormone. This condition is known as an **iodine-deficient goiter** (Figure 38.5b). In humans, the problem can be alleviated either by adding iodine to the diet or by taking thyroid hormone pills. Goiters are not unique to humans. Iodine deficiency is relatively common among vertebrates, and goiters are found frequently in reptiles and birds, particularly those that subsist on all-seed diets, which are generally low in iodine.

## Hormones of the Adrenal Glands and Pancreas Regulate the Concentrations of Energy-Yielding Molecules in the Blood

Thyroid hormone regulates an animal's metabolism. For metabolism to proceed normally, however, body cells must have adequate sources of available energy, usually in the form of glucose and fatty acids. The brain, in particular, must have a constant supply of glucose because brain cells cannot metabolize fatty acids for energy and cannot store glycogen. Regulation of energy availability to cells is in large part mediated by the hormones of the adrenal glands and the pancreas.

**Adrenal Glands** The **adrenal glands,** which sit atop the kidneys (Figure 38.6), function in glucose metabolism by producing amine and steroid hormones. The inner portion of the glands, the adrenal medulla, produces the amine hormones epinephrine and norepinephrine. Their importance in the fight-or-flight response was described in Chapter 33. If the blood glucose concentration ever decreases below normal, neurons from the sympathetic nervous system activate the adrenal medulla to release these hormones, which stimulate the liver to produce glucose that is then secreted into the blood.

The outer part of the adrenal glands—the cortex—is itself subdivided into three zones: the glomerulosa, the fasciculata, and the reticularis (see Figure 38.6b). The outermost zone, the glomerulosa, is the region that makes the mineralocorticoid aldosterone, which acts to maintain $Na^+$, $K^+$, and water balance and will be described later. The innermost cortical zone is known as the reticularis, which functions in humans to make certain androgens, but its function in other animals is not as clear. The bulk of the cortex is the middle zone, the fasciculata, which produces glucocorticoid hormones such as cortisol. **Cortisol** is called a glucocorticoid because one of its major actions is to promote gluconeogenesis (refer back to Figure 37.13b) in the liver during times of stress, thus providing glucose to the blood. Because of the combined actions of adrenal gland hormones, the blood glucose concentration rarely decreases or remains significantly below normal except in extreme circumstances. Glucocorticoids are also catabolic hormones; that is, they promote the breakdown of molecules and macromolecules. For example, they act on bone, immune, muscle, and adipose tissue to break down proteins and lipids; the breakdown products are released into the blood to provide energy for the body's organs.

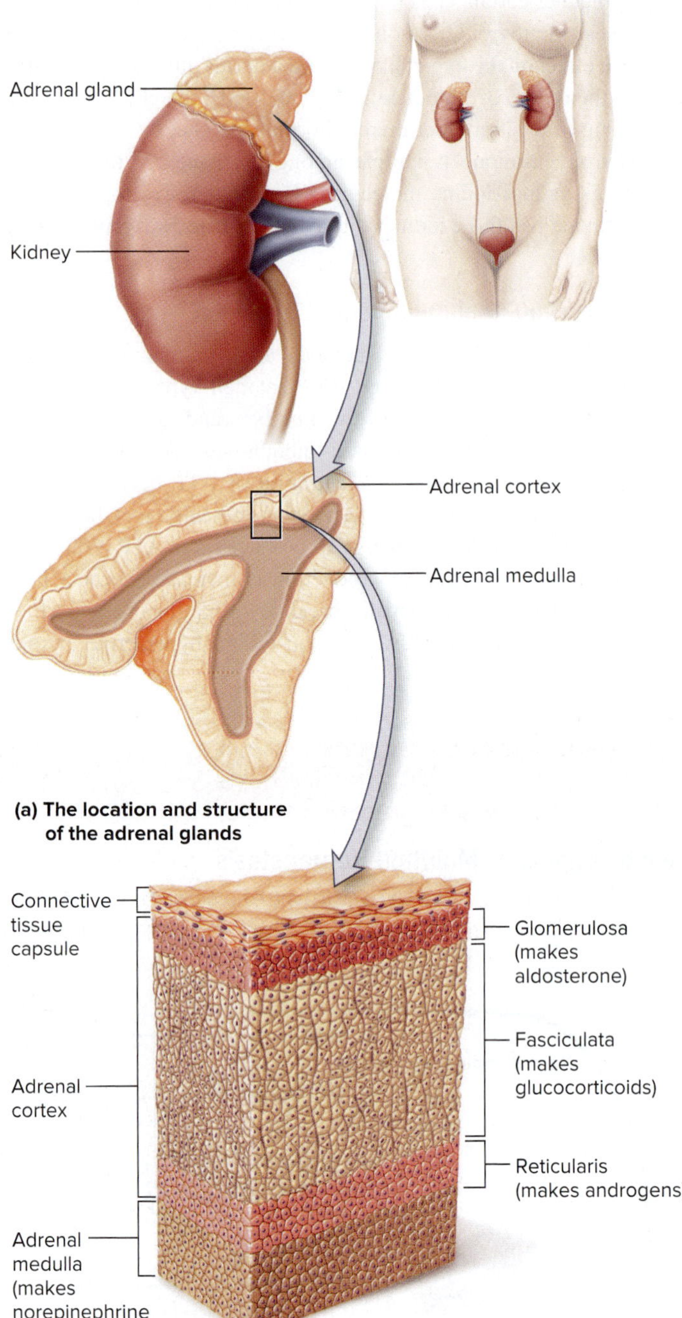


**Figure 38.6** Location, structure, and functions of the adrenal glands.

**Pancreas** The **pancreas** is a complex organ that is both an exocrine and endocrine gland (Figure 38.7). As you learned in Chapter 37, the secretions of the exocrine cells empty into the pancreatic duct and from there into the small intestine, where they aid digestion. The endocrine portion of the pancreas consists of cells that produce polypeptide hormones. Spherical clusters of endocrine cells called **islets of Langerhans** are scattered in large numbers throughout the pancreas. Within the islets are alpha cells, which make **glucagon,** and beta cells, which make **insulin.** The actions of these two hormones are antagonistic with respect to each other—insulin decreases and glucagon increases blood glucose concentrations.

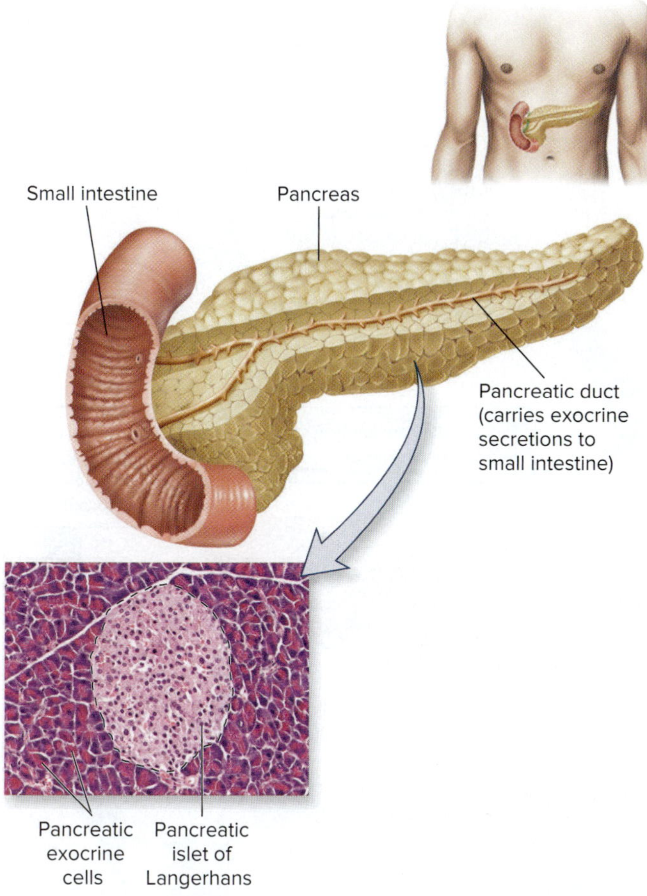


**Figure 38.7 Location, appearance, and internal structure of the mammalian pancreas.** Amidst the exocrine pancreas are scattered islets of Langerhans, which are endocrine tissue. The exocrine products are secreted into the intestine; the hormones from the islets of Langerhans are secreted into the blood.
*(bottom left)* © Cultura RM/Alamy

**BioConnections:** *The pancreas contains both exocrine and endocrine tissue. Is this structural feature ever observed in other organs? (Hint: Think about the processes described in Chapter 37 associated with digestion and absorption of food.)*

Maintaining homeostatic concentrations of glucose and other nutrients in the blood is a vital process that keeps cells functioning optimally. When an animal has not eaten for some time, its energy stores become depleted, and the blood glucose concentration begins to decrease. Under these conditions, glucagon is secreted into the blood, where it acts on the liver (Figure 38.8, steps 2b–4b). The liver contains a supply of glucose in the form of stored glycogen. Glucagon stimulates the breakdown of glycogen into many molecules of glucose, which are then secreted into the blood. This process, known as glycogenolysis (refer back to Figure 37.13a), is stimulated within seconds by glucagon. A second action of glucagon on the liver is particularly important for responses to prolonged fasting. In that case, glucagon stimulates gluconeogenesis, which also results in glucose being released into the blood.

In contrast to fasting, after an animal eats a meal, the concentrations of glucose and other nutrients in the blood become elevated. Restoring the normal blood concentrations of glucose, fats, and amino acids is almost exclusively under the control of insulin, one of the few hormones that are absolutely essential for survival in animals. The secretion of insulin is directly stimulated by an increased concentration of glucose in the blood (Figure 38.8, steps 2a–3a). Once in the blood, insulin acts on plasma membrane receptors located primarily in cells of adipose, skeletal, and cardiac muscle tissues. There, insulin acts to facilitate the transport of glucose across the plasma membrane into the cytosol. Once in the cell, glucose is metabolized for cellular functions or used in the synthesis of glycogen or triglycerides. As discussed in Chapter 37, glucose transport involves the actions of proteins called glucose transporters (GLUTs) (refer back to Figure 37.14). In the absence of insulin, most of these proteins are located in cytosolic membrane-bound vesicles. The major function of insulin is to stimulate fusion of these vesicles with the plasma membrane. Once a vesicle fuses with the plasma membrane, GLUTs begin transporting glucose from the extracellular fluid into the cell, thereby lowering the blood glucose concentration.

When the blood glucose concentration returns to normal, the stimulus for insulin secretion no longer exists, and the insulin concentration in the blood decreases. This results in a decrease in the number of GLUTs in plasma membranes, because the GLUTs are subjected to endocytosis and thereby return to membrane vesicles in the cytosol until required again. In addition, insulin also stimulates transport of amino acids into cells and promotes fat deposition in adipose tissue. In other words, the broader role of insulin is to facilitate the transfer of energy from the extracellular fluid into storage sites, primarily in muscle and fat.

In the absence of sufficient insulin, as occurs in the disease **type 1 diabetes mellitus (T1DM),** less extracellular glucose can cross plasma membranes, and consequently, glucose accumulates to a very high concentration in the blood. T1DM occurs in many vertebrates. It is caused when an animal's immune system attacks and destroys the insulin-producing cells of the islets of Langerhans. One consequence of the disease is that muscle and adipose cells cannot receive their normal amount of glucose to provide the ATP they require. In addition, the unusually large amount of glucose in the blood overwhelms the kidneys' ability to reabsorb it from the kidney filtrate, and glucose appears in the urine. Fortunately, in humans and in common house pets this form of the disease is treatable with regular monitoring of blood glucose concentration and daily administration of insulin.

The far more common form of diabetes in humans is **type 2 diabetes mellitus (T2DM).** In T2DM, the pancreas may function relatively normally and is not attacked by the immune system. The concentration of insulin in the blood may in some cases even be greater than normal. However, the cells of the body lose much of their ability to respond to insulin for reasons that are still unclear. T2DM is linked to obesity and can often be prevented or reversed with weight control. Drugs are also available that improve the ability of cells to respond to insulin.

The discovery of insulin and its application to the treatment of diabetes was one of the greatest and most influential achievements in the history of medical research, as described next.

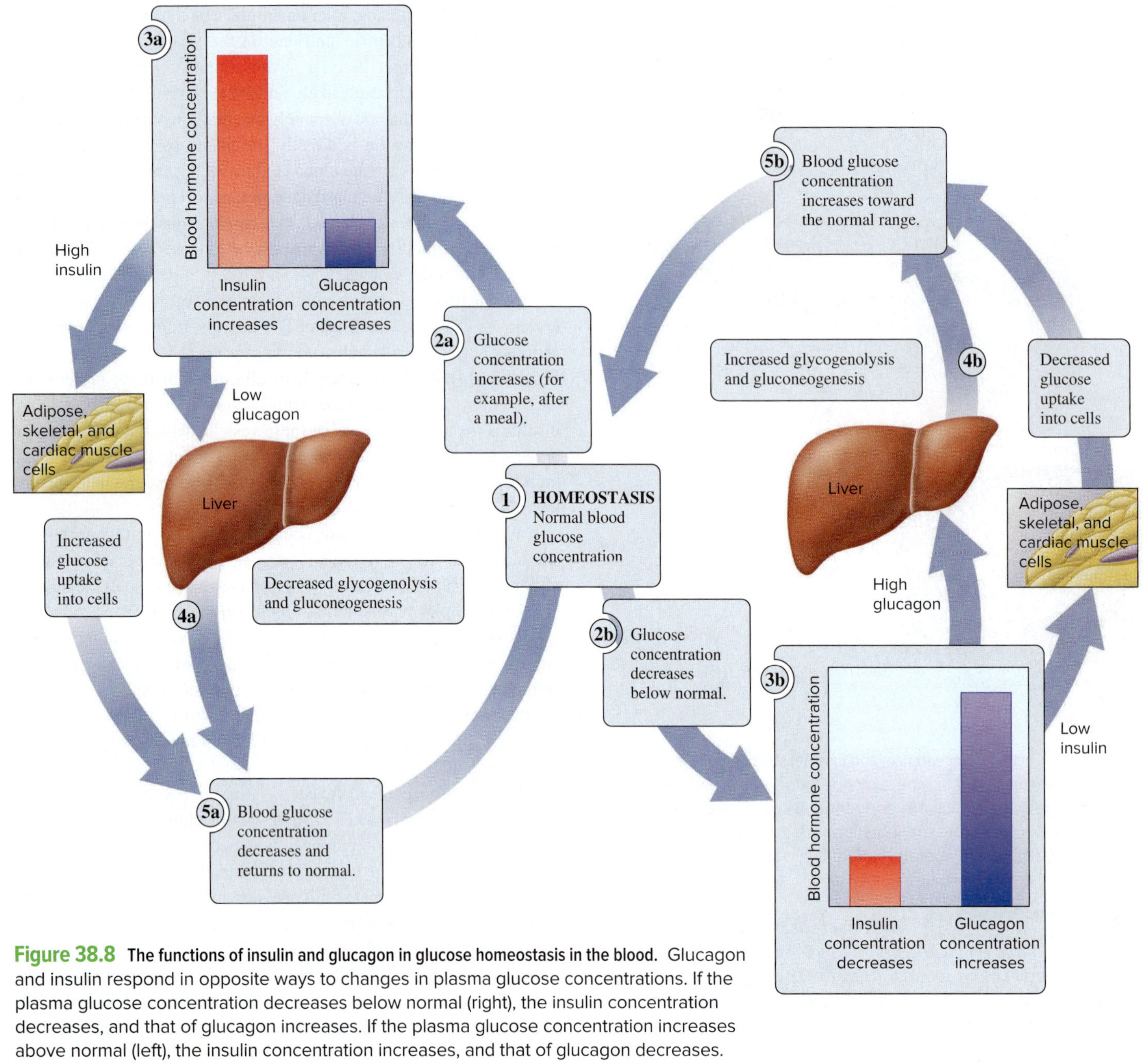


**Figure 38.8 The functions of insulin and glucagon in glucose homeostasis in the blood.** Glucagon and insulin respond in opposite ways to changes in plasma glucose concentrations. If the plasma glucose concentration decreases below normal (right), the insulin concentration decreases, and that of glucagon increases. If the plasma glucose concentration increases above normal (left), the insulin concentration increases, and that of glucagon decreases. Both of these mechanisms return the plasma glucose concentration to normal.

# FEATURE INVESTIGATION

## Banting, Best, Collip, and MacLeod Were the First to Isolate Active Insulin

By the early 20th century, scientists had discovered that the complete removal of the pancreas from experimental animals resulted immediately in T1DM. Researchers assumed, therefore, that cells of the pancreas produced a factor of some kind that prevented diabetes. The factor, however, proved impossible to isolate using the chemical purification methods available at the time. Researchers assumed that during the process of grinding up a pancreas to produce an extract, the factor was destroyed by the digestive enzymes of the exocrine parts of the pancreas.

This problem was eventually solved in 1921 by a team of Canadian scientists working at the University of Toronto: surgeon

Frederick Banting, medical student Charles Best, and biochemist James Collip. The work was performed in the laboratory of John MacLeod, a renowned expert in carbohydrate metabolism at the time. Banting had read a paper in a medical journal that described a deceased patient in whom the pancreatic duct, which carries digestive juices to the small intestine, had become clogged due to calcium deposits. The closed duct caused pressure to build up behind the blockage, which eventually caused the exocrine part of the pancreas to atrophy and die. The islets of Langerhans, however, which are not connected to ducts, survived intact. Banting hypothesized that if he were to tie off, or ligate, the pancreatic ducts of an animal, and then wait a sufficient time for the exocrine pancreas to die, he could more easily obtain an active glucose-lowering factor from the remaining islets without the problem of contamination by digestive enzymes.

Banting and Best proceeded to ligate the pancreatic ducts of several dogs (**Figure 38.9**). After waiting 7 weeks, an amount of time they previously had determined was sufficient for the exocrine part of the pancreas to atrophy, they prepared extracts of the remaining parts of the pancreas, including its islets of Langerhans. This was done by removing the atrophied pancreas from each dog and grinding it up with a mortar and pestle in an acid solution. The extract was then injected into a second group of dogs in which diabetes had previously been induced by surgically removing the

## Biology Principle

### Biology Affects Our Society

The discovery of insulin and the methods to manufacture it in large amounts has improved (and in many cases, saved) the lives of many millions of people. Today, dozens of human diseases, including diabetes, are treatable with the use of hormones.

**Figure 38.9** **The isolation of insulin by Banting, Best, Collip, and MacLeod.**

**HYPOTHESIS** Ligation of pancreatic ducts will cause atrophy of the exocrine pancreas, allowing extraction of an active glucose-lowering factor from the remaining endocrine portion of the pancreas.

**KEY MATERIALS** One group of dogs for ligation experiments; second group of dogs made diabetic by having pancreas removed.

**Experimental level** | **Conceptual level**

1 Ligate pancreatic ducts in one group of dogs by tying threads around the base of the ducts and pulling them tight.

Surgeon operating on pancreas of a dog

Ligated ducts block flow of digestive juices, which damages exocrine pancreas.

Pancreas

Islets of Langerhans

Small intestine

2 Allow 7 weeks for atrophy of pancreas.

Pancreas atrophies, but islets remain intact.

3 Remove atrophied pancreas. Prepare extract by grinding up tissue in acid. Purify by adding alcohol, filtering, removing lipids, and concentrating.

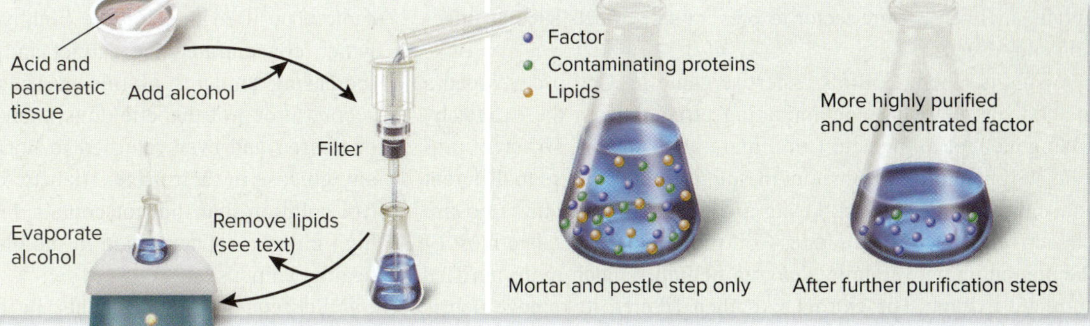

4 Remove pancreas from a second group of dogs.

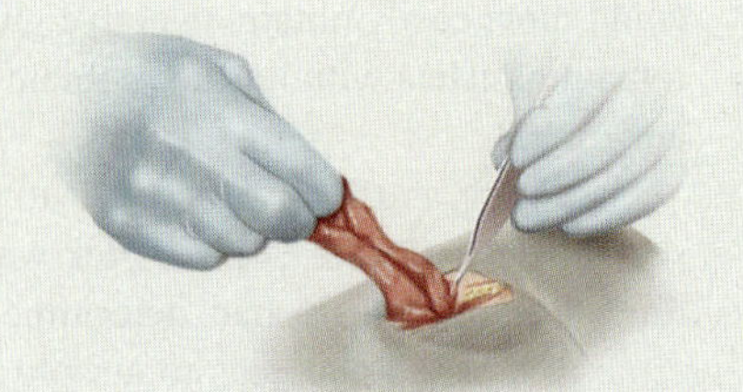

The concentration of glucose in the blood increases due to diabetes.

Glucose appears in the urine, a sign of diabetes.

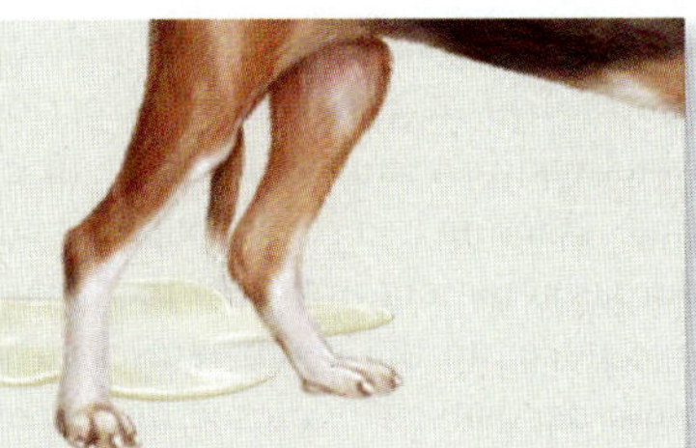

5 Inject either purified extract or control solution into the diabetic dogs. Determine the levels of glucose in the blood and in the urine.

The control solution contained all of the chemicals used to prepare the extract but did not include any material from a pancreas.

6 **THE DATA**

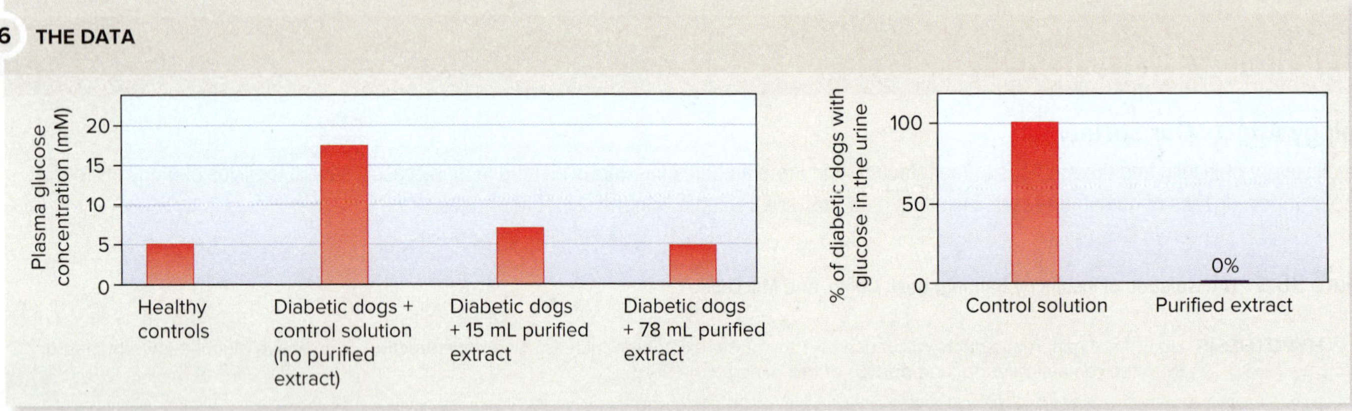


7 **CONCLUSION** Atrophy of the exocrine portion of the pancreas eliminated digestive enzymes that would have degraded the glucose-lowering factor (insulin) during its purification from the pancreas. Insulin was effective in reversing the major symptoms of diabetes in dogs.

8 **SOURCE** Banting, F. G., and Best, C. H. 1922. The Internal Secretion of the Pancreas. *Journal of Laboratory and Clinical Medicine* 7: 256–271.

entire pancreas. The researchers discovered that the extract briefly kept the diabetic dogs alive. However, they were unable to isolate sufficient quantities of the active factor in the extract to keep the experiments going longer than a day or so, nor were they able to obtain sufficiently pure factor to prevent side effects such as infection and fever.

This is where Collip's expertise came in. Collip developed a method for removing contaminating proteins from the extract by adding alcohol to the acid (see Figure 38.9, step 3). Different proteins will precipitate (come out of solution) in response to different concentrations of alcohol. Using a concentration too low to cause the active factor to precipitate, Collip was able to remove most of the contaminating proteins, leaving behind a much more purified and safe extract. The extract was then filtered to remove debris and further extracted to remove lipids. In this step, a hydrophobic solvent was added to the partially purified extract; the lipids in the extract dissolved in the solvent, which could then be removed. This was done because the researchers correctly believed that the factor was a polypeptide and, by removing lipids, they would obtain a more purified preparation. Finally, the alcohol was evaporated, which concentrated the extract and increased its potency. When increasing amounts of this purified factor were injected into a second group of diabetic dogs, their blood glucose concentrations decreased and even returned to normal, and there was no longer any glucose in their urine. Diabetic dogs that received only a control solution that did not contain the factor continued to show a high glucose concentration in the blood and urine, compared with healthy animals.

Subsequent improvements in their procedures enabled the researchers to obtain larger amounts of the factor and to more accurately assess its potency without using the ligation procedure. The first successful test in humans of the extract, which they eventually would name insulin, came in 1922 on a 14-year-old boy in Toronto, who was seriously ill from T1DM. The success of the team in rapidly isolating insulin and proving its effectiveness was so immediately significant that Banting and MacLeod were awarded the 1923 Nobel Prize in Medicine or Physiology.

*Experimental Questions*

1. How did Banting and Best propose to obtain the glucose-lowering factor produced by the pancreas?
2. What was Collip's contribution to the isolation of insulin?
3. **SCISKILLS** ▶ Propose a reason for why it was it necessary to inject insulin, as opposed to, for example, giving it in the form of an oral solution.

## 38.3 Reviewing the Concepts

- The thyroid gland makes thyroid hormone, which contains iodine. A major action of thyroid hormone in adult animals is to stimulate energy use by many different cell types (Figures 38.4, 38.5).
- The adrenal glands produce glucocorticoids, which increase the blood glucose concentration. The endocrine pancreas produces insulin and glucagon, which have opposite effects on the blood glucose concentration (Figures 38.6, 38.7, 38.8).
- Groundbreaking research by Banting, Best, Collip, and MacLeod identified and isolated insulin for therapeutic use in treating diabetes mellitus (Figure 38.9).

## 38.3 Testing Your Knowledge

1. Chronic deficiency of iodine in the diet of a person will lead to
   a. increased secretion of TRH, decreased secretion of TSH, decreased $T_3$ concentration in the blood, and a goiter.
   b. decreased secretion of TRH and TSH, decreased $T_3$ concentration in the blood, and a goiter.
   c. increased $T_3$ concentration in the blood, decreased TRH and TSH secretion, and no goiter.
   d. decreased $T_3$ concentration in the blood, increased TRH and TSH secretion, and a goiter.
   e. decreased secretion of TRH, increased secretion of TSH, decreased $T_3$ concentration in the blood, and a goiter.
2. Type _______ diabetes mellitus is the most common form, and is associated with _______ .
   a. 1, immune disease
   b. 2, immune disease
   c. 1, obesity
   d. 2, obesity

# 38.4 Hormonal Control of Mineral Balance

### Learning Outcomes

1. Describe the hormonal control of $Ca^{2+}$ homeostasis.
2. Explain the regulation in the kidneys of water, $Na^+$ and $K^+$ balance by antidiuretic hormone and aldosterone.

All animals must maintain a balance in their cells and fluids of minerals such as $Ca^{2+}$, $Na^+$, and $K^+$. These ions participate in many functions common throughout much of the animal kingdom. For example, the movement of ions across plasma membranes determines the electrical properties of neurons and muscle cells. In this section, we will see how hormones maintain a homeostatic balance of these ions in an animal's body.

## Vitamin D and Parathyroid Hormone Regulate the Concentration of $Ca^{2+}$ in the Blood

Calcium ions have critical roles in neuronal transmission, heart function, muscle contraction, and numerous other events. Therefore, the concentration of $Ca^{2+}$ in the blood is among the most tightly regulated variables in an animal's body.

Calcium is obtained from the diet and absorbed by the small intestine. Like all hydrophilic substances, $Ca^{2+}$ cannot readily cross plasma membranes, including those of the epithelial cells of the small intestine, and thus its transport must be regulated. This process is controlled by the action of a derivative of **vitamin D.** In most mammals that receive regular exposure to sunlight, skin cells produce vitamin D from a precursor called 7-dehydrocholesterol, a reaction that requires the energy of ultraviolet light (Figure 38.10). In addition, many animals obtain vitamin D from food or, in humans, from milk and other foods supplemented with the vitamin. This is especially important for people who live at latitudes that receive little sunlight for much or part of the year.

Before it can act, vitamin D must first be modified by two enzymes that each add one hydroxyl group to specific carbon atoms, first in the liver and then in the kidney. The final active product is called **1,25-dihydroxyvitamin D.** This molecule is a hormone because it is released into the blood by one organ—the kidneys—and then acts on a distant target tissue—the small intestine. The major function of 1,25-dihydroxyvitamin D in the intestine is to activate expression of a $Ca^{2+}$ active transporter in the intestinal epithelium, thereby stimulating the absorption of $Ca^{2+}$. The $Ca^{2+}$ then enters the blood, from where it is delivered to tissues for such activities as building bone and maintaining nerve, muscle, and heart functions.

When $Ca^{2+}$ is not present in the diet, or when 1,25-dihydroxyvitamin D is not formed in normal amounts because of insufficient exposure to sunlight, the blood concentration of $Ca^{2+}$ does not normally decrease significantly, because of a hormone called **parathyroid hormone (PTH).** This hormone is secreted from several small glands called parathyroid glands, which in humans are located at the back of the thyroid gland (Figure 38.11a). Cells of the parathyroid gland express plasma membrane receptors that bind extracellular $Ca^{2+}$. The

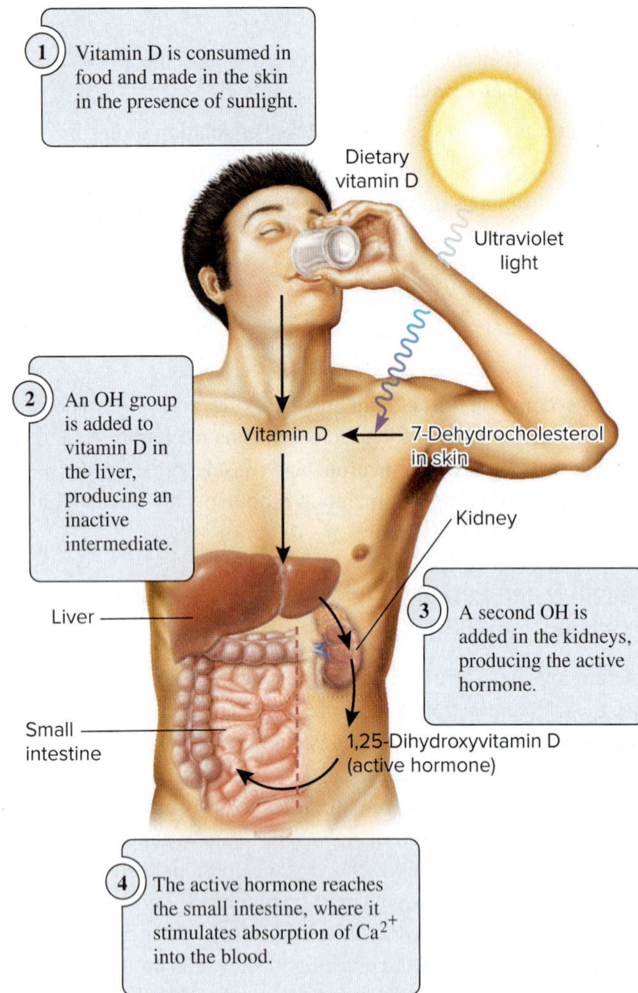


**Figure 38.10 Synthesis of the active hormone formed from vitamin D or its precursor in the skin.**

**Concept Check:** *Would you predict that all mammals synthesize the active form of vitamin D using the energy of sunlight?*

binding of $Ca^{2+}$ to these receptors inhibits the secretion of PTH. Thus, a decrease in the blood concentration of $Ca^{2+}$ stimulates the secretion of PTH. PTH acts on bone to stimulate the activity of cells that dissolve the mineral part of bone. This releases $Ca^{2+}$, which enters the blood (Figure 38.11b). Typically, only a tiny fraction of the total $Ca^{2+}$ in bone is removed in this way. Thus, besides providing a skeletal framework for the vertebrate body, bone also serves as an important reservoir of $Ca^{2+}$. PTH also acts to increase reabsorption of $Ca^{2+}$ from the filtrate in the kidneys, so less $Ca^{2+}$ is excreted in the urine. Without PTH, $Ca^{2+}$ homeostasis is not possible. Complete absence of PTH is fatal in humans and other mammals.

In addition to 1,25-dihydroxyvitamin D and PTH, another hormone called **calcitonin** has a role in $Ca^{2+}$ homeostasis in some vertebrates, notably fishes and some mammals. Calcitonin is secreted by cells in the thyroid gland. Its function is opposite in many respects to that of PTH. Calcitonin promotes excretion of $Ca^{2+}$ into the urine

## Biology Principle

### Living Organisms Maintain Homeostasis

Homeostatic control of $Ca^{2+}$ concentrations is vitally important to life in most living organisms, and in all animals. Because of the actions of PTH, the $Ca^{2+}$ concentration in the body fluids of humans rarely changes by more than a slight margin.

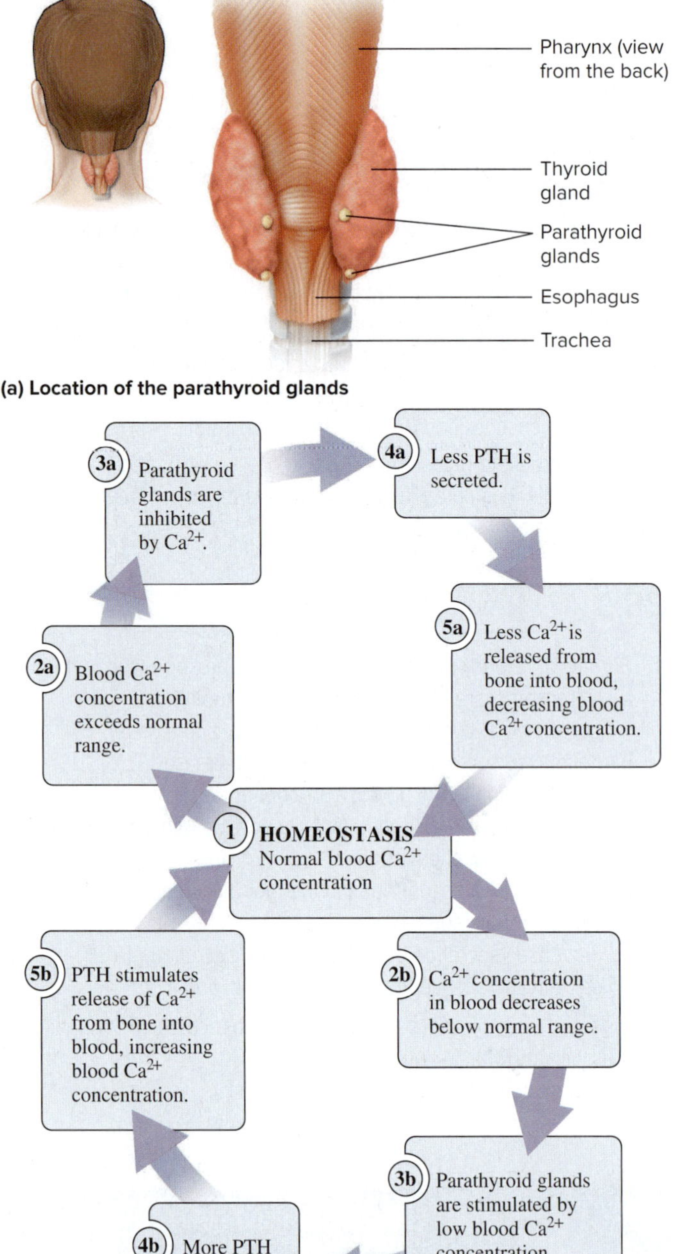


**Figure 38.11 The role of parathyroid hormone (PTH) in $Ca^{2+}$ homeostasis.** **(a)** The four parathyroid glands are located behind the thyroid. **(b)** The action of PTH: Steps 2a–5a occur when $Ca^{2+}$ is in excess; steps 2b–5b occur when $Ca^{2+}$ is below normal.

and deposition of $Ca^{2+}$ into bone, thereby decreasing the blood $Ca^{2+}$ concentration. This is especially important in marine fishes, because of the high $Ca^{2+}$ content of seawater and the entry of this ion into the body fluids of the animals.

BIO TIPS ONLINE

## Two Hormones Regulate the Extracellular Fluid Concentrations of $Na^+$ and $K^+$ in Vertebrates

Like $Ca^{2+}$, concentrations of $Na^+$ and $K^+$ in the body fluids of most animals are tightly regulated, because these ions are the basis of membrane potential formation and action potential generation (see Chapter 33), among other functions. Like $Ca^{2+}$, $Na^+$ and $K^+$ are ingested in the diet and excreted in the urine in vertebrates at rates that maintain homeostatic concentrations in the body fluids. Unlike $Ca^{2+}$, however, no large reservoirs exist in the body for $Na^+$ and $K^+$. One of the key mechanisms for regulating the concentrations of these ions in the blood is altering the rate of $Na^+$, $K^+$, and water reabsorption from the urine as it is being formed in the kidneys. This is accomplished in large part by the actions of two hormones: ADH and aldosterone (**Figure 38.12**), both of which exert their effects on the kidneys.

The vertebrate kidneys normally reabsorb most $Na^+$ and $K^+$ from the fluid that is filtered through the kidney glomeruli (refer back to Figures 37.20 through 37.22). However, dietary and other changes can alter the concentrations of these ions in the blood. When this happens, the kidneys work to restore the ions to their normal concentrations. For example, if the blood $Na^+$ concentration increases above normal, the osmolarity of the blood will increase. Osmoreceptors in the brain detect this and stimulate ADH secretion from the posterior pituitary. Recall from Chapter 37 that ADH acts on the kidneys to reabsorb water from the forming urine. By increasing the amount of ADH available to act on the kidneys, less water is excreted in the urine and more is reabsorbed into the blood. The reabsorption of water into the blood decreases the blood $Na^+$ concentration back toward normal. In addition, the synthesis and secretion of the adrenal steroid hormone **aldosterone** are directly inhibited in response to an increase in $Na^+$ concentration. Normally, aldosterone increases $Na^+$ reabsorption in the kidneys. Therefore, a decrease in aldosterone in the blood results in a greater amount of $Na^+$ excretion in the urine than normal (refer back to Figure 37.22). Aldosterone also regulates the extracellular fluid concentration of $K^+$. If the $K^+$ concentration increases above normal, aldosterone release is increased and this stimulates secretion of $K^+$ into the urine.

Aldosterone and 1,25-dihydroxyvitamin D are both steroid hormones, as is cortisol, described earlier. Their structures are similar but distinct (see Figure 38.2). These characteristics have enabled biologists to use steroid hormones as models for understanding how hormones and their receptors may have coevolved, as described next.

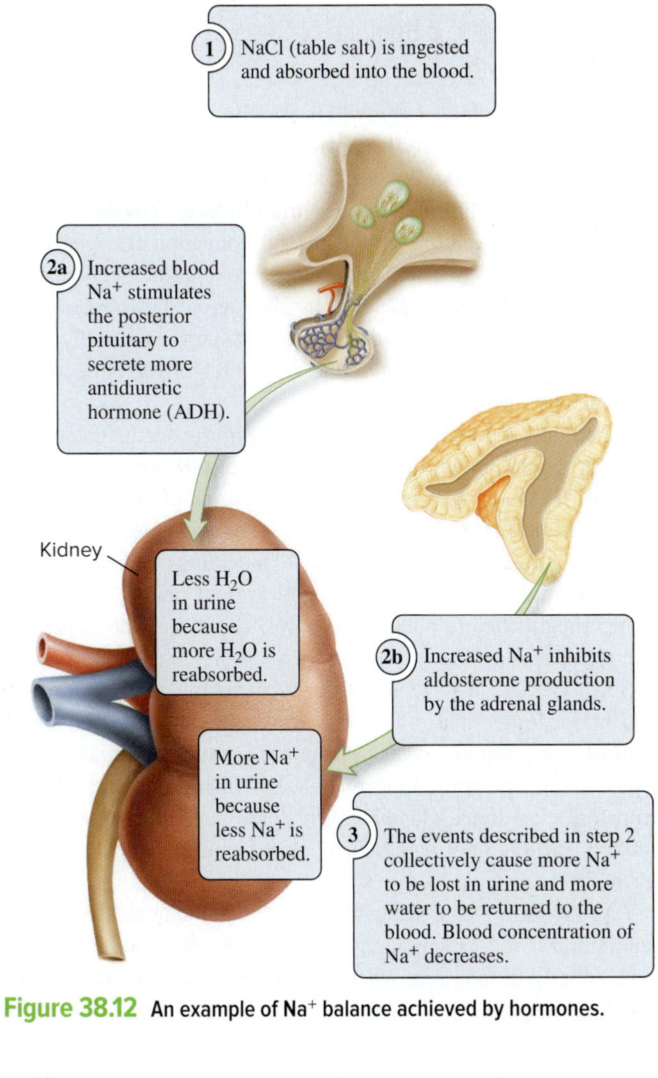

(1) NaCl (table salt) is ingested and absorbed into the blood.
(2a) Increased blood $Na^+$ stimulates the posterior pituitary to secrete more antidiuretic hormone (ADH).
Kidney
Less $H_2O$ in urine because more $H_2O$ is reabsorbed.
(2b) Increased $Na^+$ inhibits aldosterone production by the adrenal glands.
More $Na^+$ in urine because less $Na^+$ is reabsorbed.
(3) The events described in step 2 collectively cause more $Na^+$ to be lost in urine and more water to be returned to the blood. Blood concentration of $Na^+$ decreases.

**Figure 38.12** An example of $Na^+$ balance achieved by hormones.

# EVOLUTIONARY CONNECTIONS

## Hormones and Receptors Evolved as Tightly Integrated Molecular Systems

We have learned that hormones act by binding to receptor proteins located on the cell surface or in the cytosol or nucleus. This raises an intriguing question: Which evolved first, a given hormone or its receptor? There would appear to be no selection pressure to evolve one without the other. Without a receptor, a hormone has no function, and without a hormone, its receptor has no function. Hormones and their receptors function as tightly integrated molecular systems. Each molecule depends on the other for biological activity. How such systems could have evolved, or whether they evolved together, has long been a major puzzle. However, recent research has generated intriguing new hypotheses about the evolution of these signaling systems.

American biologist Joseph Thornton and colleagues recently addressed this puzzle by examining how two structurally related steroid hormones evolved separate activities and distinct receptors. Aldosterone, as noted, is a steroid hormone made by the adrenal cortex. Because it regulates the balance of two minerals ($Na^+$ and $K^+$) in the body, it is known as a mineralocorticoid. Cortisol is another steroid hormone made by the adrenal cortex and is known as a glucocorticoid because one of its major functions is to regulate glucose balance. The actions of these hormones are mediated by intracellular receptors known as the mineralocorticoid receptor (MR) and the glucocorticoid receptor (GR), respectively. Phylogenetic analysis of the gene sequences that code for these two receptors suggests that they arose at least 450 mya, after the evolution of the first vertebrates, by duplication of an ancient corticoid receptor (CR) gene.

The researchers analyzed the known sequences of the genes for the MR and GR of many vertebrate species, including the most ancient vertebrates, to deduce a theoretical sequence of an ancestral CR. They then synthesized this CR and tested its ability to bind aldosterone and cortisol. They discovered that the CR was capable of binding both hormones, particularly aldosterone. This was surprising, because aldosterone is only present in tetrapods, which arose around 300 mya, long after the proposed gene duplication event that created the MR. Therefore, it appears that a receptor with high affinity for aldosterone was present long before animals acquired the capacity to synthesize aldosterone. The receptor seems to have evolved to bind other steroids. Around 450 mya, the gene for the CR duplicated, and then the two resulting genes seem to have further mutated such that one receptor gained high affinity primarily for mineralocorticoids and the other for glucocorticoids.

These studies suggest that an animal's ability to respond to aldosterone evolved long before aldosterone did, because a high affinity receptor was already in place in animals. When aldosterone appeared in tetrapods, it used the MR derivative of the ancestral CR. In this example, the answer to "which came first" appears to be that the receptor evolved first and the hormone later.

## 38.4 Reviewing the Concepts

- The concentration of $Ca^{2+}$ in the blood is tightly regulated. 1,25-Dihydroxyvitamin D and parathyroid hormone (PTH), produced by the parathyroid glands, regulate the blood concentration of $Ca^{2+}$ (Figures 38.10, 38.11).
- The blood concentrations of $Na^+$ and $K^+$ are regulated in large part by the actions of ADH, which increases water retention in the body and therefore controls ion concentration indirectly, and aldosterone, which increases $Na^+$ reabsorption and $K^+$ secretion in the urine (Figure 38.12).
- In at least one case, it appears that a hormone's receptor evolved before the hormone itself.

## 38.4 Testing Your Knowledge

1. Which is correct?
   a. Aldosterone decreases the sodium ion concentration in the blood.
   b. 1,25-Dihydroxyvitamin D acts on intestinal cells to increase $Ca^{2+}$ transport into the blood.
   c. Parathyroid hormone acts on intestinal cells to increase $Ca^{2+}$ transport into the blood.
   d. ADH promotes loss of water from the body via the urine.
   e. Aldosterone stimulates reabsorption of $K^+$ from the urine.

## 38.5 Hormonal Control of Growth and Development

### Learning Outcome

1. Identify and describe the functions of the hormones involved in the regulation of growth and development in animals.

Hormones have a central role in controlling growth and development in animals. Growth can occur slowly over long periods or in brief spurts, with some animals exhibiting both types of growth. Although growth is determined by many factors, notably adequate nutrition, it is regulated in large part by the endocrine system.

Growth is distinguished from development, which is the process by which cells form tissues with specific functions, and tissues develop into larger and more complex structures. In this section, we will examine how both processes depend in part on the endocrine system.

### Vertebrates Require a Balance of Several Hormones for Normal Growth

Several hormones stimulate growth in vertebrates. The anterior pituitary gland produces **growth hormone (GH),** which is under the control of the hypothalamus (see Table 38.1). GH acts on the liver to produce another hormone, called **insulin-like growth factor-1 (IGF-1).** In mammals, IGF-1 stimulates the elongation of bones, especially during puberty, when mammals become reproductively mature. This growth is further accelerated by the steroid hormones of the gonads, leading to the rapid pubertal growth spurt. Eventually, however, the gonadal hormones cause the growth regions of bone to seal, preventing any further bone elongation.

In humans, if the pituitary gland fails to make adequate amounts of GH during childhood, the concentrations of GH and IGF-1 in the blood will be lower than normal. In such cases, growth is stunted, resulting in one of the major causes of short stature. Individuals with this condition can be treated with injections of recombinant human GH and will grow to relatively normal height, as long as treatment begins before puberty is completed, so that the bones can still elongate.

Hormones are also required during fetal life for development of the brain, lungs, and other organs into fully mature, functional structures. For example, cortisol from the mammalian fetal adrenal gland is vital for proper lung formation. In humans, premature babies born before the adrenal glands have matured have lungs that are not capable of inflating normally; such babies require hospitalization to survive.

Thyroid hormone influences development among all vertebrates. In amphibians, for example, thyroid hormone promotes metamorphosis, notably in tadpoles, where it promotes the resorption of the tail and development of the legs (**Figure 38.13**). This effect can be

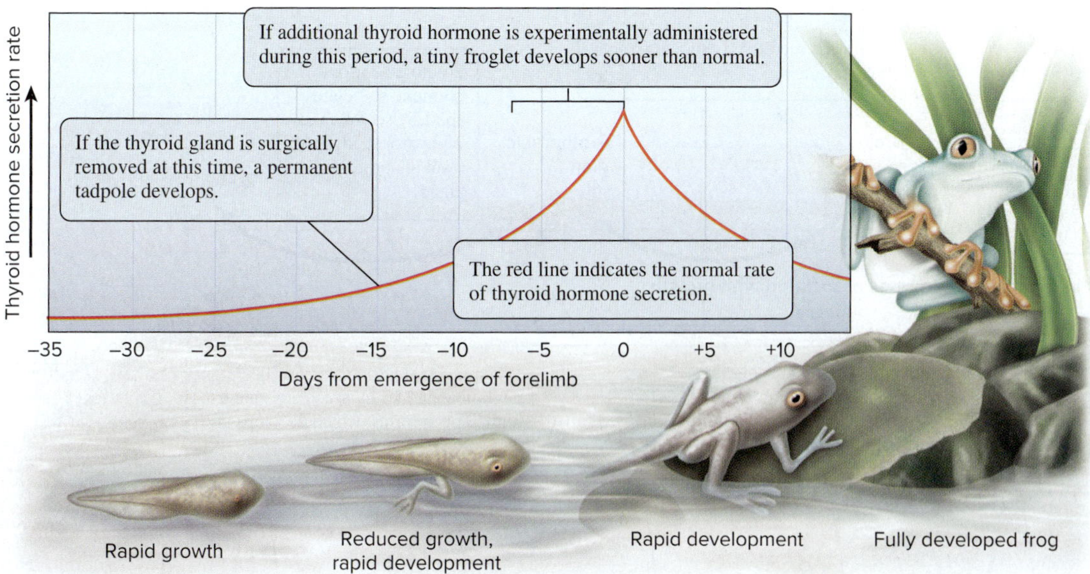


**Figure 38.13** **Effects of thyroid hormones on animal development.** Experimental manipulation of thyroid hormone concentration can slow down or accelerate development of tadpoles.

demonstrated by experimentally decreasing or increasing the concentration of thyroid hormone in the blood of a tadpole. Decreasing the thyroid hormone concentration results in a tadpole that does not transform into a frog. Increasing the thyroid hormone above normal results in a tiny froglet in which metamorphosis occurs sooner than normal.

## Many Invertebrates Grow and Develop in Spurts Under the Control of Three Hormones

Like vertebrate endocrine pathways, hormonal control systems in insects and other invertebrates often involve multiple glands and neural structures acting in concert. In insects, for example, the endocrine system is critical for the growth and development of larvae and their eventual development into pupae (Figure 38.14). In the case of larval growth and metamorphosis, specialized neurosecretory cells in the brain periodically secrete a hormone called **prothoracicotropic hormone (PTTH),** which then stimulates a pair of endocrine glands called the prothoracic glands. These glands, located in the thorax, synthesize and secrete into the hemolymph a steroid hormone called **20-hydroxyecdysone.** This hormone is secreted only periodically. In response to each burst of secretion, the larva undergoes a rapid development and molts (sheds its cuticle). It then begins a new developmental period until it molts again in response to another episode of secretion. The molting process is known as ecdysis, from which the name of the hormone is derived (refer back to Figure 26.8).

Throughout larval development, paired neurosecretory structures behind the brain, called the corpus allata, secrete another hormone, a protein called **juvenile hormone (JH).** JH determines the character of the molt that is induced by 20-hydroxyecdysone, by acting to prevent metamorphosis into an adult (hence the name juvenile, which reflects the fact that JH fosters the larval stage). As the larva ages, however, the amount of JH it produces gradually declines until its concentration is nearly zero. During this time, 20-hydroxyecdysone continues to be periodically secreted. The decline in JH below a certain concentration results in the transition from larva to pupa in response to a burst of 20-hydroxyecdysone; the near absence of JH is a prerequisite for the final step of metamorphosis into an adult.

## 38.5 Reviewing the Concepts

- Hormones play a crucial role in regulating growth and development. In vertebrates, normal growth depends on a balance between growth hormone (GH), insulin-like growth factor-1 (IGF-1), and gonadal hormones. Thyroid hormones affect development among all vertebrates (Figures 38.13).
- In insects, prothoracicotropic hormone (PTTH), 20-hydroxyecdysone, and juvenile hormone (JH) control the growth of larvae and their development into pupae (Figure 38.14).

## 38.5 Testing Your Knowledge

1. Some children with short stature can be effectively treated with
   a. cortisol.
   b. growth hormone.
   c. drugs that inhibit IFG-1 secretion.
   d. prothoracicotropic hormone.
   e. thyroid hormone.
2. ______ is a steroid hormone that ______ molting in insects.
   a. Prothoracicotropic hormone, stimulates
   b. 20-Hydroxyecdysone, stimulates
   c. Juvenile hormone, inhibits
   d. 1,25-Dihydroxyvitamin D, stimulates
   e. 20-Hydroxyecdysone, inhibits

**(a) Synthesis of invertebrate hormones associated with growth**

**(b) Effects of hormones on molting and pupation**

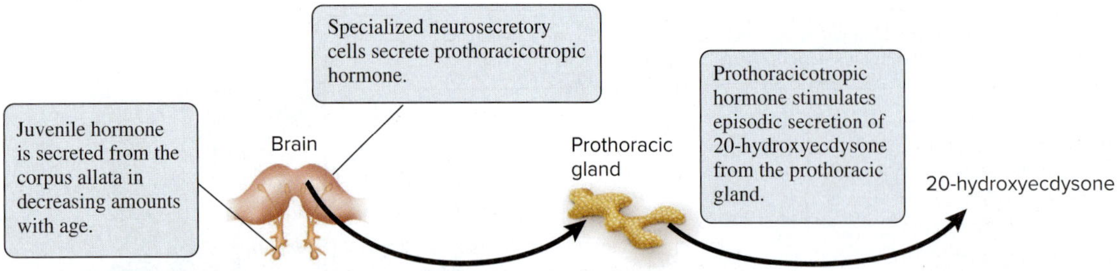

**Figure 38.14** **Hormonal control of insect development.** Development of insects requires the coordinated actions of three hormones. Note: The relative concentrations and patterns of 20-hydroxyecdysone and juvenile hormone in this figure are schematic and not representative of all insects.

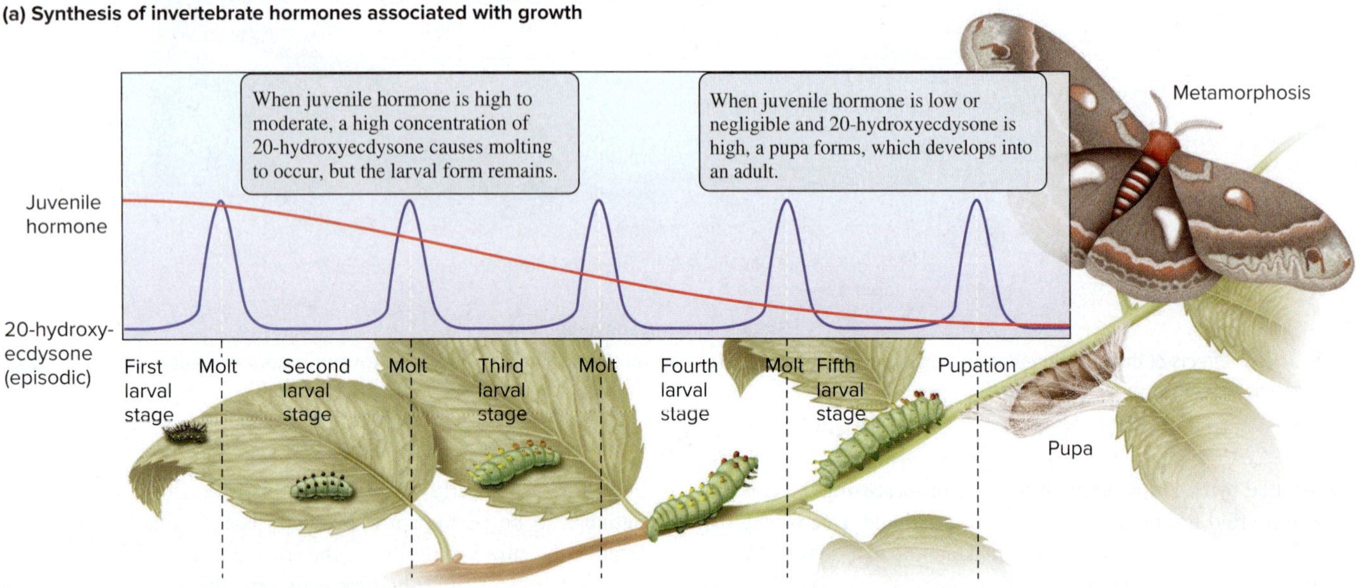

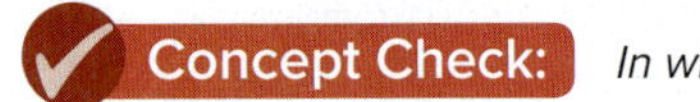

**Concept Check:** *In what part of a cell do you predict the receptor for the steroid 20-hydroxyecdysone is located?*

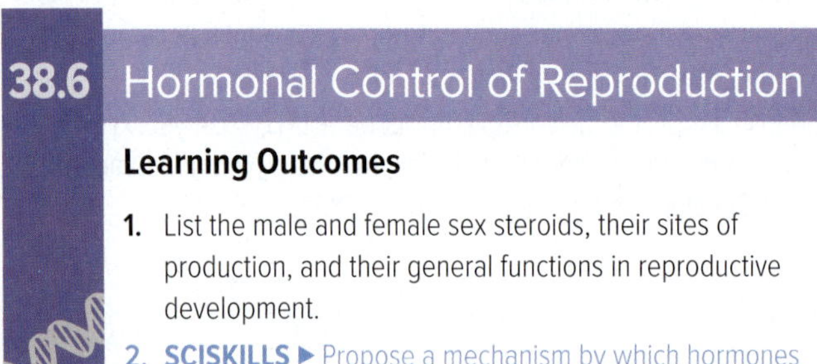

## 38.6 Hormonal Control of Reproduction

### Learning Outcomes

1. List the male and female sex steroids, their sites of production, and their general functions in reproductive development.
2. **SCISKILLS ▶** Propose a mechanism by which hormones can link nutrition with reproduction.

The topic of reproduction is covered in detail in Chapter 39. In all vertebrates and probably most invertebrates, reproduction is closely linked with endocrine function. In this section, we briefly examine the most common reproductive hormones and some of their major actions in male and female animals.

### The Gonads Secrete Sex Steroids That Influence Most Aspects of Reproduction

Hormones produced by the gonads of animals have central roles in nearly all aspects of reproduction, from reproductive behaviors to the ability to produce offspring. In males, the gonads are called testes. They produce and house the male gametes, called sperm cells. In addition, the testes produce several related steroid hormones, collectively termed **androgens.** In many vertebrates, including humans, the primary androgen is **testosterone.** The gonads of females are called the ovaries, which contain the female gametes, or egg cells. Other cells within the ovaries secrete two major steroid hormones: **progesterone** and a family of hormones called **estrogens.** The major estrogen in many animals, including humans, is **estradiol.** Collectively, the male and female gonadal hormones are referred to as the sex steroids.

The sex steroids are responsible for male- or female-specific reproductive changes associated with courtship and mating. In vertebrates, androgens from the developing testes are required for development of the male phenotype. Sex steroids are also chiefly responsible for development of secondary sex characteristics for each sex, which in humans includes growth of the appropriate external genitals, growth of the breasts in women, development of facial hair in men, and distribution and amount of fat and muscle in the body. Finally, sex steroids are required for maturation of gametes, the transition of young animals to reproductive maturity, and the ability of females to produce young.

The ability of the testes and ovaries to produce sex steroids depends on the presence of the gonadotropins secreted by the anterior pituitary, which are the same in both sexes and include **follicle-stimulating hormone (FSH)** and **luteinizing hormone (LH)** (see Table 38.1). Therefore, identical anterior pituitary gland hormones control the gonads in both male and female animals.

### Nutrition and Reproduction in Mammals Are Linked Through Hormones

An interesting feature of the endocrine control of reproduction is the observation that puberty is delayed in mammals, including humans, that are very undernourished. Similarly, fertility—the ability to produce offspring—is reduced in women and other adult female mammals under such conditions. This makes sense, because supporting the nutritional demands of a growing fetus would be difficult without adequate nutrition and energy stores. In such a case, it is more advantageous for an animal to delay reproduction until sufficient food is available.

How does the brain of a female mammal determine when sufficient energy is stored in her body to support pregnancy? The answer appears to be partly the result of hormonal signals, notably a polypeptide hormone called **leptin,** secreted by adipose cells in proportion to the amount of fat stored in the cells. For example, if the amount of adipose tissue in a woman's body decreases, so does the concentration of leptin in her blood. Leptin has been demonstrated to stimulate synthesis and secretion of reproductive hormones such as FSH and LH from the anterior pituitary gland. Consequently, undernourishment that causes a decrease in leptin also results in decreased production of reproductive hormones, thereby contributing to a loss of fertility. Therefore, leptin acts as a link between energy stores and the reproductive system.

## 38.6 Reviewing the Concepts

- Hormones produced by the gonads play vital roles in nearly all aspects of reproduction. The ability of the testes and ovaries to produce the sex steroids depends on the gonadotropins, FSH and LH. The hormone leptin links nutritional status with reproduction.

## 38.6 Testing Your Knowledge

1. Which is/are *true*?
   a. Leptin inhibits reproduction by inhibiting FSH and LH.
   b. Progesterone and estradiol are examples of androgens.
   c. FSH and LH stimulate the gonads in both males and females.
   d. Undernourishment stimulates leptin production, which in turn inhibits reproduction.
   e. Both a and d are correct.

## 38.7 Impact on Public Health

### Learning Outcomes

1. Name several hormones used therapeutically and the disease or health problem for which they are prescribed.
2. **SCISKILLS ▶** Predict the risks associated with the misuse of hormones and with exposure to endocrine disruptors in the environment.

The list of endocrine-related diseases in the human population is lengthy, ranging from relatively common disorders such as diabetes and thyroid disease to rare ones that may affect only 1 in every 100,000 or more individuals. Hormones can be used to treat many of these diseases, but they can also be misused in ways that cause health problems. Also, a wide range of industrial chemicals with hormone-like actions are thought to cause disruption to the endocrine system. In this section, we consider how hormones are used therapeutically as well as misused, and then look at an example of how environmental factors that alter the function of the endocrine system can have important public health implications.

### Hormones Are Used to Treat Millions of People

Since the advent of synthetic hormone production, doctors have been able to provide individuals with hormones that their own bodies may fail to produce. For example, millions of people self-administer insulin each day to treat diabetes or take thyroid pills to supplement an insufficient level of thyroid hormone. Many women are treated with gonadotropin hormones to increase fertility or take estrogen to control the symptoms of menopause. A growing number of men are administered androgens if the levels produced by their bodies decline significantly in later life. Other examples of therapeutic hormone use include recombinant human growth hormone to treat abnormally short stature in prepubertal children, epinephrine inhalers to treat asthma, and glucocorticoids to treat inflammation, lung disease, and skin disorders, to name just a few.

### Hormone Misuse Can Have Disastrous Consequences

The advent of hormone therapy has been of enormous benefit to society. However, the availability of hormones has also led to their misuse. Some individuals, typically those in competitive sports, self-administer hormones such as androgens. This practice may increase muscle mass and improve athletic performance, but serious side effects may occur. For instance, androgen administration has been linked to extreme aggressive behavior, and to development of cardiovascular disease and heart attacks, skin problems, and certain cancers. Women using androgens run similar health risks as men and in addition develop masculinizing traits, including increased growth of body hair and thinning scalp hair.

Another example of misuse occurs when hormones are used to boost the number of erythrocytes in order to increase the oxygen-carrying capacity of the blood. This has become common among some athletes participating in long-distance aerobic activities, such as competitive cycling and cross-country skiing. The hormone used is **erythropoietin (EPO),** which acts by stimulating the maturation of erythrocytes in the bone marrow and their release into the blood. EPO is normally made by the liver and kidneys in response to any situation where additional blood cells are required, such as following blood loss. When used abusively, however, the number of erythrocytes can become great enough that the blood becomes significantly more viscous than normal. This strains and damages the heart, which must work harder to pump the thickened blood. Since the 1990s, the international cycling community has been rocked by an alarming number of professional cyclists who died of heart attacks in the prime of their lives. These individuals had been using EPO to gain an unfair and, as it turns out, unwise advantage over their peers. Testing continues to detect injected EPO in the blood of some cyclists and other endurance athletes.

## Synthetic Compounds May Act as Endocrine Disruptors

A recent and disturbing phenomenon is the growing occurence of so-called **endocrine disruptors,** chemicals that interrupt the normal function of hormones in the body. These chemicals, which are found in lakes, streams, ocean water, and soil exposed to pollution runoff, are derived in many cases from industrial waste. They are also found in many common household products, including nearly all plastic goods. They often have molecular structures that resemble a hormone sufficiently to bind to its receptors. A common type of disruptor is recognized by estrogen receptors. If these compounds make their way into drinking water or food, they can exert estrogen-like actions or in other cases can inhibit the actions of the body's own estrogen. This can lead to dramatic consequences on fertility and on the development of embryos and fetuses. The extent of the risk from endocrine disruptors is hotly debated, but the number of mature, functional germ cells produced in animals as diverse as mollusks and human males has declined dramatically during the past 50 years in the U.S. In addition, researchers throughout the world have noted feminization of freshwater fishes downstream of wastewater facilities. For example, male fishes that were exposed to such conditions during development show increased production of proteins normally made by females bearing eggs. They also show changes in gonadal structures that resemble the female appearance. Further research is being conducted to gain more information on the consequences of these contaminants for animal and human endocrine systems.

## 38.7 Reviewing the Concepts

- Hormones are used therapeutically to treat a variety of human disorders, including diabetes, infertility, growth disorders, asthma, and inflammation.
- Androgen misuse can cause health risks such as cardiovascular disease; skin problems; cancer; infertility in men; and masculinizing traits in women. Erythropoietin misuse can seriously damage the heart.
- Endocrine disruptors may bind to hormone receptors in animals' bodies, including humans. In some cases, they may exert estrogen-like actions or inhibit the actions of the body's own estrogen.

## 38.7 Testing Your Knowledge

1. Hormones are often misused. Which of the following might result if a man injects himself with androgens in an effort to increase muscle mass?
   - a. low blood sugar
   - b. heart disease
   - c. increased fertility
   - d. immune suppression
   - e. testes enlargement due to increased LH and FSH

# Assess and Discuss

## Test Yourself

1. Which is the defining feature of hormones?
   - a. They are only produced in endocrine glands.
   - b. They are secreted into the blood, where they may reach one or more distant target cells, thereby altering cell function throughout the body.
   - c. They are released only by neurons.
   - d. They are never released by neurons.
   - e. They are secreted into ducts, where they diffuse to another, nearby gland or other structure.
2. Steroid hormones are synthesized from ______ and bind ______.
   - a. proteins, membrane receptors
   - b. fatty acids, membrane receptors
   - c. tyrosine, intracellular receptors
   - d. proteins, intracellular receptors
   - e. cholesterol, intracellular receptors
3. Which is *false* about polypeptide hormones?
   - a. They bind to receptors located on cell membranes.
   - b. Most of them are lipid-soluble.
   - c. They are the most abundant class of hormones.
   - d. They normally activate second messengers.
   - e. They bind reversibly to receptors.
4. Which is *false* about the control of hormones?
   - a. Many are regulated by negative feedback.
   - b. In some cases, they may be controlled by changes in blood concentrations of certain nutrients such as glucose.
   - c. In some cases, they may be controlled by changes in blood concentrations of certain minerals such as $Ca^{2+}$.
   - d. Their blood concentrations are controlled by their rates of synthesis and degradation.
   - e. The rate of degradation of a hormone may go up or down, but synthesis is always constant.
5. The hypothalamus and the anterior pituitary gland are physically connected by
   - a. arteries.
   - b. the infundibular stalk and portal veins.
   - c. the adrenal medulla.
   - d. the spinal cord.
   - e. the intermediate lobe.
6. Antidiuretic hormone (ADH)
   - a. increases water reabsorption in the kidneys.
   - b. regulates blood pressure by constricting arterioles.
   - c. decreases the volume of urine produced by the kidneys.
   - d. increases blood pressure when at high concentration.
   - e. does all of the above.
7. Which of the following pairs of hormones are involved in the regulation of blood $Ca^{2+}$ concentration in vertebrates?
   - a. aldosterone and prothoracicotropic hormone
   - b. insulin and glucagon
   - c. parathyroid hormone and 1,25-dihydroxyvitamin D
   - d. ADH and oxytocin
   - e. $T_3$ and TSH

8. In invertebrates, molting of larvae is stimulated by
   a. growth hormone.
   b. cortisol.
   c. juvenile hormone.
   d. 20-hydroxyecdysone.
   e. aldosterone.
9. Which of the following is *true* of the adrenal glands?
   a. They produce insulin.
   b. They produce hormones that control ion balance and glucose homeostasis.
   c. They produce only steroid hormones.
   d. They are inhibited by the hormone ACTH.
   e. They chiefly regulate calcium ion balance.
10. An animal such as a mammal has not been able to find food to eat for several days. Which hormone(s) might be present in the animal's blood in higher concentrations than normal?
    a. insulin
    b. glucagon
    c. cortisol
    d. both b and c
    e. both a and b

## Conceptual Questions

1. What is the function of leptin, and what is the benefit of an adipose-derived signaling molecule in this context?
2. Distinguish between type 1 and type 2 diabetes mellitus.
3. **PRINCIPLES** A principle of biology is that living organisms maintain homeostasis. How do the opposing actions of insulin and glucagon cooperate to maintain glucose homeostasis? When are these hormones released into the blood? What might happen to a nonfasting mammal if it were injected with a high dose of glucagon?

## Collaborative Questions

1. Discuss the role of hormones in animal development.
2. Discuss the role of the different steroid hormones and where they are produced in a mammal's body. Describe how steroid hormones differ in their structure and mechanism of action from other types of hormones.

# 39 The Production of Offspring: Reproduction and Development

© David Liebman/Pink Guppy

**Most animals reproduce by sexual reproduction.** These mud daubers, shown here mating, reproduce by sexual reproduction, which favors genetic variation in species.

## Chapter Outline



An 18-year-old woman joined her college's cross-country track team after competing for 2 years in high school. Gradually, her training increased from about 25 km/week to a very demanding 90 km/week. Recognizing the increased ability and competitiveness of college athletes relative to those at her high school, and at the urging of her track coach, she restricted her daily food intake to become as lean as possible. As her training and diet restriction progressed, she became increasingly preoccupied with monitoring what she was eating. Eventually, her weight dropped from 59 kg to 46 kg. This was followed by the cessation of her menstrual periods for several months. Alarmed that something was wrong, she visited her physician. After a battery of tests ruled out a variety of possible causes for missing her periods, she was diagnosed with secondary amenorrhea. This is a general term for the absence of menstrual cycles in a person who had previously been cycling normally. In her case, it was clear that the cause of her condition was undernourishment combined with intense exercise. She was encouraged to receive mental health counseling in conjunction with her parents and coach in order to address her eating disorder, and to work with a dietitian to tailor an appropriate meal plan to support her desired level of activity.

Secondary amenorrhea is very common among elite female athletes in sports and activities like cross-country track, gymnastics, and ballet, where physical activity is demanding and a lean body mass is encouraged. It is a warning sign that a person is overdoing things and needs to correct a nutritional imbalance. When a woman's body is stressed in this way, her reproductive function ceases. From a physiological perspective, this makes sense, as the likelihood of supporting a healthy pregnancy in such circumstances is greatly reduced and the body requires whatever energy is available just to survive.

This woman's case illustrates the delicate control of **reproduction**—the processes by which organisms produce offspring. The biological mechanisms that favor successful reproduction in the animal kingdom are extraordinarily diverse. Many of the observable differences in animal behavior and anatomy are the result of adaptations that increase an animal's chances of reproducing.

In this chapter, we will examine the diverse means of reproduction throughout the animal kingdom, including asexual and sexual reproduction. We will consider the control mechanisms of reproduction in mammals, including humans. We then will cover some of the general events in embryonic development and will conclude with a discussion of some key issues related to fertility in the human population today, including those that affected the young woman described in this introduction.

## 39.1 Overview of Sexual and Asexual Reproduction

### Learning Outcomes

1. Describe several differences between sexual and asexual reproduction.
2. **SCISKILLS** ▶ Propose an hypothesis for why most animals reproduce by sexual reproduction.

**Sexual reproduction** in animals is the production of a new individual by the joining of two haploid reproductive cells called **gametes**—typically one from each parent. This produces offspring that are genetically different from both parents. **Asexual reproduction** in animals occurs when offspring are produced from a single parent without the fusion of gametes. The offspring are clones of the parent organism. In this section, we will consider the processes of sexual and asexual reproduction, as well as their advantages and disadvantages.

### Sexual Reproduction Occurs in Most Animals

Sexual reproduction occurs in most invertebrate and vertebrate species, and involves the joining of two gametes. The gametes are spermatozoa (usually shortened to **sperm**) from the male and **egg cells,** or **ova,** (singular, ovum) from the female. (In some species, sperm and egg cells are produced by a single organism called a **hermaphrodite**). When a sperm unites with an egg—the process called **fertilization**—each haploid gamete contributes its set of chromosomes to produce a diploid cell called a fertilized egg, or **zygote.** As the zygote undergoes cell divisions and begins to develop, it is called an **embryo.**

### Asexual Reproduction Takes Several Forms

Asexual reproduction occurs in some invertebrates and in a small number of vertebrate species. These animals reproduce in one of three ways: budding, regeneration, or parthenogenesis. **Budding,** which is seen in cnidarians, occurs when a portion of the parent organism pinches off to form a complete, new individual (Figure 39.1a). In this process, cells from the parent undergo mitosis and differentiate into specific types of structures before the new individual breaks away from the parent. At any one time, a parent organism may have one, two, or multiple buds forming simultaneously. Budding continues throughout such an animal's lifetime.

Certain species of sponges, echinoderms, and worms reproduce by **regeneration,** in which a complete organism can be formed from small fragments of the animal's body. In some sea stars and starfish, for example, an arm removed by injury or predation can grow into an entirely new individual (Figure 39.1b). Similarly, a flatworm bisected into two pieces can regenerate into two new individuals.

(a) Budding

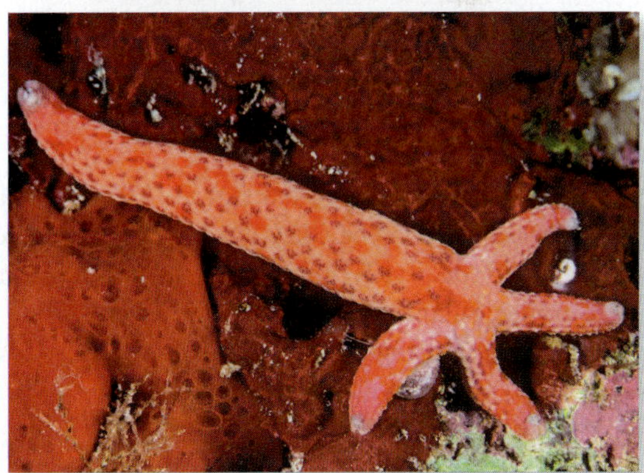

(b) Regeneration

**Figure 39.1** **Examples of asexual reproduction. (a)** *Hydra* with a single bud. **(b)** A starfish regenerating a new body from a single arm.
*(a)* © Clouds Hill Imaging Ltd./Corbis; *(b)* © WaterFrame/Alamy

**Parthenogenesis** is the development of offspring from an unfertilized egg. Animals produced by parthenogenesis are usually haploid. It occurs in several invertebrate classes and in a few species of fishes and reptiles.

### Sexual and Asexual Reproduction Have Advantages and Disadvantages

Asexual reproduction provides a relatively simple way for an organism to produce many copies of itself, whereas sexual reproduction usually requires two individuals to produce offspring. What are the advantages and disadvantages of each method?

Asexual reproduction has certain advantages over sexual reproduction. First, an animal can reproduce asexually even if it is isolated from others of its own species—for example, because the animal is nonmotile or because it rarely encounters another member of its species. Second, individuals can reproduce rapidly at any time because they need not seek out, attract, and mate with an individual of the opposite sex. Asexual reproduction, therefore, is an effective way of generating large numbers of offspring. Although many kinds of animals reproduce asexually, it is more prevalent in species that live in very stable environments, with little selection pressure for genetic diversity in the population.

Compared with asexual reproduction, sexual reproduction is associated with unique costs. Two types of gametes (sperm and eggs) must be made (often in large numbers), males and females require specialized body parts to mate with each other, and the two sexes must be able to find each other. Yet given that the vast majority of eukaryotic species reproduce sexually, this question has intrigued biologists since the time of Darwin: What is the advantage of sexual reproduction for the perpetuation of a species?

In the context of species survival, the major difference between asexual and sexual reproduction is that sexual reproduction allows for greater genetic variation due to genetic recombination. Only certain alleles from each parent are passed on, and when a set of genes from one parent mixes with a different set from the other parent, the offspring are never exactly like either of their parents. Thus, a hallmark of sexual reproduction is increased genetic variation between successive generations. One prevalent hypothesis about why sex evolved is that sexual reproduction allows more rapid adaptation to environmental changes than does asexual reproduction.

In particular, sexual reproduction allows alleles within a species to be redistributed via crossing over and independent assortment across many generations. As a result, some offspring carry combinations of alleles that promote survival and reproduction, whereas other offspring may carry less favorable combinations. As described in Chapter 19, natural selection can favor those combinations that promote greater reproductive success while eliminating offspring with lower fitness. By comparison, the alleles of asexual organisms are not reassorted from generation to generation. As a result, it is more difficult to accumulate potentially beneficial alleles within individuals of these species. Also, as described next, sexual reproduction may facilitate elimination of harmful alleles from a population.

## FEATURE INVESTIGATION

### Paland and Lynch Provided Evidence That Sexual Reproduction May Promote the Elimination of Harmful Mutations in Populations

Evolutionary biologists have suggested that the inability of asexual species to reassort alleles may be a key disadvantage compared with sexually reproducing species. To investigate this question, American researchers Susanne Paland and Michael Lynch studied the persistence of mutations in populations of *Daphnia pulex*, a freshwater organism commonly known as the water flea. The researchers chose this organism because some natural populations reproduce asexually, and others reproduce sexually.

In their experiment, shown in **Figure 39.2**, Paland and Lynch studied the sequences of several mitochondrial genes in 14 sexually reproducing

**Figure 39.2** **Paland and Lynch demonstrated the importance of sexual reproduction in reducing the frequency of harmful genetic mutations.**

**HYPOTHESIS** Sexual reproduction allows for greater mixing of alleles of different genes and thereby may prevent the accumulation of detrimental alleles in a population.

**KEY MATERIALS** The researchers collected samples of *Daphnia pulex* from many natural populations. A total of 14 sexual populations and 14 asexual populations were studied.

Experimental level | Conceptual level

1 Isolate mitochondrial DNA from members of 28 populations of *D. pulex*. This involves breaking open cells and extracting the DNA (refer back to Chapter 18).

Daphnia
DNA
Segments of mitochondrial DNA

2 Amplify regions of mitochondrial genes, using PCR. Subject the regions to DNA sequencing. The techniques of PCR and DNA sequencing are described in Chapter 18 (refer back to Figures 18.6 and 18.8).

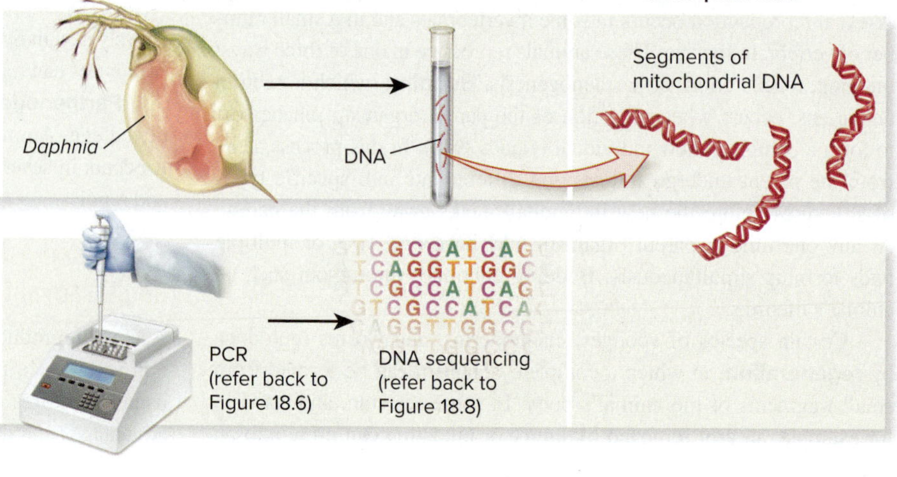

PCR (refer back to Figure 18.6)
DNA sequencing (refer back to Figure 18.8)
TCGCCATCAG
TCAGGTTGGC
TCGCCATCAG
GTCGCCATCA
AGGTTGGCC

3 Using computer technology, align the sequences and determine the number of DNA changes that would cause amino acid substitutions. These amino acid changes were categorized as those that would be highly deleterious, moderately deleterious, mildly deleterious, or neutral for protein function. Compare the sexual and asexual populations for the persistence of these types of changes.

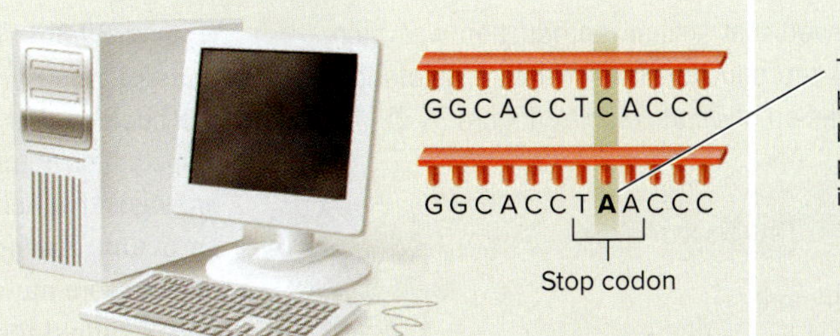

GGCACCTCACCC
GGCACCTAACCC
Stop codon
This change would be highly detrimental because it would put a stop codon into the gene.

4 **THE DATA**

Results from step 3:

| Types of amino acid substitutions (The amino acid substitutions were due to rare mutations that occurred in the natural populations of *D. pulex*.) | % of total amino acid substitutions | Allowed to persist in | |
|---|---|---|---|
| | | **Sexual populations** | **Asexual populations** |
| Highly deleterious | 73.2 | No | No |
| Moderately deleterious | 13.3 | No | Yes |
| Mildly deleterious | 4.4 | Yes | Yes |
| Neutral | 9.1 | Yes | Yes |

5 **CONCLUSION** Moderately deleterious mutations are less likely to persist in populations of animals that reproduce sexually.

6 **SOURCE** Paland, S., and Lynch, M. 2006. Transition to Asexuality Results in Excess Amino Acid Substitutions. *Science* 311: 990–992.

and 14 asexually reproducing populations of *D. pulex.* The researchers hypothesized that asexual populations would be less able to eliminate harmful mutations. As discussed in Chapter 15, random gene mutations that change the amino acid sequence of an encoded protein are much more likely to be harmful than beneficial. The alleles of sexually reproducing populations can be reassorted from generation to generation, thereby producing succeeding generations in which the detrimental alleles are lost from the population. As you can see in step 4 of Figure 39.2, the researchers discovered that both the sexual and asexual populations could eliminate highly deleterious mutations. Organisms harboring such mutations probably died and did not reproduce. In addition, the sexual and asexual populations both retained mildly deleterious and neutral mutations. However, moderately deleterious mutations were eliminated from the sexual populations but not from the asexual ones. One interpretation of these data is that sexual reproduction allowed for the reassortment of beneficial and detrimental alleles, making it easier for sexually reproducing populations to eliminate those mutations that are moderately detrimental.

### *Experimental Questions*

1. **SCISKILLS ▸** How did Paland and Lynch test the hypothesis that sexual reproduction allowed for the reduction in deleterious mutations?
2. **SCISKILLS ▸** Approximately how many times more often did deleterious mutations persist in the asexually reproducing populations compared to the sexually reproducing ones?
3. What is the proposed evolutionary benefit of sexual reproduction?

## 39.1 Reviewing the Concepts

- Asexual reproduction occurs when offspring are produced from a single parent, without the fusion of genetic material from two parents (Figure 39.1).
- Sexual reproduction is the production of a new individual by the joining of two haploid gametes: a sperm from the male and an egg from the female.

- Paland and Lynch showed that sexual reproduction was more effective than asexual reproduction in eliminating deleterious mutations from a population of *Daphnia pulex* (Figure 39.2).

## 39.1 Testing Your Knowledge

1. Choose the correct statement.
   a. The majority of animals reproduce by asexual reproduction.
   b. Budding, regeneration, and parthenogenesis account for reproduction in nearly all invertebrates.
   c. A zygote is the haploid organism produced by the union of a sperm and an egg.
   d. Potentially beneficial alleles are more likely to accumulate in species that reproduce sexually than in asexually reproducing species.
   e. In asexual reproduction, gametes from each parent join to form a zygote.

# 39.2 Gametogenesis and Fertilization

### Learning Outcomes

1. Outline the processes of spermatogenesis and oogenesis.
2. **SCISKILLS ▸** Propose reasons for why internal fertilization is advantageous.

This section covers how and where gametes are formed, and how two gametes join to form a new organism.

## Sperm and Eggs Are Produced During the Process of Gametogenesis

Male and female gametes are formed within the **gonads—**the **testes** (singular, testis) in males and the **ovaries** (singular, ovary) in females. The formation of gametes—**gametogenesis—**begins with cells called germ cells, which multiply by mitosis, resulting in diploid cells (carrying two sets of chromosomes, denoted as $2n$) called **spermatogonia** (singular, spermatogonium) in males and **oogonia** (singular, oogonium) in females (Figure 39.3). Some of these cells become **primary spermatocytes** or **primary oocytes** that may begin the process of meiosis. (See Chapter 14 for a review of the processes of mitosis and meiosis.) Until this point, the development of sperm and eggs is similar. From then on, gametogenesis differs between the two types of gamete.

**Spermatogenesis** The formation of haploid sperm from a diploid germ cell is called **spermatogenesis.** As shown in Figure 39.3a, primary spermatocytes begin this process by undergoing the first of two meiotic divisions (meiosis I). In the primary spermatocyte, meiosis I produces two haploid ($n$) cells called **secondary spermatocytes.** These cells also undergo meiosis (meiosis II), producing four haploid **spermatids** that eventually differentiate into mature haploid sperm cells. Gametogenesis in males, therefore, results in four gametes from each spermatogonium.

The most striking change in each spermatocyte as it differentiates into a sperm is the formation of a flagellum, also called the tail (Figure 39.3b). The movements of the tail require energy and make the sperm motile. The sperm also has a head region, which contains the nucleus that carries the chromosomes. At the tip of the head is a special structure called the **acrosome,** which contains proteolytic enzymes that help break down the protective outer layers surrounding an ovum. The head and tail are separated by a midpiece containing one or more mitochondria, depending on the species, that produce the ATP required for tail movements.

**Oogenesis** Whereas spermatogenesis produces four gametes from each primary spermatocyte, gametogenesis in the female, called **oogenesis,** results in the production of a single gamete from each primary oocyte (Figure 39.3c). In mammals, oogenesis begins in the female embryo: Oogonia divide by mitosis to form primary oocytes that enter meiosis I but stop the process, remaining in that arrested state after birth. Meiosis I does not resume until puberty, the time after which a mammal first becomes capable of reproducing. Beginning at puberty, meiosis I is completed in some primary oocytes, producing a haploid **secondary oocyte** plus a smaller cell, called a polar body, that eventually degenerates. Meiosis II begins in the secondary oocyte but stops at metaphase. Once the secondary oocyte is released from the ovary in the process of **ovulation,** it can become fertilized if sperm are available. Meiosis II is not complete until the oocyte is fertilized by a sperm. If the oocyte does not encounter a sperm, it never undergoes the second meiotic division. Once a haploid egg nucleus fuses with a haploid sperm nucleus, a diploid zygote is produced.

Each oocyte undergoes growth and development within an ovarian structure called a follicle. In mammals, a layer of glycoproteins called the **zona pellucida** surrounds the surface of the secondary oocyte in the follicle. In turn, a layer of cells called the cumulus mass, which provides protection and nutritive support, is outside the zona pellucida. An outer cellular layer called the theca produces hormones that control oocyte growth (Figure 39.3d). These layers have important roles in fertilization, as discussed below.

## Fertilization Involves the Union of Sperm with Egg

Fertilization is the process by which the haploid male and female gametes unite and become a diploid zygote. Several important cellular and molecular processes must occur before the nuclei of the gametes can fuse. The mechanism by which the egg and sperm make contact has been studied extensively in sea urchins, and evidence suggests that a similar process occurs in humans and other mammals. Sea urchin eggs emit chemical attractants that bind to nearby sperm. This increases cellular respiration (that is, the breakdown of nutrients to synthesize ATP) within the sperm, which helps increase sperm motility. The sperm then swim toward the egg by following the attractants' concentration gradient.

For a sperm to physically contact the egg, it must first penetrate the layers surrounding the egg's plasma membrane (Figure 39.4). In sea urchins, as in many animals, the sperm must penetrate a jelly-like layer consisting of glycoproteins and polysaccharides before contacting the plasma membrane of the egg. The sperm is able to do this because of the **acrosomal reaction,** in which proteases and other hydrolytic enzymes are released from the acrosome onto the jelly

## Biology Principle

### Structure Determines Function

The structure of a sperm cell is a good example of this principle of biology at the cellular level. Notice how each of the three major structural elements of the sperm cell is suited for a particular function. When all three elements function together, the cell is capable of fertilizing an egg.

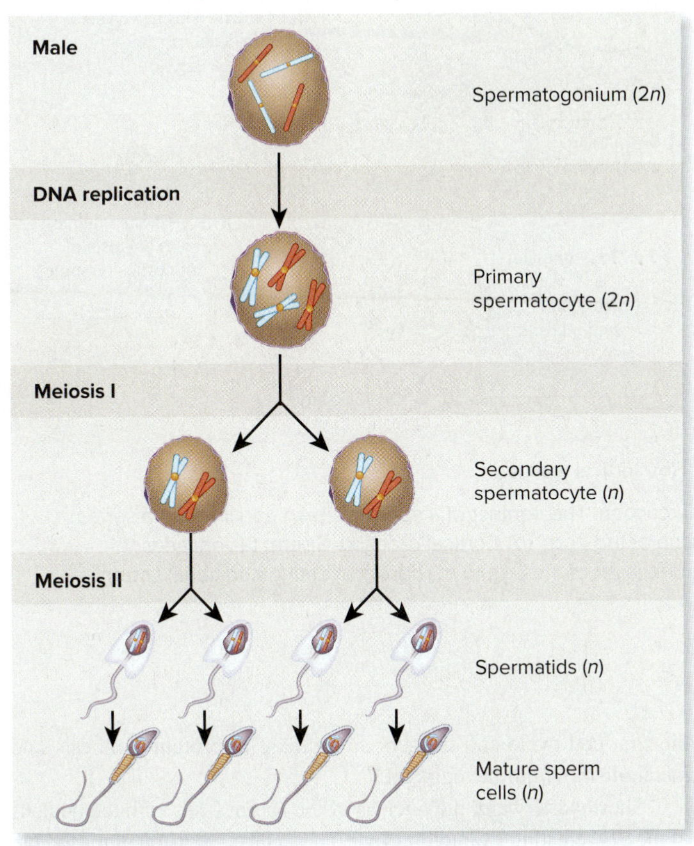


(a) Spermatogenesis

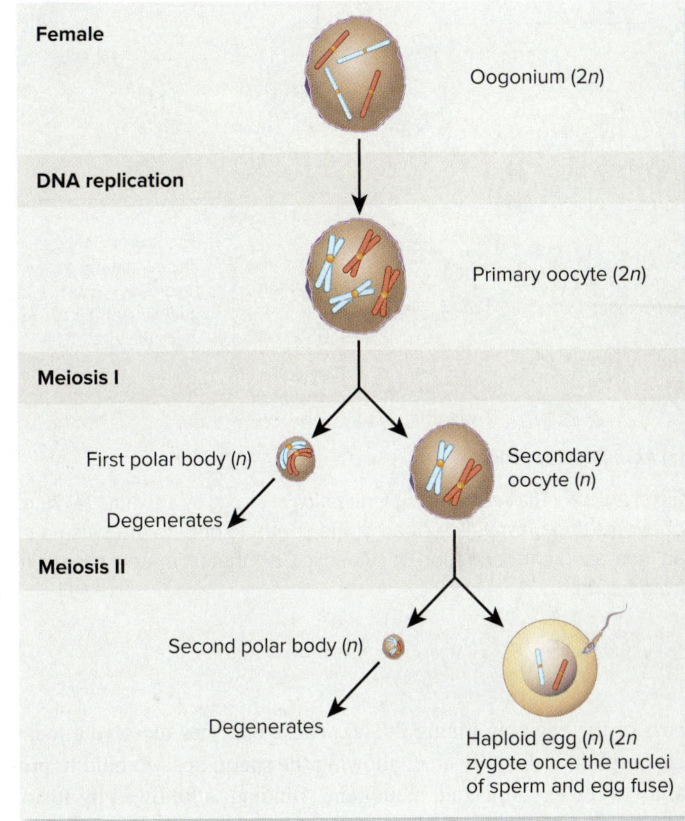


(c) Oogenesis

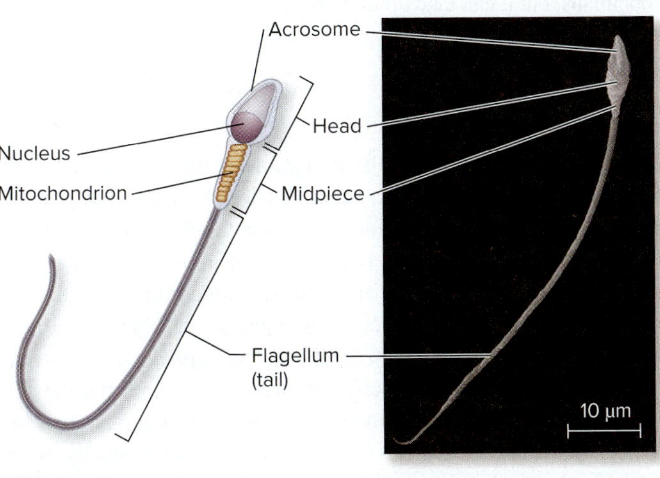


(b) Mature human sperm

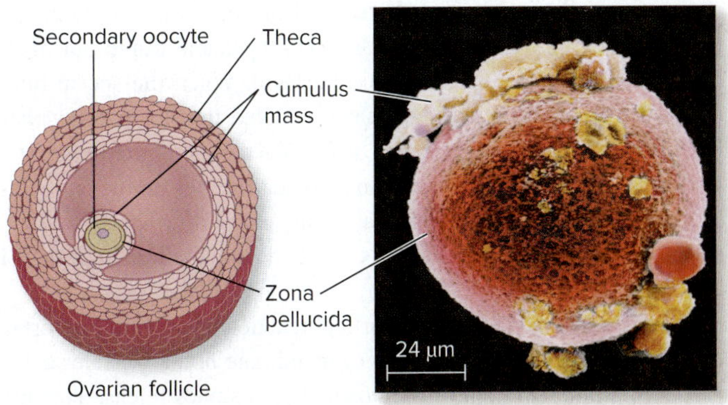


(d) Mature human follicle and oocyte

**Figure 39.3 Gametogenesis and gametes in males and females. (a)** In the process of spermatogenesis, male diploid (2*n*) germ cells undergo two meiotic divisions to produce mature haploid (*n*) sperm. **(b)** The characteristic head, midpiece, and flagellum (tail) of a mature human sperm, as seen in a drawing and accompanying scanning electron micrograph (SEM). **(c)** The process of oogenesis in females, which produces a haploid secondary oocyte that enters but does not complete meiosis II until it is fertilized. **(d)** Mature follicle and oocyte. The drawing depicts a secondary oocyte within its follicle; the SEM shows an isolated human oocyte covered by its zona pellucida and remnants of the cumulus mass.

*(b)* © Eye of Science/Science Source; *(d)* © P.M. Motta & G. Familiari/Univ. "La Sapienza"/Science Source

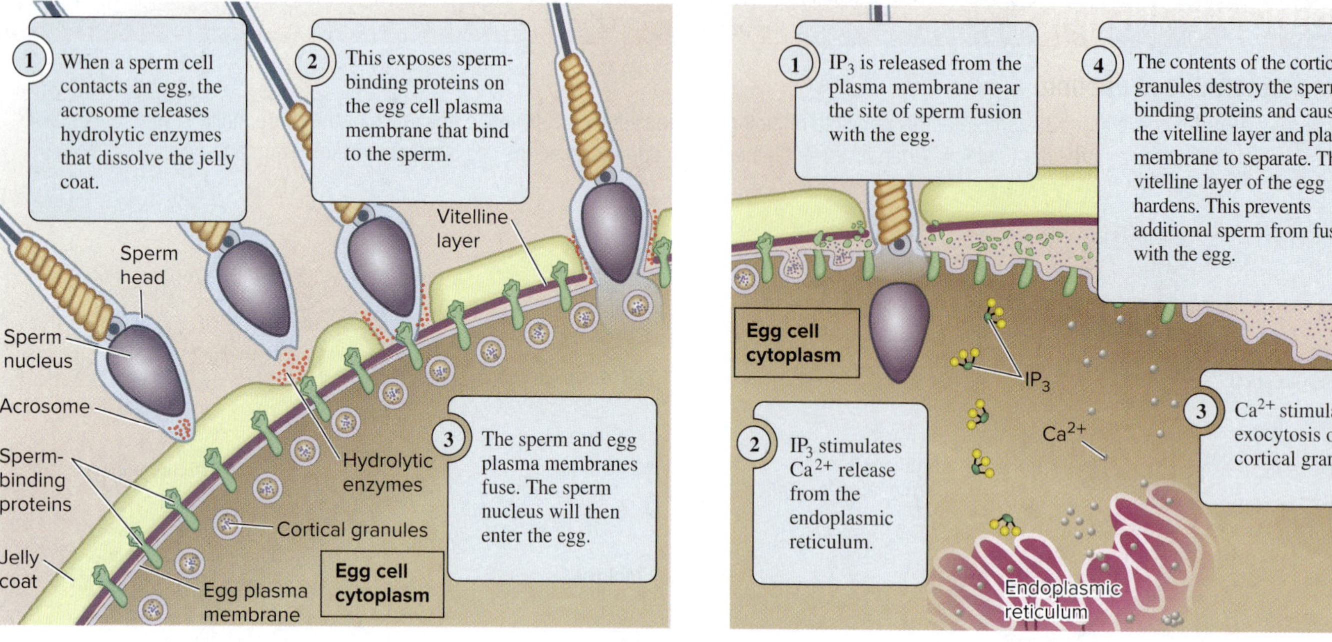

1 When a sperm cell contacts an egg, the acrosome releases hydrolytic enzymes that dissolve the jelly coat.
2 This exposes sperm-binding proteins on the egg cell plasma membrane that bind to the sperm.
Vitelline layer
Sperm head
Sperm nucleus
Acrosome
Sperm-binding proteins
Hydrolytic enzymes
3 The sperm and egg plasma membranes fuse. The sperm nucleus will then enter the egg.
Cortical granules
Jelly coat
Egg plasma membrane
Egg cell cytoplasm
1 IP3 is released from the plasma membrane near the site of sperm fusion with the egg.
4 The contents of the cortical granules destroy the sperm-binding proteins and cause the vitelline layer and plasma membrane to separate. The vitelline layer of the egg hardens. This prevents additional sperm from fusing with the egg.
Egg cell cytoplasm
IP3
2 IP3 stimulates Ca2+ release from the endoplasmic reticulum.
Ca2+
3 Ca2+ stimulates exocytosis of cortical granules.
Endoplasmic reticulum


**(a) Acrosomal reaction**

**(b) Cortical reaction**

**Figure 39.4** **The events during fertilization of an egg by a sperm. (a)** Acrosomal reaction. The contact of a sperm with an egg initiates a series of events that permits the head of the sperm to bind to the plasma membrane of the egg. **(b)** Cortical reaction. Sperm fusion leads to an increased concentration of cytosolic $Ca^{2+}$ that ultimately causes the vitelline layer of the egg to harden, preventing additional sperm from fusing.

coat of the egg (see Figure 39.4a). These enzymes dissolve a localized region of the jelly coat, allowing the sperm head to bind to proteins in the egg's plasma membrane. Binding is followed by fusion of the sperm head membrane with the egg membrane, and shortly thereafter by penetration of the sperm head and release of its nucleus into the egg.

The acrosomal reaction is followed by the **cortical reaction** (Figure 39.4b). Normally, the cytosolic $Ca^{2+}$ concentration in eggs, as in most cells, is kept low by several mechanisms, including one in which $Ca^{2+}$ is transported by an ATP-dependent pump out of the cytosol and into the endoplasmic reticulum. When the sperm binds to the egg, a small signaling molecule called inositol trisphosphate ($IP_3$) is released from the region of the plasma membrane nearest to the sperm entry point. $IP_3$ then binds to nearby sites on the endoplasmic reticulum (ER) and opens $Ca^{2+}$ channels. Within seconds after a sperm cell binds to an egg, $Ca^{2+}$ is released from the lumen of the ER and into the cytosol.

The release of $Ca^{2+}$ in the cortical reaction has several important effects. First, it causes exocytosis of membrane-bound vesicles called cortical granules in the egg's cytosol. The granules release enzymes and other substances that inactivate the sperm-binding proteins on the plasma membrane. In addition, the outer coating of the egg cell, known as the vitelline layer in sea urchins or the zona pellucida in vertebrates (see Figure 39.3), becomes hardened and begins to separate from the plasma membrane. These events create a barrier to more sperm fusing with the egg. Additionally, the burst of cytosolic $Ca^{2+}$ leads to the activation of molecular signaling pathways that initiate the first cell cycle and triggers an increase in protein synthesis and metabolism within the egg cell.

Shortly afterward, the nucleus of the sperm fuses with the nucleus of the egg, creating a diploid zygote. The first cell division of the zygote occurs approximately 90 minutes after fertilization in sea urchins and amphibians, but it can take up to 24 hours in mammals.

## In a Given Species, Fertilization Occurs Either Outside or Inside the Female

For sperm to fertilize eggs, the two gametes must physically come into contact. This can occur either outside or inside the female's body. When this occurs outside of the female, the process is called **external fertilization.** This type of fertilization occurs in aquatic environments, when eggs and sperm are released into the water close enough for fertilization to occur. The aqueous environment protects the gametes from drying out. Animals that reproduce by external fertilization show species-specific behaviors that help bring the eggs and sperm together. For instance, very soon after a female fish lays her eggs, a male deposits his sperm cells in the water nearby, where they spread over the clump of eggs. The fertilized eggs then develop outside the mother's body.

Although the aqueous environment protects against desiccation, eggs can be eaten by predators, washed away by currents, or subjected to potentially lethal changes in water temperature or other variables such as pH or oxygen. Such environmental challenges have led to selection for some species, including many aquatic or amphibious animals, to release very large numbers of eggs at once—as many as several million in some species!

In contrast to external fertilization, most terrestrial animals and some aquatic animals reproduce by **internal fertilization,** in which sperm are deposited within the reproductive tract of the female during the act called copulation, as seen in the mud daubers in the chapter-opening photo. Internal fertilization protects the gametes from environmental hazards and predation and guarantees that sperm are placed and remain in very close proximity to eggs. Once fertilization occurs within the female, the zygotes then develop into offspring.

The behaviors and anatomical structures involved in achieving internal fertilization are extremely varied among species. Typically, mating involves accessory sex organs, which are reproductive structures other than the gonads. The external accessory sex organs involved in copulation include certain of the genitalia (for example, the **penis** and the **vagina**), which are used to physically join the male and female so that sperm can be deposited directly into the female's reproductive tract. A penis or analogous structure is present in most insects, reptiles, some species of birds (ratites and waterfowl), and all mammals. However, males of other vertebrate species—including most birds—that reproduce by internal fertilization lack a structure that can be inserted into the female, instead depositing sperm in the female by cloacal contact. The cloaca is a common opening for the reproductive, digestive, and excretory systems in these animals.

## 39.2 Reviewing the Concepts

- Male and female gametes are formed within the gonads—the testes in males and the ovaries in females. Within the ovaries, each oocyte undergoes growth and development within a follicle before it leaves the ovary in a process called ovulation (Figure 39.3).
- Major events in fertilization include the acrosomal and cortical reactions (Figure 39.4).
- In external fertilization, sperm and eggs are released into an aquatic environment, where they unite. In internal fertilization, sperm are deposited within the reproductive tract of the female during copulation or by cloacal contact.

## 39.2 Testing Your Knowledge

1. In ________, gametes are released in large numbers in a watery environment.
   a. copulation
   b. internal fertilization
   c. external fertilization
   d. the cortical reaction
   e. spermatogenesis
2. In which way is gametogenesis similar in males and females?
   a. The primary spermatocyte and primary oocyte both eventually produce four gametes.
   b. The formation of gametes includes both mitosis and meiosis in both sexes.
   c. The gametes in both sexes have a flagellum.
   d. The gametes that are produced are all diploid.
   e. In both sexes, meiosis is not completed until fertilization occurs.

## 39.3 Human Reproductive Structure and Function

### Learning Outcomes

1. Describe the structure of the human male reproductive tract.
2. Outline the process of hormonal control of the male reproductive system.
3. Describe the structure of the human female reproductive tract.
4. Diagram the events of the ovarian cycle.
5. Outline the process of hormonal control of the reproductive cycle of the human female.
6. Explain how maternal hormones prepare the uterus to accept the embryo.

We turn now to a detailed look at the human reproductive system. For both sexes, we will begin with a description of the anatomy of the reproductive system, including the gonads and the accessory sex structures. We will then examine the hormones that control the production of the gametes and the preparation for and establishment of pregnancy.

### The Human Male Reproductive Tract Is Specialized for Production and Ejaculation of Sperm

The external structures of the male reproductive tract—the genitalia—consist of the penis and the scrotum, the sac that contains the testes and holds them outside the body cavity where the temperature is better suited for spermatogenesis (**Figure 39.5**).

Each testis is composed of tightly packed **seminiferous tubules** encased in connective tissue. Surrounding the tubules are Leydig cells—scattered endocrine cells that secrete the steroid hormone testosterone. Spermatogenesis begins at puberty and continues throughout life. It occurs all along the walls of the seminiferous tubules. Cells at the earliest stages of spermatogenesis—the spermatogonia—are located nearest the outer surface of the wall. Cells of more advanced stages are located progressively inward, such that the mature sperm are released into the tubule lumen. Cells within the seminiferous tubules are continuously developing from spermatogonia into spermatocytes and eventually to sperm, so at any one time, all types of cells are present along the seminiferous tubule. Support cells, called Sertoli cells, surround the developing spermatogonia and spermatocytes, providing them with nutrients and protection and having a role in their maturation into sperm.

Sperm moving out of the seminiferous tubules are emptied into the epididymis, a coiled, tubular structure located on the surface of the testis (see Figure 39.5). Here the sperm complete their differentiation by becoming motile and gaining the capacity to fertilize an egg.

Sperm leave the epididymis through the vas deferens, a muscular tube leading to the ejaculatory duct**,** which then connects to the urethra (see Figure 39.5). The urethra originates at the bladder and extends to the end of the penis. In males, the urethra not only conducts urine but also carries **semen,** a mixture containing fluid and sperm, that is released during **ejaculation**—the movement of semen through the urethra by contraction of muscles at the base of the penis.

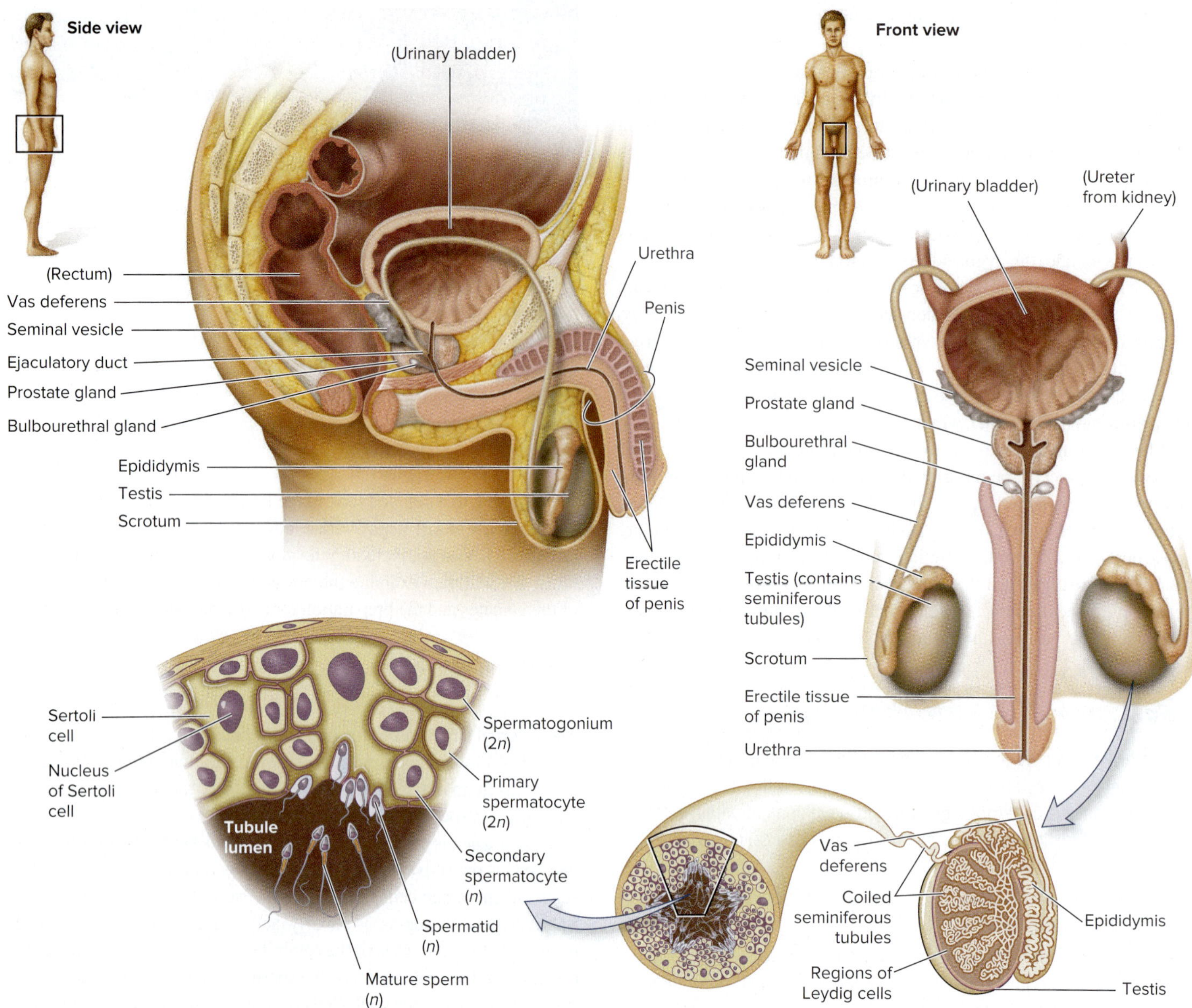


**Figure 39.5** **Male reproductive structures in humans.** Side and front views of the male reproductive system (nonreproductive structures are identified in parentheses for orientation purposes). The enlargement (at bottom) shows the internal structure of a testis and associated structures and indicates the stages of spermatogenesis.

The liquid components of semen are important for the survival and movement of sperm through the female reproductive tract. This liquid is formed by three accessory glands that secrete substances into the urethra to mix with the sperm: the seminal vesicles, the prostate gland, and the bulbourethral glands. The seminal vesicles secrete the monosaccharide fructose, the main nutrient for sperm, as well as other factors that enhance sperm motility and survival. The prostate gland and bulbourethral glands secrete a thin, alkaline fluid that protects sperm from acidic fluids in the urethra and within the female reproductive tract.

Introduction of sperm into the female reproductive system during copulation is made possible by erection of the penis. Erection occurs when blood fills spongy erectile tissue located along the length of the penis (see Figure 39.5). Sexual arousal stimulates release of the gaseous neurotransmitter nitric oxide (NO) in the penis, causing vasodilation of arteries. The pressure of the blood flowing into the penis constricts nearby veins, causing a reduction in venous drainage from the penis, engorging it with blood. After ejaculation, NO release is decreased, causing a reversal of the vascular changes responsible for erection.

## Male Reproductive Function Requires the Actions of Hormones

Recall from Chapter 38 that the hypothalamus is a structure on the ventral surface of the brain that synthesizes neurohormones, including gonadotropin-releasing hormone (GnRH) (refer back to Table 38.1).

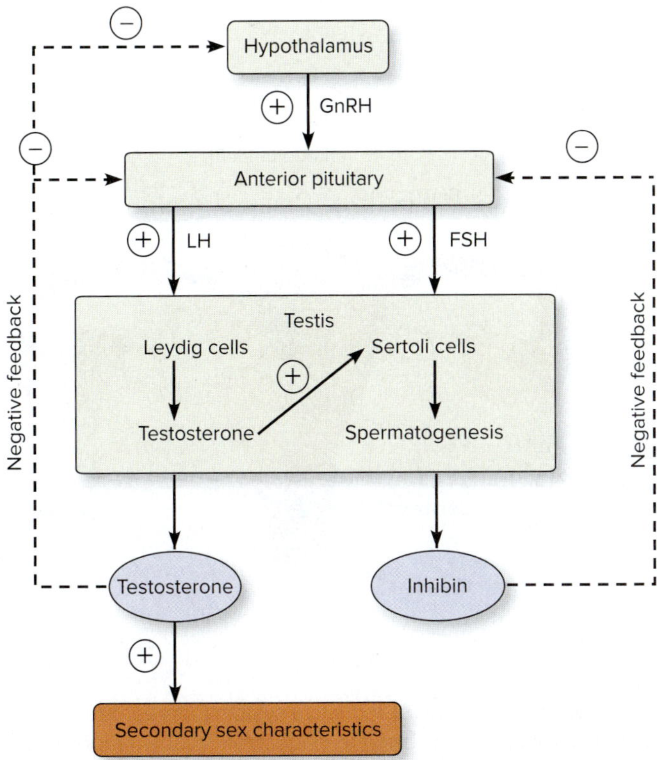


**Figure 39.6 The hormonal control of male reproduction.** In response to LH, Leydig cells in the testes secrete testosterone, which, along with FSH, acts on Sertoli cells to facilitate spermatogenesis. Negative signs indicate inhibitory effects via negative feedback, and plus signs indicate stimulatory effects.

GnRH stimulates the anterior pituitary gland to release two gonadotropins: luteinizing hormone (LH) and follicle-stimulating hormone (FSH).

In males, LH stimulates the Leydig cells of the testes to produce androgens, particularly testosterone (Figure 39.6). Testosterone stimulates the growth of the male reproductive tract and the genitalia during development and puberty, as well as the development of male secondary sex characteristics. In humans, these include facial hair growth, increased muscle mass, and deepening of the voice.

The other pituitary gonadotropin, FSH, functions along with testosterone to stimulate spermatogenesis. FSH does this by stimulating the activity of the Sertoli cells within the seminiferous tubules (see Figures 39.5 and 39.6). The Sertoli cells provide the nutritional and structural support necessary for development of the sperm. They also respond to testosterone produced by the Leydig cells, by stimulating mitosis and meiosis of the germ cells associated with them in the tubules.

Production of sperm and testosterone is kept in check by negative feedback mechanisms that control the amount of gonadotropins produced (see Figure 39.6). Testosterone at high concentrations inhibits the secretion of GnRH from the hypothalamus, so both LH and FSH are inhibited when the blood concentration of testosterone is high. Testosterone also directly inhibits LH secretion by the anterior pituitary gland. In addition, when Sertoli cells are activated by FSH, they secrete a protein hormone called inhibin, which enters the blood and inhibits further secretion of FSH. These feedback mechanisms maintain homeostatic concentrations of FSH and LH in the blood.

Before puberty, LH is not released in sufficient amounts to stimulate significant testicular production of testosterone, and the reproductive system is quiescent—spermatogenesis does not occur. Although the mechanisms that initiate puberty in mammals are still not completely understood, research has shown that increased GnRH production at that time initiates increased LH and FSH secretion from the pituitary. The testosterone induced by LH stimulates development of adult male characteristics. Testosterone is also responsible for an increased sex drive (libido) at this time.

BIO TIPS ONLINE

## The Female Reproductive Tract Is Specialized for Production and Fertilization of the Egg and Development of the Embryo

The female genitalia are composed of the larger, hair-covered outer folds called the labia majora and smaller, inner folds, called the labia minora, which surround the external opening of the reproductive tract (Figure 39.7). At the anterior part of the labia minora is the clitoris, which is erectile tissue of the same embryonic origin as the penis. Like the penis, the clitoris becomes engorged with blood during sexual arousal and is very sensitive to sexual stimulation. Unlike males, however, the openings of the reproductive tract and the urethra are separate in females. The opening of the urethra is located between the clitoris and the opening of the reproductive tract.

The external opening of the reproductive tract leads to the vagina, a tubular, smooth muscle structure into which sperm are deposited during copulation. At the end of the vagina is the cervix, the opening to the **uterus,** a pear-shaped organ that holds and nourishes the growing embryo. It consists of an inner lining of glandular and secretory cells, called the endometrium, and a thick, muscular layer. We will discuss the functions of the uterus later in the chapter.

Oocytes develop within one of the two ovaries (see Figure 39.7), which are suspended within the abdominal cavity by connective tissue. In humans, each ovary is typically a little larger than an almond. During ovulation, a secondary oocyte is released by an ovary and is quickly drawn into a thin tube, the **oviduct** (also called the Fallopian tube), by the actions of undulating fimbriae (singular, fimbria), finger-like projections of the oviduct that extend out to the ovary.

The secondary oocyte is moved down the length of the oviduct by cilia on the oviduct's inner surface. For fertilization to take place, sperm must travel through the cervix and uterus and then into the oviduct, where fertilization typically occurs. Upon contact with a sperm, the secondary oocyte completes meiosis II, and the union of sperm and the secondary oocyte creates a fertilized egg, or zygote. The zygote undergoes several cell divisions to become a **blastocyst,** a ball of approximately 32–150 cells that enters the uterus, where it will develop into an embryo, as we will examine in detail later.

## Gametogenesis in Females Is a Cyclical Process Within the Ovaries

In contrast to males, in which spermatogenesis continues throughout postpubertal life in the testes, most female mammals, including humans, appear to be born with all the primary oocytes they will ever have. At birth, each ovary in a human female has about 1 million primary oocytes, which are arrested in prophase of meiosis I. Most

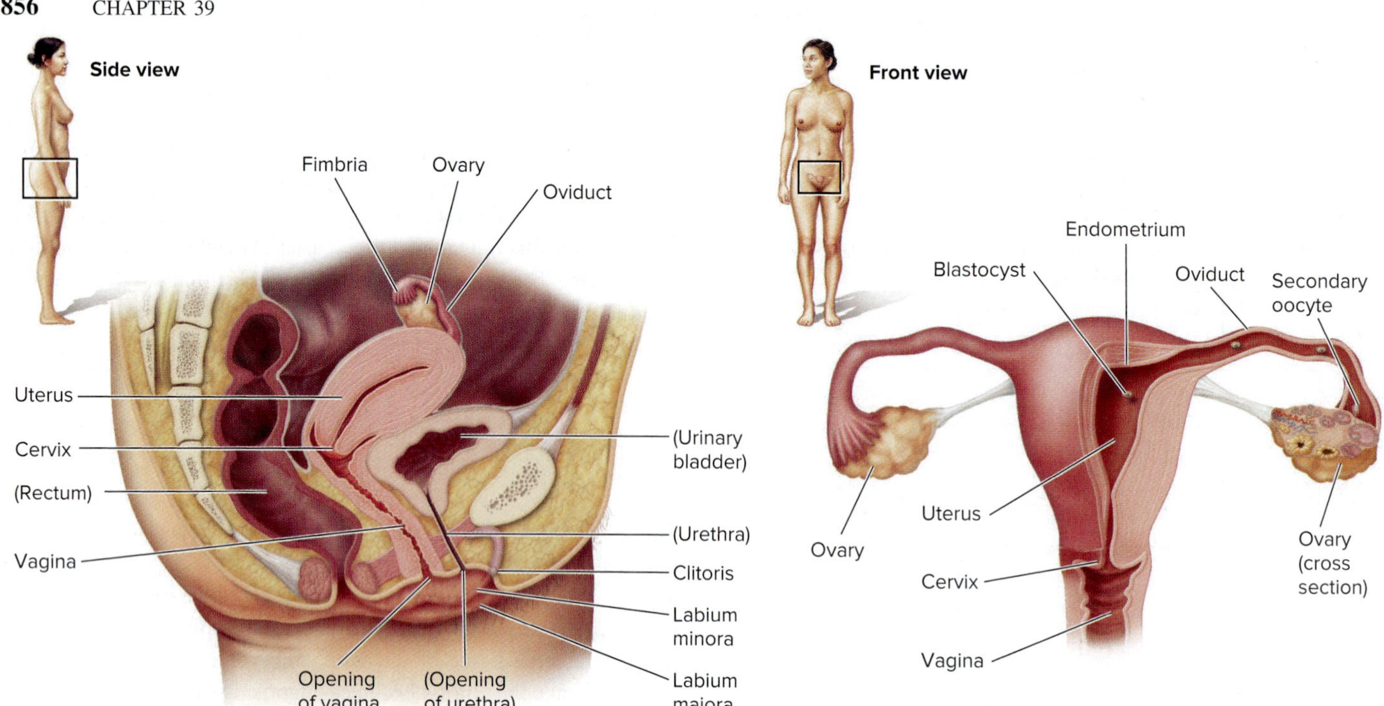


**Figure 39.7 Female reproductive structure and function in humans.** Side and front views of the female reproductive system (nonreproductive structures are identified in parentheses for orientation purposes). An oocyte moves from the ovary into the oviduct (or Fallopian tube), where it may be fertilized and develop into a blastocyst. Subsequently, the blastocyst enters the uterus, where it may implant in the endometrium, the inner lining of the uterus.

of these degenerate before the onset of puberty, when each ovary contains about 200,000 primary oocytes. Other than this degeneration, the ovaries are quiescent until puberty, when they begin to show cyclical activity.

Cells within the ovaries secrete a family of steroid hormones called estrogens. The most important estrogen is **estradiol,** which has a critical role in ovulation and influences the secondary sex characteristics of females. The secondary sex characteristics, which begin to develop at puberty, include development of breasts, widening of the pelvis (an adaptation for giving birth), and a characteristic pattern of fat deposition.

The **ovarian cycle** involves the development of an ovarian follicle, the release of a secondary oocyte, and the formation and subsequent regression of the empty follicle (Figure 39.8). During the first week of the ovarian cycle in humans, several primary oocytes that have been maturing for several months, each within their own follicle, now begin their final maturation steps (see step 1, Figure 39.8). By the beginning of the second week, all but one of these growing follicles and their primary oocytes degenerate; the single remaining follicle continues to develop and enlarge (while a new crop of immature follicles begins their slow growth period for future ovarian cycles). During that time, the primary oocyte of the enlarging follicle completes meiosis I, becomes a secondary oocyte, and begins meiosis II (see steps 2 and 3, Figure 39.8). The developing secondary oocyte is surrounded by cells of the cumulus mass and theca, which both protect and nurture it and secrete estradiol. The estradiol is secreted into the blood, where it functions to control the secretion of LH and FSH from the anterior pituitary gland, as we'll discuss shortly. Some estradiol is also secreted into the follicle, where it stimulates fluid secretion into the inner core of the follicle. As the follicle continues to grow, the fluid pressure inside the follicle increases, until it begins to form a bulge. Eventually, ovulation occurs as the follicle ruptures, and the secondary oocyte, the zona pellucida, and some surrounding supportive cells of the cumulus mass are released from the ovary (see step 3 of Figure 39.8).

Cells in the empty follicle subsequently undergo pronounced anatomical and physiological changes, developing into a structure called the **corpus luteum.** In humans, the corpus luteum is active for approximately the second half of the ovarian cycle. It is responsible for secreting hormones that stimulate the development of the uterus required for sustaining the embryo in the event of a pregnancy. If pregnancy does not occur, the corpus luteum degenerates, and a new group of follicles with their primary oocytes develops (see step 5, Figure 39.8).

In humans, a typical ovarian cycle lasts approximately 28 days. Ovulation and degeneration of additional primary oocytes continue throughout adulthood, with about 300 to 500 secondary oocytes being ovulated over a woman's 30- to 40-year reproductive lifetime. Eventually, the oocytes become nearly depleted, and a woman stops having ovarian cycles, an event called **menopause.** The average onset of menopause in the U.S. is approximately 51 years of age. After menopause, a woman is no longer capable of ovulation.

## The Ovarian Cycle Results from Changes in Hormone Secretion

We saw that in males, testosterone produced by cells in the testes exerts a negative feedback on secretion of GnRH and LH. In females, however, the situation is more complicated. Although GnRH also stimulates release of LH and FSH in females, the resulting estradiol can have both negative and positive feedback effects on the

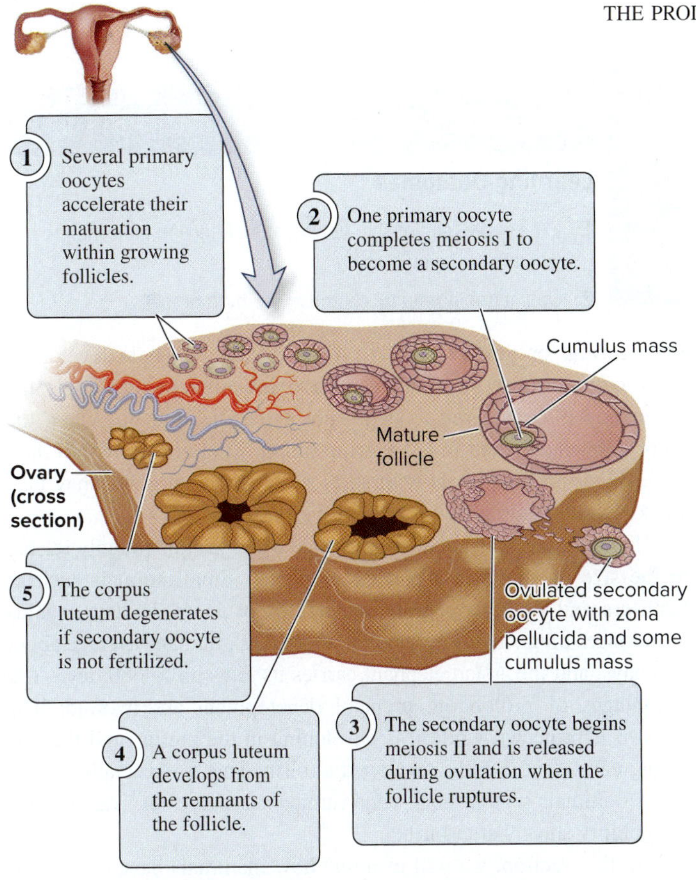


**Figure 39.8** **Follicle and oocyte development in the ovarian cycle.** Development of an oocyte and corpus luteum within the ovary are events that occur during a single ovarian cycle.

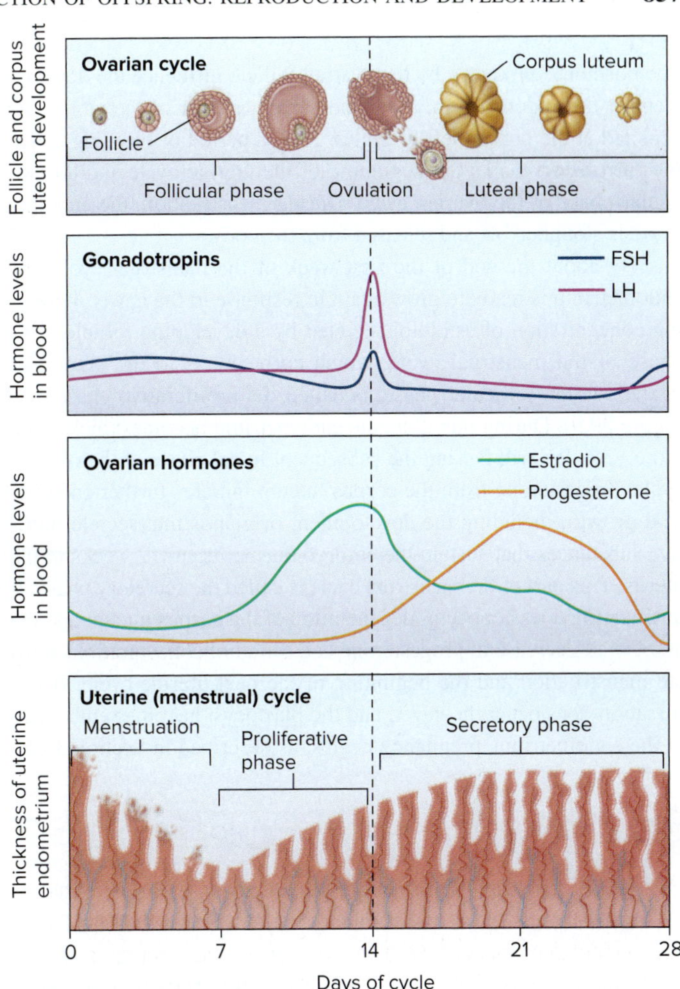


**Figure 39.9** **The ovarian and uterine cycles in a human female.** The ovarian cycle is divided into the follicular and luteal phases. The uterine cycle is divided into menstruation and the proliferative and secretory phases.

gonadotropins. To understand this, let's examine the hormone changes that occur during the ovarian cycle in a human female (Figure 39.9).

The first half of the ovarian cycle is called the **follicular phase** of the cycle, because this is when the growth and differentiation of a cohort of follicles occur. The relatively low concentration of LH that exists during follicular development is sufficient to stimulate the cells of the follicle to make estradiol. The estradiol that is produced is important for enlargement and growth of the oocytes and it is secreted into the blood, where it can influence the secretion of LH and FSH.

As these follicles develop, estradiol (and to a lesser extent, another steroid hormone called progesterone) production continues, and consequently, the concentration of estradiol in the blood begins to increase. Initially, estradiol exerts a negative feedback action on LH and FSH secretion, keeping their concentrations relatively low, and all but the largest of follicles die. When the remaining follicle is fully developed and ready for ovulation, its production of estradiol increases, such that the blood concentration of estradiol increases sharply. At that time, the feedback action of estradiol on LH and FSH switches from negative to positive, by mechanisms that involve increased GnRH secretion from the hypothalamus. This results in a sudden, sharp surge in LH and FSH in the blood. The LH released from the pituitary as a result of positive feedback by estradiol induces rupture of the follicle and ovulation.

Ovulation marks the end of the follicular phase and the beginning of the **luteal phase** of the ovarian cycle, named after the corpus luteum. During this phase, the high LH concentration initiates development of the corpus luteum. The corpus luteum secretes progesterone, the dominant ovarian hormone of the luteal phase, plus some estradiol. Progesterone inhibits LH and FSH secretion and is essential to prepare the uterus to receive and nourish the embryo. If fertilization of the secondary oocyte does not occur, the corpus luteum degenerates after 2 weeks and progesterone release decreases, allowing LH and FSH to initiate development of a new set of oocytes and their follicles. However, if fertilization does occur, the blastocyst develops a surrounding layer of cells that secrete an LH-like hormone, called **chorionic gonadotropin,** which maintains the corpus luteum and its ability to secrete progesterone. Chorionic gonadotropin is only present in pregnant women; as such, it is the hormone that is tested for by home pregnancy tests.

**Concept Check:** *Would similar surges in the blood concentrations of FSH and LH be expected to occur in human males?*

## Maternal Hormones Prepare the Uterus to Accept the Embryo

In humans, the ovarian cycle occurs in parallel with changes in the lining of the uterus during the **uterine cycle,** or **menstrual cycle.**

The hormones produced by the ovarian follicle influence the development of the endometrium, the glandular inner layer of the uterus. As depicted at the bottom left of Figure 39.9, a period of bleeding called **menstruation** marks the beginning of the uterine cycle and the follicular phase of the ovarian cycle. During menstruation, the endometrium is sloughed off and released from the body.

By about the end of the first week of the menstrual cycle, the endometrium is ready to grow again in response to the newly increasing concentration of estradiol secreted by a developing follicle. This phase of the menstrual cycle, which corresponds to the latter part of the ovarian follicular phase, is called the proliferative phase (see Figure 39.9). During this time, the endometrium becomes thicker and more vascularized. During the subsequent luteal phase of the ovarian cycle, progesterone from the corpus luteum initiates further endometrial growth, including the development of glands that secrete nutritive substances that sustain the embryo during its first 2 weeks in the uterus. This part of the menstrual cycle is called the secretory phase. If fertilization does not occur, degeneration of the corpus luteum and the associated decrease in progesterone and estradiol concentrations initiate menstruation and the beginning of the next uterine cycle. If fertilization does occur, however, and the blastocyst becomes embedded in the endometrium, pregnancy begins, as described in Section 39.4.

## 39.3 Reviewing the Concepts

- Sperm production requires testosterone. Sperm move out of the seminiferous tubules and into the epididymis, which leads into the vas deferens and the ejaculatory duct. The urethra conducts semen, a mixture containing fluid and sperm, during ejaculation. The fluid components of semen are produced in the seminal vesicles, the prostate gland, and the bulbourethral glands (Figures 39.5, 39.6).
- In the ovary, primary oocytes develop into secondary oocytes. If the secondary oocyte is fertilized by a sperm, the fertilized egg undergoes several cell divisions to become a blastocyst, a ball of cells that enters the uterus, where it will develop into an embryo (Figures 39.7, 39.8).
- In females, changes in hormone secretion produce the ovarian cycle and control the uterine cycle or menstrual cycle. The cessation of ovarian cycles is called menopause (Figure 39.9).

## 39.3 Testing Your Knowledge

1. During the secretory phase of the menstrual cycle, endometrial glands secrete
   a. hormones that increase the likelihood of pregnancy.
   b. nutritive substances that sustain an embryo during the first 2 weeks of development.
   c. hormones that prevent ovulation.
   d. waste products into the lumen of the uterus.
   e. both a and c.
2. Sperm fertilize a _________ in the _________.
   a. primary oocyte, uterus
   b. secondary oocyte, oviduct
   c. blastocyst, oviduct
   d. blastocyst, uterus
   e. secondary oocyte, ovary

## 39.4 Pregnancy and Birth in Mammals

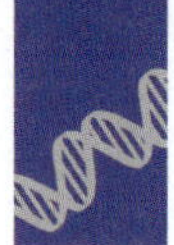

### Learning Outcomes

1. Explain the relationship between the fetal and maternal structures of the placenta.
2. Outline the hormonal control of the birth process.

**Pregnancy,** or gestation, is the time during which a developing embryo or fetus grows within the uterus of the mother. Physiologically, pregnancy is considered to begin not at fertilization but when the embryo is established in the uterine lining. This occurs within days of fertilization in animals with short gestation lengths but may take weeks in large animals with long gestations.

In mammals, gestation length varies widely and is roughly related to the size of adults in a particular species. Small animals such as hamsters and mice have gestation periods of 16 to 21 days, canids have longer pregnancies of about 50 to 75 days, humans average about 268 days, and the Asian elephant carries its fetus up to 660 days. The advantages of prolonging prenatal development are twofold: The embryo is protected while it is developing in the uterus, and the offspring can be more fully developed at birth. This is especially important for animals such as horses and ruminants, whose survival depends on mobility shortly after birth.

In this section, we will examine how mammals have evolved to retain their young in the uterus for extended periods and the role of hormones in pregnancy and birth.

### Most Mammals Retain Their Young in the Uterus and Nourish Them via a Placenta

Three types of pregnancies are found in mammals, which correspond to the three clades of mammals: monotremes, marsupials, and eutherians (see Chapter 27). Monotremes such as the platypus are the only mammals that lay fertilized eggs. In marsupials such as the kangaroo, the young are born while still extremely undeveloped. They then crawl up the mother's abdomen to her pouch, where they attach to a nipple to suckle and obtain nourishment, remaining and maturing within the pouch. Compared with marsupials, humans and other eutherian mammals retain their young within the uterus for a longer period of time and nourish them via transfer of nutrients and gases through a **placenta,** a structure that connects the developing young to the mother's uterine wall (**Figure 39.10**).

During pregnancy, many physiological changes occur in both the embryo and the mother. The first event of pregnancy in eutherian (placental) mammals is **implantation,** when the blastocyst embeds within the uterine endometrium, which typically occurs in humans around 8 to 10 days after fertilization. As mentioned earlier, the implanted blastocyst initially receives nutrients directly from endometrial glands. However, shortly after implantation, newly developing embryonic tissues merge with the endometrium to form the placenta, which remains in place and grows larger as the embryo matures into a **fetus.** (In humans, an embryo is called a fetus after the eighth week of gestation.) The placenta, therefore, has a maternal portion and a fetal portion.

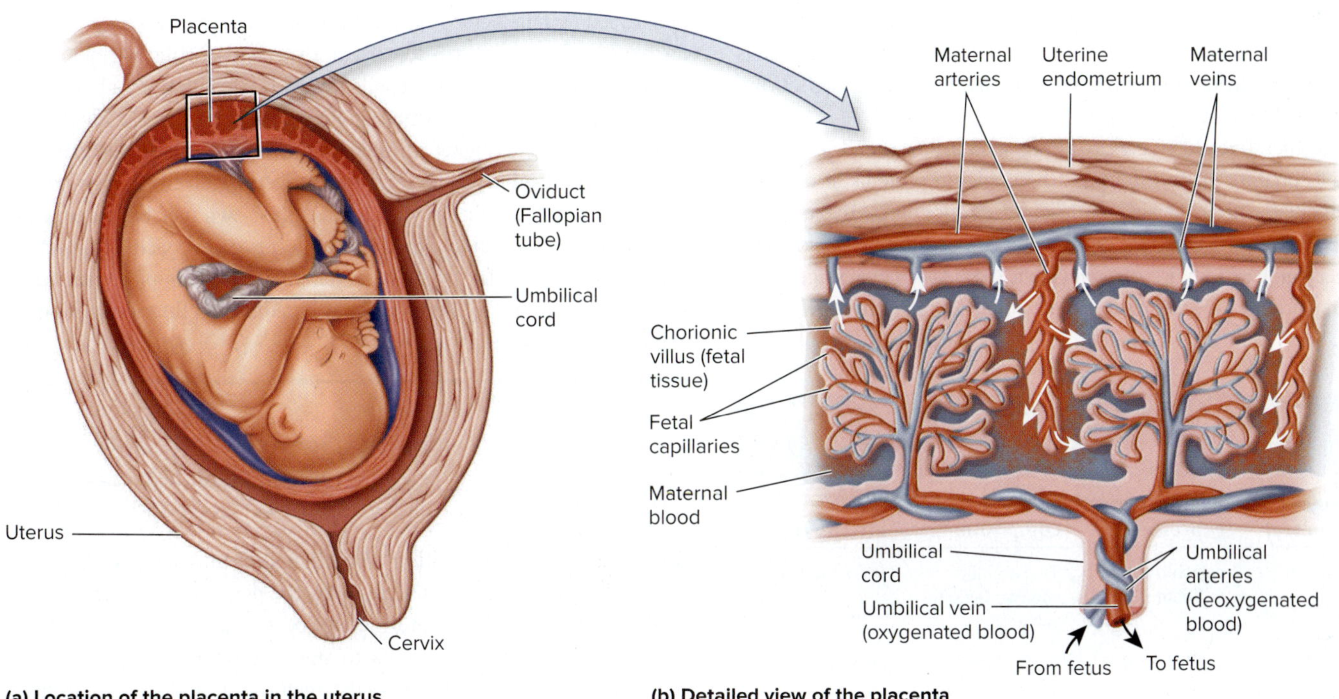


**Figure 39.10** **The structure of the placenta.** In mammals, the placenta is composed of both fetal and maternal tissues. **(a)** Overview of placental location and structure in the human. **(b)** Enlarged view of the placenta showing the relationship between fetal and maternal structures. Note that in humans, blood in the fetal and maternal circulations does not mix. The white arrows indicate diffusion of solutes such as nutrients and wastes; black arrows indicate direction of blood flow.

**BioConnections:** *Note that the umbilical arteries are shown in blue to signify that they carry deoxygenated blood, and the umbilical vein is shown in red to signify it carries oxygenated blood. Normally, arteries carry oxygenated blood, and veins carry deoxygenated blood, except in the pulmonary circulation (refer back to Figure 36.2). Why is the circulation through the placenta like that of the pulmonary circulation?*

The placenta is rich in blood vessels from both the mother and the fetus. The maternal and fetal sets of vessels lie in close proximity. The fetal portion of the placenta, called the chorion, contains convoluted structures called chorionic villi that provide a large surface area containing capillaries where nutrients, gases, and other solutes are exchanged. Nutrients and oxygen from the mother are carried through maternal arteries, where her blood pools in large areas of the fetal placenta surrounding the fetal capillaries. Solutes diffuse from the maternal blood into fetal capillaries, and from there flow into the umbilical vein, part of the fetal circulation. In turn, carbon dioxide and other waste products from the fetus are carried through the umbilical artery to the placenta, where they diffuse into the mother's circulation, from which they can be excreted. Because of this placental organization, the maternal and fetal blood do not mix.

Prenatal development in humans is generally described as having three trimesters, each of which lasts about 3 months. At the end of the first trimester, the rudiments of the organs are present, and the developing fetus is about an inch long. The second trimester is an extremely rapid phase of growth. During the third trimester, the lungs of the fetus mature so that they are ready to function as gas-exchange organs.

## EVOLUTIONARY CONNECTIONS

### The Evolution of the Globin Gene Family Has Been Important for Internal Gestation in Mammals

As discussed in Chapter 18, genes can become duplicated, which sometimes creates gene families. Gene families have been important in the evolution of complex traits because the various members of a gene family can enable the expression of complex, specialized forms and functions. An interesting example is the globin gene family in animals. Globin genes encode polypeptides that are subunits of proteins that function in oxygen binding. Hemoglobin, which is present in erythrocytes, carries oxygen throughout the body in all vertebrates and many invertebrates, delivering oxygen to all of the body's cells. In humans, the globin gene family is composed of several homologous genes that were originally derived from a single ancestral globin gene (refer back to Figure 18.13).

All of the globin polypeptides are subunits of proteins that function in oxygen binding, but the various family members tend to have specialized functions. For example, certain globin genes are expressed only during particular stages of embryonic development (refer back to Figure 12.3). This has special importance in placental mammals, because the oxygen demands of a growing embryo and fetus are quite different from the demands of its mother. These different demands are met by the differential expression of hemoglobin genes during prenatal development.

Altogether, five globin genes—designated α, β, γ, ε, and ζ—encode the major subunits that are found in hemoglobin proteins at different developmental stages. During embryonic development, the ε-globin and ζ-globin genes are turned on, resulting in embryonic hemoglobin with a very high affinity for oxygen (**Table 39.1**). At the fetal stage, these genes are turned off, and the α-globin and γ-globin genes are turned on, producing fetal hemoglobin with slightly less (but still high) affinity for oxygen. Finally, just before birth, expression of the γ-globin gene decreases, and the β-globin gene is turned on, resulting in adult hemoglobin, which has a lower affinity for oxygen than either the embryonic or fetal forms. The higher affinities of embryonic and fetal hemoglobins enable the embryo and fetus to remove oxygen from the mother's bloodstream and use that oxygen to meet their own metabolic demands. Therefore, the evolution of different globin genes each of which is expressed at particular stages of development enables placental mammals to develop in the uterus without breathing on their own or being exposed to atmospheric oxygen.

**Table 39.1 Globin Gene Expression During Mammalian Development**

| Stage of development | Globin genes expressed | Hemoglobin composition | Oxygen affinity ($P_{50}$)* |
|---|---|---|---|
| Embryo | ε-globin and ζ-globin | Two ε-globin and two ζ-globin subunits | 5–13.5 mmHg |
| Fetus | γ-globin and α-globin | Two γ-globin and two α-globin subunits | 19.5 mmHg |
| Birth to adult | β-globin and α-globin | Two β-globin and two α-globin subunits | 26.5 mmHg |

*$P_{50}$ values represent the partial pressure of oxygen required to half-saturate hemoglobin (see Chapter 36): A lower $P_{50}$ indicates a higher affinity of hemoglobin for oxygen. The value for embryos is an estimate based on in vitro experiments. All values are for human hemoglobins.

## Birth Is Dependent on Hormones That Elicit a Positive Feedback Loop

Birth—also called parturition—is initiated by the actions of several hormones and other factors secreted by the placenta and glands of the mother (**Figure 39.11**). Toward the end of pregnancy, hormones from the fetus stimulate the placenta to start secreting large amounts of estrogens such as estradiol into the maternal circulation. Estrogens have at least two major effects on uterine tissue at this time. First, they promote the formation of gap junctions between uterine smooth muscle cells, which enables coordinated uterine contractions. Second, estrogens enhance uterine sensitivity to oxytocin.

Recall from Chapter 38 that oxytocin is a posterior pituitary hormone that stimulates contraction of uterine muscle. The high concentration of estradiol in the mother's blood near the end of

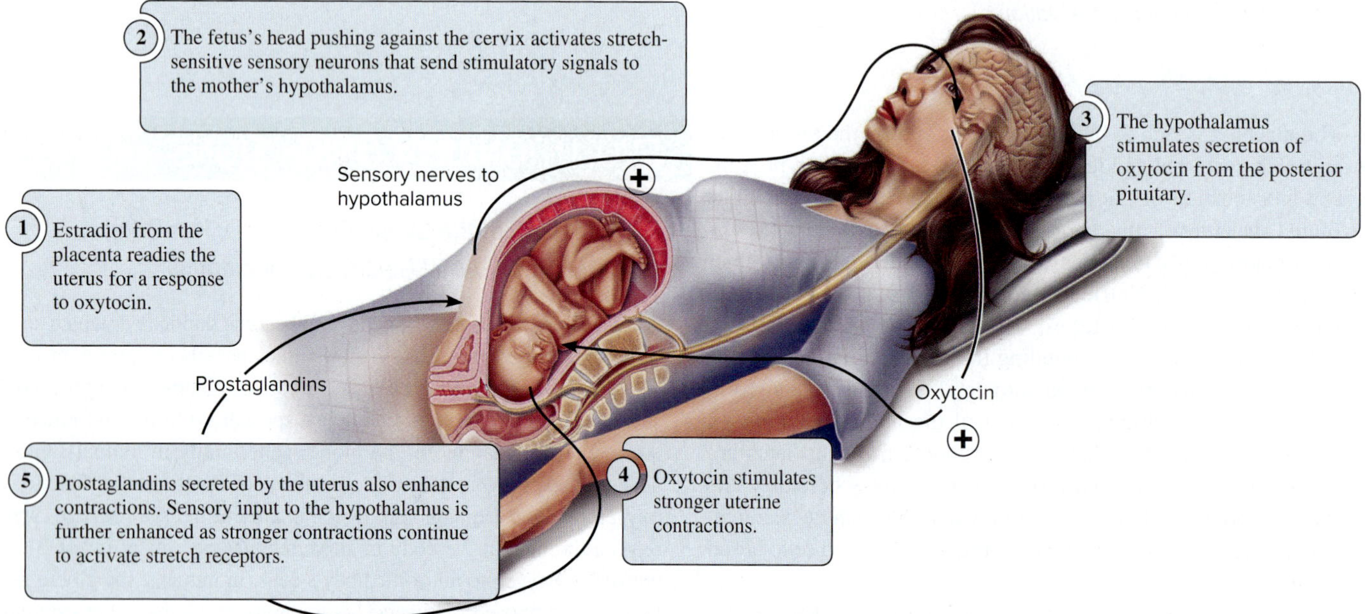

2 The fetus's head pushing against the cervix activates stretch-sensitive sensory neurons that send stimulatory signals to the mother's hypothalamus.
Sensory nerves to hypothalamus
+
3 The hypothalamus stimulates secretion of oxytocin from the posterior pituitary.
1 Estradiol from the placenta readies the uterus for a response to oxytocin.
Prostaglandins
Oxytocin
+
5 Prostaglandins secreted by the uterus also enhance contractions. Sensory input to the hypothalamus is further enhanced as stronger contractions continue to activate stretch receptors.
4 Oxytocin stimulates stronger uterine contractions.


**Figure 39.11 Hormonal control of parturition.** Birth relies on maternal hormones that act on the uterus and neural signals from the uterus. In response to sensory neural input arising from the push of the fetus on the cervix, the maternal posterior pituitary releases oxytocin into the blood, which stimulates uterine smooth muscle contractions. The secretion of prostaglandins by the uterus also increases the strength of the contractions. Sensory receptors in the uterus detect the more forceful contractions and signal the mother's posterior pituitary to secrete more oxytocin, thus completing a positive feedback loop that further strengthens the contractions.

*In what other context have you learned about positive feedback in female reproduction? See Figure 39.9 for help.*

pregnancy stimulates the production of oxytocin receptors in smooth muscle cells of the uterus, thereby making the uterus more sensitive to oxytocin. At the same time, the fetus usually positions itself with its head above the uterine cervix in preparation for birth. The pressure of the fetus's head pressing on the cervix stretches the smooth muscle of the uterus and cervix. This stretch is detected by neurons in these structures. Signals from the stretch-sensitive neurons are sent to the mother's hypothalamus, triggering the release of oxytocin into the blood.

Binding of oxytocin to its receptors initiates the strong uterine muscle contractions that are the hallmark of **labor.** In addition to its direct action on uterine muscle, oxytocin stimulates uterine secretion of small signaling molecules called prostaglandins that act with oxytocin to further increase the strength of the muscle contractions. The stronger contractions elicit ever-greater stimulation of stretch receptors, resulting in more oxytocin release from the mother's pituitary, which causes yet stronger contractions. This positive feedback loop continues until the baby is born.

Labor occurs in three stages (Figure 39.12). In stage one, dilation and thinning of the cervix occur, which makes it easier for the fetus to pass out of the uterus. In stage two, uterine contractions get stronger and more frequent. The fetus is pushed, usually headfirst, through the cervix and the vagina and out into the world. In stage three, the contractions continue for a short while. Blood vessels within the placenta and umbilical cord contract and block further blood flow, making the newborn independent from the mother. As the oxytocin-induced contractions continue for a short period, the placenta detaches from the uterine wall and is delivered a few minutes after the birth of the baby.

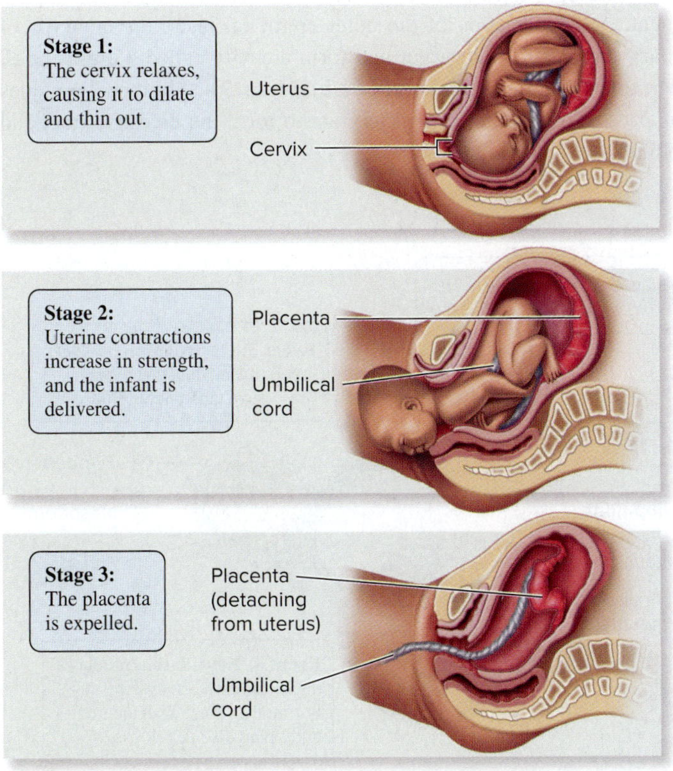

**Stage 1:** The cervix relaxes, causing it to dilate and thin out.

Uterus

Cervix

**Stage 2:** Uterine contractions increase in strength, and the infant is delivered.

Placenta

Umbilical cord

**Stage 3:** The placenta is expelled.

Placenta (detaching from uterus)

Umbilical cord

**Figure 39.12** The three stages of labor.

**Concept Check:** *Many female mammals consume the placenta after giving birth. What is the benefit of this behavior?*

## 39.4 Reviewing the Concepts

- Pregnancy is the time during which a developing embryo or fetus grow within the uterus of the mother. Implantation is when the blastocyst embeds within the uterine endometrium. Most mammals retain and nourish their young within the uterus via transfer of nutrients and gases through the placenta (Figure 39.10).
- The evolution of the globin gene family contributed to the ability of placental mammals to develop inside the mother's uterus (Table 39.1).
- Birth is initiated by hormones produced by the placenta and by the mother's endocrine system. Oxytocin stimulates the strong uterine muscle contractions that are the hallmark of the three-stage process called labor (Figures 39.11, 39.12).

## 39.4 Testing Your Knowledge

1. During the ________ stage of labor in mammals, the placenta is expelled from the uterus.
   a. first
   b. second
   c. third
   d. It is not expelled; the placenta is reabsorbed.
2. Which is *false*?
   a. Fetal hemoglobin has a higher affinity for oxygen than the hemoglobin of adult mammals.
   b. The process of birth is an example of negative feedback control.
   c. The placenta is composed of both maternal and fetal tissue.
   d. Implantation is when a blastocyst embeds in the uterine endometrium.
   e. Oxytocin and estrogens both act on the uterus in different ways to stimulate the process of birth.

# 39.5 General Events of Embryonic Development

### Learning Outcomes

1. **SCISKILLS ▶** Describe the general events of embryonic development, and identify the end result of each.
2. Describe the process of cleavage, beginning at fertilization and leading into gastrulation.
3. **SCISKILLS ▶** Predict the fates of cells in a gastrula as they differentiate into the three major germ layers.
4. Outline the early development of the nervous system during neurulation.
5. Explain the migration and fates of neural crest cells.

The process by which a fertilized egg (that is, a zygote) is transformed into an animal with distinct physiological systems and body parts is called **embryonic development.** As animals develop, cells arrange themselves in coordinated ways that lead to the establishment of a body plan. The final, adult body plan of most animals is organized along three axes: the dorsoventral axis**,** the anteroposterior axis, and the left-right axis. Along these axes are often separate sections, or body segments, each containing specific body parts such as a wing or leg.

To establish the correct body plan, each cell in a developing animal must receive information regarding its relative position within the body. Such information determines where a cell should move to, whether or not it should divide (or die), and what types of functions it will ultimately perform. This is possible because each cell receives positional information from its neighboring cells. This information is provided in a variety of ways, including intracellular and extracellular signaling molecules and by cell-to-cell contacts. A cell responds to positional information by dividing (cell division), dying (cell death, or apoptosis), or migrating from one region of the embryo to another (cell migration). Last, in the process of **cell differentiation,** different cells within a developing organism acquire specialized forms and functions, due to the expression of cell-specific genes.

Embryonic development follows a similar pattern in most animals. Most modern animals are triploblastic; that is, they develop from embryos with three distinct cell layers called the ectoderm, mesoderm, and endoderm (refer back to Figure 26.4). Triploblasts include vertebrates and most invertebrates other than sponges and cnidarians. Development in these animals can be categorized into five general events: (1) fertilization (see Figure 39.4), (2) cleavage, (3) gastrulation, (4) neurulation, and (5) organogenesis (**Figure 39.13**). In this section, we will examine the key aspects of the general events of animal development beginning with cleavage. However, it should be noted that many species also have an additional event called metamorphosis, which is a transition from a feeding larval form to an adult (refer back to Figure 26.28). Metamorphosis occurs after organogenesis and facilitates the rapid growth of young organisms into mature ones. Examples of metamorphosis include the transformation of a caterpillar into a butterfly and that of a tadpole into a frog.

## Cell Divisions Without Cell Growth Create a Cleavage-Stage Embryo

The initial cell cycles of embryos are unique because they involve repeated cell divisions without cell growth. The process by which these cell cycles occur is called **cleavage.** The embryonic cells repeatedly split in two, resulting in several generations of daughter cells that are roughly half the size of the cells that gave rise to them.

In most species in which development occurs outside the mother, where eggs can be eaten by predators, cell division during cleavage represents one of the fastest cell cycles found in nature. The cell cycle during cleavage in amphibians, for example, requires only 20 minutes. During each 20-minute cell cycle, complete genome replication, mitosis, and duplication of the nuclear envelope are followed by cytokinesis. In placental mammals, in which development occurs within the protective environment of the mother's body, cell divisions during cleavage are slower, requiring about 12 hours to complete.

The daughter cells produced during the cleavage stage of development are known as **blastomeres.** Individual blastomeres are bound together, and an outer, single-cell layer of blastomeres forms a sheet of epithelial cells that separates the embryo from its environment. After formation of the outer epithelial layer, the embryos of many animals take up water and form a cavity called a **blastocoel.** The embryo at this stage is called a **blastula.** The blastocoel provides a space into which cells will migrate to form the digestive tract and other structures of the embryo.

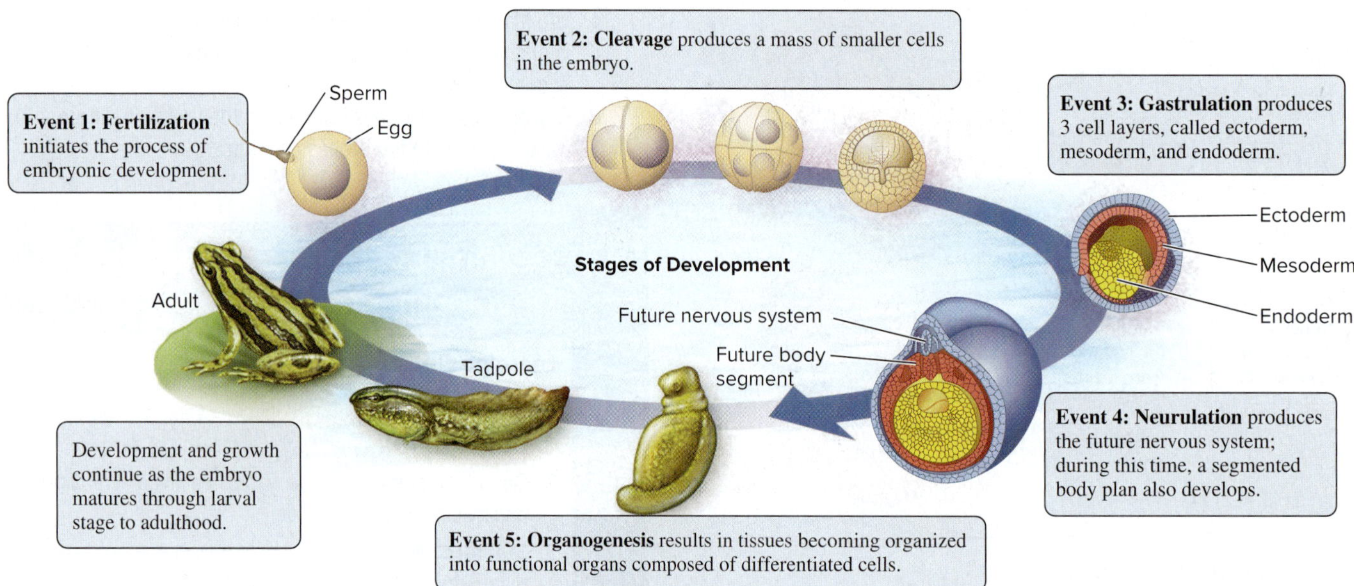

Event 2: Cleavage produces a mass of smaller cells in the embryo.
Sperm
Egg
Event 1: Fertilization initiates the process of embryonic development.
Event 3: Gastrulation produces 3 cell layers, called ectoderm, mesoderm, and endoderm.
Ectoderm
Mesoderm
Endoderm
Stages of Development
Adult
Future nervous system
Future body segment
Tadpole
Event 4: Neurulation produces the future nervous system; during this time, a segmented body plan also develops.
Development and growth continue as the embryo matures through larval stage to adulthood.
Event 5: Organogenesis results in tissues becoming organized into functional organs composed of differentiated cells.


**Figure 39.13** Overview of events of embryonic development. This figure shows the general developmental events of all vertebrate embryos, using a frog as an example.

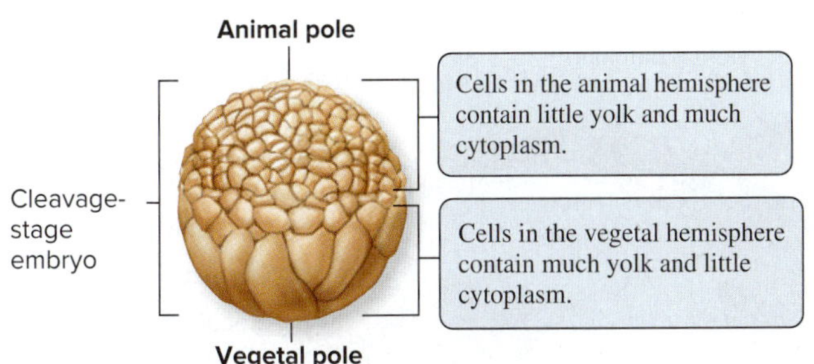


**Figure 39.14** Polarity in an amphibian cleavage-stage embryo.

**Meroblastic Cleavage: Birds, Reptiles, and Fishes** Among triploblastic organisms, cleavage-stage embryos can vary dramatically in size and appearance. This variation is in part related to whether or not the egg contains yolk and, if so, the location and amount of that yolk. Yolk is a nutrient-rich food store that is used by the developing embryo. The eggs of birds, some fishes, and some other vertebrates have large amounts of yolk. In the eggs and early embryos of these species, yolk is most concentrated toward one end—or pole—called the **vegetal pole.** Much less yolk, and much more cytoplasm, is concentrated near the opposite pole, called the **animal pole** (Figure 39.14). These poles form the apices of the vegetal and animal hemispheres, which determine in part the future anteroposterior (head-tail) and dorsoventral (back-front or top-bottom, depending on the species) axes of the embryo.

In some species that exhibit animal and vegetal poles, cleavage of the zygote is called **meroblastic cleavage,** or incomplete cleavage, because only the region of the embryo located within the animal hemisphere undergoes cell division (Figure 39.15). Instead of forming a ball of cells (a blastula), in this type of cleavage, a flattened disc of blastomeres known as a **blastoderm** develops on top of the yolk mass.

**Holoblastic Cleavage: Amphibians and Mammals** In animals whose eggs have smaller amounts of yolk, cleavage during the first cell division is complete and bisects the entire zygote into two equal-sized blastomeres. This complete cleavage, or **holoblastic cleavage,** occurs in amphibians and mammals (see Figure 39.15). In amphibians, cleavage-stage embryos form a blastula, as previously noted. In mammals, however, cleavage-stage embryos undergo a process called compaction, in which the amount of physical contact between cells is maximized. At this stage, the embryo in these species is called a **morula.** The morula then proceeds to form a blastocyst, the mammalian counterpart of a blastula.

**Cleavage and Implantation in Mammals** In mammals, the events of fertilization and cleavage occur in the oviduct (Figure 39.16). The blastocyst has a different morphological appearance than the blastula or blastoderm in nonmammalian species, and no animal-vegetal polarity that is analogous to that of other chordates exists. The blastocyst consists of an outer epithelial layer called the **trophectoderm,** which gives rise to the placenta, and an inner layer called the inner cell mass, which develops into the embryo. Upon reaching the uterus, the embryo hatches from the zona pellucida, the layer of glycoproteins that surrounds the secondary oocyte and is retained up to this time to prevent premature adhesion of the embryo to the oviduct. The

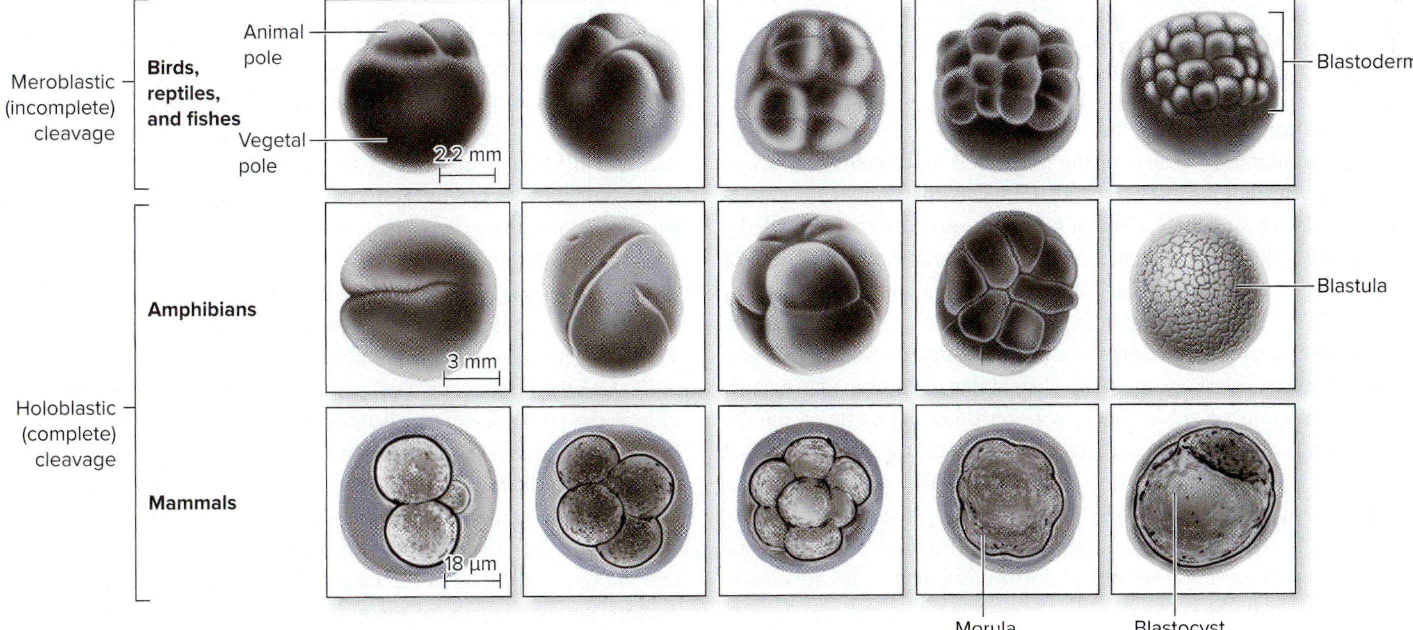


**Figure 39.15** Meroblastic and holoblastic cleavage. Early embryos of birds, reptiles, and many fishes undergo incomplete (meroblastic) cleavage, whereas most amphibian and mammalian embryos undergo complete (holoblastic) cleavage. The amount of yolk in the egg (not visible in these illustrations) contributes to many of these morphological differences observed in various species.

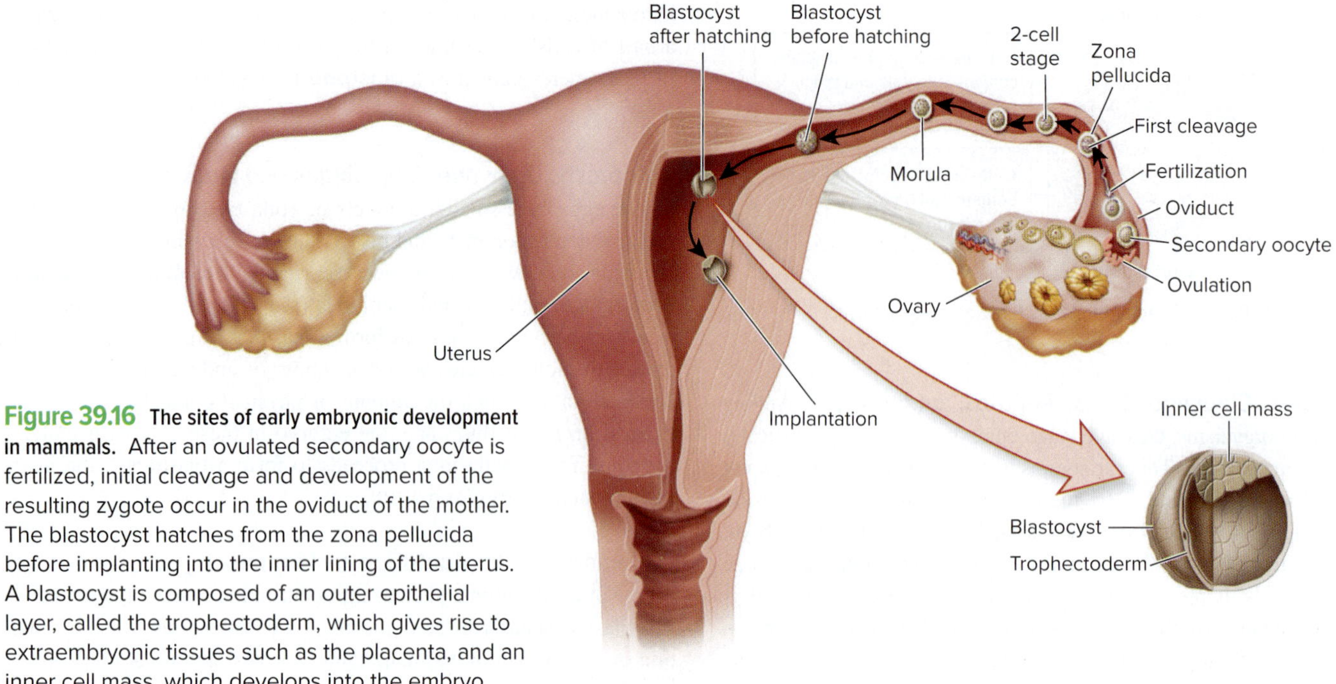


**Figure 39.16 The sites of early embryonic development in mammals.** After an ovulated secondary oocyte is fertilized, initial cleavage and development of the resulting zygote occur in the oviduct of the mother. The blastocyst hatches from the zona pellucida before implanting into the inner lining of the uterus. A blastocyst is composed of an outer epithelial layer, called the trophectoderm, which gives rise to extraembryonic tissues such as the placenta, and an inner cell mass, which develops into the embryo.

embryo then implants in the endometrium of the mother's uterus (see Figure 39.16). This entire process takes about 8 to 10 days in humans.

Toward the end of cleavage, cell cycles become less synchronous, and the embryo begins to express its own genes. The embryo's shift from existing exclusively on maternal factors to developing in response to products derived from its own genome begins 6–24 hours after fertilization in vertebrates. This is followed by the next general event of development, called gastrulation.

## Gastrulation Establishes the Three Germ Layers in the Embryo

**Gastrulation** is one of the most dramatic events of embryonic development in animals because of the major cell movements that occur. During gastrulation, the hollow ball of cells that makes up a blastula or blastocyst develops into a highly organized structure called a **gastrula** (refer back to Figure 26.4). A key event is the formation of germ layers, which are primary layers of cells that form during gastrulation. In the gastrula-stage embryo, three germ layers—**ectoderm, mesoderm,** and **endoderm**—become clearly established. These distinct germ layers are partially differentiated tissues that occupy discrete regions of the embryo, with an outer ectoderm, a middle mesoderm, and an inner endoderm layer. Each type of germ layer eventually gives rise to different structures. The organization that emerges during gastrulation is most evident by the clear establishment of the digestive tube and body axes. Gastrulation is the first time when both the anteroposterior and dorsoventral body axes are clearly evident in the embryo.

Each germ layer gives rise to different types of cells and body structures (**Figure 39.17**). The ectoderm in the gastrula forms the epidermis and nervous system in the later embryo. The mesoderm gives rise to muscles, kidneys, blood, heart, limbs, connective tissues, and notochord; the last is a key feature of all chordates, described later in this section. The endoderm becomes the epithelial lining of the pancreas, thyroid, lungs, digestive tract, liver, and urinary bladder.

Some of the most detailed descriptions of the events in gastrulation come from the study of frog and other amphibian embryos. The major events of gastrulation are depicted in **Figure 39.18** and described next.

**Invagination and Involution: Formation of Germ Layers and Archenteron** Prior to gastrulation, the blastula is enclosed in a simple, spherical epithelial cell layer. Gastrulation begins when a band of epithelial cells located at the vegetal hemisphere of the blastula—called bottle cells—invaginates (pinches in), pushing cells from the outside of the embryo to the inside (see Figure 39.18, step 1). This process creates a small opening called the **blastopore,** which defines the anteroposterior axis of the animal. The initiating site of invagination becomes what is called the dorsal lip of the blastopore. This change in morphology of only a few key cells in the embryo initiates the gastrulation process in amphibians.

Once the bottle cells change their shape and push into the interior of the embryo, other cell movements occur, and together these orchestrated movements establish the mesoderm and endoderm of the organism, including its future digestive tract. Just before invagination begins, cells of the animal hemisphere spread out and move downward. When they arrive at the blastopore, they enter the opening and subsequently migrate upward along the roof of the blastocoel, toward the animal pole of the embryo. This folding back of sheets of surface cells into the interior of the embryo is called involution (see Figure 39.18, step 2). After involution, dorsal mesodermal cells migrate toward the animal pole by crawling along the roof of the blastocoel, with endoderm following closely behind.

## Biology Principle

### Living Organisms Grow and Develop

The acquisition of specific cell types as shown here is an example of development; growth would entail an increase in the numbers and/or size of each of these cell types.

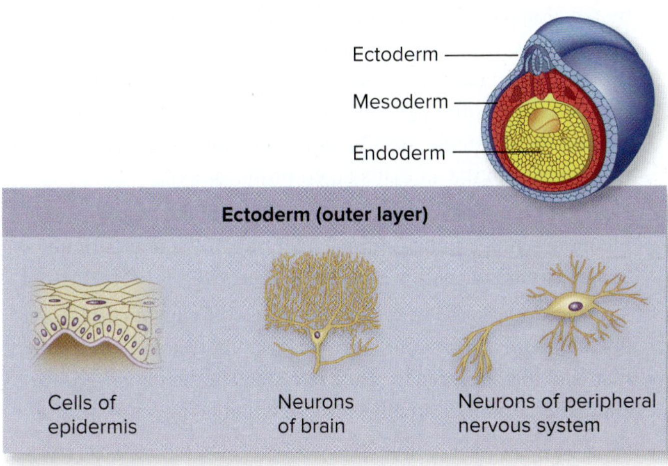


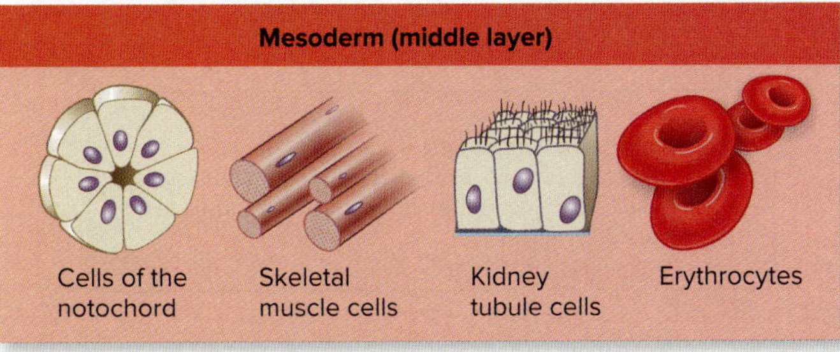


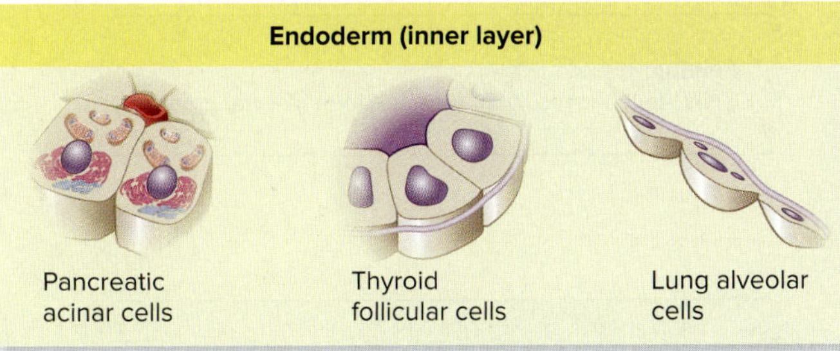


**Figure 39.17** Examples of cell types derived from ectoderm, mesoderm, and endoderm.

As the opening from the blastopore extends into the embryo, a new cavity called the **archenteron** displaces the existing blastocoel (see Figure 39.18, steps 2 and 3). The archenteron becomes the organism's future digestive tract. The blastopore opening remains sealed with a yolk-rich piece of tissue called the yolk plug until later in development. In chordates and echinoderms, the opening formed by the blastopore ultimately becomes the anus of the organism (refer back to Figure 26.5a). Meanwhile, during involution, surface cells spread from the animal hemisphere to surround the entire vegetal hemisphere to become the future ectoderm. The result of these cellular rearrangements is an embryo with three distinct germ layers.

(1) **Formation of the blastopore by invagination of bottle cells.** Gastrulation is initiated by the invagination of bottle cells, which forms a blastopore. Invagination of bottle cells forces cells behind them to involute toward the future anterior end of the embryo. The curved arrows indicate directions of cell movements.

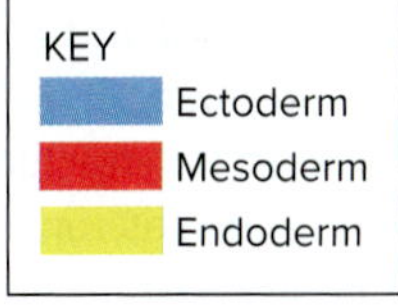


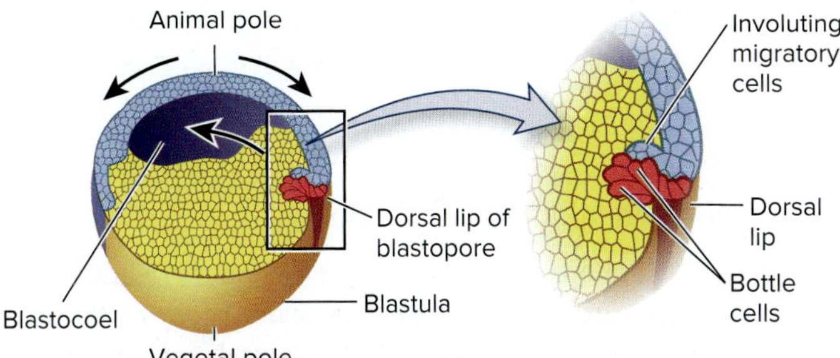


(2) **Formation of the archenteron by invagination and involution.** The cavity that begins at the blastopore expands to form the archenteron (future digestive tract). Ectoderm spreads over the embryo.

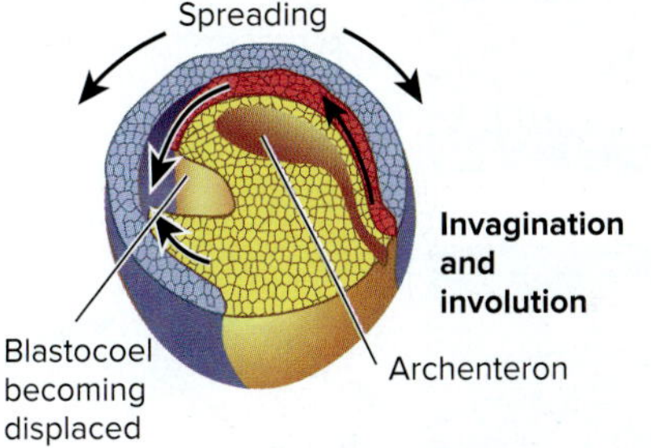


(3) **Completion of gastrulation with the beginning of notochord formation.** By the end of gastrulation, the archenteron has displaced the blastocoel and becomes closed by a yolk plug. Involution continues; some of the involuting cells become the mesoderm layer of the gastrula. The dorsal surface of the gastrula begins to thicken, and a portion of the mesoderm begins to form the notochord.

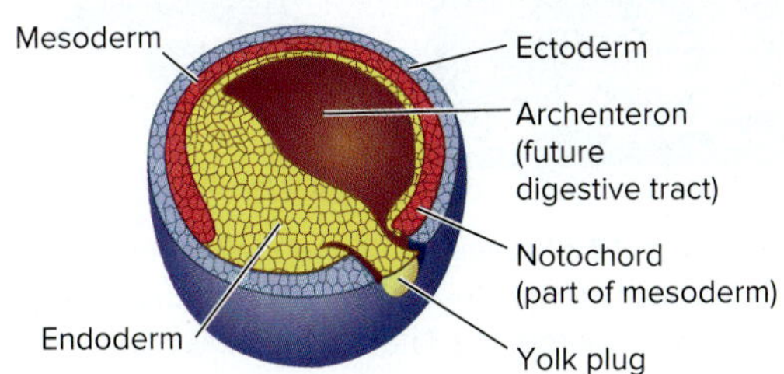


**Figure 39.18** The events of gastrulation in amphibians.

**Notochord Formation** A distinguishing anatomical feature that begins forming at the end of gastrulation in all chordates is the **notochord**—a structure derived from mesoderm that provides rigidity along the anteroposterior axis in the dorsal side of the gastrula (see Figure 39.18, step 3). The presence of a notochord defines the phylum Chordates. The notochord persists in the trunk and tail of fishes and amphibians; in birds and mammals, the notochord disappears by the time vertebrae have formed. By the time the notochord has formed, the dorsal ectoderm overlying the notochord begins to thicken, which initiates the next general event in development, called neurulation.

## Neurulation Involves Formation of the Central Nervous System and Segmentation of the Body

By studying development in several different vertebrate species, researchers are beginning to understand some of the fundamental steps in the formation of the central nervous system (CNS)—the brain and spinal cord—in vertebrates. The multistep embryological process responsible for initiating CNS formation is called **neurulation** (**Figure 39.19**). Neurulation occurs just after gastrulation and involves the formation of the **neural tube** from ectoderm located dorsal to the notochord. All neurons and their supporting cells in the CNS originate from neural precursor cells derived from the neural tube.

**Neural Tube Formation** Neurulation in vertebrates occurs in several major steps, as shown in Figure 39.19.

1. In the first step, ectoderm overlaying the notochord thickens by the elongation of cells in the dorsal region to form the neural plate, with adjacent regions that will eventually form the epidermis and a structure called the neural crest (discussed shortly). The neural plate then elongates, resulting in the formation of a single, dorsal, elongated epithelial cell layer that is aligned with the animal's anteroposterior axis.
2. Next, a column of cells along the midline of the neural plate—the medial hinge point—initiates folding of the neural plate, leading to the formation of the neural groove.
3. After folding, in the next step, bilateral columns of cells in the dorsal lateral hinge points then undergo morphological changes that lead to convergence of the two sides of the neural groove and generation of a tubelike structure that is not yet sealed on the dorsal surface.

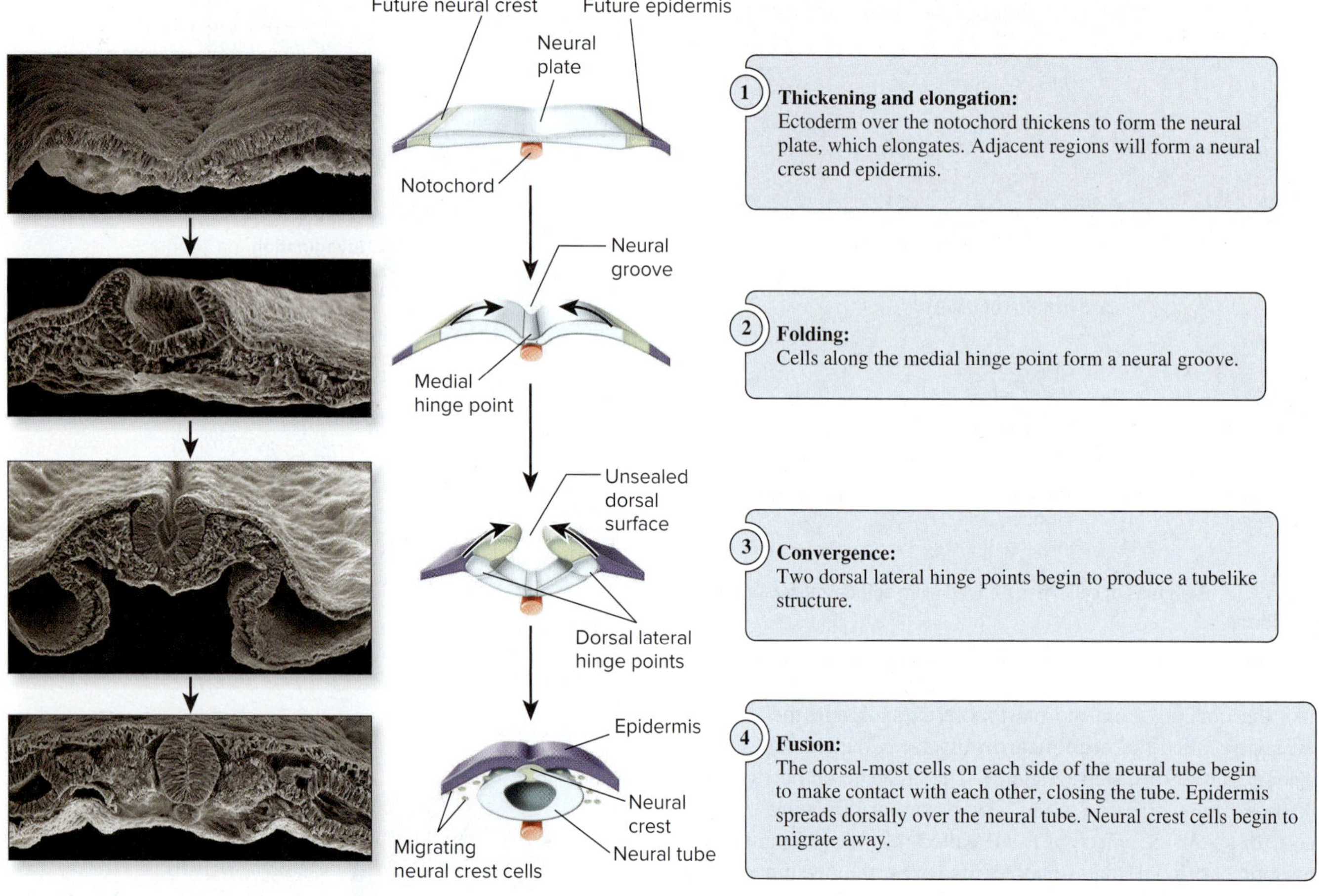


**Figure 39.19 Neurulation and the beginning of neural crest formation in vertebrates.** The major steps of neurulation are (1) thickening and elongation; (2) folding, which creates the neural groove; (3) convergence, in which the neural tube begins to take shape; and (4) fusion, in which the neural tube is completed. In a later event, cells migrate away from the neural crest to form several other structures, including the neurons of the peripheral nervous system.

*(1–4)* Courtesy Kathryn Tosney

4. In the next step of neurulation, called fusion (see Figure 39.19, step 4), the dorsal-most cells on either side of the neural tube are released from adjacent ectoderm and make contact with each other, thereby closing the neural tube. At the same time, ectoderm on either side of the neural tube moves toward the centerline, then up and over the neural tube, where it forms the dorsal epidermis of the embryo.

**Neural Crest Formation** Another important group of cells that arises during neurulation is the **neural crest,** which is unique to vertebrates. It consists of cells that originate from the ectoderm overlaying the dorsal surface of the newly formed neural tube and that migrate to other regions of the embryo (see Figure 39.19, step 4). Once these cells reach their final destination in the embryo, they differentiate into a variety of cell types different from those that arise from the rest of the ectoderm. All neurons and supporting cells of the peripheral nervous system in vertebrates are derived from neural crest cells. In addition, the neural crest gives rise to skeletal and cartilaginous structures in the head and face, melanocytes (specialized cells that provide pigmentation to the skin of vertebrates), the medulla of the adrenal glands, and connective tissue in numerous organs, notably the heart.

## Organogenesis Is the Process of Organ Formation

As described in Chapter 32, organs are specialized structures that consist of arrangements of two or more tissue types. Most organs, such as the kidney, contain all four tissue types: nervous, muscle, epithelial, and connective tissue (refer back to Figure 32.1). The developmental event in which cells and tissues form organs is called **organogenesis.** Each germ layer gives rise to particular types of cells found within different organs (see Figure 39.17).

Many organs begin to form during or just after neurulation. However, these organs become functional at different times during development. For example, the heart is the first functional organ to form in the vertebrate embryo. It begins to beat and pump blood by 2.5 days after fertilization in chicks, 9 days in mice, and about 22 days in humans. By contrast, the lungs of mammals do not acquire the ability to function until shortly before birth. As we saw in Chapter 32, the development of different organs in animals is controlled by genes in the embryo—notably the *Hox* genes. *Hox* genes are important for establishing structures along the anteroposterior axis. Many of the genes controlling the processes of gastrulation, neurulation, and organogenesis encode secreted proteins or growth factors that induce cells in their local vicinity to differentiate along a specific developmental pathway. For example, the notochord produces many signaling proteins that help establish tissue patterns in the embryo. Proteins produced within it induce segment-specific expression of the *Hox* genes in subsequent stages of development.

## 39.5 Reviewing the Concepts

- The process by which a fertilized egg is transformed into an organism with distinct physiological systems and body parts is called embryonic development. The process by which different cells within a developing organism acquire specialized forms and functions, due to the expression of cell-specific genes, is called cellular differentiation.
- Development in many animals, including vertebrates, involves five general events: fertilization, cleavage, gastrulation, neurulation, and organogenesis (Figure 39.13).
- Cleavage involves cell divisions without cell growth and results in daughter cells called blastomeres. Cleavage-stage embryos in triploblasts have animal and vegetal hemispheres. The hemispheres determine, in part, the future anteroposterior and dorsoventral axes of the embryo (Figure 39.14).
- Incomplete, or meroblastic, cleavage occurs in vertebrates whose eggs contain large amounts of yolk. Complete, or holoblastic, cleavage occurs in animals whose eggs have smaller amounts of yolk. In mammals, cleavage occurs in the oviduct and the embryo implants in the uterine wall (Figures 39.15, 39.16).
- During gastrulation, the hollow ball of cells that makes up a blastula or blastocyst develops into a gastrula, containing the three germ layers: endoderm, mesoderm, and ectoderm. Each germ layer gives rise to specific structures (Figures 39.17, 39.18).
- Neurulation is the multistep embryological process responsible for initiating formation of the nervous system. Neurulation involves the formation of the neural tube and neural crest (Figure 39.19).
- Organogenesis, the developmental event during which cells and tissues form organs, begins during or just after neurulation; however, organs become functional at different times during development.

## 39.5 Testing Your Knowledge

1. The three germ layers of triploblasts are established during
   - **a.** fertilization.
   - **b.** cleavage.
   - **c.** blastula formation.
   - **d.** gastrulation.
   - **e.** neurulation.
2. Choose the correct sequence of events.
   - **a.** cleavage→gastrulation→organogenesis→neurulation
   - **b.** neurulation→cleavage→organogenesis→gastrulation
   - **c.** cleavage→gastrulation→neurulation→organogenesis
   - **d.** organogenesis→neurulation→gastrulation→cleavage

## 39.6 Impact on Public Health

### Learning Outcomes

1. Explain the most common causes of human infertility.
2. Compare the different types of birth control and their mechanisms of pregnancy prevention.

Approximately 5–10% of individuals of reproductive age in the U.S. are infertile; that is, they cannot reproduce. In men and women alike, fertility can be decreased by a variety of factors. In this section, we discuss some of the common causes of **infertility**—the inability of a man to produce sufficient numbers or quality of sperm to impregnate a woman, or the inability of a woman to become pregnant or maintain a pregnancy. We then conclude by examining the methods in use today to prevent pregnancy.

## Infertility May Result from Disease, Developmental Disorders, Inadequate Nutrition, and Stress

As many as 75% of infertility cases have some identifiable cause, ranging from disease to toxin exposure.

**Disease** Primary among the diseases that affect fertility are sexually transmitted diseases (STDs). For example, some STDs may cause blockage in the ducts of the testes, thus preventing normal sperm transport, or can cause permanent damage to the oviducts, uterus, and surrounding tissues.

**Developmental Disorders** Developmental disorders are conditions that either are present at birth or arise during childhood and adolescence. In some developmental disorders that affect fertility, inherited mutations of genes that code for enzymes involved in the biosynthesis of reproductive hormones cause abnormal expression of those genes. The result is either too much or too little of one or more of these hormones, notably estradiol or testosterone. Other developmental disorders that compromise fertility include malformations of the cervix or oviducts.

**Inadequate Nutrition and Other Stressors** Adequate nutrition is required for normal growth and development of all parts of the body, including the reproductive system. Because the reproductive system is not essential for an individual's survival, it often becomes inactive when nutrients are chronically scarce—such as during starvation—and temporary infertility results. In this way, precious stores of energy in the body are preserved for vital functions, such as those of the brain and heart.

Malnutrition can also delay sexual development and puberty during late childhood and early adolescence. Undernourished children may begin puberty several years later than normal. The brains of all mammals, including humans, contain a center that monitors the body's fat stores. One of the triggers that initiate puberty may be a signal—such as the hormone leptin—from adipose tissue to the brain (see Chapter 38). Very low fat stores in undernourished girls, for example, result in decreased leptin in the blood; this signals the brain that the body does not contain sufficient energy to support the energetic demands of pregnancy, and thus puberty is delayed.

Starvation or poor nutrition is considered a type of stress, defined as any real or perceived threat to an animal's homeostasis. Physical and psychological forms of stress can and do affect fertility in humans. Many nonessential functions, including the maintenance of menstrual cycles in women, can be suppressed by chronic stress. The reproductive consequences of stress are much greater in females than in males, likely because only females bear the energetic cost of pregnancy. Interestingly, from a reproduction viewpoint, the human body responds to long-term strenuous physical exercise in the same way that it responds to long-term stress. This is why many young ballerinas and gymnasts experience delayed puberty and why female long-distance runners (like the one in the chapter introduction) may have menstrual cycles that are abnormal or absent (secondary amenorrhea).

**Other Causes** When the causes of infertility cannot be determined, a variety of factors come under suspicion. Among these possible causes are ingestion of toxins (for example, certain heavy metals such as cadmium), tobacco smoking, marijuana use, and injuries to the gonads.

**Treatments** Among several currently available treatments for increasing the likelihood of pregnancy in infertile couples are hormone therapy for the woman to increase egg production and a collection of procedures known as **assisted reproductive technologies (ART).** In the most common ART procedure, in vitro fertilization, sperm and eggs collected from a man and a woman are placed together in culture dishes. Once the sperm have fertilized the eggs and the resulting zygotes have undergone several cell divisions, one or more of the embryos are inserted into a woman's uterus with the goal that one will implant. When this procedure was first used in 1978, the children born as a result came to be known as "test-tube babies." Since then, several million children have been born using this technology.

## Contraception Usually Prevents Pregnancy

The use of methods to prevent fertilization or the implantation of a fertilized egg is termed **contraception.** Methods of contraception can be either permanent or temporary.

**Permanent Methods** The permanent forms of contraception surgically prevent the transport of gametes through the reproductive tract (Figure 39.20a). **Vasectomy** is a surgical procedure in men that severs the vas deferens, thereby preventing the release of sperm at ejaculation (however, semen is still released). In women, **tubal ligation** involves the cutting and sealing of the oviducts. This procedure prevents the movement of the egg from the oviduct into the uterus. These procedures are considered permanent, because it is difficult—sometimes impossible—to reverse the surgery.

**Temporary Methods** Temporary methods of preventing fertilization include barrier methods, which prevent sperm from reaching an egg (Figure 39.20b). Barrier methods include **vaginal diaphragms,** which are placed in the upper part of the vagina just prior to intercourse and block movement of sperm to the cervix, and **condoms,** which are sheathlike membranes worn over the penis that collect the ejaculate. In addition to their contraceptive function, condoms significantly reduce the risk of STDs such as HIV infection, syphilis, gonorrhea, chlamydia, and herpes. Other types of contraception do not reduce this risk.

Another temporary form of contraception involves synthetic hormones. Oral contraceptives (birth control pills) are synthetic forms of estradiol and progesterone, which are taken by mouth to prevent ovulation in women by inhibiting pituitary LH and FSH release (recall the feedback actions of estradiol and progesterone). The hormones in these pills also affect the composition of cervical mucus such that sperm cannot easily pass through it into the uterus. In addition to the oral route, hormones can be administered by injections, skin patches, and vaginal rings.

A last temporary method of contraception involves placement in the uterus of an **intrauterine device (IUD),** a small object that prevents fertilization in part by inhibiting sperm movement and survival in the uterus. Some IUDs may also induce local inflammation in the endometrium and make it less likely that a blastocyst could implant

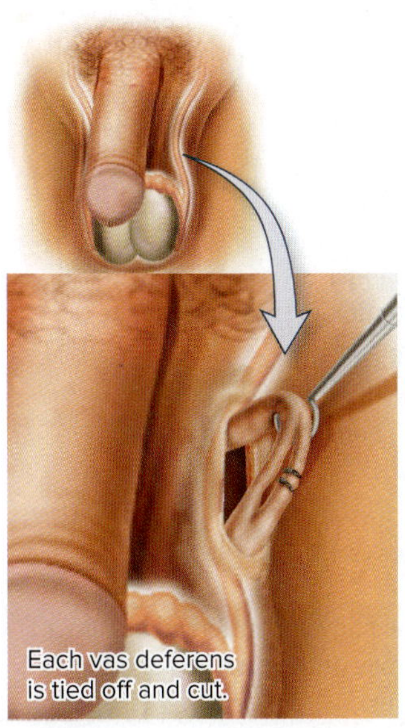

Each vas deferens
is tied off and cut.

Vasectomy (<1.0%)

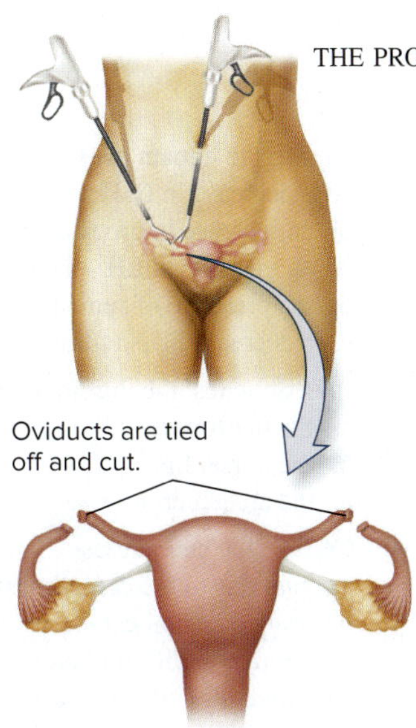

Oviducts are tied
off and cut.

Tubal ligation (<1.0%)

**(a) Permanent methods**

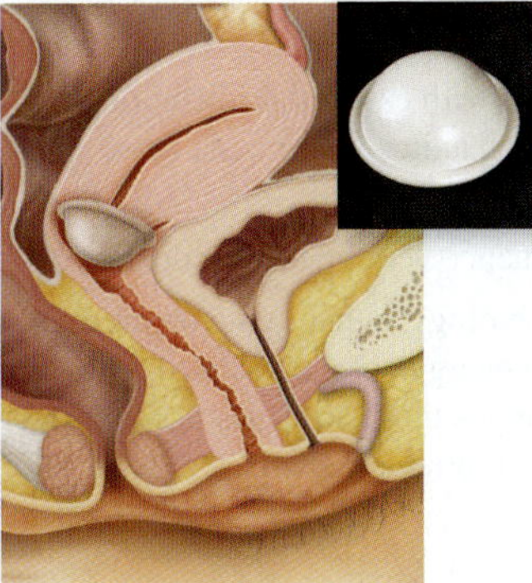

Diaphragm (5–20%)

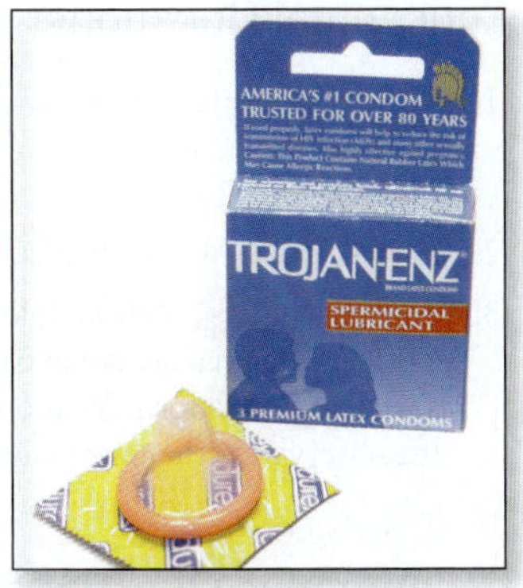


Condoms (male) (2–15%)

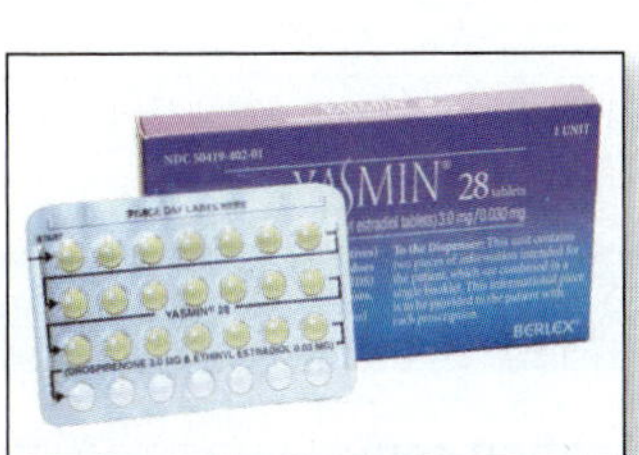


Oral contraceptive (1–2%)

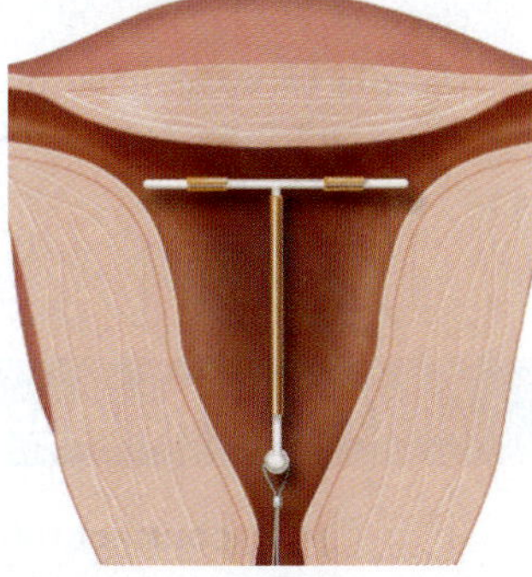

Intrauterine device (IUD) (1–2%)

**(b) Temporary methods**

**Figure 39.20** **Examples of contraceptive methods.** These methods may be used by men or women to **(a)** permanently or **(b)** temporarily prevent pregnancy. The estimated first-year failure rates for each method are given in parentheses (collected from data published by the U.S. Food and Drug Administration and other organizations). A failure rate of 10% means that 10 of every 100 women using that method of contraception will become pregnant in the first year of use. The large failure rate for use of condoms and diaphragms is due to improper use of these devices by many people. Female condoms are also available and have a failure rate of approximately 20%.

*(b diaphragm, condoms, oral contraceptive)* © McGraw-Hill Education/Jill Braaten, photographer

there. Although not as widely used as other contraceptives in the U.S., IUDs are the most commonly used means of contraception by women worldwide because of their effectiveness and simplicity of use.

In addition to the contraceptive methods used before or during intercourse, within 72 hours after intercourse, women can take one of several emergency contraception pills. These include a high dose of progestin (a synthetic form of progesterone), a high dose of combined estrogen and progestin, and a dose of ulipristal acetate, a substance that acts on the progesterone receptor. These work to prevent ovulation or to impede the ability of sperm to reach and fertilize an egg.

Used prior to the advent of modern contraception, and still in use today by individuals who prefer not to use contraceptives, is the rhythm method, which involves abstaining from sexual intercourse near the time of ovulation. Its main drawback is the difficulty in precisely pinpointing the time of ovulation; it can occur anytime within a roughly 2-week window of the 28-day cycle. Predicting the time of ovulation is also difficult in that several of the detectable changes characteristic of the midpoint of the ovarian cycle—including a small rise in body temperature and changes in the characteristics of the cervical mucus—occur only after ovulation. This explains why the rhythm method has a relatively high failure rate.

## 39.6 Reviewing the Concepts

- Infertility is the inability of a man to produce sufficient numbers or quality of sperm to impregnate a woman, or the inability of a woman to become pregnant or to maintain a pregnancy. A primary cause of infertility is STDs.
- The use of procedures to prevent fertilization or implantation of a fertilized egg is termed contraception. Methods of contraception include vasectomy and tubal ligation, vaginal diaphragms, condoms, oral contraceptives, and IUDs (Figure 39.20).

## 39.6 Testing Your Knowledge

1. Under what condition(s) might you expect temporary infertility in a woman?
   a. when she is under a great deal of emotional stress
   b. if she develops an eating disorder that greatly restricts her food intake
   c. if she embarks on a very strenuous exercise program, such as training for a marathon
   d. if she is exposed to potentially toxic compounds or chronically smokes marijuana
   e. all of the above
2. Which is *true* about the use of contraceptives?
   a. They all greatly reduce the likelihood of spreading or contracting an STD.
   b. They include barrier methods, which prevent sperm from reaching an egg.
   c. There are surgical methods available for men but not for women.
   d. The most common form of contraception worldwide is the vaginal diaphragm.
   e. They are all approximately 100% effective.

# Assess and Discuss

## Test Yourself

1. The development of offspring from unfertilized eggs is
   a. budding.
   b. cloning.
   c. fragmentation.
   d. parthenogenesis.
   e. implantation.
2. Which is considered an advantage of sexual reproduction?
   a. necessity to locate a mate
   b. increased energy expenditure in producing gametes that may not be used in reproduction
   c. increased genetic variation
   d. decreased genetic variation
   e. both a and b
3. Spermatogonia
   a. are haploid germ cells.
   b. are mature haploid cells.
   c. are mature male gametes.
   d. have flagella.
   e. are diploid germ cells.
4. Compared with external fertilization, in internal fertilization,
   a. male gametes have a higher chance of coming into close proximity to female gametes.
   b. gametes are less protected against predation or other harmful environmental factors.
   c. gametes typically become dessicated.
   d. gametes come into contact only outside the mother's reproductive tract.
   e. b and c only.
5. Which correctly describes the pathway followed by sperm in mammals?
   a. epididymis→vas deferens→seminiferous tubules→ejaculatory duct→urethra
   b. seminiferous tubules→vas deferens→epididymis→ejaculatory duct→urethra
   c. vas deferens→seminiferous tubules→epididymis→ejaculatory duct→urethra
   d. seminiferous tubules→epididymis→vas deferens→ejaculatory duct→urethra
   e. epididymis→seminiferous tubules→ejaculatory duct→vas deferens→urethra
6. The fructose in semen is secreted by
   a. the epididymis.
   b. the seminiferous tubules.
   c. the seminal vesicles.
   d. the prostate gland.
   e. the bulbourethral glands.
7. A major function of FSH is to
   a. stimulate the development of the gonads during early development.
   b. stimulate spermatogenesis in males and oocyte maturation in females.
   c. increase the secretion of testosterone by the testes.
   d. regulate the secretion of the bulbourethral glands.
   e. inhibit the activity of Sertoli cells in the testes.
8. During the human ovarian cycle, ovulation is stimulated by
   a. a decrease in FSH secretion.
   b. an increase in progesterone secretion.
   c. an increase in LH secretion.
   d. the presence of semen in the vagina.
   e. a decrease in estradiol concentration in the bloodstream.
9. In vertebrates, the digestive tract forms from
   a. the blastopore.
   b. the dorsal lip.
   c. the archenteron.
   d. the mesoderm.
   e. both a and d.
10. Cells of the neural crest
    a. give rise to the central nervous system.
    b. originate from ectoderm.
    c. migrate to different areas of the body and differentiate into a variety of cells, including neurons of the peripheral nervous system.
    d. all of the above.
    e. b and c only.

## Conceptual Questions

1. What disadvantages are associated with external fertilization, and what is one major way in which animals overcome those disadvantages?
2. How does the hypothalamus influence vertebrate reproduction?
3. **PRINCIPLES** A principle of biology is that living organisms use energy. What are some of the energy costs associated with sexual reproduction? What outweighs those costs and accounts for the observation that most animals reproduce sexually?

## Collaborative Questions

1. Describe the major events in animal development beginning with cleavage.
2. Describe the events of the ovarian and uterine cycles and their timing with respect to each other.

# Online Resource

# Immune Systems: How Animals Defend Against Pathogens and Other Dangers

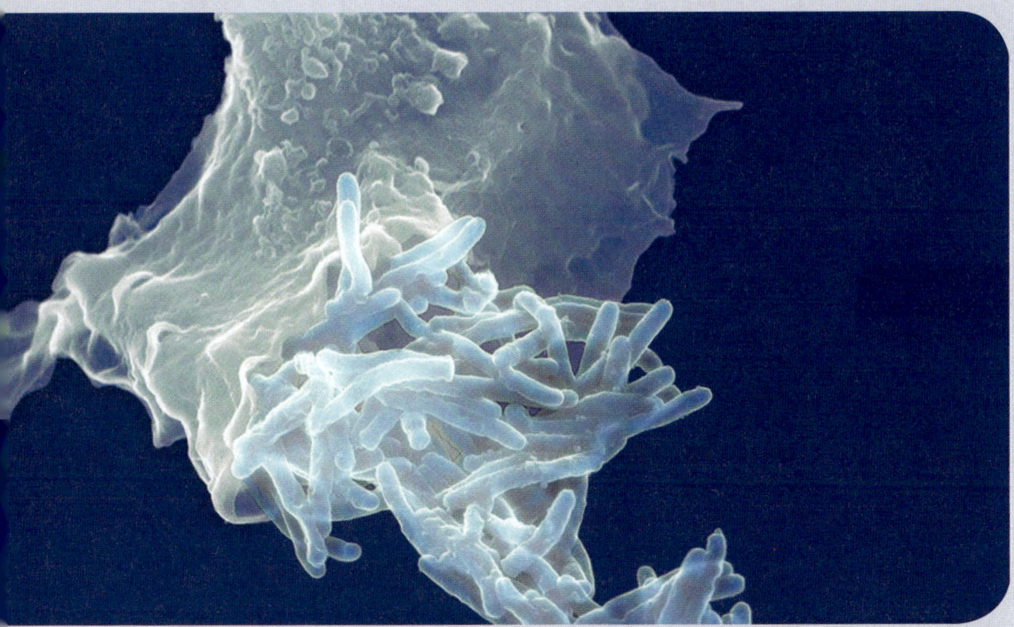

© SPL/Science Source

In an immune system response, a macrophage engulfs numerous rod-shaped bacteria (false-color SEM).

## Chapter Outline



A 50-year-old man in the northwestern U.S. was bitten while trying to remove a dead rodent from the mouth of a stray cat that his family had taken in and cared for. Within a few days, the man developed a fever, aches and pains, bleeding sores in his mouth and nose, and patches of dying, blackened skin. After he was admitted to a hospital, tests revealed the presence of the infectious bacterium *Yersinia pestis*, the organism that causes bubonic plague. This disease has caused upwards of 100 million deaths since the 14th century, and 1,000–3,000 cases still occur worldwide each year. The bacterium typically enters through the skin, normally an animal's first line of defense against infectious agents. In humans, it then enters the lymph nodes, where it multiplies rapidly and causes severe inflammation. Treatment of the disease is with a class of drugs called antibiotics, which kill bacteria. Such treatment is successful roughly 85–90% of the time, but prior to the advent of antibiotics, as many as 90% of individuals died when infected with *Y. pestis*.

All animals must defend themselves against various threats to their internal environment, including the invasion of potentially harmful microorganisms such as bacteria, the presence of foreign molecules such as the products of microorganisms, and the presence of abnormal cells such as cancer. The ability of an animal to ward off these threats—an animal's **immunity,** or immune defenses—is the subject of this chapter. The cells and organs within an animal's body that contribute to immune defenses collectively constitute an animal's **immune system.**

Immune defenses are often considered as two types: innate and acquired. **Innate immunity** refers to defenses that are present at birth and that act against foreign materials in much the same way regardless of the specific identity of the invading materials. Thus, innate immunity of animals is sometimes referred to as nonspecific immunity. An example of an innate immune response is seen in the chapter-opening photo, in which a cell known as a macrophage is engulfing numerous bacteria. All animals have innate immune defenses.

In contrast, **acquired immunity** develops only after the body is exposed to foreign substances. This type of immunity is characterized by the ability of certain cells of the immune system to recognize a particular foreign substance and initiate a response that targets that specific substance. For this reason, acquired immunity is also known as specific immunity. Another feature distinguishing acquired immunity is that repeated exposure to a foreign substance elicits greater and greater defense responses, unlike the situation for innate immunity, in which each exposure to the foreign material elicits the same defense responses. Acquired immunity has been identified in all vertebrates except for the jawless fishes and has not been unequivocally identified in invertebrates.

We begin with a brief overview of the types of infectious agents that cause disease in animals. We then consider the mechanisms that provide animals with innate and acquired defenses. We conclude with a discussion of the public health implications of some selected immunity-related conditions in humans.

## 40.1 Types of Pathogens

**Learning Outcome**

1. List the three main types of pathogens that elicit immune responses and how each damages its host organism.

All animals are susceptible to disease, often with catastrophic consequences. In recent years, outbreaks of diseases such as ebola in primates, hoof-and-mouth disease in domestic farm animals, white nose syndrome in bats, West Nile virus in birds, and white plague disease in coral have devastated large populations of these animals worldwide. To prevent or minimize disease, an animal's immune defenses must protect against a variety of foreign materials, but most important among them are disease-causing **pathogens,** which include microorganisms and viruses. Both terrestrial and aquatic animals encounter each of the three major types of pathogens: certain bacteria, viruses, and eukaryotic parasites.

### Bacteria Are Prokaryotes Responsible for Many Animal Diseases

Bacteria are single-celled prokaryotic organisms that lack a true nucleus (refer back to Figure 4.8). Bacteria can either damage tissues at the site of an open wound or release toxins that enter the blood and disrupt functions in other parts of the body. In humans, bacteria are responsible for many diseases and infections, including typhoid fever, strep throat, skin infections, middle ear infections, and food poisoning. The major ways in which bacteria gain entry into an animal's body are through direct bodily contact, open wounds, inhalation through the respiratory tract, and ingestion via fecal contamination of food or water. This last situation may arise because many infectious bacteria enter the intestines and are excreted in feces, which may be deposited near food or water sources used by animals.

### Viruses May Cause Illness or Cancer

All viruses contain nucleic acid (DNA or RNA) within a protein coat (see Chapter 17). Unlike bacteria, viruses are not living organisms and lack the metabolic machinery to synthesize the proteins they require to replicate themselves. Instead, they infect a host cell and use its biochemical and genetic machinery, including nucleotides and energy sources, to make more viruses. The viral nucleic acid directs the host cell to synthesize the proteins required for viral replication. After entering a cell, some viruses, such as the common cold virus, multiply rapidly, kill the cell, and then infect other cells. Other viruses can lie dormant within host cells before suddenly undergoing rapid replication, which causes cell damage or death. Finally, certain viruses can transform their host cells into cancerous cells. In humans, viruses are responsible for a great variety of illnesses, including some sexually transmitted diseases (refer back to Figure 17.1). Viruses typically enter an animal's body through the respiratory tract or through open wounds.

### Eukaryotic Parasites Enter an Animal's Body in Several Ways and Cause Widespread Illness

Eukaryotic parasites—whether protists, fungi, or worms—damage a host by using the host's nutrients for their own growth and reproduction or by secreting toxic chemicals. In humans, parasites account for an enormous number of cases of disease annually. Parasitic infections may enter a host through the bite of an infected insect, as in malaria; by ingestion of food or water containing parasitic organisms, such as roundworms; or in some cases by penetrating the skin, as with blood flukes.

### 40.1 Reviewing the Concepts

- Three major types of pathogens elicit an immune response: bacteria, viruses, and eukaryotic parasites. Pathogens enter the body through direct contact, open wounds, ingestion, insect bites, and the respiratory tract.

### 40.1 Testing Your Knowledge

1. Which pathogens are *not* living organisms and require a host cell's genetic machinery to reproduce themselves?
   a. bacteria
   b. roundworms
   c. eukaryotic parasites
   d. viruses
   e. the pathogens that cause such illnesses as middle ear infections and strep throat
2. Medicines that kill bacteria would *not* be effective to treat which of the following?
   a. the common cold
   b. food poisoning
   c. skin infections
   d. typhoid fever
   e. strep throat

## 40.2 Innate Immunity

### Learning Outcomes

1. Identify the general characteristics of innate defense mechanisms in animals.
2. Explain how an animal's body surface provides protection from pathogens.
3. Describe the process of phagocytosis, and explain its importance in innate immune responses.
4. List the cell types involved in innate immunity and the functions of each.
5. Outline the sequence of events in the inflammatory response.
6. Describe the role of Toll proteins and Toll-like receptors in innate immunity.

Innate immune defenses protect against foreign cells or substances without having to recognize the invaders' specific identities. This type of defense mechanism is called innate because animals inherit the ability to perform these protective functions and because this type of immunity does not require prior exposure to invaders. Instead of distinguishing among foreign materials, innate defenses recognize some general, conserved property marking the invader as foreign, such as a particular class of carbohydrate or lipid present in the cell walls of many different kinds of microbes.

In this section, we will first consider defenses at the body surfaces. We will then examine the actions of phagocytic cells, the response to injury known as inflammation, and various proteins secreted by cells of the innate immune system that facilitate the destruction of pathogens.

### The Body Surface is an Initial Line of Defense

Although not part of innate immunity, it is worth noting that an animal's initial defenses against pathogens are the cellular, anatomical, and chemical barriers provided by a surface exposed to the external environment. Very few microorganisms can penetrate the intact skin or body surface of most animals, particularly the tough, thick, or scaly skin characteristic of many vertebrates or the sturdy exoskeleton of many arthropods. In addition, glands in the body surfaces of many invertebrates and vertebrates secrete a variety of antimicrobial molecules, including mild acids and enzymes such as lysozyme that destroy bacterial cell walls.

The secretions from cells in the mucous membranes lining the respiratory and upper digestive tracts of many vertebrates also contain antimicrobial molecules. Equally important, mucus is sticky—pathogens that become stuck in it are prevented from penetrating the mucous membrane barrier and entering the internal environment of the body.

If, however, a pathogen is able to penetrate a barrier and gain entry into an animal's internal tissues and fluids, innate defense mechanisms are activated. These mechanisms are mediated by several types of cells that reside in the body fluids and tissues, as described next.

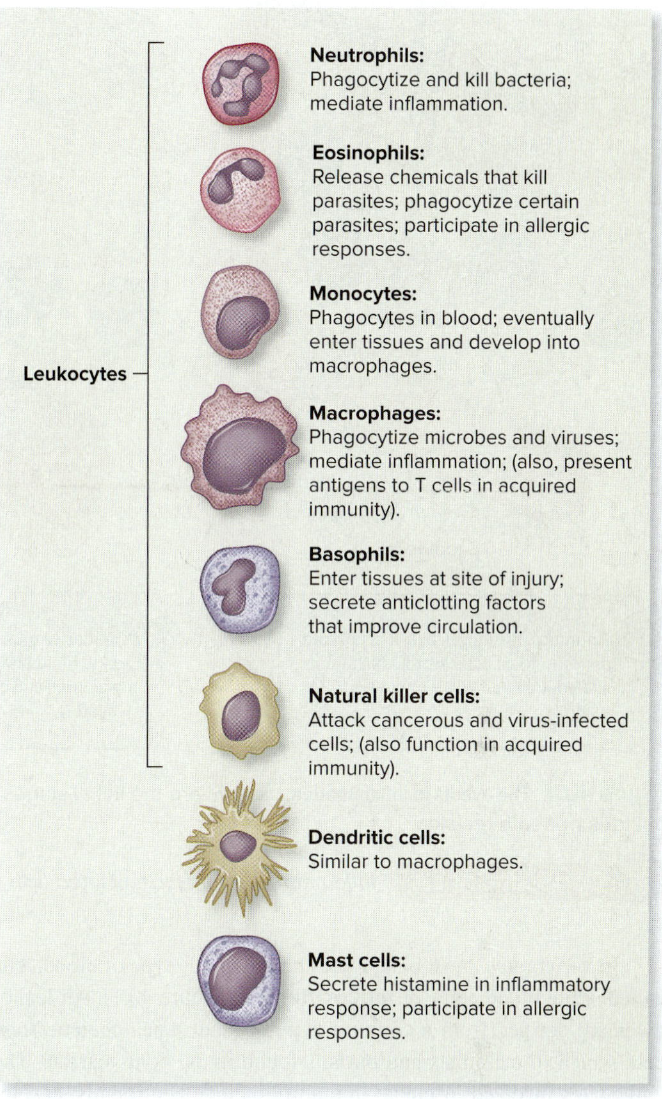


**Figure 40.1** **Cells involved in innate immunity in vertebrates.** Six of these types of cells are leukocytes, and several are phagocytes. Some of the cells also participate in acquired immunity.

**BioConnections:** *Refer back to Figure 36.4. How do the numbers of leukocytes in blood compare with those of erythrocytes?*

### In Innate Immunity, Phagocytic Cells Defend Against Pathogens That Enter the Body

Several different types of cells have key functions in innate immunity (Figure 40.1). Many of these cells are **phagocytes**—cells capable of phagocytosis. **Phagocytosis** is a type of endocytosis in which a cell engulfs a large particle, such as a bacterium, by forming an internal vesicle called a phagosome. Within the phagosome, enzymes and oxidizing compounds such as hydrogen peroxide begin the process of destroying the bacterium or other engulfed matter. Phagocytes are found in extracellular fluids and within various tissues and organs. They are present in all classes of animals and are among the most ancestral forms of immune defenses.

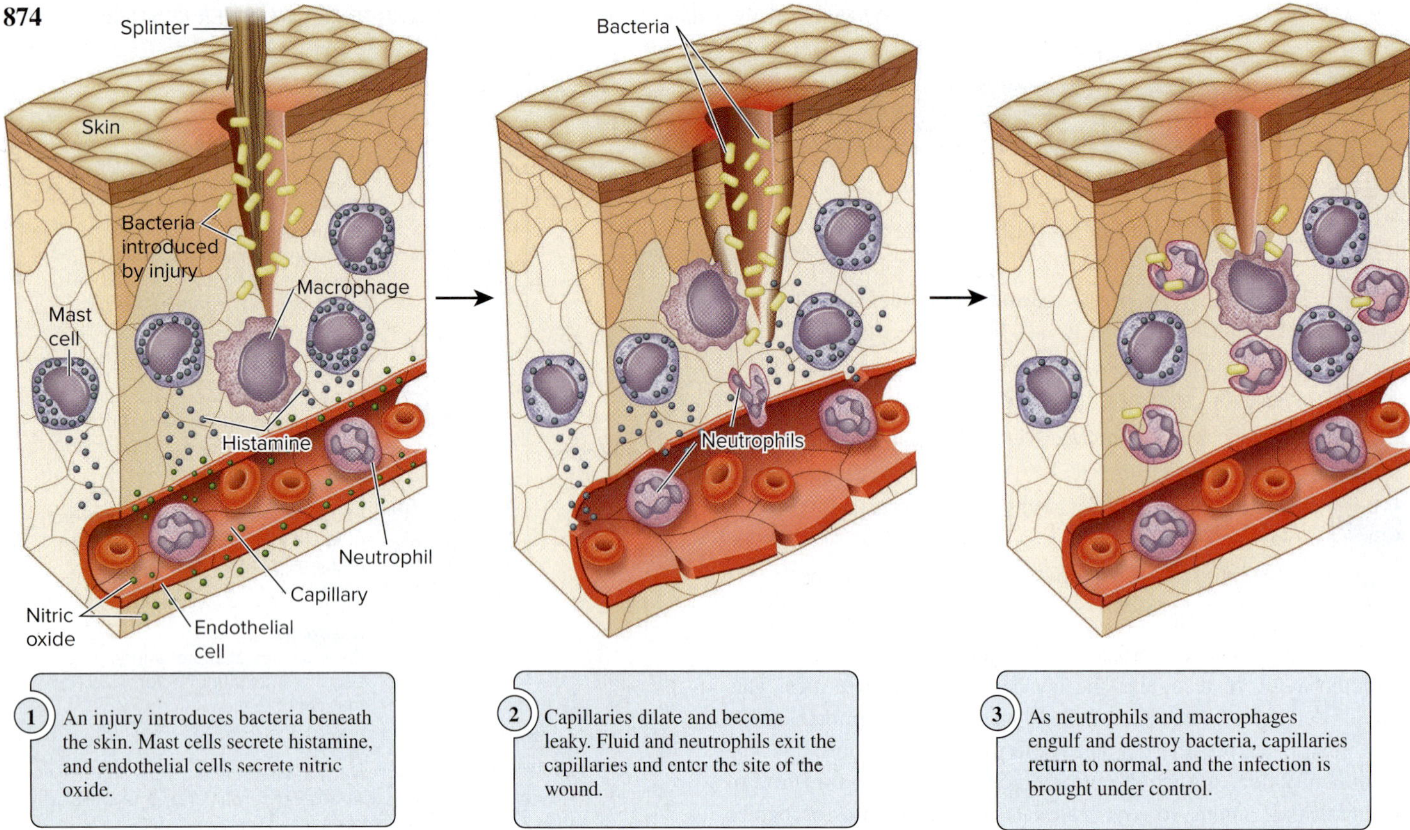


**Figure 40.2 The events in inflammation.** Shown are the initial stages of inflammation in response to a penetrating wound that introduces bacteria beneath the skin.

*Inflammation is often associated with swelling of the inflamed area. Could this swelling have an adaptive value?*

In vertebrates, most phagocytes belong to the type of blood cells called white blood cells, or **leukocytes** (see Figure 40.1). All leukocytes are derived from a common type of adult hematopoietic stem cell, which in mammals and birds is found in the bone marrow. The leukocytes involved in innate immunity include neutrophils, eosinophils, monocytes, macrophages, basophils, and natural killer (NK) cells. In addition to leukocytes, dendritic cells and mast cells are also derived from bone marrow stem cells and have important functions in innate immunity. As we will see later, some of these cells also participate in certain features of acquired immunity.

## Inflammation Is an Innate Response to Infection or Injury

**Inflammation** is an innate, local response to infection or injury. The functions of inflammation are to destroy or inactivate pathogens and to clear the infected region of dead cells and other debris. Some of the key cells involved in this process are neutrophils, macrophages, and mast cells.

The events of inflammation are induced and regulated by chemical mediators. These include a family of proteins called **cytokines** that function in both innate and acquired immune defenses. Cytokines provide a chemical communication network that synchronizes the components of the immune response. Most cytokines are secreted by more than one type of immune system cell and by certain nonimmune cells such as endothelial cells.

The sequence of events in a typical inflammatory response to a bacterial infection is summarized in Figure 40.2. A tissue injury such as that caused by a splinter begins the inflammation process, which results in the familiar signs and symptoms of local redness, swelling, heat, and pain. Substances secreted into the extracellular fluid from injured tissue cells, mast cells, and neutrophils contribute to the inflammatory response. For example, histamine from mast cells and nitric oxide (NO) from endothelial cells cause vasodilation of the small blood vessels in the infected and damaged area. This increases blood flow to the inflamed area, which accounts for the redness and heat, speeds the delivery of beneficial proteins and leukocytes, and increases local metabolism to facilitate healing. In addition, the vessels become leakier as they dilate, which helps the neutrophils and plasma proteins that participate in inflammation gain entry to the interstitial fluid. The swelling in an inflamed area also results from this increased leakiness of blood vessels.

Once neutrophils and macrophages arrive at the site of an injury, they begin the process of phagocytizing and destroying the bacteria. The neutrophils and macrophages also release substances into the extracellular fluid that can destroy bacteria even before phagocytosis occurs.

Inflammation or illness may be accompanied by **fever,** an increase in an animal's body temperature that results from an infection. In mammals, cytokines released into the circulation by activated macrophages act within the hypothalamus to raise the body's set point for temperature. The result is an increase in heat generation (for example, by shivering). Studies of animal models indicate that a small or moderate increase in temperature stimulates leukocyte activity and proliferation and provides a less hospitable environment for at least some types of pathogens.

### Defenses Against Pathogens Include Interferons and Complement Proteins

In addition to phagocytes and the inflammatory response, at least two other types of innate defenses against invading pathogens can also be involved. These defenses allow for extracellular destruction of pathogens without prior exposure to them. The first of these innate defenses is used against viruses. **Interferons** are proteins that generally inhibit viral replication inside host cells. In response to viral infection, many cell types produce interferons and secrete them into the extracellular fluid. When the interferons bind to plasma membrane receptors on the secreting cell and on other cells, each cell synthesizes a variety of proteins that interfere with the ability of the viruses to replicate. Interferons are not specific. Many kinds of viruses induce interferon synthesis, and the same interferons, in turn, can inhibit the multiplication of many different kinds of viruses.

The second type of innate defense is provided by a family of circulating proteins known as **complement.** Inactive complement proteins normally circulate in the blood at all times and are activated by contact with the surface of many different microbes. Activation of complement proteins results in a cascade of events. Among their many actions, complement proteins stimulate the release of histamine from mast cells, thereby increasing permeability of local blood vessels, as described earlier. Five of the active proteins generated in the complement cascade form a multiunit protein called the **membrane attack complex (MAC),** which, by embedding itself in a pathogen's plasma membrane or envelope, creates pore-like channels. Water and ions enter the pathogen through the channels, and the pathogen is destroyed.

## EVOLUTIONARY CONNECTIONS

### Innate Immune Responses Require Proteins That Recognize Features of Many Pathogens

As we have seen, innate immunity often depends on an immune cell's recognition of some general molecular feature common to many types of microbes or viruses. It is easy to imagine the selective advantage for animals to recognize such pathogen features, or **pathogen-associated molecular patterns (PAMPs),** as they are known. Until 1985, however, it was not known how that recognition was accomplished. At that time, German biologist Christiane Nüsslein-Volhard and American biologist Eric Wieschaus were interested in how animal embryos differentiate into adults. In the course of their studies, they discovered a protein they named Toll (now called Toll-1) that was required for the proper dorsoventral orientation of the fruit fly (*Drosophila melanogaster*). In a landmark study in 1996, however, it was discovered that Toll-1 also conferred on adult flies the ability to fight off fungal infections (see the Feature Investigation). It later became clear that a family of Toll proteins exists in animals from nematodes to mammals, including humans, and these proteins are found in the plasma membranes and endosomal membranes of macrophages and dendritic cells, among others.

In vertebrates, one function of Toll proteins is to recognize and bind to ligands containing PAMPs, such as lipopolysaccharide and other common bacterial lipids and carbohydrates, viral and bacterial nucleic acids, and a protein found in the flagellum of many bacteria. PAMPs are molecular features that are generally considered to be vital to the survival of a particular pathogen. When one of these ligands binds a Toll protein on the plasma membrane of an immune cell, second messengers are generated in the cell, triggering secretion of inflammatory mediators such as various cytokines. These in turn stimulate the activity of other leukocytes involved in the innate immune response. Some of these mediators also activate cells involved in the acquired immune response. Because many of the Toll proteins are plasma membrane-bound, bind extracellular ligands, and induce second-messenger formation, they are referred to as receptors, and the family of proteins is now known as **Toll-like receptors (TLRs).** Despite this, not all TLRs generate intracellular signals when bound to a ligand; some TLRs induce attachment of a microbe to a macrophage, for example, thereby facilitating its phagocytosis and destruction.

The importance of TLRs in mammals has been demonstrated in mice with a mutated, nonfunctional form of one member of the family called Toll-4. These mice are hypersensitive to the effects of injections of lipopolysaccharide (to mimic a bacterial infection) and are less able to ward off bacterial infection. In humans, recent studies suggest that certain naturally occurring variants in a specific TLR are associated with increased risk of certain diseases.

TLRs are currently an active area of investigation among biologists because of their role in immunity. Certain domains of these receptors have also been identified in plants, where they may also be involved in disease resistance. Therefore, TLRs may be among the first immune defense mechanisms to have evolved in living organisms. For their investigations of embryonic development and discovery of Toll proteins, Nüsslein-Volhard and Wieschaus were awarded the 1995 Nobel Prize in Physiology or Medicine.

## FEATURE INVESTIGATION

### Lemaitre and Colleagues Identify an Immune Function for Toll Protein in *Drosophila*

As you've just learned, Toll protein was initially identified as a critical molecule directing the proper dorsoventral development of *Drosophila* embryos. Interestingly, however, certain aspects of the sequence and function of Toll shared similarities with at least one important immune protein found in vertebrates. For example, both Toll and the receptor for the cytokine interleukin-1 were found to be transmembrane proteins with similar cytosolic sequences. Moreover, both proteins activated a similar intracellular signaling pathway once they had bound an extracellular ligand. This led to the hypothesis proposed by French researcher Bruno Lemaitre and coworkers, working in the laboratory of Jules Hoffman, that Toll protein may have a role in immunity in *Drosophila*.

To test this hypothesis, the researchers compared wild-type flies with mutant flies in which the *Toll* gene was underexpressed due to a mutation (**Figure 40.3**). They then infected the flies with a fungal pathogen by pricking them with a needle that had been dipped into a concentrated solution of fungal spores. Survival of the flies was monitored over the next several days and compared with the survival

**Figure 40.3** **Lemaitre and colleagues identified a function for Toll protein in immunity in *Drosophila*.**
*(top right)* © Daniel L. Geiger/SNAP/Alamy

**HYPOTHESIS** The Toll signaling pathway is required for the immune response in *Drosophila*.

**KEY MATERIALS** Wild-type *Drosophila*; mutant strain of *Drosophila* that under expresses the receptor for Toll protein; spores of the fungus *Aspergillus fumigatus*; cDNA probe to detect mRNA of antifungal polypeptide drosomycin.

1 Prick flies with a clean needle that was previously dipped in water (control) or a solution of *A. fumigatus* spores.

*A. fumigatus* is a common pathogen that *Drosophila* is normally able to defend against.

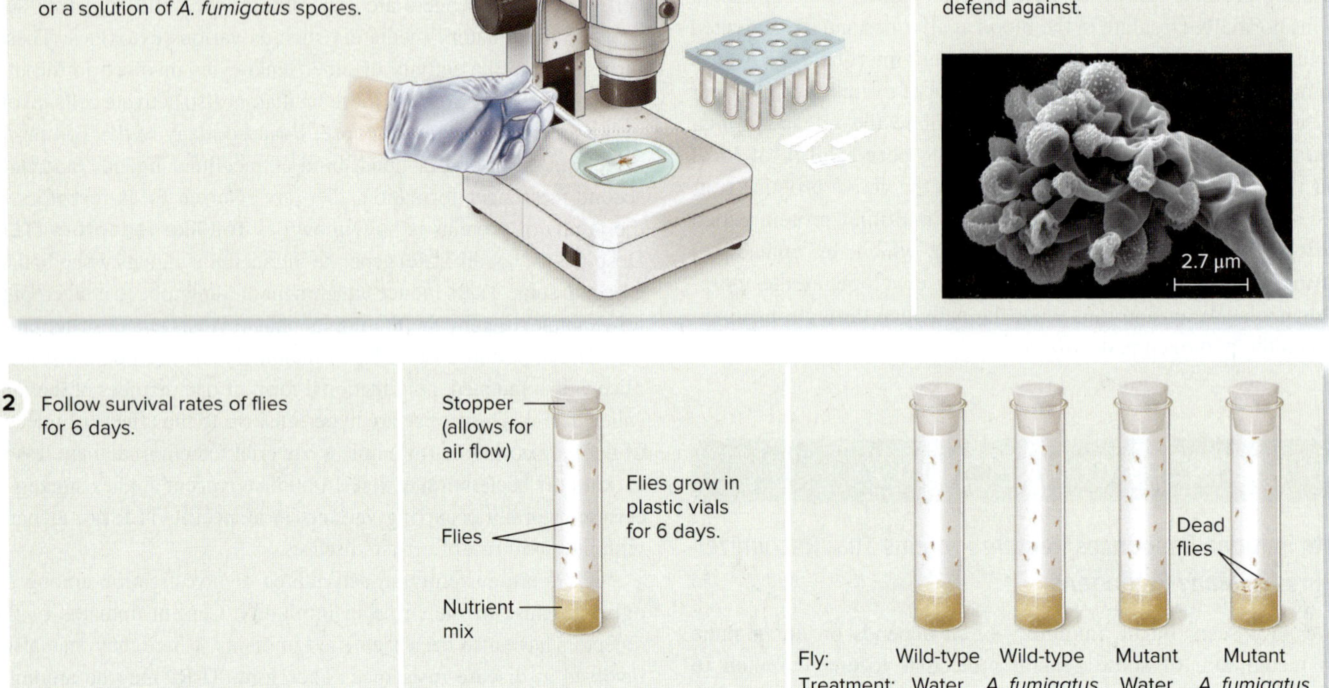


2 Follow survival rates of flies for 6 days.

3 Examine expression of antipathogen genes in wild-type and mutant flies, using Northern blot analysis:

1. Extract mRNA from flies.
2. Separate individual mRNAs by gel electrophoresis.
3. Transfer mRNAs from the gel to a nylon membrane.
4. Probe membrane with a radioactively labeled fragment of cDNA complementary to a region of the mRNA of interest.
5. Expose membrane to X-ray film.

**Experimental level**

Agarose gel

Nylon membrane

mRNA corresponding to drosomycin

Labeled cDNA probes

**Conceptual level**

All other mRNAs expressed at that time

mRNA corresponding to drosomycin, an antifungal peptide

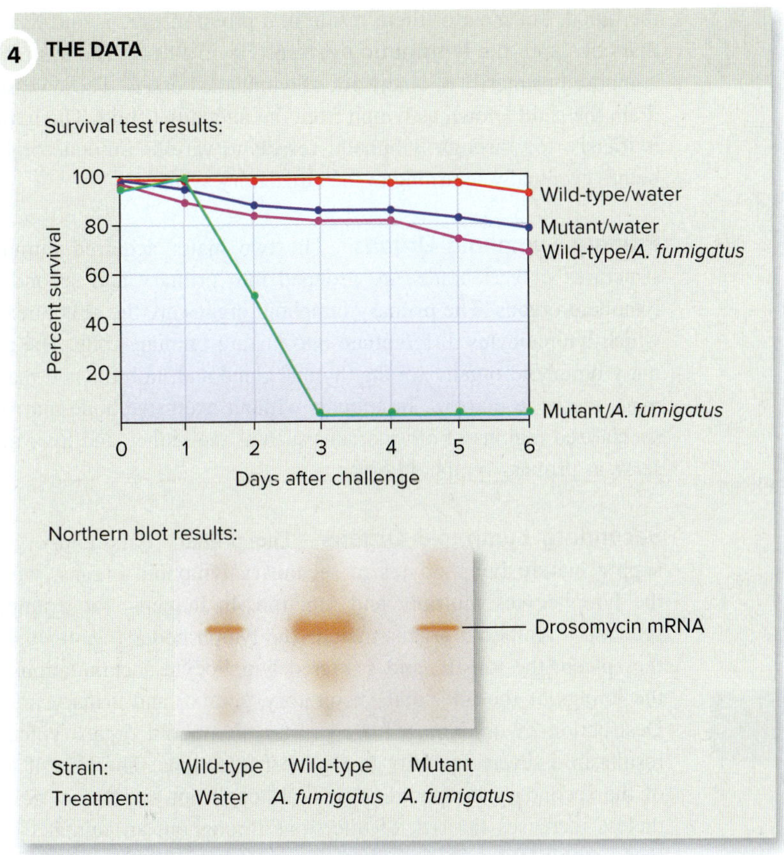


**5 CONCLUSION** Toll protein receptor is essential for survival of *Drosophila* exposed to a common fungal pathogen and for induction of antifungal genes.

**6 SOURCE** Lemaitre, B., Nicolas, E., Michaut, L., Reichhart, T., and Hoffmann, J. A. 1996. The Dorsoventral Regulatory Gene Cassette Spätzle/Toll/Cactus Controls the Potent Antifungal Response in *Drosophila* Adults. *Cell* 86: 973–983.

of wild-type flies with normal Toll protein. What they found was remarkable: Within days of being infected with the fungus *Aspergillus fumigatus*, 100% of the mutant flies died, but only about 30% of the wild-type flies died (see step 4 of Figure 40.3). The mutant flies were observed to be covered in germinating hyphae of the fungus, clearly demonstrating that the flies were unable to fight the infection. Moreover, analysis of mRNA from the treated and untreated flies revealed that the mutant flies were unable to increase expression of a key antifungal gene (known to encode a polypeptide called drosomycin) after infection with *A. fumigatus*, in contrast to the situation in wild-type animals.

Since then, these and other researchers have identified much of the mechanism by which Toll promotes immunity to certain microbes in *Drosophila* and, presumably, other insects. Activation of Toll protein induces expression of several antimicrobial genes. Unlike vertebrate TLRs, however, in *Drosophila* Toll does not directly bind PAMPs associated with microbial membranes. Rather, microbial infection induces activation of extracellular signaling molecules in *Drosophila* that, in turn, bind to Toll located on cell membranes. This binding induces intracellular signals that activate gene expression. Despite this difference, the identification of an immune function of an ancestral Toll protein paved the way for a clearer understanding of TLRs and innate immunity in all animals; in 2011, Jules Hoffman received a share of the Nobel Prize in Physiology or Medicine for this pioneering research.

*Experimental Questions*

1. What led the investigators to hypothesize that Toll protein may serve an immune function in *Drosophila*?
2. Is *Drosophila* Toll protein a receptor that binds PAMPs?
3. **SCISKILLS ▸** Based on the results of their survival studies of infected flies, was the researchers' hypothesis supported?

## 40.2 Reviewing the Concepts

- Phagocytes, many of which are leukocytes, play an important role in innate immunity (Figure 40.1).
- Inflammation is an innate, local response to infection or injury characterized by local redness, swelling, heat, and pain. The events of inflammation are induced and regulated by chemical mediators called cytokines (Figure 40.2).
- Antimicrobial proteins include interferons and complement proteins. Activation of the complement proteins results in the formation of a membrane attack complex (MAC).
- Toll-like receptors are evolutionarily ancient proteins that recognize PAMPs (Figure 40.3).

## 40.2 Testing Your Knowledge

1. An example of a PAMP is
   a. lipopolysaccharide. c. Toll protein. e. complement protein.
   b. Toll-like receptor. d. interleukin-1.
2. Place the events of inflammation following a skin puncture in the proper sequence.
   a. phagocytes destroy the pathogens→capillaries dilate→capillaries become leaky→neutrophils leave capillaries→histamine and nitric oxide are released at the wound site
   b. phagocytes destroy the pathogens→capillaries dilate→capillaries become leaky→histamine and nitric oxide are released at the wound site→neutrophils leave capillaries
   c. capillaries dilate→capillaries become leaky→phagocytes destroy the pathogens→histamine and nitric oxide are released at the wound site→neutrophils leave capillaries
   d. histamine and nitric oxide are released at the wound site→capillaries dilate→capillaries become leaky→neutrophils leave capillaries→phagocytes destroy the pathogens

# 40.3 Acquired Immunity

### Learning Outcomes

1. Describe the major classes of lymphocytes.
2. Outline the three stages of the acquired immune response.
3. Compare and contrast humoral immunity with cell-mediated immunity.
4. **SCISKILLS ▶** Explain how the structures of different immunoglobulins allow predictions to be made about their functions.
5. Explain how antibody structure and genetic rearrangement lead to the diverse set of antibodies found in mammals.
6. List the steps of a typical humoral immune response.
7. **SCISKILLS ▶** Predict the benefit of secondary immune responses, and explain the function of memory cells in these responses.

In acquired immunity, cells of the immune system first encounter and later recognize a specific foreign cell or protein to be destroyed, as opposed to recognizing some general conserved feature of pathogens. Any molecule that can trigger a specific immune response is called an **antigen.** An antigen, therefore, is any molecule that the host does not recognize as self. Most antigens are either proteins or very large polysaccharides. Antigens include the protein coats of viruses, bacterial surface proteins, specific macromolecules on pollens and other allergens, membrane proteins of cancerous cells or transplanted cells, toxins, and vaccines. We will consider acquired immune responses in the context in which they are best understood—the jawed vertebrates.

## Acquired Immune Defense Mechanisms Include Lymphoid Organs, Tissues, and Cells

The cells responsible for acquired immunity are a type of leukocyte called **lymphocytes.** Like all leukocytes, lymphocytes circulate in the blood, but most of them reside in a group of organs and tissues that constitute the **lymphatic system** (Figure 40.4a). The system is composed primarily of a network of lymphatic vessels. These vessels drain the fluid known as lymph from the interstitial fluid. The lymph is transported through lymphatic vessels to various immune organs before eventually returning to the circulatory system.

**Primary Lymphoid Organs** The two major acquired immune structures in vertebrates are grouped into primary and secondary lymphoid organs. The primary lymphoid organs are the structures in which lymphocytes differentiate into mature immune cells. The primary lymphoid organs are the thymus gland and, in birds and mammals, the bone marrow. In animals without extensive bone marrow, specialized regions of other organs such as the kidney and liver may serve as primary lymphoid organs.

**Secondary Lymphoid Organs** The primary lymphoid organs supply mature lymphocytes to secondary lymphoid organs, where the lymphocytes multiply and function. In humans, for example, secondary lymphoid organs include the lymph nodes (Figure 40.4b), the spleen, the tonsils, and scattered lymphocyte accumulations in the linings of the intestinal, respiratory, genital, and urinary tracts. Destruction of or damage to a primary lymphoid organ typically results in a severe inability to fight off infections. The loss of any of the secondary lymphoid organs, although not as serious, nevertheless increases the risk of infections throughout an animal's life. For example, the spleen must occasionally be surgically removed in humans due to injury or disease. Such individuals must be carefully monitored for the rest of their lives because of their increased vulnerability to infection.

**Recirculation of Lymphocytes** After leaving the bone marrow or thymus gland, lymphocytes circulate between the secondary lymphoid organs, blood, lymph, and all the tissues of the body. Lymphocytes from all the secondary lymphoid organs continually leave those structures and are carried to the blood. Simultaneously, some circulating lymphocytes leave venules all over the body to enter the interstitial fluid. From there, they reenter lymphatic vessels and are carried back to secondary lymphoid organs. This constant recirculation of lymphocytes increases the likelihood that any given lymphocyte will encounter an antigen it is specifically programmed to recognize.

## Lymphocytes Include B Cells and T Cells

Different kinds of lymphocytes participate in coordinated acquired immune responses (Figure 40.5). In addition to NK cells, described previously, the two major types of lymphocytes are B cells and T cells. **B cells** were first observed to mature in an avian organ called the bursa of Fabricius—and therefore were named B cells. In mammals, B cells mature within bone marrow. If stimulated by antigen, some B cells differentiate further into **plasma cells,** which synthesize and secrete **antibodies,** proteins that bind to and help destroy foreign molecules or pathogens, as described later. This type of response is called **humoral immunity.**

**T cells** get their name because they mature within the thymus gland. T cells may directly kill infected, mutated, or

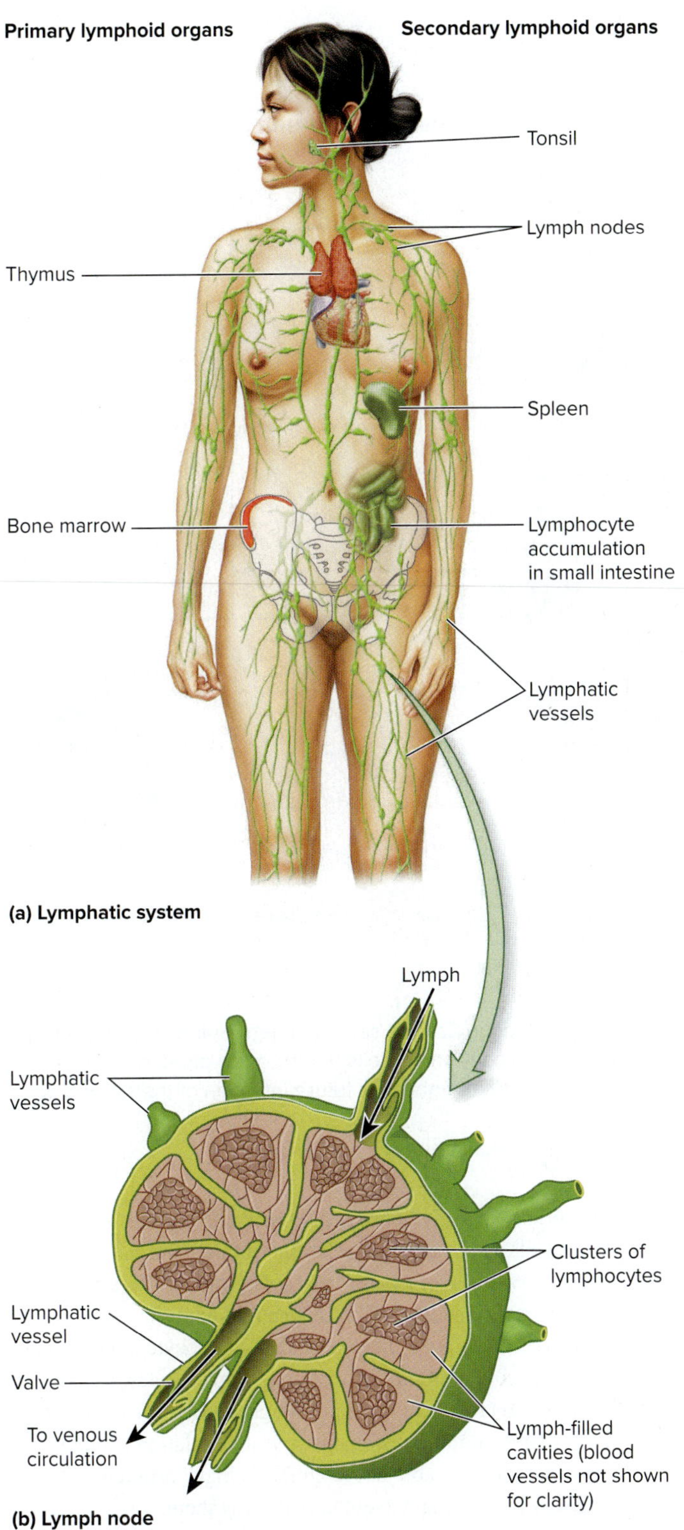


**Figure 40.4** **The lymphatic system in humans.** **(a)** The major components of the human lymphatic system. Primary lymphoid organs are shown in red, and secondary lymphoid organs are shown in green. In adult humans, the primary lymphoid organs in bone are found in the sternum, ribs, parts of the skull, small regions of the femur and humerus, and, as shown here, the pelvis. **(b)** The structure of a lymph node. Lymph nodes occur along the course of lymphatic vessels, which drain interstitial fluid from tissues and return it to the venous circulation. Within a lymph node, lymph percolates through open cavities containing clusters of lymphocytes.

*From where does lymph arise?*

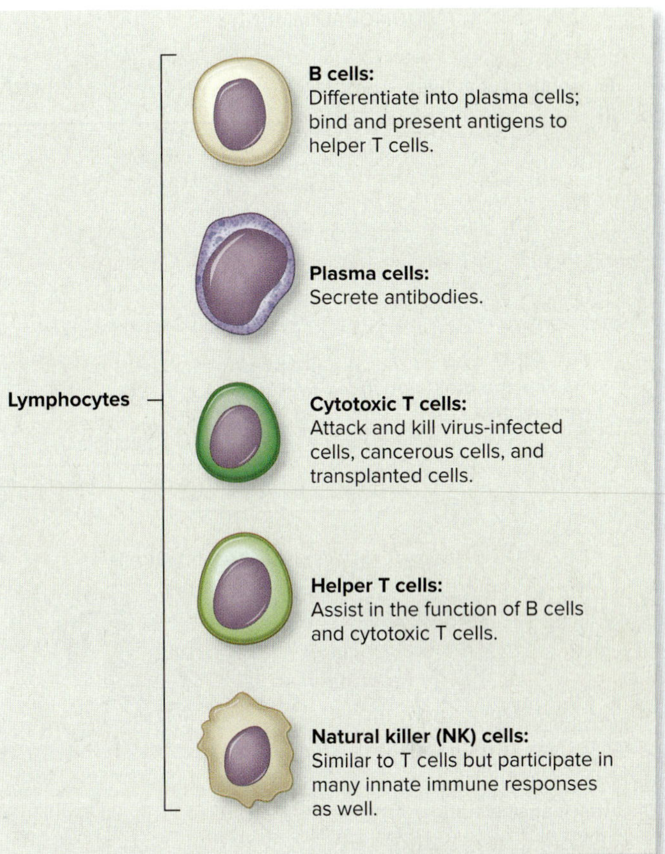

Lymphocytes
B cells:
Differentiate into plasma cells; bind and present antigens to helper T cells.
Plasma cells:
Secrete antibodies.
Cytotoxic T cells:
Attack and kill virus-infected cells, cancerous cells, and transplanted cells.
Helper T cells:
Assist in the function of B cells and cytotoxic T cells.
Natural killer (NK) cells:
Similar to T cells but participate in many innate immune responses as well.


**Figure 40.5** **Cells involved in acquired immunity in vertebrates.** Some important functions of each type of lymphocyte are included. The shape and color conventions shown in this figure will be used throughout this chapter.

transplanted cells. T cells include a few distinct types of lymphocytes, including two called cytotoxic T cells and helper T cells. **Cytotoxic T cells** travel to the location of their targets, bind to these targets by recognizing an antigen, and destroy those targets via secreted chemicals. In addition, responses mediated by cytotoxic T cells are directed against body cells that have become cancerous or infected by pathogens. (NK cells also destroy such cells by secreting toxic chemicals.) This type of immune response is called **cell-mediated immunity.**

As their name implies, **helper T cells** do not themselves function as effector cells. Instead, they assist in the activation and function of B cells and cytotoxic T cells. With only a few exceptions, B cells and cytotoxic T cells cannot function adequately unless first stimulated by cytokines secreted from helper T cells.

## Humoral and Cell-Mediated Immune Responses Occur in Three Stages

In both humoral immunity and cell-mediated immunity, an acquired immune response can usually be divided into three stages (Figure 40.6): recognition of antigen, activation and proliferation of lymphocytes, and attack against a recognized antigen.

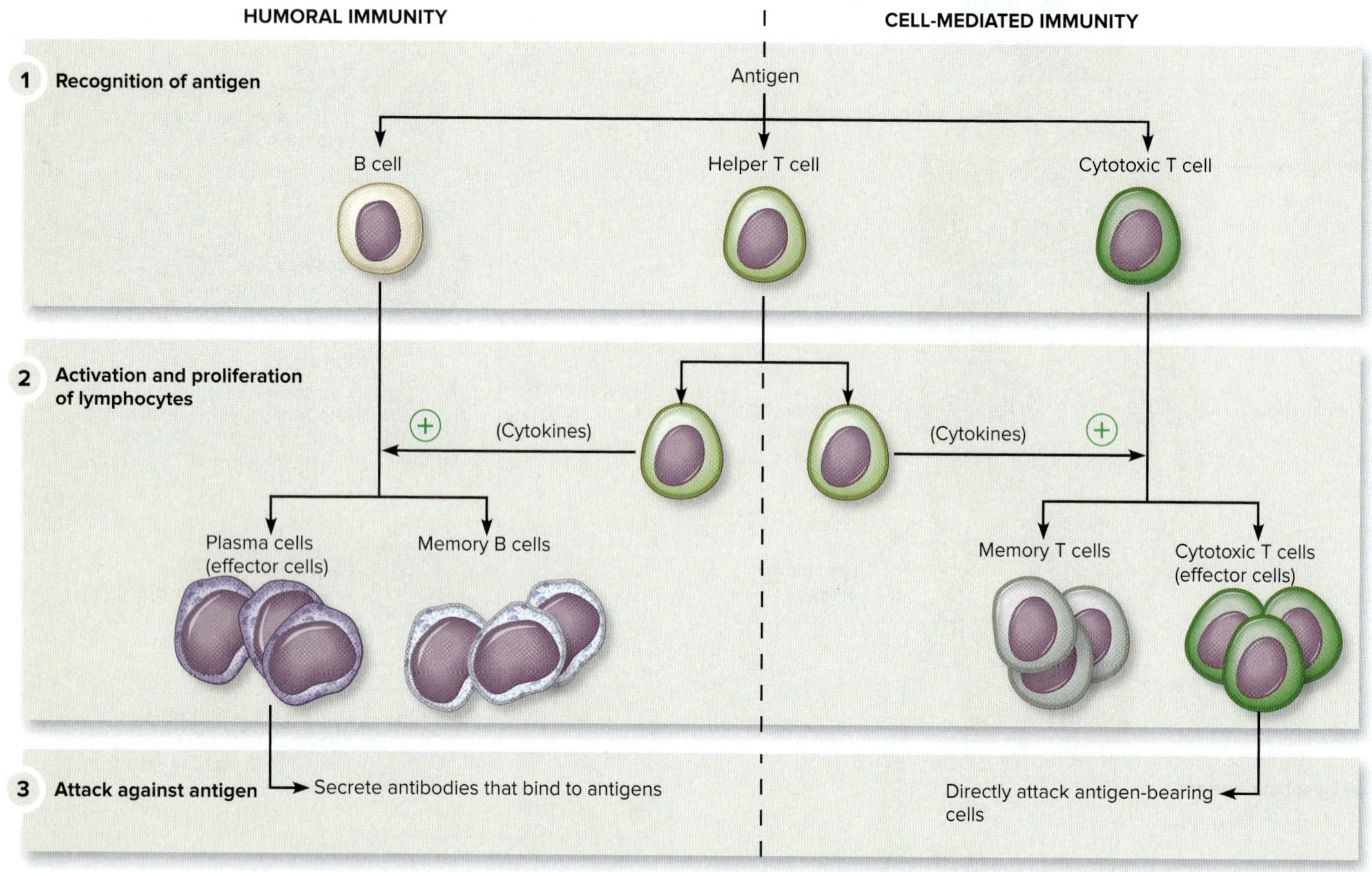

**Figure 40.6** **The three stages of an acquired immune response.** All three cell types in stage 1 recognize the same antigen. Helper T cells secrete cytokines that activate B cells and cytotoxic T cells, as indicated by the ⊕ symbols. Both B cells and T cells undergo cell division to form clones when activated, and in both cases, a portion of the cells are set aside as memory cells to fight off a future infection of the same type.

*Which cell type participates in both humoral and cell-mediated immunity?*

***Stage 1: Recognition of Antigen*** During its development, each lymphocyte synthesizes a type of membrane receptor that can bind to a specific antigen. If subsequently the lymphocyte encounters that antigen, the antigen becomes bound to the receptor. This specific binding is the meaning of the word recognize in immunology. Antigens that bind to a lymphocyte receptor are said to be recognized by the lymphocyte as being nonself molecules. The ability of lymphocytes to distinguish one antigen from another, therefore, is determined by the nature of their plasma membrane receptors. Each lymphocyte is specific for just one type of antigen.

***Stage 2: Activation and Proliferation of Lymphocytes*** In the second stage, the binding of an antigen to a receptor on a lymphocyte activates that lymphocyte. Upon activation, the lymphocyte undergoes multiple cycles of cell division. The result is the formation of many identical cells called clones that express the same receptor as the receptor that first recognized the antigen. The molecular nature of the antigen determines which individual lymphocytes will be activated to form clones. This process requires the function of helper T cells, which divide when activated and then secrete the cytokines that promote cell division. Some of the cloned lymphocytes become effector cells—the plasma cells and cytotoxic T cells that carry out the attack response—whereas others function as **memory cells,** which remain poised to recognize the antigen if it returns in the future.

In a typical human, the size of the lymphocyte population is staggering. Over 100 million different lymphocytes, each with the ability to recognize a unique antigen, are estimated to exist in a person's immune system. This vast population explains why our bodies are able to recognize so many different antigens as foreign and eventually destroy them.

***Stage 3: Attack Against Antigen*** In the third stage, the effector cells attack all antigens of the kind that initiated the immune response. In humoral immunity, plasma cells secrete antibodies, which recruit and guide other molecules and cells that perform the actual attack. In cell-mediated immunity, cytotoxic T cells directly attack and kill the cells bearing the antigens.

Once the attack is successfully completed, the great majority of plasma cells and cytotoxic T cells that participated in it die by apoptosis. However, memory cells persist even after the immune response has been successfully completed, so they can recognize and fight off any future infection bearing the same type of

## Biology Principle

### Structure Determines Function

The amino acid sequences of the variable regions of different immunoglobulins impart highly specific three-dimensional structures of immunoglobulins. This, in turn, is what allows an immunoglobulin to specifically bind a particular antigen and not others.

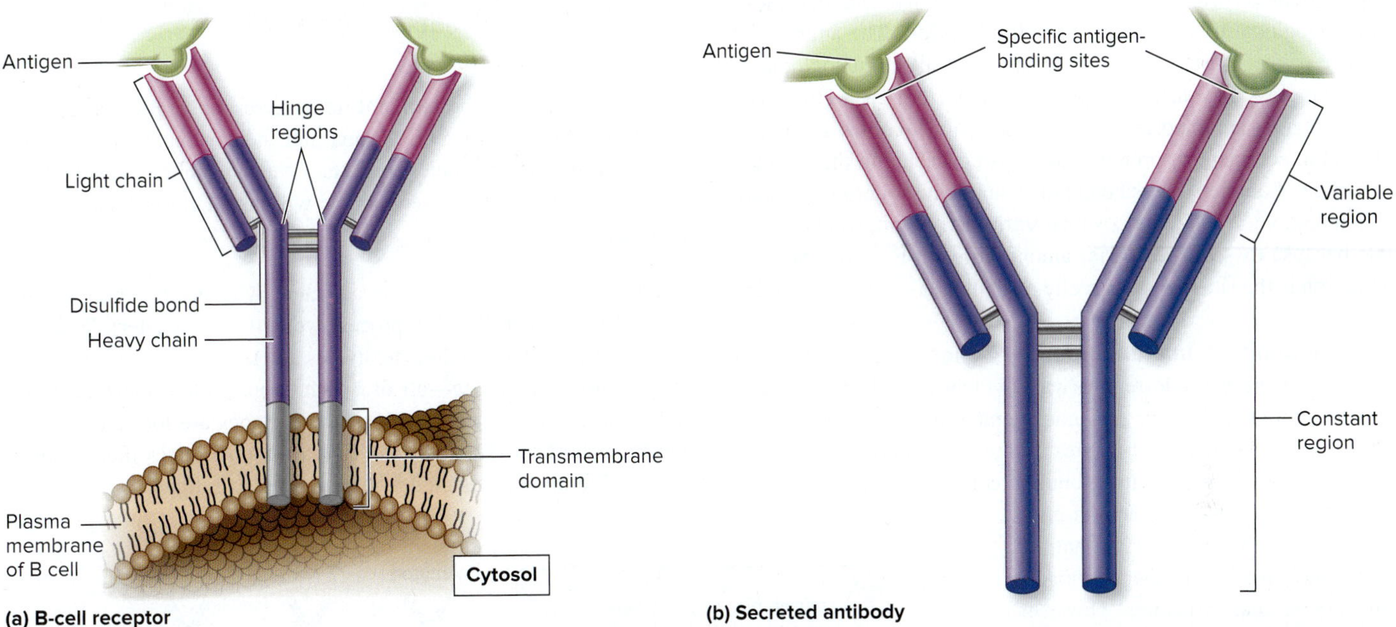


**Figure 40.7** **Immunoglobulins.** **(a)** B-cell receptor and **(b)** secreted antibody. Immunoglobulins are composed of polypeptides known as heavy chains and light chains, held together by disulfide bonds. B-cell receptors are anchored to the plasma membrane by a transmembrane domain, whereas antibodies are secreted into the extracellular fluid. Within each immunoglobulin class, the constant regions have identical amino acid sequences. In contrast, the antigen-binding sites formed by the light- and heavy-chain variable regions have unique amino acid sequences and give each receptor or antibody its specificity for a particular antigen.

antigen. Let's look at each of these three stages in more detail, first for humoral immunity and then for cell-mediated immunity.

## In Humoral Immunity, B Cells Produce Immunoglobulins That Serve as Receptors or Antibodies

In stage 1 of humoral immunity (see Figure 40.6, left side), B cells recognize antigens with the help of B-cell receptors. When B cells are activated by cytokines in stage 2, they proliferate and differentiate into plasma cells, which secrete antibodies. These proteins travel all over the body to reach and bind antigens identical to those that stimulated their production. In some cases, the binding of an antibody to an antigen simply prevents that antigen from infecting a host cell in a process called neutralization. For example, many viruses are prevented from infecting host cells in this way. In other cases, antibodies bound to an antigen guide an attack that eliminates the antigens or the cells bearing them, a process we will discuss in more detail later. Antibody-mediated responses are the major defense against bacteria, viruses, and other pathogens in the extracellular fluid, and against toxin molecules.

**B-Cell Receptors and Antibodies** B-cell receptors and the antibodies secreted by plasma cells share many structural and functional similarities (Figure 40.7). They are both members of a family of proteins called **immunoglobulins (Ig).** However, they show some differences. B-cell receptors have a transmembrane domain that anchors them in the plasma membrane of the B cell (see Figure 40.7a). Antibodies are soluble proteins that are secreted from plasma cells (see Figure 40.7b). Interestingly, B-cell receptors and the antibodies made by plasma cells are encoded by the same genes. In plasma cells, the pre-mRNA is alternatively spliced, a phenomenon described in Chapter 12 (refer back to Figure 12.17), so the transmembrane domain is not present in the protein. For this reason, the B-cell receptor in a particular B cell and secreted antibodies from the resulting plasma cells recognize the same antigen.

**Immunoglobulin Classes and Structure** Mammals have five classes of immunoglobulins, designated IgM, IgG, IgA, IgE, and IgD, which are involved in different types of functions. For example, the most abundant immunoglobulins in mammals are IgM and IgG. Together these two immunoglobulin classes provide the bulk of specific immunity against bacteria and viruses in the extracellular fluid. IgA antibodies are secreted by plasma cells in the linings of structures exposed to the environment. For example, IgA molecules secreted into saliva help keep animals' mouths relatively free of pathogens. IgE antibodies participate in defenses against multicellular eukaryotic parasites. They also attach to mast cell membranes, where they have a role in allergic responses. Finally, the functions of IgD are unclear, but they may contribute to B-cell activation.

Each immunoglobulin molecule comprises four interlinked polypeptides: two heavy chains and two light chains (see Figure 40.7b). Hinge regions provide the molecule with flexibility. One portion of an immunoglobulin is called the **constant region:** The amino acid sequence of the constant region is identical for all immunoglobulins of a given class and distinguishes the classes from each other. The constant regions are important for the binding of antibodies to immune cells and to complement proteins, and thus contribute to the eventual destruction of antigen that is bound to an antibody. A defining feature of immunoglobulins, however, is their **variable region,** which gets its name because its sequence varies among different B cells. The variable region is the site that specifically recognizes a particular antigen.

**Recombination of Immunoglobulin Genes** The amino acid sequences of the variable regions vary widely from immunoglobulin to immunoglobulin. The enormous number of variable sequences results in countless unique structures of immunoglobulins. Thus, each of the five classes of immunoglobulins contains up to millions of unique molecules, each capable of combining with only one specific antigen or, in some cases, with several antigens with very similar structures. However, the human genome contains only about 200 genes that encode immunoglobulins. This raises an intriguing question. How can millions of different immunoglobulin proteins be produced if there are only 200 immunoglobulin genes? The answer is that the 200 genes undergo a unique process involving gene rearrangements.

Along the length of a typical human immunoglobulin gene are many gene segments that code for a piece of the final immunoglobulin protein. Each segment along the length of the gene is associated with recognition sequences that bind enzymes that are expressed only in developing lymphocytes. These enzymes randomly cut out segments, which are then recombined in different ways, resulting in a new, permanent immunoglobulin gene for a given B cell.

The process of gene recombination results in each lymphocyte within an individual's body producing a unique type of immunoglobulin. Because the body makes hundreds of millions of different lymphocytes, the immune system can produce antibodies capable of recognizing an incredibly diverse array of antigens. Nearly any foreign antigen that is taken into the body is recognized by some lymphocytes in this large population.

## Activated B Cells Produce Plasma Cells That Secrete Antibodies

In stages 2 and 3 of the humoral immune response, B cells are activated and differentiate into plasma and memory cells. The plasma cells then secrete antibodies that attack the antigen that was detected (see Figure 40.6). In this section, we will take a closer look at these processes.

**Clonal Selection** B cells are activated by a specific antigen, with the aid of a helper T cell (a process we will discuss later). When an antigen-stimulated lymphocyte divides and replicates itself, the progeny of this lymphocyte—all of which express the same receptor—are clones. The process by which these clones are formed is called **clonal selection** (**Figure 40.8**). This term emphasizes that lymphocyte proliferation is selected by exposure to an antigen.

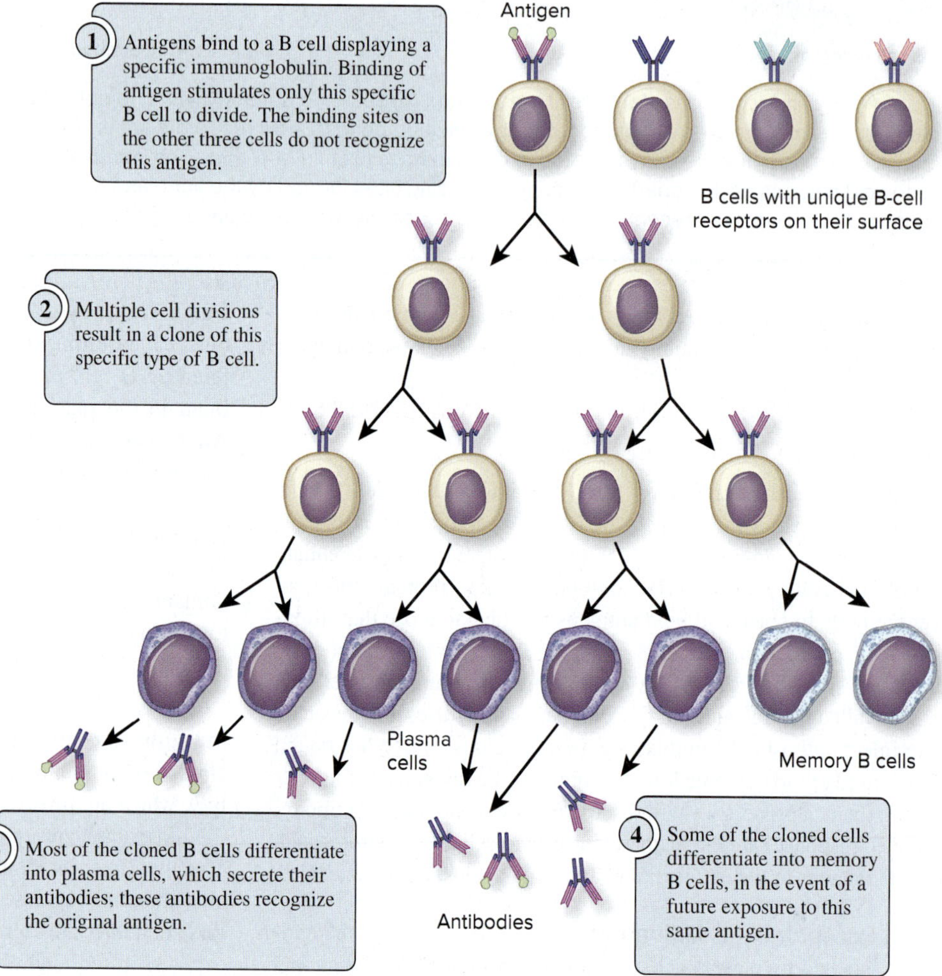

Antigen
1 Antigens bind to a B cell displaying a specific immunoglobulin. Binding of antigen stimulates only this specific B cell to divide. The binding sites on the other three cells do not recognize this antigen.
B cells with unique B-cell receptors on their surface
2 Multiple cell divisions result in a clone of this specific type of B cell.
Plasma cells
Memory B cells
3 Most of the cloned B cells differentiate into plasma cells, which secrete their antibodies; these antibodies recognize the original antigen.
Antibodies
4 Some of the cloned cells differentiate into memory B cells, in the event of a future exposure to this same antigen.


**Figure 40.8** **Clonal selection.** In this example, a B cell with a specific immunoglobulin on its surface recognizes an antigen and is stimulated to divide into a clone of identical cells.

**Antibodies Facilitate Destruction of Pathogens via Opsonization** The antibodies secreted from the plasma cells circulate through the lymphatic system and the blood. Eventually, the antibodies combine with the antigen that initiated the immune response. These antibodies then direct the attack against the pathogen to which they are now bound. Thus, immunoglobulins play two distinct roles in humoral immune responses. First, during antigen recognition, immunoglobulins (B-cell receptors) on the surface of B cells bind to antigen brought to them. Second, antibodies secreted by the resulting plasma cells bind to pathogens bearing the same antigens, marking them as the targets to be destroyed.

Instead of directly destroying the pathogens, antibodies bound to antigen on the pathogen surface inactivate the pathogens in various ways. Antibodies may physically link the pathogens to phagocytes (neutrophils and macrophages), complement proteins, or NK cells. This linkage—called **opsonization**—triggers the attack mechanism and ensures that only the pathogens, and not nearby body cells, are destroyed.

In a second mechanism, antibodies that recognize toxins produced by bacterial pathogens in the extracellular fluid bind to the toxins, thereby preventing them from harming susceptible body cells. The antibody-antigen complexes that are formed are then destroyed by phagocytes.

In a similar way, antibodies that recognize certain viral surface proteins bind to the viruses in the extracellular fluid, preventing them from attaching to the plasma membranes of potential host cells. As with bacterial toxins, the antibody-virus complexes that are formed are subsequently phagocytized.

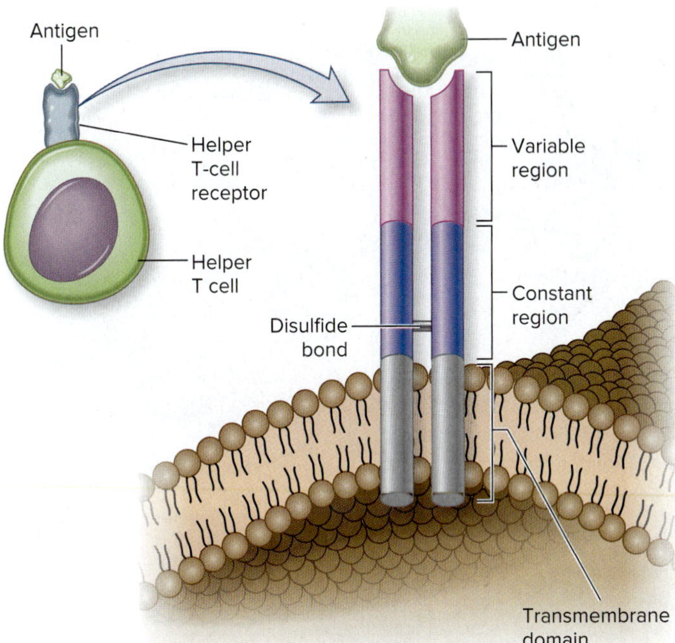


**Figure 40.9** Structure of a T-cell receptor in the plasma membrane.

**Concept Check:** *What are some structural similarities between T-cell receptors and B-cell receptors?*

## In Cell-Mediated Immunity, T Cells Recognize Antigens Complexed with Self Proteins

In cell-mediated immunity, T cells recognize and are activated by antigen (see Figure 40.6). T-cell receptors for antigens have specific regions that differ from one T cell to another. As shown in **Figure 40.9**, they are composed of two polypeptides, each with a variable and a constant region, along with a transmembrane domain. The variable regions recognize an antigen. As in B-cell development, multiple DNA rearrangements occur during T-cell maturation, leading to millions of distinct types of T cells, each with a receptor of unique specificity. For T cells, this maturation occurs as they develop in the thymus.

A T-cell receptor cannot bind to an antigen unless the antigen is already complexed with a protein that is found on the surface of a host cell, such as a macrophage. Such receptor proteins are called **major histocompatibility complex (MHC)** proteins. In humans, two major classes of MHC proteins are known. Class I MHC proteins are found on the surface of all body cells except erythrocytes (that is, all nucleated cells). Class II MHC proteins are found only on the surface of macrophages, B cells, and dendritic cells.

To recognize an antigen, the two different types of T cells have different MHC requirements. Cytotoxic T cells require antigen to be associated with class I MHC proteins, whereas helper T cells require an association with class II MHC proteins. One reason for this difference stems from the presence of different proteins on their surfaces; helper T cells can be identified by a unique membrane protein called CD4, and cytotoxic T cells are identified by a membrane protein known as CD8. CD4 binds to class II MHC proteins, whereas CD8 binds to class I MHC proteins.

**Antigen Presentation to Helper T Cells** How do antigens, which are foreign, end up complexed with MHC proteins on the surface of the body's own cells? The answer involves the mechanism known as antigen presentation. As previously noted, helper T cells can bind antigen only when the antigen appears on the plasma membrane of a host cell complexed with the cell's class II MHC proteins. Cells bearing these complexes, therefore, function as **antigen-presenting cells (APCs).** Because only macrophages, B cells, and dendritic cells express class II MHC proteins, only these cells can function as APCs for helper T cells.

Let's consider the function of macrophages as APCs for helper T cells (**Figure 40.10**). After a protein antigen has been phagocytized by a macrophage in an innate immune response, it is partially broken down into smaller peptide fragments by the macrophage's proteolytic enzymes within intracellular vesicles called endosomes. The resulting digested fragments then bind in the endosome to class II MHC proteins synthesized by the macrophage. Each fragment-MHC complex is then transported to the plasma membrane, where it is displayed on the cell surface. A specific helper T-cell receptor then binds this entire complex on the cell surface of the macrophage. The CD4 protein helps link the two cells. What is complexed to MHC proteins and presented to the helper T cells is not the intact antigen but, instead, a peptide fragment of the antigen—called an antigenic determinant, or **epitope.**

B cells process antigen and present it to helper T cells in essentially the same way as macrophages do. The ability of B cells to

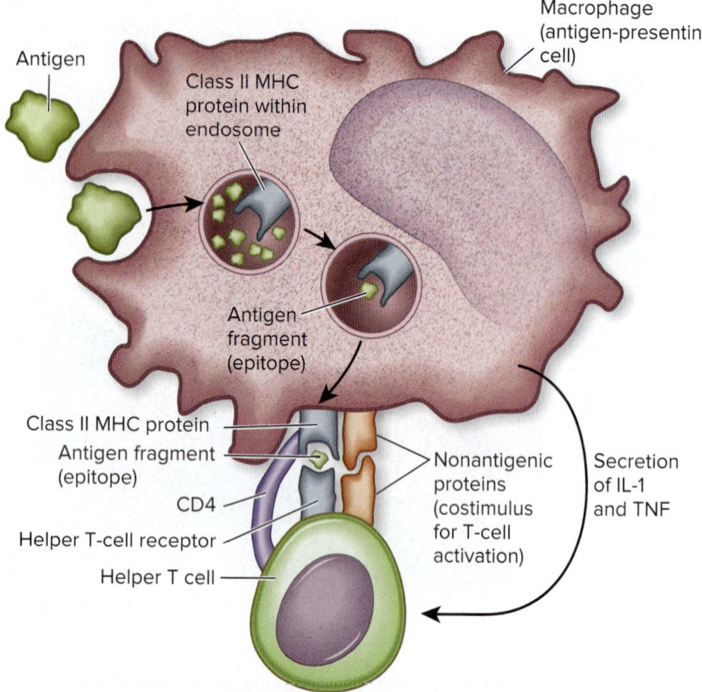


**Figure 40.10 Antigen presentation and helper T-cell activation.** In the initial events in helper T-cell activation, antigen fragments are complexed with a class II MHC protein within an antigen-presenting cell such as a macrophage. The complex is then displayed on the cell surface and binds to a helper T-cell receptor. Also required for T-cell activation are the binding of nonantigenic proteins between the APC and the attached helper T cell (a costimulus), and the actions of the cytokines interleukin 1 (IL-1) and tumor necrosis factor (TNF).

present antigen to helper T cells is a second function of B cells in response to antigenic stimulation, in addition to their differentiation into antibody-secreting plasma cells.

The binding between the helper T-cell receptor and antigen bound to class II MHC proteins on an APC is the essential antigen-specific event in helper T-cell activation. However, by itself this specific binding does not result in helper T-cell activation. In addition, interactions occur between nonantigenic proteins on the surfaces of the attached helper T cell and the APC. These interactions provide a necessary costimulus for helper T-cell activation (see Figure 40.10).

Finally, the antigen-dependent binding of the APC to the helper T cell plus the costimulus induces the APC to secrete large amounts of two cytokines—interleukin 1 (IL-1) and tumor necrosis factor (TNF). These molecules also stimulate the attached helper T cell. Thus, the APC participates in the activation of a helper T cell in three ways:

1. presenting antigen,
2. providing a costimulus, and
3. secreting cytokines.

Activated helper T cells then secrete cytokines that stimulate B cells and cytotoxic T cells.

**Helper T Cells and B-Cell Activation** Let's return to B cells to understand how they are activated by helper T cells. This process begins when a helper T cell specific for a particular antigen binds to a complex of that antigen and a class II MHC protein on an APC, thus activating the helper T cell. Along with other signals, this binding induces the activated helper T cell to divide. Some of the resulting activated helper T cells then bind to B cells that display the same antigen on their surfaces. This binding, along with additional cytokines, stimulates the B cell to go through the process of clonal selection. Thus, helper T cells are so named because their secretions help activate B cells that have bound antigen, in addition to their participation in activation of cytotoxic T cells and antigen presentation.

**Antigen Presentation to Cytotoxic T Cells** Unlike helper T cells, cytotoxic T cells require class I MHC proteins for activation. This distinction helps explain the major function of cytotoxic T cells—destruction of any of the body's own altered cells that have become cancerous or infected with viruses. The crucial point is that the antigens that complex with class I MHC proteins typically arise within body cells and are thus endogenous antigens.

How do such antigens arise? In viral infections, once a virus has entered a host cell, the expression of viral genes results in the synthesis of viral proteins, which are foreign to the cell. In cancerous cells, one or more of the cell's genes have become altered by chemicals, radiation, or other factors. Genetic changes associated with cancer lead to the production of abnormal proteins, which act as antigens.

In both virus-infected and cancerous cells, cytosolic enzymes hydrolyze some of the endogenously produced antigenic proteins into polypeptide fragments, which are transported into the endoplasmic reticulum. There the fragments are complexed with the host cell's class I MHC proteins and then shuttled by exocytosis to the plasma membrane, where a cytotoxic T cell specific for the antigen/MHC protein complex can bind to it. Once binding occurs, cytotoxic T cells release chemicals that kill the infected or cancerous cell, as discussed next.

## Activated Cytotoxic T Cells Kill Infected or Cancerous Cells

The previous sections described how immune responses provide long-term defenses against bacteria, viruses, and individual foreign molecules that enter the body's extracellular fluid. We now examine how the body's own cells that have become infected by viruses or transformed into cancerous cells are destroyed by cell-mediated immune responses (see Figure 40.6).

What is the value of destroying virus-infected host cells? First, and most importantly, such destruction prevents cells from making more viruses. Second, for cells that already are making mature viruses, it results in the release of the viruses into the extracellular fluid, where they can be neutralized by circulating antibody.

**Functions of Cytotoxic T Cells** A typical cytotoxic T-cell response triggered by viral infection of a vertebrate's body cells is summarized in Figure 40.11. The response triggered by a cancerous cell would be similar. A virus-infected cell produces foreign proteins, viral antigens that are processed and presented on the plasma membrane of the cell complexed with class I MHC proteins. Cytotoxic T cells specific for the particular antigen bind to the complex

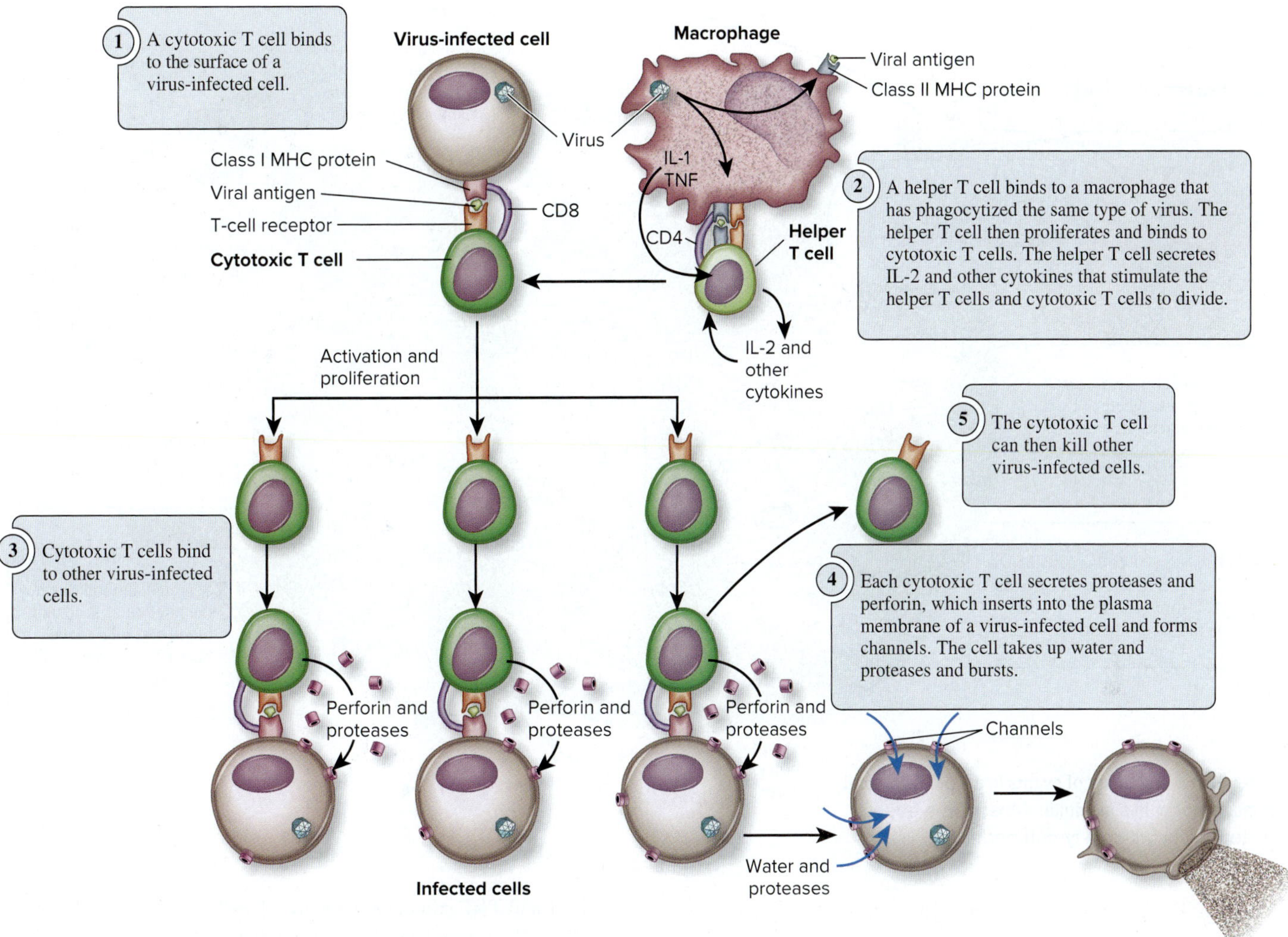


**Figure 40.11 Summary of events in the killing of virus-infected cells by cytotoxic T cells.** The sequence is similar for cancerous cells attacked by a cytotoxic T cell.

(see Figure 40.11, step 1). As with B cells, binding to antigen alone does not cause activation of the cytotoxic T cell.

Macrophages phagocytize extracellular viruses (or, in the case of cancer, antigens released from the surface of cancerous cells) and then process and present antigen, in association with class II MHC proteins, to the helper T cells (step 2). The activated helper T cells then bind to cytotoxic T cells. They also release IL-2 and other cytokines, which stimulate proliferation of helper T cells as well as cytotoxic T cells. Why is proliferation important if a cytotoxic T cell has already located and bound to its target? The answer is that there is rarely just one virus-infected or cancerous cell. By expanding the population of cytotoxic T cells capable of scanning the entire body and recognizing the particular antigen, the likelihood is greater that an appropriate cytotoxic T cell will encounter other virus-infected or cancerous cells.

The cytotoxic T cells specific for that virus then find and bind to other virus-infected cells (step 3). Each cytotoxic T cell releases the contents of its secretory vesicles directly into the extracellular space between itself and the target cell to which it is bound (thereby ensuring that other nearby host cells will not be killed). These vesicles contain proteases, and a protein called perforin, which is similar in structure to the proteins of the complement system's membrane attack complex. Perforin is inserted into the target cell's membrane and forms channels ("perforations") through the membrane (step 4). This allows proteases secreted by a cytotoxic T cell to enter the attacked cell and causes apoptosis. The cell also takes in water through the pores, which causes it to burst. The cytotoxic T cell is not harmed by this process and can then continue to kill other virus-infected cells (step 5). Target cell killing by activated cytotoxic T cells occurs by several mechanisms, but this is one of the most important.

Viruses can be eliminated from an animal's body in two ways: through the humoral actions of antibodies in body fluids and through the cell-mediated killing of virus-infected cells. Although cytotoxic T cells have an important role in the attack against such cells, they are not the only mechanisms. NK cells also destroy virus-infected and cancerous cells by secreting toxic chemicals. As mentioned earlier, NK cells can recognize general features on the surface of such cells and participate in innate immunity. In addition, in a cell-mediated immune response, NK cells can be linked to such target cells by antibodies and then can destroy them by release of toxic molecules.

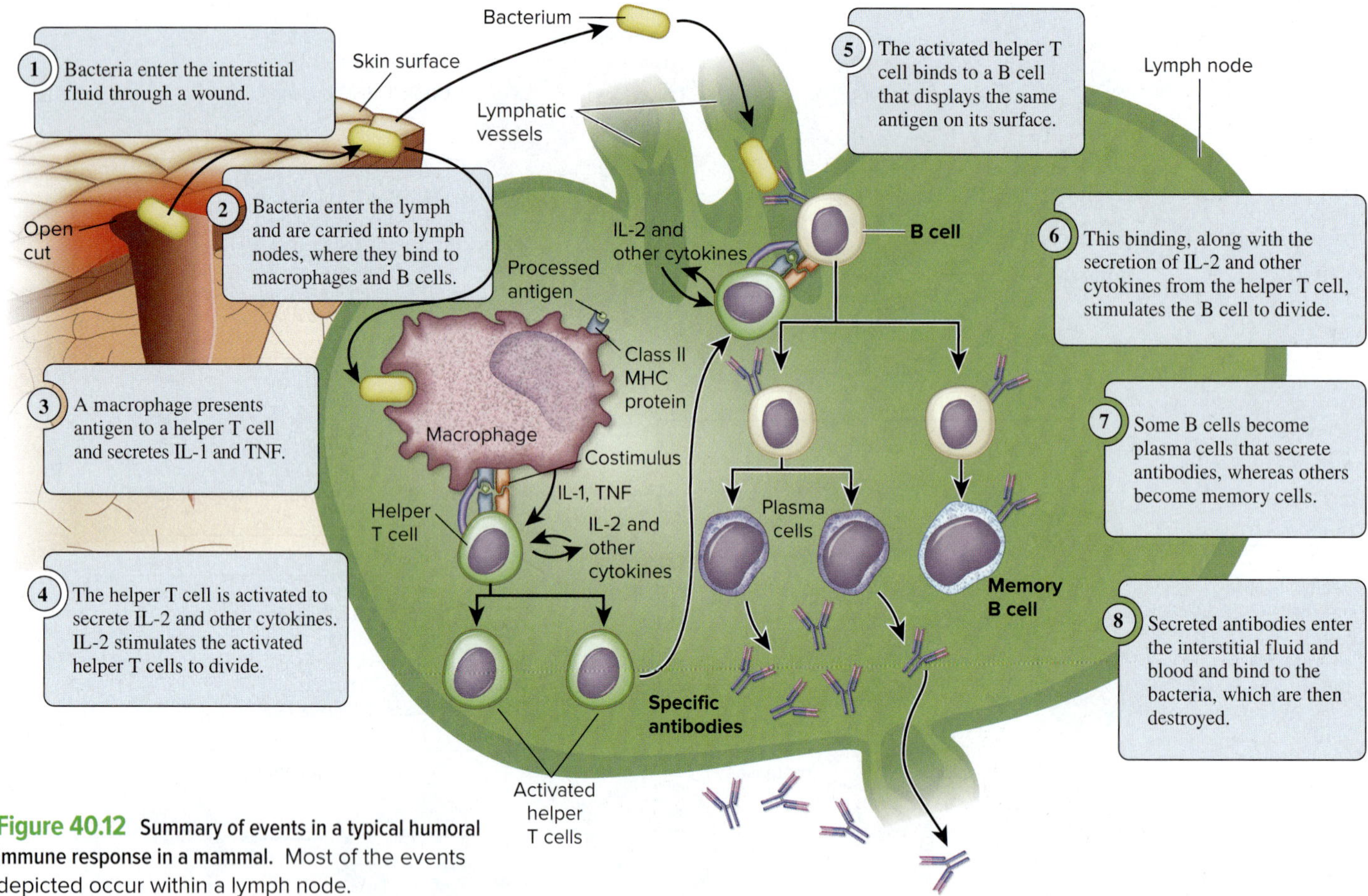


**Figure 40.12** **Summary of events in a typical humoral immune response in a mammal.** Most of the events depicted occur within a lymph node.

## Summary: Example of an Acquired Immune Response

Let's bring together our discussion of the acquired immune system by looking in detail at one example. One classic humoral immune response is that which results in the destruction of bacteria. The sequence of events, which is quite similar to the humoral response to a virus in the extracellular fluid, is summarized in **Figure 40.12**. For this example, we consider the response in mammals, in which lymph nodes are present. Many features of the response, however, are similar in other vertebrates.

This process starts the same way as for innate responses, with the bacteria penetrating one of the body's linings through an injury and entering the interstitial fluid (see Figure 40.12, step 1). The bacteria then move with lymph into the lymphatic system and are carried to lymph nodes (step 2). Within the lymph node, both a macrophage and a B cell recognize one of the bacteria as foreign and bind to it.

B-cell activation usually requires activation of helper T cells. The helper T cell binds to a complex of processed antigen and class II MHC protein on the macrophage (an APC). The macrophage phagocytizes the bacterium, hydrolyzes its proteins into peptide fragments, complexes the fragments with class II MHC proteins, and displays the complexes on its surface. Once a helper T cell specific for the complex binds to it, the helper T cell becomes activated. The macrophage helps this process in two other ways: It provides a costimulus, and it secretes the cytokines IL-1 and TNF.

IL-1 and TNF stimulate the helper T cell to secrete another cytokine, IL-2. IL-2 stimulates the activated helper T cell to divide, which leads eventually to the formation of a clone of activated helper T cells (step 4). The activated helper T cells bind to B cells and secrete IL-2 and other cytokines (step 5). Certain of these cytokines provide the additional signals that are usually required to activate nearby antigen-bound B cells to proliferate (step 6). These cells differentiate into memory cells, which help ward off possible future attacks by the same antigen, and plasma cells, which then secrete specific antibodies (step 7). The antibodies enter the blood and bind to bacterial cells, which are then destroyed (step 8).

## B Cells and T Cells That Recognize Self Molecules Must Be Killed or Inhibited

As we have seen, the lymphocytes responsible for the specific immune response in vertebrates are very capable killers of pathogens—so capable, in fact, that it raises a question: Why don't these cells attack and kill normal self cells? In other words, how does the body distinguish between self and nonself components and develop what is called **immune tolerance,** or tolerance of its own proteins and other molecules?

Recall that the huge diversity of lymphocyte receptors is ultimately the result of multiple random DNA recombination processes. It is virtually certain, therefore, that every animal possessing acquired immune defenses would have lymphocytes with receptors that could bind to that individual's own proteins. The continued existence and functioning of such lymphocytes would be disastrous, because such binding would launch an immune attack against all body cells expressing these proteins.

At least two mechanisms explain why individuals normally lack active lymphocytes that respond to self components. First, during early development in vertebrates, T cells are exposed to a wide mix of self proteins in the thymus gland. Those T cells with receptors capable of binding self proteins are destroyed by apoptosis in a process termed **clonal deletion.** The second process, termed **clonal inactivation,** occurs outside the thymus gland and causes potentially self-reacting T cells to become nonresponsive. B cells undergo similar processes. The mechanisms by which these two events occur are still under investigation. Occasionally, however, these mechanisms fail, and an animal's immune cells attack the body's own cells, resulting in an autoimmune disease. **Autoimmune diseases** are conditions in which the body's normal state of immune tolerance somehow breaks down, with the result that both humoral and cell-mediated immunity attacks are directed against the body's own cells. A growing number of human diseases, such as multiple sclerosis, rheumatoid arthritis, systemic lupus erythematosus, and type 1 diabetes mellitus, are being recognized as autoimmune in origin.

## Immunological Memory Is an Important Feature of Acquired Immunity

As we have learned, the acquired immune response to a given antigen depends on whether or not an animal's body has previously been exposed to that antigen. Consider, for example, the humoral immune response. In mammals, antibody production in response to the first contact with an antigen occurs slowly, over a few weeks. This response to an initial antigen exposure is termed a **primary immune response** (Figure 40.13). Any subsequent infection by the same pathogen elicits an immediate and heightened production of additional specific antibodies against that particular antigen, a reaction termed a **secondary immune response.**

In the case of humoral immunity, this secondary response occurs more quickly, is stronger, and lasts longer because memory B cells that were produced in response to the initial antigen exposure are quickly stimulated to multiply and differentiate into thousands of plasma cells. These cells then produce large amounts of specific antibodies. The immune system's ability to produce this secondary response is called immunological memory.

Immunological memory explains why humans and other animals are able to fight off many illnesses to which we have been previously exposed, such as many common childhood diseases. The acquired response to exposure to any type of antigen is known as **active immunity.** Active immunity not only results from natural exposure to antigens but also is the basis for the artificial exposures to antigen that occur in vaccinations. In **vaccinations,** small quantities of living, dead, or altered pathogens; small quantities of toxins; or harmless antigenic molecules derived from a microorganism or its toxin are injected into the body so as to induce a primary immune response, including the production of memory cells. Subsequent natural exposure to the immunizing antigen results in a rapid, effective response that can prevent or reduce the severity of disease.

BIO TIPS ONLINE

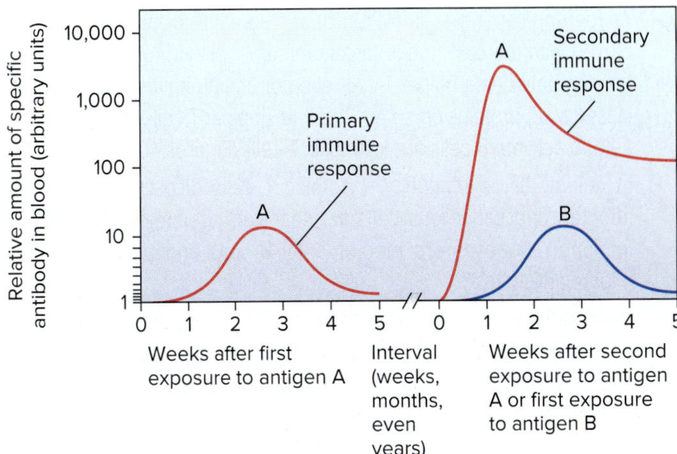


**Figure 40.13 Primary and secondary immune responses.** In a primary response, as shown on the left of this graph, an initial exposure to an antigen produces modest levels of specific antibody over a period of weeks. In a secondary response, subsequent exposure to the same antigen results in greater antibody production that occurs more rapidly and lasts longer than a primary response. (Note that the scale of the *y*-axis is logarithmic.) The secondary response is specific for that antigen. Exposure at that time to a different antigen for the first time produces the usual primary response.

**Concept Check:** *What is the advantage of a secondary immune response?*

In contrast to active immunity, another type of acquired immunity, called **passive immunity,** confers protection against disease through the direct transfer of antibodies from one individual to another. Passive immunity can occur naturally, as when a newborn mammal receives antibodies from its mother's milk. It can also occur artificially, as when a person is given an injection of IgG molecules shortly after exposure to hepatitis viruses.

## 40.3 Reviewing the Concepts

- An antigen is a foreign molecule that the host does not recognize as self and that triggers an acquired immune response.
- Lymphocytes (B and T cells) mediate acquired immune responses and reside in structures that constitute the lymphatic system (Figure 40.4). B cells differentiate into antibody-producing cells called plasma cells. T cells include cytotoxic T cells, which directly kill target cells, and helper T cells, which assist in the activation and function of B cells and cytotoxic T cells (Figure 40.5).
- Acquired immune responses occur in three stages: recognition of antigen, activation and proliferation of lymphocytes, and attack against antigen (Figure 40.6).
- B-cell receptors and antibodies are immunoglobulins; they contain a variable region that serves as the antigen-binding site (Figure 40.7). Antibodies combine with the antigen that activated the B cell and guide an attack that eliminates the antigen or the cells bearing it (Figure 40.8).

- Antigen-presenting cells (APCs) bear fragments of antigen complexed with the cell's major histocompatibility (MHC) proteins. The binding between a helper T-cell receptor and an antigen bound to class II MHC proteins on an APC activates helper T cells, which then help to activate B cells and cytotoxic T cells (Figures 40.9, 40.10).
- In cell-mediated immunity, cytotoxic T cells directly kill virus-infected and cancerous cells via secreted chemicals. Humoral immune responses are mediated by B cells and plasma cells (Figures 40.11, 40.12).
- Upon initial exposure to an antigen, the body produces a primary immune response. Subsequent exposure to the same antigen elicits a rapid and heightened response termed a secondary immune response (Figure 40.13).

## 40.3 Testing Your Knowledge

1. The structures in which lymphocytes differentiate into mature immune cells include
   a. the primary lymphoid organs.
   b. the secondary lymphoid organs.
   c. the lymph nodes and spleen.
   d. the thymus gland and bone marrow.
   e. both a and d.
2. Which cells directly recognize and kill virus-infected or cancer cells?
   a. antigen-presenting cells
   b. B cells
   c. plasma cells
   d. helper T cells
   e. cytotoxic T cells

## 40.4 Impact on Public Health

### Learning Outcomes

1. Explain the role of the immune system in allergic reactions.
2. Describe the effects of HIV on the human immune system, and describe a current method of treating HIV infection.

In this section, we will consider how the human immune system is affected by allergies in vast numbers of people, and how destruction of immune cells caused by the human immunodeficiency virus (HIV) is a continuing pandemic. Collectively, the effects of disorders of the immune system have an almost immeasurable impact on public health in terms of worker productivity, health-care resources, and the economy.

### Allergies Affect the Quality of Life of Millions of People

An **allergy** (also known as hypersensitivity) is a condition in which immune responses to environmental antigens cause inflammation and damage to body cells. Antigens that induce allergic reactions are called allergens. Common examples of allergens include ragweed pollen and animal dander. Most allergens themselves are relatively or completely harmless. It is the immune system's response to them that causes the damage. In essence, then, allergy is immunity gone awry, for the response is of inappropriate strength and duration for the stimulus. In the U.S. alone, as many as 40 million people (about 13% of the population) suffer from allergies.

For any allergy to develop, a genetically predisposed person must first be exposed to the allergen—a process called sensitization. Subsequent exposures elicit the damaging immune responses we recognize as an allergy. Hypersensitivities can be broadly classified according to the speed of the response. Allergies that take up to several days to develop are considered delayed hypersensitivities. The skin rash that appears after contact with poison ivy is an example. More common are reactions considered immediate hypersensitivities, which can develop in minutes or up to a few hours. These allergies are also called IgE-mediated hypersensitivities, because they involve IgE antibodies.

In immediate hypersensitivity, sensitization to the allergen leads to the production of specific antibodies and a clone of memory B cells. In individuals who are genetically susceptible to allergies, antigens that elicit immediate hypersensitivity reactions stimulate the production of IgE antibodies. Upon their release from plasma cells, IgE molecules circulate throughout the body and become attached to mast cells in connective tissue. When the same antigen subsequently enters the body at some future time and binds with IgE that is bound to mast cells, the mast cell is stimulated to secrete many chemicals, including histamine, which then initiate an inflammatory response.

The signs and symptoms of IgE-mediated hypersensitivity reflect both the effects of inflammatory mediators and the site in which the antigen–IgE–mast cell binding occurs. When, for example, a previously sensitized person inhales ragweed pollen, the antigen combines with the variable region of IgE, and the constant region of IgE binds to mast cells in the airways. The mast cells release their contents, which induce increased mucous secretion, increased blood flow, swelling of the epithelial lining, and contraction of the smooth muscle surrounding airways. These effects produce the congestion, runny nose, sneezing, and, in some persons, difficulty in breathing characteristic of "hay fever." Antihistamines are drugs that block the action of histamine that is released during allergic responses. These drugs prevent histamine from binding to its receptor protein on its target cells, thereby preventing or relieving some of the symptoms of allergy.

### Acquired Immunodeficiency Syndrome (AIDS) Results from HIV Infection

**Acquired immunodeficiency syndrome (AIDS)** is caused by the **human immunodeficiency virus (HIV),** a virus that incapacitates the immune response by infecting helper T cells. HIV is a retrovirus, a virus that contains RNA as its genetic material. Once HIV is inside a helper T cell, the enzyme reverse transcriptase uses the viral RNA strand to make a complementary copy of DNA, which is then integrated into the chromosomal DNA of the host's T cells (refer back to Figure 17.2). Later, viral replication within the T cell results in the death of the cell.

HIV infects helper T cells because the CD4 protein in their plasma membranes (see Figure 40.10) acts as a receptor for an HIV envelope protein. However, binding to CD4 is not sufficient to enable HIV to enter the helper T cell. Another T-cell surface protein, which normally acts as a receptor for certain cytokines, must serve as a coreceptor. Interestingly, individuals possessing a mutation in this cytokine receptor are highly resistant to HIV infection, so much research is now focused on the possible therapeutic use of chemicals that can bind to and block this coreceptor.

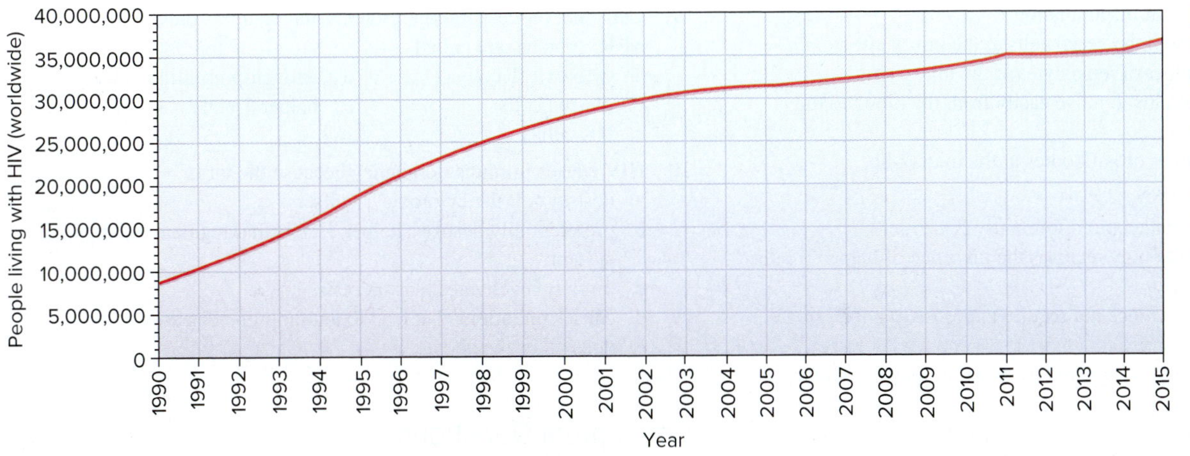


**Figure 40.14** **Worldwide incidence of people living with HIV/AIDS.** Data are from the 2016 World AIDS Day Report published by The Joint United Nations Programme on HIV/AIDS.

HIV infection leads to a reduction in the number of helper T cells. Without adequate numbers of helper T cells, neither B cells nor cytotoxic T cells can function normally. Both humoral and cell-mediated immunity are compromised. AIDS is the late stage of HIV infection, when the individual's weakened immune system has difficulty fighting certain diseases. Thus, many individuals with AIDS die from infections and cancers that ordinarily would be readily handled by a fully functional immune system.

AIDS, first described in 1981, has since reached pandemic proportions. About 33 to 35 million people worldwide are currently living with HIV infection, and an estimated 5,000–10,000 new infections occur in the world each day (Figure 40.14). The major routes of HIV transmission are (1) unprotected sexual intercourse with an infected partner; (2) transfer of contaminated blood or blood products between individuals, such as the sharing of needles among intravenous drug users or, less commonly, as a result of a blood transfusion; (3) transfer from an infected mother to her child across the placenta or during delivery; or (4) transfer via breast milk during nursing.

Treatment for HIV-infected individuals has two components: one directed against the virus itself to delay progression of the disease, and one to prevent or treat the opportunistic infections and cancers that ultimately cause death. One current antiviral approach involves administering a combination of four drugs, known as HAART (highly active antiretroviral therapy). Two of the drugs inhibit the action of reverse transcriptase in converting viral RNA into DNA within the host cell, a third drug inhibits an HIV enzyme required for assembling new viruses, and a fourth drug called a fusion inhibitor prevents the virus from entering T cells. These treatments have been effective in slowing the rate at which infection with HIV leads to AIDS. Unfortunately, however, the HAART regimen is associated with numerous side effects, including nausea, vomiting, diarrhea, metabolic disturbances, and liver damage. Much research is under way to find better treatments and ultimately to cure this disease.

## 40.4 Reviewing the Concepts

- Allergies affect millions of people and result from excessive immune responses to antigens. Acquired immunodeficiency syndrome (AIDS) is caused by the human immunodeficiency virus (HIV), which weakens the body's immunity by killing helper T cells (Figure 40.14).

## 40.4 Testing Your Knowledge

1. Which is *true* of allergies?
   a. They are autoimmune diseases.
   b. They affect roughly 40% of the U.S. population.
   c. They are an inappropriately strong immune response to antigens that are generally harmless or of little danger.
   d. They are characterized by hyposensitivity to a variety of antigens.
   e. They result in overproduction of IgG antibodies, which inhibit histamine release from mast cells.
2. Which is *correct*?
   a. AIDS sometimes, but not always, causes HIV.
   b. HIV infects helper T cells.
   c. The only ways to become infected with HIV are via unprotected sex and contamination with blood or blood products from an infected individual.
   d. HIV infects cytotoxic T cells.
   e. HIV contains DNA, which then gets integrated into the chromosomes of the host cell.

# Assess and Discuss

## Test Yourself

1. Which of the following is *not* an example of a barrier defense in animals?
   a. skin
   b. secretions from skin glands
   c. exoskeleton
   d. mucus
   e. antibodies
2. The leukocytes that are found in mucosal surfaces and that play a role in defending the body against parasitic infections are
   a. neutrophils.
   b. eosinophils.
   c. basophils.
   d. monocytes.
   e. NK cells.

3. The vascular changes of inflammation
   a. lead to an increase in bacterial cells at the injury site.
   b. decrease the number of leukocytes at the injury site.
   c. allow plasma proteins to move easily from the blood to the injury site.
   d. increase the number of antibodies at the injury site.
   e. activate lymphocytes.
4. Which statement about acquired immunity is *correct*?
   a. Acquired immunity only requires the presence of helper T cells to function properly.
   b. Acquired immunity does not require exposure to a foreign substance.
   c. Acquired immunity is triggered by exposure to a particular antigen.
   d. Acquired immunity includes inflammation.
   e. All of the above are correct.
5. Memory B cells are
   a. cloned lymphocytes that are active in subsequent infections.
   b. cloned lymphocytes that are active during a primary infection.
   c. NK cells that recognize cancer cells and destroy them.
   d. cells that produce antibodies.
   e. macrophages that have recognized self antigens.
6. The secondary immune response to a specific antigen that is characteristic of acquired immunity is primarily due to the actions of
   a. cytotoxic T cells.
   b. memory T cells.
   c. cells derived from B cells in an earlier exposure to that antigen.
   d. helper T cells.
   e. both b and c.
7. The region of an antibody that is the antigen-binding site is
   a. the constant region.
   b. the variable region.
   c. the entire heavy chain.
   d. the entire light chain.
   e. the hinge regions.
8. A major difference between the activation of B cells and cytotoxic T cells is that
   a. cytotoxic T cells must interact with antigens bound to plasma membranes.
   b. B cells interact only with free antigens.
   c. B cells are not regulated by helper T cells.
   d. cytotoxic T cells produce antibodies.
   e. only cytotoxic T cells express immunoglobulins in their membranes.
9. Cells that process foreign proteins and complex them with their MHC proteins are called
   a. cytotoxic T cells.
   b. plasma cells.
   c. NK cells.
   d. antigen-presenting cells.
   e. helper T cells.
10. HIV causes immune deficiency because the virus
   a. destroys all the cytotoxic T cells.
   b. preferentially destroys helper T cells that regulate the immune system.
   c. directly inactivates plasma cells.
   d. causes mutations that lead to autoimmune diseases.
   e. does all of the above.

## Conceptual Questions

1. Distinguish between innate and acquired immunity.
2. Explain the function of cytotoxic T cells.
3. **PRINCIPLES** A principle of biology is that living organisms interact with their environment. Such interactions include potentially threatening environmental factors such as pathogens. Discuss three types of pathogens that affect the health of animals.

## Collaborative Questions

1. Describe the basic structure of an immunoglobulin and how the structure relates to its ability to specifically recognize an antigen.
2. List the different types of lymphocytes, and briefly identify a key structure and/or function of each.

## Online Resource

**connect.mheducation.com**

SMARTBOOK® SmartBook® is the first and only adaptive reading experience designed to change the way students read and learn.

# Integrated Responses of Animal Organ Systems to a Challenge to Homeostasis

# 41

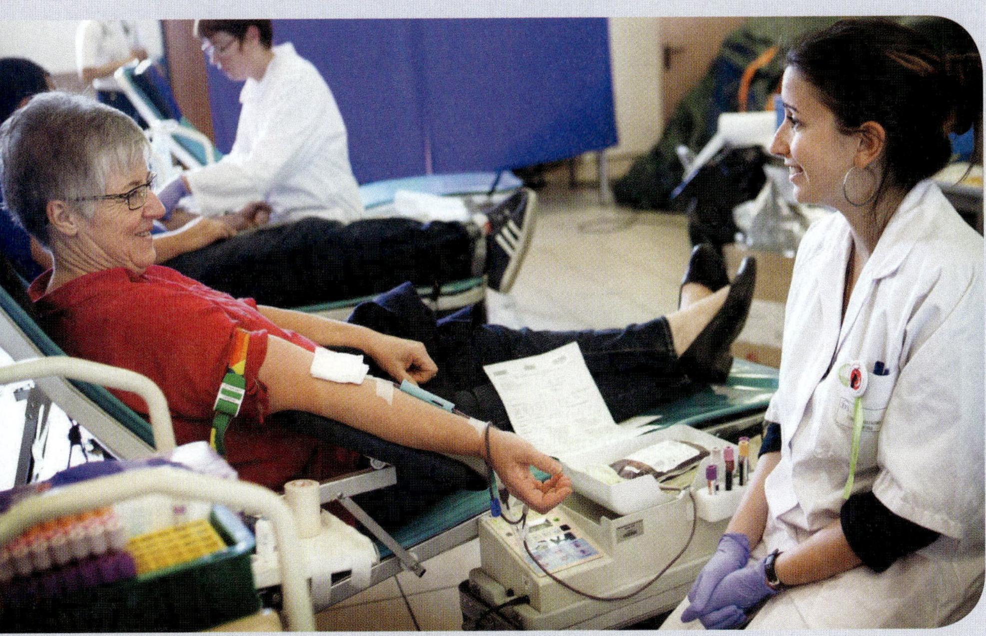

© BSIP SA/Alamy

**A person donating blood.** Donations such as these improve the chances for survival of individuals who have lost significant amounts of blood due to injury.

## Chapter Outline



Shortly into a high school soccer match, a 16-year-old girl falls to the ground after tripping over an opponent's foot. A teammate running behind the girl cannot stop herself in time and accidentally kicks the girl with a hard, cleated shoe in the lower back, near the region of the girl's left kidney. The injury is not terribly painful, and the girl is able to get up on her own and walk off the field, where she sits down and rests. By the end of the match, however, she is complaining of feeling "a little woozy." She becomes agitated yet remains alert and responsive. Nonetheless, her concerned coach has her transported by ambulance to a local hospital. A paramedic records her blood pressure as 108/66 mmHg—which, according to the girl's mother, is a little lower than usual. Her heart rate is 101 beats per minute (bpm), which is abnormally rapid. She is breathing at a rate of 20 breaths/minute, which is above normal.

The kick she received has ruptured one or more large blood vessels, causing her to bleed internally. The loss of blood through a ruptured blood vessel is called a **hemorrhage.** Without prompt treatment, a significant hemorrhage runs the risk of leading to a condition known as **shock,** in which the ability of the circulatory system to provide nutrients and oxygen to vital organs decreases to the point that those organs can no longer function properly and their cells begin to die.

One of the principles of biology that you learned in Chapter 1 is that living organisms maintain homeostasis. Throughout this unit, you have seen numerous examples of how this principle applies to animal biology. Recall from Chapter 32 that organ systems in animals do not function in isolation. That is, changes in the activity of one organ system often result in changes in other organ systems. Many times, two or more organ systems function together to control an important variable such as blood pressure. In this chapter, we will take a detailed look at how a significant challenge to homeostasis results in changes in the activities of all organ systems, which function together to meet the challenge of restoring homeostasis. Our example of a homeostatic challenge will be hemorrhage, which may be internal, as in the girl just described, or external, as when you cut yourself and bleed to the outside.

Hemorrhage can have life-threatening consequences if not compensated for by homeostatic mechanisms. A significant loss of blood decreases an animal's blood pressure, which slows the delivery of nutrients and oxygen to vital organs such as the brain and heart. The result can be catastrophic and even fatal if pressure is not corrected. The compensatory mechanisms for hemorrhage have been studied in all vertebrates but most thoroughly in mammals, and thus we will use mammals as our example in this chapter. We will begin by describing these compensatory mechanisms and then, at the end of the chapter, return to the case of the young soccer player and discover how an understanding of homeostatic control mechanisms helped her recover.

## 41.1 Effects of Hemorrhage on Blood Pressure and Organ Function

### Learning Outcomes

1. Explain why a decrease in blood pressure can be dangerous.
2. **SCISKILLS** ▶ Predict changes that may occur to an animal's blood pressure when its blood volume is increased or decreased, and explain why.

In any type of hemorrhage, no matter how small or large, one of the first responses of an animal's body is sealing the wound. This is accomplished by the clotting mechanisms described in Chapter 36 (refer back to Figure 36.5). For the purpose of our discussion, we will assume that an animal suffering a hemorrhage has normal clotting mechanisms and that they are initiated immediately following the hemorrhage and continue for some time until the wound is closed. In this section, we will focus on understanding why hemorrhage is so dangerous and how blood volume and pressure are linked.

### Decreased Blood Pressure May Lead to Cell Death

Blood circulates in the body under pressure, as described in Chapter 36. The source of that pressure is the beating of the heart. Blood flows through arteries when it leaves the heart and pushes against the inner surfaces of the vessels. This force is what we refer to when we measure a person's blood pressure using an arm cuff, for example. In most mammals, blood pressure is roughly similar to that of humans, but different animals face different challenges in getting blood to all parts of the body. Think of the distance blood must travel from the heart to the brain in a giraffe. Even in humans, the shorter distance from the heart to the brain results in a decrease in blood pressure as the blood flows upward against gravity. Giraffes have evolved an exceptionally strong heart that moves blood up their long neck. However, such adaptations are not always completely effective. You may have experienced a sense of light-headedness on occasion when standing suddenly; it takes a second or two to generate the pressure required to overcome the effect of gravity upon standing. This temporary light-headedness illustrates the importance of blood pressure. All cells require uninterrupted delivery of nutrients and oxygen to perform the metabolic and other activities required for optimal functioning and survival. Very active cells, such as those of the brain, are the first to show signs of malfunction when the rate of nutrient and oxygen delivery is decreased.

Because of its vital importance for cell function, let's focus for the moment on oxygen delivery. The rate of oxygen delivery to any tissue in an animal's body depends on at least four factors, which were described in detail in Chapter 36. These include

- the rate of blood flow to a region, which in turn is influenced by the degree to which vessels are dilated or constricted,
- the number of erythrocytes in a given volume of blood,
- the percent saturation of hemoglobin in the erythrocytes, and
- the ability of hemoglobin to release its oxygen to cells.

Following a hemorrhage, the first of these factors—blood flow—is immediately affected as blood pressure begins to decrease. Very quickly, unless compensatory events occur, this alone can be sufficient to cause widespread cell death. Adjustments to all four factors, however, can help prevent this from happening.

### Blood Volume and Blood Pressure Are Linked

Refer again to the chapter-opening photo of a person donating blood. This procedure removes about 500 mL of blood, which is roughly 10% of the total blood volume of a typical adult human. This controlled procedure has been adapted by investigators in understanding how an animal's body responds to a hemorrhage. For instance, using a thin, flexible, hollow tube (called a cannula) inserted into a vein of an experimental animal to withdraw carefully controlled amounts of blood, investigators can determine the relationship between the degree of hemorrhage and its effects on different organ systems. Then, investigators can monitor how these organ systems respond in an integrated, coordinated way—first, to prevent blood volume and pressure from continuing to decrease and, second, to help restore blood volume and pressure to normal.

The results of such experiments indicate that mammals can usually cope well with a 10% hemorrhage, as humans can when

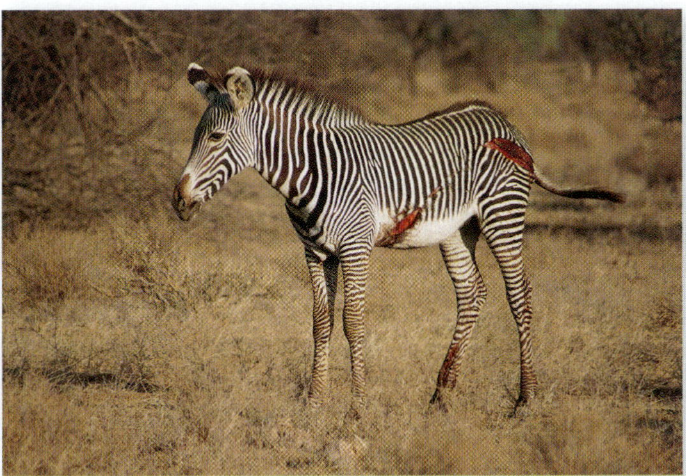

**Figure 41.1 An adult zebra that survived an attack by a lioness.** These deep wounds have begun healing. Note that there is no bleeding at this time. Nonetheless, these injuries would initially have caused considerable hemorrhaging. The animal's homeostatic responses to the loss of blood may have helped save its life.


donating blood. Symptoms of a 10% hemorrhage are relatively mild and, in some people, are barely noticeable. Blood pressure may remain normal or may decrease only slightly. Part of the reason that symptoms are minor can be attributed to the large volume of blood that is present in the large veins, such as those in the legs. Veins are expandable and can accommodate more blood than can arteries. This blood can serve as a reservoir for times when an animal becomes more active and requires greater cardiac output (refer back to Figures 36.10 and 36.13). After a hemorrhage, this reservoir can also be tapped to help maintain heart function and blood pressure. Larger hemorrhages, however—for example, 20% and greater of an animal's blood volume—do result in a significant decrease in blood pressure. Remarkably, though, animals may survive even such a large hemorrhage (Figure 41.1).

What is the link between blood volume and blood pressure? Mammals have a closed circulatory system (refer back to Figure 36.2b). In such a system, volume and pressure are closely related. Look at the artificial closed system shown in Figure 41.2. When volume is normal, so is pressure. However, a decrease in volume decreases pressure, whereas an increase in volume causes the pressure to rise.

Recall that the blood pressure of a mammal that has lost about 10% of its blood volume does not change much. What is helping to prevent this from happening? How is the venous reservoir of blood returned to the heart, and what other changes occur to improve circulatory function? As we will see next, numerous organ systems function together to help prevent a dangerous drop in blood pressure under these circumstances. These responses occur in two phases, one within seconds to minutes and one requiring hours to days or even weeks.

## 41.1 Reviewing the Concepts

- Hemorrhage is a loss of blood from a damaged blood vessel. It may decrease oxygen and nutrient delivery to cells (Figure 41.1).
- Blood volume and blood pressure are closely related; a decrease in volume due to hemorrhage can lead to a fall in pressure. Smaller hemorrhages may produce no obvious symptoms, but large hemorrhages can have serious consequences (Figure 41.2).

## 41.1 Testing Your Knowledge

1. Which is *false*?
   a. Oxygen delivery to an organ depends partly on the rate of blood flow to that organ.
   b. Blood volume and blood pressure in an animal are usually inversely related.
   c. Small hemorrhages do not usually cause a decrease in blood pressure.
   d. A typical person has about 5 L of blood.
   e. The force of gravity can influence the ability of blood to reach regions distant from the heart.

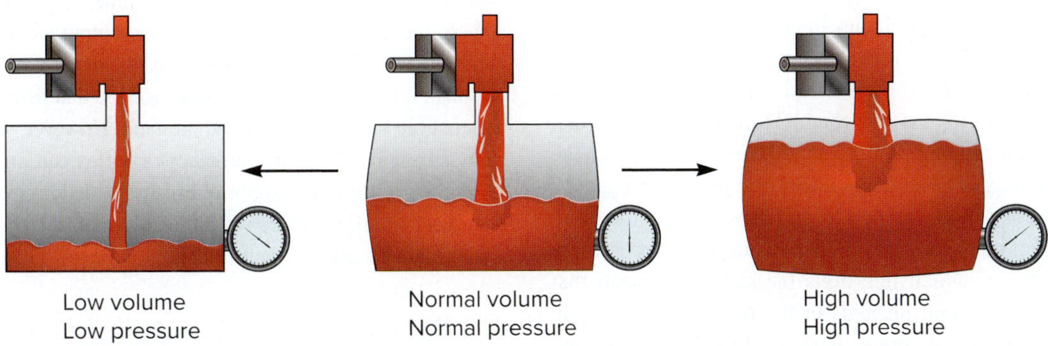


**Figure 41.2 The relationship between blood volume and blood pressure.** In a system such as the artificial one shown here, an increase in fluid volume will increase the pressure that the fluid exerts on the walls of the chamber. In a similar way, increasing blood volume will increase the pressure of the blood in vessels such as arteries. A hemorrhage will decrease volume and thus decrease pressure.

## 41.2 The Rapid Phase of the Homeostatic Response to Hemorrhage

### Learning Outcomes

1. Define baroreceptors, and describe where they are located in mammals.
2. **SCISKILLS ▶** Predict what will happen to the firing rate of baroreceptors if blood volume is increased or decreased.
3. Describe the baroreceptor reflex in mammals, its dependence on the nervous system, and the changes it induces in the function of the circulatory system.
4. Compare the distribution of blood to different organs before and after hemorrhage.
5. Explain the mechanisms that cause redistribution of blood between organs.
6. Describe how the structures of the respiratory system contribute to restoring pressure following hemorrhage.

The initial homeostatic response to a decrease in blood volume and pressure is mediated by the nervous system. When a hemorrhage occurs, the nervous system must receive information that blood pressure and volume are decreasing, then respond appropriately. Refer back to Figure 32.9, in which a typical homeostatic control system is described. Two key elements of such a system are a sensor and an integrator. For monitoring blood volume and pressure, these control elements are located in blood vessels and in the brain, respectively. In this section, we will examine how they maintain homeostasis.

### Baroreceptors Sense Changes in Blood Pressure and Initiate a Compensatory Reflex

Special pressure-sensitive regions exist within the walls of certain large arteries in all vertebrates. In mammals, these arteries include the carotid arteries in the neck, which supply blood to the brain, and the aorta, the first artery that emerges from the left ventricle. These regions contain the endings of neurons, known as **baroreceptors** (from the Greek *baros,* meaning weight or pressure), which are in constant communication with the medulla oblongata of the brainstem (**Figure 41.3a**). They mediate an important mechanism by which blood pressure is regulated in vertebrates called the **baroreceptor reflex** (**Figure 41.3b**). This response to a change in blood pressure is termed a reflex, because, like all reflexes, it is involuntary and rapid.

When the heart contracts, the carotid arteries and aorta are stretched. This stretching opens ion channels in the baroreceptors, depolarizing them and causing them to send action potentials to the brain. If the blood pressure increases above the normal range, the ion channels are opened more frequently, and consequently the baroreceptors send action potentials at a higher frequency. Nerves carry the signals to the brain, which interprets the signals to mean that blood pressure is higher than normal. In the terminology used in Chapter 32, baroreceptors are the sensors for blood pressure, and the brainstem is the integrator. The set point for blood pressure corresponds to a baseline frequency of action potentials arriving in the brainstem when blood pressure is normal. A greater frequency of action potentials than the set point is interpreted as a higher blood pressure than normal.

The effectors in this control system are nerves exiting the brain and spinal cord via the autonomic nervous system (refer back to Figure 33.19). The response to increased activation of baroreceptors is a decrease in the amount of norepinephrine released from neurons of the sympathetic nervous system onto cardiac sinoatrial node (pacemaker) cells and vascular smooth muscle cells (see Figure 41.3). Pacemaker cells control the frequency of heart contractions. In addition, there is an increase in the release of acetylcholine from neurons of the parasympathetic branch of the autonomic nervous system onto the pacemaker cells. Together, these changes result in a decreased heart rate (HR) and decreased force of contraction of the heart. The latter effect decreases the stroke volume (SV) of the heart, or the amount of blood ejected with each beat. Recall from Chapter 36 that the cardiac output (CO), or the volume of blood pumped from the heart per minute, is the product of HR and SV.

$$\underset{(\text{mL/min})}{\text{CO}} = \underset{(\text{beats/min})}{\text{HR}} \times \underset{(\text{mL/beat})}{\text{SV}}$$

Therefore, the baroreceptor reflex in this example results in a decreased CO. The relationship between CO and blood pressure (BP) was given in Chapter 36 and looks like this:

$$\text{BP} = \text{CO} \times \text{R}$$

where R is the total vascular resistance determined by the degree of vasoconstriction of blood vessels, notably the arterioles. Since CO is directly related to blood pressure, a decrease in CO decreases pressure. In addition, though, stimulation of the baroreceptors also results in vasodilation in many parts of the body, again due to decreased norepinephrine release from neurons of the sympathetic nervous system. The greater the radius of a blood vessel, the less resistance it imparts to the movement of blood (refer back to Figure 36.14). Consequently, the decreased CO and the vasodilation (decreased R) compensate for an increase in blood pressure.

By contrast, if blood pressure decreases below normal, as in hemorrhage, the walls of these arteries are stretched less than normal, and the baroreceptors send fewer action potentials to the brainstem, which interprets this as low blood pressure (see Figure 41.3b). The result is a sequence of events that are opposite to those just described. In this case, the sympathetic neurons are activated and the parasympathetic neurons are inhibited. Thus, more norepinephrine and less acetylcholine are secreted, thereby increasing cardiac output and causing vasconstriction (increased R). Together, these events increase blood pressure toward normal. When a mild hemorrhage occurs, this response is usually sufficient to restore a normal blood pressure. However, when a more serious hemorrhage happens, the baroreceptor reflex can prevent pressure from continuing to fall but cannot return it all the way back to normal. The baroreceptor reflex occurs within seconds and can be considered the first line of defense against hemorrhage-induced low blood pressure.

### The Nervous System Stimulates Redistribution of Blood to Vital Organs When Pressure is Low

As you have just learned, part of the baroreceptor reflex following a decrease in blood pressure involves vasoconstriction in blood vessels

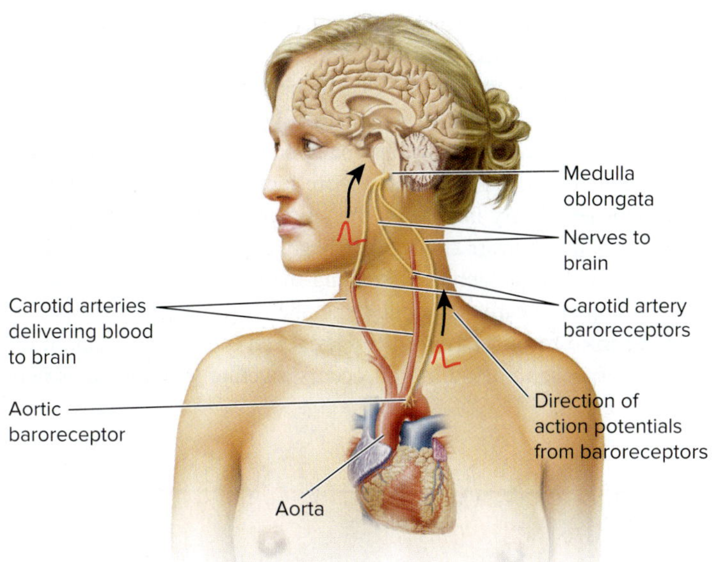


(a) Location of baroreceptors in a human

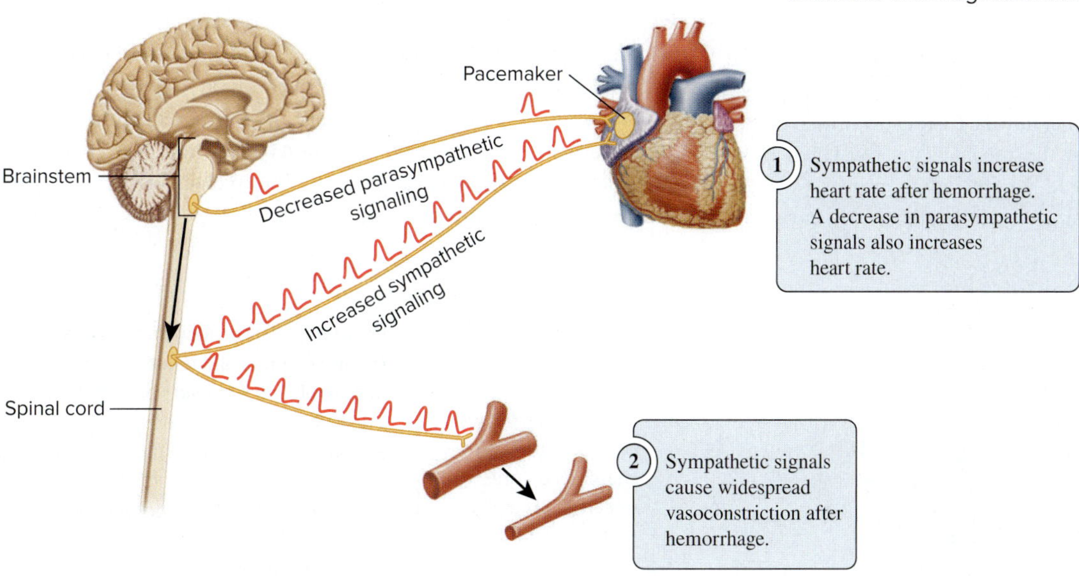


(b) The baroreceptor reflex after hemorrhage

**Figure 41.3** **Location of major baroreceptors in a human, and the baroreceptor reflex after a hemorrhage.** **(a)** Baroreceptors are neuron endings within the walls of major arteries such as the aorta and carotids. All vertebrates have baroreceptors, but the locations shown here are those of mammals. Changes in stretch of a blood vessel that contains a baroreceptor will change the frequency of action potentials generated in the baroreceptor. These signals are relayed to the brainstem, which elicits signals via the sympathetic and parasympathetic branches of the autonomic nervous system. The result is a change in heart rate and pumping strength as well as a change in peripheral vasoconstriction or vasodilation. **(b)** The baroreceptor reflex after hemorrhage. Decreased signals from the baroreceptors lead to increased signaling via nerves of the sympathetic nervous system exiting the spinal cord to the heart and blood vessels. There is also decreased signaling to the heart from nerves of the parasympathetic nervous system that exit from the brainstem. Note: Details of autonomic nervous system anatomy are omitted for simplicity.

**Bioconnections:** *In what other organs might there be stretch-sensitive receptors? Refer back to Figure 33.3 and Section 34.2 of Chapter 34 for help, and think about structures of the digestive and urinary systems (Chapter 37).*

in different parts of an animal's body. This occurs in part through the actions of the sympathetic nervous system on smooth muscle cells of arterioles. However, because vasoconstriction limits blood flow past the site of the constriction, it should not occur in parts of the body that are vital for survival, namely the heart and brain. How does the body limit vasoconstriction to some regions but not others? The answer is that the smooth muscle cells of arteries and arterioles express different types of norepinephrine receptors in different parts of the body. Binding to one type causes contraction of the muscles, resulting in vasoconstriction of the vessel. This occurs in the skin, skeletal muscles, and digestive system, among others. By contrast, when norepinephrine binds to the other type of norepinephrine receptor found in vessels of the heart and brain, it helps keep those vessels dilated to ensure sufficient blood flow to those regions (**Figure 41.4**). This redistribution of blood away from organs with less immediately vital functions to those with more important functions is an effective means of partially compensating for a decreased total blood volume.

In addition to their actions on the arterioles, the neurons of the sympathetic nervous system also have endings on the smooth muscles of many of the large veins, such as those in the legs. When stimulated by norepinephrine, these smooth muscles contract, and this helps squeeze blood up through the veins toward the heart. Recall from Chapter 36 that venous blood pressure is very low compared to arterial pressure. Compressing the veins in this way helps provide an important boost of blood from the venous reservoir.

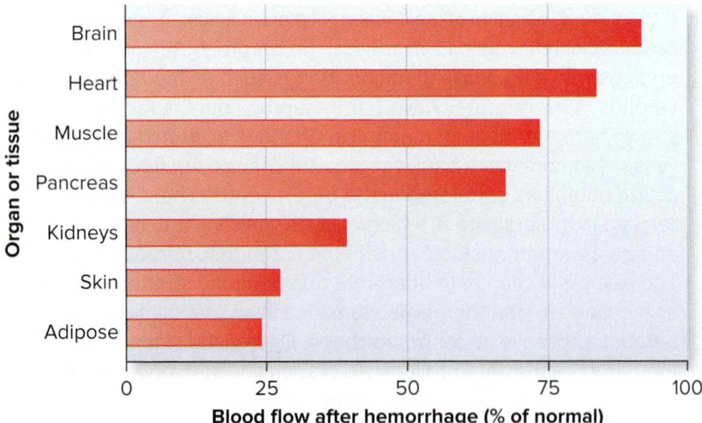


**Figure 41.4 Changes in the distribution of total blood volume after hemorrhage.** By redistributing blood away from areas that are not immediately important, the heart and brain can still receive sufficient blood despite hemorrhage. The idealized values shown here are based on research derived from numerous mammalian species.

There is an additional benefit from the increased venous blood returning to the heart. In the early 20th century, German physiologist Otto Frank and British physiologist Ernest Starling, working independently to investigate the mechanisms by which heart muscle cells contract, recognized an intriguing property of the muscle. Within limits, if the ventricles of the heart are expanded (stretched), they subsequently contract with greater force than normal. In other words, the more blood that enters the heart before a beat, the more the ventricles will be stretched, and the greater will be the force of the beat. This can be thought of as somewhat analogous to stretching a rubber band before releasing it. The more you stretch the elastic band, the more forcefully it snaps back to its original length. By increasing the amount of blood returning from the veins to the heart after hemorrhage, the contractile force of the heart chambers is increased, making them more effective in delivering a greater cardiac output. This phenomenon is now known as the **Frank-Starling Law of the Heart** and is very important in the early response to blood loss. It occurs together with the baroreceptor reflex.

## The Respiratory System Aids in Circulating Blood and Delivering Oxygen

In addition to the nervous system, the respiratory system also participates in the overall response to hemorrhage. When oxygen delivery to any region of an animal's body is decreased, the cells in that region depend more on fermentation to produce the ATP required for survival. A by-product of fermentation is lactic acid, which is released into the blood and decreases the blood pH. Recall from Chapter 36 that chemoreceptors are present in certain blood vessels of the circulatory system of animals. In mammals, the chemoreceptors are associated with the same vessels that contain the baroreceptors, but in this case they are sensitive not to pressure but to changes in the amount of oxygen, carbon dioxide, and acidity of the blood (Figure 41.5).

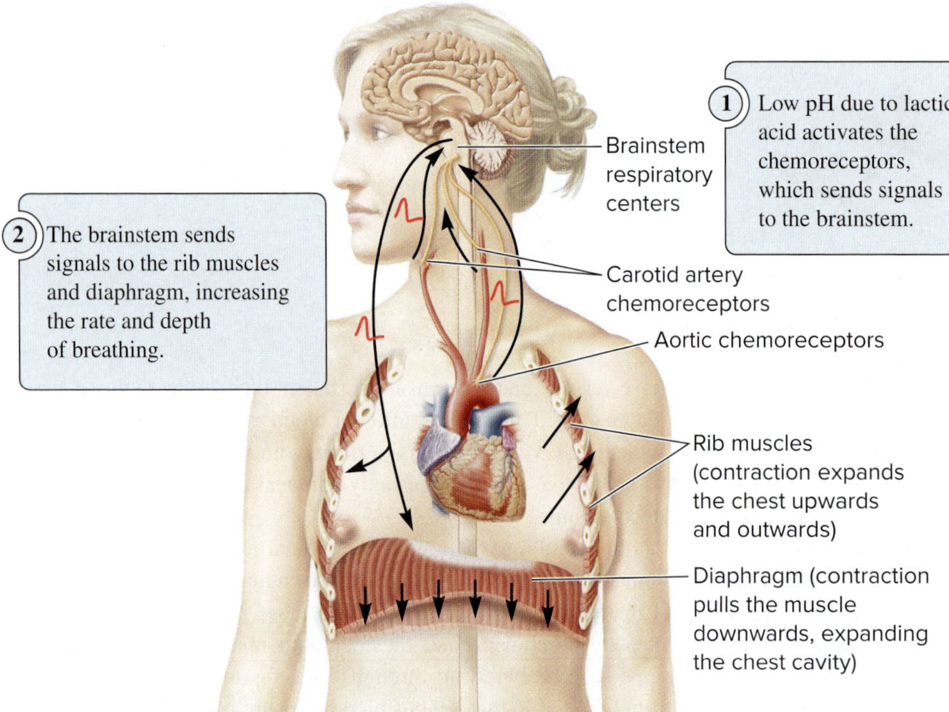


**Figure 41.5 Location of peripheral chemoreceptors and reflex activation of breathing in a human.** Nearby the baroreceptors are specialized cells sensitive to the amount of oxygen, carbon dioxide, and acid (pH) in the blood (not shown are additional chemoreceptors located within the brain). Activation of neurons associated with these cells results in signals that travel to the brainstem respiratory centers. Effector nerves leave the central nervous system and travel to the muscles of respiration, including the diaphragm and rib (intercostal) muscles, increasing the rate and depth of breathing.

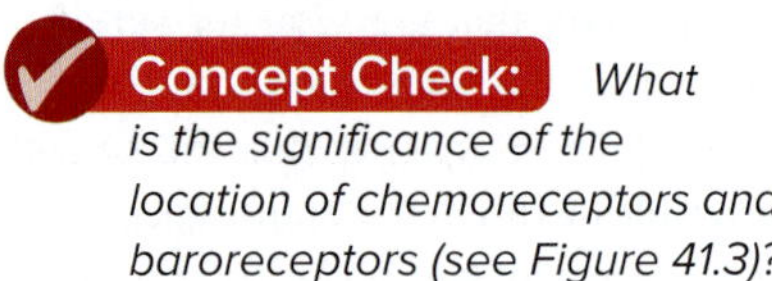

**Concept Check:** *What is the significance of the location of chemoreceptors and baroreceptors (see Figure 41.3)?*

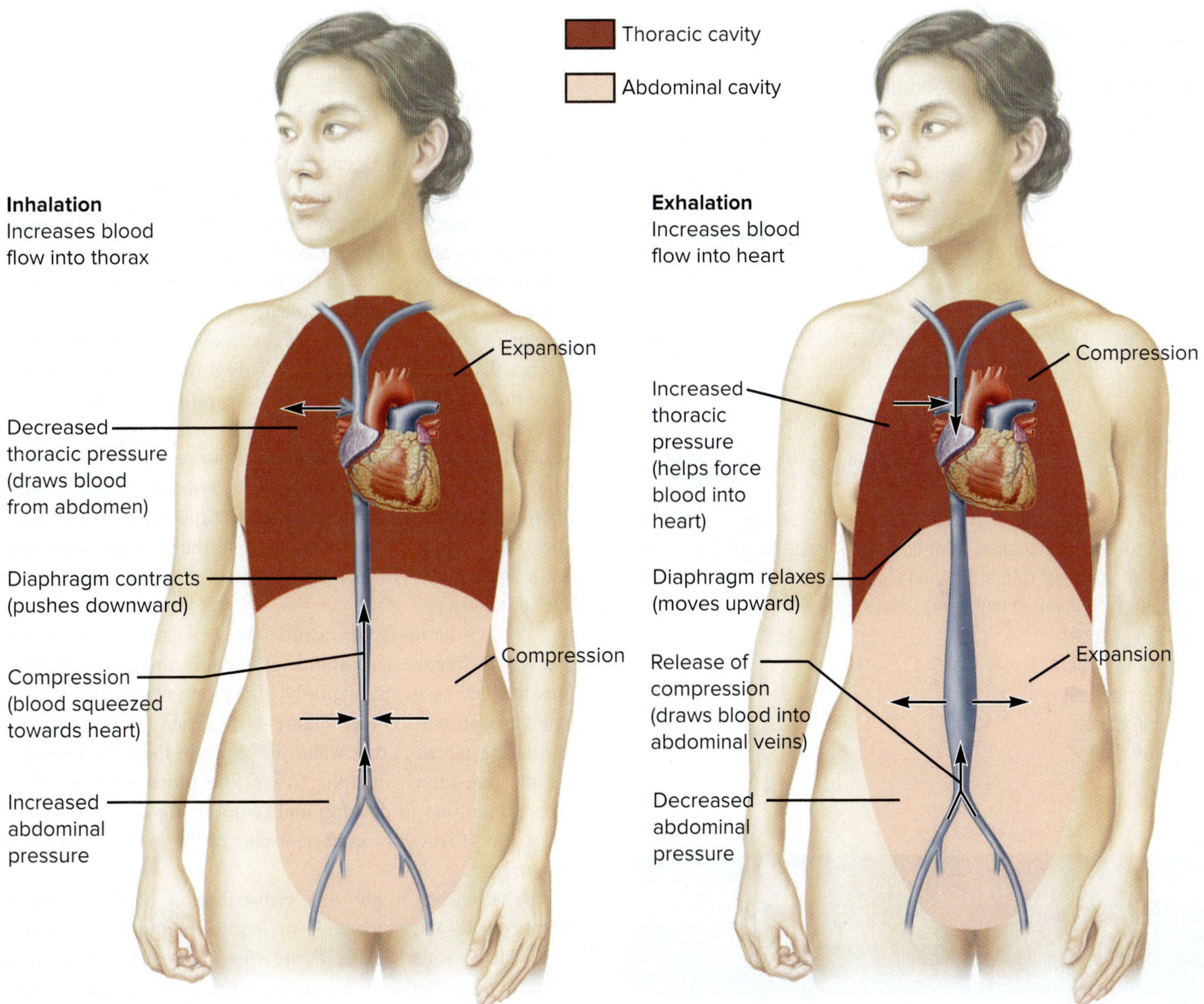


**Figure 41.6** **The thoraco-abdominal pump mechanism.** Blood is drawn toward the heart from the abdomen, through the diaphragm, and into the thoracic cavity by the actions of skeletal muscles. When an individual is inhaling, the diaphragm pushes down on the abdominal contents, raising the pressure there. Simultaneously, as the thorax expands, the pressure within it decreases. Because veins are readily collapsed or expanded, these pressure differences tend to squeeze blood from the abdomen into the thorax. Fast, deep breathing, as occurs when chemoreceptors are activated after a hemorrhage, facilitates venous return to the heart by this mechanism.

A buildup of lactic acid (decreased pH) stimulates the chemoreceptors, which then activate associated neurons that send signals to the brainstem centers that control breathing (refer back to Figure 36.24). Breaths become deeper and more rapid, thereby maximizing the saturation of hemoglobin with oxygen. In addition, the mechanical action of the chest and diaphragm creates a siphon-like effect that draws blood up from veins below the diaphragm, allowing greater venous return to the heart. This **thoraco-abdominal pump**—or respiratory pump, as it is sometimes called—is another important way in which the Frank-Starling effect is generated (**Figure 41.6**).

The acidic pH of the blood has another benefit. As you learned in Chapter 36 (refer back to Figure 36.23), a decreased pH results in a decreased affinity of hemoglobin for oxygen in (recall that this is called the Bohr effect). The lactic acid generated by poorly oxygenated tissues causes hemoglobin to unload a greater amount of oxygen, which is an important benefit that helps offset the decreased oxygen delivery to the tissues resulting from decreased blood flow.

## 41.2 Reviewing the Concepts

- Baroreceptors are pressure-sensitive neuron endings associated with certain large blood vessels. In the baroreceptor reflex, a change in blood pressure is rapidly compensated for by a change in cardiac output and in peripheral resistance in many blood vessels (Figure 41.3).
- The nervous system mediates a redistribution of blood after hemorrhage such that the vital organs continue to receive as much blood as possible (Figure 41.4).
- The respiratory system contributes to venous return via the chemoreceptors and the thoraco-abdominal pump (Figures 41.5, 41.6).

## 41.2 Testing Your Knowledge

1. After a moderate hemorrhage, an animal's cardiac output is found to have decreased from its normal value of 4.6 L/min to 3.7 L/min. Its heart rate at that time is 99 beats per minute. What is its stroke volume at that time?
   a. 366 mL/beat
   b. 0.0366 L/beat
   c. 89.1 mL/beat
   d. 3.66 L/beat
   e. 3.66 beats/L

# 41.3 The Secondary Phase of the Homeostatic Response to Hemorrhage

### Learning Outcomes

1. Describe how and why water moves between body fluid compartments after a hemorrhage.
2. Explain how the endocrine and urinary systems help restore blood volume.
3. Describe the mechanism by which hematocrit first decreases and then is returned to normal and how this depends on the endocrine and muscular-skeletal systems.
4. List some changes in the functions of the reproductive, integumentary, and digestive systems that compensate for hemorrhage.

After the baroreceptor reflex and venous return contribute to stabilizing the blood pressure, the next phase of the compensatory response to this homeostatic challenge is to restore fluid volume and erythrocytes. In this section, we will examine how this process largely requires the concerted actions of the endocrine and urinary systems, and also includes a special feature of the way in which fluid compartments in an animal's body can be redistributed without any input from other organ systems.

## Water Moves from Interstitial Fluid to the Plasma After a Hemorrhage

Recall from Figure 32.14 that water is found in three compartments of an animal's body: within cells (intracellular fluid), around cells (interstitial fluid), and in the blood (plasma). Water will move between compartments if there is an osmotic gradient, or if there is a difference in fluid pressure. Ernest Starling determined that water moves across a capillary between the plasma and interstitial fluid compartments due to the difference between these two forces (**Figure 41.7**). In mammals, plasma and interstitial fluid are very similar in composition, except that much more protein is found in plasma than in interstitial fluid. This difference creates an osmotic force that tends to draw water into a capillary. By contrast, the fluid (hydrostatic) pressure in a capillary is much higher than in the surrounding interstitial fluid, and this pressure difference tends to force fluid out of the capillary. The net effect of these forces, today called **Starling forces,** is that fluid leaves at the beginning of a capillary due to hydrostatic pressure and then, as the pressure decreases along the length of the vessel, re-enters at the other end of the capillary due to osmosis.

After a hemorrhage occurs, the delicate balance between the pressure difference and osmotic gradient is altered because the blood pressure decreases, particularly in regions where there is considerable vasoconstriction (see Figure 41.7). The net effect of the change in Starling forces is that water moves from the interstitial fluid into the plasma, thereby helping to restore blood volume. This effect takes some time to develop and is not as rapid as the baroreceptor reflex, but it is a very important part of the longer-term response to hemorrhage.

What effect might this redistribution of fluid have on the **hematocrit**—the percentage of blood volume that is occupied by erythrocytes (refer back to Figure 36.4)? For a few hours after a hemorrhage, the hematocrit does not change, because whole blood is lost from the injured vessel. There are fewer total erythrocytes, but also less total fluid in proportion, so the hematocrit is initially unchanged. However, it will eventually decrease below normal as the blood becomes diluted by the movement of fluid into capillaries. As you have learned throughout this unit, homeostatic responses to one challenge often produce new challenges that must, in turn, be addressed at a later time. In this case, helping restore blood volume so that pressure can be maintained is of more immediate importance than is the development of a decreased hemocrit. As we will see shortly, hematocrit is eventually restored during later stages of the secondary phase.

## The Endocrine and Urinary Systems Function Together to Retain Fluid in the Body

An animal that has lost significant blood would seem to benefit from decreasing the amount of fluid that is normally lost via the urine. On the other hand, a mammal's kidneys serve the important function of removing soluble wastes from the blood, which requires water for excretion. This is another example of how the response to a homeostatic challenge may result in new challenges that must be met at a later time.

The sympathetic system decreases the amount of plasma that is filtered into the kidney nephrons by mediating the vasoconstriction of

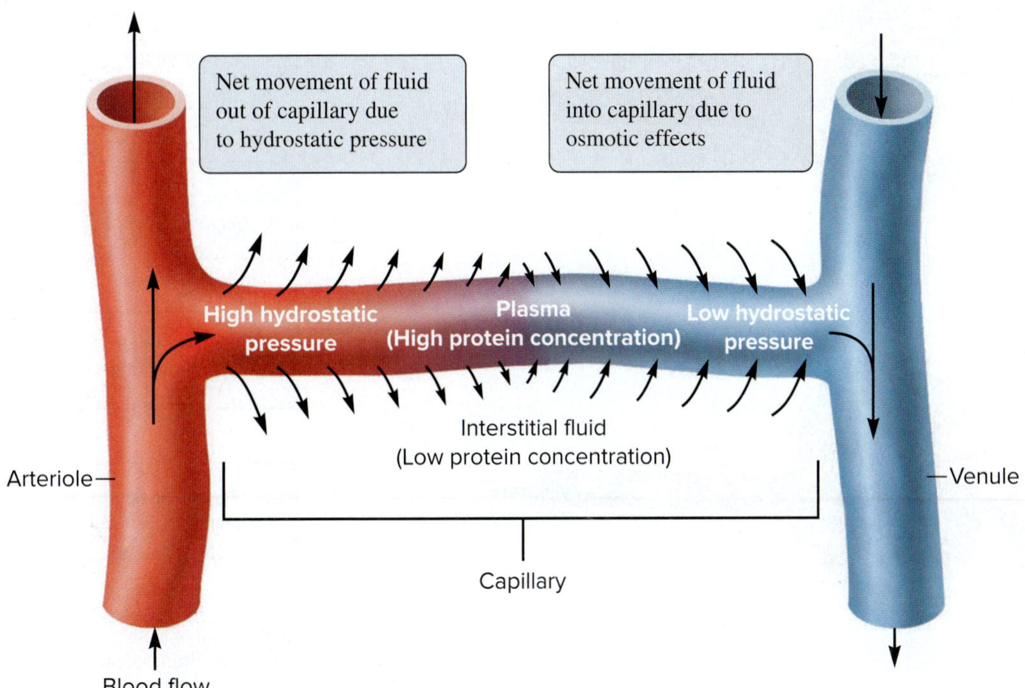


**Figure 41.7 Starling's forces in a capillary in the absence of a hemorrhage.** Capillaries are the smallest blood vessels and allow the movement of water and solutes between the plasma and interstitial fluid. Near the arteriole, the pressure of blood moving through the capillary forces plasma out of the capillary, but near the venule, the osmotic effect of proteins in the blood draws fluid into the capillary. When blood volume is lowered, as in a hemorrhage, a change in the balance between these two forces will change the net flow of fluid. Hemorrhage decreases blood pressure resulting in a greater net movement of fluid into the capillary from the interstitial fluid to the plasma. Not shown is that some fluid that exits a capillary is not recaptured but, instead, enters a nearby lymphatic vessel.

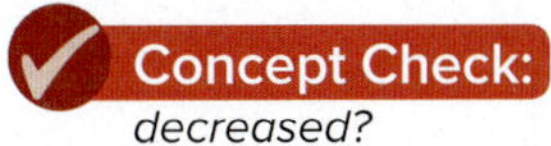


*What could happen to fluid moving across a capillary if the protein concentration of plasma was decreased?*

the renal afferent arterioles (Figure 41.8a and refer back to Figure 37.20). With less blood entering the glomerular capillaries, less fluid is filtered into the nephron tubules. This conserves fluid in the blood but also limits the ability of the kidneys to remove soluble wastes. The accumulation of wastes in the blood needs to be corrected eventually, but in the short term, the net effect of helping to minimize water loss in the urine may be lifesaving. In addition, as the limited volume of filtrate that does get formed moves through the nephron tubules, a higher amount of $Na^+$ and especially water is reabsorbed from the filtrate and returned to the blood, making the urine more concentrated and retaining more fluid in the body (Figure 41.8b).

These effects on urine and blood volume are mediated by hormones produced by the endocrine system. Aldosterone, a steroid hormone produced by the adrenal glands, is secreted whenever blood volume is decreased. It acts on distal regions of nephrons to increase $Na^+$ reabsorption, which creates an osmotic gradient that draws water from the filtrate in the tubules into the blood (refer back to Figure 37.22). The production of aldosterone is stimulated by another hormone, called angiotensin II (AII), which is formed in the blood whenever blood pressure is decreased. The signal for AII formation, in turn, is an enzyme called renin, which is released into the blood by kidney cells whenever blood pressure is low and the sympathetic nervous system is activated (Figure 41.8c). Thus, a hemorrhage elicits a sequence of events that leads to the production of aldosterone, thereby helping to minimize the volume of urine and increase the volume of the blood.

In addition to stimulating aldosterone production, AII is also a potent vasoconstrictor of arterioles and directly helps to increase blood pressure by that mechanism. In some animals, AII is also believed to stimulate thirst and thus encourages animals to seek out water to drink. This is beneficial because the ingested water will be absorbed into the blood.

Another hormone that helps retain water in the body is antidiuretic hormone (ADH) (refer back to Figure 37.23). ADH stimulates cells in the tubules of the mammalian kidneys to reabsorb water (but not ions) into the blood, concentrating the urine. The secretion of ADH is triggered by signals from the baroreceptors that activate the hypothalamic neurons that synthesize ADH. Like AII, ADH has direct actions on vascular smooth muscle cells, causing them to contract and thereby promoting vasoconstriction. Thus, AII and ADH both increase blood pressure directly by vasoconstriction and indirectly (via aldosterone, in the case of AII) by helping to retain water in the body.

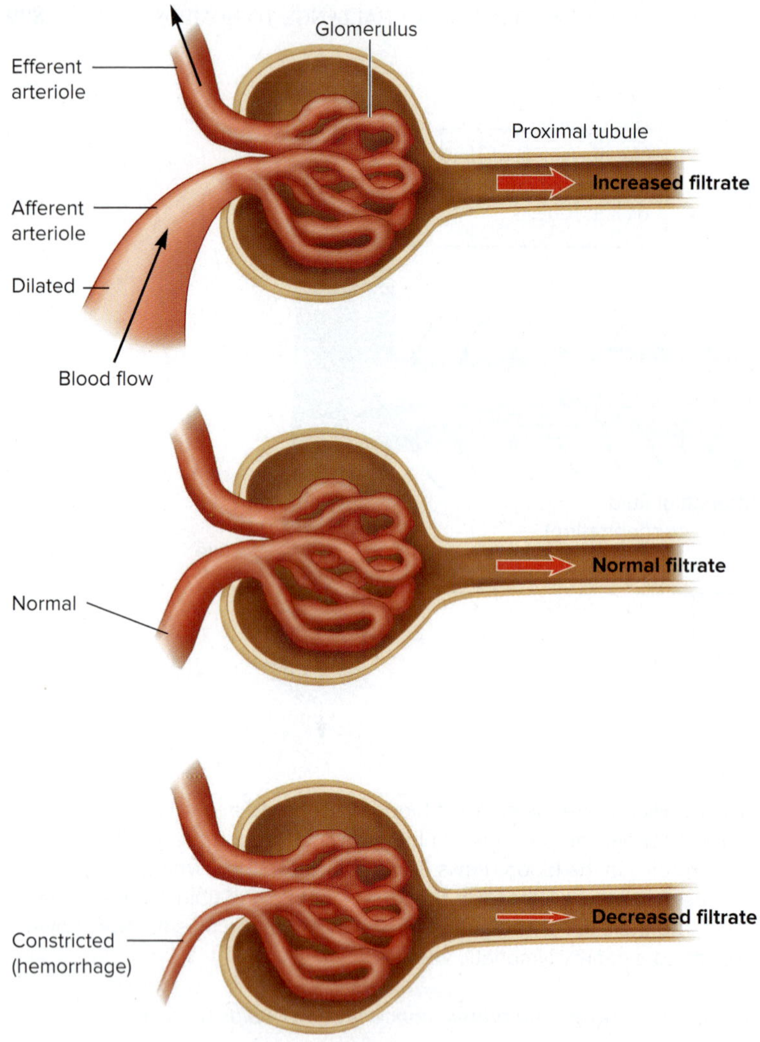


**(a) Effect of vasodilation and vasoconstriction of afferent arteriole on the amount of filtrate formed in a nephron**

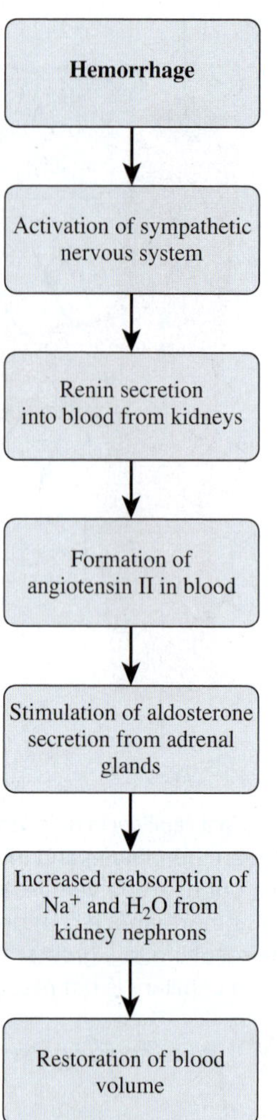


**(c) Stimulation of aldosterone secretion and its effect on blood volume**

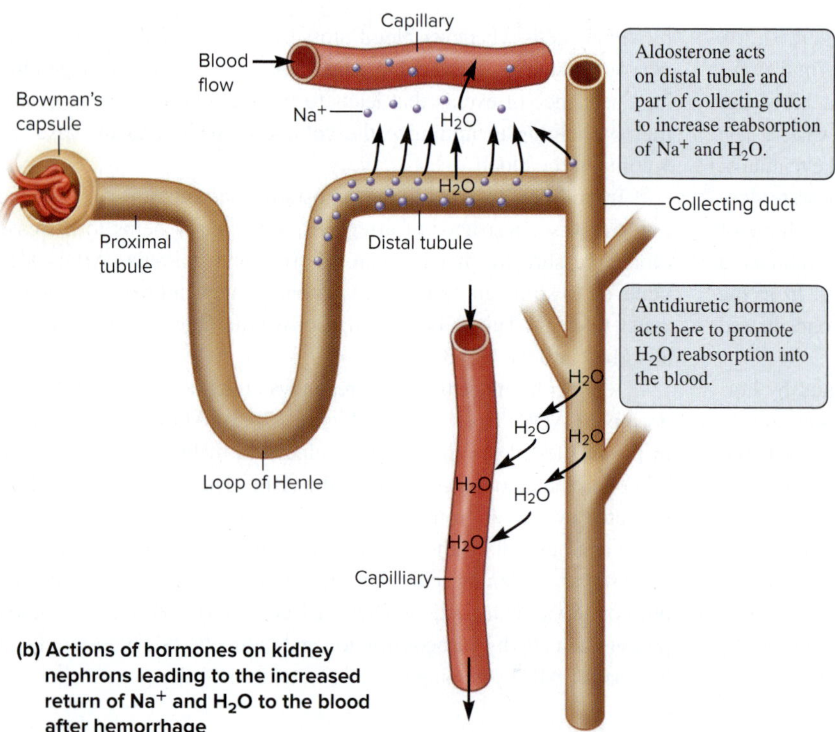


**(b) Actions of hormones on kidney nephrons leading to the increased return of $Na^+$ and $H_2O$ to the blood after hemorrhage**

**Figure 41.8 A nephron of the mammalian kidney, with its associated afferent and efferent arterioles.** **(a)** Schematic illustration of the effect of changing the diameter of afferent arterioles on the filtration of fluid out of a capillary. The afferent arteriole brings blood into a glomerulus associated with a kidney nephron. The relative amount of filtrate (red arrows) in the proximal tubule of the nephron is indicated by the thickness of the arrow. Vasoconstriction of an afferent arteriole will decrease the amount of fluid that gets filtered into the tubule, thereby preserving fluid in the blood. **(b)** Additional changes along the distal tubule of the nephron and collecting ducts promote greater reabsorption of $Na^+$ and water into the blood. The result is a decreased volume of urine that is more concentrated than normal, as well as a return of additional fluid to the circulation. **(c)** The production of the hormones angiotensin II and aldosterone, along with some major actions. Intermediate steps in AII synthesis are not shown for simplicity.

## Hematocrit Changes in a Biphasic Manner After Hemorrhage

Recall that initially hematocrit does not change when blood is lost following a hemorrhage. The total blood volume decreases, but the percentage of blood that is comprised of erythrocytes remains the same. As time goes on, the blood becomes more dilute (as just described), and the hematocrit decreases (**Figure 41.9**). This must eventually be corrected so that oxygen transport in the blood can be returned to normal. Once again, the endocrine system has an important role in this response.

Erythrocytes mature in the marrow of certain bones of the mammalian skeleton. Stem cells in the bones follow a path of differentiation characterized by the loss of the nucleus and other organelles. This differentiation is completed in the blood once the cells leave the bone marrow. The hormone erythropoietin (EPO) controls the rate of this differentiation and increases the survival of the developing cells in the marrow. EPO is made primarily by the kidneys. Whenever the total amount of hemoglobin in the blood is decreased below normal for any reason (such as a decreased hematocrit), the EPO-producing cells of the kidneys are activated to produce and secrete EPO into the blood. The effect of the hormone on the number of circulating erythrocytes is not apparent for at least several hours, and it does not fully restore hematocrit for up to several days or even weeks, depending on the magnitude of the hemorrhage.

Interestingly, some mammals such as horses, seals, and pigs contain a large reservoir of mature erythrocytes in their spleen, an organ of the immune system involved in the production of leukocytes (refer back to Figure 40.4). In response to activation by the sympathetic nervous system, smooth muscle cells in the spleen contract, causing the spleen to eject its reservoir of erythrocytes into the circulation. This happens much more quickly than the slow EPO-induced restoration of hematocrit and provides a clear survival advantage after hemorrhage. Under normal conditions, the spleen response gives such animals a boost in the oxygen-carrying capacity of the blood during periods of increased activity or, in the case of seals, when diving under water.

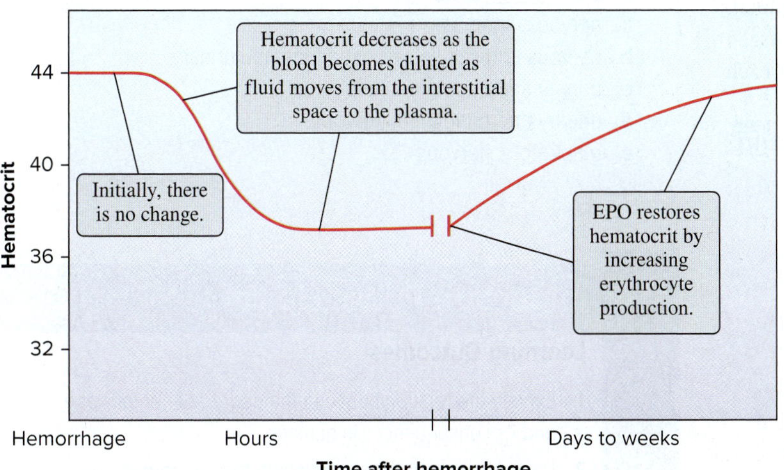


**Figure 41.9 The effect of fluid shifts on hematocrit of the blood following hemorrhage.** Note that hematocrit decreases as fluid moves from the interstitial fluid into the blood vessel. This is later compensated for by the production of more erythrocytes in the bone marrow and their release into the blood under the control of the hormone erythropoietin.

## Other Organ Systems Are Affected by Hemorrhage and May Help in the Compensatory Response

As we have seen, the circulatory, nervous, respiratory, endocrine, and urinary systems have a variety of important functions that contribute to an animal's ability to survive a serious hemorrhage. In addition, all other organ systems are affected to some degree by hemorrhage, and they all are involved at least indirectly in the overall compensatory response that restores blood volume and blood pressure.

You have already learned how the respiratory system contributes to the return of blood from the veins to the heart. The muscular-skeletal system may also assist the flow of blood from the limbs to the heart. Recall that one-way valves within limb veins help move blood against gravity toward the heart (refer back to Figure 36.13). Each time the skeletal muscles of the limbs contract, the veins coursing through those muscles are squeezed and blood is pushed through the valves. The anxiety that may be triggered by the injury and hemorrhage may encourage such muscular activity, and this has the benefit of contributing to the Frank-Starling effect.

The integumentary system also contributes indirectly as the arterioles that normally deliver blood to the skin constrict, redirecting blood to vital organs. This also happens throughout much of the alimentary canal of the digestive system. In addition, the smooth muscles of the canal relax, due to inhibitory actions of the sympathetic nervous system. This happens even if an animal has just eaten a meal prior to the hemorrhage, and it results in a considerable energy savings. In such emergency situations, energy supplies tend to be spared for the heart, the brain, and a few other organs, such as the placenta in a pregnant female.

In major homeostatic challenges such as hemorrhage, energy is also redirected away from the reproductive system. At such times, it is more important that the individual survives so that it may reproduce another day.

Finally, even the immune and lymphatic systems respond to hemorrhage. We have already learned that a major immune system organ, the spleen, may contribute to the maintenance of a normal hematocrit in certain mammals. In addition, if the injury has resulted from an external wound, part of the response to hemorrhage is to prevent infection. The cells and secretions of the innate immune response trigger a local inflammatory reaction (refer back to Figure 40.2) at the site of the wound, and the acquired immune response responds to any specific pathogens that have entered the circulation or local tissues.

**Table 41.1** summarizes some of the major events of the homeostatic response to hemorrhage in a mammal. Not all homeostatic challenges engage such widespread responses, but typically the activities of more than one organ system are affected directly or indirectly. There are many common examples of this fundamental feature of homeostasis. For example, when you are ill, you may have a fever caused by secretions of the cells of the immune system. Fever induces sweating, which in turn decreases the amount of

**Table 41.1 Summary of Key Homeostatic Responses Following a Hemorrhage**

| Variable | Rapid responses | Secondary responses |
|---|---|---|
| Rate of blood flow to organs | Decreased to most organs, but less so to heart and brain | Restored to normal as fluid shifts into the plasma and the kidneys reabsorb water |
| Heart rate (HR) | Increased | Returns to normal once pressure is capable of sustaining itself |
| Stroke volume (SV) | Decreased due to less venous return, but the decrease is minimized by increased force of contraction of the heart | Increased toward normal due to Frank-Starling effect and sympathetic nervous system (norepinephrine) |
| Cardiac output (HR × SV) | Maintained near normal by baroreceptor reflex if hemorrhage is not severe, but decreased in severe hemorhage | Slowly returned to normal or even increased above normal as venous return increases |
| Hematocrit | Not usually a rapid response except in animals with contractile spleens that release erythrocytes into the bloodstream | Decreased due to fluid shifts into blood, but eventually restored by erythropoietin |
| Breathing rate and depth | Increased (improves venous return) | Returns to normal once pressure is normal and blood pH is restored |
| Hemoglobin (Hb) saturation | Increased toward 100% by hyperventilation | Restored to the normal range |
| Unloading of oxygen from Hb | Increased by Bohr effect due to lactic acid | Restored to normal once blood flow to organs returns to normal and pH is normalized |
| Urine production | Decreased due to vasoconstriction of afferent arterioles | Further decreased due to effects of ADH and aldosterone |

fluid and alters the concentrations of $Na^+$ and other ions in the blood. The urinary and endocrine systems help to restore fluid and ion balance. If you happen to miss a meal or two, the endocrine and nervous systems together will help maintain glucose homeostasis in the blood. If you move to a high altitude or go mountain-climbing, the respiratory system will help maintain the amount of oxygen in your blood by stimulating your breathing rate. That will, however, decrease the amount of carbon dioxide in the blood, which in turn will increase the pH of the blood. The urinary system will then help restore pH to normal by excreting bicarbonate ions in the urine and reabsorbing more hydrogen ions into the blood. These and many other examples demonstrate once again that organ systems do not function in isolation but are integrated with each other to ensure homeostasis of all of an animal's physiological functions.

BIO TIPS ONLINE

## 41.3 Reviewing the Concepts

- After hemorrhage, fluid flows from the interstitial space to the plasma, which helps restore blood volume (Figure 41.7).
- Antidiuretic hormone and aldosterone (stimulated by angiotensin II) act on kidney tubules to help retain water and ions in the blood, thereby concentrating and decreasing the amount of urine produced and increasing blood volume (Figure 41.8).
- Hematocrit decreases as fluid enters the plasma due to Starling forces, then is slowly restored to normal by the action of erythropoietin on immature erythrocytes in bone marrow (Figure 41.9).
- All organ systems are affected by hemorrhage to some degree and directly or indirectly contribute to the restoration of homeostasis.

## 41.3 Testing Your Knowledge

1. The initial response to hemorrhage is mediated largely by the _______ system(s); the later responses are mediated largely by the _______ system(s).
   a. nervous, immune
   b. nervous and circulatory, endocrine and urinary
   c. urinary and endocrine, respiratory
   d. urinary and endocrine, nervous
   e. respiratory, nervous

## 41.4 Impact on Public Health

**Learning Outcomes**

1. Explain the relation between the degree of hemorrhage and its consequences in humans.
2. Describe procedures used to help restore normal circulatory function in humans after a large hemorrhage.

Let's now return to the case of the girl described at the beginning of the chapter. Her initial symptoms were light-headedness, agitation without any sign of mental confusion, rapid breathing rate, a slightly lower blood pressure than normal for her, and a very rapid heart rate. All of these suggested significant internal bleeding due to the rupture of one or more blood vessels near one of her kidneys. Once in the ambulance, therefore, she was laid on her back with her feet slightly elevated. Gravity helped to return blood from her leg veins to the right atrium of her heart. This was important because the amount of blood ejected with each heartbeat is roughly proportional to the amount of blood that fills the heart (recall the Frank-Starling Law of the Heart). She was anxious but still conscious and alert, suggesting that only a moderate hemorrhage had occurred. A decision was made to not infuse whole blood but, instead, to simply supply fluids. A cannula was placed into a vein in her hand and attached to a solution of ions, which was allowed to rapidly enter her circulation at an initial rate of about 1 liter/10 minutes. The fluid was approximately iso-osmotic compared to her blood. Recall from Chapter 32 (and refer back to Figure 5.10) that cells can be deformed and even killed when exposed to extracellular fluid with an osmolarity that is very different from normal.

After sufficient fluids were infused, the girl's blood pressure rose to 114/71 mmHg, which is within the normal range for her. Her heart rate was still somewhat elevated at 84 bpm, compared to her usual resting rate of about 72 bpm; this, however, could mainly be attributed to her continued anxiety, not to the baroreceptor reflex. Her breathing rate by this time was a steady 15 breaths per minute, normal for a girl her age, and her hematocrit was 34 (normal is about 38–46 for females). A blood sample was drawn to test for indicators of kidney damage. Later analyses of the sample, along with a CT scan of her thorax and abdomen, revealed that the kidneys and other major organs were functioning properly and undamaged, and thus it appeared that fluid replacement was sufficient in this case. Her hematocrit was low for some time, but it eventually recovered fully due to the actions of erythropoietin.

Had the hemorrhage been more severe—for example, resulting in a loss of greater than 20% of her total blood volume—the outcome might have been very different. Under such conditions, shock ensues, with a variety of symptoms, including low blood pressure, rapid heart rate, and in extreme cases a loss of consciousness and multiple organ failure. During shock, the baroreceptor reflex and endocrine events described earlier may be insufficient to prevent pressure from continuing to fall. In such situations, whole blood is sometimes infused, because the total amount of hemoglobin in the person's blood may be dangerously low. In addition, hormones such as epinephrine may be infused to facilitate heart action and vasoconstriction. Even with these treatments, however, a person who has progressed too far along in shock may not recover. One reason is that as organs begin to fail due to the profound lack of oxygen and nutrient delivery, their ability to contribute to the mechanisms compensating for the blood loss decreases. Another reason is a positive feedback cycle that can develop. As cells die in large numbers, they often release inflammatory substances that result in vasodilation, a normal response to inflammation (refer back to Figure 40.2). This can offset the vasoconstriction mediated by the sympathetic nervous system and thereby decrease blood pressure even more, which further decreases the perfusion of organs with blood, causing more cell death, and so on (**Figure 41.10**).

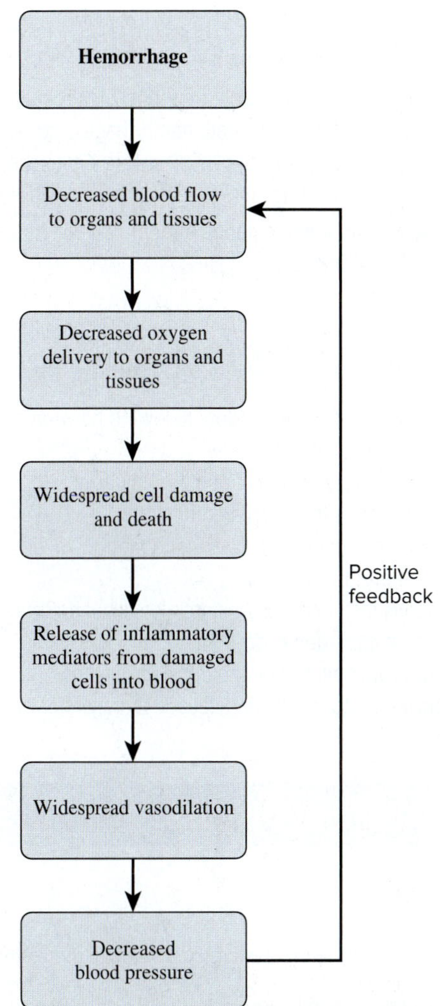


**Figure 41.10 A positive feedback cycle that worsens shock.** As blood flow to tissues is decreased, cells begin to die and release their contents, which may cause widespread inflammation. Inflammation, in turn, tends to cause small blood vessels to dilate, which decreases resistance and therefore lowers blood pressure even further.

**Bioconnections:** *In what other context have you learned about positive feedback? Refer back to Figure 39.9 for help.*

Roughly every 3 minutes, a person dies in the U.S. from traumatic injury, such as those incurred in car accidents or from operating heavy machinery. Hemorrhage is the leading cause of death due to traumatic injuries in the U.S. Each year, several hundred thousand individuals in the U.S. also suffer hemorrhaging due to gastrointestinal diseases or abnormally weakened blood vessels (aneurysms). Hemorrhage is also a very serious concern in battlefield injuries. For these reasons, investigators continue to research improved means of preventing or reversing shock. The girl in this case was fortunate that she received prompt treatment, and she was able to resume her athletic pursuits after a recovery period of nearly 4 weeks.

## 41.4 Reviewing the Concepts

- The treatment for hemorrhage depends on the magnitude of blood loss. For relatively moderate hemorrhages, iso-osmotic fluid replacement is usually sufficient. For severe hemorrhage, whole blood and vasoactive drugs may be administered (Figure 41.40).
- Hemorrhage is the leading cause of death due to traumatic injuries in the United States.

## 41.4 Testing Your Knowledge

1. Why was the girl's hematocrit low following her treatment at the hospital?
   a. She was still very anxious about her injury, and the anxiety caused changes in her spleen function.
   b. The hemorrhage depleted her bone marrow of its supply of immature erythrocytes.
   c. As blood is lost, erythrocytes are also lost, and therefore hematocrit must decrease.
   d. Her blood was diluted by the infusion of fluids.
   e. Her baroreceptor reflex had caused many of her erythrocytes to die.

# Assess and Discuss

## Test Yourself

1. How does the respiratory system contribute to the restoration of blood pressure?
   a. by increasing the saturation of hemoglobin with oxygen
   b. by helping to increase the return of blood from veins to the heart
   c. by increasing the heart rate
   d. by activating the baroreceptor reflex
   e. by the Bohr effect
2. Following hemorrhage, blood flow decreases to the _______ but is maintained as close to normal as possible in the _______.
   a. brain and heart, digestive system
   b. skin, digestive system
   c. kidneys and heart, brain and skin
   d. skin and digestive system, brain and heart
   e. digestive system and liver, skin and kidneys
3. Which of the following would increase hematocrit after hemorrhage in a person?
   a. an intravenous infusion of erythropoietin
   b. Starling forces
   c. intravenous infusion of iso-osmotic NaCl solution
   d. intravenous infusion of a hyperosmotic NaCl solution
   e. both a and b
4. The frequency of action potentials sent from baroreceptors to the brain _______ when blood volume is decreased, resulting in _______ heart rate.
   a. increases, an increased
   b. increases, a decreased
   c. decreases, an increased
   d. decreases, a decreased
5. An animal's normal stroke volume is 9 mL/beat and its normal heart rate is 125 beats/min. Immediately after a hemorrhage, its heart rate increases to 161 beats/min and its stroke volume does not change. What is its new cardiac output?
   a. 1.45 L/min
   b. 0.145 L/min
   c. 17.9 mL/min
   d. 17.9 L/min
   e. 0.055 L/min
6. In the animal in question 5, what could explain why the stroke volume remained normal?
   a. Stroke volume is not usually affected by hemorrhage.
   b. The heart muscle cannot contract more forcefully than it normally does at rest.
   c. The hemorrhage may have been so severe that venous return to the heart cannot be significantly increased.
   d. It is a compensatory response to prevent blood pressure from increasing too much.
   e. Stroke volume increases only when the heart rate does not change from normal.
7. Antidiuretic hormone
   a. causes vasodilation of blood vessels.
   b. stimulates aldosterone production by the adrenal glands.
   c. stimulates reabsorption of both $Na^+$ and water from the kidneys.
   d. is stimulated by AII.
   e. increases water but not ion reabsorption in the kidneys.
8. Which of the following act(s) as a vasoconstrictor?
   a. norepinephrine
   b. acetylcholine
   c. AII
   d. aldosterone
   e. both a and c
9. Imagine that the baroreceptor reflex was *not* functional in an animal that had received an injury causing a severe hemorrhage. What would happen?
   a. Heart rate would increase significantly, but the strength of each heartbeat would not.
   b. The initial phase of recovery would be normal, but the delayed responses would be absent.
   c. The animal would probably develop shock very quickly as its pressure rapidly fell.
   d. Heart rate would decrease immediately after the hemorrhage but then slowly return to normal.
   e. There would be little significant effect on the overall compensatory response to hemorrhage.
10. A day after a moderate hemorrhage, what would be *true*?
    a. An animal's blood pressure would be normal, but its stamina (ability to be active for long periods) would be decreased.
    b. The concentration of erythropoietin in its blood would be increasing.
    c. Its respiration rate would be increased above normal.
    d. The flow of blood to its vital organs would be decreased.
    e. Both a and b would be true.

## Conceptual Questions

1. Try to imagine the integrated responses of an animal's body to other major homeostatic challenges. Don't limit yourself to mammals. For example, what are some of the challenges a bird might encounter while flying at high altitudes over a mountain range in a cold climate?
2. Some people get momentarily light-headed after donating blood and getting up from the chair or cot on which they were reclining. Why? Despite the light-headedness, few people actually faint after this procedure. What prevents them from fainting?
3. **PRINCIPLES** A principle of biology is that living organisms use energy. In what ways does homeostasis require energy, and how might that be applied to the integrated response to hemorrhage?

## Collaborative Questions

1. Identify some secondary problems that arise as a result of the integrated response to hemorrhage in a mammal.
2. Describe the nature of the baroreceptor reflex that occurs if blood pressure is *increased*.

## Online Resource

# Unit VIII
# ECOLOGY

**Ecology** is the study of interactions among organisms and between organisms and their environments. Interactions among organisms and the living environment are called **biotic** interactions, and those between organisms and the nonliving environment are termed **abiotic** interactions. These interactions, in turn, govern the numbers of species in an area and their population densities. Ecology ranges in scale from the study of the behavior of individual organisms to the study of populations, communities, ecosystems, and biomes.

In this unit, we will explore each of the broad areas of behavioral ecology, population growth and species interactions, community and ecosystem ecology, and biomes and the physical environment. Toward the end of the unit, we will consider how knowledge of these ecological disciplines facilitates conservation of life on Earth, and examine the broader impacts of humans on ecological systems. Chapter 42 discusses animal behavior and how behavior contributes to the survival and reproductive success of organisms. Chapter 43 examines population growth and the constraints on growth provided by competitors and natural enemies. This includes the demographic tools needed to study population growth and simple mathematical models of growth. In Chapter 44, we consider the factors that influence the number of species in a community and examine different measures of diversity. Chapter 44 also addresses the flow of energy through the living and nonliving components of the environment. At the largest scales, variation in temperature and moisture create distinct large-scale habitats, called biomes, which we discuss in Chapter 45. In Chapter 46, we will consider how human populations have continued to grow and impact natural systems. These impacts include global warming and climate change, pollution, habitat destruction, overexploitation and invasive species. Finally, in Chapter 47 we turn our attention to the conservation of life on Earth and the various strategies biologists use to protect biodiversity.

Ecological studies have important implications in the real world, as will be amply illustrated by examples discussed throughout the unit. However, there is a distinction between ecology and **environmental science**—the application of ecology to real-world problems. To use an analogy, ecology is to environmental science as physics is to engineering. Both physics and ecology provide the theoretical framework on which to pursue more applied studies. Engineers rely on the principles of physics to build bridges. Similarly, environmental scientists rely on the principles of ecology to solve environmental problems.

***The following biology principles will be emphasized in this unit:***

- ***Living organisms use energy.*** *In Chapter 44, we discuss the flow of energy in ecosystems.*
- ***Living organisms interact with their environment.*** *Chapter 45 provides examples of the influence of temperature and water on the distribution and abundance of organisms at large scales called biomes.*
- ***Biology is an experimental science.*** *Throughout the unit, we will examine numerous examples of experiments that ecologists have used to investigate how ecological systems function.*
- ***Biology affects our society.*** *Humans rely on natural systems for survival, as described in Chapter 46, but at the same time natural systems are frequently degraded by the actions of humans. In Chapter 47, the last chapter of the unit, we discuss some of the conservation efforts currently under way to save and secure life on Earth.*

42 43 44 45 46 47

# 42 Behavioral Ecology: The Struggle to Find Food and Mates and to Pass On Genes

© Takafumi Minamimoto, Ph.D.

Monkey "working" in Barry Richmond's lab.

## Chapter Outline

In 2004, American neuroscientist Barry Richmond and colleagues trained four monkeys to release a lever at the exact moment a spot on a computer screen changed color from red to green. The monkeys had to complete this task three times, but only on the third trial did they receive a food reward, regardless of how they performed on the first two trials. As a result, the monkeys made fewer errors in the third trial than in the first two trials.

Next, the team injected a short strand of DNA into the rhinal cortex that decreased the binding of a receptor known to be involved in processing reward signals. The effects were only temporary, 10–12 weeks, but during that time the monkeys were unable to determine how many trials were left before the reward was given. As a result, the monkeys worked vigilantly to receive the reward on every trial, making few errors even on trials one and two. The researchers believe that this research may eventually provide insight into how human motivation works and how it might be disrupted in behavioral disorders.

**Behavior** is the observable response of organisms to external or internal stimuli. In the early 20th century, scientific studies of animal behavior, termed **ethology** (from the Greek *ethos*, meaning habit or manner), focused on specific genetic and physiological mechanisms of behavior. For example, we could hypothesize that male deer rut or fight with other males in the fall because a change in day length stimulates the eyes, brain, and pituitary gland and triggers hormonal changes in their bodies. These factors, which directly affect behavior, are called **proximate causes.** However, we could also hypothesize that male deer fight to determine which deer get to mate with the most female deer and pass on their genes. This hypothesis leads to a different answer from the one that is concerned with changes in day length. This answer focuses on the adaptive significance of fighting to the deer—that is, on the effect of a particular behavior on reproductive success. These factors are called **ultimate causes** of behavior.

In this chapter, we will explore the role of both proximate and ultimate causes of behavior. We begin the chapter by exploring how behavior is determined, which involves the roles of both genetics and the environment. We will consider how different behaviors include movement, communication, and group living. Finally, we will examine whether an organism can truly behave in a way that benefits others at a cost to itself, and how behavior shapes different mating systems.

## 42.1 The Influence of Genetics and Learning on Behavior

### Learning Outcomes

1. Outline the differences among innate behavior, classical conditioning, and operant conditioning.
2. Describe how memory is used in local movement.

As we will see, behavior is controlled by both genetics and the environment, and in this section, we will discuss the influence of both. Genetics allows the evolutionary responses of ancestors to selection to be incorporated into current behavior, while responses to the environment permit animals the opportunity to change their behavior over the short term.

### Innate Behavior Patterns Are Genetically Programmed

Behaviors that appear to be genetically programmed are referred to as **innate** (also called instinctual). For example, a spider will spin a specific web without ever seeing a member of its own species build one. A classic example of innate behavior is the egg-rolling response in geese (Figure 42.1). If an incubating goose notices an egg out of the nest, she will extend her neck toward the egg, get up, and then roll the egg back to the nest using her beak. Such behavior functions to improve the survival of offspring. Eggs that roll out of the nest can get cold and fail to hatch. Geese that fail to exhibit the egg-rolling response would pass on fewer of their genes to future generations. Interestingly, any round object, from a wooden egg to a volleyball, can elicit the egg-rolling response. Such objects are called **releasers.**

### Learning Occurs When Animals Modify Their Behavior Based on Past Experience

Although many of the behavioral patterns exhibited by animals are innate, animals can make modifications to their behavior based on previous experience, a process that involves learning. Perhaps the simplest form of learning is **habituation,** in which an organism learns to ignore a repeated stimulus. For example, animals in African safari parks become habituated to the presence of vehicles containing tourists. Habituation can be a problem at airports, where birds eventually ignore the alarm calls designed to scare them away from the runways.

Habituation is a form of nonassociative learning, without a positive or negative reinforcement. A positive or negative association between a stimulus and a response is termed **associative learning.** The two main types of associative learning are classical conditioning and operant conditioning.

In **classical conditioning,** an involuntary response comes to be associated with a stimulus that did not originally elicit the response. This type of learning is generally associated with the Russian psychologist Ivan Pavlov. In his original experiments in the 1920s, Pavlov restrained a hungry dog in a harness and presented small portions of food at regular intervals. The dog salivated whenever it smelled the food. Pavlov then began to sound a metronome when presenting the food. Eventually, the dog salivated at the sound of the metronome whether or not the food was present. Classical conditioning is widely observed in animals. For example, many insects quickly learn to associate certain flower odors with nectar rewards. In some humans, the sound of a dentist's drill is enough to produce a feeling of uneasiness, tension, and sweaty palms.

In **operant conditioning,** an animal's behavior is reinforced by a consequence—either a reward, as observed in the monkey described at the beginning of the chapter, or it can be reinforced by a punishment. Operant conditioning, also called trial-and-error learning, is common in animals. Often it is associated with negative rather than positive reinforcement. For example, toads eventually refuse to strike at insects that sting, such as wasps and bees, and birds will learn to avoid bad-tasting butterflies (Figure 42.2).

Behavior is also affected by spatial learning in which specific landmarks are learned to construct cognitive maps, mental representations

(1) The female goose extends her neck toward the egg.
(2) The goose gets up from the nest and approaches the egg.
(3) The goose places her neck above the egg.
(4) The goose rolls the egg back to the nest with her beak and neck.


**Figure 42.1 Innate behavior.** Female geese retrieve eggs that have rolled outside the nest through a set sequence of movements. The goose completes the entire sequence even if a researcher takes the egg away before the goose has rolled it back to the nest.

(a) Blue jay eating monarch

(b) Vomiting reaction

**Figure 42.2** **Operant conditioning, also known as trial-and-error learning.** **(a)** A young blue jay will eat a monarch butterfly, not knowing that it is noxious. **(b)** After the first experience of eating a monarch and vomiting, a blue jay will avoid the insects in the future.
*(a–b)* © L.P. Brower, Sweet Briar College

**Concept Check:** *What's the difference between operant conditioning and classical conditioning?*

**BioConnections:** *Look ahead to Figure 43.12. How might operant conditioning relate to the similar appearance of king snakes and coral snakes?*

of the spatial relationships of objects, to aid in local movements. For example, in the fall, some birds known as nutcrackers (genus *Nucifraga*) store thousands of seeds in many underground caches over a 35-km$^2$ area. The birds relocate many of these caches in the winter but leave alone the caches of other nutcrackers. As long ago as the 1930s, Dutch-born ethologist Niko Tinbergen showed how the female digger wasp uses landmarks to relocate her nests, as described next.

# FEATURE INVESTIGATION

## Tinbergen's Experiments Show That Digger Wasps Learn the Positions of Landmarks to Find Their Nests

In the sandy, dry soils of Europe, the solitary female digger wasp (*Philanthus triangulum*) digs four to five nests in which to lay her eggs. Each nest stretches obliquely down into the ground for 40–80 cm. After digging the nesting holes, the wasp performs a sequence of apparently genetically programmed events. She catches and stings a honeybee, which paralyzes it; returns to the nest; drags the bee into the nest; and lays an egg on it. The egg hatches into a larva, which then feeds on the paralyzed bee. However, the larva needs to ingest five or six bees before it is fully developed. This means the wasp must catch and sting four or five more bees for each larva. She can carry only one bee at a time. After each visit, the wasp must seal the nest with soil to prevent the bee from being stolen, find another bee, relocate the nest, open it, and add the new bee. How does the wasp relocate the nest after spending considerable time away? Niko Tinbergen observed the wasps hover and fly around the nest each time they took off. He hypothesized that they were learning the nest position by creating a cognitive map of the landmarks in the area.

To test his hypothesis, Tinbergen experimentally adjusted the landmarks around the burrow that the wasps might be using as cues (**Figure 42.3**). First, he put a ring of pinecones around the nest entrance to train the wasp to associate the pinecones with the nest. Then, when the wasp was out hunting, he moved the circle of pinecones a distance from the real nest and constructed a sham nest, making a slight depression in the sand and mimicking the covered entrance of the burrow. On returning, the wasp flew straight to the sham nest and tried to locate the entrance. Tinbergen chased it away. When it returned, it again flew to the sham nest. Tinbergen repeated this nine times, and every time the wasp chose the sham nest. Tinbergen got the same result with 16 other wasps, and not once did any of them choose the real nest.

Next Tinbergen experimented with the type of stimulus that might be eliciting the learning. He hypothesized that the wasps could be responding to the distinctive scent of the pinecones rather than their appearance. He trained the wasps by placing a circle of pinecones that had no scent and two small pieces of cardboard coated in pine oil around the real nest. He then moved the cones to surround a sham nest and left the scented cardboard around the real nest. The returning wasps again ignored the real nest with the scented cardboard and flew to the sham nest. He concluded that for the wasps, sight was apparently more important than smell in determining landmarks.

**Figure 42.3** **How Niko Tinbergen discovered the digger wasp's nest-locating behavior.**

**Concept Check:** *How would you test what type of spatial landmarks are used by female digger wasps?*

**HYPOTHESIS** Digger wasps (*Philanthus triangulum*) use visual landmarks to locate their nests.

**STARTING LOCATION** The female digger wasp excavates an underground nest, to which she returns daily, bringing food to the larvae located inside.

**Experimental level**

1 Place a ring of pinecones around the nest to train the wasp to associate pinecones with the nest.

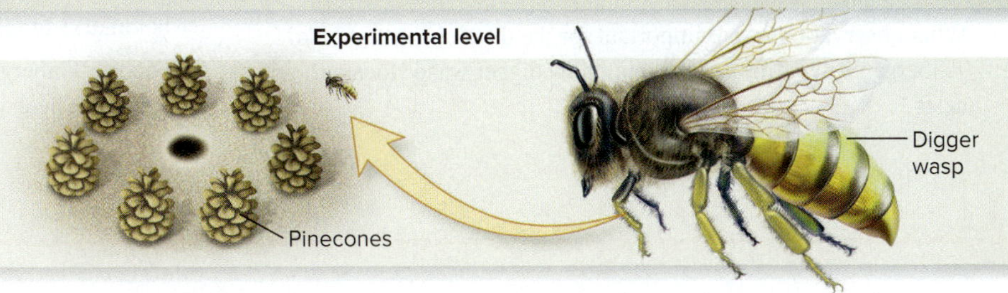

2 After the wasp leaves the nest to hunt, move the pinecones 30 cm from the real nest. The wasp returns and flies to the center of the pinecone circle instead of the real nest. Repeated experiments yield similar results (see data), indicating that the wasp uses landmarks as visual cues.

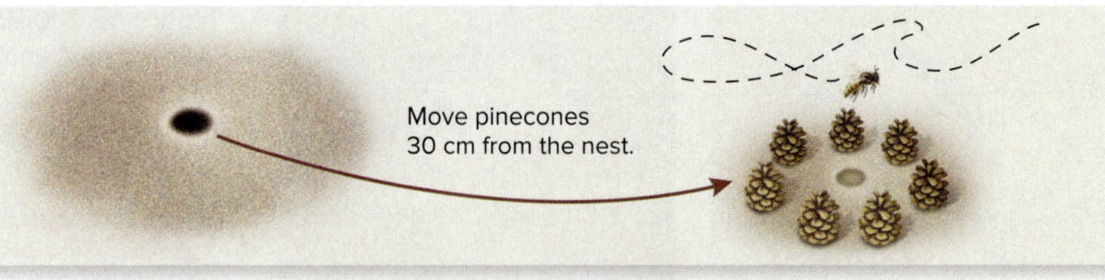


3 To test whether it is the shape or the smell of the pinecones that elicits the response, perform the same experiment as above, except use pinecones with no scent and add 2 small pieces of cardboard coated with pine oil.

4 After the wasp leaves the nest, move the pinecones 30 cm from the nest, but leave the scented cardboard at the nest. The wasps again fly to the pinecone nest (see data), indicating that it is the arrangement of cones, not their smell, that elicits the learning.

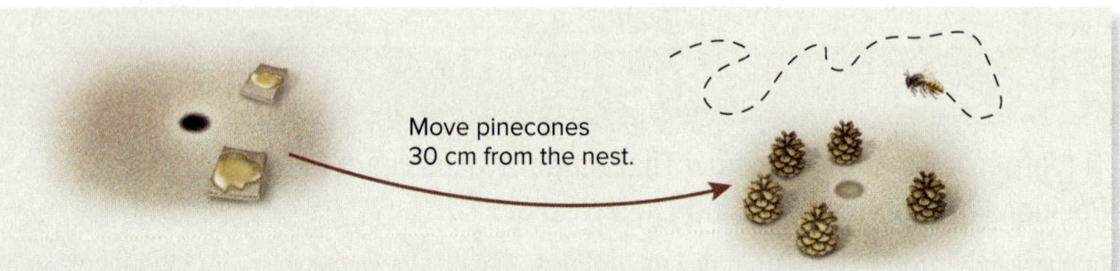


5 **THE DATA***

**Results from steps 1 and 2:**

| Wasp # | Number of return visits per wasp to real nest without pinecones | Number of return visits per wasp to sham nest with pinecones |
|---|---|---|
| 1–17 | 0 | ~9 |

**Results from steps 3 and 4:**

| Wasp # | Number of return visits per wasp to real nest with scented cardboard | Number of return visits per wasp to sham nest with pinecones |
|---|---|---|
| 18–22 | 0 | ~6 |

*Seventeen wasps, numbered 1–17, were studied as described in steps 1 and 2. Five wasps, numbered 18–22, were studied as described in steps 3 and 4.

6 **CONCLUSION** Digger wasps remember the positions of visual landmarks and use them as aids in local movements.

7 **SOURCE** Tinbergen, N. 1951. *The Study of Instinct*. Clarendon Press, Oxford.

*Experimental Questions*

1. What observations were important for the development of Niko Tinbergen's hypothesis explaining how digger wasps locate their nests?
2. How did Tinbergen test the hypothesis that the wasps were using landmarks to relocate the nest? What were the results?
3. Did the Tinbergen experiment rule out any other cue the wasps may have been using besides the sight of pinecones?

## Both Genetics and Learning Influence Most Behaviors

Much of the behavior we have discussed so far has been presented as either innate (genetic) or learned, but the behavior we observe in nature is usually a mixture of both. Bird songs present a good example. Many birds learn their songs as juveniles, when they hear their parents sing. For example, if juvenile white-crowned sparrows are raised in isolation, their adult songs do not resemble the typical species-specific song. If they hear only the song of a different species, such as the song sparrow, they again sing a poorly developed adult song. However, if they hear the song of the white-crowned sparrow, they will learn to sing a fully developed white-crowned sparrow song. The birds are genetically programmed to learn, but they will sing the correct song only if the appropriate instructive program is in place to guide learning.

Another example of how innate behavior interacts with learning can occur during a limited time period of development, called a **critical period.** At this time, many animals develop irreversible species-specific behavior patterns. This process is called **imprinting.** One of the best examples of imprinting was demonstrated by the Austrian ethologist Konrad Lorenz in the 1930s. Lorenz noted that young birds of some species imprint on their mother during a critical period that is usually within a few hours after hatching. This behavior serves them well, because in many species of ducks and geese, it would be hard for the mother to keep track of all her offspring as they walk or swim. After imprinting takes place, the offspring keep track of the mother.

The survival of the young birds requires that they quickly learn to follow their mother's movements. Lorenz raised greylag geese from eggs, and soon after they hatched, he used himself as the model for imprinting. As a result, the young goslings imprinted on Lorenz and followed him around (Figure 42.4). For the rest of their life, they preferred the company of Lorenz and other humans to geese. Studies have shown that even an object as foreign as a black box, watering can, or flashing light can be imprinted on if it is the first moving object the chick sees during the critical period.

**Figure 42.4 Konrad Lorenz being followed by his imprinted geese.** Newborn geese imprint on the first object they see after hatching and later will follow that particular object only. They normally imprint on their mother but can imprint on humans. The first thing these young geese saw after hatching was ethologist Konrad Lorenz.



## 42.1 Reviewing the Concepts

- Behavior is usually due to the interaction of an organism's genes and the environment.
- Genetically programmed behaviors are termed innate and often involve a releaser that initiates the behavior (Figure 42.1).
- Organisms can often make modifications to their behavior based on previous experience, a process called learning. Some forms of learning include habituation, classical conditioning, operant conditioning, and spatial learning (Figures 42.2, 42.3).
- Much behavior is a mixture of innate and learned behaviors. A good example of this occurs in a process called imprinting, in which animals develop a strong attachment that influences subsequent behavior (Figure 42.4).

## 42.1 Testing Your Knowledge

1. Birds can come to ignore the presence of a scarecrow in a field of crops. This behavior is
   a. learned.
   b. habituation.
   c. operant conditioning.
   d. classical conditioning.
   e. innate.
2. After their young hatch, black-headed gulls will remove the old egg shells, which, unlike the camouflaged outsides, are white inside and may attract predators. This behavior is likely
   a. learned.
   b. habituation.
   c. operant conditioning.
   d. classical conditioning.
   e. innate.
3. Green sea turtles feed off the coast of Brazil yet swim east for 2,300 km (1,429 miles) to lay their eggs on Ascension Island, an 8-km-wide island in the center of the Atlantic Ocean between Brazil and Africa. What could be the ultimate reason turtles return to this tiny island to lay their eggs?
   a. They use chemical cues.
   b. They use magnetic cues.
   c. Fewer predators exist here than on other beaches.
   d. They use navigation.
   e. They learn underwater landmarks.

## 42.2 Communication

### Learning Outcome

1. Give examples of how animals use chemical, auditory, visual, and tactile communication.

Much animal behavior is directed at individuals of the same species—for example, in courtship, fighting over mates, or defense of territories or food sources. **Communication** is the use of specially designed signals or displays to modify the behavior of others. The use of different forms of communication between organisms depends on the environment in which they live. For example, while diurnal animals use a variety of gestures and displays to communicate with each other, visual communication plays little role in the signals of nocturnal animals. Similarly, for animals in dense forests, sounds are of prime importance. Sound, however, is a temporary signal. Scent can last longer and is often used to mark the large territories of some mammals. Many animals use tactile communication, or touch, to communicate with others of their same species. In this section, we outline the various types of communication—chemical, auditory, visual, and tactile—that occur among animals.

## Chemical Communication Is Often Used to Mark Territories or Attract Mates

The chemical marking of territories is common among animals, especially among members of the canine and feline families, which urinate on prominent markers at the boundaries of their territories. Scent trails are often used by social insects to recruit workers to help bring prey to the nest. Fire ants (genus *Solenopsis*) attack large, living prey, and many ants are needed to drag the prey back to the nest. The scout that finds the prey lays down a scent trail from the prey back to the nest. The scent excites other workers, which follow the trail to the prey. The scent marker evaporates easily, and the trail effectively disappears in a few minutes to avoid confusion over old trails.

Animals also use chemicals to attract mates. Female moths attract males by powerful chemical attractants called **pheromones.** Male moths have receptors that can detect as little as a single molecule. Among social organisms, some individuals use pheromones to manipulate the behavior of others. For example, a queen bee releases pheromones that suppress the reproductive system of workers, which ensures that she is the only reproductive female in the hive.

## Auditory Communication Is Often Used to Attract Mates and to Deter Competitors

Many organisms communicate by making sound. Because the ground can absorb sound waves, sound travels farther in the air, which is why many birds and insects perch on branches or leaves when singing. Air is, on average, 14 times less turbulent at dawn and dusk than during the rest of the day, so sound carries farther then, which helps explain the preference of most animals for calling at these times. Some insects utilize the very plants on which they feed as a medium of song transmission. Many male leafhopper and planthopper insects vibrate their abdomens on leaves and create species-specific courtship songs that are transmitted by adjacent vegetation and are picked up by nearby females of the same species.

Although many males use auditory communication to attract females, some females use calls to attract the attention of males. Female elephant seals scream loudly when approached by a nondominant male. This attracts the attention of the dominant male, which drives the nondominant male away. In this way, the female is guaranteed a mating with the strongest male. Sound production can attract predators as well as mates. Some bats listen for the mating calls of male frogs to find their prey. Parasitic flies detect and locate chirping male crickets and then deposit larvae on or near them. Sound may also be used by males during competition over females. In many animals, lower-pitched sounds come from larger males, so by calling to one another, males can gauge the size of their opponents and decide whether it is worth fighting.

## Visual Communication Is Often Used in Courtship and Aggressive Displays

In courtship, animals use a vast number of visual signals to identify and select potential mates. Competition among males for the most impressive displays to attract females has led to elaborate coloration and extensive ornamentation in some species. For example, peacocks and males of many other bird species have developed elaborate plumage to attract females. Male fireflies have developed light flashes that are species specific with regard to number and duration of flashes (**Figure 42.5**). Females respond with a flash of their own.

Visual signals are also used to resolve disputes over territories or mates. Deer and antelope have antlers or horns that they use to display and spar over territory and females. Most of these matches never develop into outright fights, because the males assess their opponent's strength by the size of these ornaments and concede to a stronger rival so as to avoid injury.

## Tactile Communication Strengthens Social Bonds and Conveys Information About Food

Animals often use tactile communication to establish bonds between group members. Primates frequently groom one another, and canines and felines may nuzzle and lick each other. Many insects use tactile communication to convey information on the whereabouts of food.

A particularly fascinating example of tactile communication among animals is the dance of the honeybee, studied by German ethologist Karl von Frisch in the 1940s. Bees commonly live in large hives; in the case of the European honeybee (*Apis mellifera*), the hive consists of 30,000–40,000 individuals. The flowering plants on which the bees forage can be located miles from the hive and store more nectar and pollen than an individual bee can carry back to the nest. The scout bee that locates the plants returns to the hive and recruits more workers to join her (**Figure 42.6a**). The scout dances on the vertical side of a honeycomb, and the dance is monitored by other bees, which follow and touch her to interpret the message. If the food is relatively close to the hive, less than 50 m away, the scout performs a round dance, rapidly moving in a circle, first in one direction and then the other (**Figure 42.6b**). If the food is more than 50 m away, the scout performs a different type of dance, called a waggle dance. In this dance, the scout traces a figure 8, in the middle of which she waggles her abdomen and produces bursts of sound. Again, the other bees maintain contact with her. The amazing part of the waggle dance is that the angle at which the central part of the figure 8 deviates from the vertical direction of the comb represents the same angle at which the food source deviates from the point at which the Sun hits the horizon (**Figure 42.6c**). The direction is always up-to-date, because the bee adjusts the dance as the Sun moves across the sky.

## Biology Principle

### Structure Determines Function

In fireflies, a morphological feature influences animal behavior.

(a) Firefly flashing

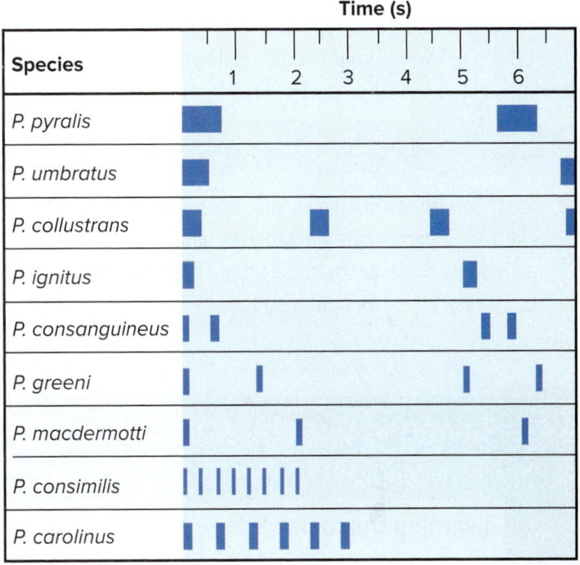


(b) Male flash patterns

**Figure 42.5** **Visual communication in fireflies.** **(a)** Communication between fireflies is conducted by species-specific light flashes emitted by organs on the underside of the abdomen. **(b)** Different species have different flash frequencies and durations. For example, *Photinus pyralis* and *P. umbratus* make relatively long flashes with long intervals between them, whereas *P. consimilis* and *P. carolinus* flashes are short but rapid.

*(a)* © Darwin Dale/Science Source

---

(a) Bees clustering around a recently returned scout, seen in the center, vibrating her abdomen.

(b) Round dance

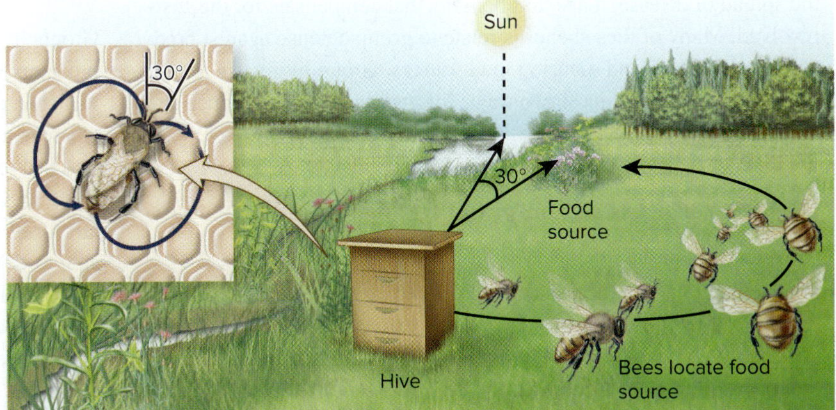


(c) Waggle dance: The angle of the waggle to the vertical orientation of the honeycomb corresponds to the angle of the food source from the Sun.

**Figure 42.6** **Tactile communication among honeybees regarding food sources.** **(a)** Bees gather around a newly returned scout to receive information about nearby food sources. **(b)** If the food is less than 50 m away, the scout performs a round dance. **(c)** If the food is more than 50 m away, the scout performs a waggle dance, which conveys information about its location. If the dance is performed at a 30° angle to the right of the hive's vertical plane, then the food source is located at a 30° angle to the right of the Sun.

*(a)* © Scott Camazine/Alamy

## 42.2 Reviewing the Concepts

- Communication is a form of behavior. The use of different forms of communication between organisms depends on the environment in which they live.
- Chemical communication often involves marking territories; auditory and visual forms of communication are used to attract or defend mates; and tactile communication is used to establish bonds between group members (Figures 42.5, 42.6).

## 42.2 Testing Your Knowledge

1. Most terrestrial mammals are nocturnal. This means they are likely to communicate primarily via what means?
   a. visual  c. chemical  e. both b and c
   b. auditory  d. both a and b

# 42.3 Living in Groups and Optimality Theory

### Learning Outcomes

1. Detail the costs and benefits of living in groups.
2. Outline the costs and benefits of defending a territory.
3. **SCISKILLS** ▶ Explain how game theory is used to determine winners between contestants.

Some of the more complex animal behavior occurs when individuals live together in groups such as flocks or herds. When is it best to live alone and when does it benefit an individual to live in a group? One way to approach this question is to assess the costs and benefits involved. Although group living increases competition for food and the spread of disease, it also has benefits that compensate for the costs involved. Many of these benefits relate to group defense against predators and locating and securing food sources. Group living can reduce predator success in at least two ways: through increased vigilance and through protection in numbers.

## Living in Large Groups May Reduce the Risk of Predation Because of Increased Vigilance

For many predators, success depends on the element of surprise. If an individual prey is alerted to an attack, the predator's chance of success is lowered. A woodpigeon (*Columba palumbus*) in a flock takes to the air when it spots a goshawk (*Accipiter gentilis*). Once one pigeon takes flight, the other members of the flock are alerted and follow suit. If each individual in a group occasionally scans the environment for predators, the larger the group, the less time an individual forager needs to devote to vigilance and the more time it can spend feeding. This is referred to as the **many-eyes hypothesis** (Figure 42.7). Of course, cheating is a possibility, because some birds might never look up, relying on others to keep watch while they keep feeding. However, the individual that happens to be scanning when a predator approaches is most likely to escape, whereas cheaters are more likely to get caught and not pass their genes to future generations.

**Figure 42.7** **Living in groups and the many-eyes hypothesis.** The larger the number of woodpigeons, the less likely an attack will be successful.

**Concept Check:** *What other advantages are there to large groups of individuals when being attacked by a predator?*

## Living in Groups Offers Protection by the Selfish Herd

Group living also provides protection in sheer numbers. Typically, predators take one prey animal per attack. In any given attack, an individual antelope in a herd of 100 has a 1 in 100 chance of being selected, whereas a single individual has a 1 in 1 chance. Large herds may be attacked more frequently than a solitary individual, but a herd is unlikely to attract 100 times more attacks than an individual, often because of the territorial nature of predators. Furthermore, large numbers of prey are able to defend themselves better than single individuals, which usually choose to flee. For example, groups of nesting black-headed gulls mob a crow, thereby reducing the crow's ability to steal the gulls' eggs.

Research has shown that within a group, each individual can minimize the danger to itself by choosing the location that is as close to the center of the group as possible. This was the subject of a famous paper, "The Geometry of the Selfish Herd," by the British evolutionary biologist W. D. Hamilton. The explanation of this type of defense is that predators are likely to attack prey on the periphery because they are easier to isolate visually. Many animals in herds tend to bunch close together when they are under attack, making it physically difficult for the predator to get to the center of the herd.

## Predator Success May Be Increased in Groups

Group living may allow predators such as lions and wolves the opportunity to locate and take down prey of a disproportionately large size, which a single predator would be unable to handle. Furthermore, as we saw with honeybees, when animals search for food within a group, each individual may be able to exploit the discovery of others.

Interestingly, the actual size of predator groups may be different from the optimal size. Imagine that a solitary killer whale hunting for a seal has a given payoff for its efforts (Figure 42.8). Here, payoff is defined as the amount of energy gained by eating the prey minus the amount of energy expended in capturing it. If another killer whale joins in the hunt, the seal is more easily caught, and, despite having to share the prey, payoff for either individual increases. A third killer whale increases the payoff even more. Three whales is the optimal group size, for which the payoff to each individual is maximized. When the group size becomes four or five, then the average payoff to each individual decreases, because the food has to be shared among more individuals. However, living in a group of four or five is still a better option than hunting alone, so whales continue to join groups of two, three, or four other whales. When the group size is six or more, a whale would do better on its own than in the group. In this example, the most commonly observed or stable group size is five individuals. Paradoxically, the optimal group size of three, which affords the greatest payoff to each whale, is not equivalent to the stable group size.

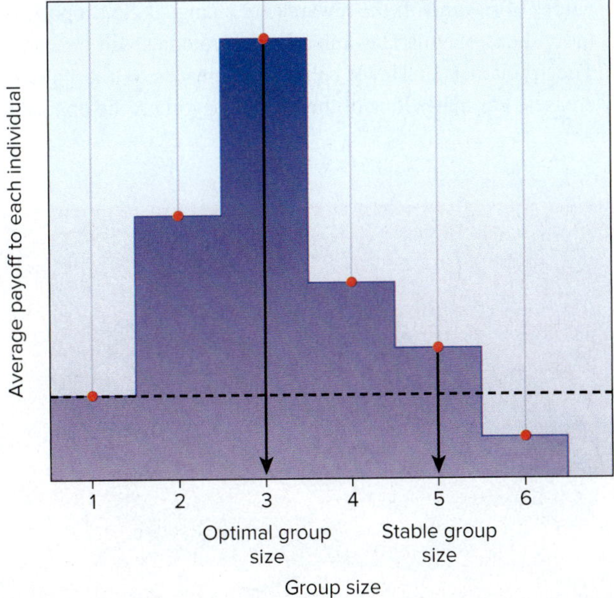


**Figure 42.8 Optimal and stable group size.** The payoff to a group of two or three predators is greater than for a solitary predator. The highest payoff is for a group size of three, which is the optimal size. However, other solitary individuals would benefit from joining the group, and group size increases to five individuals. In a group size of six, any one individual would do better on its own. The stable group size is then five.

## Some Animals Defend a Territory

Some animals prefer living a solitary existence to living in a herd. Many of these animals actively defend a **territory,** a fixed area in which an individual excludes other members of its own species, and sometimes other species, by aggressive behavior or territory marking. The primary benefit of a territory is that it provides exclusive access to a particular resource, whether it be food, mates, or sheltered nesting sites. Large territories provide more of a resource but may be costly to defend, whereas small territories that are less costly to defend may not provide enough resources.

The best size of a territory may be predicted by **optimality theory,** which predicts that an animal should behave in a way that maximizes the benefits of a behavior minus its costs. In this case, the benefits are the nutritional or caloric value of the food items within the territory, and the costs are patrolling and defending the territory. When the difference between the energetic benefits of food gathering and the energetic costs of maintaining a territory is maximized, an organism is optimizing its behavior. Optimality theory can also be used in decisions about food gathering and when it may be profitable to remain at a patch to look for food versus when is better to look for a completely new patch.

In studies of the territorial behavior of the golden-winged sunbird (*Nectarinia reichenowi*) in East Africa, American ornithologists Frank Gill and Larry Wolf measured the energy content of nectar as the benefit of maintaining a territory and compared it with the energy costs of activities such as perching, flying, and fighting (Figure 42.9). Defending the territory ensured that other sunbirds did not take nectar from available flowers, thus increasing the amount of nectar in each flower. In defending a territory, the sunbird gained 780 kilocalories (kcal) a day in extra nectar content. However, the sunbird also spent 728 kcal in defense of the territory, yielding a net gain of 52 kcal a day and making territorial defense advantageous.

Defending a territory can be dangerous, and biologists have long been interested in determining the outcome of contests between individuals. American mathematician John Nash developed **game theory** to establish the outcome of contests in which an individual's choice of strategy depends on the actions of other individuals. Originally developed to understand human economic behavior, game theory has been used to help understand the behavior of animals in competitive situations, as described next.

**Figure 42.9 The golden-winged sunbird of East Africa (*Nectarinia reichenowi*) defends its territory.**

# Quantitative Analysis

## GAME THEORY ESTABLISHES WHETHER INDIVIDUALS FIGHT OR FLEE

Imagine a game in which individuals of the same species engage in contests over resources. The possible behaviors for any given individual are to display, presenting itself as a contestant; to fight to win, attacking until the opponent retreats; and to retreat, giving up the resource to the opponent. Some individuals, the "Hawks," always fight to win, risking injury to themselves in the process. "Doves" initially perform aggressive displays when encountering another Dove, but always retreat before injury occurs. When a Hawk encounters a Dove, the Dove does not display, but instead assumes a submissive posture (**Figure 42.10**).

Let's assign the winner of a contest +50 and the loser 0. The cost of injury is −100 because the injured player may not be able to compete again for a long time or may die, and the cost of spending time in a display is −10. It is assumed that Hawks and Doves reproduce in proportion to their scores. What happens when Hawks fight other Hawks or Doves, or when Doves fight Doves? We can analyze such a contest by constructing a two-by-two matrix with the average rewards for four possible types of encounter (**Table 42.1**). In a population of all Hawks, the chance of winning is 50%. Half the population wins and gets a reward of 50, and half the population loses and gets −100. The average is $^1/_2(50) + {}^1/_2(-100) = -25$. Thus, any individual playing Dove would do better because, when a Dove meets a Hawk, it gets 0 in every encounter, since it retreats immediately, which is not very good but still better than −25. Therefore, Hawk is not always the best strategy. In a population of all Doves, the chance of winning is 50%. Half of the population wins and gets a reward of 40 (the reward minus the cost of displaying), and half of the population loses and gets −10 (the cost of display); the score, on average is $^1/_2(50 - 10) + {}^1/_2(-10) = +15$. Even though the average score is positive, Dove is not always the best strategy, either. In this population, any individual Hawk would do very well, and the Hawk strategy would soon spread because when a Hawk meets a Dove, it gets +50 in every encounter.

Although all individuals in a population will not always adopt a Hawk or Dove strategy, a mixture of Hawk and Dove strategies is more likely. In such a population, a stable equilibrium is reached in which the average reward for a Hawk is equal to the average reward for a Dove. For the rewards we have specified, the stable mixture can be calculated as follows. If $h$ is the proportion of Hawks in the population, the proportion of Doves is $(1 - h)$. The average reward for a Hawk, H, is the reward for each type of fight multiplied by the probability of meeting each type of contestant.

Therefore, the average reward for a Hawk will be

$$\mathrm{H} = -25h + 50(1 - h)$$

For Doves, the average reward will be

$$\mathrm{D} = 0h + 15(1 - h)$$

At the stable point, H is equal to D. When H = D, the proportion of Hawks $(h) = 7/12$, and the proportion of Doves $(1 - h) = 5/12$. You can verify this as follows:

$$\mathrm{H} = -25(7/12) + 50(5/12) = -175/12 + 250/12 = 75/12$$
$$\mathrm{D} = 0(7/12) + 15(5/12) = 0 + 75/12 = 75/12.$$

This stable point could be achieved in two ways. First, the population could consist of 7/12 of individuals who always played Hawk and 5/12 who always played Dove. Alternatively, the population could consist of individuals who all adopted a mixed strategy, playing Hawk in 7/12 contests and Dove in the other 5/12.

Despite the simplicity of these models, some important conclusions can be drawn from them:

1. Fighting strategy is frequency dependent: It depends on what other animals are doing. A Hawk strategy is good for an individual in a population of Doves but not in one of Hawks. Territory owners may be more likely to behave like Hawks than challengers.
2. The strategies employed by individuals are dependent on the values of rewards. If the rewards are changed, the proportion of individuals playing Hawk and Dove strategies will change.
3. The frequency of Hawk behavior increases as the payoff increases. For males, one of the biggest payoffs is the opportunity

**Table 42.1** Average Rewards to the Attacker in the Game Between Hawk and Dove

| | Opponent | |
|---|---|---|
| **Attacker** | **Hawk** | **Dove** |
| **Hawk** | $^1/_2(50) + {}^1/_2(-100) = -25$ | +50 |
| **Dove** | 0 | $^1/_2(50 - 10) + {}^1/_2(-10) = +15$ |

Costs and rewards: Winner +50 Injury −100

Loser 0 Display −10

**Figure 42.10** **Hawk and Dove strategies.** In these dueling meerkats, the individual on the left is escalating the fight, behaving in a Hawklike manner, and the individual on the right is in a submissive posture, a Dovelike strategy.

to mate with a female because failure to do so is equivalent to genetic death. Fights over females are therefore commonly severe, sometimes fatal, and Hawk strategies dominate.

In the real world, animals also adjust their strategies according to the vigor of their opponents. It would be pointless to adopt a Hawk strategy against a bigger opponent, even for a territory owner. Players may use a strategy that follows the rule "if larger, behave like a Hawk; if smaller, behave like a Dove; if equally matched, adopt the Hawk strategy if a territory owner."

**Crunching the Numbers:** What is the average reward to a Hawk in a Hawk-versus-Hawk confrontation, where the reward to the winner is +30 and all other rewards are the same as in Table 42.1?

## 42.3 Reviewing the Concepts

- Many benefits of group living relate to defense against predators, offering protection through sheer numbers. Groups of predators may enjoy greater success hunting in a pack than alone (Figures 42.7, 42.8).
- Some animals prefer a solitary existence over group living and may maintain exclusive territories. The size of a territory, a fixed area in which an individual or a group excludes other members of its own species, tends to be optimized according to the costs and benefits involved (Figure 42.9).
- Game theory can be used to determine the strategies used by animals in contests over territories, food, or mates (Figure 42.10; Table 42.1).

## 42.3 Testing Your Knowledge

1. In nature, the most commonly observed number of individuals in a group is generally
   a. <100.
   b. >10.
   c. the optimal group size.
   d. the stable group size.
   e. the selfish herd size.
2. What is the main advantage of defending a territory?
   a. exclusive access to mates
   b. exclusive use of resources
   c. reduced conflict with others
   d. reduced transmission of disease
   e. reduced predation pressure
3. In any given population, the most commonly observed fighting strategy is
   a. Hawk.
   b. Dove.
   c. a mixture of Hawks and Doves.
   d. if smaller, behave like a Hawk.
   e. if a territory owner, behave like a Dove.

## 42.4 Altruism

### Learning Outcomes

1. **SCISKILLS** ▶ Evaluate the arguments for and against the concept of group selection.
2. Describe how the concept of kin selection can explain altruistic behavior.
3. **SCISKILLS** ▶ Predict which types of relatives are more or less likely to be the recipients of apparent altruistic acts.
4. Explain how eusociality can be seen as an example of altruism.

As discussed in Chapter 19, each species exhibits adaptations that favor reproductive success. Via natural selection, organisms that have favorable adaptations are more likely to reproduce and pass their genes to future generations. However, ecologists have identified many examples in which some individuals forgo reproduction or exhibit behaviors that put them at risk, apparently for the benefit of the group. **Altruism** is a behavior that appears to benefit others at a cost to oneself. In this section, we will consider why altruistic behavior takes place.

### Kin Selection Is Explained by Hamilton's Rule

We are not surprised if parents forgo food for themselves while passing food on to their offspring because all offspring have copies of their parents' genes. In taking care of their young, parents are actually caring for copies of their own genes. Such behavior is known as **kin selection,** selection for behavior that lowers an individual's own survival and reproductive capabilities but enhances the survival and reproductive success of a relative. As we will see later, kin selection can operate via many different relatives, not just parents and their offspring.

The probability that any two individuals will share a copy of a particular gene is a quantity, $r$, called the **coefficient of relatedness.** During meiosis in a diploid species, any given copy of a gene has a 50% chance of segregating into an egg or sperm. A mother and father are, on average, related to their children by an amount $r = 0.5$, because half of a child's genes come from its mother and half from its father. By similar reasoning, brothers or sisters are related by an amount $r = 0.5$ (they share half their mother's genes and half their father's); grandchildren and grandparents, by 0.25; and cousins, by 0.125 (Figure 42.11). In 1964, British evolutionary biologist W. D. Hamilton realized the implication of the coefficient of relatedness for the evolution of altruism. An organism can pass on its genes not only through having offspring but also through ensuring the survival of siblings, nieces, nephews, and cousins. This means an organism has a vested interest in protecting its brothers and sisters, and even the offspring of its parents' siblings.

The term **inclusive fitness** is used to designate the total number of copies of genes passed on through one's relatives, as well as one's own reproductive output. Hamilton proposed that an altruistic gene is favored by natural selection when

$$rB > C$$

where $r$ is the coefficient of relatedness of donor (the altruist) to the recipient, $B$ is the benefit received by the recipient of the altruism, and $C$ is the cost incurred by the donor. This is known as **Hamilton's rule.**

Let's examine a situation involving altruism within a group of animals. Many insect larvae, especially caterpillars, are soft-bodied

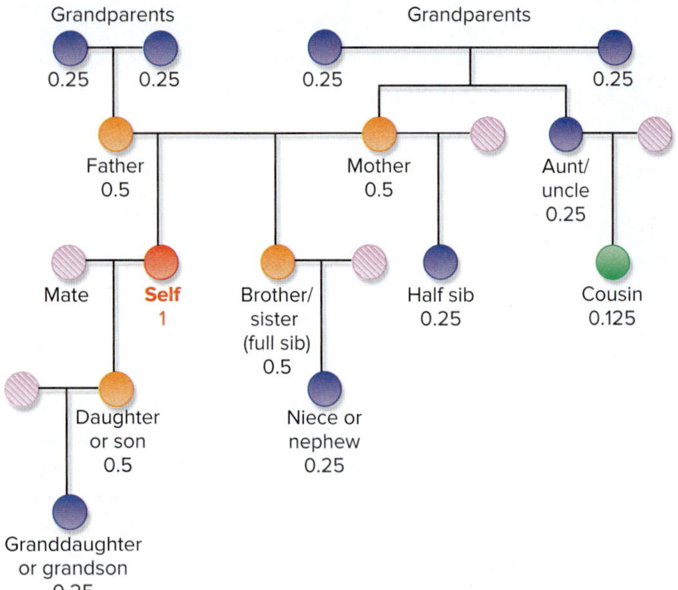


**Figure 42.11** **Degree of genetic relatedness to self in a diploid organism.** Pink hatched circles represent completely unrelated individuals.

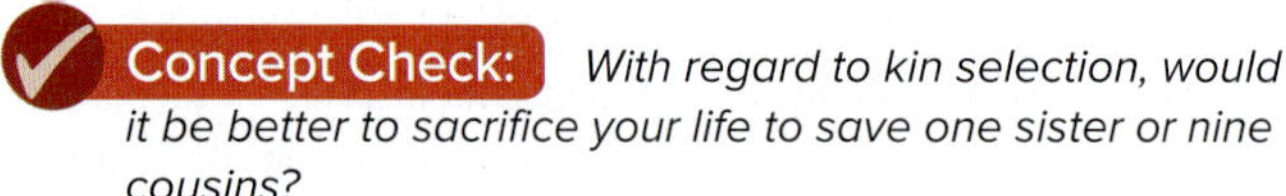

**Concept Check:** *With regard to kin selection, would it be better to sacrifice your life to save one sister or nine cousins?*

creatures. They rely on possessing a bad taste or toxin to deter predators and may advertise this condition with bright warning colors. For example, noxious *Datana ministra* caterpillars, which feed on oaks and other trees, have bright red and yellow stripes and adopt a specific posture with head and tail ends upturned when threatened (**Figure 42.12**). Unless it is born with an innate avoidance of this prey type, a predator has to kill and eat one of the caterpillars in order to learn to avoid similar individuals in the future. It is of no personal benefit to a single unlucky caterpillar to be killed. However, animals with warning colors often aggregate in kin groups because they hatch from the same egg mass. In this case, the death of one individual is likely to benefit its siblings, which are less likely to be attacked by the same predator in the future, and thus its genes will be preserved. This explains why the genes for bright color and a warning posture are successfully passed on from generation to generation. In a case where $r = 0.5$, $B$ might be 50, and $C = 1$, the benefit of 25 is greater than 1, so the genes for this behavior are favored by natural selection.

A common example of altruism in social animals occurs when a sentry raises an alarm call in the presence of a predator. This behavior has been observed in Belding's ground squirrels (*Spermophilus beldingi*). The squirrels feed in groups, with certain individuals acting as sentries and watching for predators. As a predator approaches, the sentry typically gives an alarm call, and the group members retreat into their burrows. Similar behavior occurs in prairie dogs (*Cynomys* spp.) (**Figure 42.13**). In drawing attention to itself, the caller is at a higher risk of being attacked by the predator. However, in many groups, those closest to the sentry are most likely to be offspring or brothers or sisters; thus, the altruistic act of alarm calling is reasoned to be favored by kin selection.

BIO TIPS ONLINE

**Figure 42.12** **Altruistic behavior.** *Datana ministra* caterpillars exhibit a bright, striped warning pattern to advertise their bad taste to predators.

© Peter Stiling

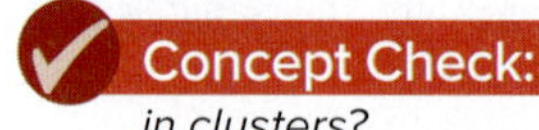

**Concept Check:** *Why do these caterpillars congregate in clusters?*

## Biology Principle

### The Genetic Material Provides a Blueprint for Reproduction

The similarities in DNA between kin promote behavior in which some animals act to save the lives of their close relatives.

**Figure 42.13** **Alarm calling, an example of kin selection.** This prairie dog sentry is emitting an alarm call to warn other individuals, which are often close kin, of the presence of a predator. It is believed that by doing so, the sentry draws attention away from the others but becomes an easier target itself.

© Danita Delimont/Alamy

## Altruism in Eusocial Animals Arises Partly from Genetics and Partly from Lifestyle

Perhaps the most extreme form of altruism is the evolution of sterile castes in social animals, in which the vast majority of females, known as workers, rarely reproduce themselves but instead help one reproductive female (the queen) to raise offspring, a phenomenon called **eusociality.** In insects, the explanation of eusociality lies partly in the particular genetics of most social insect reproduction. Females develop from fertilized eggs and are diploid, the product of fertilization of an egg by a sperm. Males develop from unfertilized eggs and are haploid. Such a system of sex determination is called the **haplodiploid system** (refer back to Figure 15.16). If they have the same parents, each daughter receives an identical set of genes from her haploid father. The other half of a female's genes comes from her diploid mother, so the coefficient of relatedness ($r$) of sisters is 0.50 (from father) + 0.25 (from mother) = 0.75. The result is that females are more closely related to their sisters (0.75) than they would be to their own offspring (0.50). This suggests that it is evolutionarily more advantageous for females to stay in the nest or hive and care for other female offspring of the queen, which are their full sisters.

Elegant though these types of explanations are, they do not explain the whole picture. Large eusocial colonies of termites exist, but termites are diploid, not haplodiploid. How do we account for the existence of eusociality? For termites, the coefficient of relatedness among sisters is 0.5 rather than 0.75, so it shouldn't matter to females whether they rear siblings or daughters of their own. Rather than proposing kin selection as a simple explanation, some ecologists have suggested that particular lifestyles of some animals, as well as genetics, promotes eusociality. In the 1970s, a eusocial mammal, the naked mole rat (*Heterocephalus glaber*) was found living in arid areas of Africa in large underground colonies where only one female, the queen, mates with different males and produces offspring (Figure 42.14). Like termites, the species is diploid. The large nests, or burrows, of the mole rat are enclosed, making it difficult for individuals to escape and form their own colony. Many females are forced to live together, making it possible for the queen to suppress reproduction in other females. She does this by producing a pheromone in her urine that is passed around the colony by grooming.

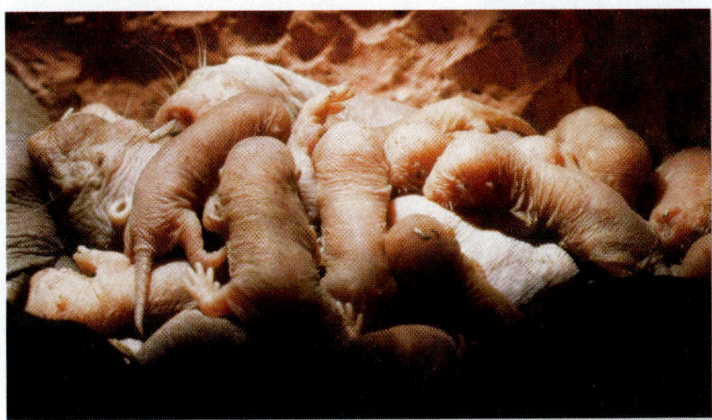

**Figure 42.14** **A naked mole rat colony (*Heterocephalus glaber*).** In this mammalian species, most females do not reproduce; only the queen (shown underneath her babies) has offspring.


## 42.4 Reviewing the Concepts

- Altruism is behavior that benefits others at a cost to oneself. Most apparently altruistic acts are associated with outcomes beneficial to those most closely related to the individual, a concept termed kin selection (Figures 42.11, 42.12, 42.13).
- In eusociality, a phenomenon in which most individuals of a species do not reproduce, altruism may arise partly from the unique genetics of the animals and partly from lifestyle (Figure 42.14).

## 42.4 Testing Your Knowledge

1. In diploid organisms, a mother is related to her daughter by $r$ =
   a. 0. b. 0.25. c. 0.5. d. 0.75. e. 1.0.
2. The existence of worker ants that do not mate can be explained by
   a. eusociality.
   b. a haplodiploid mating system.
   c. the enclosed nature of their nests.
   d. both a and b.
   e. both b and c.

## 42.5 Mating Behavior

### Learning Outcomes

1. Compare and contrast promiscuous, monogamous, polygynous, and polyandrous mating systems.
2. **SCISKILLS ▸** Predict the circumstances that favor polygyny.

In nature, males produce millions of sperm, but females produce far fewer eggs. It would seem that the majority of males are superfluous because one male could easily fertilize all the females in a local area. If one male can mate with many females, why in most species does the sex ratio remain at approximately 1 to 1? The answer lies with natural selection. Let's consider a hypothetical population that contains 10 females to every male; each male mates, on average, with 10 females. A parent whose $F_1$ offspring were exclusively sons could expect to have 10 times the number of $F_2$ offspring than a parent with the same number of daughters. Under such conditions, natural selection would favor the spread of genes for male-producing tendencies, and males would become prevalent in the population. If the population were mainly males, females would be at a premium, and natural selection would favor the spread of genes for female-producing tendencies. Such constraints operate on the numbers of both male and female offspring, keeping the sex ratio at about 1:1. This concept was developed in 1930 by the British geneticist Ronald Fisher and has come to be known as Fisher's principle.

Even though the sex ratio is fairly even in most species, that doesn't mean that one female always mates with one male or vice versa. Four different types of mating systems occur in nature (Figure 42.15). In some species, mating is promiscuous, with each female and each male mating with many partners within a breeding season. In monogamy, each individual mates exclusively with one partner over at least a single breeding cycle and sometimes for longer. In contrast, polygamy describes a mating pattern in which either males or females mate with more than one partner in a breeding season. There are two types of

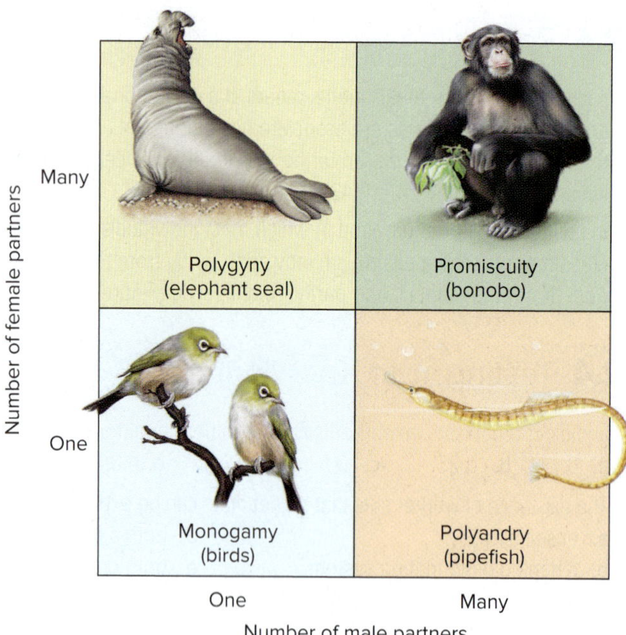


**Figure 42.15** The four different animal mating strategies.

polygamy. In polygyny (Greek, for many females), one male mates with more than one female, but females mate only with one male. In polyandry (Greek, for many males), one female mates with several males, but males mate with only one female.

## In Promiscuous Mating Systems, Each Male or Female Mates with Multiple Partners

Chimpanzees and bonobos are somewhat **promiscuous;** each male mates with many females and vice versa. In these species, sex has a social as well as a reproductive function, allowing for the alleviation of conflict within the social group. However, females that maximize the genetic diversity of their offspring are also more likely to have at least some offspring that will survive in a changing world. Intertidal and terrestrial mollusks are also usually promiscuous. Individuals copulate with several partners and eggs are fertilized with sperm from several individuals. These mollusks are slow moving and they risk desiccation when searching for a mate. The risk of not finding a mate is believed to promote promiscuous mating.

## In Monogamous Mating Systems, Males and Females Are Paired for at Least One Reproductive Season

In **monogamy,** each individual mates exclusively with one partner over at least a single breeding cycle and sometimes for longer. Males and females do not exhibit much **sexual dimorphism—**a pronounced difference in the morphologies of the two sexes within a species—and are generally similar in body size and structure (Figure 42.16a). Several hypotheses have been developed to explain the existence of monogamy. The first is the **mate-guarding hypothesis,** which suggests that males stay with a female to protect her from being fertilized by other males. Such a strategy may be advantageous when receptive females are widely scattered and difficult to find.

The **male-assistance hypothesis** maintains that males remain with females to help them rear their offspring. Monogamy is common among birds, where about 70% of the pairings remain intact during at least one breeding season. According to the male-assistance hypothesis, monogamy is prevalent in birds because eggs and chicks take a considerable amount of parental care. Most eggs need to be incubated continuously if they are to hatch, and chicks require almost continual feeding. It is therefore in the male's best interest to help raise his young, because he would have few surviving offspring if he did not.

The **female-enforced monogamy hypothesis** suggests that females stop their male partners from being polygynous. Male and female burying beetles (*Nicrophorus defodiens*) work together to bury small, dead animals, which provide a food resource for their developing offspring. Males release pheromones to attract other females to the site. However, although an additional female might increase the male's fitness, the additional developing offspring might compete with

**(a) Monogamous species**

**(b) Polygynous species**

**(c) Polyandrous species**

**Figure 42.16** Sexual dimorphism in body size and mating system. **(a)** In monogamous species, such as these Manchurian cranes (*Grus japonensis*), males and females do not exhibit pronounced sexual dimorphism and appear very similar. **(b)** In polygynous species, such as white-tailed deer, (*Odocoileus virginianus*), males are bigger than females and have large horns with which they engage other males in combat over females. **(c)** In polyandrous species, females are usually bigger, as with these golden silk spiders (*Nephila clavipes*).

the offspring of the first female, decreasing her fitness. As a result, on sensing these pheromones, the first female interferes with the male's attempts at signaling, preserving the monogamous relationship.

## In Polygynous Mating Systems, One Male Mates with Many Females

In **polygyny** (Greek, meaning many females), one male mates with more than one female in a single breeding season. Physiological constraints often dictate that female organisms must care for the young. Because of these constraints, at least in many organisms with internal fertilization, such as mammals and some fishes, males are able to mate with and then desert several females. Polygynous systems are therefore associated with uniparental care of young, with males contributing little. Sexual dimorphism is typical in polygynous mating systems, with males developing a larger body size to boost success in competition over mates (Figure 42.16b). Sexual selection is prevalent in polygynous mating systems, with females choosing the most attractive males (intersexual selection) or mating with the most competitively superior males (the "winners" of intrasexual selection) (refer back to Figure 19.18). Several types of polygyny are known, including resource-based polygyny, harem mating, and communal courting.

**Resource-Based Polygyny** When some critical resource is patchily distributed and in short supply, certain males may dominate the resource and breed with more than one visiting female. The major source of nestling death in the lark bunting (*Calamospiza melanocorys*), which lives in North American grasslands, is overheating from too much exposure to the Sun. Prime territories are therefore those with abundant shade, and some males with shaded territories attract two females, even though the second female can expect no help from the male in the process of rearing young. Males in some exposed territories remain bachelors for the season. From the dominant male's point of view, polygyny is advantageous; from the female's point of view, there may be costs. Although by choosing dominant males, a female may be gaining access to good resources, she may have to share these resources with other females.

**Harem Mating** Sometimes males defend a group of females without commanding a resource-based territory. This pattern is more common where females naturally congregate in groups or herds, perhaps to avoid predation, as with horses, zebras, and some deer, and where space is limited, as with southern elephant seals, which mate on beaches. Usually, the largest and strongest males command most of the matings, but defending the harem is usually so exhausting that males may manage to remain the strongest male for only a year or two.

**Communal Courting** Polygynous mating can occur where neither resources nor groups of females are defended. In some instances, particularly in birds and mammals, males display in designated communal courting areas called **leks** (Figure 42.17). Females come to these areas specifically to find a mate, and they choose a prospective mate after the males have performed elaborate displays. Most females seek to mate with males possessing the most desirable traits, so a few of the flashiest males perform the vast majority of the matings. At a lek of the white-bearded manakin (*Manacus manacus*) of South America, one male accounted for 75% of the 438 matings when there were as many as 10 males. A second male mated 56 times (13% of matings), but six others mated a total of only 10 times.

**Figure 42.17** Male birds at a lek. Black grouse (*Tetrao tetrix*) congregate at a moorland lek in Scotland in April. Females visit the leks, and males display to them.



## In Polyandrous Mating Systems, One Female Mates with Many Males

**Polyandry** (Greek, meaning many males), in which one female mates with several males, is rarer than polygyny. Nevertheless, it occurs in some species of birds, fishes, and insects. Sexual dimorphism is present, with the females being the larger of the sexes (Figure 42.16c). In the Arctic tundra, the summer season is short but very productive, providing a bonanza of insect food for 2 months. The productivity of the breeding grounds of the spotted sandpiper (*Actitis macularia*) is so high that the female becomes rather like an egg factory, laying up to five clutches of four eggs each in 40 days. Her reproductive success is limited not by food but by the number of males she can find to incubate the eggs, and females compete for males, defending territories where the males sit.

Polyandry is also seen in some species where egg predation is high and males are needed to guard the nests. For example, in the pipefish (*Syngnathus typhle*), males have brood pouches that provide eggs with safety and a supply of oxygen- and nutrient-rich water. Females produce enough eggs to fill the brood pouches of two males and may mate with more than one male.

Ultimately, as we have seen, most behaviors have evolved to maximize an individual's reproductive output. In a successful group of individuals, this leads to population growth. But we are not knee deep in black grouse or spotted sandpipers, so there must be some constraints on reproductive output. In Chapter 43, we will turn to the topic of population ecology and explore how populations grow and the factors that limit their growth.

## 42.5 Reviewing the Concepts

- Four types of mating systems are found among animals: promiscuity, monogamy, polygyny, and polyandry. Relative body size of males and females depends on mating system (Figures 42.15, 42.16).
- Polygynous mating can often occur in situations when males dominate a resource, defend groups of females, or display in common courting areas called leks (Figure 42.17).

## 42.5 Testing Your Knowledge

1. Fisher's principle suggests the sex ratio of males to females in a population should tend toward
   a. 100:1. c. 1:1. e. 1:100.
   b. 10:1. d. 1:10.
2. In polyandrous mating systems,
   a. males are usually bigger than females.
   b. females are usually bigger than males.
   c. males and females are of roughly equal size.

# Assess and Discuss

## Test Yourself

1. What is the proximate cause of male deer fighting over females?
   a. to determine their supremacy over other males
   b. to injure other males so that these other males cannot mate with females
   c. to maximize the number of genes they pass on
   d. because changes in day length stimulate this behavior
   e. because fighting helps rid the herd of weaker individuals
2. Certain behaviors seem to have very little environmental influence. Such behaviors are the same in all individuals regardless of the environment and are referred to as ________ behaviors.
   a. genetically programmed d. all of the above
   b. instinctual e. b and c only
   c. innate
3. Patrick has decided to teach his puppy a few new tricks. Each time the puppy responds correctly to Patrick's command, the puppy is given a treat. This is an example of
   a. habituation. d. imprinting.
   b. classical conditioning. e. orientation.
   c. operant conditioning.
4. Nutcrackers are able to relocate caches of seeds by means of
   a. innate behavior. d. operant conditioning.
   b. imprinting. e. cognitive maps.
   c. classical conditioning.
5. For group living to have evolved, the benefits of living in a group must be greater than the costs associated with it. Which of the following is an example of a benefit of living in a group?
   a. reduced spread of disease and/or parasites
   b. increased food availability
   c. reduced competition for mates
   d. decreased risk of predation
   e. all of the above
6. The modification of behavior based on prior experience is called
   a. habituation. d. adjustment behavior.
   b. learning. e. innate.
   c. association.
7. When an individual behaves in a way that reduces its own fitness but increases the fitness of others, the organism is exhibiting
   a. kin selection. d. selfishness.
   b. group selection. e. ignorance.
   c. altruism.
8. In naked mole rats, mothers are related to daughters by $r =$
   a. 0. d. 0.5.
   b. 0.125. e. 0.75.
   c. 0.25.
9. In ants, which employ a haplodiploid mating system, fathers are related to sons by $r =$
   a. 0. c. 0.25. e. 0.75.
   b. 0.125. d. 0.5.
10. In a polygynous mating system,
    a. one male mates with one female.
    b. one female mates with many different males.
    c. one male mates with many different females.
    d. many different females mate with many different males.

## Conceptual Questions

1. Some male spiders are eaten by the females after copulation. How can this act be seen to benefit the males?
2. Male parental care occurs in only 7% of fishes and amphibian families with internal fertilization but in 69% of amphibian families with external fertilization. Propose an explanation for why this is so.
3. **PRINCIPLES** A principle of biology is that new properties emerge from complex interactions. Male brown bears (*Ursus arctos*) can be infanticidal, killing existing cubs when they move into a new territory. Explain why bear hunting may have severe consequences for bear populations.

## Collaborative Questions

1. Discuss ways in which organisms communicate with each other, giving examples of each.
2. If a bumblebee queen mates with two males, instead of one, how would this influence eusociality?

# Online Resource

# Population Growth and Species Interactions

43

A wolf in Yellowstone National Park.

© Barrett Hedges/National Geographic/Corbis

## Chapter Outline



The last wild wolves in Montana, Wyoming, and Idaho were killed in the early 1900s as part of a government-funded wolf extermination project. At the time, it was the policy of the federal government to exterminate wolves everywhere, even in national parks. The last wolves in Yellowstone National Park were killed in 1924, when two pups were killed at Soda Butte Creek in the northeast corner of the park. In the 1960s and 1970s, however, public attitudes changed, and wolves received legal protection under the Endangered Species Act of 1973. Wolves from Canada slowly began to filter into northern Montana in the 1980s. After a long, heated debate, wolf reintroductions were started in the continental United States a decade later. Between 1995 and 1996, 66 wolves were captured near Alberta, Canada, and released in Yellowstone Park, Wyoming, and central Idaho.

The combined population of wolves in these states plus Montana showed dramatic growth. In 2011, wolves in Idaho were removed from the endangered species list. Tags for hunting and trapping wolves went on sale May 5, 2011, in Idaho for the fall hunt. Surprisingly, a population that was federally endangered increased so fast that wolves became a game species that could be harvested by humans.

For sexually reproducing species such as wolves, a **population** can be defined as a group of interbreeding individuals of the same species occupying the same area at the same time. This chapter explores **population ecology,** the study of the factors that affect population size and how these factors change over space and time. To study populations, we need to employ some of the tools of **demography**—the study of birth rates, death rates, age distributions, and the sizes of populations. We begin our discussion by examining the methods that ecologists use to measure population density. We will explore characteristics of populations and how growth rates are determined by the number of reproductive individuals in the population and their fertility rate. These data are used to construct simple mathematical models that allow us to analyze population growth, such as that of the reintroduced wolves, and predict future growth. We will also look at the factors that limit the growth of populations, such as competition, predation, and parasitism.

# 43.1 Measuring Population Size and Density

## Learning Outcomes

1. List the different techniques ecologists use to measure population density.
2. **SCISKILLS** ▶ Calculate population size using mark-recapture data, and explain the limitations of this technique.

Within their areas of distribution, organisms occur in varying numbers. We recognize this pattern by saying a plant or an animal is "rare" in one place and "common" in another. For more precision, ecologists quantify distribution further and talk in terms of population **density**—the numbers of organisms in a given unit area or volume. Population growth affects population size and population density, and knowledge of both can help us make decisions about the management of species. How long will it take for a population of an endangered species to recover to a healthy level if we protect it from its most serious threats?

The simplest method for measuring population size is to visually count the number of organisms in a population. We can reasonably do this only if the area they live in is small and the organisms are relatively large. For example, we can readily determine the number of gumbo limbo trees (*Bursera simaruba*) on small islands in the Florida Keys. Normally, however, population ecologists calculate the density of plants or animals in a small area and use these data to estimate the total abundance over a larger area. For plants, algae, or other sessile organisms such as intertidal animals, it is fairly easy to count numbers of individuals per square meter or, for larger organisms such as trees, numbers per hectare (an area of land equivalent to 2.471 acres). However, many plants are clonal; that is, they grow in patches of genetically identical individuals, so rather than count individuals, we can also use the amount of ground covered by plants as an estimate of vegetation density.

### Quadrats and Transects Are Used to Determine the Densities of Plants and Sessile Animals

Plant ecologists use a sampling device called a **quadrat,** a square frame that often measures 50 × 50 cm and encloses an area of 0.25 m$^2$ (Figure 43.1a). They then count the numbers of plants of a given species inside the quadrat and repeat this several times to obtain a density estimate per square meter. For example, if you counted densities of 20, 35, 30, and 15 plants in four quadrats, you could reliably say that the average density of this species was 25 individuals per 0.25 m$^2$, or 100/m$^2$. For larger plants, such as trees, a quadrat would be ineffective. To count such organisms, many ecologists perform a **line transect,** in which a long piece of string is stretched out and any tree along its length is counted. For example, to count tree species on larger islands in the Florida Keys, we could lay out a 100-m line transect and count all the trees within 1 m on either side of the transect. In effect, this transect is little more than a long, thin quadrat encompassing 200 m$^2$. By performing five such transects, we could obtain estimates of tree density per 1,000 m$^2$ and then extrapolate that to a number per hectare or per island.

Several different sampling methods exist for quantifying the density of animal populations, which are more mobile than plants. Pitfall traps set into the ground can catch species such as spiders, lizards, or beetles wandering over the surface (Figure 43.1b). Sweep nets can be passed over grassland vegetation to dislodge and capture the insects feeding there. Mist nets—very fine netting spread between trees—can entangle flying birds and bats (Figure 43.1c). Baited live traps can capture terrestrial animals (Figure 43.1d). Population density can thus be estimated as the number of animals caught per trap or per unit area where a given number of traps are set—for example, 10 traps per 100 m$^2$ of habitat. For some larger terrestrial or marine species, captured animals can be fitted with radio collars and followed remotely, using an antennal tracking device. Their home ranges can be determined and population estimates developed based on the area of available habitat.

(a) Quadrat

(b) Pitfall trap

(c) Mist net

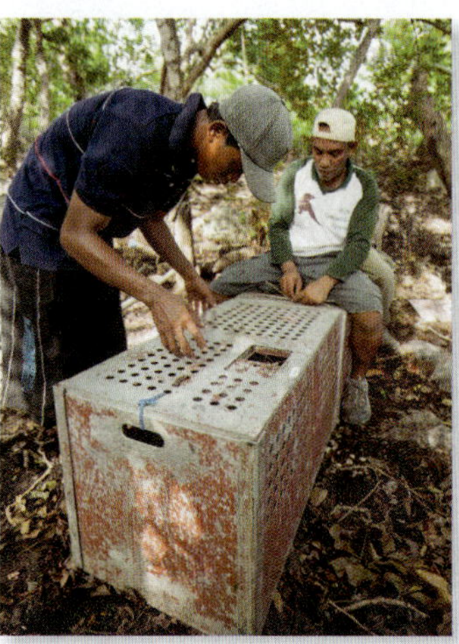
(d) Live mammal trap

**Figure 43.1** **Sampling techniques.** **(a)** Quadrats are frequently used to count the number of plants per unit area. **(b)** Pitfall traps set into the ground catch wandering species such as beetles and spiders. **(c)** Mist nets consist of very fine mesh to entangle birds or bats. **(d)** Baited live traps catch terrestrial animals, including Komodo dragons, shown here on Rinca Island, Indonesia.

In the case of very mobile animals, such as fish in the ocean or large terrestrial animals with a large range, it may not be feasible to measure population density. As an alternative, ecologists may use different methods to estimate population size. For example, we can estimate relative population size by examining catch per unit effort, which is especially valuable in commercial fisheries. We can't easily expect to count the number of fishes in an area of ocean, but we can count the number caught—say, per 100 hours of trawling.

## Quantitative Analysis

### MARK-RECAPTURE CAN BE USED TO ESTIMATE POPULATION SIZE

Sometimes, population biologists capture animals and then tag and release them (Figure 43.2). The rationale behind the **mark-recapture technique** is that after the tagged animals are released, they mix freely with unmarked individuals and within a short time are randomly mixed within the population. The population is then resampled, and the numbers of marked and unmarked individuals are recorded. We assume that the ratio of marked to unmarked individuals in the second sample is the same as the ratio of marked individuals in the first sample to the total population size. Thus,

$$\frac{\text{Number of individuals marked in first catch}}{\text{Total population size, } N} = \frac{\text{Number of marked recaptures in second catch}}{\text{Total number of second catch}}$$

Let's say we catch 50 largemouth bass in a lake and mark them with colored fin tags. A week later, we return to the lake and catch 40 fish and 5 of them were previously tagged fish. If we assume no immigration or emigration has occurred, which is quite likely in a closed system like a lake, and we assume there have been no births or deaths of fish, then the total population size is given by rearranging the equation:

$$\text{Total population size, } N = \frac{\text{Number of marked individuals in first catch} \times \text{Total number of second catch}}{\text{Number of marked recaptures in second catch}}$$

Using our data,

$$N = \frac{50 \times 40}{5} = 400$$

From this equation, we estimate that the lake has a total population size of 400 largemouth bass. This might be useful information for game and fish personnel who wish to know the total size of a fish population in order to set catch limits.

However, the mark-recapture technique can have drawbacks. Some animals that have been marked may learn to avoid the traps. Recapture rates will then be low, resulting in an overestimate of population size. Imagine that instead of 5 tagged fish out of 40 recaptured fish, we get only 2 tagged fish. Now our population size estimate is 2,000/2 = 1,000, a dramatic increase in our estimate. On the other hand, some animals can become "trap-happy," particularly if the traps are baited with food. This would result in an underestimate of the population size.

**Figure 43.2 The mark-recapture technique for estimating population size.** An ear tag identifies this Rocky Mountain goat (*Oreamnos americanus*) in Olympic National Park, Washington. Recapture of such marked animals permits estimates of population size.


**Concept Check:** *If we mark 110 Rocky Mountain goats and recapture 100 goats, 20 of which have ear tags, what is the estimate of the total population size?*

**Crunching the Numbers:** In a study of rodent populations, you catch 60 mice, mark them with plastic ear tags, and recatch 40, of which 12 are marked. What is the population size? You later realize that half of the ear tags fell off the mice prior to being recaptured. By how much have you misestimated the population size?

## 43.1 Reviewing the Concepts

- Population ecology studies how populations grow and what factors promote and limit growth. Ecologists measure population density, the numbers of organisms in a given unit area, in many ways (Figure 43.1).
- Similarly, a variety of techniques can be used to determine population size, including the mark-recapture technique (Figure 43.2).

## 43.1 Testing Your Knowledge

1. What method would you use to estimate the population density of beetles?
   a. quadrats
   b. line transects
   c. pitfall traps
   d. mist nets
   e. all of the above

## 43.2 Demography

### Learning Outcomes

1. Describe the difference between information summarized in a life table and information in a survivorship curve.
2. Differentiate among type I, II, and III survivorship curves, and give examples of organisms that exhibit each of those survivorship curves.
3. **SCISKILLS ▶** Use survivorship and age-specific fertility data to predict a population's growth.

Populations are not usually constant, and ecologists are often interested in how population size and structure changes over time. **Demography** is the statistical study of populations that focuses on various factors such as births, deaths, fertility, and age structure. One way to determine how a population will change is to examine a cohort of individuals from birth to death. For most animals and plants, this involves marking a group of individuals in a population as soon as they are born or germinate and following their fate through their lifetime. For some long-lived organisms, such as tortoises, elephants, or trees, this is impractical, so a snapshot approach is used, in which researchers examine the age structure of a population at one point in time. Recording the presence of juveniles and mature individuals, researchers use this information to construct a **life table**—a table that provides data on the number of individuals alive in each age class. Age classes can be created for any time period, but they often represent 1 year. Generally, only females and their female offspring are included in these tables, because only females produce offspring. The resulting numbers are usually similar to those if males and females were counted together. In this section, we will examine how to construct life tables and plot survivorship curves, which show at a glance the general pattern of population survival over time.

### Life Tables and Survivorship Curves Summarize Survival Patterns

Let's consider a life table for the North American beaver (*Castor canadensis*). Prized for their pelts, by the mid-19th century, these animals had been hunted and trapped to near extinction. Beavers began to be protected by laws in the 20th century, and populations recovered in many areas, often growing to what some considered to be nuisance status. In Newfoundland, Canada, legislation supported trapping as a management technique. From 1964 to 1971, trappers provided mandibles from which teeth were extracted for age classification. If many mandibles were obtained from, say, 1-year-old beavers, then such animals were probably common in the population. If the number of mandibles from 2-year-old beavers was low, then we know there was high mortality for the 1-year-old age class. From the mandible data, researchers constructed a life table (**Table 43.1**). The number of individuals alive at the start of the time period (in this case, a year) is referred to as $n_x$, where $n$ is the number, and $x$ refers to the particular age class. By subtracting the value of $n_x$ from the number alive at the start of the previous year, we can calculate the number dying in a given age class or year, $d_x$. Thus, $d_x = n_x - n_{x+1}$. For example, in Table 43.1, 273 beavers were alive at the start of their sixth year ($n_5$), and only 205 were alive at the start of the seventh year ($n_6$); thus, 68 died during the sixth year: $d_5 = n_5 - n_6$, or $d_5 = 273 - 205 = 68$.

**Table 43.1** Life Table for the Beaver (*Castor canadensis*) in Newfoundland, Canada

| Age (years), $x$ | Number alive at start of year, $n_x$ | Number dying during year, $d_x$ | Age-specific survivorship rate, $l_x$ | Age-specific fertility, $m_x$ | $l_x m_x$ |
|---|---|---|---|---|---|
| 0–1 | 3,695 | 1,995 | 1.000 | 0.000 | 0 |
| 1–2 | 1,700 | 684 | 0.460 | 0.315 | 0.145 |
| 2–3 | 1,016 | 359 | 0.275 | 0.400 | 0.110 |
| 3–4 | 657 | 286 | 0.178 | 0.895 | 0.159 |
| 4–5 | 371 | 98 | 0.100 | 1.244 | 0.124 |
| 5–6 | 273 | 68 | 0.074 | 1.440 | 0.107 |
| 6–7 | 205 | 40 | 0.055 | 1.282 | 0.071 |
| 7–8 | 165 | 38 | 0.045 | 1.280 | 0.058 |
| 8–9 | 127 | 14 | 0.034 | 1.387 | 0.047 |
| 9–10 | 113 | 26 | 0.031 | 1.080 | 0.033 |
| 10–11 | 87 | 37 | 0.024 | 1.800 | 0.043 |
| 11–12 | 50 | 4 | 0.014 | 1.080 | 0.015 |
| 12–13 | 46 | 17 | 0.012 | 1.440 | 0.017 |
| 13–14 | 29 | 7 | 0.007 | 0.720 | 0.005 |
| 14+ | 22 | 22 | 0.006 | 0.720 | 0.004 |

Net reproductive rate, $R_0 = \sum l_x m_x = 0.938$

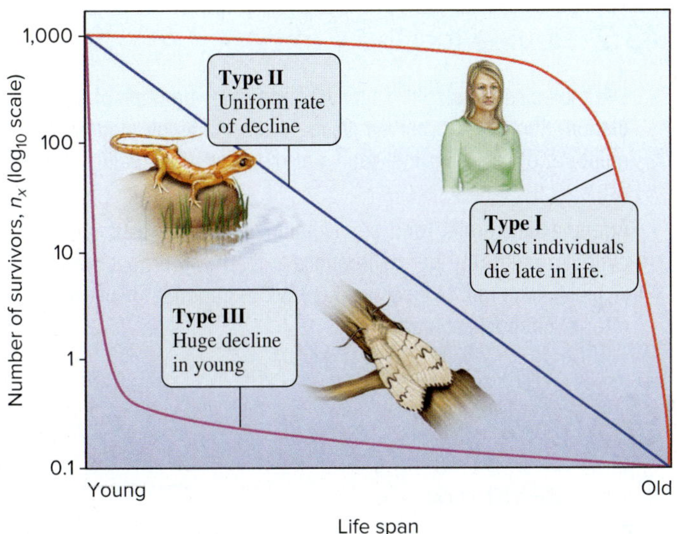


**Figure 43.3** Idealized survivorship curves.

**BioConnections:** *Refer back to Figure 26.16. Which type of survivorship curve would you expect in trematodes?*

A simple and informative exercise is to plot numbers of surviving individuals ($n_x$) at each age, creating a **survivorship curve.** The value of $n_x$ is typically expressed on a logarithmic scale. Survivorship curves generally fall into one of three patterns (Figure 43.3). In a type I curve, the rate of loss for juveniles is relatively low, and most individuals are lost later in life, as they become older and more prone to sickness and predators. Organisms that exhibit type I survivorship have relatively few offspring but invest much time and resources in raising their young. Many large mammals, including humans, exhibit type I curves. At the other end of the scale is a type III curve, in which the rate of loss for juveniles is relatively high, and the survivorship curve flattens out for those organisms that have avoided early death. Many fishes and marine invertebrates fit this pattern. Most of the juveniles die or are eaten, but a few reach a favorable habitat and thrive. For example, once they find a suitable rock face on which to attach themselves, barnacles grow and survive very well. Many insects and plants also fit the type III survivorship curve, because they lay many eggs or release hundreds of seeds, respectively. Type II curves represent a middle ground, with fairly uniform death rates over time. Species with type II survivorship curves include many birds, small mammals, reptiles, and some annual plants.

## Age-Specific Fertility Data Are Needed to Predict Population Growth

To calculate how a population grows, we need information on birth rates as well as mortality and survivorship rates. For any given age, we can determine the per female rate of female offspring production or **age-specific fertility rate** ($m_x$). For example, if 100 females of a given age produce 75 female offspring, $m_x = 0.75$. An examination of the beaver age-specific fertility rates in Table 43.1 illustrates a couple of general points. First, for this beaver population in particular, and for many organisms in general, no babies are born to very young females. As females mature sexually, age-specific fertility goes up, and it remains fairly high until later in life, when females reach postreproductive age.

With the addition of age-specific fertility data, we can calculate the growth rate of a population. First, we use the survivorship data to find the proportion of individuals alive at the start of any given age class. This **age-specific survivorship rate,** termed $l_x$, equals $n_x/n_0$, where $n_0$ is the number alive at time 0, the start of the study, and $n_x$ is the number alive at the beginning of age class $x$. Let's examine the beaver life table in Table 43.1. The proportion of the original beaver population still alive at the start of the sixth age class, $l_5$, equals $n_5/n_0 = 273/3{,}695$, or 0.074. This means that 7.4% of the original beaver population survived to age 5. Next we multiply $l_x$ by $m_x$, the age-specific fertility rate, to give us $l_xm_x$, an average number of female offspring per female in that age class. This column represents the contribution of each age class to the overall population growth rate. In our example, the average number of female offspring per female for beavers aged 5–6 is 1.440.

The number of offspring born to females of any given age class depends on two things: the number of females in that age class and their age-specific fertility rate. Thus, although fertility of young beavers is very low, there are so many females in the age class that $l_xm_x$ for 1-year-olds is quite high. Age-specific fertility for older beavers is much higher, but the fewer females in these age classes cause $l_xm_x$ to be low. Maximum values of $l_xm_x$ occur for females of an intermediate age, 3–4 years old in the case of the beaver. The overall growth rate per generation is the number of offspring born to all females of all ages, where a generation is defined as the mean period between birth of females and birth of their offspring. Therefore, to calculate the generational growth rate, we sum all the values of $l_xm_x$—that is, $\sum l_xm_x$, where the $\sum$ symbol means "sum of." This summed value, $R_0$, is called the **net reproductive rate.**

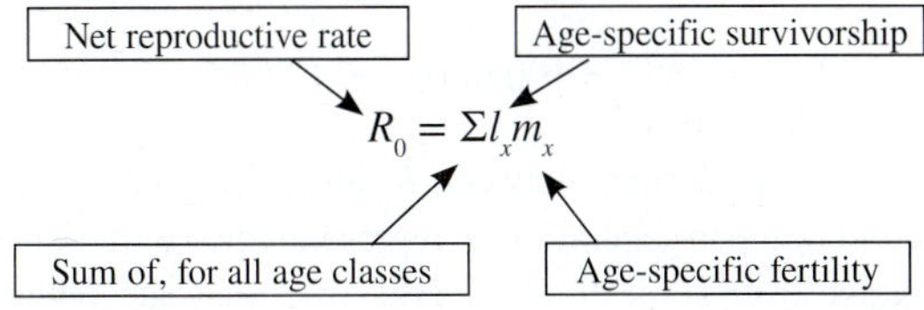


To estimate the size of a population in the next generation, $N_{t+1}$, we multiply the number of individuals in the population at time $t$ by the net reproductive rate, $R_0$:

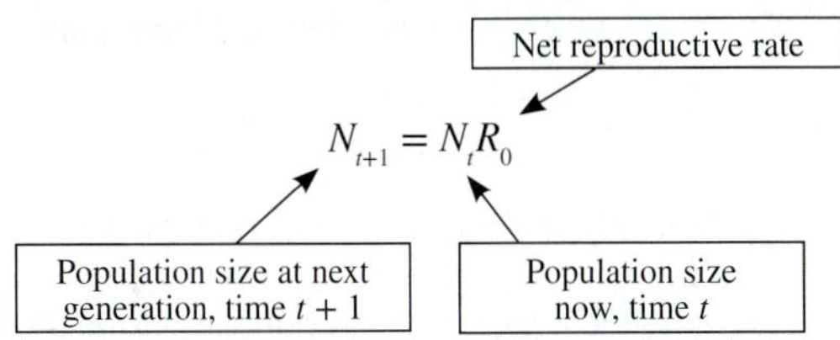

Let's consider an example in which the number of beavers alive now, $N_t$, is 1,000, and $R_0 = 1.1$. The size of the population in the next generation, $N_{t+1}$, is given by

$$N_{t+1} = N_t R_0$$
$$N_{t+1} = 1{,}000 \times 1.1$$
$$= 1{,}100$$

Therefore, the number of beavers in the next generation is 1,100 and the population will have increased. This means the beaver population is reproducing at a rate that is 10% greater than simply replacing itself.

In estimating future population size, much depends on the value of $R_0$. If $R_0 > 1$, the population is increasing. If $R_0 < 1$, the population is in decline. If $R_0 = 1$, the population size stays the same, and we say it is at **equilibrium.** In the case of the beavers, Table 43.1 reveals that $R_0 = 0.938$, which is less than 1, and therefore the population is declining. This is valuable information, because it tells us that at that time, the beaver population in Newfoundland needed some form of protection (perhaps bans on trapping and hunting) in order to attain a population level at equilibrium.

Because of the effort involved in calculating $R_0$, ecologists sometimes use a shortcut to predict population growth. Imagine a bird species that breeds annually. To measure population growth, ecologists first count the number of birds in a starting population, $N_0$. Let's say $N_0 = 100$. The next year, ecologists count 105 birds in the same population, so $N_1 = 105$. The **finite rate of increase,** $\lambda$, is the ratio of the population size from one year to the next, calculated as

$$\lambda = N_1/N_0$$

In this case, $\lambda = 1.05$, or 105%. In other words, the population increases 5% in a given year. $\lambda$ is often given as percent annual growth, and $t$ is a number of years. Let's consider a population of birds growing at a rate of 5% per year. To calculate the size of the population after 5 years, we substitute $\lambda$ for $R_0$:

$$N_t = 100, \lambda = 1.05, \text{ and } t = 5$$

therefore,

$$N_{t+5} = 100\,(1.05)^5 = 127.6$$

What's the difference between $R_0$ and $\lambda$? $R_0$ represents the net reproductive rate per generation. $\lambda$ represents the finite rate of population change over some time interval, often a year. When species are annual breeders that live 1 year, such as annual plants, $R_0 = \lambda$. For species that breed for multiple years, $R_0 \neq \lambda$. The size of a given population can be predicted in two ways:

$$N_t = N_0 R_0^{\,t}, \text{ where } t = \text{a number of generations}$$

or

$$N_t = N_0 \lambda^t, \text{ where } t = \text{a number of time intervals}$$

Populations grow when $R_0$ or $\lambda > 1$; populations decline when $R_0$ or $\lambda < 1$; and they are at equilibrium when $R_0$ or $\lambda = 1$.

## 43.2 Reviewing the Concepts

- Life tables summarize the survival and fertility patterns of a population. Survivorship curves illustrate life tables by plotting the numbers of surviving individuals at different ages (Figure 43.3; Table 43.1).
- Age-specific fertility and survivorship data help estimate the overall growth rate per generation, or the net reproductive rate ($R_0$). The finite rate of increase, $\lambda$, provides information on growth rates over specific time periods.

## 43.2 Testing Your Knowledge

1. In a survivorship curve,
   a. $l_x$ is plotted against time.
   b. survivorship is plotted against $n_x$.
   c. log $n_x$ is plotted against time.
   d. $n_x$ is plotted against log time.
   e. log $n_x$ is plotted against log time.
2. Plot the survivorship curve for American beavers using the data in Table 43.1. What type of curve results?
   a. type I
   b. type II
   c. type III
   d. type IV
   e. none of the above
3. If $\Sigma\, l_x m_x = 1.0$, which of the following statements is *true?*
   a. There are few females of reproductive age.
   b. The population is increasing.
   c. The survivorship of females is low.
   d. Over a 1-year interval, $\lambda = 1$.
   e. Population size will be unchanged from one generation to the next.

## 43.3 How Populations Grow

### Learning Outcomes

1. **SCISKILLS ▶** Calculate the population growth of continuously breeding organisms using the per capita growth rate ($r$).
2. Distinguish between exponential growth and logistic growth.
3. Describe how density-dependent factors can regulate population size.

Life tables provide estimates about how populations can grow from generation to generation. As we have seen, other population growth models can provide valuable insights into how populations grow over time periods shorter than whole generations. In this section, we will contrast two different models of population growth. The first model assumes resources are not limiting, and it results in prodigious growth. The second, and perhaps more biologically realistic, model assumes resources are limiting, and it results in limits to growth and eventual stable population sizes.

## Knowing the per Capita Growth Rate Helps Predict How Populations Will Grow

The change in population size over any time period can be written as the number of births per unit time interval minus the number of deaths per unit time interval.

For example, if in a population of 1,000 rabbits, there were 100 births and 50 deaths over the course of 1 year, then the population would grow in size to 1,050 the next year. We can write this formula mathematically as

$$\frac{\text{Change in numbers}}{\text{Change in time}} = \text{Births} - \text{Deaths}$$

or

$$\frac{\Delta N}{\Delta t} = B - D$$

The Greek letter delta, $\Delta$, indicates change, so that $\Delta N$ is the change in number, and $\Delta t$ is the change in time; $B$ is the number of births per time unit; and $D$ is the number of deaths per time unit.

The numbers of births and deaths can be expressed per individual in the population, so the birth of 100 rabbits to a population of 1,000 would represent a per capita birth rate, $b$, of 100/1,000, or 0.10. Similarly, the death of 50 rabbits in a population of 1,000 would be a per capita death rate, $d$, of 50/1,000, or 0.05. Now we can rewrite our equation giving the rate of change in a population.

$$\frac{\Delta N}{\Delta t} = bN - dN$$

For our rabbit example,

$$\frac{\Delta N}{\Delta t} = 0.10 \times 1{,}000 - 0.05 \times 1{,}000 = 50$$

so if $\Delta t = 1$ year, the rabbit population increases by 50 individuals in a year.

Ecologists often simplify this formula by representing $b - d$ as $r$, the **per capita rate of increase.** Thus, $bN - dN$ can be written as $rN$. Because ecologists are interested in population growth rates over very short time intervals, so-called instantaneous growth rates, instead of writing

$$\frac{\Delta N}{\Delta t}$$

they write

$$\frac{dN}{dt}$$

which is the notation of differential calculus. The equations essentially mean the same thing, except that $dN/dt$ reflects very short time intervals. Thus,

$$\frac{dN}{dt} = rN = (0.10 - 0.05)N = 50$$

## Exponential Growth Occurs When the per Capita Rate of Increase Remains Above Zero

How do populations grow? Clearly, much depends on the value of the per capita rate of increase rate, $r$. When $r < 0$, the population decreases; when $r = 0$, the population remains constant; and when $r > 0$, the population increases. When $r = 0$, the population is again referred to as being at equilibrium, where no changes in population size will occur and there is zero population growth.

Even if $r$ is only fractionally above 0, population increase is rapid, and when plotted graphically a J-shaped curve results (Figure 43.4). We refer to this type of population growth as **exponential growth.** How do field data fit this simple model for exponential growth? In a new and expanding population when resources are not limited, exponential growth is often observed. The wolf populations in Montana, Wyoming, and Idaho that we discussed at the beginning of the chapter provide a good example. Wolves began dispersing into Montana from Canada in the early 1980s, but the release of wolves in nearby Yellowstone National Park and neighboring Idaho and their subsequent population growth caused an exponential increase in wolf numbers in these areas (Figure 43.5a).

The Burmese python, *Python molorus,* is now well established in an area of south Florida called the Everglades. At up to 7 m (22 feet) long and 90 kg (200 lb) in weight, the snakes have a large appetite and can eat prey as large as white-tailed deer and alligators. Most likely, a small number of adult snakes, once pets, were released prior to 1985, and since then their numbers have increased dramatically. Researchers fitted an exponential growth curve to python abundance based on capture numbers (Figure 43.5b). From a relatively small number of pythons likely released prior to 1985, population densities in the Everglades National Park were estimated conservatively at 30,000 individuals in 2007 based on similar density estimates of this species in its native India.

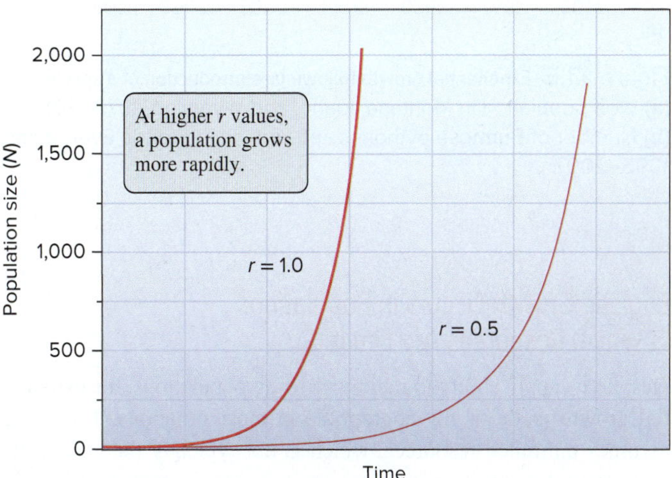


**Figure 43.4 Exponential population growth.** At higher $r$ values, the slope of the curve gets steeper. In theory, a population with unlimited resources could grow indefinitely.

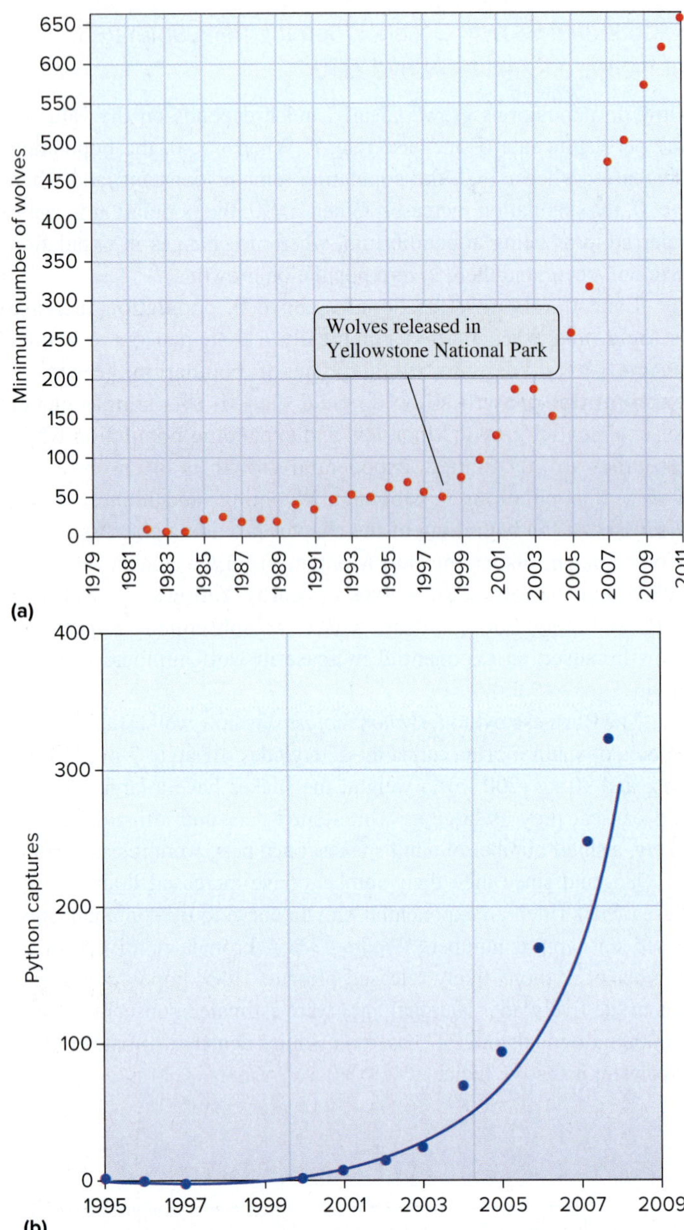


**Figure 43.5** **Exponential growth following reintroduction of a species.** **(a)** Wolf population in Montana, Idaho, and Wyoming, 1979–2011. **(b)** Number of Burmese pythons captured in the Florida Everglades, 1995–2007.

## Logistic Growth Occurs in Populations in Which Resources Are Limited

Despite its applicability to rapidly growing populations, the exponential growth model is not appropriate in most situations. The model assumes unlimited resources, which is not typically the case in the real world. For most species, resources become limiting as populations grow, and the per capita rate of increase decreases. The maximum population size that can be sustained by an environment is known as the **carrying capacity (*K*).** A more realistic equation to explain population growth, one that takes into account the amount of available resources, is

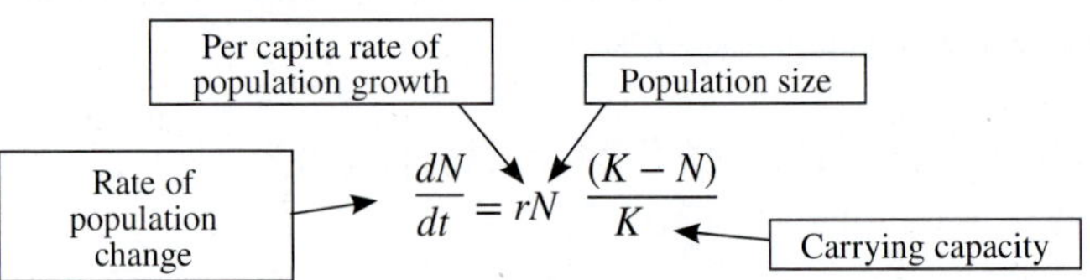


$$\frac{dN}{dt} = rN\frac{(K-N)}{K}$$

where $(K - N)/K$ represents the proportion of the carrying capacity that is unused by the population. This equation is called the **logistic equation.**

As the population size, $N$, grows, it moves closer to the carrying capacity, $K$, with fewer available resources for population growth. If $K = 1{,}000$, $N = 500$, and $r = 0.1$, then population growth is 25 individuals per unit of time. At larger values of $N$, the value of $(K - N)/K$ becomes small, and population growth is smaller. If $K = 1{,}000$, $N = 900$, and $r = 0.1$, then

$$\frac{dN}{dt} = (0.1)(900) \times \frac{(1{,}000 - 900)}{1{,}000}$$

$$\frac{dN}{dt} = 9$$

In this instance, population growth is only 9 individuals per unit of time.

Let's consider how an ecologist would use the logistic equation. First, the value of $K$ would come from intense field and laboratory work in which researchers would determine the amount of resources, such as food, needed by each individual and then determine the amount of available food in the wild. Field censuses determine $N$, and field censuses of births and deaths per unit time provide $r$. When this type of population growth is plotted over time, an S-shaped growth

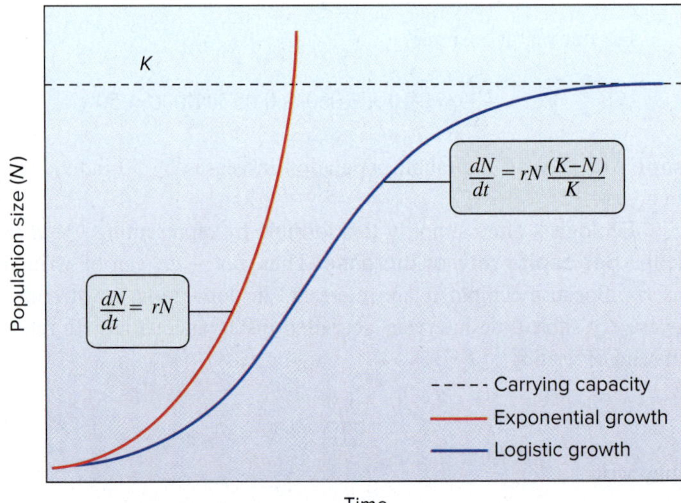


**Figure 43.6** **Exponential versus logistic growth.** Exponential (J-shaped) growth occurs in an environment with unlimited resources, whereas logistic (S-shaped) growth occurs in an environment with limited resources.

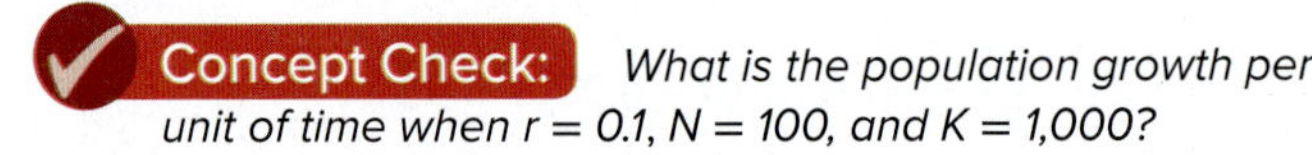

*What is the population growth per unit of time when r = 0.1, N = 100, and K = 1,000?*

## Biology Principle

### Biology Is an Experimental Science

Tests of population growth models are more accurately performed using manipulative experiments than by simple field observations.

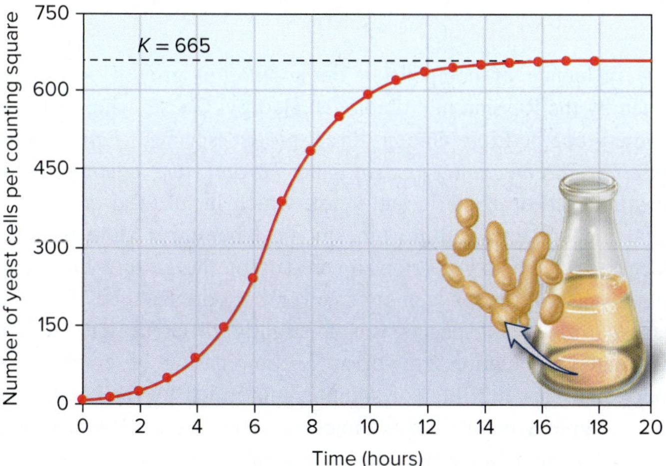


**Figure 43.7 Logistic growth of yeast cells in culture.** Early tests of the logistic growth curve were validated by growth of yeast cells in laboratory cultures. These populations showed the typical S-shaped growth curve.

curve results (**Figure 43.6**). This pattern, in which the growth of a population slows down as it approaches $K$, is called **logistic growth.**

Does the logistic growth model provide a better fit to growth patterns of plants and animals in the wild than the exponential model? In some instances, such as laboratory cultures of bacteria and yeasts, the logistic growth model provides a very good fit (**Figure 43.7**). In nature, however, variations in temperature, rainfall, or resources can cause frequent changes in carrying capacity and thus in population size. Therefore, the logistic growth model is rarely an exact fit to population growth in the field. Is the logistic model therefore of little value? Not really. It is a useful starting point for thinking about how populations grow, and it seems intuitively correct. However, the carrying capacity is a difficult feature of the environment to identify for most species, and it varies with time and local climate patterns. Also, populations are affected by interactions with other species. In Section 43.4, we will see how predators, parasites, and competitors can affect population densities. As described later, such population limitations are often influenced by a process known as density dependence.

## 43.3 Reviewing the Concepts

- The per capita rate of increase ($r$) helps determine how populations grow over any time period. When $r$ is $> 0$, exponential (J-shaped) growth occurs. Exponential growth can be observed in an environment where resources are not limited (Figures 43.4, 43.5).
- Logistic (S-shaped) growth takes into account the upper boundary for a population, called carrying capacity, and occurs in an environment where resources are limited (Figures 43.6, 43.7).

## 43.3 Testing Your Knowledge

1. Under what conditions will a population grow?
   a. $R_0 = 0.5$
   b. $r = 0.4$
   c. $\lambda = 0.8$
   d. a and b only
   e. b and c only
2. Suppose you have a pond that is being overgrown by an aggressive weed that doubles in number every day. If unchecked, it would cover the pond in 30 days. Weed growth is very slow, almost negligible, to begin with, and you decide not to remove the weed until the pond is half covered. When will that be?
   a. day 1 b. day 10 c. day 15 d. day 20 e. day 29

## 43.4 Species Interactions

### Learning Outcomes

1. Describe the four general types of species interactions.
2. **SCISKILLS ▶** Predict how the competitive exclusion principle can lead to resource partitioning among species.
3. List and describe strategies plants and animals use to avoid being eaten, and give examples of the effects of natural enemies on prey populations.

In this section, we turn from considering populations on their own to investigating how they interact with populations of other species that live in the same locality. Such species interactions can take a variety of forms (**Table 43.2**). **Competition** can be defined as an interaction that affects both species negatively (–/–), as both species compete over food or other resources. Sometimes an interaction is quite one-sided, being detrimental to one species and neutral to the other; such an

**Table 43.2 Summary of the Types of Species Interactions**

| Nature of interaction | Species 1* | Species 2* |
|---|---|---|
| Competition | – | – |
| Amensalism | – | 0 |
| Mutualism | + | + |
| Commensalism | + | 0 |
| Predation, herbivory, parasitism, parasitoidism | + | – |

*+ = positive effect; – = negative effect; 0 = no effect.

interaction is called **amensalism** (−/0). **Mutualism** is an interaction in which both species benefit (+/+), whereas **commensalism** benefits one species and is neutral to the other (+/0). Finally, four types of interactions are positive for one species but detrimental to the other (+/−):

- **Predation** A predator feeds on prey and causes their rapid death.
- **Herbivory** An animal feeds on a plant.
- **Parasitism** A parasite establishes a relatively long-term relationship with its host, which may or may not lead to the host's death.
- **Parasitoidism** An organism lays eggs in a host and the resulting larvae remain in the host and usually kill it.

To illustrate how species interact in nature, let's consider a rabbit population in a woodland community (Figure 43.8). To determine what factors influence the size and density of the rabbit population, we need to understand each of its possible species interactions. For example, the rabbit population could be limited by the quality or availability of food. It is also likely that other species, such as deer, use the same resource and thus compete with the rabbits. The rabbit population could be limited by predation from foxes or by the virus that causes the disease myxomatosis, which is usually spread by fleas and mosquitoes. Finally, gut microbes help the rabbit to digest cellulose-rich plant materials and receive nutrients for themselves.

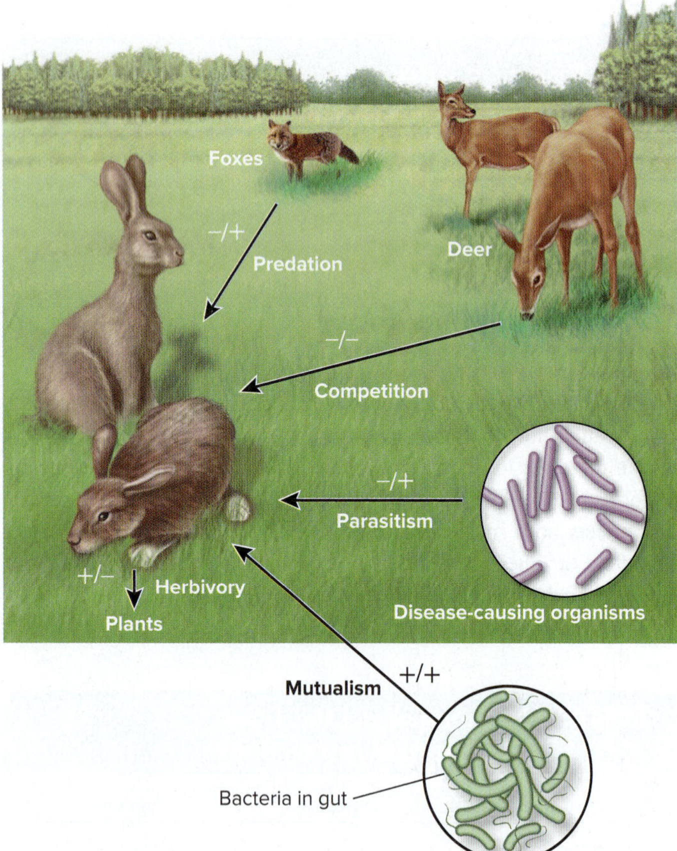


**Figure 43.8** **Species interactions.** These rabbits can interact with a variety of species, experiencing predation by foxes, competition with deer for food, and parasitism from various disease-causing organisms. Herbivory occurs when rabbits feed on various plants. Finally, mutualism occurs between the rabbit and gut microbes that aid in the digestion of cellulose-rich plant material. The effects of each species on the other are shown by the terms assigned to the arrows, as discussed in the text.

## Competition Can Negatively Affect the Population Growth of Competing Species

The influence of competition on population growth was illustrated by the Russian microbiologist Georgyi Gause, who, in 1934, studied competition among three protist species, *Paramecium aurelia*, *Paramecium bursaria*, and *Paramecium caudatum*, all of which fed on bacteria and yeast, which in turn fed on an oatmeal medium in a culture tube in the laboratory. More bacteria occurred in the oxygen-rich upper part of the culture tube, and more yeast in the oxygen-poor lower part of the tube. Because each *Paramecium* species was a slightly different size, Gause calculated population growth as a combination of numbers of individuals per milliliter of solution multiplied by their unit volume to give a population volume for each species. When grown separately, population volume of all three *Paramecium* species followed a logistic growth pattern (Figure 43.9a). When Gause cultured *P. caudatum* and *P. aurelia* together, *P. caudatum* went extinct (Figure 43.9b). Both species utilized bacteria as food, but *P. aurelia* grew at a rate six times faster than *P. caudatum.*

However, when Gause cultured *P. caudatum* and *P. bursaria* together, neither went extinct (Figure 43.9c). The population volume of each was much less than when they were grown alone, because some competition occurred between them. Gause discovered, however, that *P. bursaria* was better able to utilize the yeast in the lower part of the culture tubes. *P. bursaria* have tiny green algae inside them, which produce oxygen and allow *P. bursaria* to thrive in the lower oxygen levels at the bottom of the tubes. From these experiments, Gause concluded that two species with exactly the same requirements cannot live together in the same place and use the same resources—that is, occupy the same **niche.** This became known as the **competitive exclusion principle.** The term **resource partitioning** describes the differentiation of niches, both in space and in time, that enables similar species to coexist in a community, just like the two species of *Paramecium* feeding in different parts of the culture tube. We can think of resource partitioning as reflecting the results of past competition, in which competition led the inferior competitor to eventually occupy a different niche.

By reviewing studies that have investigated competition in nature, we can see how frequently it occurs and in what particular circumstances it is most important. In a 1983 review of field studies by American ecologist Joseph Connell, competition was found in 55% of 215 species surveyed, demonstrating that it is frequent in nature. Generally in studies of single pairs of species utilizing the same resource, competition is almost always reported, whereas in studies involving more species, the frequency of competition decreases. Why should this be the case? Imagine a resource such as a series of different-sized grains with four types of species—ants,

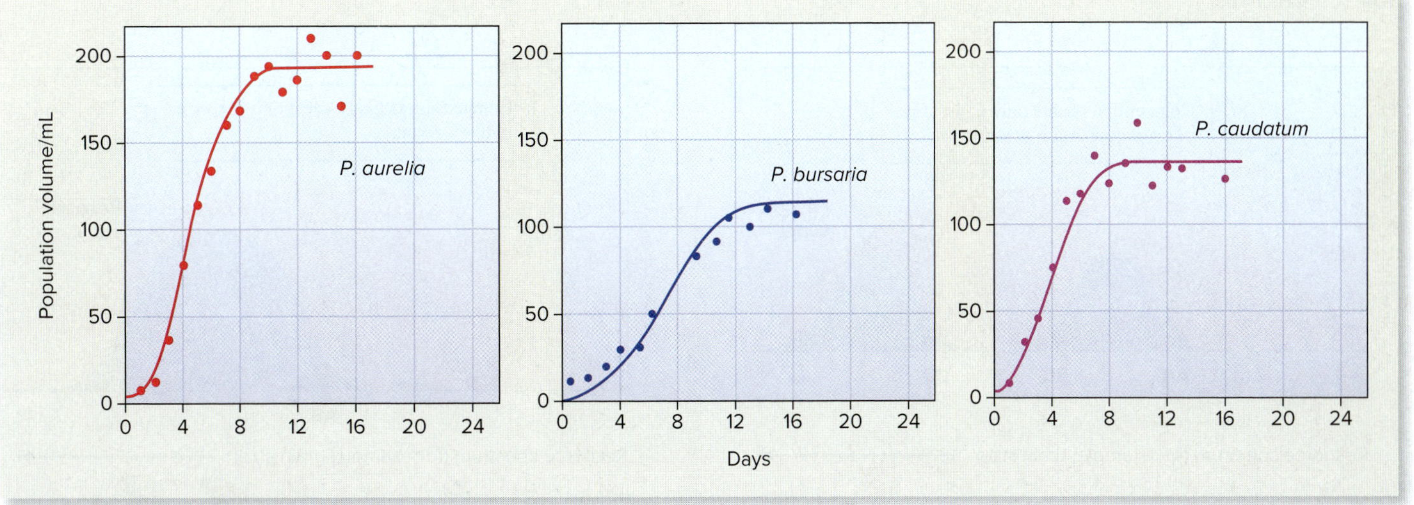


**(a) Each *Paramecium* species grown alone**

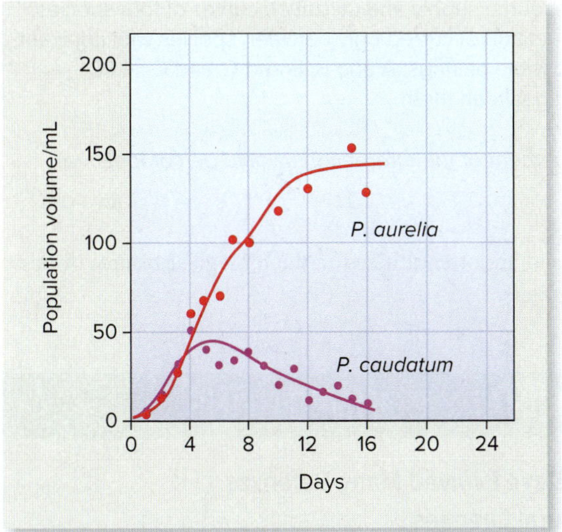


**(b) Competition between *P. aurelia* and *P. caudatum***

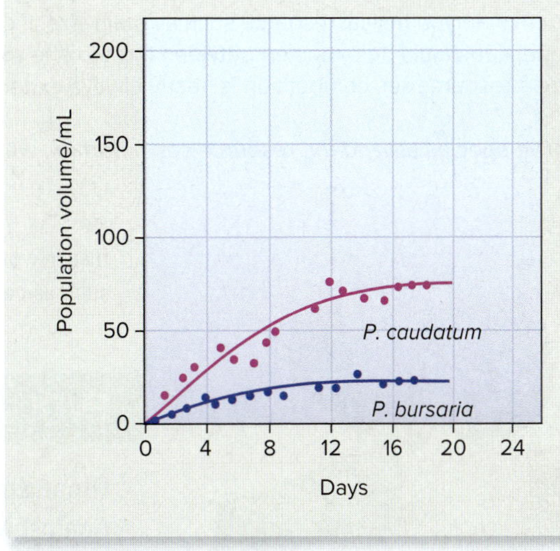


**(c) Competition between *P. caudatum* and *P. bursaria***

**Figure 43.9 Competition among *Paramecium* species.** **(a)** When grown alone, each of three species, *Paramecium aurelia, Paramecium bursaria,* and *Paramecium caudatum,* grows according to the logistic model. **(b)** When *P. aurelia* is grown with *P. caudatum,* the density of *P. aurelia* is lower than when grown alone, and *P. caudatum* goes extinct. **(c)** When *P. caudatum* is grown with *P. bursaria,* the population densities of both are lowered, but the species coexist.

beetles, mice, and birds—feeding on it (Figure 43.10a). The ants feed on the smallest grain, the beetles and mice on the intermediate sizes, and the birds on the largest. If only adjacent species competed with each other, competition would be expected only between the ant–beetle, beetle–mouse, and mouse–bird. Thus, competition would be found in only three out of the six possible species pairs (50%). Naturally, the percentage would vary according to the number of species on the resource spectrum. If only three species occurred along the spectrum, we would expect competition in two of the three pairs (67%). If just two species utilized the resource spectrum, however, we would expect competition in almost 100% of the cases (Figure 43.10b).

Some other general patterns were evident from Connell's review. Plants showed a high degree of competition, perhaps because they are rooted in the ground and cannot easily escape or perhaps because they are competing for the same set of limiting nutrients—water, light, and minerals. Marine organisms tended to compete more than terrestrial ones, most likely because many of the species studied lived in the intertidal zone and were attached to the rock face, in a manner similar to the rooting of plants. Because the area of the rock face is limited, competition for space is quite important.

## Predators, Herbivores, Parasites, and Parasitoids Interact with Their Prey or Host Species in Different Ways

Predation, herbivory, parasitism, and parasitoidism are interactions that have a positive effect for one species and a negative effect for the other. These categories of species interactions can be classified according to how lethal they are for the prey and the length of association between the consumer and prey (Figure 43.11). Each has particular characteristics that set it apart. Herbivory usually involves nonlethal feeding on plants, whereas predation generally results in the death of the prey. Parasitism, like herbivory, is typically nonlethal and differs from predation in that the adult parasite typically lives and reproduces for long periods in or on the living host (refer back to Figure 26.16). Parasitoids are organisms that lay eggs in a host that the resulting larvae remain with and almost always kill. For example, a parasitoid female wasp inserts its eggs into a host, such as a caterpillar, and the

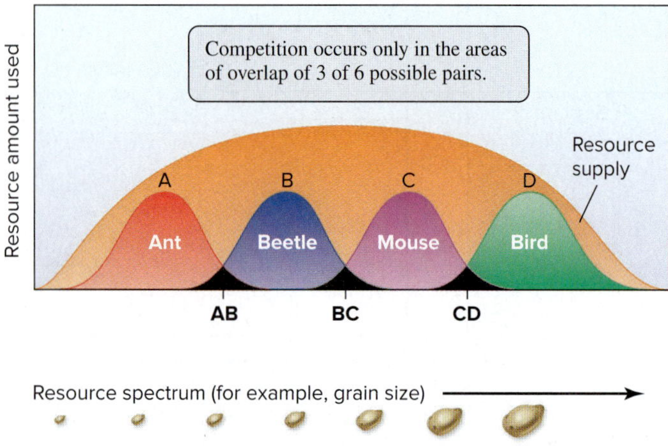


(a) Competition among 4 species for a resource

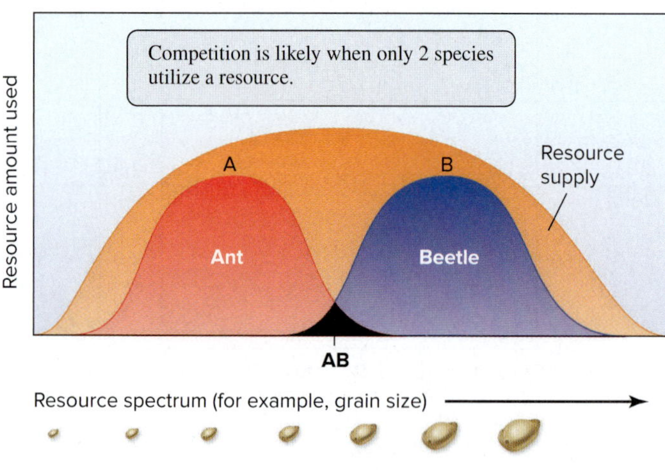


(b) Competition among 2 species for a resource

**Figure 43.10 The frequency of competition according to the number of species involved.** **(a)** Resource supply and utilization curves of four species, A, B, C, and D, along the spectrum of a hypothetical resource such as grain size. If competition occurred only between species with adjacent resource utilization curves, competition would be expected between three of the six possible pairings: A and B, B and C, and C and D. **(b)** When only two species utilize a resource set, competition is nearly always expected between them.

**Concept Check:** *If five species utilized the resource set in part (a), what percent of the interactions would be competitive?*

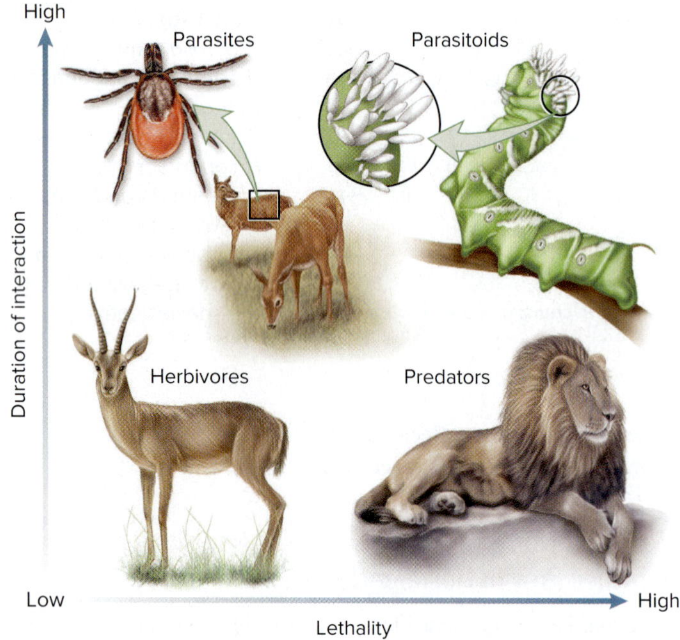


**Figure 43.11 Possible interactions between populations.** Lethality represents the probability that an interaction results in the death of the prey. Duration represents the length of the interaction between the consumer and the prey.

**Concept Check:** *Where might omnivores fit in this figure?*

resulting larvae consume the insides of the caterpillar and eventually kill it. The larvae emerge to create white pupae on the outside of the caterpillar. They thus have features in common with parasites (a long association) and predators (lethality). Parasitoids are common in the insect world and are often utilized in the biological control of pests, such as caterpillars.

## EVOLUTIONARY CONNECTIONS

### Organisms Have Evolved Many Defenses Against Natural Enemies

The variety of strategies that organisms have evolved to avoid being eaten suggests that predators, herbivores, parasites, and parasitoids, also known as **natural enemies,** are a strong selective force. Common strategies include chemical and mechanical defenses, camouflage, displays of intimidation, and armor and weaponry.

**Chemical and Mechanical Defenses** A great many animal species have evolved chemical defenses against predation. One of the classic examples of a chemical defense involves the bombardier beetle (*Stenaptinus insignis*), which has been studied by entomologist Tom Eisner and coworkers. These beetles possess a reservoir of hydroquinone and hydrogen peroxide in their abdomen. When threatened, they eject the chemicals into an "explosion chamber," where the subsequent release of oxygen causes the whole mixture to be violently ejected as a hot spray (about 88°C, or 190°F) that can be directed at the beetle's attackers (Figure 43.12a). Many other arthropods, such as millipedes, also have chemical sprays, and the phenomenon is also found in vertebrates, as anyone who has had a close encounter with a skunk can testify. Often associated with a chemical defense is an **aposematic coloration,** or warning coloration, which advertises an organism's unpalatable taste or its poisonous nature. For instance, many tropical frogs have bright warning coloration that calls attention to their harmful effects due to toxins in their skin (Figure 43.12b).

(a) This bombardier beetle (*Stenaptinus insignis*) directs its hot, stinging spray at a forceps "attacker."

(b) Aposematic coloration advertises the poisonous nature of this blue poison arrow frog (*Dendrobates azureus*) from South America.

(c) Camouflage allows this Pygmy sea horse (*Hippocampus bargibanti*) from Bali to blend in with its background.

(d) In a display of intimidation, this porcupine fish (*Diodon hystrix*) puffs itself up to look threatening to its predators.

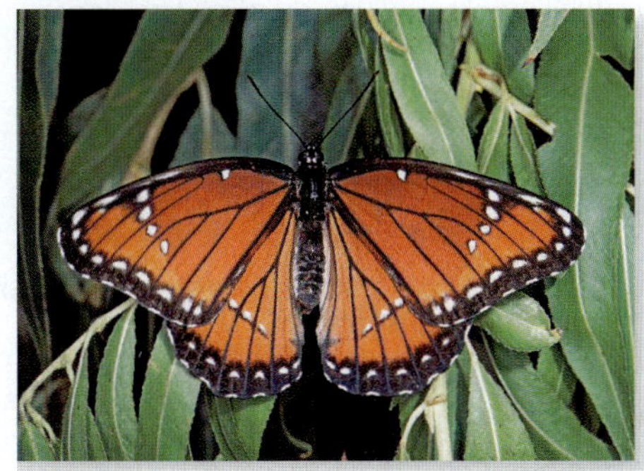

(e) Müllerian mimicry. Viceroy (left) and monarch (right) butterflies are both noxious and have similar color patterns.

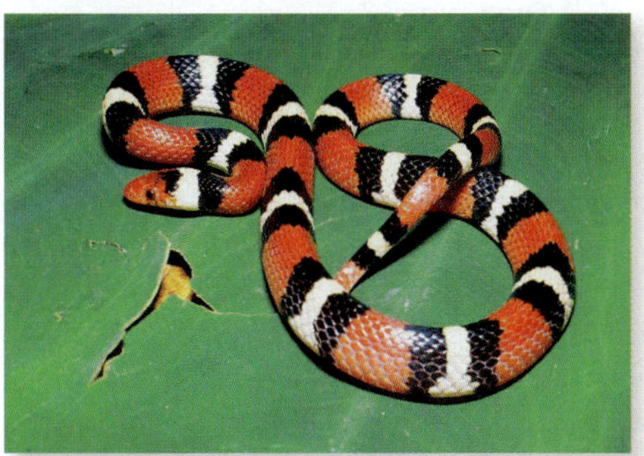

(f) In this example of Batesian mimicry, an innocuous scarlet king snake (*Lampropeltis elapsoides*) (left) mimics the venomous eastern coral snake (*Micrurus fulvius*) (right).

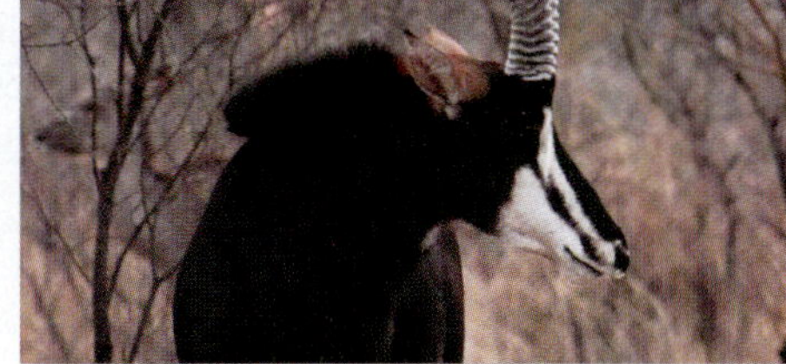

(g) Sable antelope have horns that may be used as predator defense.

**Figure 43.12** Antipredator adaptations.

*(a)* © CB2/ZOB/Supplied by WENN/Newscom; *(b)* © Hans D. Dossenbach/ardea.com; *(c)* © Thomas Aichinger/V&W/The Image Works; *(d)* © Paul Springett 01/Alamy RF; *(e left)* © Rick & Nora Bowers/Alamy; *(e right)* © Eric Carr/Alamy; *(f left)* © Suzanne L. & Joseph T. Collins/Science Source; *(f right)* © Jason Ondreicka/Alamy RF; *(g)* © cd123/123RF.com

**BioConnections:** *Refer back to Figure 42.2. Do predators learn to avoid bad-tasting prey by habituation, classical conditioning, or operant conditioning?*

Many unusual and powerful chemicals are present in plants such as nicotine in tobacco, morphine in poppies, cocaine in coca, and caffeine in coffee. These compounds are not part of the primary metabolic pathway that plants use to obtain energy and are therefore referred to as **secondary metabolites.** Most of these chemicals are bitter tasting or toxic, thereby deterring herbivores from feeding. The staggering variety of secondary metabolites in plants, over 25,000 identified so far, may be testament to the large number of organisms that feed on plants. In an interesting twist, many of these compounds have medicinal properties that have proved to be beneficial to humans.

In addition to containing chemical compounds, many plants have an array of mechanical defenses, such as thorns, hairs, and spines.

**Camouflage** Camouflage is the blending of an organism with the background of its habitat and is a common method of avoiding detection by predators (refer back to Figure 27.17b). For example, many grasshoppers are green and blend in with the foliage on which they feed. Stick insects mimic branches and twigs with their long, slender bodies. In most cases, these animals stay perfectly still when threatened, because movement alerts a predator. Camouflage is prevalent in the vertebrate world, too. Many sea horses adopt a body shape and color pattern similar to the environment in which they are found (**Figure 43.12c**).

**Displays of Intimidation** Some animals put on displays of intimidation in an attempt to discourage predators. For example, a cat arches its back, a frilled lizard extends it collar, and a porcupine fish inflates itself when threatened in order to appear larger (**Figure 43.12d**). All of these animals have evolved displays to deceive potential predators about the ease with which they can be eaten.

**Mimicry** There are two major types of mimicry. In **Müllerian mimicry,** two or more toxic species converge to look the same, thus reinforcing the basic distasteful design. The viceroy butterfly (*Limenitis archippus*) and the monarch butterfly (*Danaus plexippus*) are examples of Müllerian mimicry. Both species are unpalatable and look similar, but the viceroy can be distinguished from the monarch by a black line that crosses its hindwings (**Figure 43.12e**).

**Batesian mimicry** is the mimicry of an unpalatable species (the model) by a palatable one (the mimic). Some of the best examples involve flies, especially hoverflies of the family Syrphidae, which are striped black and yellow and resemble stinging bees and wasps but are themselves harmless. Among vertebrates, the nonvenomous scarlet king snake (*Lampropeltis elapsoides*) mimics the venomous eastern coral snake (*Micrurus fulvius*), thereby gaining protection from would-be predators (**Figure 43.12f**).

**Armor and Weaponry** The shells of tortoises and turtles are a strong means of defense against most predators, as are the quills of porcupines (refer back to Figure 27.17c). Though many animals developed horns and antlers for sexual selection, these projections can also be used in defense against predators (**Figure 43.12g**). Invertebrate species often have powerful claws, pincers, or, in the case of scorpions, venomous stingers that can be used in defense as well as offense.

## Natural Enemies Can Have a Dramatic Effect on Prey Populations

Despite this impressive array of defenses, many studies have shown a strong effect of natural enemies on their prey populations. Invasive species, which are species introduced into a new area usually by humans and tend to spread rapidly, provide particularly striking examples of the effects of natural enemies.

### Biology Principle

**Biology Affects Our Society**

Control of prickly pear by the cactus moth cleared hundreds of thousands of hectares of cacti, allowing sheep to graze the area and farming to thrive.

**(a) Before biological control**

**(b) After biological control**

**Figure 43.13** **Successful biological control of prickly pear cactus.** The prickly pear cactus (*Opuntia stricta*) in Chinchilla, Australia, **(a)** before and **(b)** after control by the cactus moth (*Cactoblastis cactorum*).

*(a–b)* Source: State Library of Queensland, neg. no.'s API-101-01-0001r, API-101-01-0002r

**Concept Check:** *Why can invasive predators have such strong effects on native prey?*

The prickly pear cactus (*Opuntia stricta*) was imported from South America into Australia in the 19th century as a food source for an insect that was used to make a bright red dye for the British. The cactus quickly became invasive and established itself as a major pest of rangeland. The small cactus moth (*Cactoblastis cactorum*), which feeds on *Opuntia* cactus, was introduced from South America in the 1920s and within a short time had controlled cactus populations. This reuniting of an invasive plant with its natural enemy successfully saved hundreds of thousands of acres of valuable rangeland from being overrun by the cacti (**Figure 43.13**).

## Mutualism and Commensalism May Increase the Population Growth of Interacting Species

Not all species interactions have negative effects on one of the species. In mutualism, both species gain from the interaction. For example, in mutualistic pollination systems, such as occurs between plants and insects such as bees, the plant benefits by the transfer of pollen and the bee typically gains a nectar meal. In commensalism, one species benefits, and the other remains unaffected. For example, in some forms of seed dispersal, barbed seeds are transported to new germination sites in the fur of mammals. The seeds benefit, but the mammals are generally unaffected.

## Density-Dependent Factors May Regulate Population Sizes

Parasitism, parasitoidism, herbivory, predation, and competition are some of the many factors that may reduce the population densities of living organisms and stabilize them at equilibrium levels. **Density-dependent factors** are those in which the effect of the factors depends on the density of the population. For example, a pathogen that kills more of a population when densities are higher and less of a population when densities are lower is acting in a density-dependent manner. Many predators develop a visual search image for a particular prey. When a prey is rare, predators tend to ignore it and kill relatively few. When a prey is common, predators key in on it and kill relatively more. In England, for example, predatory shrews kill proportionately more moth pupae in leaf litter when the pupae are common than when they are rare. Density-dependent mortality may also occur as population densities increase and competition for scarce resources increases, reducing offspring production or survival. Parasitism may also act in a density-dependent manner. Parasites are able to pass from host to host more easily as the host's densities increase.

Density dependence can be detected by plotting mortality, expressed as a percentage, against population density (Figure 43.14). If a positive slope results and mortality increases with density, the factor tends to have a greater effect on dense populations than on sparse ones and is clearly acting in a density-dependent manner.

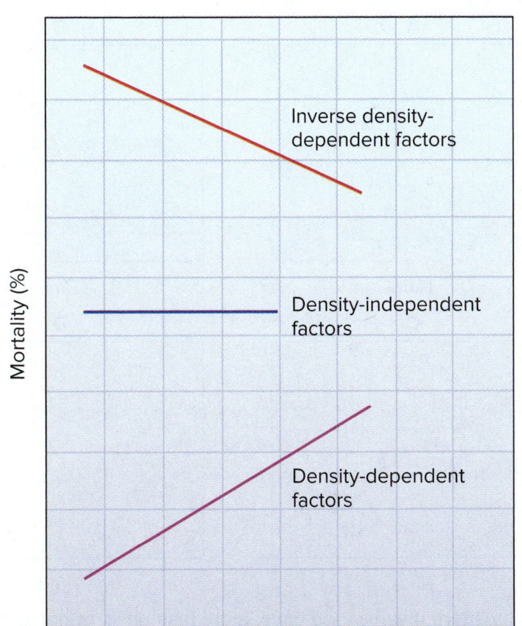


**Figure 43.14** **Three ways that factors affect mortality in response to changes in population density.** For a density-dependent factor, mortality increases with population density; for a density-independent factor, mortality remains unchanged. For an inverse density-dependent factor, mortality decreases as a population increases in size.

**Concept Check:** *Which types of factors tend to stabilize populations at equilibrium levels?*

A **density-independent factor** is a mortality factor whose influence is not affected by changes in population size or density. When mortality is plotted against density, a flat line results. In general, density-independent factors are physical factors, including weather, drought, freezes, floods, and disturbances such as fire. For example, in hard freezes, the same proportions of organisms such as birds or plants are usually killed, no matter how large the population size.

Finally, a mortality factor that decreases with increasing population size is considered an **inverse density-dependent factor.** In this case, a negative slope results when mortality is plotted against density. For example, if a territorial predator such as a lion always kills the same number of wildebeest prey, regardless of wildebeest density, it is acting in an inverse density-dependent manner, because it is taking a smaller proportion of the population at higher density. Some mammalian predators, being highly territorial, often act in this manner on herbivore density.

Determining which factors act in a density-dependent or density-independent fashion has large practical implications. Foresters, game managers, and conservation biologists alike are interested in learning how to maintain populations. For example, if a specific disease were to act in a density-dependent manner on elk, there wouldn't be much point in game managers attempting to kill off predators such as wolves to increase herd sizes for hunters, because proportionately more elk would be killed by disease. Species interactions can clearly be very important in influencing both the growth of individual populations and the structure of communities—groups of species living in a particular area. What factors determine the numbers of species in a given area? What factors influence the stability of a community, and what are the effects of disturbances on community structure? In Chapter 44, on community and ecosystem ecology, we will explore these and other questions.

## 43.4 Reviewing the Concepts

- Species interactions can take a variety of forms that differ based on their effect on the species involved from competition, predation, herbivory, parasitism, and parasitoidism to mutualism and commensalism (Figure 43.8; Table 43.2).
- Surveys have shown that competition occurs frequently in nature. Research has shown that two species with the same resource requirements cannot occupy the same niche. Resource partitioning between species allows them to coexist (Figures 43.9, 43.10).
- Natural enemies include predators, herbivores, parasites, and parasitoids. The most common defenses against natural enemies are chemical defense and aposematic coloration, camouflage, mimicry, and physical defenses such as spines or armor (Figures 43.11, 43.12).
- Natural enemies can have a dramatic effect on prey populations, such as invasive species (Figures 43.13).

- Density-dependent factors are mortality factors whose influence varies with population density. Density-independent factors are those whose influence does not vary with density (Figure 43.14).

## 43.4 Testing Your Knowledge

1. A species interaction in which one species benefits but the other species is unaffected is called
   a. mutualism. c. parasitism. e. mimicry.
   b. amensalism. d. commensalism.
2. According to the competitive exclusion hypothesis,
   a. two species that use exactly the same resource show very little competition.
   b. two species with the same niche cannot coexist.
   c. one species that competes with several different species for resources will be excluded from the community.
   d. all competition between species results in the extinction of at least one of the species.
   e. none of the above are correct.
3. Tapeworms, which are a type of human parasite, have
   a. low lethality and low duration of interaction.
   b. low lethality and high duration of interaction.
   c. high lethality and low duration of interaction.
   d. high lethality and high duration of interaction.
   e. none of the above.

# Assess and Discuss

## Test Yourself

1. A student decides to conduct a mark-recapture experiment to estimate the population size of mosquitofish, a species of freshwater fish, in a small pond near his home. In the first catch, he marked 45 individuals. Two weeks later, he captured 62 individuals, of which 8 were marked. What is the estimated size of the population based on these data?
   a. 134 c. 558 e. 22,320
   b. 349 d. 1,016

Questions 2–5 refer to the following table:

| Age | $n_x$ | $d_x$ | $l_x$ | $m_x$ | $l_x m_x$ |
|---|---|---|---|---|---|
| 0 | 100 | 35 | 1.00 | 0 | 0 |
| 1 | 65 | ? | 0.65 | 0 | 0 |
| 2 | 45 | 15 | ? | 3 | 1.35 |
| 3 | 30 | 20 | 0.30 | 1 | ? |
| 4 | 10 | 10 | 0.10 | 1 | 0.10 |
| 5 | 0 | 0 | 0.00 | 1 | 0.0 |

2. How many individuals die between their first and second birthdays?
   a. 65 c. 35 e. 20
   b. 45 d. 25
3. What proportion of newborns survive to age 2?
   a. 0.55 c. 0.35 e. 0.15
   b. 0.45 d. 0.20
4. What is the net reproductive rate?
   a. 5 c. 1.75 e. 0.80
   b. 2.5 d. 1.45
5. This population is
   a. increasing.
   b. decreasing.
   c. at equilibrium.
   d. It can't be determined from this information.
6. _______________ survivorship curves are usually associated with organisms that have high mortality rates in the early stages of life.
   a. Type I c. Type III e. Types II and III
   b. Type II d. Types I and II
7. If the net reproductive rate ($R_0$) is equal to 0.5, what assumptions can we make about the population?
   a. This population is essentially not changing in numbers.
   b. This population is in decline.
   c. This population is growing.
   d. This population is in equilibrium.
   e. None of the above are true.
8. The maximum number of individuals a certain area can sustain is known as
   a. the intrinsic rate of growth. d. the logistic equation.
   b. the resource limit. e. the equilibrium size.
   c. the carrying capacity.

Questions 9 and 10 refer to the following generalized growth patterns as plotted on arithmetic scales. Match the following descriptions with the patterns indicated below.

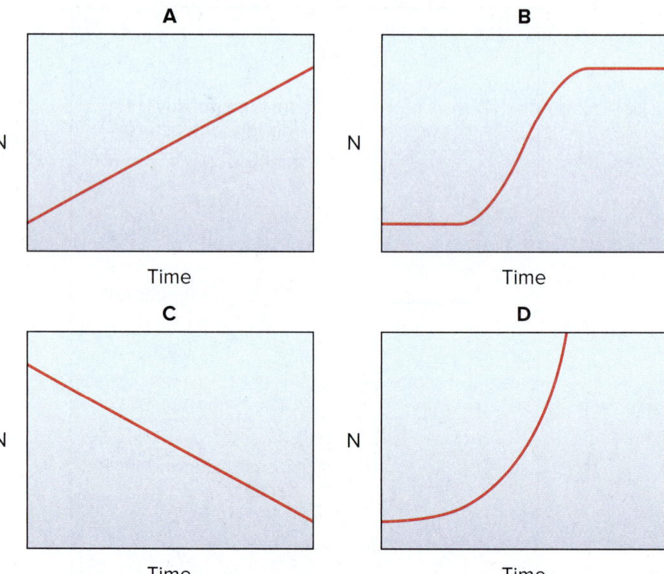


Each pattern may be used once, more than once, or not at all.

9. Which pattern is found when a population exhibits a constant per capita rate of increase?
   a. A d. D
   b. B e. none of the above
   c. C
10. Which pattern is found when a population is heading toward extinction?
   a. A d. D
   b. B e. none of the above
   c. C

## Conceptual Questions

1. As a researcher, you are using the mark-recapture procedure (see Figure 43.2) with mice. Say you recapture twice as many mice with ear tags as you should, because they are enticed to come to traps baited with peanut butter. How does this influence your estimate of population size? Does it increase or decrease, and by how much?
2. Using the logistic equation, calculate population growth when $K = 1{,}000$, $N = 500$, and $r = 0.1$ and when $K = 1{,}000$, $N = 100$, and $r = 0.1$. Compare the results with those shown in Section 43.3, where $K = 1{,}000$, $N = 900$, and $r = 0.1$. At which time is population growth highest as a percentage of the population?
3. **PRINCIPLES** A principle of biology is that biology affects our society. How is it possible that mutualisms between humans and other organisms have produced some of the most drastic environmental changes on Earth?

## Collaborative Questions

1. Discuss the different types of antipredator strategies that plants and animals have evolved.
2. You survey an annually breeding butterfly population and discover 1,000 females. The next year, you count 1,200 females in the same area. What is the finite rate of increase, $x$? What will the population be 5 years from now if the rate of increase is the same each year?

## Online Resource

# 44

# Communities and Ecosystems: Ecological Organization at Large Scales

© Gary Braasch/Corbis

**Mt. St. Helens erupting in 1980. Ecologists have been monitoring gradual change in the species composition of the area since the disturbance.**

## Chapter Outline



At 8:32 a.m. on May 18, 1980, Mount St. Helens, in the Washington Cascades, erupted. The blast felled trees over a 600-$km^2$ area, and the landslide that followed destroyed everything in its path, killing nearly 60 people. For more than 30 years, ecologists have studied the recovery of plant and animal species and how the appearance of some species facilitates the recovery of others.

The assemblage of many populations that live in the same place at the same time is known as a **community. Community ecology** explores the factors that influence the number and abundance of species in a community, In this chapter, we begin by considering patterns of species richness and diversity. We will examine why, on a global scale, the number of species is usually greatest in the tropics and declines toward the poles. Next, we discuss why the recovery of communities following a disturbance such as a fire or a volcanic eruption tends to occur in a predictable sequence—a process termed succession—which may be determined by the balance between the rates of immigration and extinction.

The term **ecosystem** was first used in 1935 by the British plant ecologist A. G. Tansley to describe the system formed by the interaction between a community of organisms and its physical environment. **Ecosystem ecology** addresses the flow of energy and the production of biomass within ecosystems. In Section 44.5, we first explore energy flow, the movement of energy through an ecosystem. In examining energy flow, our main task will be to consider the complex networks of feeding relationships between species and how these are represented by food webs. Finally, we will focus on the measurement of biomass, the total mass of living matter in a given area, usually measured in grams or kilograms per square meter.

## 44.1 Patterns of Species Richness and Species Diversity

### Learning Outcomes

1. Identify the latitudinal gradient of species richness.
2. List and describe three hypotheses for observed patterns of species richness.
3. **SCISKILLS ▸** Calculate the Shannon diversity index.
4. **SCISKILLS ▸** Calculate the effective number of species.

Community ecology addresses which factors influence the number of different species in a community, or **species richness.** Globally, the number of species of most taxa varies along a latitudinal gradient, generally increasing from polar to temperate areas and reaching a maximum in the tropics. For example, the species richness of North American birds increases from Arctic Canada to Panama (Figure 44.1). A similar pattern exists for mammals, amphibians, reptiles, and plants. Although the latitudinal gradient of species richness is an important pattern, species richness is also influenced by topographical variation. More mountains mean more hilltops, valleys, and differing habitats; thus, the number of birds is greater in the U.S. mountainous West. Species richness is also reduced by the peninsular effect, in which the number of species decreases as a function of distance from the main body of land.

Many hypotheses for the latitudinal gradient in species richness have been advanced. We will consider three hypotheses for patterns of species richness. Although they are discussed separately here, these hypotheses are not mutually exclusive. All can potentially contribute to patterns of species richness.

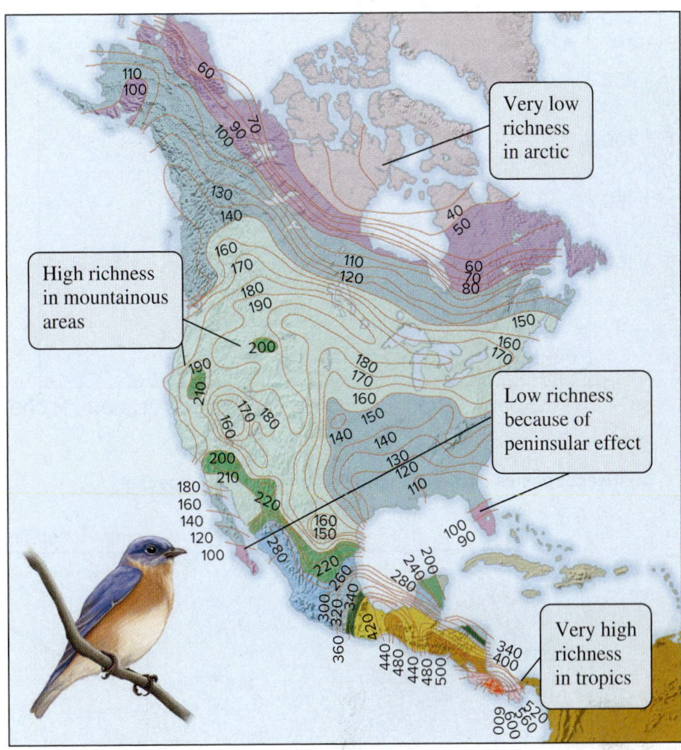


**Figure 44.1 Species richness of birds in North America.** The values indicate the numbers of different species in a given area. Contour lines show equal numbers of bird species, with colors indicating incremental changes. Note the pronounced latitudinal gradient toward the tropics and the high diversity in California and northern Mexico, regions of considerable topographical variation and habitat diversity.

### The Species-Time Hypothesis Suggests Communities Diversify with Age

Many ecologists argue that communities diversify, or gain species, with time. Tropical communities are usually older than temperate communities, because the species in temperate regions are periodically wiped out by glaciers. The **species-time hypothesis** proposes that temperate regions have less rich communities than tropical ones because they are younger. According to this idea, Ice Ages have driven many species extinct in temperate regions and it takes time for remaining species to evolve and diversify in those regions after the glaciers have retreated. More than half of the families of flowering plants have no temperate representatives.

In support of the species-time hypothesis, British ecologist H. John Birks found a significant correlation between the numbers of species of insects on various British trees and the evolutionary ages of these tree species in Britain (Figure 44.2a). Many of the tree species in Britain are relatively recent colonists, having appeared following the departure of the glaciers that covered most of the islands. Birks used radiocarbon dating of pollen collected from deep lake sediments to estimate the length of time a tree species had been present in Britain. No tree species had been present in Britain for longer than 13,000 years. He then gathered information on numbers of insect species present on trees from lists provided by other experts who had been examining the insect fauna of trees in Britain for many years. The significant relationship between pollen age and the total number of insects indicated that older tree species support more insect species.

However, ecologists recognize drawbacks to the species-time hypothesis. For example, this hypothesis may help explain variations in the species richness of terrestrial organisms, but it has limited applicability to marine organisms. Although we might not expect terrestrial species, particularly plants, to redistribute themselves quickly following a glaciation—especially if there is a physical barrier, like the English Channel, to overcome—there seems to be no reason that marine organisms couldn't relatively easily shift their distribution patterns during glaciations, yet the latitudinal gradient of species richness still exists in marine habitats.

### The Species-Area Hypothesis Suggests That Large Areas Support More Species

The **species-area hypothesis** proposes that larger areas contain more species than smaller areas because they can support larger populations and a greater range of habitats. Much evidence supports the area hypothesis. For example, in 1974, American ecologist Donald Strong showed that insect species richness on tree species in Britain was better correlated with the area over which a tree species could be found than with time of habitation since the last Ice Age (Figure 44.2b). The points cluster more tightly around the line of

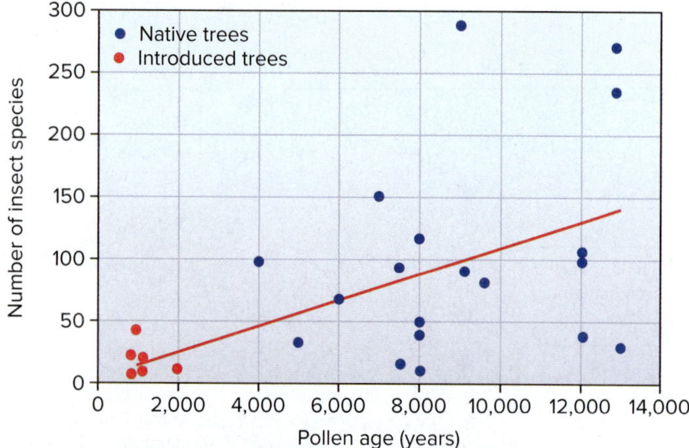


**(a) Insect species richness increases on older tree species**

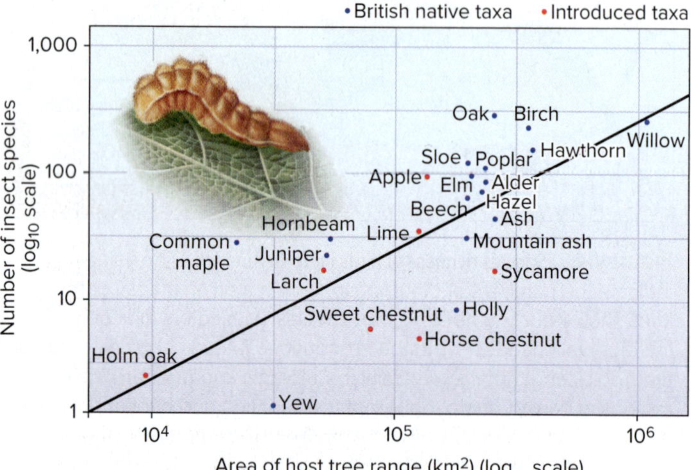


**(b) Insect species richness increases on more widely occurring tree species**

**Figure 44.2 Relationship between species richness on British host trees and both evolutionary time and area.** **(a)** Insect species richness is greater on evolutionary older tree species, which supports the species-time hypothesis. **(b)** A positive correlation is also found between insect species richness and the area of the host tree's range, in square kilometers (km²), which supports the species-area hypothesis. Note the log scales in part **(b).**

best fit. Even relatively newly introduced species, such as apple trees, supported a large number of insect herbivores if they were planted over a wide area. The observation that the number of species tends to increase with increasing area is called the **species-area effect.**

The large, climatically similar area of the tropics has been proposed as a reason that the tropics have high species richness. However, the area hypothesis seems unable to explain why, if increased richness is linked to increased area, more species are not found in certain regions such as the vast contiguous landmass of Asia. Furthermore, although tundra may be the world's largest biome in terms of landmass, it has low species richness. Finally, the largest marine system, the open ocean, which has the greatest volume of any habitat, has fewer species than tropical nearshore waters, which have a relatively small volume.

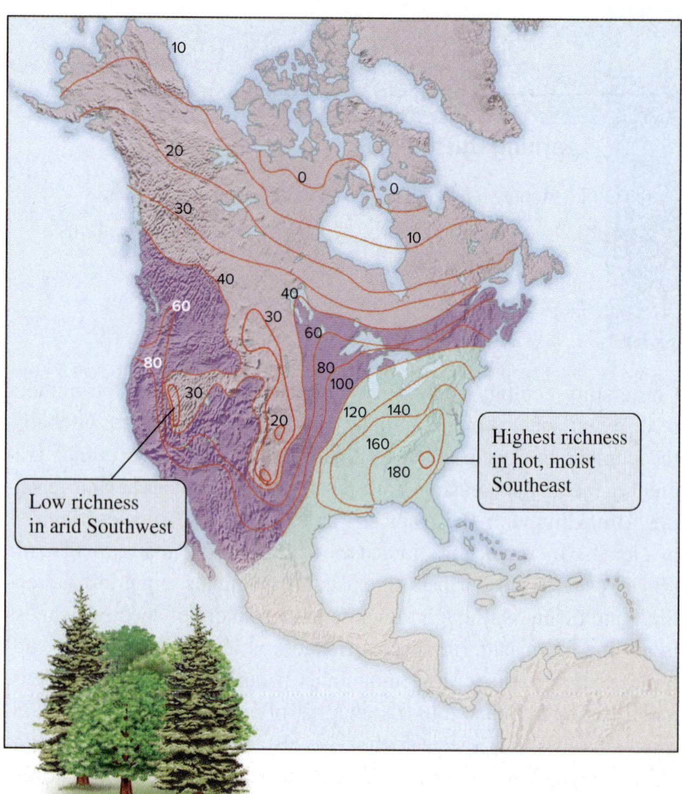


**Figure 44.3 Tree species richness in North America.** Contour lines show equal numbers of tree species, with colors indicating incremental changes. Tree species richness and evapotranspiration rates are highest in the Southeast.

**Concept Check:** *Why doesn't the species richness of trees increase in mountainous areas of the West, as it does for birds?*

## The Species-Productivity Hypothesis Suggests That Greater Productivity Results in More Species

The **species-productivity hypothesis** proposes that greater production by plants results in greater overall species richness. An increase in plant productivity, the total mass of plant material produced over time, leads to an increase in the number of herbivores and hence an increase in the number of predator, parasite, and scavenger species. Productivity itself is influenced by factors such as temperature and rainfall, because many plants grow better where it is warm and wet. For example, in 1987, Canadian biologist David Currie and colleagues showed that the species richness of trees in North America is best predicted by the **evapotranspiration rate,** the rate at which water moves into the atmosphere through the processes of evaporation from the soil and transpiration of plants, both of which are influenced by the amount of solar energy (Figure 44.3).

Once again, however, exceptions are observed. In 1993, American researchers Robert Latham and Robert Ricklefs showed that although patterns of tree richness in North America support the productivity hypothesis, the pattern does not hold for broad comparisons between continents. For example, the temperate forests of eastern Asia support substantially higher numbers of tree species (729) than do climatically similar areas of North America (253) or Europe (124). These three areas have different evolutionary histories and different neighboring areas from which species might have invaded. In summary, although there is no one simple cause of the polar-to-tropical species richness gradients, the factors of evolutionary time, area, and productivity each influence species richness to some degree.

## Quantitative Analysis

### CALCULATING SPECIES DIVERSITY

So far, we have discussed communities in terms of variations in species richness. However, ecologists need to take into account not only the number of species in a community but also their frequency of occurrence, or **relative abundance.** For example, consider two hypothetical communities, A and B, both with two species and 100 total individuals.

| | Number of individuals of species 1 | Number of individuals of species 2 |
|---|---|---|
| Community A | 99 | 1 |
| Community B | 50 | 50 |

The species richness of community B equals that of community A, because they both contain two species. However, community B is considered more diverse than A because the distribution of individuals between species is more even. One would be much more likely to encounter both species in community B than in community A, where one species dominates. **Species diversity** is a measure of the diversity of an ecological community that incorporates both species number and relative abundance.

To measure the species diversity of a community, ecologists calculate what is known as a diversity index. Although many different indices are available, the most widely used is the **Shannon diversity index ($H_S$),** which is calculated as

$$H_S = -\sum p_i \ln p_i$$

where $p_i$ is the proportion of individuals belonging to species $i$ in a community, ln is the natural logarithm, and $\sum$ indicates summation. For example, for a species in which there are 50 individuals out of a total of 100 in the community, $p_i$ is 50/100, or 0.5. The natural log of 0.5 is −0.693. For this species, $p_i \ln p_i$ is then 0.5(−0.693) = −0.347. For a hypothetical community with 5 species and 100 total individuals, the Shannon diversity index is calculated as follows:

| | Species | Abundance | $p_i$ | $\ln p_i$ | $p_i \ln p_i$ |
|---|---|---|---|---|---|
| | 1 | 50 | 0.5 | −0.693 | −0.347 |
| | 2 | 30 | 0.3 | −1.204 | −0.361 |
| | 3 | 10 | 0.1 | −2.302 | −0.230 |
| | 4 | 9 | 0.09 | −2.408 | −0.217 |
| | 5 | 1 | 0.01 | −4.605 | −0.046 |
| Total | 5 | 100 | 1.00 | $\Sigma p_i \ln p_i$ | =−1.201 |

Remember that in the equation, the negative sign in front of the summation changes the summed value to positive, so the index actually becomes 1.201, not −1.201.

Values of the Shannon diversity index for real communities often fall between 1.5 and 3.5, with the higher the value, the greater the diversity. Table 44.1 calculates the diversity of two bird communities in Indonesia with similar species richness but differing species abundance. The bird communities were surveyed in a pristine unlogged forest or in a selectively logged lowland forest. To document diversity, British biologist Stuart Marsden established census stations in the two forests and recorded the type and number of all bird species for a number of 10-minute periods. Although a greater number of individual birds was seen in the logged areas (2,345) than in the unlogged ones (1,824), a high proportion of the individuals in the logged areas (0.386) belonged to just one species, *Nectarinia jugularis*. Although only one more bird species was found in the unlogged area than in the logged area, calculation of the Shannon diversity index showed a higher diversity of birds in the unlogged area, 2.284 versus 2.037, which is a considerable difference, considering the logarithmic nature of the index.

A problem with diversity indices is that the results are not easy to compare. For example, imagine our hypothetical community has 5 species each with 20 individuals ($n$ = 100). The Shannon diversity index is 1.609, which is 34% greater than the value of 1.201 from the other hypothetical community. A better comparison would be to calculate the **effective number of species,** which converts values from species diversity indices into equivalent numbers of species. For the Shannon diversity index, we take the exponential of 1.609, $e^{1.609}$, so that our value of 1.609 becomes 5.0, which is the same as the actual number of species. For our value of 1.201, the effective number of species is 3.323, which means that this community is 50.5% less diverse than the community with an effective number of species of 5.0.

**Crunching the Numbers:** **(a)** Community A has 2 species each with 50 individuals. Community B has 3 species, 1 with 80 individuals and the other 2 with 10 individuals each. Calculate the Shannon diversity index for both communities. What can you conclude? **(b)** Calculate the effective number of species for the bird communities in the logged and unlogged areas of the Indonesian forests mentioned above. What is the percentage difference in the effective number of species?

**Table 44.1** Shannon Diversity Index of Bird Species on Logged and Unlogged Sites in Indonesia

| | Unlogged | | | Logged | | |
|---|---|---|---|---|---|---|
| **Species** | $N$ | $p_i$ | $p_i \ln p_i$ | $N$ | $p_i$ | $p_i \ln p_i$ |
| *Nectarinia jugularis*, olive-backed sunbird | 410 | 0.225 | −0.336 | 910 | 0.386 | −0.367 |
| *Ducula bicolor*, pied imperial pigeon | 230 | 0.126 | −0.261 | 220 | 0.093 | −0.221 |
| *Philemon subcorniculatus*, grey-necked friarbird | 210 | 0.115 | −0.249 | 240 | 0.102 | −0.233 |
| *Nectarinia aspasia*, black sunbird | 190 | 0.104 | −0.235 | 120 | 0.051 | −0.152 |
| *Dicaeum vulneratum*, ashy flowerpecker | 185 | 0.101 | −0.232 | 280 | 0.119 | −0.253 |
| *Ducula perspicillata*, white-eyed imperial pigeon | 170 | 0.093 | −0.221 | 180 | 0.076 | −0.196 |
| *Phylloscopus borealis*, arctic warbler | 160 | 0.088 | −0.214 | 140 | 0.059 | −0.167 |
| *Eos bornea*, red lory | 88 | 0.048 | −0.146 | 73 | 0.031 | −0.108 |
| *Ixos affinis*, golden bulbul | 76 | 0.042 | −0.133 | 31 | 0.013 | −0.056 |
| *Geoffroyus geoffroyi*, red-cheeked parrot | 44 | 0.024 | −0.089 | 54 | 0.023 | −0.087 |
| *Rhyticeros plicatus*, Papuan hornbill | 24 | 0.013 | −0.056 | 27 | 0.011 | −0.050 |
| *Cacatua moluccensis*, Moluccan cockatoo | 12 | 0.007 | −0.035 | 1 | 0.001 | −0.007 |
| *Tanygnathus megalorynchos*, great-billed parrot | 9 | 0.005 | −0.026 | 11 | 0.005 | −0.026 |
| *Eclectus roratus*, electus parrot | 7 | 0.004 | −0.022 | 0 | 0 | 0 |
| *Macropygia amboinensis*, brown cuckoo-dove | 6 | 0.003 | −0.017 | 7 | 0.003 | −0.017 |
| *Cacomantis sepulcralis*, ruby-breasted cuckoo | 3 | 0.002 | −0.012 | 0 | 0 | 0 |
| *Trichoglossus haematodus*, rainbow lorikeet | 0 | 0 | 0 | 64 | 0.027 | −0.097 |
| **Total** | **1,824** | **1.0** | | **2,345** | **1.0** | |
| **Shannon diversity index** | | | **2.284** | | | **2.037** |

## 44.1 Reviewing the Concepts

- The number of species of most taxa varies according to geographic location, generally increasing from polar areas to tropical areas (Figure 44.1).
- Different hypotheses for the variation in species richness have been advanced, including the species-time hypothesis, the species-area hypothesis, and the species-productivity hypothesis (Figures 44.2, 44.3).
- Species diversity takes into account both species richness and species abundance. The most widely used measure of species diversity is called the Shannon diversity index (Table 44.1), but the most easily interpreted measure is called the effective number of species.

## 44.1 Testing Your Knowledge

1. In North America, the highest numbers of bird species are generally found
   **a.** at the tip of peninsulas, such as Florida and Baja California.
   **b.** in mountainous areas of the U.S. Southwest.
   **c.** in wet areas of the U.S. Southeast.
   **d.** in Central America.
   **e.** in high-productivity areas of the Arctic.
2. Lake Baikal in Siberia is an ancient, unglaciated temperate lake and contains 580 species of bottom-dwelling invertebrates. Great Slave Lake, a comparably sized lake that was once glaciated at the same latitude in northern Canada, contains only four species in the same zone. This supports which of the following hypotheses?
   **a.** species-time **c.** species-productivity **e.** both b and c
   **b.** species-area **d.** both a and b
3. A community of birds contains 8 individuals of species A, 6 of B, 4 of C, and 2 of D. What is the value of the Shannon diversity index?
   **a.** 4 **c.** 0.30 **e.** 1.280
   **b.** 1 **d.** 0.366

## 44.2 Species Diversity and Community Stability

**Learning Outcome**

1. Describe the diversity-stability hypothesis, and explain the evidence supporting it.

In this section, we consider the relationship between species diversity and community stability. A community is often seen as stable when little to no change can be detected in the number of species and their abundance over a given time period. The community may then be said to be in equilibrium. Community stability is an important consideration to ecologists. A decrease in the stability of a community over time may alert ecologists to a possible problem. For example, an invasive species such as the zebra mussel, *Dreissena polymorpha,* in the Great Lakes can cause changes in the community of other invertebrates living there.

In this section, we explore the question of whether communities with higher species diversity are more stable than communities with lower diversity, using evidence from a field experiment.

### The Diversity-Stability Hypothesis States That Species-Rich Communities Are More Stable Than Those with Fewer Species

The link between species richness and stability was first explicitly proposed by English ecologist Charles Elton in the 1950s. He suggested that a disturbance in a species-rich community would be cushioned by large numbers of interacting species and would not produce as drastic an effect as it would on a species-poor community. Thus, an introduced predator or parasite could cause extinctions in a species-poor community but probably not in a more diverse community, where its effects would be buffered by interactions with more species. Elton argued that outbreaks of pests are often found on cultivated land or land disturbed by humans, both of which are species-poor communities with few naturally occurring species. His argument became known as the **diversity-stability hypothesis.**

However, some ecologists began to challenge Elton's association of diversity with stability. Ecologists pointed out many examples of introduced species that have assumed pest proportions in species-rich areas, including rabbits in Australia and pigs in North America. They noted that disturbed or cultivated land may suffer from pest outbreaks not because of its simple nature but because individual species, including introduced species, often have no natural enemies in the new environment, in contrast to the long associations between native species and their natural enemies. For example, in Europe, coevolved predators, such as foxes, prevent rabbit populations from increasing to pest proportions. Research was needed to determine if a link existed between diversity and stability.

### Field Studies Have Linked Stability to Diversity

In 1996, American ecologist David Tilman reported the relationship between species diversity and stability from an 11-year study of 207 grassland plots in Minnesota that varied in their species richness. He measured the biomass of every species of plant, in each plot, at the end of every year and obtained the average species biomass. He then calculated how much this biomass varied from year to year through a statistical measure called the coefficient of variation. Less variation in biomass signified community stability. Year-to-year variation in plant community biomass was significantly lower in plots with greater plant species richness (**Figure 44.4**). The results showed that greater diversity enhances community stability.

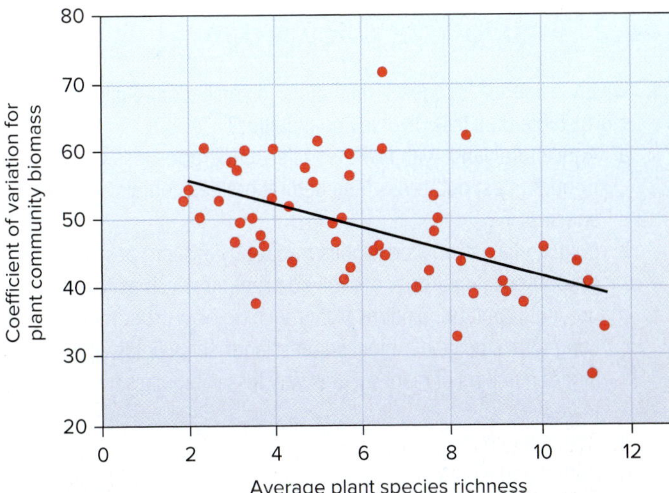


**Figure 44.4** **Biomass variation and species richness.** Tilman's 11-year study of grassland plots in Minnesota revealed that year-to-year variability in community biomass was lower in species-rich plots. Each dot represents an individual plot. Only the plots from one field are graphed.

Tilman suggested that more diverse communities are more likely to contain disturbance-resistant species that, in the event of a disturbance, could grow and compensate for the loss of disturbance-sensitive species. For example, when a change in climate such as drought decreased the abundance of competitively dominant species that thrived in normal conditions, unharmed drought-resistant species increased in mass and replaced them. Such declines in the number of sensitive species and compensatory increases in other species acted to stabilize total community biomass.

Although ecologists recognize a link between species diversity and community stability, they are also aware that, over long periods of time, communities may experience severe disturbances. The change in the composition and structure of communities that follows occurs in a predictable way termed succession, which is described next.

### 44.2 Reviewing the Concepts

- Community stability is an important concept in ecology. The diversity-stability hypothesis maintains that species-rich communities are more stable than communities with fewer species. Tilman's field studies, which showed that year-to-year variation in plant biomass decreased with increasing species diversity, established a link between diversity and stability (Figure 44.4).

## 44.2 Testing Your Knowledge

1. Which evidence suggests that more diverse communities are more stable than less diverse communities?
   a. Agricultural land, with fewer species, undergoes less frequent pest outbreaks than natural prairies containing more species.
   b. In Australia, introduced rabbits frequently assume pest proportions. In Europe, coevolved predators such as foxes prevent rabbit populations from reaching pest proportions.
   c. Long-term studies of American grasslands show fields with higher numbers of plant species vary less in biomass from year to year.
   d. Both a and b are correct.
   e. Both b and c are correct.
2. Ecologists began to question Elton's link of increased stability to increased diversity because
   a. cultivated land undergoes few outbreaks of pests.
   b. highly disturbed areas have high numbers of species.
   c. pest outbreaks are caused by lack of long associations with natural enemies, not because they occur in simple systems.
   d. all of the above.

## 44.3 Succession: Community Change

### Learning Outcomes

1. Distinguish between primary and secondary succession.
2. Compare and contrast facilitation, inhibition, and tolerance as mechanisms of succession.

Ecologists use different terms to describe how change in a community occurs over time. The term **succession** describes the gradual and continuous change in species composition of a community following a disturbance. **Primary succession** refers to succession on a newly exposed site that has no biological legacy in terms of plants, animals, or microbes, such as bare ground caused by a volcanic eruption or the sediment created by the retreat of glaciers. In primary succession on land, the plants must often build up the soil, and thus a long time—even hundreds of years—may be required for the process. Only a tiny proportion of the Earth's surface is currently undergoing primary succession, including the area around Mount St. Helens and on new lava flows around the volcanoes in Hawaii and off the coast of Iceland, and behind retreating glaciers in Alaska and Canada.

**Secondary succession** refers to succession on a site that has already supported life but has undergone a disturbance such as a fire, tornado, hurricane, or flood. In terrestrial areas, soil is already present. Clearing a natural forest and farming the land for several years is an example of a severe forest disturbance that does not kill all native species. Some plants and many soil bacteria, nematodes, and insects are still present. Secondary succession occurs if farming is ended. The secondary succession in abandoned farmlands (also called old fields) can lead to a pattern of vegetation quite different from one that develops after primary succession following glacial retreat. For example, the plowing and added fertilizers, herbicides, and pesticides may have caused substantial changes in the soil of an old field, allowing species that require a lot of nitrogen to colonize. These species would not be present for many years in newly created glacial soils.

American plant ecologist Frederic Clements is often viewed as the founder of successional theory. His work in the early 20th century emphasized succession as proceeding through several stages to a distinct end point, or **climax community.** Although disturbance can return a community from a later stage to an earlier stage, generally the community progresses in one direction. Clements's depiction of succession focused on a process termed facilitation, but two other mechanisms of succession—inhibition and tolerance—have since been described. Let's examine the evidence for each of them.

### Facilitation Assumes Each Invading Species Creates a More Favorable Habitat for Succeeding Species

A key assumption of Clements's view of succession was that each colonizing species makes the environment a little different, such as a little shadier or a little richer in soil nitrogen, so that it becomes more suitable for other species, which then invade and outcompete the earlier residents. This process, known as **facilitation,** continues until the most competitively dominant species have colonized, when the community is at climax. The composition of the climax community for any given region is determined by climate, soil condition, and frequency of disturbance.

Succession following the gradual retreat of Alaskan glaciers is often used as a specific example of facilitation as a mechanism of succession. Over the past 200 years, the glaciers in Glacier Bay have undergone a dramatic retreat of nearly 100 km (Figure 44.5). Succession in Glacier Bay has followed a distinct pattern of vegetation. As glaciers retreat, they leave moraines—deposits of stones, pulverized rock, and debris that serve as soil. In Alaska, the bare soil has a low nitrogen content and little organic matter. In the pioneer stage, the soil is first colonized by a black crust of cyanobacteria, mosses, lichens, horsetails (*Equisetum variegatum*), and the occasional river beauty (*Epilobium latifolium*) (Figure 44.6a). Because the cyanobacteria are nitrogen fixers, the soil nitrogen increases a little, but soil depth and litterfall (fallen leaves, twigs, and other plant material) are still minimal. At this stage, there may be a few seeds and seedlings of dwarf shrubs of the rose family, commonly called mountain avens (*Dryas drummondii*); alders (*Alnus sinuata*); and spruce but they are rare. After about 40 years, mountain avens dominate the landscape (Figure 44.6b). Soil nitrogen increases, as does soil depth and litterfall, and alder trees begin to invade.

At about 60 years, alders form dense, close thickets (Figure 44.6c). Alders have nitrogen-fixing bacteria that live mutualistically in their roots and convert nitrogen from the air into a biologically useful form. Soil nitrogen dramatically increases, as does litterfall. Sitka spruce trees (*Picea sitchensis*) begin to invade at about this time. After about 75 to 100 years, the spruce trees begin to overtop the alders, shading them out. The litterfall is still high, and the large volume of needles

**Figure 44.5** **The degree of glacier retreat at Glacier Bay, Alaska, since 1794.** **(a)** Primary succession begins on the bare rock and soil evident at the edges of the retreating glacier. **(b)** The lines reflect the position of the glacier in 1794 and its subsequent retreat northward.

*(a)* © Charles D. Winters/Science Source

**Concept Check:** *Why might ecologists think of walking the coastline of Glacier Bay as the equivalent of walking back in time?*

turns the soil acidic. The shade causes competitive exclusion of many of the original understory species, including alder, and only mosses carpet the ground. At this stage, seedlings of western hemlock (*Tsuga heterophylla*) and mountain hemlock (*Tsuga mertensiana*) may also occur. After 200 years, a mixed spruce-hemlock climax forest results (**Figure 44.6d**).

What other evidence is there of facilitation? Experimental studies of early primary succession on Mount St. Helens, which show that decomposition of fungi allows mosses and other fungi to colonize the soil, provide evidence of facilitation. Succession on sand dunes also supports the facilitation model, in that pioneer plant species stabilize the sand dunes and facilitate the establishment of subsequent plant species. The foredunes, those nearest the shoreline, are the most frequently disturbed and are maintained in a state of early succession, whereas more stable communities develop farther away from the shoreline.

Succession also occurs in aquatic communities. Although soils do not develop in marine environments, facilitation may still be encountered when one species enhances the quality of settling and establishment sites for another species. When experimental test plates used to measure settling rates of marine organisms were placed in the Delaware Bay, researchers discovered that certain cnidarians enhanced the attachment of tunicates, and both facilitated the attachment of mussels, the dominant species in the community. In this experiment, the smooth surface of the test plates prevented many species from colonizing, but once the surface became rougher, because of the presence of the cnidarians, many other species were able to colonize. In a similar fashion, early colonizing bacteria, which create biofilms on rock surfaces, can facilitate succession of other organisms.

## Inhibition Implies That Early Colonists Can Prevent Later Arrivals from Replacing Them

Although data on succession in some communities fit the facilitation model, researchers have proposed alternative hypotheses of how succession may operate. In the process known as **inhibition,** early colonists prevent colonization by other species. For example, removing the litter of *Setaria faberi*, an early successional plant species in New Jersey old fields, causes an increase in the biomass of a later species, *Erigeron annuus*. The release of toxic compounds from decomposing *Setaria* litter or physical obstruction by the litter itself blocks the establishment of *Erigeron*. Without the litter present, however, *Erigeron* dominates and reduces the biomass of *Setaria*.

Inhibition has been seen as the primary method of succession in the marine intertidal zone, where space is limited. In this habitat, early successional species are at a great advantage in maintaining possession of valuable space. In 1974, American ecologist Wayne Sousa created an environment for testing how succession works in the intertidal zone by scraping rock faces clean of all algae or putting out fresh boulders or concrete blocks. The first colonists of these areas were the green algae *Ulva*. By removing *Ulva* from the substrate, Sousa showed that the large red alga *Chondracanthus canaliculatus* was able to colonize more quickly (**Figure 44.7**). The results of Sousa's study indicate that early colonists can inhibit rather than facilitate the invasion of subsequent colonists. Succession may eventually occur because early colonizing species, such as *Ulva*, are more susceptible than later successional species, such as *Chondracanthus*, to the rigors of the physical environment and to attacks by herbivores, such as crabs (*Pachygrapsus crassipes*).

## Tolerance Suggests That Early Colonists Neither Facilitate nor Inhibit Later Colonists

In 1977, researchers Joseph Connell and Ralph Slatyer proposed a third mechanism of succession, which they termed **tolerance.** In this process, any species can start the succession, but the eventual

Stage | Pioneer | Dryas | Alder | Spruce-hemlock
Time (years) since glacial retreat | 5 | 40 | 60 | 200
Soil depth (cm) | 5.2 | 7.0 | 8.8 | 15.1
Soil N (g/m2) | 3.8 | 5.3 | 21.8 | 53.3
Soil pH | 7.2 | 7.3 | 6.8 | 3.6
Litterfall (g/m2/yr) | 1.5 | 2.8 | 277 | 261
Cyanobacteria
Moss
Lichens
Mountain avens
(Dryas drummondii)
Alder
(Alnus sinuata)
Spruce
(Picea sitchensis)
Western hemlock
(Tsuga heterophylla)
(a) Pioneer stage
(b) Dryas stage
(c) Alder stage
(d) Spruce-hemlock stage


**Figure 44.6** **The pattern of primary succession at Glacier Bay, Alaska.** **(a)** The first species to colonize the bare ground following retreat of the glaciers are small species such as cyanobacteria, moss, and lichens. **(b)** Mountain avens (*Dryas drummondii*) is a flower common in the *Dryas* stage. **(c)** Soil nitrogen and litterfall increase rapidly as alder (*Alnus sinuata*) invades. Note also the appearance of a few spruce trees higher up the valley. **(d)** Sitka spruce (*Picea sitchensis*) and hemlock (*Tsuga heterophylla*) trees make up a climax spruce-hemlock forest at Glacier Bay, with moss carpeting the ground. Two hundred years ago, glaciers occupied this spot.

*(a)* © Leon Werdinger/Alamy; *(b)* © James Hager/agefotostock; *(c)* © Howie Garber/AccentAlaska.com; *(d)* © Craig Lovell/Eagle Visions P/Newscom

climax community is reached in a somewhat orderly fashion. The species that establish and remain do not change the environment in ways that either facilitate or inhibit subsequent colonists. Species have differing tolerances to the intensity of competition as more species accumulate. Relatively competition-intolerant species are more successful early in succession, when the intensity of competition is low and resources are abundant. Relatively competition-tolerant species appear later in succession and at climax. Connell and Slatyer found the best evidence for the tolerance model in American plant ecologist Frank Egler's earlier work on floral succession. In the 1950s, Egler showed that succession in plant communities is determined largely by species that already exist in the ground as buried seeds or old roots. Whichever species germinates first or regenerates from roots initiates the succession sequence. Germination or root regeneration, in turn, depends on the timing of a disturbance. For example, an early-season tree fall would promote early-germinating species to grow in the subsequent light gap, whereas a late-season tree fall would promote the growth of late-germinating species. As succession proceeds, earlier germinating or regenerating species may be outcompeted by different species.

The key distinction between the three models is in the manner in which succession proceeds. In the facilitation model, species replacement is facilitated by previous colonists; in the inhibition model, it is inhibited by the action of previous colonists; and in the tolerance model, species may be affected by previous colonists, but they do not require them (**Figure 44.8**).

## 44.3 Reviewing the Concepts

- Succession describes the gradual and continuous change in community structure over time. Primary succession refers to succession on a newly exposed site with no prior biological legacy; secondary succession refers to succession on a site that has supported life but has undergone a disturbance (Figures 44.5, 44.6).
- Three mechanisms have been proposed for succession. In facilitation, each species facilitates or makes the environment more suitable for subsequent species. In inhibition, initial species inhibit later colonists. In tolerance, any species can start the succession, and species replacement is unaffected by previous colonists (Figures 44.7, 44.8).

## Biology Principle

### Biology Is an Experimental Science

The results of Sousa's study indicate that early colonists can inhibit rather than facilitate the invasion of subsequent colonists.

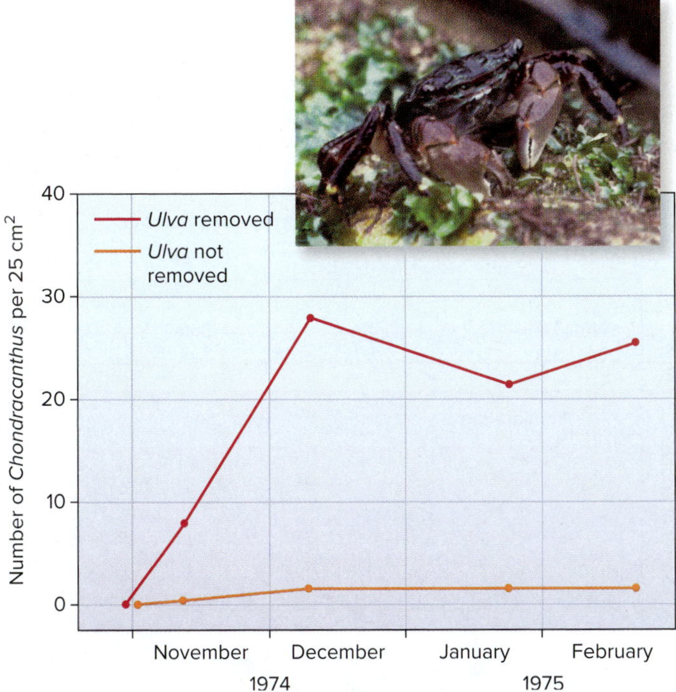


**Figure 44.7 Inhibition as a method of succession in the marine intertidal zone.** Removing *Ulva* from intertidal rock faces allowed colonization by *Chondracanthus*. The inset shows *Ulva* on a rock face with the striped shore crab *Pachygrapsus crassipes,* a herbivore.



## 44.3 Testing Your Knowledge

1. Which of the following are examples of secondary succession?
   a. plants growing in cracks on the pavement of a quiet street
   b. the recovery of vegetation following the 2004 Indonesian tsunami
   c. the colonization of new sand by beach plants
   d. the recovery of forests following a wildfire
   e. both b and d
2. In New England salt marshes, *Spartina* grass stabilizes the substrate and reduces water velocity, which enables other seedlings to emerge. This is an example of
   a. primary succession.
   b. facilitation.
   c. tolerance.
   d. inhibition.
   e. a climax community.

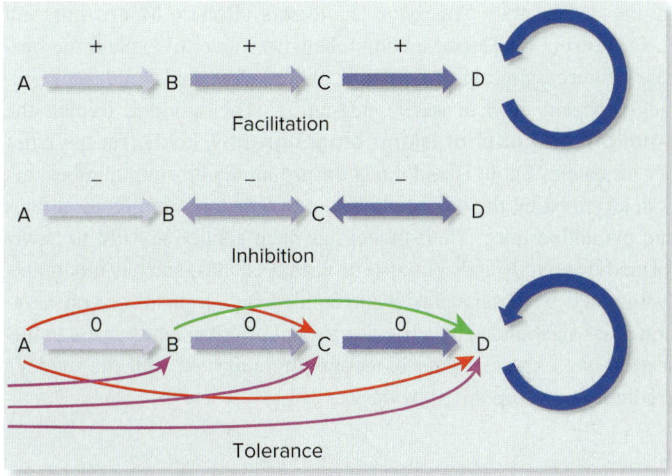


**Figure 44.8 Three models of succession.** A, B, C, and D represent four stages. D represents the climax community. An arrow indicates "is replaced by," and + = facilitation, – = inhibition, and 0 = no effect. The facilitation model is the classic model of succession. In the inhibition model, early-arriving species outcompete later-arriving species. The tolerance model depends on which species gets there first. The colored arrows show that succession may bypass some stages in the tolerance model.

**Concept Check:** *Inhibition implies that competition exists between species, with early-arriving species tending to outcompete later arrivals, at least for a while. Does competition or mutualism feature more prominently in facilitation?*

3. A tree falls in a forest in spring, and flowers germinate in the light gap. Following a tree fall in autumn, different species of flowers germinate in the light gaps. This illustrates the principle of
   a. facilitation.
   b. tolerance.
   c. inhibition.
   d. primary succession.
   e. climax communities.

## 44.4 Island Biogeography

### Learning Outcomes

1. **SCISKILLS ▶** Use a graph to illustrate the equilibrium model of island biogeography.
2. List the predictions of the model, and discuss how well the evidence supports them.

In some newly formed habitats such as volcanic islands, succession may be affected not only by facilitation, inhibition, or tolerance but also by the ability of species to colonize isolated areas. In these cases, species richness is affected by distance of habitats from a source pool of colonists and the size of the areas to be colonized.

In the 1960s, two American ecologists, Robert MacArthur and E. O. Wilson, developed a comprehensive model to explain the process of succession on new islands, where a gradual buildup of species proceeds from a sterile beginning. Their model, termed the **equilibrium model of island biogeography,** holds that the number of species on an island tends toward an equilibrium number that is determined by the balance between two factors: immigration rates and extinction rates. Their model has been applied not just to newly formed oceanic islands but also to virtual islands, such as mountains surrounded by deserts, lakes surrounded by dry land, or conservation areas surrounded by agricultural land or urban landscapes. In this section, we explore island biogeography and how well the model's predictions are supported by data.

## The Island Biogeography Model Suggests That During Succession, Gains in Immigration Are Balanced by Losses from Extinction

MacArthur and Wilson's model of island biogeography suggests that species repeatedly arrive on an island and either thrive or become extinct. The rate of immigration of new species is highest when no species are present on the island. As the number of species accumulates, the immigration rate decreases, since subsequent immigrants are more likely to represent species already present on the island. The rate of extinction is low at the time of first colonization, because few species are present and many have large populations. With the addition of new species, the populations of some species diminish, so the probability of extinction increases. Over time, the number of species tends toward an equilibrium, $\hat{S}$, in which the rates of immigration and extinction are equal. Species may continue to arrive and go extinct, but the number of species on the island remains approximately the same.

MacArthur and Wilson reasoned that when plotted graphically, both the immigration and extinction lines would be curved, for several reasons (Figure 44.9a). First, species arrive on islands at different rates. Some organisms, including plants with seed-dispersal mechanisms and winged animals, are more mobile than others and arrive quickly. Other organisms arrive more slowly. This pattern causes the immigration curve to start off steep but get progressively shallower. On the other hand, extinctions rise at accelerating rates, because as later species arrive, competition increases and more species are likely to go extinct. The strength of the island biogeography model was that it generated several testable predictions:

1. The number of species should increase with increasing island size (area), a concept known as the species-area relationship (see Figure 44.2b). Extinction rates would be lower on larger islands because population sizes would be larger and less susceptible to extinction (Figure 44.9b).
2. The number of species should decrease with increasing distance of the island from the mainland, or the **source pool,** the pool of potential species available to colonize the island. Immigration rates would be greater on islands near the source pool because species do not have as far to travel (see Figure 44.9b).

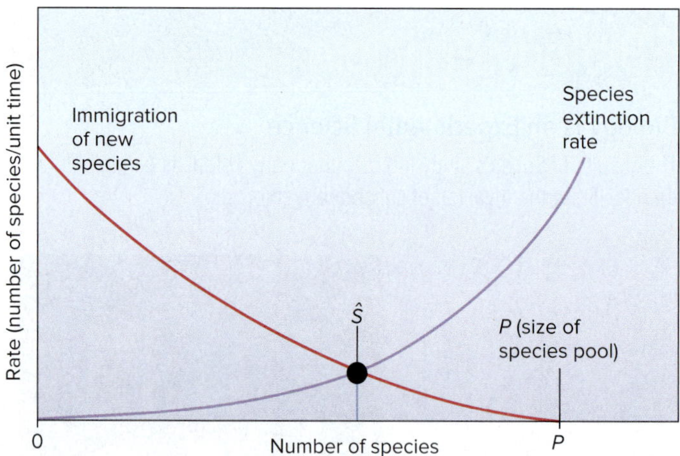

(a) Effects of immigration and extinction on species number

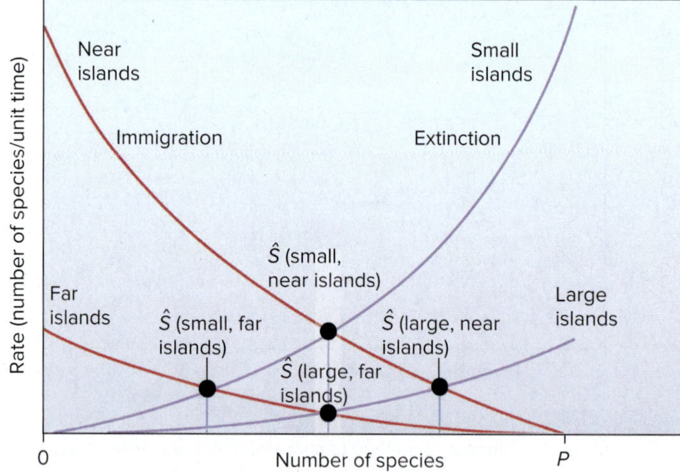

(b) Added effects of island size and proximity to the mainland on species number

**Figure 44.9 MacArthur and Wilson's equilibrium model of island biogeography. (a)** The interaction of immigration rate and extinction rate produces an equilibrium number of species on an island, $\hat{S}$. $\hat{S}$ can vary from 0 species to $P$ species, the total number of species available to colonize. **(b)** $\hat{S}$ varies according to the island's size and distance from the mainland. An increase in distance (near to far) lowers the immigration rate. An increase in island area (small to large) lowers the extinction rate.

**Concept Check:** *Can you think of a scenario where there would be large numbers of species on a small island?*

**BioConnections:** *Look ahead to Figure 47.6. How might the model of island biogeography be useful in the design of nature reserves?*

3. The turnover of species should be considerable. The number of species on an island might remain relatively constant, but the composition of the species should vary over time as new species colonize the island and others become extinct.

**(a) Lesser Antilles Islands**

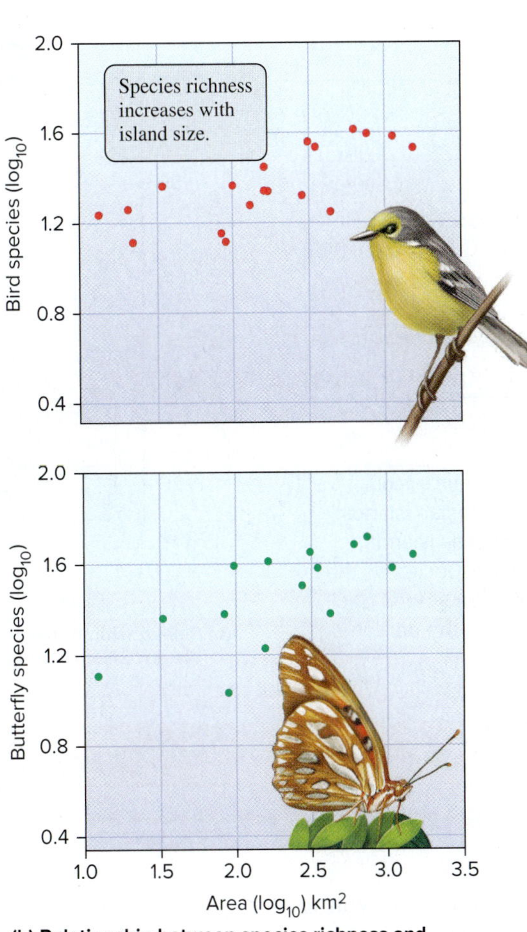


**(b) Relationship between species richness and island size**

**Figure 44.10 Species richness and island size. (a)** The Lesser Antilles extend from Anguilla in the north to Grenada in the south. **(b)** On these islands, the number of bird and butterfly species increases with the area of an island. Note that these relationships are traditionally plotted on a double logarithmic scale, a so-called log-log plot, in which the horizontal axis is the logarithm to the base 10 of the area and the vertical axis is the logarithm to the base 10 of the number of species. A linear plot of the area versus the number of species would be difficult to produce because of the wide range of area and richness of species involved. Logarithmic scales condense this variation to manageable limits.

**Concept Check:** *Calculate the approximate change in bird species richness across islands in the Lesser Antilles.*

Let's examine the three predictions of the island biogeography model and see how well the data support each one.

**Species-Area Relationships** The West Indies has traditionally been a key location for ecologists studying island biogeography. The physical geography and the plant and animal life of the islands are well known. Furthermore, the Lesser Antilles, from Anguilla in the north to Grenada in the south, enjoy a similar climate and are surrounded by deep water (**Figure 44.10a**). In 1999, Robert Ricklefs and Irby Lovette summarized the available data on the richness of species of four groups of animals—birds, bats, reptiles and amphibians, and butterflies—across 19 islands that varied in area over two orders of magnitude (13–1,510 km$^2$). In each case, a positive correlation occurred between area and species richness (**Figure 44.10b**).

**Species-Distance Relationships** In studies of the numbers of lowland forest bird species in Polynesia, MacArthur and Wilson found that the number of species decreased with the distance from the source pool of New Guinea (**Figure 44.11**). They expressed the richness of bird species on the islands as a percentage of the number of bird species found on New Guinea. A significant decline in this percentage was observed with increasing distance. More-distant islands contained lower numbers of species than nearer islands. This research substantiated the prediction of species richness declining with increasing distance from the source pool.

**Species Turnover** Studies involving species turnover on islands are difficult to perform because detailed and complete species lists are needed over long periods of time, usually many years and often decades. The lists that do exist are often compiled in a casual way and are not usually suitable for comparison with more modern data. In 1980, British researcher Francis Gilbert reviewed 25 investigations carried out to demonstrate turnover and found a lack of this type of rigor in nearly all of them. Furthermore, most of the observed turnover in these studies, usually less than 1% per year, or less than one species per year, appeared to be due to immigrants that never became established rather than to the extinction of well-established species. More recent studies have revealed similar findings, suggesting that the rates of turnover are low rather than high, giving little conclusive support to the third prediction of the equilibrium model of island biogeography. Even the most rigorous study, by E. O. Wilson and his student Dan Simberloff, showed negligible turnover, as described next.

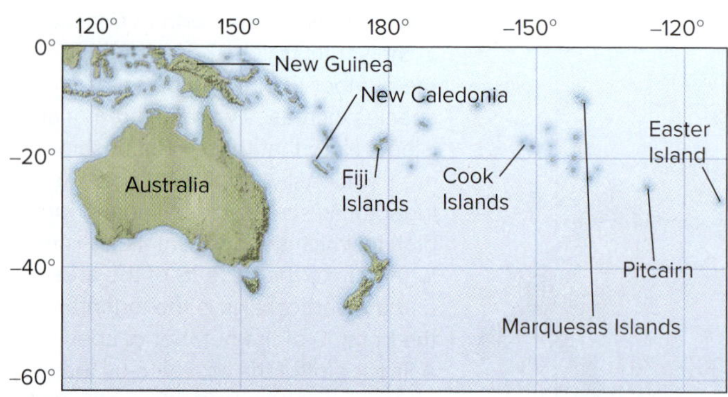


**(a) New Guinea and neighboring islands**

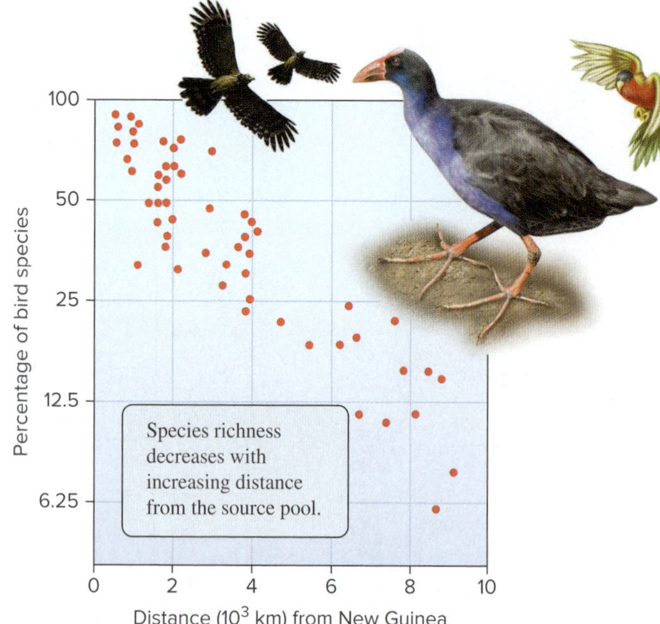


**(b) Relationship between species richness and distance from source**

**Figure 44.11** **Species richness and distance from the source pool.** **(a)** Map of Australia, New Guinea, and these Polynesian Islands: New Caledonia, Fiji Islands, Cook Islands, Marquesas Islands, Pitcairn, and Easter Island. **(b)** The number of bird species on the islands decreases with increasing distance from the source pool, New Guinea. The species richness is expressed as the percentage of bird species on New Guinea.

## FEATURE INVESTIGATION

### Simberloff and Wilson's Experiments Tested the Predictions of the Equilibrium Model of Island Biogeography

In the 1960s, American ecologists Daniel Simberloff and E. Wilson conducted possibly the most rigorous test of the equilibrium model of island biogeography ever performed, using islands in the Florida Keys. First, they surveyed small red mangrove (*Rhizophora mangle*) islands, 11–25 m in diameter, taking a census of the numbers of all their terrestrial arthropods. Then they enclosed each island with a plastic tent and had the islands fumigated with methyl bromide, a short-acting insecticide, to kill all arthropods on them. The tents were removed, and periodically thereafter Wilson and Simberloff surveyed the islands to examine recolonization rates. At each survey, they counted all the species present, noting any species not there at the previous census and the absence of others that were previously there but had presumably gone extinct (see the data of **Figure 44.12**). In this way, they estimated turnover of species on islands.

After 250 days, all but one of the islands had a number of arthropod species similar to the number before fumigation, even though population densities were still low. The data indicated that recolonization rates were higher on islands nearer to the mainland

**Figure 44.12** **Simberloff and Wilson's experiments on the equilibrium model of biogeography.**
*(photos)* Courtesy Dr. D. Simberloff, University of Tennessee

**HYPOTHESIS** The island biogeography model predicts higher species richness for islands closer to the mainland and significant turnover of species on islands.

**STARTING LOCATION** Mangrove islands in the Florida Keys.

**Experimental level** | **Conceptual level**

1 Take initial census of all terrestrial arthropods on 4 mangrove islands. Erect a framework over each mangrove island.

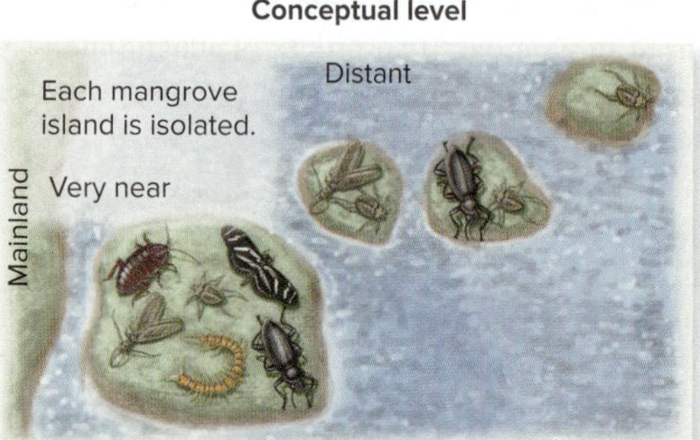

**2** Cover the framework with tents and fumigate with methyl bromide to kill all arthropod species.

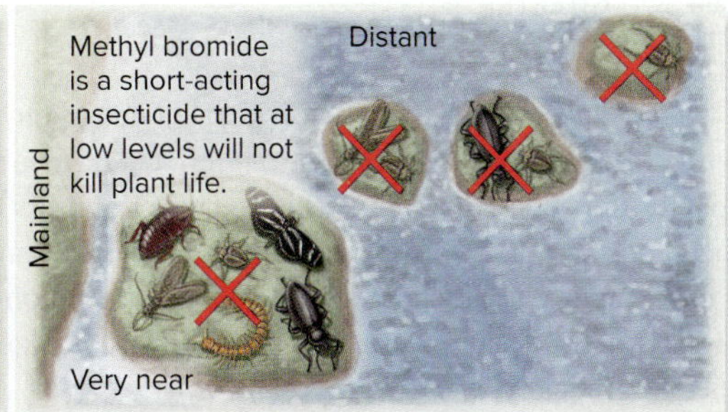

Methyl bromide is a short-acting insecticide that at low levels will not kill plant life.
Distant
Mainland
Very near

**3** Remove the tents and conduct censuses every month to monitor recolonization of arthropods and to determine extinction rates.

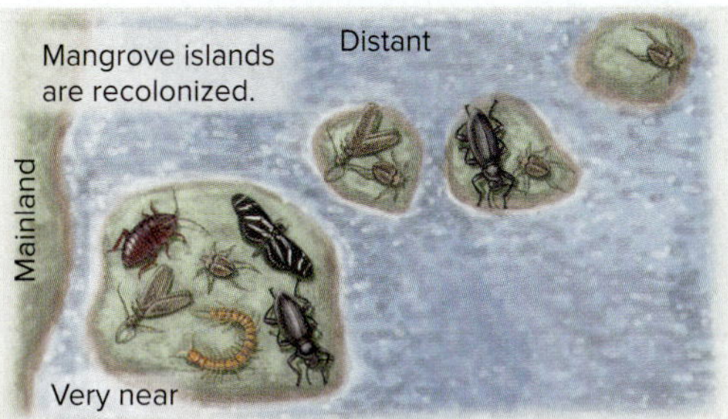


**4 THE DATA** Island E2 was closest to the mainland and supported the highest number of species both before and after fumigation. E3 and ST2 were at an intermediate distance from the mainland, and E1 was the most distant.

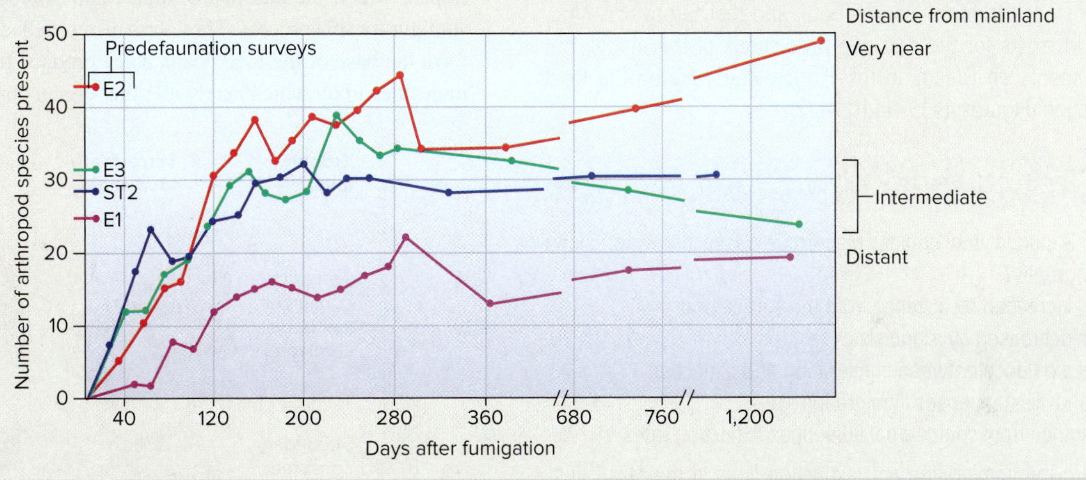


**5 CONCLUSION** Island distance from the mainland influences species richness on mangrove islands in the Florida Keys. However, species turnover is minimal, and species richness changes little following initial recolonization.

**6 SOURCE** Simberloff, D. S. 1978. Colonization of Islands by Insects: Immigration, Extinction and Diversity, pp. 139–153 in L. A. Mound and N. Waloff (eds.), *Diversity of Insect Faunas*. Blackwell Scientific Publications, Oxford, U.K.

than on far islands—as the island biogeography model predicts. However, the data, which consisted of lists of species on islands before and after extinctions, provided little support for the prediction of substantial turnover. Rates of turnover were low, only 1.5 extinctions per year, compared with the 15–40 species found on the islands within a year. Simberloff and Wilson concluded that turnover probably involves only a small subset of transient or less important species, with the more important species remaining permanent after colonization.

*Experimental Questions*

1. What was the purpose of Simberloff and Wilson's study?
2. Why did the researchers conduct a thorough species survey of arthropods before experimental removal of all the arthropod species?
3. **SCISKILLS** ▶ What did the researchers conclude about the relationship between island proximity to the mainland and species richness and turnover?

The equilibrium model of island biogeography has stimulated much research to confirm the strong effects of area and distance on species richness. However, species turnover appears to be low rather than considerable, which suggests that succession on most islands is a fairly orderly process. This means that colonization is not a random process and that the same species seem to colonize first and other species gradually appear in the same order.

The principles of island biogeography have been applied to wildlife preserves, which are essentially islands in a sea of developed land consisting of agricultural fields or urban sprawl. Conservationists have therefore utilized the model of island biogeography in the design of nature preserves, a topic we will return to in Chapter 47.

## 44.4 Reviewing the Concepts

- In the equilibrium model of island biogeography, the number of species on an island tends toward an equilibrium number determined by the balance between immigration and extinction rates (Figure 44.9).
- The model predicts that the number of species increases with increasing island size, that the number of species decreases with distance from the source pool, and that turnover is high. Support exists for the first two predictions of the model, but experiments on islands in the Florida Keys refuted the third prediction (Figures 44.10, 44.11, 44.12).

## 44.4 Testing Your Knowledge

1. Which is part of the original MacArthur-Wilson theory of island biogeography?
   - **a.** $\hat{S}$ is increased by distance from the source pool.
   - **b.** $\hat{S}$ is decreased by island size.
   - **c.** $\hat{S}$ is a balance between immigration and extinction.
   - **d.** Island size influences immigration rates.
   - **e.** Distance from source pool influences extinction rates.
2. Why are the immigration and extinction lines in the MacArthur-Wilson model both curved?
   - **a.** Species arrive at different rates.
   - **b.** Some organisms are more mobile than others.
   - **c.** Competition increases as more species arrive.
   - **d.** Later-arriving species tend to be better competitors.
   - **e.** All of the above are true.

## 44.5 Food Webs and Energy Flow

### Learning Outcomes

1. Compare and contrast food chains and food webs, and distinguish primary producers and primary, secondary, and tertiary consumers.
2. List and describe the different types of ecological pyramids.

We now turn our attention to ecosystem ecology, which concerns the movement of energy and materials through organisms and their communities. A key factor that affects species richness is the amount of available energy and the feeding relationships among organisms. These feeding relationships can be characterized by an unbranched **food chain**—a linear depiction of energy flow, with each organism feeding on and deriving energy from the preceding organism. In this section, we will consider the unidirectional flow of energy in a food chain and examine a food web, a more complex model of interconnected food chains. We will then explore two of the most important features of food webs—chain length and the pyramid of numbers—and learn how the passage of nutrients through food webs can result in the accumulation of harmful chemicals in the tissues of organisms at higher trophic levels.

### The Main Trophic Levels Within Food Chains Consist of Primary Producers, Primary Consumers, and Secondary Consumers

Each level in a food chain is called a **trophic level** (from the Greek *trophos*, meaning feeder), with different species feeding at different levels. In a food-chain diagram, an arrow connects each trophic level with the one above it (**Figure 44.13**). Food chains typically consist of organisms that obtain energy in different ways. **Autotrophs** harvest light or chemical energy and store that energy in carbon compounds. Most autotrophs, which include plants, algae, and photosynthetic bacteria, use sunlight for this process. These organisms, called **primary producers,** form the base of the food chain. They produce the energy-rich organic molecules upon which nearly all other organisms depend.

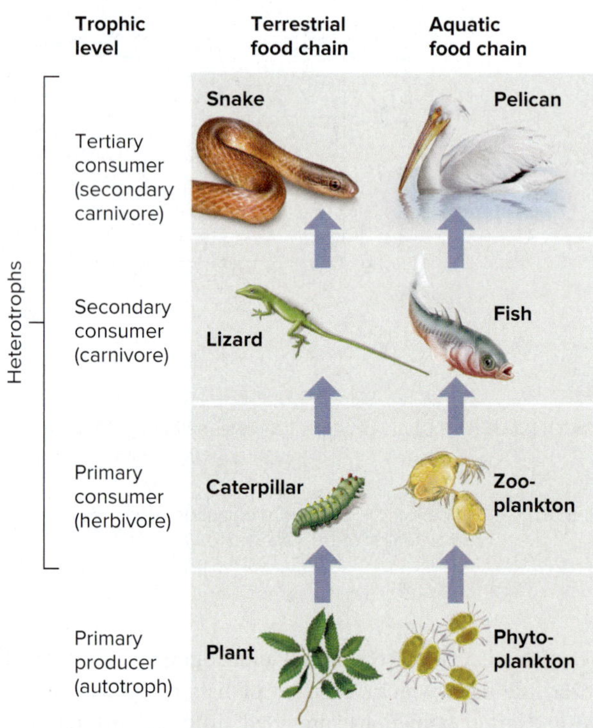


**Figure 44.13** **Food chains.** Two examples of the flow of food energy up the trophic levels: a terrestrial food chain and an aquatic food chain.

**BioConnections:** *In these two food chains, plants and protists (phytoplankton) are the producers. Refer back to Section 23.4. What other organisms are producers and could also support food chains?*

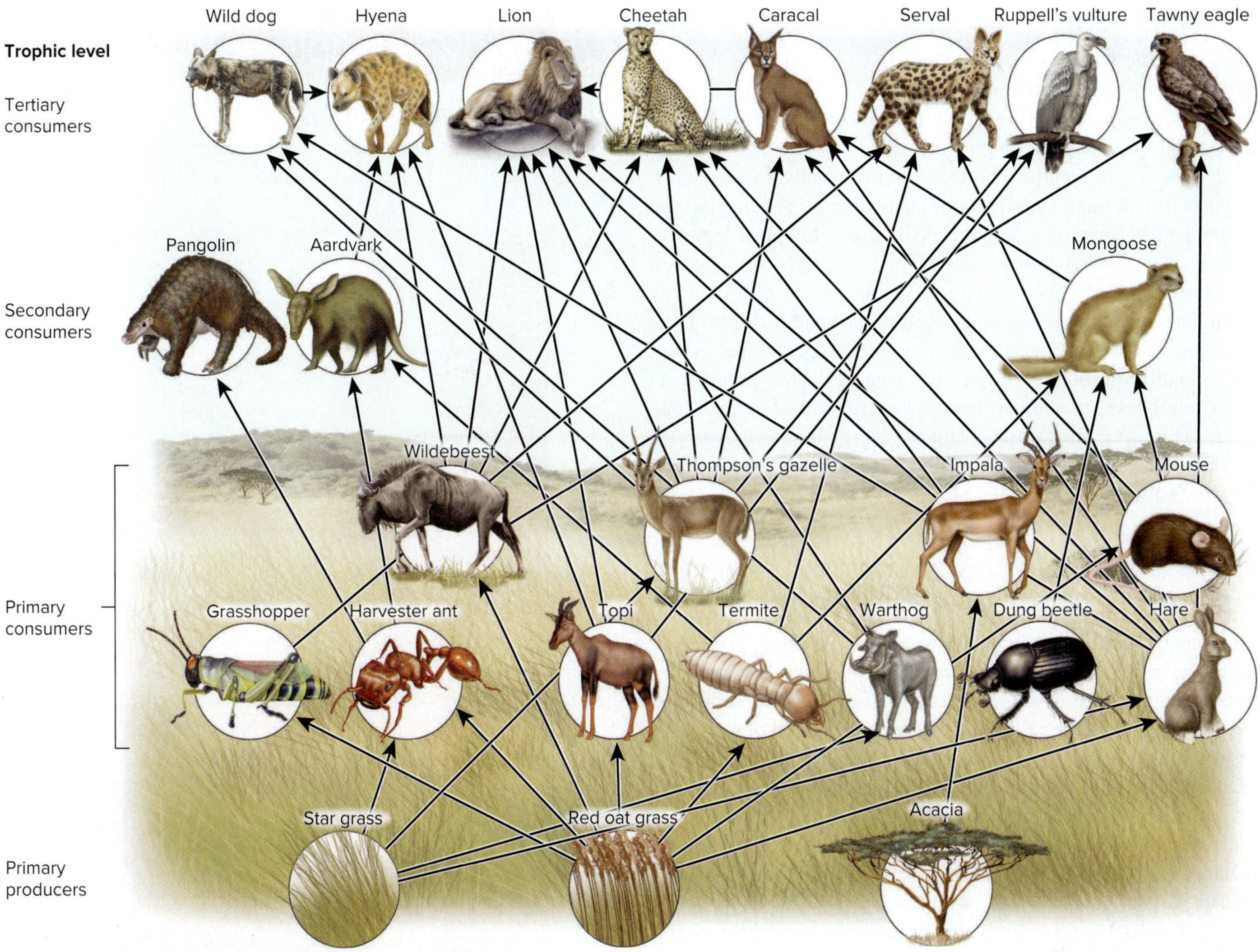

**Figure 44.14 A food web from an African savanna ecosystem.** Each trophic level is occupied by different species. Generally, each species feeds on, or is fed upon by, more than one species.

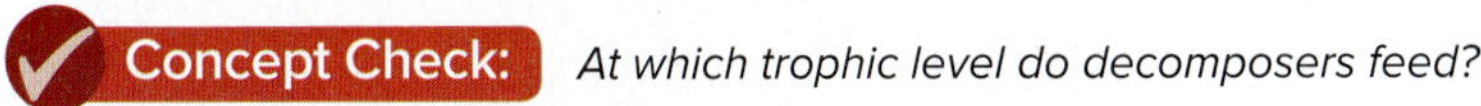

*At which trophic level do decomposers feed?*

Organisms that consume organic molecules from their environment to sustain life and thus receive their nutrition by eating other organisms are termed **heterotrophs.** Heterotrophs that obtain their food by consuming primary producers are termed **primary consumers** (also called herbivores) and include most protists, most animals, and even some plants such as mistletoe, which is parasitic on other plants. Organisms that eat primary consumers are **secondary consumers** (also called carnivores). Organisms that feed on secondary consumers are **tertiary consumers,** and so on. Thus, energy enters a food chain through primary producers, via photosynthesis, and is passed up the food chain to primary, secondary, and tertiary consumers (see Figure 44.13).

At each trophic level, many organisms die before they are eaten. Most energy from the first trophic level, such as the plants, goes unconsumed by herbivores. Instead, unconsumed plants die and decompose in place. This material, along with dead remains of animals and waste products, is called **detritus.** Consumers that get their energy from detritus, called **detritivores,** break down dead organisms from all trophic levels. For example, carrion beetles feed on the dead bodies of other animals. Some detritivores, which do not ingest their food but feed by absorbing it on a molecular scale, are known as decomposers. Prominent here are fungi and bacteria. In terrestrial systems, detritivores probably consume 80–90% of plant matter, with different groups such as earthworms and fungi working in concert to extract most of the energy. Detritivores may, in turn, support a community of predators that feed on them.

In nature, branching of food chains occurs at all trophic levels. For example, many different herbivore species may feed on the same plant species. Also, each species of herbivore may feed on several different plant species. For instance, on the African savanna, cheetahs, lions, and hyenas all eat a variety of prey, including wildebeest, impala, and Thompson's gazelle. These, in turn, eat a variety of trees and grasses. It is more correct, then, to draw relationships between these plants and animals not as a simple chain but as a **food web,** a complex model of interconnected food chains in which there are multiple links among species (Figure 44.14).

## In Most Food Webs, Chain Lengths Are Short

Let's examine some of the characteristics of food webs in more detail. The concept of chain length refers to the number of links between the trophic levels involved. For example, if a lion feeds on a zebra, and a zebra feeds on grass, the chain length is two. In many food webs, chain lengths tend to be short, usually five or fewer, because of two main factors. First, many organisms cannot digest all their prey. They take only the easily digestible plant leaves or animal tissue such as muscles and internal organs, leaving the hard wood or energy-rich bones behind. Second, much of the energy assimilated by animals is used in maintenance and is lost from the organism as heat. Both of these factors acting together means that, on average, only about 10% of available energy is transferred from one trophic level to another. Because energy is lost at each link, after a few links, most of the available energy has been expended and relatively little energy is available for higher trophic levels (**Figure 44.15**).

## Ecological Pyramids Describe the Distribution of Numbers, Biomass, or Energy Between Trophic Levels

The abundance of organisms, biomass, or available energy at each trophic level of a food web can be expressed graphically as an ecological pyramid. One of the best-known ecological pyramids, described by British ecologist Charles Elton in 1927, is the **pyramid of numbers,** in which the number of individuals decreases at each trophic level, with a large number of individuals at the base and fewer individuals at the top. For example, in a grassland, there may be hundreds of individual plants per square meter, dozens of insects that feed on the plants, a few spiders feeding on the insects, and birds that feed on the spiders (**Figure 44.16a**).

Ecologists have, however, discovered some exceptions to this pyramid. One single producer such as an oak tree can support hundreds of herbivorous beetles, caterpillars, and other primary consumers, which in turn may support thousands of insect predators. This is called an inverted pyramid of numbers.

One way to reconcile this apparent exception is to weigh the organisms in each trophic level, creating a **pyramid of biomass.** For example, an oak tree weighs more than all of its herbivores and predators combined. American ecologist Howard Odum measured the pyramid of biomass for a freshwater ecosystem, Silver Springs, in Florida (**Figure 44.16b**). Beds of eelgrass (genus *Sagittaria*) and attached algae make up most of the producers. Insects, snails, herbivorous fishes, and turtles eat the producers. Other fishes form the secondary and tertiary consumers. Odum also noted the presence of fungi and bacteria, which were involved in decomposition on all trophic levels.

Another way is to express the pyramid in terms of production rate. The **pyramid of energy** shows the rate of energy production rather than biomass (**Figure 44.16c**). The laws of thermodynamics ensure that the highest amounts of energy are found at the lowest trophic levels. The energy pyramid for Silver Springs is also very accurate in that it shows that large amounts of energy pass through decomposers, despite their relatively small biomass.

## Biology Principle

### Living Organisms Use Energy

Within trophic levels, energy is lost to maintenance, and between trophic levels, energy is lost to imperfect efficiency of transfer.

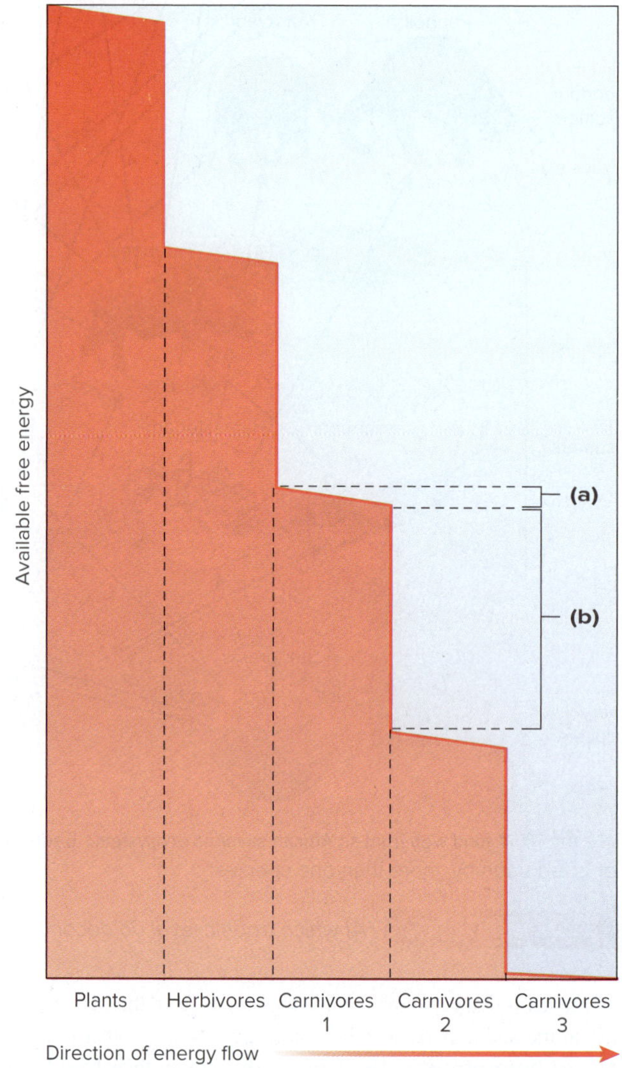


**Figure 44.15** **Energy flow through a food web.** In this graph of energy flow through a food web, there are five trophic levels and four links between the trophic levels. **(a)** Energy lost as heat in a single trophic level. **(b)** Energy lost in the conversion from one trophic level to another.

## 44.5 Reviewing the Concepts

- Ecosystem ecology concerns the movement of energy and materials through organisms and their communities. Organisms that obtain energy from light or chemicals are primary producers (or autotrophs). Organisms that feed on primary producers

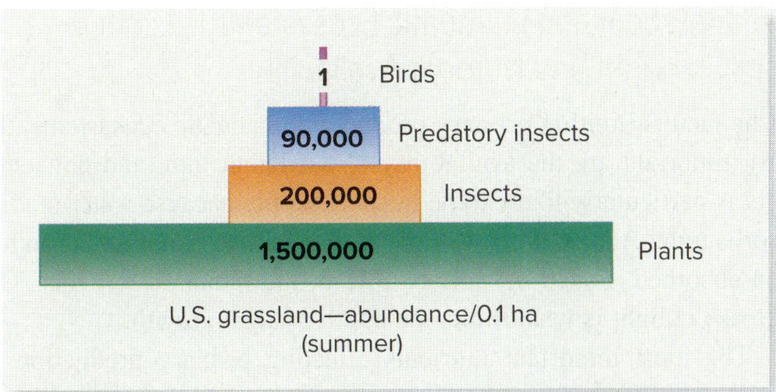


**(a) Pyramid of numbers**

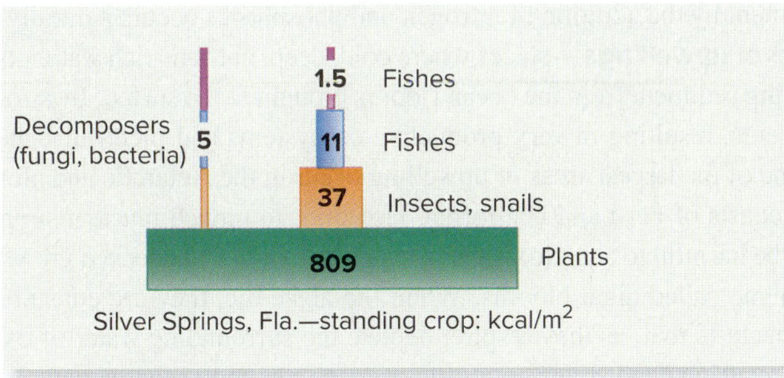


**(b) Pyramid of biomass**

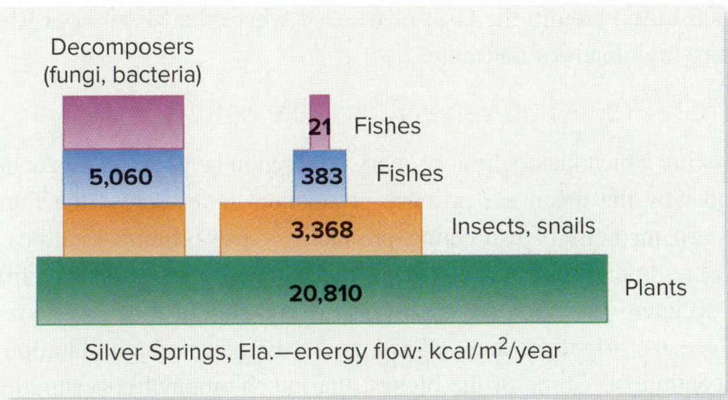


**(c) Pyramid of energy**

**Figure 44.16** **Ecological pyramids in food webs.** **(a)** In this pyramid of numbers, the abundance of species in an American grassland decreases with increasing trophic level. **(b)** In a pyramid of biomass, the amount of biological material is used instead of numbers of individuals. Note the presence of decomposers that decompose material at all trophic levels. **(c)** A pyramid of energy for Silver Springs, Florida. Note the large energy production of decomposers, despite their small biomass.

are called primary consumers (or herbivores). Organisms that feed on primary consumers are called secondary consumers (or carnivores). Consumers that get their energy from the remains and waste products of organisms are called detritivores (Figures 44.13, 44.14).

- Food webs are complex models of interconnected food chains in which multiple links occur between species. Food webs tend to have five or fewer links between top and bottom trophic levels. Energy conversions are not 100% efficient, and usable energy is lost within each trophic level and from one trophic level to the next (Figures 44.15, 44.16).

## 44.5 Testing Your Knowledge

1. As we learned in Chapter 23, some bacteria and archaea are able to use energy from the oxidation of sulfur, iron, or hydrogen. These bacteria can be classified as
   **a.** heterotrophs. **c.** producers. **e.** both b and c.
   **b.** autotrophs. **d.** both a and b.
2. Detritivores that feed on the dung of herbivores feed at which trophic level?
   **a.** 1 **b.** 2 **c.** 3 **d.** 4 **e.** 5
3. When considering the average food chain, which of the following statements is *true?*
   **a.** Secondary consumers are the most abundant organisms in an ecosystem.
   **b.** Most plant biomass is eaten by herbivores.
   **c.** Biomass decreases as you move up the food chain.
   **d.** The trophic level with the highest species abundance is usually the primary producers.
   **e.** All of the above are true.

## 44.6 Biomass Production in Ecosystems

### Learning Outcomes

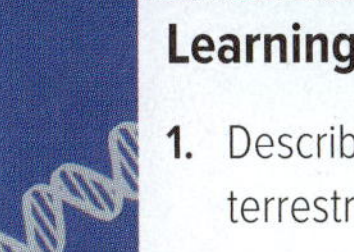

1. Describe the factors that limit primary production in terrestrial and aquatic ecosystems.
2. Explain the fate of most primary production.

In this section, we will take a closer look at biomass production in ecosystems. Because the bulk of the Earth's biosphere, 99.9% by mass, consists of primary producers, when we measure ecosystem biomass production, we are primarily interested in plants, algae, or cyanobacteria. Their production is called **gross primary production (GPP).** Gross primary production is equivalent to the carbon fixed during photosynthesis. **Net primary production (NPP)** is GPP minus the energy used during cellular respiration (R) of photosynthetic organisms.

$$NPP = GPP - R$$

NPP is thus the amount of energy available to primary consumers. Unless otherwise noted, the term **primary production** refers to NPP.

## Primary Production Is Influenced in Terrestrial Ecosystems by Water, Temperature, and Nutrient Availability

In terrestrial ecosystems, water is a major determinant of primary production, and primary production shows an almost linear increase with annual precipitation, at least in arid regions. Likewise, temperature, which affects production primarily by slowing or accelerating plant metabolic rates, is also important. A lack of **nutrients,** key elements in usable form, particularly nitrogen and phosphorus, can also limit primary production in terrestrial ecosystems, as farmers know only too well. Fertilizers are commonly used to boost the production of annual crops.

In 1984, Susan Cargill and Robert Jefferies showed how a lack of both nitrogen and phosphorus limited production in salt marsh sedges and grasses in Hudson Bay, Canada (Figure 44.17). Of the two nutrients, nitrogen was the **limiting factor**—the one in the shortest supply for growth; without it, the addition of phosphorus did not increase production. However, once nitrogen was added and was no longer limiting, phosphorus became the limiting factor. The addition of nitrogen and phosphorus together increased production the most. This result supports a principle known as **Liebig's law of the minimum,** named for Justus von Liebig, a 19th-century German chemist, which states that species biomass or abundance is limited by the scarcest factor. This factor can change, as the Hudson Bay experiment showed. When sufficient nitrogen is available, phosphorus becomes the limiting factor. Once phosphorus becomes abundant, then productivity will be limited by another nutrient.

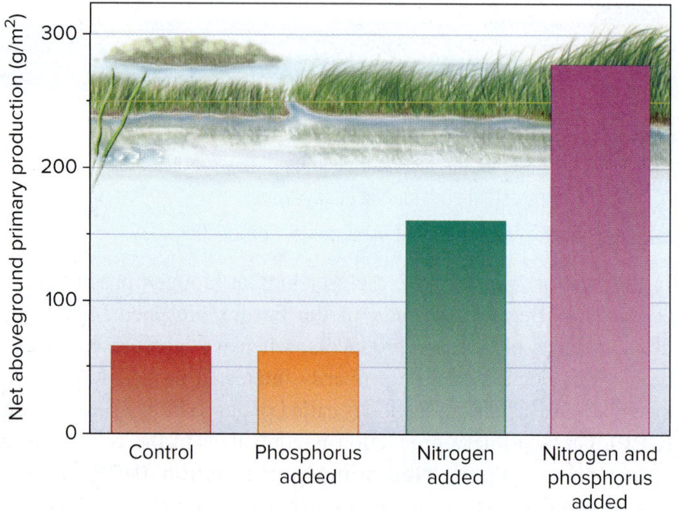


**Figure 44.17 Limitation of primary production by nitrogen and phosphorus.** Net aboveground primary production of a salt marsh sedge (*Carex subspathacea*) in response to nutrient addition. Nitrogen is the limiting factor. After nitrogen is added, phosphorus becomes the limiting factor.

## Primary Production in Aquatic Ecosystems Is Limited Mainly by Light and Nutrient Availability

Of the factors limiting primary production in aquatic ecosystems, the most important are the availability of sufficient light and nutrients. Light is particularly likely to be in short supply because water readily absorbs light. At a depth of 1 m, more than half the solar radiation has been absorbed. By 20 m, only 5–10% of the radiation remains. The decrease in light is what limits the depth of algal growth.

The most important nutrients affecting primary production in aquatic systems are nitrogen and phosphorus, because they occur in very low concentrations. Whereas soil contains about 0.5% nitrogen, seawater contains only 0.00005% nitrogen. Enrichment of the aquatic environment by the addition of nitrogen and phosphorus occurs naturally in areas of **upwellings**—places where cold, deep, nutrient-rich water containing sediment from the ocean floor is brought to the surface by strong currents, resulting in very productive ecosystems and plentiful fishes. Some of the largest areas of upwelling occur in the Antarctic and along the coasts of Peru and California. However, too much nutrient supply can be harmful to aquatic systems, resulting in large, unchecked growths of algae called algal blooms. When the algae die, they are consumed by bacteria that, as they respire, deplete the surrounding water of oxygen, causing dead zones with little oxygen to support other aquatic life. Such dead zones are prominent along coastal areas where fertilizer-rich rivers discharge into the oceans. The largest of these is the 22,000-km$^2$ (8,500-mile$^2$) area in the Gulf of Mexico where the Mississippi River dumps high loads of nutrients.

## Primary Production Varies Across the Earth

Knowing which factors limit primary production helps ecologists understand why the mean net primary production varies across the Earth. Modern methods of estimating productivity use orbiting satellites to measure differences in the electromagnetic radiation reflected back from the different vegetation types on Earth (Figure 44.18). When we look at the oceans, bright greens, yellows, and reds indicate high chlorophyll concentrations. Some of the highest marine chlorophyll concentrations occur at continental margins, where river nutrients pour into the oceans. Upwellings along coasts also bring nutrient-rich water to the surface. Northern oceans, and to a lesser extent southern oceans, are also very productive, because seasonal storms and temperature changes allow vertical mixing of water, bringing nutrient-rich water to the surface. In the spring, the increased amount of light and nutrients permit rapid phytoplankton growth until the nutrients are all used up. Many other marine areas, including tropical oceans, are highly unproductive.

Over land, the productivity of forests in all parts of the world, from the tropics to northern and southern temperate areas, is high, but production is often higher in temperate rather than tropical forests. This matches the pattern of productivity observed in the oceans. Although tropical forests enjoy warm temperatures and abundant rainfall, such conditions weather soils rapidly. Tropical soils are low in available forms of most plant nutrients because of loss through leaching. In contrast, temperate soils tend to have much greater concentrations of essential nutrients because of lower rates of nutrient loss and more frequent grinding of fresh minerals by the cycles of continental glaciations over the past 3 million years. Prairies and savannas are also highly productive because their plant biomass usually dies and

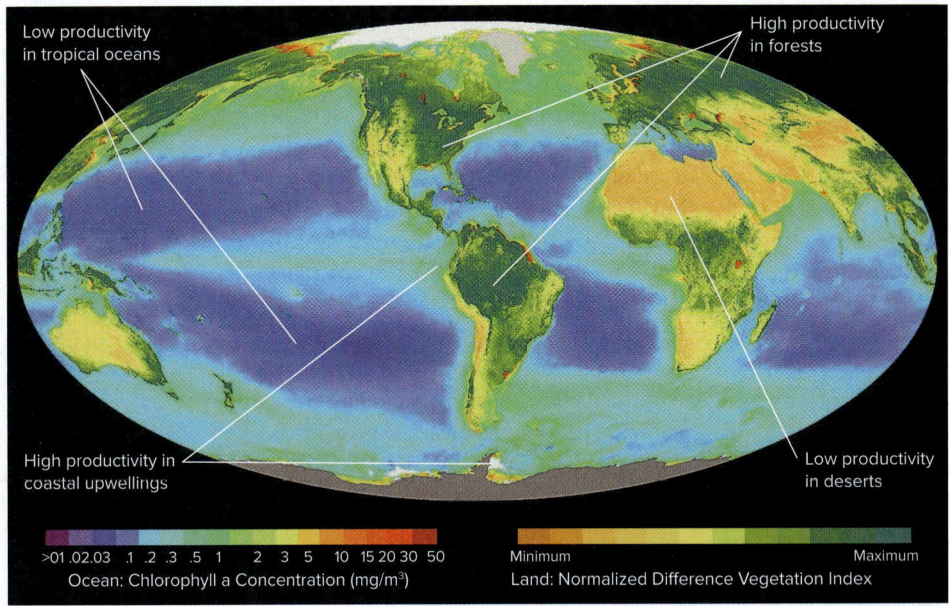


**Figure 44.18 Primary productivity measured by satellite imagery.** Ocean chlorophyll concentrations and the Normalized Difference Vegetation Index (NDVI) on land provide good data on marine and terrestrial productivity, respectively.

Source: Provided by the SeaWiFS Project, NASA/Goddard Space Flight Center, and ORBIMAGE

decomposes each year, returning a portion of the nutrients to the soil, and temperatures and rainfall are not limiting. Deserts and tundra have low productivity because of a lack of water and low temperatures, respectively. Wetlands tend to be extremely productive, primarily because water is not limiting and nutrient levels are high.

## Most Primary Production Is Eaten by Detritivores

A strong relationship exists between primary production and secondary production, usually measured as the biomass of herbivores. This means that more plant biomass, and thus more primary production, leads to an increased biomass of consumers. However, it has been shown that in ecosystems as diverse as forests and salt marshes most primary production goes to detritivores, not herbivores.

In 1962, American ecologist John Teal examined energy flow in a Georgia salt marsh (**Figure 44.19**). In salt marshes, most of the energy from the Sun goes to two types of organisms: *Spartina* plants and marine algae. The *Spartina* plants are rooted in the ground, whereas the algae float on the water's surface or live on the mud

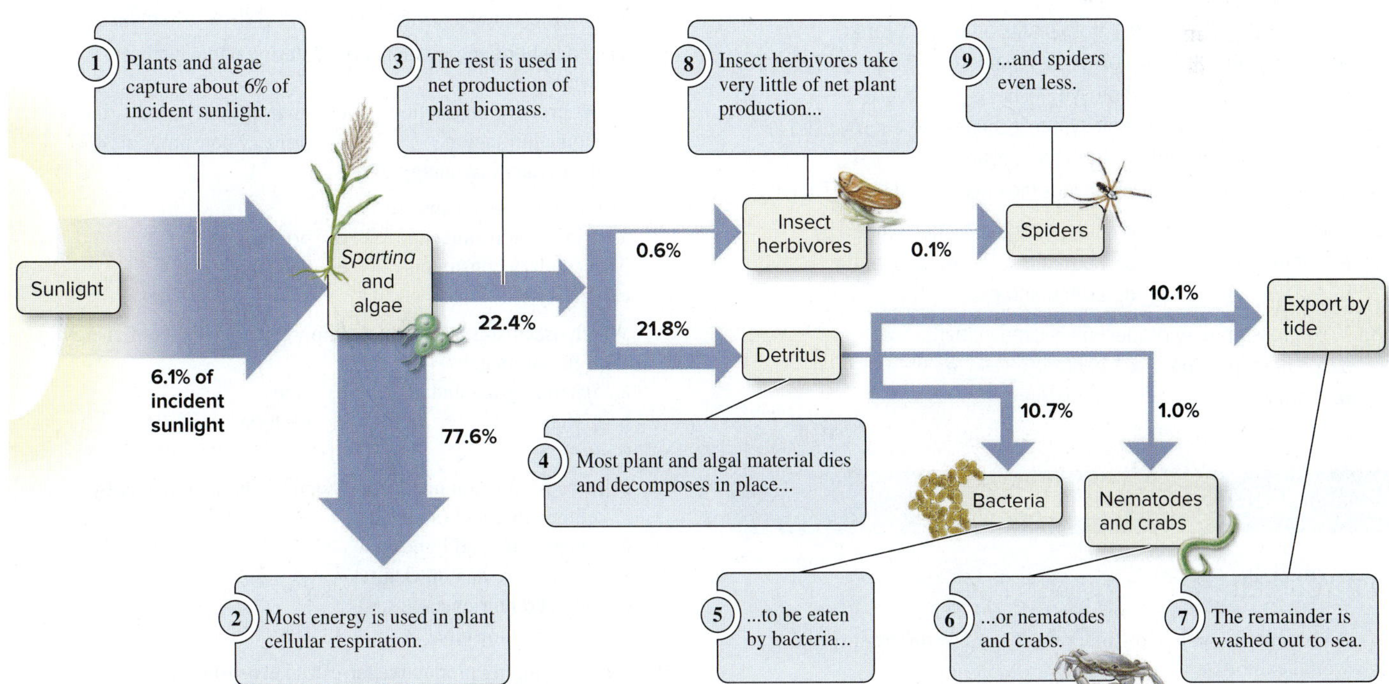


**Figure 44.19 Energy-flow diagram for a Georgia salt marsh.** Numbers represent the percentage of gross primary production that flows into different trophic levels or is used in plant respiration.

or on *Spartina* leaves at low tide. These photosynthetic organisms absorb about 6% of the sunlight. Most of the plant energy, 77.6%, is used in plant and algal cellular respiration. Of the energy that is accumulated in plant biomass, 22.4%, most dies in place and rots on the muddy ground, to be consumed by bacteria. Bacteria are the major detritivores in this system, followed distantly by nematodes and crabs, which feed on tiny food particles as they sift through the mud. Some of this dead material is also removed from the system (exported) by the tide. The herbivores take very little of the plant production, around 0.6%, eating only a small proportion of the *Spartina* and none of the algae. A fraction of herbivore biomass is then consumed by spiders. Overall, if we view the species in ecosystems as transformers of energy, then plants, algae, and cyanobacteria are by far the most important organisms on the planet, other species of bacteria are next, and animals are a distant third.

## 44.6 Reviewing the Concepts

- Net primary production (NPP) is gross primary production minus the energy released during respiration via photosynthetic organisms. NPP in terrestrial ecosystems is limited primarily by temperature and the availability of water and nutrients. In aquatic ecosystems, it is limited mainly by the availability of light and nutrients (Figures 44.17, 44.18).
- Secondary production is limited by available primary production, but most primary production goes to detritivores (Figure 44.19).

## 44.6 Testing Your Knowledge

1. Net primary production is
   a. the energy that passes from plants on to herbivores.
   b. gross primary production minus the energy used by herbivores.
   c. the energy fixed in photosynthesis.
   d. the energy fixed in photosynthesis minus the energy used in respiration of photosynthetic organisms.
   e. gross primary production minus the energy used by consumers.
2. Primary production in terrestrial systems is often limited by
   a. temperature. c. nutrients e. a, b, and c.
   b. water. d. both b and c.
3. Most gross primary production is used in (by)
   a. plant respiration. c. bacteria. e. both c and d.
   b. herbivores. d. nematodes.

# Assess and Discuss

## Test Yourself

1. A community with many individuals but few different species would exhibit
   a. low abundance and high species complexity.
   b. high stability.
   c. low species richness and high abundance.
   d. high species diversity.
   e. high abundance and high species richness.
2. Which of the following statements best represents the productivity hypothesis regarding species richness?
   a. The larger the area, the greater the number of species that will be found there.
   b. Temperate regions have a lower species richness due to the lack of time available for migration after the last Ice Age.
   c. The number of species in a particular community is directly related to the amount of available energy.
   d. As invertebrate productivity increases, species richness will increase.
   e. Species richness is not related to primary productivity.
3. The process of primary succession occurs
   a. around a recently erupted volcano.
   b. on a newly plowed field.
   c. on a hillside that has suffered a mudslide.
   d. on a recently flooded riverbank.
   e. on none of the above.
4. The mechanism of succession in which early colonizers exclude subsequent colonists from moving into a community is referred to as
   a. facilitation. d. inhibition.
   b. competitive exclusion. e. natural selection.
   c. secondary succession.
5. On which types of islands would you expect species richness to be greatest?
   a. small, near mainland
   b. small, distant from mainland
   c. large, near mainland
   d. large, distant from mainland
   e. species richness is equal on all these types of islands.
6. The amount of energy that is fixed during photosynthesis is known as
   a. net primary production. d. gross primary production.
   b. biomagnification. e. primary consumption.
   c. the pyramid of energy.
7. Autotrophic organisms are
   a. primary consumers. d. primary producers.
   b. secondary consumers. e. decomposers.
   c. tertiary consumers.
8. Which organisms are the most important consumers of energy in a Georgia salt marsh?
   a. *Spartina* grass and algae d. crabs
   b. insects e. bacteria
   c. spiders
9. Primary production in aquatic systems is limited mainly by
   a. temperature and moisture.
   b. temperature and light.
   c. temperature and nutrients.
   d. light and nutrients.
   e. light and moisture.
10. The most highly productive terrestrial ecosystems are
    a. deserts. c. forests. e. tundra.
    b. prairies. d. savannas.

## Conceptual Questions

1. Forest A has 5 tree species with 100 individuals and forest B has 5 tree species with 10 individuals. What is the Shannon diversity index for both forests? Which forest has the highest diversity? What does this exercise tell you about the limitations of the Shannon diversity index?
2. At what trophic level does a carrion beetle feed?
3. **PRINCIPLES** A principle of biology is that biology is an experimental science. In the nutrient-poor heathlands of Europe, scotch heather (*Calluna vulgaris*) and cross-leaved heath (*Erica tetralix*) are gradually replaced by variegated purple moor grass (*Molinia caerulea*) and wavy hair grass (*Deschampsia flexuosa*). Adding *Calluna* litter or nitrogen fertilizer speeds up this process. Explain this phenomenon and which mechanism of succession is supported.

## Collaborative Questions

1. List some possible ecological disturbances, their likely frequency in natural communities, and the severity of their effects.
2. Calculate the species diversity of the following four communities. Which community has the highest diversity? What is the maximum diversity each community could have?

| | Relative abundance of species | | | | Maximum possible |
|---|---|---|---|---|---|
| **Community** | **Species 1** | **Species 2** | **Species 3** | **$H_S$** | **diversity** |
| 1 | 90 | 10 | — | | |
| 2 | 50 | 50 | — | | |
| 3 | 80 | 10 | 10 | | |
| 4 | 33.3 | 33.3 | 33.3 | | |

## Online Resource

# 45

# How Climate Affects the Distribution of Species on Earth

© AGF/UIG/Getty Images

African savannah, Masai Mara, Kenya.

## Chapter Outline



For most areas of the world, we need only see a picture for a few seconds to know where we are: a tropical rain forest, a hot desert, or the African savannah. This is because the plants and animals in these large-scale communities, called **biomes,** are characteristic of the climate in which they live. In this chapter, we examine how and why climate varies across the Earth and the global distribution of biome types. First, we discuss how temperature differences on Earth arise and how these differences affect air circulation and precipitation patterns. We see how the tilt of the Earth causes seasonality and how elevation and proximity to water can affect local climate. In section 45.2 we examine how variations in temperature and rainfall patterns produce different biome types.

Humans are fundamentally changing the appearance of traditional biomes on Earth as they plant crops, cut down old-growth forests, and urbanize areas. Some recent studies have suggested that human-dominated areas now cover more of the Earth's land surface than natural ones. In order to understand the extent of agriculture, deforestation, and urbanization at large scales, we need to develop an understanding of what naturally limits the distribution of certain large-scale communities, such as tropical forests or temperate prairies.

## 45.1 Climate

### Learning Outcomes

1. Explain how differences in global temperatures drive atmospheric circulation.
2. Discuss how the tilt and rotation of the Earth influences weather patterns.
3. Explain how elevation and proximity to large bodies of water can change local temperature and precipitation patterns.
4. Explain how waves and tides are generated in marine systems.

Temperature and precipitation are the most important components of **climate**, the prevailing weather pattern in a given region. The distribution and abundance of organisms are primarily influenced by these factors. Therefore, to understand the patterns of abundance of life on Earth, ecologists need to study the global climate. In this chapter, we examine global climate patterns, focusing on how temperature variation drives atmospheric circulation and how features such as elevation and landmass can alter these patterns. Knowing how and why climate changes around the world enable us to understand and predict the occurrence of biomes, large-scale terrestrial communities such as tropical rain forests and hot deserts.

### Atmospheric Circulation Is Driven by DIfferences in Global Temperatures

Substantial differences in temperature occur over the Earth, mainly due to latitudinal variations in the incoming solar radiation. In higher latitudes, such as northern Canada and Russia, the Sun's rays hit the Earth obliquely and are spread out over more of the planet's surface than they are in equatorial areas (**Figure 45.1**). More heat is also lost in the atmosphere of higher latitudes because the Sun's rays travel a greater distance through the atmosphere, allowing more heat to be dissipated by cloud cover. The result is that 40% less solar energy strikes polar latitudes than equatorial areas. Generally, temperatures increase as the amount of solar radiation increases (**Figure 45.2**). However, at the tropics, both cloudiness and rain reduce average temperature, so temperatures do not continue to increase toward the equator.

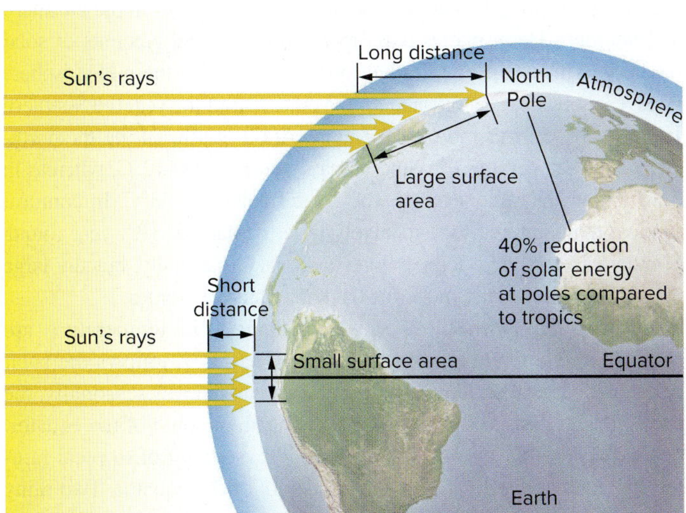


**Figure 45.1 The intensity of solar radiation at different latitudes.** In polar areas, the Sun's rays strike the Earth at an oblique angle and deliver less energy than at tropical locations. In tropical areas, the energy is concentrated over a smaller surface and travels a shorter distance through the atmosphere.

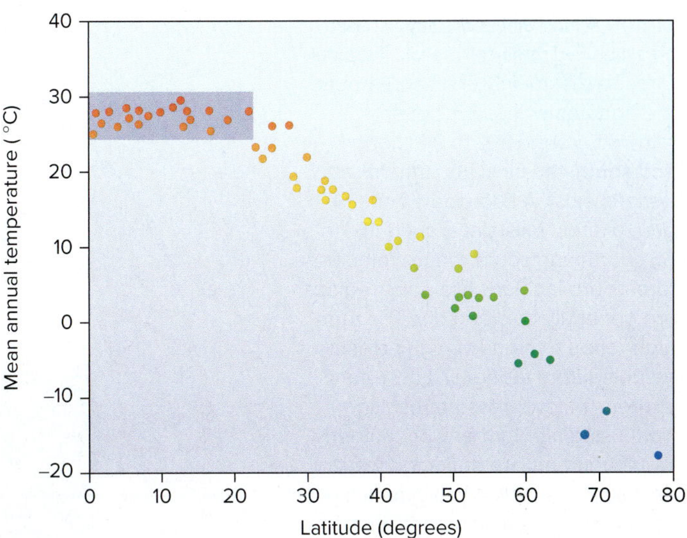


**Figure 45.2 Variation of the Earth's temperature.** The temperatures shown in this figure were measured at moderately moist continental locations of low elevation.

**Concept Check:** *Why is there a wide band of similar temperatures at the tropics? (See the orange dots highlighted in the purple rectangle.)*

Global patterns of atmospheric circulation and precipitation are influenced by solar energy. In 1735, English meteorologist George Hadley made the initial contribution to a model of general atmospheric circulation. In his model, high temperatures at the equator cause the surface equatorial air to heat up and rise vertically into the atmosphere. As the warm air rises away from its source of heat, it cools and becomes less buoyant, but the cool air does not sink back to the surface because of the warm air behind it. The warm air rising near the equator forms towers of cumulus clouds that provide rainfall, which, in turn, maintains the lush vegetation of the equatorial rain forests. As the upper flow of air moves toward the poles, it begins to subside, or fall back to Earth, at about 20–30° north and south of the equator. These **subsidence zones** are areas of high pressure and are the sites of the world's tropical deserts, because the subsiding air is relatively dry, having released all of its moisture over the equator (**Figure 45.3**).

From the center of the subsidence zones, the surface flow splits into two directions, one of which flows toward the poles and the other toward the equator. The equatorial flow from both hemispheres meets near the equator in a region called the intertropical convergence zone (ITCZ). In the three-cell model, the circulation between 30° and 60° latitude is opposite that of the cell nearest the equator because the net surface flow is poleward. Additional zones of high precipitation

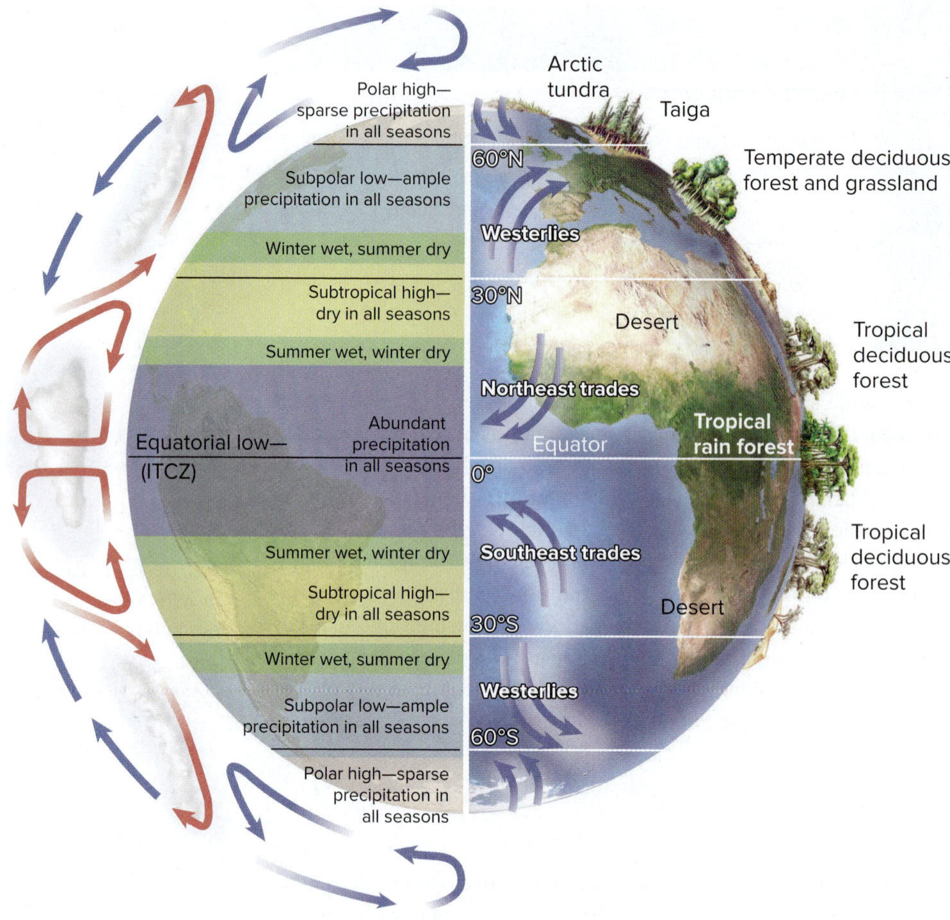


**Figure 45.3 Global circulation based on a modified three-cell model.** Tropical forests exist mainly in a band around the equator, where it is hot and rainy. At around 20° to 30° north and south, the air is hot and dry, and deserts exist. A secondary zone of precipitation exists at around 45° to 55° north and south, where temperate forests are located. The polar regions are generally cold and dry. The term high refers to areas of high pressure resulting from falling air. Low refers to areas of low pressure resulting from rising air. Northerly or southerly air movements are deflected west or east and are shown as westerlies or so-called northeast/southeast trade winds, which allowed sailing ships to explore the world.

occur in this cell, usually between 45° and 55°. In the final circulation cell, at the poles, the air has cooled and descends, but it has little moisture left, explaining why many high-latitude regions are desert-like in condition. The distributions of the major biomes are largely determined by temperature differences and the rain patterns they generate. Hot, tropical forest blankets the tropics, where rainfall is high. At about 20–30° latitude, the air cools and descends, but it is without moisture, so the hot deserts occur around that latitude. The middle cell of the circulation model shows us that at about 45° to 55° latitude, the air has warmed and gained moisture, so it ascends, dropping rainfall over the wet, temperate forests of the Pacific Northwest and western Europe in the Northern Hemisphere and New Zealand and Chile in the Southern Hemisphere. In the high-latitude regions of the Northern and Southern Hemispheres (above 60° latitude), only short tundra vegetation can be supported.

## The Tilt and Rotation of the Earth Also Affect Climate

The three-cell model provides a good understanding of global circulation, but it is oversimplified. Overlaid on these patterns is a seasonality caused by the tilt of the Earth. The Earth's axis of rotation is tilted at 23.5° from the vertical for the full 365 days it takes to orbit the Sun (**Figure 45.4**). The solar equator, the area receiving the most solar energy, varies seasonally and reaches 23.5° north on June 21 and 23.5° south on December 21. In the Northern Hemisphere, these dates are called the summer and winter solstices, from the Latin *sol,* sun, and *sistit,* stands. For several days before and after each solstice, the noontime elevation of the Sun appears in the same place. On March 21 and September 22, the so-called spring and autumn equinoxes, all locations in the Northern and Southern Hemispheres have approximately equal amounts of solar radiation. This means that for half of the year the Northern Hemisphere receives more solar energy, and for the other half of the year the Southern Hemisphere receives more solar energy. At 60° north, during the northern winter, temperatures in Siberia may average only –12°C, whereas in the summer they may average 16°C, a difference of 28°C. In contrast, tropical temperatures vary relatively little, perhaps 2–3°C, year round. Southern Hemisphere temperatures also vary seasonally, but the large expanses of open water moderate the temperature extremes.

Associated with these seasonal changes in temperature are changes in day length and precipitation. The ITCZ follows the solar equator (**Figure 45.5**). Thus, as the solar equator moves seasonally, the band of rainfall associated with it drifts north or south of the equator, depending on the season. The result is a seasonality of tropical rainfall between about 20° north and 20° south of the equator. Two rainy seasons of equal length exist at the equator, both about 3 to 4 months long, as the ITCZ drifts north or south. At the solar equator, there is just one rainy season of about 4–5 months. As a result of changes in latitude, there is much global variation in precipitation patterns.

In addition to seasonal changes in temperatures, the wind direction is deflected based on the rotation of the Earth, a phenomenon

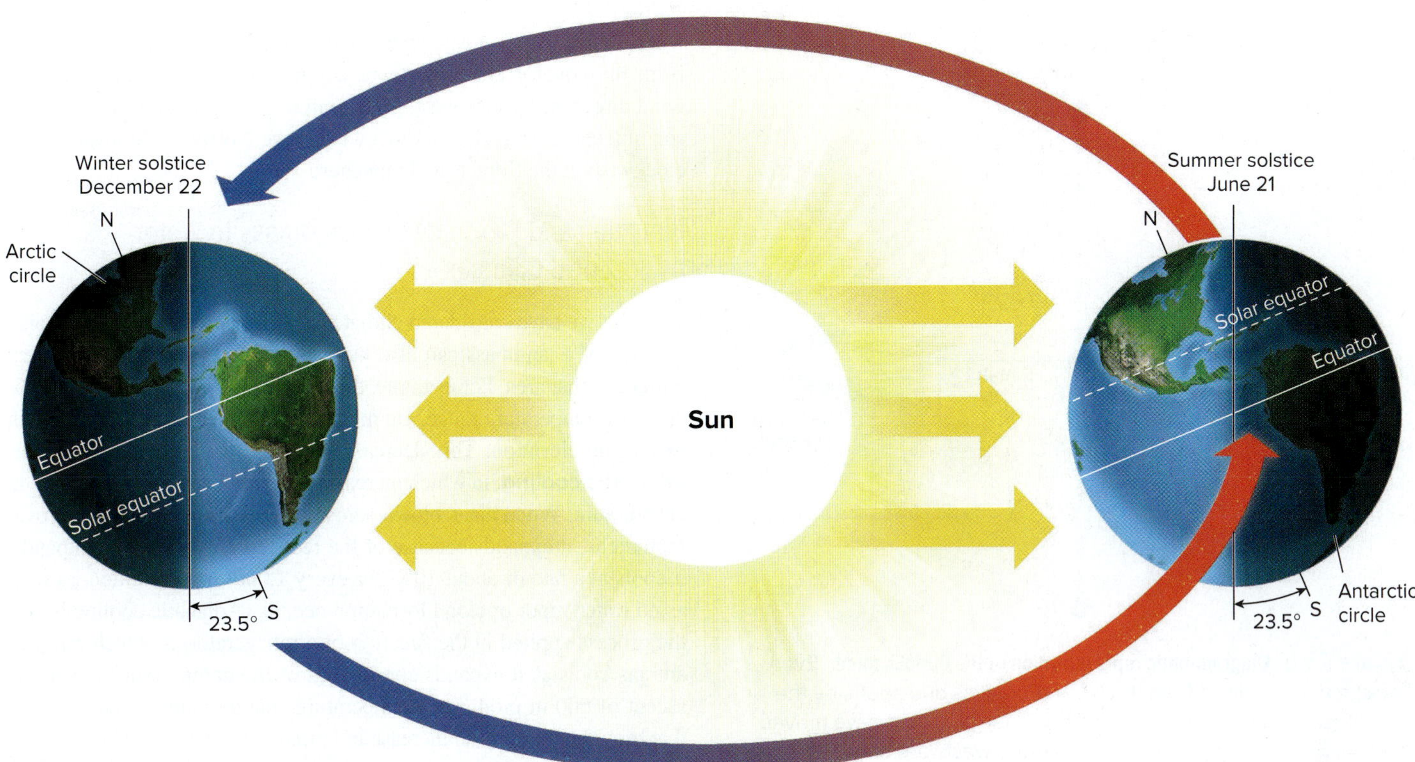


**Figure 45.4** **Seasonality.** The constant 23.5° tilt of the Earth's axis causes changes in the solar energy striking the Earth during its 365-day procession around the Sun.

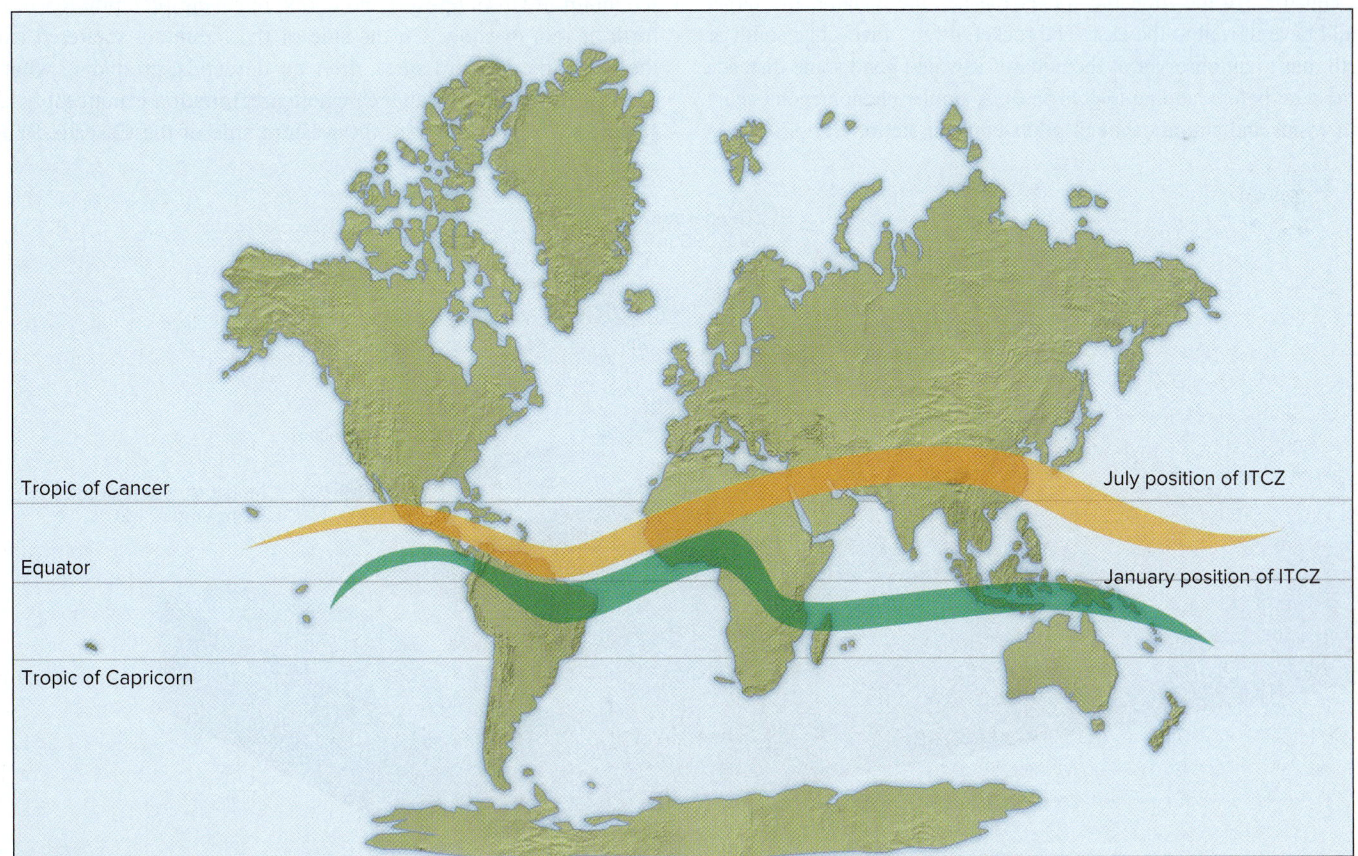


**Figure 45.5** **Seasonal variations in the intertropical convergence zone.**

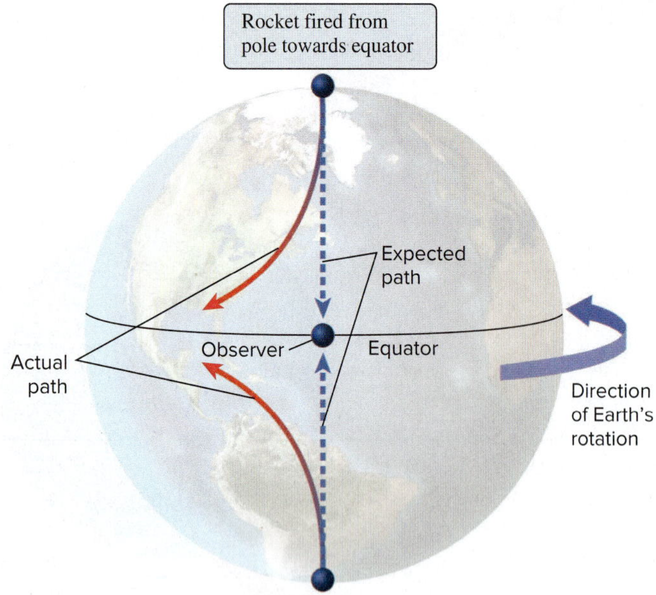


**Figure 45.6 Diagrammatic representation of the Coriolis force.** Even though a rocket fired from the North Pole flies due south, by the time it reaches the equator its intended target would have moved and the landing point would be at a more westward spot.

called the Coriolis force, after the French scientist Gaspard Gustave de Coriolis. Imagine a rocket fired from the North or South Pole toward the equator. By the time the rocket reaches the equator, the target would have moved to the east. The rocket always travels due south or north, but to an observer at the equator it would bend some distance to the west before landing (Figure 45.6). A similar phenomenon occurs with winds and aircraft, though pilots compensate for Coriolis forces when flying. The opposite phenomenon happens when a rocket is fired from the equator either north or south. The Coriolis force deflects wind directions east or west, gives spin to storm systems, and is a reason hurricanes spin counterclockwise in the Northern Hemisphere but clockwise in the Southern Hemisphere.

## Elevation and Proximity of a Landmass to Water Affect Local Climates

Thus far, we have considered climate on a global scale. The geographic features of a landmass can also have an important effect on the local climate in that area. For example, the elevation of a region greatly influences its temperature range. On mountains, temperatures decrease with increasing elevation. This decrease is a result of a process known as **adiabatic cooling**, in which increasing elevation leads to a decrease in air pressure. When air is blown across the Earth's surface and up over mountains, it expands because of the reduced pressure. As it expands, it cools at a rate of about 10°C for every 1,000 m in elevation, as long as no water vapor or cloud formation occurs. (Adiabatic cooling is also the process applied in the function of a refrigerator, in which refrigerant gas cools as it expands coming out of the compressor.) A vertical ascent of 600 m produces a temperature change roughly equivalent to that brought about by an increase in latitude of 1,000 km. This explains why mountaintop vegetation, even in tropical areas, can have the characteristics of a colder biome.

Mountains can also influence patterns of precipitation. For example, when warm, moist air encounters the windward side of a mountain, it flows upward and cools, releasing precipitation in the form of rain or snow. On the side of the mountain sheltered from the wind (the leeward side), drier air descends, producing what is called a **rain shadow**, an area where precipitation is noticeably less (Figure 45.7). For example, the western side of the Cascade Range

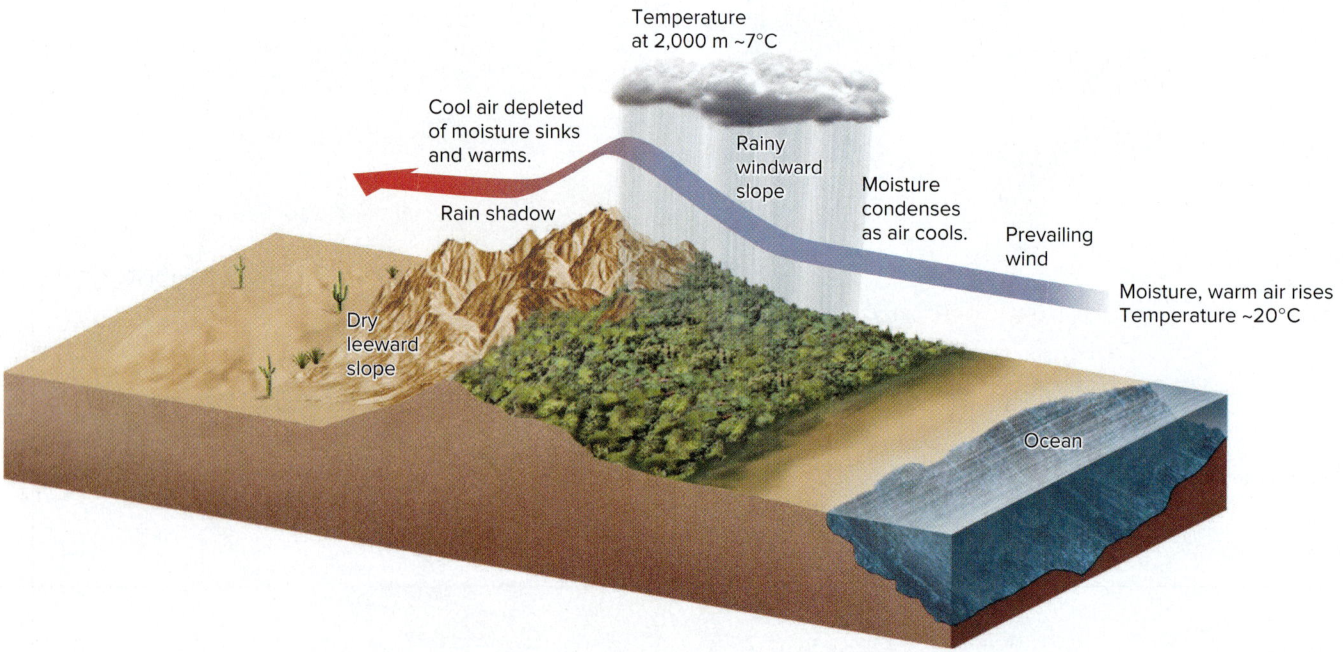


**Figure 45.7 The influence of elevation on climate.**

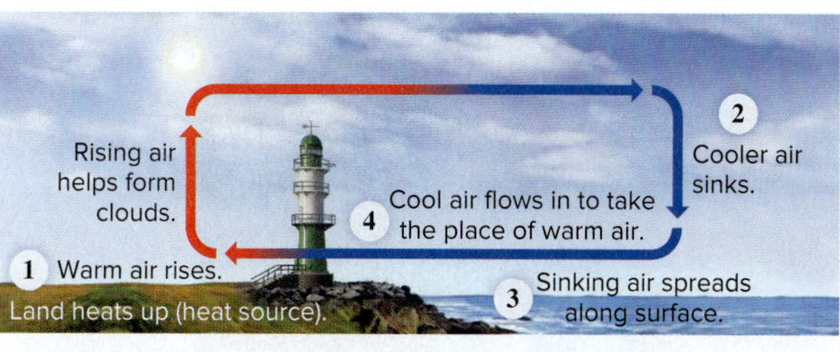


(a) Sea breeze

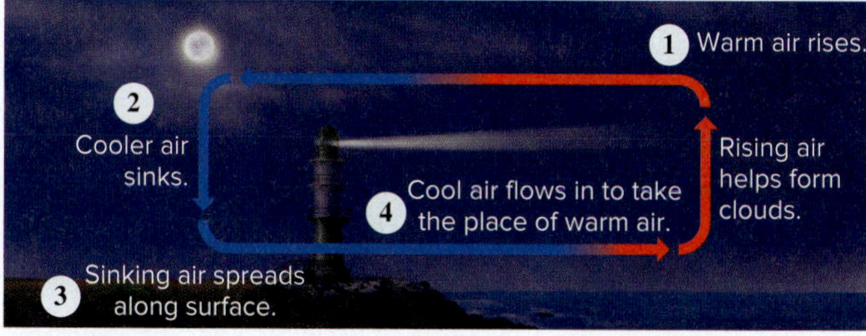


(b) Land breeze

**Figure 45.8 Proximity to water affects climate. (a)** When the landmass is warmer than the ocean, a sea breeze is created, bringing in cooler, moisture-laden air and, sometimes, fog. **(b)** When the sea is warmer than the land, a land breeze results.

in Washington State receives more than 500 cm of annual precipitation, whereas the eastern side receives only 50 cm. The Gobi Desert in central Asia lies in the rain shadow of the Himalayas, and the Atacama Desert in Chile and Peru is in the rain shadow of the Andes.

The proximity of a landmass to a large body of water can affect climate because land heats and cools more quickly than large bodies of water. Recall from Chapter 2 that water has a very high specific heat—the amount of energy required to raise the temperature of 1 gram of a substance by 1°C. The specific heat of the land is much lower than that of the water, allowing the land to warm quicker than water. During the day, the warmed air rises and cooler air flows in to replace it. This pattern creates the familiar onshore sea breezes in coastal areas (Figure 45.8). At night, the land cools quicker than the sea, and so the pattern is reversed, creating offshore breezes. The sea, therefore, has a moderating effect on the temperatures of coastal regions and especially islands. The climates of coastal regions may differ markedly from those of their climatic zones. Many never experience frost, and fog is often evident. Thus, along coastal areas, vegetation patterns may differ from those in areas farther inland. In fact, some areas of the U.S., including Florida, would be deserts were it not for the warm water of the sea and the moisture-laden clouds that form above them.

## Ocean Currents, Waves, and Tides Also Affect Climate

Together with the rotation of the Earth, winds also create ocean currents. The major ocean currents act as "pinwheels" between continents, running clockwise in the ocean basins of the Northern Hemisphere and counterclockwise in those of the Southern Hemisphere (Figure 45.9). The Gulf Stream, equivalent in flow to 50 times the world's major rivers combined, brings warm water from the Caribbean and the U.S. coasts across the Atlantic Ocean, where it combines with the North Atlantic Drift to moderate the climate of Europe. The Humboldt Current brings cool conditions to the western coast of South America and almost to

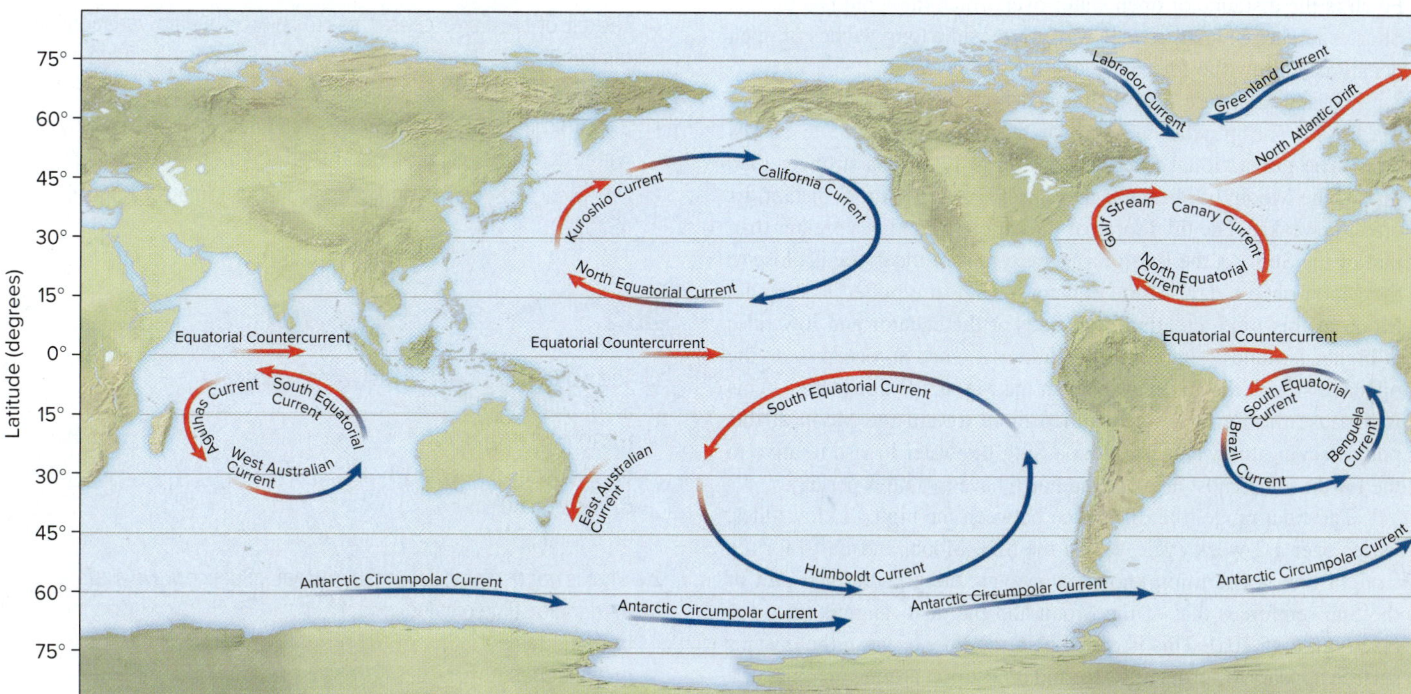


**Figure 45.9 Major ocean currents of the world.** The red arrows represent warm water; the blue arrows, cold water.

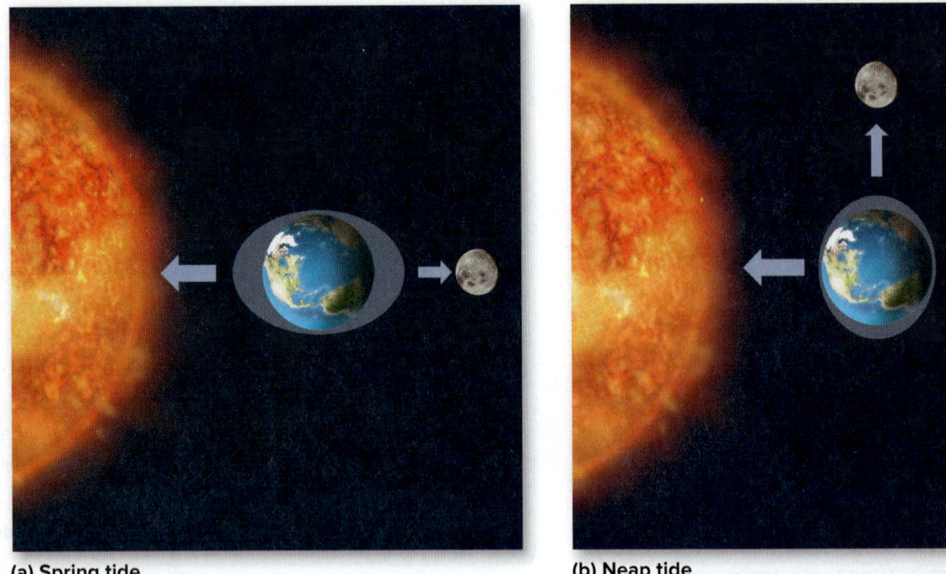


**Figure 45.10** **Tide formation.** **(a)** Spring tides form when the Sun, Earth, and Moon are in alignment. **(b)** Neap tides result when the Sun, Moon, and Earth are at a 90° angle.

the equator, and the California Current brings cooler climate to the Hawaiian Islands.

In addition to currents, wind also creates waves, which can range in size from small ripples to huge swells. As the wind blows, the friction between the air and the water creates small ripples. Once the ripples have formed, the wind has something to push against and the waves may increase in size. Four factors influence wave size: wind speed, fetch, duration of time of wind, and water depth. Fetch is the distance of open water over which the wind has blown. Shorter fetches in lakes reduce wave size, while long fetches of open ocean may allow the formation of 4- to 5-m waves.

Coastal areas are also influenced by tides, the rise and fall of sea levels. The gravitational pull of the Moon and the Sun causes tides. The gravitational force of the Sun is 179 times stronger than that of the Moon, but because the Sun is 389 times more distant, its effect is weaker. The tidal force of the Moon is 2.2 times greater than that of the Sun. As the Earth turns, each area of the globe is close to the Moon once a day. At the equator, oceans are pulled toward the Moon at this time, creating high tides at the equator and low tides at higher latitudes (**Figure 45.10**). Similarly, when an ocean is on the opposite side of the Earth away from the Moon, the tide is high. This is because the Earth is itself pulled more toward the Moon at this point, leaving the water behind, causing the water to rise relative to the Earth. Thus, most areas of the Earth have two tides per day.

The tidal range, the difference between the high and low tides, varies over a 2-week cycle. When the Sun, Moon, and Earth form a straight line, a condition known as syzygy, the gravitational pull of the Sun reinforces that of the Moon and the tidal range is maximal (see Figure 45.10a). This is called a spring tide, because the ocean's surface appears to spring upward, and it occurs at around the time of a new or full Moon. When the Sun, Earth, and Moon are at a 90° angle, the gravitational forces of the Sun partially cancel those of the Moon, and the tidal range is minimal (see Figure 45.10b). This is termed a neap tide, and it occurs when the Moon is in its first or third quarter.

## 45.1 Reviewing the Concepts

- DIfferences in global temperatures are caused by variations in incoming solar radiation and patterns of atmospheric circulation (Figures 45.1, 45.2, 45.3).
- The tilt of the Earth causes seasonality, including variations in temperature and rainfall. The rotation of the Earth affects wind direction (Figures 45.4, 45.5, 45.6).
- Elevation and the proximity of a landmass to a large body of water can affect climate (Figures 45.7, 45.8).
- Ocean currents, waves, and tides also affect climate (Figures 45.9, 45.10).

## 45.1 Testing Your Knowledge

1. The world's major subsidence zones occur at
   - a. 0° and 60°.
   - b. 30° and the poles.
   - c. 0° and the poles.
   - d. 30° and 60°.
   - e. 0° and 30°.
2. What is *not* one of the four factors that influence wave size?
   - a. duration time of wind
   - b. water temperature
   - c. water depth
   - d. length of fetch
   - e. wind speed

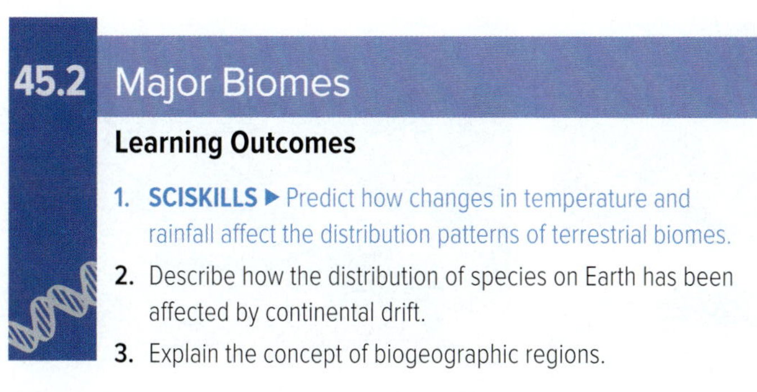

## 45.2 Major Biomes

### Learning Outcomes

1. **SCISKILLS ▶** Predict how changes in temperature and rainfall affect the distribution patterns of terrestrial biomes.
2. Describe how the distribution of species on Earth has been affected by continental drift.
3. Explain the concept of biogeographic regions.

As mentioned at the beginning of this chapter, a biome is a large-scale community characteristic of the climate in which it is found. Many types of classification schemes are used for mapping the geographic locations of terrestrial biomes, but one of the most useful was developed by the American ecologist Robert Whittaker, who classified biomes according to the physical factors of average annual precipitation and temperature (**Figure 45.11**). This classification scheme recognizes 10 terrestrial biomes distributed widely across the Earth (**Figures 45.12, 45.13**).

Although broad terrestrial biomes are a useful way of defining the main large-scale communities on Earth, ecologists acknowledge that not all communities fit neatly into 1 of these 10 major biome types. One biome type often grades into another at biome boundaries, as seen on mountain ranges. Soil conditions can also influence biome vegetation. In California, temperate grassland with isolated trees is abundant, except on serpentine soils, which are dry and nutrient-poor, and support only sparse vegetation that is adapted to this extreme environment. In the eastern U.S., most of New Jersey's coastal plain, called the Pine Barrens, consists of sandy, nutrient-poor soil that cannot support the surrounding deciduous forest and instead contains grasses and low shrubs growing among open stands of pygmy pitch pine and oak trees.

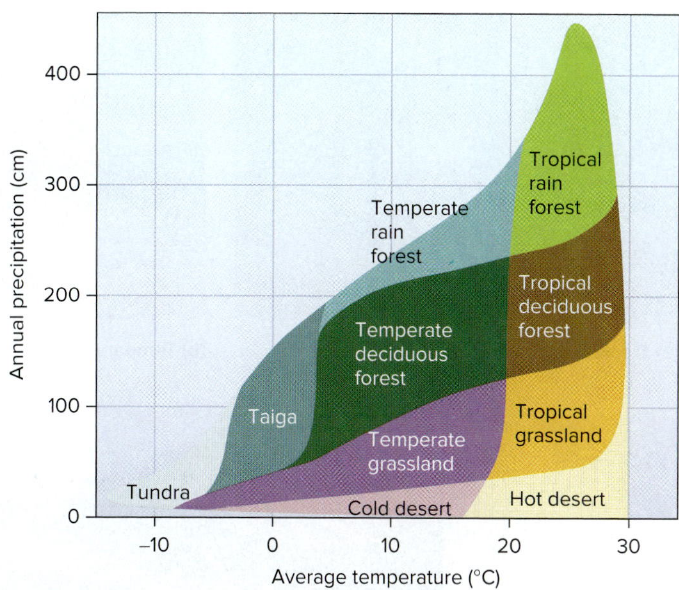


**Figure 45.11** **The relationship between the world's terrestrial biome types and temperature and precipitation patterns.**

Concept Check: *What other factors may influence biome types?*

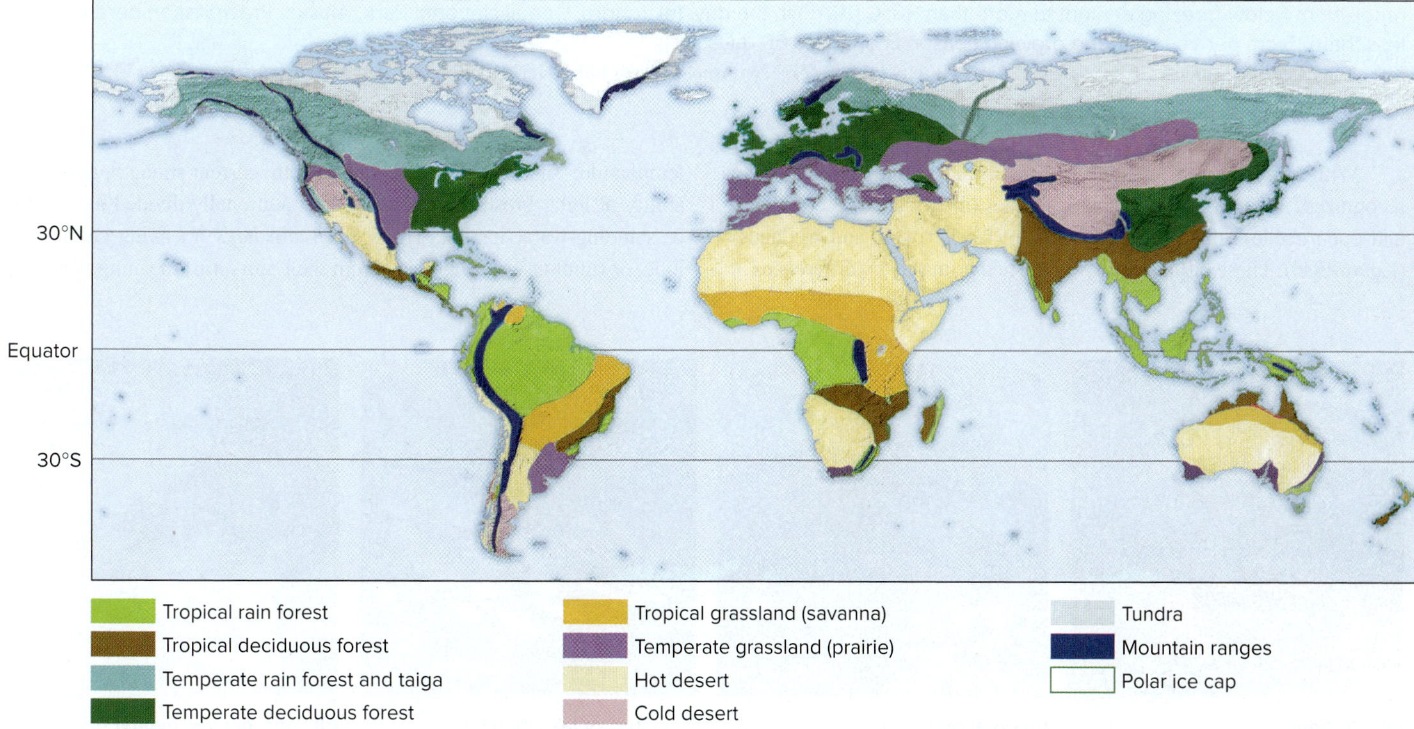


**Figure 45.12** **Geographic location of terrestrial biomes.** The distribution patterns of taiga and temperate rain forest are combined because of their similarity in tree species and because temperate rain forest is actually limited to a very small area.

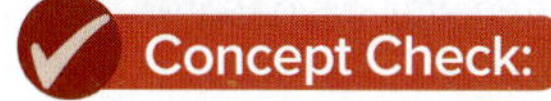
Concept Check: *In a globally warmed world, what biome might expand into areas currently occupied by tundra?*

**(a) Tropical deciduous forest**

**(b) Temperate coniferous forest, or taiga**

**(c) Tropical grassland, or savannah**

**(d) Hot desert**

**(e) Tundra**

**Figure 45.13** **Examples of terrestrial biomes.** **(a)** Tiger in tropical deciduous forest, India. Tropical deciduous forests have distinct wet and dry seasons. In the dry season trees often shed their leaves. **(b)** Temperate coniferous forest, or taiga, Canada. Taiga consists of coniferous trees with a conical shape to prevent bough breakage under heavy snow. Mammals are often heavily furred. **(c)** Tropical grassland, or savannah, Kenya. Savannah contains the greatest assemblages of large mammals on Earth. Rainfall is insufficient to support large trees but sufficient to support large swaths of grassland. **(d)** Hot desert, Namibia. Here rainfall is less than 30 cm per year, and temperatures can range from below freezing at night to more than 50°C (112°F) in the day. **(e)** Tundra, Denali National Park, Alaska. Precipitation here is scant, less than 25 cm per year, and the growing season is only 50–60 days long.

*(a)* © Theo Allofs/theoallofs.com; *(b)* © John E. Marriot/Getty Images; *(c)* © Jon Arnold Images Ltd./Alamy RF; *(d)* © Eye Ubiquitous/Getty Images; *(e)* © Bruce Coleman/Photoshot

Within aquatic environments, several different biome types are also recognized, including marine aquatic biomes (intertidal zone, coral reef, and open ocean) and freshwater habitats (lakes, rivers, and wetlands) (Figure 45.14). These biomes are distinguished primarily by differences in temperature, salinity, oxygen content, depth, current strength, and availability of light. Freshwater habitats are traditionally divided into lentic, or standing-water habitats (from the Latin *lenis,* meaning calm), and lotic, or running-water habitats (from the Latin *lotus,* meaning washed).

**(a) Intertidal**

**(b) Coral reef**

**(c) Lentic, or lake, habitat**

**(d) Lotic, or river, habitat**

**Figure 45.14** **Aquatic biomes.** **(a)** Intertidal, Washington State. The intertidal is alternately submerged and exposed by a daily cycle of tides. **(b)** Coral reef, Caribbean Sea. Coral reefs exist in warm water of at least 20°C but less than 30°C. **(c)** Lentic, or lake, habitat, Florida. Lentic habitats consist of still, often deep water. **(d)** Lotic, or river, habitat, Pacific Northwest. Lotic habitats have flowing, often well-oxygenated water and are able to support fish such as trout.

*(a)* © Nature Picture Library/Alamy; *(b)* © adokon/Getty Images RF; *(c)* © imageBROKER/Alamy; *(d)* © Ron Crabtree/Getty Images RF

## EVOLUTIONARY CONNECTIONS

### Continental Drift and Biogeography Help Explain Species Distributions

Knowledge of biomes does not tell us everything about the distribution of plant and animal life on Earth. An understanding of evolution and geological change over large areas and time periods helps explain some of the patterns we see. For example, South America, Africa, and Australia all have similar biomes, ranging from tropical to temperate, yet each continent has distinctive animal life. South America is inhabited by sloths, anteaters, armadillos, and monkeys with prehensile tails. Africa possesses a wide variety of antelopes, zebras, giraffes, lions, baboons, the okapi, and the aardvark. Australia, which has no native placental mammals except bats, is home to a variety of marsupials such as kangaroos, koala bears, Tasmanian devils, and wombats, as well as the egg-laying monotremes—namely, the duck-billed platypus and echidnas. Most continents also have distinct species of plants; for example, eucalyptus trees are native only in Australia. In South American deserts, succulent plants belong to the family Cactaceae, the cacti. In Africa, they belong to the genus Euphorbia, the spurges.

One explanation for these species distributions is that abiotic factors are of paramount importance and that each region supports the fauna best adapted to it. However, the spread of introduced species has proved this explanation incorrect: European rabbits introduced into Australia proliferated rapidly, and eucalyptus from Australia grows well in California. A better explanation is that different flora and fauna are the result of the independent evolution of separate, unconnected populations, which have generated different species in different places. A knowledge of continental drift and evolution is therefore vital to understanding modern distributions of species.

The relative location of the landmasses on Earth has changed enormously over time as a result of continental drift, the slow shifting of the Earth's landmasses over time (refer back to Figure 22.5). The current distribution of the essentially flightless bird family, the ratites, in the Southern Hemisphere is the result of continental drift. The common ancestor of these birds occurred in a supercontinent that included what are now South America, Africa, and Australia. As the continent split apart, genera evolved separately so that today we have ostriches in Africa, emus in Australia, and rheas in South America.

Continental drift is not the only mechanism that creates widely separate populations of closely related species. The distributions of many present-day species are relics of once much broader distributions. For example, there are currently four living species of tapir: three in Central and South America and one in Malaysia (**Figure 45.15**). Fossil records reveal a much more widespread distribution over much of Europe, Asia, and North America. The oldest fossils come from Europe, making it likely that this was the center of origin of tapirs. The migrations of tapirs resulted in a more widespread distribution. However, global cooling resulted in the extinction of tapirs in all areas except the two tropical locations.

## Biology Principle

### All Life Is Related by an Evolutionary History

Knowing that all tapir species share a common European ancestor helps to explain their current distribution pattern.

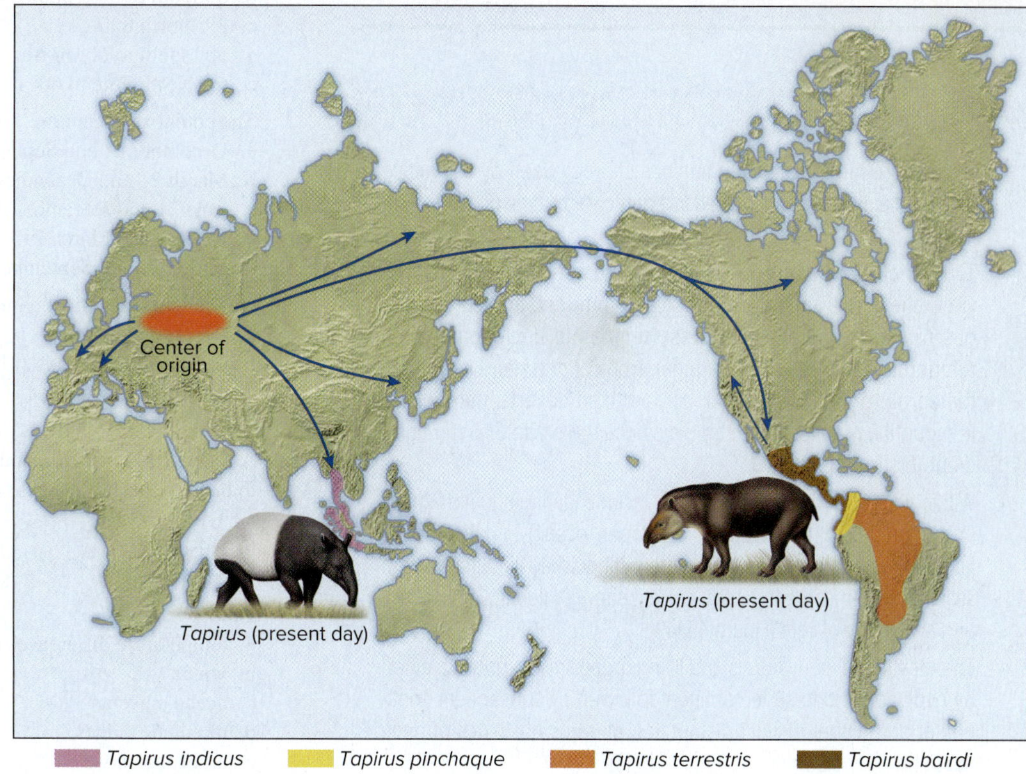


**Figure 45.15 Tapir distribution.** This map shows the geographic ranges of four modern tapir species (see key below the map). Fossil evidence suggests a European origin and a more widespread distribution, with tapirs dying out in other regions.

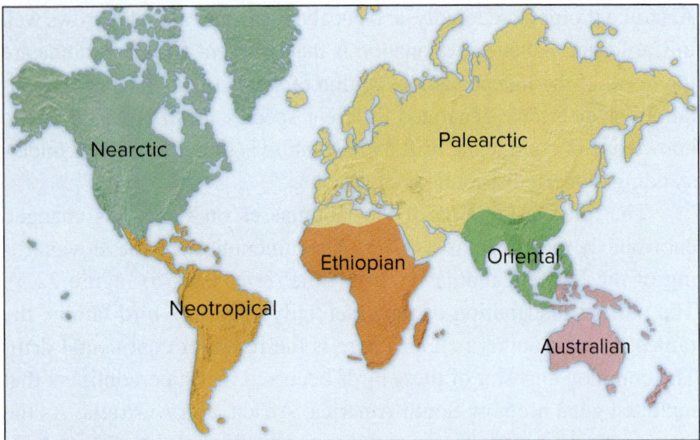


**Figure 45.16** **The biogeographic regions proposed by A. R. Wallace.** Note that the borders do not always demarcate continents.

Alfred Russel Wallace was one of the earliest scientists to realize that certain plant and animal taxa were restricted to certain geographic areas of the Earth (see Chapter 19). For example, the distribution patterns of guinea pigs, anteaters, and other groups are confined to Central and South America, from central Mexico southward. Based on this and other observations, Wallace divided the world's biota into six major **biogeographic regions:** Nearctic, Palearctic, Neotropical, Ethiopian, Oriental, and Australian (**Figure 45.16**). These regions are still widely accepted today, though debate continues about the exact location of the boundary lines. Biogeographic regions correspond largely to continents but more exactly to areas bounded by major barriers to dispersal, such as oceans, deserts, and mountain ranges.

## 45.2 Reviewing the Concepts

- Biomes are large-scale communities characterized by distinctive plant and animal life and located in a particular climate (Figures 45.11, 45.12).
- Terrestrial biomes are generally named for their climate and vegetation type and include tropical rain forest, tropical deciduous forest, temperate rain forest, temperate deciduous forest, temperate coniferous forest (taiga), tropical grassland (savanna), temperate grassland (prairie), hot and cold deserts, and tundra. In mountain ranges, biome type may change on an elevation gradient (Figure 45.13).
- Within aquatic environments, biomes include marine aquatic biomes (the intertidal zone, coral reef, and open ocean) and freshwater lakes, rivers, and wetlands. These are distinguished by differences in salinity, oxygen content, depth, current strength (lentic versus lotic), and availability of light (Figure 45.14).
- The location of many fossils and living organisms can be explained by evolution in one supercontinent followed by subsequent continental drift. Other distinct distribution patterns are relics of once much broader distributions (Figure 45.15).
- Six major biogeographic regions—Nearctic, Palearctic, Neotropical, Ethiopian, Oriental, and Australian—each of which contains distinct fauna and flora, are recognized today (Figure 45.16).

## 45.2 Test Your Knowledge

1. Which terrestrial biome type is present where rainfall is substantial, temperatures are hot year round, but there is a distinct dry season?
   a. tropical rain forest
   b. tropical deciduous forest
   c. temperate forest
   d. savanna
   e. prairie
2. The uniquely marsupial fauna of Australia is largely a result of
   a. adiabatic cooling.
   b. climate.
   c. continental drift and evolution.
   d. rain shadows.
   e. biogeographic regions.

# Assess and Discuss

## Test Yourself

1. The northeast trade winds blow from
   a. 60° north to 30° north.
   b. 0° south to 30° south.
   c. 30° north to 0°.
   d. 30° south to 60° north.
   e. the North Pole to 60° north.
2. The equinoxes occur on
   a. December 21 and September 22.
   b. March 21 and December 21.
   c. June 21 and December 21.
   d. June 21 and March 21.
   e. March 21 and September 22.
3. On the leeward side of a mountain, air
   a. warms and ascends.
   b. warms and descends.
   c. cools and descends.
   d. cools and ascends.
4. A rocket fired from the South Pole toward the equator would appear to bend to the east because of the Coriolis force.
   a. True
   b. False
5. What is the driving force that determines the circulation of the atmospheric air?
   a. temperature differences of the Earth
   b. winds
   c. ocean currents
   d. mountain ridges
   e. all of the above

6. What characteristics are commonly used to identify the terrestrial biomes of the Earth?
   a. temperature
   b. precipitation
   c. vegetation
   d. all of the above
   e. a and b only
7. The most common biome type, by area occupied, is the
   a. open ocean.
   b. tropical rainforest.
   c. tundra.
   d. hot desert.
   e. lentic habitats.
8. In this biome, rainfall is between about 25 cm and 100 cm and average temperatures are between –10°C and 20°C. Where are you?
   a. tropical rain forest
   b. tropical deciduous forest
   c. tropical grassland
   d. temperate grassland
   e. temperate deciduous forest
9. Tidal range is maximal in
   a. a neap tide.
   b. a spring tide.
   c. syzygy.
   d. a and c.
   e. b and c.
10. Aquatic biomes are distinguished primarily by differences in
    a. temperature
    b. salinity
    c. current strength
    d. depth
    e. all of the above

## Conceptual Questions

1. If mountains are closer to the Sun than valleys, why aren't they hotter?
2. Why are tides affected more by the gravitational effects of the Moon, when the Sun is so much bigger?
3. **PRINCIPLES** A principle of biology is that living organisms interact with their environment. In most locations on Earth at about 20–30° latitude, air cools and descends, and hot deserts occur. Florida is situated between 31° north and 24° north. Why does it not support a desert biome?

## Collaborative Questions

1. If the Earth were not tilted 23.5° on its axis, how would climate and life, be affected?
2. Based on your knowledge of biomes, identify the biome in which you live. In your discussion, list and describe the organisms that you have observed in your biome. Why might your observations not fit the biome predicted to occur from temperature/precipitation profiles?

## Online Resource

# 46 The Age of Humans

Source: Photo courtesy of National Park Service, Everglades National Park

**Burmese python in the Florida Everglades.** This snake was introduced into the Florida Everglades by humans and has dramatically decreased the populations of several native species.

## Chapter Outline



In January of 2016, British geoscientist Colin Waters and colleagues proposed formal recognition of a new geological epoch, the Anthropocene, the last and latest epoch of the quaternary period (refer back to Figure 22.4). Some scientists have argued that the Anthropocene is a pop-culture term and a political statement rather than a new epoch. But support is growing to embrace the Anthropocene, a term coined in the 1980s by American ecologist Eugene Stoermer, and later popularized by Dutch atmospheric chemist Paul Crutzen. In the 1970s and '80s, Crutzen had shown how human pollutants can damage the ozone layer, a discovery for which he won the Nobel Prize in 1995, along with Mexican chemist Mario Molina and American chemist Frank Rowland.

While agreement is widespread that humans have a profound effect on the global environment, the formal recognition of a new geological time unit requires formal criteria to be met. Global-scale changes must be recorded in geological stratigraphic (layered) material, usually rock, and in ice or marine sediments. Waters and colleagues noted the appearance of manufactured items in these sediments, including plastics, concrete, aluminum, and particulates from fossil-fuel combustion. Perhaps the most distinctive new identifiable items in sediments were radioactive elements from the 500 aboveground nuclear blasts that occurred between 1945 and 1963, which created globe-encircling debris.

Whether the Anthropocene gains wide acceptance in the scientific literature, there is no disputing the immense impacts of humans on natural systems. In this chapter, we will examine some of these impacts, which include global warming, effects on biogeochemical cycles, habitat destruction such as deforestation, the overexploitation of various species via overfishing and overhunting, and the introduction of invasive species to all corners of the globe (see chapter-opening photo). We will begin the chapter addressing the huge numbers of humans on the planet and our prodigious rate of growth.

## 46.1 Human Population Growth

**Learning Outcomes**

1. Describe how differences in age structure and human fertility across different countries affect human population growth.
2. Explain the concept of an ecological footprint.

In 2015, the world's population was estimated to be increasing at the rate of 150 people every minute: about 2 per minute in developed nations and 148 in less-developed nations. In this section, we examine human population growth trends in more detail and discuss how knowledge of the human population's age structure and fertility levels can help predict its future growth. We then investigate the carrying capacity of the Earth for humans and explore how the concept of an ecological footprint, which measures human resource use, can help us determine this carrying capacity.

### Human Populations Show Extreme Recent Growth

Until the beginning of agriculture and the domestication of animals, about 10,000 B.C.E., the average rate of human population growth was very low. With the establishment of agriculture, the world's population grew to about 300 million by 1 C.E. and to 800 million by the year 1750. Between 1750 and 1998, a relatively short period of human history, the world's human population surged from 800 million to 6 billion (Figure 46.1). In 2016, the number of humans was estimated at 7.5 billion. Scientists are very interested in determining when and at what size the human population will level off.

### Knowledge of a Population's Age Structure Can Help Predict Its Future Growth

The age structure of a population can help to predict future population growth. In all populations, **age structure** refers to the relative numbers of individuals of each defined age group. This information is commonly displayed as a population pyramid. In West Africa, for example, children younger than age 15 make up nearly half of the population, creating a pyramid with a wide base and narrow top (Figure 46.2a).

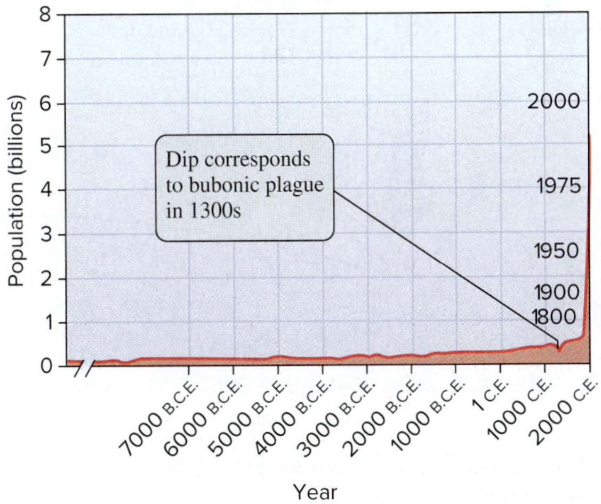


**Figure 46.1** The growth pattern of the human population through history.

Even if fertility rates decline, the population is expected to significantly increase as these young people move into childbearing age. The age structure of Western Europe is much more uniform (Figure 46.2b). Even if the fertility rate of young women in Western Europe increases to a level higher than that of their mothers, the annual numbers of births will still be relatively low because of the low number of women of childbearing age.

### Human Population Fertility Rates Vary Widely Around the World

Global population growth can be estimated by determining the **total fertility rate (TFR),** the average number of live births a woman has during her lifetime (Figure 46.3). The total fertility rate differs considerably from one geographic area to another. In Africa, the total fertility rate of 4.6 in 2010 declined substantially since the 1970s, when it was around 6.7 children per woman. In Latin America and Southeast Asia, the rates declined even more from the 1970s and are now at around 2.3. Canada and most countries in Europe have a TFR of less than 2.0 (the TFR was about 1.86 in the U.S. in 2013); in Russia, fertility rates have dropped to 1.34. In China, although the TFR is only 1.7, the population there will still continue to increase until at least 2025 because of the large number of women of reproductive age. Although the global TFR declined from 4.47 in the 1970s to 2.36 in 2015, this is still greater than the average of 2.3 needed for zero population growth. The replacement rate is slightly higher than 2.0, to replace mother and father, due to natural mortality prior to reproduction. The replacement rate varies globally, from 2.1 in developed countries to between 2.5 and 3.3 in less-developed countries.

The wide variation in fertility rates makes it difficult to predict future population growth. The 2010 United Nations report shows world population projections to the year 2100 for three different growth scenarios: low, medium, and high (Figure 46.4). The three scenarios are based on three different assumptions about fertility rate. Using a low fertility rate estimate of only 1.5 children per woman, the population would reach a maximum of about 8 billion people by 2050. A more realistic assumption may be to use the fertility rate estimate of 2.0 or even 2.5, in which case the population would continue to rise to 10 or 16 billion, respectively.

### The Concept of an Ecological Footprint Helps Estimate Carrying Capacity

Recall from Chapter 43 that carrying capacity refers to the maximum population size that can be sustained by an environment. What is the Earth's carrying capacity for the human population, and when will it be reached? Estimates vary widely. Much of the speculation on the upper boundary of the world's population size centers on lifestyle. To use a simplistic example, if everyone on the planet ate meat extensively and drove large cars, then the carrying capacity would be a lot less than if people were vegetarians and used bicycles as their main means of transportation.

In the 1990s, Swiss researcher Mathis Wackernagel and his coworkers calculated how much land is needed to support each person on Earth. Everybody has an effect on the Earth, because we consume the land's resources, including crops, wood, fossil fuels, minerals, and so on. Thus, each person has an **ecological footprint**—the aggregate

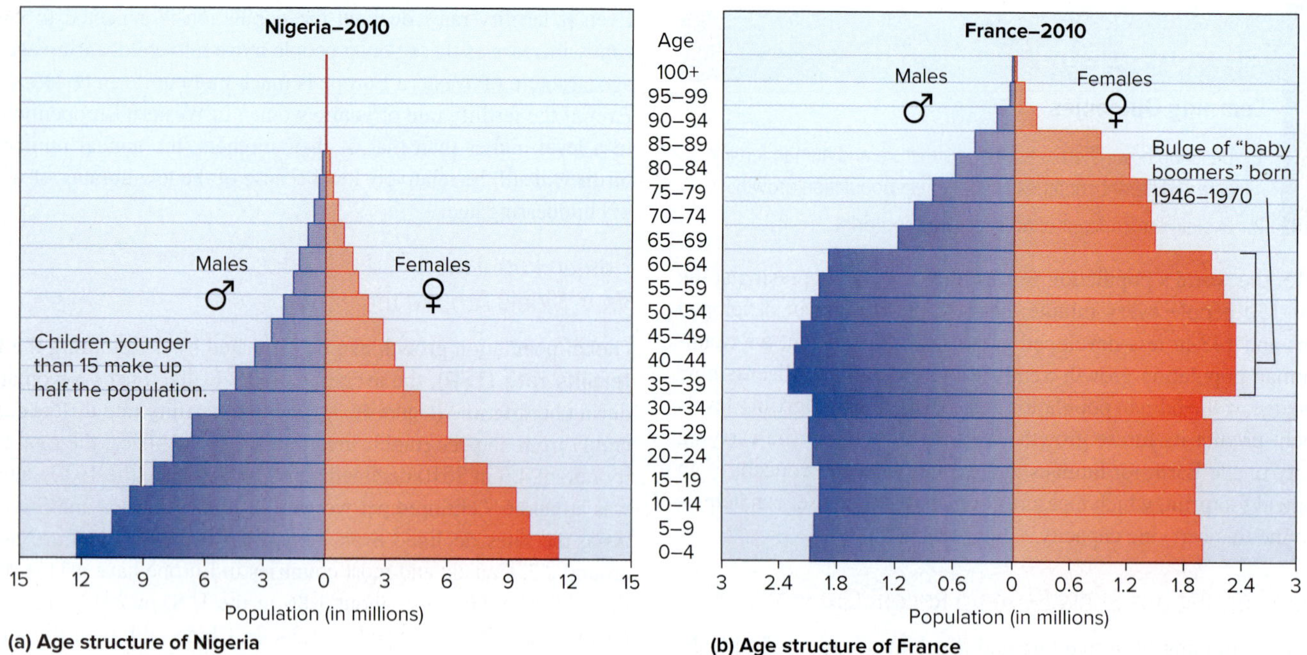


**(a) Age structure of Nigeria** **(b) Age structure of France**

**Figure 46.2 The age structure of human populations in Nigeria and France, as of 2010. (a)** In developing areas of the world such as Nigeria, children are the most abundant age group. Population growth is rapid. **(b)** In the developed countries of Western Europe, the age structure is more evenly distributed. The bulge represents those born in the post–World War II "baby boom," when birth rates climbed due to stabilization of political and economic conditions. Population growth is close to zero.

*If the population pyramid in (a) was inverted, what would you conclude about the age structure of the population?*

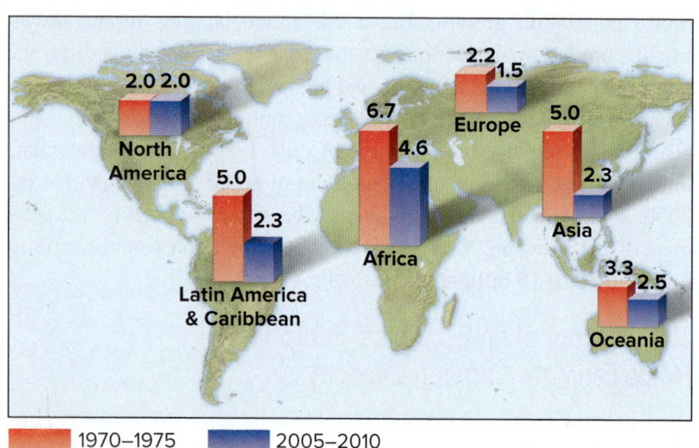


**Figure 46.3 Total fertility rates (TFRs) among major regions of the world.** Data refer to the average number of children born to a woman during her lifetime.

total of productive land needed for survival in a sustainable world. The average footprint size for everyone on the planet is about 3 hectares (1 ha = 10,000 m$^2$), but wide variation is found around the globe (**Figure 46.5**). The ecological footprint of the average Canadian is 7.5 hectares and about 10 hectares for the average American.

In most developed countries, the largest component of land is for energy, followed by food and then forestry. Much of the land needed

## Biology Principle

### Biology Affects Our Society

How TFR is calculated has a great influence on assumptions about how human global population size will change over the next 80 years.

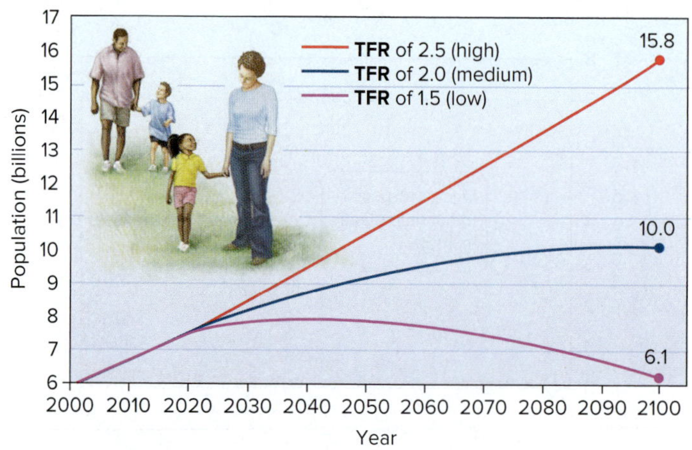


**Figure 46.4 Population predictions for 2000–2100, using three different total fertility rates (TFRs).**

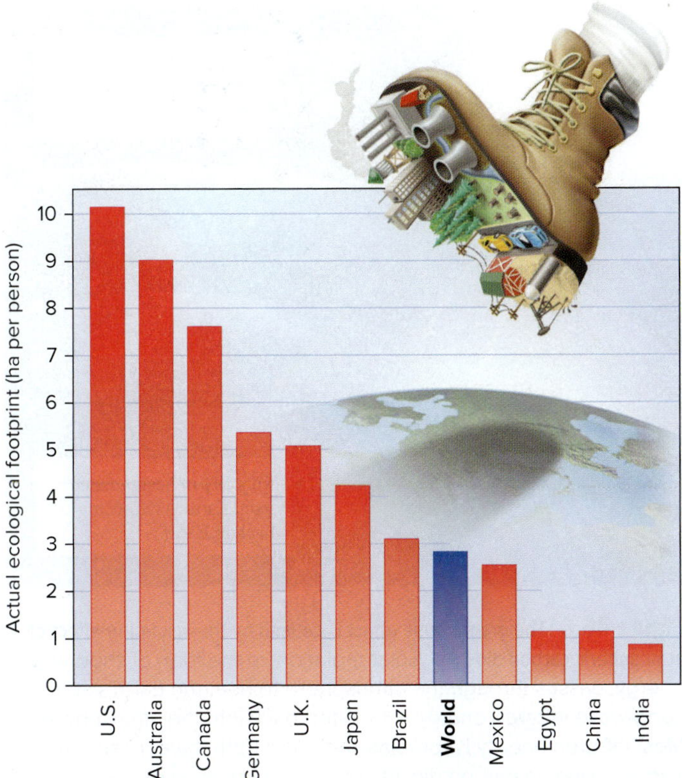


**Figure 46.5 Ecological footprints of different countries.** The term ecological footprint refers to the amount of productive land needed to support the average individual of that country.

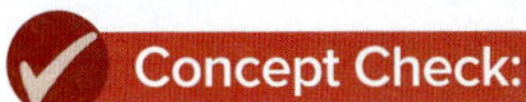

*What is your ecological footprint?*

for energy serves to absorb the $CO_2$ emitted by the use of fossil fuels. If everyone required 10 hectares, as the average American does, we would need three Earths to provide us with the needed resources. Many people in less-developed countries use far fewer resources. However, globally we are already beyond the Earth's carrying capacity for humans if we were to live in a sustainable manner. This has happened because many people currently live in an unsustainable manner, using more resources than can be regenerated in any given year.

What's your personal ecological footprint? Several different calculations are available on the Internet that you can use to find out. A rapidly growing human population combined with an increasingly large per capita ecological footprint makes it increasingly difficult to preserve other species on the planet, a subject we will examine further in our discussion of conservation biology (Chapter 47).

## 46.1 Reviewing the Concepts

- Human population growth has skyrocketed over the last 250 years (Figure 46.1).
- Differences in the age structure of a population, the numbers of individuals in each age group, can influence future patterns of population growth (Figure 46.2).
- Although they have been declining worldwide, total fertility rates (TFRs) differ markedly between developed and less-developed countries. Predicting the growth of the human population depends on the total fertility rate that is projected (Figures 46.3, 46.4).
- The ecological footprint refers to the amount of productive land needed to support each person on Earth. Because many people live in a nonsustainable manner, globally we are already in an ecological deficit (Figure 46.5).

## 46.1 Testing Your Knowledge

1. The average total fertility rate needed for zero population change across the world is
   a. 1.7. c. 2.0. e. 2.5.
   b. 1.9. d. 2.3.
2. The total fertility rate is highest in
   a. Latin America.
   b. Southeast Asia.
   c. Europe.
   d. Africa.
   e. North America.

# 46.2 Global Warming and Climate Change

### Learning Outcomes

1. Describe how the greenhouse effect contributes to global warming.
2. Explain how global warming may affect sea levels and precipitation patterns.
3. Predict how climate change will change species distribution patterns.

**Global warming** refers to an increase in Earth's average surface temperature. In recent years, ecologists have been investigating how human activities have contributed to global warming. A key impact of global warming is **climate change,** which is a long-term change in Earth's climate or a change in climate in a particular region. In most regions on Earth, global warming is expected to increase surface temperatures. However, because global warming can affect ocean currents and the circulation of the atmosphere, some regions may actually become colder or experience seasonal changes such as colder winters.

To understand how global warming causes climate change, we will begin by exploring how Earth's atmosphere acts like the glass panes in a greenhouse, trapping heat inside. By understanding this process, ecologists are better able to predict the future effects of a warming Earth. We will explore how human activities have contributed to global warming and consider some of the potential consequences of climate change, which include rising sea levels, changes in precipitation, and changes in the distributions of species.

## The Greenhouse Effect Causes the Earth's Temperature to Rise

The Earth is warmed by the greenhouse effect. In a greenhouse, sunlight penetrates the glass and heats the plants inside. The heat is reradiated but the glass acts to trap the heat inside. Similarly, solar radiation in the form of short-wave energy passes through the atmosphere to heat the surface of the Earth. This energy is then reradiated from the Earth's warmed surface into the atmosphere, but in the form of long-wave infrared radiation. Atmospheric gases absorb much of this infrared energy and radiate it a second time to the Earth's surface, causing its temperature to rise further (**Figure 46.6**). The greenhouse effect is important to life on Earth. Without the greenhouse effect, global temperatures would be much lower than they are, perhaps averaging only –17°C compared with the existing average of +15°C.

The greenhouse effect is caused by a small group of gases, mainly water vapor, that together make up less than 1% of the total volume of the atmosphere. After water vapor, the four most significant greenhouse gases are carbon dioxide, methane, nitrous oxide, and chlorofluorocarbons (**Table 46.1**).

Ecologists are concerned that human activities are increasing the strength of the greenhouse effect, resulting in the gradual elevation of the Earth's surface temperature. In particular, the burning of fossil fuels appears to be the main culprit of global warming. Fossil fuels such as oil, coal, and natural gas are high in carbon and release carbon dioxide ($CO_2$) into the atmosphere when burned. The atmospheric concentrations of most greenhouse gases have increased since the industrial revolution over 200 years ago. Of those increasing, the one that is the greatest contributor to global warming is $CO_2$. As Table 46.1 shows, $CO_2$ has a lower global warming potential per unit of gas (relative absorption) than any other major greenhouse gas, but its concentration in the atmosphere is much higher. An analysis of $CO_2$ trapped in glaciers shows that concentrations of atmospheric $CO_2$ increased from about 280 ppm (parts per million) in the preindustrial 19th century to 400 ppm in 2013 (**Figure 46.7a**). Since 1957, air samples have been collected directly at Mauna Loa, Hawaii, a relatively unpolluted site. The data show a 25% increase in atmospheric $CO_2$ levels in just 57 years, from 313 ppm to 400 ppm during 1957–2015 (**Figure 46.7b**).

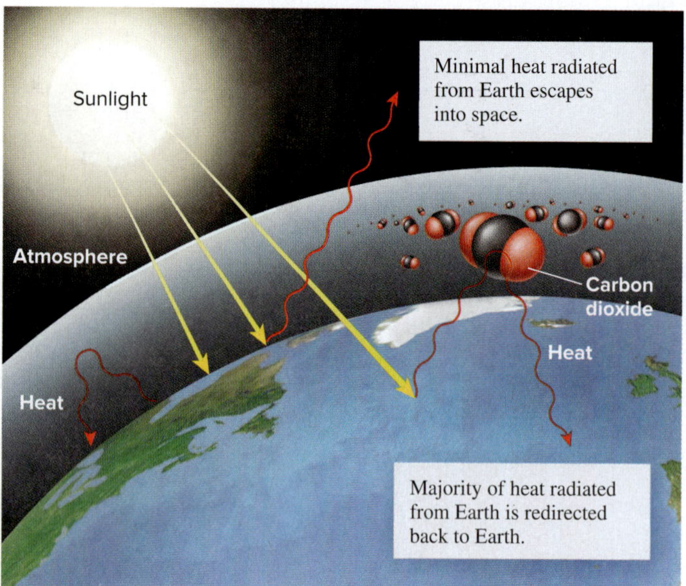


**Figure 46.6 The greenhouse effect is caused by the insulating effect of atmospheric carbon dioxide.** Solar radiation, in the form of short-wave energy, passes through the atmosphere to heat the Earth's surface. Long-wave infrared energy is radiated back into the atmosphere. Most infrared energy is reradiated back to Earth by atmospheric gases, including carbon dioxide molecules, causing global temperatures to rise.

In addition, seasonal oscillations in $CO_2$ also occur, as seen in the orange line of Figure 46.7b. The Northern Hemisphere has greater land area and plant biomass than the Southern Hemisphere. In the northern summer, more $CO_2$ is used up by plants and atmospheric $CO_2$ level declines slightly. In the northern winter, less $CO_2$ is absorbed and atmospheric $CO_2$ levels increase.

To predict the effects of global warming, most scientists focus on a future point, about 2100, when the concentration of atmospheric $CO_2$

**Table 46.1 The Major Greenhouse Gases and Their Contribution to Global Warming**

| | Carbon dioxide ($CO_2$) | Methane ($CH_4$) | Nitrous oxide ($N_2O$) | Chlorofluorocarbons (CFCs) |
|---|---|---|---|---|
| Relative absorption in ppm of increase* | 1 | 21 | 310 | 10,000 |
| Atmospheric concentration (ppm†) | 400 | 1.75 | 0.315 | 0.0005 |
| Contribution to global warming | 73% | 7% | 19% | 1% |
| Percent from natural sources; type of source | 20–30%; volcanoes | 70–90%; swamps, gas from termites and ruminants | 90–100%; soils | 0% |
| Major human-made sources | Fossil-fuel use, deforestation | Rice paddies, landfills, biomass burning, coal and gas exploitation | Cultivated soil, fossil-fuel use, automobiles, industry | Previously manufactured products (for example, aerosol propellants) but now banned in the U.S. and the E.U. |

*Relative absorption is the warming potential per unit of gas.
†ppm = parts per million

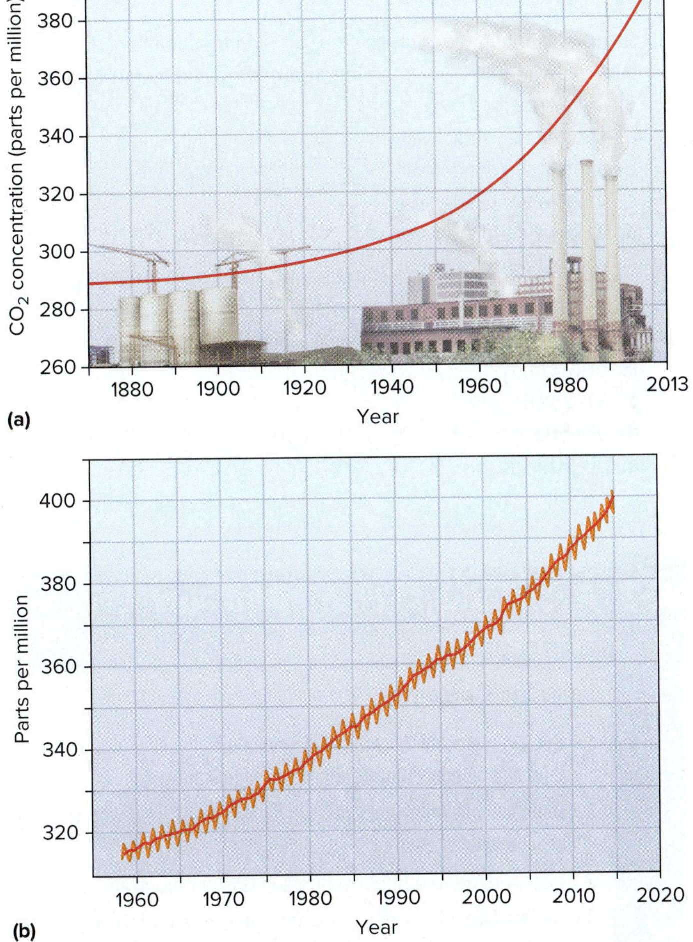


**Figure 46.7** **The increase of atmospheric carbon dioxide.** **(a)** Increases in the burning of fossil fuels have caused an increase in atmospheric $CO_2$ since the late 19th century. **(b)** Since 1957, $CO_2$ levels have shown consistent increases in air taken directly from Mauna Loa, Hawaii.

**Concept Check:** *Why does the amount of $CO_2$ fluctuate seasonally in the graph?*

will have doubled—that is, increased to about 700 ppm compared with the late-20th-century level of 350 ppm. Ecologists hypothesize that at that time, average global temperatures will be about 1–6°C (about 2–10°F) warmer than present and will increase an additional 0.5°C each decade. This increase in heat might not seem like much, but it is comparable to the warming that ended the last Ice Age. As discussed next, one consequence of global warming is its effect on sea levels.

## Global Warming Is Expected to Cause Sea Levels to Rise

The term **sea level** refers to the average level for the surface of one or more of Earth's oceans. The 2007 report of the Intergovernmental Panel on Climate Change (IPCC) suggested that a business-as-usual scenario with the same level of increasing greenhouse gas emissions would lead to a 2°C warming by about the middle of this century. With a 2–3°C warming, sea levels are expected to rise by several meters due to the melting of glaciers. Sea levels have already risen 10–25 cm over the past century, and seawater entry into coastal forests has killed trees in many areas.

An alarming problem is the long time lag between warming temperatures and the time it takes for ice to melt. Even if future temperature increases were prevented, glaciers would continue to melt for many years. For example, the ice sheets of Greenland will take hundreds of years to melt, even though the increase in temperature needed to melt them could be reached in 50–100 years. You might envision this as being similar to an ice cube on a kitchen counter—it takes a while for the ice cube to melt, even though room temperature is considerably above freezing. Therefore, even if no greenhouse gases were added to the atmosphere after 2020, we would have already committed to at least a quarter of the sea-level rise of 80–120 cm that could be expected by 2100.

## One Effect of Global Warming on Climate Change Is to Alter Precipitation Patterns on Land

An increase in global temperatures is also expected to promote climate change and alter global precipitation. Higher temperatures increase evaporation, and warmer air holds more water vapor than cooler air. Global warming will generally lead to more frequent and heavier precipitation. Over the period 1986–2000, the average increase in precipitation over land was 3.5 mm per year. However, over normally dry areas, global warming is expected to increase evaporation of water from Earth's surface, but the level of moisture in the atmosphere over dry regions may not be high enough to facilitate rain. Therefore, the increased temperatures will likely lead to widespread droughts in desert areas. We can make the analogy that the atmosphere is like a sponge; warm, moist air allows the sponge to absorb more water, but in dry conditions nothing more can be rung out of the sponge. Northern and southern temperate areas have already shown precipitation increases, as have tropical and subtropical areas. However, desert areas such as the Sahel region south of the Sahara have become drier. In essence, the wet get wetter and the dry get drier.

## Climate Change Is Expected to Alter the Distribution of Species

Assuming that the previous scenario of climate change is accurate, scientists are beginning to consider the consequences on natural and human-made ecosystems. As most regions on Earth become warmer and a few regions become colder, most plant species cannot easily disperse and move north or south into newly created climatic regions that will be suitable for them. Many tree species take hundreds, even thousands, of years for seed dispersal.

Paleobotanist Margaret Davis was among the first to investigate how species' ranges would be changed in a globally warmed world. She predicted that in the event of a $CO_2$ doubling, many tree species in North America would suffer range contractions because their southern ranges would retreat northward (**Figure 46.8**). Of course, this contraction in the trees' distribution could be offset by the creation of new favorable habitats in Canada. However, most scientists hypothesize that the climatic zones would shift toward the poles faster than the tree species could either migrate via seed dispersal or evolve to become better adapted to their altered climate. Therefore, climate change is expected to cause the extinction of many species. In addition, changed precipitation patterns will alter the distribution patterns and densities of organisms still further. When Margaret Davis mapped

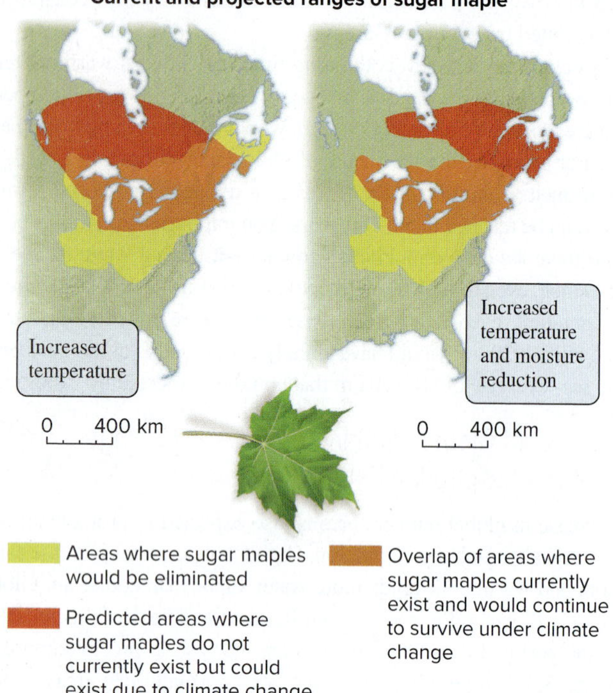


**Figure 46.8 Possible changes in the ranges of sugar maples due to climate change.** The map on the left takes into account only temperature changes, whereas the one on the right takes into account temperature and precipitation changes. The yellow shading indicates areas where sugar maples now exist but would not survive due to climate changes associated with a doubling of atmospheric $CO_2$ levels. The red shading indicates areas where sugar maples do not currently exist but would have a favorable climate to exist if $CO_2$ levels doubled. The orange shading indicates the overlap where sugar maples currently exist and where they could also exist due to climate change.

out the future distribution of the sugar maple, *Acer saccharum,* in a globally warmed world and took into account changed precipitation patterns, the predicted future area of distribution was considerably less than the prediction taking into account only temperature changes.

## 46.2 Reviewing the Concepts

- Global warming is an increase in the surface temperature on Earth. It is expected to result in climate change.
- The greenhouse effect is the process by which short-wave solar radiation passes through the atmosphere to warm the Earth and is reradiated back into the atmosphere as long-wave infrared radiation. Much of this radiation is radiated back to Earth's surface a second time, causing its temperature to rise (Figures 46.6).
- An increase in greenhouse gases, particularly of $CO_2$, is intensifying the greenhouse effect, causing climate change (Table 46.1; Figure 46.7).
- Two consequences of climate change are rising sea levels and changes in precipitation patterns.
- Ecologists predict that climate change will have a large impact on the distribution of the world's organisms (Figure 46.8).

## 46.2 Testing Your Knowledge

1. Solar radiation in the form of ________ energy heats the Earth's surface, whereas ________ infrared energy is radiated back to the atmosphere.
   a. short-wave, short-wave
   b. short-wave, long-wave
   c. long-wave, short-wave
   d. long-wave, long-wave
2. The rise in sea levels due to global warming over the past century is thought to be in the range of
   a. 1–5 cm.
   b. 5–10 cm.
   c. 10–25 cm.
   d. 25–50 cm.
   e. 50–100 cm.

## 46.3 Pollution and Human Influences on Biogeochemical Cycles

### Learning Outcomes

1. Outline the steps of the carbon cycle and the environmental effects of elevated atmospheric concentrations of $CO_2$.
2. Describe the processes of the water cycle and how they are affected by humans.
3. Describe the role of phosphorus in lake eutrophication.
4. List the five main steps of the nitrogen cycle and the human influences on it.

A unit of energy moves through an ecosystem only once, passing through the trophic levels of a food web from producer to consumer and dissipating as heat. In contrast, chemical elements such as carbon or nitrogen follow a cycle, moving from the physical environment to organisms and back to the environment, where the cycle begins again. Because the movements of chemicals through ecosystems involve biological, geological, and chemical transport mechanisms, they are termed **biogeochemical cycles.** Biological mechanisms involve the absorption of chemicals by living organisms and their subsequent release back into the environment. Geological mechanisms include weathering and erosion of rocks, and elements transported by surface and subsurface drainage. Chemical transport mechanisms include dissolved matter in rain and snow, atmospheric gases, and dust blown by the wind.

In addition to the basic building blocks of carbon, hydrogen, and oxygen, the elements required in the greatest amounts by living organisms are phosphorus and nitrogen. In this section, we take a detailed look at the cycles of these nutrients, which can be divided into two broad types: (1) local cycles, such as the phosphorus cycle, which involve elements with no atmospheric mechanism for long-distance transfer, and (2) global cycles, which involve an interchange between the atmosphere and the ecosystem. In our discussion of biogeochemical cycles, we will take a close look at how these cycles are affected by human activities, such as the burning of fossil fuels and the use of fertilizers.

## The Burning of Fossil Fuels Alters the Carbon Cycle

The movement of carbon from the atmosphere into organisms and back again is known as the carbon cycle (**Figure 46.9**). Carbon dioxide ($CO_2$) is present in the atmosphere at a level of about 400 parts per million (ppm), or about 0.04%. Phototrophs, which include plants, algae, and cyanobacteria, acquire $CO_2$ from the atmosphere and incorporate it into the organic matter of their own biomass via photosynthesis. Each year, plants, algae, and cyanobacteria remove approximately one-seventh of the $CO_2$ from the atmosphere. At the same time, respiration and the decomposition of plants recycle a similar amount of carbon back into the atmosphere as $CO_2$. Much material from primary producers is also transformed into deposits of coal, gas, and oil, which are collectively known as **fossil fuels.** Herbivores can return some $CO_2$ to the atmosphere, eating plants and giving off $CO_2$, but the amount flowing through this part of the cycle is minimal.

Over time, much carbon is also incorporated into the shells of marine organisms, which eventually form huge limestone deposits on the ocean floor or in terrestrial rocks, where turnover is extremely low. As a result, rocks and fossil fuels contain the largest reserves of carbon. Natural sources of $CO_2$ such as volcanoes, hot springs, and fires release large amounts of $CO_2$ into the atmosphere. Human activities, primarily the burning of fossil fuels, are increasingly causing large amounts of $CO_2$ to enter the atmosphere. In addition, deforestation contributes to elevation of atmospheric $CO_2$ because there is less vegetation to absorb $CO_2$ from the atmosphere. Direct measurements over the past five decades show a steady rise in atmospheric $CO_2$ (see Figure 46.7), a pattern that shows no sign of slowing. As discussed in Section 46.2, elevated levels of atmospheric $CO_2$ are the primary cause of global warming but also have other dramatic environmental effects, boosting plant growth but lowering the amount of herbivory (see the Feature Investigation).

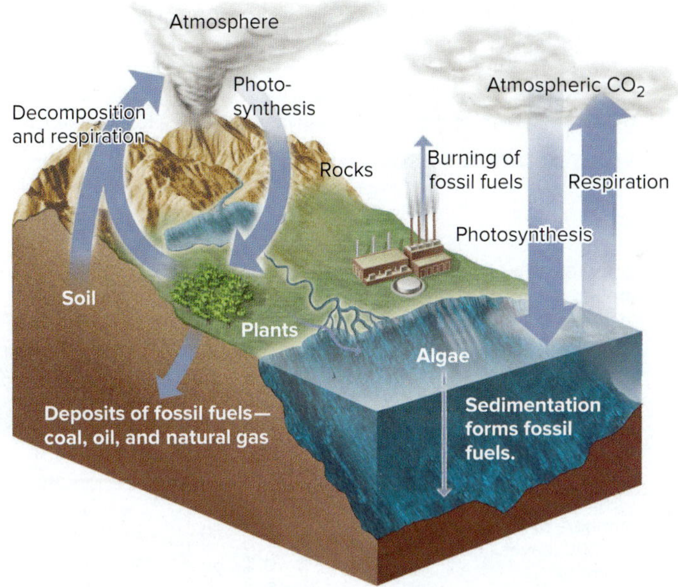


**Figure 46.9** **The carbon cycle.** Each year, plants and algae remove about one-seventh of the $CO_2$ in the atmosphere. Animal respiration is so small it is not represented. The width of the arrows indicates the relative contribution of each process to the cycle.

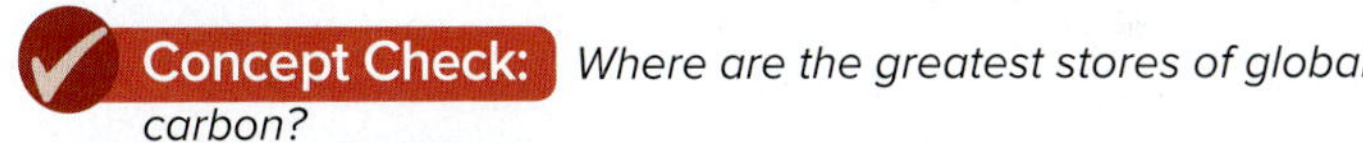

**Concept Check:** *Where are the greatest stores of global carbon?*

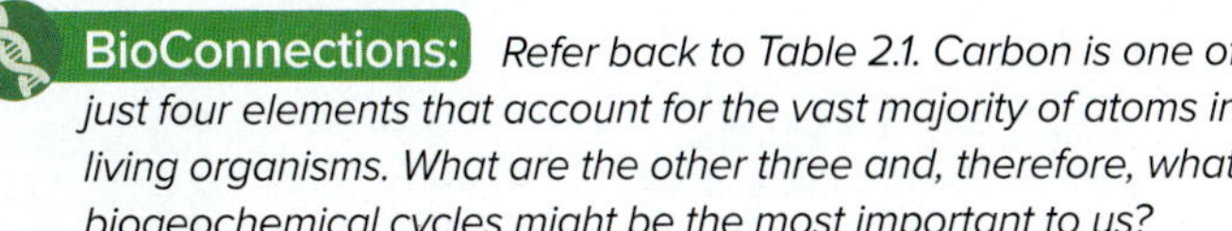

**BioConnections:** *Refer back to Table 2.1. Carbon is one of just four elements that account for the vast majority of atoms in living organisms. What are the other three and, therefore, what biogeochemical cycles might be the most important to us?*

# FEATURE INVESTIGATION

## Stiling and Drake's Experiments with Elevated $CO_2$ Showed an Increase in Plant Growth but a Decrease in Herbivory

How will forests of the future respond to elevated $CO_2$? To begin to answer such a question, ecologists ideally would enclose large areas of forests with chambers, increase the $CO_2$ content within the chambers, and measure the responses. This has proved to be difficult for two reasons. First, it is hard to enclose large trees in chambers, and second, it is expensive to increase $CO_2$ levels over such a large area. However, in a discovery-based investigation, ecologists Peter Stiling and Bert Drake were able to increase $CO_2$ levels around patches of forest at the Kennedy Space Center in Cape Canaveral, Florida. In much of Florida's forests, trees are small, only 3–5 m high at maturity, because frequent lightning-initiated fires prevent the growth of larger trees. In the 1990s, Stiling and Drake teamed up with NASA engineers to create 16 circular, open-topped chambers (**Figure 46.10**). In eight of these, they increased atmospheric $CO_2$ to double their ambient levels, from around 360 ppm to 720 ppm, the latter of which is the atmospheric concentration predicted by the end of the 21st century. The experiments commenced in 1996 and lasted until 2007. Not surprisingly, plants produced more biomass in elevated $CO_2$, because $CO_2$ is limiting to plant growth, but the data revealed much more.

Because the chambers were open-topped, insect herbivores could come and go. Insect herbivores cause the largest amount of herbivory in North American forests, because most vertebrate herbivores cannot access the high foliage. Censuses were conducted of all damaged leaves but focused on leaves damaged by leaf miners, the most common type of herbivore at this site. Leaf miners are small moths whose larvae are small enough to burrow between the surfaces of plant leaves, creating brown areas as shown in Figure 46.10, step three. The condition of the leaf mine reveals whether the larva survived and emerged, was eaten by predators, or was attacked by parasitoids.

In the chambers with elevated $CO_2$, although biomass increased, densities of damaged leaves, including those damaged by leaf miners, were lower in every year studied. Part of the reason for the decline was

**Figure 46.10** The effects of elevated atmospheric $CO_2$ on insect herbivory.
*(photos)* © Peter Stiling

**GOAL** To determine the effects of elevated $CO_2$ on a forest ecosystem; effects on herbivory are highlighted here.

**STUDY LOCATION** Patches of forest at the Kennedy Space Center in Cape Canaveral, Florida.

| | Experimental level | Conceptual level |
|---|---|---|
| **1** Erect 16 open-top chambers around native vegetation. Increase $CO_2$ levels from 360 ppm to 720 ppm in half of them. |  | Expected atmospheric $CO_2$ level is 720 ppm by end of the 21st century. Open-top chambers allow movement of herbivores in and out of chambers. |
| **2** Conduct a yearly count of numbers of insect herbivores per 200 leaves in each chamber. | 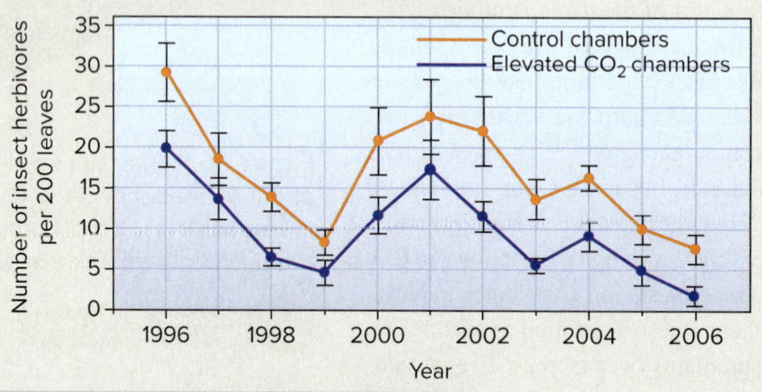 | |
| **3** Count number of herbivores that died due to nutritional inadequacy. Monitor attack rates on insect herbivores by natural enemies such as predators and parasitoids. | 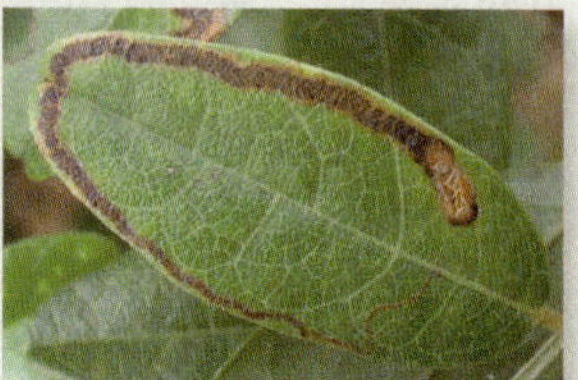 | Elevated $CO_2$ reduces foliar nitrogen, inhibits normal insect development, and prolongs the feeding time of herbivores, allowing natural enemies greater opportunities to attack them. |

**4 THE DATA**

| Source of mortality* | Elevated $CO_2$ (% mortality) | Control (% mortality) |
|---|---|---|
| Nutritional inadequacy | 10.2 | 5.0 |
| Predators | 2.4 | 2.0 |
| Parasitoids | 10.0 | 3.2 |

*Data refer only to mortality of larvae within leaves and do not sum to 100%. Mortality of eggs on leaves, pupae in the soil, and flying adults is unknown.

**5 CONCLUSION** Elevated $CO_2$ decreases insect herbivory in a Florida forest.

**6 SOURCE** Stiling, P., and Cornelissen, T. 2007. How Does Elevated Carbon Dioxide ($CO_2$) Affect Plant-Herbivore Interactions? A Field Experiment and Meta-Analysis of $CO_2$-Mediated Changes on Plant Chemistry and Herbivore Performance. *Global Change Biology* 13: 1823–1842.

that even though plants increased in mass, the existing soil nitrogen was diluted over a greater volume of plant material, so the nitrogen level in leaves decreased. This resulted in increased insect mortality by two means. First, poorer leaf quality directly increased insect death because leaf nitrogen levels may have been too low to support the normal development of the leaf miners. Second, lower leaf quality increased the amount of time insects needed to feed to gain sufficient nitrogen. Increased feeding times, in turn, led to increased exposure to natural enemies, such as predatory spiders and ants and parasitoids (see Figure 43.11), so mortality from natural enemies also increased (see step 4 of Figure 46.10). Thus, in a world of elevated $CO_2$, plant growth may increase, and herbivory could decrease.

*Experimental Questions*

1. What was the hypothesis of Stiling and Drake's experiment?
2. **SCISKILLS ▸** Explain the purpose of increasing the $CO_2$ levels in only half of the chambers in the experiment and not all of the chambers.
3. **SCISKILLS ▸** Imagine the percent mortality from parasitoids in the eight elevated $CO_2$ chambers was 6, 11, 9, 8, 13, 12, 14 and 7, and that in the ambient chambers was 1, 4, 5, 2, 3, 6, 1.6, and 3. Perform a statistical test to see if mortality from parasitoids was different between elevated and ambient $CO_2$ chambers.

## Global Warming and the Human Interruption of Water Flow Affect the Water Cycle

The water cycle, also called the hydrological cycle, differs from the cycles of other nutrients in that very little of the water that cycles through ecosystems is chemically changed by any of the cycle's components (**Figure 46.11**). It is a physical process, fueled by the Sun's energy, rather than a chemical one, because it consists of essentially two phenomena: evaporation and precipitation. Even so, the water cycle has important biological components. Over land, 90% of the water that reaches the atmosphere is moisture that has passed through plants and exited from the leaves via evapotranspiration. Only about 2% of the total volume of Earth's water is found in the bodies of organisms or is held frozen or in the soil. The rest cycles from bodies of water, to the atmosphere, and then to the land and back again.

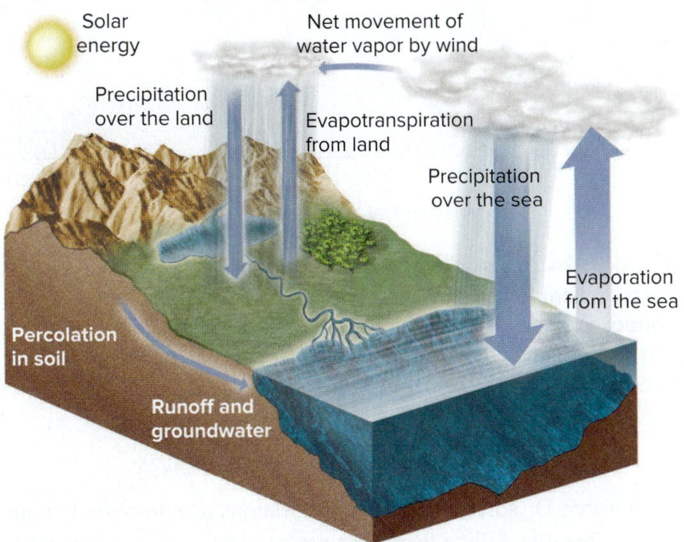


**Figure 46.11 The water cycle.** This cycle is primarily a physical process, not a chemical one. Solar energy drives the water cycle, causing evaporation of water from the ocean and evapotranspiration from the land. This is followed by condensation of water vapor into clouds and precipitation. The width of the arrows indicates the relative contribution of each process to the cycle.

Human activities have altered the water cycle in different ways. As discussed in Section 46.2, climate change may have the greatest effect on the water cycle by altering precipitation patterns around the world. On a smaller scale, humans have built structures that affect the flow of moving bodies of water. To increase the amount of available water and to create hydroelectric power, humans have interrupted the water cycle in many ways, most prominently through the use of dams to create reservoirs. Such dams, such as those on the Columbia River in Washington State (**Figure 46.12**), can greatly interfere with the migration of fishes such as salmon and affect their ability to reproduce and survive. Other activities, such as tapping into underground water supplies, or **aquifers,** for drinking water removes more water than is put back by rainfall and can cause shallow ponds and lakes to dry up and sinkholes to develop, exacerbating local shortages.

## Fertilizers and Other Pollutants Affect the Phosphorus Cycle

All living organisms require phosphorus, which becomes incorporated into ATP, DNA, and RNA, and it is an essential mineral that helps maintain a strong, healthy skeleton.

The phosphorus cycle is relatively simple (**Figure 46.13**). Phosphorus has no gaseous phase and thus no atmospheric component; that is, it is not moved by wind. As a result, phosphorus cycles only locally. The Earth's crust is the main storehouse for this element. Weathering and erosion of rocks release phosphorus into the soil. Plants have the metabolic means to absorb dissolved ionized forms of phosphorus, the most important of which occurs as phosphate ($HPO_4^{2-}$ or $H_2PO_4^-$). Herbivores obtain their phosphorus only from eating plants, and carnivores obtain it by eating herbivores. When plants and animals excrete wastes or die, the phosphorus becomes available to decomposers, which release it back to the soil.

Leaching and runoff eventually wash much phosphate into aquatic systems, where plants and algae utilize it. Phosphate that is not taken up into the food chain settles to the ocean floor or lake bottom, forming sedimentary rock.

Phosphorus can be particularly limiting in aquatic ecosystems. When more phosphorus is added, the growth of algae and aquatic

# Biology Principle

## Biology Affects Our Society

Interruptions in biogeochemical cycles, such as changes in the water cycle, can have severe repercussions for human societies, such as depletion of fish stocks.

**Figure 46.12 A dam on the Columbia River, Washington State.** Dams can increase water supplies and provide hydroelectric power but can interfere with migration of fishes, such as salmon and trout.


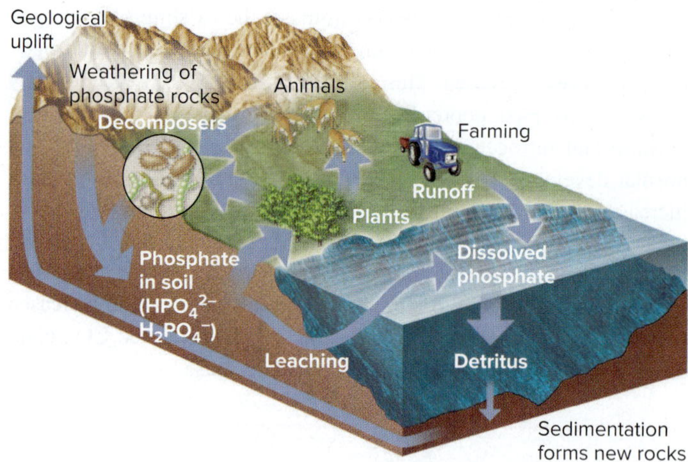


**Figure 46.13 The phosphorus cycle.** Unlike other major biogeochemical cycles, the phosphorus cycle does not have an atmospheric component and thus cycles only locally. The widths of the arrows indicate the relative contribution of each process to the cycle.

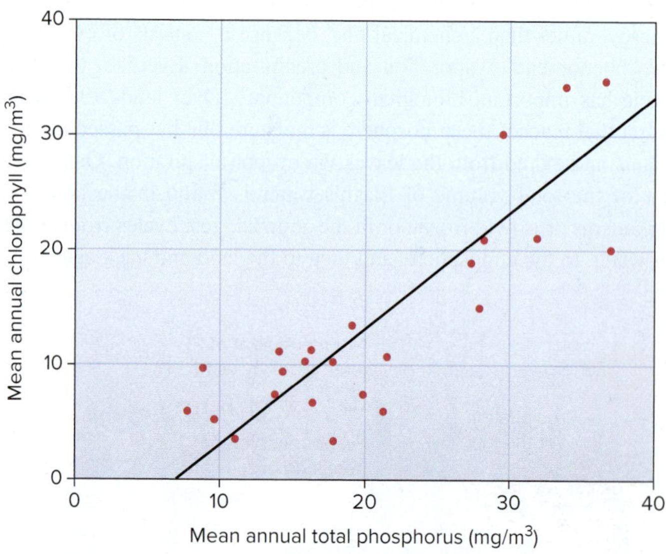


**Figure 46.14 The relationship between primary production and total phosphorus concentration.** As shown in this graph, primary production (measured by chlorophyll concentration) increases linearly with an increase in phosphorus. Each dot represents a different lake.

plants can dramatically increase. In lakes, primary production and total phosphorus concentration appear to follow a linear relationship (**Figure 46.14**). What is the consequence of high primary production? When the algae and plants die, they sink to the bottom, where bacteria decompose them and consume the dissolved oxygen in the water. Dissolved oxygen concentrations can drop too low for fishes to breathe, killing them. The process by which elevated nutrient levels lead to an overgrowth of algae and the subsequent depletion of water oxygen concentrations is known as **eutrophication.** Eutrophication is frequently due to the enrichment of water with nutrients derived from human activities, such as fertilizer use and sewage dumping.

Lake Erie became eutrophic in the 1960s due to the runoff of fertilizer rich in phosphorus from farms and to the industrial and domestic pollutants released from the many cities along its shores (**Figure 46.15a**). Fish species such as white fish and lake trout became severely depleted. The U.S. and Canada teamed together to reduce the levels of discharge by 80%, primarily through eliminating phosphorus in laundry detergents and maintaining strict controls on the phosphorus content of wastewater from sewage treatment plants. Fortunately, lake systems have great potential for recovery after phosphorus inputs are reduced, and Lake Erie has experienced fewer algal blooms, clearer water, and a restoration of fish populations (**Figure 46.15b**).

**(a) Polluted (eutrophication)**

**(b) Cleared up**

**Figure 46.15** **Phosphorus pollution in Lake Erie.** The lake **(a)** in the 1960s, when eutrophic and polluted by industrial effluent and fertilizer runoff, and **(b)** in 2007, after eutrophication was reversed by pollution control laws.

*(a)* © Bill Brooks/Alamy RF; *(b)* © Rolf Hicker Photography/Alamy

## Fertilizers and Industrial Pollutants Also Affect the Nitrogen Cycle

Nitrogen is an essential component of proteins, nucleic acids, and chlorophyll. Because 78% of the Earth's atmosphere consists of nitrogen gas ($N_2$), it may seem that nitrogen should not be in short supply for organisms. However, nitrogen is often a limiting factor in ecosystems because $N_2$ molecules must be broken apart before the individual nitrogen atoms can combine with other elements. Because of its triple bond, $N_2$ is very stable, and only certain bacteria can break it apart into usable forms such as ammonia ($NH_3$). This process, called nitrogen fixation, is a critical component of the five-part nitrogen cycle (**Figure 46.16**):

1. A few species of bacteria can accomplish **nitrogen fixation,** that is, convert atmospheric $N_2$ to forms usable by other organisms. The bacteria that fix nitrogen are fulfilling their own metabolic needs, but in the process, they release ammonia ($NH_3$) or ammonium ($NH_4^+$), which can be used by some plants. Cyanobacteria are important nitrogen fixers in terrestrial and aquatic systems (refer back to Figure 30.10).
2. In the process of **nitrification,** soil bacteria convert $NH_3$ or $NH_4^+$ to nitrate ($NO_3^-$), a form of nitrogen commonly used by plants. The bacteria *Nitrosomonas* and *Nitrococcus* first oxidize the forms of ammonia to nitrite ($NO_2^-$), after which the bacterium *Nitrobacter* converts $NO_2^-$ to $NO_3^-$.
3. **Assimilation** is the process by which inorganic substances are incorporated into organic molecules. In the nitrogen cycle, organisms assimilate nitrogen by taking up $NH_3$, $NH_4^+$, and $NO_3^-$ formed through nitrogen fixation and nitrification and incorporating them into other molecules. Plant roots take up these forms of nitrogen through their roots, and animals assimilate nitrogen from the plant tissues they ingest.
4. Ammonia can also be formed in the soil through the decomposition of plants and animals and the release of animal waste. **Ammonification** is the conversion of organic nitrogen to $NH_3$ and $NH_4^+$. This process is carried out by bacteria and fungi. Most soils are slightly acidic and, because of an excess of $H^+$, the $NH_3$ rapidly gains an additional $H^+$ to form $NH_4^+$. Because many soils lack nitrifying bacteria,

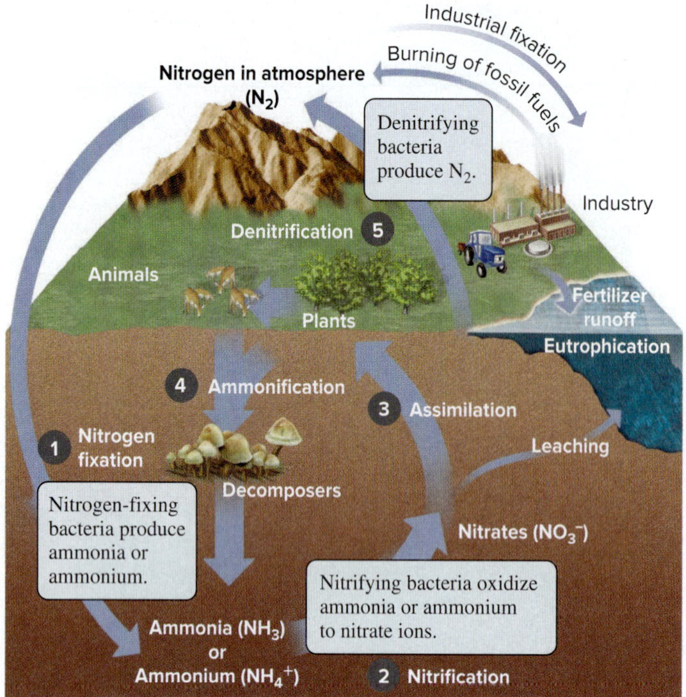


**Figure 46.16** **The nitrogen cycle.** The five main parts of the nitrogen cycle are (1) nitrogen fixation, (2) nitrification, (3) assimilation, (4) ammonification, and (5) denitrification. The recycling of nitrogen from dead plants and animals into the soil and then back into plants is of paramount importance because this is the main pathway for nitrogen to enter the soil. The width of the arrows indicates the relative contribution of each process to the cycle.

ammonification is the most common pathway for nitrogen to enter the soil.

5. **Denitrification** is the reduction of $NO_3^-$ to $N_2$. Denitrifying bacteria, which are anaerobic and use $NO_3^-$ in their metabolism instead of $O_2$, perform the reverse of their nitrogen-fixing counterparts by delivering $N_2$ to the atmosphere. This process delivers only a relatively small amount of nitrogen to the atmosphere.

Human activities have approximately doubled the rate of nitrogen input to the nitrogen cycle. Industrial fixation of nitrogen for the production of fertilizer makes a major contribution to the pool of nitrogen-containing material in the soils and waters of agricultural regions. As with phosphorus, fertilizer runoff can cause eutrophication of rivers and lakes, and, as the resultant algae die, decomposition by bacteria depletes the oxygen level of the water, resulting in fish kills. Excess $NO_3^-$ in surface or groundwater systems used for drinking water is also a health hazard, particularly for infants. In the body, $NO_3^-$ is converted to $NO_2^-$, which then combines with hemoglobin to form methemoglobin, a type of hemoglobin that does not carry oxygen. In infants, the production of large amounts of $NO_2^-$ can cause methemoglobinemia, a dangerous condition in which the level of $O_2$ carried through the body decreases. Finally, burning fossil fuels releases not only carbon but also nitrogen in the form of nitrous oxide ($N_2O$), which contributes to air pollution. $N_2O$ can also react with rainwater to form nitric acid ($HNO_3$), a component of acid rain, which decreases the pH of lakes and streams and increases fish mortality.

## 46.3 Reviewing the Concepts

- Elements such as carbon, phosphorus, and nitrogen recycle from the physical environment to organisms and back in what are called biogeochemical cycles.
- In the carbon cycle, phototrophs incorporate $CO_2$ from the atmosphere into their biomass; decomposition of plants and respiration recycle most of this $CO_2$ back to the atmosphere. Human activities, primarily the burning of fossil fuels, are causing increased amounts of $CO_2$ to enter the atmosphere (Figures 46.9, 46.10).
- The water cycle is a physical rather than a chemical process because it consists of essentially two phenomena: evaporation and precipitation. Climate change is altering precipitation patterns. In addition, alteration of the water cycle by dams can greatly disrupt migration of fishes such as salmon and trout (Figures 46.11, 46.12).
- The phosphorus cycle lacks an atmospheric component and thus is a local cycle. An overabundance of phosphorus can cause the overgrowth of algae and subsequent depletion of oxygen levels, called eutrophication (Figures 46.13, 46.14, 46.15).
- The nitrogen cycle has five parts: nitrogen fixation, nitrification, assimilation, ammonification, and denitrification. In the nitrogen cycle, atmospheric nitrogen is unavailable for use by most organisms and must be converted to usable forms by certain bacteria. The activities of humans, including the production and use of fertilizers, have altered the nitrogen cycle (Figure 46.16).

## 46.3 Testing Your Knowledge

1. Industrial fixation of nitrogen for the production of fertilizer makes a significant contribution to the pool of nitrogen in
   a. soils.
   b. water.
   c. the atmosphere.
   d. a and b.
   e. all of the above.
2. The increase of atmospheric $CO_2$ levels is caused by
   a. the burning of fossil fuels.
   b. deforestation.
   c. acidification.
   d. a and b.
   e. all of the above.

## 46.4 Pollution and Biomagnification

### Learning Outcome

1. Define biomagnfication and explain its relationship to pollution.

In Section 46.3, we considered how pollution generated by humans can affect biogeochemical cycles. Certain types of pollutants can also accumulate within the bodies of animals and plants. The tendency of certain chemicals to concentrate in higher trophic levels in food chains is called **biomagnification,** and it presents a problem for certain organisms. The passage of dichlorodiphenyltrichloroethane (DDT), an insecticide used against mosquitoes and agricultural pests, in food chains provides a startling example. DDT was first synthesized by chemists in 1874. In 1939, its insecticidal properties were recognized by Paul Müller, a Swiss scientist who won the 1948 Nobel Prize in Physiology or Medicine for his discovery and subsequent research on the uses of the chemical. The first important application of DDT was in human health programs during and after World War II, particularly as a means of controlling mosquito-borne malaria; at that time, its use in agriculture also began. The global production of DDT peaked in 1970, when 175 million kilograms of the insecticide was manufactured.

DDT has several chemical and physical properties that profoundly influence the nature of its ecological effect. First, DDT is persistent in the environment. It is not rapidly degraded to other, less toxic chemicals by microorganisms or by physical agents such as light and heat. The typical persistence in soil of DDT is about 10 years, which is two to three times longer than the persistence of many other insecticides. Another important characteristic of DDT is its low solubility in water and its high solubility in fats, or lipids. In the environment, most lipids are present in living tissue. Therefore, because of its high lipid solubility, DDT tends to concentrate in biological tissues.

Because biomagnification occurs at each step of the food chain, organisms at higher trophic levels can accumulate especially high concentrations of DDT in their lipids. A typical pattern of biomagnification is illustrated in **Figure 46.17**, which shows the relative amounts of DDT found in a Lake Michigan food chain. The highest concentration of the insecticide was found in gulls, tertiary consumers that feed on fishes, which are the secondary consumers that eat small insects. An unanticipated effect of DDT on bird species was its interference with the metabolic process of eggshell formation. The result was thin-shelled eggs that often broke under the weight of incubating birds (**Figure 46.18**). DDT was responsible for a dramatic decrease in the populations of many birds due to failed reproduction. Relatively high levels of the chemical were also found to be present in some game fishes, which, as a result, became unfit for human consumption.

Because of growing awareness of the adverse effects of DDT, most industrialized countries, including the U.S., banned the use of the chemical by the early 1970s. The good news is that following the outlawing of DDT, populations of the most severely affected bird species have recovered. However, had scientists initially possessed a more thorough knowledge of how DDT accumulates in food chains, some of the damage to the bird populations might have been prevented.

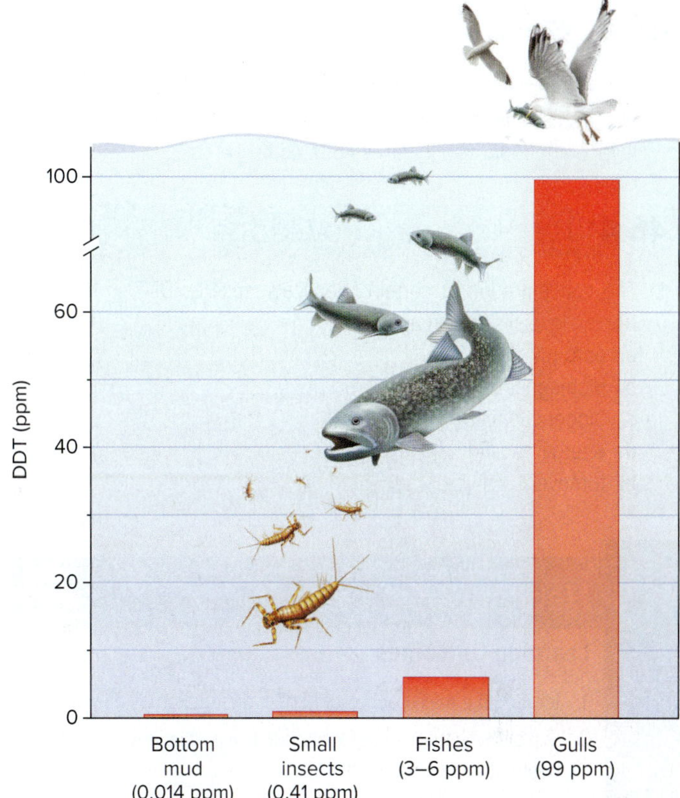


**Figure 46.17** **Biomagnification in a Lake Michigan food chain.** The DDT tissue concentration in gulls, a tertiary consumer, was about 240 times that in the small insects sharing the same environment. The biomagnification of DDT in lipids causes its concentration to increase at each successive link in the food chain. PPM is the equivalent of 1 mg of DDT per kg of biological material.

Cl
Cl
H
C
$CCl_3$
DDT (dichlorodiphenyltrichloroethane)
• Persists in environment
• High solubility in lipids
• Found in high concentrations at higher trophic levels


**Figure 46.18** **Thinning of eggshells caused by DDT.** Ibis eggs, Texas Gulf Coast.

## 46.4 Reviewing the Concepts

- The increase in the concentration of certain chemicals in living organisms, called biomagnification, can occur at each trophic level of the food web (Figures 46.17, 46.18).

## 46.4 Testing Your Knowledge

1. The concentration of certain chemicals, such as DDT, in higher trophic levels is known as
   a. eutrophication.
   b. biomagnification.
   c. biogeochemical cycling.
   d. energy transfer.
   e. turnover.

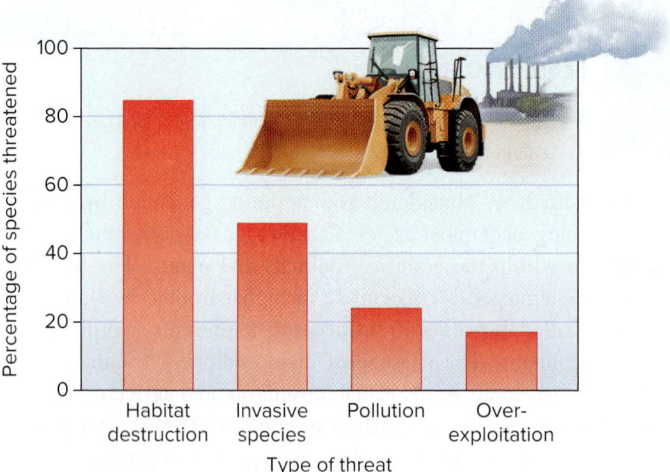


**Figure 46.19** **Percentages of plant and animal species threatened by various causes in the U.S.** Species can suffer from multiple threats, so categories do not sum to 100%.

## 46.5 Habitat Destruction

### Learning Outcomes

1. Identify the main causes of habitat destruction by humans.
2. Explain why tropical deforestation is a particularly destructive form of habitat loss.
3. Describe the impact of agriculture on ecological change.

**Habitat destruction** is usually a human-driven process in which natural habitat is altered in a way that prevents it from supporting the species that were originally present. Organisms that previously occupied the habitat are displaced or unable to survive, thereby reducing biodiversity. Habitat destruction is the primary cause of species extinction, and a high percentage of existing species are threatened by this process (**Figure 46.19**). Other human-driven practices that are threatening the survival of many species include pollution, which has already been discussed in this chapter, and overexploitation and invasive species, which are described later.

Habitat destruction includes deforestation, conversion of habitat to agricultural land, urbanization, strip mining, quarrying, and many other forms of land modification. Urbanization, the development of cities on previously natural or agricultural areas, is the most human-dominated and fastest-growing type of land use worldwide, and it devastates the land more severely than nearly any other form of habitat destruction. Freshwater habitats have also suffered via dam construction and river channelization. Wetlands have been drained for agricultural purposes and have been filled in for urban or industrial development. In the U.S., as much as 90% of the freshwater marshes have disappeared in states such as Iowa and California, though the national average is approximately 53%. In this section, we will examine the two most widespread and interrelated types of habitat destruction, at least in terms of their influence on species extinctions, which are deforestation and the conversion of land for agricultural purposes.

### Deforestation by Humans Threatens the Existence of Many Species

**Deforestation,** the conversion of forested areas to nonforested land, is a prime cause of the extinction of species (**Figure 46.20**). About one-third of the world's land surface is covered with forests, and much of this area is at risk of deforestation.

**Figure 46.20** **Deforestation.** Cascade Mountains near Seattle, Washington, 1906.
Source: Library of Congress Prints and Photographs Division [LC-USZ62-67666]

**Figure 46.21 Habitat destruction as an important cause of species extinction.** The ivory-billed woodpecker, the third-largest woodpecker in the world, was thought to have gone extinct in the southeastern U.S. because of habitat destruction.


Many species live in forests. For example, among North American terrestrial wildlife, about one-quarter of the bird species (272 species) and more than 10% of mammal species (49 species) have an obligatory relationship with forest cover, meaning that they depend on trees for food and nesting sites. In terms of wildlife use, oaks are among the most valuable trees in North America. At least 100 species of birds and mammals include acorns in their diets, and for many species of wildlife, the annual acorn crop is a major determinant of their abundance. Most woodpeckers, as well as many other types of birds, nest in holes that they excavate in trees, and their food usually consists of insects collected on or in trees. The ivory-billed woodpecker, *Campephilus principalis*, the largest woodpecker in North America and an inhabitant of wetlands and forests of the southeastern United States, was widely assumed to have gone extinct in the 1950s due to destruction of its habitat by heavy logging (Figure 46.21).

Tropical forests, primarily rain forests but also deciduous forests, are among the most species-rich terrestrial habitats and exist primarily in three areas of the globe: Africa, Asia, and Latin America. Latin America contains more than the other two areas combined, with Brazil containing the greatest amount of tropical forest of any nation. Although tropical deforestation occurs for different reasons, clearing land for agriculture has been identified as the prime cause of forest loss. Tropical forests, which once covered 14% of the Earth's dry land, now cover only about 7%. Current rates of tropical deforestation increased from 0.67% per year in the 1990s to 0.84% annually between 2000 and 2005.

Logging in excess of regrowth is also a significant cause of loss, particularly in Asian forests. Fuelwood collection can also be important but generally is more of a problem in lightly wooded areas. Finally, the construction of mines, dams, and oil installations is a minor cause of direct deforestation, but indirectly the discharge of chemicals and silt into rivers can cause much damage. The roads built into these regions often open them up to further development.

Conservation of tropical forests would save many rare species from extinction and has other benefits as well. The Amazon rainforest has been termed "the lungs of the planet" because it produces more than 20% of the world's oxygen. Many of the world's crops, including oranges, lemons, bananas, cacao (chocolate), coffee, and vanilla, evolved in tropical rain forests. Rain forests remain a depository of genetic variation that could be used in future breeding of these crops. Rain forests are particularly rich in plants with unusual chemicals that they use in defense (refer back to Chapter 43). Many of these chemicals also have medicinal value. Although many prescription drugs sold worldwide come from plant-derived sources, less than 1% of tropical plants have been tested for their medicinal properties. One estimate has valued tropical forests at $2,400 an acre if renewable and sustainable resources such as fruit, rubber, and nuts are harvested. In comparison, forests are worth only $400 an acre for timber and $60 an acre when converted to agricultural grazing land. For these reasons, it seems to make good economic and ecological sense to slow the rate of tropical deforestation.

## The Development of Land for Agriculture Results in Many Types of Ecological Changes

More land has been converted to agriculture since 1945 than in the 18th and 19th centuries combined. The destruction of habitat to plant agricultural crops and provide pasture for grazing livestock can create soil erosion, increased flooding, declining soil fertility, silting of rivers, and desertification as well as remove valuable wildlife habitat. While the average area of land under crop cultivation worldwide averages is 12%, with an additional 26% given over to rangeland for grazing, this amount varies substantially between regions (Table 46.2).

The planting of crops and the grazing of livestock have produced perhaps the most far-reaching ecological effects on Earth (Figure 46.22). Most introduced crops will always remain dependent on people for their survival, requiring that water and fertilizer be added or competing weeds and insect herbivores be removed. The population size of humans is increased by the cultivation of these crops and domesticated animals, and similarly, populations of crops and domesticated animals are increased by the efforts of humans. Excluding Antarctica, a staggering 37.3% of the world's land area is given over to crops and grazing pastures for animals such as cattle and sheep and the number of

**Table 46.2 Percentage of Land Area Used for Agricultural Purposes**

| Continental area | % cropland | % pastures | Total |
|---|---|---|---|
| World | 11.5 | 25.8 | 37.3 |
| Asia | 20.0 | 33.0 | 53.0 |
| Central America and the Caribbean | 15.6 | 36.7 | 52.3 |
| Europe | 13.1 | 7.9 | 21.0 |
| Middle East and North Africa | 7.6 | 28.1 | 35.6 |
| North America | 11.4 | 12.5 | 23.9 |
| Oceania | 6.2 | 48.2 | 54.4 |
| South America | 6.8 | 28.8 | 35.6 |
| Sub-Sahara and Africa | 8.1 | 33.9 | 42.0 |

**Figure 46.22** **Modern agriculture and its impact on the environment.** **(a)** This wheat crop could not exist without human help, and at the same time, crops sustain human populations. **(b)** The grazing of livestock has radically altered the ecological landscape.
*(a)* Source: USDA; *(b)* © Martial Colomb/Photographer's Choice/Getty Images RF

livestock in the world runs into the billions (Table 46.3). No other human endeavor on Earth requires as much land as agriculture. Unfortunately, the ramifications for natural systems don't stop simply with the acreage of land used. As we have seen, runoff water from agricultural land often contains high levels of nitrogen and phosphorus from fertilizers as well as residual pesticides and manure, all of which contaminate streams. Some 4 billion tons of topsoil are washed into U.S. waterways each year, severely impacting the organisms that live there. Topsoil can also be whipped away by wind.

**Table 46.3** **Numbers of Livestock in the World in 2005**

| Livestock | Millions |
|---|---|
| Chickens | 19,600 |
| Cattle | 1,400 |
| Sheep | 1,081 |
| Ducks | 1,046 |
| Pigs | 980 |
| Goats | 807 |
| Geese | 301 |
| Turkeys | 279 |
| Buffaloes | 174 |
| Horses | 58 |
| Donkeys | 44 |
| Camels | 19 |
| Mules | 12 |

## 46.5 Reviewing the Concepts

- Habitat destruction involves the conversion of habitat to agricultural land, urbanization, the draining of swamps, strip mining, dam destruction, and many other forms of land modification. It is the primary cause of species extinction and is threatening many current species (Figure 46.19).
- Deforestation—in particular, tropical deforestation—threatens many species with extinction because of the high numbers of species that live in forests and the high rate of forest loss (Figures 46.20, 46.21).
- The planting of crops and the grazing of livestock are among the most important drivers of ecological change. Habitat destruction has changed huge areas of the planet into agricultural areas and rangelands for livestock (Figure 46.22; Tables 46.2, 46.3).

## 46.5 Testing Your Knowledge

1. Habitat destruction includes
   a. deforestation.
   b. conversion of habitat to agricultural land.
   c. urbanization.
   d. the draining of wetlands.
   e. all the above.

## 46.6 Overexploitation

### Learning Outcomes

1. Outline the main causes of overexploitation of plant and animal populations.
2. Describe examples of the overexploitation of animals and plants.
3. **SCISKILLS ▶** Calculate the maximum sustainable yield of a population based on its carrying capacity and per capita rate of increase.

In ecology, **overexploitation** is the practice in which humans harvest a particular species at a rate that is unsustainable, based on its natural rate of mortality and capacity for reproduction. Overexploitation, particularly the hunting of animals, has been the cause of many extinctions in the past. In this section, we will consider several examples of overexploitation and then examine how ecologists make calculations to determine if overexploitation is occurring.

## Many Species Have Gone Extinct or Are Currently Threatened Due to Overexploitation

With regard to animals, hunting and fishing have been the main practices that facilitate overexploitation. For other groups, such as plants, the ability to identify valuable species and remove them from their native habitats has led to overexploitation. In 2015, a study by Canadian conservation biologist Chris Darimont and colleagues examined the effects of humans as hunters of terrestrial mammals and fishers of marine fishes. They compared 2,125 estimates of exploited animal populations and showed that humans kill adult prey at rates up to 14 times higher than other predators (Figure 46.23). Darimont and colleagues termed humans super predators and suggested that, in the long run, this level of hunting and fishing would not be sustainable. As seen in Figure 46.25, humans are predators of species that are considered the top predators in their native environment. For example, humans have greatly decreased the populations of wolves, which are successful predators in many different habitats.

Let's now consider a few examples in which humans have overexploited animals and plants.

**Land Mammals** The list of mammals threatened with extinction or actually driven extinct by hunting is long. In the prairies of North America, hunting diminished populations of buffalo from around 70 million in the 18th century to 1,150 by 1899. Since then, buffalo numbers have rebounded to about 350,000. In Russia, another inhabitant of the prairies, the Eurasian wild horse also known as the tarpan, *Equus ferus*, had been hunted to extinction by the 1860s.

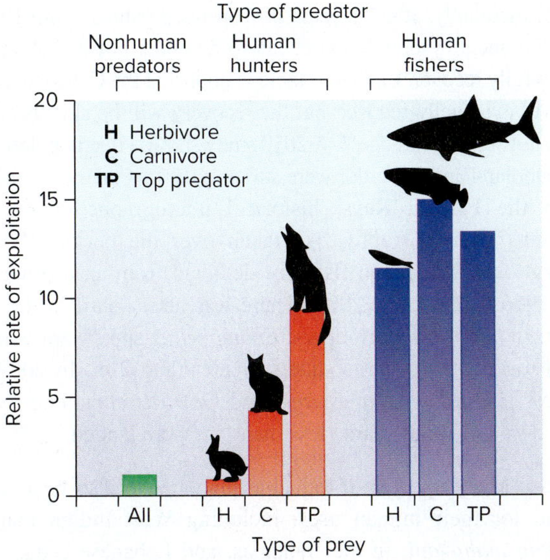


**Figure 46.23** **Humans as hunters.** The rates at which humans exploit adult land mammals and marine fish vastly exceed the impacts of other predators. Marine fish of all trophic levels are similarly affected. In contrast, land predators are exploited at much higher rates than herbivores. The term top predator refers to the most successful predator in a given habitat, which also can be a prey to humans.

Even today, poachers threaten many land mammals. Rhinoceroses and elephants both have horns that some people believe will cure everything from cancer to hangovers. The black market value for rhino horn in Southeast Asia is about $65,0000 per kilogram, making it more expensive by weight than gold, diamonds, or cocaine. Roughly 100,000 African elephants were poached across the continent between 2010 and 2012. In 2011 alone, poachers killed roughly 1 in every 12 African elephants. Many experts believe that the last of the great woolly mammoth populations were hunted to extinction at the end of the last glacial period about 12,000 years ago.

**Whales** The question as to whether whales should be exploited has been the subject of vigorous and worldwide debate since at least the 1960s. The level of popular interest in this question probably exceeds that of any other group of exploited animals. The history of whaling in general, and Antarctic whaling in particular, has been characterized by a progression from more valuable or more easily caught species to less attractive ones, as populations of the original targets were depleted (Figure 46.24). In the Antarctic, blue whales, *Balaenoptera musculus*, dominated the catches through the 1930s, but by the middle 1950s, few were being taken, although the species was not legally protected until 1965. As the populations of blue whales diminished, attention turned to the fin whale, *B. physalus*, which was originally the most abundant of all whales in the Southern Ocean. By the 1960s, numbers of this species had diminished rapidly. Humpback whales, *Megaptera novaeangliae*, though never very numerous, were valuable because of their high oil yield and the ease with which they could be caught. Catches were never very great, but the populations in most areas collapsed dramatically in the early 1960s. Sei whales, *B. borealis*, were almost ignored by whalers until the bigger species were no longer available. They were hardly taken at all until about 1958, but then catches increased rapidly and reached a peak of about 20,000 in 1964–1965. Catches declined rapidly thereafter, this time due to the introduction of 10 catch limits. Then the relatively small minke whales, *B. acutorostrata*, which were ignored in the Southern Ocean until 1971–1972, began to be taken. Since that time, minkes have been the largest component of the southern baleen whale catch.

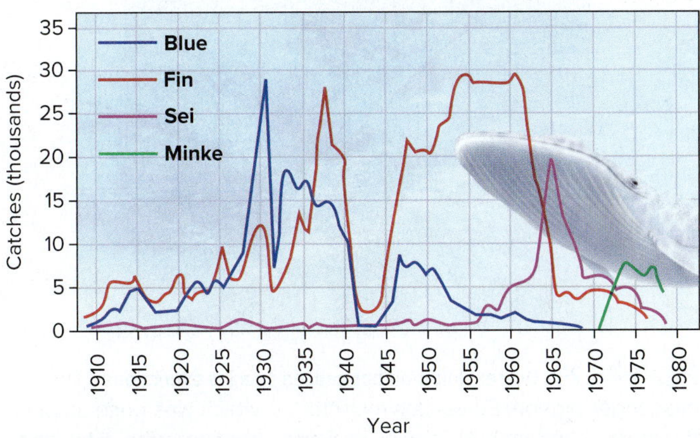


**Figure 46.24** **Sequential decline of whale catches in the Antarctic, showing the strong effect of human predators.**

The story was similar in the Northern Hemisphere, although the populations were much smaller, about 20% of that in Antarctic waters.

In 1982, the International Whaling Commission (IWC) voted for a moratorium on all commercial whaling. The IWC proposed that commercial whaling be ended in 1985–1986, a proposal that did not actually take effect until 1988. The good news is that following the moratorium, the populations of some whales have increased. Blue whales appear to have quadrupled their numbers off the California coast during the 1980s, showing the impact that protection from a predator can have. Steve Palumbi's laboratory estimated past population sizes of North Atlantic whales by examining mitochondrial DNA sequence variation. Population size estimates for fin and humpback whales suggested that pre-whaling populations were 6–20 times higher than present-day population estimates and that full recovery of whale populations will take another 70–100 years.

**Birds** A poignant example of overexploitation was the dodo, *Raphus cucullatus*, a flightless bird native only to the island of Mauritius that had no known predators. A combination of overexploitation and introduced species led to its extinction within 200 years of the arrival of humans. Sailors hunted it for its meat, and the rats and pigs they brought to the island, the latter as a food source, destroyed the dodos' eggs and chicks in their ground nests.

Two abundant species of North American birds, the passenger pigeon and the Carolina parakeet, had been hunted to extinction by the early 20th century. The passenger pigeon, *Ectopistes migratorius*, was once the most common bird in North America, probably accounting for over 40% of the entire number of birds (**Figure 46.25**). Flock sizes were estimated to be over 1 billion birds. It may seem improbable that the most common bird on the continent could be hunted to extinction for its meat, but that is just what happened. The flocking behavior of the birds made them relatively easy targets for hunters, who used special firearms to harvest the birds in quantity. In 1876, in Michigan alone, over 1.6 million birds were killed and sent to markets in the eastern U.S. Similarly, the Carolina parakeet, *Conuropsis carolinensis*, the only species of parrot native to the eastern U.S., had been similarly hunted to extinction by the early 1900s.

**Figure 46.25 Overexploitation has caused species extinctions.** The passenger pigeon, *Ectopistes migratorius*, which was once among the most abundant bird species on Earth, was hunted to extinction for its meat.



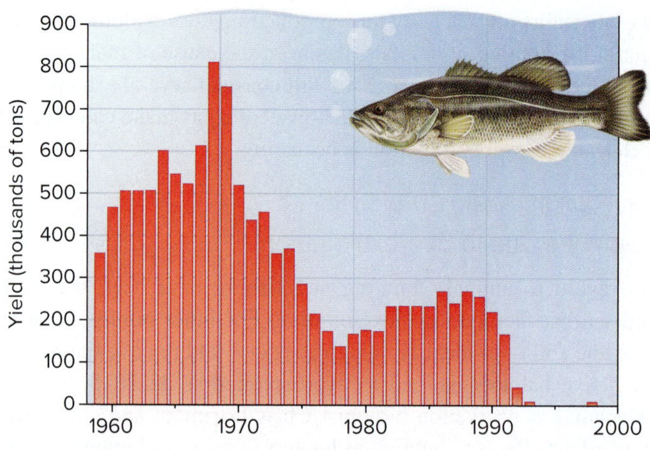


**Figure 46.26 Commercial cod fisheries.** Changes in Canadian cod yields from 1960 to 2000.

**Fishes** Many species of fish are harvested to the degree that the rate of removal exceeds the rate of reproduction, the fish populations crash, and the fishery collapses. In the case of the Canadian cod fishery, overfishing and collapse came in the early 1990s after hundreds of years of fishing (**Figure 46.26**). Since that time, the species has not been economically fished.

One of the most famous examples of overfishing is the Peruvian anchovy industry, which collapsed in the 1970s. This fishing industry was the largest of its kind in the world until 1972, when it collapsed due to overfishing and a severe El Niño event. In the El Niño, the coastal waters warmed, stopping cold, nutrient-rich water from coming to the surface. Incredibly, after 16 years of suspended fishing, from 1972 until 1988, the anchovy population began to rebound, but it took another 5 years to fully recover. Likewise, there is hope that the Canadian cod fisheries will eventually recover, but the recovery will be slow because the fish do not spawn until age 7. A 2010 study showed that populations near Newfoundland and Labrador were still only 10% of their original sizes.

In the Florida Keys, historical photographs document the reduction of large trophy fish taken over the period 1956–2007 (**Figure 46.27**). The mean fish size declined from an estimated 19.9 kg in 1956 to 2.3 kg in 2007! There was also a shift in the species caught. In 1956, large groupers, *Epinephellus* spp., were commonly landed together with large sharks greater than 2 m. In 2007, small snappers, *Lutjanus campechanus* and *Ocyurus chrysurus*—average length 34.4 cm—and sharks less than 1 m were landed.

**Plants** Many species of valuable plants have also been severely reduced for their human uses, including West Indian mahogany, *Swietenia mahogani*, in the Bahamas, and Lebanese cedar, *Cedrus libani*, which in Lebanon has been reduced to a few scattered forest remnants. Rare cacti and orchids have also been threatened by collectors, who seek to own a rare organism or to profit from its sale. Rare orchids are very valuable. A single stem of Rothschild's orchid, *Paphiopedilum rothschildianum,* also known as the Gold of Kinabalu, is reported to sell for \$5,000 (**Figure 46.28**). No wonder that after its

**Figure 46.27** **Change in the sizes of trophy fish caught on Key West charter boats.** **(a)** 1957, **(b)** early 1980s, and **(c)** 2007.

*(a)* Source: From the archives of Monroe County Public Library, Key West, Florida. Photo by Wil-Art Studio/Art Stickel; *(b)* Source: From the archives of Monroe County Public Library, Key West, Florida. Photo by Dale McDonald; *(c)* Photo courtesy Loren McClenachan, Ph.D.

**Figure 46.28** **Rothschild's orchid, *Paphiopedilum rothschildianum*.** Overcollection in northern Borneo has pushed this species close to extinction in the wild.



discovery, this orchid was stripped from the wild in northern Borneo by orchid smugglers, pushing it close to extinction.

Plants may also be collected for their medicinal value, real or supposed. In the U.S., American ginseng, *Panax quinquefolicus*, native to eastern North America, has been sought after to treat the common cold. Plants are so enthusiastically collected by "sang hunters," who also sell it to Chinese traders, that ginseng has become threatened or endangered in some states. The Botanic Gardens Conservation International Organization has stated that over 400 medical plant species are at risk of extinction because of overcollection and deforestation, including many species of yew tree, *Taxus* spp., whose bark is used for cancer drugs.

## Quantitative Analysis

### ECOLOGISTS MAKE CALCULATIONS TO DETERMINE IF OVEREXPLOITATION IS OCCURRING

For a population that is being harvested by humans, the **yield** is the number of individuals harvested in a given unit of time. **Maximum sustainable yield (MSY)** is the largest number of individuals that can be removed without causing long-term decreases in the population. If the maximum sustainable yield is exceeded on a consistent basis, overexploitation is occurring.

According to the logistic equation described in Chapter 43, population increase, $dN/dt$, occurs according to the following equation.

$$\frac{dN}{dt} = rN\,\frac{(K - N)}{K}$$

Where

$N$ is the number of individuals
$t$ is a unit of time
$r$ is the per capita rate of population growth
$K$ is the carrying capacity

The greatest population growth, and thus the maximum sustainable yield, occurs at the midpoint of the logistic curve, at $K/2$ (refer back to Figure 43.6).

MSY can thus be estimated as

$$\text{MSY} = \frac{rK}{2}$$

As an example, if $K/2 = 10{,}000$ and $r = 0.14$, then MSY = 1,400.

MSY has been extensively used for fisheries management. However, the use of the model has some drawbacks. For example, biologists cannot easily go under water and count the number of available individuals. The model also ignores the age and reproductive status of the individuals being harvested. As an alternative, simple models based on data collection from fisheries may be used, which compare costs and revenues and consider MSY (**Figure 46.29**). The maximum profit (red line) is higher than the profit obtained at MSY (blue line).

Why is MSY important to fisheries (and the harvesting of other species)? Let's assume that total economic costs increase linearly as fishing effort increases. Total revenue also increases with fishing effort, up to a point. After MSY is reached, any increase in effort would result in a decrease in revenue as the fish populations became overfished. The maximum profit is equal to the biggest difference between the cost and revenue curves, and as the red line in Figure 46.29 shows, this occurs below the effort needed to reach the maximum sustainable yield.

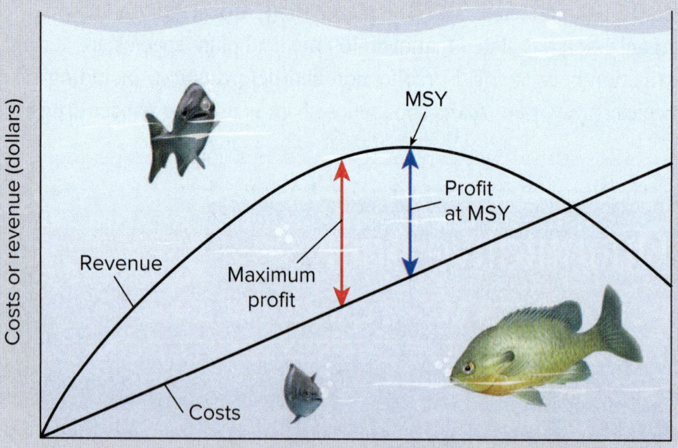


**Figure 46.29 Economic fisheries model based on revenue and costs.** The model assumes that costs increase linearly with effort but revenue reaches a maximum at maximum sustainable yield (MSY). Maximum profit, however, occurs below MSY.

Why do fisheries become overfished? Most estimates of MSY are nearly always too high, because of overestimates of population size. Furthermore, incremental improvements are made to fishing gear over the years and the same level of effort results in increased catches. Finally, as fish populations decline, the prices inch upward, making it more likely for commercial fishing to keep harvesting them. In the case of the cod industry, the introduction of radar and sonar allowed crews to pursue fish over huge areas. The crews also caught enormous numbers of noncommercial fish, which were discarded. Among these were important prey species for cod, such as capelin. Finally, although undersized cod, which cannot spawn, were returned to the ocean, such discards do not always survive. As the cod populations went into a tailspin, increased effort resulted in more undersized fish being caught and more discarded.

**Crunching the Numbers:** In a study of red snapper, *Lutjanus campechanus*, in the Gulf of Mexico, the carrying capacity of a particular population is estimated at 100,000 and the per capita rate of increase is 0.05. What is the maximum sustainable yield?

## 46.6 Reviewing the Concepts

- Human overexploitation of plants and animals occurs via hunting, fishing, and removing species from their native habitat. On average, humans decrease prey populations much more than natural predators (Figure 46.23).
- Human overexploitation has decreased the populations of many species and driven some of them to extinction. Examples include whales, the passenger pigeon, and many species of fishes and plants (Figures 46.24–46.28).
- The maximum sustainable yield is the largest number of individuals that can be harvested without causing long-term decreases in the population. Simple economic models suggest that populations should be harvested at levels of maximum profit, which is well below the maximum sustainable yield (Figure 46.29).

## 46.6 Testing Your Knowledge

1. Which was the first group of whales to be overharvested in the Southern Hemisphere?
   a. blue
   b. fin
   c. minke
   d. sei
   e. humpback
2. The maximum yield occurs at which point of the logistic curve?
   a. beginning
   b. middle
   c. upper end

## 46.7 Invasive Species

### Learning Outcomes

1. Define introduced species and invasive species.
2. Compare and contrast invasive species as competitors, as predators, and as parasites.

**Introduced species** are those species moved by humans from their native habitat to another location. Most often, the species are introduced for agricultural or landscaping purposes or as sources of timber, meat, or wool, and they may need humans for their continued survival. In some cases, such as plants, insects, or aquatic species, organisms are unintentionally transported via the movement of cargo by ships or planes. Regardless of the way they have been transported, some introduced species become **invasive species,** spreading naturally and impacting native species. Why do invasive species become successful in their new habitat? In most cases, they have one or more of the following characteristics.

- Invasive species may reproduce rapidly.
- They may not have natural enemies in their new habitat.
- Certain native species may lack defenses against the invasive species.
- Invasive species may compete aggressively for resources.
- They may tolerate a wide variety of habitat conditions.

In the U.S. alone, over 4,500 invasive species have been identified, and 15% cause severe ecological or economic harm. Of introduced vertebrates, 142 species have self-sustaining populations in the wild. These include ring-necked pheasants, *Phasianus colchicus,* which were brought over by hunters, and Burmese pythons, *Python molorus,* which were introduced by pet owners. Of the 300 most invasive weeds in the U.S., over half were brought in for gardening, horticulture, or landscape purposes. These include purple loosestrife, *Lythrum salicaria,* and Japanese honeysuckle, *Lonicera japonica,* in the Northeast; kudzu, *Pueraria lobata,* in the Southeast; Chinese tallow, *Sapium sebiferum,* in the South; and leafy spurge, *Euphorbia esula,* in the Great Plains. In this section, we will examine how the interactions between invasive and native species may involve competition, predation, and parasitism.

### Invasive Species May Compete Against Native Species

On Anticosti Island, a large island in the St. Lawrence Seaway off the coast of Quebec, the black bear, *Ursus americanus,* was driven extinct by introduced white-tailed deer, *Odocoileus virginianus*. Historical records indicate that black bears were once numerous there. In 1896, 220 white-tailed deer were introduced onto the island, and by 1934, in the absence of predators, their numbers had increased to 50,000. Feeding deer reduced the abundance and sizes of many shrub species, with the result that berry density decreased to only 0.28 berry/$m^2$. This was 235 times lower than the minimum 66 berries/$m^2$ necessary for black bears to maintain body mass. Bears need to build up energy reserves for the winter, and without the berries the bears became extinct.

Placental mammals have been found to be particularly effective competitors in Australia, where introduced dingoes, *Canis lupus,* ssp. *dingo,* are thought to have outcompeted the thylacine, *Thylacine cynocephalus,* a native marsupial, wolflike animal, in mainland Australia (Figure 46.30a,b). Sheep introduced for the wool industry are thought to compete with a variety of kangaroo species, especially the brush-tailed rock wallaby and larger species, such as the red kangaroo and the western gray kangaroo.

Introduced vertebrates are a problem in aquatic environments, too. In California, 48 of 137 species of freshwater fish are nonnative. Of these, 26 have been well studied and 24 are known to have a negative impact on native fish. After construction of the Welland Canal linking the Atlantic Ocean with the Great Lakes, many native fish populations were reduced in abundance through competition for food by the alewife, *Alosa pseudoharengus,* which invaded from the St. Lawrence Seaway.

Invasive plants are major competitors of native plants throughout the world.

- In the U.S. Northeast, purple loosestrife is a strong competitor, choking out other plant species in wetlands (Figure 46.30c). This, in turn, has endangered several species of ducks and a species of turtle that depends on the native plants for food and shelter.
- In some cases, invasive plants have become so successful that they have produced near monocultures—areas where they exist as the sole plant species. Examples include Chinese tallow trees, *Sapium sebiferum,* in the southern U.S. and garlic mustard, *Alliaria petiolata,* in the eastern and central U.S.
- In California the introduced species barb goatgrass, *Aegilops triuncialis,* is prolific on serpentine soils. These soils have low nutrients and low moisture content and may contain toxic metals that harm some species of native plants. Because barb goatgrass can tolerate these conditions, it can use the limited nutrients and water that are largely untapped by other plant species. The result is that barb goatgrass thrives and can form dense stands, outcompeting other species.
- In south Florida, three invasives—Brazilian pepper, *Schinus terebinthifolius*, Australian pine, *Casuarina equisetifolia,* and, especially punk-tree, *Melaleuca quinquenervia*—are outcompeting native vegetation in the Everglades. In north Florida and southeastern states, kudzu vine, *Pueraria lobata,* imported from Japan to stabilize construction areas, outcompetes native plants in large areas of native habitat.

### Invasive Species May Act as Predators

Many striking examples of the powerful effects of invasive species have been provided by predator introductions. The brown tree snake, *Boiga irregularis,* was inadvertently introduced by humans to the island of Guam in Micronesia, shortly after World War II. It probably arrived as a stowaway in U.S. military transport ships from the Admiralty Islands, near Papua New Guinea, during World War II. The growth and spread of its population over the next 40 years closely coincided with a precipitous decline in the island's forest birds. On Guam, the brown tree snake has no natural predators to control it. Because the bird species on Guam did not evolve with this species of

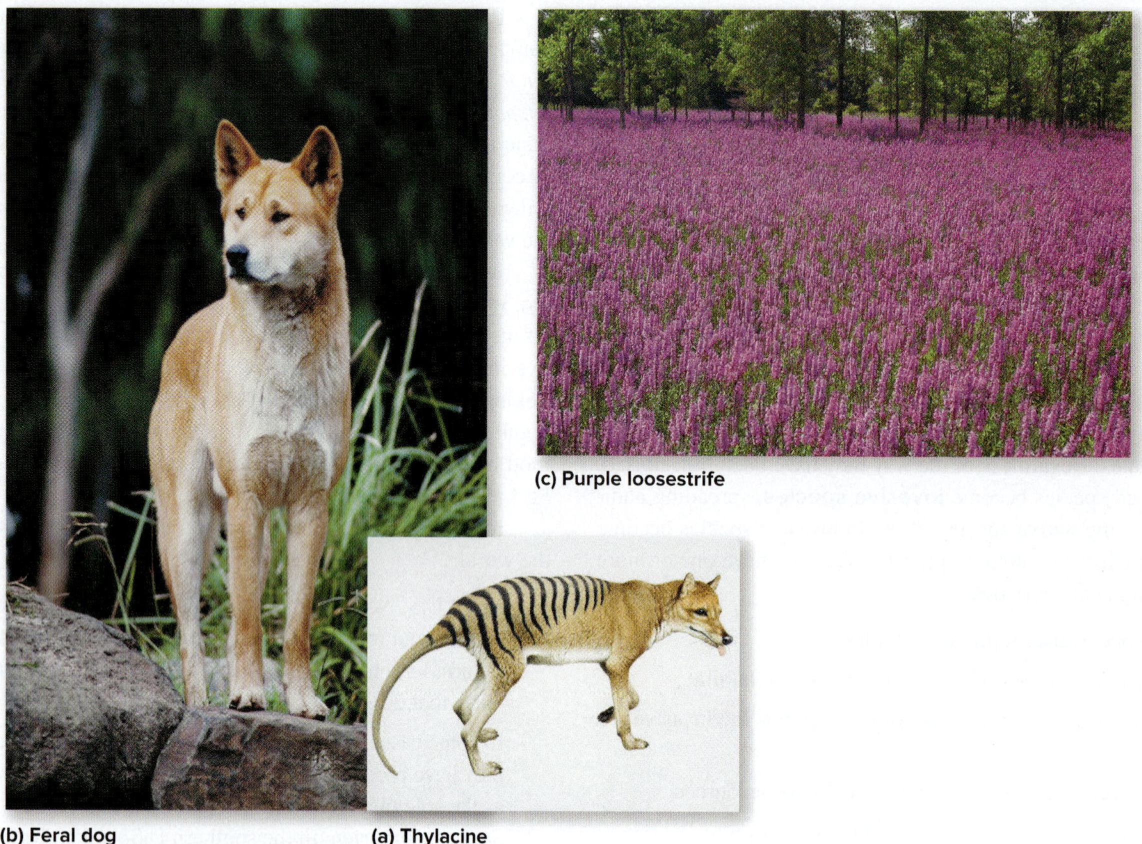


**Figure 46.30** **Introduced species as competitors.** **(a)** The thylacine in Australia may have been outcompeted by **(b)** feral dogs called dingoes. **(c)** Introduced plants can choke out native vegetation, as happens with purple loosestrife in the U.S. Northeast.
*(a)* © Kenneth Lilly/Getty Images RF; *(b)* © Royalty-Free/Corbis; *(c)* © Reimar Gaertner/Alamy RF

snake, there was no selective pressure to evolve defenses against it. Eight of the island's 11 native species of forest birds had gone extinct by the 1980s!

The U.S. government is attempting control measures, but trapping brown tree snakes is extremely time and labor intensive. However, scientists have a new weapon in their arsenal, tablets of acetaminophen, the active ingredient in Tylenol, packed into dead mice. In humans, acetaminophen gives pain relief, but in snakes it disrupts the oxygen-carrying ability of the snake's blood proteins. Only 80 milligrams, equivalent to a child's dose, is lethal. Furthermore, the brown tree snake is one of only a few snake species that will feed on dead animals as well as animals it has killed. However, placing acetaminophen-laden dead mice on the forest floor could impact other wildlife. A solution was to attach long, 1- to 2-m (4-ft) paper streamers to the mice and airdrop them onto the forest canopy, where they lodge in the foliage and only the snakes can find them. A small subset of the airdropped mice are fitted with radio transmitters. If a signal is found to move, it is because the mouse has very likely been eaten by a snake. The first airdrop was made in 2010, and it is too early to see how effectively this control measure is working.

Sea lampreys, *Petromyzon marinus,* are primitive, jawless fish that feed by rasping holes in the sides of larger fish species and feeding on their blood (Figure 46.31a). They spawn in freshwater streams, and the juveniles return to salt water (or one of the Great Lakes as a substitute) to develop. Such fish have modifications of their kidneys that allow them to exist in fresh water or salt water. Sea lampreys found their way into Lake Ontario in the mid-1800s by way of the Erie Canal (Figure 46.31b). Improvements to the Welland Canal allowed lamprey to bypass Niagara Falls and colonize the rest of the Great Lakes. In 1921, lampreys were found in Lake Erie and by 1938 they had been found all the way up into Lake Superior. Lake trout, *Salvelinus namaycush,* were the preferred prey in the Great Lakes. The lake trout fishing industry declined somewhat in the 1940s due to overfishing, but the decline was hastened by the arrival of lampreys (Figure 46.31c). By the 1960s, lake trout catches had been reduced by over 90% of their historic averages, and the fishery collapsed.

In the late 1940s, the state of Michigan began looking at control measures for lamprey. Barriers were erected across the mouths of streams to prevent the return of spawning adults, but they were difficult to maintain and were not 100% effective. Attention turned to chemical control of juvenile lampreys, which spend 3–5 years buried in stream sediments, filter feeding. The chemical TFM (trifluoromethyl nitrophenol) proved effective in controlling juvenile lampreys in streams and was first applied in 1958 to streams feeding into Lake Superior. Treatments reduced lamprey densities by 90%,

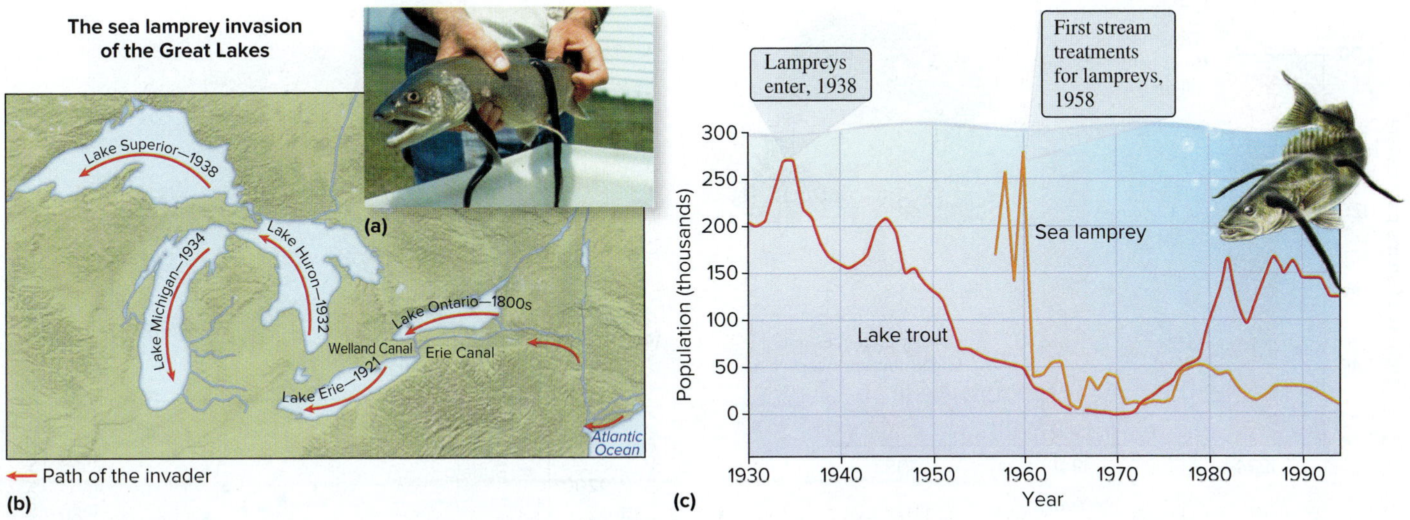


**Figure 46.31 Effects of invasive sea lampreys on the lake trout population in the Great Lakes.** **(a)** Sea lampreys attached to a lake trout. **(b)** Historical passage of lampreys into the Great Lakes from the Atlantic. **(c)** Effects of lampreys on the trout population in Lake Superior. Note that detailed records of sea lamprey populations were not available before 1956.

*(a)* Source: U.S. Fish and Wildlife Service, Bugwood.org

and the population of lake trout increased in the 1980s, aided by an active program of restocking. To maintain lamprey control, streams are treated every 3–5 years. Removal of invasive species is time consuming and costly but may lead to recovery of native prey.

Finally, let's return to the Florida Everglades (see chapter-opening photo) to see the effect of the Burmese pythons on native vertebrates. Pythons were likely introduced prior to 1985 and the population has grown exponentially since (refer back to Figure 43.5b). Pythons in Florida consume a wide range of vertebrate prey, including endangered species such as the wood stork, *Mycteria americana,* and the Key Largo woodrat, *Neotoma floridana.* Survey data gathered by counting animals and roadkills showed that raccoons, opossums, and rabbits were often observed during the period 1993–1999, before pythons became common. In the period 2003–2011, when pythons were common, the decline in observations was 99.3% for raccoons, 98.9% for opossums, 94.1% for white-tailed deer, and 87.5% for bobcats (**Figure 46.32**). No rabbits or foxes were even observed. In south Florida today, these species exist only in locations peripheral to the Everglades, where pythons are rare. Perhaps python removal would permit wildlife to recover, but because constrictors feed only on prey they have killed, it would not be feasible to control them with acetaminophen, as done with the brown tree snake.

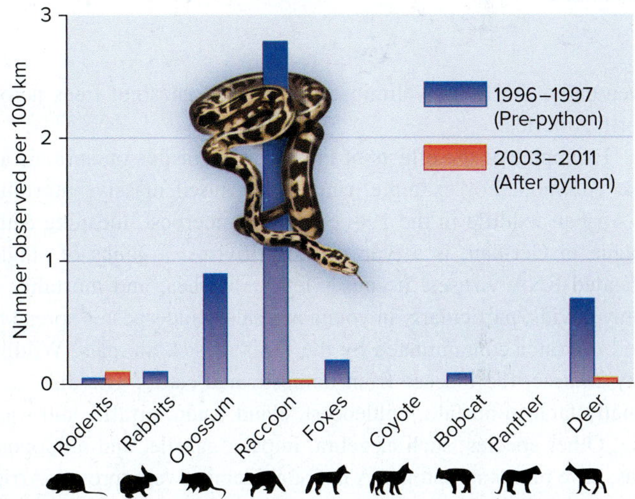


**Figure 46.32 Effect of invasive Burmese pythons on wildlife abundance in the Everglades National Park, Florida.** According to roadside surveys, a decline in mammal abundance occurred after pythons became common (2003–2011).

**BioConnections:** *Based on models of population growth, when do scientists believe the first Burmese pythons were released into the Everglades?*

## Some Invasive Species Are Parasites

Chestnut blight, *Cryphonectria parasitica,* is a fungus from Asia that was accidentally introduced to New York around 1904 from imported Asian chestnut trees, *Castanea crenata.* At that time, the American chestnut tree, *Castanea dentata,* was one of the most common trees in the eastern U.S. It was said that a squirrel could jump from one chestnut to another all the way from Maine to Georgia without touching the ground. The densest populations of chestnuts occurred in the Appalachian Mountains. Here, humans and wildlife alike feasted on the bountiful supply of nuts. The tree was ingrained in American culture, as in "chestnuts roasting on an open fire." By the 1950s, however, chestnut blight had significantly reduced the density of American chestnut trees in all areas of the United States, including North Carolina (**Figure 46.33**). Because at least one tree in four in these forests was a chestnut, the effect was very noticeable, although oaks and hickories replaced chestnuts in the canopy.

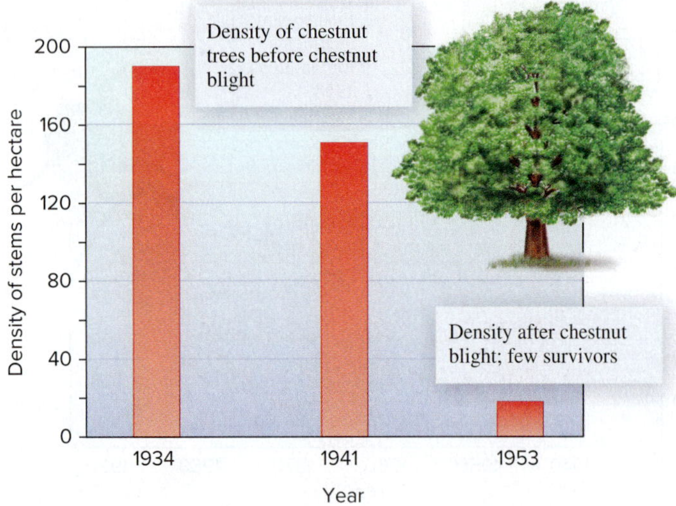


**Figure 46.33 Effects of introduced parasites on American chestnut trees.** The reduction in density of American chestnut trees in North Carolina following the introduction of chestnut blight disease from Asia shows the severe effect that parasites can have on their hosts. By the 1950s, this once prevalent species had been virtually eliminated.

Eventually, the fungus eliminated nearly all chestnut trees across North America.

Examples of the effects of introduced parasites on animals are also common. For example, rinderpest caused massive mortality in African wildlife in the 19th century. Rinderpest, meaning cattle plague in German, is a type of morbillivirus, a genus of single-stranded RNA viruses. It causes fever, diarrhea, and mortality in many bovids, particularly in young animals. Rinderpest is spread by food or water contaminated by the feces of sick animals. Wildlife may contract the disease from domesticated cattle. The disease is usually fatal in buffalo, wildebeest, eland, kudu, giraffe, and warthog. Other species, such as zebra, impala, gazelle, and hippopotamus, have minor symptoms. A major epidemic swept through Africa in the 1890s, probably from infected cattle imported into Somalia from India by an Italian military expedition. In the breeds of cattle found in India, the disease caused only mild infections, but across Africa more than 80% of cattle died. The disease spread 5,000 km in 8 years, arriving in North Africa in 1889 and reaching the Cape of Good Hope in 1897.

In the Serengeti of Tanzania, the common bovids are the wildebeest, *Connochoetes taurinus*, and the buffalo, *Syncerus cafter*; both were susceptible to rinderpest, and their numbers were reduced to about 200,000 and 30,000, respectively, following the rinderpest outbreak. In the 1960s, British veterinary scientist Walter Plowright developed an effective vaccine against rinderpest. The disease was brought under control and eventually eradicated. After the control of rinderpest in the 1960s, populations of wildebeest and buffalo increased over a 20-year period (Figure 46.34). The number of zebra, *Equus burchelli*, was also about 200,000 when detailed wildlife surveys began in the 1950s, but it was not susceptible to rinderpest, so its numbers changed little over the 1960s and 1970s.

(a)

(b)

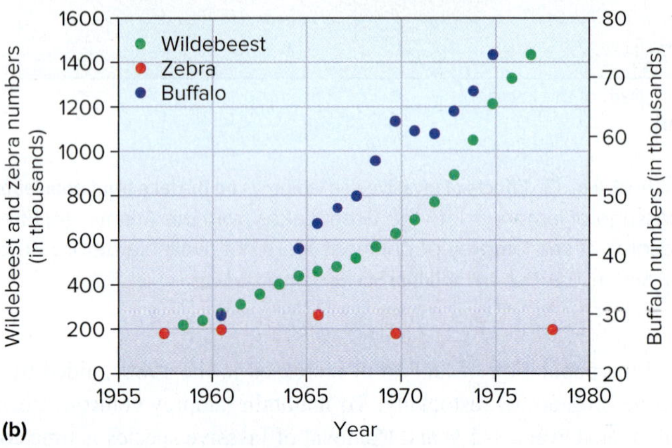


**Figure 46.34 Rebound of wildebeest and buffalo populations after the elimination of rinderpest. (a)** Cattle herds were decimated by rinderpest in the late 19th century. **(b)** Populations of wildebeest and buffalo, which are usually killed by this virus, increased dramatically after rinderpest was eliminated in the 1960s. Numbers of zebra, which are not killed by this virus, remained steady.

*(a)* Source: Photo courtesy of Texas A & M University College of Veterinary Medicine & Biomedical Sciences

**BioConnections:** *What type of population growth was exhibited by wildebeest and buffalo from 1958 to 1975, following the reduction of rinderpest?*

## 46.7 Reviewing the Concepts

- Invasive species are those species introduced into new geographic areas by humans and spread on their own without human support. Invasive species may impact native species via competition, predation, and parasitism (Figures 46.30–46.34).

## 46.7 Testing Your Knowledge

1. Why are invasive species introduced into areas of the world where they are not normally found?
   a. for agriculture
   b. for timber
   c. for meat
   d. by accident
   e. all of the above

**2.** Burmese pythons
  **a.** were released in Guam, where they have killed many species of birds.
  **b.** were released in the Great Lakes area to control invasive lampreys.
  **c.** are invasive in south Florida, where they have severely impacted raccoons, opossums, and rabbits.
  **d.** were important in the spread of chestnut blight in the U.S.
  **e.** none of the above.

# Assess and Discuss

## Test Yourself

1. The amount of land necessary for survival for each person in a sustainable world is known as
   a. the sustainability level.
   b. an ecological impact.
   c. an ecological footprint.
   d. survival needs.
   e. all of the above.
2. Which anthropogenic gas contributes most to global warming?
   a. methane
   b. chlorofluorocarbons
   c. carbon dioxide
   d. nitrogen oxides
   e. sulfur dioxide
3. The data show that atmospheric carbon dioxide levels have increased by what percent over the past 57 years?
   a. 1
   b. 5
   c. 12
   d. 25
   e. 52
4. Eutrophication is
   a. caused by an overabundance of nitrogen, which leads to a decrease in bacteria populations.
   b. caused by an overabundance of nutrients, which leads to an increase in algal populations.
   c. usually caused by oil spills.
   d. normally seen in dry, hot regions of the world.
   e. accelerated by habitat loss.
5. Human production of fertilizers first impacts which stage of the nitrogen cycle?
   a. nitrogen fixation
   b. nitrification
   c. assimilation
   d. ammonification
   e. denitrification
6. Which is *not* an example of habitat destruction?
   a. deforestation
   b. agriculture
   c. urbanization
   d. draining of wetlands
   e. overharvesting
7. The passenger pigeon, once the most common bird in North America, was driven to extinction by
   a. pollution.
   b. overexploitation.
   c. habitat destruction.
   d. invasive species.
   e. all of the above.
8. Most recorded extinctions have been caused by
   a. invasive species.
   b. habitat destruction.
   c. overexploitation.
   d. a and b equally.
   e. a, b, and c equally.
9. Invasive species come from
   a. agricultural introductions.
   b. accidental transportation via ships.
   c. landscape plants and their pests.
   d. a and b.
   e. a, b, and c.
10. Invasive species commonly
    a. outcompete native species.
    b. prey on native species.
    c. parasitize native species.
    d. a and b.
    e. a, b, and c.

## Conceptual Questions

1. The Earth's atmosphere consists of 78% nitrogen. Why is nitrogen a limiting nutrient, and how can we increase the supply of nitrogen to plants?
2. Why does maximum sustainable yield occur at the midpoint of the logistic curve and not where the population is at carrying capacity?
3. **PRINCIPLES** A principle of biology is that biology affects our society. In one family, parents, who were born in 1900, have twins at age 20 but then have no more children. Their children, grandchildren, and so on behave in the same way. In another family, parents, who were also born in 1900, delay reproduction until age 33 but have triplets. Their children and grandchildren behave in the same way. Which family has the most descendants by 2000? What can you conclude?

## Collaborative Questions

1. Discuss what might limit human population growth in the future.
2. As a group, try to predict the atmospheric concentration of $CO_2$ in 2100. Discuss what effects this might have on the environment.

# Online Resource

# 47 Biodiversity and Conservation Biology

Spix's macaw (*Cyanopsitta spixii*). Less than 100 individuals of this species are known to exist in the rain forests of Brazil.

## Chapter Outline

In 2009, Jeff Corwin, an American conservationist and host for programs on Animal Planet and other television networks, published a book entitled *100 Heartbeats: The Race to Save the Earth's Most Endangered Species*. The Hundred Heartbeat Club had been created earlier by biologist E. O. Wilson to highlight the plight of animal species, such as Spix's macaw (*Cyanopsitta spixii*) in Brazil (see chapter-opening photo), the Chinese river dolphin (*Lipotes vexillifer*), and the Philippine eagle (*Pithecophaga jefferyi*), that have 100 or fewer individuals left alive (and hence that number of heartbeats away from extinction). Saving species from extinction is important in its own right, but, as we will see, conservation of biological diversity also has great economic and social value to humankind.

Biological diversity, or **biodiversity,** encompasses the genetic diversity of species, the variety of species, and the different ecosystems they form. The field of **conservation biology** uses principles and knowledge from molecular biology, genetics, and ecology to protect and sustain biological diversity. Because it draws from nearly all chapters of this textbook, a discussion of conservation biology is a fitting way to conclude our study. In this chapter, we begin by examining the questions of what biodiversity is and why it should be conserved and exploring how much diversity is needed for ecosystems to function properly.

Later, we consider what is being done to help conserve the world's endangered plant and animal life. This includes identifying global areas rich in species and establishing parks and refuges of the appropriate size, number, and connectivity. We also discuss conservation of particularly important types of species and how to restore damaged habitats to a more natural condition. We then examine how captive-breeding programs help to build populations of rare species prior to their release back into the wild. Some programs have also used modern genetic techniques such as cloning to help breed and perhaps eventually increase populations of endangered species.

## 47.1 Genetic, Species, and Ecosystem Diversity

### Learning Outcome

1. List and describe the three levels of biodiversity.

Biodiversity can be examined on three levels: genetic diversity, species diversity, and ecosystem diversity. Each level of biodiversity provides valuable benefits to humanity.

Genetic diversity consists of the amount of genetic variation occurring within and between populations. Without such variation, populations cannot respond so quickly to changes in environmental conditions, resulting in population decline and even extinction. Maintaining genetic variation in the wild relatives of crops may be vital to the continued success of crop-breeding programs. For example, the café marron (*Ramosmania rodriguesii*), a wild relative of the coffee plant that is native to a tiny island off the coast of Mauritius, was assumed to be extinct until 1979, when one surviving tree was identified. Today, cuttings from the tree are being cultured in London's Kew Gardens (Figure 47.1). The plant may contain genes that would allow coffee to be grown in a wider range of soils and elevations.

The second level of biodiversity concerns **species diversity,** the number and relative abundance of species in a community (refer back to Section 44.1). In 1973, the U.S. Endangered Species Act (ESA) was enacted, which was designed to protect both endangered and threatened species. **Endangered species** are those species that are in danger of extinction throughout all or a significant portion of their range. **Threatened species** are those species likely to become endangered in the foreseeable future. Many species are currently threatened. According to the International Union for Conservation of Nature and Natural Resources (IUCN), more than 25% of the fish species that live on coral reefs and 22% of all mammals, 12% of birds, and 31% of amphibians are threatened with extinction. In 2000, the World Wildlife Foundation placed Atlantic cod, *Gadus morhua*, on the endangered species list as a result of overfishing. Nine of 17 populations of commercially important Chinook salmon, *Oncorhynchus tshawytscha*, in California and Oregon are listed as endangered or threatened.

**Figure 47.1 Café marron being cultured in London's Kew Gardens.** These plants are derived from just one surviving individual found in Mauritius.



The last level of biodiversity is ecosystem diversity, the diversity of structure and function within an ecosystem. Conservation has largely focused attention on species-rich ecosystems such as tropical rain forests. Over 120 prescription drugs used to treat malaria, cancer, and other diseases were developed from rain forest plants, yet less than 1% of such plants have been tested for medicinal properties. However, some ecologists have argued that relatively species-poor ecosystems such as prairies are also highly threatened and in equal need of conservation. More than 99% of the original tallgrass prairie in the United States has been converted to agricultural land.

### 47.1 Reviewing the Concepts

- Biodiversity represents diversity at three levels: genetic diversity, species diversity, and ecosystem diversity. Conservation biology uses knowledge from molecular biology, genetics, and ecology to protect biological diversity (Figure 47.1).

### 47.1 Testing Your Knowledge

1. In 1977, Rafael Guzman, a Mexican biologist, discovered a previously unknown wild relative of corn that is resistant to many of the viral diseases that infect domestic corn. Agriculturalists believe crossbreeding could improve current corn crops. In this case, biodiversity is important at which level?
   a. ecosystem
   b. species
   c. genetic
   d. community
   e. both a and b

## 47.2 Value of Biodiversity to Human Welfare

### Learning Outcomes

1. Outline the benefits of biological diversity to human welfare.
2. Provide graphical representations of possible relationships between biodiversity level and ecosystem function.
3. Describe experimental evidence that shows how species diversity and ecosystem function are linked.

Why should biodiversity be a concern? American biologists Paul Ehrlich and E. O. Wilson have suggested that the loss of biodiversity should be an area of great concern for at least three reasons:

1. Humans depend on plants, animals, and microorganisms for a wide range of food, medicine, and industrial products.
2. Ecosystems provide an array of essential services, such as clean air and water.

3. Humans have an ethical responsibility to protect what are our only known living companions in the universe.

In this section, we examine some of the primary reasons that preserving biodiversity matters and explore the link between biodiversity and ecosystem function.

## Society Benefits Economically from Biodiversity

The pharmaceutical industry is heavily dependent on plant and fungal products for source material. An estimated 50,000–70,000 plant species are used in traditional and modern medicine. About 25% of the prescription drugs in the U.S. alone are derived from plants, and the 2015 market value of such drugs was estimated to be $374 billion, accounting for a little less than half the global pharmaceutical market. Rapamycin was developed from a soil fungus on Easter Island and suppresses immune reactions. It is used to prevent organ rejection in transplants and as a coating on heart stents. Worldwide prescription drug sales are forecast to be $1 trillion in 2020. Many medicines come from plants found only in tropical rain forests. These include quinine, a drug from the bark of the Cinchona tree (*Cinchona officinalis*), which is used for treating malaria, and vincristine, a drug derived from rosy periwinkle (*Catharanthus roseus*), which is a treatment for leukemia and Hodgkin disease. Many chemicals of therapeutic importance are likely to be found in the numerous rain forest plant species that have not yet been fully analyzed. As mentioned in Chapter 46, the continued destruction of rain forests could mean the loss of potential lifesaving medical treatments.

Animal products have also been important to the pharmaceutical industry. The blood of the horseshoe crab (*Limulus polyphemus*) clots when exposed to toxins produced by some bacteria. Pharmaceutical companies use the blood enzyme responsible for this clotting to ensure that their products are free of bacterial contamination. The venom of gila monsters, *Heloderma suspectum,* one of only two venomous lizards in the world, is being used to treat people who are resistant to conventional treatment for type 2 diabetes, a disease that may affect 30% of Americans at some point in their life. A protein from the South American pit viper, *Bothrops jacara,* may help control human blood pressure. Tarantula venom may be helpful in treating neurological disorders such as Parkinson's disease.

Although humans depend on only about 20 plant species to provide 90% of the world's food, wild relatives of these crops provide a useful reservoir of genetic material for developing pest-resistant varieties or strains that can grow in marginal areas. In the 1970s, infusion of genetic material from wild corn in Mexico was used to protect U.S. commercial corn from a leaf fungus, which had killed 15% of the crop. The Lake Placid mint, *Dicerandra frutescens,* known only from central Florida, produces a powerful insect-repelling chemical that may have benefits for crop protection. Another endangered species, buffalo clover, *Trifolium stoloniferum,* has high protein content and is a perennial, making it of high potential value as a forage crop. The guayule, *Parthenium argentartum,* is of great potential value to industry. It has high amounts of natural rubber and grows in the deserts of the Southwest U.S., adding economic value to marginal lands. Animals, too, have great commercial value. Salmon fishing in the Pacific Northwest supports over 60,000 jobs and injects over $1 billion into the economy. In the U.K., the sea fish catch is worth over $500 million annually.

## Natural Ecosystems Provide Essential Services to Humans

Beyond the direct economic gains from biodiversity, humans benefit enormously from the essential services that natural ecosystems provide (**Table 47.1**). For example, forests soak up carbon dioxide, maintain soil fertility, and retain water, helping to prevent or minimize flooding; estuaries provide water filtration and protect rivers and coastal shores from excessive erosion. *Prochlorococcus,* an abundant ocean-dwelling genus of cyanobacteria was discovered only in 1986, yet it is estimated to produce about 20% of the oxygen we breathe. Other ecosystem functions include the maintenance of populations of natural predators to regulate pest outbreaks and reservoirs of pollinators to pollinate crops and other plants. The loss of biodiversity can disrupt an ecosystem's ability to carry out such functions.

In the 1990s, farmers in India began using the anti-inflammatory drug diclofenac to reduce pain and fever in their livestock. They could hardly anticipate that vultures scavenging on dead carcasses would accumulate large doses of the drug and die of renal failure. But the consequences did not stop there. Following a 97% reduction in vulture numbers over a 14-year period, the population of feral dogs exploded, buoyed by the availability of uneaten carcasses. The incidence of rabies in humans increased, with estimates of an additional 48,000 people dying over the 14-year time span. The loss of the scavenging services of these vultures was estimated to cost India $24 billion.

**Table 47.1 Examples of the World's Ecosystem Services**

| Service | Example |
|---|---|
| Atmospheric gas supply | Regulation of carbon dioxide, ozone, and oxygen levels |
| Climate regulation | Regulation of carbon dioxide, nitrogen dioxide, and methane levels |
| Disturbance regulation | Storm protection; flood control |
| Water regulation | Regulation of hydrological flows' |
| Water supply | Irrigation; water for industry |
| Erosion control | Retention of topsoil; reduction of accumulation of sediments in lakes |
| Soil formation | Soil formation processes |
| Nutrient cycling | Nitrogen, phosphorus, and carbon cycles |
| Waste treatment | Sewage purification |
| Pollination | Pollination of crops |
| Biological control | Pest population regulation |
| Wilderness and refuges | Habitat for wildlife |
| Food production | Crops; livestock |
| Raw materials | Fossil fuels; timber |
| Genetic resources | Medicines; genes for plant resistance |
| Recreation | Ecotourism; outdoor recreation |
| Cultural | Aesthetic and educational value |

A 2014 paper in the journal *Global Environmental Change* by economist Robert Costanza and colleagues made an attempt to calculate the monetary value of ecosystems to various economies (in 2007 dollars). They came to the conclusion that, at the time, the world's ecosystems were worth more than \$124 trillion a year, nearly twice the gross national product of the world's economies combined (\$75.2 trillion in 2007 dollars).

## There Are Ethical Reasons for Conserving Biodiversity

Arguments can also be made against the loss of biodiversity on ethical grounds. As only one of many species on Earth, it has been argued that humans have no right to destroy other species and the environment around us. American philosopher Tom Regan suggests that animals should be treated with respect because they have a life of their own and therefore have value apart from anyone else's interests. American law professor Christopher Stone, in an influential 1972 article titled "Should Trees Have Standing?" has argued that entities such as nonhuman natural objects, like trees or lakes, should be given legal rights just as corporations are treated as individuals for certain purposes. As E.O. Wilson proposed in a 1984 concept known as biophilia, humans have innate attachments with species and natural habitats because of our close association for over millions of years.

## Ecologists Have Described Several Relationships Between Ecosystem Function and Biodiversity

Because biodiversity affects the health of ecosystems, ecologists have explored the question of how much diversity is needed for ecosystems to function properly. In doing so, they have described several possible relationships between biodiversity and ecosystem function. In the 1950s, ecologist Charles Elton proposed in the **diversity-stability hypothesis** that species-rich communities are more stable than those with fewer species (refer back to Section 44.2). If we use stability as a measure of ecosystem function, Elton's hypothesis suggests a linear correlation between diversity and ecosystem function; as diversity increases, ecosystem function increases proportionately (Figure 47.2a). Australian ecologist Brian Walker proposed an alternative to this idea, termed the **redundancy hypothesis** (Figure 47.2b). According to this hypothesis, ecosystem function increases rapidly at fairly low levels of diversity, but then levels off because most additional species are functionally redundant. Two other alternatives relating species richness and ecosystem services have been proposed. The **keystone hypothesis** (Figure 47.2c) proposes ecosystem function dramatically rises as biodiversity approaches its natural levels. Finally, the **idiosyncratic hypothesis** addresses the possibility that although ecosystem function can change as the number of species increases or decreases, the amount and direction of change are unpredictable (Figure 47.2d).

Determining which model is most correct is very important, as our understanding of the effect of species loss on ecosystem function can greatly affect the way we manage our environment.

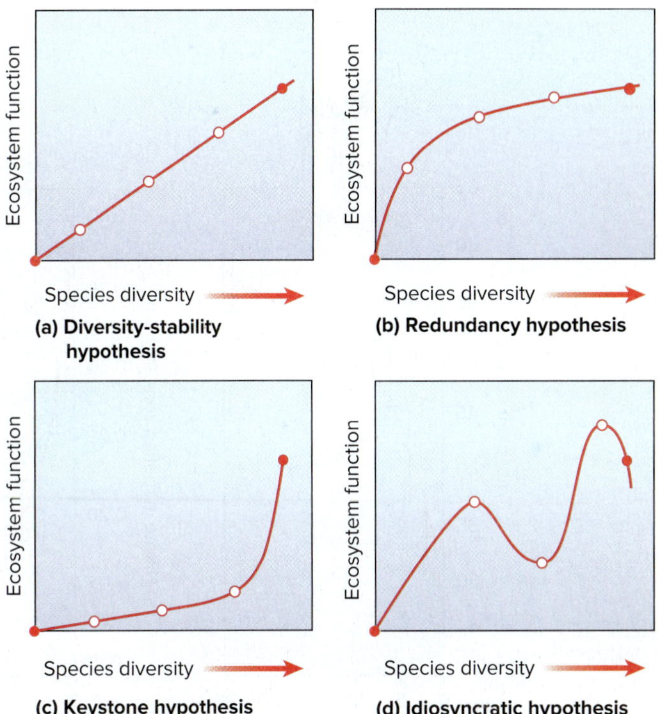


**Figure 47.2 Graphical representations of possible relationships between ecosystem function and biodiversity.** The two solid dots represent the end points of a continuum of species richness. The first dot is at the origin, where there are no species and no community services. The second dot represents natural levels of species diversity. The relationship is strongest in **(a)** and weakest in **(d)**.

## Field Experiments Have Been Used to Determine How Much Diversity Is Needed for Normal Ecosystem Function

In the mid-1990s, David Tilman and colleagues performed experiments in the field to determine how much biodiversity was necessary for proper ecosystem functioning. Tilman's previous experiments had suggested that species-rich grasslands were more stable (that is, they were more resistant to the ravages of drought and recovered from drought more quickly) than species-poor grasslands (refer back to Figure 44.4). In the subsequent experiments, Tilman's group sowed multiple plots, each 3 m × 3 m, and on comparable soils, with seeds of 1, 2, 4, 6, 8, 12, or 24 species of prairie plants. Exactly which species were sown into each plot was determined randomly from a pool of 24 native species. The treatments were replicated 21 times, for a total of 147 plots. The results showed that plots with more species (more diverse plots) had increased productivity, expressed as a percentage of plant cover (the amount of ground covered by leaves of plants) than plots with fewer species (less diverse plots). This occurred because of a greater variety of plant growth forms that could utilize light at different levels of the canopy. More diverse plots also used more nutrients, such as nitrate ($NO_3^-$), than less diverse plots because a greater variety of plant root lengths could utilize nutrients at different levels of the soil (Figure 47.3a,b). Furthermore, the frequency of invasive plant species (species not originally planted in the plots) decreased with increased plant species richness (Figure 47.3c).

Although Tilman's experiments show a relationship between species diversity and ecosystem function, they also suggest that most of the advantages of increasing diversity come with the first 5–10 species, beyond which adding more species appears to have little to no effect. This supports the redundancy hypothesis (compare Figure 47.3a with Figure 47.2b).

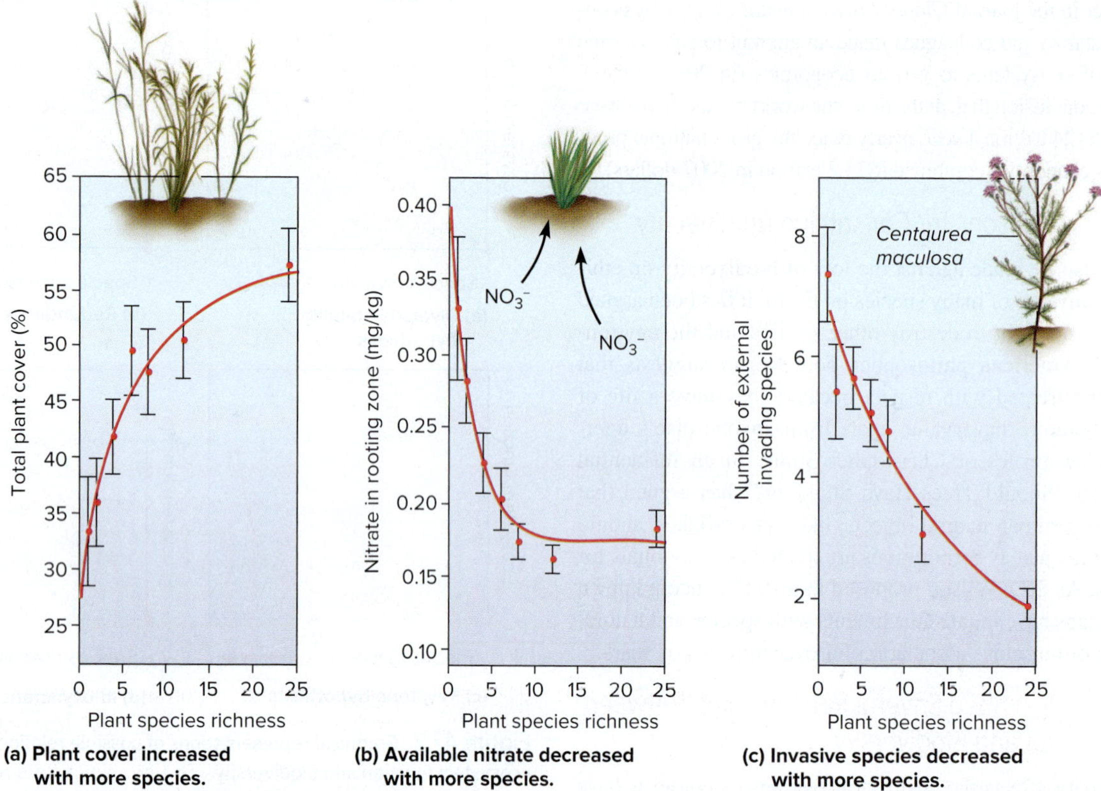


**Figure 47.3** The relationship of species richness to ecosystem function.

This is also observed on a larger scale. The productivity of temperate forests on different continents is roughly the same, despite different numbers of tree species being present—729 in East Asia, 253 in North America, and 124 in Europe. The presence of more tree species may ensure a supply of "backups," should some of the most-productive species die off from insect attack or disease. This can happen, as was seen in the demise of the American chestnut tree. Diseases devastated this species, and its presence in forests dramatically decreased by the mid-20th century (refer back to Figure 46.33). The forests filled in with other species and continued to function as before in terms of nutrient cycling and gas exchange. However, although the forests continued to function without the American chestnuts, some important changes occurred. For example, the loss of chestnuts deprived bears and other animals of an important source of food and may have affected their reproductive capacity and hence the size of their populations.

## 47.2 Reviewing the Concepts

- The preservation of biodiversity has been justified because of its economic value, because of the value of ecosystem services, and on ethical grounds (Table 47.1).
- Four models describe the relationship between biodiversity and ecosystem function: the diversity-stability, redundancy, keystone, and idiosyncratic hypotheses (Figure 47.2).
- Field experiments have shown that increased biodiversity results in increased ecosystem function and support the redundancy hypothesis (Figure 47.3).

## 47.2 Testing Your Knowledge

1. The idea that there is a linear correlation between diversity and ecosystem function is known as the __________ hypothesis.
   a. keystone
   b. diversity-stability
   c. linear
   d. redundancy
   e. idiosyncratic
2. Experimental evidence suggests that
   a. ecosystem function levels off at fairly low levels of diversity.
   b. ecosystem function plummets as soon as biodiversity decreases from natural levels.
   c. there is a linear correlation between diversity and ecosystem function.
   d. there is an unpredictable relationship between diversity and ecosystem function.
   e. there is no relationship between diversity and ecosystem function.

## 47.3 Conservation Strategies

### Learning Outcomes

1. Detail the criteria that conservation biologists use to target areas for protection.
2. Explain how the principles of island biogeography and landscape ecology are used to create nature preserves.
3. Describe different approaches conservation biologists use to protect individual species.
4. Define restoration ecology and the approaches used to restore degraded ecosystems and populations of species.

In their efforts to maintain the diversity of life on Earth, conservation biologists are active on many fronts. How do they decide on which areas or species to focus and which strategies to employ to protect diversity? We begin this section by discussing how conservation biologists identify the global habitats most in need of conservation. Next, we explore the concept of nature preserves and how they should be constructed. These issues are within the realm of landscape ecology, which studies the spatial arrangement of communities and ecosystems in a geographic area. Next we discuss how conservation efforts often focus on certain species that can have a disproportionate influence on their ecosystem. We will also examine the field of restoration ecology, studying how wildlife habitats can be established from degraded areas and how captive breeding programs have been used to reestablish populations of threatened species in the wild.

### Conservation Seeks to Establish Protected Areas

Currently, about 12.85% of global land area is under some form of environmental protection. More than 147,000 separate areas are protected, with more being added daily. Conservation biologists often must make decisions regarding which habitats should be protected. Many conservation efforts have focused on saving habitats in so-called megadiversity countries, because they often have the greatest number of species. However, more recent strategies have promoted preservation of certain key areas with the highest levels of unique species or the preservation of representative areas of all types of habitat, even relatively species-poor areas.

**Megadiversity Countries** One method of targeting areas for conservation is to identify **megadiversity countries,** those countries with the greatest numbers of species. Using the number of plants, vertebrates, and selected groups of insects as criteria, American biologist Russell Mittermeier and colleagues determined that just 17 countries are home to nearly 70% of all known species. Brazil, Indonesia, and Colombia top the list, followed by Australia, Peru, Mexico, Madagascar, China, and nine other countries (Figure 47.4a). The megadiversity country approach suggests that conservation efforts should be focused on the most biologically rich countries. However, although megadiversity areas may contain the most species, they do not necessarily contain the greatest number of unique species. The mammal species list for Peru is 344, and for Ecuador, it is 271; of these, however, 208 species are common to both.

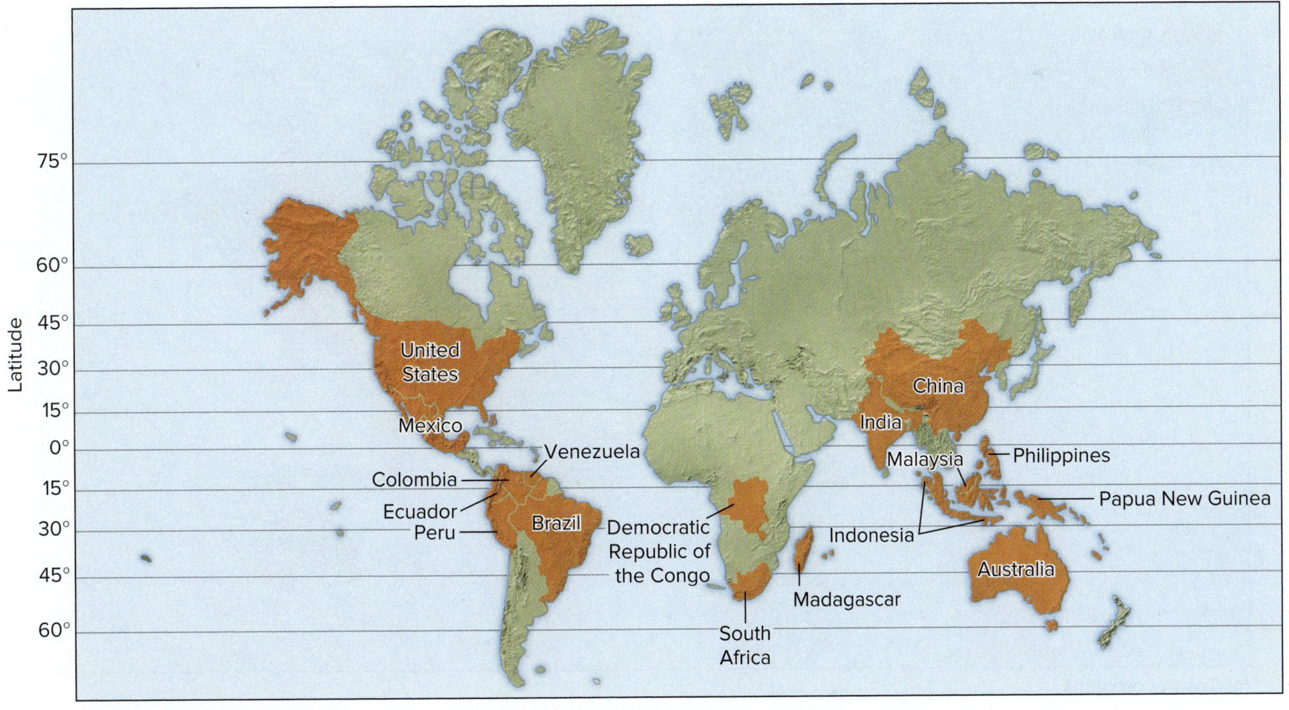


(a) Megadiversity countries

**Figure 47.4 Maps of four global biodiversity conservation priority approaches.** **(a)** Megadiversity countries. The top 17 countries are shown in orange. **(b)** Biodiversity hot spots. Hot spots have high numbers of endemic species. Different colors distinguish biodiversity hot spots. **(c)** Crisis ecoregions. **(d)** "Last of the wild," which are shown in orange.

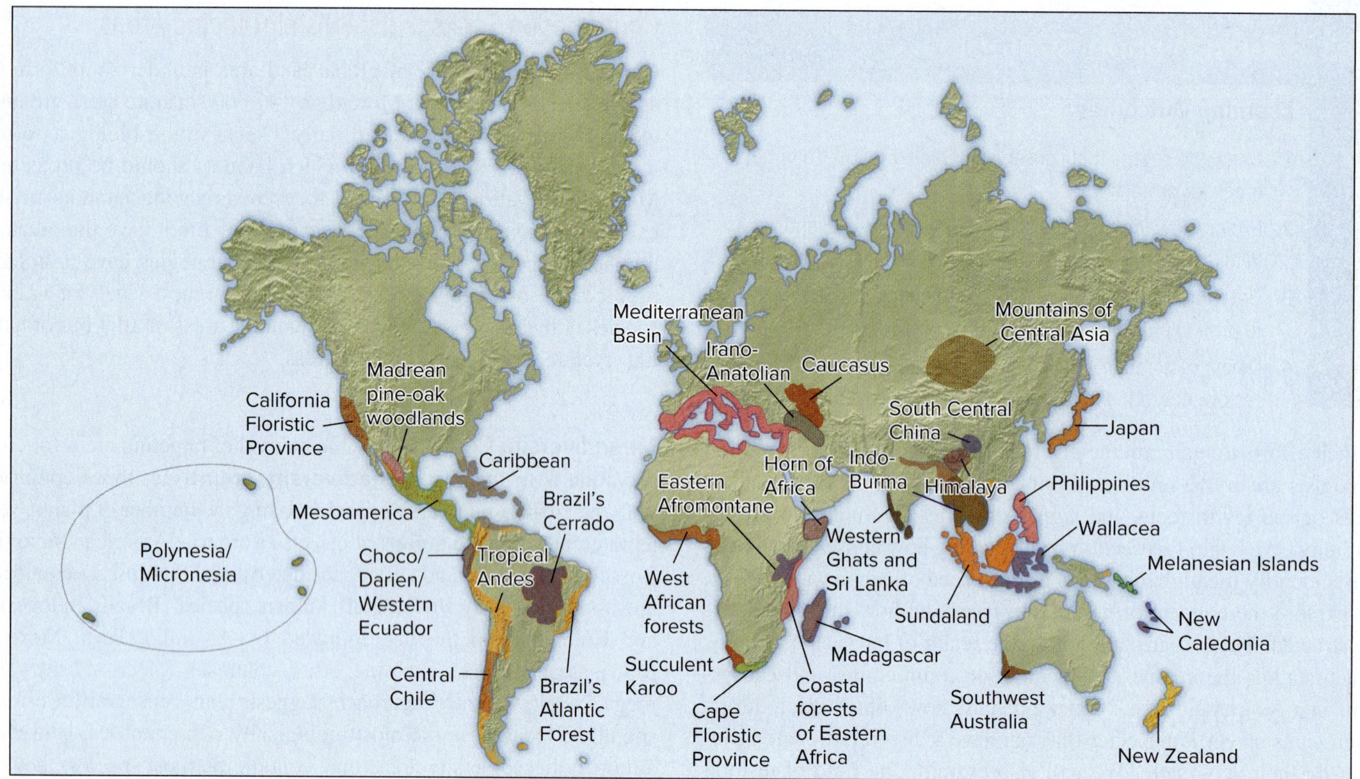


**(b) Biodiversity hot spots**

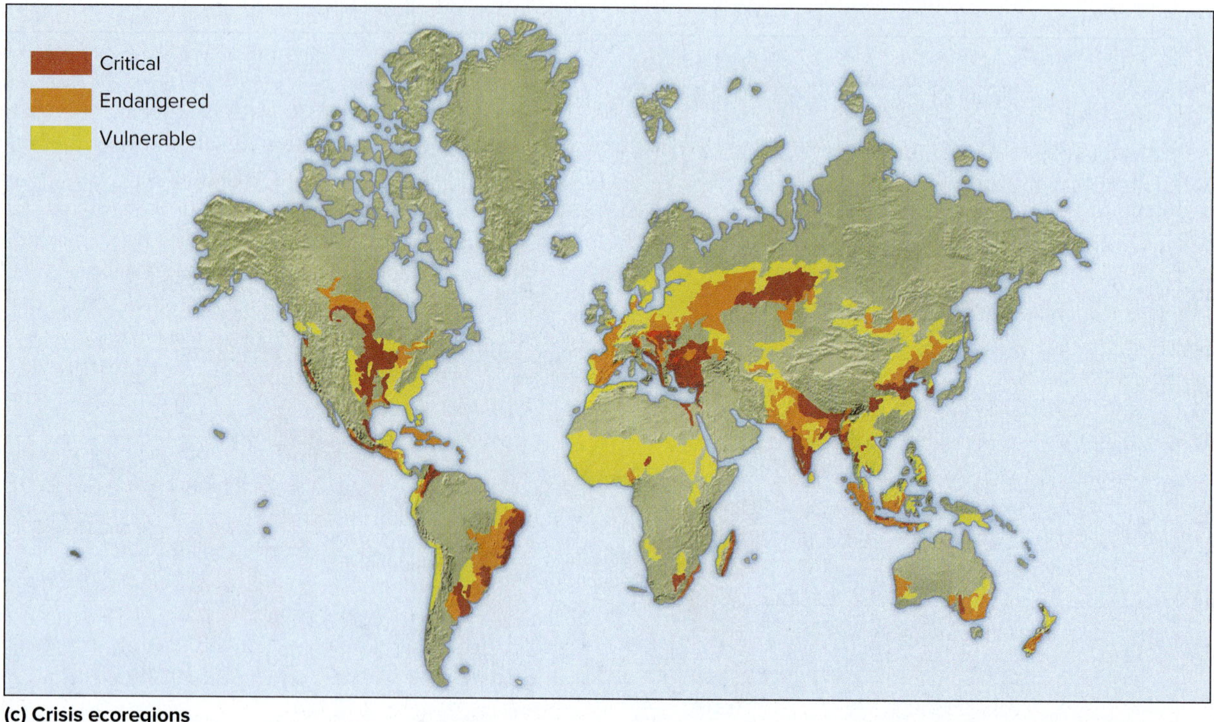


**(c) Crisis ecoregions**

**Figure 47.4** *(Continued).*

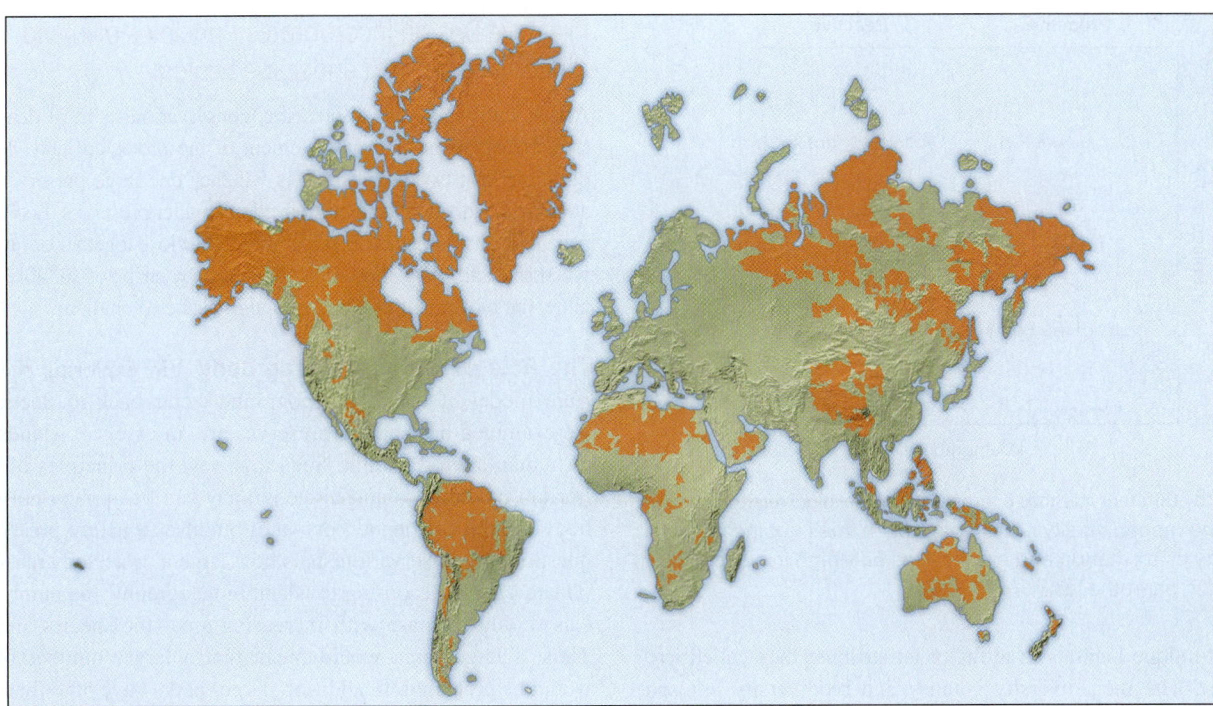

(d) "Last of the wild"

**Figure 47.4** *(Continued).*

**Areas Rich in Endemic Species** Another method of setting conservation priorities—one adopted by the organization Conservation International—takes into account the number of species that are **endemic;** that is, they are found only in a particular place or region. This approach suggests that conservationists focus their efforts on **biodiversity hot spots,** regions that are biologically diverse and under threat of destruction. To qualify as a biodiversity hot spot, a region must meet two criteria: (1) It must contain at least 1,500 species of vascular plants as endemic species and (2) it must have lost at least 70% of its original habitat. Vascular plants were chosen as the primary group of organisms to determine whether or not an area qualifies as a hot spot, mainly because most other terrestrial organisms depend on them to some extent.

Conservationists Norman Myers, Russell Mittermeier, and colleagues identified 34 hot spots that together occupy a mere 2.3% of the Earth's surface but contain 150,000 endemic plant species, or 50% of the world's total (Figure 47.4b). This approach proposes that protecting geographic hot spots will prevent the extinction of a larger number of endemic species than would protecting areas of a similar size elsewhere. The main argument against using hot spots as the criterion for targeting conservation efforts is that the areas richest in endemic species—tropical forests—would receive the majority of attention and funding, perhaps at the expense of protecting other areas.

**Representative Habitats** In a third approach to prioritizing areas for conservation, scientists have recently argued that we need to conserve representatives of all major habitats. Prairies are a case in point. An example is the Pampas region of South America, which is arguably the most threatened habitat on the continent because of rapid conversion of its natural grasslands to ranch land and agriculture. The Pampas does not compare well in richness or endemics with the rain forests, but it is a unique area that without preservation could disappear. By selecting habitats that are most distinct from those already preserved, many areas that are threatened but not biologically rich may be preserved in addition to the less immediately threatened, but richer, tropical forests.

Within these areas, some regions were seen as more critical than others. Habitat loss has been most extensive in temperate grasslands and tropical and temperate deciduous forests, where, in each case, over 50% of the land has been converted to other uses. At the same time, very little, less than 10%, of these agriculturally rich areas has been protected. Such habitats are thus termed crisis ecoregions (Figure 47.4c). Within these crisis ecoregions, designations vary from vulnerable, with conversion rates to other land uses of greater than 20%; endangered, with conversion rates of greater than 40%; and critical, with conversion rates of greater than 50%.

**Last of the Wild** The final approach to conservation involves preservation of regions of the world relatively untouched by humans. Scientists have mapped out the extent of the human footprint on the globe. The areas of the Earth that fall within the lowest 10% of the human-affected areas have been termed the "last of the wild" (Figure 47.4d). Such areas, because they are relatively pristine, offer a great opportunity for conservationists because of their relatively intact communities. Such areas include tundra and boreal forests of Russia and Canada as well as some desert biomes and tropical forests.

**Comparing the Four Conservation Approaches** In 2006, conservationists T.M. Brooks and colleagues contrasted these four conservation strategies (Figure 47.5). They suggested that many conservation strategies weigh the value of endemic species, taxonomically unique

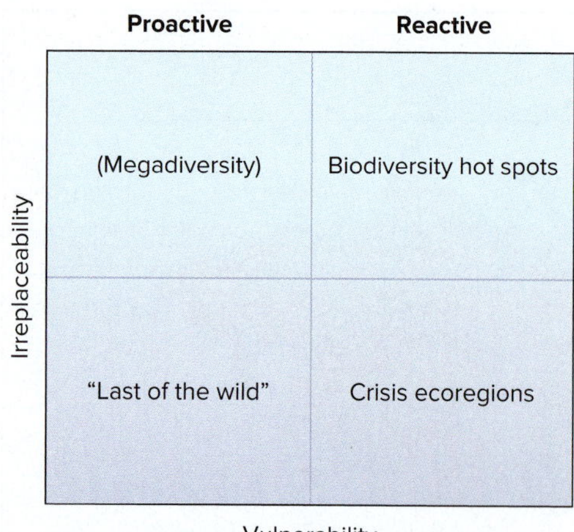


**Figure 47.5 Different methods of preserving global biodiversity.** Because the megadiversity country approach does not incorporate vulnerability as a criterion and emphasizes only high irreplaceability, its name is in parentheses.

species, or unique habitats in terms of an attribute they called irreplaceability. The megadiversity country and biodiversity hot spot approaches rate high on the attribute of irreplaceability. Other strategies place greater value on the actual vulnerability of biomes that are currently being threatened. This includes the crisis ecoregion and biodiversity hot spot approaches.

They also ranked the strategies in terms of being proactive or reactive in nature. The "last of the wild" strategy is the only proactive approach to conservation, preserving wild areas before they have been severely impacted. In contrast, the other three approaches are reactive, seeking to preserve areas that have already been much impacted. The best strategy of identifying areas for conservation efforts might be one that seeks to preserve a combination of different areas containing those with high species richness, large numbers of endemic species, various representative habitat types, and relatively untouched areas.

## Preserve Design Incorporates Principles of Island Biogeography and Landscape Ecology

After identifying areas to preserve, conservationists must determine the size, arrangement, and management of the protected land. Among the questions conservationists ask is whether one large preserve is preferable to an equivalent area composed of smaller preserves. Ecologists also need to determine whether parks should be close together or far apart and whether or not they should be connected by strips of suitable habitat to allow the movement of plants and animals between them.

**The Role of Island Biogeography** In exploring the equilibrium model of island biogeography (refer back to Section 44.4), we examined how nature preserves are, in essence, islands in a sea of human-altered habitat. Seen this way, the principles of the equilibrium model of island biogeography can be applied not only to a body of land surrounded by water but also to nature preserves. One question for conservationists is how large a protected area should be (Figure 47.6a). According to island biogeography, the number of species should increase with increasing area (the species-area effect). Thus, a larger area would mean that a larger number of species would be protected. In addition, larger parks have other benefits. For example, they are beneficial for organisms that require large spaces, including migrating species and species with extensive territories, such as lions and tigers.

A related question is whether it is preferable to protect a single, large preserve or several smaller ones (Figure 47.6b). This is called the **SLOSS debate** (for single large or several small). Proponents of the single, large preserve claim that a larger preserve is better able to protect more and larger populations than an equal area divided into small areas. According to island biogeography, a larger block of habitat should support more species than several smaller blocks.

However, many empirical studies suggest that multiple small sites of equivalent area will contain more species, because a series of small sites is more likely to contain a broader variety of habitats

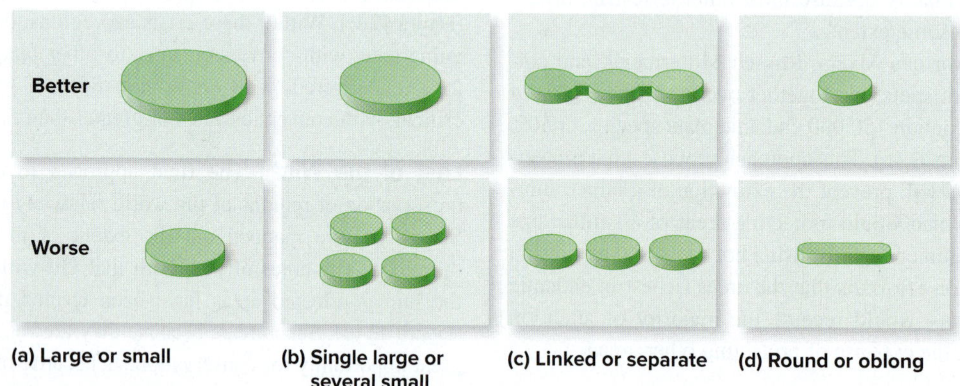


**Figure 47.6 The theoretical design of nature preserves.** **(a)** A larger preserve holds more species and has low extinction rates. **(b)** A given area should be fragmented into as few pieces as possible. **(c)** Maintaining or creating corridors between fragments may also enhance dispersal. **(d)** Circular-shaped areas minimize the amount of edge effects. The labels better and worse refer to theoretical principles generated by the equilibrium model of island biogeography, but empirical data have not supported all the predictions.

Concept Check: *What are some of the potential risks in connecting areas via movement corridors?*

than one large site. Looking at a variety of sites, American conservationists Jim Quinn and Susan Harrison concluded that animal life was richer in collections of small parks than in a smaller number of larger parks. In their study, having more habitat types outweighed the effect of area on biodiversity. In addition, another benefit of a series of smaller parks is a reduction of extinction risk by a single event such as a wildfire or the spread of disease.

**The Role of Landscape Ecology** Landscape ecology is an area of ecology that examines the spatial arrangement of communities and ecosystems in a geographic area. Landscape ecologists have suggested that small preserves should be linked together by **movement corridors,** thin strips of land that permit the movement of species between patches (Figure 47.6c). Such corridors ideally facilitate movements of organisms that are vulnerable to predation outside their natural habitat or have poor powers of dispersal between habitat patches. In this way, if a population in one small preserve experiences a disaster, immigrants from neighboring populations can more easily recolonize it. This avoids the need for humans to physically move new plants or animals into an area.

Several types of habitat function as movement corridors, including hedgerows in Europe, which facilitate movement and dispersal of species between forest fragments (Figure 47.7). In China, corridors of habitat have been established to link small, adjacent populations of giant pandas. However, disadvantages are associated with movement corridors. Corridors also can facilitate the spread of disease, invasive species, and fire between small preserves.

**Figure 47.7** **Movement corridors.**
© Andrew Parker/Alamy

**Concept Check:** *How do these European hedgerows act as movement corridors?*

Finally, parks are often designed to minimize **edge effects,** the special physical conditions that exist at the boundaries, or edges, of ecosystems. Habitat edges, particularly those between a natural habitat such as a forest and developed land, are often different in physical characteristics from the habitat core. For example, the center of a forest is shaded by trees and has less wind and light than the forest edge, which is unprotected. Many forest-adapted species therefore shy away from forest edges and prefer forest centers. For this reason, circular parks are generally preferable to oblong parks, because the amount of edge is minimized (Figure 47.6d).

## The Single-Species Approach Focuses Conservation Efforts on Particular Types of Species

The single-species approach to conservation focuses on saving species that are deemed particularly important. As with habitat conservation, different approaches are used to identify which species to focus effort on.

**Indicator Species** Some conservation biologists have suggested that certain organisms can be used as **indicator species,** those species whose status provides information on the overall health of an ecosystem. For example, corals are good indicators of marine processes such as siltation—the accumulation of sediments transported by water. Because siltation reduces the availability of light, the abundance of many marine organisms decreases in such situations, with corals among the first to display a decline in health. Coral bleaching is also an indicator of climate change.

Polar bears (*Ursus maritimus*), shown on the textbook cover and described in Chapter 1, are thought to be a mammalian indicator species of global climate change (Figure 47.8a). Most scientists are in agreement that global warming is causing the ice in the Arctic to melt earlier in the spring than in the past. Because polar bears rely on the ice to hunt for seals, the earlier breakup of the ice is leaving the bears less time to feed and build the fat that enables them to sustain themselves and their young. A U.S. Geological Survey study concluded that future reduction of arctic ice could result in a loss of two-thirds of the world's polar bear population within 50 years. In May 2008, the polar bear was listed as a threatened species under the U.S. Endangered Species Act (ESA).

**Umbrella Species** **Umbrella species** are those whose habitat requirements are so large that protecting them would protect many other species existing in the same habitat. The Northern spotted owl (*Strix occidentalis*) of the Pacific Northwest is considered to be an important umbrella species (Figure 47.8b). A pair of birds needs at least 800 hectares of old-growth forest for survival and reproduction, so maintaining healthy owl populations is thought to help ensure survival of many other forest-dwelling species. In the southeast area of the U.S., the red-cockaded woodpecker (*Picoides borealis*) is often seen as the equivalent of the spotted owl because it requires large tracts of old-growth long-leaf pine (*Pinus palustris*), including old diseased trees in which it can excavate its nests.

**Flagship Species** In the past, conservation resources were often allocated to a **flagship species,** a single large or instantly recognizable species. Such species were typically chosen because they were attractive and thus more readily engendered support from the public

(a) Indicator species: Polar bear

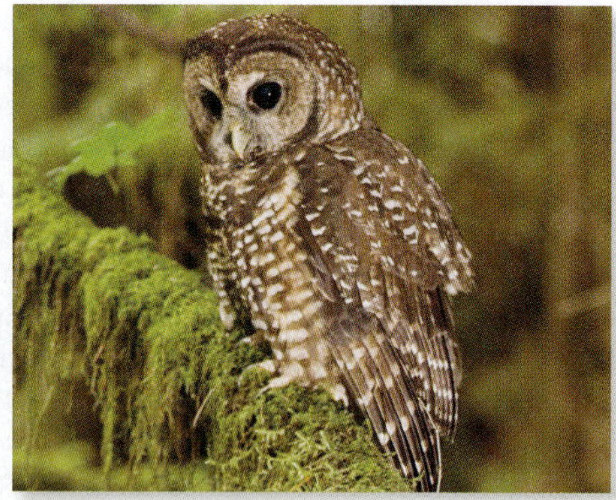

(b) Umbrella species: Northern spotted owl

(c) Flagship species: Florida panther

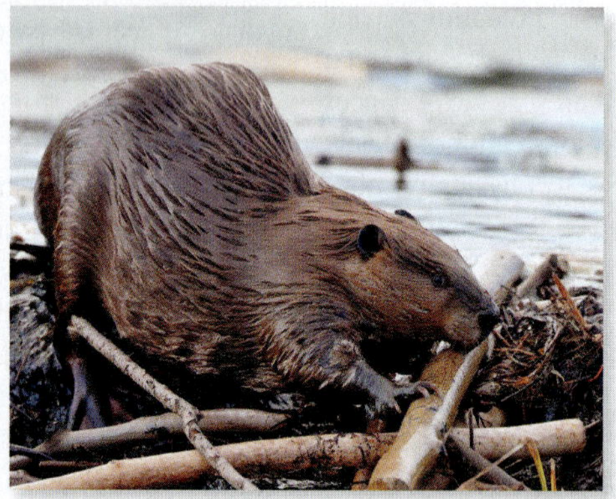

(d) Keystone species: American beaver

**Figure 47.8 Indicator, umbrella, flagship, and keystone species. (a)** Polar bears have been called an indicator species of global climate change. **(b)** The Northern spotted owl is considered an umbrella species for the old-growth forest in the Pacific Northwest. **(c)** The Florida panther has become a flagship species for Florida. **(d)** The American beaver, a keystone species, creates large dams across streams, and the resultant lakes provide habitats for a great diversity of species.

*(a)* © Robert E. Barber/Alamy; *(b)* © Rick A. Brown/rick@moosephoto.com; *(c)* © T. Kitchin & V. Hurst/NHPA/Photoshot; *(d)* © Robert McGouey/Wildlife/Alamy

for their conservation. The concept of the flagship species, typically a charismatic vertebrate such as the American buffalo (*Bison bison*), has often been used to raise awareness for conservation in general. The giant panda (*Ailuropoda melanoleuca*) is the World Wildlife Fund's emblem for endangered species, and the Florida panther (*Puma concolor*) has become a symbol of the state's conservation campaign (Figure 47.8c).

**Keystone Species** A different conservation strategy focuses on **keystone species,** species within a community that have a role out of proportion to their abundance or biomass. The beaver, a relatively small animal, can completely alter the composition of a community by building a dam and flooding an entire river valley (Figure 47.8d). The resultant lake may become a home to fish species, wildfowl, and aquatic vegetation. A decline in the number of beavers could have serious ramifications for the remaining community members, promoting fish die-offs, waterfowl loss, and the death of vegetation adapted to waterlogged soil.

American ecologist John Terborgh considers tropical palm nuts and figs to be keystone plant species because they produce fruit during otherwise fruitless times of the year and are thus critical resources for tropical forest fruit-eating animals, including primates, rodents, and many birds. Together, these fruit eaters account for as much as three-quarters of the tropical forest animal biomass. Without the fruit trees, widespread extinction of these animals could occur.

## Restoration Ecology Attempts to Rehabilitate Degraded Ecosystems and Populations

Another approach to conservation is to rehabilitate previously degraded habitat. **Restoration ecology** is the full or partial repair or replacement of biological habitats and/or their populations that have been degraded or destroyed. It can focus on restoring or rehabilitating a habitat, or it can involve returning species to the wild following captive breeding.

**Habitat Restoration** The three basic approaches to habitat restoration are complete restoration, rehabilitation, and ecosystem replacement. In complete restoration, conservationists attempt to return a habitat to its composition and condition prior to the disturbance. Under the leadership of American ecologist Aldo Leopold, the University of Wisconsin pioneered the restoration of prairie habitats as early as 1935, converting agricultural land back to species-rich prairies (Figure 47.9a).

The second approach aims to return the habitat to something similar to, but a little less than, full restoration, a goal called rehabilitation. In Florida, phosphate mining involves removing a layer of topsoil, mining the phosphate-rich layers, returning the topsoil, and then replanting the area. Unfortunately, species such as cogongrass (*Imperata cylindrica*), an invasive Southeast Asian species, may invade these disturbed areas. Even so, the restoration serves to revegetate the area (Figure 47.9b).

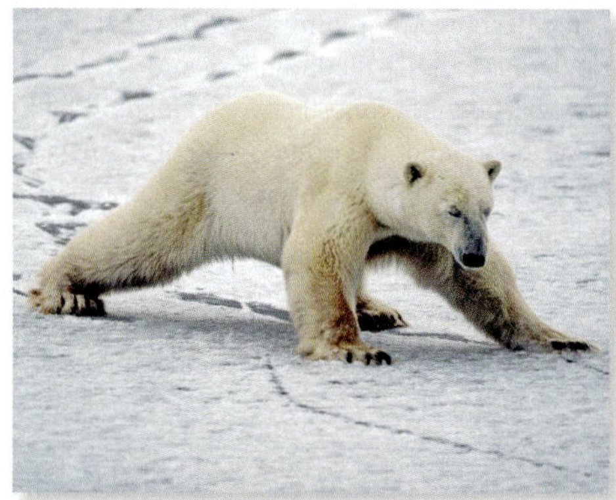

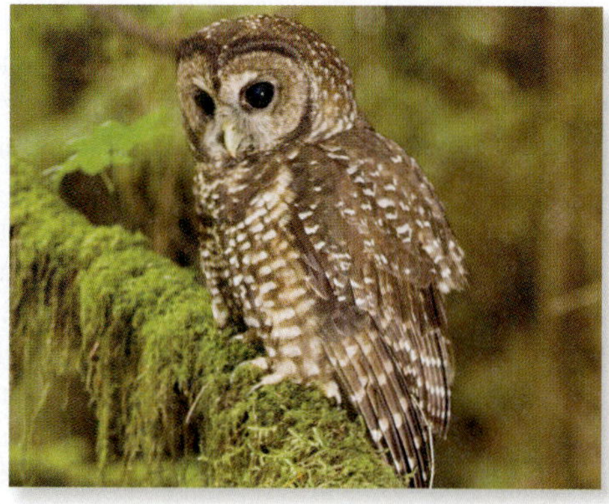

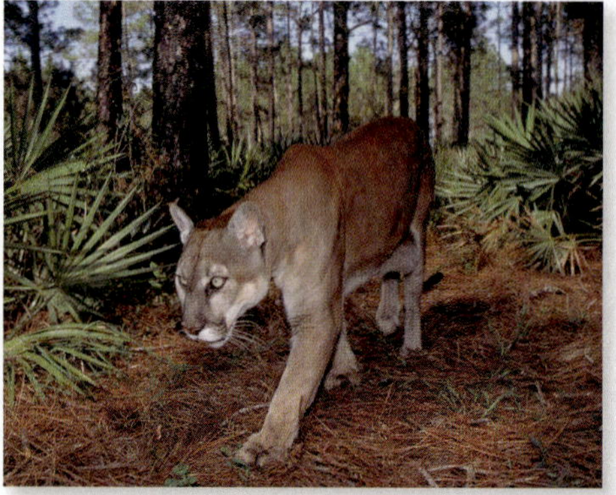

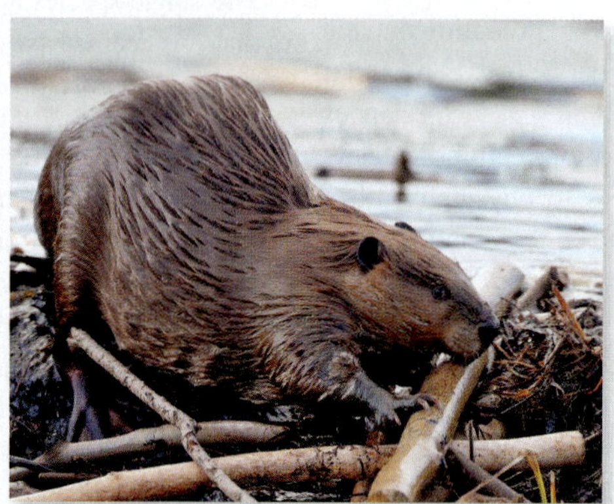

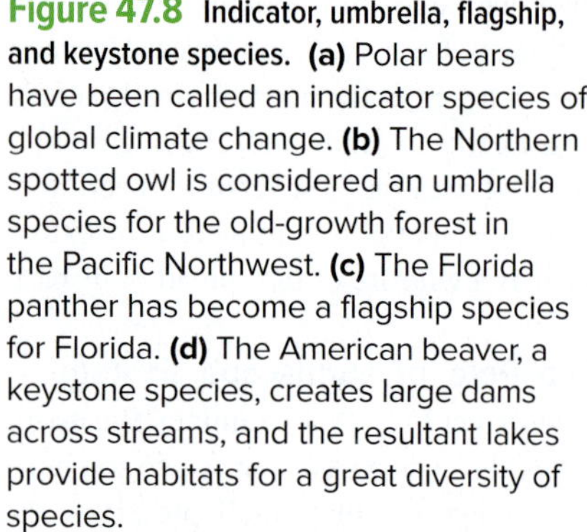

## Biology Principle

### Biology Affects Our Society

Restoration of human-degraded habitats can lead to recovery of habitat and increased biodiversity.

**(a) Complete restoration**

**(b) Rehabilitation**

**(c) Ecosystem replacement**

**Figure 47.9** **Habitat restoration.** **(a)** The University of Wisconsin pioneered the practice of complete restoration of agricultural land to native prairies. **(b)** In Florida, complete restoration after phosphate mining is not usually possible. After topsoil is replaced, invasive species such as cogongrass often grow, resulting in habitat rehabilitation rather than complete restoration. **(c)** These old open-pit mines in Middlesex, England, have been converted to valuable freshwater habitats, replacing the wooded ecosystem that was originally present.

*(a)* © University of Wisconsin-Madison Arboretum and University Boards of Regents; *(b)* Courtesy of D.L. Rockwood, School of Forest Resources and Conservation, University of Florida, Gainesville, FL; *(c)* © Sally A. Morgan/Ecoscene/Corbis

The third approach, termed replacement, makes no attempt to restore what was originally present but instead replaces the original ecosystem with a different one. Ecosystem replacement is particularly useful for places in which the terrain has been substantially altered by past human activities. It would be nearly impossible to re-create the original landscape of an area that was mined for stone or gravel. In these situations, however, wetlands or lakes may be created in the open pits (Figure 47.9c).

**Reintroductions and Captive Breeding** Reintroducing species to areas where they previously existed is a valuable conservation strategy that can reestablish populations into areas where they once occurred (see Chapter 43 opening photograph). Captive breeding, the propagation of animals and plants outside their natural habitat to produce stock for subsequent release into the wild, has proved valuable in reestablishing breeding populations following extinction or near extinction. Zoos, aquariums, and botanical gardens often play a key role in captive breeding, propagating species that are highly threatened in the wild.

Several classic programs illustrate the value of reintroduction and captive breeding. The peregrine falcon (*Falco peregrinus*) had become extinct in nearly all of the eastern U.S. by the mid-1940s, a decline that was linked to the effects of DDT (refer back to Figure 46.17). In 1970, American biologist Tom Cade gathered falcons from other parts of the country to start a captive breeding program at Cornell University. Since then, the program has released thousands of birds into the wild, and in 1999, the peregrine falcon was removed from the list of endangered species. A captive breeding program is also helping to save the California condor (*Gymnogyps californicus*) from extinction. At a cost of $35 million, this is the most expensive species conservation project ever undertaken in the U.S. In the 1980s, there were only 22 known condors, some in captivity and some in the wild. Scientists made the decision to capture the remaining wild birds in order to protect and breed them (Figure 47.10a). By 2016, the captive population numbered 167 individuals, and 268 birds were living in the coastal mountains of California; northern Baja California, Mexico; the Grand Canyon area of Arizona; and Zion National Park, Utah (Figure 47.10b). A milestone was reached in 2003, when a pair of captive-reared California condors bred in the wild.

Because the number of individuals in any captive breeding program is initially small, care must be exercised to avoid inbreeding. Matings are usually carefully arranged to maximize resultant genetic variation in offspring. The use of genetic engineering to clone endangered species is a new area that may eventually help bolster populations of captive-bred species.

## Can Cloning Save Endangered Species?

In 1997, geneticist Ian Wilmut and colleagues at Scotland's Roslin Institute announced to the world that they had cloned a now-famous sheep, Dolly, from mammary cells of an adult ewe. Since then, interest has arisen among conservation biologists about whether the same technology might be used to save species on the verge of extinction. Scientists were encouraged that in January 2001, an Iowa farm cow called Bessie gave birth to a cloned Asian gaur (*Bos gaurus*), an endangered species. The gaur, an oxlike animal native to the jungles of India and Burma, was cloned from a single skin cell taken from a dead animal. To clone the gaur, scientists removed the nucleus from a cow's egg and replaced it with a nucleus from the gaur's cell. The treated egg was then placed into the cow's uterus. Unfortunately, the gaur died

(a) A condor chick being fed using a puppet

(b) A released captive-bred condor

**Figure 47.10** **Captive breeding programs.** The California condor (*Gymnogyps californicus*), the largest bird in the U.S., with a wingspan of nearly 3 m, has been bred in captivity in California. **(a)** A researcher at the San Diego Wild Animal Park feeds a chick with a puppet so that the birds will not become habituated to the presence of humans. **(b)** This captive-bred condor soars over the Grand Canyon. Note the tag on the underside of its wing.
*(a)* © Corbis; *(b)* © Simpson/Photri Images/Alamy

from dysentery 2 days after birth, although scientists believe this was unrelated to the cloning procedure. In 2003, another type of endangered wild cattle, the Javan banteng (*Bos javanicus*), was successfully cloned (**Figure 47.11**). In 2005, clones of the African wildcat (*Felis libyca*) successfully produced wildcat kittens. This is the first time that clones of a wild species have bred. In 2009, a cloned Pyrenean ibex (*Capra pyrenaica*) was born but lived only 7 minutes due to physical defects in the lungs. The last wild Pyrenean ibex had died in Spain in 2000. Other candidates for cloning include the Sumatran tiger (*Panthera tigris*) and the giant panda. Brazil plans to clone eight of its endangered species. Cloning extinct animals such as the woolly mammoth (*Mammuthus primigenius*) or Tasmanian tiger (*Thylacinus cynocephalus*) would be more difficult due to a lack of fully preserved DNA.

## Biology Principle

### The Genetic Material Provides a Blueprint for Reproduction

Genetic cloning utilizes a somatic cell, which contains the complete DNA or genetic blueprint of the animal that is to be cloned.

**Figure 47.11** **Cloning an endangered species.** In 2004, this 8-month-old cloned Javan banteng (*Bos javanicus*) made its public debut at the San Diego Zoo.
© Ken Bohn/UPI/Newscom

**BioConnections:** *Livestock and even pets have been cloned as well as endangered species (refer back to the chapter-opening photograph of Chapter 18). What are some of the arguments in favor and against genetic cloning of species?*

Despite the promise of cloning, a number of issues remain unresolved:

1. Scientists would have to develop an intimate knowledge of different species' reproductive cycles. For sheep and cows, this was routine, based on the vast experience in breeding these species, but eggs of different species, even if they could be harvested, often require different nutritive media in laboratory cultures.
2. Because it is desirable to use natural mothers available for breeding, scientists will have to identify surrogate females of similar but more common species that can carry the fetus to term.
3. Some argue that cloning does not address the root causes of species loss, such as habitat fragmentation or poaching, and that resources would be better spent elsewhere—for example, in preserving the remaining habitat of endangered species.
4. Cloning might not be able to increase the genetic variability of the population. However, if it were possible to use cells from deceased animals—for example, from their skin—these clones could theoretically reintroduce lost genes back into the population.

Many biologists believe that while cloning may have a role in conservation, it is only part of the solution and we need to address what made the species go extinct before attempting to restore it.

Conservation is clearly a matter of great importance, and a failure to value and protect our natural resources adequately could be a grave mistake. Some authors, most recently the American ecologist and geographer Jared Diamond, have investigated why many societies of the past—including Angkor Wat, Easter Island, and the Mayans—collapsed or vanished, leaving behind monumental ruins. Diamond has concluded that the collapse of these societies occurred partly because people inadvertently destroyed the ecological resources on which their societies depended. Modern nations such as Rwanda face similar issues. The country's population density is the highest in Africa, and it has a limited amount of land that can be used for growing crops. By the late 1980s, the need to feed a growing population had led to the wholesale clearing of Rwanda's forests and wetlands, with the result that little additional land was available to farm. Increased population pressure, along with food shortages fueled by environmental scarcity, were likely contributing factors in igniting the genocide of 1994.

As we hope you have seen throughout this textbook, an understanding of biology is vital to learning and helping to solve many of society's problems. The study of biology has a huge potential for improving people's lives and society at large. Biology offers the opportunity to unlock new diagnoses and treatments for diseases, to improve nutrition and food production, and to maintain biological diversity.

## 47.3 Reviewing the Concepts

- Habitat conservation strategies commonly target megadiversity countries, countries with the largest number of species; biodiversity hot spots, areas with the largest number of endemic species, those unique to the area; representative habitats including crisis ecoregions, areas that represent the major habitats; and "last of the wild," pristine areas that have not been severely affected by human activities (Figures 47.4, 47.5).
- Conservation biologists employ many strategies in protecting biodiversity. Principles of the equilibrium model of island biogeography and landscape ecology are used in the theory and practice of park preserve design (Figure 47.6, 47.7).
- The single-species approach focuses conservation efforts on indicator species, umbrella species, flagship species, and keystone species (Figure 47.8).
- Restoration ecology seeks to repair or replace populations and their habitats. Three basic approaches to habitat restoration are complete restoration, rehabilitation, and ecosystem replacement (Figure 47.9).
- Captive breeding is the propagation of animals outside their natural habitat and reintroducing them to the wild. Cloning of endangered species has been accomplished on a very small scale and despite its limitations may have a role in conservation biology (Figures 47.10, 47.11).

## 47.3 Testing Your Knowledge

**1.** According to the equilibrium model of island biogeography, which type of park would contain most species?
   **a.** one large park
   **b.** several small parks of combined area equal to a large park
   **c.** several small parks connected by a movement corridor
   **d.** a circular park
   **e.** all of the above

**2.** Over time, dark forms of the peppered moth (*Biston betularia*) became more common in polluted environments because predators were less able to detect them on trees darkened by soot. These moths are regarded by many as
   **a.** keystone species.
   **b.** indicator species.
   **c.** umbrella species.
   **d.** flagship species.
   **e.** endangered species.

**3.** After being used for mining, what was once a deciduous forest is replaced by grassland to be used for public recreation. This process is known as
   **a.** complete restoration.
   **b.** rehabilitation.
   **c.** ecosystem replacement.
   **d.** bioremediation.
   **e.** habitat repair.

# Assess and Discuss

## Test Yourself

1. Which of the following best describes an endangered species?
   a. a species that is likely to become extinct in a portion of its range
   b. a species that has disappeared in a particular community but is present in other natural environments
   c. a species that is extinct
   d. a species that is in danger of becoming extinct throughout all or a significant portion of its range
   e. both b and d
2. Biological diversity is important and should be preserved because
   a. food, medicines, and industrial products are all benefits of biodiversity.
   b. ecosystems provide valuable services to us in many ways.
   c. many species can be used as valuable research tools.
   d. we have an ethical responsibility to protect our environment.
   e. all of the above are correct.
3. The idea that humans have an intimate love of life, coined by E. O. Wilson, is known as
   a. biodiversity
   b. biophilia
   c. the call of the wild
   d. biotheology
   e. the "last of the wild"
4. The research conducted by Tilman and colleagues demonstrated that
   a. as diversity increases, productivity increases.
   b. as diversity decreases, productivity increases.
   c. areas with higher diversity demonstrate less efficient use of nutrients.
   d. species-richness increases lead to an increase in invasive species.
   e. increased diversity results in increased susceptibility to disease.
5. Which hypothesis best describes the idea that a small decline in species diversity results in a large drop in ecosystem function?
   a. diversity-stability
   b. redundancy
   c. keystone
   d. idiosyncractic
   e. community
6. The acronym SLOSS stands for
   a. several large or single small.
   b. single large or several small.

c. single large or single small.
d. several large or several small.
e. both c and d.

7. Saving endangered habitats, such as the Argentine pampas, focuses on
a. saving genetic diversity.
b. saving keystone species.
c. conservation in a megadiversity country.
d. preserving an area rich in endemic species.
e. preserving a representative habitat.

8. Biodiversity hot spots are those areas rich in
a. species.
b. habitats.
c. rare species.
d. biodiversity.
e. endemic species.

9. Small strips of land that connect and allow organisms to move between small patches of natural habitat are called
a. biological conduits.
b. edge effects.
c. movement corridors.
d. migration pathways.
e. landscape breaks.

10. The American buffalo, *Bison bison*, is considered a
a. keystone species.
b. flagship species.
c. umbrella species.
d. indicator species.
e. extinct species.

## Conceptual Questions

1. **PRINCIPLES** A principle of biology is that biology affects our society. What is the value of increased biodiversity for human society?
2. Distinguish between the following conservation strategies to conserve biodiversity: Megadiversity countries, biodiversity hot spots, crisis ecoregions, and "last of the wild."
3. Describe how the concepts of vulnerability, irreplaceability, and proactive and reactive strategies are useful in conservation.

## Collaborative Questions

1. Discuss the differences between indicator species, umbrella species, flagship species, and keystone species.
2. You are called upon to design a park to maximize biodiversity in a tropical country. What are your recommendations?

## Online Resource

**connect.mheducation.com**

SMARTBOOK® SmartBook® is the first and only adaptive reading experience designed to change the way students read and learn.

# APPENDIX A

## Periodic Table of the Elements

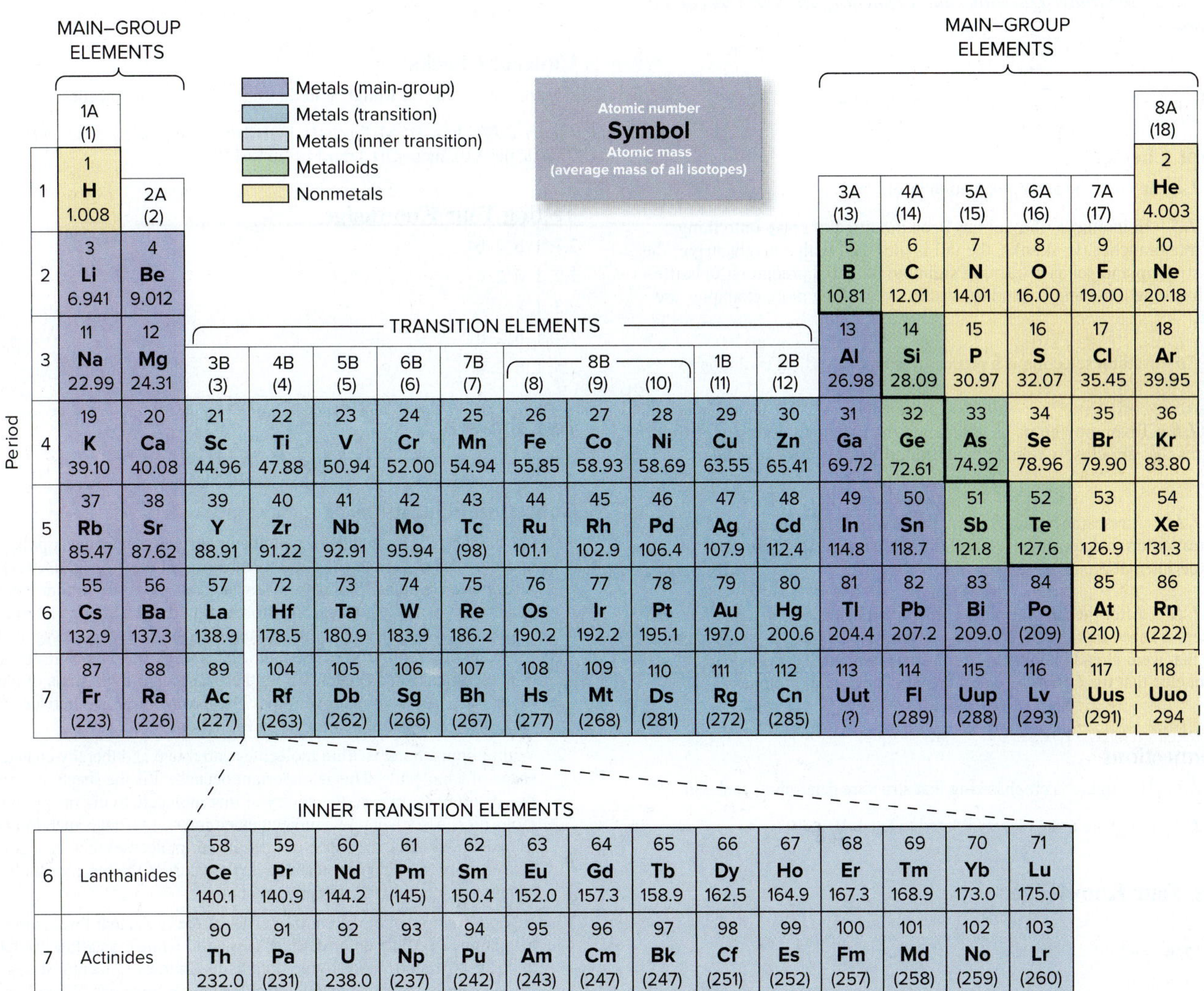


| Period | 1A (1) | 2A (2) | 3B (3) | 4B (4) | 5B (5) | 6B (6) | 7B (7) | 8B (8) | 8B (9) | 8B (10) | 1B (11) | 2B (12) | 3A (13) | 4A (14) | 5A (15) | 6A (16) | 7A (17) | 8A (18) |
|---|---|---|---|---|---|---|---|---|---|---|---|---|---|---|---|---|---|---|
| 1 | 1 **H** 1.008 | | | | | | | | | | | | | | | | | 2 **He** 4.003 |
| 2 | 3 **Li** 6.941 | 4 **Be** 9.012 | | | | | | | | | | | 5 **B** 10.81 | 6 **C** 12.01 | 7 **N** 14.01 | 8 **O** 16.00 | 9 **F** 19.00 | 10 **Ne** 20.18 |
| 3 | 11 **Na** 22.99 | 12 **Mg** 24.31 | | | | | | | | | | | 13 **Al** 26.98 | 14 **Si** 28.09 | 15 **P** 30.97 | 16 **S** 32.07 | 17 **Cl** 35.45 | 18 **Ar** 39.95 |
| 4 | 19 **K** 39.10 | 20 **Ca** 40.08 | 21 **Sc** 44.96 | 22 **Ti** 47.88 | 23 **V** 50.94 | 24 **Cr** 52.00 | 25 **Mn** 54.94 | 26 **Fe** 55.85 | 27 **Co** 58.93 | 28 **Ni** 58.69 | 29 **Cu** 63.55 | 30 **Zn** 65.41 | 31 **Ga** 69.72 | 32 **Ge** 72.61 | 33 **As** 74.92 | 34 **Se** 78.96 | 35 **Br** 79.90 | 36 **Kr** 83.80 |
| 5 | 37 **Rb** 85.47 | 38 **Sr** 87.62 | 39 **Y** 88.91 | 40 **Zr** 91.22 | 41 **Nb** 92.91 | 42 **Mo** 95.94 | 43 **Tc** (98) | 44 **Ru** 101.1 | 45 **Rh** 102.9 | 46 **Pd** 106.4 | 47 **Ag** 107.9 | 48 **Cd** 112.4 | 49 **In** 114.8 | 50 **Sn** 118.7 | 51 **Sb** 121.8 | 52 **Te** 127.6 | 53 **I** 126.9 | 54 **Xe** 131.3 |
| 6 | 55 **Cs** 132.9 | 56 **Ba** 137.3 | 57 **La** 138.9 | 72 **Hf** 178.5 | 73 **Ta** 180.9 | 74 **W** 183.9 | 75 **Re** 186.2 | 76 **Os** 190.2 | 77 **Ir** 192.2 | 78 **Pt** 195.1 | 79 **Au** 197.0 | 80 **Hg** 200.6 | 81 **Tl** 204.4 | 82 **Pb** 207.2 | 83 **Bi** 209.0 | 84 **Po** (209) | 85 **At** (210) | 86 **Rn** (222) |
| 7 | 87 **Fr** (223) | 88 **Ra** (226) | 89 **Ac** (227) | 104 **Rf** (263) | 105 **Db** (262) | 106 **Sg** (266) | 107 **Bh** (267) | 108 **Hs** (277) | 109 **Mt** (268) | 110 **Ds** (281) | 111 **Rg** (272) | 112 **Cn** (285) | 113 **Uut** (?) | 114 **Fl** (289) | 115 **Uup** (288) | 116 **Lv** (293) | 117 **Uus** (291) | 118 **Uuo** 294 |

INNER TRANSITION ELEMENTS

| | | | | | | | | | | | | | | | |
|---|---|---|---|---|---|---|---|---|---|---|---|---|---|---|---|
| 6 | Lanthanides | 58 **Ce** 140.1 | 59 **Pr** 140.9 | 60 **Nd** 144.2 | 61 **Pm** (145) | 62 **Sm** 150.4 | 63 **Eu** 152.0 | 64 **Gd** 157.3 | 65 **Tb** 158.9 | 66 **Dy** 162.5 | 67 **Ho** 164.9 | 68 **Er** 167.3 | 69 **Tm** 168.9 | 70 **Yb** 173.0 | 71 **Lu** 175.0 |
| 7 | Actinides | 90 **Th** 232.0 | 91 **Pa** (231) | 92 **U** 238.0 | 93 **Np** (237) | 94 **Pu** (242) | 95 **Am** (243) | 96 **Cm** (247) | 97 **Bk** (247) | 98 **Cf** (251) | 99 **Es** (252) | 100 **Fm** (257) | 101 **Md** (258) | 102 **No** (259) | 103 **Lr** (260) |

**The complete Periodic Table of the Elements.** In some cases, the average atomic mass has been rounded to one or two decimal places, and in others only an estimate is given in parentheses due to the short-lived nature or rarity of those elements. The symbols and names of some of the elements between 113–118 are temporary until the chemical characteristics of these elements become better defined.

# APPENDIX B

# Answers to In-Chapter and End-of-Chapter Questions

***Answers to Collaborative Questions and Crunching the Numbers can be found online.***

## Chapter 1

### Concept Checks

***Figure 1.4*** It would be at the population level.

***Figure 1.5*** In monkeys, the tail has been modified to grasp onto things, such as tree branches. In skunks, the tail is modified with a bright stripe; the tail can stick up and act as a warning signal to potential predators. In cattle, the tail has long hairs and is used to swat insects. Many more examples are possible.

***Figure 1.6*** Natural selection is a process that causes vertical evolution to happen.

***Figure 1.8*** Taxonomy helps us appreciate the unity and diversity of life. Organisms that are closely related evolutionarily are placed in smaller groups.

***Figure 1.11*** A researcher can compare the results in the experimental group and control group to determine if a single variable is causing a particular outcome in the experimental group.

***Figure 1.12*** After the *CFTR* gene was identified by discovery-based science, researchers realized that the *CFTR* gene was similar to other genes that encoded proteins that were already known to be transport proteins. This provided an important clue that the *CFTR* gene also encodes a transport protein.

### BioConnections

***Figure 1.3*** This figure is emphasizing that structure determines function.

***Figure 1.7*** Fungi are more closely related to animals.

### Testing Your Knowledge

1.1: 1. c 2. c

1.2: 1. c 2. b

1.3: 1. e 2. e 3. b

### Test Yourself

1. d 2. a 3. a 4. c 5. d 6. b 7. d 8. d 9. a 10. b

### Conceptual Questions

1. Principles (a) through (f) apply to individuals whereas (g) and (h) apply to populations.
2. The unity among different species occurs because modern species have evolved from a group of related ancestors. Some of the traits in those ancestors are also found in modern species, which thereby unites them. The diversity is due to the variety of environments on the Earth. Each species has evolved to occupy its own unique environment. For every species, many traits are evolutionary adaptations to survival in a specific environment. For this reason, evolution also promotes diversity.
3. The principles are outlined in Figure 1.3. Students can rephrase these principles in their own words.

## Chapter 2

### Concept Checks

***Figure 2.11*** The oil would be in the center of the soap micelles.

***Figure 2.14*** It is $10^{-6}$ M. Since $[H^+][OH^-]$ always equals $10^{-14}$ M, if $[H^+] = 10^{-8}$ M (that is, pH 8.0), then $[OH^-]$ must be $10^{-6}$ M.

### Testing Your Knowledge

2.1: 1. b 2. b

2.2: 1. d 2. c

2.3: 1. a

2.4: 1. b 2. d

2.5: 1. b 2. b

### Test Yourself

1. b 2. b 3. b 4. d 5. c 6. e 7. b 8. c 9. c 10. e

### Conceptual Questions

1. Covalent bonds are bonds in which atoms share electrons. A polar covalent bond is formed between two atoms of different electronegativities, whereas a nonpolar covalent bond is one between two atoms of similar electronegativities, such as two carbon atoms. A hydrogen bond is an interaction between a hydrogen and an electronegative atom, such as oxygen. The van der Waal dispersion forces are temporary, weak bonds, resulting from random electrical forces generated by the changing distributions of electrons in the outer shells of nearby atoms. The strong attraction between two oppositely charged atoms forms an ionic bond.
2. Within limits, bonds within molecules can rotate and thereby change the shape of a molecule. This is important because it is the shape of a molecule that determines, in part, the ability of that molecule to interact with other molecules. Also, when two molecules do interact through such forces as hydrogen bonds, the shape of one or both molecules may change as a consequence. The change in shape is often part of the mechanism by which signals are sent within and between cells.
3. A good example of emergent properties at the molecular level is that of the formation of sodium chloride (NaCl), a solid white crystalline compound that is very important for living organisms. In their elemental states, sodium is a soft, highly reactive metal and chlorine is a toxic gas. When they combine through ionic bonds, the two elements produce a completely new and harmless substance found in all the world's oceans and soils. Another example described in this chapter is water, a liquid that is vital for all life but which is formed from two gases, hydrogen and oxygen, with very different properties.

## Chapter 3

### Concept Checks

***Figure 3.5*** Recall that the reverse of a dehydration reaction is called a hydrolysis reaction, in which a molecule of water is added to the molecule being broken down, resulting in the formation of monomers.

***Figure 3.9*** The phospholipids would be oriented such that their polar regions dissolved in the water layer and the nonpolar regions dissolved in the oil. Thus, the phospholipids would form a layer at the interface between the water and oil.

*Figure 3.12* 71; one less than the number of amino acids in the polypeptide.

*Figure 3.18* Yes. The opposite strand must be complementary to the first strand, because pairs can form only between A and T, and G and C. For instance, if a portion of the first strand is AATGCA, the opposite strand along that region would be TTACGT.

### BioConnections

*Figure 3.6* Cellulose is believed to be the most abundant organic molecule on Earth. In addition to being part of plant cells, it is also found in many other organisms, including many protists.

### Feature Investigation Questions

1. Anfinsen was testing the hypothesis that the information necessary for determining the three-dimensional shape of a protein is contained within the protein itself. In other words, the chemical characteristics of the amino acids that make up a protein determines the three-dimensional shape.
2. The urea disrupts hydrogen bonds and ionic interactions that are necessary for protein folding. The mercaptoethanol disrupted the S—S bonds that also form between certain amino acids of the same polypeptide. Both substances cause the polypeptide to unfold, disrupting the three-dimensional shape.
3. Anfinsen removed the urea and mercaptoethanol from the protein solution by size-exclusion chromatography. After removing the urea and mercaptoethanol, Anfinsen discovered that the protein refolded into its proper three-dimensional shape and became functional again. This was important because the solution contained only the protein and lacked any other cellular material that could possibly assist in protein folding. This demonstrated that the protein could refold into the functional conformation without the need for any cellular factors.

### Testing Your Knowledge

3.1: 1. b 2. e

3.2: 1. e

3.3: 1. e

3.4: 1. e 2. c

3.5: 1. b 2. a

3.6: 1. b 2. d

3.7: 1. a 2. d

### Test Yourself

1. b 2. b 3. e 4. e 5. d 6. b 7. e 8. d 9. b 10. b

### Conceptual Questions

1. If the amino acid sequence of a protein is changed, as might occur in a mutation (as you will learn in later chapters), the arrangement of bonds between amino acids may change. This can create local regions of altered tertiary structure, and in some cases can significantly alter the shape of the entire protein.
2. Saturated fatty acids are saturated with hydrogen and have only single (C—C) bonds, whereas unsaturated fatty acids have one or more double (C=C) bonds. The double bonds in unsaturated fatty acids alters their shape, resulting in a kink in the structure. Saturated fatty acids are unkinked and are better able to stack tightly together. Fats containing saturated fatty acids have a higher melting point than those containing mostly unsaturated fatty acids; consequently, saturated fats tend to be solids at room temperatures, and unsaturated fatty acids are usually liquids at room temperature.
3. The structures of macromolecules in all cases determine their function. For example, the structure of a protein determines its three-dimensional shape. This, in turn, allows a protein to interact specifically with certain other molecules. Certain intracellular signaling molecules, for example, have shapes that are determined by the structural arrangement of various protein domains; when combined in a precise way they create a functional protein. Likewise, the structure of different lipids determines such functional characteristics as male/female differences, cellular membrane formation, and energy storage. The different structures of polysaccharides determine their usefulness as energy stores, or as components of plant cell walls.

## Chapter 4

### Concept Checks

*Figure 4.1* Organic molecules form the chemical foundation for the structure and function of living organisms. Modern organisms can synthesize organic molecules. However, to explain how life got started, biologists need to understand how organic molecules were made prior to the existence of living cells.

*Figure 4.2* These vents release hot gaseous substances from the interior of the Earth. Organic molecules can form in the temperature gradient between the extremely hot vent water and the cold water that surrounds the vent.

*Figure 4.3* A liposome is more similar to real cells, which are surrounded by a membrane that is composed of a phospholipid bilayer.

*Figure 4.4* Certain chemicals, such as RNA molecules, have properties that provide advantages and therefore cause them to increase in number relative to other molecules.

*Figure 4.5* You would use transmission electron microscopy. The other methods do not have good enough resolution.

*Figure 4.7* The primary advantage is that it gives an image of the 3-D surface of a material.

*Figure 4.11* Centrioles: Not found in plant cells; their role is not entirely clear, but they are found in the centrosome, which is where microtubules are anchored. Lysosome: Not found in plant cells; site where macromolecules are degraded. Chloroplasts: Not found in animal cells; function in photosynthesis. Cell wall: Not found in animal cells; important in cell shape. Central vacuole: Not found in animal cells; site that provides storage and regulates cell volume.

*Figure 4.16* Both dynein and microtubules are anchored in place. Using ATP as a source of energy, dynein tugs on microtubules. Because the microtubules are anchored, they bend in response to the force exerted by dynein.

*Figure 4.19* The nuclear lamina organizes the nuclear envelope and also helps to organize/anchor the chromosomes. The nuclear matrix is inside the nucleus and helps to organize the chromosomes into chromosome territories.

*Figure 4.22* A secreted protein begins its synthesis in the cytosol and then is further synthesized into the ER. It travels via a vesicle to the *cis-*, *medial-*, and *trans-*Golgi and then is secreted when a vesicle fuses with the plasma membrane.

*Figure 4.26* It increases the surface area where ATP synthesis takes place, thereby making it possible to increase the amount of ATP synthesis.

*Figure 4.29* Bacteria and mitochondria are similar in size; they both have circular chromosomes; they both divide by binary fission; and they both make ATP. Bacterial chromosomes are larger, and they make all of their own cellular proteins. Mitochondria chromosomes are smaller, and they import most of their proteins from the cytosol.

*Figure 4.31* The four functions of the ECM in animals are strength, structural support, organization, and cell signaling.

*Figure 4.32* The proteins would become more linear, and the fiber would come apart.

*Figure 4.33* GAGs are highly negatively charged molecules that tend to attract positively charged ions and water. Their high water content gives GAGs a gel-like character, which makes them difficult to compress.

*Figure 4.34* Because the secondary cell wall is usually rigid, it prevents cell growth. If it were made too soon, it might prevent a cell from attaining its proper size.

### BioConnections

*Figure 4.3* Phospholipids are amphipathic molecules; they have a polar end (the head groups) and a nonpolar end (the two fatty acyl tails). Phospholipids form a bilayer such that the heads interact with water, whereas the tails are shielded from the water. This is an energetically favorable structure.

*Figure 4.10* Alternative splicing produces proteins with slightly different structures, because they have certain regions that have different amino acid sequences. The functions of such proteins are often similar, but specialized for the cell type in which they are expressed.

*Figure 4.12* The surfaces of cells involved with gas exchange are highly convoluted. This provides a much greater surface area, thereby facilitating the movement of gases across the membrane.

*Figure 4.15* The type of movement shown in part (b) occurs during muscle contraction.

*Figure 4.17* Cilia carry out a variety of functions, including motility, moving food particles into a feeding groove, and dispersing across moist surfaces.

*Figure 4.20* During cell division, the chromosomes condense and form more compact structures.

*Figure 4.28* These processes are very similar in that DNA replication occurs and then mitochondria or bacterial cells split in two. They are different in that bacteria have cell walls and septa must form between the two daughter cells, which does not occur when mitochondria divide.

### Testing Your Knowledge

4.1: 1. c 2. c 3. b

4.2: 1. b 2. d

4.3: 1. c 2. d 3. b

4.4: 1. b 2. b 3. e

4:5: 1. a 2. b 3. e

4.6: 1. c 2. d

4.7: 1. d

4.8: 1. e 2. c 3. b

4.9: 1. c 2. b

### Test Yourself

1. d 2. d 3. a 4. b 5. e 6. d 7. e 8. e 9. d 10. c

### Conceptual Questions

1. There are a lot of possibilities. The interactions between a motor protein (dynein) and cytoskeletal filaments (microtubules) cause a flagellum to bend. Tubulin proteins bind to each other to form microtubules.
2. If the motor is bound to a cargo and the motor can walk along a filament that is fixed in place, this will cause the movement of the cargo when the motor is activated. If the motor is fixed in place and the filament is free to move, this will cause the filament to move when the motor is activated. If both the motor and filament are fixed in place, the activation of the motor will cause the filament to bend.
3. ATP synthesis occurs along the inner mitochondrial membrane. The invaginations of this membrane greatly increase its surface area, thereby allowing for a greater amount of ATP synthesis.

## Chapter 5

### Concept Checks

*Figure 5.3* More double bonds and shorter fatty acyl tails make the membrane more fluid. Changing the cholesterol concentration can also affect fluidity, but that depends on the starting level of cholesterol. If cholesterol was at a level that maximized stability, increasing the cholesterol concentration would probably increase fluidity.

*Figure 5.4* The low temperature prevents lateral diffusion of membrane proteins. Therefore, after fusion, all of the mouse proteins would stay on one side of the fused cell, and all of the human proteins would remain on the other.

*Figure 5.7* Although both of these molecules penetrate the bilayer fairly quickly, methanol has a polar —OH group and therefore crosses a bilayer more slowly than methane.

*Figure 5.10* Water will move from outside to inside, from the hypotonic medium into the hypertonic medium.

*Figure 5.12* The purpose of gating is to regulate the function of channels.

*Figure 5.20* The function of the protein coat is to promote the formation of a vesicle.

*Figure 5.22* Adherens junctions and desmosomes are cell-to-cell junctions, whereas hemidesmosomes and focal adhesions are cell-to-ECM junctions.

*Figure 5.23* Tight junctions in your skin prevent harmful things like toxins and viruses from entering your body. They also prevent materials like nutrients from leaking out of your body.

### BioConnections

*Figure 5.7* Leucine would cross an artificial membrane more easily because it is more hydrophobic than lysine.

*Figure 5.8* Gradients of sodium and potassium ions are important for the conduction of action potentials.

*Figure 5.19* Plasmodesmata facilitate the movement of nutrients in a cell-to-cell manner. This is called symplastic transport.

*Figure 5.23* If tight junctions did not exist, substances in the lumen of your intestine might directly enter your blood. This could be potentially harmful if you consumed something with a toxic molecule in it. Likewise, materials from blood could be lost by diffusing into the lumen of your small intestine.

### Feature Investigation Questions

1. Most cells allow movement of water across the cell membrane by passive diffusion. However, it was noted that certain cell types had a much higher rate of water movement, indicating that water channels might be made in these cells.
2. The researchers identified water channels by characterizing proteins that are present in red blood cells and kidney cells but not other types of cells. Red blood cells and kidney cells have a faster rate of water movement across the membrane than other cell types. These cells are more likely to have water channels. By identifying proteins that are found in both of these types of cells but not in other cells, the researchers were identifying possible candidate proteins that function as water channels. In addition, CHIP28 had a structure that resembled other known channel proteins. Agre and his associates experimentally created multiple copies of the gene that produces the CHIP28 protein and then artificially transcribed the genes to produce many mRNAs. The mRNAs were injected into frog oocytes where they could be translated to make the CHIP28 proteins. After altering the frog oocytes by introducing the CHIP28 mRNAs, they compared the rate of water transport in the altered oocytes versus normal frog oocytes. This procedure allowed them to introduce the candidate protein to a cell type that normally does not have the protein present.
3. After artificially introducing the candidate protein into the frog oocytes, the researchers found that the experimental oocytes took up water at a much faster rate in a hypotonic solution as compared to the control oocytes. The results indicated that the presence of the CHIP28 protein did increase water transport into cells.

### Testing Your Knowledge

5.1: 1. c 2. c

5.2: 1. c 2. b

5.3: 1. e 2. c

5.4: 1. d 2. b

5.5: 1. c

5.6: 1. c

5.7: 1. e

### Test Yourself

1. c 2. c 3. b 4. d 5. c 6. e 7. d 8. e 9. e 10. c

### Conceptual Questions

1. See Figure 5.1 for the type of drawing you should have made. The membrane is considered a mosaic of lipid, protein, and carbohydrate molecules. The membrane exhibits properties that resemble a fluid because lipids and proteins can move relative to each other within the membrane.
2. Integral membrane proteins can contain transmembrane segments that cross the membrane, or they may contain lipid anchors. Peripheral membrane proteins are noncovalently bound to integral membrane proteins or to the polar heads of phospholipids.

3. Lipid bilayers, channels, and transporters cause the plasma membrane to be selectively permeable. This allows a cell to take up needed nutrients from its extracellular environment and to export waste products into the environment. Cell junctions between cells and the ECM allow cells to sense changes in the ECM and to respond to such changes.

## Chapter 6

### Concept Checks

***Figure 6.2*** The solution of dissolved $Na^+$ and $Cl^-$ has more entropy. A salt crystal is very ordered, whereas the ions in solution are much more disordered.

***Figure 6.4*** It speeds up the rate. When the activation energy is lower, it takes less time for reactants to reach a transition state where a chemical reaction can occur. It does not affect the direction of a reaction.

***Figure 6.5*** The activation energy is lowered during the second step when the substrates undergo induced fit.

***Figure 6.6*** At a substrate concentration of 0.5 mM, enzyme A would have a higher velocity. Enzyme A would be very near its $V_{max}$, whereas enzyme B would be well below its $V_{max}$.

***Figure 6.9*** The oxidized form is $NAD^+$.

***Figure 6.12*** The first phase is named the energy investment phase because some ATP is used up. The second phase is called the cleavage phase because a 6-carbon molecule is broken down into two 3-carbon molecules. The energy liberation phase is so named because NADH and ATP are made.

***Figure 6.13*** The molecules that donate phosphates are 1,3-bisphosphoglycerate and phosphoenolpyruvate.

***Figure 6.16*** The main products are 2 $CO_2$, 3 NADH, 1 $FADH_2$, and 1 ATP.

***Figure 6.18*** It is called cytochrome oxidase because it removes electrons from (oxidizes) cytochrome *c*. Another possible name would be oxygen reductase because it reduces oxygen.

***Figure 6.19*** The $\gamma$ subunit turns clockwise, when viewed from the intermembrane space. The $\beta$ subunit in the back right is in conformation 3, and the one on the left is in conformation 1.

***Figure 6.22*** The advantage is that the cell can use the same enzymes to metabolize different kinds of organic molecules. This saves the cell energy because it is costly to make a lot of different enzymes, which are composed of proteins.

### BioConnections

***Figure 6.12*** Glycolytic muscle fibers rely on glycolysis for their ATP needs. Because glycolysis does not require oxygen, such muscle fibers can function without oxygen.

### Feature Investigation Questions

1. The researchers attached an actin filament to the $\gamma$ subunit of ATP synthase. The actin filament was fluorescently labeled so the researchers could determine if the actin filament moved when viewed under the fluorescence microscope.
2. When functioning in the hydrolysis of ATP, the actin filament was seen to rotate. The actin filament was attached to the $\gamma$ subunit of ATP synthase. The rotational movement of the filament was the result of the rotational movement of the enzyme. In the control experiment, no ATP was added to stimulate enzyme activity. In the absence of ATP, no movement was observed.
3. No, the observation of counterclockwise rotation is the opposite of what would be expected inside the mitochondria. During the experiment, the enzyme was not functioning in ATP synthesis but instead was running backwards and hydrolyzing ATP.

### Testing Your Knowledge

6.1: 1. e 2. b

6.2: 1. a 2. a

6.3: 1. e 2. d 3. c

6.4: 1. c 2. d

6.5: 1. a

6.6: 1. e

6.7: 1. e 2. d 3. e

6.8: 1. d

### Test Yourself

1. d 2. e 3. b 4. e 5. c 6. a 7. b 8. a 9. d 10. b

### Conceptual Questions

1. During feedback inhibition, the product of a metabolic pathway binds to an allosteric site on an enzyme that acts earlier in the pathway. The product inhibits this enzyme, thereby preventing the overaccumulation of the product.
2. The movement of $H^+$ through the *c* subunits causes the $\gamma$ subunit to rotate. As it rotates, it sequentially alters the conformation of the subunits, where ATP is made. This causes (1) ADP and $P_i$ to bind with moderate affinity, (2) ADP and $P_i$ to bind very tightly such that ATP is made, and (3) ATP to be released.
3. As discussed in this chapter, the phases of glucose metabolism are regulated in a variety of ways. For example, key enzymes in glycolysis and the citric acid cycle are regulated by the availability of substrates and by feedback inhibition. The electron transport chain is regulated by the ATP/ADP ratio. Such regulation ensures that a cell does not waste energy making ATP when it is in sufficient supply. Also, the production of too much NADH is potentially harmful because at high levels it has the potential to haphazardly donate its electrons to other molecules and promote the formation of free radicals, highly reactive chemicals that damage DNA and cellular proteins.

## Chapter 7

### Concept Checks

***Figure 7.1*** Both heterotrophs and autotrophs carry out cellular respiration.

***Figure 7.3*** The Calvin cycle can occur in the dark as long as there is sufficient $CO_2$, ATP, and NADPH.

***Figure 7.4*** Gamma rays have higher energy than radio waves.

***Figure 7.5*** To drop down to a lower orbital at a lower energy level and become more stable, an electron could release energy in the form of heat, release energy in the form of light, or transfer energy to another electron by resonance energy transfer.

***Figure 7.7*** By having different pigment molecules, plants can absorb a wider range of wavelengths of light.

***Figure 7.8*** ATP and NADPH are made in the stroma. $O_2$ is made in the thylakoid lumen.

***Figure 7.9*** Noncyclic electron flow produces equal amounts of ATP and NADPH. However, plants usually need more ATP than NADPH. Cyclic photophosphorylation allows plants to make just ATP, thereby increasing the relative amount of ATP.

***Figure 7.11*** An electron has its highest amount of energy just after it has been boosted by light in PSI.

***Figure 7.12*** NADPH reduces organic molecules and makes them more able to form C—C and C—H bonds.

***Figure 7.15*** The arrangement of cells in $C_4$ plants makes the level of $CO_2$ high and the level of $O_2$ low in the bundle sheath cells.

***Figure 7.16*** When there is plenty of moisture and it is not too hot, $C_3$ plants are more efficient. However, under hot and dry conditions, $C_4$ and CAM plants have the advantage because they lose less water and avoid photorespiration.

### BioConnections

***Figure 7.2*** Two guard cells make up one stoma.

***Figure 7.15*** Water is taken up by the roots of plants and moves via the vascular system to the leaves.

### Feature Investigation Questions

1. The researchers were attempting to determine the biochemical pathway of the process of carbohydrate synthesis in plants. The researchers wanted to identify different molecules produced in plants over time to determine the steps of the biochemical pathway.
2. The purpose for using $^{14}C$ was to label the different carbon molecules produced during the biochemical pathway. The researchers could "follow" the carbon molecules from $CO_2$ that were incorporated into the organic molecules during photosynthesis. The radioactive isotope provided the researchers with a method of labeling the different molecules. The purpose of the experiment was to determine the steps in the biochemical pathway of photosynthesis. By examining samples from different times after the introduction of the labeled carbon source, the researchers would be able to determine which molecules were produced first and, thus, products of the earlier steps of the pathway versus products of later steps of the pathway. The researchers used two-dimensional paper chromatography to separate the different molecules from each other. Afterward, the different molecules were identified by different chemical methods. The text describes the method of comparing two-dimensional paper chromatography results of unknown molecules to known molecules and identifying the unknown with the known molecule it matched.
3. The researchers were able to determine the biochemical process that plants use to incorporate $CO_2$ into organic molecules. The researchers were able to identify the biochemical steps and the molecules produced at these steps in what is now called the Calvin cycle.

### Testing Your Knowledge

7.1: 1. e 2. d 3. e

7.2: 1. b 2. b 3. a

7.3: 1. c 2. d

7.4: 1. b 2. a

7.5: 1. a 2. e

### Test Yourself

1. c 2. c 3. c 4. a 5. b 6. b 7. c 8. b 9. e 10. c

### Conceptual Questions

1. The two stages of photosynthesis are the light reactions and the Calvin cycle. The key products of the light reactions are ATP, NADPH, and $O_2$. The key product of the Calvin cycle is carbohydrate. The initial product is G3P, which is used to make sugars and other organic molecules.
2. NADPH is used during the reduction phase of the Calvin cycle. It donates its electrons to 1,3-BPG.
3. At the level of the biosphere, the role of photosynthesis is to incorporate carbon dioxide into organic molecules. These organic molecules can then be broken down, by autotrophs and by heterotrophs, to make ATP. The organic molecules made during photosynthesis are also used as starting materials to synthesize a wide variety of organic molecules and macromolecules that are made by cells.

## Chapter 8

### Concept Checks

***Figure 8.1*** It is glucose.

***Figure 8.3*** Endocrine signals are more likely to exist for a longer period of time. This is necessary because endocrine signals called hormones travel relatively long distances to reach their target cells. Therefore, the hormone must exist long enough to reach its target cells.

***Figure 8.4*** The effect of a signaling molecule is to cause a cellular response. Most signaling molecules do not enter the cell. Therefore, to exert an effect, they must alter the conformation of a receptor protein, which, in turn, stimulates an intracellular signal transduction pathway that leads to a cellular response.

***Figure 8.6*** Phosphorylation of a protein via a kinase involves ATP hydrolysis, which is an exergonic reaction. The energy from this reaction usually alters the conformation of the phosphorylated protein, thereby influencing its function. Phosphorylation is used to regulate protein function.

***Figure 8.7*** The α subunit has to hydrolyze its GTP to GDP and $P_i$. This changes the conformation of the α subunit so that it can reassociate with the β and γ subunits.

***Figure 8.13*** The signal transduction pathway begins with the G protein and ends with protein kinase A being activated. The cellular response involves the phosphorylation of target proteins. The phosphorylation of target proteins will change their function in some way, which is how the cell is responding.

***Figure 8.14*** Depending on the protein involved, phosphorylation can activate or inhibit protein function. Phosphorylation of phosphorylase kinase and glycogen phosphorylase activates their function, whereas it inhibits glycogen synthase.

***Figure 8.15*** Signal amplification allows a single signaling molecule to affect many proteins within a cell, thereby amplifying a cellular response.

### BioConnections

***Figure 8.5*** Most receptors and enzymes bind their ligands noncovalently and with high specificity. Enzymes, however, convert their ligands (which are reactants) into products, whereas receptors undergo a conformation change after a ligand binds.

***Figure 8.10*** The GTP-bound form of Ras is active and promotes cell division. To turn the pathway off, Ras hydrolyzes GTP to GDP and $P_i$. If this cannot occur due to a mutation, the pathway will be continuously on, and uncontrolled cell division will result.

### Testing Your Knowledge

8.1: 1. e 2. e 3. b

8.2: 1. c 2. c

8.3: 1. e 2. d

8.4: 1. c

8.5: 1. a 2. d

8.6: 1. b 2. e

### Test Yourself

1. d 2. c 3. d 4. e 5. d 6. a 7. a 8. b 9. d 10. e

### Conceptual Questions

1. Cells need to respond to a changing environment, and cells need to communicate with each other.
2. In the first stage, a signaling molecule binds to a receptor, causing receptor activation. In the second stage, one type of signal is transduced or converted to a different signal inside the cell. In the third stage, the cell responds in some way to the signal, possibly by altering the activity of enzymes, structural proteins, or transcription factors. When the estrogen receptor is activated, the second stage, signal transduction, is not needed because the estrogen receptor is an intracellular receptor that directly activates the transcription of genes to elicit a cellular response.
3. Cell signaling allows cells to respond to environmental changes. For example, if a yeast cell is exposed to glucose, cell signaling will allow it to adapt to that change and utilize glucose more readily. Likewise, cell signaling allows plants to grow toward light. In addition, cells in a multicellular organism respond to changes in signaling molecules, such as hormones, and thereby coordinate their activities.

## Chapter 9

### Concept Checks

***Figure 9.4*** Cytosine is found in both DNA and RNA.

***Figure 9.5*** A phosphoester bond is a single covalent bond between a phosphorus atom and an oxygen atom. A phosphodiester linkage involves two phosphoester bonds. This linkage occurs along the backbone of DNA and RNA strands.

***Figure 9.6*** Because it is antiparallel and obeys the AT/GC rule, it would be 3′–CTAAGCAAG–5′.

***Figure 9.11*** It would be 1/8 half-heavy and 7/8 light.

***Figure 9.17*** The lagging strand is made discontinuously in the direction opposite to the movement of the replication fork.

***Figure 9.21*** Proteins hold the bottoms of the loops in place.

***Figure 9.22*** Proteins that compact the radial loop domains are primarily responsible for the X shape.

### BioConnections

***Figure 9.1*** When a bacterium dies, it may release some of its DNA into the environment. Such DNA can be taken up via transformation by living bacteria, even bacteria of other species. If the DNA that is taken up encodes an antibiotic resistance gene, such a gene may be incorporated into the genome of the living bacterium and make it resistant to an antibiotic.

***Figure 9.22*** If chromosomes did not become compact, they might get tangled up with others during cell division, which would prevent the even segregation of chromosomes into the two daughter cells.

### Feature Investigation Questions

1. Previous studies had indicated that mixing different strains could lead to transformation or the changing of a strain into a different one. Griffith had shown that mixing heat-killed type S with living type R would result in the transformation of the type R to type S. Though mutations could cause the changing of the identity of certain strains, the type R to type S transformation was not due to mutation but was more likely due to the transmission of a biochemical substance between the two strains. Griffith recognized this and referred to the biochemical substance as the "transformation principle." If Avery, MacLeod, and McCarty could determine the biochemical identity of this "transformation principle," they could identify the genetic material for this organism.
2. A DNA extract contains DNA that has been purified from a sample of cells.
3. The researchers could not verify that the DNA extract was completely pure and did not have small amounts of contaminating molecules, such as proteins and RNA. The researchers were able to treat the extract with enzymes to remove proteins (using protease), RNA (using RNase), or DNA (using DNase). Removing the proteins or RNA did not alter the transformation of the type R to type S strains. Only the enzymatic removal of DNA disrupted the transformation, indicating that DNA is the genetic material.

### Testing Your Knowledge

9.1: 1. b

9.2: 1. a 2. b 3. c

9.3: 1. c 2. d

9.4: 1. b 2. c

9.5: 1. a 2. c 3. a

9.6: 1. d 2. c

### Test Yourself

1. d 2. b 3. d 4. d 5. b 6. c 7. b 8. d 9. d 10. c

### Conceptual Questions

1. The genetic material must contain the information necessary to construct an entire organism. The genetic material must be accurately copied and transmitted from parent to offspring and from cell to cell during cell division in multicellular organisms. The genetic material must contain variation that can account for the known variation within each species and among different species. Griffith discovered something called the transformation principle, and his experiments showed the existence of biochemical genetic information. In addition, he showed that this genetic information can move from one individual to another of the same species. In his experiments, Griffith took heat-killed type S bacteria and mixed them with living type R bacteria and injected them into a live mouse, which died after the injection. By themselves, these two strains would not kill the mouse, but when they were put together, the genetic information from the heat-killed type S bacteria was transferred into the living type R bacteria, thus transforming the type R bacteria into type S.
2. In a DNA double helix, the two strands hydrogen-bond with each other according to the AT/GC rule. This provides the basis for DNA replication. In addition, as described in later chapters, hydrogen bonding between complementary bases is the basis for the transcription of RNA, which is needed for gene expression.
3. As discussed in the answer to number 2, the structure of DNA provides a way for it to be replicated. In addition, the sequences of bases within genes store information to make polypeptides with a defined amino acid sequence.

## Chapter 10

### Concept Checks

***Figure 10.1*** The usual direction of flow of genetic information is from DNA to RNA to protein, though exceptions occur.

***Figure 10.2*** If a terminator was removed, transcription would occur beyond the normal stopping point. Eventually, RNA polymerase would encounter a terminator from an adjacent gene, and transcription would end.

***Figure 10.5*** The ends of protein-encoding genes do not have a poly T region that acts as a template for the synthesis of a poly A tail. Instead, the poly A tail is added after the pre-mRNA is made by an enzyme that attaches many adenine nucleotides in a row.

***Figure 10.7*** A protein-encoding gene would still be transcribed into RNA if the start codon was missing. However, it would not be translated properly into a polypeptide.

***Figure 10.8*** It would bind to a 5′–UGG–3′ codon, and it would carry tryptophan.

***Figure 10.10*** The function of the anticodon in tRNA is to recognize a codon in an mRNA.

***Figure 10.13*** Each mammal is closely related to the other mammals, and *E. coli* and *Serratia marcescens* are also closely related. The mammals are relatively distantly related to the bacterial species.

***Figure 10.14*** A region near the 5′ end of the mRNA is complementary to a region of rRNA in the small subunit. These complementary regions hydrogen-bond with each other to promote the binding of the mRNA to the small ribosomal subunit.

### BioConnections

***Figure 10.3*** Both DNA and RNA polymerase use DNA strands as a template and connect nucleotides to each other in a 5′ to 3′ direction based on the complementarity of base pairing. One difference is that DNA polymerase needs a preexisting strand, such as an RNA primer, to begin DNA replication, whereas RNA polymerase can begin the synthesis of RNA on a bare template strand. Another key difference is that DNA polymerase connects deoxyribonucleotides, whereas RNA polymerase connects ribonucleotides.

***Figure 10.11*** The attachment of an amino acid to a tRNA is an endergonic reaction. ATP provides the energy to catalyze this reaction.

### Feature Investigation Questions

1. A triplet mimics mRNA because it can cause a specific tRNA to bind to the ribosome. This was useful to Nirenberg and Leder because it allowed them to correlate the binding of a tRNA carrying a specific amino acid with a triplet sequence.
2. The researchers were attempting to match codons with appropriate amino acids. By labeling one amino acid in each of the 20 tubes for each codon, the researchers were able to identify the correct relationship by detecting which tube resulted in radioactivity on the filter.
3. The AUG triplet would have shown radioactivity in the methionine test tube. Even though AUG acts as the start codon, it also codes for the amino acid

methionine. The other three codons act as stop codons and do not code for an amino acid. In these cases, the researchers would not have found radioactivity trapped on filters.

### Testing Your Knowledge

10.1: 1.b 2. d 3. d

10.2: 1. d 2. a

10.3: 1. b 2. d

10.4: 1. d 2. d

10.5: 1. d 2. d 3. c

10.6: 1. d 2. c

### Test Yourself

1. b 2. a 3. d 4. e 5. e 6. e 7. d 8. d 9. d 10. b

### Conceptual Questions

1. This mutation would not affect transcription. RNA polymerase would still transcribe the gene. However, it would affect translation. A polypeptide would be made that was only 5 amino acids long.
2. Each of these 20 enzymes catalyzes the attachment of a specific amino acid to a specific tRNA molecule.
3. During transcription, a DNA strand is used as a template for the synthesis of RNA. Most genes encode mRNAs, which contain the information to make polypeptides. During translation, an mRNA binds to a ribosome and a polypeptide is made, which becomes a unit within a functional protein.

## Chapter 11

### Concept Checks

***Figure 11.1*** The binding of an ncRNA to DNA or another RNA could be inhibited by the formation of a stem-loop, because it would inhibit base-pairing.

***Figure 11.5*** Telomerase uses the ncRNA called TERC as a template to make DNA.

***Figure 11.6*** HOTAIR binds to the target gene because the RNA in HOTAIR is complementary to the target gene.

***Figure 11.8*** RISC binds to an mRNA because the miRNA is complementary to the mRNA. The miRNA and mRNA hydrogen bond to each other.

***Figure 11.11*** The crRNA directly recognizes the bacteriophage DNA.

### BioConnections

***Figure 11.4*** DNA polymerase cannot copy the 3' end of a DNA strand because it needs a primer to make DNA and it synthesizes DNA only in the 5' to 3' direction.

***Figure 11.10*** In eukaryotes, proteins that are destined for the ER, Golgi, vacuoles, lysosomes, plasma membrane, and secretion need SRP to reach their correct destination. In bacteria and archaea, proteins that are secreted need SRP.

### Feature Investigation Questions

1. The *mex-3* mRNA corresponds to the sense strand.
2. The researchers began with a copy of the *mex-3* gene in a plasmid. When RNA polymerase and nucleotides were added, the sense strand of *mex-3* mRNA was made. They also modified this plasmid by placing the promoter on the opposite side of the *mex-3* coding sequence. When RNA polymerase and nucleotides were added, the antisense strand of *mex-3* mRNA was made.
3. The mixture of sense and antisense *mex-3* mRNA was the most effective at causing the degradation of *mex-3* mRNA.

### Testing Your Knowledge

11.1: 1. e 2. b

11.2: 1. e

11.3: 1. e

11.4: 1. e 2. e

11.5: 1. e

11.6: 1. d 2. c

11.7: 1. a

### Test Yourself

1. e 2. a 3. b 4. c 5. d 6. d 7. a 8. a 9. c 10. d

### Conceptual Questions

1. HOTAIR: scaffold, guide
   RNA of SRP: scaffold, alters protein function
   microRNA: guide
   crRNA: guide
2. RNA interference is the phenomenon in which an miRNA or an siRNA silences the expression of an mRNA. The double-stranded pre-miRNA or pre-siRNA is cleaved by dicer into a smaller double-stranded RNA that associates with RISC. One of the RNA strands is then degraded, so that only a short single-stranded miRNA or siRNA that recognizes an mRNA is found within RISC. This miRNA or siRNA guides RISC to the mRNA and causes it to be silenced.
3. The structure of HOTAIR provides a scaffold for the binding of two histone-modifying complexes. Its structure also is complementary to GA-rich sequences next to certain target genes, such as the *HoxD* gene, and thereby guides those histone-modifying complexes to those genes. Such genes are then covalently modified, which causes them to be repressed.

## Chapter 12

### Concept Checks

***Figure 12.2*** Gene regulation causes each cell type to express its own unique set of proteins, which, in turn, are largely responsible for the morphologies and functions of cells.

***Figure 12.6*** The *lacZ*, *lacY*, and *lacA* genes are under the control of the *lac* promoter.

***Figure 12.7*** Negative control refers to the action of a repressor protein, which inhibits transcription when it binds to the DNA. Inducible refers to the action of a small effector molecule. When it is present, it promotes transcription.

***Figure 12.9*** In this case, the repressor keeps the *lac* operon turned off unless lactose is present in the environment. The activator allows the bacterium to choose between glucose and lactose.

***Figure 12.14*** Some histone modifications may alter chromatin structure in a way that promotes transcription, whereas others may inhibit transcription.

***Figure 12.17*** The advantage of alternative splicing is that it allows a single gene to encode two or more polypeptides. This enables organisms to have smaller genomes, which is more efficient and easier to package into a cell.

***Figure 12.18*** When iron levels rise in the cell, the iron binds to IRP and removes it from the mRNA that encodes ferritin. This results in the rapid translation of ferritin protein, which can store excess iron. Unfortunately, ferritin storage does have limits, so iron poisoning can occur if too much is ingested.

### BioConnections

***Figure 12.8*** In eukaryotic cells, cAMP acts as a second messenger in signal transduction pathways.

***Figure 12.15*** A nucleosome is composed of DNA wrapped around an octamer of histone proteins.

### Testing Your Knowledge

12.1: 1. d 2. a

12.2: 1. b

12.3: 1. c 2. e

12.4: 1. d 2. b

12.5: 1. d 2. a

### Test Yourself

1. d 2. b 3. c 4. c 5. c 6. d 7. c 8. d 9. c 10. d

### Conceptual Questions

1. In an inducible operon, the presence of a small effector molecule causes transcription to occur. In repressible operons, a small effector molecule inhibits transcription. The effects of these small molecules are mediated through regulatory proteins that bind to the DNA. Repressible operons usually encode anabolic enzymes, and inducible operons encode catabolic enzymes.
2. a. regulatory protein; b. small effector molecule; c. segment of DNA; d. small effector molecule; and e. regulatory protein.
3. Gene regulation offers key advantages such as (1) proteins are made only when they are needed; (2) proteins are made in the correct cell type; and (3) proteins are made at the correct stage of development. These advantages are important for reproduction and sustaining life.

## Chapter 13

### Concept Checks

***Figure 13.1*** Only germ-line cells give rise to gametes (sperm or egg cells). A somatic cell cannot give rise to a gamete and therefore cannot be passed to offspring.

***Figure 13.2*** This is a mutation that occurred in a somatic cell, so it cannot be transmitted to an offspring.

***Figure 13.5*** A thymine dimer is harmful because it can cause errors in DNA replication.

***Figure 13.6*** First, we subtract the number of colonies formed in the absence of mutagen (2) from those formed in the presence of mutagen (44) and obtain a value of 42 for the number of mutations caused by the mutagen. If we divide 42 by 2 million, the rate is $2.1 \times 10^{-5}$.

***Figure 13.7*** UvrC and UvrD are responsible for removing the damaged DNA. UvrC makes cuts on both sides of the damage, and then UvrD removes the damaged region.

***Figure 13.9*** Growth factors turn on a signaling pathway that ultimately leads to cell division.

***Figure 13.12*** The type of cancer associated with this fusion is leukemia, which is a cancer of blood cells. The gene fusion produces a chimeric gene that is expressed in blood cells because it has the *bcr* promoter. The abnormal fusion protein promotes cancer in these cells.

***Figure 13.13*** Checkpoints prevent cell division if a genetic abnormality is detected. This helps to properly maintain the genome, thereby minimizing the possibility that a cell harboring a mutation will divide to produce two daughter cells.

***Figure 13.14*** Cancer would not occur if both copies of the *Rb* gene and both copies of the *E2F* gene were rendered inactive due to mutations. An active copy of the *E2F* gene is needed to promote cell division.

### BioConnections

***Figure 13.9*** Drugs that inhibit protein kinases may be used to combat cancer if they target the protein kinases that are overactive in certain forms of cancer.

### Feature Investigation Questions

1. Some biologists believed that heritable traits may be altered by physiological events. This suggests that mutations may be stimulated by certain needs of the organism. Others believed that mutations are random. If a mutation had a beneficial effect that improved survival and/or reproductive success, these mutations would be maintained in the population through natural selection.
2. The Lederbergs were testing the hypothesis that mutations are random events. By subjecting the bacteria to some type of environmental stress, the bacteriophage, the researchers would be able to see if the stress induced mutations or if mutations occurred randomly.
3. When looking at the number and location of colonies that were resistant to viral infection, the pattern was consistent among the secondary plates. This indicates that the mutation that allowed the colonies to be resistant to viral infection occurred on the master plate. The secondary plates introduced the selective agent that allowed the resistant bacteria colonies to survive and reproduce while the other colonies were destroyed. Thus, mutations occurred randomly in the absence of any selective agent.

### Testing Your Knowledge

13.1: 1. b 2. d

13.2: 1. b 2. d

13.3: 1. e

13.4: 1. d 2. d

### Test Yourself

1. d 2. d 3. d 4. e 5. d 6. b 7. a 8. b 9. c 10. d

### Conceptual Questions

1. Random mutations are more likely to be harmful than beneficial. The genes within each species have evolved to work properly. They have functional promoters, coding sequences, terminators, and so on, that allow the genes to be expressed. Mutations are more likely to disrupt these sequences. For example, mutations within the coding sequence may produce early stop codons, frameshift mutations, and missense mutations that result in a nonfunctional polypeptide. On rare occasions, however, mutations are beneficial; they may produce a gene that is expressed better than the original gene or produce a polypeptide that functions better.
2. A spontaneous mutation originates within a living cell. It may be due to spontaneous changes in nucleotide structure, errors in DNA replication, or products of normal metabolism that may alter the structure of DNA. The causes of induced mutations originate from outside the cell. They may be physical agents, such as UV light or X-rays, or chemicals that act as mutagens. Both spontaneous and induced mutations may cause a harmful phenotype such as a cancer. In many cases, induced mutations are avoidable if the individual can prevent exposure to the environmental agent that acts as a mutagen.
3. Mutations may alter the expression of a gene and/or alter the function of a protein encoded by a gene. In many cases, such changes are harmful because a gene may not be expressed at the correct level, or the protein may not function as well as the normal (nonmutant) protein.

## Chapter 14

### Concept Checks

***Figure 14.1*** Chromosomes are readily seen when they are compacted in a dividing cell. By adding such a drug, you increase the percentage of cells that are actively dividing.

***Figure 14.2*** Interphase consists of the $G_1$, S, and $G_2$ phases of the cell cycle.

***Figure 14.4*** As shown in the inset, each object is a pair of sister chromatids.

***Figure 14.5*** The astral microtubules, which extend away from the chromosomes, are important for positioning the spindle apparatus within the cell. The polar microtubules project into the region between the two poles. Polar microtubules that overlap with each other play a role in the separation of the two poles. Kinetochore microtubules are attached to kinetochores at the centromeres and are needed to sort the chromosomes.

***Figure 14.7*** The mother cell (in $G_1$ phase) and the daughter cells have the same chromosome composition. They are genetically identical.

***Figure 14.8*** Cytokinesis in both animal and plant cells separates a mother cell into two daughter cells. In animal cells, cytokinesis involves the formation of a cleavage furrow, which constricts like a drawstring to separate the cells. In plants, the two daughter cells are separated by the formation of a cell plate, which forms a cell wall between the two daughter cells.

*Figure 14.10* The mother cell is diploid with two sets of chromosomes, whereas the four resulting cells are haploid with one set of chromosomes.

*Figure 14.11* The reason for meiosis in animals is to produce gametes. These gametes combine during fertilization to produce a diploid organism. Following fertilization, the purpose of mitosis is to produce a multicellular organism.

*Figure 14.13* Inversions and the translocations shown here do not affect the total amount of genetic material.

### BioConnections

*Figure 14.3* Checkpoints prevent cancer by checking the integrity of the genome. If abnormalities in DNA structure are detected or if a chromosome is not properly attached to the spindle, the checkpoint will delay cell division until the problem is fixed. If it cannot be fixed, the checkpoint will initiate the process of apoptosis, thereby killing a cell that may harbor mutations. This prevents the proliferation of cells that have the potential to be cancerous.

### Testing Your Knowledge

14.1: 1. d 2. b 3. d

14.2: 1. c 2. b 3. d

14.3: 1. a 2. a 3. c

14.4: 1. d 2. e 3. c

### Test Yourself

1. b 2. e 3. b 4. e 5. c 6. a 7. c 8. d 9. b 10. c

### Conceptual Questions

1. In diploid species, chromosomes are present in pairs, one from each parent, and contain similar gene arrangements. Such chromosomes are homologous. When DNA is replicated, two identical copies are created and are sister chromatids.
2. There are four copies. A karyotype shows homologous chromosomes that come in pairs. Each member of the pair has replicated to form a pair of sister chromatids. Therefore, four copies of each gene are present. See the inset to Figure 14.1.
3. Mitosis is a process that produces two daughter cells with the same genetic material as the original daughter cell. In the case of plants and animals, this allows a fertilized egg to develop into a multicellular organism composed of many genetically identical cells.

## Chapter 15

### Concept Checks

*Figure 15.2* Having blue eyes is a variant (also called a trait). A character is a more general term, which in this case would refer to eye color.

*Figure 15.4* In this procedure, stamens are removed from the purple flower to prevent self-fertilization.

*Figure 15.5* The reason why offspring of the $F_1$ generation exhibit only one variant of each character is because one trait is dominant over the other.

*Figure 15.6* The ratio of alleles (*T* to *t*) is 1:1. The reason why the phenotypic ratio is 3:1 is because *T* is dominant to *t*.

*Figure 15.7* If the linked hypothesis had been correct, the ratio would have been 3 round, yellow to 1 wrinkled, green.

*Figure 15.10* There would be four possible ways of aligning the chromosomes, and eight different types of gametes (*ABC, abc, ABc, abC, Abc, aBC, AbC, aBc*) could be produced.

*Figure 15.11* No. If two parents are affected with the disease, they must be homozygous for the mutant allele if it's recessive. Two homozygous parents would have to produce all affected offspring. If they don't, then the inheritance pattern is not recessive.

*Figure 15.12* In a dominant pattern of inheritance, all affected offspring have at least one affected parent.

*Figure 15.15* No. You need a genetically homogenous population to study the norm of reaction. A wild population of squirrels is not genetically homogenous, so it could not be used.

*Figure 15.16* The person would be a female. The presence of the Y chromosome determines maleness in humans.

### BioConnections

*Figure 15.8* Sexual reproduction is the process in which two haploid gametes (for example, sperm and egg) combine with each other to begin the life of a new individual. Each gamete contributes one set of chromosomes. The resulting zygote has chromosomes that occur in pairs (one from each parent). The members of each pair are called homologs of each other; they carry the same types of genes.

*Figure 15.9* Alleles segregate from each other during the process of meiosis. Meiosis begins with a diploid mother cell that has pairs of genes, which may be found in different alleles. During meiosis, these pairs of genes separate and end up in different haploid cells. Therefore, each haploid cell has only one copy of each gene. In other words, each haploid cell has only one allele of a given gene.

### Testing Your Knowledge

15.1: 1. c 2. d

15.2: 1. b 2. b

15.3: 1. b

15.4: 1. b 2. d

15.5: 1. b 2. d

### Test Yourself

1. e 2. c. 3. b 4. d 5. a 6. c 7. c 8. d 9. d 10. d

### Conceptual Questions

1. The genotype is the genes (alleles) that an individual carries, whereas the phenotype is the characteristics that they display. It is possible for two individuals to have the same phenotype and different genotypes. For example, *TT* and *Tt* pea plants have different genotypes, but they have same phenotype (tall).
2. Though the patterns have some similarities, there are certain features to distinguish them. If a female showing the recessive trait is crossed to a male showing the dominant trait, all of the male offspring would show the recessive trait if it were X-linked. If it was not X-linked, half or all of the male offspring would show the dominant trait if the father was heterozygous or homozygous, respectively. If the father showed the dominant trait and the mother was heterozygous, half of the male offspring would show the recessive trait if it was X-linked, but none of the females would.
3. The environment is needed so that genes can be expressed. For example, organic molecules and energy are needed for transcription and translation. In addition, environmental factors influence the outcome of traits. For example, sunlight can cause a tanning response, thereby affecting the darkness of the skin.

## Chapter 16

### Concept Checks

*Figure 16.1* Only the allele inherited from the father would be expressed in the offspring. Because he is heterozygous, half of the offspring would be normal size and half would be dwarf.

*Figure 16.3* The Barr body is much more compact than the other Xchromosome in the cell. This compaction prevents most of the genes on the Barr body from being expressed.

*Figure 16.7* The gene is located in the chloroplast DNA. In this species, chloroplasts are transmitted from parent to offspring via eggs but not via sperm.

*Figure 16.10* Crossing over occurred during oogenesis in the heterozygous female of the $F_1$ generation to produce the recombinant offspring of the $F_2$ generation.

### BioConnections

*Figure 16.6* The evolutionary origin of these organelles is an ancient endosymbiotic relationship. Mitochondria are derived from purple bacteria, and chloroplasts are derived from cyanobacteria.

*Figure 16.10* Crossing over occurs during prophase of meiosis I.

### Feature Investigation Questions

1. Bateson and Punnett were testing the hypothesis that the gene pairs that influence flower color and pollen shape would assort independently of each other. The two traits were expected to show a pattern consistent with Mendel's law of independent assortment.
2. The expected results were a phenotypic ratio of 9:3:3:1. The researchers expected 9/16 of the offspring would have purple flowers and long pollen, 3/16 of the offspring would have purple flowers and round pollen, 3/16 of the offspring would have red flowers and long pollen, and 1/16 of the offspring would have red flowers and round pollen.
3. In this problem, you hypothesize that the genes independently assort, because this allows you to calculate the expected values as shown in the data of Figure 16.9. The calculated chi square value is very high and you would reject the hypothesis that the genes independently assort. As an alternative, you would accept the hypothesis that the genes are linked.

### Testing Your Knowledge

16.1: 1. e

16.2: 1. a 2. c

16.3: 1. a 2. c

16.4: 1. a 2. d

16.5: 1. a 2. d

16.6: 1. c 2. b

### Test Yourself

1. e 2. d. 3. c 4. a 5. a 6. c 7. e 8. d 9. c 10. c

### Conceptual Questions

1. Epigenetics is the study of mechanisms that lead to changes in gene expression that can be passed from cell to cell and are reversible, but do not involve a change in the sequence of DNA. Not all epigenetic changes are passed from parent to offspring. For example, a cigarette smoker could acquire an epigenetic change in a lung cell; this change would not be passed to offspring.
2. A Barr body is an X chromosome in the somatic cells of female mammals that has undergone X-chromosome inactivation. It is highly compacted, and most of its genes are not expressed.
3. Epigenetics alters the way that genes are expressed, and thereby can affect an individual's traits. Such traits, in turn, may affect reproduction.

## Chapter 17

### Concept Checks

*Figure 17.1* Viruses vary with regard to their structure and their genomes. Genome variation is described in Table 17.1.

*Figure 17.4* There appears to be two nucleoids in this bacterial cell.

*Figure 17.5* The loop domains are held in place by proteins that bind to the DNA at the bases of the loops. The proteins also bind to each other.

*Figure 17.6* Bacterial chromosomes and plasmids are similar in that they typically contain circular DNA molecules. However, bacterial chromosomes are usually much longer than plasmids and carry many more genes. Also, bacterial chromosomes tend to be more compacted due to the formation of loop domains and supercoiling.

*Figure 17.7* 16 hours is the same as 32 doublings. So, $2^{32} = 4{,}294{,}967{,}296$. (The actual number would be much less because the cells would deplete the growth media and grow more slowly than the maximal rate.)

*Figure 17.9* During conjugation, only one strand of the DNA from an F factor is transferred from the donor to the recipient cell. The single-stranded DNA in both cells is then used as a template to create double-stranded F factor DNA in both cells.

*Figure 17.11* Transduction is not a normal part of the phage life cycle. It is a mistake in which a piece of the bacterial chromosome is packaged into a phage coat and is then transferred to another bacterial cell.

### BioConnections

*Figure 17.2* Viral release occurs as a budding process in which a membrane vesicle is formed that surrounds the capsid. Similarly, exocytosis involves the formation of a membrane vesicle that encloses some type of cargo.

*Figure 17.4* A nucleoid is not a membrane-bound organelle. It is simply a region where a bacterial chromosome is found. A cell nucleus in a eukaryotic cell has an envelope with a double membrane.

*Figure 17.10* Griffiths was able to show that genetic material was transferred to type R bacteria, which converted them to type S. This occurred via transformation. Later, Avery, MacLeod, and McCarty determined that DNA was the material that was being transferred.

### Testing Your Knowledge

17.1: 1. d 2. a

17.2: 1. c 2. d 3. e

17.3: 1. b 2. a

### Test Yourself

1. c 2. e 3. c 4. b 5. b 6. a 7. d 8. d 9. b 10. c

### Conceptual Questions

1. Viruses are similar to living cells in that they contain a genetic material that provides a blueprint to make new viruses. However, viruses are not composed of cells, and by themselves, they do not carry out metabolism, use energy, maintain homeostasis, or even reproduce. A virus or its genetic material must be taken up by a living cell to replicate.
2. Conjugation—The process involves a direct physical contact between two bacterial cells in which a donor cell transfers a strand of DNA to a recipient cell. Transformation—This occurs when a living bacterium takes up genetic information that has been released from a dead bacterium. Transduction—When a virus infects a donor cell, it incorporates a fragment of bacterial chromosomal DNA into a newly made virus particle. The virus then transfers this fragment of DNA to a recipient cell.
3. Horizontal gene transfer is the transfer of genes from one organism to another organism that is not its offspring. These acquired genes sometimes increase survival and therefore may have an evolutionary advantage. Such genes may even promote the formation of new species. From a medical perspective, an important example of horizontal gene transfer is when one bacterium acquires antibiotic resistance from another bacterium and then itself becomes resistant to that antibiotic. This phenomenon is making it increasingly difficult to treat a wide variety of bacterial diseases.

## Chapter 18

### Concept Checks

*Figure 18.2* No. A recombinant vector has been made, but it has not been cloned. In other words, many copies of the recombinant vector have not been made yet.

*Figure 18.3* The insertion of chromosomal DNA into the vector disrupts the *lacZ* gene, thereby preventing the expression of β-galactosidase. The functionality of *lacZ* can be determined by providing the growth medium with a colorless compound, X-Gal, which is cleaved by β-galactosidase into a blue dye. Bacterial colonies containing recircularized vectors form blue colonies, whereas colonies containing recombinant vectors carrying a segment of chromosomal DNA will be white.

*Figure 18.5* The 600-bp piece would be closer to the bottom. Smaller pieces travel faster through the gel.

*Figure 18.6* The primers are complementary to sequences at each end of the DNA region to be amplified.

*Figure 18.8* If a dideoxynucleotide ddNTP is added to a growing DNA strand, the strand can no longer grow because the 3'—OH group, the site of attachment for the next nucleotide, is missing.

*Figure 18.9* A fluorescent spot identifies a cDNA that is complementary to a particular DNA sequence. Because the cDNA was generated from mRNA, this technique identifies a gene that has been transcribed in a particular cell type under a given set of conditions.

*Figure 18.10* The sgRNA is composed of two different components of the bacterial defense system, crRNA and tracrRNA, which have been linked together.

*Figure 18.11* One reason is that more complex species tend to have more genes. A second reason is that species vary with regard to the amount of repetitive DNA present in their genome.

*Figure 18.16* For DNA transposons, inverted repeats are recognized by transposase, which cleaves the DNA and inserts the transposon into a new location.

*Figure 18.17* Retrotransposons. A single retrotransposon can be transcribed into multiple copies of RNA, which can be converted to DNA by reverse transcriptase, and inserted into multiple sites in the genome.

### BioConnections

*Figure 18.2* Plasmids are small, circular molecules of DNA that exist independently of the bacterial chromosome. They have their own origin of replication. Many plasmids carry genes that convey some type of selective advantage to the host cell, such as antibiotic resistance.

*Figure 18.6* Primers are needed in a PCR experiment because DNA polymerase cannot begin DNA replication on a bare template strand.

*Figure 18.13* The family members that are expressed at early stages of development (embryonic and fetal stages) have a higher affinity for oxygen than the adult form. This allows the embryo and fetus to obtain oxygen from its mother's bloodstream.

### Testing Your Knowledge

18.1: 1. d 2. b 3. b

18.2: 1. e 2. c

18.3: 1. e

18.4: 1. d 2. d

18.5: 1. d 2. c

### Test Yourself

1. d 2. b 3. b 4. b 5. c 6. e 7. e 8. a 9. c 10. e

### Conceptual Questions

1. A ddNTP is missing an oxygen at the 3' position. This prevents the further growth of a DNA strand, thereby causing chain termination.
2. a. yes

   b. No, it's only one chromosome in the nuclear genome.

   c. yes

   d. yes
3. The genome contains the information for the production of cellular proteins; it is a blueprint. The production of proteins is largely responsible for determining cellular characteristics, which, in turn, determine an organism's traits and allow it to reproduce.

## Chapter 19

### Concept Checks

*Figure 19.2* A single organism does not evolve. Populations may evolve from one generation to the next.

*Figure 19.5* Due to a changing global climate, the island fox became isolated from the mainland species. Over time, natural selection resulted in adaptations for the population on the island and eventually resulted in a new species with characteristics that are somewhat different from the mainland species.

*Figure 19.6* Many answers are possible. One example is the wing of a bird and the wing of a bat.

*Figure 19.9* Rhesus and green monkeys = 0, Congo puffer fish and European flounder = 2, and Rhesus monkey and Congo puffer fish = 10. Pairs that are closely related evolutionarily have fewer differences than do pairs that are more distantly related.

*Figure 19.11* If $C^R$ is 0.4, then $C^W$ must be 0.6, because the allele frequencies add up to 1.0. The heterozygote ($2pq$) equals 2(0.4)(0.6), which equals 0.48, or 48%.

*Figure 19.12* Over the short run, alleles that confer better fitness would be favored and increase in frequency, perhaps enhancing diversity. Over the long run, however, an allele that confers high fitness in the homozygous state may become monomorphic, thereby reducing genetic diversity.

*Figure 19.13* Stabilizing selection eliminates alleles that give phenotypes that deviate significantly from the average phenotype. For this reason, it tends to decrease genetic diversity.

*Figure 19.15* If malaria was eradicated, there would be no selective advantage for the heterozygote. The $H^S$ allele would eventually be eliminated because the $H^SH^S$ homozygote has a lower fitness. Directional selection would occur.

*Figure 19.18* This is likely to be a form of intersexual selection. Such traits are likely to be involved in mate choice.

*Figure 19.20* The bottleneck effect decreases genetic diversity. This may eliminate adaptations that promote survival and reproductive success. Therefore, the bottleneck effect makes it more difficult for a population to survive.

*Figure 19.22* Migration promotes gene flow, which tends to make the allele frequencies in neighboring populations more similar to each other. It also promotes genetic diversity by introducing new alleles into populations.

### BioConnections

*Figure 19.2* Both natural selection and chemical selection involve processes in which the relative proportions of something in a population increase compared with something else. In natural selection, it is the relative proportions of individuals with certain traits that increases. In chemical selection, molecules with certain characteristics increase their relative numbers compared with other molecules.

*Figure 19.21* There are lots of possibilities. The idea is that you are changing one codon to another codon that specifies the same amino acid. For example, changing a codon from GGA to GGG is likely to be neutral because both codons specify glycine.

*Figure 19.23* Inbreeding favors homozygotes. Initially, inbreeding would result in more homozygotes in a population. Over the long run, however, if a homozygote has a lower fitness, inbreeding would accelerate the elimination of the allele from the population.

### Feature Investigation Questions

1. The island has a moderate level of isolation but is located near enough to the mainland to have some migrants. The island is an undisturbed habitat, so the

researchers would not have to consider the effects of human activity on the study. Finally, the island had an existing population of ground finches that would serve as the study organism over many generations.

2. First, the researchers were able to show that beak depth is a genetic trait that has variation in the population. Second, the depth of the beak is an indicator of the types of seeds the birds can eat. The birds with larger beaks can eat larger and drier seeds; therefore, changes in the types of seeds available could act as a selective force on the bird population. During the study period, annual changes in rainfall occurred, which affected the seed sizes produced by the plants on the island. In the drier year, fewer small seeds were produced, so the birds would have to eat larger, drier seeds.
3. The researchers found that following the drought in 1978, the average beak depth in the finch population increased. This indicated that birds with larger beaks were better able to adapt to the environmental changes due to the drought and produce more offspring. This is direct evidence of the phenomenon of natural selection.

### Testing Your Knowledge

19.1: 1. e 2. e

19.2: 1. e 2. b

19.3: 1. e 2. e

19.4: 1. c 2. a 3. a

19.5: 1. b 2. b

19.6: 1. e 2. a

### Test Yourself

1. d 2. b 3. b 4. b 5. d 6. d 7. c 8. e 9. b 10. d

### Conceptual Questions

1. In convergent evolution, two different species from different lineages show similar characteristics because they occupy similar environments. An example is the long snout and tongue of both the giant anteater, found in South America, and the echidna, found in Australia. This enables these animals to feed on ants, but the two structures evolved independently. These observations support the idea that evolution results in adaptations to particular environments.
2. The frequency of the disease is a genotype frequency because it represents individuals with the disease. If we let $q^2$ represent the genotype frequency, then $q$ equals the square root of 0.04, which is 0.2. If $q = 0.2$, then $p = 1 - q$, which is 0.8. The frequency of heterozygous carriers is $2pq$, which is $2(0.8)(0.2) = 0.32$, or 32%.
3. Homologous structures are two or more structures that are similar because they are derived from a common ancestor. An example is the same set of bones that is found in the human arm, turtle arm, bat wing, and whale flipper. The forearms in these species have been modified to perform different functions. This supports the idea that all of these animals evolved from a common ancestor by descent with modification.

## Chapter 20

### Concept Checks

***Figure 20.1*** There are a lot of possibilities. Certain grass species look quite similar. Elephant species look very similar. And so on.

***Figure 20.3*** Temporal isolation is an example of a prezygotic isolating mechanism. Because the species breed at different times of the year, hybrid zygotes are not formed between the two species.

***Figure 20.5*** Hybrid sterility is a type of postzygotic isolating mechanism. A hybrid forms between the two species, but it is sterile.

***Figure 20.10*** The insects on different host plants would tend to breed with each other, and natural selection would favor the development of traits that are an advantage for feeding on that host. Over time, the accumulation of genetic changes may lead to reproductive isolation between the populations of insects.

***Figure 20.11*** If the *Gremlin* gene was underexpressed, less gremlin protein would be produced. Because gremlin protein inhibits apoptosis, more cell death would occur, and the result would probably be smaller feet, and maybe they would not be webbed.

***Figure 20.12*** By comparing the number of *Hox* genes in many different animal species, a general trend is observed that animals with more complex body structures have a greater number of *Hox* genes.

### BioConnections

***Figure 20.2*** Female choice is a prezygotic isolating mechanism.

***Figure 20.7*** The Hawaiian Islands have many different ecological niches that can be occupied by birds. The first founding bird inhabitants evolved to occupy those niches, thereby evolving into many different species.

### Feature Investigation Questions

1. Podos hypothesized that the morphological changes in the beak would also affect the birds' songs. A bird's song is an important component for mate choice. If changes in the beak alter the song of the bird, reproductive ability would be affected. Podos suggested that changes in the beak morphology could thus lead to reproductive isolation among the birds.
2. Podos first caught male birds in the field and collected data on beak size. The birds were banded for identification and released. Later, the banded birds' songs were recorded and analyzed for range of frequencies and trill rates. The results were then compared with similar data from other species of birds to determine if beak size constrained the frequency range and trill rate of the song.
3. The results of the study did indicate that natural selection on beak size due to changes in diet could lead to changes in song. Considering the importance of bird song to mate choice, the changes in the song could also lead to reproductive isolation. The phrase "by-product of adaptation" refers to changes in the phenotype that are not directly acted on by natural selection. In the case of the Galápagos finches, the changes in beak size were directly related to diet; however, as a consequence of that selection, the song pattern was also altered. The change in song pattern was a by-product.

### Testing Your Knowledge

20.1: 1. d 2. a

20.2: 1. b 2. b

20.3: 1. a 2. e

20.4: 1. d

### Test Yourself

1. b 2. c 3. e 4. d 5. c 6. a 7. b 8. b 9. c 10. b

### Conceptual Questions

1. Prezygotic isolating mechanisms prevent the formation of the zygote. An example is mechanical isolation, the incompatibility of genitalia. Postzygotic isolating mechanisms act after the formation of the zygote. An example is inviability of the hybrid that is formed. (Other examples shown in Figure 20.2 would also be correct.) Postzygotic mechanisms are more costly because some energy is spent in the formation of a zygote and its subsequent growth.
2. Allopatric speciation and sympatric speciation are distinguished by the level of geographic isolation. In allopatric species, two groups of the same species become very isolated from each other and then develop into separate species. This typically involves the accumulation of many small genetic changes that leads to adaptations and reproductive isolation. In sympatric speciation, there may not be a physical separation. For example, polyploidy can cause reproductive isolation and result in new species. Also, species may become adapted to local environments and develop into new species.

3. One example involves the *Hox* genes, which control morphological features along the anteroposterior axis in animals. An increase in the number of *Hox* genes during evolution is associated with an increase in body complexity and may have spawned many different animal species.

## Chapter 21

### Concept Checks

*Figure 21.2* A phylum is broader than a family.

*Figure 21.3* Yes. They can have many common ancestors, depending on how far back you go in the tree. For example, dogs and cats have a common ancestor that gave rise to mammals, and an older common ancestor that gave rise to vertebrates. The most recent common ancestor is the point at which two species diverged from each other.

*Figure 21.4* An order is a smaller taxon that would have a more recent common ancestor.

*Figure 21.7* A hinged jaw is the character common to the salmon, lizard, and rabbit, but not to the lamprey.

*Figure 21.8* Changing the second G to an A is common to species A, B, and C, but not to species G.

*Figure 21.11* Gorillas and humans would be expected to have fewer genetic differences because their common ancestor (named C) is more recent than that of orangutans and gorillas, which is ancestor B.

*Figure 21.12* Monophyletic groups are based on the concept that a particular group of species descended from a common ancestor. When horizontal gene transfer occurs, not all of the genes in a species were inherited from the common ancestor, so this muddles the concept of monophyletic groups.

### BioConnections

*Figure 21.1* The domains Bacteria and Archaea have organisms with prokaryotic cells.

*Figure 21.10* There are lots of possibilities. The idea is that you are changing one codon to another codon that specifies the same amino acid. For example, changing a codon from GGA to GGG is likely to be neutral because both codons specify glycine.

### Testing Your Knowledge

21.1: 1. c 2. b

21.2: 1. d 2. a 3. b

21.3: 1. b 2. c

21.4: 1. a 2. d

21.5: 1. c

### Test Yourself

1. c 2. d 3. e 4. d 5. b 6. d 7. b 8. b 9. c 10. e

### Conceptual Questions

1. The scientific name of every species has two parts, which are the genus name and the species epithet. The genus name is always capitalized, but the species name is not. Both names are italicized. An example is *Canis lupus*.
2. If neutral mutations occur at a relatively constant rate, they act as a molecular clock on which to measure evolutionary time. Genetic diversity between species that is due to neutral mutation gives an estimate of the time elapsed since the last common ancestor. A molecular clock can provide a timescale to a phylogenetic tree.
3. Morphological analysis focuses on morphological features of extinct and modern species. Many traits are analyzed to obtain a comprehensive picture of two species' relatedness. Convergent evolution leads to similar traits that arise independently in different species as they adapt to similar environments. Convergent evolution can, therefore, cause errors if a researcher assumes that a particular trait arose only once and that all species having the trait have the same common ancestor.

## Chapter 22

### Concept Checks

*Figure 22.2* In a sedimentary rock formation, the layer at the bottom is usually the oldest.

*Figure 22.3* For this time frame, you would analyze the relative amounts of the rubidium-87 and strontium-87 isotopes.

*Figure 22.9* Most animal species, including fruit flies, fishes, and humans, exhibit bilateral symmetry.

*Figure 22.15* Defining features of primates are grasping hands; eyes situated on the front of the head to facilitate binocular vision; a large brain; digits with flat nails instead of claws; and complex social behavior including well-developed parental care.

### BioConnections

*Figure 22.7* First, the process of membrane invagination created the nuclear envelope. Second, endocytosis may have enabled an ancient archaeon to take up a bacterial cell. Over time, bacterial genes were transferred to the nucleus, which gave rise to the eukaryotic nuclear genome. An engulfed bacterial cell eventually became a mitochondrion, and an engulfed cyanobacterial cell became a chloroplast in algae and plants.

*Figure 22.12* Two key features are mammary glands and hair. They also have specialized teeth, external ears, and enlarged skulls that harbor highly developed brains. Mammals are typically endothermic.

### Testing Your Knowledge

22.1: 1. b 2. e

22.2: 1. c 2. a 3. b

22.3: 1. e 2. a

### Test Yourself

1. c 2. e 3. a 4. b 5. c 6. e 7. a 8. b 9. d 10. d

### Conceptual Questions

1. The relative ages of fossils can be determined by the locations in sedimentary rock formation. Older fossils are found in lower layers. A common way to determine the ages of fossils is via radioisotope dating, which is often conducted using a piece of igneous rock from the vicinity of the fossil. A radioisotope is an unstable isotope of an element that decays spontaneously, releasing radiation at a constant rate. The half-life is the length of time required for a radioisotope to decay to exactly one-half of its initial value. To determine the age of a rock (and that of a nearby fossil), scientists can measure the amount of a given radioisotope as well as the amount of the decay product.
2. The process of membrane invagination created the nuclear envelope. In addition, endocytosis may have enabled an ancient archaeon to take up a bacterial cell. Over time, bacterial genes were transferred to the nucleus, which gave rise to the eukaryotic nuclear genome. An engulfed bacterial cell eventually became a mitochondrion, and an engulfed cyanobacterial cell became a chloroplast in algae and plants.
3. Several examples are described in this chapter. In some cases, catastrophic events like volcanic eruptions and glaciers caused mass extinctions, which allowed new species to evolve and flourish. In other cases, changing environmental conditions (for example, changes in temperature and moisture) played key roles. One interesting example is adaptation to terrestrial environments. Plant species evolved seeds that are desiccation resistant, whereas animal species evolved eggs with shells. Mammalian species evolved internal gestation.

## Chapter 23

### Concept Checks

***Figure 23.13*** The motion of the stiff filament of a prokaryotic flagellum is more like that of a propeller shaft than the flexible arms of a human swimmer.

***Figure 23.8*** Endospores allow bacterial cells to survive treatments and environmental conditions that would kill ordinary cells.

***Figure 23.22*** Gametes of *Plasmodium falciparum* undergo fusion to produce zygotes while in the mosquito host.

***Figure 23.31*** Fungal hyphae growing into a substrate having a much higher solute concentration will tend to lose cell water to the substrate, a process that could inhibit fungal growth. This process explains how salting or drying foods helps to protect them from fungal degradation and thus are common preservation techniques.

***Figure 23.33*** You might filter the air entering the patient's room and limit the entry of visitors.

### BioConnections

***Figure 23.35*** The amanitin toxin, by interfering with the function of RNA polymerase II, inhibits transcription in eukaryotic cells.

### Testing Your Knowledge

23.1: 1. a 2. e

23.2: 1. d 2. c

23.3: 1. b 2. c

23.4: 1. a

23.5: 1. c 2. a 3. b 4. d

23.6: 1. a

23.7: 1. c 2. a 3. b 4. d

### Test Yourself

1. b 2. c 3. b 4. d 5. c 6. a 7. c 8. d 9. e

### Conceptual Questions

1. Small cell size and simple division processes allow many bacteria to divide much more rapidly than eukaryotes. This helps to explain why food can spoil so quickly and why infections can spread very rapidly within the body. Other factors also influence these rates.
2. Many bacterial cells are capable of motility; bacterial cells that enter a body can be moved via the circulatory systems of animals or the internal transport systems of plants.
3. Endospores that enter the human body or food supplies may germinate, thereby producing progeny that produce toxins.

## Chapter 24

### Concept Checks

***Figure 24.13*** Mycorrhizal fungi provide their plant partners with water and minerals absorbed from a much larger area of soil than plant roots can exploit on their own.

### Feature Investigation Questions

1. Investigators fed the experimental mice sterile food so that food microbes would not be present, allowing the effects of fecal transplants to be more easily detected. In this case, the investigators removed a potentially confounding factor from their experiment.
2. Microbes from healthy children enabled mice receiving these microbes to grow larger than mice receiving microbes from stunted children. This was a first clue that the microbes directly affected child growth, a connection that was established by additional studies.

### Testing Your Knowledge

24.1: 1. c 2. c

24.2: 1. b

24.3: 1. b 2. b

24.4: 1. b

### Test Yourself

1. c 2. d 3. c 4. c 5. b 6. e 7. e 8. b 9. c 10. d

### Conceptual Questions

1. Ice microbiomes include colored surface films of algae that absorb solar radiation, thereby reducing the amount reflected into space. This has a warming effect on the physical environment, but could also influence other organisms present. Soil microbiomes contain diverse prokaryotic and eukaryotic species, including nitrogen-fixing bacteria and mycorrhizal fungi that aid plant growth.
2. One example of a host is the common, abundant nearshore green alga *Cladophora*, whose microbiota include nitrogen-fixing bacteria that aid host growth and obtain oxygen and organic molecules from the host.
3. Mycorrhizal fungi, which are components of many plant root microbiomes, mobilize and transport phosphorus and other minerals to plant roots. In this way, the host plant changes the mineral content of the soil.

## Chapter 25

### Concept Checks

***Figure 25.3*** Liverworts, mosses, and hornworts grow very close to surfaces such as soil or tree trunks. Raising their sporophytes off the surface helps to disperse spores into air currents.

***Figure 25.9*** The polyester cutin found in cuticle, sporopollenin on spore walls, and lignin on water-conducting tracheids of vascular tissues are resistant to decay and thus help plants fossilize.

***Figure 25.10*** During the Carboniferous period (Coal Age), atmospheric oxygen levels reached historic high levels that were able to supply the large needs of giant insects, which obtain oxygen by diffusion.

***Figure 25.22*** Importantly, ears of modern *Zea mays* do not readily shatter when the fruits are mature, as do those of teosinte. This feature enables human harvesting of the fruits (known as grains).

### Feature Investigation Questions

1. The investigators obtained many samples from around the world because they wanted to increase their chances of finding as many species as possible.
2. Although cannabinoids are produced in glandular hairs that cover the plant surface, these compounds are most abundant on leaves near the flowers. Collecting such leaves reduces the chances that compounds might be missed by the analysis.

### Testing Your Knowledge

25.1: 1. a 2. c

25.2: 1. d 2. e

25.3: 1. e 2. e

25.4: 1. d 2. b

25.5: 1. d

### Test Yourself

1. c 2. e 3. d 4. a 5. c 6. d 7. a 8. d 9. d 10. c

### Conceptual Questions

1. Flowering plants display many coevolutionary associations with animals; these associations enhance plant pollination and seed dispersal and also benefit the animals. Flowering plants have also taken advantage of wind (and sometimes water) to transport pollen and seeds effectively. These reproductive adaptations have greatly enhanced evolutionary diversification of the flowering plants.
2. Apple, strawberry, and cherry plants coevolved with animals that use the fleshy, sweet portion of the fruits as food and excrete the seeds, thereby dispersing them. Humans have sensory systems similar to those of the target animals and likewise are attracted by the same colors, odors, and tastes.
3. Vascular tissues allow tracheophytes to effectively conduct water from roots to stems and to leaves. Waxy cuticle helps prevent loss of water by evaporation through plant surfaces. Stomata allow plants to achieve gas exchange under moist conditions and help them avoid losing excess water under arid conditions.

## Chapter 26

### Concept Checks

***Figure 26.6*** The coelom functions as a hydrostatic skeleton, which aids in movement. This feature permitted increased burrowing activity and contributed to the development of a profusion of wormlike body shapes.

***Figure 26.8*** The main members of the Ecdysozoa are the arthropods (insects, spiders, and crustaceans) and the nematodes.

***Figure 26.11*** Sponges aren't eaten by other organisms because they produce toxic chemicals and contain needle-like silica spicules that are hard to digest.

***Figure 26.12*** The dominant life stages are jellyfish: medusa; sea anemone: polyp; Portuguese man-of-war: polyp (in a large floating colony).

***Figure 26.14*** Having no specialized respiratory or circulatory system, flatworms obtain oxygen by diffusion. A flattened shape ensures no cells are too far from the body surface.

***Figure 26.18*** (1) A ciliary feeding device, and (2) a respiratory device are the two main functions of the lophophore.

***Figure 26.21*** Some advantages of segmentation are organ duplication, minimization of body distortion during movement, and specialization of some segments.

***Figure 26.23*** An annelid is segmented and possesses a true coelom whereas a nematode is unsegmented and has a pseudocoelom. In addition, nematodes molt but annelids do not.

***Figure 26.26*** All arachnids have a body consisting of two tagmata: a cephalothorax and an abdomen. Insects have three tagmata: a head, thorax, and abdomen.

***Figure 26.28*** Some key insect adaptations are the development of wings, an exoskeleton that reduced water loss and aided in the colonization of land, and the development of a variety of mouthparts.

***Figure 26.31*** In embryonic development, deuterostomes have radial cleavage, indeterminate cleavage, and the blastopore becomes the anus. (In protostomes, cleavage is spiral and determinate, and the blastopore becomes the mouth.)

### BioConnections

***Figure 26.4*** A shared derived character.

***Figure 26.20*** Mollusks arose in the Cambrian period, 543–490 mya. Three hundred million years later ammonites flourished, yet none are alive today.

***Figure 26.29*** These organs, called statocysts, are located at the base of the antennules.

***Figure 26.33*** Notochord.

### Testing Your Knowledge

26.1: 1. b

26.2: 1. a 2. c 3. c

26.3: 1. d

26.4: 1. d 2. b

26.5: 1. c

26.6: 1. c 2. c 3. a

26.7: 1. b 2. d 3. a

26.8: 1. c 2. e 3. c

### Test Yourself

1. e 2. e 3. d 4. d 5. d 6. d 7. d 8. b 9. c 10. c

### Conceptual Questions

1. (1) Type of body symmetry. (2) Presence of germ layers. (3) Patterns of embryonic development.
2. The five main feeding methods used by animals are (1) suspension feeding, (2) decomposition, (3) herbivory, (4) predation, and (5) parasitism. Suspension feeding is usually used to filter out food particles from the water column. A great many phyla are filter feeders, including sponges, rotifers, lophophorates, some mollusks, echinoderms, and tunicates. Decomposers usually feed on dead material such as animal carcasses or dead leaves. For example, many fly and beetle larvae feed on dead animals, and earthworms consume dead leaves from the surface of the Earth. Earthworms and crabs also sift through soil or mud, eating the substrate and digesting the soil-dwelling bacteria, protists, and dead organic material. Herbivores eat plants or algae and are especially common in the arthropoda. Adult moths and butterflies also consume nectar. Snails are also common plant feeders. Predators feed on other animals, killing their prey, and may be active hunters or sit-and-wait predators. Many scorpions and spiders actively pursue their prey, whereas web-spinning spiders ambush their prey using webs. Parasites also feed on other animals but do not normally kill their hosts. Endoparasites live inside their hosts and include flukes, tapeworms, and nematodes. Ectoparasites live on the outside of their hosts and include ticks and lice.
3. Polyphyletic.

## Chapter 27

### Concept Checks

***Figure 27.1*** Vertebrates (but not invertebrates) usually possess a (1) notochord; (2) dorsal hollow nerve chord; (3) pharyngeal slits; (4) postanal tail, exhibited by all chordates; (5) vertebral column; (6) cranium; and (7) endoskeleton of cartilage or bone.

***Figure 27.6*** Ray-finned fishes (but not sharks) have a (1) bony skeleton; (2) mucus-covered skin; (3) swim bladder; and (4) operculum covering the gills.

***Figure 27.7*** Both lungfishes and coelocanths are Sarcopterygians, having lobe fins.

***Figure 27.9*** The advantages to animals that moved onto land included an oxygen-rich environment and a bonanza of food in the form of terrestrial plants and the insects that fed on them.

***Figure 27.10*** No. Caecilians and some salamanders give birth to live young.

***Figure 27.11*** Besides the amniotic egg, other critical innovations in amniotes are thoracic breathing; internal fertilization; a thicker, less permeable skin; and more efficient kidneys.

***Figure 27.12*** Snakes evolved from tetrapod ancestors but subsequently lost their limbs. Some species have tiny vesitgial limbs.

***Figure 27.15*** Adaptations in birds to reduce body weight for flight include a lightweight skull; reduction of organ size; and a reduction of organs outside of breeding season. Also female birds have one ovary and relatively few eggs, and no urinary bladder.

### BioConnections

***Figure 27.5*** Collagen-secreting cells. Cartilage is not mineralized and is softer and more flexible than bone.

***Figure 27.14*** Both classes have four-chambered hearts and care for their young.

### Feature Investigation Questions

1. The researchers were interested in determining the method in which *Hox* genes controlled limb development.
2. The researchers bred mice that were homozygous for certain mutations in specific *Hox* genes. This allowed the researchers to determine the function of individual genes.
3. The researchers found that homozygous mutants would develop limbs of shorter lengths compared to the wild-type mice. The reduced length was due to the lack of development of particular bones in the limb, specifically, the radius, ulna, and some carpels. These results indicated that simple mutations in a few genes could lead to dramatic changes in limb development.

### Testing Your Knowledge

27.1: 1. b 2. e

27.2: 1. c 2. d 3. e

27.3: 1. e 2. d

27.4: 1. c 2. d 3. c 4. e

27.5: 1. b

### Test Yourself

1. a 2. d 3. a 4. d 5. c 6. d 7. c 8. d 9. a 10. c

### Conceptual Questions

1. Both taxa have external limbs that move when the attached muscles contract or relax. The difference is that arthropods have external skeletons with the muscles attached internally, whereas vertebrates have internal skeletons with the muscles attached externally.
2. Endothermy (warm-bloodedness) probably evolved independently in both birds and mammals. If the common ancestor of reptiles and birds were endothermic, the chances are that all reptiles would be endothermic.
3. Possibly. Both birds and reptiles lay amniotic eggs and possess scales, though these only cover the legs in birds. Birds and crocodilians also share a four-chambered heart. Finally, birds share many skeletal similarities with certain dinosaurs.

## Chapter 28

### Concept Checks

***Figure 28.6*** As in the case of shoots, the capacity to divide the root into two equal pieces by means of a line drawn from the circular edges through the center would indicate that a root has superficial radial symmetry. In order to determine that an organ has radial symmetry at the cellular level, you would have to compare the microscopic views of randomly chosen, wedge-shaped pieces of cross slices. If the structure of the wedges is similar, the organ has radial symmetry at the microscopic level.

***Figure 28.8*** Locating stomata on the darker and cooler lower leaf surface helps reduce water loss from the leaf.

***Figure 28.12*** A twig having five sets of bud scale scars is likely to be approximately 6 years old.

***Figure 28.16*** Cactus stems are green and photosynthetic, playing the role served by the leaves of most plants.

***Figure 28.20*** A woody stem builds up a thicker layer of wood than inner bark, in part because older tracheids and vessel element walls are not lost during shedding of bark, which is the case for secondary phloem. In addition, plants typically produce a greater volume of xylem than phloem tissue per year, in part because vessel elements are relatively wide. A large volume of water-conducting tissue helps plants maintain a large amount of internal water.

***Figure 28.23*** Lateral roots are produced from internal meristematic tissue because roots do not produce axillary buds like those from which shoot branches develop. Internal production of branch roots helps to prevent them from shearing off as the root tip grows through abrasive soil.

### BioConnections

***Figure 28.1*** The flowering plant life cycle is like that of other plants in that a haploid gametophyte generation alternates with a diploid sporophyte generation, but the size of flowering plant gametophytes is smaller than those of moss and fern.

### Feature Investigation Questions

1. The advantages of using natural plants include the opportunity to avoid influencing plants with unnatural environmental factors, such as artificial light, and the exposure of all experimental plants to similar growth conditions. In addition, the investigators studied the leaves of some large trees, which would be hard to accommodate in a greenhouse.
2. Pinnately-veined leaves were splinted to prevent their breaking, since they were cut at the single main vein, which has both support and conducting functions.
3. Sack and associates measured leaf water conduction at two or more places on each leaf because the effect of cutting a vein might have affected some portions of leaves more than others.

### Testing Your Knowledge

28.1: 1. d 2. a

28.2: 1. e 2. b

28.3: 1. d 2. e

28.4: 1. d 2. c

### Test Yourself

1. d 2. b 3. a 4. c 5. a 6. d 7. e 8. b

### Conceptual Questions

1. If overall plant architecture were bilaterally symmetrical, plants would be shaped like higher animals, with a distinct front (ventral surface) and back (dorsal surface). By comparison with radially symmetrical organisms, bilaterally symmetrical plants would have reduced ability to deploy branches and leaves in a way that would fill available lighted space and would thus be unable to take optimal photosynthetic advantage of their habitats.
2. If leaves were generally radially symmetrical (shaped like spheres or cylinders), leaves would not have maximal ability to absorb sunlight, and they would not be able to optimally disperse excess heat from their surfaces.
3. Although tall herbaceous plants exist (palms and bamboo are examples), the additional support and water-conducting capacity that are provided by secondary xylem allow woody plants to grow tall.

## Chapter 29

### Concept Checks

***Figure 29.4*** Auxin efflux carriers could be located on the upper sides of root cells, thereby allowing auxin to move upward in roots.

***Figure 29.6*** The triple response of dicot seedlings to internally produced ethylene allows them to protect the delicate apical meristem from damage as the seedling emerges through the soil.

***Figure 29.7*** The active conformation of phytochrome absorbs far-red light. Such absorption causes the active conformation of phytochrome to change to the inactive conformation and to move out of the nucleus and into the cytosol.

*Figure 29.8* The inactive conformation of phytochrome would absorb the red portion of sunlight, thereby converting phytochrome into the active conformation.

*Figure 29.9* Exposing plants to brief periods of darkness during the daytime will have no effect on flowering because flowering is determined by night length.

*Figure 29.11* Yes, just as shoots exhibit negative gravitropism in upward growth, roots are capable of using negative phototropism to grow downward, because light decreases with depth in the soil.

*Figure 29.13* Predators are more likely to be able to find their prey if the latter are concentrated and exposed while feeding on plants. Plants benefit when predation removes herbivores, a process that lessens damage to plants.

*Figure 29.15* Similar suites of protective plant hormones, such as jasmonic acid, are used in both types of defenses.

### BioConnections

*Figure 29.3* The defense hormone/plasma membrane receptor is the plant signaling system most similar to the general diagram shown in Figure 8.4.

### Testing Your Knowledge

29.1: 1. a

29.2: 1. d 2. b 3. a 4. c 5. e 6. d

29.3: 1. a 2. c 3. b 4. c

29.4: 1. c

29.5: 1. e

### Test Yourself

1. c 2. c 3. a 4. e 5. d 6. c 7. d 8. d

### Conceptual Questions

1. Behavior is defined as the responses of living things to a stimulus. Therefore, because plants display many kinds of responses to diverse stimuli, they display behavior.
2. Many kinds of disease-causing bacteria and fungi occur in nature, and these organisms evolve very quickly, producing diverse elicitors. Thus, plants must maintain a stock of resistance genes, each having many alleles.
3. Talking implies a conversation with "listeners" who detect a message and respond to it. Thus, plants that exude volatile compounds that attract enemies of herbivores could be interpreted as "talking" to those enemies. The message is "Hey, you guys, there's food for you over here." In addition, research has revealed that some plants near those under attack respond to volatile compounds by building up defenses. "Talking" to other plants does not enhance the "talker's" fitness. But the ability to "listen" enhances the "listener's" fitness, because it can take preemptive actions to prevent attack.

## Chapter 30

### Concept Checks

*Figure 30.8* Mineral leaching occurs more readily from sandy soils than from clay soils.

*Figure 30.10* Soil crusts containing nitrogen-fixing cyanobacteria increase soil fertility, fostering the growth of larger plants that stabilize soils against erosion and provide forage for animals.

*Figure 30.16* Your drawing should show that equivalent amounts of water will move into and out of a turgid cell having a water potential of 0, when placed into pure water. When placed in pure water, a plasmolyzed cell having a water potential of –1.0 MPa will gain water. Your drawing should show that more water moves into the cell than out of it. When placed in pure water, a flaccid cell having a water potential of –0.5 MPa will gain water. This is because water moves from a region of higher water potential to a region of lower water potential, and 0 is greater than –0.5. Your drawing should show that more water moves into the flaccid cell than out of it.

*Figure 30.22* The evaporation of water has a powerful cooling effect because it disperses heat so effectively. Water has the highest heat of vaporization of any known liquid.

*Figure 30.25* In its desert habitat, times of drought and contrasting availability of water sufficient to support the development and photosynthetic function of leaves do not occur at predictable times, as is the case for temperate forests. For this reason, ocotillo leaf abscission is not amenable to the evolution of genetic mechanisms that allow leaf drop to be precisely timed in anticipation of the onset of drought.

### BioConnections

*Figure 30.19* Tight junctions of intestinal epithelium and Casparian strips of endodermal cells of plant roots both incorporate materials that form a tight seal, preventing movement of materials from one location to another.

### Testing Your Knowledge

30.1: 1. e 2. b

30.2: 1. c 2. c

30.3: 1. c 2. d

30.4: 1. d 2. d

30.5: 1. d 2. c 3. d

### Test Yourself

1. e 2. a 3. b 4. d 5. a 6. e 7. a 8. c

### Conceptual Questions

1. In the case of plant fertilizers, more is not better, because the ion concentration of overfertilized soil may become so high as to draw water from plant cells. In this case, the cells would be bathed in a hypertonic solution and would likely lose water to the solution. If plant cells lose too much water, they will die.
2. Agricultural experts are concerned that adding excess fertilizer to crop fields increases the costs of crop production. Ecologists are concerned that excess fertilizers will wash from crop fields into natural waters and cause harmful overgrowths of cyanobacteria, algae, and aquatic plants. Methods for closely monitoring crop nutrient needs so that only the appropriate amount of fertilizer is applied would help to allay both groups' concerns.
3. When the natural vegetation is removed, transpiration stops, so water is not transported from the ground to the atmosphere, where it may be an important contributor to local rainfall. Extensive removal of plants actually changes local climate in ways that reduce agricultural productivity and human survival.

## Chapter 31

### Concept Checks

*Figure 31.2* Because gametophytes are haploid, they lack the potential for allele variation at each gene locus that is present in diploid sporophytes. Hence, gametophytes are more vulnerable to environmental stresses. By living within the diploid tissues of flowers, flowering plant gametophytes are protected to some extent, and the plant does not lose its gamete-producing life cycle stage.

*Figure 31.3* Some flowers lack some of the major flower parts.

*Figure 31.4* By clustering its stamens around the pistil, the hibiscus flower increases the chance that a pollinator will both pick up pollen and deliver pollen from another hibiscus flower on the same trip.

*Figure 31.6* The absence of showy petals often correlates with wind pollination, because large petals would interfere with the shedding of pollen in the wind.

*Figure 31.9* The maximum number of cells in a mature male gametophyte of a flowering plant is three: a tube cell and two sperm cells.

*Figure 31.10* Female gametophytes are not photosynthetic and cannot produce their own food. Enclosed within ovules, female gametophytes lack direct access to the outside environment. Carpels contain veins of vascular tissue that bring nutrients from sporophytic tissue to ovules.

*Figure 31.13* During their maturation, the cotyledons of eudicot seeds absorb the nutrients originally present in endosperm.

## BioConnections

*Figure 31.1* The female gametophytes of a flowering plant are produced within ovules that lie within ovaries of flowers. The male gametophytes of flowering plants are produced by pollen grains that originate within the stamen anthers of a flower.

*Figure 31.11* The plant pollen tube is analogous in function to a human penis in that both structures accomplish internal fertilization. The plant pollen tube grows long enough to deposit sperm at the micropyle within the body of the female gametophyte, much as an animal penis deposits sperm within the female's body. In both cases, sperm are more likely to survive and accomplish fertilization than if they were deposited outside the female body.

## Feature Investigation Questions

1. The large flowers of this lily enabled investigators to more easily mark petals and record the positions of marks over time.
2. Time-lapse video reduced the amount of time investigators would have to spend recording changes in the positions of petal marks.
3. Results obtained by using mathematical models can be compared with actual measurements to assess the accuracy of the models. An accurate mathematical model indicates a relatively full understanding of the physical processes involved. If models are sufficiently accurate, they can be utilized in other situations.

## Testing Your Knowledge

31.1: 1. d 2. a

31.2: 1. c

31.3: 1. b

31.4: 1. b

31.5: 1. d

## Test Yourself

1. a 2. b 3. d 4. b 5. d 6. e 7. b 8. c 9. c

## Conceptual Questions

1. Pollen grains are vulnerable to mechanical damage and microbial attack during the journey through the air from the anthers of a flower to a stigma. Sporopollenin is an extremely tough polymer that helps to protect pollen cells from these dangers. The function of the beautiful sculptured patterns of sporopollenin on pollen surfaces is unclear.
2. The embryos within seeds are vulnerable to mechanical damage and microbial attack after they are dispersed. Seed coats protect embryos from these dangers and also help to prevent seeds from germinating until conditions are favorable for seedling survival and growth.
3. Flower diversity is an evolutionary response to diverse pollination circumstances. For example, plants such as oak and corn that are wind-pollinated produce flowers having a poorly developed perianth. If such wind-pollinated flowers had large, showy perianths, they would get in the way of pollen dispersal or acquisition. On the other hand, flowers that are pollinated by animals often have diverse shapes and attractive petals of differing colors or fragrances that have coevolved with different types of animal pollinators.

# Chapter 32

## Concept Checks

*Figure 32.6* No. The brain, for example, does not contain muscle tissue (although the blood vessels supplying the brain do contain smooth muscle).

*Figure 32.7* Surface area is important to any living organism that needs to exchange materials with the environment. A good example of a high surface area/volume ratio is that of most tree leaves. This makes leaves ideally suited for such processes as light absorption (required for photosynthesis) and the exchange of gases and water with the environment.

*Figure 32.8* No, not necessarily. Body temperature, for example, is maintained at different set points in birds and mammals. Other vertebrates and most invertebrates do not have temperature set points; their body temperature simply conforms close to that of the environment. As another example, a giraffe has a set point for blood pressure that is higher than that of a human being, because a giraffe's circulatory system must generate enough pressure to pump blood up its long neck to its brain.

*Figure 32.9* Humans are endotherms and homeotherms.

*Figure 32.10* The spider is a heterotherm. Its body temperature changes with that of the environment.

*Figure 32.12* Heat moves from the warmer artery to the cooler veins. This prevents heat from reaching the extremities and being lost to the environment.

*Figure 32.13* Blood, including plasma and blood cells, would leak out of the blood vessel into the interstitial space. The fluid level of the bloodstream would decrease, and that of the interstitial space near the site of the injury would increase. Eventually the blood that entered the interstitial space would be degraded by enzymes, resulting in the characteristic skin appearance of a bruise. If the injury were very severe, the fluid level in the blood could decrease to a point where the various tissues and organs of the body would not receive sufficient nutrients and oxygen to function normally.

*Figure 32.14* No, obligatory exchanges must always occur, but animals can minimize obligatory losses through modifications in behavior. For example, terrestrial animals that seek shade on a hot, sunny day reduce evaporative water loss. As another example, reducing activity minimizes water loss due to respiration.

## BioConnections

*Figure 32.6* Note that both the stomach and the intestine depicted in Figures 32.6 and 37.6 contain layers of muscle wrapped around the lumen. Although you will learn later that the stomach and intestine have many different functions, this similarity in anatomy suggests that both of these organs may perform the similar activity of mechanically breaking apart chunks of food, and propelling the contents from one region to another.

*Figure 32.8* Negative feedback is a common way in which the product of a chemical reaction inhibits the rate of that reaction, thus maintaining homeostatic levels of the product in a cell.

## Feature Investigation Questions

1. Symptoms of prolonged, heavy exercise include fatigue, muscle cramps, and even occasionally seizures. Fatigue results from the reduction in blood flow to muscles and other organs. Muscle cramps and seizures are the result of imbalances in plasma electrolyte levels. Cade and his colleagues hypothesized that maintaining proper water and electrolyte levels would prevent these problems, and that if water and electrolyte levels were maintained, athletic performance should not decrease as rapidly with prolonged exercise.
2. To test their hypothesis, the researchers created a drink that would restore the correct proportions of lost water and electrolytes within the athletes. If the athletes consumed the drink during exercise, they should not experience as much fatigue or muscle cramping, and thus their performance should be enhanced compared with a control group of athletes that drank only water.
3. A drink containing electrolytes such as those in sports drinks would be of little or no benefit to a person who was not exercising with at least some effort since perspiration (and thus loss of ions and water) would usually be minor under such conditions.

## Testing Your Knowledge

32.1: 1. b 2. (See following text)

| | |
|---|---|
| Supports the body and produces locomotion | Musculoskeletal system |
| Absorbs nutrients into the body | Digestive system |
| Regulates growth, metabolism, reproduction, mineral balance | Endocrine system |

| | |
|---|---|
| Distributes solutes throughout the body | Circulatory system |
| Provides a barrier against pathogens entering the body | Integumentary system |

32.2: 1. d

32.3: 1. a 2. c

32.4: 1. c 2. e

32.5: 1. e 2. e

## Test Yourself

1. c 2. c 3. d 4. b 5. c 6. c 7. d 8. b 9. d 10. c

## Conceptual Questions

1. Structure and function are related in that the function of a given organ, for example, depends in part on the organ's size, shape, and cellular and tissue arrangement. Clues about a physical structure's function can often be obtained by examining the structure's form. For example, the extensive surface area of a moth's antennae suggests that the antennae are important in detecting the presence of airborne chemicals. Likewise, any structure that contains a large surface area for its volume is likely involved in some aspect of signal detection, cell-to-cell communication, or transport of materials within the animal or between the animal and the environment. Surface area increases by a power of 2, and volume increases by a power of 3 as an object enlarges; this means that in order to greatly increase surface area of a structure such as an antenna, without occupying enormous volumes, specializations must be present (such as folds) to package the structure in a small space.
2. Homeostasis is the ability of animals to maintain a stable internal environment by adjusting physiological processes, despite changes in the external environment. Examples include maintenance of ion and water balance, pH of body fluids, and body temperature. Some animals conform to their external environment to achieve homeostasis, but others regulate their internal environment themselves. Ectotherms depend on the environment to warm themselves, while endotherms generate their own internal heat. Heterotherms have body temperatures that may vary widely; homeotherms maintain a relatively stable body temperature. Body temperature influences most chemical reactions in an animal's body and is therefore closely linked with homeostasis.
3. Maintaining homeostasis requires continual supplies of energy. Animals consume food, and the energy from that food helps sustain activities that maintain variables such as body temperature and pH, and synthesis of complex molecules. Without this energy, it would be difficult or impossible for animals to maintain many important biological processes within a narrow range despite changes in the environment.

# Chapter 33

## Concept Checks

***Figure 33.3*** Many reflexes, such as the knee-jerk reflex, cannot be prevented once started. Others, however, can be controlled to an extent. Open your eyes widely and gently touch your eyelashes. A reflex that protects your eye will tend to make you close your eyelid. However, you can overcome this reflex with a bit of difficulty if you need to, for example, when you are putting in contact lenses.

***Figure 33.5*** Yes, the flow of $K^+$ down its concentration gradient does create an electrical gradient because $K^+$ is electrically charged. The net flow of $K^+$ will stop when the concentration gradient balances the electrical gradient. This occurs at the equilibrium potential.

***Figure 33.8*** When the $K^+$ channels open (at 1 msec), the $Na^+$ channels would still be opened, so the part of the curve that slopes downward would not occur as rapidly, and perhaps the cell would not be able to restore its resting potential.

***Figure 33.10*** The action potential can move faster down an axon. This is especially important for long axons, such as those that carry signals from the spinal cord to distant muscles.

***Figure 33.12*** In the absence of such enzymes, neurotransmitters would remain in the synapse for too long, and the postsynaptic cell could become overstimulated. In addition, the ability of the postsynaptic cell to respond to multiple, discrete inputs from the presynaptic cell would be compromised.

***Figure 33.20*** No, these divisions are present in all vertebrates.

***Figure 33.21*** The occipital lobe is part of the cerebral cortex, and therefore also part of the cerebrum.

***Figure 33.23*** Thinking requires energy! Even daydreaming requires energy; imagine how much energy the brain uses when you concentrate for 60 minutes on a difficult exam. In fact, you just expended energy thinking about this question!

## BioConnections

***Figure 33.5*** Electrochemical gradients are used in transporting solutes across a cell membrane. They are also important in the production of ATP in mitochondria, in muscle contraction, and osmotic regulation of cell volume.

***Figure 33.6*** Water molecules move through membrane channels called aquaporins.

***Figure 33.14*** Animals are multicellular heterotrophs (cannot make their own food) whose cells lack a cell wall. Most animals have a nervous system, muscles, the ability to move about during at least some phase of their life cycle, and to reproduce sexually.

## Feature Investigation Questions

1. Gaser and Schlaug hypothesized that repeated exposure to musical training would increase the size of certain areas of the brain associated with motor, auditory, and visual skills. All three skills are commonly used in reading and performing musical pieces. The researchers used MRI to examine the areas of the brain associated with motor, auditory, and visual skills in three groups of individuals: professional musicians, amateur musicians, and nonmusicians. The researchers found that certain areas of the brain were larger in the professional musicians compared to the other groups, and larger in the amateur musicians compared to the nonmusicians.
2. Schmithorst and Holland found that, when exposed to music, certain regions of the brains of musicians were activated differently compared with the brains of nonmusicians. This study supports the hypothesis that there is a difference in the brains of musicians versus nonmusicians. The experiment conducted by Gaser and Schlaug compared the size of certain regions of the brain among professional musicians, amateur musicians, and nonmusicians. Schmithorst and Holland, however, were also able to detect functional differences between musicians and nonmusicians.
3. A trained athlete might be expected to have greater development of brain areas devoted to motor functions. For example, athletes who engage in repetitive activities such as tennis and badminton have been found to have greater volumes of motor and balance regions of the brain including the cerebellum. Someone deaf from birth might be predicted to have decreased development of auditory regions such as those in the temporal lobes. However, it is not necessarily this simple. For example, MRI studies sometimes but not always reveal differences in cell number of the temporal lobes of deaf versus hearing subjects, but there is generally significantly less development of fiber tracts (white matter) in this region in the brains of deaf persons. Therefore, the size of a brain structure (that is, the number of neurons) by itself may not always be predictive of function.

## Testing Your Knowledge

33.1: 1. d 2. c

33.2: 1. a 2. e

33.3: 1. d 2. a

33.4: 1. a 2. c

33.5: 1. (see following text)

nerve net—cnidarian

two ventral nerve cords—*Planaria*

nerve ring and radial nerves—sea star (echinoderm)

brain with several well-defined subdivisions—*Drosophila*

33.5: 2. c

33.6: 1. c

33.6: 2. (see following text)

thalamus—filtering incoming sensory information

hypothalamus—controlling body temperature and circadian rhythms

basal nuclei—fine tuning, starting, or inhibiting movement

cerebral cortex—decision making and exercising judgment

pons—controlling breathing/relay between cerebellum and rest of brain

33.7: 1. d 2. b

### Test Yourself

1. c 2. d 3. d 4. e 5. b 6. d 7. b 8. c 9. a 10. b

### Conceptual Questions

1. In a graded potential, a weak stimulus causes a small change in the membrane potential, whereas a strong stimulus produces a greater change. Graded potentials occur along the dendrites and cell body. If a graded potential reaches the threshold potential at the axon hillock, an action potential results. This is a change in the membrane potential that is of a constant value and is propagated from the axon hillock to the axon terminal.
2. An increase in extracellular $Na^+$ concentration would slightly depolarize neurons, thereby changing the resting membrane potential. This effect would be minimal, however, because the resting membrane is not very permeable to $Na^+$. However, the shape of the action potentials in such neurons would be a little steeper, and the peak a little higher, because the electrochemical gradient favoring $Na^+$ entry into the cell through voltage-gated channels would be greater.
3. The activities of animal nervous systems are replete with examples of new properties emerging from complex interactions. For example, you learned about reflexes in this chapter, which are behaviors that emerge from interactions between individual neurons that form communication circuits between the peripheral and central nervous systems. You also learned about such "higher" properties as conscious thought, which also emerges from the interactions between many individual cells, each of which is in communication with up to hundreds of thousands of other cells. Individually, the cells cannot "think," but networked together in elaborate ways, a person like yourself can think, remember, plan ahead, and interpret your environment.

## Chapter 34

### Concept Checks

***Figure 34.8*** This orientation permits animals to detect circular or angular movement of the head in three different planes. The fluid in a canal that is oriented in the same plane as the plane of movement will respond maximally to the movement. For example, the canal that is oriented horizontally would respond greatest to horizontal movements, while the other two canals would not. Overall, by comparing the signals from the three canals, the brain can interpret the motion in three dimensions.

***Figure 34.9*** The advantage of having heat-sensing organs on either side of the head is that it provides information about the source of the heat, much like our ability to detect the origin of a sound based on differences in arrival times to our two ears.

***Figure 34.14*** Because red-green color blindness is a sex-linked recessive gene, males require only a single defective allele on an X chromosome, whereas females require two defective alleles, one on each X chromosome.

***Figure 34.17*** Photoreceptors (rods and cones) synapse with bipolar cells which in turn synapse with ganglion cells.

***Figure 34.20*** The receptor proteins are located in the membranes of microvilli extending from the sensory receptor cells.

### BioConnections

***Figure 34.1*** The term sensory receptor refers to a type of cell that can respond to a particular type of stimulus. The term membrane receptor refers to a protein within a cell membrane that binds a ligand, thereby generating signals that initiate a cellular response.

***Figure 34.3*** Cilia are cell extensions that contain, in their internal structure, microtubules and motor proteins that cause the cilia to beat, or move, in a coordinated fashion. Stereocilia are membrane projections that are not motile, but instead are deformed by the movements of surrounding fluids.

***Figure 34.7*** Statoliths are also found in the roots and shoots of plants. They serve as a gravity-detection mechanism that results in roots growing downward, and shoots upward.

### Feature Investigation Questions

1. One possibility is that many different types of odor molecules might bind to one or just a few types of receptor proteins, with the brain responding differently depending on the number or distribution of the activated receptors. The second hypothesis is that organisms can make a large number of receptor proteins, each type binding a particular odor molecule or group of odor molecules. According to this hypothesis, it is the *type* of receptor protein, and not the number or distribution of receptors, that is important for olfactory sensing. The researchers extracted RNA molecules from the olfactory receptor cells of the nasal epithelium. They then used this RNA to identify genes that encoded G-protein-coupled receptor proteins.
2. The results of the experiment conducted by Buck and Axel support the hypothesis that animals discriminate between different odors based on having a variety of receptor proteins that recognize different odor molecules. Current research suggests that each olfactory receptor cell has a single type of receptor protein that is specific to particular odor molecules. Because most odors are due to multiple chemicals that activate many different types of odor receptor proteins, the brain detects odors based on the combination of the activated receptor proteins. Odor seems to be discriminated by many ollfactory receptor proteins, which are in the membrane of separate olfactory receptor cells.
3. Among other things, animals use odors to detect potential sources of food, shelter, mates, and danger. Discerning between the odors of food versus those of potentially toxic material is key to the survival of many animals (think, for example, how we recognize fresh milk and spoiled milk). Many animals also use olfaction in finding shelter; some birds, for example can distinguish the odors produced by particular beneficial plants that they use to build their nests. These plants have natural pathogen-killing properties and help maintain a healthier nest site. Odor preference also plays an important role in animals finding mates via the release of pheromones, particularly animals that live in dark environments such as nocturnal animals. The ability to discern the odors of different predators even when the predator may not be close enough to hear, or may not be in sight, is a huge survival advantage for many animals, who generally can differentiate such odors from those of nonthreatening species.

### Testing Your Knowledge

34.1: 1. b

34.2: 1. c 2. c

34.3: 1. c

34.4: 1. e 2. c

34.5: 1. c 2. c

34.6: 1. b 2. c

### Test Yourself

1. e 2. b 3. a 4. b 5. d 6. b 7. c 8. b 9. d 10. b

### Conceptual Questions

1. Sensory transduction—The process by which incoming stimuli are converted into neural signals. An example would be the signals generated in the retina when a photon of light strikes a photoreceptor. Perception—An awareness of the sensations that are experienced. An example would be an awareness of what a particular visual image is.
2. The organ of Corti contains the hair cells and sensory neurons that initiate signaling. The hair cells sit on top of the basilar membrane, and their stereocilia are embedded in the tectorial membrane at the top of the organ of

Corti. Pressure waves of different frequencies cause the basilar membrane to vibrate at particular sites. This bends the stereocilia of hair cells back and forth, sending oscillating signals to the sensory neurons. Consequently, the sensory neurons send intermittent action potentials to the CNS via the auditory nerve. Hair cells at the end of the basilar membrane closest to the oval window respond to high-pitched sounds, and lower-pitched sounds trigger hair cell movement farther along the basilar membrane.

3. Of the various senses, the sense of olfaction (smell) is least important for the survival of humans. As diurnal animals, we rely largely on our visual sense. Sounds are a critical way to learn about impending danger, such as a car horn, but also are our major means of communication. Other senses, such as the ability to sense pain, have acutely important functions from time to time. Olfaction, though often a pleasurable sense and at times a protective one (think of the smell of spoiled food), nonetheless provides little survival advantage to us. In fact, many people spend much of their lives with greatly diminished olfactory abilities, whether from chronic allergies or other problems, and are not hindered in any significant way. The story is very different for animals such as nocturnal mammals, which rely very heavily on olfaction to find food, locate mates, and avoid predators.

## Chapter 35

### Concept Checks

***Figure 35.1*** In addition to not having a requirement to shed their skeletons periodically, animals with endoskeletons can use their skin as an efficient means of heat transfer (and, to an extent in amphibians, water transfer). In addition, the body surface of such animals is often a highly sensitive sensory organ.

***Figure 35.3*** If a tendon is torn, its ability to link a muscle to bone is reduced or lost. Therefore, when a muscle such as the one shown in this illustration contracts, it will not be able to move the bone from which the tendon has become dislodged.

***Figure 35.8*** As cellular ATP concentrations begin to decline in dying cells, bound cross-bridges are unable to detach from thin filaments. With time, however, the muscle tissues begin to degrade and the connections are lost.

***Figure 35.11*** $Na^+$ enters the muscle cell because all cells have an electrochemical gradient for $Na^+$ that favors diffusion of $Na^+$ from extracellular to intracellular fluid (see Chapter 33). This is because cells have a negative membrane potential and because $Na^+$ concentrations are higher in the extracellular fluid. The acetylcholine receptor on skeletal muscle cells is also a ligand-gated ion channel; when acetylcholine binds the receptor, it induces a shape change that opens the channel. This allows the entry of $Na^+$ into the cell.

### BioConnections

***Figure 35.10*** Voltage-gated $Ca^{2+}$ channels exist in the terminals of all axons that communicate by chemical signaling (neurotransmitter release). In those cases, depolarization of the axon terminal opens $Ca^{2+}$ channels, allowing $Ca^{2+}$ to enter the terminal and trigger exocytosis of stored vesicles containing neurotransmitter molecules.

***Figure 35.11*** Myelin is a lipid-rich membrane sheath that speeds up conduction of action potentials along an axon. Action potentials are regenerated at discrete lengths along the axon wherever the myelin sheath is interrupted by a node of Ranvier. This is known as saltatory conduction.

### Testing Your Knowledge

35.1: 1. b 2. b

35.2: 1. e 2. d

35.3: 1. c

35.4: 1. e 2. c

### Test Yourself

1. d 2. e 3. d 4. c 5. e 6. d 7. a 8. e 9. c 10. e

### Conceptual Questions

1. Electrical events precede mechanical (contractile) events in skeletal muscle. Acetylcholine released to the synapse of a neuromuscular junction binds to its receptor on the postsynaptic cell. It is the electrical activity of the action potential that triggers an increase in cytosolic $Ca^{2+}$ and this, in turn, activates the sliding-filament mechanism.
2. a. The cycle begins with the binding of an energized myosin cross-bridge to an actin molecule on a thin filament.
   b. The cross-bridge moves, and the thin filaments slide past the thick filaments.
   c. The ATP binds to myosin, causing the cross-bridge to detach.
   d. The ATP bound to myosin is hydrolyzed by ATPase, re-forming the energized state of myosin.
3. The use of energy released by the hydrolysis of ATP is fundamental to muscle function and locomotion. Recall that ATP must be hydrolyzed during the cross-bridge cycle for skeletal muscle cells to shorten. Energy is also used to maintain calcium ion balance in the sarcoplasmic reticulum and is used in all forms of locomotion. The amount of energy expended by an animal during locomotion reflects how well they are adapted to the environment in which they must move.

## Chapter 36

### Concept Checks

***Figure 36.1*** Open circulatory systems evolved prior to closed systems. However, this does not mean that open systems are in some way "primitive" compared to closed circulatory systems. It is better to think of open systems as being ideally suited to the needs of those animals that have them. Arthropods are an incredibly successful order of animals, with the greatest number of species, and inhabiting virtually every ecological niche on the planet. Clearly, their type of circulatory system has not prevented arthropods from achieving their great success.

***Figure 36.6*** The aorta and all arteries branching from it carry oxygenated blood.

***Figure 36.8*** The left and right ventricles pump blood through the semilunar valves into the aorta and the pulmonary trunk, respectively.

***Figure 36.9*** Body fluids, both extracellular and intracellular, contain large amounts of charged ions, which are capable of conducting electricity. The slight electric currents generated by the beating heart muscle cells are conducted through the surrounding body fluids by the movements of ions in those fluids. This is recorded by the surface electrodes and amplified by the recording machine.

***Figure 36.11*** No, erythrocytes never leave the blood vessels unless a vessel is cut.

***Figure 36.13*** The valves open toward the heart. When the head is upright, the valves are open, and blood drains from the head to the right atrium by gravity. When the giraffe lowers its head to drink, however, gravity would prevent the venous blood from reaching the heart; instead, blood would pool in the head and could raise pressure in the head and brain. The valves in the neck veins work the same way as those in the legs of other animals, helping to propel blood against gravity to the heart.

***Figure 36.16*** Imagine holding several thin sheets of a wet substance, such as paper. If you wave them in the air, what happens? The sheets stick to one another because of surface tension and other properties of moist surfaces. This is what happens to the lamellae in gills when they are in air. When the lamellae stick to each other, the surface area available for gas exchange is reduced and the fish suffocates.

***Figure 36.17*** Several factors probably limit insect body size, but the respiratory system most likely is one such factor. If an insect grew to the size of a human, for example, the trachea and tracheoles would be so large and extensive that there would be little room for any other internal organs in the body! Also, the mass of the animal's body and the forces generated during locomotion would probably collapse the tracheoles. Finally, diffusion of oxygen from the surface of the body to the deepest regions of a human-sized insect would take far too long to support the metabolic demands of internal structures.

***Figure 36.18*** As the lungs expand, the pressure within them decreases, as defined by Boyle's law. This permits air to flow into the lungs.

*Figure 36.23* An increase in the blood concentration of $HCO_3^-$ would favor the reaction $HCO_3^- + H^+ \rightarrow H_2CO_3 \rightarrow CO_2 + H_2O$. This would reduce the $H^+$ concentration of the blood, thereby raising the pH; the $CO_2$ formed as a result would be exhaled. These changes would shift the hemoglobin curve to the left of the usual position.

### BioConnections

*Figure 36.4* Immune defenses are found in most living organisms. Many bacteria produce antibacterial secretions that kill other bacteria. Plants, as shown in Figure 28.14, have a wide array of pathogen-fighting mechanisms.

*Figure 36.16* Countercurrent exchange is an efficient means of heat transfer between arteries and veins, such as those near the skin surface of the legs of a wading bird. Heat from the descending arteries is transferred to surrounding veins, which return the warm blood to the heart, preventing heat loss through the skin to the water.

*Figure 36.24* The brainstem includes the midbrain, pons, and medulla oblongata. See Figure 33.20 for an illustration of the major parts of the human brain.

### Testing Your Knowledge

36.1: 1. c 2. a

36.2: 1. e 2. a

36.3: 1. e 2. b

36.4: 1. c 2. e

36.5: 1. c 2. e

36.6: 1. b

36.7: 1. b 2. b

36.8: 1. c 2. e

36.9: 1. b 2. b

36.10: 1. d

36.11: 1. e 2. c 3. a

### Test Yourself

1. b 2. a 3. c 4. b 5. a 6. a 7. a 8. d 9. b 10. b

### Conceptual Questions

1. ***Closed circulatory system***—In a closed circulatory system, the blood and interstitial fluid are contained within tubes called blood vessels and are transported by a pump called the heart. All of the nutrients and oxygen that tissues require are delivered directly to them by the blood vessels. Advantages of closed circulatory systems are that different parts of an animal's body can receive blood flow in proportion to that body part's metabolic requirements at any given time. Due to its efficiency, a closed circulatory system allows organisms to become larger. ***Open circulatory system***—In an open circulatory system, the organs are bathed in hemolymph that ebbs and flows into and out of the heart(s) and body cavity, rather than blood being directed to all cells. Like a closed circulatory system, there are a pump and blood vessels, but these two structures are less developed and less complex compared to a closed circulatory system. Partly as a result, organisms such as mollusks and arthropods are generally limited to being relatively small, although exceptions do exist.
2. Carbon dioxide, hydrogen ions, and heat are produced by metabolism; the more active a cell is, the more these products are generated. The products, in turn, reduce the ability of hemoglobin to bind oxygen. In this way, more active regions of an animal's body obtain more oxygen in proportion to the metabolic demand at that time.
3. Hemoglobin is a protein with quaternary structure (see Chapter 3) in which the different subunits cooperate to bind up to a total of four oxygen molecules. The association of iron and a polypeptide forms a new molecule with complex ligand-binding properties. It is the structure of the subunits and their relationship to each other that contributes to their ability to bind $O_2$ and to the nonlinear relationship of the oxygen-hemoglobin dissociation curve. In addition, however, interactions of hemoglobin with other molecules, such as $CO_2$, change the structure of hemoglobin in such a way that its properties change. Under such conditions, hemoglobin is less able to bind $O_2$ and consequently it releases the gas. Any molecule that binds to hemoglobin will alter its structure and change its properties; these revert to the original state once the bound molecules are released. A particularly dramatic example of the relationship between the structure and function of hemoglobin is that which occurs in sickle cell disease, due to a mutation that alters the shape of hemoglobin and renders it less functional.

## Chapter 37

### Concept Checks

*Figure 37.2* Extracellular digestion protects the interior of cells from enzyme activity, and allows for the intake of larger amounts of food that can be slowly digested and absorbed gradually.

*Figure 37.8* Active transport requires energy provided by ATP. Thus, absorption of nutrients by this mechanism is an energy-requiring event, and some portion of an animal's regular nutrient consumption is used to provide the energy required to absorb the nutrients.

*Figure 37.11* By resynthesizing triglycerides from absorbed fatty acids and monoglycerides, a steep diffusion gradient is maintained for further diffusion of the smaller molecules into the cell.

*Figure 37.14* The time required for the vesicles to move to the plasma membrane and fuse with it is much shorter than the time required for new GLUTs to be synthesized by activation of GLUT genes. Thus, the action of insulin on cells is very quick, because the GLUTs are already synthesized.

*Figure 37.15* The glycerol and fatty acids used to make glucose are the breakdown products of triglycerides that were stored in adipose tissue during the absorptive period. The amino acids used to make glucose are derived from the breakdown of protein in muscle and other tissue.

*Figure 37.17* Secretion of substances into excretory organ tubules is advantageous because it increases the amount of a substance that gets removed from the body by the excretory organs. This is important, because many substances that get secreted are potentially toxic. Filtration, though efficient, is limited by the volume of fluid that can leave the capillaries and enter the excretory tubule.

*Figure 37.18* Insects remove soluble wastes by the process of secretion, unlike many other animals such as the annelid shown in part (a), which removes soluble wastes by filtration.

### BioConnections

*Figure 37.8* Transmembrane transport processes are not unique to animals, and one or more types are found in virtually all cells.

*Figure 37.14* Exocytosis, a feature characteristic of animal cells, involves the fusion of intracellular vesicles with the plasma membrane, resulting in the release of the vesicle contents into the extracellular fluid. See Figure 5.22 for a general description and Figure 32.12 for a specific example unique to animal cells.

*Figure 37.23* Insulin-sensitive cells such as those of adipose tissue contain glucose transporters (GLUTs) in the membranes of cytosolic vesicles. When an adipose cell is stimulated by insulin, the GLUT-containing vesicles fuse with the cell's plasma membrane, making it possible for the cell to transport glucose from the extracellular fluid into the cell cytosol.

*Figure 37.26* Countercurrent exchange promotes heat retention in the extremities of some animals, and also facilitates gas exchange across the gills of fishes.

### Feature Investigation Questions

1. The surprising observation that some people with gastritis or ulcers have living bacteria (*H. pylori*) in their stomachs, and that administering bacteria-killing compounds provided some relief from the symptoms, led to the hypothesis that *H. pylori* infection is a cause of ulcers in humans.
2. The results did support the hypothesis; however, the results also clearly indicated that not all ulcers are due to *H. pylori* infection.
3. A combined treatment with bismuth and an antibiotic is the most effective treatment. It is apparent, however, that even in individuals with continued

*H. pylori* infection, some ulcers will heal on their own. In the absence of bismuth/antibiotic therapy, though, the likelihood of a recurrence of a new ulcer is much greater.

### Testing Your Knowledge

37.1: 1. c 2. b

37.2: 1. c

37.3: 1. a 2. b 3. d

37.4: 1. c 2. a

37.5: 1. a 2. b

37.6: 1. a 2. c

37.7: 1. e 2. c

37.8: 1. d 2. d

### Test Yourself

1. d 2. b 3. a 4. e 5. e 6. e 7. d 8. e 9. e 10. a

### Conceptual Questions

1. The crop is a dilation of the esophagus, which stores and softens food. The gizzard contains swallowed pebbles that help pulverize food. Both of these functions are adaptations that assist digestion in birds, which do not have teeth and therefore do not chew food. Humans, like many animals, can chew food before swallowing. Sauropod dinosaurs were herbivores that probably contained a gizzard-type stomach in which stones helped to grind coarse vegetation. Such stones would have become smooth after months or even years of rumbling around in the gizzard. Some of these sauropods are known to have lacked the sort of grinding teeth characteristic of modern mammalian herbivores, and thus a gizzard would have aided in their digestion much as it does in modern birds.
2. During filtration, an organ acts like a sieve or filter, removing some of the water and its small solutes from the blood, interstitial fluid, or hemolymph, while retaining blood cells and large solutes such as proteins. Reabsorption is the process whereby epithelial cells of an excretory organ recapture useful solutes that were filtered. Secretion is the process whereby epithelial cells of an excretory organ transport unneeded or harmful solutes from the blood to the excretory tubules for elimination. Some substances such as glucose and amino acids are reabsorbed but not secreted, while some other substances such as toxic compounds are not reabsorbed and are secreted. Still other substances, namely proteins, are not filtered at all.
3. By functioning in fluid and ion balance, the kidneys help regulate blood pressure, solute composition (and thereby electrical activity of nervous and muscle tissue), the pH of body fluids (which in turn is vital for many homeostatic processes), and osmolarity (important for maintaining cell shape and function, among other things). Healthy kidneys are important for all the organ systems of vertebrates.

## Chapter 38

### Concept Checks

***Figure 38.10*** Not all mammals use the energy of sunlight to synthesize vitamin D. Many animals, such as those that inhabit caves or that are strictly nocturnal, rarely are exposed to sunlight. Some of these animals get their vitamin D from dietary sources. How others maintain calcium balance without dietary or sunlight-derived active vitamin D remains uncertain.

***Figure 38.14*** Because 20-hydroxyecdysone is a steroid hormone, you would predict that its receptor would be intracellular. All steroid hormones interact with receptors located either in the cytosol or, more commonly, in the nucleus. The hormone-receptor complex then acts to promote or inhibit transcription of one or more genes. The receptor for 20-hydroxyecdysone is indeed found in cell nuclei.

### BioConnections

***Figure 38.3*** When dopamine is secreted from an axon terminal into a synapse where it diffuses to a postsynaptic cell, it is considered a neurotransmitter. When it is secreted from an axon terminal into the extracellular fluid, from where it diffuses into the blood, it is considered a hormone.

***Figure 38.7*** In addition to the pancreas, certain other organs in an animal's body may contain both exocrine and endocrine tissue or cells. For example, you learned in Chapter 37 that the vertebrate alimentary canal is composed of several types of secretory cells. Some of these cells release hormones into the blood that regulate the activities of the pancreas and other structures, such as the gallbladder. Other cells of the alimentary canal secrete exocrine products such as acids or mucus into the gut lumen that directly aid in digestion or act as a protective coating, respectively.

### Feature Investigation Questions

1. Banting and Best based their procedure on a medical condition that results when pancreatic ducts are blocked. The exocrine cells will deteriorate in a pancreas that has obstructed ducts; however, the islet cells are not affected. The researchers proposed to experimentally replicate the condition to isolate the cells suspected of secreting the glucose-lowering factor. From these cells, they assumed they would be able to extract the substance of interest without contamination or degradation due to exocrine products.
2. The extracts obtained by Banting and Best did contain insulin, the glucose-lowering factor, but were of low strength and purity. Collip developed a procedure to obtain a more purified extract with higher concentrations of insulin.
3. Because insulin is a polypeptide, it cannot be absorbed across the intestines; moreover, acid and digestive enzymes would degrade it in the stomach and small intestine. Thus, oral administration of insulin would be ineffective.

### Testing Your Knowledge

38.1: 1. c 2. b

38.2: 1. a 2. a

38.3: 1. d 2. d

38.4: 1. b

38.5: 1. b 2. b

38.6: 1. c

38.7: 1. b

### Test Yourself

1. b 2. e 3. b 4. e 5. b 6. e 7. c 8. d 9. b 10. d

### Conceptual Questions

1. Leptin is produced by adipose cells in proportion to the amount of stored triglyceride. Leptin stimulates the secretion of reproductive hormones such as LH and FSH; when adipose stores are low, leptin secretion decreases and this removes some of the stimulation to LH and FSH. The benefit of this relationship is that it helps ensure that fertility is linked with adequate energy reserves in females.
2. Type 1 DM is characterized by insufficient production of insulin due to the immune system destroying the insulin-producing cells of the pancreas. In type 2 DM, insulin is still produced by the pancreas, but adipose and muscle cells do not respond normally to insulin.
3. Insulin acts to lower blood glucose concentrations, for example, after a meal, whereas glucagon elevates blood glucose, for example, during fasting. Insulin acts by stimulating the insertion of glucose transporter proteins into the cell membrane of muscle and fat cells. Glucagon acts by stimulating glycogenolysis in the liver. If a high dose of glucagon were injected into an animal, including humans, the blood concentration of glucagon would increase rapidly. This would stimulate increased glycogenolysis, resulting in blood glucose concentrations that were above normal.

## Chapter 39

### Concept Checks

***Figure 39.9*** FSH and LH concentrations do not surge in males, but instead remain fairly steady, because the testes do not show cyclical activity. Sperm production in males is constant throughout life after puberty.

***Figure 39.12*** Pregnancy and subsequent lactation require considerable energy and, therefore, nutrient ingestion. Consuming the placenta provides the female with a rich source of protein and other important nutrients.

### BioConnections

***Figure 39.10*** In addition to its other functions, the placenta must serve the function of the lungs for the fetus, because the fetus's lungs are not breathing air during this time. Arteries always carry blood away from the heart; veins carry blood to the heart. Consequently, blood leaving the heart of the fetus and traveling through arteries to the placenta is deoxygenated. As blood leaves the placenta and returns to the heart, the blood has become oxygenated as oxygen diffuses from the maternal blood into fetal blood. That oxygenated blood then gets pumped from the fetal heart through other arteries to the rest of the fetus's body.

***Figure 39.11*** Positive feedback also occurs during ovulation in the ovarian cycle (see Figure 39.9). Stimulation of an ovarian follicle by LH causes growth of the follicle and the release of estradiol, which further stimulates LH, which causes more follicle activity, and so on until ovulation occurs.

### Feature Investigation Questions

1. Using *Daphnia*, Paland and Lynch compared the accumulation of mitochondrial mutations between sexually reproducing populations and asexually reproducing populations.
2. Of all the mutations observed in the populations, 17.7% were either moderately or mildly deleterious and all of these persisted in the asexually reproducing species, whereas only 4.4% persisted in sexually reproducing species. Thus, asexually reproducing species were about four times more likely to retain a deleterious mutation.
3. Sexual reproduction allows for mixing of the different alleles of genes with each generation, thereby increasing genetic variation within the population. This could prevent the accumulation of deleterious alleles in the population.

### Testing Your Knowledge

39.1: 1. d

39.2: 1. c 2. b

39.3: 1. b 2. b

39.4: 1. c 2. b

39.5: 1. d 2. c

39.6: 1. e 2. b

### Test Yourself

1. d 2. c 3. e 4. a 5. d 6. c 7. b 8. c 9. c 10. e

### Conceptual Questions

1. External fertilization results in exposure of gametes to predation and other environmental dangers. Many such animals have evolved the ability to lay enormous numbers of eggs to compensate for these dangers.
2. Cells of the hypothalamus produce two important hormones that regulate reproduction. GnRH stimulates the anterior pituitary gland to release two gonadotropic hormones, LH and FSH. These two hormones regulate the production of gonadal hormones and development of gametes in both sexes. In addition, increased secretion of GnRH contributes to the initiation of puberty. The hypothalamus also produces oxytocin, a hormone that is stored in the posterior pituitary gland and that acts to stimulate milk release during lactation.
3. Sexual reproduction requires that males and females of a species produce different gametes and that these gametes come into contact with each other. This requires males and females to expend energy to locate mates. It also may require the production of very large numbers of gametes to increase the likelihood that the eggs are fertilized. These costs are outweighed by the genetic diversity afforded by sexual reproduction.

## Chapter 40

### Concept Checks

***Figure 40.2*** Although swelling is one of the most obvious manifestations of inflammation, it has no significant adaptive value of its own. It is a consequence of fluid leaking out of blood vessels into the interstitial space. It can, however, contribute to pain sensations, because the buildup of fluid may cause distortion of connective tissue structures such as tendons and ligaments. Pain, while obviously unpleasant, is an important signal that alerts many animals to the injury and serves as a reminder to protect the injured site.

***Figure 40.4*** Recall from Chapter 36 that as blood circulates, a portion of the plasma—the fluid part of blood—exits venules and capillaries and enters the interstitial fluid. Most of the plasma is reabsorbed back into the capillaries, but a portion gets left behind. That excess fluid is drained away by lymph vessels and becomes lymph. Without lymph vessels, fluid would accumulate outside of the blood, in the interstitial fluid.

***Figure 40.6*** Helper T cells contribute to both humoral and cellular immunity.

***Figure 40.9*** Both B- and T-cell receptors have transmembrane domains, a constant region, and a variable region that binds a specific antigen.

***Figure 40.13*** Because an animal may encounter the same type of pathogen many times during its life, having a secondary immune response means that future infections will be fought off much more efficiently.

### BioConnections

***Figure 40.1*** There are far more erythrocytes in blood than all leukocytes combined (in humans, for example, there are at least 1,000 times more erythrocytes).

### Feature Investigation Questions

1. The amino acid sequence of Toll protein shared similarities with a portion of a protein known to be involved in immune responses in vertebrates. In addition, activation of Toll protein and the vertebrate immune protein (a cytokine receptor) resulted in the generation of some of the same intracellular signals. This suggested that in addition to its characterized role in embryonic development, Toll may also be important in immune functions in flies.
2. No, Toll protein is not a receptor that recognizes pathogen-associated molecular patterns (PAMPs) expressed on microbial surfaces, and thus it is distinguishable from Toll-like receptors in vertebrates. Toll is, however, a transmembrane protein that binds to extracellular signals; these signals arise, however, not from the microbes themselves but rather from proteins that are endogenous to flies and that are generated during infections.
3. Yes, the results of the survival study clearly implicated Toll as a protein required for the induction of antimicrobial proteins and the ability to withstand fungal infection. Thus, the investigators' hypothesis was supported.

### Testing Your Knowledge

40.1: 1. d 2. a

40.2: 1. a 2. d

40.3: 1. e 2. e

40.4: 1. c 2. b

### Test Yourself

1. e 2. b 3. c 4. c 5. a 6. e 7. b 8. a 9. d 10. b

### Conceptual Questions

1. Innate immunity is present at birth and is found in all animals. These defenses recognize general, conserved features common to a wide array of pathogens and include external barriers, such as the skin, and internal defenses involving phagocytes and other cells. Acquired immunity develops *after* an animal has been exposed to a *particular* antigen. The responses include humoral and cell-mediated defenses. Acquired immunity appears to be largely restricted to vertebrates. Unlike innate immunity, in acquired immunity, the response to an antigen is greatly increased if an animal is exposed to that antigen again at some future time.
2. Cytotoxic T cells are "attack" cells that are responsible for cell-mediated immunity. Once activated, they migrate to the location of their targets, bind to the targets by combining with an antigen on them, and directly kill the targets via secreted chemicals.
3. Pathogens are disease-causing viruses and microorganisms including certain bacteria, and eukaryotic parasites such as certain protists, fungi, and small

worms. Bacteria are single-celled prokaryotes that lack a true nucleus but are capable of reproducing on their own, whereas viruses are nucleic acids packaged in a protein coat and require a host cell to reproduce.

## Chapter 41

### Concept Checks

***Figure 41.5*** Note that the receptors are located in or associated with blood vessels supplying the brain, a vital organ that among other functions controls many of the compensatory responses to changes in blood pressure or blood oxygen levels. Other receptors are located in the aorta, the first major vessel to leave the heart. Thus, blood pressure and gases are monitored in the general circulation and also specifically in the circulation entering the brain.

***Figure 41.7*** If the plasma protein concentration decreases, Starling forces would favor a net outflow of fluid from a capillary because there would be less osmotic force favoring fluid movement into the capillary.

### Bioconnections

***Figure 41.3*** Stretch-sensitive receptors are widespread in animal bodies. The familiar knee-jerk response is a reflex triggered by the stretch of receptors located in tendons in the knee. Other examples include stretch receptors in muscles of crustaceans that provide feedback information on the animal's posture and movement, and receptors in the stomach that relay a sense of fullness when the stomach is stretched after eating.

***Figure 41.10*** Positive feedback also occurs during the ovarian cycle in mammals at the time of ovulation (see Figure 39.9), and during the process of birth in mammals (see Figure 39.11).

### Testing Your Knowledge

41.1: 1. b

41.2: 1. b

41.3: 1. b

41.4: 1. d

### Test Yourself

1. b 2. d 3. a 4. c 5. a 6. c 7. e 8. e 9. c 10. e

### Conceptual Questions

1. Animals in nature are confronted with many types of homeostatic challenges that often require integrated responses in multiple organ systems. For example, a bird flying at high altitudes such as over a mountain range faces the challenge of obtaining sufficient oxygen from the environment. You might predict that in such situations a bird may increase its breathing rate, adjust its cardiac output, or both; indeed, both of these changes and several others do occur. Likewise, imagine a fish that migrates between fresh and salt water, such as a salmon. These animals alter the function of their respiratory and urinary systems to help compensate for the changes in ion and water movement across the gills when moving from one environment to another. Yet another common example is starvation, during which the nervous, endocrine, urinary, and digestive systems help maintain glucose homeostasis by processes such as gluconeogenesis.
2. Light-headedness can occur in some people when donating blood, which is essentially a carefully controlled hemorrhage. Initially, as the homeostatic processes described in this chapter are just beginning, there is a period of instability with respect to blood pressure control. While lying or sitting down, this is rarely noticeable, but upon standing gravity counteracts the movement of blood through limb veins back to the heart, causing a sudden decrease in pressure. Fainting does not usually occur, however, because the baroreceptor reflex responds immediately to this sudden change in pressure, and within seconds the heart rate and cardiac output are increased due to the actions of the sympathetic nervous system. This phenomenon can actually happen anytime, even under ordinary circumstances, but is more noticeable when a person's blood volume is reduced such as when donating blood.
3. All of the homeostatic responses to hemorrhage described in this chapter require energy. Increasing the activity of any muscle, including that of the heart and the respiratory muscles, requires a considerable increase in expenditure of energy (ATP). Activity of the nervous system, so vital to the compensatory response to hemorrhage, requires continual ATP production and hydrolysis to maintain ion concentration gradients across the plasma membrane of neurons; without this, there could be no flow of current along a neuron. Many of the transport processes in the kidneys also require ATP to drive the ion pumps that move ions across membranes and that create osmotic gradients for the movement of water.

## Chapter 42

### Concept Checks

***Figure 42.2*** In classical conditioning, an involuntary response comes to be associated with a stimulus that did not originally elicit the response, as with Pavlov's dogs salivating at the sound of a metronome.

***Figure 42.3*** Tinbergen manipulated pinecones, but not all digger wasp nests are surrounded by pinecones. You could manipulate branches, twigs, stones, and leaves to determine the necessary size and dimensions of objects that digger wasps use as landmarks.

***Figure 42.7*** The individuals in the center of the group are less likely to be attacked than those on the edge of the group. This is referred to as the geometry of the selfish herd.

***Figure 42.11*** Because of the genetic benefit, the answer is nine cousins. Consider Hamilton's rule, expressed in the formula $rB > C$. Using cousins, $B = 9$, $r = 0.125$, and $C = 1$, and $1.125 > 1$. Using sisters, $B = 1$, $r = 0.5$, and $C = 0.5$. Because $rB$ would not be greater than $C$, there would be no net genetic benefit in self-sacrifice.

***Figure 42.12*** All the larvae in the group are likely to be the progeny of one egg mass from one adult female moth. The death of the one caterpillar teaches a predator to avoid the pattern and benefits the caterpillar's close kin.

### BioConnections

***Figure 42.2*** Toxic or bad-tasting prey species converge on the same color patterns to reinforce the basic distasteful design.

### Feature Investigation Questions

1. Tinbergen observed the activity of digger wasps as they prepared to leave the nest. Each time, the wasp hovered and flew around the nest for a period of time before leaving. Tinbergen suggested that during this time, the wasp was making a mental map of the nest site. He hypothesized that the wasp was using characteristics of the nest site, particularly landmarks, to help relocate it.
2. Tinbergen placed pinecones around the nest of the wasps. When the wasps left the nest, he removed the pinecones from the nest site and set them up in the same pattern a distance away, constructing a sham nest. For each trial, the wasps would go directly to the sham nest, which had the pinecones around it. This indicated to Tinbergen that the wasps identified the nest based on the pinecone landmarks.
3. Yes. Tinbergen also conducted an experiment to determine if the wasps were responding to the visual cue of the pinecones or the chemical cue of the pinecone scent. The results of this experiment indicated that the wasps responded to the visual cue of the pinecones and not their scent.

### Testing Your Knowledge

42.1: 1. b 2. e 3. c

42.2: 1. e

42.3: 1. d 2. b 3. c

42.4: 1. c 2. e

42.5: 1. c 2. b

### Test Yourself

1. d 2. d 3. c 4. e 5. d 6. b 7. c 8. d 9. a 10. c

### Conceptual Questions

1. The donation of the male's body to the female is the ultimate nuptial gift. It is possible that this meal enables the females to produce more eggs. In this way, the male's genes will be passed on to future generations.
2. Certainty of paternity influences degree of parental care. With internal fertilization, certainty of paternity is relatively low. With external fertilization, eggs and sperm are deposited together, and paternity is more certain. This explains why males of some species, such as mouth-breeding cichlid fish, are more likely to engage in parental care.
3. As male bears are killed by hunters, new males move in to a territory and kill existing cubs. Thus, not only are bears killed directly by hunters, but population growth is also slowed as cubs are killed and population recovery is prolonged.

## Chapter 43

### Concept Checks

***Figure 43.2*** The total population size, $N$, would be estimated to be $110 \times 100/20$, or 550.

***Figure 43.6*** $dN/dt = 0.1 \times 100\,(1000 - 100)/1000 = 9$.

***Figure 43.10*** There would be 10 possible pairings (AB, AC, AD, AE, BC, BD, BE, CD, CE, DE), of which only neighboring species (AB, BC, CD, DE) competed. Therefore, competition would be expected in 4/10 pairings, or 40% of the cases.

***Figure 43.11*** Omnivores, such as bears, can feed on both plant material, such as berries, and animals, such as salmon. As such, omnivores may act as both predators and herbivores depending on what they are feeding on.

***Figure 43.13*** Because there is no evolutionary history between invasive predators and native prey, the native prey often have no defenses against these predators and are very easily caught and eaten.

***Figure 43.14*** Only density-dependent factors operate in this way.

### BioConnections

***Figure 43.3*** Type III.

***Figure 43.12*** Operant conditioning.

### Testing Your Knowledge

43.1: 1. c

43.2: 1. c 2. b 3. e

43.3: 1. b 2. e

43.4: 1. d 2. b 3. b

### Test Yourself

1. b 2. e 3. b 4. c 5. a 6. c 7. b 8. c 9. d 10. c

### Conceptual Questions

1. Decrease, by 50%.
2. At medium values of $N$, $(K - N)/K$ is closer to a value of 1, and population growth is relatively large. If $K = 1{,}000$, $N = 500$, and $r = 0.1$, then

$$\frac{dN}{dt} = (0.1)(500) \times \frac{(1{,}000 - 100)}{1{,}000}$$

$$\frac{dN}{dt} = 25$$

However, if population sizes are low ($N = 100$), $(K - N)/K$ is so small that growth is low.

$$\frac{dN}{dt} = (0.1)(100) \times \frac{(1{,}000 - 100)}{1{,}000}$$

$$\frac{dN}{dt} = 9$$

By comparing these two examples with that shown in Section 43.3, we see that growth is small at high and low values of $N$ and is greatest at intermediate values of $N$. Growth is greatest when $N = K/2$. However, when expressed as a percentage, growth is greatest at low population sizes. Where $N = 100$, percentage growth = 9/100 = 9%. Where $N = 500$, percentage growth = 25/500 = 5%, and where $N = 900$, percentage growth = 9/100 = 1%.

3. Mutualisms between humans and crops and humans and livestock have had far reaching effects. Excluding Antarctica, a staggering 37.3% of the world's land area is given over to crops and rangeland for animals such as cattle and sheep. Most of these species remain dependent on humans for their survival.

## Chapter 44

### Concept Checks

***Figure 44.3*** Species richness of trees doesn't increase in the mountainous areas of the U.S. Southwest because rainfall in the western United States is low compared to that in the east.

***Figure 44.5*** As we walk forward from the edge of the glacier to the mouth of the inlet, we are walking backward in ecological time to communities that originated hundreds of years ago.

***Figure 44.8*** Competition features more prominently. Although early colonists tend to make the habitat more favorable for later colonists, it is the later colonists who outcompete the earlier ones, and this fuels species change.

***Figure 44.9*** If a small island was extremely close to the mainland, it could continually receive migrating species from the source pool. Even though these species could not complete their life cycle on such a small island, extinctions would rarely be recorded because of this continual immigration.

***Figure 44.10*** At first glance, the change looks small, but the data are plotted on a log scale. On this scale, an increase in bird richness from 1.2 to 1.6 equals an increase from 16 to 40 species, a change of over 100%.

***Figure 44.14*** It depends on the trophic level of their food, whether dead vegetation or dead animals. Many decomposers feed at multiple trophic levels.

### BioConnections:

***Figure 44.9*** The model helps conservationists design the best shaped and optimally placed nature reserves in a "sea" of developed land.

***Figure 44.13*** Cyanobacteria.

### Feature Investigation Questions

1. Simberloff and Wilson were testing the three predictions of the theory of island biogeography. One prediction suggested that the number of species should increase with increasing island size. Another prediction suggested that the number of species should decrease with increasing distance of the island from the source pool. Finally, the researchers were testing the prediction that the turnover of species on islands should be considerable.
2. Simberloff and Wilson used the information gathered from the species survey to determine whether the same types of species recolonized the islands or if colonizing species were random.
3. The data suggested that species richness did increase with island size. Also, the researchers found that in all but one of the islands, the number of species was similar to the number of species before fumigation.

### Testing Your Knowledge

44.1: 1. d 2. a 3. e

44.2: 1. c 2. d

44.3: 1. e 2. b 3. b

44.4: 1. c 2. e

44.5: 1. e 2. b 3. d

44.6: 1. d 2. e 3. a

### Test Yourself

1. c 2. c 3. a 4. d 5. c 6. d 7. d 8. a 9. d 10. c

### Conceptual Questions

1. The value of the Shannon Index is 1.609 for both forests. By this measure, diversity is equal between both forests. The Shannon Index is unable to discriminate between communities that have different species abundance but the same relative proportions of species. An observer would be more likely to encounter a variety of trees in forest A than in forest B.
2. Carrion beetles are decomposers. They feed on dead animals such as mice, at trophic level 3 or 4. Mice generally feed on vegetative material (trophic level 1) or crawling arthropods (trophic level 2), so mice themselves feed at trophic level 2 or 3.
3. Facilitation. *Calluna* litter enriches the soil with nitrogen, facilitating the growth of the grasses. Adding fertilizer also increases soil nitrogen.

## Chapter 45

### Concept Checks

***Figure 45.2*** This occurs because increasing cloudiness and rain at the tropics maintain fairly constant temperatures across a wide latitudinal range.

***Figure 45.11*** Soil conditions can also influence biome type. Nutrient-poor soils, for example, may support vegetation different from that of the surrounding area.

***Figure 45.12*** Taiga.

### Testing Your Knowledge

45.1: 1. b 2. b

45.2: 1. b 2. c

### Test Yourself

1. c 2. e 3. b 4. b 5. a 6. d 7. a 8. d 9. e 10. e

### Conceptual Questions

1. Mountains are cooler than valleys because of adiabatic cooling. Air at higher altitudes expands because of decreased pressure. As it expands, air cools, at a rate of 10°C for every 1,000 m in elevation. As a result, mountain tops can be much cooler than the plains or valleys that surround them.
2. The Sun may be much bigger than the Moon, but it is also much further away from the Earth, so its effects on tides are much less than those of the smaller but closer Moon.
3. Florida is a peninsula that is surrounded by the Atlantic Ocean and the Gulf of Mexico. Differential heating between the land and the sea creates onshore sea breezes on both the east and west coasts. These breezes often drift across the whole peninsula, bringing heavy rain.

## Chapter 46

### Concept Checks

***Figure 46.2*** There were very few juveniles in the population and many mature adults. The population would be in decline.

***Figure 46.5*** Many different ecological footprint calculators are available on the Internet. Does altering inputs such as type of transportation, amount of meat eaten, or amount of waste generated make a difference?

***Figure 46.7*** The Northern Hemisphere has greater land area and plant biomass than the Southern Hemisphere, so in the northern summer more $CO_2$ is used up by plants and atmospheric $CO_2$ level declines slightly. In the northern winter, less $CO_2$ is absorbed and atmospheric $CO_2$ levels increase.

***Figure 46.9*** The greatest stores are in rocks and fossil fuels.

### BioConnections

***Figure 46.9*** Oxygen, hydrogen, and nitrogen. Therefore, the water and nitrogen cycles would be very important.

***Figure 46.32*** The first introductions of Burmese pythons into the Everglades were likely made prior to 1985.

***Figure 46.34*** Exponential growth.

### Feature Investigation Questions

1. The researchers were testing the effects of increased carbon dioxide levels on the forest ecosystem. The researchers were testing the effects of increased carbon dioxide levels on primary production as well as other trophic levels in the ecosystem.
2. By increasing the carbon dioxide levels in only half of the chambers, the researchers were maintaining the control treatments necessary for all scientific studies. By maintaining equal numbers of control and experimental treatments, the researchers could compare data to determine what effects the experimental treatment had on the ecosystem.
3. $t_{14} = 5.667$, $P<0.001$. $x_1 = 10.00$, s.d = 2.93; $x_2 = 3.20$, s.d = 1.72

### Testing Your Knowledge

46.1: 1. d 2. d

46.2: 1. b 2. c

46.3: 1. d 2. d

46.4: 1. b

46.5: 1. e

46.6: 1. a 2. b

46.7: 1. e 2. c

### Test Yourself

1. c 2. c 3. d 4. b 5. a 6. e 7. b 8. b 9. e 10. e

### Conceptual Questions

1. Nitrogen molecules have a triple bond, making them hard to break apart. Only a few species of bacteria can break apart atmospheric nitrogen or fix nitrogen. The excess ammonia, $NH_3$, or ammonium, $NH_4^+$, gradually accumulates and can be used by plants.
2. Maximum sustainable yield represents the number of individuals that can be removed from a population without affecting population growth. This is rather like removing the interest from a bank account and not touching the principal. Maximal sustainable yield occurs at the steepest point of the growth curve and this occurs at the midpoint of the logistic curve.
3. The family who has triplets has 27 offspring, compared to 32 for the family who has twins. Delaying reproduction can reduce population growth.

## Chapter 47

### Concept Checks

***Figure 47.6*** Corridors might also promote the movement of invasive species or the spread of fire between areas.

***Figure 47.7*** They act as habitat corridors because they permit movement of species between forest fragments.

### BioConnections:

***Figure 47.11*** Genetic cloning could be used to save threatened species or even to resurrect recently extinct species. Cloning may theoretically be able to increase genetic variability of populations if it were possible to use cells from deceased animals. However, cloning is not a panacea because habitat loss, poaching or invasive species may still prevent re-introductions of species back into the wild.

## Testing Your Knowledge

47.1: 1. c

47.2: 1. b 2. a

47.3: 1. a 2. b 3. c

## Test Yourself

1. d 2. e 3. b 4. a 5. c 6. b 7. e 8. e 9. c 10. b

## Conceptual Questions

1. Increased species diversity increases ecosystem function. Ecosystem functions such as nutrient cycling, regulation of atmospheric gases, pollination of crops, pest regulation, water purity, storm protection, and sewage purification are all likely to be increased by increased species diversity. In addition, increased plant species diversity increases likely availability of new medicines for humans.
2. Megadiversity countries are those with the greatest number of species. Biodiversity hot spots conserve the greatest numbers of endemic species. Crisis ecoregions are those areas of the Earth which represent distinct biome types, such as temperate grasslands and tropical deciduous forests, but have undergone substantial habitat loss. "Last of the wild" areas are relatively pristine areas such as much tundra and taiga, deserts, and some tropical rainforests.
3. Habitats or species are considered vulnerable if they are currently being threatened such that habitats may be lost or species may go extinct. There is a clear need to protect such entities. However, species and habitats may also be unique and should also be protected, even if they are not currently threatened. Examples may include rare species, which are restricted to a small area, such as panda bears, or rare habitats such as prairie remnants, many of which, are now found only in cemetery plots and have been spared from the plough. Most conservation strategies are reactive, seeking to preserve the last few remaining individuals of a species or areas of habitat. Only the "last of the wild" strategy is proactive, identifying areas for conservation before they have become severely impacted.

# INDEX

Note: Page numbers followed by f refer to figures; page numbers followed by t refer to tables.

## A

## B

## C

## D

## E

## F

## G

## I

## J

## M

## O

## P

## Q



## R

## S

## T

## U



## V



## W

## X



## Y



## Z